W9-AVL-743

BIOLOGY

The Unity and Diversity of Life

BIOLOGY

The Unity and Diversity of Life

NINTH EDITION

CECIE STARR / RALPH TAGGART

LISA STARR
Biological Illustrator

Brooks/Cole
Thomson Learning

Australia • Canada • Mexico • Singapore • Spain • United Kingdom • United States

BIOLOGY PUBLISHER: Jack C. Carey

PROJECT DEVELOPMENT EDITOR: Kristin Milotich

MEDIA PROJECT MANAGER: Pat Waldo

EDITORIAL ASSISTANT: Daniel Lombardino

DEVELOPMENTAL EDITOR: Mary Arbogast

MARKETING TEAM: Rachel Alvelais, Mandie Houghan, Laura Hubrich, Carla Martin-Falcone

PRODUCTION EDITOR: Jamie Sue Brooks

PRODUCTION DIRECTOR: Mary Douglas, Rogue Valley Publications

EDITORIAL PRODUCTION: Karen Stough, Jamee Rae

PERMISSIONS EDITOR: Mary Kay Hancharick

TEXT AND COVER DESIGN, ART DIRECTION: Gary Head, Gary Head Design

COVER PHOTO: *Snow leopard, a currently endangered species (Section 28.2).* © Art Wolfe / PhotoResearchers

ART EDITOR: Mary Douglas, Jamie Sue Brooks

INTERIOR ILLUSTRATION: Lisa Starr, Raychel Ciemma, Precision Graphics, Preface Inc., Robert Demarest, Jan Flessner, Darwen Hennings, Vally Hennings, Betsy Palay, Nadine Sokol, Kevin Somerville, Lloyd Townsend

PHOTO RESEARCHER: Myrna Engler

PRINT BUYER: Vena M. Dyer

COMPOSITION: Preface, Inc. (Angela Harris)

COLOR PROCESSING: H&S Graphics (Tom Anderson, Michelle Kessel, Laurie Riggle, and John Deady)

COVER PRINTING, PRINTING AND BINDING: Von Hoffmann Press

BOOKS IN THE BROOKS/COLE BIOLOGY SERIES

Biology: The Unity and Diversity of Life, Ninth, Starr/Taggart
Biology: Concepts and Applications, Fourth, Starr
Laboratory Manual for Biology, Perry/Morton/Perry
Human Biology, Fourth, Starr/McMillan
Perspectives in Human Biology, Knapp
Human Physiology, Fourth, Sherwood
Fundamentals of Physiology, Second, Sherwood
Psychobiology: The Neuron and Behavior, Hoyenga/Hoyenga

Introduction to Cell and Molecular Biology, Wolfe
Molecular and Cellular Biology, Wolfe

Introduction to Microbiology, Second, Ingraham/Ingraham

Genetics: The Continuity of Life, Fairbanks
Human Heredity, Fifth, Cummings
Introduction to Biotechnology, Barnum
Sex, Evolution, and Behavior, Second, Daly/Wilson

Plant Biology, Rost et al.
Plant Physiology, Fourth, Salisbury/Ross
Plant Physiology Laboratory Manual, Ross

General Ecology, Second, Krohne
Living in the Environment, Eleventh, Miller
Environmental Science, Eighth, Miller
Sustaining the Earth, Fourth, Miller
Environment: Problems and Solutions, Miller
Environmental Science, Fifth, Chiras

Oceanography: An Invitation to Marine Science, Third, Garrison
Essentials of Oceanography, Second, Garrison
Introduction to Ocean Sciences, Segar
Oceanography: An Introduction, Fifth, Ingmanson/Wallace
Marine Life and the Sea, Milne

Printed in United States of America

10 9 8 7 6 5 4 3 2

Library of Congress Cataloging-in-Publication Data
Starr, Cecie.
Biology: the unity and diversity of life/Cecie Starr, Ralph Taggart.—9th ed.
p. cm.
Includes bibliographical references (p.).
ISBN 0-534-57546-3
1. Biology. I. Taggart, Ralph. II. Title.
QH308.2 .S72 2000
570—dc21 00-040359

Annotated Instructor's Edition: 0-534-37794-7

For more information about this or any other Brooks/Cole products, contact:
BROOKS/COLE
511 Forest Lodge Road
Pacific Grove, CA 93950 USA
www.brookscole.com
1-800-423-0563 (Thomson Learning Academic Resource Center)

CONTENTS IN BRIEF

DETAILED CONTENTS

INTRODUCTION

I PRINCIPLES OF CELLULAR LIFE

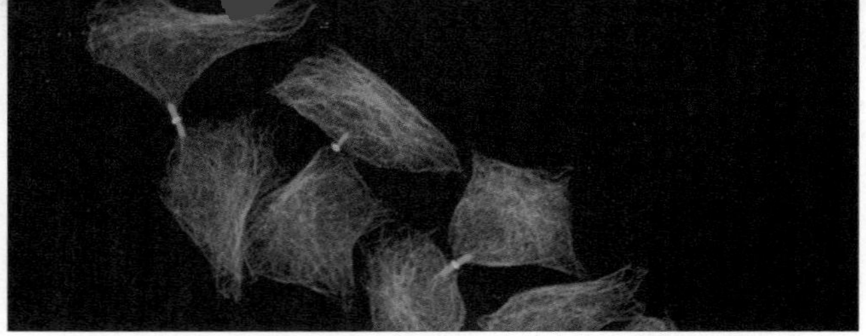

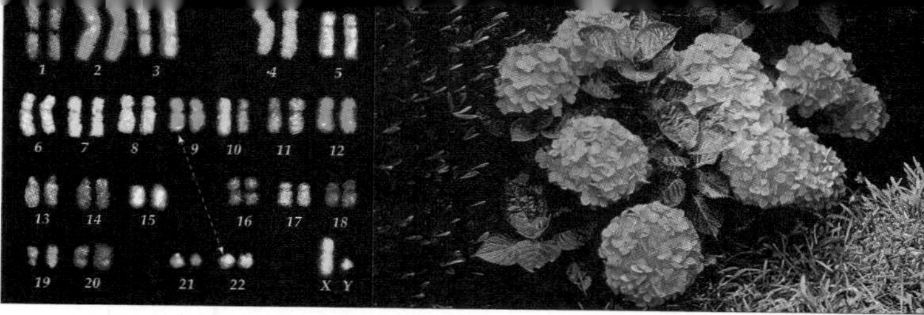
1
2
3
4
5
6
7
8
9
10
11
12
13
14
15
16
17
18
19
20
21
22
X Y

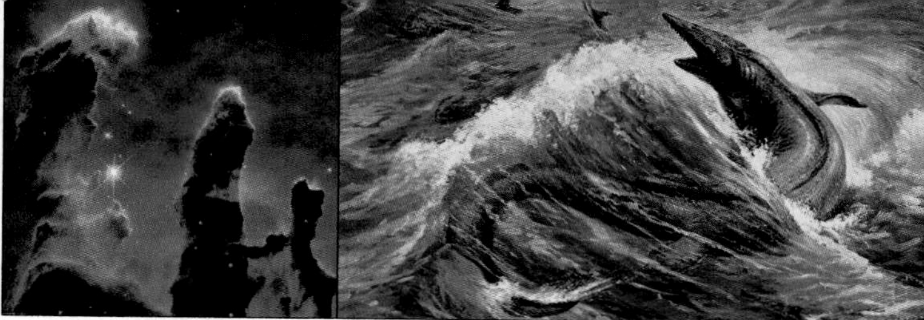

26 ANIMALS: THE INVERTEBRATES

27 ANIMALS: THE VERTEBRATES

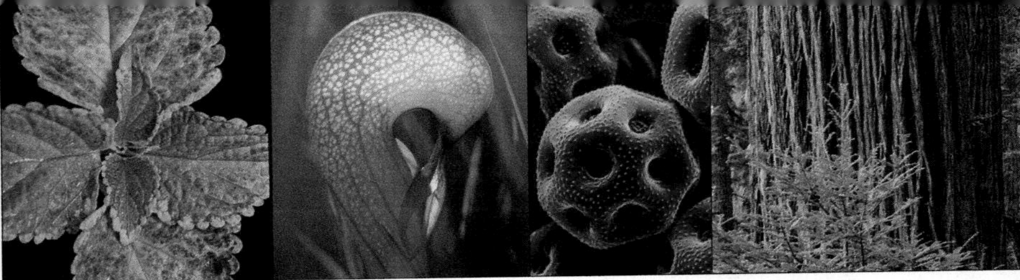

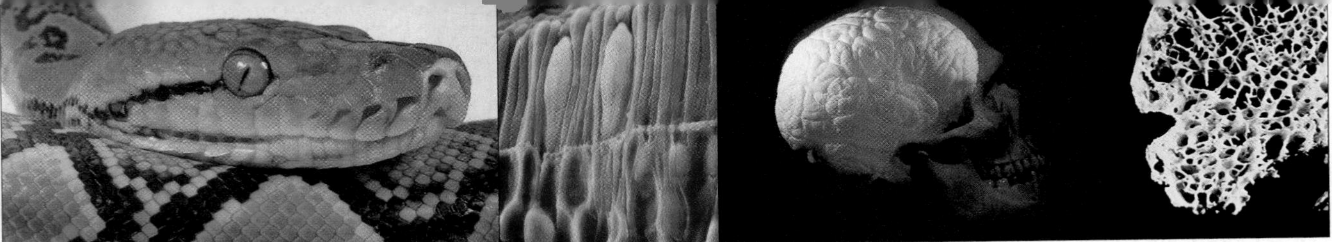

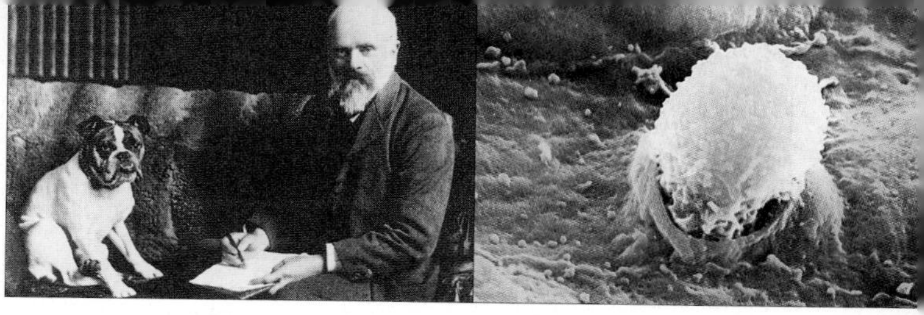

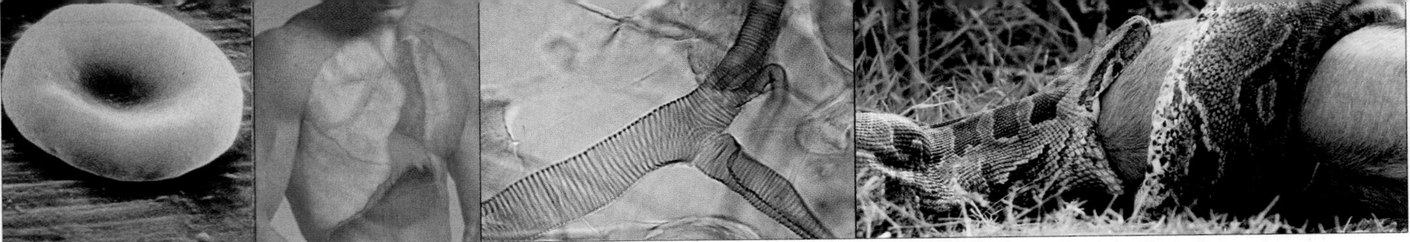

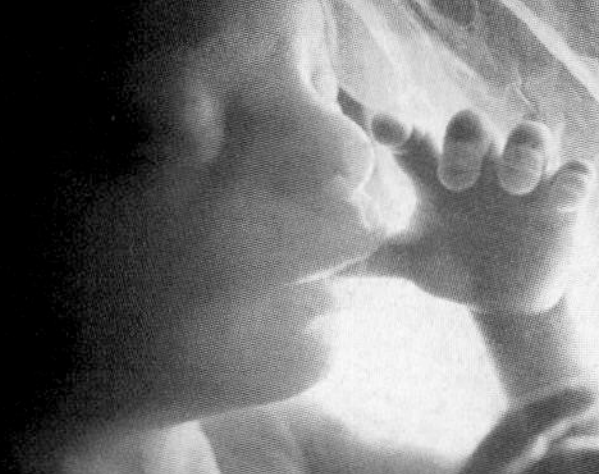

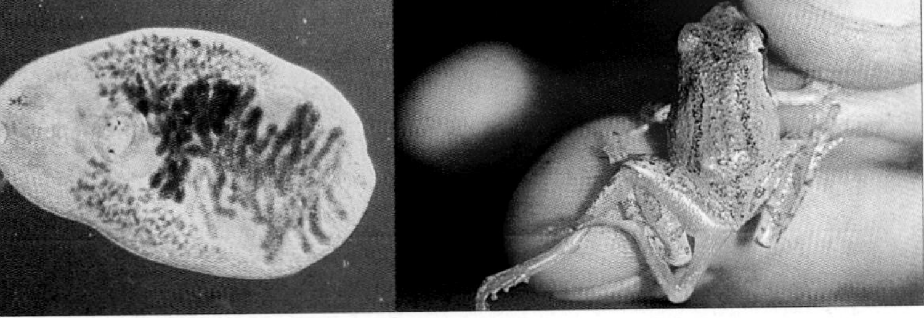

VII ECOLOGY AND BEHAVIOR

Preface

Teachers of introductory biology know all about the Red Queen effect, whereby one runs as fast as one can to stay in the same place. New and modified information from hundreds of fields of inquiry piles up daily, and somehow teachers are expected to distill it into Biology Lite, a one-course zip through the high points that still manages to help students deepen their understanding of a world of unbelievable richness.

Restricting textbook content runs the risk of splintering understanding of that world, as when an emphasis on human biology inadvertently reinforces archaic notions that everything on Earth is here in the service of Us; or when a molecular focus excludes knowledge of whole organisms. (Here I am reminded of the well intentioned but lethal blanketing of habitats with DDT and of Jacques Monod's first clueless encounters with *E. coli*.)

We offer this book as a coherent account of the sweep of life's diversity and its underlying unity. Through its examples of problem solving and experiments, it shows the power of thinking critically about the natural world. It highlights key concepts, current understandings, and research trends for major fields of biological inquiry. It explains the structure and function of a broad sampling of organisms in enough detail so students can develop a working vocabulary about life's parts and processes.

The book starts with an overview of the basic concepts and scientific methods. Three units on the principles of biochemistry, inheritance, and evolution follow. They are the conceptual framework for exploring life's unity and diversity, starting with an evolutionary survey of each kingdom. Units on comparative anatomy and physiology of plants, then animals, are next. The last unit focuses on patterns and consequences of organisms interacting with one another and with the environment. This conceptual organization parallels the levels of biological organization.

TOPIC SPREADS Ongoing feedback from teachers of more than three million students helped us refine our approach to writing. We keep the story line in focus for students by subscribing to the question "How do you eat an elephant?" and its answer, "One bite at a time." We organized descriptions, art, and supporting evidence for each topic on two facing pages, at most. Each *topic spread* starts with a numbered tab and concludes with boldface, summary statements of key points (see below). Students can use the statements to check whether they understand one topic spread's concepts before starting another.

By clearly organizing topics within chapters, we offer teachers flexibility in assigning text material to fit course requirements. For example, those who spend little time on photosynthesis might choose to bypass topic spreads with details of the properties of light or the chemiosmotic theory of ATP formation. They might or might not assign the Focus essays (one on the impact of photosynthesis on

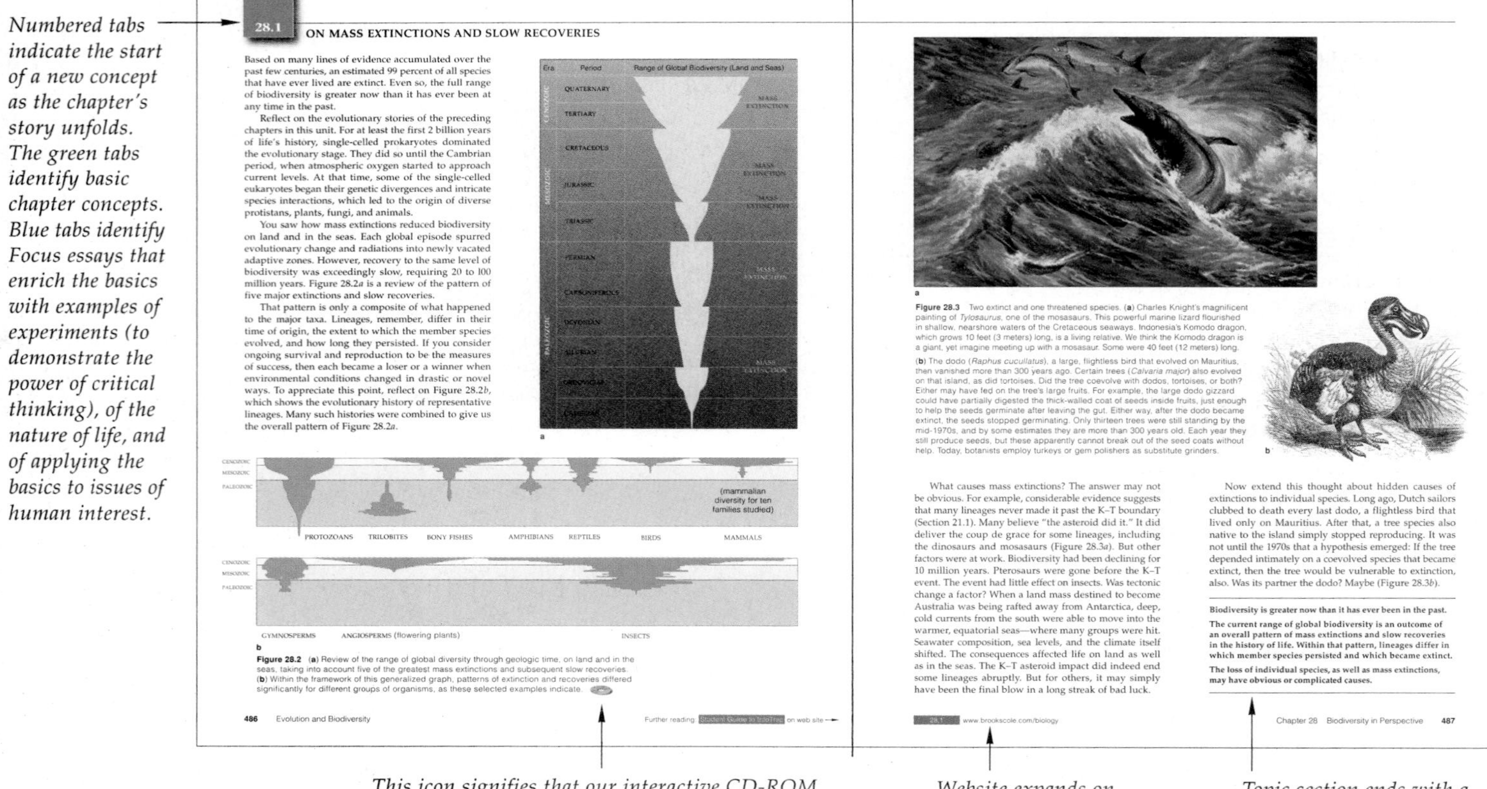

Numbered tabs indicate the start of a new concept as the chapter's story unfolds. The green tabs identify basic chapter concepts. Blue tabs identify Focus essays that enrich the basics with examples of experiments (to demonstrate the power of critical thinking), of the nature of life, and of applying the basics to issues of human interest.
This icon signifies that our interactive CD-ROM further explores the concept being illustrated.
Website expands on the section's topic.
Topic section ends with a summary of key concepts.

28.1 ON MASS EXTINCTIONS AND SLOW RECOVERIES

Based on many lines of evidence accumulated over the past few centuries, an estimated 99 percent of all species that have ever lived are extinct. Even so, the full range of biodiversity is greater now than it has ever been at any time in the past.

Reflect on the evolutionary stories of the preceding chapters in this unit. For at least the first 2 billion years of life's history, single-celled prokaryotes dominated the evolutionary stage. They did so until the Cambrian period, when atmospheric oxygen started to approach current levels. At that time, some of the single-celled eukaryotes began their genetic divergences and intricate species interactions, which led to the origin of diverse protistans, plants, fungi, and animals.

You saw how mass extinctions reduced biodiversity on land and in the seas. Each global episode spurred evolutionary change and radiations into newly vacated adaptive zones. However, recovery to the same level of biodiversity was exceedingly slow, requiring 20 to 100 million years. Figure 28.2*a* is a review of the pattern of five major extinctions and slow recoveries.

That pattern is only a composite of what happened to the major taxa. Lineages, remember, differ in their time of origin, the extent to which the member species evolved, and how long they persisted. If you consider ongoing survival and reproduction to be the measures of success, then each became a loser or a winner when environmental conditions changed in drastic or novel ways. To appreciate this point, reflect on Figure 28.2*b*, which shows the evolutionary history of representative lineages. Many such histories were combined to give us the overall pattern of Figure 28.2*a*.

Era
Period
Range of Global Biodiversity (Land and Seas)
CENOZOIC
MESOZOIC
PALEOZOIC
QUATERNARY
TERTIARY
CRETACEOUS
JURASSIC
MASS EXTINCTION
a
CENOZOIC
MESOZOIC
PALEOZOIC
(mammalian diversity for ten families studied)
PROTOZOANS
TRILOBITES
BONY FISHES
AMPHIBIANS
REPTILES
BIRDS
MAMMALS
GYMNOSPERMS
ANGIOSPERMS (flowering plants)
INSECTS
b

Figure 28.2 (**a**) Review of the range of global diversity through geologic time, on land and in the seas, taking into account five of the greatest mass extinctions and subsequent slow recoveries. (**b**) Within the framework of this generalized graph, patterns of extinction and recoveries differed significantly for different groups of organisms, as these selected examples indicate.

486 Evolution and Biodiversity

Further reading: Student Guide to InfoTrac on web site

Figure 28.3 Two extinct and one threatened species. (**a**) Charles Knight's magnificent painting of *Tylosaurus*, one of the mosasaurs. This powerful marine lizard flourished in shallow, nearshore waters of the Cretaceous seaways. Indonesia's Komodo dragon, which grows 10 feet (3 meters) long, is a living relative. We think the Komodo dragon is a giant, yet imagine meeting up with a mosasaur. Some were 40 feet (12 meters) long.

(**b**) The dodo (*Raphus cucullatus*), a large, flightless bird that evolved on Mauritius, then vanished more than 300 years ago. Certain trees (*Calvaria major*) also evolved on that island, as did tortoises. Did the tree coevolve with dodos, tortoises, or both? Either may have fed on the tree's large fruits. For example, the large dodo gizzard could have partially digested the thick-walled coat of seeds inside fruits, just enough to help the seeds germinate after leaving the gut. Either way, after the dodo became extinct, the seeds stopped germinating. Only thirteen trees were still standing by the mid-1970s, and by some estimates they are more than 300 years old. Each year they still produce seeds, but these apparently cannot break out of the seed coats without help. Today, botanists employ turkeys or gem polishers as substitute grinders.

What causes mass extinctions? The answer may not be obvious. For example, considerable evidence suggests that many lineages never made it past the K–T boundary (Section 21.1). Many believe "the asteroid did it." It did deliver the coup de grace for some lineages, including the dinosaurs and mosasaurs (Figure 28.3*a*). But other factors were at work. Biodiversity had been declining for 10 million years. Pterosaurs were gone before the K–T event. The event had little effect on insects. Was tectonic change a factor? When a land mass destined to become Australia was being rafted away from Antarctica, deep, cold currents from the south were able to move into the warmer, equatorial seas—where many groups were hit. Seawater composition, sea levels, and the climate itself shifted. The consequences affected life on land as well as in the seas. The K–T asteroid impact did indeed end some lineages abruptly. But for others, it may simply have been the final blow in a long streak of bad luck.

Now extend this thought about hidden causes of extinctions to individual species. Long ago, Dutch sailors clubbed to death every last dodo, a flightless bird that lived only on Mauritius. After that, a tree species also native to the island simply stopped reproducing. It was not until the 1970s that a hypothesis emerged: If the tree depended intimately on a coevolved species that became extinct, then the tree would be vulnerable to extinction, also. Was its partner the dodo? Maybe (Figure 28.3*b*).

Biodiversity is greater now than it has ever been in the past.

The current range of global biodiversity is an outcome of an overall pattern of mass extinctions and slow recoveries in the history of life. Within that pattern, lineages differ in which member species persisted and which became extinct.

The loss of individual species, as well as mass extinctions, may have obvious or complicated causes.

28.1 www.brookscole.com/biology

Chapter 28 Biodiversity in Perspective 487

the biosphere; the other on a recent, novel idea about how photosynthesis got started in the first place). All topic spreads of the chapter flow as parts of the same story, but some clearly offer depth that may be treated as optional.

The topic spreads are not gimmicks. Ongoing feedback guided decisions about when to add depth and when to loosen core material with applications. Within the spreads, headings and subheadings help students keep track of the hierarchy of information. Transitions between spreads help them keep the greater story in focus and discourage memorization for its own sake. To avoid disrupting the basic story line while still attending to interested students, we include some enriching details in optional illustrations.

The clear organization helps students find assigned topics easily and has translated into improved test scores. This is a tangible outcome, but we are more pleased that the clarity helps give students enough confidence to dig deeper into biological science. We also are happy to hear that the story has lured many of them into reading far more than they planned to do.

BALANCING CONCEPTS WITH APPLICATIONS Each chapter starts with a lively or sobering application and an adjoining list of key concepts, the chapter's advance organizer. Strategically placed examples of applications parallel core material, not so many as to be distracting but enough to keep students interested in continuing with the basics. Brief applications are integrated in the text. Focus essays afford more depth on many medical, environmental, and social issues without interrupting the conceptual flow. The book's last four pages index all applications separately for fast reference.

FOUNDATIONS FOR CRITICAL THINKING To help students increase their capacity for critical thinking, we walk them through experiments that yielded evidence in favor of or against hypotheses. The main index lists all of the selected experimental tests and observational tests (index entries *Experiment* and *Test, observational*).

We use certain chapter introductions as well as entire chapters to show students some productive results of critical thinking. The introductions to Mendelian genetics (Chapter 11), DNA structure and function (13), speciation (19), immunology (40), and behavior (47) are examples. Also, each chapter has a set of *Critical Thinking* questions. Katherine Denniston developed most of these thought-provoking questions. Daniel Fairbanks developed many of the *Genetics Problems*, which help students grasp the principles of inheritance (Chapters 11 and 12).

VISUAL OVERVIEWS OF CONCEPTS We simultaneously develop text and art as inseparable parts of the same story. We give visual learners a means to work their way through a visual overview of major processes before reading the corresponding (and possibly intimidating) text. Students repeatedly let us know how much they appreciate this art. Overview illustrations have step-by-step descriptions of biological parts and processes. Instead of "wordless" diagrams, we break down information into a series of illustrated callouts. For example, in Figure 14.14, callouts integrated with the art walk students through the stages by which a mature mRNA transcript becomes translated.

Many anatomical drawings are integrated overviews of structure/function. Students need not jump back and forth from text, to tables, to illustrations, and back again to see how an organ system is put together and what its parts do. We hierarchically arrange descriptions of parts to reflect a system's structural and functional organization.

ZOOM SEQUENCES Many illustrations progress from macroscopic to microscopic views of a system or process. Figure 7.3, for example, starts with a plant leaf and ends with reaction sites in the chloroplast. Figures 38.20 and 38.21 start with a ballerina's biceps and move down through levels of skeletal muscle contraction.

COLOR CODES Consistent use of colors for molecules, cell structures, and processes helps students track what is going on. We use these colors throughout the book:

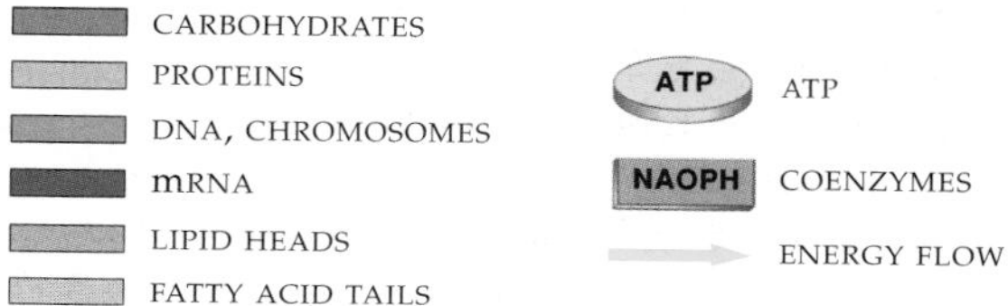
CARBOHYDRATES
PROTEINS
DNA, CHROMOSOMES
mRNA
LIPID HEADS
FATTY ACID TAILS
ATP
ATP
NAOPH
COENZYMES
ENERGY FLOW

ICONS Small, simple diagrams next to an illustration help students relate a topic to the big picture. For instance, a simple diagram of a cell reminds them of the location of the plasma membrane relative to the cytoplasm. Other icons relate reactions and processes to certain locations and to how they tie in with one another. Others remind students of evolutionary relationships among organisms, as in Chapters 26 and 27. A multimedia icon directs them to art in the CD-ROM packaged in its own envelope at the back of their book. Another directs them to supplemental material on the Web and a third, to InfoTrac.

A COMMUNITY EFFORT

Each new edition starts the same way. I go back through banks of past reviews from a network of more than 2,000 teachers and reviewers. The cumulative wisdom about educational pitfalls and promises is just astounding, and it keeps me humble. Rereading pivotal journal articles follows, as do phone calls and e-mail flurries with my special advisors, contributors, and new reviewers for the next two years. Each edition, I again acknowledge those individuals whose contributions continue to shape our collective thinking. There is no way to describe their thoughtful assistance. I can only salute their commitment to quality in education.

This time Daniel Fairbanks, ongoing advisor for the genetics unit, wrote a new chapter on recombinant DNA technology and genetic engineering. Students will benefit from the years of research Dr. Fairbanks conducted while writing a genetics textbook. Stephen Wolfe, well-known author in cell biology, advised on biochemistry, genetics, and evolution. Paul Hertz and Tyler Miller checked the evolution and ecology units; Paul made sketches for new biogeochemical cycles. Lauralee Sherwood rewrote critical passages on animal physiology. Besides continuing with his ongoing support, Tom Garrison graciously allowed us to adapt the Easter Island story from his exceptional oceanography textbook.

Ron Hoham's reviews helped us refine the protistan pages. Eugene Kozloff checked the invertebrate material. Nancy Dengler, Richard Falk, John Jackson, and Thomas Rost carefully reviewed chapters on plants. Elizabeth Moore–Landecker guided us through the fungi. Alan Mann and Mark Weiss helped update the sections on human evolution. Linda Barham and John Jackson refined and compiled answers for end-of-chapter Self-Quizzes. David Goodin created original structural models of molecules.

Bruce Reid, Michael Renfroe, Robin Tyser, and Mattie Roig were especially helpful with detailed challenges. Lisa Starr applied her talents in biochemistry, biomedical research, computer graphics, and fine arts to create the stunning new illustrations for this edition.

Mary Douglas and Gary Head worked with me every step of the way on the book's page-by-page design and production. Mary efficiently keeps a few dozen people, including me, focused and on schedule. Gary also designs our covers and creates such photographic gems as Fred-and-Ginger, the streetwise snail (Figure 19.1). They are the best of the best, and I can't do books without them.

Pat Waldo, force of nature, and her talented colleagues Chris Evers, Amanda Kaufman, Veronica Oliva, and Steve Bolinger keep us at the creative forefront of multimedia.

We have utmost appreciation for the daily important contributions of our teammates Kristin Milotich, Mary Arbogast, Jamie Sue Brooks, Daniel Lombardino, and Rachel Alvelais. Myrna Engler's excellent sleuthing turned up our fine new photographs. Karen Stough and Angela Harris helped make production a breeze. We also acknowledge the valuable support of Tami Cueny, Kelly Fielding, Michelle Kessel, Carole Lawson, Dixie Conner, Kevin Howard, Wes Webber, Verbal Clark, and Rick Cook.

By trusting me to participate fully in the development of this book all the way through production, Tim McEwen, Susan Badger, Geoffrey Burn, Gary Carlson, and Kathie Head contributed to its high quality.

So many years have passed since Jack Carey convinced me to write my first book. He remains close counselor and abiding friend. And nothing, in all that time, has shaken our shared belief in the intrinsic capacity of biology to enrich the lives of each new generation of students.

CECIE STARR
May 2000

General Advisors/Contributors

JOHN ALCOCK *Arizona State University*
AARON BAUER *University of Chicago*
ROBERT COLWELL *University of Connecticut*
JERRY COYNE *University of Chicago*
GEORGE COX *San Diego State University*
KATHERINE DENNISTON *Towson State University*
DANIEL FAIRBANKS *Brigham Young University*
DAVID GOODIN *The Scripps Research Institute*
PAUL HERTZ *Barnard College*
RONALD HOHAM *Colgate University*
JOHN JACKSON *North Hennipin Community College*
EUGENE KOZLOFF *University of Washington*
ROBERT LAPEN *Central Washington University*
WILLIAM PARSON *University of Washington*
CLEON ROSS *Colorado State University*
LAURALEE SHERWOOD *West Virginia University*
STEPHEN WOLFE *University of California, Davis*

Advisors/Contributors to the Ninth Edition

LINDA BARHAM *Meridian Community College*
A. KENT CHRISTIANSEN *University of Michigan*
FRED DELCOMYN *University of Illinois, Urbana*
NANCY DENGLER *University of California, Davis*
CATHY DONALD-WHITNEY *Collin County Community College*
RICHARD FALK *University of California, Davis*
TOM GARRISON *Orange Coast College*
ANN LUMSDEN *Florida State University*
ANNE MCNABB *Virginia Polytechnic Institute and State University*
ALAN MANN *University of Pennsylvania*
G. TYLER MILLER, JR. *Wilmington, North Carolina*
PETER V. MINORKSY *Western Connecticut State University*
ELIZABETH MOORE–LANDECKER *Glassboro State University*
ALLISON MORRISON–SHETLER *Georgia State University*
DAVID MORTON *Frostburg State University*
RICHARD MURPHY *University of Virginia Medical School*
DAVID NORRIS *University of Colorado*
MICHAEL RENFROE *James Madison University*
BRUCE REID *Kean College of New Jersey*
DAVID REZNICK *University of California, Riverside*
MATTIE ROIG *Broward Community College*
THOMAS ROST *University of California, Davis*
JUDY SCHNEIDEWENT *Milwaukee Area Technical College*
ROGER SLOBODA *Dartmouth College*
IAN TIZARD *Texas A&M University*
ROBIN TYSER *University of Wisconsin, LaCrosse*
MARK WEISS *Wayne State University*

Contributors of Influential Reviews

ALDRIDGE, DAVID, *North Carolina Agricultural and Technical State University*
ARMSTRONG, PETER, *University of California at Davis*
BAJER, ANDREW, *University of Oregon*
BAKKEN, AIMEE, *University of Washington*
BARBOUR, MICHAEL, *University of California, Davis*
BARKWORTH, MARY, *Utah State University*
BELL, ROBERT A., *University of Wisconsin, Stevens Point*
BINKLEY, DAN, *Colorado State University*
BLEEKMAN, GEORGE, *American River College*
BRENGELMANN, GEORGE, *University of Washington*
BRINSON, MARK, *East Carolina University*
BROWN, ARTHUR, *University of Arkansas*
BUCKNER, VIRGINIA, *Johnson County Community College*
CALVIN, CLYDE, *Portland State University*
CASE, CHRISTINE, *Skyline College*
CASE, TED, *University of California, San Diego*
COLAVITO, MARY, *Santa Monica College*
CONKEY, JIM, *Truckee Meadows Community College*
CROWCROFT, PETER, *University of Texas at Austin*
DANIELS, JUDY, *Washtenau Community College*
DAVIS, JERRY, *University of Wisconsin, La Crosse*
DEMMANS, DANA, *Finger Lakes Community College*
DEMPSEY, JEROME, *University of Wisconsin*
DESAIX, JEAN, *University of North Carolina*
DETHIER, MEGAN, *University of Washington*
DEWALT, R. EDWARD, *Louisiana State University*
DIBARTOLOMEIS, SUSAN, *Millersville University of Pennsylvania*
DIEHL, FRED, *University of Virginia*
DLUZEN, DEAN, *Northeastern Ohio Universities College of Medicine*
DOYLE, PATRICK, *Middle Tennessee State University*
DUKE, STANLEY H., *University of Wisconsin, Madison*
DYER, BETSEY, *Wheaton College*
EDLIN, GORDON, *University of Hawaii, Manoa*
EDWARDS, JOAN, *Williams College*
ELMORE, HAROLD W., *Marshall University*
ENDLER, JOHN, *University of California, Santa Barbara*
ENGLISH, DARREL, *Northern Arizona University*
ERWIN, CINDY, *City College of San Francisco*
EWALD, PAUL, *Amherst College*
FISHER, DAVID, *University of Hawaii, Manoa*
FISHER, DONALD, *Washington State University*
FLESSA, KARL, *University of Arizona*
FONDACARO, JOSEPH, *Hoechst Marion Roussel, Inc.*
FRAILEY, CARL, *Johnson County Community College*
FRISBIE, MALCOLM, *Eastern Kentucky University*
FROEHLICH, JEFFREY, *University of New Mexico*
FULCHER, THERESA, *Pellissippi State Technical Community College*
GAGLIARDI, GRACE S., *Bucks County Community College*
GENUTH, SAUL M., *Mt. Sinai Medical Center*
GHOLZ, HENRY, *University of Florida*
GIBSON, THOMAS, *San Diego State University*
GOODMAN, H. MAURICE, *University of Massachusetts Medical School*
GOSZ, JAMES, *University of New Mexico*
GREGG, KATHERINE, *West Virginia Wesleyan College*
HARRIS, JAMES, *Utah Valley Community College*
HARTNEY, KRISTINE BEHRENTS, *California State University, Fullerton*
HASSAN, ASLAM, *Univeristy of Illinois College of Veterinary Medicine*
HELGESON, JEAN, *Collin County Community College*
HESS, WILFORD M., *Brigham Young University*
HUFFMAN, DAVID, *Southwest Texas State University*
INGRAHAM, JOHN L., *University of California, Davis*
JENSEN, STEVEN, *Southwest Missouri State University*
JOHNSON, LEONARD R., *University of Tennessee College of Medicine*
JUILLERAT, FLORENCE, *Indiana University, Purdue University*
KAREIVA, PETER, *University of Washington*
KAUFMAN, JUDY, *Monroe Community College*
KAYE, GORDON I., *Albany Medical College*
KAYNE, MARLENE, *Trenton State College*
KENDRICK, BRYCE, *University of Waterloo*
KEYES, JACK L., *Linfield College, Portland Campus*
KILLIAN, JOELLA C., *Mary Washington College*
KIRKPATRICK, LEE A., *Glendale Community College*
KREBS, CHARLES, *University of British Columbia*
KREBS, JULIA E., *Francis Marion University*
KUTCHAI, HOWARD, *University of Virginia Medical School*
LANZA, JANET, *University of Arkansas, Little Rock*
LASSITER, WILLIAM, *University of North Carolina*
LEVY, MATTHEW, *Mt. Sinai Medical Center*
LEWIS, LARRY, *Bradford University*
LITTLE, ROBERT, *Medical College of Georgia*
LOHMEIER, LYNNE, *Mississippi Gulf Coast Community College*
LOPO, ALINA C., *University of California-Los Angeles Medical Center*
MACKLIN, MONICA, *Northeastern State University*
MARTIN, JAMES, *Reynolds Community College*
MARTIN, TERRY, *Kishwaukee College*
MATSON, RONALD, *Kennesaw State College*
MATTHEWS, ROBERT, *University of Georgia*
MAXWELL, JOYCE, *California State University, Northridge*
MCCLURE, JERRY, *Miami University*
MCEDWARD, LARRY, *University of Florida*
MCKEAN, HEATHER, *Eastern Washington State University*
MCKEE, DOROTHY, *Auburn University, Montgomery*
MCNABB, ANNE, *Virginia Polytechnic Institute & State University*
MILLER, TYLER, *St. Andrews Presbyterian College*
MOISES, HYLAN C., *University of Michigan Medical School*
MYRES, BRIAN, *Cypress College*
NAGARKATTI, PRAKASH, *Virginia Polytechnic Institute & State University*
NELSON, RILEY, *University of Texas at Austin*
PEARCE, FRANK, *West Valley College*
PECHENIK, JAN, *Tufts University*
PECK, JAMES H., *University of Arkansas, Little Rock*
PERRY, JAMES, *University of Wisconsin, Center-Fox Valley*
PETERSON, GARY, *South Dakota State University*
PIPERBERG, JOEL, *Millersville University*
PLETT, HAROLD, *Fullerton College*
REEVE, MARIAN, *Emeritus, Merritt Community College*
RICKETT, JOHN, *University of Arkansas, Little Rock*
ROBBINS, ROBERT, *National Science Foundation*
ROSE, GREIG, *West Valley College*
RUIBAL, RODOLFO, *University of California, Riverside*
SALISBURY, FRANK, *Utah State University*
SCHAPIRO, HARRIET, *San Diego State University*
SCHLESINGER, WILLIAM, *Duke University*
SCHNERMANN, JURGEN, *University of Michigan School of Medicine*
SHEGAL, PREM, *East Carolina University*
SHONTZ, NANCY, *Grand Valley State University*
SELLERS, LARRY, *Louisiana Tech University*
SHOPPER, MARILYN, *Johnson County Community College*
SMITH, JERRY, *St. Petersburg Junior College, Clearwater Campus*
SMITH, MICHAEL E., *Valdosta State College*
SMITH, ROBERT L., *West Virginia University*
SOLOMON, NANCY, *Miami University*
STEARNS, DONALD, *Rutgers University*
STEELE, KELLY P., *Appalachian State University*
STEINERT, KATHLEEN, *Bellevue Community College*
SUMMERS, GERALD, *University of Missouri*
SUNDBERG, MARSHALL D., *Louisiana State University*
SWANSON, ROBERT, *North Hennepin Community College*
SWEET, SAMUEL, *University of California, Santa Barbara*
TAYLOR, JANE, *Northern Virginia Community College*
TERHUNE, JERRY, *Jefferson Community College, University of Kentucky*
TROUT, RICHARD E., *Oklahoma City Community College*
WAALAND, ROBERT, *University of Washington*
WAHLERT, JOHN, *City University of New York, Baruch College*
WALSH, BRUCE, *University of Arizona*
WARING, RICHARD, *Oregon State University*
WARNER, MARGARET R., *Purdue University*
WEBB, JACQUELINE F., *Villanova University*
WEIGL, ANN, *Winston-Salem State University*
WELKIE, GEORGE W., *Utah State University*
WENDEROTH, MARY PAT, *University of Washington*
WHITE, EVELYN, *Alabama State University*
WHITENBERG, DAVID, *Southwest Texas State University*
WINICUR, SANDRA, *Indiana University, South Bend*
YONENAKA, SHANNA, *San Francisco State University*

Current configurations of Earth's oceans and land masses—the geologic stage upon which life's drama continues to unfold. Thousands of separate images were pieced together to create this remarkable, true-color image of our planet.

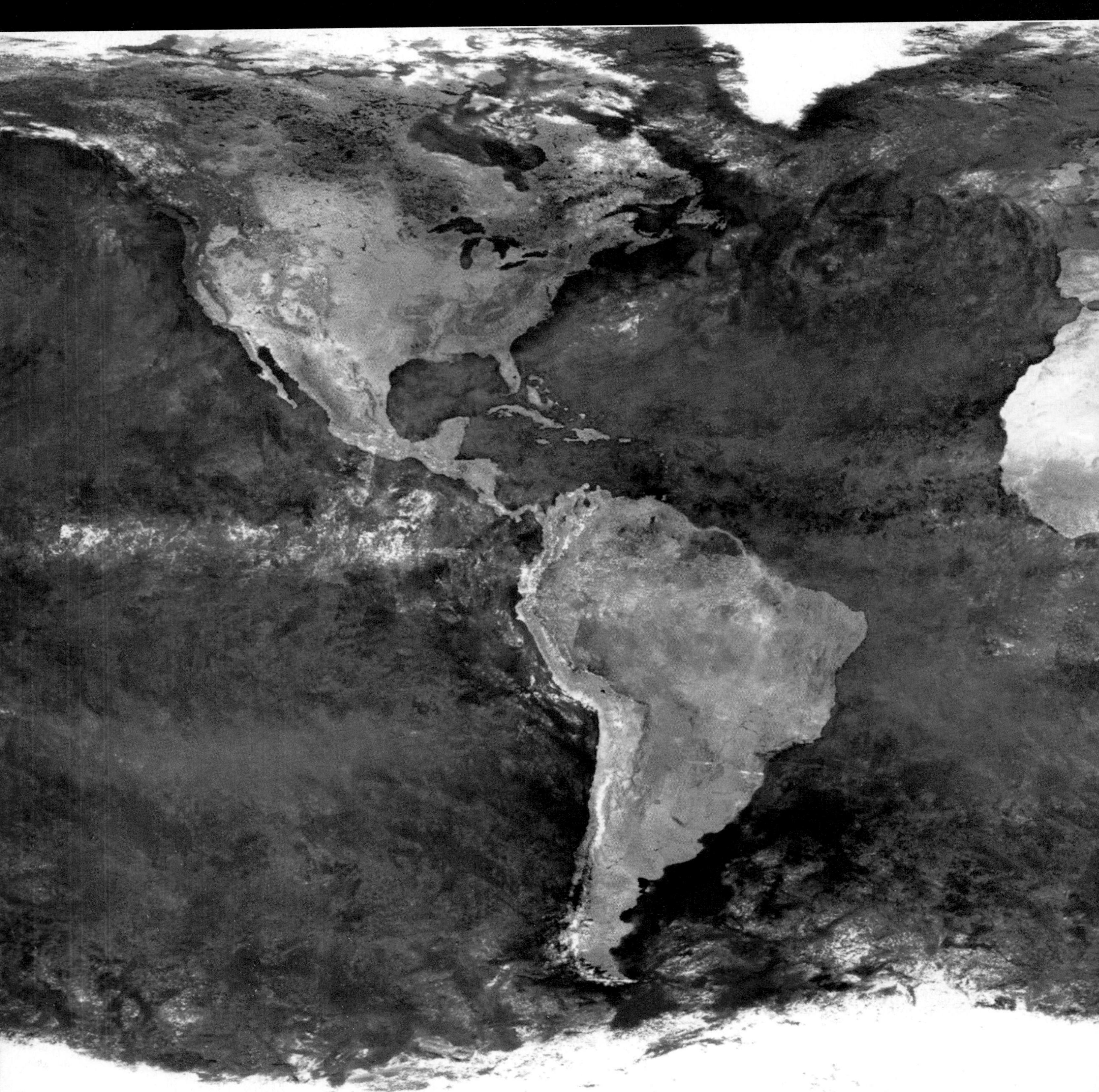

INTRODUCTION

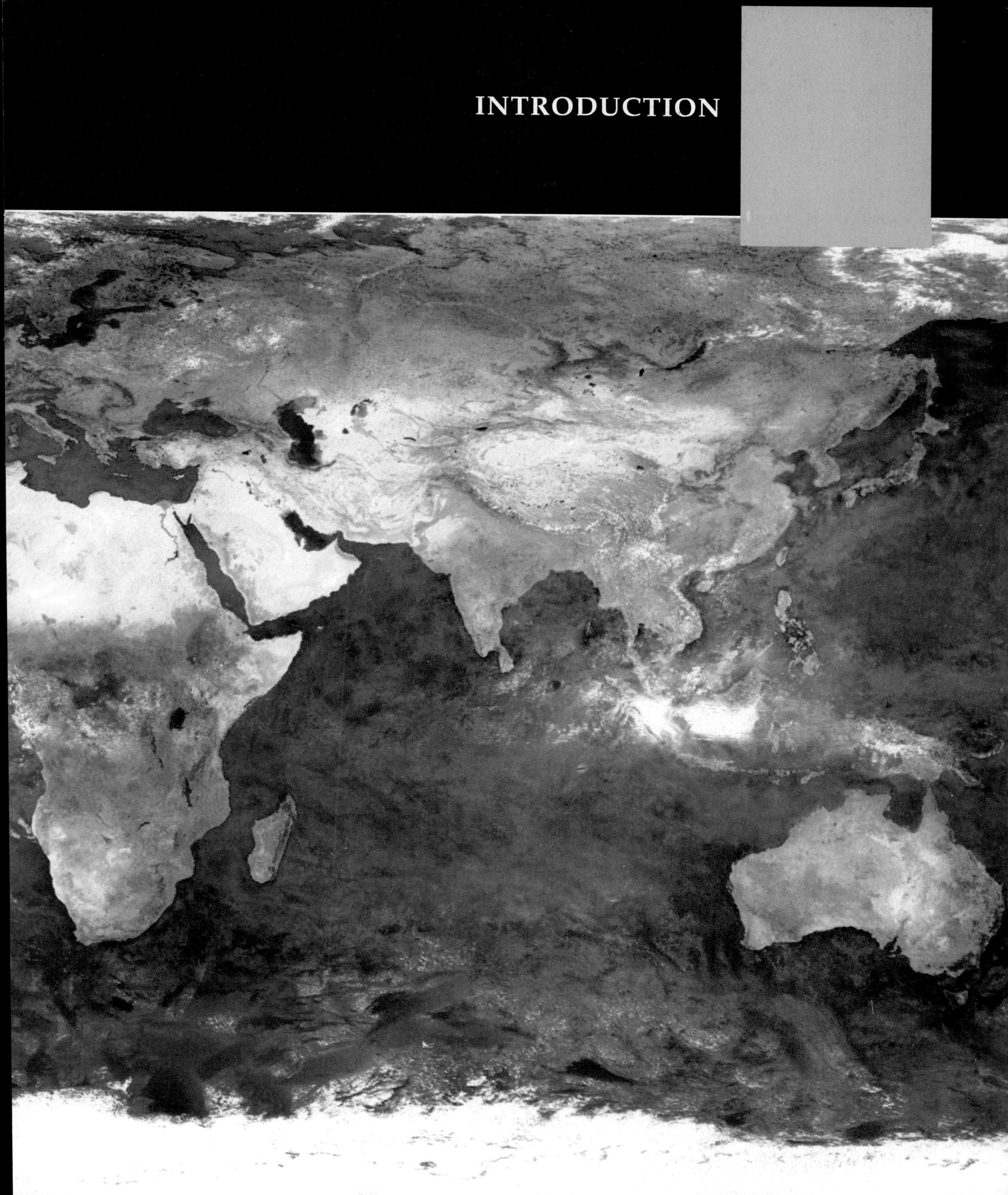

1 CONCEPTS AND METHODS IN BIOLOGY

Biology Revisited

Buried somewhere in that mass of tissue just above and behind your eyes are memories of first encounters with the living world. In that brain are early memories of discovering your hands and feet, your family, friends, the change of seasons, the scent of rain-drenched earth and grass. Still in residence are memories of your early introductions to a great disorganized parade of plants and animals, mostly living, sometimes dead. There, too, are memories of questions—*"What is life?"* and, inevitably, *"What is death?"* There also are memories of answers, some satisfying, others less so.

By making observations, asking questions, and accumulating answers, you have gradually built up a store of knowledge about life. Experience and education have been refining the questions, and no doubt some answers are difficult to come by.

Think of the world's forests—once vast, now astonishingly diminished by logging, conversion for agriculture, urban expansion, and other activities that help keep many people sheltered, warm, fed, and alive. In 1999, the human population surpassed 6 billion; it is still growing. With few forests remaining, where will we find more usable land and forest products?

Think of a college student, twenty years old, whose motorcycle skidded into a truck. Now he is comatose, and doctors say his brain is functionally dead. His breathing, heart rate, and other basic functions will continue only as long as he stays hooked up to

Figure 1.1 Think back on all you have ever known and seen. This is a foundation for your deeper probes into life.

a respirator and other artificial support systems. Would you say the unfortunate student is still "alive"?

Or think of a human egg, recently penetrated by a sperm inside a woman's body. At first the fertilized egg does not get bigger, but a series of programmed cuts divides it into a cluster of a few dozen tiny cells. Would you call the microscopically small mass a *human* life? If you learn more about how embryos actually develop, will the knowledge influence your thoughts about such incendiary issues as birth control and abortion?

If questions like these have ever crossed your mind, your thoughts about life obviously run deep. And you can approach this course in **biology**—the scientific study of life—with confidence, because you have been studying life ever since information started penetrating your brain. You simply are *revisiting* biology, in ways that might help carry your thoughts to deeper, more organized levels of understanding.

Return to the question, *What is life?* Offhandedly, you might respond that you know it when you see it. However, as you will see later in the book, the question opens up a story that has been unfolding in countless directions for about 3.8 billion years!

From the biological perspective, "life" is an outcome of ancient events by which nonliving matter—atoms and molecules—became assembled into the first living cells. "Life" is a way of capturing and using energy and raw materials. "Life" is a way of sensing and responding to changes in the environment. "Life" is a capacity to reproduce, grow, and develop. And "life" evolves, meaning that the traits characterizing the individuals of a population can change over the generations. Even so, this short list only hints at the meaning of life. Deeper insight requires wide-ranging study of life's characteristics.

Throughout this book, you will come across many diverse examples of how organisms are constructed, how they function, where they live, and what they do. The examples support certain concepts which, when taken together, will give you a sense of what "life" is.

This chapter introduces the basic concepts. It also sets the stage for forthcoming descriptions of scientific observations, experiments, and tests that help show how you can develop, modify, and refine your views of life. As you continue with your reading, you may find it useful to return occasionally to this simple overview as a way to reinforce your grasp of the details.

KEY CONCEPTS

1. Unity underlies the world of life, for all organisms are alike in key respects. They consist of one or more cells made of the same kinds of substances, put together in the same basic ways. Their activities require inputs of energy, which they must get from their surroundings. All organisms sense and respond to changing conditions in their environment. They all have a capacity to grow and reproduce, based on instructions contained in DNA.

2. The world of life shows immense diversity. Many millions of different kinds of organisms, or species, now inhabit the Earth, and many millions more lived in the past. Each one of those species is unique in some of its traits—that is, in some aspects of its body plan, body functioning, and behavior.

3. Theories of evolution, especially a theory of evolution by natural selection as formulated by Charles Darwin, help explain the meaning of life's diversity.

4. Biology, like other branches of science, is based on systematic observations, hypotheses, predictions, and observational and experimental tests. The external world, not internal conviction, is the testing ground for scientific theories.

1.1 DNA, ENERGY, AND LIFE

Nothing Lives Without DNA

DNA AND THE MOLECULES OF LIFE Picture a frog on a rock, busily croaking. Without even thinking about it, you know the frog is alive and the rock is not. Would you be able to explain why? At a fundamental level, both are no more than concentrations of the same units of matter, called protons, electrons, and neutrons. The units are building blocks of atoms, which are building blocks of larger bits of matter called molecules. And it is at the molecular level that differences between living and nonliving things start to emerge.

You will never, ever find a rock made of nucleic acids, proteins, carbohydrates, and lipids. In nature, only cells build these molecules, which you will read about later. All living things consist of one or more of cells, which are the smallest units of matter having a capacity for life. The signature molecule of cells is a nucleic acid known as **DNA**. No chunk of granite or quartz has it.

Encoded in DNA's structure are the instructions for assembling a dazzling array of proteins from a limited number of smaller building blocks, the amino acids. By analogy, if you follow suitable instructions and invest energy in the task, you might organize a heap of a few kinds of ceramic tiles (representing amino acids) into diverse patterns (representing proteins), as in Figure 1.2.

Among the proteins are enzymes. When these worker molecules get an energy boost, they swiftly build, split, and rearrange the molecules of life. Some enzymes work with a class of nucleic acids called RNAs in carrying out DNA's protein-building instructions. Think of this as a flow of information, from *DNA to RNA to protein*. As you will see from Chapter 14, this molecular trinity is central to our understanding of life.

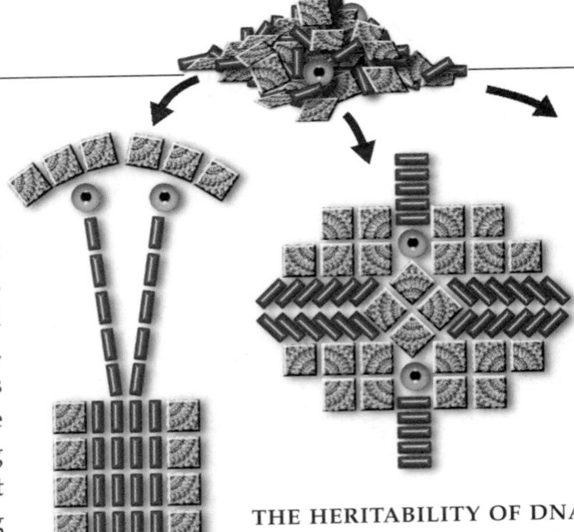

Figure 1.2 Examples of objects built from the same materials but with different assembly instructions.

THE HERITABILITY OF DNA We humans tend to think we enter the world abruptly and leave it the same way. But we are much more than this. *We and all other organisms are part of a journey that began about 3.8 billion years ago, starting with the origin of the first living cells.*

Under present-day conditions on Earth, new cells arise only from cells that already exist. They do so by **inheritance**, which is the acquisition of traits by way of transmission of DNA from parent to offspring. Why do baby storks look like storks and not pelicans? Because they inherited stork DNA, which isn't exactly the same as pelican DNA in its molecular details.

Reproduction refers to actual mechanisms by which a cell or an organism produces offspring. Often it starts when a sperm fertilizes an egg. But the fertilized egg never could form if the sperm and egg had not formed earlier, according to DNA instructions passed on from cell to cell through countless generations according to the principles of inheritance and reproduction.

For frogs and humans and other large organisms, DNA also guides **development**, the transformation of a fertilized egg into a multicelled adult with cells, tissues, and organs specialized for certain tasks. Development proceeds through a series of stages. As one example, a moth is only the adult stage of a winged insect (Figure 1.3). First, a fertilized egg develops into an immature larval stage—a caterpillar that eats leaves and grows

Figure 1.3 "The insect"—a series of stages of development. Different adaptive properties emerge at each stage. Shown here, a silkworm moth, from the egg (**a**), to a larval stage (**b**), to a pupal stage (**c**), and on to the winged form of the adult (**d**,**e**).

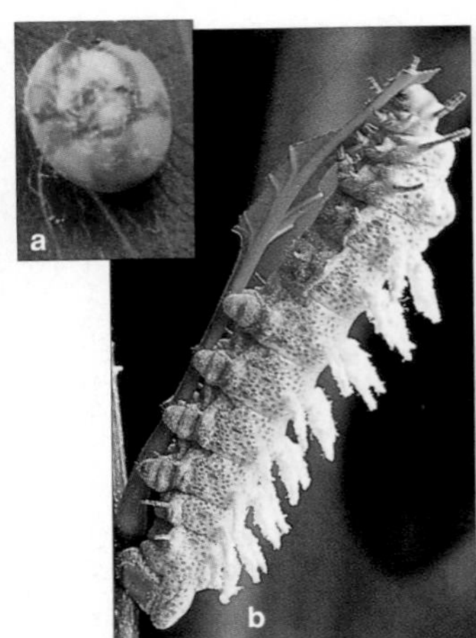
a
b
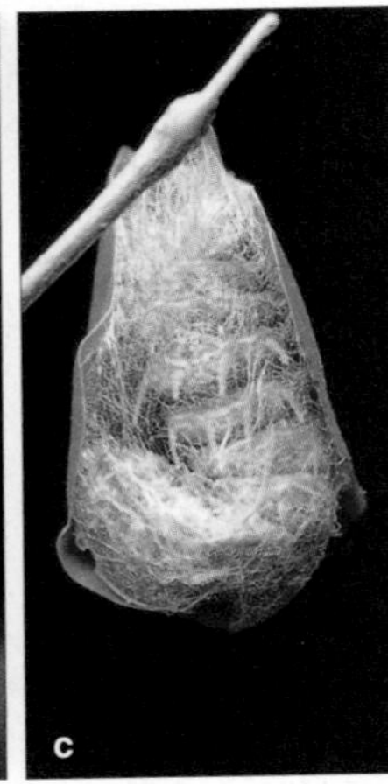
c
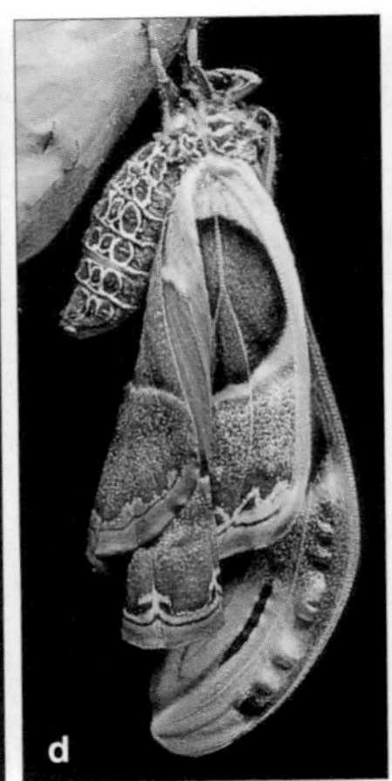
d
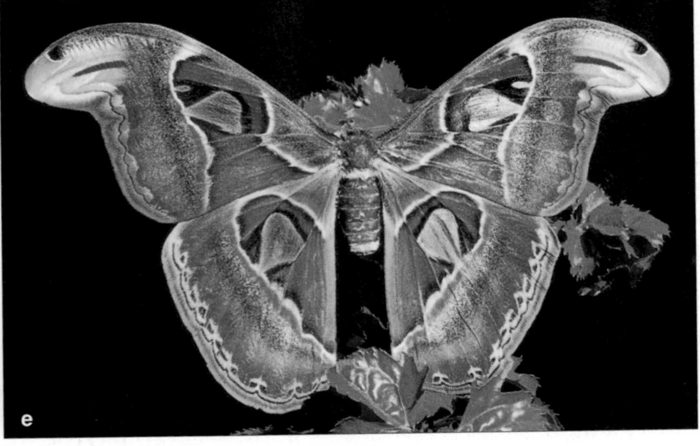
e

Further reading: Student Guide to InfoTrac on web site →

Figure 1.4 Response to signals from pain receptors, activated by a lion cub flirting with disaster.

rapidly until an internal alarm clock goes off. Then its tissues are remodeled into a different stage—a pupa. In time, an adult emerges that is adapted to reproduce. It produces sperm or eggs. Its wing color, patterns, and fluttering frequency are adapted for attracting a mate.

And so "the insect" is a series of organized stages. Its development from egg to adult will not proceed properly unless each stage is completed before the next begins. Instructions for each stage were written into moth DNA long before each moment of reproduction—and so the ancient moth story continues.

Nothing Lives Without Energy

ENERGY DEFINED Everything in the entire universe has some amount of **energy**, which is most simply defined as the capacity to do work. And nothing—absolutely nothing—happens in the universe without a *transfer* of energy. For instance, a single, undisturbed atom can do nothing except vibrate incessantly with its own energy. Suppose it absorbs extra energy from the sun and starts vibrating faster. Now some energy is on the move. That energy can do work by getting transferred elsewhere. If by chance the atom collides with a neighboring atom, one may give up, grab, or share energy with the other. Molecules form, become rearranged, and are split apart by such energy transfers. When this kind of molecular work is done, cells stay alive, grow, and reproduce.

METABOLISM DEFINED Each living cell has the capacity to (1) obtain and convert energy from its surroundings and (2) use energy to maintain itself, grow, and make more cells. We call this capacity **metabolism**. Think of a cell in a leaf that produces food by photosynthesis. It intercepts sunlight energy and converts it to chemical energy, in the form of ATP molecules. ATP is an energy carrier that helps drive hundreds of activities. It easily transfers some energy to metabolic workers—in this case, enzymes that assemble sugar molecules. ATP also forms by aerobic respiration. This process can release energy that cells have tucked away in sugars and other kinds of molecules.

SENSING AND RESPONDING TO ENERGY It is often said that only organisms respond to the environment. Yet even a rock shows responsiveness, as when it yields to the force of gravity and tumbles down a hill or changes its shape slowly under the repeated battering of wind, rain, or tides. The difference is this: *Organisms sense changes in their surroundings, then they make controlled, compensatory responses to them.* How? Every organism has **receptors**, which are molecules and structures that detect stimuli. A **stimulus** is a specific form of energy detected by receptors. Examples are sunlight energy, heat energy, a hormone molecule's chemical energy, and the mechanical energy of a bite (Figure 1.4).

Cells adjust metabolic activities in response to signals from receptors. Each cell (and organism) can withstand only so much heat or cold. It must rid itself of harmful substances. It requires certain foods, in certain amounts. Yet temperatures do shift, harmful substances might be encountered, and food is sometimes plentiful or scarce.

For example, after you finish a snack, simple sugars leave your gut and enter your blood. Blood is part of your *internal* environment (the other part is tissue fluid that bathes your cells). Over the long term, too much or too little sugar in the blood can cause problems, such as diabetes. When the sugar level rises, a glandular organ, the pancreas, normally steps up its secretion of insulin. Most of your cells have receptors for this hormone, which stimulates cells to take up sugar. When enough cells do so, the sugar level in blood returns to normal.

Organisms respond so exquisitely to energy changes that their internal operating conditions remain within tolerable limits. We call this a state of **homeostasis**. It is one of the key defining features of life.

All organisms consist of one or more cells, the smallest units of life. Under present-day conditions, new cells form only through the reproduction of cells that already exist.

DNA, the molecule of inheritance, encodes protein-building instructions, which RNAs help carry out. Many proteins are enzymes, the metabolic workers necessary to construct DNA and all other complex molecules of life.

Cells live only for as long as they engage in metabolism. They acquire and transfer energy that is used to assemble, break down, stockpile, and dispose of materials in ways that promote survival and reproduction.

Single cells and multicelled organisms sense and respond to environmental conditions in ways that help maintain their internal operating conditions.

1.2 ENERGY AND LIFE'S ORGANIZATION

Levels of Biological Organization

Taken as a whole, the metabolic activities of single cells and multicelled organisms maintain the great pattern of organization in nature, as sketched out in Figure 1.5. Consider the hierarchy. Life's properties emerge when DNA and other molecules become organized into cells. The **cell** is the smallest unit of organization having a capacity to survive and reproduce on its own, given DNA instructions, suitable conditions, building blocks, and energy inputs. Free-living, single cells such as an amoeba fit the definition. Does the definition of cells hold for **multicelled organisms**, which typically consist of specialized, interdependent cells organized as tissues and organs? Yes. You might find this a strange answer. After all, your own cells could never live alone in nature, because body fluids must continually bathe them. Yet even isolated human cells stay alive under controlled conditions in laboratories around the world. Investigators routinely maintain isolated human cells for use in important experiments, as in cancer studies.

ecosystem (community together with its physical environment)
populations of shrubs and trees
community (all of the populations living in the same area)
population of zebras
populations of grasses
former multicelled individual, about to revert to molecules and atoms
multicelled individual

BIOSPHERE
All those regions of Earth's waters, crust, and atmosphere in which organisms can exist

ECOSYSTEM
A community and its physical environment

COMMUNITY
The populations of all species occupying the same area

POPULATION
A group of individuals of the same kind (that is, the same species) occupying a given area

MULTICELLED ORGANISM
An individual composed of specialized, interdependent cells most often organized in tissues, organs, and organ systems

ORGAN SYSTEM
Two or more organs interacting chemically, physically, or both in ways that contribute to survival of the whole organism

ORGAN
A structural unit in which a number of tissues, combined in specific amounts and patterns, perform a common task

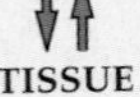

TISSUE
An organized group of cells and surrounding substances functioning together in a specialized activity

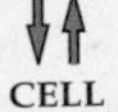

CELL
Smallest unit having the capacity to live and reproduce, independently or as part of a multicelled organism

ORGANELLE
Inside all cells except bacteria, a membrane-bound sac or compartment for a separate, specialized task

MOLECULE
A unit in which two or more atoms of the same element or different ones are bonded together

ATOM
Smallest unit of an element (a fundamental substance) that still retains the properties of that element

SUBATOMIC PARTICLE
An electron, proton, or neutron; one of the three major particles of which atoms are composed

Figure 1.5 Levels of organization in nature.

You typically find cells and multicelled organisms as part of a **population**, defined as a group of organisms of the same kind (such as a herd of zebras). The next level of organization is the **community**—all the populations of all species living in the same area (such as the African savanna's bacteria, grasses, trees, zebras, lions, and so on). The next level, the **ecosystem**, is the community *and*

Further reading: Student Guide to InfoTrac on web site

Producers trap, convert, and use or store some energy from the sun.
PRODUCERS
NUTRIENT CYCLING
CONSUMERS, DECOMPOSERS
ONE-WAY FLOW OF ENERGY
Energy is transferred from one organism to another; in time, all flows back to the environment.
a
b
beetle larva
c

Figure 1.6 Example of the one-way flow of energy and the cycling of materials through the biosphere.

(**a**) Plants of a warm, dry grassland called the African savanna capture energy from the sun and use it to build plant parts. Some of the energy ends up inside plant-eating organisms, including this adult male elephant. He eats huge quantities of plants to maintain his eight-ton self and produces great piles of solid wastes—dung—that still contain some unused nutrients. Although most organisms might not recognize it as such, elephant dung is an exploitable food source.

(**b**) And so we next have little dung beetles rushing to the scene almost simultaneously with the uplifting of an elephant tail. Working rapidly, they carve fragments of moist dung into round balls, which they roll off and bury in burrows. In those balls the beetles lay eggs—a reproductive behavior that will help assure their forthcoming offspring (**c**) of a compact food supply.

Thanks to beetles, dung does not pile up and dry out into rock-hard mounds in the intense heat of the day. Instead, the surface of the land is tidied up, the beetle offspring get fed, and the leftover dung accumulates in beetle burrows—there to enrich the soil that nourishes the plants that sustain (among others) the elephants.

its physical and chemical environment. The **biosphere** includes all regions of the Earth's atmosphere, waters, and crust in which organisms live. Astoundingly, *this globe-spanning organization begins with the convergence of energy, certain materials, and DNA in tiny, individual cells.*

Interdependencies Among Organisms

A great flow of energy into the world of life starts with **producers**, which are plants and other organisms that make their own food. Animals are **consumers**. Directly or indirectly, they depend on energy that became stored in the tissues of producers. For example, some energy is transferred to zebras after they browse on plants. It gets transferred again when lions devour a baby zebra that wandered away from its herd. And it is transferred again when fungal and bacterial decomposers feed on the tissues and remains of lions, elephants, or any other organism. **Decomposers** break down sugars and other biological molecules to simpler materials—which may be cycled back to producers. In time, all of the energy that the plants captured from the sun's rays returns to the environment, but that's another story.

For now, keep in mind that organisms connect with one another by a one-way flow of energy *through* them and a cycling of materials *among* them, as in Figure 1.6. Their interconnectedness affects the structure, size, and composition of populations and communities. It affects ecosystems, even the biosphere. Understand the extent of their interactions and you will gain insight into the environmental effects of acid rain, amplification of the greenhouse effect, and other modern-day problems.

Nature shows levels of organization. The characteristics of life emerge at the level of single cells and extend through populations, communities, ecosystems, and the biosphere.

A one-way flow of energy through organisms and a cycling of materials among them organizes life in the biosphere. In nearly all cases, energy flow starts with energy from the sun.

1.3 IF SO MUCH UNITY, WHY SO MANY SPECIES?

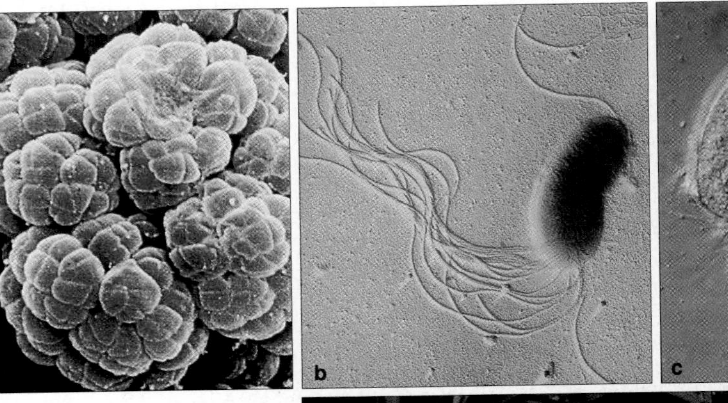

a
b
c

So far, we have focused on life's unity, on the characteristics that all living things have in common. Think of it! They are put together from the same "lifeless" materials. They remain alive by metabolism —by ongoing energy transfers at the cellular level. They interact in their requirements for energy and for raw materials. They have the capacity to sense and respond to the environment in highly specific ways. They all have a capacity to reproduce, based on instructions encoded in DNA. And they all inherited their molecules of DNA from individuals of a preceding generation.

Superimposed on the common heritage is immense diversity. You share the planet with many millions of different kinds of organisms, or **species**. Many millions more preceded you during the past 3.8 billion years, but their lineages vanished; they are extinct. Centuries ago, scholars tried to make sense of the confounding diversity. One of them, Carolus Linneaus, came up with a classification scheme that assigns a two-part name to each newly identified species. The first part designates the **genus** (plural, genera). Each genus encompasses all species that seem closely related by way of their recent descent from a common ancestor. The second part of the name designates a particular species within that genus.

For example, *Quercus alba* is the scientific name for the white oak, and *Q. rubra* is the name for the red oak. As this example suggests, once you spell out the name of the genus in a document, you may abbreviate the name wherever else it appears in that document.

Biologists also classify life's diversity by assigning species to groups at more encompassing levels. Among other things, they group genera that apparently share a common ancestor into the same *family*, related families into the same *order*, related orders into the same *class*, and related classes into the same *phylum* (plural, phyla). At a higher level, they assign related phyla to the same *kingdom*. They are still refining the groups. For instance, scholars once recognized only two kingdoms (animals and plants). Later, most biologists came to accept five kingdoms. Today, compelling evidence suggests there should be at least six: the **Archaebacteria**, **Eubacteria**, **Protista**, **Fungi**, **Plantae**, and **Animalia** (Figure 1.7). We use a six-kingdom scheme for this book. At this point in your reading, it is enough simply to become familiar with a few of the defining features of their members.

All of the world's bacteria (singular, bacterium) are single cells, of a type known as *prokaryotic*. The word means they don't have a nucleus, a membrane-bound sac that otherwise might keep their DNA separate from the

d

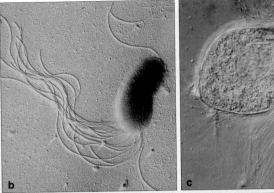

Figure 1.7 A few representatives of life's diversity.

KINGDOM ARCHAEBACTERIA. (**a**) From the muck of an anaerobic (oxygen-free) habitat, a colony of cells (*Methanosarcina*).

KINGDOM EUBACTERIA. (**b**) A eubacterium sporting a number of bacterial flagella. Flagella are used for motility.

KINGDOM PROTISTA. (**c**) A trichomonad that lives in a termite's gut. Compared to bacteria, most protistans are much larger and have greater internal complexity. They range from microscopically small single cells, such as this one, to giant seaweeds.

KINGDOM FUNGI. (**d**) A stinkhorn fungus. Some species of fungi are parasites and some cause diseases, but the vast majority are decomposers. Without decomposers, communities would gradually become buried in their own wastes.

KINGDOM PLANTAE. (**e**) Trunk of a redwood growing near the coast of California. Like nearly all plants, redwoods produce their own food by photosynthesis. (**f**) Flower of a plant from the family of composites. Its colors and patterning guide bees to nectar. The bees get food. The plant gets help reproducing. Like many other organisms, they interact in a mutually beneficial way.

KINGDOM ANIMALIA. (**g**,**h**) Male bighorn sheep competing for females and so displaying a characteristic of the kingdom— they actively move about in their environment.

Further reading: Student Guide to InfoTrac on web site →

rest of the cell interior. Different bacterial species are producers, consumers, and decomposers. Of all kingdoms, theirs shows the greatest metabolic diversity.

We find diverse archaebacteria in extreme environments, much like the forms that are thought to have prevailed soon after life originated. Eubacteria are much more successful in their world distribution. Different kinds live in about every place you might imagine. Eubacteria inside your gut and on your skin outnumber the trillions of cells making up your body.

Generalizing about protistans is not easy. Like plants, fungi, and animals, they are *eukaryotic*, meaning their DNA is located inside a nucleus. Most types are larger and show far more internal complexity than bacteria. A spectacular variety of microscopically small, single-celled producers and consumers are protistans. So are multicelled brown algae and other "seaweeds," some of which are giants of the underwater world.

Most fungi, including the common field mushrooms sold in grocery stores, are multicelled. These eukaryotic decomposers and consumers feed in a distinctive way. They secrete enzymes that digest food outside of the fungal body, then their cells absorb the digested bits.

Both plants and animals are multicelled, eukaryotic organisms. Nearly all plants are photosynthetic. Along with other producers, they make up the all-important food base for communities, especially on land. Animals are a type of consumer that ingests other organisms or their tissues. They include herbivores (which graze on photosynthesizers), carnivores (meat eaters), parasites, and scavengers. Unlike plants, animals actively move about during at least some stage of their life.

Pulling this all together, you start to get a sense of what it means when someone says life shows unity *and* diversity.

Unity threads through the world of life, for all organisms are alike in important ways. They are composed of the same substances, which are assembled in the same basic ways. They engage in metabolism, and they sense and respond to their environment. They all have a capacity to reproduce, based on heritable instructions encoded in their DNA.

Immense diversity threads through the world of life, for organisms differ enormously in body form, in the functions of their body parts, and in their behavior.

To make the study of life's diversity more manageable, we group organisms into six kingdoms—the archaebacteria, eubacteria, protistans, fungi, plants, and animals.

AN EVOLUTIONARY VIEW OF DIVERSITY

Given that organisms are so much alike, what could account for their great diversity? One key explanation is called evolution by means of natural selection. A few simple examples will be enough to introduce you to its premises, which build on the simple observation that the individuals of a population vary in their traits.

Mutation—Original Source of Variation

DNA has two striking qualities. Its instructions work to ensure that offspring will resemble their parents, yet they also permit variations in the details of most traits. As an example, having five fingers on each hand is a human trait. Yet some humans are born with six fingers on each hand instead of five. This is an outcome of a **mutation**, a molecular change in the DNA. Mutations are the original source of variations in heritable traits.

Many mutations are harmful. A change in even a bit of DNA may be enough to sabotage the body's growth, development, or functioning. One such mutation causes *hemophilia A*. If a person affected by this blood-clotting disorder gets even a small cut or bruise, an abnormally lengthy time passes before a clot forms and stops the bleeding. Yet some variations are harmless or beneficial. A classic case is a mutation in light-colored moths that results in dark offspring. Moths fly at night and rest in the day, when birds that eat them are active. Birds tend to miss light-colored moths that rest on light tree trunks. Those moths are camouflaged; they are "hiding in the open," as shown in Figure 1.8.

Suppose that people build coal-burning factories nearby. In time, smoke laden with soot darkens the tree trunks. Now the dark moths are much less conspicuous to bird predators, so they have a better chance to survive and reproduce. Where soot prevails, the variant (darker) form of the trait is more adaptive than the common form. An **adaptive trait** is any form of a trait that gives the individual an advantage, in terms of surviving and reproducing, under a given set of environmental conditions.

a
b

Figure 1.8 Example of different forms of the same trait (body surface coloration), adaptive to two different environmental conditions. (**a**) On a light-colored tree trunk, light-colored moths (*Biston betularia*) are hidden from predators, but dark ones stand out. (**b**) The dark color is more adaptive in places where tree trunks are darkened with soot.

Figure 1.9 *Facing page*: Some of the 300+ varieties of domesticated pigeons, an outcome of artificial selection practices. Breeders began with variant forms of traits in captive populations of wild rock doves.

WILD
ROCK
DOVE

Evolution Defined

Now imagine a population of light-colored moths in a sooty forest. At some point, a DNA mutation arose in the population, and it resulted in a moth of a darker color. When the mutated individual reproduced, some of its offspring inherited the trait. Birds saw and ate many light-colored moths, but most of the dark ones escaped detection and lived long enough to reproduce. So did their dark offspring, and so did *their* offspring. Over the generations, the frequency of the dark form of the trait increased and that of the light form decreased. As more time passes, the dark form of the trait might even become the more common, and people might end up referring to "the population of dark-colored moths." **Evolution** is under way. In biology, the word means genetically based change in a line of descent over time. Like moths, individuals of most populations typically show different forms of many (or most) of their traits. And the frequencies of those different forms relative to one another can change over successive generations.

Natural Selection Defined

Long ago, the naturalist Charles Darwin used pigeons to explain a conceptual connection between evolution and variation in traits. Domesticated pigeons display great variation in size, feather color, and other traits (Figure 1.9). As Darwin knew, pigeon breeders select certain forms of traits. For example, if breeders prefer tail feathers that are black with curly edges, they will allow only those individual pigeons having the most black and the most curl in their tail feathers to mate and produce offspring. Over time, "black" and "curly" will become the most common forms of tail feathers in the captive population, and different forms of the two traits will become less common or will be eliminated.

Pigeon breeding is a case of **artificial selection**, for the selection among different forms of a trait is taking place in an artificial environment—under contrived, manipulated conditions. Yet Darwin saw the practice as a simple model for *natural* selection, a favoring of some forms of traits over others in nature. Whereas breeders are "selective agents" that promote the reproduction of some individuals over others in captive populations, a pigeon-eating peregrine falcon is one of many selective agents that operate across the range of variation among

Further reading: Student Guide to InfoTrac on web site

pigeons in the wild. Generation after generation, the swifter or more effectively camouflaged pigeons have a better chance of living long enough to reproduce than the not-so-swift or too-conspicuous ones among them. What Darwin identified as **natural selection** simply is a difference in *which* individuals of each generation survive and reproduce, the difference being an outcome of which have more adaptive forms of traits.

Unless you happen to be a pigeon breeder, Darwin's pigeon example may not be firing rockets through your imagination. So think about an example closer to home. Certain bacteria and fungi make *antibiotics*, metabolic products that kill bacterial competitors for nutrients in soil. Starting in the 1940s, we learned to use antibiotics to control diseases that result after bacteria invade the body and use its tissues for nutrients. Doctors routinely prescribed these "wonder drugs" for mild infections as well as serious ones. Some manufacturers even added an antibiotic to toothpaste and chewing gum.

As it turned out, antibiotics are powerful agents of natural selection. Over time, some bacteria have mutated in ways that help them resist the antibiotics produced by their bacterial neighbors. For example, streptomycin is an antibiotic that binds with some essential bacterial proteins and inhibits their activity. In certain variant strains of bacteria, mutations slightly changed the form of the proteins, so streptomycin is unable to bind with them. The mutant bacteria escape streptomycin's effects.

In infected patients, an antibiotic acts against bacteria that are susceptible to its action—but it actually favors variant strains that have resistance to it! Presently, such antibiotic-resistant strains are making it difficult to treat typhoid, tuberculosis, gonorrhea, staph (*Staphylococcus*) infections, and some other bacterial diseases. In a few patients, the "superbugs" responsible for tuberculosis cannot be successfully eliminated.

When resistance to antibiotics evolves by selection processes, the antibiotics must also evolve if they are to overcome the defenses. For example, drug companies have now modified parts of the streptomycin molecule. Such molecular changes in the laboratory produce more effective antibiotics—until new generations of more resistant superbugs enter the deadly evolutionary competition for nutrients.

Later in the book, we will consider the mechanisms by which populations of moths, pigeons, bacteria, and all other organisms evolve. Meanwhile, keep in mind the following points about natural selection. They are central to biological inquiry, for they have consistently proved useful in explaining a great deal about nature.

1. Individuals of a population vary in form, function, and behavior. Much of the variation is heritable; it can be transmitted from parents to offspring.

2. Some forms of heritable traits are more adaptive to prevailing environmental conditions. They improve an individual's chance of surviving and reproducing, as by helping it secure food, a mate, hiding places, and so on.

3. Natural selection is the outcome of differences in survival and reproduction among individuals that show variation in one or more traits.

4. Natural selection leads to a better fit with prevailing environmental conditions. Adaptive forms of traits tend to become more common and other forms less so. The population changes in its characteristics; it evolves.

In short, in the evolutionary view, *life's diversity is the sum total of variations in traits that have accumulated in different lines of descent generation after generation, as by natural selection and other processes of change.*

Mutations in DNA introduce variations in heritable traits.

Although many mutations are harmful, some give rise to variations in form, function, or behavior that are adaptive under prevailing environmental conditions.

Natural selection is a result of differences in survival and reproduction among individuals of a population that vary in one or more heritable traits. The process helps explain evolution—changes in lines of descent over the generations.

1.5 THE NATURE OF BIOLOGICAL INQUIRY

The preceding sections sketched out major concepts in biology. Now consider approaching this or any other collection of "facts" with a critical attitude. *"Why should I accept that they have merit?"* The answer requires insight into how biologists make inferences about observations and then test the predictive power of their inferences against actual experiences in nature or the laboratory.

Observations, Hypotheses, and Tests

To get a sense of "how to do science," start by following some practices that are pervasive in scientific research:

1. Observe some aspect of nature, carefully check what others have found out about it, and then frame a question or identify a problem related to your observation.

2. Develop **hypotheses**, or educated guesses, about possible answers to questions or solutions to problems.

3. Using hypotheses as a guide, make a **prediction**—that is, a statement of what you should observe in the natural world if you were to go looking for it. This is often called the "if–then" process. (*If* gravity does not pull objects toward Earth, *then* it should be possible to observe apples falling up, not down, from a tree.)

4. Devise ways to **test** the accuracy of your predictions, as by making systematic observations, building models, and conducting experiments. **Models** are theoretical, detailed descriptions or analogies that help us visualize something that has not yet been directly observed.

5. If the tests do not confirm the prediction, check to see what might have gone wrong. For example, maybe you overlooked a factor that influenced the test results. Or maybe the hypothesis is not a good one.

6. Repeat the tests or devise new ones—the more the better, for hypotheses that withstand many tests are likely to have a higher probability of being useful.

7. Objectively analyze and report the test results and the conclusions you have drawn from them.

You might hear someone refer to these practices as "the scientific method," as if all scientists march to the drumbeat of an absolute, fixed procedure. They do not. Many observe, describe, and report on some subject, then leave it to others to hypothesize about it. Some are lucky; they stumble onto information they are not even looking for, although chance does favor the prepared mind. It is not one single method they have in common. It is a critical attitude about being shown rather than told, and taking a logical approach to problem solving.

Logic encompasses thought patterns by which an individual draws a conclusion that does not contradict the evidence used to support it. Lick a cut lemon and you notice it is mouth-puckeringly sour. Lick ten more.

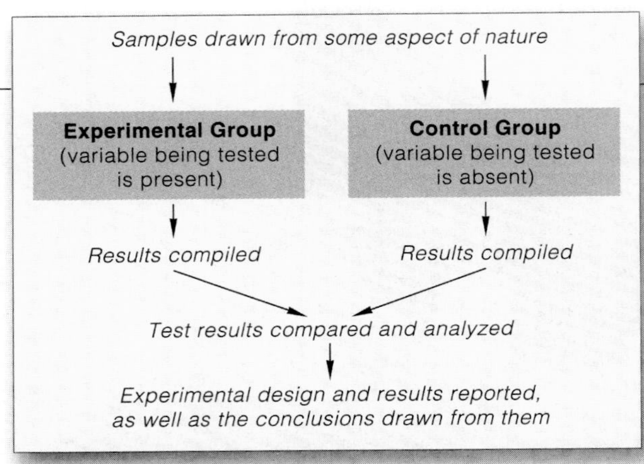

Samples drawn from some aspect of nature
Experimental Group (variable being tested is present)
Control Group (variable being tested is absent)
Results compiled
Results compiled
Test results compared and analyzed
Experimental design and results reported, as well as the conclusions drawn from them

Figure 1.10 Generalized sequence of steps involved in an experimental test of a prediction based on a hypothesis.

You notice the same thing each time, so you conclude all lemons are mouth-puckeringly sour. You correlated one specific (lemon) with another (sour). By this pattern of thinking, called **inductive logic**, an individual derives a general statement from specific observations.

Express the generalization in "if–then" terms, and you have a hypothesis: "If you lick any lemon, then you will get an extremely sour taste in your mouth." By this pattern of thinking, called **deductive logic**, an individual makes inferences about specific consequences or specific predictions that must follow from a hypothesis.

You decide to test the hypothesis by tracking down and sampling all the varieties of lemons in the vicinity. One variety, the Meyer lemon, is actually mellow, for a lemon. You also discover that some people cannot taste anything. So you must modify the original hypothesis: "If most people lick any lemon *except* the Meyer lemon, they will get an extremely sour taste in their mouth." Suppose, after sampling all the known lemon varieties in the world, you conclude the modified hypothesis is a good one. You can never prove it beyond all shadow of a doubt, because there might be lemon trees growing in places people don't even know about. You *can* say the hypothesis has a high probability of not being wrong.

Comprehensive observations are a logical means to test the predictions that flow from hypotheses. So are **experiments**. These tests simplify observation in nature or the laboratory by manipulating and controlling the conditions under which observations are made. When suitably designed, observational and experimental tests allow you to predict that something will happen if a hypothesis isn't wrong (or won't happen if it *is* wrong). Figure 1.10 gives a general idea of the steps involved.

AN ASSUMPTION OF CAUSE AND EFFECT Experiments start from the premise that any aspect of nature has one or more underlying causes. With this premise, science is distinct from faith in the supernatural (meaning "beyond nature"). Experiments deal with potentially falsifiable hypotheses. Hypotheses of this sort can be tested in the natural world in ways that might disprove them.

Further reading: Student Guide to InfoTrac on web site →

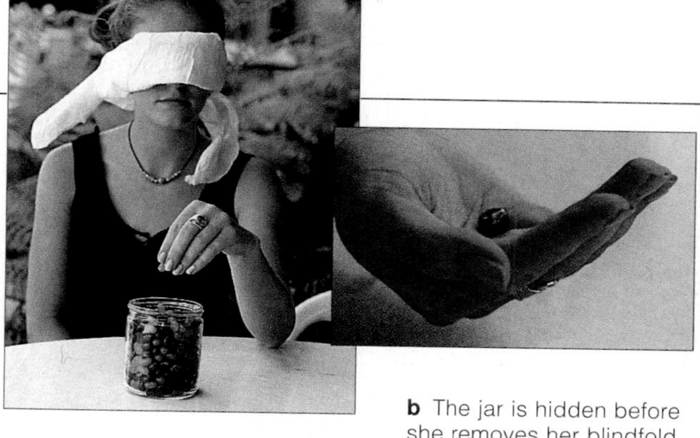

a Natalie, blindfolded, randomly plucks a jellybean from a jar of 120 green and 280 black jellybeans. That is a ratio of 30 to 70 percent.

b The jar is hidden before she removes her blindfold. She observes only a single green jellybean in her hand and assumes the jar holds only green jellybeans.

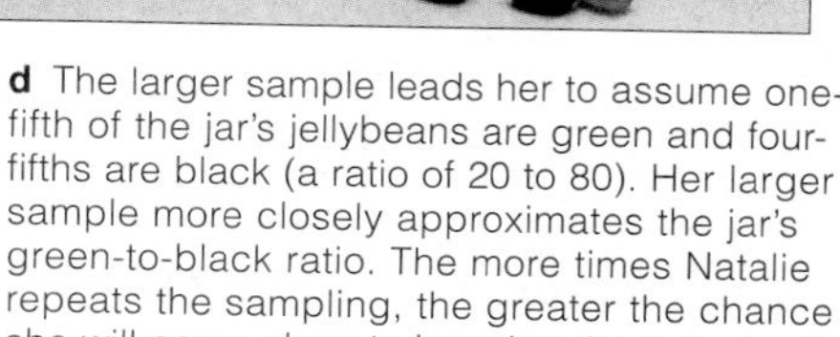

c Still blindfolded, Natalie randomly plucks 50 jellybeans from the jar and ends up with 10 green and 40 black ones.

d The larger sample leads her to assume one-fifth of the jar's jellybeans are green and four-fifths are black (a ratio of 20 to 80). Her larger sample more closely approximates the jar's green-to-black ratio. The more times Natalie repeats the sampling, the greater the chance she will come close to knowing the actual ratio.

Figure 1.11 A simple demonstration of sampling error. 

EXPERIMENTAL DESIGN To get conclusive test results, experimenters rely on certain practices. They refine test designs by searching the literature for information that may relate to their inquiry. They design experiments to test one prediction of a hypothesis at a time. Each time, they set up a **control group**: a standard for comparison with one or more experimental groups. Ideally, their control group is identical with an experimental group in all respects *except* for the one variable being studied. **Variables** are specific aspects of objects or events that may differ or change over time and among individuals.

Section 1.6 shows the design of a recent experiment. As you will see, the experimenters directly manipulated a single variable in an attempt to support or disprove a prediction. They also tried to hold constant any other variables that could influence the results.

SAMPLING ERROR Rarely can experimenters observe *all* individuals of a group. Rather, they use large-enough samples to avoid risking tests with groups that are not representative of the whole. They usually must rely on samples (or subsets) of populations, events, and other aspects of nature. In general, the larger their sampling, the less likely it will be that any differences among the individuals will distort results (Figure 1.11).

About the Word "Theory"

Suppose no one has disproved a hypothesis after years of rigorous tests. Suppose scientists use it to explain more data or observations, which could involve more hypotheses. When a hypothesis meets these criteria, it may become accepted as a **scientific theory**.

You may hear someone apply the word "theory" to a speculative idea, as in the expression "It's only a theory." However, a scientific theory differs from speculation for a simple reason: *Many researchers have tested its predictive power many times, in many ways, and have yet to find evidence that disproves it*. This is why Darwin's view of natural selection is a respected theory. We use it successfully to explain many diverse issues, such as the origin of life, the relationship between plant toxins and plant-eating animals, the sexual advantages of strongly colored or patterned wings or feathers, the reason that certain cancers run in families, or why antibiotics that doctors often prescribe may no longer be effective. By yielding reasoned evidence that life evolved in the past, the theory even influenced views of Earth history.

An exhaustively tested theory might be as close to the truth as scientists can get with the evidence at hand. For example, Darwin's theory stands, with only minor modification, after more than a century's worth of many thousands of different tests. We cannot show that the theory holds under all possible conditions; an infinite number of tests would be required to do so. As for any theory, we can only say it has a *high or low probability* of being a good one. So far, biologists have not found any evidence that refutes Darwin's theory. Yet they still keep their eyes open for any new information and new ways of testing that might disprove its premises.

And this point gets us back to the value of thinking critically. Scientists must keep asking themselves: *Will observations or experiments show that a hypothesis is false?* They expect one another to put aside pride or bias by testing ideas, even in ways that may prove them wrong. Even if an individual doesn't or won't do this, others will—for science proceeds as a community that is both cooperative and competitive. Ideally, its practitioners share their ideas, with the understanding that it is just as important to expose errors as it is to applaud insights. Individuals can and often do change their minds when presented with contradictory evidence. As you will see, this is a strength of science, not a weakness.

A scientific approach to studying nature is based on asking questions, formulating hypotheses, making predictions, devising tests, and objectively reporting the results.

A scientific theory is a testable explanation about the cause or causes of a broad range of related phenomena. It remains open to tests, revision, and tentative acceptance or rejection.

1.6 THE POWER OF EXPERIMENTAL TESTS

BIOLOGICAL THERAPY EXPERIMENTS If you worry about the increasing frequencies of strains of pathogenic (disease-causing) bacteria that resist antibiotics, you are not alone. Bruce Levin of Emory University and Jim Bull of the University of Texas are among the investigators who search for possible alternatives to antibiotic therapy. They had been studying literature on a *biological therapy* that enlists bacteriophages—the "bacteria eaters"—to fight infections. Bacteriophages, a class of viruses, attack a narrow range of bacterial strains. When they contact a target, they typically inject a few enzymes and genetic material into it. What happens next is a hostile takeover of the cell's metabolic machinery, and the cell itself makes many new viral particles. The cell dies after viral enzymes rupture its outer membrane. Virus particles slip through the ruptured membrane and typically infect new cells.

The biologists wondered, as others had decades ago, whether injections of bacteriophages could help people resist or fight off bacterial infections. The discovery of antibiotics in the 1940s had diverted attention away from that idea, at least in Western countries. Now, with so many lethal pathogens breaching the antibiotic arsenal, the bacteria eaters were starting to look good again.

Levin and Bull focused on promising phage therapy experiments that were conducted in 1982 by two British researchers, H. Williams Smith and Michael Huggins. They started their research by testing whether Smith and Huggins's results could be duplicated.

They selected 018:K1:H7, a strain of *Escherichia coli* originally isolated from a human patient with meningitis. Like a harmless *E. coli* strain that lives in the intestines of humans and other mammals (Figure 1.12), this one departs from the body in feces. Bacteriophages used in laboratory work typically ignore 018:K1:H7, so Levin and Bull looked for some *E. coli* killers in samples from an Atlanta sewage treatment plant. They successfully isolated two kinds of bacteriophages and named them *H* (for Hero, an effective killer) and *W* (for the less effective Wimp).

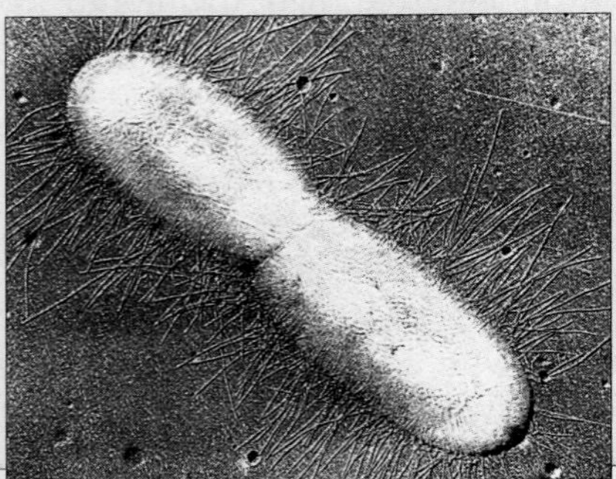

Figure 1.12 Two experiments to compare the effectiveness of bacteriophage injections against antibiotics for treating a bacterial infection. To the left, a micrograph shows a harmless *Escherichia coli* cell in the process of dividing in two.

Hypothesis: If bacteriophages specifically target and destroy cells of *E. coli* 018:K1:H7 in petri dishes, then they will do the same in laboratory mice that have been infected by that strain.

Prediction: Laboratory mice injected with a preparation that contains more than 10^7 particles of *H* bacteriophage will not die following an injection of *E. coli* strain 018:K1:H7.

Experimental test of the prediction:

Researchers establish large populations of H bacteriophage and of E. coli 018:K1:H7 from which to draw their samples, and select a specific strain of laboratory mice.

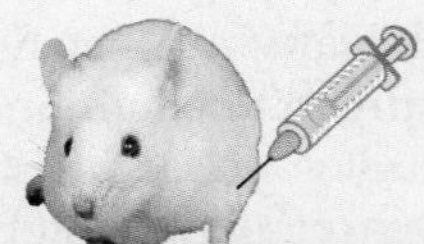

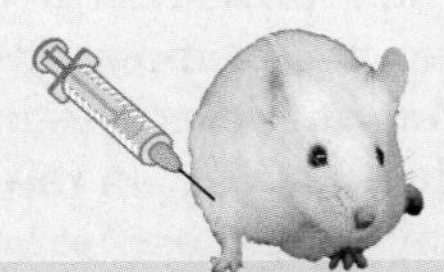

EXPERIMENTAL GROUP	CONTROL GROUP
E. coli injected into right thigh of 15 of the mice; bacteriophage injected into their left thigh.	*E. coli injected into 15 other mice; no bacteriophage injected into this sampling.*

Test results:

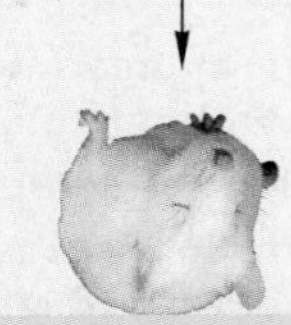

All mice survive.	*All mice die within 32 hours.*

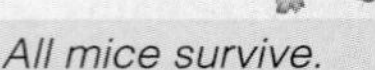

a

Another prediction derived from the same hypothesis: *H* bacteriophage that target and destroy 018:K1:H7 will be more effective than single doses of streptomycin in treating mice infected with 018:K1:H7.

Experimental test of the new prediction:

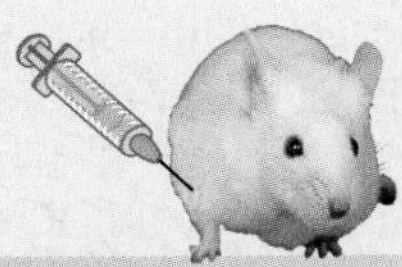

Researchers inject 48 laboratory mice with 018:K1:H7, then divide them into four groups of 12 each. Eight hours later . . .

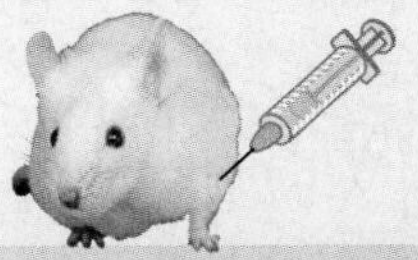

	Test results:
EXPERIMENTAL GROUP 1 *. . . 12 mice receive a single injection of H bacteriophage.*	*11 of 12 mice survive.*
EXPERIMENTAL GROUP 2 *. . . 12 mice receive a single dose of 60 micrograms/gram streptomycin.*	*5 of 12 mice survive.*
EXPERIMENTAL GROUP 3 *. . . 12 mice receive a single dose of 100 micrograms/gram streptomycin.*	*3 of 12 mice survive.*
CONTROL GROUP *. . . 12 mice receive an injection of saline solution only.*	*All control mice die.*

b

Further reading: Student Guide to InfoTrac on web site →

With graduate student Terry DeRouin and laboratory technician Nina Moore Walker, the biologists grew a large population of 018:K1:H7 in a culture flask. They selected a specific strain of laboratory mice, all females of the same age. Fifteen mice of one experimental group were each injected with 018:K1:H7 in one thigh and more than 10^7 particles of *H* bacteriophage in the other thigh. All mice survived. Fifteen mice of a control group were injected with 018:K1:H7 only. None survived. Figure 1.12*a* outlines this experimental test. Figure 1.12*b* outlines a second test of another prediction derived from the same hypothesis. The results reinforced Smith and Huggins's conclusion that certain bacteriophages can be as good as or better than antibiotics at stopping specific bacterial infections.

Each type of bacteriophage targets specific strains of one or at most a few bacterial species. At present, it takes too long for clinicians to discover which specific bacterium is infecting a patient, so the right bacteriophage might not be enlisted until it is too late to do the patient any good. On the bright side, procedures are now being developed that can dramatically accelerate the identification process.

IDENTIFYING IMPORTANT VARIABLES In nature, many factors can influence the outcome of an infection. For example, genetic differences among infected individuals lead to differences in how their immune system responds to the invasion. Age, nutrition, and health at the time of infection influence the outcome. Some pathogens are deadlier than others. And so on. That is why researchers try to simplify and control variables in their experiments. Variables, recall, are specific aspects of objects or events that may differ over time and among individuals. All of the *E. coli* cells in Levin and Bull's experiments were descended from the same parent cell and raised on the same nutrients at the same temperature in a flask. They could be expected to respond in the same way to the bacteriophage attack. All the mice were the same age and sex, and were raised under identical laboratory conditions. Each mouse in each experimental group received the same amount of bacteriophages or streptomycin. And each mouse in a given control group received the same injection of saline solution. Thus the focus was on *one variable*—a specific bacteriophage versus a specific antibiotic—in a simple, controlled, artificial situation.

BIAS IN REPORTING THE RESULTS Whether intentional or not, experimenters run the risk of interpreting data in terms of what they want to prove or dismiss. A few have even been known to fake measurements or nudge findings to reinforce their bias. That is why science emphasizes presenting test results in quantitative terms—that is, with actual counts or some other precise form. Doing so allows other experimenters to check or test the results readily and systematically, as Levin and Bull did. At this writing, they are assembling a detailed report of their own experiments, for publication in a science journal.

1.7 THE LIMITS OF SCIENCE

The call for objective testing strengthens the theories that emerge from scientific studies. It also puts limits on the kinds of studies that can be carried out. Beyond the realm of science, some events remain unexplained. Why do we exist, for what purpose? Why does any one of us have to die at a particular moment? Such questions lead to *subjective* answers, which come from within, as an outcome of all the experiences and mental connections shaping human consciousness. Because people differ vastly in this regard, subjective answers do not readily lend themselves to scientific analysis and experiments.

This is not to say subjective answers are without value. No human society can function for long unless its members share a commitment to certain standards for making judgments, even subjective ones. The moral, aesthetic, philosophical, and economic standards vary from one society to the next. But they all guide their members in deciding what is important and good, and what is not. All attempt to give meaning to what we do.

Every so often, scientists stir up controversy when they happen to explain some part of the world that was considered to be beyond natural explanation—that is, as belonging to the "supernatural." This is often the case when a society's moral codes are interwoven with religious narratives. Exploring some longstanding view of the world from the perspective of science might be misinterpreted as questioning morality even though the two are not the same thing.

As one example, centuries ago in Europe, Nicolaus Copernicus studied the planets and concluded the Earth circles the sun. Today this seems obvious enough. Back then it was heresy. The prevailing belief was that the Creator made the Earth (and, by extension, humans) the immovable center of the universe. Later a respected scholar, Galileo Galilei, studied the Copernican model of the solar system, thought it was a good one, and said so. He was forced to retract his statement publicly, on his knees, and put the Earth back as the fixed center of things. (Word has it that when he stood up he muttered, "Even so, it *does* move.") Later still, Darwin's theory of evolution ran up against the same prevailing belief.

Today, as then, society has sets of standards. Those standards might be questioned when some new, natural explanation runs counter to supernatural beliefs. This doesn't mean that the scientists who raise questions are less moral, less lawful, less sensitive, or less caring than anyone else. It simply means one more standard guides their work: *The external world, not internal conviction, must be the testing ground for scientific beliefs.*

Systematic observations, hypotheses, predictions, tests—in all these ways, science differs from systems of belief that are based on faith, force, or simple consensus.

1.8 SUMMARY

1. There is unity in the living world, for all organisms have these characteristics in common: They all consist of one or more cells. They are assembled from the same kinds of atoms and molecules according to the same laws of energy. They survive by metabolism, and by sensing and responding to specific conditions in the environment. Organisms have the capacity to survive and reproduce, based upon the heritable instructions encoded in the molecular structure of their DNA. Table 1.1 lists these characteristics.

2. The characteristics of life extend from cells, through multicelled organisms, then populations, communities, ecosystems, and the biosphere.

3. Many millions of species (kinds of organisms) exist. Many millions more were alive in the past and became extinct. Classification schemes place all known species in ever more inclusive groupings, from genus on up through family, order, class, phylum, and kingdom.

4. Life's diversity arises through mutations (changes in the structure of DNA molecules). These molecular changes are the basis for variation in heritable traits. These are traits that parents bestow on their offspring, including most details of body form and functioning.

5. Darwin's theory of evolution by natural selection is a cornerstone of biological inquiry. Its key premises are:

a. Individuals of a population differ in the details of their shared heritable traits. Variant forms of traits may affect the ability to survive and reproduce.

b. Natural selection is the outcome of differences in survival and reproduction among individuals that differ in one or more traits. Adaptive forms of a trait tend to become more common and less adaptive ones become less common or disappear. Thus a population's defining traits may change over successive generations; the population can evolve.

6. There are diverse methods of scientific inquiry. The following terms are important aspects of those methods:

a. Theory: Explanation of a broad range of related phenomena, supported by many tests. An example is Darwin's theory of evolution by natural selection.

b. Hypothesis: A proposed explanation of a specific phenomenon. Sometimes called an educated guess.

c. Prediction: A claim about what can be expected in nature, based on premises of a theory or hypothesis.

d. Test: An attempt to produce actual observations that match predicted or expected observations.

e. Conclusion: A statement about whether a theory or hypothesis should be accepted, modified, or rejected, based on tests of the predictions derived from it.

7. Logic is a pattern of thought by which an individual draws a conclusion that does not contradict evidence used to support the conclusion. Inductive logic means an individual derives a general statement from specific observations. Deductive logic means individuals make inferences about particular consequences or predictions that must follow from a hypothesis. Such a pattern of thinking is often expressed in "if–then" terms.

8. Predictions that flow from hypotheses can be tested by comprehensive observations or by experiments in nature or in the laboratory.

9. Experimental tests simplify observations in nature or the laboratory because conditions under which the observations are made are manipulated and controlled. They are based on the premise that any aspect of nature has one or more underlying causes, whether obvious or not. With such an assumption of cause and effect, only those hypotheses that can be tested in ways that might disprove them are scientific.

10. A control group is a standard against which one or more experimental groups (test groups) are compared. Ideally, it is the same as each experimental group in all variables except the one variable being investigated.

11. A variable is a specific aspect of an object or event that might differ over time and between individuals. Experimenters directly manipulate the variable they are studying to support or disprove their prediction.

12. Test results might be distorted by sampling error —chance differences between a population, event, or some other aspect of nature and the samples chosen to represent it. Test results are less likely to be distorted when samplings are large and when they are repeated.

13. Systematic observations, hypotheses, predictions, and experimental tests are the foundation of scientific theories. The external world, not internal conviction, is the testing ground for those theories.

Table 1.1 Summary of Life's Key Characteristics

SHARED CHARACTERISTICS THAT REFLECT LIFE'S UNITY

1. Organisms consist of one or more cells.
2. Organisms are constructed of the same kinds of atoms and molecules according to the same laws of energy.
3. Organisms engage in metabolism; they acquire and use energy and materials to grow, maintain themselves, and reproduce.
4. Organisms sense and make controlled responses to internal and external conditions.
5. Heritable instructions encoded in DNA give organisms their capacity to grow and reproduce. DNA instructions also guide the development of complex multicelled organisms.

FOUNDATIONS FOR LIFE'S DIVERSITY

1. Mutations (changes in the molecular structure of DNA) give rise to variation in heritable traits, including most details of body form, functioning, and behavior.
2. Diversity is the sum total of variations that accumulated in different lines of descent over the past 3.8 billion years, as by natural selection and other processes of change.

Review Questions

For this chapter and subsequent chapters, *italics* after a review question identify the section where you can find answers. They include section numbers and *CI* (for Chapter Introduction).

1. Why is it difficult to formulate a simple definition of life? *CI*
2. Name the molecule of inheritance in cells. *1.1*
3. Write out simple definitions of the following terms: *1.1*
   a. cell c. metabolism
   b. energy d. ATP
4. How do organisms sense changes in their surroundings? *1.1*
5. Study Figure 1.5. Then, on your own, arrange and define the levels of biological organization. *1.2*
6. Study Figure 1.6. Then, on your own, make a sketch of the one-way flow of energy and the cycling of materials through the biosphere. To the side of the sketch, write out definitions of producers, consumers, and decomposers. *1.2*
7. List the shared characteristics of life. *CI, 1.3*
8. What are the two parts of the scientific name for each kind of organism? *1.3*
9. List the six kingdoms of species as outlined in this chapter, and name some of their general characteristics. *1.3*
10. Define mutation and adaptive trait. Explain the connection between mutations and the immense diversity of life. *1.4*
11. Write brief definitions of evolution, artificial selection, and natural selection. *1.4*
12. Define and distinguish between: *1.5*
    a. hypothesis and prediction
    b. observational test and experimental test
    c. inductive and deductive logic
    d. speculation and scientific theory
13. With respect to experimental tests, define variable, control group, and experimental group. *1.5*
14. What does sampling error mean? *1.5*

Self-Quiz (*Answers in Appendix III*)

1. _______ is the capacity of cells to extract energy from sources in their environment, and to transform and use energy to grow, maintain themselves, and reproduce.
2. _______ is a state in which the internal environment is being maintained within tolerable limits.
3. The _______ is the smallest unit of life.
4. If a form of a trait improves chances for surviving and reproducing in a given environment, it is a(n) _______ trait.
5. The capacity to evolve is based on variations in heritable traits, which originally arise through _______ .
6. You have some number of traits that also were present in your great-great-great-great-grandmothers and -grandfathers. This is an example of _______ .
   a. metabolism c. a control group
   b. homeostasis d. inheritance
7. DNA molecules _______ .
   a. contain instructions for traits
   b. undergo mutation
   c. are transmitted from parents to offspring
   d. all of the above
8. For many years in a row, a dairy farmer allowed his best milk-producing cows but not the poor producers to mate. Over many generations, milk production increased. This outcome is an example of _______ .
   a. natural selection c. evolution
   b. artificial selection d. both b and c

Figure 1.13 A spider (*Dolomedes*) and its prey: a tiny minnow. The spider delivered paralyzing venom as well as digestive enzymes into its captive and is now sucking predigested juices from it. Such spiders can move about below the water's surface. Hairs on their body trap oxygen (for aerobic respiration) during their hunting expeditions.

9. A control group is _______ .
   a. a standard against which experimental groups are compared
   b. identical to experimental groups except for one variable
   c. a standard with several variables against which an experimental group is compared
   d. both a and b are correct
10. A specific aspect of an object or event that may change over time or change among individuals is a _______ .
    a. control group c. variable
    b. experimental group d. sampling error
11. The fewer the individuals from a population that are chosen at random for an experimental group, _______ .
    a. the greater the chance of sampling error
    b. the smaller the chance of sampling error
    c. the less likely differences among them will distort the test results
12. Match the terms with the most suitable descriptions.

_____ adaptive trait	a. statement of what you should find in nature if you were to go looking for it
_____ natural selection	b. educated guess
_____ theory	c. improves chance of surviving and reproducing in environment
_____ hypothesis	d. related set of hypotheses that form a broad, applicable, testable explanation
_____ prediction	e. outcome of differences in survival and reproduction among individuals that differ in details of one or more traits

Critical Thinking

1. Some spiders (*Dolomedes*) that feed on insects around ponds occasionally capture tadpoles and small fishes, as in Figure 1.13, and isn't that fun to think about? While they are immature, the female spiders confine themselves to a small patch of vegetation next to the pond. When sexually mature, they mate and store sperm that fertilize the eggs. Only then do they move out and occupy larger areas around the pond. Develop hypotheses to explain what might cause the spiders to live in different places at different times. Design an experiment to test each hypothesis.

2. A scientific theory about some aspect of nature rests upon inductive logic. The assumption is that, because an outcome of some event has been observed to happen with great regularity, it will happen again. However, we cannot know this for certain, because there is no way to account for all possible variables that may affect the outcome. To illustrate this point, Garvin McCain and Erwin Segal offer a parable:

> Once there was a highly intelligent turkey. It lived in a pen, attended by a kind, thoughtful master, and had nothing to do but reflect on the world's wonders and regularities. It observed some major regularities. Morning always began with the sky getting light, followed by the clop, clop, clop of its master's friendly footsteps, then by the appearance of delicious food. Other things varied—sometimes the morning was warm and sometimes cold—but food always followed footsteps. The sequence of events was so predictable, it became the basis of the turkey's theory about the goodness of the world. One morning, after more than one hundred confirmations of the theory, the turkey listened for the clop, clop, clop, heard it, and had its head chopped off.

The turkey learned the hard way that explanations about the world only have a high or low probability of not being wrong. Today, some people take this uncertainty to mean that "facts are irrelevant—facts change." If that is so, should we just stop doing scientific research? Why or why not?

3. Witnesses in a court of law are asked to "swear to tell the truth, the whole truth, and nothing but the truth." What are some of the problems inherent in the question? Can you think of a better alternative?

4. Many popular magazines publish an astounding number of articles on diet, exercise, and other health-related topics. Some authors recommend a specific diet or dietary supplement. What kinds of evidence do you think the articles should include so that you can decide whether to accept their recommendations?

5. Although scientific information often is used when making a decision, it cannot tell an individual what is "right" or "wrong." Give an example of this from your own experience.

6. As the old saying goes, Everybody complains about the weather, but nobody does anything about it. Maybe you can start thinking about how we humans might change at least one aspect of local weather conditions. Two researchers at Arizona State University found that, at least for their investigation of the northeast coast of North America, it really does rain more on weekends, when we'd rather be out having fun!

As R. Cerveny and R. Balling, Jr., reasoned, if the amount of rain falling each day of the week is a random event, then rules of probability should apply. (*Probability* means the chance that each possible outcome of an event will occur is proportional to the number of ways it can be reached.) Thus, each day of the week should get one-seventh (14.3 percent) of the total rainfall for the week. But it turns out Monday is driest (13.1 percent). Days of the week are wetter, and Saturday is the wettest of all (with 16 percent of the total). Sunday is a bit above average.

What causes this effect? As the researchers hypothesized, if *air pollution* is greater during weekdays than on weekends, then cyclic human activities may influence regional patterns of rainfall. Monday through Friday, coal-burning factories and gas-powered vehicles release quantities of tiny particles into the air. The particles can promote updrafts and act as "platforms" for the formation of water droplets. Air pollution must build up during the week and carry over into Saturday. Over the weekend, fewer factories operate and fewer commuters are on the road, so by Monday, the air must be cleaner—and drier.

What kind of evidence do you suppose Cerveny and Balling gathered to test the hypothesis? Jot down a few ideas and check them against the researchers' article in the 6 August 1998 issue of *Nature*. Whether or not you continue in biology, this exercise will be good practice for searching through scientific literature for actual data that you can evaluate on topics of interest to you.

7. Scientists devised experiments to shed light on whether different fish species of the same genus compete in their natural habitat. They constructed twelve ponds, identical in chemical composition and physical characteristics. Then they released the following individuals in each pond:

Ponds 1, 2, 3:	Species A	(300 individuals each pond)
Ponds 4, 5, 6:	Species B	(300 individuals each pond)
Ponds 7, 8, 9:	Species C	(300 individuals each pond)
Ponds 10, 11, 12:	Species A, B, C	(300 individuals of each species in each pond)

Does this experimental design take into consideration all factors that can affect the outcome? If not, how would you modify it?

Selected Key Terms

For this chapter and subsequent chapters, these are the **boldface** terms that appear in the text, in the sections indicated here by *italic* numbers (or *CI*, short for Chapter Introduction). As a study aid, make a list of the terms, write a definition for each, and check it against the one in the text. You will use these terms later on. Becoming familiar with each one will help give you a foundation for understanding the material in later chapters.

adaptive trait *1.4*
Animalia *1.3*
Archaebacteria *1.3*
artificial selection *1.4*
biology *CI*
biosphere *1.2*
cell *1.2*
community *1.2*
consumer *1.2*
control group *1.5*
decomposer *1.2*
deductive logic *1.5*
development *1.1*
DNA *1.1*
ecosystem *1.2*
energy *1.1*
Eubacteria *1.3*
evolution *1.4*
experiment *1.5*
Fungi *1.3*
genus *1.3*
homeostasis *1.1*
hypothesis *1.5*
inductive logic *1.5*
inheritance *1.1*
metabolism *1.1*
model *1.5*
multicelled organism *1.2*
mutation *1.4*
natural selection *1.4*
Plantae *1.3*
population *1.2*
prediction *1.5*
producer *1.2*
Protista *1.3*
receptor *1.1*
reproduction *1.1*
species *1.3*
stimulus *1.1*
test, scientific *1.5*
theory, scientific *1.5*
variable *1.5*

Readings *See also www.infotrac-college.com*

Carey, S. 1994. *A Beginner's Guide to the Scientific Method.* Belmont, California: Wadsworth. Paperback.

Committee on the Conduct of Science. 1989. *On Being a Scientist.* Washington, D.C.: National Academy of Sciences. Paperback.

McCain, G., and E. Segal. 1988. *The Game of Science.* Fifth edition. Pacific Grove, California: Brooks/Cole. Paperback.

Moore, J. 1993. *Science as a Way of Knowing—The Foundations of Modern Biology.* Cambridge, Massachusetts: Harvard University Press.

FACING PAGE: *Living cells of a plant (Elodea), observed with the aid of a microscope. Each rectangular cell contains efficient chemical factories called chloroplasts (the round, bright green parts inside).*

PRINCIPLES OF CELLULAR LIFE I

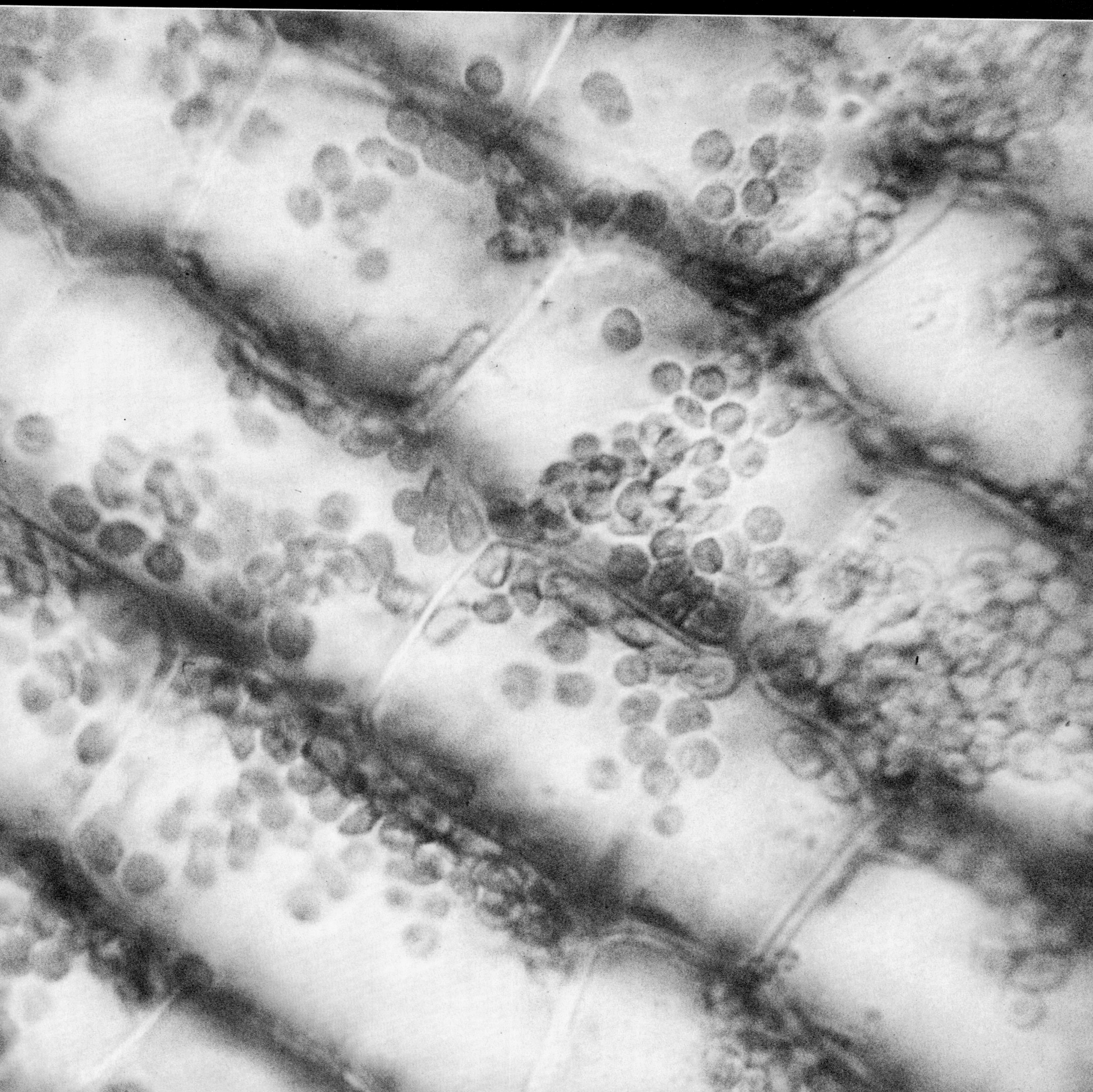

2

CHEMICAL FOUNDATIONS FOR CELLS

Checking Out Leafy Clean-Up Crews

Right now you are breathing in oxygen. You would die without it. Two centuries ago, no one had a clue to what oxygen is, where it comes from, and how it helps keep people alive. Then researchers started unlocking the secrets of this chemical substance and others. As their understanding of chemistry deepened, they began to conjure up such amazing things as nuclear power, synthetic fertilizers, nylons, lipsticks, fabric cleaners, aspirin, antibiotics, and plastic parts of refrigerators, computers, television sets, jet planes, and cars.

Today, our chemical "magic" brings us benefits *and* problems. For example, the scale of agriculture required to sustain the human population, which now surpasses 6 billion, is astounding. Synthetic fertilizers boost crop yields by providing plants with nitrogen, phosphorus, and other growth-enhancing nutrients. The upside is that fertilizer-fed crops help keep more people than you might imagine from starving to death. The downside is that plants do not take up every last bit of fertilizer. Nutrients in runoff from the fields enter lakes, rivers, and seas—where they "feed" such organisms as the protistans that cause huge fish kills (Section 23.10).

What about diverting the runoff to holding ponds, where water can evaporate? Farmers commonly do this. However, evaporation ponds are like magnets that pull selenium from the soil. High concentrations of selenium are toxic to grazing animals and waterfowl.

In 1996, Norman Terry and his coworkers grew cattails and other grasses in ten experimental plots that had become contaminated by runoff from agricultural fields (Figure 2.1). As they knew, plants can incorporate selenium into their tissues and convert some of it into dimethyl selenide—a gas about 600 times less toxic than selenium. They measured how much selenium settled in the mud and how much became incorporated into plant tissues or escaped into the air. As they predicted, before runoff trickled away from the plots, the plants significantly reduced selenium levels.

Terry's research is an example of **bioremediation**—the use of living organisms to withdraw harmful substances from the environment. Another example is the use of sunflowers to clean a pond at Chernobyl, in the Ukraine. Following a total meltdown at a nuclear power plant, strontium 90, cesium 137, and other extremely nasty radioactive elements contaminated the surroundings, including the pond (Section 51.8).

Fertilizers, wetlands, people, pumpkins, the air you breathe—*everything in and around you is "chemistry."* Every solid, liquid, or gaseous substance you care to think about is a collection of one or more elements. Think of the **elements** as fundamental forms of matter that have mass and take up space. Here on Earth, we can't break an element apart into something else except in physics laboratories. Break a chunk of copper into smaller and smaller bits, and the smallest bit you end up with is still copper.

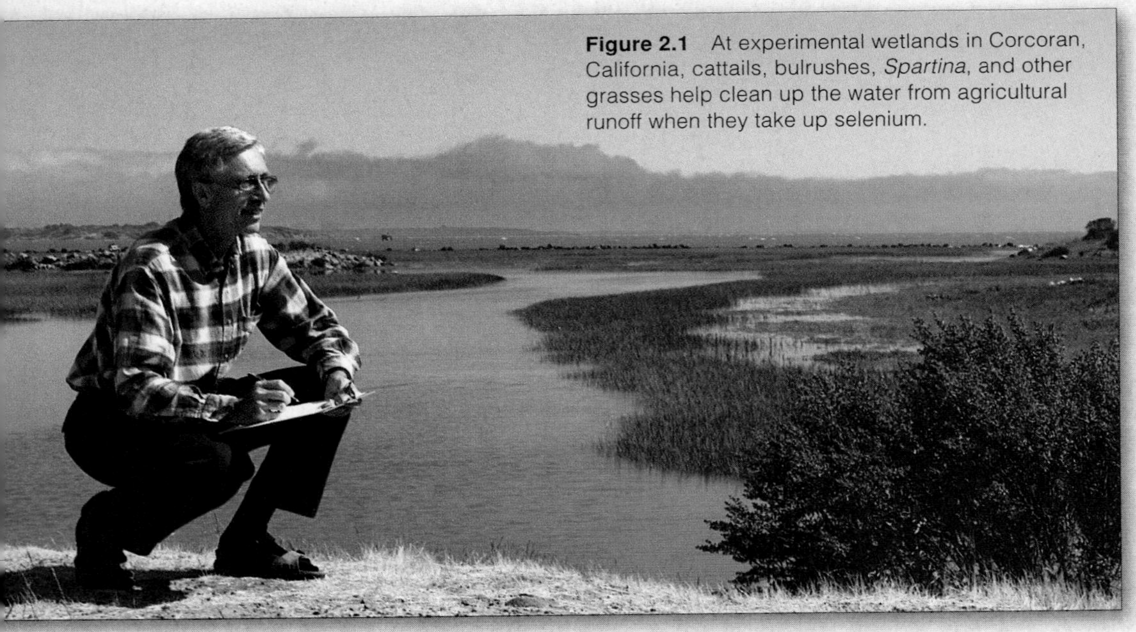

Figure 2.1 At experimental wetlands in Corcoran, California, cattails, bulrushes, *Spartina*, and other grasses help clean up the water from agricultural runoff when they take up selenium.

Ninety-two elements occur naturally on Earth. Four of these—oxygen, hydrogen, carbon, and nitrogen—are the most abundant elements in your body, as they are in all other living organisms (Figure 2.2). Your body also contains some phosphorus, potassium, sulfur, calcium, sodium, and chlorine, plus a number of trace elements such as iodine. A *trace* element simply is one that represents less than 0.01 percent of body weight.

The structure and function of each organism depend on the availability of certain kinds and amounts of elements. For example, without daily intakes of magnesium, your muscles will become sore and weak, and your brain won't work properly. Without magnesium, older leaves of plants droop, turn yellow, and die. The herbicide 2,4-D stimulates weeds to grow faster than vital elements can be absorbed, so the weeds literally grow themselves to death.

As you might deduce from this story, safeguarding the environment, our food supplies, and our health depends on knowledge of chemistry. So do efforts to minimize side effects of its applications on biological systems. You owe it to yourself and others to gain insight into the structure and behavior of chemical substances. By demystifying chemistry's "magic," you will be better equipped to assess its benefits and risks.

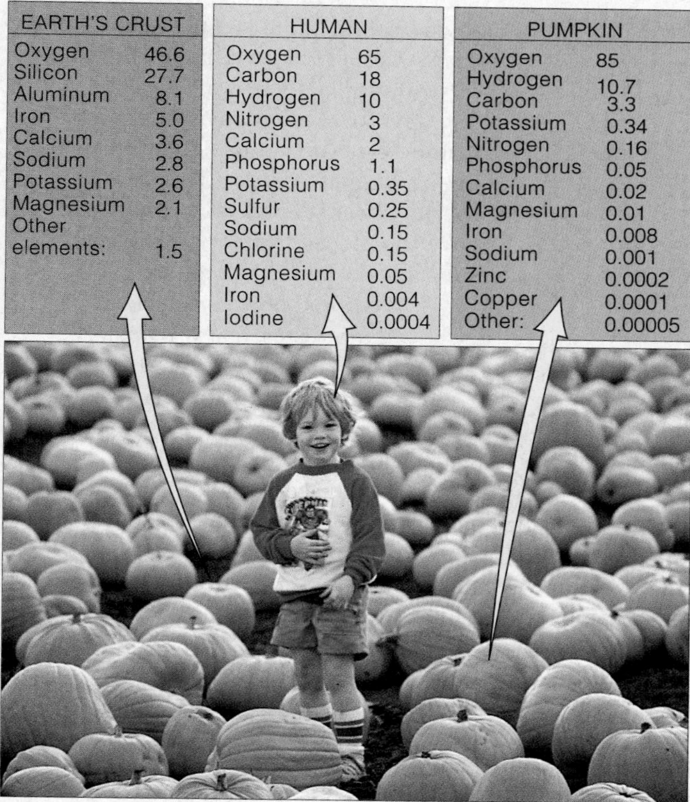
EARTH'S CRUST
Oxygen 46.6
Silicon 27.7
Aluminum 8.1
Iron 5.0
Calcium 3.6
Sodium 2.8
Potassium 2.6
Magnesium 2.1
Other elements: 1.5
HUMAN
Oxygen 65
Carbon 18
Hydrogen 10
Nitrogen 3
Calcium 2
Phosphorus 1.1
Potassium 0.35
Sulfur 0.25
Sodium 0.15
Chlorine 0.15
Magnesium 0.05
Iron 0.004
Iodine 0.0004
PUMPKIN
Oxygen 85
Hydrogen 10.7
Carbon 3.3
Potassium 0.34
Nitrogen 0.16
Phosphorus 0.05
Calcium 0.02
Magnesium 0.01
Iron 0.008
Sodium 0.001
Zinc 0.0002
Copper 0.0001
Other: 0.00005

Figure 2.2 Proportions of elements in a human body and the fruit of pumpkin plants, compared to proportions of elements in the materials of the Earth's crust. How are the proportions similar? How do they differ?

KEY CONCEPTS

1. All substances consist of one or more elements, such as hydrogen, oxygen, and carbon. Each element is composed of atoms, which are the smallest units of matter that still display the element's properties. The atoms of elements are composed of protons, electrons, and (except for the hydrogen atom) neutrons.

2. The atoms that make up each kind of element have the same number of protons and electrons, but they may vary slightly from one another in their number of neutrons. Variant forms of an element's atoms are called isotopes.

3. Atoms have no overall electric charge unless they become ionized—that is, unless they lose electrons or acquire more of them. An ion is an atom or molecule that has lost or gained one or more electrons and thereby has acquired an overall positive or negative charge.

4. Whether a given atom will interact with other atoms depends on how many electrons it has and how they are structurally arranged within the atom. When energetic interactions unite two or more atoms, this is a chemical bond.

5. The molecular organization and activities of living things arise largely from ionic, covalent, and hydrogen bonds between atoms.

6. Life probably originated in water and is exquisitely adapted to its properties. Foremost among those properties are its temperature-stabilizing effects, cohesiveness, and capacity to dissolve or repel a variety of substances.

7. All organisms depend on the controlled formation, use, and disposal of hydrogen ions (H^+). The pH scale is a measure of the concentration of H^+ ions in solutions.

2.1 REGARDING THE ATOMS

Structure of Atoms

What are the smallest particles that retain the properties of an element? **Atoms**. A line of about a million of them would fit in the period ending this sentence. Small as atoms are, physicists have split them into more than a hundred kinds of smaller particles. The only subatomic particles you will need to consider in this book are the ones called **protons**, **electrons**, and **neutrons**.

All atoms have one or more protons, which carry a positive electric charge (p^+). Except for hydrogen, atoms also have one or more neutrons, which are uncharged. Protons and neutrons make up the atom's core region, or atomic nucleus (Figure 2.3). Zipping about the nucleus and occupying most of the atom's volume are one or more electrons, which carry a negative charge (e^-). Each atom has just as many electrons as protons. This means that an atom carries no net charge, overall.

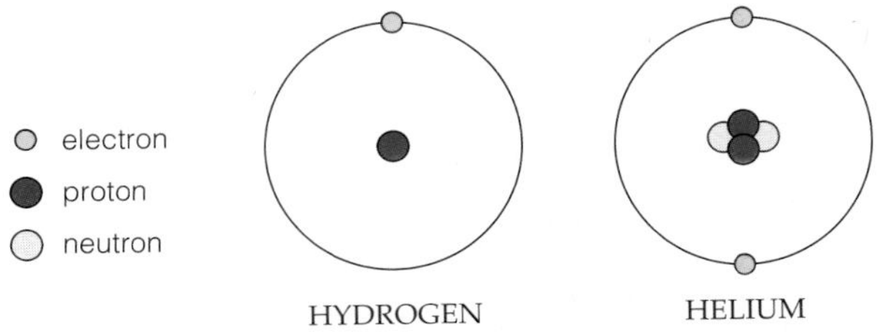

electron
proton
neutron

HYDROGEN

HELIUM

Figure 2.3 The hydrogen atom and helium atom according to one model of atomic structure. The model is highly simplified; the nucleus of these two representative atoms really would be an invisible speck at the scale employed here.

Each element has a unique *atomic* number, which refers to the number of protons in its atoms. To give examples, the atomic number is 1 for the hydrogen atom, which has a single proton. And it is 6 for the carbon atom, which has six protons (Table 2.1).

Also, each element has a *mass* number, which is the combined number of protons and neutrons in the atomic nucleus. A carbon atom, with six protons and six neutrons, has a mass number of 12.

Why bother with atomic numbers and mass numbers? They can give you an idea of whether and how substances will interact. *That knowledge may help you predict how substances might behave in individual cells, in multicelled organisms, and in the environment, under many conditions.*

Table 2.1 Atomic Number and Mass Number of Elements Common in Living Things

Element	Symbol	Atomic Number	Most Common Mass Number
Hydrogen	H	1	1
Carbon	C	6	12
Nitrogen	N	7	14
Oxygen	O	8	16
Sodium	Na	11	23
Magnesium	Mg	12	24
Phosphorus	P	15	31
Sulfur	S	16	32
Chlorine	Cl	17	35
Potassium	K	19	39
Calcium	Ca	20	40
Iron	Fe	26	56
Iodine	I	53	127

Isotopes—Variant Forms of Atoms

Samples of most naturally occurring elements contain two or more **isotopes**. The word indicates that, even though all atoms of an element have the same number of protons, they don't all have the same number of neutrons. Carbon has three isotopes, nitrogen has two, and so on. If the element's symbol has a superscript number to the left of it, this signifies which isotope is being discussed. For example, carbon's three isotopes are abbreviated ^{12}C (six protons and six neutrons, this being the most common form), ^{13}C (six protons, seven neutrons), and ^{14}C (six protons, eight neutrons). We also write these out as carbon 12, carbon 13, and carbon 14.

All isotopes of an element interact with other atoms in the same way. As you will see, this means cells are able to use any isotope of an element for metabolic activities.

Have you heard of radioactive isotopes (radioisotopes)? A physicist, Henri Becquerel, discovered them in 1896, after he had placed a heavily wrapped rock on top of an unexposed photographic plate located in a desk drawer. The rock contained isotopes of uranium, which emit energy. A few days after the plate was exposed to those energetic emissions, a faint image of that rock appeared on it. Marie Curie, Becquerel's coworker, gave the name "radioactivity" to the substance's chemical behavior.

As we now know, a **radioisotope** is an isotope that has an unstable nucleus and that stabilizes itself by spontaneously emitting energy and particles. (These particles are much smaller than protons, electrons, and neutrons.) This process, radioactive decay, transforms a radioisotope into an atom of a different element at a known rate. For instance, over a predictable time span, carbon 14 becomes nitrogen 14. You will look at some uses of this radioisotope and others in Unit III.

Elements are forms of matter that occupy space, have mass, and cannot be degraded to something else by ordinary means.

Atoms, the smallest particles that are unique to each element, have one or more positively charged protons, negatively charged electrons, and (except for hydrogen) neutrons.

Most elements have two or more isotopes (atoms that differ in the number of neutrons). A radioisotope has an unstable nucleus and stabilizes itself by spontaneously emitting particles and energy at a known rate.

Further reading: Student Guide to InfoTrac on web site

USING RADIOISOTOPES TO TRACK CHEMICALS AND SAVE LIVES

Radioisotopes make splendid tracers. A **tracer** is any substance with a radioisotope attached to it, rather like a shipping label, that researchers can track after they deliver it into a cell, a body, an ecosystem, or some other system. Laboratory devices can detect emissions from the tracer and precisely follow its movement through a pathway or pinpoint its final destination.

Melvin Calvin and some other botanists gave us a classic example. They used tracers to figure out the steps by which plants synthesize carbohydrates during photosynthesis. As they knew, all isotopes of an element have the same number of electrons, so all the isotopes must interact with other atoms the same way. Plant cells, they hypothesized, should be able to use any isotope of carbon when they build carbon compounds. By putting plant cells in a medium enriched with a tracer (^{14}C instead of ^{12}C), they were able to track the uptake of carbon through each reaction step leading to the formation of sugars and starches.

As another example, botanists use a radioisotope of phosphorus (^{32}P) as a tracer to identify how plants take up and use soil nutrients and synthetic fertilizers. Such findings are used to improve crop yields.

Besides being research tools, radioisotopes also are diagnostic tools in medicine. For safety considerations, clinicians use only the kinds that rapidly decay into harmless elements. Consider how clinicians analyze a human thyroid. This gland of ours, located in front of the windpipe, is the only one that takes up iodine. Iodine is a building block for thyroid hormones, which have great influence over growth and metabolism. If a patient's symptoms point to abnormal outputs of the thyroid hormones, clinicians may inject a trace amount of an iodine radioisotope (^{123}I) into the patient's blood. Then they use a photographic imaging device to scan the gland. Figure 2.4 shows examples of the images.

Radioisotopes also have uses in *PET* (for *P*ositron-*E*mission *T*omography). With this device, clinicians obtain images of particular body tissues. Suppose clinicians attach a tracer to glucose (or some other molecule). They inject the labeled glucose into the patient, who is then moved into a PET scanner (Figure 2.5*a*). Because all cells require glucose, cells throughout the patient take up the labeled glucose. Uptake is greater in some tissues than in others, depending on the tissue's metabolic activity at the time of examination. Laboratory devices can detect the radioactive emissions and use them to form an image, such as the one in Figure 2.5*b,c*. Such images can reveal variations or abnormalities in metabolic activity.

Radioisotopes that are energetic enough to destroy cells have uses in some therapies. For example, emissions from plutonium 238 drive artificial pacemakers, which smooth out irregular heartbeats. The radioisotope is sealed in a case before the pacemaker is inserted into a patient so its dangerous emissions won't damage body cells.

By contrast, with *radiation therapy*, the idea is to allow radioisotopes to destroy or impair the activity of targeted living cells that are not functioning properly. For example, bombardment with emissions from a source of radium 226 or cobalt 60 may destroy small, localized cancers.

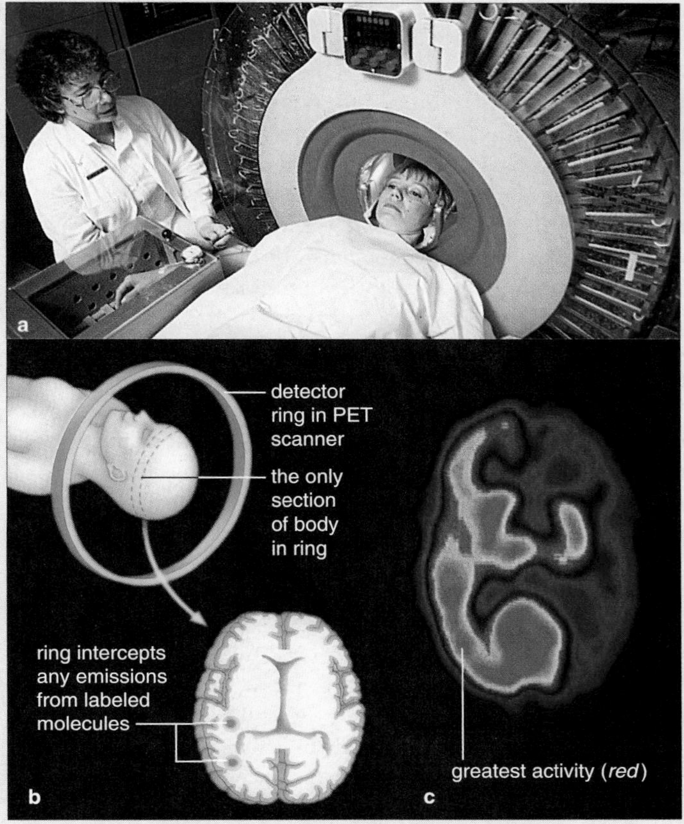
a
detector ring in PET scanner
the only section of body in ring
ring intercepts any emissions from labeled molecules
b
greatest activity (red)
c

Figure 2.5 (**a**) Patient being moved into a PET scanner. (**b**) Inside, a ring of detectors intercepts the radioactive emissions from labeled molecules that were injected into the patient. Computers analyze and color-code the number of emissions from each location in the scanned body region.

(**c**) Brain scan of a child who has a neurological disorder. Different colors in a brain scan signify differences in metabolic activity. The right half of this brain shows very little activity. By comparison, cells of the left half absorbed and used the labeled molecule at expected rates.

normal

enlarged

cancerous

Figure 2.4 Scans of the thyroid gland from three patients.

2.3

WHAT HAPPENS WHEN ATOM BONDS WITH ATOM?

Electrons and Energy Levels

Cells stay alive because energy inherent in all electrons makes things happen (Figure 2.6). Countless atoms in cells acquire electrons, share them, and donate them to other atoms. Atoms of certain elements do this easily, but others do not. What determines whether one atom will interact with another in such ways? *The outcome depends on the number and arrangement of their electrons.*

Tinker with magnets and you can get a sense of the attractive force between unlike charges (+ –) as well as the repulsive force between like charges (++ or – –). Electrons carry a negative charge. In an atom, they repel each other but are attracted to the positive charge of protons. They spend as much time as possible near the protons and far away from each other by moving about in different orbitals (Figure 2.7). Orbitals are volumes of space around the atomic nucleus in which electrons are likely to be at any instant. As a rough analogy, picture three preschoolers circling a cookie jar, not yet expert in the art of sharing. Each is drawn inexorably to the cookies but dreads being shoved away by the others. Two of them might bob and weave about on opposite sides of the jar to avoid a direct hit. But all three never, ever will occupy the same space at the same time.

An orbital can house one or at most two electrons. Because atoms differ in their number of electrons, they also differ in how many occupied orbitals they have.

Hydrogen is the simplest atom. It has a lone electron in a spherical orbital, closest to the nucleus. The orbital corresponds to the *lowest available energy level*. In every other atom, two electrons fill that first orbital. And two more electrons are occupying a second spherical orbital around the first one. Larger atoms have even more electrons. These are occupying orbitals farther from the nucleus, at *higher energy levels*.

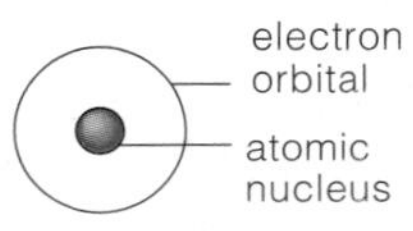
electron
orbital
atomic
nucleus

a Hydrogen atom

The **shell model** is a simple way to think about how electrons are distributed in atoms. By this model, a series of "shells" enclose all orbitals available to electrons (Figure 2.8). The first orbital, which is spherical, is inside the first shell. A second shell (at a higher energy level) encloses the first shell. The next four available orbitals fit in it. More orbitals fit in a third shell, a fourth shell, and so on through the large, complex atoms of heavier elements (Appendix VI).

Figure 2.6 How even a lone electron can make things happen. Physicists confined electrons inside a bubble of liquid helium, then used fluctuating sound waves to pop the bubble. When an electron escaped, it made the white-centered red flash shown here.

How can you predict whether atoms will interact? Check for electron vacancies in their outermost shell. Vacancies mean an atom might give up, gain, or share electrons under suitable conditions, which of course will change the distribution or number of its electrons. You can use the shell model to visualize what goes on here.

In Figure 2.8, which shows the electron distribution for several atoms, electrons are assigned to an energy level (a circle, or shell). Count the electron vacancies—that is, one or more unfilled orbitals in the outermost shell of each atom. As the figure shows, helium is one of the atoms with no vacancies. It is an *inert* atom, which shows little tendency to enter chemical reactions.

Now look again at Figure 2.2, which lists the most abundant of the elements making up a typical organism (a human). These elements include hydrogen, oxygen, carbon, and nitrogen. Atoms of all four elements have electron vacancies. Because of this, *they tend to fill the vacancies by forming bonds with other atoms.*

From Atoms to Molecules

Each **chemical bond** is a union between the electron structures of atoms. Take a moment to study Figure 2.9.

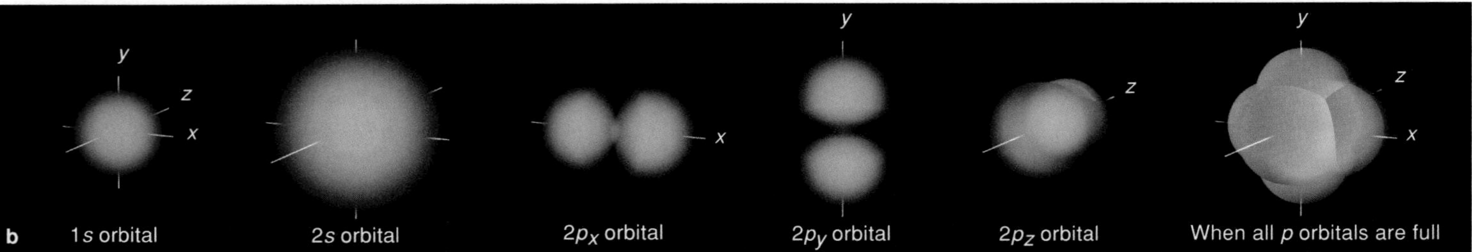
b
y
z
x
1*s* orbital
2*s* orbital
x
2*p*x orbital
y
2*p*y orbital
z
2*p*z orbital
y
z
x
When all *p* orbitals are full

Figure 2.7 Electron arrangements in atoms, which are viewed in terms of three axes. Each axis (*x*, *y*, or *z*) is perpendicular to the other two. (**a**) A hydrogen atom's electron occupies a spherical 1*s* orbital, at the lowest energy level. (**b**) Every atom has a 1*s* orbital, occupied by one or two electrons. The 2*s* orbital and the *p* orbitals at the second energy level can each hold two electrons.

Further reading: Student Guide to InfoTrac on web site →

Figure 2.8 Examples of the shell model of how electrons are distributed in atoms. Successive shells correspond to higher energy levels from the nucleus. Hydrogen, carbon, and other atoms having electron vacancies (unfilled orbitals) inside their outermost shell tend to give up, accept, or share electrons. Helium and other atoms that have no electron vacancies in their outermost shell are inert; they show little if any tendency to interact with other atoms.

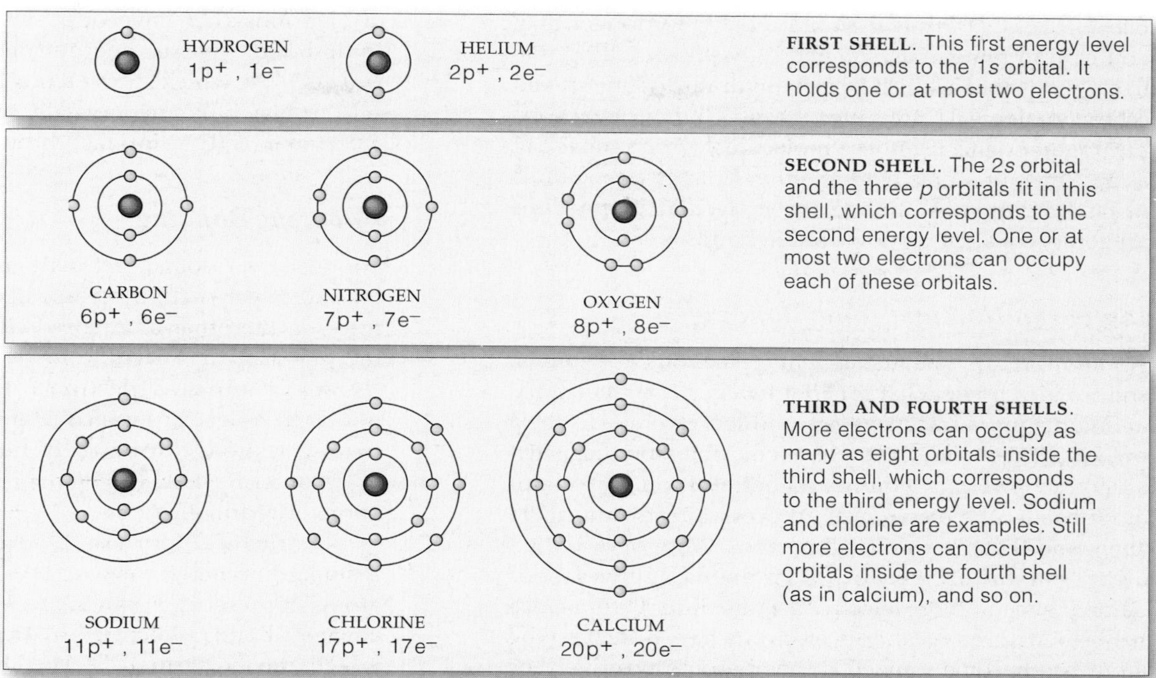

HYDROGEN
1p+ , 1e–
HELIUM
2p+ , 2e–
FIRST SHELL. This first energy level corresponds to the 1s orbital. It holds one or at most two electrons.
CARBON
6p+ , 6e–
NITROGEN
7p+ , 7e–
OXYGEN
8p+ , 8e–
SECOND SHELL. The 2s orbital and the three p orbitals fit in this shell, which corresponds to the second energy level. One or at most two electrons can occupy each of these orbitals.
SODIUM
11p+ , 11e–
CHLORINE
17p+ , 17e–
CALCIUM
20p+ , 20e–
THIRD AND FOURTH SHELLS. More electrons can occupy as many as eight orbitals inside the third shell, which corresponds to the third energy level. Sodium and chlorine are examples. Still more electrons can occupy orbitals inside the fourth shell (as in calcium), and so on.

Figure 2.9 Chemical bookkeeping. We use symbols for elements when writing *formulas*, which identify the composition of compounds. For example, water has the formula H_2O. The subscript indicates two hydrogen (H) atoms are present for every oxygen (O) atom. We use such symbols and formulas when writing *chemical equations*, which are representations of the reactions among atoms and molecules. The substances entering a reaction (reactants) are to the left of the reaction arrow, and the products are to the right, as shown by the following chemical equation for photosynthesis:

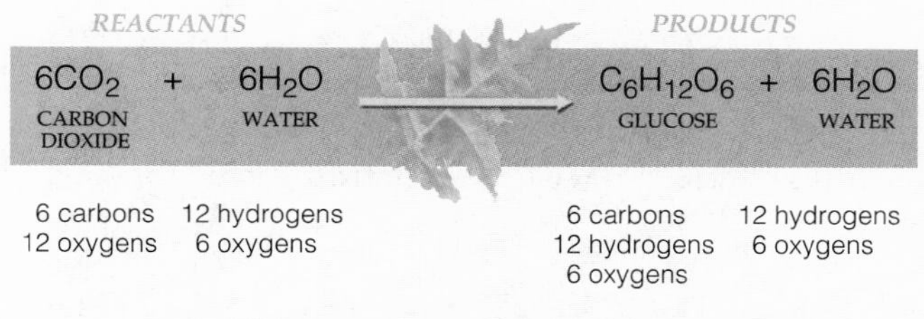

REACTANTS
PRODUCTS
6CO2 + 6H2O
C6H12O6 + 6H2O
CARBON DIOXIDE
WATER
GLUCOSE
WATER
6 carbons
12 oxygens
12 hydrogens
6 oxygens
6 carbons
12 hydrogens
6 oxygens
12 hydrogens
6 oxygens

You may read about reactions for which reactants and products are expressed in moles. A *mole* is a certain number of atoms or molecules of any substance, just as "a dozen" can refer to any twelve cats, roses, and so forth. Molar weight, in grams, equals the total atomic weight of all atoms making up that substance.

For example, the atomic weight of carbon is 12, so one mole of carbon weighs 12 grams. A mole of oxygen (atomic weight 16) weighs 16 grams. Can you state why a mole of water (H_2O) weighs 18 grams, and why a mole of glucose ($C_6H_{12}O_6$) weighs 180 grams?

It summarizes some of the conventions used to describe metabolic reactions. When two or more atoms bond, a **molecule** results. Some molecules have one element only. Molecular nitrogen (N_2), with its two nitrogen atoms, is like this. So is molecular oxygen (O_2).

The molecules of **compounds** consist of two or more different elements in proportions that never vary. Water is an example. In each molecule of water you find one oxygen atom bonded to two hydrogen atoms. Water molecules in rainclouds, the ocean, a Siberian lake, your bathtub, the petals of a leaf or flower, or anywhere else always have twice as many hydrogen as oxygen atoms.

By contrast, in a **mixture**, two or more elements are simply intermingling in proportions that can vary (and usually do). Swirl together some water and the sugar sucrose, which is a compound of carbon, hydrogen, and oxygen, and you get a mixture.

In an atom, electrons occupy orbitals, which are volumes of space around the nucleus. By a simplified model, orbitals are arranged as a series of shells that surround the nucleus. The successive shells correspond to levels of energy, which become greater with distance from the nucleus.

One or two electrons at most occupy any orbital. Atoms having unfilled orbitals in their outermost shell tend to interact with other atoms; those with no vacancies do not.

In molecules of an element, all of the atoms are of the same kind. In molecules of a compound, atoms of two or more elements are bonded together, in unvarying proportions.

2.4 IMPORTANT BONDS IN BIOLOGICAL MOLECULES

Eat your peas! Drink your milk! Probably for longer than you care to remember, somebody has been telling you to eat foods that are rich in carbohydrates, proteins, and other "biological molecules." Only living organisms put together and use these molecules, which consist of a few kinds of atoms held together by only a few kinds of bonds. Foremost among the molecular interactions are ionic, covalent, and hydrogen bonds.

Ion Formation and Ionic Bonding

An atom, recall, has just as many electrons as protons, so it carries no net charge. That balance can change for atoms having a vacancy—an unfilled orbital—in their outermost shell. For example, a chlorine atom has such a vacancy and can acquire another electron. A sodium atom has a lone electron in an orbital in its outermost shell, and that electron can be knocked out of or pulled away from the orbital. Any atom that has either lost or gained one or more electrons is an **ion**. The balance between its protons and its electrons has shifted, so the atom has become ionized; it has become positively or negatively charged (Figure 2.10*a*).

In living cells, neighboring atoms commonly accept or donate electrons among one another. When one atom loses an electron and one gains, both become ionized. Depending on cellular conditions, the two ions may not separate; they may remain together as a result of the mutual attraction of opposite charges. An association of two ions that have opposing charges is called an **ionic bond**. You see one outcome of ionic bonding in Figure 2.10*b*, which shows a portion of a crystal of table salt, or NaCl. In such crystals, sodium ions (Na^+) and chloride ions (Cl^-) interact through ionic bonds.

Covalent Bonding

Suppose two atoms, each with an unpaired electron in its outermost shell, meet up. Each exerts an attractive force on the other's unpaired electron but not enough to yank it away. Each atom becomes more stable by *sharing* its unpaired electron with the other. A sharing of a pair of electrons is a **covalent bond**. For example, a hydrogen atom can partially fill the electron vacancy in its outermost shell when it is covalently bonded to another hydrogen atom.

MOLECULAR HYDROGEN (H_2)

In structural formulas, a single line between two atoms represents a *single* covalent bond. Molecular hydrogen has such a bond, which can be written as H—H. In a *double* covalent bond, two atoms share two pairs of electrons. Molecular oxygen (O=O) is like this. In a *triple* covalent bond, two atoms share three pairs of electrons. Molecular nitrogen (N≡N) is like this. All three examples happen to be gaseous molecules. Each time you breathe in some air, you draw a stupendous number of H_2, O_2, and N_2 molecules into your nose.

Covalent bonds are nonpolar or polar. In a *nonpolar* covalent bond, participating atoms exert the same pull on the electrons and both share them equally. "Nonpolar" implies that there is no difference in charge

SODIUM ATOM 11 p⁺ 11 e⁻ — electron transfer — CHLORINE ATOM 17 p⁺ 17 e⁻

SODIUM ION 11 p⁺ 10 e⁻ — CHLORIDE ION 17 p⁺ 18 e⁻

a Formation of sodium and chloride ions

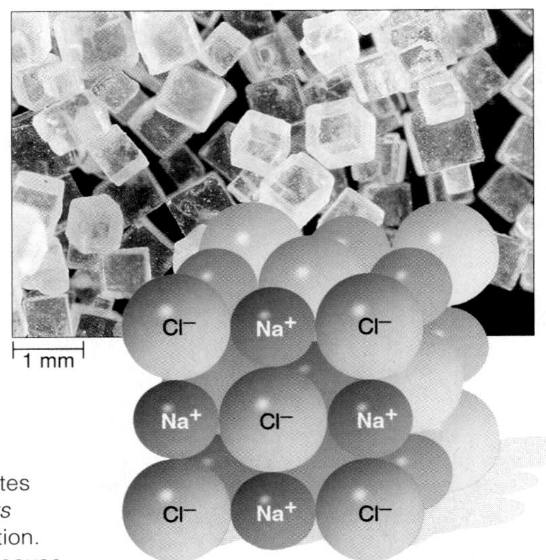

1 mm
Cl⁻
Na⁺
Cl⁻
Na⁺
Cl⁻
Na⁺
Cl⁻
Na⁺
Cl⁻

b Crystals of sodium chloride (NaCl)

Figure 2.10 (**a**) Ionization by way of an electron transfer. In this case, a sodium atom donates the single electron in its outermost shell to a chlorine atom, which has an unfilled orbital in *its* outermost shell. A sodium ion (Na^+) and a chloride ion (Cl^-) are the outcome of this interaction. (**b**) In each crystal of table salt, or NaCl, many sodium and chloride ions remain together because of the mutual attraction of opposite charges. Their interaction is a case of ionic bonding.

Further reading: Student Guide to InfoTrac on web site

between two ends of the bond (that is, at its two poles). Molecular hydrogen is a simple example of a nonpolar bond. Its two H atoms, each with one proton, attract the shared electrons equally.

In a *polar* covalent bond, atoms of different elements (which have different numbers of protons) do not exert the same pull on shared electrons. The more attractive atom ends up with a slight negative charge; the atom is "electronegative." Its effect is balanced out by the other atom, which ends up with a slight positive charge. In other words, taken together, the atoms interacting in a polar covalent bond have no *net* charge, but the charge is distributed unevenly between the bond's two ends.

As an example, a water molecule has two polar covalent bonds: H—O—H. In this molecule, electrons are less attracted to the hydrogens than to the oxygen, which has more protons. A water molecule carries no *net* charge, but you will see shortly that its polarity can weakly attract neighboring polar molecules and ions.

Hydrogen Bonding

The patterns of electron sharing in covalent bonds hold atoms together in specific arrangements in molecules. Some of the patterns also give rise to weak attractions and repulsions between charged functional groups of molecules, as well as between molecules and ions. Like interacting skydivers, such interactions break and form easily (Figure 2.11). Yet they have important roles in the structure and functioning of biological molecules.

For example, a **hydrogen bond** is a weak attraction between an electronegative atom (such as an oxygen or nitrogen atom taking part in a polar covalent bond) and a hydrogen atom taking part in a second polar covalent bond. Hydrogen's slight positive charge weakly attracts the atom with the slight negative charge (Figure 2.12).

Figure 2.11 Like skydivers who briefly clasp hands to form an orderly pattern, weak attractions within and between molecules (and between ions and molecules) can form and break easily.

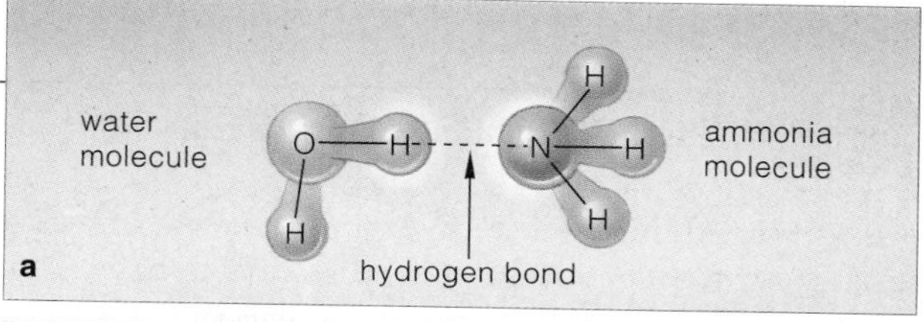
water molecule
O
H
H
N
H
H
H
ammonia molecule
a
hydrogen bond

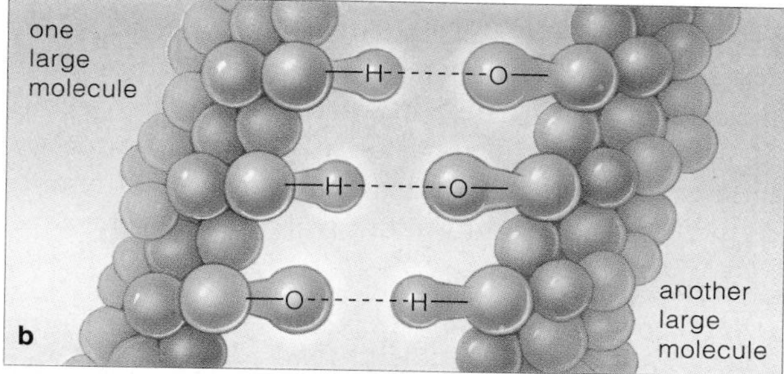
one large molecule
H
O
H
O
O
H
another large molecule
b

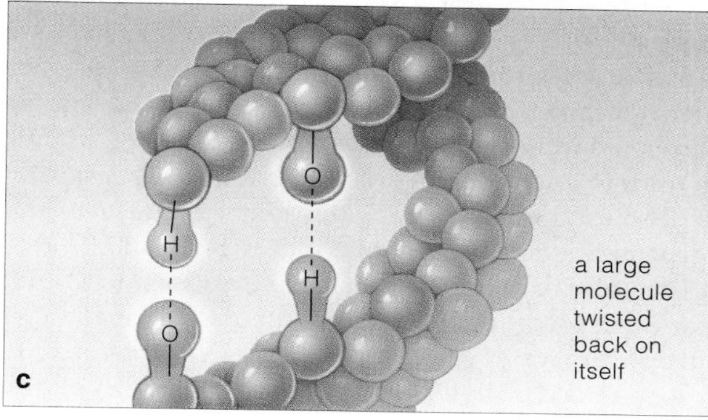
O
H
H
O
H
a large molecule twisted back on itself
c

Figure 2.12 Three examples of hydrogen bonds. Compared to covalent bonds, a hydrogen bond is easier to break. Collectively, however, extensive hydrogen bonding has a major role in water, DNA, proteins, and many other substances.

Hydrogen bonds may form between two or more molecules. They also may form in different parts of the same molecule where it twists and folds back on itself. For example, many such bonds form between the two strands of a DNA molecule. Individually, the hydrogen bonds break easily. Collectively, they stabilize DNA's structure. Similarly, hydrogen bonds form between the molecules that make up water. As you will read next, they contribute to water's life-sustaining properties.

In an ionic bond, two ions of opposite charge attract each other and stay together. Ions form when atoms gain or lose electrons and so acquire a net positive or negative charge.

In a covalent bond, atoms share a pair of electrons. When atoms share the electrons equally, the bond is nonpolar. When sharing is not equal, the bond itself is polar—slightly positive at one end, slightly negative at the other.

In a hydrogen bond, a covalently bound atom showing a slight negative charge weakly interacts with a covalently bound hydrogen atom showing a slight positive charge.

2.5 PROPERTIES OF WATER

No sprint through basic chemistry is complete unless it leads us to the collection of molecules called water. Life originated in water. Many organisms still live in it. The ones that don't cart water around with them, in cells and tissue spaces. Many metabolic reactions require water as a reactant. Cell shape and internal structure depend on it. These topics will repeatedly occupy our attention in the book, so you may find it useful to become familiar with the following points about water's properties.

Polarity of the Water Molecule

A water molecule, remember, has no net charge, but the charge that it does carry is unevenly distributed. As a result of its electron arrangements and bond angles, the water molecule's oxygen "end" is a bit negative and its hydrogen end is a bit positive. Figure 2.13*a* is a simple way to think about this charge distribution. Because of the resulting polarity, one water molecule attracts and hydrogen-bonds with others (Figures 2.13*b* and 2.14).

The polarity of the water molecule also attracts other polar molecules, including the sugars. We call these **hydrophilic** (water-loving) **substances**, for they readily hydrogen-bond with water. By contrast, water's polarity repels nonpolar molecules, including oils. We call these **hydrophobic** (water-dreading) **substances.** Observe this for yourself by shaking a bottle that contains water and salad oil, then setting it on a table. Not long afterward, new hydrogen bonds replace the ones that broke apart when you shook the bottle. As the water molecules reunite, they push molecules of oil aside, forcing them to cluster as droplets or as a film at the water's surface.

Life depends on hydrophobic interactions. For example, a thin, oily membrane separates the cell's watery surroundings and watery interior. Membrane organization starts with countless hydrophobic interactions (Section 5.1).

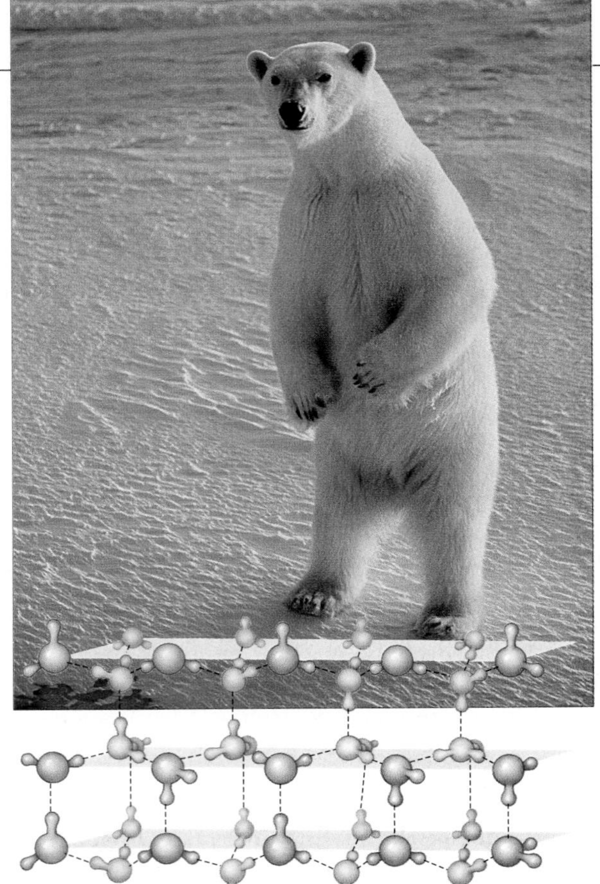

Figure 2.14 Hydrogen bonding pattern of ice, which in vast quantities blankets the habitat of choice for polar bears. Below 0°C, each water molecule is hydrogen-bonded to four others in a three-dimensional lattice. The molecules are spaced farther apart than they would be in liquid water at room temperature, when molecular motion is greater and not as many bonds form. That is why ice floats on water; it has fewer molecules than the same volume of liquid water (the lattice is less dense).

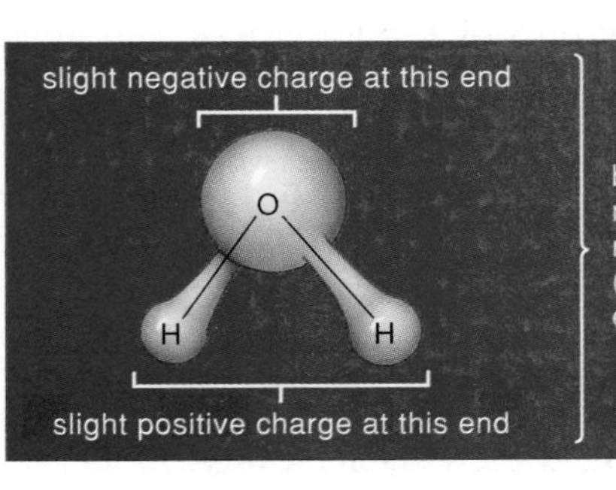
a
slight negative charge at this end
O
H
H
slight positive charge at this end
but the whole molecule has no net charge (+ and – balance each other)

b
H
O
H
O
H
H

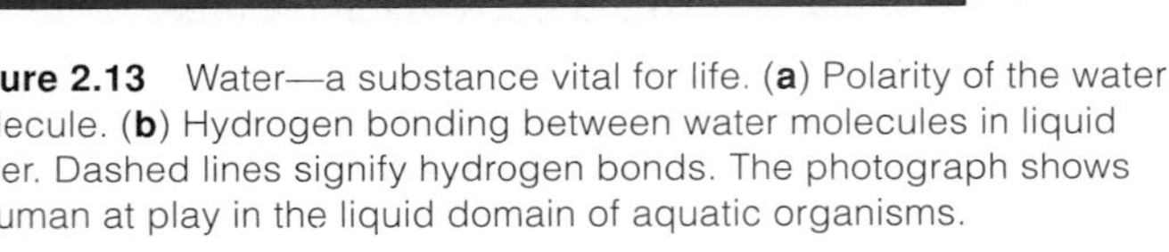

Figure 2.13 Water—a substance vital for life. (**a**) Polarity of the water molecule. (**b**) Hydrogen bonding between water molecules in liquid water. Dashed lines signify hydrogen bonds. The photograph shows a human at play in the liquid domain of aquatic organisms.

Further reading: Student Guide to InfoTrac on web site →

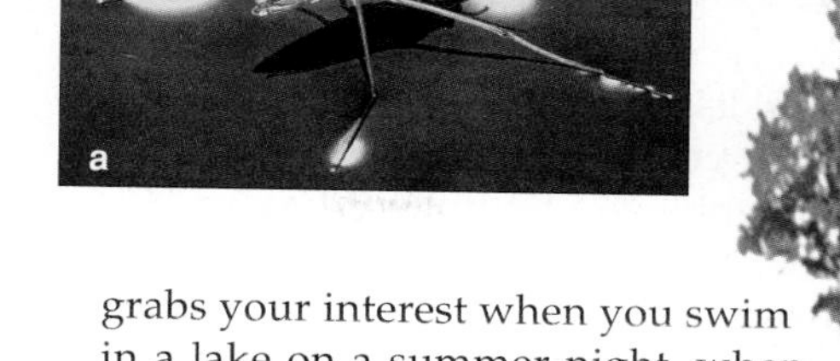
a

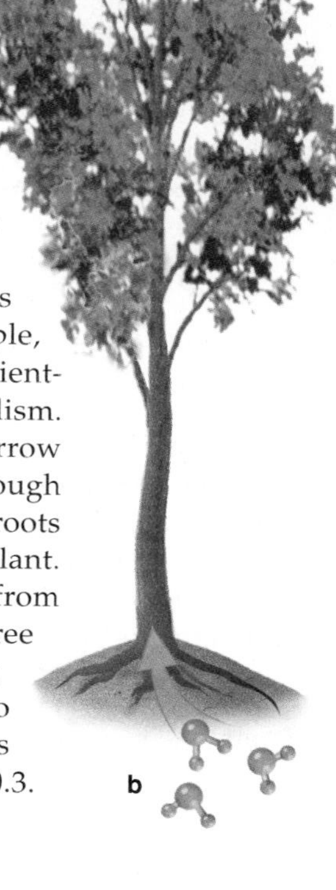
b

Figure 2.15 Water's cohesion. (**a**) Water strider (*Gerris*). This long-legged bug feeds on insects that land or fall on water. Because of its fine, water-resistant leg hairs and water's high surface tension, the bug scoots across the surface. (**b**) Water's cohesion, combined with its evaporation from leaves, pulls water up to the tops of trees.

Water's Temperature-Stabilizing Effects

Cells consist mostly of water, and they release a great deal of heat energy during metabolism. If it were not for hydrogen bonds in liquid water, cells might cook in their own juices. To see why this is so, start with these observations: Every molecule vibrates incessantly, and its motion increases when it absorbs heat. **Temperature** is simply a measure of the molecular motion of a given substance. Compared to most other fluids, water can absorb more heat energy before its temperature rises measurably. Why? Much of the added energy disrupts hydrogen bonding between neighboring molecules of water rather than causing an increase in the motion of individual molecules. In liquid water, the stupendous number of hydrogen bonds can buffer large swings in temperature. Such bonds help stabilize the temperature of multicelled organisms and of aquatic habitats.

Even when the temperature of liquid water is not shifting much, hydrogen bonds are constantly breaking. But they also are forming again just as fast. By contrast, a large energy input can increase molecular motion so much that hydrogen bonds stay broken, and individual molecules at the water's surface escape into the air. By this process, called **evaporation**, heat energy converts liquid water to the gaseous state. As large numbers of molecules break free and depart, they carry away some energy and lower the water's surface temperature.

Evaporative water loss can help cool you and some other mammals when you work up a sweat on hot, dry days. Under such conditions, sweat—which is about 99 percent water—evaporates from your skin.

Below 0°C, hydrogen bonds resist breaking and lock water molecules in the latticelike bonding pattern of ice (Figure 2.14). Ice is less dense than water. During winter freezes, ice sheets may form near the surface of ponds, lakes, and streams. Like a blanket, ice "insulates" the liquid water beneath it and helps protect many fishes, frogs, and other aquatic organisms against freezing.

Water's Cohesion

Life also depends on water's cohesion. **Cohesion** means something has a capacity to resist rupturing when placed under tension—that is, stretched—as by the weight of a bug's legs (Figure 2.15*a*). Think of a lake, pool, or some other body of liquid water. Uncountable episodes of hydrogen bonding exert a continual inward pulling on water molecules at or near the surface. The hydrogen bonding results in a high surface tension. That tension grabs your interest when you swim in a lake on a summer night, when too many night-flying insects splat against water and float on it.

Cohesion works inside organisms as well as on the outside. For example, trees and other plants require nutrient-laden water for growth and metabolism. Largely because of the cohesion, narrow columns of liquid water move through pipelines of vascular tissues, from roots to leaves and all other parts of the plant. On sunny days, water evaporates from leaves; individual molecules break free (Figure 2.15*b*). Hydrogen bonding "pulls up" more water molecules into leaf cells as replacements, in ways that you will read about in Section 30.3.

Water's Solvent Properties

Finally, water is a fine solvent; ions and polar molecules easily dissolve in it. Any dissolved substance is called a **solute**. In general, we say a substance is *dissolved* after water molecules cluster around its ions or molecules and keep them dispersed in fluid. Such clusters are "spheres of hydration." This is what happens to solutes in cellular fluid, in the sap of maple trees, in blood, in fluid traveling through your gut, and in every other fluid associated with life.

Watch this happen when you pour some table salt (NaCl) into a cup of water. After a while, the salt crystals separate into Na^+ and Cl^-. Each Na^+ attracts the negative end of some of the water molecules at the same time Cl^- attracts the positive end of others (Figure 2.16). The spheres of hydration keep the ions dispersed in the fluid.

Na+
Cl−

Figure 2.16 Two spheres of hydration.

A water molecule has no net charge, but it shows polarity. Polarity allows water molecules to hydrogen-bond to each other and to other polar (hydrophilic) substances. Water molecules tend to repel nonpolar (hydrophobic) substances.

Water has temperature-stabilizing effects, internal cohesion, and a capacity to dissolve many substances. These properties influence the structure and functioning of organisms.

2.6 ACIDS, BASES, AND BUFFERS

A great variety of ions dissolved in the fluids inside and outside cells influence cell structure and functioning. Among the most influential are **hydrogen ions**, or H^+. Hydrogen ions are the same thing as free (unbound) protons. They have far-reaching effects largely because they are chemically active and there are so many of them.

The pH Scale

At any instant in liquid water, some water molecules break apart into hydrogen ions and **hydroxide ions** (OH^-). This ionization of water is the basis of the **pH scale**, as in Figure 2.17. Biologists use this scale when measuring the H^+ concentration of seawater, tree sap, blood, and other fluids. Pure water (not rainwater or tapwater) always contains just as many H^+ as OH^- ions. This condition also may occur in other fluids, and it signifies neutrality. We assign neutrality a value of 7 at the midpoint of the pH scale, which ranges from 0 (the highest H^+ concentration) to 14 (the lowest). *The greater the H^+ concentration, the lower the pH.*

Starting at neutrality, each change by one unit of the pH scale corresponds to a tenfold increase or decrease in H^+ concentration. An easy way to sense the differences is to dissolve a bit of baking soda (pH 9) on your tongue, then water (7), then lemon juice (2.3).

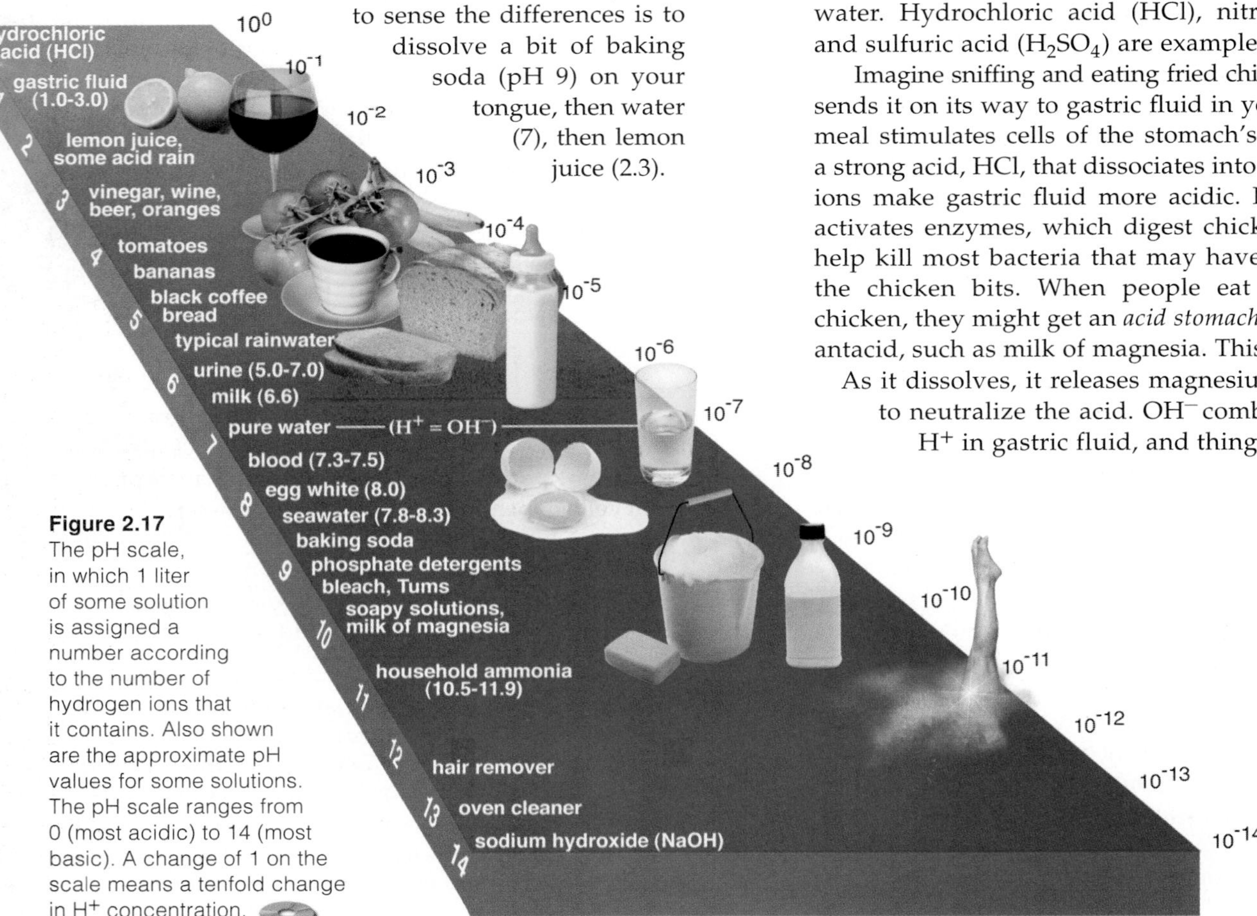

10^0
10^{-1}
10^{-2}
10^{-3}
10^{-4}
10^{-5}
10^{-6}
10^{-7}
10^{-8}
10^{-9}
10^{-10}
10^{-11}
10^{-12}
10^{-13}
10^{-14}
0 hydrochloric acid (HCl)
1 gastric fluid (1.0-3.0)
2 lemon juice, some acid rain
3 vinegar, wine, beer, oranges
4 tomatoes
bananas
black coffee
5 bread
typical rainwater
6 urine (5.0-7.0)
milk (6.6)
pure water ($H^+ = OH^-$)
7 blood (7.3-7.5)
egg white (8.0)
8 seawater (7.8-8.3)
baking soda
9 phosphate detergents
bleach, Tums
soapy solutions, milk of magnesia
10
household ammonia (10.5-11.9)
11
12 hair remover
13 oven cleaner
14 sodium hydroxide (NaOH)

Figure 2.17 The pH scale, in which 1 liter of some solution is assigned a number according to the number of hydrogen ions that it contains. Also shown are the approximate pH values for some solutions. The pH scale ranges from 0 (most acidic) to 14 (most basic). A change of 1 on the scale means a tenfold change in H^+ concentration.

How Do Acids Differ From Bases?

When they dissolve in water, substances categorized as **acids** *donate* protons (H^+) to other solutes or to water molecules. By contrast, the substances we categorize as **bases** *accept* H^+ when dissolved in water, and OH^- forms directly or indirectly after they do this. *Acidic* solutions, such as lemon juice, gastric fluid, and coffee, release more H^+ than OH^-; their pH is below 7. *Basic* solutions, such as seawater, baking soda, and egg white, release more OH^- than H^+. Such solutions are also called "alkaline" fluids; they have a pH above 7.

The fluid inside most human cells is about 7 on the pH scale. The pH values of most fluids bathing these cells are slightly higher; they range between 7.3 and 7.5. This also is the case for the fluid portion of human blood. By contrast, seawater is more alkaline than the body fluids of organisms that live in it.

Think of most acids as being either weak or strong. Weak ones such as carbonic acid (H_2CO_3) are reluctant H^+ donors. Depending on the pH, they just as easily accept H^+ after giving it up, so they alternate between acting as an acid and acting as a base. By contrast, strong acids completely give up H^+ when they dissociate in water. Hydrochloric acid (HCl), nitric acid (HNO_3), and sulfuric acid (H_2SO_4) are examples.

Imagine sniffing and eating fried chicken. Swallowing sends it on its way to gastric fluid in your stomach. The meal stimulates cells of the stomach's lining to secrete a strong acid, HCl, that dissociates into H^+ and Cl^-. The ions make gastric fluid more acidic. Increased acidity activates enzymes, which digest chicken proteins and help kill most bacteria that may have lurked in or on the chicken bits. When people eat too much fried chicken, they might get an *acid stomach* and reach for an antacid, such as milk of magnesia. This is a strong base. As it dissolves, it releases magnesium ions and OH^- to neutralize the acid. OH^- combines with excess H^+ in gastric fluid, and things calm down.

Further reading: Student Guide to InfoTrac on web site

High concentrations of strong acids or bases also can disrupt the external environment and pose dangers to life. Read the labels on bottles of ammonia, drain cleaner, and other products that are often stored in households. Many can cause severe *chemical burns*. So can sulfuric acid in car batteries. Smoke from fossil fuels, exhaust from motor vehicles, and nitrogen fertilizers release strong acids, which alter the pH of rain (Figure 2.18). Some regions are sensitive to the pH of this *acid rain*, owing to their soil type and vegetation cover. The altered chemistry of habitats in such regions drastically affects the functioning of organisms. We will return to this topic in Section 51.1.

Figure 2.18 Sulfur dioxide emissions from a coal-burning power plant. Camera lens filters revealed the otherwise invisible emissions. Sulfur dioxide and other airborne pollutants dissolve in water vapor to form acidic solutions. They are a major component of acid rain.

Buffers Against Shifts in pH

Metabolic reactions are sensitive to even slight shifts in pH, for H^+ and OH^- can combine with many different molecules and alter their functions. Normally, control mechanisms minimize unsuitable shifts in pH, as they do when HCl enters the stomach in response to a meal. Many of the controls involve buffer systems.

A **buffer system** is a partnership between a weak acid and the base that forms when the acid dissolves in water. The two work as a pair to counter slight shifts in pH. Remember, when a strong base enters a fluid, the OH^- level rises. But a weak acid neutralizes part of the added OH^- by combining with it. By this interaction, the weak acid's partner forms. Later, if a strong acid floods in, the base will accept H^+ and thereby become its partner in the system.

Bear in mind, the action of a buffer system cannot make new hydrogen ions or eliminate ones that already are present. It can only bind or release them.

In all complex, multicelled organisms, diverse buffer systems operate in the internal environment—in blood and tissue fluids. For example, metabolic reactions in the vertebrate lungs and kidneys help control the acid–base balance of this environment, at levels suitable for life (Sections 41.4, 43.4, and 43.6). For now, simply think of what happens when the blood level of H^+ decreases and the blood is not as acidic as it should be. At such times, carbonic acid that is dissolved in blood releases H^+ and so becomes the partner base, bicarbonate:

$$\underset{\text{CARBONIC ACID}}{H_2CO_3} \longrightarrow \underset{\text{BICARBONATE}}{HCO_3^-} + H^+$$

When blood becomes more acidic, more H^+ becomes bound to the base, thus forming the partner acid:

$$\underset{\text{BICARBONATE}}{HCO_3^-} + H^+ \longrightarrow \underset{\text{CARBONIC ACID}}{H_2CO_3}$$

Uncontrolled shifts in pH have drastic outcomes. If blood's pH (7.3–7.5) declines even to 7, an individual will enter into a *coma*, a sometimes irreversible state of unconsciousness. An increase to 7.8 can lead to *tetany*, a potentially lethal condition in which skeletal muscles enter a state of uncontrollable contraction. In *acidosis*, carbon dioxide builds up in blood, too much carbonic acid forms, and blood pH severely decreases. *Alkalosis* is an uncorrected increase in blood pH. Both conditions weaken the body and can be lethal.

Salts

Salts are compounds that release ions *other than* H^+ and OH^- in solutions. Salts and water often form when a strong acid and strong base interact. Depending on a solution's pH value, salts can form and dissolve easily. Consider how sodium chloride forms, then dissolves:

$$\underset{\text{HYDROCHLORIC ACID}}{HCl\ (\text{acid})} + \underset{\text{SODIUM HYDROXIDE}}{NaOH\ (\text{base})} \longrightarrow \underset{\text{SODIUM CHLORIDE}}{NaCl\ (\text{salt})} + H_2O$$

$$NaCl \rightarrow Na^+ \quad Cl^- \ (\text{ionization})$$

Many salts dissolve into ions that serve key functions in cells. For example, nerve cell activity depends on ions of sodium, potassium, and calcium. Muscles contract with the help of calcium ions. And water absorption by plant cells depends largely on potassium ions.

Hydrogen ions (H^+) and other ions dissolved in the fluids inside and outside cells affect cell structure and function.

When dissolved in water, acidic substances release H^+, and basic (alkaline) substances accept them. Certain acid–base interactions, as in buffer systems, help maintain the pH value of a fluid—that is, its H^+ concentration.

A buffer system counters slight shifts in pH by releasing hydrogen ions when their concentration is too low or by combining with them when the concentration is too high.

Salts are compounds that release ions other than H^+ and OH^-, and many of those ions have key roles in cell functions.

2.7 SUMMARY

1. Chemistry helps us understand the nature of all the substances that make up cells, organisms, and the Earth, its waters, and the atmosphere. Each substance consists of one or more elements. Of the ninety-two naturally occurring elements, the most common in organisms are oxygen, carbon, hydrogen, and nitrogen. Organisms also have lesser amounts of many other elements, such as calcium, phosphorus, potassium, and sulfur.

2. Table 2.2 summarizes some key chemical terms that you will encounter throughout this book.

3. Elements consist of atoms. An atom has one or more positively charged protons, an equal number of negatively charged electrons, and (except for hydrogen atoms) one or more uncharged neutrons. The protons and neutrons occupy the core region, the atomic nucleus. Most elements have isotopes: two or more forms of atoms with the same number of protons but different numbers of neutrons.

4. An atom has no net charge. But it may lose or gain one or more electrons and so become an ion, which has an overall positive or negative charge.

5. Whether an atom interacts with others depends on the number and arrangement of its electrons, which occupy orbitals (volumes of space) inside a series of shells around the atomic nucleus. Atoms with unfilled orbitals in the outermost shell tend to bond with other elements.

6. Generally, a chemical bond is a union between the electron structures of atoms.

a. In an ionic bond, a positive ion and negative ion stay together because of a mutual attraction of their opposite charges.

b. Various atoms often share one or more pairs of electrons in single, double, or triple covalent bonds. Such electron sharing is equal in nonpolar covalent bonds, and it is unequal in polar covalent bonds. The interacting atoms have no net charge, but the bond is slightly negative at one end and slightly positive at the other.

c. In a hydrogen bond, one covalently bonded atom (such as oxygen or hydrogen) that displays a slight negative charge is weakly attracted to the slight positive charge of a hydrogen atom that is taking part in a different polar covalent bond.

7. Polar covalent bonds join together three atoms in water molecules (two hydrogens, one oxygen). The polarity of a water molecule invites hydrogen bonding between water molecules. Such hydrogen bonding is the basis of liquid water's ability to resist temperature changes more than other fluids do, display internal cohesion, and easily dissolve polar or ionic substances. These properties greatly influence the metabolic activity, structure, shape, and internal organization of cells.

8. By the pH scale, a solution is assigned a number that reflects its H^+ concentration. This ranges from 0 (highest concentration) to 14 (lowest). At pH 7, the H^+ and OH^- concentrations are equal. Acids release H^+ in water, and bases combine with them. Buffer systems help maintain pH values of blood, tissue fluids, and the fluid inside cells.

Table 2.2 Summary of Key Players in the Chemical Basis of Life

ELEMENT	Fundamental form of matter that occupies space, has mass, and cannot be broken apart into a different form of matter by ordinary physical or chemical means.
ATOM	Smallest unit of an element that still retains the characteristic properties of that element.
Proton (p^+)	Positively charged particle of the atomic nucleus. All atoms of an element have the same number of protons, which is the atomic number. A proton without an electron zipping around it is a hydrogen ion (H^+).
Electron (e^-)	Negatively charged particle that can occupy a volume of space (orbital) around an atomic nucleus. All atoms of an element have the same number of electrons. Electrons can be shared or transferred among atoms.
Neutron	Uncharged particle of the nucleus of all atoms except hydrogen. For a given element, the mass number is the number of protons and neutrons in the nucleus.
MOLECULE	Unit of matter in which two or more atoms of the same element, or different ones, are bonded together.
Compound	Molecule composed of two or more different elements in unvarying proportions. Water is an example.
Mixture	Intermingling of two or more elements in proportions that can and usually do vary.
ISOTOPE	One of two or more forms of atoms of an element that differ in their number of neutrons.
Radioisotope	Unstable isotope, having an unbalanced number of protons and neutrons, that emits particles and energy.
Tracer	Molecule of a substance to which a radioisotope is attached. In conjunction with tracking devices, it can be used to follow the movement or destination of that substance in a metabolic pathway, the body, etc.
ION	Atom that has gained or lost one or more electrons, thus becoming positively or negatively charged.
SOLUTE	Any molecule or ion dissolved in some solvent.
Hydrophilic substance	Polar molecule or molecular region that can readily dissolve in water.
Hydrophobic substance	Nonpolar molecule or molecular region that strongly resists dissolving in water.
ACID	Substance that donates H^+ when dissolved in water.
BASE	Substance that accepts H^+ when dissolved in water; OH^- forms directly or indirectly afterward.
SALT	Compound that releases ions other than H^+ or OH^- when dissolved in water.

Review Questions

1. What is an element? Name four elements (and their symbols) that make up more than 95 percent of the body weight of all living organisms. *CI*

2. Define atom, isotope, and radioisotope. *2.1*

3. How many electrons can occupy each orbital around an atomic nucleus? Using the shell model, explain how the orbitals available to electrons are distributed in an atom. *2.3*

4. Define molecule, compound, and mixture. *2.3*

5. Distinguish between:
   a. ionic and hydrogen bonds *2.4*
   b. polar and nonpolar covalent bonds *2.4*
   c. hydrophilic and hydrophobic interactions *2.5*

6. If a water molecule has no net charge, then why does it attract polar molecules and repel nonpolar ones? *2.5*

7. Label the atoms in each water molecule in the sketch below. Indicate which parts of each molecule carry a slight positive charge (+) and which carry a slight negative charge (−). *2.5*

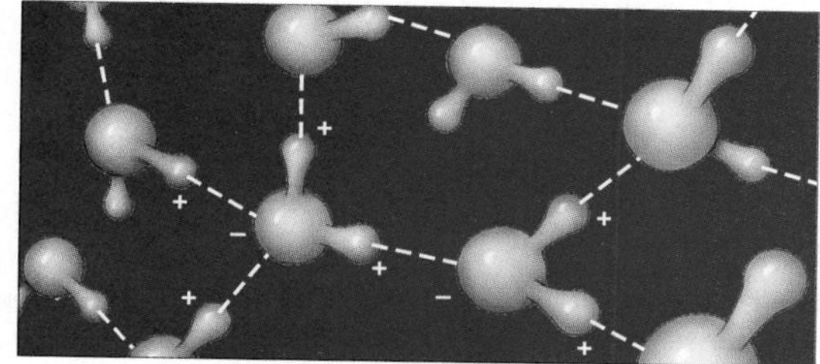
+
+
−
+
+
+
−
+

8. Define acid and base. Then describe the behavior of a weak acid in solutions having a high or low pH value. *2.6*

Self-Quiz (*Answers in Appendix III*)

1. Electrons carry (a) ______ charge.
   a. positive b. negative c. zero

2. Atoms share electrons unequally in a(n) ______ bond.
   a. ionic c. polar covalent
   b. nonpolar covalent d. hydrogen

3. A water molecule shows ______ .
   a. polarity d. solvency
   b. hydrogen-bonding capacity e. a and b
   c. heat resistance f. all of the above

4. In liquid water, spheres of hydration form around ______ .
   a. nonpolar molecules d. solvents
   b. polar molecules e. b and c
   c. ions f. all of the above

5. Hydrogen ions (H^+) are ______ .
   a. the basis of pH values d. dissolved in blood
   b. unbound protons e. both a and b
   c. targets of certain buffers f. all of the above

6. When dissolved in water, a(n) ______ donates H^+; however, a(n) ______ accepts H^+.

7. Match the terms with the most suitable descriptions.

______ trace element	a. weak acid and its partner base work as a pair to counter pH shifts
______ buffer system	b. union between electron structures of two atoms
______ chemical bond	c. less than 0.01% of body weight
______ temperature	d. measure of molecular motion in some defined region

Critical Thinking

1. An ionic compound forms when calcium combines with chlorine. Referring to Figure 2.8, give the compound's formula. (Hint: Be sure the outermost shell of each atom is filled.)

2. David, an inquisitive three-year-old, poked his fingers in the water inside a metal pan on the stove and discovered that it was warm. Then he touched the pan itself and got a nasty burn. Devise a hypothesis to explain why water in a metal pan heats up far more slowly than the pan itself.

3. When molecules absorb microwaves, which are a form of electromagnetic radiation, they move more rapidly. Explain why a microwave oven can heat foods.

4. From what you know about cohesion, devise a hypothesis to explain why water forms droplets.

5. Edward is trying to study a chemical reaction that an enzyme catalyzes (speeds up). H^+ forms in the reaction, but the enzyme is destroyed at low pH. What can he include in his reaction mix to protect the enzyme while he studies the reaction? Explain how your suggestion might solve the problem.

6. Many reactions occur on molecular parts of enzymes and other proteins. Cells must have access to those parts. Through interactions with water and ions, a soluble protein can become dispersed in cellular fluid rather than settling against some cell structure. An electrically charged cushion around it makes this happen. Using the sketch below as a guide, explain the chemical interactions by which such a cushion forms, starting with major bonds in the protein itself.

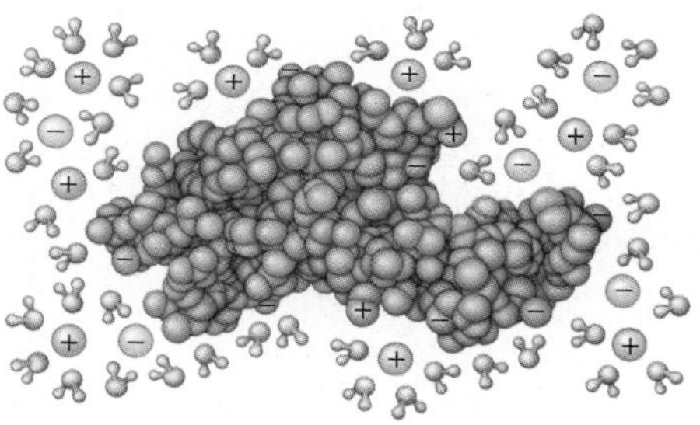

Selected Key Terms

acid *2.6*
atom *2.1*
base *2.6*
bioremediation *CI*
buffer system *2.6*
chemical bond *2.3*
cohesion *2.5*
compound *2.3*
covalent bond *2.4*
electron *2.1*
element *CI*
evaporation *2.5*
hydrogen bond *2.4*
hydrogen ion (H^+) *2.6*
hydrophilic substance *2.5*
hydrophobic substance *2.5*
hydroxide ion (OH^-) *2.6*
ion *2.4*
ionic bond *2.4*
isotope *2.1*
mixture *2.3*
molecule *2.3*
neutron *2.1*
pH scale *2.6*
proton *2.1*
radioisotope *2.1*
salt *2.6*
shell model *2.3*
solute *2.5*
temperature *2.5*
tracer *2.2*

Readings *See also www.infotrac-college.com*

Ritter, P. 1996. *Biochemistry: A Foundation*. Pacific Grove, California: Brooks/Cole. Good, easy-to-read introduction to biochemistry, with plenty of human-interest applications.

3

CARBON COMPOUNDS IN CELLS

Carbon, Carbon, in the Sky—Are You Swinging Low and High?

High in the mountains of the Pacific Northwest, vast forests of conifers have endured another murderously cold winter (Figure 3.1). Like all other living organisms, these evergreen, cone-bearing trees cannot grow or reproduce in the absence of liquid water. Yet through the winter months, water in their habitat is locked away from them, in the form of snow and ice.

The trees get by anyway. Their needle-shaped leaves have an epidermis—a layer of interconnected cells with a thick, waxy wall that restricts water loss. About the only way precious water can escape is through small gaps across the epidermis that can close or open in response to changing conditions. During the cool, dry days of autumn, the trees enter dormancy. Metabolic activities idle and growth ceases, but water conserved inside them is enough to keep their leaf cells alive.

With the arrival of spring, rising temperatures and water from melting snow stimulate renewed growth. Tree roots soak up mineral-laden water. And carbon dioxide, a gaseous molecule of one carbon atom and two oxygen atoms, moves in from the air, through gaps in the leaf epidermis. With their photosynthetic magic, the conifers turn those simple materials into sugars, starches, and other carbon-based compounds. They are premier producers of the northern forests.

Producers, recall, are organisms that use self-made organic compounds as their structural materials and as packets of energy. One way or another, compounds made by producers of forests and other ecosystems all over the world nourish every consumer and decomposer.

Plants of the great prevailing forests and plains at northern latitudes of Canada, the United States, and other parts of the Northern Hemisphere are now breaking dormancy sooner than they did just two decades ago. Why? No one knows for sure, but this change in their life cycles may be one outcome of long-term change in the global climate.

Figure 3.1 Conifers beneath the first snows of winter on Silver Star Mountain, Washington. As is the case for all other organisms, the structure, activities, and very survival of these trees start with the carbon atom and its diverse molecular partners in organic compounds.

Researchers have looked at atmospheric concentrations of carbon dioxide since the early 1950s. Among other things, they found that the concentration shifts with the seasons. It declines during spring and summer, when photosynthesizers take up huge amounts of carbon dioxide from their surroundings. It rises at other times of the year, when huge populations of decomposers that release the gas as a metabolic by-product undergo rapid increases in their population size.

According to the researchers' measurements, the spring decline starts a full week earlier than it did in the mid-1970s. Besides this, the seasonal swings are becoming more pronounced—by as much as 20 percent in Hawaii and a whopping 40 percent in Alaska.

Swings in the atmospheric concentration of carbon dioxide were greatest in 1981 and 1990. Intriguingly, temperatures of the lower atmosphere are rising also, and in 1981 and 1990, they were uncommonly high. Are we in the midst of a long-term, worldwide rise in atmospheric temperature? And is this *global warming* promoting a longer growing season, hence the wider seasonal swings?

The picture gets more intricate. We humans burn great quantities of coal, gasoline, and other fossil fuels for energy. Fossil fuels are rich in carbon, which is released (as carbon dioxide) by the burning processes. The released carbon may be contributing to the global warming, in ways you will read about in later chapters.

For now, the point to keep in mind is this: *Carbon permeates the world of life—from the energy-requiring activities and structural organization of individual cells, to physical and chemical conditions that span the globe and that influence life everywhere.*

With this chapter, we turn to life-giving properties that emerge out of the molecular structure of carbon-rich compounds. Study the chapter well, especially the summary in Table 3.1. It will serve as your foundation for understanding how different organisms put such compounds together, how they use them, and how the effects of those uses ripple through the biosphere.

KEY CONCEPTS

1. Organic compounds have a backbone of one or more carbon atoms to which hydrogen, oxygen, nitrogen, and other atoms are attached. We define cells partly by their capacity to assemble the organic compounds known as carbohydrates, lipids, proteins, and nucleic acids.

2. Cells put together large biological molecules from their pools of smaller organic compounds, which include simple sugars, fatty acids, amino acids, and nucleotides.

3. Glucose and other simple sugars are carbohydrates. So are organic compounds composed of two or more sugar units, of one or more types, that are covalently bonded together. The most complex carbohydrates that cells assemble are polysaccharides, many of which consist of hundreds or thousands of sugar units.

4. Lipids are greasy or oily compounds that show little tendency to dissolve in water but dissolve in nonpolar compounds, including other lipids. They include the neutral fats (triglycerides), phospholipids, waxes, and sterols.

5. Cells use carbohydrates and lipids as building blocks and as their major sources of energy.

6. Proteins have truly diverse roles. Many are structural materials. Many are enzymes, a type of molecule that enormously increases the rate of specific metabolic reactions. Other kinds transport cell substances, contribute to cell movements, trigger changes in cell activities, and defend the body against injury and disease.

7. For living organisms, ATP and other nucleotides are crucial players in metabolism. DNA and RNA, strandlike nucleic acids assembled from nucleotide subunits, are the basis of inheritance and reproduction.

3.1 PROPERTIES OF ORGANIC COMPOUNDS

The Molecules of Life

Under present-day conditions in nature, *only living cells synthesize carbohydrates, lipids, proteins, and nucleic acids.* These are the molecules characteristic of life. Different classes of biological molecules act as the cells' packets of instantly available energy, energy stores, structural materials, metabolic workers, libraries of hereditary information, and cell-to-cell signals.

The molecules of life are **organic compounds**, with hydrogen and often other elements covalently bonded to carbon atoms. The term is a holdover from a time when chemists thought "organic" substances were the ones they got from animals and vegetables, as opposed to "inorganic" substances they got from minerals. The term persists, even though researchers now synthesize organic compounds in laboratories. And it persists even though there are reasons to believe organic compounds were present on Earth *before* organisms were.

Carbon's Bonding Behavior

By far, organisms consist mainly of oxygen, hydrogen, and carbon (Figure 2.2). Much of the oxygen and the hydrogen is in the form of water. Remove the water, and carbon makes up more than half of what's left.

Carbon's importance in life arises from its versatile bonding behavior. As shown in the sketch below, *each carbon atom can share pairs of electrons with as many as four other atoms.* Each covalent bond formed in this way is relatively stable. Such bonds join carbon atoms together in chains. These form a backbone to which many other elements, including hydrogen, oxygen, and nitrogen, become attached.

single covalent bond
carbon atom

The bonding arrangement is the start of wonderful three-dimensional shapes of organic compounds. A chain of carbon atoms, bonded covalently one after another, forms a backbone from which other atoms can project:

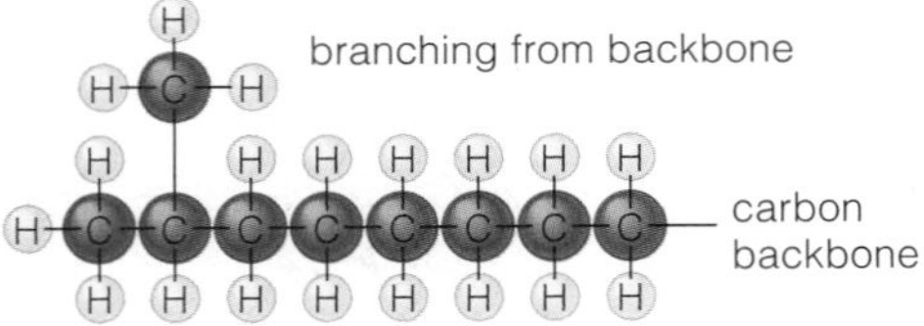
branching from backbone
carbon backbone

The backbone can coil back on itself in a ring structure, which we can diagram in such ways as:

or
carbon rings

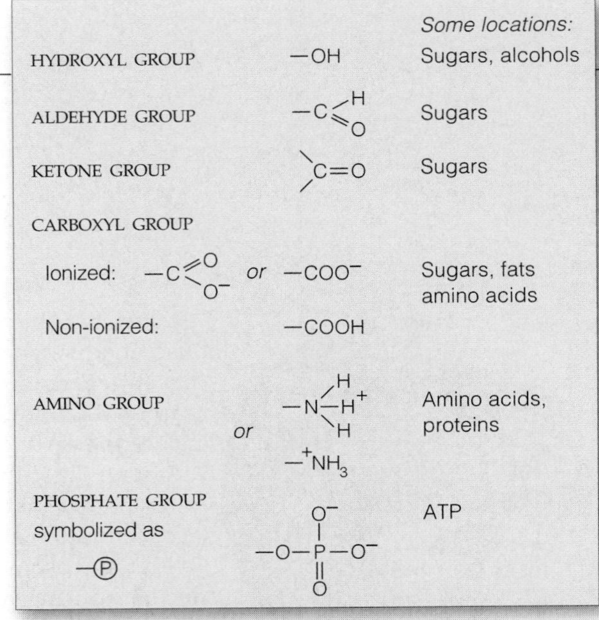
Some locations:
HYDROXYL GROUP —OH Sugars, alcohols
ALDEHYDE GROUP Sugars
KETONE GROUP Sugars
CARBOXYL GROUP
Ionized: or —COO⁻ Sugars, fats amino acids
Non-ionized: —COOH
AMINO GROUP or Amino acids, proteins
PHOSPHATE GROUP symbolized as —Ⓟ ATP

Figure 3.2 Examples of functional groups. 

Functional Groups

A carbon backbone with only hydrogen atoms bonded covalently to it is a hydrocarbon, which is a very stable structure. Besides hydrogen atoms, biological molecules also have **functional groups**: various kinds of atoms or clusters of them covalently bonded to the backbone.

To get a sense of their importance, consider a few of the functional groups shown in Figure 3.2. Sugars and other organic compounds classified as **alcohols** have one or more hydroxyl groups (—OH). Alcohols dissolve quickly in water, because water molecules easily form hydrogen bonds with —OH groups. The backbone of a protein forms by reactions between amino groups and carboxyl groups. As you will see, the backbone is the start of bonding patterns that produce the protein's three-dimensional structure. As other examples, amino groups can combine with H^+ and act as buffers against decreases in pH. And the functional groups shown in Figure 3.3 are a molecular starting point for differences between the males and females of many species.

How Do Cells Build Organic Compounds?

Using carbon (from carbon dioxide), water, and sunlight (as an energy source), the photosynthetic cells you read about earlier put together simple sugar molecules. Like all living cells, they also use such molecules as a starting point for assembling other small molecules, especially the fatty acids, amino acids, and nucleotides. As you will read shortly, cells use some assortment of these four classes of small organic compounds as subunits for building all the organic compounds they require for their structure and functioning.

How do they do it? It will take more than one chapter to sketch out answers (and best guesses) to the question.

Further reading: Student Guide to InfoTrac on web site

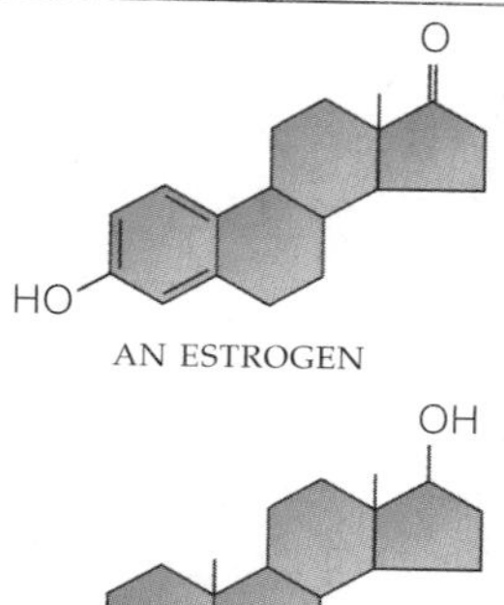

O
HO
AN ESTROGEN
OH
O
TESTOSTERONE

FEMALE WOOD DUCK
MALE WOOD DUCK

Figure 3.3 Notable differences in traits between male and female wood ducks (*Aix sponsa*). Two different sex hormones have key roles in the development of feather color and other traits that help males and females recognize each other (and therefore influence reproductive success). Both of the hormones, testosterone and one of the estrogens, have the same carbon ring structure. As you can see, however, the ring structures have different functional groups attached to them.

Figure 3.4 Examples of metabolic reactions by which most biological molecules are put together, rearranged, and broken apart.

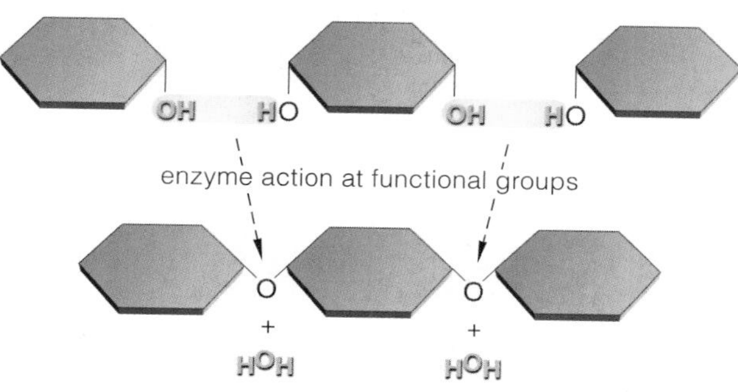

OH
HO
OH
HO
enzyme action at functional groups
O
+
HOH
O
+
HOH

a Two condensation reactions. Enzymes remove an —OH group and H atom from two molecules, which covalently bond as a larger molecule. Two water molecules form.

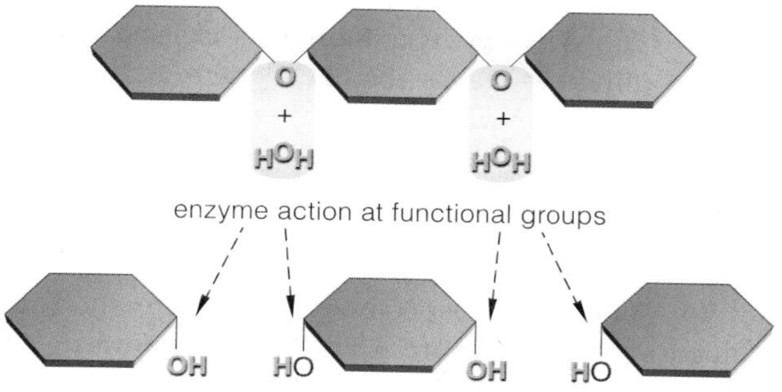

O
+
HOH
O
+
HOH
enzyme action at functional groups
OH
HO
OH
HO

b Hydrolysis, a water-requiring cleavage reaction. Enzyme action splits a molecule into three parts, then attaches an H atom and an —OH group derived from a water molecule to each exposed site.

At this point in your reading, simply become aware that the reactions by which cells build organic compounds, and even rearrange them and break them apart, require more than an energy input. They also require the class of proteins called **enzymes**, which make specific metabolic reactions proceed faster than they would on their own. Different enzymes mediate different kinds of reactions. Most of the reactions fall into five categories, which you will encounter in chapters to come:

1. *Functional-group transfer*. One molecule gives up a functional group, which another molecule accepts.

2. *Electron transfer*. One or more electrons stripped from one molecule are donated to another molecule.

3. *Rearrangement*. A juggling of its internal bonds converts one type of organic compound into another.

4. *Condensation*. Through covalent bonding, two molecules combine to form a larger molecule.

5. *Cleavage*. A molecule splits into two smaller ones.

To get a sense of what goes on, consider two examples of these events. First, in many **condensation reactions**, enzymes remove a hydroxyl group from one molecule and an H atom from another, then speed the formation of a covalent bond between the two molecules at their exposed sites (Figure 3.4*a*). As a typical but incidental outcome of the reaction, the discarded atoms join to form a molecule of water. A series of condensation reactions can produce starches and other polymers. A polymer is a large molecule having three to millions of subunits, which may or may not be identical. Biologists call the subunits monomers, as in the sugar monomers of starch.

As the second example, a cleavage reaction called **hydrolysis** is like condensation in reverse (Figure 3.4*b*). Enzymes that recognize specific functional groups split molecules into two or more parts, then attach an —OH group and a hydrogen atom derived from a molecule of water to the exposed sites. With hydrolysis, cells can cleave large polymers such as starch into smaller units when these are required for building blocks or energy.

Organic compounds have diverse, three-dimensional shapes and functions. The diversity starts at their carbon backbones and with bonding arrangements that arise from it.

Functional groups that are covalently bonded to the carbon backbone add enormously to the structural and functional diversity of organic compounds.

Carbohydrates, lipids, proteins, and nucleic acids are the main biological molecules, the organic compounds that only living cells can assemble under conditions that now occur in nature.

Cells assemble, rearrange, and degrade organic compounds mainly through enzyme-mediated reactions involving the transfer of functional groups or electrons, rearrangement of internal bonds, and a combining or splitting of molecules.

3.2 CARBOHYDRATES

Consider first the carbohydrates—the most abundant of all biological molecules, which cells use as structural materials and transportable or storage forms of energy. Most **carbohydrates** consist of carbon, hydrogen, and oxygen in a 1:2:1 ratio, which also may be written as $(CH_2O)_n$. Three classes are called the **monosaccharides**, **oligosaccharides**, and **polysaccharides**.

The Simple Sugars

"Saccharide" comes from a Greek word meaning sugar. A *mono*saccharide, meaning "one monomer of sugar," is the simplest carbohydrate. It has at least two —OH groups joined to the carbon backbone plus an aldehyde or a ketone group. Most monosaccharides are sweet tasting and readily dissolve in water. The most common have a backbone of five or six carbon atoms that tends to form a ring structure when dissolved in cells or body fluids. Ribose and deoxyribose, the sugar components of RNA and DNA, respectively, have five carbon atoms. Glucose has six (Figure 3.5*a*). Besides being the main energy source for most organisms, glucose is a precursor (parent molecule) of many compounds and a building block for larger carbohydrates. Vitamin C (a sugar acid) and glycerol (an alcohol with three —OH groups) are other examples of compounds having sugar monomers.

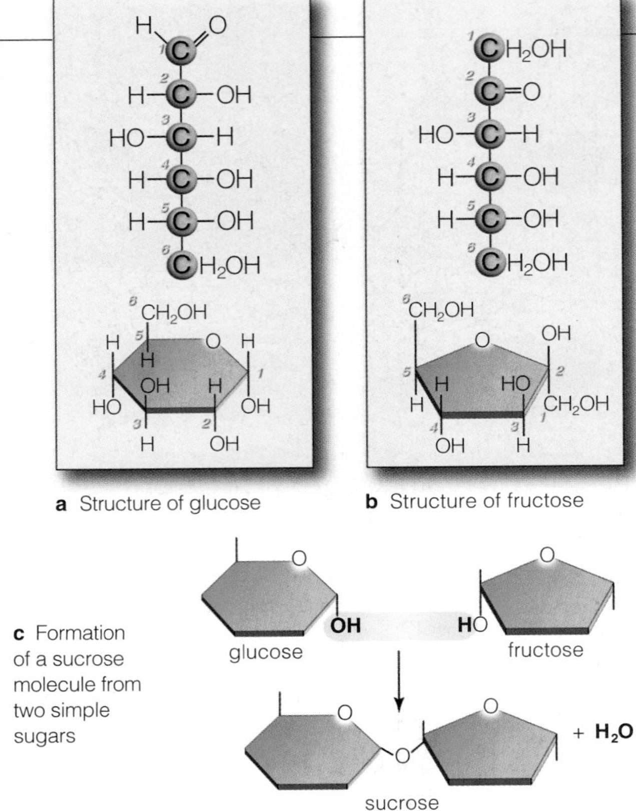

a Structure of glucose
b Structure of fructose
c Formation of a sucrose molecule from two simple sugars
glucose
fructose
OH
HO
sucrose
+ H_2O

Figure 3.5 Straight-chain and ring forms of (**a**) glucose and (**b**) fructose. For reference purposes, carbon atoms of simple sugars are commonly numbered in sequence, starting at the end closest to the molecule's aldehyde or ketone group. (**c**) Condensation of two monosaccharides into a disaccharide.

Short-Chain Carbohydrates

Unlike the simple sugars, an *oligo*saccharide is a short chain of two or more sugar monomers that are bonded covalently. (*Oligo-* means "a few.") The type known as *di*saccharides consists of just two sugar units. Lactose, sucrose, and maltose are examples. Lactose (a glucose and a galactose unit) is a milk sugar. Sucrose, the most plentiful sugar in nature, consists of one glucose and one fructose unit, as in Figure 3.5*c*. Plants convert their stores of carbohydrates to sucrose, which can be easily transported through their leaves, stems, and roots. Table sugar is sucrose crystallized from sugarcane and sugar beets. Proteins and other large molecules often have oligosaccharides attached as side chains to the carbon backbone. Some chains take part in the body's defenses against disease, others in cell membrane functions.

Complex Carbohydrates

The "complex" carbohydrates, or *poly*saccharides, are straight or branched chains of many sugar monomers (often hundreds or thousands) of the same or different types. Cellulose, starch, and glycogen, the most common polysaccharides, consist only of glucose. In cell walls of plants, cellulose is a structural material (Figure 3.6*a*). Like steel rods in reinforced concrete, fibers of cellulose are tough and insoluble; they withstand considerable weight and stress. Plants store their photosynthetically produced sugars as large starch molecules (Figure 3.6*b*). Enzymes can readily hydrolyze starch to glucose units.

If both cellulose and starch consist of glucose, why are their properties so different? The answer starts with differences in covalent bonding patterns between their monomers, which in both are bonded together in chains.

In cellulose, many glucose chains stretch out side by side and hydrogen-bond to one another at —OH groups (Figure 3.7*a*). This bonding arrangement stabilizes the chains into a tightly bundled pattern that resists being digested, at least by most enzymes.

In starch, the pattern of covalent bonding positions each sugar monomer at an angle relative to the next monomer in line. The chain ends up coiling like a spiral staircase (Figure 3.7*b*). The coils are not particularly stable. In the kinds of starches with branched chains, they are even less stable. A great many —OH groups project outward from the coiled chains, and this makes them readily accessible to enzymes.

In animals, glycogen is the sugar-storage equivalent of starch in plants. Muscle and liver cells house large stores of it. When the level of sugar in blood decreases, liver cells degrade glycogen, so glucose is released and

Further reading: Student Guide to InfoTrac on web site —

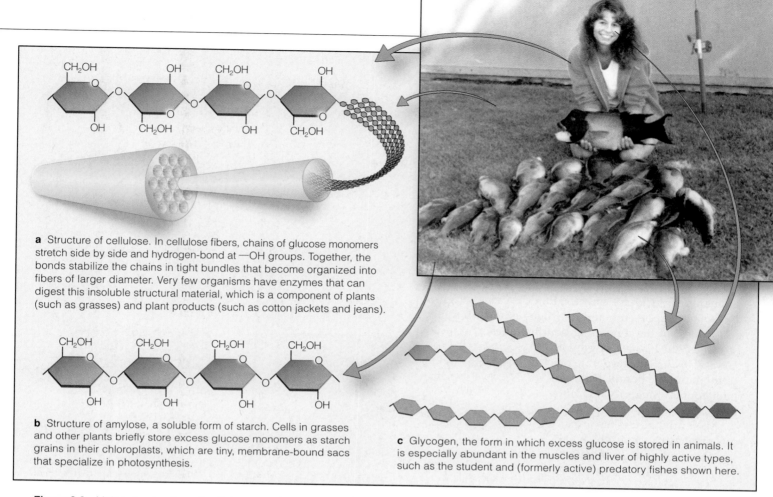

a Structure of cellulose. In cellulose fibers, chains of glucose monomers stretch side by side and hydrogen-bond at —OH groups. Together, the bonds stabilize the chains in tight bundles that become organized into fibers of larger diameter. Very few organisms have enzymes that can digest this insoluble structural material, which is a component of plants (such as grasses) and plant products (such as cotton jackets and jeans).

b Structure of amylose, a soluble form of starch. Cells in grasses and other plants briefly store excess glucose monomers as starch grains in their chloroplasts, which are tiny, membrane-bound sacs that specialize in photosynthesis.

c Glycogen, the form in which excess glucose is stored in animals. It is especially abundant in the muscles and liver of highly active types, such as the student and (formerly active) predatory fishes shown here.

Figure 3.6 Molecular structure of cellulose, starch, and glycogen, and their typical locations in a few organisms. All three carbohydrates consist only of glucose monomers.

Figure 3.7 Comparison of bonding patterns between glucose monomers in cellulose and in starch. (**a**) In cellulose, bonds form between monomers of neighboring glucose chains. This bonding pattern stabilizes the chains and allows them to become tightly bundled. (**b**) In amylose, a form of starch, the bonding pattern between monomers causes the chains to coil. The coiling orients the bonds in a way that is accessible to enzymes.

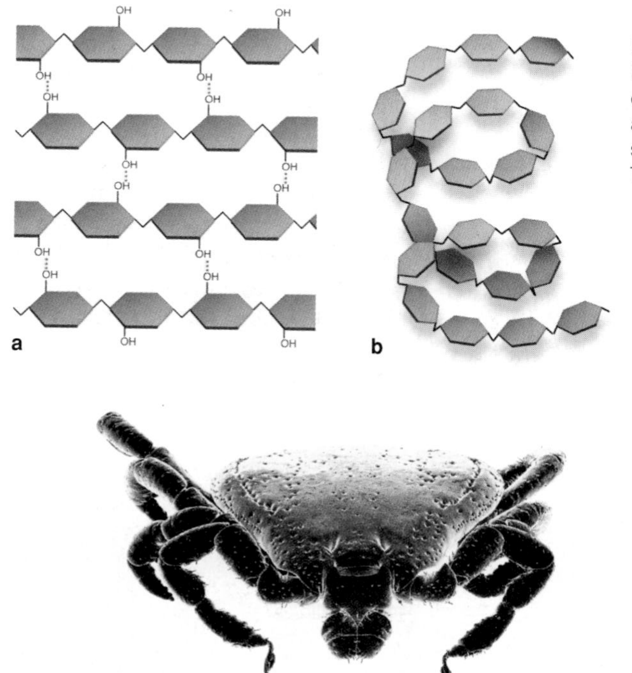
a
b

Figure 3.8 Scanning electron micrograph of a tick. Its body covering is a protective cuticle reinforced with chitin.

enters the bloodstream. When you exercise strenuously but briefly, your muscle cells tap their glycogen stores for a rapid burst of energy. Figure 3.6*c* represents a few of glycogen's many branchings.

The polysaccharide chitin has nitrogen-containing groups attached to its glucose monomers. It is the main structural material for the external skeletons and other hard body parts of crabs, earthworms, insects, ticks, and many other animals (Figure 3.8). Chitin also is the main structural material in the cell walls of many fungi.

The simple sugars (such as glucose), oligosaccharides, and polysaccharides (such as starch) are carbohydrates. Every cell requires carbohydrates as structural materials, stored forms of energy, and transportable packets of energy.

3.3 LIPIDS

Being mostly hydrocarbon, **lipids** show little tendency to dissolve in water, although they readily dissolve in nonpolar substances. All are greasy or oily to the touch. Cells use different kinds as their main energy reservoirs, as structural materials (for example, in cell membranes and surface coatings), and as signaling molecules. Let's look first at the kinds with fatty acid components—the fats, phospholipids, and waxes. We also will consider the sterols, each with a backbone of four carbon rings.

Fats and Fatty Acids

Lipids called **fats** have one, two, or three fatty acids attached to glycerol. Each **fatty acid** has a backbone of as many as thirty-six carbon atoms, a carboxyl group (—COOH) at one end, and hydrogen atoms occupying most or all of the remaining bonding sites. It typically stretches out like a flexible tail. *Unsaturated* tails incorporate one or more double bonds. *Saturated* tails contain single bonds only. Figure 3.9*a* shows examples.

Most animal fats have many saturated fatty acids, which pack together by weak interactions. They stay solid at temperatures that keep most plant fats liquid, as "vegetable oils." The packing interactions in plant fats are not as stable because of rigid kinks in their fatty acid tails. That is why vegetable oils flow freely.

Butter, lard, vegetable oils, and other natural fats consist mostly of **triglycerides**: neutral fats having three fatty acid tails attached to glycerol (Figure 3.9*b*). Triglycerides are the body's most abundant lipids and its richest energy source. Gram for gram, they yield more than twice as much energy when broken down, compared to starches and other complex carbohydrates. Notable quantities of triglycerides are stored as droplets in the cells of body fat (that is, adipose tissue) in vertebrates. A thick layer of triglycerides under the skin has survival value for some animals, including the penguins shown in Figure 3.10. It helps insulate their body against near-freezing temperatures of their habitats.

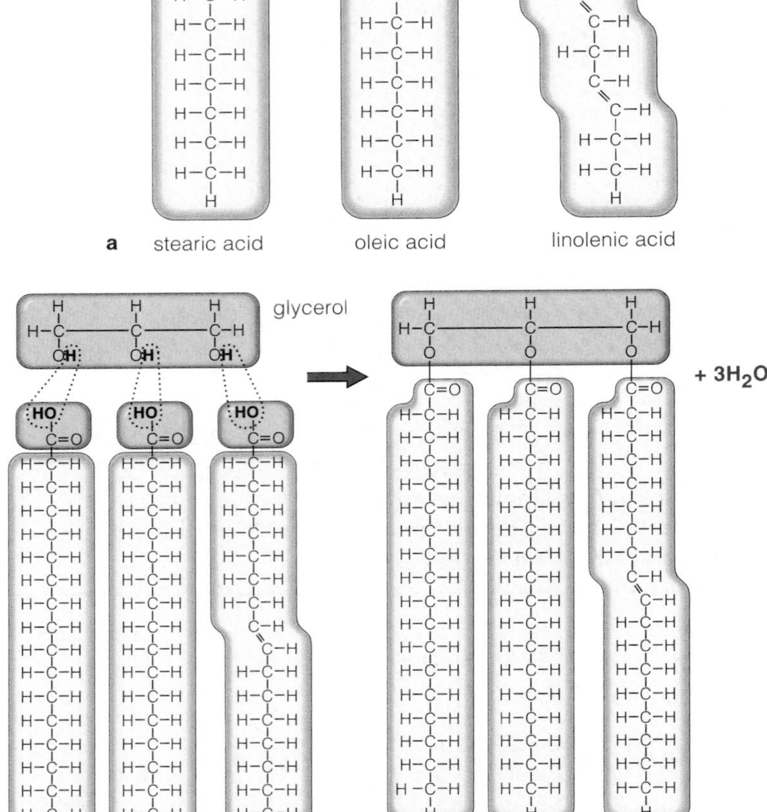

a
stearic acid
oleic acid
linolenic acid
glycerol
+ $3H_2O$
triglyceride
b
three fatty acids

Figure 3.9 (**a**) Structural formulas for three fatty acids. In stearic acid, the carbon backbone is fully saturated with hydrogen atoms. Oleic acid, with a double bond in its backbone, is an unsaturated fatty acid. Linolenic acid, with its three double bonds, is a "polyunsaturated" fatty acid. (**b**) Condensation of fatty acids and glycerol into a triglyceride.

Figure 3.10 Triglyceride-protected penguins taking the plunge.

Further reading: Student Guide to InfoTrac on web site →

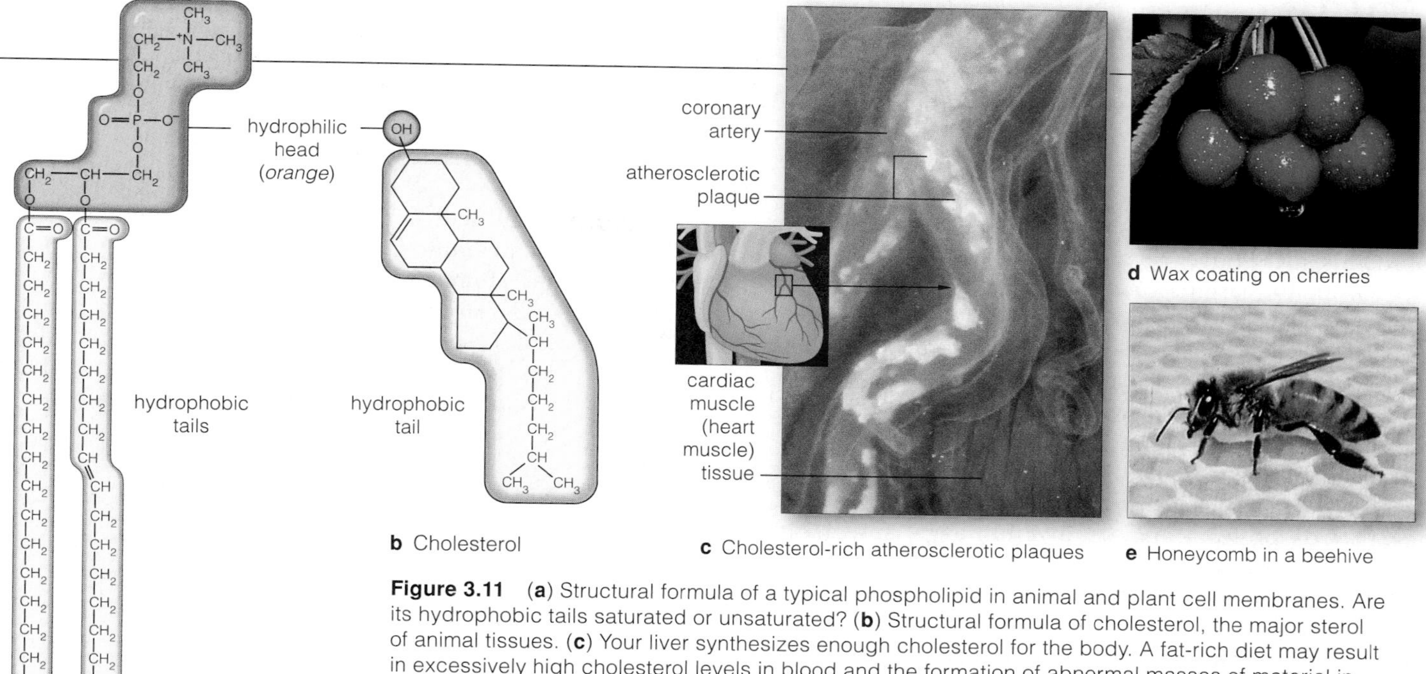
hydrophilic
head
(orange)
hydrophobic
tails
a One of the
phospholipids
hydrophobic
tail
b Cholesterol
coronary
artery
atherosclerotic
plaque
cardiac
muscle
(heart
muscle)
tissue
c Cholesterol-rich atherosclerotic plaques
d Wax coating on cherries
e Honeycomb in a beehive

Figure 3.11 (**a**) Structural formula of a typical phospholipid in animal and plant cell membranes. Are its hydrophobic tails saturated or unsaturated? (**b**) Structural formula of cholesterol, the major sterol of animal tissues. (**c**) Your liver synthesizes enough cholesterol for the body. A fat-rich diet may result in excessively high cholesterol levels in blood and the formation of abnormal masses of material in certain blood vessels (arteries). Such *atherosclerotic plaques* may clog the arteries that deliver blood to the heart. (**d**) Demonstration of the water-repelling attribute of a cherry cuticle. (**e**) Honeycomb, a structural material constructed of a firm, water-repellent, waxy secretion called beeswax.

One fatty acid is a precursor of eicosanoids, a class of local signaling molecules that include prostaglandins. Different eicosanoids bind to receptors on cells and help regulate many physiological processes, such as muscle contraction, message transmission through the nervous system, inflammation, and immune responses.

Phospholipids

A **phospholipid** has a glycerol backbone, two fatty acid tails, and a hydrophilic "head" with a phosphate group and another polar group (Figure 3.11*a*). Phospholipids are the main materials of cell membranes, which have two layers of lipids. Heads of one layer are dissolved in the cell's fluid interior, and heads of the other layer are dissolved in the surroundings. Sandwiched between the two are all the fatty acid tails, which are hydrophobic.

Sterols and Their Derivatives

Sterols are among the many lipids that have no fatty acids. Sterols differ in the number, position, and type of their functional groups, but all have a rigid backbone of four fused-together carbon rings:

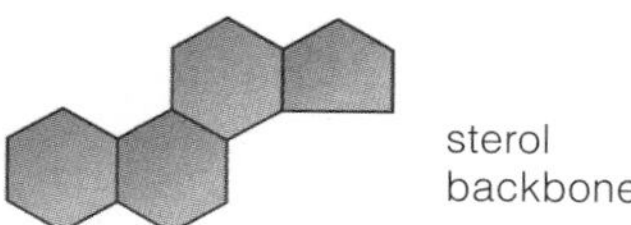
sterol
backbone

Sterols occur in eukaryotic cell membranes. Cholesterol is the most common type in tissues of animals (Figure 3.11*b*,*c*). Cells also remodel cholesterol into compounds such as vitamin D (required for good bones and teeth), steroids, and bile salts. Steroids include sex hormones, such as estrogens and testosterone, that govern sexual traits and gamete formation. Bile salts have roles in the digestion of fats in the small intestine.

Waxes

The lipids called **waxes** have long-chain fatty acids tightly packed and linked to long-chain alcohols or to carbon rings. They have a firm consistency and repel water. Waxes and another lipid, cutin, make up most of the cuticle that covers the aboveground plant parts. The covering helps plants conserve water and fend off some parasites. A waxy cherry cuticle is an example (Figure 3.11*d*). In many animals, waxy secretions from cells are incorporated in coatings that protect, lubricate, and impart pliability to skin or hair. Among waterfowl and other birds, wax secretions help keep feathers dry. As another example, beeswax is the material of choice when bees construct their honeycombs (Figure 3.11*e*).

Being largely hydrocarbon, lipids can dissolve in nonpolar substances but resist dissolving in water.

Triglycerides (neutral fats), which have a glycerol head and three fatty acid tails, are the body's main energy reservoirs. Phospholipids are the main components of cell membranes.

Sterols such as cholesterol are membrane components as well as precursors of steroid hormones and other compounds. Waxes are firm yet pliable components of water-repelling and lubricating substances.

AMINO ACIDS AND THE PRIMARY STRUCTURE OF PROTEINS

Have you ever wondered how permanent waves work (Figure 3.12)? The characteristics of many mammalian body parts—including hair—start with the structure of proteins, just as they do for all other organisms.

Of all the large biological molecules, **proteins** are the most diverse. Proteins of the class called enzymes make metabolic events proceed much faster than they otherwise would. Structural proteins are the stuff of spider webs, butterfly wings, feathers, cartilage and bone, and a dizzying array of other body parts and products. Transport proteins move molecules and ions across cell membranes and cart them about through body fluids. Nutritious proteins abound in milk, eggs, and many seeds. Protein hormones and other regulatory types are signals for change in cell activities. Many proteins act as weapons against disease-causing bacteria and other invaders. Amazingly, cells build diverse proteins from their pools of only twenty kinds of amino acids!

a

b

Figure 3.12 Appearance of the hair of actress Nicole Kidman (**a**) before and (**b**) after a permanent wave. The difference, as you will see from this section and the next, starts with strings of amino acids in polypeptide chains.

Amino Acid Structure

Every **amino acid** is a small organic compound that consists of an amino group, a carboxyl group (an acid), a hydrogen atom, and one or more atoms known as its R group. As you can see from the structural formula in Figure 3.13, these parts generally are covalently bonded to the same carbon atom. Figure 3.14 shows a number of amino acids that we will consider later in the book.

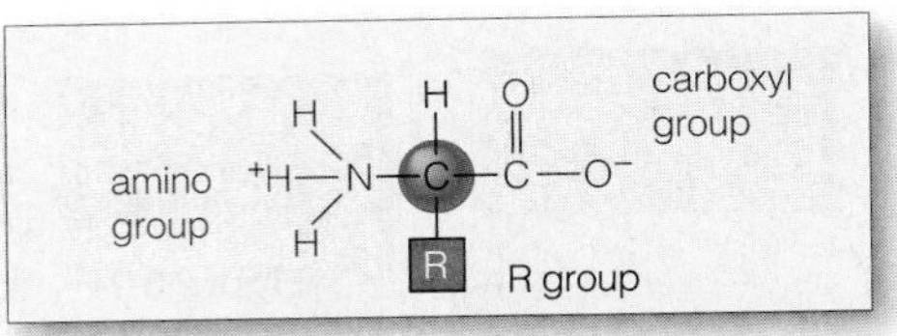
carboxyl group
amino group
R group

Figure 3.13 Generalized structural formula for amino acids.

What Is A Protein's Primary Structure?

When a cell synthesizes a protein, amino acids become linked, one after the other, by peptide bonds. As Figure 3.15 shows, this is the type of covalent bond that forms between one amino acid's amino group (NH_3^+) and the carboxyl group ($-COO^-$) of the next amino acid.

When peptide bonds join two amino acids together, we have a dipeptide. When they join three or more, we have a **polypeptide chain**. In such chains, the carbon

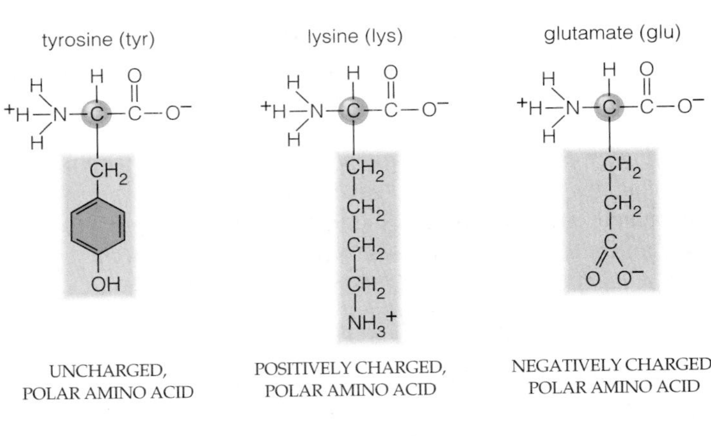
tyrosine (tyr)
lysine (lys)
glutamate (glu)
UNCHARGED, POLAR AMINO ACID
POSITIVELY CHARGED, POLAR AMINO ACID
NEGATIVELY CHARGED, POLAR AMINO ACID

Figure 3.14 Structural formulas for eight of the twenty common amino acids. *Green* boxes highlight the R groups, which are side chains with functional groups. Each type of side chain contributes in a major way to the distinctive properties of each amino acid.

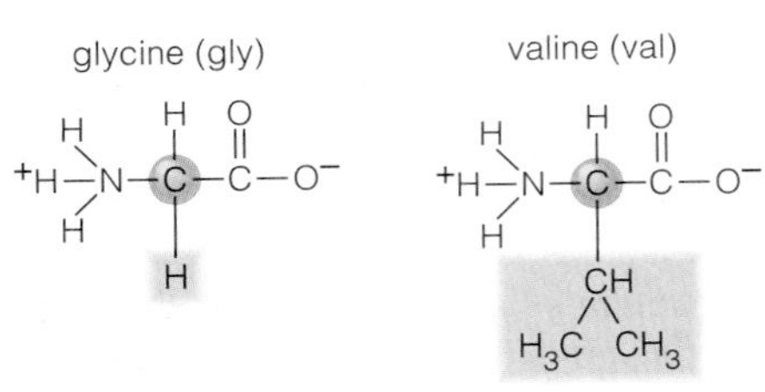
glycine (gly)
valine (val)

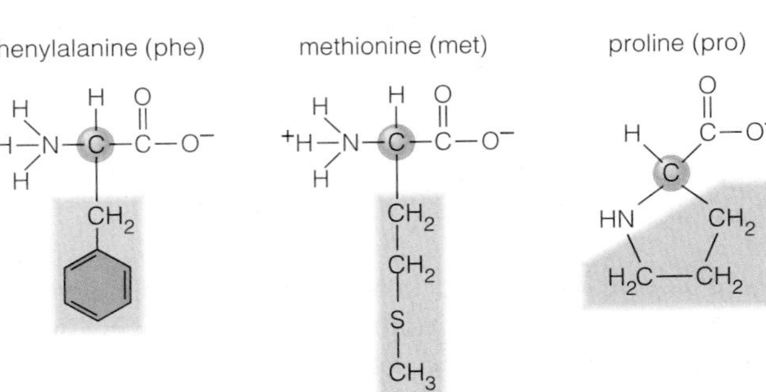
phenylalanine (phe)
methionine (met)
proline (pro)

FIVE OF THE NONPOLAR AMINO ACIDS

Further reading: Student Guide to InfoTrac on web site

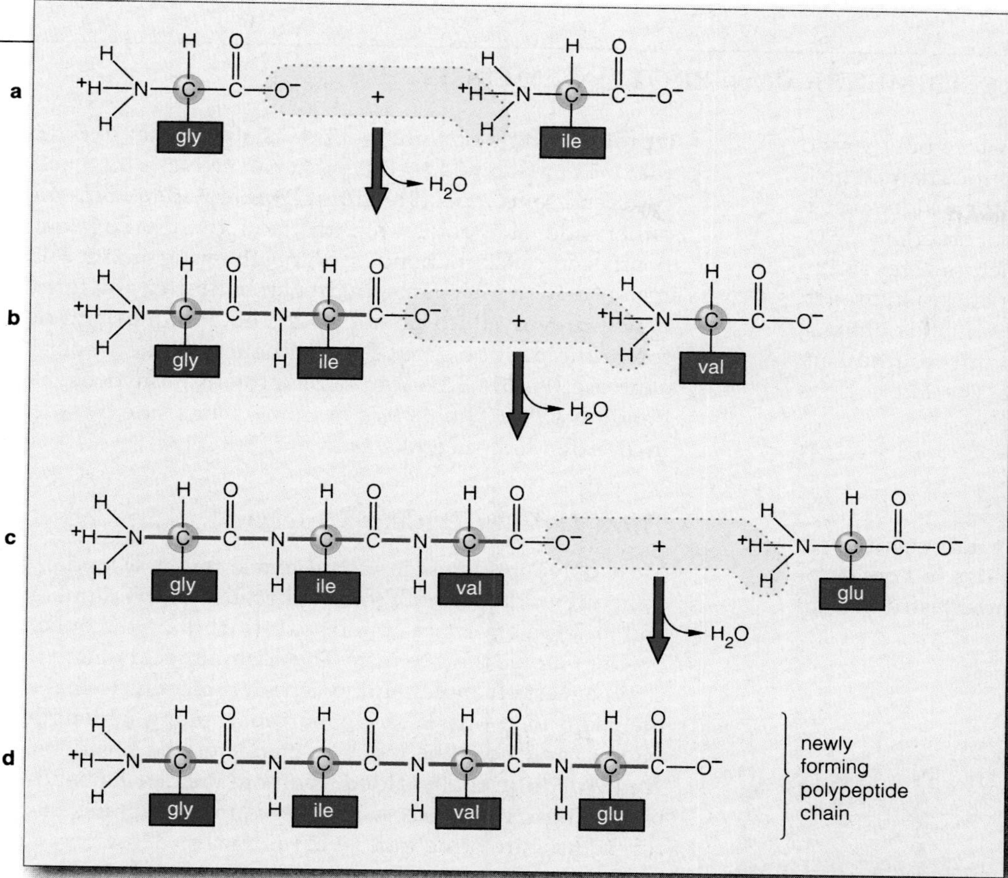
a
gly
ile
H_2O
b
gly
ile
val
H_2O
c
gly
ile
val
glu
H_2O
d
gly
ile
val
glu
newly forming polypeptide chain

Figure 3.15 Peptide bond formation during protein synthesis.

(**a**) The first two amino acids shown are glycine (gly) and isoleucine (ile). They are at the start of the sequence for one of two polypeptide chains that make up the protein insulin in cattle.

(**b**) Through a condensation reaction, the isoleucine becomes joined to the glycine by a peptide bond. A water molecule forms as a by-product of the reaction.

(**c**) A peptide bond forms between isoleucine and valine (val), another amino acid, and water again forms.

(**d**) Remember, DNA specifies the order in which the different kinds of amino acids follow one another in a growing polypeptide chain. In this case, glutamate (glu) is the fourth amino acid specified. Ultimately, the result is one of the two polypeptide chains of an insulin molecule, as in Figure 3.16. Chapter 14 has more details on the steps that lead from DNA instructions to proteins.

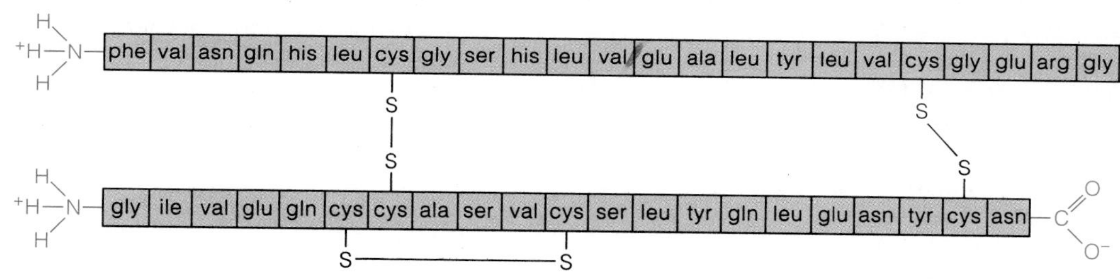
phe val asn gln his leu cys gly ser his leu val glu ala leu tyr leu val cys gly glu arg gly phe phe tyr thr pro lys ala
gly ile val glu gln cys cys ala ser val cys ser leu tyr gln leu glu asn tyr cys asn
S
S

Figure 3.16 Diagram of the amino acid sequence for an insulin molecule (in cattle), as deduced by Frederick Sanger. This protein, which functions as a hormone, is composed of two polypeptide chains. Two disulfide bridges (—S—S—), each formed by a condensation reaction at two R groups (sulfhydryl groups), link the chains together.

backbone has nitrogen atoms positioned in this regular pattern: —N—C—C—N—C—C—.

For each particular kind of protein, different amino acid units are selected one at a time from the twenty kinds available. Their orderly progression is prescribed by the cell's DNA. Overall, the resulting sequence of amino acids is unique for each kind of protein, and it is the protein's *primary* structure (Figures 3.15 and 3.16).

Now consider this: Different cells make thousands of different proteins. Many of the proteins are fibrous, with polypeptide chains organized as strands or sheets. Collectively, many such molecules contribute to the shape and internal organization of cells. Other kinds of proteins are globular, with one or more polypeptide chains folded in compact, rather rounded shapes. Most enzymes are globular proteins. So are actin and other proteins that contribute to cell movement.

Regardless of the type of protein, its shape and its function arise from the primary structure—that is, from information built into its amino acid sequence. As you will see next, that information dictates which parts of a polypeptide chain will coil, bend, or interact with other chains nearby. And the type and arrangement of atoms in coiled, stretched-out, or folded regions determine whether that protein will function as, say, an enzyme, a transporter, a receptor, or even an inadvertent target for a bacterium or virus.

A protein consists of one or more polypeptide chains in which amino acids are strung together. The amino acid sequence (which kind of amino acid follows another in the chain) is unique for each kind of protein and gives rise to its unique structure, chemical behavior, and function.

3.5 HOW DOES A THREE-DIMENSIONAL PROTEIN EMERGE?

The preceding section gave you a sense of how amino acids are strung together in a polypeptide chain, which is a protein's primary structure. Now consider just a few examples of the protein shapes that emerge.

For the most part, the primary structure gives rise to a protein's shape in two ways. First, it allows hydrogen bonds to form between different amino acids along a polypeptide chain's length. Second, it puts R groups in positions that allow them to interact. The interactions force the chain to bend and twist.

Second Level of Protein Structure

Hydrogen bonds form at regular, short intervals along a new polypeptide chain, and they give rise to a coiled or extended pattern known as the secondary structure of a protein. Think of a polypeptide chain as a set of rigid playing cards joined by links that can swivel a bit. Each card represents a peptide group (Figure 3.17). Atoms on either side of it rotate slightly around their covalent bonds and can form bonds with neighboring atoms. For instance, in many chains, hydrogen bonds readily form between every third amino acid. The bonding pattern forces the peptide groups to coil helically, like a spiral staircase (Figure 3.17*a*). In other proteins, a hydrogen-bonding pattern holds two or more chains side by side in an extended, sheetlike array (Figure 3.17*b*).

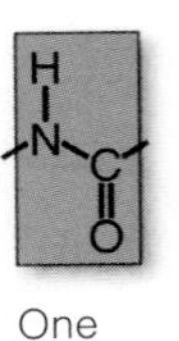

One peptide group

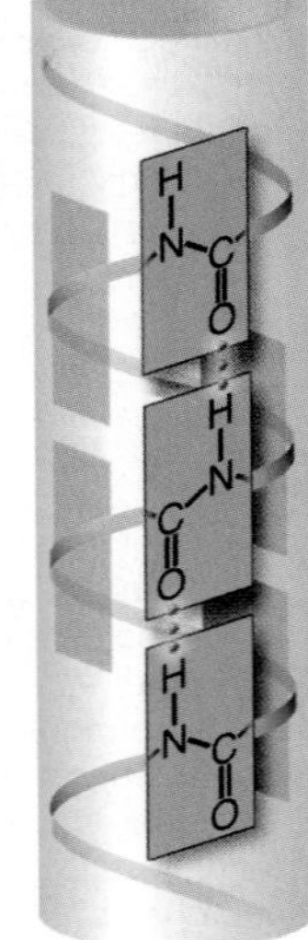

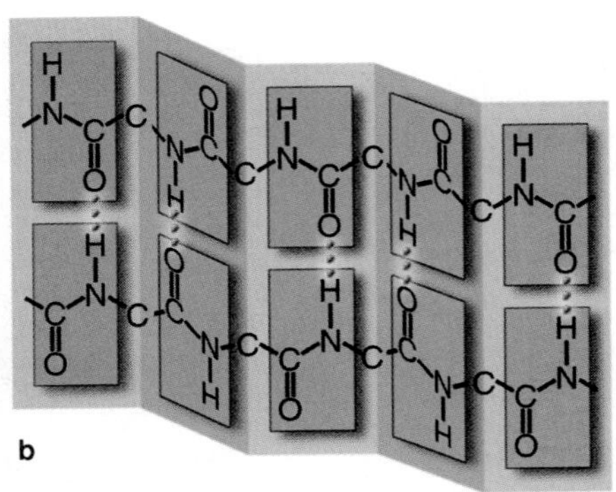

Figure 3.17 Bonding patterns between peptide groups of polypeptide chains. Extensive hydrogen bonding (*dotted* lines) can bring about (**a**) coiling of one chain or (**b**) sheetlike arrays of two or more chains.

Third Level of Protein Structure

Most polypeptide chains that have a coiled secondary structure undergo more folding because of the number and location of certain amino acids (including the bulky proline) along their length. These amino acids bend a chain at certain angles and in certain directions to make the chain loop out. R groups far apart along its length interact and hold the loops in characteristic positions. A polypeptide chain folded as an outcome of its bend-producing amino acids and its R-group interactions has reached a third structural level (tertiary structure).

Figure 3.18*a* shows how a polypeptide chain became folded into the compact, tertiary structure of the protein globin. Hydrogen bonds between particular functional groups along the chain's length caused the folding.

Fourth Level of Protein Structure

Imagine that bonds form between *four* molecules of globin and that an iron-containing functional group, a heme group, is positioned near the center of each. The outcome is **hemoglobin**, an oxygen-transporting protein

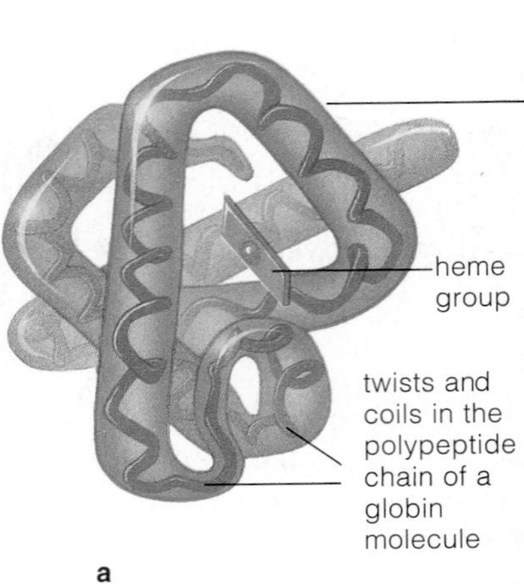

heme group
twists and coils in the polypeptide chain of a globin molecule
a

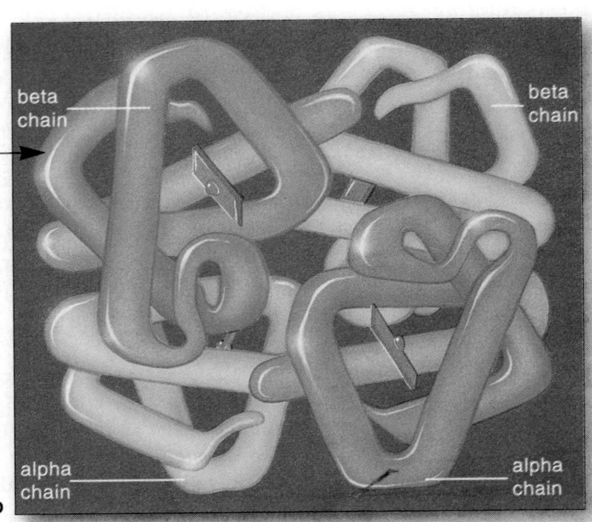

beta chain
beta chain
alpha chain
alpha chain
b

Figure 3.18 (**a**) Globin molecule. This coiled polypeptide chain associates with a heme group, an iron-containing group that strongly attracts oxygen. There is a whole family of globin molecules, identical in most parts of their amino acid sequences but unique in other parts. In this diagram, the artist drew a transparent "green noodle" around each chain to help you visualize how it folds in three dimensions.

(**b**) Hemoglobin, a protein that transports oxygen in blood, consists of four polypeptide chains (globin molecules) and four heme groups. Two chains, designated *alpha*, differ slightly from the other two (the *beta* chains) in their amino acid sequence. To keep this sketch from looking like a tangle of noodles, the two chains in the background are tinted differently from those in the foreground.

Further reading: Student Guide to InfoTrac on web site

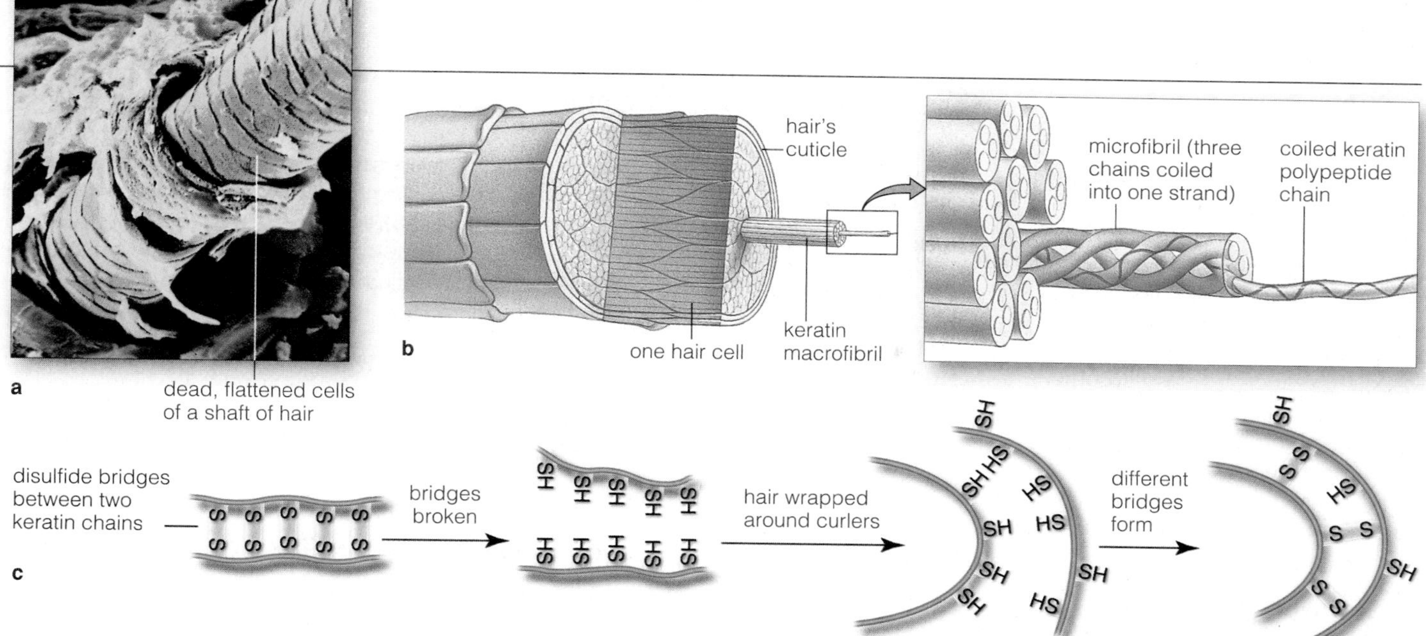
a
dead, flattened cells of a shaft of hair
b
hair's cuticle
one hair cell
keratin macrofibril
microfibril (three chains coiled into one strand)
coiled keratin polypeptide chain
c
disulfide bridges between two keratin chains
bridges broken
hair wrapped around curlers
different bridges form

Figure 3.19 Structure of hair. (**a**,**b**) Hair cells develop from modified skin cells. They synthesize polypeptide chains of the protein keratin. Disulfide bridges link three chains together as fine fibers, which get bundled into larger, cable-like fibers. The larger fibers almost fill the cells, which eventually die off. Dead, flattened cells form a tubelike cuticle around the developing hair shaft.

(**c**) For a permanent wave, hair is exposed to chemicals that break the disulfide bridges. When hairs wrap around curlers, their polypeptide chains are held in new positions. Now exposure to a different chemical causes new disulfide bridges to form. But they form between different sulfur-bearing amino acids than before. The displaced bonding locks the hair in curled positions (compare Figure 3.12).

(Figure 3.18*b*). As you read this, each of your mature red blood cells is transporting a billion molecules of oxygen, bound to 250 million hemoglobin molecules.

Hemoglobin is at the fourth (quaternary) level of protein structure. In every protein at this level, two or more polypeptide chains are joined by numerous weak interactions (such as hydrogen bonds) and sometimes by covalent bonds between sulfur atoms of R groups.

Proteins with quaternary structure are globular or fibrous. Hemoglobin is one of the **globular proteins**, which have one or more polypeptide chains folded in a rounded shape. Most enzymes are globular proteins, also. The **fibrous proteins** are long strands or sheets of polypeptide chains. Examples are keratin (Figure 3.19) and collagen, the most common animal protein. Skin, bone, corneas, arteries, and many other animal body parts depend on the strength inherent in collagen.

Glycoproteins and Lipoproteins

Some proteins have other organic compounds attached to their polypeptide chains. For example, lipoproteins form when certain proteins circulating in blood combine with cholesterol, triglycerides, and phospholipids that were absorbed from the gut after a meal. Similarly, most glycoproteins have linear or branched oligosaccharides bonded to them. Nearly all the proteins at the surface of animal cells are glycoproteins. So are most protein secretions from cells and many proteins in blood.

Structural Changes by Denaturation

Breaking weak bonds of a protein or any other large molecule disrupts its three-dimensional shape, an event called **denaturation**. For example, weak hydrogen bonds are sensitive to increases or decreases in temperature and pH. If the temperature or pH exceeds a protein's range of tolerance, its polypeptide chains will unwind or change shape, and the protein will lose its function. Consider the protein albumin, concentrated in the "egg white" of uncooked chicken eggs. When you cook eggs, the heat does not disrupt the strong covalent bonds of albumin's primary structure. But it destroys weaker bonds contributing to the three-dimensional shape. For some proteins, denaturation might be reversed when normal conditions are restored—but albumin isn't one of them. There is no way to uncook a cooked egg.

Proteins have a primary structure, which is the sequence of different kinds of amino acids along a polypeptide chain. An individual's DNA specifies that sequence.

Proteins have a secondary structure: a coiled pattern or an extended, sheetlike pattern that arises by hydrogen bonding at short, regular intervals along a polypeptide chain.

At the third level of protein structure, bonding at certain amino acids makes a coiled chain bend and loop at certain angles, in certain directions. Interactions among R groups in the chain hold the loops in characteristic positions.

At the fourth level of protein structure, numerous hydrogen bonds and other interactions join two or more polypeptide chains. Many proteins are this structurally complex.

3.6 FOOD PRODUCTION AND A CHEMICAL ARMS RACE

The next time you shop for groceries, reflect on what it takes to provide you with your daily supply of organic compounds. For example, those heads of lettuce typically grew in fertilized cropland. Possibly they competed with weeds, which didn't know the nutrients in the fertilizer were meant for the lettuce. Leaf-chewing insects didn't know the lettuce wasn't meant to be their salad bar. Each year, those food pirates and others ruin or gobble up nearly half of what people all over the world try to grow.

Most plants aren't entirely defenseless. They evolved under the intense selection pressures of attacks by insects, fungi, and other organisms, and in natural settings they often can repel the attackers with toxins. A **toxin** is an organic compound, a normal metabolic product of one species, but its chemical effects can harm or kill individuals of a different species that come in contact with it. Humans, too, encounter traces of natural toxins in most of what they eat, even in such familiar edibles as hot peppers, potatoes, figs, celery, rhubarb, and alfalfa sprouts. Still, we do not die in droves from these natural toxins, so apparently our bodies have chemical defenses against them.

Well over 2,500 years ago, farmers realized that sulfur kills insects. About a thousand years later, desperate but misguided farmers were dispensing arsenic, lead, and mercury through croplands. Toxic metal applications continued until the late 1920s, when someone made the connection between the toxins and human poisonings and deaths. These toxins resist breakdown, and we still find traces of them in vegetables and other plants grown on contaminated lands.

a

b Atrazine. Farmers spray this herbicide directly on foliage of corn and other crop plants. It kills weeds within a few days but does not affect the plants. Other herbicides with low persistence include glyphosate (Roundup), daminozide (Alar), and alachlor (Lasso).

c DDT (dichlorodiphenyltrichloroethane). This nerve cell poison does not break down until two to fifteen years after its application. Other high-persistence insecticides include chlordane.

d Malathion. Like other organophosphates, it is a nerve cell poison that persists for one to twelve weeks only. Although organophosphates break down more rapidly than chlorinated hydrocarbons, they are much more toxic. About four dozen organophosphates are cheap and effective. In the United States they account for half of all insecticide uses. The Environmental Protection Agency wants to minimize pesticide residues on food. At this writing methyl parathion has been banned for food crops; azenphos methyl applications must end at least three weeks before harvest time. Farmers are contesting these actions and want the EPA to consider economic and trade issues as well as human health.

Figure 3.20 (**a**) Lettuce, a common crop plant, and a low-flying crop duster with its rain of pesticides.(**b–d**) A few of the major pesticides, some more toxic than others. (**e**) And think about *that* the next time you sidle up to a salad bar.

Further reading: Student Guide to InfoTrac on web site →

During the 1600s, farmers took a cue from plants and started using an *organic* compound extracted from tobacco leaves as a natural pesticide. By the mid-1800s, toxins from chrysanthemum flowers and from the roots of some tropical forest plants were added to the arsenal. Then, in 1945, scientists started developing synthetic toxins to protect crops, food stores, freshwater supplies, pets, and ornamental plants. And they started to understand the mechanisms by which toxins exert their effects.

The *herbicides* kill weeds by disrupting metabolism and growth. Most of the *insecticides* clog a target insect's airways, disrupt functioning of its nerves and muscles, or block its reproduction. *Fungicides* work against harmful fungi, including a mold that makes aflatoxin, one of the deadliest poisons. By 1995, people in the United States were spraying or spreading more than 1.25 billion pounds of toxins through fields, gardens, homes, and commercial and industrial sites (Figure 3.20).

e

There are downsides to pesticide applications. Right along with the pests, some toxic organic compounds kill birds and other predators which, in natural settings, help control pest population sizes. Besides, targeted pests have been developing resistance to the chemical arsenal for reasons that you read about earlier, in Chapter 1.

Also, pesticides cannot be released haphazardly into the environment because they can be inhaled, ingested with food, or absorbed through the skin. Different types stay active for weeks or years. Some trigger rashes, hives, sickening headaches, asthma, and joint pain in millions of people. And some trigger life-threatening allergic reactions in abnormally sensitive individuals.

Presently, DDT and other long-lived pesticides are banned in the United States. Even rapidly degradable ones, such as malathion and other organophosphates, are subjected to rigorous application standards and ongoing safety tests. With so many pests around us, however, many farmers resist limits on applications (Figure 3.20*d*).

3.7 NUCLEOTIDES

Nucleotides are small organic compounds with major roles in metabolism. Each consists of a sugar, at least one phosphate group, and a base. The sugar is either ribose or deoxyribose. Both sugars have a five-carbon ring structure. The only difference is that ribose has an oxygen atom attached to carbon 2 in its ring structure and deoxyribose does not. The bases have a single or double carbon ring structure that incorporates nitrogen.

One nucleotide, **ATP** (adenosine triphosphate), has a string of three phosphate groups attached to its sugar component (Figure 3.21*a*). ATP can readily transfer a phosphate group to many other molecules inside cells, and the acceptor molecules become energized enough to enter a reaction. Although an ATP molecule acquires its phosphate groups at certain reaction sites, it is able to deliver them to nearly all other reaction sites in a cell. In this respect ATP is central to metabolism.

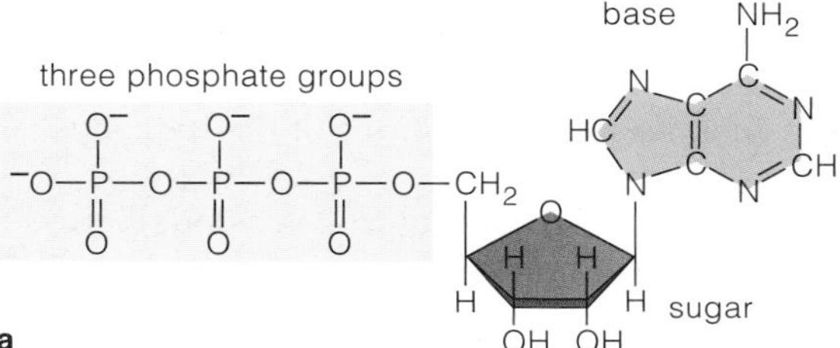

three phosphate groups
base
NH_2
sugar
a

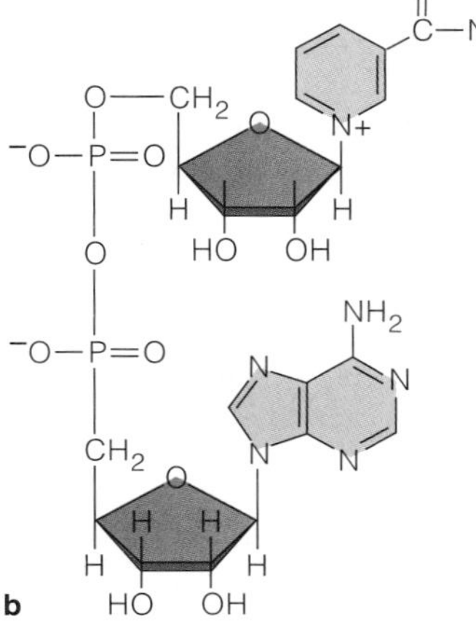

b

Figure 3.21 Structural formulas for (**a**) ATP and (**b**) NAD^+. Both of these nucleotides have central roles in cell metabolism.

Other nucleotides are subunits of **coenzymes**, enzyme helpers that can accept hydrogen atoms plus electrons stripped away from molecules at a reaction site and then transfer them elsewhere. One coenzyme is NAD^+ (short for nicotinamide adenine dinucleotide; see Figure 3.21*b*). FAD (flavin adenine dinucleotide) is another kind. Still other nucleotides function as chemical messengers between cells and within the cell interior. You will read about one (cyclic adenosine monophosphate, or cAMP) later in the book. As you will see, nucleotides also are building blocks for the large, single- or double-stranded molecules known as nucleic acids.

Nucleotides serve as energy carriers, chemical messengers, and subunits for coenzymes and for nucleic acids.

3.8 NUCLEIC ACIDS

As you read in the preceding section, some nucleotides carry out specific tasks for metabolic reactions. A few kinds have a different—and absolutely vital—function in all cells. These particular nucleotides serve as the monomers for single- or double-stranded molecules classified as **nucleic acids**. In such strands, a covalent bond connects the sugar component of one nucleotide with the phosphate group of the next nucleotide in the sequence (Figure 3.22*a*).

All cells start out life and then maintain themselves by way of instructions they inherited in some number of double-stranded molecules of deoxyribonucleic acid, or **DNA**. This nucleic acid consists of four different kinds of nucleotides, the structural formulas for which are shown in Figure 3.22*b*. The four differ only in their component base, which is adenine, guanine, thymine, or cytosine. The sequence of bases encodes heritable information about how to synthesize all of the diverse proteins that give each new cell the potential to grow, maintain itself, and reproduce. The particular bases that occur in at least some portions of the sequence are unique to each species.

Figure 3.22*c* shows how hydrogen bonds between bases join the two strands together along the length of a DNA molecule. Think of each "base pair" as one rung of a ladder, and the two sugar–phosphate backbones as the ladder's two posts. The ladder twists spirally, in the shape of a double helix.

Like DNA, the **RNAs** (ribonucleic acids) consist of four kinds of nucleotide monomers. Unlike DNA, the bases are adenine, guanine, cytosine, and uracil. Also

Figure 3.22 (**a**) Simplified example of how a series of nucleotides is covalently bonded in a single strand of DNA or RNA. (**b**) The four kinds of nucleotides that can serve as monomers for DNA molecules. Two nucleotide bases, adenine and guanine, have a double ring structure. The other two nucleotide bases, thymine and cytosine, have a single ring structure. (**c**) DNA double helix.

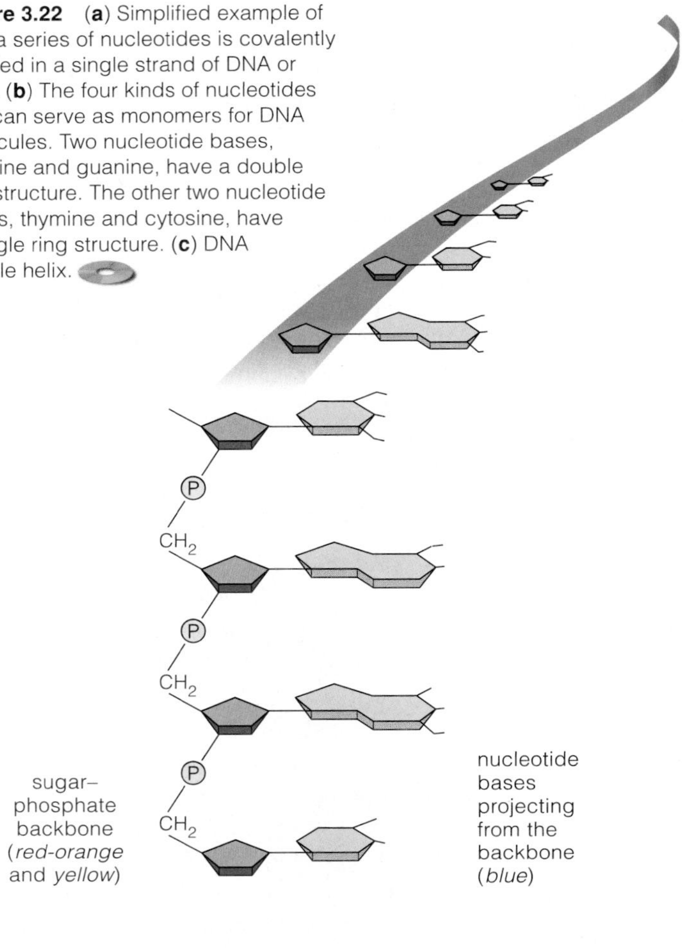

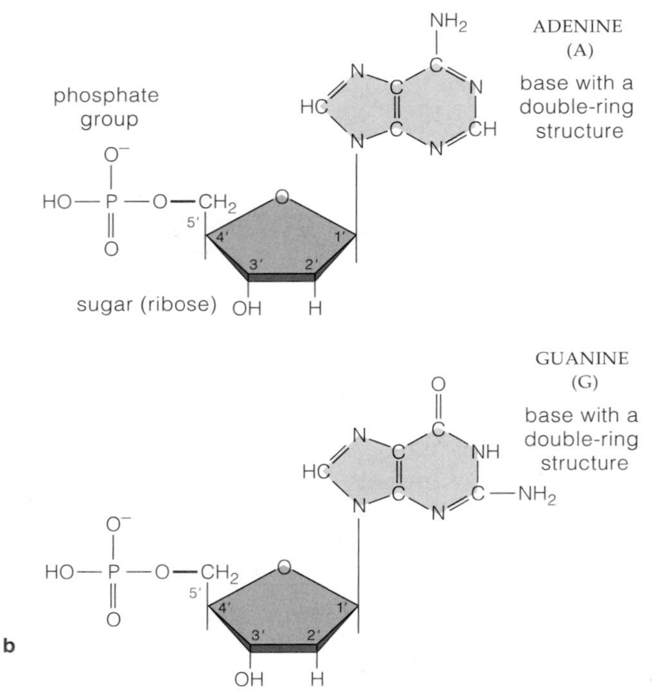

P
CH2
sugar–phosphate backbone (*red-orange* and *yellow*)
nucleotide bases projecting from the backbone (*blue*)
a
phosphate group
ADENINE (A)
base with a double-ring structure
sugar (ribose)
THYMINE (T)
base with a single-ring structure
GUANINE (G)
base with a double-ring structure
CYTOSINE (C)
base with a single-ring structure
b

Further reading: Student Guide to InfoTrac on web site

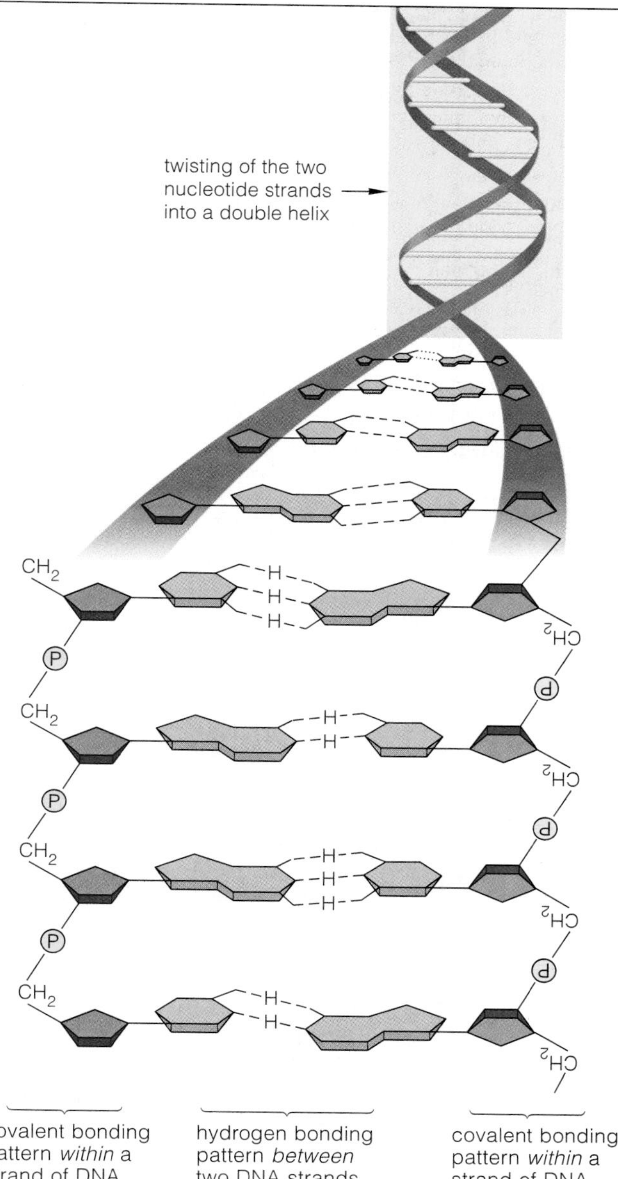
twisting of the two nucleotide strands into a double helix
CH_2
P
covalent bonding pattern *within* a strand of DNA
hydrogen bonding pattern *between* two DNA strands
covalent bonding pattern *within* a strand of DNA
c

unlike DNA, RNA molecules are usually single strands of nucleotides. RNAs do not encode protein-building instructions, but different classes are the key players in carrying out those instructions in cells. Ever since cells first appeared on Earth, RNAs have served in processes by which genetic information is used to build proteins.

Nucleic acids consist of one or two strands of nucleotides joined in strandlike array by covalent bonds.

DNA is a double-stranded nucleic acid. Encoded in its nucleotide sequence is heritable information about all proteins a new cell requires to survive and reproduce.

The RNAs are single-stranded nucleic acids with roles in processes by which a cell uses genetic information in DNA to build proteins.

3.9 SUMMARY

1. Organic compounds consist of one or more elements covalently bonded to carbon atoms that commonly form linear and ring-shaped backbones. The properties that are characteristic of life emerge out of the molecular structure of carbon-rich compounds.

2. The central role of carbon in the world of life arises from its versatile bonding behavior. Each carbon atom can form relatively stable covalent bonds with as many as four other atoms. The bonding arrangement is the basis of carbon backbones, which can extend as straight chains or coil back on themselves as ring structures.

3. A carbon backbone that has only hydrogen atoms bonded to it is a hydrocarbon, a highly stable structure. Besides hydrogen, biological molecules have different functional groups: a variety of atoms or clusters of them covalently bonded to the carbon backbone. Functional groups impart diverse properties to organic compounds. Hydroxyl (—OH) and phosphate groups are examples.

4. In nature, only living cells can assemble the large organic compounds called biological molecules—the complex carbohydrates and lipids, proteins, and nucleic acids. To do so, they draw from small pools of organic compounds, which include simple sugars, fatty acids, amino acids, and nucleotides.

5. Cells assemble, rearrange, and break down organic compounds mainly through enzyme-mediated reactions involving the transfer of functional groups or electrons, the rearrangement of internal bonds, and a combining or splitting of molecules. As an example, in condensation reactions, enzymes remove a hydroxyl group from one molecule and an H atom from another, then speed the formation of a covalent bond between the two molecules at their exposed sites. A molecule of water usually forms also. Hydrolysis is like condensation in reverse.

6. Most carbohydrates consist of carbon, hydrogen, and oxygen in a 1:2:1 ratio. They are simple sugars, short-chain oligosaccharides, and polysaccharides.

a. Cells use the simple sugars (such as glucose) and oligosaccharides (such as the disaccharide sucrose) as building blocks and transportable forms of energy.

b. Complex carbohydrates (e.g., cellulose and starch) are structural materials and forms of energy storage.

7. Lipids are hydrocarbons that show little tendency to dissolve in water, but they readily dissolve in nonpolar substances. Cells use lipids (including triglycerides) as energy storage forms, as structural components of cell membranes, and as precursors of other compounds.

a. Fats are lipids with one, two, or three fatty acids attached to glycerol. Saturated tails have single covalent bonds only. Unsaturated tails have some double bonds. Triglycerides (three fatty acid tails) are the body's most abundant lipids and its richest energy source.

Table 3.1 Summary of the Main Organic Compounds in Living Things			
Category	Main Subcategories	Some Examples and Their Functions	
CARBOHYDRATES ... contain an aldehyde or a ketone group, and one or more hydroxyl groups	**Monosaccharides** (simple sugars)	Glucose	Energy source
	Oligosaccharides	Sucrose (a disaccharide)	Most common form of sugar transported through plants
	Polysaccharides (complex carbohydrates)	Starch, glycogen Cellulose	Energy storage Structural roles
LIPIDS ... are largely hydrocarbon; generally do not dissolve in water but dissolve in nonpolar substances, such as other lipids	**Lipids with fatty acids**		
	Glycerides: one, two, or three fatty acid tails attached to glycerol backbone	Fats (e.g., butter), oils (e.g., corn oil)	Energy storage
	Phospholipids: phosphate group, one other polar group, and (often) two fatty acids attached to glycerol backbone	Phosphatidylcholine	Key component of cell membranes
	Waxes: long-chain fatty acid tails attached to alcohol	Waxes in cutin	Water conservation at aboveground surfaces of plants
	Lipids with no fatty acids		
	Sterols: four carbon rings; the number, position, and type of functional groups differ among various sterols	Cholesterol	Component of animal cell membranes; precursor of many steroids and of vitamin D
PROTEINS ... are one or more polypeptide chains, each with as many as several thousand covalently linked amino acids	**Fibrous proteins** Long strands or sheets of polypeptide chains; often tough, water-insoluble	Keratin Collagen	Structural element of hair, nails Structural element of bone, cartilage
	Globular proteins One or more polypeptide chains folded and linked into globular shapes; many roles in cell activities	Enzymes Hemoglobin Insulin Antibodies	Great increase in rates of reactions Oxygen transport Control of glucose metabolism Tissue defense
NUCLEIC ACIDS (AND NUCLEOTIDES) ... are chains of units (or individual units) that each consist of a five-carbon sugar, phosphate, and a nitrogen-containing base	**Adenosine phosphates**	ATP cAMP (Section 37.2)	Energy carrier Messenger in hormone regulation
	Nucleotide coenzymes	NAD^+, $NADP^+$, FAD	Transport of protons (H^+), electrons from one reaction site to another
	Nucleic acids Chains of thousands to millions of nucleotides	DNA, RNAs	Storage, transmission, translation of genetic information

b. Phospholipids are the main components of cell membranes. Sterols such as cholesterol are components of cell membranes and precursors of steroid hormones and other compounds. Waxes are pliable components of water-repelling, protective, and lubricating substances.

8. Proteins are the most diverse biological molecules. They function in enzyme activity, structural support, transport, cell movements, changes in cell shape and activities, and defense against disease.

a. Each protein consists of one or more polypeptide chains in which amino acids are joined one after another by peptide bonds. The amino acid sequence, which is unique for each kind of protein, gives rise to its unique structure, chemical behavior, and function.

b. The amino acid sequence is the protein's primary structure. An individual's DNA specifies the sequence.

c. A protein has a secondary structure (sheetlike or coiled) arising from hydrogen bonding along the chain. A third-level structure arises from interactions among amino acids that hold the chain in bends and loops. Proteins with a fourth-level structure consist of two or more polypeptide chains bonded together.

9. Nucleotides serve as energy carriers (mostly ATP), coenzymes (such as NAD^+), and monomers for strands of DNA and RNA, which are the basis of inheritance and reproduction.

10. Table 3.1 summarizes the main biological molecules.

Review Questions

1. Define organic compound. Name the type of chemical bond that predominates in the backbone of such a compound. *3.1*

2. Define hydrocarbon. Also define functional group and give an example. *3.1*

3. Name the "molecules of life." Do they break apart most easily at their hydrocarbon portion or at functional groups? *3.1*

4. Describe the difference between a condensation reaction and hydrolysis. *3.1*

5. Select one of the carbohydrates, lipids, proteins, or nucleic acids described in this chapter. Speculate on how its functional groups and bonds between the carbon atoms in its backbone contribute to its final shape and function. *3.2 through 3.8*

6. Which item listed includes all of the other items listed? *3.3*
   a. triglyceride c. wax e. lipid
   b. fatty acid d. sterol f. phospholipid

7. Explain how a hemoglobin molecule's three-dimensional shape arises, starting with its primary structure. *3.4, 3.5*

Self-Quiz (*Answers in Appendix III*)

1. Each carbon atom can share pairs of electrons with as many as ______ other atoms.
   a. one b. two c. three d. four

2. Hydrolysis is a ______ reaction.
   a. functional group transfer d. condensation
   b. electron transfer e. cleavage
   c. rearrangement f. both b and d

3. ______ is a simple sugar (monosaccharide).
   a. glucose c. ribose e. both a and b
   b. sucrose d. chitin f. both a and c

4. In unsaturated fats, fatty acid tails have one or more ______
   a. single covalent bonds b. double covalent bonds

5. ______ are to proteins as ______ are to nucleic acids.
   a. sugars; lipids c. amino acids; hydrogen bonds
   b. sugars; proteins d. amino acids; nucleotides

6. A denatured protein or DNA molecule has lost its ______ .
   a. hydrogen bonds c. function
   b. shape d. all of the above

7. Nucleotides include ______ .
   a. ATP c. RNA subunits
   b. DNA subunits d. all of the above

8. Match each molecule with the most suitable description.
   ___ long sequence of amino acids a. carbohydrate
   ___ the main energy carrier b. phospholipid
   ___ glycerol, fatty acids, phosphate c. protein
   ___ two strands of nucleotides d. DNA
   ___ one or more sugar monomers e. ATP

Critical Thinking

1. Jack decided to celebrate summer by making a crab salad and a peach pie for some friends. He had such a good time, he forgot to put the cap on the bottle of olive oil after making the salad. He also didn't tighten the lid on the shortening can after making the pie crust. A few weeks later, the oil had turned rancid but the shortening still smelled okay. Both substances are fats. Both were stored in a dark cupboard, at room temperature. Yet one spoiled and the other stayed fresh. Speculate on which molecular bonds in these substances were the starting point for the difference.

2. In the following list, identify which is the carbohydrate, fatty acid, amino acid, and polypeptide:
   a. $^{+}NH_3$—CHR—COO^- c. $(glycine)_{20}$
   b. $C_6H_{12}O_6$ d. $CH_3(CH_2)_{16}COOH$

3. A clerk in a health-food store tells you that certain "natural" vitamin C tablets extracted from rose hips are better for you than synthetic vitamin C tablets. Given your understanding of the structure of organic compounds, what would be your response? Design an experiment to test whether the vitamins differ.

4. Rabbits that eat green, leafy vegetables containing xanthophyll accumulate this yellow pigment molecule in body fat but not in muscles. What chemical properties of the molecule might cause this selective accumulation?

5. Cows can digest grasses, but people cannot. The cow's four-chambered stomach houses populations of certain microorganisms that are not normal residents of the human stomach. What kind of metabolic reactions do you think the microorganisms carry out? What do you think might happen to a cow undergoing treatment with an antibiotic that killed the microorganisms?

6. A Gary Larsen cartoon states that sheep with steel wool have no natural enemies. You might laugh at the thought of such a preposterous animal without knowing what wool is. Wool is the keratin-rich, soft undercoat of sheep, Angora goats, llamas, and other hairy mammals. Keratin is a strandlike, fibrous protein. Many hydrogen bonds hold it in a helical shape. Visualize three such chains coiled together, with many disulfide bridges in the end regions stabilizing the trio in a tight, ropelike array. Each hair has linear aggregates of the ropelike fibers, which resist stretching (Figure 3.19). Given keratin's structure, speculate why, when you run a wool sweater through the hot cycle of a dryer, it shrinks pathetically and permanently.

7. Reflect on the component atoms of the different classes of organic compounds. Now look at the structural formulas for the pesticides shown in Figure 3.20. Which one is a glycoprotein? An organophosphate? A chlorinated hydrocarbon?

8. Many farmers believe crop losses will be disastrous if pesticides are banned without sufficient research and effective alternatives. Many environmental activists are pushing for rapid bans to protect people and the environment. What do you think?

Selected Key Terms

alcohol *3.1*
amino acid *3.4*
ATP *3.7*
carbohydrate *3.2*
coenzyme *3.7*
condensation reaction *3.1*
denaturation *3.5*
DNA *3.8*
enzyme *3.1*
fat *3.3*
fatty acid *3.3*
fibrous protein *3.5*
functional group *3.1*
globular protein *3.5*
hemoglobin *3.5*
hydrolysis *3.1*
lipid *3.3*
monosaccharide *3.2*
nucleic acid *3.8*
nucleotide *3.7*
oligosaccharide *3.2*
organic compound *3.1*
phospholipid *3.3*
polypeptide chain *3.4*
polysaccharide *3.2*
protein *3.4*
RNA *3.8*
sterol *3.3*
toxin *3.6*
triglyceride *3.3*
wax *3.3*

Readings See also *www.infotrac-college.com*

Atkins, P. 1987. *Molecules.* New York: Scientific American Library. A molecular "glossary."

Ritter, P. 1996. *Biochemistry: An Introduction.* Pacific Grove, California: Brooks/Cole. Great human applications.

Wolfe, S. 1995. *Introduction to Molecular and Cellular Biology.* Belmont, California: Wadsworth.

4

CELL STRUCTURE AND FUNCTION

Animalcules and Cells Fill'd With Juices

Early in the seventeenth century, a scholar by the name of Galileo Galilei arranged two glass lenses within a cylinder. With this instrument he happened to look at an insect, and later he described the stunning geometric patterns of its tiny eyes. Thus Galileo, who was not a biologist, was among the first to record a biological observation made through a microscope. The study of the cellular basis of life was about to begin. First in Italy, then in France and England, scholars set out to explore a world whose existence had not even been suspected.

At midcentury Robert Hooke, Curator of Instruments for the Royal Society of England, was at the forefront of these studies. When Hooke first turned a microscope to thinly sliced cork from a mature tree, he observed tiny compartments (Figure 4.1*c*). He gave them the Latin name *cellulae*, meaning small rooms—hence the origin of the biological term "cell." They actually were the interconnecting walls of dead plant cells, which is what cork is made of, but Hooke did not think of them as being dead because neither he nor anyone else at the time knew that cells could be alive. In still other plant tissues, he observed cells "fill'd with juices" but could not imagine what they represented.

Given the simplicity of their instruments, it is just amazing that the pioneers in microscopy observed as much as they did. Antony van Leeuwenhoek, a Dutch shopkeeper, had exceptional skill in constructing lenses and possibly the keenest vision of all (Figure 4.1*a*). By the late 1600s, he was discovering natural wonders everywhere, including "many very small animalcules, the motions of which were very pleasing to behold," in scrapings of tartar from his own teeth. Elsewhere he observed a variety of protistans, sperm, and even a bacterium—an organism so small that it would not be seen again for another two centuries!

a
b

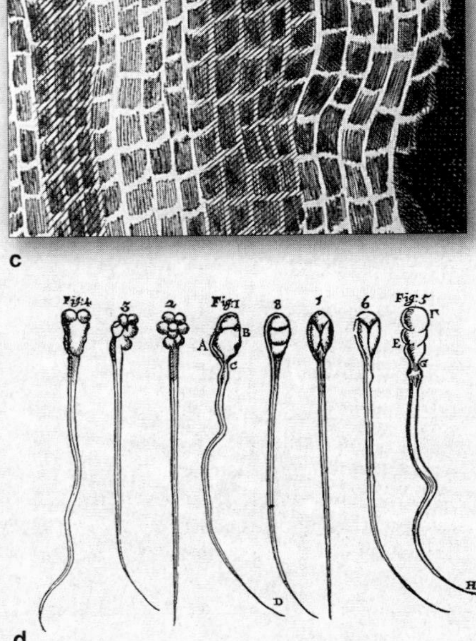

c
d

Figure 4.1 Early glimpses into the world of cells. (**a**) Antony van Leeuwenhoek, with microscope in hand. (**b**) Robert Hooke's compound microscope and (**c**) his drawing of cell walls from cork tissue. (**d**) One of van Leeuwenhoek's early sketches of sperm cells. (**e**) Cartoon evidence of the startling impact of microscopic observations on nineteenth-century London.

In the 1820s, improvements in lenses brought cells into sharper focus. Robert Brown, a botanist, was noticing an opaque spot in a variety of cells. He called it a nucleus. In 1838 still another botanist, Matthias Schleiden, wondered if the nucleus had something to do with a cell's development. As he hypothesized, each plant cell must develop as an independent unit even though it is part of the plant.

By 1839, after years of studying animal tissues, the zoologist Theodor Schwann had this to say: Animals as well as plants consist of cells and cell products—and even though the cells are part of a whole organism, to some extent they have an individual life of their own.

A decade later a question remained: Where do cells come from? Rudolf Virchow, a physiologist, completed his own studies of a cell's growth and reproduction—that is, its division into two daughter cells. Every cell, he reasoned, comes from a cell that already exists.

And so, by the middle of the nineteenth century, microscopic analysis had yielded three generalizations, which together constitute the **cell theory**. *First, every organism is composed of one or more cells. Second, the cell is the smallest unit having the properties of life. Third, the continuity of life arises directly from the growth and division of single cells.* All three insights still hold true.

This chapter is not meant for memorization. Read it simply to gain an overview of current understandings of cell structure and function. In later chapters, you can refer back to it as a road map through the details. With its images from microscopy, this chapter and others in the book can transport you into spectacular worlds of juice-fill'd cells and animalcules.

MICROCOSM dedicated to the London Water Companies
MONSTER SOUP commonly called THAMES WATER, being a correct representation of that precious stuff doled out to us

e

KEY CONCEPTS

1. All organisms consist of one or more cells. The cell is the smallest unit that still retains the characteristics of life. Since the origin of life, each new cell has arisen from a preexisting cell. These are the three generalizations of the cell theory.

2. All cells have an outermost, double-layered membrane. This plasma membrane separates the cell interior from the surroundings. In addition, cells contain cytoplasm, an organized internal region where energy conversions, protein synthesis, movements, and other activities necessary for survival proceed. In eukaryotic cells only, DNA is enclosed within a membrane-bound nucleus. In prokaryotic cells (bacteria), DNA is concentrated in part of the cell interior.

3. The plasma membrane and internal cell membranes consist largely of phospholipid and protein molecules. The phospholipids form two adjacent layers that give the membrane its basic structure and prevent water-soluble substances from freely crossing it. Proteins embedded in cell membranes or positioned at their surfaces perform most membrane functions.

4. The nucleus is one of many organelles. Organelles are membrane-bound compartments inside eukaryotic cells. They physically separate different metabolic reactions and allow them to proceed in orderly fashion. Bacteria do not have comparable organelles.

5. When cells grow, they increase faster in volume than in surface area. This physical constraint on increases in size influences cell size and shape.

6. Different microscopes modify light rays or accelerated beams of electrons in ways that allow us to form images of incredibly small specimens. They are the foundation for our current understanding of cell structure and function.

4.1 BASIC ASPECTS OF CELL STRUCTURE AND FUNCTION

Inside your body and at its moist surfaces, trillions of cells live in interdependency. In northern forests, four-cell structures—pollen grains—escape from pine trees. In scummy pondwater, a single-celled amoeba moves freely, thriving on its own. For humans, pines, amoebas, and all other organisms, the **cell** is the smallest unit that retains the properties of life (Figure 4.2). Each living cell can survive on its own or has the potential to do so. Its structure is highly organized for metabolism. It senses and responds to its environment. And, based on inherited instructions in its DNA, it has the potential to reproduce.

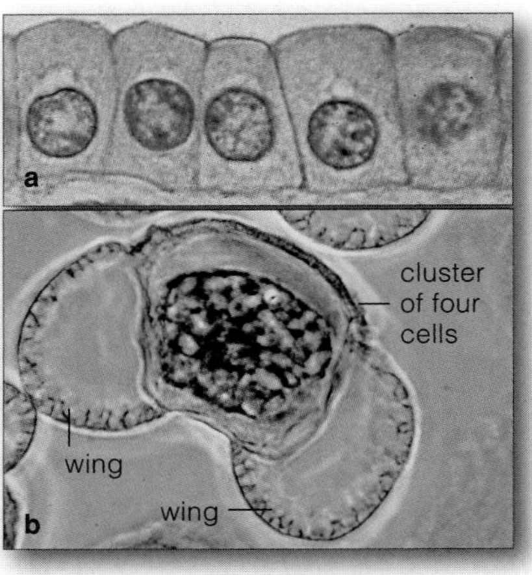

a
cluster of four cells
wing
wing
b

Figure 4.2 (**a**) A row of cells at the surface of a tiny tube inside a kidney. Each has a nucleus (the sphere stained darker red). (**b**) A winged pollen grain. One of its cells will give rise to a sperm that may meet up with an egg and help start a new pine tree.

Structural Organization of Cells

Cells differ enormously in size, shape, and activities, as you might gather by comparing a tiny bacterium with one of your relatively giant liver cells. Yet they are alike in three respects. All cells start out life with a plasma membrane, a region of DNA, and a region of cytoplasm:

1. **Plasma membrane**. This thin, outermost membrane maintains the cell as a distinct entity. By doing so, it allows metabolic events to proceed apart from random events in the environment. A plasma membrane does not isolate the cell interior. Substances and signals continually move across it in highly controlled ways.

2. **Nucleus** or **nucleoid**. Depending on the species, the DNA occupies a membranous sac in the cell interior (nucleus) or simply an interior region (nucleoid).

3. **Cytoplasm**. Cytoplasm is everything between the plasma membrane and region of DNA. It consists of a semifluid matrix and other components, such as **ribosomes** (structures on which proteins are built).

This chapter introduces two fundamentally different kinds of cells. The cytoplasm of **eukaryotic cells** includes organelles, which are tiny sacs and other compartments bounded by membranes. One of these is the nucleus, the defining feature of eukaryotic cells. **Prokaryotic cells** do not have a nucleus. No cell membranes intervene between the nucleoid (where DNA is located) and the cytoplasm surrounding it. The only prokaryotic cells are bacteria. Beyond the bacterial realm, all other organisms —from amoebas to peach trees to puffball mushrooms to zebras—consist of one to trillions of eukaryotic cells.

Lipid Bilayer of Cell Membranes

All cell membranes have the same structural framework of two sheets of lipid molecules. Figure 4.3 shows this **lipid bilayer** arrangement for the plasma membrane, the continuous boundary that bars the free passage of water-soluble substances into and out of the cell. Within eukaryotic cells, membranes subdivide the cytoplasm into specific zones in which substances are synthesized, processed, stockpiled, or degraded.

Diverse proteins embedded in the lipid bilayer or positioned at one of its surfaces carry out most of the

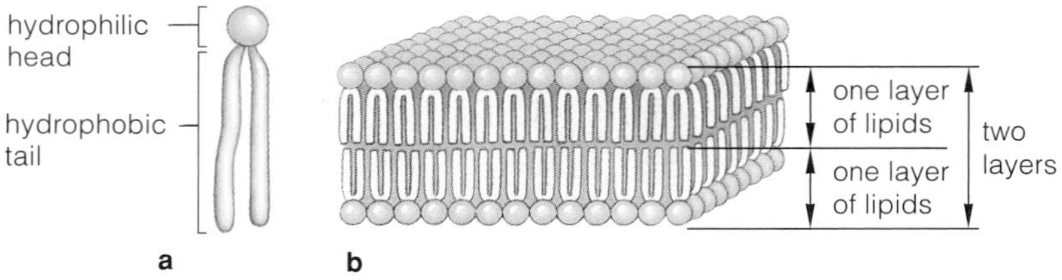

hydrophilic head
hydrophobic tail
one layer of lipids
one layer of lipids
two layers
a
b

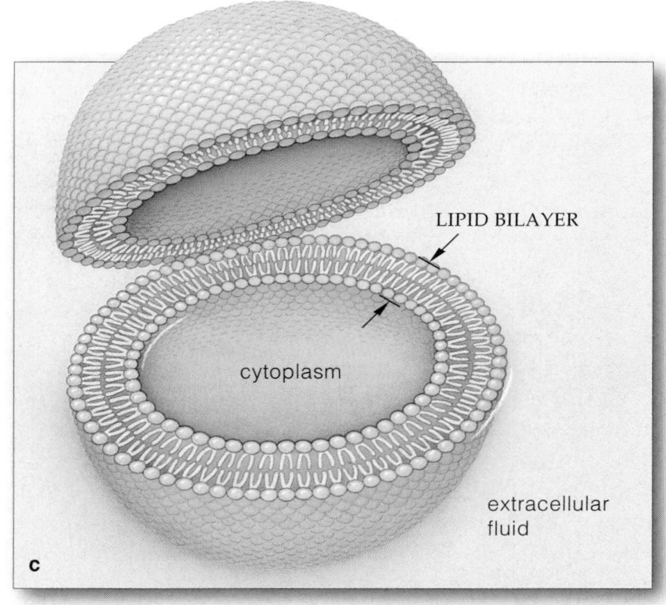

LIPID BILAYER
cytoplasm
extracellular fluid
c

Figure 4.3 Organization of the lipid bilayer of cell membranes. (**a**) Icon for phospholipid molecules, the most abundant components of cell membranes. These and other lipids are arranged as two layers. (**b**) The hydrophobic tails of the molecules are sandwiched between their hydrophilic heads, which are dissolved in cytoplasm on one side of the bilayer and in extracellular fluid on the other (**c**).

Diameter (cm):	0.5	1.0	1.5
Surface area (cm^2):	0.79	3.14	7.07
Volume (cm^3):	0.06	0.52	1.77
Surface-to-volume ratio:	13.17:1	6.04:1	3.99:1

a

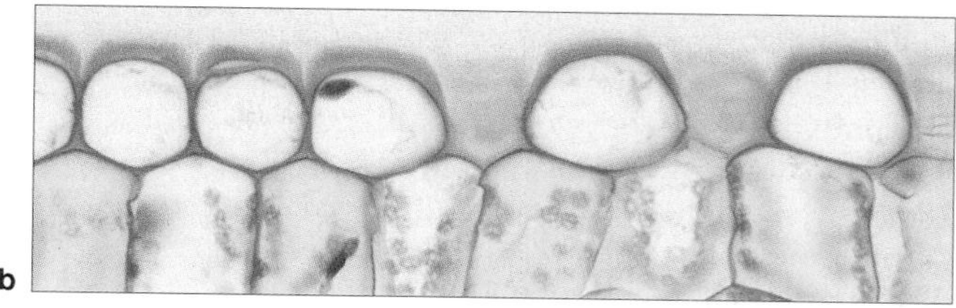
b

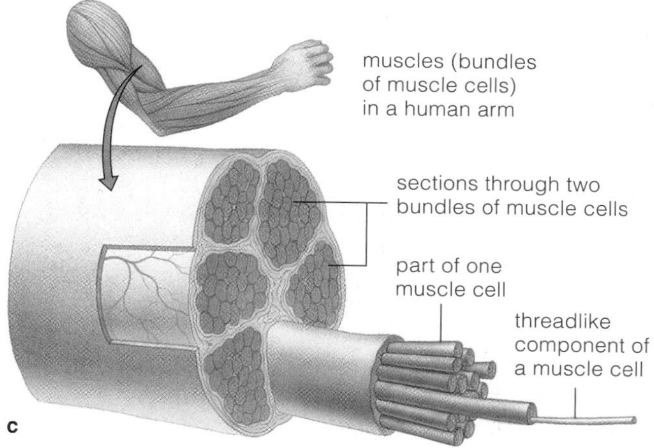
muscles (bundles of muscle cells) in a human arm
sections through two bundles of muscle cells
part of one muscle cell
threadlike component of a muscle cell
c

Figure 4.4 (**a**) An example of the surface-to-volume ratio. This physical relationship between increases in volume and in surface area imposes restrictions on the size and shape of cells, including the ones near the surface of leaves (**b**) and in skeletal muscles (**c**).

membrane functions. For example, some proteins act as passive channels for water-soluble substances. Others pump substances across the bilayer. Still others serve as receptors, which can latch onto hormones and other types of signaling molecules that trigger changes in cell activities. You will read more about membrane proteins in the chapter to follow.

Cell Size and Cell Shape

You may be wondering how small cells really are. Can any be observed with the unaided human eye? There are a few, including the "yolks" of bird eggs, cells in the red part of watermelons, and the fish eggs we call caviar. However, most cells cannot be observed without microscopes. To give you a sense of cell sizes, red blood cells are about 8 millionths of a meter across. You could fit a string of 2,000 of them across your thumbnail!

Why are most cells so small? A physical relationship called the **surface-to-volume ratio** constrains increases in a cell's size. By this relationship, an object's volume increases with the cube of the diameter, but its surface area increases only with the square. Figure 4.4*a* gives an example of this. Simply put, *if a cell expands in diameter during growth, its volume will increase more rapidly than its surface area will.*

Suppose you figure out a way to make a round cell grow four times wider than it normally would. Its volume increases 64 times (4^3), but its surface area only increases 16 times (4^2). As a result, each unit of the cell's plasma membrane must now serve four times as much cytoplasm as it did previously! Moreover, past a certain point, the inward flow of nutrients and outward flow of wastes will not be fast enough, and the cell will die.

A large, round cell also would have trouble moving materials *through* its cytoplasm. In small cells, such as those shown in Figure 4.4*b*, the random, tiny motions of molecules can easily distribute materials. If a cell is not small, you usually can expect it to be long and thin or to have outfoldings and infoldings that increase its surface relative to its volume. *The smaller or narrower or more frilly-surfaced the cell, the more efficiently materials cross its surface and become distributed through the interior.*

Multicelled body plans show evidence of surface-to-volume constraints, also. For example, cells attach end to end in strandlike algae, and each one interacts directly with its surroundings. Your skeletal muscle cells are very thin, but each one is as long as a biceps or some other muscle of which it is part (Figure 4.4*c*). Your circulatory system delivers materials to muscle cells and trillions of others, and it also sweeps away their metabolic wastes. Its many "highways" cut through the volume of body tissues, and by doing so they shrink the distance to and from individual cells.

All cells have an outermost plasma membrane, an internal region of cytoplasm, and an internal region of DNA.

Besides the plasma membrane, eukaryotic cells have internal, membrane-bound compartments, including a nucleus.

Each cell membrane has a lipid bilayer structure, consisting largely of phospholipid molecules. The hydrophobic parts of the lipids are sandwiched between the hydrophilic parts, which are dissolved in the fluid surroundings.

A lipid bilayer imparts structure to the cell membrane and serves as a barrier to water-soluble substances.

Proteins embedded in the bilayer or positioned at one of its surfaces carry out most membrane functions.

During their growth, cells increase faster in volume than they do in surface area. This physical constraint on increases in size influences the size and shape of cells. It also influences the body plans of multicelled organisms.

4.2 MICROSCOPES—GATEWAYS TO CELLS

Modern microscopes are gateways to astounding worlds. Some even afford glimpses into the structure of molecules. Different kinds use wavelengths of light or accelerated electrons (Figure 4.5). The micrographs in Figures 4.6 and 4.7 hint at the details now being observed. A **micrograph** simply is a photograph of an image that came into view with the aid of a microscope.

LIGHT MICROSCOPES Picture a series of waves moving across an ocean. Each **wavelength** is the distance from one wave's peak to the peak of the wave behind it. Light also travels as waves from sources such as the sun and illuminated specimens. In a *compound light microscope*, two or more sets of glass lenses bend light emanating from a cell or some other specimen in ways that form an enlarged image of it (Figure 4.5*a*,*b*).

A living cell must be small or thin enough for light to pass through. It would help if cell parts differed in color and density from the surroundings, but most are nearly colorless and appear uniformly dense. To get around this problem, microscopists stain cells (expose them to dyes that react with some parts of a specimen but not others). Staining may alter and kill cells. Dead cells break down fast, so cells typically are pickled or preserved before staining.

Suppose you use the best glass lens system. When you magnify the diameter of a specimen by 2,000 times or more, you discover that cell parts appear larger but are not clearer. To understand why, think about the distance between the two crests of a wavelength of light. To give two examples, that distance is about 750 nanometers for red light and 400 nanometers for violet (wavelengths of all other colors fall in between). If a cell structure is less than one-half of a wavelength long, that structure will not be able to disturb the rays of light streaming past, and it will not be visible.

ELECTRON MICROSCOPES Better resolution of very fine details is possible with the assistance of electrons. Remember, electrons are particles of matter, but they also behave like waves. In electron microscopy, streams of electrons are accelerated to wavelengths of about 0.005 nanometer—about 100,000 times shorter than wavelengths of visible light. The electrons cannot pass through glass lenses, but a magnetic field can bend them from their path and focus them.

In a *transmission electron microscope*, a magnetic field is the "lens." Accelerated electrons are directed through a specimen, focused into an image, and magnified. With *scanning electron microscopes*, a narrow beam of electrons moves back and forth across a specimen to which a thin coat of metal was applied. The metal responds by emitting some of its own electrons. A detector tied into electronic circuitry then transforms the energy of electrons into an image of the specimen's surface on a television screen. Most of the scanning images have fantastic depth (Figure 4.7*d*).

c

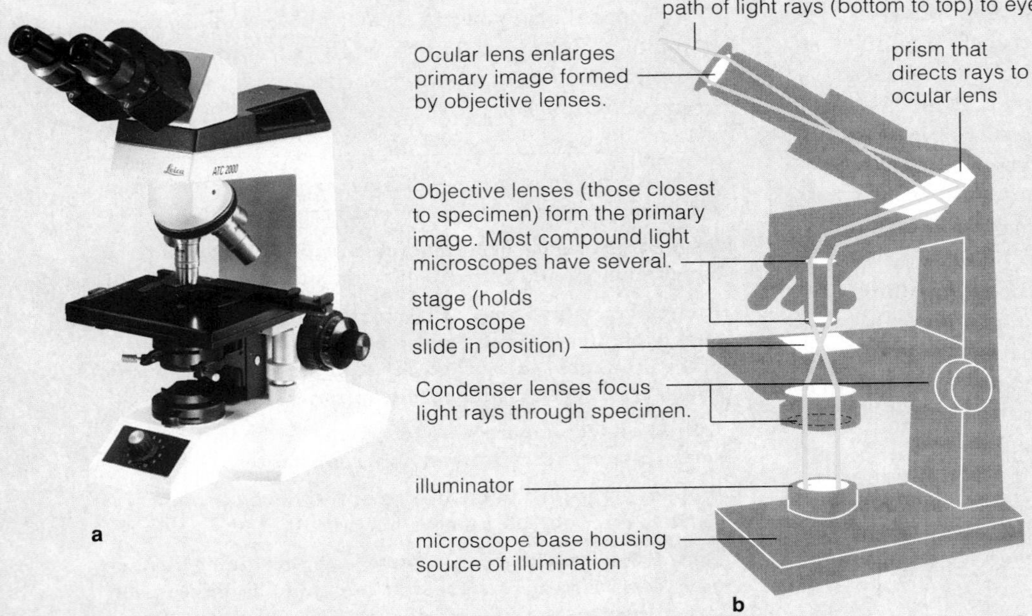
path of light rays (bottom to top) to eye
Ocular lens enlarges primary image formed by objective lenses.
prism that directs rays to ocular lens
Objective lenses (those closest to specimen) form the primary image. Most compound light microscopes have several.
stage (holds microscope slide in position)
Condenser lenses focus light rays through specimen.
illuminator
microscope base housing source of illumination
a
b

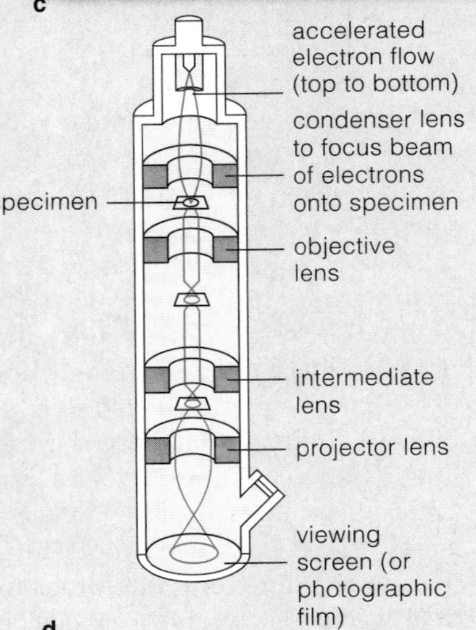
accelerated electron flow (top to bottom)
condenser lens to focus beam of electrons onto specimen
specimen
objective lens
intermediate lens
projector lens
viewing screen (or photographic film)
d

Figure 4.5 (**a**) Exterior view and (**b**) diagram of a compound light microscope. Exterior view (**c**) and diagram (**d**) of an electron microscope.

Further reading: Student Guide to InfoTrac on web site

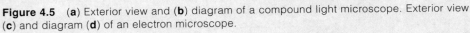

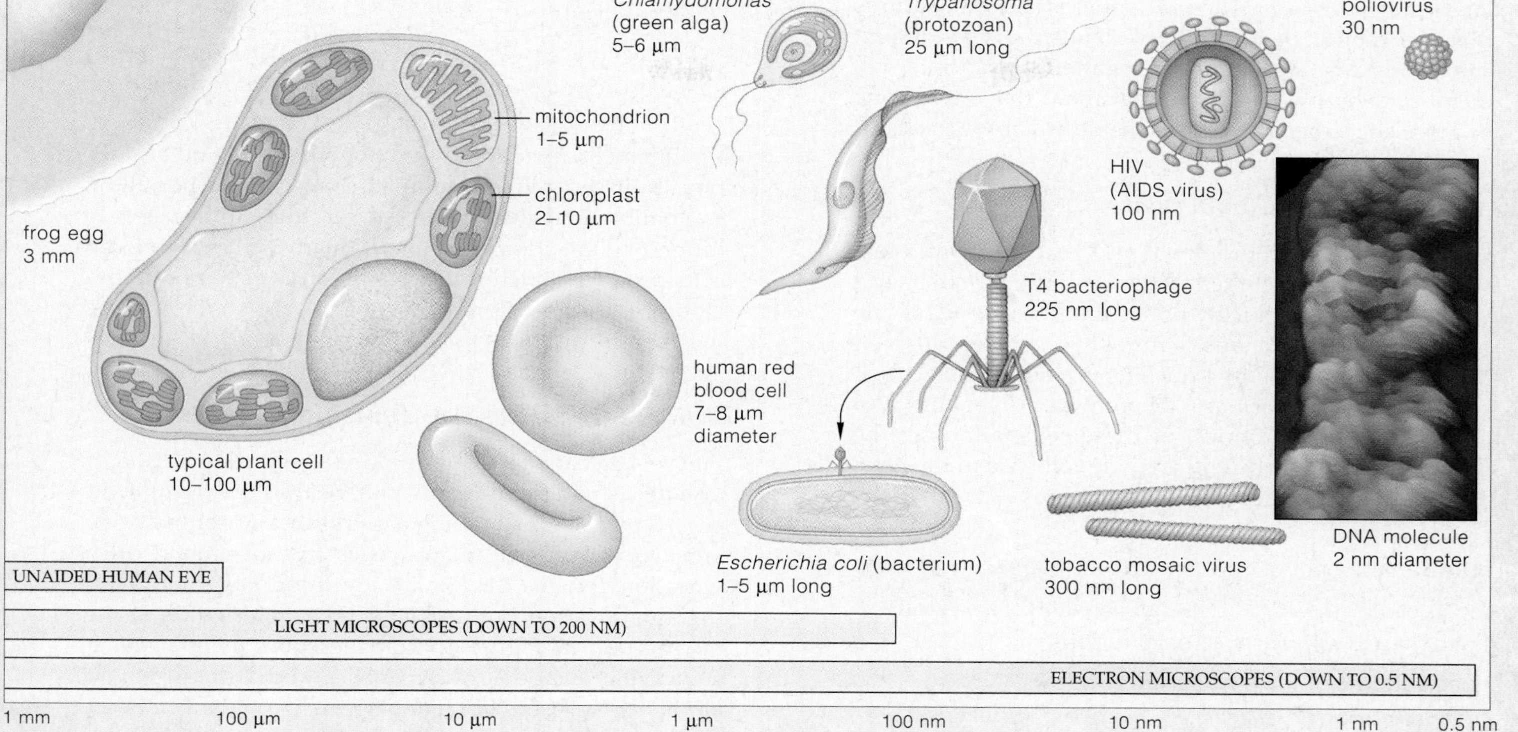
Chlamydomonas
(green alga)
5–6 µm
Trypanosoma
(protozoan)
25 µm long
poliovirus
30 nm
mitochondrion
1–5 µm
chloroplast
2–10 µm
HIV
(AIDS virus)
100 nm
frog egg
3 mm
T4 bacteriophage
225 nm long
human red
blood cell
7–8 µm
diameter
typical plant cell
10–100 µm
Escherichia coli (bacterium)
1–5 µm long
tobacco mosaic virus
300 nm long
DNA molecule
2 nm diameter
UNAIDED HUMAN EYE
LIGHT MICROSCOPES (DOWN TO 200 NM)
ELECTRON MICROSCOPES (DOWN TO 0.5 NM)
1 mm
100 µm
10 µm
1 µm
100 nm
10 nm
1 nm
0.5 nm

Figure 4.6 Units of measure used in microscopy. A *scanning tunneling microscope*, which can provide magnifications up to 100 million, gave us the photomicrograph of part of a DNA molecule. The scope's needlelike probe has a single atom at its tip. When voltage is applied between the tip and an atom at a specimen's surface, it tunnels into the electron orbitals. As the tip moves over a specimen's contours, a computer analyzes the tunneling motion and creates a three-dimensional view of the surface atoms.

1 centimeter	(cm)	= 1/100 meter, or 0.4 inch
1 millimeter	(mm)	= 1/1,000 meter
1 micrometer	(µm)	= 1/1,000,000 meter
1 nanometer	(nm)	= 1/1,000,000,000 meter

1 meter = 10^2 cm = 10^3 mm = 10^6 µm = 10^9 nm

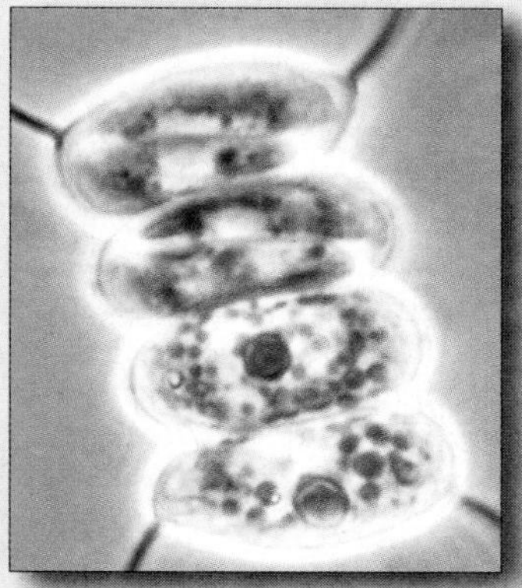

a Light micrograph (phase-contrast process)

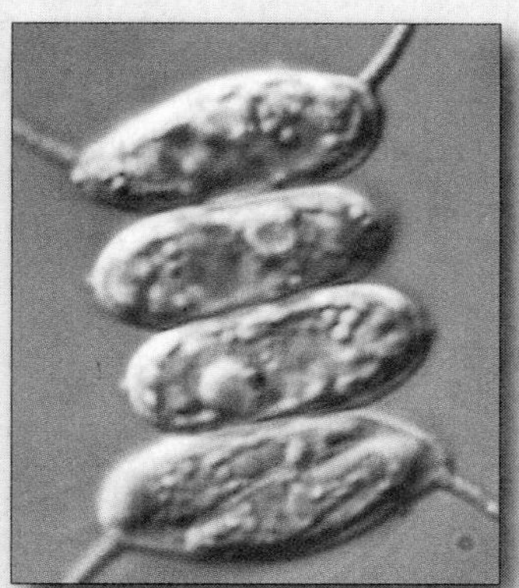

b Light micrograph (Nomarski process)

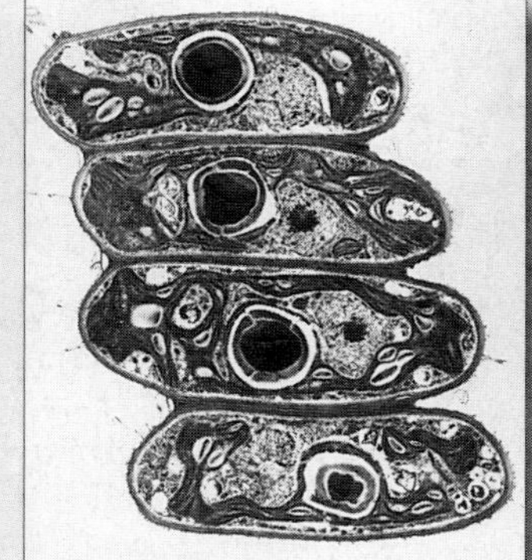

c Transmission electron micrograph, thin section

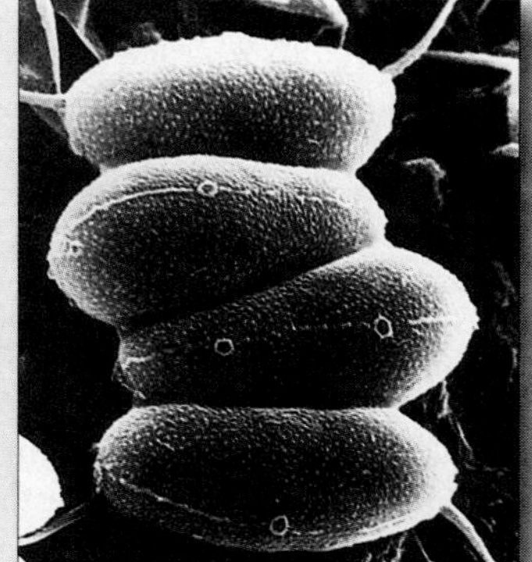

d Scanning electron micrograph

10 µm

Figure 4.7 How different microscopes can reveal different aspects of the same organism—in this case, a green alga (*Scenedesmus*). The images of all four specimens are at the same magnification. The phase-contrast and Nomarski processes mentioned in (**a**) and (**b**) can create optical contrasts without staining the cells. Both processes enhance the usefulness of light micrographs. As for other micrographs in the book, the short horizontal bar below the micrograph in (**d**) provides you with a visual reference for size. A micrometer (µm) is 1/1,000,000 of a meter. Using the scale bar, can you estimate the length and width of *Scenedesmus*?

4.3 DEFINING FEATURES OF EUKARYOTIC CELLS

We turn now to organelles and other structural features that are typical of the cells of plants, animals, fungi, and protistans. We define an **organelle** as an internal, membrane-bound sac or compartment that serves one or more specialized functions inside eukaryotic cells.

Major Cellular Components

Observe some micrographs of a typical eukaryotic cell, such as the ones included in the preceding section, and you probably will quickly notice that the nucleus is one of the most conspicuous features. Remember, any cell that starts out life with a nucleus is a eukaryotic cell; it contains a "true nucleus." Many other organelles and structures also are typical of these cells, although the numbers and kinds differ from one cell type to the next. Table 4.1 lists the most common features.

Table 4.1 Common Features of Eukaryotic Cells

ORGANELLES AND THEIR MAIN FUNCTIONS	
Nucleus	Localizing the cell's DNA
Endoplasmic reticulum	Routing and modifying newly formed polypeptide chains; also, synthesizing lipids
Golgi body	Modifying polypeptide chains into mature proteins; sorting and shipping proteins and lipids for secretion or for use inside cell
Various vesicles	Transporting or storing a variety of substances; digesting substances and structures in the cell; other functions
Mitochondria	Producing many ATP molecules in highly efficient fashion
NON-MEMBRANOUS STRUCTURES AND THEIR FUNCTIONS	
Ribosomes	Assembling polypeptide chains
Cytoskeleton	Imparting overall shape and internal organization to the cell; moving the cell and its internal structures

Think about the list, and you might find yourself asking: What is the advantage of partitioning the cell interior with such organelles? *The compartmentalization allows a large number of activities to occur simultaneously in very limited space.* Consider a photosynthetic cell in a leaf. It can put together starch molecules by one set of reactions and break them apart by another set. Yet the cell would get nothing if the synthesis and breakdown reactions proceeded at the exact same time on the same starch molecule. Without membranes of organelles, the balance of diverse chemical activities that helps keep eukaryotic cells alive would spiral out of control.

Figure 4.8 *Facing page*: Generalized sketches showing some of the features that are typical of (**a**) many plant cells and (**b**) many animal cells.

Organelle membranes have another function besides physically separating incompatible reactions. They allow compatible and interconnected reactions to proceed at different times. For instance, a plant's photosynthetic cells produce starch molecules in an organelle called a chloroplast, then store and later release starch for use in different reactions inside the same organelle.

Which Organelles Are Typical of Plants?

Figure 4.8*a* can start you thinking about the location of organelles in a typical plant cell. Bear in mind, calling a cell "typical" is like calling a cactus or a water lily or an elm tree a "typical" plant. As is true of animal cells, variations on the basic plan are mind-boggling. With this qualification in mind, also take a close look at the micrograph in Figure 4.9, on the subsequent page. It shows the locations of organelles and structures you are likely to observe in many specialized plant cells.

Which Organelles Are Typical of Animals?

Now start thinking about organelles of a typical animal cell, such as the one shown in Figures 4.8*b* and 4.10. Like plant cells, it has a nucleus, many mitochondria, and the other features listed in Table 4.1. *The structural similarities point to basic functions that are necessary for survival, regardless of the cell type.* We will return to this concept throughout the book.

Comparing Figures 4.8 through 4.10 also will give you an initial idea of how plant and animal cells differ in their structure. For example, you will never observe an animal cell surrounded by a cell wall. (You might see many different kinds of fungal and protistan cells with one, however.) What other differences can you identify?

Eukaryotic cells contain a number of organelles, which are internal, membrane-bound sacs and compartments that serve specific metabolic functions.

Organelles physically separate chemical reactions, many of which are incompatible.

Organelles separate different reactions in time, as when certain molecules are put together, stored, then used later in other reaction sequences.

All eukaryotic cells contain certain organelles (such as the nucleus) and structures (such as ribosomes) that perform functions essential for survival. Specialized cells also may incorporate additional kinds of organelles and structures.

Further reading: Student Guide to InfoTrac on web site —

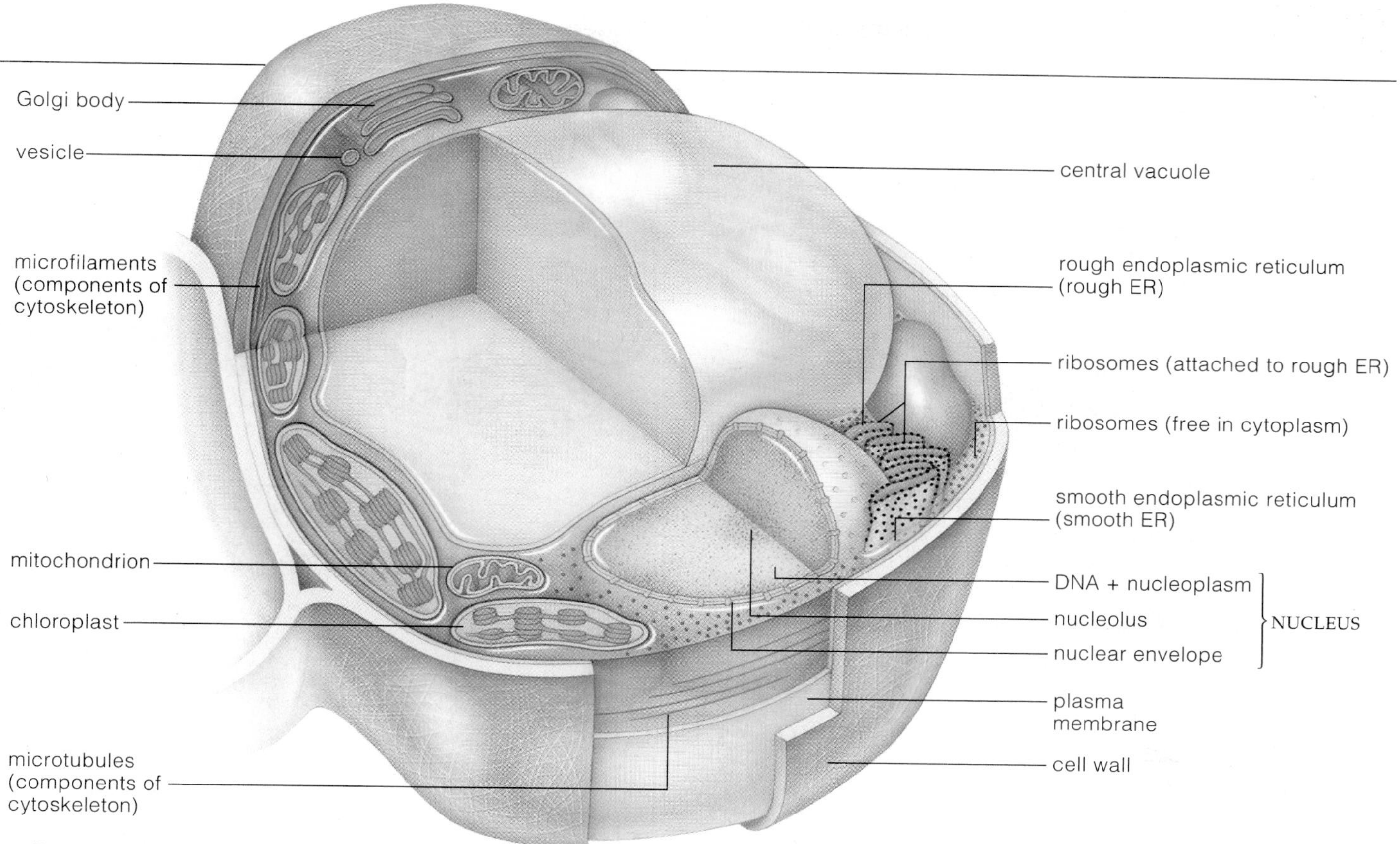

Golgi body
vesicle
central vacuole
microfilaments (components of cytoskeleton)
rough endoplasmic reticulum (rough ER)
ribosomes (attached to rough ER)
ribosomes (free in cytoplasm)
smooth endoplasmic reticulum (smooth ER)
mitochondrion
DNA + nucleoplasm
chloroplast
nucleolus
nuclear envelope
NUCLEUS
plasma membrane
cell wall
microtubules (components of cytoskeleton)

a Components of a typical plant cell.

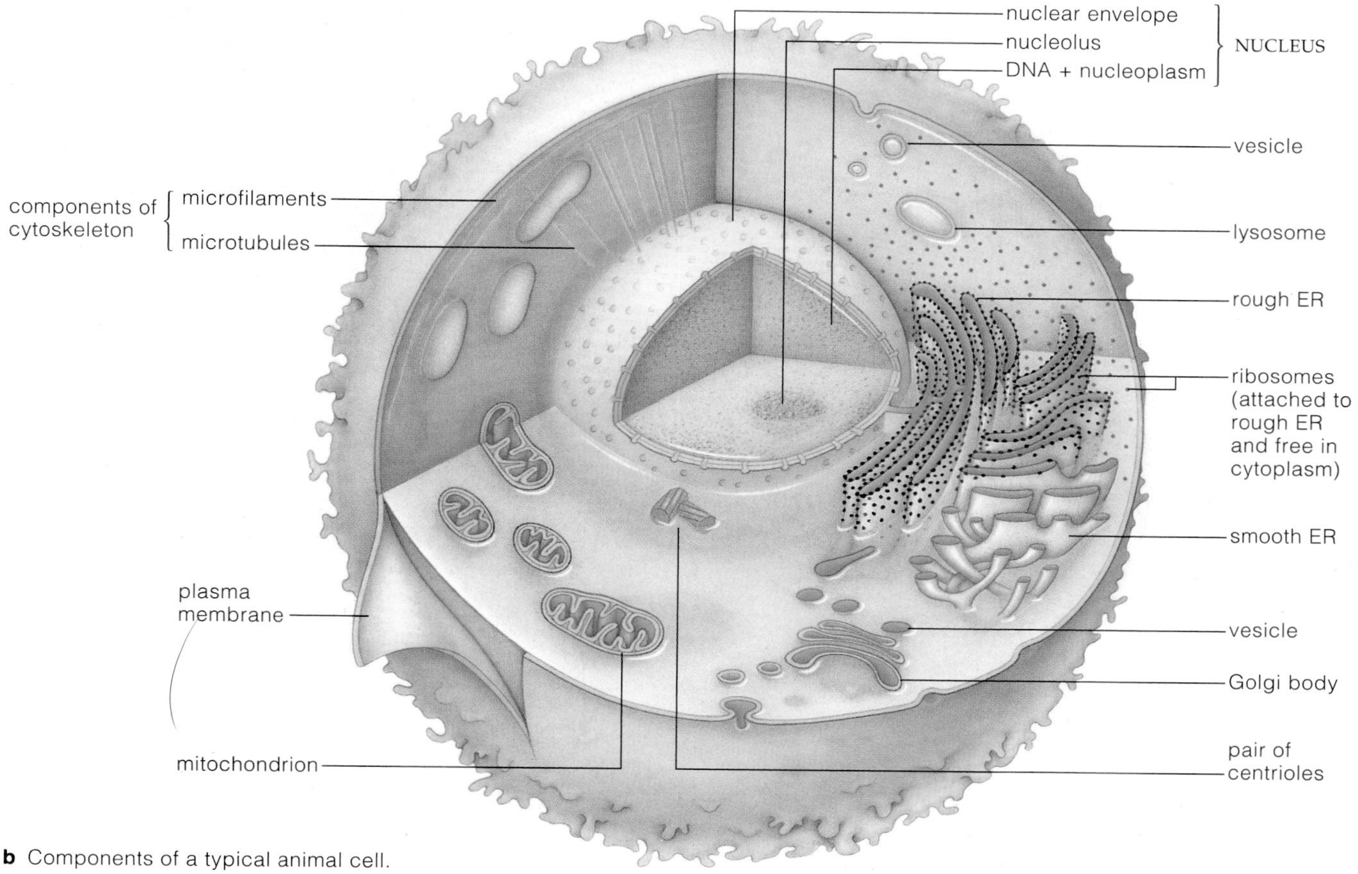

nuclear envelope
nucleolus
DNA + nucleoplasm
NUCLEUS
vesicle
components of cytoskeleton: microfilaments, microtubules
lysosome
rough ER
ribosomes (attached to rough ER and free in cytoplasm)
smooth ER
plasma membrane
vesicle
Golgi body
mitochondrion
pair of centrioles

b Components of a typical animal cell.

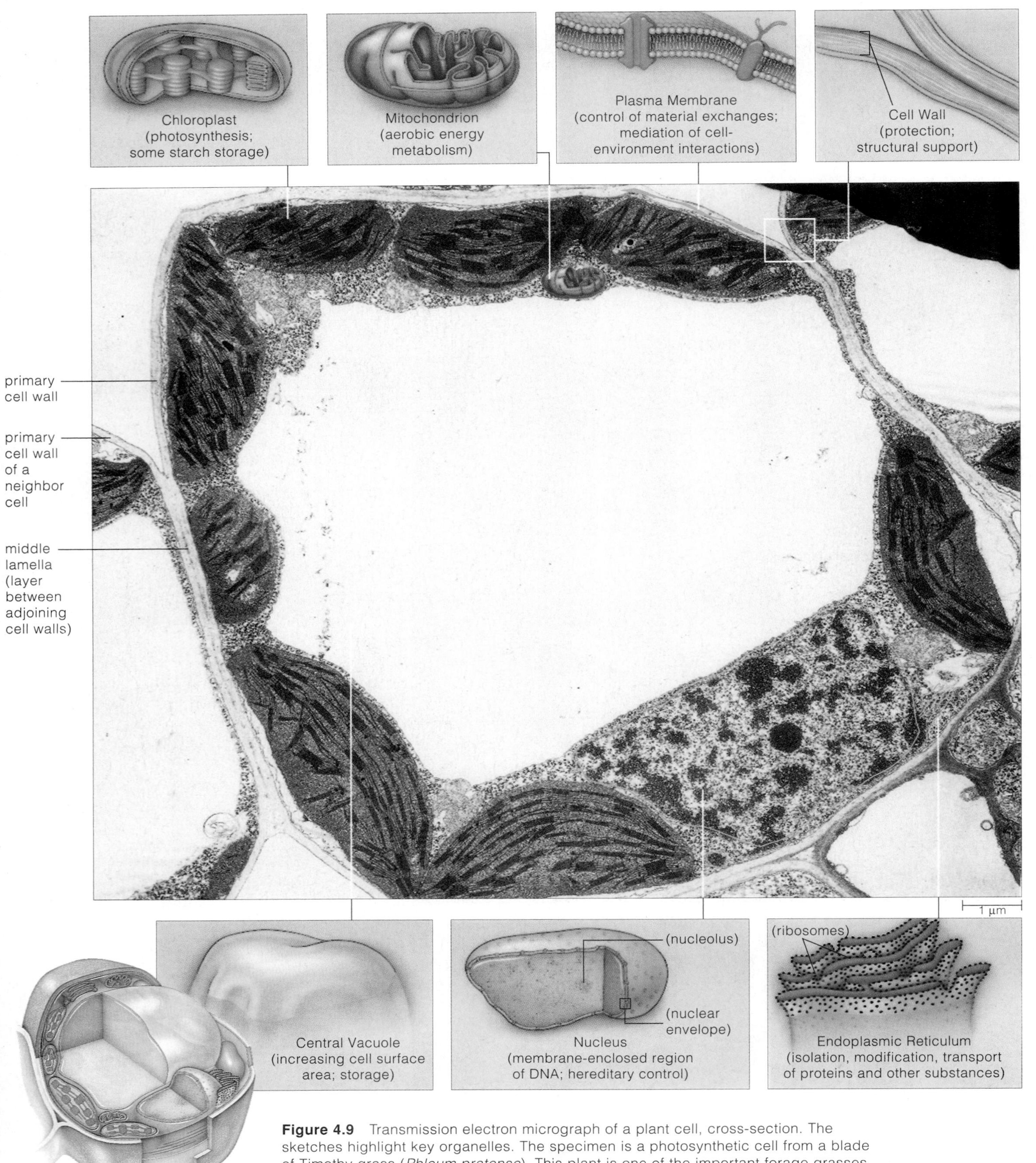

Chloroplast
(photosynthesis;
some starch storage)
Mitochondrion
(aerobic energy
metabolism)
Plasma Membrane
(control of material exchanges;
mediation of cell-
environment interactions)
Cell Wall
(protection;
structural support)
primary
cell wall
primary
cell wall
of a
neighbor
cell
middle
lamella
(layer
between
adjoining
cell walls)
1 µm
Central Vacuole
(increasing cell surface
area; storage)
(nucleolus)
(nuclear
envelope)
Nucleus
(membrane-enclosed region
of DNA; hereditary control)
(ribosomes)
Endoplasmic Reticulum
(isolation, modification, transport
of proteins and other substances)

Figure 4.9 Transmission electron micrograph of a plant cell, cross-section. The sketches highlight key organelles. The specimen is a photosynthetic cell from a blade of Timothy grass (*Phleum pratense*). This plant is one of the important forage grasses that ranchers grow on lands that are suitable for grazing but not for agriculture.

Further reading: Student Guide to InfoTrac on web site

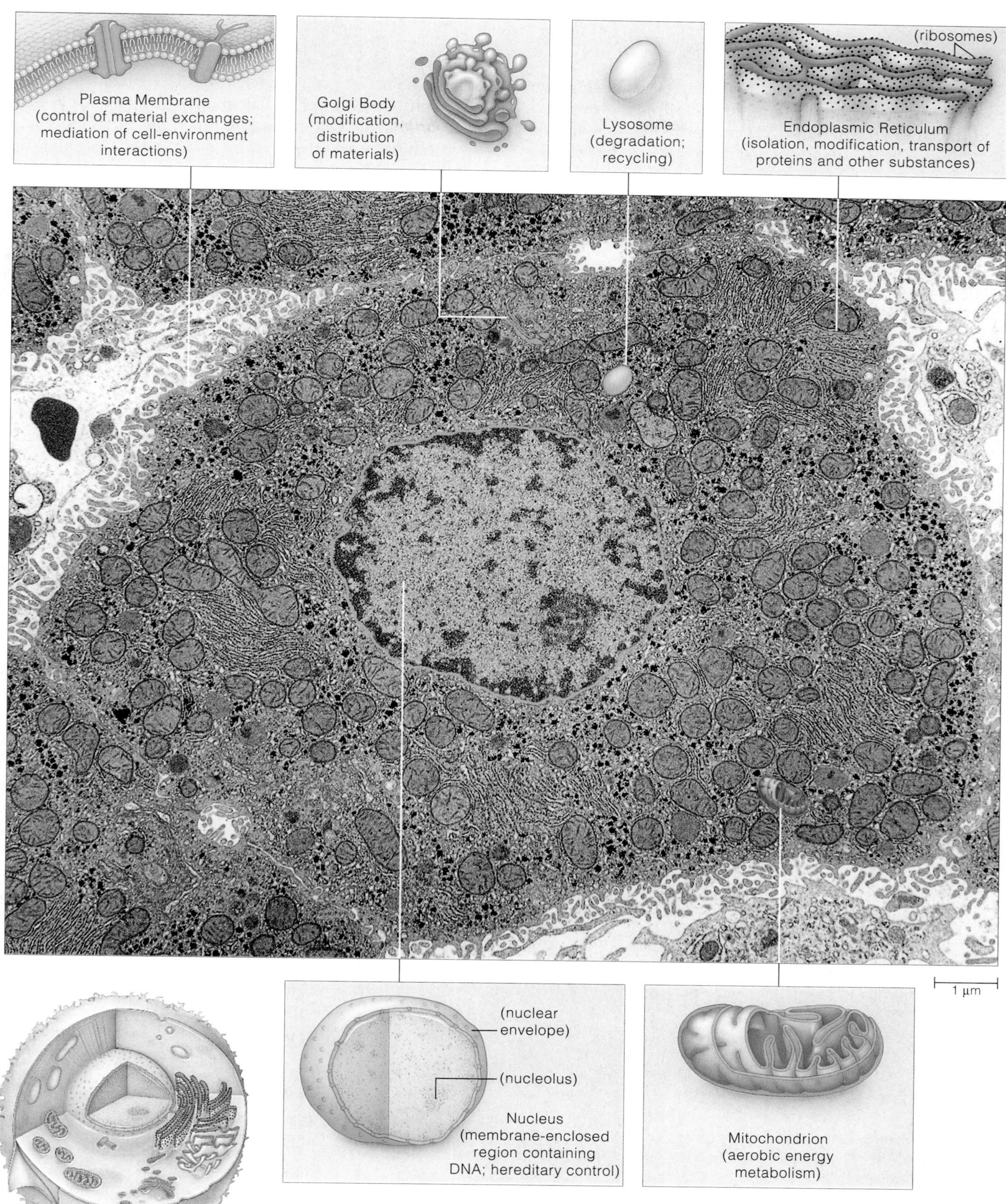

Plasma Membrane
(control of material exchanges; mediation of cell-environment interactions)
Golgi Body
(modification, distribution of materials)
Lysosome
(degradation; recycling)
(ribosomes)
Endoplasmic Reticulum
(isolation, modification, transport of proteins and other substances)
1 µm
(nuclear envelope)
(nucleolus)
Nucleus
(membrane-enclosed region containing DNA; hereditary control)
Mitochondrion
(aerobic energy metabolism)

Figure 4.10 Transmission electron micrograph of an animal cell, cross-section. The sketches highlight key organelles. The specimen is a cell from the liver of a rat.

4.4 THE NUCLEUS

Constructing, operating, and reproducing cells simply cannot be done without carbohydrates, lipids, proteins, and nucleic acids. It takes a class of proteins—enzymes—to build and use these molecules. Said another way, a cell's structure and function begin with proteins. *And instructions for building proteins are contained in DNA.*

Unlike bacteria, eukaryotic cells have their genetic instructions distributed among several to many DNA molecules of different lengths. For example, each human body cell has forty-six molecules of DNA. Stretch them end to end, and they would be about a meter long. Similarly, stretch the twenty-six DNA molecules in a frog cell end to end and they would extend ten meters. We have no idea what frogs are doing with so much of it, but that's a lot of DNA!

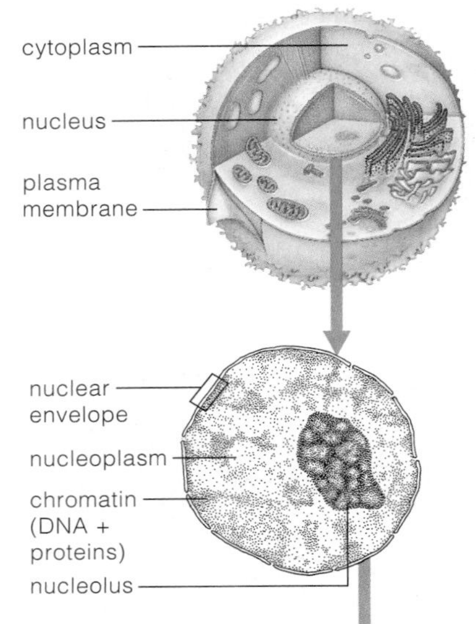

cytoplasm
nucleus
plasma membrane
nuclear envelope
nucleoplasm
chromatin (DNA + proteins)
nucleolus

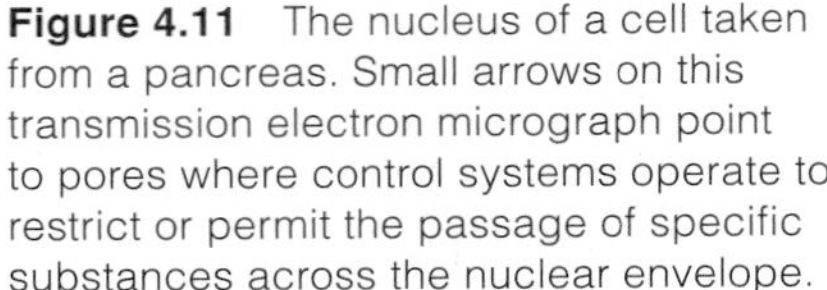

Figure 4.11 The nucleus of a cell taken from a pancreas. Small arrows on this transmission electron micrograph point to pores where control systems operate to restrict or permit the passage of specific substances across the nuclear envelope.

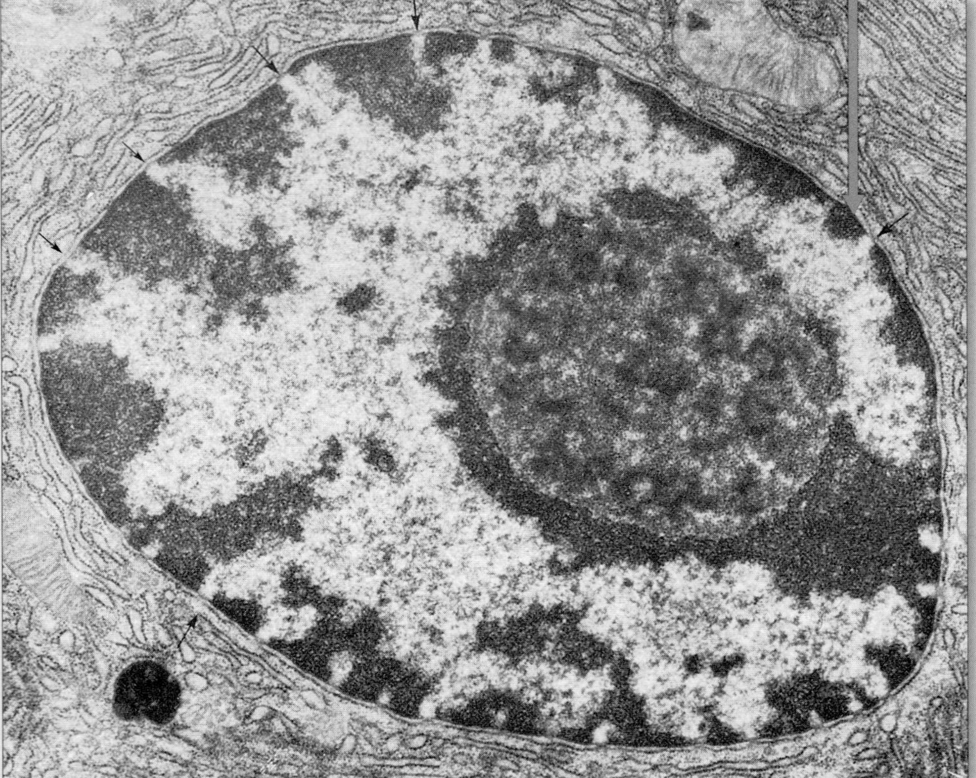

Eukaryotic cells protect their DNA inside a **nucleus**. Figure 4.11 shows the distinctive structure of this type of organelle, which has two functions. *First*, the nucleus physically tucks away all the DNA molecules, apart from the complex metabolic machinery of the cytoplasm. This localization of DNA makes it easier for parent cells to copy their genetic instructions before the time comes for them to divide. DNA molecules can be sorted out into parcels—one for each daughter cell that forms. *Second*, outer membranes of the nucleus form a boundary where cells can control the passage of substances and signals to and from the cytoplasm.

Nuclear Envelope

Unlike the cell itself, a nucleus has two outer membranes, one wrapped around the other. This double-membrane system is called a **nuclear envelope**. It consists of two lipid bilayers in which numerous protein molecules are embedded (Figure 4.12). It surrounds the fluid portion of the nucleus, which is the nucleoplasm.

Stitched on the innermost surface of the nuclear envelope are attachment sites for protein filaments. These anchor the molecules of DNA to the envelope and help keep them organized. On the outer surface of the envelope is a profusion of ribosomes. All proteins are built either on such membrane-bound ribosomes or on ribosomes located in the cytoplasm.

As is the case for all cell membranes, a nuclear envelope's lipid bilayers keep water-soluble substances from moving freely into and out of the nucleus. But many pores, each composed of clusters of proteins, span both bilayers. Ions and small, water-soluble molecules cross the nuclear envelope at the pores. As you will see, large molecules (including the subunits of ribosomes) cross the bilayers at pores in highly controlled ways.

Nucleolus

Take a look at the nucleus in Figure 4.11. Inside is a rather round, dense mass of material. What is it? While eukaryotic cells are growing, one or more of these masses form inside its nucleus. Each is a **nucleolus** (plural, nucleoli). Here, a large number of protein and RNA molecules

Further reading: Student Guide to InfoTrac on web site →

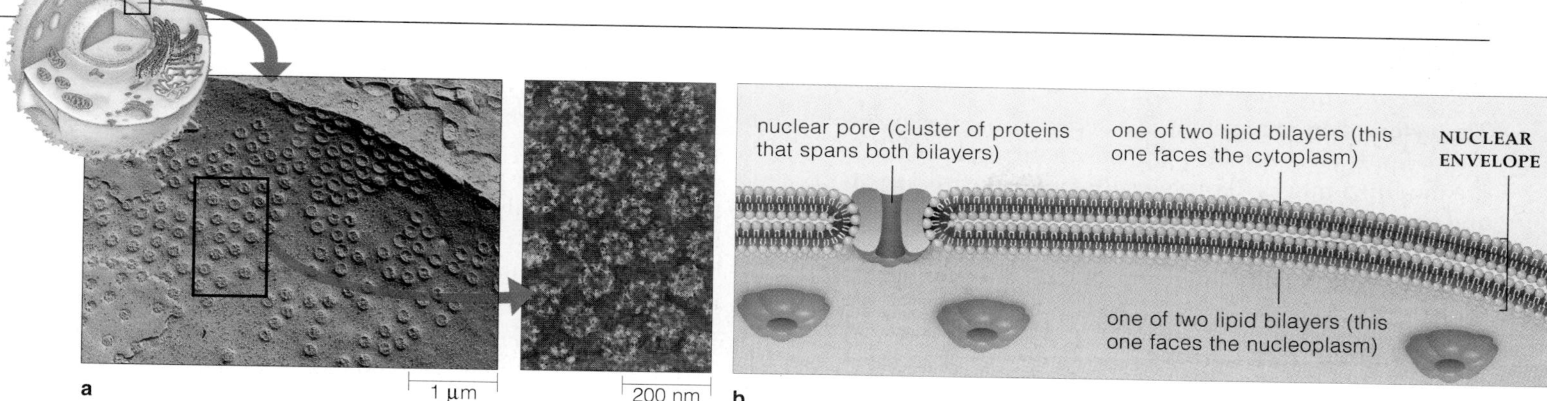
a
1 µm
200 nm
b
nuclear pore (cluster of proteins that spans both bilayers)
one of two lipid bilayers (this one faces the cytoplasm)
NUCLEAR ENVELOPE
one of two lipid bilayers (this one faces the nucleoplasm)

Figure 4.12 (**a**) Part of the outer surface of a nuclear envelope. *Left*: This specimen was fractured to reveal the intimate layering of its two lipid bilayers. *Right*: Nuclear pores. Each pore across the envelope is an organized array of membrane proteins. It permits the selective transport of substances into and out of the nucleus. (**b**) Sketch of the nuclear envelope's structure.

are being constructed. These particular materials are subunits from which ribosomes are built. The subunits pass through nuclear pores and reach the cytoplasm. At times of protein synthesis, intact ribosomes form (each from two subunits) in the cytoplasm.

Chromosomes

When eukaryotic cells are not busy dividing, their DNA looks like thin threads inside the nucleus. Many protein molecules are attached to the threads, a bit like beads on a string. Except at extreme magnification, the beaded threads have a grainy appearance, as in Figure 4.11. However, when the cell is preparing to divide, it duplicates its DNA molecules so each of its daughter cells will get all of the required hereditary instructions. In addition, each DNA molecule becomes folded and twisted into a condensed structure, proteins and all.

Early microscopists bestowed the name *chromatin* on the seemingly grainy substance and *chromosomes* on the condensed structures. We now define **chromatin** as the cell's collection of DNA, together with all proteins associated with it (Table 4.2). Each **chromosome** is one DNA molecule and its associated proteins, regardless of whether it is in threadlike or condensed form:

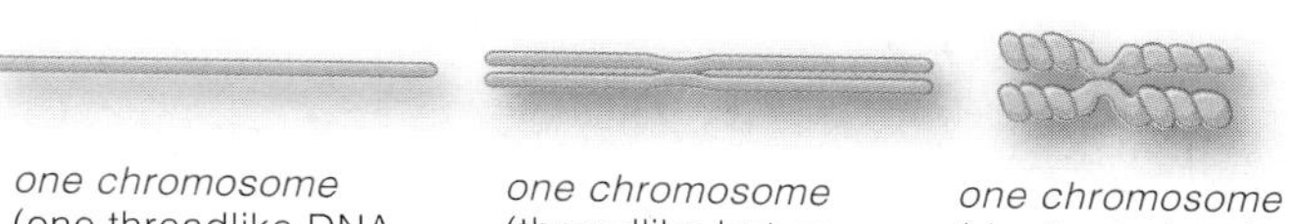

one chromosome (one threadlike DNA molecule + proteins; not duplicated)	*one chromosome* (threadlike but now duplicated; two DNA molecules + proteins)	*one chromosome* (duplicated and also condensed)

In other words, the appearance of "the chromosome" changes over the life of a eukaryotic cell. In chapters to come, you will look at different aspects of chromosomes, and you may find it useful to remember this point.

Table 4.2 Summary of Components of the Nucleus	
Nuclear envelope	Pore-riddled double-membrane system that selectively controls the passage of various substances into and out of the nucleus
Nucleoplasm	Fluid interior portion of the nucleus
Nucleolus	Dense cluster of RNA and proteins that will be assembled into subunits of ribosomes
Chromosome	One DNA molecule and many proteins that are intimately associated with it
Chromatin	Total collection of all DNA molecules and their associated proteins in the nucleus

What Happens to the Proteins Specified by DNA?

Outside the nucleus, polypeptide chains for proteins are assembled on ribosomes. What happens to the new chains? Many become stockpiled in the cytoplasm or get used at once. Many others enter a cytomembrane system. As described in the next section, this system consists of different organelles, including endoplasmic reticulum, Golgi bodies, and vesicles.

Thanks to DNA's instructions, many proteins take on particular, final forms in the cytomembrane system. Lipids also are packaged and assembled in the system by enzymes and other proteins (which also were built according to DNA's instructions). As you will see next, vesicles deliver the proteins and lipids to specific sites within the cell or to the plasma membrane, for export.

The nucleus, an organelle with two outer membranes, keeps the DNA molecules of eukaryotic cells separated from the metabolic machinery of the cytoplasm.

The localization makes it easier to organize the DNA and to copy it before a parent cell divides into daughter cells.

Pores across the nuclear envelope help control the passage of many substances between the nucleus and cytoplasm.

4.5 THE CYTOMEMBRANE SYSTEM

The **cytomembrane system** is a series of organelles in which lipids are assembled and new polypeptide chains are modified into final proteins. Its products are sorted and shipped to different destinations. Figure 4.13 shows how its organelles—the ER, Golgi bodies, and various vesicles—functionally interconnect with one another.

Endoplasmic Reticulum

The functions of the cytomembrane system begin with **endoplasmic reticulum**, or **ER**. In animal cells, the ER is continuous with the nuclear envelope and extends through cytoplasm. Its membrane regions appear rough or smooth, depending mainly on whether ribosomes are attached to the membrane facing the cytoplasm.

We typically observe *rough* ER arranged into stacks of flattened sacs with many ribosomes attached (Figure 4.14*a*). Every new polypeptide chain is synthesized on ribosomes. But only the newly forming chains having a built-in signal can enter the space within rough ER or become incorporated into ER membranes. (The signal is a string of fifteen to twenty specific amino acids.) Once the chains are in rough ER, enzymes may attach oligosaccharides and other side chains to them. Many specialized cells secrete the final proteins. Rough ER is abundant in such cells. For example, in your pancreas, ER-rich gland cells make and secrete enzymes that end up in the small intestine and help digest your meals.

Smooth ER is free of ribosomes and curves through cytoplasm like connecting pipes (Figure 4.14*b*). Many cells assemble most lipids inside the pipes. Smooth ER is well developed in seeds. In liver cells, some drugs and toxic metabolic wastes are inactivated in it. Sarcoplasmic reticulum, a type of smooth ER in skeletal muscle cells, functions in muscle contraction.

Golgi Bodies

In **Golgi bodies**, enzymes put the finishing touches on proteins and lipids, sort them out, and package them inside vesicles for shipment to specific locations. For example, an enzyme in one Golgi region might attach a phosphate group to a new protein, thereby giving it a mailing tag to its proper destination.

Commonly, a Golgi body looks vaguely like a stack of pancakes; it is composed of a series of flattened membrane-bound sacs (Figure 4.15). In functional terms, the last portion of a Golgi body corresponds to the top pancake. Here, vesicles form as patches of the membrane bulge out, then break away into the cytoplasm.

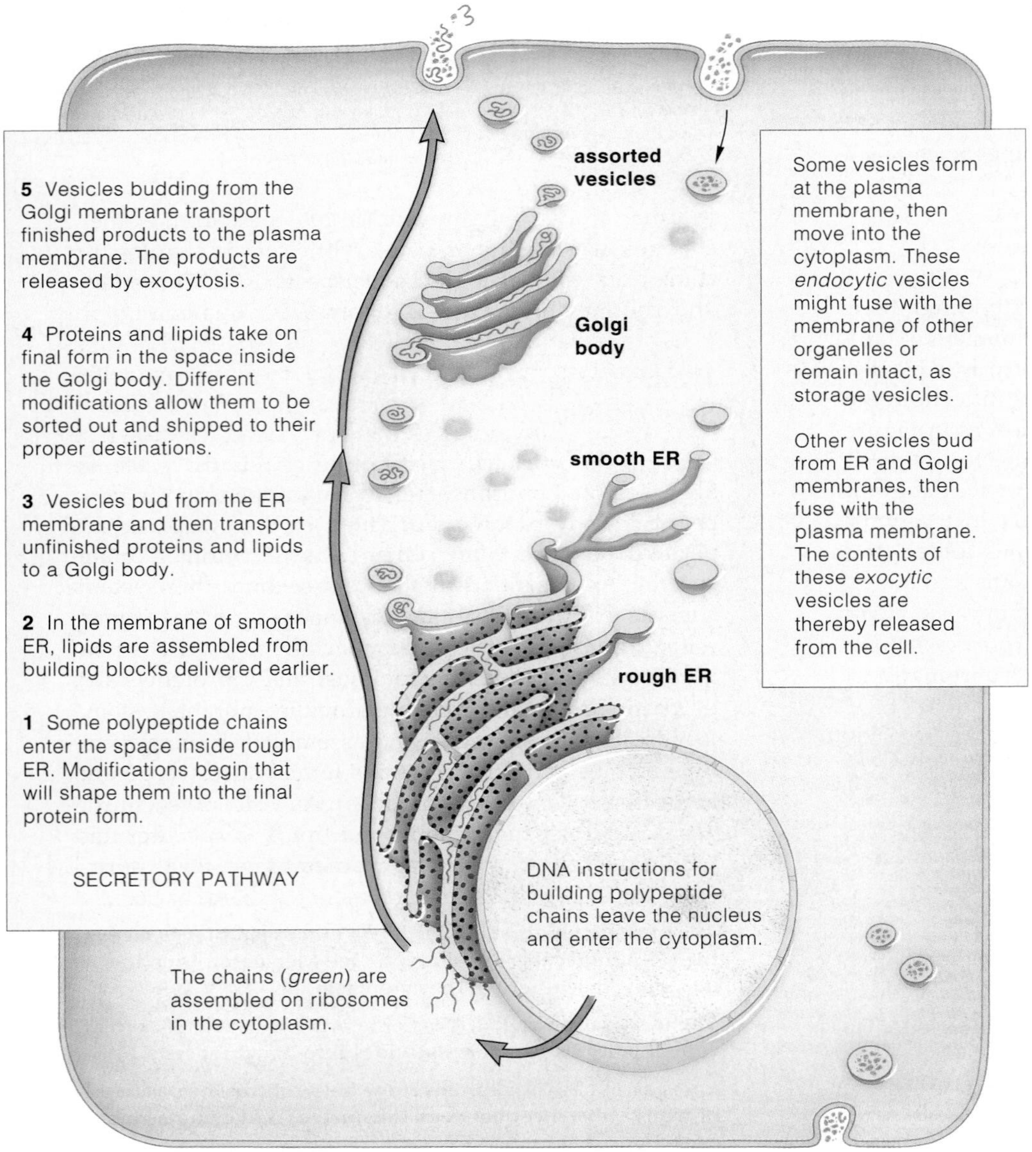

5 Vesicles budding from the Golgi membrane transport finished products to the plasma membrane. The products are released by exocytosis.
4 Proteins and lipids take on final form in the space inside the Golgi body. Different modifications allow them to be sorted out and shipped to their proper destinations.
3 Vesicles bud from the ER membrane and then transport unfinished proteins and lipids to a Golgi body.
2 In the membrane of smooth ER, lipids are assembled from building blocks delivered earlier.
1 Some polypeptide chains enter the space inside rough ER. Modifications begin that will shape them into the final protein form.
SECRETORY PATHWAY
assorted vesicles
Golgi body
smooth ER
rough ER
DNA instructions for building polypeptide chains leave the nucleus and enter the cytoplasm.
The chains (green) are assembled on ribosomes in the cytoplasm.
Some vesicles form at the plasma membrane, then move into the cytoplasm. These endocytic vesicles might fuse with the membrane of other organelles or remain intact, as storage vesicles.
Other vesicles bud from ER and Golgi membranes, then fuse with the plasma membrane. The contents of these exocytic vesicles are thereby released from the cell.

Figure 4.13 Cytomembrane system, a membrane system in the cytoplasm that assembles, modifies, packages, and ships proteins and lipids. *Green* arrows highlight a secretory pathway by which certain proteins and lipids are packaged and released from many types of cells, including gland cells that secrete mucus, sweat, and digestive enzymes.

Further reading: Student Guide to InfoTrac on web site

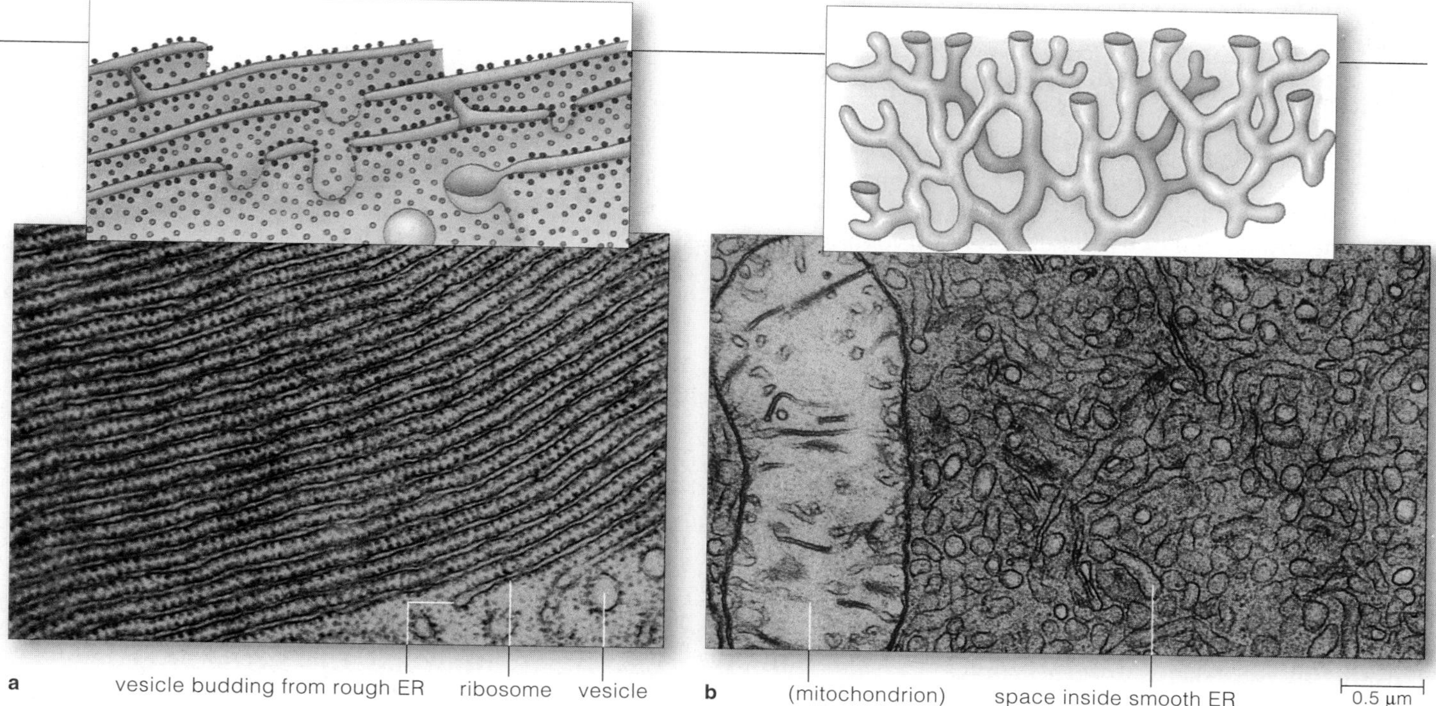
a
vesicle budding from rough ER
ribosome
vesicle
b
(mitochondrion)
space inside smooth ER
0.5 µm

Figure 4.14 Transmission electron micrographs and sketches of endoplasmic reticulum. (**a**) Many ribosomes dot the flattened surfaces of rough ER that face the cytoplasm. (**b**) This section shows the diameters of the many interconnected, pipelike regions of smooth ER.

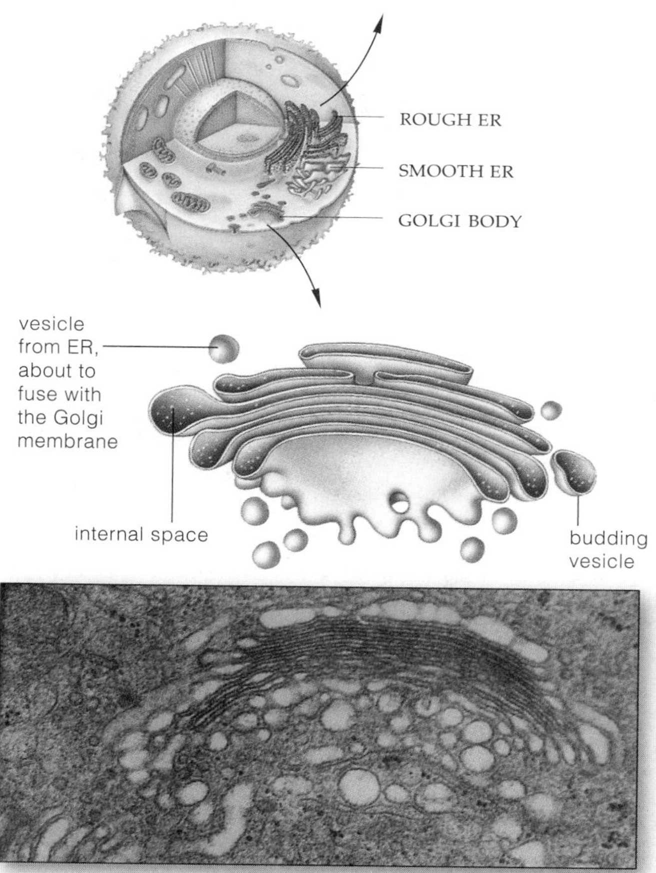
ROUGH ER
SMOOTH ER
GOLGI BODY
vesicle from ER, about to fuse with the Golgi membrane
internal space
budding vesicle
0.25 µm

Figure 4.15 Sketch and micrograph of a Golgi body from an animal cell.

A Variety of Vesicles

Vesicles are tiny, membranous sacs that move through the cytoplasm or take up positions in it. A common type, the lysosome, buds from Golgi membranes of animal cells and certain fungal cells. Lysosomes are organelles of intracellular digestion. They contain a potent brew, rich with diverse enzymes that speed the breakdown of proteins, complex carbohydrates, nucleic acids, and some lipids. Often, lysosomes fuse with vesicles that formed at the plasma membrane. The vesicles typically contain molecules, bacteria, or other items that docked at the plasma membrane. Lysosomes even digest whole cells or cell parts. For example, as a tadpole is developing into an adult frog, its tail slowly disappears. Lysosomal enzymes are responding to developmental signals and are helping to destroy cells that make up the tail.

Peroxisomes, another type, are tiny sacs of enzymes that break down fatty acids and amino acids. Hydrogen peroxide, a potentially harmful product, forms during the reactions. Enzyme action converts it to water and oxygen or channels it into reactions that break down alcohol. After someone drinks alcohol, nearly half of it is degraded in peroxisomes of liver and kidney cells.

Many proteins take on final form and lipids are synthesized in the ER and Golgi bodies of the cytomembrane system.

Lipids, proteins (such as enzymes), and other items become packaged in vesicles destined for export, storage, membrane building, intracellular digestion, and other cell activities.

4.6 MITOCHONDRIA

Recall, from Section 3.7, that ATP molecules are premier energy carriers. Energy associated with their phosphate groups is delivered to nearly all reaction sites, where it drives nearly all cell activities. Many ATP molecules form when organic compounds are completely broken down to carbon dioxide and water in a **mitochondrion** (plural, mitochondria).

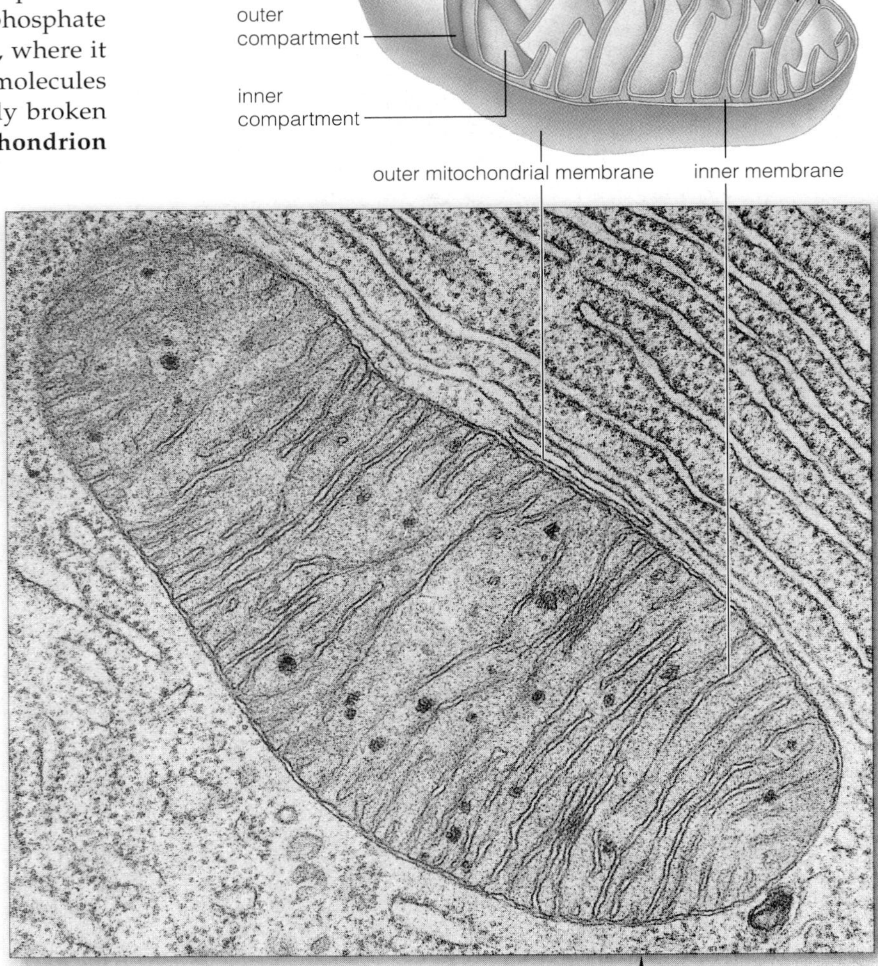

cristae
outer compartment
inner compartment
outer mitochondrial membrane
inner membrane
0.5 µm

Figure 4.16 Sketch and transmission electron micrograph, thin section, of a typical mitochondrion. This organelle specializes in the production of large quantities of ATP, the main energy-carrying molecule between different reaction sites in cells. The process that produces ATP in mitochondria cannot proceed without free oxygen.

Only eukaryotic cells contain these organelles. The example in Figure 4.16 will give you an idea of their structure. The kind of ATP-forming reactions that proceed in mitochondria extract far more energy from organic compounds than can be done by any other means. They cannot run to completion without plenty of oxygen. Like all other land-dwelling vertebrates, every time you breathe in, you take in oxygen mainly for mitochondria in cells—in your case, many trillions of individual cells.

Each mitochondrion has a double-membrane system. As you can tell from Figure 4.8, the outermost membrane faces the cytoplasm. Most commonly, the inner membrane repeatedly folds back on itself. Each inner fold is a crista (plural, cristae).

What is the function of the intricate membrane system? It forms two distinct compartments within a mitochondrion. Enzymes and other proteins stockpile hydrogen ions in the outer compartment. Electron transfers drive the stockpiling, and oxygen helps keep the machinery running by binding and thus removing the spent electrons. Hydrogen ions flow out of the compartment in controlled ways. Energy inherent in the flow drives ATP formation, as Chapter 8 describes.

All eukaryotic cells contain one or more mitochondria. You may find only one in a single-celled yeast. You might find a thousand or more in energy-demanding cells, such as those of muscles. Take a look at the profusion of mitochondria in Figure 4.10, which is a micrograph of merely one thin slice from a liver cell. It alone tells you that the liver is an exceptionally active, energy-demanding organ.

In terms of size and biochemistry, mitochondria resemble bacteria. They even have their own DNA and some ribosomes, and they divide on their own. Perhaps they evolved from ancient bacteria that were engulfed by a predatory, amoebalike cell yet managed to escape digestion. Perhaps they were able to reproduce inside the cell and continued doing so in the descendant cells. If they became permanent, protected residents, they might have lost structures and functions required for independent life while they were becoming mitochondria. We return to this topic in Section 21.3.

The organelles called mitochondria are the ATP-producing powerhouses of all eukaryotic cells.

Energy-releasing reactions proceed at the compartmented, internal membrane system of mitochondria. The reactions, which require oxygen, produce far more ATP than can be made by any other cellular reactions.

Further reading: Student Guide to InfoTrac on web site

SPECIALIZED PLANT ORGANELLES

Chloroplasts and Other Plastids

Many plant cells contain plastids, a general category of organelles that specialize in photosynthesis or function in storage. Three types are common in different parts of plants. They are the chloroplasts, chromoplasts, and amyloplasts.

Of all eukaryotic cells, only the photosynthetic ones have **chloroplasts**. These organelles convert sunlight energy into the chemical energy of ATP, which is used to make sugars and other organic compounds. Chloroplasts commonly are oval or disk-shaped. Their semifluid interior, the stroma, is enclosed by two outer membrane layers. In the stroma is a third membrane called the thylakoid membrane. It is folded into a system of interconnecting, disk-shaped compartments. In many chloroplasts, these compartments stack, one atop the other, as in Figure 4.17. Each stack is a granum (plural, grana).

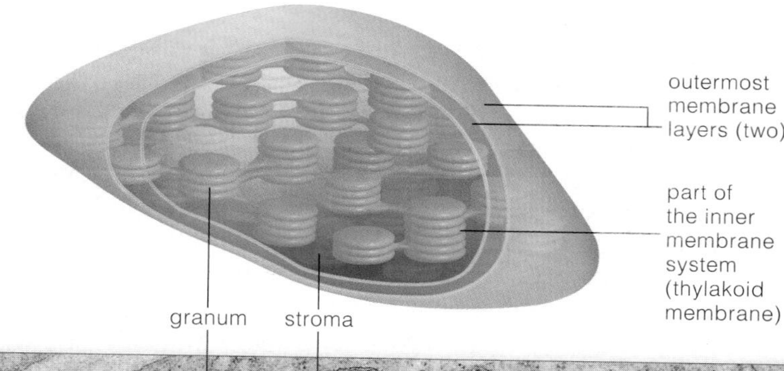
outermost membrane layers (two)
part of the inner membrane system (thylakoid membrane)
granum
stroma

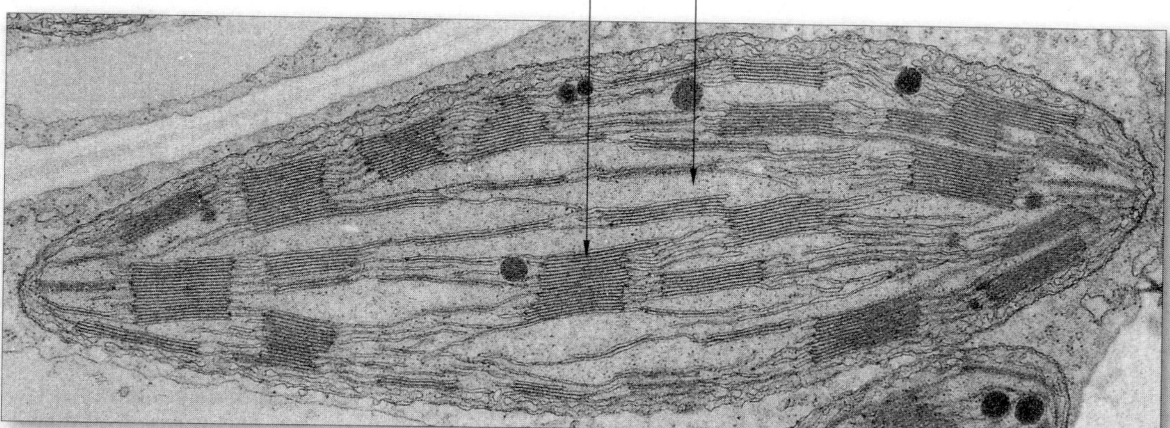
0.5 μm

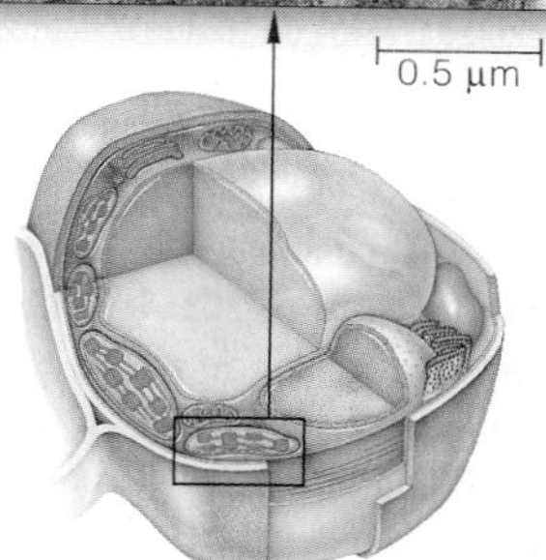

Figure 4.17 Generalized sketch of a chloroplast, the key defining feature of every photosynthetic eukaryotic cell. The transmission electron micrograph shows a thin section of a chloroplast from a photosynthetic corn cell.

The first stage of photosynthesis starts and ends at a thylakoid membrane. Many light-trapping pigments, enzymes, and other proteins carry out the reactions. They work together to absorb light energy and "store" it in the form of ATP. Inside the stroma, ATP energy is used to make sugars, and then starch and other organic compounds, from carbon dioxide and water. Clusters of the new starch molecules (starch grains) may briefly accumulate inside the stroma.

The most abundant photosynthetic pigments are chlorophylls, which reflect or transmit green light. Others include carotenoids, which reflect or transmit yellow, orange, and red light. The relative abundances of the different pigments influence the colors of plant parts.

In many ways, chloroplasts resemble photosynthetic bacteria. Like mitochondria, they might have evolved from bacteria that were engulfed by predatory cells but escaped digestion and became permanent residents in them. We return to this idea in Section 21.3.

Unlike chloroplasts, chromoplasts lack chlorophylls but have an abundance of carotenoids. They are the source of red-to-yellow colors of many flowers, autumn leaves, ripening fruits, and carrots and other roots. The pigment colors commonly attract animals that pollinate plants or disperse seeds. Amyloplasts lack pigments. Often they store starch grains and are abundant in cells of stems, potato tubers (underground stems), and seeds.

Central Vacuole

Many mature, living plant cells have a **central vacuole** (Figure 4.8). This fluid-filled organelle stores amino acids, sugars, ions, and toxic wastes. As it enlarges, it causes fluid pressure to build up inside the cell and so forces the cell's still-pliable cell wall to enlarge. Hence the cell itself enlarges. As the cell surface area increases, so does the rate at which water and other substances can be absorbed across the plasma membrane.

In most cases, the central vacuole increases so much in volume that it takes up 50 to 90 percent of the cell's interior. The cytoplasm ends up as a very narrow zone between the central vacuole and plasma membrane.

Photosynthetic eukaryotic cells contain chloroplasts and other plastids that function in food production and storage.

Many plant cells have a central vacuole. When this storage vacuole enlarges during growth, cells are forced to enlarge, and this increases the surface area available for absorption.

4.8 COMPONENTS OF THE CYTOSKELETON

An interconnected system of fibers, threads, and lattices extends between the nucleus and plasma membrane of eukaryotic cells. This system, the **cytoskeleton**, gives cells their internal organization, shape, and capacity to move. Some elements reinforce the plasma membrane and nuclear envelope. Others form scaffolds that hold protein clusters in membranes or cytoplasmic regions. Others are like railroad tracks upon which organelles are shipped from one site to another! Many elements are permanent; others appear only at certain times in a cell's life. Before a cell divides, for instance, many new microtubules form a "spindle" structure that moves its chromosomes, then disassemble when the task is done.

Figure 4.18 shows the cytoskeleton of an animal cell, isolated from its home tissue, as it was stretching out from left to right across a glass slide. Researchers were studying its **microtubules** and **microfilaments**. These two classes of cytoskeletal elements, acting singly or collectively, are responsible for nearly all movements of eukaryotic cells. In addition, some animal cells have **intermediate filaments**, which are ropelike cytoskeletal elements that impart mechanical strength to cells and tissues. All three classes grow through polymerization, by which many monomers become joined together.

Microtubules

A microtubule is a long, hollow cylinder, twenty-five nanometers wide, made of tubulin monomers (Figure 4.19*a*). It is the largest cytoskeletal element. Tubulin is a protein made of two chemically distinct polypeptide chains, each folded into a globular shape. In a growing microtubule, all tubulin subunits become oriented in the same direction, which puts slightly different chemical and electrical properties at opposite ends of the cylinder. Thus a microtubule shows polarity. One end (the *minus* end) tends to lose monomers. Usually it is stabilized by becoming anchored in a centrosome, a type of MTOC. (MTOCs, short for *M*icrotubule *O*rganizing *C*enters, are sites of dense material that give rise to microtubules.) The cylinder's other end (the *plus* end) is free to grow rapidly through the cytoplasm.

Each microtubule shows dynamic instability: it can suddenly shorten (depolymerize) as well as lengthen. By one model, GTP becomes bound to free tubulin. (GTP is an organic compound made of guanine, ribose, and three phosphate groups.) Binding puts tubulin in a shape that fits right in at the plus end, like a brick being added to a wall. Then hydrolysis removes one phosphate group from GTP to form GDP. Regions of the cylinder that contain GDP are not stable. However, when "bricks" stack up at the growing end faster than hydrolysis is occurring, they act as a cap that stabilizes the cylinder. When conditions trigger loss of the GTP cap, the cylinder curves, weakens, and disassembles.

Microtubules govern the division of cells and some aspects of their shape as well as many cell movements. Cells cannot do much without them. As you might well imagine, microtubules have become prime targets in the chemical warfare between vulnerable species and their attackers. For example, autumn crocus and other plants of the genus *Colchicum* synthesize colchicine, a poison that inhibits the assembly and promotes disassembly of microtubules, with dire effects on browsing animals. The plant cells themselves have an evolved insensitivity to colchicine, which cannot bind well to the type of tubulin monomers they produce.

Taxol, a poison offered up by the western yew (*Taxus brevifolia*), can stabilize assembled microtubules and so prevent formation of new ones that a cell might require for other tasks. For example, like colchicine and some other microtubule poisons, taxol is able to block cell division by interfering with the formation of

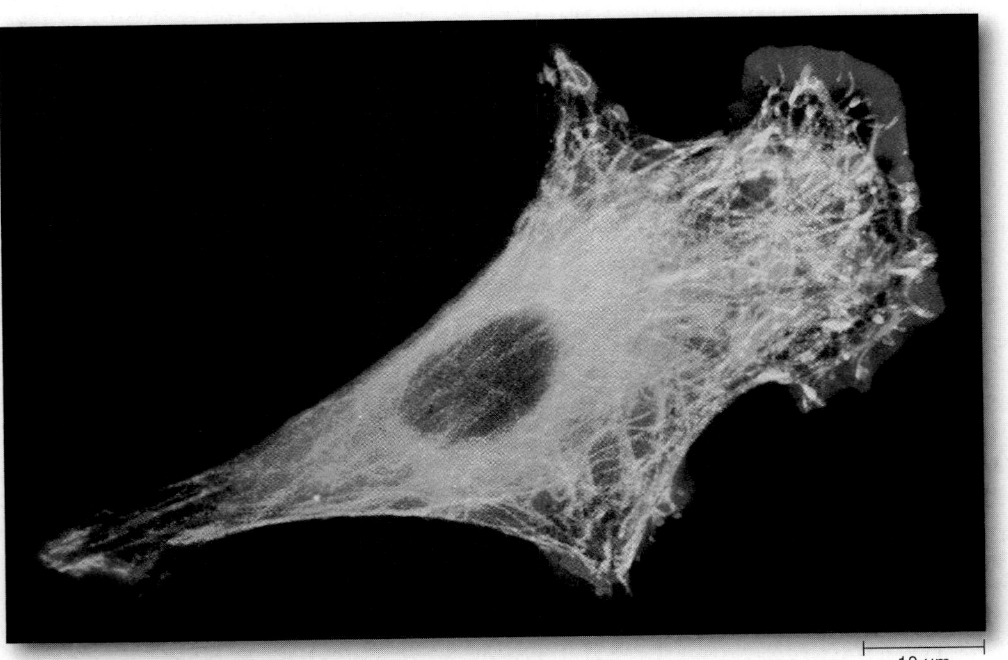

10 μm

Figure 4.18 Locations of three types of structural elements in the cytoskeleton of a fibroblast, a type of animal cell that gives rise to certain animal tissues. This composite of three micrographic images reveals the presence of many microtubules (tinted *green*). Two kinds of microfilaments are tinted *blue* and *red*.

Further reading: Student Guide to InfoTrac on web site

microtubular spindles. Doctors have used it to decrease the uncontrolled cell divisions that underlie the growth of some benign and malignant tumors.

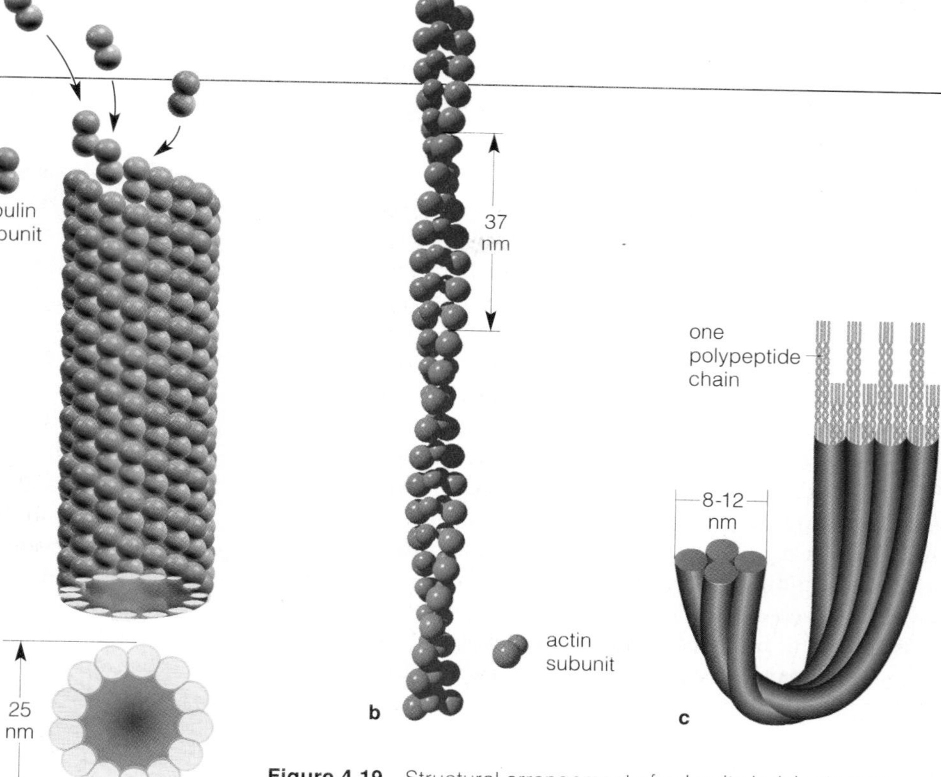

tubulin subunit
25 nm
a
37 nm
b
actin subunit
one polypeptide chain
8-12 nm
c

Figure 4.19 Structural arrangement of subunits in (a) microtubules, (b) microfilaments, and (c) one of the intermediate filaments.

Microfilaments

Microfilaments, the thinnest of the cytoskeletal elements, are five to seven nanometers wide (Figure 4.19*b*). Each is made of two polypeptide chains that are helically twisted together. The chains are assembled from monomers of the protein actin.

Like microtubules, the actin filaments are polar and have a dynamic structure. Monomers tend to be added to one end and removed from the other end. Unlike microtubules, however, they generally are organized in bundles or networks. As explained in several chapters in the book, microfilaments take part in a great variety of movements, especially the kinds that affect the cell surface. They also contribute to the development and maintenance of animal cell shapes.

Myosin and Other Accessory Proteins

Both tubulin and actin have been highly conserved over evolutionary time; the microtubules and microfilaments of all eukaryotic species are assembled from them. In spite of the uniformity, monomers of a variety of other proteins become attached to them. Thus embellished, microtubules and microfilaments that were assembled according to the same basic pattern can serve different functions in different regions of the cell.

For example, monomers of myosin, dynein, or some other **motor protein** typically attach to the surface of microtubules and microfilaments in ways that cause cell movement. Myosin is abundant in muscle cells, and it is part of the machinery by which they contract. "Crosslinking proteins" splice adjacent microfilaments together. Some kinds take part in the formation of the **cell cortex**—an extensive, three-dimensional network of microfilaments and other proteins just beneath the plasma membrane. The cortex reinforces the cell surface and facilitates movements as well as changes in shape. Other crosslinking proteins splice microfilaments in a stable network having the properties of a gel. Spectrin, another accessory protein, attaches microfilaments to the plasma membrane. As one more example, the integrins span this outermost membrane, attach to microfilaments inside the cell, and connect them with proteins outside.

Intermediate Filaments

Intermediate filaments, the most stable elements of the cytoskeleton, are between eight and twelve nanometers wide (Figure 4.19*c*). The six known groups mechanically strengthen cells or cell parts and help maintain their shape. For example, desmins and vimentins help hold the contractile units of muscle cells in organized arrays. Lamins help form a scaffold that reinforces the nucleus. Diverse cytokeratins structurally reinforce the cells that give rise to nails, claws, horns, and hairs.

Unlike the other two classes of cytoskeletal elements, intermediate filaments are present only in animal cells of specific tissues. Moreover, because each cell usually has only one or sometimes two kinds, researchers can use intermediate filaments to identify cell type. Such typing has proved to be a useful tool in diagnosing the tissue origin of different forms of cancer.

Every eukaryotic cell has a cytoskeleton, the diverse elements of which are the basis of its shape, internal structure, and capacity for movement.

Microtubules are key organizers of the cytoskeleton and help move certain cell structures. Microfilaments take part in diverse movements and in the formation and maintenance of cell shape. Intermediate filaments structurally reinforce certain animal cells and internal cell structures.

4.9 THE STRUCTURAL BASIS OF CELL MOTILITY

If a cell lives, it moves. This is true of free-living single cells, such as the sperm and ciliated cells in Figure 4.20. It is true also of cells with fixed positions in the tissues of multicelled organisms. Eukaryotic cells rearrange or shunt organelles and chromosomes, contract (shorten), thrust out long or wide lobes of cytoplasm, stir fluids, or bend a tail and propel the cell body forward. These and all other movements of cell structures, the cell itself, and entire multicelled organisms are forms of motility.

Section 4.8 introduced the major players in cellular motility—the microtubules, microfilaments, and other proteins that interact with them. Energy inputs, as from ATP, trigger motions at the molecular level. Many such motions combine to produce movement at the cellular level, as when a muscle cell contracts. The coordinated movements of many cells cause movements of tissues or organs, as when bundles of skeletal muscle cells interact with bones to make a thumb and finger turn a page.

Mechanisms of Cell Movements

Microfilaments, microtubules, or both take part in most aspects of motility. They do so by three mechanisms.

First, *the length of a microtubule or microfilament can grow or diminish by the controlled assembly or disassembly of its subunits.* When either lengthens or shortens at one end, a chromosome or some other structure attached to the other end is pushed or dragged through cytoplasm.

Some cells crawl about on protrusions of the body surface that form by microfilament assembly. *Amoeba proteus*, a soft-bodied protistan, has **pseudopods** ("false feet"): temporary, lobelike protrusions from the body. Inside each lobe, microfilaments grew rapidly in length, and the attached plasma membrane was dragged along with them. Similarly, the animal cell in Figure 4.18 was migrating on sheetlike extensions of microfilaments.

Second, *parallel rows of microfilaments or microtubules actively slide in specific directions.* For instance, a muscle cell has a series of contractile units. Microfilaments of actin are attached to one end or the other of each unit, and many myosin strands lie parallel with them. ATP energizes the myosin, which has oarlike projections that repeatedly bind and release microfilament neighbors. The short, repeated strokes make the microfilaments slide over the myosin strands, toward the center of the contractile unit—which shortens (contracts) as a result.

A similar sliding mechanism might be operating as cells crawl about. If the microfilament network beneath the plasma membrane at a cell's trailing end contracts, then some cytoplasmic gel would be squeezed into the leading end, which would bulge forward in response.

Third, *microtubules or microfilaments shunt organelles or parts of the cytoplasm from one location to another.* For example, chloroplasts move to new light-intercepting positions in response to the changing angle of the sun's overhead position. How? They are attached to myosin monomers that are "walking" over microfilaments and carrying chloroplasts with them. This shunt mechanism also causes cytoplasmic streaming: a dynamic and often rapid flowing of certain components in the cytoplasmic gel. Observe a living plant cell with a light microscope, and you may see pronounced streaming.

Flagella and Cilia

To biologists, the **flagellum** (plural, flagella) and **cilium** (plural, cilia) are classic examples of structures for cell motility. Both motile structures have a ring of nine pairs of microtubules and a central pair. A system of spokes and links stabilizes this "9 + 2 array." The array arises from a **centriole**, a barrel-shaped structure that is one type of microtubule-producing center. A centriole remains at the base of the completed array, as in Figure 4.21. In that location it is known as a **basal body**.

How do flagella and cilia differ? Flagella typically are longer and less profuse than cilia. Sperm and many other free-living cells use flagella as whiplike tails when swimming (Figure 4.20*a*). In many multicelled species, ciliated epithelial cells stir air or fluid. Many thousands line the airways in your chest (Figure 4.20*b*). When they beat, they direct mucus-trapped particles away from the lungs.

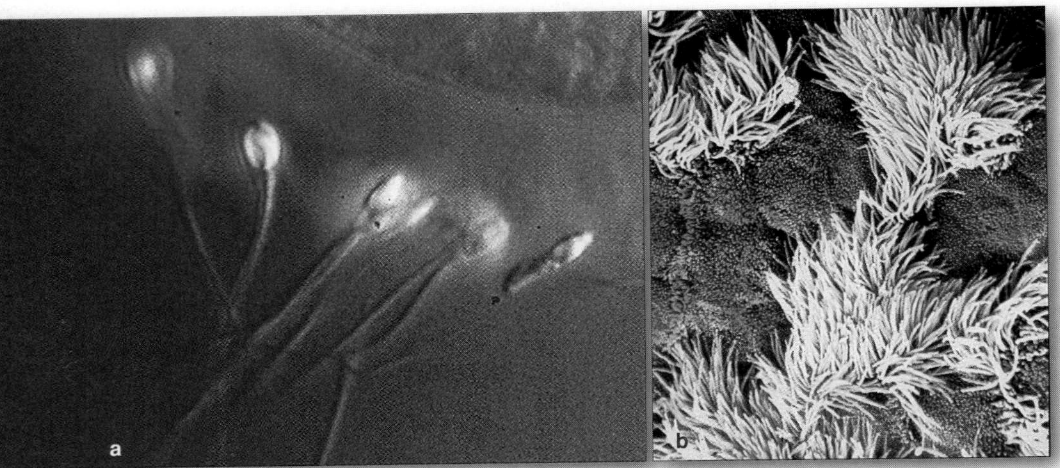
a
b

Figure 4.20 Flagella and cilia. (**a**) Human sperm consisting of a DNA-packed head, mitochondria, and a flagellum. (**b**) From the lining of an airway to the lungs, a view of the free surface of mucus-secreting cells (*brown*) and ciliated cells (*yellow*).

Further reading: Student Guide to InfoTrac on web site

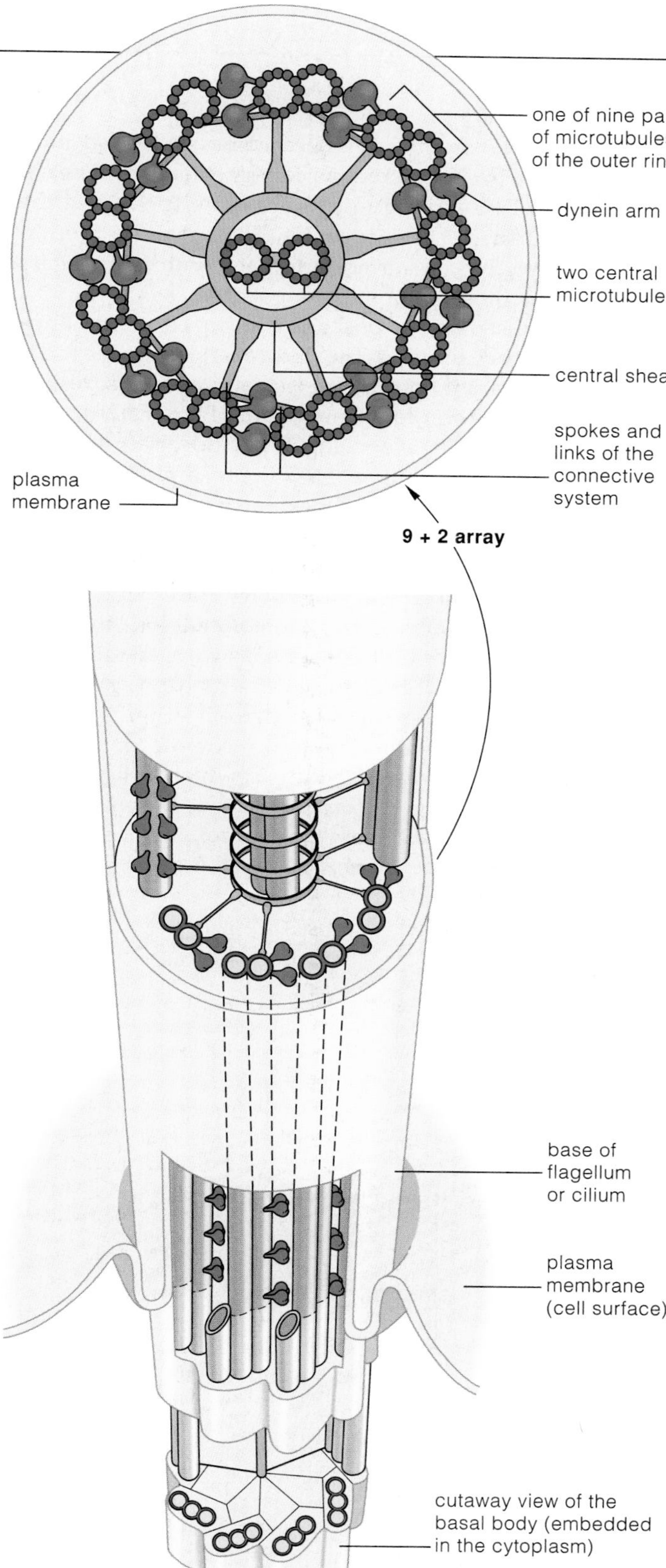
one of nine pairs of microtubules of the outer ring
dynein arm
two central microtubules
central sheath
spokes and links of the connective system
plasma membrane
9 + 2 array
base of flagellum or cilium
plasma membrane (cell surface)
cutaway view of the basal body (embedded in the cytoplasm)

Figure 4.21 Internal organization of flagella and cilia. Both motile structures have a system of microtubules and a connective system of spokes and linking elements. Nine pairs (doublets) of the microtubules are arranged as an outer ring around two central microtubules. This is a 9 + 2 array.

Figure 4.22 Model of a sliding mechanism responsible for the beating of flagella and cilia, as proposed by cell biologists K. Summers and R. Gibbons. Inside an unbent flagellum (or cilium), all microtubule doublets extend the same distance into the tip. When the flagellum bends, doublets on the side that is bending the most are being displaced farthest from the tip, as shown here:

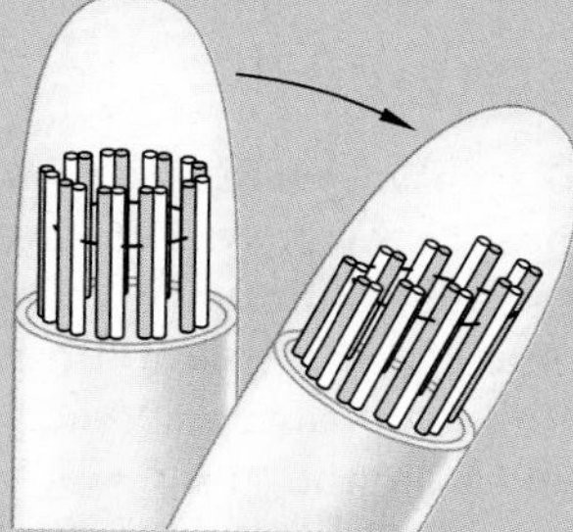

A sliding mechanism operates between the microtubule doublets of the outer ring. A series of motor proteins (dynein) form short arms that extend from each doublet toward the next doublet in the ring. Inputs of ATP energy cause the arms of one doublet to attach to the doublet in front of them, tilt in a short power stroke that pulls on the attached doublet, then release their hold. Repeated power strokes and cross-bridgings force the doublet to slide in a direction toward the base of the flagellum. As the attached doublet moves, *its* arms force the *next* doublet in line to slide down a bit, which forces the *next* doublet to slide down a bit also, and so on in sequence.

The system of interconnecting spokes and links extends through the length of the flagellum and prevents doublets from sliding totally out of the 9 + 2 array. This restriction forces the flagellum to bend and thus accommodate the internal displacement of the sliding doublets.

In short, in the 9 + 2 array of flagella and cilia, energized motor proteins make the nine doublets slide in sequence, and restrictions imposed by a system of spokes and links convert the doublet sliding into a bending motion.

Flagella and cilia beat by a sliding mechanism, as outlined in Figure 4.22. Extending from each pair of microtubules in the outer ring are short arms of motor proteins (dynein). When energized by ATP, the arms attach to the pair in front of them, tilt in a downward-directed short stroke, then release their hold. Repeated strokes make the pairs slide down in sequence. Spokes and links connect all the pairs. The collective downward displacement makes the flagellum or cilium bend.

Cell contractions and migrations, chromosome movements, and all other forms of cell motility arise at organized arrays of microtubules, microfilaments, and accessory proteins.

Different mechanisms cause these cytoskeletal elements to assemble or disassemble, slide past one another, and shunt structures to new locations.

4.10 CELL SURFACE SPECIALIZATIONS

This survey of eukaryotic cells concludes with a look at cell walls and some other specialized surface structures. Many of these architectural marvels are constructed of various secretions from the cells themselves. Others are cytoplasmic bridges or sets of membrane proteins that connect neighboring cells and allow them to interact.

Eukaryotic Cell Walls

Single-celled eukaryotic species are directly exposed to their surroundings. Many have a **cell wall**, a structural component that wraps continuously around the plasma membrane. A cell wall protects and physically supports its owner. The wall is porous, so water and solutes can easily move to and from the plasma membrane. A great variety of protistans have a cell wall. (Figure 4.23 shows one.) So do plant cells and many types of fungal cells.

For instance, young plant cells in actively growing regions secrete gluelike polysaccharides (such as pectin), glycoproteins, and cellulose. The cellulose molecules join together into ropelike strands, which become embedded in the gluey matrix. The secretions combine as a **primary wall** (Figure 4.24*a*). Primary walls are quite sticky, and they cement adjacent cells together. They also are thin and pliable, so the cell surface area can continue to enlarge under the pressure of incoming water.

Figure 4.23 From a freshwater habitat, one of the single-celled, walled protistans (*Ceratium*, a dinoflagellate).

At cell surfaces exposed to air, waxes and other cell secretions accumulate. The deposits form a cuticle. This semitransparent, protective surface covering restricts evaporative water loss from plants (Figure 4.25*a*).

Many plant cells develop only a thin wall. These are the cells that retain the capacity to divide or change shape during growth and development. As other plant cells mature, they stop enlarging and start secreting material on the primary wall's inner surface. The deposits combine to form a rigid, **secondary wall** that reinforces cell shape (Figure 4.24*e*). Whereas cellulose makes up less than 25 percent of the primary wall, the additional deposits now contribute more to structural support.

In woody plants, up to 25 percent of the secondary wall consists of lignin. Lignin is a complex molecule; it has a six-carbon ring structure to which a three-carbon chain and an oxygen atom are attached. Lignin makes plant parts stronger, more waterproof, and less inviting to insects and other plant-attacking organisms.

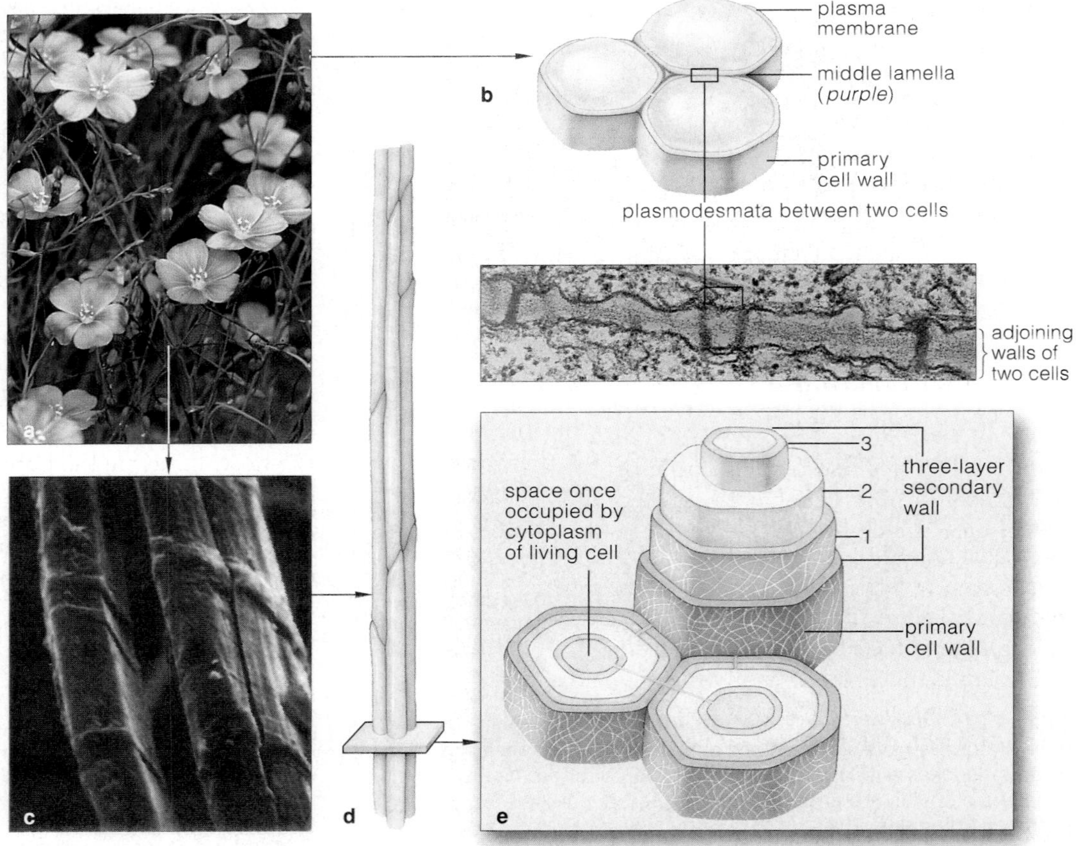

a
b
plasma membrane
middle lamella (*purple*)
primary cell wall
plasmodesmata between two cells
adjoining walls of two cells
c
d
e
space once occupied by cytoplasm of living cell
3
2
1
three-layer secondary wall
primary cell wall

Figure 4.24 Examples of cell walls in flax plants (**a**). (**b**) Primary cell wall of young cells in flower petals. Cell secretions form the middle lamella, a layer between the walls of adjoining cells. The layer is thickest in adjoining corners. Plasmodesmata, membrane-lined channels across the adjacent walls, connect the cytoplasm of neighboring cells. (**c**,**d**) Part of the lustrous fibers in a cell from a flax stem. We make linen from such fibers, which are three times stronger than cotton fibers. (**e**) In flax fibers, as in many other types of plant cells, more layers become deposited inside the primary wall. The layers stiffen the wall and help maintain its shape. Later, the cell dies, leaving the stiffened walls behind. This also happens in water-conducting pipelines that thread through most plant tissues. Interconnected, stiffened walls of dead cells form the tubes.

Further reading: Student Guide to InfoTrac on web site →

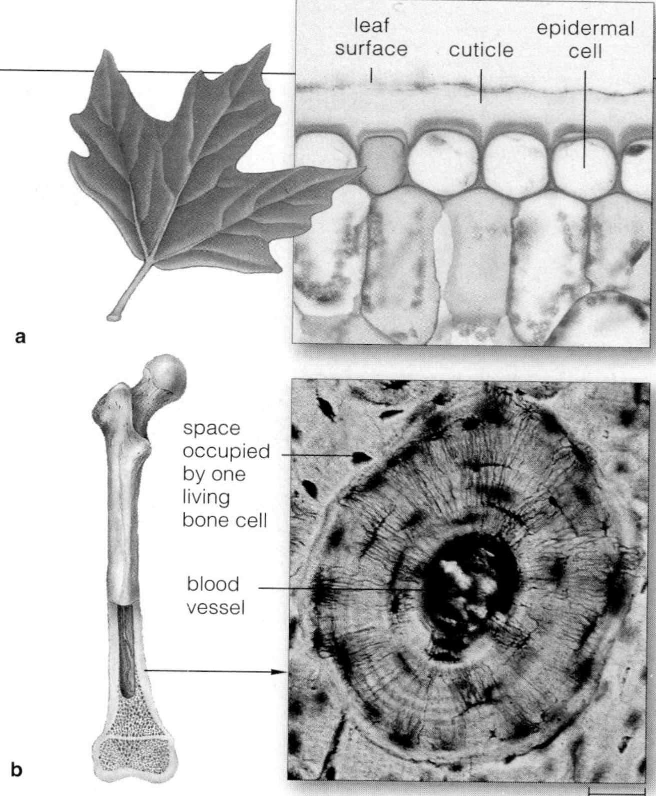
leaf surface
cuticle
epidermal cell
a
space occupied by one living bone cell
blood vessel
b
75 µm

Figure 4.25 (**a**) Section through a plant cuticle, a surface layer composed of cell secretions. (**b**) Section through compact bone tissue, stained for microscopy.

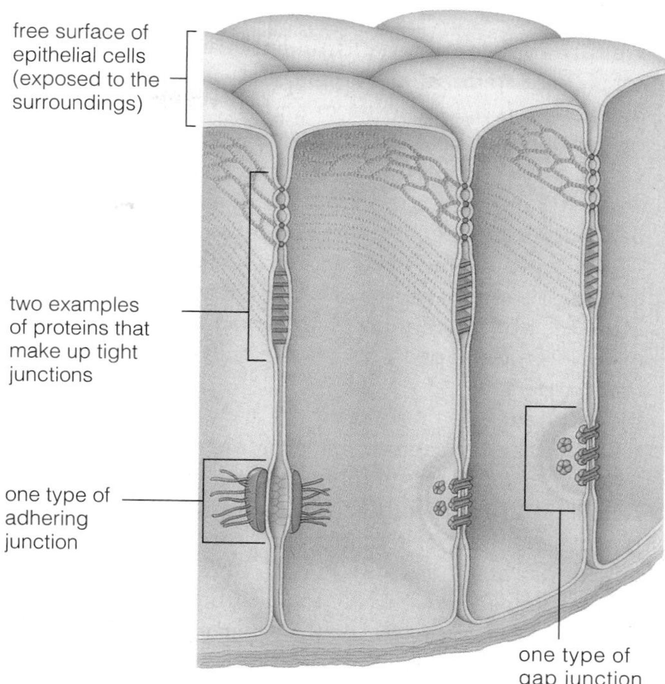
free surface of epithelial cells (exposed to the surroundings)
two examples of proteins that make up tight junctions
one type of adhering junction
one type of gap junction

Figure 4.26 Composite drawing of the most common types of cell junctions in animals. These are cells from epithelial tissue.

Matrixes Between Animal Cells

Although animal cells have no walls, diverse matrixes composed of cell secretions and even materials drawn from the surroundings intervene between many of them. Think of cartilage at the knobby ends of your leg bones. Cartilage consists of scattered cells and their secretions, which form collagen or elastin fibers embedded in a "ground substance" of modified polysaccharides. As another example, an extensive matrix widely separates the living bone cells in bone tissue (Figure 4.25*b*).

Cell-to-Cell Junctions

Even when a wall or some other structure imprisons a cell in its own secretions, the only contact that cell has with the outside world is *through* its plasma membrane. In multicelled species, membrane components project into adjacent cells as well as the surrounding medium. Among the components are junctions where the cell sends and receives diverse signals and materials, where it recognizes and cements itself to cells of the same type.

In plants, for instance, many tiny channels cross the adjacent primary walls of living cells and interconnect their cytoplasm. Figure 4.24*b* shows a few. Each channel is a plasmodesma (plural, plasmodesmata). The plasma membranes of adjoining cells have merged and fully line the channels, so there can be an uninterrupted flow of substances between cells. Thus, living cells inside the plant body have the potential to exchange substances.

In most animal tissues, three categories of cell-to-cell junctions are common (Figure 4.26). *Tight* junctions link the cells of epithelial tissues, which line the body's outer surface and internal cavities and organs. They seal adjoining cells together; water-soluble substances can't leak between them. Thus gastric fluid cannot leak across the stomach's lining and damage surrounding tissues. *Adhering* junctions join cells in tissues of the skin, heart, and other organs subjected to stretching. *Gap* junctions link the cytoplasm of neighboring cells. They are open channels for a rapid flow of signals and substances.

We will be returning to the cell walls, intercellular substances, and cell-to-cell interactions in later chapters. For now, these are the points to remember:

A variety of protistan, plant, and fungal cells have a porous but protective wall that surrounds the plasma membrane. The cells themselves secrete wall-forming materials.

Cell secretions form a cuticle at the surfaces of plants and some animals, extracellular matrixes in tissues, and other specialized structures.

In multicelled organisms, coordinated cell activities depend on cell-to-cell junctions, which are protein complexes or cytoplasmic bridges that serve as physical links and sites of communication between cells.

4.11 PROKARYOTIC CELLS—THE BACTERIA

We turn now to bacteria. Unlike the cells you have considered so far, all bacterial cells are prokaryotic; their DNA is *not* enclosed in a nucleus. *Prokaryotic* means "before the nucleus." Biologists selected the word as a reminder that bacterial cells already had appeared on the Earth before the nucleus evolved in the forerunners of eukaryotic cells.

As a group, the bacteria are the smallest cells, although a rare exception was recently discovered. Most are not much more than a micrometer wide; even the rod-shaped species are only a few micrometers in length. In terms of their structure, bacteria are the simplest kinds of cells to think about. Most species have a semirigid or rigid cell wall that wraps around the plasma membrane, structurally supports the cell, and imparts shape to it (Figure 4.27*a*). Dissolved substances can move freely to and from the plasma membrane because the wall is permeable. Often, sticky polysaccharides surround the cell wall. They help the bacterium attach to interesting surfaces, such as river rocks, teeth, and the vagina. In many of the disease-causing (pathogenic) bacteria, the polysaccharides form a thick, jellylike capsule that surrounds and helps protect the wall.

Like eukaryotic cells, bacteria have a plasma membrane that helps to control the movement of substances to and from the cytoplasm. A bacterial plasma membrane, too, has proteins that serve as channels, transporters, and receptors for signals and substances. It incorporates built-in machinery for metabolic reactions, such as the breakdown of energy-rich compounds. In photosynthetic types, clusters of some membrane proteins harness light energy and convert it to chemical energy in ATP.

Bacterial cells are too small to contain more than a small volume of cytoplasm, but they have many ribosomes upon which polypeptide chains are assembled. Apparently, these cells are small enough and so internally simple that they do not require a cytoskeleton.

The cytoplasm of a bacterium is distinct from that of eukaryotic cells. It is continuous with an irregularly shaped region of DNA. Membranes do not surround the region, which is named a nucleoid (Figure 4.27*c*). A circular molecule of DNA, also called the bacterial chromosome, occupies this region.

Extending from the surface of many bacterial cells are one or more threadlike motile structures known as bacterial flagella (singular, flagellum). These are not the same as eukaryotic flagella, because the 9 + 2 array of microtubules is absent. Bacterial flagella help a cell move rapidly in its fluid surroundings. Other surface projections include pili (singular, pilus). These are the main protein filaments that help many kinds of bacteria attach to various surfaces, even to one another.

There are two kingdoms of prokaryotic cells: the Archaebacteria and Eubacteria (Section 1.3). Together, they contain the most metabolically diverse organisms. Different kinds have managed to exploit energy and raw materials in just about every kind of environment.

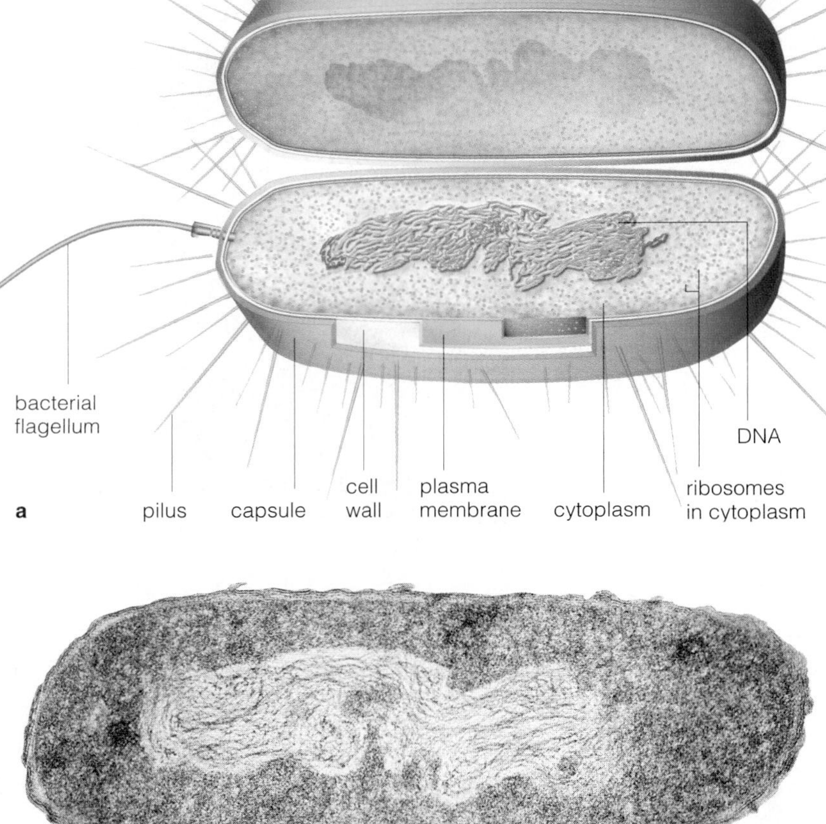

a
bacterial flagellum
pilus
capsule
cell wall
plasma membrane
cytoplasm
DNA
ribosomes in cytoplasm
b
0.5 μm

Figure 4.27 (**a**) Generalized sketch of a typical prokaryotic body plan. (**b**) Micrograph of the bacterium *Escherichia coli*.

Your own gut is home to a large population of a normally harmless strain of *E. coli*. A harmful strain has repeatedly contaminated large quantities of meat sold commercially. The same strain also has contaminated hard apple cider sold at a few roadside stands. Cooking the meat thoroughly or boiling the cider would have killed the bacterial cells. Where this was not done, people who ate the meat or drank the cider became quite sick. Some died.

Facing page: (**c**) Researchers manipulated this *E. coli* cell to release its single, circular molecule of DNA. (**d**) Cells of different bacterial species are shaped like balls, rods, or corkscrews. The ball-shaped cells of *Nostoc*, a photosynthetic bacterium, stick together inside a thick, gelatin-like sheath of their own secretions. Chapter 22 gives other fine examples. (**e**) Like this *Pseudomonas marginalis* cell, many species have one or more bacterial flagella, motile structures that propel cells through fluid environments.

Further reading: Student Guide to InfoTrac on web site →

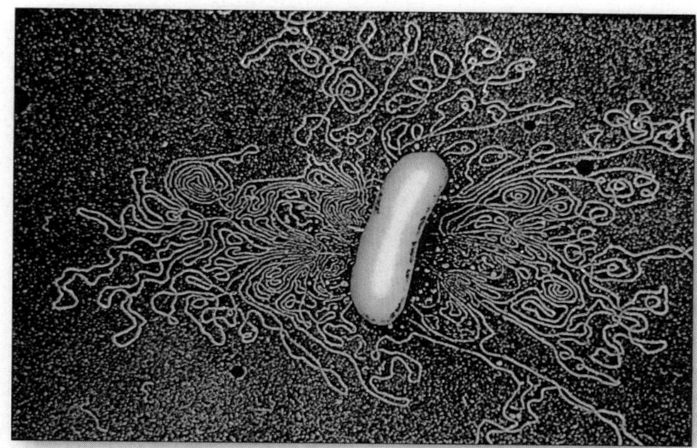
c

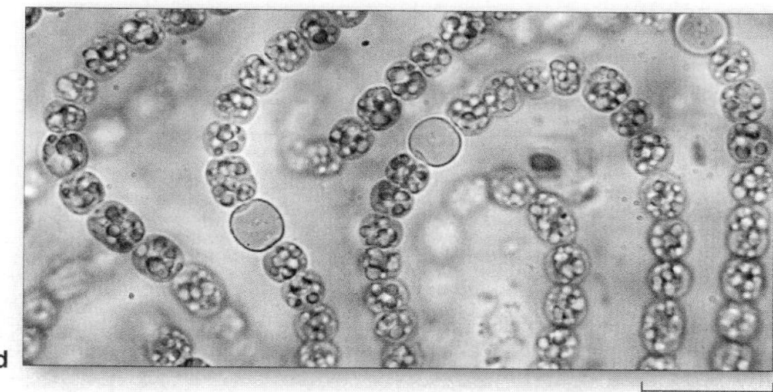
d

1 μm

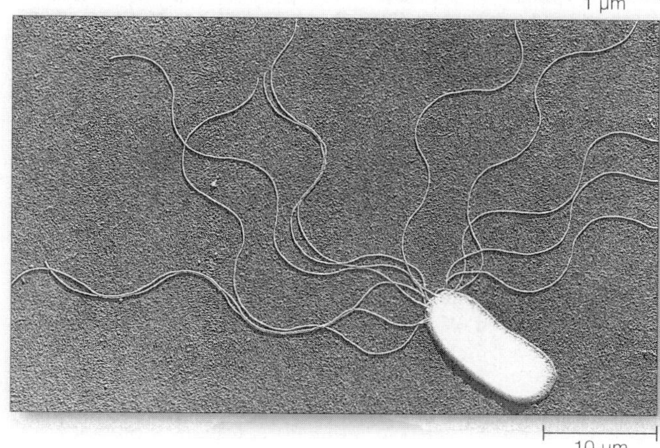
e

10 μm

In addition, ancient prokaryotic cells gave rise to all the protistans, plants, fungi, and animals ever to appear on Earth. The evolution, structure, and functions of these remarkable cells are topics of later chapters.

Bacteria alone are prokaryotic cells; their DNA is not housed inside a nucleus. Most species have a cell wall around the plasma membrane. Generally, their cytoplasm does not have any organelles comparable to those of eukaryotic cells.

Bacteria are the simplest cells, but as a group they show the most metabolic diversity. Their metabolic activities proceed at the plasma membrane and within the cytoplasm.

4.12 SUMMARY

1. Three generalizations constitute the cell theory:
   a. All living things are composed of one or more cells.
   b. The cell is the smallest entity that still retains the properties of life. That is, it either lives independently or has a built-in, genetic capacity to do so.
   c. New cells arise only from cells that already exist.

2. At the minimum, a newly formed cell has a plasma membrane, a region of cytoplasm, and a region of DNA.
   a. The plasma membrane (a thin, outer membrane) maintains the cell as a distinct, separate entity. It allows metabolic events to proceed apart from random events in the environment. Many substances and signals are continually moving across it in highly controlled ways.
   b. Cytoplasm is all the fluids, ribosomes, structural elements, and (in eukaryotic cells) organelles between the plasma membrane and the region of DNA.

3. Membranes are vital to cell structure and function. They consist of lipids (phospholipids, for the most part) and proteins. The lipids are arrayed as two layers, with all the hydrophobic tails of both layers sandwiched in between all the hydrophilic heads. This lipid bilayer imparts structure to the membrane and bars passage of water-soluble substances across it. Diverse proteins are embedded in the bilayer or attached to its surfaces.

4. Proteins carry out most cell membrane functions. For example, many serve as channels or pumps that allow or promote passage of water-soluble substances across the lipid bilayer. Others are receptors for extracellular substances that trigger changes in cell activities.

5. Cell membranes divide the cytoplasm of eukaryotic cells into functional compartments called organelles. Prokaryotic cells do not have comparable organelles.

6. Organelle membranes separate metabolic reactions in the space of the cytoplasm and allow different kinds to proceed in orderly fashion. (In bacteria, many similar reactions proceed at the plasma membrane.)
   a. The nuclear envelope functionally separates the DNA from the metabolic machinery of the cytoplasm.
   b. The cytomembrane system includes the ER, Golgi bodies, and vesicles. Many new proteins are modified into final form and lipids are assembled in this system. Finished products are packaged and then shipped off to destinations inside or outside the cell.
   c. Mitochondria are specialists in oxygen-requiring reactions that produce many ATP molecules.
   d. Chloroplasts trap sunlight energy and produce organic compounds in photosynthetic eukaryotic cells.

7. The cytoskeleton of eukaryotic cells functions in cell shape, internal organization, and movements.

8. Table 4.3 on the next page summarizes the defining features of both prokaryotic and eukaryotic cells.

Table 4.3 Summary of Typical Components of Prokaryotic and Eukaryotic Cells

		PROKARYOTIC	EUKARYOTIC			
Cell Component	Function	Archaebacteria, Eubacteria	Protistans	Fungi	Plants	Animals
Cell wall	Protection, structural support	✓*	✓*	✓	✓	*None*
Plasma membrane	Control of substances moving into and out of cell	✓	✓	✓	✓	✓
Nucleus	Physical separation and organization of DNA	*None*	✓	✓	✓	✓
DNA	Encoding of hereditary information	✓	✓	✓	✓	✓
RNA	Transcription, translation of DNA messages into polypeptide chains of specific proteins	✓	✓	✓	✓	✓
Nucleolus	Assembly of subunits of ribosomes	*None*	✓	✓	✓	✓
Ribosome	Protein synthesis	✓	✓	✓	✓	✓
Endoplasmic reticulum (ER)	Initial modification of many of the newly forming polypeptide chains of proteins; lipid synthesis	*None*	✓	✓	✓	✓
Golgi body	Final modification of proteins, lipids; sorting and packaging them for use inside cell or for export	*None*	✓	✓	✓	✓
Lysosome	Intracellular digestion	*None*	✓	✓*	✓*	✓
Mitochondrion	ATP formation	**	✓	✓	✓	✓
Photosynthetic pigments	Light–energy conversion	✓*	✓*	*None*	✓	*None*
Chloroplast	Photosynthesis; some starch storage	*None*	✓*	*None*	✓	*None*
Central vacuole	Increasing cell surface area; storage	*None*	*None*	✓*	✓	*None*
Bacterial flagellum	Locomotion through fluid surroundings	✓*	*None*	*None*	*None*	*None*
Flagellum or cilium with 9 + 2 microtubular array	Locomotion through or motion within fluid surroundings	*None*	✓*	✓*	✓*	✓
Cytoskeleton	Cell shape; internal organization; basis of cell movement and, in many cells, locomotion	*None*	✓*	✓*	✓*	✓

* Known to be present in cells of at least some groups.
** Many groups use oxygen-requiring (aerobic) pathways of ATP formation, but mitochondria are not involved.

Review Questions

1. State the three key points of the cell theory. *CI*

2. Suppose you wish to observe the three-dimensional surface of an insect's eye. Would you see more details with the aid of a compound light microscope, transmission electron microscope, or scanning electron microscope? *4.2*

3. Label the organelles in the two diagrams below of a plant cell and an animal cell. What are the main differences between the two kinds of cells? *4.3*

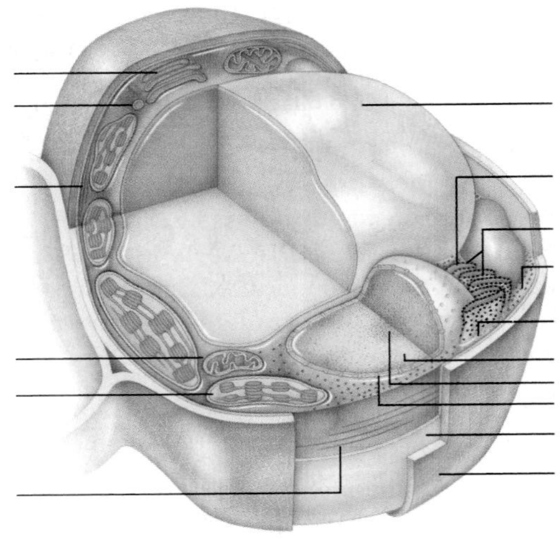

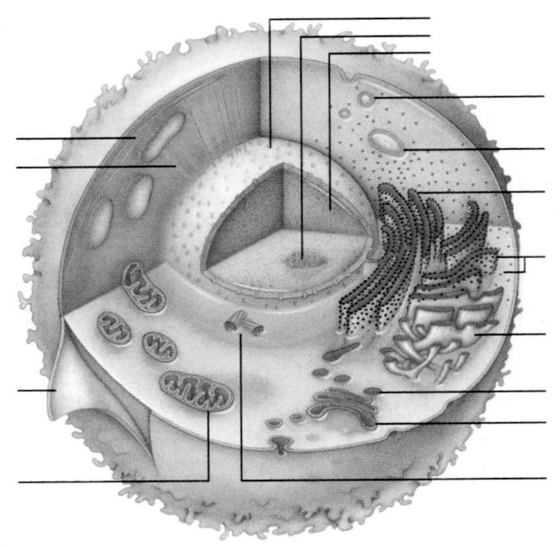

4. Label the parts of this bacterial cell. *4.11*

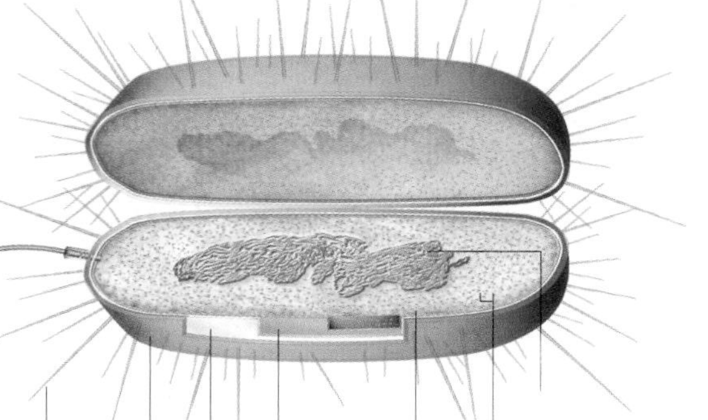

Figure 4.28 Daily chest thumping for a child who is affected by cystic fibrosis.

5. Describe three features that all cells have in common. After reviewing Table 4.3, write a paragraph on the key differences between prokaryotic and eukaryotic cells. *4.1, 4.3, 4.11*

6. Briefly characterize the structure and function of the cell nucleus, nuclear envelope, and nucleolus. *4.4*

7. Define chromosome and chromatin. Do chromosomes always have the same appearance during a cell's life? *4.4*

8. Which organelles are part of the cytomembrane system? *4.5*

9. Is this statement true or false: Plant cells have chloroplasts, but not mitochondria. Explain your answer. *4.6, 4.7*

10. What are the functions of the central vacuole? *4.7*

11. Define cytoskeleton. How does it aid in cell functioning? *4.8*

12. What gives rise to the 9 + 2 array of cilia and flagella? *4.9*

13. Cell walls are typical of which organisms: bacteria, protistans, fungi, plants, animals? Are the walls impermeable? *4.10*

14. In certain plant cells, is a secondary wall deposited inside or outside the surface of the primary wall? *4.10*

15. In multicelled organisms, coordinated interactions depend on linkages and communications between cells. What types of junctions occur between adjacent animal cells? Plant cells? *4.10*

Self-Quiz (*Answers in Appendix III*)

1. Cell membranes consist mainly of a ______ .
   a. carbohydrate bilayer and proteins
   b. protein bilayer and phospholipids
   c. lipid bilayer and proteins
   d. none of the above

2. Organelles ______ .
   a. are membrane-bound compartments
   b. are typical of eukaryotic cells, not prokaryotic cells
   c. separate chemical reactions in time and space
   d. all of the above are features of the organelles

3. Cells of many protistans, plants, and fungi, but not animals, commonly have ______ .
   a. mitochondria c. ribosomes
   b. a plasma membrane d. a cell wall

4. Is this statement true or false: The plasma membrane is the outermost component of all cells. Explain your answer.

5. Unlike eukaryotic cells, prokaryotic cells ______ .
   a. lack a plasma membrane c. do not have a nucleus
   b. have RNA, not DNA d. all of the above

6. Match each cell component with its function.

___ mitochondrion	a. synthesis of polypeptide chains
___ chloroplast	b. initial modification of new polypeptide chains
___ ribosome	c. final modification of proteins; sorting, shipping tasks
___ rough ER	d. photosynthesis
___ Golgi body	e. formation of many ATP

Critical Thinking

1. Why is it likely that you will never encounter a predatory two-ton living cell on the sidewalk?

2. In compound light microscopes having blue filters, the lens transmits only blue light. Think about the spectrum of visible light (as in Figure 7.4). Then speculate on why blue light is efficient for viewing objects at high magnification.

3. Your professor shows you an electron micrograph of a cell with large numbers of mitochondria and Golgi bodies. You notice that this particular cell also contains a great deal of rough endoplasmic reticulum. What kinds of cellular activities would require such an abundance of the three kinds of organelles?

4. *Cystic fibrosis* is a fatal genetic disorder. Affected glands secrete far more than they should, with far-reaching effects. In time, digestive enzymes clog a duct between the pancreas and small intestine, food can't be digested properly, and even if food intake increases, malnutrition results. Cysts form in the pancreas, which degenerates and becomes fibrous (hence the disorder's name). Thick mucus builds up in the respiratory tract; affected people have trouble expelling airborne bacteria and particles that enter lungs (Figure 4.28). The disorder may arise from a defective protein in the plasma membrane of gland cells that secrete mucus, digestive enzymes, and sweat. Review Section 4.5, then name the organelles involved in the secretory pathway in those cells.

Selected Key Terms

basal body *4.9*
cell *4.1*
cell cortex *4.8*
cell theory *CI*
cell wall *4.10*
central vacuole *4.7*
centriole *4.9*
chloroplast *4.7*
chromatin *4.4*
chromosome *4.4*
cilium *4.9*
cytomembrane system *4.5*
cytoplasm *4.1*
cytoskeleton *4.8*
ER (endoplasmic reticulum) *4.5*
eukaryotic cell *4.1*
flagellum *4.9*
Golgi body *4.5*
intermediate filament *4.8*
lipid bilayer *4.1*
microfilament *4.8*
micrograph *4.2*
microtubule *4.8*
mitochondrion *4.6*
motor protein *4.8*
nuclear envelope *4.4*
nucleoid *4.1*
nucleolus *4.4*
nucleus *4.1, 4.4*
organelle *4.3*
plasma membrane *4.1*
primary wall *4.10*
prokaryotic cell *4.1*
pseudopod *4.9*
ribosome *4.1*
secondary wall *4.10*
surface-to-volume ratio *4.1*
vesicle *4.5*
wavelength *4.2*

Readings See also *www.infotrac-college.com*

Alberts, B. et al. 1994. *Molecular Biology of the Cell.* Third edition. New York: Garland. See Chapter 16: The Cytoskeleton.

deDuve, C. 1985. *A Guided Tour of the Living Cell.* New York: Freeman. Beautiful introduction to the cell; two short volumes.

5 A CLOSER LOOK AT CELL MEMBRANES

It Isn't Easy Being Single

As small as it may be, a cell is a living thing engaged in the risky business of survival. Think of a single-celled amoeba and how something as ordinary as water can challenge its very existence. Water bathes the amoeba inside and out, donates its individual molecules to many metabolic reactions, and dissolves the ions that are necessary for cell functioning. If all goes well, the amoeba holds onto enough water and dissolved ions—not too little, not too much—to survive. But who is to say that life consistently goes well?

Next, think of the cells making up a goose barnacle. Not long ago, this marine organism attached itself to the submerged side of a log near a coastline. Now the log is drifting offshore, at the mercy of ocean currents. At feeding time, the barnacle opens its hinged shell and extends many featherlike appendages, which trap bacteria and other bits of food suspended in the water (Figure 5.1*a*).

The fluids bathing each cell of the barnacle body are salty, rather like the salt composition of seawater. And seawater normally is in balance with the salty fluid inside the barnacle's cells.

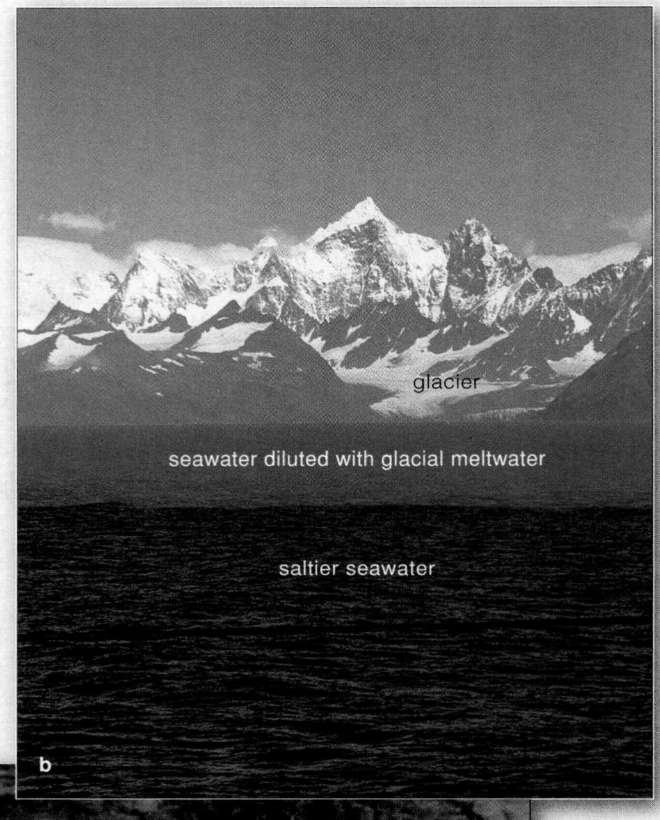
glacier
seawater diluted with glacial meltwater
saltier seawater
b

a

However, suppose the log drifts into a part of the ocean where meltwater from a glacier has diluted the water (Figure 5.1*b*). For reasons you will explore in this chapter, salts inevitably move out of the barnacle's body—and out of its cells. The previously exquisite salt–water balance gradually spirals out of control, and cells die. So, in time, does the barnacle.

The same thing happens to burrowing worms and other soft-bodied organisms that live between the high and low tide marks along a rocky shore. For instance, after an unpredictably fierce storm along the coast, seawater becomes highly dilute with runoff from the land. Dilution upsets the balance between body fluids and the water moving in with the tides. At such times, the functions of cells, tissues, and organs are disrupted, and the resulting death toll in the intertidal zone can be catastrophic.

With these examples we begin to see the cell for what it is: a tiny, organized bit of life in a world that is, by comparison, unorganized and sometimes harsh. How finely adapted a living cell must be to its environment! It must be built in such a way that it can bring in certain substances, release or keep out other substances, and conduct its internal activities with great precision.

For this bit of life, precision begins at the plasma membrane—a flimsy bilayer of lipids, dotted with diverse proteins, that surrounds the cytoplasm. Across the membrane, the cell exchanges substances with its surroundings in highly selective ways.

For eukaryotic cells, precision continues at the membranes of internal compartments called organelles. Aerobic respiration, photosynthesis, and many other metabolic processes depend on the selective movement of substances across those internal boundary layers. Gain insight into the structure and function of cell membranes, and you will gain insight as well into survival at life's most fundamental level.

Figure 5.1 (**a**) Goose barnacles, a type of marine animal that glues itself to logs and other floating objects in the sea. As is the case for other organisms, drastic changes in salt concentrations can threaten their body cells. Such changes would occur if the barnacles were to accidentally end up in glacial meltwaters.

(**b**) The dark-blue seawater in this photograph is quite salty. The lighter blue water has been diluted by meltwater from glaciers, including the one in the background. It is much, much lower in salts than seawater.

KEY CONCEPTS

1. Cell membranes consist mainly of lipids, organized as a double layer. The lipid "bilayer" gives the membrane its basic structure and prevents the haphazard movement of water-soluble substances across it. Proteins embedded in the bilayer or associated with one of its surfaces carry out most membrane functions.

2. The plasma membrane has receptor proteins (which receive chemical signals that can trigger changes in cell activities) and recognition proteins (which identify a cell as being of a certain type). Spanning all cell membranes are transport proteins through which specific water-soluble substances cross the bilayer.

3. A concentration gradient is a difference in the number of molecules (or ions) of a substance between two regions. The molecules tend to show a net movement down the gradient, to the region where they are less concentrated. This behavior is called diffusion.

4. Metabolism depends on concentration gradients that drive the directional movements of substances. Cells have mechanisms for increasing or decreasing water and solute concentrations across the plasma membrane and internal cell membranes.

5. With passive transport, a solute crosses a membrane simply by diffusing through the interior of transport proteins. With active transport, membrane proteins pump a solute across a membrane, against its concentration gradient. Active transport requires an energy input.

6. By a molecular behavior called osmosis, water diffuses across any selectively permeable membrane to a region where its concentration is lower.

7. Larger packets of material move across the plasma membrane by processes of endocytosis and exocytosis.

5.1 MEMBRANE STRUCTURE AND FUNCTION

Earlier chapters provided you with a brief look at the structure of cell membranes and the general functions of their component parts. Here, we incorporate some of the background information in a more detailed picture.

The Lipid Bilayer of Cell Membranes

Fluid bathes the two surfaces of a cell membrane and is vital for its functioning. The membrane, too, has a fluid quality; it is not a solid, static wall between cytoplasmic and extracellular fluids. For instance, puncture a cell with a fine needle, and its cytoplasm will not ooze out. The membrane will flow over the puncture site and seal it!

How does a fluid membrane remain distinct from its fluid surroundings? To arrive at the answer, start by reviewing what we have already learned about its most abundant components, the phospholipids. Recall that a **phospholipid** has a phosphate-containing head and two fatty acid tails attached to a glycerol backbone (Figure 5.2*a*). The head is hydrophilic; it easily dissolves in water. Its tails are hydrophobic; water repels them. Immerse a number of phospholipid molecules in water, and they interact with water molecules and with one another until they spontaneously cluster in a sheet or film at the water's surface. Their jostlings may even force them to become organized in two layers, with all fatty acid tails sandwiched between all hydrophilic heads. This **lipid bilayer** arrangement, remember, is the structural basis of cell membranes (Section 4.1 and Figure 5.2*c*).

The organization of each lipid bilayer minimizes the total number of hydrophobic groups exposed to water, so the fatty acid tails do not have to spend a lot of energy fighting water molecules, so to speak. A "punctured" membrane exhibits sealing behavior precisely because a puncture is energetically unfavorable. It leaves far too many hydrophobic groups exposed to the surrounding fluid.

Ordinarily, few cells get jabbed by fine needles. But the self-sealing behavior of membrane phospholipids is good for more than damage control. Among other things, it functions in vesicle formation. For example, as vesicles bud away from ER or Golgi membranes, phospholipids interact hydrophobically with cytoplasmic water. They get pushed together, and the rupture seals. You will read more about vesicle formation later in the chapter.

Fluid Mosaic Model of Membrane Structure

Figure 5.3 shows a bit of membrane that corresponds to the **fluid mosaic model**. By this model, cell membranes are a mixed composition—a *mosaic*—of phospholipids, glycolipids, sterols, and proteins. The phospholipid heads as well as the length and saturation of the tails are not all the same. (Recall that unsaturated fatty acids have one or more double bonds in their backbone and fully saturated ones have none.) The glycolipids are structurally similar to phospholipids, but their head incorporates one or more sugars. In animal cell membranes, cholesterol is the most abundant sterol (Figure 5.2*b*). Phytosterols are their equivalent in plant cell membranes.

Also by this model, the membrane is *fluid* as a result of the motions and interactions of its component parts.

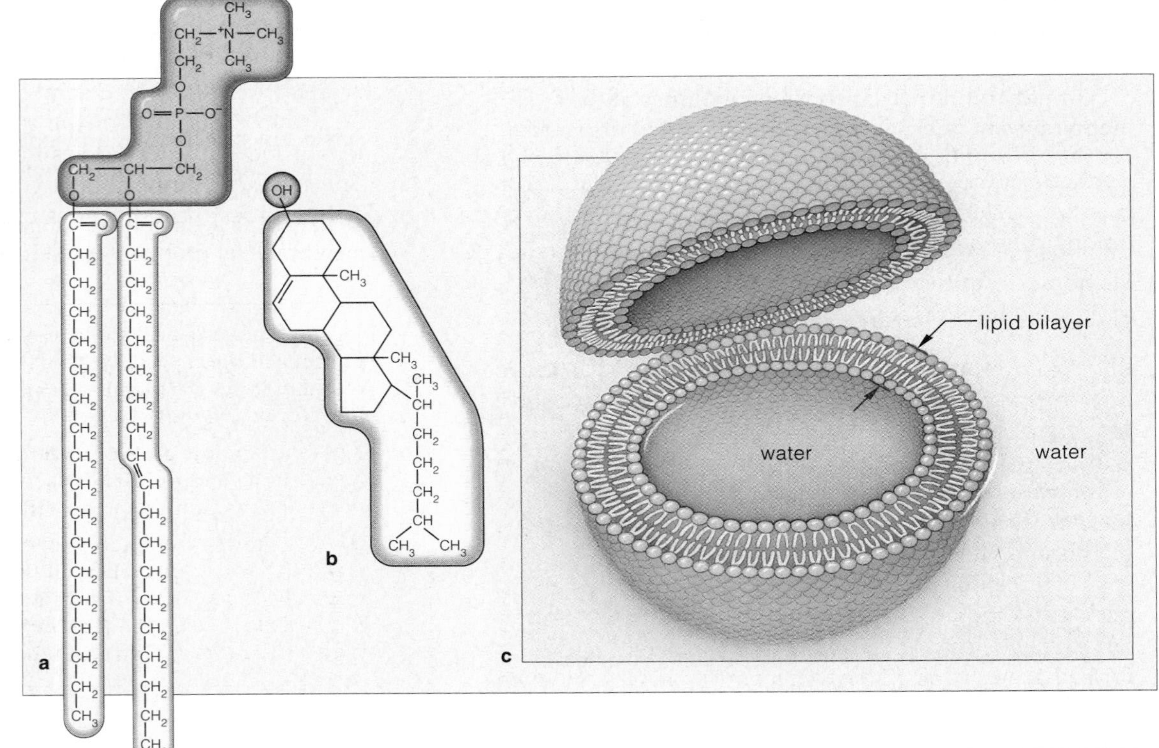

a
b
c
lipid bilayer
water
water

Figure 5.2 (**a**) The structural formula for phosphatidylcholine, a phospholipid that is one of the most common components of the membranes of animal cells. *Orange* indicates its hydrophilic head; *yellow* indicates its hydrophobic tails.

(**b**) Structural formula for cholesterol, the major sterol in animal tissues.

(**c**) Diagram of how lipids spontaneously organize themselves into a bilayer structure when placed in liquid water.

Further reading: Student Guide to InfoTrac on web site

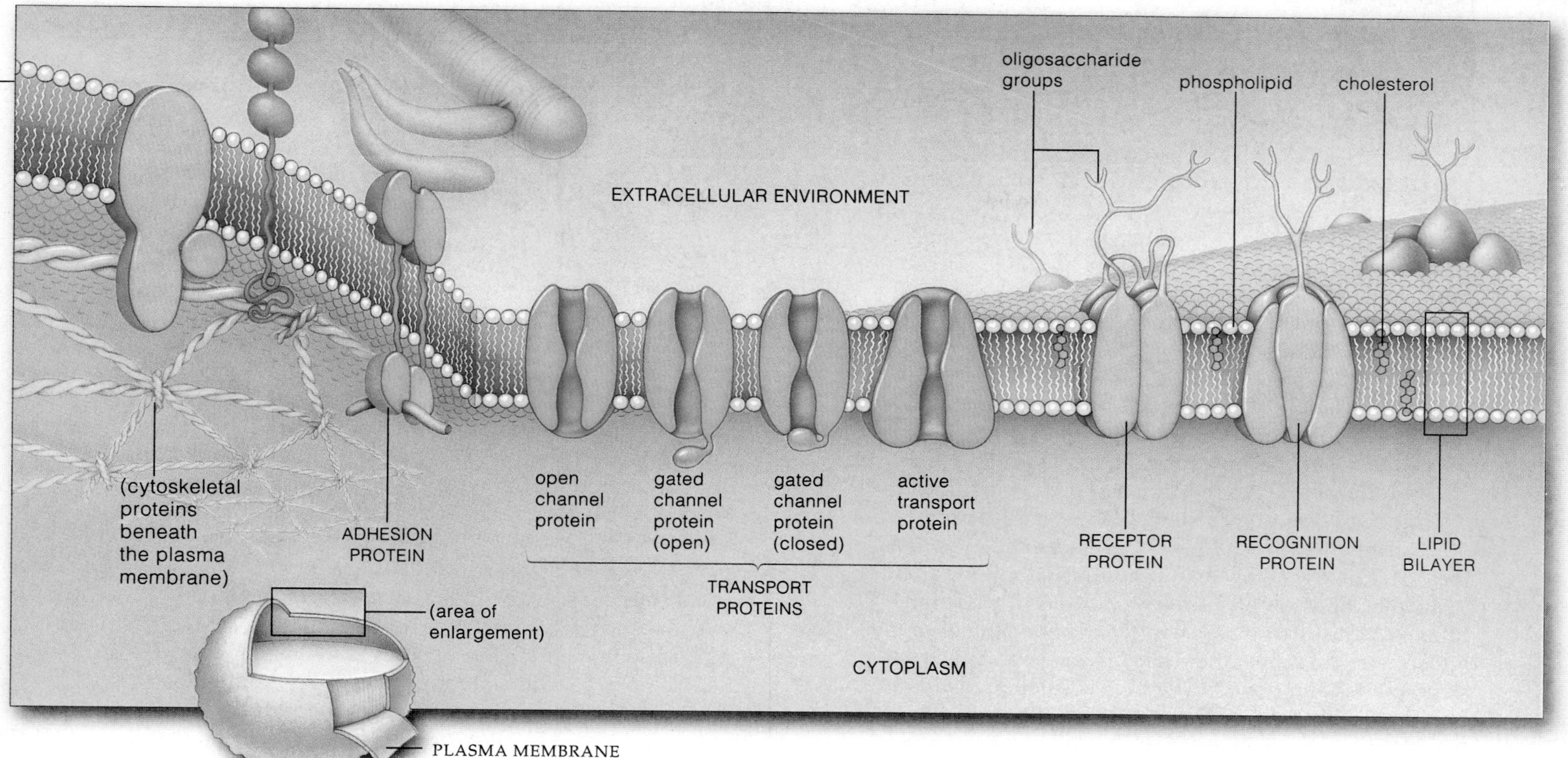
oligosaccharide groups
phospholipid
cholesterol
EXTRACELLULAR ENVIRONMENT
(cytoskeletal proteins beneath the plasma membrane)
ADHESION PROTEIN
open channel protein
gated channel protein (open)
gated channel protein (closed)
active transport protein
TRANSPORT PROTEINS
RECEPTOR PROTEIN
RECOGNITION PROTEIN
LIPID BILAYER
(area of enlargement)
CYTOPLASM
PLASMA MEMBRANE

Figure 5.3 Cutaway view of part of a plasma membrane, based on the fluid mosaic model. In addition to the specialized proteins shown, enzymes also are associated with cell membranes.

The hydrophobic interactions that give rise to most of a membrane's structure are weaker than covalent bonds. This means most phospholipids and some proteins are free to drift sideways. Also, the phospholipids can spin about their long axis and flex their tails, which keeps neighboring molecules from packing together in a solid layer. Short or kinked (unsaturated) hydrophobic tails contribute to the membrane fluidity.

The fluid mosaic model is a good starting point for exploring membranes. But bear in mind, cell membranes differ in composition and molecular arrangements. They are not even the same on both surfaces of their bilayer. For example, oligosaccharides and other carbohydrates are covalently bonded to protein and lipid components of a plasma membrane, but only on its outward-facing surface (Figure 5.3). Also, these carbohydrates differ in number and kind from one species to the next, even among different cells of the same individual.

Overview of Membrane Proteins

The proteins embedded in a lipid bilayer or attached to one of its surfaces carry out most membrane functions. Many are enzyme components of metabolic machinery. Others are **transport proteins** that span the bilayer and allow water-soluble substances to move through their interior. They bind molecules or ions on one side of the membrane, then release them on the other side.

The **receptor proteins** bind extracellular substances, such as hormones, that trigger changes in cell activities. For example, certain enzymes that crank up machinery for cell growth and division become switched on when somatotropin, a hormone, binds with receptors for it. Different cells have different combinations of receptors.

Diverse **recognition proteins** at the cell surface are like molecular fingerprints; their oligosaccharide chains identify a cell as being of a specific type. For example, the plasma membrane of your cells bristles with "self" proteins. Certain white blood cells chemically recognize self proteins and leave your cells alone, but they attack invading bacterial cells bearing "nonself" proteins at their surface. Finally, **adhesion proteins** of multicelled organisms help cells of the same type locate and stick to one another and stay positioned in the proper tissues. They are glycoproteins, with oligosaccharides attached. After tissues form, the sites of adhesion may become a type of cell junction, as described in Section 4.10.

A cell membrane has two layers composed mainly of lipids, phospholipids especially. This lipid bilayer is the structural foundation for the membrane and also serves as a barrier to water-soluble substances.

The hydrophobic parts of membrane lipids are sandwiched between the hydrophilic parts, which are dissolved either in cytoplasmic fluid or extracellular fluid.

Diverse proteins associated with the bilayer carry out most membrane functions. Many of the membrane proteins are enzymes, transporters of substances across the bilayer, or receptors for extracellular substances. Other types function in cell-to-cell recognition or adhesion.

5.2 TESTING IDEAS ABOUT CELL MEMBRANES

TESTING MEMBRANE MODELS Recall, from Chapter 1, that scientific methods are used to reveal nature's secrets. To gain insight into how researchers discovered some structural details of cell membranes, imagine yourself duplicating some of their test methods. You can start by attempting to identify the molecular components of the plasma membrane.

Your first challenge is to secure a membrane sample that is large enough to study and not contaminated with organelle membranes. Using red blood cells will simplify the task. These cells are abundant, easy to collect, and structurally simple. As they are maturing, their nucleus disintegrates. Ribosomes, hemoglobin, and just a few other components remain, but these are enough to keep the cells functional for their brief, four-month life span.

By placing a sample of red blood cells in a test tube filled with distilled water, you can separate the plasma membranes from the other cell components. Cells have more solutes and fewer water molecules than distilled water does. For reasons that you will read about shortly, the difference in solute concentrations between the two regions causes water to move across a plasma membrane, into the cells. These particular cells have no mechanisms for actively expelling the excess water, so they swell. In time they burst, and their contents spill out.

Now you have a mixture of membranes, hemoglobin, and other components in water. How do you separate all the bits of membrane? You place a tube that holds a solution of the cell parts in a **centrifuge**, a motor-driven rotary device that can spin test tubes at very high speed (Figure 5.4). Structures and molecules move in response to the centrifugal force. How far they move depends on their mass, density, and shape—as well as on the mass, density, and fluidity of the solution in the tubes. If the bits of membrane have greater mass and density than the solution, they will move downward in the tube. If they have less mass or density, they will remain near the top of the tube.

Each cell component has a molecular composition that gives it a characteristic density. As the centrifuge spins at an appropriate speed, the components having the greatest density move toward the bottom of the tube. Other cell components take up positions in layers above them, according to their relative densities.

You perform centrifugation properly, so one layer in the solution consists only of membrane shreds. You draw off this layer carefully and examine it with a microscope to check whether the membrane sample is contaminated.

Afterward, by using standard procedures of chemical analysis, you discover that the plasma membrane of red blood cells consists of lipids and proteins.

In the past, two competing structural models guided research into membranes. According to the "protein coat" model, a cell membrane consists of a lipid bilayer that has a layer of proteins coating both of its outward-facing surfaces. According to the "fluid mosaic" model, proteins are largely embedded within the bilayer.

You decide to test the protein coat model. First you calculate how much protein it would take to coat the inner and outer surfaces of a known number of red blood cells. Then you separate a membrane sample into its lipid and protein fractions and measure the amount of each. In this way, you can compare the *observed* ratio of proteins to lipids against the ratio *predicted* on the basis of the model.

Such calculations were performed with membrane samples. There are indeed enough lipids for a bilayer arrangement. And there *might* have been enough proteins to cover both of its surfaces *if* all the membrane proteins had a fibrous (not globular) structure. That possibility was discarded for two reasons. First, it became clear that proteins stretched out in a thin layer over lipids would be an unfavorable arrangement, in terms of the energy cost required to maintain it. Second, biochemical analysis showed that the proteins are globular. Such evidence does not favor the protein coat model.

Now you perform an observational test of the fluid mosaic model. You prepare cell samples for microscopy by two research methods, called *freeze–fracturing* and *freeze–etching*. First you immerse the cell sample in liquid nitrogen, an extremely cold fluid. The cells freeze instantly. Then you strike them with an exceedingly fine blade, as in Figure 5.5. A suitably directed blow can fracture cells in such a way that one lipid layer of the plasma membrane will separate from the other. Such preparations can be observed under extremely high magnifications.

If the protein coat model were correct, then you would observe a perfectly smooth, pure lipid layer. But the freeze–fractured cell membranes you observe reveal many bumps and other irregularities in the

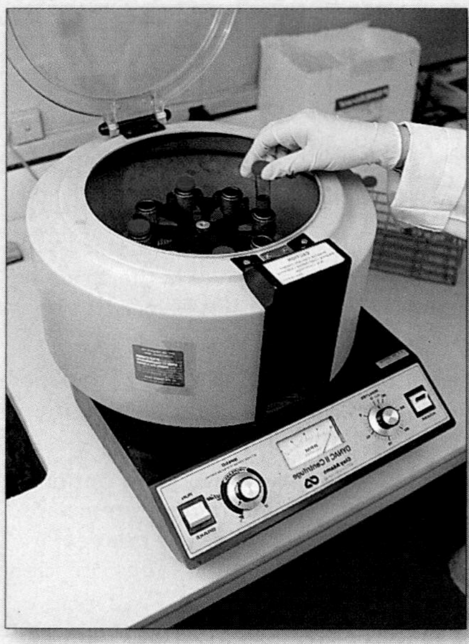

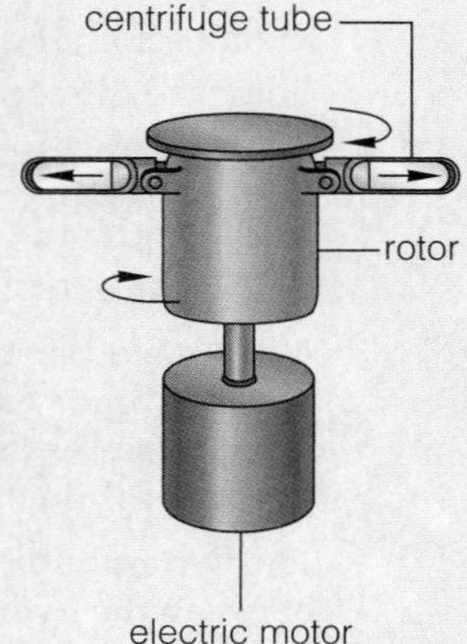
centrifuge tube
rotor
electric motor

Figure 5.4 One example of a centrifuge. The sketch shows the arrangement of the tubes and the rotor of this high-speed spinning device.

Further reading: Student Guide to InfoTrac on web site

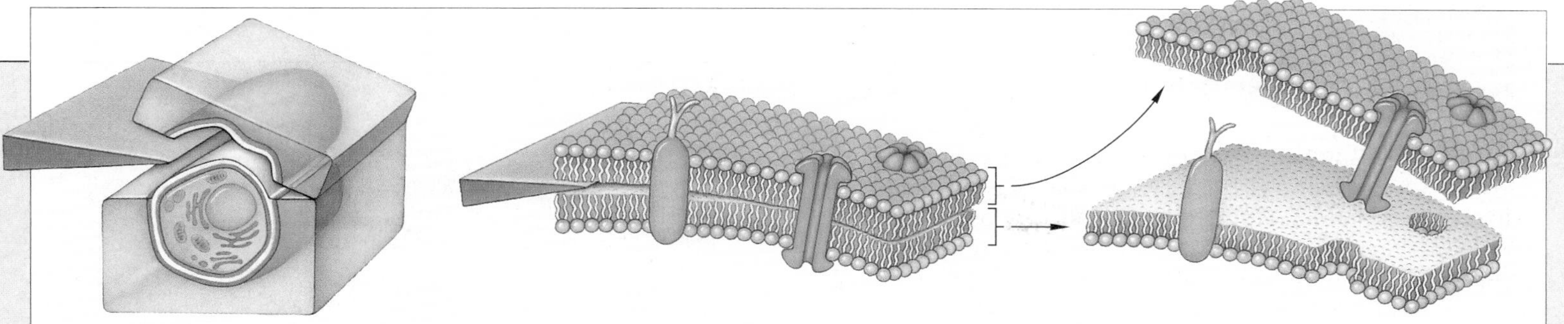

a With freeze–fracturing, specimens being prepared for electron microscopy are rapidly frozen, then fractured by a sharp blow from the edge of an ultrafine blade.

b A fractured cell membrane often splits down the middle of its lipid bilayer. Typically, many particles and depressions dot the inner surface of one exposed layer, and a complementary pattern of particles and depressions dots the other. The particles are membrane proteins. Depressions are sites where they projected into the facing layer.

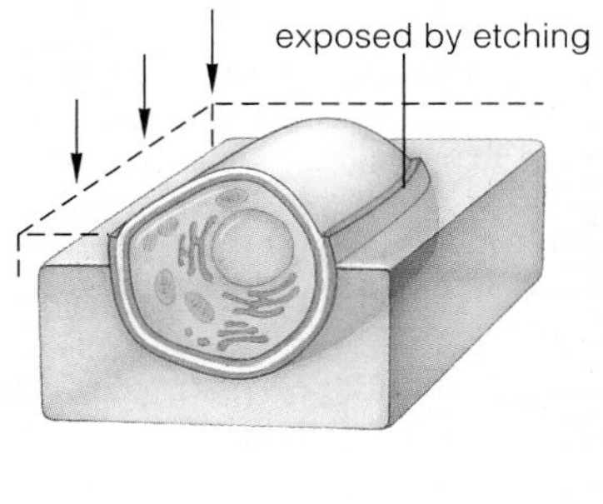
exposed by etching

c Commonly, the specimens are also freeze–etched. With this method, more ice is forced to evaporate away from the fracture face, so that a portion of the outward-facing surface of the cell membrane is exposed.

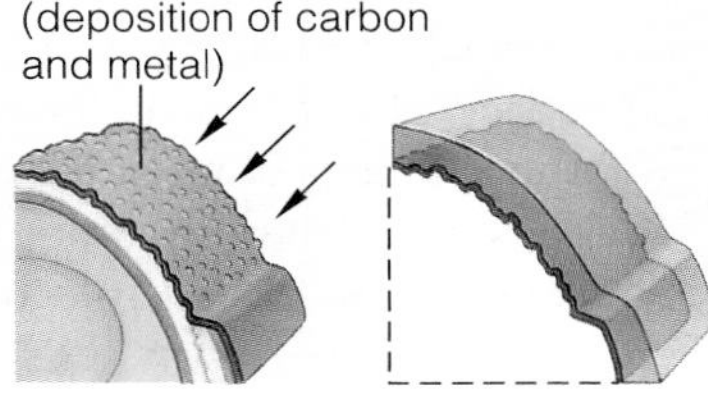
(deposition of carbon and metal)

d By metal-shadowing methods, the fractured surface is coated with a layer of carbon and heavy metal, such as platinum. The coat is thin enough to replicate details of the exposed specimen surface. The metal replica, not the specimen, is used to make micrographs of the details.

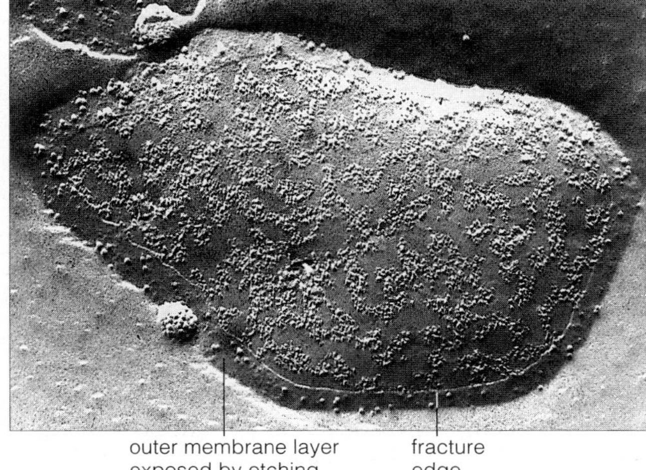
outer membrane layer exposed by etching
fracture edge

Figure 5.5 Freeze–fracturing and freeze–etching methods. The micrograph shows part of a replica of a red blood cell that was prepared by the techniques described in the text.

separated layers of lipids. The "bumps" are proteins, incorporated directly in the bilayer—just as the fluid mosaic model predicts.

OBSERVING MEMBRANE FLUIDITY Consider now an experiment that yielded evidence of the fluid motion of membrane proteins. Researchers induced an isolated human cell and an isolated mouse cell to fuse. The plasma membranes of the cells from the two species merged to form a single, continuous membrane in the new, hybrid cell. Analysis showed that, in less than one hour, most of the membrane proteins were mixed together. They had been free to drift laterally through the hybrid membrane (Figure 5.6).

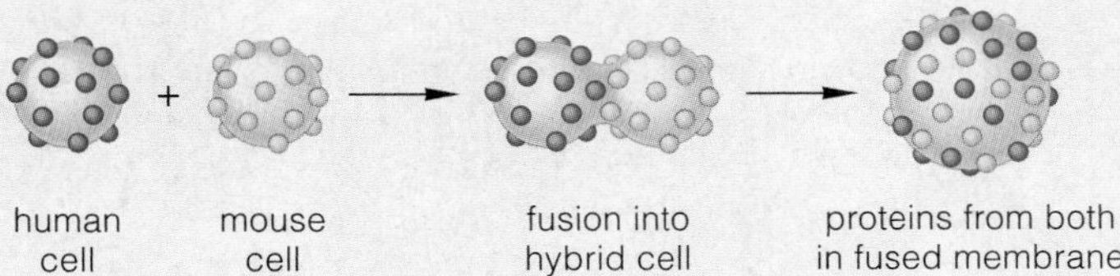
+
human cell
mouse cell
fusion into hybrid cell
proteins from both in fused membrane

Figure 5.6 Result of an experiment in which the plasma membranes from cells of different species were induced to fuse. Membrane proteins drifted laterally and became mixed.

Through observational tests, we also know that the fluidity of a membrane is subject to change, depending on environmental temperatures—and that at least some cells can respond to the danger.

For example, after some researchers decreased the temperature of a solution that contained live bacterial or yeast cells, the membranes of those cells started to stiffen. Such stiffening can disrupt the functioning of membrane proteins. However, the cells responded by rapidly synthesizing unsaturated fatty acids, and the infusion of those kinked lipids into the cell membranes countered the stiffening effect. However, past a certain low temperature, the membranes did solidify. The temperature at which this occurs varies, depending on the lipid composition of a given membrane.

CROSSING SELECTIVELY PERMEABLE MEMBRANES

All molecules contain energy in the form of chemical bonds. Cell membranes help make that energy available for cell functioning. How? They selectively concentrate reactants and products in amounts most favorable for specific reactions.

Think of the water bathing both sides of the bilayer of a cell membrane. Plenty of substances are dissolved in it, but the kinds and amounts are not the same near the two surfaces of the bilayer. The membrane itself is establishing and maintaining the differences, which are vital for metabolism. How? Like all cell membranes, it shows **selective permeability**: Owing to its molecular structure, the membrane permits some substances but not others to cross it in certain ways, at certain times.

For instance, being largely nonpolar, the hydrocarbon chains of the lipid bilayer passively let carbon dioxide, molecular oxygen, and other small, nonpolar molecules cross the membrane. Some water molecules cross, also. Even though they are polar, they slip through gaps that open up when the chains flex and bend (Figure 5.7*a*).

By contrast, large, polar molecules such as glucose almost never move freely across the bilayer. Neither do ions (Figure 5.7*b*). Such water-soluble substances cross the bilayer passively or actively, through the interior of transport proteins. The water molecules in which they are dissolved cross with them.

Membrane structure also allows cells to import and export substances in bulk across the plasma membrane. The processes are called endocytosis and exocytosis.

At any time in a cell's life, all of these mechanisms are operating simultaneously. They help cells increase, decrease, and maintain concentrations of the molecules and ions that are crucial for metabolism.

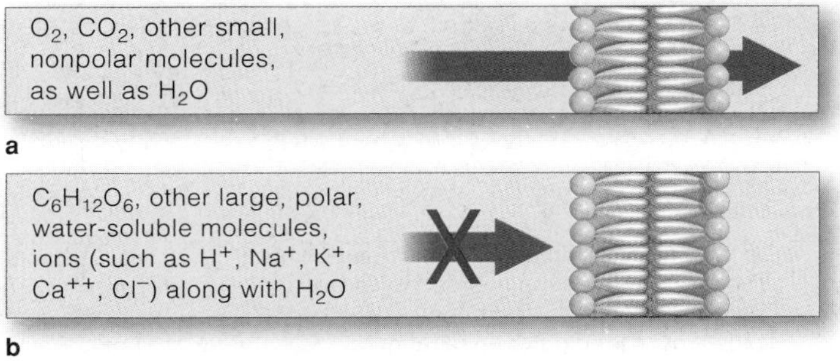

O_2, CO_2, other small, nonpolar molecules, as well as H_2O
a
$C_6H_{12}O_6$, other large, polar, water-soluble molecules, ions (such as H^+, Na^+, K^+, Ca^{++}, Cl^-) along with H_2O
b

Figure 5.7 Selective permeability of cell membranes. (**a**) Small, nonpolar molecules and some water molecules cross the lipid bilayer. (**b**) Ions and large, polar, water-soluble molecules and the water dissolving them cannot cross the bilayer; transport proteins spanning the bilayer must help them across.

Concentration Gradients and Diffusion

Concentration refers to the number of molecules or ions of a substance in a specified region, as in a volume of fluid or air. *Gradient* means the number in one region is not the same as it is in another. Thus, a **concentration gradient** is a difference in the number of molecules or ions of a given substance between adjoining regions.

In the absence of other forces, a substance moves from a region where it is more concentrated to a region where it is less concentrated. *The energy inherent in its individual molecules, which keeps them in constant motion, drives the directional movement.* Although the molecules collide randomly and careen back and forth millions of times a second, the *net* movement is away from the place of greater concentration (and the most collisions).

Diffusion is the name for the net movement of like molecules or ions down a concentration gradient. It is a key factor in the movement of substances across cell membranes and through cytoplasmic fluid. In multicelled organisms, diffusion also moves substances from one body region to another, and between the body and its environment. For example, when oxygen builds up in photosynthetic leaf cells, it diffuses across their plasma membrane, into air inside the leaf, then into air outside the leaf—where the oxygen concentration is lower.

Like other substances, oxygen tends to diffuse in a direction established by its *own* concentration gradient, not those of any other substances dissolved in the same fluid. You see the outcome of this tendency when you squeeze a drop of dye into water. Molecules of the dye diffuse to the region where they are less concentrated. And the water molecules move to the region where *they* are not as concentrated (Figure 5.8).

Several factors affect the rate at which molecules or ions move down a concentration gradient. A gradient's steepness is a factor. So are molecular size, temperature, and electric or pressure gradients that may be present.

Diffusion is faster when a gradient is steep. Then, far more molecules are moving outward from the region of greatest concentration compared to the number that are moving in. The difference in how many molecules are moving in either direction declines as the gradient decreases. If the gradient vanishes, individual molecules will still be in motion. But

Figure 5.8 Example of diffusion. A drop of dye enters a bowl of water. The dye molecules *and* the water molecules slowly become evenly dispersed, because each substance shows a net movement down its own concentration gradient.

Further reading: Student Guide to InfoTrac on web site

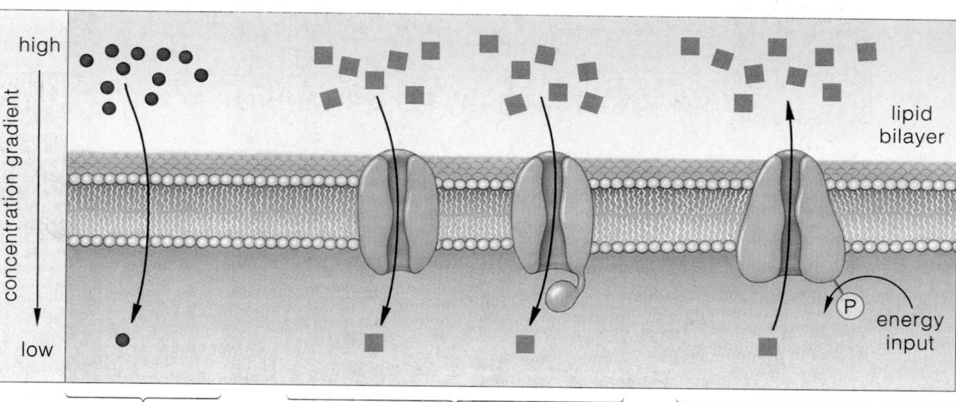
high
low
concentration gradient
lipid bilayer
P
energy input

DIFFUSION ACROSS LIPID BILAYER
Lipid-soluble substances as well as water diffuse across.

PASSIVE TRANSPORT
Water-soluble substances, and water, diffuse through interior of transport proteins. No energy boost required. Also called facilitated diffusion.

ACTIVE TRANSPORT
Specific solutes are pumped through interior of transport proteins. Requires energy boost.

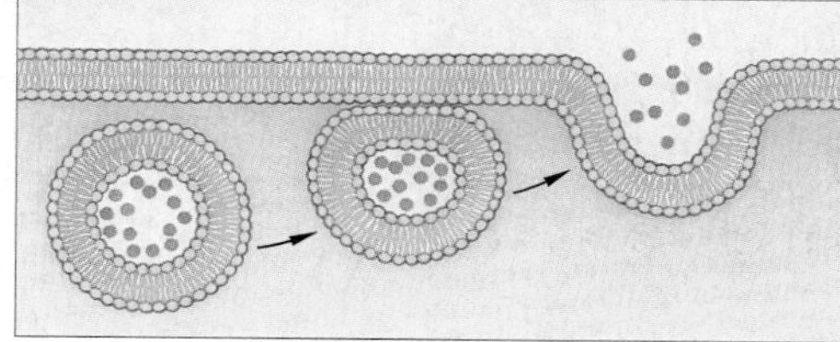

EXOCYTOSIS
Vesicle in cytoplasm moves to plasma membrane, fuses with it; contents released to the outside.

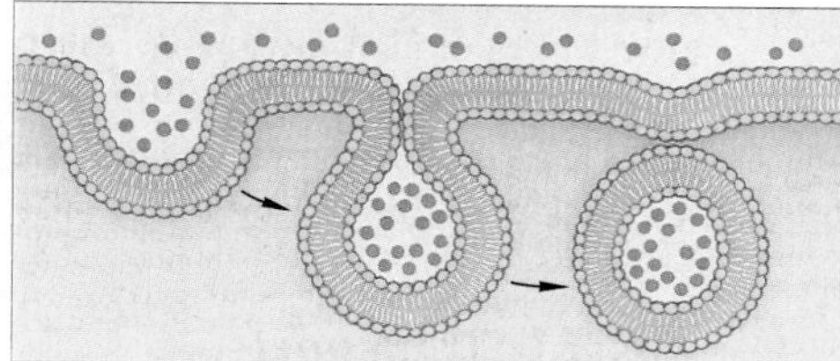

ENDOCYTOSIS
Vesicle forms from a patch of inward-sinking plasma membrane, enters cytoplasm.

Figure 5.9 Overview of major mechanisms by which solutes can cross cell membranes. Exocytosis and endocytosis proceed only at the plasma membrane.

then the total number moving one way or the other during a given interval will be about the same. A net distribution of molecules that is nearly uniform through two adjoining regions is called "dynamic equilibrium."

In addition, the rates of diffusion are faster at higher temperatures. More heat energy causes molecules to move faster and therefore to collide more frequently. Hence, diffusion is more rapid than in a cooler adjoining region.

Molecular size also affects diffusion rates. Generally, small molecules tend to move down their concentration gradient more rapidly than larger ones do.

Besides this, the rate and direction of diffusion might be under the influence of an electric gradient. An **electric gradient** simply is a difference in electric charges of adjoining regions. For example, in the fluid bathing the surface of each cell membrane are many kinds of dissolved ions. Each ion is contributing to the fluid's overall electric charge. Opposite charges attract. So the fluid having the more negative charge, overall, tends to exert the greatest pull on positively charged substances, such as sodium ions. Many biological events, including information flow through nervous systems, involve the combined force of electric and concentration gradients across membranes.

Finally, as you will see shortly, the presence of a pressure gradient also may affect the rate and direction of diffusion. A **pressure gradient** is a difference in the pressure being exerted in adjoining regions.

Overview of Membrane Crossing Mechanisms

Before taking a closer look at how substances cross membranes, study Figure 5.9, which is an overview of the key mechanisms. Together, these mechanisms help supply cells and organelles with raw materials and rid them of wastes, at controlled rates. They contribute to maintaining the volume and pH of a cell or organelle within functional ranges.

Again, small nonpolar molecules (such as oxygen) and water can simply diffuse across the lipid bilayer. Polar molecules cross by diffusing through the interior of transport proteins that span the bilayer. Some kinds of proteins help the substance follow its concentration gradient simply by functioning as a channel across the membrane. They are the basis of *passive transport*, or "facilitated" diffusion. Other proteins engage in *active transport*. They, too, allow polar molecules to cross the membrane through their interior, but the net direction of movement is against the concentration gradient. Active transport only operates when transport proteins become energized, as by a phosphate-group transfer from ATP.

Certain mechanisms also move substances in bulk across a plasma membrane. *Exocytosis* involves fusion between the plasma membrane and a membrane-bound vesicle that formed inside the cytoplasm. *Endocytosis* involves an inward sinking of a small patch of plasma membrane, which seals back on itself to form a vesicle in the cytoplasm.

Molecules or ions of a substance constantly collide because of their inherent energy of motion. The collisions result in diffusion, a net outward movement of a substance from one region into an adjoining region where it is less concentrated.

A concentration gradient is a form of energy. It can drive the directional movement of a substance across a cell membrane. The steepness of the gradient, temperature, molecular size, electric gradients, and pressure gradients influence the rates of diffusion.

Metabolic reactions depend on the chemical energy inherent in concentrated amounts of molecules and ions. Cells have mechanisms for increasing or decreasing those concentrations across the plasma membrane and internal cell membranes.

5.4 PROTEIN-MEDIATED TRANSPORT

Let's now take a look at how water-soluble substances diffuse into and out of cells or organelles. Whether by passive or active transport mechanisms, a great variety of proteins have roles in moving molecules and ions across cell membranes. These proteins span the lipid bilayer, and their interior is able to open on both sides of it (Figure 5.10).

To understand how such protein molecules work, you must know they are not rigid blobs of atoms. When such a protein interacts with a solute, it changes from one shape to another, then back again. The changes start when a solute weakly binds to a site in the protein's interior. Part of the protein closes in behind the bound solute, and part of the interior opens up on the other side of the cell membrane. Then the solute dissociates (separates) from the binding site. Think of it as hopping onto the transport protein on one side of the membrane and hopping off on the other side.

Passive Transport

Many transport proteins permit ions and other solutes to diffuse freely across a membrane. **Passive transport** is the name for a flow of solutes through the interior of transport proteins, down their concentration gradients. Energetically, the flow costs only what the cell already spent to produce and maintain the gradients. Passive transport itself adds nothing more to the energy cost.

Transport proteins allow solutes to move both ways across a cell membrane. In cases of passive transport, the *net* direction of movement during a given interval depends on how many molecules or ions of the solute are making random contact with vacant binding sites in the interior of the proteins (Figure 5.11). The binding and transport simply proceed more often on the side of the membrane where the solute is more concentrated. Because there are more molecules around, the random encounters with binding sites are more frequent than they are on the other side of the membrane.

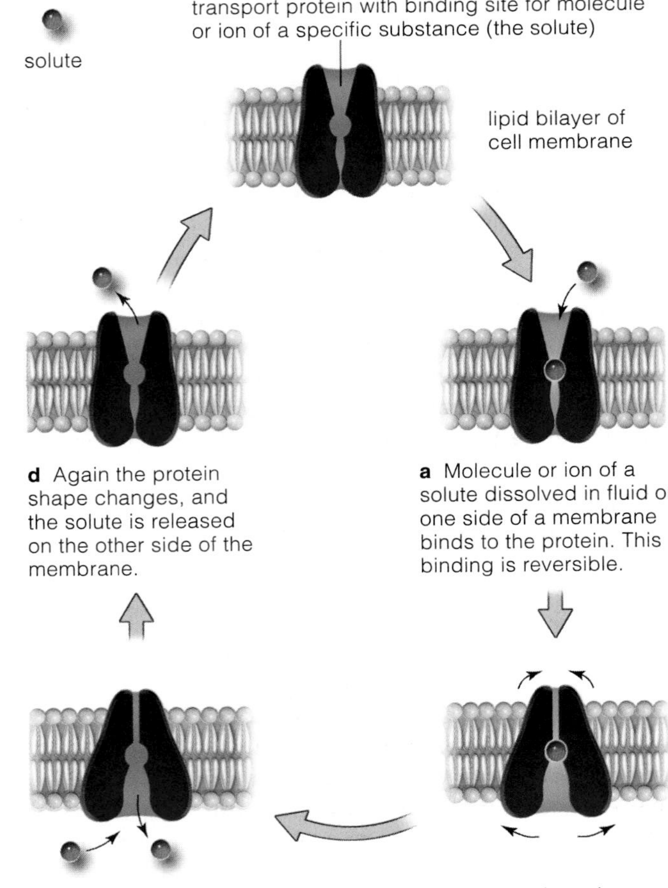
solute
transport protein with binding site for molecule or ion of a specific substance (the solute)
lipid bilayer of cell membrane
a Molecule or ion of a solute dissolved in fluid on one side of a membrane binds to the protein. This binding is reversible.
b Binding alters the shape of the protein. Part of the protein molecule closes in behind the solute and part opens in front of it.
c Being exposed to fluid on this side of the membrane, the binding site releases the solute. The newly vacated binding site attracts another molecule or ion of the solute.
d Again the protein shape changes, and the solute is released on the other side of the membrane.

Figure 5.11 Passive transport across a cell membrane. Solutes are able to move in both directions through transport proteins. In passive transport, *net* movement will be down the concentration gradient (from higher to lower) until concentrations are the same on both sides of the membrane.

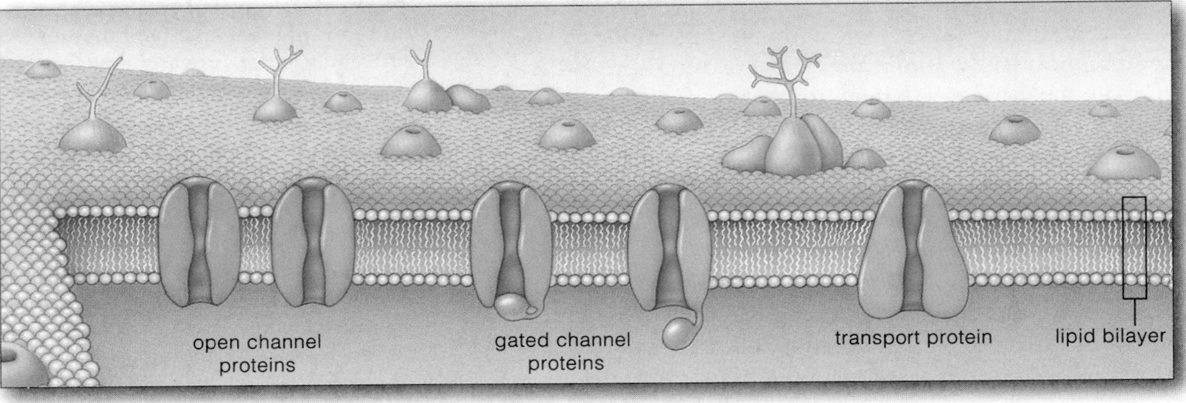
open channel proteins
gated channel proteins
transport protein
lipid bilayer

Figure 5.10 Sketch of transport proteins that span the lipid bilayer of a plasma membrane, cutaway view. Such proteins passively allow or actively assist ions and large polar molecules such as glucose to move through their interior, from one side of the cell membrane to the other.

Further reading: Student Guide to InfoTrac on web site →

Figure 5.12 Active transport across a cell membrane. ATP transfers a phosphate group to a transport protein. The transfer sets in motion reversible changes in the protein's shape that result in a greater net movement of solute particles *against* the concentration gradient.

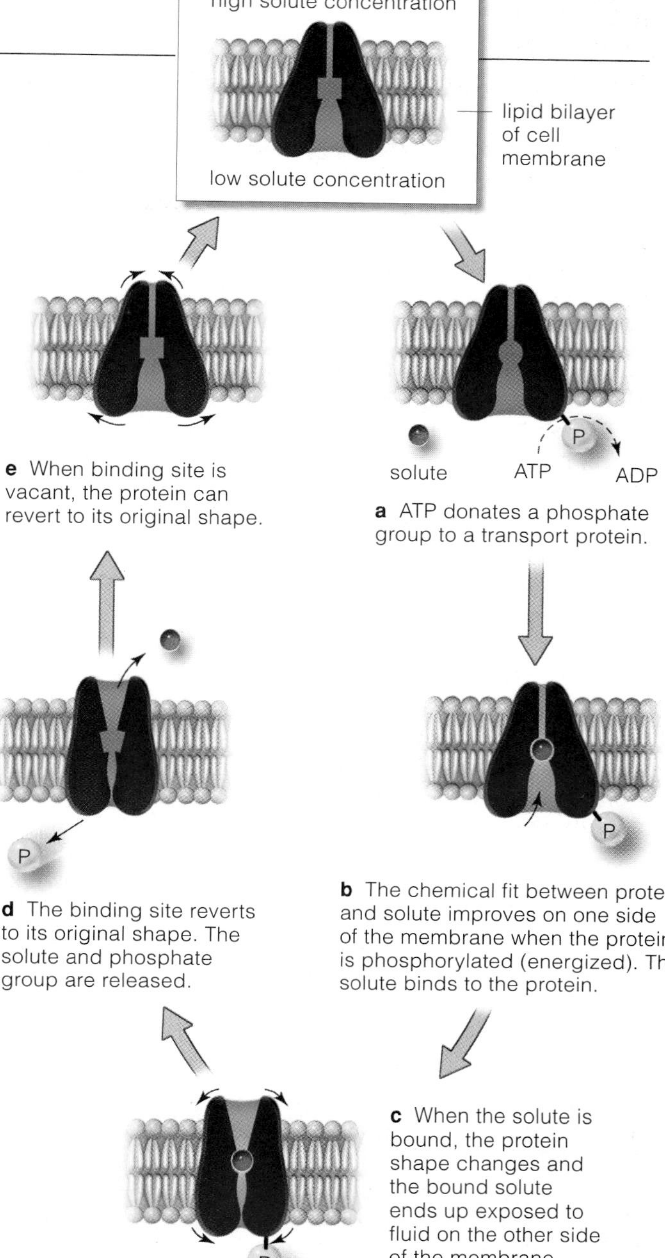

high solute concentration
lipid bilayer of cell membrane
low solute concentration
solute
ATP
ADP
P
a ATP donates a phosphate group to a transport protein.
b The chemical fit between protein and solute improves on one side of the membrane when the protein is phosphorylated (energized). The solute binds to the protein.
c When the solute is bound, the protein shape changes and the bound solute ends up exposed to fluid on the other side of the membrane.
d The binding site reverts to its original shape. The solute and phosphate group are released.
e When binding site is vacant, the protein can revert to its original shape.

If nothing else were going on, the passive two-way transport would proceed until a solute's concentrations were equal across the membrane. But other processes usually affect the outcome. For example, blood delivers glucose to all body tissues, where nearly all cells use it as an energy source and as a building block. Cells can rapidly take up glucose when its blood concentration is high, and still the gradient is maintained. How? As fast as some glucose molecules diffuse into a cell, others are entering metabolic reactions. Thus, when cells quickly use glucose, they help maintain a concentration gradient that favors the uptake of *more* glucose.

Active Transport

Only in dead cells do solute concentrations become equal on both sides of membranes. Living cells never stop expending energy to pump potassium and other solutes to and from their interior. In **active transport**, energy-driven mechanisms called "membrane pumps" make solutes cross membranes *against* concentration gradients. ATP provides most of the energy to do this.

When ATP gives up one of its phosphate groups to a transport protein, the chemical fit between the solute and the protein binding site improves on one side of the membrane (Figure 5.12*a,b*). After a solute particle binds at the site, the protein's folded shape changes in such a way that the bound solute becomes exposed to fluid bathing the other side of the membrane (Figure 5.12*c*). Now the binding site reverts to its less attractive state, and the solute is released. A less attractive site means fewer molecules or ions of the solute can make the return trip. The *net* movement is to the side of the membrane where the solute is more concentrated.

One type of active transport system, the **calcium pump**, helps keep the calcium concentration in a cell at least a thousand times lower than outside. Another, the **sodium–potassium pump**, moves potassium ions (K^+) across the plasma membrane. Activation of this protein facilitates binding of a sodium ion (Na^+) on one side of a cell membrane. After the Na^+ makes the crossing, it is released. Its release facilitates the binding of K^+ at a different binding site on the protein, which reverts to its original shape after K^+ is delivered to the other side.

Through operation of such active transport systems, concentration and electric gradients can be maintained across membranes. The gradients are vital to many cell activities and physiological processes, including muscle contraction and nerve cell (neuron) function. We return to the mechanisms of active transport in later chapters. For now, keep these points in mind:

Transport proteins span the lipid bilayer of cell membranes. When they bind a solute on one side of a membrane, they undergo a reversible change in shape that shunts the solute through their interior, to the other side.

With passive transport, a solute simply diffuses through the protein; its net movement is down its concentration gradient.

With active transport, the net diffusion of a solute is uphill, against its concentration gradient. The transporting protein must be activated, as by ATP energy, to counter the energy inherent in the gradient.

5.5 MOVEMENT OF WATER ACROSS MEMBRANES

By far, more water diffuses across cell membranes than any other substance, so the key factors that influence its directional movement deserve special attention.

Osmosis

Turn on a faucet or watch a waterfall, and the moving water provides a demonstration of bulk flow. **Bulk flow** is the mass movement of one or more substances in response to pressure, gravity, or some other external force. It accounts for some movement of water through complex plants and animals. With each beat, your heart creates fluid pressure that drives a volume of blood, which is mainly water, through interconnected blood vessels. Sap runs inside conducting tissues that thread through maple trees, and this, too, is bulk flow.

What about the movement of water into and out of cells or organelles? A membrane intervening between two regions allows the small, polar water molecules to cross but restricts the passage of ions and large polar molecules. **Osmosis** is the name for the diffusion of water molecules in response to a water concentration gradient between two regions separated by a selectively permeable membrane.

Concentrations of solutes in water on both sides of a membrane influence osmosis, as you can see from Figure

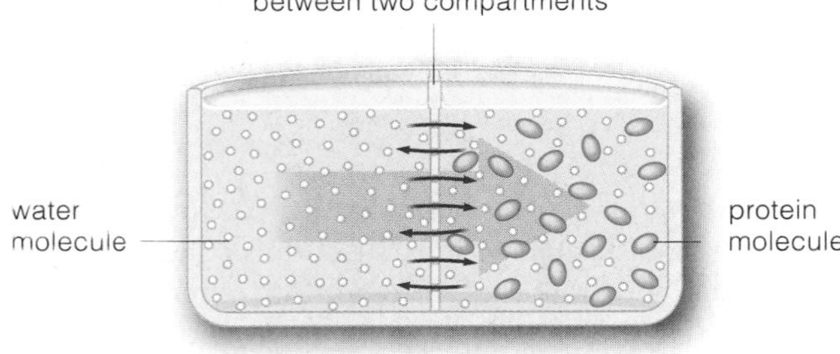

selectively permeable membrane between two compartments
water molecule
protein molecule

Figure 5.13 Demonstration of how a solute concentration gradient influences osmotic movement. Start with a container divided by a membrane that water but not proteins can cross. Pour water into the left compartment. Pour the same volume of a protein-rich solution into the right compartment. There, proteins occupy some of the space. The net diffusion of water in this example is from left to right (large *gray* arrow).

5.13. *The side that has more solute particles has a lower concentration of water.* To gain insight into why this is so, imagine dissolving a small amount of glucose in water. Compared to an equivalent volume of water, the glucose solution has fewer water molecules—because each molecule of glucose occupies some of the space formerly occupied by molecules of water.

It is mainly the *total number* of molecules or ions, not the type of solute, that dictates the concentration of water. Dissolve one mole of an amino acid or urea in 1 liter of water, and the water concentration decreases about as much as it did in the glucose solution. Add one mole of sodium chloride (NaCl) to 1 liter of water, and it dissociates into equal numbers of sodium ions and chloride ions. There are now two moles of solute particles—twice as many as in the glucose solution—so the water concentration has decreased proportionately.

Effects of Tonicity

Given that water molecules tend to move osmotically to a region where water is less concentrated, the direction of their movement tends to be toward a region where solutes are more concentrated. Figure 5.14 illustrates this tendency. Suppose you decide to make a simple observational test of this statement. You construct three sacs out of a membrane that water but not sucrose can

2M sucrose solution

1 liter of distilled water

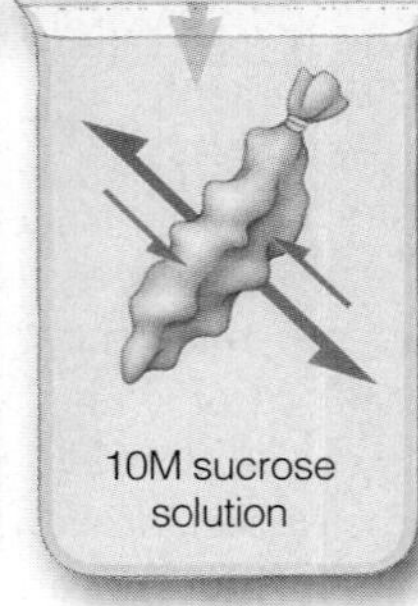

10M sucrose solution

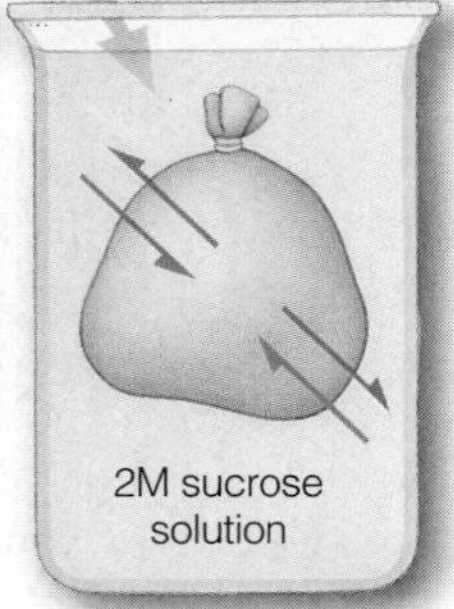

2M sucrose solution
a

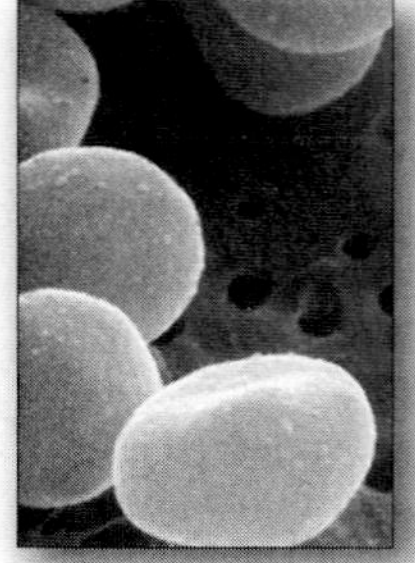

HYPOTONIC CONDITIONS

Water diffuses into red blood cells, which swell up

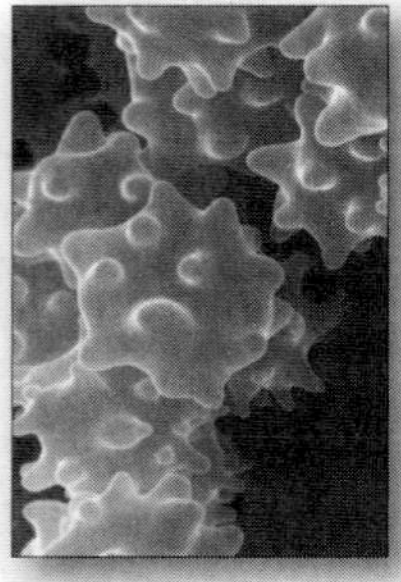

HYPERTONIC CONDITIONS

Water diffuses out of the cells, which shrink

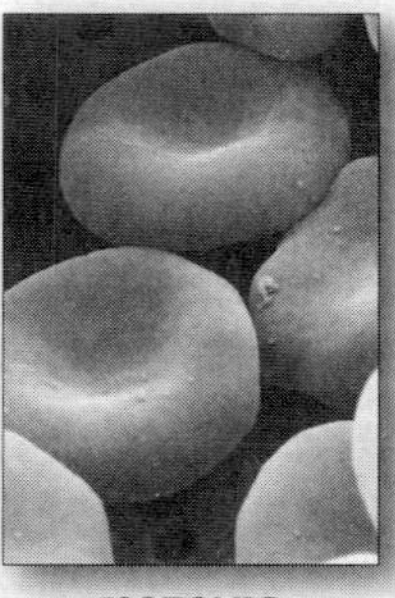

ISOTONIC CONDITIONS

No net movement of water, no change in cell size or shape

b

Figure 5.14 Effect of tonicity on water movement. (**a**) The arrow widths show the direction and relative amounts of water movement. (**b**) The micrographs correspond to the sketches. They show shapes that human red blood cells assume when you place them in fluids of higher, lower, and equal solute concentrations. Normally, solutions inside and outside of red blood cells are in balance. This type of cell does not have any built-in mechanisms that would help it adjust to drastic changes in solute levels in its fluid surroundings.

Further reading: Student Guide to InfoTrac on web site

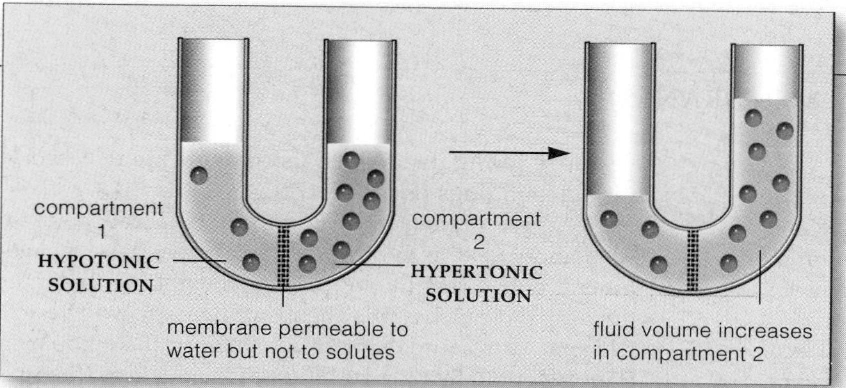
compartment
1
HYPOTONIC
SOLUTION
compartment
2
HYPERTONIC
SOLUTION
membrane permeable to
water but not to solutes
fluid volume increases
in compartment 2

Figure 5.15 Increase in fluid volume owing to osmosis. In time, the net diffusion across a membrane separating two compartments is equal. Then, the fluid volume in compartment 2 is greater because the membrane is impermeable to solutes.

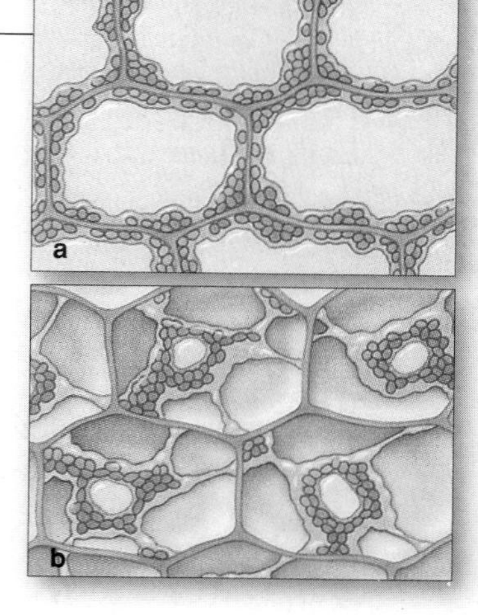
a
b

Figure 5.16 An osmotically induced loss of internal fluid pressure (called plasmolysis) from young plant cells such as those in (**a**). The cytoplasm and central vacuole shrink; the plasma membrane moves away from the wall, as in (**b**).

cross, and you fill each with a 2M sucrose solution. (*M* stands for *M*olarity, the number of moles of a solute in 1 liter of fluid.) Next you immerse one of the sacs in 1 liter of distilled water (which has no solutes), one in a 10M sucrose solution, and the third in a 2M sucrose solution. In each case, the extent and direction of water movement are dictated by tonicity (Figure 5.14).

Tonicity refers to the relative solute concentrations of two fluids. When two fluids on opposing sides of a membrane differ in solute concentration, the one having fewer solutes is called the **hypotonic solution**, and the one with more is the **hypertonic solution**. Water tends to diffuse from hypotonic to hypertonic fluids. **Isotonic solutions** have the same solute concentrations, so water shows no net osmotic movement from one to the other.

Normally, the fluid inside your cells and the tissue fluid bathing them are isotonic. If a tissue fluid became drastically hypotonic, so much water would diffuse into the cells that they would burst. If the fluid became too hypertonic, an outward diffusion of water would shrivel the cells. Most cells have built-in mechanisms that adjust to shifts in tonicity. Red blood cells do not; Figure 5.14 shows what happened to them during a demonstration of the effects of tonicity differences. That is why patients who are severely dehydrated are given infusions of a solution isotonic with blood. Such solutions move by bulk flow from a bottle positioned above the patient, through a tube, and directly into an incised vein.

Effects of Fluid Pressure

Animal cells generally can avoid bursting by engaging in the ongoing selective transport of solutes across the plasma membrane. Cells of plants and many protistans, fungi, and bacteria also avoid that unpleasant prospect with the help of pressure exerted on their cell walls.

Pressure differences as well as solute concentrations influence the osmotic movement of water. Take a look at Figure 5.15. It shows how water continues to diffuse across a membrane between a hypotonic solution and a hypertonic solution until the water concentration is the same on both sides. As you can see, the *volume* of the formerly hypertonic solution has increased (because its solutes cannot diffuse out). Any volume of fluid exerts **hydrostatic pressure**, or a force directed against a wall, membrane, or some other structure that encloses the fluid. The greater the solute concentration of the fluid, the greater will be the hydrostatic pressure it exerts.

Living cells cannot increase in volume indefinitely (Section 4.1). At some point, hydrostatic pressure that develops in a cell counters the inward diffusion of water. That point is the **osmotic pressure**, the amount of force that prevents further increase in a solution's volume.

Think of a young plant cell, with its pliable primary wall. As it grows, water diffuses into it and hydrostatic pressure increases against the wall. The wall expands, and the cell volume increases. The thin walls are strong enough for the cell's internal fluid pressure to develop to the point where it counterbalances water uptake.

Plant cells are vulnerable to water loss, which can happen when soil dries or becomes too salty. Water stops diffusing in, the cells lose water, and internal fluid pressure drops. Such osmotically induced shrinkage of cytoplasm is called plasmolysis (Figure 5.16). Plants can adjust somewhat to the loss of pressure, as when they actively take up potassium ions against a concentration gradient by mechanisms outlined in the next section.

As you will read in Chapters 39 and 41, hydrostatic and osmotic pressure also influence the distribution of water in the blood, tissue fluid, and cells of animals.

Osmosis is the net diffusion of water between two solutions that differ in water concentration and that are separated by a selectively permeable membrane. The greater the number of molecules and ions dissolved in a solution, the lower its water concentration will be.

Water tends to move osmotically to regions of greater solute concentration (from hypotonic to hypertonic solutions). There is no net diffusion between isotonic solutions.

The fluid pressure that a solution exerts against a membrane or wall also influences the osmotic movement of water.

BULK TRANSPORT ACROSS MEMBRANES

Exocytosis and Endocytosis

Transport proteins can move only small molecules and ions into or out of cells. When it comes to taking in or expelling large molecules or particles, cells use vesicles that form through exocytosis and endocytosis.

By **exocytosis**, a vesicle moves to the cell surface, and the protein-studded lipid bilayer of its membrane fuses with the plasma membrane. While this exocytic vesicle is losing its identity, its contents are released to the surroundings (Figure 5.17*a*).

By three pathways of **endocytosis**, a cell takes in substances next to its surface. In all three cases, a small indentation forms at the plasma membrane, balloons inward, and pinches off. The resulting endocytic vesicle transports its contents or stores them in the cytoplasm (Figure 5.17*b*). By *receptor-mediated* endocytosis, the first pathway, membrane receptors chemically recognize and bind specific substances, such as lipoproteins, vitamins, iron, peptide hormones, growth factors, and antibodies. The receptors become concentrated in tiny indentations in the plasma membrane (Figure 5.18). Each of these pits looks like a woven basket on its cytoplasmic side. The basket consists of protein filaments (clathrin) that are interlocked into stable, geometric patterns. When the pit sinks in the cytoplasm, the basket closes back on itself and becomes the vesicle's structural framework.

The second pathway, *bulk-phase* endocytosis, is less selective. An endocytic vesicle forms around a small volume of extracellular fluid regardless of what kinds of substances happen to be dissolved in it. Bulk-phase endocytosis operates at a fairly constant rate in nearly all eukaryotic cells. By continually pulling patches of plasma membrane into the cytoplasm, this pathway compensates for membrane that steadily departs from the cytoplasm in the form of exocytic vesicles.

The third pathway, **phagocytosis**, is an active form of endocytosis by which a cell engulfs microorganisms, large edible particles, and cellular debris. (Phagocytosis literally means "cell eating.") Amoebas and some other protistans get food this way. In multicelled organisms, macrophages and some other white blood cells engage in phagocytosis when they defend the body against

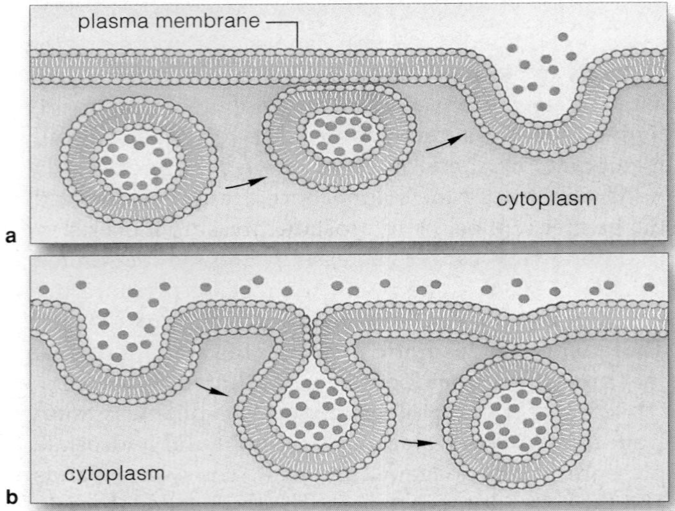
plasma membrane
cytoplasm
a
cytoplasm
b

Figure 5.17 (**a**) Exocytosis. Cells release substances when an exocytic vesicle's membrane fuses with the plasma membrane. (**b**) Endocytosis. A bit of plasma membrane balloons inward beneath water and solutes outside, then it pinches off as an endocytic vesicle that moves into the cytoplasm.

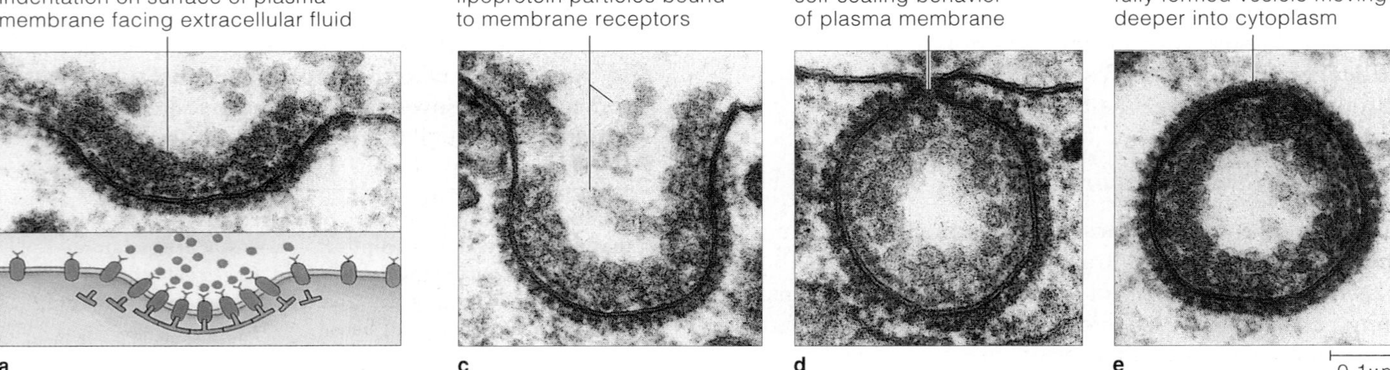
indentation on surface of plasma membrane facing extracellular fluid
lipoprotein particles bound to membrane receptors
self-sealing behavior of plasma membrane
fully formed vesicle moving deeper into cytoplasm
a
c
d
e
0.1µm
b
clathrin filaments of coated pit

Figure 5.18 Example of receptor-mediated endocytosis, at the plasma membrane of an immature chicken egg. (**a**) This shallow indentation is a coated pit. (**b**) The cytoplasmic surface of each pit has a basketlike array of clathrin filaments. (**c**) Receptor proteins at the pit's outer surface preferentially bind lipoprotein particles. (**d**) The pit deepens and rounds out. (**e**) The formed endocytic vesicle contains lipoproteins that the cell will use or store.

Further reading: Student Guide to InfoTrac on web site —

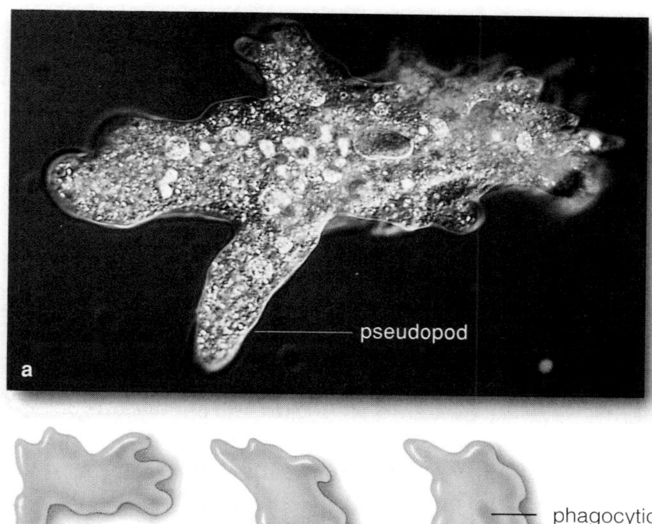
pseudopod
a
edible
bacterium
phagocytic
vesicle
b

Figure 5.19 Phagocytosis in amoebas (including the *Amoeba proteus* cell in **a**), certain white blood cells such as macrophages, and some other cells. (**b**) Lobes of the amoeba's cytoplasm extend outward and surround the target. The plasma membrane of the extensions fuses together, thereby forming a phagocytic vesicle. Such vesicles move deeper into the cytoplasm and then fuse with lysosomes. Their contents are digested and, along with the vesicle's membrane components, are recycled elsewhere.

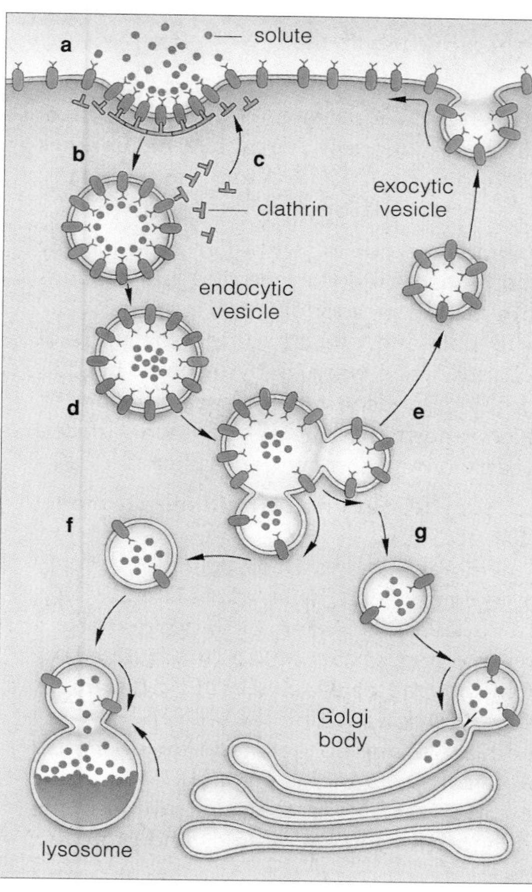
solute
a
b
c
clathrin
exocytic
vesicle
endocytic
vesicle
d
e
f
g
Golgi
body
lysosome

a Molecules get concentrated inside coated pits of plasma membrane.

b Endocytic vesicles form from the pits.

c Vesicles lose molecules of clathrin, which then return to the plasma membrane.

d Enclosed molecules are sorted and often released from receptors.

e Many sorted molecules are cycled back to the plasma membrane.

f,g Many other sorted molecules are delivered to lysosomes and stay there. Still others are routed to spaces in the nuclear envelope and inside ER membranes, and still others to Golgi bodies.

Figure 5.20 Cycling of membrane lipids and proteins. This example starts with receptor-mediated endocytosis. The plasma membrane gives up small patches of itself to endocytic vesicles that form from coated pits. It gets membrane back from exocytic vesicles that budded from ER membranes and Golgi bodies. The membrane initially appropriated by endocytic vesicles will cycle receptor proteins and lipids back to the plasma membrane.

invasions of harmful viruses, bacteria, cancerous body cells, and other threats to health.

A phagocytic cell gets busy after a target binds with certain receptors that bristle from its plasma membrane. Binding sends signals into the cell. The signals trigger a directional assembly and crosslinking of microfilaments into a dynamic, ATP-requiring network just beneath the plasma membrane. The network contracts in ways that squeeze some cytoplasm toward the cell margins, thus forming lobes called pseudopods (Figure 5.19). The pseudopods flow over the target and fuse at their tips. The result is a phagocytic vesicle, which sinks into the cytoplasm. There it fuses with lysosomes, the organelles of intracellular digestion in which trapped items are digested to fragments and smaller, reusable molecules.

Membrane Cycling

As long as a cell stays alive, exocytosis and endocytosis continually replace and withdraw patches of its plasma membrane. And they apparently do so at rates that can maintain the plasma membrane's total surface area.

As an example, neurons release neurotransmitters in bursts of exocytosis. Each neurotransmitter is a type of signaling molecule released from one cell that acts on neighboring cells. Cell biologist John Heuser and T. S. Reese documented an intense burst of endocytosis that immediately followed an intense episode of exocytosis in neurons—and that counterbalanced it. Figure 5.20 gives more examples of the ways in which cells cycle their membrane lipids and proteins.

Whereas transport proteins in a cell membrane deal only with ions and small molecules, exocytosis and endocytosis move larger packets of material across the plasma membrane.

By exocytosis, a cytoplasmic vesicle fuses with the plasma membrane, so that its contents are released outside the cell. By endocytosis, a small patch of the plasma membrane sinks inward and seals back on itself, forming a vesicle inside the cytoplasm. Membrane receptors often mediate this process.

5.7

SUMMARY

Membrane Structure and Function

1. The plasma membrane is a structural and functional boundary between the cytoplasm and the surroundings of all cells. In eukaryotic cells, organelle membranes subdivide the fluid portion of the cytoplasm into many functionally diverse compartments.

2. A cell membrane consists of two water-impermeable layers of lipids (phospholipids especially) and proteins associated with the layers, as shown in Figure 5.21.

a. Fatty acid tails and other hydrophobic parts of the lipids are sandwiched between hydrophilic heads.

b. Many different kinds of proteins are embedded in the lipid bilayer or positioned at one of its two surfaces. The proteins carry out most membrane functions.

3. These are the key features of the fluid mosaic model of membrane structure:

a. A cell membrane shows fluid behavior, mainly because its lipid components twist, move laterally, and flex hydrocarbon tails. Also, some of these lipids have ring structures and many have kinked (unsaturated) or short fatty acid tails, all of which disrupt what might otherwise be tight packing within the bilayer.

b. A membrane is a mosaic, or composite, of diverse lipids and proteins. The proteins are embedded in the bilayer and are at its surface. Its two layers differ in the number, kind, and arrangement of lipids and proteins.

4. All cell membranes include transport proteins and proteins that structurally reinforce the membrane. The plasma membrane also has proteins that serve in signal reception, cell recognition, and adhesion. Differences in the number and types of proteins among cells affect metabolism, cell volume, pH, and responsiveness to substances that make contact with the membrane.

a. Transport proteins help water-soluble substances cross the membrane by passing through their interior, which opens to both sides of the membrane.

b. Receptor proteins bind extracellular substances, and binding triggers alterations in metabolic activities. Recognition proteins are like molecular fingerprints; they identify cells as being of a given type. Adhesion proteins help cells of tissues adhere to one another and to form cell junctions.

Movement of Substances Into and Out of Cells

1. Molecules or ions of a substance tend to move from a region of higher to lower concentration. Movement in response to a concentration gradient is called diffusion.

a. Diffusion rates are influenced by the steepness of the concentration gradient, temperature, and molecular size, as well as by gradients in electrical charge and pressure that may occur between two regions.

b. Cells have built-in mechanisms that work with or against gradients to move solutes across membranes.

c. Metabolism requires chemical energy inherent in concentration and electric gradients across cell membranes.

2. Oxygen, carbon dioxide, and other small nonpolar molecules diffuse across a membrane's lipid bilayer. Ions and large, polar molecules such as glucose cross it with the passive or active help of transport proteins. Water moves through proteins and through the bilayer.

3. Transport proteins bind specific solutes on one side of a cell membrane, and they shunt solutes to the other side as an outcome of reversible changes in their shape.

a. Passive transport does not require energy inputs; the protein allows a solute simply to diffuse through its interior in the direction of the concentration gradient.

b. Active transport requires energy boosts, as from ATP. The protein pumps a solute across the membrane against its concentration gradient.

4. Osmosis is the diffusion of water across a selectively permeable membrane in response to a concentration gradient.

5. By exocytosis, vesicles in the cytoplasm move to the plasma membrane. When their membrane fuses with it, their contents are automatically released to the outside.

6. By endocytosis, a small patch of plasma membrane sinks into the cytoplasm and seals back on itself to form a vesicle. Three pathways are called receptor-mediated endocytosis (requires recognition of specific solutes), bulk-phase endocytosis (indiscriminate uptake of some extracellular fluid), and phagocytosis (active uptake of large particles, cell parts, or whole cells).

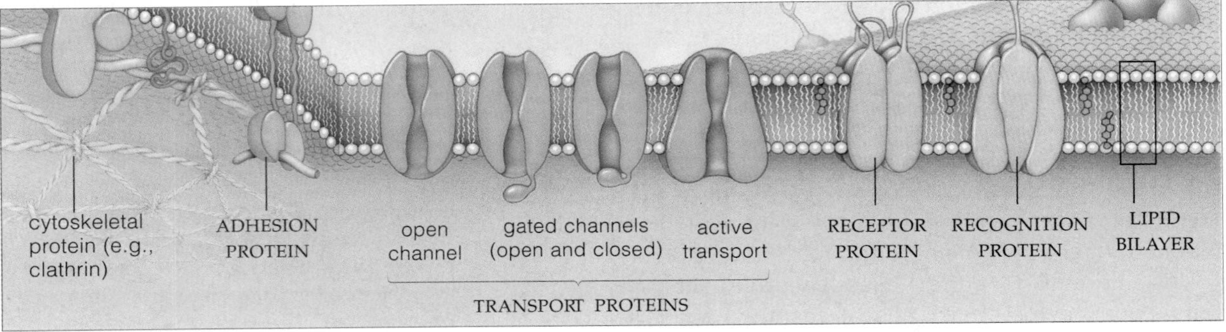

cytoskeletal protein (e.g., clathrin)
ADHESION PROTEIN
open channel
gated channels (open and closed)
active transport
TRANSPORT PROTEINS
RECEPTOR PROTEIN
RECOGNITION PROTEIN
LIPID BILAYER

Figure 5.21 Summary of the main types of proteins associated with plasma membranes.

Review Questions

1. Describe the fluid mosaic model of cell membranes. What imparts fluidity to the membrane? What makes it a mosaic? *5.1*

2. State the functions of transport proteins, receptor proteins, recognition proteins, and adhesion proteins. *5.1*

3. Define diffusion. Does diffusion occur in response to a solute concentration gradient, an electric gradient, a pressure gradient, or some combination of these? *5.3*

4. If all transport proteins work by changing shape, then how do passive transporters differ from active transporters? *5.4*

5. Define osmosis. Explain how solute concentrations of two solutions on either side of a membrane influence the osmotic movement of water down the water concentration gradient. *5.5*

6. Define hypertonic, hypotonic, and isotonic solutions. Does each term refer to a property inherent in a given solution? Or are the terms used only when comparing solutions? *5.5*

7. Define exocytosis and endocytosis. Describe the main features of the three pathways of endocytosis. *5.6*

Self-Quiz (*Answers in Appendix III*)

1. Cell membranes consist mainly of a ______ .
   a. carbohydrate bilayer and proteins
   b. protein bilayer and phospholipids
   c. lipid bilayer and proteins

2. In a lipid bilayer, ______ of lipid molecules are sandwiched between ______ .
   a. hydrophilic tails; hydrophobic heads
   b. hydrophilic heads; hydrophilic tails
   c. hydrophobic tails; hydrophilic heads
   d. hydrophobic heads; hydrophilic tails

3. Most membrane functions are carried out by ______ .
   a. proteins c. nucleic acids
   b. phospholipids d. hormones

4. All cell membranes incorporate ______ .
   a. transport proteins c. recognition proteins
   b. adhesion proteins d. all of the above

5. Immerse a living cell in a hypotonic solution, and water will tend to ______ .
   a. move into the cell c. show no net movement
   b. move out of the cell d. move in by endocytosis

6. A ______ is a device that spins test tubes and separates cell components according to their relative densities.

7. ______ can readily diffuse across a lipid bilayer.
   a. Glucose c. Carbon dioxide
   b. Oxygen d. b and c

8. Sodium ions cross a membrane through transport proteins that receive an energy boost. This is an example of ______ .
   a. passive transport c. facilitated diffusion
   b. active transport d. a and c

Critical Thinking

1. The bacterium *Vibrio cholerae* causes a disease, *cholera*. Infected people have severe diarrhea and may lose up to twenty liters of fluid a day. *V. cholerae* enters the body when someone drinks contaminated water, then it adheres to the intestinal lining. It secretes a metabolic product that is toxic to cells of the lining, and they in turn start secreting chloride ions (Cl^-). Sodium ions (Na^+) follow the chloride ions into the intestinal fluid. Explain how the series of ion movements causes the massive fluid loss.

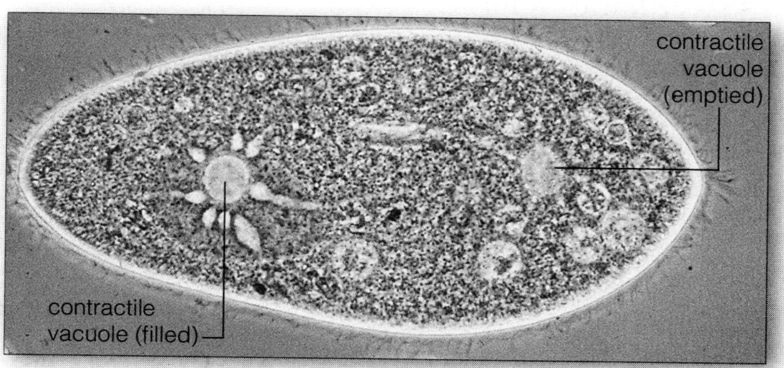

contractile
vacuole
(emptied)
contractile
vacuole (filled)

Figure 5.22 Contractile vacuoles of *Paramecium*, a protistan.

2. Many cultivated fields in California require heavy irrigation. Over the years, most of the water has evaporated from the soil, leaving behind all of the irrigation water's solutes. What kinds of problems might the altered soil conditions cause for plants?

3. Imagine that you are a juvenile shrimp living in an *estuary*, where freshwater draining from the land mixes with saltwater from the sea. Many people who own homes around a large lake want boat access to the sea, so they ask their city for permission to build a canal between the lake and estuary. If they succeed, what might happen to you and other estuary inhabitants?

4. Water moves osmotically into *Paramecium*, a single-celled protistan of aquatic habitats. If unchecked, the influx would bloat the cell and rupture its plasma membrane. An energy-requiring mechanism involving contractile vacuoles expels the excess (Figure 5.22). Water enters tubelike extensions of this organelle and collects in a central space in the vacuole. When full, the vacuole contracts and squirts excess water out of a pore that opens to the outside. Are the fluid surroundings hypotonic, hypertonic, or isotonic relative to *Paramecium*'s cytoplasm?

5. Certain species of bacteria thrive in environments where the temperatures approach the boiling point of water—for example, in the steam-venting fissures of volcanoes and in hot springs of Yellowstone National Park. Assume that the lipid bilayer of the bacterial cell membranes consists mainly of phospholipids. What features might the fatty acid tails of the phospholipids have that help stabilize the membranes at such extreme temperatures?

Selected Key Terms

active transport *5.4*
adhesion protein *5.1*
bulk flow *5.5*
calcium pump *5.4*
centrifuge *5.2*
concentration gradient *5.3*
diffusion *5.3*
electric gradient *5.3*
endocytosis *5.6*
exocytosis *5.6*
fluid mosaic model *5.1*
hydrostatic pressure *5.5*
hypertonic solution *5.5*
hypotonic solution *5.5*
isotonic solution *5.5*
lipid bilayer *5.1*
osmosis *5.5*
osmotic pressure *5.5*
passive transport *5.4*
phagocytosis *5.6*
phospholipid *5.1*
pressure gradient *5.3*
receptor protein *5.1*
recognition protein *5.1*
selective permeability *5.3*
sodium–potassium pump *5.4*
transport protein *5.1*

Readings *See also www.infotrac-college.com*

Alberts, B. et al. 1994. *Molecular Biology of the Cell*. Third edition. New York: Garland.

6

GROUND RULES OF METABOLISM

You Light Up My Life

Find yourself out and about at night or snorkeling in the ocean, and you might see fireflies and other insects, squids, or certain other organisms emitting light. Figure 6.1*a* shows fireflies (actually a type of beetle) engaging in social displays in a tropical forest. Different varieties, known locally as kittyboos, light up the night with green, yellow-green, yellow, or orange flashes.

Kittyboos emit light when enzymes—luciferases—convert chemical energy to light energy. The reactions get under way when a phosphate group from ATP is transferred to luciferin, a highly fluorescent substance. Luciferases, with a little help from oxygen, convert the activated luciferin to a different molecule. Their action boosts electrons of the molecule to a higher energy level. As the excited electrons quickly return to a lower energy level, they release energy as *fluorescent* light. Fluorescent light is emitted when a destabilized molecule reverts to a stable configuration. When organisms flash with it, this is called **bioluminescence**.

Imaginative biologists learned how to borrow light from such flashers to make *bioluminescent gene transfers*. They now insert copies of the genes for bioluminescence into bacteria, plants, and other organisms (Figure 6.1*b*)! Besides being fun to think about, these transfers have practical applications.

Each year, for example, 3 million people die from a lung disease caused by *Mycobacterium tuberculosis*. No one antibiotic is effective against all the different strains of this bacterium, so an infected person cannot receive effective treatment until the strain causing the infection is identified. A fast way to do this is to expose bacterial cells in samples taken from a patient to luciferase genes. The genes typically slip into the bacterial DNA of some cells. Clinicians isolate those cells, then expose colonies of the descendants to different antibiotics. If a particular antibiotic doesn't work, the colonies glow. Their cells have churned out gene products—including luciferase. If the colonies don't glow, the antibiotic works.

Christopher and Pamela Contag, two postdoctoral students at Stanford University, wanted to light up bacteria that cause *Salmonella* infections in laboratory mice. Why? Researchers of viral or bacterial diseases typically infect dozens to hundreds of laboratory mice for experiments. The only option has been to kill infected mice and examine tissues to find out whether infection occurred—a costly, tediously painstaking practice that also requires dispatching the experimental animals.

The Contags approached David Benaron, a medical imaging researcher at Stanford, with this hypothesis: If live, infectious bacteria were made bioluminescent, then flashes would shine through the tissues of infected animals. In a preliminary test of this novel idea, the researchers put glowing *Salmonella* cells into a thawed chicken breast from a market. A glow showed through.

A kittyboo's bioluminescent organs, where luciferin and luciferases are kept separated until signals from the nervous system command them to mix it up
a

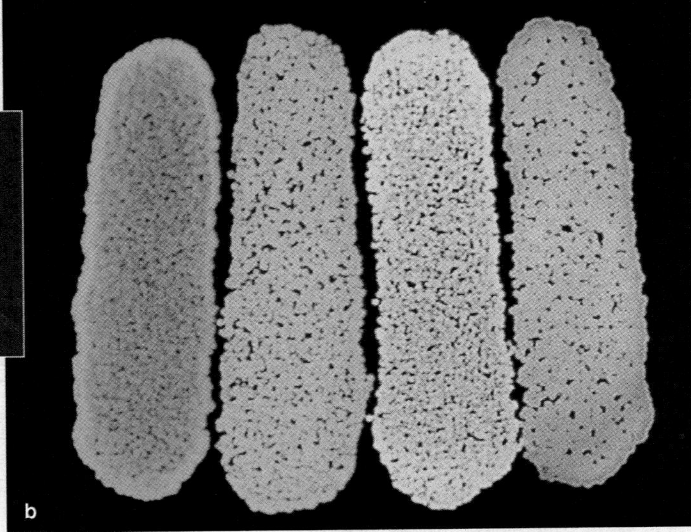
b

Figure 6.1 (**a**) Jamaican fireflies, known locally as kittyboos (*Pyrophorus noctilucus*). This type of beetle lights up the night with brief bioluminescent flashes, which help potential mates find each other in the dark. (**b**) Micrograph of four colonies of bacterial cells. Each colony started with a parent bacterial cell that had taken up a kittyboo gene for a glowing color.

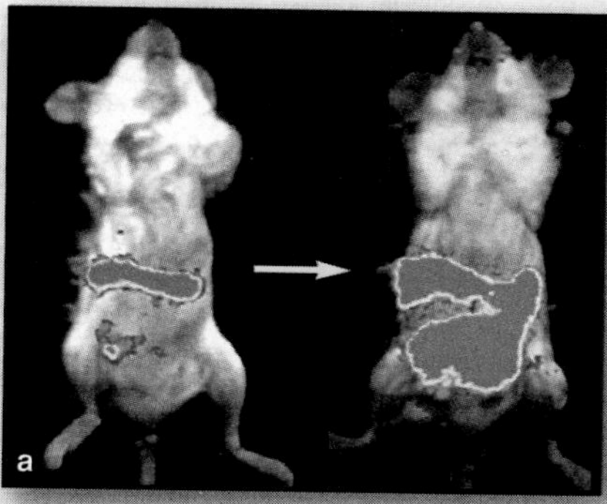
a

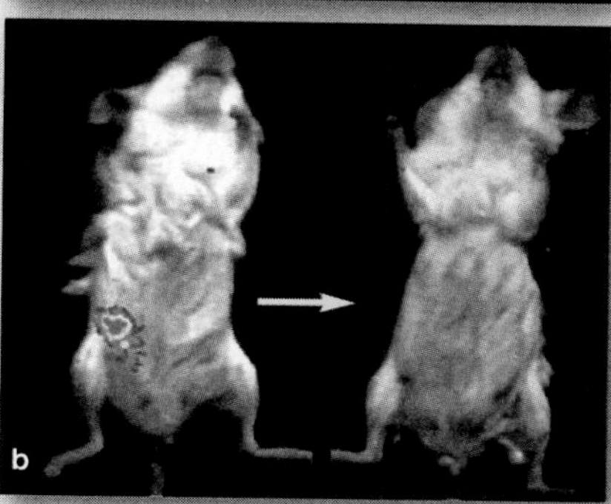
b

Figure 6.2 Using bioluminescent bacterial cells to chart the location of infectious bacteria inside living laboratory mice and their spread through body tissues. (**a**) False-color images in this pair of photographs show how the infection spread in a control group that had not been given a dose of antibiotics. (**b**) This pair shows how antibiotics had killed most of the infectious bacterial cells.

Next the Contags transferred bioluminescence genes into three strains of *Salmonella*. Then they injected the strains into mice of three experimental groups and used a digital imaging camera to track the infection in each group. The first strain was weak; the mice were able to fight off the infection in less than six days and did not glow. The second strain was not as weak but could not spread through the mouse body; it remained localized. The third strain was dangerous. It spread very rapidly through the mouse gut—and the entire gut glowed.

Thus bioluminescent gene transfer, combined with imaging of enzyme activity, can be used to track the course of infection and to evaluate the effectiveness of drugs in living organisms (Figure 6.2). It may also have uses in gene therapy, whereby copies of functional genes replace defective or cancer-causing genes in patients.

Why use bioluminescent organisms to introduce a chapter? They give us visible signs of **metabolism**—of the cell's capacity to acquire energy and use it to build, break apart, store, and release substances in controlled ways. Each flash reminds us that living cells are taking in energy-rich solutes, building membranes, storing things, replenishing enzymes, and checking out their DNA. A constant supply of energy drives all of these activities. The story of metabolism starts with ways in which cells get energy and channel it into the reactions by which they stay alive, grow, and reproduce.

KEY CONCEPTS

1. Cells engage in metabolism, or chemical work. That is, they use energy to stockpile, build, rearrange, and break apart substances. Cells also use energy for mechanical work, as when they move cell structures such as flagella. They also channel energy into electrochemical work, as when they move charged substances into or out of the cytoplasm or an organelle compartment.

2. All organisms are adapted to secure energy from their environment, such as energy from the sun and from inorganic or organic molecules. Their cells couple the energy inputs to thousands of energy-requiring reactions.

3. ATP is the main carrier of energy in cells. It couples reactions that release energy from the sun's ray's or from molecules with reactions that require energy.

4. ATP forms when energy inputs lead to the transfer of a phosphate group or inorganic phosphate to a molecule called ADP. Later, ATP transfers a phosphate group to enzymes, glucose, and other molecules, which primes them to enter a reaction (activates them).

5. Many metabolic reactions involve electron transfers from one substance to another. Such transfers, or oxidation–reduction reactions, occur singly or in a series of small steps through electron transport systems.

6. Thousands of reactions proceed in cells in orderly, enzyme-mediated sequences called metabolic pathways. Operation of the pathways is coordinated in ways that maintain, increase, or decrease the relative amounts of various substances in cells.

7. Chemical reactions proceed far too slowly on their own to sustain life. In living organisms, the action of specific enzymes greatly increases the rate of specific reactions. Control of enzyme activity is central to metabolism.

ENERGY AND THE UNDERLYING ORGANIZATION OF LIFE

Defining Energy

If you have ever watched a house cat stalking a mouse, you know it can "freeze" its position to avoid detection before springing at its unsuspecting prey. Like anything else in the universe that is stationary, the cat has a store of **potential energy**—a capacity to do work, simply owing to its position in space and the arrangement of its parts. As a cat springs, some of its potential energy is transformed into **kinetic energy**, the energy of motion.

Energy on the move does work when it imparts motion to other things. In skeletal muscle cells inside the cat, ATP gave up some of its potential energy to molecules of contractile units and set them in motion. The combined motions in many muscle cells resulted in the movement of whole muscles. The transfer of energy from ATP also resulted in the release of another form of energy called **heat**, or *thermal* energy.

The potential energy of molecules has its own name: **chemical energy**. It is measurable, as in kilocalories. A kilocalorie is the same thing as 1,000 calories, which is the amount of energy it takes to heat 1,000 grams of water from 14.5°C to 15.5°C at standard pressure.

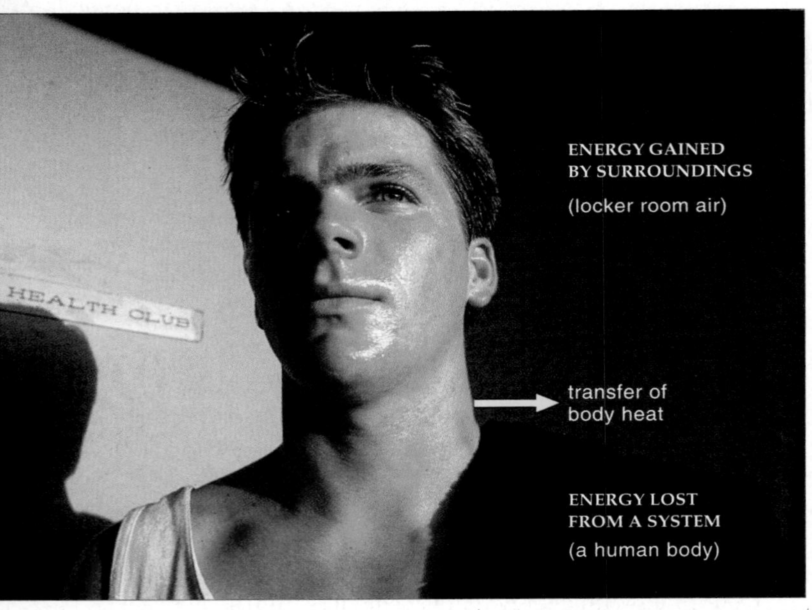
HEALTH CLUB
ENERGY GAINED BY SURROUNDINGS
(locker room air)
transfer of body heat
ENERGY LOST FROM A SYSTEM
(a human body)
NET ENERGY CHANGE = 0

Figure 6.3 Example of how the total energy content of any system *together with its surroundings* remains constant.

"System" means all matter in a specific region, such as a human body, a plant, a DNA molecule, or a galaxy. "Surroundings" can be a small region in contact with the system or as vast as the entire universe. The system shown (a human male) is giving off heat to the surroundings (a locker room) by evaporative water loss from sweat. What one region loses, the other region gains, so the total energy content of both does not change.

What Can Cells Do With Energy?

All organisms have specific adaptations for securing energy from their environment. Some harness energy from the sun, and others extract energy from inorganic or organic substances in the environment. Regardless of the source, energy inputs become *coupled* to thousands of energy-requiring processes in cells. Cells use energy for *chemical* work—to stockpile, build, rearrange, and break apart substances. They channel it into *mechanical* work—to move flagella and other cell structures and (in multicelled species) the whole body or portions of it. They channel it into *electrochemical* work—to move charged substances into or out of the cytoplasm or an organelle compartment.

How Much Energy Is Available?

Like single cells, we cannot create energy from scratch; we must get it from someplace else. Why? According to the **first law of thermodynamics**, the total amount of energy in the universe remains constant. More energy cannot be created; existing energy cannot vanish. It can only be converted from one form to some other form.

Think about what the law means. The universe has only so much energy, distributed in a variety of forms. One form can be converted to another, as when corn plants absorb energy from the sun and convert it to the chemical energy of starch. After you eat and digest corn, your cells extract energy from starch and convert it to other forms, such as kinetic energy for moving about.

With each metabolic conversion, some of the energy escapes to the surroundings, as heat. Even when you "do nothing," your body gives off about as much heat as a 100-watt lightbulb because of conversions in your cells. The energy being released is transferred to atoms and molecules making up the air, and the conversion of thermal to kinetic energy "heats up" the surroundings (Figure 6.3). The kinetic energy increases the number of ongoing, random collisions among molecules in the air. And with each collision, a bit more energy is released as heat. However, none of the energy ever vanishes.

The One-Way Flow of Energy

Energy available for conversions in cells resides mainly in covalent bonds. Glucose, glycogen, starches, fatty acids, and other organic compounds have organized arrangements of many of these bonds and are said to have a high energy content. When the compounds enter metabolic reactions, specific bonds break or become rearranged. During that molecular commotion, some amount of heat energy is lost to the surroundings. In general, cells cannot recapture energy lost as heat.

Further reading: Student Guide to InfoTrac on web site

Figure 6.4 An example of the one-way flow of energy into the world of life that compensates for the one-way flow of energy out of it. The sun continuously loses energy, much of it in the form of wavelengths of light (Section 7.2). Living cells intercept some of the energy and convert it to useful forms of energy, stored in bonds of organic compounds. Each time a metabolic reaction proceeds in cells, stored energy is released—and some inevitably is lost to the surroundings, mostly as heat.

The lower photograph shows green, water-dwelling, photosynthetic cells (*Volvox*). They live in tiny, spherical colonies. The orange cells function in reproduction. They form new colonies inside the parent sphere.

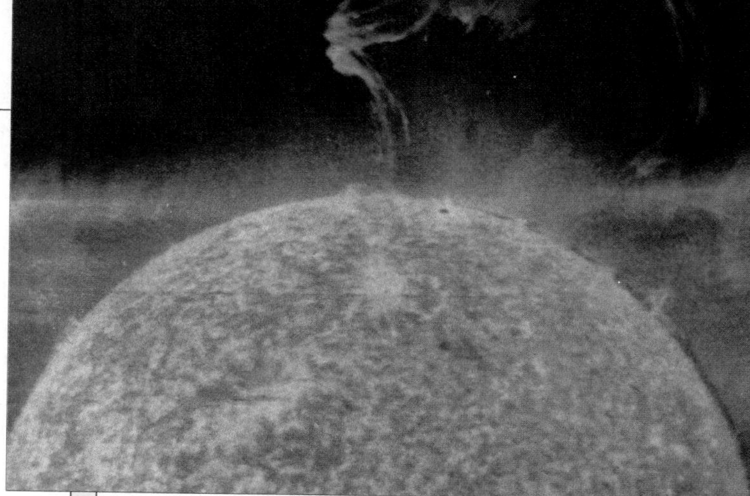

ENERGY LOST
one-way flow of energy from sun to Earth's environment

ENERGY GAINED
one-way flow of energy from environment to organisms

Producer organisms harness sun's energy, use it to build organic compounds from simple raw materials available in their environment.

All organisms tap potential energy stored in organic compounds to drive energy conversions that keep them alive. Some energy is lost with each conversion.

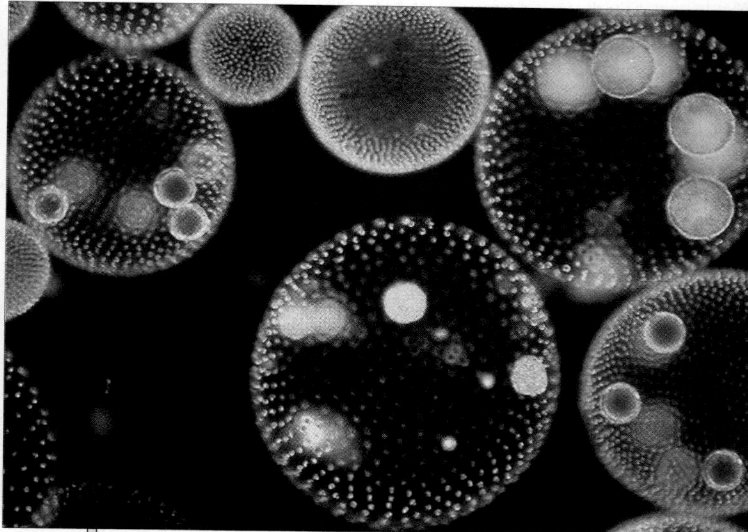

ENERGY LOST
one-way flow of energy from organisms back to the environment

For example, your cells release usable energy from glucose by breaking all of its covalent bonds. After many steps, six molecules of carbon dioxide and six of water remain. Compared with glucose, those leftovers have more stable arrangements of atoms, but chemical energy in their bonds is much less than the total chemical energy of glucose. Why? *Some energy was lost at each step leading to their formation*. Said another way, glucose is a better source of usable energy.

What about the heat that was transferred from cells to their surroundings when carbon dioxide formed? It is not at all useful. Cells cannot convert it to other forms, so it cannot be used to do work.

Bad news for cells of the remote future: The amount of "low-quality" energy in the universe is increasing. No energy conversion can ever be 100 percent efficient—even highly efficient ones lose heat—so the total amount of energy in the universe is spontaneously flowing from forms rich in energy to forms having less of it. Billions of years from now, all of the energy available for conversions will be dissipated.

Without energy inputs to maintain it, any organized system tends to become more and more disorganized over time. **Entropy** is a measure of the degree of a system's disorder. Think of the Egyptian pyramids—originally organized, presently crumbling, and many thousands of years from now, dust. It seems that the ultimate destination of those pyramids and everything else in the universe is a state of maximum entropy. That, basically, is the point to remember about the **second law of thermodynamics**.

Can life be one glorious pocket of resistance to the depressing flow toward maximum entropy? After all, in each new organism, new bonds form and hold atoms together in precise arrays. So molecules become more organized and have a richer store of energy, not poorer!

Yet a simple example will show that the second law does indeed apply to life on Earth. The primary energy source for life is the sun, which has been releasing energy since it first formed. Plants can capture sunlight energy, convert it to other forms, then lose energy to other organisms that feed, directly or indirectly, on the plants. At each energy transfer, some energy is lost as heat that joins the universal pool. *Overall, energy still flows in one direction*. The world of life maintains its amazing degree of organization only because it is being resupplied with energy that is being lost from someplace else (Figure 6.4).

The amount of energy in the universe remains constant. Energy can undergo conversions from one form to another, but it cannot be created out of nothing or destroyed.

The total amount of energy in the universe is spontaneously flowing from usable to nonusable forms.

A steady flow of sunlight energy into the interconnected web of life compensates for the steady flow of energy leaving it.

ENERGY CHANGES AND CELLULAR WORK

Energy Inputs, Energy Outputs

When cells convert one form of energy to another, there is a change in the amount of potential energy available to them. The greater the initial amount of potential energy a cell taps into, the larger the energy change can be—and the more work can be done.

Imagine a Martian who is not happy that the NASA Rover is inching around her planet. She decides to push it to the top of a rocky hill (Figure 6.5*a*). To do this, she converts some potential energy stored in her muscles to kinetic energy. Once the Rover is precariously perched on the hill, it has potential energy (owing to its position) and it tends to roll down on its own, spontaneously. The higher up the Rover has been pushed relative to its final position at the base of the hill, the greater the energy change and the more work done—in this case, a bigger impact and more broken parts (Figure 6.5*b*).

The same ground rule applies to the cellular world, in which temperature and pressure remain constant. *Energy changes in cells tend to proceed spontaneously in the direction that results in a decrease in usable energy.*

For instance, glucose ($C_6H_{12}O_6$) is built from carbon dioxide ($6CO_2$) and water ($12H_2O$). Each substance has potential energy in chemical bonds. But the bond energy of glucose exceeds those of the other two substances combined. Thus, the assembly of glucose from carbon dioxide and water is not something that will proceed spontaneously. Visualize carbon dioxide and water at the base of an energy hill. On their own, they simply don't have enough energy for an uphill run to make glucose.

In photosynthetic cells, *energy inputs* from the sun drive reactions that synthesize glucose. The outcome is a net increase in usable energy for the cell (Figure 6.5*c*). Said another way, the reaction sequence by which glucose forms is **endergonic** (meaning "energy in").

Now picture the reactions running in reverse, from glucose (at the top of the energy hill) to carbon dioxide and water (at the base). Energetically, a downhill run is favorable. It proceeds spontaneously and ends with a net loss in energy. So the reaction sequence that breaks down glucose is **exergonic**, which means "energy out" (Figure 6.5*d*). All cells, including photosynthetic ones, release energy from glucose and other large molecules.

Energy input required to push Rover uphill
a

Potential energy released by the downhill run
b

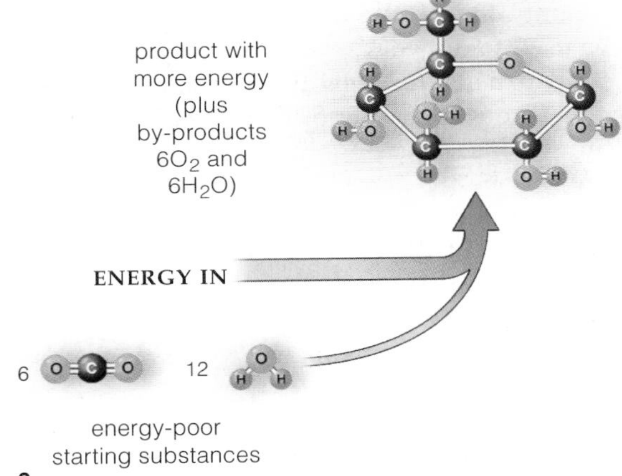
product with more energy (plus by-products $6O_2$ and $6H_2O$)
ENERGY IN
6
12
energy-poor starting substances
c

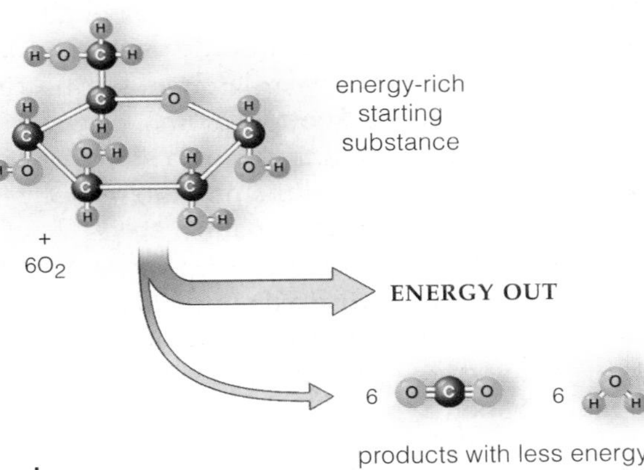
energy-rich starting substance
+ $6O_2$
ENERGY OUT
6
6
products with less energy
d

Figure 6.5 Examples of energy changes involved in (**a**,**b**) mechanical work and (**c**,**d**) chemical work.

Further reading: Student Guide to InfoTrac on web site

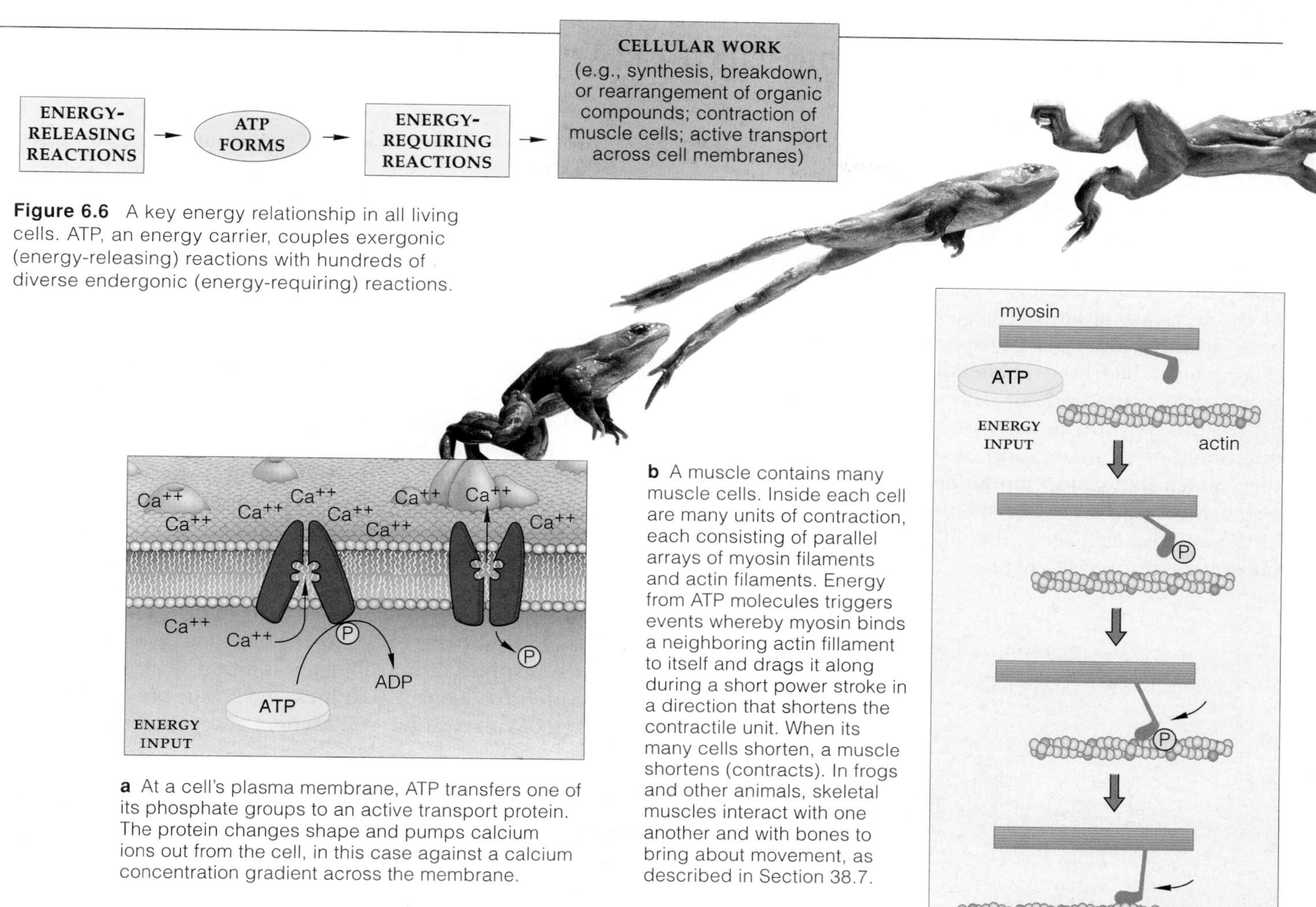

ENERGY-RELEASING REACTIONS
ATP FORMS
ENERGY-REQUIRING REACTIONS
CELLULAR WORK
(e.g., synthesis, breakdown, or rearrangement of organic compounds; contraction of muscle cells; active transport across cell membranes)
myosin
ATP
ENERGY INPUT
actin
P
Ca++
ADP
ATP
ENERGY INPUT

Figure 6.6 A key energy relationship in all living cells. ATP, an energy carrier, couples exergonic (energy-releasing) reactions with hundreds of diverse endergonic (energy-requiring) reactions.

a At a cell's plasma membrane, ATP transfers one of its phosphate groups to an active transport protein. The protein changes shape and pumps calcium ions out from the cell, in this case against a calcium concentration gradient across the membrane.

b A muscle contains many muscle cells. Inside each cell are many units of contraction, each consisting of parallel arrays of myosin filaments and actin filaments. Energy from ATP molecules triggers events whereby myosin binds a neighboring actin fillament to itself and drags it along during a short power stroke in a direction that shortens the contractile unit. When its many cells shorten, a muscle shortens (contracts). In frogs and other animals, skeletal muscles interact with one another and with bones to bring about movement, as described in Section 38.7.

Figure 6.7 Examples of cellular work initiated when ATP (symbolized as a yellow coin) gives up energy to molecules that participate in specific metabolic reactions.

The Role of ATP

Rays of sunlight are a form of kinetic energy, released during stupendous exergonic reactions at the surface of the sun. Here on Earth, photoautotrophs can convert some of that energy into chemical bond energy of ATP. Also, all organisms can make ATP through exergonic reactions during aerobic respiration and other glucose-degrading processes. Regardless of how it forms, ATP can activate many different molecules by transferring one of its phosphate groups to them. Section 6.3 explains this process, which is called phosphorylation.

ATP's role is like currency in an economy: cells earn it by exergonic reactions and can spend it in endergonic reactions that drive hundreds of cellular activities, such as those shown in Figures 6.6 and 6.7. That is why we often use a cartoon coin to symbolize ATP.

The Role of Electron Transfers

Think back on earlier chapters that describe the making and breaking of chemical bonds. Remember how atoms, ions, and molecules can be made to accept or give up electrons? Electrons are transferred in virtually every reaction that harnesses energy for use in the formation of ATP. For example, in plant cells, energy from the sun drives electrons from water molecules. Being attracted to the oppositely charged electrons, hydrogen ions go along for the ride—and both join up with carbon and oxygen to form new organic compounds. In all cells, ATP jump-starts the release of electrons from glucose to form more ATP, then oxygen or some other substance picks up the "spent" electrons. Section 6.4 takes a closer look at the nature of electron transfers in cells.

On their own, energy changes in cells spontaneously run in the direction that results in a decrease in usable energy.

ATP, an energy carrier in all living cells, couples energy-releasing reactions with energy-requiring reactions.

Electron transfers are involved in all of the reactions that harness the energy necessary to make ATP.

A LOOK AT TYPICAL PHOSPHATE-GROUP TRANSFERS

How Does ATP Give Up Energy?

To gain insight into how energy is transferred during metabolic reactions, let's take a closer look at the ATP molecule. As you know, **ATP** is short for adenosine triphosphate, which is one of the organic compounds called nucleotides (Section 3.7). Each molecule consists of the five-carbon sugar ribose to which a nucleotide base (adenine) and three phosphate groups are attached (Figure 6.8*a*). Its triphosphate tail is where the action is, so to speak. Hundreds of different kinds of enzymes can readily hydrolyze the covalent bond between the outermost phosphate group of the molecule's tail and then attach that group to another substance. Enzymes also can break the next covalent bond in the tail and transfer still another phosphate group elsewhere.

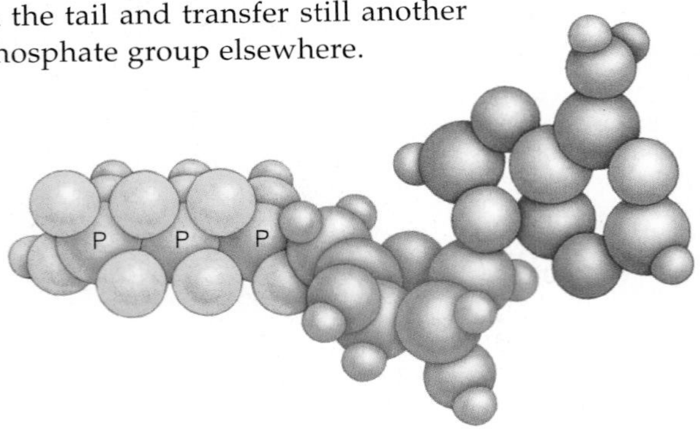
P
P
P
a

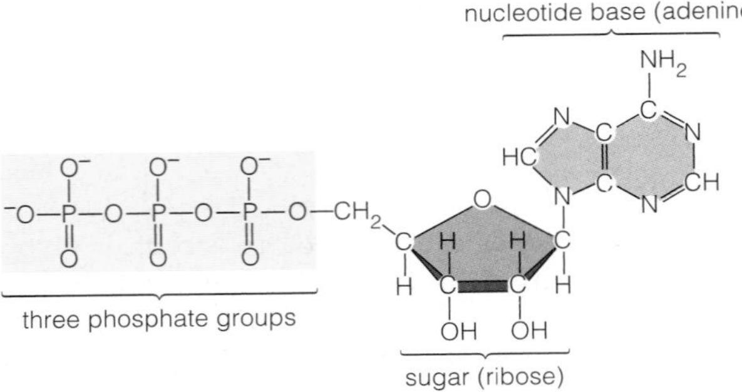
nucleotide base (adenine)
three phosphate groups
sugar (ribose)
b

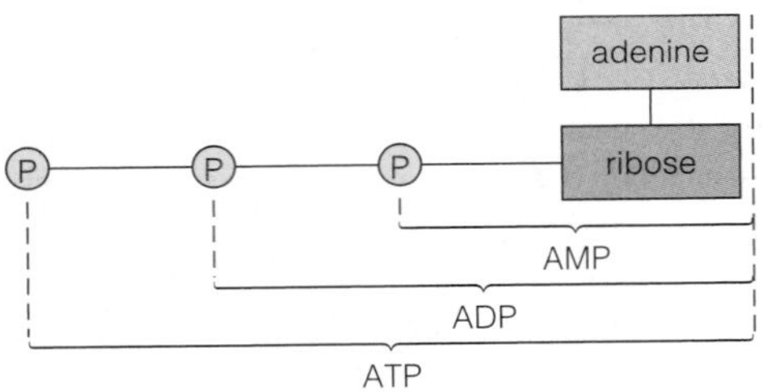
adenine
ribose
P
AMP
ADP
ATP
c

Figure 6.8 (**a**) Three-dimensional model of an ATP molecule showing its component atoms. (**b**) Structural formula for ATP. (**c**) Molecular relationship between ATP and two structurally similar nucleotides. When ATP gives up one of its phosphate groups, adenosine diphosphate (ADP) forms. Two phosphate-group transfers leave adenosine monophosphate (AMP).

Why are those bonds so readily broken? After all, in itself, a covalent bond is a stable interaction between atoms, and this same type of bond holds all the other component parts of ATP together. However, the three phosphate groups are clustered together. By studying Figure 6.8, you see that the clustering puts a number of negative charges in proximity. Remember, like charges repel each other. In this case, the repulsions destabilize the molecule's tail.

Getting rid of a phosphate group helps stabilize the molecule and results in more stable products of much lower energy. The products are adenosine diphosphate, or **ADP**, and a free inorganic phosphate atom, which is abbreviated $\mathbf{P_i}$ (Figure 6.8*c*). In other words, hydrolysis of ATP is an energetically favorable reaction, one that releases a large quantity of usable energy. This is the energy that can drive endergonic reactions. It can put glucose, enzymes, and other molecules in an activated state, meaning they become primed to react.

Any transfer of a phosphate group to a molecule is a **phosphorylation**. Phosphorylation is at the heart of many diverse cell activities, including photosynthesis, aerobic respiration, contraction, and active transport.

Renewing Supplies of ATP

ATP is the main energy carrier for so many metabolic reactions, you might speculate that cells must have a way to renew it, and you would be right. After ATP gives up a phosphate group to become ADP, enzymes can use available energy to attach phosphate to ADP, as shown in Figure 6.9.

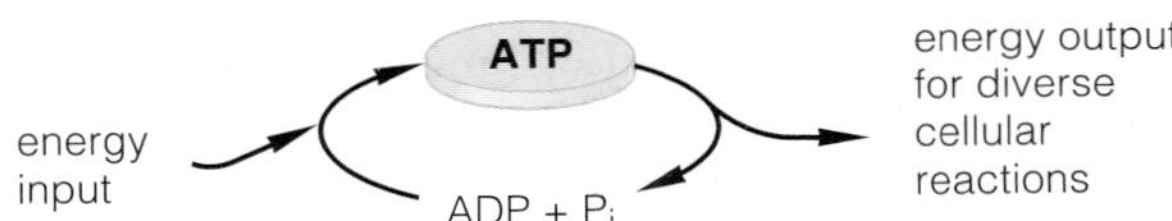
ATP
energy input
ADP + P_i
energy output for diverse cellular reactions

Figure 6.9 The ATP/ADP cycle, by which an ATP molecule that gives up a phosphate group is regenerated.

Regenerating ATP this way is called the **ATP/ADP cycle**. It is an important aspect of metabolism.

Phosphorylation is the transfer of a phosphate group from one molecule, such as ATP, to another molecule.

The transfer releases enough usable energy to activate an acceptor molecule, meaning the molecule becomes primed to enter a reaction.

Because the covalent bonds between its phosphate groups lend themselves to hydrolysis by so many enzymes, ATP is the main renewable energy carrier in cells.

Further reading: Student Guide to InfoTrac on web site —

6.4 A LOOK AT TYPICAL ELECTRON TRANSFERS

Recall, from Section 6.2, that many cellular activities involve electron transfers from one substance to another. An electron transfer is also called an **oxidation–reduction reaction**. "Oxidation" simply refers to the removal of electrons from substances. "Reduction" refers to the addition of electrons to substances.

As a molecule is being reduced, it often acquires a hydrogen ion (H^+) at the same time. The ions, which are abundant in fluids bathing cell membranes, are attracted to the opposite charge of the electrons and go along for the ride. Such reductions are known as hydrogenations. Every saturated fat is fully hydrogenated. When its fatty acid tails were being synthesized, electron transfers caused two hydrogen ions to become attached to every carbon atom in their backbone.

Often electron transfers occur singly, as when ATP phosphorylates a protein dealing with active transport and thereby primes it for reaction. Electron transfers also proceed one after another, in a series of tiny steps.

For example, this happens when cells are running low on energy and oxidize glucose, fatty acids, and other organic compounds. Remember, a molecule such as glucose is energy-rich yet unstable, compared with the most stable products (CO_2 and H_2O) of its complete breakdown. Atmospheric oxygen is so abundant that CO_2 is the most stable form of carbon. H_2O the most stable form of hydrogen. Cells extract the most energy from glucose when its carbon and hydrogen atoms combine with oxygen in air.

The more stable forms cannot be reached all at once in cells. Imagine throwing some glucose into a wood-burning fire. The carbon atoms and hydrogen atoms of the glucose molecules would very quickly let go of one another and combine with oxygen in air—but all of the released energy would be lost as heat. Compared to the wood-burning fire, cells are able to release energy from glucose far more efficiently. How? They strip electrons from it and send them on through transport systems built into cell membranes.

An **electron transport system** consists of enzymes, coenzymes, and other molecules organized for electron transfers at a cell membrane. As one molecule becomes oxidized, the next in line becomes reduced. Electrons entering the system are at a higher energy level than ones leaving. Think of the electrons as dropping down a staircase and releasing energy at each step (Figure 6.10). At certain steps, energy is used to do work, as when it leads to the formation of H^+ concentration and electric gradients. Such gradients are used to produce ATP, in ways described in later chapters.

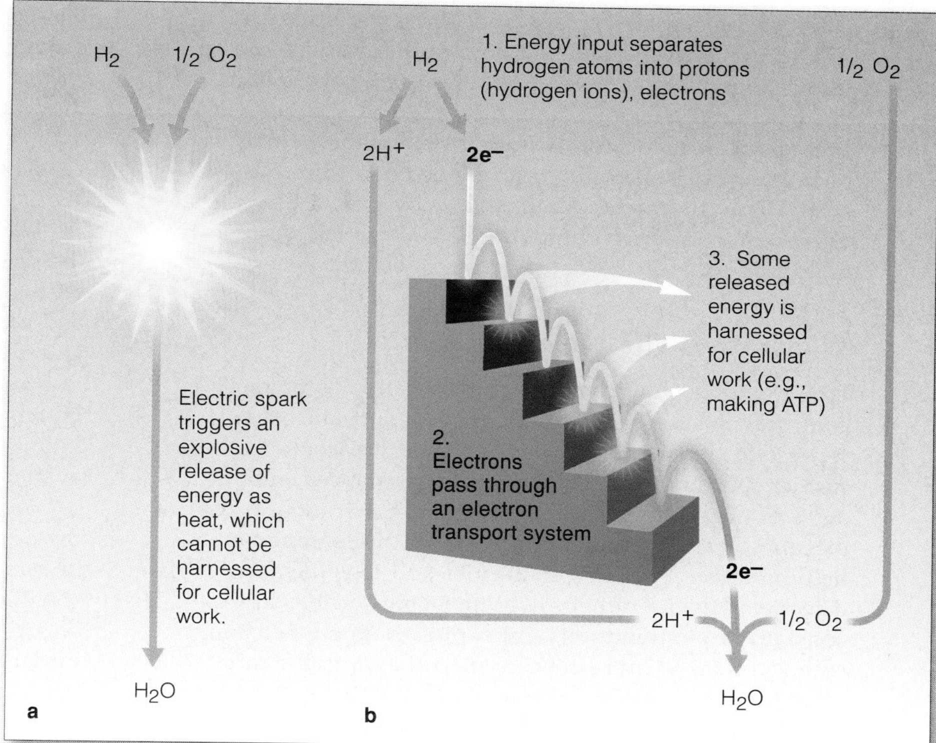
H2
1/2 O2
Electric spark triggers an explosive release of energy as heat, which cannot be harnessed for cellular work.
H2O
a
H2
1. Energy input separates hydrogen atoms into protons (hydrogen ions), electrons
1/2 O2
2H+
2e–
3. Some released energy is harnessed for cellular work (e.g., making ATP)
2. Electrons pass through an electron transport system
2e–
2H+
1/2 O2
H2O
b

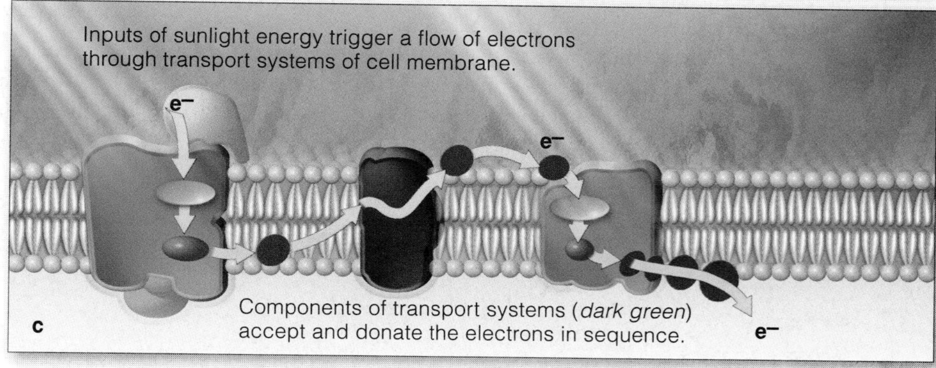
Inputs of sunlight energy trigger a flow of electrons through transport systems of cell membrane.
e–
e–
Components of transport systems (dark green) accept and donate the electrons in sequence.
e–
c

Figure 6.10 Examples of uncontrolled and controlled energy release. (**a**) Hydrogen and oxygen exposed to an electric spark react and release energy all at once. (**b**) Electron transport systems allow the same reaction to occur in small steps that let cells harness some of the released energy. (**c**) Electron-donating complexes and electron transport systems of a chloroplast's thylakoid membrane. The transport systems accept electrons that were excited to a higher energy level by sunlight.

Many electron transfers, or oxidation–reduction reactions, occur singly. Also, when cells degrade organic compounds as fuel, electron transfers proceed as a series of small steps to make more efficient use of the energy being released.

6.5 THE DIRECTIONAL NATURE OF METABOLISM

Later chapters will explain how energy inputs drive ATP formation, especially in photosynthesis and aerobic respiration. Photosynthesis is an example of **anabolism**, the metabolic reactions by which the cell synthesizes energy-rich organic compounds from smaller, energy-poor precursors. Aerobic respiration is an example of **catabolism**, the metabolic reactions by which the cell degrades energy-rich compounds into smaller, simpler molecules to release usable energy for cellular work.

Which Way Will a Reaction Run?

From the preceding chapter, you have an idea that cells control their internal concentrations of substances with respect to the surroundings, and that eukaryotic cells further control the concentrations of substances on both sides of organelle membranes. At any time, thousands of concentration gradients across cell membranes are helping to drive specific molecules and ions in specific directions. Even on their own, molecules or ions are in constant random motion, which puts them on collision courses. But the more concentrated they are, the more often they collide. Energy associated with the collisions might be enough to cause a **chemical reaction**—that is, to make a molecule combine with something else, split into smaller parts, or change its shape.

Nearly all chemical reactions in cells are reversible. In other words, they might start out in the "forward" direction, from starting substances to products. But they also can run in "reverse," with the products being converted back to starting substances. When you see a chemical equation with opposing arrows, this signifies the reaction is reversible. Each arrow means *yields*:

$$\underset{\text{STARTING SUBSTANCES}}{A + B} \rightleftharpoons \underset{\text{PRODUCT}}{C}$$

Which way such a reaction runs depends partly on the energy content of the participants. It also depends on the reactant-to-product ratio. When the energy level and concentration of reactant molecules is high, this is an energetically favorable state, and the reaction tends to proceed spontaneously and strongly in the forward direction. However, when the product concentration is high enough, more molecules or ions of product are available to revert spontaneously to reactants.

Any reversible reaction tends to run spontaneously toward **chemical equilibrium**, the time at which it will be running at about the same pace in both directions (Figure 6.11). Almost always, the *amounts* of the reactant and product molecules are not the same at that time. Picture a party with just as many people drifting in as drifting out of two rooms. The number in each room stays the same overall—say, thirty in one and ten in the other—even as the mix of people in each room changes.

Each reaction has a characteristic ratio of reactant to product molecules at equilibrium. For instance, glucose–1–phosphate can form from glucose–6–phosphate, which is nineteen times higher in energy. (As Figure 6.12 shows, both are "glucose," but their phosphate group is attached to a different carbon atom.) Depending on the reactant-to-product ratio at the outset, this reaction runs in the forward or the reverse direction. In time it will run both ways at the same rate, but it will do so only when there are nineteen glucose–6–phosphate molecules for each glucose–1–phosphate molecule. For this reaction, the equilibrium ratio is 19:1.

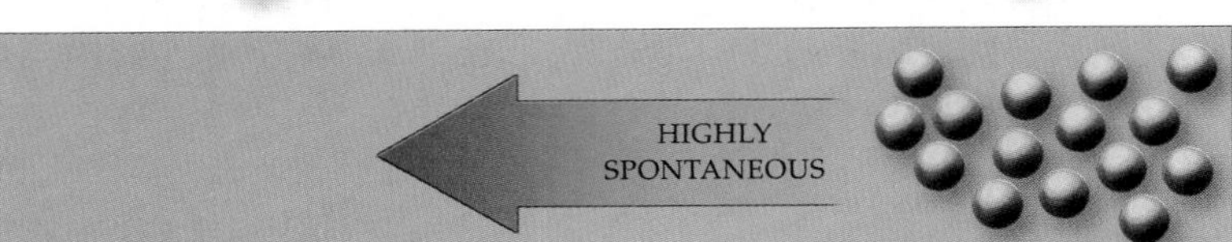
RELATIVE CONCENTRATION OF REACTANT
RELATIVE CONCENTRATION OF PRODUCT
HIGHLY SPONTANEOUS
EQUILIBRIUM
HIGHLY SPONTANEOUS

Figure 6.11 Chemical equilibrium. When the concentration of reactant molecules is high, a reaction runs most strongly in the forward direction (to products). When the concentration of product molecules is high, it runs most strongly in reverse. At equilibrium, the rates of the forward and reverse reactions are the same.

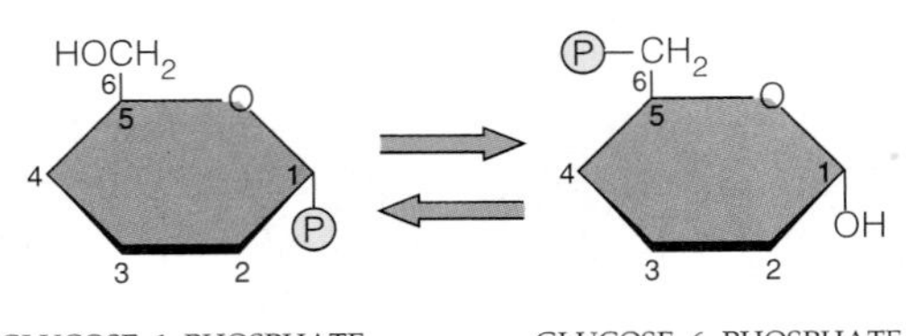
HOCH2
P–CH2
OH
GLUCOSE–1–PHOSPHATE
GLUCOSE–6–PHOSPHATE

Figure 6.12 A reversible reaction. Glucose is primed to enter reactions when a phosphate group becomes attached to it. When there is a high concentration of glucose–1–phosphate, the reaction tends to run in the forward direction. With a high glucose–6–phosphate concentration, it runs in reverse. (The 1 and 6 of these names simply identify which particular carbon atom of the glucose ring has a phosphate group attached to it.)

Further reading: Student Guide to InfoTrac on web site

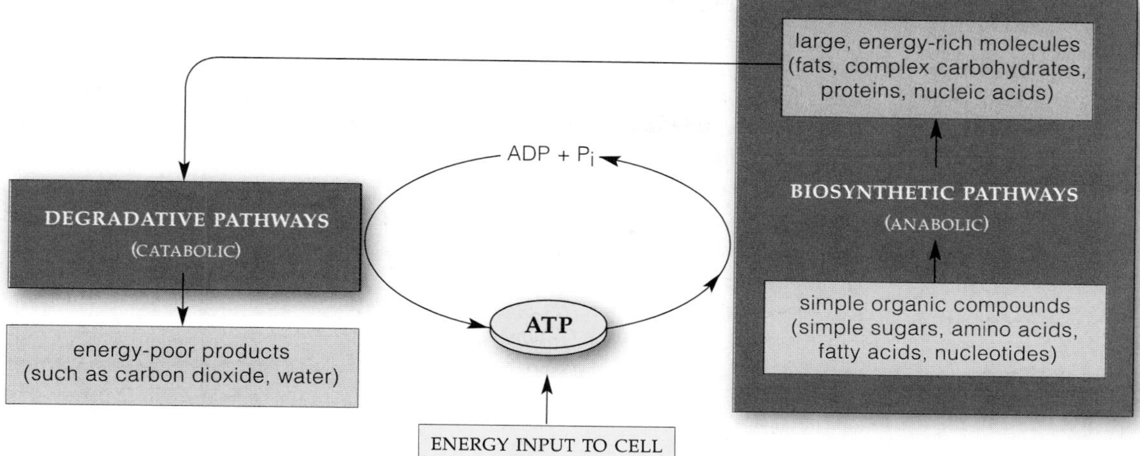
large, energy-rich molecules (fats, complex carbohydrates, proteins, nucleic acids)
BIOSYNTHETIC PATHWAYS (ANABOLIC)
simple organic compounds (simple sugars, amino acids, fatty acids, nucleotides)
ADP + Pi
ATP
DEGRADATIVE PATHWAYS (CATABOLIC)
energy-poor products (such as carbon dioxide, water)
ENERGY INPUT TO CELL

Figure 6.13 Energy relationship between the main biosynthetic and degradative pathways in cells. ATP forms by inputs into the cell and by catabolic pathways. ATP drives synthesis of large, energy-rich biological molecules from simple precursors by anabolic pathways.

No Vanishing Atoms at the End of the Run

When metabolic reactions run in either the forward or reverse direction, they rearrange atoms, but they never destroy them. By the **law of conservation of mass**, the total mass of all substances entering a reaction equals the total mass of all the products. When you study any chemical equation, count up the individual atoms of each reactant and product molecule. There should be as many atoms of each element to the right of the arrow as there are to the left, even though they are combined in different forms. When you write out equations for metabolic reactions, they must balance this way.

Metabolic Pathways

Energy inputs drive reactions that involve thousands of substances within the confines of a cell. Most of the reactions occur in orderly, enzyme-mediated sequences called **metabolic pathways**.

In the *biosynthetic* pathways, small molecules are assembled into molecules having higher bond energies, such as complex carbohydrates, lipids, and proteins. Such pathways require energy inputs (Figure 6.13). In *degradative* pathways, large molecules are broken down to products with lower bond energies. Such pathways yield energy. Photosynthesis is the main biosynthetic pathway in the world of life. Aerobic respiration is the main degradative pathway.

The participants of metabolic pathways go by these names: **Substrates** are substances that enter a reaction. They are also called reactants or precursors. Between the start and conclusion of a pathway, any substance that forms is an **intermediate**. Those remaining at the end of a reaction or a pathway are **end products**. ATP and a few other compounds are **energy carriers**; they activate enzymes and other molecules with phosphate-group transfers. ATP is the coupling agent between the main biosynthetic and degradative pathways. Most **enzymes** are proteins that speed specific reactions. (A few RNAs also display enzyme activity.) **Cofactors** are coenzymes (NAD^+ and some other organic compounds) and metal ions. They either assist enzymes or pick up electrons, atoms, or functional groups from one reaction site and taxi them to a different site. **Transport proteins**, recall, help substances cross membranes in controlled ways. Controls over these proteins help adjust concentrations on both sides of a cell membrane and thereby influence metabolic reactions.

Many pathways advance step by step in a straight line from substrates to end products. Many others are cyclic; the steps proceed in a circle, with end products becoming reactants (Figure 6.14). As you will see soon enough, intermediates or end products of one pathway also can enter different metabolic pathways.

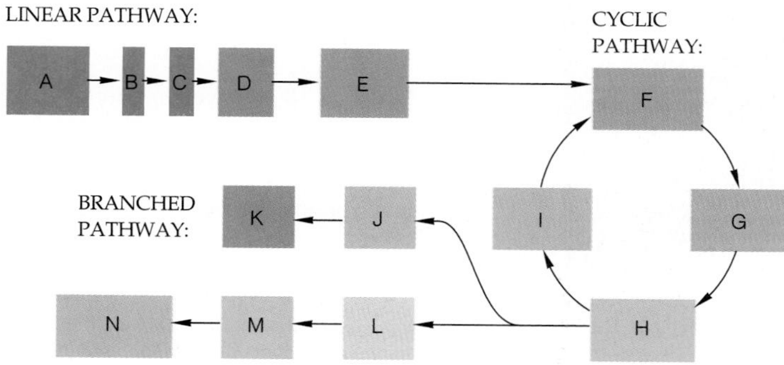
LINEAR PATHWAY:
A
B
C
D
E
CYCLIC PATHWAY:
F
G
H
I
BRANCHED PATHWAY:
K
J
N
M
L

Figure 6.14 Types of reaction sequences of metabolic pathways.

A cell can simultaneously increase, decrease, and maintain the concentrations of thousands of different substances by coordinating thousands of metabolic reactions.

The coupling of energy-releasing reactions with energy-requiring reactions, as by ATP, is central to metabolism. Metabolic pathways are enzyme-mediated reaction sequences from substrates and intermediates to end products.

6.6 ENZYME STRUCTURE AND FUNCTION

Without enzymes, the dynamically steady state called "you" would quickly cease to exist. Reactions simply wouldn't proceed fast enough for your body to process food, build and tear down hemoglobin and other vital molecules, send signals to and from brain cells, make muscles contract, and do everything else to stay alive.

To see how enzymes work, start with this concept: *Cells control their internal concentrations of substances with respect to the surroundings, and eukaryotic cells also control the concentrations across membranes of organelles.* Remember, molecules or ions of any substance are in constant, random motion that puts them on collision courses. The more concentrated they are, the more often they collide. And energy associated with the collisions might be enough to cause a metabolic reaction—that is, to make a molecule combine with something else, split into smaller parts, or change its shape.

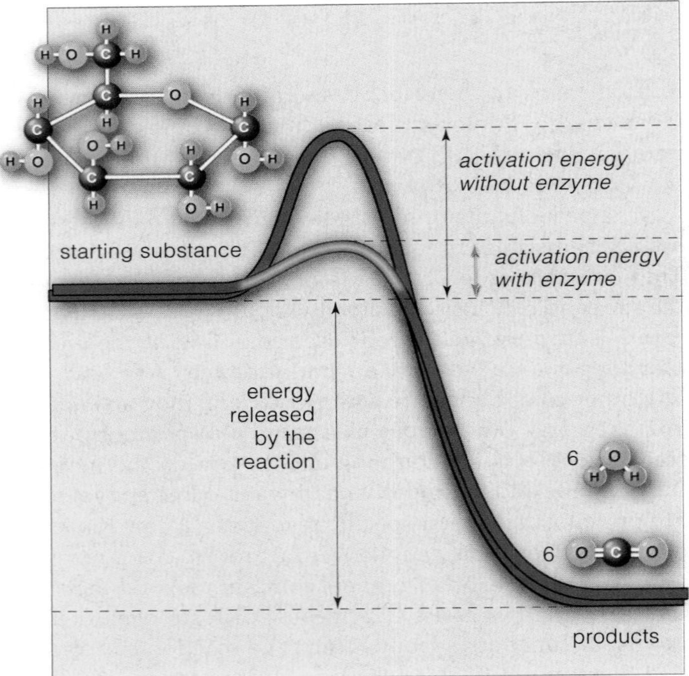

starting substance
activation energy without enzyme
activation energy with enzyme
energy released by the reaction
6
6
products

Four Features of Enzymes

By definition, enzymes are catalytic molecules; *they speed the rate at which reactions approach equilibrium*. Again, nearly all enzymes are proteins. All share four features. First, enzymes do not make anything happen that could not happen on its own, but they usually make it happen hundreds to millions of times faster. Second, reactions do not permanently alter or use up enzyme molecules; the same enzyme may act repeatedly. Third, the same type of enzyme usually works for the forward and the reverse directions of a reaction. Fourth, each type of enzyme is very picky about its substrates. Its substrates are specific substances that it can chemically recognize, bind, and modify in certain ways. For instance, thrombin is one of the enzymes necessary to clot blood. It only recognizes a side-by-side arrangement of arginine and glycine (two amino acids) in a protein molecule. When it does so, it cleaves the peptide bond between them.

Enzyme–Substrate Interactions

Take a look at Figure 6.15. A metabolic reaction occurs when participating molecules collide—provided they collide with some minimum amount of energy called the **activation energy**. Collision can be spontaneous or enzymes can promote it; this makes no difference. The activation energy is like a hill, an energy barrier, that must be surmounted *one way or another* before a reaction will proceed.

Enzymes make the energy barrier smaller, so to speak. How? Every enzyme has one

ENZYME!
b

Figure 6.15 Activation energy. (**a**) Before reactants can enter a metabolic reaction, an energy input must activate them. Only then can they spontaneously proceed to end products. (**b**) An enzyme enhances the rate of a specific reaction by lowering the amount of activation energy required to boost reactants to the transition state. Going back to the analogy in Figure 6.5, they lower the hill (energy barrier).

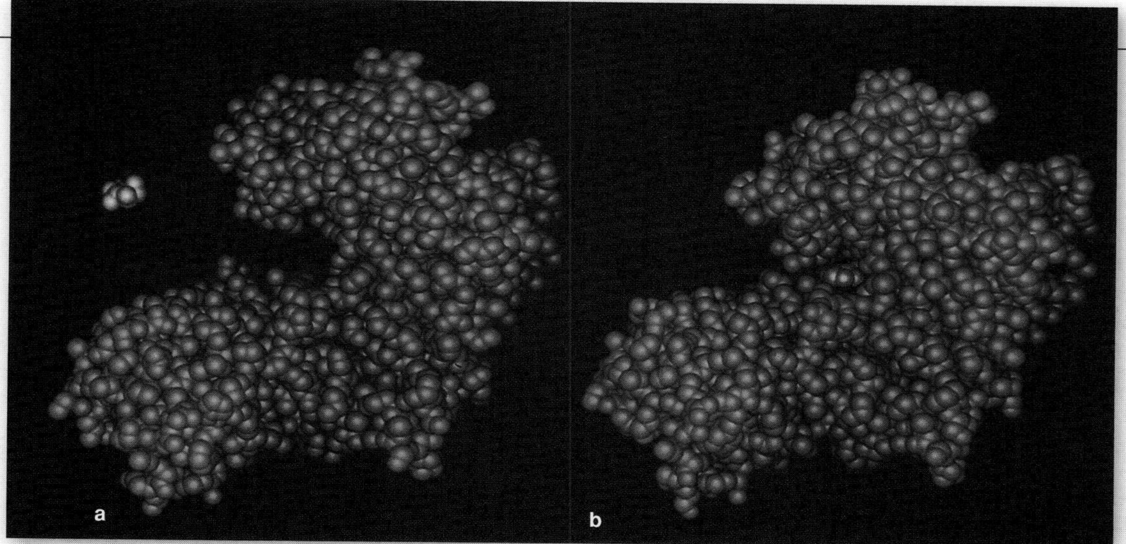

a
b
two substrate molecules
substrates contacting active site of enzyme
active site
TRANSITION STATE
(tightest binding but least stable)
end product
enzyme unchanged by the reaction
c

Figure 6.16 Model of the enzyme hexokinase. Hexokinase catalyzes the phosphorylation of glucose. (**a**) A glucose molecule (color-coded *red*) is heading toward the active site, a cleft in the enzyme (*green*). (**b**) When the glucose molecule makes contact with the site, parts of the enzyme briefly close in around it and prod the molecule to enter into the reaction.

(**c**) Induced-fit model of enzyme–substrate interactions. Only when the substrate is bound in place is an enzyme's active site complementary to it. The fit is most precise during the transition state of a reaction. An enzyme–substrate complex is short-lived, for the attractive forces holding it together are usually weak.

or more **active sites**. At these crevices in the molecule's surface, the enzyme interacts with its substrates and catalyzes a reaction. Figure 6.16 shows the active site of one enzyme, which activates glucose by catalyzing the attachment of a phosphate group to it.

According to Daniel Koshland's **induced-fit model**, a surface region of each substrate has chemical groups that are almost but not quite complementary to chemical groups in an active site. When substrates first settle into the site, the contact strains some of their bonds. Strained bonds are easier to break, so they promote the formation of new bonds (in the products). Also in the active site, interactions among charged or polar groups often shift the electric charge in substrates. And that redistribution primes substrates for conversion to an activated state.

When substrates fit most precisely in the active site of an enzyme, they are in an activated, *transition* state and will now react (Figure 6.16*c*). And so the reaction must proceed, just as NASA's Rover must roll down the Martian hill if something pushes it over the crest (Figure 6.5).

What induces the transition state or gets substrates over the energy barrier once that state is reached? The following are among the mechanisms involved:

1. *Helping substrates get together*. Substrate molecules rarely collide if their concentrations are low. Binding at an active site is like a localized boost in concentration. The boost increases the rate by 10,000 to 10,000,000 times, depending on the particular reaction.

2. *Orienting substrates in positions favoring reaction*. On their own, substrates collide from random directions. By contrast, weak but extensive bonding at an active site puts reactive chemical groups on precise collision courses much more frequently.

3. *Promoting acid–base reactions.* In many active sites, acidic or basic side groups of amino acids are poised to donate or accept hydrogen atoms from substrates. The loss or addition destabilizes covalent bonds in a substrate and makes them easier to break. Hydrolysis works this way.

4. *Shutting out water.* Some active sites bind substrates so tightly that some or all of the water molecules that bathed the site are shut out. A nonpolar environment lowers the activation energy for certain reactions, such as the attachment of a carboxyl group ($-COO^-$) to a molecule, by as much as 500,000 times.

Depending on the enzyme, such mechanisms work alone or in combination with one another to bring about the transition state.

Enzymes catalyze (speed) the rate at which specific reactions reach equilibrium. They do so by lowering the amount of activation energy necessary to make substrates react.

Enzymes change the rate, not the outcome, of a reaction. They only act on specific substrates. And they may catalyze the same reaction repeatedly, as long as substrates are available.

6.7 FACTORS INFLUENCING ENZYME ACTIVITY

You probably don't get much done when you feel too hot or cold or out of sorts because you ate too many sour plums or salty potato chips. When the cupboard is bare, you focus on food. Maybe you call a friend to go shopping with you, and if you drive too fast to the grocery store, police tend to slow you down. In such respects, you have a lot in common with the enzymes. They, too, respond to shifts in temperature, pH, and salinity, and to the relative abundances of particular substances. Many even engage helpers for specific tasks. And all normal enzymes respond to metabolic police.

Enzymes and the Environment

Temperature, recall, is a measure of molecular motion. You may think increases in temperature must increase the rate of enzyme-mediated reactions by making the substrates collide more frequently with active sites. That is so, but only until some point on the temperature scale. Past that point—which differs among enzymes—the increased molecular motion disrupts weak bonds that hold the enzyme in its three-dimensional shape. Substrates can no longer bind to the active site, so the reaction rate declines sharply, as in Figure 6.17*a*.

Expose an organism to temperatures that are far higher than it normally encounters. Its enzymes will change in shape and thereby throw metabolic activities into turmoil. This is what happens when sick people develop dangerously high fevers. They usually die when their internal temperature reaches 44°C (112°F).

Similarly, pH values that rise or sink beyond each enzyme's range of tolerance disrupt enzyme structure and function (Figure 6.18). Most enzymes work best when the pH is between 6 and 8. For example, trypsin is active in a mammal's small intestine, where the pH is 8 or so. Pepsin, a protein-digesting enzyme, is one of the exceptions. It functions in gastric fluid, a highly acidic fluid (pH of about 1–2) that denatures most enzymes.

Enzyme activity also will suffer if the environment gets far saltier than is normally encountered. Extremely high ion concentrations disrupt interactions that help hold most enzymes in their three-dimensional shapes.

How Is Enzyme Action Controlled?

Each cell controls its enzyme activity. By coordinating control mechanisms, it maintains, lowers, or raises the concentrations of substances. Controls that adjust how fast enzyme molecules are synthesized affect how many are available for a metabolic pathway. Other controls boost or slow the action of enzyme molecules that were synthesized earlier. For example, enzymes are activated or inhibited through **allosteric control** when a specific substance combines with them at a binding site *other than* the active site. (*Allo-* means different, *steric* means structure, or state.) Figure 6.19 shows two models of the binding, which is reversible.

Picture a bacterium, busily synthesizing tryptophan (and other amino acids) used to construct its proteins. After a bit, protein synthesis slows, so tryptophan is no longer required. But the tryptophan pathway is in full swing. The concentration of its end product (tryptophan) continues to rise. Now **feedback inhibition** kicks in: a cellular change, caused by a specific activity, *shuts down*

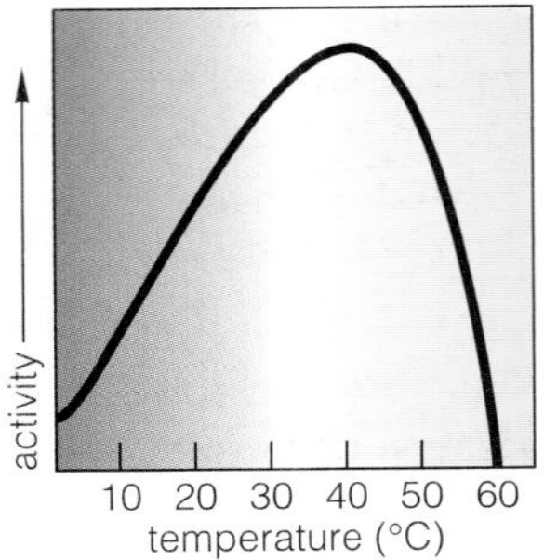
activity
10 20 30 40 50 60
temperature (°C)
a

b

Figure 6.17 (**a**) Example of how temperatures outside the range of tolerance for one enzyme influence its activity. (**b**) Siamese cats show observable effects of such changes. Fur on the ears and paws has more of a dark brown pigment, melanin, than the rest of the body does. A heat-sensitive enzyme controlling melanin production is less active in warmer parts of the body, which end up with lighter fur.

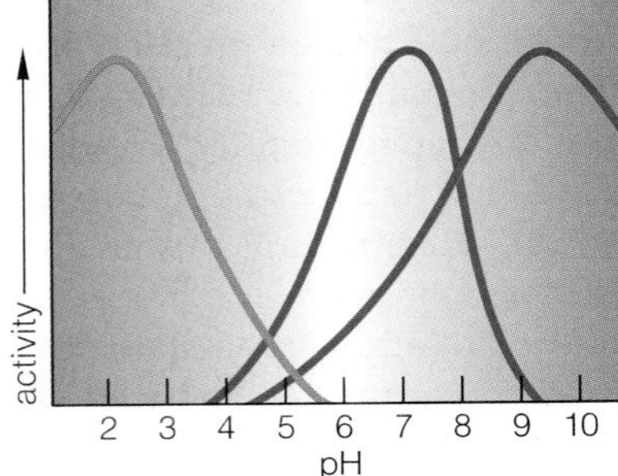
activity
2 3 4 5 6 7 8 9 10
pH

Figure 6.18 Diagram comparing changes in activity of three kinds of enzymes at different pH values. One of the enzymes functions best in neutral solutions, as indicated by the *brown* line. (Remember, 7 is neutrality on the pH scale.) Another enzyme (*red* line) functions best in basic solutions, and another (*purple* line) in acidic solutions.

Further reading: Student Guide to InfoTrac on web site —

the activity that brought it about. In this case, a feedback loop starts and ends with a specific allosteric enzyme in the pathway. When tryptophan molecules accumulate, the unused ones bind with the allosteric site. This shuts down the enzyme and blocks the pathway.

By contrast, if tryptophan molecules are scarce when the demand for them increases, then the enzyme will remain free of inhibition, so the production of tryptophan will increase. Such feedback loops quickly adjust the concentrations of many substances (Figure 6.20).

In humans and other multicelled organisms, control of enzyme activity is just amazing. Cells not only work to keep themselves alive, they work with other cells in ways that can benefit the whole body! As an example, this vast enterprise relies on hormones, a type of signaling molecule. Specialized cells release hormones into the surrounding tissue. Any cell having receptors for a given hormone responds to it, then its program for building a protein or some other activity changes. The hormone trips cell controls into action—and the activities of specific enzymes change.

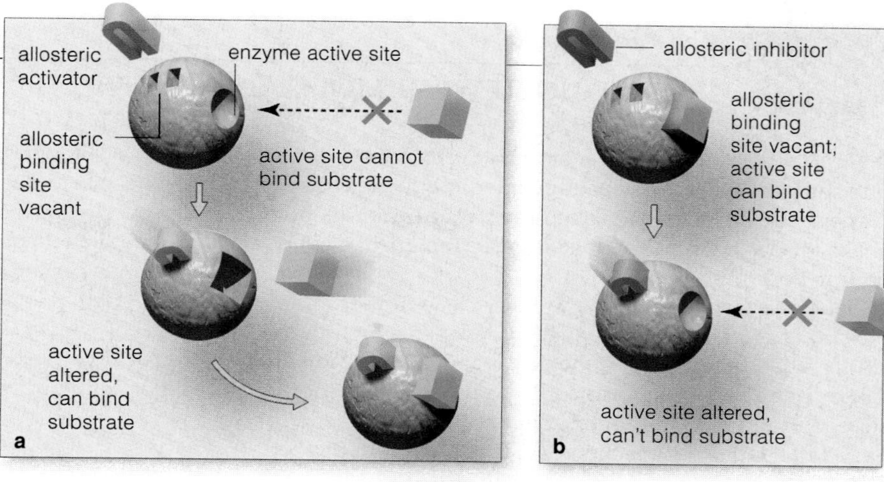
allosteric activator
enzyme active site
allosteric binding site vacant
active site cannot bind substrate
active site altered, can bind substrate
a
allosteric inhibitor
allosteric binding site vacant; active site can bind substrate
active site altered, can't bind substrate
b

Figure 6.19 Example of allosteric control of enzymes. (**a**) Binding of an activator protein to a vacant allosteric site causes a change in the active site's shape that allows it to bind substrate. (**b**) Binding of an inhibitor protein to a vacant allosteric site causes a change in the active site's shape that shuts out substrate.

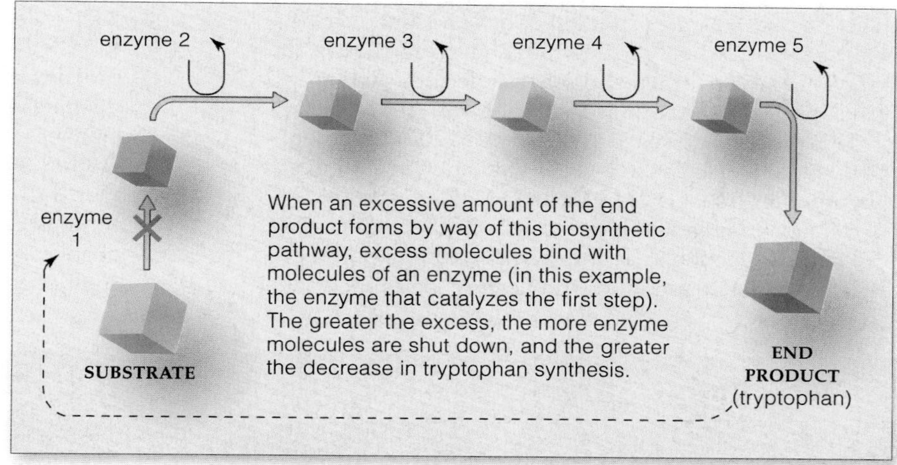
enzyme 2
enzyme 3
enzyme 4
enzyme 5
enzyme 1
SUBSTRATE
When an excessive amount of the end product forms by way of this biosynthetic pathway, excess molecules bind with molecules of an enzyme (in this example, the enzyme that catalyzes the first step). The greater the excess, the more enzyme molecules are shut down, and the greater the decrease in tryptophan synthesis.
END PRODUCT (tryptophan)

Figure 6.20 Example of feedback inhibition of a metabolic pathway. Five kinds of enzymes act in sequence to convert a substrate to the end product, tryptophan. When end product accumulates, some of the excess molecules bind to molecules of the first enzyme and block the entire pathway.

Enzyme Helpers

In many metabolic reactions, enzymes enormously hasten the transfer of one or more electrons, atoms, or functional groups from one substrate to another. Cofactors either assist the reactions or briefly act as the transferring agents. Cofactors include complex organic compounds called **coenzymes** as well as metal ions that associate with the enzyme molecule.

NAD^+ (for nicotinamide adenine dinucleotide) and FAD (for flavin adenine dinucleotide) are examples of coenzymes that are derived from vitamins. Both accept electrons and hydrogen atoms released during many degradative reactions, such as glucose breakdown, and then transfer the electrons to other reaction sites. Being attracted to the electrons, the unbound protons (H^+) go along for the ride. When reduced—that is, after they've accepted electrons (and hydrogen)—NAD^+ and FAD are abbreviated NADH and $FADH_2$, respectively. Another coenzyme, $NADP^+$ (nicotinamide adenine dinucleotide phosphate), is a key player in photosynthesis and other synthesis reactions. Like NAD^+, it is derived from the vitamin niacin. When reduced, it is abbreviated NADPH.

Ferrous iron (Fe^{++}), one of the metal ions that serve as cofactors, is a component of cytochrome molecules. Cytochromes are proteins that show enzyme activity. They are embedded in cell membranes, including the membranes of chloroplasts and mitochondria.

Enzymes function best when the cellular environment stays within limited ranges of temperature, pH, and salinity. The actual ranges depend on the type of enzyme.

Control mechanisms govern the synthesis of new enzymes and stimulate or inhibit the activity of existing enzymes. By controlling enzymes, cells control the concentrations and kinds of substances available to them.

Some enzymes require cofactors (coenzymes and metal ions), which help catalyze reactions or transfer electrons, atoms, and functional groups from one substrate to another.

GROWING OLD WITH MOLECULAR MAYHEM

Somewhere in those slender strands of DNA in your cells are snippets of instructions for constructing two enzymes: superoxide dismutase and catalase (Figure 6.21*a*,*b*). Both kinds of enzymes may help keep you from growing old before your time.

The two enzymes help your body clean house, so to speak. Together, they use oxygen (O_2) to strip hydrogen from substrate molecules, thereby helping to neutralize many potentially toxic wastes. The same reactions occur in other organisms, ranging from bacteria to plants.

At the end of the reactions, O_2 is supposed to pick up electrons. Sometimes it picks up only one—which is not enough to complete the reaction but is enough to give the oxygen a negative charge (O_2^-).

Like other unbound molecular fragments that have the wrong number of electrons, O_2^- is a **free radical**. Free radicals can slip away from a variety of enzyme-catalyzed reactions, such as the oxidation of fats and amino acids. They can form when ionizing radiation (gamma rays and x-rays) bombards water and other molecules. And they can escape from electron transport chains.

Free radicals are so reactive, they can even become attached to molecules that usually do not take part in just any reaction. These molecules include DNA, the lipids of cell membranes, and membrane receptors that trigger cell suicide (Section 15.6). Free radicals can disrupt the structure of such molecules and destroy their function.

Enter superoxide dismutase. Under its biochemical prodding, two rogue oxygen molecules will combine with hydrogen ions. Hydrogen peroxide (H_2O_2) and O_2 are the outcome. Hydrogen peroxide is a normal by-product of certain aerobic reactions, and its accumulation can be lethal to cells.

Enter catalase. Under *its* prodding, two molecules of hydrogen peroxide react and split into ordinary water and ordinary oxygen:

$$2H_2O_2 \longrightarrow 2H_2O + O_2$$

This crucial reaction is supposed to occur before H_2O_2 can do major damage.

Thus both superoxide dismutase and catalase belong to a class of molecules called **antioxidants**, which protect the body by scavenging for free radicals. Vitamins E and C, as well as beta-carotene, serve similar protective functions.

When people age, their capacity to make functional proteins starts to falter. Among the proteins are superoxide dismutase and catalase. Cells synthesize copies of both enzymes in ever diminishing numbers, in crippled form, or both. When that happens, free radicals and hydrogen peroxide can accumulate. Like loose cannons, they careen through cells as tiny blasts at the structural integrity of proteins, DNA, membrane lipids, and other essential components.

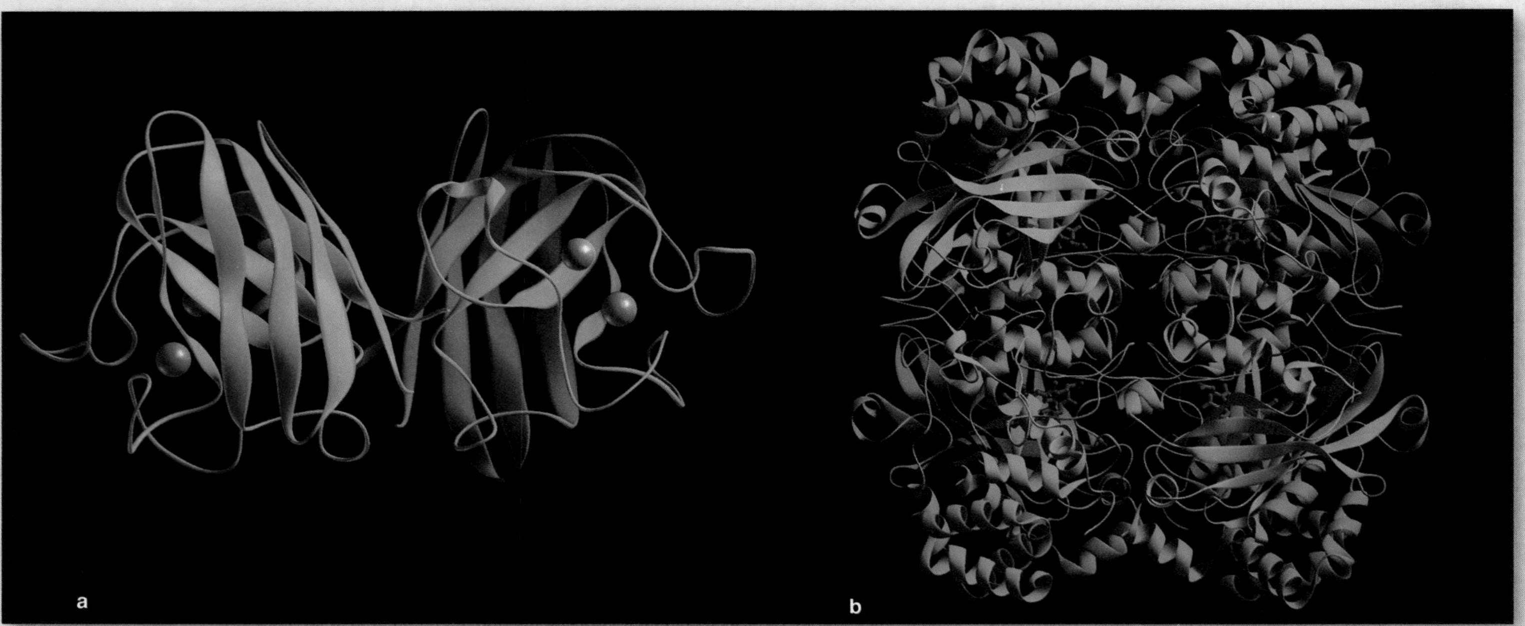
a
b

SUPEROXIDE DISMUTASE CATALASE

Figure 6.21 Computer models of the molecular structure of (**a**) superoxide dismutase and (**b**) catalase. (**c**) Adult owner of skin with a spattering of age spots, visible evidence of free radicals on the loose. At one time he, like the boy shown in (**d**), had of a good supply of smoothly functioning molecules of superoxide dismutase and catalase. Both enzymes help keep free radicals inside the body in check.

Further reading: Student Guide to InfoTrac on web site –

Cells under attack suffer or die outright. Those brown "age spots" you may have noticed on an older person's skin are evidence of assaults by free radicals. The irregular dark spots on the skin of the adult male in Figure 6.21*c* are examples. Each age spot is a mass of brownish-black pigment molecules that build up in cells whenever free radicals take over—all for the want of specific enzymes.

And so with this chapter, we have considered the kinds of activities that help keep all cells alive and functioning smoothly. At times they might have seemed remote from your interests. But these activities help define who *you* are and who you will become, age spots and all.

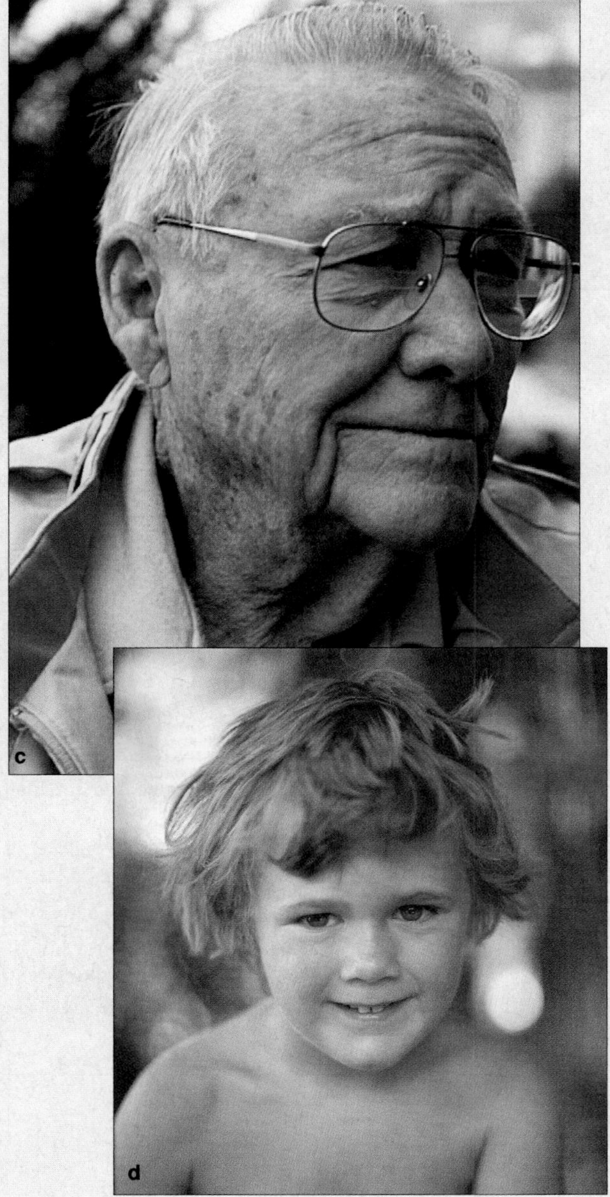

c
d

6.9 SUMMARY

1. Cells store, break down, and dispose of substances by acquiring and using energy and raw materials from outside sources. Metabolism, the sum of these energy-driven activities, underlies the survival of organisms.

2. Two laws of thermodynamics affect life. *First*, energy undergoes conversion from one form to another, but its total amount never increases or decreases as a result of conversions. Thus, the total amount of energy in the universe holds constant. *Second*, energy spontaneously flows in one direction, from usable forms to forms that are less and less usable.

a. All matter has some amount of potential energy (as measured by the capacity to do work) by virtue of its position in space and the arrangement of its parts.

b. Potential energy may be transformed into kinetic energy, the energy of motion. Mechanical movements and heat, which correspond to the degree of molecular motion, are two common forms of kinetic energy.

c. Chemical energy (the potential energy inherent in molecular bonds) is often measured in kilocalories.

3. Like all organized systems, the cell tends to become disorganized without energy. It inevitably loses some of its chemical potential energy during every metabolic reaction, mainly in the form of heat. It stays organized and alive as long as it counters its energy expenditures (outputs) with energy replacements (inputs).

4. The sun is life's primary energy source. In plants and other photosynthetic organisms, cells trap energy from the sun and convert it to chemical bond energy of organic compounds. Plants, then organisms that feed on plants and one another, use energy that was stored in these organic compounds to do cellular work.

5. Exergonic (energy out) reactions end with a net loss in energy. Endergonic (energy in) reactions end with a net gain in energy. Cells conserve energy by coupling energy-releasing reactions with energy-requiring ones.

6. ATP is the main energy carrier in every cell. It forms when a phosphate group or inorganic phosphate is attached to ADP. ATP gives up energy at many reaction sites when it phosphorylates reactants or intermediates, which often become primed to enter specific reactions.

7. ATP couples energy-releasing and energy-requiring reactions in cells.

8. Much of the cellular work leading to ATP formation involves electron transfers between substances. We call such transfers oxidation–reduction reactions.

a. Molecules that lose electrons are oxidized; those that gain electrons are reduced.

b. Hydrogen ions often are attracted to free electrons and are simultaneously transferred to molecules along with them. These are called hydrogenation reactions.

c. Electron transfers may occur singly or in a series of small steps through electron transport systems.

9. Metabolic reactions may release energy when they run toward equilibrium, or they may require energy inputs to drive them away from equilibrium.

10. Most metabolic reactions proceed strongly in the forward direction (to products) when concentrations of starting substances are high. They also run strongly in reverse (to starting substances) when concentrations of products are high.

a. When left to themselves, reversible reactions run toward chemical equilibrium, at which time they are proceeding at about the same rate in both directions.

b. Cellular mechanisms work with and against this tendency. During a given interval, reversible reactions help increase, decrease, and maintain concentrations of the thousands of different substances that are required for specific reactions.

c. Regardless of the direction of a metabolic reaction or the extent of the molecular cleavages, combinations, and rearrangements, the total number of atoms that end up in all the products will equal the total number of atoms that were present in the starting substances.

11. Metabolic pathways are orderly, stepwise sequences of enzyme-mediated reactions. Table 6.1 summarizes the main participants.

a. A substrate (or reactant) is a substance that enters a reaction or pathway. Intermediates are the substances that form between the reactants and end products.

b. By the end of a biosynthetic pathway, energy-rich organic compounds have been assembled from smaller molecules of lower energy content.

c. By the end of a degradative pathway, energy-rich molecules have been broken down to smaller ones of lower energy content.

12. Overall, the main metabolic pathways are anabolic (biosynthetic) and catabolic (degradative).

a. Cells assemble energy-rich organic compounds from smaller molecules of lower energy content in the biosynthetic pathways, such as photosynthesis.

b. Cells break down energy-rich molecules to small ones of lower energy content in degradative pathways, such as aerobic respiration.

c. ATP is the main coupling agent between energy-releasing and energy-requiring pathways.

13. Enzymes are catalysts; they greatly enhance the rate of a reaction that involves specific substrates but do not change the outcome. Nearly all enzymes are proteins, although some RNAs also show catalytic activity.

a. Enzymes lower the activation energy required to start a reaction. They bind substrates at an active site, strain its bonds, and make the bonds easier to break.

b. Each kind of enzyme functions best within limited ranges of temperature, pH, and salinity.

c. Cofactors assist enzymes in speeding a reaction, or they carry electrons, hydrogen, or functional groups stripped from substrates to other reaction sites. They include coenzymes (such as NAD^+) and metal ions.

d. Controls stimulate or inhibit enzyme activity at key steps in metabolic pathways. They help coordinate the kinds and amounts of substances available at any given time in the cell.

Table 6.1 Main Participants in Metabolic Pathways

SUBSTRATE	Substance that enters a metabolic reaction or pathway; also called a reactant
INTERMEDIATE	Substance formed between reactants and end products of a pathway
END PRODUCT	Substance remaining at end of reaction or pathway
ENZYME	Usually a protein that enhances reaction rates
COFACTOR	Coenzyme (such as NAD^+) or metal ion; assists enzymes or taxis electrons, hydrogen, or functional groups between reaction sites
ATP	Main energy carrier in cells; couples energy-releasing reactions with energy-requiring ones
TRANSPORT PROTEIN	Protein that passively assists substances across a cell membrane or actively pumps them across

Review Questions

1. State the first and second laws of thermodynamics. Does life violate the second law? *6.1*
2. Define and give a few examples of potential energy and kinetic energy. What is the name for the potential energy of molecules? *6.1*
3. Define mechanical work, chemical work, and electrical work as accomplished by cells. *6.1*
4. Give examples of a change in potential energy involved in (a) mechanical work and (b) chemical work. *6.2*
5. Make a simple diagram of the ATP molecule. Highlight which parts of an ATP molecule can be transferred to another molecule and then later replaced. *6.3*
6. What is an oxidation–reduction reaction? *6.4*
7. Define and describe four key features of enzymes. *6.6*
8. Define activation energy, then state four ways in which enzymes may lower it. *6.6*
9. Briefly describe the induced-fit model of enzyme–substrate interactions. *6.6*
10. Define feedback inhibition as it relates to the activity of an allosteric enzyme. *6.7*
11. Define free radical. How does it relate to aging? *6.8*
12. Fill in the blanks in the sketch on the facing page. *6.5*

Self-Quiz *(Answers in Appendix III)*

1. ______ is life's primary source of energy.
   a. Food   c. Sunlight
   b. Water   d. ATP

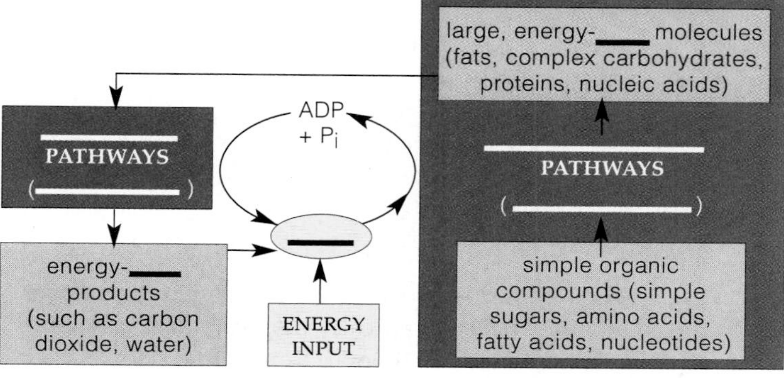
large, energy-___ molecules (fats, complex carbohydrates, proteins, nucleic acids)
ADP + P_i
PATHWAYS
(______)
PATHWAYS
(______)
energy-___ products (such as carbon dioxide, water)
ENERGY INPUT
simple organic compounds (simple sugars, amino acids, fatty acids, nucleotides)

2. If we liken chemical equilibrium to the bottom of an energy hill, then an ______ reaction is an uphill run.
   a. endergonic
   b. exergonic
   c. ATP-assisted
   d. both a and c

3. Phosphate-group transfers from ATP to another molecule are a ______ mechanism for delivering energy.
   a. rapid
   b. renewable
   c. near-universal
   d. all of the above

4. Electron transport systems involve ______ .
   a. enzymes and cofactors
   b. electron transfers
   c. cell membranes
   d. all of the above

5. Which statement is *not* correct? A metabolic pathway ______ .
   a. has an orderly sequence of reaction steps
   b. is mediated by one enzyme, which initiates the reactions
   c. may be biosynthetic or degradative, overall
   d. all of the above

6. Which is *not* true of chemical equilibrium?
   a. Product and reactant concentrations are always equal.
   b. The rates of the forward and reverse reactions are the same.
   c. There is no further net change in product and reactant concentrations.

7. Enzymes are ______ .
   a. enhancers of reaction rates
   b. influenced by temperature
   c. influenced by pH
   d. not influenced by salinity
   e. a through c
   f. all of the above

8. All enzymes incorporate a(n) ______ .
   a. active site
   b. coenzyme
   c. metal ion
   d. all of the above

9. The coenzymes NAD^+, FAD, and $NADP^+$ are ______ .
   a. cofactors
   b. allosteric enzymes
   c. metal ions
   d. both a and b

10. Match each substance with the most suitable description.

___ coenzyme or metal ion	a. reactant or substrate
___ adjusts gradients at membrane	b. enzyme
___ substance entering a reaction	c. cofactor
___ substance formed while a reaction is proceeding	d. intermediate
___ substance at end of reaction	e. end product
___ enhances reaction rate	f. energy carrier
___ mainly ATP	g. transport protein

Critical Thinking

1. When cyanide, a toxic organic compound, binds to an enzyme that is a component of electron transport systems, the outcome is *cyanide poisoning*. Binding prevents the enzyme from donating electrons to a nearby acceptor molecule in the system. What effect will this have on ATP production? From what you know of ATP function, what effect will this have on a person's health?

2. AZT, or azidothymidine, is a drug used to alleviate symptoms of *AIDS* (acquired immunodeficiency syndrome). AZT is very similar in molecular structure to thymidine, one of the DNA nucleotides. AZT also can fit into the active site of an enzyme produced by the virus that causes AIDS. Infection puts the viral enzyme and viral genetic material (RNA) into a cell. There, the RNA is used as a template (structural pattern) for joining nucleotides to form a strand of DNA. The viral enzyme takes part in the assembly reactions. Propose a model to explain how AZT might inhibit replication of the virus inside cells.

3. The bacterium *Clostridium botulinum* is an obligate anaerobe, meaning it dies quickly in the presence of oxygen. It lives in oxygen-free pockets in soil, and it can enter a metabolically inactive, resting state by forming spores. The spores may end up on the surfaces of garden vegetables. When the picked vegetables are being canned, the spores must be destroyed. If they are not destroyed, *C. botulinum* may grow and produce botulinum, a toxin. When tainted canned foods are not cooked properly, the toxin remains active and causes a type of food poisoning called *botulism*. This bacterium is not able to produce either superoxide dismutase or catalase. Develop a hypothesis to explain how the absence of these enzymes is related to its anaerobic life-style.

4. *Pyrococcus furiosus* thrives at 100°C, the boiling point of water. This species of bacterium was discovered growing in a volcanic vent in Italy. Enzymes isolated from *P. furiosus* cells do not function well below 100°C. What is it about the structure of these enzymes that allows them to remain stable and active at such high temperatures? (Hint: Review Section 3.5, which summarizes the interactions that maintain protein structure.)

5. Hydrogen peroxide is a free radical that attacks cell structure and function by reacting with and disrupting nearly all molecules it contacts. Some research indicates its cumulative effects over time may contribute to the gradual deterioration associated with aging. For example, decreases in fruit fly longevity are directly proportional to a decrease in a specific type of enzyme in a specific organelle. Reflect on the organelles introduced earlier, in Chapter 4, then identify the enzyme and the organelle.

Selected Key Terms

Readings *See also www.infotrac-college.com*

Fenn, J. *Engines, Energy, and Entropy*. 1982. New York: Freeman. Deceptively simple paperback.

7 HOW CELLS ACQUIRE ENERGY

Sunlight and Survival

Think about the last time you were hungry and craved a bit of apple, maybe, or lettuce, chicken, pizza, bread, or any other kind of food. Where did it come from? For the answer, look past the refrigerator, the market or restaurant, or even the farm. Look to individual plants—the starting point for nearly all of the food you put into your mouth. Plants use environmental sources of energy and raw materials to build glucose and all other organic compounds necessary for survival. Organic compounds, recall, are built on a framework of carbon atoms. So the questions become these:

1. *Where does the carbon come from in the first place?*

2. *Where does the energy come from to drive the synthesis of carbon-based compounds?*

The answers vary according to an organism's mode of nutrition.

Plants generally are "self-nourishing" organisms, or **autotrophs**. As their carbon source, they use carbon dioxide (CO_2), a gaseous compound present in the air and dissolved in aquatic habitats. Plants, some bacteria, and many protistans are *photo*autotrophs, which means they capture sunlight energy to drive a metabolic process called **photosynthesis**. By this process, energy from the sun is converted to chemical bond energy in ATP, then ATP gives up energy at sites where glucose and other organic compounds are synthesized. An enzyme helper, the coenzyme $NADP^+$, typically picks up electrons and hydrogen, then delivers them to those same sites.

Many other organisms are **heterotrophs**, meaning they must feed on autotrophs, one another, and organic wastes. (*Hetero-* means other, as in "being nourished by other organisms.") That is how most bacteria, many protistans, and all fungi and animals stay alive. Unlike plants, heterotrophs cannot nourish themselves with sunlight and raw materials from their environment.

How do we know such things? We didn't have a clue until observational and experimental tests began in the mid-seventeenth century. Before then, most people assumed that plants got the raw materials they needed to make food from the soil they grew in. By 1882 a few chemists had an inkling that plants use sunlight, water, and something in the air to make food. T. Englemann, a botanist, was curious: What parts of sunlight do plants favor? As he already knew, when plants and algae are photosynthesizing, they release oxygen. He also knew that, like many other organisms, certain free-living bacterial cells require oxygen for aerobic respiration. Those cells move toward places where conditions favor their activities and away from unfavorable conditions.

Englemann hypothesized: If bacterial cells require oxygen, then they will move toward places where photosynthesis is most effective. He put a strand of a green alga, *Spirogyra,* in a water droplet that contained such bacteria. He mounted the strand on a microscope slide, then used a crystal prism to break up a beam of sunlight and to cast a spectrum of colors across it. As Figure 7.1 shows, bacteria congregated mainly where violet and red light fell on the algal strand. Englemann concluded that algal cells were releasing more oxygen in the area illuminated by light of those colors—which is the most effective light for photosynthesis.

Such observations yielded insight into the process of photosynthesis. Ultimately, they also helped reveal a great pattern in nature, because nearly all organisms depend on photosynthesis—*the main pathway by which carbon and energy enter the web of life.* Once a cell makes or takes up organic compounds, it uses or stores them. *All* autotrophs and heterotrophs store energy in organic compounds, and that energy can be released by other processes. Aerobic respiration, the most common energy-releasing process, requires oxygen to run to completion. Use Figure 7.2 as your preview of the chemical links between photosynthesis and aerobic respiration—the focus of this chapter and the next.

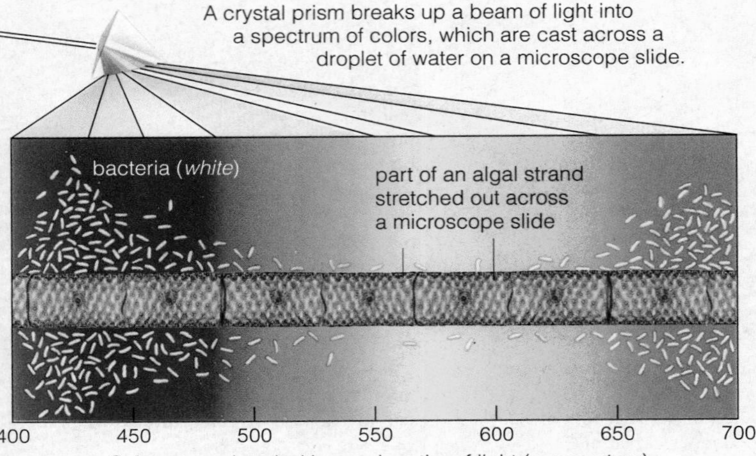

A crystal prism breaks up a beam of light into a spectrum of colors, which are cast across a droplet of water on a microscope slide.
bacteria (white)
part of an algal strand stretched out across a microscope slide
400
450
500
550
600
650
700
Colors associated with wavelengths of light (nanometers)

Figure 7.1 Results from T. Englemann's observational test that correlated portions of visible light with photosynthesis in *Spirogyra*, a strandlike green alga. Many oxygen-requiring bacterial cells moved to the colors where algal cells released the most oxygen, which is a by-product of their photosynthetic activity.

Plants capture energy from the sun, which drives photosynthesis
PHOTOSYNTHESIS
The main energy-acquiring process
1. Sunlight energy is converted to chemical bond energy in ATP. Typically, NADPH also forms.
2. ATP and NADPH are used in reactions that form glucose, other energy-rich organic compounds.
Carbon dioxide, water are required
Oxygen is released
AEROBIC RESPIRATION
The main energy-releasing process
1. Usable energy is released when cells break down glucose and other organic compounds.
2. Released energy is coupled to electron transfers that bring about the formation of many ATP molecules.
Carbon dioxide, water are released
Oxygen is required
ATP is available to drive cellular tasks

Figure 7.2 Links between photosynthesis—the main energy-requiring process in the world of life—and aerobic respiration, the main energy-releasing process.

At times, such processes might seem far removed from your daily interests. But remember this: The food that nourishes you and most other organisms cannot be produced or used without them. You will be returning to this point in later chapters. It provides perspective on many important issues, including nutrition and dieting, agriculture and human population growth, genetic engineering, as well as the impact of pollution on our sources of food—hence on our survival.

KEY CONCEPTS

1. Plants, some bacteria, and many protistans use sunlight energy, carbon dioxide, and water to produce glucose and other organic compounds, which have a backbone of carbon atoms. The metabolic process by which they accomplish this is called photosynthesis.

2. Photosynthesis is the main route by which carbon and energy enter the web of life.

3. In the first stage of photosynthesis, sunlight energy is trapped and converted to chemical bond energy in ATP molecules. Typically, water molecules are split, their electrons and hydrogen atoms are picked up by $NADP^+$ to form NADPH, and their oxygen atoms are released as a by-product.

4. In the second stage of photosynthesis, ATP delivers energy to reaction sites where glucose is synthesized. Carbon dioxide provides carbon and oxygen for the reactions, and NADPH provides the electrons and hydrogen.

5. In the photosynthetic cells of plants and in many protistans, both stages of the reactions proceed inside organelles called chloroplasts.

6. Often we summarize photosynthesis this way:

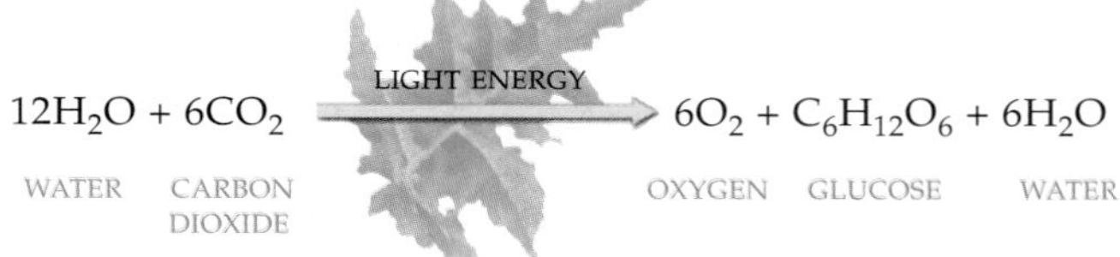
LIGHT ENERGY
WATER
CARBON DIOXIDE
OXYGEN
GLUCOSE
WATER

$$12H_2O + 6CO_2 \xrightarrow{\text{LIGHT ENERGY}} 6O_2 + C_6H_{12}O_6 + 6H_2O$$

7.1 PHOTOSYNTHESIS—AN OVERVIEW

Where the Reactions Take Place

Let's start with a look at **chloroplasts**, the organelles of photosynthesis in plants and in the protistans called algae. Chloroplasts, recall, specialize in producing and briefly storing food (Section 4.7). Figure 7.3 shows the structure and functional zones of one type. All chloroplasts have two outermost membranes that surround a largely fluid interior, the **stroma**. Still another membrane weaves through the stroma; we call it a thylakoid membrane system. In many species, parts of this system are often folded repeatedly into disk-shaped sacs, or **thylakoids**, stacked one on top of the other as in Figure 7.3*d*, *e*. Each stack of disks is a granum (plural, grana). Membranous channels interconnect the stacks.

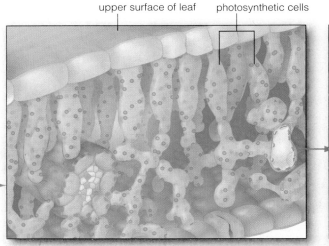

Two stages of photosynthesis—the *light-dependent* and *light-independent* reactions—proceed in the interior of a chloroplast. The first stage occurs at the thylakoid membrane system (Figure 7.3*e,f*). The interconnecting spaces within all thylakoids and channels form a single compartment in which hydrogen ions (H^+) accumulate. ATP forms when these ions flow across the membrane. The second stage of photosynthesis, the set of reactions by which sugars are assembled, occurs in the stroma.

a Part of a leaf

upper surface of leaf

photosynthetic cells

b Cutaway view of a small section from the leaf. Its upper and lower surfaces enclose many photosynthetic cells.

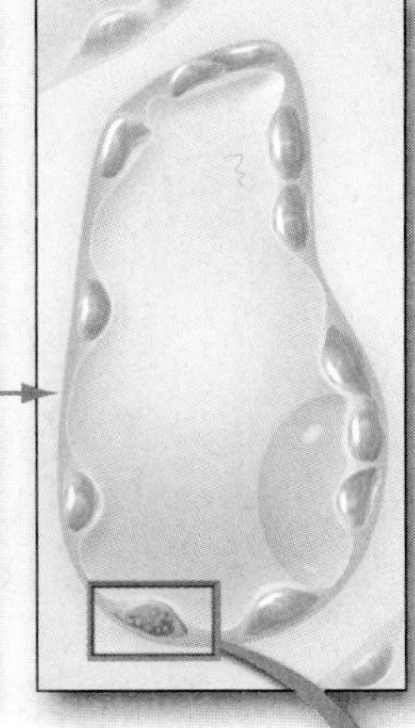

c One of the photosynthetic cells

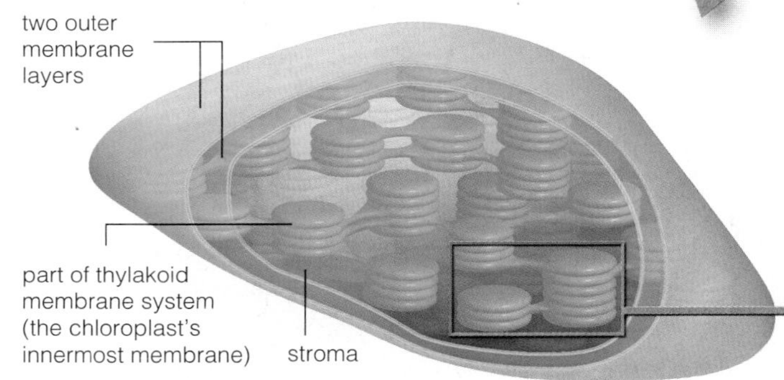
two outer membrane layers
part of thylakoid membrane system (the chloroplast's innermost membrane)
stroma

d Cutaway view of one of the chloroplasts inside the photosynthetic cell shown in (**c**).

Figure 7.3 Zooming in on sites of photosynthesis inside one of the leaves of a typical plant.

Energy and Materials for the Reactions

In the light-dependent reactions, absorption of energy from sunlight drives the formation of ATP from the chloroplast's pool of ADP and inorganic phosphate (P_i). In this way, energy from the sun becomes temporarily stored as chemical bond energy of ATP. Typically, water molecules are split and a coenzyme, $NADP^+$, picks up their electrons and hydrogen atoms to form NADPH.

In the light-independent reactions, the ATP donates energy to sites where carbon, hydrogen, and oxygen are assembled into glucose ($C_6H_{12}O_6$). NADPH delivers the electrons and hydrogen atoms. And carbon dioxide (CO_2) provides the carbon and oxygen atoms.

Overall, the reactions of photosynthesis are often summarized as a simple equation:

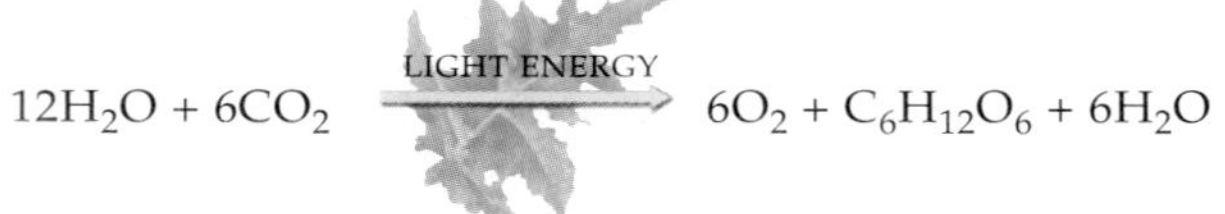
LIGHT ENERGY

$$12H_2O + 6CO_2 \xrightarrow{\text{LIGHT ENERGY}} 6O_2 + C_6H_{12}O_6 + 6H_2O$$

Remember the description of tracers in Section 2.2? By attaching radioisotopes to atoms of the two kinds of reactants given in the summary equation, researchers showed where each atom ends up in the three products:

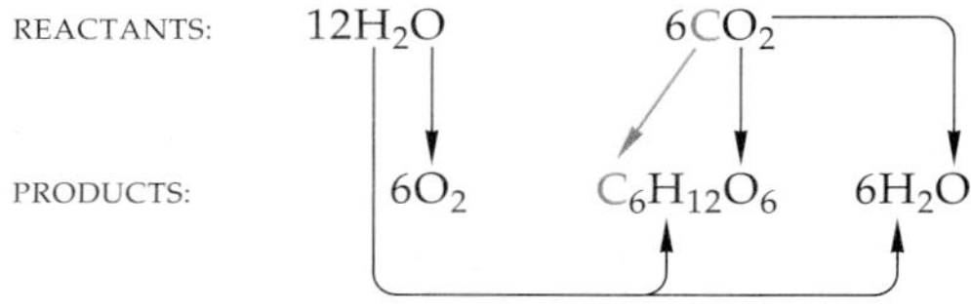
REACTANTS:
12H2O
6CO2
PRODUCTS:
6O2
C6H12O6
6H2O

This summary equation shows *glucose* as the carbon-rich end product of the reactions, to keep the chemical bookkeeping simple. But you will find very little glucose in a chloroplast. Each newly formed glucose molecule has an attached phosphate group; it is primed to react

Further reading: Student Guide to InfoTrac on web site

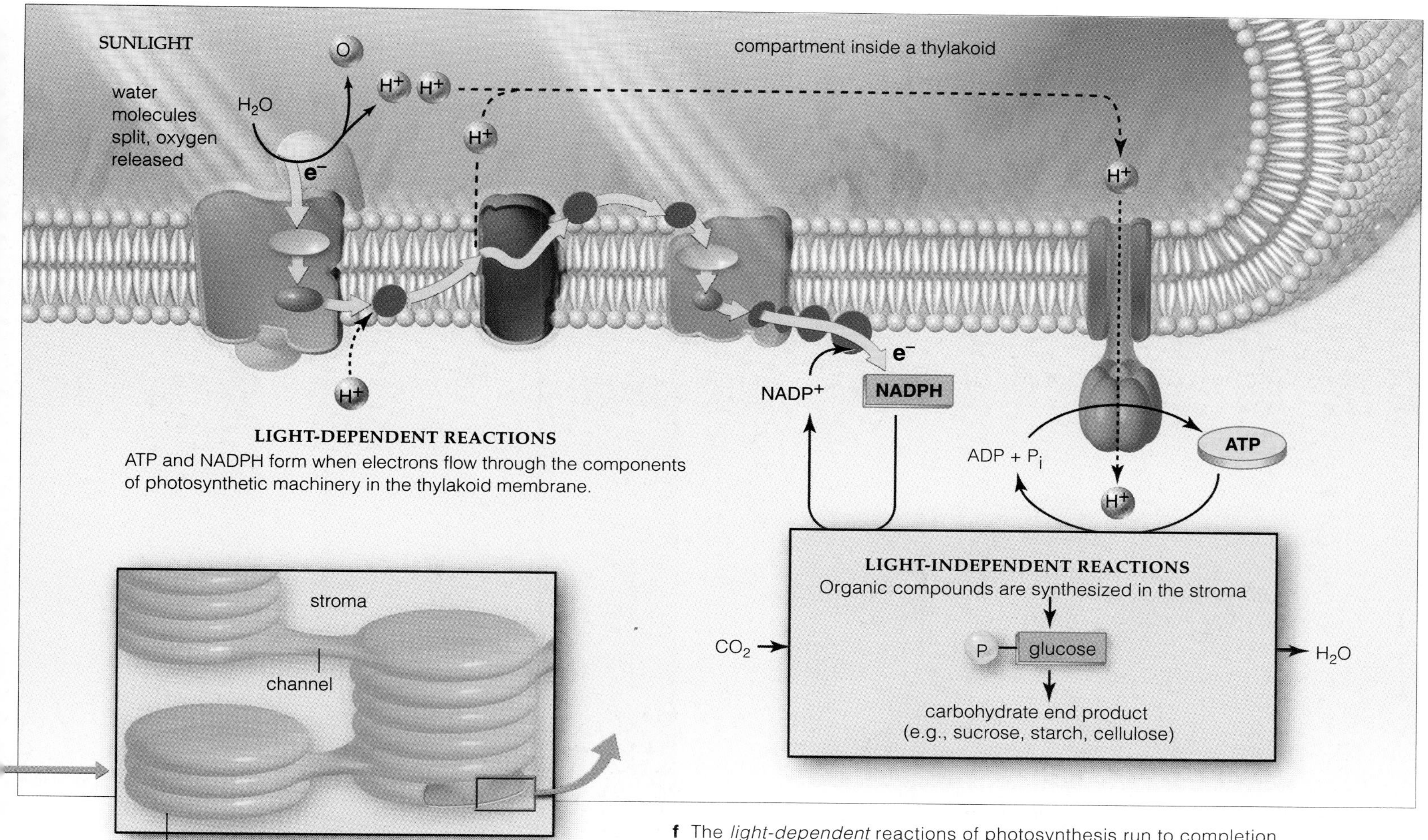

LIGHT-DEPENDENT REACTIONS
ATP and NADPH form when electrons flow through the components of photosynthetic machinery in the thylakoid membrane.

LIGHT-INDEPENDENT REACTIONS
Organic compounds are synthesized in the stroma

e One granum (a stack of disks, or thylakoids, that are part of the thylakoid membrane system)

f The *light-dependent* reactions of photosynthesis run to completion at thylakoids. The *light-independent* reactions proceed in the stroma.

with something else. Almost always, it is used at once in reactions that form sucrose, starch, and cellulose. Just keep in mind that these are the main end products of photosynthesis even though we "stop" with glucose.

The diagram below is a simplified version of Figure 7.3. Where you see it repeated in sections to follow, use it as a reminder to refer back to Figure 7.3 to reinforce your grasp of how the details fit into the big picture:

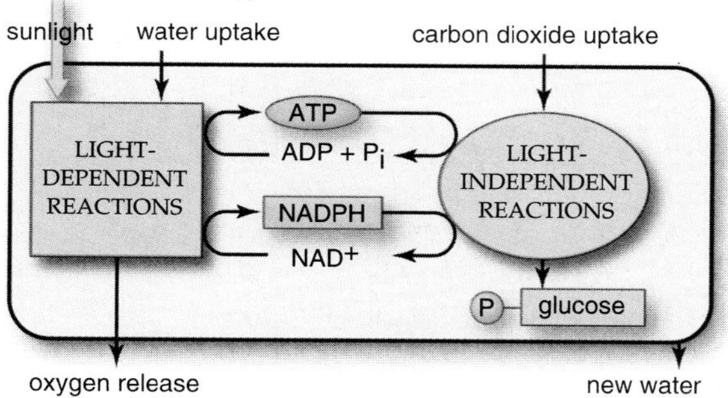

sunlight
water uptake
carbon dioxide uptake
LIGHT-DEPENDENT REACTIONS
ATP
ADP + Pi
NADPH
NAD+
LIGHT-INDEPENDENT REACTIONS
P
glucose
oxygen release
new water

We turn next to the details of photosynthesis. Before we do, think about this: Two thousand chloroplasts, lined up single file, would be no wider than a dime. Imagine all of the chloroplasts in just one corn or rice plant—each a small factory for producing sugars and starch—and you may get an inkling of the magnitude of the metabolic events required to feed you and every other organism on this planet.

Chloroplasts, the organelles of photosynthesis in plants and many protistans, specialize in food production.

In the first stage of photosynthesis, sunlight energy drives the formation of ATP and NADPH, and oxygen is released. In chloroplasts, this stage occurs at thylakoids.

The second stage occurs in the stroma, where energy from ATP drives glucose formation. For these synthesis reactions, carbon dioxide provides the carbon and oxygen atoms, and NADPH provides the electrons and hydrogen atoms.

SUNLIGHT AS AN ENERGY SOURCE

Each second, more than 2 million metric tons of the sun's mass enter thermonuclear reactions that release stupendous amounts of energy. It takes a mere eight minutes for a fraction of the released energy to reach the Earth's atmosphere, 160 million kilometers away. Each day, that fraction averages 7,000 kilocalories per square meter of the Earth as a whole. Almost a third is reflected back into space. Of the amount of sunlight energy that does reach the Earth's surface, only about 1 percent is intercepted by photoautotrophs—which are the entry point for a one-way flow of energy through nearly all webs of life (Figure 7.4*a*).

Properties of Light

Photosynthesis starts with energy from the sun that has radiated across space in undulating motion, a bit like waves crossing a sea. The horizontal distance between crests of every two successive waves is a **wavelength** (Figure 7.4*b*). There are many different wavelengths of radiant energy. The entire range of all the wavelengths represents the **electromagnetic spectrum** (Figure 7.4*c*).

Collectively, different kinds of photoautotrophs can absorb wavelengths between 380 and 750 nanometers. That is the range of *visible* light, which we humans and

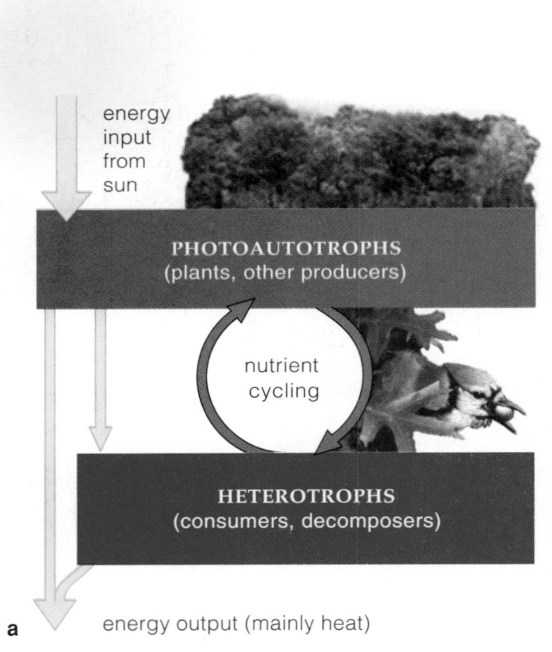
energy input from sun
PHOTOAUTOTROPHS
(plants, other producers)
nutrient cycling
HETEROTROPHS
(consumers, decomposers)
a
energy output (mainly heat)

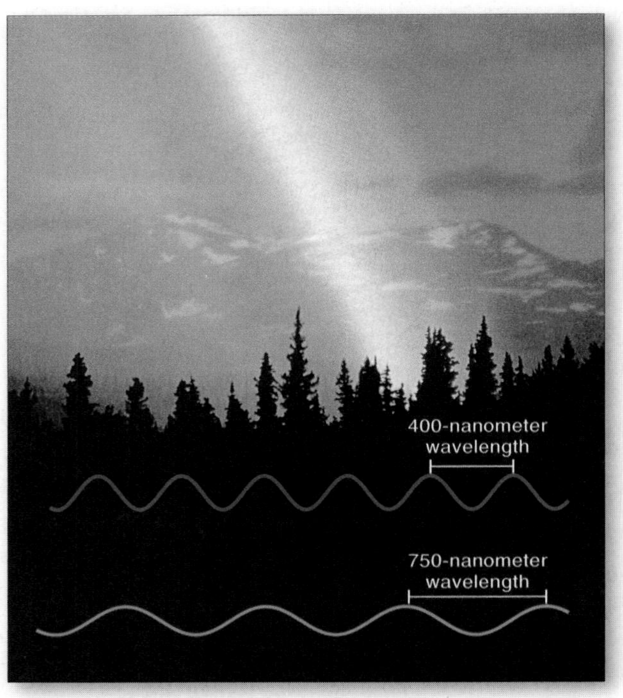
400-nanometer wavelength
750-nanometer wavelength
b

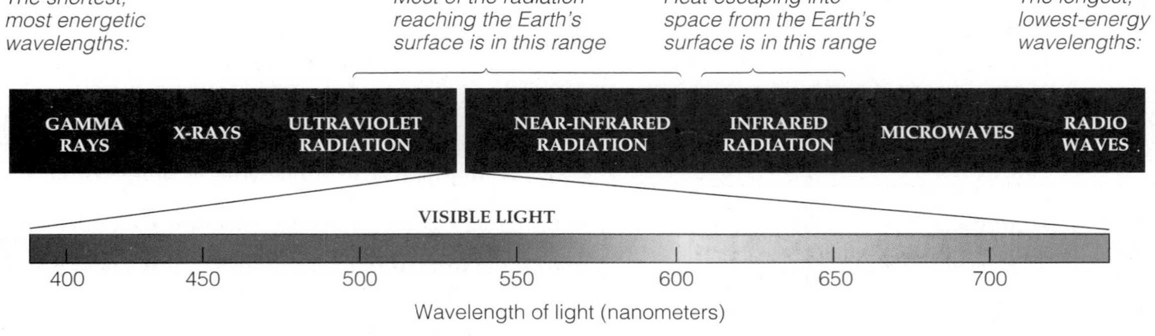
The shortest, most energetic wavelengths:
Most of the radiation reaching the Earth's surface is in this range
Heat escaping into space from the Earth's surface is in this range
The longest, lowest-energy wavelengths:
GAMMA RAYS
X-RAYS
ULTRAVIOLET RADIATION
NEAR-INFRARED RADIATION
INFRARED RADIATION
MICROWAVES
RADIO WAVES
VISIBLE LIGHT
400
450
500
550
600
650
700
c
Wavelength of light (nanometers)

Figure 7.4 (**a**) Model of how photoautotrophs are the basis of a one-way flow of energy through nearly all webs of life, starting with energy inputs from the sun. (**b**) Two examples of wavelengths, the horizontal distance between crests of successive waves. (**c**) The electromagnetic spectrum. Visible light is one of many forms of electromagnetic radiation in the spectrum.

Further reading: Student Guide to InfoTrac on web site

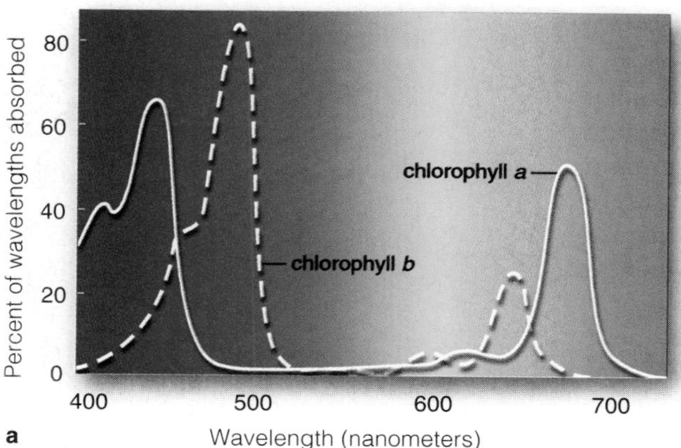
Percent of wavelengths absorbed
80
60
40
20
0
chlorophyll a
chlorophyll b
400
500
600
700
a
Wavelength (nanometers)

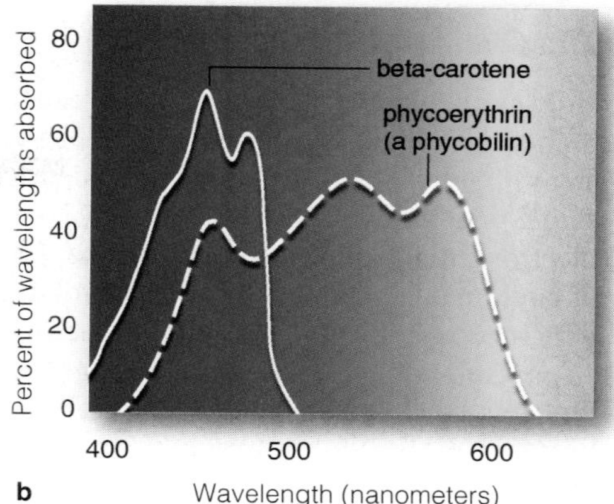
Percent of wavelengths absorbed
80
60
40
20
0
beta-carotene
phycoerythrin
(a phycobilin)
400
500
600
b
Wavelength (nanometers)

Figure 7.5 (**a**) Two absorption spectra that indicate the efficiency with which chlorophylls *a* and *b* respond to different wavelengths. (**b**) Absorption spectra for a beta-carotene (a carotenoid) and for a phycobilin.

many other organisms perceive as different colors. Wavelengths shorter than this, such as ultraviolet (UV) radiation, are energetic enough to break the bonds of organic compounds—hence to kill cells. Hundreds of millions of years ago, before a layer of ozone (O_3) gradually accumulated in the upper atmosphere, only water shielded the surface of the Earth from UV radiation. The lethal bombardment kept photoautotrophs below the surface of the seas.

When absorbed by matter, the energy of visible light can be measured as if it were organized in packets, which we call **photons**. Each type of photon has a fixed amount of energy. Those having the most energy travel as the shortest wavelengths, which correspond to blue-violet light. Photons having the least energy travel as long wavelengths that correspond to red light (Figure 7.4*b*).

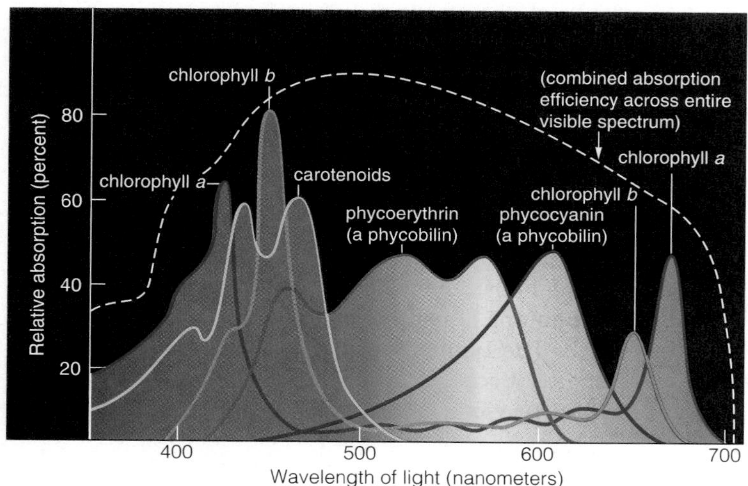
Relative absorption (percent)
80
60
40
20
chlorophyll b
(combined absorption efficiency across entire visible spectrum)
chlorophyll a
carotenoids
chlorophyll a
chlorophyll b
phycoerythrin
(a phycobilin)
phycocyanin
(a phycobilin)
400
500
600
700
Wavelength of light (nanometers)

Figure 7.6 Combined energy absorption efficiency, indicated by the dashed line, of certain chlorophylls (*green* lines), carotenoids (*yellow-orange* lines), and phycobilins (*purple and blue* lines).

Pigments—Molecular Bridge From Sunlight to Photosynthesis

Collectively, wavelengths of visible light look white to us. A crystal prism intercepting white light can sort it out into its component colors by bending the different wavelengths by different degrees. In the 1800s, recall, T. Englemann used a prism as part of an experiment to identify which wavelengths drive photosynthesis in a green alga (Figure 7.1). But molecular biology was far in the future, and Englemann was not able to identify the actual bridge between sunlight and photosynthetic activity.

Pigments are the molecular bridge. **Pigments** are different molecules that absorb wavelengths of light, and organisms put them to many uses. Most pigments absorb only some wavelengths and transmit the rest. A few, such as the melanins in animals, absorb so many wavelengths they appear dark or black.

An **absorption spectrum** is a diagram that shows how effectively a pigment molecule absorbs different wavelengths in the spectrum of visible light. Consider chlorophylls, the main pigments of photosynthesis. As Figure 7.5*a* shows, chlorophylls absorb all wavelengths except very little of the green and yellow-green ones, which they mainly transmit. (That is why plant parts with an abundance of chlorophylls look green to us.) Other, *accessory* pigments that impart different colors also occur in photoautotrophs. Figure 7.5*a* and 7.6 show how their assistance extends the range of wavelengths that drive photosynthesis. As you can see, the combined absorption efficiency is high indeed.

Radiation from the sun travels in waves that differ in length and energy content. Short wavelengths pack the most energy.

We perceive wavelengths of visible light as different colors and measure their energy content in packets called photons.

Chlorophylls and certain other pigments absorb specific wavelengths of visible light. They are the molecular bridge between the sun's energy and photosynthetic activity.

The Chemical Basis of Color

How is it that pigment molecules can impart color to plants and other organisms? Each has a light-catching array of atoms, which often are joined by alternating single and double bonds. Figure 7.7 shows examples. Electrons that are distributed around the atomic nuclei of such arrays can absorb photons of specific energies, which correspond to specific colors of light.

Remember, when an atom's electrons absorb energy, they move to a higher energy level (Section 2.3). In a pigment molecule, an input of energy destabilizes the distribution of electrons inside the light-catching array. Within 10^{-15} of a second, excited electrons return to a lower energy level, the electron distribution stabilizes, and energy is emitted in the form of light. When any destabilized molecule emits light as it reverts to its more stable configuration, this is called a **fluorescence**.

Excitation occurs only when the quantity of energy of an incoming photon matches the amount of energy required to boost an electron to a higher energy level. Suppose a sunbeam strikes a pigment. If photons of its red or orange wavelengths match the amount of energy necessary for the boost, the pigment will absorb them. However, the photons of its blue or violet wavelengths are a mismatch. The pigment molecule cannot absorb these wavelengths. Because it transmits (reflects) them, it takes on a blue or violet color.

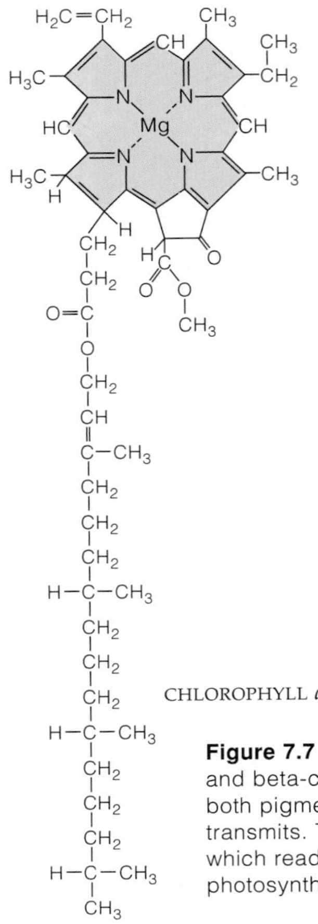

CHLOROPHYLL *a*

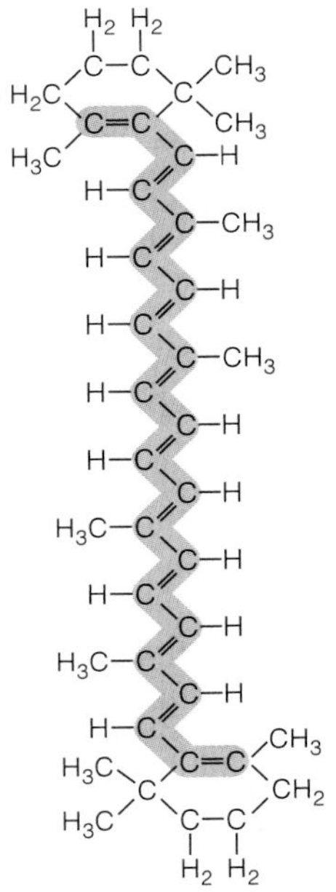

BETA-CAROTENE

Figure 7.7 Structural formulas for chlorophyll *a* and beta-carotene. The light-catching array of both pigments is shaded in the color of light it transmits. The backbone is pure hydrocarbon, which readily dissolves in the lipid bilayer of photosynthetic cell membranes.

On the Variety of Photosynthetic Pigments

Most pigments respond to only part of the rainbow of visible light. If acquiring energy is so vital for life, why doesn't each photosynthetic pigment go after the whole rainbow? In other words, why isn't each one black?

If the first photoautotrophs evolved in the seas, then so did their pigments. Ultraviolet and red wavelengths do not penetrate water as deeply as green and blue wavelengths do. Possibly natural selection favored the evolution of diverse pigments at different depths. Many red algae live in deep water, and indeed they are nearly black. Green algae live in shallow water, where their chlorophylls absorb red wavelengths. Their accessory pigments harvest more wavelengths, and some even function as shields against ultraviolet radiation.

Today, **chlorophylls** are the main pigments in all but one marginal group of photoautotrophs. Remember, photosynthetic pigments respond best to red and blue-to-violet light. Chlorophyll *a*, a grass-green pigment that absorbs blue-violet and red wavelengths, is a key player in the light-dependent reactions (Figure 7.7). Chlorophyll *b* absorbs blue and red-orange wavelengths. In chloroplasts, it is one of several accessory pigments, busily harvesting wavelengths that chlorophyll *a* misses. Chlorophyll *b*, a bluish-green pigment, occurs in plants, green algae, and a few photoautotrophic bacteria.

All photoautotrophs also contain **carotenoids**. These accessory pigments absorb blue-violet and blue-green wavelengths that chlorophylls miss. They reflect red, orange, and yellow wavelengths. You can expect flowers, fruits, and vegetables in this color range to have an abundance of carotenoids.

Usually the backbone of a carotenoid molecule has a carbon ring at each end. Many types, including beta-carotene, are pure hydrocarbons (Figure 7.7). The type called xanthophylls also contain oxygen. Some of the carotenoids have proteins attached, in which case they appear purple, violet, blue, green, brown, or black.

Carotenoids are less abundant than the chlorophylls in green leaves, but in many plants they become visible in autumn (Figure 7.8). Each year, tourists spend about a billion dollars just to watch the three-week demise of chlorophyll in the deciduous trees of New England.

Other accessory pigments include the red and blue **anthocyanins** and **phycobilins**. The anthocyanins are pigments in many flowers, and the phycobilins are the signature pigments of red algae and cyanobacteria.

Further reading: Student Guide to InfoTrac on web site

Figure 7.8 Leaf color. Inside intensely green leaves, photosynthetic cells are continually synthesizing chlorophyll molecules. The chlorophylls mask the presence of carotenoids and other accessory pigments. In autumn, however, chlorophyll synthesis lags behind chlorophyll breakdown in many species. When that happens, the other pigments in the leaf are unmasked, and more colors show through.

Also in autumn, anthocyanins accumulate in leaf cells. These water-soluble pigments appear red if fluids moving through plants are slightly acidic, blue if the fluids are basic (alkaline), or colors in between if the fluids are of intermediate pH. Soil conditions contribute to the pH values.

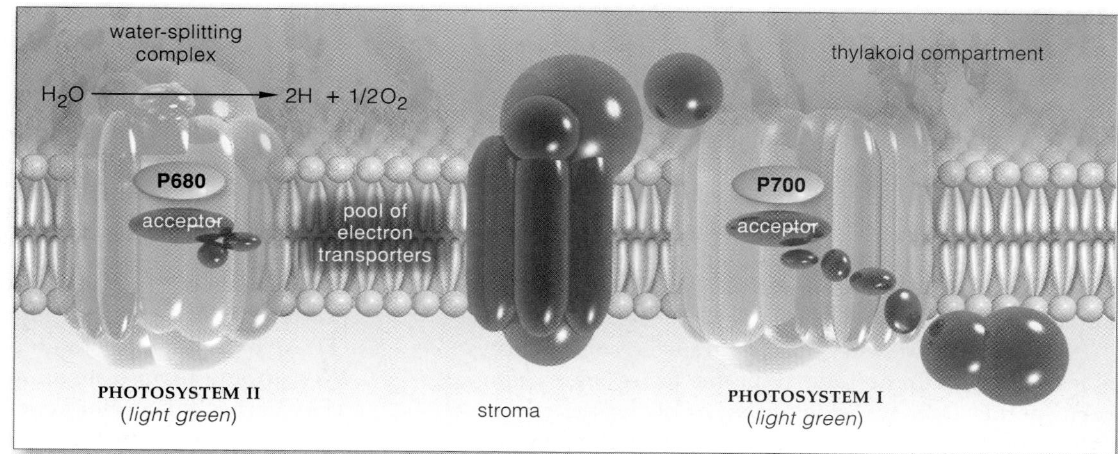
water-splitting complex
H_2O → $2H + 1/2O_2$
thylakoid compartment
P680
acceptor
pool of electron transporters
P700
acceptor
PHOTOSYSTEM II
(light green)
stroma
PHOTOSYSTEM I
(light green)

Figure 7.9 How two kinds of photosystems, designated I and II, are arranged in the thylakoid membrane system of chloroplasts. Components of neighboring electron transport systems are coded *dark green*. One transport system extends from photosystem II to P700 of photosystem I. A second transport system that extends from P700 gives up electrons and hydrogen to a coenzyme ($NADP^+$).

Where Are Photosynthetic Pigments Located?

We find photosynthetic pigments among some bacteria of ancient lineages. For example, the archaebacterium *Halobacterium halobium* has bacterorhodopsin (a purple pigment) in its plasma membrane. Cyanobacteria have chlorophyll *a* and other pigments embedded in internal foldings of the plasma membrane.

The thylakoid membrane system of chloroplasts has pigments organized in clusters called **photosystems**. In some plant species, the membrane has many thousands of these clusters. Each photosystem consists of proteins and 200 to 300 pigment molecules. As Figure 7.9 shows, it is located next to its functional partner—an electron transport system. How the two systems work together is the topic of the next two sections.

About Those Roving Pigments

There are only about eight classes of pigments, but this limited group gets around in the world. For example, animals synthesize some pigments (such as melanin) but not carotenoids. These originate with photoautotrophs and move up through food webs, as when tiny aquatic snails feed on green algae and flamingos subsequently eat the snails. Flamingos modify ingested carotenoids in diverse ways. For instance, their cells can split beta-carotene molecules to form two molecules of vitamin A. Vitamin A is the precursor of retinol, a visual pigment that transduces light to electric signals in the flamingo's eyes. Beta-carotene molecules that become dissolved in fat reservoirs under the skin are taken up by skin cells that differentiate and give rise to bright pink feathers.

The chlorophylls are the main photosynthetic pigments in photoautotrophs. Carotenoids and other accessory pigments enhance their light-harvesting function.

Photosystems (pigment clusters) are the functional partners of electron transport systems in the thylakoid membrane.

7.4 THE LIGHT-DEPENDENT REACTIONS

We are now ready to look more closely at the first stage of photosynthesis, the **light-dependent reactions**, by tracking three events that proceed in the chloroplast. *First*, pigments absorb light energy and give up excited electrons, which enter electron transport systems. *Second*, water molecules are split, ATP and NADPH form, and oxygen is released. *Third*, the pigments that gave up electrons in the first place get electron replacements.

What Happens to the Absorbed Energy?

Photon absorption, recall, can boost electrons of atoms of a photosynthetic pigment to a higher energy level, but the electrons quickly emit the excitation energy as they return to a lower energy level. If nothing else were around to intercept it, all of the emitted energy would simply escape as fluorescent light and heat. However, photosynthetic pigments are not isolated.

As you read earlier, hundreds of pigment molecules are clustered together in each of the photosystems that are embedded in the thylakoid membrane. Most of the pigments of each cluster only harvest the energy from photons. Instead of releasing the excitation energy as a fluorescent afterglow, a harvester directly transfers it to another pigment molecule, which randomly passes it on to a neighboring pigment molecule, and so on. Figure 7.10 illustrates the "random walk" of excitation energy among the pigments of a photosystem.

Each time a harvester makes a transfer, some energy is lost (as heat). In no time at all, the energy remaining corresponds to a wavelength that only a specialized chlorophyll *a* can trap. That chlorophyll is the **reaction center** for the photosystem. It accepts excitation energy,

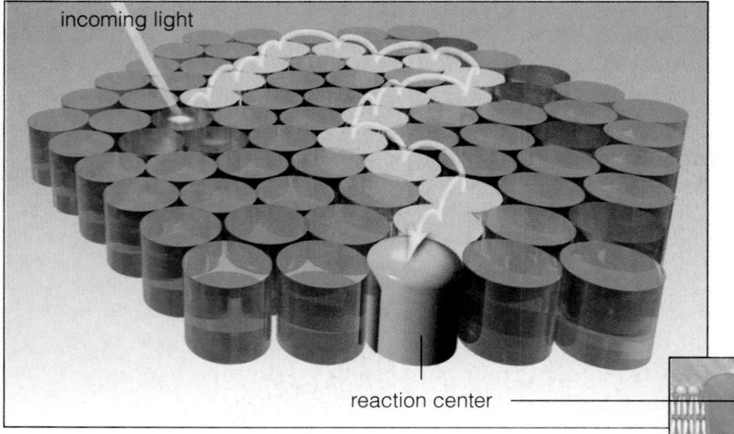
incoming light
reaction center
PHOTOSYSTEM

Figure 7.10 A small sampling of harvester pigments of a photosystem. Notice the random flow of energy from photons among them. The energy of excitation quickly reaches a reaction center which, when suitably activated, releases electrons for photosynthesis.

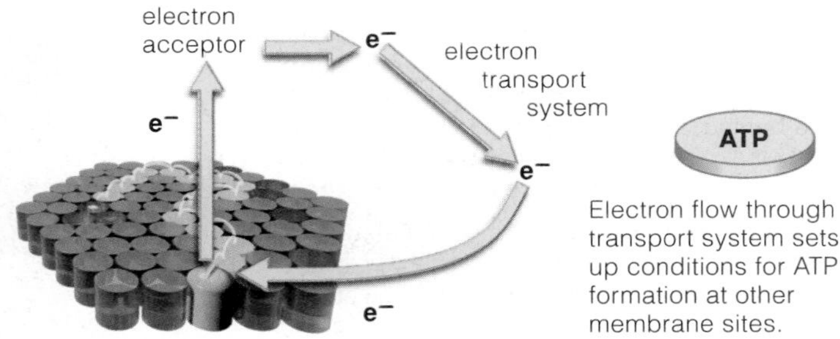
electron acceptor
e$^-$
electron transport system
e$^-$
e$^-$
e$^-$
ATP
Electron flow through transport system sets up conditions for ATP formation at other membrane sites.

Figure 7.11 Cyclic pathway of ATP formation.

but it does not pass energy on to another pigment. A suitably activated reaction center donates electrons to an acceptor molecule near an electron transport system.

Electron transport systems are organized arrays of enzymes, coenzymes, and other proteins associated with a cell membrane (Section 6.4). Electrons are transferred step-by-step through the array, and some energy escapes at each step. In chloroplasts, much of the energy drives the machinery that produces ATP and NADPH.

Cyclic and Noncyclic Electron Flow

In ancient bacteria, photosynthesis evolved as a way to produce ATP, not to synthesize organic compounds. (Those bacteria got enough electrons to make NADPH by stripping them from simple inorganic compounds.) They harnessed sunlight energy to cycle electrons from a photosystem, to a transport system, then back to the photosystem (Figure 7.11). This *cyclic* pathway still runs in all photoautotrophs. It requires a *type I* photosystem, which has a reaction center designated P700.

The electron flow from P700 does not pack enough punch to make NADPH, too. But more than 2 billion years ago, it seems that the energy for electron transfers increased as if two batteries had been hooked together. Another photosystem, *type II*, entered the picture. And the two photosystems operating together could harness enough energy to strip electrons from water molecules.

Take a look at Figure 7.12*a*. As you see, the newer pathway of electron flow is *noncyclic*. There is a linear flow of electrons from water to photosystem II, through a transport system to photosystem I, then on through a transport system that delivers the electrons to NADP$^+$.

The machinery works when photosystem II absorbs enough photon energy to make its designated reaction center—P680—give up electrons. The energy input also triggers **photolysis**. By this reaction sequence, water molecules split into oxygen, hydrogen ions (H$^+$), and electrons. When P680 releases the excited electrons, new electrons released from water replace them (Figure 7.12*a*).

Further reading: Student Guide to InfoTrac on web site

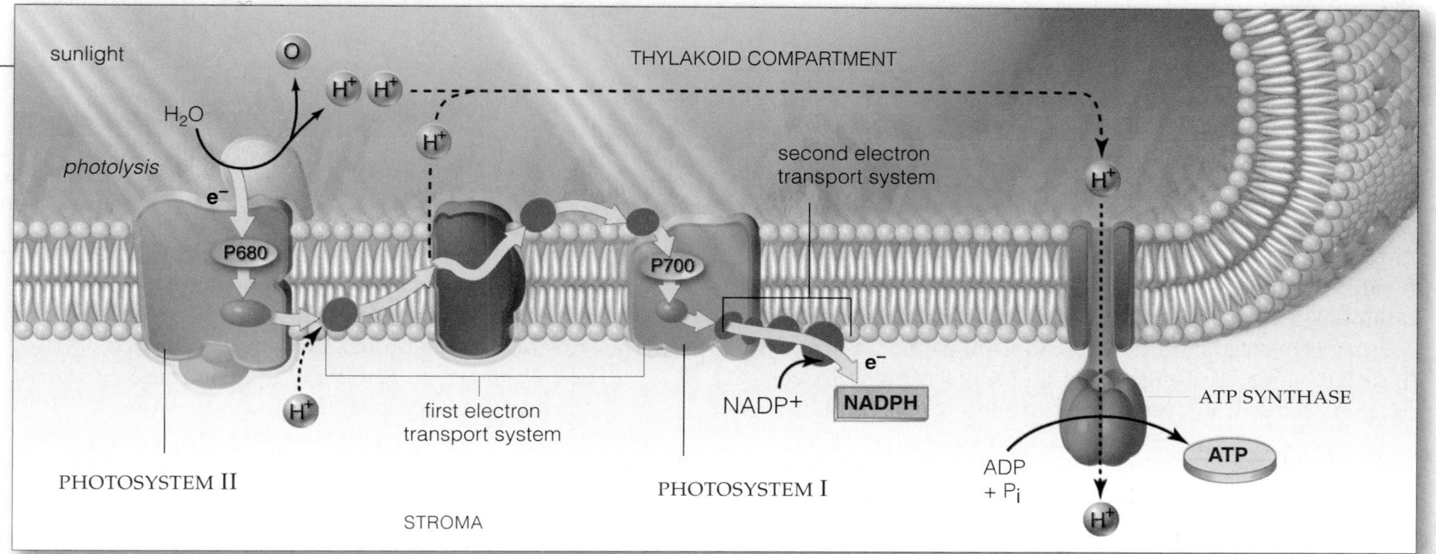
sunlight
THYLAKOID COMPARTMENT
H_2O
photolysis
P680
P700
second electron transport system
first electron transport system
$NADP^+$
NADPH
ATP SYNTHASE
ATP
ADP + P_i
PHOTOSYSTEM II
PHOTOSYSTEM I
STROMA

Figure 7.12 (**a**) Filling in the details for Figure 7.3—the noncyclic pathway of photosynthesis, by which ATP and NADPH form. *Yellow* arrows show electron flow. Electrons released from water molecules split by photolysis move through two photosystems (*light green*) and two transport systems (*dark green*). The joint operation of the two photosystems boosts electrons to an energy level high enough to drive NADPH formation. Also, as you will read in Section 7.5, hydrogen ions move across the thylakoid membrane in ways that lead to ATP formation at proteins called ATP synthases.

(**b**) Diagram the energy changes associated with the noncyclic pathway, and you will end up with this "Z scheme."

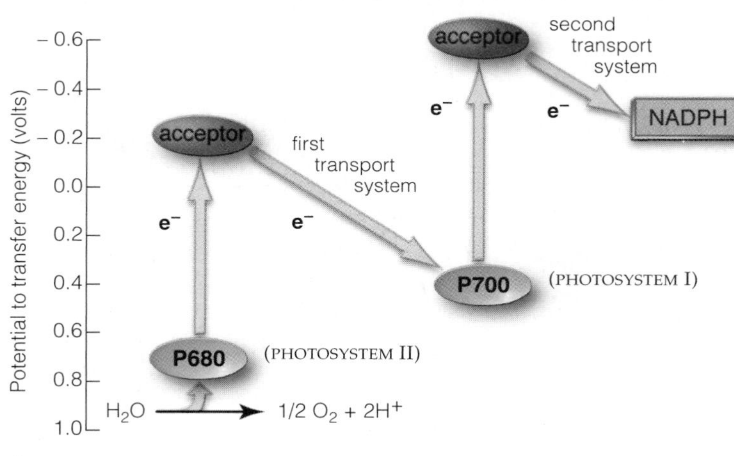
Potential to transfer energy (volts)
– 0.6
– 0.4
– 0.2
0.0
0.2
0.4
0.6
0.8
1.0
acceptor
first transport system
acceptor
second transport system
NADPH
P700
(PHOTOSYSTEM I)
P680
(PHOTOSYSTEM II)
H_2O
$1/2\ O_2 + 2H^+$
b

The excited electrons move on to a transport system, then to P700 of photosystem I. They still have a bit of extra energy. And they get more upon their arrival, for photons are also bombarding photosystem I. The energy boost puts electrons at a higher energy level that allows them to enter a transport system. There, $NADP^+$ accepts two electrons and a hydrogen ion to become NADPH.

Today, both the cyclic and noncyclic pathways of electron flow operate in all photosynthetic organisms. Which one dominates at a given time depends on the organism's metabolic demands for ATP and NADPH.

The Legacy—A New Atmosphere

On sunny days, on the surfaces of aquatic plants, you see bubbles of oxygen, a by-product of the noncyclic pathway (Figure 7.13). When the pathway first evolved, the oxygen being released dissolved in seawater, wet mud, and similar bacterial habitats. By 1.5 billion years ago, enormous quantities of dissolved oxygen were escaping to what had been an oxygen-free atmosphere. The accumulation of oxygen changed the atmosphere forever. That vast, global change made possible aerobic respiration, which became the most efficient pathway for releasing usable energy from organic compounds. Ultimately, the emergence of the noncyclic pathway allowed you and every other animal to be around today, breathing the oxygen that helps keep your cells alive.

Figure 7.13 Visible evidence of photosynthesis—bubbles of oxygen escaping from the leaves of *Elodea*, a plant of freshwater habitats.

In the light-dependent reactions, sunlight energy drives the release of electrons from photosystems in thylakoid membranes. The electron flow through nearby transport systems results in the formation of ATP, NADPH, or both.

ATP can form when electrons cycle from and back to a type I photosystem. Both ATP and NADPH can form by a noncyclic pathway in which electrons flow from water, through two photosystems (types II and I), and finally to $NADP^+$.

All photosynthetic species employ one or both pathways in response to the cell's changing needs for ATP and NADPH.

Oxygen, a by-product of the noncyclic pathway, changed the early atmosphere and made aerobic respiration possible.

7.5 A CLOSER LOOK AT ATP FORMATION IN CHLOROPLASTS

As you probably have noticed, we saved the trickiest question for last. Exactly how does ATP form during the noncyclic pathway of photosynthesis? To arrive at the answer, let's walk through Figure 7.14.

Inside the chloroplast, photon absorption triggers photolysis, the pathway's first step. At this step, enzyme activity repeatedly splits water molecules into oxygen, hydrogen ions, and electrons. O_2 forms from the free oxygen atoms. It diffuses out of the chloroplast and out of the cell. The hydrogen ions (H^+) are left behind. And they accumulate in the fluid inside the thylakoid compartment (Figure 7.14*a*).

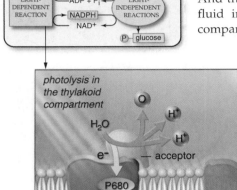

LIGHT-DEPENDENT REACTION
ATP
ADP + P_i
NADPH
NAD^+
LIGHT-INDEPENDENT REACTIONS
P
glucose

Therefore, *by a combination of photolysis and electron transport*, the concentration of hydrogen ions becomes greater inside the thylakoid compartment, compared to the stroma. The unequal distribution of these positively charged ions also creates a difference in electric charge across the membrane. An electric gradient, as well as a concentration gradient, has become established.

The combined force of the H^+ concentration gradient and electric gradient propels hydrogen ions through the interior of ATP synthases, a type of transport protein that spans the thylakoid membrane (Figure 7.14*c*). In other words, the ions flow out from the compartment, through ATP synthases, into the stroma. ATP synthases have built-in enzymatic machinery. And the flow of ions through them drives the machinery, which catalyzes the attachment of unbound phosphate to a molecule of ADP. In this way, ATP forms.

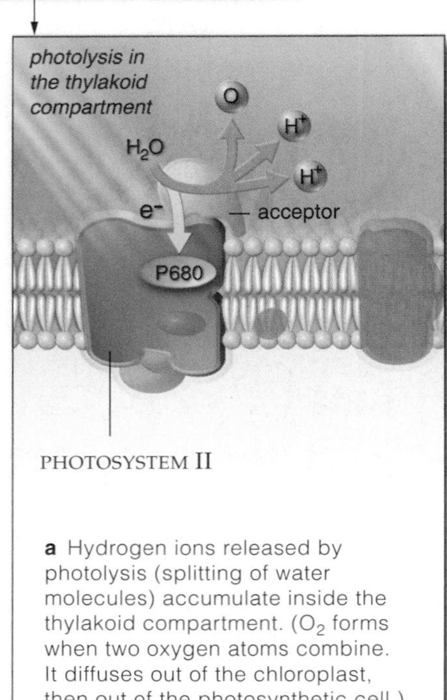

photolysis in the thylakoid compartment
O
H^+
H_2O
H^+
e^-
acceptor
P680
PHOTOSYSTEM II

a Hydrogen ions released by photolysis (splitting of water molecules) accumulate inside the thylakoid compartment. (O_2 forms when two oxygen atoms combine. It diffuses out of the chloroplast, then out of the photosynthetic cell.)

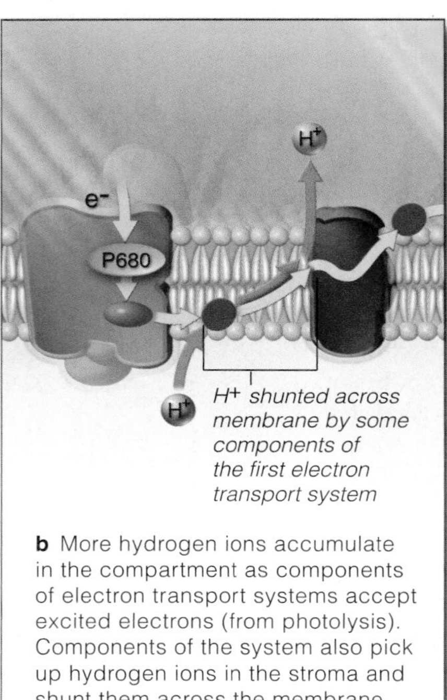

e^-
H^+
P680
H^+
H^+ shunted across membrane by some components of the first electron transport system

b More hydrogen ions accumulate in the compartment as components of electron transport systems accept excited electrons (from photolysis). Components of the system also pick up hydrogen ions in the stroma and shunt them across the membrane.

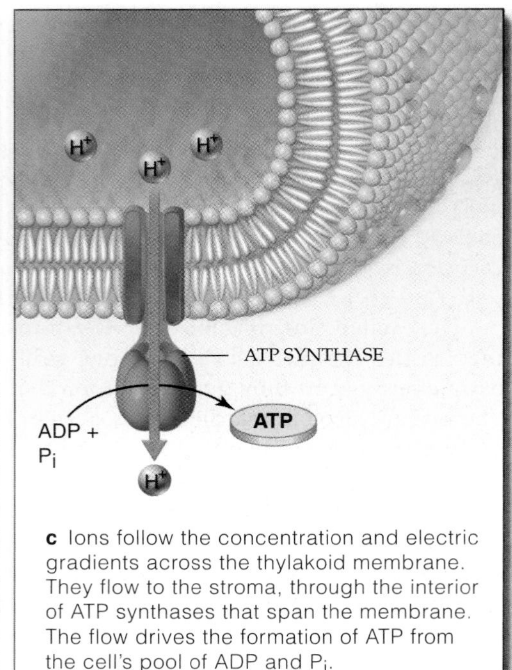

H^+
H^+
H^+
ATP SYNTHASE
ADP + P_i
ATP
H^+

c Ions follow the concentration and electric gradients across the thylakoid membrane. They flow to the stroma, through the interior of ATP synthases that span the membrane. The flow drives the formation of ATP from the cell's pool of ADP and P_i.

Figure 7.14 How ATP forms in chloroplasts during the noncyclic pathway of photosynthesis. 

The released electrons are picked up by a primary acceptor molecule, which transfers them to the electron transport system positioned next to photosystem I in the thylakoid membrane. More hydrogen ions flow into the thylakoid compartment while the electron transport system is operating. This is the case for both the cyclic and noncyclic pathways. The thylakoid membrane has many photosystems and many transport systems. At the same time that certain components of the transport systems accept electrons, they also pick up hydrogen ions from the stroma. They immediately shunt those ions across the membrane, as shown in Figure 7.14*b*.

The sequence of events just described is called the chemiosmotic model of ATP formation in chloroplasts. As you will see in the next chapter, the same model applies to ATP formation in mitochondria.

In chloroplasts, H^+ concentration and electric gradients form across the thylakoid membrane. The flow of ions from the thylakoid compartment to the stroma drives ATP formation.

Further reading: Student Guide to InfoTrac on web site

7.6 THE LIGHT-INDEPENDENT REACTIONS

The **light-independent reactions** are the "synthesis" part of photosynthesis. They occur as a cyclic pathway called the **Calvin–Benson cycle**. The reactions require energy from ATP, hydrogen and electrons from NADPH, and carbon and oxygen from carbon dioxide (CO_2), which is present in the air (or water) that surrounds photosynthetic cells. We say these reactions are light-independent because they do not depend directly on sunlight. They can proceed just as well in the dark, as long as ATP and NADPH are available.

How Do Plants Capture Carbon?

Let's track a CO_2 molecule that diffuses into air spaces inside a leaf and ends up next to a photosynthetic cell. It diffuses into the cell, then into a chloroplast's stroma. An enzyme attaches the carbon atom of CO_2 to **RuBP** (ribulose bisphosphate), a compound with a backbone of five carbon atoms. **Rubisco** (for RuBP carboxylase) is the enzyme's name. Its action produces an unstable six-carbon intermediate that splits into two molecules of **PGA** (phosphoglycerate). PGA is a stable molecule with a three-carbon backbone. Incorporating a carbon atom from CO_2 into a stable organic compound is called **carbon fixation**. Quite simply, food can't be produced without this first step of the Calvin–Benson cycle.

The cycle yields phosphorylated glucose, and it also regenerates the RuBP. For our purposes, we can focus on the carbon atoms of the substrates, intermediates, and end products, as in Figure 7.15.

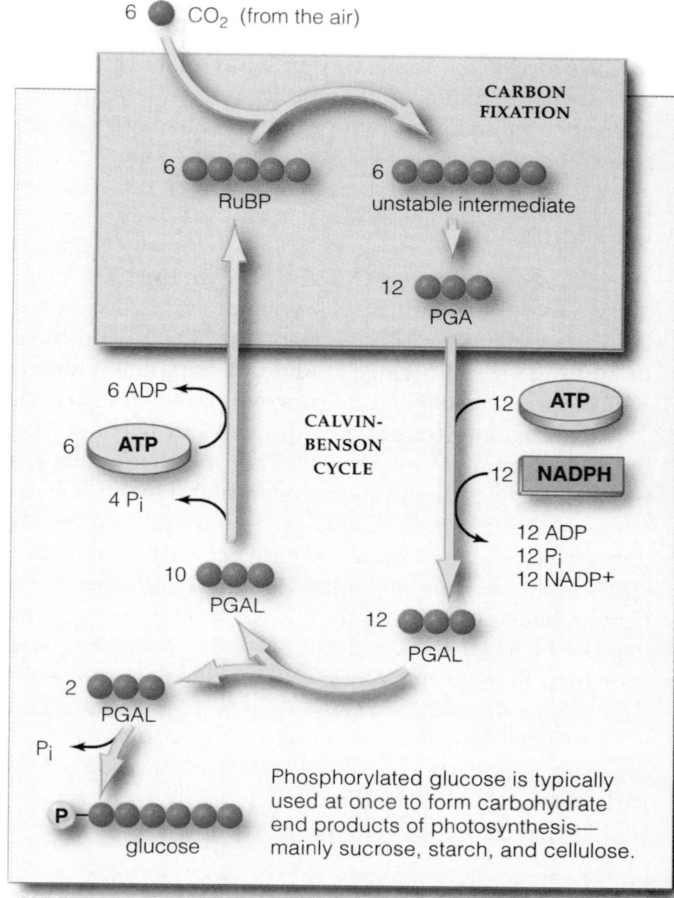

6 CO_2 (from the air)
CARBON FIXATION
6 RuBP
6 unstable intermediate
12 PGA
6 ADP
6 ATP
4 P_i
CALVIN-BENSON CYCLE
12 ATP
12 NADPH
12 ADP
12 P_i
12 NADP+
10 PGAL
12 PGAL
2 PGAL
P_i
P
glucose
Phosphorylated glucose is typically used at once to form carbohydrate end products of photosynthesis—mainly sucrose, starch, and cellulose.

Figure 7.15 Summary of the light-independent reactions of photosynthesis. *Red* circles are carbon atoms of key molecules. All intermediates have one or two phosphate groups attached, but to keep things simple, we show only the phosphate group on the resulting glucose. Also, many water molecules that formed in the light-dependent reactions enter this pathway, and six remain at its conclusion. Appendix V (Figure C) shows details.

How Do Plants Build Glucose?

Each PGA accepts a phosphate group from ATP, plus hydrogen and electrons from NADPH. The resulting intermediate is called **PGAL** (phosphoglyceraldehyde). To build *one* six-carbon sugar phosphate, carbon atoms from six CO_2 must be fixed and twelve PGAL must form. Most of the PGAL becomes rearranged to form new RuBP, which can be used to fix more carbon. But two of the PGAL combine, thus forming glucose with a phosphate group attached to its six-carbon backbone.

When phosphorylated this way, glucose is primed to enter other reactions. Plants use it as a building block for their main carbohydrates—sucrose, cellulose, and starch. *The synthesis of these organic compounds by other pathways marks the end of the light-independent reactions.*

With six turns of the cycle, enough RuBP molecules form to replace the ones used in carbon fixation. The ADP, $NADP^+$, and phosphate remaining diffuse through the stroma, to sites of the light-dependent reactions. There they are converted back to NADPH and ATP.

Photosynthetic cells convert phosphorylated glucose to sucrose or starch during daylight hours. Of all plant carbohydrates, sucrose is the most easily transported. Starch is the most common storage form. The cells also convert excess PGAL to starch. They briefly store starch (as starch grains) inside the stroma. After the sun goes down, they convert starch to sucrose for export to other living cells in leaves, stems, and roots. Ultimately, the products and intermediates of photosynthesis end up as energy sources and as building blocks for all of the lipids, amino acids, and other organic compounds that are required for growth, survival, and reproduction.

In the light-independent reactions (the Calvin–Benson cycle), carbon is "captured" from carbon dioxide, glucose forms during reactions that require ATP and NADPH, and RuBP (necessary to capture the carbon) is regenerated.

7.7 FIXING CARBON—SO NEAR, YET SO FAR

If sunlight intensity, air temperature, rainfall, and soil composition were the same everywhere, photosynthesis might proceed the same way in every plant. However, environments differ—and so do photosynthetic details. A brief comparison of two carbon-fixing adaptations to stressful environments will illustrate this point.

C4 Plants

All plants must take up CO_2 for growth. But CO_2 is not always plentiful inside leaves, which have a waxy cover that helps plants conserve water. The CO_2 must diffuse in and the O_2 out mainly at **stomata** (singular, stoma), which are microscopic openings across the leaf surface.

Stomata close on hot, dry days. Water is conserved but CO_2 cannot get into leaves. Photosynthetic cells are busy, so oxygen accumulates. A high O_2 level in leaves triggers *photorespiration*. This process wastes fixed CO_2 and lowers a plant's sugar-making capacity. Remember rubisco, the enzyme that attaches carbon to RuBP in the Calvin–Benson cycle? Rubisco also can attach *oxygen* to it if the O_2 level rises and the CO_2 level falls. Only one (not two) PGA forms, along with a glycolate molecule that is later degraded to CO_2.

Kentucky bluegrass is one of many **C3 plants.** "C3" refers to *three*-carbon PGA, the first intermediate of its carbon-fixing pathway (Figure 7.16*a*). By contrast, the first intermediate formed when corn and many other plants fix carbon is the *four*-carbon oxaloacetate. Hence *their* name, **C4 plants**.

When C4 plants close stomata on hot, dry days, they still maintain enough CO_2 in leaves by *fixing carbon twice*, in two types of photosynthetic cells. Mesophyll cells, the first type, use CO_2 to form oxaloacetate—which is transferred to bundle-sheath cells at leaf veins (Figure 7.16*b*). There, CO_2 is released and its local concentration rises, so rubisco cannot use oxygen. In the mesophyll cells, CO_2 is fixed again during the Calvin–Benson cycle. With this pathway, C4 plants get by with tinier

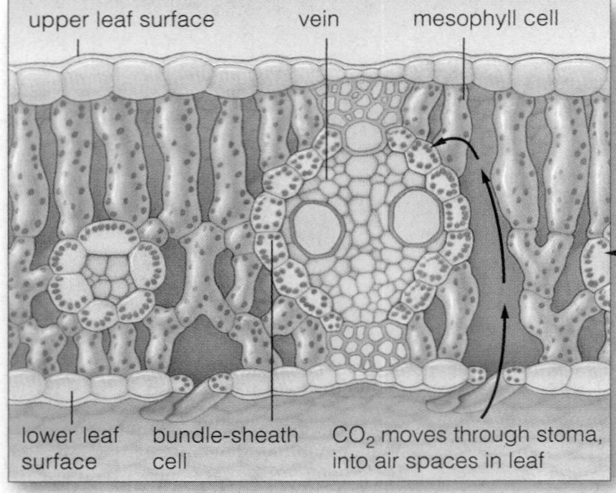

upper leaf surface
vein
mesophyll cell
lower leaf surface
bundle-sheath cell
CO_2 moves through stoma, into air spaces in leaf
Typical C4 plant leaf, in cross-section

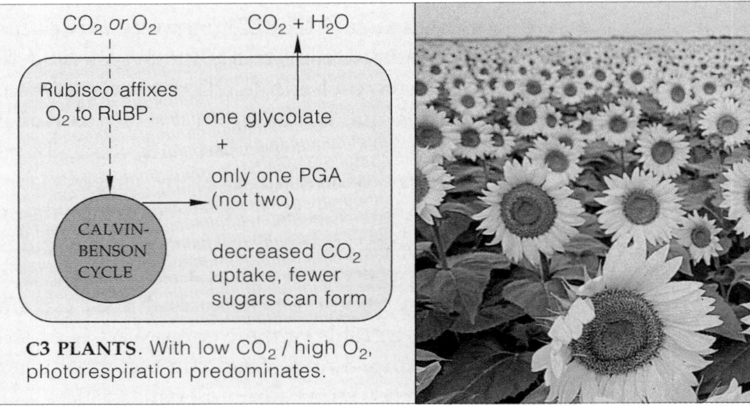

CO_2 or O_2
CO_2 + H_2O
Rubisco affixes O_2 to RuBP.
one glycolate
+
only one PGA (not two)
CALVIN-BENSON CYCLE
decreased CO_2 uptake, fewer sugars can form

C3 PLANTS. With low CO_2 / high O_2, photorespiration predominates.

a

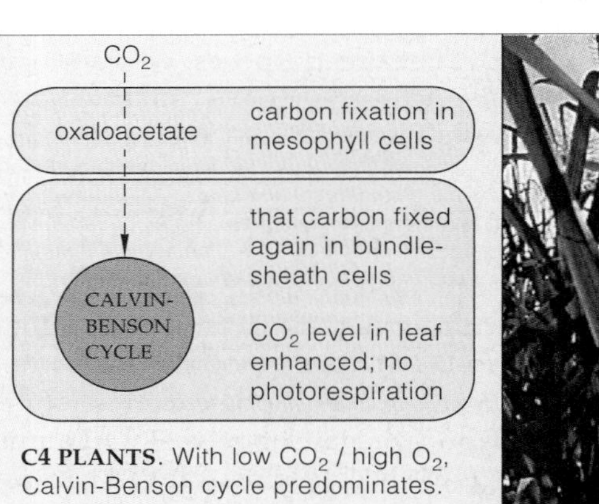

CO_2
oxaloacetate
carbon fixation in mesophyll cells
that carbon fixed again in bundle-sheath cells
CALVIN-BENSON CYCLE
CO_2 level in leaf enhanced; no photorespiration

C4 PLANTS. With low CO_2 / high O_2, Calvin-Benson cycle predominates.

b

CO_2
stomata open at night; CO_2 uptake but no water loss
stomata close during day; CO_2 in leaf used
CALVIN-BENSON CYCLE

CAM PLANTS. With low CO_2 / high O_2, Calvin-Benson cycle predominates.

c

Figure 7.16 Three ways of fixing carbon in hot, dry weather, when the CO_2 level is low and the O_2 level is high in leaves. (**a**) The C3 pathway is common among evergreen trees and shrubs as well as many nonwoody plants of temperate zones, such as sunflowers. (**b**) The C4 pathway is common among grasses and other plants that evolved in the tropics and fix CO_2 twice. Corn, crabgrass, and the sugarcane shown here are examples. (**c**) CAM plants open stomata and fix carbon at night. They include pineapple, cacti and many other succulents (plants with a low surface-to-volume ratio), and orchids.

Further reading: Student Guide to InfoTrac on web site

stomata, lose less water, and make more glucose than C3 plants can when conditions are hot, bright, and dry.

Experiments show that photorespiration lowers the photosynthetic efficiency of many C3 crop plants. For example, when hothouse tomatoes were grown at CO_2 levels high enough to block photorespiration, growth rates increased by as much as five times. If the process is so wasteful, why hasn't natural selection eliminated it? The answer may lie with rubisco, the enzyme that sets the whole process in motion. That switch-hitting enzyme evolved long ago, when atmospheric levels of oxygen were still low and carbon dioxide levels high. Maybe a gene coding for its structure cannot mutate without adversely affecting the carbon-fixing activity. Or maybe the pathway that degrades glycolate to carbon dioxide has proved so adaptive that it cannot be eliminated. Glycolate can be toxic at high concentrations.

The C4 pathway evolved separately in many lineages of flowering plants over the past 50 million to 60 million years. Before then, atmospheric CO_2 levels were higher, and they gave C3 plants a selective advantage.

Which pathway will be most adaptive in years to come? Atmospheric CO_2 levels have been increasing for decades. Some ecologists predict that they will double over the next fifty years. If that happens, photorespiration will decline again—and many vital crops might benefit.

CAM Plants

We see a carbon-fixing adaptation to desert conditions in cacti. A cactus is one of the succulents; it has juicy, water-storing tissues and thick surface layers that restrict water loss. It cannot open stomata on hot days without losing precious water. It opens them and fixes CO_2 *at night*. Its cells store the resulting intermediate in their central vacuoles, then use it for photosynthesis the next day when stomata are closed. Many plants are adapted this way. They are **CAM plants** (short for Crassulacean Acid Metabolism). Unlike C4 species, CAM plants do not fix carbon twice in different types of cells. They fix it in the same cells but at different times (Figure 7.16*c*).

Many plants die during prolonged droughts, but some CAM plants survive by keeping stomata closed even at night. They repeatedly fix the CO_2 that forms by aerobic respiration. Not much forms, but it is enough to allow these plants to maintain very low rates of metabolism. CAM plants grow slowly. Try growing a cactus plant in Seattle or some other place with a mild climate, and it will compete poorly with C3 and C4 plants.

C4 plants and CAM plants both have modified ways of fixing carbon for photosynthesis. The modifications counter the stress imposed by hot, dry conditions in their environments.

Focus on Science

7.8 LIGHT IN THE DEEP DARK SEA?

During the late 1980s a graduate student, Cindy Lee Van Dover, was puzzling over an eyeless shrimp. Divers had discovered it at a hydrothermal vent deep in the Atlantic Ocean. **Hydrothermal vents** are fissures in the seafloor where molten rock rises and mixes with cold seawater. Mineral-rich water, superheated by 650 degrees, spews into the perpetually dark surroundings. The sides of the vents teem with life—including shrimps and bacteria.

When Van Dover saw videotapes of shrimps clinging to the sides of the vent, she noticed a pair of bright strips on their back. By examining specimens in the lab, she saw that the strips connected to a nerve. The eyeless shrimps, it seemed, were equipped with a novel sensory organ.

EYELESS SHRIMP

Steven Chamberlain, a neuroscientist, confirmed the strips are a sensory organ with photoreceptors, or light-absorbing pigmented cells. Later, Ete Szuts, an expert on pigments, isolated the pigment. Its absorption spectrum is the same as that for rhodopsin, a visual pigment in eyes as complex as yours.

Van Dover had a hunch: If the strips serve as an eye, then we should find *light* on the seafloor. On later dives, special cameras confirmed her hypothesis. Like coils in a toaster oven, vents release heat (infrared radiation). They also release faint radiation at the low end of the visible spectrum—light up to nineteen times brighter than expected for infrared. A billion billion photons per square inch per second strike a typical sunlit tree. Only a trillion photons per square inch per second reach photoautotrophic bacteria living 72 meters (240 feet) below the surface of the Black Sea. *But just as many photons are available to organisms at hydrothermal vents.*

About this time Van Dover casually asked a colleague, "Hey, what if there's enough light for photosynthesis?" The response was, "What a stupid idea."

Euan Nisbet, who studies ancient environments, thought about this. The first cells arose about 3.8 billion years ago. As Nisbet and Van Dover hypothesized, What if they arose at hydrothermal vents? They could have used inorganic compounds such as hydrogen sulfide (for their hydrogen and electrons) and carbon dioxide (for carbon). Survival probably depended on being able to move away from light at vents and thereby avoid being boiled alive. Millions of years later, some bacterial descendants were evolving in hot springs near the ocean's surface. In time they used their light-detecting machinery to absorb light from the sun. They were adapted to a new energy source; they had become photosynthetic.

Some observations support the intriguing hypothesis that light-sensing machinery of deep-sea bacteria became modified for shallow-water photosynthesis. For example, absorption spectra for the chlorophyll in evolutionarily ancient photosynthetic bacteria correspond to the light measured at hydrothermal vents. Also, photosynthetic machinery contains iron, sulfur, manganese, and other minerals—which are abundant at hydrothermal vents.

7.9 AUTOTROPHS, HUMANS, AND THE BIOSPHERE

We conclude this chapter with a story that reinforces how photosynthesizers and other autotrophs fit in the world of living things. It is a story of mind-boggling numbers of single-celled and multicelled species on land and in the sunlit waters of the Earth.

Each spring, you sense renewed growth of autotrophs on land as leaves unfurl and lawns and fields turn green. You might not be aware that uncountable numbers of single-celled autotrophs also drift through the surface waters of the world ocean. You can't see them without a microscope; a row of 7 million cells of one aquatic species would be less than a quarter-inch long. In some regions, a cupful of seawater might hold 24 million cells of one species, and that wouldn't include cells of all the other aquatic species.

Nearly all of the drifters are photoautotrophic bacteria and protistans. Together they are the "pastures of the seas," the producers that ultimately feed all other marine organisms. The pastures "bloom" in the spring, when nutrient inputs sustain rapid reproduction. At that time seawater becomes warmer and enriched with nutrients churned up from the deep by winter currents.

Until NASA gathered data from satellites in space, we had no idea of their numbers and distribution. Figure 7.17*a* shows visual evidence of their activities one winter in the surface waters of the North Atlantic Ocean. Figure 7.17*b* shows a springtime bloom stretching from North Carolina all the way past Spain!

Collectively, the cells help shape the global climate, for they deal in staggering numbers of reactant and product molecules. For instance, they sponge up nearly half the carbon dioxide that we humans release each year, as when we burn fossil fuels or burn vast tracts of forests to clear land for farming. Without aquatic photoautotrophs, atmospheric carbon dioxide would accumulate more rapidly and possibly contribute to global warming, as described in Section 49.9. If the atmosphere warms by only a few degrees, then sea levels will rise and all lowlands near the coasts of continents and islands will be submerged.

Even though such global change is a real possibility, tons of industrial wastes, raw sewage, and fertilizers in runoff from croplands drain into the ocean each day. The pollutants seriously alter the chemical composition of seawater. How long will marine photoautotrophs be able to function in this new chemical brew? The answer will have impact on your own life in more ways than one.

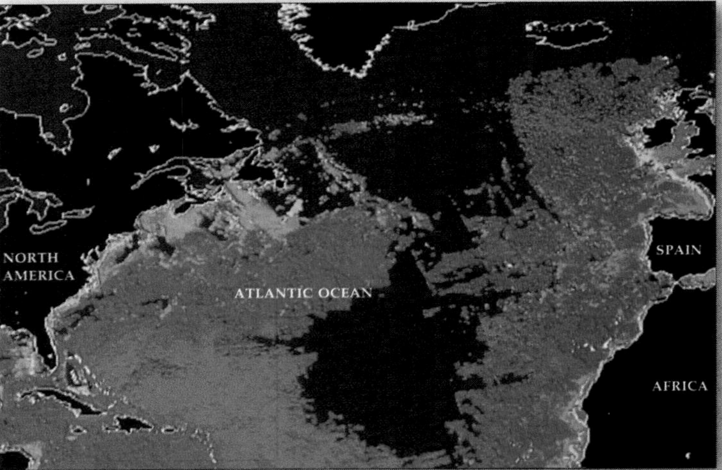
NORTH AMERICA
ATLANTIC OCEAN
SPAIN
AFRICA

a Photosynthetic activity in winter.

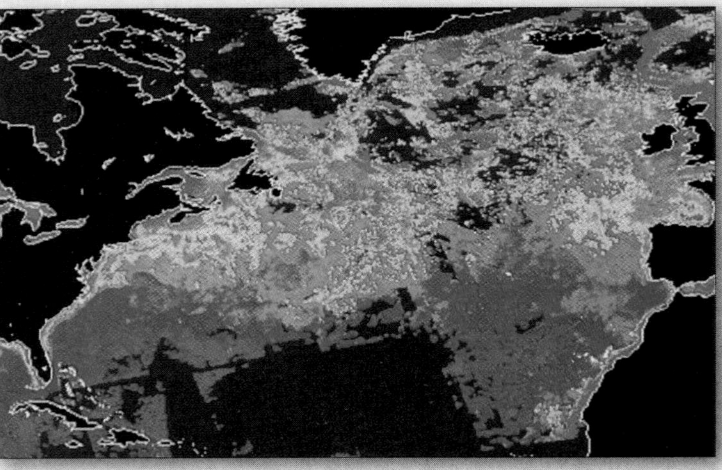

b Photosynthetic activity in spring.

Figure 7.17 Two satellite images that help convey the magnitude of photosynthetic activity during springtime in the North Atlantic portion of the world ocean. In these color-enhanced images, *red-orange* shows where chlorophyll is most concentrated.

Other autotrophs also influence your life in ways you might not expect. At hydrothermal vents, in hot springs, even in waste heaps of coal mines are diverse bacteria classified as *chemo*autotrophs. They, too, use carbon dioxide as a carbon source. But they use inorganic compounds in their environment as energy sources.

For example, some of those bacteria living on the sides of hydrothermal vents strip hydrogen and electrons from hydrogen sulfide (Section 7.8). Other chemoautotrophs living in soil obtain energy from nitrogen-containing wastes and remains of animals and other organisms.

Although they, too, are microscopically small, the chemoautotrophs also exist in monumental numbers. They influence the cycling of nitrogen, phosphorus, and other elements through the biosphere. We will be returning to their impact on the environment. In this unit, we turn next to pathways by which cells release the chemical bond energy of glucose and other biological molecules —the chemical legacy of autotrophs everywhere.

Further reading: Student Guide to InfoTrac on web site

7.10 SUMMARY

1. Cell structure and function are based on organic compounds, the synthesis of which depends on sources of carbon and energy. Plants and other autotrophs use carbon dioxide as their source of carbon. Some types use sunlight as the energy source. Others use inorganic compounds. Animals and other heterotrophs are not self-nourishing; they must get carbon and energy from organic compounds already synthesized by autotrophs.

2. Photosynthesis is the main process by which carbon and energy enter the web of life. It has two stages: the light-dependent and light-independent reactions. Key reactants, intermediates, and products are summarized in Figure 7.18 and in this simplified equation:

$$12H_2O + 6CO_2 \xrightarrow{\text{LIGHT ENERGY}} 6O_2 + C_6H_{12}O_6 + 6H_2O$$

WATER CARBON DIOXIDE → OXYGEN GLUCOSE WATER

3. In plants and algae, photosynthesis proceeds inside the organelles called chloroplasts.

a. Two outermost membranes enclose a chloroplast's semifluid interior. A third membrane extends through the stroma as a system of interconnected channels and disks (which are often arranged in stacks).

b. Light-dependent reactions proceed at thylakoids: disk-shaped sacs of the inner membrane system. ATP and NADPH form. Oxygen is released as a by-product.

c. Light-independent reactions occur in the stroma. Glucose molecules form. Each has a phosphate group attached and typically is used at once for the assembly of sucrose, starch, and cellulose. Those three organic compounds are the main end products of photosynthesis.

4. The light-dependent reactions start after pigments absorb wavelengths of visible light from the sun.

a. Chlorophyll *a*, the main photosynthetic pigment, absorbs wavelengths that are most efficient for driving photosynthesis. Like other chlorophylls, it absorbs nearly all wavelengths of visible light except for some yellow-green and green ones, which it reflects or transmits.

b. Accessory pigments can absorb wavelengths that chlorophyll *a* cannot absorb. They include chlorophyll *b* and the carotenoids (such as beta-carotene).

5. In chloroplasts, photosynthetic pigments are part of photosystems. Each photosystem has a cluster of 200 to 300 pigments, most of which are harvesters of energy. Light absorption raises some electrons of the harvester pigments to a higher energy level. Then the energy of excitation passes randomly among the pigments until it reaches a chlorophyll *a* that acts as a reaction center. When suitably activated, that chlorophyll alone releases excited electrons to a nearby acceptor molecule, which is poised at the start of an electron transport system.

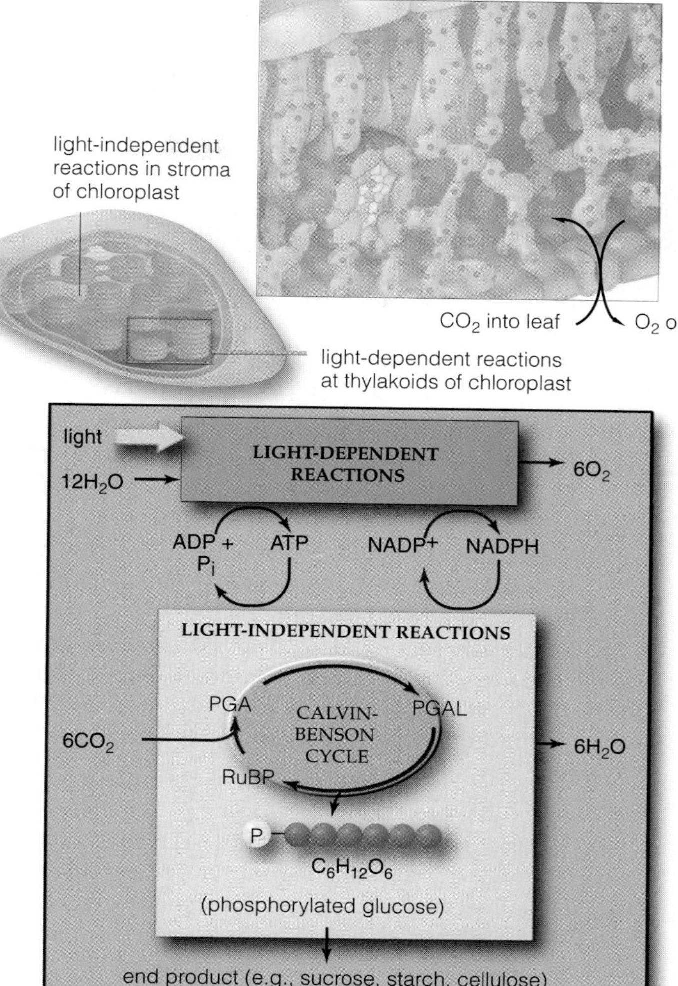

light-independent reactions in stroma of chloroplast
CO_2 into leaf
O_2 out
light-dependent reactions at thylakoids of chloroplast
light
$12H_2O$
LIGHT-DEPENDENT REACTIONS
$6O_2$
ADP + P_i
ATP
NADP+
NADPH
LIGHT-INDEPENDENT REACTIONS
PGA
CALVIN-BENSON CYCLE
PGAL
$6CO_2$
RuBP
$6H_2O$
P
$C_6H_{12}O_6$
(phosphorylated glucose)
end product (e.g., sucrose, starch, cellulose)

Figure 7.18 Summary of photosynthesis. In the light-dependent reactions, sunlight energy is converted to chemical bond energy of ATP. With a noncyclic pathway, water molecules are split by photolysis. $NADP^+$, a coenzyme, accepts released electrons and hydrogen to form NADPH. Oxygen is released as a by-product. With the light-independent reactions, a phosphorylated glucose molecule forms in the Calvin–Benson cycle, the RuBP on which the cycle turns is regenerated, and new water molecules form. *Each turn* of the cycle requires one CO_2, three ATP, and two NADPH. Because each glucose molecule has a backbone of six carbon atoms, its formation requires six turns of the cycle.

6. The thylakoid membrane incorporates two types of photosystems: I and II. An ancient pathway of electron flow leading to ATP formation uses only photosystem I. A chlorophyll *a* molecule designated P700 is at this photosystem's reaction center. Photosystems I and II both operate in a noncyclic pathway by which ATP and NADPH form. The reaction center of photosystem II has a chlorophyll *a* designated P680.

7. In the cyclic pathway, sunlight energy is converted to chemical bond energy of ATP when excited electrons

cycle from P700, through a transport system, then back to photosystem I. New electrons are not required; the electrons released from P700 simply return to P700.

8. Noncyclic electron flow has these features:

a. There is a *linear* flow of excited electrons from water to P680 (photosystem II), into a transport system to P700 (photosystem I), then on through a transport system that delivers electrons to $NADP^+$.

b. Water molecules give up the electrons when they are split into oxygen and hydrogen (photolysis).

c. By 1.5 billion years ago, oxygen released by the noncyclic pathway had accumulated in the atmosphere. It ultimately made possible aerobic respiration.

9. With the light-dependent reactions, ATP forms from the cell's pools of ADP and inorganic phosphate.

a. While components of the transport systems accept electrons, they also pick up hydrogen ions (H^+) from the stroma and shunt them into a thylakoid compartment.

b. H^+ from water molecules split during photolysis also collects in the compartment.

c. The activity sets up H^+ concentration and electric gradients across the thylakoid membrane. So H^+ flows out through ATP synthases (into the stroma), and the energy behind this flow drives the formation of ATP.

10. Here are the key points about the light-independent reactions, which proceed in the stroma:

a. The reactions require ATP and NADPH. The ATP delivers energy (by phosphate-group transfers), and the NADPH delivers hydrogen and electrons to reaction sites. Glucose forms and is used in the assembly of starch, cellulose, and other end products of photosynthesis.

b. The reactions are steps of the Calvin–Benson cycle. They start when an enzyme, rubisco, affixes carbon from CO_2 to five-carbon RuBP. The intermediate formed splits into two PGA. ATP phosphorylates each one. NADPH gives electrons and H^+ to the formation of two PGAL.

c. For every six carbon atoms that enter the cycle by way of carbon fixation, twelve PGAL form. Two PGAL are used to produce a six-carbon sugar phosphate. The remainder are used to regenerate the RuBP.

11. Rubisco, a carbon-fixing enzyme, evolved when the atmosphere had far more CO_2 and far less oxygen. Today, when the O_2 level is higher than the CO_2 level in leaves, it attaches oxygen rather than carbon to RuBP. This results in the formation of only one PGA (not two) and glycolate, a compound that cannot be used to form sugars but instead is degraded to carbon dioxide and water. This wasteful process is called photorespiration.

12. Photorespiration predominates in C3 plants such as sunflowers during hot, dry conditions. Then, stomata close and O_2 from photosynthesis builds up inside the leaf to levels higher than the CO_2 levels. Sugarcane and other C4 plants can raise the CO_2 level by fixing carbon twice, in two cell types. CAM plants such as cacti raise it by fixing carbon at night, when stomata are open.

Review Questions

1. A cat eats a bird, which earlier ate a caterpillar that chewed on a weed. Which organisms are autotrophs? Heterotrophs? *CI*

2. Summarize the photosynthesis reactions as an equation. Name where each stage takes place inside a chloroplast. *7.1*

3. Which of the following pigments are most visible in a maple leaf in summer? Which become the most visible in autumn? *7.3*

a. chlorophylls	c. anthocyanins
b. phycobilins	d. carotenoids

4. How does chlorophyll *a* differ in function from accessory pigments during the light-dependent reactions? *7.3*

5. Fill in Figure 7.19 for the light-dependent reactions. *7.4*

6. With respect to the light-dependent reactions, how do the cyclic and noncyclic pathways of electron flow differ? *7.4*

7. What substance does *not* take part in the Calvin–Benson cycle: ATP, NADPH, RuBP, carotenoids, O_2, CO_2, or enzymes? *7.6*

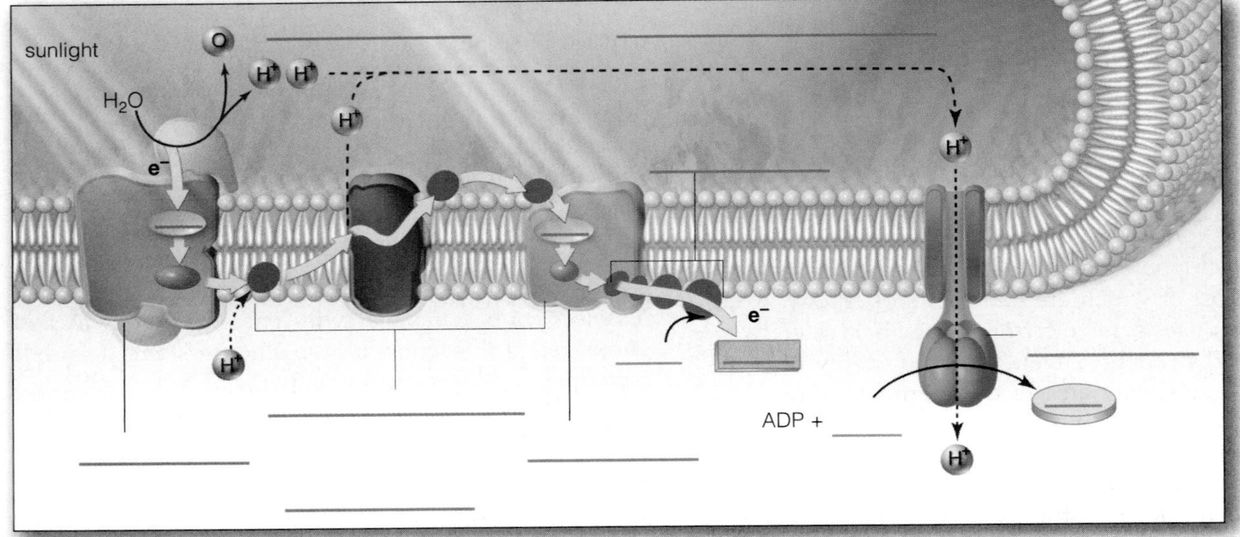

sunlight
H₂O
O
H⁺
H⁺
e⁻
H⁺
H⁺
e⁻
ADP +
H⁺
H⁺

Figure 7.19 Review Figure 7.12, which shows the light-dependent reactions of photosynthesis. Then, on your own, fill in the blanks (*red* lines) with the names of key components and activities.

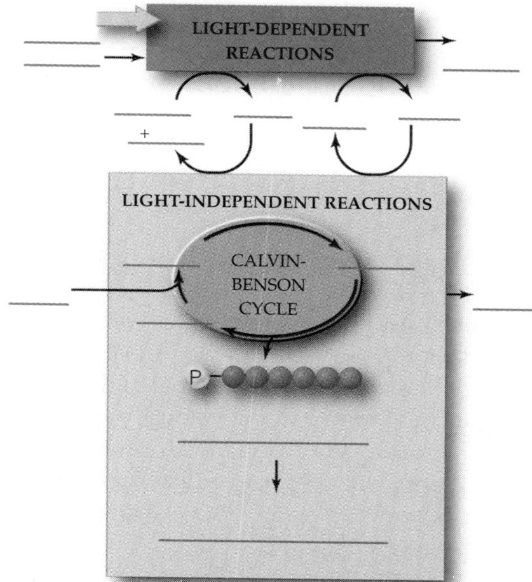
LIGHT-DEPENDENT REACTIONS
+
LIGHT-INDEPENDENT REACTIONS
CALVIN-BENSON CYCLE
P

Figure 7.20 Review Figure 7.18. Then, on your own, fill in the blanks (*red* lines) with the names of the key reactants, intermediates, and products of photosynthesis.

8. Fill in the blanks for Figure 7.20. Which substances are the original sources of carbon atoms and hydrogen atoms used in the synthesis of glucose in the Calvin–Benson cycle? *7.1, 7.6*

9. How many carbon atoms from CO_2 must enter the Calvin–Benson cycle to produce one glucose molecule? Why? *7.6*

Self-Quiz (*Answers in Appendix III*)

1. Photosynthetic autotrophs use ______ from the air as a carbon source and ______ as their energy source.

2. In plants, light-dependent reactions proceed at the ______.
   a. cytoplasm c. stroma
   b. plasma membrane d. thylakoid membrane

3. In the light-*dependent* reactions, ______.
   a. carbon dioxide is fixed c. CO_2 accepts electrons
   b. ATP and NADPH form d. sugar phosphates form

4. Identify which of the following substances accumulates inside the thylakoid compartment of chloroplasts during the light-dependent reactions:
   a. glucose c. chlorophyll e. hydrogen ions
   b. carotenoids d. fatty acids

5. When a photosystem absorbs light, ______.
   a. sugar phosphates are produced
   b. electrons are transferred to ATP
   c. RuBP accepts electrons
   d. light-dependent reactions begin

6. The light-*independent* reactions proceed in the ______.
   a. cytoplasm b. plasma membrane c. stroma d. grana

7. The Calvin–Benson cycle starts when ______.
   a. light is available
   b. light is not available
   c. carbon dioxide is attached to RuBP
   d. electrons leave a photosystem

8. ATP phosphorylates ____ in the light-independent reactions.
   a. RuBP b. $NADP^+$ c. PGA d. PGAL

9. Match each event with its most suitable description.
   ____ RuBP used; PGA forms a. photon absorption
   ____ ATP and NADPH used b. noncyclic pathway
   ____ NADPH forms c. CO_2 fixation
   ____ ATP and NADPH form d. PGAL forms
   ____ only ATP forms e. H^+ and e^- to $NADP^+$
   ____ energy matches amount needed to excite electrons f. cyclic pathway

Critical Thinking

1. About 200 years ago, Jan Baptista van Helmont performed an experiment on the nature of photosynthesis. He wanted to know where growing plants acquire the raw materials necessary to increase in size. For his experiment, he planted a tree seedling weighing 5 pounds in a barrel filled with 200 pounds of soil. He watered the tree regularly. Five years passed. Van Helmont again weighed the tree and the soil. The tree weighed 169 pounds, 3 ounces. The soil weighed 199 pounds, 14 ounces. Because tree weight had increased so much and soil weight had decreased so little, he concluded the tree had gained weight after absorbing the water he had added to the barrel.

Given what you know about the composition of biological molecules, why was van Helmont's conclusion misguided? Reflect on the current model of photosynthesis and then give a more plausible explanation of his results.

2. Like other accessory pigments, carotenoids extend the effective range of light absorption beyond chlorophyll *a*, the main pigment of photosynthesis. They also protect plants from *photo-oxidation*. This destructive process begins when the excitation energy of chlorophylls drives the conversion of oxygen into free radicals, which affect carotenoid synthesis. In fact, plants damaged by photo-oxidation cannot make cartenoids. Grow them in light and they will bleach white and die. Given this result, which molecules in the plant cells are among the first to go?

3. Suppose a garden in your neighborhood is filled with red, white, and blue petunias. Explain the floral colors in terms of which wavelengths of light they are absorbing and reflecting.

4. A busily photosynthesizing plant takes up molecules of CO_2 that have incorporated radioactively labeled carbon atoms ($^{14}CO_2$). Identify the compound in which the labeled carbon will appear first: NADPH, PGAL, pyruvate, or PGA.

Selected Key Terms

absorption spectrum *7.2*
anthocyanin *7.3*
autotroph *CI*
C3 plant *7.7*
C4 plant *7.7*
Calvin–Benson cycle (light-independent reactions) *7.6*
CAM plant *7.7*
carbon fixation *7.6*
carotenoid *7.3*
chlorophyll *7.3*
chloroplast *7.1*
electromagnetic spectrum *7.2*
electron transport system *7.4*
fluorescence *7.3*
heterotroph *CI*
hydrothermal vent *7.8*
light-dependent reactions *7.4*
PGA *7.6*
PGAL *7.6*
photolysis *7.4*
photon *7.2*
photosynthesis *CI*
photosystem *7.3*
phycobilin *7.3*
pigment *7.2*
reaction center *7.4*
rubisco *7.6*
RuBP *7.6*
stoma (stomata) *7.7*
stroma *7.1*
thylakoid *7.1*
wavelength *7.2*

Readings *See also www.infotrac-college.com*

Bazzaz, F., and E. Jajer. January 1992. "Plant Life in a CO_2-Rich World." *Scientific American*, 266:68–74.

Zimmer, C. November 1996. "The Light at the Bottom of the Sea." *Discover*, 63–73.

8 HOW CELLS RELEASE STORED ENERGY

The Killers Are Coming! The Killers Are Coming!

In 1990, thanks to selective breeding experiments gone wrong, descendants of "killer" bees that flew out of South America a few decades earlier buzzed across the border between Mexico and Texas. By 1995, they had invaded 13,287 square kilometers of southern California and were busily setting up colonies. By 1998, when nectar-rich desert flowers bloomed profusely after heavy El Niño storms, the invasion extended even farther west and north than scientists had predicted.

When provoked, the bees behave in a terrifying way. For example, thousands flew into action simply because a construction worker started up a tractor a few hundred yards away from their hive. Agitated bees entered a nearby subway station and started stinging passengers on the platform and in trains. They killed one person and injured a hundred others.

Where did these bees come from? Some queen bees were shipped from Africa to Brazil for breeding experiments in the 1950s. Why? As it happens, honeybees are big business. Besides being a source of nutritious honey, bees are rented to commercial orchards, where their collective pollinating activities may significantly enhance fruit production. For example, enclose a blossoming orchard tree in a pollinator-excluding cage, and less than 1 percent of the tree's flowers will set fruit. But put a hive of honeybees inside the cage with the tree and 40 percent of the flowers will set fruit. Compared to their relatives in Africa, bees in Brazil are rather sluggish pollinators and honey producers. By cross-breeding the two, researchers thought they might come up with a strain of mild-mannered but zippier bees. So they put local bees and imported ones together in netted enclosures, complete with artificial hives. Then they let nature take its course.

Figure 8.1 One of the mild-mannered honeybees buzzing in for a landing on a flower, wings beating with energy provided by ATP. If this were one of its Africanized relatives protecting a hive, possibly you would not stay around to watch the landing. Both kinds of bees look alike. How can we tell them apart? From our own biased perspective, Africanized bees are the ones with an attitude problem.

Twenty-six African queen bees escaped. That was bad enough. Then beekeepers got wind of preliminary experimental results. After learning that the first few generations of offspring were more energetic but not overly aggressive, they imported hundreds of African queens and encouraged them to mate with the locals. And they set off a genetic time bomb.

Before long, African bees became established in commercial hives—and in wild bee populations. Their traits became dominant. The "Africanized" bees do everything other bees do, but they do more of it faster. Their eggs develop into adults more quickly. Adults fly more rapidly, outcompete other bees for nectar, and even die sooner.

When something disturbs their hives or swarms, Africanized bees become extremely agitated. They can remain that way for as long as eight hours. Whereas a mild-mannered honeybee might chase an intruding animal fifty yards or so, a squadron of Africanized bees will chase it a quarter of a mile. If they catch up to it, they collectively can sting it to death.

Doing things faster means having a continuous supply of energy and efficient ways of using it. An Africanized bee's stomach can hold thirty milligrams of sugar-rich nectar—which is enough fuel to fly sixty kilometers. That's more than thirty-five miles! Besides this, compared to other kinds of bees, an Africanized bee's flight muscle cells have larger mitochondria. These organelles specialize in releasing a great deal of energy from sugars and other organic compounds, then converting it to the energy of ATP.

Whenever they tap into the stored energy of organic compounds, Africanized bees reveal their biochemical connection with other organisms. Study a primrose or puppy, a mold growing on stale bread, an amoeba in pondwater, or a bacterium living on your skin, and you will discover that their energy-releasing pathways differ in some details. But all of the pathways require characteristic starting materials. They yield predictable products and by-products. And they yield the universal energy currency of life—ATP.

In fact, throughout the biosphere, organisms put energy and raw materials to use in amazingly similar ways. *At the biochemical level, we find undeniable unity among all forms of life.* We will return to this idea in the concluding section of the chapter.

KEY CONCEPTS

1. The cells of all organisms can release energy stored in glucose and other organic compounds, then use it in ATP production. The energy-releasing pathways differ from one another. But the main types all start with the breakdown of glucose to pyruvate.

2. The initial breakdown reactions, known as glycolysis, can proceed in the presence of oxygen or in its absence. Said another way, these reactions can be the first stage of either aerobic or anaerobic pathways.

3. Two kinds of energy-releasing pathways are entirely anaerobic, from start to finish. We call them fermentation and anaerobic electron transport. They proceed only in the cytoplasm, and none yields more than a small amount of ATP for each glucose molecule metabolized.

4. Another pathway, aerobic respiration, also starts in the cytoplasm. But this one runs to completion in organelles called mitochondria. Compared with the other pathways, aerobic respiration releases far more energy from glucose.

5. Aerobic respiration has three stages. First, pyruvate forms from glucose (during glycolysis). Second, different reactions break down the pyruvate to carbon dioxide. These reactions liberate electrons and hydrogen, which coenzymes deliver to an electron transport system. Third, stepwise electron transfers through the system help set up the conditions that favor ATP formation. Free oxygen accepts the electrons at the end of the line and combines with hydrogen, thereby forming water.

6. Over evolutionary time, photosynthesis and aerobic respiration became linked on a global scale. The oxygen-rich atmosphere, a long-term outcome of photosynthetic activity, sustains aerobic respiration, which has become the dominant energy-releasing pathway. And most kinds of photosynthesizers use carbon dioxide and water from aerobic respiration as raw materials when they synthesize organic compounds:

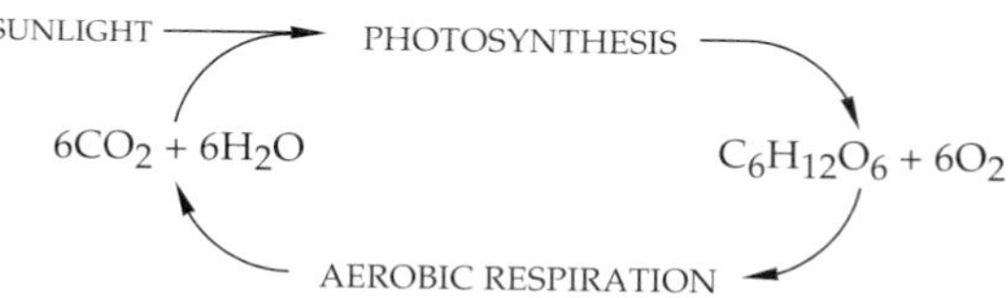
SUNLIGHT
PHOTOSYNTHESIS
$6CO_2 + 6H_2O$
$C_6H_{12}O_6 + 6O_2$
AEROBIC RESPIRATION

8.1 HOW DO CELLS MAKE ATP?

Organisms stay alive by taking in energy. Plants and all other organisms that engage in photosynthesis get energy from the sun. Animals get energy secondhand, thirdhand, and so on, by eating plants and one another. Regardless of its source, energy must be in a form that can drive thousands of life-sustaining reactions. Energy that becomes converted into the chemical bond energy of adenosine triphosphate—ATP—serves that function.

Plants make ATP during photosynthesis, which they then use to produce glucose and other carbohydrates. Plants and all other organisms also can make ATP by breaking down carbohydrates (glucose especially), fats, and proteins. During the breakdown reactions, electrons are stripped from intermediates, then energy associated with the liberated electrons drives the formation of ATP. Electron transfers of the sort outlined in Section 6.4 are central to these energy-releasing pathways.

Comparison of the Main Types of Energy-Releasing Pathways

The first energy-releasing pathways evolved about 3.8 billion years ago, when conditions were very different on Earth. Because the atmosphere had little free oxygen, the pathways must have been *anaerobic*, which means they could run to completion without utilizing oxygen. Many bacteria and protistans still live in places where oxygen is absent or not always available. They make ATP by anaerobic routes, mainly fermentation pathways and anaerobic electron transport. Some cells in your own body use an anaerobic route for short periods, but only when they are not receiving enough oxygen. Your cells, like most others, rely primarily on **aerobic respiration**, an oxygen-dependent pathway of ATP formation. With each breath you take, you are providing your actively respiring cells with a fresh supply of oxygen.

Make note of this point: *The main energy-releasing pathways all start with the same reactions in the cytoplasm.* During this initial stage of reactions, called **glycolysis**, enzymes cleave and rearrange a glucose molecule into two molecules of **pyruvate**, which has a backbone of three carbon atoms. Once glycolysis is completed, the energy-releasing pathways differ. Most importantly, only the aerobic pathway continues inside a mitochondrion (Figure 8.2). There, oxygen serves as the final acceptor of electrons used during the reactions. The anaerobic pathways start and end in the cytoplasm. A substance other than oxygen is the final electron acceptor.

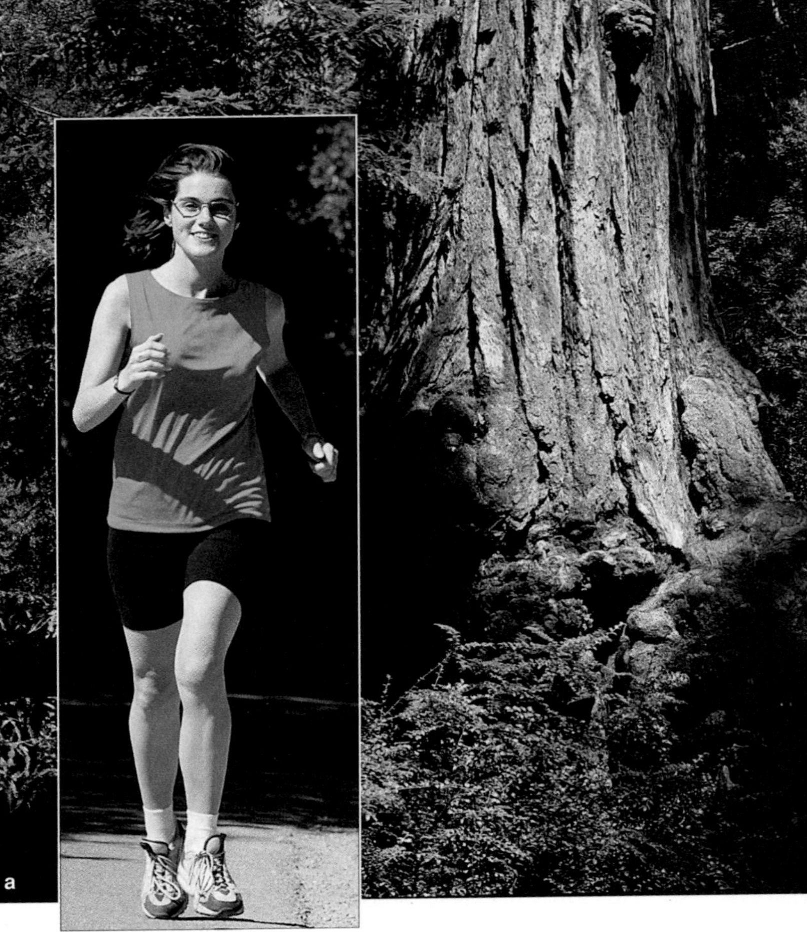
a

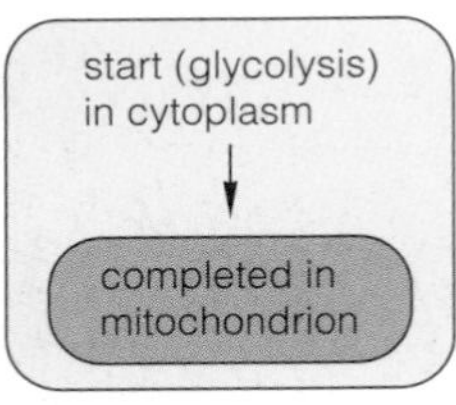
start (glycolysis) in cytoplasm
completed in mitochondrion
AEROBIC RESPIRATION

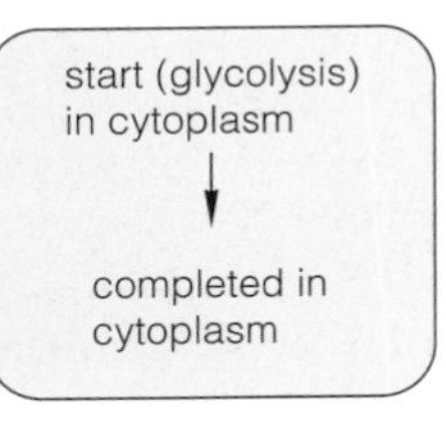
start (glycolysis) in cytoplasm
completed in cytoplasm
ANAEROBIC ENERGY-RELEASING PATHWAYS

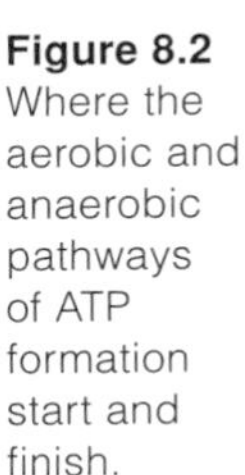

Figure 8.2 Where the aerobic and anaerobic pathways of ATP formation start and finish.

As you examine the energy-releasing pathways in sections to follow, keep in mind that the reaction steps do not proceed by themselves. Enzymes catalyze each step, and the intermediate molecules formed at one step serve as substrates for the next enzyme in the pathway.

Overview of Aerobic Respiration

Of all energy-releasing pathways, aerobic respiration gets the most ATP for each glucose molecule. Whereas anaerobic routes typically have a net yield of two ATP molecules, the aerobic route commonly yields thirty-six or more. If you were a bacterium, you wouldn't require much ATP. Being far larger, more complex, and highly active, you rely absolutely on the aerobic route's high yield. When a glucose molecule is the starting material, aerobic respiration can be summarized this way:

$$\underset{\text{GLUCOSE}}{C_6H_{12}O_6} + \underset{\text{OXYGEN}}{6O_2} \longrightarrow \underset{\text{CARBON DIOXIDE}}{6CO_2} + \underset{\text{WATER}}{6H_2O}$$

However, as you can see, the summary equation only tells us what the substances are at the start and finish of the pathway. In between are three reaction stages.

Further reading: Student Guide to InfoTrac on web site —

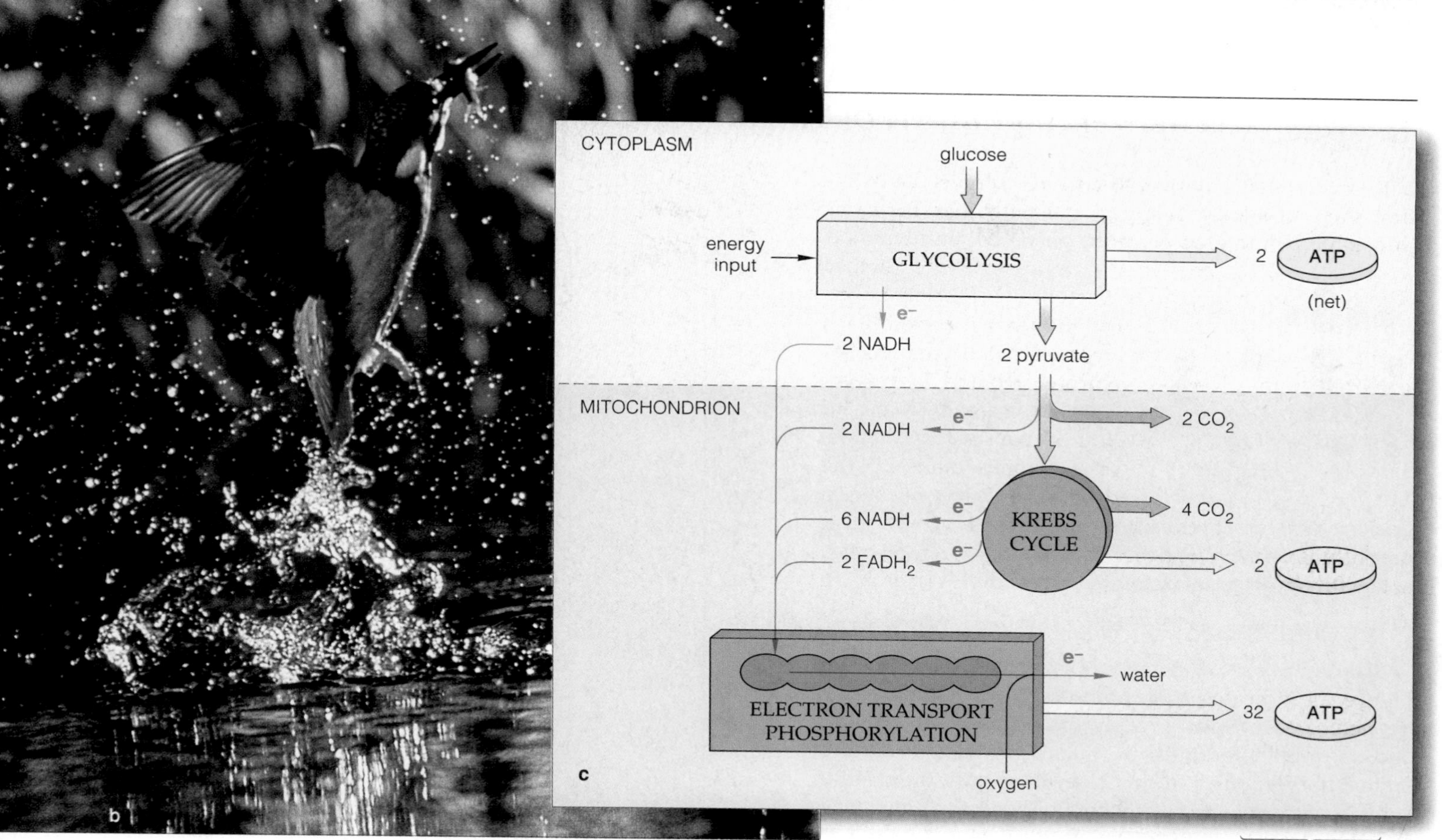

CYTOPLASM
glucose
energy input
GLYCOLYSIS
2 ATP
(net)
e⁻
2 NADH
2 pyruvate
MITOCHONDRION
2 NADH
e⁻
2 CO2
6 NADH
e⁻
KREBS CYCLE
4 CO2
2 FADH2
e⁻
2 ATP
e⁻
water
ELECTRON TRANSPORT PHOSPHORYLATION
32 ATP
oxygen
c
b
Typical Energy Yield: 36 ATP

Figure 8.3 Overview of aerobic respiration, which has three stages. From start to finish, the typical net energy yield from each glucose molecule is thirty-six ATP. Only this pathway delivers enough ATP to construct and maintain giant redwoods and all other large, multicelled organisms. It alone delivers enough ATP for highly active animals, such as bees, humans (**a**), and kingfishers (**b**) and other birds.

(**c**) In the first stage (glycolysis), enzymes partially break down glucose to pyruvate. In the second stage, which is mainly the Krebs cycle, enzymes completely break down pyruvate to carbon dioxide. NAD^+ and FAD pick up electrons and hydrogen stripped from intermediates at both stages. In the final stage (electron transport phosphorylation), those reduced coenzymes (NADH and $FADH_2$) give up electrons and hydrogen to a transport system. Energy released during the flow of electrons through the system drives ATP formation. Oxygen accepts the electrons at the end of the third stage.

Let's use Figure 8.3 and the following descriptions as a brief overview of these reactions. The initial stage, again, is glycolysis. Most of the second stage consists of a cyclic pathway, the **Krebs cycle**, when enzymes break down pyruvate to carbon dioxide and water.

NAD^+ (nicotinamide adenine dinucleotide) and **FAD** (flavin adenine dinucleotide) take part in glycolysis and the Krebs cycle. These organic compounds, derived from vitamins, are *coenzymes* (Section 6.7). They help enzymes by accepting electrons (e^-) and hydrogen removed from intermediates, then transferring the electrons elsewhere. Unbound hydrogen atoms, recall, are naked protons, or hydrogen ions (H^+). So they tag along with the oppositely charged electrons. In their reduced form, the coenzymes are abbreviated NADH and $FADH_2$.

Not much ATP forms during glycolysis or the Krebs cycle. The large energy harvest comes *after* coenzymes deliver their cargo to an electron transport system.

In the third stage, the transport system functions as machinery for **electron transport phosphorylation**. It sets up H^+ concentration and electric gradients, which drive ATP formation at nearby membrane proteins. It is during this final stage that so many ATP molecules are produced. As it ends, oxygen inside the mitochondrion accepts the "spent" electrons from the last component of the transport system. Oxygen picks up hydrogen at the same time and thereby forms water.

Cells drive nearly all metabolic activities by releasing energy from glucose and other organic compounds and converting it to chemical bond energy of ATP.

All of the main energy-releasing pathways start inside the cytoplasm with glycolysis, a stage of reactions that break down glucose to pyruvate.

The most common anaerobic pathways, which include the fermentation routes, end in the cytoplasm. Each has a net energy yield of two ATP.

Aerobic respiration, an oxygen-dependent pathway, runs to completion in the mitochondrion. From start (glycolysis) to finish, it commonly has a net energy yield of thirty-six ATP.

8.2 GLYCOLYSIS: FIRST STAGE OF ENERGY-RELEASING PATHWAYS

Let's track what happens to a glucose molecule in the first stage of aerobic respiration. Remember, the same things happen to glucose in the anaerobic pathways.

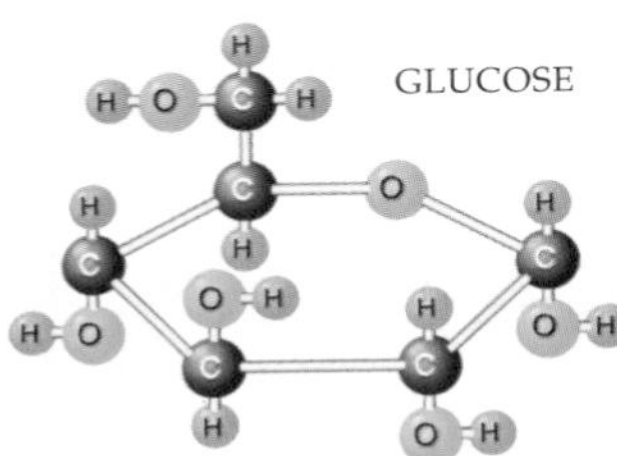

GLUCOSE

As described earlier, in Section 3.2, glucose is one of the simple sugars. Each molecule of it contains six carbon, twelve hydrogen, and six oxygen atoms, all joined by covalent bonds. The carbons make up the backbone. With glycolysis, glucose or some other carbohydrate in the cytoplasm is partially broken down, the result being two molecules of the three-carbon compound pyruvate:

glucose → glucose–6–phosphate → 2 pyruvate

The first steps of glycolysis are *energy-requiring*. As Figure 8.4 indicates, they proceed only when two ATP molecules each transfer a phosphate group to glucose and so donate energy to it. Such transfers, recall, are "phosphorylations." In this case, they raise the energy content of glucose to a level high enough to allow entry into the *energy-releasing* steps of glycolysis.

The first energy-releasing step cleaves the activated glucose into two molecules. We can call each of these PGAL (phosphoglyceraldehyde). Each PGAL becomes converted to an unstable intermediate that allows ATP to form by giving up a phosphate group to ADP. The next intermediate in the sequence does the same thing.

Thus, a total of four ATP form by **substrate-level phosphorylation**. We define this metabolic event as the direct transfer of a phosphate group from a substrate of a reaction to some other molecule—in this case, ADP. Remember, though, two ATP were invested to jump-start the reactions. So the *net* energy yield is only two ATP.

Meanwhile, the coenzyme NAD^+ picks up electrons and hydrogen atoms liberated from each PGAL, thus becoming NADH. When the NADH gives up its cargo at a different reaction site, it reverts to NAD^+. Said another way, like other coenzymes, NAD^+ is reusable.

In sum, glycolysis converts energy stored in glucose to a transportable form of energy, in ATP. NAD^+ picks up electrons and hydrogen stripped from glucose. These have roles in the next stage of reactions. So do the end products of glycolysis—the two molecules of pyruvate.

Glycolysis is an energy-releasing stage of reactions in which glucose or some other carbohydrate is partially broken down to two molecules of pyruvate.

Two NADH and four ATP form. However, when we subtract the two ATP required to start the reactions, the *net* energy yield of glycolysis is two ATP per glucose molecule.

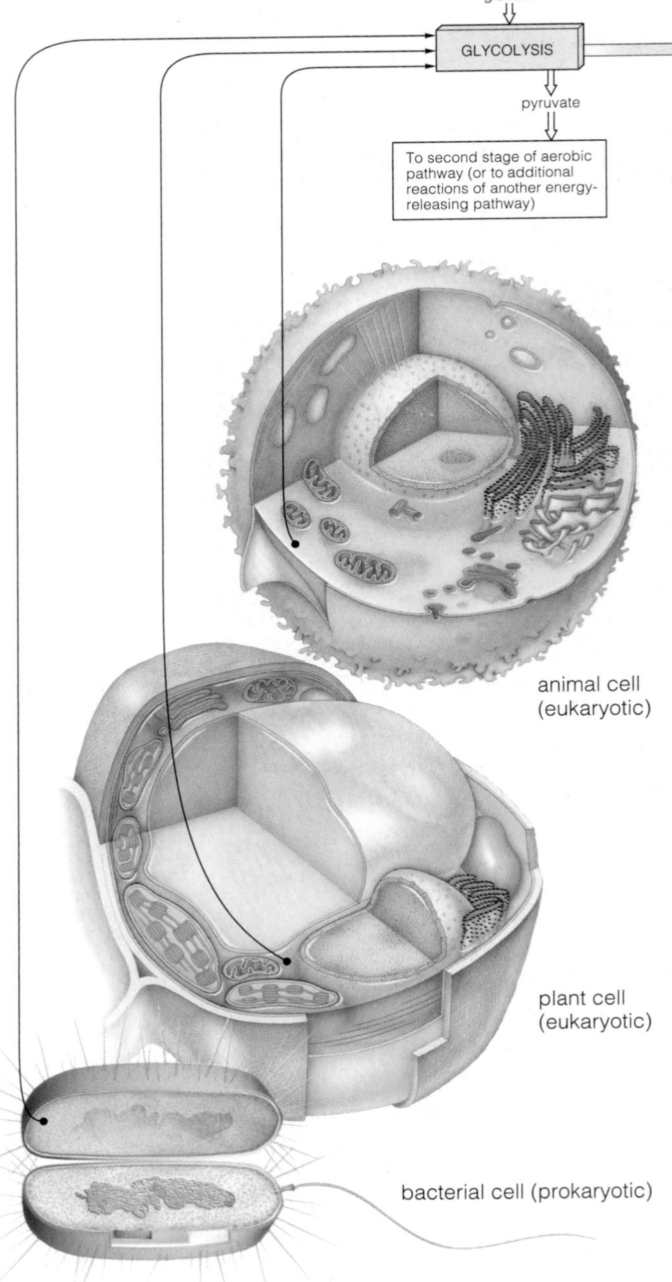

glucose
GLYCOLYSIS
pyruvate
To second stage of aerobic pathway (or to additional reactions of another energy-releasing pathway)
animal cell (eukaryotic)
plant cell (eukaryotic)
bacterial cell (prokaryotic)

Figure 8.4 Glycolysis, first stage of the main energy-releasing pathways. The reaction steps proceed inside the cytoplasm of every living prokaryotic and eukaryotic cell. In this example, glucose is the starting material. By the time the reactions end, two pyruvate, two NADH, and four ATP have been produced. Cells invest two ATP to start glycolysis, however, so the *net* energy yield of glycolysis is two ATP. For an expanded picture of the reactions, refer to Appendix V (Figure A).

Depending on the type of cell and on environmental conditions, the pyruvate may enter the second set of reactions of the aerobic pathway, which includes the Krebs cycle. Or it may be used in other reactions, such as those of fermentation pathways.

Further reading: Student Guide to InfoTrac on web site —

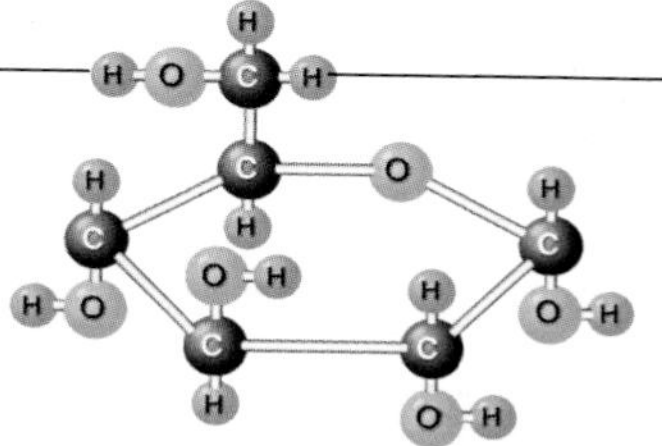
H
H
O
C
H
H
C
O
H
C
H
O
O
H
H
C
H
O
H
H
C
C
H
O
H

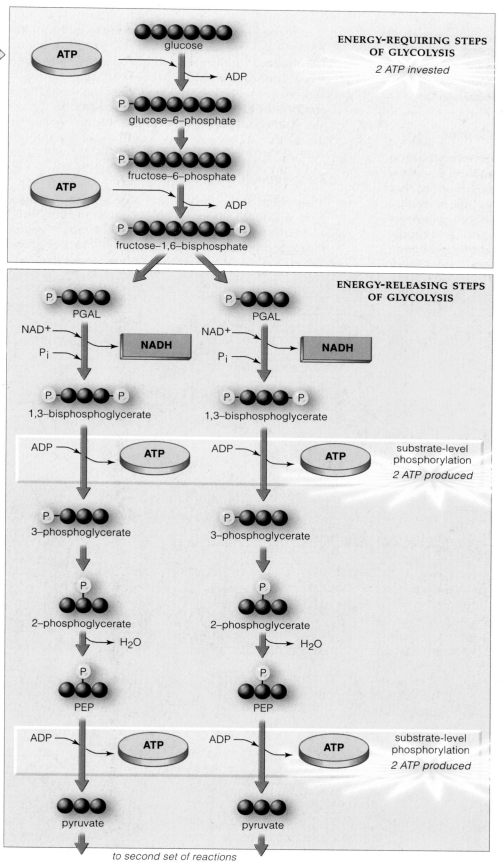

ENERGY-REQUIRING STEPS OF GLYCOLYSIS

2 ATP invested

ATP → glucose → ADP

P-glucose-6-phosphate

P-fructose-6-phosphate

ATP → ADP

P-fructose-1,6-bisphosphate-P

ENERGY-RELEASING STEPS OF GLYCOLYSIS

P-PGAL | P-PGAL

NAD^+, P_i → NADH | NAD^+, P_i → NADH

P-1,3-bisphosphoglycerate-P | P-1,3-bisphosphoglycerate-P

ADP → ATP | ADP → ATP

substrate-level phosphorylation

2 ATP produced

P-3-phosphoglycerate | P-3-phosphoglycerate

P 2-phosphoglycerate | P 2-phosphoglycerate

H_2O | H_2O

P PEP | P PEP

ADP → ATP | ADP → ATP

substrate-level phosphorylation

2 ATP produced

pyruvate | pyruvate

to second set of reactions

a This diagram tracks the fate of the six carbon atoms (*red* spheres) of a glucose molecule, shown above. Glycolysis starts with an energy investment of two ATP.

b A phosphate-group transfer from one of the ATP molecules to glucose causes atoms in glucose to undergo rearrangements.

c A phosphate-group transfer from another ATP causes rearrangements that form the intermediate fructose–1,6–bisphosphate.

d The intermediate splits at once into two molecules, each with a three-carbon backbone. We can call these two PGAL.

e Each PGAL gives up two electrons and a hydrogen atom to NAD^+, thus forming two NADH.

f Each PGAL also combines with inorganic phosphate (P_i) in the cytoplasm and then makes a phosphate-group transfer to ADP.

g *Thus two ATP have formed by the direct transfer of phosphate from two intermediate molecules that served as substrates for the reactions.* Formation of two ATP means the original energy investment of two ATP has been paid off.

h During the next two enzyme-mediated reactions, the two intermediates each release a hydrogen atom and an —OH group, which combine to form water.

i Two 3–phosphoenolpyruvate (PEP) molecules result. Each PEP makes a phosphate-group transfer to ADP.

j *Once again, two ATP have formed by substrate-level phosphorylation.*

k In sum, the net energy yield from glycolysis is two ATP for each glucose molecule entering the reactions. Two molecules of pyruvate, the end product, may enter the next set of reactions in an energy-releasing pathway.

SECOND STAGE OF THE AEROBIC PATHWAY

Suppose two pyruvate molecules formed by glycolysis leave the cytoplasm and enter a **mitochondrion** (plural, mitochondria). In this organelle, both the second and third stages of the aerobic pathway run to completion. Figure 8.5 shows its structure and functional zones.

Preparatory Steps and the Krebs Cycle

During the second stage, a bit more ATP forms. Carbon atoms depart from the pyruvate, in the form of carbon dioxide. And coenzymes latch onto the electrons and hydrogen stripped from intermediates of the reactions.

In a few preparatory reactions, an enzyme removes a carbon atom from each pyruvate molecule. An enzyme helper, coenzyme A, becomes **acetyl–CoA** by combining with the remaining two-carbon fragment. The fragment is transferred to **oxaloacetate**, the entry point for the Krebs cycle. The name of this cyclic pathway honors Hans Krebs, who began working out its details in the 1930s. It also is called the citric acid cycle. Notice, in Figure 8.6, that *six* carbon atoms enter the second stage of reactions (three in each pyruvate backbone). Notice also that *six* depart, in six carbon dioxide molecules, during the preparatory steps and the Krebs cycle.

Functions of the Second Stage

The second stage of reactions serves three functions. First, it loads electrons and hydrogen onto both NAD^+ and FAD, which results in NADH and $FADH_2$. Second, through substrate-level phosphorylations, it produces two ATP molecules. And third, it rearranges the Krebs

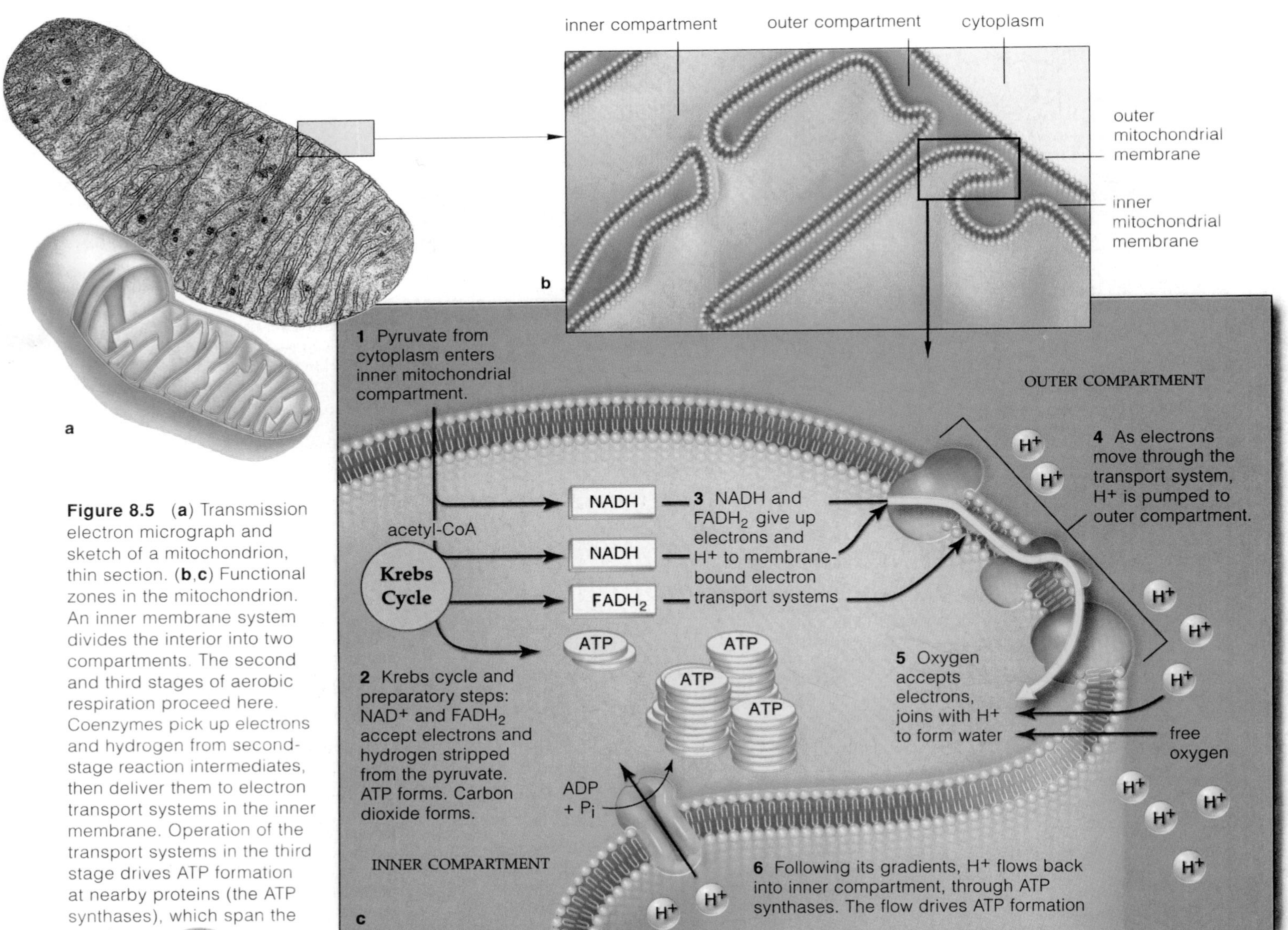

inner compartment
outer compartment
cytoplasm
outer mitochondrial membrane
inner mitochondrial membrane
a
b
c
1 Pyruvate from cytoplasm enters inner mitochondrial compartment.
OUTER COMPARTMENT
acetyl-CoA
Krebs Cycle
NADH
NADH
$FADH_2$
3 NADH and $FADH_2$ give up electrons and H^+ to membrane-bound electron transport systems
4 As electrons move through the transport system, H^+ is pumped to outer compartment.
H^+
ATP
2 Krebs cycle and preparatory steps: NAD^+ and $FADH_2$ accept electrons and hydrogen stripped from the pyruvate. ATP forms. Carbon dioxide forms.
5 Oxygen accepts electrons, joins with H^+ to form water
free oxygen
ADP + P_i
INNER COMPARTMENT
6 Following its gradients, H^+ flows back into inner compartment, through ATP synthases. The flow drives ATP formation

Figure 8.5 (**a**) Transmission electron micrograph and sketch of a mitochondrion, thin section. (**b**,**c**) Functional zones in the mitochondrion. An inner membrane system divides the interior into two compartments. The second and third stages of aerobic respiration proceed here. Coenzymes pick up electrons and hydrogen from second-stage reaction intermediates, then deliver them to electron transport systems in the inner membrane. Operation of the transport systems in the third stage drives ATP formation at nearby proteins (the ATP synthases), which span the membrane.

Further reading: Student Guide to InfoTrac on web site

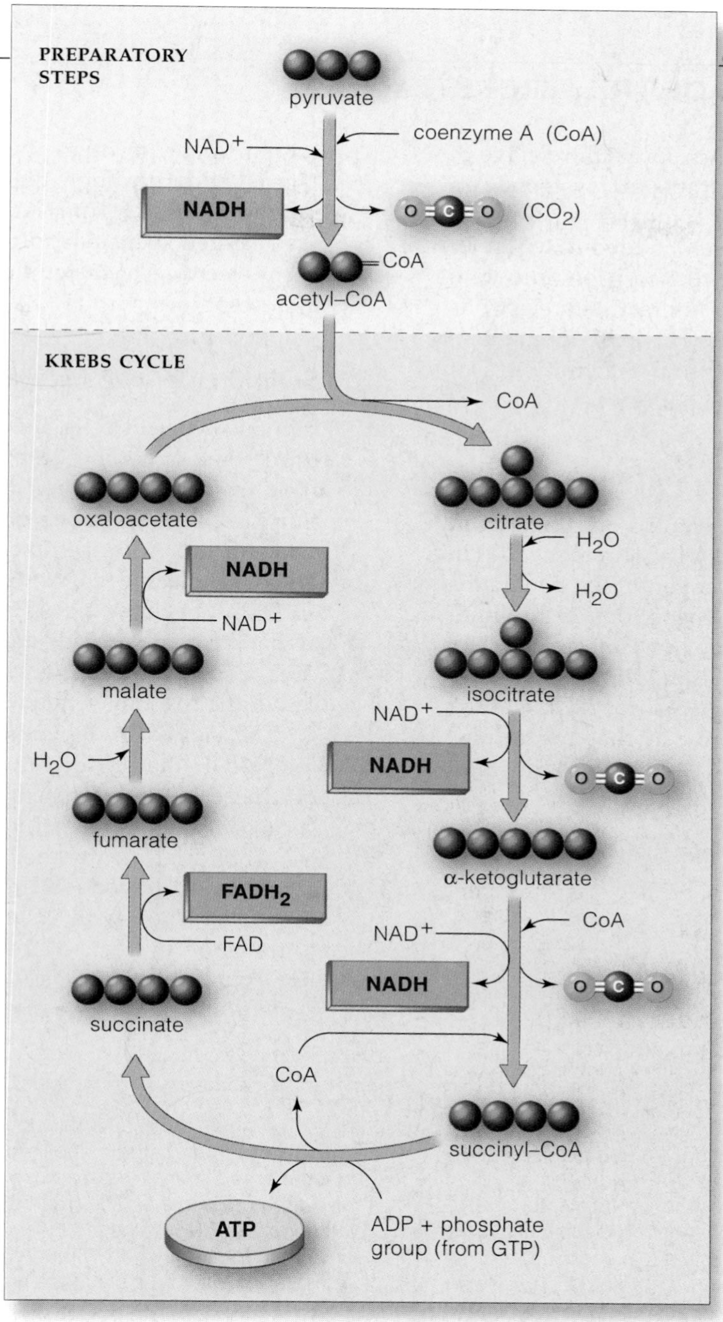

PREPARATORY STEPS

pyruvate
NAD^+
coenzyme A (CoA)
NADH
(CO_2)
CoA
acetyl–CoA

KREBS CYCLE

CoA
oxaloacetate
citrate
H_2O
H_2O
isocitrate
NAD^+
NADH
α-ketoglutarate
NAD^+
NADH
CoA
succinyl–CoA
CoA
ADP + phosphate group (from GTP)
ATP
succinate
FAD
$FADH_2$
fumarate
H_2O
malate
NAD^+
NADH

a Pyruvate from glucose enters a mitochondrion. It undergoes initial conversions before entering cyclic reactions (the Krebs cycle).

b Pyruvate is stripped of a carboxyl group (COO^-), which departs as CO_2. It also gives up hydrogen and electrons to NAD^+, forming NADH. Then a coenzyme joins with the remaining two-carbon fragment, forming acetyl–CoA.

c The acetyl–CoA transfers its two-carbon group to oxaloacetate, a four-carbon compound that is the entry point into the Krebs cycle. The result is citrate, with a six-carbon backbone. Addition and then removal of H_2O changes citrate to isocitrate.

d Isocitrate enters conversion reactions, and a COO^- group departs (as CO_2). Hydrogen and electrons are transferred to NAD^+, forming NADH.

e Another COO^- group departs (as CO_2) and another NADH forms. The resulting intermediate attaches to a coenzyme A molecule, forming succinyl–CoA. *At this point, three carbon atoms have been released, balancing out the three that entered the mitochondrion (in pyruvate).*

f Now the attached coenzyme is replaced by a phosphate group (donated by the substrate GTP). That phosphate group becomes attached to ADP. Thus, for each turn of the cycle, one ATP forms by substrate-level phosphorylation.

g Electrons and hydrogen are transferred to FAD, forming $FADH_2$.

h In the final conversion reactions, the oxaloacetate is regenerated, and hydrogen and electrons are transferred to NAD^+, forming NADH.

Figure 8.6 Second stage of aerobic respiration: the Krebs cycle and reaction steps that precede it. For each three-carbon pyruvate molecule entering the cycle, three CO_2, one ATP, four NADH, and one $FADH_2$ molecules form. The steps shown proceed *twice.* Why? Glucose, remember, was broken down earlier to *two* pyruvate molecules.

cycle intermediates into oxaloacetate. Cells have only so much oxaloacetate, which must be regenerated to keep the cyclic reactions going.

The two ATP that form in the second stage do not add much to the small yield from glycolysis. However, for each glucose molecule, *many* coenzymes pick up electrons and hydrogen for transport to the sites of the third and final stage of the aerobic pathway:

Glycolysis:	2 NADH
Pyruvate conversion before Krebs cycle:	2 NADH
Krebs cycle:	2 $FADH_2$ + 6 NADH
Coenzymes sent to third stage:	2 $FADH_2$ + 10 NADH

Overall, these are the key points to remember about the second stage of aerobic respiration:

During the second stage of aerobic respiration, two pyruvate molecules from glycolysis enter a mitochondrion.

Each pyruvate molecule gives up a carbon atom, then its two-carbon remnant enters the Krebs cycle. All of the carbon atoms of pyruvate eventually end up in carbon dioxide.

The second stage yields only two ATP. But the reactions regenerate oxaloacetate, the entry point for the Krebs cycle. And many coenzymes pick up electrons and hydrogen that were stripped from substrates, for delivery to the final stage of the pathway.

THIRD STAGE OF THE AEROBIC PATHWAY

ATP production goes into high gear in the third stage of the aerobic pathway. Electron transport systems and neighboring proteins called ATP synthases serve as the production machinery. They are embedded in the inner membrane that divides the mitochondrion into two compartments (Figure 8.7). They interact with electrons and unbound hydrogen—that is, H^+ ions. Remember, coenzymes deliver this bounty from reaction sites of the first two stages of the aerobic pathway.

Electron Transport Phosphorylation

Briefly, during the final stage, electrons get transferred from one molecule of each transport system to the next in line. When certain molecules accept and then donate the electrons, they also pick up hydrogen ions in the inner compartment. Shortly afterward, they release them to the outer compartment. Their shuttling action sets up H^+ concentration and electric gradients across the inner mitochondrial membrane. Nearby in that membrane, the ions follow the gradients and flow back to the inner compartment, through the interior of ATP synthases. The H^+ flow through these transport proteins drives formation of ATP from ADP and unbound phosphate. Free oxygen keeps ATP production going. It withdraws spent electrons at the end of the transport systems and then combines with H^+. Water is the result.

Summary of the Energy Harvest

In many types of cells, thirty-two ATP form during the third stage of aerobic respiration. Add these to the net yield from the preceding stages, and the total harvest is thirty-six ATP from one glucose molecule (Figure 8.8). That's a lot! An anaerobic pathway may use eighteen glucose molecules to produce the same amount of ATP.

Think of thirty-six ATP as a typical yield only. The actual amount depends on cellular conditions, as when cells require a given intermediate elsewhere and pull it out of the reaction sequence.

The yield also depends on how particular cells use the NADH that formed during glycolysis. Any NADH produced in the cytoplasm can't enter a mitochondrion. It can only deliver electrons and hydrogen *to* the outer mitochondrial membrane, where proteins shuttle them across to NAD^+ or FAD molecules already inside this organelle. Then both coenzymes deliver the electrons to transport systems of the inner membrane. However, FAD puts them at a *lower* entry point in the transport system, so *its* deliveries produce less ATP (Figure 8.8).

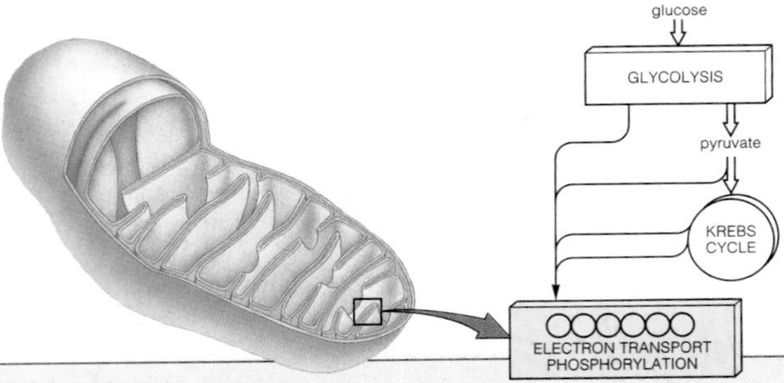
glucose
GLYCOLYSIS
pyruvate
KREBS CYCLE
ELECTRON TRANSPORT PHOSPHORYLATION

Figure 8.7 Electron transport phosphorylation, the third and final stage of aerobic respiration. The reactions proceed at electron transport systems and at ATP synthases, a type of transport protein, in the inner mitochondrial membrane. Each electron transport system consists of specific enzymes, cytochromes, and other proteins that act in sequence.

The inner membrane functionally divides the mitochondrion into two compartments. The third-stage reactions start in the inner compartment, when NADH and $FADH_2$ give up electrons and hydrogen to transport systems. Electrons are transferred *through* the system, but electron transport proteins drive unbound hydrogen (H^+) *to* the outer compartment:

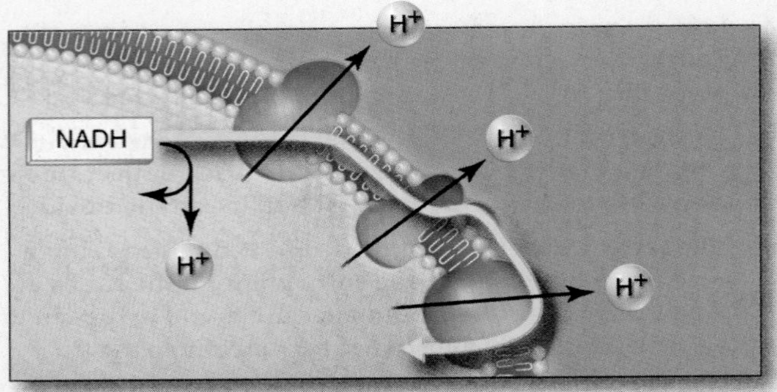
NADH
H+
H+
H+
H+
H+

In short order, there is a higher concentration of H^+ in the outer compartment compared to the inner one. Concentration and electric gradients now exist across the membrane. The ions follow the gradients and flow across the membrane, through the interior of the ATP synthases. Energy associated with the flow drives the formation of ATP from ADP and unbound phosphate (P_i). Hence the name, electron transport *phosphorylation*:

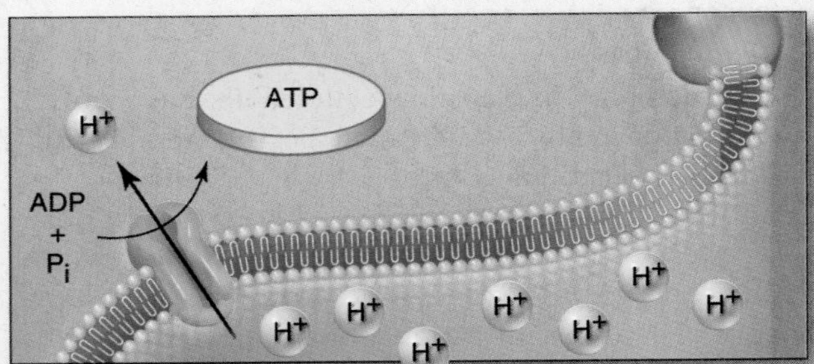
ATP
H+
ADP
+
Pi
H+
H+
H+
H+
H+
H+
H+

Do these metabolic events sound familiar? They should. Recall, from Section 7.5, that ATP forms in much the same way inside chloroplasts. By the *chemiosmotic* model, H^+ concentration and electric gradients across a cell membrane drive ATP formation. The model applies also to mitochondria, although the ions flow in the opposite direction compared to chloroplasts.

Further reading: Student Guide to InfoTrac on web site

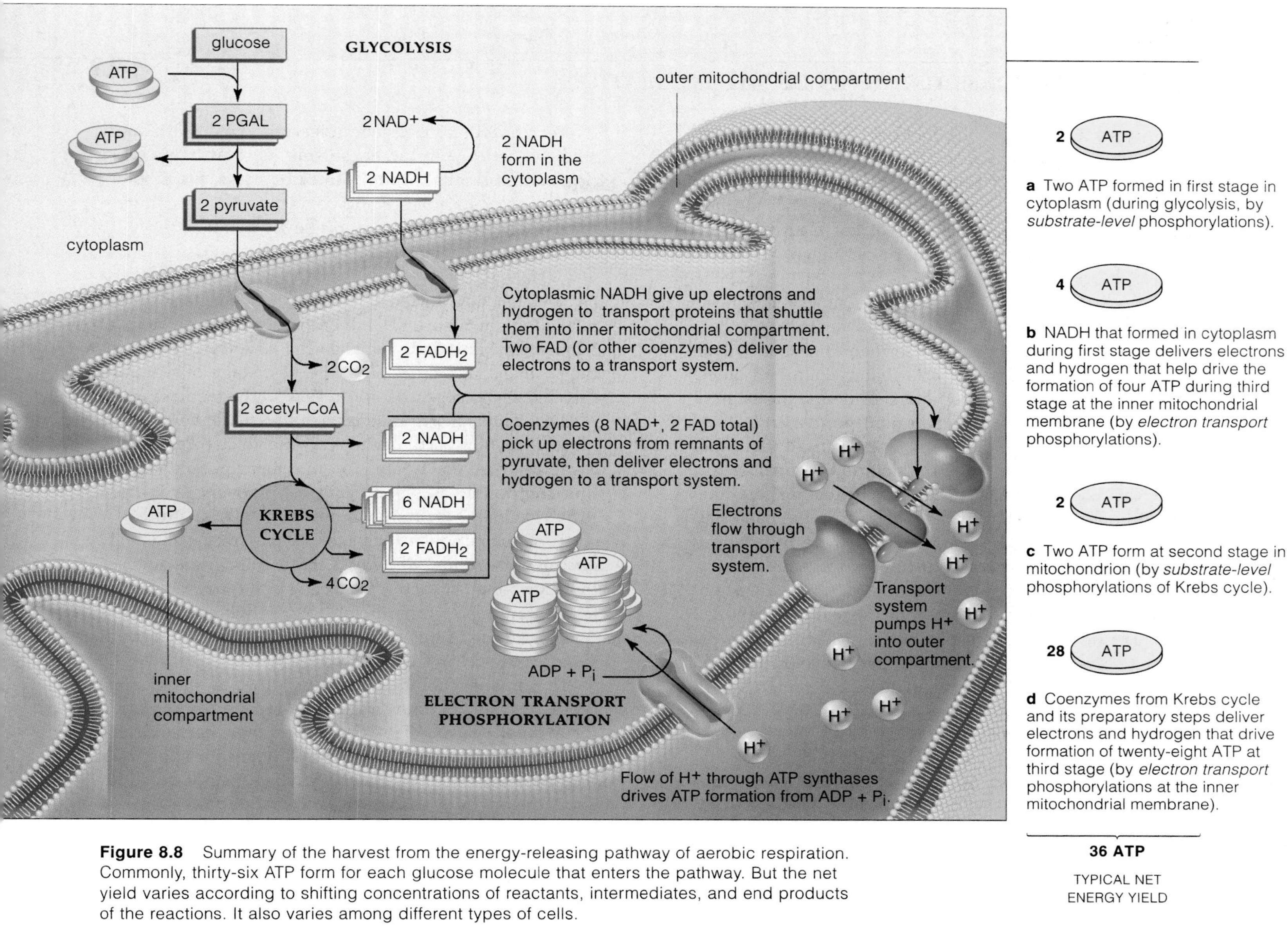

glucose
GLYCOLYSIS
ATP
2 PGAL
ATP
2NAD+
2 NADH
2 NADH form in the cytoplasm
outer mitochondrial compartment
2 pyruvate
cytoplasm
Cytoplasmic NADH give up electrons and hydrogen to transport proteins that shuttle them into inner mitochondrial compartment. Two FAD (or other coenzymes) deliver the electrons to a transport system.
2 FADH2
2CO2
2 acetyl-CoA
2 NADH
Coenzymes (8 NAD+, 2 FAD total) pick up electrons from remnants of pyruvate, then deliver electrons and hydrogen to a transport system.
H+
ATP
KREBS CYCLE
6 NADH
2 FADH2
4CO2
ATP
ATP
ATP
Electrons flow through transport system.
Transport system pumps H+ into outer compartment.
inner mitochondrial compartment
ADP + Pi
ELECTRON TRANSPORT PHOSPHORYLATION
Flow of H+ through ATP synthases drives ATP formation from ADP + Pi.

2 

a Two ATP formed in first stage in cytoplasm (during glycolysis, by *substrate-level* phosphorylations).

4 

b NADH that formed in cytoplasm during first stage delivers electrons and hydrogen that help drive the formation of four ATP during third stage at the inner mitochondrial membrane (by *electron transport* phosphorylations).

2 

c Two ATP form at second stage in mitochondrion (by *substrate-level* phosphorylations of Krebs cycle).

28 

d Coenzymes from Krebs cycle and its preparatory steps deliver electrons and hydrogen that drive formation of twenty-eight ATP at third stage (by *electron transport* phosphorylations at the inner mitochondrial membrane).

36 ATP

TYPICAL NET ENERGY YIELD

Figure 8.8 Summary of the harvest from the energy-releasing pathway of aerobic respiration. Commonly, thirty-six ATP form for each glucose molecule that enters the pathway. But the net yield varies according to shifting concentrations of reactants, intermediates, and end products of the reactions. It also varies among different types of cells.

As you read earlier, cells differ in how they use the NADH from glycolysis, which cannot enter mitochondria. At the outer mitochondrial membrane, these NADH give up electrons and hydrogen to transport proteins, which shuttle the electrons and hydrogen across the membrane. NAD^+ or FAD already inside the mitochondrion accept them, thus forming NADH or $FADH_2$.

Any NADH inside the mitochondrion delivers electrons to the highest possible entry point into a transport system. When it does, enough H^+ is pumped across the inner membrane to make *three* ATP. By contrast, any $FADH_2$ delivers them to a lower entry point. Fewer hydrogen ions can be pumped, so only *two* ATP can form.

In liver, heart, and kidney cells, for example, electrons and hydrogen from glycolysis enter the highest entry point of transport systems, so the energy harvest is thirty-eight ATP. More commonly, as in skeletal muscle and brain cells, they are transferred to FAD—so the harvest is thirty-six ATP.

One final point. Glucose, recall, has more energy (stored in more covalent bonds) than carbon dioxide or water. About 686 kilocalories of energy are released when glucose is broken down to those more stable end products. Much of this energy escapes (as heat), but about 7.5 kilocalories are conserved in every mole of ATP. So when 36 ATP form through breakdown of a glucose molecule, the energy-conserving efficiency of aerobic respiration is $36 \times 7.5 / 686 \times 100$, or 39 percent.

In the final stage of the aerobic pathway, coenzymes deliver electrons to transport systems of the inner mitochondrial membrane. As electrons move through the system, they set up H^+ gradients that drive ATP formation at nearby proteins in the membrane. Oxygen is the final electron acceptor.

Again, from start (glycolysis in the cytoplasm) to finish (in mitochondria), the pathway commonly has a net yield of thirty-six ATP for every glucose molecule metabolized.

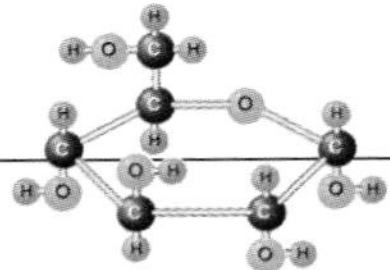

ANAEROBIC ROUTES OF ATP FORMATION

So far, we have tracked the fate of a glucose molecule through the pathway of aerobic respiration. We turn now to its use as a substrate for fermentation pathways. Remember, these are anaerobic pathways; they do *not* use oxygen as the final acceptor of the electrons that ultimately drive the ATP-forming machinery.

Fermentation Pathways

Diverse kinds of organisms use fermentation pathways. Many are bacteria and protistans that make their homes in marshes, bogs, mud, deep-sea sediments, the animal gut, canned foods, sewage treatment ponds, and other oxygen-free settings. Some fermenters actually will die if they are exposed to oxygen. Bacteria responsible for many diseases, including botulism and tetanus, are like this. Other kinds of fermenters, including the bacterial "employees" of yogurt manufacturers, are indifferent to the presence of oxygen. Still others can use oxygen, but they also might use a fermentation pathway when oxygen becomes scarce. Even your muscle cells do this.

As is true of aerobic respiration, glycolysis serves as the first stage of the fermentation pathways. Here also, enzymes split glucose and rearrange the fragments into two pyruvate molecules. Here again, two NADH form, and the net energy yield is two ATP.

However, fermentation reactions do not completely break down glucose to carbon dioxide and water, and they produce no more ATP beyond the tiny yield from glycolysis. *The final steps of fermentation serve only to regenerate NAD$^+$, the coenzyme with a central role in the breakdown reactions.*

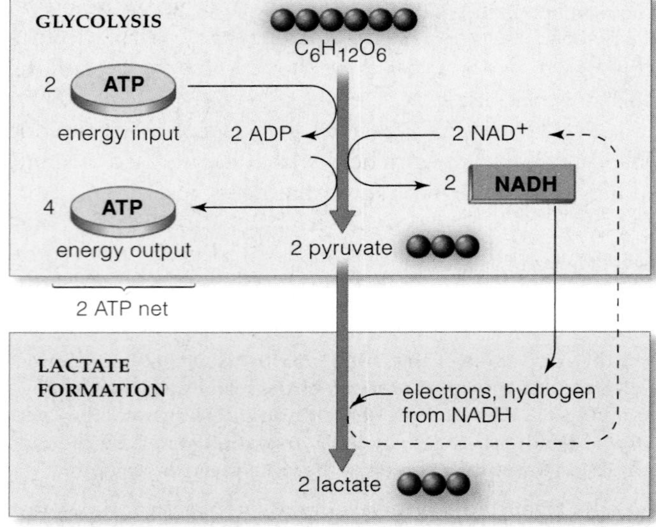
GLYCOLYSIS
$C_6H_{12}O_6$
2 ATP
energy input
2 ADP
2 NAD+
2 NADH
4 ATP
energy output
2 pyruvate
2 ATP net
LACTATE FORMATION
electrons, hydrogen from NADH
2 lactate

Figure 8.9 Lactate fermentation. In this anaerobic pathway, electrons end up in lactate, the reaction product.

Fermentation yields enough energy to sustain many single-celled anaerobic organisms. It even helps carry some aerobic cells through times of stress. But it is not enough to sustain large, active, multicelled organisms, this being one reason why you never will meet up with an anaerobic elephant.

LACTATE FERMENTATION With these points in mind, take a look at Figure 8.9, which tracks the main steps of **lactate fermentation**. During this anaerobic pathway, the *pyruvate* molecules from the first stage of reactions (glycolysis) accept the hydrogen and electrons from NADH. The transfer regenerates the NAD$^+$ and, at the same time, converts each pyruvate to a three-carbon compound called lactate. You may hear people refer to this compound as "lactic acid." However, its ionized form (lactate) is far more common in cellular fluids.

Some bacteria, such as *Lactobacillus*, rely exclusively on this anaerobic pathway. Left to their own devices, their fermentation activities often spoil food. Yet certain fermenters have commercial uses, as when they break down glucose in huge vats where cheeses, yogurt, and sauerkraut are produced.

In humans, rabbits, and many other animals, some types of cells also may switch to lactate fermentation for a quick fix of ATP. When your own demands for energy are intense but brief—say, during a short race—muscle cells use this pathway. They cannot do so for long; they would throw away too much of glucose's stored energy for too little ATP. When their stores of glucose become depleted, muscles fatigue and lose their ability to contract.

ALCOHOLIC FERMENTATION In the anaerobic route called **alcoholic fermentation**, enzymes convert each pyruvate molecule that formed during glycolysis to an intermediate form: acetaldehyde. The NADH transfers electrons and hydrogen to this form and so converts it to an alcoholic end product—ethanol (Figure 8.10).

Certain species of single-celled fungi called yeasts are renowned for their use of this pathway. One type, *Saccharomyces cerevisiae,* makes bread dough rise. Bakers mix the yeast with sugar, then blend both into dough. When yeast cells degrade the sugar, they release carbon dioxide. Bubbles of the gas expand the dough (make it rise). Oven heat forces the gas out of the dough, and a porous product remains.

Beer and wine producers use yeasts on a large scale. Vintners use wild yeasts living on grapes and cultivated strains of *S. ellipsoideus,* which remain active until the alcohol concentration in wine vats exceeds 14 percent. (Wild yeast cells die when the concentration exceeds 4 percent.) That is why some birds get drunk on naturally fermenting berries. That is why landscapers don't plant

Further reading: Student Guide to InfoTrac on web site

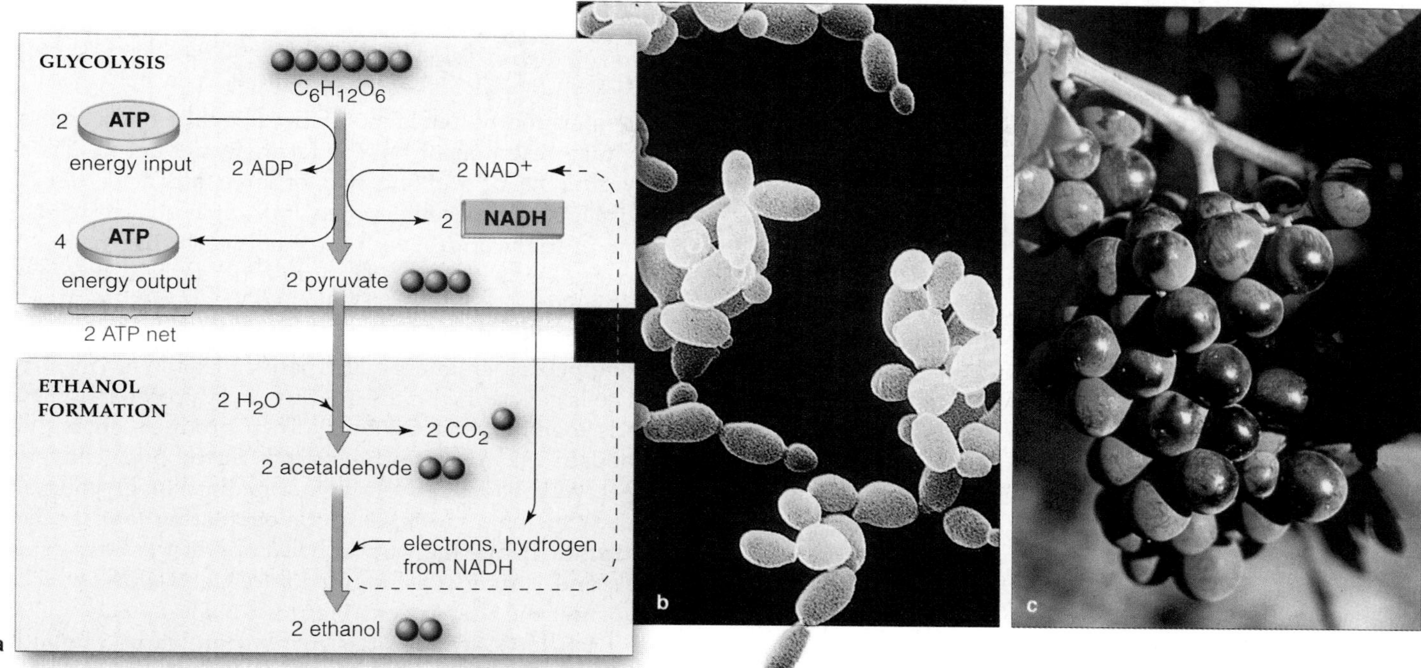
GLYCOLYSIS
$C_6H_{12}O_6$
2 ATP
energy input
2 ADP
2 NAD+
2 NADH
4 ATP
energy output
2 pyruvate
2 ATP net
ETHANOL FORMATION
2 H_2O
2 CO_2
2 acetaldehyde
electrons, hydrogen from NADH
2 ethanol
a
b
c

Figure 8.10 (**a**) Alcoholic fermentation, one of the anaerobic pathways. Acetaldehyde, an intermediate of the reactions, is the final acceptor of electrons. Ethanol is the end product. Yeasts, single-celled organisms, use this pathway. (**b**) Activity of one species of *Saccharomyces* makes bread dough rise. (**c**) Another species lives on sugar-rich tissues of ripened grapes.

prodigious berry-producing shrubbery near highways; drunk birds doodle into windshields (Figure 8.11). Wild turkeys similarly have been known to get tipsy when they gobble fermenting apples in untended orchards.

Anaerobic Electron Transport

Especially among the bacteria, we find less common energy-releasing pathways, some of which are topics of later chapters in the book. For example, many bacterial species have key roles in the global cycling of sulfur, nitrogen, and other crucial elements. Collectively, their metabolic activities influence nutrient availability for organisms everywhere.

Figure 8.11 Robin feasting on the fermented berries of a pyracantha bush.

For example, certain bacteria use **anaerobic electron transport**. Electrons stripped from some type of organic compound move on through transport systems of their plasma membrane. Commonly, an inorganic compound in the environment serves as the final electron acceptor. The net energy yield varies, but it is always small.

Even as you read this, some anaerobic bacteria that live in waterlogged soil are stripping electrons from a variety of compounds. They dump electrons on sulfate. Hydrogen sulfide, a putrid-smelling gas, is the result. The sulfate-reducing bacteria also live in many aquatic habitats that are enriched with decomposed organic material. They even live on the deep ocean floor, near hydrothermal vents. As described in Section 50.11, they form the food production base for unique communities.

In fermentation pathways, an organic substance that forms during the reactions serves as the final acceptor of electrons from glycolysis. The reactions regenerate NAD+, which is required to keep the pathway operational.

In anaerobic electron transport, an inorganic substance (but not oxygen) usually serves as the final electron acceptor.

For each glucose molecule metabolized, anaerobic pathways typically have a net energy yield of two ATP molecules, which only form during glycolysis.

8.6 ALTERNATIVE ENERGY SOURCES IN THE HUMAN BODY

So far, you have looked at what happens after a lone glucose molecule enters an energy-releasing pathway. Now you can start thinking about what cells do when they have too many or too few molecules of glucose.

Carbohydrate Breakdown in Perspective

THE FATE OF GLUCOSE AT MEALTIME Consider what happens to you or any other mammal during a meal. Glucose and certain other small organic molecules are being absorbed across the gut lining, then the blood transports them through the body. A rise in the glucose level in blood prompts the pancreas to release insulin, a hormone that stimulates cells to take up glucose at a faster rate. The cells convert the windfall of glucose to glucose–6–phosphate and so "trap" it in the cytoplasm. (When phosphorylated, glucose cannot be transported back out, across the plasma membrane.) Look again at Figure 8.4, and you see that glucose–6–phosphate is the first activated intermediate of glycolysis.

If your glucose intake exceeds cellular demands for energy, ATP-producing machinery goes into high gear. Unless a cell is rapidly using ATP, its concentration of ATP can rise to high levels. Then, glucose–6–phosphate is diverted into a biosynthesis pathway that assembles glucose units into glycogen, a storage polysaccharide (Section 3.3). This is especially the case for muscle and liver cells, which maintain the largest glycogen stores.

THE FATE OF GLUCOSE BETWEEN MEALS When you are not eating, glucose is not entering your bloodstream, and its level in the blood declines. If the decline were not countered, that would be bad news for the brain, your body's glucose hog. The brain constantly takes up more than two-thirds of the freely circulating glucose because its many hundreds of millions of cells simply use this sugar alone as their preferred energy source.

The pancreas responds to the decline by secreting glucagon, a hormone that prompts liver cells to convert glycogen back to glucose and send it back to the blood. Only liver cells do this; muscle cells won't give it up. The blood glucose level rises, and brain cells keep on functioning. Thus, *hormones control whether the body's cells use free glucose as an energy source or tuck it away.*

A word of caution: Don't let the preceding examples lead you to believe cells squirrel away large amounts of glycogen. In adult humans, glycogen makes up merely 1 percent or so of the body's total energy reserves, the energy equivalent of two cups of cooked pasta. Unless you eat on a regular basis, you will deplete the liver's small glycogen stores in less than twelve hours. Of the total energy reserves in, say, a typical adult American, 78 percent (about 10,000 kilocalories) is concentrated in body fat and 21 percent in proteins.

Energy From Fats

The question becomes this: How does the body access its huge reservoir of fats? A fat molecule, recall, has a glycerol "head" and one, two, or three fatty acid "tails." Most fats that become stored in your body are in the form of triglycerides, with three tails each. Triglycerides accumulate in fat cells of adipose tissue, which forms at buttocks and other strategic places beneath the skin.

When blood glucose levels decline, triglycerides can be tapped as an energy alternative. Then, enzymes in fat cells cleave the bonds holding the glycerol and fatty acids together, and the breakdown products enter the bloodstream. Afterward, enzymes in the liver convert the glycerol to PGAL—an intermediate of glycolysis. Nearly all cells can take up the circulating fatty acids. Enzymes cleave the carbon backbone of the fatty acid tails and convert the fragments to acetyl–CoA, which can enter the Krebs cycle (Figures 8.6 and 8.12).

Each fatty acid tail has many more carbon-bound hydrogen atoms than glucose, so its breakdown yields much more ATP. In between meals or during sustained exercise, fatty acid conversions supply about half of the ATP that muscle, liver, and kidney cells require.

What happens if you eat too many carbohydrates? Exceed the glycogen-storing capacity of your liver and muscle cells, and the excess gets converted to fats. *Too much glucose ends up as excess fat.* Worse yet, 25 percent of the people in the United States are blessed with a combination of genes that allows them to eat as much as they like without gaining weight—but a diet far too rich in carbohydrates keeps the other 75 percent fat. For them, insulin levels remain elevated, which "tells" the body to store fat rather than use it for energy. We will return to this topic in Sections 37.7 and 42.10.

Energy From Proteins

Eat more proteins than your body requires to grow and maintain itself, and its cells won't store them. Enzymes split these proteins into amino acid units. Then they remove the amino group ($-NH_3^+$) from each unit, and ammonia (NH_3) forms. What happens to the leftover carbon backbones? Depending on conditions in the cell, the outcome varies. The backbones can be converted to carbohydrates or fats. Or they may enter the Krebs cycle, as shown in Figure 8.12, where coenzymes can pick up hydrogen and electrons stripped away from the carbon atoms. The ammonia that forms undergoes conversions to become urea. This nitrogen-containing waste product would be toxic if it accumulated to high concentrations. Normally your body excretes urea, in urine.

As this brief discussion makes clear, maintaining and accessing the body's energy reserves is complicated

Further reading: Student Guide to InfoTrac on web site

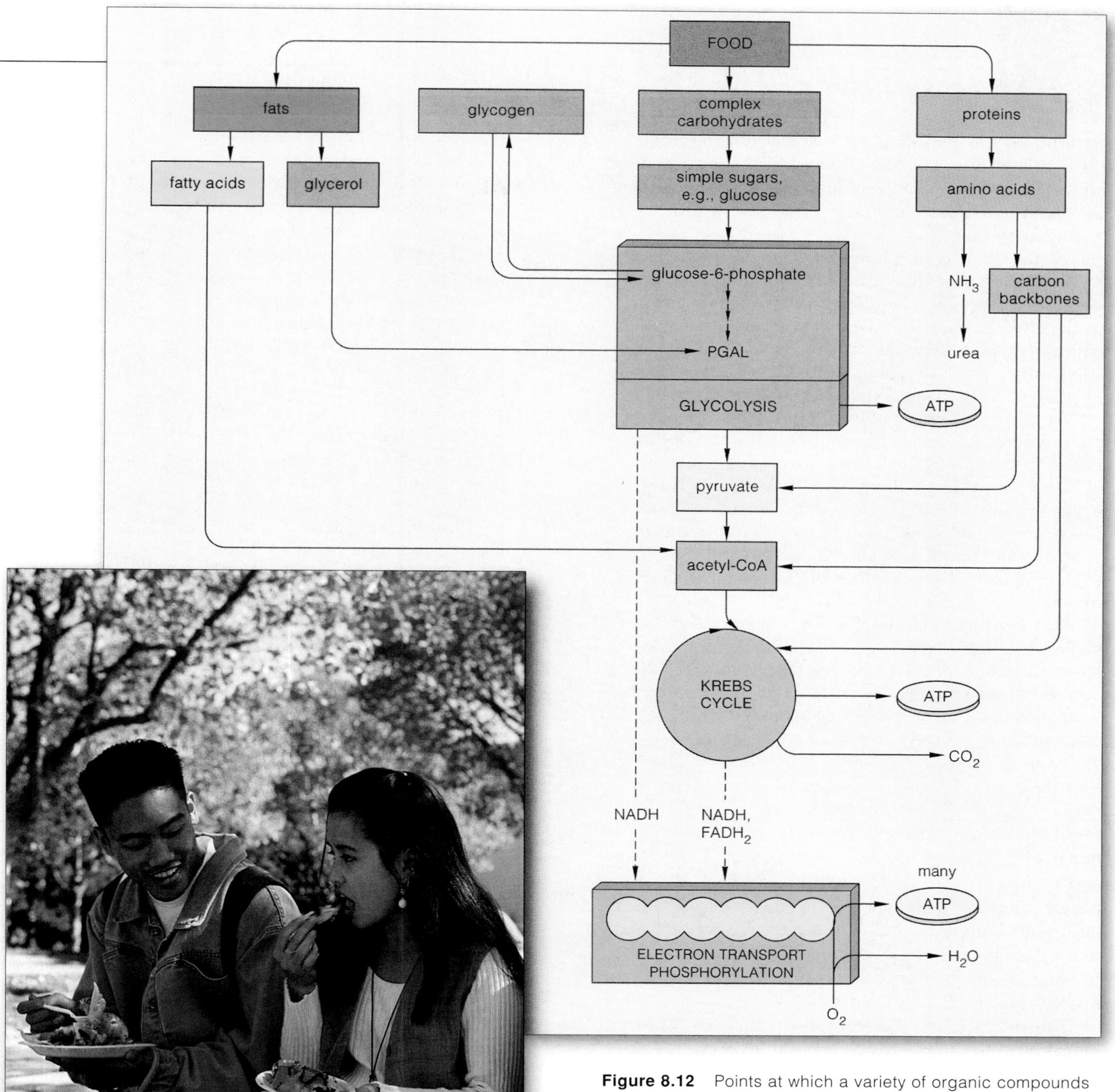
FOOD
fats
glycogen
complex carbohydrates
proteins
fatty acids
glycerol
simple sugars, e.g., glucose
amino acids
glucose-6-phosphate
NH_3
carbon backbones
PGAL
urea
GLYCOLYSIS
ATP
pyruvate
acetyl-CoA
KREBS CYCLE
ATP
CO_2
NADH
NADH, $FADH_2$
many
ATP
ELECTRON TRANSPORT PHOSPHORYLATION
H_2O
O_2

Figure 8.12 Points at which a variety of organic compounds can enter the reaction stages of aerobic respiration. Complex carbohydrates, fats, and proteins cannot enter the aerobic pathway directly. In humans and other mammals, the digestive system as well as individual cells must first break apart these large molecules into simpler, degradable subunits.

business. Hormonal controls over the disposition of glucose are special only because glucose is the fuel of choice for the all-important brain. However, as you will see in later chapters, providing all of your cells, organs, and organ systems with energy starts with the kinds and proportions of food you put in your mouth.

This concludes our look at aerobic respiration and other energy-releasing pathways. The section to follow may help you get a sense of how they fit into the larger picture of life's evolution and interconnectedness.

In humans and other mammals, the entrance of glucose or other organic compounds into an energy-releasing pathway depends on the kinds and proportions of carbohydrates, fats, and proteins in the diet as well as on the type of cell.

PERSPECTIVE ON LIFE

In this unit, you read about photosynthesis and aerobic respiration—the main pathways by which cells trap, store, and release energy. What you might not know is that the two pathways became linked, on a grand scale, over evolutionary time.

When life originated more than 3.8 billion years ago, the Earth's atmosphere had little free oxygen. The earliest single-celled organisms probably used reactions similar to glycolysis to make ATP. Without oxygen, fermentation pathways must have dominated. By about 1.5 billion years later, oxygen-producing photosynthetic cells had emerged. They irrevocably changed the course of evolution.

Oxygen, a by-product of the noncyclic pathway of photosynthesis, began to accumulate in the atmosphere. Probably through mutations that affected the proteins of electron transport systems, some cells started using oxygen as an electron acceptor. At some point in the past, descendants of those fledgling aerobic cells abandoned photosynthesis entirely. Among them were the forerunners of animals and all other organisms that engage in aerobic respiration.

With aerobic respiration, the flow of carbon, hydrogen, and oxygen through the metabolic pathways of living organisms came full circle. For the final products of this aerobic pathway—carbon dioxide and water—are the same materials that are necessary to build organic compounds in photosynthesis:

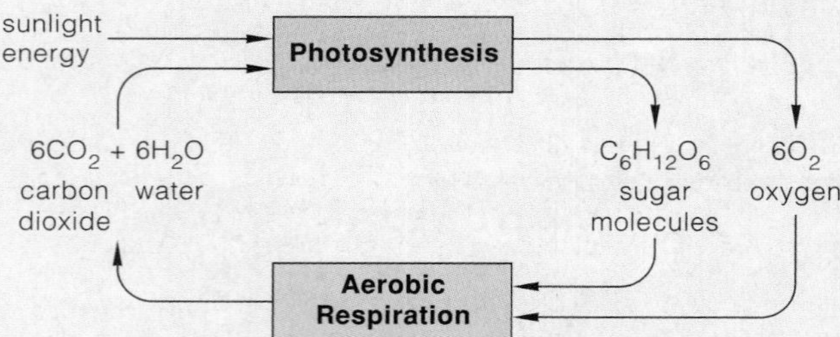
sunlight energy
Photosynthesis
$6CO_2 + 6H_2O$
carbon dioxide
water
$C_6H_{12}O_6$
sugar molecules
$6O_2$
oxygen
Aerobic Respiration

Perhaps you have difficulty fathoming the connection between yourself—an intelligent being—and such remote-sounding events as energy flow and the cycling of carbon, hydrogen, and oxygen. Is this really the stuff of humanity?

Think back, for a moment, on the structure of a water molecule. Two hydrogen atoms sharing electrons with an oxygen atom may not seem close to your daily life. And yet, through that sharing, water molecules show polarity—and they hydrogen-bond with one another. Their chemical behavior is a beginning for the organization of lifeless matter that leads to the organization of all living things.

For now you can imagine other molecules interspersed through water. The nonpolar kinds resist interaction with water; the polar kinds dissolve in it. On their own, the phospholipids among them assemble into a two-layered film. Such lipid bilayers, remember, serve as the very framework of all cell membranes, hence all cells. From the beginning, the cell has been the fundamental *living* unit.

The essence of life is not some mysterious force. It is metabolic control. With a cell membrane to contain them, reactions *can* be controlled. With mechanisms built into their membranes, cells can respond to energy changes and shifting concentrations of substances in the environment. The response mechanisms operate by "telling" proteins—enzymes—when and what to build or tear down.

And it is not some mysterious force that creates the proteins themselves. DNA, the slender double-stranded treasurehouse of inheritance, has the chemical structure—*the chemical message*—that allows molecule to reproduce molecule, one generation after the next. In your own body, DNA strands tell trillions of cells how countless molecules must be built or torn apart for their stored energy.

So yes, carbon, hydrogen, oxygen, and other atoms of organic molecules represent the stuff of you, and us, and all of life. But it takes more than molecules to complete the picture. Life exists as long as an unbroken flow of energy sustains its organization. Molecules are assembled into cells, cells into organisms, organisms into communities, and so on up through the biosphere. It takes energy inputs from the sun to maintain these levels of organization. And energy flows through time in one direction—from organized to less organized forms. Only as long as energy continues to flow into the web of life can life continue in all its rich diversity.

In short, life is no more *and no less* than a marvelously complex system of prolonging order. Sustained by energy transfusions from the sun, life continues onward, through its capacity for self-reproduction. For with the hereditary instructions contained in DNA, energy and materials can be organized, generation after generation. Even with the death of individuals, life elsewhere is prolonged. With each death, molecules are released and may be cycled once more, as raw materials for new generations.

In this flow of energy and cycling of material through time, each birth is affirmation of our ongoing capacity for organization, each death a renewal.

Further reading: Student Guide to InfoTrac on web site —

8.8 SUMMARY

1. Phosphate-group transfers from ATP are central to metabolism. Autotrophic cells alone can tap energy from the environment to make ATP. But *all* cells are able to make ATP by pathways that release chemical energy from organic compounds, such as glucose.

2. After glucose enters these pathways, enzymes strip electrons and hydrogen from intermediates that form along the way. Coenzymes pick these up and deliver them to other reaction sites, where the pathway ends. NAD^+ is the main coenzyme; the aerobic route also uses FAD. The reduced forms of these coenzymes are designated NADH and $FADH_2$.

3. All of the main energy-releasing pathways start with glycolysis, a stage of reactions that begin and end in the cytoplasm. Glycolytic reactions can be completed either in the presence of oxygen or in its absence.

a. During glycolysis, enzymes break down a glucose molecule to two pyruvate molecules. Two NADH and four ATP form during the reactions.

b. The net energy yield is two ATP (because two ATP had to be invested up front to get the reactions going).

4. Aerobic respiration continues on through two more stages: (1) the Krebs cycle and a few steps preceding it, and (2) electron transport phosphorylation. The stages occur only in organelles called mitochondria, which are present only in eukaryotic cells.

5. The second stage of the aerobic pathway starts when an enzyme strips a carbon atom from each pyruvate. Coenzyme A binds the remaining two-carbon fragment (to form acetyl–CoA), then transfers it to oxaloacetate, the entry point of the Krebs cycle. The cyclic reactions, along with the steps immediately preceding them, load up ten coenzymes with electrons and hydrogen (eight NADH and two $FADH_2$). Two ATP form. Three carbon dioxide molecules are released for each pyruvate that entered this second stage.

6. The third stage of the aerobic pathway proceeds at a membrane that divides the interior of a mitochondrion into two compartments. Electron transport systems and ATP synthases are embedded in this inner membrane.

a. Coenzymes deliver electrons from the first two stages to transport systems. In the outer compartment, hydrogen ions accumulate; therefore, concentration and electric gradients form across the membrane.

b. Hydrogen ions follow the gradients and flow from the outer to the inner compartment, through the interior of ATP synthases. Energy released during the ion flow drives the formation of ATP from ADP and unbound phosphate.

c. Oxygen withdraws electrons from the transport system and, at the same time, combines with hydrogen ions to form water molecules. The oxygen is the final acceptor of electrons that initially resided in glucose.

7. Aerobic respiration has a typical net energy yield of thirty-six ATP for each glucose molecule metabolized. Yields vary, according to cell type and cell conditions.

8. The fermentation pathways and anaerobic electron transport also start with glycolysis, but they do not use oxygen. They are anaerobic, start to finish.

a. Lactate fermentation has a net energy yield of two ATP, which form in glycolysis. The remaining reaction regenerates NAD^+. The two NADH from glycolysis transfer electrons and hydrogen to two pyruvate from glycolysis. Two lactate molecules are the end products.

b. Alcoholic fermentation has a net energy yield of two ATP from glycolysis, and its remaining reactions serve to regenerate NAD^+. Enzymes convert pyruvate from glycolysis to acetaldehyde, and carbon dioxide is released. The NADH from glycolysis transfer electrons and hydrogen to the two acetaldehyde molecules, thus forming two ethanol molecules, the end products.

c. Certain bacteria use anaerobic electron transport. Electrons are stripped from various organic compounds and travel through transport systems in the bacterial cell's plasma membrane. An inorganic compound in the environment often serves as the final electron acceptor.

9. In humans and other mammals, simple sugars such as glucose from carbohydrates, glycerol and fatty acids from fats, and carbon backbones of amino acids from proteins can enter ATP-producing pathways.

Review Questions

1. Is this true or false: Aerobic respiration occurs in animals but not plants, which make ATP only by photosynthesis. *8.1*

2. For this diagram of the aerobic pathway, fill in all blanks and write in the number of molecules of pyruvate, coenzymes, and end products. Also write in the net ATP formed in each stage, and the net ATP formed from start (glycolysis) to finish. *8.1*

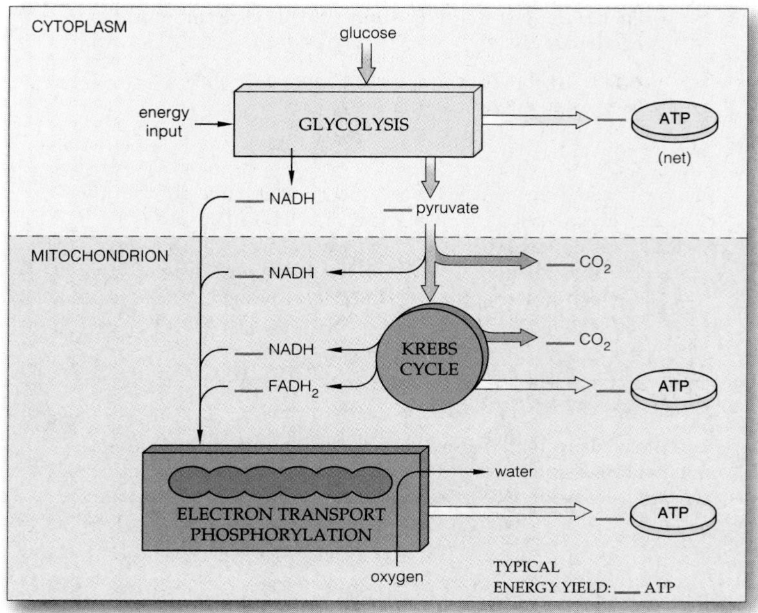

CYTOPLASM
glucose
energy input
GLYCOLYSIS
ATP
(net)
__ NADH
__ pyruvate
MITOCHONDRION
__ NADH
__ CO_2
__ NADH
KREBS CYCLE
__ CO_2
__ $FADH_2$
ATP
water
ELECTRON TRANSPORT PHOSPHORYLATION
ATP
oxygen
TYPICAL ENERGY YIELD: __ ATP

3. Is glycolysis energy-*requiring* or energy-*releasing*? Or do both kinds of reactions occur during glycolysis? *8.2*

4. In what respect does *electron transport* phosphorylation differ from *substrate-level* phosphorylation? *8.2, Figure 8.7*

5. Sketch the double-membrane system of the mitochondrion and show where transport systems and ATP synthases are located. *8.3*

6. Name the compound that is the entry point for the Krebs cycle, and state whether it directly accepts the pyruvate from glycolysis. For each glucose molecule, how many carbon atoms enter the Krebs cycle? How many depart from it, and in what form? *8.3*

7. Is this statement true or false: Muscle cells cannot contract at all when deprived of oxygen. If true, explain why. If false, name the alternative(s) available to them. *8.5*

Self-Quiz (*Answers in Appendix III*)

1. Glycolysis starts and ends in the ______.
   a. nucleus c. plasma membrane
   b. mitochondrion d. cytoplasm

2. Which of the following does *not* form during glycolysis?
   a. NADH c. $FADH_2$
   b. pyruvate d. ATP

3. Aerobic respiration is completed in the ______.
   a. nucleus c. plasma membrane
   b. mitochondrion d. cytoplasm

4. In the last stage of aerobic respiration, ______ is the final acceptor of electrons that originally resided in glucose.
   a. water c. oxygen
   b. hydrogen d. NADH

5. ______ engage in lactate fermentation.
   a. *Lactobacillus* cells c. Sulfate-reducing bacteria
   b. Muscle cells d. a and b

6. In alcoholic fermentation, ______ is the final acceptor of the electrons stripped from glucose.
   a. oxygen c. acetaldehyde
   b. pyruvate d. sulfate

7. The fermentation pathways produce no more ATP beyond the small yield from glycolysis, but the remaining reactions ______.
   a. regenerate ADP c. dump electrons on an inorganic substance (not oxygen)
   b. regenerate NAD^+

8. In certain organisms and under certain conditions, ______ can be used as an energy alternative to glucose.
   a. fatty acids c. amino acids
   b. glycerol d. all of the above

9. Match the event with its most suitable metabolic description.
   ___ glycolysis a. ATP, NADH, $FADH_2$, CO_2, and water form
   ___ fermentation b. glucose to two pyruvate
   ___ Krebs cycle c. NAD^+ regenerated, two ATP net
   ___ electron transport phosphorylation d. H^+ flows through ATP synthases

Critical Thinking

1. Diana suspects that a visit to her family doctor is in order. After eating carbohydrate-rich food, she always experiences sensations of being intoxicated and becomes nearly incapacitated, as if she had been drinking alcohol. She even wakes up with a hangover the next day. Having completed a course in freshman biology, Diana has an idea that something is affecting the way her body is metabolizing glucose. Explain why.

2. The cells of your body absolutely do not use nucleic acids as alternative energy sources. Suggest why.

3. The body's energy needs and its programs for growth depend on balancing the levels of amino acids in blood with proteins in cells. Cells of the liver, kidneys, and intestinal lining are especially important in this balancing act. When the levels of amino acids in blood decline, lysozymes in cells can rapidly digest some of their proteins (structural and contractile proteins are spared, except in cases of malnutrition). The amino acids released this way enter the blood and thereby help maintain the required levels.

Suppose you embark on a body-building program. You already eat plenty of carbohydrates, but a nutritionist advises a protein-rich diet that includes protein supplements. Speculate on how the extra dietary proteins will be put to use, and in which tissues.

4. Each year, Canada geese lift off in precise formation from their northern breeding grounds. They head south to spend the winter months in warmer climates, and then they make the return trip in spring. As is true of other migratory birds, their flight muscle cells are efficient at using fatty acids as an energy source. (Remember, the carbon backbone of fatty acids can be cleaved into fragments that can be converted to acetyl–CoA for entry into the Krebs cycle.)

Suppose a lesser Canada goose from Alaska's Point Barrow has been steadily flapping along for three thousand kilometers and is nearing Klamath Falls, Oregon. It looks down and notices a rabbit sprinting like the wind from a coyote with a taste for rabbit. With a stunning burst of speed, the rabbit reaches the safety of its burrow.

Which energy-releasing pathway predominated in muscle cells in the rabbit's legs? Why was the Canada goose relying on a different pathway for most of its journey? And why wouldn't the pathway of choice in goose flight muscle cells be much good for a rabbit making a mad dash from its enemy?

5. Reflect on this chapter's introduction, then question 4. Now speculate on which energy-releasing pathway is predominating in agitated Africanized bees that are chasing a farmer through a cornfield.

Selected Key Terms

acetyl–CoA *8.3*
aerobic respiration *8.1*
alcoholic fermentation *8.5*
anaerobic electron transport *8.5*
electron transport phosphorylation *8.1*
FAD *8.1*
glycolysis *8.1*
Krebs cycle *8.1*
lactate fermentation *8.5*
mitochondrion *8.3*
NAD^+ *8.1*
oxaloacetate *8.3*
pyruvate *8.1*
substrate-level phosphorylation *8.2*

Readings *See also www.infotrac-college.com*

Levi, P. October 1984. "Travels with C." *The Sciences.* Journey of a carbon atom through the world of life.

Wolfe, S. 1995. *An Introduction to Molecular and Cellular Biology.* Belmont, California: Wadsworth. Exceptional reference text.

FACING PAGE: *Human sperm, one of which will penetrate this mature egg and so set the stage for the development of a new individual in the image of its parents.*

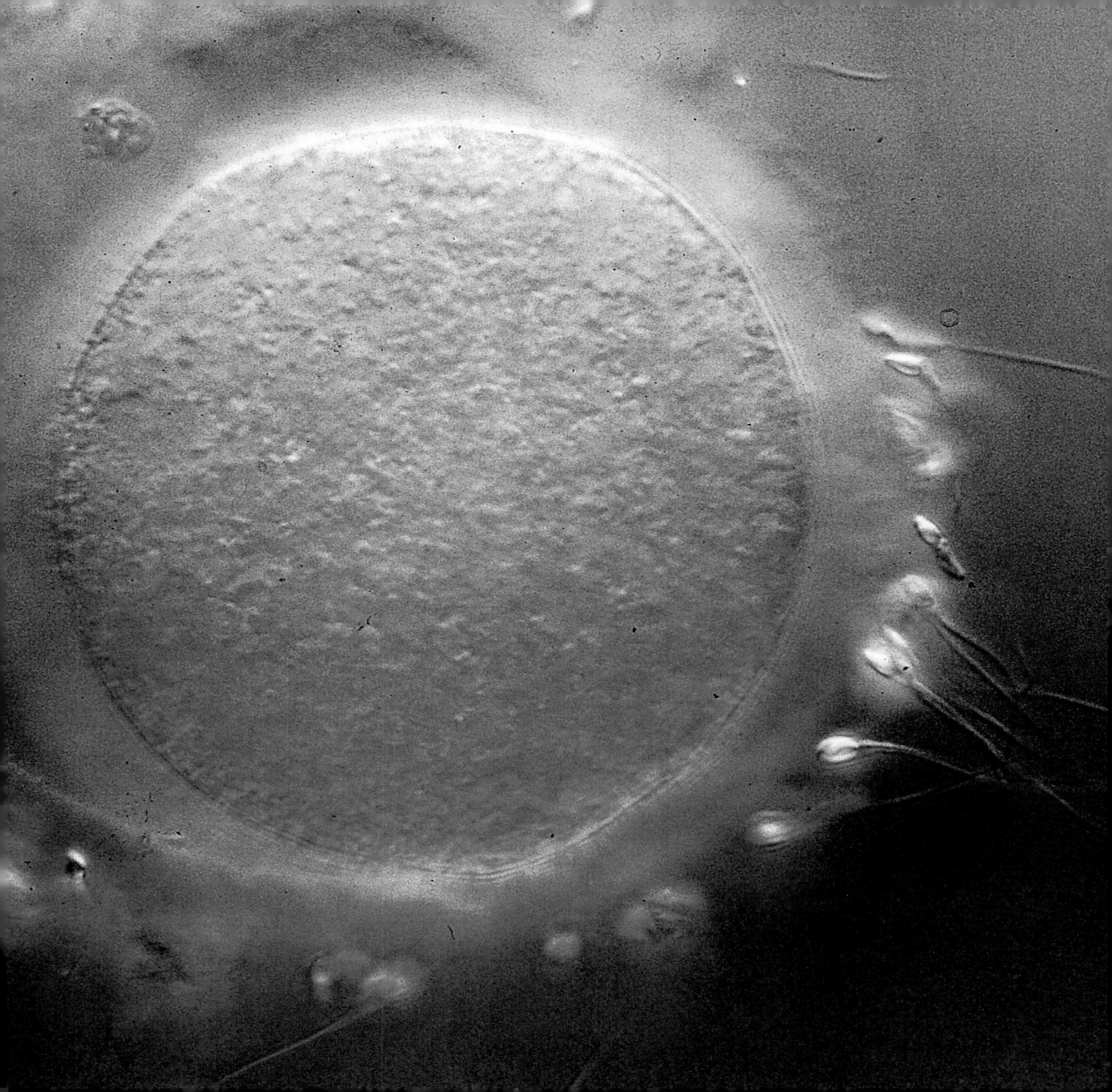

9

CELL DIVISION AND MITOSIS

Silver In the Stream of Time

Five o'clock, and the first rays from the sun dance over the wild Alagnak River of the Alaskan tundra. It is September, and life is ending and beginning in the clear, cold waters. By the thousands, mature silver salmon have returned from the open ocean to spawn in their shallow native home. The females are tinged with red, the color of spawners, and they are dying.

This morning you observe a female salmon releasing translucent pink eggs into a shallow "nest," which her fins hollowed out in the gravel riverbed (Figure 9.1). Within moments a male sheds a cloud of sperm, and fertilization follows. Trout and other predators eat most of the eggs. But if you wait around, you will find that some eggs survive and give rise to a new generation.

Within three years, the pea-sized eggs have become streamlined salmon, fashioned from billions of cells. A few of their cells will develop into eggs or sperm. In time, on some September morning, they will take part in an ongoing story of birth, growth, death, and rebirth.

For you, as for salmon and every other multicelled species, growth as well as reproduction depends on *cell division*. Inside your mother, a fertilized egg divided in two, then the two into four, and so on until billions of cells were growing, developing in specialized ways, and dividing at different times to produce all of your genetically prescribed body parts. Your body now has roughly 65 trillion living cells—and many of them are still dividing. Every five days, for instance, divisions replace the tissue that lines your small intestine.

Understanding cell division—and, ultimately, how new individuals are put together in the image of their parents—begins with answers to three questions. *First*, what instructions are necessary for inheritance? *Second*, how are those instructions duplicated for distribution into daughter cells? *Third*, by what mechanisms are the duplicated instructions parceled out to daughter cells? We will require more than one chapter to consider the nature of cell reproduction and other mechanisms of inheritance. Even so, the points made early in this chapter can help you keep the overall picture in focus.

Begin with the word **reproduction**. In biology, this means that parents produce a new generation of cells or multicelled individuals like themselves. The process starts in cells that are programmed to divide. And the

sexually mature
female salmon

Figure 9.1 The last of one generation and the first of the next in Alaska's Alagnak River.

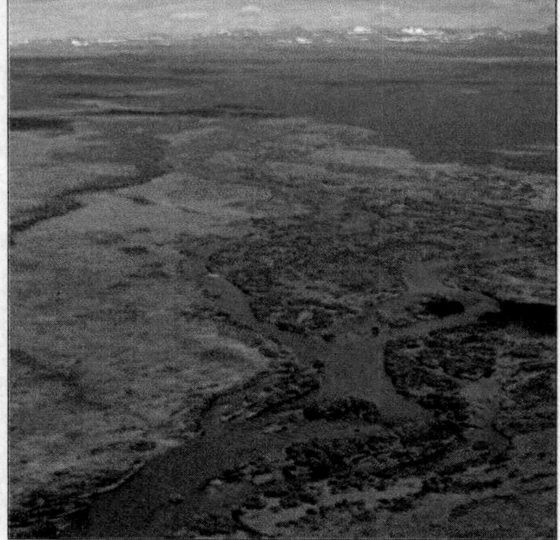

ground rule for division is this: *Parent cells must provide their daughter cells with hereditary instructions, encoded in DNA, and enough metabolic machinery to start up their own operation.*

DNA, recall, contains instructions for synthesizing proteins. Some proteins are structural materials. Many are enzymes that speed the assembly of specific organic compounds, such as the lipids that cells use as building blocks and sources of energy. Unless a daughter cell receives the necessary instructions for making proteins, it simply will not be able to grow or function properly.

Also, the cytoplasm of a parent cell already contains enzymes, organelles, and other operating machinery. When a daughter cell inherits what looks like a blob of cytoplasm, it really is getting start-up machinery—which will keep that cell operating until it can use its inherited DNA for growing and developing on its own.

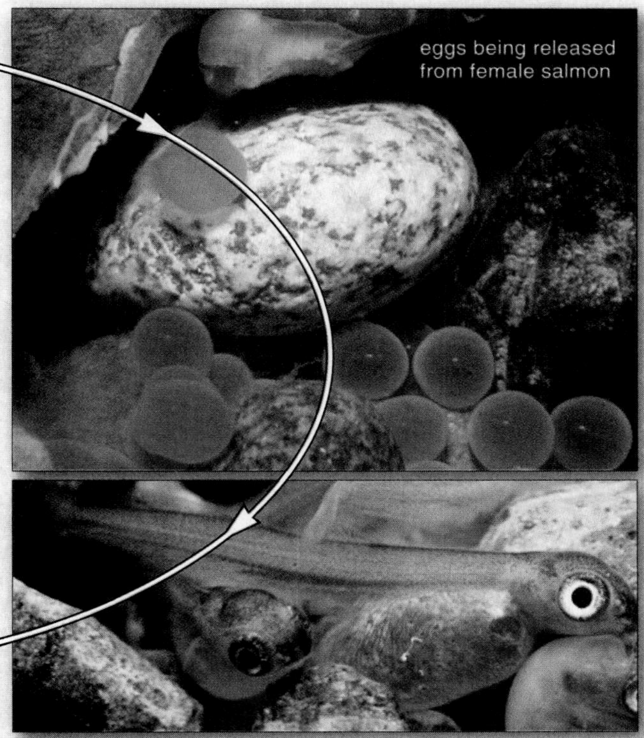
eggs being released
from female salmon

Fingerlings—young fishes growing, from fertilized eggs, by way of mitotic cell divisions

KEY CONCEPTS

1. The continuity of life depends on reproduction, by which parents produce a new generation of cells or multicelled individuals like themselves. Cell division is the bridge between generations.

2. When a cell divides, its two daughter cells must each receive a required number of DNA molecules and some cytoplasm. For eukaryotic cells, a division mechanism called mitosis sorts out the DNA into two new nuclei. A separate mechanism divides the cytoplasm.

3. Mitosis is one part of the cell cycle. The other part is interphase, an interval when each new cell formed by mitosis and cytoplasmic division increases in mass, increases the number of its components, then duplicates its DNA. The cycle ends when that cell divides.

4. In eukaryotic cells, many proteins with structural and functional roles are attached to the DNA. Each DNA molecule, with its attached proteins, is a chromosome.

5. Members of the same species have a characteristic number of chromosomes in their cells. The chromosomes differ from one another in length and shape, and they carry different portions of the hereditary instructions.

6. The body cells of humans and many other organisms have a diploid chromosome number; they contain two of each type of chromosome characteristic of the species.

7. Mitosis keeps the chromosome number constant, from one cell generation to the next. Therefore, if a parent cell is diploid, the daughter cells also will be diploid.

8. Mitotic cell division is the basis of growth and tissue repair in multicelled eukaryotes. It also is the means by which single-celled eukaryotes and many multicelled eukaryotes reproduce asexually.

9.1 DIVIDING CELLS: THE BRIDGE BETWEEN GENERATIONS

Overview of Division Mechanisms

In plants, animals, and all other eukaryotic organisms, hereditary instructions are distributed among a number of DNA molecules. Before the cells of such organisms reproduce, they must undergo *nuclear* division. **Mitosis** and **meiosis** are two nuclear division mechanisms. Both sort out and then package a parent cell's DNA into new nuclei for their forthcoming daughter cells. A separate mechanism splits the cytoplasm into daughter cells.

Multicelled organisms grow, replace dead or worn-out cells, and repair tissues by way of mitosis and the cytoplasmic division of body cells, which are called **somatic cells**. (Cut yourself peeling a potato and mitotic cell divisions will replace the cells that the knife sliced away.) Also, many protistans, fungi, plants, and some animals reproduce asexually by mitotic cell division.

By contrast, meiosis occurs only in **germ cells**, a cell lineage set aside for the formation of gametes (such as sperm and eggs) and sexual reproduction. As you will read in the next chapter, meiosis has much in common with mitosis, but the end result is different.

Prokaryotic cells—bacteria—reproduce asexually by a different mechanism, called prokaryotic fission. We will consider the bacteria later, in Section 22.2.

Some Key Points About Chromosomes

In a nondividing cell, the DNA molecules are stretched out like thin threads, with many proteins attached to them. Each DNA molecule, with its attached proteins, is a **chromosome**. When a cell prepares for mitosis, each threadlike chromosome gets duplicated. It now consists of two DNA molecules, which will stay together until late in mitosis. As long as they remain attached, the two are called **sister chromatids** of the chromosome.

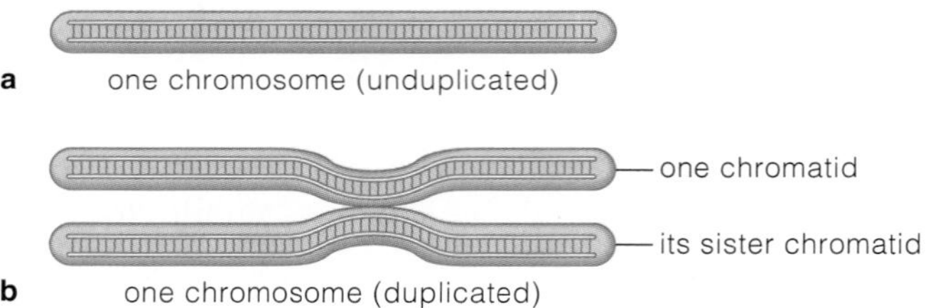
a
one chromosome (unduplicated)
one chromatid
its sister chromatid
b
one chromosome (duplicated)

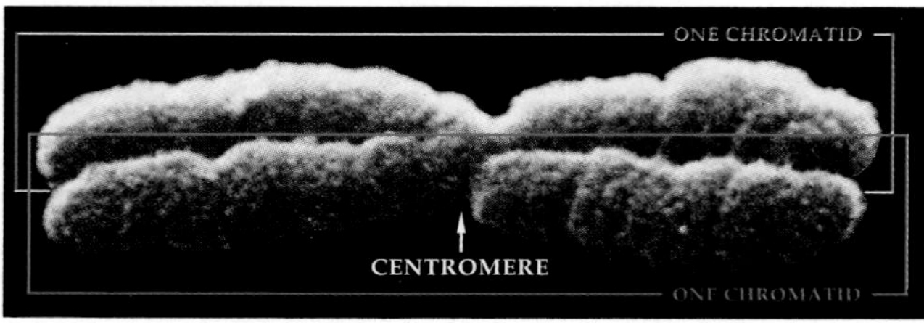
ONE CHROMATID
CENTROMERE
ONE CHROMATID
c

Figure 9.2 (**a**,**b**) Sketches of a chromosome in the unduplicated and duplicated states. Chromosomes are duplicated before cell division. (**c**) This scanning electron micrograph shows a human chromosome in the duplicated state; it consists of two sister chromatids attached at the centromere.

Figure 9.2 illustrates a eukaryotic chromosome in the unduplicated and duplicated states. Notice how the duplicated chromosome narrows down in one section along its length. This is the **centromere**, a small region with attachment sites for microtubules that move the chromosome during nuclear division. Bear in mind, the sketch in Figure 9.2 is highly simplified. For example, the centromere's location differs among chromosomes. And the two strands of a DNA molecule do not look like a ladder; they twist rather like a spiral staircase and are much longer than can be shown here.

Mitosis and the Chromosome Number

Each species has a characteristic **chromosome number**, which refers to the sum total of chromosomes in cells of a given type. Human somatic cells have 46, those of gorillas have 48, and those of pea plants have 14.

Actually, your 46 chromosomes are like volumes of two sets of books. Each set is numbered, from 1 to 23. For example, you have two "volumes" of chromosome 22—that is, *a pair of them*. Generally, both members of each pair have the same length and shape, and they carry the same portion of hereditary instructions for the same traits. Think of them as two sets of books on how to build a house. Your father gave you one set. Your mother had her own ideas about storage, plumbing, and so on, so she gave you an alternate edition. Her set covers the same topics but says slightly different things about many of them.

We say the chromosome number is **diploid**, or $2n$, if a cell has two of each type of chromosome characteristic of the species. The body cells of humans, gorillas, pea plants, and a great many other organisms are like this. (By contrast, as Chapter 10 describes, eggs and sperm of these organisms have a *haploid* chromosome number, or n. This means they contain only one of each type of chromosome characteristic of the species.)

With mitosis, a diploid parent cell can produce two diploid daughter cells. This doesn't mean each merely gets forty-six or forty-eight or fourteen chromosomes. If only the total mattered, one cell might get, say, two pairs of chromosome 22 and no pairs whatsoever of chromosome 9. Neither would be able to function like the parent cell *without two of each type of chromosome.*

Mitosis keeps the chromosome number constant, division after division, from one cell generation to the next. Thus, if a parent cell is diploid, its daughter cells will be diploid also.

Further reading: Student Guide to InfoTrac on web site

9.2 THE CELL CYCLE

Mitosis is only one phase of the **cell cycle**. Such cycles start each time new cells form, and they end when those cells complete their own division. The cycle starts again for each new daughter cell, as in Figure 9.3. Usually, the longest phase of the cell cycle is **interphase**, which has three parts. During interphase, a cell increases its mass, roughly doubles its number of cytoplasmic components, then duplicates its DNA. Biologists abbreviate the four parts of a cell cycle this way:

G1 Of interphase, a "Gap" (interval) of cell growth and functioning before the onset of DNA replication

S Of interphase, a time of "Synthesis" (DNA replication)

G2 Of interphase, a second "Gap" (interval) after DNA replication, when the cell prepares for division

M *Mitosis*; nuclear division only, usually followed by cytoplasmic division

The cell cycle lasts about the same length of time for cells of the same type. Its duration differs among cells of different types. As examples, all neurons (nerve cells) in your brain are arrested at interphase and usually do not divide again. All red blood cells form (and replace your worn-out ones) at an average rate of 2–3 million every second. Early in the development of a sea urchin embryo, the number of cells doubles every two hours.

Adverse conditions may disrupt a cell cycle. When deprived of a vital nutrient, for instance, the free-living cells called amoebas do not leave interphase. Even so, if any cell proceeds past a certain point in interphase, the cycle normally will continue regardless of outside conditions owing to built-in controls over its duration.

We turn now to mitosis and to how it maintains the chromosome number through turn after turn of the cell cycle. Figure 9.4 only hints at the divisional ballet that begins when a cell leaves interphase.

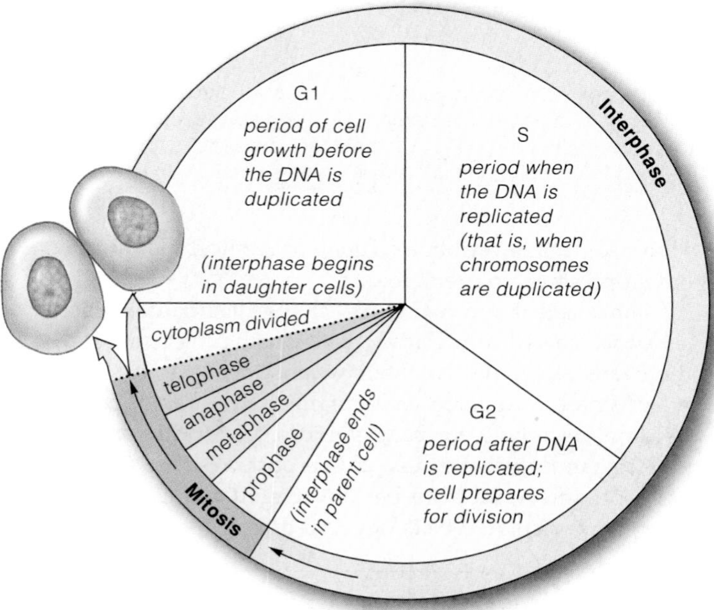
G1
period of cell growth before the DNA is duplicated
(interphase begins in daughter cells)
S
period when the DNA is replicated (that is, when chromosomes are duplicated)
Interphase
G2
period after DNA is replicated; cell prepares for division
(interphase ends in parent cell)
prophase
metaphase
anaphase
telophase
cytoplasm divided
Mitosis

Figure 9.3 Eukaryotic cell cycle, generalized. The length of each part differs among different cell types.

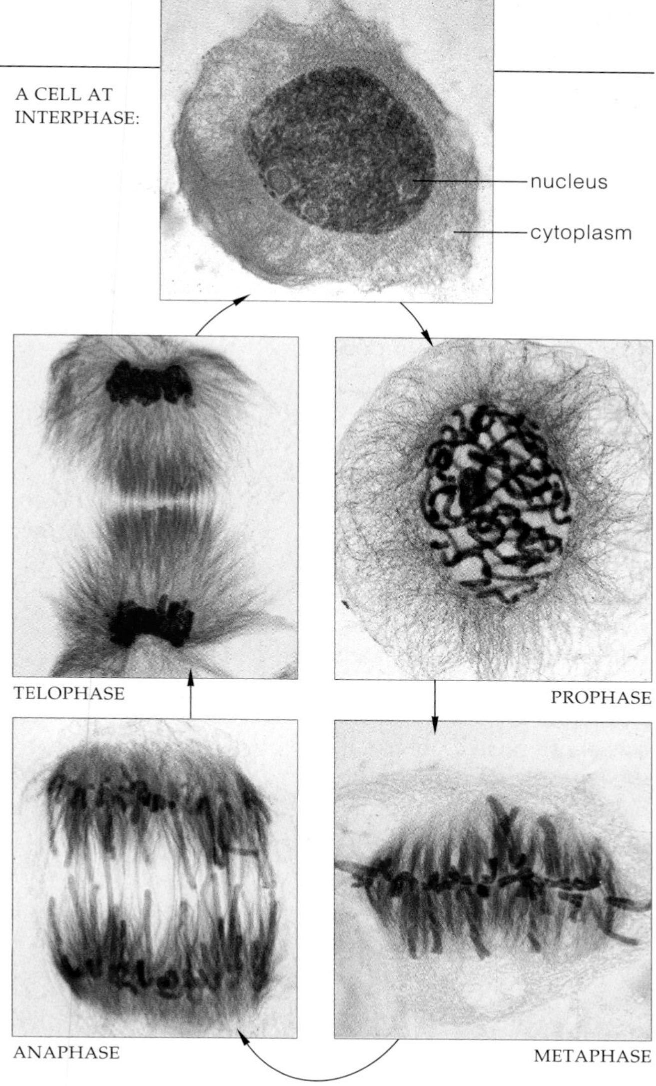
A CELL AT INTERPHASE:
nucleus
cytoplasm
TELOPHASE
PROPHASE
ANAPHASE
METAPHASE

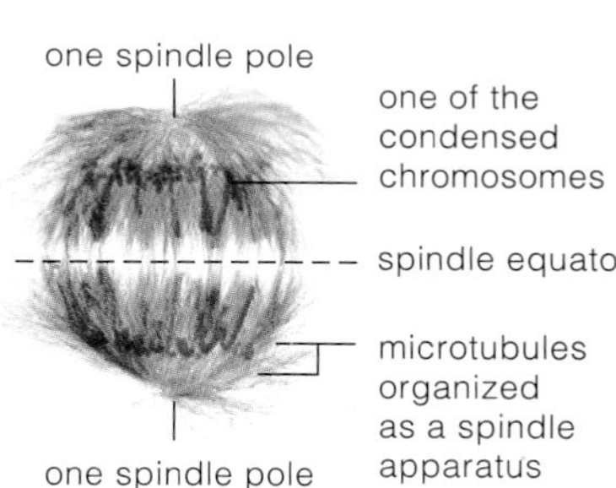
one spindle pole
one of the condensed chromosomes
spindle equator
microtubules organized as a spindle apparatus
one spindle pole

Figure 9.4 Mitosis in a cell from the African blood lily, *Haemanthus*. The chromosomes are stained *blue*, and the many microtubules are stained *red*. Before reading further, take a moment to become familiar with the labels on the micrographs.

A cell cycle starts at interphase, when a new cell (formed by mitosis and cytoplasmic division) increases its mass and the number of its cytoplasmic components, then duplicates its chromosomes. The cycle ends when the cell divides.

9.3 STAGES OF MITOSIS—AN OVERVIEW

When a cell makes the transition from interphase to mitosis, it has stopped constructing new cell parts, and the DNA has been replicated. Within that cell, major changes will now proceed smoothly, one after the other, through four stages. The sequential stages of mitosis are **prophase**, **metaphase**, **anaphase**, and **telophase**.

Figure 9.5 shows mitosis in an animal cell. By comparing its series of photographs with those of the plant cell in Figure 9.4, you see the chromosomes are changing positions. They are not doing so on their own. A **spindle apparatus** is moving them. Each spindle consists of microtubules organized as two distinct sets. Each set extends from one of the two poles (end points) of the spindle. The two sets overlap each other a bit at the spindle equator, midway between the poles. Formation of this bipolar, microtubular spindle will establish the final destinations of chromosomes during mitosis, as you will see shortly.

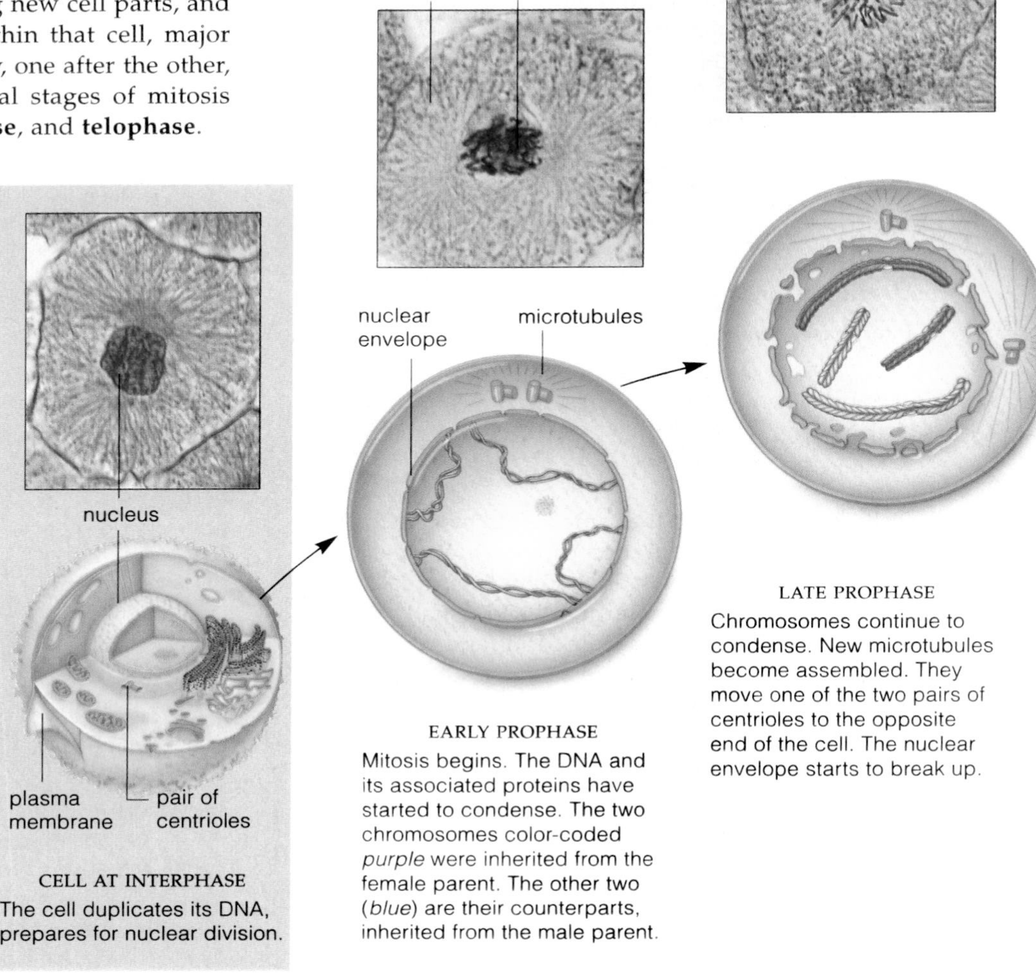

cell
chromosomes
nucleus
plasma membrane
pair of centrioles
CELL AT INTERPHASE
The cell duplicates its DNA, prepares for nuclear division.
nuclear envelope
microtubules
EARLY PROPHASE
Mitosis begins. The DNA and its associated proteins have started to condense. The two chromosomes color-coded *purple* were inherited from the female parent. The other two (*blue*) are their counterparts, inherited from the male parent.
LATE PROPHASE
Chromosomes continue to condense. New microtubules become assembled. They move one of the two pairs of centrioles to the opposite end of the cell. The nuclear envelope starts to break up.

Figure 9.5 Mitosis in a generalized animal cell. This nuclear division mechanism ensures that each daughter cell will have the same chromosome number as the parent cell. For clarity, this diagram shows only two pairs of chromosomes from a diploid ($2n$) cell. With only rare exceptions, the picture is more complicated, as suggested by the micrographs of mitosis in a whitefish cell.

Prophase: Mitosis Begins

We know a cell is in prophase when its chromosomes become visible in the light microscope as threadlike forms. ("Mitosis" comes from the Greek *mitos*, for thread.) Each chromosome was duplicated earlier, during interphase, so each is now two sister chromatids joined at the centromere. During early prophase, sister chromatids of each chromosome twist and fold into a more compact form. By late prophase, all of the chromosomes will be condensed into thicker, rod-shaped forms.

Meanwhile, in the cytoplasm, most microtubules of the cytoskeleton are breaking down to tubulin subunits (Section 4.8). The subunits reassemble near the nucleus as *new* microtubules of the spindle. Many of the new microtubules will extend from one spindle pole or the other to a centromere of a chromosome. The remainder will not interact with chromosomes. Instead, they will extend from the poles and overlap each other.

While new microtubules are assembling, the nuclear envelope is a physical barrier that prevents them from interacting with the chromosomes inside the nucleus. However, when prophase draws to a close, the nuclear envelope starts to break up.

Many cells have two barrel-shaped **centrioles**. Each centriole started duplicating itself during interphase, so there are two pairs of them when prophase is under way. Microtubules start moving one pair to the opposite pole of the newly forming spindle. Centrioles, recall, give rise to flagella or cilia. If you observe them in cells of an organism, you can bet that flagellated cells (such as sperm) or ciliated cells develop during its life cycle.

Transition to Metaphase

So much happens between prophase and metaphase that researchers give this transitional period its own

Further reading: Student Guide to InfoTrac on web site

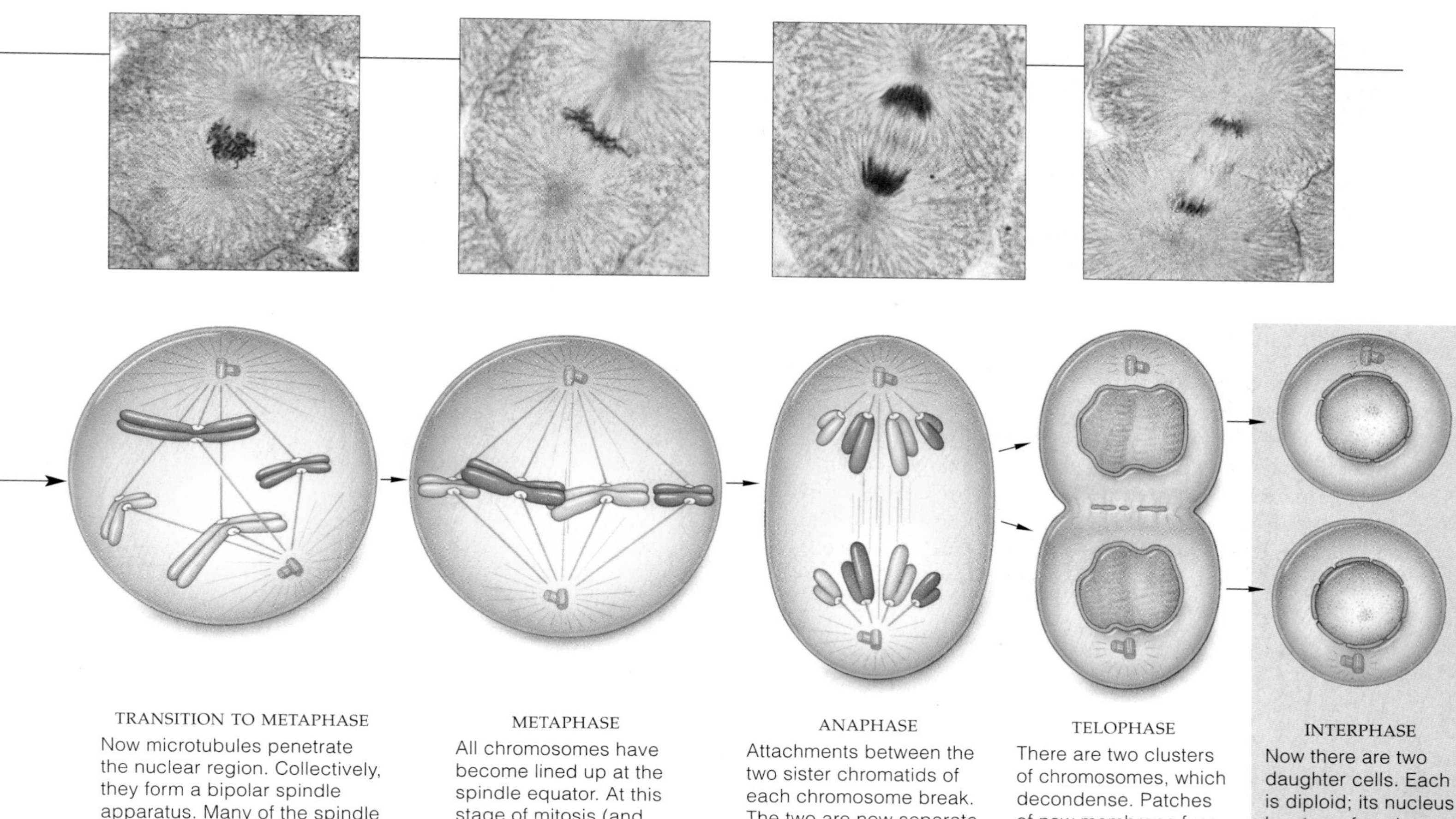

TRANSITION TO METAPHASE
Now microtubules penetrate the nuclear region. Collectively, they form a bipolar spindle apparatus. Many of the spindle microtubules become attached to the two sister chromatids of each chromosome.

METAPHASE
All chromosomes have become lined up at the spindle equator. At this stage of mitosis (and of the cell cycle), they are in their most tightly condensed form.

ANAPHASE
Attachments between the two sister chromatids of each chromosome break. The two are now separate chromosomes, which microtubules move to opposite spindle poles.

TELOPHASE
There are two clusters of chromosomes, which decondense. Patches of new membrane fuse to form a new nuclear envelope. Mitosis is completed.

INTERPHASE
Now there are two daughter cells. Each is diploid; its nucleus has two of each type of chromosome, just like the parent cell.

name, "prometaphase." The nuclear envelope breaks up completely into numerous tiny, flattened vesicles. Now the chromosomes are free to interact with microtubules that are extending toward them, from the poles of the forming spindle. Microtubules from both poles harness each chromosome and start pulling on it. The two-way pulling orients the chromosome's two sister chromatids toward opposite poles. Meanwhile, overlapping spindle microtubules ratchet past each other and push the poles of the spindle apart. The push–pull forces are balanced when the chromosomes reach the spindle's midpoint.

When all of the duplicated chromosomes are aligned midway between the poles of a completed spindle, we call this metaphase (*meta-* means "midway between"). The alignment is crucial for the next stage of mitosis.

From Anaphase Through Telophase

At anaphase, the sister chromatids of each chromosome separate from each other and move to opposite poles by two mechanisms. First, microtubules attached to the centromere regions shorten and *pull* the chromosomes to the poles. Second, the spindle elongates as overlapping microtubules continue to ratchet past each other and *push* the two spindle poles even farther apart. Once each chromatid is separated from its sister, it has a new name. It is a separate chromosome in its own right.

Telophase gets under way as soon as each of two clusters of chromosomes arrives at a spindle pole. The chromosomes, no longer harnessed to the microtubules, return to threadlike form. Vesicles derived from the old nuclear envelope fuse and form patches of membrane around the chromosomes. Patch joins with patch, and soon a new nuclear envelope separates each cluster of chromosomes from the cytoplasm. If the parent cell was diploid, each cluster contains two chromosomes of each type. With mitosis, remember, each new nucleus has the same chromosome number as the parent nucleus. Once two nuclei form, telophase is over—and so is mitosis.

Prior to mitosis, each chromosome in a cell's nucleus is duplicated, so that it consists of two sister chromatids.

Mitosis proceeds through four consecutive stages called prophase, metaphase, anaphase, and telophase.

A microtubular spindle moves sister chromatids of every chromosome apart, to opposite spindle poles. Around each of two clusters of chromosomes, new nuclear envelope forms. Both daughter nuclei formed this way have the same chromosome number as the parent cell's nucleus.

9.4 DIVISION OF THE CYTOPLASM

The cytoplasm usually divides at some time between late anaphase and the end of telophase. As you might gather by comparing Figures 9.6 and 9.7, the actual mechanism of **cytoplasmic division** (or cytokinesis, as it is often called) differs among organisms.

Cell Plate Formation in Plants

As described in Section 4.10, most plant cells are walled, which means their cytoplasm cannot be pinched in two. Cytoplasmic division of such cells involves **cell plate formation**, as shown in Figure 9.6. By this mechanism, vesicles packed with wall-building materials fuse with one another and with remnants from the microtubular spindle. Together, they form a disklike structure—a cell plate. At this location, deposits of cellulose accumulate.

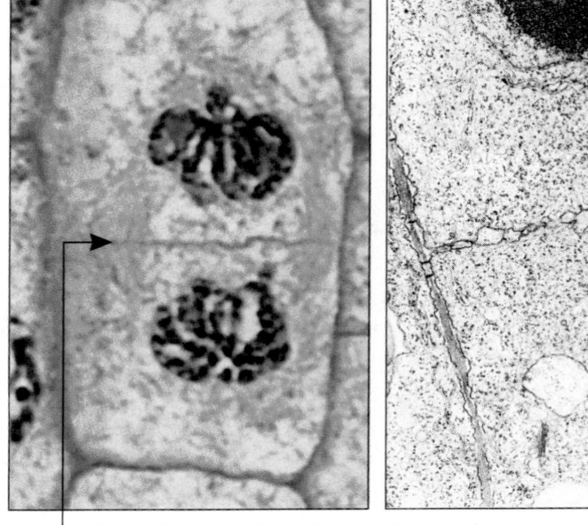

light micrograph and transmission electron micrograph showing cell plate formation in a dividing plant cell

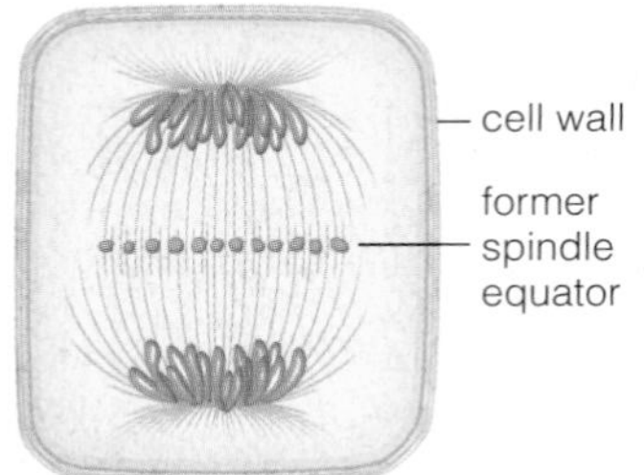
cell wall
former spindle equator

a As mitosis ends, vesicles converge at the spindle equator. They contain cementing materials and structural materials for a new primary cell wall.

vesicles converging

b A cell plate starts forming as membranes of the vesicles fuse. Materials inside the vesicles get sandwiched between two new membranes that elongate along the plane of the cell plate.

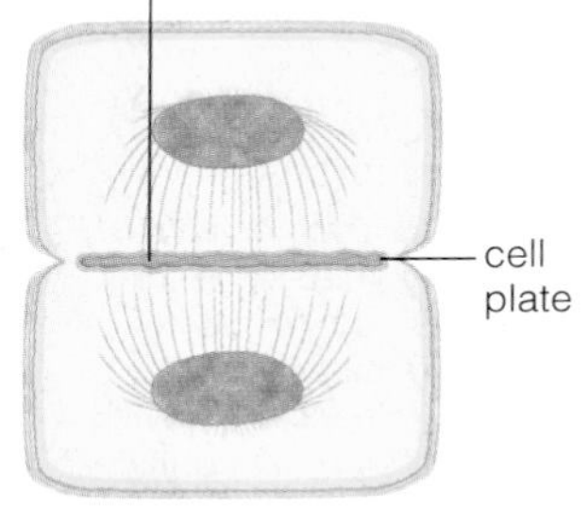
cell plate

c Cellulose is deposited on the inside of the "sandwich." (In time, deposits will form two cell walls. Other deposits will form a middle lamella and cement the walls together; refer to Section 4.10.)

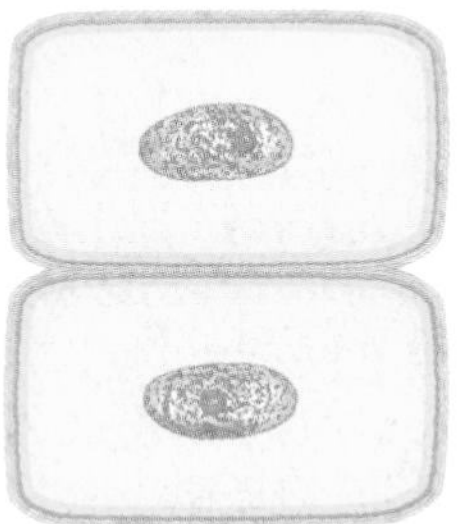

d A cell plate grows at its margins until it fuses with the parent cell's plasma membrane. During growth, when new plant cells expand and their walls are still thin, new material is deposited on the old primary wall.

Figure 9.6 Cytoplasmic division of a plant cell, as brought about by cell plate formation. 

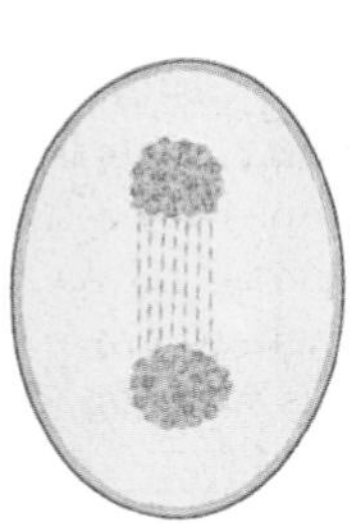

a Mitosis is over, and the spindle is now disassembling.

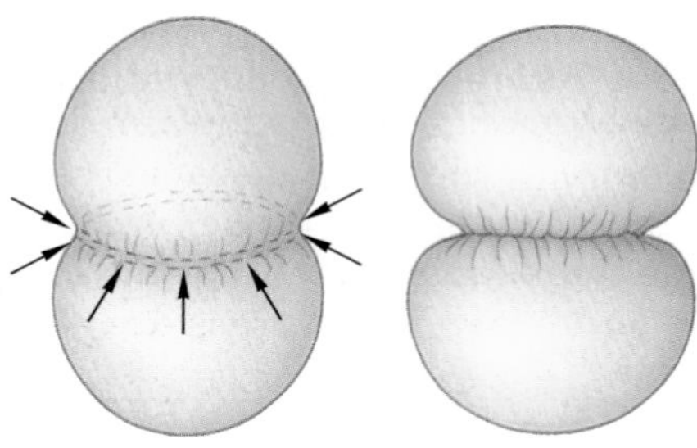

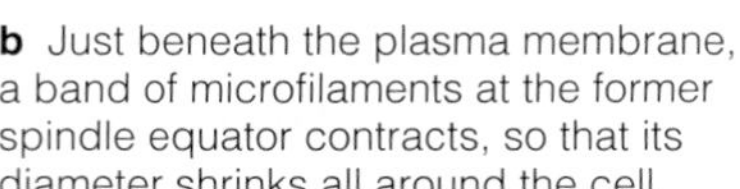

b Just beneath the plasma membrane, a band of microfilaments at the former spindle equator contracts, so that its diameter shrinks all around the cell.

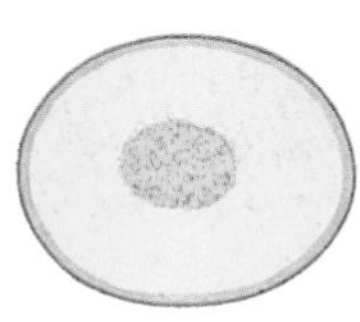

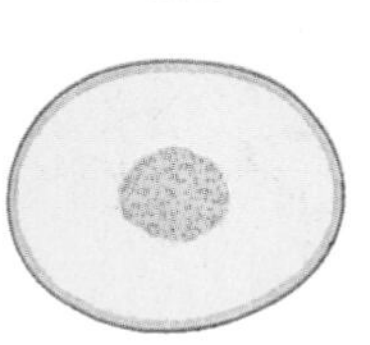

c The contractions continue and cut the cell in two.

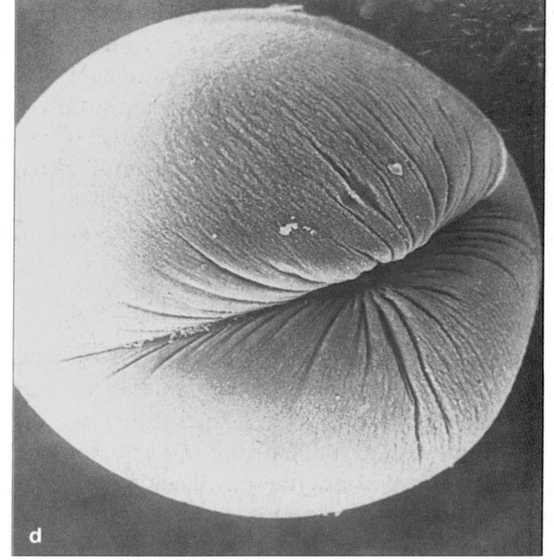
d

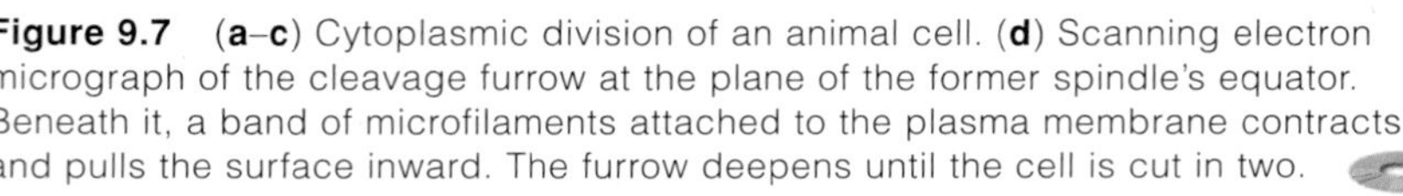

Figure 9.7 (**a**–**c**) Cytoplasmic division of an animal cell. (**d**) Scanning electron micrograph of the cleavage furrow at the plane of the former spindle's equator. Beneath it, a band of microfilaments attached to the plasma membrane contracts and pulls the surface inward. The furrow deepens until the cell is cut in two.

Further reading: Student Guide to InfoTrac on web site

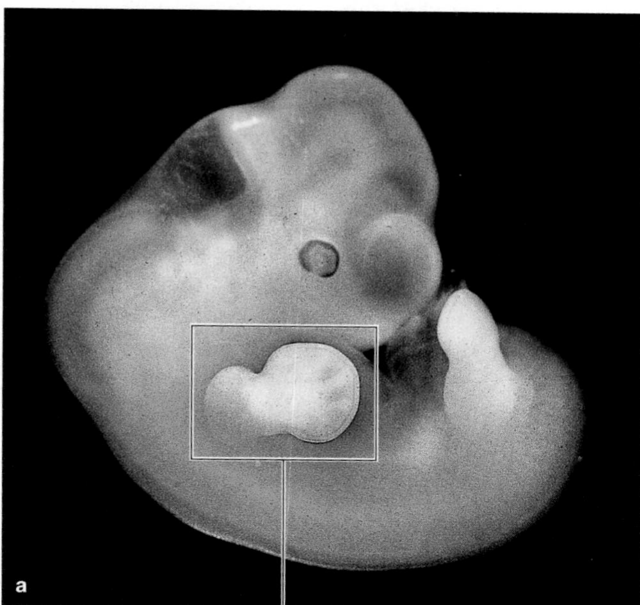
a
future arm and hand of embryo, five weeks old

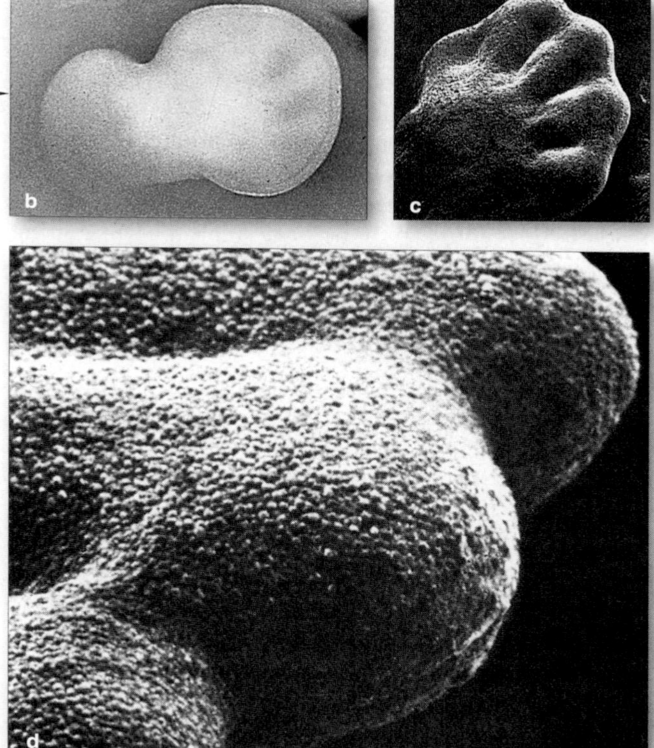
b
c
d

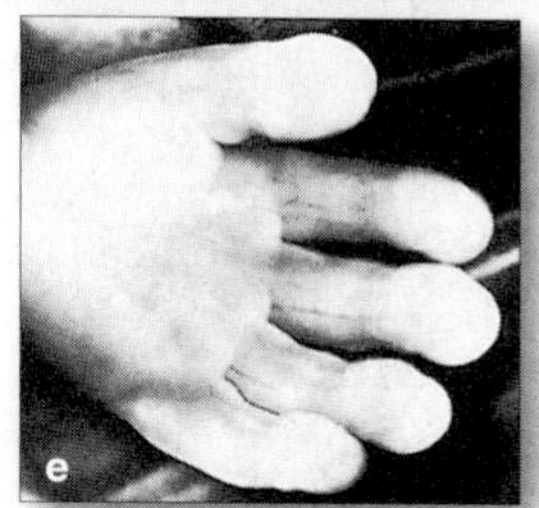
e

Figure 9.8 A few moments in the emergence of the human hand by way of mitosis, cytoplasmic divisions, and other developmental processes. Many individual cells produced by mitotic cell divisions are visible in (**d**). The photograph in (**e**) shows the digits at a later stage of development.

In time, the cellulose deposits are thick enough to form a crosswall. The new crosswall bridges the cytoplasm and divides the parent cell into two daughter cells.

Cytoplasmic Division of Animal Cells

Unlike plant cells, an animal cell is not confined within a cell wall, and its cytoplasm typically "pinches in two." Look at Figure 9.7, a surface view of a newly fertilized animal egg. An indentation is forming about midway between the egg's two poles ("ends"). The indentation in its plasma membrane is a **cleavage furrow**. Such a furrow is the first visible sign that an animal cell is undergoing cytoplasmic division. It will extend around the cell and continue to deepen along the plane of the former spindle's midpoint until the cell is cut in two.

A band of microfilaments beneath the cell's plasma membrane generates the force for cytoplasmic division. Microfilaments, remember, are threadlike cytoskeletal elements. These particular ones are organized so that they slide past one another (Section 4.9). When they do, they pull the plasma membrane inward until the two daughter nuclei are cut off in separate cells, each with its own cytoplasm and plasma membrane.

This concludes our picture of mitotic cell division. Look now at your two hands and try to visualize all of the cells in your palms, thumbs, and fingers. Imagine the divisions that produced all the generations of cells that preceded them when you were developing early on, inside your mother (Figure 9.8). And be grateful for the astonishing precision of the mechanisms that led to their formation at certain times, in certain numbers, for the alternatives can be terrible indeed. Why? Good health, and survival itself, depends absolutely on the proper timing and the completion of cell cycle events, including mitosis. Some genetic disorders arise from mistakes in the duplication or distribution of even one chromosome. Also, when normal controls that prevent cells from dividing are lost, unchecked cell divisions may destroy surrounding tissues and, ultimately, the organism. Section 9.6 touches on a landmark case of such losses, which we will explore further in Section 15.6.

Following mitosis, a separate mechanism cuts the cytoplasm into two daughter cells, each with a daughter nucleus.

In plants, cytoplasmic division often involves the formation of a cell plate and a crosswall between the adjoining, new plasma membranes of daughter cells.

Cytoplasmic division in animals may involve cleavage. Rings of microfilaments around a parent cell's midsection slide past one another in a way that pinches the cytoplasm in two.

9.5 A CLOSER LOOK AT THE CELL CYCLE

Close this book for a moment. Be sure you have a clear picture of the flow of events in interphase, mitosis, and on through cytoplasmic division. How easily you will get through many later chapters depends on how well you understand the cell division story. If parts of the picture are still not clear, it may be worth your time to read the preceding sections once again before getting into the details presented here.

The Wonder of Interphase

If you could coax the DNA molecules from just one of your somatic cells to stretch in a single line, one after another, that line would extend past the fingertips of your outstretched arms. Salamander DNA is even more amazing. A single line of it would extend ten meters! The wonder is, enzymes and other proteins in the cell selectively scan all the DNA, switch protein-building instructions on and off, and make base-by-base copies of each DNA molecule—all during interphase.

G1, S, and G2 of interphase have distinct patterns of biosynthesis. Most of your cells remain in G1, when they assemble most of the carbohydrates, lipids, and proteins they use or export. The cells destined to divide enter S, when they copy the DNA and the histones and other proteins associated with it. During G2, these cells produce proteins that will drive mitosis to completion.

Once S begins, events normally proceed at about the same rate in all cells of a species and continue through mitosis. Given this observation, you may well assume the cycle has built-in molecular brakes. Apply brakes that operate in G1, and the cycle stalls in G1. Lift the brakes and the cycle runs to completion. Said another way, *control mechanisms govern the rate of cell division.*

Imagine a car losing its brakes just as it starts down a steep mountain road. As you will read later on in the book, that is how cancer starts. Controls over division are lost, and the cell cycle cannot stop turning.

On Chromosomes and Spindles

The precision with which mitosis parcels out DNA for forthcoming daughter cells is impressive. That precision depends on chromosome organization and interactions among microtubules and motor proteins.

ORGANIZATION OF METAPHASE CHROMOSOMES Even during interphase, eukaryotic DNA has many proteins bound tightly to it. **Histones** are among them. Many histones are like spools for winding up small stretches of DNA. Each histone–DNA spool is a single structural unit called a **nucleosome**. Other histones stabilize the spools. During mitosis (and meiosis also), interactions between histones and DNA make the chromosome coil back on itself repeatedly. The coiling greatly increases the chromosome's diameter. Other proteins besides the histones form a structural scaffold when the DNA folds even more, possibly into a series of loops (Figure 9.9).

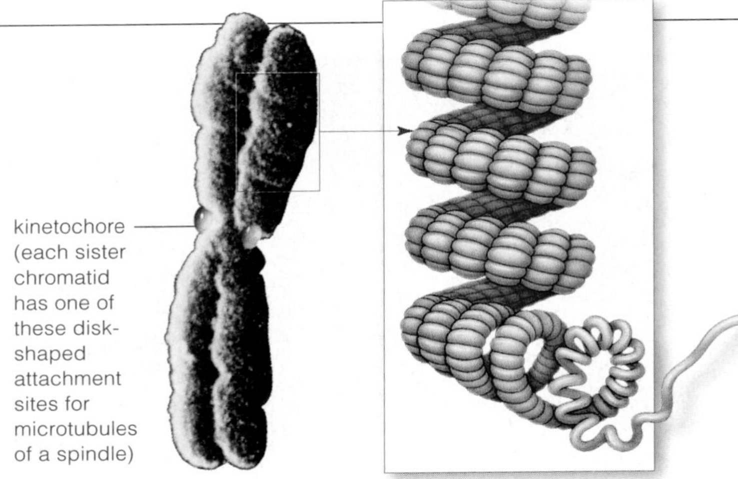
kinetochore (each sister chromatid has one of these disk-shaped attachment sites for microtubules of a spindle)

a A duplicated human chromosome at metaphase, at its most condensed. Interactions among some chromosomal proteins may keep loops of DNA tightly packed in a "supercoiled" array.

Figure 9.9 One model of the levels of organization in a human chromosome at metaphase.

A chromosome acquires its distinct shape and size late in prophase, when condensation is nearly complete. By then, each of its sister chromatids has at least one constricted region, the most prominent of which is the centromere. As you will read next, small, disk-shaped structures at the surface of centromeres serve as the docking sites for spindle microtubules (Figure 9.9*a*). We call them **kinetochores**.

When threadlike molecules of DNA are condensing into such compact chromosome structures, why don't they get tangled up? Actually, it appears that they do, but an enzyme called DNA topoisomerase chemically recognizes such tangling and puts things right. When researchers deliberately interfered with this enzyme's activity, the tangles persisted. Later, sister chromatids failed to separate at anaphase.

SPINDLES COME, SPINDLES GO All through anaphase, the spindle microtubules attached to chromosomes do not change position, yet the distance shrinks between these microtubules and the spindle poles. How? When a kinetochore slides over them, the microtubules shorten by disassembling (Figure 9.10). A kinetochore is like a train chugging along a railroad track—except the track falls apart after the train has passed over it. It contains two motor proteins, dynein and kinesin, that may drive the sliding motion.

Motor proteins, remember, are accessory proteins for cytoskeletal elements. As mentioned in Section 4.8,

Further reading: Student Guide to InfoTrac on web site

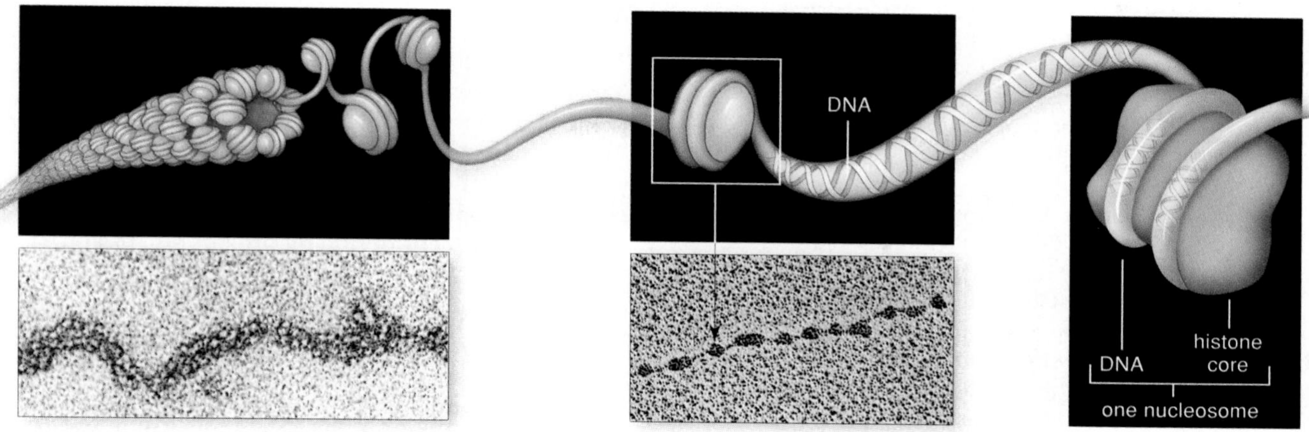
DNA
histone
core
DNA
one nucleosome

b At a deeper level of structural organization, the chromosomal proteins and the DNA are arranged as a cylindrical fiber (solenoid) thirty nanometers in diameter.

c Immerse a chromosome in saltwater and it loosens up to a beads-on-a-string organization. The "string" is one DNA molecule. Each "bead" is a nucleosome.

d A nucleosome consists of a double loop of DNA around a core of eight histones. Other histones stabilize the structural array.

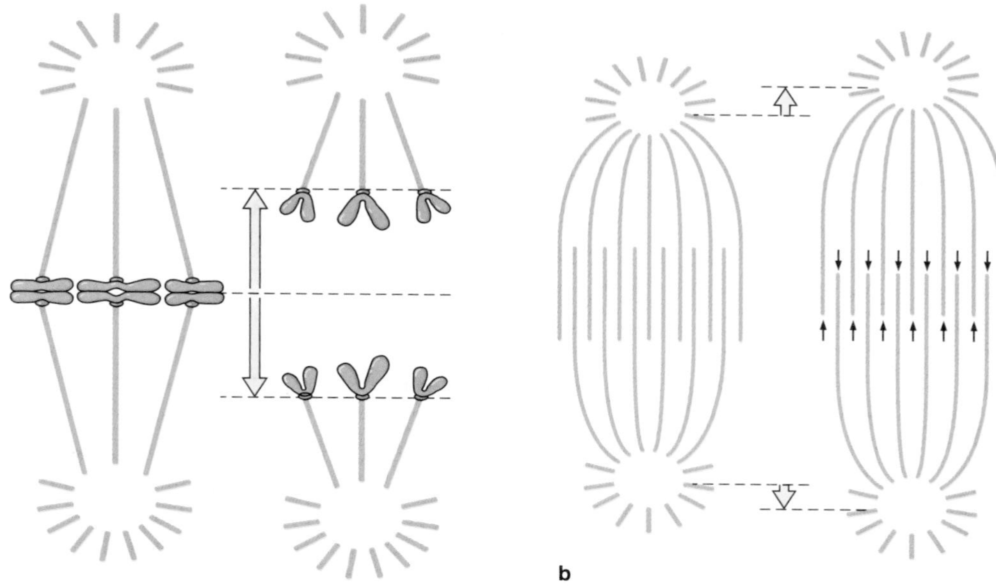
a
b

Figure 9.10 Models of two of the mechanisms that separate sister chromatids of a chromosome from each other at anaphase. In (**a**), the microtubules attached to the chromatids *shorten* and thereby decrease the distance between the kinetochores and spindle poles. In (**b**), overlapping microtubules ratchet past each other. As they do this, they move the spindle poles apart and thereby increase the distance between the sister chromatids of each chromosome.

they are attached to and project from microtubules and microfilaments that take part in cell movements.

And what about the spindle microtubules extending from both poles but not attached to kinetochores? They, too, incorporate dynein and kinesin. They actively slide past one another where they overlap in a spindle.

A final point about spindle structure and function: Over evolutionary time, spindle microtubules became targets in the chemical warfare between certain edible plants and animals that browse on them. For instance, plants of the genus *Colchicum* make a poison that blocks assembly and promotes disassembly of microtubules. The poison, colchicine, is a favorite of those who study cancer and other expressions of cell division. A *Critical Thinking* question at the chapter's end mentions another kind. Spindles in cells will disassemble within seconds or minutes after exposure to such microtubule poisons.

Once the S stage of interphase begins, the cell cycle turns at about the same rate in all cells of a given type, all the way through mitosis. Molecular mechanisms control whether a cell enters S and thereby control the rate of cell division.

The condensed form of metaphase chromosomes arises by interactions among DNA and structural proteins, including histones, that associate with DNA throughout the cell cycle.

During mitosis, motor proteins act on microtubules of the spindle to produce chromosomal movements.

Focus on Science

9.6 HENRIETTA'S IMMORTAL CELLS

Each human starts out as a single fertilized egg. By the time of birth, mitotic cell divisions and other processes have resulted in a human body of about a trillion cells. Even in adults, billions of cells still divide. For example, cells of the stomach's lining divide every day. Liver cells usually do not divide, but if part of the liver becomes injured or diseased, repeated cell divisions will yield more new cells until the damaged part is finally replaced.

In 1951, George and Margaret Gey of Johns Hopkins University were trying to develop a way to keep human cells dividing *outside* the body. With such isolated cells, these researchers and others could investigate basic life processes. They also could conduct studies of cancer and other diseases without having to experiment directly on patients and gamble with human lives. The Geys used normal and diseased human cells, which local physicians had sent them. But they couldn't stop the descendants of those precious cells from dying out within a few weeks.

Mary Kubicek, a laboratory assistant, worked with the Geys in their efforts to start a self-perpetuating lineage of cultured human cells. She was about to give up after dozens of failed attempts. Even so, in 1951 she decided to prepare one more sample of cancer cells for culture. She gave the sample the code name **HeLa cells**, for the first two letters of the patient's first and last names.

The HeLa cells began to divide. And divide. And divide again! By the fourth day there were so many cells that Kubicek subdivided them into more culture tubes. Unfortunately, tumor cells in the patient were just as vigorous. Six months after she was diagnosed as having cancer, tumor cells had spread to tissues throughout her body. Only eight months after the diagnosis, Henrietta Lacks, a young woman from Baltimore, was dead.

Although Henrietta passed away, some of her cells lived on in the Geys' laboratory as the first successful human cell culture. In time, HeLa cells were shipped to research laboratories all over the world. Some journeyed into space for experiments on the *Discoverer XVII* satellite. Each year hundreds of scientific papers describe work that was based on HeLa cells.

Henrietta was thirty-one when runaway cell divisions killed her. And now, decades later, her legacy is benefitting humans everywhere—in her cellular descendants that are still alive and dividing, day after day after day.

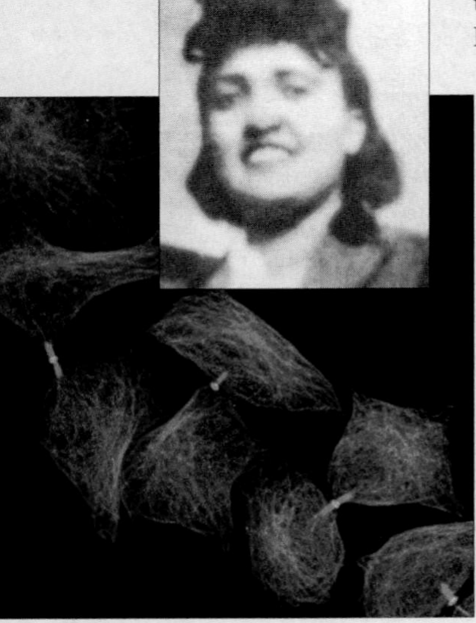

Figure 9.11 *Above:* Henrietta Lacks, a casualty of cancer whose cellular contribution to science is still helping others. *Below:* Dividing HeLa cells.

9.7 SUMMARY

1. Through specific division mechanisms, summarized in Table 9.1, a parent cell provides each daughter cell with the hereditary instructions (DNA) and cytoplasmic machinery necessary to start up its own operation.

a. In eukaryotic cells, the nucleus divides by mitosis or meiosis. Cytoplasmic division typically follows.

b. Prokaryotic cells divide by prokaryotic fission.

2. Each eukaryotic chromosome is one DNA molecule with numerous proteins attached. Chromosomes in a given cell differ in length, shape, and which portion of the hereditary instructions they carry.

a. "Chromosome number" refers to the sum total of chromosomes in cells of a given type. Cells having a diploid chromosome number ($2n$) contain two of each kind of chromosome.

b. Mitosis divides the nucleus into two equivalent nuclei, each with the same chromosome number as the parent cell. It maintains the chromosome number from one cell generation to the next.

c. Mitosis is the basis of growth, tissue repair, and cell replacements among multicelled eukaryotes. It is the basis of asexual reproduction in many single-celled eukaryotes. (Meiosis occurs only in germ cells.)

3. A cell cycle starts when a new cell forms. It proceeds through interphase and ends when the cell reproduces by mitosis and cytoplasmic division. In interphase, a cell carries out its functions. If it is to divide again, the cell increases in its mass and cytoplasmic components and duplicates its chromosomes in preparation for division.

4. A duplicated chromosome has two DNA molecules attached at the centromere. While the two stay attached to each other, they are called sister chromatids.

5. Mitosis proceeds through four continuous stages:

a. Prophase. Duplicated, threadlike chromosomes start to condense. A spindle apparatus starts to form. The nuclear envelope starts to break up and, during the *transition* to metaphase (prometaphase), its remnants form vesicles, and microtubules from opposite poles of

Table 9.1 Summary of Cell Division Mechanisms

Mechanisms	Functions
MITOSIS, CYTOPLASMIC DIVISION	In all multicelled eukaryotes, the basis of increases in body size during growth, tissue repair, and cell replacements. Also the basis of *asexual* reproduction in single-celled and many multicelled eukaryotes.
MEIOSIS, CYTOPLASMIC DIVISION	In single-celled and multicelled eukaryotes, the basis of gamete formation and of *sexual* reproduction.
PROKARYOTIC FISSION	In bacterial cells only, the basis of *asexual* reproduction.

Further reading: Student Guide to InfoTrac on web site →

the developing spindle attach to only one of two sister chromatids of each chromosome. *At* metaphase, all the chromosomes are aligned at the spindle equator.

b. Anaphase. Microtubules pull sister chromatids of each chromosome away from each other, to opposite spindle poles. Now each type of parental chromosome is represented by a daughter chromosome at both poles:

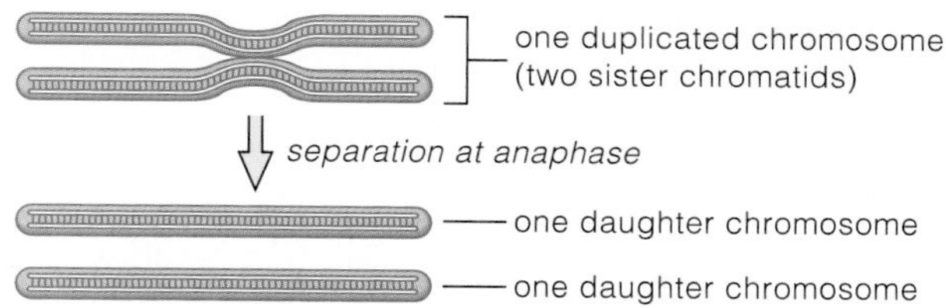
one duplicated chromosome
(two sister chromatids)
separation at anaphase
one daughter chromosome
one daughter chromosome

c. Telophase. Chromosomes decondense to threadlike form. A new nuclear envelope forms around them. Each nucleus has the parental chromosome number.

6. Separate mechanisms divide the cytoplasm near the end of nuclear division or afterward.

Review Questions

1. Define mitosis and meiosis, two mechanisms that operate in eukaryotic cells. Does either one divide the cytoplasm? *9.1*

2. Define somatic cell and germ cell. *9.1*

3. What is a chromosome called when it is in the unduplicated state? In the duplicated state (with two sister chromatids)? *9.1*

4. Describe the microtubular spindle and its functions. *9.3*

5. Using Figure 9.5 as a guide, name and describe the key features of the stages of mitosis. *9.3*

6. Briefly explain how cytoplasmic division differs in typical plant and animal cells. *9.4*

Self-Quiz (*Answers in Appendix III*)

1. A somatic cell having two of each type of chromosome has a(n) ______ chromosome number.
   a. diploid b. haploid c. tetraploid d. abnormal

2. A duplicated chromosome has ______ chromatid(s).
   a. one b. two c. three d. four

3. In a chromosome, a ______ is a constricted region with attachment sites for microtubules.
   a. chromatid b. cell plate c. centromere d. cleavage

4. Interphase is the part of the cell cycle when ______.
   a. a cell ceases to function
   b. a germ cell forms its spindle apparatus
   c. a cell grows and duplicates its DNA
   d. mitosis proceeds

5. After mitosis, the chromosome number of a daughter cell is ____ the parent cell's.
   a. the same as c. rearranged compared to
   b. one-half d. doubled compared to

6. Mitosis and cytoplasmic division function in ______.
   a. asexual reproduction of single-celled eukaryotes
   b. growth, tissue repair, and sometimes asexual reproduction in many multicelled eukaryotes
   c. gamete formation in prokaryotes
   d. both a and b

7. Only ____ is not a stage of mitosis.
   a. prophase b. interphase c. metaphase d. anaphase

8. Match each stage with the events listed.
   ____ metaphase a. sister chromatids move apart
   ____ prophase b. chromosomes start to condense
   ____ telophase c. chromosomes decondense and daughter nuclei form
   ____ anaphase d. all duplicated chromosomes are aligned at spindle equator

Critical Thinking

1. Suppose you have a way to measure the amount of DNA in a single cell during the cell cycle. You first measure the amount at the G1 phase. At what points during the rest of the cell cycle would you predict changes in the amount of DNA per cell?

2. The cervix is part of the uterus, in which embryos develop. A screening procedure, the *Pap smear,* detects the earliest stages of *cervical cancer,* when chances for survival are greatest. Freezing precancerous cells or hitting them with a laser beam kills them. A hysterectomy (removing the uterus) rids the uterus of cancer. If the cancer is caught early, treatment is 90+ percent effective. If it spreads, survival chances plummet to less than 9 percent.

Most cervical cancers develop slowly. Unsafe sex increases the risk (Section 45.14). Why? A major risk factor is infection by human papillomaviruses that cause genital warts. In 93 percent of all cases, viral genes coding for tumor-inducing proteins had been inserted into the DNA of previously normal cervical cells.

Not all women request Pap smears. Many wrongly believe the procedure is costly. Many do not recognize the importance of abstinence or "safe" sex. Others simply do not want to know whether they have cancer. Knowing what you have learned so far about the cell cycle and cancer, what would you say to a woman who falls into one or more of these groups?

3. Pacific yews (*Taxus brevifolius*) face extinction. People started stripping its bark and killing the trees when they heard that *taxol,* a chemical extracted from the bark, might be useful for treating breast and ovarian cancer. (Synthesizing taxol in the laboratory may save the species.) Taxol can prevent microtubules from disassembling into tubulin subunits. What does this tell you about its potential as an anticancer drug?

4. X-rays and gamma rays emitted from some radioisotopes chemically damage DNA, especially in cells engaged in DNA replication. High-level exposure can result in *radiation poisoning.* Hair loss and damage to the lining of the stomach and intestines are two early symptoms. Speculate why. Also speculate on why highly focused radiation therapy is used against some cancers.

Selected Key Terms

anaphase *9.3*
cell cycle *9.2*
cell plate formation *9.4*
centriole *9.3*
centromere *9.1*
chromosome *9.1*
chromosome number *9.1*
cleavage furrow *9.4*
cytoplasmic division *9.4*
diploid (chromosome number) *9.1*
germ cell *9.1*
HeLa cell *9.6*
histone *9.5*
interphase *9.2*
kinetochore *9.5*
meiosis *9.1*
metaphase *9.3*
mitosis *9.1*
motor protein *9.5*
nucleosome *9.5*
prophase *9.3*
reproduction *CI*
sister chromatid *9.1*
somatic cell *9.1*
spindle apparatus *9.3*
telophase *9.3*

Readings *See also www.infotrac-college.com*

Murray, A., and M. Kirschner. March 1991. "What Controls the Cell Cycle?" *Scientific American* 264(3): 56–63.

10

MEIOSIS

Octopus Sex and Other Stories

The couple clearly are interested in each other. First he caresses her with one tentacle, then another—and then another and another. She reciprocates with a hug here, a squeeze there. This goes on for hours. Finally the male reaches under his mantle, a fold of tissue that drapes around most of his body. He removes a packet of sperm from a reproductive organ and inserts it into an egg chamber underneath the female's mantle. For every sperm that successfully fertilizes an egg, a new octopus may grow and develop.

Unlike the one-to-one coupling between a male and female octopus, sex for the slipper limpet is a group enterprise. Slipper limpets are marine animals, relatives of the familiar snails on land. Before becoming transformed into a sexually mature adult, a slipper limpet must pass on through a free-living stage of development called a larva. When a limpet larva is about to undergo its programmed transformation, it settles on a rock or pebble or shell. If it settles down all by itself, it will become a female. Then, if another larva settles and develops on the first limpet, the second limpet will function right off as a male. However, if *another* limpet develops into a male on top of it, the second limpet will switch gears and develop into a female. Later, the third limpet will also become a female if still another limpet develops into a male on top of *it*—and so on amongst ten or more limpets!

Slipper limpets typically live in such piles, with the bottom one always being the oldest female and the uppermost one being the youngest male (Figure 10.1*a*).

a

b

Figure 10.1 Examples of variations in reproductive modes of eukaryotic organisms. (**a**) Slipper limpets, busily perpetuating the species by group participation in sexual reproduction. The tiny crab in the foreground is merely a passerby. (**b**) Live birth of an aphid, a type of insect that reproduces sexually in autumn but can switch to an asexual mode in summer.

Until they make the gender switch, the male limpets release sperm, which fertilize a female's eggs, which grow up to become males and then, most likely, females—and so it goes, from one sexually flexible generation to the next.

Limpets are not alone in having unusual variations in their mode of reproduction. For example, sexual reproduction is common in many life cycles—and so are asexual episodes that are based on mitotic cell divisions. Orchids, dandelions, and many other plants reproduce very well with or without engaging in sex. Aquatic animals called flatworms can engage in sex or can split their small body into two roughly equivalent parts, which each grow into a new flatworm.

And what about aphids! In summertime, nearly all aphids are females, which produce more females from *unfertilized* egg cells (Figure 10.1*b*). Only when autumn approaches do male aphids finally develop and do their part in the sexual phase of the life cycle. Even then, females that manage to survive through the winter can do without males. Come summer, the females begin another round of producing offspring all by themselves.

These examples only hint at the immense variation in reproductive modes among eukaryotic organisms. And yet, despite the variation, *sexual* reproduction dominates nearly all of the life cycles, and it always involves certain events. Briefly, before the time of cell division, chromosomes are duplicated in cells that are set aside for reproduction. For instance, the immature reproductive cells called **germ cells** develop in male and female animals. Germ cells undergo meiosis and cytoplasmic division. In time, the cellular descendants of germ cells mature and become **gametes**, or sex cells. When gametes manage to get together at fertilization, they form the first cell of a new individual.

With this chapter, we turn to the kinds of cells that serve as the bridge between generations of organisms. Specialized phases of reproduction and development, including asexual episodes, loop out from the basic life cycle of many eukaryotic species. Regardless of the specialized details, all of the life cycles turn on three events: *meiosis, the formation of gametes, and fertilization*. These three interconnected events are the hallmarks of sexual reproduction. As you will see in many chapters throughout the book, they have contributed to the diversity of life.

KEY CONCEPTS

1. Sexual reproduction proceeds through three events: meiosis, gamete formation, and fertilization. Sperm and eggs are familiar gametes.

2. Meiosis, a nuclear division mechanism, occurs only in cells that are set aside for sexual reproduction. The immature germ cells of male and female animals are examples. Meiosis sorts out a germ cell's chromosomes into four new nuclei. After meiosis, gametes form by way of cytoplasmic division and other events.

3. Cells with a diploid chromosome number contain two of each type of chromosome characteristic of the species. The two function as a pair during meiosis. Commonly, one chromosome of the pair is maternal, with hereditary instructions from a female parent. The other is paternal, with the same categories of hereditary instructions from a male parent.

4. Meiosis divides the chromosome number by half for each forthcoming gamete. Thus, if both parents have a diploid chromosome number ($2n$), the gametes that form will be haploid (n). Later, union of two gametes at fertilization restores the diploid number in the new individual ($n + n = 2n$).

5. During meiosis, each pair of chromosomes may swap segments. Each time they do, they exchange hereditary information about certain traits. Also, meiosis assigns one of each pair of chromosomes to a forthcoming gamete—but *which* gamete is its destination is a matter of chance. Hereditary instructions are further shuffled at fertilization. All three reproductive events lead to variations in traits among offspring.

6. In most plants, spore formation and other events intervene between meiosis and gamete formation.

10.1 COMPARING SEXUAL WITH ASEXUAL REPRODUCTION

When an orchid, flatworm, or aphid reproduces all by itself, what sort of offspring does it get? By the process of **asexual reproduction**, one parent alone produces offspring, and each offspring inherits the same number and kinds of genes as its parent. **Genes** are particular stretches of chromosomes—that is, of DNA molecules. Taken together, the genes for every species contain all of the heritable bits of information that are necessary to produce new individuals. Rare mutations aside, this means asexually produced offspring can only be clones: genetically identical copies of the parent.

Inheritance gets much more interesting with **sexual reproduction**. This process involves meiosis, gamete formation, and fertilization (union of the nuclei of two gametes). In humans and other sexually reproducing species, the first cell of a new individual has *pairs of genes* on pairs of homologous chromosomes. Typically, one of each pair is maternal and the other paternal in origin.

If instructions encoded in every pair of genes were identical down to the last detail, sexual reproduction would produce clones, also. Just imagine—you, every single person you know, the entire human population might be a clone, with everybody looking alike. But the two genes of a pair may *not* be identical. Why not? The molecular structure of genes can change; this is what we mean by mutation. Depending on their structure, two genes that happen to be paired in a person's cells may "say" slightly different things about a trait. Each unique molecular form of the same gene is called an **allele**.

Such tiny differences affect thousands of traits. For example, whether your chin has a dimple depends on which pair of alleles you inherited at one chromosome location. One kind of allele at that location says "put a dimple in the chin," another kind says "no dimple." This leads us to a key reason why members of sexually reproducing species don't all look alike. *Through sexual reproduction, offspring inherit new combinations of alleles, which lead to variations in the details of their traits.*

This chapter describes the cellular basis of sexual reproduction. More importantly, it starts you thinking about far-reaching effects of gene shufflings at different stages of the process. The process introduces variations in traits among offspring that are typically acted upon by agents of natural selection. Thus, *variation in traits is a foundation for evolutionary change.*

Asexual reproduction produces genetically identical copies of the parent. Sexual reproduction introduces variations in the details of traits among offspring.

Sexual reproduction dominates the life cycles of eukaryotic species. Meiosis, formation of gametes, and fertilization are the basic events of this process.

10.2 HOW MEIOSIS HALVES THE CHROMOSOME NUMBER

Think "Homologues"

Think back on the preceding chapter and its focus on mitotic cell division. Unlike mitosis, **meiosis** divides chromosomes into separate parcels not once *but twice* prior to cell division. Unlike mitosis, it is the first step leading to the formation of gametes.

Gametes, recall, are sex cells such as sperm or eggs. In nearly all multicelled eukaryotic organisms, gametes develop from cells that arise in specialized reproductive structures or organs. Figure 10.2 shows a few examples of where gametes form.

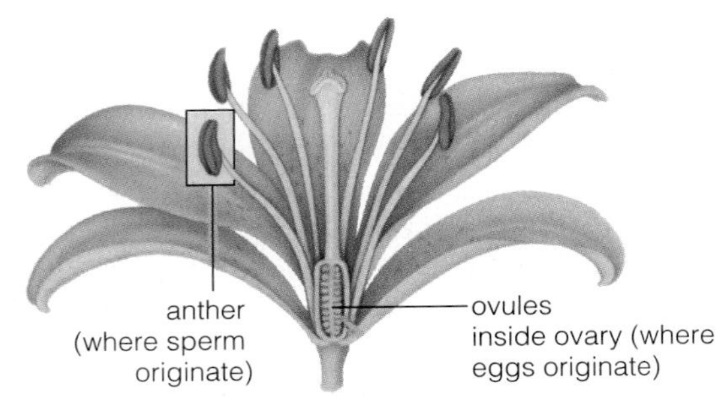

anther
(where sperm
originate)
ovules
inside ovary (where
eggs originate)

a FLOWERING PLANT

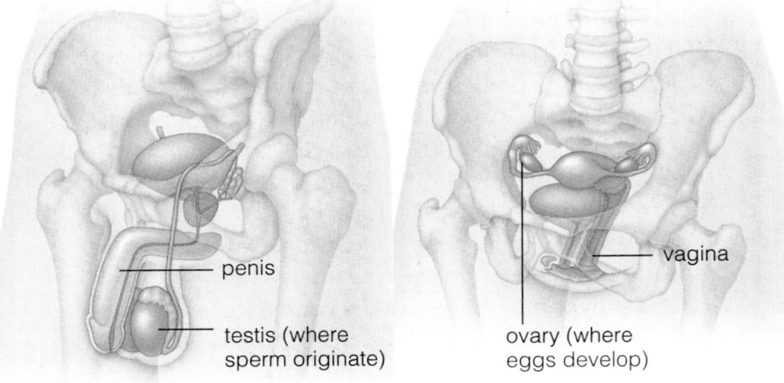

penis
testis (where
sperm originate)
vagina
ovary (where
eggs develop)

b HUMAN MALE

c HUMAN FEMALE

Figure 10.2 Examples of gamete-producing structures.

As you know, the **chromosome number** is the sum total of chromosomes in cells of a given type (Section 9.1). Germ cells start out with the same chromosome number as somatic cells (the rest of the body's cells). If a cell has a **diploid number** ($2n$), it has a *pair* of each type of chromosome, often from two parents. In general, the chromosomes of each pair have the same length and shape. Their genes deal with the same traits. And they line up with each other during meiosis. We call them **homologous chromosomes** (*hom-* means alike).

Further reading: Student Guide to InfoTrac on web site —

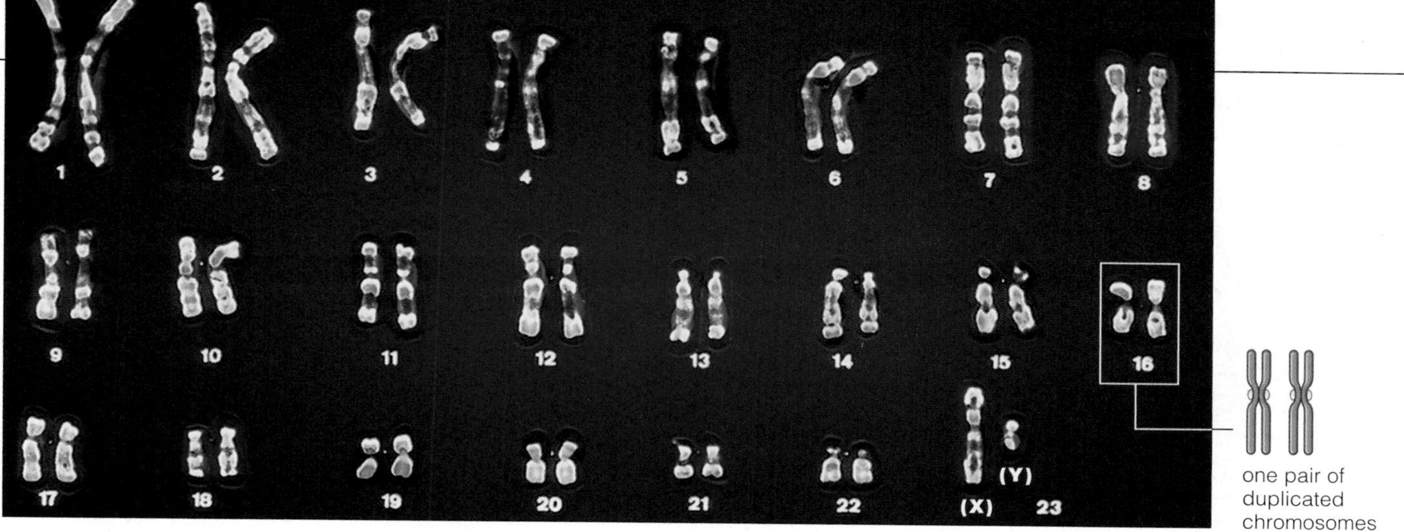

1
2
3
4
5
6
7
8
9
10
11
12
13
14
15
16
17
18
19
20
21
22
(X)
(Y)
23
one pair of duplicated chromosomes

Figure 10.3 From a diploid cell of a human male, twenty-three pairs of homologous chromosomes. The sketch to the right of the photograph corresponds to two chromosomes (in the white box).

As you can probably deduce from Figure 10.3, your own germ cells have 23 + 23 homologous chromosomes. After meiosis, 23 chromosomes—one of each type—end up in gametes. Thus meiosis halves the chromosome number, so that gametes have a **haploid number** (n).

Two Divisions, Not One

Meiosis resembles mitosis in some respects, even though the outcome is different. Before interphase gives way to meiosis, a germ cell duplicates its DNA. Each duplicated chromosome now consists of two DNA molecules. These remain attached at a narrowed-down region called the centromere. For as long as the two stay attached, they are called **sister chromatids** of the chromosome:

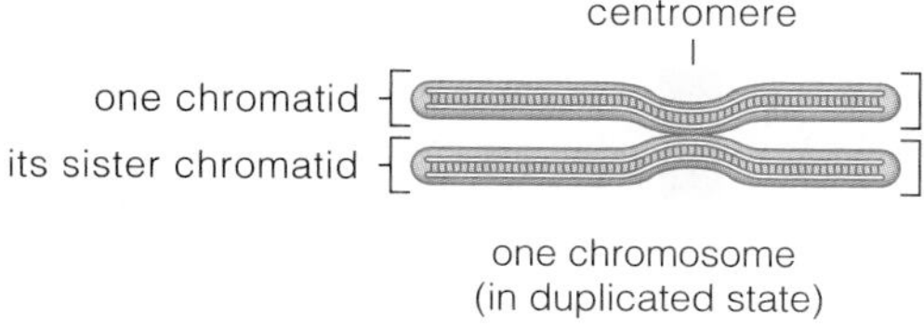

centromere
one chromatid
its sister chromatid
one chromosome
(in duplicated state)

As in mitosis, the microtubules of a spindle apparatus move the chromosomes in prescribed directions.

With meiosis alone, *chromosomes proceed through two consecutive divisions, which conclude with the formation of four haploid nuclei.* We call the two nuclear divisions meiosis I and meiosis II:

DNA is replicated during interphase	MEIOSIS I	*DNA is not replicated between divisions*	MEIOSIS II
	PROPHASE I		PROPHASE II
	METAPHASE I		METAPHASE II
	ANAPHASE I		ANAPHASE II
	TELOPHASE I		TELOPHASE II

During meiosis I, each duplicated chromosome lines up with its partner, *homologue to homologue;* then the two partners are moved apart from each other:

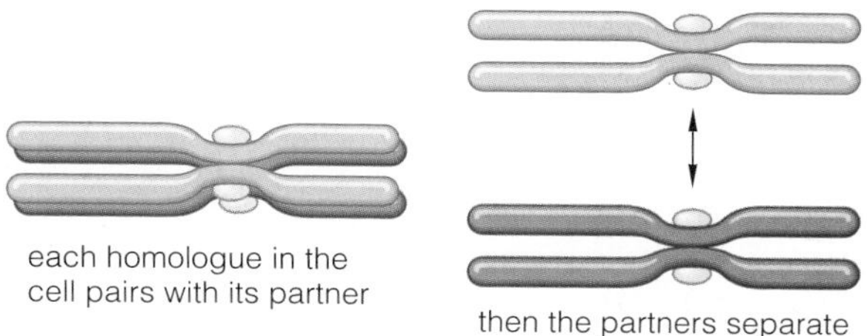

each homologue in the cell pairs with its partner
then the partners separate

The cytoplasm typically starts to divide at some point after homologues have been separated from each other. The cytoplasmic division results in two daughter cells. Each daughter cell is haploid; it has only one of each type of chromosome. But remember, the chromosomes are still in the duplicated state.

Next, during meiosis II, *the two sister chromatids of each chromosome are separated from each other:*

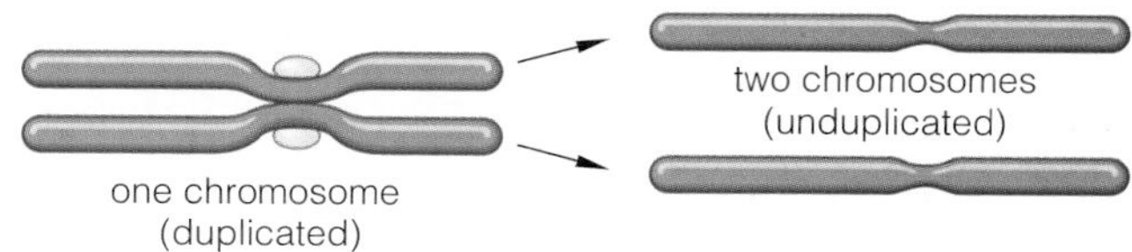

one chromosome
(duplicated)
two chromosomes
(unduplicated)

Each sister chromatid is now a chromosome in its own right. Four nuclei now form. Often the cytoplasm divides once more. The final outcome is four haploid cells.

Figure 10.4 on the next two pages illustrates the key events of meiosis I and II.

Meiosis is a type of nuclear division mechanism that reduces the parental chromosome number by half, to the haploid number (n).

Meiosis proceeds only in immature cells, such as germ cells, that are set aside for sexual reproduction. It is the first step leading to the formation of gametes.

10.3 VISUAL TOUR OF THE STAGES OF MEIOSIS

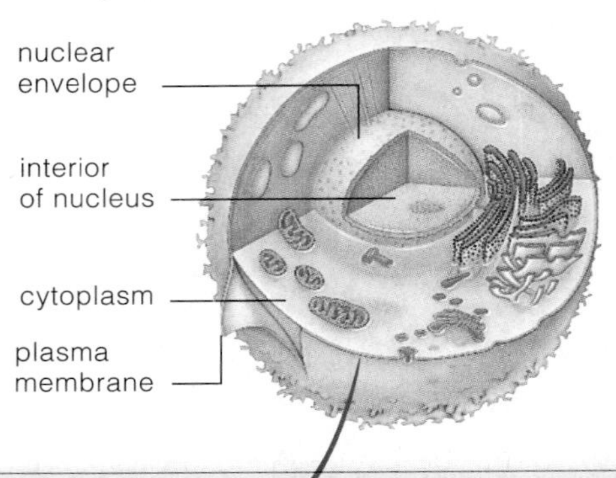
nuclear envelope
interior of nucleus
cytoplasm
plasma membrane

Figure 10.4 Meiosis in a generalized animal cell. This nuclear division mechanism reduces the chromosome number in an immature reproductive cell by half (to the haploid number) for forthcoming gametes. To keep things simple, only two pairs of homologous chromosomes are shown. Maternal chromosomes are shaded *purple* and paternal chromosomes, *blue*.

A GERM CELL AT INTERPHASE

A germ cell with a diploid chromosome number (2*n*) is about to leave interphase and undergo the first division of meiosis. The cell's DNA is already duplicated, so each chromosome is in the duplicated state; it consists of two sister chromatids.

MEIOSIS I

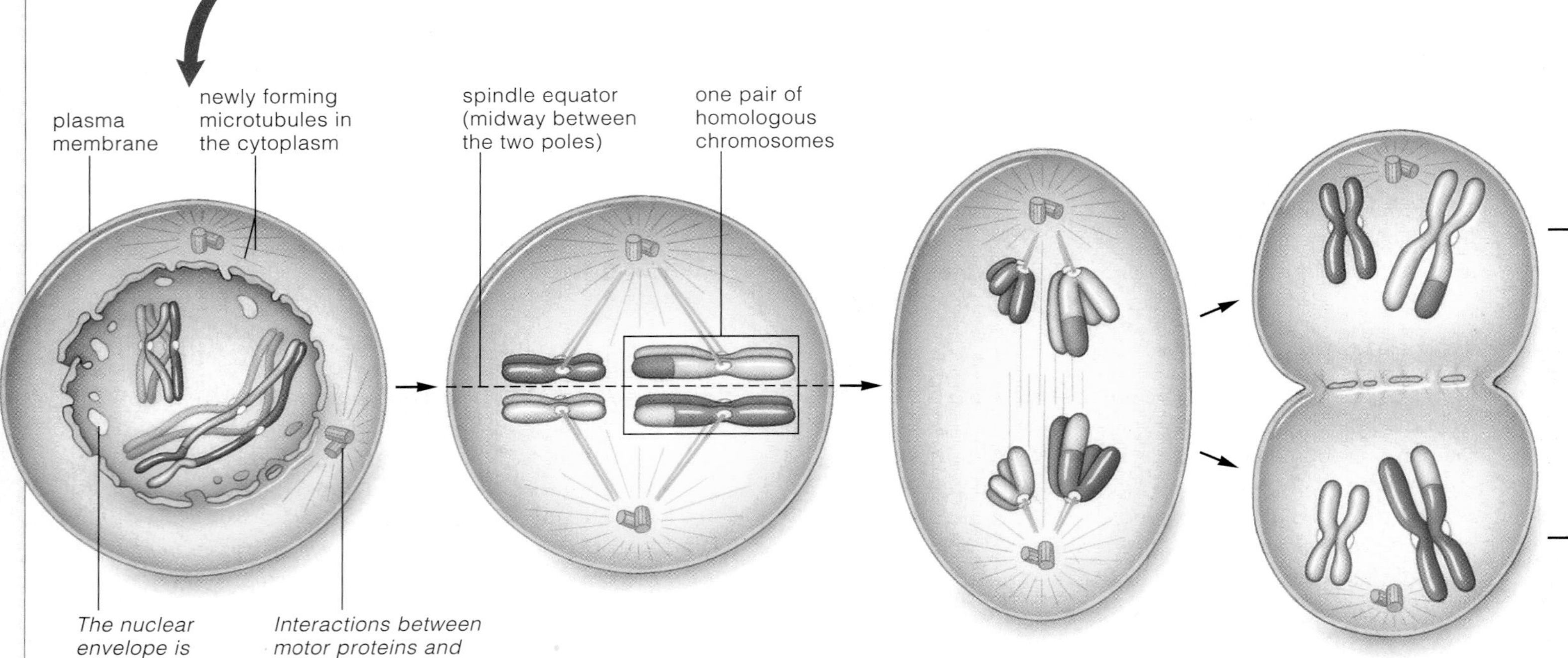
plasma membrane
newly forming microtubules in the cytoplasm
spindle equator (midway between the two poles)
one pair of homologous chromosomes
The nuclear envelope is breaking apart; microtubules will be able to penetrate the nuclear region.
Interactions between motor proteins and microtubules are moving one of two pairs of centrioles toward the opposite spindle pole.

PROPHASE I

Each duplicated chromosome is in threadlike form, but now it starts to twist and fold into more condensed form. It pairs with its homologue, and the two typically swap segments. The swapping, called crossing over, is indicated by the break in color on the pair of larger chromosomes. Each chromosome becomes attached to some microtubules of a newly forming spindle.

METAPHASE I

Motor proteins have been driving the movement of microtubules that became attached to the kinetochores of chromosomes. As a result, the chromosomes have been pushed and pulled into position midway between the spindle poles. Now the spindle is fully formed, owing to dynamic interactions of motor proteins, microtubules, and the chromosomes themselves.

ANAPHASE I

Microtubules extending from the poles and overlapping at the spindle equator *lengthen* and push the poles apart. At the same time, other microtubules extending from the poles to the kinetochores of chromosomes *shorten*, and each chromosome is thereby pulled away from its homologous partner. These motions move the homologous partners to opposite poles.

TELOPHASE I

At some point, the cytoplasm of the germ cell divides. Two cells, each with a haploid chromosome number (*n*), result. That is, the cells have one of each type of chromosome that was present in the parent (2*n*) cell. All chromosomes are still in the duplicated state.

Further reading: Student Guide to InfoTrac on web site →

Of the four haploid cells that form by way of meiosis and cytoplasmic divisions, one or all may proceed to develop into gametes and function in sexual reproduction.

(In plants, the cells that form after meiosis is completed may develop into spores, which take part in a stage of the life cycle that precedes gamete formation.)

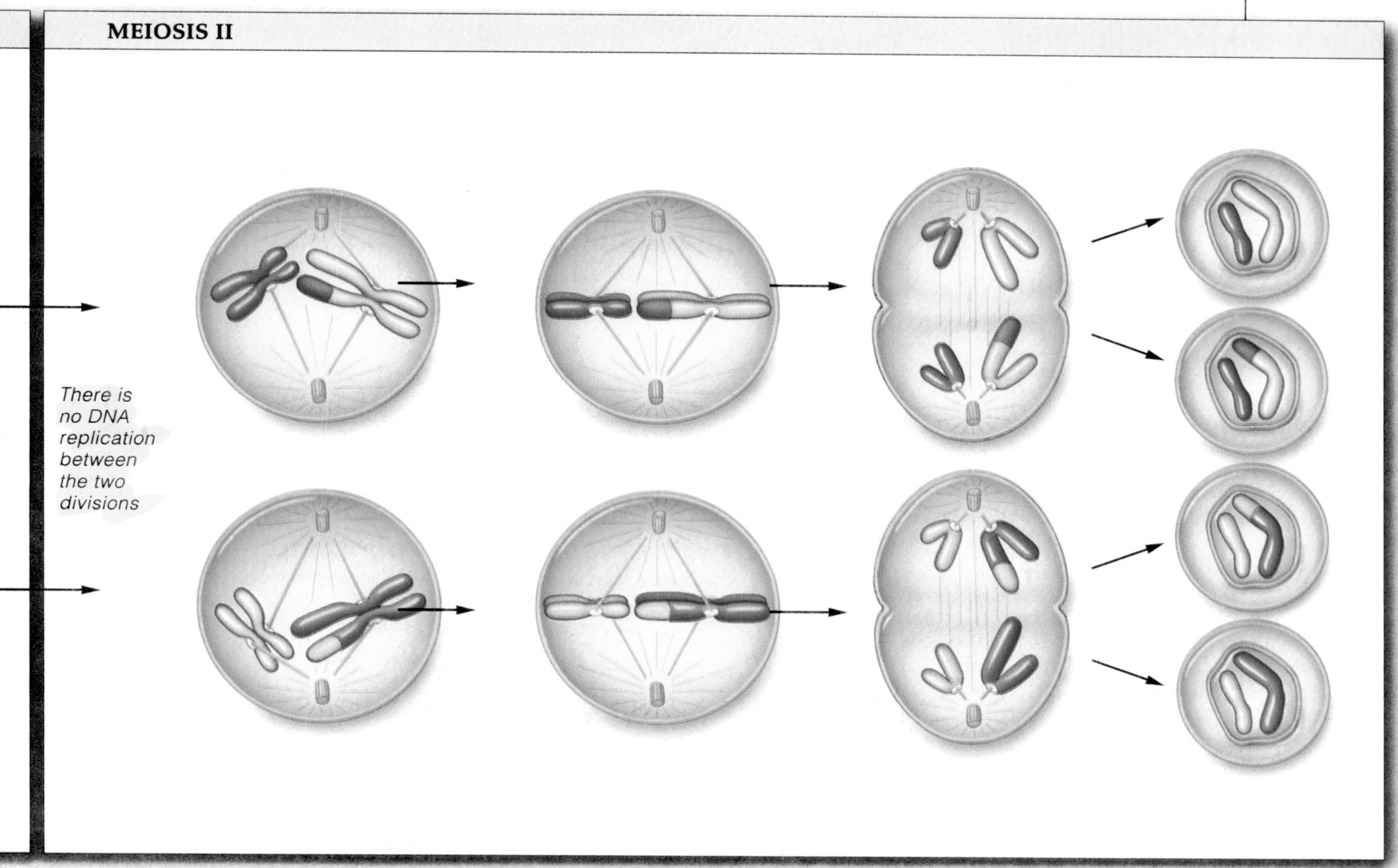
MEIOSIS II
There is no DNA replication between the two divisions

PROPHASE II

In each of the two daughter cells, microtubules have already moved one member of the centriole pair to the opposite pole of the spindle during the transition to prophase II. Now, at prophase II, microtubules attach to the kinetochores of chromosomes, and motor proteins drive the movement of chromosomes toward the spindle's equator.

METAPHASE II

Now, in each daughter cell, interactions among motor proteins, spindle microtubules, and each duplicated chromosome have moved all of the chromosomes so that they are positioned at the spindle equator, midway between the two poles.

ANAPHASE II

The attachment between the two chromatids of each chromosome breaks. Once that happens, each of the former "sister chromatids" is a chromosome in its own right. Motor proteins interact with kinetochore microtubules to move the separated chromosomes to opposite poles of the spindle.

TELOPHASE II

By the time telophase II is completed, there will be four daughter nuclei. Also, at the time when division of the cytoplasm is completed, each new, daughter cell will have a haploid chromosome number (n). All of those chromosomes will now be in the unduplicated state.

10.4 A CLOSER LOOK AT KEY EVENTS OF MEIOSIS I

The preceding overview, in Sections 10.2 and 10.3, is enough to convey the overriding function of meiosis—that is, *the reduction of the chromosome number by half for forthcoming gametes*. However, as you will now read, two other events that take place during prophase and metaphase of meiosis I contribute greatly to the adaptive advantage of sexual reproduction.

That advantage, recall, is production of offspring with new combinations of alleles. During growth and development, those combinations are translated into a new generation of individuals that differ in the details of some number of traits.

What Goes On in Prophase I?

Prophase I of meiosis is a time of major gene shufflings. Reflect on Figure 10.5*a*, which shows two chromosomes condensed to threadlike form. All chromosomes in a germ cell condense this way. As they do, each is drawn close to its homologue. Molecular interactions stitch homologues together point by point along their length, with little space between. The intimate, parallel array favors **crossing over**, a molecular interaction between two of the *non*sister chromatids of a pair of homologous chromosomes. Nonsister chromatids break at the same

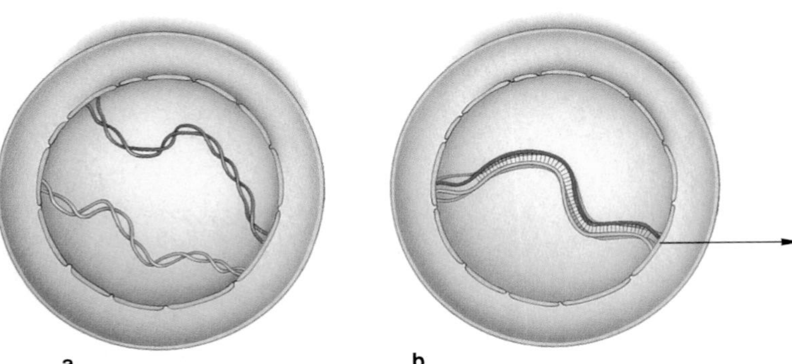

a
b

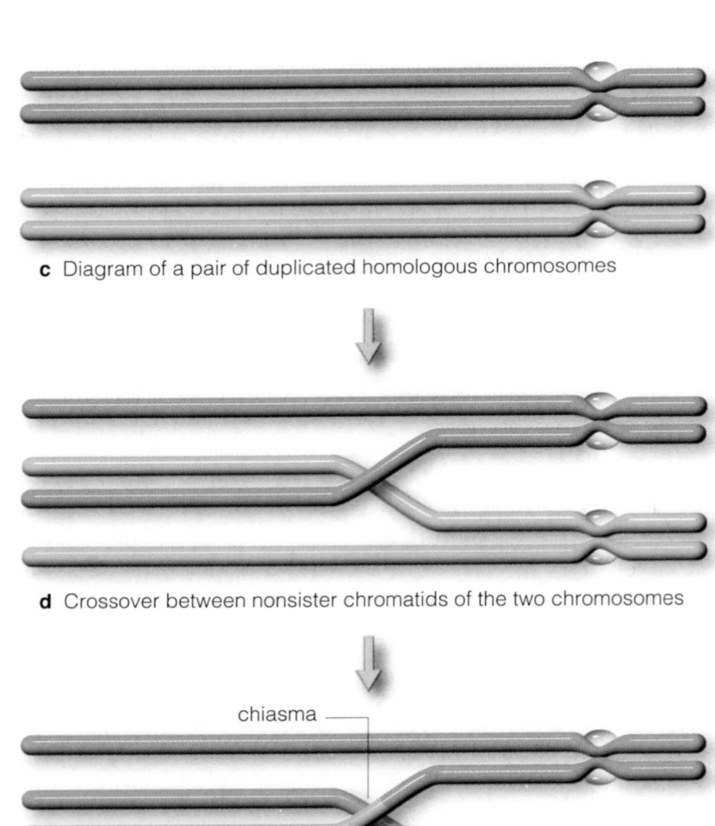

c Diagram of a pair of duplicated homologous chromosomes
d Crossover between nonsister chromatids of the two chromosomes
chiasma
e Nonsister chromatids exchange segments

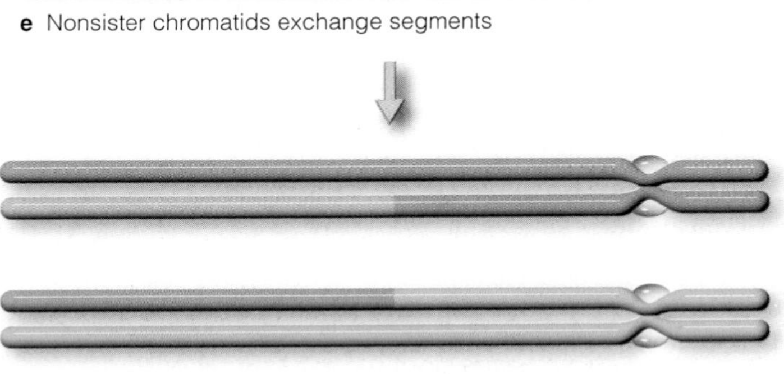

f Homologues have new combinations of alleles

Figure 10.5 Key events during prophase I, the first stage of meiosis. For clarity, this diagram of a cell shows only one pair of homologous chromosomes and one crossover event, although more than one typically occur. *Blue* signifies the paternal chromosome, and *purple* signifies its maternal homologue.

(**a**) Both of the chromosomes were duplicated earlier, during interphase. Early in prophase I, the two sister chromatids of each duplicated chromosome are in thin, threadlike form. They are all positioned so closely together, they look like a single thread.

(**b**) Each chromosome becomes zippered to its homologue, so all four chromatids are intimately aligned. When the two sex chromosomes have different forms (such as X paired with Y) they still get zippered together, although only in a small region along their length.

(**c**,**d**) We show the pair of chromosomes as if they were already condensed, then teased apart so that you can visualize what goes on. Bear in mind, their double-stranded DNA molecules are still tightly aligned at this stage. The intimate contact allows one (and usually more) crossover to occur at intervals along the length of nonsister chromatids.

(**e**) Nonsister chromatids exchange segments. As prophase I ends, the chromosomes continue to condense into thicker, rodlike forms. Then they unzipper from each other except at places where they physically cross each other. Those places are called chiasmata (singular, chiasma, meaning "cross"). Each chiasma does not last long, but it is indirect evidence that a crossover occurred at some place in the chromosomes.

(**f**) What is the function of crossing over? It breaks up old combinations of alleles and puts new ones together in pairs of homologous chromosomes.

Further reading: Student Guide to InfoTrac on web site

places along their length. At these break points, they exchange corresponding segments—that is, genes.

Gene swapping would be rather pointless if each type of gene never varied from one chromosome to the next. But remember, a gene can have slightly different forms: alleles. You can safely bet that some number of alleles on one chromosome will *not* be identical to their allelic partners on the homologue. Thus each crossover represents a chance to swap slightly different versions of hereditary instructions for particular traits.

We will look at the mechanism of crossing over in later chapters. For now, it is enough to remember this: *Crossing over leads to genetic recombination, which in turn leads to variation in the traits of offspring.*

Metaphase I Alignments

Major shufflings of whole chromosomes begin during the transition from prophase I to metaphase I, which is the second stage of meiosis. Suppose the shufflings are proceeding at this moment in one of your germ cells. By now, crossovers have made genetic mosaics of the chromosomes, but put this aside in order to simplify tracking. Just call the twenty-three chromosomes you inherited from your mother the *maternal* chromosomes and their twenty-three homologues from your father the *paternal* chromosomes.

Kinetochore microtubules have already oriented one chromosome of each pair toward one spindle pole and its homologue toward the other pole (refer to Section 10.3). Now they are moving all the chromosomes, which soon will become positioned at the spindle's equator.

Have all maternal chromosomes become attached to one spindle pole and all paternal chromosomes to the other pole? Maybe, but probably not. Remember, the initial contacts between kinetochore microtubules and the chromosomes are random. Because of the random grabs, the eventual positioning of maternal or paternal chromosomes at the spindle's equator at metaphase I follows no particular pattern. Now carry this thought one step further. *Either one* of each pair of homologous chromosomes can end up at either pole of the spindle after they move apart at anaphase I.

Think about the possibilities when you are tracking merely three pairs of homologues. As you can see from Figure 10.6, by metaphase I, three pairs of homologues may be arranged in any one of four possible positions. Here, eight combinations (2^3) of maternal and paternal chromosomes are possible for forthcoming gametes.

Of course, a human germ cell has twenty-three pairs of homologous chromosomes, not just three. So a grand total of 2^{23}—or *8,388,608*—combinations of maternal and paternal chromosomes is possible every time a human germ cell gives rise to sperm or eggs!

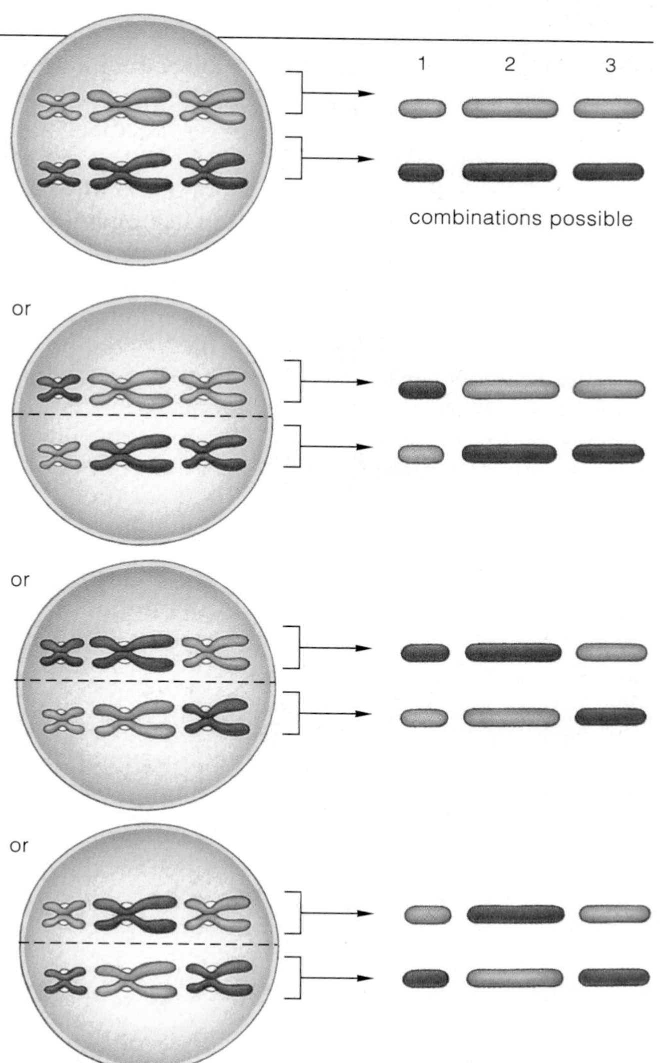
1
2
3
combinations possible
or
or
or

Figure 10.6 Possible outcomes of the random alignment of only three pairs of homologous chromosomes at metaphase I. Three types of chromosomes are labeled 1, 2, and 3. Maternal chromosomes are *purple;* paternal ones are *blue.* With merely four possible alignments, eight combinations of maternal and paternal chromosomes are possible in forthcoming gametes.

Moreover, in each sperm or egg, many hundreds of alleles inherited from the mother might not "say" the exact same thing about hundreds of different traits as the alleles inherited from the father. Are you beginning to get an idea of why such fascinating combinations of traits show up even in the same family?

Crossing over is an interaction between a pair of homologous chromosomes. It breaks up old combinations of alleles and puts new ones together during prophase I of meiosis.

The random attachment and subsequent positioning of each pair of maternal and paternal chromosomes at metaphase I lead to different combinations of maternal and paternal traits in each generation of offspring.

10.5 FROM GAMETES TO OFFSPRING

The gametes that form following meiosis are not all the same in their details. For example, human sperm have one tail, opossum sperm have two, and roundworm sperm have none. Crayfish sperm look like pinwheels. Most eggs are microscopic in size, yet an ostrich egg tucked inside its shell is as large as a baseball. From its appearance alone, you might not believe that a plant gamete is even remotely like an animal's.

Later chapters contain details of how gametes form in the life cycles of representative organisms, including humans. Figure 10.7 and the following points may help you keep the details in perspective.

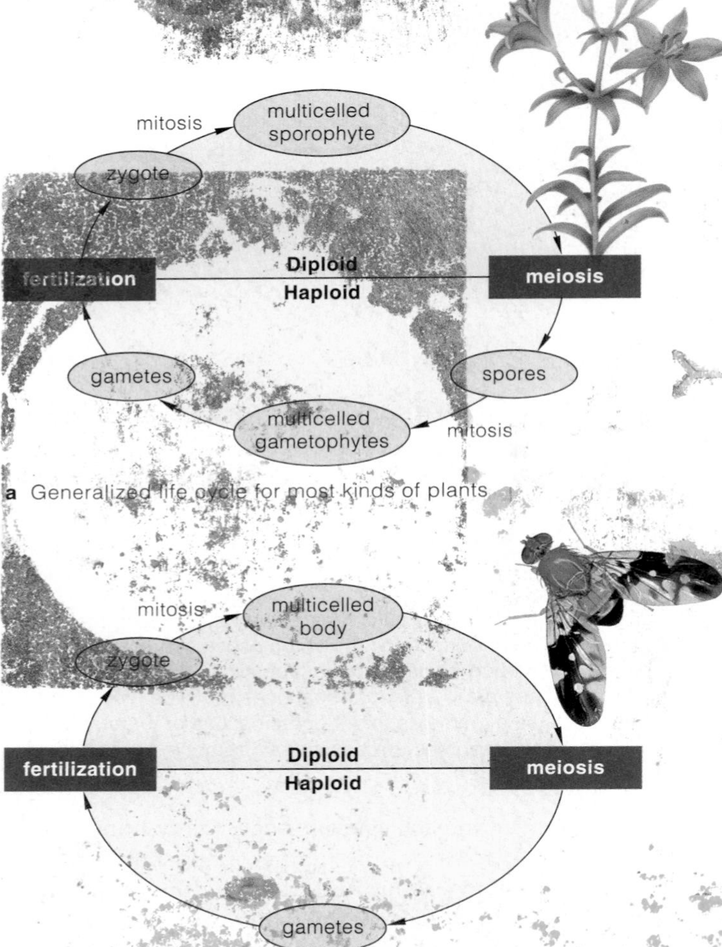

mitosis
multicelled sporophyte
zygote
fertilization
Diploid
Haploid
meiosis
gametes
spores
multicelled gametophytes
mitosis
a Generalized life cycle for most kinds of plants
mitosis
multicelled body
zygote
fertilization
Diploid
Haploid
meiosis
gametes
b Generalized life cycle for animals

Figure 10.7 Generalized life cycles for (**a**) most plants and (**b**) animals. The zygote is the first cell that forms when the nuclei of two gametes fuse together at fertilization.

For plants, a sporophyte (spore-producing body) develops by way of mitotic cell divisions, from the zygote. After meiosis, gametophytes (gamete-producing bodies) form. A lily plant is a sporophyte. Gametophytes form in parts of its flowers.

Chapters 22 through 26, 31, and 44 include specific examples of life cycles of representative organisms.

Gamete Formation in Plants

For pine trees, apple trees, roses, dandelions, corn, and nearly all other familiar plants, certain events intervene between the times of meiosis and gamete formation. Among other things, spores form.

Spores are haploid resting cells, often walled, that are good at resisting drought, cold, and other adverse environmental conditions. When favorable conditions return, spores germinate (resume growth) and develop into a haploid body or structure that produces gametes. So *gamete*-producing bodies and *spore*-producing bodies develop during the life cycle of most kinds of plants. Figure 10.7*a* is a generalized diagram of these events.

Gamete Formation in Animals

In male animals, gametes form by a process known as spermatogenesis. Inside the male reproductive system, a diploid germ cell grows in size. It becomes a primary spermatocyte, a large immature cell that enters meiosis and cytoplasmic divisions. Four haploid cells result and develop into spermatids (Figure 10.8). These immature cells change in form and develop a tail, thus becoming a **sperm**, a common type of mature male gamete.

In female animals, gametes form by a process that is called oogenesis. In human females, for example, a diploid germ cell develops into an **oocyte**, or immature egg. Unlike a sperm cell, an oocyte accumulates many cytoplasmic components. Also, as Figure 10.9 indicates, its daughter cells differ in size and function. When the oocyte undergoes cytoplasmic division after meiosis I, one cell—the secondary oocyte—gets nearly all of the cytoplasm. The other cell, called the first polar body, is quite small. Some time afterward, both cells undergo meiosis II and then cytoplasmic division. One daughter cell of the secondary oocyte develops into a second polar body. The other gets most of the cytoplasm and develops into the gamete. A mature female gamete is called an ovum (plural, ova) or, more commonly, an **egg**.

Thus, one egg and three polar bodies have formed. Polar bodies don't have much in the way of nutrients or metabolic machinery and don't function as gametes. In time, they degenerate. But polar body formation allows the egg to end up with a suitable (haploid) number of chromosomes. Also, by getting most of the cytoplasm, the egg receives enough start-up machinery to support the new individual right after fertilization.

More Shufflings at Fertilization

The chromosome number characteristic of the parents is restored at **fertilization**, the time when a male gamete unites with a female gamete and their haploid nuclei

Further reading: Student Guide to InfoTrac on web site →

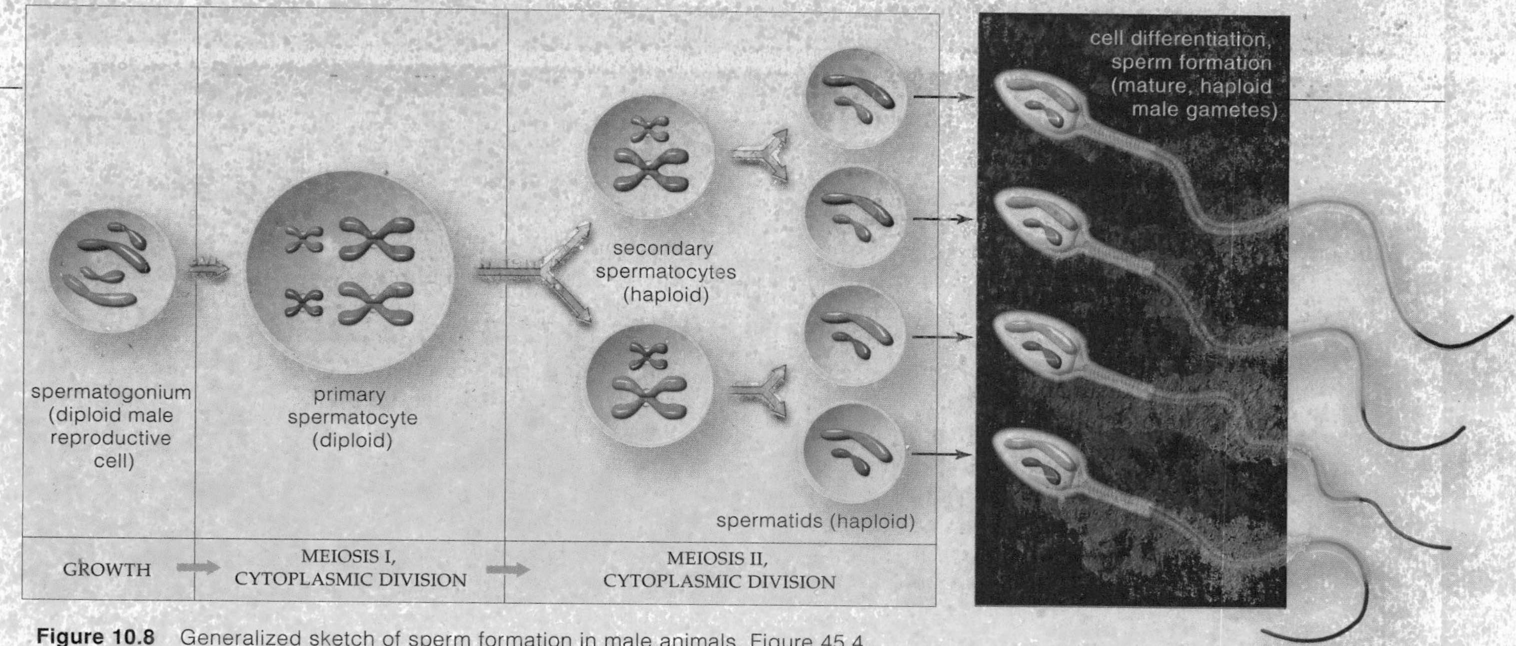
cell differentiation, sperm formation (mature, haploid male gametes)
secondary spermatocytes (haploid)
spermatogonium (diploid male reproductive cell)
primary spermatocyte (diploid)
spermatids (haploid)
GROWTH
MEIOSIS I, CYTOPLASMIC DIVISION
MEIOSIS II, CYTOPLASMIC DIVISION

Figure 10.8 Generalized sketch of sperm formation in male animals. Figure 45.4 shows a specific example (how sperm form in human males).

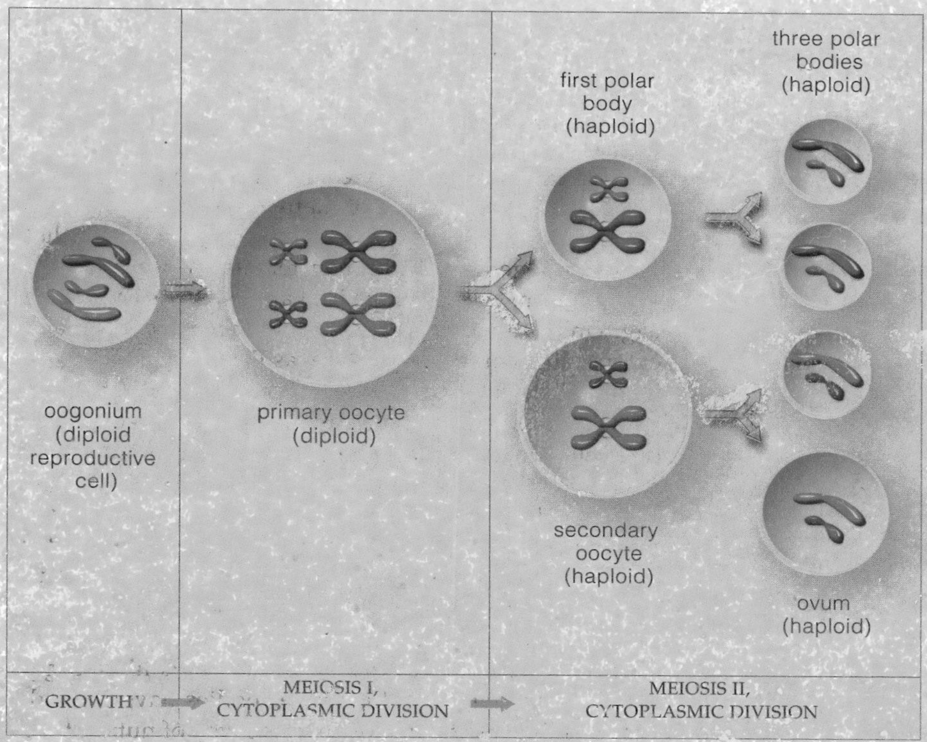
three polar bodies (haploid)
first polar body (haploid)
oogonium (diploid reproductive cell)
primary oocyte (diploid)
secondary oocyte (haploid)
ovum (haploid)
GROWTH
MEIOSIS I, CYTOPLASMIC DIVISION
MEIOSIS II, CYTOPLASMIC DIVISION

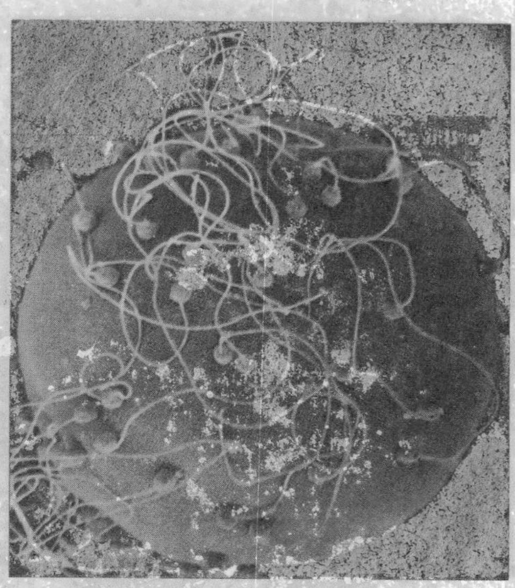

Figure 10.9 Egg formation in female animals. In animals, eggs are far larger than sperm, as the micrograph of sea urchin gametes (*above*) indicates. In addition, the three polar bodies are much smaller than an egg. Figure 45.7 shows a specific example (how eggs form in human females).

fuse. If meiosis did not precede it, fertilization would double the chromosome number each new generation. Such changes in chromosome number would disrupt the hereditary instructions encoded in chromosomes, usually for the worse. Why? Those instructions operate as a complex, fine-tuned package in each individual.

Fertilization contributes to variation in offspring. Reflect on the possibilities for humans alone. During prophase I, an average of two or three crossovers takes place in each human chromosome. Even without the crossovers, the random positioning of pairs of paternal and maternal chromosomes at metaphase I results in one of millions of possible chromosome combinations in each gamete. And of all the male and female gametes that are produced, which two actually get together is a matter of chance. The sheer number of combinations that can exist at fertilization is staggering!

Cumulatively, crossing over, the distribution of random mixes of homologous chromosomes into gametes, and fertilization contribute to variation in the traits of offspring.

10.6

MEIOSIS AND MITOSIS COMPARED

In this unit our focus has been on two nuclear division mechanisms. Single-celled eukaryotic species reproduce asexually by way of mitosis, followed by cytoplasmic division. Many multicelled eukaryotic species depend on mitosis and cytoplasmic division during episodes of asexual reproduction in their life cycle. All depend on it for growth and tissue repair. By contrast, meiosis occurs only in reproductive cells, such as germ cells that give rise to the gametes used in sexual reproduction. Figure 10.10 summarizes the basic similarities and differences between the two nuclear division mechanisms.

The end results of the two mechanisms differ in a crucial way. *Mitotic cell division only produces clones*—genetically identical copies of a parent cell. *But meiotic cell division, in conjunction with fertilization, promotes variation in traits among offspring*. First, crossing over at prophase I of meiosis puts new combinations of alleles in chromosomes. Second, the random assignment of either member of a pair of homologous chromosomes to either pole of the spindle at metaphase I affects gametes, which end up with mixes of maternal and paternal alleles. And third, different combinations of alleles are brought together simply by chance during fertilization. In later chapters, you will be reading about the ways in which both meiosis and fertilization contributed to the truly stunning diversity and evolution of sexually reproducing organisms.

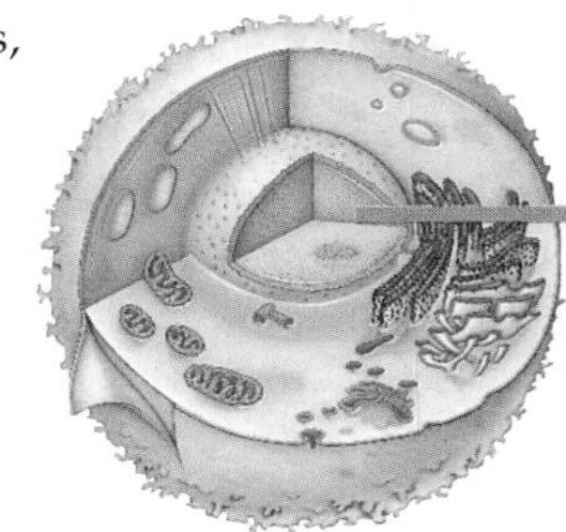

A *somatic cell* with a diploid chromosome number (2*n*) is at interphase. Before mitotic division begins, its DNA is replicated (all chromosomes are duplicated).

Figure 10.10 Summary of mitosis and meiosis. Both diagrams use a diploid (2*n*) animal cell as the example. They are arranged to help you compare similarities and differences between the division mechanisms. Maternal chromosomes are coded *purple*, and paternal chromosomes are coded *blue*.

MEIOSIS I

A *germ cell* with a diploid chromosome number (2*n*) is at interphase. Before mitotic division begins, its DNA is replicated (all chromosomes are duplicated).

PROPHASE I

Each duplicated chromosome (consisting of two sister chromatids) condenses to threadlike form, then rodlike form. *Crossing over* occurs. Each chromosome unzips from its homologue. Each gets attached to the spindle in transition to metaphase.

METAPHASE I

All chromosomes are now positioned at the spindle's equator.

ANAPHASE I

Each chromosome is separated from its homologue. They are moved to opposite poles of the spindle.

TELOPHASE I

When the cytoplasm divides, there are two cells. Each has a haploid (*n*) number of chromosomes, but these are still in the duplicated state.

Further reading: Student Guide to InfoTrac on web site

MITOSIS

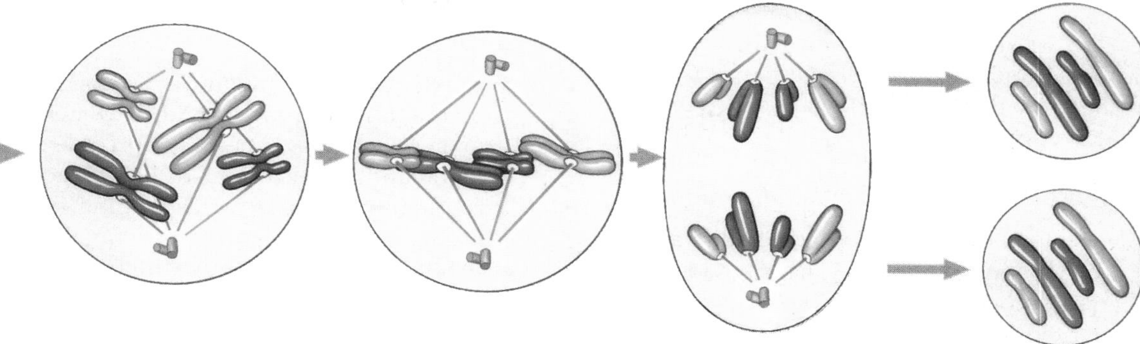

PROPHASE

Each duplicated chromosome (consisting of two sister chromatids) condenses from threadlike form to rodlike form. Each gets attached to the spindle during the transition to metaphase.

METAPHASE

All chromosomes are now positioned at the spindle's equator.

ANAPHASE

Sister chromatids of each chromosome are separated from each other. These new, daughter chromosomes are moved to opposite poles of the spindle.

TELOPHASE

When the cytoplasm divides, there are two cells. Each is diploid (2*n*)—*it has the same chromosome number as the parent cell.*

MEIOSIS II

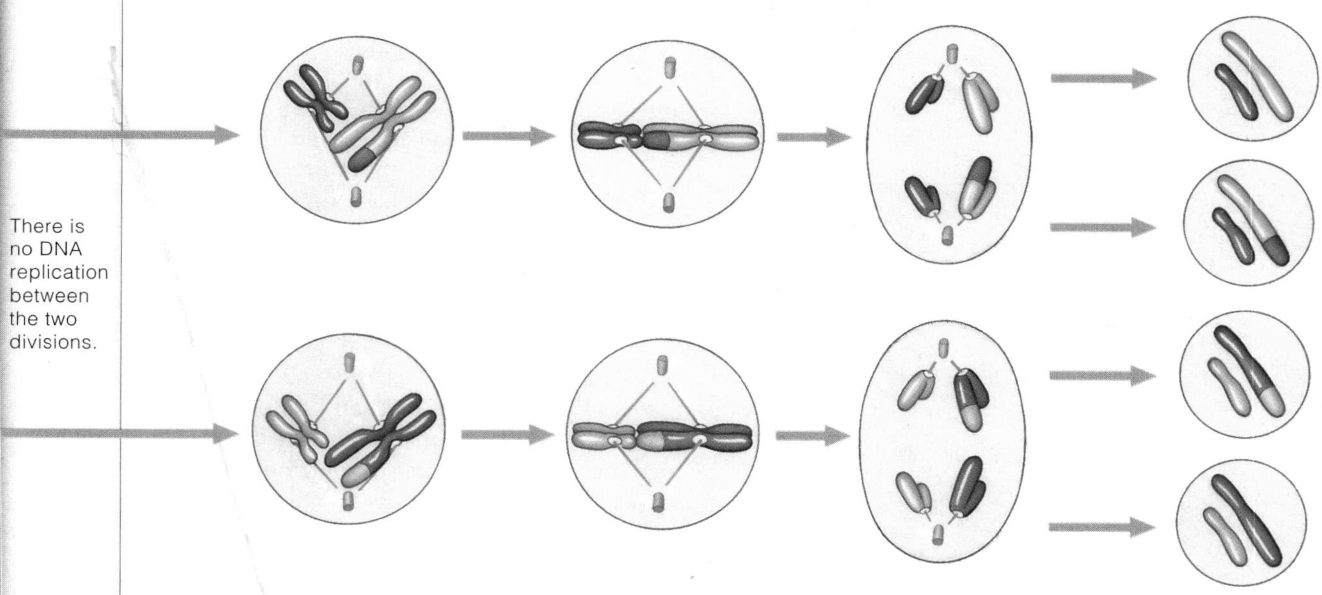

There is no DNA replication between the two divisions.

PROPHASE II

Before prophase II, the two centrioles in each new cell were moved apart and a new spindle formed. Now, each chromosome becomes attached to the spindle and starts moving toward its equator.

METAPHASE II

All chromosomes are now positioned at the spindle's equator.

ANAPHASE II

Sister chromatids of each chromosome are separated from each other. These new, daughter chromosomes are moved to opposite poles of the spindle.

TELOPHASE II

Four daughter nuclei form. When the cytoplasm divides, each new cell is haploid (*n*). *The original chromosome number has been reduced by half.* One or all of these cells may become gametes.

10.7

SUMMARY

1. The life cycle of each sexually reproducing species includes meiosis, gamete formation, and fertilization.

a. Meiosis, a nuclear division mechanism, reduces the chromosome number of a parent germ cell by half. It precedes the formation of haploid gametes (typically, sperm in males, eggs in females).

b. At fertilization, a sperm and egg nuclei fuse. This event restores the chromosome number (Figure 10.11).

2. A germ cell with a diploid chromosome number ($2n$) has *two* of each type of chromosome characteristic of its species. Commonly, one of each pair of chromosomes is maternal (inherited from a female parent) and the other is paternal (from a male parent).

3. Each pair of maternal and paternal chromosomes shows homology, meaning the two are alike. Generally the two have the same length, same shape, and same sequence of genes. And they interact during meiosis.

4. Chromosomes become duplicated during interphase. Each consists of two DNA molecules that will remain attached, as sister chromatids, during mitosis.

5. Meiosis consists of two consecutive divisions, which both require a microtubular spindle apparatus.

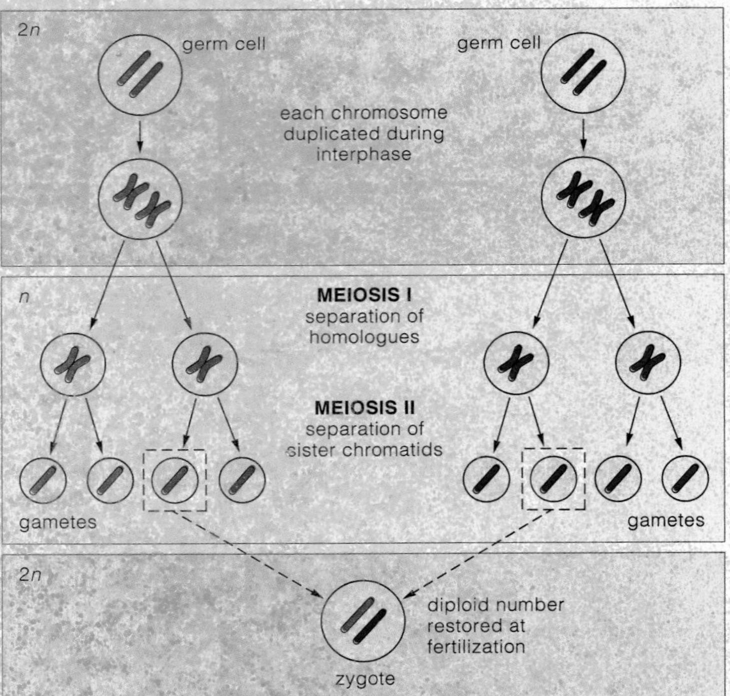
2n
germ cell
germ cell
each chromosome duplicated during interphase
n
MEIOSIS I
separation of homologues
MEIOSIS II
separation of sister chromatids
gametes
gametes
2n
diploid number restored at fertilization
zygote

Figure 10.11 Summary of changes in chromosome number at different stages of sexual reproduction, using diploid ($2n$) germ cells as the example. Meiosis reduces the chromosome number by half (n). Union of haploid nuclei of two gametes at fertilization restores the diploid number.

a. In meiosis I, kinetochores (of chromosomes) and motor proteins (projecting from microtubules) interact to move each duplicated chromosome away from its partner, the homologous chromosome.

b. In meiosis II, similar interactions move the sister chromatids of each chromosome away from each other.

6. Meiosis I, the first nuclear division, is characterized by the following events and outcomes:

a. Crossing over occurs in prophase I. Two nonsister chromatids of each pair of homologous chromosomes commonly break at corresponding sites and exchange segments. And this puts new combinations of alleles together. Alleles (slightly different molecular forms of the same gene) specify different forms of the same trait.

b. Different combinations of alleles lead to variation in the details of a given trait among offspring.

c. Also in prophase I, a microtubular spindle forms outside the nucleus, and the nuclear envelope starts to break up. In cells with duplicated pairs of centrioles, one pair starts moving to the opposite spindle pole.

d. All the pairs of homologous chromosomes have become positioned at the spindle equator at metaphase I. For each pair, either the maternal chromosome or its homologue can be oriented toward either pole.

e. In anaphase I, spindle microtubules interact with the kinetochores to move each duplicated chromosome away from its homologue, to opposite spindle poles.

7. Meiosis II (second nuclear division) is characterized by these events and outcomes:

a. At metaphase II, all the duplicated chromosomes are positioned at the spindle equator.

b. Sister chromatids are moved apart in anaphase II. Each is now a separate, unduplicated chromosome.

c. By the end of telophase II, four nuclei that have a haploid chromosome number (n) have been formed.

8. When the cytoplasm divides, there are four haploid cells. One or all of these may function as gametes (or as plant spores that give rise to gamete-producing bodies).

9. Crossing over, the chance allocation of different mixes of pairs of maternal and paternal chromosomes to different gametes, and the chance of any two gametes meeting at fertilization all contribute to the immense variation in details of traits among offspring.

Review Questions

1. The diploid chromosome numbers for the somatic cells of a few organisms are listed at right. How many chromosomes will end up in the gametes of each organism? *10.2*

Organism	Diploid number
Fruit fly, *Drosophila melanogaster*	8
Garden pea, *Pisum sativum*	14
Corn, *Zea mays*	20
Frog, *Rana pipiens*	26
Earthworm, *Lumbricus terrestris*	36
Human, *Homo sapiens*	46
Chimpanzee, *Pan troglodytes*	48
Amoeba, *Amoeba*	50
Horsetail, *Equisetum*	216

2. A diploid germ cell has four pairs of homologous chromosomes, designated AA, BB, CC, and DD. How would the chromosomes of the gametes be designated? *10.2*

3. Look at the chromosomes in the germ cell in the diagram at the right. Is this cell at anaphase I or anaphase II? *10.3, 10.6*

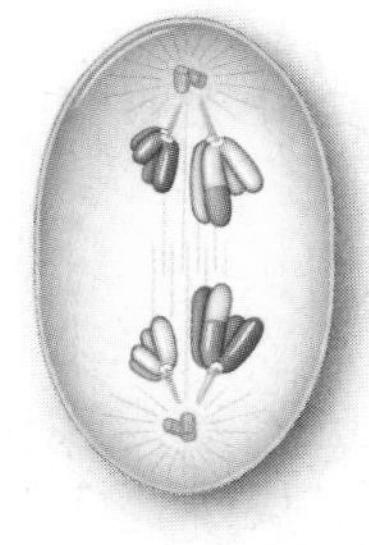

The cell is at anaphase ___ rather than anaphase ___. I know this because:

___

___

___

___

4. Define meiosis and describe its main stages. In what key respects is meiosis *not* like mitosis? *10.3, 10.6*

5. Actor Michael Douglas (Figure 10.12*a*) inherited a gene from each parent that influences the chin dimple trait. One form of the gene called for a dimple and the other didn't, but one is all it takes for this particular trait. Figure 10.12*b* shows what the chin of Mr. Douglas might have looked like if he had inherited two ordinary forms of the gene instead. What is the name for the alternative forms of the same gene? *10.1, 10.4*

6. Outline the main steps by which gametes form in plants. Do the same for gamete formation in animals. *10.5*

7. Genetically speaking, what is the key difference between the outcomes of sexual and asexual reproduction? *10.1, 10.6*

Self-Quiz (*Answers in Appendix III*)

1. Sexual reproduction requires ___.
   a. meiosis c. fertilization
   b. gamete formation d. all of the above

2. Meiosis is a division mechanism that produces ___.
   a. two cells c. eight cells
   b. two nuclei d. four nuclei

3. An animal cell having two rather than one of each type of chromosome has a ___ chromosome number.
   a. diploid c. normal gamete
   b. haploid d. both b and c

4. Meiosis ___ the parental chromosome number.
   a. doubles c. maintains
   b. reduces d. corrupts

5. Generally, a pair of homologous chromosomes ___.
   a. carry the same genes c. interact at meiosis
   b. are the same length, shape d. all of the above

6. Before the onset of meiosis, all chromosomes are ___.
   a. condensed c. duplicated
   b. released from protein d. both b and c

7. Each chromosome moves away from its homologue and ends up at the opposite spindle pole during ___.
   a. prophase I c. anaphase I
   b. prophase II d. anaphase II

8. Sister chromatids of each chromosome move apart and end up at opposite spindle poles during ___.
   a. prophase I c. anaphase I
   b. prophase II d. anaphase II

9. Match each term with its description.
   ___ chromosome number a. different molecular forms of the same gene
   ___ alleles b. none between meiosis I and II
   ___ metaphase I c. pairs of homologous chromosomes are now aligned at the spindle equator
   ___ interphase d. sum total of chromosomes in all cells of a given type

a
b

Figure 10.12 Example of the chin dimple trait (actually a fissure in the chin surface).

Critical Thinking

1. Assume you can measure the amount of DNA in a primary oocyte, then in a primary spermatocyte, which gives you a mass *m*. What mass of DNA would you expect to find in each mature gamete (egg and sperm) that forms after meiosis? What mass of DNA would you expect to find (1) in an egg fertilized by one of the sperm and (2) in that egg after the first DNA duplication?

2. Adam has a pair of alleles that influence whether a person is right- or left-handed. One allele says "left," and its partner says "right." Visualize one of his germ cells, in which chromosomes are being duplicated prior to meiosis. Visualize what happens to the chromosomes during anaphase I and II. (It might help to use index cards as models of the sister chromatids of each chromosome.) What fraction of Adam's sperm will carry the gene for right-handedness? For left-handedness?

3. Adam also has one allele for long eyelashes, and a partner allele (on the homologous chromosome) for short eyelashes. What fraction of his sperm will have these gene combinations:

right-handed, long eyelashes *left-handed, long eyelashes*
right-handed, short eyelashes *left-handed, short eyelashes*

Selected Key Terms

allele *10.1*
asexual reproduction *10.1*
chromosome number *10.2*
crossing over *10.4*
diploid number *10.2*
egg (ovum) *10.5*
fertilization *10.5*
gamete *CI*
gene *10.1*
germ cell *CI*
haploid number *10.2*
homologous chromosome *10.2*
meiosis *10.2*
oocyte *10.5*
sexual reproduction *10.1*
sister chromatid *10.2*
sperm *10.5*
spore *10.5*

Readings *See also www.infotrac-college.com*

Klug, W., and M. Cummings. 1994. *Concepts of Genetics*. Fourth edition. New York: Macmillan.

Wolfe, S. 1995. *Introduction to Molecular and Cellular Biology*. Belmont, California: Wadsworth.

11 OBSERVABLE PATTERNS OF INHERITANCE

A Smorgasbord of Ears and Other Traits

Basketball ace Charles Barkley has them. So does actor Tom Cruise. Actress Joan Chen doesn't, and neither did a monk named Gregor Mendel. To see how you fit in with these folks, use a mirror to check out your ears. Is the fleshy lobe at the base of each ear attached to the side of your head? If so, you and Barkley and Cruise have something in common. Or is the fleshy lobe not attached, so that you can flap it back and forth? If so, you are like Chen and Mendel (Figure 11.1).

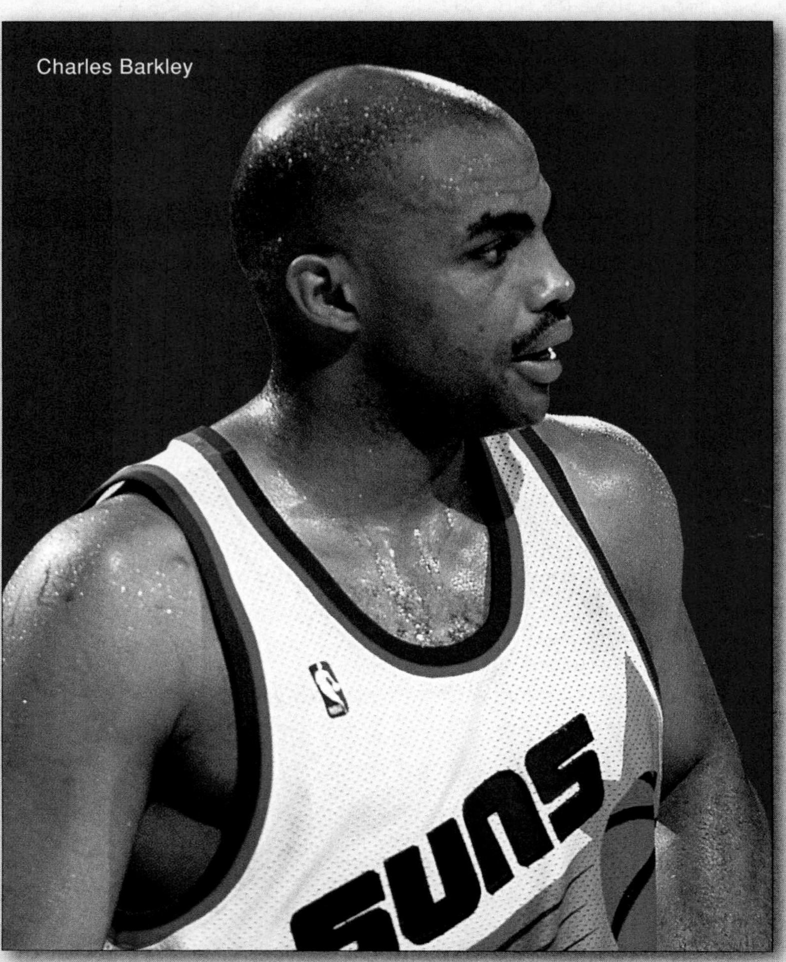
Charles Barkley
SUNS

Tom Cruise

Figure 11.1 Attached and detached earlobes of a few representative humans. This sampling provides observable evidence of a trait governed by a certain gene, which exists in different molecular forms in the human population. Do you have one or the other version of the trait? It depends on which molecular forms of the gene you inherited from your mother and father. As Gregor Mendel perceived, such easily observable traits can be used to identify patterns of inheritance from one generation to the next.

Whether a person is born with attached or detached earlobes depends on a single kind of gene. That gene comes in slightly different molecular forms—alleles. Only one form has information about detached lobes. The information is put to use while a human body is developing inside the mother. It calls for a death signal, which is sent to all the cells positioned between the newly forming lobes and the head. Without the signal, the cells don't die and earlobes don't detach.

We all have genes for thousands of traits, including earlobes, cheeks, lashes, and eyeballs. Most traits vary in their details from one person to the next. Remember, we inherit pairs of genes, on pairs of chromosomes. In some pairs, one allele has strong effects and overwhelms the other allele's contribution. The outgunned allele is said to be recessive to the dominant one. If you have detached earlobes, dimpled cheeks, long lashes, or large eyeballs, you have at least one and quite possibly two dominant alleles that affect the trait in a particular way.

When both alleles of a pair are recessive, nothing masks their effect on a trait. You get *attached* earlobes with one pair of recessive alleles (and *flat* feet with another, a *straight* nose with another, and so on).

How did we discover such remarkable things about our genes? It started with Mendel. By analyzing garden

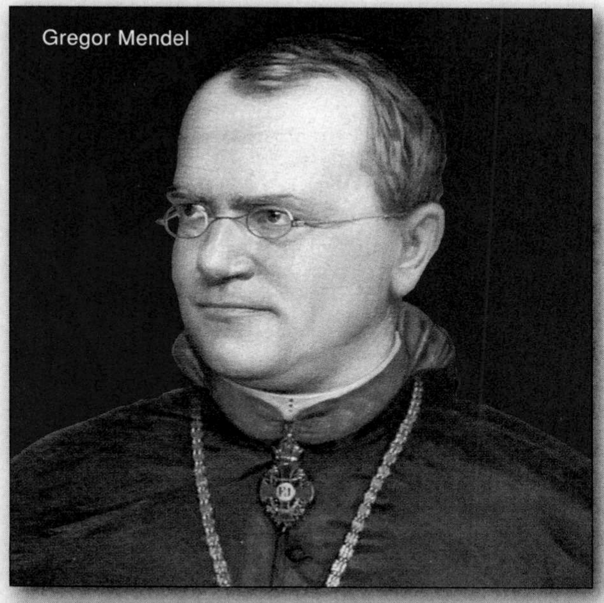
Gregor Mendel

Joan Chen

pea plants generation after generation, Mendel found indirect but *observable* evidence of how parents bestow units of hereditary information—genes—on offspring. This chapter focuses on both the methods and the results of Mendel's experiments. They remain a classic example of how a scientific approach can pry open important secrets about the natural world. And to this day, they serve as the foundation for modern genetics.

KEY CONCEPTS

1. Genes are units of information about heritable traits. Alleles, which are slightly different molecular forms of a gene, specify different versions of the same trait.

2. Each gene has a particular location on a particular chromosome of a species. Humans, pea plants, and other organisms with a diploid chromosome number inherit *pairs* of genes, at equivalent locations on pairs of homologous chromosomes.

3. When the two members of each pair of homologous chromosomes are moved apart from each other during meiosis, their pairs of genes are moved apart, also, and end up in different gametes. Gregor Mendel found indirect evidence of this gene segregation when he crossbred pea plants showing different versions of the same trait, such as purple or white flowers.

4. Each pair of homologous chromosomes in a germ cell is sorted out for distribution into one gamete or another independently of how the other pairs of homologous chromosomes are assorted. Mendel discovered indirect evidence of this when he tracked many plants having observable differences in two traits, such as flower color and plant height.

5. The contrasting forms of traits that Mendel happened to study were specified by nonidentical alleles. One allele was dominant, in that its effect on a trait masked the effect of a recessive allele paired with it.

6. Not all traits have such clearly dominant or recessive forms. One allele of a pair may be fully or partially dominant over its partner or codominant with it. Two or more gene pairs often influence the same trait, and some single genes influence many traits. Besides this, environmental factors induce further variation in traits.

11.1 MENDEL'S INSIGHT INTO INHERITANCE PATTERNS

More than a century ago, people wondered about the basis of inheritance. It was common knowledge that sperm and eggs both transmit information about traits to offspring. But few suspected that the information is organized in units (genes). Instead, the idea was that a father's blob of information "blended" with the mother's blob, like cream into coffee, at fertilization.

However, carried to its logical conclusion, blending would slowly dilute a population's shared pool of hereditary information until there was only a single version left of each trait. If that were so, why did, say, freckles keep showing up among the children of nonfreckled parents through the generations? Why weren't all the descendants of a herd of white stallions and black mares gray? The blending theory could scarcely explain the obvious variation in traits that people could observe with their own eyes. Nevertheless, few disputed the theory.

Charles Darwin was among the scholarly dissidents. According to the key premise of his theory of natural selection, individuals of a population show variation in heritable traits. Through the generations, the variations that improve the chance of surviving and reproducing show up with greater frequency than those that do not. Less advantageous variations may persist among fewer individuals, or they may even disappear. It is not that separate versions of a given trait are "blended out" of the population. Rather, *each version of a trait may persist in a population, at frequencies that can change over time.*

Just before Darwin presented his theory, someone was gathering evidence that eventually would support his key premise. A monk, Gregor Mendel, had already guessed that sperm and eggs carry distinct "units" of information about heritable traits. By carefully analyzing traits of pea plants generation after generation, Mendel found indirect but *observable* evidence of how parents transmit genes to offspring.

Mendel's Experimental Approach

Mendel spent most of his adult life in a monastery in Brno, a city near Vienna that has since become part of the Czech Republic. However, Mendel was not a man of narrow interests who accidentally stumbled onto principles of great import. The monastery of St. Thomas was close to European capitals that were the centers of scientific inquiry.

Having been raised on a farm, Mendel was aware of agricultural principles and their applications. He kept abreast of the breeding experiments and developments described in the available literature. He was a member of the regional agricultural society. He also won several awards for developing improved varieties of vegetables and fruits. Shortly after entering the monastery, he spent two years studying mathematics, physics, and botany at the University of Vienna. Few scholars of his time had combined talents in plant breeding and mathematics.

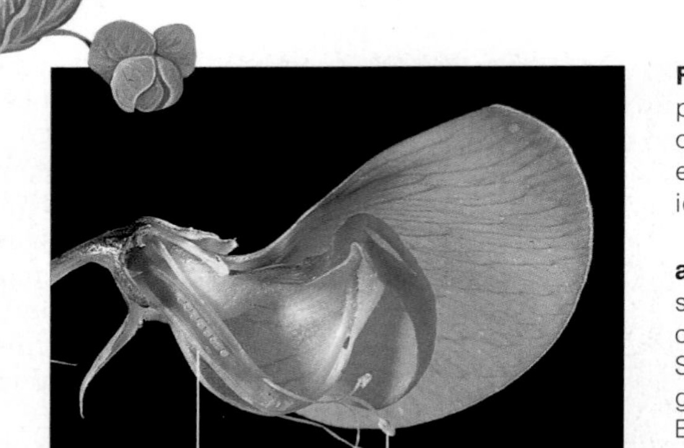
carpel
stamen

Figure 11.2 Garden pea plant (*Pisum sativum*), the organism Mendel chose for experimental tests of his ideas about inheritance.

a Garden pea flower. The section shows the location of its stamens and carpel. Sperm-producing pollen grains form in stamens. Eggs develop, fertilization takes place, and seeds mature inside the carpel.

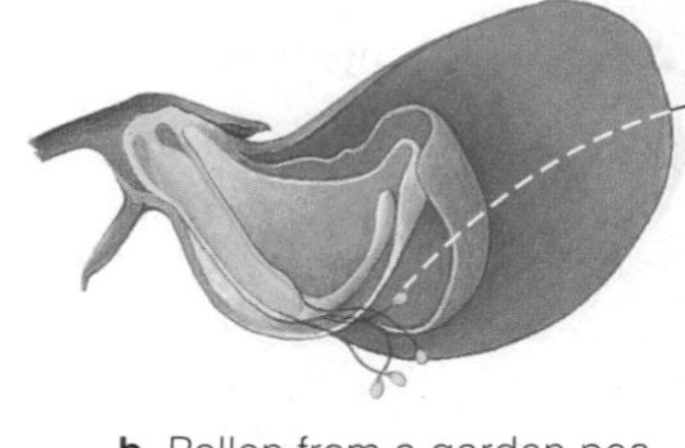

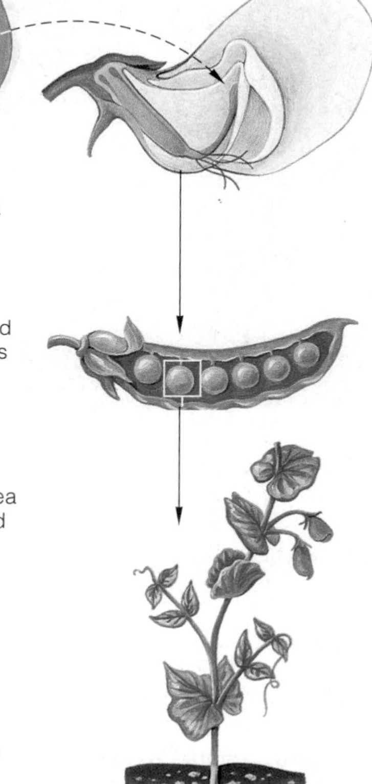

b Pollen from a garden pea plant that breeds true for purple flowers is brushed onto a floral bud of a plant that breeds true for white flowers. The white flower had its stamens snipped off. This is one way to assure cross-fertilization of plants.

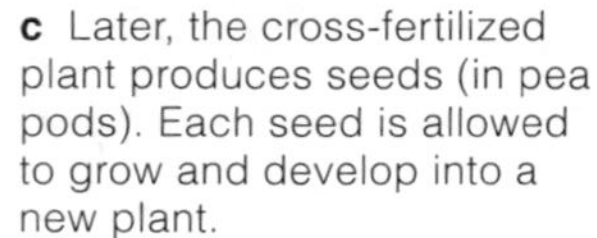

c Later, the cross-fertilized plant produces seeds (in pea pods). Each seed is allowed to grow and develop into a new plant.

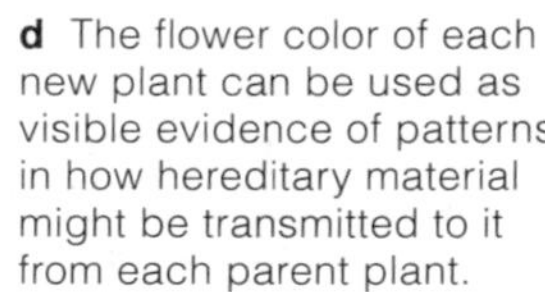

d The flower color of each new plant can be used as visible evidence of patterns in how hereditary material might be transmitted to it from each parent plant.

Further reading: Student Guide to InfoTrac on web site

Shortly after his university training, Mendel began experimenting with the garden pea plant, *Pisum sativum* (Figure 11.2). This plant is self-fertilizing. Male as well as female gametes (call them sperm and eggs) develop in different parts of the same flower, where fertilization takes place. Nearly all the plants breed true for certain traits. In other words, successive generations are just like their parents in one or more traits, as when all of the offspring grown from seeds of self-fertilized, white-flowered parent plants have white flowers.

As Mendel knew, we can cross-fertilize pea plants by transferring pollen from one plant's flower to the flower of another plant. For his experimental studies, he could open flower buds of a plant that bred true for a trait—say, white flowers—and snip out the stamens. (Stamens bear pollen grains in which sperm develop.) Then he could brush buds without stamens with pollen from a plant that bred true for a *different* version of the same trait: purple flowers. As Mendel hypothesized, he could use such clearly observable differences to track a trait through many generations. If there were patterns to the trait's inheritance, *then those patterns might tell him something about heredity itself.*

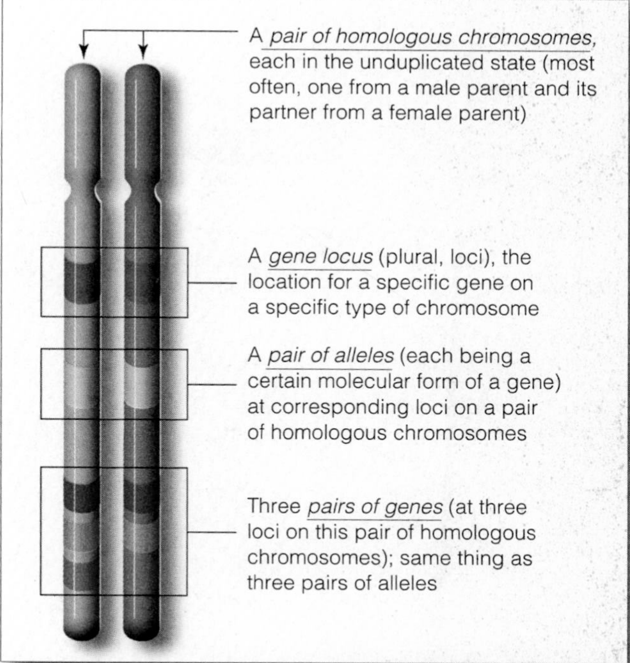

A pair of homologous chromosomes, each in the unduplicated state (most often, one from a male parent and its partner from a female parent)
A gene locus (plural, loci), the location for a specific gene on a specific type of chromosome
A pair of alleles (each being a certain molecular form of a gene) at corresponding loci on a pair of homologous chromosomes
Three pairs of genes (at three loci on this pair of homologous chromosomes); same thing as three pairs of alleles

Figure 11.3 A few genetic terms illustrated. Diploid organisms have pairs of genes, on pairs of homologous chromosomes. For example, you inherited one chromosome of each pair from your mother and the other, homologous chromosome from your father.

Most genes can have slightly different molecular forms, called alleles. Different alleles specify different versions of the same trait. An allele at one location on a chromosome may or may not be identical to its partner on the homologous chromosome.

Some Terms Used in Genetics

Having read the chapter on meiosis, you already have insight into the mechanisms of sexual reproduction, which is more than Mendel had. Neither he nor anyone else of his era knew about chromosomes. So he could not have known that a chromosome number is reduced by half in gametes, then restored when gametes meet at fertilization. Even so, Mendel sensed what was going on. As we follow his thinking, let's simplify the story by substituting a few of the modern terms used in studies of inheritance (see also Figure 11.3):

1. **Genes** are units of information about specific traits, and they are passed from parents to offspring. Each gene has a specific location (locus) on a chromosome.

2. Cells with a diploid chromosome number ($2n$) have pairs of genes, on pairs of homologous chromosomes.

3. Mutation can alter a gene's molecular structure. The alteration may change the gene's information about a trait (as when the gene for flower color specifies purple and a mutated version specifies white). All the different molecular forms of the same gene are called **alleles**.

4. When offspring of genetic crosses inherit a pair of *identical* alleles for a trait, generation after generation, they are a **true-breeding lineage**. By contrast, when offspring of a genetic cross inherit a pair of *nonidentical* alleles for a trait, they are **hybrid offspring**.

5. When both alleles of a pair are identical, this is a *homozygous* condition. When the two are not identical, this is a *heterozygous* condition.

6. An allele is said to be *dominant* when its effect on a trait masks that of any *recessive* allele paired with it. We use capital letters for dominant alleles and lowercase letters for recessive ones. *A* and *a* are examples.

7. Putting this all together, a **homozygous dominant** individual has a pair of dominant alleles (*AA*) for a trait under study. A **homozygous recessive** individual has a pair of recessive alleles (*aa*). And a **heterozygous** individual has a pair of nonidentical alleles (*Aa*).

8. Two terms help keep the distinction clear between genes and the traits they specify. **Genotype** refers to the particular genes an individual carries. **Phenotype** refers to an individual's observable traits.

9. When tracking the inheritance of traits through generations of offspring, these abbreviations apply:

P	parental generation
F_1	first-generation offspring
F_2	second-generation offspring

11.2 MENDEL'S THEORY OF SEGREGATION

Mendel had an idea that in every generation, a plant inherits two "units" (genes) of information about a trait, one from each parent. To test this idea, he performed what is now known as **monohybrid crosses**. These are experimental crosses between two parents that breed true (are homozygous) for different versions of a single trait. The F_1 offspring are hybrids; each inherits a pair of nonidentical alleles (is heterozygous) for that trait.

Predicting Outcomes of Monohybrid Crosses

Mendel tracked many individual traits through two generations. For instance, in one series of experiments, he crossed true-breeding, purple-flowered plants and true-breeding, white-flowered ones. All plants grown from the seeds that resulted from this cross had purple flowers. Mendel allowed these plants to self-fertilize. Some plants grown from the seeds had white flowers!

If Mendel's hypothesis were correct—if each plant had inherited two units of information about flower color—then the unit for "purple" had to be dominant, because it had masked the unit for "white" in F_1 plants.

Let's rephrase his thinking. Germ cells of pea plants are diploid, with pairs of homologous chromosomes. Assume one parent is homozygous dominant (AA) and the other is homozygous recessive (aa) for flower color. Following meiosis, a sperm or egg carries one allele for flower color (Figure 11.4). Thus, when a sperm fertilizes an egg, only one outcome is possible: $A + a = Aa$.

Before continuing, you should know Mendel crossed hundreds of plants and tracked thousands of offspring. Besides this, he counted and recorded the number of plants showing dominance or recessiveness. As you can see from Figure 11.5, an intriguing ratio emerged. On average, three of every four F_2 plants had the dominant phenotype, and one had the recessive phenotype.

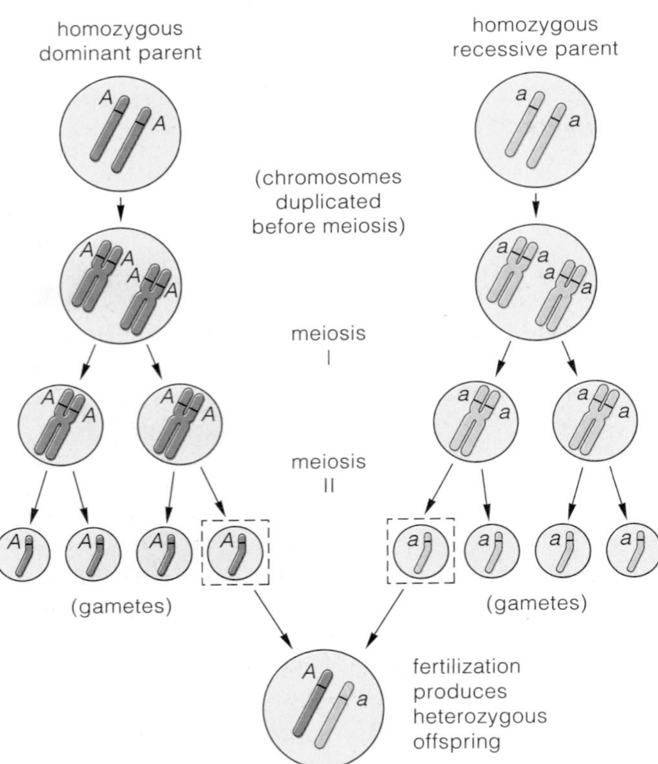
homozygous dominant parent
homozygous recessive parent
(chromosomes duplicated before meiosis)
meiosis I
meiosis II
(gametes)
(gametes)
fertilization produces heterozygous offspring

Figure 11.4 Monohybrid cross, showing how one gene of a pair segregates from the other gene. Two parents that breed true for two versions of a trait produce only heterozygous offspring.

Figure 11.5 *Right*: Numerical results from Mendel's monohybrid cross experiments with the garden pea plant (*P. sativum*). The numbers are his counts of the F_2 plants that carried dominant or recessive hereditary "units" (alleles) for the trait. On average, the dominant-to-recessive ratio was 3:1.

Trait Studied	Dominant Form	Recessive Form	F_2 Dominant-to-Recessive Ratio
SEED SHAPE	5,474 round	1,850 wrinkled	2.96:1
SEED COLOR	6,022 yellow	2,001 green	3.01:1
POD SHAPE	882 inflated	299 wrinkled	2.95:1
POD COLOR	428 green	152 yellow	2.82:1
FLOWER COLOR	705 purple	224 white	3.15:1
FLOWER POSITION	651 along stem	207 at tip	3.14:1
STEM LENGTH	787 tall	277 dwarf	2.84:1

Average ratio for all traits studied: **3:1**

Further reading: Student Guide to InfoTrac on web site →

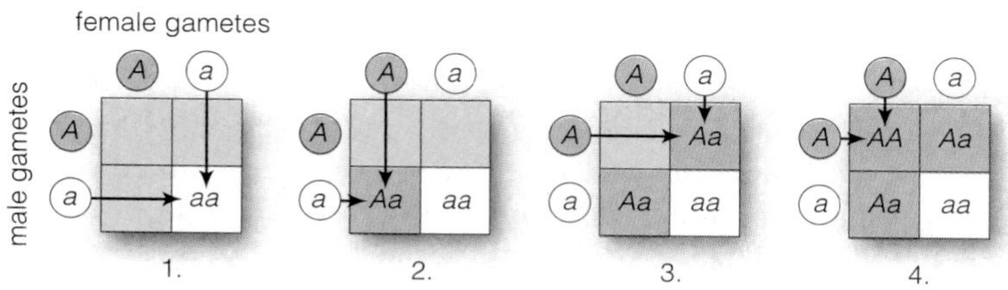
female gametes
male gametes
A
a
aa
Aa
AA
1.
2.
3.
4.

Figure 11.6 Punnett-square method of predicting the probable outcome of a genetic cross. Circles represent gametes. *Italic* letters on gametes represent dominant or recessive alleles. The squares show the different genotypes possible among offspring. In this case, gametes are from a self-fertilizing heterozygous (*Aa*) plant.

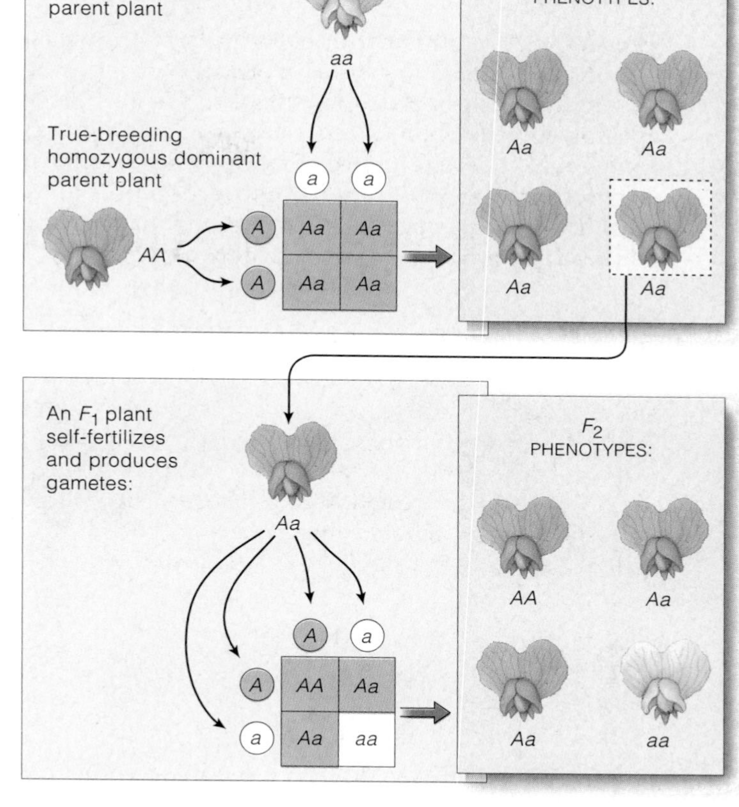
True-breeding homozygous recessive parent plant
aa
True-breeding homozygous dominant parent plant
AA
F_1 PHENOTYPES:
Aa
An F_1 plant self-fertilizes and produces gametes:
Aa
AA
F_2 PHENOTYPES:
aa

Figure 11.7 *Right*: Results from one of Mendel's monohybrid crosses. On average, the dominant-to-recessive ratio among the second-generation (F_2) plants was 3:1.

To Mendel, the ratio suggested that fertilization is a chance event, with a number of possible outcomes. And he had an understanding of probability, which applies to chance events *and therefore could help him predict the possible outcomes of genetic crosses.* **Probability** simply means this: The chance that each outcome of a given event will occur is proportional to the number of ways in which that event can be reached.

The **Punnett-square method**, explained in Figure 11.6 and applied in Figure 11.7, may help you visualize the possibilities. As you can see, if half of a plant's sperm (or eggs) were *a* and half were *A*, then four outcomes would be possible each time a sperm fertilized an egg:

POSSIBLE EVENT:	PROBABLE OUTCOME:
sperm *A* meets egg *A*	1/4 *AA* offspring
sperm *A* meets egg *a*	1/4 *Aa*
sperm *a* meets egg *A*	1/4 *Aa*
sperm *a* meets egg *a*	1/4 *aa*

By this prediction, an F_2 plant has three chances in four of getting at least one dominant allele (purple flowers). It has one chance in four of getting two recessive alleles (white flowers). That is a probable phenotypic ratio of three purple to one white, or 3:1.

Mendel's observed ratios were not *exactly* 3:1. You can see this for yourself by looking at the numerical results listed in Figure 11.5. Why did Mendel put aside the deviations? To understand why, flip a coin several times. As we all know, a coin is just as likely to end up heads as tails. But often a coin ends up heads, or tails, several times in a row. So if you flip the coin only a few times, the observed ratio might differ greatly from the predicted ratio of 1:1. Flip the coin many, many times, and you are more likely to come close to the predicted ratio. Mendel understood the rules of probability—and performed a large number of crosses. Almost certainly, this kept him from being confused by minor deviations from the predicted results of the experimental crosses.

Testcrosses

By running **testcrosses**, Mendel gained support for his prediction. In this type of experimental test, an organism shows dominance for a specified trait but its genotype is unknown, so it is crossed to a known homozygous recessive individual. Test results may reveal whether the organism is homozygous dominant or heterozygous.

Regarding the monohybrid crosses just described, Mendel tested his prediction that the purple-flowered F_1 offspring were heterozygous by crossing them with true-breeding, white-flowered plants. If they were all homozygous dominant, then all the F_2 offspring would show the dominant form of the trait. If heterozygous, there would be about as many dominant as recessive plants. Sure enough, when old enough to flower, about half the F_2 plants had purple flowers (*Aa*) and half had white (*aa*). Can you construct two Punnett squares that show the possible outcomes of this testcross?

The results from Mendel's monohybrid crosses and testcrosses became the basis of a theory of **segregation**, which we state here in modern terms:

MENDEL'S THEORY OF SEGREGATION. **Diploid cells have pairs of genes, on pairs of homologous chromosomes. The two genes of each pair are separated from each other during meiosis, so they end up in different gametes.**

Predicting Outcomes of Dihybrid Crosses

By another series of experiments, Mendel attempted to explain how *two* pairs of genes might be assorted into gametes. He selected true-breeding plants that differed in two traits, including flower color and plant height. In such **dihybrid crosses**, F_1 offspring inherit two gene pairs, each consisting of two nonidentical alleles.

Let's diagram one of Mendel's dihybrid crosses. We can use *A* for flower color and *B* for height as dominant alleles and, as their recessive counterparts, *a* and *b*:

TRUE-BREEDING PARENTS: purple flowers, tall × white flowers, dwarf

AABB   *aabb*

GAMETES: *AB* *AB* *ab* *ab*

F_1 HYBRID OFFSPRING: *AaBb*

As Mendel would have predicted, the F_1 offspring from this cross are all purple-flowered and tall (*AaBb*).

When those F_1 plants reproduce, how will the two gene pairs be assorted into gametes? The answer partly depends on the chromosomal locations of those pairs. Assume one pair of homologous chromosomes carries the *Aa* alleles and a different pair carries the *Bb* alleles. Next, think of how all chromosomes become positioned at the spindle equator during metaphase I of meiosis (Figures 10.4 and 11.8). The chromosome with the *A* allele might be positioned to move to either spindle pole (then into one of four gametes). The same is true of its homologue. And the same is true of the chromosomes with the *B* and *b* alleles. Thus, after meiosis and gamete formation, four combinations of alleles are possible in the sperm or eggs: 1/4 *AB*, 1/4 *Ab*, 1/4 *aB*, and 1/4 *ab*.

Given the alternative alignments of chromosomes at metaphase I, several allelic combinations are possible at fertilization. Simple multiplication (four kinds of sperm times four kinds of eggs) tells us sixteen combinations of gametes are possible in the F_2 offspring of a dihybrid cross. Use the Punnett-square method to diagram the probabilities (Figure 11.9). Now add up all the possible phenotypes and you get 9/16 tall purple-flowered, 3/16 dwarf purple-flowered, 3/16 tall white-flowered, and 1/16 dwarf white-flowered plants. That is a probable phenotypic ratio of 9:3:3:1. Results from one dihybrid cross that Mendel described were close to this ratio.

Figure 11.8 Independent assortment. This example tracks two pairs of homologous chromosomes. An allele at one locus on a chromosome may or may not be identical with its partner allele on the homologous chromosome. At meiosis, either chromosome of a pair may become attached to either pole of the spindle. Thus, in this case, two different lineups are possible at metaphase I.

Nucleus of a diploid (2*n*) reproductive cell with only two pairs of homologous chromosomes

OR

Possible alignments of the homologous chromosomes at metaphase I of meiosis, as shown by two diagrams:

The resulting alignments of chromosomes at metaphase II:

The combinations of alleles possible in the forthcoming gametes:

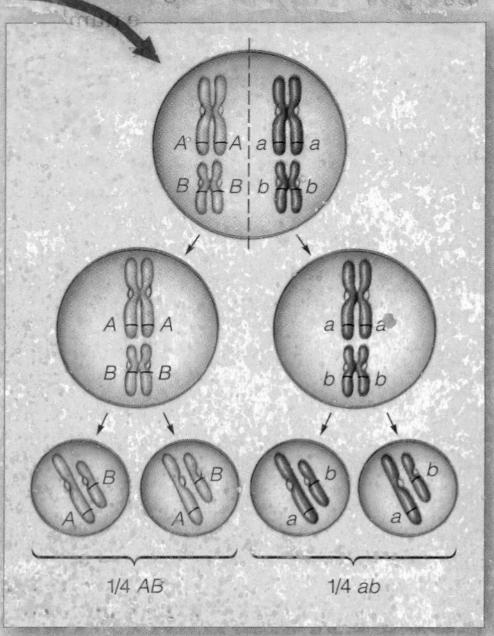

A A a a
B B b b
1/4 AB
1/4 ab

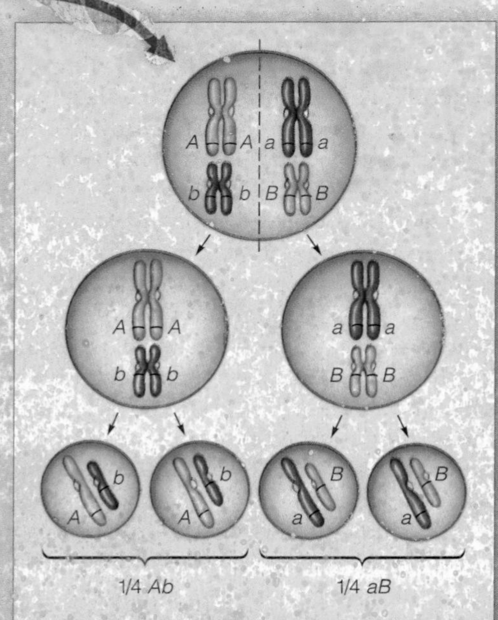

A A a a
b b B B
1/4 Ab
1/4 aB

Further reading: Student Guide to InfoTrac on web site →

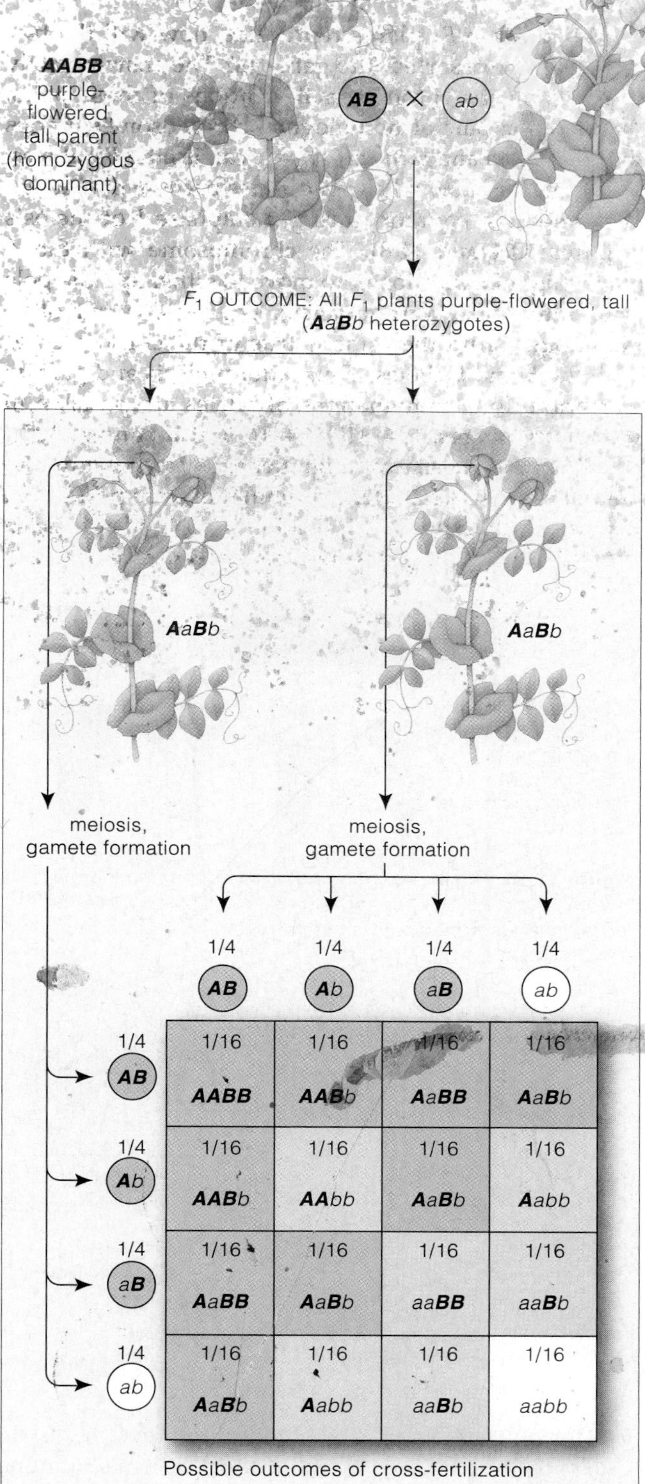
AABB
purple-flowered, tall parent (homozygous dominant)
AB
×
ab
aabb
white-flowered, dwarf parent (homozygous recessive)
F_1 OUTCOME: All F_1 plants purple-flowered, tall (AaBb heterozygotes)
AaBb
AaBb
meiosis, gamete formation
meiosis, gamete formation
1/4 AB
1/4 Ab
1/4 aB
1/4 ab
1/16 AABB
1/16 AABb
1/16 AaBB
1/16 AaBb
1/16 AABb
1/16 AAbb
1/16 AaBb
1/16 Aabb
1/16 AaBB
1/16 AaBb
1/16 aaBB
1/16 aaBb
1/16 AaBb
1/16 Aabb
1/16 aaBb
1/16 aabb
Possible outcomes of cross-fertilization
ADDING UP THE F_2 COMBINATIONS POSSIBLE:
9/16 or 9 purple-flowered, tall
3/16 or 3 purple-flowered, dwarf
3/16 or 3 white-flowered, tall
1/16 or 1 white-flowered, dwarf

Figure 11.9 Results from Mendel's dihybrid cross between parent plants that bred true for different versions of two traits: flower color and plant height. *A* and *a* represent dominant and recessive alleles for flower color. *B* and *b* represent dominant and recessive alleles for plant height. As the Punnett square indicates, the probabilities of certain combinations of phenotypes among the F_2 offspring occur in a 9:3:3:1 ratio, on the average.

The Theory in Modern Form

Mendel could do no more than analyze the numerical results from his dihybrid crosses, because he didn't know that seven pairs of homologous chromosomes carry the pea plant's "units" of inheritance. It just seemed to him that the two units for the first trait he was tracking had been assorted into gametes independently of the two units for the other trait. In time his interpretation became known as the theory of **independent assortment**, which we state here in modern terms: By the end of meiosis, each pair of homologous chromosomes—and the genes they carry—have been sorted for shipment into gametes independently of how the other pairs were sorted out.

Independent assortment and hybrid crossing lead to stupendous genetic variation. In a monohybrid cross involving only a single gene pair, three genotypes are possible: *AA*, *Aa*, and *aa*. We can represent this as 3^n, where n is the number of gene pairs. When we consider more gene pairs, the number of possible combinations increases dramatically. Even if parents differ in merely ten pairs of genes, almost 60,000 genotypes are possible among their offspring. If they differ in twenty pairs of genes, the number approaches 3.5 billion!

In 1865 Mendel presented his ideas to the Brünn Natural History Society. His ideas had little impact. The next year his paper was published, and apparently it was read by few and understood by no one. In 1871 he became abbot of the monastery, and his experiments gradually gave way to administrative tasks. He died in 1884, never to know his experiments would become the starting point for the development of modern genetics.

Today, Mendel's theory of segregation still stands. Hereditary material is indeed organized in units (genes) that retain their identity and are segregated from each other for distribution into different gametes. But the theory of independent assortment does not apply to all gene combinations, as you will see in the next chapter.

MENDEL'S THEORY OF INDEPENDENT ASSORTMENT. By the end of meiosis, genes on pairs of homologous chromosomes have been sorted out for distribution into one gamete or another independently of gene pairs of other chromosomes.

11.4 DOMINANCE RELATIONS

For the most part, Mendel studied traits having clearly dominant or recessive forms. As the remaining sections of this chapter will make clear, however, the expression of other traits is not as straightforward.

Incomplete Dominance

In **incomplete dominance**, one allele of a pair isn't fully dominant over its partner, so a heterozygous phenotype *somewhere in between* the two homozygous phenotypes emerges. Cross a true-breeding red snapdragon and a true-breeding white one. All F_1 offspring will have pink flowers. Cross two F_1 plants and expect red, pink, and white snapdragons in a predictable ratio (Figure 11.10). What causes this inheritance pattern? Red snapdragons have two alleles that allow them to make an abundance of red pigment molecules. White snapdragons have two mutant alleles that render them pigment-free. The pink ones are heterozygous. Their one red allele can specify enough pigment to make flowers pink but not red.

homozygous parent
×
homozygous parent

All F_1 offspring are heterozygous for flower color:

Cross two of the F_1 plants, and the F_2 offspring will show three phenotypes in a 1:2:1 ratio:

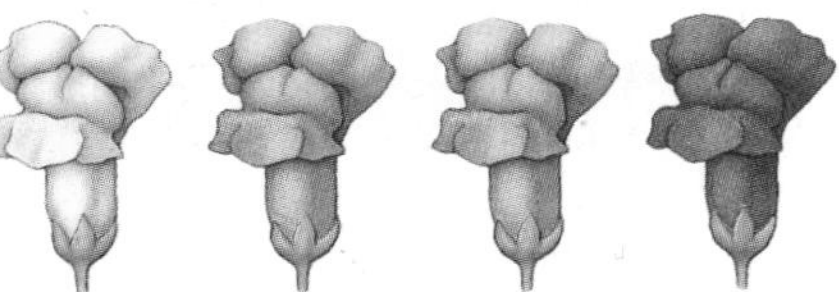

Figure 11.10 Visible evidence of incomplete dominance in heterozygous (pink) snapdragons, in which an allele for red pigment is paired with a "white" allele.

ABO Blood Types: A Case of Codominance

In **codominance**, a pair of nonidentical alleles specify two phenotypes, which are both expressed at the same time in heterozygotes. As an example, think about one of the glycolipids projecting from the plasma membrane of your red blood cells. It helps give red blood cells their unique identity, although it comes in slightly different molecular forms. A method of analysis known as *ABO blood typing* reveals which form a person has.

In humans, an enzyme dictates the glycolipid's final structure. The gene for that enzyme has three alleles. Two alleles, I^A and I^B, are codominant when paired. The third, i, is recessive; a pairing with either I^A or I^B masks its effect. Together, they are a **multiple allele system**, which we define as the presence of three or more alleles of a gene among individuals of a population.

Before each glycolipid molecule became positioned at the cell surface, it was modified in the cytomembrane system (Section 4.5). First an oligosaccharide chain was attached to a lipid molecule, then a sugar was attached to the end of that chain. Alleles I^A and I^B specify two slightly different versions of the enzyme catalyzing that final step. The two attach different sugars, which give the glycolipid its identity—either A or B.

Which alleles do you have? With either I^AI^A or I^Ai, you have type A blood. With I^BI^B or I^Bi, your blood is type B. With codominant alleles I^AI^B, it is AB—meaning you have both versions of the sugar-attaching enzyme. But if you are homozygous recessive (ii), the molecules never did get an additional sugar attached to them. Your blood type is neither A nor B; that is what type "O" means. Figure 11.11 summarizes the possibilities.

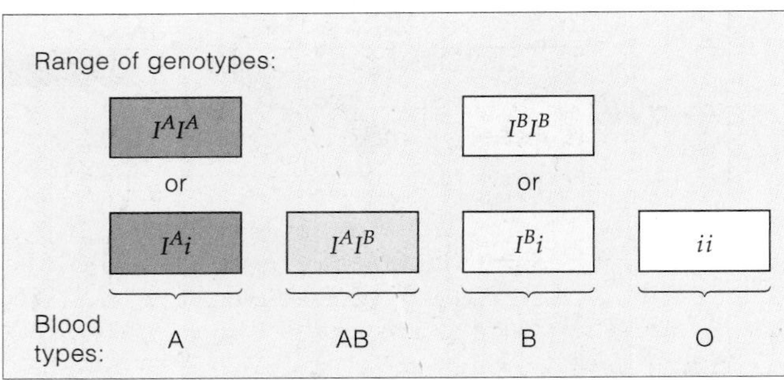
Range of genotypes:
I^AI^A
or
I^Ai
I^AI^B
I^BI^B
or
I^Bi
ii
Blood types:
A
AB
B
O

Figure 11.11 Allelic combinations that are related to ABO blood typing.

During *transfusions*, the blood of two people mixes. Unless their blood is compatible, the recipient's immune system perceives the red blood cells from the donor as "nonself." It will act against those cells and may cause death (Section 39.4).

One allele may be fully dominant, incompletely dominant, or codominant with its partner on the homologous chromosome.

Further reading: Student Guide to InfoTrac on web site

11.5 MULTIPLE EFFECTS OF SINGLE GENES

Expression of the alleles at just a single location on a chromosome may have positive or negative effects on two or more traits. This phenotypic outcome of a single gene's activity is known as **pleiotropy** (after the Greek *pleio–*, meaning more, and *–tropic*, meaning to change).

The genetic disorder *sickle-cell anemia* is a classic example of how the alleles at a single locus can have pleiotropic effects. The disorder arises from a mutated gene for beta-globin, one of two kinds of polypeptide chains in the hemoglobin molecule. Hemoglobin, recall, is the oxygen-transporting protein in red blood cells. We designate the mutant allele as Hb^S instead of Hb^A. Heterozygotes (Hb^A/Hb^S) usually show few symptoms of the disorder. Their red blood cells are able to produce enough normal hemoglobin molecules to compensate for the abnormal ones. In homozygotes (Hb^S/Hb^S), the red blood cells can only produce abnormal hemoglobin. This one abnormality may have drastic repercussions throughout the body, for it disrupts the concentration of oxygen in the bloodstream.

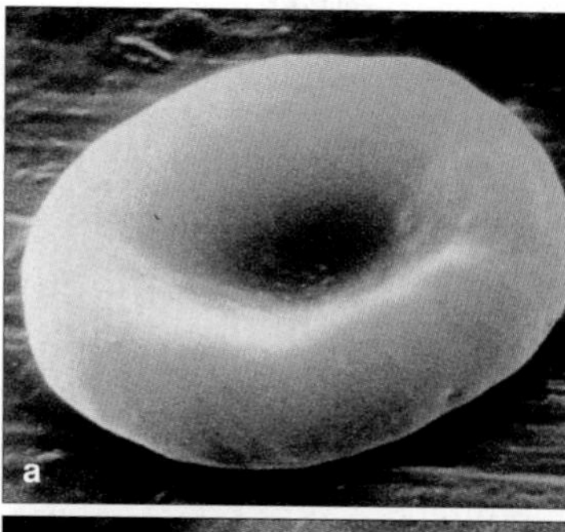

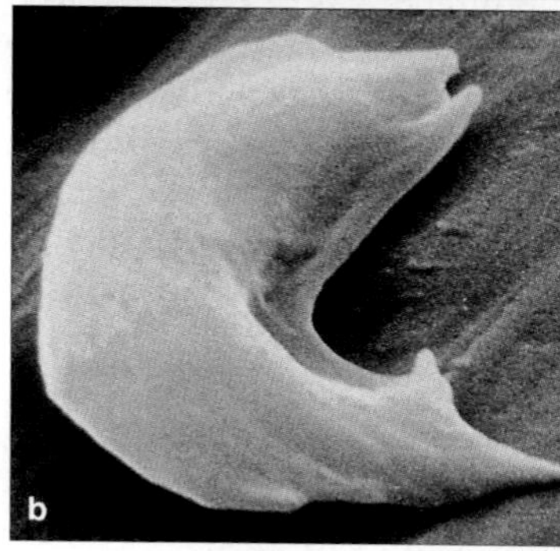

Figure 11.12 (**a**) Red blood cell from a person affected by the genetic disorder sickle-cell anemia. This scanning electron micrograph shows the surface appearance of the affected individual's red blood cells when the blood is adequately oxygenated. (**b**) This scanning electron micrograph shows the sickle shape that red blood cells assume when the concentration of oxygen in blood is low. (**c**) The diagram tracks the wide range of symptoms characteristic of an individual who is homozygous recessive for the disorder.

Humans, like most organisms, depend on the intake of oxygen for aerobic respiration. Oxygen in air flows into the lungs, then diffuses into the blood. There it binds to hemoglobin, which transports it through arteries, arterioles, and then small-diameter, thin-walled capillaries threading past all living cells in the body. A steep oxygen concentration gradient exists between blood and the fluid within the surrounding tissues, so most of the oxygen diffuses into the tissues and on into cells. With this extensive cellular uptake, the concentration of oxygen in blood declines. The decline is most pronounced at high altitudes and during strenuous activity.

In the red blood cells of people who bear the mutant gene, abnormal hemoglobin molecules stick together in rod-shaped arrangements. The rods distort cells into a sickle shape, as in Figure 11.12*b*. (A sickle is a farm tool with a long, crescent-shaped blade.) The distorted cells rupture easily, and their remnants clog and rupture capillaries. When these oxygen transporters are rapidly destroyed, the cells in affected tissues become starved for oxygen. Also, their clumping effect leads to local failures in the capacity of the circulatory system to deliver oxygen and to carry away carbon dioxide and other metabolic wastes.

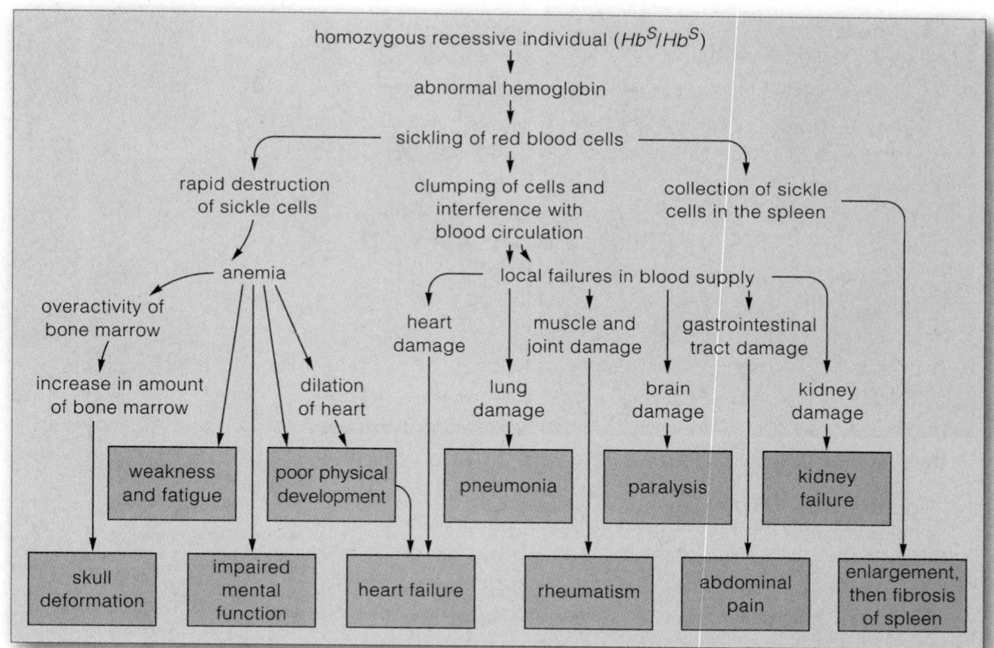
homozygous recessive individual (Hb^S/Hb^S)
abnormal hemoglobin
sickling of red blood cells
rapid destruction of sickle cells
clumping of cells and interference with blood circulation
collection of sickle cells in the spleen
anemia
local failures in blood supply
overactivity of bone marrow
increase in amount of bone marrow
dilation of heart
heart damage
muscle and joint damage
gastrointestinal tract damage
lung damage
brain damage
kidney damage
weakness and fatigue
poor physical development
pneumonia
paralysis
kidney failure
skull deformation
impaired mental function
heart failure
rheumatism
abdominal pain
enlargement, then fibrosis of spleen
c

Over time, ongoing expression of the mutant gene may damage tissues and organs throughout the body. Figure 11.12*c* tracks how the successive alterations in phenotype may proceed. In chapters to come, you will read about other aspects of this genetic disorder.

The alleles at a single gene location may have positive or negative effects on two or more traits.

The effects may not be simultaneous. Rather, they may have repercussions over time. The gene may lead to an alteration in one trait. That change may alter another trait, and so on.

11.6 INTERACTIONS BETWEEN GENE PAIRS

Often a trait results from interactions among products of two or more gene pairs. For example, two alleles of a gene may mask expression of another gene's alleles, so some expected phenotypes may not appear at all. Such interactions between the product of pairs of genes are called **epistasis** (meaning the act of stopping).

Hair Color in Mammals

Epistasis is common among the gene pairs responsible for skin or fur color in mammals. Consider the black, brown, or yellow fur of Labrador retrievers (Figure 11.13). The different colors arise from variations in the amount and distribution of melanin, a brownish black pigment. A variety of enzymes and other products of many gene pairs affect different steps in the production of melanin and its deposition in certain body regions.

The alleles of one gene specify an enzyme required to produce melanin. Expression of allele *B* (black) has a more pronounced effect and is dominant to *b* (brown). Alleles of a different gene control the extent to which molecules of melanin will be deposited in a retriever's hairs. Allele *E* permits full deposition. Two recessive alleles (*ee*) reduce deposition, and fur will be yellow.

In some individuals, those two gene pairs are not able to interact, owing to a certain allelic combination at still another gene locus. There, a gene (*C*) calls for tyrosinase, the first of several enzymes in a melanin-producing pathway. An individual bearing one or two dominant alleles (*CC* or *Cc*) can make the functional enzyme. An individual bearing two recessive alleles (*cc*) cannot. When the biosynthetic pathway for melanin production gets blocked, then *albinism*—the absence of melanin—is the resulting phenotype (Figure 11.14).

a BLACK LABRADOR

b YELLOW LABRADOR

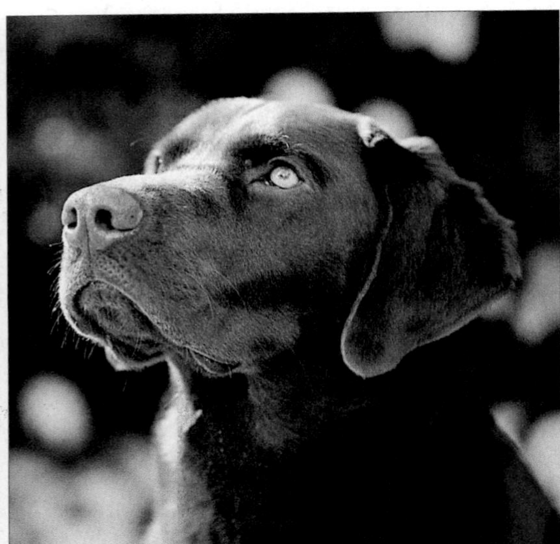

c CHOCOLATE LABRADOR

Figure 11.13 The heritable basis of coat color among Labrador retrievers. The trait arises through interactions among the alleles of two pairs of genes.

One kind of gene is involved in melanin production. Allele *B* (black) of this gene is dominant to allele *b* (brown). A different kind of gene influences the deposition of melanin pigment in individual hairs. Allele *E* of this gene promotes deposition, but a pairing of recessive alleles (*ee*) of the gene blocks deposition, and a yellow coat results.

F_1 offspring of a dihybrid cross produce F_2 offspring in a 9:3:4 ratio, as the Punnett-square diagram at *right* indicates.

The yellow Labrador in photograph (**b**) probably has genotype *BBee*, because it can produce melanin but cannot deposit pigment in hairs. After studying the photograph, can you say why?

HOMOZYGOUS PARENTS:
BBEE × bbee
F1 PUPPIES:
BbEe

ALLELIC COMBINATIONS POSSIBLE AMONG F_2 PUPPIES:

	BE	Be	bE	be
BE	BBEE	BBEe	BbEE	BbEe
Be	BBEe	BBee	BbEe	Bbee
bE	BbEE	BbEe	bbEE	bbEe
be	BbEe	Bbee	bbEe	bbee

RESULTING PHENOTYPES:

- 9/16 or 9 black
- 3/16 or 3 brown
- 4/16 or 4 yellow

Further reading: Student Guide to InfoTrac on web site

Figure 11.14 A rare albino rattlesnake. Like other animals that cannot produce melanin, it has pink eyes and its body surface is white, overall. (Eyes look pink because the absence of melanin from a tissue layer in the eyeball allows red light to be reflected from blood vessels in the eyes.) In birds and mammals, surface coloration results from pigments in feathers, fur, or skin. In fishes, amphibians, and reptiles, color-bearing cells give skin its surface coloration. Some of the cells contain melanin pigments or yellow-to-red pigments. Others contain crystals that reflect light and alter the effect of other pigments present.

The mutation affecting melanin production in the snake shown here had no effect on the production of yellow-to-red pigments and of light-reflecting crystals. Therefore, the snake's skin appears to be iridescent yellow as well as white.

P:
RRpp
(rose comb)
×
rrPP
(pea comb)
F_1:
RrPp
(all walnut comb)
RrPp × RrPp
F_2:
9/16 walnut
(RRPP, RRPp, RrPP, or RrPp)
3/16 rose
(RRpp or Rrpp)
3/16 pea
(rrPP or rrPp)
1/16 single
(rrpp)
comb
a WALNUT COMB
b ROSE COMB
c PEA COMB
d SINGLE COMB

Figure 11.15 Interaction between two genes that affect the same trait in domestic chicken breeds. The initial cross is between a Wyandotte (with a rose comb, **b**, on the crest of its head) and brahma (pea comb, **c**). With complete dominance at the locus for pea comb and at the locus for rose comb, products of the two gene pairs interact and give rise to a walnut comb (**a**). With full recessiveness at both gene loci, products interact and give rise to a single comb (**d**).

Comb Shape in Poultry

In some cases, interaction between two gene pairs results in a phenotype that neither pair can produce alone. The geneticists W. Bateson and R. Punnett identified two interacting gene pairs (*R* and *P*) that affect comb shape in chickens. Allelic combinations of *rr* at one gene locus and *pp* at the other locus result in the least common phenotype, the single comb. The presence of dominant alleles (*R*, *P*, or both) results in varied phenotypes.

Take a look at Figure 11.15. The diagram shows the combinations of alleles that interact to specify the rose, pea, and walnut combs shown in the photographs.

Genes often interact, as when alleles of one gene mask the expression of another gene, and when some expected phenotypes may not appear at all.

11.7 HOW CAN WE EXPLAIN LESS PREDICTABLE VARIATIONS?

Regarding the Unexpected Phenotype

As Mendel demonstrated, the phenotypic effects of one or two pairs of certain genes show up in predictable ratios when you track them from one generation to the next. Besides this, certain interactions among two or more gene pairs can produce phenotypes in predictable ratios, as the example of Labrador coat color clearly demonstrated in Section 11.6. However, even when you decide to track a single gene through the generations, you may discover that the resulting phenotypes are not quite what you had expected.

Consider *camptodactyly*—a rare genetic abnormality that affects both the shape and the movement of fingers. Certain people who carry the mutant allele for this heritable trait develop immobile, bent fingers on both hands. Other people develop immobile, bent fingers on the left or right hand only. Others who carry the mutant allele develop fingers that are not affected either way.

What is the source of such confounding variation? Recall that most organic compounds are synthesized by a series of metabolic steps, *and different enzymes—each a gene product—regulate different steps.* Maybe one gene has mutated in one of a number of different ways. Maybe the product of another gene blocks the pathway or causes it to run nonstop or not long enough. Or maybe poor nutrition or another factor that is variable in the individual's environment affects a crucial enzyme in the pathway. These are the sorts of variable factors that commonly introduce far less predictable variations in the phenotypes resulting from gene expression.

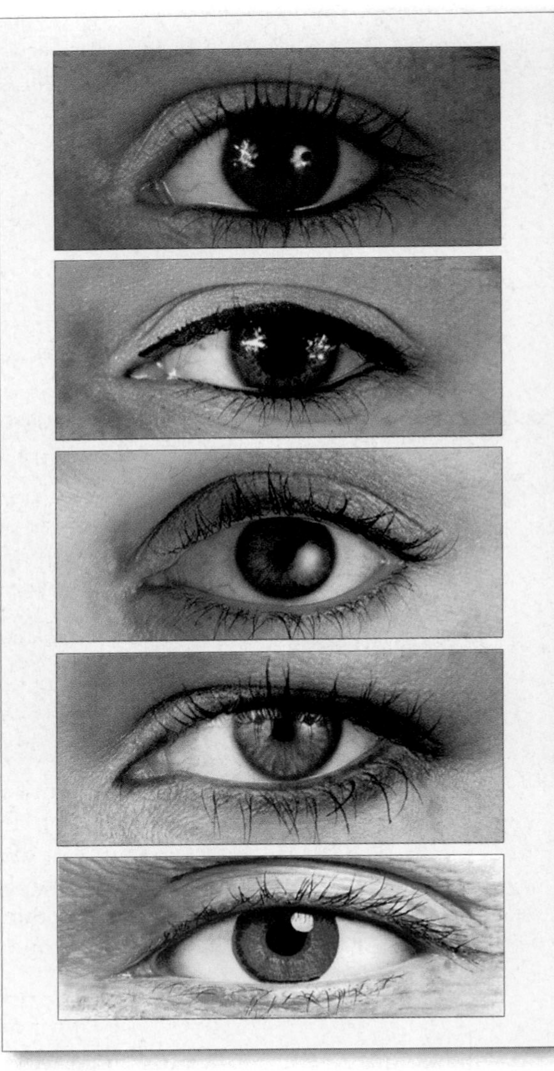

Figure 11.16 Samples from a range of continuous variation in human eye color. Different pairs of genes interact to produce and deposit melanin. Among other things, this pigment helps color the eye's iris. Different combinations of alleles result in small differences in eye color. Therefore, the frequency distribution for the eye-color trait appears to be continuous over a range from black to light blue.

Continuous Variation in Populations

Generally, the individuals of a population display a range of small differences in most traits. This characteristic of populations, which is known as **continuous variation**, is primarily an outcome of the number of genes affecting a trait and the number of environmental factors that influence their expression. In most cases, the greater the number of genes and environmental factors, the more continuous will be the expected distribution of all the versions of that trait.

Think about your own eye color. The colored part is the iris, a doughnut-shaped, pigmented structure just beneath the cornea. As is true of all humans, the color of your iris is the cumulative outcome of a number of gene products. Some products take part in the stepwise production and distribution of melanin, the same light-absorbing pigment that influences coat color in mammals. Dark eyes that seem to be almost black have abundant deposits of melanin molecules in the iris—so much so that most of the light that enters gets absorbed. Dark brown eyes don't have as many deposits of melanin, and some light that is not absorbed is reflected out from the iris. Light brown or hazel eyes have even less (Figure 11.16).

Green, gray, or blue eyes do not contain green, gray, or blue pigments. In these cases, the iris does incorporate different amounts of melanin, but not very much of it. As a result, many or most of the blue wavelengths of light that do enter the eye are reflected out.

How might you describe the continuous variation of some trait within a group, such as the college students in Figure 11.17*a*? They range from very short to very tall, with average heights much more common than either extreme. You can start out by dividing the full range of the different phenotypes into measurable categories. Next, count all the individual students in each category. This will give you the relative frequencies of all phenotypes distributed across the range of measurable values.

The bar chart in Figure 11.17*c* plots the proportion of students in each category against the range of the measured phenotypes. The vertical bars that are the shortest

Further reading: Student Guide to InfoTrac on web site

1	4	8	10	16	16	15	15	14	13	13	11	9	8	8	5	1	2
Number of individuals																	
60	61	62	63	64	65	66	67	68	69	70	71	72	73	74	75	76	77
Height (inches)																	

a Students organized according to height, as an example of continuous variation

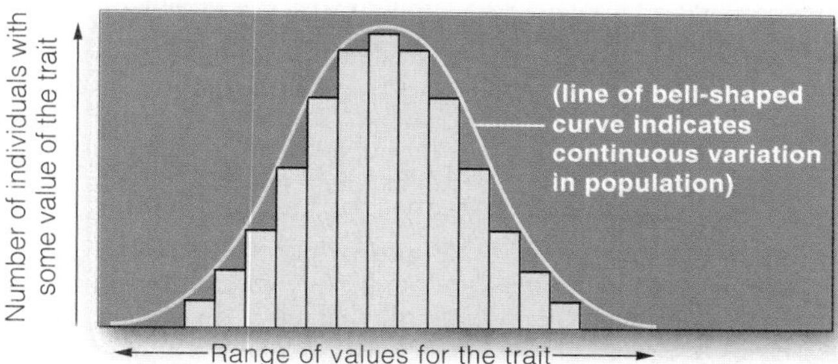
Number of individuals with some value of the trait
(line of bell-shaped curve indicates continuous variation in population)
Range of values for the trait

b Idealized bell-shaped curve for a population that displays continuous variation in some trait

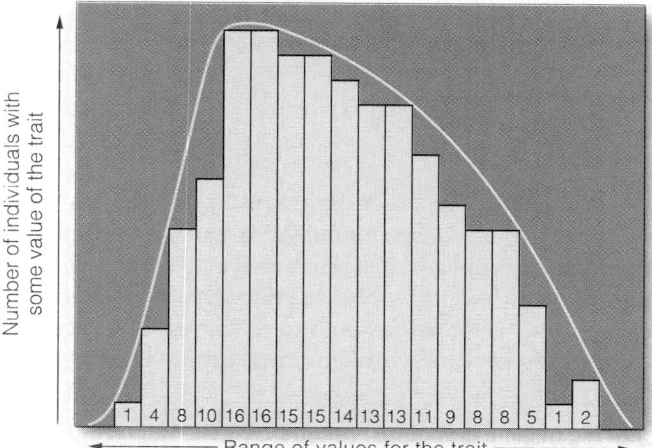
Number of individuals with some value of the trait
1 4 8 10 16 16 15 15 14 13 13 11 9 8 8 5 1 2
Range of values for the trait

c Specific bell-shaped curve corresponding to the distribution of the trait (height) illustrated by the photograph in (**a**)

Figure 11.17 Continuous variation in body height, a trait that is one of the characteristics of the human population.

(**a**) Suppose you want to find the frequency distribution for height in a group of 169 biology students at Brigham Young University. You decide on how finely the range of possible heights should be divided. After this, you measure each student and assign her or him to the appropriate category. Finally, you divide the number in each category by the total number of all students in all categories.

(**b**) Commonly, a bar graph is used to depict continuous variation in a population. In such graphs, the proportion of individuals in each category is plotted against the range of measured phenotypes. The curved line above this particular set of bars is an idealized example of the kind of "bell-shaped" curve that emerges for populations showing continuous variation in a trait. The bell-shaped curve in (**c**) is a specific example of this type of diagram.

represent categories with the least number of students. The bar that is tallest represents the category with the greatest number of students. Finally, draw a graph line around all of the bars and you end up with a "bell-shaped" curve. Such curves are typical of populations that show continuous variation in a trait.

Enzymes and other products of genes regulate each step of most metabolic pathways. Mutations, gene interactions, and environmental conditions may affect one or more of the steps, and this leads to variations in phenotypes.

For most traits, the individuals of a population display continuous variation—that is, a range of small differences.

The greater the number of genes and environmental factors that can influence a trait, the more continuous will be the expected distribution of all the versions of that trait.

11.8 EXAMPLES OF ENVIRONMENTAL EFFECTS ON PHENOTYPE

We have mentioned, in passing, that environmental conditions often contribute to variable gene expression among individuals of a population. Before leaving this chapter, consider just two examples of environmentally induced variations in phenotype.

Possibly you have observed a Himalayan rabbit and probably a Siamese cat. Both of these furry mammals carry an allele that specifies a heat-sensitive version of one of the enzymes necessary for melanin production. At the surface of warm body regions, the enzyme is less active. Fur growing there is lighter in color than fur in cooler regions, which include the ears and other body parts that project away from the main body mass. The experiment shown in Figure 11.18 provided observable evidence of an environmental effect on gene expression.

The environment also influences genes that govern phenotypes of plants. You may have observed the color variation in the floral clusters of *Hydrangea macrophylla*, a species widely favored in home gardens (Figure 11.19). In this type of plant, the action of genes responsible for floral color produces different phenotypes, depending on the acidity of the soil in which the plant is growing.

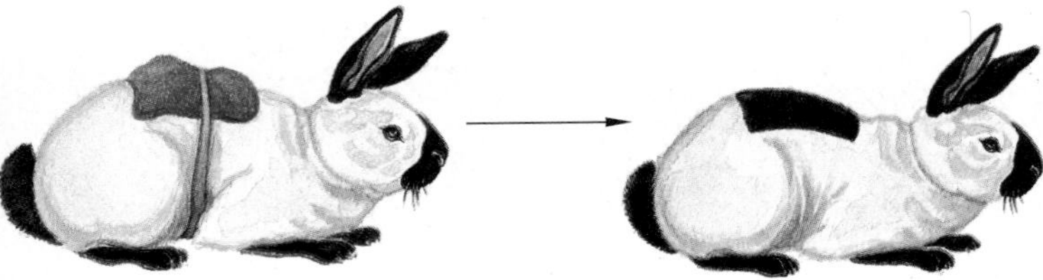
Icepack is strapped to hair-free patch.
New hair growing in patch exposed to cold is black.

Figure 11.18 Observable effect of differences in environmental conditions on gene expression in animals. A Himalayan rabbit normally has black hair only on its long ears, nose, tail, and lower leg limbs. For one experiment, a patch of a rabbit's white fur was plucked clean, then an icepack was secured over the hairless patch. Where the colder temperature had been maintained, the hairs that grew back were black.

Himalayan rabbits are homozygous for the *ch* allele of the gene for tyrosinase, an enzyme required to produce melanin. The allele specifies a heat-sensitive version of the enzyme, which is able to function only when the air temperature is below about 33°C. When cells that give rise to hairs grow under warmer conditions, they cannot produce melanin and hairs appear light. This happens in body regions that are massive enough to conserve a fair amount of metabolic heat. Ears and other slender extremities are cooler because they tend to lose metabolic heat more rapidly.

Figure 11.19 Effect of environmental conditions on gene expression in a favorite garden plant (*Hydrangea macrophylla*). Even plants that carry the same alleles have floral colors ranging from pink to blue, depending on the acidity of the soil in which they happen to be growing.

And so we conclude this chapter, which has dealt with heritable and environmental factors that give rise to variations in phenotype. What is the take-home lesson? Simply this: An individual's phenotype is an outcome of complex interactions among its genes, enzymes and other gene products, and environmental factors.

Owing to gene mutations, cumulative gene interactions, and environmental effects on genes, individuals of a population show degrees of variation for many traits.

Further reading: Student Guide to InfoTrac on web site —

11.9 SUMMARY

1. A gene is a unit of information about a heritable trait. Alleles of a gene are different molecular versions of that information. Through experimental crosses with pea plants, Mendel gathered indirect evidence that diploid organisms have two genes for each trait and that genes retain their identity when transmitted to offspring.

2. *Monohybrid* crosses are experimental crosses between two individuals that are homozygous (true-breeding) for different versions of a trait; offspring are heterozygous (they inherit a pair of nonidentical alleles) for the trait. Mendel's monohybrid crosses with garden pea plants provided indirect evidence that some forms of a gene may be dominant over other, recessive forms.

3. A homozygous dominant individual has inherited two dominant alleles (*AA*) for the trait being studied. A homozygous recessive has two recessive alleles (*aa*), and a heterozygote has two nonidentical alleles (*Aa*).

4. In Mendel's monohybrid crosses ($AA \times aa$), all of the F_1 offspring were *Aa*. Then crosses between F_1 plants resulted in these combinations of alleles in F_2 offspring:

	A	*a*
A	*AA*	*Aa*
a	*Aa*	*aa*

AA (dominant)
Aa (dominant)
Aa (dominant)
aa (recessive)
} the expected phenotypic ratio of 3:1

5. Results from Mendel's monohybrid crosses led to the formulation of a theory of segregation. In modern terms, diploid organisms have pairs of genes, on pairs of homologous chromosomes. The two genes of each pair segregate from each other at meiosis, so that each gamete formed ends up with one or the other gene.

6. *Dihybrid* crosses are experimental crosses between two individuals that breed true (are homozygous) for different versions of two traits; so the F_1 offspring are heterozygous (inherit two nonidentical alleles) for both traits. The phenotypes of the F_2 offspring from Mendel's dihybrid crosses were close to a 9:3:3:1 ratio:

9 dominant for both traits
3 dominant for *A*, recessive for *b*
3 dominant for *B*, recessive for *a*
1 recessive for both traits

7. Mendel's dihybrid crosses led to the formulation of a theory of independent assortment. In modern terms, by the end of meiosis, the gene pairs of two homologous chromosomes have been sorted out for distribution into one gamete or another, independently of how the gene pairs of other chromosomes were sorted out.

8. Four factors commonly influence gene expression:

a. Degrees of dominance may occur between some pairs of genes.

b. The products of pairs of genes may interact to influence the same trait.

c. One gene may have positive or negative effects on two or more traits, a condition called pleiotropy.

d. Environmental conditions to which an individual is subjected may affect gene expression.

Review Questions

1. Distinguish between these terms: *11.1*
   a. gene and allele
   b. dominant allele and recessive allele
   c. homozygote and heterozygote
   d. genotype and phenotype
2. Define a true-breeding lineage. What is a hybrid? *11.1*
3. Distinguish between monohybrid and dihybrid crosses. What is a testcross, and why is it useful in genetic analysis? *11.2, 11.3*
4. Do segregation and independent assortment proceed during mitosis, meiosis, or both? *11.2, 11.3*
5. What do the vertical and horizontal arrows of the diagram at right represent? What do the bars and the curved line represent? *11.7*

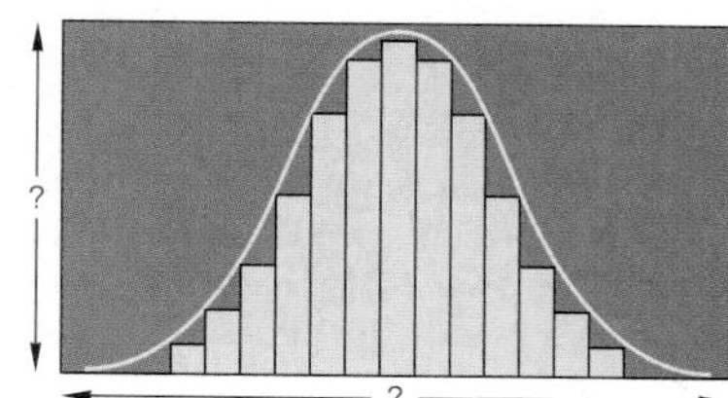
?
?

Self-Quiz (*Answers in Appendix III*)

1. Alleles are ______.
   a. different molecular forms of a gene
   b. different phenotypes
   c. self-fertilizing, true-breeding homozygotes
2. A heterozygote has a ______ for the trait being studied.
   a. pair of identical alleles
   b. pair of nonidentical alleles
   c. haploid condition, in genetic terms
   d. a and c
3. The observable traits of an organism are its ______.
   a. phenotype c. genotype
   b. sociobiology d. pedigree
4. F_1 offspring of the monohybrid cross $AA \times aa$ are ______.
   a. all *AA* c. all *Aa*
   b. all *aa* d. 1/2 *AA* and 1/2 *aa*
5. Second-generation offspring from a cross are the ______.
   a. F_1 generation c. hybrid generation
   b. F_2 generation d. none of the above
6. Assuming complete dominance will occur, the offspring of the cross $Aa \times Aa$ will show a phenotypic ratio of ______.
   a. 3:1 c. 1:2:1
   b. 9:1 d. 9:3:3:1
7. Crosses between F_1 individuals resulting from the cross $AABB \times aabb$ lead to F_2 phenotypic ratios close to ______.
   a. 1:2:1 c. 1:1:1:1
   b. 3:1 d. 9:3:3:1
8. Match each example with the most suitable description.
   ____ dihybrid cross a. *bb*
   ____ monohybrid cross b. $AaBb \times AaBb$
   ____ homozygous condition c. *Aa*
   ____ heterozygous condition d. $Aa \times Aa$

Critical Thinking—Genetics Problems

(Answers in Appendix IV)

1. One gene has alleles *A* and *a*. Another has alleles *B* and *b*. For each genotype listed, what type(s) of gametes will be produced? (Assume independent assortment occurs before gametes form.)

a. *AABB*   c. *Aabb*
b. *AaBB*   d. *AaBb*

2. Still referring to Problem 1, what will be the genotypes of the offspring from the following matings? Indicate the frequencies of each genotype among them.

a. *AABB* × *aaBB*   c. *AaBb* × *aabb*
b. *AaBB* × *AABb*   d. *AaBb* × *AaBb*

3. In one experiment, Mendel crossed a pea plant that bred true for green pods with one that bred true for yellow pods. All the F_1 plants had green pods. Which form of the trait (green or yellow pods) is recessive? Explain how you arrived at your conclusion.

4. Return to Problem 1, and assume you now study a third gene having alleles *C* and *c*. For each genotype listed, what type(s) of gametes will be produced?

a. *AABBCC*   c. *Aa BB Cc*
b. *Aa BB cc*   d. *Aa Bb Cc*

5. Mendel crossed a true-breeding tall, purple-flowered pea plant with a true-breeding dwarf, white-flowered plant. All F_1 plants were tall and had purple flowers. If an F_1 plant self-fertilizes, then what is the probability that a randomly selected F_2 offspring will be heterozygous for the genes specifying height and flower color?

6. At a gene location on a human chromosome, a dominant allele controls *tongue rolling*, an ability to curl up the two sides of the tongue (Figure 11.20). People who are homozygous for a recessive allele at that locus can't roll the tongue. At a different gene locus, a dominant allele controls whether the earlobes will be attached or detached (refer to Figure 11.1). These two pairs of genes assort independently. Suppose a tongue-rolling, detached-earlobed woman marries a man who has attached earlobes and cannot roll his tongue. Their first child has the phenotype of the father. Given this outcome,

a. What are the genotypes of the mother, father, and child?

b. What is the probability that a second child of theirs will have detached earlobes and won't be a tongue roller?

7. Bill and his wife, Marie, are hoping to have children. Both have notably flat feet and long eyelashes, and they tend to sneeze a lot (hence the name of the *achoo syndrome*). Dominant alleles give rise to these traits: *A* (foot arch), *E* (eyelash length), and *S* (chronic sneezing). Bill is heterozygous for all three dominant alleles and Marie is homozygous.

a. What is Bill's genotype? What is Marie's genotype?

b. Marie becomes pregnant four times. What is the probability each child will have the dominant phenotypes for all three traits?

c. What is the probability that each child will have short lashes, high arches, and no chronic tendency to sneeze?

8. *DNA fingerprinting* is a method of identifying individuals by locating unique base sequences in their DNA molecules (Section 16.3). Before researchers refined the method, attorneys often relied on the ABO blood-typing system to settle disputes over paternity. Suppose that you, as a geneticist, are asked to testify during a paternity case in which the mother has type A blood, the child has type O blood, and the alleged father has type B blood. How would you respond to the following statements?

a. Attorney of the alleged father: "The mother's blood is type A, so the child's type O blood must have come from the father. Because my client has type B blood, he could not be the father."

b. Mother's attorney: "Because further tests prove this man is heterozygous, he must be the father."

9. Suppose you identify a new gene in mice. One of its alleles specifies white fur color. A second allele specifies brown fur color. You want to determine whether the relationship between the two alleles is one of simple dominance or incomplete dominance. What sorts of genetic crosses would give you the answer? On what types of observations would you base your conclusions?

10. Your sister moves away and gives you her purebred Labrador retriever, a female named Dandelion. Suppose you decide to breed Dandelion and sell puppies to help pay for your college tuition. Then you discover that two of her four brothers and sisters show *hip dysplasia*, a heritable disorder arising from a number of gene interactions. If Dandelion mates with a male Labrador known to be free of the harmful genes, can you guarantee to a buyer that puppies will not develop the disorder? Explain your answer.

11. A dominant allele *W* confers black fur on guinea pigs. If a guinea pig is homozygous recessive (*ww*), it has white fur. Fred would like to know whether his pet black-furred guinea pig is homozygous dominant (*WW*) or heterozygous (*Ww*). How might he determine his pet's genotype?

12. Red-flowering snapdragons are homozygous for allele R^1. White-flowering snapdragons are homozygous for a different allele (R^2). Heterozygous plants (R^1R^2) bear pink flowers. What phenotypes should appear among F_1 offspring of the crosses listed? What are the expected proportions for each phenotype?

a. $R^1R^1 \times R^1R^2$ c. $R^1R^2 \times R^1R^2$
b. $R^1R^1 \times R^2R^2$ d. $R^1R^2 \times R^2R^2$

Notice, in Problem 12, that in cases of incomplete dominance it is inappropriate to refer to either allele of a pair as dominant or recessive. When the phenotype of a heterozygous individual is halfway between those of the two homozygotes, then there is no dominance. Such alleles are usually designated by superscript numerals, as shown here, rather than by uppercase letters for dominance and lowercase letters for recessiveness.

Figure 11.20 A student at San Diego State University exhibiting the tongue-rolling trait for the benefit of a tongue-roll-challenged student.

13. Two pairs of genes affect comb type in chickens (Figure 11.15). When both genes are recessive, a chicken has a single comb. A dominant allele of one gene, P, gives rise to a pea comb. Yet a dominant allele of the other (R) gives rise to a rose comb. An epistatic interaction occurs when a chicken has at least one of both dominants, P— R —, which gives rise to a walnut comb. Predict the F_1 ratios resulting from a cross between two walnut-combed chickens that are heterozygous for both genes ($PpRr$).

14. As described in Section 11.5, a single mutant allele gives rise to an abnormal form of hemoglobin (Hb^S instead of Hb^A). Homozygotes (Hb^SHb^S) develop sickle-cell anemia. But the heterozygotes (Hb^AHb^S) show few outward symptoms.

Suppose a woman's mother is homozygous for the Hb^A allele and her father is homozygous for the Hb^S allele. She marries a male who is heterozygous for the allele, and they plan to have children. For *each* of her pregnancies, state the probability that this couple will have a child who is:

a. homozygous for the Hb^S allele
b. homozygous for the Hb^A allele
c. heterozygous Hb^AHb^S

15. Certain dominant alleles are so vital for normal development that an individual who is homozygous recessive for a mutant recessive form of the allele cannot survive. Such recessive, *lethal alleles* can be perpetuated by heterozygotes.

Consider the Manx allele (M^L) in cats. Homozygous cats (M^LM^L) die when they are still embryos inside the mother cat. In heterozygotes (M^LM), the spine develops abnormally, and the cats end up with no tail whatsoever (Figure 11.21).

Two M^LM cats mate. Among their *surviving* progeny, what is the probability that any one kitten will be heterozygous?

16. A recessive allele a is responsible for *albinism*, an inability to produce or deposit melanin in tissues. Humans and some other organisms can have this phenotype (Figure 11.22). In each of the following cases, what are the possible genotypes of the father, of the mother, and of their children?

a. Both parents have normal phenotypes; some of their children are albino and others are unaffected.
b. Both parents are albino and have only albino children.
c. The woman is unaffected, the man is albino, and they have one albino child and three unaffected children.

Figure 11.21 Manx cat.

Figure 11.22 An albino male in India.

17. Kernel color in wheat plants is determined by two pairs of genes. Alleles of one pair show incomplete dominance over alleles of the other pair.

For the gene pair at one locus on the chromosome, allele A^1 imparts one dose of red color to the kernel, whereas allele A^2 does not. At the second locus, allele B^1 gives one dose of red color to the kernel, whereas allele B^2 does not. One kernel with genotype $A^1A^1B^1B^1$ is dark red. A different kernel with genotype $A^2A^2B^2B^2$ is white. All other genotypes have kernel colors in between the two extremes.

a. Suppose you cross a plant grown from a dark red kernel with a plant grown from a white kernel. What genotypes and what phenotypes would you expect among the offspring?
b. If a plant with genotype $A^1A^1B^1B^2$ self-fertilizes, what genotypes and what phenotypes would be expected among the offspring? In what proportions?

Selected Key Terms

allele *11.1*
codominance *11.4*
continuous variation *11.7*
dihybrid cross *11.3*
epistasis *11.6*
F_1 *11.1*
F_2 *11.1*
gene *11.1*
genotype *11.1*
heterozygous *11.1*
homozygous dominant *11.1*
homozygous recessive *11.1*
hybrid offspring *11.1*
incomplete dominance *11.4*
independent assortment *11.3*
monohybrid cross *11.2*
multiple allele system *11.4*
phenotype *11.1*
pleiotropy *11.5*
probability *11.2*
Punnett-square method *11.2*
segregation *11.2*
testcross *11.2*
true-breeding lineage *11.1*

Readings *See also www.infotrac-college.com*

Fairbanks, D. J., and W. R. Andersen. 1999. *Genetics: The Continuity of Life*. Monterey, California: Brooks-Cole.

Orel, V. 1996. *Gregor Mendel: The First Geneticist*. New York: Oxford University Press.

12 CHROMOSOMES AND HUMAN GENETICS

The Philadelphia Story

Positioned at strategic locations in chromosomes are genes that work together to bring about orderly cell growth and division. Some specify enzymes and other proteins that perform these tasks, and neighboring genes control when, whether, and how fast the tasks proceed. If something disturbs the neighborhood, cell growth and division can spiral out of control and lead to cancer.

The first abnormal chromosome to be associated with cancer was named the *Philadelphia chromosome* after the city in which it was discovered. It shows up in cells of humans affected by a type of leukemia.

Leukemias arise when stem cells in bone marrow overproduce the white blood cells in charge of the body's housekeeping and defense. (All stem cells are unspecialized and retain the capacity for mitotic cell division. Their descendants also divide, and a portion become specialized cell types.) Leukemic cells infiltrate bone marrow and often crowd out the stem cells that give rise to red blood cells and platelets. Without red blood cells, anemia develops. Without platelets, the body starts to bleed internally. Leukemic cells also infiltrate the blood and organs, including the lymph nodes, spleen, and liver, and skew their functioning.

Untreated patients with acute leukemia often die within weeks or months. Those with chronic leukemia live longer but often die after infections overwhelm their body's weakened defenses. The Philadelphia chromosome is associated with a chronic leukemia.

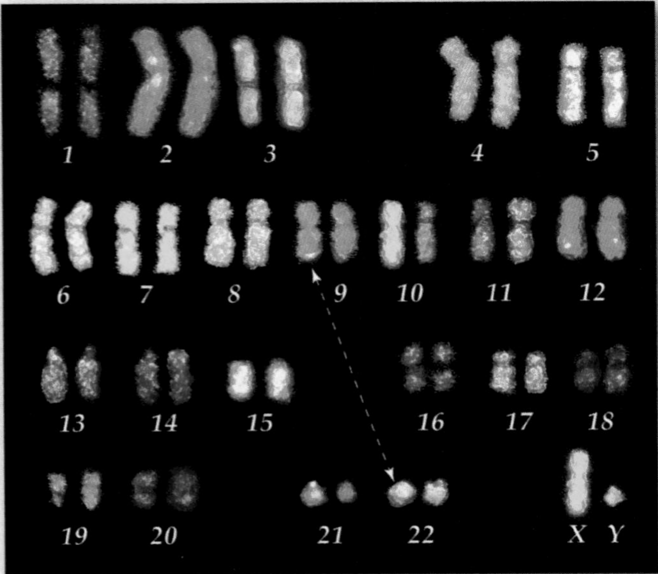
1
2
3
4
5
6
7
8
9
10
11
12
13
14
15
16
17
18
19
20
21
22
X
Y

No one knew about the Philadelphia chromosome until microscopists learned how to identify its physical appearance. Chromosomes, recall, are most highly condensed at metaphase of mitosis. In that state, their size, length, and centromere location are easiest to identify. The chromosomes of many species also have distinct patterns of bands when stained a certain way. A preparation of metaphase chromosomes based on their defining features is a **karyotype**. Section 12.2 shows you how to construct a karyotype diagram from a photograph of metaphase chromosomes, just like microscopists do. The idea is to line up all the chromosomes, from largest to smallest, and position the sex chromosomes last in the lineup.

As it turned out, the Philadelphia chromosome is physically longer than its normal counterpart, human chromosome 9. The extra length is actually a piece of chromosome 22! What happened?

By chance, both chromosomes broke inside a stem cell in bone marrow. Then, in an instance of reciprocal translocation, each broken piece became reattached to the wrong chromosome—and a gene located at the end of chromosome 9 fused with a gene in chromosome 22. The altered gene region specifies an abnormal protein. In some way, that protein stimulates the unrestrained division of white blood cells.

A Philadelphia chromosome is not easy to identify in standard karyotypes. Things should change with *spectral* karyotyping, a newer research and diagnostic tool that artificially colors human chromosomes in an unambiguous way. Figure 12.1 is an example.

You began this unit of the book by looking at cell division, the starting point of inheritance. You thought about how chromosomes—and the genes they carry—are shuffled during meiosis and at fertilization. In this chapter you will delve more deeply into patterns of chromosomal inheritance. As the Philadelphia story suggests, the described methods of analysis are not remote from the world of your interests. An inherited collection of bits of information in chromosomal DNA gives rise to traits that, for better or worse, define each organism, young and old alike (Figure 12.2).

Figure 12.1 Pinpointing a killer—how a reciprocal translocation between human chromosome 9 and chromosome 22 might appear with the help of a new imaging technique called spectral karyotyping. One outcome of this translocation, the Philadelphia chromosome, is a risk factor in chronic myelogenous leukemia, a type of cancer.

Figure 12.2 The DNA icon for an Internet site, *Rare Genetic Diseases in Children*, which allows interested individuals to seek information about specific genetic diseases and disorders. At this writing, the Philadelphia chromosome is on the site's topic board.

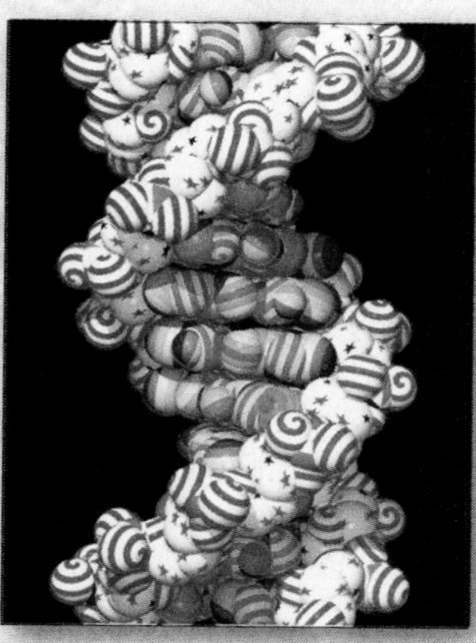

Insights into the chromosomal basis of inheritance started to accumulate after the rediscovery of Gregor Mendel's work. In 1884 Mendel had just passed away, and his paper on pea plants had been gathering dust in a hundred libraries for nearly two decades. Then improvements in the resolving power of microscopes rekindled efforts to find the hereditary material in cells. By 1882, Walther Flemming had seen threadlike bodies—chromosomes—inside the nuclear region of dividing cells. By 1884, a question was taking shape: Could chromosomes be the hereditary material?

Microscopists soon realized each gamete has half the number of chromosomes of a fertilized egg. In 1887, August Weismann hypothesized that a special division process must reduce the chromosome number by half before gametes form. Sure enough, in that same year meiosis was discovered. Weismann began to promote another hypothesis: If a chromosome number is halved during meiosis and restored at fertilization, then the cell's hereditary material must be half paternal and half maternal in origin. This view of heredity was hotly debated, and it prompted a flurry of experimental crosses—just like the ones Mendel had carried out.

Finally, in 1900, researchers came across Mendel's paper while checking literature related to their own genetic crosses. To their surprise, their experimental results confirmed what Mendel's results had already suggested: Diploid cells have two units (genes) for each heritable trait, and the two units are segregated from each other before gametes form.

In the decades that followed, researchers learned a great deal more about chromosomes. Turn now to a few high points of their work, which will be our starting point for exploring human inheritance.

KEY CONCEPTS

1. Cells of many organisms contain pairs of homologous chromosomes, which interact during meiosis. Typically, one chromosome of each pair is maternal in origin, and its homologue is paternal in origin.

2. One gene follows another in sequence along the length of a chromosome. Each kind of gene has its own position, or locus, in that sequence. Different molecular forms of the gene (alleles) may occupy the locus.

3. The combination of alleles in a chromosome does not necessarily remain intact throughout meiosis and gamete formation. During an event called crossing over, some alleles in the sequence swap places with their partner on the homologous chromosome. And alleles that swap places may or may not be identical.

4. Allelic recombinations contribute to variations in the phenotypes of offspring.

5. A chromosome may change structurally, as when a segment of it is deleted, duplicated, inverted, or moved to a new location. Also, the chromosome number of an individual's cells may change as a result of an improper separation of duplicated chromosomes during meiosis or mitosis.

6. Chromosome structure and the parental chromosome number rarely change. When changes do occur, they often give rise to genetic abnormalities or disorders.

12.1 THE CHROMOSOMAL BASIS OF INHERITANCE—AN OVERVIEW

Genes and Their Chromosome Locations

Earlier chapters provided you with a general sense of the structure of chromosomes and of what happens to them during meiosis. To refresh your memory and get a glimpse of where you are going from here with your reading, take a moment to study the following list:

1. **Genes** are units of information about heritable traits. The genes of eukaryotic species are distributed among a number of chromosomes, and each has its own location—a gene locus—in a particular type of chromosome.

2. A cell with a diploid chromosome number ($2n$) has inherited pairs of **homologous chromosomes**. All but one pair of homologous chromosomes are identical in their length, shape, and gene sequence. The single exception is a pairing of nonidentical sex chromosomes, such as X with Y. During meiosis, the two members of a pair of homologous chromosomes interact and then segregate from each other.

3. Compare the gene at one locus on a chromosome with its partner on a homologous chromosome, and you find that their molecular form may be the same or slightly different. Although many different forms of that gene may occur among members of a population, each diploid cell has only a pair of them.

4. The different molecular forms of a gene that are possible at a given locus are called **alleles**. They arise through mutation.

5. A *wild-type* allele is the most common form of a gene, either in a natural population or in standard, laboratory-bred strains of a species. Any other, less common form of a gene is called a *mutant* allele.

6. Genes on the same chromosome are physically linked together. The farther apart two linked genes are, the more vulnerable they are to **crossing over**, an event by which homologous chromosomes exchange corresponding segments. Crossing over results in **genetic recombination**: nonparental combinations of alleles in gametes, then in offspring.

7. Independent assortment (random alignment of each pair of homologous chromosomes during metaphase I of meiosis) results in nonparental combinations of alleles in gametes, then in offspring.

8. Abnormal occurrences during meiosis or mitosis occasionally change the structure of chromosomes and the parental chromosome number.

Autosomes and Sex Chromosomes

In all but one case, a pair of homologous chromosomes are exactly like each other in their length, shape, and gene sequence. Microscopists discovered the exception during the late 1800s. For many organisms, a distinctive chromosome occurs in female *or* male individuals, but not both. For example, a diploid cell of a human male has one **X chromosome** and one **Y chromosome** (XY), and that of a human female has two X chromosomes (XX). This inheritance pattern is common among many organisms, including all mammals and fruit flies.

A different pattern prevails for butterflies, moths, some birds, and certain fishes. For them, inheriting two identical sex chromosomes results in a male. Inheriting two nonidentical sex chromosomes results in a female.

Figure 12.1 shows examples of the human X and Y chromosomes. Notice that they are physically different; one is much shorter than the other. Besides this, they do not carry the same genes. Despite the differences, the two can synapse (become zippered together briefly) in a small region along their length, and this allows them to function as homologues during meiosis.

The human X and Y chromosomes fall in the more general category of **sex chromosomes**. The term refers to distinctive types of chromosomes which, in certain combinations, determine a new individual's sex—that is, whether a male or a female will develop. All other chromosomes in an individual's cells are the same in both sexes; they are called **autosomes**.

Karyotype Analysis

Today, microscopists can routinely analyze the physical appearance of sex chromosomes and autosomes taken from a cell at metaphase of mitosis. A preparation of metaphase chromosomes can be sorted out by their defining features and used to construct a karyotype diagram. You read about a new method of karyotyping in this chapter's introduction. You can use Section 12.2 to walk through a simple procedure for constructing your own karyotype diagrams.

Diploid cells have pairs of genes, on pairs of homologous chromosomes. At each gene locus, the alleles (alternative forms of a gene) may be identical or nonidentical.

As a result of crossing over and other events during meiosis, new combinations of alleles and parental chromosomes end up in offspring. Abnormal events during meiosis or mitosis also can change the structure and number of chromosomes.

Autosomes are the pairs of chromosomes that are the same in males and females of a species. One other pair, the sex chromosomes, govern the sex of a new individual.

Further reading: Student Guide to InfoTrac on web site

Focus on Science

12.2 KARYOTYPING MADE EASY

Karyotype diagrams help answer questions about an individual's chromosomes. Chromosomes are the most condensed and easiest to identify in cells that are going through metaphase of mitosis. Technicians don't count on finding a cell that happens to be dividing in the body when they go looking for it. Instead, they culture cells **in vitro** (literally, "in glass"). They put a sample of cells, usually from blood, into a glass container. The container holds a solution that stimulates cell growth and mitotic cell divisions.

Dividing cells can be arrested at metaphase by adding colchicine to a culture medium. As mentioned in Section 9.5, technicians and researchers use colchicine to block spindle formation. (Colchicine is an extract of autumn crocus and other plants of the genus *Colchicum*.) When a spindle can't form, the sister chromatids of duplicated chromosomes can't move to opposite spindle poles during nuclear division. As you can imagine, by using suitable colchicine concentrations and exposure times, technicians can stockpile metaphase cells, and this increases the chance of finding candidates for karyotype diagrams.

Following colchicine treatment, the culture medium is transferred to the tubes of a centrifuge, a motor-driven spinning device (Figure 12.3*a*). The cells have greater mass and density than the solution bathing them. As a result, the spinning force moves them farthest from the center of rotation, to the bottom of the attached tubes. Separation in response to a spinning force is called centrifugation.

Afterward, the cells are transferred to a saline solution. When immersed in this hypotonic fluid, they swell (by osmosis) and move apart. The metaphase chromosomes thus move apart also. The cells are ready to be mounted on a microscope slide, fixed as by air-drying, and stained.

Chromosomes take up some stains uniformly along their length, and this allows identification of chromosome size and shape. With other staining procedures, horizontal bands show up along the length of the chromosomes of certain species. If researchers direct a ray of ultraviolet light at the chromosomes, the bands will fluoresce. You can see an example of this in Figure 10.3.

In the final steps of karyotype preparation, the chromosomes are photographed through the microscope, and the image is enlarged. Next, the photograph itself is cut apart, one chromosome at a time, and individual cutouts are arranged according to size, shape, and length of arms. Then all the pairs of homologous chromosomes are horizontally aligned by centromeres. Figure 12.3*f* shows an example of a karyotype diagram prepared in this manner.

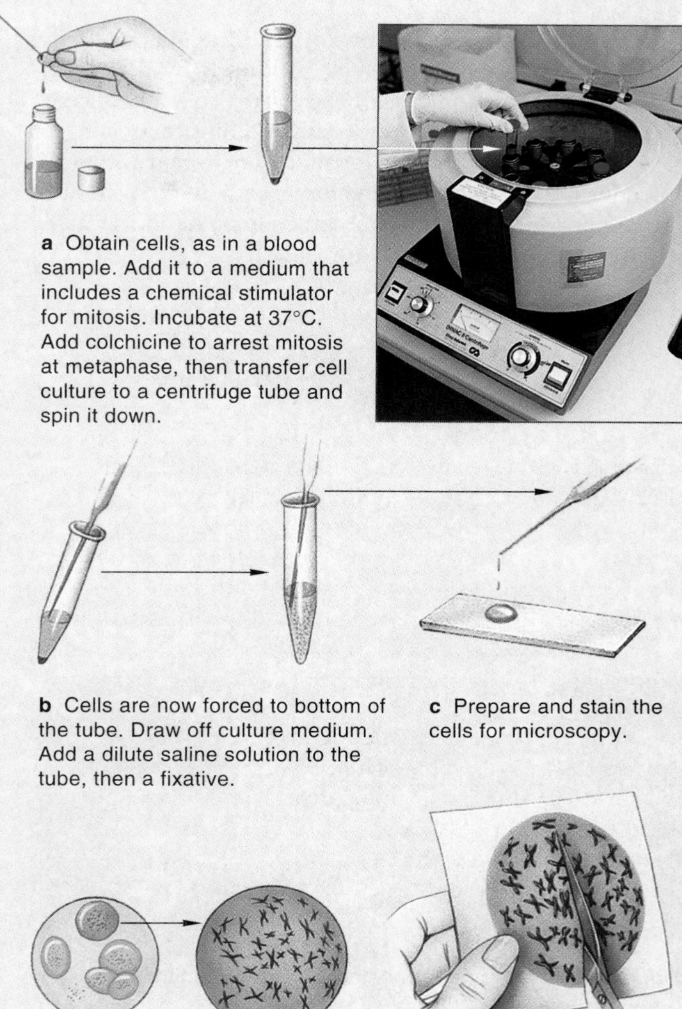

a Obtain cells, as in a blood sample. Add it to a medium that includes a chemical stimulator for mitosis. Incubate at 37°C. Add colchicine to arrest mitosis at metaphase, then transfer cell culture to a centrifuge tube and spin it down.

b Cells are now forced to bottom of the tube. Draw off culture medium. Add a dilute saline solution to the tube, then a fixative.

c Prepare and stain the cells for microscopy.

d Place cells on microscope slide. Observe.

e Photograph one cell through the microscope. Enlarge the image of its chromosomes. Cut the image apart. Arrange chromosomes as a set.

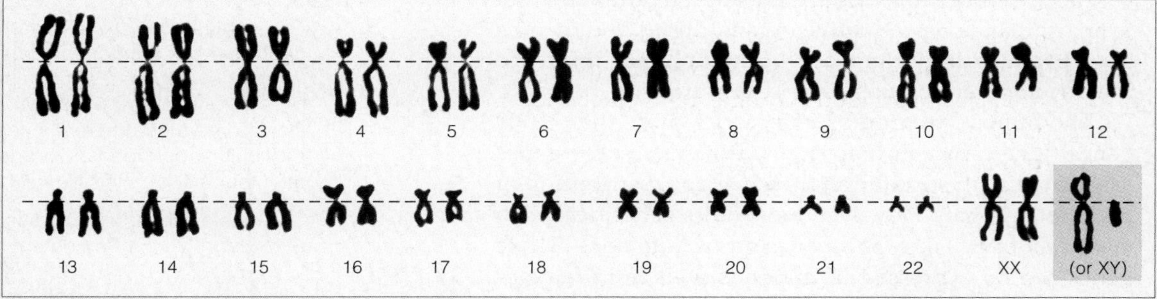
1
2
3
4
5
6
7
8
9
10
11
12
13
14
15
16
17
18
19
20
21
22
XX
(or XY)
f

Figure 12.3 (**a–e**) Karyotype preparation. (**f**) Human karyotype. Human somatic cells have 22 pairs of autosomes and 1 pair of sex chromosomes (XX or XY). That is a diploid number of 46. These are metaphase chromosomes; each is in the duplicated state, with two sister chromatids joined at the centromere.

12.3 SEX DETERMINATION IN HUMANS

Analyzing human cells, as by karyotyping, has yielded evidence that each normal egg produced by a female has one X chromosome. Half the sperm cells produced by a male carry an X chromosome and half carry a Y.

If an X-bearing sperm fertilizes an X-bearing egg, the new individual will develop into a female. If the sperm happens to carry a Y chromosome, the individual will develop into a male (Figure 12.4).

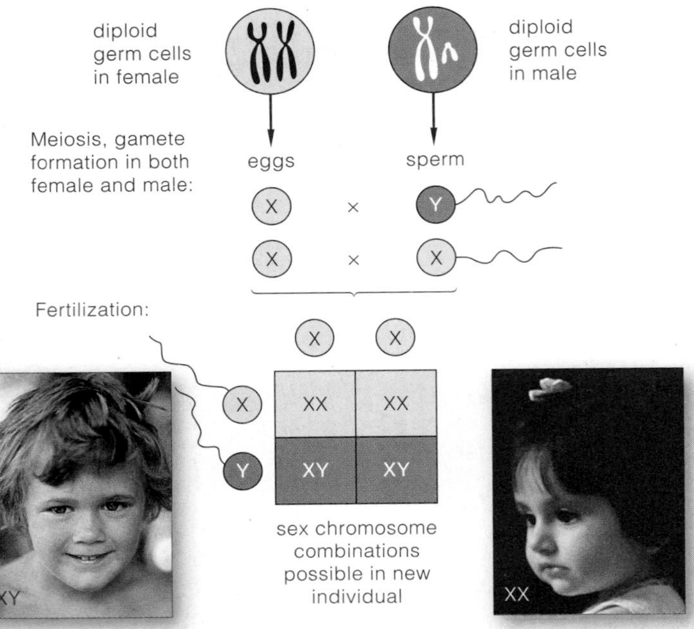
diploid germ cells in female
diploid germ cells in male
Meiosis, gamete formation in both female and male:
eggs
sperm
X
Y
X
X
Fertilization:
X
X
X
Y
XX
XX
XY
XY
sex chromosome combinations possible in new individual
XY
XX

Figure 12.4 Pattern of sex determination in humans.

At this writing, fewer than two dozen genes have been identified on the human Y chromosome. One of them is the master gene for male sex determination. Its expression leads to the formation of testes, the primary male reproductive organs (Figure 12.5). In that gene's absence, ovaries form automatically. Ovaries are the primary female reproductive organs. Testes and ovaries both produce important sex hormones that influence the development of particular sexual traits.

The human X chromosome carries more than 2,300 genes. Like other chromosomes, it carries some genes associated with sexual traits, such as the distribution of body fat and hair. However, most of its genes deal with *nonsexual* traits, such as blood-clotting functions. These genes can be expressed in males as well as in females (males, remember, also carry one X chromosome).

A certain gene on the human Y chromosome dictates that a new individual will develop into a male. In the absence of the Y chromosome (and the gene), a female develops.

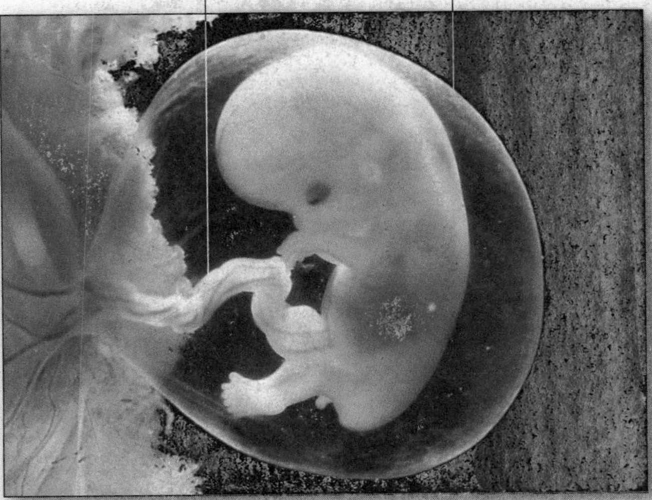
umbilical cord (lifeline between the embryo and the mother's tissues)
amnion (a protective, fluid-filled sac surrounding and cushioning the embryo)

a A human embryo, eight weeks old and about an inch long. The mass at left is part of the placenta, an organ that forms from maternal and embryonic tissues.

Figure 12.5 Boys, girls, and the Y chromosome.

For about the first four weeks of its existence, a human embryo has neither male nor female traits, even though it normally carries either XY or XX chromosomes. Soon enough, however, ducts and other internal structures that can develop either way start forming.

(**a**–**c**) In an XX embryo, the ovaries (primary female reproductive organs) start to form automatically—*in the absence of a Y chromosome.* By contrast, in an XY embryo, the testes (primary male reproductive organs) start to form during the next four to six weeks. Apparently, a gene region on the Y chromosome governs a fork in the developmental road that can lead to maleness.

The newly forming testes start to produce testosterone and other sex hormones. These hormones are crucial for the development of a male reproductive system. By contrast, in an XX embryo, newly forming ovaries start to produce different kinds of sex hormones, particularly estrogens. Estrogens are crucial for the development of a female reproductive system.

The master gene for male sex determination is named *SRY* (short for the *S*ex-determining *R*egion of the *Y* chromosome). The same gene has been identified in DNA from male humans, chimpanzees, mice, rabbits, pigs, horses, cattle, and tigers, among others. None of the females tested had the gene. Tests with mice indicate that the gene region becomes active about the time that testes are starting to develop.

Further reading: Student Guide to InfoTrac on web site

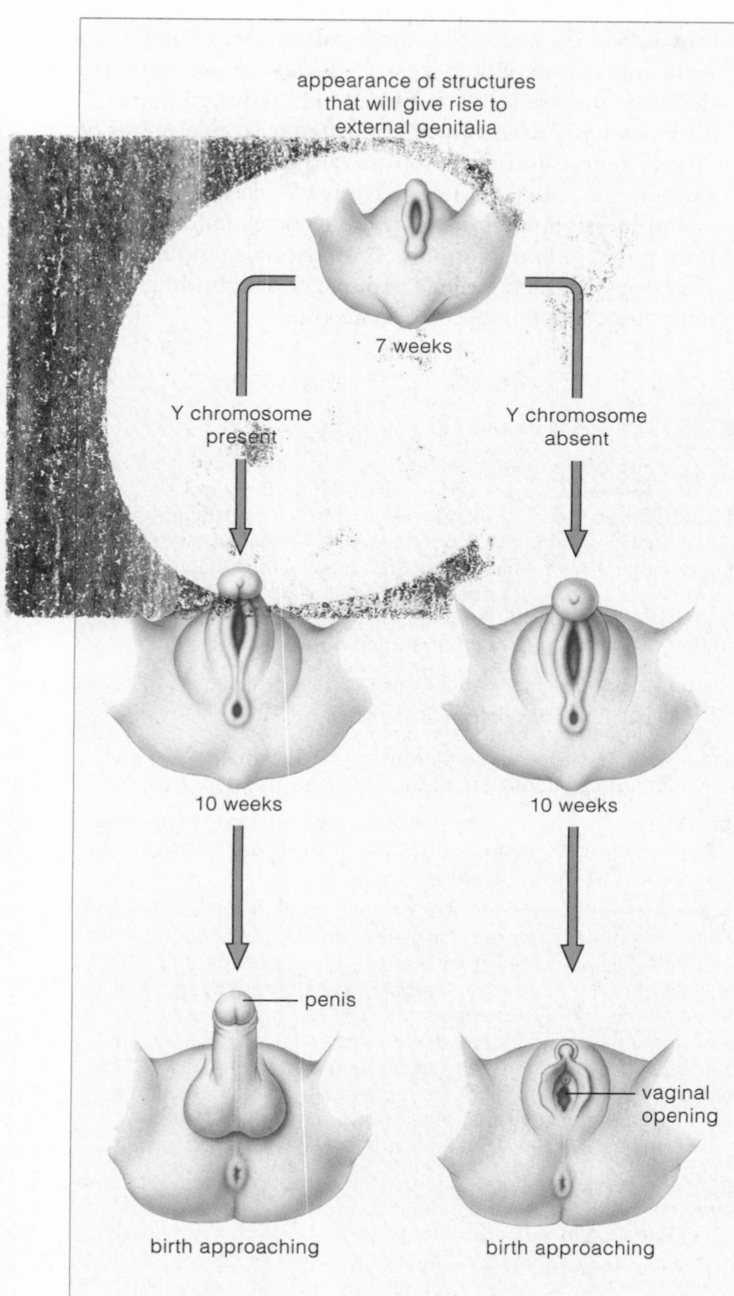
appearance of structures
that will give rise to
external genitalia
7 weeks
Y chromosome
present
Y chromosome
absent
10 weeks
10 weeks
penis
vaginal
opening
birth approaching
birth approaching

b External appearance of the newly forming reproductive organs in human embryos.

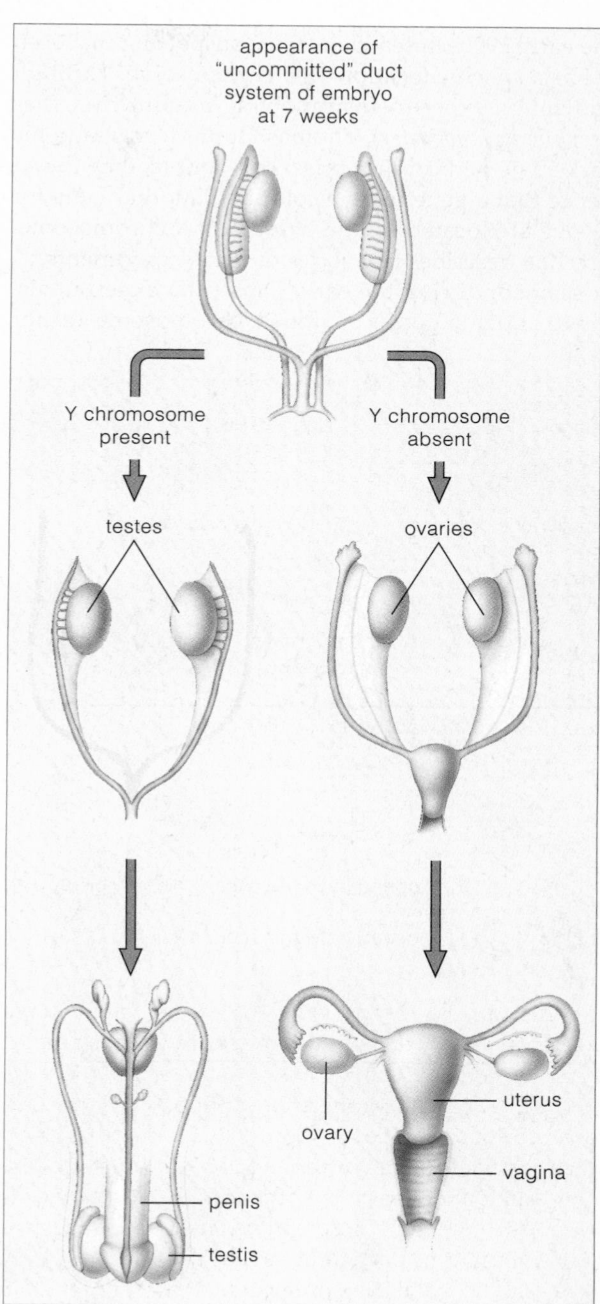
appearance of
"uncommitted" duct
system of embryo
at 7 weeks
Y chromosome
present
Y chromosome
absent
testes
ovaries
penis
testis
ovary
uterus
vagina

c A duct system that forms early on in the human embryo. The same internal system may develop into the male primary reproductive organs *or* the female primary reproductive organs. The presence or absence of the *SRY* gene in the embryo is the major factor that determines which developmental pathway will be followed.

The *SRY* gene resembles DNA regions that specify regulatory proteins. As described in Section 15.1, such proteins bind with certain parts of DNA and thereby turn genes on and off. It seems that the *SRY* gene product regulates a cascade of reactions that are necessary for male sex determination.

12.4 EARLY QUESTIONS ABOUT GENE LOCATIONS

Linked Genes—Clues to Inheritance Patterns

By the early 1900s, researchers were suspecting that each gene has a specific location on a chromosome. Through hybridization experiments involving mutant fruit flies (*Drosophila melanogaster*), Thomas Hunt Morgan and his coworkers helped confirm this. For example, they found evidence that a gene for eye color and another gene for wing size are located on the *Drosophila* X chromosome. Figure 12.6 describes one series of their experiments.

It seemed, during the early *Drosophila* experiments, that two mutant genes on the X chromosome (*w* for white eyes and *m* for miniature wings) were "linked." That is, they traveled together during meiosis, so that they ended up inside the same gamete. For a time, they were called "sex-linked genes." Today, researchers use the more precise terms **X-linked** and **Y-linked genes**.

Eventually, researchers identified a large number of linked genes, and those on each type of chromosome came to be called a **linkage group**. *D. melanogaster*, for example, has four linkage groups, corresponding to its four pairs of homologous chromosomes. Indian corn (*Zea mays*) has ten linkage groups, corresponding to its ten pairs of homologous chromosomes.

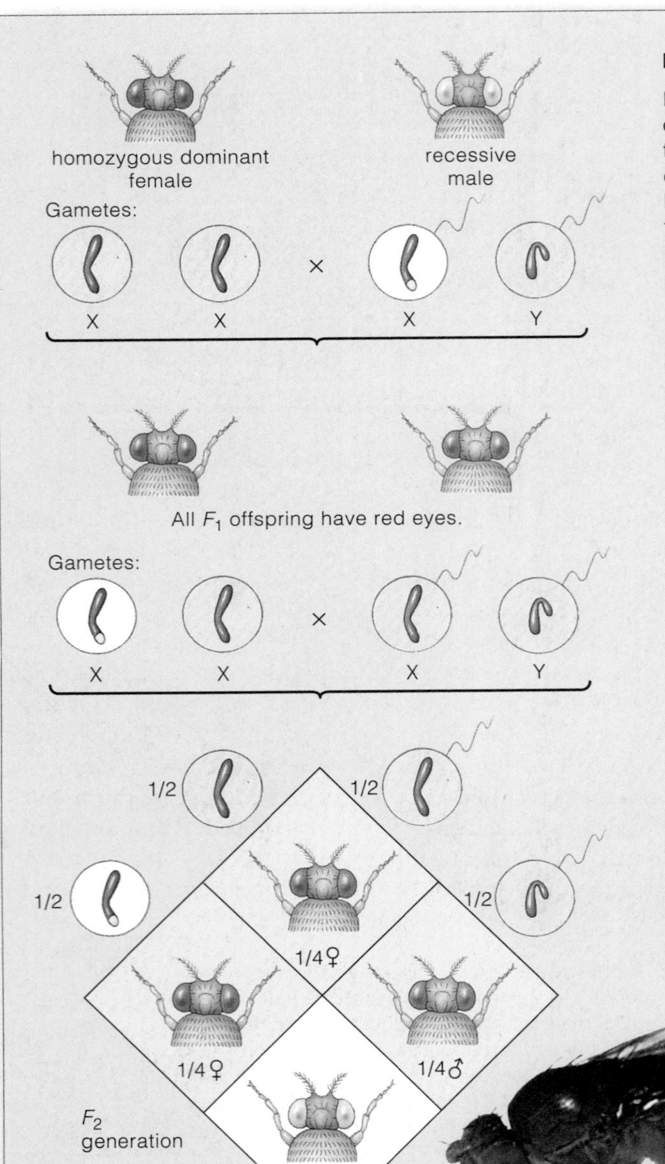

homozygous dominant female
recessive male
Gametes:
X
X
×
X
Y
All F_1 offspring have red eyes.
Gametes:
X
X
×
X
Y
1/2
1/2
1/2
1/2
1/4♀
1/4♀
1/4♂
1/4♂
F_2 generation

Figure 12.6 X-linked genes as clues to inheritance patterns.

In the early 1900s, the embryologist Thomas Morgan was studying patterns of inheritance. He and his coworkers discovered an apparent genetic basis for connections between sex determination and some nonsexual traits. For example, human males and females both have blood-clotting factors. Yet blood-clotting disorders (hemophilias) show up most often in males, not females, of a family lineage. Sex-specific outcomes were not like anything Mendel saw in his hybrid crosses of pea plants. With respect to phenotype, it made no difference which parent plant carried a recessive allele.

Morgan studied eye color and other nonsexual traits of fruit flies (*Drosophila melanogaster*). The flies can live in bottles on cornmeal, molasses, agar, and yeast. A female lays hundreds of eggs in a few days, and the offspring can reproduce in less than two weeks. So Morgan could track hereditary traits through nearly thirty generations of thousands of flies in a year's time.

At first, all flies were wild type for eye color; they had brick-red eyes. Then, as a result of an apparent mutation in a gene controlling eye color, a white-eyed male appeared in one of the bottles.

Morgan established true-breeding strains of white-eyed males and females. Then he performed paired, **reciprocal crosses**. (In the first of such paired crosses, one parent displays the trait of interest. In the second, the other parent displays it.) For the first cross, Morgan allowed white-eyed males to mate with homozygous red-eyed females. All F_1 offspring had red eyes. Of the F_2 offspring, however, only some of the males had white eyes. In the second cross, white-eyed females were mated with true-breeding red-eyed males. Half of the F_1 offspring were red-eyed females and half were white-eyed males. And of the F_2 offspring, 1/4 were red-eyed females, 1/4 white-eyed females, 1/4 red-eyed males, and 1/4 white-eyed males!

The seemingly odd results implied a relationship between an eye-color gene and sex determination. Probably the gene locus was on a sex chromosome. But which one? Because females (XX) could be white-eyed, the recessive allele would have to be on one of their X chromosomes. Suppose white-eyed males (XY) also carry the recessive allele on their X chromosome. Suppose there is no corresponding eye-color allele on the Y chromosome. If that were so, then the males would have white eyes, for they have no dominant allele to mask the effect of the recessive one.

The diagram at left illustrates the expected results when Morgan's idea of an X-linked gene is combined with Mendel's concept of segregation. By proposing that one particular gene is located on an X chromosome but not on the Y, Morgan was able to explain the outcome of his reciprocal crosses. His experimental results matched the predicted outcomes.

Further reading: Student Guide to InfoTrac on web site

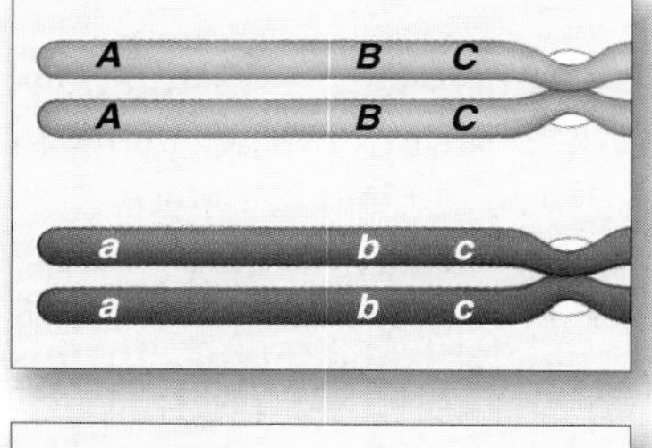
A
B
C
A
B
C
a
b
c
a
b
c

a One pair of homologous chromosomes in the duplicated state (each consists of two sister chromatids). One is shaded *blue*, the other *purple*. Three different genes are shown. Alleles at all three gene loci are nonidentical (*A* with *a*, *B* with *b*, and *C* with *c*).

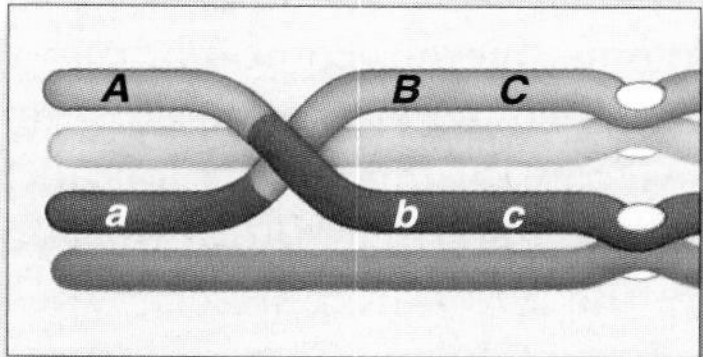
A
B
C
a
b
c

b In prophase I of meiosis, two nonsister chromatids exchange segments. The exchange represents one crossover event.

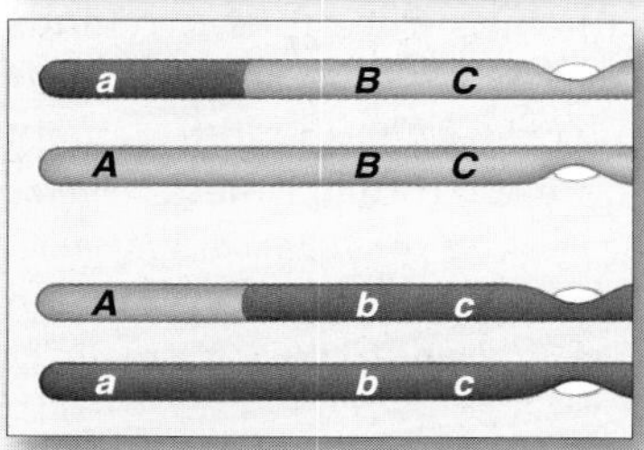
a
B
C
A
B
C
A
b
c
a
b
c

c This is the outcome of the crossover: genetic recombination between nonsister chromatids (which are shown after meiosis, as unduplicated, separate chromosomes).

Figure 12.7 Review of crossing over. As explained earlier in Section 10.4, this event occurs in prophase I of meiosis.

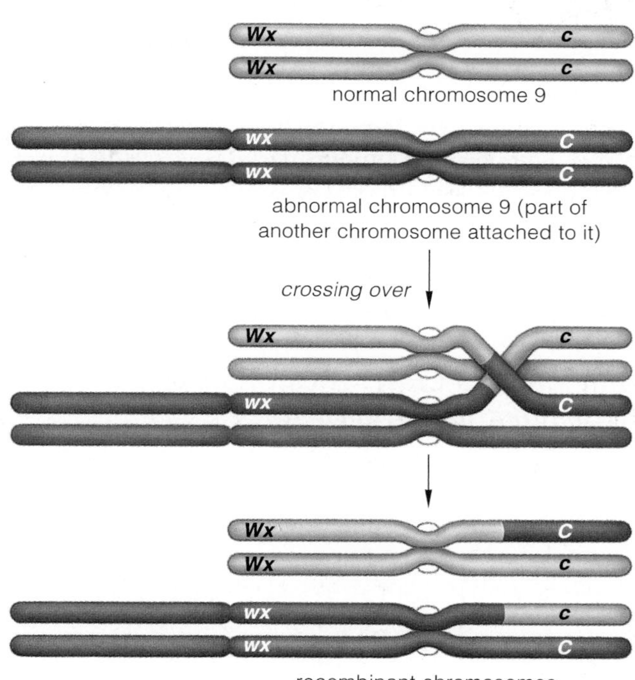
Wx
c
Wx
c
normal chromosome 9
wx
C
wx
C
abnormal chromosome 9 (part of another chromosome attached to it)
crossing over
Wx
c
wx
C
Wx
C
Wx
c
wx
c
wx
C
recombinant chromosomes

Figure 12.8 Harriet Creighton and Barbara McClintock's correlation of a cytological difference with a genetic difference in chromosome 9 from a strain of Indian corn (*Zea mays*).

Crossing Over and Genetic Recombination

If linked genes always stayed together through meiosis, then a dihybrid cross between true-breeding parents should always have a predictable outcome. Specifically, the most frequent phenotypes among the F_2 offspring should be those of the original parents. (Here you may wish to review Section 10.4 and Figure 12.7.) However, a number of puzzling results from the early *Drosophila* experiments did not match this expectation.

For example, Morgan crossed a *Drosophila* female that was recessive for white eyes and miniature wings with a wild-type male (red eyes and long wings). As expected, all F_1 males had white eyes and miniature wings, and all females were wild type. After crossing F_1 flies, Morgan analyzed 2,441 of the F_2 offspring. A significant number had white eyes and long wings or red eyes and miniature wings! As you will see in the next section, 900 (or 36.9 percent) were recombinants. As Morgan hypothesized, physical exchanges must have occurred between X chromosomes during meiosis.

It was not until 1931 that two genetic researchers, Harriet Creighton and Barbara McClintock, discovered evidence of such exchanges. They were experimenting with two *Z. mays* chromosomes that differed physically in a way that could be distinguished with a microscope. Such distinguishable features are **cytological markers** for the genes being studied.

The researchers used corn that was heterozygous for two genes on chromosome 9. Two alleles at one gene locus specified colored (C) or colorless (*c*) seeds. One allele at the other gene locus specified the synthesis of two forms of starch (*Wx*); its partner specified only one (*wx*). In one experiment, one of the chromosomes 9 had genotype *cWx* and a normal appearance (Figure 12.8). Another chromosome 9 had genotype *Cwx*, and it was longer; an abnormal event caused a piece of a different chromosome to become attached to one of its ends.

During meiosis, crossing over sometimes occurred between the two gene loci on a pair of the cytologically different chromosomes. The outcome was two kinds of genetic recombinants: *CWx* and *cwx*. As Creighton and McClintock realized, when genetic recombination had occurred, cytological features of the chromosomes also had changed. When *wx* ended up with *c* instead of *C*, so did the extra length. They observed no such correlation in F_1 generations that retained the parental genotype.

Genes at different loci on the same chromosome belong to the same linkage group and do not assort independently during meiosis.

Crossing over between homologous chromosomes disrupts gene linkages and results in nonparental combinations of genes in chromosomes. Correlations between specific genes and cytological markers provide evidence of recombination.

12.5 RECOMBINATION PATTERNS AND CHROMOSOME MAPPING

As we now know, crossing over is not a rare event. In fact, for humans and most other eukaryotic species, meiosis cannot even be completed properly unless each pair of homologous chromosomes takes part in at least one crossover. The preceding section briefly described experimental evidence of this remarkable event. Let us now consider a few specific examples of the ways in which crossing over disrupts gene linkages and what researchers do with this information.

How Close Is Close? A Question of Recombination Frequencies

Figure 12.9 shows Morgan's cross between a wild-type *Drosophila* male and a female that was recessive for white eyes and for miniature wings. Again, the two parental genotypes showed up equally among the F_1 offspring (50 percent white-eyed, miniature-winged males and 50 percent wild-type females). Of the 2,441 F_2 offspring analyzed, 36.9 percent were recombinants.

Morgan's group also extended their investigations to other X-linked genes. In one series of experiments, a true-breeding white-eyed, yellow-bodied mutant female was crossed with a wild-type male (with red eyes and a gray body). As expected, 50 percent of the F_1 offspring showed one or the other parental phenotype. In this case, however, only 129 of 2,205 of the F_2 offspring that were analyzed—or 1.3 percent—were recombinants.

From the results of many such experimental crosses, it appeared that the genes controlling eye color and wing size were not as "tightly linked" as those for eye color and body color. Also, the two parental genotypes were represented in approximately equal numbers of the F_2 offspring, and the same was true of the two kinds of recombinant genotypes. What was going on here? As Morgan hypothesized, certain alleles tend to remain together during meiosis more often than others *because they are positioned closer together on the same chromosome.*

Imagine any two genes at two different locations on the same chromosome. *The probability that a crossover*

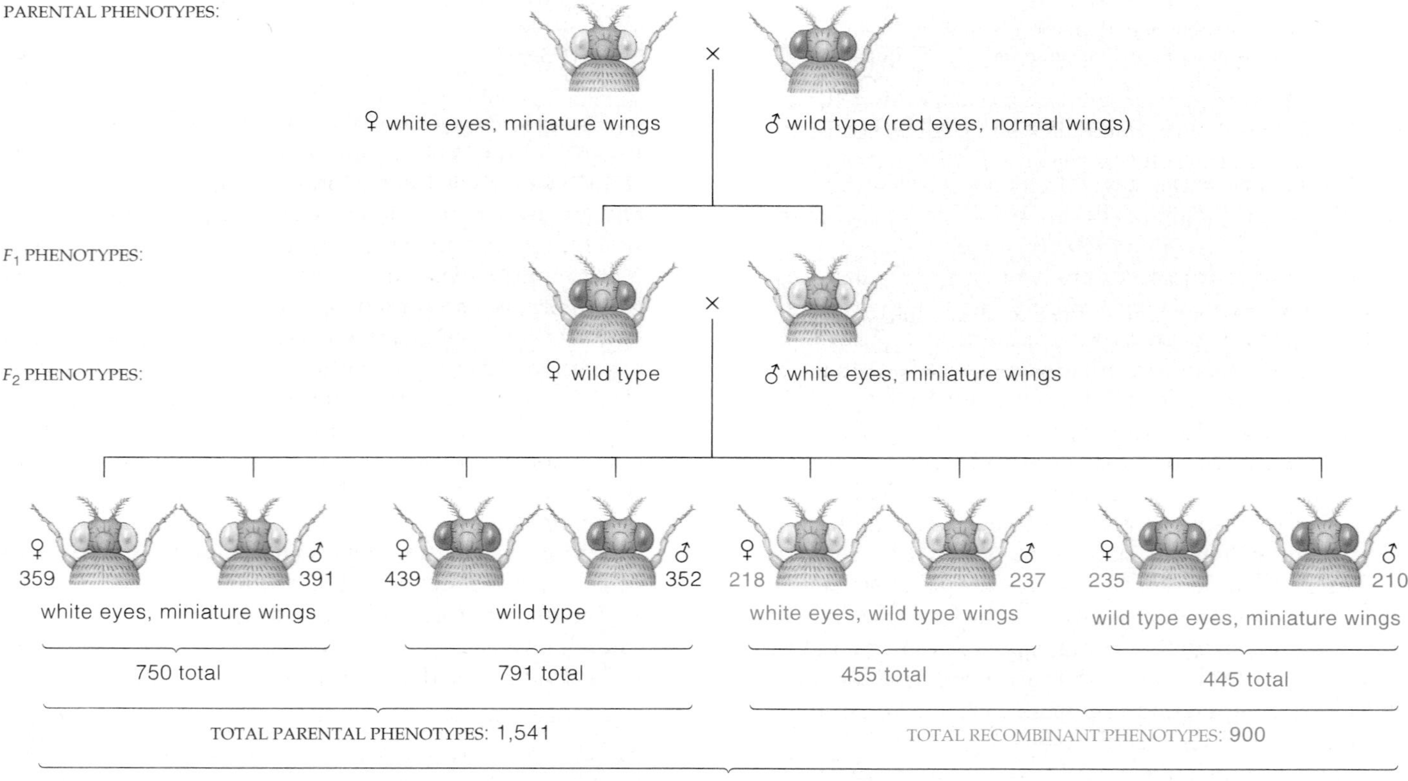

PARENTAL PHENOTYPES:
♀ white eyes, miniature wings
♂ wild type (red eyes, normal wings)
F_1 PHENOTYPES:
F_2 PHENOTYPES:
♀ wild type
♂ white eyes, miniature wings
♀ 359
♂ 391
white eyes, miniature wings
750 total
♀ 439
♂ 352
wild type
791 total
♀ 218
♂ 237
white eyes, wild type wings
455 total
♀ 235
♂ 210
wild type eyes, miniature wings
445 total
TOTAL PARENTAL PHENOTYPES: 1,541
TOTAL RECOMBINANT PHENOTYPES: 900
Total Offspring: 1,541 + 900 = 2,441
PERCENT RECOMBINANTS: 900/2,441 × 100 = 36.9%

Figure 12.9 Experimental crosses with *Drosophila melanogaster* that provided indirect evidence of how crossing over can disrupt linkage groups. In this example, Morgan's group tracked a mutant gene for white eyes and a different mutant gene for miniature wings. As you probably know, the symbol ♀ signifies female and the symbol ♂ signifies male.

Further reading: Student Guide to InfoTrac on web site —

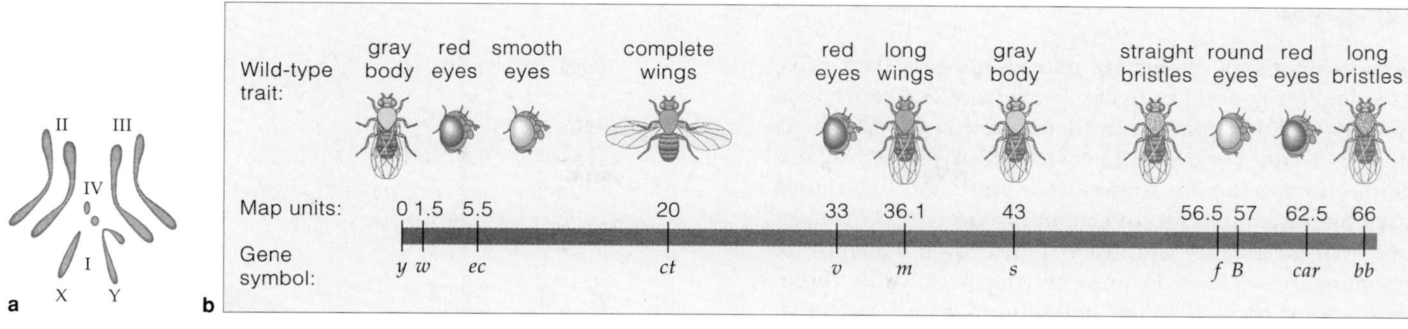
II
III
IV
I
X
Y
a
b
Wild-type trait:
gray body
red eyes
smooth eyes
complete wings
red eyes
long wings
gray body
straight bristles
round eyes
red eyes
long bristles
Map units:
0 1.5 5.5
20
33
36.1
43
56.5 57
62.5
66
Gene symbol:
y w
ec
ct
v
m
s
f B
car
bb

Figure 12.10 (**a**) *Drosophila melanogaster* chromosomes. (**b**) A linkage map for a number of genes on chromosome I (here, the X chromosome). As you can see, more than one gene can influence the same trait, such as eye color.

will disrupt their linkage is proportional to the distance that separates the two loci. Suppose genes *A* and *B* are twice as far apart as two other genes, *C* and *D*:

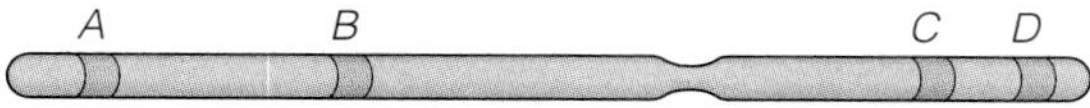
A
B
C
D

We would expect crossing over to disrupt the linkage between *A* and *B* much more often.

Two genes are very closely linked when the distance between their loci is small; their allelic combinations nearly always end up in the same gamete. Linkage is more vulnerable to crossover when the distance between the two loci is greater. When two gene loci are very far apart, crossing over disrupts linkage so often that those genes assort independently of each other into gametes.

Linkage Mapping

After using testcrosses to analyze crossover patterns, Alfred Sturtevant, one of Morgan's students, proposed that the percentage of recombinants in gametes might be a quantitative measure of the relative positions of genes along a chromosome. This investigative approach is now called **linkage mapping**.

The relative linear distance between any two linked genes on a genetic map is expressed in map units. One map unit corresponds to a crossover frequency of 1 percent, twenty map units correspond to a crossover frequency of 20 percent, and so on. For instance, take a look at Figure 12.10*b*, which is a linkage map for one of the *D. melanogaster* chromosomes. The amount of gene recombination to be expected between "smooth eyes" and "long wings" would be 30.6 percent (5.5 map units subtracted from 36.1 map units).

Linkage maps do not show *actual* physical distances between gene loci. The most accurate approximations have been calculated only for closely linked genes. Why? As map distance increases, multiple crossovers occur and skew recombination frequencies. Even so, the *map distance* between two gene loci as estimated by genetic crosses generally correlates with the *physical distance*, or length of DNA between them, as calculated by other methods. (Exceptions to this generalization occur in the centromere region.)

Of the several thousand known genes in the four types of *Drosophila* chromosomes, the positions of about a thousand have been mapped. What about the 50,000 to 100,000 genes on the twenty-three types of human chromosomes? Unlike fruit flies, humans do not lend themselves to experimental crosses. Nevertheless, some tight gene linkages have been identified by tracking the resulting phenotypes, one generation after another, in certain families.

For example, recessive alleles at two gene loci on the X chromosome cause *color blindness* and *hemophilia* (to be described shortly). One female carried both alleles, but she herself was symptom-free. So was her father, so she must have inherited a normal X chromosome from him (remember, males have only one X and one Y). The X chromosome that she inherited from her mother must have carried both mutant alleles. The female gave birth to six sons. Three sons developed color blindness and hemophilia; two were unaffected. Here was phenotypic evidence that recombination had not occurred. But her sixth son started life as a fertilized, recombinant egg; he was color-blind only. Thus, for the two mutant alleles in this family, the recombination frequency is 1/6 (or 0.167 percent).

Many affected families would have to be examined to get a good estimate of the map distance between the two gene loci. Generally speaking, these genes do not have high recombination frequencies because they are very closely linked at one end of the X chromosome.

The farther apart two genes are on a chromosome, the greater will be the frequency of crossing over and therefore of genetic recombination between them.

Linkage mapping is a method of measuring the relative linear distances between genes on the same chromosome. Such maps roughly correspond to actual physical distances, or lengths of DNA, as calculated by other methods.

12.6 HUMAN GENETIC ANALYSIS

Some organisms, including pea plants and fruit flies, are ideal for genetic analysis. They grow and reproduce rapidly in small spaces, under controlled conditions. It does not take very long to track a trait through many generations. Humans are another story. We live under variable conditions in diverse environments. We select our own mates and reproduce if and when we want to. Humans live as long as the geneticists who study them, so tracking traits through generations can be tedious. Most human families are so small, there are not enough offspring for easy inferences about inheritance.

Constructing Pedigrees

To get around the problems associated with analyzing human inheritance, geneticists put together pedigrees. A **pedigree** is a chart that shows genetic connections among individuals. When genetic researchers construct them, they use standardized methods and definitions, as well as standardized symbols to represent individuals, as shown in Figure 12.11*a*.

When they analyze pedigrees, geneticists rely on their knowledge of probability and Mendelian inheritance patterns, which may yield clues to the genetic basis for a trait. For example, clues might suggest that an allele responsible for a disorder is dominant or recessive or that it is located on a particular autosome or sex chromosome.

Gathering a great many family pedigrees increases the numerical base for analysis. When any trait follows a simple Mendelian inheritance pattern, a geneticist has greater confidence in predicting the probability of its occurrence among children of prospective parents. We will return to this topic later.

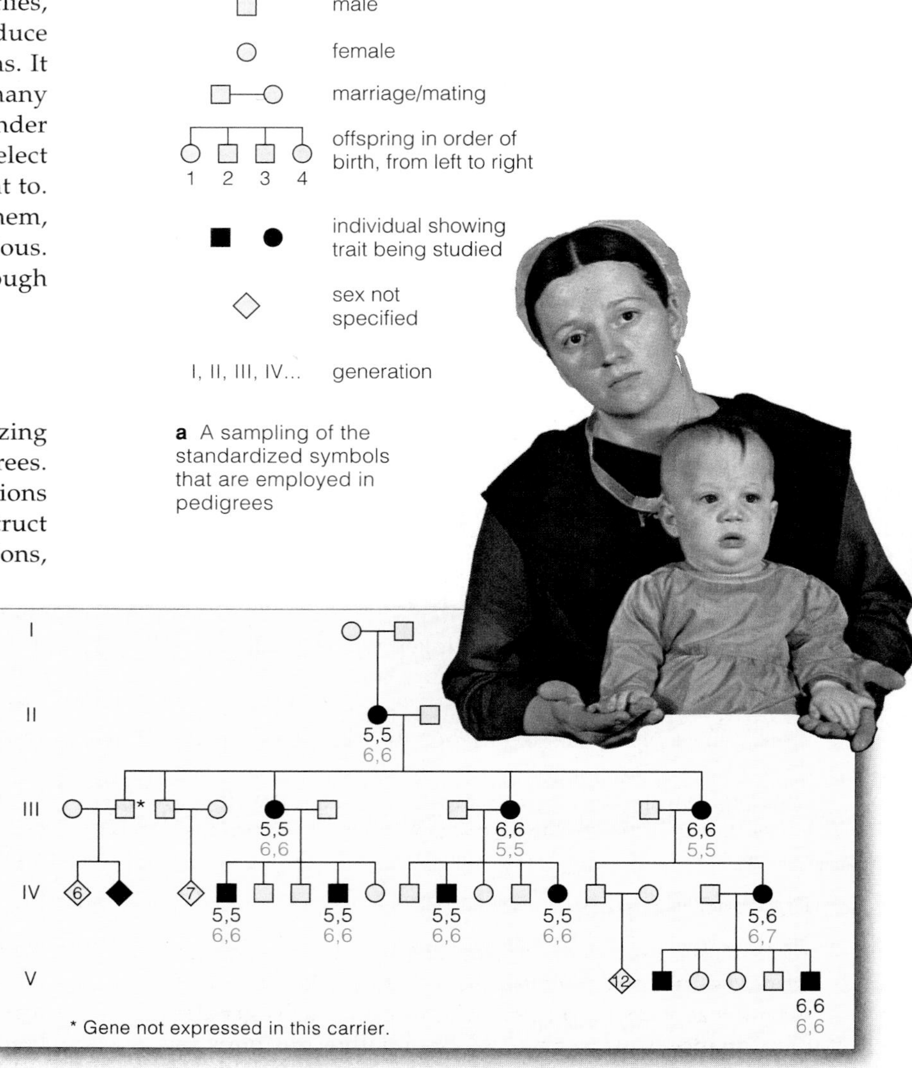
male
female
marriage/mating
offspring in order of birth, from left to right
1 2 3 4
individual showing trait being studied
sex not specified
I, II, III, IV...
generation
I
II
III
IV
V
5,5
6,6
*
5,5
6,6
6,6
5,5
6,6
5,5
6
7
5,5
6,6
5,5
6,6
5,5
6,6
5,5
6,6
5,6
6,7
12
6,6
6,6
* Gene not expressed in this carrier.

a A sampling of the standardized symbols that are employed in pedigrees

b Pedigree for a family in which polydactyly recurs

c Nancy Wexler and a pedigree she constructed to track Huntington disorder

Figure 12.11 (**a**) Some standardized symbols used in constructing pedigree diagrams. (**b**) Example of a pedigree for *polydactyly*, a condition in which an individual has extra fingers, extra toes, or both. Expression of the gene for this trait can vary from one individual to the next. *Black* numerals signify the number of fingers on each hand where data were available; *blue* numerals signify the number of toes on each foot.

(**c**) From human genetic researcher Nancy Wexler, a pedigree for *Huntington disorder*, by which the human nervous system progressively degenerates. Wexler and her team constructed an extended family tree for nearly 10,000 people in Venezuela. As analysis of affected and unaffected individuals revealed, a dominant allele on human chromosome 4 is the genetic culprit. Wexler has a special interest in Huntington disorder; she herself has a 50 percent chance of developing it.

Further reading: Student Guide to InfoTrac on web site

Table 12.1 Examples of Human Genetic Disorders and Genetic Abnormalities

Disorder or Abnormality*	Main Consequences	Disorder or Abnormality*	Main Consequences
AUTOSOMAL RECESSIVE INHERITANCE		**X-LINKED RECESSIVE INHERITANCE**	
Albinism *11.6; 11.9 CT*	Absence of pigmentation	Color blindness *12.7*	Inability to distinguish among some or all colors
Blue offspring *18.12 CT*	Bright blue skin coloration	Fragile X syndrome *12.7*	Mental retardation
Cystic fibrosis *4.12 CT*	Excessive glandular secretions leading to tissue, organ damage	Hemophilia *12.7; 12.12 CT*	Impaired blood-clotting ability
Ellis–van Creveld syndrome *18.11*	Extra fingers, toes, short limbs	Muscular dystrophies *12.12 CT; 15.7 CT*	Progressive loss of muscle function
Galactosemia *12.7*	Brain, liver, eye damage	Testicular feminization syndrome *37.2*	XY individual but having some female traits; sterility
Phenylketonuria (PKU) *12.11*	Mental retardation	X-linked anhidrotic dysplasia *15.4*	Mosaic skin (patches with or without sweat glands); other effects
Sickle-cell anemia *11.5; 14.1; 14.5; 18.6*	Adverse pleiotropic effects on organs throughout body	**CHANGES IN CHROMOSOME NUMBER**	
AUTOSOMAL DOMINANT INHERITANCE		Down syndrome *12.10; 45.4*	Mental retardation; heart defects
Achondroplasia *12.7*	One form of dwarfism	Klinefelter syndrome *12.10*	Sterility; retardation
Achoo syndrome *11.9 CT*	Chronic sneezing	Turner syndrome *12.10*	Sterility; abnormal ovaries, abnormal sexual traits
Camptodactyly *11.7*	Rigid, bent little fingers	XYY condition *12.10*	Mild retardation or free of symptoms
Familial cholesterolemia *16 CI*	High cholesterol levels in blood; eventually clogged arteries	**CHANGES IN CHROMOSOME STRUCTURE**	
Huntington disorder *12.6; 12.7*	Nervous system degenerates progressively, irreversibly	Chronic myelogenous leukemia *12 CI*	Overproduction of white blood cells in bone marrow; organ malfunctions
Marfan syndrome *12.12 CT*	Abnormal or no connective tissue	Cri-du-chat syndrome *12.9*	Mental retardation; abnormally shaped larynx
Polydactyly *12.6*	Extra fingers, toes, or both		
Progeria *12.8*	Drastic premature aging		
Neurofibromatosis *14.7*	Tumors of nervous system, skin		

* *Italic* numbers indicate sections in which a disorder is described. *CI* signifies *C*hapter *I*ntroduction. *CT* signifies an end-of-chapter *C*ritical *T*hinking question.

Regarding Human Genetic Disorders

Table 12.1 is a list of some heritable traits that have been studied in detail. A few are abnormalities, or deviations from the average condition. Said another way, a **genetic abnormality** is nothing more than a rare, uncommon version of a trait, as when a person is born with six toes on each foot instead of five. Whether an individual or society at large views an abnormal trait as disfiguring or merely interesting is subjective. As the classic novel *The Hunchback of Notre Dame* suggests, there is nothing inherently life-threatening or even ugly about it.

By comparison, a **genetic disorder** is an inherited condition that sooner or later will cause mild to severe medical problems. A **syndrome** is a recognized set of symptoms that characterize a given disorder.

Because alleles underlying severe genetic disorders put people at great risk, they are rare in populations. Why, then, don't they disappear entirely? There are two reasons. First, rare mutations introduce new copies of the alleles into the population. Second, in heterozygotes, the harmful allele is paired with a normal one that may cover its functions, so it still can be passed to offspring.

You may hear someone refer to a genetic disorder as a disease, but the terms are not always interchangeable. A disease also is an abnormal alteration in the way the body functions, and it, too, is characterized by a set of symptoms. But **disease** is illness caused by infectious, dietary, or environmental factors, not by inheritance of mutant genes. If an individual's previously workable genes get altered in a way that disrupts body functions, the illness might be called a *genetic* disease.

With these qualifications in mind, we turn next to examples of inheritance in the human population. As you will see, genetic analyses of family pedigrees have often revealed simple Mendelian inheritance patterns for certain traits. Researchers have traced many of the traits to dominant or recessive alleles on an autosome or X chromosome. They have traced others to changes in the structure or number of chromosomes.

For many genes, pedigree analysis might reveal simple Mendelian inheritance patterns that will allow inferences about the probability of their transmission to children.

A genetic abnormality is a rare or less common version of an inherited trait. A genetic disorder is an inherited condition that results in mild to severe medical problems.

12.7 INHERITANCE PATTERNS

Autosomal Recessive Inheritance

For some traits, inheritance patterns reveal two clues that point to a recessive allele on an autosome. *First,* if both parents are heterozygous, any child of theirs will have a 50 percent chance of being heterozygous and a 25 percent chance of being homozygous recessive, as Figure 12.12 indicates. *Second,* if the parents are both homozygous recessive, any child of theirs will be, also.

On average, 1 in 100,000 newborns is homozygous for a recessive allele that causes *galactosemia*. It cannot produce functional molecules of an enzyme that stops a product of lactose breakdown from accumulating to toxic levels. Lactose normally is converted to glucose and galactose, then to glucose–1–phosphate (which is broken down by glycolysis or converted to glycogen). In affected persons, the full conversion is blocked:

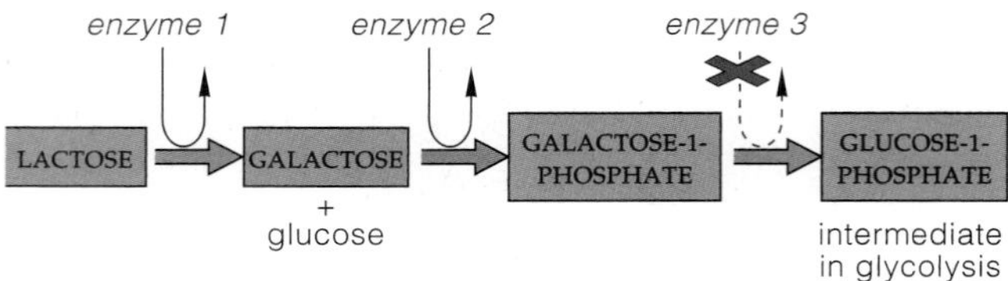

enzyme 1
enzyme 2
enzyme 3
LACTOSE
GALACTOSE
+ glucose
GALACTOSE-1-PHOSPHATE
GLUCOSE-1-PHOSPHATE
intermediate in glycolysis

A high blood level of galactose damages the eyes, liver, and brain. This telling symptom of the disorder can be detected in urine samples. Malnutrition, diarrhea, and vomiting are other symptoms. Untreated galactosemics often die in childhood. If affected individuals are quickly placed on a restricted diet that excludes dairy products, however, they can grow up symptom-free.

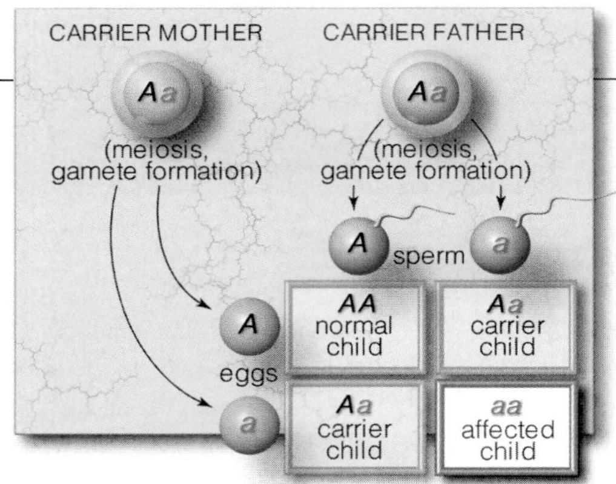

CARRIER MOTHER
CARRIER FATHER
Aa
Aa
(meiosis, gamete formation)
(meiosis, gamete formation)
A
sperm
a
A
eggs
a
AA normal child
Aa carrier child
Aa carrier child
aa affected child

Figure 12.12 A pattern for autosomal recessive inheritance. In this case, both parents are heterozygous carriers of the recessive allele (shown in *red*).

Autosomal Dominant Inheritance

Two clues of a different sort indicate that an autosomal dominant allele is responsible for a trait. *First,* the trait typically appears in each generation, for the allele is usually expressed, even in heterozygotes. *Second,* if one parent is heterozygous and the other is homozygous recessive, there is a 50 percent chance that any child of theirs will be heterozygous (Figure 12.13*a*).

A few dominant alleles persist in populations even though they cause severe genetic disorders. Some persist by spontaneous mutations. For others, expression of the dominant allele may not interfere with reproduction or affected people reproduce before symptoms are severe.

For example, *Huntington disorder* is characterized by progressive involuntary movements and deterioration of the nervous system, and eventual death. Symptoms may not even start to show up until an affected person is past age thirty. Most people have already reproduced by then. Affected individuals usually die during their forties or fifties, perhaps before they realize that they have transmitted the mutant allele to their children.

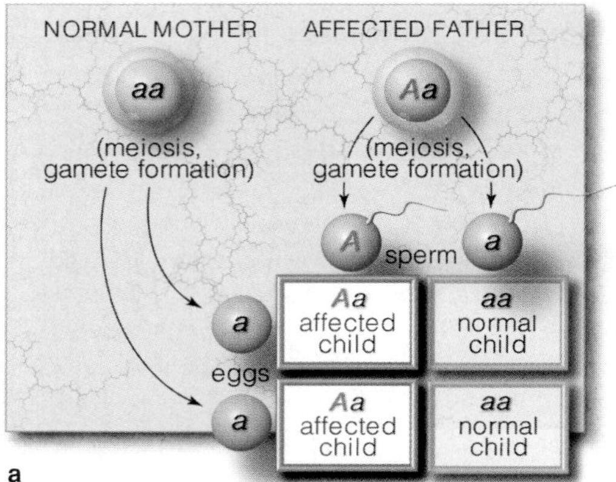

NORMAL MOTHER
AFFECTED FATHER
aa
Aa
(meiosis, gamete formation)
(meiosis, gamete formation)
A
sperm
a
a
eggs
a
Aa affected child
aa normal child
Aa affected child
aa normal child
a

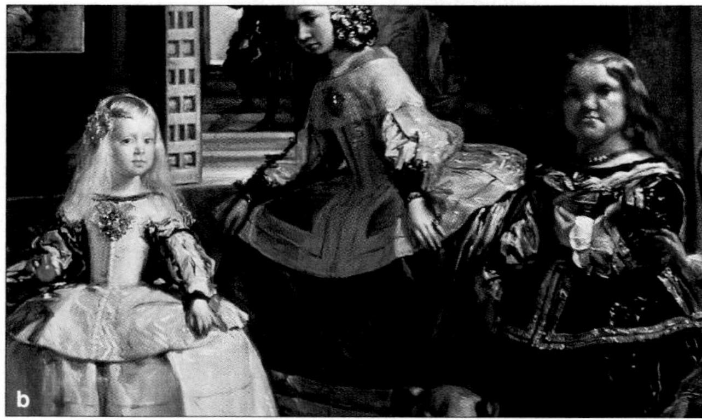

b

Figure 12.13 (**a**) A pattern for autosomal dominant inheritance. This dominant allele (*red*) is fully expressed in the carriers. (**b**) Infanta Margarita Teresa of the Spanish court and her maids, including the achondroplasic woman at the far right.

As another example, *achondroplasia* affects about 1 in 10,000 people. Commonly, the homozygous dominant condition leads to stillbirth, but heterozygotes are able to reproduce. While they are young, cartilage components of their skeleton form improperly. At maturity, affected people have abnormally short arms and legs relative to other body parts (Figure 12.13*b*). Adult achondroplasics are less than 4 feet, 4 inches tall. Often the dominant allele has no other phenotypic effects.

Further reading: Student Guide to InfoTrac on web site

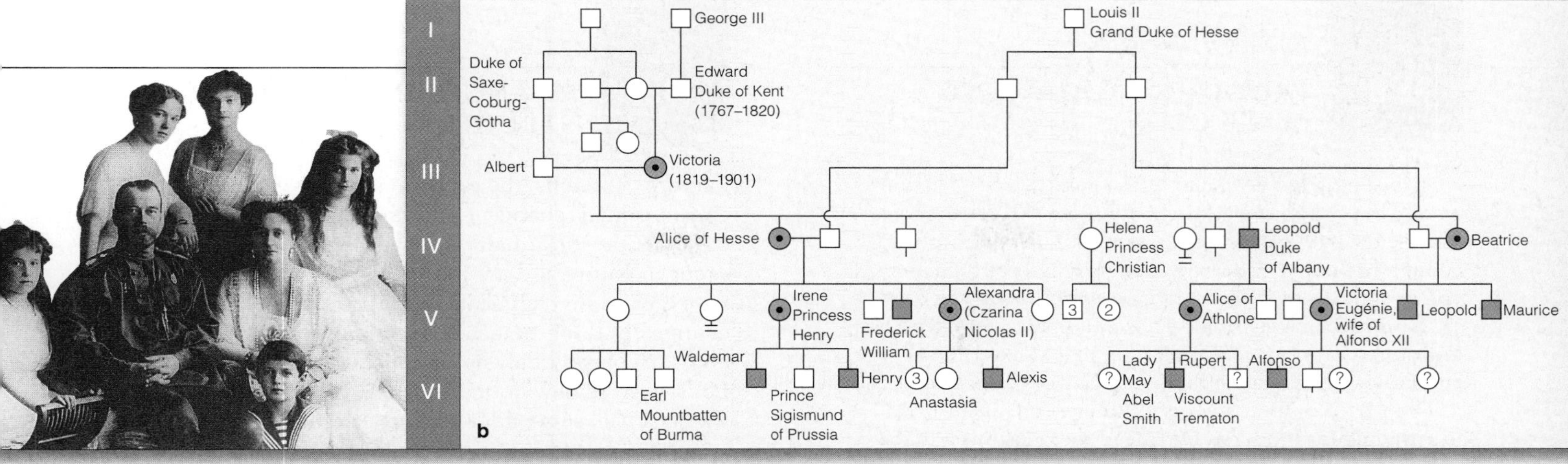
George III
Louis II
Grand Duke of Hesse
Duke of Saxe-Coburg-Gotha
Edward
Duke of Kent
(1767–1820)
Albert
Victoria
(1819–1901)
Alice of Hesse
Helena
Princess
Christian
Leopold
Duke
of Albany
Beatrice
Irene
Princess
Henry
Alexandra
(Czarina
Nicolas II)
Frederick
William
Alice of
Athlone
Victoria
Eugénie,
wife of
Alfonso XII
Leopold
Maurice
Waldemar
Earl
Mountbatten
of Burma
Prince
Sigismund
of Prussia
Henry
Anastasia
Alexis
Lady
May
Abel
Smith
Rupert
Viscount
Trematon
Alfonso
b

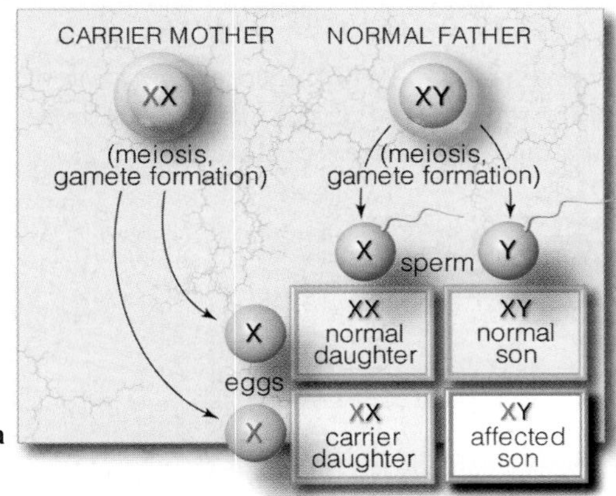
CARRIER MOTHER
NORMAL FATHER
XX
XY
(meiosis,
gamete formation)
(meiosis,
gamete formation)
X
sperm
Y
X
eggs
X
XX
normal
daughter
XY
normal
son
XX
carrier
daughter
XY
affected
son
a

Figure 12.14 (**a**) One pattern for X-linked inheritance. In this example, assume the mother carries the recessive allele on one of her X chromosomes (*red*).

(**b**) *Above:* Partial pedigree for Queen Victoria's descendants, showing carriers and affected males who carried the X allele for hemophilia A. Of the Russian royal family members in the photograph, the mother was a carrier and Crown Prince Alexis was hemophilic. He was a focus of political intrigue that helped trigger the Russian Revolution of 1917.

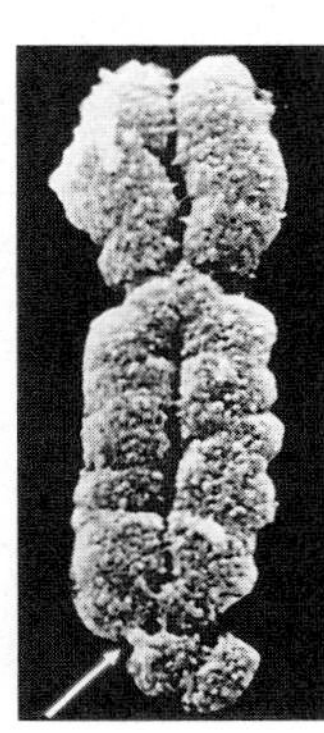

Figure 12.15 Scanning electron micrograph of the fragile X chromosome from a cultured cell. The arrow points to the fragile site.

X-Linked Recessive Inheritance

Distinctive clues often show up when a recessive allele on an X chromosome causes a genetic disorder. *First,* males show the recessive phenotype more often than females. The recessive allele can be masked in females, who may inherit a dominant allele on their other X chromosome. The allele is not masked in males, who have only one X chromosome (Figure 12.14*a*). *Second,* a son cannot inherit the recessive allele from his father. A daughter can. If a daughter is heterozygous, there is a 50 percent chance each son of hers will inherit the allele.

Color blindness, an inability to distinguish among some or all colors, is a common X-linked recessive trait. For instance, in red–green color blindess, the individual lacks some or all of the sensory receptors that normally respond to visible light of red or green wavelengths.

Hemophilia A, a blood-clotting disorder, is a case of X-linked recessive inheritance. In most people, a blood-clotting mechanism quickly stops bleeding from minor injuries. Clot formation requires the products of several genes, some of which are on the X chromosome. If any of the X-linked genes is mutated in a male, the absence of its functional product prolongs bleeding. About 1 in 7,000 males inherits the mutant allele for hemophilia A. Clotting is close to normal in heterozygous females.

The frequency of hemophilia A was unusually high in royal families of nineteenth-century Europe. Queen Victoria of England was a carrier (Figure 12.14*b*). At one time, the recessive allele was present in eighteen of her sixty-nine descendants.

Fragile X syndrome, an X-linked recessive disorder that causes mental retardation, affects 1 in 1,500 males in the United States. In cultured cells, an X chromosome carrying the mutant allele is constricted near the end of its long arm (Figure 12.15). This constriction is called a fragile site because the end of the chromosome tends to break away. The term is misleading, for the end breaks away only in cultured cells, not in cells in the body.

A mutant gene, not breakage, causes the syndrome. It specifies a protein required for normal development of brain cells. Within that gene, a segment of DNA is repeated several times. Certain mutations result in the addition of many more repeats in the DNA; hence their name, *expansion* mutations. The outcome is a mutant allele that cannot function properly. Because that allele is recessive, males who inherit it and females who are homozygous for it have fragile X syndrome. Expansion mutations are now known to be the cause of Huntington disorder and some other genetic disorders.

Genetic analyses of family pedigrees have revealed simple Mendelian inheritance patterns for certain traits, as well as for many genetic disorders that arise from expression of specific alleles on an autosome or X chromosome.

12.8 PROGERIA—TOO YOUNG TO BE OLD

Imagine being ten years old with a mind trapped inside a body that is rapidly getting a bit more shriveled, more frail—*old*—with each passing day. You are just barely tall enough to peer over the top of the kitchen counter, and you weigh less than thirty-five pounds. Already you are bald and have a crinkled nose. Possibly you have a few more years to live. Could you, like Mickey Hayes and Fransie Geringer, still laugh with your friends?

Of every 8 million newborn humans, one is destined to grow old far too soon. On one of its autosomes, that rare individual carries a mutant gene that gives rise to *Hutchinson–Gilford progeria syndrome*. Through hundreds, thousands, then many billions of DNA replications and mitotic cell divisions, terrible information encoded in that gene was systematically distributed to every cell in the growing embryo, and later in the newborn. Its legacy will be accelerated aging and a greatly reduced life span. The photograph of Mickie Hayes and Fransie Geringer in Figure 12.16 shows some of the symptoms.

The mutation causes gross disruptions in interactions among genes that bring about the body's growth and development. Observable symptoms start before age two. Skin that should be plump and resilient starts to thin. Skeletal muscles weaken. Tissues in limb bones that should lengthen and grow stronger start to soften. Hair loss is pronounced; extremely premature baldness is inevitable. There are no documented cases of progeria running in families, so it seems likely the gene mutates spontaneously, at random. It appears to be dominant over its normal partner on the homologous chromosome.

Most progeriacs can expect to die in their early teens as a result of strokes or heart attacks. These final insults are brought on by a hardening of the walls of arteries, a condition typical of advanced age. When Mickey Hayes turned eighteen, he was the oldest living progeriac. Fransie was seventeen when he died.

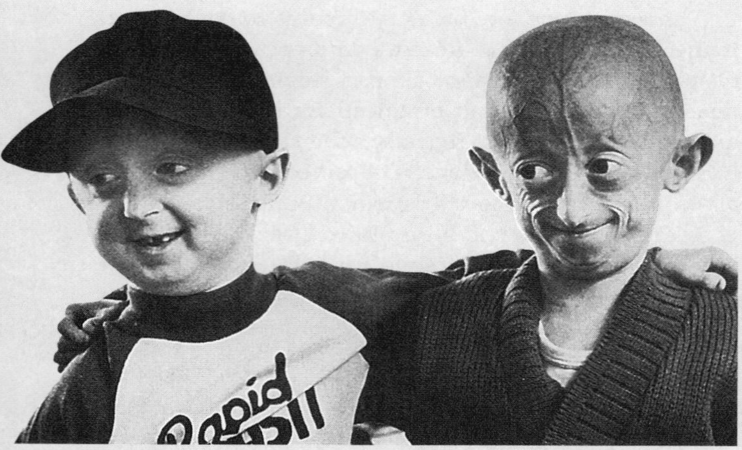

Figure 12.16 Two boys who met at a gathering of progeriacs at Disneyland, California, when they were not yet ten years old.

12.9 CHANGES IN CHROMOSOME STRUCTURE

On rare occasions, the physical structure of one or more chromosomes changes, and the outcome is a genetic disorder or abnormality. Such changes spontaneously occur in nature and are induced in research laboratories by exposure to chemicals or by irradiation. Either way, changes may be detected by microscopic examination and karyotype analysis of cells at mitosis or meiosis. Let's now review four kinds of structural change. As you will see, some have serious or lethal consequences.

Major Categories of Structural Change

DUPLICATION Even normal chromosomes have gene sequences that are repeated several to many hundreds or thousands of times. These are **duplications**:

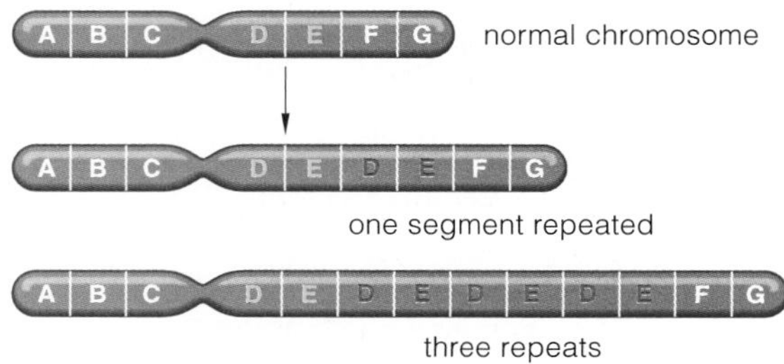
A B C D E F G
normal chromosome
A B C D E D E F G
one segment repeated
A B C D E D E D E D E F G
three repeats

INVERSION With an **inversion**, a linear stretch of DNA within the chromosome becomes oriented in the reverse direction, with no molecular loss:

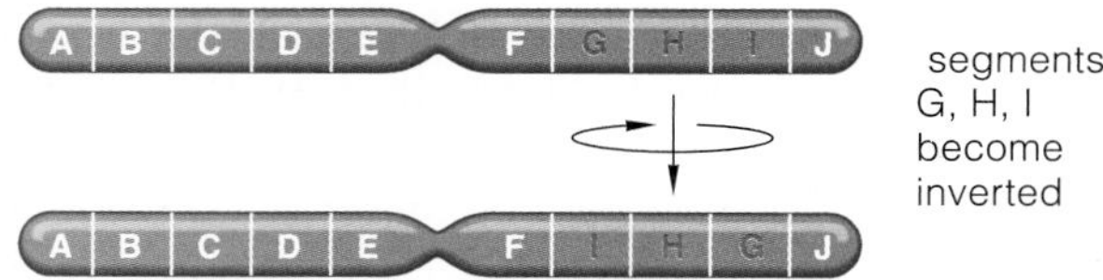
A B C D E F G H I J
segments G, H, I become inverted
A B C D E F I H G J

TRANSLOCATION As the chapter introduction showed, the Philadelphia chromosome that has been linked to a form of leukemia results from **translocation**, whereby a broken part of a chromosome becomes attached to a *non*homologous chromosome. Most translocations are reciprocal (both chromosomes exchange broken parts):

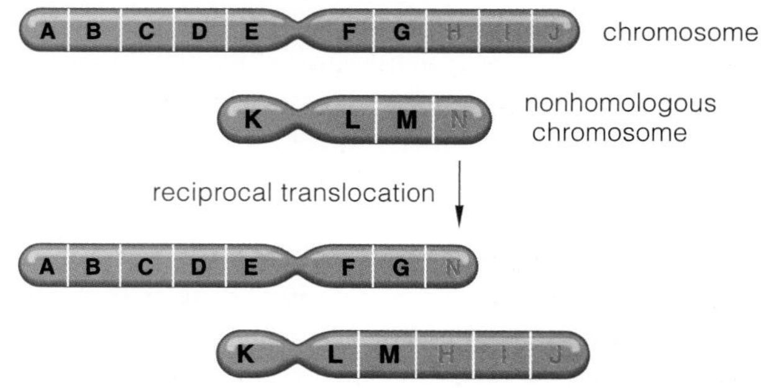
A B C D E F G H I J
chromosome
K L M N
nonhomologous chromosome
reciprocal translocation
A B C D E F G N
K L M H I J

Further reading: Student Guide to InfoTrac on web site

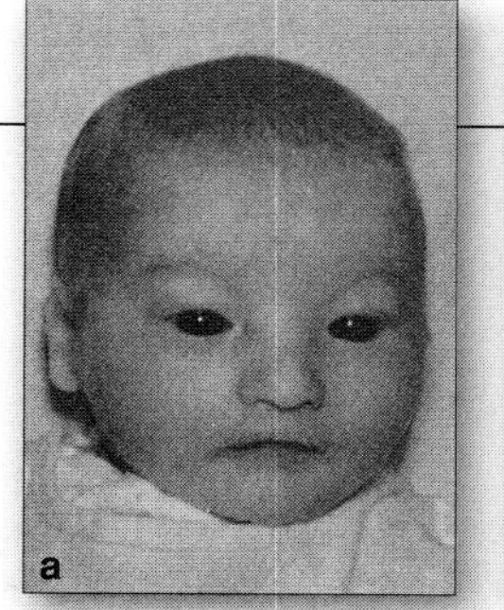
a
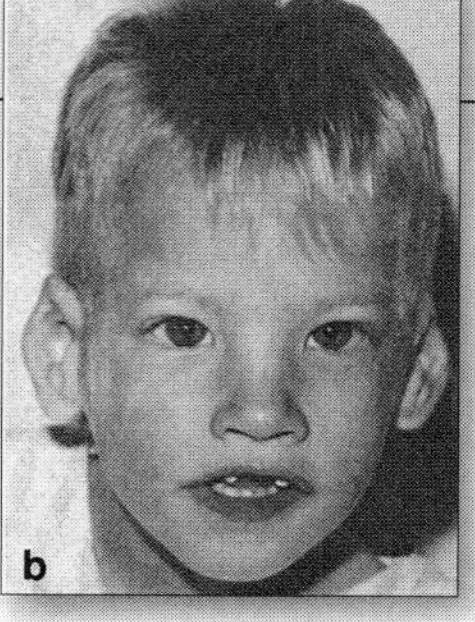
b

Figure 12.17 (**a**) A male infant who later developed cri-du-chat syndrome. The ears are positioned low on the side of the head relative to the eyes. (**b**) The same boy, four years later. By this age, affected humans no longer make mewing sounds typical of the syndrome.

DELETION Viral attacks, irradiation (ionizing radiation especially), chemical assaults, or other environmental factors may trigger a **deletion**, the loss of some segment of a chromosome:

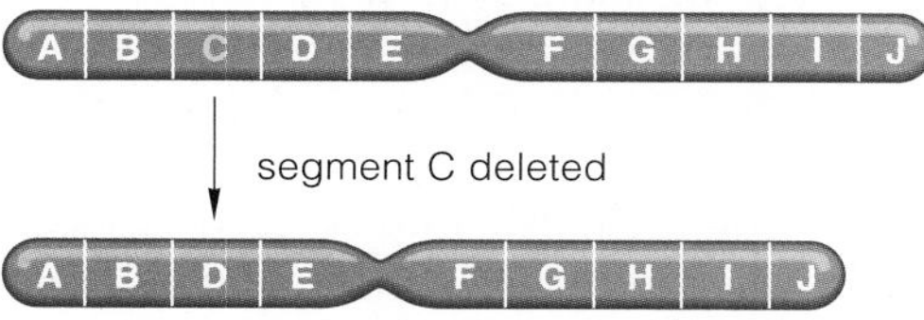
A B C D E F G H I J
segment C deleted
A B D E F G H I J

Don't species with diploid cells have an advantage after a deletion occurs? That is, wouldn't genes on the homologous chromosome cover the loss? Maybe, but the sword of chance cuts both ways. If the remaining segment happens to carry a harmful recessive allele, nothing will mask or compensate for *its* effects.

Most deletions are lethal or cause serious disorders in mammals, for they disrupt normal gene interactions that underlie a program of growth, development, and maintenance activities. For example, one deletion from human chromosome 5 results in mental retardation and the development of an abnormally shaped larynx. When affected infants cry, they produce sounds rather like a cat's meow. Hence *cri-du-chat* (cat-cry), the name of this disorder. Figure 12.17 shows an affected child.

Does Chromosome Structure Evolve?

Changes in chromosome structure tend to be selected against rather than conserved over evolutionary time. Even so, for many species, we have interesting signs of changes that occurred in the past. To give one example, many duplications have done their bearers no harm. Although duplications are relatively rare events, they have accumulated over millions, even billions, of years and are now built into the DNA of all species.

Biologists speculate that duplicates of some gene sequences with neutral effects could have an adaptive advantage. In effect, a copy could free up a gene for chance mutations that might turn out to be useful, and the normal gene would still issue the required product. One or more duplicated gene sequences could become slightly modified, and then the products of those genes could function in slightly different or new ways.

Several kinds of duplicated, modified genes seem to have had pivotal roles in evolution. Consider the gene regions for the polypeptide chains of hemoglobin, as shown in Section 3.5. In humans and other primates, gene regions for the polypeptide chains of hemoglobin contain multiple sequences that are remarkably similar. The sequences specify whole families of chains, which have slight structural differences. Each of the resulting hemoglobin molecules transports oxygen with slightly different efficiencies, depending on cellular conditions.

Certain duplications, inversions, and translocations probably helped put primate ancestors of humans on a unique evolutionary road. They apparently contributed to divergences that led to the modern apes and humans. Of our twenty-three pairs of chromosomes, eighteen are nearly identical to their counterparts in chimpanzees and gorillas. The other five pairs differ at inverted and translocated regions. You can observe such similarities for yourself by comparing chromosomes of a human with those of a gibbon, one of the apes (Figure 12.18).

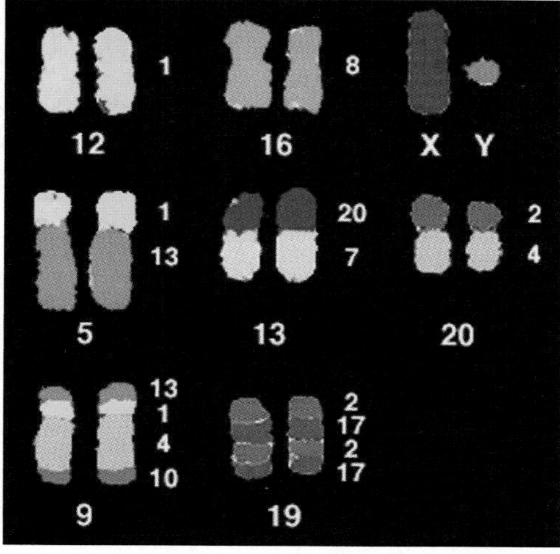
1
12
8
16
X Y
1
13
5
20
7
13
2
4
20
13
1
4
10
9
2
17
2
17
19

Gibbon chromosomes 12, 16, X, and Y, which are structurally identical to human chromosomes 12, 16, X, and Y (compare Figure 12.1).

Gibbon chromosomes 5, 13, and 20 show translocations that correspond to human chromosomes 1/13, 20/7, and 2/4.

Gibbon chromosome 9 corresponds to parts of human chromosomes. Gibbon chromosome 19 duplications correspond to human chromosomes 2 and 17.

Figure 12.18 Spectral karyotype of duplicated chromosomes from an ape. Colors identify which parts are structurally identical in human chromosomes.

On rare occasions, a segment of a chromosome may get lost, inverted, moved to a new location, or duplicated.

Most chromosome changes are harmful or lethal, especially when they alter gene interactions that underlie growth, development, and maintenance activities.

Over evolutionary time, many changes have been conserved; they confer adaptive advantages or have had neutral effects.

12.10 CHANGES IN CHROMOSOME NUMBER

Occasionally, abnormal events occur before or during cell division, then gametes and new individuals end up with the wrong chromosome number. The consequences range from minor to lethal physical changes.

Categories and Mechanisms of Change

With **aneuploidy**, individuals usually have one extra or one less chromosome. This condition is a major cause of human reproductive failure. Possibly it affects about one-half of all fertilized eggs. As autopsies reveal, most *miscarriages* (spontaneous aborting of embryos before pregnancy reaches full term) are aneuploids.

With **polyploidy**, individuals have three or more of each type of chromosome. About one-half of all species of flowering plants are polyploid (Section 19.4). Often, researchers can induce polyploidy in undifferentiated plant cells by exposing them to colchicine (Section 12.2). Some species of insects, fishes, and other animals are polyploids. However, polyploidy is lethal for humans. All but about 1 percent of human polyploids die before birth, and the rare newborns die soon after birth.

Chromosome numbers can change during mitotic or meiotic cell divisions. Suppose a cell cycle proceeds through DNA duplication and mitosis but is arrested before the cytoplasm divides. The cell is now *tetra*ploid, with four of each type of chromosome. Suppose one or more pairs of chromosomes fail to separate in mitosis or meiosis, an event called **nondisjunction**. Some or all of the forthcoming cells will have too many or too few chromosomes, as in the Figure 12.19 example.

The chromosome number also may get changed at fertilization. Visualize a normal gamete that unites by chance with an $n + 1$ gamete (one extra chromosome). The new individual will be "trisomic" ($2n + 1$); it will have three of one type of chromosome and two of every other type. What if an $n - 1$ gamete unites with a normal n gamete? Then, the new individual will turn out to be "monosomic" ($2n - 1$).

Change in the Number of Autosomes

Nearly all changes in the number of autosomes arise by nondisjunction during gamete formation. Let's look at one of the most common of the resulting disorders. A trisomic 21 newborn, with three chromosomes 21, will develop *Down syndrome*. The symptoms vary greatly, but most affected individuals show moderate to severe mental impairment. About 40 percent develop heart defects. Abnormal development of the skeleton means older children have shortened body parts, loose joints, and poorly aligned bones of the hips, fingers, and toes. Muscles and muscle reflexes are weaker than normal, and speech and other motor skills develop slowly. With special training, trisomic 21 individuals often take part in normal activities; Figure 12.20 gives examples. As a group, they tend to be cheerful and affectionate, and they derive great pleasure from socializing.

Down syndrome is one of many disorders that can be detected before birth. Before detection procedures were widespread, about 1 in 700 newborns of all ethnic groups was trisomic 21. The number is now closer to 1 in 1,100, for some women elect to terminate pregnancy when the condition is detected (Section 12.11). The risk is greater if pregnant women are more than thirty-five years old, as indicated by the diagram in Figure 12.20*b*.

Change in the Number of Sex Chromosomes

Most sex chromosome abnormalities arise as a result of nondisjunction during meiosis and gamete formation. Let's look at a few phenotypic outcomes.

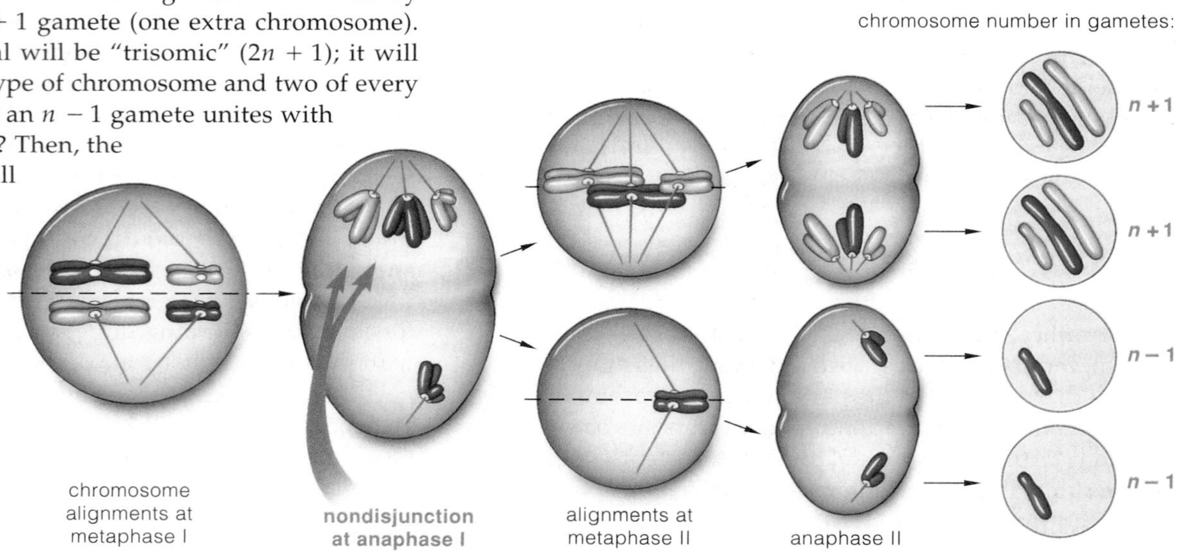

chromosome number in gametes:
n + 1
n + 1
n − 1
n − 1
chromosome alignments at metaphase I
nondisjunction at anaphase I
alignments at metaphase II
anaphase II

Figure 12.19 Example of nondisjunction. Of two pairs of homologous chromosomes shown, one pair fails to separate at anaphase I of meiosis. The chromosome number changes in the gametes. (Make a sketch of nondisjunction at anaphase II. What will the chromosome numbers be in gametes?)

Further reading: Student Guide to InfoTrac on web site

1
2
3
4
5
6
7
8
9
10
11
12
13
14
15
16
17
18
19
20
21
22
23
a

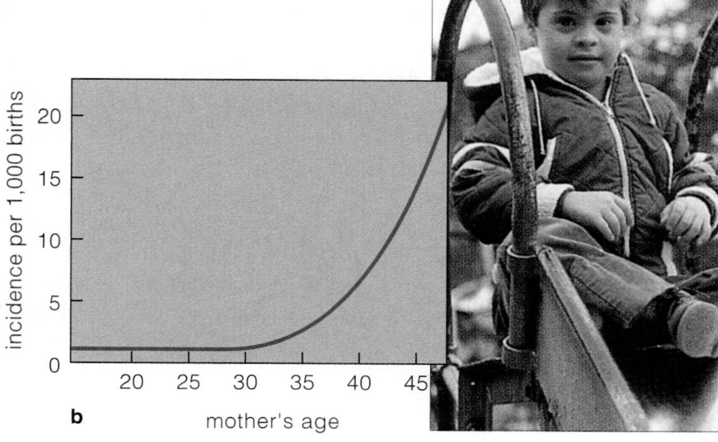
incidence per 1,000 births
0
5
10
15
20
20
25
30
35
40
45
b
mother's age

Figure 12.20 Down syndrome. (**a**) Karyotype revealing a trisomic 21 condition (*arrows*). (**b**) Relationship between the frequency of Down syndrome and mother's age at the time of childbirth. Results are from a study of 1,119 affected children who were born in Victoria, Australia, between 1942 and 1957. The young lady above was a lively participant in the Special Olympics, held annually in San Mateo, California.

TURNER SYNDROME Inheriting one X chromosome with no corresponding X or Y chromosome gives rise to *Turner syndrome*, which affects 1 in 2,500 to 10,000 or so newborn girls. Nondisjunction affecting sperm accounts for 75 percent of the cases. We see fewer people with Turner syndrome compared to people having other sex chromosome abnormalities. The likely reason is that at least 98 percent of all X0 zygotes spontaneously abort early in pregnancy. Approximately 20 percent of all the spontaneously aborted embryos in which chromosome abnormalities were detected have been X0.

Despite the near-lethality, X0 survivors are not as disadvantaged as other aneuploids. They grow up well proportioned, albeit short—4 feet, 8 inches tall, on the average. Generally, their behavior is normal during childhood. But most Turner females are infertile. They do not have functional ovaries and so cannot produce eggs or sex hormones. Without sex hormones, breast enlargement and the development of other secondary sexual traits are reduced. Possibly as a result of their arrested sexual development and small size, X0 females often become passive and are easily intimidated by peers during their teens. Some patients have benefited from hormone therapy and corrective surgery.

KLINEFELTER SYNDROME Of every 500 to 2,000 liveborn males, one has inherited two X chromosomes and one Y chromosome. This XXY condition results mainly from nondisjunction in the mother (about 67 percent of the time, compared to 33 percent in the father). Symptoms of the resulting *Klinefelter syndrome* develop after the onset of puberty. XXY males are taller than average and are sterile or nearly so. Their testes usually are much smaller than average; the penis and scrotum are not. Often facial hair is sparse, and breasts are somewhat enlarged. Injections of the hormone testosterone can reverse the feminized traits but not the low fertility. Some XXY males display mild mental impairment, but many fall in the normal range for intelligence. Except for their low fertility, many affected individuals show no outward symptoms at all.

XYY CONDITION About 1 in 1,000 males has one X and two Y chromosomes. *XYY males* tend to be taller than average. Some may be mildly retarded, but most are phenotypically normal. At one time, XYY males were thought to be genetically predisposed to become criminals. The erroneous conclusion was based on small numbers of cases in narrowly selected groups, such as prison inmates. Investigators often knew who the XYY males were, and this may have biased their evaluations. There were no **double-blind studies**, by which different investigators gather data independently of one another and then match them up only after both sets of data are completed. In this case, the same investigators gathered the karyotypes and the personal histories. Fanning the stereotype was a sensationalized report in 1968 that a mass-murderer of young nurses was XYY. He wasn't.

In 1976, a Danish geneticist issued a report on a large-scale study based on records of 4,139 tall males, twenty-six years old, who had reported to their draft board. Besides giving results of physical examinations and intelligence testing, the records provided clues to social and economic status, educational history, and any criminal convictions. Only twelve of the males were XYY, which left more than 4,000 for the control group. The only finding of significance was that tall, mentally impaired males who engage in criminal activity are just more likely to get caught—irrespective of karyotype.

Most changes in chromosome number arise as an outcome of nondisjunction during meiosis and gamete formation.

12.11 PROSPECTS IN HUMAN GENETICS

With the arrival of their newborn, parents typically ask, "Is our baby normal?" Quite naturally, they want their baby to be free of genetic disorders, and most of the time it is. But what are the options when it is not?

We do not approach heritable disorders and diseases the same way. We attack diseases with antibiotics, surgery, and other weapons. But how do we attack a heritable "enemy" that can be transmitted to offspring? Do we institute regional, national, or global programs to identify people who might be carrying harmful alleles? Do we tell them they are "defective" and run a risk of bestowing a disorder on their children? Who decides which alleles are "harmful"? Should society bear the cost of treating genetic disorders before and after birth? If so, should society also have a say in whether an affected embryo will be born at all, or whether it should be aborted? An **abortion** is the expulsion of a pre-term embryo or fetus from the uterus.

Questions such as these are only the tip of an ethical iceberg. And we do not have anwers that are universally acceptable throughout our society.

PHENOTYPIC TREATMENTS Often, the symptoms of genetic disorders can be either minimized or suppressed by (1) exerting dietary controls, (2) making adjustments to specific environmental conditions, and (3) intervening surgically or by way of hormone replacement therapy.

For example, dietary control works in *phenylketonuria*, or PKU. A certain gene specifies an enzyme that converts one amino acid to another—phenylalanine to tyrosine. If an individual is homozygous recessive for a mutated form of the gene, the first of these amino acids accumulates inside the body. If excess amounts are diverted into other pathways, then phenylpyruvate and other compounds may form. High levels of phenylpyruvate in the blood can impair the functioning of the brain. When affected people restrict their intake of phenylalanine, they are not required to dispose of excess amounts, so they can lead normal lives. Among other things, they can avoid soft drinks and other food products that are sweetened with aspartame, a compound that contains phenylalanine.

Environmental adjustments help counter or minimize the symptoms of some disorders, as when albinos avoid exposure to direct sunlight. Surgical reconstructions also minimize many physical problems. For example, surgeons close up a form of *cleft lip* in which a vertical fissure cuts through the lip and extends into the roof of the mouth.

GENETIC SCREENING Through large-scale screening programs in the general population, affected persons or carriers of a harmful allele often can be detected early enough to start preventive measures before symptoms develop. For example, most hospitals in the United States routinely screen newborns for PKU, so today it is less common to see people with symptoms of this disorder.

GENETIC COUNSELING If a first child or close relative has a severe heritable problem, prospective parents may worry about their next child. They may request help in evaluating their options from a qualified professional counseler. *Genetic counseling* often includes diagnosis of parental genotypes, detailed pedigrees, and genetic testing for hundreds of known metabolic disorders. Geneticists, too, may be contacted to help predict risks for genetic disorders. During genetic counseling, the prospective parents must be reminded that the same risk usually applies to each pregnancy.

PRENATAL DIAGNOSIS Methods of *prenatal diagnosis* can be used to determine the sex as well as more than a hundred genetic conditions of the embryo or fetus. (*Prenatal* means "before birth." The term *embryo* applies until eight weeks after fertilization, after which the term *fetus* is appropriate.)

For example, suppose a woman who is forty-five years old becomes pregnant. She worries that her forthcoming child may develop Down syndrome. She might request prenatal diagnosis by *amniocentesis* (Figure 12.21). With this diagnostic procedure, a clinician withdraws a tiny sample of the fluid inside the amnion, the membranous sac surrounding the fetus. Some cells that the fetus has sloughed

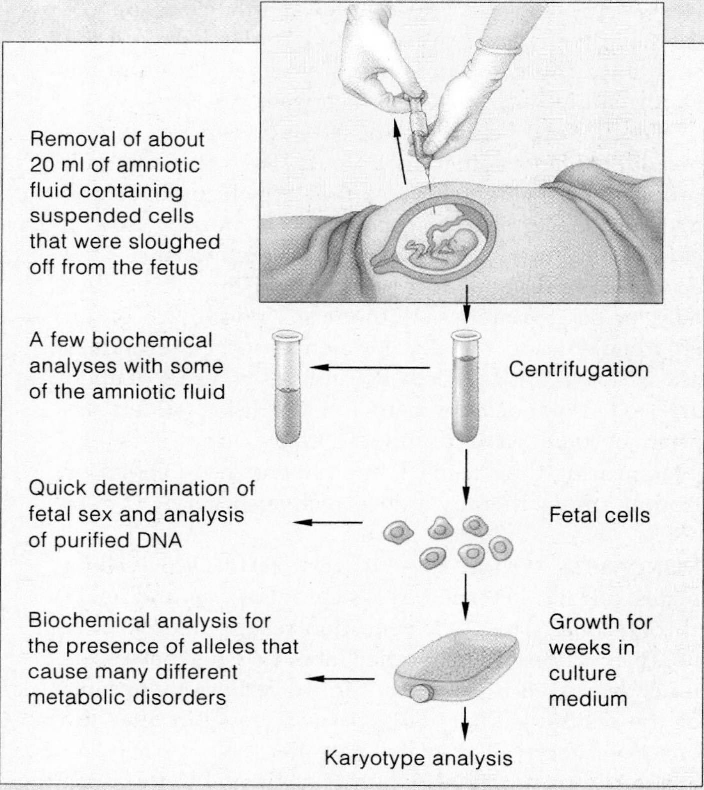
Removal of about 20 ml of amniotic fluid containing suspended cells that were sloughed off from the fetus
A few biochemical analyses with some of the amniotic fluid
Centrifugation
Quick determination of fetal sex and analysis of purified DNA
Fetal cells
Biochemical analysis for the presence of alleles that cause many different metabolic disorders
Growth for weeks in culture medium
Karyotype analysis

Figure 12.21 Steps in amniocentesis, a prenatal diagnostic tool.

Further reading: Student Guide to InfoTrac on web site →

off are suspended in the sample. The cells are cultured and then analyzed.

Chorionic villi sampling (CVS) is a different diagnostic procedure. A clinician withdraws cells from the chorion, a fluid-filled, membranous sac that surrounds the amnion. CVS can be perfomed weeks before amniocentesis. It can yield results as early as the ninth week of pregnancy.

Direct visualization of the developing fetus is possible with *fetoscopy*. A fiber-optic device, an endoscope, uses pulsed sound waves to scan the uterus and visually locate particular parts of the fetus, umbilical cord, or placenta (Figure 12.22). Fetoscopy has been used to diagnose blood cell disorders, such as sickle-cell anemia and hemophilia.

All three procedures may accidentally cause infection or puncture the fetus. Occasionally the punctured amnion does not reseal itself quickly, and an excessive amount of amniotic fluid may leak out and cause problems for the fetus. A mother-to-be who requests amniocentesis runs a 1 to 2 percent greater risk of miscarriage. For CVS, she runs a 0.3 percent risk that her future child will have missing or underdeveloped fingers and toes. Fetoscopy increases the risk of a miscarriage by 2 to 10 percent.

Parents-to-be probably should seek counseling from their doctor to help them weigh the risks and benefits of such procedures as applied to their own circumstances. They may wish to (1) ask about the small overall risk of 3 percent that any child will have some kind of birth defect, (2) ask about the severity of genetic disorder that a child might be at risk of developing, and (3) consider how old the woman is at the time of pregnancy.

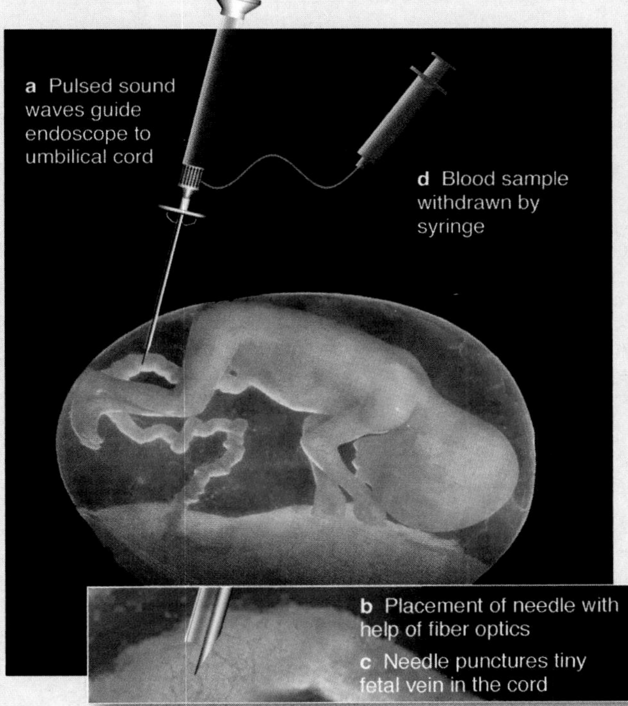
a Pulsed sound waves guide endoscope to umbilical cord
d Blood sample withdrawn by syringe
b Placement of needle with help of fiber optics
c Needle punctures tiny fetal vein in the cord

Figure 12.22 Fetoscopy for prenatal diagnosis.

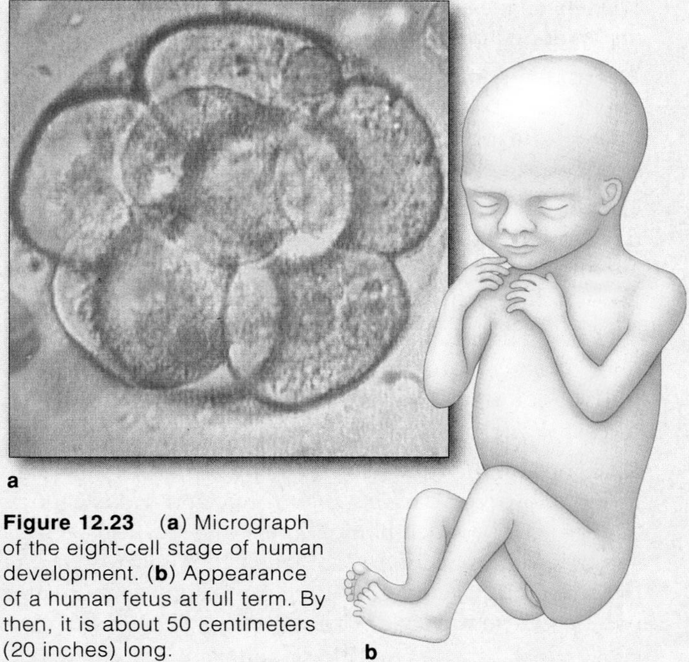
a
b

Figure 12.23 (**a**) Micrograph of the eight-cell stage of human development. (**b**) Appearance of a human fetus at full term. By then, it is about 50 centimeters (20 inches) long.

REGARDING ABORTION What happens when prenatal diagnosis does reveal a serious problem? Do prospective parents opt for induced abortion? We can only say here that they must weigh their awareness of the severity of the genetic disorder against ethical and religious beliefs. Worse, they must play out their personal tragedy on a larger stage, dominated by a nationwide battle between fiercely vocal "pro-life" and "pro-choice" factions. We return to this volatile issue in Section 45.13.

PREIMPLANTATION DIAGNOSIS Another procedure, *preimplantation diagnosis*, relies on **in-vitro fertilization**. In vitro, recall, means "in glass." Sperm and eggs taken from prospective parents are quickly transferred to an enriched medium in a glass petri dish. One or more eggs may become fertilized. In two days, mitotic cell divisions may convert a fertilized egg into a ball of eight cells, such as the one shown in Figure 12.23*a*.

According to one view, the tiny, free-floating ball is a *pre*-pregnancy stage. Like the unfertilized eggs discarded monthly from a woman, the ball is not attached to the uterus. All cells in the ball have the same genes and are not yet committed to giving rise to specialized cells of a heart, lungs, and other organs. Doctors take one of the undifferentiated cells and analyze its genes for suspected disorders. If the cell has no detectable genetic defects, the ball is inserted into the uterus.

Some couples who are at risk of passing on muscular dystrophy, cystic fibrosis, and other disorders have opted for the procedure. Many *"test-tube" babies* have been born in good health and are free of the mutant alleles.

12.12 SUMMARY

1. Genes, the units of instruction for heritable traits, are arranged one after the other along chromosomes.

2. Human somatic cells are diploid ($2n$). They contain twenty-three pairs of homologous chromosomes that interact during meiosis. Each pair has the same length, shape, and gene sequence (except for the XY pairing).

3. Human females have two X chromosomes. Males have one X paired with one Y. All other chromosomes are autosomes (the same in both females and males). A gene on the Y chromosome determines sex.

4. Pedigrees, or charts of genetic connections through lines of descent, often provide clues to inheritance of a trait. Certain patterns are characteristic of dominant or recessive alleles on autosomes or the X chromosome.

5. Genes on the same chromosome represent a linkage group. However, crossing over (breakage and exchange of segments between homologues) disrupts linkages, as summarized in Section 12.4. The farther apart two gene loci are along the length of a chromosome, the greater will be the frequency of crossovers between them.

6. A chromosome's structure may become altered on rare occasions. A segment might be deleted, inverted, moved to a new location (translocated), or duplicated.

7. The chromosome number can change. Gametes and offspring might get one more or one less chromosome than their parents (aneuploidy). They might also get three or more of each type of chromosome (polyploidy). Nondisjunction at meiosis (prior to gamete formation) accounts for most of these chromosome abnormalities.

8. Crossing over adds to potentially adaptive variation in traits in a population. By contrast, most changes in chromosome number or structure are harmful or lethal. Over evolutionary time, however, changes occasionally establish themselves in chromosomes of the species.

Review Questions

1. What is a gene? What are alleles? *12.1*
2. Distinguish between: *CI, 12.1*
   a. homologous and nonhomologous chromosomes
   b. sex chromosomes and autosomes
   c. karyotype and karyotype diagram
3. Define genetic recombination, and describe how crossing over can bring it about. *12.1, 12.4*
4. Define pedigree, genetic abnormality, and genetic disorder. *12.6*
5. Contrast a typical pattern of autosomal recessive inheritance with that of autosomal dominant inheritance. *12.7*
6. Describe two clues that often show up when a recessive allele on an X chromosome causes a genetic disorder. *12.7*
7. Distinguish among a chromosomal deletion, duplication, inversion, and translocation. *12.9*
8. Define aneuploidy and polyploidy. Make a sketch of an example of nondisjunction. *12.10*

Parents: *AB* × *ab*

F_1 offspring: All *AaBb*

meiosis, gamete formation

Equal ratios of two types of gametes: 50% *AB* 50% *ab*

a Full linkage between two genes (no crossovers): half of the gametes have one parental genotype and half have the other.

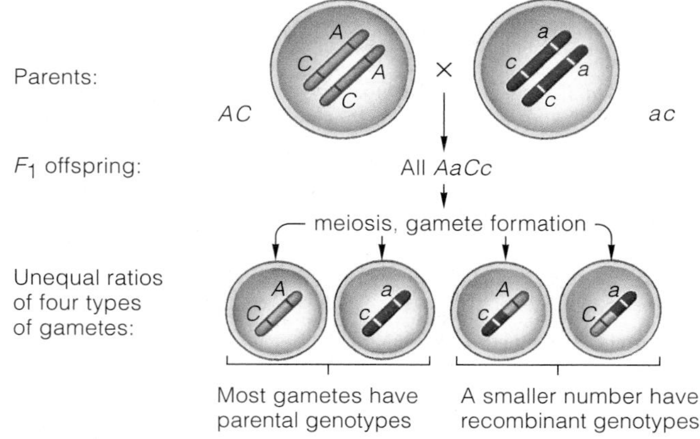

Parents:
AC
ac
F1 offspring:
All AaCc
meiosis, gamete formation
Unequal ratios of four types of gametes:
Most gametes have parental genotypes
A smaller number have recombinant genotypes

b Incomplete linkage; crossing over affected the outcome.

Figure 12.24 Summary examples of how (**a**) complete and (**b**) incomplete gene linkage can affect a dihybrid cross.

Self-Quiz (*Answers in Appendix III*)

1. ______ segregate during ______.
   a. Homologues; mitosis
   b. Genes on nonhomologous chromosomes; meiosis
   c. Homologues; meiosis
   d. Genes on one chromosome; mitosis
2. The probability of a crossover occurring between two genes on the same chromosome is ______.
   a. unrelated to the distance between them
   b. increased if they are closer together on the chromosome
   c. increased if they are farther apart on the chromosome
3. Chromosome structure can be altered by ______.
   a. deletions c. inversions e. all of the above
   b. duplications d. translocations
4. Nondisjunction can be caused by ______.
   a. crossing over in mitosis
   b. segregation in meiosis
   c. failure of chromosomes to separate during meiosis
   d. multiple independent assortments
5. A gamete affected by nondisjunction would have ______.
   a. a change from the normal chromosome number
   b. one extra or one missing chromosome
   c. the potential for a genetic disorder
   d. all of the above
6. Genetic disorders can be caused by ______.
   a. altered chromosome number c. mutation
   b. altered chromosome structure d. all of the above

Figure 12.25 Mutant fruit fly with vestigial wings

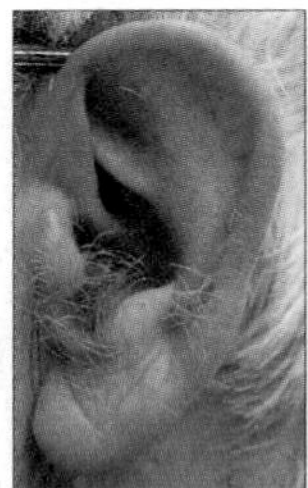

Figure 12.26 Hairy pinnae

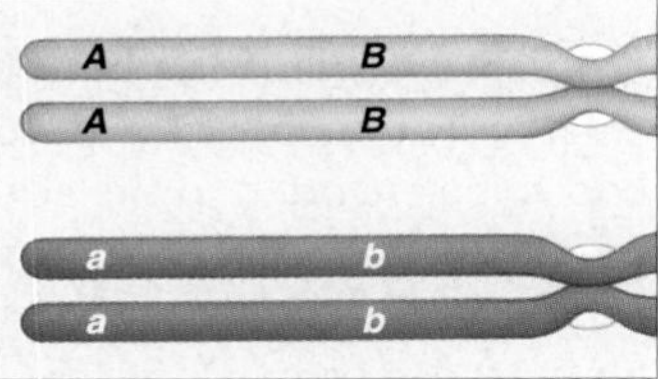

A B
A B
a b
a b

Figure 12.28 Two of your linked genes, on a pair of homologous chromosomes.

7. Is this statement true or false: Independent assortment, crossing over, and changes in chromosome structure and number contribute to phenotypic variation in a population.

8. Match the chromosome terms appropriately.

____ crossing over	a. number and defining features of an individual's metaphase chromosomes
____ deletion	b. chromosome segment moves to a nonhomologous chromosome
____ nondisjunction	c. disrupts gene linkages at meiosis
____ translocation	d. causes gametes to have abnormal chromosome numbers
____ karyotype	e. loss of a chromosome segment

Critical Thinking—Genetics Problems

(*Answers in Appendix IV*)

1. Human females are XX and males are XY.
   a. Does a male inherit the X from his mother or father?
   b. With respect to X-linked alleles, how many different types of gametes can a male produce?
   c. If a female is homozygous for an X-linked allele, how many types of gametes can she produce with respect to that allele?
   d. If a female is heterozygous for an X-linked allele, how many types of gametes can she produce with respect to that allele?

2. Expression of a dominant allele at a gene locus governing wing length in *D. melanogaster* results in long wings. Figure 12.25 shows how homozygosity for a recessive allele results in formation of vestigial (short) wings. You cross a homozygous dominant, long-winged fly with a homozygous recessive, vestigial-winged fly. Then you ask a technician to expose the fertilized eggs to a level of x-rays known to induce mutation and deletions. Later, when the irradiated eggs develop into adults, most flies are heterozygous and have long wings. A few have vestigial wings. What might explain these results?

3. Suppose expression of one allele of a Y-linked gene results in nonhairy ears in males. Expression of another allele results in rather long hairs, a condition called *hairy pinnae* (Figure 12.26).
   a. Why would you *not* expect females to have hairy pinnae?
   b. Any son of a hairy-eared male will also be hairy-eared, but no daughter will be. Explain why.

4. *Marfan syndrome* is a genetic disorder with pleiotropic effects. Expression of a mutant allele leads to the absence or abnormal formation of a connective tissue (fibrillin) in many organs. Often affected persons are tall and thin, with double-jointed fingers. Arms, fingers, and lower limbs are disproportionately long. The spine is abnormally curved. Eye lenses are easy to dislocate. Thin tissue flaps in the heart that act like valves to direct blood flow may billow the wrong way as the heart contracts. The heart beats irregularly. The aorta, the main artery carrying blood away from the heart, is fragile. Its diameter widens; its wall may tear.
   An estimated 40,000 people in the United States are affected, including a few prominent athletes. Spontaneous mutation gives rise to the allele. Genetic analysis shows it follows a pattern of autosomal dominant inheritance. What is the chance any child will inherit the allele if one parent is heterozygous for it?

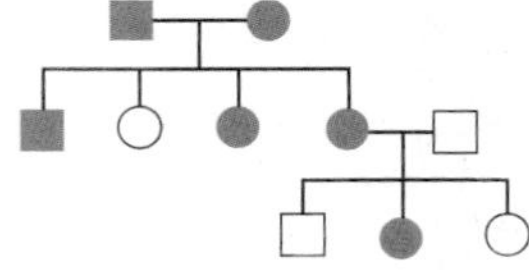

Figure 12.27 Identify the mystery pedigree.

5. In the Figure 12.27 pedigree, does the phenotype indicated by *red* circles and squares follow a Mendelian inheritance pattern that is autosomal dominant, autosomal recessive, or X-linked?

6. One kind of *muscular dystrophy*, a genetic disorder, is due to a recessive X-linked allele. Usually, symptoms start in childhood. Over time, a slowly progressing loss of muscle function leads to death, usually by age twenty or so. Unlike color blindness, this disorder is nearly always restricted to males. Suggest why.

7. Suppose you carry two linked genes with alleles *Aa* and *Bb*, respectively, as in Figure 12.28. If the crossover frequency between these two genes is zero, what genotypes would be expected among the gametes you produce, and with what frequencies?

8. Say you have alleles for lefthandedness and straight hair on one chromosome and alleles for righthandedness and curly hair on the homologous chromosome. If the two loci for the alleles are very close together along the chromosome, how likely is it that crossing over will occur between them? If they are distant from each other, is a crossover more or less likely to occur?

9. Individuals showing *Down syndrome* usually have an extra chromosome 21, so their body cells contain 47 chromosomes.
   a. At which stages of meiosis I and II could a mistake occur that could result in the altered chromosome number?
   b. In a few cases, 46 chromosomes are present, including two normal-appearing chromosomes 21 and a longer-than-normal chromosome 14. Explain how this situation can arise.

10. In the human population, mutation of two different genes on the X chromosome causes two types of X-linked *hemophilia* (types A and B). In a few known cases, a woman is heterozygous for both mutant alleles (one on each of her two X chromosomes). All her sons should have either hemophilia A or B. Yet, on very rare occasions, such a woman gives birth to a son who does not have hemophilia, and his one X chromosome does not have either mutant allele. Explain how such an X chromosome could arise.

Selected Key Terms

abortion *12.11*
allele *12.1*
aneuploidy *12.10*
autosome *12.1*
crossing over *12.1*
cytological marker *12.4*
deletion *12.9*
disease *12.6*
double-blind study *12.10*
duplication *12.9*
gene *12.1*
genetic abnormality *12.6*
genetic disorder *12.6*
genetic recombination *12.1*
homologous chromosome *12.1*
inversion *12.9*
in vitro *12.2*
in-vitro fertilization *12.11*
karyotype *CI*
linkage group *12.4*
linkage mapping *12.5*
nondisjunction *12.10*
pedigree *12.6*
polyploidy *12.10*
reciprocal cross *12.4*
sex chromosome *12.1*
syndrome *12.6*
translocation *12.9*
X chromosome *12.1*
X-linked gene *12.4*
Y chromosome *12.1*
Y-linked gene *12.4*

Readings

Fairbanks, D., and W. R. Andersen. 1999. *Genetics: The Continuity of Life*. Monterey, California: Brooks-Cole.

13

DNA STRUCTURE AND FUNCTION

Cardboard Atoms and Bent-Wire Bonds

One might have wondered, in the spring of 1868, why Johann Friedrich Miescher was collecting cells from the pus of open wounds and, later, from the sperm of a fish. Miescher, a physician, wanted to identify the chemical composition of the nucleus. These particular cells have very little cytoplasm, which makes it easier to isolate the nuclear material for analysis.

Miescher finally succeeded in isolating an organic compound with the properties of an acid. Unlike other substances in cells, it incorporated a notable amount of phosphorus. Miescher called the substance nuclein. He had discovered what came to be known many years later as **deoxyribonucleic acid**, or **DNA**.

The discovery did not cause even a ripple through the scientific community. At the time, no one really knew much about the physical basis of inheritance—that is, *which chemical substance encodes the instructions for reproducing parental traits in offspring*. Few even suspected that the cell nucleus might hold the answer. For a time, researchers generally believed hereditary instructions had to be encoded in the structure of some unknown class of proteins. After all, heritable traits are spectacularly diverse. Surely the molecules encoding information about those traits were structurally diverse also. Proteins are put together from potentially limitless combinations of twenty different amino acids, so the thinking was that they could function as the sentences (genes) in each cell's book of inheritance.

By the early 1950s, however, the results of many ingenious experiments clearly indicated that DNA was the substance of inheritance. Moreover, in 1951, Linus Pauling did something that no one had done before. Through his training in biochemistry, a talent for model building, and a few great educated guesses, he deduced the three-dimensional structure of the protein collagen. Pauling's discovery was truly electrifying. If someone could pry open the secrets of proteins, then why not assume the same might be done for DNA? And once the structural details of the DNA molecule were understood, wouldn't they provide clues to its biological functions? *Who would go down in history as having discovered the very secrets of inheritance?*

Scientists around the world started scrambling after that ultimate prize. Among them were James Watson, a young postdoctoral student from Indiana University, and Francis Crick, an unflappably exuberant researcher working at Cambridge University. Exactly how could DNA, a molecule that consisted of only four kinds of subunits, hold genetic information? Watson and Crick spent long hours arguing over everything they had read about the size, shape, and bonding requirements of the subunits of DNA. They fiddled with cardboard cutouts of the subunits. They even badgered chemists to help them identify any potential bonds they might have overlooked. Then they assembled models from bits of metal, held together with wire "bonds" bent at seemingly suitable angles.

In 1953, Watson and Crick put together a model that fit all the pertinent biochemical rules and all the facts about DNA that they had gleaned from other sources. They had discovered the structure of DNA (Figures 13.1 and 13.2). And the breathtaking simplicity of that structure enabled them to solve another long-standing riddle—*how the world of life can show such unity at the molecular level and yet show such spectacular diversity at the level of whole organisms.*

Figure 13.1 James Watson and Francis Crick posing in 1953 by their newly unveiled structural model of DNA.

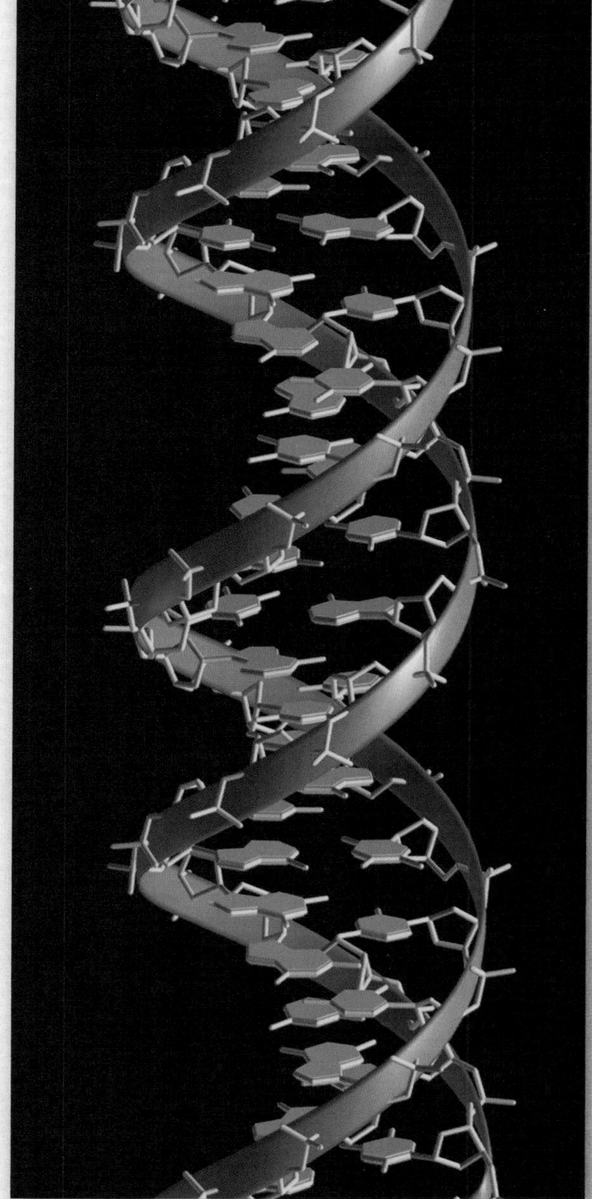

Figure 13.2 A recent computer-generated model of DNA, corresponding to the prototype that Watson and Crick put together decades ago.

With this chapter, we turn to investigations that led to our current understanding of DNA. The story is more than a march through details of its structure and function. *It also is revealing of how ideas are generated in science.* On the one hand, having a shot at fame and fortune quickens the pulse of men and women in any profession, and scientists are no exception. On the other hand, science proceeds as a community effort, with individuals sharing not only what they can explain but also what they do not understand. Even when an experiment fails to produce the anticipated results, it might turn up information that others can use or lead to questions that others can answer. Unexpected results, too, might be clues to something important about the natural world.

KEY CONCEPTS

1. In all living cells, DNA molecules are the storehouses of information about heritable traits.

2. In a DNA molecule, two strands of nucleotides twist together, like a spiral stairway. Each strand consists of four kinds of nucleotides that are the same except for one component—a nitrogen-containing base. The four bases are adenine, guanine, thymine, and cytosine.

3. Great numbers of nucleotides are arranged one after another in each strand of the DNA molecule. In at least some regions, the order in which one kind of nucleotide follows another is unique for each species. Hereditary information is encoded in the particular sequence of nucleotide bases.

4. Hydrogen bonds connect the bases of one strand of the DNA molecule to bases of the other strand. As a rule, adenine pairs (hydrogen-bonds) with thymine, and guanine with cytosine.

5. Before a cell divides, its DNA is replicated with the assistance of enzymes and other proteins. Each double-stranded DNA molecule starts unwinding. As it does so, a new, complementary strand is assembled bit by bit on the exposed bases of each parent strand, according to the base-pairing rule stated above.

13.1 DISCOVERY OF DNA FUNCTION

Early and Puzzling Clues

The year was 1928. Frederick Griffith, an army medical officer, was attempting to develop a vaccine against *Streptococcus pneumoniae*, a bacterium that is one cause of the lung disease pneumonia. (When introduced into the body, vaccines mobilize internal defenses against a real attack. Many vaccines are preparations of either killed or weakened bacterial cells.) Griffith never did develop a vaccine. But his work unexpectedly opened a door to the molecular world of heredity.

Griffith isolated and cultured two different strains of the bacterium. He noticed that colonies of one strain had a rough surface appearance, but those of the other strain appeared smooth. He designated the two strains *R* and *S* and used them in a series of four experiments:

1. Laboratory mice were injected with live R cells. The mice did not develop pneumonia, as indicated in Figure 13.3. *The R strain was harmless.*

2. Other mice were injected with live S cells. The mice died. Blood samples taken from them teemed with live S cells. *The S strain was pathogenic* (disease-causing).

3. S cells were killed by exposure to high temperature. Mice injected with these cells did not die.

4. Live R cells were mixed with heat-killed S cells and injected into mice. The mice died—and blood samples from them teemed with *live* S cells!

What was going on in the fourth experiment? Maybe heat-killed S cells in the mixture weren't really dead. But if that were true, then mice injected with heat-killed S cells alone (experiment 3) would have died. Maybe harmless R cells in the mixture had mutated into a killer form. But if that were true, then mice injected with the R cells alone (experiment 1) would have died.

The simplest explanation was as follows: *Heat killed the S cells but did not destroy their hereditary material—including the part that specified "how to cause infection."* Somehow, that material had been transferred from the dead S cells to living R cells, which put it to use.

Further experiments showed the harmless cells had indeed picked up information on causing infections and were permanently transformed into pathogens. After a few hundreds of generations, descendants of those transformed bacterial cells were still infectious!

The unexpected results of Griffith's experiments intrigued Oswald Avery and his fellow biochemists. In time they were even able to transform harmless bacterial cells with *extracts* of killed pathogenic cells. Finally in 1944, after rigorous chemical analyses, they felt confident in reporting that the hereditary substance in their extracts probably was DNA—not proteins, as was then widely believed. To give experimental evidence for their conclusion, they reported that they had added certain protein-digesting enzymes to some extracts, but cells exposed to those extracts were transformed anyway. To other extracts, they had added an enzyme that digests DNA but not proteins. Doing so blocked hereditary transformation.

Despite these impressive experimental results, many biochemists refused to give up on the proteins. Avery's findings, they said, probably applied only to bacteria.

genetic material
viral coat
sheath
base plate
tail fiber
a

Confirmation of DNA Function

By the early 1950s molecular detectives, including Max Delbrück, Alfred Hershey, Martha Chase, and Salvador Luria, were using viruses as experimental subjects. The viruses they had selected, called **bacteriophages**, infect *Escherichia coli* and other bacteria.

Viruses are biochemically simple infectious agents. Although they are not alive, they do contain hereditary information about building more new virus particles. At some point after a virus has infected a host cell, viral enzymes take over that cell's metabolic machinery—which starts churning out substances that are necessary to construct new virus particles.

1 Mice injected with live cells of harmless strain R.

Mice do not die. No live R cells in their blood.

2 Mice injected with live cells of killer strain S.

Mice die. Live S cells in their blood.

3 Mice injected with heat-killed S cells.

Mice do not die. No live S cells in their blood.

4 Mice injected with live R cells *plus* heat-killed S cells.

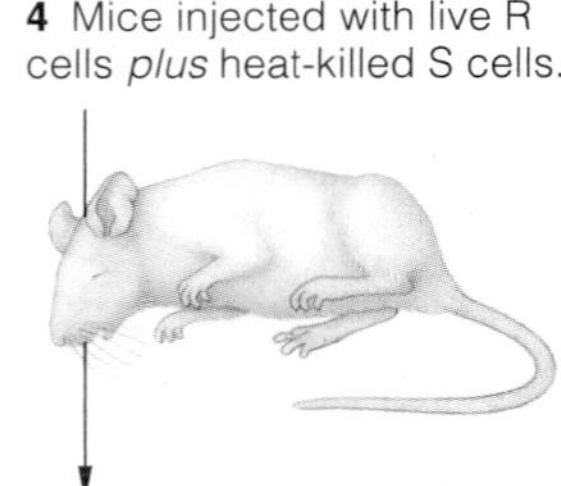

Mice die. Live S cells in their blood.

Figure 13.3 Results of Griffith's experiments with a harmless and a pathogenic strain of *Streptococcus pneumoniae*, as described in the text above.

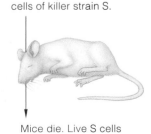

Further reading: Student Guide to InfoTrac on web site

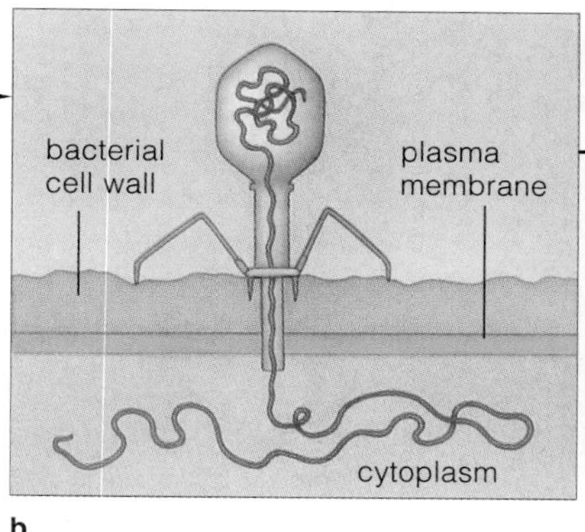
bacterial cell wall
plasma membrane
cytoplasm

b

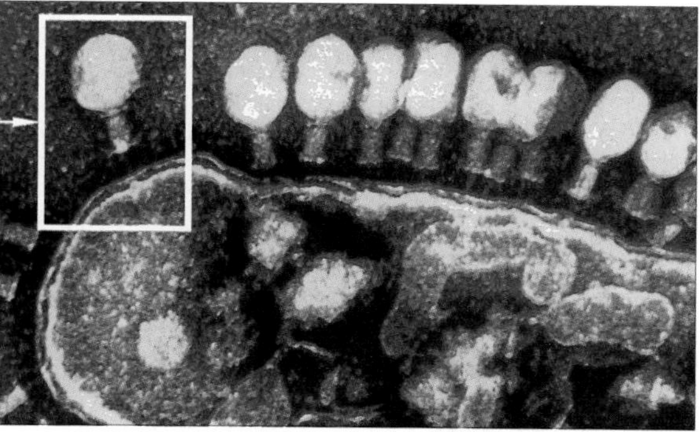

c

Figure 13.4 (**a**,**b**) *Far left*: Structural organization of a T4 bacteriophage. The diagram shows the genetic material of this type of virus being injected into the cytoplasm of a host cell. For this bacteriophage, it is DNA (the *blue*, threadlike strand). (**c**) Electron micrograph of T4 virus particles infecting a bacterium (*Escherichia coli*), which has just become an unwilling host.

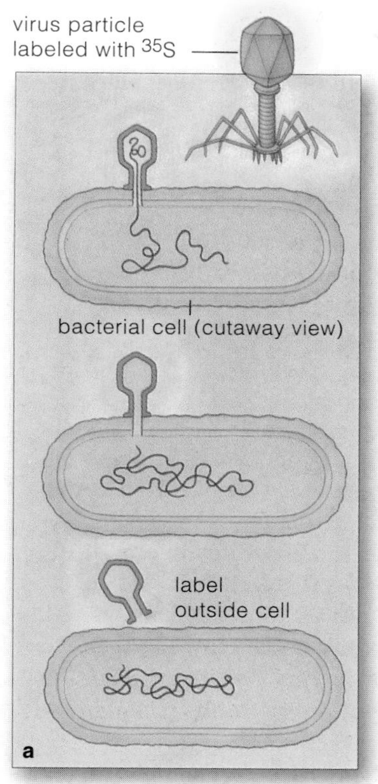
virus particle labeled with ^{35}S
bacterial cell (cutaway view)
label outside cell
a

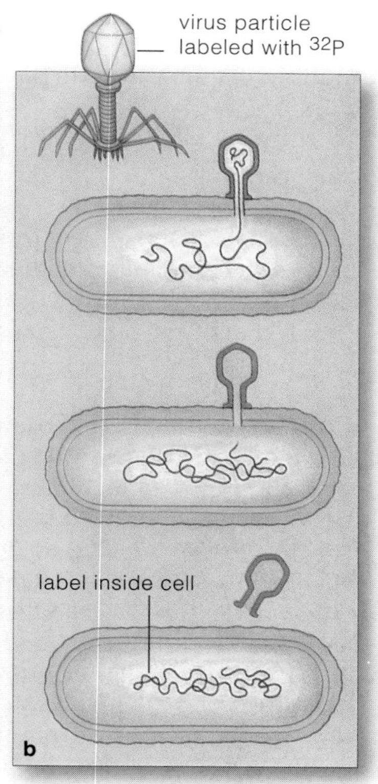
virus particle labeled with ^{32}P
label inside cell
b

Figure 13.5 Examples of the landmark experiments that pointed to DNA as the substance of heredity. In the 1940s, Alfred Hershey and his colleague, Martha Chase, were studying the biochemical basis of inheritance. They were aware that certain bacteriophages consist of proteins and DNA. *Did the proteins, DNA, or both contain viral genetic information?*

To find a possible answer, Hershey and Chase started with two known biochemical facts to design two experiments: First, the proteins of bacteriophages incorporate sulfur (S) but not phosphorus (P). Second, the DNA of bacteriophages contains phosphorus but not sulfur.

(**a**) In one experiment, some bacterial cells were grown on a culture medium that included a radioisotope of sulfur, ^{35}S. When the bacterial cells synthesized proteins, they had to take up that radioisotope—which the researchers used as a tracer. (Here you may wish to review Section 2.2.)

After the cells were labeled with the tracer, bacteriophages were allowed to infect them. As the infection ran its course, viral proteins were synthesized inside the host cells. These proteins also became labeled with ^{35}S. And so did the new generation of virus particles.

Next, the labeled bacteriophages were allowed to infect a new batch of unlabeled bacteria that were suspended in a fluid culture medium. Afterward, Hershey and Chase whirred the fluid in a kitchen blender. Whirring dislodged the viral protein coats from the cells. The particles became suspended in the fluid medium. Chemical analysis revealed the presence of labeled protein in the fluid. But there was no evidence of labeled protein *inside* the bacterial cells.

(**b**) For the second experiment, Hershey and Chase cultured more bacterial cells. The phosphorus that was available to them for synthesizing DNA included the radioisotope ^{32}P. Later, bacteriophages were allowed to infect the cells.

As predicted, the viral DNA synthesized inside infected cells became labeled, as did the new generation of virus particles. Next, the labeled particles were allowed to infect bacteria that were suspended in a fluid medium. Then they were dislodged from the host cells.

Analysis showed the labeled viral DNA was not in the fluid. DNA stayed *inside* host cells, where its hereditary instructions had to be used to make more virus particles. Here was evidence that DNA is the genetic material of this type of virus.

By 1952, researchers knew that some bacteriophages consist only of DNA and a protein coat. Also, electron micrographs revealed that the main part of the viruses remains *outside* the cells they are infecting (Figure 13.4). Possibly, such viruses were injecting genetic material alone *into* host cells. If that were true, was the material DNA, protein, or both? Through many experiments, researchers accumulated strong evidence that DNA, not proteins, serves as the molecule of inheritance. Figure 13.5 describes two of these landmark experiments.

Information for producing the heritable traits of single-celled and multicelled organisms is encoded in DNA.

13.2 DNA STRUCTURE

What Are the Components of DNA?

Long before the bacteriophage studies were under way, biochemists knew that DNA contains only four types of nucleotides that are the building blocks of nucleic acids. Each **nucleotide** consists of a five-carbon sugar (which, in DNA, is deoxyribose), a phosphate group, and one of the following nitrogen-containing bases:

adenine (A)	guanine (G)	thymine (T)	cytosine (C)

As you can see from Figure 13.6, all four types of nucleotides in DNA have their component parts joined together in much the same way. However, T and C are pyrimidines, which are single-ring structures. A and G are purines, which are larger, bulkier molecules; they have double-ring structures.

By 1949, Erwin Chargaff, a biochemist, had shared two crucial insights into the composition of DNA with the scientific community. First, the amount of adenine relative to guanine differs from one species to the next. Second, the amount of adenine in DNA always equals that of thymine, and the amount of guanine always equals that of cytosine. We may show this as:

A = T and **G = C**

The proportions of those four kinds of nucleotides relative to each other were tantalizing clues. In some way, the proportions almost certainly were related to the arrangement of the nucleotides in a DNA molecule.

The first convincing evidence of that arrangement emerged from Maurice Wilkins's research laboratory in England. Rosalind Franklin, one of Wilkins's colleagues, had obtained especially good **x-ray diffraction images** of DNA fibers. (Maybe for the reasons sketched out in Section 13.3, Franklin's contribution has only recently been acknowledged.) X-ray diffraction images can be made by directing a beam of x-rays at a molecule. The molecule scatters the beam in patterns that are captured on film. The pattern consists only of dots and streaks; it alone does not reveal molecular structure. However, researchers use photographic images of those patterns to calculate the positions of the molecule's atoms.

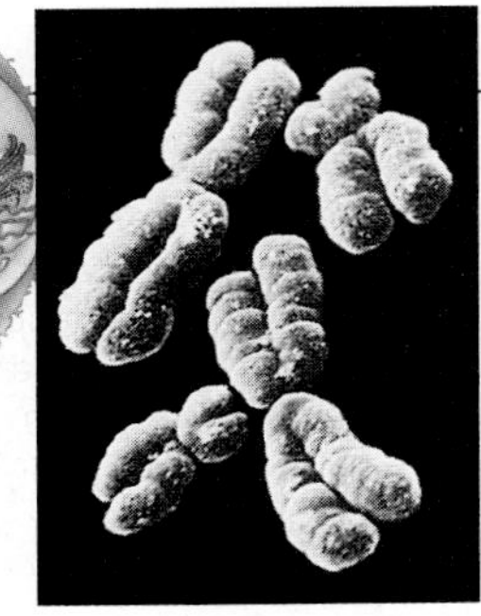

All chromosomes in a cell contain DNA. What does DNA contain? Four kinds of nucleotides, A, G, C, and T. Here are the structural formulas for those nucleotides:

ADENINE (A)
base with a double-ring structure

phosphate group

sugar (ribose)

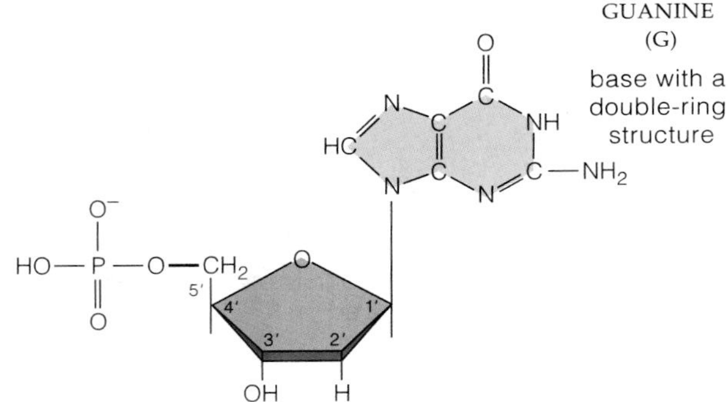
GUANINE
(G)
base with a double-ring structure

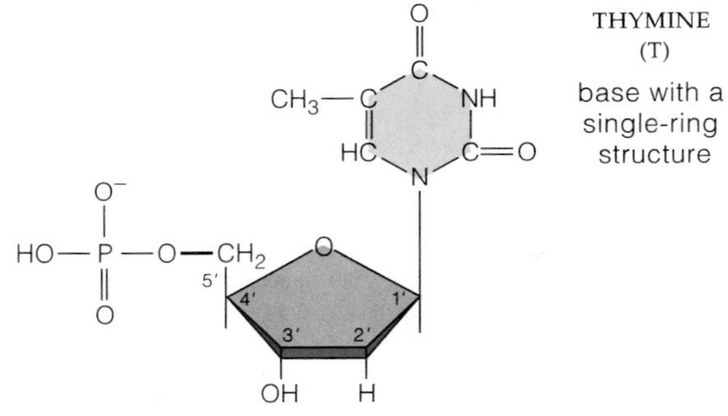
THYMINE
(T)
base with a single-ring structure

CYTOSINE (C)
base with a single-ring structure

Figure 13.6 Four kinds of nucleotides that serve as building blocks for DNA. The small numerals on the structural formulas identify the carbon atoms to which other parts of the molecule are attached.

As you can see, each nucleotide in a DNA molecule has a five-carbon sugar (shaded *red*), which has a phosphate group attached to the fifth carbon atom of its carbon ring structure. A nucleotide also has one of four kinds of nitrogen-containing bases (*blue*), which is attached to the first carbon atom. The four kinds of nucleotides in DNA differ only in which base they have: adenine, guanine, thymine, or cytosine.

Further reading: Student Guide to InfoTrac on web site

Figure 13.7 Representations of a DNA double helix. Notice how the two sugar–phosphate backbones run in *opposing* directions. Think of the sugar (deoxyribose) units of one strand as being upside down.

By comparing the numerals used to identify each carbon atom of the deoxyribose molecule (1′, 2′, 3′, and so on), you see that one strand runs in the 5′ → 3′ direction and the other runs in the 3′ → 5′ direction.

2-nanometer diameter, overall

distance between each pair of bases = 0.34 nanometer

In all these respects, the Watson–Crick model of DNA structure is consistent with the known biochemical and x-ray diffraction data.

each full twist of the DNA double helix = 3.4 nanometers

The pattern of base pairing (A only with T, and G only with C) is consistent with the known composition of DNA (A = T and G = C).

DNA does not readily lend itself to x-ray diffraction. However, researchers could rapidly spin a suspension of DNA molecules, spool them onto a rod, and gently pull them into gossamer fibers, like cotton candy. If the atoms in DNA were arranged in a regular order, x-rays directed at a fiber should scatter in a regular pattern that could be captured on film.

Calculations based on Franklin's images strongly indicated that the DNA molecule had to be long and thin, with a 2-nanometer diameter. Some molecular configuration was being repeated every 0.34 nanometer along its length, and another one every 3.4 nanometers.

Could the sequence of nucleotide bases be twisting, like a circular stairway? Certainly Pauling thought so. After all, he discovered the helical shape of collagen. He and everybody else—including Wilkins, Watson, and Crick—were thinking "helix." Watson later wrote, "We thought, why not try it on DNA? We were worried that *Pauling* would say, why not try it on DNA? Certainly he was a very clever man. He was a hero of mine. But we beat him at his own game. I still can't figure out why."

Pauling, it turned out, made a big chemical mistake. His model had hydrogen bonds at phosphate groups holding DNA's structure together. That does happen in highly acidic solutions. It doesn't happen in cells.

Patterns of Base Pairing

As Watson and Crick perceived, DNA consists of *two* strands of nucleotides, held together at their bases by hydrogen bonds. The bonds form when the two strands run in opposing directions and twist together into a double helix (Figure 13.7). Two kinds of base pairings form along the length of the molecule: A—T and G—C. This bonding pattern permits variation in the order of bases in any given strand. For example, in even a tiny stretch of DNA from a rose, gorilla, human, or any other organism, the sequence might be:

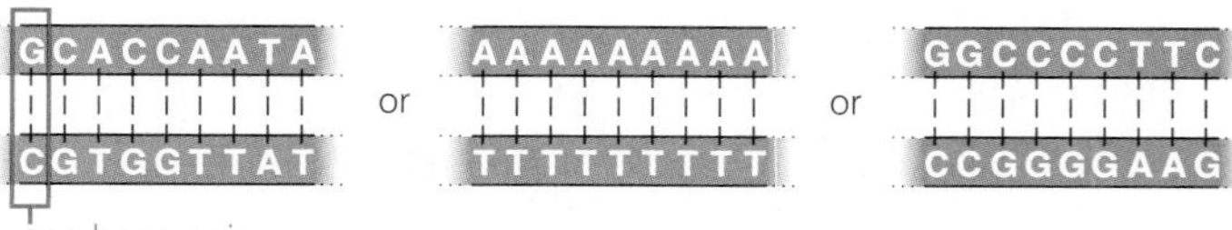
GCACCAATA
CGTGGTTAT
or
AAAAAAAAA
TTTTTTTTT
or
GGCCCCTTC
CCGGGGAAG
one base pair

In fact, even though all DNA molecules show the same bonding pattern, each species has unique base sequences in its DNA. *This molecular constancy and variation among species is the foundation for the unity and diversity of life.*

The pattern of base pairing between the two strands in DNA is constant for all species—A with T, and G with C. However, the DNA molecules of each species show unique differences in the sequence of base pairs along their length.

13.3 ROSALIND'S STORY

In 1951, Rosalind Franklin arrived at King's Laboratory of Cambridge University with impressive credentials. Earlier, in Paris, she had refined existing procedures for x-ray diffraction while studying the structure of coal. She also had devised a new mathematical approach to interpreting x-ray diffraction images and had built three-dimensional models of molecules, as Pauling had done. Now she had been asked to run an x-ray crystallography laboratory, which she would create with state-of-the-art equipment. Her assignment? Investigate the structure of DNA.

No one bothered to tell her that, down the hall, Maurice Wilkins was already working on the puzzle. Even the graduate student assigned to assist her failed to mention it. And no one bothered to tell Wilkins about Franklin's assignment, so he assumed she was just a technician hired to do his x-ray crystallography work because he could not do it himself. And so began a poisonous clash. To Franklin, Wilkins seemed inexplicably prickly; to Wilkins, Franklin displayed an appalling lack of deference that technicians usually show to researchers.

Figure 13.8 Rosalind Franklin.

Wilkins had a prized cache of crystalline fibers of DNA, each having parallel arrays of hundreds of millions of DNA molecules—and these he gave to his "technician."

Five months later, Franklin gave a talk on what she had learned so far. DNA, she said, might have two, three, or four parallel chains twisted into a helix, with phosphate groups projecting outward. She had measured DNA's density and assigned DNA fibers to 1 of 230 categories of crystals, based on the symmetry of their parallel chains.

With his background in crystallography, Crick would have recognized the significance of that symmetry *if* he had been present. (To wit, *paired* chains running in opposite directions would look the same even if flipped 180°. Two paired chains? No. DNA's density ruled that out. But *one pair* of chains? Yes!) Watson was in the audience, but he didn't have a clue to what Franklin was talking about.

Later, Franklin created an outstanding x-ray diffraction image of wet DNA fibers that fairly screamed *Helix!* She also worked out DNA's length and diameter. But she had been working with dry fibers for so long, she did not dwell on her new data. Wilkins, however, did. In 1953, he let Watson see Franklin's exceptional x-ray diffraction image and reminded him of what she had reported fourteen months earlier. And when Watson and Crick finally did focus on her data, they had the final bits of information necessary to start building a DNA model—one with two helically twisted chains running in opposing directions.

Not until ten years after Franklin's untimely death in 1958 did Watson acknowledge her pivotal discoveries.

13.4 DNA REPLICATION AND REPAIR

How Is a DNA Molecule Duplicated?

The discovery of DNA structure was a turning point in studies of inheritance. Until then, no one could explain **DNA replication**, or how the molecule of inheritance is duplicated before the cell divides. Once Watson and Crick had assembled their model, Crick understood at once how this might be done.

As he knew, enzymes can easily break the hydrogen bonds between the two nucleotide strands of a DNA molecule. When these enzymes and other proteins act on the molecule, one strand can unwind from the other, thereby exposing stretches of nucleotide bases. Cells have stockpiles of free nucleotides, and these can pair with the exposed bases.

Each parent strand remains intact, and a companion strand is assembled on each one according to this base-pairing rule: A to T, and G to C. As soon as a stretch of a new, partner strand forms on a stretch of the parent strand, the two twist together into a double helix, in the manner shown in Figure 13.9. Because the parent DNA strand is conserved during the replication process, half

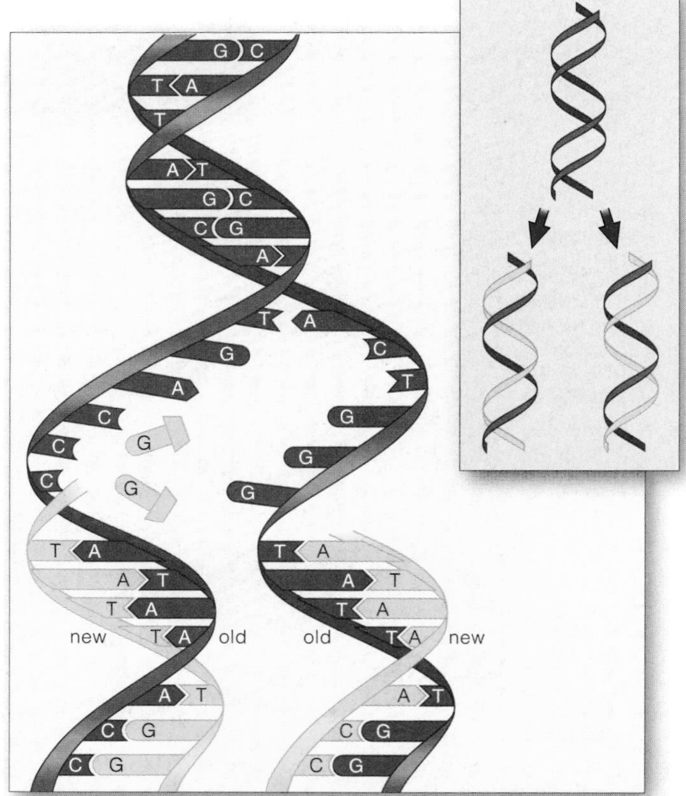

G C
T A
A T
G C
C G
new
old
old
new

Figure 13.9 Overview of the semiconservative nature of DNA replication. The original two-stranded DNA molecule is shown in *blue*. Each parent strand remains intact, and a new strand (*yellow*) is assembled on each one.

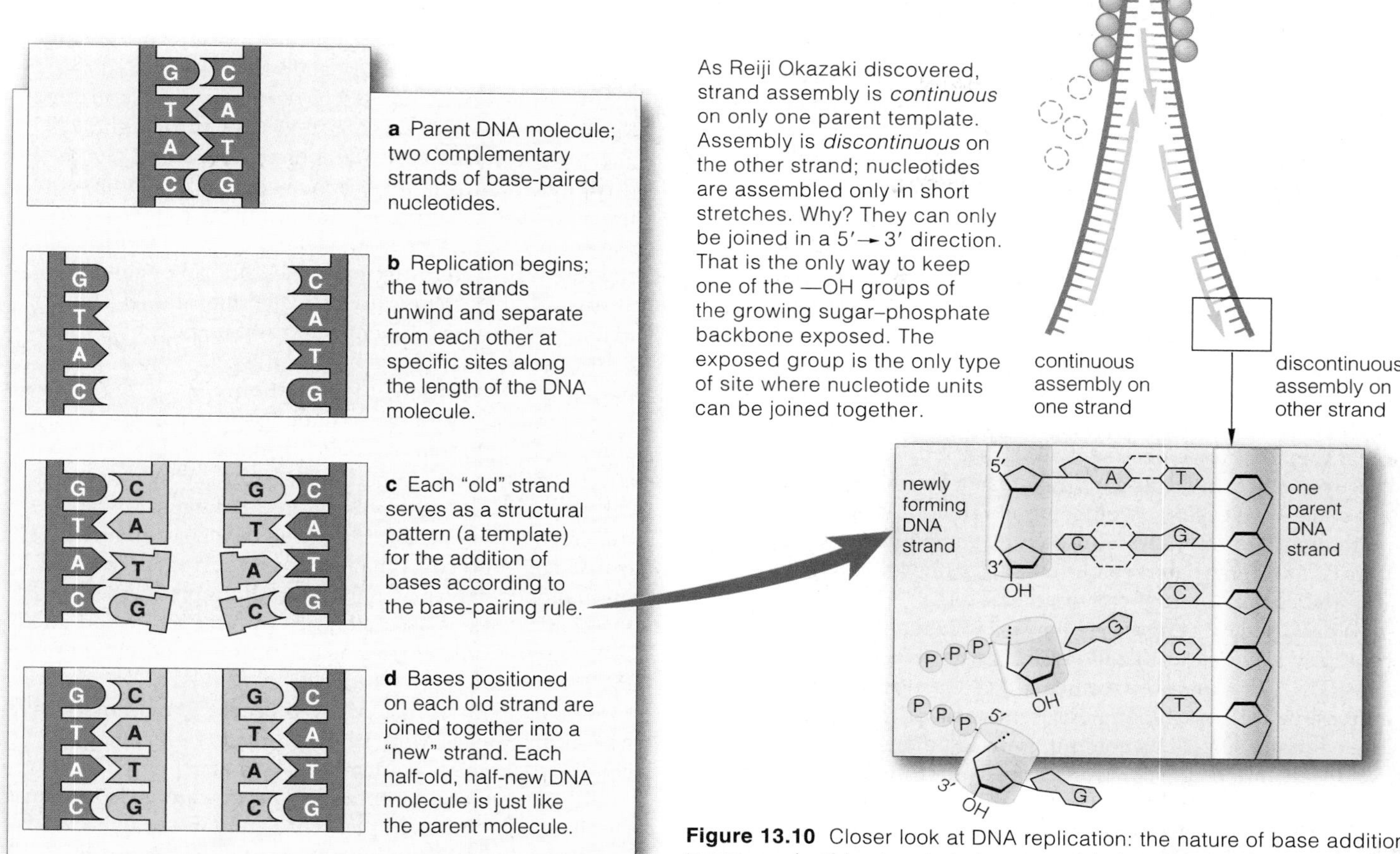
G C
T A
A T
C G
a Parent DNA molecule; two complementary strands of base-paired nucleotides.
b Replication begins; the two strands unwind and separate from each other at specific sites along the length of the DNA molecule.
c Each "old" strand serves as a structural pattern (a template) for the addition of bases according to the base-pairing rule.
d Bases positioned on each old strand are joined together into a "new" strand. Each half-old, half-new DNA molecule is just like the parent molecule.
As Reiji Okazaki discovered, strand assembly is continuous on only one parent template. Assembly is discontinuous on the other strand; nucleotides are assembled only in short stretches. Why? They can only be joined in a 5′→3′ direction. That is the only way to keep one of the —OH groups of the growing sugar–phosphate backbone exposed. The exposed group is the only type of site where nucleotide units can be joined together.
continuous assembly on one strand
discontinuous assembly on other strand
newly forming DNA strand
one parent DNA strand
5′
3′
OH
P-P-P
5′
3′
OH

Figure 13.10 Closer look at DNA replication: the nature of base additions to new nucleotide strands being assembled on a parent template.

of every double-stranded DNA molecule is "old" and half is "new" (Figure 13.9). That is why biologists refer to the process as *semiconservative* replication.

DNA replication uses a team of molecular workers. In response to cellular signals, the replication enzymes become active along the length of the DNA molecule. Together with other proteins, some enzymes unwind the strands in both directions and prevent them from rewinding. Enzyme action jump-starts the unwinding but is not required to unzip hydrogen bonds between the strands; hydrogen bonds are individually weak.

Now enzymes called **DNA polymerases** can attach short stretches of free nucleotides to unwound portions of the parent template (Figure 13.10). Free nucleotides brought up for strand assembly supply the energy that drives the replication process. Each has three phosphate groups attached. DNA polymerase splits away two of the groups, and some energy released by the reaction is used to attach the nucleotide to a growing strand.

DNA ligases fill in tiny gaps between the new short stretches to form a continuous strand. Then enzymes wind the template strand and complementary strand together to form a DNA double helix.

As you will read in Section 16.2, some replication enzymes are used in recombinant DNA technology.

Monitoring and Fixing the DNA

DNA polymerases, DNA ligases, and other enzymes also engage in **DNA repair**. By this process, enzymes excise and repair altered parts of the base sequence in one strand of a double helix. Suppose a few bases are lost from one strand during replication. DNA polymerases can "read" the complementary sequence on the other strand. With assistance from other repair enzymes, they restore the original sequence. And remember crossing over, whereby nonsister chromatids of homologous chromosomes break and exchange segments at meiosis? The same enzymes that bring about that recombination also repair double-strand breaks. They initiate strand exchange with homologous DNA. Section 14.5 gives you an idea of what happens when mutation or some other factor compromises the excision–repair function.

DNA is replicated prior to cell division. Enzymes unwind its two strands. Each strand remains intact throughout the process—it is conserved—and enzymes assemble a new, complementary strand on each one.

Enzymes involved in replication also repair the DNA where base-pairing errors have crept into the nucleotide sequence.

13.5 DOLLY, DAISIES, AND DNA

Imagine the possibility of making a genetically identical copy—a clone—of yourself. Is the image that far-fetched? Consider this: Researchers have been cloning complex animals for more than a decade. For example, some use in vitro fertilization methods to grow cattle embryos in petri dishes. After cleavage is under way in a new embryo, they split the tiny cluster of embryonic cells, and the two tinier clusters continue to develop as two identical-twin cattle embryos. These are implanted in surrogate mothers, and they are born as cloned calves. Figure 13.11*a* shows cloned Holsteins that are prized for their milk production.

A researcher who clones farm animals derived from embryonic cells has to wait for the clones to grow up to see if they display a desired trait. Using a differentiated cell from a prized adult would be faster, for the prized genotype would already be demonstrated and could be maintained indefinitely. However, it seemed impossible to trick a differentiated cell into using its DNA in a new way to become the first cell of an embryo.

What does "differentiated" mean? When an embryo first grows from a fertilized egg, all of its cells have the same DNA and are pretty much alike. Then different embryonic cells start using different parts of their DNA. Their unique selections commit them to being liver cells, heart cells, brain cells, and other specialists in structure, composition, and function (Section 15.3).

Then, in 1997 in Scotland, a research group led by Ian Wilmut coaxed a differentiated cell from a sheep udder into becoming an "uncommitted" first cell of an embryo. Thus a lamb named Dolly was born a clone of her mother without help from a father. Wilmut's group had managed to transfer nuclei from differentiated cells into enucleated unfertilized eggs, which they had removed earlier from a pregnant ewe. (*Enucleated* means that the nucleus was surgically removed from a cell with a microscopically small needle.) After hundreds of attempts, one egg proceeded to grow, by mitotic cell divisions, into a whole sheep. Dolly developed into a healthy adult and gave birth to a lamb of her own (Figure 13.11*b*).

In science, extraordinary claims call for extraordinary proof—in this case, successful repeats of the experiment. Resounding success came in 1998. Ryozo Yanagimachi and his coworkers at the University of Hawaii cloned three generations of mice (*Mus musculus*). They quickly transferred nuclei from mature cumulus cells from an ovary into unfertilized, enucleated eggs. (Cumulus cells provide nutritional support to neighboring eggs.) Shortly afterward, they chemically activated the eggs as a way to jump-start development into fully formed mice. The resting period may have helped the cells adjust to the unconventional mode of fertilization.

By the year's end, experimenters in Japan gave us a new way of looking at Daisy the Cow. Yukio Tsunoda and his coworkers slipped nuclei from a cow's cumulus cells and oviduct cells into enucleated eggs. Of 240 transfers, 67 embryos formed. Ten of the survivors were transferred into cows that served as surrogate mothers. Eight cloned calves were born. Only four survived, but the results suggest that cloning cows from differentiated cells may become at least as efficient as producing them by in vitro fertilization.

In short, individuals of three genera of mammals have now been cloned. More importantly, the efficiency of cloning techniques is increasing at a startling pace.

All of which raises some far-out questions. Is human cloning next? For example, will a human Daisy be able to reproduce copies of herself without involving men? Will men be able to pay to have their DNA inserted into an enucleated cell, have the cell implanted in a surrogate mother, and make a little repeat of themselves?

Recently Lee Silver, a molecular biologist at Princeton University, said he knows of two qualified specialists in human fertility who are interested in cloning humans. He suggested that anyone who thinks the technology will move slowly is being naive. Ethically speaking, should such an attempt be made at all? At this writing, scientists and nonscientists alike are actively debating the issue.

Figure 13.11 (**a**) A clone of Holsteins resulting from in vitro fertilization. (**b**) Dolly, a cloned sheep. She started her life as a differentiated cell that was extracted from an adult ewe, then induced to undergo mitotic cell divisions. She is shown with her first lamb. She clearly is able to breed normally and reproduce the old-fashioned way. However, her DNA may be deteriorating (aging) faster than it should.

Further reading: Student Guide to InfoTrac on web site

13.6 SUMMARY

1. The hereditary information of cells and multicelled organisms is encoded in DNA (deoxyribonucleic acid).

2. DNA consists of nucleotide subunits. Each of these has a five-carbon sugar (deoxyribose), one phosphate group, and one of four kinds of nitrogen-containing bases (adenine, thymine, guanine, or cytosine).

3. A DNA molecule consists of two nucleotide strands twisted together into a double helix. The bases of one strand pair (hydrogen-bond) with bases of the other.

4. The bases of the two strands in a DNA double helix pair in constant fashion. Adenine pairs with thymine (A to T), and guanine with cytosine (G to C). *Which* base pair follows the next (A–T, T–A, G–C, or C–G) varies along the length of the strands.

5. Overall, the DNA of one species includes a number of unique stretches of base pairs that set it apart from the DNA of all other species.

6. During DNA replication, enzymes unwind the two strands of a double helix and assemble a new strand of complementary sequence on each parent strand. Two double-stranded molecules result. One strand of each molecule is "old" (it is conserved); the other is "new."

7. Some of the enzymes involved in DNA replication also repair DNA where base-pairing errors have been introduced into the nucleotide sequence.

Review Questions

1. Name the three molecular parts of a nucleotide in DNA. Also name the four different bases in these nucleotides. *13.2*

2. What kind of bond joins two DNA strands in a double helix? Which nucleotide base-pairs with adenine? With guanine? *13.2*

3. Explain how DNA molecules can show both constancy and variation from one species to the next. *13.2*

Self-Quiz (*Answers in Appendix III*)

1. Which is *not* a nucleotide base in DNA?
   a. adenine c. uracil e. cytosine
   b. guanine d. thymine

2. What are the base-pairing rules for DNA?
   a. A–G, T–C c. A–U, C–G
   b. A–C, T–G d. A–T, G–C

3. A DNA strand having the sequence C–G–A–T–T–G would be complementary to the sequence ______.
   a. C–G–A–T–T–G c. T–A–G–C–C–T
   b. G–C–T–A–A–G d. G–C–T–A–A–C

4. One species' DNA differs from others in its ______.
   a. sugars c. base sequence
   b. phosphate groups d. all of the above

5. When DNA replication begins, ______.
   a. the two DNA strands unwind from each other
   b. the two DNA strands condense for base transfers
   c. two DNA molecules bond
   d. old strands move to find new strands

6. DNA replication requires ______.
   a. free nucleotides c. many enzymes
   b. new hydrogen bonds d. all of the above

7. Match the DNA terms appropriately.
   ____ DNA polymerase a. two nucleotide strands that are twisted together
   ____ constancy in base pairing b. A with T, G with C
   ____ replication c. hereditary material duplicated
   ____ DNA double helix d. replication enzyme

Critical Thinking

1. Chargaff's data suggested that adenine pairs with thymine, and guanine pairs with cytosine. What other data available to Watson and Crick suggested that adenine-guanine and cytosine-thymine pairs normally do not form?

2. One of Matthew Meselson and Frank Stahl's experiments supported the semiconservative model of DNA replication. The researchers made "heavy" DNA by growing *Escherichia coli* in a medium enriched with ^{15}N, a heavy isotope of nitrogen. They prepared "light" DNA by growing *E. coli* in the presence of ^{14}N, the more common isotope. An available technique helped them identify which replicated molecules were heavy, light, or hybrid (one heavy strand, one light). Use two pencils of two different colors, one for heavy strands and one for light strands. Starting with a DNA molecule having two heavy strands, sketch the daughter molecules that would form after one replication in a ^{14}N-containing medium. Now sketch the four DNA molecules that would result if these daughter molecules were replicated a second time in the ^{14}N medium.

3. Mutations (permanent changes in base sequences of genes) are the original source of genetic variation. This variation is the raw material of evolution. Yet how can both statements be true, given that cells have efficient mechanisms to repair DNA before mutations can become established?

4. As indicated in Section 4.11, a pathogenic strain of *E. coli* has acquired an ability to produce a dangerous toxin that has caused medical problems and fatalities. This is especially the case for young children who have ingested undercooked, contaminated beef. Develop a hypothesis to explain how a normally harmless bacterium such as *E. coli* can become a pathogen.

5. In October 1999, scientists announced the amazing discovery of a woolly mammoth that had been frozen in glacial ice for the past 20,000 years. Its soft organs are intact. They plan to carefully use hair dryers to thaw it. They want to minimize disruption of its cells, and its DNA—which they plan to use to clone a woolly mammoth. Consider Section 13.5, then speculate on the pros and cons of cloning an extinct animal.

Selected Key Terms

adenine (A) *13.2*
bacteriophage *13.1*
cytosine (C) *13.2*
deoxyribonucleic acid (DNA) *CI*
DNA ligase *13.4*
DNA polymerase *13.4*
DNA repair *13.4*
DNA replication *13.4*
guanine (G) *13.2*
nucleotide *13.2*
thymine (T) *13.2*
x-ray diffraction image *13.2*

Readings *See also www.infotrac-college.com*

Watson, J. 1978. *The Double Helix.* New York: Atheneum. Highly personal view of scientists and their methods, interwoven into an account of how DNA structure was discovered.

Wolfe, S. 1995. *Introduction to Molecular and Cellular Biology.* Belmont, California: Wadsworth. Comprehensive and accessible.

14 FROM DNA TO PROTEINS

Beyond Byssus

Picture a mussel, of the sort shown in Figure 14.1. Hard-shelled but soft of body, it is using its muscular foot to probe a wave-scoured rock. At any moment, pounding waves can whack the mussel into the water, hurl it repeatedly against the rock with shell-shattering force, and so offer up a gooey lunch for gulls.

By chance, the mussel's foot comes across a crevice in the rock. The foot moves, broomlike, and sweeps the crevice clean. It presses down, forcing air out from underneath it, then arches up. The result is a vacuum-sealed chamber, rather like the one that forms when a plumber's rubber plunger is being squished down and up to unclog a drain. Into this vacuum chamber the mussel spews a fluid, consisting of keratin and other proteins, which bubbles into a sticky foam. Now, by curling its foot into a small tubular shape and pumping the foam through it, the mussel produces sticky threads about as wide as a human whisker. As a final touch, it varnishes the threads with another type of protein and ends up with an adhesive called byssus, which anchors the mussel to the rock.

Byssus is the world's premier underwater adhesive. Nothing that humans have manufactured even comes close. (Sooner or later, water chemically degrades or deforms synthetic adhesives.) Byssus truly fascinates biochemists, dentists, and surgeons looking for better ways to do tissue grafts and to rejoin severed nerves. Genetic engineers insert mussel DNA into yeast cells,

Figure 14.1 Mussels (*Mytilus californianus*) busily demonstrating the importance of proteins for survival. When mussels come across a suitable anchoring site, they use their muscular foot rather like a plumber's plunger to create a vacuum chamber. In this chamber they manufacture the world's best underwater adhesive from a mix of proteins. The adhesive anchors them to substrates in their wave-swept habitat.

which go on to reproduce in large numbers and serve as "factories" for translating mussel genes into useful quantities of proteins. This exciting work, like the mussel's own byssus-building efforts, starts with one of life's universal precepts: *Every protein is synthesized in accordance with instructions contained in DNA.*

You are about to trace the steps leading from DNA to proteins. Many enzymes are players in this pathway. So is another kind of nucleic acid besides DNA. The same steps produce *all* proteins, from mussel-inspired adhesives to the keratin in your hair and fingernails to the insect-digesting enzymes of a Venus flytrap.

Start out by thinking of each cell's DNA as a book of protein-building instructions. The alphabet used to create the book is simple enough: A, T, G, and C (for the nucleotide bases adenine, thymine, guanine, and cytosine). How do you get from that alphabet to a protein? The answer starts with DNA's structure.

DNA, recall, is a double-stranded molecule. Which kind of nucleotide base follows the next along the length of a strand—that is, the **base sequence**—differs from one kind of organism to the next. As you read in the preceding chapter, before a cell divides, its DNA is replicated as the two strands unwind entirely from each other. However, at other times in a cell's life, the two strands unwind only in certain regions to expose particular base sequences—genes. Most of the genes contain instructions for building proteins.

It takes two steps, **transcription** and **translation**, to carry out a gene's protein-building instructions. In eukaryotic cells, transcription proceeds in the nucleus. At this step, a selected base sequence in DNA serves as a structural pattern—a template—for assembling a strand of **ribonucleic acid** (RNA) from the cell's pool of free nucleotides. Afterward, the RNA moves into the cytoplasm, where translation proceeds. At this second step, RNA directs the assembly of amino acids into polypeptide chains. The newly formed chains become folded into the three-dimensional shapes of proteins.

In short, DNA guides the synthesis of RNA, then RNA guides the synthesis of proteins:

DNA —*transcription*→ RNA —*translation*→ PROTEIN

The newly synthesized proteins will play structural and functional roles in cells. And some even will have roles in synthesizing more DNA, RNA, and proteins.

KEY CONCEPTS

1. Life cannot exist without enzymes and other proteins. Proteins consist of polypeptide chains, which consist of amino acids. The sequence of amino acids corresponds to a gene, which is a sequence of nucleotide bases in a DNA molecule.

2. The path leading from genes to proteins consists of two steps, called transcription and translation.

3. In transcription, the double-stranded DNA molecule is unwound at a gene region, then an RNA molecule is assembled on the exposed bases of one of the strands.

4. In translation, a certain type of RNA directs the linkage of one amino acid after another, in the sequence required to produce the specified polypeptide chain.

5. With few exceptions, the genetic "code words" by which DNA's instructions are translated into proteins are the same in all species of organisms.

6. A mutation is a permanent alteration in a gene's base sequence. Mutations are the original source of genetic variation in populations.

7. Mutations cause changes in protein structure, protein function, or both. The changes may lead to small or large differences in traits among individuals of a population.

14.1 CONNECTING GENES WITH PROTEINS

GARROD'S HYPOTHESIS Early in the 1900s Archibald Garrod, a physician, was puzzling over certain illnesses. They appeared to be heritable, for they kept recurring in the same families. Those illnesses also appeared to be metabolic disorders. In each case, blood or urine samples from affected patients contained abnormally high levels of a substance known to be produced at a certain step in a metabolic pathway.

Most likely, the enzyme that operated at the *next* step in the metabolic pathway was defective. Something was preventing it from chemically recognizing or interacting properly with the substance that is supposed to be its substrate. If Garrod's reasoning were correct, then the metabolic pathway would be blocked from that step onward. Figure 14.2 illustrates this concept.

Garrod's hypothesis could explain why molecules of a particular substance were accumulating in excess amounts in the body fluids of affected individuals. He suspected that his patients differed from unaffected individuals in a key aspect of their metabolism: Each one had inherited a single metabolic defect. Garrod concluded that specific "units" of inheritance (genes) are expressed through the synthesis of specific enzymes of metabolic pathways.

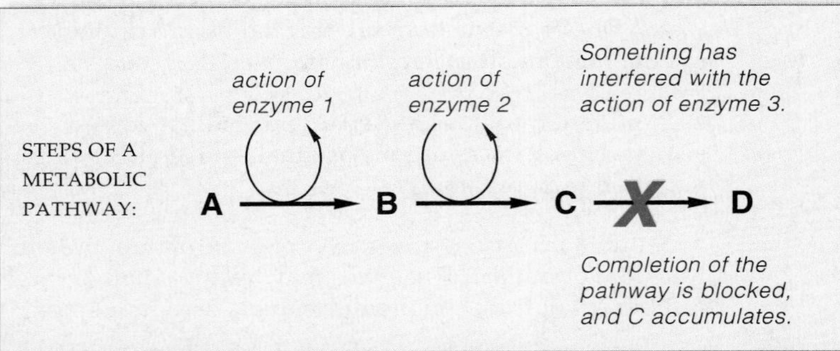
action of enzyme 1
action of enzyme 2
Something has interfered with the action of enzyme 3.
STEPS OF A METABOLIC PATHWAY:
A
B
C
X
D
Completion of the pathway is blocked, and C accumulates.

Figure 14.2 Example of how a defective enzyme can block completion of a metabolic pathway.

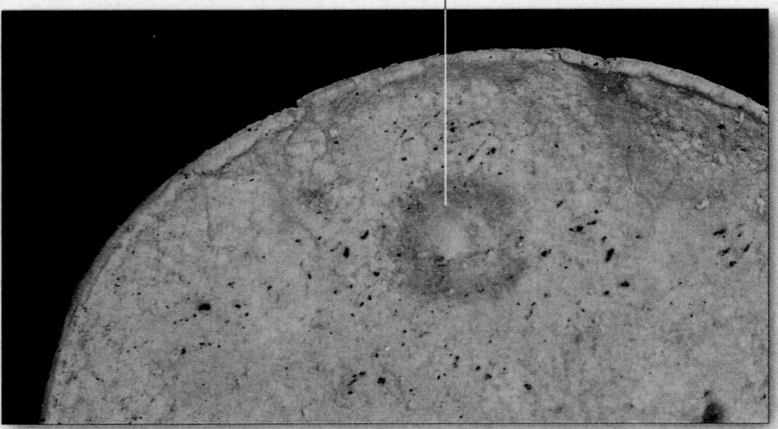
fungal colony on a tortilla, and isn't that appetizing

Figure 14.3 Colony of the red bread mold (*Neurospora crassa*) and other fungal species on a stale tortilla.

BEADLE, TATUM, AND A BREAD MOLD Thirty-three years later two researchers, George Beadle and Edward Tatum, were experimenting with the red bread mold (*Neurospora crassa*). This fungus is a common spoiler of baked goods (Figure 14.3). But it also lends itself to genetic experiments and has become an organism of choice in the laboratory. *N. crassa* can be grown easily on an inexpensive culture medium that contains only sucrose, mineral salts, and biotin, which is one of the B vitamins. The fungal cells can synthesize all of the other nutrients they require, including other vitamins.

Suppose a fungal enzyme that takes part in a synthesis pathway is defective as a result of a gene mutation. The researchers suspected that this had happened in some of the strains of *N. crassa* that they were studying. One strain grew only when it was supplied with vitamin B_6, another with vitamin B_1, and so on.

Chemical analysis of cell extracts revealed a different defective enzyme in each mutant strain. In other words, *each inherited mutation corresponded to a defective enzyme*. Here was evidence favoring the "one-gene, one-enzyme" hypothesis.

CLUES FROM GEL ELECTROPHORESIS Later on, the one-gene, one-enzyme hypothesis was refined as a result of investigations into the genetic basis of *sickle-cell anemia*. This heritable disorder arises from the presence of an abnormal version of a protein, hemoglobin, in the red blood cells of affected individuals. The abnormal hemoglobin is designated HbS instead of HbA (Section 11.5).

In 1949, the biochemists Linus Pauling and Harvey Itano subjected molecules of HbS and HbA to **gel electrophoresis**. This laboratory procedure uses an electric field to move molecules through a viscous gel and separate them according to their size, shape, and net surface charge. Often the gel is sandwiched between glass or plastic plates to form a viscous slab. The two ends of the slab are suspended in two salt solutions that are connected by electrodes to a power source (Figure 14.4). When voltage is applied to the apparatus, the molecules present in the gel migrate through the electric field according to their individual charge, and they move away from one another in the gel. Later on, the molecules can be pinpointed by staining the gel after a predetermined period of electrophoresis.

Pauling and Itano carefully layered a mixture of HbS and HbA molecules on the gel at the top of the slab. As the molecules gradually migrated down through the gel, they separated into distinct bands. The band that moved fastest carried the greatest

Further reading: Student Guide to InfoTrac on web site

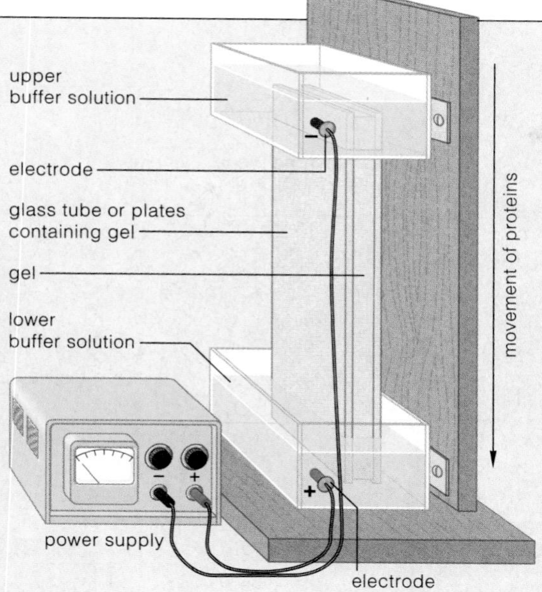

upper buffer solution
electrode
glass tube or plates containing gel
gel
lower buffer solution
power supply
electrode
movement of proteins

Figure 14.4 Diagram of one type of apparatus that is used for gel electrophoresis studies.

surface charge, and this turned out to be composed of HbA molecules. HbS molecules moved more slowly.

ONE GENE, ONE POLYPEPTIDE Later, Vernon Ingram pinpointed the biochemical difference between HbS and HbA. Hemoglobin, recall, has four polypeptide chains (Figure 14.5). Two of the chains are designated alpha and the other two, beta. An abnormal HbS chain arises from a gene mutation that affects protein synthesis. The mutation causes valine instead of glutamate to be added as the sixth amino acid of the beta chain (Figure 14.5*c*). Whereas glutamate carries an overall negative charge, valine has no net charge. That is the reason why HbS chains behaved differently in the electrophoresis studies.

Because of that one mutation, HbS hemoglobin has a "sticky" (hydrophobic) patch. In blood capillaries, where oxygen concentrations are at their lowest, hemoglobin molecules interact at the sticky patches. They aggregate into rods and distort red blood cells, and the consequences adversely affect organs throughout the body (Section 11.5).

The discovery of the genetic difference between alpha and beta chains of hemoglobin suggested that *two* genes code for hemoglobin—one for each kind of polypeptide chain. More importantly, it provided further evidence that genes code for proteins in general, not just for enzymes.

And so a more precise hypothesis emerged: *The amino acid sequences of polypeptide chains—the structural units of proteins—are encoded in genes.*

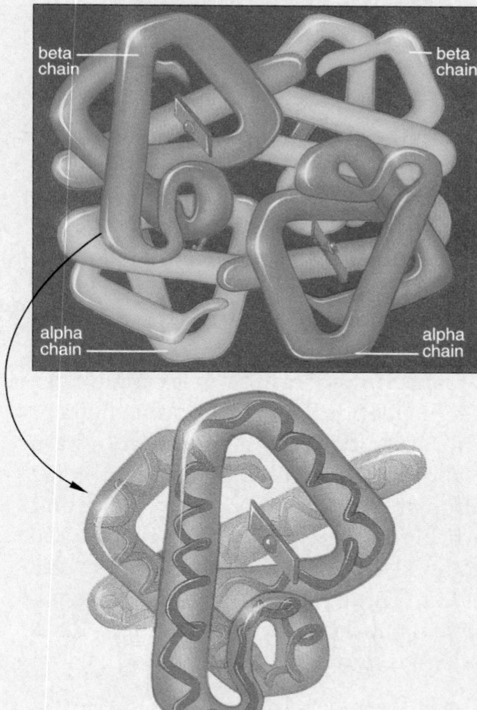

beta chain
beta chain
alpha chain
alpha chain

a *Above*: Arrangement of the four polypeptide chains and heme groups of a hemoglobin molecule. *Below*: Closer view of one of the beta chains.

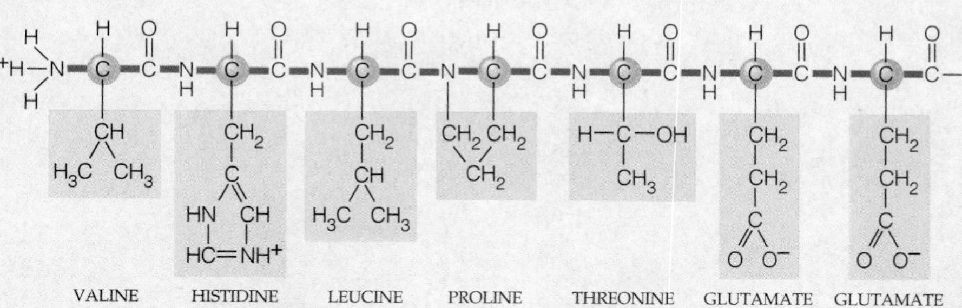

VALINE
HISTIDINE
LEUCINE
PROLINE
THREONINE
GLUTAMATE
GLUTAMATE

b Normal sequence of amino acids at the start of the beta chain that is characteristic of HbA molecules.

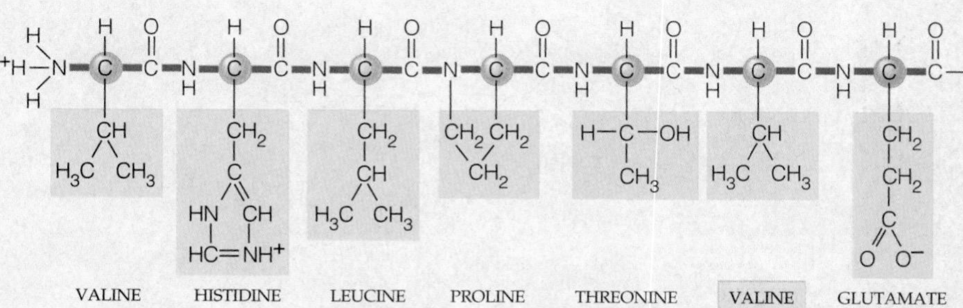

VALINE
HISTIDINE
LEUCINE
PROLINE
THREONINE
VALINE
GLUTAMATE

c The single amino acid substitution (*yellow*) that results in the abnormal beta chain that is characteristic of HbS molecules.

Figure 14.5 A single amino acid substitution that starts with a gene mutation and ends with the symptoms of sickle-cell anemia, which are described in Section 11.5.

14.2 HOW IS DNA TRANSCRIBED INTO RNA?

The Three Classes of RNA

Before turning to the details of protein synthesis, let's clarify one point. The chapter introduction might have left you with the impression that synthesis of proteins requires only one class of RNA molecules. Actually, it requires three. Transcription of most genes produces **messenger RNA**, or **mRNA**—the only class of RNA that carries *protein-building* instructions. Transcription of some other genes produces **ribosomal RNA**, or **rRNA**, a major component of ribosomes. Ribosomes, recall, are the structural units upon which polypeptide chains are assembled. Transcription of still other genes produces **transfer RNA**, or **tRNA**, which delivers amino acids one by one to a ribosome in the order specified by mRNA.

The Nature of Transcription

An RNA molecule is almost but not quite like a single strand of DNA. RNA, too, consists of only four types of nucleotides. Each nucleotide has a five-carbon sugar, ribose (not DNA's deoxyribose), a phosphate group, and a base. Three types of bases—adenine, cytosine, and guanine—are the same in RNA and DNA. But in RNA, the fourth type of base is **uracil**, not thymine (Figure 14.6). Like thymine, uracil can pair with adenine. This means a new RNA strand can be put together on a DNA region according to base-pairing rules (Figure 14.7).

Transcription resembles DNA replication in another respect. Enzymes add nucleotides to a growing RNA strand one at a time, in the 5′ → 3′ direction. Here you might wish to refer to the simple explanation of strand assembly in Section 13.4 (Figure 13.10.)

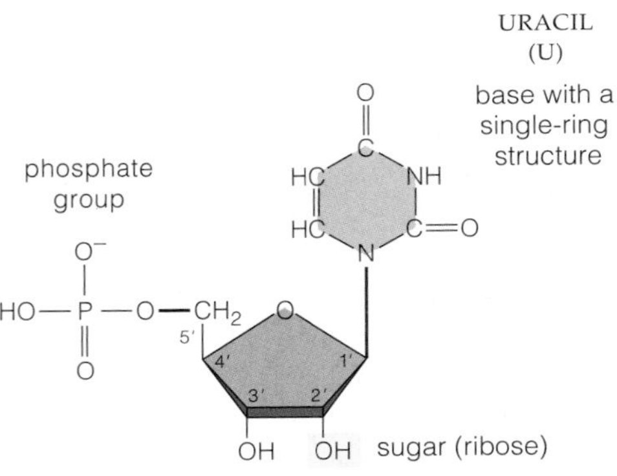
URACIL
(U)
base with a single-ring structure
phosphate group
O⁻
HO—P—O—CH₂
O
HC
HC
C
NH
N
C=O
5′
4′
1′
3′
2′
OH
OH
sugar (ribose)

Figure 14.6 Structural formula for one of the four types of RNA nucleotides. The three others have a different base (adenine, guanine, or cytosine instead of the uracil shown here). Compare Section 13.2, which shows DNA's four nucleotides. Notice that the sugars of DNA and RNA differ at one group only (*yellow*).

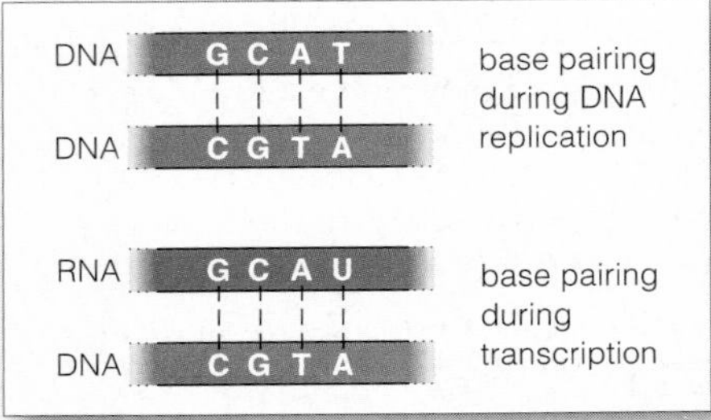
DNA
G C A T
DNA
C G T A
base pairing during DNA replication
RNA
G C A U
DNA
C G T A
base pairing during transcription

Figure 14.7 An example of base pairing of RNA with DNA during transcription, as compared to base pairing during DNA replication.

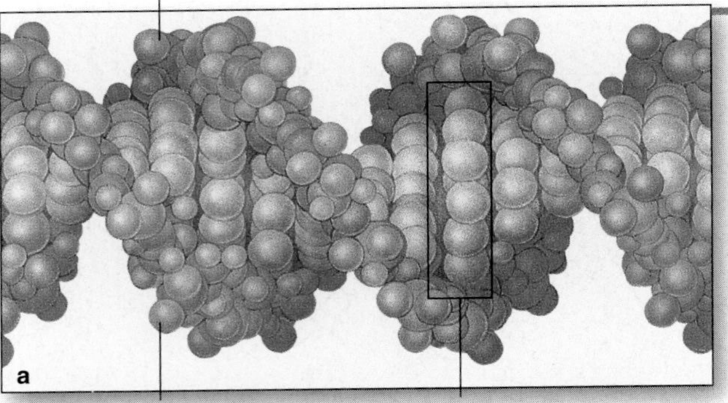
sugar-phosphate backbone of one strand of nucleotides in a DNA double helix
a
sugar-phosphate backbone of the other strand of nucleotides
part of the sequence of base pairs in DNA

Figure 14.8 The process of gene transcription, by which an RNA molecule is assembled on a DNA template. The sketch in (**a**) shows a gene region in part of a DNA double helix. In this region, the base sequence of one of the two nucleotide strands (not both) is about to be transcribed into an RNA molecule, in the manner shown in (**b**) through (**e**).

Transcription *differs* from DNA replication in three key respects. First, only a selected stretch of one DNA strand, not the whole molecule, serves as the template. Second, instead of DNA polymerase, a different enzyme, **RNA polymerase**, catalyzes the addition of nucleotides to the 3′ end of a growing strand of RNA. Third, transcription results in a single, free strand of RNA nucleotides, not in a double helix.

Transcription is initiated at a **promoter**, a base sequence in DNA that signals the start of a gene. Proteins position an RNA polymerase on DNA and thereby help it bind to the promoter. The enzyme moves along the DNA strand, joining nucleotides one after another (Figure 14.8). When that enzyme reaches a particular base, however, the new RNA molecule is released as a free transcript.

Finishing Touches on mRNA Transcripts

In eukaryotic cells alone, a new mRNA molecule is unfinished. That "pre-mRNA" must be modified before its protein-building instructions can be put to use. Just as a dressmaker might snip off some threads or add bows on a dress before it leaves the shop, so do eukaryotic cells tailor their pre-mRNA.

For example, enzymes attach a cap to the 5′ end of a pre-mRNA molecule. The cap is a nucleotide

Further reading: Student Guide to InfoTrac on web site

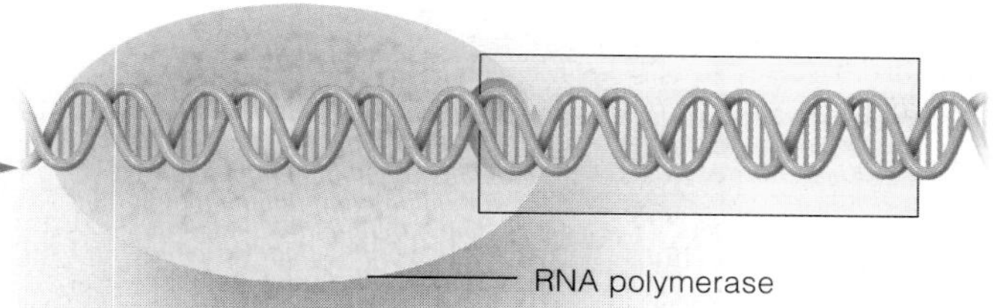
RNA polymerase

b A molecule of RNA polymerase binds with a promoter region in the DNA. It will recognize the base sequence located downstream from that site as a template for linking together the nucleotides adenine, cytosine, guanine, and uracil into a strand of RNA.

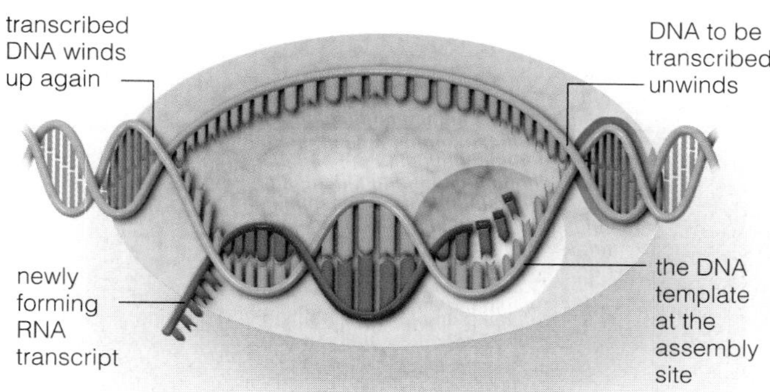
transcribed DNA winds up again
DNA to be transcribed unwinds
newly forming RNA transcript
the DNA template at the assembly site

c All through transcription, the DNA double helix is unwound in front of the RNA polymerase. Short lengths of the newly forming RNA strand briefly wind up with its DNA template strand. New stretches of RNA unwind from the template (and the two DNA strands wind up again).

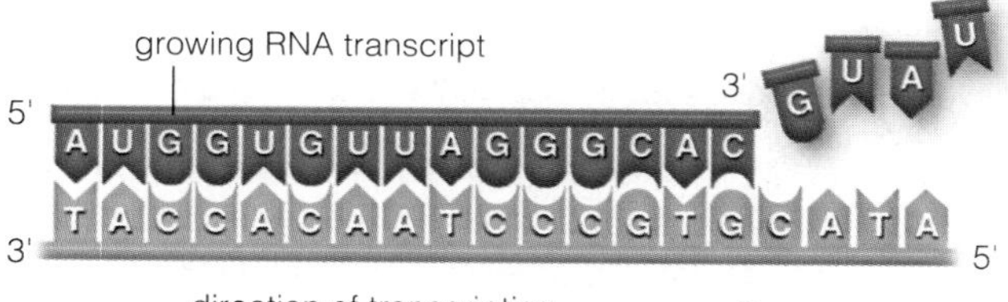
growing RNA transcript
5'
3'
A U G G U G U U A G G G C A C
G U A U
T A C C A C A A T C C C G T G C A T A
3'
5'
direction of transcription

d What happened at the assembly site? RNA polymerase catalyzed the base-pairing of RNA nucleotides, one after another, with exposed bases on the DNA template strand.

5' A U G G U G U U A G G G C A C G U A U 3'

e At the end of the gene region, the last stretch of the new mRNA transcript is unwound and released from the DNA.

that has a methyl group and phosphate groups bonded with it. Also, enzymes attach a tail of about 100 to 200 adenine-containing nucleotides to the 3′ end of most pre-mRNA transcripts. Hence the name, "poly-A tail" (for multiple adenine units). The tail gets wound up with proteins. Later, in the cytoplasm, the cap will assist the binding of mRNA to a ribosome. Also, enzymes will slowly destroy the wound-up tail from the tip on back. Such tails "pace" the access of enzymes to mRNA and thereby dictate how long an mRNA molecule will last. Apparently, they help keep protein-building messages intact for as long as the cell requires them.

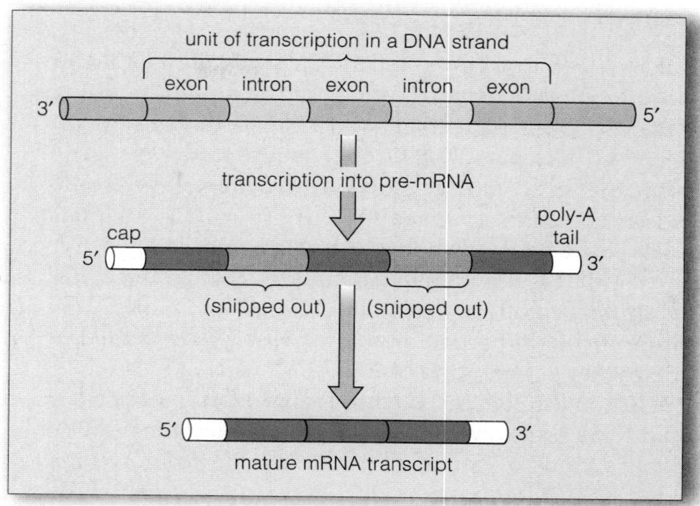
unit of transcription in a DNA strand
exon intron exon intron exon
3′
5′
transcription into pre-mRNA
cap
poly-A tail
5′
3′
(snipped out) (snipped out)
5′
3′
mature mRNA transcript

Figure 14.9 Transcription and modification of new mRNA in the nucleus of eukaryotic cells. Its cap is a nucleotide with functional groups attached. Its tail is a string of adenine nucleotides.

Besides these alterations, the pre-mRNA itself gets modified. Most eukaryotic genes contain one or more **introns**, base sequences that must be removed before a pre-mRNA molecule can be translated. The introns intervene between **exons**, the parts that remain in the mRNA when it gets translated into protein. As Figure 14.9 shows, the introns are transcribed right along with the exons, but they are snipped out before the mRNA leaves the nucleus in mature form.

It could be that some introns are evolutionary junk, the leftovers of past mutations that led nowhere. Yet many introns are sites where instructions for building a particular protein can be snipped apart and spliced back together in various ways. The alternative splicing allows different cells in your body to use the same gene to make different versions of a pre-mRNA transcript, and therefore different versions of the resulting protein. We return to this topic in Section 15.3.

During gene transcription, a sequence of exposed bases in one of the two strands of a DNA molecule serves as the template for assembling a single strand of RNA. The assembly follows these base-pairing rules: adenine only with uracil, and cytosine only with guanine.

Before leaving the nucleus, each new mRNA transcript, or pre-mRNA, undergoes modification into final form.

14.3 DECIPHERING THE mRNA TRANSCRIPTS

What Is the Genetic Code?

Like a strand of DNA, an mRNA molecule is a linear sequence of nucleotides. What are the protein-building "words" encoded in that sequence? Gobind Khorana, Marshall Nirenberg, and other investigators came up with the answer. They showed that ribosomes "read" nucleotide bases *three at a time*, as triplets. Base triplets in an mRNA strand were given this name: **codons**.

Figure 14.10 will give you an idea of how the order of different codons in an mRNA strand dictates the order in which particular amino acids will be added to a growing polypeptide chain.

Count the codons listed in Figure 14.11, and you see that there are sixty-four kinds. Notice how most of the twenty kinds of amino acids correspond to more than one codon. Glutamate corresponds to the code words GAA *and* GAG, for example. Also notice how AUG has dual functions. It codes for the amino acid methionine, and it also is an initiation codon, a START signal for translating an mRNA transcript at a ribosome. That is, "three-bases-at-a-time" selections start at a particular

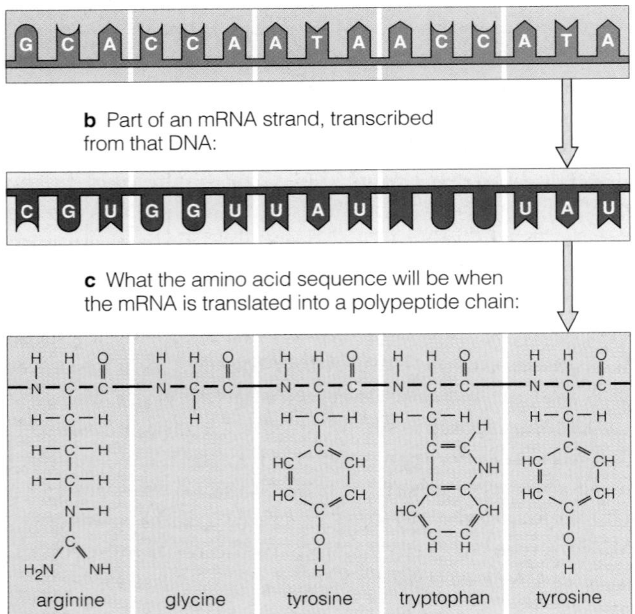

a Base sequence of a gene region in DNA:
G C A C C A A T A A C C A T A
b Part of an mRNA strand, transcribed from that DNA:
C G U G G U U A U U A U
c What the amino acid sequence will be when the mRNA is translated into a polypeptide chain:
arginine
glycine
tyrosine
tryptophan
tyrosine

Figure 14.10 The steps from genes to proteins. (**a**) This diagram represents a region of a DNA double helix that was unwound during transcription. (**b**) The exposed bases on one DNA strand served as a template for assembling an mRNA strand. In the new mRNA transcript, every three nucleotide bases equaled one codon. Each codon calls for one amino acid in a polypeptide chain. (**c**) Referring to Figure 14.11, can you fill in the blank codon for tryptophan in the chain?

Amino acids that correspond to base triplets:

FIRST BASE	SECOND BASE OF A CODON: U	C	A	G	THIRD BASE
U	phenylalanine	serine	tyrosine	cysteine	U
U	phenylalanine	serine	tyrosine	cysteine	C
U	leucine	serine	STOP	STOP	A
U	leucine	serine	STOP	tryptophan	G
C	leucine	proline	histidine	arginine	U
C	leucine	proline	histidine	arginine	C
C	leucine	proline	glutamine	arginine	A
C	leucine	proline	glutamine	arginine	G
A	isoleucine	threonine	asparagine	serine	U
A	isoleucine	threonine	asparagine	serine	C
A	isoleucine	threonine	lysine	arginine	A
A	methionine (or START)	threonine	lysine	arginine	G
G	valine	alanine	aspartate	glycine	U
G	valine	alanine	aspartate	glycine	C
G	valine	alanine	glutamate	glycine	A
G	valine	alanine	glutamate	glycine	G

Figure 14.11 The genetic code. The codons in mRNA are nucleotide bases, "read" in blocks of three. Sixty-one of the base triplets correspond to specific amino acids. Three others serve as signals that stop translation. The left column of the diagram shows the first of the three nucleotides in each codon in mRNA. The middle columns show the second nucleotide. The right column shows the third. Reading from left to right, for instance, the triplet UGG corresponds to tryptophan. Both UUU and UUC correspond to phenylalanine.

AUG in the transcript's nucleotide sequence. Codons UAA, UAG, and UGA do not correspond to amino acids. They serve as STOP signals, which prevent further addition of amino acids to a new polypeptide chain.

The set of sixty-four different codons is the **genetic code**. It is the basis of protein synthesis in all organisms.

Structure and Function of tRNA and rRNA

In a cell's cytoplasm are pools of free amino acids and free tRNA molecules. The tRNAs each have a molecular "hook," an attachment site for amino acids. They also have an **anticodon**, a nucleotide triplet that can base-pair with a codon (Figure 14.12). When tRNAs bind to

Further reading: Student Guide to InfoTrac on web site

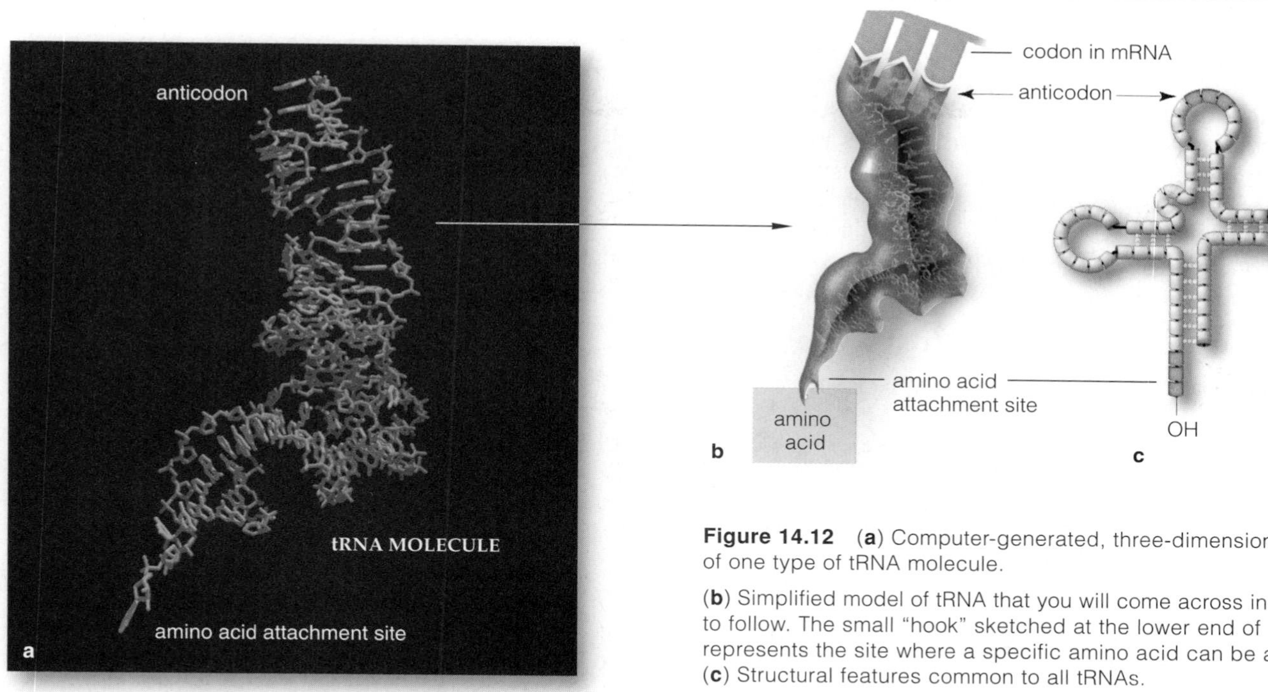
anticodon
tRNA MOLECULE
amino acid attachment site
a
codon in mRNA
anticodon
amino acid attachment site
amino acid
b
OH
c

Figure 14.12 (**a**) Computer-generated, three-dimensional model of one type of tRNA molecule.

(**b**) Simplified model of tRNA that you will come across in illustrations to follow. The small "hook" sketched at the lower end of this model represents the site where a specific amino acid can be attached. (**c**) Structural features common to all tRNAs.

a A small ribosomal subunit. (The *red* arrow points to a platform for chain assembly on the surface)

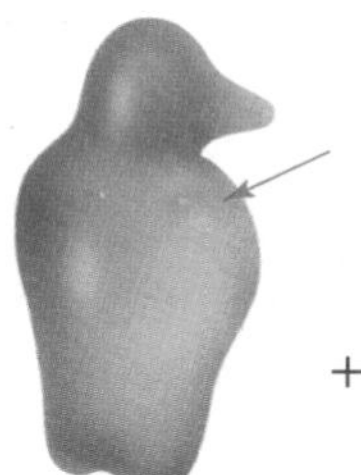

b A large ribosomal subunit, with a tunnel through parts of its interior

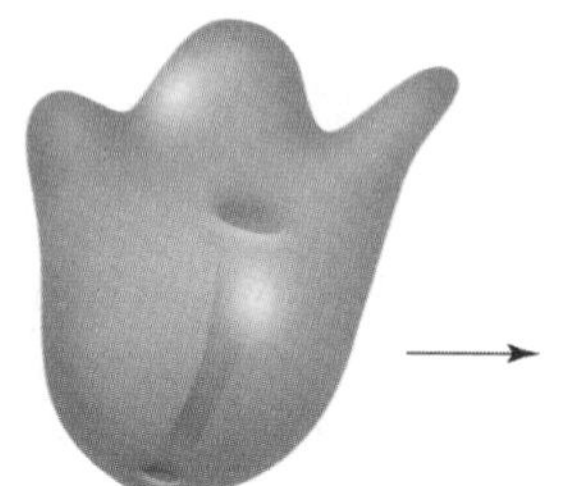

c Side view of an intact ribosome, showing how the platform and tunnel are aligned.

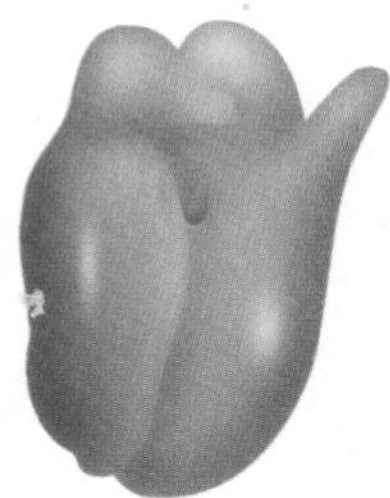

Figure 14.13 Model of eukaryotic ribosomes. Polypeptide chains are assembled on the platform of the small ribosomal subunit. Newly forming chains may move through the tunnel of the large ribosomal subunit.

the codons, they automatically position their attached amino acids in the order specified by mRNA.

A cell has a cytoplasmic pool of sixty-four kinds of codons, but it is able to utilize fewer kinds of tRNAs. How do tRNAs match up with more than one type of codon? According to base-pairing rules, adenine must pair with uracil, and cytosine with guanine. For codon–anticodon interactions, however, the rules loosen up for the first and third bases. To give one example, AUU, AUC, and AUA specify isoleucine. All three of these codons can pair with a single type of tRNA that carries isoleucine. Such freedom in codon–anticodon pairing at a base is known as the "wobble effect."

Even before anticodons interact with the codons of an mRNA strand, that strand must bind to specific parts of the surface of ribosomes. As shown in Figure 14.13, each ribosome has two subunits. These are assembled inside the nucleus from rRNA and protein components, some of which show enzyme activity.

At some point the subunits are shipped separately to the cytoplasm. There, intact, functional ribosomes are put together, each from two subunits—but only when messages encoded in mRNA are to be translated.

The nucleotide sequence of both DNA and mRNA encodes protein-building instructions. The genetic code is a set of sixty-four base triplets, which are nucleotide bases read in blocks of three. A codon is a base triplet in mRNA.

Different combinations of codons specify the amino acid sequence of different polypeptide chains, start to finish.

mRNAs are the only molecules that carry protein-building instructions from DNA into the cytoplasm.

tRNAs deliver amino acids to ribosomes and base-pair with codons in the order specified by mRNA. Their action translates mRNA into a sequence of amino acids.

rRNAs are components of ribosomes, the structures upon which amino acids are assembled into polypeptide chains.

14.4 HOW IS mRNA TRANSLATED?

Stages of Translation

The protein-building code built into mRNA transcripts of DNA becomes translated at intact ribosomes in the cytoplasm. Translation proceeds through three stages: initiation, elongation, and termination.

During the stage called *initiation*, a particular tRNA that can start transcription and an mRNA transcript are both loaded onto a ribosome. First, the initiator tRNA binds with the small ribosomal subunit. AUG, the start codon for the transcript, matches up with this tRNA's anticodon. At the same time, the AUG binds with the small subunit. Second, a large ribosomal subunit binds with the small subunit. When joined together this way, the three form an initiation complex (Figure 14.14*b*). Now the next stage can begin.

In the *elongation* stage of translation, a polypeptide chain is assembled as the mRNA passes between two ribosomal subunits, like a thread being moved through the eye of a needle. Some components of the ribosomes are enzymes. They join individual amino acids together in a sequence dictated by the sequence of codons in the mRNA molecule. Figure 14.14*f–i* shows how a peptide bond forms between the most recently attached amino acid of a growing polypeptide chain and the next amino acid delivered to the intact ribosome. Here you might wish to refer to the description and sketch of peptide bond formation in Section 3.4 (Figure 3.15).

During the last stage of translation, *termination*, a STOP codon in the mRNA moves onto the platform, and no tRNA has a corresponding anticodon. Now proteins called release factors bind to the ribosome. They trigger enzyme activity that detaches the mRNA *and* the chain from the ribosome (Figure 14.14*j–l*).

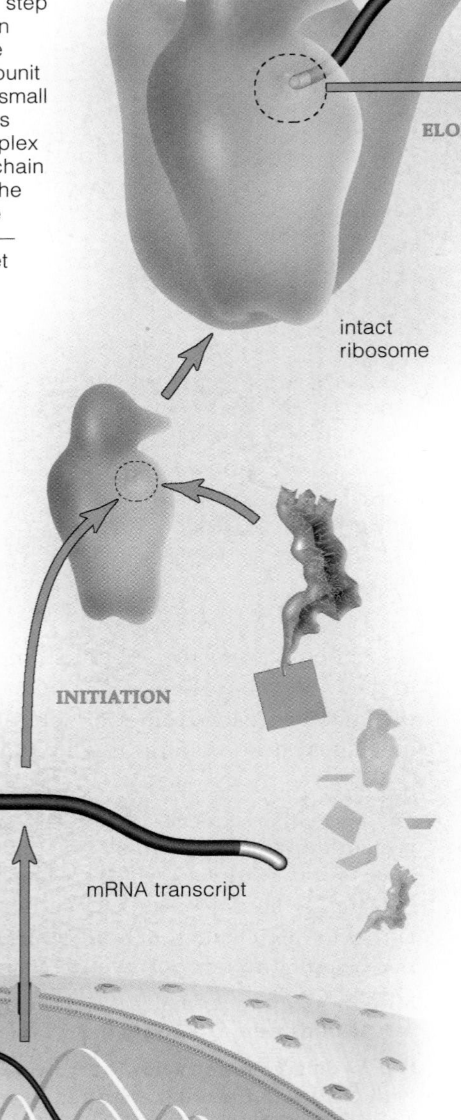
ELONGATION
intact ribosome
INITIATION
mRNA transcript

a A mature mRNA transcript leaves the nucleus by passing through pores across the nuclear envelope. Thus it enters the cytoplasm, which contains pools of many free amino acids, tRNAs, and ribosomal subunits.

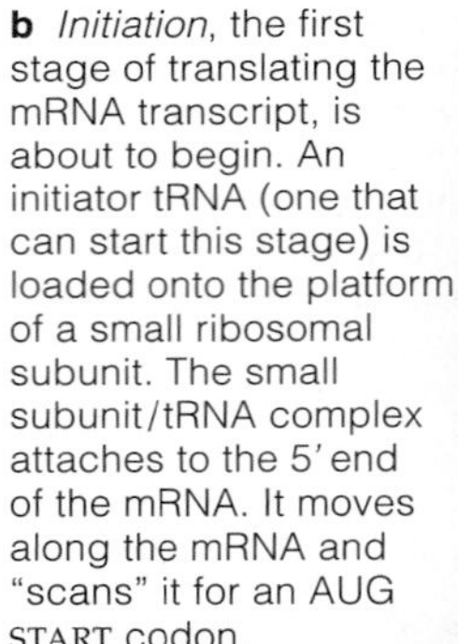

b *Initiation*, the first stage of translating the mRNA transcript, is about to begin. An initiator tRNA (one that can start this stage) is loaded onto the platform of a small ribosomal subunit. The small subunit/tRNA complex attaches to the 5′ end of the mRNA. It moves along the mRNA and "scans" it for an AUG START codon.

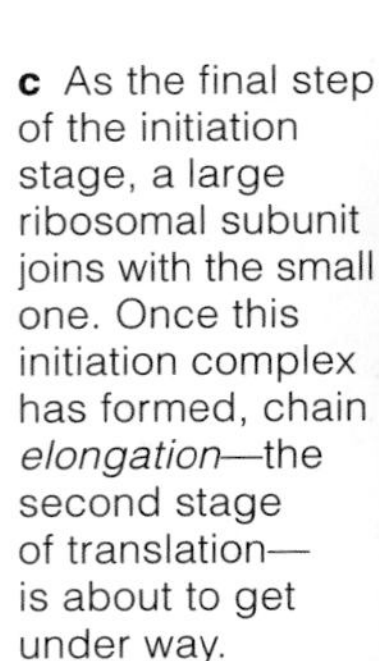

c As the final step of the initiation stage, a large ribosomal subunit joins with the small one. Once this initiation complex has formed, chain *elongation*—the second stage of translation—is about to get under way.

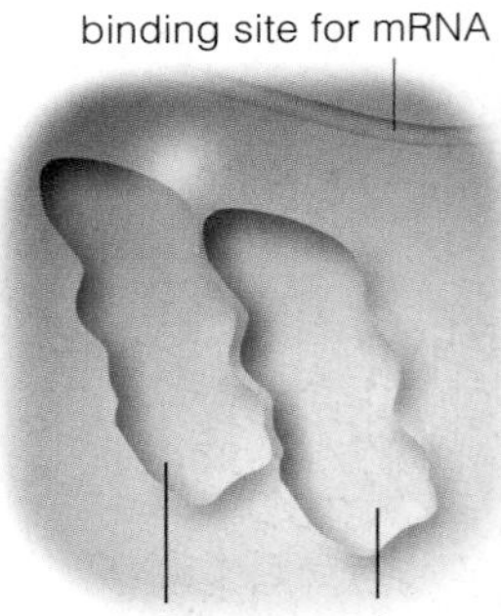
binding site for mRNA
P (first binding site for tRNA)
A (second binding site for tRNA)

d This close-up of the small ribosomal subunit's platform shows the relative positions of binding sites for an mRNA transcript and for tRNAs that deliver amino acids to the intact ribosome.

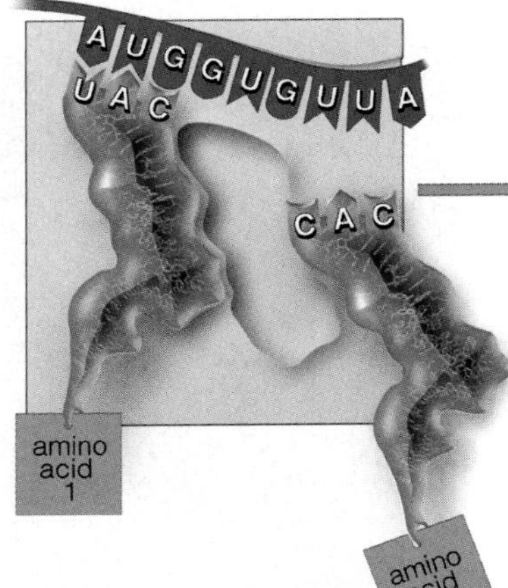
A U G G U G U U A
U A C
C A C
amino acid 1
amino acid 2

e The initiator tRNA has become positioned in the first tRNA binding site (designated *P*) on the ribosome platform. Its anticodon matches up with the START codon (AUG) of the mRNA, which also has become positioned in *its* binding site. Another tRNA is about to move into the platform's second tRNA binding site (designated *A*). It is one that can bind with the codon following the START codon.

Figure 14.14 Translation, the second step of protein synthesis. 

Further reading: Student Guide to InfoTrac on web site

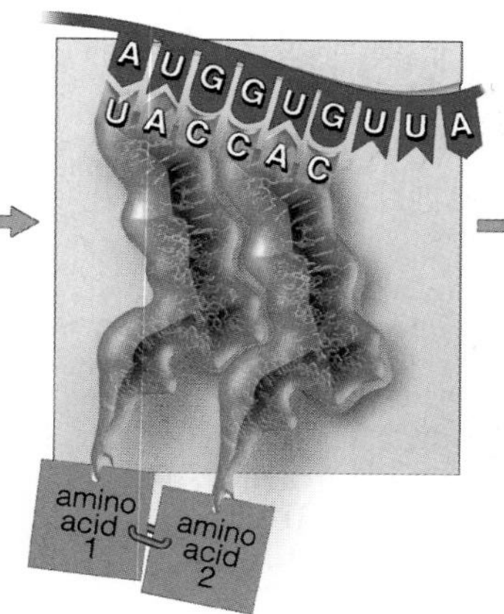
A U G G U G U U A
U A C C A C
amino acid 1
amino acid 2

f Enzyme action breaks the bond between the initiator tRNA and the amino acid hooked to it. At the same time, enzyme action catalyzes the formation of a peptide bond between that amino acid and the one hooked to the second tRNA. Then the initiator tRNA is released from the ribosome.

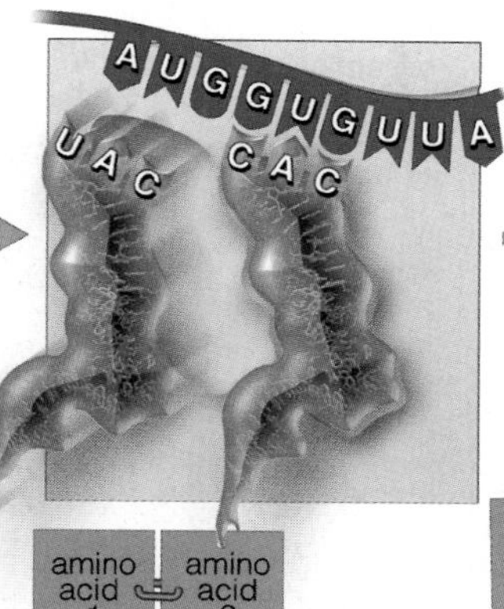
A U G G U G U U A
U A C
C A C
amino acid 1
amino acid 2

g Now the first amino acid is attached only to the second one—which is still hooked to the second tRNA. This tRNA is about to move into the ribosomal platform's *P* site and slide the mRNA along with it by one codon. This will align the third codon in the *A* site.

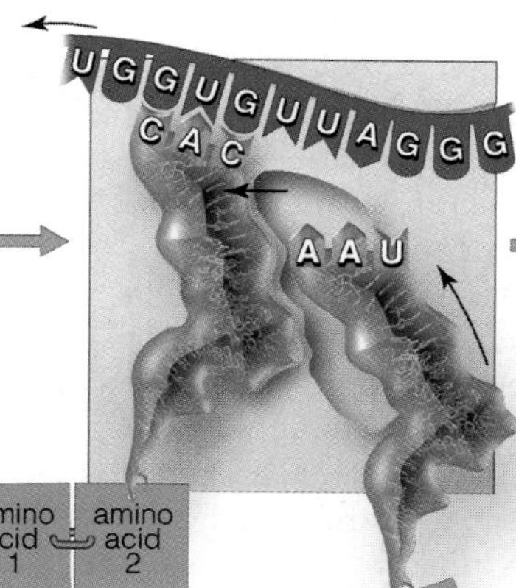
U G G U G U U A G G G
C A C
A A U
amino acid 1
amino acid 2
amino acid 3

h A third tRNA is about to move into the vacated *A* site. Its anticodon is able to base-pair with the third codon of the mRNA transcript. Enzymes will now catalyze the formation of a peptide bond between amino acids 2 and 3.

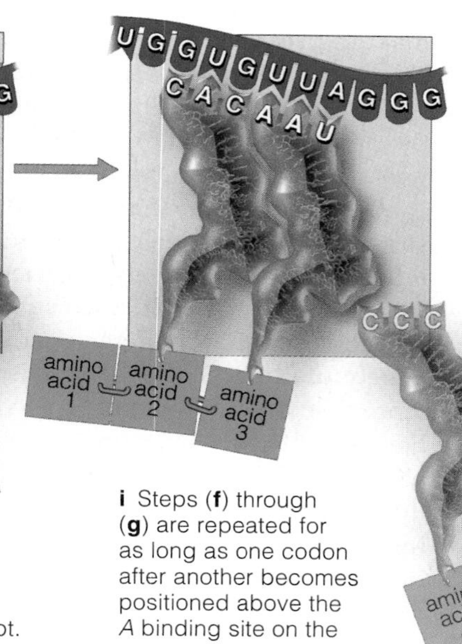
U G G U G U U A G G G
C A C A A U
C C C
amino acid 1
amino acid 2
amino acid 3
amino acid 4

i Steps (**f**) through (**g**) are repeated for as long as one codon after another becomes positioned above the *A* binding site on the ribosomal platform.

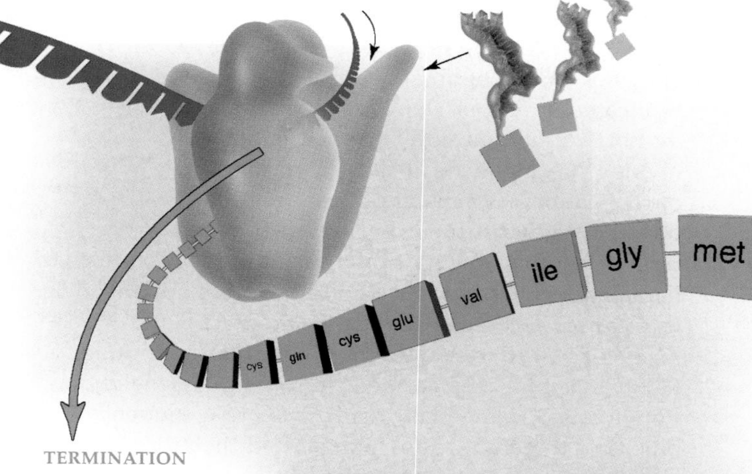
cys
gln
cys
glu
val
ile
gly
met
TERMINATION

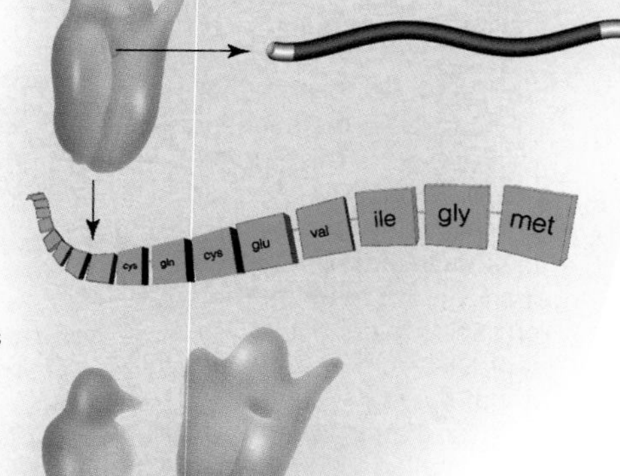
cys
gln
cys
glu
val
ile
gly
met

j A STOP codon moves onto the ribosomal assembly platform. It is the signal to release the mRNA transcript from the ribosome.

k The newly formed polypeptide chain also is released from the ribosome. It is free to join the pool of proteins in the cytoplasm or to enter rough ER of the cytomembrane system.

l The two ribosomal subunits separate.

What Happens to the New Polypeptides?

Unfertilized eggs and other cells that will be called upon to rapidly synthesize many copies of different proteins usually stockpile mRNA transcripts in their cytoplasm. In cells that are rapidly using or secreting proteins, you often observe polysomes. Each polysome is a cluster of many ribosomes translating one mRNA transcript at the same time. The transcript threads through all of them, one after another, like the thread of a pearl necklace.

After new polypeptide chains are synthesized, many join the cytoplasmic pool of free proteins. Many others enter the ribosome-studded, flattened sacs of rough ER, which is part of the cytomembrane system (Section 4.4). There they take on final form before they are shipped to their ultimate destinations inside or outside the cell.

Translation is initiated when a small ribosomal subunit and an initiator tRNA arrive at an mRNA's START codon and a large ribosomal subunit binds to them.

tRNAs deliver amino acids to the ribosome in the order dictated by the sequence of mRNA codons, to which the tRNA anticodons base-pair. A polypeptide chain lengthens as peptide bonds form between the amino acids.

Translation ends when a STOP codon triggers events that cause the chain and mRNA to detach from the ribosome.

14.5 DO MUTATIONS AFFECT PROTEIN SYNTHESIS?

Whenever a cell puts its genetic code into action, it is making precisely those proteins that it requires for its structure and functions. If something changes a gene's code words, the resulting protein may change, also. If the protein is central to cell architecture or metabolism, we can expect the outcome to be an abnormal cell.

Gene sequences do change. Sometimes one base gets substituted for another in the nucleotide sequence. At other times, an extra base is inserted into the sequence or a base is lost from it. Such small-scale changes in the nucleotide sequence of genes in the DNA molecule are **gene mutations**. There is some leeway here; remember, more than one codon may specify the same amino acid. For example, if a mutation were to change UCU to UCC, it probably would not have dire effects, for both codons specify serine. More often, however, mutations give rise to proteins with altered or lost functions.

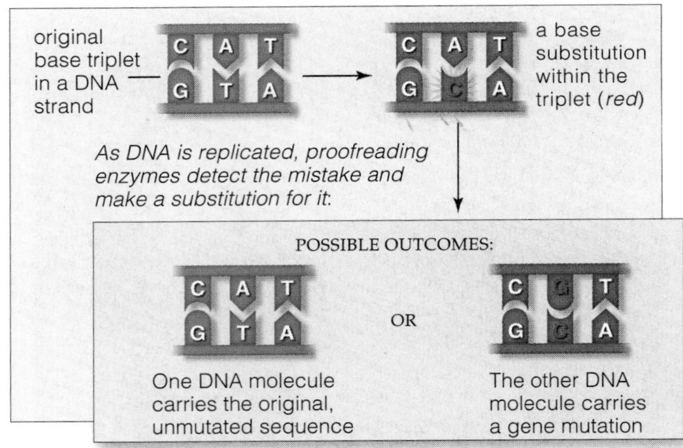

original base triplet in a DNA strand
C A T
G T A
C A T
G C A
a base substitution within the triplet (*red*)
As DNA is replicated, proofreading enzymes detect the mistake and make a substitution for it:
POSSIBLE OUTCOMES:
C A T
G T A
OR
C G T
G C A
One DNA molecule carries the original, unmutated sequence
The other DNA molecule carries a gene mutation

a Example of a base-pair substitution

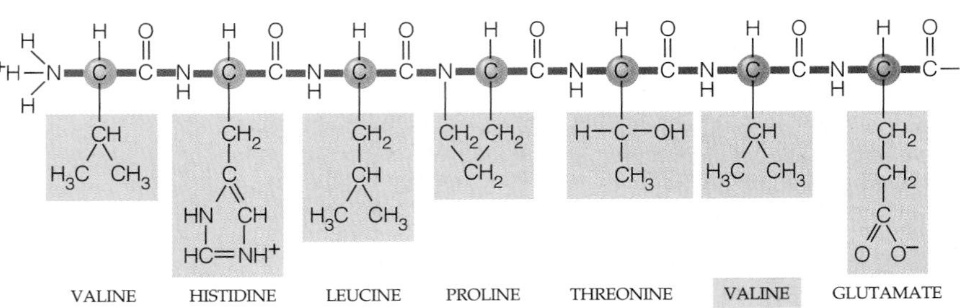

VALINE
HISTIDINE
LEUCINE
PROLINE
THREONINE
VALINE
GLUTAMATE

b Outcome of the base-pair substitution

Figure 14.15 Common types of mutations. (**a**) One example of a base-pair substitution. (**b**) Remember Figure 14.5? The base-pair substitution shown here is the type of molecular change that caused a different amino acid (valine, not glutamate) to be substituted in beta chains of hemoglobin.

Common Gene Mutations and Their Sources

Figure 14.15 gives an example of a common gene mutation. Here, one base (adenine) is wrongly paired with another base (cytosine) when DNA is being replicated. Proofreading enzymes can recognize an error in a newly replicated strand of DNA and fix it. If they do not, a mutation will become established in one DNA molecule during the next round of replication. As a result of this mutation, a **base-pair substitution**, one amino acid can replace another during protein synthesis. This is what happens, recall, in people who carry Hb^S, a mutant allele that causes sickle-cell anemia (Sections 11.5 and 14.1).

Figure 14.16 has an example of a different mutation. Here, an *extra* base became inserted into a gene region. Polymerases, remember, read nucleotide sequences in blocks of three. This insertion shifted the "three-bases-at-a-time" reading frame by one base; hence the name, *frameshift* mutation. The affected gene has a different message, and an altered version of the protein will be synthesized. Frameshift mutations fall within broader categories of gene mutation known as **insertions** and **deletions**. In such cases, one to several base pairs are inserted into a DNA molecule or deleted from it.

As a final example, Barbara McClintock discovered that mutations may result when transposable elements,

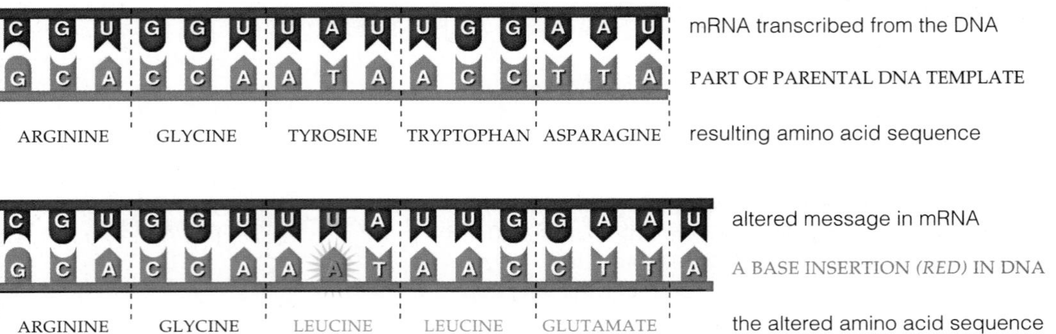

C G U G G U U A U U G G A A U
G C A C C A A T A A C C T T A
mRNA transcribed from the DNA
PART OF PARENTAL DNA TEMPLATE
ARGININE
GLYCINE
TYROSINE
TRYPTOPHAN
ASPARAGINE
resulting amino acid sequence
C G U G G U U U A U U G G A A U
G C A C C A A A T A A C C T T A
altered message in mRNA
A BASE INSERTION *(RED)* IN DNA
ARGININE
GLYCINE
LEUCINE
LEUCINE
GLUTAMATE
the altered amino acid sequence

Figure 14.16 Example of an insertion, a mutation in which an extra base gets inserted into a gene region of DNA. This insertion has caused a *frameshift*; it has changed the reading frame for base triplets in the DNA and in the mRNA transcript of that region. As a result, the wrong amino acids will be called up when the mRNA transcript becomes translated into protein.

Further reading: Student Guide to InfoTrac on web site

Figure 14.17 Barbara McClintock, who won a Nobel Prize for her meticulous research and eventual insight that some DNA segments can slip into and out of different locations in DNA molecules. We call these segments transposons. In her hands is an ear of Indian corn (*Zea mays*). The curiously nonuniform coloration of its kernels sent her on the road to discovery.

Each corn kernel is a seed, which has the potential to grow into a new corn plant. All of its cells have the same pigment-coding genes. However, some of the kernels are colorless or spottily colored. In the ancestor of the plant from which this ear of corn was plucked, a gene in a germ cell left its position in a DNA molecule, invaded another DNA molecule, and shut down a pigment-encoding gene.

The plant inherited the mutation. As cell divisions proceeded in the growing plant, none of the mutated cell's descendants was able to synthesize pigment molecules. Each gave rise to colorless kernel tissue. Later, in some cells, the movable DNA slipped out of the pigment-encoding gene. All the descendants of *those* cells produced pigment—and colored kernel tissue.

or **transposons**, are on the move. These DNA segments move spontaneously from one location to another in the same DNA molecule or to a different one. Often they inactivate genes into which they become inserted. Their unpredictability can cause interesting variations in traits. Figure 14.17 and *Critical Thinking* question 5 at the end of this chapter provide two examples.

Causes of Gene Mutations

Many mutations arise spontaneously as DNA is being replicated. This should not come as a surprise, given the swift pace of replication and the huge pools of free nucleotides concentrated near the growing DNA strands. Proofreading and repair enzymes detect and fix most of them. However, a low number of mistakes do slip past those enzymes with predictable frequency.

Each gene has a characteristic **mutation rate**, which is the probability it will mutate spontaneously during a specified interval, such as each DNA replication cycle. (This is not the same as mutation *frequency*, the number of times a gene mutation has occurred in a population, as in 1 million gametes that produced 500,000 people.)

Not all mutations are spontaneous. Many result after exposure to mutagens (mutation-causing agents in the environment). Ultraviolet radiation, especially the 260-nanometer wavelength in sunlight, is mutagenic. This is the wavelength DNA absorbs most strongly, and it may induce crosslinks to form between two pyrimidine neighbors on the same DNA strand. Skin cancers are one outcome. Other mutagens are gamma rays and x-rays. They can ionize water and other molecules around the DNA, so free radicals form. Free radicals are molecular fragments having an unpaired electron—and they can attack DNA's structure. The next section takes a closer look at the effects of such **ionizing radiation**.

Natural and synthetic chemicals in the environment can accelerate the rate of spontaneous mutations. For example, substances called **alkylating agents** transfer methyl or ethyl groups to reactive sites on the bases or phosphate groups of DNA. At an alkylated site, DNA becomes more susceptible to base-pair disruptions that invite mutation. Many **carcinogens**, which are cancer-causing agents, operate by alkylating the DNA.

The Proof Is in the Protein

Spontaneous mutations are rare in terms of a human life. (The rate for eukaryotes ranges between 10^{-4} and 10^{-6} per gene per generation.) If one arises in a somatic cell, any good or bad consequences will not endure, for it cannot be passed on to offspring. If the mutation arises in a germ cell or gamete, however, it may enter the evolutionary arena. The same can happen with a mutation in an asexually reproducing organism or cell.

In all such cases, nature's test is this: *A protein that is specified by a heritable mutation may have harmful, neutral, or beneficial effects on an individual's ability to function in the prevailing environment.* As you will read in the next unit of the book, the outcomes of gene mutations can have powerful evolutionary consequences.

A gene mutation is an alteration in one to several bases in the nucleotide sequence of DNA. The most common are base-pair substitutions, base insertions, and base deletions.

Each gene has a spontaneous and characteristic mutation rate, which may be accelerated by exposure to harmful radiation and certain chemicals in the environment.

A protein specified by a mutated gene may have harmful, neutral, or beneficial effects on the ability of an individual to function in the prevailing environment.

Focus on Health

14.6 MUTAGENIC RADIATION

Each day, radiation bombards the human body. However, only two categories of radiation are mutagenic; they can cause mutations.

Protons, neutrons, x-rays, and gamma rays are among the forms of *ionizing* radiation. Their wavelengths are highly energetic, and they can damage DNA that takes a direct hit. More commonly, their effects are indirect. When such high-energy rays strike molecules—including water molecules that bathe the cell's organic compounds—they often strip its atoms of an electron. In this way the atom becomes an ion (hence the name, ionizing radiation). The unpaired electron left behind is highly reactive, and its molecular owner has become a free radical.

When free radicals collide with other molecules, they can create new free radicals. When they collide with the phosphate bonds in DNA, they can break them. A break in one strand of double-stranded DNA doesn't amount to much, because repair enzymes easily patch things up. When both strands break, repair enzymes often fail to restore the original base sequence. It is not the ionizing radiation itself that directly causes mutation. Rather, its agents are the free radicals that it induces to form.

Ionizing radiation is potentially quite dangerous. It can deeply penetrate living tissue, in which case a trail of free radicals marks its passage. Think about this when your doctor or dentist recommends an x-ray examination. X-rays are especially useful for diagnosing medical and dental problems. Because they pass more easily through some tissues than others, they can be used to produce an image of the body's interior on photographic film. The patient's exposure time is usually brief, and the dosage is extremely low to minimize the risk of mutations and radiation-induced disease.

But ionizing radiation has a cumulative mutagenic effect. Repeated exposure to low levels over many years can cause problems. That is why x-ray technicians use lead aprons and shielding booths. With respect to long-term exposure, there may be no safe level of radiation.

Unlike ionizing radiation, *nonionizing* radiation is not energetic enough to invite the formation of free radicals. Rather, it simply excites electrons enough to boost them to a higher energy level. Nonionizing radiation can only penetrate single-celled organisms (bacteria and protistans) and the outermost layers of living cells of multicelled organisms.

Ultraviolet (UV) light is the most significant form of mutagenic nonionizing radiation. The nucleotides of DNA easily absorb it. Maximum absorption—and maximum mutagenesis—occurs at a wavelength of 254 nanometers. Cytosine and thymine are particularly vulnerable to excitation; they may be converted to lesions with altered base-pairing properties. The introduction to Chapter 15 describes one such lesion—a pyrimidine dimer. If left unrepaired, such lesions can cause mutations in a newly synthesized DNA strand during replication.

14.7 SUMMARY

1. Cells, and multicelled organisms, cannot stay alive without enzymes and other proteins. A protein consists of one or more polypeptide chains, each of which is composed of a linear sequence of amino acids.

a. An amino acid sequence of a polypeptide chain corresponds to a gene region in a DNA molecule. Each gene is a sequence of nucleotide bases in one of the DNA molecule's two strands. The bases are A, T, G, and C (adenine, thymine, guanine, and cytosine).

b. For most genes, that sequence corresponds to a linear sequence of specific amino acids for a particular polypeptide chain. (Some genes specify tRNA or rRNA, not the mRNA that is translated into proteins.)

c. Understanding of the connection between genes and proteins started with studies of gene mutations that affected enzymes known to catalyze steps in metabolic pathways. Comparisons between normal and abnormal proteins, hemoglobin especially, led to the hypothesis that the amino acid sequence of polypeptide chains is encoded in genes.

2. The biochemical path from genes to proteins has two steps, called transcription and translation:

DNA —*transcription*→ RNA —*translation*→ PROTEIN

a. During transcription, the double-stranded DNA is unwound at a gene region. Enzymes use its exposed bases as a template, or a structural pattern, to assemble a strand of ribonucleic acid (RNA) from the cell's pool of free nucleotides.

b. During translation, three classes of RNAs interact in the synthesis of polypeptide chains, which later will twist, fold, and often become modified into the final, three-dimensional shape of the protein.

c. Figure 14.18 is a visual summary of this flow of genetic information from DNA to proteins, as it occurs in eukaryotic cells. The DNA is transcribed into RNA in the nucleus, but RNA is translated in the cytoplasm. Prokaryotic cells (bacteria) lack a nucleus; transcription and translation proceed in their cytoplasm.

3. Here are the key points concerning transcription:

a. When RNA is transcribed from exposed bases of DNA, base-pairing rules govern its assembly. Guanine pairs with cytosine, as in DNA replication, but uracil (not thymine) pairs with adenine in RNA:

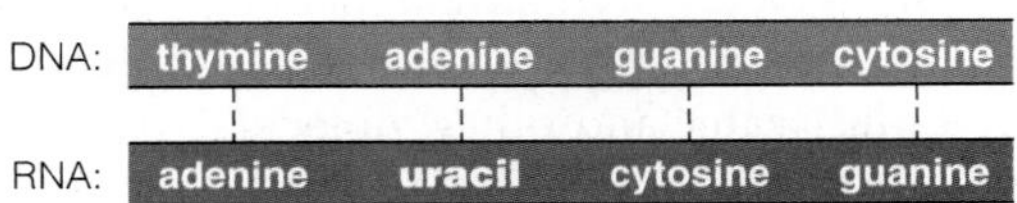

DNA:
thymine
adenine
guanine
cytosine
RNA:
adenine
uracil
cytosine
guanine

b. Different gene regions in DNA serve as templates for assembling different RNA molecules.

Further reading: Student Guide to InfoTrac on web site

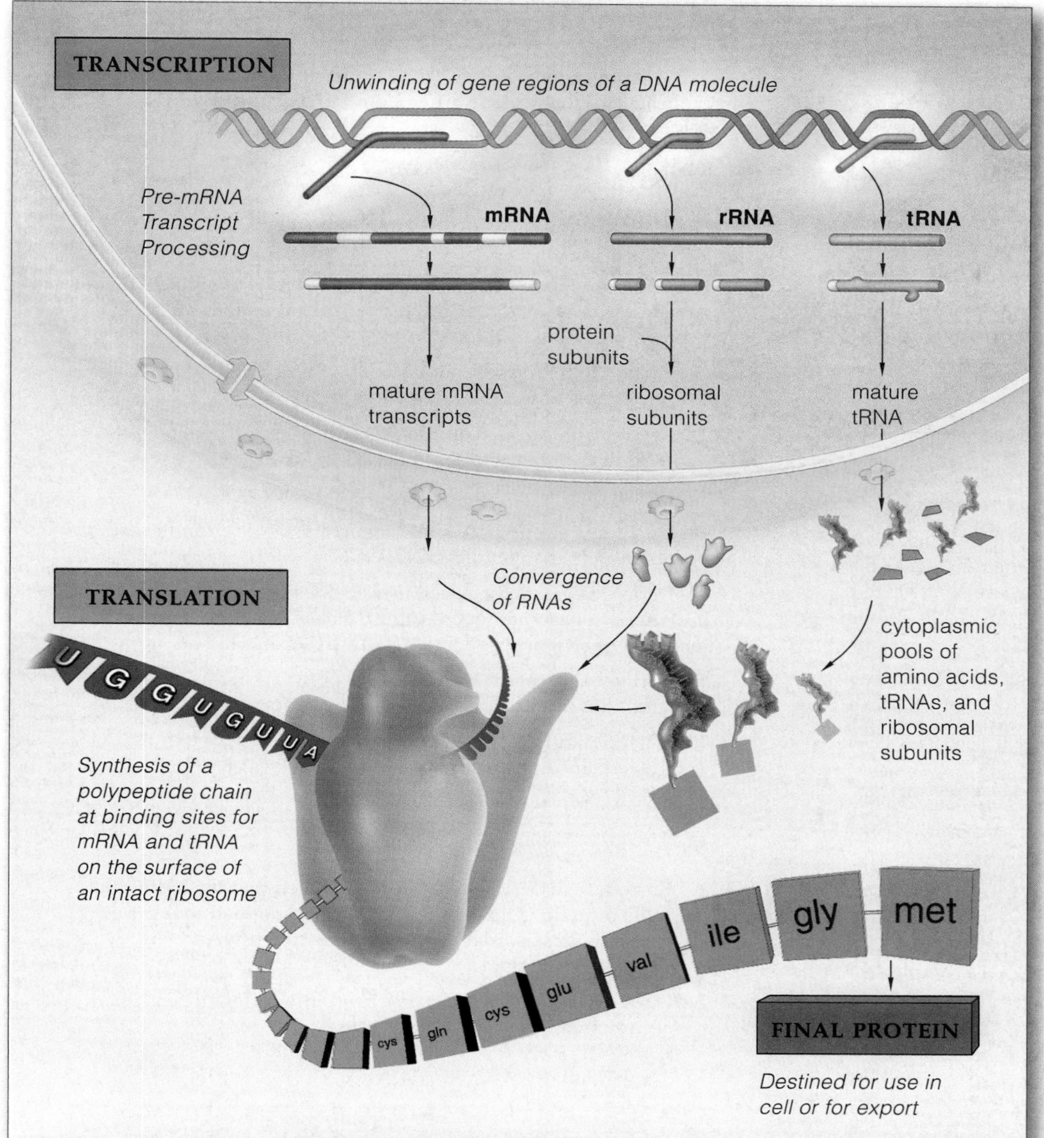
TRANSCRIPTION
Unwinding of gene regions of a DNA molecule
Pre-mRNA Transcript Processing
mRNA
rRNA
tRNA
protein subunits
mature mRNA transcripts
ribosomal subunits
mature tRNA
TRANSLATION
Convergence of RNAs
cytoplasmic pools of amino acids, tRNAs, and ribosomal subunits
U G G U G U U A
Synthesis of a polypeptide chain at binding sites for mRNA and tRNA on the surface of an intact ribosome
met
gly
ile
val
glu
cys
gln
cys
FINAL PROTEIN
Destined for use in cell or for export

Figure 14.18 Visual summary of protein synthesis—transcription and translation—as it proceeds in eukaryotic cells. All three classes of RNA are assembled in the nucleus and shipped to the cytoplasm. In prokaryotic cells, transcription as well as translation proceeds in the cytoplasm.

c. Messenger RNA (mRNA) is the only class of RNA that carries protein-building instructions.

d. A ribosome has two subunits made of ribosomal RNA (rRNA) and other factors. Each is a physical site where polypeptide chains are assembled. Its subunits are built in the nucleus (or, in prokaryotic cells, in the cytoplasm). They meet up only in the cytoplasm.

e. Transfer RNA (tRNA) is the vehicle of translation; it will latch on to free amino acids in the cytoplasm and deliver them to ribosomes. It will do so in a sequence that corresponds to the sequential message of mRNA.

f. RNA transcripts of eukaryotic cells are processed into final form before being shipped from the nucleus. For example, mRNA's noncoding portions (introns) are excised, so only its coding portions (exons) get spliced together. Only mature mRNA transcripts are translated into protein.

4. Here are the key points about translation:

a. mRNA interacts with many tRNAs and ribosomes in ways that assemble amino acids in a sequence that produces a specific kind of polypeptide chain.

b. Translation is based on the genetic code: a set of sixty-four base triplets. A triplet is a series of nucleotide bases that ribosomal proteins "read" in blocks of three.

c. One base triplet in mRNA is called a codon. An anticodon is a complementary triplet in tRNA. Some combination of different codons specifies what the amino acid sequence of a polypeptide chain will be, start to finish.

5. Translation proceeds through the following three stages:

a. Initiation stage. One small ribosomal subunit, one initiator tRNA, and one mRNA transcript converge. The small subunit binds reversibly with a large ribosomal subunit.

b. Chain elongation stage. A number of tRNAs deliver many amino acids to the ribosome. Their anticodons base-pair with codons in mRNA. The amino acids become joined one after another by peptide bonds to form a new polypeptide chain.

c. Chain termination. A STOP codon in mRNA causes the chain and the mRNA to detach from the ribosome.

6. Gene mutations are potentially heritable, small-scale alterations in the nucleotide sequence of DNA.

a. Many gene mutations arise spontaneously during DNA replication. Others arise after the DNA is exposed to mutagens, such as ultraviolet radiation and other mutation-causing agents in the environment.

b. A base-pair substitution (one base replaces a base of a different kind) changes only one codon, but this may cause an amino acid substitution that alters protein function. Insertions of one or more bases into a gene or deletions from it can shift the reading frame to specify different amino acids. Transposons (movable elements) may inactivate genes into which they become inserted.

c. Each gene has a characteristic mutation rate: the probability that it will spontaneously mutate in some specified time interval, such as a DNA replication cycle.

7. A protein specified by a mutant gene might have harmful, neutral, or beneficial effects on the individual. The outcome depends on prevailing conditions in the internal and external environments. Somatic mutations affect individuals only. Mutations in reproductive cells are heritable and can enter the evolutionary arena.

Review Questions

1. Are polypeptide chains assembled on DNA? If so, state how. If not, state how and where they are assembled. *CI, 14.1*

2. Define gene transcription and translation, the two stages of events by which polypeptide chains of proteins are synthesized. Both stages proceed in the cytoplasm of prokaryotic cells. Where does each stage proceed in eukaryotic cells? *CI*

3. Briefly state how gel electrophoresis, a common laboratory procedure, works. Explain how it provided a clue that small differences in normal and abnormal versions of the same protein may lead to big differences in how the proteins function. *14.1*

4. Name the three classes of RNA and briefly describe their functions. *14.2, 14.3*

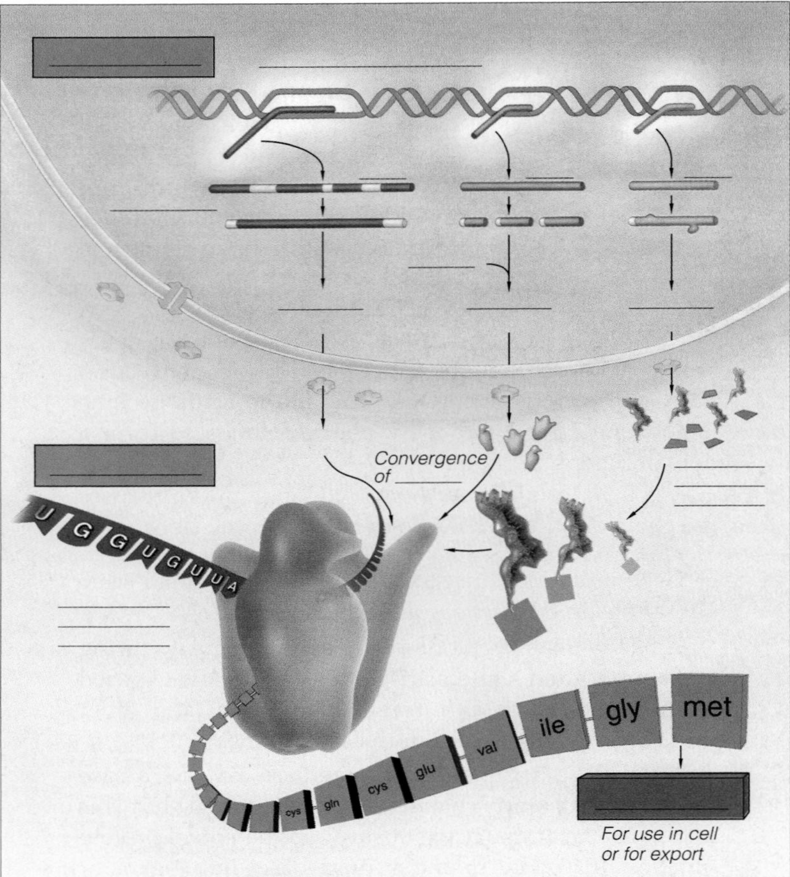

Convergence of ______
U G G U G U U A
met
gly
ile
val
glu
cys
gln
cys
For use in cell or for export

5. In what key respect does the sequence of nucleotide bases in RNA differ from those in DNA? *14.2*

6. How does the process of gene transcription resemble DNA replication? How does it differ from DNA replication? *14.2*

7. Pre-mRNA transcripts of eukaryotic cells contain introns and exons. Are the introns or exons snipped out before the transcript leaves the nucleus? *14.2*

8. Distinguish between codon and anticodon. *14.3*

9. Cells use the set of sixty-four codons in the genetic code to build polypeptide chains from twenty kinds of amino acids. Are some amino acids specified by different codons? If so, in what respect do the codons differ? *14.3*

10. Name the three stages of translation and briefly describe the key events of each one. *14.4*

11. Review Figure 14.18 on the preceding page. Then, on your own, fill in the blanks of the diagram at lower left.

12. Define gene mutation. *14.5*

13. Do all mutations arise spontaneously? Do environmental agents trigger change in each case? *14.5*

14. Define and state the possible outcomes of the following mutations: a base-pair substitution, a base insertion, and an insertion of a transposon at a new DNA location. *14.5*

15. Define and explain the difference between mutation rate and mutation frequency. *14.5*

16. What determines whether an altered product of a mutation will have helpful, neutral, or harmful effects? *14.5*

Self-Quiz (*Answers in Appendix III*)

1. Garrod, then Beadle and Tatum, deduced that the individual, heritable mutations they investigated corresponded to ______ .
   a. metabolic disorders
   b. defective enzymes
   c. defective tRNAs
   d. both a and b

2. DNA's many genes are transcribed into different ______ .
   a. proteins
   b. mRNAs only
   c. mRNAs, tRNAs, and rRNAs
   d. All are correct.

3. An RNA molecule is ______ .
   a. a double helix
   b. usually single-stranded
   c. always double-stranded
   d. usually double-stranded

4. An mRNA molecule is produced by ______ .
   a. replication
   b. duplication
   c. transcription
   d. translation

5. Each codon calls for a specific ______ .
   a. protein
   b. polypeptide
   c. amino acid
   d. carbohydrate

6. Referring to Figure 14.11, use the genetic code to translate the mRNA sequence UAUCGCACCUCAGGAGACUAG. Notice that the first codon in the frame is UAU. Which of the following three amino acid sequences is being specified?

   a. TYR—ARG—THR—SER—GLY—ASP—

   b. TYR—ARG—THR—SER—GLY

   c. TYR—ARG—TYR—SER—GLY—ASP—

7. Anticodons pair with ______ .
   a. mRNA codons   c. tRNA anticodons
   b. DNA codons   d. amino acids

8. Match the terms with the suitable description.

___ alkylating agent	a. part remaining in mRNA transcript
___ chain elongation	b. base triplet coding for amino acid
___ exon	c. second stage of translation
___ genetic code	d. base triplet that pairs with codon
___ anticodon	e. one environmental agent that induces mutation in DNA
___ intron	f. set of sixty-four codons for mRNA
___ codon	g. part removed from a pre-mRNA transcript

Critical Thinking

1. Sandra discovered a tRNA with a mutation in DNA that encodes the anticodon 3′-AAU instead of 3′-AUU. In cells with the mutated tRNA, what will be the effect on protein synthesis?

2. A DNA polymerase made an error during the replication of an important gene region of DNA. None of the DNA repair enzymes detected or repaired the damage. A portion of the DNA strand with the error is shown here:

...AATTCC ACTCCTATGG
...TTAAGG TGAGGATACC

After the DNA molecule is replicated and two daughter cells have formed, one cell is carrying a mutation and the other cell is normal. Develop a hypothesis to explain this observation.

3. In the bacterium *E. coli*, the end product (E) of the following metabolic pathway is absolutely essential for life:

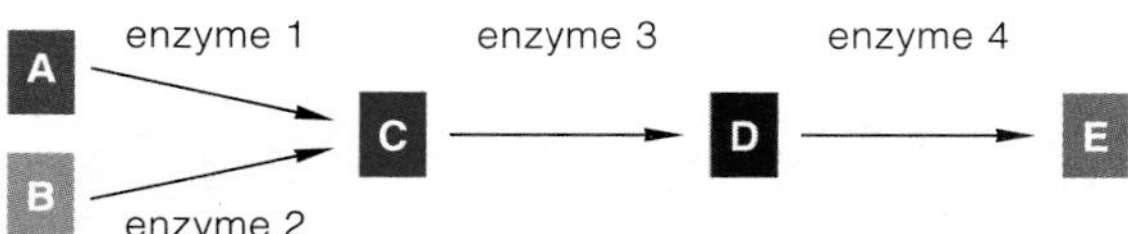
A
B
enzyme 1
enzyme 2
C
enzyme 3
D
enzyme 4
E

Ginetta, a geneticist, is attempting to isolate mutations in the genes for the four enzymes of this pathway. She has been able to isolate mutations in the genes for enzymes 1 and 2. In each case, the mutant *E. coli* cells synthesize a reduced amount of the pathway's end product E. However, Ginetta has not been able to isolate *E. coli* cells that have mutations in the genes for enzymes 3 and 4. Develop a hypothesis to explain why.

4. List all possible codons in the genetic code that could be changed into a STOP codon by a single nucleotide substitution.

5. *Neurofibromatosis* is a human autosomal dominant disorder caused by mutations in the *NF1* gene. It is characterized by soft, fibrous tumors in the peripheral nervous system and skin as well as abnormalities in muscles, bones, and internal organs (Figure 14.19).

Because the gene is dominant, an affected child usually has an affected parent. Yet in 1991, scientists reported on a boy who had neurofibromatosis yet whose parents did not. When they examined both copies of his *NF1* gene, they found the copy he had inherited from his father contained a transposon, a DNA segment that had moved to a new location. Neither the father nor the mother had a transposon in any of the copies of their own *NF1* genes. Speculate on the cause of neurofibromatosis in the human body and how it may have arisen.

6. The genome of a species is all the DNA in a haploid number of its chromosomes. The human genome has about 3.2 billion base pairs. But half of that may consist of transposons, which some researchers call *parasitic DNA*.

Some transposons contain instructions for synthesizing an enzyme that cuts the transposon out of the DNA and inserts it into a new location. *Retrotransposons* are more complex. These gene segments contain instructions for a reverse transcriptase, a type of enzyme that uses RNA as a template to make DNA. That DNA is a movable duplicate of the retrotransposon. Its mode of replication is uncannily similar to that of retroviruses (Section 22.8). Many biologists suspect that the first retroviruses were retrotransposons that left one species for a new one, maybe after inserting themselves into the DNA of a virus.

Either way, parasitic DNA can be duplicated hundreds of times and passed on through generations. As it moves in and out of a genome, it can switch genes on or off, maybe at unsuitable times or in the wrong cells. John McDonald of the University of Georgia argues that their destabilizing effects represent threats to survival. Over evolutionary time, natural selection may have worked at the molecular level to silence these genetic parasites.

Think about how eukaryotic chromosomes have many more genes and greater structural complexity than bacterial DNA. In the distant past, did some chromosomal proteins start to interact to isolate parasitic DNA from vital genes? Was this the impetus for the evolution of controls over complex genomes?

Or think about how vertebrate DNA is highly methylated. Methyl groups (alkylating agents) tend to shut down genes, and they are pervasive in vertebrate genomes. Did methylation first evolve as a way to control invading transposons? Was that control later extended to control vital genes?

If you find these ideas provocative, you may opt to continue your reading on this subject. Start with Ayala Ochert's highly readable article "Transposons" (*Discover*, December 1999).

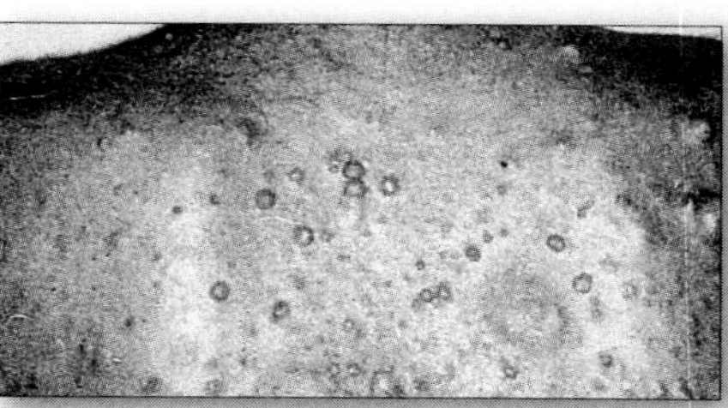

Figure 14.19 Soft skin tumors on a person with neurofibromatosis, a type of genetic disorder.

Selected Key Terms

alkylating agent *14.5*
anticodon *14.3*
base sequence *CI*
base-pair substitution *14.5*
carcinogen *14.5*
codon *14.3*
deletion (of base) *14.5*
exon *14.2*
gel electrophoresis *14.1*
gene mutation *14.5*
genetic code *14.3*
insertion (of base) *14.5*
intron *14.2*
ionizing radiation *14.5*
mRNA (messenger RNA) *14.2*
mutation rate *14.5*
promoter (RNA) *14.2*
ribonucleic acid (RNA) *CI*
RNA polymerase *14.2*
rRNA (ribosomal RNA) *14.2*
transcription *CI*
translation *CI*
transposon *14.5*
tRNA (transfer RNA) *14.2*
uracil *14.2*

Readings *See also www.infotrac-college.com*

Crick, F. 1988. *What Mad Pursuit: A Personal View of Scientific Discovery*. New York: HarperCollins. Crick's autobiography.

Fairbanks, F., and W. R. Andersen. 1999. *Genetics: The Continuity of Life*. Monterey, California: Brooks/Cole.

15 CONTROLS OVER GENES

When DNA Can't Be Fixed

1992 was an unforgettable year for Laurie Campbell. She finally turned eighteen. In that same year she just happened to notice a suspiciously black mole on her skin. It had an odd lumpiness about it, a ragged border, and an encrusted surface. No fool, Laurie quickly made an appointment with her family doctor, who quickly ordered a biopsy. The mole turned out to be a *malignant melanoma*, the deadliest form of skin cancer.

Laurie was lucky. She detected a cancer in its earliest stage, before it could spread through her body. Now she regularly checks out the appearance of other moles on her skin. She is aware of having become a statistic—one of 500,000 people in the United States who develop skin cancer in any given year and one of the 23,000 with malignant melanoma. She knows now that, of 7,500 who will die each year from skin cancer, 5,600 have malignant melanoma.

Laurie is smart. She plotted out the position of every mole on her body. Once a month, that body map is her guide for a quick but thorough self-examination. Figure 15.1 shows examples of what she looks for. Laurie also schedules a medical examination every six months.

Changes in DNA trigger skin cancer. The ultraviolet wavelengths in rays from the sun, tanning lamps, and other sources of nonionizing radiation can cause the bad molecular changes. For example, they can promote covalent bonding between two adjacent thymine bases in a nucleotide strand. The two nucleotides to which the bases belong become an abnormal, bulky structure—a thymine dimer—within the DNA. Normally, at least seven gene products interact as a DNA repair mechanism to remove such bulky lesions. But mutation in one or more of the genes can disrupt the repair machinery. Thymine dimers can accumulate in skin cells and trigger cancers and other lesions.

C
T
T
C
thymine dimer

The risk of skin cancer is greater for some than others. At one extreme are people affected by the genetic disorder *xeroderma pigmentosum*. Even brief exposure to sunlight puts them at risk of developing skin tumors and dying early from cancer.

You are at risk if you have moles that are chronically irritated, as by shaving or abrasive clothing. You are at risk if your skin, including the lip surface, is chronically chapped, cracked, or sore. You are at risk if your family has a history of cancer or if you have had radiation therapy. And, like Laurie, you are at risk if you burn

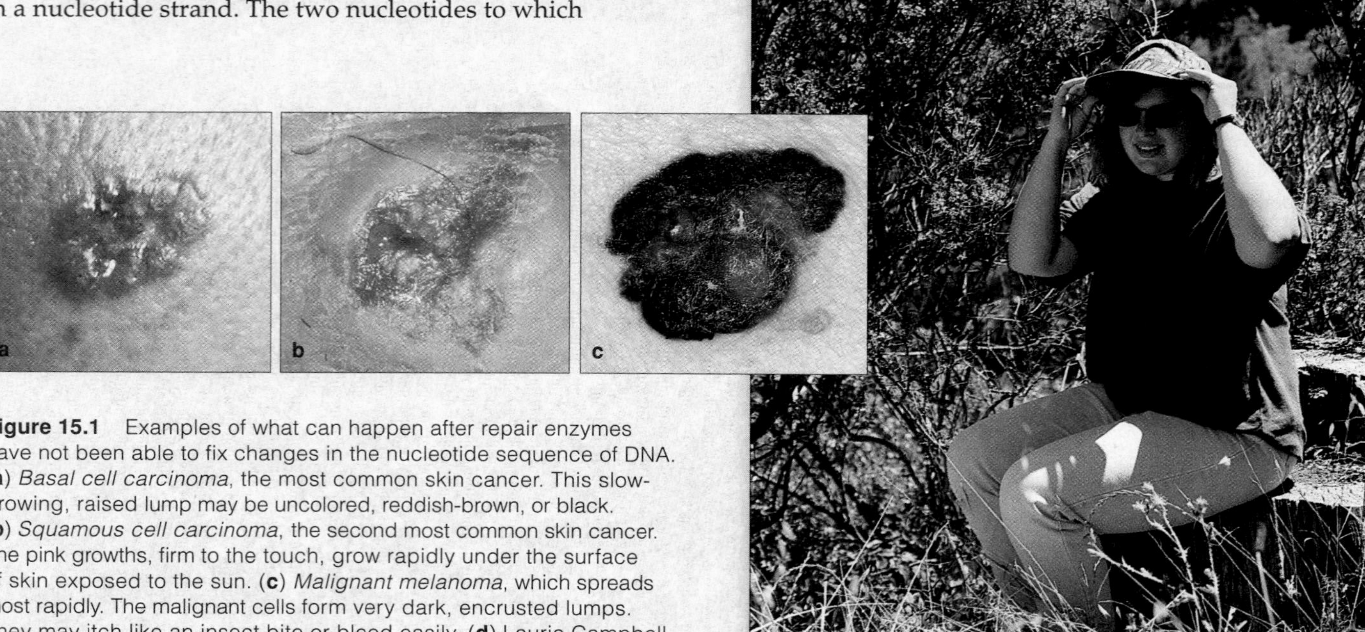
a
b
c
d

Figure 15.1 Examples of what can happen after repair enzymes have not been able to fix changes in the nucleotide sequence of DNA. (**a**) *Basal cell carcinoma*, the most common skin cancer. This slow-growing, raised lump may be uncolored, reddish-brown, or black. (**b**) *Squamous cell carcinoma*, the second most common skin cancer. The pink growths, firm to the touch, grow rapidly under the surface of skin exposed to the sun. (**c**) *Malignant melanoma*, which spreads most rapidly. The malignant cells form very dark, encrusted lumps. They may itch like an insect bite or bleed easily. (**d**) Laurie Campbell avoiding the sun—and melanoma.

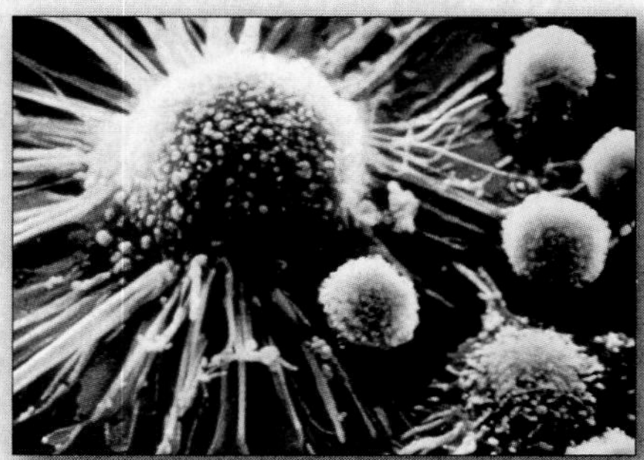

Figure 15.2 A patrol of white blood cells in an encounter with a body cell that has undergone cancerous transformation.

easily. Damaged DNA is the reality for all of us in spite of the ill-advised, socially promoted allure of tanning and staying out unprotected under the sun.

By definition, all **cancers** are malignant forms of tumors, which are tissue masses that arise through mutations in the genes that govern cell growth and division. Tumor cells don't respond to normal controls; they go on dividing as long as conditions for growth remain favorable. Cells of common skin moles and other *benign* tumors grow in an unprogrammed way. But they grow slowly, and they still have the surface recognition proteins that hold them together in their home tissue. Most benign tumors are left alone unless they become overly large or irritating.

In a *malignant* tumor, the abnormal cells grow and divide more rapidly, with destructive physical and metabolic effects on surrounding tissues. The cells are grossly disfigured (Figure 15.2). They cannot construct a normal cytoskeleton or plasma membrane, and they cannot synthesize normal recognition proteins. Also, malignant cells can break loose from their home tissue, enter lymph or blood vessels, travel along, then slip out and invade other parts of the body where they do not belong. And there they may start growing as new tumors. **Metastasis** (meh-TAH-stu-SIS) is the name for the process of abnormal cell migration and tissue invasion.

This chapter is your invitation to learn about the controls that govern when and how fast genes will be transcribed and translated, and whether gene products will be activated or shut down. By starting out with cancerous transformations, it invites you to reflect on how lucky we are when proper gene controls are in place and cells are operating as they should. Each year in the developed countries alone, 15 to 20 percent of all deaths result from cancer. It is not just a human problem. Researchers also have observed cancer in most of the animal species they have studied to date.

KEY CONCEPTS

1. In cells, a variety of controls govern when, how, and to what extent genes are expressed. The control elements operate in response to changing chemical conditions and to reception of external signals.

2. Control is exerted by way of regulatory proteins and other molecules that operate before, during, or after gene transcription. Different controls interact with DNA, with RNA that has been transcribed from the DNA, or with gene products—that is, with the resulting polypeptide chains or final proteins.

3. Prokaryotic cells depend on rapid control over short-term shifts in nutrient availability and other aspects of their surrounding environment. They commonly rely on regulatory proteins that help make quick adjustments in rates of gene transcription.

4. All eukaryotic cells depend on controls over short-term shifts in diet and levels of activity. In complex multicelled species, they also depend on controls over an intricate, long-term program of growth and development.

5. Controls over eukaryotic cells come into play when new cells contact one another in developing tissues. They also come into play when those cells start interacting with their neighbors by way of hormones and other signaling molecules.

6. Although all cells of a multicelled organism inherit the same genes, different cell types activate or suppress many of those genes in different ways. The controlled, selective use of genes leads to the synthesis of the proteins that give each type of cell its distinctive structure, function, and products.

15.1 OVERVIEW OF GENE CONTROL

At this very moment, bacteria are feeding on nutrients in your gut. Red blood cells are binding, transporting, or giving up oxygen, and great numbers of epithelial cells in your skin are busily synthesizing the protein keratin. Like cells everywhere, they are functioning by virtue of the protein products of genes.

Cells are selective about which gene products they make or require. Different kinds express certain genes just once, only at certain times, all the time, or not at all. *Which genes are being expressed depends on the type of cell, its moment-by-moment adjustments to changing chemical conditions, which external signals it happens to be receiving —and its built-in control systems.*

For example, availability of nutrients shifts rapidly and often for enteric bacteria (those living in intestines). Like other prokaryotes, such bacteria rapidly transcribe genes for nutrient-digesting enzymes. They make many copies of those enzymes whenever nutrients are moving through the intestines. They cut back on synthesis of the enzymes when nutrients are scarce. By comparison, your cells do not encounter drastic shifts in the solute concentrations and composition of the fluid that bathes them, and few exhibit rapid shifts in transcription.

The "systems" that control the expression of genes consist of molecules. For instance, **regulatory proteins** influence transcription, translation, and gene products. As you will see, such molecules exert their effects by interacting with DNA, RNA, new polypeptide chains, or final proteins (such as enzymes). Some components are activated or inhibited by signaling molecules, such as hormones. Others operate in response to changing concentrations of substances outside or inside the cell.

Negative control systems block a particular activity in a cell, and **positive control systems** promote it. For instance, one regulatory protein inhibits transcription of a gene when it binds with DNA, but the action of a different regulatory protein enhances its transcription. Usually, regulatory proteins do not act alone. To give an example, some types bind to DNA when required to do so, then release their grip on the binding site when a different control element interacts with them.

Summing up, these are the main concepts to keep in mind as you read through the rest of this chapter:

Cells exert selective control over when, how, and to what extent each of their genes is expressed.

The expression of a given gene depends on the type of cell and its functions, on chemical conditions, and on signals from the outside environment.

Many regulatory proteins and other molecules exert control over gene expression through their interactions with DNA or RNA. Others exert control through their interaction with gene products—new polypeptide chains or final proteins.

15.2 CONTROL IN BACTERIAL CELLS

Let's first consider some examples of gene control in prokaryotic cells—that is, bacteria. When nutrients are plentiful and other environmental conditions also favor growth, bacteria tend to grow and divide indefinitely. Gene controls promote the rapid synthesis of enzymes that have roles in nutrient digestion and other growth-related activities. Transcription is fast, and translation is initiated even before mRNA transcripts are finished. Remember, bacteria have no nucleus; nothing separates their DNA from ribosomes in the cytoplasm.

When a nutrient-degrading pathway utilizes several enzymes, all genes for those enzymes are transcribed, often as a continuous mRNA molecule. The genes are not transcribed when conditions turn unfavorable. The rest of this section gives two cases of how controls can adjust transcription rates downward or upward.

Negative Control of Transcription

The enteric bacterium *Escherichia coli* inhabits the gut of all mammals. It survives on glucose, lactose (a sugar in milk), and other ingested nutrients. Like other adult mammals, you probably do not drink milk around the clock. When you do, however, *E. coli* cells in your gut rapidly transcribe three genes for enzymes that have roles in breakdown reactions that begin with lactose.

A promoter precedes the three genes, which are next to one another. A **promoter**, recall, is a base sequence that signals the start of a gene. A different sequence, an **operator**, intervenes between a promoter and bacterial genes. It is a binding site for a **repressor**, a regulatory protein that can block transcription (Figure 15.3). An arrangement in which a promoter and operator service more than one gene is an **operon**. Elsewhere in *E. coli* DNA, a different gene codes for this repressor, which can bind with the operator *or* with a lactose molecule.

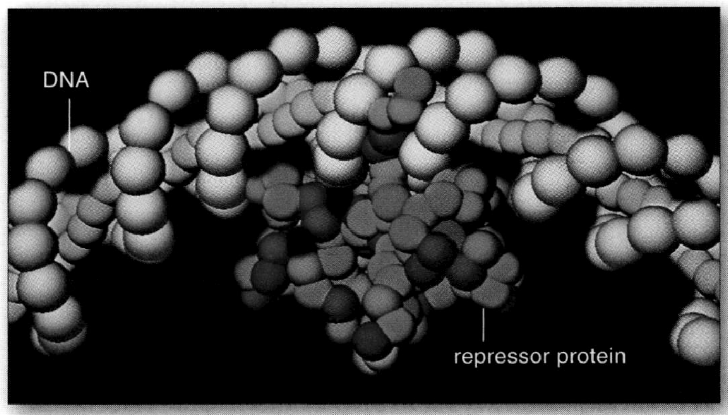
DNA
repressor protein

Figure 15.3 Computer model of one type of repressor protein binding to an operator at a site in a bacterial DNA molecule.

Further reading: Student Guide to InfoTrac on web site —

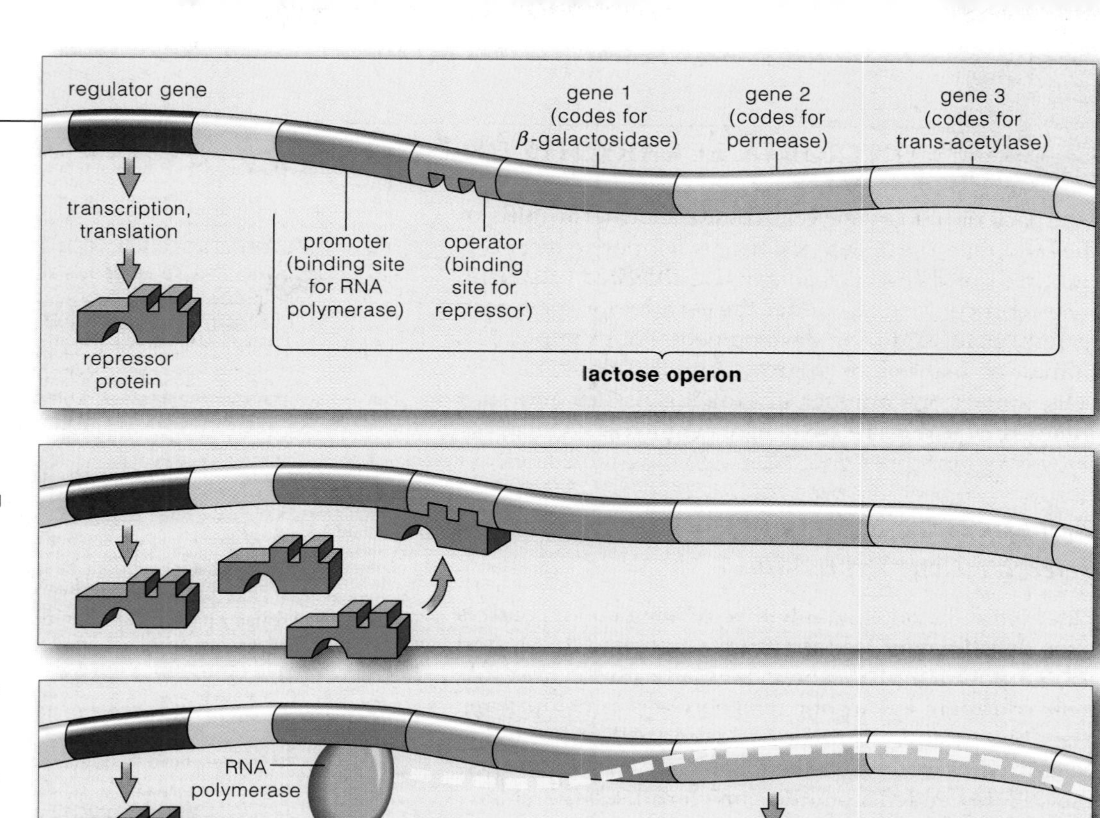

regulator gene
gene 1 (codes for β-galactosidase)
gene 2 (codes for permease)
gene 3 (codes for trans-acetylase)
transcription, translation
promoter (binding site for RNA polymerase)
operator (binding site for repressor)
repressor protein
lactose operon
RNA polymerase
mRNA transcript
translation into three polypeptide chains for three different enzymes
lactose
β-galactosidase
permease
trans-acetylase

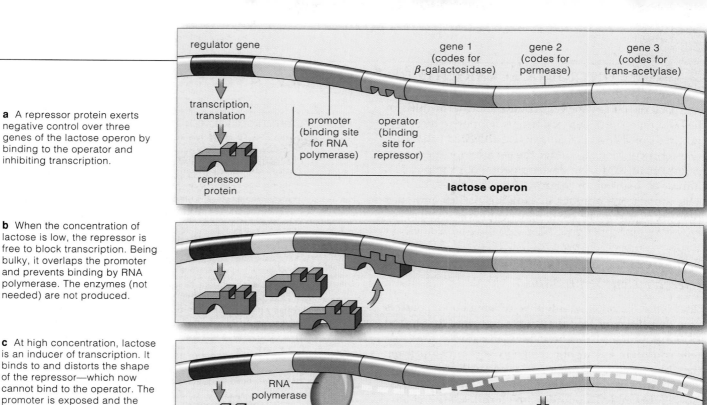

a A repressor protein exerts negative control over three genes of the lactose operon by binding to the operator and inhibiting transcription.

b When the concentration of lactose is low, the repressor is free to block transcription. Being bulky, it overlaps the promoter and prevents binding by RNA polymerase. The enzymes (not needed) are not produced.

c At high concentration, lactose is an inducer of transcription. It binds to and distorts the shape of the repressor—which now cannot bind to the operator. The promoter is exposed and the genes can be transcribed.

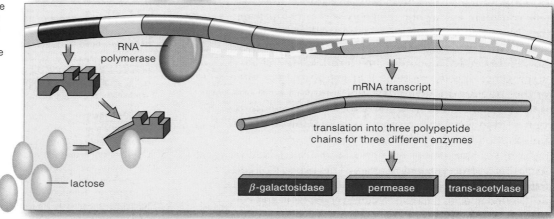

Figure 15.4 Negative control of the lactose operon. The operon's first gene codes for an enzyme that splits lactose, a disaccharide, into two subunits (glucose and galactose). The second enzyme codes for an enzyme that transports lactose into cells. The third enzyme functions in metabolizing certain sugars.

When the lactose concentration is low, a repressor binds with the operator, as in Figure 15.4. Being a large molecule, it overlaps the promoter, so transcription is blocked. Thus, *lactose-degrading enzymes are not built if they are not required.* When many lactose molecules are around, chances are greater that one of them will bind with the repressor. Binding alters the repressor's shape, so it cannot bind with the operator. RNA polymerase is free to transcribe the genes. Therefore, *lactose-degrading enzymes are synthesized only when required.*

Positive Control of Transcription

E. coli cells pay far more attention to glucose than to lactose. They transcribe genes for glucose breakdown faster, and continuously. Even if lactose is present, the lactose operon isn't used much *unless glucose is absent.* At such times, CAP, a regulatory protein, acts on the operon. CAP is one of the **activator proteins**, which are part of positive control systems. To understand how it works, you have to know the lactose operon's promoter is not good at binding RNA polymerase. It does better when CAP adheres to it first. But CAP will not do this until it is activated by a small molecule called cAMP.

Among other things, cAMP forms from ATP, which *E. coli* makes by breaking down glucose (in glycolysis). Not much cAMP is available in the cell when glucose is plentiful and glycolysis is running full bore. When such conditions prevail, CAP does not get primed to adhere to the promoter, so transcription of the lactose operon genes slows almost to a standstill.

Suppose glucose is scarce but lactose is available. cAMP accumulates, and CAP–cAMP complexes form. Now the lactose operon genes are transcribed rapidly, and the cell uses lactose as an alternative energy source.

Prokaryotic cells, which must respond rapidly to changing conditions, commonly rely on a small number of regulatory proteins that exert rapid, on-off control of transcription.

15.3 CONTROL IN EUKARYOTIC CELLS

Like bacteria, eukaryotic cells control short-term shifts in diet and in levels of activity. If cells happen to be one of hundreds or trillions of cells in a multicelled organism, long-term controls also enter the picture, for their gene activity changes during development. For example, as a human or plant embryo grows and develops, its new cells contact one another in growing tissues and start interacting with their particular neighbors. They do so by way of hormones and other signaling molecules.

Cell Differentiation and Selective Gene Expression

Later on in the book, you will be reading about controls over development, especially in Chapters 32, 44, and 45. For now, simply become familiar with the idea that gene control in eukaryotic cells becomes quite intricate. Start by considering this basic point: All of the living cells in your body inherited the same genes (because they all descended from the same fertilized egg). Many of the genes specify proteins that are absolutely vital for the structure and everyday functioning of each one of those cells. That is why genes for those proteins are controlled in ways that promote ongoing, low levels of transcription.

Even so, *nearly all of your cells became specialized in composition, structure, and function*. This process of **cell differentiation** proceeds during the development of all multicelled species. It arises as embryonic cells and cell lineages descended from them activate and suppress a small fraction of their genes in unique, selective ways. As examples, only immature red blood cells use the genes for making hemoglobin. Only certain white blood cells use the genes for making weapons called antibodies.

Controls Before and After Transcription

Many of the genes that govern the housekeeping tasks in all eukaryotic cells are under positive controls that promote continuous, low levels of transcription. In the case of many others, we see fluctuating adjustments of transcription. Why? In every multicelled organism, the internal environment (tissue fluids and blood) affords much more stable operating conditions for the body's individual cells. Most often, gene transcription rates rise or fall by small degrees in response to slight shifts in concentrations of signaling molecules, nutrients, and products in that environment.

As the examples in Figure 15.5*a* indicate, some gene sequences are repeatedly duplicated or rearranged in genetically programmed fashion prior to transcription. Besides this, programmed chemical modifications often shut down many genes. So does the orderly packaging of DNA by histones and other chromosomal proteins, as you may realize after reflecting on the organization of eukaryotic chromosomes (Section 9.5).

Controls also come into play after genes have been transcribed. As Figure 15.5*b–d* indicates, many controls govern RNA transcript processing, transport of mature RNAs from the nucleus, and rates of translation in the cytoplasm. Other controls deal with the modification of new polypeptide chains. Others deal with activating, inhibiting, and breaking down existing proteins. As one example, consider enzymes, the proteins that catalyze

a CONTROLS RELATED TO TRANSCRIPTION. At any given time, most genes of a multicelled organism are shut down, either permanently or temporarily. The genes necessary for a cell's everyday tasks are under positive controls that promote ongoing, low levels of transcription. With the help of these controls, a cell is assured of having enough enzymes and other proteins to carry out its most basic functions. Transcription of many other genes often shifts only slightly. In this case, controls work to assure chemical responsiveness even when concentrations of specific substances rise or fall only slightly.

Also, even before some genes are transcribed, a portion of their base sequences may be amplified, rearranged, or chemically modified in temporary or reversible ways. These changes are not mutations. They are heritable, programmed events that affect the manner in which particular genes will be expressed, if at all.

1. *Gene amplification.* Some cells that require enormous numbers of certain molecules briefly increase the number of the required genes. Sometimes multiple rounds of DNA replication produce hundreds or thousands of gene copies prior to transcription! This happens in immature amphibian eggs and glandular cells of some insect larvae (Sections 15.4 and 15.5).

2. *DNA rearrangements.* In a few cell types, many base sequences are alternative "choices" for parts of a gene. Prior to transcription, they are snipped out of the DNA. Combinations of the snippets are spliced together as the gene's final base sequence. For instance, this happens when B lymphocytes, a special class of white blood cells, are forming (Section 40.9). Different cells transcribe and translate their uniquely rearranged DNA into staggeringly diverse versions of protein weapons called antibodies, which act specifically against one of staggeringly diverse kinds of pathogens and other foreign agents in the body.

3. *Chemical modification.* Numerous histones and other proteins interact with eukaryotic DNA in highly organized fashion. The DNA–protein packaging, as well as chemical modifications to particular sequences, influences gene activity. Usually, only a small fraction of a cell's genes are available for transcription. A dramatic shutdown occurs in female mammals. As Section 15.4 describes, this event is called X chromosome inactivation.

Figure 15.5 Examples of the levels of control over gene expression in eukaryotes.

Further reading: Student Guide to InfoTrac on web site

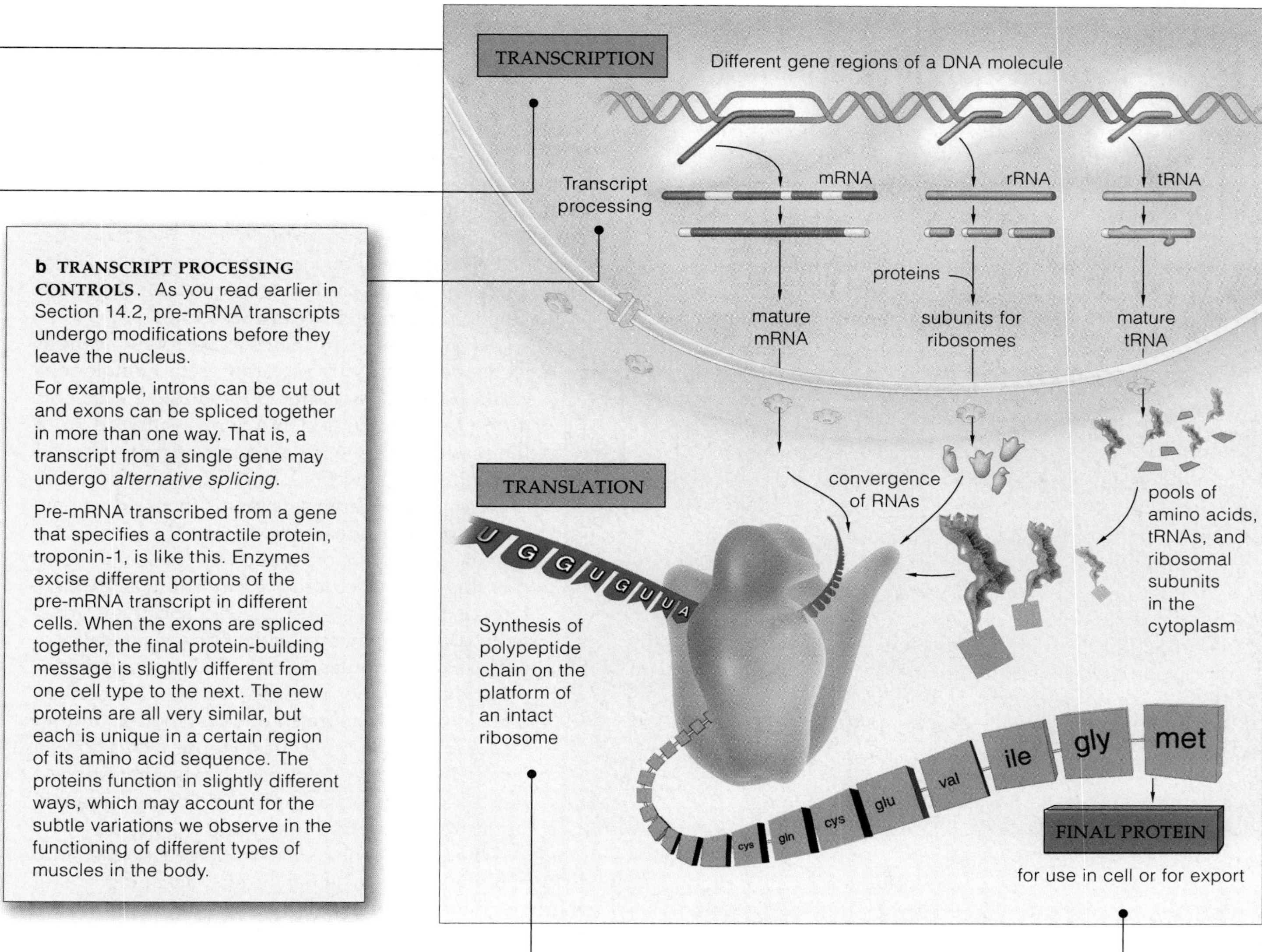
TRANSCRIPTION
Different gene regions of a DNA molecule
Transcript processing
mRNA
rRNA
tRNA
proteins
mature mRNA
subunits for ribosomes
mature tRNA
TRANSLATION
convergence of RNAs
pools of amino acids, tRNAs, and ribosomal subunits in the cytoplasm
U G G U G U U A
Synthesis of polypeptide chain on the platform of an intact ribosome
cys
gln
cys
glu
val
ile
gly
met
FINAL PROTEIN
for use in cell or for export

b TRANSCRIPT PROCESSING CONTROLS. As you read earlier in Section 14.2, pre-mRNA transcripts undergo modifications before they leave the nucleus.

For example, introns can be cut out and exons can be spliced together in more than one way. That is, a transcript from a single gene may undergo *alternative splicing.*

Pre-mRNA transcribed from a gene that specifies a contractile protein, troponin-1, is like this. Enzymes excise different portions of the pre-mRNA transcript in different cells. When the exons are spliced together, the final protein-building message is slightly different from one cell type to the next. The new proteins are all very similar, but each is unique in a certain region of its amino acid sequence. The proteins function in slightly different ways, which may account for the subtle variations we observe in the functioning of different types of muscles in the body.

c CONTROLS OVER TRANSLATION. Diverse controls govern when, how rapidly, and how often a given mRNA transcript will be translated. Sections 37.2, and 44.3 provide elegant examples.

The stability of a transcript influences the number of protein molecules that can be produced from it. Enzymes destroy transcripts from the poly-A tail on up (Section 14.2). The tail's length and its attached proteins affect the pace of degradation. Also, after leaving the nucleus, some transcripts are temporarily or permanently inactivated. For example, in unfertilized eggs, many transcripts are inactivated and stored in the cytoplasm. These "masked messengers" will not be available for translation until after fertilization, when great numbers of protein molecules will be required for the early cell divisions of the new individual.

d CONTROLS FOLLOWING TRANSLATION. Before they can become fully functional, many polypeptide chains must pass through the cytomembrane system (Section 4.5). There they undergo modification, as when enzymes attach specific oligosaccharides or phosphate groups to them.

Also, a variety of control mechanisms govern the activation, inhibition, and stability of the enzymes and other molecules that are involved in protein synthesis. Allosteric control of tryptophan synthesis, as described in Section 6.7, is an example.

nearly all metabolic reactions. In addition to selectively transcribing and translating genes for enzymes, diverse control systems activate and inhibit the molecules of enzymes that have already been synthesized in the cell.

Just imagine the coordination necessary to govern which of the cell's thousands of types of enzymes are to be stockpiled, deployed, or degraded in a specified interval. *That coordination governs all short-term and long-term aspects of cell structure and function.*

In multicelled species, gene controls guide the moment-by-moment activities of cells. They also guide intricate, long-term patterns of bodily growth and development.

All cells of complex organisms inherit the same genes, yet most become specialized in composition, structure, and function. This process of cell differentiation arises when different populations of cells activate and suppress some of their genes in highly selective, unique ways.

15.4

EVIDENCE OF GENE CONTROL

By some estimates, the cells of a multicelled organism rarely use more than 5 to 10 percent of their genes at a given time. One way or another, control mechanisms are keeping most of the genes repressed. Besides this, *which* genes are being expressed varies, depending on the stage of growth and development the organism is passing through. The following examples only hint at the kinds of controls operating at different stages.

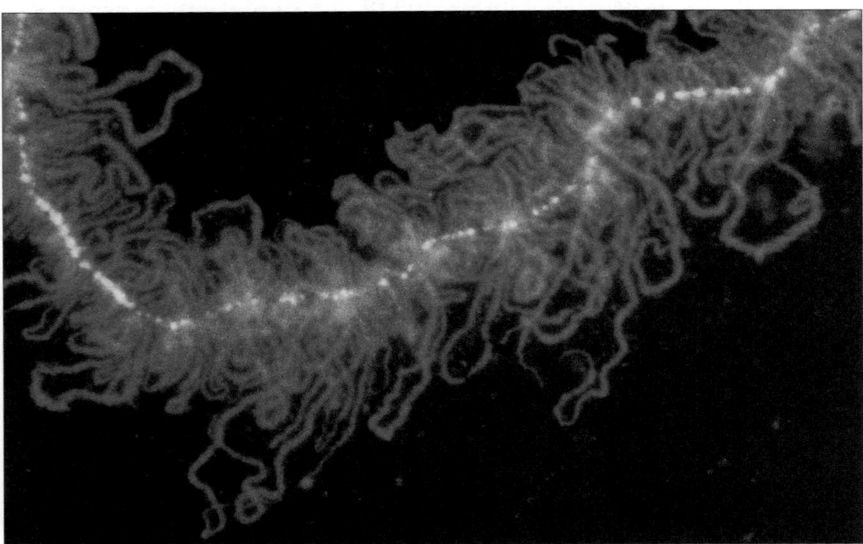

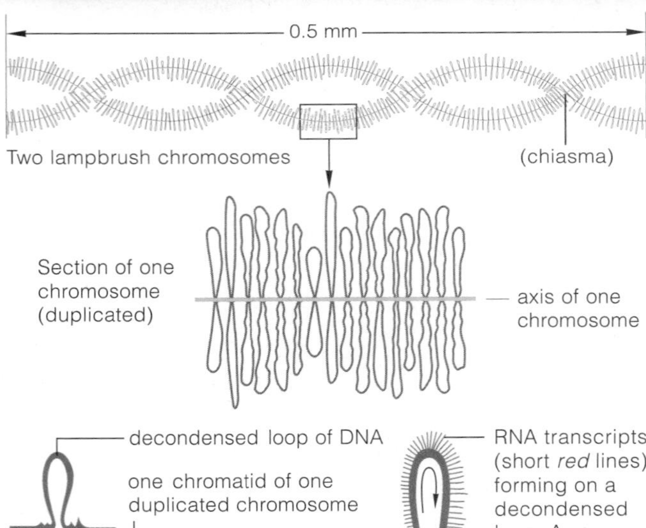

0.5 mm
Two lampbrush chromosomes
(chiasma)
Section of one chromosome (duplicated)
axis of one chromosome
decondensed loop of DNA
one chromatid of one duplicated chromosome
sister chromatid
RNA transcripts (short *red* lines) forming on a decondensed loop. Arrow shows direction of transcription.

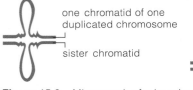

Figure 15.6 Micrograph of a lampbrush chromosome from a germ cell of a newt (*Notophthalmus viridiscens*). During prophase I, gene regions of the duplicated chromosomes decondensed and looped outward. A *red* fluorescent dye labeled a component of RNA–protein complexes, the presence of which is evidence of gene activity. (A *white* fluorescent dye labeled the DNA making up the axis of each duplicated chromosome's structural framework.)

Transcription in Lampbrush Chromosomes

Amphibian life cycles are such that the unfertilized eggs of females grow extremely fast while they are maturing. Growth requires numerous copies of enzymes and other proteins. Each germ cell that gives rise to eggs stockpiles large numbers of RNAs and ribosomes, all of which will participate in rapid protein synthesis.

At prophase I of meiosis, a protein scaffold forms between pairs of duplicated, homologous chromosomes. The sister chromatids of each pair decondense, and the DNA loops outward from the scaffold. The loops are profuse and stiffened with a coat of new RNA transcripts and proteins. The overall appearance is bristly, hence the name **lampbrush chromosomes** (Figure 15.6).

What is going on here? Histones and other proteins that structurally organize the DNA have loosened their grip, so the genes specifying RNA molecules are now accessible. Histones, recall, are part of nucleosomes, the unit of organization in eukaryotic chromosomes. As Section 9.5 describes, a nucleosome is a stretch of DNA looped around a core of histone molecules. Figure 15.7*b* shows a model based on photomicrographs that suggest how nucleosome packaging might be loosened up during transcription.

The chromosomes are packaged further into distinct domains. This higher order packaging is still poorly understood, but it is known to have a crucial role in controlling transcription. There are about 10,000 loops in many amphibian eggs. Each corresponds to a specific DNA sequence. Yet most of the DNA remains condensed in regions called chromomeres, which for the most part are *not* transcribed. Controls keep some parts of the chromosomes extended for transcription even as they keep other parts condensed and thereby inactive. The domains are large, and they are precise.

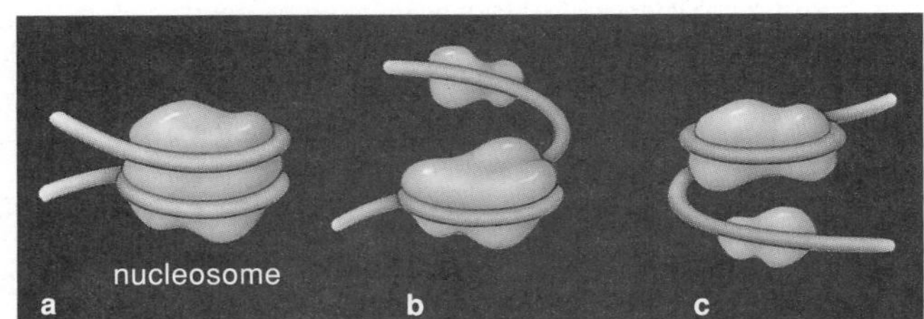

nucleosome
a
b
c

Figure 15.7 Model of the kinds of changes in the structural organization of chromosomal DNA that might promote gene transcription. The tight DNA–histone packing in nucleosomes (**a**) might loosen up at gene sequences that are about to be transcribed in the manner shown in (**b**–**c**). Eukaryotic genes are known to be transcribed primarily in regions where the packing of chromosomal DNA is most relaxed.

Further reading: Student Guide to InfoTrac on web site

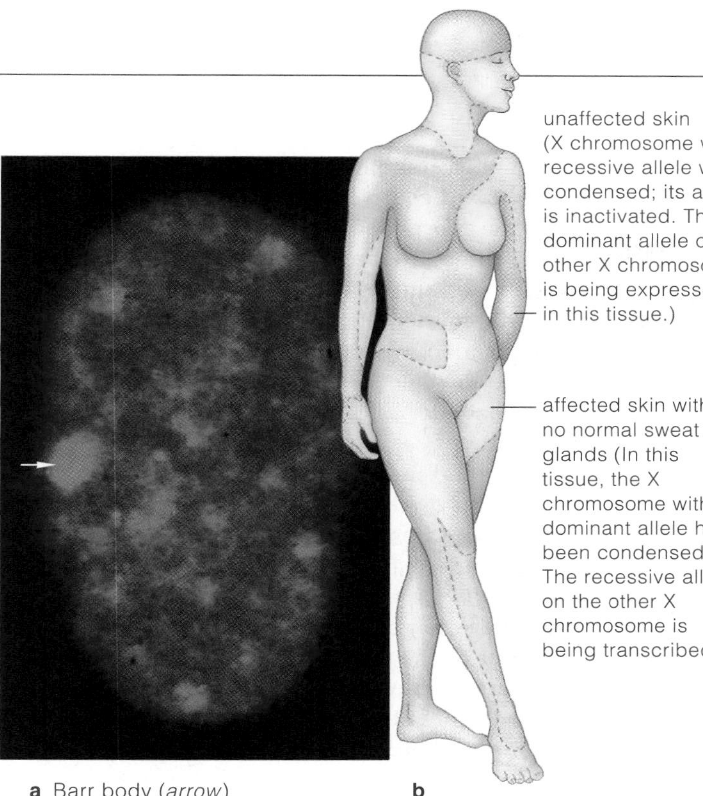
unaffected skin (X chromosome with recessive allele was condensed; its allele is inactivated. The dominant allele on other X chromosome is being expressed in this tissue.)
affected skin with no normal sweat glands (In this tissue, the X chromosome with dominant allele has been condensed. The recessive allele on the other X chromosome is being transcribed.)
a Barr body (*arrow*)
b

Figure 15.8 (**a**) Micrograph of an inactivated X chromosome, called a Barr body, as it appears in a human female's somatic cell during interphase. The X chromosome is not condensed this way in a human male's cells. (**b**) Anhidrotic ectodermal dysplasia, a mosaic pattern of gene expression. The condition arises as a result of random X chromosome inactivation.

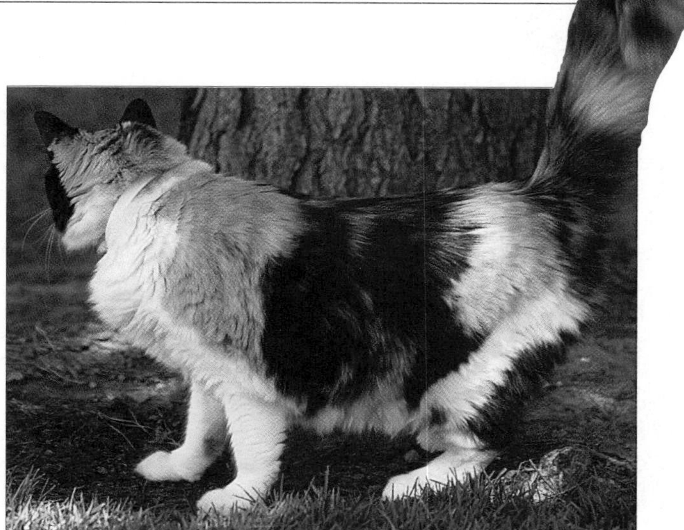

Figure 15.9 Why is this female calico cat "calico"? In her cells, one X chromosome carries a dominant allele for the brownish-black pigment melanin. The allele on her other X chromosome specifies yellow fur. At an early stage of the cat's embryonic development, one of the two X chromosomes was inactivated at random in each cell that had formed by then.

In all descendants of those cells, the same chromosome also became inactivated, which left them with only one functional allele for the coat-color trait. We see patches of different colors, depending on which allele was inactivated in cells that formed a given tissue region. (The white patches result from a gene interaction involving the "spotting gene," which blocks melanin synthesis entirely.)

X Chromosome Inactivation

As you know, proteins impart structural organization to eukaryotic DNA. In female mammals, the structure of one of the two X chromosomes in each body cell is chemically modified in a way that completely closes off access to its genes.

A mammalian embryo destined to become female inherits one X chromosome from the mother and another from the father. One of the two condenses in each cell and is inactivated. Condensation is part of the program of development, but the outcome is random. *One or the other chromosome may be inactivated.* We observe such a condensed X chromosome in the interphase nucleus; it shows up as a dense spot (Figure 15.8*a*). The condensed X chromosome was named a **Barr body** (after Murray Barr, its discoverer).

Whenever a maternal (or paternal) X chromosome is inactivated in a cell, it also becomes inactivated in all descendants of that cell. Thus, when development ends, the female is a "mosaic" for those X chromosomes. *She bears patches of tissue where the genes of the maternal X chromosome are expressed—and patches of tissues in which the genes of the paternal X chromosome are expressed.* Any pair of alleles on those two chromosomes may or may not be identical. That is why tissues such as skin may have different features in different body regions. Mary Lyon discovered this **mosaic tissue effect**, which arises from random X chromosome inactivation.

We see the effect in human females who have one X chromosome with a recessive allele that blocks sweat gland formation. The absence of sweat glands in such heterozygotes is one symptom of *anhidrotic ectodermal dysplasia*. In skin patches without the glands, only genes on the X chromosome with the mutant allele are being transcribed; the X chromosome with the dominant allele is condensed. In patches with the glands, the dominant allele is active (Figure 15.8*b*). The same effect shows up in female calico cats, which are heterozygous for black and yellow coat-color alleles on their X chromosomes. The coat color in a given body region depends on which X chromosome's genes are transcribed (Figure 15.9).

In a typical cell of multicelled organisms, controls repress all but about 5 to 10 percent of the genes at a given time.

The remaining fraction of genes is selectively expressed at different stages of growth and development. Lampbrush chromosome formation, X chromosome inactivation, and other cellular events provide evidence of this.

15.5 EXAMPLES OF SIGNALING MECHANISMS

The story that opened this chapter introduced the idea that a great variety of signals influence gene activity. The following examples from animals and plants will give you a sense of their effects at the molecular level.

Hormonal Signals

Hormones, a major category of signaling molecules, can stimulate or inhibit gene activity in target cells. Any cell with receptors for a given hormone is a target. Animal cells secrete hormones into tissue fluid. Most hormone molecules are picked up by the bloodstream, which distributes them to cells some distance away. In plants, hormones do not travel far from cells that secrete them.

Certain hormones bind to membrane receptors at the surface of target cells. Others enter target cells, bind to regulatory proteins, and so help initiate transcription. In many cases, a hormone molecule must first combine briefly with an enhancer. **Enhancers** are base sequences that serve as binding sites for suitable activator proteins. They may or may not be adjacent to a promoter in the same DNA molecule. When some distance separates the two, a loop forms in the DNA, and looping unites the enhancer with the promoter. RNA polymerase can bind avidly to the bound complex at the base of the loop.

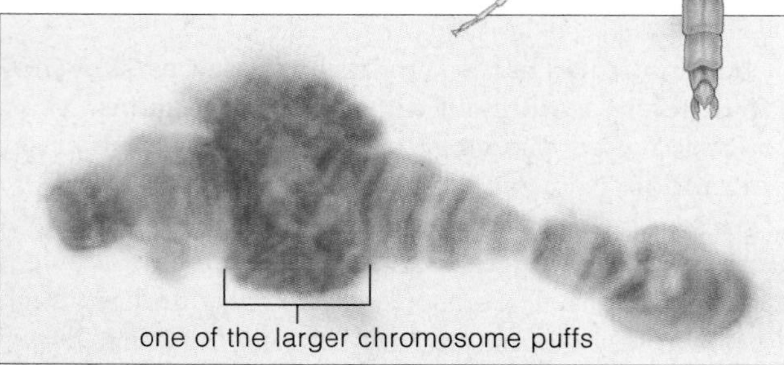
one of the larger chromosome puffs

Figure 15.10 Visible evidence of transcription in a polytene chromosome from a larva of a midge (*Chironomus*). Midges are flies, one to ten millimeters long. The short-lived, winged adults often congregate in great swarms, which helps them find mates in a hurry. They might seem to be pests merely because of their large numbers. They also look somewhat like mosquitoes, but they don't bite.

Most midge larvae develop in aquatic habitats. To sustain their rapid growth, they feed continually, as on decaying organic material. They require a lot of saliva, and they must continuously transcribe genes for saliva's protein components. Those genes have undergone amplification. Ecdysone, a hormone, serves as a regulatory protein that helps promote their transcription. Midge chromosomes loosen and puff out in regions where the genes are being transcribed in response to the hormonal signal. The puffs become large and appear quite diffuse when transcription is most intense. Staining techniques reveal banding patterns in the chromosomes, as evident in the micrograph.

Consider the effect of **ecdysone**, a hormone with key roles in many insect life cycles. Immature stages called larvae grow rapidly during the cycles. They continually feed on organic matter, such as decaying leaves, and require copious amounts of saliva to prepare food for digestion. DNA in their salivary gland cells has been replicated repeatedly. The copies of DNA molecules have remained together in parallel array, forming what is known as a **polytene chromosome**. When ecdysone binds to a receptor on the gland cells, its signal triggers very rapid transcription of multiple copies of genes in the DNA. The gene regions affected by this hormonal signal puff out during transcription, as in Figure 15.10. Afterward, translation of mRNA transcripts made from the genes produces the protein components of saliva.

In vertebrates, certain hormones have widespread effects on gene expression because many types of cells have receptors for them. As one example, the pituitary gland secretes somatotropin, or growth hormone. This hormonal signal stimulates synthesis of all the proteins required for cell division and, ultimately, the body's growth. Most cells have receptors for somatotropin.

Other vertebrate hormones signal only certain cells at certain times. Prolactin, a secretion from the pituitary gland, is like this. A few days after a female mammal gives birth, prolactin can be detected in her blood. This hormone activates genes in mammary gland cells that have receptors for it. Those genes have responsibility for milk production. Liver cells and heart cells have the same genes but do not have the receptors necessary to respond to signals from prolactin.

Explaining hormonal control of gene activity is like explaining a full symphony orchestra to someone who has never seen one or heard it perform. Many separate parts must be defined before their interactions can be understood! We will return to this topic later on. As you will especially see in Chapter 32, 37, 44, and 45, some of the most elegant examples of hormonal controls are drawn from studies of plant and animal reproduction and development.

Sunlight as a Signal

Plant a few seeds from a corn or bean plant in a pot that contains moist, nutrient-rich soil. Next, let the seeds germinate, but keep them in total darkness. After eight days have passed, they will develop into seedlings that are spindly and notably pale, owing to the absence of chlorophyll (Figure 15.11). Now expose the seedlings to a single burst of dim light from a flashlight. Within ten minutes, they will start converting stockpiles of certain molecules to the activated forms of chlorophylls, the

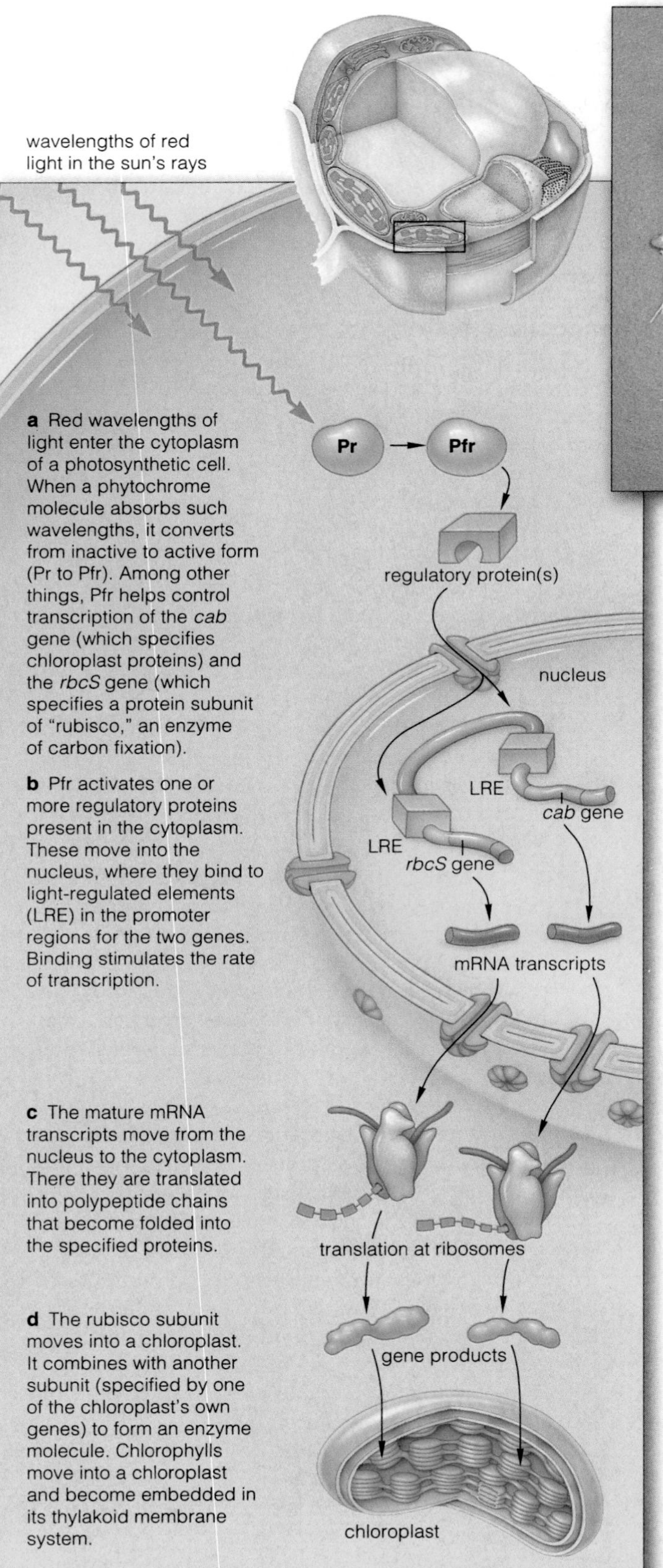
wavelengths of red
light in the sun's rays
a Red wavelengths of light enter the cytoplasm of a photosynthetic cell. When a phytochrome molecule absorbs such wavelengths, it converts from inactive to active form (Pr to Pfr). Among other things, Pfr helps control transcription of the *cab* gene (which specifies chloroplast proteins) and the *rbcS* gene (which specifies a protein subunit of "rubisco," an enzyme of carbon fixation).
b Pfr activates one or more regulatory proteins present in the cytoplasm. These move into the nucleus, where they bind to light-regulated elements (LRE) in the promoter regions for the two genes. Binding stimulates the rate of transcription.
c The mature mRNA transcripts move from the nucleus to the cytoplasm. There they are translated into polypeptide chains that become folded into the specified proteins.
d The rubisco subunit moves into a chloroplast. It combines with another subunit (specified by one of the chloroplast's own genes) to form an enzyme molecule. Chlorophylls move into a chloroplast and become embedded in its thylakoid membrane system.
Pr
Pfr
regulatory protein(s)
nucleus
LRE
cab gene
LRE
rbcS gene
mRNA transcripts
translation at ribosomes
gene products
chloroplast

Figure 15.11 Sunlight as a signal for gene activity. The photograph shows the effect of an absence of light on corn seedlings. The two seedlings at *left* were the control group; they were grown inside a sunlit greenhouse. The two seedlings positioned next to them were grown in total darkness for eight days. The dark-grown plants were not able to convert their stockpiled precursors of chlorophyll molecules to active form, and they never did green up. The diagram illustrates one model for the mechanism by which phytochrome might help to control gene transcription in plants. Red wavelengths can convert the phytochrome molecule from inactive form (here designated Pr) to active form (Pfr). In this form, the phytochrome can serve as a regulator of transcription.

light-trapping pigment molecules that donate electrons for the reactions of photosynthesis.

Phytochrome, a blue-green pigment, is a signaling molecule that helps plants adapt over the short term to changes in light conditions. Section 32.4 takes a close look at this molecule. For now, simply be aware that it alternates between active and inactive forms. At sunset, at night, or in shade, far-red wavelengths predominate. At such times, the phytochrome in cells is inactive. It is activated at sunrise, when red wavelengths dominate the sky. Also, the amount of red or far-red wavelengths that a plant intercepts varies from day to night and as seasons change. Such variations act as control signals over phytochrome activity. That activity influences the transcription of certain genes at certain times of day and year. The genes specify a number of enzymes and other proteins that help seeds to germinate, stems to lengthen and then branch, and leaves to grow. Gene products also help flowers, fruits, and seeds to form.

Elaine Tobin and her coworkers at the University of California, Los Angeles, performed experiments that provided evidence in favor of phytochrome control. Using dark-grown seedlings of duckweed (*Lemna*), they discovered a marked increase in the number of certain mRNA transcripts after a one-minute exposure to red light. Exposure had enhanced transcription of the genes for proteins that bind chlorophylls and for rubisco, an enzyme that mediates carbon fixation (Sections 7.6 and 7.7). In the absence of the proteins, chloroplasts do not develop properly and they don't turn green.

Hormones and other signaling molecules, including diverse kinds that respond to environmental stimuli, have profound influence on gene expression.

LOST CONTROLS AND CANCER

THE CELL CYCLE REVISITED Every second, millions of cells in your skin, bone marrow, gut lining, liver, and elsewhere divide and replace their worn-out, dead, and dying predecessors. They do not divide willy-nilly. Controls govern the expression of genes that specify enzymes and other proteins required for cell growth, DNA replication, chromosome movements, and division of the cytoplasm. Diverse regulatory proteins control the synthesis and use of those gene products. And they control when the division machinery is put to rest.

Controls over the cell cycle are extensive. At various interrelated points, gene activity can be stepped up or slowed down, and gene products can advance, delay, or block the cycle. Some genes directly regulate passage through the cycle and are its primary controllers. Their products include **protein kinases**, a class of enzymes that attach phosphate groups to proteins. They operate at the boundary between G1 and S of the cell cycle, and during the transition from G2 to mitosis (Section 9.2).

Other genes encode proteins that modify the activity of the primary control genes or their products. Among them are genes for **growth factors**: transcriptional signals sent by one cell to trigger growth in other cells. For example, when epidermal growth factor (EGF) binds to its receptor on target cells, it triggers tyrosine kinase activity—a signal for mitosis. Other genes specify the receptors for growth factors.

Certain gene products exert indirect control over the cell cycle. For example, some are enzymes that help maintain DNA replication. Others take part in stockpiling weapons to be released when the cell is scheduled to die.

If genes specifying any of these products become mutated, the cell may be deprived of crucial enzymes or proteins. Cancer may follow.

CHARACTERISTICS OF CANCER At the least, four features characterize all cancer cells. First, *their plasma membrane and cytoplasm change profoundly.* The membrane becomes more permeable; its proteins are lost or altered, and abnormal ones form. The cytoskeleton becomes disorganized, shrinks, or both. Enzyme action shifts, as in amplified reliance on glycolysis. Second, *cancer cells grow and divide abnormally.* Controls that prevent overcrowding in tissues are lost. Cell populations reach high densities. Proteins trigger abnormal increases in the number of small blood vessels that can service the growing cell mass. Third, *cancer cells have a weakened capacity for adhesion.* Recognition proteins are altered or lost, so cells cannot stay anchored in proper tissues. Thus they can break away and establish colonies in distant tissues (Figure 15.12). Fourth, *cancer cells are lethal.* Unless eradicated, they will put the individual on a painful road to death.

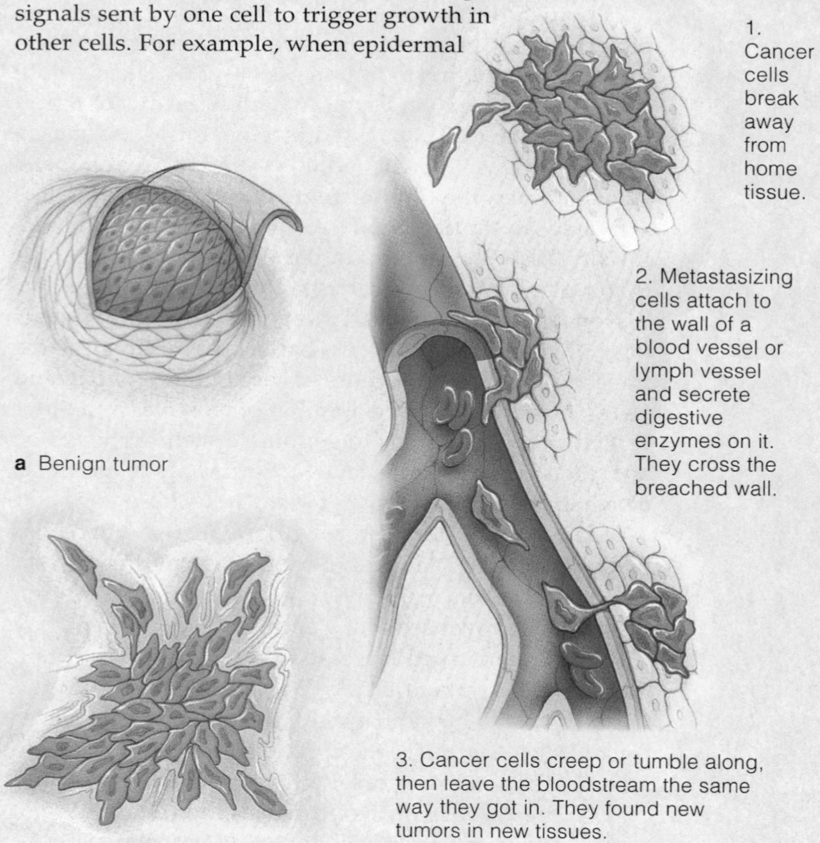

a Benign tumor
b Malignant tumor
1. Cancer cells break away from home tissue.
2. Metastasizing cells attach to the wall of a blood vessel or lymph vessel and secrete digestive enzymes on it. They cross the breached wall.
3. Cancer cells creep or tumble along, then leave the bloodstream the same way they got in. They found new tumors in new tissues.
c Metastasis

Figure 15.12 Evidence of loss of gene control: (**a**) A benign tumor, a mass of cells with an outwardly normal appearance that are enclosed in a capsule of connective tissue. (**b**) A malignant tumor, a disorganized collection of cancer cells. (**c**) Steps in metastasis.

ONCOGENES Any gene having the potential to induce cancerous transformation is an **oncogene**. Oncogenes were first identified in retroviruses (a class of RNA viruses). They are mutated forms of normal genes that specify the growth factors and other proteins that are necessary for normal cell functioning. Think of the normal forms as *proto*-oncogenes.

One category of oncogenes specifies proteins that induce cell proliferation. To give examples, *erbB* specifies the receptor for epidermal growth factor, *src* specifies tyrosine kinase, and *myc* specifies a factor necessary for transcription.

Another category specifies proteins that are *anti*-oncogenes; they inhibit cell proliferation. The product of one such oncogene, *p53*, prevents some cancers. To date, mutation in *p53* has been implicated in 90 percent of the reported cases of small-cell lung cancer and more than 50 percent of breast and colon cancer.

A third category of oncogenes suppresses or triggers programmed cell death. One of these, *bcl-21*, helps keep normal body cells from dying before their time.

Oncogene expression is highly controlled in all normal cells. Typically, mitosis is induced

Further reading: Student Guide to InfoTrac on web site —

Signal
to die
docks at
receptor.
Signal
causes
activation
of ICE-like
proteases.

Figure 15.13 An artist's representation of weapons of cell death being unleashed inside a cell.

when one cell type secretes a growth factor, which acts on the cells bearing the receptor for that factor. Mitosis is suppressed when one cell type secretes anti-oncogenes. Uncontrolled activation of any oncogene may spark tumor formation. Inactivation of any anti-oncogene releases one of the brakes on tumor formation.

Oncogenes can form following certain insertions of viral DNA into cellular DNA. They may also form after carcinogens (cancer-inducing agents) change the DNA. As you read earlier, ultraviolet radiation and other forms of nonionizing radiation (such as x-rays) are carcinogens. So are many natural and synthetic compounds, including asbestos and certain components of tobacco smoke.

Remember those chromosome alterations and gene mutations? Some cancers arise when base substitutions or deletions alter a proto-oncogene or one of the controls over its transcription. Others arise by translocations, as described in the introduction to Chapter 12 and Section 12.9. Destabilizing transposons also cause some cancers.

HERE'S TO SUICIDAL CELLS! We conclude with a case study of a gene and its cancerous transformation.

The first cell of a new multicelled individual contains marching orders that will guide its descendants along a program of growth, development, and reproduction, then on to death. As part of that program, many cells heed calls to self-destruct when they complete a prescribed function. If they become altered in ways that might pose a threat to the body as a whole, as by infection or cancer, they can execute themselves. **Apoptosis** (app-oh-TOE-sis) is the name for this form of cell death. It starts with molecular signals that activate and unleash lethal weapons of self-destruction, which were stockpiled earlier within the cell.

Protein-cleaving enzymes called **ICE-like proteases** are such weapons.Think of them as folded pocketknives or lethal Ninja weapons. When popped open, they chop apart structural proteins, including the building blocks of cytoskeletal elements and nucleosomes that organize the DNA (Figure 15.13).

A body cell in the act of suicide shrinks away from its neighbors. Its cytoplasm seems to roil, and its surface repeatedly bubbles outward and inward. No longer are its chromosomes extended through the nucleoplasm; they are bunched up near the nuclear envelope. The nucleus, then the cell, breaks apart. Phagocytic white blood cells patrol tissues. If they encounter suicidal cells or remnants of them, the patrolling cells will swiftly engulf them.

The timing of cell death is predictable in some cells, such as pigment-packed keratinocytes. These form the densely packed sheets of dead cells that are continually sloughed off and replaced at the skin surface. They have a three-week life span, more or less. Keratinocytes and other body cells, even the kinds that are supposed to last for a lifetime, can be induced to die ahead of schedule. All it takes is sensitivity to signals that can activate ICE-like proteases and other enzymes of death.

The knives remain sheathed in cancer cells, which are supposed to—but don't—commit suicide on cue. Maybe they have lost normal receptors that would allow contact with their signaling neighbors. Maybe they are receiving abnormal signals about when to grow, divide, or cease dividing. In some cancers, for instance, the anti-oncogene *bcl-2* either has been tampered with or it has been shut down. Apoptosis cannot occur.

As you might deduce from all of this, cancer is a multistep process, involving more than one oncogene. Researchers have much to learn. But they have already identified many of the genes in the process. Through their work, it may be possible to diagnose carriers early and frequently. Cancer that can be detected early enough might still be curable, as by surgery.

15.7 SUMMARY

1. Cells are equipped with controls that govern gene expression; that is, which gene products appear, when, and in what amounts. When control mechanisms come into play depends on cell type, on prevailing chemical conditions, and on signals from other cell types that can change a target cell's activities.

2. Regulatory proteins, enzymes, hormones, and other molecules are components of control systems. Various kinds interact with one another, with DNA and RNA, and with products of genes. Control systems operate before, during, and after transcription and translation.

3. In all cell types, two of the most common categories of control systems operate to inhibit or enhance gene transcription. Their effects are reversible.

a. With negative control systems, a regulatory protein binds at a specific DNA sequence and prevents one or more genes from being transcribed.

b. With positive control systems, a regulatory protein binds to DNA and promotes transcription.

4. Unlike eukaryotic cells, most bacteria (prokaryotic cells) do not require many genes to live, grow, and reproduce. Most of their control systems affect rates of transcription. Control of operons (groupings of related bacterial genes and control elements) is an example.

5. Promoters are binding sites in DNA that signal the start of a gene. In all cells, activator proteins bind next to promoters, and the complex promotes transcription. In bacterial cells only, operators serve as binding sites for regulatory proteins that can inhibit transcription.

6. Eukaryotic cells use more complex gene controls. Gene activity must change rapidly in response to short-term shifts in the surroundings, as in prokaryotic cells. However, for all multicelled species, it also must be intricately adjusted during growth and development, when great numbers of cells grow and divide, make physical contact with one another in diverse tissues, and interact chemically over the long term.

7. Being descended from the same cell, all of the cells in a multicelled organism inherit the same assortment of genes. However, different types of cells activate and suppress some fraction of the genes in different ways. This behavior is called selective gene expression.

8. Cell differentiation is one outcome of selective gene expression. By this process, different lineages of cells in multicelled organisms become specialized in function, appearance, and composition.

9. Genes associated with the cell cycle are extensively controlled. Gene activity can be stepped up or slowed. Gene products can advance, delay, or block the cycle. Products of the primary control genes include protein receptors for control signals. Other gene products, such as diverse growth factors and protein kinases, regulate primary control genes or their products. Cancer may follow mutation of any of these genes.

10. Cancer involves loss of normal controls over the cell cycle and mechanisms of programmed cell death.

a. In cancer cells, the structure and function of both the plasma membrane and the cytoplasm are severely compromised. These cells grow and divide abnormally. They have a weakened capacity for adhesion to their home tissue. Unless eradicated, they are lethal.

b. Growth factors, enzymes, and other products of certain genes normally induce cell proliferation, inhibit cell proliferation, or suppress or trigger programmed cell death (apoptosis). Oncogenes are mutated forms of those essential genes, and they have the potential to induce cancer.

Review Questions

1. Briefly describe the general characteristics of normal cells. Then explain the difference between a benign tumor and a malignant tumor. *CI, 15.6*

2. In what fundamental way do negative and positive controls of transcription differ? Is the effect of one or the other form of control (or both) reversible? *15.2*

3. Distinguish between: *15.2*
   a. repressor protein and activator protein
   b. promoter and operator
   c. repressor and enhancer

4. Describe one type of control of transcription for the lactose operon in *E. coli*, a prokaryotic cell. *15.2*

5. A plant, fungus, or animal consists of diverse cell types. How might the diversity arise, given that body cells in each of these organisms inherit the same set of genetic instructions? As part of your answer, define cell differentiation and the general way that selective gene expression brings it about. *15.3*

6. Using the diagram on the facing page, define at least three types of gene controls in eukaryotic cells and indicate the levels at which they take effect. *15.3*

7. What is a Barr body? Does it appear in the cells of males, females, or both? Explain your answer. *15.4*

Self-Quiz (*Answers in Appendix III*)

1. The expression of a given gene depends on the _______.
   a. type of cell and its functions
   b. chemical conditions
   c. environmental signals
   d. all of the above

2. Regulatory proteins interact with _______.
   a. DNA   c. gene products
   b. RNA   d. all of the above

3. In prokaryotic cells but not eukaryotic cells, a(n) _______ precedes the genes of an operon.
   a. lactose molecule   c. operator
   b. promoter   d. both b and c

4. A base sequence signaling the start of a gene is a(n) _______.
   a. promoter   c. enhancer
   b. operator   d. activator protein

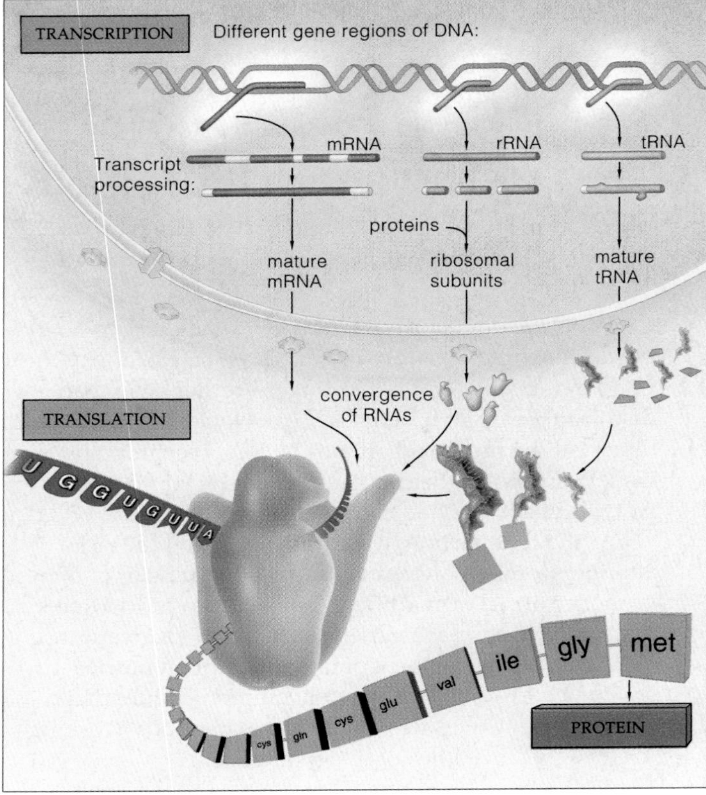
TRANSCRIPTION
Different gene regions of DNA:
mRNA
rRNA
tRNA
Transcript processing:
proteins
mature mRNA
ribosomal subunits
mature tRNA
convergence of RNAs
TRANSLATION
U G G U G U U A
ile
gly
met
val
glu
cys
gln
cys
PROTEIN

5. An operon most typically governs ______ .
   a. bacterial genes
   b. a eukaryotic gene
   c. genes of all types
   d. DNA replication

6. Eukaryotic genes guide ______ .
   a. rapid short-term activities
   b. overall growth
   c. development
   d. all of the above

7. Cell differentiation ______ .
   a. occurs in all complex multicelled organisms
   b. requires different genes in different cells
   c. involves selective gene expression
   d. both a and c
   e. all of the above

8. X chromosome inactivation may result in a ______ for some traits.
   a. male phenotype
   b. uniform tissue effect
   c. mosaic tissue effect
   d. patterned tissue effect

9. Hormones interact with ______ .
   a. membrane receptors
   b. regulatory proteins
   c. enhancers
   d. all of the above

10. Apoptosis is ______ .
   a. cell division after severe tissue damage
   b. programmed cell death by suicide
   c. a popping sound in mutated toes

11. ICE-like proteases are ______ .
   a. structural proteins
   b. lethal weapons
   c. environmental signals
   d. low-temperature enzymes

12. Match the terms with their most suitable descriptions.
   ______ phytochrome
   ______ Barr body
   ______ oncogene
   ______ repressor
   ______ lampbrush chromosome

   a. inhibits gene transcription
   b. gene that induces cancerous transformation
   c. intense, controlled transcription in selected domains of DNA
   d. helps plants adapt to daily and seasonal changes in light
   e. inactivated X chromosome

Critical Thinking

1. Geraldo isolated a strain of *E. coli* in which a mutation has affected the capacity of CAP to bind to a region of the lactose operon, as it would do normally. State how the mutation affects transcription of the lactose operon when cells of this bacterial strain are subjected to these conditions:
   a. Lactose and glucose are available.
   b. Lactose is available but glucose is not.
   c. Both lactose and glucose are absent.

2. *Duchenne muscular dystrophy*, a genetic disorder, affects boys almost exclusively. Early in childhood, muscles begin to atrophy (waste away) in affected individuals, who typically die in their teens or early twenties as a result of respiratory failure. Muscle biopsies of women who carry a gene associated with the disorder reveal some regions of atrophied muscle tissue. Yet the muscle tissue adjacent to these regions was normal or even larger and more chemically active, as if to compensate for the weakness of the adjoining region. How can you explain these observations?

3. Unlike most rodents, guinea pigs are already well developed at the time of birth. Within a few days, they are able to eat grass, vegetables, and other plant material. Suppose a breeder decides to separate the baby guinea pigs from their mothers after three weeks. He wants to keep the males and females in different cages, but it is quite difficult to identify the sex of guinea pigs when they are so young. Suggest a simple test that the breeder can perform to identify their sex.

4. Individuals affected by *pituitary dwarfism* cannot synthesize somatotropin (also called growth hormone). Children with this genetic abnormality will not grow to normal height. Develop a hypothesis to explain why therapy that is based on somatotropin injections may be effective.

5. Transcription is generally controlled by the binding of a protein to a DNA sequence "upstream" from a gene, rather than by modification of RNA polymerase. Develop a hypothesis to explain why this is so.

6. The closer a mammalian species is to humans in its genetic makeup, the more useful information it may yield in laboratory studies of the mechanisms of cancer. Do you support the use of any mammal for cancer research? Why or why not?

Selected Key Terms

activator protein *15.2*
apoptosis *15.6*
Barr body *15.4*
cancer *CI*
cell differentiation *15.3*
ecdysone *15.5*
enhancer *15.5*
growth factor *15.6*
hormone *15.5*
ICE-like protease *15.6*
lampbrush chromosome *15.4*
metastasis *CI*
mosaic tissue effect *15.4*
negative control system *15.1*
oncogene *15.6*
operator *15.2*
operon *15.2*
phytochrome *15.5*
polytene chromosome *15.5*
positive control system *15.1*
promoter *15.2*
protein kinase *15.6*
regulatory protein *15.1*
repressor *15.2*

Readings *See also www.infotrac-college.com*

Duke, R., D. Ojcius, and J. Ding-E Young. December 1996. "Cell Suicide in Health and Disease." *Scientific American*, 80–87.

Murray, A., and M. Kirschner. March 1991. "What Controls the Cell Cycle?" *Scientific American*, 264(3): 56–63.

Tijan, R. February 1995. "Molecular Machines That Control Genes." *Scientific American*, 54–61.

16 RECOMBINANT DNA AND GENETIC ENGINEERING

Mom, Dad, and Clogged Arteries

Butter! Bacon! Eggs! Ice cream! Cheesecake! Possibly you think of such foods as enticing, off-limits, or both. After all, who among us doesn't know about animal fats and the dreaded cholesterol?

Soon after you feast on those fatty foods, cholesterol enters the bloodstream. Cholesterol is important. It is a structural component of animal cell membranes, and without membranes, there would be no cells. Cells also remodel cholesterol into a variety of molecules, such as the vitamin D necessary for the development of good bones and teeth. Normally, however, the liver itself synthesizes enough cholesterol for your cells.

Some proteins circulating in blood combine with cholesterol and other substances to form lipoprotein particles. The high-density lipoproteins, or *HDLs*, transport cholesterol to the liver, where it can be metabolized. The low-density lipoproteins (*LDLs*) normally end up in cells that store or use cholesterol.

Sometimes too many LDLs form. The excess infiltrates the elastic walls of arteries, where it contributes to the formation of atherosclerotic plaques (Figure 16.1). These abnormal masses interfere with blood flow and narrow the arterial diameter. If the plaques clog one of the tiny coronary arteries that deliver blood to the heart, chest pains or a heart attack may result.

How well you handle dietary cholesterol depends on genes you inherited from your parents. One gene specifies a protein that serves as the cell's receptor for LDLs. Inherit two "good" alleles of that gene, and your blood level of cholesterol will tend to remain so low that your arteries may never clog up, even with a high-fat diet. Inherit two copies of a certain mutated allele, however, and you are destined to develop a rare genetic disorder called *familial cholesterolemia*. Levels of cholesterol become abnormally high. Many affected people die of heart attacks in childhood or their teens.

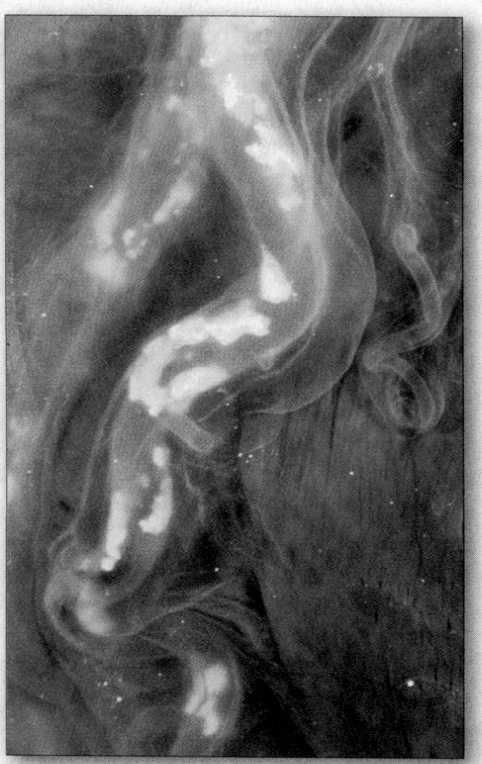

Figure 16.1 Life-threatening plaques (*bright yellow-white*) in one of the coronary arteries, small vessels that deliver blood to the heart. Abnormally high levels of cholesterol contribute to plaque formation.

In 1992 a woman from Quebec, Canada, became a milestone in the history of genetics. She was thirty years old. Like two of her younger brothers who had died from heart attacks in their early twenties, she inherited the mutant allele for the LDL receptor. She herself survived a heart attack when she was sixteen. At twenty-six, she had coronary bypass surgery.

At the time, people were hotly debating the risks and the promises of **gene therapy**—the transfer of one or more normal or modified genes into an individual's body cells to correct a genetic defect or boost resistance to disease. Even so, the woman opted for an untried, physically wrenching procedure that was designed to give her body working copies of the good gene.

Surgeons removed about 15 percent of the woman's liver. Researchers put cells from it in a nutrient-rich medium to promote cell growth and division. *And they spliced the functional allele for the LDL receptor into the genetic material of a harmless virus*. That modified virus served roughly the same function as a hypodermic needle. It was allowed to infect the cultured liver cells and insert copies of the good gene into them.

Later, about a billion modified cells were infused into the woman's portal vein, the main blood vessel that leads directly to the liver. There, some cells took up residence and started to make the missing cholesterol receptor. Two years after that, a fraction of the woman's liver cells were behaving normally and sponging up cholesterol from the blood. Her blood levels of LDLs had declined nearly 20 percent. Also, scans of her arteries showed no evidence of the progressive clogging that had nearly killed her.

Her cholesterol levels remain more than twice as high as normal, and it is too soon to know whether the gene therapy will prolong her life. Yet the intervention does provide strong evidence that gene therapy may be able to help some patients.

As you might gather from this pioneering clinical application, recombinant DNA technology has great potential for medicine. It also has staggering potential

Figure 16.2 One outcome of genetic manipulation by way of artificial selection practices: a large kernel from a modern strain of corn next to the tiny kernels of an ancestral species, which was recovered from a prehistoric cave in Mexico.

for agriculture and industry. Think of it! For thousands of years, we humans have been changing genetically based traits of species. By artificial selection practices, we produced new crop plants and breeds of cattle, cats, dogs, and birds from wild ancestral stocks. We were selective agents for meatier turkeys, sweeter oranges, seedless watermelons, flamboyant ornamental roses, and big juicy corn kernels (Figure 16.2). We produced splendid hybrid organisms, including the mule (horse × donkey) and the plants that produce the tangelo (tangerine × grapefruit).

Of course, we have to remember we are newcomers on the evolutionary stage. During the 3.8 billion years before we even made our entrance, nature conducted uncountable numbers of genetic experiments by way of mutation, crossing over, and other events that introduce changes in genetic messages. Those countless changes are the source of life's rich diversity.

But the striking thing about human-directed changes is that the pace has picked up. Today, researchers use **recombinant DNA technology** to analyze genes. They cut and recombine DNA from different species, then insert it into bacterial, yeast, or mammalian cells that replicate their DNA and divide at a rapid pace. The cells copy the foreign DNA as if it were their own and churn out useful quantities of recombinant DNA molecules. The technology also is the basis of **genetic engineering**, by which genes are isolated, modified, and inserted back into the same organism or into a different one. The protein products of those modified genes might cover the functions of their missing or malfunctioning counterparts.

The new technology does not come without risks. With this chapter, we consider its basic aspects. At the chapter's end, we also address some ecological, social, and ethical questions related to its application.

KEY CONCEPTS

1. Genetic experiments have been proceeding in nature for billions of years, through gene mutations, crossing over, recombination, and other natural events.

2. Humans are now purposefully bringing about genetic changes by way of recombinant DNA technology. Such manipulations are called genetic engineering.

3. With this technology, researchers isolate, cut, and splice together gene regions from different species, then greatly amplify the number of copies of the genes that interest them. The genes, and in some cases the proteins they specify, are produced in quantities that are large enough to use for research and for practical applications.

4. Three activities are at the heart of recombinant DNA technology. First, procedures based on specific types of enzymes are used to cut DNA molecules into fragments. Second, the fragments are inserted into cloning tools, such as plasmids. Third, the fragments containing the genes of interest are identified, then copied rapidly and repeatedly.

5. Genetic engineering involves isolating, modifying, and inserting genes back into the same organism or into a different one. The goal is to beneficially modify traits that the genes influence. Human gene therapy, which focuses on controlling or curing genetic disorders, is an example.

6. The new technology raises social, legal, ecological, and ethical questions regarding its benefits and risks.

16.1 A TOOLKIT FOR MAKING RECOMBINANT DNA

Restriction Enzymes

In the 1950s, the scientific community was agog over the discovery of DNA's structure. The excitement gave way to frustration, for no one could figure out the sequence of nucleotides—the order of genes and gene regions along a chromosome. Robert Holley and his colleagues managed to sequence a relatively small tRNA by using digestive enzymes that could break the molecule, step by step, into fragments small enough to characterize chemically. But DNA? DNA molecules are much longer than RNA. No one knew how to cut one into fragments long enough to have unique, and therefore analyzable, sequences. Digestive enzymes can cut DNA, but they chop away in no particular order. So there was no telling how the fragments had been arranged in the molecule.

Then, by accident, Hamilton Smith discovered that *Haemophilus influenzae*, a bacterium, swiftly responds to a bacteriophage attack by chopping the foreign DNA. Wonderfully, extracts from *H. influenzae* cells held an enzyme that restricts its activity to a specific kind of site in DNA. It was suitably named a **restriction enzyme**.

In time, several hundred strains of bacteria offered up a toolkit of restriction enzymes that recognize and cut specific sequences of four to eight bases in DNA. Table 16.1 lists a few. They are among many that make *staggered* cuts, which leave single-stranded "tails" on the end of a DNA fragment. Tails made by *Taq*I are two bases long (CG). *Eco*RI makes tails four bases long (AATT).

How many cuts do restriction enzymes make? That depends partly on the molecule. To give one example, the eight-base sequence recognized by *Not*I is rare in the DNA of mammals. Usually the resulting fragments are tens of thousands of base pairs long. That is long enough to be useful for studying the organization of a **genome**, which is all the DNA in a haploid number of chromosomes for each species. For the human species, the genome is about 3.2 billion base pairs long.

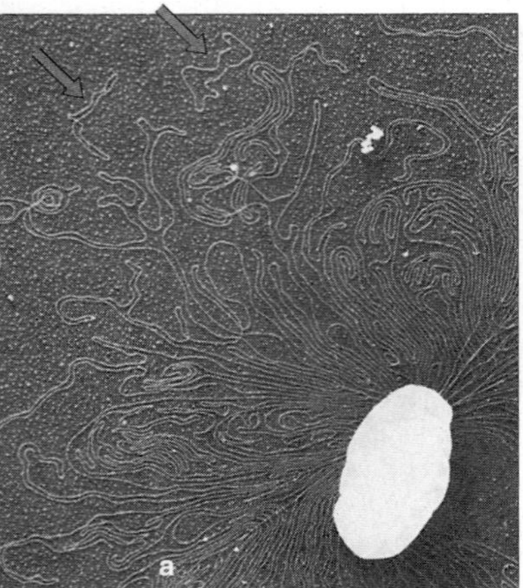

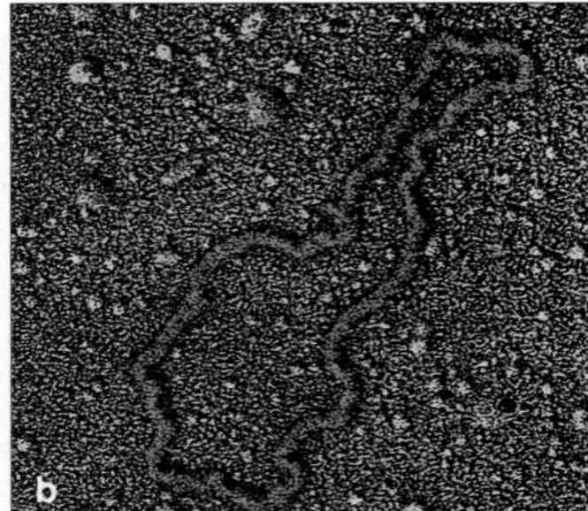

Figure 16.3 (**a**) Plasmids (*blue* arrows) released from a ruptured *Escherichia coli* cell. (**b**) One plasmid at higher magnification.

Table 16.1 Examples of Restriction Enzymes

Bacterial Source	Enzyme's Abbreviation	Its Specific Cut
Thermus aquaticus	*Taq*I	5' T C G A 3' 3' A G C T 5'
Escherichia coli	*Eco*RI	5' G A A T T C 3' 3' C T T A A G 5'
Nocardia otitidus caviarum	*Not*I	5' G C G G C C G C 3' 3' C G C C G G C G 5'

Modification Enzymes

DNA fragments with staggered cuts have sticky ends. "Sticky" means a restriction fragment's single-stranded tail has a capacity to base-pair with a complementary tail of any other DNA fragment or molecule cut by the same restriction enzyme. *Mix together DNA fragments cut by the same restriction enzyme, and the sticky ends of any two fragments having complementary base sequences will base-pair and form a recombinant DNA molecule:*

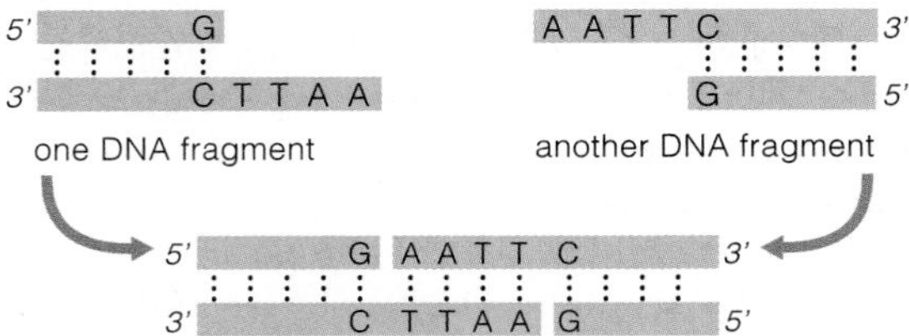

5' G
3' C T T A A
one DNA fragment
A A T T C 3'
G 5'
another DNA fragment
5' G A A T T C 3'
3' C T T A A G 5'

Notice the nicks where such fragments base-pair. **DNA ligase**, a modification enzyme, can seal these nicks:

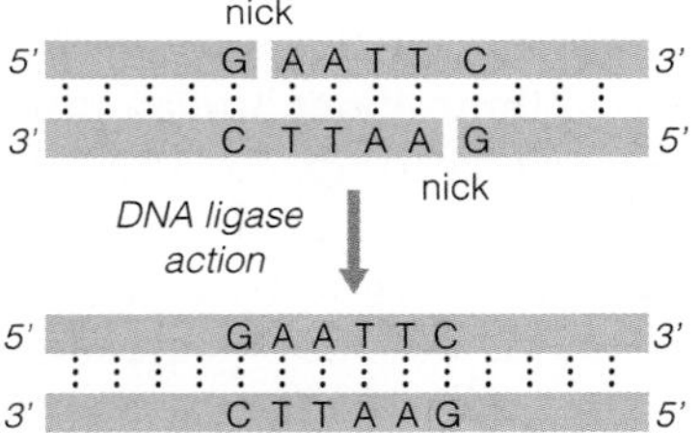

nick
5' G A A T T C 3'
3' C T T A A G 5'
nick
DNA ligase action
5' G A A T T C 3'
3' C T T A A G 5'

Cloning Vectors for Amplifying DNA

Restriction and modification enzymes made it possible to insert foreign DNA into bacterial cells that have the capacity to replicate it. As you know, a bacterial cell has a single chromosome, a circular DNA molecule with all the genes that it requires to grow and reproduce. Many species of bacteria also may inherit plasmids. As Figure 16.3 shows, a **plasmid** is a small circle of extra DNA. It carries a few genes, and it gets replicated, also. Bacterial cells usually survive without plasmids, although some

Further reading: Student Guide to InfoTrac on web site —

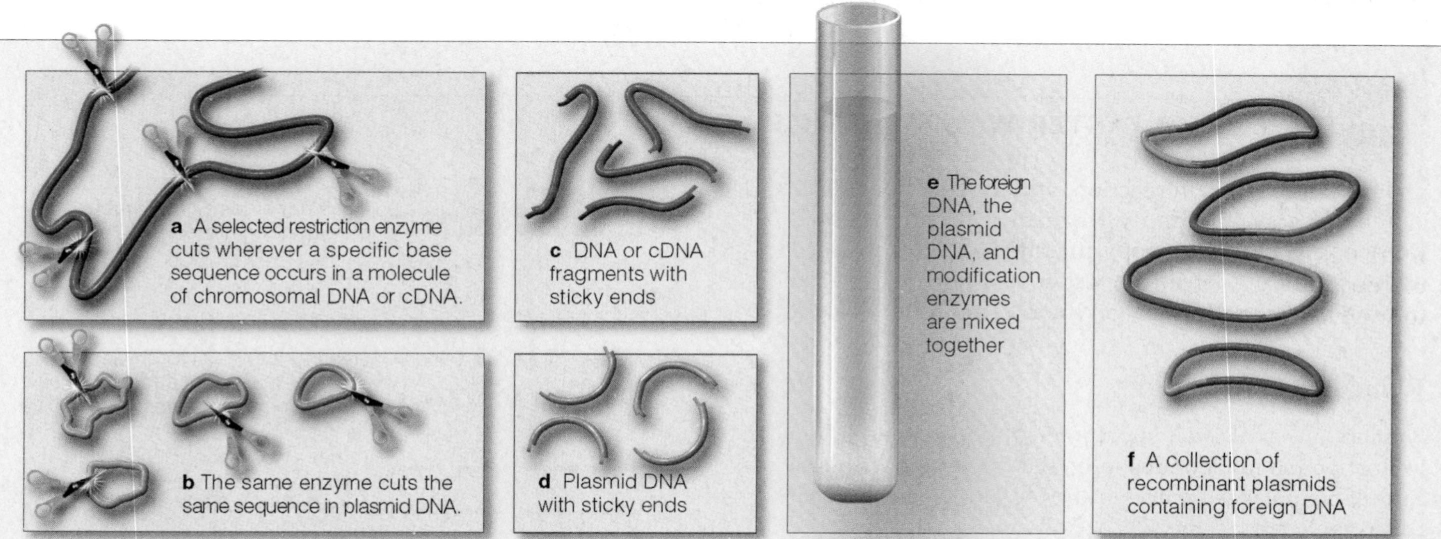
a A selected restriction enzyme cuts wherever a specific base sequence occurs in a molecule of chromosomal DNA or cDNA.
b The same enzyme cuts the same sequence in plasmid DNA.
c DNA or cDNA fragments with sticky ends
d Plasmid DNA with sticky ends
e The foreign DNA, the plasmid DNA, and modification enzymes are mixed together
f A collection of recombinant plasmids containing foreign DNA

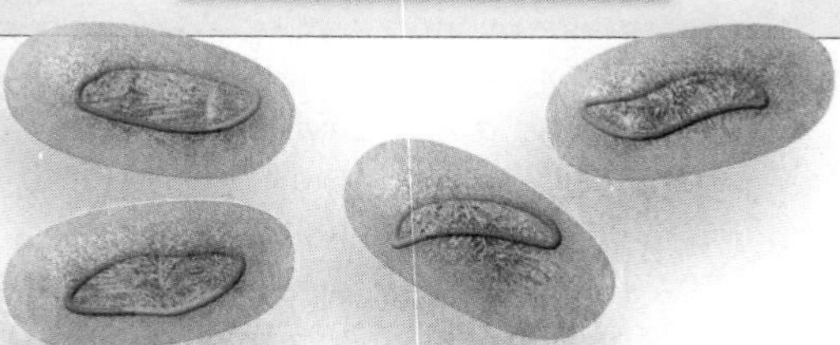
g Host cells able to divide rapidly take up recombinant plasmids

Figure 16.4 (**a–f**) Formation of recombinant DNA—in this case, a collection of either chromosomal DNA fragments or cDNA sealed into bacterial plasmids. (**g**) Recombinant plasmids are inserted into host cells that can rapidly amplify the foreign DNA of interest.

plasmid genes do offer special benefits (as when they confer resistance to antibiotics).

Under favorable conditions, bacteria divide rapidly and often—every thirty minutes, for some species. In short order, huge populations of genetically identical cells form. Before every division, each cell's replication enzymes duplicate the bacterial chromosome. They also replicate the plasmid, sometimes repeatedly, so a cell can end up with many identical copies of foreign DNA.

In research laboratories, such foreign DNA can be inserted into a plasmid for replication. The outcome is called a **DNA clone**, because bacterial cells make many identical, "cloned" copies of it.

A modified plasmid that can accept foreign DNA is called a **cloning vector**. It serves as a taxi for delivering foreign DNA into a bacterium, yeast, or some other cell that starts a "cloning factory"—a population of rapidly dividing descendants, all with identical copies of the foreign DNA (Figure 16.4).

Reverse Transcriptase To Make cDNA

Researchers analyze genes to unlock secrets about gene products and how they are put to use. However, even when a host cell takes up a gene, it might not be able to synthesize the protein. For example, most eukaryotic genes incorporate noncoding sequences (introns). New mRNA transcripts of those genes cannot be translated until introns are snipped out and coding regions (exons) spliced together. Bacterial cells can't remove introns, so they often can't translate human genes into proteins.

Researchers get around this problem with **cDNA**, a DNA strand "copied" from a mature mRNA transcript. **Reverse transcriptase**, an enzyme from RNA viruses, catalyzes transcription in reverse. That is, it assembles a complementary DNA strand on an mRNA transcript (Figure 16.5). Other enzymes remove the RNA from the hybrid mRNA–cDNA molecule, make a complementary DNA strand, and thus make double-stranded cDNA.

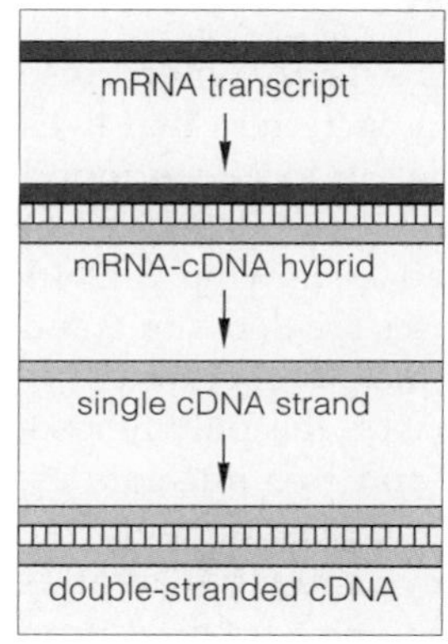
mRNA transcript
mRNA-cDNA hybrid
single cDNA strand
double-stranded cDNA

a A mature mRNA transcript of a gene of interest is used as a template to assemble a single strand of DNA. The enzyme reverse transcriptase catalyzes the assembly. An mRNA–cDNA hybrid molecule is the result.

b By enzyme action, the mRNA is removed and a second strand of DNA is assembled on the DNA strand remaining. The outcome is double-stranded cDNA,"copied" from an mRNA template.

Figure 16.5 Formation of cDNA from an mRNA transcript.

Double-stranded cDNA can be further modified, as by attaching signals for transcription and translation. The modified cDNA can be inserted into a plasmid for amplification. Then the recombinant plasmids can be inserted into bacterial cells, which may use the cDNA instructions for synthesizing a protein of interest.

Restriction enzymes and modification enzymes can cut apart chromosomal DNA or cDNA and splice it into plasmids and other cloning vectors. Then the recombinant plasmids can be inserted into bacteria or other rapidly dividing cells that can make multiple, identical quantities of the foreign DNA.

16.2 PCR—A FASTER WAY TO AMPLIFY DNA

The polymerase chain reaction, widely known as **PCR**, is another way to amplify fragments of chromosomal DNA or cDNA. These copy-making reactions proceed with astounding speed in test tubes, not in bacterial cloning factories. Primers get them going.

What Are Primers?

Primers are synthetic, short nucleotide sequences, ten to thirty or so nucleotides long, that base-pair with any complementary sequences in DNA. The workhorses of DNA replication—the DNA polymerases—chemically recognize primers as START tags. Following a computer program, machines synthesize a primer one step at a time. How do researchers decide on the order of those nucleotides? First, they must identify short nucleotide sequences located just before and just after the DNA region from a cell that interests them. Then they build primers that have the complementary sequences.

What Are the Reaction Steps?

For PCR, the enzyme of choice is a DNA polymerase extracted from *Thermus aquaticus*, a bacterium that lives in superheated water of hot springs. It has been found even in water heaters. The enzyme is not destroyed at the elevated temperatures required to unwind a DNA double helix. Most DNA polymerases are denatured and permanently lose their activity at such temperatures.

Researchers mix together primers, the polymerase, cellular DNA from an organism, and free nucleotides. Next, they expose the mixture to precise temperature cycles. During each temperature cycle, the two strands of all the DNA molecules in the mixture unwind from each other.

Primers become positioned on exposed nucleotides at targeted sites according to base-pairing rules (Figure 16.6). Each round of reactions doubles the number of DNA molecules amplified from the target site. Thus, if there are 10 such molecules in the test tube, there soon will be 20, then 40, 80, 160, 320, 640, 1,280, and so on. Thus a targeted region from a single DNA molecule can be rapidly amplified to billions of molecules.

In short, *PCR amplifies samples that contain even tiny amounts of DNA*. At this writing, it is the amplification procedure of choice in thousands of laboratories all over the world. As you will see in the next section, such samples can be obtained even from a single hair follicle or drop of blood left at the scene of a crime.

PCR is a method of amplifying chromosomal DNA or cDNA inside test tubes. Compared to cloning methods, PCR is far more rapid.

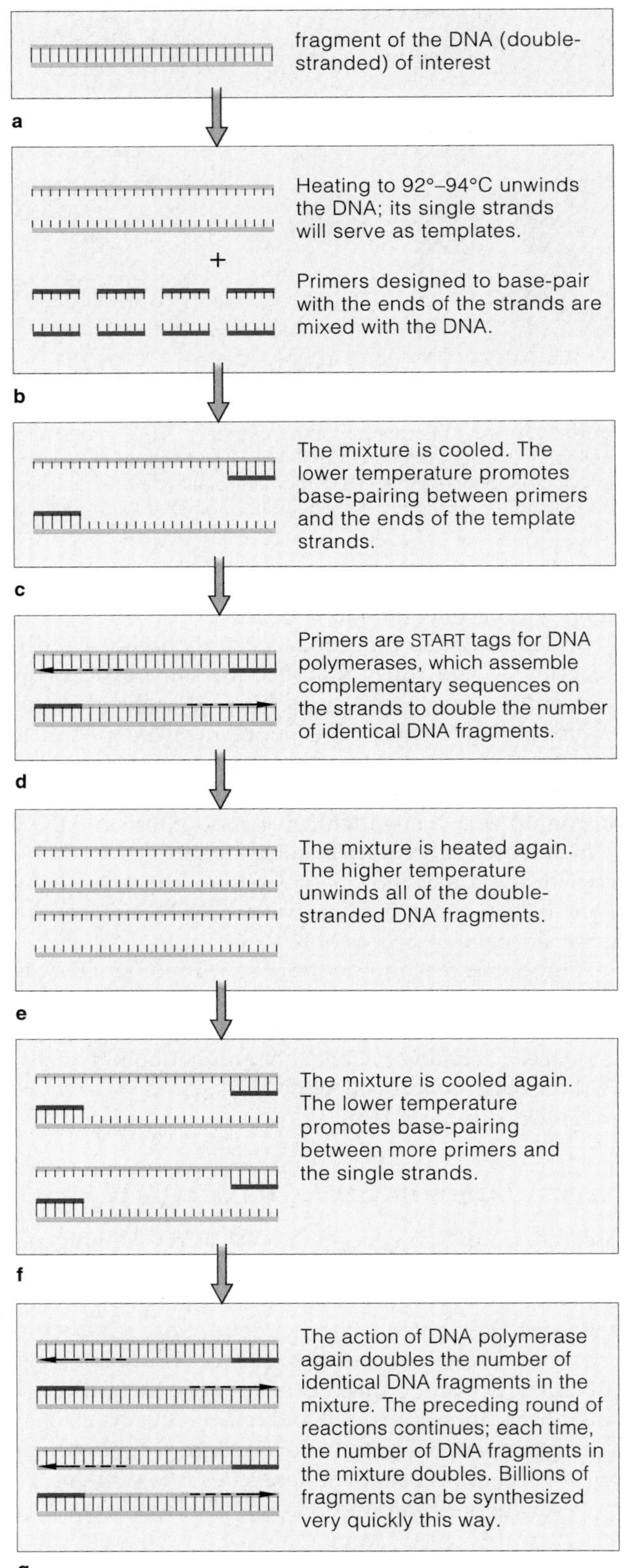

a
fragment of the DNA (double-stranded) of interest
b
Heating to 92°–94°C unwinds the DNA; its single strands will serve as templates.
+
Primers designed to base-pair with the ends of the strands are mixed with the DNA.
c
The mixture is cooled. The lower temperature promotes base-pairing between primers and the ends of the template strands.
d
Primers are START tags for DNA polymerases, which assemble complementary sequences on the strands to double the number of identical DNA fragments.
e
The mixture is heated again. The higher temperature unwinds all of the double-stranded DNA fragments.
f
The mixture is cooled again. The lower temperature promotes base-pairing between more primers and the single strands.
g
The action of DNA polymerase again doubles the number of identical DNA fragments in the mixture. The preceding round of reactions continues; each time, the number of DNA fragments in the mixture doubles. Billions of fragments can be synthesized very quickly this way.

Figure 16.6 The polymerase chain reaction (PCR).

Further reading: Student Guide to InfoTrac on web site

Focus on Science

16.3 DNA FINGERPRINTS

Except for identical twins, no two people have exactly the same sequence of bases in their DNA. By detecting the differences in DNA sequences, scientists can distinguish one person from another. As you know, each human has a unique set of fingerprints, a marker of his or her identity. Like all other sexually reproducing species, each human also has a **DNA fingerprint**, which is a unique array of DNA fragments that were inherited from each parent in a Mendelian pattern. DNA fingerprints are so accurate that even full siblings are readily distinguished from one another.

More than 99 percent of the DNA is exactly the same in all humans. But DNA fingerprinting focuses only on the part that tends to differ from one person to the next. Throughout the human genome are **tandem repeats**—short regions of repeated DNA—that differ substantially among people. For example, the five bases TTTTC are repeated anywhere from four to fifteen times in tandem in different people, and three bases (CGG) are repeated five to fifty times in tandem.

The number of repeats may increase or decrease during DNA replication. Mutation rates in tandem repeats are much higher than the rates at most other sites in DNA. Because such mutations occurred over many generations in many different family lineages, each person is usually heterozygous for a repeat number at any given tandem-repeat locus. By examining many tandem-repeat sites, researchers found out that each person carries a unique combination of repeat numbers.

Researchers detect differences at tandem-repeat sites with **gel electrophoresis**, a laboratory technique that uses an electric field to force molecules through a viscous gel (Section 14.1, Figure 14.4). In this case, it separates DNA fragments according to their length. Size alone dictates how far each fragment moves through the gel, so tandem repeats of different sizes migrate at different rates.

A gel is immersed in a buffered solution, then DNA fragments from individuals are added to the gel. When an electric current is applied to the solution, one end of the gel takes on a negative charge, and the other end a positive charge. DNA molecules carry a negative charge (because of the negatively charged phosphate groups), so they migrate through the gel toward the positively charged pole. They do so at different rates, and so they separate into bands according to length. The smaller the fragment, the farther it will migrate. After a set time, researchers can identify fragments of different lengths by staining the gel or by specifically highlighting fragments that contain tandem repeats.

Figure 16.7 shows a series of tandem-repeat DNA fragments that were separated by gel electrophoresis. They are DNA fingerprints from seven individuals and from blood collected at a crime scene. Notice how much their DNA fingerprints differ. Can you identify which one of the patterns exactly matches the pattern from the crime scene? DNA fingerprints help forensic scientists identify criminals and victims, and exclude innocent suspects. A few drops of blood, semen, or cells from a hair follicle at a crime scene or on a suspect's clothing often yield enough to do the trick. Such an analysis even confirmed that human bones exhumed from a shallow pit in Siberia belonged to five members of the Russian imperial family, all shot to death in secrecy in 1918. In addition, because DNA fragments that make up a DNA fingerprint are inherited, scientists use them to resolve questions of paternity.

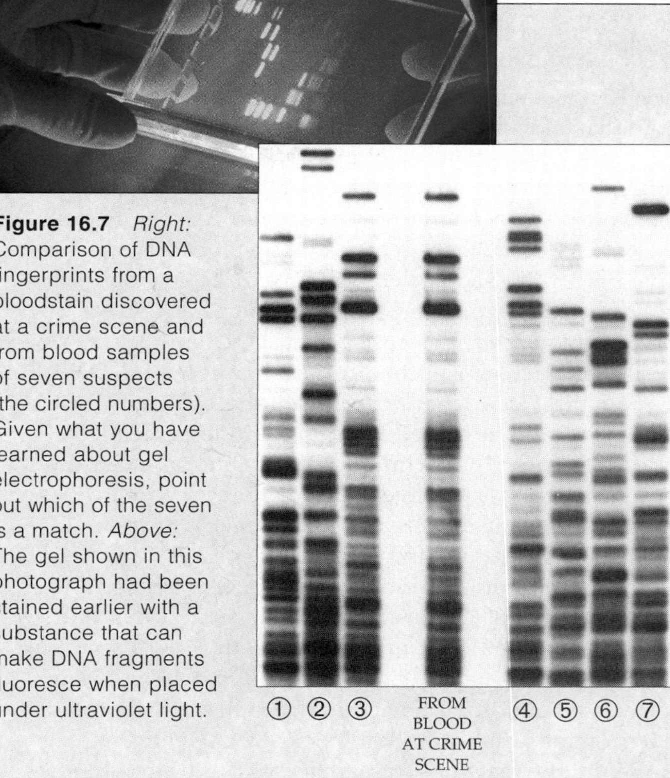

① ② ③
FROM BLOOD AT CRIME SCENE
④ ⑤ ⑥ ⑦

Figure 16.7 *Right:* Comparison of DNA fingerprints from a bloodstain discovered at a crime scene and from blood samples of seven suspects (the circled numbers). Given what you have learned about gel electrophoresis, point out which of the seven is a match. *Above:* The gel shown in this photograph had been stained earlier with a substance that can make DNA fragments fluoresce when placed under ultraviolet light.

When DNA fingerprinting was first used as evidence in court, attorneys challenged conclusions based on it. But DNA fingerprinting is now firmly established as an accurate and unambiguous way to identify individuals and paternity. It is widely used to convict the guilty and exonerate innocent suspects.

For example, in 1998, an eleven-year-old girl was raped, stabbed, and strangled to death in Elisabethfehn, Germany. In the largest mass screening to date, police requested blood samples from 16,400 males in and around the vicinity. A single DNA fingerprint from a thirty-year-old mechanic matched the evidence, and he confessed to the crime.

The variation in tandem repeats also can be detected as restriction fragment length polymorphisms, or RFLPs (DNA fragments of different sizes that had been cleaved by restriction enzymes). In the case of tandem repeats, a restriction enzyme cleaves the DNA flanking the repeat. Alternatively, researchers might use PCR to amplify the tandem repeat region. Either way, differences in the size of the fragments, which reveal genetic differences, can be detected with electrophoresis.

16.4 HOW IS DNA SEQUENCED?

In 1995, researchers accomplished a feat that was little more than a dream a few decades ago. They determined the entire DNA sequence of an organism—*Haemophilus influenzae*, a bacterium that causes respiratory infections in humans. The genome of this species has 1.8 million nucleotides. Since then, genomes of several other species have been fully sequenced. By current projections, we will know the sequence of all 3.2 billion nucleotides of the human genome before the year 2002.

The molecular sleuths are using **automated DNA sequencing**. This laboratory technique can reveal the sequence of either cloned DNA or PCR-amplified DNA in a few hours. The technique has all but replaced the more laborious, expensive methods that preceded it.

Researchers use the four standard nucleotides (T, C, A, and G) for automated DNA sequencing. They also use four modified versions, which we can represent as T*, C*, A*, and G*. Each modified version has been labeled. Attached to it is a molecule that fluoresces a particular color when it passes through a laser beam. Each time a modified nucleotide becomes incorporated in a growing DNA strand, it blocks DNA synthesis.

Before the reactions, researchers mix the eight kinds of nucleotides together. To that mixture they add the millions of copies of the DNA to be sequenced, a type of primer, and DNA polymerase. Then they separate the DNA into single strands, and the reactions begin.

The primer binds with its complementary sequence on one of the strands. DNA polymerase synthesizes a new DNA strand right behind the primer. One by one, it adds nucleotides in the order dictated by the exposed sequence. Each time, one of the standard nucleotides *or* one of the modified versions may be attached.

Suppose that DNA polymerase encounters a T in a template strand. It will catalyze the base-pairing of either A or A* to it. If A is added to the new strand, replication will continue. If A* is added, replication will stop; the modified nucleotide will *block* addition of more nucleotides to that DNA strand. The same thing happens at each nucleotide in a template strand. If a standard nucleotide is added, replication will continue; if a modified version is added, replication stops.

Remember, the starting mixture contained millions of identical copies of the DNA sequence. Because either a standard or modified nucleotide could be added at each exposed base, the new strands ended at different locations in the sequence. The mixture now contains millions of copies of tagged fragments having different lengths. These can be separated by length into *sets* of fragments. And each set corresponds to just one of the nucleotides in the entire base sequence.

The automated DNA sequencer is a machine that separates the sets of fragments by gel electrophoresis. The set having the shortest fragments migrates fastest

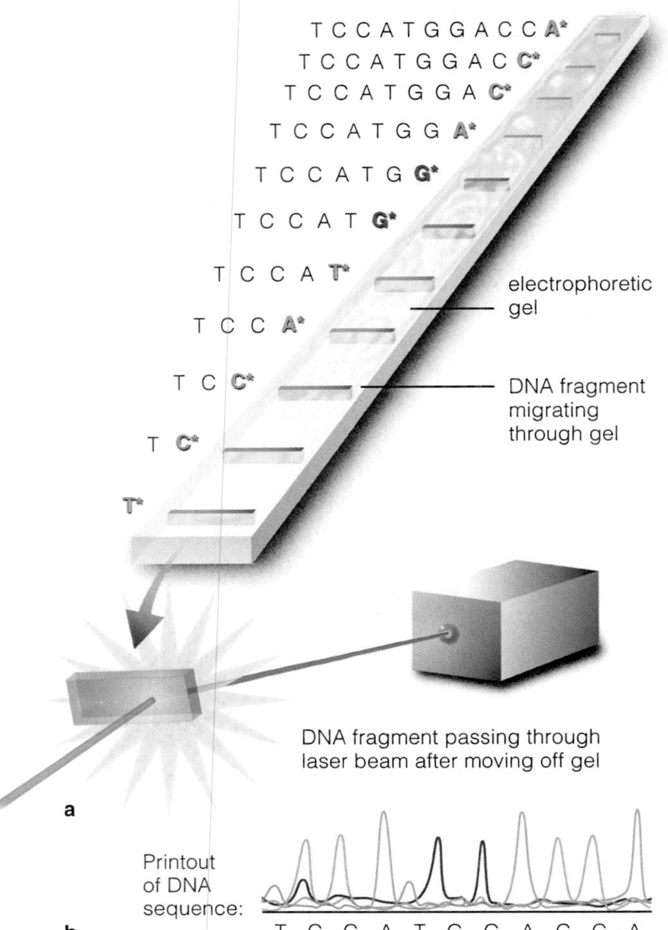

T C C A T G G A C C A*
T C C A T G G A C C*
T C C A T G G A C*
T C C A T G G A*
T C C A T G G*
T C C A T G*
T C C A T*
T C C A*
T C C*
T C*
T*
electrophoretic gel
DNA fragment migrating through gel
DNA fragment passing through laser beam after moving off gel
a
Printout of DNA sequence:
T C C A T G G A C C A
b

Figure 16.8 Automated DNA sequencing. (**a**) DNA fragments from an organism's genome become labeled at their end with a modified nucleotide that fluoresces a certain color. (**b**) Printout of the DNA sequence used in this example. Each peak indicates absorbance by a particular labeled nucleotide.

through the gel and peels away from it first. The last set to peel away has the longest fragments. Because its fragments have a modified nucleotide at the 3′ end, the set fluoresces a particular color as it passes through a laser beam (Figure 16.8*a*). The automated sequencer detects the color and thus determines which nucleotide is on the end of the fragments in each set. It assembles the information from all the nucleotides in the sample and reveals the entire DNA sequence.

Figure 16.8*b* shows the printout from an automated DNA sequencer. Each peak in the tracing represents the detection of a particular color as the sets of fragments reached the end of the gel.

With automated DNA sequencing, the order of nucleotides in a cloned or amplified DNA fragment can be determined.

Further reading: Student Guide to InfoTrac on web site

16.5 FROM HAYSTACKS TO NEEDLES—ISOLATING GENES OF INTEREST

Any genome consists of thousands of genes. There are about 4,000 in the *E. coli* genome, for example, and over 100,000 in the human genome. What if you wanted to learn about or modify the structure of any one of those genes? First you would have to isolate that gene from all others in the genome.

There are several ways to do this. If part of the gene sequence is already known, you can design primers to amplify the whole gene or part of it with PCR. Often, though, researchers must isolate and clone a gene. First they make a **gene library**. This is a mixed collection of bacteria that contain different cloned DNA fragments, one of which is the gene of interest. A *genomic* library contains cloned DNA fragments from an entire genome. A *cDNA* library contains DNA derived from mRNA. A cDNA library is usually the most useful, for it is free of introns. Even then, the gene of interest is like a needle in a haystack. How can you isolate it from all the others in the library? One way is to use a nucleic acid probe.

What Are Probes?

A **probe** is a very short stretch of DNA labeled with a radioisotope so that it can be distinguished from other DNA molecules in a given sample. Part of the probe must be able to base-pair with some portion of the gene of interest. Any base-pairing that takes place between sequences of DNA (or RNA) from different sources is called **nucleic acid hybridization**.

How do you acquire a suitable probe? Sometimes part of the gene or a closely related gene has already been cloned, in which case you can use it as a probe. If the gene's structure is a mystery, you still may be able to work backward from the amino acid sequence of its protein product, assuming the protein is available. By using the genetic code as a guide (Section 14.3), maybe you can build a DNA probe that is more or less similar to the gene of interest.

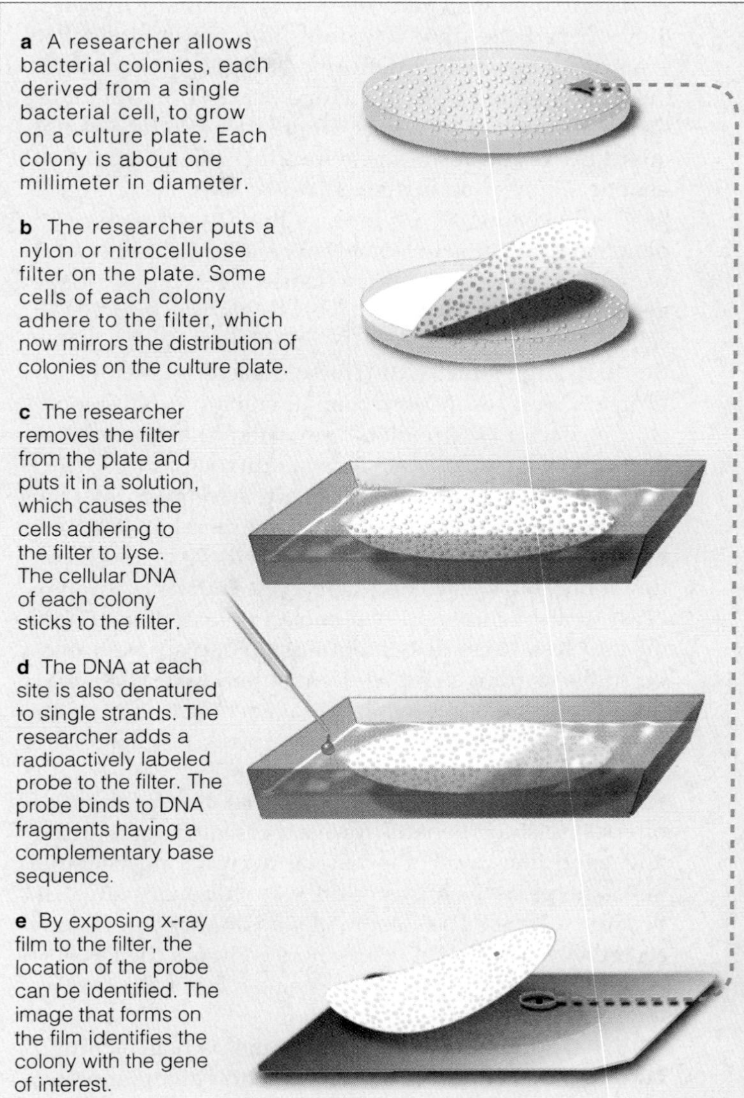
a A researcher allows bacterial colonies, each derived from a single bacterial cell, to grow on a culture plate. Each colony is about one millimeter in diameter.
b The researcher puts a nylon or nitrocellulose filter on the plate. Some cells of each colony adhere to the filter, which now mirrors the distribution of colonies on the culture plate.
c The researcher removes the filter from the plate and puts it in a solution, which causes the cells adhering to the filter to lyse. The cellular DNA of each colony sticks to the filter.
d The DNA at each site is also denatured to single strands. The researcher adds a radioactively labeled probe to the filter. The probe binds to DNA fragments having a complementary base sequence.
e By exposing x-ray film to the filter, the location of the probe can be identified. The image that forms on the film identifies the colony with the gene of interest.

Figure 16.9 Use of a probe to identify bacterial colonies that have taken up a gene library.

Screening for Genes

Once you have a gene library and a suitable probe, you are ready to hunt down the gene. Figure 16.9 shows the steps of one isolation method. The first step is to take bacterial cells of the library and spread them apart on the surface of a gelled growth medium in a petri plate. When spread out sufficiently, individual cells undergo division. Each cell starts a colony of genetically identical cells. The bacterial colonies appear as hundreds of tiny white spots on the surface of a culture medium.

After colonies appear, you lay a nylon filter on top of the colonies. Some cells stick to the filter at locations that mirror the locations of the original colonies. You use solutions to rupture the cells, and the DNA so released sticks to the filter. You make the DNA unwind to single strands and then add the probes. The probes hybridize only with DNA from the colony that took up the gene of interest. If you expose the probe-hybridized DNA to x-ray film, the pattern formed by the radioactivity will identify that colony. With this information you can now culture cells from that one colony, which will replicate only the cloned gene.

Probes may be used to identify one particular gene among many in gene libraries. Bacterial colonies that have taken up the library can be cultured to isolate the gene.

16.6 USING THE GENETIC SCRIPTS

As researchers decoded the genetic scripts of species, they opened the door to astonishing possibilities. For example, genetically engineered bacteria now produce medically valued proteins. Huge bacterial populations thrive in stainless steel vats, where they produce useful quantities of the desired gene products. Beneficiaries of genetic engineering include *diabetics*, who must receive insulin injections for as long as they live. Insulin is a pancreatic hormone. At one time, medical supplies of it were extracted from pigs and cattle. Later on, synthetic genes for human insulin were transferred into *E. coli* cells. Those cells gave rise to populations that became the first large-scale, cost-effective bacterial factory for proteins. Besides insulin, human somatotropin, blood-clotting factors, hemoglobin, interferon, and a variety of drugs and vaccines are also manufactured this way.

Genetically engineered bacteria also hold potential for industry and for cleaning up environmental messes. For instance, as you know, many microorganisms break down organic wastes and help cycle nutrients through ecosystems. Some modified bacteria break down crude oil into less harmful compounds. When sprayed onto oil spills, as from a shipwrecked supertanker, they may help avert an environmental disaster. Other bacteria are genetically engineered to sponge up excess phosphates or heavy metals from the environment.

In addition, bacterial species that house plasmids offer astonishing benefits for basic research, agriculture, and gene therapy. For example, deciphering bacterial genes helps us reconstruct life's evolutionary story. In Sections 20.4 and 21.6, you will read about evolutionary secrets revealed when researchers compare the DNA or RNA of different organisms. The next two sections give a few more examples.

What about the "bad" bunch—the pathogenic fungi, bacteria, and viruses? Natural selection favors mutated genes that improve a pathogen's chances of evading a host organism's natural defenses. Mutation is frequent among pathogens that reproduce rapidly, so designing new antibiotics and other defenses against new gene products is a constant challenge. But if we learn about the genes, we may have advance warning of the "plan of attack." The story of HIV, a rapidly mutating virus that causes AIDS, gives insight into the magnitude of the problem (Section 40.11).

Knowing about genes allows us to genetically engineer beneficial microorganisms for uses in medicine, industry, agriculture, and environmental remediation.

Knowing about genes can help us reconstruct evolutionary histories of species.

Knowing about genes can help us devise counterattacks against rapidly mutating pathogens.

16.7 DESIGNER PLANTS

Regenerating Plants From Cultured Cells

Many years ago, Frederick Steward and his coworkers cultured cells from carrot plants and induced some of them to develop into small embryos. As you will read in Section 31.7, some embryos grew into whole plants. Today, researchers routinely regenerate crop plants and many other plant species from cultured cells. They use various methods to pinpoint a gene in a culture that contains, say, millions of cells. Suppose the researcher includes a toxic product of a pathogen in the culture medium. If only a few cells carry a gene that confers resistance to the toxin, they will be the only ones left.

Once whole plants are regenerated from preselected cultured cells, researchers can hybridize them with other varieties. The idea is to transfer desired genes to a plant lineage to improve herbicide resistance, pest resistance, and other traits that can benefit crops and gardens.

To protect the food supply for most of the human population, botanists are combing the world for seeds of the wild ancestors of potatoes, corn, and other crop plants. They send their prizes—seeds that carry genes of a plant's lineage—to seed banks. These safe storage facilities preserve the genetic diversity. The supply is vulnerable; farmers focus mainly on genetically similar varieties of high-yield crop plants. Genetic uniformity makes our food crops dangerously vulnerable to many kinds of pathogenic fungi, viruses, and bacteria.

To give an example, a new fungal strain of *Southern corn leaf blight* destroyed much of the 1970 corn crop in the United States. This widespread epidemic happened because all of the plants were genetically similar; they all carried the same gene that conferred susceptibility to the disease. Since then, seed companies have become more attentive to offering corn with genetic diversity. To introduce genetic diversity into existing varieties, plant breeders must tap into the seeds in seed storage.

The search for beneficial genes is a race against time. Farmers throughout the world want high yields, which many of the genetically uniform varieties provide. They discard older, more diverse varieties. By doing so, they also discard the genetic diversity that is so essential for the future of our food supply.

Figure 16.10 Crown gall tumor on a woody plant, an abnormal tissue growth triggered by a gene carried by the Ti plasmid.

Further reading: Student Guide to InfoTrac on web site

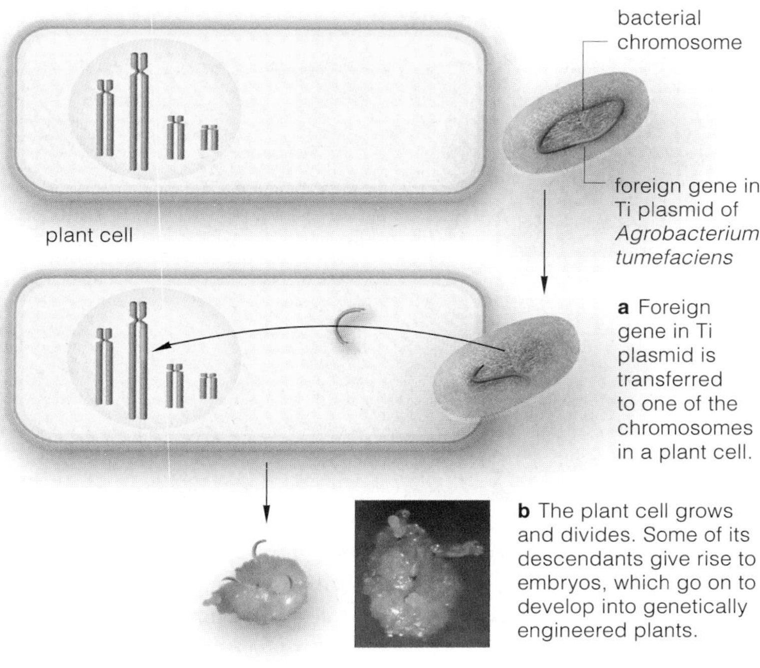
bacterial chromosome
foreign gene in Ti plasmid of *Agrobacterium tumefaciens*
plant cell
a Foreign gene in Ti plasmid is transferred to one of the chromosomes in a plant cell.
b The plant cell grows and divides. Some of its descendants give rise to embryos, which go on to develop into genetically engineered plants.

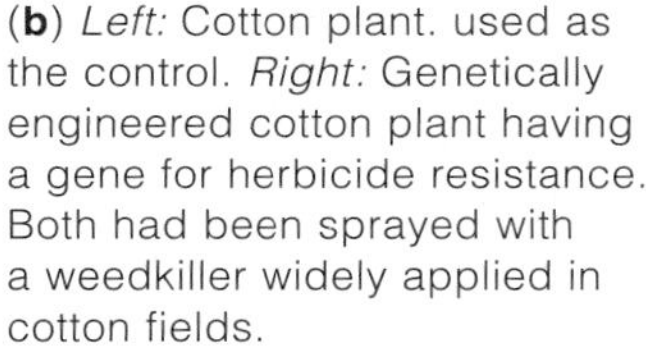

Figure 16.11 Gene transfer from the bacterium *Agrobacterium tumefasciens* to a plant cell, by way of the Ti plasmid gene.

a

b

Figure 16.13 (**a**) Control plant (*left*) and three genetically engineered aspen seedlings. Vincent Chiang and coworkers suppressed a regulatory gene of a lignin biosynthetic pathway. The modified plants synthesized normal lignin, but not as much. Lignin production decreased by as much as 45 percent—yet cellulose production increased 15 percent. Root, stem, and leaf growth were greatly enhanced. Plant structure did not suffer. Wood harvested from such trees might make it easier to manufacture paper and some clean-burning fuels, such as ethanol. (Lignin, a tough polymer, reinforces secondary cell walls of plants. It must be chemically extracted from the wood used to make paper.)

(**b**) *Left:* Cotton plant. used as the control. *Right:* Genetically engineered cotton plant having a gene for herbicide resistance. Both had been sprayed with a weedkiller widely applied in cotton fields.

How Are Genes Transferred Into Plants?

A simple example gives insight into how genes can be transferred into a plant. Genetic engineers often insert new or modified genes into the Ti (*T*umor-*I*nducing) plasmid from *Agrobacterium tumefaciens*. This bacterium infects many species of flowering plants. Some plasmid genes invade a plant's DNA, then induce the formation of abnormal tissue masses known as crown gall tumors (Figure 16.10). Before introducing the Ti plasmid into plant cells, researchers first remove the tumor-inducing genes and insert a desired gene into it. They place plant cells in a culture of the modified bacteria. When some cells take up the gene, whole plants may be regenerated from the cellular descendants, as in Figure 16.11. The expression of foreign genes in plants can sometimes be quite dramatic, as shown in Figure 16.12, at left.

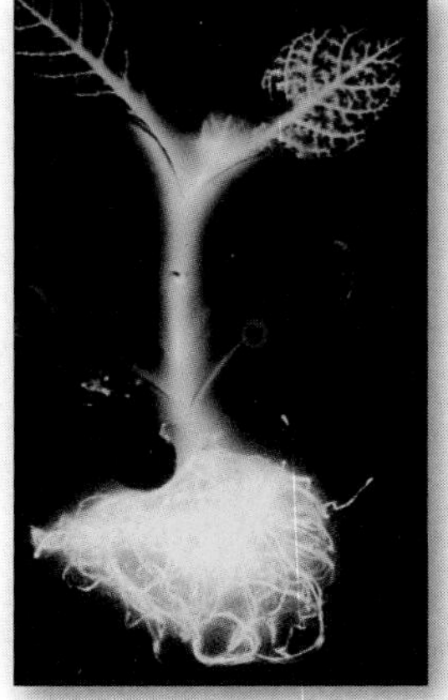

Figure 16.12 A modified plant that glows in the dark owing to a gene transfer. A firefly gene was inserted into the plant's DNA and is being expressed. As described in the Chapter 6 introduction, the gene's product is luciferase, an enzyme required for bioluminescence.

In nature, *A. tumefaciens* infects the plants called dicots, which include beans, peas, potatoes and other essential crops. Geneticists are working to modify the bacterium so it also can infect monocots that are vital food crops, including wheat, corn, and rice. In some cases, they use electric shocks or chemicals to deliver modified genes into plant cells. As another delivery trick, they blast microscopic particles coated with DNA into the cells.

Despite many obstacles, improved varieties of crop plants have been engineered or are in the works. For example, genetically engineered cotton plants display resistance to an herbicide (Figure 16.13). Farmers spray the herbicide in a field of the modified cotton plants and kill the weeds, and the plants remain unaffected.

Also on the horizon are engineered plants that can serve as factories for pharmaceuticals. A few years ago, genetically engineered tobacco plants that make human hemoglobin and other proteins were planted in a field in North Carolina. Ecologists later found no trace of foreign genes or proteins in the soil or in other plants or animals in the vicinity. As a final example, mustard plant cells grown in a Stanford University laboratory have synthesized biodegradable plastic beads that may be used for manufacturing plastics.

Genetically diverse plant species are essential to protecting our vulnerable food supply. Genetic engineers can design plants with new beneficial traits.

16.8 GENE TRANSFERS IN ANIMALS

Supermice and Biotech Barnyards

The first mammals enlisted for experiments in genetic engineering were laboratory mice. Consider an example of this work. R. Hammer, R. Palmiter, and R. Brinster corrected a hormone deficiency that leads to dwarfism in mice. Insufficient levels of somatotropin (or growth hormone) cause the abnormality. The researchers used a microneedle to inject the gene for rat somatotropin into fertilized mouse eggs, which they implanted in an adult female. The gene was successfully integrated into mouse DNA, and the eggs developed into mice. Young mice in which that foreign gene was expressed were 1–1/2 times larger than the dwarf littermates. In other experiments, researchers transferred the gene for human somatotropin into a mouse embryo. There, it became integrated into mouse DNA. The genetically modified embryo developed into a "supermouse" (Figure 16.14).

Today, human genes are being inserted into mouse embryos as part of ongoing research into the molecular basis of genetic disorders, such as Alzheimer's disease. In addition to microneedles, microscopic laser beams are used to penetrate the plasma membrane of cultured cells. The new methods have varying degrees of success.

Animals of "biotech barnyards" are competing with bacterial factories as genetically engineered sources of proteins. For example, goats produce CFTR protein (for treating cystic fibrosis) and TPA (which may diminish the severity of heart attacks). Cattle may soon produce human collagen for repairing cartilage, bone, and skin.

In 1997 researchers reported the first case of cloning a mammal from an adult cell. They extracted a nucleus from a ewe's mammary gland cell and inserted it into an egg (from another ewe) from which the nucleus had been removed. Signals from the egg cytoplasm sparked the development of an embryo, which was implanted into a surrogate mother. As described in Section 13.5, the result was the clone *Dolly*.

Although animals have been routinely cloned from embryonic tissue for years, cloning from adult animal cells has advantages. An adult animal already displays a version of some trait that may be sought after, and cloning might maintain its genotype indefinitely.

The experiment that produced Dolly opened up the possibility of making genetically engineered clones of sheep, cattle, and other farm animals from adults with proven genetic potential. (It also raised the jarring idea that human clones might be just around the corner. The United States promptly imposed a ban on the federal funding of research into human cloning.)

One year later, Steve Stice and his colleagues used similar cloning procedures and went further down the road to designer cattle. Starting with a cell culture from cattle, they induced the growth and development of six genetically identical calves. More importantly, they also engineered specific, targeted alterations in the cloning cell lineage. Their goal is to genetically modify cattle for resistance to bovine spongiform encephalopathy. (You may know this condition as mad cow disease. Humans who eat infected meat are vulnerable to it; Section 22.9.) Their technique also is being used to insert the gene for human serum albumin into the chromosomes of dairy cows. This protein is medically useful because it helps control blood pressure. At present, it must be separated from large quantities of donated human blood. Imagine how much easier it would be to obtain large quantities of the protein from bountiful supplies of milk.

Figure 16.14 Evidence of a successful gene transfer. Two ten-week-old mouse littermates. *Left:* This one weighs 29 grams. *Right:* This one weighs 44 grams. It grew from a fertilized egg into which a gene for human somatotropin had been inserted.

Mapping and Using the Human Genome

In the early 1980s, biologists were in an uproar over a costly proposal before the National Institutes of Health (NIH) to map the entire human genome. Many argued that benefits would be incalculable—not only for pure research, but also for efforts to cure genetic disorders and combat diseases. To others, it was a boondoggle; such mapping could not be done, and it would divert NIH funding from worthy endeavors. But then Leroy Hood, a molecular biologist, invented automated DNA sequencing (Section 16.4), and the race was on.

By 1990, the Human Genome Initiative was under way. The international effort came with a projected price tag of 3 billion dollars—about 92 cents for each base pair in the heritable script for human life. For the first phase, biologists were to pinpoint the location of the stretches of nucleotides that were already sequenced. Once those "reference markers" were assigned to the chromosomes, biologists could fill in the long spans between them.

By 1991, not even 2,000 genes had been sequenced. But the pace quickened after Craig Ventner realized that even a tiny bit of cDNA could be used as a molecular hook to drag the whole sequence of its parent gene out

Further reading: Student Guide to InfoTrac on web site

of a cDNA library. He named the hooks ESTs (*E*xpressed *S*equence *T*ags). By tapping commercial sources, Ventner and his colleague, Mark Adams, put together a cDNA library that had been produced from various neurons, the communication cells of nervous systems. Within a few months, they had sequenced 2,000 more genes. By 1994, about 35,000 genes had been decoded.

Later, Ventner and Hamilton Smith used a software program, the TIGR Assembler, to work out the first full genome of an organism—the bacterium *H. influenzae* you read about in Section 16.1. Through application of advanced technologies, the Human Genome Initiative will be completed ahead of schedule.

Even when the whole human genome is sequenced, it will not be easy to manipulate it to advantage. Put yourself in the position of some genetic researcher. You have modified a human gene and you have to figure out how to get it into a host cell of a particular tissue. That's not all. You have to be sure the modified gene gets inserted at a suitable site—say, at or very near the locus for its abnormal counterpart in a chromosome. You must also make sure the cell will synthesize the specified protein at suitable times, in suitable amounts.

Today, most experimenters employ stripped-down viruses as vectors to put genes into cultured human cells, which may incorporate the foreign genes into their own DNA. However, the viral genetic material can undergo rearrangements, deletions, and other alterations, which can shut down or disrupt gene expression. Synthetic, streamlined versions of human chromosomes might be developed that will be recognized by a cell's machinery for DNA replication and protein synthesis.

In some cases of gene therapy, simply inserting a number of genetically modified cells into a tissue does the individual good, even if the cells make as little as 10 to 15 percent of the required protein. The introduction to this chapter is a case in point. So far, however, no one can predict where new genes will end up. The danger is that a gene insertion may disrupt the function of some other gene, including the genes controlling cell growth and cell division. One-for-one gene swaps by way of a process called homologous recombination are possible. But Oliver Smithies, one of the best at this, currently puts genes right where they are supposed to go, once every 100,000 or so tries. The body may be somewhat forgiving of a small load of cells bearing near-misses. But mistakes in germ cells—which give rise to all cells of a new individual—probably would have unexpected and disastrous effects on the body's development.

Although the technical details are still being worked out, modified genes are being transferred into cells of humans and other mammals during experimental and clinical trials.

Focus on Bioethics

16.9 WHO GETS ENHANCED?

This chapter opened with a historic, inspiring case, the first proof that gene therapy may help save human lives. It closes with questions that invite you to consider some social and ethical issues surrounding the application of recombinant DNA technology to the human genome.

To most of us, human gene therapy to correct genetic abnormalities seems like a socially acceptable goal. Let's take this idea one step further. Is it also socially desirable or acceptable to change certain genes of a normal human individual (or sperm or egg) to alter or enhance traits?

The idea of selecting desirable human traits is called *eugenic engineering*. Yet who decides which forms of a trait are most "desirable"? For example, would it be okay to engineer taller or blue-eyed or fair-skinned boys and girls? Would it be okay to engineer "superhumans" with amazing strength or intelligence? Are there some people narcissistic enough to commission a clone of themselves, one with a few genetically engineered "enhancements"?

Recently, more than 40 percent of Americans surveyed say gene therapy would be okay to make their offspring smarter or better looking. A poll of British parents found 18 percent willing to use genetic enhancement to prevent children from being aggressive and 10 percent to keep them from growing up to be homosexual.

There are those who say the DNA of any organism must never be altered. Put aside the fact that nature itself alters DNA much of the time and has done so for nearly all of life's history. The concern is that *we* do not have the wisdom to bring about beneficial changes without causing great harm to ourselves or to the environment. To be sure, when it comes to altering human genes, one is reminded of our very human tendency to leap before we look. And yet, when it comes to restricting genetic modifications of any sort, one also is reminded of another old saying: "If God had wanted us to fly, he would have given us wings." Something about the human experience did give us a capacity to imagine wings of our own making—and that capacity carried us to the frontiers of space.

Where are we going from here? To gain perspective on the question, spend some time reading the history of the human species. It is a history of survival in the face of challenges, threats, bumblings, and sometimes disasters on a grand scale. It is also a story of our connectedness with the environment and with one another.

The basic questions confronting you are these: Should we be more cautious, believing the risk takers may go too far? And what do we as a species stand to lose if risks are not taken? At this writing, a young man with a genetic disorder died after gene therapy that used an adenovirus as a vector. Yet two *SCID-X1* infants with a mutant gene that disabled their immune system may be cured. A viral vector put copies of the nonmutated gene into stem cells taken from their bone marrow. Months after the modified cells were reinfused into their bone marrow, both "bubble boys" left their isolation tents. Their immune system is working normally; they are healthy and living at home.

16.10 SAFETY ISSUES

Many years have passed since foreign DNA was first transferred into a plasmid. That gene transfer ignited a debate that will continue well into the next century. The issue is this: *Do potential benefits of gene modifications and gene transfers outweigh potential dangers?*

Genetically engineered bacteria are "designed" so they cannot survive except in the laboratory. As added precautions, "fail-safe" genes are built into the foreign DNA in case they escape. These genes are silent unless the captives are exposed to environmental conditions—whereupon the genes get activated, with lethal results for their owner. Say the package includes a *hok* gene next to a promoter of the lactose operon (Section 15.2). Sugars are plentiful in the environment. If they were to activate the *hok* gene, the protein product of that gene will destroy membrane function and the wayward cell.

What about a worst-case scenario? Remember how retroviruses are now being used to insert genes into cultured cells? If they escape from laboratory isolation and infect organisms, what might be the consequences?

And what about genetically engineered plants and animals released into the environment? For example, Steven Lindlow thought about how frost destroys many crops. Knowing that a surface protein of a bacterium promotes formation of ice crystals, he excised the "ice-forming" gene from bacterial cells. As he hypothesized, spraying "ice-minus bacteria" on strawberry plants in an isolated field prior to a frost would help plants resist freezing. He actually had deleted a *harmful* gene from a species, yet a bitter legal battle ensued. The courts ruled in his favor. His coworkers sprayed a strawberry patch; nothing bad happened.

Then there was the potato plant designed to kill the insects that attack it. It also was too toxic for people to eat. Or think of how crop plants compete poorly with weeds for nutrients. Many have been designed to resist weedkillers so that farmers can spray for weeds and not worry about killing their crops. Herbicides do have toxic effects on more than their targets (Section 3.6). If crop plants offer herbicide resistance, will farmers be less or more apt to spread them about?

And what if engineered plants or animals transfer modified genes to organisms in the wild? Think of how the advantage in acquiring resources would tilt toward vigorous weeds blessed with herbicide-resistant genes. Such possibilities are why standards for rigorous and extended safety tests must be in place *before* modified organisms enter the environment.

For more on safety issues, turn to *Critical Thinking* question 8 at the chapter's end.

Rigorous safety tests are carried out before genetically modified organisms are released into the environment.

16.11 SUMMARY

1. Uncountable numbers of gene mutations and other forms of genetic "experiments" have been proceeding in nature for at least 3 billion years.

2. Through artificial selection practices, humans have manipulated the genetic character of many species for many thousands of years. Currently, recombinant DNA technology has enormously expanded our capacity to genetically modify organisms.

3. A genome is all the DNA in the haploid chromosome number for a species. By recombinant DNA technology, a genome can be cut into fragments, then the fragments can be amplified (copied over and over again) to make useful quantities that permit analysis of the nucleotide sequence of the genome or a specific portion of it.

4. Researchers work with chromosomal DNA or with cDNA (a DNA strand transcribed by the enzyme reverse transcriptase from a mature mRNA transcript). The first choice is better for questions about DNA regions that control gene expression or that contain introns. cDNA is a better choice for questions about the amino acid sequence of a protein. Either way, researchers use restriction enzymes, which cut the DNA into fragments.

5. The use of plasmids or other cloning vectors is one way to amplify DNA fragments. Many bacteria contain plasmids, which are small circles of DNA with a few genes in addition to those of the bacterial chromosome.

a. Certain restriction enzymes make staggered cuts that leave the fragments with single-stranded tails. Such tails base-pair with complementary tails of any other DNA cut by the same enzyme, such as plasmid DNA.

b. DNA ligase, a modification enzyme, seals base-pairing sites between plasmid DNA and foreign DNA. A plasmid modified to accept foreign DNA is a cloning vector; it can deliver foreign DNA into a bacterium or some other cell that can start a population of rapidly dividing descendant cells. All of the descendants have identical copies of the foreign DNA. Collectively, all of the identical copies are a DNA clone.

6. Currently, the polymerase chain reaction (PCR) is the fastest way to amplify fragments of chromosomal DNA or cDNA. Bacterial factories are not needed; the reactions occur in test tubes. The reactions use a supply of nucleotide building blocks and primers: synthetic, short nucleotide sequences that will base-pair with any complementary DNA sequence and that enzymes (DNA polymerases) recognize to start replication. Each round of replication doubles the number of fragments.

7. For sexually reproducing species, no two individuals have exactly the same DNA base sequence (except for identical twins). When restriction enzymes are used to cut an individual's DNA, the result is a unique array of restriction fragments called a DNA fingerprint.

Further reading: Student Guide to InfoTrac on web site —

a. The DNA in all humans is more than 99 percent identical except at tandem repeats (short stretches of repeated base sequences, such as TTTC). The number and combination of tandem repeats is unique in each individual and can be detected by gel electrophoresis.

b. DNA fingerprinting has uses in forensic science and in resolving paternity suits.

8. Automated DNA sequencing can rapidly reveal the base sequence of cloned DNA or PCR-amplified DNA fragments. The method labels each fragment with one of four modifed nucleotides, then separates them according to length by gel electrophoresis. Each label fluoresces a certain color under a laser beam. A machine reads each as it peels off the gel and assembles the whole sequence.

9. A gene library is a mixed collection of bacterial cells that took up different cloned DNA or cDNA fragments. A gene may be isolated from the library with use of a probe, a very short stretch of radioactively labeled DNA that is known or suspected to be similar to or identical with part of that gene and can base-pair with it. A base pairing between nucleotide sequences from different sources is called nucleic acid hybridization.

10. In genetic engineering, specific genes are modified and inserted into the same organism or a different one. In gene therapy, copies of normal or modified genes are inserted into individuals in order to correct a genetic defect or boost resistance to disease.

11. Generally, recombinant DNA technology and genetic engineering have enormous potential for research and applications in medicine and agriculture, in the home and industry. As with any new technology, however, the potential benefits must be weighed against the potential risks, including ecological and social repercussions.

Review Questions

1. Distinguish between recombinant DNA technology and genetic engineering. *CI*

2. Distinguish these terms from one another: *16.1*
   a. chromosomal (genomic) DNA and cDNA
   b. cloning vector and DNA clone

3. Define PCR. Can fragments of chromosomal DNA, cDNA, or both be amplified by PCR? *16.2*

4. Define DNA fingerprint. Briefly describe which portions of the DNA are used in DNA fingerprinting. *16.3*

5. Outline the steps of automated DNA sequencing. *16.4*

6. Define cDNA library, then briefly explain how a gene can be isolated from it. Define probe and nucleic acid hybridization as part of your answer. *16.5*

7. Give three examples of applications that can be derived from knowledge of an organism's genome. *CI, 16.6–16.8*

8. Name one of the ways in which modified genes have been inserted into mammalian cells. *16.8*

9. Define gene therapy. Once the human genome has been fully sequenced, why will it be difficult to manipulate its genes to advantage? *CI, 16.8*

Self-Quiz (*Answers in Appendix III*)

1. ______ is the transfer of normal genes into body cells to correct a genetic defect.
   a. Reverse transcription
   b. Nucleic acid hybridization
   c. Gene mutation
   d. Gene therapy

2. DNA fragments result when ______ cut DNA molecules at specific sites.
   a. DNA polymerases
   b. DNA probes
   c. restriction enzymes
   d. RFLPs

3. ______ are small circles of bacterial DNA that are separate from the circular bacterial chromosome.

4. Foreign DNA that was inserted into a plasmid and then replicated many times in a population of bacteria is a ______ .
   a. DNA clone
   b. gene library
   c. DNA probe
   d. gene map

5. By reverse transcription, ______ is assembled on ______ .
   a. mRNA; DNA
   b. cDNA; mRNA
   c. DNA; enzymes
   d. DNA; agar

6. PCR stands for ______ .
   a. polymerase chain reaction
   b. polyploid chromosome restrictions
   c. polygraphed criminal rating
   d. politically correct research

7. By gel electrophoresis, fragments of a gene library can be separated according to ______.
   a. shape
   b. length
   c. species

8. Automated DNA sequencing relies on ______ .
   a. supplies of standard and labeled nucleotides
   b. primers and DNA polymerases
   c. gel electrophoresis and a laser beam
   d. all of the above

9. Match the terms with the most suitable description.

____ DNA fingerprint	a. selecting "desirable" traits
____ Ti plasmid	b. deciphering 3.2 billion base pairs of 23 human chromosomes
____ nature's genetic experiments	c. used in some gene transfers
____ nucleic acid hybridization	d. unique array of DNA fragments inherited in Mendelian pattern from each of two parents
____ Human Genome Initiative	e. base pairing of nucleotide sequences from different DNA or RNA sources
____ eugenic engineering	f. mutations, crossovers

Critical Thinking

1. In the following diagram, which restriction enzyme made the cuts (indicated in *red*) in part of a DNA molecule from two different organisms?

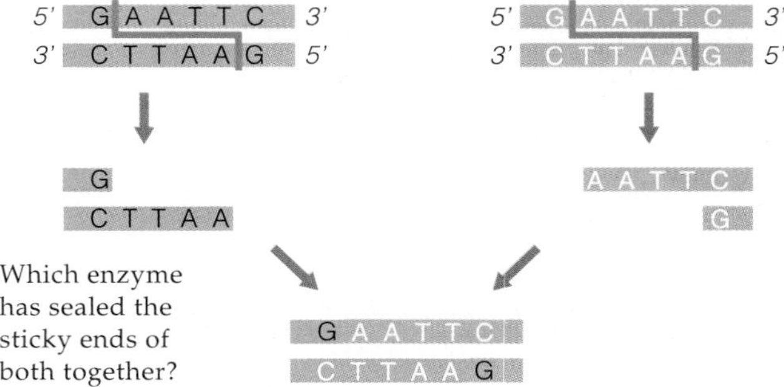
5' G A A T T C 3'
3' C T T A A G 5'
G
C T T A A
A A T T C
G
G A A T T C
C T T A A G

Which enzyme has sealed the sticky ends of both together?

2. Ryan, a forensic scientist, obtained a very small DNA sample from a crime scene. In order to examine the sample by DNA fingerprinting, he must amplify the sample by the polymerase chain reaction. He estimates there are 50,000 copies of the DNA in his sample. Calculate the number of copies he will have after fifteen cycles of PCR.

3. A game warden in Africa confiscated eight ivory tusks from elephants. Some tissue is still attached to the tusks. Now he must determine whether the tusks were taken illegally from northern populations of endangered elephants or from other populations of elephants to the south that can be hunted legally. How can he use DNA fingerprinting to find the answer?

4. The Human Genome Initiative is not yet completed, yet knowledge about a number of the newly discovered genes is already being used to detect genetic disorders. Ask yourself: What will be done with genetic information about individuals? Will insurance companies and potential employers request it? At this writing, many women have already refused to take advantage of genetic screening for a gene associated with the development of breast cancer. Should medical records about people participating in genetic research and genetic clinical services be made available to other individuals? If not, how could such information be protected?

5. Lunardi's Market put out a bin of tomatoes having splendid vine-ripened redness, flavor, and texture. The sign posted above the bin identified them as genetically engineered produce. Most shoppers selected unmodified tomatoes in the adjacent bin, even though those tomatoes were pale pink, mealy-textured, and tasteless. Which ones would you pick? Why?

6. Avoiding genetically engineered food is probably impossible in the United States. At least 45 percent of its cotton, 38 percent of its soybean, and 25 percent of its corn crops are engineered to withstand weedkillers or make their own pesticides. For years, the corn and soybeans have found their way into breakfast cereals, tofu, soy sauce, vegetable oils, beer, soft drinks, and other food products. They are fed to farm animals.

By contrast, public resistance to genetically engineered food is high in Europe, especially Britain. Many people, including the Prince of Wales, speak out against what the tabloids call "Frankenfood." Protesters routinely vandalize crops (Figure 16.15). Worries abound that such foods may be more toxic, have lower nutritional value, and promote natural selection for antibiotic resistance. Some people worry that designer plants will cross-pollinate with wild plants to produce "superweeds."

Biotechnologists envision a new Green Revolution. They argue that designer plants can hold down food production costs, reduce dependence on pesticides and herbicides, enhance crop yields, and offer improved flavor, nutritional value, even salt tolerance and drought tolerance. Yet the chorus of critics in Europe may provoke a trade war with the United States. The issue is not small potatoes, so to speak. In 1998, the value of agricultural exports reached about 50 billion dollars. Flattery, threats, and bullying are rampant on both sides of the Atlantic. Restrictions on genetic engineering will profoundly impact United States agriculture, and inevitably the impact will trickle down to what you eat and how much you pay for it.

All of which invites you to read up on scientific research on this issue and form your own opinions. The alternatives are being swayed either by catchy, frightful phrases (Frankenfood, for example) or by biased reports from groups (such as chemical manufacturers) with their own agendas.

Possibly start with Christopher Bond's article, "Politics, Misinformation, and Biotechnology" in the 18 February 2000 *Science* (287: 1201). Dodd argues that biotechnology attempts to solve real-world problems of sickness, hunger, and dwindling resources; and that . . ."hysteria and unworkable propositions advanced by those who can afford to take their next meal for granted have little currency among those who are hungry."

7. What if it were possible to create life in test tubes? This is the question behind attempts to model and eventually create *minimal organisms*, which we define as living cells having the smallest set of genes necessary to survive and reproduce.

As recent experiments by Craig Ventner and Claire Fraser revealed, *Mycoplasma genitalium*, a bacterium with 517 genes (and 2,209 transposons) might be a candidate. By disabling its genes one at a time in the laboratory, they discovered that it may have only 265–350 essential protein-coding genes.

What if those genes were to be synthesized one at a time and inserted into an engineered cell consisting only of a plasma membrane and cytoplasm? Would the cell come to life? The possibility that it might prompted Ventner and Fraser to seek advise from a panel of bioethicists and theologians. As Arthur Caplan, a bioethicist at the University of Pennsylvania reported, no one on the panel objected to synthetic life research. They felt that much good might come of it, provided scientists didn't claim to have found the "secret of life." The 10 December 1999 issue of *Science* includes an essay from the panel and an article on *M. genitalium* research. Read both, then write down your thoughts about creating life in a test tube.

Figure 16.15 Activists ripping some genetically modified crop plants from a field in Great Britain.

Selected Key Terms

automated DNA sequencing *16.4*
cDNA *16.1*
cloning vector *16.1*
DNA clone *16.1*
DNA fingerprint *16.3*
DNA ligase *16.1*
gel electrophoresis *16.3*
gene library *16.5*
gene therapy *CI*
genetic engineering *CI*
genome *16.1*
nucleic acid hybridization *16.5*
PCR *16.2*
plasmid *16.1*
primer *16.2*
probe (nucleic acid) *16.5*
recombinant DNA technology *CI*
restriction enzyme *16.1*
reverse transcriptase *16.1*
tandem repeats *16.3*

Readings *See also www.infotrac-college.com*

Cho, M. et al. 10 December 1999. "Ethical Considerations in Synthesizing a Minimal Genome." *Science* 286:2087–2090.

Fairbanks, D., and W. R. Andersen. 1999. *Genetics: The Continuity of Life*. Monterey, California: Brooks-Cole.

Joyce, G. December 1992. "Directed Molecular Evolution." *Scientific American* 267(6): 90–97.

Watson, J. D. et al., 1992. *Recombinant DNA*. Second edition. New York: Scientific American Books.

FACING PAGE: *Millions of years ago, a bony fish died, and sediments gradually buried it. Today its fossilized remains are studied as one more piece of the evolutionary puzzle.*

PRINCIPLES OF EVOLUTION

17 EMERGENCE OF EVOLUTIONARY THOUGHT

Legends of the Flood

With this unit, we turn to evolutionary theories and to ways in which they can be used to interpret the past and present, even to predict possible futures for the natural world. Today the theories are routinely used in science and widely accepted in society at large. This was not the case when early evolutionists in Europe started to work out the details. They rankled more than a few people with their astounding theories, which revealed connections between the geologic record, the fossil record, and the sweep of biological diversity.

As evidence accumulated in support of the theories, many zoologists, geologists, and others came to accept them as useful guides for research. Others were unable to reconcile the theories with the Bible, the premier book of the Western world. That book had helped many generations cope with the uncertainties of life, death, and the unpredictability of nature. It included stories of a great flood and other catastrophes. Catastrophes did indeed occur, but they had been interpreted at an early time in human history when people obviously had a limited understanding of geologic processes.

For example, William Ryan and Walter Pitman drew from marine geology, genetics, archaeology, botany, and linguistics to pinpoint where the biblical flood may have occurred. They zeroed in on the Black Sea (Figure 17.1). This landlocked sea is 1,800 meters (6,000 feet) deep, is freshened by rainfall and large rivers, and loses excess water through the Bosporus and Dardanelles straits.

What Ryan and Pitman discovered coincides with the biblical account of the great flood. It also coincides with stone tablets inscribed in Babylon long before the Bible was written, and with the epic of Gilgamesh and many other myths from Greece, Mesopotamia, and the Levant. What follows is the picture they pieced together.

The most recent ice age set in motion events that helped shape human history. Starting 120,000 years ago, water evaporated from the sea and fell as snow over the northern half of Europe, Asia, and North America. So much water evaporated that the sea level was 120 meters (400 feet) lower than today. Snow accumulated as vast ice sheets, some 3 kilometers (2 miles) thick. However, about 20,000 years ago, the global climate warmed and a colossal meltdown began. Meltwater drained to the sea, but it also helped form immense freshwater lakes on land. One of the lakes was destined to become the Black Sea. It rose high enough to connect with a narrow outlet to the Mediterranean Sea (Figure 17.2*a*).

Between 8,200 and 7,800 years ago, glacial conditions returned. Then the climate warmed and the meltdown continued. The ice receded far north, and nearly all of the meltwater was diverted to the sea. Also, rainfall was scarce and evaporation rates were high. And so the lake shrank. By 7,600 years ago, its surface was 150 meters (500 feet) below its outlet to the sea.

When the ice sheets prevailed, cool, dry deserts and grasslands formed to the south, and modern humans (*Homo sapiens*) witnessed the change. Hunter–gatherers moved into fertile valleys and river deltas in Europe and Asia. Where water, wild game, and edible plants were available, they founded settlements—only to abandon them when the climate changed and resources vanished. The lake beckoned the displaced more than once. Being so far below sea level, the lake's edge was warmer than the surrounding cold, arid mountains and plateaus. It welcomed the thirsty when springs, ponds, and other lakes dried up. In the moist, fertile soil of that Edenlike oasis, wild wheat and barley took root. So did complex trading cultures and early attempts at farming.

All the while, the sea had been rising. By 7,600 years ago it was high enough to lap over the crest of the Bosporus valley, which plunged to the lake far below. At first, trickles cut into the soft earth. The cut deepened until water thundered through at 200 times the flow volume of Niagara Falls (Figure 17.2*c*). For

Atlantic Ocean
Black Sea
Caspian Sea
Mediterranean Sea
Aegean Sea
Levant
Mesopotamia
Tigris
Euphrates

Figure 17.1 The geologic region where global climate changes led to a catastrophe that altered the course of human history.

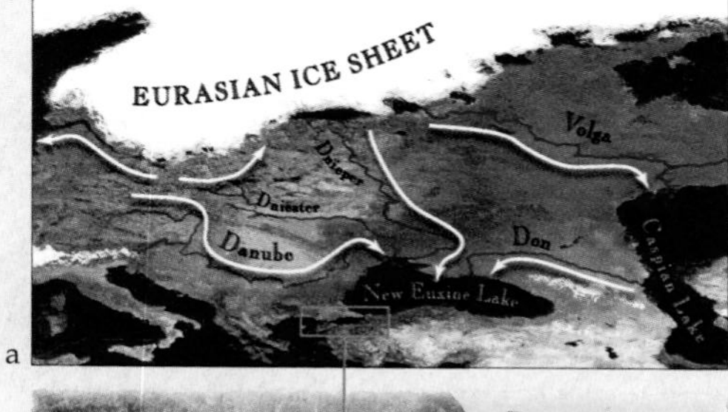
EURASIAN ICE SHEET
Volga
Dnieper
Dniester
Danube
Don
Caspian Lake
New Euxine Lake

a

Meltwater flow from Eurasian ice sheet

Freshwater Lake
Sea of Marmara
(fresh)
Aegean Sea

b

Freshwater lake (future Black Sea) connects with Mediterranean 14,500 years ago

(marine)
Dardanelles

c

Flow of seawater in same region before flood 7,600 years ago

Black Sea
(marine)
Bosporus
(marine)

d

Present-day exchange of seawater and freshwater between the Mediterranean and Black seas

Figure 17.2 Historical connection between the Mediterranean and the ice age freshwater lake that became the Black Sea.

at least the next ten months, water quickly advanced through deltas and valleys, permanently submerging villages and driving humans and wildlife before it.

These were farmers and fishermen, not nomads who would simply pack up and move on. They built houses and boats; they traded obsidian, leather, herbs, and finely incised pottery. They could not comprehend the force behind their mass dispersal. In time they settled in the Levant, Mesopotamia, Egypt, the Persian Gulf, central Asia, and western Europe. They brought with them precious seeds, farm animals, and fragments of languages, ideas, and technologies. And they carried with them a terrifying story of a cataclysmic flood.

The story endured. It influenced investigations into processes of nature, as when sixteenth-century biblical scholars argued that the Earth was about 6,000 years old. To them, that was plenty of time for man's creation, fall from grace, and salvation after a punishing flood. The only remembered geologic changes that made it into oral and written traditions were catastrophes so immense that they must have been divinely invoked. Otherwise the Earth was just an unchanging stage for the human drama. This was a core hypothesis of what came to be known as theories of **catastrophism**.

Biological science emerged within the framework of such influential beliefs, and so did awareness of changes in the Earth and its creatures. Understand this, and you will gain insight into why acceptance of the very idea of biological evolution was so long in coming.

KEY CONCEPTS

1. In the past, cultural beliefs and limited knowledge of geologic forces shaped scholarly explanations of devastating floods, volcanic eruptions, and other natural catastrophes. Evolutionary theories gave scholars a new way to interpret and investigate the natural world.

2. Today, we define biological evolution as heritable changes in lines of descent, or lineages. Evidence of evolution comes from many different investigations that began centuries ago.

3. From the fifteenth century onward, explorers found puzzling differences in the world distribution of species. By the eighteenth century, anatomists were identifying similarities and differences in the body structure and patterning among the embryos and adult forms of major groups of animals. In sequences of sedimentary rocks, geologists were finding sequences of fossils. The deeper the sedimentary layers, the simpler were the fossils.

4. Charles Darwin came of age at a time when scholars were attempting to reconcile the new findings from global explorations, comparative morphology, and geology with prevailing cultural beliefs.

5. As Darwin and Alfred Wallace perceived, populations evolve by way of natural selection. By this process, differences in survival and reproduction arise among individuals that differ in one or more heritable traits.

6. Here are the key premises of the theory of evolution by natural selection: Individuals of a population share certain traits, most of which are heritable. Those traits vary in their details from one individual to the next. The forms of traits that prove to be most adaptive under prevailing environmental conditions tend to increase in frequency through successive generations. Other forms decrease in frequency or disappear. Thus, the traits that characterize a population can change over time; the population can evolve.

17.1 EARLY BELIEFS, CONFOUNDING DISCOVERIES

The Great Chain of Being

Our story begins more than two thousand years ago, when the seeds of biological inquiry were beginning to take hold among the ancient Greeks. At the time, popular belief held that supernatural beings intervened directly in human affairs. For example, people "knew" that angry gods inflicted a common ailment known as the sacred disease. And yet, from one physician of the school of Hippocrates, these thoughts come down to us:

It seems to me the disease called sacred . . . has a natural cause, as other diseases have. Men think it divine merely because they do not understand it. But if they called everything divine that they did not understand, there would be no end of divine things! . . . If you watch these fellows treating the disease, you see them use all kinds of incantations and magic—but they are also careful in regulating diet. Now if food makes the disease better or worse, how can they say it is the gods who do this? . . . It does not really matter whether you call such things divine or not. In Nature, all things are alike in this, in that they can be traced to preceding causes.

— ON THE SACRED DISEASE (400 B.C.)

He perceived that all aspects of nature have underlying causes—a key premise of modern science (Section 1.5).

Aristotle was foremost among the early naturalists, and he described the world around him in great detail. He had no reference books or instruments to guide him, for biological science in the Western world began with the great thinkers of this age. Yet here was a man who was no mere collector of random tidbits of information. In his thoughtful descriptions, we find evidence of a mind perceiving connections between observations and attempting to explain the order of things.

Aristotle believed (as did others) that each kind of organism was distinct from all the rest. Nevertheless, he wondered about organisms that seemed to have rather blurred positions in nature. For example, even though some sponges look very much like plants, they do not make their own food, as plants do. They capture and digest it, as animals do. In time, Aristotle came to view nature as a continuum of organization, from lifeless matter through complex forms of plant and animal life.

By the fourteenth century, Aristotle's idea had been transformed into a rigid view of life. A Chain of Being was seen to extend from the "lowest" forms of life to humans and on up to spiritual beings. Each kind of being, or "species" as it was called, was a separate link in the great chain. All of those links were designed and forged at the same time, at the same center of creation, and had not changed since. Scholars thought that once they had discovered, named, and described all of the links, the meaning of life would be revealed to them.

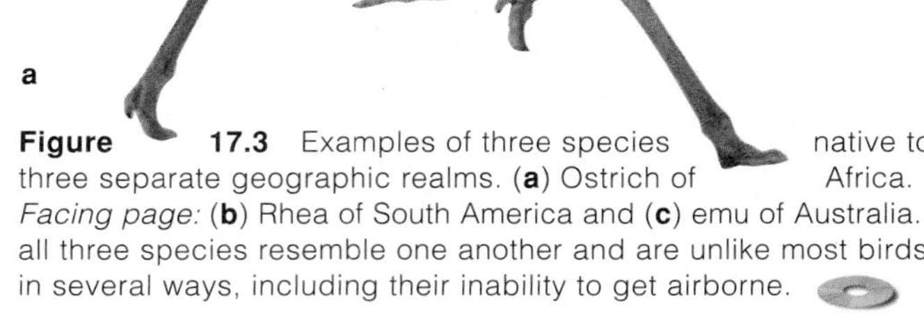
a

Figure 17.3 Examples of three species native to three separate geographic realms. (**a**) Ostrich of Africa. *Facing page:* (**b**) Rhea of South America and (**c**) emu of Australia. all three species resemble one another and are unlike most birds in several ways, including their inability to get airborne.

Questions From Biogeography

Until the fifteenth century, naturalists were not aware that the world is a great deal bigger than Europe, so the task of locating and describing all species seemed manageable. Then global explorations began, and the known world expanded enormously. Naturalists were soon overwhelmed by descriptions of tens of thousands of plants and animals that explorers were discovering in Asia, Africa, the Pacific Islands, and the New World.

In 1590 the naturalist Thomas Moufet attempted to sort through the bewildering array. He simply gave up and wrote such gems as this description of locusts and grasshoppers: "Some are green, some black, some blue. Some fly with one pair of wings; others with more; those that have no wings they leap; those that cannot fly or leap they walk; some have long shanks, some shorter. Some there are that sing, others are silent." It was not a work of subtle distinctions.

Even so, a few scholars began to examine the world distribution of organisms, a discipline now known as **biogeography** (see the Chapter 50 introduction). They soon realized that many plants and animals are unique to isolated places, such as remote oceanic islands. But certain species separated by great distances resemble one another (Figure 17.3).

How did so many species get from one center of creation to oceanic islands and other isolated locations? And what did the similarities and differences among them mean?

Questions From Comparative Morphology

By the eighteenth century, many scholars were engaged in **comparative morphology**, the systematic study of similarities and differences in the body plans between major groups, such as different kinds of vertebrates. Think of the bones of a human arm, whale flipper, and

Further reading: Student Guide to InfoTrac on web site →

b

c

bat wing. They differ in size, shape, and function. Yet all have similar locations in the body. They consist of the same tissues, arranged in the same overall patterns. They develop in similar ways in embryos. Comparative morphologists who deduced all of this wondered: Why are some animals that are so different in some features so much alike in others?

By one hypothesis, basic body plans were so perfect that there was no need to come up with a totally new one for each organism at the time of their creation. Yet if that were true, then how could there be parts with no function? For example, some snakes have bones that correspond to a pelvic girdle—a set of bones to which hind legs attach (Figure 17.4). Snakes don't have legs. Why the bones? Humans have bones that correspond to a few tailbones of many other mammals, but they don't have a tail. Why do they have parts of one? (After reading Chapter 20, you may have a good idea.)

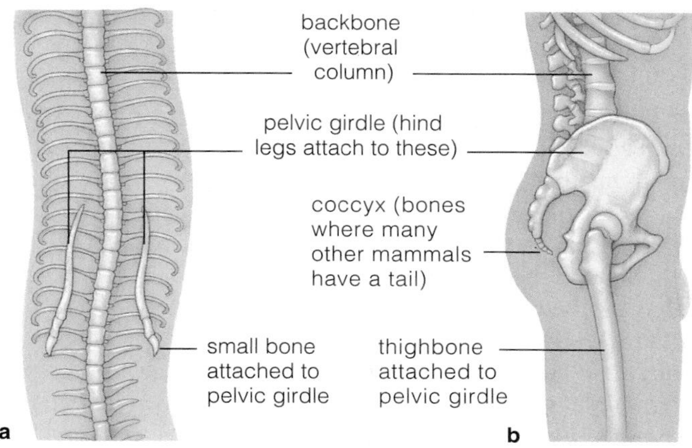
backbone (vertebral column)
pelvic girdle (hind legs attach to these)
coccyx (bones where many other mammals have a tail)
small bone attached to pelvic girdle
thighbone attached to pelvic girdle
a
b

Figure 17.4 (**a**) Python bones corresponding to the pelvic girdle of other vertebrates, including humans (**b**). Small "hind limbs" protrude through the skin on the underside of the snake.

Questions About Fossils

From the late 1600s on, geologists added to the growing confusion. They began mapping layers of rocks at sites where erosion or quarrying had cut deep into the earth. They found similar layers around the world. (Such beds consist of distinct, multiple layers of sedimentary rock. Section 20.10 shows an example.) Most agreed that sand and other sediments had been deposited at different times and had slowly compacted into rock layers. They realized that certain layers contained distinctive **fossils**, which were thought to be stone-hard evidence of life in ancient times.

For example, deep layers contained fossils of simple marine organisms. Some fossils in the overlying layers were similar but structurally more complex. The fossils in layers above these closely resembled living marine organisms. What did increasing structural complexity among fossils of a given type represent? Were these *sequences* of fossils, separated in time? Another puzzle: many fossils were unique in some traits yet similar to certain existing species in other traits! Could species so similar yet so far apart in time be *related* in some way?

Taken as a whole, the findings from biogeography, comparative morphology, and geology did not fit with prevailing beliefs. Georges-Louis Leclerc de Buffon and a few others started to formulate novel hypotheses. If dispersal of all species from a center of creation was not possible, given the vast oceans and other barriers, *then perhaps species had originated in more than one place*. And if they were not created in a perfect state—and fossil sequences and the presence of "useless" body parts in certain organisms suggested they were not—*then perhaps species had been modified over time*. Awareness of change in lines of descent—**evolution**—was in the wind.

Awareness of biological evolution emerged over centuries, through the cumulative observations of many naturalists, biogeographers, comparative anatomists, and geologists.

17.2 A FLURRY OF NEW THEORIES

Squeezing New Evidence Into Old Beliefs

In the nineteenth century, naturalists tried to reconcile the growing evidence of change in lines of descent with a traditional conceptual framework that did not allow for change. Foremost among them was Georges Cuvier, a respected anatomist. For years he had compared body plans of fossils and living organisms. He acknowledged the abrupt changes in the fossil record, corresponding to discontinuities between certain layers of sedimentary beds. Was the record evidence of changing populations of organisms that lived in those ancient environments? Cuvier thought so. And he was right, as you will see from the evolutionary story (Sections 19.1 and 20.10).

Based on that inference, Cuvier constructed his own theory of catastrophism. There was one time of creation, he said, that populated the world with every species. A global catastrophe destroyed many of them. Survivors repopulated the world. It was not that survivors were *new* species; naturalists simply hadn't yet found fossils of them that would date to the time of creation. Later catastrophes wiped out more species, and repopulation by the survivors followed, as recorded by fossils.

Many scholars accepted his theory, but others kept at the puzzle. One hypothesis—inheritance of *acquired* characteristics—was pushed by Jean Lamarck. During each individual's life, thought Lamarck, environmental pressures and internal "needs" bring about permanent changes in body form and functioning, then offspring inherit the necessary changes. And so life, created long ago in a simple state, gradually improved. The force for change was a drive toward perfection, up the Chain of Being. The drive was centered in nerves that directed an unknown "fluida" to body parts in need of change.

Apply his hypothesis to modern giraffes. Say they had a short-necked ancestor. Pressed by a need to find food, it kept stretching its neck to browse upon leaves beyond the reach of other animals. Stretching directed fluida to its neck, which lengthened permanently. The longer neck was inherited by offspring, which stretched their necks, also. So generations of animals desiring to reach ever loftier leaves led to the modern giraffe.

As Lamarck correctly inferred, the environment *is* a factor in evolution. His hypothesis, however, like others proposed at the time, has not been supported by any tests carried out since then. No one has found evidence that the environment can modify traits of an existing individual in a way that can be passed on to offspring.

Voyage of the Beagle

In 1831, in the midst of the confusion, Charles Darwin was twenty-two years old and wondering what to do with his life. Ever since he was eight, he had wanted to hunt, fish, collect shells, or simply watch insects and birds—anything but sit in school. Later, at his father's insistence, he did try to study medicine in college. The crude, painful procedures used on patients at that time sickened him. His by-now exasperated father urged him to become a clergyman, so he packed for Cambridge. His grades were good enough to earn him a degree in theology. But he spent most of his time among faculty members with leanings toward natural history.

a
b

Figure 17.5 (**a**) Charles Darwin and (**b**) a blue-footed booby, one of the species he observed during his five-year voyage around the world on the *Beagle*. A replica of this ship is shown in (**c**), sailing off the coast of South America. During stops along Argentina's coast, Darwin ventured into the Andes. He saw fossils of marine organisms embedded in rocks 3.6 kilometers above sea level. (**d**,**e**) Galápagos Islands, isolated in the ocean far to the west of Ecuador. They arose through volcanic action about 5 million years ago, so organisms could not have originated there. Winds or ocean currents must have carried them to the new islands.

John Henslow, a botanist, perceived Darwin's real interests. He arranged for Darwin to function as ship's naturalist aboard H.M.S. *Beagle*. The *Beagle* was about to embark on a five-year voyage that would take Darwin around the world (Figure 17.5). Abruptly, the young man who hated schoolwork and had no formal training as a naturalist started to work with enthusiasm.

The *Beagle* sailed first to South America to complete work on mapping the long coastline. During the Atlantic crossing, Darwin collected and examined marine life. He also read Henslow's parting gift, the first volume of Charles Lyell's *Principles of Geology*. During stops along the coast and at various islands, he observed diverse species in environments ranging from sandy shores to

Further reading: Student Guide to InfoTrac on web site

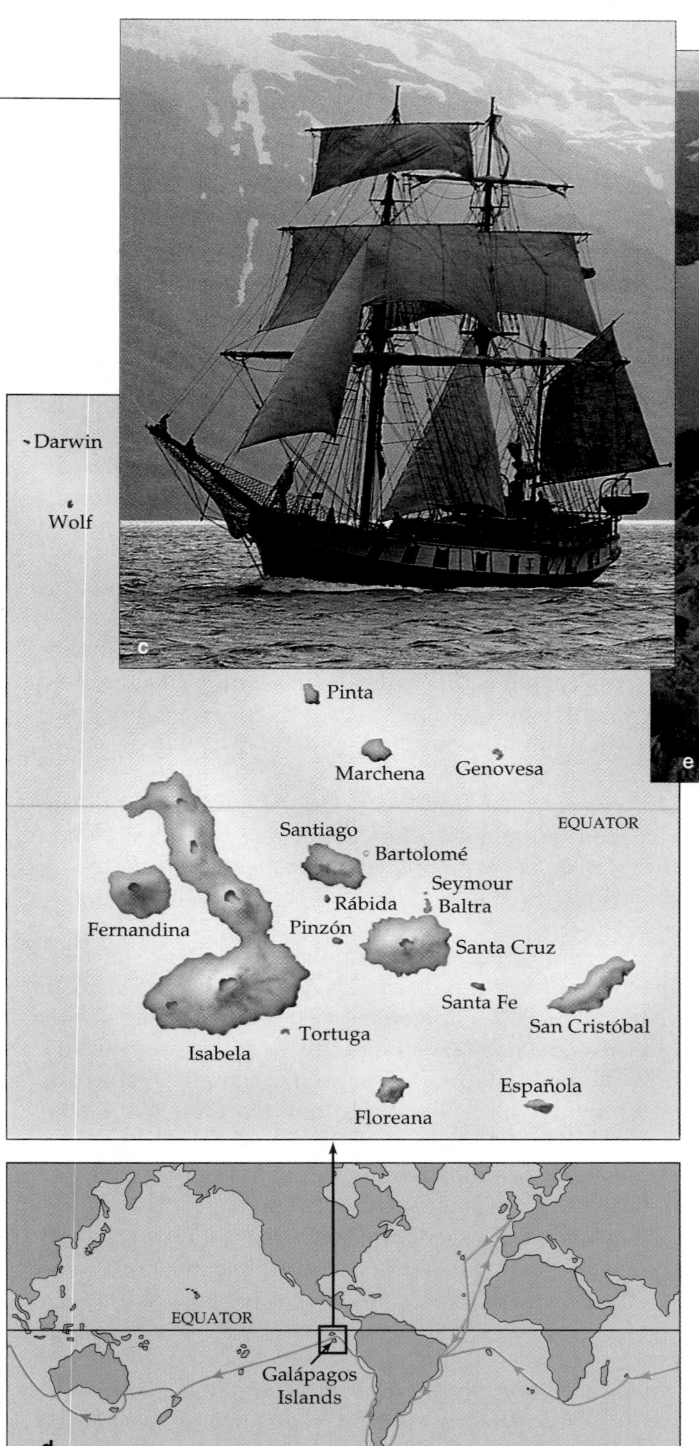
Darwin
Wolf
c
Pinta
Marchena
Genovesa
EQUATOR
Santiago
Bartolomé
Seymour
Rábida
Baltra
Fernandina
Pinzón
Santa Cruz
Santa Fe
San Cristóbal
Tortuga
Isabela
Española
Floreana
EQUATOR
Galápagos
Islands
d

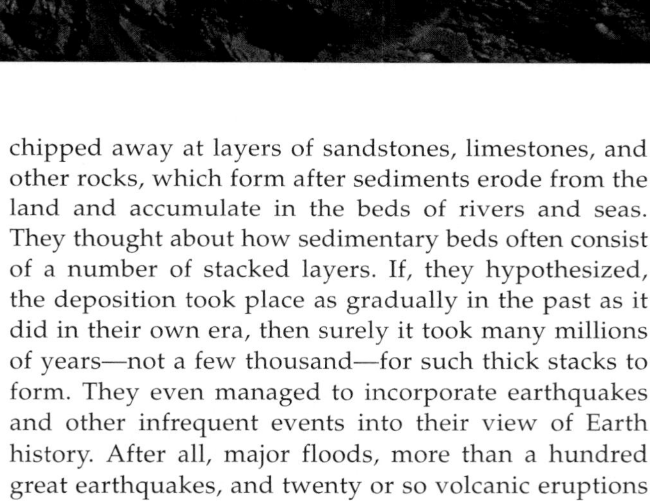
e

high mountains. And he started circling the question of evolving life, which was now on the minds of many respected individuals, including his own grandfather.

Darwin started mulling over a rather radical theory that Lyell was advancing in his book. Lyell and other geologists were arguing that catastrophes had no more and possibly less effect on Earth history than did subtle processes of change. They had thought about how long it takes for rains, the pounding surf, and other forces of nature to sculpt the landscape. For years geologists had chipped away at layers of sandstones, limestones, and other rocks, which form after sediments erode from the land and accumulate in the beds of rivers and seas. They thought about how sedimentary beds often consist of a number of stacked layers. If, they hypothesized, the deposition took place as gradually in the past as it did in their own era, then surely it took many millions of years—not a few thousand—for such thick stacks to form. They even managed to incorporate earthquakes and other infrequent events into their view of Earth history. After all, major floods, more than a hundred great earthquakes, and twenty or so volcanic eruptions happen every year, so catastrophes are not unusual.

Their view of gradual, uniformly repetitive change became the **theory of uniformity**. It directly challenged prevailing views of the age of the Earth.

The theory bothered scholars who firmly believed the Earth was only about 6,000 years old. They thought people had recorded everything that happened during those thousands of years, and in all that time no one ever mentioned seeing a species evolve. Yet by Lyell's calculations, it must have taken millions of years to mold the present landscape. *Wasn't that enough time for species to evolve in diverse ways?* Later, Darwin thought so. But exactly *how* did they evolve? He would end up devoting the rest of his life to that burning question.

Prevailing beliefs can influence how we interpret clues to natural processes and their observable outcomes.

Darwin's observations during a global voyage helped him think about these processes in a novel way.

17.3 DARWIN'S THEORY TAKES FORM

Old Bones and Armadillos

After Darwin returned to England in 1836, he talked with other naturalists about possible evidence that life evolves. By carefully studying all of the notes from his journey, he came up with some possibilities.

In Argentina, for example, he had observed fossils of glyptodonts, which are now extinct. Of all animals on Earth, only living armadillos are like glyptodonts (Figure 17.6). And of all places on Earth, armadillos live only in the same places where glyptodonts once lived. If the two kinds of animals had been created at the same time, lived in the same place, and were so much alike, why is only one still alive? Wouldn't it be reasonable to assume glyptodonts were early relatives of armadillos? Many of their shared traits might have been retained through countless generations. Other traits might have been modified in the armadillo branch of a family tree. Descent with modification—it seemed possible. What, then, was the driving force for evolution?

A Key Insight—Variation in Traits

While Darwin assessed his notes, an influential essay by Thomas Malthus, a clergyman and economist, made him look closely at a topic of social interest. Malthus had correlated population size with famine, disease, and war. Humans, he claimed, run out of food, living space, and other resources because they reproduce too much. The larger a population gets, the more people there are to reproduce in each generation. Population size burgeons, resources dwindle, and the struggle to live intensifies. Many people starve, get sick, and engage in war and other forms of competition for the remaining resources.

After Darwin reflected on his personal observations, he suspected that any population has the capacity to produce more individuals than the environment is able to support. To give one example, a single sea star can release 2,500,000 eggs per year, but the seas obviously do not fill up with sea stars. In each generation, nearly all of the eggs and larvae end up inside the bellies of predators. Many of the survivors starve or succumb to disease or some other environmental insult.

Assume that the environment restricts the number of reproducing individuals. Which individuals will be the winners and losers? What influences the outcome? Darwin thought about the populations he had observed during his voyage. As he recalled, individuals were not alike down to the last detail. They showed variations in size, coloration, and other traits. *And it dawned on him that variations in traits might affect an individual's ability to secure resources—and therefore to survive and reproduce in particular environments.*

Did the Galápagos Islands show evidence of this? Between these volcanic islands and the South American coastline are 900 kilometers of open ocean. The islands offer diverse habitats along their rocky shores, deserts, and mountain flanks. Nearly all of their inhabitants live nowhere else—although they resemble species on the mainland. Were those remote islands colonized by species that flew, floated, or were blown over from the mainland? If so, then in the different island habitats, island-hopping descendants of the colonizers might have undergone modifications, over time, as adaptive responses to local conditions.

As Darwin eventually learned from other naturalists in England, thirteen species of finches were distributed throughout the Galápagos Islands. He himself had collected specimens of some of these birds, and now he attempted to correlate variations in their traits with environmental challenges.

Imagine yourself in his place. The finches of one population, you notice, have a large, strong bill suitable for cracking seeds (Figure 17.7). Yet a few individuals with a slightly stronger bill crack seeds that are too tough for their neighbors. If most of the seeds being produced in a particular habitat during a given interval have hard coats, then a strong-billed bird will have a competitive edge. It will have a better chance than the other birds of surviving and producing offspring. Assuming the trait is heritable, the same will be true of that bird's strong-billed descendants.

a

b

Figure 17.6 (**a**) A living armadillo. (**b**) Painting of a glyptodont (extinct). Although separated in time, these animals share unusual traits and a restricted distribution. To Darwin, they were a clue that helped him develop a theory of evolution by natural selection.

Further reading: Student Guide to InfoTrac on web site

Figure 17.7 Four species of finches that live on the Galápagos Islands. (**a**) *Geospiza conirostris* and (**b**) *G. scandens*, both with a bill adapted for eating cactus flowers and fruits. Other finches have thick, strong bills that crush cactus seeds. (**c**) *Certhidea olivacea*, a tree-dwelling finch, resembles warblers in song and behavior. It uses its slender beak to probe for insects. (**d**) *Camarhynchus pallidus* feeds on wood-boring insects such as termites. It has learned to break cactus spines and twigs to suitable lengths, and then hold the "tools" and use them to probe bark for insects hidden from its view.

Figure 17.8 Alfred Wallace. Darwin worked out a theory of natural selection long before Wallace did. However, Wallace was first to report it; he circulated a brief letter describing his views about the process.

Take these thoughts one step further. If factors in the environment continue to "select" the most adaptive version of a trait, then the population will become one of mostly strong-billed birds. *And a population is evolving if forms of heritable traits are changing over the generations.*

Recall, from Section 1.4, that Darwin offered pigeon breeding and other *artificial* selection practices as a way to explain *natural* selection of traits in the wild. When breeders favor pigeons with, say, black-feathered tails, they encourage black-tailed pigeons of each generation to mate but not white-tailed ones. It was an easy way to show how selection could lead to an increase in the frequency of one form of a trait in a captive population.

Anticipating that his view would be controversial, Darwin waited to announce it and searched for flaws in his reasoning. He waited too long. More than a decade after he wrote up but never published an essay on his theory, another respected naturalist—Alfred Wallace—sent him a letter (Figure 17.8). Wallace had developed the same theory, then quickly wrote up and circulated his ideas. Colleagues insisted Darwin present a formal paper at the same time Wallace presented his. (Wallace also believed Darwin deserved most of the credit.) The next year, in 1859, Darwin did publish *On the Origin of Species*, his detailed evidence in support of the theory.

Although you may have heard that Darwin's book fanned an intellectual firestorm, the idea that diversity is the product of evolution was accepted almost at once by most naturalists. But Darwin's specific explanation, of gradual evolution by natural selection, was fiercely debated. Nearly seventy years passed before advances in a new field, genetics, led to widespread acceptance of his explanation. Until that happened, people generally associated Darwin's name mainly with the premise that life evolves—something others had suggested before him.

DARWIN'S THEORY OF EVOLUTION BY NATURAL SELECTION. A population can evolve (change over time) when individuals differ in one or more heritable traits that are responsible for differences in the ability to survive and reproduce.

REGARDING THE FIRST OF THE "MISSING LINKS"

At the time when Darwin formally presented his theory of evolution by natural selection, it faced a serious challenge. If scholars subscribed to the theory, then they would be tentatively accepting that all species are related, by way of descent, to ever more ancient species. Darwin saw with his own eyes that artificial selection practices by pigeon breeders and others could mold the traits of a population in no time at all. So he argued it was entirely possible for one species to evolve gradually into a separate species, with one or more traits that were uniquely its own, over hundreds or thousands of generations. But if that were so, then where were the "missing links"? That is, if each species evolved from others, then where were the fossils of all the *transitional forms*? Where were the species with traits that bridged two major groups of organisms?

More than a century would pass before more fossil finds as well as the application of molecular biology and genetic analysis would yield plausible answers to that question. In Darwin's time, though, the presumed absence of transitional forms was a major point of contention in discussions of the theory. Ironically, a fossil of just such a transitional form had already been unearthed at the Solnhofen limestones of Bavaria, in southern Germany.

The fossilized skeleton initially was labeled a small theropod (meat-eating) dinosaur, about the size of a pigeon. The specimen's shape said "dinosaur." Like the theropods, it had a long, bony tail, clawed fingers, and a heavy jaw with short, spiky teeth (Figure 17.9).

Later, diggers unearthed another fossil of the same type. Later still, someone noticed the feathers. If those fossils were of dinosaurs, whatever were they doing with *feathers*? Close examination revealed that the feathers were like those of modern birds. The specimen type was named *Archaeopteryx* (meaning "ancient winged one").

Between 1860 and 1988, six specimens of *Archaeopteryx* and a fossilized feather were found. Anti-evolutionists tried to dismiss them as forgeries. Someone, they asserted, had pressed bones and feathers of existing birds against wet plaster. After the imprinted plaster casts dried, they merely looked like fossils. But microscopic examination confirmed that the fossils are real. Further confirmation of their antiquity is found in the remains of clearly ancient species of worms, jellyfishes, and many other kinds of organisms preserved in the same limestone layers.

Through radiometric dating methods, we now know that *Archaeopteryx* lived 150 million years ago. When you examine the photograph in Figure 17.9, you might wonder: How could the remains of those winged creatures be so well preserved after so much time? *Archaeopteryx* happened to live in tropical forests next to a large lagoon. The lagoon was warm, still, and stagnant, because coral reefs barred inputs of oxygenated water from the sea. So it was not a favorable habitat for scavenging animals that might otherwise have feasted on—and obliterated the remains of—*Archaeopteryx* and other organisms that fell from the skies or drifted offshore. With each tropical storm surge, though, fine sediments were swept over the reefs. They gently buried the carcasses littered at the bottom of the lagoon. Over time, the soft mud compacted and hardened to become the exceptionally fine limestone tombs for more than 600 species, including *Archaeopteryx*.

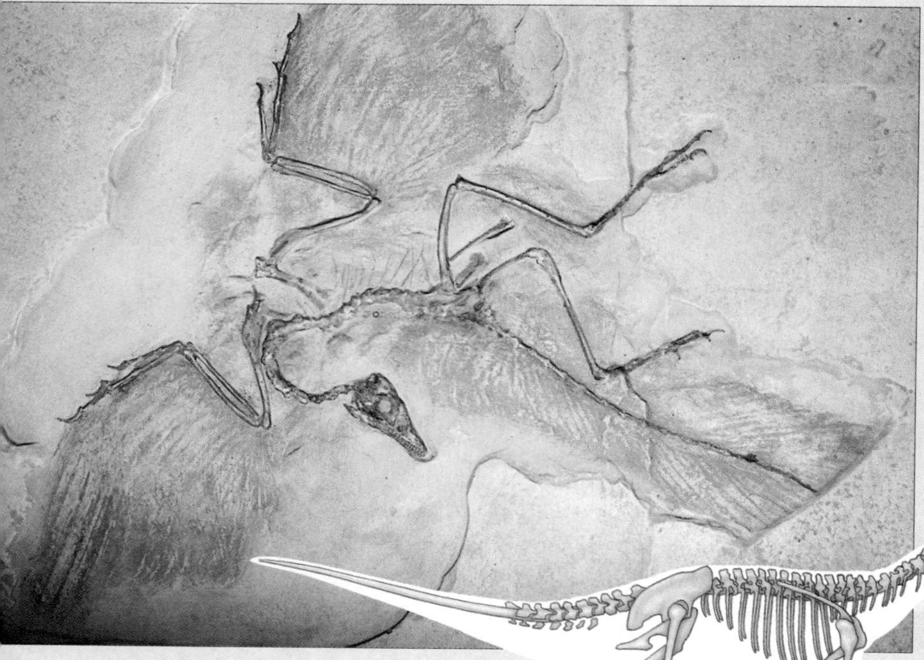

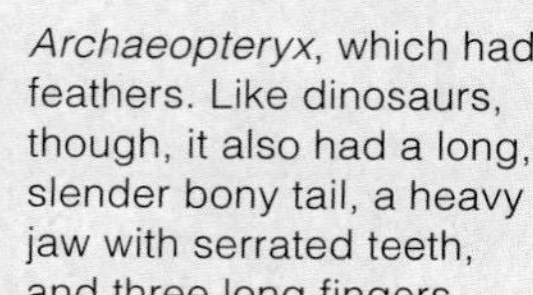

Figure 17.9 *Above:* A photograph of one of the *Archaeopteryx* fossils. *Right:* Comparison of the skeletons of a dinosaur (a type of reptile) and of *Archaeopteryx*, which was on or very near the evolutionary road leading from reptiles to birds.

Dromaeosaurus, a small, two-legged dinosaur that had the most traits in common with birds. Like birds, it had long, slender hind legs, three toes forward and one toe pointing in reverse, and long forelimbs. Unlike birds, it had no feathers.

Archaeopteryx, which had feathers. Like dinosaurs, though, it also had a long, slender bony tail, a heavy jaw with serrated teeth, and three long fingers.

Further reading: Student Guide to InfoTrac on web site

17.5 SUMMARY

1. Awareness of evolution—changes in lines of descent over time—emerged through comparisons of the body structure and patterning for major groups of animals (comparative morphology), questions about the world distribution of plants and animals (biogeography), and observations of fossils of different types in different layers of sedimentary rocks.

2. Here are the key points of the theory of evolution by natural selection, as proposed by Darwin and Wallace:

a. Individuals of all populations have a capacity to produce more offspring than the environment is able to support, so individuals must compete for resources.

b. The individuals of a population vary in size, form, and other traits. Variant forms of a trait may be more or less adaptive under prevailing conditions.

c. When some form of a trait proves adaptive under prevailing conditions, and when it has a heritable basis, its bearers tend to survive and then reproduce more frequently than individuals with less adaptive forms of the trait. Through successive generations, the adaptive version becomes more common in the population.

d. Natural selection is a difference in survival and reproduction among individuals of a population that differ from one another in one or more traits.

e. Natural selection results in modifications of traits within a line of descent. Over time, it may bring about the evolution of a new species, with an array of traits that is uniquely its own relative to other species.

3. One "test" of Darwin's theory would be evidence of one major kind of organism changing into another kind. *Archaeopteryx*, a now-extinct, transitional form between reptiles and birds, provided early evidence. Not until decades later did studies at many levels of biological organization offer a large body of evidence in support of the theory of evolution by natural selection.

Review Questions

1. Define and contrast the theories of catastrophism and of uniformity. *CI, 17.2*

2. Define biogeography and comparative anatomy. How did studies in both disciplines contradict the idea that species have remained unchanged since the time of creation? *17.1*

3. Cuvier and Lamarck interpreted the fossil record differently. Briefly state how their interpretations differed. *17.2*

4. Define evolution. Define evolution by natural selection. Can an individual evolve? *17.3*

Self-Quiz *(Answers in Appendix III)*

1. Ancient scholars interpreted natural disasters in terms of ______.
   a. cultural beliefs
   b. natural selection
   c. understanding of geologic forces shaping the Earth

2. Biologists define evolution as ______.
   a. the origin of a species
   b. heritable change in a line of descent over generations
   c. inheritance of characteristics acquired by the individual

3. Darwin saw that populations of Galápagos finches ______.
   a. show variation in traits
   b. resemble birds in South America
   c. are adapted to different island habitats
   d. all of the above

4. ______ is a difference in survival and reproduction among individuals of a population that differ in traits.
   a. Natural selection
   b. Artificial selection
   c. both a and b

5. Match the following individuals with their ideas.

____ Cuvier	a. theory of natural selection
____ Lamarck	b. populations outgrow resources
____ Lyell	c. catastrophism
____ Malthus	d. inheritance of traits that were acquired through environmental pressures and internal desires for change
____ Darwin, Wallace	e. geologic evidence that the Earth is extremely ancient

Critical Thinking

1. Darwin was not aware of Gregor Mendel's investigations into patterns of inheritance, as described in Chapter 11. How might knowledge of Mendel's results have helped him when he was developing his theory of natural selection?

2. In 1996, after reflecting on evidence of evolution that has been accumulating for more than a hundred years, Pope John Paul II acknowledged that the theory of evolution ". . . has been progressively accepted by researchers, following a series of discoveries in various fields of knowledge. The convergence, neither sought nor fabricated, of the results of work that was conducted independently is in itself a significant argument in favor of this theory."

His words infuriated individuals who believe in a strict, literal interpretation of the biblical account of creation. Some of these individuals have been demanding that biology instructors treat evolution as "just a theory" and give equal time to the biblical interpretation.

Refer to Section 1.5 on the nature of scientific inquiry. Do you equate scientific and religious explanations of life's origin and history as alternative theories? Why or why not?

Selected Key Terms

biogeography *17.1*
catastrophism (theory of) *CI*
comparative morphology *17.1*
evolution *17.1*
fossil *17.1*
natural selection (theory of) *17.3*
uniformity (theory of) *17.2*

Readings *See also www.infotrac-college.com*

Darwin, C. 1957. *Voyage of the Beagle.* New York: Dutton. An account of what Darwin saw and thought about during his voyage, in his own words.

Ryan, W., and W. Pitman. 1998. *Noah's Flood: The New Scientific Discoveries About the Event That Changed History.* New York: Simon & Schuster.

18 MICROEVOLUTION

Designer Dogs

We humans have tinkered rather ruthlessly with the modern descendants of a long and distinguished lineage. The lineage originated some 40 million years ago with the appearance of tree-dwelling carnivores that looked rather like small weasels. Their descendants evolved along separate branchings that now include the weasels, badgers, otters, skunks, bears, pandas, raccoons—and dogs.

About 50,000 years ago, we began domesticating wild dogs. No doubt the advantages of doing so were important. Times were tough in the days before police protection and supermarkets. Dogs welcomed to the hearth could guard people and their possessions. They could corner, kill, eat, and thus dispose of big rats and other unwelcome vermin.

By 14,000 years ago, we started to develop different varieties (breeds) through artificial selection. Individual dogs having desired forms of traits were selected from each new litter and, later, encouraged to breed. Those having undesired forms of traits were passed over.

After favoring the pick of the litter over hundreds or thousands of generations, we ended up with sheep-herding border collies, badger-hunting dachshunds, wily retrievers, and sled-pulling huskies. And at some point we began to delight in the odd, extraordinary dog. Imagine! In practically no time at all, evolutionarily speaking, we picked our way through the pool of variant dog alleles and came up with such extremes as Great Danes and chihuahuas (Figure 18.1).

Sometimes our canine designs exceeded the limits of biological common sense. For example, how long would a tiny, finicky-eating, nearly hairless, nearly defenseless chihuahua last in the wild? Not long. What about the English bulldog, bred for a very short snout and a compressed face? Long ago, breeders thought these particular traits would allow the dogs to get a better grip on the nose of a bull. (Why they wanted dogs to bite bulls is a story in itself.) So now the roof of the bulldog mouth is ridiculously wide and it is often flabby, so bulldogs have trouble breathing. They sometimes get so short of air, they pass out.

Through our centuries-old fascination with artificial selection, we produced thousands of varieties of crop plants, ornamental plants, cats, cattle, and birds as well as dogs. With the currently available technologies of genetic engineering, we are now mixing the genes of many different species and producing incredible new varieties, including tobacco plants that produce useful quantities of hemoglobin and mustard plants that produce plastic.

So, when you hear someone wonder about whether "evolution" takes place, remind yourself that **evolution** simply means genetic change in a line of descent over the generations. Selective breeding practices provide abundant, tangible evidence that heritable changes do, indeed, occur. The actual mechanisms that bring about change are the focus of this chapter. Later chapters will explain their role in the evolution of new species from parent species.

Figure 18.1 Two designer dogs. About 50,000 years ago, humans began domesticating wild dogs. From that ancestral stock, artificial selection resulted in startling diversity among rather closely related breeds, such as the Great Dane (*legs, left*) and the chihuahua (*possibly fearful of being stepped on, right*).

KEY CONCEPTS

1. All individuals of a population generally have the same number and kinds of genes. Those genes give rise to an array of traits that characterize the population.

2. Mutations may result in two or more slightly different molecular forms of a gene, called alleles, which influence a trait in different ways. Individuals of the population may or may not inherit the same combinations of alleles, so they may show variation. That is, they may not be exactly alike in the details of shared traits.

3. Any allele at a given gene locus may become more or less common in the population, relative to the other kinds, or it may disappear. *Microevolution* means that the allele frequencies of a population have changed over time.

4. Allele frequencies change as an outcome of mutation, gene flow, genetic drift, and natural selection. Mutation alone produces new alleles. Gene flow, genetic drift, and natural selection shuffle existing alleles into, through, or out of populations.

5. Natural selection is not a purposeful search for the "best" individuals in a given population. It simply is the outcome of differences in shared traits that in some way influence which individuals of each generation reproduce.

6. Natural selection contributes to the ongoing adaptation of a species to its environment.

18.1 INDIVIDUALS DON'T EVOLVE—POPULATIONS DO

Examples of Variation in Populations

As Charles Darwin once perceived, individuals don't evolve; populations do. By definition, a **population** is a group of individuals of the same species occupying a given area. To understand how a population evolves, start with the variation in traits among its individuals.

Certain features characterize a population. All of its members have the same body plan, as when jays have wings, feathers, feet with three toes forward and one toe back, and so on. These are *morphological* traits (*morpho-* means form). Cells and body parts of all individuals in the population operate much the same way during short-term metabolic tasks, growth, and reproduction. These are *physiological* traits, which relate to the body's functioning. Also, individuals respond the same way to basic stimuli, as when babies instinctively imitate adult facial expressions. These are *behavioral* traits.

Especially in sexually reproducing species, the details of most traits vary among individuals. Pigeon feathers and snail shells vary in patterning or coloration within a population (Section 1.4 and Figure 18.2). Some individuals of a frog population might be more sensitive to winter cold or better at attracting a mate than others. Humans differ in the distribution, color, texture, and amount of hair. These examples only hint at the stunning variation within populations; almost every trait of every species is variable.

Many traits, such as those of the pea plants studied by Gregor Mendel, vary in *qualitatively different* ways. They come in two or more distinct forms (morphs) in a population, as when feathers are yellow or white. Such qualitative variation is known as **polymorphism**. Other traits, such as eye color and height, show continuous, *quantitatively different* variation. That is, individuals of a population show small, incremental differences in traits that may be quantified, as described in Section 11.7.

The "Gene Pool"

Information about heritable traits occurs in genes, which are specific regions of DNA molecules. In general, all of the individuals of a population have the same number and kinds of genes. We say "in general" because the males and females of sexually reproducing populations differ in some genes on the sex chromosomes.

Think of all the genes in the entire population as a **gene pool**—a pool of genetic resources that, in theory at least, is shared by all members of a population and is passed on to the next generation. Each kind of gene in the pool usually is present in two or more slightly different molecular forms, called **alleles**.

Figure 18.2 *Facing page*: From islands of the Caribbean, variation in shell color and banding patterns in populations of the same snail species. Different individuals carry different alleles for the genes that specify most traits. The shells differ because their owners had different combinations of alleles at particular gene locations along their chromosomes.

Individuals inherit different combinations of alleles. This leads to variations in phenotype—to differences in the details of traits. For example, whether your hair is black, brown, red, or blond depends on which alleles of certain genes you inherited from your two parents. And don't forget: *Offspring inherit genes, not phenotypes*. Environmental conditions often modify expression of a gene, but the phenotypic variations resulting from their effects last no longer than the individual (Section 11.8).

Which alleles end up in a given gamete and, later, in a new individual? The outcome depends on five events, as described in earlier chapters and summarized here:

1. Gene mutation (produces new alleles)
2. Crossing over at meiosis I (puts novel combinations of alleles in chromosomes)
3. Independent assortment at meiosis I (puts mixes of maternal and paternal chromosomes in gametes)
4. Fertilization (combines alleles from two parents)
5. Change in chromosome number or structure (leads to the loss, duplication, or repositioning of genes)

Of the five events just listed, only mutation *creates* new alleles. The other four only shuffle *existing* alleles into different combinations. But what a shuffle! Consider this: Each human gamete will end up with one of 10^{600} possible combinations of alleles. Not even 10^{10} humans are alive today. So unless you are an identical twin, it is extremely unlikely that another person with your exact genetic makeup has ever lived, or ever will.

Stability and Change in Allele Frequencies

Imagine yourself in a large garden in summer, watching butterflies flitting about. They all look the same except in wing color. A few wings are white. Most are blue. Most likely, you muse, the "blue" allele must be more common than the "white" allele. With genetic analysis, you could identify **allele frequencies**, the abundance of each kind of allele in the population. You could track the rate of genetic change over the generations.

Suppose you start with the "Hardy–Weinberg rule," as given in Section 18.2, to set up a theoretical reference point for measuring patterns of change. At that point, called **genetic equilibrium**, the frequencies of alleles at a given gene locus remain stable one generation after

Further reading: Student Guide to InfoTrac on web site —

you know, gene interactions underlie the phenotypes of complex organisms. A mutation that has severe effects on phenotype usually will result in death; it is a **lethal mutation**.

By comparison, a **neutral mutation** does not help *or* harm an individual. Natural selection can neither increase nor decrease the frequency of neutral mutations in a population, for these do not have any discernible effect on the individual's chances of surviving and reproducing. For example, if you carry a mutant gene that resulted in attached earlobes instead of detached ones, this alone should not stop you from surviving and reproducing just as well as anybody else.

the next. The population is *not* evolving with respect to that gene, for five conditions are being met. First, there have been no gene mutations. Second, the population is very large. Third, it is isolated from other populations of the species. Fourth, the gene has no effect at all on survival or reproduction. Finally, all mating is random.

Rarely, if ever, do all five conditions prevail at the same time in nature. Gene mutation is infrequent but inevitable. Three occurrences—*natural selection, gene flow,* and *genetic drift*—may drive the population away from genetic equilibrium, even in a few generations. The term **microevolution** refers to small-scale changes in allele frequencies, as brought about by mutation, natural selection, gene flow, and genetic drift.

Mutations Revisited

Mutations, remember, are heritable changes in DNA that typically give rise to altered gene products. They are the only source of new alleles. We cannot predict exactly when or in which individual they will appear. Yet each gene has its own **mutation rate**, which is the probability of its mutating during or in between DNA replications. On average, the rate is between 10^{-5} and 10^{-6} per gene locus per gamete, each generation. In a single reproductive season, about 1 gamete in 100,000 to 1,000,000 has a new mutation at any given locus.

Mutations often give rise to structural, functional, or behavioral modifications that decrease the individual's chances of surviving and reproducing. Even a single biochemical change can have devastating effects.

For example, skin, bones, tendons, and many other vertebrate organs cannot develop without collagen, a structural protein. If the gene specifying the molecular form of collagen mutates, drastic changes in the lungs, arteries, skeleton, and other body parts may follow. As

Finally, every so often, a gene mutation bestows an advantage. For instance, a product of a mutant gene that influences growth might make a corn plant grow larger or faster and so give it the best access to sunlight and nutrients. Or maybe a neutral mutation turns out to be advantageous after conditions in the environment change. Even if the advantage is small, chance events or natural selection might preserve that mutant gene and favor its representation in the next generation.

Mutations are so rare, they usually have little or no immediate effect on allele frequencies of a population. But beneficial mutations, and neutral ones, have been accumulating in different lineages for billions of years. Through all that time, they have been the raw material for evolutionary change—the basis for the staggering range of biological diversity, past and present.

In the evolutionary view, then, the reason you don't look like a bacterium or a grass plant or an earthworm or even neighbors down the street began with different mutations that originated at different times in the past, in different lines of descent.

Certain morphological, physiological, and behavioral traits characterize a population. The traits differ in their details from one individual to the next.

Differences in the combinations of alleles that individuals carry give rise to variations in phenotype. By phenotypic variation, we mean differences in the details of structural, functional, and behavioral traits that the individuals of a population have in common.

For sexually reproducing species, a population's individuals represent a pool of genetic resources—that is, a gene pool.

Mutation alone creates *new* alleles. Natural selection, gene flow, and genetic drift change the *frequencies* of alleles in the gene pool. The evolutionary story starts with these changes.

Focus on Science

18.2 WHEN IS A POPULATION *NOT* EVOLVING?

Look again at Figure 11.1, which shows the earlobes of Tom Cruise, Gregor Mendel, and other celebrated people. Now ask this: Is the number of individuals with *detached* earlobes staying the same, one generation after the next, in the whole human population? What about the number of individuals who have *attached* earlobes? Stated more broadly, how do we know whether or not a population is evolving with respect to earlobes or any other trait?

Almost a hundred years ago, a mathematician and a doctor came up with the answer. Working independently, they thought about what it would take to maintain the frequencies of alleles of an idealized population. They came up with what is now called the **Hardy–Weinberg rule**, in their honor.

They started out with this formula: In a population at genetic equilibrium, the proportions of genotypes at one gene locus for which there are two kinds of alleles are

$$p^2\ AA + 2pq\ Aa + q^2\ aa = 1$$

where p is the frequency of allele A, and q is the frequency of allele a. Their rule is this: *Allele frequencies will stay the same through successive generations if there is no mutation, if the population is infinitely large and is isolated from other populations of the same species, if mating is random regarding the alleles, and if all individuals survive and reproduce equally.* Thus, the Hardy–Weinberg rule presents a hypothesis for determining whether any of the assumptions just listed currently apply to the population.

To test whether the allele frequencies of a population will remain constant from one generation to the next in the absence of evolutionary forces, you decide to track a pair of alleles through a population of butterflies. These sexually reproducing organisms have pairs of genes, on pairs of homologous chromosomes. You start by assuming the population is currently at genetic equilibrium. The pair of alleles you are interested in govern wing color. Allele A specifies dark-blue wings. Allele a is associated with white wings. The heterozygous (Aa) condition results in medium-blue wings.

In the population as a whole, the frequencies of A and a must add up to 1. For example, if A occupies half of all the loci for this gene in the population, then a must occupy the other half (0.5 + 0.5 = 1). If A occupies 90 percent of all the loci, then a must occupy 10 percent (0.9 + 0.1 = 1). No matter what the proportions of the two kinds of alleles,

$$p + q = 1$$

During meiosis in germ cells, each allele segregates from its partner, and the two end up in separate gametes. Therefore, p is also the proportion of gametes carrying the A allele, and q is the proportion with the a allele. To find the expected frequencies of the three genotypes (AA, Aa, and aa) that are possible in the next generation, you construct a Punnett square:

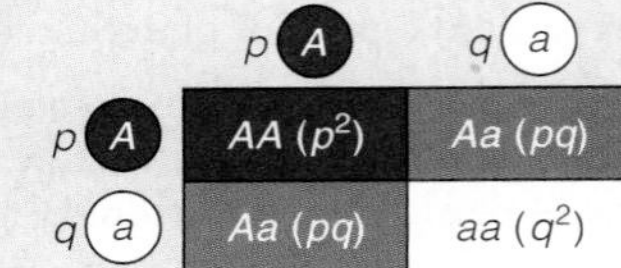
p A
q a
p A
q a
AA (p²)
Aa (pq)
Aa (pq)
aa (q²)

The frequencies of the genotypes add up to 1:

$$p^2 + 2pq + q^2 = 1$$

To see whether the allele frequencies and genotypic frequencies will remain the same through the generations, you work through an example. You find the population has 1,000 butterflies, each of which produces two gametes:

490 AA individuals produce 980 A gametes
420 Aa individuals produce 420 A and 420 a gametes
90 aa individuals produce 180 a gametes

You notice the frequency of A among the 2,000 gametes is

$$(980 + 420)/2{,}000 = 0.7$$

Also,

$$q = (420 + 180)/2{,}000 = 0.3$$

At fertilization, the gametes combine at random and give rise to the next generation, as given in the Punnett square. Assuming the population remains constant at 1,000 individuals, you now have

$p^2\ AA = 0.7 \times 0.7 = 0.49$ — 490 AA individuals
$2pq\ Aa = 2 \times 0.7 \times 0.3 = 0.42$ *or* 420 Aa individuals
$q^2\ aa = 0.3 \times 0.3 = 0.09$ — 90 aa individuals

and

$$p^2 + 2pq + q^2 = 0.49 + 0.42 + 0.09 = 1$$

The allele frequencies have not changed:

$$A = \frac{2 \times 490 + 420}{2{,}000 \text{ alleles}} = \frac{1{,}400}{2{,}000} = 0.7 = p$$

$$a = \frac{2 \times 90 + 420}{2{,}000 \text{ alleles}} = \frac{600}{2{,}000} = 0.3 = q$$

You notice the genotype frequencies have not changed, either. And as long as the five assumptions of the Hardy–Weinberg rule are being met, frequencies should stay the same through the generations. To test this, you calculate allele frequencies in the gametes of the *next* generation:

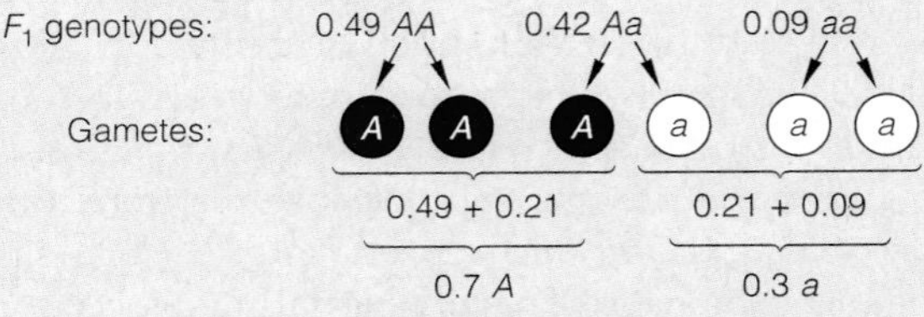
F₁ genotypes:
0.49 AA
0.42 Aa
0.09 aa
Gametes:
A A A a a a
0.49 + 0.21
0.21 + 0.09
0.7 A
0.3 a

which is back where you started from. Because the allele frequencies for dark-blue, medium-blue, and white wings are the same as they were in the original gametes, they will yield the same phenotypic frequencies as you saw in the second generation.

Further reading: Student Guide to InfoTrac on web site

You could do similar calculations for additional wing colors. You could go on with your calculations until you ran out of paper (or patience). However, as long as the five assumptions continue to hold, the allele frequencies and the range of values for the wing-color trait will not change, as you can see from Figure 18.3.

Therefore, when genotypes and phenotypes do *not* show up in the proportions that you predicted on the basis of the Hardy–Weinberg rule, this tells you that one or more conditions of the rule are being violated. And the hunt can begin for the specific evolutionary force, or forces, driving the change.

Starting generation:
490 *AA* butterflies (dark-blue wings)
420 *Aa* butterflies (medium-blue wings)
90 *aa* butterflies (white wings)
The next generation:
490 *AA* butterflies
420 *Aa* butterflies
90 *aa* butterflies
(NO CHANGE)
The next generation:
490 *AA* butterflies
420 *Aa* butterflies
90 *aa* butterflies
(NO CHANGE)

Figure 18.3 A hypothetical population of butterflies at genetic equilibrium.

18.3 NATURAL SELECTION REVISITED

We now turn from our idealized population that never changes to the real-world processes of change. Of these processes, natural selection might account for most of the morphological and physiological changes that have accrued throughout the history of life.

Darwin, recall, was able to explain natural selection after correlating his understanding of inheritance with certain features of populations and their environments. Before we consider the modes of natural selection, let's restate his correlations in modern terms:

1. *Observation*: All populations in nature have the reproductive capacity to increase in numbers over the generations.

2. *Observation*: No population is able to increase indefinitely, for its individuals will run out of food, living space, and other resources that sustain it.

3. *Inference*: Sooner or later, the individuals of a population will end up competing for resources.

4. *Observation*: All of the individuals have the same genes, which specify the same assortment of traits. Collectively, their genes represent a pool of heritable information.

5. *Observation*: Most, if not all, kinds of genes occur in slightly different molecular forms (alleles), which give rise to differences in phenotypic details.

6. *Inferences*: Some phenotypes are better than others at helping the individual compete for resources, and therefore to survive and reproduce. Thus, the alleles for those phenotypes increase in the population, and other alleles decrease. Over time, the genetic change leads to increased **fitness**—that is, to an increase in adaptation to the environment.

7. *Conclusions*: **Natural selection** simply is the result of differences in survival and reproduction among individuals that differ in heritable traits. Adaptation is one outcome of this microevolutionary process.

Evolutionary biologists have been documenting the results of natural selection in thousands of field studies of populations of all kinds of organisms. They find that this microevolutionary process has different results. As you will see in sections to follow, sometimes the result is a shift in the range of values for a given trait in some direction. At other times, the result may be stabilization or disruption of an existing range of values.

As Darwin perceived, natural selection is the outcome of variations in shared traits that influence which individuals of a population survive and reproduce in each generation. Natural selection can lead to increased fitness—that is, to an increase in adaptation to the environment.

18.4 DIRECTIONAL CHANGE IN THE RANGE OF VARIATION

What Is Directional Selection?

In cases of **directional selection**, allele frequencies that underlie some range of variation in phenotypes shift in a consistent direction. Such shifts occur in response to a directional change in the environment or to one or more new environmental conditions. They also occur when a mutation appears and proves to be adaptive. As Figure 18.4 shows, the forms at one end of the range for some trait become more common than midrange forms. Let's look at a few documented cases of this outcome.

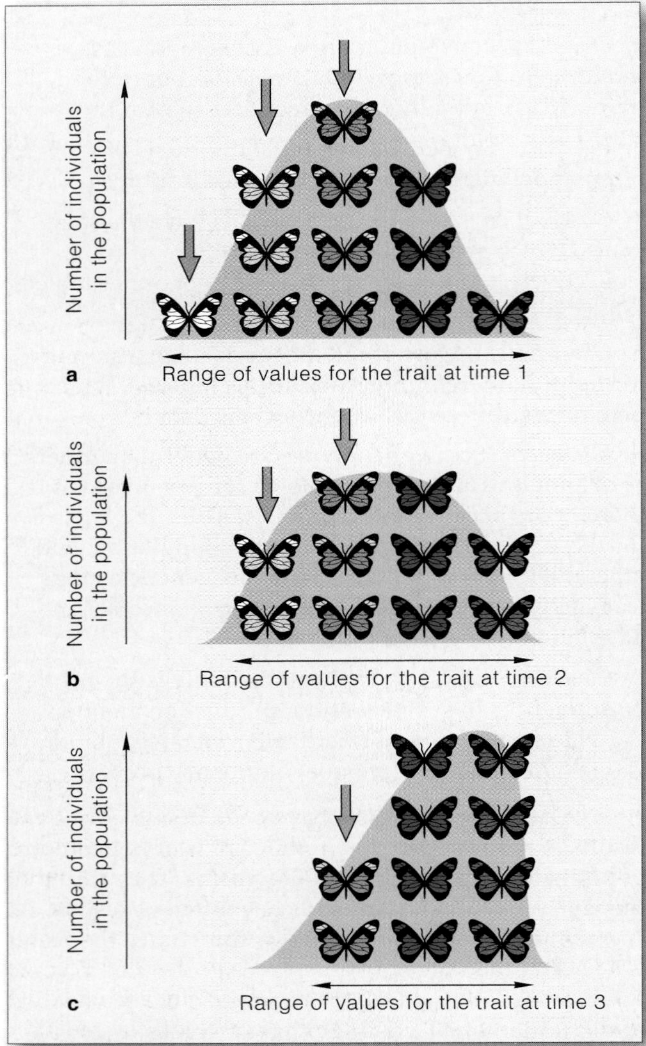
Number of individuals in the population
a
Range of values for the trait at time 1
Number of individuals in the population
b
Range of values for the trait at time 2
Number of individuals in the population
c
Range of values for the trait at time 3

Figure 18.4 Directional selection, using phenotypic variation within a population of butterflies as the example. A bell-shaped curve (*brown*) represents the range of continuous variation in wing color. The most common forms (*medium blue*) are between extreme forms of the trait (*white* at one end of the curve, *deep purple* at the other). *Orange* arrows signify which forms are being selected against over time.

The Case of the Peppered Moths

In England, biologists tracked directional selection in about a hundred moth species, including the peppered moth (*Biston betularia*). Peppered moths feed and mate at night. During the day, they rest motionless on birches and other trees. Their wings and body have a mottled pattern, in shades that range from light gray to nearly black. Apparently their behavior, coloration, and wing patterns help camouflage them from moth-eating birds, which actively hunt for food during the day.

The English industrial revolution began during the 1850s. Outpourings of sooty smoke altered conditions in many parts of the surrounding countryside. Before then, the light moths were the most common form, and a dark form was rare. Also before conditions changed, light-gray speckled lichens grew in profusion on tree trunks. Lichens can camouflage light moths that rest on them, but not dark ones (Figure 18.5*a*).

Lichens are sensitive to air pollution. Between 1848 and 1898, the soot and other pollutants from factories started to kill the lichens and darken tree trunks. Now the rare form of the moth was better camouflaged, as Figure 18.5*b* indicates. The moth collectors hypothesized that conditions had previously favored light moths—but that the changed conditions would favor dark ones.

In the 1950s, H. B. Kettlewell used a *mark–release–recapture* method to test the prediction. He bred both forms of the moth in captivity, then marked hundreds so that they could be identified. He released moths near heavily industrialized areas around Birmingham and in Dorset, an unpolluted area. After a time, he recaptured as many moths as he could. As Table 18.1 shows, more dark moths were recaptured in the polluted area and more light ones in the pollution-free area. By stationing observers in blinds near groups of moths that had been tethered to trees, Kettlewell and his coworkers gathered direct evidence of birds capturing more of the light moths around Birmingham and more of the dark ones around Dorset. Directional selection was operating.

In 1952, strict pollution controls went into effect. Lichens made a comeback. Tree trunks became free of soot, for the most part. As you might have predicted, natural selection had phenotypes shifting in the reverse direction. Where levels of pollution have declined, the frequency of dark moths has been declining, also.

Pesticide Resistance

Widespread use of chemical pesticides in agriculture has resulted in directional selection. Initial applications kill most of the insects, worms, or other pests, but some individuals usually manage to survive. Some aspect of their structure, physiology, or behavior allows them to

Further reading: Student Guide to InfoTrac on web site

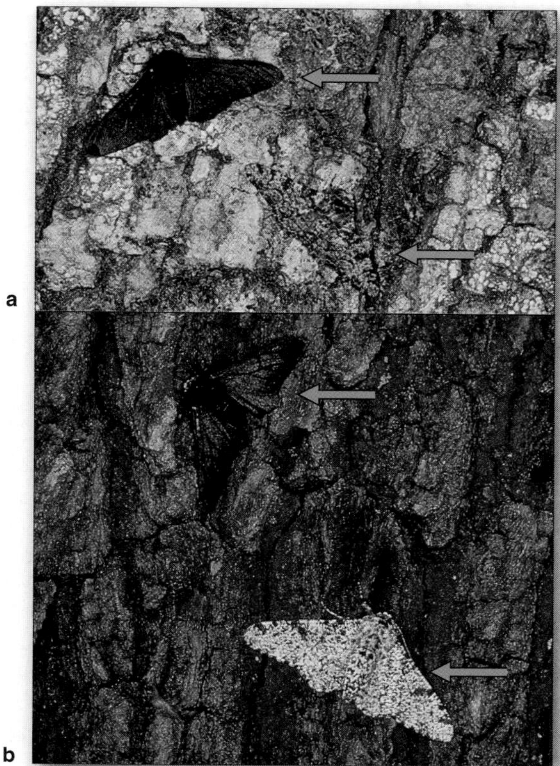
a
b

Figure 18.5 Individuals of a population of peppered moths (*Biston betularia*) that has undergone directional selection in response to changes in the environment. Light-winged and dark-winged individuals are resting on a lichen-covered tree trunk in (**a**) and on a soot-darkened tree trunk in (**b**).

Table 18.1 Marked *Biston betularia* Recaptured in Polluted Versus Nonpolluted Areas

	Near Birmingham (pollution high)	Near Dorset (pollution low)
LIGHT-GRAY MOTHS:		
Released	64	393
Recaptured	16 (25%)	54 (13.7%)
DARK-GRAY MOTHS:		
Released	154	406
Recaptured	82 (53%)	19 (4.7%)

Data after H. B. Kettlewell.

resist the chemical effects. When their resistance has a heritable basis, it becomes more common in the next generation, and the next, and the next. The chemicals are agents of selection, which favor the most resistant forms! Today, 450 different species are resistant to one or more pesticides. Worse, the pesticides also kill the natural predators of the pests. When freed from natural constraints, the populations of resistant pests burgeon, and crop damage is greater than ever. This outcome of directional selection is called *pest resurgence*.

Maybe pesticide use will lessen in fields of plants that are genetically engineered for pesticide resistance. These plants will not escape the coevolutionary arms race, but they might help keep our food supplies one step ahead of the pests. However, many consumers are leery of genetically engineered food.

What about using *biological controls*? By this practice, natural enemies of pests, including parasitic wasps and predatory beetles, are raised in commercial insectaries in great numbers and then released at preselected sites. The practice has an advantage, in that a control species can coevolve with the pests. But farmers must replace the ones that migrate from the fields or are destroyed at harvest time, and they pass on the cost to consumers.

Antibiotic Resistance

When your grandparents were young, up to one-fourth of the annual deaths in the United States alone were caused by bacterial agents of tuberculosis, pneumonia, and scarlet fever. In the 1940s, we started treating such bacterial-induced diseases with antibiotics.

Remember, certain microorganisms that live in soil produce diverse antibiotics, which can destroy bacterial competitors for nutrients (Section 1.4). Streptomycins, for example, block protein synthesis in target cells. The penicillins disrupt the formation of covalent bonds that hold bacterial cell walls together. Penicillin derivatives cause the wall to weaken until it ruptures.

Antibiotics should be prescribed with restraint and care. Why? Besides performing their intended function, they commonly disrupt the balances among bacterial populations that normally compete for resources in the mammalian intestines and of yeast cells in the vaginal canal. Such disruptions lead to secondary infections.

Worse yet, antibiotics have been overprescribed in the human population. Too frequently they have been used for simple infections that many individuals could have overcome successfully on their own. Disturbingly, antibiotics have lost their punch. Over time, they did destroy the most susceptible cells of target populations. But they also favored their replacement by much more resistant cells. Millions of people around the world are now dying each year of tuberculosis, cholera, and other bacterial infections. Even vancomycin, held in reserve as "the antibiotic of last resort," is no longer effective against certain pathogenic strains of enteric bacteria. In 1996 the World Health Organization announced that, in the race for supremacy, pathogens are sprinting ahead.

With directional selection, allele frequencies underlying a range of variation tend to shift in a consistent direction in response to directional change in the environment.

18.5 SELECTION AGAINST OR IN FAVOR OF EXTREME PHENOTYPES

As you have seen, natural selection can bring about a directional shift in a population's range of phenotypic variation. Depending on the prevailing environmental conditions, the process also may favor either the most common or the most extreme phenotypes in that range.

Stabilizing Selection

In **stabilizing selection**, intermediate forms of a trait are favored and alleles that specify extreme forms are eliminated from a population (Figure 18.6). This mode of selection tends to counteract the effects of mutation, gene flow, and genetic drift and to preserve the most common phenotypes.

Consider the selection pressures on the gallmaking fly *Eurosta solidaginis*. After a fly larva develops from an egg, it bores into a stem of a tall goldenrod (*Solidago altissima*). There, it feeds upon plant tissues as it grows into an adult. In response, plant cells multiply rapidly and enclose the invader in a gall, a type of tumorous mass. Genetic analysis reveals that flies with different phenotypes induce formation of galls of different sizes.

A wasp that parasitizes the larvae can breach small galls, not the large ones (Figure 18.7). Thus it promotes selection in favor of flies that induce formation of large galls. However, the downy woodpecker and some other insect-eating birds can reach larvae in the large galls. By their action, both kinds of natural enemies influence allele frequencies in the fly population.

Researchers conducted studies in Pennsylvania. As they discovered, flies that caused *intermediate-size* galls to form had the highest survival rate and fitness. The offspring of flies associated with *small* galls and *large* galls were more vulnerable to wasp and bird attacks.

In sum, parasitic wasps are operating against one phenotypic extreme in the fly population, and predatory birds are operating against the other. The intermediate phenotypes are being favored.

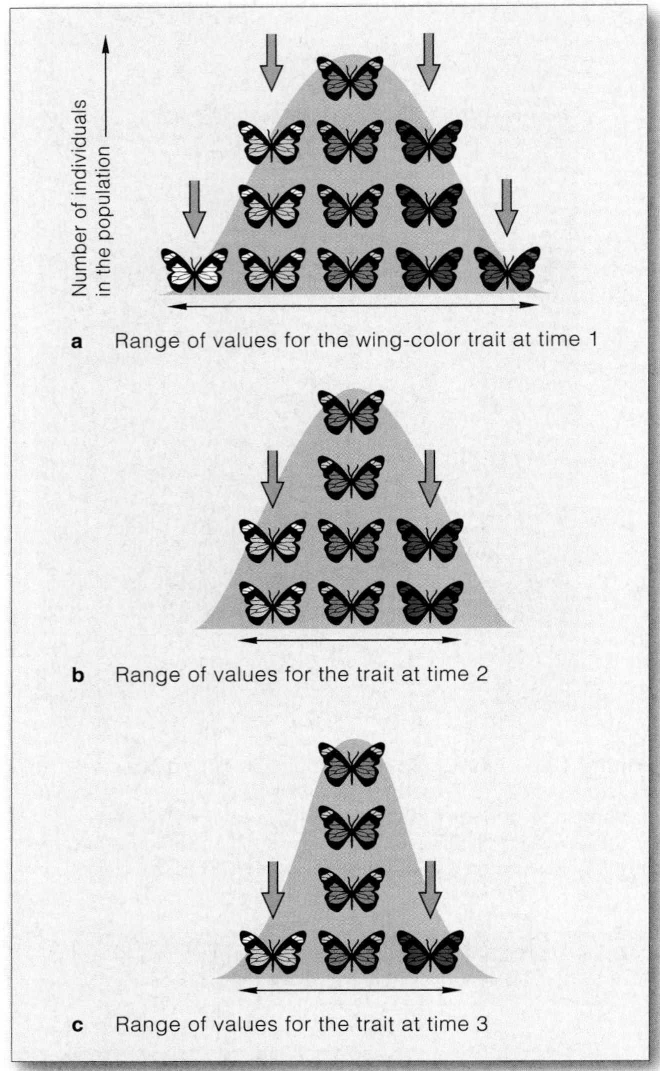

Number of individuals in the population
a Range of values for the wing-color trait at time 1
b Range of values for the trait at time 2
c Range of values for the trait at time 3

Figure 18.6 Stabilizing selection, using phenotypic variation within a population of butterflies as the example.

gall
a

b

c
d

Figure 18.7 Example of stabilizing selection. (**a**,**b**) Larvae of the fly *Eurosta solidaginis* induce formation of tumors (galls) on goldenrod stems. (**c**) Downy woodpeckers (*Dendrocopus pubescens*) and other birds prefer to chisel into large-size galls and eat the larvae. (**d**) The egg-laying device of a parasitic wasp (*Eurytoma gigantea*) can penetrate only the thin wall of small galls. Its eggs develop into larvae, then *its* larvae eat fly larvae.

Warren Abrahamson and his coworkers monitored twenty populations of *Eurosta* in Pennsylvania. They found that the larvae in small and large galls have low relative fitnesses, and larvae in intermediate-size galls have relatively high fitnesses. Thus, there is a stabilizing component to selection pressures created by the fly's natural enemies.

Further reading: Student Guide to InfoTrac on web site

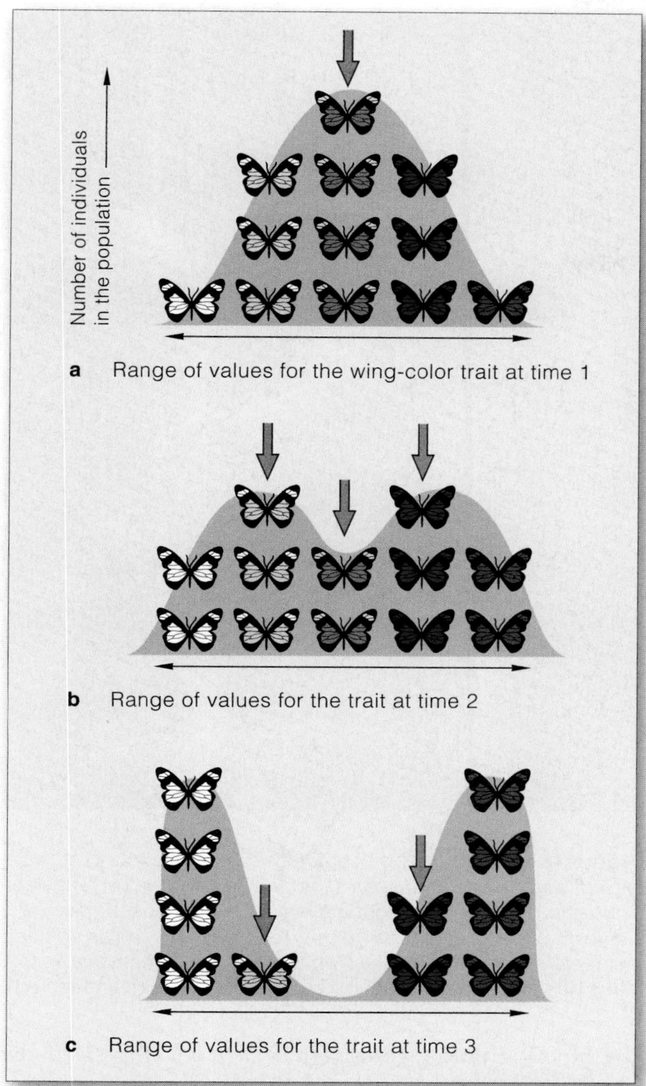

Number of individuals in the population
a Range of values for the wing-color trait at time 1
b Range of values for the trait at time 2
c Range of values for the trait at time 3

Figure 18.8 Disruptive selection, using phenotypic variation within a population of butterflies as the example.

Disruptive Selection

In **disruptive selection**, forms at both ends of the range of variation are favored and the intermediate forms are selected against (Figure 18.8). Thomas Smith found a splendid example of this in a remote rain forest in Cameroon, West Africa. Smith had read about unusual variation in bill size in populations of the black-bellied seedcracker (*Pyrenestes ostrinus*). These African finches have large or small bills—but no sizes in between. The pattern holds for females and males, throughout their geographic range. (This is simply remarkable; imagine every person in Texas being four feet *or* six feet tall, with no intermediates.) If the bill pattern is unrelated to gender or geography, what causes it?

a
b

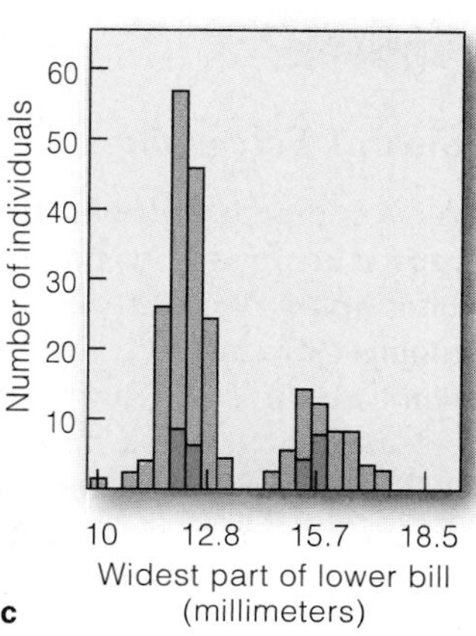

Number of individuals
60
50
40
30
20
10
10
12.8
15.7
18.5
Widest part of lower bill (millimeters)
c

Figure 18.9 Disruptive selection among African finches. Feeding trials showed that birds with large bills are better users of hard seeds. Small-billed birds are better at using soft seeds, not hard ones. (**a**,**b**) Two individuals displaying small and large bill sizes.

(**c**) Graph of survival of juvenile birds during the dry season, when competition for resources is most intense. Individuals with *very* small, *very* large, or intermediate-size bills cannot feed efficiently on either type of seed; they survive poorly. For this graph, the *tan* bars show the number of nestlings; *orange* bars show the only survivors among them. The findings are based on measurements of 2,700 netted individuals.

If, as Smith hypothesized, the persistence of only two sizes of bills relates to seed-cracking ability (which directly affects survival), then disruptive selection may be eliminating birds with intermediate-size bills. But what factors cause the disruptive selection pressure on feeding performance? Cameroon's swamp forests are flooded during the wet season, and lightning-sparked fires burn during the dry season. Two species of sedge (fire-resistant, grasslike plants) dominate these forests. One has hard seeds; the other has soft. When finches reproduce, both hard and soft seeds are abundant.

All birds prefer soft seeds for as long as they can get them. Birds with small bills are better at using soft seeds; large-billed birds are better at using hard ones. When the dry season peaks, soft seeds and other food supplies dwindle. Small-billed birds, and the youngest ones, are at a competitive disadvantage. Many do not survive (Figure 18.9).

Smith also performed experimental crosses between large- and small-billed birds. All offspring had large *or* small bills. Along with other data, the crosses suggest two alleles at a single autosomal gene locus control bill size and feeding performance.

With stabilizing selection, intermediate phenotypes are favored and extreme phenotypes at both ends of the range of variation are eliminated.

With disruptive selection, intermediate forms are selected against; extreme forms in the range of variation are favored.

18.6 SPECIAL TYPES OF SELECTION

Sexual Selection

As you may have noticed, individuals of most sexually reproducing species have a distinctively male or female phenotype. We call this sexual dimorphism (*dimorphos* means "having two forms"). How does this condition come about, and what maintains it? Here again, natural selection is at work. In this case, it is **sexual selection**. The traits being favored are advantageous, with respect to survival and reproduction, simply because males or females prefer them. Through nonrandom mating, the alleles for preferred traits prevail over the generations.

Sexual dimorphism is particularly striking among many mammals and birds, including the pair in Figure 18.10. Often males are larger, have flashier coloration and patterning, and are much more aggressive than the females. Remember those male bighorn sheep butting heads (Section 1.3)? Such fights waste time and energy, and they might cause serious injuries. Why, then, do alleles that contribute to aggressive behavior persist in a population? The increased chance of mating offsets the costs. Male bighorn sheep fight only to control areas where receptive females gather during a winter rutting season. Winners mate often, with a number of females. Losers will not challenge a stronger, larger male.

Females are the main agents of selection. They exert direct control over reproductive success by choosing their mates. We return to this topic in Chapter 47.

Figure 18.10 One outcome of sexual selection. This male bird of paradise (*Paradisaea raggiana*) is engaged in a flashy courtship display. He caught the eye (and, perhaps, the sexual interest) of the smaller, less colorful female. Males of this species compete fiercely for females, which serve as selective agents. (Why do you suppose drab-colored females have been favored?)

Maintaining Two or More Alleles

Balancing selection includes all forms of selection that maintain two or more alleles for a trait in a population. When this type of genetic variation persists over time, we call it **balanced polymorphism** (after *polymorphos*, "having many forms"). A population is in a state of balanced polymorphism when nonidentical alleles for a trait are being maintained at frequencies greater than 1 percent. Frequencies may shift slightly, but over time they often will bounce back to the same values. Smith's work with African finches is only one example of the effect of balancing selection through the generations.

Sickle-Cell Anemia—Lesser of Two Evils?

We sometimes find cases of balanced polymorphism where environmental conditions favor heterozygotes (which carry nonidentical alleles for a specified trait), not the homozygotes (which carry identical alleles for the trait). Said another way, *the heterozygote has a higher fitness than either homozygote.*

Let's look at the environmental pressures that favor an Hb^A/Hb^S pairing in humans. Hb^S specifies a mutant form of hemoglobin, an oxygen-transporting protein in the blood. Homozygotes (Hb^S/Hb^S) develop *sickle-cell anemia*, a genetic disorder that has serious pleiotropic effects on phenotype (Section 11.5). The frequency of Hb^S is high in tropical and subtropical regions of Asia and Africa. Often, Hb^S/Hb^S homozygotes die in their early teens or early twenties. However, in those same regions, heterozygotes (Hb^A/Hb^S) make up nearly one-third of the human population! Why is this combination of alleles maintained at such high frequency?

The balancing act, an outcome of natural selection, is most pronounced in areas with the highest incidence of *malaria* (Figure 18.11 and Section 23.7). A mosquito transmits *Plasmodium*, the parasite agent of the disease, to humans. The parasite multiplies in the liver and then in red blood cells. These cells rupture and release new parasites during severe, recurring bouts of infection.

Who is more likely to survive recurring infections? Hb^A/Hb^S heterozygotes. That allelic combination gives them two forms of hemoglobin, with interesting results. They have enough nonmutated molecules to maintain body functions. And the *altered* molecules distort the

Further reading: Student Guide to InfoTrac on web site

a

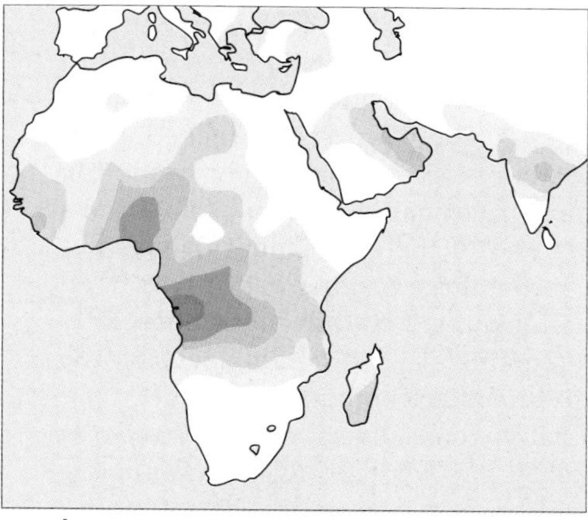

b

less than 1 in 1,600
1 in 400–1,600
1 in 180–400
1 in 100–180
1 in 64–100
more than 1 in 64

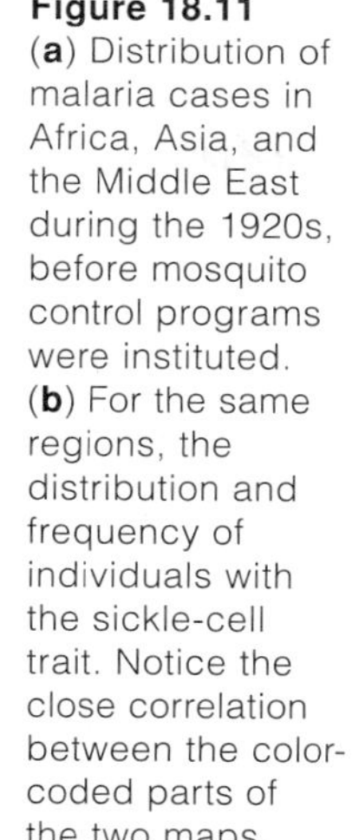

Figure 18.11 (**a**) Distribution of malaria cases in Africa, Asia, and the Middle East during the 1920s, before mosquito control programs were instituted. (**b**) For the same regions, the distribution and frequency of individuals with the sickle-cell trait. Notice the close correlation between the color-coded parts of the two maps.

shape of red blood cells in a way that interferes with circulation. The distorted cells are destroyed when they circulate through the spleen. But the slowdown in blood flow is a factor in survival because it interferes with the parasite's ability to rapidly infect new cells.

Thus the persistence of the "harmful" Hb^S allele is a matter of relative evils. Natural selection has favored one allelic combination, Hb^A/Hb^S, because its bearers show greatest fitness in environments where malaria is the most prevalent. In those places, Hb^A/Hb^S has a higher fitness than either Hb^S/Hb^S or Hb^A/Hb^A.

In tropical and subtropical habitats of Asia and the Middle East, malaria has been a selective force for more than 2,000 years. Although sickle-cell anemia occurs at high frequencies in these regions, its symptoms are much less severe than they are in Central Africa, where the Hb^S allele became established much later in time. Most likely, other gene products are modifying some of the widespread effects of the Hb^S allele in ways that minimize symptoms of the disorder.

With sexual selection, some version of a trait simply gives the individual an advantage in reproductive success. Sexual dimorphism is one outcome of sexual selection.

Balanced polymorphism is a state in which natural selection is maintaining two or more alleles over the generations at frequencies greater than 1 percent.

18.7 GENE FLOW

When individuals of the same species move about over time, they may move their alleles between populations. Alleles are lost from a population when individuals permanently leave it, an act called *emigration*. Alleles enter a population when new individuals move in, an act called *immigration*.

The physical flow of alleles, or **gene flow**, tends to counter genetic differences that arise through mutation, natural selection, and genetic drift. And it helps keep separated populations genetically similar.

Think of the acorns that blue jays disperse when they store nuts for the winter. Each fall the jays may make hundreds of round trips from acorn-bearing oak trees to bury acorns in the soil of their home territories, which may be up to a mile away (Figure 18.12). The alleles flowing in with "immigrant acorns" help reduce genetic differences that might otherwise arise among neighboring stands of oaks. Gene flow apparently is operating among peppered moths, also, for the form that doesn't match the prevailing background color is being maintained at higher than expected frequencies.

Figure 18.12 Gene flow among oak populations, courtesy of feathered travel agents. Blue jays hoard acorns in their home territory, but they might shop at nut-bearing trees up to a mile away. Some acorns contribute to the allele pool of an oak population some distance away from the parent tree.

Or think of the millions of people from economically bankrupt, politically explosive countries who seek more stable homes. The sheer scale of their movements is unprecedented, but hardly unique. Throughout human history, immigrations may have minimized many of the genetic differences that otherwise would have built up among geographically separated groups of people.

Gene flow is the physical movement of alleles into and out of a population, through immigration and emigration.

18.8 GENETIC DRIFT

Chance Events and Population Size

Genetic drift is a random change in allele frequencies over the generations, brought about by chance alone. The magnitude of its effect on genetic diversity and on the range of phenotypes relates to population size. Its impact tends to be minor or insignificant in very large populations but significant in small ones.

Sampling error, a rule of probability, helps explain the difference. By this rule, the fewer times a chance event occurs, the greater will be the variance from the expected outcome of that occurrence. You saw a simple demonstration of this in Section 1.5.

Also think back on the coin-flipping example given earlier, in Section 11.2. Each time you flip a coin, there is a 50 percent chance it will turn up heads or tails. With, say, only ten flips, the odds are great that the coin will turn up heads 7 times and tails 3 times. But with a thousand flips, the odds that it will turn up heads 700 times and tails 300 times are virtually nil. Similarly, sampling error applies every time random mating and fertilization take place in a population.

Bear in mind, genetic drift has nothing to do with how a population got small in the first place. *Genetic drift simply increases the chance of a given allele becoming more or less prevalent when the number of individuals in a population is small.*

Figure 18.13 is a computer simulation of the effect of genetic drift in two populations: one large, the other small. The outcomes parallel the findings of an actual experiment with beetles (*Tribolium*) that mate with one another at random. Researchers grouped 1,320 beetles into twelve populations of 10 beetles and twelve of 100. Which beetles ended up in which group was a matter of chance. At the outset, the frequency of a wild-type allele (call it *A*) was 0.5. The researchers tracked allele *A* for twenty generations. Every time, they randomly removed some offspring to maintain each population's original size. At the end of the experiment, *A* was not the only allele left in the large groups, but it was fixed in seven of the small groups. **Fixation** means that only one kind of allele remains at a particular locus in the population; all individuals are homozygous for it.

Thus, *in the absence of other forces, random change in allele frequencies leads to the homozygous condition and a loss of genetic diversity over the generations.* This happens in all populations; it just happens faster in small ones. Once alleles inherited from an original population are fixed, their frequencies will not change again unless mutation or gene flow introduce new alleles.

Bottlenecks and the Founder Effect

Genetic drift is pronounced when very few individuals rebuild a population or found a new one. A **bottleneck** is a severe reduction in population size brought about by intense selection pressure or some natural calamity. Suppose contagious disease, habitat loss, hunting, or a massive volcanic blast destroys much of a population. Even if a moderate number do survive the bottleneck, the allele frequencies will have been altered at random.

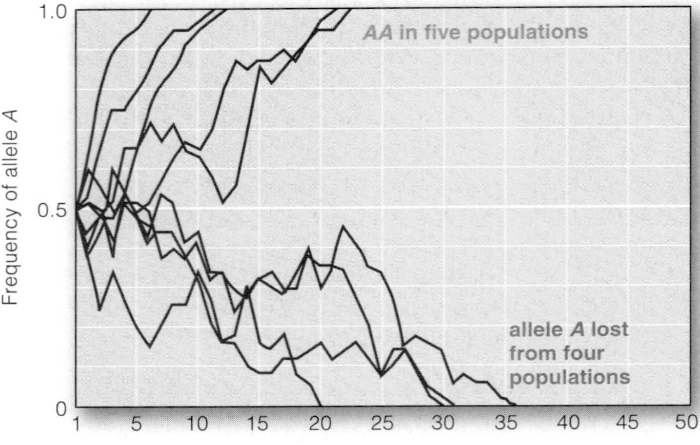

1.0
0.5
0
Frequency of allele A
AA in five populations
allele A lost from four populations
1
5
10
15
20
25
30
35
40
45
50
Generation (25 stoneflies at the start of each)
a

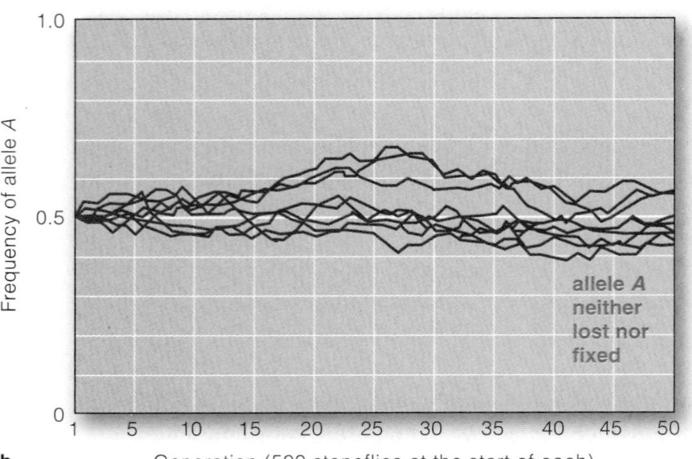

1.0
0.5
0
Frequency of allele A
allele A neither lost nor fixed
1
5
10
15
20
25
30
35
40
45
50
Generation (500 stoneflies at the start of each)
b

Figure 18.13 Computer simulation of the effect of genetic drift on one allele's frequency in small populations and in large populations. Equal fitness is assumed for three simulations (*AA* = 1, *Aa* = 1, and *aa* = 1).

(**a**) The size of nine populations of a species (stoneflies, in this case) was maintained at 25 breeding individuals in each generation, through fifty generations. The five graph lines reaching the top of the diagram tell you that allele *A* became fixed in five of the small populations. The four lines plummeting off the bottom of the diagram tell you it was lost from four of them. As you can see, *alleles can be fixed or lost even in the absence of selection.*

(**b**) The size of nine other populations was maintained at 500 individuals in each generation, through fifty generations. Allele *A* did not become fixed in any of these large populations. The magnitude of genetic drift was much less in every generation than in the small populations.

Further reading: Student Guide to InfoTrac on web site

Figure 18.14 Example of the founder effect. A seabird carries a few seeds, stuck to its feathers, from the mainland to a remote oceanic island. By chance, most of the seeds carry an allele for orange flowers that was not common in the original population. In the absence of further gene flow or selection for flower color, genetic drift will fix the allele in the island population.

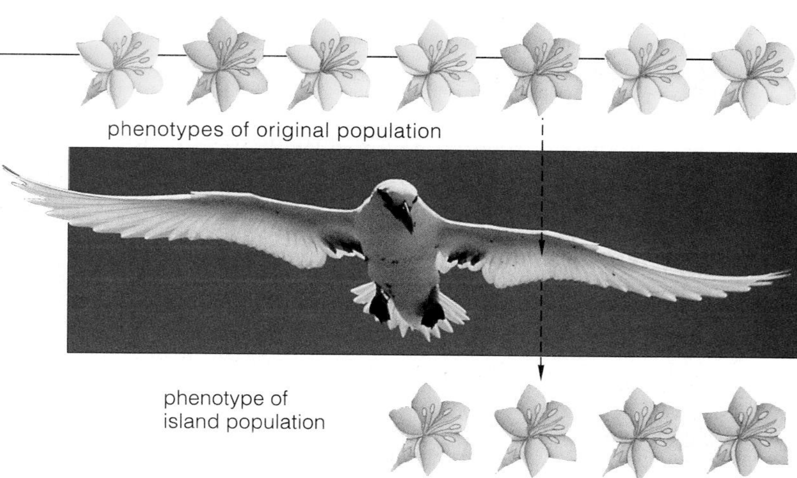
phenotypes of original population
phenotype of
island population

In the 1890s, hunters killed all but twenty of a large population of northern elephant seals. The population recovered. When its size passed 30,000, electrophoretic analysis of a sample of twenty-four genes revealed no variation in the population. A number of alleles had been lost in the bottleneck.

The genetic outcome can be similarly dicey after a few individuals leave a population and establish a new one elsewhere. This form of bottlenecking is called a **founder effect**. By chance, the allele frequencies of the founders may not be the same as those of the original population. In the absence of further gene flow, natural selection will influence those frequencies in drastically different ways because of its interaction with genetic drift. As you might deduce, the founder effect is quite pronounced on isolated islands (Figure 18.14).

Genetic Drift and Inbred Populations

Inbreeding refers to nonrandom mating among closely related individuals, which have many identical alleles in common. Inbreeding is a form of genetic drift in a small population—that is, within the group of relatives that are preferentially interbreeding. Like genetic drift, inbreeding leads to the homozygous condition. It also can lower fitness when the alleles that are increasing in frequency are recessive and have harmful effects.

Most human societies forbid or discourage incest (inbreeding between parents and children or between siblings). But inbreeding among other close relatives is common in small communities that are geographically or culturally isolated from a larger population. The Old Order Amish of Pennsylvania, for instance, is a highly inbred group having distinct genotypes. One outcome of their inbreeding is a high frequency of a recessive allele that causes *Ellis–van Creveld syndrome*. Affected individuals have extra fingers or toes and short limbs. Section 12.6 shows one affected individual. The allele probably was rare when the small number of founders immigrated to Pennsylvania. At present, about 1 in 8 are heterozygous and 1 in 200 are homozygous for it.

Figure 18.15 A few of the remaining cheetahs, which have some bad alleles that made it through a severe bottleneck.

Bottlenecks and inbreeding are a bad combination for endangered species, the populations of which have become small and vulnerable to extinction. Cheetahs, for instance, apparently survived a drastic bottleneck in the nineteenth century. Survivors mated with their own offspring when no other options were available.

Inbreeding among survivors and their descendants resulted in strikingly similar alleles among the 20,000 existing cheetahs (Figure 18.15). One, a mutated allele, affects fertility. Most male cheetahs have a low sperm count, and 70 percent of the sperm are abnormal. Other shared alleles result in far lower resistance to disease. Therefore, infections that are seldom life-threatening to other cat species can be devastating to cheetahs. In one outbreak of *feline infectious peritonitis* in a wild animal park, the viral pathogen had very little effect on captive lions, but it killed many cheetahs. For them, infection triggered uncontrollable inflammation. Fluid filled the body cavity that houses the heart and other organs, and the cats died in agony. There is no vaccine.

The Florida panther also is highly inbred. Pressure from hunting and urban sprawl has helped put this cat on the endangered species list. Only fifty remain.

Genetic drift is the random change in allele frequencies over the generations, brought about by chance alone. The magnitude of its effect is greatest in small populations, such as the ones that make it through a bottleneck.

Barring mutation, selection, and gene flow, the chance losses and increases of the various alleles at a given locus lead to the homozygous condition and a loss of genetic diversity.

18.9 SUMMARY

1. Individuals of a population generally have the same number and kind of genes. But genes come in different allelic forms, and this leads to variations in their traits.

2. A population is evolving when some forms of a trait (and the alleles specifying them) are becoming more or less common with respect to the other kinds, over the generations. This happens as a result of mutation, gene flow, genetic drift, and natural selection (Table 18.2).

3. Genetic equilibrium, a state in which a population is not evolving, is used as a baseline to measure change. The Hardy–Weinberg rule is this: Genetic equilibrium occurs only if there is no mutation, if the population is very large and isolated from other populations of the same species, and if there is no selection (all members survive and reproduce equally by random mating).

4. Gene mutations, heritable changes in DNA, are the only source of *new* alleles. New combinations of existing alleles occur by crossing over, independent assortment at meiosis, and mixing of alleles at fertilization.

5. Natural selection simply is a difference in survival and reproduction among individuals of a population that differ in the details of their heritable traits. Over time, natural selection leads to an increase in fitness (increased adaptation to the environment).

a. Selection pressure may shift a range of variation for a trait in one direction (*directional* selection), favor both extremes (*disruptive* selection), or favor eliminate both extremes and favor intermediate forms (*stabilizing* selection).

b. Most species are adapted to fairly stable habitats, so extreme phenotypes are often lost. Thus stabilizing selection may be the most common mode.

c. Selection can result in balanced polymorphism. A population is in this state when nonidentical alleles for a given trait are being maintained over the generations at frequencies greater than 1 percent.

d. Sexual selection, by one sex or the other, leads to forms of traits that offer an advantage in reproductive success. Sexual dimorphism, which is the persistence of phenotypic differences between males and females of a species, is one outcome of sexual selection.

6. Gene flow is a change in allele frequencies brought about by the physical movement of alleles into and out of a population (by immigration and emigration).

7. Genetic drift is a change in allele frequencies over the generations due to chance events alone. Because of sampling error, the magnitude of its effect is greater in small populations than in large ones.

a. In the absence of mutation, natural selection, and gene flow, sooner or later genetic drift can result in loss of diversity.

Table 18.2 Summary of Microevolutionary Processes

MUTATION	A heritable change in DNA
NATURAL SELECTION	Change or stabilization of allele frequencies owing to differences in survival and reproduction among variant individuals of a population
GENETIC DRIFT	Random fluctuation in allele frequencies over time due to chance occurrences alone
GENE FLOW	Change in allele frequencies as individuals leave or enter a population

b. Drift has greatest impact following a bottleneck (a severe reduction in population size from which the population recovers). The alleles that make it through a bottleneck might give rise to differences in the range of phenotypes, compared to the original population. In the type of bottlenecking known as the founder effect, a few emigrants from one population establish a small subpopulation in a new environment.

Review Questions

1. Name three broad categories of heritable traits that help characterize a population. *18.1*

2. Explain the difference between continuous variation and polymorphism. *18.1*

3. How do lethal mutations and neutral mutations differ? *18.1*

4. Define genetic equilibrium. Which occurrences can drive allele frequencies away from genetic equilibrium? *18.1, 18.2*

5. Define fitness, with respect to phenotype. *18.3*

6. Identify the mode of selection (stabilizing, directional, or disruptive) for each of the following diagrams: *18.4, 18.5*

7. Explain how sexual dimorphism might arise. *18.6*

8. Why has natural selection apparently favored the allelic combination Hb^A/Hb^S? *18.6*

9. Define bottleneck and the founder effect. Are these cases of genetic drift, or do they set the stage for it? *18.8*

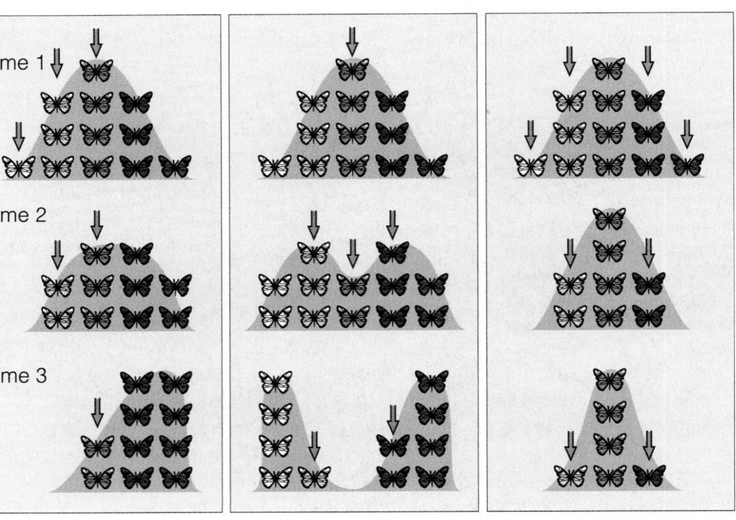
time 1
time 2
time 3

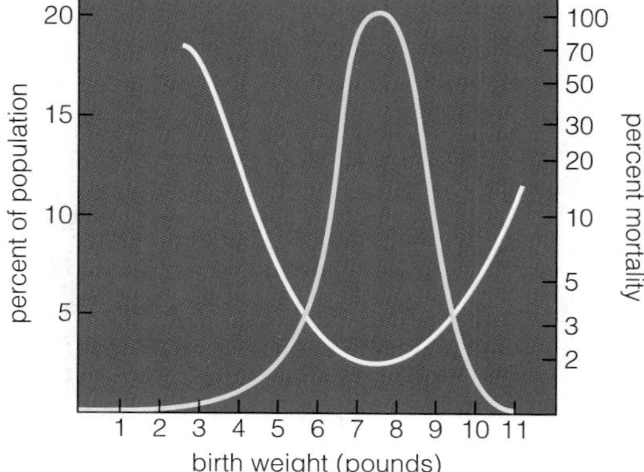

20
15
10
5
percent of population
100
70
50
30
20
10
5
3
2
percent mortality
1 2 3 4 5 6 7 8 9 10 11
birth weight (pounds)

Figure 18.16 The weight distribution for 13,730 human newborns (*yellow* curve) correlated with mortality rate (*white* curve).

Self-Quiz (*Answers in Appendix III*)

1. Individuals don't evolve; ________ do.

2. Genetic variation gives rise to variation in ________ traits.
   a. morphological
   b. physiological
   c. behavioral
   d. all of the above

3. Sickle-cell anemia first appeared in Asia, the Middle East, and Africa. The causative allele entered the U.S. population when people were forcibly brought over from Africa prior to the Civil War. In microevolutionary terms, this is a case of ________ .
   a. mutation
   b. genetic drift
   c. gene flow
   d. natural selection

4. Natural selection may occur when there are ________ .
   a. differences in the adaptiveness of forms of traits to prevailing environmental conditions
   b. differences in survival and reproduction among individuals that differ in one or more traits
   c. both a and b

5. Directional selection ________ .
   a. eliminates uncommon forms of alleles
   b. shifts allele frequencies in a steady, consistent direction
   c. favors intermediate forms of a trait
   d. works against adaptive traits

6. Disruptive selection ________ .
   a. eliminates uncommon forms of alleles
   b. shifts allele frequencies in a steady, consistent direction
   c. doesn't favor intermediate forms of a trait
   d. both b and c

7. Match the evolution concepts.
   ____ gene flow
   ____ natural selection
   ____ mutation
   ____ genetic drift
   a. source of new alleles
   b. changes in a population's allele frequencies due to chance alone
   c. immigration, emigration change allele frequencies
   d. differences in survival and reproduction among variant individuals

Critical Thinking

1. As long-term studies show, prospects for human newborns of very high or very low birth weight are not good. Being born too small increases the risk of stillbirth and early infant death (Figure 18.16). Also, pre-term instead of full-term pregnancies increase the risks. Is this an outcome of directional, disruptive, or stabilizing selection?

2. Consider the brilliantly hued male sugarbird and subdued-hued female in Figure 18.17. In the evolutionary view, what is maintaining this morphological difference over time?

3. For centuries *tuberculosis* (TB) has caused many deaths around the world. A bacterium (*Mycobacterium tuberculosis*) causes this

Figure 18.17 Sugarbirds.

contagious lung disease. TB declined steadily in the United States, where public health officials predicted there would be no more cases by the year 2000. But caseloads are increasing. The AIDS epidemic, an influx of immigrants from regions where TB is still common, and severe overcrowding in tenements and homeless shelters contribute to the increase. At one time, antibiotics could cure TB within six to nine months. But now antibiotic-resistant strains of *M. tuberculosis* have evolved. Currently, two or even three different antibiotics must be used simultaneously to block different metabolic pathways of this bacterium. How might this agressive approach help circumvent the problem of antibiotic resistance?

4. A few families in a remote region of Kentucky show a high frequency of *blue offspring*, an autosomal recessive disorder. The skin of affected individuals has a bright blue appearance. Homozygous recessives lack the enzyme diaphorase. The enzyme catalyzes reactions that help maintain hemoglobin (the oxygen-carrying red pigment in blood) in its normal molecular form. Without the enzyme, a blue form of hemoglobin accumulates in blood. Skin and its blood capillaries are transparent, so the color of pigments in blood shows through and contributes to skin coloration. This gives the skin of nonmutated individuals a pinkish cast—and blue skin its color.

Formulate a hypothesis to explain why the blue-offspring trait is rather common among a cluster of families but rare in the human population at large.

Selected Key Terms

allele *18.1*
allele frequency *18.1*
balanced polymorphism *18.6*
bottleneck *18.8*
directional selection *18.4*
disruptive selection *18.5*
evolution *CI*
fitness *18.3*
fixation *18.8*
founder effect *18.8*
gene flow *18.7*
gene pool *18.1*
genetic drift *18.8*
genetic equilibrium *18.1*
Hardy–Weinberg rule *18.2*
inbreeding *18.8*
lethal mutation *18.1*
microevolution *18.1*
mutation rate *18.1*
natural selection *18.3*
neutral mutation *18.1*
polymorphism *18.1*
population *18.1*
sampling error *18.8*
sexual selection *18.6*
stabilizing selection *18.5*

Readings *See also www.infotrac-college.com*

Darwin, C. 1957. *Voyage of the Beagle*. New York: Dutton.

Gould, S. J. 1977. *Ever Since Darwin*. New York: Norton.

Smith, T. B. January 1991. "A Double-Billed Dilemma." *Natural History*, 14–21.

Weiner, J. 1994. *The Beak of the Finch: Evolution in Real Time*. New York: Alfred A. Knopf.

19

SPECIATION

The Case of the Road-Killed Snails

If you happen to be a snail living in a garden in Bryan, Texas, it doesn't take much to keep your genes away from snails in a backyard across the street. By day, the sunbaked asphalt would be about as inviting to a snail as a desert would be to a catfish. Besides, day or night, a street-traversing snail is vulnerable to cars, trucks, skateboards, and bicycles (Figure 19.1*a*). Whatever else it might be, that strip of asphalt is a formidable barrier to the flow of genes between populations. For snails, that is.

Whether any physical barrier deters gene flow depends in large part on the organism's mode of locomotion or dispersal. It also depends on how fast and how long an organism *can* move in response to environmental factors or its own hormonal signals.

A snail is not swift, and it does not roam very far from its home population. Compare it with the wild black duck, leg-banded in Virginia in 1969, that turned up eight years later in Korea. Compare it to the wandering albatross, one of the supreme barrier busters. After lifting off from Kerguelen Island in the Indian Ocean, one of these birds soared westward past the southern tip of Africa, across the Atlantic, and around South America's Cape Horn. After it had traveled thirteen thousand kilometers, it landed in Chile. With its lightweight body and wingspan of 3.65 meters (12 feet across), that bird was able to exploit the great prevailing winds of the Southern Hemisphere.

And yet, in 1859, some snails did cross an ocean. Humans had transported garden-variety snails (*Helix aspersa*) all the way from France to California, then turned them loose. The idea was that the snails would mate, multiply, and meet the demand for that French delicacy, *escargots aux fines herbes*. It was bad enough that the importers brought over a less tasty species by mistake. Worse, the exotic species exceeded expectations and became

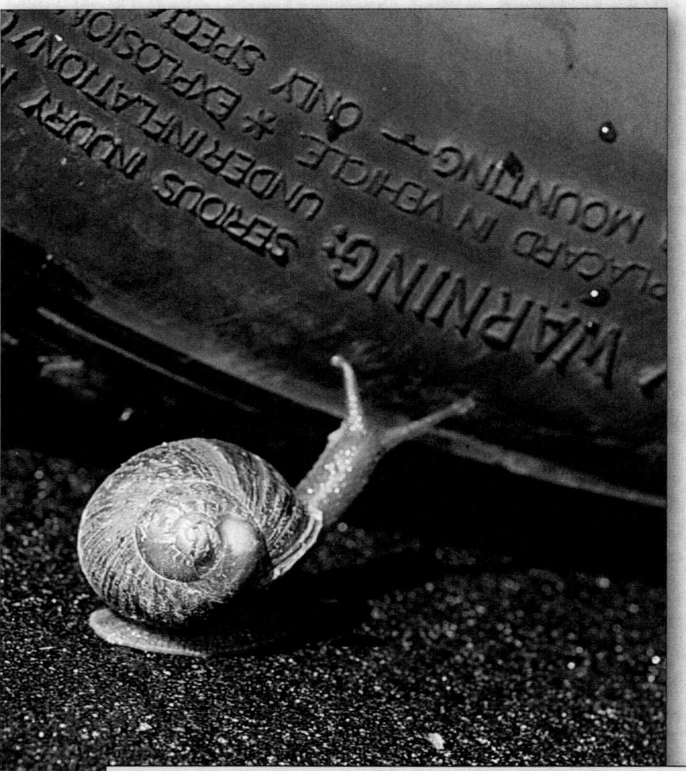
a
WARNING
SERIOUS INJURY
UNDERINFLATION
PLACARD IN VEHICLE
MOUNTING

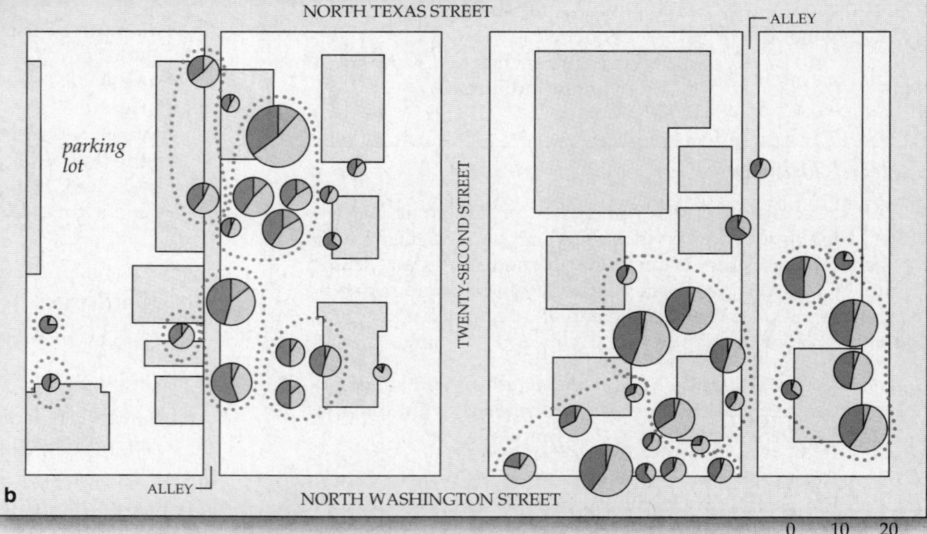
NORTH TEXAS STREET
ALLEY
parking lot
TWENTY-SECOND STREET
ALLEY
NORTH WASHINGTON STREET
b
0 10 20
meters

Figure 19.1 (**a**) A snail (*Helix aspersa*) encountering a major barrier to gene flow. It only appears to be reading the warning label. (**b**) Results from a study of neighboring populations of snails, all descended from founders that ended up in a small town in Texas in the 1930s. For each population, a circle represents relative abundances of three alleles (color-coded *pink*, *orange*, and *brown*) for an enzyme, leucine aminopeptidase. Genetic variation is greater between populations living on opposite sides of Twenty-Second Street. For example, notice the higher frequencies of the allele coded *orange* in the block to the west of the street.

an absolute nuisance in gardens and nurseries through much of the southwestern United States.

By the 1930s, *H. aspersa* had hitched rides, possibly as eggs in the soil of plant containers, to Bryan, Texas. They founded small colonies in the local vegetation. Forty years later Robert Selander, now of Pennsylvania State University, was down on his hands and knees with a few graduate students, scouring patches of vegetation on two adjacent city blocks. Why? Selander wanted to determine the effect of genetic drift on the introduced populations. He and his students collected every single snail—2,218 of them—from fourteen local populations. For each population, they determined the allele frequencies for five different genes.

For each of the five genes, results from their analysis pointed to some genetic variation among the colonies on the same block—and to major differences in allele frequencies *between* different blocks. Figure 19.1*b* shows the results for one of the genes studied.

Assume the genetic differences between populations will continue to increase, as through natural selection or genetic drift. Will the time come when snails from opposite sides of the street can no longer interbreed successfully even if they do manage to get together? In other words, *will they become members of separate species?* Or will stabilizing selection work against significant genetic divergence by eliminating extreme phenotypes from the neighboring populations? After all, how much can a workable package of *H. aspersa* genes evolve in adjacent patches of vegetation—and under very similar environmental pressures—in a small town in Texas?

And with these questions, we arrive at the topic of **speciation**—that is, to changes in allele frequencies that are significant enough to mark the formation of daughter species from a parent species. As you read in the preceding chapter, Charles Darwin proposed long ago that natural selection might lead to speciation but was not sure how this actually happened. Obviously, no one was around to watch species originate during the past. No one lives long enough to know whether many existing populations are at intermediate stages leading to speciation. However, evidence is accumulating that supports certain models of speciation.

As you read through this chapter, it is important for you to recognize that speciation is not the same thing as natural selection. Evolutionary biologists now view speciation as a *potential consequence* of natural selection —and of other microevolutionary processes that may work in concert with natural selection.

KEY CONCEPTS

1. A species consists of one or more populations of individuals that can interbreed under natural conditions and produce fertile offspring and that are reproductively isolated from other such populations. This definition is restricted to sexually reproducing species.

2. Populations of the same species have a shared genetic history, they are maintaining genetic contact over time, and they are evolving independently of other species.

3. Speciation is the process by which daughter species evolve from a parent species.

4. By one model, speciation starts when a geographic barrier arises between populations or subpopulations of a species. Thereafter, mutation, natural selection, and genetic drift operate independently in each population and lead to irreversible genetic divergence of one from the other.

5. Such populations may come to differ in certain alleles that affect morphological, physiological, or behavioral traits associated with reproduction. When the genetic differences affecting reproduction become great enough, the populations can no longer interbreed even if they later coexist in the same area. Speciation is completed.

6. The model of speciation just described, which applies to geographically separated populations, is known as allopatric speciation. This might be the way most species originate in nature. Sympatric speciation and parapatric speciation might be less prevalent routes.

7. By sympatric speciation, species form within the home range of their parent species. By parapatric speciation, adjacent populations become distinct species while still maintaining contact along a common border between their home ranges.

8. The timing, rate, and direction of speciation vary within and between lineages. The extinction of some number of species is inevitable for all lineages.

19.1 ON THE ROAD TO SPECIATION

What Is a Species?

MORPHOLOGICAL SPECIES CONCEPT "If it looks like a duck, walks like a duck, and quacks like a duck, then probably it's a duck." Let's use this familiar saying as a starting point for defining what is and what is not a species. **Species** is a Latin word. Generally, it simply means "kind," as in "a particular kind of duck."

Based on the appearance of the body alone—that is, on morphological traits—few of us would have trouble distinguishing a duck from, say, a chicken. At the very least, chickens don't have a flattened bill and webbed feet. And a chicken's body is not adapted to tipping bottoms up in a pond, head down, to strain food from the mud or water. But think about this: There are forty known species of ducks around the world. Especially in the interior of the Northern Hemisphere's continents, ducks that show great morphological diversity gather in large numbers to nest in marshlands. How do we know a particular duck belongs to one species and not another? More to the point, how does the *duck* know it?

Even when early naturalists were defining species on the basis of morphological traits, common sense told them other factors must also be considered. Sometimes individuals of the same species have strikingly different appearances only because they responded to different environmental conditions. Those arrowhead plants in Figure 19.2 are a fine example. Sometimes pronounced differences in form and coloration emerge as part of the life history of a species, so that the young do not look at all like the adults. You have only to think about the wriggly larval beginnings of adult butterflies to know that this is so. And sometimes sexual dimorphism can be extreme. Thus, at first glance, you might well dismiss a busily mating male anglerfish as a puny extension of the female's body (Figure 19.3).

Variation in morphology is not the only source of potential confusion. Individuals that outwardly appear to be almost identical may belong to genetically distinct species. Consider the gray treefrogs *Hyla versicolor* and *H. chrysoscelis*. We find both frog species in parts of the central and eastern United States. Both have warty skin and prominent sticky pads on their toes. Both have a white belly, yellow or yellow-orange coloration on the inside of the hind legs, and a large white spot beneath each eye. Both also live in the same forest habitats and prefer the same kind of prey. It took careful behavioral and cytological research to identify *H. versicolor* and *H. chrysoscelis* as separate species. During the breeding season, the distinctive call of each type of male frog attracts only females of the same species (Figure 19.4). Also, the two species show a chromosomal difference; one is diploid and the other is tetraploid.

BIOLOGICAL SPECIES CONCEPT Morphological details can vary so enormously, perhaps we should extend our consideration past the details, down to a basic function that unites members of a species and isolates them from all other species.

For example, using *reproduction* as a basic, defining function is central to the **biological species concept.** The evolutionary biologist Ernst Mayr has phrased the concept this way: "Species are groups of interbreeding natural populations that are reproductively isolated from other such groups." No matter how extensive the phenotypic variation, populations belong to the same

a

b

Figure 19.2 Pronounced morphological differences between two plants of the same species. Shown are mature leaves of arrowheads (*Sagittaria sagittifolia*) growing (**a**) on land and (**b**) in water. The difference is attributable to adaptive responses to different environmental conditions, not to genetic differences.

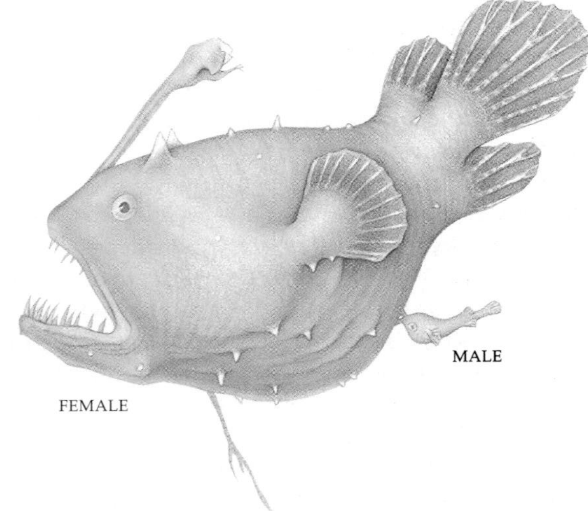
MALE
FEMALE

Figure 19.3 Pronounced sexual dimorphism in the deep-sea anglerfish *Linophryne*, also known as the net devil. The female is about ten centimeters long. The tiny male is attached to her.

Further reading: Student Guide to InfoTrac on web site

a *Hyla versicolor* **b** *H. chrysoscelis*

Figure 19.4 Which frog is shown here—*Hyla versicolor* or *H. chrysoscelis*? You cannot tell the two species apart on the basis of external appearance. Identification depends on other clues, such as the female-attracting calls made by the male frogs during the breeding season. (**a**, **b**) Sound spectrograms (visual records of each note's frequency, or pitch) reveal a behavioral difference between the two species.

species for as long as their individuals have the form, physiology, and behavior that allow them to interbreed and produce fertile offspring. Their capacity to contribute to a shared gene pool is qualification for membership. Mayr's concept doesn't apply to asexually reproducing organisms, such as bacteria. But it is a useful guide for research into the factors that define the vast majority of species, which do reproduce sexually.

If we subscribe to Mayr's concept, then speciation is the attainment of reproductive isolation. Bear in mind, reproductive isolation does not evolve purposefully to promote formation of a species or maintain its identity. Rather, *any structural, functional, or behavioral difference that favors reproductive isolation simply is a by-product of genetic change*.

Recall that genetic changes between populations of the same species can only be countered by **gene flow**, the movement of alleles into and out of populations by immigration and emigration. Gene flow may exert its homogenizing effect even when two populations are geographically separate, as long as some gene exchange continues between them. This microevolutionary process helps maintain their common reservoir of alleles.

What if some geographic barrier arises and prevents an intermingling of genes between populations or even subpopulations of a species? Then, each will embark on its own evolutionary road. Through **genetic divergence**, differences will build up between the gene pools of the genetically separate populations, for mutation, natural selection, and genetic drift will now be free to operate independently in each one. Figure 19.5 is a simple way to think about genetic divergence. Each horizontal line

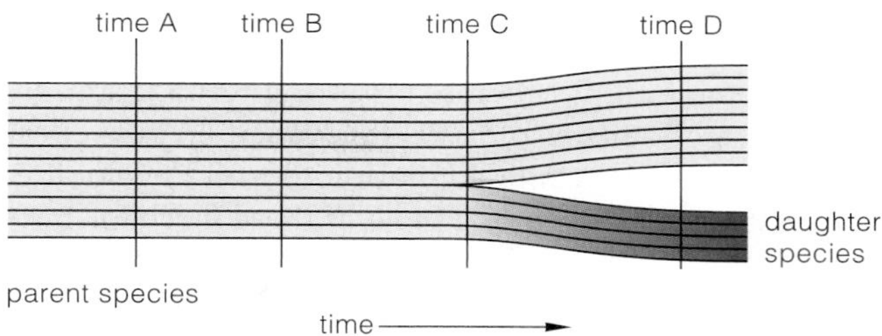

time A
time B
time C
time D
daughter species
parent species
time

Figure 19.5 Simplified diagram of genetic divergence. Each horizontal line represents a different population.

of the sketch represents a population. At times A and B, gene flow helped to keep them in genetic contact, and they belonged to the same species. After a geographic barrier arose at time C, divergences were under way.

Most species apparently arose by gradual genetic divergence. Yet, as you will see, many flowering plants and other species arose far more rapidly. Said another way, *the process of speciation can vary in its duration*.

Depending on how the affected populations interact and on their patterns of distribution, *speciation also may vary in its details*. The preceding paragraphs started you thinking about what is probably the main route, which is called allopatric speciation. Later in the chapter you will be taking a closer look at this route and two others.

Reproductive Isolating Mechanisms

Let us now define **reproductive isolating mechanisms** as any heritable feature of body form, functioning, or behavior that prevents interbreeding between one or more genetically divergent populations. Some prevent successful mating or pollination between individuals of the divergent populations so hybrid zygotes cannot form. Let's look next at some of these mechanisms.

THE BIOLOGICAL SPECIES CONCEPT. A species is one or more populations of individuals that (1) are interbreeding under natural conditions and producing fertile offspring, and (2) are reproductively isolated from other such populations.

Speciation is the process by which a daughter species forms from a population or subpopulation of the parent species. The process can vary in its details and in the length of time it takes before reproductive isolation is complete.

By perhaps the most common mechanism of speciation, a geographic barrier separates populations of a species. Then differences build up in their gene pools because mutation, natural selection, and genetic drift operate independently in each. This outcome is called genetic divergence.

Reproductive isolating mechanisms may evolve simply as by-products of the genetic changes. They are heritable traits that, one way or another, prevent interbreeding.

19.2 REPRODUCTIVE ISOLATING MECHANISMS

Let's first consider mechanisms of *prezygotic* isolation, which come into play before or during fertilization. By contrast, mechanisms of *postzygotic* isolation take effect after fertilization, while an embryo is still developing. Table 19.1 serves as an overview of these mechanisms.

Table 19.1 Categories of Isolating Mechanisms

PREZYGOTIC ISOLATION (*Mating or zygote formation is blocked*)	
Ecological isolation	Potential mates occupy different local habitats within the same area
Temporal isolation	Potential mates occupy overlapping ranges but reproduce at different times
Behavioral isolation	Potential mates meet but cannot figure out what to do about it
Mechanical isolation	Potential mates attempt engagement, but sperm cannot be successfully transferred
Gametic mortality	Sperm is transferred, but egg is not fertilized (gametes die or gametes are incompatible)
POSTZYGOTIC ISOLATION (*Hybrids don't work*)	
Zygotic mortality	Egg is fertilized, but zygote or embryo dies
Hybrid inviability	First-generation hybrid forms but shows very low fitness
Hybrid infertility	Hybrid is sterile or partially so

After Alan Templeton, Jerry Coyne, and others.

Prezygotic Isolation

ECOLOGICAL ISOLATION Populations that are adapted to different microenvironments in the same habitat may be isolated ecologically. In the seasonally dry foothills of the Sierra Nevada are populations of two manzanita species. One grows at elevations between 600 and 1,850 meters in open forests of conifers; the other grows at elevations between 750 and 3,350 meters. The two rarely hybridize where their ranges overlap. Like all manzanitas, both species have built-in, physiological mechanisms that can promote conservation of water during the dry season. But one is adapted to the more sheltered sites, where water stress is not as intense as it is on more exposed (and drier) rocky hillsides—which the other species prefers. Cross-pollination is unlikely because of the ecological differences (Figure 19.6).

TEMPORAL ISOLATION What if individuals of divergent populations still have the potential to interbreed? If they reproduce at different times, this makes no difference. Almost all animals mate and most plants get pollinated quickly, sometimes in less than a day, so there is not much chance of overlap with others. Extreme temporal isolation occurs among periodical cicadas (*Magicicada*), which live in the eastern United States. These insects mature below the soil surface, where they feed on juices of tree roots. Three species that differ in size, color, and song emerge and reproduce every 17 years (Figure 19.7). Each of the three has a "sibling species," which is just about the same, morphologically. But its sibling species emerges every 13 years. Only once every 221 years do the two species release gametes at the same time!

Figure 19.6 Two species of manzanita: (**a**) *Arctostaphylos patula* and (**b**) *A. viscida*. The two species are ecologically isolated in the same forests in the Sierra Nevada.

Figure 19.7 An adult *Magicicada septendecim*, a periodical cicada that matures underground. It emerges every seventeen years to reproduce. Often its populations overlap the habitats of its sibling species (*M. tredecim*), which reproduces every thirteen years.

Adults live but a few weeks. They are a taste thrill to birds, which often gorge themselves so frenziedly that they vomit; which makes one suspect why periodical cicadas stay underground for so long.

Further reading: Student Guide to InfoTrac on web site

Figure 19.8 A sampling of courtship displays that precede copulation (sexual union) between a male and female albatross. Members of the same species recognize the visual, acoustical, and tactile components of such displays. Section 47.5 gives other examples.

BEHAVIORAL ISOLATION Behavioral differences are key barriers to gene flow among related species that live in the same territory. For instance, before male and female birds copulate, they often engage in intricate courtship rituals (Figure 19.8). A female is genetically equipped to recognize singing, head bobbing, wing spreading, or prancing of a male of the same species as an overture to sex. Females of different species usually ignore it.

MECHANICAL ISOLATION Incompatibility between body parts of potential mates or pollinators is an example of mechanical isolation. *Salvia apiana* and *S. mellifera* are two sage species. A nectar cup in their flowers attracts pollinators that use some petals as a landing platform. *S. apiana*'s pollen-bearing stamens extend from the cup, above a platform of petals large enough to hold large pollinators. *S. mellifera* has a smaller landing platform for small pollinators. When small bees visit *S. apiana*, the larger flower's stamens often do not brush against them, so they don't transport pollen to flowers of the other species. Similarly, large pollinators of *S. apiana* cannot land on *S. mellifera* and cross-pollinate it. Figure 19.9 shows a similar example of mechanical isolation.

GAMETIC MORTALITY Gametes of different species may have evolved incompatibilities at the molecular level. For example, when the pollen of one species lands on a plant of a different species, it usually does not recognize the molecular signals that are supposed to trigger its growth down through the plant's tissues to the egg.

Postzygotic Isolation

What about the isolating mechanisms that take effect *after* zygotes form? Some postzygotic mechanisms take effect after fertilization, when an embryo is developing. Unsuitable interactions among some of the embryo's genes or gene products lead to early death, sterility, or hybrids of low fitness. Hybrid offspring commonly are weak, with low survival rates. A few types are sturdy but sterile. Mules, which result from a cross between a female horse and a male donkey, are strong but sterile.

Figure 19.9 Mechanical isolation, as demonstrated by the precise fit between a zebra orchid and a wasp species that is one of its few pollinators. The flower releases a chemical attractant that mimics a sex pheromone produced by female wasps. This sex signal induces the wasp to attempt to mate. The plant benefits by becoming pollinated.

Mechanisms of reproductive isolation that evolve between genetically divergent populations take effect before, during, or after fertilization.

Prezygotic mechanisms block mating or pollination between two populations; hybrid zygotes cannot form. Postzygotic mechanisms take effect after hybrid zygotes formed.

19.3 SPECIATION IN GEOGRAPHICALLY ISOLATED POPULATIONS

Allopatric Speciation Defined

If we assume physical separation between populations promotes the genetic changes that are necessary for speciation, then allopatry may be the main speciation route. By the model for **allopatric speciation**, some type of physical barrier arises and prevents gene flow between populations or subpopulations of a species. (*Allo-* means different, and *patria* can be taken to mean homeland.) Reproductive isolating mechanisms arise in the genetically diverging populations. The process of speciation is completed when individuals of the two populations no longer will interbreed even if changing circumstances put them back together in the same area.

Whether a geographic barrier proves to be effective at blocking gene flow between populations depends on an organism's means of travel, how fast it can travel, and whether it is compelled or behaviorally inclined to disperse. You have only to reflect on the garden snails and the wandering albatross described at the start of this chapter to conclude that this is so.

The Pace of Geographic Isolation

Some measurable distance separates the populations of most species, and gene flow among them is more of an intermittent trickle than a steady stream. In addition, barriers can rapidly arise and shut off the trickles. For example, in the 1800s, a monstrous earthquake changed the course of the Mississippi River. As one outcome, the change separated populations of insects that could neither swim nor fly, and it effectively cut off gene flow between populations now living along opposite shores.

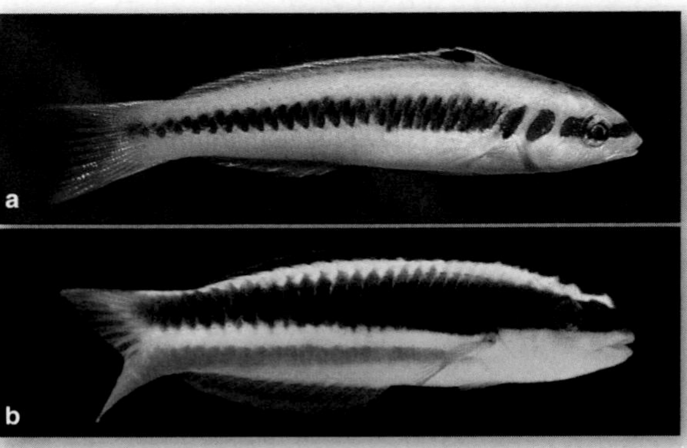

a
b

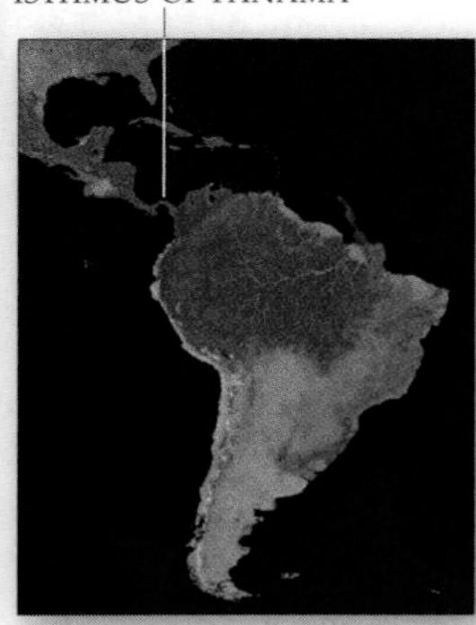

ISTHMUS OF PANAMA

Figure 19.10 (**a**) Blue-headed wrasse (*Thalassoma bifasciatum*) from the Atlantic Ocean near the Isthmus of Panama. (**b**) Cortez rainbow wrasse (*T. lucasanum*) from the Pacific Ocean near the isthmus. The species may be related by descent from a shared ancestral population that split when geologic forces created the isthmus. You might see individuals with body coloration and patterning that differ from these specimens. Such variation is common among reef fishes.

Geographic isolation also can proceed slowly, over great spans of time. We find evidence of such extended events in the fossil record, which affords glimpses into the breakup of formerly continuous environments.

For example, during past ice ages, glaciers advanced down through North America and Europe, and they gradually cut off populations from one another. When the glaciers retreated, the descendant plants, animals, and other organisms came in contact. Some groups that had descended from the same parent population were no longer reproductively compatible; they had evolved into separate species. In other groups, however, genetic divergences did not proceed far, and the descendants are still interbreeding. For them, reproductive isolation was not completed. Speciation did not occur.

Another example: As you might know, the Earth's crust is fractured into gigantic plates, somewhat like a cracked eggshell. In the past, imperceptibly slow but colossal movements of the plates caused land masses to collide and to break up. Such geologic movements uplifted part of the seafloor in the region now called the Isthmus of Panama. The uplifting divided an ancient ocean basin, and it set the stage for allopatric speciation among populations of fishes and other marine species.

In the 1980s, John Graves compared four enzymes from the muscle cells of two related species of isthmus fishes (Figure 19.10). Both species are strong swimmers. As Graves knew, temperature has different effects on the activity of different enzymes. He also knew that seawater on the Pacific side of the isthmus is cooler by about 2°–3°C than it is on the Atlantic side and that it varies more with the changing seasons. Graves found that all four "Pacific" enzymes function better at lower temperatures, compared with the four "Atlantic" enzymes. Analysis by gel electrophoresis revealed slight differences in electric charge between two of the four pairs of enzyme molecules as a result of slight differences in their amino acid sequences.

Graves drew these tentative conclusions: First, selection pressures that are imposed by even small differences in environmental conditions may have caused divergences in the molecular structure of certain enzymes. Second, it appears that closely related fish populations on opposite sides of the isthmus are diverging from each other. Why? The alternative molecular forms of the enzymes that Graves analyzed are already displaying detectable differences in catalytic activity.

How might we interpret results from his study of geographically isolated populations? For the isthmus fishes, a gradual accumulation of gene mutations that are neutral or adaptive might be a microevolutionary "foot in the door." In other words, genetic divergences here might be evidence of speciation in progress.

Further reading: Student Guide to InfoTrac on web site

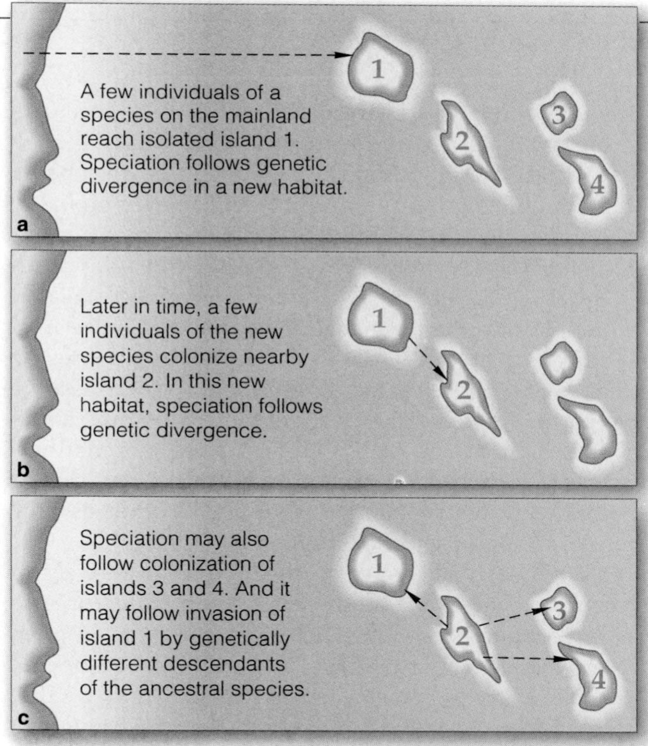
A few individuals of a species on the mainland reach isolated island 1. Speciation follows genetic divergence in a new habitat.
a
Later in time, a few individuals of the new species colonize nearby island 2. In this new habitat, speciation follows genetic divergence.
b
Speciation may also follow colonization of islands 3 and 4. And it may follow invasion of island 1 by genetically different descendants of the ancestral species.
c
1
2
3
4

Figure 19.11 Sketch of allopatric speciation on an isolated archipelago. (Can you envision other possibilities?)

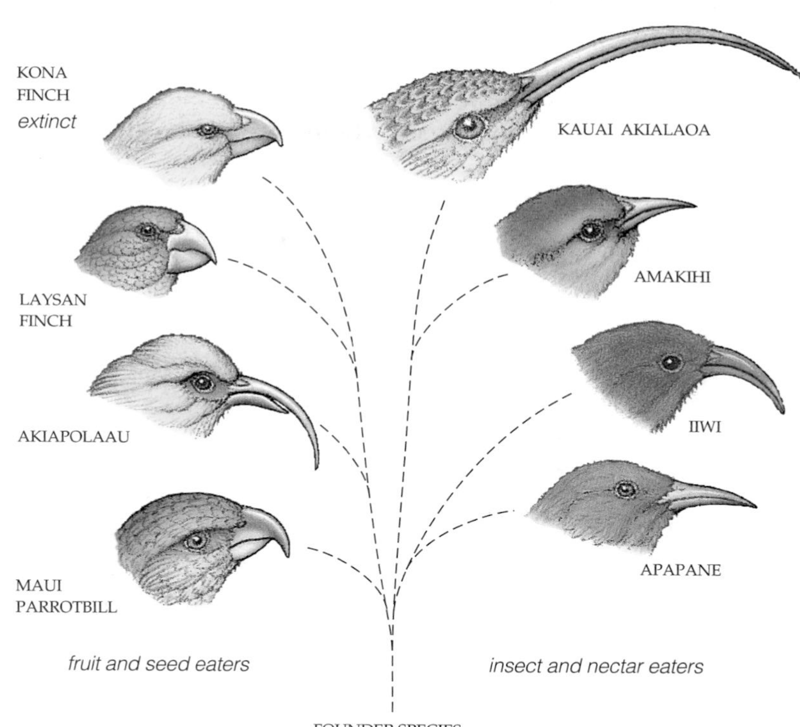
KONA FINCH
extinct
KAUAI AKIALAOA
AMAKIHI
LAYSAN FINCH
IIWI
AKIAPOLAAU
APAPANE
MAUI PARROTBILL
fruit and seed eaters
insect and nectar eaters
FOUNDER SPECIES

Figure 19.12 A few Hawaiian honeycreepers, a fine example of how a new arrival in species-poor habitats on an isolated archipelago can be the start of a flurry of allopatric speciation.

Allopatric Speciation on Archipelagos

An **archipelago** is an island chain some distance away from a continent. Some archipelagos, such as the Florida Keys, are so near the mainland that gene flow remains more or less unimpeded, which means there is little, if any, speciation. Others are so isolated that they are like laboratories for the study of evolution (Figure 19.11). Foremost among these are the Hawaiian Archipelago, nearly 4,000 kilometers from the California coast, and the Galápagos Islands, about 900 kilometers from the South American coast. Islands of both chains are only the tops of great volcanoes, some still active, that rise from a deep seafloor. When each volcano first broke the surface of the sea, it was devoid of life.

Remember those Galápagos finches (Section 17.3)? A few finches from the mainland apparently evolved in isolation on one Galápagos island. Later, some of their descendants reached other islands in the chain. In those new and unoccupied habitats—even in different parts of the same habitats—conditions varied, and the island hoppers were subject to a variety of selection pressures. In time, genetic divergences within and between the islands paved the way for new episodes of allopatric speciation. Later on, some island-hopping new species even invaded the island of the ancestors. The distances between these islands are enough to foster divergence but not enough to stop occasional invasions.

Bursts of speciation in the Hawaiian Archipelago have been so dramatic, they are a premier example of adaptive radiation in a species-poor environment, as described in Section 19.5. The youngest island, Hawaii, is less than a million years old. Here alone we find a great diversity of habitats, ranging from rain forests to alpine grasslands below high, snow-capped volcanoes. When the ancestors of Hawaiian honeycreepers arrived, they found a veritable buffet of fruits, seeds, nectars, and tasty insects—and not many competitors for them. The near-absence of competition from other bird species fanned allopatric speciations. Figure 19.12 only hints at the resulting variation that arose among the different species of Hawaiian honeycreepers. Today, these and thousands of other species of animals as well as plants that evolved in the archipelago are found nowhere else in the world. As another example of the potential for speciation, the Hawaiian Islands represent less than 2 percent of the world's land mass, yet they are home to 40 percent of all species of fruit flies (*Drosophila*).

ALLOPATRIC SPECIATION. **Some type of physical barrier intervenes between populations or subpopulations of a species and prevents gene flow among them. By doing so, it favors genetic divergence and speciation.**

19.4 MODELS FOR OTHER SPECIATION ROUTES

Geographic isolation might not always be required for reproductive isolation. By the models of sympatric and parapatric speciation, genetically divergent populations that are *not* separated by geographic barriers and are in contact may become reproductively isolated, also.

Sympatric Speciation

By the model for **sympatric speciation**, a species may form *within* the home range of an existing species, in the absence of a physical barrier. (*Sym-* means together with, as in "together with others in the homeland.")

EVIDENCE FROM CICHLIDS IN AFRICA In 1994, Ulrich Schliewen and his coworkers found evidence in favor of the sympatric speciation model. They had studied fishes called cichlids that live together in two lakes in Cameroon, West Africa. These lake basins are collapsed cones of volcanoes, and each is very small (Figure 19.13). Yet eleven cichlid species coexist in one basin and nine coexist in the other.

The researchers analyzed mitochondrial DNA from all the species in one lake. They did the same for all the species in the other lake as a basis for comparison. In nucleotide sequences, the species of each lake are like each other and *not* like related species in nearby lakes and rivers. For example, nine species in one lake are the only ones that carry an unusual base-pair substitution in the gene that specifies the protein cytochrome *b*.

Physical and chemical conditions are too uniform in the lakes to foster geographic separation. For instance, the uniform shorelines do not present even small-scale topographic barriers. Therefore, allopatric populations could not have formed even on a small scale. The crater rim isolates the lakes from all but tiny creeks trickling in from higher elevations. Thus there has been no gene flow from outside.

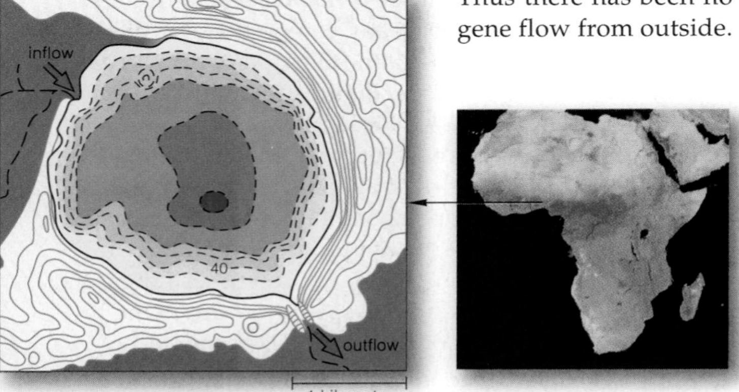
inflow
40
outflow
1 kilometer

Figure 19.13 Topographical map of Lake Barombi Mbo in Cameroon, West Africa. Apparently, nine kinds of cichlids evolved by sympatric speciation in this small, isolated crater lake. Even microgeographic separation is absent from the lake. There is separation by feeding preferences, but all species breed near the lake bottom, in sympatry.

The lakes must have been colonized before connections with a river system on the outside were severed.

Cichlids are not sluggish swimmers. Besides being highly mobile, individuals of different species probably encounter one another often in such small crater lakes. Said another way, they all must be living in sympatry.

Because clusters of species are more closely related to one another than to species anywhere else, we may assume they are all descended from a single ancestor. Because of their restricted distribution, we may assume they formed in the same lake. Yet there is nothing about these lakes that can isolate different fish populations. Apparently, then, ancestors of each present-day cichlid species were never isolated from one another as they were evolving in separate directions.

Within each lake, species do show small degrees of *ecological* separation that arises through differences in feeding preferences. Some cichlid species feed in open waters and others at the lake bottom. Even so, they all *breed* close to the bottom, in sympatry. It may be that ecological separation on a small scale was enough to influence sexual selection among potential mates and, over generations, to provide the reproductive isolation that can lead to speciation.

SPECIATION BY WAY OF POLYPLOIDY Quite possibly, sympatric speciation has been a prevalent evolutionary event among flowering plants. Consider that about half of all known species of flowering plants are polyploid. **Polyploidy**, remember, is a change in the chromosome number; offspring inherit three or more of each type of chromosome characteristic of the parental stock. Such changes arise when chromosomes separate improperly during meiosis or mitosis. They also arise when a germ cell replicates its DNA but fails to divide, then goes on to function as a gamete (Section 12.10).

Speciation may have been rapid for many flowering plants that engage in self-fertilization or some asexual reproductive mode. Suppose they produced polyploid offspring. If the extra chromosomes paired with each other during meiosis, maybe the extra set of genes did no harm. Common bread wheat is one of the species that might have arisen by polyploidy, although in this case cross-fertilization also was involved (Figure 19.14).

Allen Orr has evidence that polyploid animals are rare because of failed **dosage compensation**. By this normal event, genes on sex chromosomes are expressed at the same levels in females *and* males. For instance, X chromosome inactivation in female mammals means the cells of both males and females have a single active X chromosome (Section 15.4). Polyploidy skews dosage compensation, with bad or fatal effects. The mechanism is absent in plants, so their chromosome doublings are not as problematic as they are in animals.

Further reading: Student Guide to InfoTrac on web site —

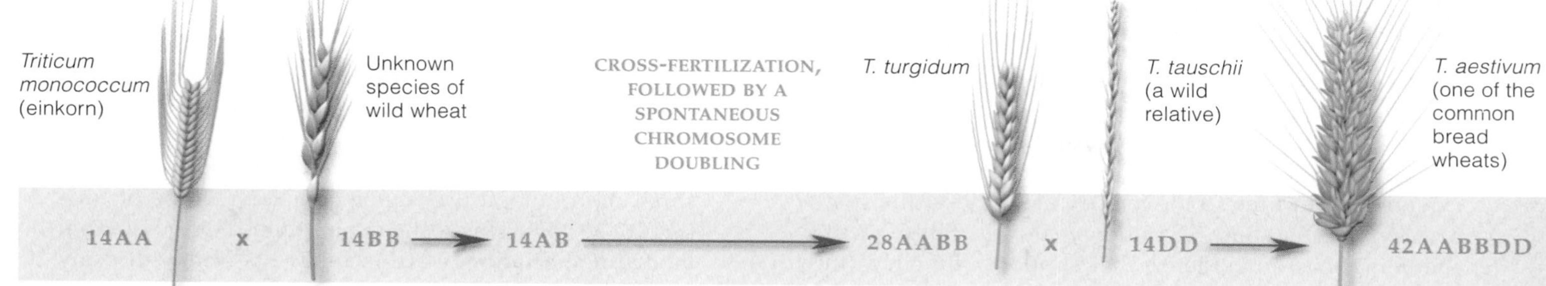
Triticum monococcum (einkorn)
Unknown species of wild wheat
CROSS-FERTILIZATION, FOLLOWED BY A SPONTANEOUS CHROMOSOME DOUBLING
T. turgidum
T. tauschii (a wild relative)
T. aestivum (one of the common bread wheats)
14AA
x
14BB
14AB
28AABB
x
14DD
42AABBDD

a By 11,000 years ago, humans had started to cultivate wild wheats. Einkorn (*Triticum monococcum*) is still around. It has diploid chromosome number of 14 (two sets of 7 chromosomes, shown above as 14AA). Long ago, einkorn wheat probably hybridized with another wild wheat species having the same chromosome number.

b If the AB hybrid offspring were sterile but self-fertilizing, an interbreeding population of AB plants could have arisen by asexual reproduction. About 8,000 years ago, polyploidy did indeed arise in such a population. Wild emmer (*T. turgidum*) plants are tetraploid (AABB), with a chromosome number of 28 (two sets of 14). They are fertile; at meiosis, the A chromosomes pair with each other, and the B chromosomes pair with each other.

c Later, an AABB plant probably hybridized with *T. tauschii*, a wild relative of the wild emmer. Its diploid chromosome number must have been 14 (two sets of 7 DD). Populations of the hybrid descendants now include common bread wheats, such as *T. aestivum*, which have a chromosome number of 42 (six sets of 7 AABBDD).

Figure 19.14 Presumed sympatric speciation in wheat by polyploidy and hybridizations over time. Wheat grains 11,000 years old were found in the Near East. Diploid wild wheats still grow there.

BULLOCK'S ORIOLE
BALTIMORE ORIOLE
HYBRID ZONE
BULLOCK'S ORIOLE
BALTIMORE ORIOLE

Figure 19.15 A possible setting for parapatric speciation? Hybrids arise along the common border of the ranges of two species of orioles. To the east are Baltimore orioles; to the west are Bullock's orioles. Hybridization was once so common that they were considered subspecies. In 1997, the American Ornithologist Union decided they are separate species.

Parapatric Speciation

By the model for **parapatric speciation**, neighboring populations become distinct species while maintaining contact along a common border. (*Para-* means near, as in "near another homeland.") Interbreeding individuals produce hybrid offspring in this region, which is called a **hybrid zone**. Evidence that might support parapatric speciation is sketchy. Why? In many cases, it is difficult to determine whether or not geographically separated populations are only *subspecies* (that is, geographically distinct populations of the same species).

For example, Bullock's orioles differ from Baltimore orioles in the color patterning of most of their feathers. The male orioles have different territorial songs, and they generally mate with their own kind. As K. Corbin determined, these birds also differ in the frequencies of alleles for several enzymes. Their geographic ranges do differ, but the ranges overlap in the American Midwest (Figure 19.15). Interbreeding was once common in the hybrid zone, although it is now becoming less frequent. Is this an example of parapatric speciation in progress? Maybe. But it might also be an example of "secondary contact." In other words, previously isolated subspecies that diverged recently from a common ancestor may have been getting together again.

SYMPATRIC SPECIATION. **Daughter species arise from a group of individuals within an existing population. This seems to have been a common speciation event among the polyploid flowering plants.**

PARAPATRIC SPECIATION. **Adjacent populations evolve into distinct species even while maintaining contact along their common border.**

19.5 PATTERNS OF SPECIATION

Branching and Unbranched Evolution

All species—past and present—are related by descent. They share genetic connections through lineages that extend back in time to the molecular origin of the first prototypic cells, some 3.8 billion years ago. Subsequent chapters focus on evidence that supports this view. In anticipation of those chapters, let's start thinking about ways to interpret the large-scale histories of species.

The fossil record provides evidence of two patterns of evolutionary change in lineages—one branching and the other unbranched. The first is called **cladogenesis** (from the Greek *klados*, meaning branch; and *genesis*, meaning origin). This is the pattern by which a lineage splits, with populations becoming genetically isolated and then diverging in different evolutionary directions. This is the pattern of speciation described earlier.

By the second pattern, **anagenesis**, changes in allele frequencies and in morphology accumulate within an unbranched line of descent. (In this context, *ana-* means renewed.) Directional changes are confined within one lineage, and gene flow never does cease among all of its populations. At some point in time, allele frequencies and morphology have changed so much that we assign a separate name to the descendants. Such changes are occurring in the moth populations you read about earlier.

Evolutionary Trees and Rates of Change

Evolutionary trees summarize information about the continuity of relationship among species. Figure 19.16 is a simple way to start thinking about how tree diagrams are constructed. Each *branch* represents a single line of descent from a common ancestor. And each *branch point* represents a time of genetic divergence and speciation, as brought about by microevolutionary processes.

Figure 19.16 also shows how an evolutionary tree diagram conveys *rates of change*, the estimated length of time between speciation events. Branches with slight angles imply that the species emerged through many small morphological changes over long spans of time (Figure 19.16*a*). This is the main premise of the **gradual model of speciation**, which actually fits well with many fossil sequences. As one example, in many sedimentary rock layers, we find sequences of intricately perforated shells of foraminiferans, a type of protistan. Observable characteristics of fossils in the sequences give evidence of slow change.

Alternatively, evolutionary tree diagrams for some lineages are constructed with short, horizontal branches that make an abrupt 90-degree turn, as in Figure 19.16*b*. The **punctuation model of speciation** is consistent with such diagrams. By this model, most of the changes in morphology are compressed into a brief period when populations are starting to diverge—say, within merely hundreds or thousands of years. Bottlenecks, founder effects, strong directional selection, or a combination of these bring about rapid speciation. The daughter species recover quickly from the adaptive wrenching and then change little over the next 2 million to 6 million years or so. Proponents of the model argue that reproductive cohesion prevailed for about 99 percent of the history of most lineages. They find evidence of abrupt change in many parts of the fossil record.

Apparently, changes in lines of descent have been gradual, abrupt, or both. Species originated at different times, and they differed in how long they persisted on the evolutionary stage. Remember, some lineages have endured without much change, producing a species here and losing a species there, often over millions of years. Other lineages have branched bushily, and sometimes spectacularly, during times of adaptive radiation.

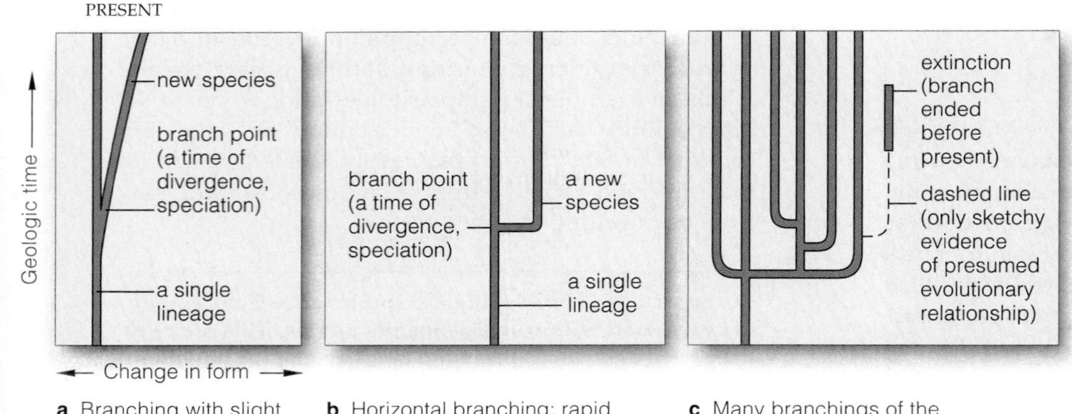
PRESENT
Geologic time
new species
branch point (a time of divergence, speciation)
a single lineage
Change in form
branch point (a time of divergence, speciation)
a new species
a single lineage
extinction (branch ended before present)
dashed line (only sketchy evidence of presumed evolutionary relationship)

a Branching with slight angle; new species formed through gradual changes in traits over geologic time.

b Horizontal branching; rapid change in traits around time of speciation. Vertical continuation of branch means traits of new species did not change much thereafter.

c Many branchings of the same lineage at or near the same point in geologic time; an adaptive radiation has taken place.

Figure 19.16 How to read evolutionary tree diagrams.

Adaptive Radiations

An **adaptive radiation** is a burst of divergences from a single lineage that give rise to many new species, each adapted to an unoccupied or a new habitat or to using a novel resource. Figure 19.17 provides an example. In the past, lineages often diversified in such a way when the member species found themselves in vacant **adaptive zones**. Think of

Further reading: Student Guide to InfoTrac on web site

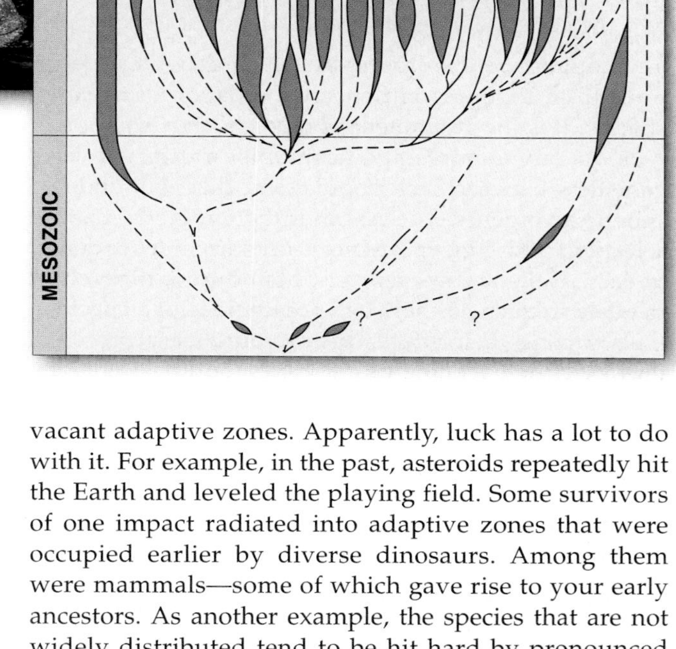
marsupials (kangaroos, etc.)
rabbits
rodents
primates
bats
order Insectivora
order Carnivora
whales
even-toed hoofed mammals
odd-toed hoofed mammals
elephants
monotremes (platypus, etc.)
CENOZOIC
MESOZOIC
?
?

Figure 19.17 Example of adaptive radiation, as shown by an evolutionary tree diagram. This one started about 65 million years ago, at the start of a geologic era called the Cenozoic, and it led to mammals as different as the opossum and walrus. Variations in the width of the branches correspond to the range of species diversity (the number of groups) within the lineage as represented at different points in time. The wider the branch, the greater the diversity.

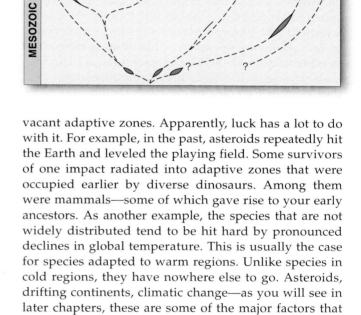

adaptive zones as ways of life, such as "burrowing in seafloor sediments" or "catching winged insects in the air at night." A lineage may radiate into such zones if it has physical, evolutionary, or ecological access to them.

Physical access means a lineage happens to be there when adaptive zones open up. For example, mammals were once distributed through uniform tropical regions of a single continent. That continent split into several land masses. Habitats and resources changed in many different ways on the separate masses and set the stage for independent radiations.

Evolutionary access means that modification of an existing structure or function will permit a lineage to exploit the environment in a new or more efficient way. Such modifications are known as **key innovations**. For example, the forelimbs of certain five-toed vertebrates evolved into wings. That innovation opened up novel adaptive zones for the ancestors of birds and bats.

Ecological access means that the lineage can enter an unoccupied adaptive zone or displace resident species. Sections 48.7 through 48.10 provide detailed examples.

Extinctions—End of the Line

Extinction is the irrevocable loss of a species. The fossil record shows twenty or more **mass extinctions**—large, catastrophic events in which entire families or other major groups disappeared during the same phase or at the same point in geologic time. For example, 95 percent of all known species abruptly disappeared 250 million years ago. In other times, only certain groups or certain regions were extinguished.

By studying the statistical distribution of extinctions, David Raup found that most were clustered in time, even though they differed in size. Like George Simpson before him, he realized that times of reduced diversity follow extinction events, then new species arise and fill vacant adaptive zones. Apparently, luck has a lot to do with it. For example, in the past, asteroids repeatedly hit the Earth and leveled the playing field. Some survivors of one impact radiated into adaptive zones that were occupied earlier by diverse dinosaurs. Among them were mammals—some of which gave rise to your early ancestors. As another example, the species that are not widely distributed tend to be hit hard by pronounced declines in global temperature. This is usually the case for species adapted to warm regions. Unlike species in cold regions, they have nowhere else to go. Asteroids, drifting continents, climatic change—as you will see in later chapters, these are some of the major factors that have contributed to a general pattern of extinctions.

Lineages have changed gradually, abruptly, or both. Their member species originated at different times and differ in how long they have persisted.

Adaptive radiations are bursts of divergences from a single lineage that gave rise to many new species, each adapted to a vacant or new habitat or to using a novel resource.

Repeated and often large extinctions have occurred in the past. After times of reduced diversity, new species formed and occupied the new or vacated adaptive zones.

Taken together, the persistence, branchings, and extinctions of species account for the full range of biological diversity at any point in geologic time.

19.6 SUMMARY

1. A species is a single kind of organism, recognized partly in terms of its morphology.

2. By the biological species concept, a species is one or more populations of individuals that are interbreeding and producing fertile offspring under natural conditions and that are reproductively isolated from other such populations. The concept identifies a species mainly in terms of the portion of alleles that promote or maintain reproductive isolation. Also, it applies only to sexually reproducing organisms.

3. The populations of a species have a shared genetic history, they are maintaining genetic contact over time, and they are evolving independently of other species.

4. Speciation is the process by which daughter species form from a population or subpopulation of a parent species. It is the attainment of reproductive isolation.

a. As one example, if a geographic barrier prevents gene flow between such populations, they will undergo genetic divergence. The reason is that mutation, natural selection, and genetic drift can operate independently in each. Genetic divergence is a buildup of differences in allele frequencies between populations of a species.

b. Among their effects, microevolutionary processes may by chance alone give rise to reproductive isolating mechanisms that prevent interbreeding. Isolation may then result in irreversible genetic differences between the populations.

c. Speciation proceeds gradually, by way of genetic divergence. It also may proceed instantaneously, as by polyploidy or other changes in chromosome number.

5. Prezygotic isolating mechanisms prevent mating or pollination between individuals of populations. They include differences in reproductive timing or behavior, incompatibilities in reproductive structures or gametes, and occupation of different microenvironments in the same area. Postzygotic mechanisms lead to early death, sterility, or unfit hybrid offspring. They take effect after fertilization, while the embryo develops.

6. There are three current models of speciation:

a. Allopatric speciation. Geographic barriers prevent gene flow between the populations of a species, which genetically diverge in such a way that interbreeding will not occur in nature even if their individuals make contact with each other later on. Allopatric speciation might be the most prevalent speciation route.

b. Sympatric speciation. Reproductive isolation of individuals that occupy the same home range leads to speciation. Speciation by polyploidy is an example.

c. Parapatric speciation. Adjacent populations give rise to a new species while maintaining contact along a border between their home ranges.

7. Lineages differ in their time, rates, and direction of speciation. Speciation can proceed gradually, rapidly, or both. Extensive branching of a lineage (divergences of one or more of its populations) in the same geologic time span is an adaptive radiation. Often this occurs as member species of a lineage become adapted to new or unoccupied habitats or to using novel resources.

8. Evolutionary tree diagrams indicate the relationships among groups of species. Each branch of such trees represents a line of descent (lineage). Branch points are speciation events, as brought about by natural selection, genetic drift, and other microevolutionary processes.

9. Extinctions are losses of species. Most are clustered in time, and their size varies greatly. Asteroid impacts, volcanism, continental drift, and other environmental changes have apparently brought about most of them.

10. Persistences, branchings, and extinctions of species account for the full range of biological diversity through geologic time (Table 19.2).

Table 19.2 Summary of Processes and Patterns of Evolution

MICROEVOLUTIONARY PROCESSES		
Mutation	Original source of alleles	Stability or change in a species is the outcome of balances or imbalances among all of these processes, the effects of which are influenced by population size and by the prevailing environmental conditions.
Gene flow	Preserves species cohesion	
Genetic drift	Erodes species cohesion	
Natural selection	Preserves or erodes species cohesion, depending on environmental pressures	
MACROEVOLUTIONARY PROCESSES		
Genetic persistence	Basis of the unity of life. The biochemical and molecular basis of inheritance extends from the origin of first cells through all subsequent lines of descent.	
Genetic divergence	Basis of life's diversity, as brought about by adaptive shifts, branchings, and radiations. Rates and times of change have varied within and between lineages.	
Genetic disconnect	Extinction. End of the line for a species. Extinctions tend to be clustered in time as discrete events; many are simultaneous, mass extinctions of many species.	

Review Questions

1. How does the biological species concept differ from a definition of species based on morphological traits alone? *19.1*

2. Define speciation and describe a speciation model. *CI, 19.3–19.5*

3. Give examples of reproductive isolating mechanisms. *19.2*

4. Interpret the following features of evolutionary tree diagrams: *19.5*
   a. a single line
   b. soft-angled branching of line
   c. horizontal branching
   d. vertical continuation of branch
   e. many branchings of a line
   f. dashed line
   g. branch ending before present

5. Define and give an example of an adaptive radiation. *19.5*

6. Define extinction. What kinds of events might bring about extinctions? *19.5*

Self-Quiz (*Answers in Appendix III*)

1. Sexually reproducing individuals of a species ______.
   a. can interbreed under natural conditions
   b. can produce fertile offspring
   c. have a shared genetic history
   d. all of the above

2. Reproductive isolating mechanisms ______.
   a. prevent interbreeding c. reinforce genetic divergence
   b. prevent gene flow d. all of the above

3. When potential mates occupy overlapping ranges but reproduce at different times, this is a case of ______ isolation.
   a. postzygotic c. temporal
   b. mechanical d. gametic

4. In an evolutionary tree diagram, a branch point represents ______, and a branch that ends represents ______.
   a. a single species; incomplete data on lineage
   b. a single species; a time of extinction
   c. a time of divergence; extinction
   d. a time of divergence; speciation complete

5. An evolutionary tree diagram with horizontal branches that abruptly become vertical is consistent with the ______.
   a. gradual model of speciation
   b. punctuation model of speciation
   c. idea of small changes in form over long spans of time
   d. both a and c

6. Match each term with its most suitable description.
   ____ cladogenesis a. burst of microevolutionary activity within a lineage
   ____ anagenesis b. catastrophic disappearance of major groups of organisms
   ____ adaptive radiation c. branching lineages
   ____ extinction d. a species lost from a lineage
   ____ mass extinction e. genetic and morphological change in unbranched lineage

Critical Thinking

1. You notice several duck species in the same lake habitat. The females of different species look very similar to one another, but the males of each particular species have feathers with distinctive patterns and colors. Speculate on which forms of reproductive isolation may be keeping each species distinct. How does the appearance of the male ducks provide a clue to the answer?

2. One family of mammals includes true horses (*Equus*), zebras (*Hippotrigris* and *Dolichohippus*), and donkeys and asses (*Asinus*). Zebroids are hybrid offspring of wild zebras and domesticated horses that were confined to the same pasture (Figure 19.18). The unnatural confinement breached the reproductive barriers between the two lineages. Those barriers have been in place since a divergence more than 3 million years ago. What does the breach say about the number of genetic changes required to attain reproductive isolation in nature?

3. Suppose a small population that was founded by only a few individuals is cut off from gene exchange with the main population of their species. By Mayr's view of the *founder effect*, the frequencies of alleles for a few gene products change as a result of genetic drift, and the change triggers a number of changes in other genes that are affected by those alleles. If that is so, then genetic changes in newly founded populations may be so great that speciation occurs rapidly—so rapidly that we will seldom be able to document it through the fossil record. Does this view correspond to the gradual model or punctuated model of speciation?

Figure 19.18 A mixed herd of zebroids and horses.

4. A key innovation, recall, is some modification in structure or function that permits a species to exploit the environment in a more efficient or novel way, compared to the ancestral species. Identify a key innovation of an existing species, such as humans. Then describe how that innovation might be the basis of an adaptive radiation in environments of the distant future.

5. Richard Lenski uses bacterial populations in culture tubes to develop model systems for studying evolution. He is fond of saying that such a population is the equivalent of the entire human population, that he can replicate it several times over, that bacteria produce several generations in a day, and that he can store them in the deep freeze, then bring them back to active form, unaltered, to directly compare ancestors and descendants. Are bacterial models relevant to evolutionary studies of sexually reproducing organisms? Before you answer, read a short article by P. Raine and M. Travisano entitled "Adaptive Radiation in a Heterogenous Environment" (*Nature*, 2 July 1998, 69–72).

Selected Key Terms

adaptive radiation *19.5*
adaptive zone *19.5*
allopatric speciation *19.3*
anagenesis *19.5*
archipelago *19.3*
biological species concept *19.1*
cladogenesis *19.5*
dosage compensation *19.4*
evolutionary tree *19.5*
extinction *19.5*
gene flow *19.1*
genetic divergence *19.1*
gradual model, speciation *19.5*
hybrid zone *19.4*
key innovation *19.5*
mass extinction *19.5*
parapatric speciation *19.4*
polyploidy *19.4*
punctuation model, speciation *19.5*
reproductive isolating mechanism *19.1*
speciation *CI*
species *19.1*
sympatric speciation *19.4*

Readings *See also www.infotrac-college.com*

Futuyma, D. 1998. *Evolutionary Biology*. Third edition. Sunderland, Massachusetts: Sinauer.

Mayr, E. 1976. *Evolution and the Diversity of Life*. Cambridge, Massachusetts: Belknap Press of Harvard University Press.

20

THE MACROEVOLUTIONARY PUZZLE

He Sees Seashells . . . So Where's the Sea?

About 500 years ago, Leonardo da Vinci was brooding about seashells entombed in layered rocks of northern Italy's high mountains, hundreds of kilometers from the sea. How did they get there? If he accepted the traditional explanation, he would have to agree that stupendous floodwaters deposited those shells in the mountains during the great Deluge (Figure 20.1). But many shells were thin and fragile—and intact. Surely they would have been battered to bits if they had been swept across such distances, then up the mountains.

Da Vinci also brooded about the rocks. They were stacked like cake layers, and some contained shells but others had none. Then he remembered how large rivers swollen with spring floodwaters deposit silt in the sea. *Did the layers slowly accumulate long ago, as a series of silt deposits?* If so, then shells in the mountains would be evidence of a progression of communities of organisms that once lived in the seas! Da Vinci did not announce his novel idea, perhaps knowing that it would be met with deafening silence, imprisonment, or worse.

By the 1700s, fossils were accepted as evidence of past life. They were still being interpreted through the prism of prevailing cultural beliefs, as when a Swiss naturalist excitedly unveiled the remains of a giant salamander and announced they were the skeleton of a man who had drowned in the Deluge.

By midcentury, however, scholars began to question such interpretations. Extensive mining, quarrying, and canal excavations were under way. The diggers were finding similar rock layers and similar fossil sequences in distant places, such as the cliffs on both sides of the English Channel. More than a few scholars began to view the findings as evidence of connections between Earth history and the history of life.

Ever since, fossils have been analyzed in increasingly refined ways. Together with biochemical studies and other modern sources of information, they yield good evidence of evolution through vast spans of time.

What does the evidence tell us? *Only populations that already exist can evolve.* As a result of mutations, natural selection, and genetic drift, each species is a mosaic of ancestral and novel traits that emerged earlier, along unbranched or branching evolutionary roads that began with the first populations of the very first species on

Earth. Thus, *all species that ever evolved are related to one another, by way of descent*. This principle of evolution guides efforts to make sense of often puzzling scraps of evidence of past life. It guides the task of identifying and sorting out the many lines of descent, or **lineages**, that connect all species, past and present.

Ultimately, then, life is a story of *species*—of how and when each kind of organism originated, whether its defining traits persisted or were modified, and whether it vanished or endured. Life also is a story of **macroevolution**—of large-scale patterns, trends, and rates of change among families and other more inclusive groups of species. We turn to this larger picture in the next unit of the book. Here we begin with the nature of the evidence that supports it.

Figure 20.1 Two interpretations of the past. *Left:* From the Sistine Chapel, Michelangelo's painting of the onset of the great Deluge. As you read in the Chapter 17 introduction, that catastrophic flood became integrated into oral and written traditions. *Below:* A modern-day photographer captures a pattern of fossilized shells of ammonites. About 65 million years ago, all of these marine animals abruptly perished together with many other major groups of organisms. That mass extinction is one intriguing piece of the evolutionary puzzle.

KEY CONCEPTS

1. In the evolutionary view, all species that have ever lived are related—some closely, others remotely so. Their relatedness is based on this premise: Each new species evolved from variant individuals of species that already existed, starting with the first living cells to appear on Earth.

2. The term *macroevolution* refers to the patterns, trends, and rates of change among lineages over geologic time.

3. The fossil record, the geologic record, and radiometric dating of rocks yield evidence of macroevolution.

4. Morphological comparisons help us understand and reconstruct patterns of change through time. Among the most revealing aspects of morphology are cases of similar, homologous structures in the adult forms or embryos of different lineages.

5. Homology implies descent from a common ancestor, with evolution in different lineages proceeding through conservative modifications to the shared body plan.

6. Biochemical comparisons within and between major lineages provide strong evidence of macroevolution.

7. Stunning diversity characterizes the distribution of species through time and through the global environment. Biological systematics attempts to discern patterns in life's diversity through taxonomy, phylogenetic reconstruction, and classification.

8. Taxonomy is concerned with identifying and naming new species. Phylogenetic reconstruction is concerned with working out evolutionary connections. Classification is an attempt to organize information about species into retrieval systems.

FOSSILS—EVIDENCE OF ANCIENT LIFE

Fossilization

Fossil comes from a Latin word for something that has been "dug up." In general, **fossils** are the recognizable, stone-hard evidence of ancient life, as in Figure 20.2.

Most of the fossils discovered so far are bones, teeth, shells, seeds, capsules of spores, and other hard parts. (Soft parts usually are the first to decompose when an organism dies.) Imprints of leaves, stems, tracks, trails, burrows, and other *trace* fossils yield indirect evidence of past life. Even fossilized feces, or coprolites, contain residual evidence of which species were being eaten—and therefore were present—in ancient environments.

Fossilization is a very slow process that starts when an organism, or traces of it, becomes buried in volcanic ash or in sediments at the bottom of some lake, lagoon, or sea. Sooner or later, water infiltrates the organic remains, which become infused with dissolved metal ions and other inorganic compounds. More and more sediments gradually accumulate above the burial site, and they exert ever increasing pressure on the remains. Over great spans of time, the pressure and the chemical changes transform those remains to stony hardness.

Preservation is favored when organisms are buried rapidly, in the absence of oxygen. Gentle entombment by volcanic ash or anaerobic mud is best. Preservation also is favored when a burial site stays undisturbed. Most often, however, erosion and other geologic insults crush, deform, break, or scatter the fossils.

Figure 20.2 (**a**) A fossil hunter's dream: a complete skeleton of a bat that lived 50 million years ago. Erosion and other forces of nature have left few ancient burial sites undisturbed, so intact fossils are rare. Even jumbled parts of the ducklike birds in (**b**) are a good find. It will take hours of preparation and analysis to identify the species. (**c**) Fossilized parts of the oldest known land plant (*Cooksonia*). Its stems were not even seven centimeters tall. (**d**) Another complete fossil: skeletal remains of an ichthyosaur, a dolphin-like marine reptile that lived 200 million years ago.

Further reading: Student Guide to InfoTrac on web site —

Figure 20.3 The geologic time scale. The major boundaries mark times of mass extinctions. Radiometric dating methods (Section 20.11) allowed researchers to assign absolute dates. Life originated during the Archean.

The time spans are not to scale. If they were, the Archean and Proterozoic portions would run off the page. Think of the spans as minutes on a clock that runs from midnight to noon. If we say that life originated at midnight, the Paleozoic began at 10:04 A.M., the Mesozoic at 11:09 A.M., and the Cenozoic at 11:47 A.M. The Recent epoch of the Cenozoic started in the last 0.1 second before noon.

Eon	Era	Period	Epoch	Millions of Years Ago (mya)
PHANEROZOIC	CENOZOIC	QUATERNARY	Recent	0.01–
			Pleistocene	1.55
		TERTIARY	Pliocene	5
			Miocene	25
			Oligocene	38
			Eocene	54
			Paleocene	65
	MESOZOIC	CRETACEOUS	Late	100
			Early	138
		JURASSIC		205
		TRIASSIC		240
	PALEOZOIC	PERMIAN		290
		CARBONIFEROUS		360
		DEVONIAN		410
		SILURIAN		435
		ORDOVICIAN		505
		CAMBRIAN		570
PROTEROZOIC				2,500
ARCHEAN				4,600

Interpreting the Geologic Tombs

We find similar fossil-containing layers of sedimentary rock over vast areas, even on different continents. Such layers formed long ago through the gradual deposition of volcanic ash, silt, and other materials, one above the other. This layering of sedimentary deposits is called **stratification**. In general, the deepest layers were the first to form. Layers closest to the surface formed last. Because particles tend to settle in response to gravity, we can assume that most sedimentary layers must have formed horizontally. Where they are tilted or ruptured, this is evidence of subsequent geologic disturbance.

Understand how rock layers form, and you realize that fossils in a particular layer are from a similar age in Earth history. Specifically, *the older the layer, the older the fossils*. Given that rock layers formed in sequence, then their fossil assemblages are unique to sequential ages. That is why fossils can be used to assign relative dates to the record of the rocks.

Early geologists identified four abrupt transitions in the sequence of fossil assemblages. They found them in the fossil record from regions around the world. They used the transitions as boundaries between four great intervals in time. The oldest fossils they knew about were said to be from the first interval, the Proterozoic. These were followed by fossils from the Paleozoic, the Mesozoic, and finally the "modern" era, the Cenozoic.

We still use their boundaries between four intervals in a chronological chart of Earth history. Figure 20.3 is a modern version of this **geologic time scale**. As we now know, the boundaries correlate with extinction events. For example, the boundary between the Mesozoic and Cenozoic marks the mass extinction of the dinosaurs and other groups of reptiles, which was followed by a major adaptive radiation of mammals (Section 19.5).

Interpreting the Fossil Record

The fossils for about 250,000 known species are clues to evolutionary history. Judging from the current range of diversity, there must have been many, many millions of ancient, now-extinct species. We never will be able to recover fossils for most of them, so our "record" of past life is incomplete, with built-in biases. Why is this so?

Most importantly, colossal movements in the Earth's crust obliterated evidence from crucial intervals in the past. Besides this, most species of ancient communities simply weren't preserved. Consider that bony fishes and hard-shelled mollusks are well represented in the fossil record. Soft-bodied worms and jellyfishes are not, even though they may have been just as common or more so. Population density and body size also skew the record. For example, a population of plants may have released millions of spores in a single growing season, whereas the earliest humans lived in small groups and produced few offspring. What are the chances of finding even one fossilized bone of an early human, compared to finding spores of plants that lived at the same time?

Also, the fossil record is heavily biased toward some environments. Most species we know about lived either on land or in shallow seas which, as a result of geologic uplifting, became part of continents. We have recovered precious few fossils from the Southern Hemisphere and from sediments beneath the ocean—which covers nearly three-fourths of the Earth's surface! Why? Less time has been spent searching for fossils in those environments.

Fossils, the stone-hard physical evidence of ancient life, are present in layers of sedimentary rock. The deeper the layers, the older the fossils. The geologic time scale is based on sequences of fossils in sedimentary rocks.

The completeness of the fossil record varies as a function of the kinds of organisms represented, where they lived, and the stability of their burial sites.

20.2 EVIDENCE FROM COMPARATIVE MORPHOLOGY

Comparing body forms and structures of major lineages yields strong evidence of evolution. This field of inquiry is known as **comparative morphology**. A few examples of how limb bones of diverse vertebrates evolved give insight into one of this field's guiding principles: *When it comes to introducing changes in morphology, evolution tends to follow the line of least resistance.*

Morphological Divergence and Homologous Structures

Recall, from Chapter 18, that populations of the same species start to diverge genetically after gene flow ceases between them. Over long time spans, populations also may diverge considerably in the morphological traits that help characterize their species. Change from the body form of a common ancestor is a pattern of macroevolution. We call it **morphological divergence**. (*Morpho-* is Greek for body form.)

Even when related species diverge greatly with respect to morphology, they remain alike in other ways. Remember, they started out with the same body form and same structures, and they evolved through modifications to that shared plan. Look carefully enough, and it is possible to identify the underlying similarities.

An example: All land-dwelling vertebrates are descended from the first amphibians that ventured onto land. Divergences led to reptiles, then birds and mammals. Knowledge of the stem reptile (the one ancestral to all other reptiles and to birds and mammals) is based on fossilized, five-toed limb bones of a species that crouched low to the ground (Figure 20.4*a*). Its descendants expanded into new habitats on land; a few even returned to the seas when the environment changed (Figure 20.4*b–g*).

That five-toed limb was a key innovation for pterosaurs, birds, and bats, the evolutionary clay that became molded into different kinds of limbs with different functions. In lineages that gave rise to penguins and porpoises, it evolved into flippers that assisted in swimming. In a lineage leading to the modern horse, it evolved into long, one-toed limbs suitable for running fast. Among moles, it became quite stubby, which facilitates burrowing into the earth. Among elephants, it became strong and pillarlike, suitable for supporting considerable weight. The five-toed limb also became modified into the human arm and hand, in which a thumb is positioned in opposition to four fingers as the basis of grasping and precision movements.

What we have been describing is an example of **homology**—a similarity in one or more body

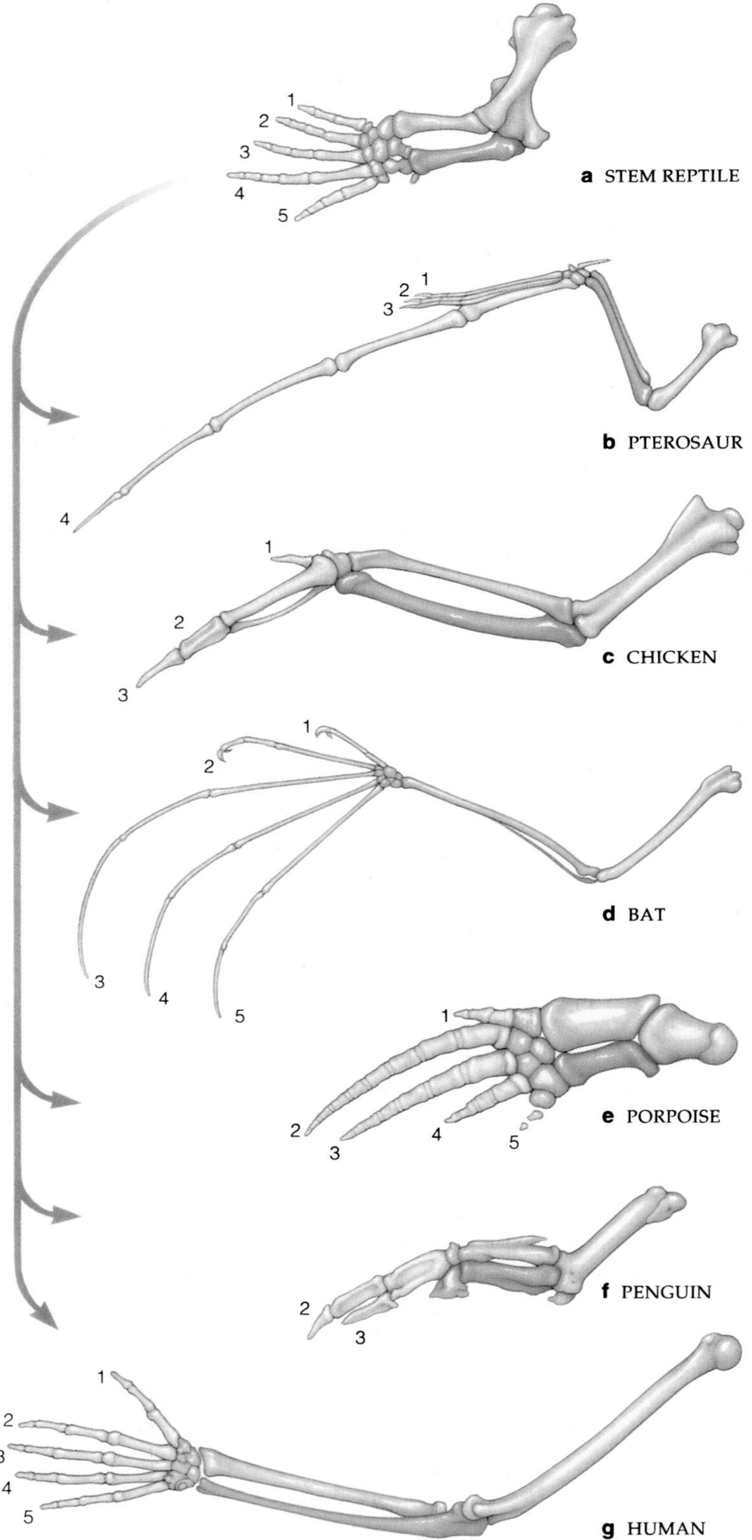
a STEM REPTILE
b PTEROSAUR
c CHICKEN
d BAT
e PORPOISE
f PENGUIN
g HUMAN

Figure 20.4 Morphological divergence in the vertebrate forelimb, starting with the generalized form of a stem reptile. Diverse forms evolved, and similarities in the number and position of bones were preserved. The drawings are not to the same scale.

Further reading: Student Guide to InfoTrac on web site

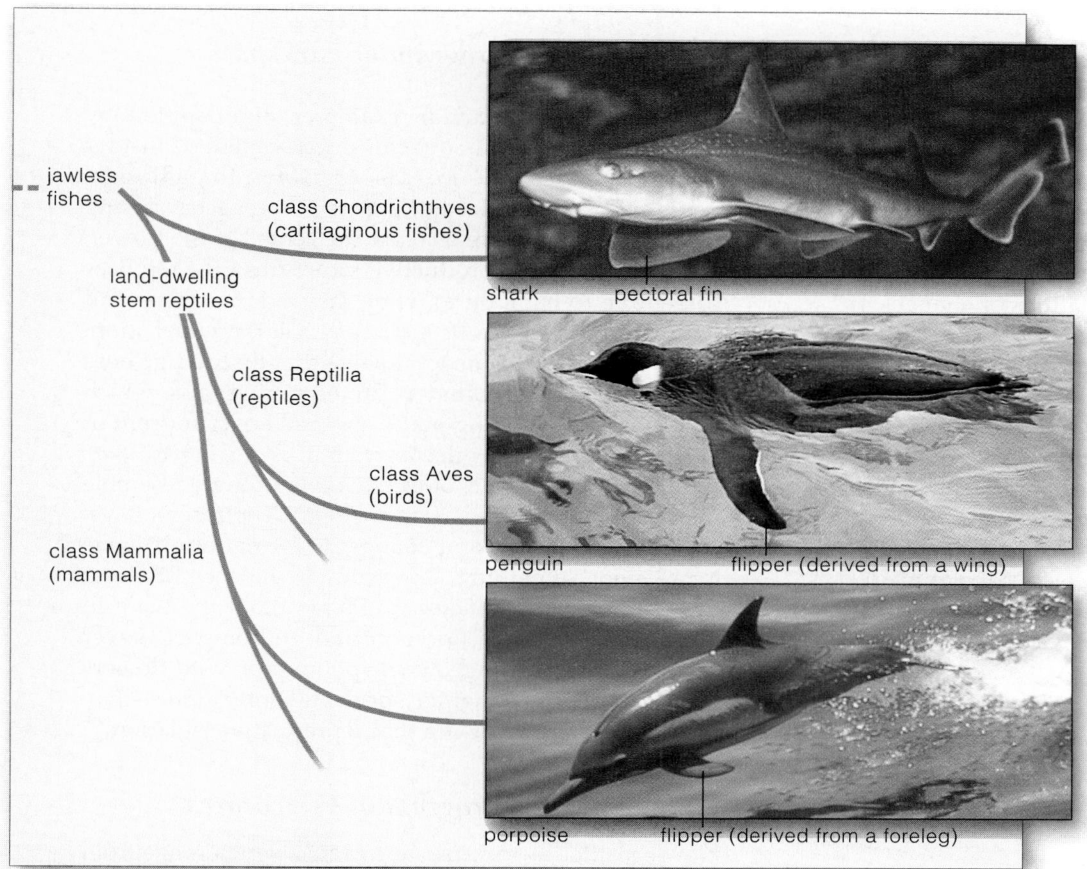
jawless fishes
class Chondrichthyes (cartilaginous fishes)
land-dwelling stem reptiles
class Reptilia (reptiles)
class Aves (birds)
class Mammalia (mammals)
shark
pectoral fin
penguin
flipper (derived from a wing)
porpoise
flipper (derived from a foreleg)

Figure 20.5 Morphological convergence among three lineages that diverged greatly over evolutionary time. In the lineages that led to the modern sharks, penguins, and porpoises, it appears that natural selection favored adaptations for rapid swimming. As one outcome, sharks, penguins, and porpoises have similar body forms even though the last shared vertebrate ancestor lived in the remote past.

parts in different organisms that we can attribute to descent from a common ancestor, even though they may or may not serve the same functions. (*Homo-* means the same.) Such similarities have been major clues for biologists who are searching for connections among pieces of the macroevolutionary puzzle.

Potential Confusion From Analogous Structures

Body parts with similar form and functions in different lineages do not always signify homology. Rather, major differences in form may have evolved *independently* in lineages having only remote evolutionary connections. This apparently happened when individuals of different lineages encountered similar environmental pressures. If those individuals responded to the pressures by using comparable body parts in much the same way, it seems likely that, by natural selection, the parts would have become modified in similar ways. In time they could end up resembling each other.

Even if their evolutionary connections are remote, lineages may thus evolve in similar ways under similar environmental pressures. We call this macroevolutionary pattern **morphological convergence**.

For example, like the sharks, penguins and porpoises are both fast-swimming predators of the seas. Their lineages diverged quite early in the evolution of vertebrates, and their flippers are very similar in shape to the pectoral fin of sharks (Figure 20.5). All three body parts help stabilize the body in water, and yet they are not homologous structures.

Sharks never left the water, and shark fins haven't changed much from the ancestral form. By contrast, aquatic ancestors of penguins gave rise to four-legged land-dwellers that evolved into birds. Later on, ancestral birds became adapted to life in water, not in air. Penguin flippers are modified *wings* that evolved from fins. Also, after the ancestors of porpoises left the water, their early land-dwelling descendants evolved into four-legged mammals, the descendants of which returned to the seas. Porpoise flippers are *modified front legs*, not modified fins or wings. Both kinds of flippers simply "converged" on the pectoral fins of sharks.

The body parts just mentioned are an example of **analogy**. The word refers to body parts that were once quite different in evolutionarily distant lineages, but then converged in structure and function because those lineages responded to similar environmental pressures. (The Greek *analogos* means similar to one another.)

Homologous structures are considered strong evidence of morphological divergence. These are the same body parts that became modified in different ways in separate lines of descent from a common ancestor.

Analogous structures indicate morphological convergence. Body parts that differed in evolutionarily distant lineages became quite similar in structure and function when those lineages responded to similar environmental pressures.

20.3 EVIDENCE FROM PATTERNS OF DEVELOPMENT

Comparing the patterns by which groups of plants or animals undergo growth and development also yields evidence of evolution. Remember, every new plant or animal follows an intricate, long-term pattern of bodily growth and development (Sections 15.3, 32.2, and 44.2). The new individual proceeds through a series of stages of changes in form, and each stage must be successfully completed before the next can begin.

As you might deduce, gene mutations or changes in chromosome structure tend to be selected against when they disrupt a key stage of development. That is when many constraints on evolution come into play. Every so often, however, such disruptions to a DNA sequence do not disrupt development. Rather, they shift one of the steps in ways that natural selection favors. This might happen, for example, when a transposon inserts itself in a different location in a chromosome and elevates or reduces expression of a gene that affects development.

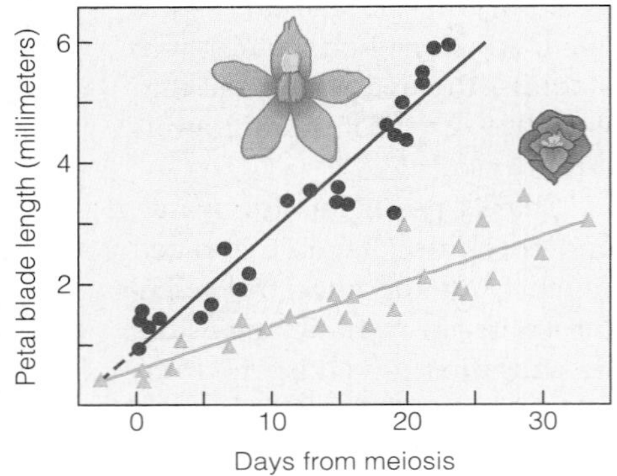

Petal blade length (millimeters)
6
4
2
0
10
20
30
Days from meiosis

a Lengthening of *D. decorum* petals (the *dark blue* graph line) plotted against pollen and egg development in other floral structures, starting with meiosis. The *light-blue* graph line plots the same events over the same span of time for *D. nudicaule*. Petals of this species develop at a slower rate.

b Bud and flower (side and front view) of *D. decorum*   **c** Flower of *D. nudicaule*

Figure 20.6 From comparative morphology, evidence of evolutionary relationship between two larkspurs (*Delphinium*). The time required for flowers to develop and mature is similar in both species. The *rate* of development is slower for most floral structures of *D. nudicaule*. Changes in petal shape are not nearly as great as they are in the other species, *D. decorum*.

Developmental Program of Larkspurs

Flowering plants provide examples of evolution by way of changes in the rates of development. Consider flowers of *Delphinium decorum*, a larkspur in California. A ring of petals guides honeybees to the nectar-storing tube, shown in Figure 20.6. At the center of the flower, outward-bulging reproductive structures give the bees something to hang on to while they draw nectar—and pollinate the flower. *D. nudicaule*, a larkspur of more recent origin, has compact flowers that discourage bees from landing. Its pollinators are hummingbirds, which don't require a landing platform; they hover in front of a flower as they draw nectar from it.

The mature flowers of *D. nudicaule* strongly resemble the buds of *D. decorum* (Figure 20.6). The resemblance probably is a result of a change in the pattern of gene expression; something slows down the rate of floral development in *D. nudicaule*. Think about the plant in which the slowdown first occurred. Its compact flower would have discouraged cross-pollination with flowers of *D. decorum*. A daughter species—*D. nudicaule*—may have originated following that reproductive isolation.

Developmental Program of Vertebrates

Now consider the diverse vertebrates, which range from fishes to amphibians, reptiles, birds, and mammals. By comparing the ways in which their embryos develop, we can find compelling evidence of their evolutionary connection with one another.

The life cycles of all vertebrate lineages include stages of embryonic development. The early stages are strikingly similar. Look at Figure 20.7*a*. Without the labels, would you know which embryo is an early stage of a fish, lizard, chicken, or human?

The early embryos of vertebrates strongly resemble one another because they inherited the same ancient plan for development. According to that plan, tissues form only when cells divide in certain patterns and also interact in prescribed ways. The gut, heart, bones, skeletal muscles, and other parts grow and develop in prescribed ways. But they do so only if each developmental step is properly completed before the onset of the next step.

During the evolution of all vertebrates, most DNA alterations that disrupted pivotal steps of embryonic development probably had lethal effects on later stages. Embryos of different groups have remained similar because alterations that disrupted the early steps were strongly selected against.

Further reading: Student Guide to InfoTrac on web site

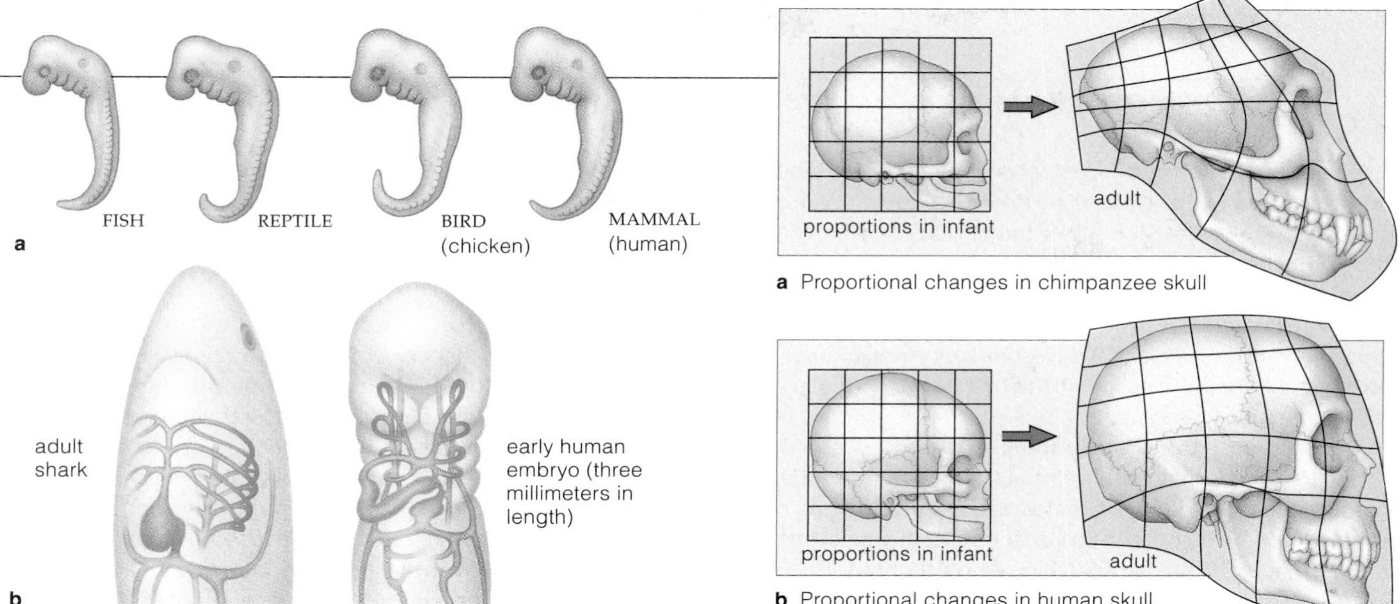
FISH
REPTILE
BIRD (chicken)
MAMMAL (human)
a
adult shark
early human embryo (three millimeters in length)
b
proportions in infant
adult
a Proportional changes in chimpanzee skull
proportions in infant
adult
b Proportional changes in human skull

Figure 20.7 From comparative embryology, examples of evidence of evolutionary relationship among vertebrates.

(**a**) Adult vertebrates show great diversity, yet their very early embryos retain striking similarities. This is evidence of change in a shared program of development. (**b**) Fishlike structures still form in early embryos of reptiles, birds, and mammals. For example, a two-chambered heart (*orange*), certain veins (*blue*), and parts of arteries called aortic arches (*red*) develop in fish embryos—and they persist in adult fishes. The same structures form in an early human embryo.

Figure 20.8 Pronounced morphological differences between two lines of descent, thought to be an outcome of changes in the timing of developmental steps. This example compares proportional changes in the skull bones of a human and a chimpanzee. The skulls are quite similar in infants.

Think of the representations of infant skulls as paintings on a blue rubber sheet divided into a grid. Stretching the sheet deforms the grid's squares. For the adult skulls, differences in size and shape within corresponding grid sections reflect differences in growth patterns.

How, then, did adults of different groups get to be so different? At least some differences probably came about by heritable changes in the onset, rate, or time of completion of developmental steps. Such changes could increase or reduce the relative sizes of tissues or organs and thereby modify the form or structure of body parts. (These are called allometric changes.) They also could lead to adult forms that retain some juvenile features.

Molecular biologists Roy Britten, Wanda Reynolds, and others hypothesize that transposons, not single gene mutations, brought about most of the variation among lineages. Remember enhancers and promoters? These short sequences in eukaryotic DNA help switch genes on and off (Sections 14.2 and 15.5). Transposons have their own enhancers and promoters. When they change locations and insert themselves next to a gene, they can have a powerful regulatory effect on it.

The "rubber grid" in Figure 20.8 indicates how a change in growth rates might underlie the pronounced proportional differences in skull bones of two primates: chimpanzees and humans. From infants to adults, the grid for humans stays fairly intact in the cranial region that encloses the brain. For chimps, facial bone growth outstrips the growth of cranial bones; the grid becomes extremely distorted. Adult skulls for the two lineages end up with significant proportional differences.

Only primates carry the *Alu* transposon, and they have been doing so for at least 30 million years. The Alu enhancer binds with several hormones—including estrogen and thyroid hormones—that have important roles in the timing of many developmental events. Was this transposon pivotal in primate evolution?

Think about this: Between 6 million and 4 million years ago, the forerunners of humans and chimpanzees diverged from the same ancestral stock. More than 98 percent of human DNA still is identical to chimpanzee DNA. How can the remainder account for the large morphological differences between the two primates? Humans happen to carry about a million copies of the Alu transposon, which make up more than 5 percent of our genome. At least some of those copies might have fanned or restricted gene expression in certain tissues. Reflect again on the proportional changes in developing skull bones between humans and chimpanzees. Among the early ancestors of humans, the rates of bone growth apparently changed in a heritable way. Ever since, the rapid growth characteristic of chimpanzee skull bones has been reduced in humans.

Similarities in patterns of development may be clues to evolutionary relationship among plant and animal lineages.

Heritable changes that alter key steps in a developmental program may be enough to bring about major differences in the adult forms of related lineages. Transposons as well as single gene mutations may bring about such changes.

20.4 EVIDENCE FROM COMPARATIVE BIOCHEMISTRY

All species have a mix of ancestral and novel traits. The kinds and numbers of traits they do or do not share are clues to how closely they are related. This is also the case for biochemical traits. Remember, the DNA of each species contains instructions for making the RNAs and proteins necessary to produce new individuals. Thus, *comparisons of the DNA, RNA, or proteins from different species are additional ways of evaluating the evolutionary relationships among species.*

For example, from morphological studies alone, you might already have an idea that monkeys, humans, chimpanzees, and other primates are related. You can test your idea by searching for differences in the amino acid sequence of a protein such as hemoglobin, which all three primates synthesize. Similarly, you could test it by determining whether the nucleotide sequences in their DNA match closely or not much at all. Logically, the species that are most similar in their biochemistry are the most closely related. This premise is central to comparative analysis at the molecular level. Let's look at just two of the methods for deducing relationships.

Protein Comparisons

Suppose two species have the same conserved gene. The amino acid sequences of the gene's product are the same or nearly so. The absence of mutation implies that the two species are close relatives. What if the sequences differ quite a bit? If so, many neutral mutations must have accumulated in them. A very long time must have passed since the species shared a common ancestor.

Consider the highly conserved gene that specifies cytochrome *c*, a protein component of electron transport chains. Species ranging from aerobic bacteria to humans synthesize it. In humans it has a primary structure of 104 amino acids. Figure 20.9 shows how the amino acid sequences for cytochrome *c* from a fungus, a plant, and an animal show striking similarity. Now think about this: Compared to human cytochrome *c*, the *entire* sequence is identical in chimpanzees. It differs by 1 amino acid in rhesus monkeys, 18 in chickens, 19 in turtles, and 56 in yeasts. On the basis of this biochemical information, would you assume humans are more closely related to a chimpanzee or a rhesus monkey? A chicken or a turtle?

Nucleic Acid Comparisons

Typically, structural alterations that resulted from gene mutations are dispersed through nucleotide sequences of DNA and RNA molecules. Some number of unique alterations have accumulated in each lineage. Suppose you isolate a single strand of DNA (or RNA) from two species. *The extent to which the strand from one species base-pairs with a comparable strand from the other is a rough measure of the evolutionary distance between them.*

That is the thinking behind comparisons based on **nucleic acid hybridization**. As mentioned in Section 16.5, this refers to any base-pairing that occurs between DNA or RNA sequences from two different sources. For instance, researchers induce the two strands of a double-stranded DNA molecule to unwind from each other. They also induce a DNA molecule from another species to unwind, then recombine the single strands into "hybrid" molecules. Finally, they subject the two double-stranded, hybridized molecules to heating, which breaks the hydrogen bonds holding them together.

The amount of heat energy required to pull apart a hybrid DNA molecule is a measure of the similarity between its two strands. It takes more heat energy to disrupt the hybridized DNA of closely related species. Figure 20.10 describes one application of this method.

Molecular Clocks

In addition to using molecular comparisons to identify the sequence of evolutionary divergences, researchers use them to estimate the timing of divergences. To do so, they must make assumptions about the constancy of evolutionary change at the molecular level.

Think about our own species. When compared to our close primate relatives, we have some unique traits, the result of mutations that accumulated after divergence from the ancestral primate stock. We also share many genes with other species. For example, the last shared ancestor of certain yeasts and humans lived hundreds of millions of years ago. Yet the overall structure of less than 10 percent of the known gene products (proteins) of yeasts has changed; the rest are highly conserved.

Even so, neutral mutations have introduced slight structural differences in the highly conserved genes of different lineages. **Neutral mutations**, recall, have little

$^{+}NH_3$-gly asp val glu lys gly lys lys ile phe ile met lys cys ser gln cys his thr val glu lys gly gly lys his lys thr gly pro asn leu his gly leu phe gly arg lys thr gly gln ala pro gly tyr ser

$^{+}NH_3$-ala ser phe ser glu ala pro pro gly asn pro asp ala gly ala lys ile phe lys thr lys cys ala gln cys his thr val asp ala gly ala gly his lys gln gly pro asn leu his gly leu phe gly arg gln ser gly thr thr ala gly tyr ser

$^{+}NH_3$-thr glu phe lys ala gly ser ala lys lys gly ala thr leu phe lys thr arg cys leu gln cys his thr val glu lys gly gly pro his lys val gly pro asn leu his gly ile phe gly arg his ser gly gln ala glu gly tyr ser

Figure 20.9 Primary structure of three versions of cytochrome *c*, a major protein component of electron transport systems in cells. The amino acid sequence is highly conserved, even in lineages that are evolutionarily distant. The three sequences are from a representative yeast (*top row*), wheat (*middle row*), and primate (*bottom row*). The *gold* sequences highlight amino acids that are identical in all three species. The probability that this pronounced molecular resemblance resulted by chance alone is extremely low.

Further reading: Student Guide to InfoTrac on web site

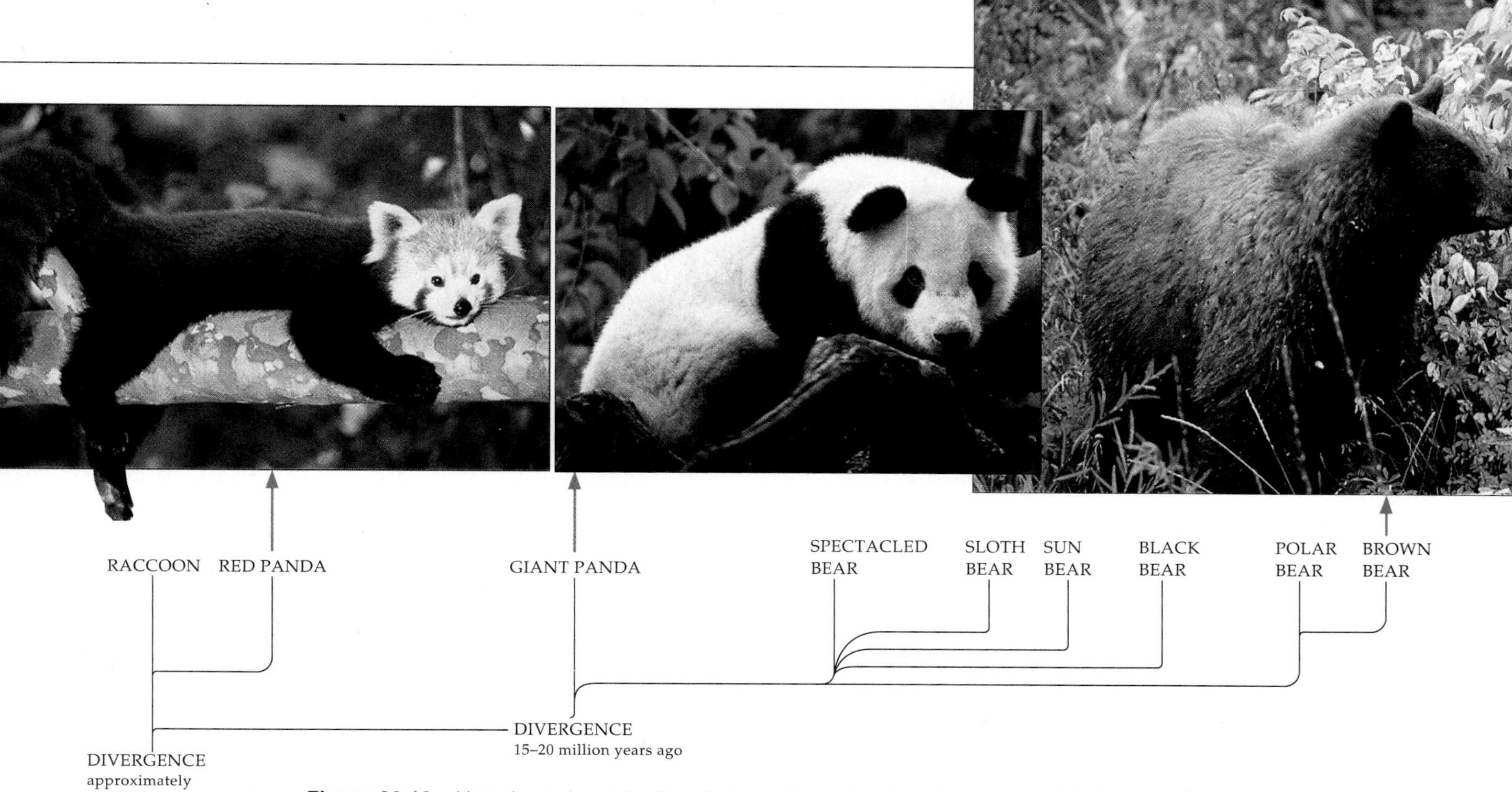
RACCOON
RED PANDA
GIANT PANDA
SPECTACLED BEAR
SLOTH BEAR
SUN BEAR
BLACK BEAR
POLAR BEAR
BROWN BEAR
DIVERGENCE
15–20 million years ago
DIVERGENCE
approximately
40 million years ago

Figure 20.10 How the red panda, the giant panda, and various bears are related, according to DNA–DNA hybridization studies.

A giant panda is a plant-eating mammal. Unlike deer, cattle, and other ungulates (plant-eating mammals with hooves), it cannot efficiently digest tough cellulose fibers of plants. Ungulates pummel plant material inside multiple stomach chambers (Section 42.1). Permanent bacterial residents in the chambers produce cellulose-digesting enzymes. By contrast, a panda has a *meat-eating* mammal's gut. Its diet consists of bamboo which, pound for pound, is not nearly as nutrient-packed as meat. The panda must grasp bamboo and eat quantities of it to get enough proteins, fats, and carbohydrates. Its rounded, short-toed paws are not good at holding and stripping leaves from bamboo stalks. But each front paw has a thumblike bony digit. When a panda uses a "thumb" in opposition to the five toes on a paw, it can grip a bamboo meal.

The giant panda certainly looks like a bear. It also resembles the red panda, which lives in the same general area in China and eats bamboo. Then again, neither bears nor the red panda has thumbs. Is the giant panda more closely related to red pandas or bears?

Researchers studied the extent to which single-stranded DNA from a giant panda will hybridize with DNA from a red panda and from bears. The more mismatched nucleotide bases, the greater the evolutionary distance between these mammals. Results from DNA–DNA hybridization studies support an earlier hypothesis. The three are related by descent from a shared ancestor that lived more than 40 million years ago. A divergence took place among the descendants. One branching led to modern raccoons and to the red panda. Another led to bears. Between 20 million and 15 million years ago, a split from the bear lineage gave rise to ancestors of the giant panda. Thus the giant panda appears to be much more closely related to bears than to the red panda.

ala ala asn lys asn lys gly ile ile trp gly glu asp thr leu met glu tyr leu glu asn pro lys lys tyr ile pro gly thr lys met ile phe val gly ile lys lys lys glu glu arg ala asp leu ile ala tyr leu lys lys ala thr asn glu-COO^-
ala ala asn lys asn lys ala val glu trp glu glu asn thr leu tyr asp tyr leu leu asn pro lys lys tyr ile pro gly thr lys met val phe pro gly leu lys lys pro gln asp arg ala asp leu ile ala tyr leu lys lys ala thr ser ser-COO^-
asp ala asn ile lys lys asn val leu trp asp glu asn asn met ser glu tyr leu thr asn pro lys lys tyr ile pro gly thr lys met ala phe gly gly leu lys lys glu lys asp arg asn asp leu ile thr tyr leu lys lys ala cys glu-COO^-

or no effect on survival or reproduction. They can slip past agents of selection and accumulate in the DNA.

By some calculations, neutral mutations in highly conserved genes accumulated at a regular rate. Think of the accumulation of neutral mutations in a lineage as a series of predictable ticks of a **molecular clock**. Then turn the hands of the clock back, so that the total number of ticks will "unwind" down through the great geologic intervals of the past. Where the last tick stops, that is roughly the time of origin for the lineage.

Biochemical similarities are greatest among the most closely related species and weakest among the most distantly related.

20.5 IDENTIFYING SPECIES, PAST AND PRESENT

So far in this book, we have mentioned only a few of the kinds of organisms that originated and disappeared over the past 3.8 billion years. Many millions of species traveled on that long evolutionary road. At the present time, the surviving, descendant species also number in the millions.

Back in the 1500s, European naturalists were content simply to identify and describe all the new specimens that were being discovered during an exciting period of global exploration (Section 17.1). Previously unknown species have been turning up ever since. For example, almost on a weekly basis, biologists in tropical forests of Brazil and elsewhere find new examples of diversity, especially among insects. Recently, dozens of unknown species were discovered living in complete darkness in a cave 24 meters (80 feet) underground in southwestern Romania. Their ancestors had been trapped inside 5.5 million years ago and had been nourished, directly or indirectly, by chemoautotrophic bacteria that lived in warm water, rich in hydrogen sulfide, on the cave floor.

Even this brief look at the magnitude of diversity is enough to convey the challenges facing **taxonomy**. This field of biology is concerned with identifying, naming, and classifying species. When they make their judgment calls, taxonomists attempt to be objective, but available information about species can be interpreted differently and arranged in different ways.

Assigning Names to Species

Let's start with something easy: the system of naming species, which all biologists agree upon. Corn plants, vanilla orchids, houseflies, humans—these and all other known species have a scientific name that is recognized immediately throughout the world (Figure 20.11). Each name was assigned in accordance with a system that Carl von Linné devised in the eighteenth century. You may have heard this Swedish naturalist referred to by his more prestigious, latinized name, Linnaeus.

Linnaeus was a naturalist whose enthusiasm knew no bounds. He sent ill-prepared students around the world to gather specimens of plants and animals, and is said to have lost a third of his collectors to the rigors of their expeditions. Although perhaps not commendable as a student adviser, Linnaeus did go on to develop a **binomial system** by which species are assigned a two-part Latin name.

The first part of the species name is generic; it is descriptive of similar species that are thought to be the same type of organism and thus are grouped together. Such a grouping is called a **genus** (plural, genera). The second part is the **specific name** which, in combination with the generic name, refers to one kind of organism only. Such scientific names provide an unambiguous way for people anywhere to discuss the same organism.

KINGDOM
PHYLUM
CLASS
ORDER
FAMILY
GENUS
SPECIES
Plantae
Anthophyta (flowering plants)
Monocotyledonae (monocots)
Commelinales
Poaceae
Zea
Z. mays
Plantae
Anthophyta
Monocotyledonae
Orchidales
Orchidaceae
Vanilla
V. planifolia
Animalia
Arthropoda
Insecta
Diptera
Muscidae
Musca
M. domestica
Animalia
Chordata
Mammalia
Primates
Hominidae
Homo
H. sapiens
COMMON NAME:
corn
vanilla orchid
housefly
human

Figure 20.11 Taxonomic classification of four representative organisms. Each is assigned to seven ever more inclusive categories, or taxa: species, genus, family, order, class, phylum, kingdom.

Further reading: Student Guide to InfoTrac on web site

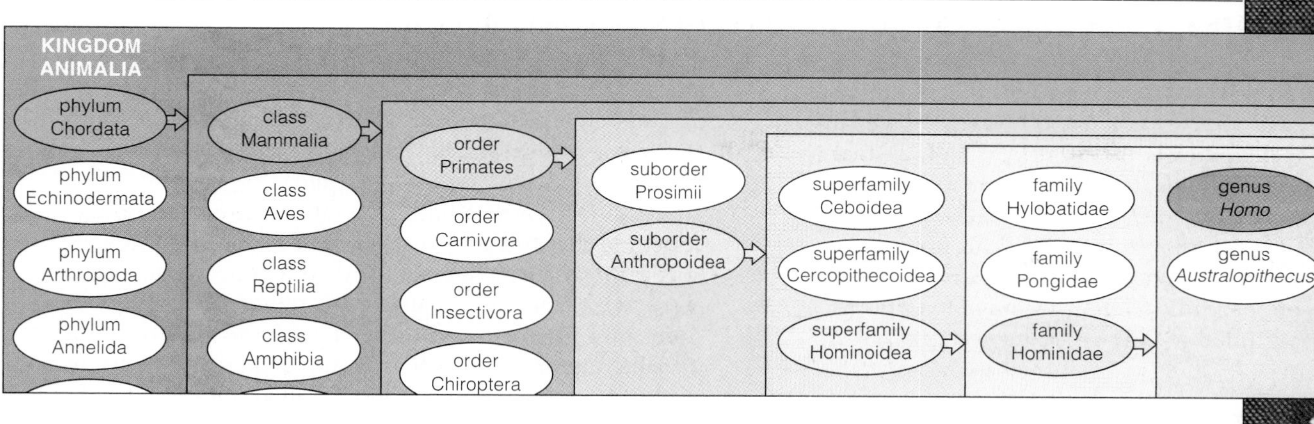
KINGDOM ANIMALIA
phylum Chordata
phylum Echinodermata
phylum Arthropoda
phylum Annelida
class Mammalia
class Aves
class Reptilia
class Amphibia
order Primates
order Carnivora
order Insectivora
order Chiroptera
suborder Prosimii
suborder Anthropoidea
superfamily Ceboidea
superfamily Cercopithecoidea
superfamily Hominoidea
family Hylobatidae
family Pongidae
family Hominidae
genus Homo
genus Australopithecus
species
Homo sapiens (only living species of this genus)

Figure 20.12 Classification of one species, *Homo sapiens* (modern humans), of the animal kingdom. The species itself is represented by Linnaeus, here outfitted with equipment suitable at the time for exploration. Appendix I includes additional taxa.

For example, there is only one *Ursus maritimus*, or polar bear. Other bears include *Ursus arctos* (the brown bear) and *Ursus americanus* (the black bear). Notice that the first letter of the generic name is capitalized and that the second (specific) name is not. The specific name is never used without the full or abbreviated generic name preceding it, for it also can be the second name of a species in a different group. Thus, *U. americanus* does indeed mean black bear, but *Homarus americanus* means Atlantic lobster, and *Bufo americanus* means American toad. (Hence one would not order *americanus* for dinner unless one is willing to take what one gets.)

Groupings of Species—The Higher Taxa

Shortly after Linnaeus formulated the binomial system, it became the core organizing unit for a classification scheme. **Classification schemes** are organized ways of retrieving information about particular species. The Linnaean scheme was based on perceived similarities or differences in morphological traits, such as the number of legs, body size, wings or no wings, coloration, and so forth. Later, biologists invented ever more inclusive groupings that are meant to reflect relationships among species. Such groupings of species are known as **higher taxa** (singular, taxon). Some familiar higher taxa include family, order, class, phylum, and kingdom.

As you might deduce from Figure 20.11, ranking by taxa alone does not yield information about patterns of relationships. In themselves, the higher taxa reflect only relative degrees of morphological similarity. A corn plant and a vanilla orchid are closely related (both are monocots) but are grouped in separate orders. A human and a housefly are both animals but are only distantly related; they are not even in the same phylum. Humans belong to a superfamily designated Hominoidea, not to superfamily Ceboidea. But they are closer to Ceboidea than to suborder Prosimii, as Figure 20.12 indicates.

Taxonomists now attempt to assign species to higher taxa in terms of **phylogeny**. We define this word as the scientific study of evolutionary relationships among species. The connectedness starts with what researchers perceive to be the most ancestral forms of life. It continues on through all of the genetic divergences that led to all of the descendant species.

Putting Classification Schemes To Work

Classifying organisms has its own rewards for anyone who burns with curiosity about life's diversity. But it has its practical applications as well. For example, how might you go about deciding which habitats should be protected to help conserve a rare, poorly understood species? You might start with species that are classified as closely related to the rare species. If the relatives are better understood, then you can make inferences on the basis of *their* needs for particular habitats. As another example, how might you test the effectiveness of a new antibiotic against a pathogenic bacterium, such as the one causing Legionnaires' disease? Perhaps you could conduct tests on a close primate relative of humans, the assumption being that its physiological responses to infection and to the antibiotic will be similar, if not the same. In short, the information that becomes organized in classification schemes can touch our lives in many different ways, at many different levels.

Each species is assigned a two-part scientific name, which consists of its genus and specific name. It also is assigned to ever more inclusive groupings of species, the higher taxa.

Classification schemes organize information about species and simplify its retrieval. The newer, phylogenetic schemes attempt to reflect evolutionary relationships among species.

FINDING EVOLUTIONARY RELATIONSHIPS AMONG SPECIES

The Systematics Approach

Evolutionary systematics is the branch of biology that applies evolutionary theory to the task of identifying patterns of diversity over time and in the environment. It approaches the patterns through work in taxonomy, classification, and phylogenetic reconstruction. Reflect, for a moment, on the kinds of clues we have to patterns of diversity, as outlined in the preceding sections:

The fossil record
The geologic record
Comparative embryology
Homologous structures
Comparative biochemistry

Such clues are interpreted through an understanding of microevolutionary processes. For example, suppose you start with the assumption that the forerunners of living reptiles evolved by gene mutation, natural selection, and genetic drift. If this is true, then lizards, dinosaurs, and other reptilian groups arose from their ancestral stock by genetic divergences and speciation events. To identify the similarities and differences in morphology, life-styles, and habitat pressures of such groups, you study reptiles in a Costa Rican rain forest, and then in the Gila Desert. You search the literature on biochemical studies for clues to how closely or distantly the groups are related. You compare all the findings against both the fossil record and information on environments to which earlier reptiles were adapted.

After analyzing all of the available information, you conclude that turtles are at one end of the evolutionary spectrum, and dinosaurs and crocodiles are at the other. You also suspect birds share a common ancestor with dinosaurs. To portray your perceptions of the pattern of diversity among the reptiles, you decide to construct an evolutionary tree diagram. How will you do it?

Actually, There Is No Single Systematics Approach

Within systematics there are different schools of thought on how to represent patterns of diversity. With *classical* taxonomy, classification schemes and evolutionary tree diagrams are constructed to reflect the perceived degree of morphological divergences among major lineages. In Figure 20.13*a*, for example, birds are not grouped with dinosaurs and other reptiles. Feathers alone are seen to be a key innovation that allowed the ancestors of birds to diverge so far from dinosaurs that their descendants are a distinct group.

With *cladistic* taxonomy, groups are arranged by branch points in an evolutionary tree diagram (*clad–* is from a Greek word for branch). Only species that share derived traits are grouped past the branch point that represents the last shared common ancestor. A **derived trait** is a novel feature that evolved only once and is shared only among descendants of the ancestral species in which it evolved. Figure 20.13*b* is a cladistic scheme.

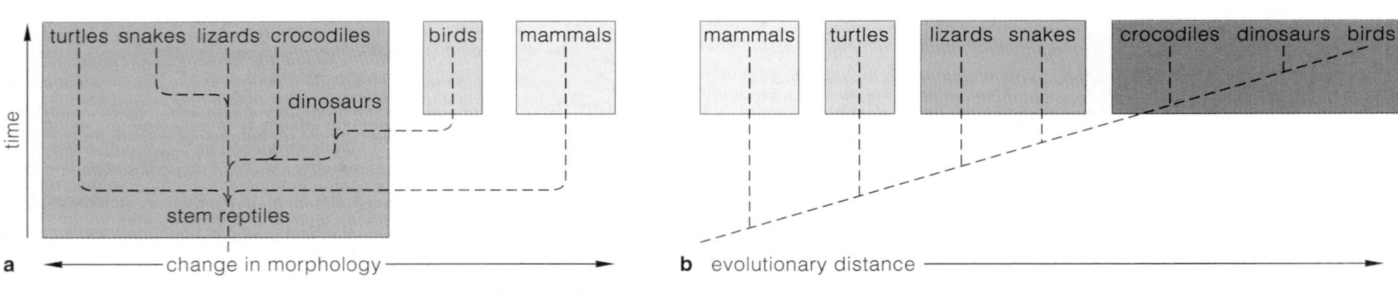

turtles snakes lizards crocodiles
dinosaurs
birds
mammals
time
stem reptiles
a
change in morphology
mammals
turtles
lizards snakes
crocodiles dinosaurs birds
b
evolutionary distance

Figure 20.13 Reconstructions of the evolutionary history of a few groups of vertebrates, according to two schools of systematics. (**a**) By classical systematics, each group is defined by a number of traditionally accepted morphological traits, such as the presence or absence of scales, feathers, and fur. (**b**) In a cladistic scheme, they are grouped on the basis of derived traits.

GREEN PYTHON

NILE CROCODILE

Further reading: Student Guide to InfoTrac on web site

OWL

DINOSAURS

SEA TURTLE

CHAMELEON

As you can see by comparing Figure 20.13*a* and *b*, these two approaches to assessing patterns of diversity among organisms can translate into large differences in how taxonomists end up grouping them. For example, by using branchings as the cut-off points, so to speak, crocodiles and dinosaurs have far more in common with birds than they do with lizards and snakes.

Evolutionary tree diagrams based upon a cladistic approach are known as **cladograms**. Section 20.7 shows how they are constructed. Bear in mind, cladograms do not directly convey "who came from whom." They only portray *relative* relationships among groups. The groups that are closer together on a cladogram simply share a more recent common ancestor than those farther apart.

Cladograms might be a more objective—hence less treacherous—way to reconstruct evolutionary history. When classical taxonomists assign more importance to one morphological divergence over another, they make judgment calls, and some of the calls may be subjective.

Why bother thinking about the difference between these two approaches? Which approach you subscribe to may influence more than how you look at crocodiles and birds. It may influence how you decide to assign relative importance to the similarities and differences among, say, regional human populations. It may affect how you interpret the kinds and distributions of fossils that are presumably on or near the evolutionary road to modern humans. As you will see shortly, it may also shape your view of the entire sweep of life's evolution. Especially with the recent developments in comparative biochemistry, we have startling new insights into the first major divergences from the first organisms ever to appear on Earth.

Reconstructing the evolutionary history of a given lineage must be based on detailed understanding of the morphology, life-styles, and habitats of the groups represented, and on biochemical comparisons of its existing members.

With classical taxonomy, taxa are grouped by assigning relative importance to morphological divergences.

With cladistic taxonomy, taxa are grouped by derived traits that reflect how recently they shared a common ancestor.

Focus on Science

20.7 CONSTRUCTING A CLADOGRAM

To see how phylogenetic reconstruction works in practice, consider how you might go about constructing and then interpreting a cladogram. Briefly, you will be analyzing an organism from different groups in terms of a number of morphological, physiological, and behavioral traits (also called characters) that clearly differ among them. Your choice of organisms to study represents the *ingroup*, and it is based on at least some understanding that they are likely to be related. You also select a different organism, one not too distantly related, that can serve as your point of reference for evaluating the evolutionary distances within the ingroup.

Let's suppose you are interested in the evolution of vertebrates, which are all animals with a backbone. You select seven of them: a lamprey, trout, lungfish, turtle, cat, gorilla, and human. For each vertebrate, you know investigators have recorded the six different traits shown in Figure 20.14*a*. To keep things simple, you focus only on the presence (+) or absence (−) of these traits. From information that is presented in Chapter 27, you decide that the lamprey is only distantly related to the other vertebrates listed. For example, it lacks jaws and paired appendages. This means you can use the lamprey as the outgroup. Compared to the ingroup, an *outgroup* is the organism exhibiting the fewest of the shared traits. The idea is to interpret any deviation from an outgroup as being a derived trait, which suggests a morphological divergence and a branching in the evolutionary tree.

LAMPREY
TURTLE

CAT
GORILLA

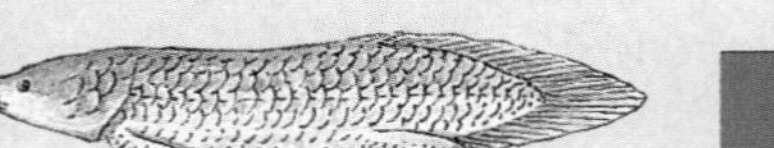
LUNGFISH

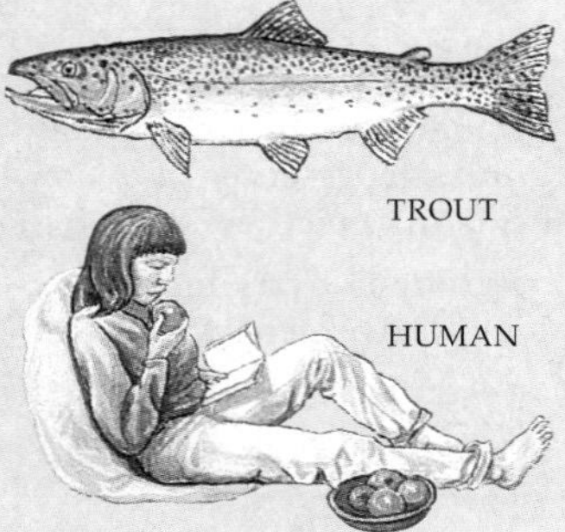
TROUT
HUMAN

Figure 20.14 Selection of seven vertebrates, which are about to be arranged in the type of evolutionary tree diagram known as a cladogram.

Taxon	Jaws	Limbs	Hair	Lungs	Tail	Shell
Lamprey	−	−	−	−	+	−
Turtle	+	+	−	+	+	+
Cat	+	+	+	+	+	−
Gorilla	+	+	+	+	−	−
Lungfish	+	−	−	+	+	−
Trout	+	−	−	−	+	−
Human	+	+	+	+	−	−

a Tabulation of traits present or absent

Taxon	Jaws	Limbs	Hair	Lungs	Tail	Shell
Lamprey	0	0	0	0	0	0
Turtle	1	1	0	1	0	1
Cat	1	1	1	1	0	0
Gorilla	1	1	1	1	1	0
Lungfish	1	0	0	1	0	0
Trout	1	0	0	0	0	0
Human	1	1	1	1	1	0

b Tabulation of outgroup and ingroup conditions

In Figure 20.14*b*, each zero (0) across the columns of traits for each vertebrate indicates an outgroup condition. Each one (1) means the vertebrate shows the derived trait.

As the next step, you look for the derived traits that your selected groups of vertebrates do or do not share. For example, you see that the human and the gorilla share five derived traits, whereas the human and the cat share only four. For so few traits, you can now sketch out a simple cladogram. In practice, systematists may resort to computers. They often use a great many traits to examine many taxa, and it takes a computer to find the pattern that is best supported by so much information.

Figure 20.15*a* shows how you would sketch information based on the jaw trait. All of the vertebrates *other than the outgroup* have jaws. The jaw is a derived trait. Figure 20.15*b* shows what happens to the cladogram when you add information based on lungs. All the vertebrates except the lamprey and the trout have lungs. The lungfish, human, gorilla, turtle, and cat appear to be a *monophyletic* group, meaning they have a common evolutionary heritage. (They share two derived traits with respect to the lamprey and one with respect to the trout.)

Figure 20.15*c* through *e* shows the resulting changes in the cladogram as you keep adding information about other derived traits to it.

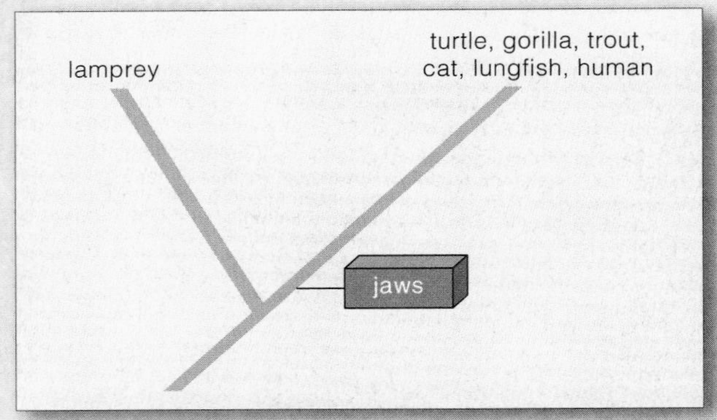
lamprey
turtle, gorilla, trout,
cat, lungfish, human
jaws
a

Figure 20.15 Step-by-step construction of a cladogram, an evolutionary tree diagram that groups taxa by derived traits.

Further reading: Student Guide to InfoTrac on web site —

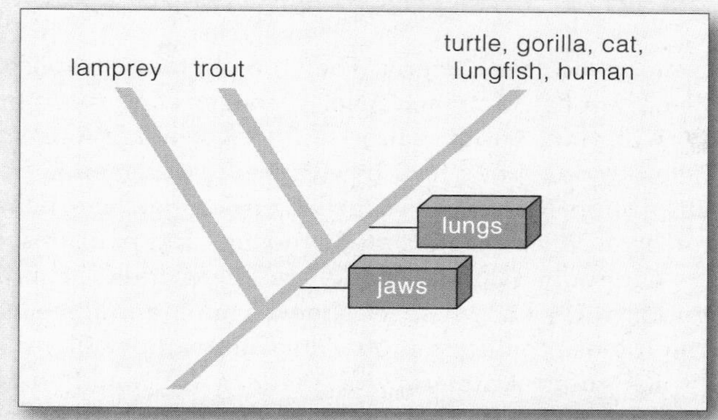
lamprey
trout
turtle, gorilla, cat, lungfish, human
lungs
jaws
b

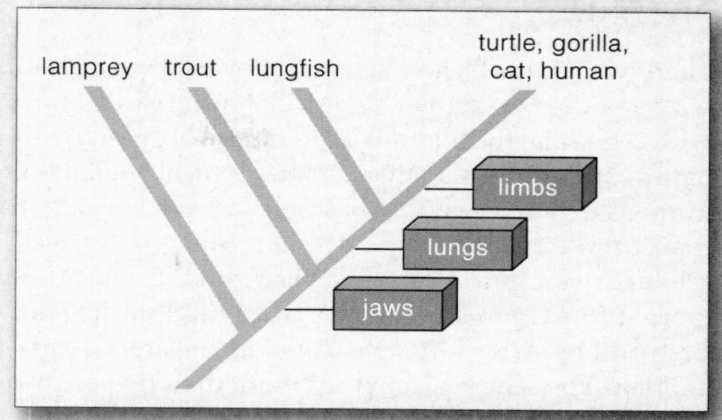
lamprey
trout
lungfish
turtle, gorilla, cat, human
limbs
lungs
jaws
c

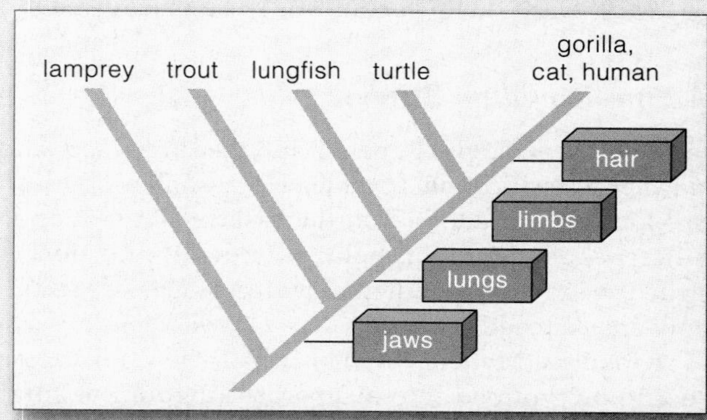
lamprey
trout
lungfish
turtle
gorilla, cat, human
hair
limbs
lungs
jaws
d

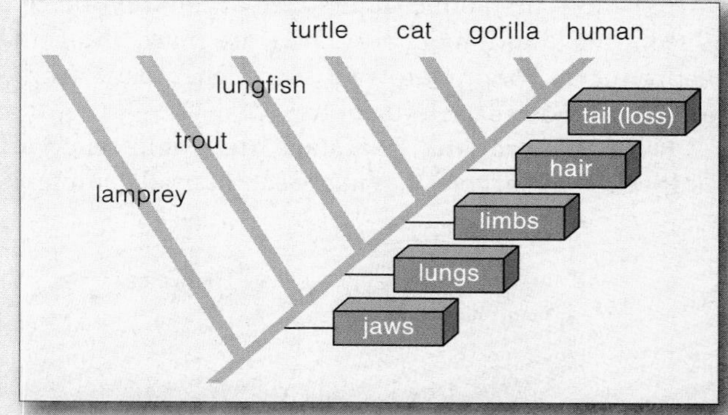
turtle
cat
gorilla
human
lungfish
trout
lamprey
tail (loss)
hair
limbs
lungs
jaws
e

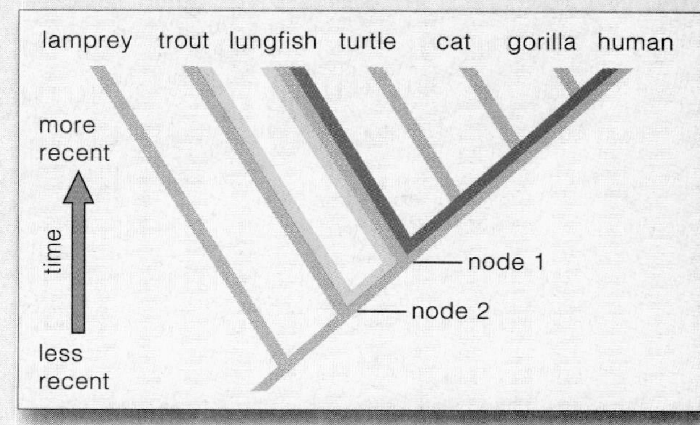
lamprey
trout
lungfish
turtle
cat
gorilla
human
more recent
time
less recent
node 1
node 2
f

Wisely, you did not use the presence of a hard shell as a trait. Only the turtle has a shell. Although the shell is a derived trait relative to the outgroup, it is not shared with any of the other animals under consideration. Therefore, it is not useful in reconstructing this phylogeny.

Every species has many such characters. The traits may be useful in identifying the species but not in determining relationships. In this example, none of the traits provides information that is contradictory to other traits. In reality, such contradictions often show up. Why? All organisms are a mixture of evolutionarily ancient traits and derived traits. Also, not all of the traits you initially regarded as homologous may actually be homologous. Morphological convergences may have occurred (Section 20.2).

How will you "read" the final cladogram? Remember, it indicates relative relatedness, not ancestors. As you can see, a human is more closely related to a gorilla than to a cat. The gorilla isn't the *ancestor* of humans (it, too, is a modern organism), but both organisms shared a common ancestor more recently than either did with the cat. The cat, human, and gorilla as a group are more closely related to one another than they are to the turtle. These are easy relationships to perceive. Perhaps less obviously, your cladogram also tells you a lungfish is more closely related to a human than to a trout. Follow the lines or branches of the cladogram from the human and lungfish back to the intersection (or node) where they meet (Figure 20.15*f*, node 1). Next, trace back the branches of the lungfish and trout to their intersection (Figure 20.15*f*, node 2). As you can see, the trout–lungfish intersection is nearer to the base of the tree, compared with the lungfish–human intersection. The higher up the evolutionary tree diagram, the more derived traits are shared by the taxa. The lower they are on the tree, the fewer traits are shared.

Another way to think about this is to regard the axis of the cladogram as a time bar, but one without absolute dates. *The lower the position of the branch point between two groups on a cladogram, the more distant is the most recent common ancestor of the taxa being compared.*

Finally, even though the axis of the cladogram can be viewed as a time line, bear in mind that the ordering of taxa from left to right is not important. No matter how the taxa are arranged across the top of the cladogram, follow the branches back to the lungfish–human and the trout–lungfish intersections and you will get the same result. Also remember this: All of the species positioned at the top of the cladogram are modern, *existing* species; they are neither ancestors nor descendants of one another.

20.8 HOW MANY KINGDOMS?

There was a time when just about everything in nature was classified as animal, vegetable, or mineral. Starting with Aristotle, individuals who paid closer attention to nature started to perceive its hierarchical organization in more detailed ways. It has been a long investigative road from Aristotle to the present. Many classification schemes were put forth, then discarded or modified as more information emerged. A two-kingdom scheme as first proposed by Linnaeus divided life into plants or animals. His scheme lasted for more than two centuries even though it had nowhere to put tens of thousands of bacterial species, which continue to be discovered.

A Five-Kingdom Scheme

In 1969, the ecologist Robert Whittaker proposed a *five-kingdom* classification scheme. He assigned species to one kingdom or another on the basis of similarities and differences in their morphology, modes of nutrition, cell structure, and developmental features. The key features only are sketched out here:

Kingdom Monera. In Whittaker's view, all bacteria should be grouped into a separate kingdom of single-celled organisms that display little internal complexity but far more metabolic diversity than eukaryotic cells. Within this kingdom's boundaries are many thousands of producers (photoautotrophs and chemoautotrophs) and decomposers (heterotrophs).

Kingdom Protista. Protistans are single-celled and multicelled eukaryotes having great internal complexity, compared to the bacteria. They are photoautotrophs or heterotrophs. Many are major pathogens and parasites.

Kingdom Fungi. These multicelled eukaryotes are heterotrophs. They feed by extracellular digestion and absorption. Many species are major decomposers, hence nutrient cyclers. Others are pathogens and parasites.

Kingdom Plantae. Plants are eukaryotes. Of more than 295,000 known species, nearly all are multicelled, chloroplast-equipped producers (photoautotrophs).

Kingdom Animalia. All are eukaryotic, and all are multicelled heterotrophs ranging from the microscopic to the giant. We know of more than a million species, including herbivores, carnivores, and parasites.

Figure 20.16 has a modified version of Whittaker's scheme, which enjoyed wide acceptance until recently.

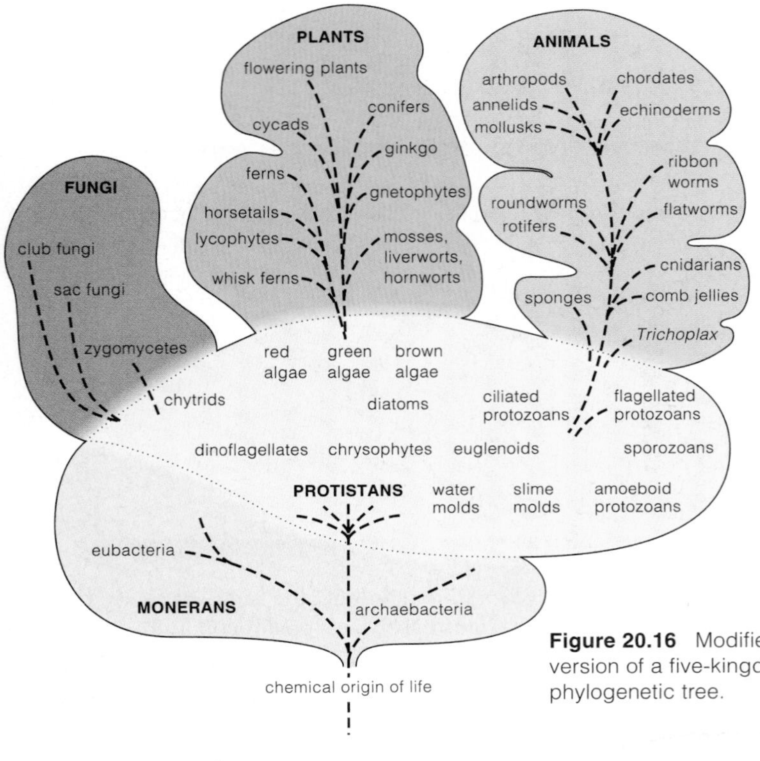

Figure 20.16 Modified version of a five-kingdom phylogenetic tree.

Along Came the Archaebacteria

For years, biologists knew about archaebacteria. These single-celled prokaryotes thrive in harsh habitats, such as salt lakes and hydrothermal vents. Compared with more common bacteria, such as *Escherichia coli*, their cell wall, plasma membrane, and chemical composition have unique features. Even so, the differences did not seem to be significant enough to pull the archaebacteria out of the kingdom Monera.

Then, in the 1970s, the microbiologist Carl Woese and other researchers at the University of Illinois took advantage of new nucleic acid hybridization and DNA sequencing methods (Chapter 16). After comparing the base sequences of rRNAs from a variety of organisms, they proposed that bacteria should be divided into two major taxa. Archaebacteria, they said, are as different from eubacteria as they are from eukaryotic cells.

Evidence supporting their conclusion came in 1996. Carol Bult and her colleagues sequenced all 1.7 million base pairs of DNA from the bacterium *Methanococcus jannaschii*. Woese and Bult worked together to decipher the sequence. More than half the genes had never been found before. Some are closer to the genes of humans and other eukaryotes than to the ones in eubacteria.

A consensus is growing among biologists to replace the five-kingdom scheme with one that better reflects the evolutionary history of life. By this newer scheme, the first great divergence led to the **Eubacteria** and the **Archaebacteria**. Soon after, the next major branching —from the archaebacteria—gave rise to the ancestors of all single-celled eukaryotes. Those early eukaryotic species later gave rise to all multicelled forms of life.

A THREE-DOMAIN SCHEME Evidence of this primordial branching has given rise to a *three-domain* scheme, as in Figure 20.17. As you might suspect, microbiologists (who know the most about bacteria) are its fiercest

Further reading: Student Guide to InfoTrac on web site

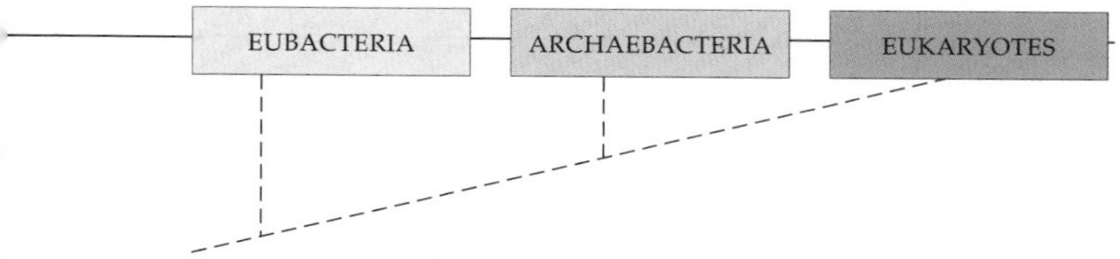

EUBACTERIA ARCHAEBACTERIA EUKARYOTES

Figure 20.17 A three-domain classification scheme, especially favored by microbiologists.

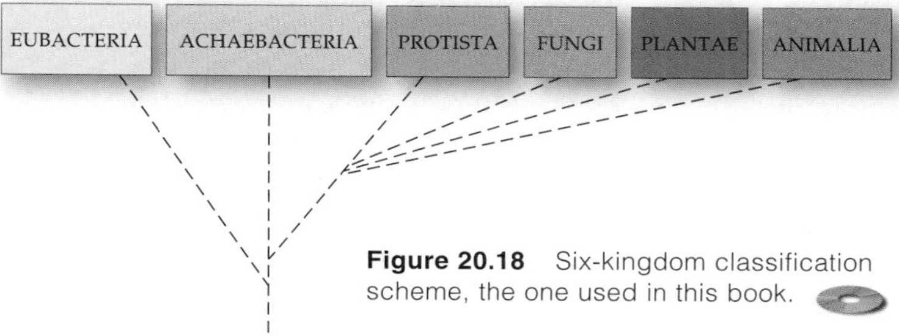
EUBACTERIA
ACHAEBACTERIA
PROTISTA
FUNGI
PLANTAE
ANIMALIA

Figure 20.18 Six-kingdom classification scheme, the one used in this book.

advocates. Those showing the most reluctance to adopt it argue that it does not give suitable weight to the far greater biodiversity among eukaryotes.

A SIX-KINGDOM SCHEME For our introductory book, it is enough to recognize a **six-kingdom classification scheme**. By this scheme, the archaebacteria and the eubacteria are each accorded the rank of kingdom, like protistans, fungi, plants, and animals (Figure 20.18).

Our understanding of evolutionary relationships is itself evolving, but don't lose sleep over this. Simply know that *species* are the real entities in such schemes. Higher taxa—the groupings of "twigs, branches, and limbs" of evolutionary trees—are convenient analogies.

As you read through the next unit, think about the genetic continuity that threads through the history of life. The boundaries of kingdoms and other taxa enclose and cut across lines of descent that have continued, unbroken, through time. The boundaries have not been assigned arbitrarily, but neither are they absolute. All classification schemes, including the most recent ones, are subject to ongoing modifications as more pieces of the macroevolutionary puzzle come together.

Each species is assigned a two-part scientific name (genus and specific name). It is classified in ever more inclusive groupings of species, the higher taxa.

Classification schemes organize and simplify the retrieval of information about species. Phylogenetic schemes attempt to reflect evolutionary relationships among species.

A widely accepted classification scheme groups organisms into five kingdoms: Monera, Protista, Fungi, Plantae, and Animalia. The groupings are based on morphology, modes of nutrition, cell structure, and other traits.

Recent evidence from comparative biochemistry strongly favors a six-kingdom scheme, which divides the Monera into two kingdoms: the Archaebacteria and Eubacteria.

Focus on Science

20.9 ALONG CAME DEEP GREEN

In 1999, during the XVI International Botanical Congress, botanists weighed in with their own version of the tree of life. They recognize five kingdoms of multicelled species within the domain of eukaryotes: animals, fungi, green plants, red plants, and one more taxon (stramenoplies) that includes brown plants. In their view, lumping all traditional plants into one kingdom detracts from the powerful predictive value of phylogenetic classification.

For example, *Deep Green* (the Green Plant Phylogeny Research Coordination Group) is attempting to make evolutionary sense of more than 500,000 species of green plants. So far, their research points to two major lineages, one with most of the green algae, the other with some green algae and all land plants. If the common ancestor of the two lineages were single celled, then multicellularity arose at least twice—once along the diverging road to *Ulva*, *Codium*, and other green seaweeds, and once along the road leading to land plants. Further, even if land plants did have a distant marine ancestor, after the time of divergence, all descendants that evolved in this second great lineage have been freshwater species.

This is no pie-in-the-sky classification scheme. In 1994, Deep Green formed because of advances in technological, theoretical, and computational analysis. Considerable data were already available, but no mechanism was in place to study green plants in a coordinated way. Some plant groups were overstudied; others were still obscure. The data sets from studies of different molecules and of morphological characters were not being compared; they rarely encompassed the same taxa. Software for data storage, retrieval, and analysis had to be improved before it could handle analyses of such size and complexity.

Starting in 1994, Brent Mishler, Russell Chapman, Pamela Soltis, Mark Buchheim, Charles Delwiche, and Elizabeth Zimmer led a five-year effort involving 200 botanists from twelve countries. Participants discussed standards for maintaining and adding to phylogenetic databases. They implemented findings through a series of meetings, workshops, and massive collaborative analyses.

Beyond the intellectual satisfaction of revealing the evolutionary history of plant lineages, as in Figure 20.19, the work of Deep Green may provide a rigorous basis for predicting new sources of biochemicals, biological control agents, and genetic material for crop improvements. Their basic research should prove to be a wellspring for many practical applications.

Figure 20.19 *Amborella*, which grows wild only in New Caledonia, a South Pacific island. Comparative genetic analyses suggests it is a living representative of a divergence that started the flowering plant lineage (compare Section 25.1).

20.10 EVIDENCE OF A CHANGING EARTH

We turn now to the impact of the Earth's evolution on the history of life. While early geologists were busily discovering fossils and mapping the record of the rocks, Charles Darwin was so taken with one of their theories of Earth history that it helped shape his view of life. By the **theory of uniformity**, recall, mountain building and erosion had repeatedly worked over the Earth's surface in precisely the same ways through time (Section 17.2). Great stacks of sedimentary layers, as in Figure 20.20, were evidence of this. Yet, the more geologists clinked their hammers against the rocks, the more they realized repetitive change was only part of the picture. Like life, the Earth had changed irreversibly; it evolved.

An Outrageous Hypothesis

For example, geologists noticed that the coastlines of continents on both sides of the southern Atlantic looked like jigsaw puzzle pieces that could fit together. Were the continents once joined, and did they drift apart? This idea came to be known as continental drift. Similarly, they wondered whether mountain ranges that parallel the coasts formed when huge, drifting masses collided.

Figure 20.21 shows an early attempt to sketch all of the continents as one landmass. Later, Alfred Wegener proposed a refined model of the supercontinent, which he named Pangea. He used clues from glacial deposits to infer climate zones of past times and clues from the world distribution of fossils and existing species.

Figure 20.20 A splendid slice through time: the Grand Canyon of the American Southwest, once part of an ocean basin. Its sedimentary rock layers formed slowly, over hundreds of millions of years. Later, geologic forces lifted the ancient stacked layers above sea level. Later still, the erosive force of rivers carved the deep canyon walls and exposed the layers.

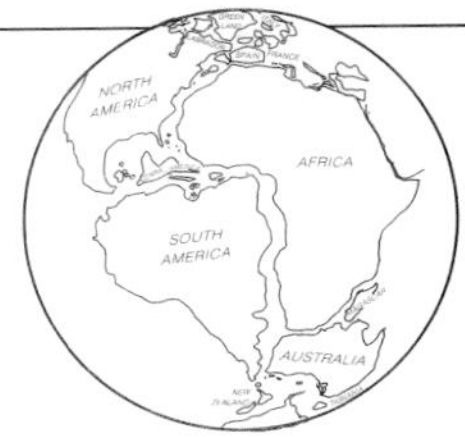
NORTH AMERICA
AFRICA
SOUTH AMERICA
AUSTRALIA

Figure 20.21 Wishful reconstruction of the ancient supercontinent Pangea by H. Baker, who conveniently squished the continents together and rotated islands and peninsulas to improve the fit.

Most scholars could not accept the hypothesis. They knew the continents were thinner and less dense than the underlying mantle. Continents drifting under their own power seemed about as plausible as a facial tissue plowing on its own across the surface of the water in a bathtub. The theory of uniformity was so powerfully entrenched, it was shutting out alternative ideas.

Then scientists made an amazing discovery. When iron-rich rocks first form, their internal structure takes on a north–south orientation in response to the Earth's magnetism. Yet the orientations in ancient rocks were *not* aligned with the magnetic poles. Either the poles moved—or continents did. Using the magnetic record of 200-million-year-old rocks from North America and Europe as a guide, scientists made sketches. In their sketches, they rotated the rocks to the inferred, original north–south alignment. The alignments dovetailed only when they joined the two continents together.

More puzzles! Deep-sea probes had revealed evidence of seafloor spreading. Scientists found that molten rock erupts from great, continuous ridges on the seafloor, flows laterally in both directions, then hardens to form new crust. As the seafloor spreads, it forces older crust down into deep trenches elsewhere in the seafloor. As it turned out, the ridges and trenches are edges of huge, relatively thin plates that are somewhat like pieces of a cracked eggshell (Figure 20.22). When the crustal plates slowly move, they raft the continents to new positions. The drift hypothesis could now be incorporated into a broader theory of crustal movements.

In time, researchers did demonstrate the predictive power of the new theory, which is now known as **plate tectonics**. Earlier, in South America, Africa, India, and Australia, fossils of the *same* plants (*Glossopteris*, Figure 20.23) and of a mammal-like reptile had been found in the *same* succession of glacial deposits, coal seams, and basalt. The plant's seeds were much too heavy and the animals were too small for dispersal across the empty vastness of the open ocean. They must have evolved on the same land mass. That land mass was once part of an early supercontinent, named Gondwana.

In the 1930s, A. du Toit predicted that *Glossopteris* fossils, the mammal-like reptile, and the succession of glacial deposits, coal seams, and basalt on the southern continents would also be discovered on Antarctica. It was not until the 1960s that evidence gathered during Antarctic expeditions supported the prediction—and the plate tectonics theory.

Further reading: Student Guide to InfoTrac on web site —

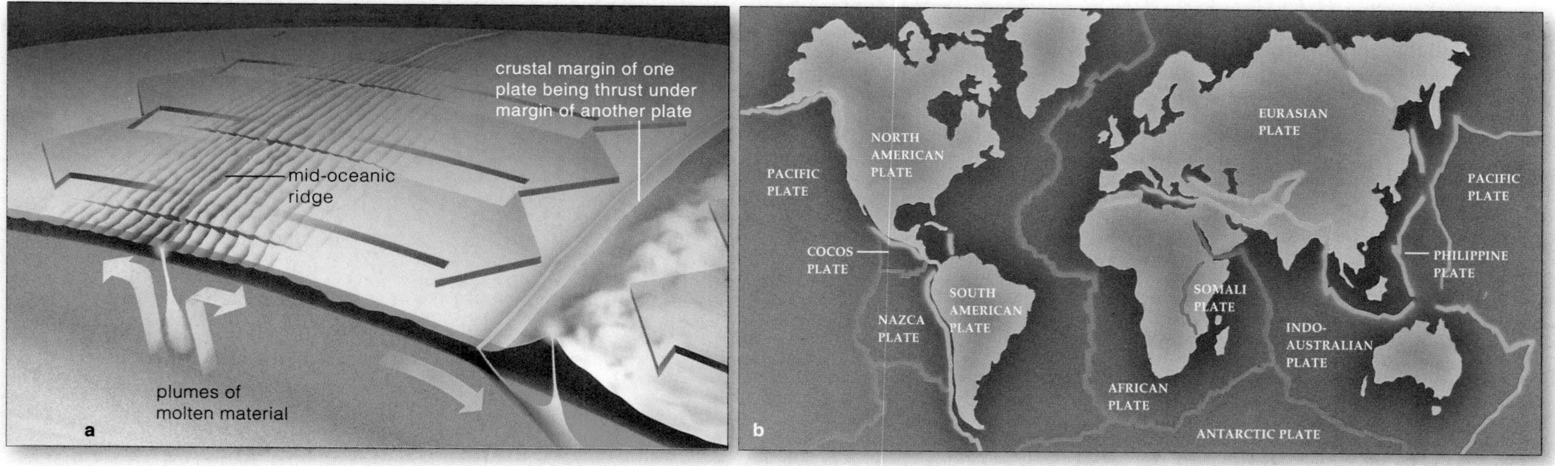

crustal margin of one plate being thrust under margin of another plate
mid-oceanic ridge
plumes of molten material
a
NORTH AMERICAN PLATE
EURASIAN PLATE
PACIFIC PLATE
PACIFIC PLATE
COCOS PLATE
PHILIPPINE PLATE
SOUTH AMERICAN PLATE
NAZCA PLATE
SOMALI PLATE
INDO-AUSTRALIAN PLATE
AFRICAN PLATE
ANTARCTIC PLATE
b

Figure 20.22 Some forces of geologic change. (**a**) The Earth's crust is fractured into rigid plates that slowly split apart, drift, and collide. Driving the movements are huge upwelling plumes from the interior that spread laterally beneath the crust. Plumes rupture the crust at ridges on the ocean floor. Plumes also cause deep rifting and splitting within continents. Such rifting is proceeding in Missouri, at Lake Baikal in Russia, and in eastern Africa. Long ago, *superplumes* violently ruptured the crust. Volcanoes still form at these "hot spots," as they do today in the Hawaiian Archipelago and elsewhere.

At mid-oceanic ridges, molten material seeps out, cools, and slowly forces seafloor away from the rupture. Seafloor spreading displaces plates away from these ridges. Often, displacement forces other parts of one plate under an adjoining plate and uplifts it. The Cascades, Andes, and other great mountain ranges that parallel the coasts of continents formed this way. (**b**) Current configuration of the crustal plates.

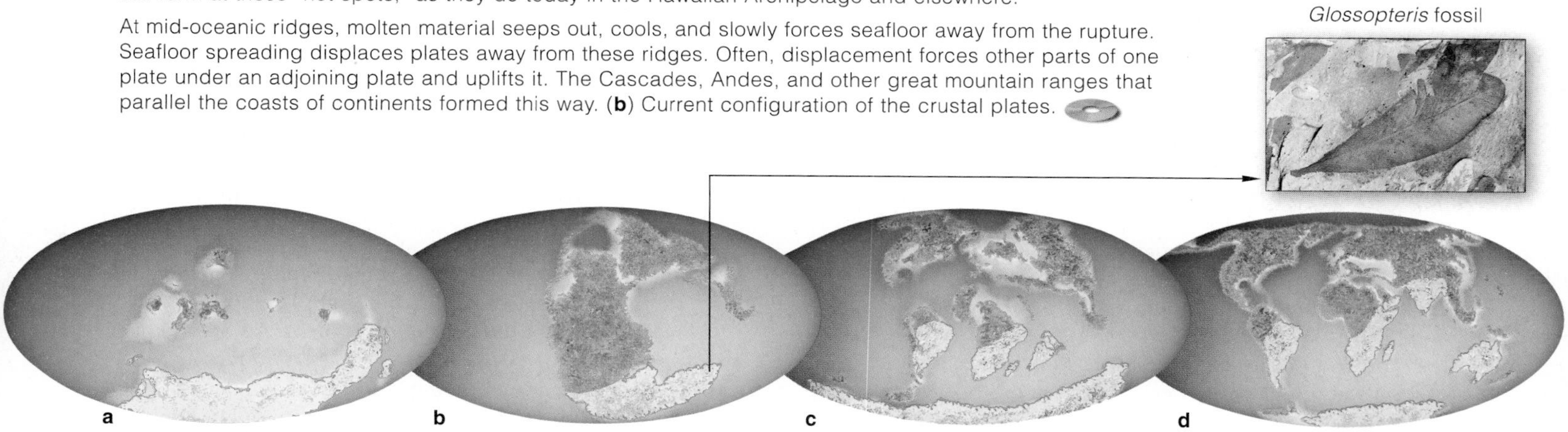

Glossopteris fossil

a b c d

Figure 20.23 Reconstructions of drifting continents. (**a**) Gondwana (*yellow*), 420 million years ago. (**b**) All land masses collided to form Pangea. About 260 million years ago, certain seed ferns (*Glossopteris*) and other plants lived on the Gondwana portion of this supercontinent but nowhere else. (**c**) Position of fragments from Pangea's breakup 65 million years ago, when the last dinosaurs disappeared. (**d**) The fragments 10 million years ago, closing in on their present positions.

Drifting Continents, Changing Seas

Take a good look at Figure 20.23. In the remote past, plate movements put enormous land masses on collision courses. In time the land masses converged and formed supercontinents, which later split apart at deep rifts to become new ocean basins. Gondwana, one of the early supercontinents, drifted south from the tropics, across the south polar region, then north until it crunched into other land masses to form a single world continent. That supercontinent—Pangea—stretched from pole to pole, and an immense ocean spanned the rest of the globe! All the while, slow erosive forces of water and wind were resculpting the surface of the land through time. As if this were not enough, asteroids and meteorites bombarded the crust. The impacts and their aftermath had long-term effects on global temperature and climate.

All of those changes in the land, oceans, and atmosphere profoundly influenced the evolution of life. Think of early life flourishing in the warm, shallow waters along the shorelines of continents. When continents collided, the shorelines vanished. Such changes were devastating for many lineages. Yet, even as old habitats vanished, new ones opened for the survivors—and evolution took off in new directions. As you will see in the unit to follow, the histories of the Earth and life are inseparable.

Over the past 3.8 billion years, changes in the Earth's crust, the atmosphere, and the oceans profoundly affected the evolution of life.

Focus on Science

20.11 DATING PIECES OF THE MACROEVOLUTIONARY PUZZLE

For interested students, we leave this chapter with a look at a powerful tool for dating fossils and gaining insights into the history of life. Remember, probably for as long as they have been digging up the earth, people have been turning up fossilized leaves, shells, skeletons, and other stone-hard evidence of past life. Figure 20.24 shows two examples. *How do we know how old they are?*

a
b

Figure 20.24 (**a**) Fossilized frond of a tree fern that unfurled its leaves more than 250 million years ago. (**b**) A fossilized leaf that dropped from a sycamore tree 50 million years ago.

As Section 20.10 described, some time ago, geologists hypothesized that if newly formed rocks accumulate on top of older rocks, then fossils in older rock layers must be more ancient than fossils in more recently deposited layers. They used their perception to count backwards through great numbers of rock layers and to construct a chronology of Earth history—a geologic time scale. They defined the boundaries of successive spans by sequences of fossils and other clues buried in those layers. However, *no one was able to assign firm dates along the length of the scale until the discovery of radioactive decay, which made it possible to convert relative ages into absolute ones.*

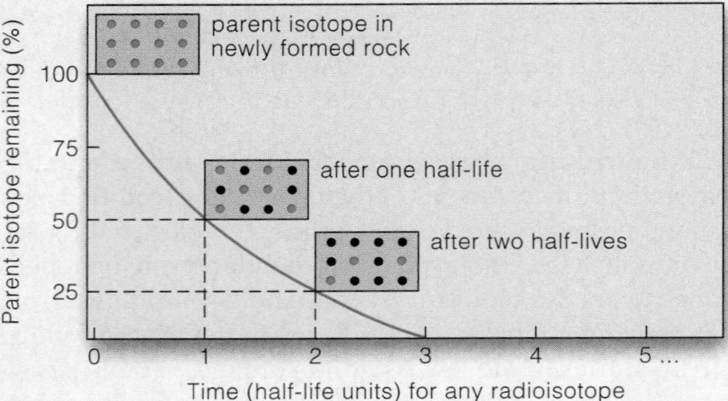
parent isotope in newly formed rock
after one half-life
after two half-lives
Parent isotope remaining (%)
100
75
50
25
0
1
2
3
4
5 ...
Time (half-life units) for any radioisotope

Figure 20.25 Generalized diagram of radioactive decay. Compare the proportion of the parent radioisotope in a rock sample over time, to the proportion of each decay product. During *each* half-life unit, the amount of the parent radioisotope will diminish by half.

By a method called **radiometric dating,** researchers now measure proportions of (1) a radioisotope in a mineral that became trapped long ago in a new rock and (2) a daughter isotope that formed from it in the same rock. Remember, the atomic nucleus of a radioactive element has an *unstable* number of protons and neutrons. The nucleus spontaneously decays—it gives up energy and one or more particles of itself—until it reaches a more stable configuration.

Half-life is the time it takes for half of a given quantity of any radioisotope to decay into a different, daughter isotope that is less unstable (Figure 20.25). The rate of decay is constant; changes in pressure, temperature, or chemical state cannot alter it. For example, uranium 238 is present in zircon, which is a mineral in most volcanic rocks. Its nuclei decay into thorium 234, a still-unstable daughter isotope that in turn decays into something else, and so on through a series of intermediate daughter isotopes to lead 206—the final, most stable configuration for this series (Table 20.1). By using uranium 238, with its half-life of 4.5 billion years, researchers realized that the Earth formed more than 4.6 billion years ago.

Radiometric dating works for volcanic rocks or ashes. However, most fossil-containing rocks formed by the compaction of sand and other sediments. Thus, the only way to date most fossil-containing rocks is to determine their position relative to volcanic rocks in the same area.

You may have heard of dating methods that rely on carbon 14 (^{14}C). The greatest numbers of this radioisotope form continually in the upper atmosphere, where they combine with oxygen to form carbon dioxide. Together with more stable isotopes of carbon, small amounts of ^{14}C enter the web of life by photosynthesis. All organisms incorporate it. When organisms die, the amount of ^{14}C starts to decrease as an outcome of radioactive decay. ^{14}C has a half-life of 5,730 years. Thus, amounts in bone, wood, shells, and other organic substances are too small to detect by direct analysis. Instead, researchers employ scintillation counters to monitor emissions from a given sample. The older that sample is, the fewer emissions (counts) will be recorded in a specified period.

Table 20.1 Radioisotopes Commonly Used in Radiometric Dating

Radioisotope (unstable)		More Stable Product	Half-Life (years)	Useful Range (years)
Samarium 147	⟶	Neodymium 143	106 billion	>100 million*
Rubidium 87	⟶	Strontium 87	48.8 billion	>100 million
Thorium 232	⟶	Lead 208	14 billion	>200 million
Uranium 238	⟶	Lead 206	4.5 billion	>100 million
Potassium 40	⟶	Argon 40	1.25 billion	>100,000
Uranium 235	⟶	Lead 207	700 million	>100 million
Carbon 14	⟶	Nitrogen 14	5,730	0–60,000

* The symbol > means greater than.

20.12 SUMMARY

1. All species, past and present, are related by way of descent from their common ancestors, starting with the origin of the first cells on Earth.

2. Macroevolution refers to patterns, trends, and rates of change among groupings of species over long spans of time. Those clues to the past provide us with insight into the sweeping continuity of relationship in nature.

3. The fossil record, the geologic record, comparative morphology, and comparative biochemistry have yielded extensive evidence of evolution. The evidence is based on similarities and differences in body form, functions, behavior, and biochemistry.

4. Fossils are recognizable, physical evidence of life in the distant past. They start to form after an organism or traces of it become buried in volcanic ash or sediments. The organic remains become infused with mineral-rich water. As sediments accumulate above the burial site, they impose increasing pressure which, with chemical changes, transforms the remains to stony hardness.

a. Fossils are present in layers of sedimentary rock. The deepest layers accumulate first; they are the oldest. Thus, the older the layers, the older the fossils.

b. Scientists use abrupt transitions in the sequence of fossil assemblages as boundaries for intervals of the geologic time scale. In the modern version of the scale, life originated during the Archean eon. Oldest to most recent fossils extend from the Proterozoic era through the Paleozoic, Mesozoic, and Cenozoic eras.

c. The completeness of the fossil record is variable in terms of the species represented, where they lived, and the stability of their tombs since fossilization occurred.

5. Comparative morphology often reveals similarities that imply evolutionary relationship among groups.

a. *Homology* refers to similarity in one or more body parts between different groups of organisms that imply descent from a shared ancestor. Homologous structures suggest morphological divergence: modification of the same body parts in different ways in different lines of descent from the common ancestor.

b. By contrast, *analogy* refers to body parts that once differed in evolutionarily remote lineages but that later converged in structure and function as those lineages responded to similar environmental pressures. They are evidence of morphological convergence.

c. The patterns by which the plants and animals develop show underlying similarities. Changes in DNA that lead to modifications in the onset, rate, or time of the developmental steps probably brought about most morphological differences between related lineages.

6. Comparative biochemistry can identify similarities and differences among species at the molecular level.

a. Nucleic acid comparisons, as by hybridization of the DNA from different species, give strong evidence of evolutionary relationships. The strength with which single-stranded DNA or RNA from one species base-pairs with a single strand from a different species is a rough measure of evolutionary distance between them.

b. The DNA of all species contains an accumulation of neutral mutations in highly conserved genes. Such mutations are like ticks of a molecular clock; they help date divergences from a common ancestor.

7. Taxonomists identify, name, and classify species. By the Linnaean binomial system, each kind of organism can be assigned a two-part Latin name. The first part (genus) identifies species that have shared similarities and presumably were derived from common ancestors. In conjunction with the second part of the name (species epithet), it is the name of the particular species.

8. Classification systems use higher taxa, or ever more inclusive groupings, from species to genera, families, orders, classes, phyla, and kingdoms. Such groupings reflect phylogeny (presumed evolutionary relatedness).

9. This book uses a scheme that groups organisms into six great kingdoms: Archaebacteria, Eubacteria, Protista, Plantae, Fungi, and Animalia. It is based on perceived evolutionary relationships; it is a phylogenetic scheme.

10. Like life, the Earth has evolved. Some of its features change repetitively, as when mountains rise and erode slowly by the same processes. Other features changed irreversibly, as explained by plate tectonics theory:

a. The Earth's crust is fractured into huge, thin, rigid plates that slowly split apart, drift, and collide with one another, rafting the land masses along with them.

b. Great plumes of molten material welling up from Earth's interior drive the movements. At mid-oceanic ridges, material seeps out, cools, hardens, and laterally displaces older seafloor. Seafloor spreading has forced older crust down elsewhere, into deep trenches. Great mountain ranges that parallel coasts formed when one plate thrust under another, which uplifted as a result.

c. Large-scale, long-term changes in the Earth's land masses have changed the ocean and atmosphere, and they have profoundly influenced the evolution of life.

Review Questions

1. Distinguish macroevolution from microevolution. *CI* (*Also compare 19.6*)

2. Will biologists ever be able to read a complete fossil record? Why or why not? *20.1*

3. Name the three eras of the geologic time scale. One of these, the Proterozoic, was so vast that it has since been divided into great eons. In which eon did life originate? *20.1*

4. Define the difference between: *20.2*
   a. homologous and analogous structures
   b. morphological divergence and convergence

5. Comparative morphology refers to comparisons of body form and structures, embryonic and adult, for major lineages. Give an example of such a comparison. *20.2, 20.3*

6. Give reasons why two organisms that are quite different in outward appearances may belong to the same lineage. *20.2, 20.3*

7. Name a protein specified by a gene that has been highly conserved in organisms ranging from bacteria to humans. *20.4*

8. Why do evolutionary biologists apply heat energy to hybrid molecules that contain DNA from two species? *20.4*

9. What type of mutation is the basis of a molecular clock? What does the last tick of a molecular clock signify? *20.4*

10. What type of evidence favors putting archaebacteria in their own kingdom? *20.8*

11. What is the basis of continental drift? Of seafloor spreading? In general, how did such changes in the Earth influence the evolution of life? *20.10*

12. How do biologists determine the age of a fossil? *20.1, 20.11*

13. Define radiometric dating. What does half-life mean? *20.11*

Self-Quiz (*Answers in Appendix III*)

1. Heritable changes in DNA underlying morphological differences between lineages ______.
   a. may have been caused mostly by transposons
   b. affected the onset, rate, and time of development steps
   c. both a and b

2. Morphological convergences may lead to ______.
   a. analogous structures
   b. homologous structures
   c. divergent structures
   d. both a and c

3. A classification system that is ______ is based on presumed evolutionary relationship.
   a. epigenetic
   b. credited to Linnaeus
   c. phylogenetic
   d. both b and c

4. *Pinus banksiana, Pinus strobus,* and *Pinus radiata* are ______.
   a. three families of pine trees
   b. three different names for the same organism
   c. several species belonging to the same genus
   d. both a and c

5. Increasingly inclusive taxa range from ______ to ______.
   a. kingdom; species
   b. kingdom; genera
   c. genera; kingdom
   d. species; kingdom

6. Match these terms suitably.
   ___ phylogeny
   ___ fossil
   ___ stratification
   ___ homology
   ___ molecular clock
   ___ analogy

   a. accumulation of neutral mutations
   b. stone-hard evidence of life in past
   c. similar body parts in different lineages owing to common descent
   d. e.g., shark fins, penguin flippers
   e. evolutionary relationship among species, ancestors to descendants
   f. layers of sedimentary rock

Critical Thinking

1. At the end of your backbone are several small, fused bones, called the coccyx. Is the coccyx a vestigial structure—all that is left of a tail that was a feature of the evolutionarily distant vertebrate (and primate) ancestors of humans? Or is it the start of a newly evolving structure with an as-yet-undetermined function? Take an educated guess, then describe some ways in which you might test whether your answer is plausible.

2. Protein comparisons and nucleic acid hybridization studies help us estimate evolutionary relationship and approximate times for divergences from ancestral stocks. DNA sequence comparisons yield even more accurate estimates. Reflect on the genetic code (Section 14.3). Then suggest why such a code may be a more accurate measure of mutations, mutation rates, and biochemical relatedness.

3. Shannon thinks there are too many kingdoms and sees no good reason to make another one for something as small as archaebacteria. "Keep them with the other prokaryotes!" she says. Taxonomists would call her a "lumper." By contrast, Geoffrey is a "splitter." He sees no good reason to withhold kingdom status from archaebacteria simply because they are part of a microscopic world that not many people know about. Which may be the more useful choice: more or fewer boundaries between groups of organisms? Explain your answer.

4. When walking along a path cut into the side of a steep mountain, you pass rocky layers that are fractured and folded back on themselves in bizarre patterns. You look closely and discover a fossilized shell in the "highest" layer in the series. How would you use the theories of continental drift and plate tectonics to help you decide whether the fossil is of recent or ancient origin?

5. "Missing links" are undiscovered transitional forms between major groups of organisms. *Archaeopteryx*, described earlier in Section 17.4, is one example. Drawing information from Sections 20.1 and 20.10, what are some possible reasons for apparent gaps in the fossil record?

Selected Key Terms

analogy *20.2*
Animalia *20.8*
Archaebacteria *20.8*
binomial system *20.5*
cladogram *20.6*
classification scheme *20.5*
comparative morphology *20.2*
derived trait *20.6*
Eubacteria *20.8*
evolutionary systematics *20.6*
fossil *20.1*
fossilization *20.1*
Fungi *20.8*
genus *20.5*
geologic time scale *20.1*
half-life *20.11*
higher taxon (taxa) *20.2*
homology *20.2*
lineage *CI*
macroevolution *CI*
molecular clock *20.4*
Monera *20.8*
morphological convergence *20.2*
morphological divergence *20.2*
neutral mutation *20.4*
nucleic acid hybridization *20.4*
phylogeny *20.5*
Plantae *20.8*
plate tectonics theory *20.10*
Protista *20.8*
radiometric dating *20.11*
six-kingdom classification scheme *20.8*
specific name *20.5*
stratification *20.1*
taxonomy *20.5*
theory of uniformity *20.10*

Readings *See also www.infotrac-college.com*

Brooks, D. R., and D. A. McLennan. 1991. *Phylogeny, Ecology, and Behavior*. Chicago: University of Chicago Press.

Dott, R., Jr., and R. Batten. 1998. *Evolution of the Earth*. Fourth edition. New York: McGraw-Hill. Good historical perspective on correlations between biological and geologic evolution.

Morell, V. April 1997. "On the Origin of (Amazonian) Species." *Discover* 18(4): 56–64. How evolutionary biologists are working to correlate the geologic history of the Amazon basin, which is astoundingly rich in biodiversity, with patterns of speciation.

Ochert, A., December 1999. "Transposons." *Discover* 59–66.

FACING PAGE: *Patterns of diversity in nature, here represented by different species of plants and fungi.*

EVOLUTION AND BIODIVERSITY IV

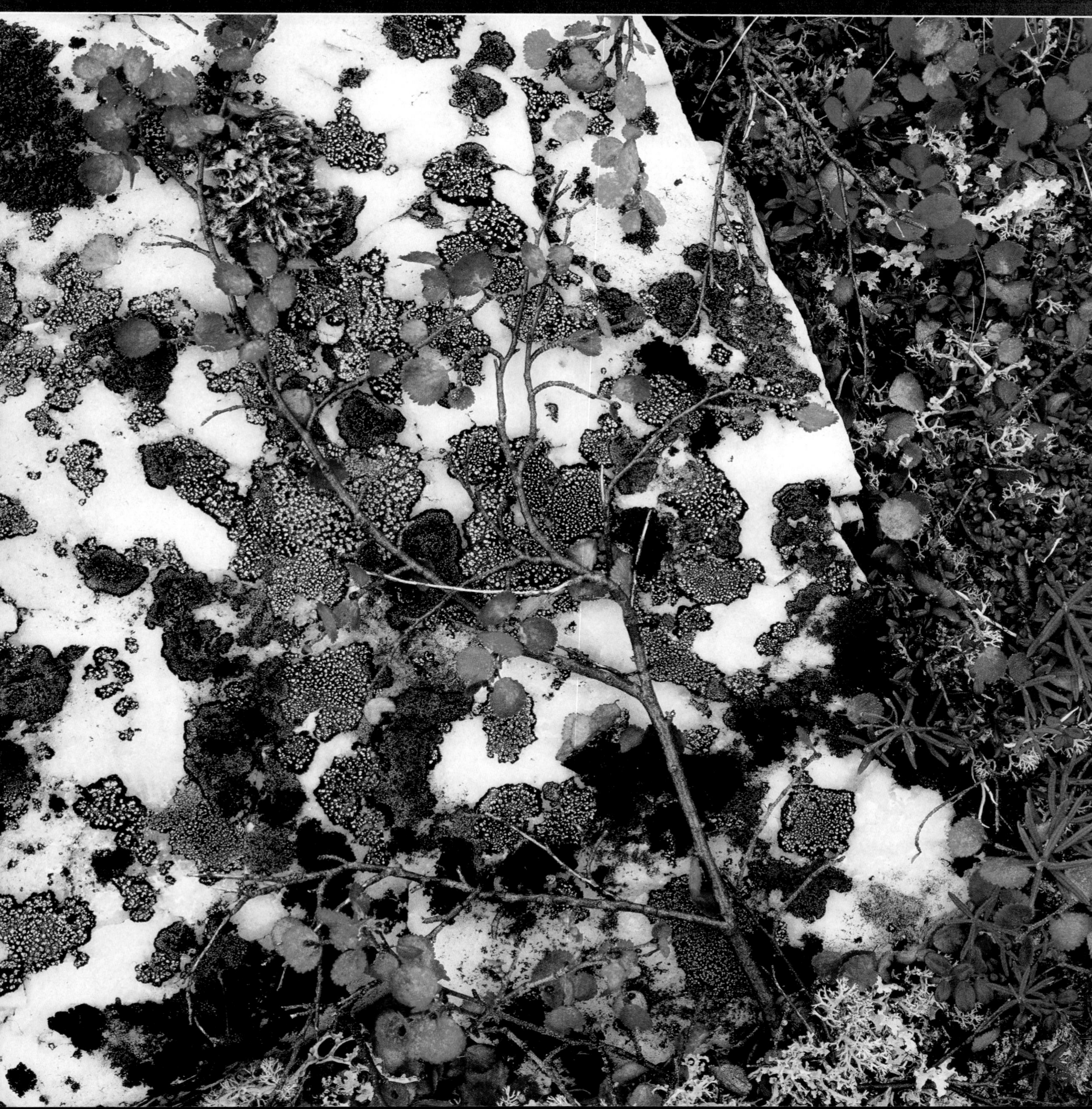

21

THE ORIGIN AND EVOLUTION OF LIFE

In The Beginning . . .

Some clear evening, watch the moon as it rises from the horizon and think of the 380,000 kilometers between it and you. *Five billion trillion times* farther away from you are galaxies—systems of stars—at the boundary of the known universe. Wavelengths traveling through space move faster than anything else—millions of meters per second—yet long wavelengths that originated from faraway galaxies many billions of years ago are only now reaching the Earth.

If we are to accept all known measures, all of the near and distant galaxies in the vast space of the universe are moving away from one another, which means the universe must be expanding. And the prevailing view of how the colossal expansion came about may account for every bit of matter in every living thing.

Figure 21.1 Part of the great Eagle nebula, a hotbed of star formation 7,000 light-years from Earth, in the constellation Serpens. (The Latin *nebula* means mist.) Not shown in this image are a few huge, young stars above the pillars. For the past few million years, intense ultraviolet radiation from the stars has been eroding the less dense surface of the pillars. (By analogy, visualize a strong prevailing wind blowing away sand in a desert and exposing rocks.) Globules of denser gases and dust that have resisted erosion are visible at the surface. Each pillar is wider than our solar system—more than 10 billion miles across! New stars are hatching from the protruding globules; some are shining brightly on the tips of gaseous streamers.

Think about how you rewind a videotape on a VCR, then imagine "rewinding" the universe. As you do this, the galaxies start moving back together. After 12 to 15 billion years of rewinding, all galaxies, all matter, and all of space are compressed into a hot, dense volume about the size of the sun. You have arrived at time zero.

That incredibly hot, dense state lasted only for an instant. What happened next is known as the **big bang,** a stupendous, nearly instantaneous distribution of matter and energy throughout the known universe. About a minute later, temperatures dropped to a billion degrees. Fusion reactions produced most of the light elements, including helium, which are still the most abundant elements in the universe. Radio telescopes detect cooled, diluted background radiation that is a relic of the big bang, left over from the beginning of time.

Over the next billion years, uncountable numbers of gaseous particles collided and condensed under gravity's force to become the first stars. When the stars became massive enough, nuclear reactions were ignited in their core, and they gave off tremendous light and heat. Massive stars continued to contract, and many became dense enough to promote the formation of heavier elements.

All stars have a life history, from birth to an often spectacularly explosive death. In what might be called the original stardust memories, the heavier elements released during the explosions became swept up in the gravitational contraction of new stars, and they became raw materials for the formation of even heavier elements. Even as you read this page, the Hubble space telescope is revealing astounding glimpses of such star-forming activity, as in the dust clouds of Orion, Serpens, and other constellations (Figure 21.1).

Now imagine a time long ago, when explosions of dying stars ripped through our galaxy and left behind a dense cloud of dust and gas that extended trillions of kilometers in space. As the cloud cooled, countless bits of matter gravitated toward one another. By 4.6 billion years ago, the cloud had flattened out into a slowly rotating disk. At the dense, hot center of that great disk, the shining star of our solar system—the sun—was born.

The remainder of this chapter is a sweeping slice through time, one that cuts back to the formation of the Earth and the chemical origins of life. It is the starting point for the next seven chapters, which will take us along lines of descent that led to the present range of species diversity.

The story is not complete. Even so, the available evidence from many avenues of research points to a principle that can help us organize bits of information about the past: *Life is a magnificent continuation of the physical and chemical evolution of the universe, of galaxies and stars, and of the planet Earth.*

KEY CONCEPTS

1. We have evidence that life originated about 3.8 billion years ago. The origin and subsequent evolution of life have been correlated with the physical and chemical evolution of the universe, the stars, and the planet Earth.

2. All inorganic and organic compounds necessary for self-replication, membrane assembly, and metabolism—for the structure and functions of living cells—could have formed spontaneously under conditions that apparently prevailed on the early Earth.

3. The history of life, from its chemical beginnings to the present, spans five intervals of geologic time. It extends through two great eons—the Archean and Proterozoic—and the Paleozoic, Mesozoic, and Cenozoic eras.

4. Not long after life originated, divergences led to two great prokaryotic lineages called the archaebacteria and eubacteria. Soon afterward, the ancestors of eukaryotes diverged from the archaebacterial lineage.

5. Archaebacteria and eubacteria dominated the Archean and Proterozoic eons. Eukaryotic cells originated late in the Proterozoic and became spectacularly diverse. A theory of endosymbiosis helps explain the profusion of specialized organelles that evolved in eukaryotic cells.

6. All six kingdoms of life have a history of persistences, extinctions, and radiations of different lineages.

7. Throughout the history of life, asteroid impacts, drifting and colliding continents, and other environmental insults have had profound impact on the direction of evolution.

21.1 CONDITIONS ON THE EARLY EARTH

Origin of the Earth

Figure 21.1 gave you a view of one of the vast clouds in the universe. Such clouds consist mostly of hydrogen gas, along with water, iron, silicates, hydrogen cyanide, ammonia, methane, formaldehyde, and other simple inorganic and organic substances. The contracting cloud that became our solar system probably was similar in composition. We assume the edges of that cloud cooled between 4.6 and 4.5 billion years ago. Mineral grains and ice orbiting the new sun started clumping together as a result of electrostatic attraction and gravity's pull (Figure 21.2). In time, larger, faster clumps collided and shattered. Some became more massive by sweeping up asteroids, meteorites, and the other rocky remnants of collisions, and gradually they evolved into planets.

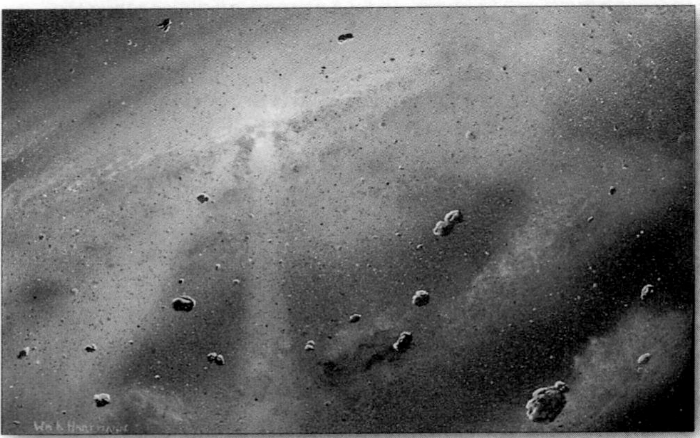

Figure 21.2 Representation of the cloud of dust, gases, and clumps of rock and ice around the early sun.

As the Earth was forming, much of its inner rocky material melted. Asteroid impacts and the Earth's own internal compression and radioactive decay of minerals could have generated the heat necessary to do this. As rocks melted, nickel, iron, and other heavy materials moved to the Earth's interior; lighter ones floated to the surface. The process produced a crust, mantle, and core. The **crust** is an outer region of basalt, granite, and other low-density rocks. It envelops the intermediate-density rocks of the **mantle**, which wraps around a core of very high-density, partially molten nickel and iron.

Four billion years ago, the Earth was a thin-crusted inferno (Figure 21.3*a*). Within 200 million years, life had originated on its surface! We have no record of the event. As far as we know, movements in the mantle and crust, volcanic activity, and erosion obliterated all traces of it. Still, we can put together a plausible explanation of how life originated by considering three questions:

First, can we identify physical and chemical conditions that prevailed on the Earth when life originated?

Second, do known physical, chemical, and evolutionary principles support the hypothesis that large organic molecules formed spontaneously and evolved into molecular systems displaying the fundamental properties of life?

Third, can we devise experiments to test the hypothesis that living systems emerged by chemical evolution?

The First Atmosphere

When the first patches of crust were forming, hot gases blanketed the Earth. Probably this first atmosphere was a mix of gaseous hydrogen (H_2), nitrogen (N_2), carbon monoxide (CO), and carbon dioxide (CO_2). Did it hold gaseous oxygen (O_2)? Probably not. Rocks subjected to intense heat, as happens during volcanic eruptions, do release oxygen, but not much. Also, free oxygen would have reacted at once with other elements. An oxygen atom has an electron vacancy in its outermost shell and tends to fill it by bonding with other atoms (Section 2.3).

If the early atmosphere had not been relatively free of oxygen, organic compounds necessary to assemble cells in the first place would not have been able to form spontaneously—*on their own*. Any oxygen would have attacked their structure and disrupted their functioning.

What about liquid water? Dense clouds blanketed the early Earth, but water reaching the molten surface would have evaporated at once. After the crust cooled and solidified, however, rains fell on the parched rocks. For millions of years, the runoff eroded mineral salts and other compounds from the rocks. Salt-laden waters collected in the depressions in the crust and formed the first seas. If liquid water had not accumulated, then cell membranes—which take on their bilayer organization only in water—could not have formed. No membrane, no cell. Life at its most basic level *is* the cell, which has a capacity to survive and reproduce on its own.

Synthesis of Organic Compounds

Reduce a cell to its lowest common denominator and all that remains are proteins, complex carbohydrates and lipids, and nucleic acids. Existing cells assemble these molecules from smaller organic compounds: the simple sugars, fatty acids, amino acids, and nucleotides. Energy from the environment drives the synthesis reactions. Were small organic compounds also present on the early Earth? Were there sources of energy that spontaneously drove their assembly into the large molecules of life?

Mars, meteorites, the Earth's moon, and the Earth all formed at the same time, from the same cosmic cloud. Rocks collected from Mars, meteorites, and the moon contain precursors of biological molecules, so the same precursors must have been present on the Earth. If this were the case, *then energy from the sun, from lightning, or*

Further reading: Student Guide to InfoTrac on web site

a
to
vacuum
pump
electrodes
spark
discharge
CH_4
NH_3
H_2O
H_2
gases
water out
condenser
water in
water droplets
water containing
organic compounds
boiling water
liquid water in trap
b

Figure 21.3 (**a**) Representation of the Earth during its formation, when the moon's orbit was much closer than it is today. If the Earth had condensed into a planet of smaller diameter, its gravitational mass would not have been great enough to hold on to an atmosphere. If it had settled into an orbit closer to the sun, water would have evaporated from its hot surface. If the Earth's orbit had been more distant from the sun, its surface would have been far colder, and water would have been locked up as ice. Without liquid water, life as we know it never would have originated on Earth.

(**b**) Stanley Miller's experimental apparatus, used to study the synthesis of organic compounds under conditions that presumably existed on the early Earth. The condenser cools circulating steam so that water droplets form.

even from heat being vented from the crust could have been enough to drive their combination into organic molecules.

Stanley Miller conducted the first experimental test of that prediction. First he mixed methane, hydrogen, ammonia, and water inside a reaction chamber of the sort depicted in Figure 21.3*b*. Then he kept the mixture circulating and bombarded it with a spark discharge to simulate lightning. In less than one week, amino acids and other small organic compounds had formed.

In other experiments that simulated conditions on the early Earth, glucose, ribose, deoxyribose, and other sugars formed spontaneously from formaldehyde, and adenine from hydrogen cyanide. Ribose and adenine occur in ATP, NAD^+, and other nucleotides vital to cells.

However, if *complex* organic compounds had formed directly in seawater, could they have lasted long? The spontaneous direction of the required reactions would have been toward hydrolysis, not condensation, in water. How did more lasting bonds form?

By one hypothesis, clay in the rhythmically drained muck of tidal flats and estuaries served as templates (structural patterns) for the spontaneous assembly of proteins and other complex organic compounds. Clay consists of thin, stacked layers of aluminosilicates with metal ions at their surfaces, which attract amino acids. Expose amino acids to some clay, warm the clay with rays from the sun, then alternately moisten and dry it. Condensation reactions will proceed at its surfaces and yield proteins and other complex organic compounds.

By another hypothesis, complex organic compounds formed spontaneously near hydrothermal vents on the seafloor, where species of archaebacteria are thriving today. As experimental tests by Sidney Fox and others show, when amino acids are heated and then placed in water, they spontaneously order themselves into small protein-like molecules, which Fox calls "proteinoids."

However the first proteins formed, their molecular structure dictated how they could interact with other compounds. Suppose some proteins had the structure to function as weak enzymes, and that they hastened bond formation between amino acids. Such enzyme-directed synthesis of proteins would have had selective advantage. In the chemical competition for available amino acids, protein configurations that could promote reactions would win. Proteins would have been favored in still another way—for proteins have the capacity to bind metal ions and other agents of metabolism.

As you will read next, the evolution of metabolism must have been based on such chemical modification. For now, simply reflect on the possibility that selection was at work before the origin of living cells, favoring the chemical evolution of enzymes and other complex organic compounds.

Many diverse experiments yield indirect evidence that the complex organic molecules characteristic of life could have formed under conditions that existed on the early Earth.

21.2 EMERGENCE OF THE FIRST LIVING CELLS

Origin of Agents of Metabolism

A defining characteristic of life is *metabolism*. The word refers to all the reactions by which cells harness energy and use it to drive their activities, such as biosynthesis. During the first 600 million years or so of Earth history, enzymes, ATP, and other organic compounds may have assembled spontaneously, perhaps in the same physical locations. If they did so, their close association would have naturally promoted chemical interactions and the beginning of metabolic pathways.

Imagine an ancient estuary, rich in clay deposits. It is a coastal region where seawater mixes with mineral-rich water being drained from the land. There, beneath the sun's rays, countless aggregations of organic molecules stick to the clay. At first there are quantities of an amino acid; call it *D*. Throughout the estuary, *D* molecules get incorporated into new proteins—until the supply of *D* dwindles. However, suppose a protein that is weakly catalytic also is present in the estuary. It can promote the formation of *D* by acting on an abundant, simpler substance—call it *C*. By chance, some aggregations of organic molecules include that particular enzyme-like protein, and so they have an advantage in the chemical competition for starting materials.

In time, *C* molecules become scarce. At that point, the advantage tilts to aggregations that can promote formation of *C* from even simpler substances *B* and *A*. Assume *B* and *A* are carbon dioxide and water. As you know, carbon dioxide and water occur in essentially unlimited amounts in the atmosphere and in the seas. Chemical selection has favored a synthetic pathway:

$$A + B \rightarrow C \rightarrow D$$

Finally, suppose some aggregations are better than others at absorbing and using energy. Which molecules could bestow such an advantage? Think of the energy-trapping pathway that now dominates the world of life: photosynthesis. It starts at pigments called chlorophylls. The portion of a chlorophyll molecule that absorbs light and gives up electrons is a porphyrin ring structure. Porphyrins also are present in cytochromes, which are part of electron transport systems in all photosynthetic and aerobically respiring cells. Porphyrin molecules can spontaneously assemble from formaldehyde—one of the molecular legacies of cosmic clouds (Figure 21.4). Was porphyrin a major electron transporter of some of the early metabolic pathways? Perhaps.

Origin of Self-Replicating Systems

Another defining characteristic of life is a capacity for reproduction, which now starts with protein-building instructions in DNA. The DNA molecule is fairly stable, and it is easily replicated before each cell division. As you know from earlier chapters, arrays of enzymes and RNA molecules operate together and carry out DNA's encoded instructions.

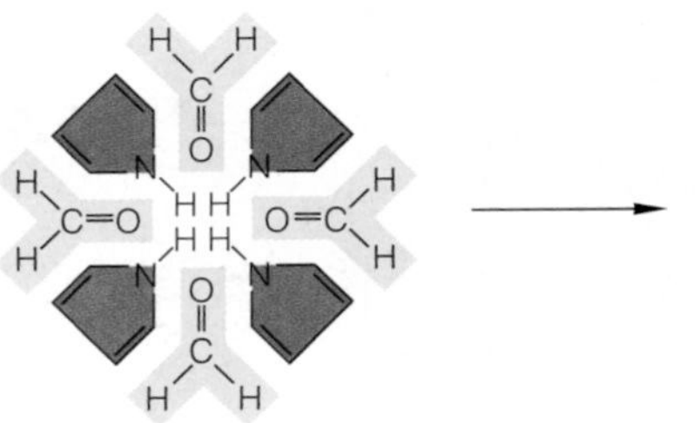

four pyrrole rings plus four formaldehyde molecules

6H

porphyrin ring system

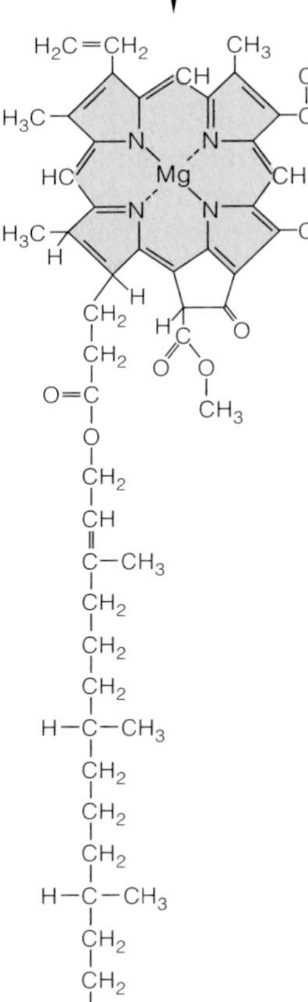

chlorophyll *a*, one of the light-trapping pigments of photosynthetic plant cells

Figure 21.4 One hypothetical sequence by which formaldehyde, an organic compound, underwent chemical evolution into porphyrin. Formaldehyde was present when the Earth formed. Porphyrin is the light-trapping and electron-donating component of all existing chlorophyll molecules. It also is a component of cytochrome, which is a protein component of electron transport systems that are part of many metabolic pathways.

Most existing enzymes get assistance from small organic molecules or metal ions called coenzymes. Intriguingly, some categories of coenzymes have a structure identical to that of RNA nucleotides. Another clue: Mix together and then heat up precursors of ribonucleotides and short chains of phosphate groups, and they self-assemble into single strands of RNA. On the early Earth, energy from the sun's rays or from geothermal events could have driven the spontaneous formation of RNA from such starting materials.

Very simple self-replicating systems of RNA, enzymes, and coenzymes have been created in some laboratories. Did RNA later become the information-

Further reading: Student Guide to InfoTrac on web site

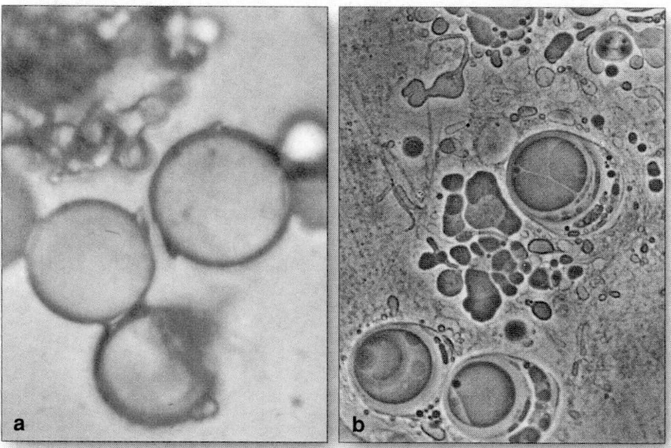
a
b

Figure 21.5 Microscopic spheres of (**a**) proteins and (**b**) lipids that self-assembled under abiotic conditions.

storing templates upon which simple proteins could be synthesized? Perhaps RNA did so initially. We can only speculate at present, because the RNA molecules that we know about are too chemically fragile to serve as templates for protein synthesis.

Yet the use of RNA templates might have set up an **RNA world** that preceded DNA's dominance as the main informational molecule. Whatever the case, DNA eventually assumed this function, probably because it can form long nucleotide chains in more stable fashion.

We still don't know how DNA entered the picture. Until we identify the likely chemical ancestors of RNA and DNA, the story of life's origin will be incomplete. Filling in the details will require imaginative sleuthing. For instance, researchers ran a computer program that incorporated information about natural energy sources and simple inorganic compounds of the sort thought to have been present on the early Earth. They asked their advanced computer to subject the chosen compounds to random chemical competition and natural selection as might have occurred untold billions of times in the past. They ran the program repeatedly. And always the outcome was the same: Simple precursors inevitably evolved—and they organized themselves as interacting systems of large, complex molecules.

Origin of the First Plasma Membranes

Experimental tests are more revealing of the origin of the plasma membrane—the outermost membrane of all living cells. This cellular component consists of a lipid bilayer, studded with proteins that carry out diverse functions. Its main role is to control which substances move into and out of the cell. Without control, cells can neither exist nor reproduce.

Probably, one avenue of molecular evolution led to **proto-cells**: simple membrane sacs that surrounded and protected information-storing templates and metabolic agents from the environment. We do know that simple membrane sacs can form spontaneously. For example, for one experiment, Fox heated amino acids until they formed protein-like chains, which he then placed in hot water. After cooling, the chains assembled into small, stable spheres (Figure 21.5*a*). Like membranes of cells, these proteinoid spheres were selectively permeable to various substances. In other experiments, the spheres picked up free lipid molecules from their surroundings, and a lipid–protein film formed at their surface.

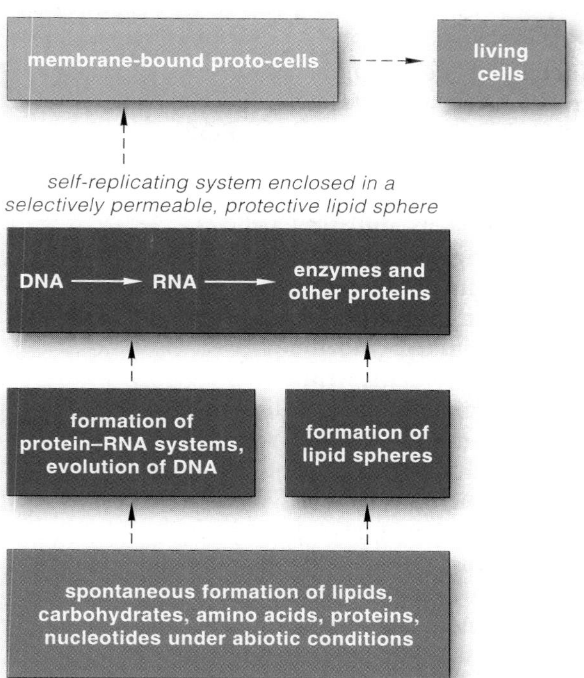
membrane-bound proto-cells
living cells
self-replicating system enclosed in a selectively permeable, protective lipid sphere
DNA
RNA
enzymes and other proteins
formation of protein–RNA systems, evolution of DNA
formation of lipid spheres
spontaneous formation of lipids, carbohydrates, amino acids, proteins, nucleotides under abiotic conditions

Figure 21.6 One possible sequence of events that led to the first self-replicating systems, then to the first living cells.

In still other experiments, by David Deamer and his coworkers, fatty acids and glycerol combined to form long-tail lipid molecules under laboratory conditions that simulated evaporating tidepools. The lipids self-assembled into small, water-filled sacs. Many were like cell membranes (Figure 21.5*b*).

In short, there are major gaps in the story of life's origins. But there also is strong experimental evidence that chemical evolution probably led to the molecules and structures that are characteristic of life. Figure 21.6 summarizes the milestones in that chemical evolution, which preceded the first cells.

Although the story is not yet complete, many laboratory experiments and computer simulations indirectly show that chemical and molecular evolution could have given rise to proto-cells.

21.3 ORIGIN OF PROKARYOTIC AND EUKARYOTIC CELLS

The first living cells originated in the **Archean** eon, which lasted from 3.9 billion to 2.5 billion years ago. Those cells emerged as molecular extensions of the evolving universe, of our solar system and the Earth. Maybe they originated in tidal flats or seafloor sediments (Section 7.8). Fossils indicate they were like the existing bacteria. Specifically, they were **prokaryotic cells**, without a nucleus. They may have been little more than membrane-bound, self-replicating sacs of DNA and other complex organic molecules. Given the absence of free oxygen, they must have secured energy through anaerobic pathways—fermentation, most likely. Energy was plentiful. Geologic processes already enriched the seas with organic compounds. So "food" was available, predators were absent, and cellular structures were free from oxygen attacks.

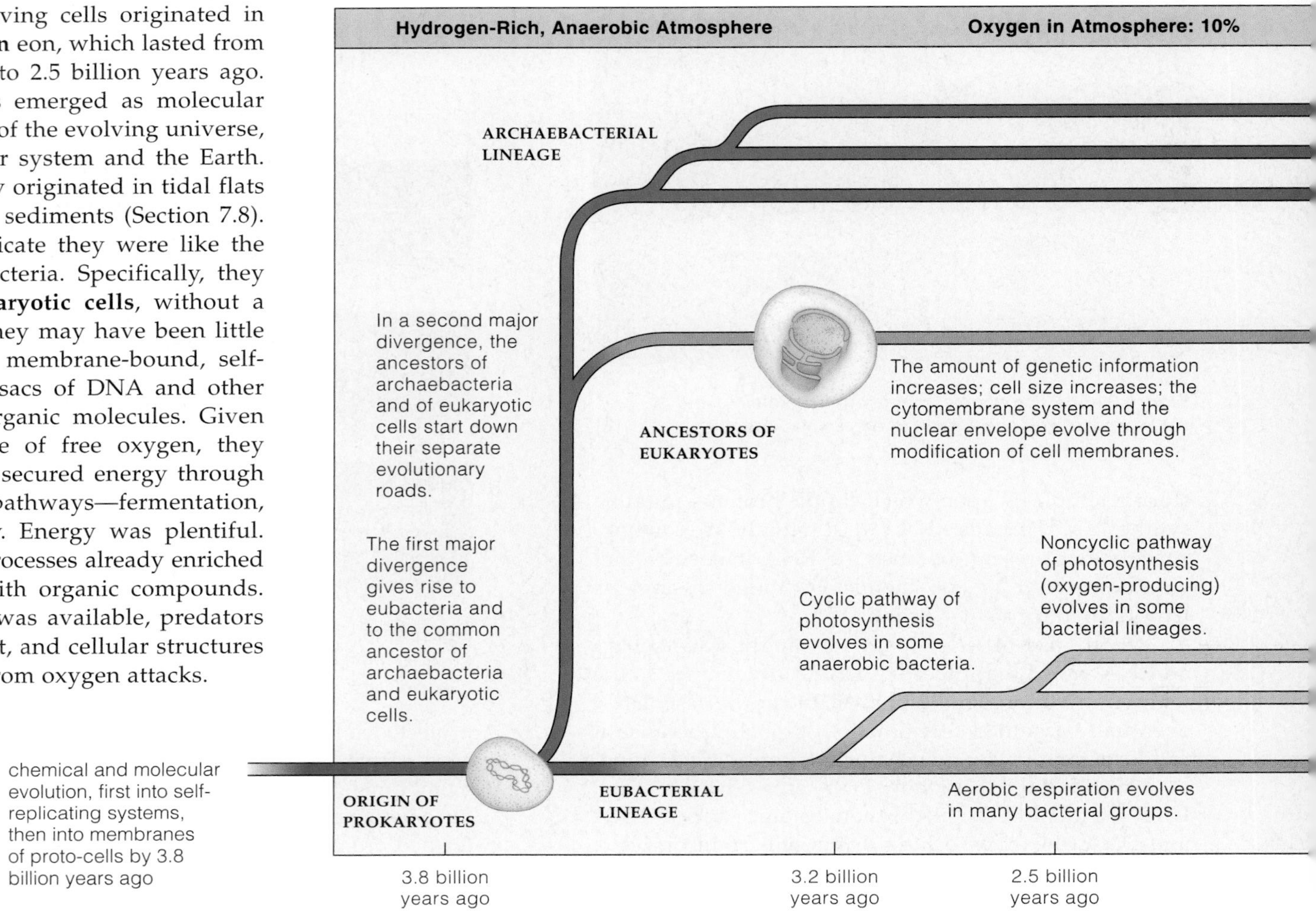

Hydrogen-Rich, Anaerobic Atmosphere

Oxygen in Atmosphere: 10%

ARCHAEBACTERIAL LINEAGE

In a second major divergence, the ancestors of archaebacteria and of eukaryotic cells start down their separate evolutionary roads.

ANCESTORS OF EUKARYOTES

The amount of genetic information increases; cell size increases; the cytomembrane system and the nuclear envelope evolve through modification of cell membranes.

The first major divergence gives rise to eubacteria and to the common ancestor of archaebacteria and eukaryotic cells.

Cyclic pathway of photosynthesis evolves in some anaerobic bacteria.

Noncyclic pathway of photosynthesis (oxygen-producing) evolves in some bacterial lineages.

chemical and molecular evolution, first into self-replicating systems, then into membranes of proto-cells by 3.8 billion years ago

ORIGIN OF PROKARYOTES

EUBACTERIAL LINEAGE

Aerobic respiration evolves in many bacterial groups.

3.8 billion years ago

3.2 billion years ago

2.5 billion years ago

Figure 21.7 An evolutionary tree of life that reflects mainstream thinking about the connections among major lineages. The diagram incorporates ideas about the origins of some eukaryotic organelles.

Some populations of those first prokaryotic cells apparently diverged in two major directions shortly after the time of origin. One lineage gave rise to the **eubacteria**. The other lineage gave rise to the common ancestor of **archaebacteria** and **eukaryotic cells** (Figure 21.7).

Between 3.5 and 3.2 billion years ago, light-trapping pigments, electron transport systems, and other metabolic machinery evolved in some anaerobic eubacteria. With these innovations, the cells became the earliest practitioners of the cyclic pathway of photosynthesis. (You read about this ATP-forming pathway in Section 7.4). Sunlight, an unlimited source of energy, had been tapped. For nearly 2 billion years, photosynthetic descendants of those cells dominated the living world. Their tiny but numerous populations formed very large mats in which sediments had collected. The mats built up, one above the other. Calcium deposits hardened and preserved the mats, which came to be called **stromatolites** (Figure 21.8).

Figure 21.8 Stromatolites exposed at low tide in western Australia's Shark Bay. These mounds started forming 2,000 years ago. Structurally, they are identical with stromatolites that formed more than 3 billion years ago.

Further reading: Student Guide to InfoTrac on web site

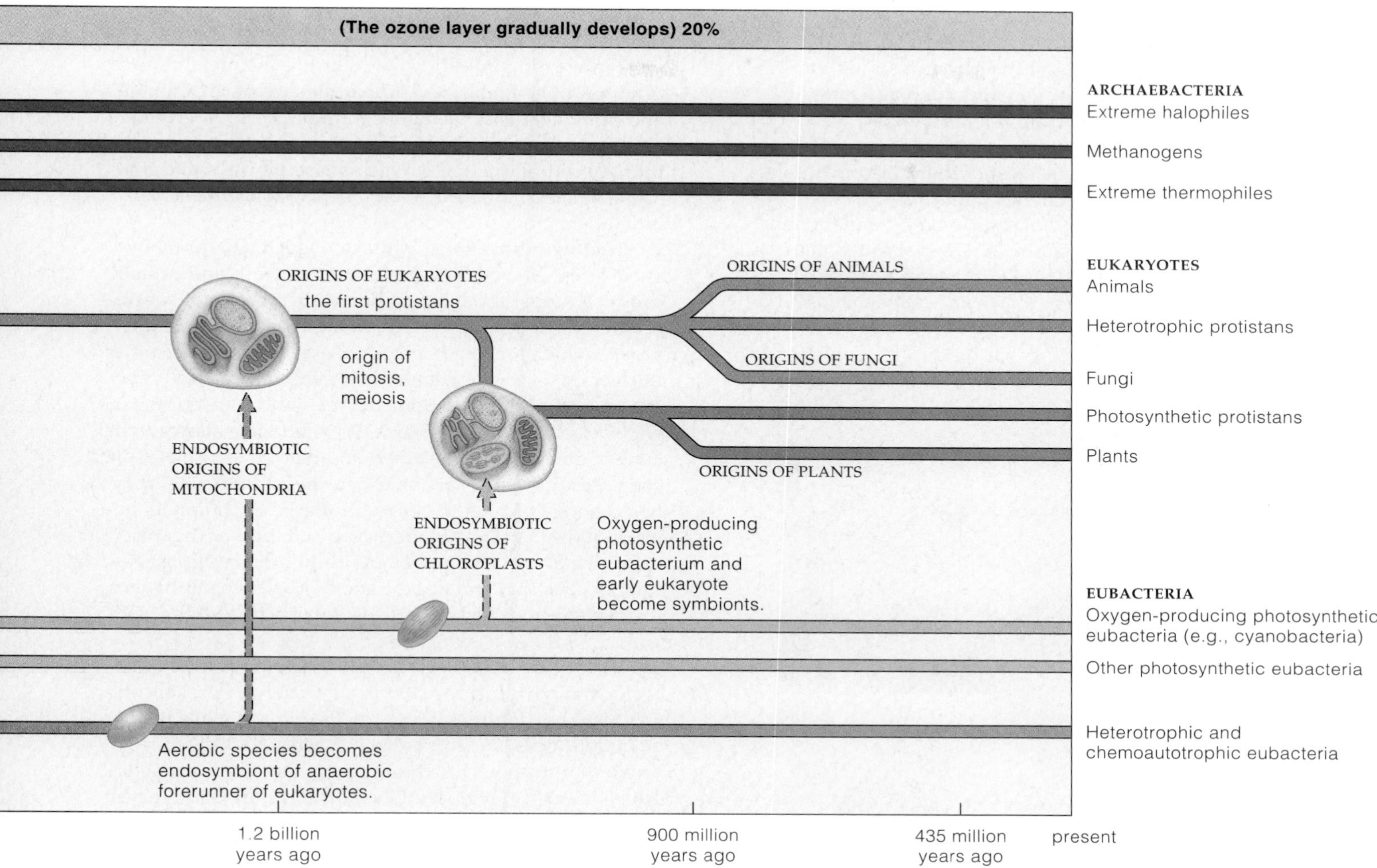
(The ozone layer gradually develops) 20%
ARCHAEBACTERIA
Extreme halophiles
Methanogens
Extreme thermophiles
EUKARYOTES
Animals
Heterotrophic protistans
Fungi
Photosynthetic protistans
Plants
ORIGINS OF EUKARYOTES
the first protistans
origin of
mitosis,
meiosis
ORIGINS OF ANIMALS
ORIGINS OF FUNGI
ORIGINS OF PLANTS
ENDOSYMBIOTIC
ORIGINS OF
MITOCHONDRIA
ENDOSYMBIOTIC
ORIGINS OF
CHLOROPLASTS
Oxygen-producing
photosynthetic
eubacterium and
early eukaryote
become symbionts.
EUBACTERIA
Oxygen-producing photosynthetic
eubacteria (e.g., cyanobacteria)
Other photosynthetic eubacteria
Heterotrophic and
chemoautotrophic eubacteria
Aerobic species becomes
endosymbiont of anaerobic
forerunner of eukaryotes.
1.2 billion
years ago
900 million
years ago
435 million
years ago
present

By the dawn of the **Proterozoic** eon, 2.5 billion years ago, photosynthetic machinery had become altered in some eubacterial species, and the noncyclic pathway of photosynthesis emerged. Oxygen, one of the pathway's by-products, started to accumulate. In time, this had two irreversible effects. First, *an oxygen-rich atmosphere stopped the further chemical origin of living cells.* Except in a few anaerobic habitats, complex organic compounds could no longer form spontaneously and resist attack. Second, *aerobic respiration became the dominant energy-releasing pathway.* In many prokaryotic lineages, selection favored metabolic equipment that "neutralized" oxygen by using it as an electron acceptor! This key innovation contributed to the rise of multicelled eukaryotes and their invasion of far-flung environments (Section 8.7).

Eukaryotic cells evolved in the Proterozoic, possibly before 1.2 billion years ago. We have fossils, 900 million years old, of complex algae, fungi, and plant spores. As you know, organelles are the premier defining features of eukaryotic cells. *Where did they come from?* The next section describes a few plausible hypotheses.

By about 800 million years ago, stromatolites along the shores of Laurentia, an early supercontinent, were declining dramatically. Perhaps newly evolved, bacteria-eating animals were using them as a concentrated food source. Fossilized embryos found in China (clusters of cells no wider than a pin) give evidence that the first animals were soft-bodied forerunners of today's sponges and marine worms. Before 570 million years ago, in "Precambrian" times, some of their small descendants started the first adaptive radiation of animals.

The first cells evolved by about 3.8 billion years ago, during the Archean. All were prokaryotic cells, and most, if not all, probably made ATP by fermentation routes.

Early on, ancestors of archaebacteria and eukaryotic cells diverged from the lineage that led to modern eubacteria.

Oxygen-releasing photosynthetic bacteria evolved. In time, the oxygen-enriched atmosphere put an end to the further spontaneous chemical evolution of life. That atmosphere was a key selection pressure in the evolution of eukaryotic cells.

Focus on Science

21.4 WHERE DID ORGANELLES COME FROM?

Thanks to Andrew Knoll, William Schopf, Jr., and other globe-hopping microfossil hunters, we have tantalizing evidence of early life. One fossil treasure is a strand of bacterial cells that lived 3.5 billion years ago, not long after life originated. Others, from the Proterozoic, were eukaryotic cells with a few membrane-bound organelles in the cytoplasm (Figure 21.9). Most of their descendants now have a profusion of organelles (Figure 21.10).

Where did eukaryotic organelles come from? Speculations abound. Some organelles probably evolved through gene mutations and natural selection. For others, researchers make a good case for evolution by way of endosymbiosis.

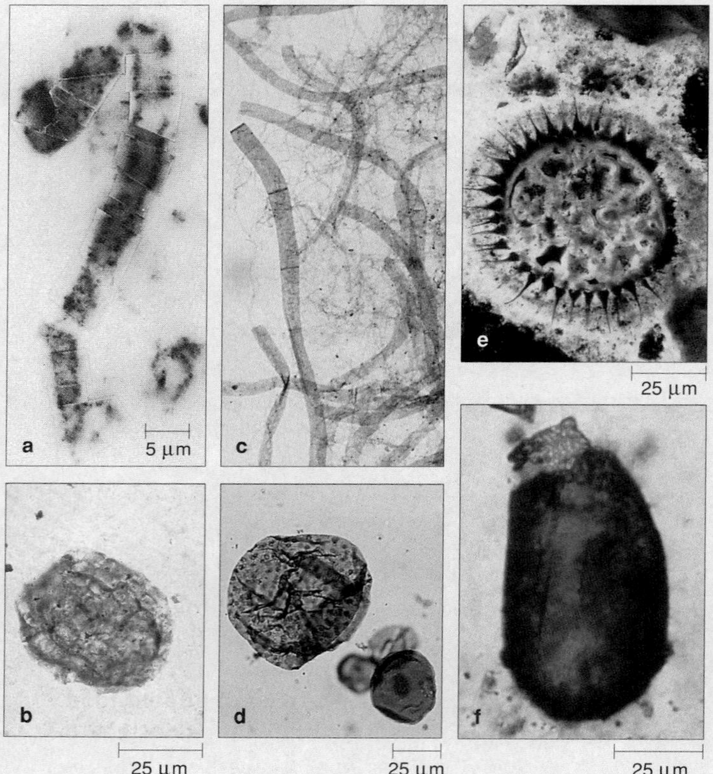
a
5 μm
c
e
25 μm
b
d
f
25 μm
25 μm
25 μm

Figure 21.9 From Australia, (**a**) a strand of walled prokaryotic cells 3.5 billion years old and (**b**) one of the oldest known eukaryotes—a protistan 1.4 billion years old. From Siberia, (**c**) a multicelled alga 900 million to 1 billion years old and (**d**) eukaryotic microplankton 850 million years old. (**e**) From China, a eukaryotic cell that lived 560 million years ago. (**f**) From Spitsbergen, Norway, a protistan that was alive 50 million years before the dawn of the Cambrian.

ORIGIN OF THE NUCLEUS AND ER Prokaryotic cells do not have an abundance of organelles, but some species have interesting infoldings of their plasma membrane (Figure 21.11). Embedded in that membrane are enzymes and other agents of metabolism. In the early forerunners of eukaryotic cells, similar infoldings may have extended into the cytoplasm and served as passageways to the surface. They may have evolved into ER channels and into an envelope around the DNA.

What would be the advantage of such membranous enclosures? Maybe they protected the genes and protein products from "foreigners." Remember, bacterial species often transfer plasmid DNA among themselves. So do yeasts, which are very simple eukaryotic cells. At first, a nuclear envelope may have been favored because it got the cell's genes, replication enzymes, and transcription enzymes out of the cytoplasm. It would have allowed vital genetic messages to be copied and read, free of metabolic competition from what could become an unmanageable hodgepodge of foreign genes. Similarly, ER channels might have kept important proteins and other organic compounds away from metabolically hungry "guests"—foreign cells that one way or another became permanent residents inside the host cell, as described next.

A THEORY OF ENDOSYMBIOSIS It appears likely that accidental partnerships between a variety of prokaryotic species arose countless times on the evolutionary road to eukaryotic cells. Perhaps some partnerships resulted in the origin of mitochondria, chloroplasts, and other organelles. This is a story of endosymbiosis, as developed in greatest

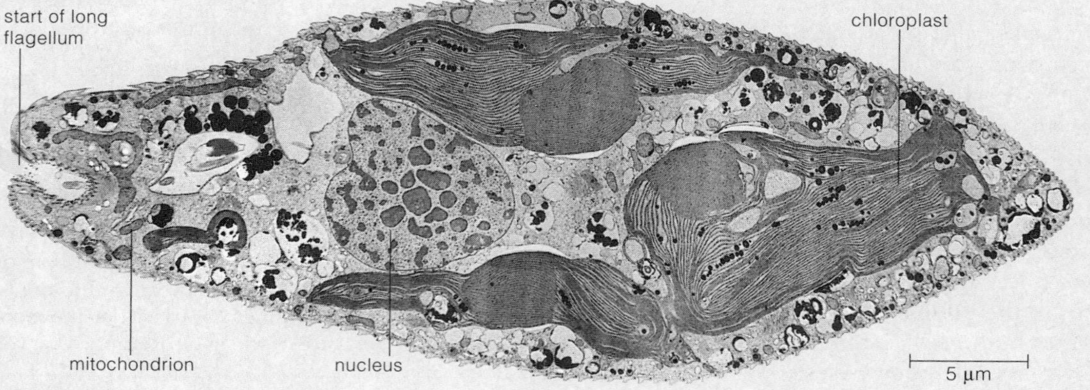
start of long flagellum
chloroplast
mitochondrion
nucleus
5 μm

Figure 21.10 A fine example of the profusion of diverse organelles that are hallmarks of eukaryotic cells: *Euglena*, a single-celled protistan, sliced lengthwise for this micrograph. It also has a long flagellum, which could not fit in the image area at this magnification.

Further reading: Student Guide to InfoTrac on web site →

Figure 21.11 (**a**) Sketch of one idea concerning the origin of endoplasmic reticulum and the nuclear envelope. In prokaryotic ancestors of eukaryotic cells, infoldings of the plasma membrane might have given rise to both cell components. (**b**) Such infoldings are present in the cytoplasm of many kinds of existing bacteria, including *Nitrobacter*, sketched here in cutaway view.

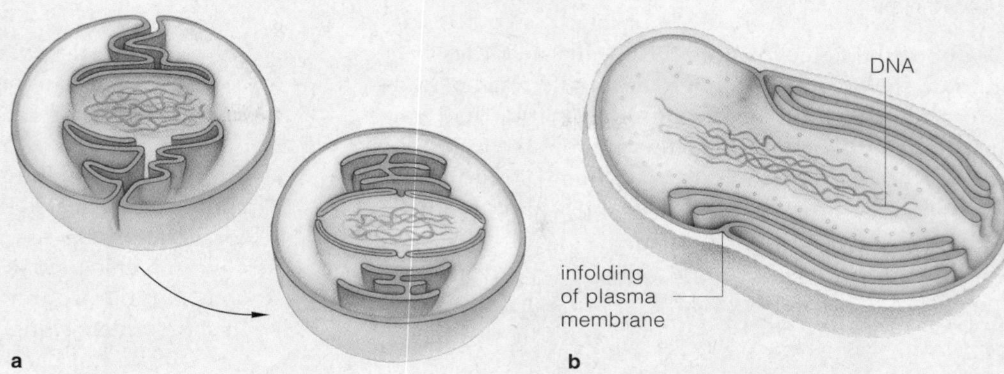
DNA
infolding
of plasma
membrane
a
b

detail by Lynn Margulis. *Endo-* means within; *symbiosis* means living together. In cases of **endosymbiosis**, one species (the guest) lives permanently inside another species (the host), and the interaction benefits both.

According to one theory, eukaryotic cells evolved by endosymbiosis long after the noncyclic pathway of photosynthesis emerged and oxygen had accumulated to significant levels in the atmosphere. In some bacterial groups, certain electron transport systems in the plasma membrane had become modified and included "extra" cytochromes. The cytochromes could donate electrons to oxygen. Thus the bacteria could extract energy from organic compounds by aerobic respiration.

By 1.2 billion years ago and possibly much earlier, the forerunners of eukaryotes were engulfing aerobic bacteria. Maybe they were like existing soft-bodied, amoebalike cells that weakly tolerate free oxygen. They would have entrapped food by sending out cytoplasmic extensions. Endocytic vesicles could have formed around food and could have delivered it to the cytoplasm for digestion.

A key point of the theory is that some aerobic bacteria resisted digestion and thrived in the protected, nutrient-rich environment. In time, they were releasing extra ATP, which their hosts came to depend on for their growth, increased activity, and assembly of hard parts and other structures. How did the guests benefit? They no longer had to seek food or duplicate the metabolic functions that hosts performed for them. The anaerobic and aerobic cells were now incapable of independent life. The guests had become mitochondria, supreme suppliers of ATP.

EVIDENCE OF ENDOSYMBIOSIS Strong evidence favors Margulis's theory. There are plenty of examples of nature continuing to tinker with endosymbionts, including the cell in Figure 21.12. Its mitochondria are like bacteria in size and structure. The inner mitochondrial membrane is like a bacterial plasma membrane. Each mitochondrion replicates its own DNA and divides independently of the host cell's division. A few genetic code words in its DNA and mRNA have unique meanings. They are translated into a few proteins required for special mitochondrial tasks. Compared to the genetic code of living cells, the "mitochondrial code" has a few distinct differences.

Chloroplasts, too, may be stripped-down descendants of oxygen-evolving, photosynthetic bacteria. Perhaps predatory aerobic bacteria engulfed such photosynthetic cells, which escaped digestion, absorbed nutrients from their host's cytoplasm, and continued to function. By providing aerobically respiring hosts with oxygen, they would have promoted their endosymbiotic existence.

In their metabolism and overall nucleic acid sequence, chloroplasts resemble some eubacteria. Their DNA is self-replicating, and they divide independently of the cell's division. Chloroplasts vary in their shape and array of light-absorbing pigments, just as photosynthetic eubacteria do. They might have originated a number of times in a number of different lineages.

Adding to the intrigue are ciliated protistans and marine slugs that "enslave" chloroplasts! The slugs eat algae but retain algal chloroplasts in their gut cells. The chloroplasts draw nutrients from the host tissues, and they continue to photosynthesize and release oxygen for weeks.

However they arose, new kinds of cells did appear on the evolutionary stage. They had become equipped with a nucleus, cytomembranes, and mitochondria, chloroplasts, or both. They were eukaryotic cells, the first **protistans**. With their efficient metabolic strategies, early protistans underwent rapid divergences and adaptive radiations. In no time at all, evolutionarily speaking, some of their descendants gave rise to the great kingdoms of plants, fungi, and animals, as sketched out earlier in Figure 21.7.

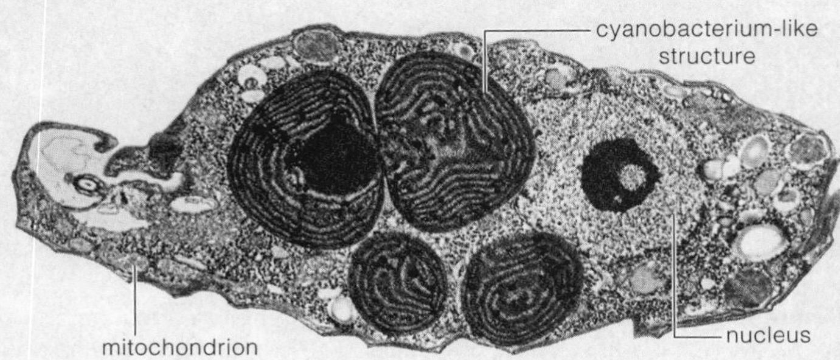
cyanobacterium-like
structure
mitochondrion
nucleus

Figure 21.12 *Cyanophora paradoxa*, a protistan. Its mitochondria resemble bacteria. Its photosynthetic structures look like spherical cyanobacteria (which are photosynthetic) without the cell wall.

21.5 LIFE IN THE PALEOZOIC ERA

We divide the **Paleozoic** into the Cambrian, Ordovician, Silurian, Devonian, Carboniferous, and Permian periods. Before the dawn of that era, gradual rifting had split the supercontinent Laurentia apart. From the Cambrian on into the Silurian, the great fragments straddled the equator, and warm, shallow seas lapped their margins:

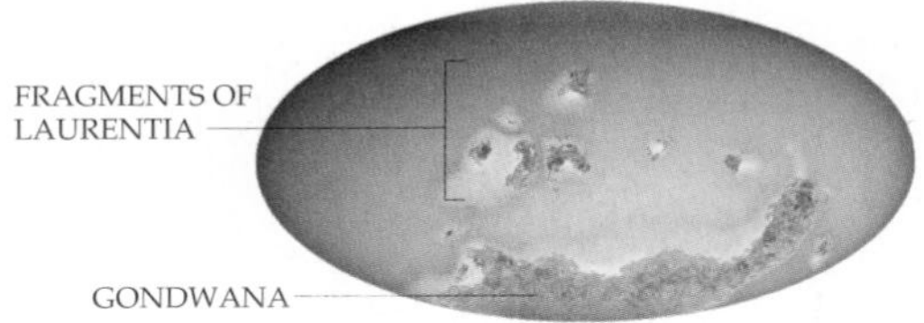
FRAGMENTS OF
LAURENTIA
GONDWANA

The global conditions restricted pronounced seasonal changes in prevailing winds, ocean currents, and the upward churning of nutrient deposits on the seafloor. As a result, supplies of nutrients along shorelines at or near the equator were stable but limited.

Most major animal phyla had evolved earlier, in the Precambrian seas. Possibly some of their ancestors were among the **Ediacarans**, peculiar organisms shaped like fronds, disks, and blobs that nearly defy classification (Figure 21.13*a,b*). Like Ediacarans, the early Cambrian animals had flattened bodies, with a good surface-to-volume ratio for taking up nutrients (Figure 21.13*c*). Most existed on or in seafloor sediments, where dead organisms and organic debris settled. They ranged from sponges to simple vertebrates, and they were diverse.

How could so much diversity arise? Possibly genes governing early growth and development were far less intertwined than they are at present, so there might not have been as much selection against mutant alleles and novel traits. Also, warm, shallow waters near the new coasts were vacant adaptive zones with many enticing opportunities for new ways to secure food.

Entombed in sedimentary beds from the Cambrian are fossilized organisms that have punctures, missing chunks, and healed wounds. These are not artifacts of fossilization; the animals were injured while they were alive. Diverse predators and prey with armorlike shells, spines, mouths, and novel feeding structures evolved in short order. Things were starting to get lively!

Later in the Cambrian, temperatures in the shallow seas changed drastically. Trilobites (Figure 21.13*d*), one of the most common invertebrates, almost vanished. In the Ordovician, the supercontinent Gondwana had been drifting south, and parts became submerged in shallow seas. The emergence of vast new marine environments promoted adaptive radiations. Many new reef organisms evolved. Among them were the swift, shelled predators called nautiloids. Their surviving descendants include the chambered nautiluses (Section 26.13).

Later on, Gondwana straddled the South Pole, and huge glaciers formed across its surface. When enormous volumes of water were locked up as ice, shallow seas throughout the world were drained. This was the first ice age that we know about. It may have triggered the first global mass extinction. At the Ordovician–Silurian boundary, reef life everywhere collapsed.

Gondwana drifted north during the Silurian and on into the Devonian. This was a pivotal time of evolution. Reef communities recovered. Armor-plated fishes with massive jaws diversified. And in the wet lowlands, small stalked plants appeared (Figure 21.14*b,c*). So did fungi and many invertebrates, such as segmented worms. In

Figure 21.13 Representatives from the Precambrian and Cambrian seas. Two Ediacarans, about 600 million years old: (**a**) *Spriggina* and (**b**) *Dickensonia*. The oldest known Ediacarans lived 610 million years ago and the most recent in Cambrian times, 510 million years ago. (**c**) From British Columbia's Burgess Shale, a fossilized marine worm. (**d**) A beautifully preserved fossil of one of the earliest trilobites.

Further reading: Student Guide to InfoTrac on web site

Figure 21.14 (**a**) Life in the Silurian seas. Some of these shelled animals (nautiloids) were twelve feet long. (**b**) Silurian swamp, dominated by forerunners of modern ferns and club mosses. (**c**) Fossils of a Devonian plant (*Psilophyton*), possibly one of the earliest ancestors of conifers and other seed-bearing plants. (**d**) Reptiles (*Dimetrodon*) of a mostly hotter and drier time—the Permian. Some fossils of these carnivores were found in Texas. Giant club mosses and horsetails had declined, and conifers, cycads, and other gymnosperms replaced them.

Devonian times, fishes that would become ancestors to amphibians invaded the land. The fishes sported lobed fins, the forerunners of legs and other limbs. They had simple lungs. As described in Section 27.6, lobed fins and lungs were key innovations that would prove most advantageous for life out of water, in dry land habitats.

Then, as they say, something bad happened. At the boundary between the Devonian and Carboniferous, sea levels swung catastrophically for unknown reasons and caused another mass extinction. Afterward, plants and insects embarked on adaptive radiations on land.

Throughout Carboniferous times, land masses were gradually submerged and drained many times. Organic debris piled up. Then it was compacted and converted to coal, in the manner described in Section 25.4.

Insects, amphibians, and early reptiles flourished in vast swamp forests of Permian times (Figure 21.14*d*). Ancestors of seed-producing plants, including cycads, ginkgos, and conifers, dominated those forests.

As the Permian drew to a close, something caused the greatest of all mass extinctions. More than 50 percent of all families disappeared, and only 5 percent of the known species survived. At the time, all land masses were colliding together. Eventually they formed Pangea, a supercontinent that extended all the way from the North to the South Pole. A single world ocean lapped its margins:

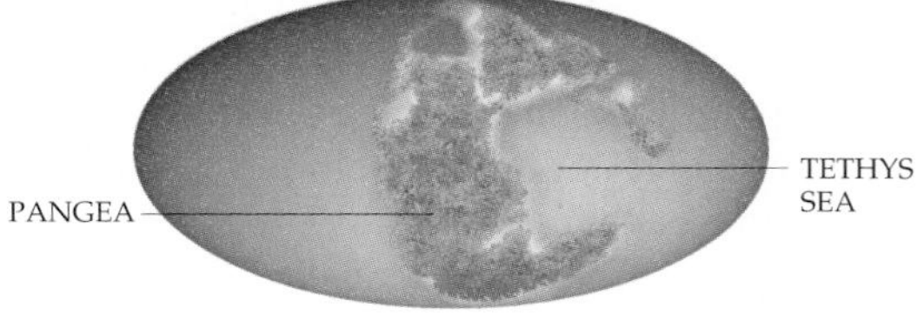

PANGEA
TETHYS SEA

As you will see, the new distribution of oceans, land masses, and land elevations had catastrophic effects on global climates—and on the course of life's evolution.

Early in the Paleozoic era, organisms of all six kingdoms were flourishing in the seas. By the end of the era, many lineages had successfully invaded the land, including the wet lowlands of the supercontinent Pangea.

21.6 LIFE IN THE MESOZOIC ERA

Speciation on a Grand Scale

We divide the **Mesozoic** into the Triassic, Jurassic, and Cretaceous periods. It lasted about 175 million years. Early on in the Cretaceous, the supercontinent Pangea started to break up. Its huge fragments slowly drifted apart, and we can assume that the resulting geographic isolation favored divergences and speciation:

This was an era of spectacular expansion in the range of global diversity. In the seas, invertebrates and fishes underwent adaptive radiations. On land, conifers and other seed-bearing **gymnosperms,** insects, and reptiles became the most visibly dominant lineages. Flowering plants—the **angiosperms**—emerged in the late Jurassic or early Cretaceous. It would take them less than 40 million years to displace conifers and related plants in most environments (Figure 21.15 and Section 25.8).

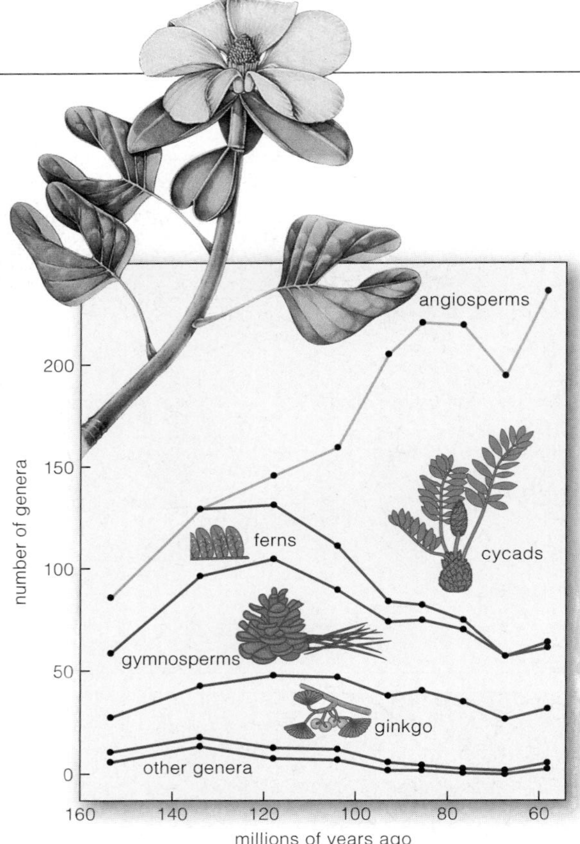
angiosperms
ferns
cycads
gymnosperms
ginkgo
other genera
number of genera
0
50
100
150
200
160
140
120
100
80
60
millions of years ago

Figure 21.15 Range of diversity among vascular plants of the Jurassic and Cretaceous. Conifers and other gymnosperms were dominant, then started to decline even before flowering plants began a great adaptive radiation that continued into the Cenozoic. *Upper left:* From the Cretaceous, a floral shoot of *Archaeanthus linnenbergeri.* In many traits, this now-extinct flowering plant resembled living magnolias.

Rise of the Ruling Reptiles

Early in the Triassic, the first **dinosaurs** evolved from a reptilian lineage. They were not much bigger than a turkey. Possibly most species had high metabolic rates, and maybe they were warm-blooded. Many sprinted about on two legs. Dinosaurs weren't the dominant land animals. Instead, center stage belonged to *Lystrosaurus* and other plant-eating, mammal-like reptiles that were too large to be bothered by most predators.

Adaptive zones may have opened up for dinosaurs in a frightening way. There is a string of five craters in France, Quebec, Manitoba, and North Dakota; one is the size of Rhode Island. About 214 million years ago, according to radioisotope dating, fragments from an asteroid or a comet apparently fell one after another as the Earth spun beneath them, just as a string of huge fragments from comet Shoemaker–Levy 9 hit Jupiter in 1994. The resulting blast waves, global firestorm, lava flows, and earthquakes must have been stupendous. Most of the animals lucky enough to survive that time of mass extinction (as well as later ones) were smaller, had higher rates of metabolism, and were less vulnerable than others to drastic changes in outside temperatures.

Descendants of the surviving dinosaurs became the ruling reptiles; they endured for 140 million years. Some species reached monstrous proportions. Among them were the ultrasaurs, fifteen meters tall.

Many dinosaurs perished in another mass extinction at the end of the Jurassic, then in a pulse of extinctions in the Cretaceous. Perhaps plumes of molten material ruptured the crust and triggered changes in the global climate. Perhaps asteroids or comets inflicted the blows. Not all lineages survived those episodes. Yet some did recover, and new ones evolved. Duckbilled dinosaurs appeared in forests and swamps. Tanklike *Triceratops* and other plant eaters flourished in open habitats. They were prey for the fearsomely toothed, agile, and swift *Velociraptor* of motion picture fame.

About 120 million years ago, global temperatures skyrocketed 25 degrees. By one theory, plumes spread out beneath the crust and "greased" the crustal plates into moving twice as fast. A superplume or a rash of them broke through the crust. The crust in what is now the South Atlantic opened like a zipper. Basalt and lava poured from the fissures; volcanoes spewed nutrient-rich ashes. Simultaneously, the plumes released great amounts of carbon dioxide, one of the "greenhouse" gases that absorb some of the heat radiating from the Earth before it escapes into outer space. The nutrient-enriched planet warmed—and remained warm for 20

Further reading: Student Guide to InfoTrac on web site

Figure 21.16 If dinosaurs of this sort had not disappeared at the end of the Cretaceous, would then-tiny mammals ever have ventured out from under the shrubbery? Would *you* even be here today?

million years. Photosynthetic organisms flourished on land and in shallow seas. Their remains were slowly buried and converted into the world's vast oil reserves.

About 65 million years ago, the last dinosaurs and many marine organisms vanished in a mass extinction (Figure 21.16). As described in the next section, their disappearance apparently coincided with a direct hit by an asteroid the size of Mount Everest. Over time, the impact site slowly drifted into a position that we now call the northern Yucatán peninsula.

The Mesozoic was a time of major adaptive radiations and of a mass extinction in which the last dinosaurs and many marine organisms disappeared.

21.7 HORRENDOUS END TO DOMINANCE

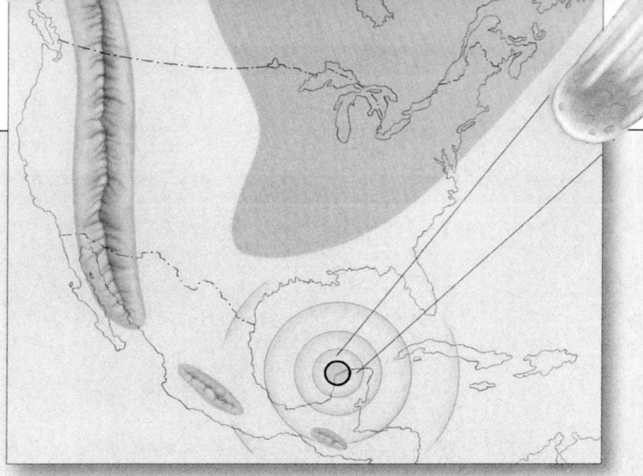

Figure 21.18 Artist's interpretation of what might have happened in the last few minutes of the Cretaceous.

It has only been about 55,000 years since the first modern humans walked the Earth. Since then, how many times have people, puffed up with self-importance, set out to conquer neighbors, lands, and seas? Think about it. Then think about the dinosaurs. Were they good at reigning supreme? No question about it; their lineage dominated the land for 140 million years. In the end, did it matter? Not a bit. Sixty-five million years ago, at the Cretaceous–Tertiary (K–T) boundary, nearly all remaining members of their most excellent lineage perished. Why? Bad luck.

A thin layer of iridium-rich rock distributed around the world dates precisely to the K–T boundary. Iridium is rare here but common in asteroids. The **asteroids** are rocky, metallic bodies, 1,000 kilometers to a few meters in diameter, that are hurtling through space. When planets of our solar system were forming, their gravitational pull swept most asteroids from the sky. At least 6,000 still orbit the sun in a belt between Mars and Jupiter (Figure 21.17). The orbits of dozens of others take them across Earth's orbit, like Russian roulette on a cosmic scale.

By analyzing iridium levels in soils, gravity maps, and other evidence, Walter Alvarez and Luis Alvarez hypothesized that an asteroid impact caused the K–T mass extinction. Later, researchers identified the impact site. Massive movements in the crust transported the site to what is now the northern Yucatán peninsula of Mexico (Figure 21.18). The impact crater is 9.6 kilometers deep and 300 kilometers across—wider than Connecticut. This crater as well as other evidence strongly supports what is now called the **K–T asteroid impact theory**.

To make a crater that big, the asteroid had to hit the Earth at 160,000 kilometers (100,000 miles) per hour. At least 200,000 cubic kilometers of debris and dense gases were blasted skyward. The crust itself heaved violently. Monstrous waves, 120 meters high, raced across the ocean, obliterating life on islands and then slamming into coasts of continents. Researchers had hypothesized that atmospheric debris blocked out sunlight for months. If so, plants and other producers that sustained the web of life would have withered and died; animals and other consumers would have starved to death. The hypothesis had problems. By some calculations, the volume of debris blasted aloft would not have been enough to have had such significant consequences.

Figure 21.17 What one of the asteroids looks like, courtesy of the *Galileo* spacecraft that flew past it on the way to Jupiter. This is only a small asteroid; it would extend halfway between Baltimore and Washington, D.C.

Then comet Shoemaker–Levy 9 slammed into Jupiter. Particles blasted into the Jovian atmosphere triggered an intense heating of an area larger than the Earth. That event supports a **global broiling hypothesis**, proposed first by H. J. Melosh and his colleagues. Briefly stated, energy released at the K–T impact site was equivalent to detonating 100 million nuclear bombs. Trillions of tons of vaporized debris rose in a colossal fireball, then rapidly condensed into particles the size of sand grains. Seconds later, a cooler fireball made of steam, carbon dioxide, and unmelted rock formed. When the debris fell to the Earth, it raised the atmosphere's temperature by thousands of degrees. The sky above the planet must have glowed with heat ten times more intense than the noonday sun above Death Valley in summer. In one horrific hour, nearly all plants erupted in flames and every animal out in the open was broiled alive.

Things haven't settled down much since dinosaurs disappeared. For instance, about 2.3 million years ago, a huge chunk from outer space apparently hit the Pacific Ocean. At about the same time, vast ice sheets started forming abruptly in the Northern Hemisphere. Long-term shifts in climate may have been ushering in this most recent ice age, but the global impact would have accelerated it. Water vaporized by the impact could have contributed to a global cloud cover that limited the amount of sunlight reaching the surface. Ancestors of humans were around when all this happened. The extreme climate shift surely tested their adaptability.

Almost certainly, severe environmental tests await all existing lineages. When we become too smitten with our importance in the world of life, we would do well to step back, from time to time, and reflect on what is going on above and beneath the Earth's surface. Asteroids and superplumes do have a way of leveling the playing field, as they did for gigantic dinosaurs and tiny mammals.

Further reading: Student Guide to InfoTrac on web site →

21.8 LIFE IN THE CENOZOIC ERA

The breakup of Pangea triggered events that continued into the present era, the **Cenozoic**. At the dawn of the Cenozoic, land masses were on collision courses:

Coastlines fractured. The Cascades, Andes, Himalayas, and Alps rose through volcanic activity, uplifting, and other events at crustal rifts and plate boundaries. These geologic changes brought about major shifts in climate that influenced the further evolution of life.

During the Paleocene epoch, climates were warmer and wetter. Tropical and subtropical forests extended farther north and south than they do today. Woodlands and forests spread even into polar regions. Although their key traits developed before the dinosaurs left the scene, mammals now began a major adaptive radiation.

The global climate warmed even more in the Eocene epoch, and subtropical forests extended north into the polar regions. A variety of mammals, including primates, bats, rhinos, hippos, elephants, horses, and assorted carnivores, emerged. By the late Eocene, climates became cooler and drier, and seasonal changes became pronounced. Woodlands and dry grasslands dominated vast tracts of land. Patterns of vegetational growth changed so much that many mammals were driven to extinction.

From the Oligocene through the Pliocene, an abundance of grazing and browsing animals flourished in the woodlands and grasslands. Among them were camels and the "giraffe rhinoceros," along with some fearsome carnivores that stalked them (Figure 21.19).

At present, the distribution of land masses favors biodiversity. The richest ecosystems are tropical forests of South America, Madagascar, and Southeast Asia, as well as the marine ecosystems of tropical archipelagos in the Pacific. Yet we are in the midst of what may be one of the greatest of all mass extinctions. About 50,000 years ago, nomadic humans followed migrating herds of wild animals around the Northern Hemisphere. By a few thousand years later, major groups of mammals were extinct. The pace of extinction has since accelerated as humans hunt for food, fur, feathers, or fun, and as they destroy habitats to clear land for farm animals or crops (Chapters 28 and 51).

Figure 21.19 Examples of Cenozoic species. (**a**) In the early Paleocene, in what is now Wyoming, dense forests of sequoia, laurel, and other plants were home to diverse mammals. On the ground, raccoonlike *Chriacus* faces a tree-climbing rodent (*Ptilodus*). Higher up is *Peradectes*, a marsupial. (**b**) *Indricotherium*, the "giraffe rhinoceros." At 15 tons and 5.5 meters tall at the shoulder, it was the largest mammal known. It browsed in late Eocene and early Miocene woodlands. (**c**) Saber-tooth cats (*Smilodon*) and small horses from the Pleistocene. Fossils were recovered from pitch pools in Rancho LaBrea, California.

Major geologic changes during the Cenozoic triggered shifts in climate. The great adaptive radiation of mammals began, first in tropical forests, then in woodlands and grasslands.

21.9 SUMMARY OF EARTH AND LIFE HISTORY

We conclude this overview chapter with an illustration that correlates milestones in the evolution of life with the evolution of the Earth. As you study Figure 21.20, keep in mind that it is only a generalized summary. For example, it charts the five greatest mass extinctions yet there were numerous others in between. The illustration's diagram of the full range of biodiversity conveys the shrinking and expansion of species through time. It is a combined range for *all* of the major groups on land and in the seas. Remember, each major group has its own evolutionary history. Some of its member species may have persisted to the present; some or all may be extinct. Some of the groups may be represented by only one or a few species; others may be represented by hundreds or thousands or a million.

With these qualifications in mind, you are ready to turn to the next chapters in this unit. They will provide you with far richer detail of the history and the current range of biodiversity for all six kingdoms of life.

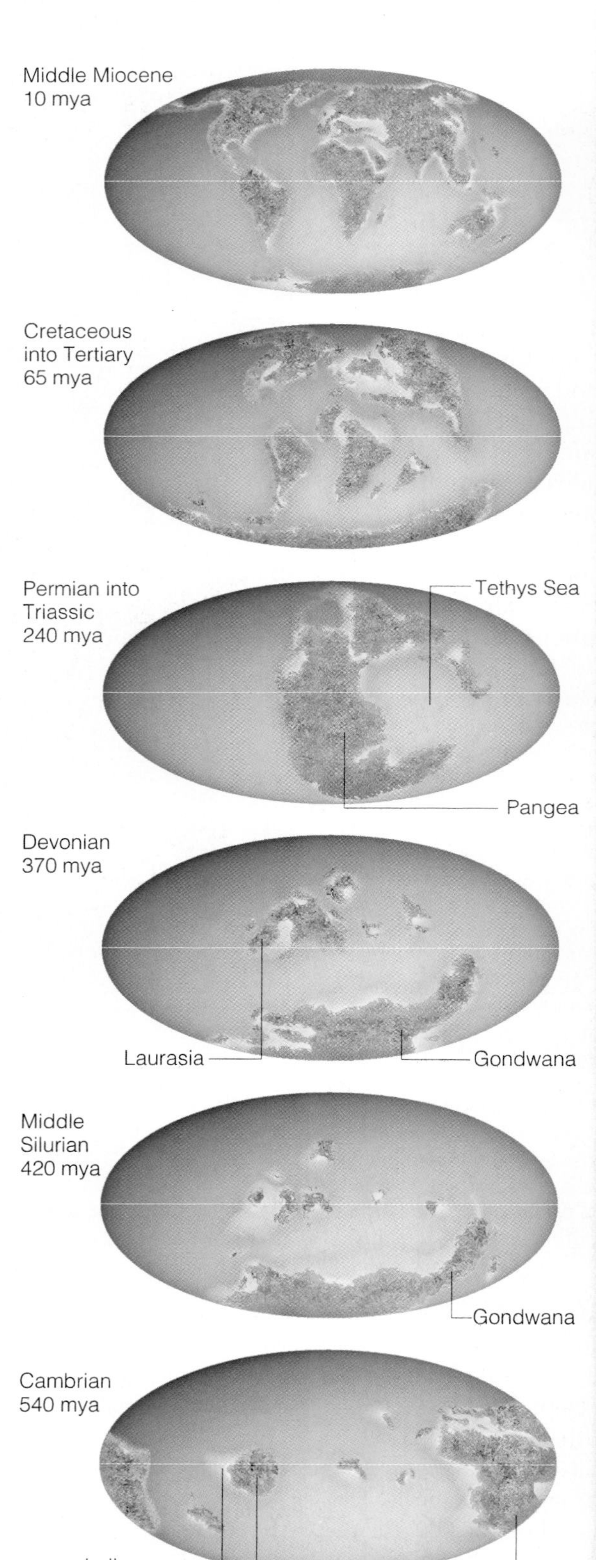

Middle Miocene
10 mya
Cretaceous into Tertiary
65 mya
Permian into Triassic
240 mya
Tethys Sea
Pangea
Devonian
370 mya
Laurasia
Gondwana
Middle Silurian
420 mya
Gondwana

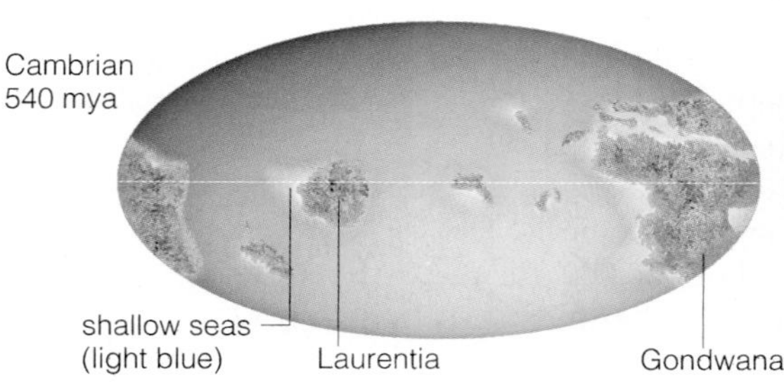

Cambrian
540 mya
shallow seas (light blue)
Laurentia
Gondwana

Figure 21.20 Summary of major events in the evolution of the Earth and of life. As you read through the chapters to follow, you may wish to return to this illustration now and then to remind yourself of how the details fit into the greater evolutionary picture.

	Period	Epoch
CENOZOIC ERA	Quaternary	Recent
		Pleistocene
	Tertiary	Pliocene
		Miocene
		Oligocene
		Eocene
		Paleocene
MESOZOIC ERA	Cretaceous	Late
		Early
	Jurassic	
	Triassic	
PALEOZOIC ERA	Permian	
	Carboniferous	
	Devonian	
	Silurian	
	Ordovician	
	Cambrian	
PROTEROZOIC EON		
ARCHEAN EON AND EARLIER		

Further reading: Student Guide to InfoTrac on web site

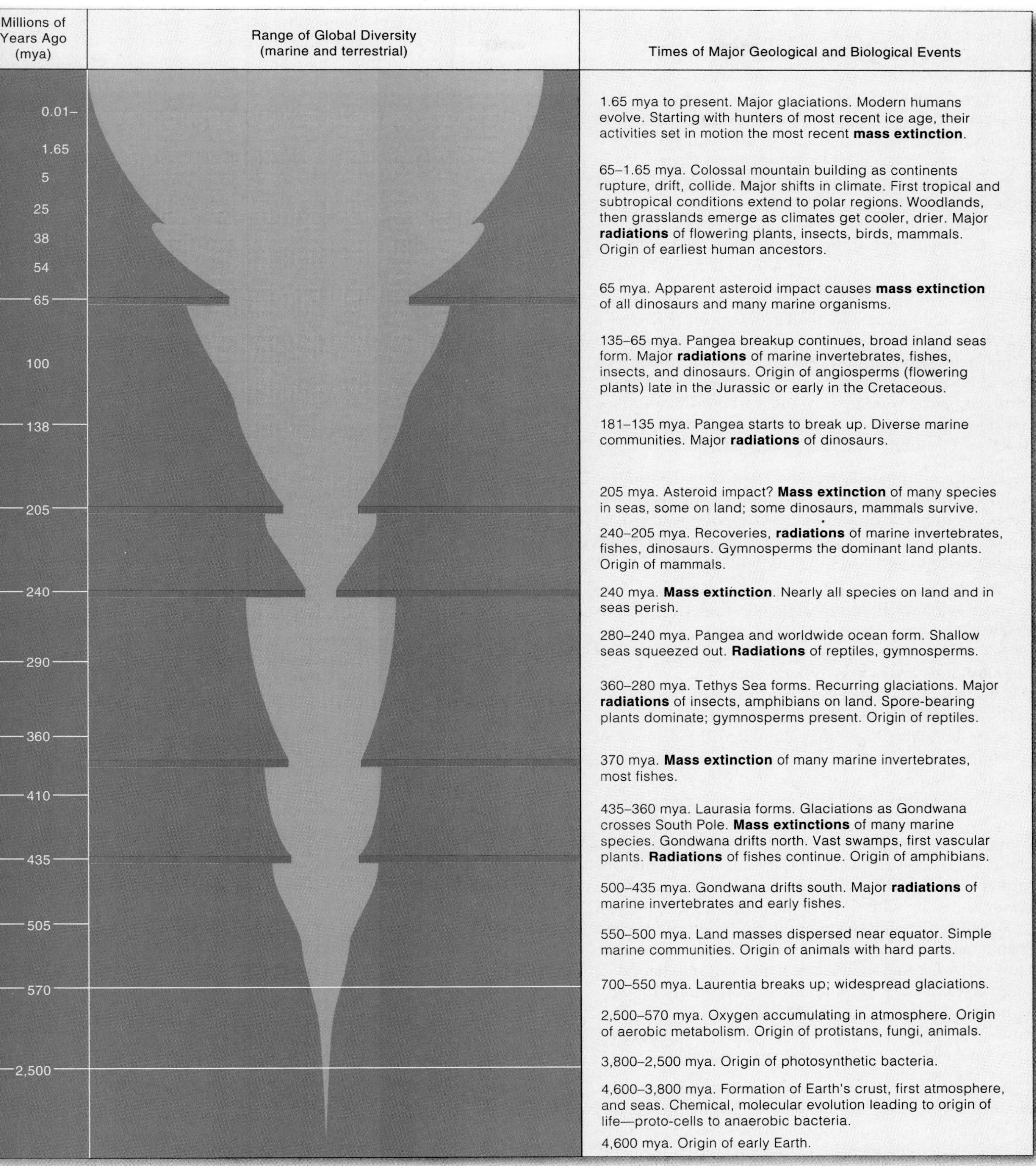

Millions of Years Ago (mya)
Range of Global Diversity (marine and terrestrial)
Times of Major Geological and Biological Events
0.01–
1.65
5
25
38
54
65
100
138
205
240
290
360
410
435
505
570
2,500
1.65 mya to present. Major glaciations. Modern humans evolve. Starting with hunters of most recent ice age, their activities set in motion the most recent **mass extinction**.
65–1.65 mya. Colossal mountain building as continents rupture, drift, collide. Major shifts in climate. First tropical and subtropical conditions extend to polar regions. Woodlands, then grasslands emerge as climates get cooler, drier. Major **radiations** of flowering plants, insects, birds, mammals. Origin of earliest human ancestors.
65 mya. Apparent asteroid impact causes **mass extinction** of all dinosaurs and many marine organisms.
135–65 mya. Pangea breakup continues, broad inland seas form. Major **radiations** of marine invertebrates, fishes, insects, and dinosaurs. Origin of angiosperms (flowering plants) late in the Jurassic or early in the Cretaceous.
181–135 mya. Pangea starts to break up. Diverse marine communities. Major **radiations** of dinosaurs.
205 mya. Asteroid impact? **Mass extinction** of many species in seas, some on land; some dinosaurs, mammals survive.
240–205 mya. Recoveries, **radiations** of marine invertebrates, fishes, dinosaurs. Gymnosperms the dominant land plants. Origin of mammals.
240 mya. **Mass extinction**. Nearly all species on land and in seas perish.
280–240 mya. Pangea and worldwide ocean form. Shallow seas squeezed out. **Radiations** of reptiles, gymnosperms.
360–280 mya. Tethys Sea forms. Recurring glaciations. Major **radiations** of insects, amphibians on land. Spore-bearing plants dominate; gymnosperms present. Origin of reptiles.
370 mya. **Mass extinction** of many marine invertebrates, most fishes.
435–360 mya. Laurasia forms. Glaciations as Gondwana crosses South Pole. **Mass extinctions** of many marine species. Gondwana drifts north. Vast swamps, first vascular plants. **Radiations** of fishes continue. Origin of amphibians.
500–435 mya. Gondwana drifts south. Major **radiations** of marine invertebrates and early fishes.
550–500 mya. Land masses dispersed near equator. Simple marine communities. Origin of animals with hard parts.
700–550 mya. Laurentia breaks up; widespread glaciations.
2,500–570 mya. Oxygen accumulating in atmosphere. Origin of aerobic metabolism. Origin of protistans, fungi, animals.
3,800–2,500 mya. Origin of photosynthetic bacteria.
4,600–3,800 mya. Formation of Earth's crust, first atmosphere, and seas. Chemical, molecular evolution leading to origin of life—proto-cells to anaerobic bacteria.
4,600 mya. Origin of early Earth.

21.10 SUMMARY

1. The evolutionary story of life begins with the "big bang," a model for the origin of the universe.

a. By this model, all matter and all of space were once compressed in a fleeting state of enormous heat and density. Time began with the near-instantaneous distribution of matter and energy through the known universe, which has been expanding ever since.

b. Helium and the other light elements formed right after the big bang. Heavier elements originated during the formation, evolution, and death of stars.

c. Every element of the solar system, the Earth, and life itself are products of the physical and chemical evolution of the universe and its stars.

2. Four billion years ago, the Earth was organized as a high-density core, a mantle of intermediate density, and a thin, extremely unstable crust of low-density rocks. Its first atmosphere probably contained gaseous hydrogen, nitrogen, carbon monoxide, and carbon dioxide. Free oxygen and liquid water could not have accumulated at the surface under the prevailing conditions.

3. Water accumulated after the crust cooled off. Over hundreds of millions of years, runoff from rains carried dissolved mineral salts and other compounds to crustal depressions, where the early seas formed. Life could not have originated without salty liquid water.

4. Many diverse studies and experiments have yielded indirect evidence that life originated under conditions that presumably existed on the early Earth.

a. Comparative investigations of the composition of cosmic clouds, rocks from other planets, and rocks from the Earth's moon suggest that precursors of complex molecules associated with life were available.

b. In laboratory tests that simulated the primordial conditions, including the absence of free oxygen, those precursors spontaneously assembled into sugars (such as glucose), amino acids, and other organic compounds.

c. Known chemical principles as well as advanced computer simulations indicate that metabolic pathways could have evolved through chemical competition for limited supplies of organic molecules (which could have accumulated by natural geologic processes in the seas).

d. Self-replicating systems of RNA, enzymes, and coenzymes have been synthesized in the laboratory. How DNA entered the picture is not yet understood.

e. In laboratory simulations of conditions thought to have prevailed on the early Earth, lipids as well as lipid–protein membranes having some of the properties of cell membranes have formed spontaneously.

5. Since life originated some 3.8 billion years ago, it has been influenced by major changes in the Earth's crust, atmosphere, and oceans. Forces of change have included asteroid impacts, volcanism, and movements of crustal plates, which altered the distribution of land masses and oceans as well as global and local climates. Another force of change: the activities of organisms—especially the oxygen-releasing photosynthesizers and, currently, the human species.

6. Abrupt discontinuities in the fossil record mark the times of global mass extinctions. They are the boundary markers for five great intervals in a geologic time scale. Radiometric dating has allowed us to assign absolute dates to this time scale:

a. Archean: 3.9 billion to 2.5 billion years ago
b. Proterozoic: 2.5 billion to 570 million years ago
c. Paleozoic: 570 to 240 million years ago
d. Mesozoic: 240 to 65 million years ago
e. Cenozoic: 65 million years ago to the present

7. The first living cells were prokaryotic (bacteria). Not long after they arose in the Archean, the first divergence led to eubacteria, and to a shared prokaryotic ancestor of both archaebacteria and eukaryotes. Some eubacteria used a cyclic pathway of photosynthesis.

8. During the Proterozoic, the noncyclic pathway of photosynthesis evolved in some lineages of eubacteria. Oxygen, a by-product of the pathway, slowly started to accumulate in the atmosphere.

a. Eventually, the atmospheric concentration of free oxygen blocked the further spontaneous formation of organic molecules. From that time on, the spontaneous origin of life was no longer possible on Earth.

b. The abundance of atmospheric oxygen became a selective pressure that brought about the evolution of aerobic respiration. Aerobic respiration was a key step toward the origin of the first eukaryotic cells.

c. Mitochondria and chloroplasts, both important eukaryotic organelles, probably evolved as an outcome of endosymbiosis between certain aerobic bacteria and the anaerobic bacterial forerunners of eukaryotes.

d. The oxygen-rich atmosphere promoted formation of a layer of ozone (O_3). In time, that atmospheric shield against destructive ultraviolet radiation allowed some lineages to move out of the seas, into low wetlands.

9. By the early Paleozoic, diverse organisms of all six lineages had become established in the seas. By its end, the invasion of land was under way. From that time on, there have been pulses of mass extinctions and adaptive radiations.

Review Questions

1. Compare the presumed chemical and physical conditions that are thought to have prevailed on the Earth 4 billion years ago with conditions that are now prevailing. *21.1*

2. Describe examples of the kinds of experimental evidence for the spontaneous origin of (a) large organic molecules, (b) the self-assembly of proteins, and (c) the formation of organic membranes and spheres, under laboratory conditions similar to those of the early Earth. *21.1, 21.2*

3. Summarize the key points of the theory of endosymbiotic origins for mitochondria and chloroplasts. Cite evidence that favors this theory. *21.4*

4. Describe the prevailing conditions that probably favored the Cambrian "explosion" of diversity among marine animals, as evidenced by the fossil record. *21.5*

5. During which geologic time spans did plants, fungi, and insects invade the land? What kind of vertebrates first invaded the land, and when? *21.5*

6. What were global conditions like when gymnosperms and dinosaurs originated? *21.6*

7. Briefly explain how an asteroid impact and "global broiling" may have caused the mass extinctions at the K–T boundary. *21.7*

8. Would you expect the Paleozoic, Mesozoic, or Cenozoic to be called "the age of mammals"? As part of your answer, explain the differences between global conditions in each era. *21.8*

Self-Quiz (*Answers in Appendix III*)

1. Life originated by ______ .
   a. 4.6 billion years ago c. 3.8 billion years ago
   b. 2.8 million years ago d. 3.8 million years ago

2. Through study of the geologic record, we know that the evolution of life has been profoundly influenced by ______ .
   a. tectonic movements of the Earth's crust
   b. bombardment of the Earth by celestial objects
   c. profound shifts in land masses, shorelines, and oceans
   d. physical and chemical evolution of the Earth
   e. all of the above

3. ______ was the first to obtain indirect evidence that organic molecules could have been formed on the early Earth.
   a. Darwin c. Fox
   b. Miller d. Margulis

4. An abundance of ______ was conspicuously absent from the Earth's atmosphere 4 billion years ago.
   a. hydrogen c. carbon monoxide
   b. nitrogen d. free oxygen

5. Which of the following statements is false?
   a. The first living cells were prokaryotic.
   b. The cyclic pathway of photosynthesis first appeared in some eubacterial species.
   c. Oxygen began accumulating in the atmosphere after the noncyclic pathway of photosynthesis evolved.
   d. In the Proterozoic, increasing levels of atmospheric oxygen enhanced the spontaneous formation of organic molecules.
   e. All are correct.

6. The first eukaryotic cells emerged during the ______ .
   a. Paleozoic c. Archean e. Cenozoic
   b. Mesozoic d. Proterozoic

7. Match the geologic time interval with the events listed.
   ____ Archean
   ____ Proterozoic
   ____ Paleozoic
   ____ Mesozoic
   ____ Cenozoic
   a. major radiations of dinosaurs, origin of flowering plants and mammals
   b. chemical evolution, origin of life
   c. major radiations of flowering plants, insects, birds, mammals; emergence of human forms
   d. oxygen present; origin of aerobic metabolism, protistans, fungi, animals
   e. rise of early plants on land, origin of amphibians, and origin of reptiles

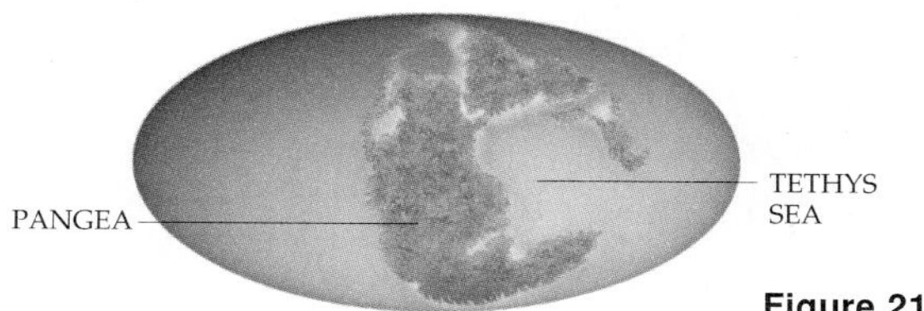
PANGEA
TETHYS SEA

Figure 21.21

Critical Thinking

1. Briefly explain, in terms of hydrophilic and hydrophobic interactions, how proto-cells might have formed in water from aggregations of lipids, proteins, and nucleic acids.

2. Reflect on Figure 21.21. Now reflect on this: the Atlantic Ocean is gradually widening, and the Pacific Ocean and Indian Ocean are closing in on each other. Many millions of years from now, continents will collide and form a second Pangea. Write a short essay on what environmental conditions might be like on that future supercontinent and on what types of species might survive there.

3. According to one estimate, there is a chance that about 10^{20} planets have formed in the universe that are capable of sustaining life—but there is only one chance at intelligent life per planet. Given your knowledge of molecular biology and evolutionary processes, do you agree with this estimate? If so, speculate on why the odds are so low.

4. We know of a number of large asteroids that may intersect Earth's orbit in the distant future. There probably are a number we don't know about. Would you use this as an excuse not to worry about polluting the environment, not to take care of your physical health (as by avoiding drugs), and not to care about our cultural evolution? Why or why not?

Selected Key Terms

angiosperm *21.6*
archaebacterium *21.3*
Archean *21.3*
asteroid *21.7*
big bang *CI*
Cenozoic *21.8*
crust, of Earth *21.1*
dinosaur *21.6*
Ediacaran *21.5*
endosymbiosis (theory of) *21.4*
eubacterium *21.3*
eukaryotic cell *21.3*
global broiling hypothesis *21.7*
gymnosperm *21.6*
K–T asteroid impact theory *21.7*
mantle, of Earth *21.1*
Mesozoic *21.6*
Paleozoic *21.5*
prokaryotic cell *21.3*
Proterozoic *21.3*
protistan *21.4*
proto-cell *21.2*
RNA world *21.2*
stromatolite *21.3*

Readings *See also www.infotrac-college.com*

de Duve, C. September–October 1995. "The Beginnings of Life on Earth." *American Scientist* 83: 428–437.

———. April 1996. "The Birth of Complex Cells." *Scientific American* 274(4): 50–57.

Dott, R., Jr., and R. Batten. 1994. *Evolution of the Earth.* Fifth edition. New York: McGraw-Hill.

Horgan, J. February 1991. "Trends in Evolution: In the Beginning . . . ." *Scientific American* 264(2): 116–125.

Impey, C. and W. Hartman. 2000. *The Universe Revealed.* Belmont, California: Wadsworth. Paperback

Wright, K. March 1997. "When Life Was Odd." *Discover* 18(3): 52–61. Update on the puzzling Ediacarans.

22

BACTERIA AND VIRUSES

The Unseen Multitudes

Did a friend ever mention that you are nearly 1/1,000 of a mile tall? Probably not. What would be the point of measuring people in units as big as miles? Even so, we think this way, in reverse, whenever we measure microorganisms. For the most part, **microorganisms** are single-celled organisms that are too small to be seen without the aid of a microscope.

The bacterial cells shown in Figure 22.1 are a case in point. To measure them, you would have to divide one meter into a thousand units, or millimeters. Next, you would have to divide one of the millimeters into a thousand smaller units, or micrometers. To give you a sense of how small that is, a single millimeter would be about as small as the dot of this "i." And a *thousand* bacteria would fit side by side on top of the dot!

To be sure, viruses are smaller still, as you might deduce after looking at Figure 22.2. We measure them in nanometers (billionths of a meter). But viruses are not alive. We consider them in this chapter because one kind or another infects just about every organism.

Bacteria generally are the smallest microorganisms, but they vastly outnumber the individuals in all other kingdoms combined. Their reproductive potential is staggering. Under ideal conditions, some divide about every twenty minutes. If that rate of reproduction were to hold constant, a single bacterium would have nearly a billion descendants in ten hours!

So why don't the unseen multitudes take over the world? Sooner or later, their burgeoning populations use up available nutrients and pollute the surroundings with their own metabolic wastes. In other words, they alter the very conditions that initially favored their reproduction. Besides this, certain kinds of viruses attack just about every type of bacterium and help keep their population sizes in check. So do seasonal changes in living conditions.

Of course, you probably do not find much comfort in this when you serve as host for one of the pathogenic types. **Pathogens** are infectious, disease-causing agents that invade target organisms and multiply inside or on

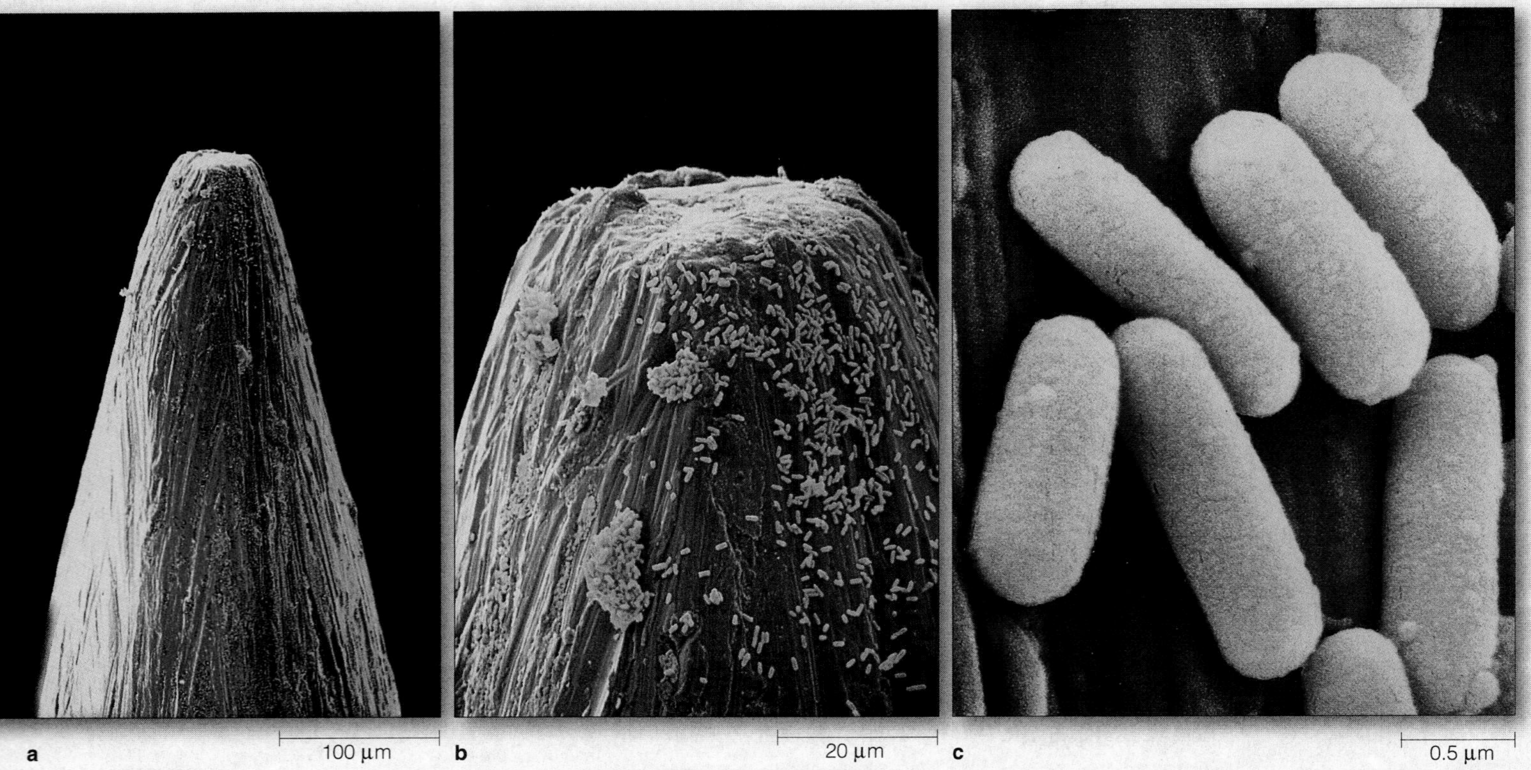

a
100 µm
b
20 µm
c
0.5 µm

Figure 22.1 (**a–c**) How small are bacteria? Shown here, *Bacillus* cells peppering the tip of a pin. The cells in (**c**) are magnified more than 16,600 times.

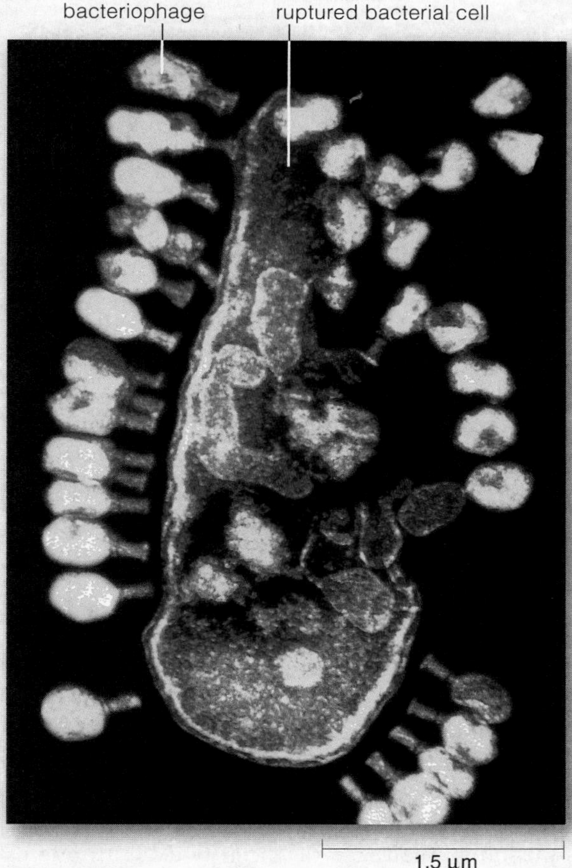
bacteriophage
ruptured bacterial cell
1.5 µm

Figure 22.2 How small are the viruses? Shown here, bacteriophage particles, each about 225 nanometers tall, that infected a bacterial cell. The cell has ruptured open.

them. Disease follows when the metabolic activities of pathogenic cells damage the tissues and interfere with the body's normal functioning.

Certain pathogens can indeed make you suffer, but they should not give every microorganism a bad name. For example, think back on the uncountable numbers of photosynthetic bacterial cells in the seas (Section 7.9). Collectively, they help provide food and oxygen for entire communities and have major roles in the global cycling of carbon. Or think about the kinds of bacterial species that feed on organic debris. Together with other decomposers, they help cycle nutrients that sustain entire communities.

From the human perspective, microorganisms are good or bad, even dangerous. Basically, however, they are simply surviving and reproducing like the rest of us, in ways that are the topics of this chapter.

KEY CONCEPTS

1. In structural terms, we find the simplest forms of life among the bacteria. Most bacteria are microscopically small. Smaller still are the viruses.

2. Bacteria alone are prokaryotic cells. They do not have a profusion of internal, membrane-bound organelles, as eukaryotic cells do. As a group, bacteria show great metabolic diversity, and many kinds display complex behavior.

3. Most bacteria reproduce by prokaryotic fission. This cell division mechanism follows DNA replication and results in two genetically equivalent daughter cells.

4. Bacteria were the first living organisms on Earth. Not long after they originated, they diverged into lineages that gave rise to eubacteria and to the common ancestor of archaebacteria and eukaryotic cells.

5. A virus, a noncellular infectious particle, consists of nucleic acid (either DNA or RNA), a protein coat, and sometimes an outer envelope. It cannot replicate without pirating the metabolic machinery of a specific type of host cell.

6. Nearly all viral multiplication cycles proceed through five steps: attachment to a host cell, penetration of its plasma membrane, replication of viral DNA or RNA and synthesis of viral proteins, then assembly of new viral particles, and finally release from the infected cell.

7. Most of us tend to judge microorganisms through the prism of human interests. Yet their lineages are the most ancient, their adaptations are diverse, and they are simply surviving and reproducing like the rest of us.

22.1 CHARACTERISTICS OF BACTERIA

Of all organisms, bacteria are the most abundant and far-flung. Thousands of species inhabit diverse places, such as deserts, hot springs, glaciers, and seas. Some have been carrying on for millions of years 2,780 meters (9,121 feet) below the Earth's surface! Billions may live in a handful of rich soil. The ones in your gut and on your skin outnumber your own cells. Bacteria have the longest evolutionary history. Trace any lineage back far enough and you come across bacterial ancestors. From *Escherichia coli* to amoebas, elephants, clams, and coast redwoods, all organisms interconnect, regardless of size, numbers, and evolutionary distance. Table 22.1 and Figure 22.3 introduce features that help characterize the remarkable bacteria.

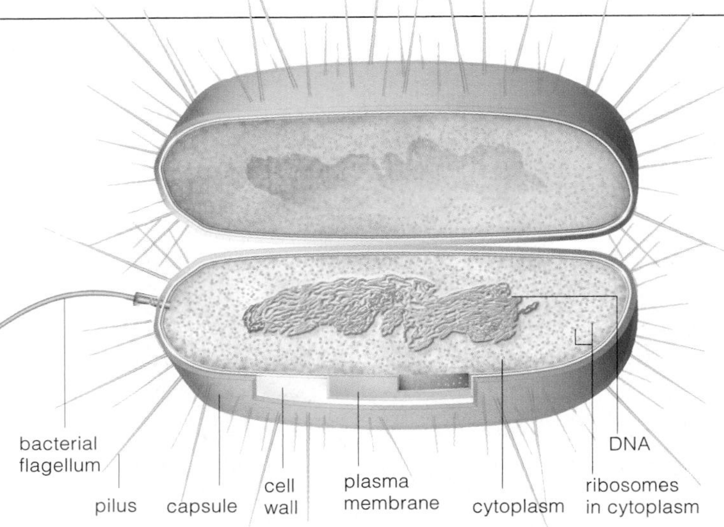
bacterial flagellum
pilus
capsule
cell wall
plasma membrane
cytoplasm
ribosomes in cytoplasm
DNA

Figure 22.3 Generalized body plan of a bacterium. 

Splendid Metabolic Diversity

All organisms take in energy and carbon to meet their nutritional requirements. Compared with other species, however, bacteria display the greatest diversity in their means of securing resources.

Like plants, *photoautotrophic* bacteria build organic compounds by photosynthesis; they are "self-feeders." They tap sunlight for energy and use carbon dioxide as their carbon source. Their plasma membrane contains the photosynthetic machinery. Some photosynthesizers use electrons and hydrogen from water molecules for the synthesis reactions, and release oxygen. Anaerobic photosynthesizers (which die in the presence of oxygen) get electrons and hydrogen from inorganic compounds, such as gaseous hydrogen and hydrogen sulfide.

Chemoautotrophic bacteria are self-feeders. Carbon dioxide is the usual carbon source. Some species strip organic compounds for electrons and hydrogen. Others use inorganic substances, such as gaseous hydrogen, sulfur, nitrogen compounds, and a form of iron.

Photoheterotrophic bacteria are not self-feeders. They use energy from the sun for photosynthesis, but their carbon sources are fatty acids, complex carbohydrates, and other compounds that various organisms produce.

Table 22.1 Characteristics of Bacterial Cells
1. All bacterial cells are prokaryotic; they do not have a membrane-bound nucleus.
2. Bacterial cells in general have a single chromosome (a circular DNA molecule); many species also have plasmids.
3. Most bacteria have a cell wall of peptidoglycan.
4. Most bacteria reproduce by prokaryotic fission.
5. Collectively, bacteria show great metabolic diversity.

Chemoheterotrophic bacteria are parasites or saprobes, not self-feeders. The parasites live on or in a living host and draw glucose and other nutrients from it. Saprobic types get nutrients from the organic products, wastes, or remains of other organisms.

Bacterial Sizes and Shapes

So far, you have a general sense of the microscopically small sizes of bacteria. The width and length of these cells typically fall between 1 and 10 micrometers.

Three basic shapes are common among bacteria. A spherical shape is a **coccus** (plural, cocci; from a word that means berries). A rod shape is a **bacillus** (plural, bacilli, meaning small staffs). A cell body with one or more twists is a **spirillum** (plural, spirilla):

coccus
bacillus
spirillum

Don't let the simple categories fool you. Cocci may also be oval or flattened. Bacilli may be skinny (like straws) or tapered (like cigars). Surface extensions make some bacteria look starlike. Square bacteria live in salt ponds in Egypt. Also, after daughter cells divide, they may stick together in chains, sheets, and other aggregations, as in Figure 22.4. Some spiral species are curved like a comma, and others are flexible or like stiff corkscrews.

Structural Features

Bacteria alone are **prokaryotic cells**, meaning they were around before the evolution of the nucleated cell (*pro-*, before; *karyon*, nucleus). Few have membrane-bounded compartments of any sort for isolating metabolic events. Reactions take place in the cytoplasm or at the plasma membrane. For example, protein synthesis proceeds at

Further reading: Student Guide to InfoTrac on web site

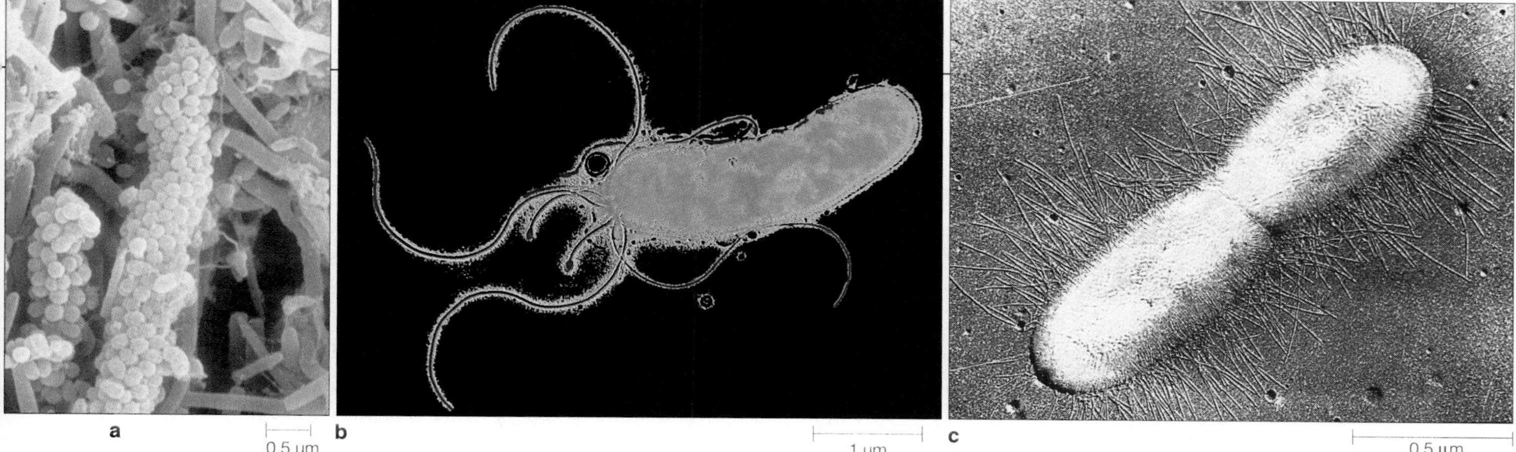
a
0.5 µm
b
1 µm
c
0.5 µm

Figure 22.4 (**a**) Surface view of numerous bacilli and cocci attached to a human tooth. (**b**) *Helicobacter pylori* cell, with its tuft of flagella. This pathogen can colonize the stomach lining and trigger inflammation. If untreated, an infection can lead to gastritis, peptic ulcers, and possibly stomach cancer. *H. pylori* may contaminate water and food, especially unpasteurized milk. Currently, a combination of antibiotics and an antacid rids the body of the pathogen, after which ulcers heal. (**c**) Example of pili, filamentous structures that project from the surface of many bacterial cells. This *Escherichia coli* cell is dividing in two.

Figure 22.5 Scanning electron micrographs of bacterial cell walls. (**a**) Dividing *Bacillus subtilis* cell. Its wall has a smooth surface texture. (**b**) A dividing *E. coli* cell. Its wrinkled cell wall is characteristic of Gram-negative bacteria.

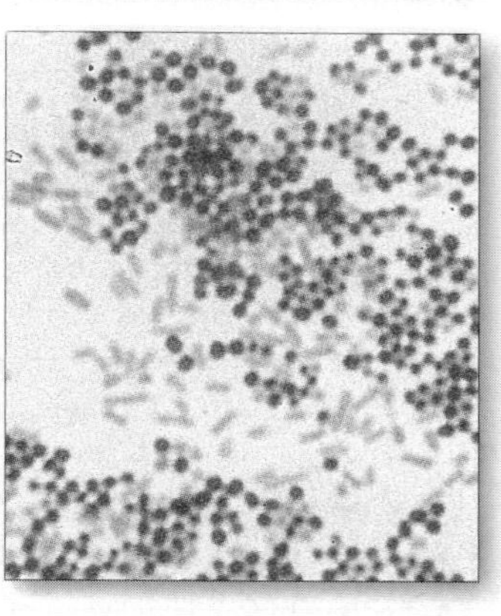

Figure 22.6 Gram staining. Cocci and rods smeared on a slide are stained with a purple dye (such as crystal violet), washed off, then stained with iodine. All cells are now purple. The slide is washed with alcohol, which renders the Gram-negative cells colorless. Now the slide is counterstained (with safranin), washed, and dried. Gram-positive cells (*Staphylococcus aureus* in this case) remain purple. The counterstain, however, colors the Gram-negative cells (*Escherichia coli*) pink.

- stain with purple dye
- stain with iodine
- wash with alcohol
- counterstain with safranin

ribosomes that are distributed through the cytoplasm or attached to the side of the plasma membrane facing the interior. This does not mean bacteria are somehow inferior to eukaryotic cells. Being tiny, fast reproducers, they do not require great internal complexity.

A wall usually surrounds the plasma membrane. A **cell wall** is a semirigid, permeable structure that helps a cell maintain shape and resist rupturing when internal fluid pressure rises (Figure 22.5 and Section 5.5). The cell wall of eubacteria is composed of peptidoglycan molecules. In such molecules, peptide groups crosslink numerous polysaccharide strands to one another.

Clinicians can identify many bacterial species on the basis of their wall structure and composition. Consider a staining reaction called the **Gram stain**. A sample of unknown bacterial cells is exposed to a purple dye, then to iodine, an alcohol wash, and a counterstain. The cell wall of *Gram-positive* species remains purple. The wall of *Gram-negative* species loses color after the wash, but the counterstain turns it pink (Figure 22.6).

A sticky mesh, or **glycocalyx**, often encloses the cell wall. It consists of polysaccharides, polypeptides, or both. When highly organized and attached firmly to the wall, we call the mesh a capsule. When less organized and loosely attached to the wall, we call it a slime layer. The mesh helps a bacterial cell attach to teeth, mucous membranes, rocks in streambeds, and other interesting surfaces (Figure 22.4*a*). It also helps some encapsulated species avoid being engulfed by phagocytic, infection-fighting cells of their host organism.

Some bacteria have one or more **bacterial flagella** (singular, flagellum), which are used in motility. These don't have the same structure as a eukaryotic flagellum, and they do not operate the same way. They move the cell by rotating like a propeller. Many species also have **pili** (singular, pilus). These short, filamentous proteins project above the cell wall, as in Figure 22.4*c*. Some pili help cells adhere to surfaces. Others help them attach to one another as a prelude to conjugation, an interaction that is described in the next section.

Bacteria are microscopic, prokaryotic cells. They generally have one circular bacterial chromosome, and often they have extra DNA in the form of plasmids.

Nearly all bacteria have a wall around the plasma membrane. Many have a capsule or slime layer around the cell wall.

BACTERIAL GROWTH AND REPRODUCTION

The Nature of Bacterial Growth

Between cell divisions, bacteria grow through increases in their component parts. We measure the growth of a large, multicelled organism in terms of increases in size, but doing this for a microscopically small bacterium would be a bit pointless. Instead, we measure bacterial growth as an increase in the number of cells in a given population. Under ideal conditions, each cell divides in two, the division of two cells results in four, four result in eight, and so forth. Many types can divide every half hour; a few can do so every ten minutes. Such rates of increase lead to large population sizes in short order.

Natural conditions that promote the growth of some bacterial populations might not sustain the growth of others. Most organisms could not become established in the most extreme environments—say, Antarctica, the Negev Desert, or deep inside the Earth. There, the few species that do endure, inside rocks, rarely reproduce. And few can live in the highly acidic wastewater from mining operations. K-12, the strain of *E. coli* originally isolated from the human gut, has been cultivated for such a long time in the laboratory that it no longer is able to grow when reintroduced into the gut. Through microevolutionary processes, it became adapted to the conditions in its artificial environment.

Prokaryotic Fission

When a bacterium has nearly doubled in size, it divides in two. Each daughter cell inherits a single **bacterial chromosome**, which is a circularized, double-stranded DNA molecule that has only a few proteins attached to it. In some species, the daughter cell merely buds from the parent cell. Most often, however, bacteria reproduce by a division mechanism called **prokaryotic fission**.

As prokaryotic fission begins, a parent cell replicates its DNA (Figure 22.7). The two molecules of DNA that result are attached to the plasma membrane at adjacent sites. The cell synthesizes lipid and protein molecules, which become incorporated in the membrane between the attachment sites. Membrane growth moves the two DNA molecules apart. New wall material is deposited above the membrane. The membrane and wall grow through the cell midsection and divide the cytoplasm. The result is two genetically equivalent daughter cells.

Especially in microbiology, you may hear someone refer to this division mechanism as *binary* fission, but such usage can cause confusion. The same term applies to a form of asexual reproduction among flatworms and some other multicelled animals. It refers to growth by way of mitotic cell divisions, then division of the whole body into two parts of the same or different sizes.

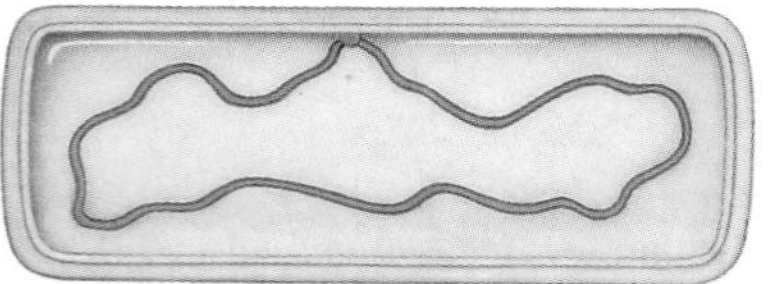

a Bacterium (cutaway view) before DNA replication. The bacterial chromosome is attached to the plasma membrane.

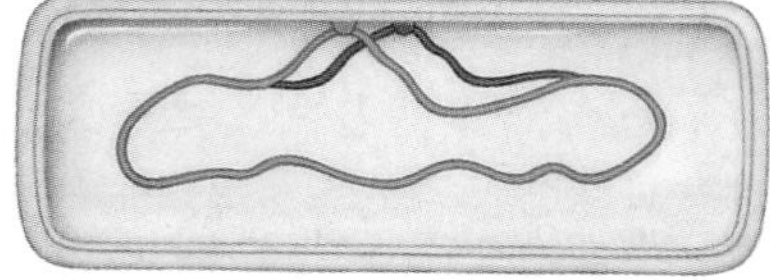

b DNA replication starts. It proceeds in two directions away from the same site in the bacterial chromosome.

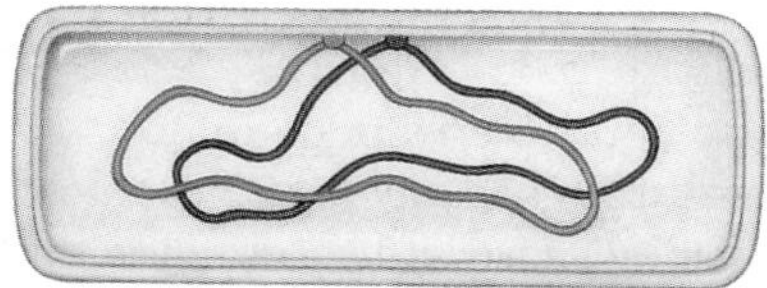

c The new copy of DNA is attached at a membrane site near the attachment site of the parent DNA molecule.

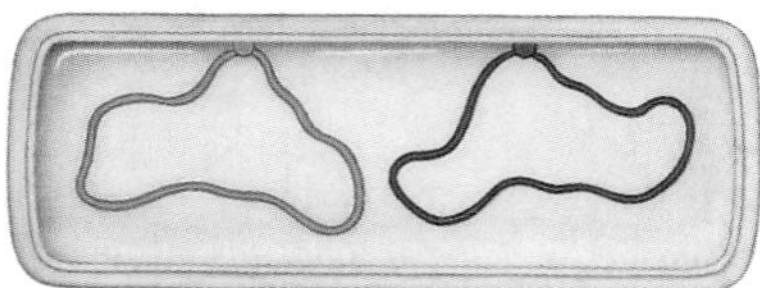

d New membrane grows between the two attachment sites. As it increases, it moves the two DNA molecules apart.

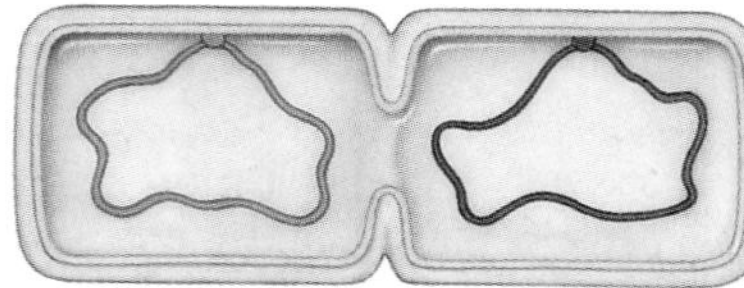

e At the cell midsection, deposits of new membrane and new wall material extend down into the cytoplasm.

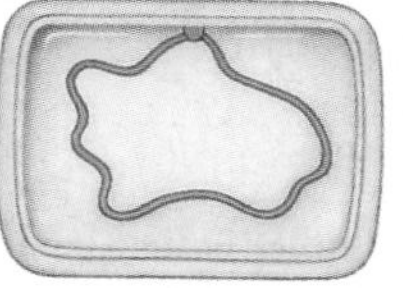

f The ongoing, organized deposition of membrane and wall material at the cell midsection divides the cell in two.

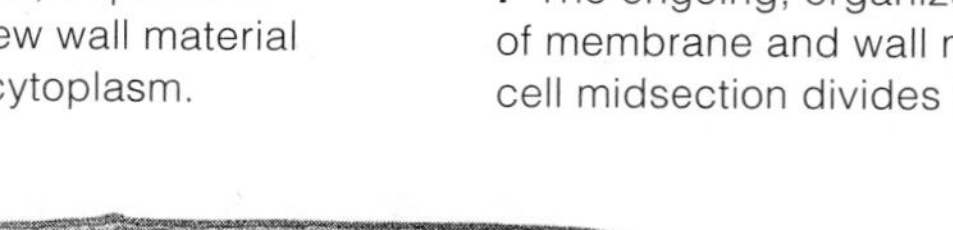

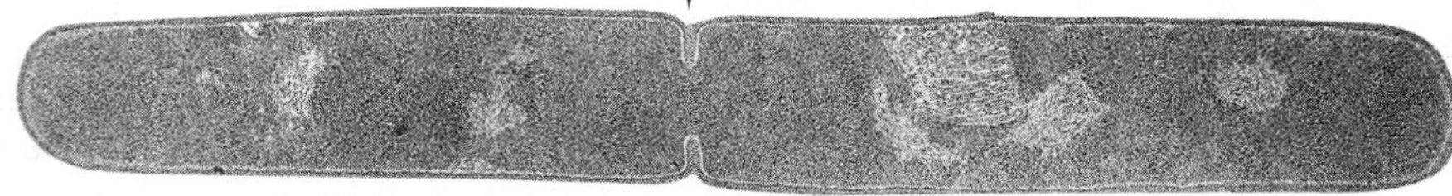

Figure 22.7 Bacterial reproduction by way of prokaryotic fission, a cell division mechanism. The micrograph shows cytoplasmic division of *Bacillus cereus*, as brought about by the formation of new membrane and wall material. This cell has been magnified 13,000 times its actual size.

Further reading: Student Guide to InfoTrac on web site

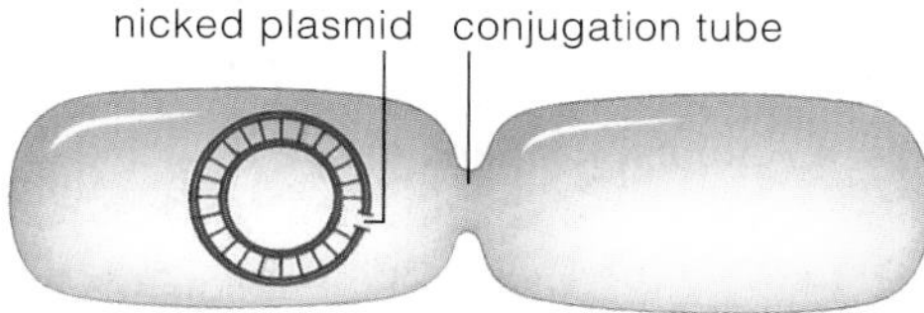
nicked plasmid
conjugation tube

a A conjugation tube has already formed between a donor and a recipient cell. An enzyme has nicked the donor's plasmid.

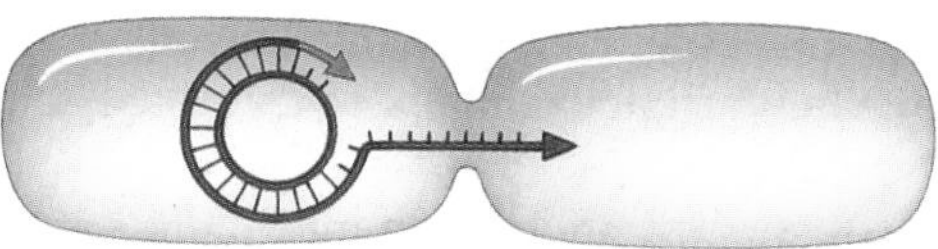

b DNA replication starts on the nicked plasmid. The displaced DNA strand moves through the tube and enters the recipient cell.

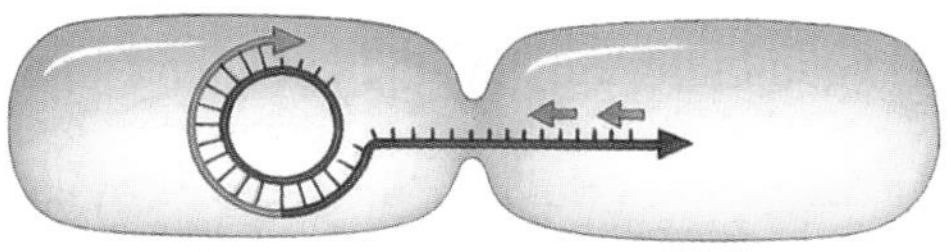

c In the recipient cell, replication starts on the transferred DNA.

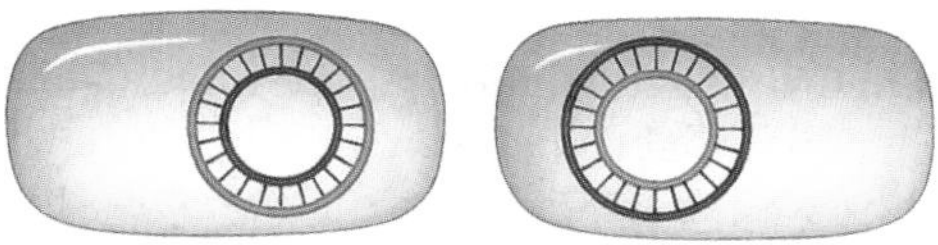

d The cells separate from each other; the plasmids circularize.

Figure 22.8 Simplified sketch of bacterial conjugation. For clarity, the plasmid's size is greatly increased and the bacterial chromosome is not shown.

Bacterial Conjugation

Daughter bacterial cells also may inherit one or more plasmids. A **plasmid**, recall, is a small, self-replicating circle of extra DNA that has a few genes (Section 16.1). The F (Fertility) plasmid has certain genes that confer the means to engage in **bacterial conjugation**. By this mechanism, a donor cell transfers plasmid DNA to a recipient cell. The transfers are known to occur among many bacterial species, such as *Salmonella*, *Streptococcus*, and *E. coli*—even between *E. coli* and yeast cells, in the laboratory. The F plasmid carries genetic instructions for synthesizing a structure called a sex pilus. Sex pili at the surface of a donor cell can hook onto a recipient cell and pull it right next to the donor. Shortly after the two cells make contact, a conjugation tube develops between them. After this, plasmid DNA is transferred through the tube, in the manner shown in Figure 22.8.

Most bacteria reproduce by way of prokaryotic fission, a cell division mechanism that follows DNA replication. Each daughter cell inherits a single bacterial chromosome (one DNA molecule). Many species also transfer plasmid DNA.

22.3 BACTERIAL CLASSIFICATION

Just a few decades ago, reconstructing the evolutionary history of bacteria appeared to be an impossible task. Except for stromatolites (Section 21.3), the most ancient groups of bacteria are not well represented in the fossil record. Most groups are not represented at all.

Given their elusive histories, the many thousands of known species of prokaryotic cells traditionally have been classified by **numerical taxonomy**. By this practice, traits of an unidentified cell are compared with those of a known bacterial group. Such traits typically include cell shape, motility, staining attributes of the cell wall, nutritional requirements, metabolic patterns, and the presence or absence of endospores. The greater the total number of traits that the cell has in common with the known group, the closer is the inferred relatedness.

Since the 1970s, nucleic acid hybridization studies, gene sequencing, and other methods of comparative biochemistry have been revealing compelling evidence of bacterial phylogenies (Section 20.4). Comparisons of ribosomal RNAs are notably useful. Remember, rRNAs are indispensable to protein synthesis, and they cannot undergo drastic changes in their overall base sequence without loss of function. The numerous small changes that accumulated in the rRNAs of different lineages can

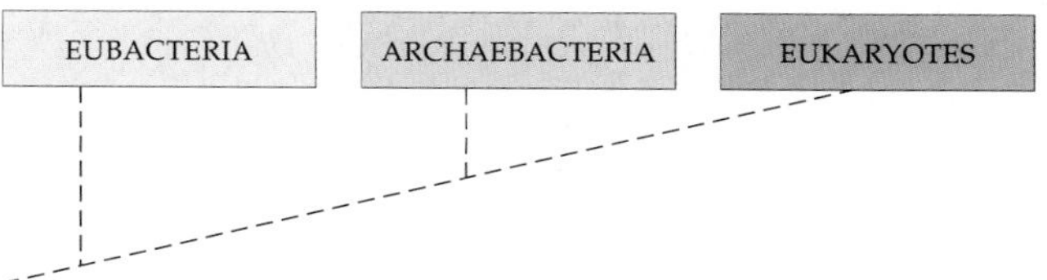
EUBACTERIA
ARCHAEBACTERIA
EUKARYOTES

Figure 22.9 Evolutionary tree diagram showing a relationship between archaebacteria, eubacteria, and eukaryotes, based on evidence from comparative biochemistry.

be directly measured. Surprisingly, the measurements are uniting some groups that did not even appear to be related on the basis of other tests. For example, it now seems a key divergence began shortly after prokaryotic cells first appeared on Earth (Section 21.3). One branch led to **eubacteria**, the most common prokaryotic cells. (Here, the *eu*– signifies "typical.") The other branch led to the common ancestor of both the **archaebacteria** and the first eukaryotic cells. Figure 22.9 is a tree diagram of these evolutionary relationships.

Prokaryotic cells are classified by numerical taxonomy: the total percentage of observable traits they have in common with a known bacterial group. They also are classified more directly by comparisons at the biochemical level.

Prokaryotic cells are now assigned to one of two lineages, called the eubacteria and archaebacteria.

22.4 MAJOR BACTERIAL GROUPS

By now, you probably have sensed that *species* is the basic unit in bacterial classification schemes. However, the definition of species that fits sexually reproducing organisms does not fit bacteria—which do not form reproductively isolated populations of interbreeding individuals. Each bacterial cell generally does its own thing; genetic recombination is infrequent. Besides this, bacteria do not display spectacular variation in traits, and variations that do occur are dictated by relatively few genes. If two types show only minor differences, one may be classified as a **strain**, not a new species.

These are just a few of the problems we face when attempting to bring more order to our understanding of bacterial kingdoms. Until evolutionary relationships are sorted out, the bacteria will continue to be grouped mainly according to numerical taxonomy, as described earlier. Table 22.2 summarizes some major groupings that are touched upon in this book. Appendix I lists their habitats and characteristics.

Table 22.2 Representative Archaebacteria and Eubacteria

Some Major Groups	Representatives
ARCHAEBACTERIA	
Methanogens	*Methanobacterium*
Extreme halophiles	*Halobacterium, Halococcus*
Extreme thermophiles	*Sulfolobus, Thermoplasma*
EUBACTERIA	
Photoautotrophs:	
Cyanobacteria, green sulfur bacteria, and purple sulfur bacteria	*Anabaena, Nostoc, Rhodopseudomonas, Chloroflexus*
Photoheterotrophs:	
Purple nonsulfur and green nonsulfur bacteria	*Rhodospirillum, Chlorobium*
Chemoautotrophs:	
Nitrifying, sulfur-oxidizing, and iron-oxidizing bacteria	*Nitrosomonas, Nitrobacter, Thiobacillus*
Chemoheterotrophs:	
Spirochetes	*Spirochaeta, Treponema*
Gram-negative aerobic rods and cocci	*Pseudomonas, Neisseria, Rhizobium, Agrobacterium*
Gram-negative facultative anaerobic rods	*Salmonella, Escherichia, Proteus, Photobacterium*
Rickettsias and chlamydias	*Rickettsia, Chlamydia*
Myxobacteria	*Myxococcus*
Gram-positive cocci	*Staphylococcus, Streptococcus, Deinococcus*
Endospore-forming rods and cocci	*Bacillus, Clostridium*
Gram-positive nonsporulating rods	*Lactobacillus, Listeria*
Actinomycetes	*Actinomyces, Streptomyces*

22.5 ARCHAEBACTERIA

We divide the kingdom of archaebacteria into three major groups: methanogens, extreme halophiles, and extreme thermophiles. In many respects, archaebacteria are unique in their composition, structure, metabolism, and nucleic acid sequences. They differ as much from other bacteria as they do from eukaryotic cells. Many investigators believe that the existing archaebacteria, which can withstand conditions as hostile as those on the early Earth, resemble the first living cells. Hence the name of the kingdom (*archae–* means "beginning").

METHANOGENS The **methanogens** (methane makers) live in swamps, stockyards, the termite and mammalian gut, and other oxygen-free habitats (Figure 22.10). They are strict anaerobes; they die in the presence of oxygen.

Methanogens produce ATP by anaerobic electron transport, a pathway described in Section 8.5. Usually, they use hydrogen gas (H_2) as their source of electrons, although some groups get them from ethanol and other alcohols. Nearly all use carbon dioxide as their carbon source and as a final electron acceptor for the reactions, which end with the formation of methane (CH_4). As a group, the methanogens produce about 2 billion tons of methane every year. Most of the methane production is released from wetlands, termites, ruminants, landfills, and rice paddies. Collectively their activities influence the levels of carbon dioxide in the atmosphere and the cycling of carbon through ecosystems.

Long ago, huge methane deposits accumulated on the seafloor. For example, geologists recently found 35

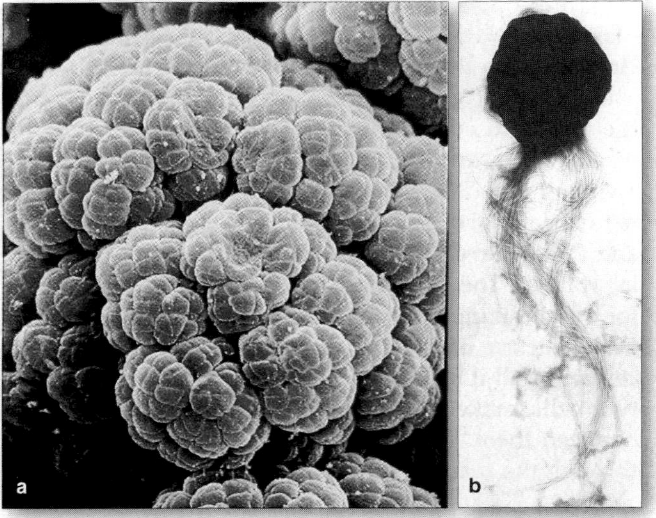
a
b

Figure 22.10 Representative archaebacteria. (**a**) Scanning electron micrograph of a dense population of *Methanosarcina* cells. Each of these methanogens has a thick polysaccharide wall. (**b**) Transmission electron micrograph of *Methanococcus jannaschii*, a heat-loving methane producer.

Further reading: Student Guide to InfoTrac on web site —

Figure 22.11 (**a**) Representative habitat for methanogens. The stomach chambers of cattle and other ruminants house methanogen populations. Cattle belch—a lot—and release quantities of methane. You might have noticed the resulting exceptionally pungent air around stockyards.

Representative habitats for extreme halophiles. (**b**) Great Salt Lake, Utah. Halophilic bacteria make purplish pigments and a green alga (*Dunaliella salina*) makes beta-carotene pigments when salinity and light intensity are high, and pH and nutrient levels are low. The pigments impart a reddish-purple color to the water. (**c**) Commercial seawater evaporating ponds, San Francisco Bay, another home of *Halobacterium* and *D. salina*. Both organisms thrive in hypersaline conditions, where salt levels greatly exceed that of seawater (for example, 300 versus 35 grams per liter).

(**d**) Representative habitat for extreme thermophiles: hot, sulfur-rich water in Emerald Pool, Yellowstone National Park. Microbes with an abundance of carotenoids impart an orange color to the rim of this hot spring.

billion tons of it 400 kilometers off the South Carolina coast. If the ocean circulation were to change, as it has in the past, then all of that gas could move abruptly and explosively to the surface. The rapid release of so much methane would drastically change the atmosphere and therefore the global climate (Section 49.9).

EXTREME HALOPHILES There are many salty habitats around the world, but **extreme halophiles** (salt lovers) thrive in exceptionally salty ones. Great Salt Lake, the Dead Sea, and seawater evaporation ponds are like this (Figure 22.11*b*,*c*). Extreme halophiles spoil commercial sea salt, salted fish, and animal hides. Most make ATP by aerobic pathways. When oxygen levels are low, some make ATP by photosynthesis that starts with a unique pigment, bacteriorhodopsin, in the plasma membrane. Absorption of light energy triggers a metabolic process that leads to an increase in an H^+ gradient across the plasma membrane. ATP forms when these ions follow the gradient and flow through the interior of proteins that span the membrane (compare Section 7.5).

EXTREME THERMOPHILES Geothermally heated soils, sulfur-rich hot springs, and wastes from coal mines are habitats of the "heat lovers," or **extreme thermophiles** (Figure 22.11*d*). These archaebacteria are the most heat-tolerant prokaryotes known. Some populations grow at above-boiling temperatures; all do best at temperatures above 80°C! Nearly all are strict anaerobes that require sulfur as an electron acceptor or electron donor.

Solfolobus, the first extreme thermophile discovered, grows in acidic hot springs. So does *Thermus aquaticus*, which biotechnologists employ as a source of extremely heat-stable DNA polymerases (Section 16.2).

Certain extreme thermophiles live in shallow water around volcanoes. We find some of these off the coast of Italy where geothermally heated water spews from openings in the sediments. Other species are the basis of food webs in sediments around hydrothermal vents, where the surrounding water can exceed 110°C. They use hydrogen sulfide spewing from these vents as a source of electrons. Their existence at vents is cited as evidence that life may have originated on the seafloor.

Like the first cells on Earth, archaebacteria live in extremely inhospitable habitats. In their properties, they differ as much from eubacteria as from eukaryotic cells.

22.6 EUBACTERIA

We recognize more than 400 genera of prokaryotes. By far, most of them are eubacteria. Unlike other bacterial cells, eubacteria have fatty acids incorporated into their plasma membrane. When a species has a cell wall—as nearly all do—the wall incorporates peptidoglycan. We still have not learned enough about the evolutionary histories of eubacteria to move much beyond taxonomic classification. Here we focus on modes of nutrition to give you a sense of the biodiversity in this kingdom.

A Sampling of Biodiversity

PHOTOAUTOTROPHIC EUBACTERIA The cyanobacteria, once called blue-green algae, are the classic example of photoautotrophic eubacteria. They are also among the most common photoautotrophs. All cyanobacteria are aerobic cells that engage in photosynthesis. Most types live in ponds and other freshwater habitats. They may grow as mucus-sheathed chains of cells, which often form thick, dense, slimy mats at the surface of nutrient-enriched water (Figure 22.12*a*).

Anabaena and other types also convert nitrogen gas (N_2) to ammonia, which they use in biosynthesis. When nitrogen compounds are scarce, some cells develop into **heterocysts**. These modified cells make a nitrogen-fixing enzyme (Figure 22.12*b*,*c*). They produce and then share nitrogen compounds with photosynthetic cells; in return they receive carbohydrates. The shared substances move freely across cytoplasmic junctions between cells.

Anaerobic photoautotrophs, such as green bacteria, get electrons from hydrogen sulfide and hydrogen gas, not from water. They may resemble anaerobic bacteria in which the cyclic pathway of photosynthesis evolved.

CHEMOAUTOTROPHIC EUBACTERIA Nearly all eubacteria in this category influence the global cycling of nitrogen, sulfur, and other nutrients. For example, as you know, nitrogen is a major building block for amino acids and proteins. Without it, there would be no life. Nitrifying bacteria in soil strip electrons from ammonia. Plants use the end product, nitrate, as a nitrogen source.

CHEMOHETEROTROPHIC EUBACTERIA Nearly all bacteria fall in this category. Many, including pseudomonads, are decomposers; their enzymes can break down organic compounds and even pesticides in soil. Other "good" species, at least by our standards, are *Lactobacillus* (used in the manufacture of pickles, sauerkraut, buttermilk, and yogurt) and actinomycetes (sources of antibiotics). *E. coli* produces vitamin K and compounds that help us digest fat, it helps newborn mammals digest milk, and its activities help stop many food-borne pathogens from colonizing the gut. Sugarcane and corn benefit from a symbiont, the nitrogen-fixing spirochete *Azospirillum*.

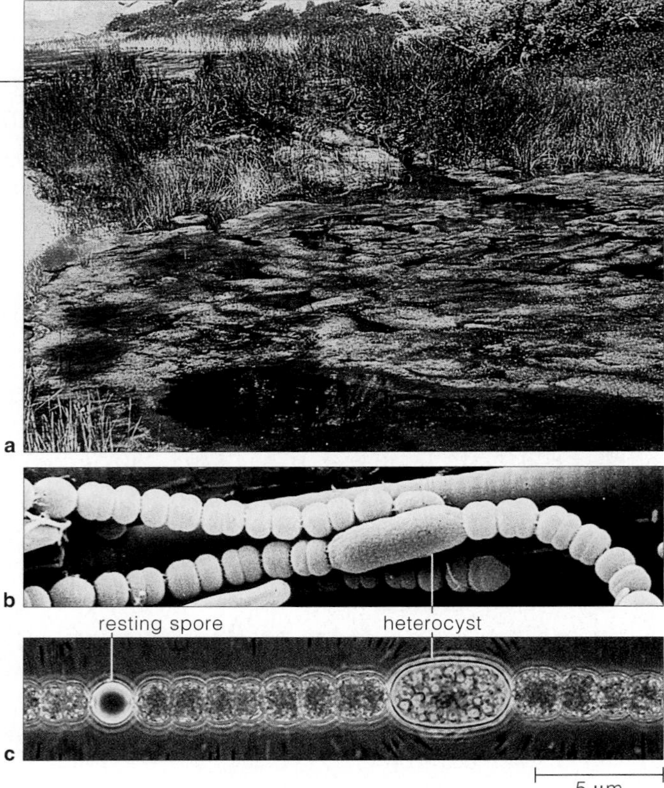
a
b
resting spore
heterocyst
c
5 µm

Figure 22.12 Cyanobacteria—common photoautotrophs. (**a**) A cyanobacterial population near the surface of a nutrient-enriched pond. (**b**,**c**) Resting spores form when conditions do not favor growth. A nitrogen-fixing heterocyst is also shown.

The plants use some of the nitrogen initially fixed by the bacterium and give up some sugars to it. *Rhizobium* is a fine symbiont with peas, beans, and other legumes. *Deinococcus radiodurans* resists high radiation doses that would kill other organisms. A genetically engineered strain holds promise for bioremediation; its *E. coli* genes allow it to break down toxic nuclear wastes, including extremely nasty mercury compounds. At present, there is no other mechanism in place to rid the environment of the alarming accumulation of nuclear wastes.

Also in this category are most pathogenic bacteria. We admire pseudomonads as decomposers in soil, not when they grow on our soaps, antiseptics, and other carbon-rich goods. They are especially bad because they can transfer plasmids with antibiotic-resistance genes.

Some *E. coli* strains cause a form of diarrhea that is the main cause of infant death in developing countries. *Clostridium botulinum* can taint fermented grain as well as food in improperly sterilized or sealed cans and jars. Its toxins cause *botulism*, a form of poisoning that can disrupt breathing and lead to death. *C. tetani*, one of its relatives, causes the disease *tetanus* (Section 34.5).

Like many other bacteria that typically live in soil, *C. tetani* can form an **endospore**. This resting structure forms inside the cell, around a copy of the bacterial chromosome and part of the cytoplasm (Figure 22.13). Endospores form when a depletion of nitrogen or some other nutrient arrests cell growth. They are released as

Further reading: Student Guide to InfoTrac on web site —

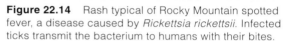

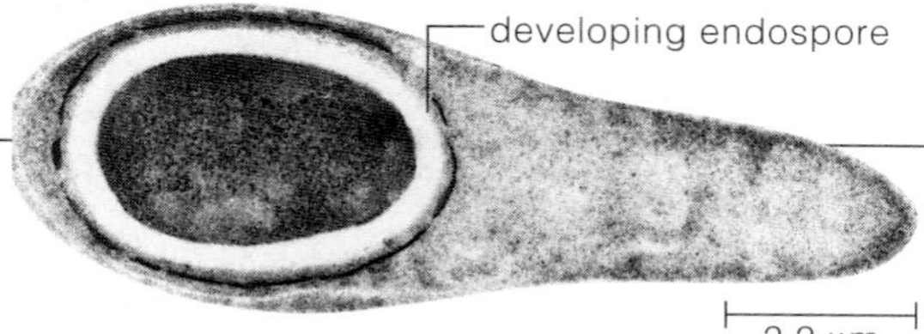
developing endospore
2.2 μm

Figure 22.13 Developing endospore in *Clostridium tetani*, one of the dangerous pathogens.

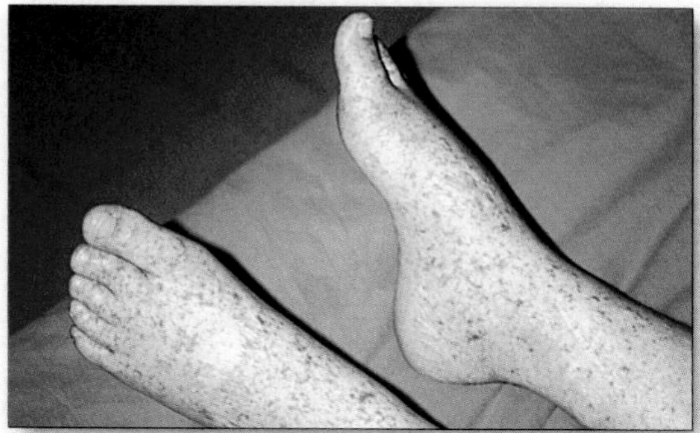
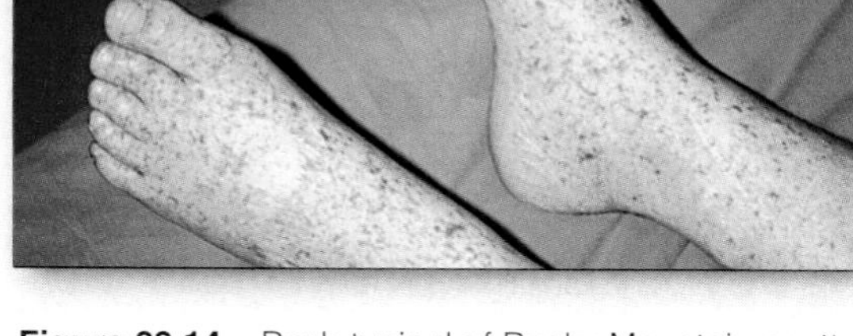

Figure 22.14 Rash typical of Rocky Mountain spotted fever, a disease caused by *Rickettsia rickettsii*. Infected ticks transmit the bacterium to humans with their bites.

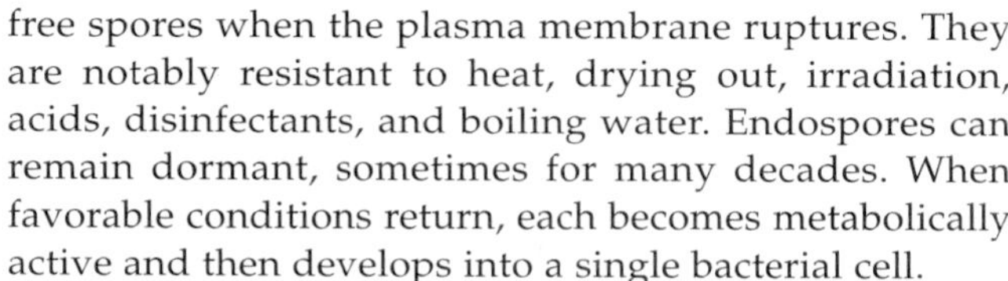
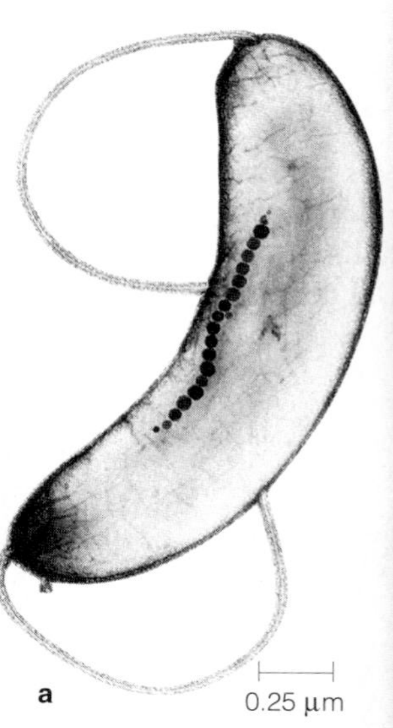
a
0.25 μm

b

Figure 22.15 (**a**) Micrograph of a magnetotactic bacterium. Inside the cytoplasm is a chain of magnetite particles that acts like a compass when the bacterium moves about. (**b**) A gathering of myxobacteria (*Chondromyces crocatus*). Cells of this species aggregate and form spore-bearing structures.

free spores when the plasma membrane ruptures. They are notably resistant to heat, drying out, irradiation, acids, disinfectants, and boiling water. Endospores can remain dormant, sometimes for many decades. When favorable conditions return, each becomes metabolically active and then develops into a single bacterial cell.

Many chemoautotrophic eubacteria taxi from host to host inside the gut of insects. For example, bites from infected ticks transmit *Borrelia burgdorferi* from deer and some other wild animals to humans, who develop *Lyme disease*. A "bull's-eye" rash often develops around the bite (Section 26.20). Severe headaches, backaches, chills, and fatigue follow. Without prompt treatment, the symptoms worsen. Another tick-traveling pathogen, *Rickettsia rickettsii*, causes *Rocky Mountain spotted fever*. After the bacterium enters a host through a tick bite, it penetrates the cytoplasm and the nucleus of host cells. Three to twelve days after this, a high fever and severe headache develop. Three to five days after that, a rash develops on the hands and feet (Figure 22.14). Diarrhea and gastrointestinal cramps are common. If untreated, the symptoms can persist for more than two weeks.

Regarding the "Simple" Bacteria

Bacteria are small. Their insides are not elaborate. *But bacteria are not simple.* A brief look at their behavior will reinforce this point. Bacteria move toward nutrient-rich regions. Aerobes move toward oxygen; anaerobes avoid it. Photosynthetic types move into light and away from light that is too intense. Many species can tumble away from toxins. Such behaviors often start with membrane receptors that are stimulated by changes in chemical conditions or in the intensity of light coming in from a particular direction. Such a change triggers a shift in metabolic activities inside the cell, which might then lead to an adjustment in the direction of movement.

Magnetotactic bacteria contain a chain of magnetite particles that serves as a tiny compass (Figure 22.15*a*). The compass helps them sense which way is north and also down. These bacteria swim toward the bottom of a body of water, where oxygen concentrations are lower and therefore more suitable for their growth.

Some species even show *collective* behavior, as when millions of *Myxococcus xanthus* cells form a "predatory" colony. These cells secrete enzymes that digest "prey," such as cyanobacteria, that become stuck to the colony. Then the cells absorb the breakdown products. What's more, the cells migrate, change direction, and move as a single unit toward what may be food!

Many myxobacteria colonies form **fruiting bodies** (spore-bearing structures). Under suitable conditions, some cells in the colony differentiate and form a slime stalk, others form branchings, and others form clusters of spores (Figure 22.15*b*). The spores disperse when a cluster bursts open; each may give rise to a new colony. As you will see in the next chapter, certain eukaryotic species also form spore-bearing structures.

Eubacteria are the most common and diverse prokaryotic cells. They are adapted to nearly all environments.

22.7 THE VIRUSES

Defining Characteristics

In ancient Rome, *virus* meant "poison" or "venomous secretion." In the late 1800s, this rather nasty word was bestowed on newly discovered pathogens, smaller than the bacteria being studied by Louis Pasteur and others.

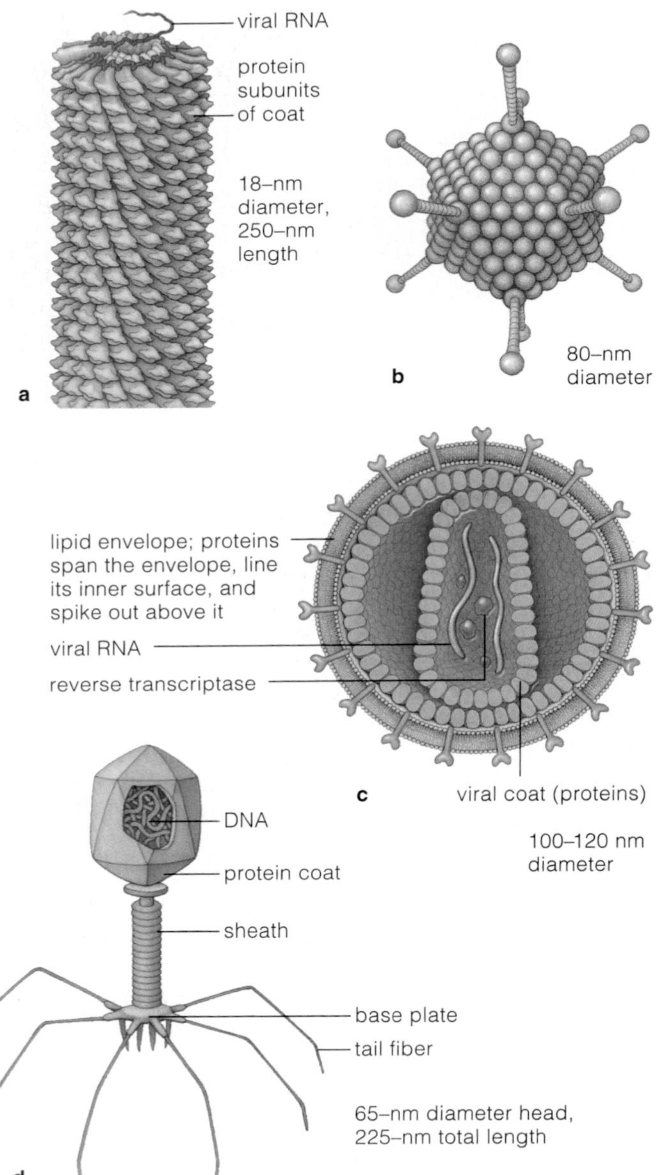

viral RNA
protein subunits of coat
18–nm diameter, 250–nm length
a
b
80–nm diameter
lipid envelope; proteins span the envelope, line its inner surface, and spike out above it
viral RNA
reverse transcriptase
c
viral coat (proteins)
100–120 nm diameter
DNA
protein coat
sheath
base plate
tail fiber
65–nm diameter head, 225–nm total length
d

Figure 22.16 Body plans of viruses. (**a**) *Helical* viruses have a rod-shaped coat of protein subunits, coiled helically around their nucleic acid. The upper subunits of the coat have been removed from this portion of a tobacco mosaic virus to reveal the RNA. (**b**) *Polyhedral* viruses, including this adenovirus, have a many-sided coat. (**c**) *Enveloped* viruses, including HIV, have an envelope around a helical or polyhedral coat. (**d**) *Complex* viruses, such as T-even bacteriophages, have additional structures attached to the coat.

Many viruses deserve the name. They attack humans, cats, cattle, birds, insects, plants, fungi, protistans, and bacteria. You name it, there are viruses that can infect it.

Today we define a **virus** as a noncellular infectious agent that has two characteristics. First, a viral particle consists of a protein coat wrapped around a nucleic acid core—that is, around its genetic material. Second, the virus cannot reproduce itself. It can be reproduced only after its genetic material and a few enzymes enter a host cell and subvert the cell's biosynthetic machinery.

The genetic material of a virus is DNA *or* RNA. The coat consists of one or more types of protein subunits organized into a rodlike or polyhedral (many-sided) shape, as in Figure 22.16. The coat protects the genetic material during the journey to a new host cell. It also contains proteins that can bind with specific receptors on host cells. An envelope, made mostly of membrane remnants from a previously infected cell, encloses the coat of some viruses. The envelope bristles with spikes of glycoproteins. Coats of complex viruses have sheaths, tail fibers, and other accessory structures attached.

The vertebrate immune system detects certain viral proteins. The problem is, genes for many viral proteins mutate at high frequencies, so a virus often can elude the immune fighters. For example, people susceptible to lung infections get new "flu shots" each year because envelope spikes on influenza viruses keep changing.

Examples of Viruses

Each kind of virus can multiply only in certain hosts. It cannot be studied easily unless the investigator cultures living host cells. This is why much of our knowledge of viruses comes from **bacteriophages**, a group of viruses that infect bacterial cells. Unlike cells of humans and other complex, multicelled species, bacterial cells can be cultured easily and rapidly. This is also why bacteria and bacteriophages were used in early experiments to determine the function of DNA (Section 13.1). They are still used as research tools in genetic engineering.

Table 22.3 lists some major groups of animal viruses. These viruses contain double- or single-stranded DNA or RNA, which is replicated in various ways. Animal viruses range in size from parvoviruses (18 nanometers) to brick-shaped poxviruses (350 nanometers). Many kinds cause diseases, such as the common cold, certain cancers, warts, herpes, and influenza (Figure 22.17). One, HIV, is the trigger for *AIDS*. By attacking certain white blood cells, HIV weakens the immune system's ability to fight infections that might not otherwise be life threatening. Researchers attempting to develop drugs against HIV and other diseases use HeLa cells and other immortal cell lineages for early experiments (Section 9.6). Later they must use laboratory animals,

Further reading: Student Guide to InfoTrac on web site —

Table 22.3 Classification of Some Major Animal Viruses

DNA VIRUSES	Some Diseases/Consequences
Parvoviruses	Gastroenteritis; roseola (fever, rash) in small children; aggravation of symptoms of sickle-cell anemia
Adenoviruses	Respiratory infections (fever, cough, sore throat, rash), diarrhea in infants, conjunctivitis (inflamed, pebbly eye membranes); some cause tumors
Papovaviruses	Benign and malignant warts
Orthopoxviruses	Smallpox, cowpox, monkeypox
Herpesviruses:	
H. simplex type I	Oral herpes, cold sores
H. simplex type II	Genital herpes (Section 45.14)
Varicella–zoster	Chicken pox, shingles
Epstein–Barr	Infectious mononucleosis; cancers of skin, liver, cervix, pharynx; Burkitt's lymphoma (malignant tumor of jaw, face)
Cytomegalovirus	Hearing loss, mental retardation
Hepadnavirus	Hepatitis B (severe liver infection)

RNA VIRUSES	Some Diseases/Consequences
Picornaviruses:	
Enteroviruses	Polio, hemorrhagic eye disease, hepatitis A (infectious hepatitis)
Rhinoviruses	Common cold
Hepatitis A virus	Inflammation of liver, kidneys, spleen
Togaviruses	Forms of encephalitis (inflammation of the brain), rubella
Flaviviruses	Yellow fever (fever, chills, jaundice), dengue (fever, severe muscle pain) St. Louis encephalitis
Coronaviruses	Upper respiratory infections, colds
Rhabdoviruses	Rabies, other animal diseases
Filoviruses	Hemorrhagic fevers, as by *Ebola* virus (Section 22.10)
Paramyxoviruses	Measles, mumps, respiratory ailments
Orthomyxoviruses	Influenza
Bunyaviruses	
Bunyamwera virus	California encephalitis
Phlebovirus	Hemorrhagic fever, encephalitis
Hantavirus	Hemorrhagic fever, kidney failure
Arenaviruses	Hemorrhagic fevers
Retroviruses:	
HTLV-I, HTLV-II*	Adult T-cell leukemia
HIV	AIDS
Reoviruses	Respiratory and intestinal infections

* Human T-cell leukemia virus.

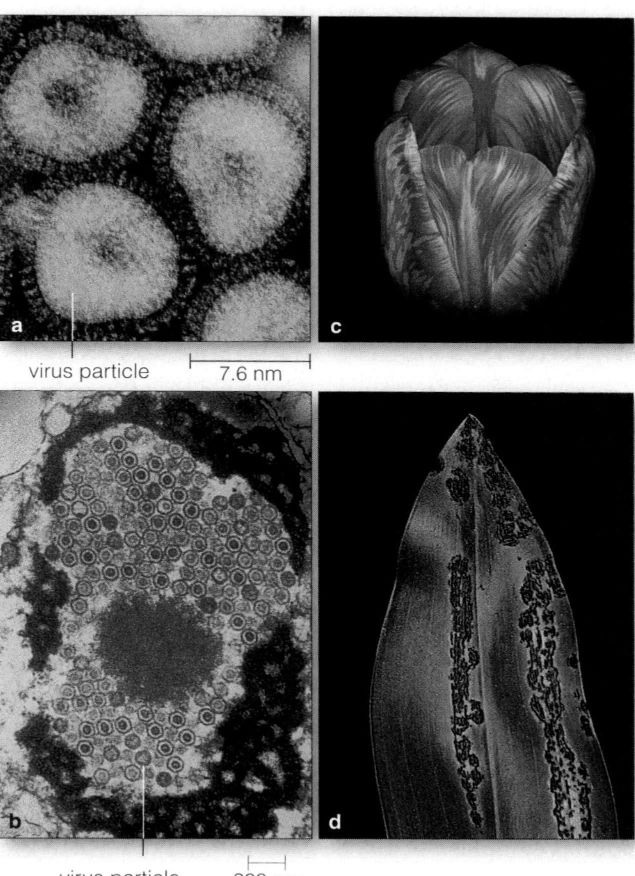

a
c
virus particle
7.6 nm
b
d
virus particle
300 nm

Figure 22.17 Some viruses and their effects. (**a**) Particles of a DNA virus that causes a herpes infection in humans. (**b**) Particles of an enveloped RNA virus that causes influenza in humans. Spikes project from the lipid envelope. (**c**) Streaking of a tulip blossom. A harmless virus infected pigment-forming cells in the colorless parts. (**d**) An orchid leaf infected by a rhabdovirus.

then human volunteers, to test a new drug for toxicity and effectiveness. Why? It takes a functioning immune system to test responses.

Plant viruses must breach plant cell walls to cause diseases. They typically hitch rides on the piercing or sucking devices of insects that feed on plant juices. Some RNA viruses infect tobacco plants (the tobacco mosaic virus), barley, potatoes, and other major crop plants. Certain DNA viruses infect such valuable crops as cauliflower and corn. Figure 22.17*c,d* shows visible effects of two viral infections.

A virus is a nonliving infectious particle that consists of nucleic acid enclosed in a protein coat and sometimes an outer envelope. It cannot multiply without pirating the metabolic machinery of a specific type of host cell.

Diverse viruses infect organisms in all kingdoms.

22.8 VIRAL MULTIPLICATION CYCLES

Viruses multiply in a variety of ways. Even so, nearly all of their multiplication cycles proceed through five basic steps, as outlined here:

1. *Attachment.* A virus attaches to a host cell. Any cell is a suitable host if molecular groups on a virus particle are able to chemically recognize and lock on to specific molecular groups at the cell surface.

2. *Penetration.* Either the whole virus or its genetic material alone penetrates the cell's cytoplasm.

3. *Replication and synthesis.* In an act of molecular piracy, the viral DNA or RNA directs the host cell into producing many copies of viral nucleic acids and proteins, including enzymes.

4. *Assembly.* The viral nucleic acids and viral proteins are put together to form new infectious particles.

5. *Release.* New virus particles are released from the cell.

Consider this list with respect to some bacteriophages. Lytic and lysogenic pathways are common among their replication cycles. In a **lytic pathway**, steps 1 through 4 proceed rapidly, and new particles are released when a host cell undergoes lysis (Figure 22.18). **Lysis** means the plasma membrane, cell wall, or both are damaged, so that cytoplasm leaks out and the cell dies. Late into most lytic pathways, a viral enzyme that causes swift destruction of the bacterial cell wall is synthesized.

In a **lysogenic pathway**, a latent period extends the cycle. The virus does not kill its host outright. Instead, a viral enzyme cuts a host chromosome, then integrates viral genes into it. Thus, when an infected cell prepares to divide, it replicates the recombinant molecule. As a result of that single instance of genetic recombination, miniature time bombs are passed on to all of that cell's descendants. Later on, a molecular signal or some other stimulus may reactivate the cycle.

Latency occurs in the multiplication cycles of many viruses, not just among bacteriophages. Type I *Herpes simplex*, which causes *cold sores*, is an example. Nearly everybody harbors this virus. It remains latent inside a ganglion (an aggregation of neuron cell bodies) in facial tissue. Stress factors such as sunburn can reactivate the virus. Then, virus particles move down the neurons to their tips near the skin. There they infect epithelial cells and cause painful skin eruptions.

Like other enveloped viruses, the herpesviruses can enter a host cell by their own version of endocytosis, then leave it by budding from the plasma membrane. Figure 22.19 shows how they accomplish this.

The multiplication cycle of the RNA viruses has an interesting twist to it. In the host cell's cytoplasm, their

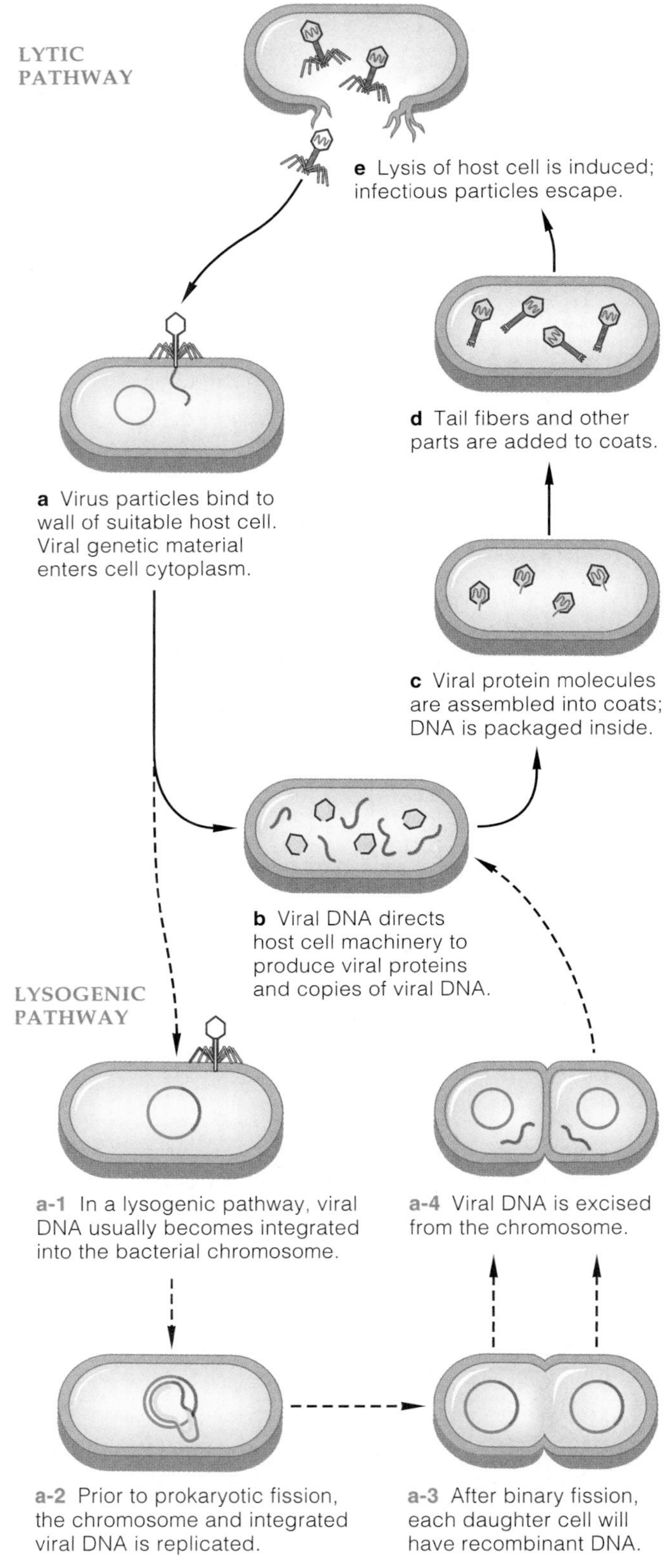

LYTIC PATHWAY
e Lysis of host cell is induced; infectious particles escape.
d Tail fibers and other parts are added to coats.
a Virus particles bind to wall of suitable host cell. Viral genetic material enters cell cytoplasm.
c Viral protein molecules are assembled into coats; DNA is packaged inside.
b Viral DNA directs host cell machinery to produce viral proteins and copies of viral DNA.
LYSOGENIC PATHWAY
a-1 In a lysogenic pathway, viral DNA usually becomes integrated into the bacterial chromosome.
a-4 Viral DNA is excised from the chromosome.
a-2 Prior to prokaryotic fission, the chromosome and integrated viral DNA is replicated.
a-3 After binary fission, each daughter cell will have recombinant DNA.

Figure 22.18 Generalized multiplication cycle for some bacteriophages. New viral particles may be produced and released by a lytic pathway. For certain viruses, the lytic pathway may expand to include a lysogenic pathway.

Further reading: Student Guide to InfoTrac on web site —

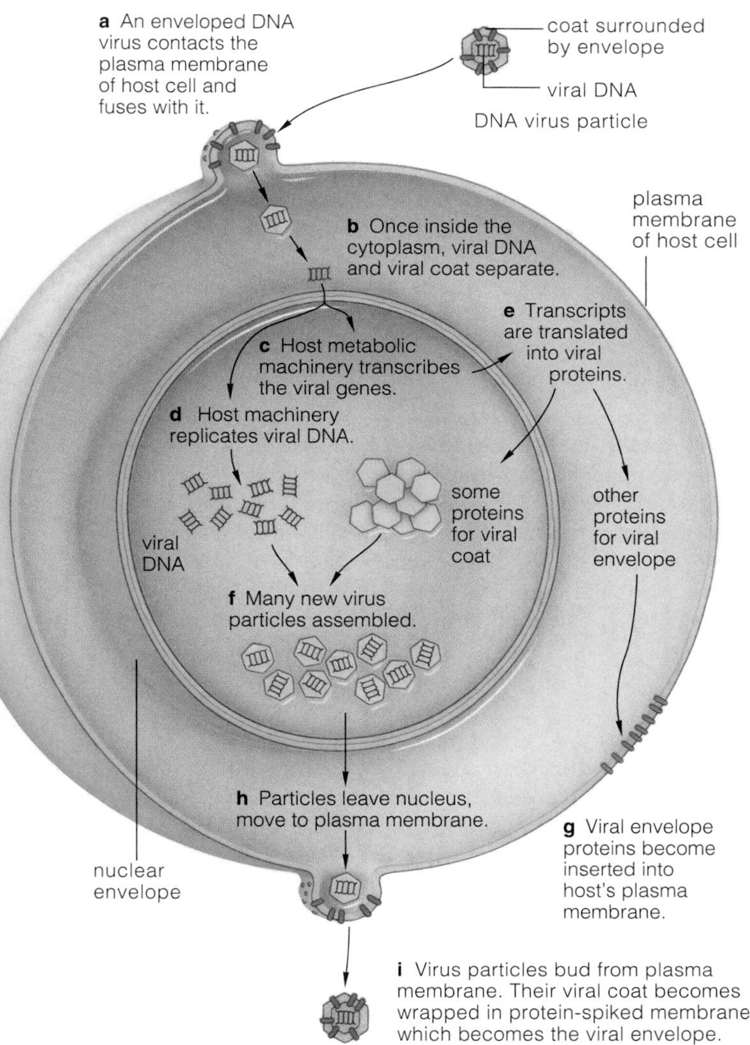

a An enveloped DNA virus contacts the plasma membrane of host cell and fuses with it.
coat surrounded by envelope
viral DNA
DNA virus particle
plasma membrane of host cell
b Once inside the cytoplasm, viral DNA and viral coat separate.
c Host metabolic machinery transcribes the viral genes.
e Transcripts are translated into viral proteins.
d Host machinery replicates viral DNA.
viral DNA
some proteins for viral coat
other proteins for viral envelope
f Many new virus particles assembled.
h Particles leave nucleus, move to plasma membrane.
g Viral envelope proteins become inserted into host's plasma membrane.
nuclear envelope
i Virus particles bud from plasma membrane. Their viral coat becomes wrapped in protein-spiked membrane, which becomes the viral envelope.
j The finished particle is equipped to infect a new potential host cell.

Figure 22.19 Multiplication cycle of one type of enveloped DNA virus infecting an animal cell.

RNA serves as a template for synthesizing either DNA or mRNA. For example, HIV, a retrovirus, carts its own enzymes into cells. These assemble DNA on viral RNA by reverse transcription (Sections 16.1 and 40.11).

The multiplication cycles of nearly all viruses include five basic steps: attachment to a suitable host cell, penetration into the cell, viral DNA or RNA replication and synthesis of viral proteins, assembly of new viral particles, and then release from the infected cell.

Bacteriophage replication cycles commonly follow a rapid, lytic pathway and an extended, lysogenic pathway.

Replication cycles of RNA viruses involve the use of viral RNA as a template for synthesizing DNA or mRNA.

Focus on Health

22.9 INFECTIOUS PARTICLES TINIER THAN VIRUSES

VIROIDS You may have trouble visualizing this, but some infectious agents are more stripped down than viruses. **Viroids** are tightly folded strands or circles of RNA, smaller than anything in viruses. They contain no protein-coding genes whatsoever. In this they resemble the noncoding portions of eukaryotic DNA called introns (Section 14.2). They might be self-splicing introns that escaped from DNA; they might even be remnants left over from an RNA world (Section 21.2). Viroids have no protein coat, but their tight folding might help protect them from a host's enzymes. These bits of "naked" RNA are known to cause a number of diseases in plants. Each year, they destroy many millions of dollars' worth of potatoes, citrus, and other important crop plants.

PRIONS Eight rare, fatal degenerative diseases of the nervous system are linked to small proteins known as **prions**. The proteins are altered products of a gene that is present in unaffected as well as infected individuals. Unaltered and altered forms of the protein molecule are found at the surface of neurons, the communication cells of the nervous system. You may have heard of *kuru* and *Creutzfeldt–Jakob* diseases (CJD). They slowly destroy muscle coordination and brain functions in humans. *Scrapie*, a disease of sheep, is so named because infected animals rub against trees or posts until they scrape off most of their wool.

In 1996, more than 150,000 cattle in Great Britain staggered about, drooled, and showed other symptoms of BSE (bovine spongiform encephalopathy), or *mad cow disease*. Prions had caused abnormal changes in proteins of neurons, and spongy holes and amyloid deposits had formed in the cattle brains. BSE is fatal.

How did it happen? Ground-up tissues of sheep that had died of scrapie had been used as a supplement in cattle feed. The practice was banned in 1988, but prions were already in the food chain.

Medical researchers already knew that 1 in 1 million people around the world develops Creutzfeldt–Jakob disease each year. Prions also are linked to this fatal condition. Symptoms, which include loss of vision and speech, rapid mental deterioration, and spastic paralysis, may take decades to develop. At the same time as the major outbreak of BSE in Great Britain, ten people were diagnosed with a variant of CJD; four had already died, probably as a result of eating meat of infected animals. Genetic analysis and interviews linked all the cases to exposure to BSE.

Ten other countries also had reported cases of BSE, possibly as a result of importing cattle feed from Great Britain. The incidence of BSE in Great Britain is declining since strict precautionary measures have been instituted. In the United States, importation of live cattle or meat products from BSE-affected countries has been banned since 1989.

22.10 THE NATURE OF INFECTIOUS DISEASES

CATEGORIES OF DISEASES Just by being human, you are a potential host for a great many pathogenic bacteria, viruses, fungi, protozoans, and parasitic worms. When a pathogen has invaded your body and is multiplying in host cells and tissues, this is an **infection**. Its outcome, **disease**, results if the body's defenses cannot be mobilized fast enough to prevent the pathogen's activities from interfering with normal body functions. You may have heard of *contagious* diseases. This simply means that the pathogenic agents can be transmitted by direct contact with body fluids secreted or otherwise released from infected individuals, as by explosive wet sneezes.

During an **epidemic**, a disease spreads rapidly through part of a population for a limited time, then the outbreak subsides. What happens when an epidemic breaks out in several countries around the world at the same time? We call this a **pandemic**. AIDS, an incurable disease, is an example. Millions are infected by the causative agent, HIV (short for the *H*uman *I*mmunodeficiency *V*irus).

Sporadic diseases such as whooping cough break out irregularly and affect just a few people. *Endemic* diseases pop up more or less continuously, but they don't spread far in large populations. Tuberculosis is like this. So is impetigo, a highly contagious bacterial infection that often spreads no further than, say, a single day-care center.

AN EVOLUTIONARY VIEW Now look at disease in terms of a pathogen's prospects for survival. Like you, *a pathogen lives only for as long as it has access to outside sources of energy and raw materials*. To a microscopic organism or viral particle, a human host is a veritable treasurehouse of both. With such a bountiful supply of resources, the pathogen can multiply or replicate itself to amazing population sizes. And, evolutionarily speaking, the ones that leave the most descendants win.

There are two significant barriers to complete world dominance by pathogens. First, any species with a history of being attacked by a particular pathogen has coevolved with it and has built-in defenses against it. The premier example is the vertebrate immune system, as described in Chapter 40. Second, if a pathogen kills its host too quickly, it may vanish before having time to infect a new one. This is one reason why most pathogens have less-than-fatal effects on a host. After all, an infected individual that lives longer spreads more of the next generation of a pathogen and contributes to its reproductive or replicative success. Usually, a host dies only if a pathogen enters its body in overwhelming numbers, if it is a novel host (one with no coevolved defenses against that pathogen), or if a mutant strain of the pathogen emerges and can surmount the host's current array of defenses.

Being equipped with the evolutionary perspective, you probably can work out the numbers on your own. To wit: *The greater the population density of host individuals, the greater will be the kinds and frequencies of infectious diseases transmitted among them.* This brings us to the bad news.

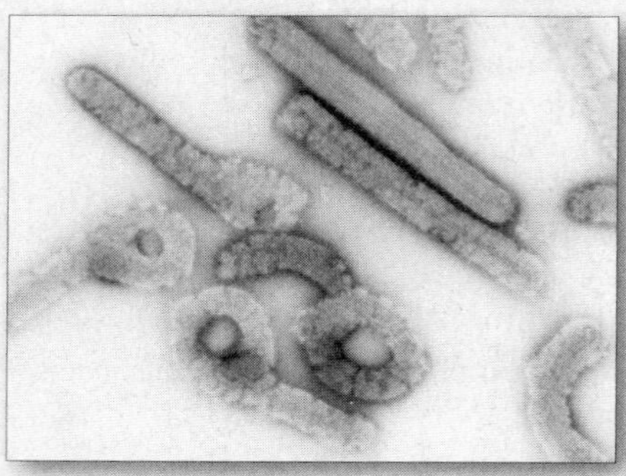

Figure 22.20 Transmission electron micrograph of *Ebola* virus particles.

***EBOLA* AND OTHER EMERGING PATHOGENS** Thanks to planes, trains, and automobiles, people travel often and in droves all around the world. Among their more exotic destinations are virgin tropical forests and other remote environments where the human body was an infrequent (or nonexistent) opportunity for pathogens. At present, strange and often dangerous pathogens are opportunistic about the novel, two-legged treasurehouses of metabolic machinery and nutrients entering their habitats. Human travelers can become infected within hours and taxi such pathogens far away, and eventually back home.

We know little about the deadly, **emerging pathogens**. Some have been around for a long time and only now are taking great advantage of the increased presence of the novel human hosts. Others are newly mutated strains of existing species.

Consider *Ebola*, one of the viruses that cause a deadly, hemorrhagic fever. This RNA virus appears filamentous or circular in electron micrographs (Figure 22.20). It may have coevolved with monkeys in Africa's tropical forests. By 1976, it was infecting humans. It kills between 70 and 90 percent of its victims. There is no vaccine or treatment for the disease, which starts with high fever and flu-like aches. Within a few days, nausea, vomiting, and diarrhea begin. Blood vessels are destroyed. Blood seeps from the circulatory system, into the surrounding tissues and out through all the body's orifices. The liver and kidneys may rapidly turn to mush. Patients often become deranged and soon die of circulatory shock. There have been four *Ebola* epidemics since 1976. Each time, governments mobilized agencies around the world; quarantine procedures were implemented to limit the spread of the disease.

Or consider the *monkeypox* virus, a relative of the smallpox virus. After infecting a host, this DNA virus replicates in the lymph nodes, spleen, and bone marrow. The bloodstream delivers new viral particles to the skin, where they produce large, hard, painful, pus-filled sores. When the virus overwhelms the immune system, death follows from secondary infections, especially pneumonia.

Like smallpox, one of the most deadly human diseases, monkeypox kills as many as one in ten. Unlike smallpox, which is highly contagious, monkeypox was not much of a threat at first. Its few victims were mainly children who

Further reading: Student Guide to InfoTrac on web site —

Table 22.4 The World's Eight Most Deadly Infectious Diseases

Disease	Cause	Estimated New Cases per Year	Estimated Deaths per Year
Acute respiratory infections*	Bacteria, viruses	1 billion	4.7 million
Diarrheas**	Bacteria, viruses, parasites	1.8 billion	3.1 million
Tuberculosis	Bacteria	9 million	3.1 million
Malaria	Sporozoans	110 million	2.5–2.7 million
AIDS	HIV	5.6 million	2.6 million
Measles	Viruses	200 million	1 million
Hepatitis B	Virus	200 million	1 million
Tetanus	Bacteria	1 million	500,000

* Includes pneumonia, influenza, and whooping cough.

** Includes amoebic dysentery, cryptosporidiosis, and gastroenteritis.

had trapped, skinned, and eaten infected monkeys and other small animals in the African forests. Virus particles slipped into the body through small cracks in the hands.

A mutant monkeypox virus appears to have entered villages. More than *Ebola* and other attention-grabbing pathogens, it has the potential to break out in pandemics. The virus easily infects and replicates itself in human hosts, and it resists breakdown in the environment. Paul Ewald, an evolutionary biologist, says it has the right starting material for making a very nasty pathogen.

The smallpox vaccine works against monkeypox virus. But the World Health Organization essentially eradicated smallpox decades ago. There are few stores of the vaccine, and there are no plans to manufacture more.

ABOUT THE DRUG-RESISTANT STRAINS There is an old saying that, when you attack nature, it comes back at you with a pitchfork. We have already considered antibiotic resistance in several contexts, in Sections 1.4, 1.6, and 18.4. Here we reinforce the consequences in terms of the upsurge in infectious diseases.

One of the resistant pathogens is having a field day close to home. Because of economic pressures over the past few decades, the number of working mothers has skyrocketed in the United States. So has the number of preschoolers enrolled in day-care centers. In effect, each center is a population of hosts whose immune systems are still developing and vulnerable to contagious diseases. Now think of *Streptococcus pneumoniae*. This bacterium causes pneumonia, meningitis, and middle-ear infections in people of all ages; about 40,000 to 50,000 cases end in death every year. The risk of infection is 36 times greater in large day-care centers than it is for children cared for at home. As recently as 1988, drug-resistant strains of *S. pneumoniae* were practically unheard of in the United States. They are becoming the rule, not the exception.

A mere twenty-five years ago, antibiotics and vaccines were considered invincible, and the Surgeon General of the United States announced we could "close the book on infectious diseases." There are more than 6 billion people.

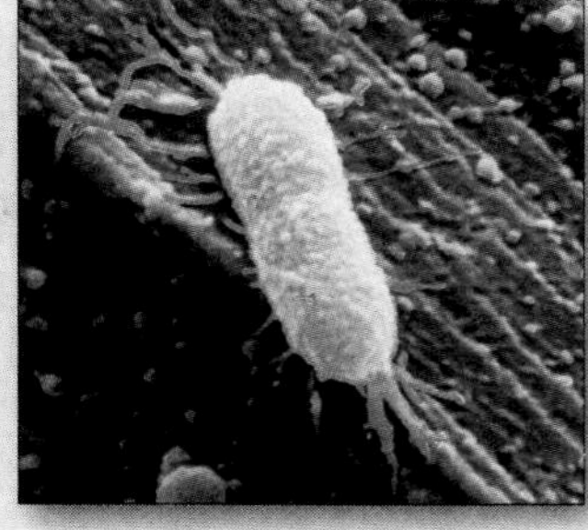

Figure 22.21 Bits of food and *Pseudomonas* on a kitchen knife blade. The cell attached itself to the blade by pili at its surface. Think about *that* the next time you use an unwashed kitchen knife previously used on raw poultry, beef, or seafood.

Most live in crowded cities. During any specified interval, as many as 50 million people are on the move within and between countries in search of a better life (Section 46.7). Even putting aside emerging pathogens, is it any wonder that the pathogens responsible for cholera, tuberculosis, and other familiar diseases are striking with a vengeance? Table 22.4 lists some sobering numbers.

PATHOGENS LURKING IN THE KITCHEN Each year, food poisoning hits more than 80 million people in the United States alone. It kills about 9,000 of them, mainly the very young, the very old, or those with compromised immune systems. Yet the cases reported to health agencies number only in the thousands; people often dismiss their misery as "the 24-hour flu." *Salmonella enteriditis* as well as certain strains of *E. coli* are prominent among the culprits. They typically are ingested with undercooked beef or poultry, contaminated water, unpasteurized milk or cider, and vegetables grown in fields fertilized with manure.

A few examples: In 1966, *S. enteriditis* slipped into ingredients used to make a popular ice cream. In a single outbreak, possibly as many as 224,000 people suffered from stomach cramps, diarrhea, and other symptoms of the disease *salmonella*. In 1993 in Washington State, more than 500 people became sick and three children died after eating restaurant hamburgers tainted with a pathogenic strain of *E. coli*. A recent outbreak of food poisoning was traced to unpasteurized apple juice.

Outbreaks such as these certainly grab our attention, yet pathogens lurking in the kitchen probably account for at least half of all cases of food poisoning. Examine wood or plastic cutting boards or sleek, stainless steel knives with electron microscopes and you see nooks and crannies where microbes can lurk (Figure 22.21). Carlos Enriquez and his colleagues at the University of Arizona sampled 75 dishrags and 325 sponges in several homes. Most harbored colonies of *Salmonella* and *E. coli*, as well as *Pseudomonas* and *Staphylococcus*. Bacteria can live two weeks in a wet sponge. Use the sponge to wipe down the kitchen, and you spread them about. Antibacterial soaps, detergents, and weak solutions of household bleach get rid of them. You can sanitize sponges simply by putting them through a dishwasher cycle.

Do you think food poisoning is of small concern? The annual cost of treating the infections known to be caused by food-borne pathogens ranges between 5 billion and 22 billion dollars.

22.11 SUMMARY

1. A major divergence occurred soon after the origin of life. One lineage gave rise to eubacteria, the other to common ancestors of archaebacteria and eukaryotes. Archaebacteria (methanogens, extreme halophiles, and extreme thermophiles) live in extreme environments, like those in which life probably originated. They are unique in wall structure and other features, and they share some features with eukaryotic cells. By far, the most common existing prokaryotic cells are eubacteria.

2. All bacteria are prokaryotic cells; they do not have a nucleus. Some species have membrane infoldings and other structures in the cytoplasm. None has a profusion of organelles, which is typical of eukaryotic cells.

a. Three basic bacterial shapes are common: cocci (spheres), bacilli (rods), and spirilla (spirals).

b. Nearly all eubacteria have a cell wall consisting of peptidoglycan. It protects the plasma membrane and helps it resist rupturing. The composition and structure of the cell wall help identify particular bacterial species.

c. A sticky mesh of polysaccharides (a glycocalyx) may surround the wall as a capsule or slime layer. It helps the bacterium attach to substrates and sometimes resist a host organism's infection-fighting mechanisms.

d. Some species have one or more bacterial flagella, structures that rotate a bit like a propeller and serve in motility. Many have pili, filamentous proteins that help cells adhere to a surface or facilitate conjugation.

3. Prokaryotic fission is a cell division mechanism used only by bacteria. The mechanism involves replication of the bacterial chromosome and division of a parent cell into two genetically equivalent daughter cells.

4. Many species have plasmids, small circles of DNA that are replicated independently of the single, circular bacterial chromosome. Plasmids may be transmitted to daughter cells and may be transferred to cells of the same species or a different species by bacterial conjugation.

5. Bacteria as a group show great metabolic diversity, as in their modes of acquiring energy and carbon.

a. *Photoautotrophs* use sunlight and carbon dioxide during photosynthesis. They include cyanobacteria and other oxygen-releasing species of eubacteria. They also include green nonsulfur and purple nonsulfur bacteria that do not produce oxygen; these get electrons from sulfur and other inorganic compounds, not from water.

b. *Photoheterotrophs* use sunlight energy and organic compounds as sources of carbon. They include certain archaebacteria and eubacteria.

c. *Chemoautotrophs*, including the nitrifying bacteria, use carbon dioxide but not sunlight. Different kinds get energy by stripping electrons from a variety of organic or inorganic substances.

d. *Chemoheterotrophs* are parasites (which get carbon and energy from living hosts) or saprobes (which feed on the organic products, wastes, or remains of other organisms). Most bacterial species are in this category. They include major decomposers and pathogens.

6. Bacteria make behavioral responses to stimuli, as when they move toward regions with more nutrients.

7. Viruses are nonliving, noncellular agents that infect particular species of nearly all organisms.

a. Each virus particle consists of a core of DNA or RNA and a protein coat that sometimes is enclosed in a lipid envelope. Many glycoproteins project, spikelike, from the envelopes. The coats of complex viruses have sheaths, tail fibers, and other accessory structures.

b. A virus particle cannot reproduce on its own. Its genetic material must enter a host cell and direct the cellular machinery to synthesize materials necessary to produce new virus particles.

8. Nearly all viral multiplication cycles include five steps: attachment to a suitable host cell, penetration of it, viral DNA or RNA replication and protein synthesis, assembly of new viral particles, and release.

9. Two pathways are common in the multiplication cycle of bacteriophages (bacteria-infecting viruses). In a lytic pathway, multiplication is rapid and new viral particles are released by lysis. In a lysogenic pathway, the infection enters a latent period. The host cell is not killed outright, and the viral nucleic acid may undergo genetic recombination with a host cell chromosome.

10. Multiplication cycles of viruses are diverse. They are rapid or they enter a latent phase. Penetration and release of most enveloped types occur by endocytosis and budding. DNA viruses spend part of the cycle in the nucleus of a host cell. RNA viruses complete the cycle in the cytoplasm. The viral RNA is the template for mRNA synthesis and for protein synthesis.

Review Questions

1. Describe the key metabolic and structural features of bacteria. Make sketches of the three basic shapes of bacterial cells. *22.1*

2. Compared to your own bodily growth, how is bacterial growth measured? *22.2*

3. With respect to bacterial classification, what are some of the pitfalls of numerical taxonomy? How are comparisons of rRNA sequences assisting classification efforts? *22.3*

4. Name a few of the photoautotrophic, chemoautotrophic, and chemoheterotrophic eubacteria. Describe some that are likely to give humans the most trouble, medically speaking. *22.4, 22.6*

5. What is a virus? Why is a virus considered to be no more alive than a chromosome? *22.7*

6. Distinguish between:
   a. microorganism and pathogen *CI*
   b. infection and disease *22.10*
   c. epidemic and pandemic *22.10*

7. Label the components of the following viruses: *22.7*

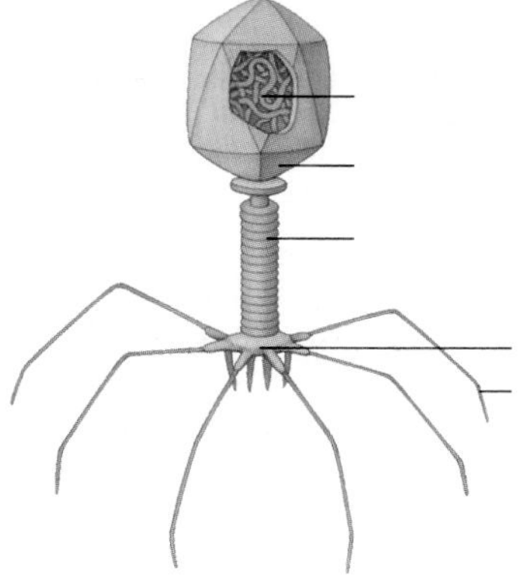

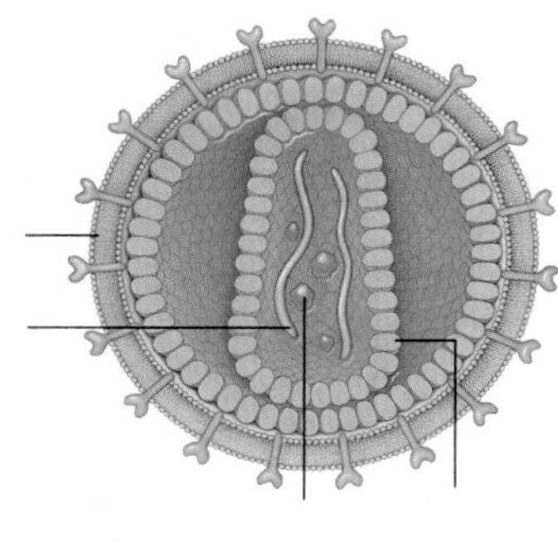

Self-Quiz (*Answers in Appendix III*)

1. Nondividing bacteria have ______ chromosome(s) and may have extra circles of ______ called plasmids.
   a. one; RNA
   b. two; RNA
   c. one; DNA
   d. two; DNA

2. ______ live in habitats much like those of the early Earth.
   a. Cyanobacteria
   b. Eubacteria
   c. Archaebacteria
   d. Protozoans

3. Which of the following is not grouped with archaebacteria?
   a. halophiles
   b. cyanobacteria
   c. thermophiles
   d. methanogens

4. Bacteria reproduce by ______ .
   a. mitosis
   b. meiosis
   c. prokaryotic fission
   d. longitudinal fission

5. Eubacterial cell walls are composed of ______ ; and a sticky mesh of polysaccharides, a ______ , often surrounds the wall.
   a. peptidoglycan; plasma membrane
   b. cellulose; glycocalyx
   c. cellulose; plasma membrane
   d. peptidoglycan; glycocalyx

6. Most bacteria are ______ and include major decomposers and pathogens.
   a. photoautotrophs
   b. photoheterotrophs
   c. chemoautotrophs
   d. chemoheterotrophs

7. Viruses are ______ .
   a. the simplest living organisms
   b. infectious particles
   c. nonliving
   d. both a and b
   e. both b and c

8. Viruses have a ______ and a ______ .
   a. DNA core; carbohydrate coat
   b. DNA or RNA core; plasma membrane
   c. DNA-containing nucleus; lipid envelope
   d. DNA or RNA core; protein coat

9. Match the terms with their most suitable description.

____ archaebacteria	a. infectious small protein
____ eubacteria	b. nonliving infectious particle with nucleic acid core and protein coat
____ virus	c. at home in stockyards
____ plasmid	d. methanogens, extreme halophiles, extreme thermophiles
____ prokaryotic fission	e. most common prokaryotic cells
____ methanogen	f. small circle of bacterial DNA
____ prion	g. bacterial cell division mechanism

Critical Thinking

1. *Salmonella* bacteria, which cause a form of food poisoning, often live in poultry and eggs, but not in newly hatched chicks until chicks eat bacteria-laden feces of healthy adult chickens. Harmless bacteria ingested this way colonize the surface of intestinal cells, leaving no place for *Salmonella* to take hold. Some farmers raise thousands of chicks in confined quarters with no adult chickens. Should they consider feeding the chicks a known mixture of bacteria from a lab or a mixture of unknown bacteria from healthy adult chickens? Devise an experiment to test which approach may be more effective.

2. Reflect on the description of mad cow disease in Section 22.9. The FDA is moving to prohibit all protein supplements for sheep, cattle, and other ruminants. Investigate the extent of this practice in the United States and how it has been monitored.

3. Water seeping from the earth at boiling springs is above 100°C; at hot springs (Figure 22.11*d*) it approaches the boiling point. The water gradually cools as it spills out through shallow channels. Using the following list as a guide, predict which organisms you might find (a) in boiling spring water, (b) near the edges of a hot spring, and (c) along the edges of an outflow channel where the upper temperature is 72°C. (Not all species of the groups listed grow at these temperatures.)

Fishes 38°C	Fungi 60°–62°C
Insects 45°–50°C	Cyanobacteria 70°–74°C
Vascular plants 45°C	Methanogens 110°C
Algae 55–60°C	Extreme thermophiles 113°C

Selected Key Terms

archaebacterium *22.3*
bacillus (bacilli) *22.1*
bacterial chromosome *22.2*
bacterial conjugation *22.2*
bacterial flagellum *22.1*
bacteriophage *22.7*
cell wall *22.1*
coccus (cocci) *22.1*
disease *22.10*
emerging pathogen *22.10*
endospore *22.6*
epidemic *22.10*
eubacterium *22.3*
extreme thermophile *22.5*
fruiting body *22.6*
glycocalyx *22.1*
Gram stain *22.1*
halophile *22.5*
heterocyst *22.6*
infection *22.10*
lysis *22.8*
lysogenic pathway *22.8*
lytic pathway *22.8*
methanogen *22.5*
microorganism *CI*
numerical taxonomy *22.3*
pandemic *22.10*
pathogen *CI*
pilus (pili) *22.1*
plasmid *22.2*
prion *22.9*
prokaryotic cell *22.1*
prokaryotic fission *22.2*
spirillum (spirilla) *22.1*
strain (bacterial) *22.4*
viroid *22.9*
virus *22.7*

Readings *See also www.infotrac-college.com*

Brock, T., M. Madigan, J. Martinko, and J. Parker. 1997. *Biology of Microorganisms*. Eighth edition. Englewood Cliffs, New Jersey: Prentice-Hall. Exceptionally lucid, well-illustrated survey.

Daszak, P., A. Cunningham, and A. Hyatt. 21 January 2000. "Emerging Infectious Diseases of Wildlife—Threats to Biodiversity and Human Health." *Science*, 287:443–449

Hively, W. May 1997. "Looking for Life in All the Wrong Places." *Discover*, 76–85. Account of microbes that live in the most extreme environments on and in the earth.

Oernt, W. October 1999. "Killer Pox in the Congo." *Discover*, 74–79.

23

PROTISTANS

Kingdom at the Crossroads

More than 2 billion years ago, in tidal flats and soils, in estuaries, lagoons, lakes, and streams, prokaryotic cells were inconspicuously changing the world. Ever since the time of life's origin, Earth's atmosphere had been free of oxygen, and anaerobic bacteria had reigned supreme. Their realm was about to shrink, drastically.

An oxygen-releasing pathway of photosynthesis was operating in vast populations of bacterial cells. At first the free oxygen combined with iron in rocks. When iron-rich deposits were rusted out—oxidized—free oxygen started to accumulate. With nothing much to combine with, it slowly became more concentrated in water, then in air.

The oxygen-enriched atmosphere was a selection pressure of global dimensions. Thereafter, anaerobic species that could not neutralize the highly reactive, potentially lethal gas were restricted to black muds, stagnant waters, and other anaerobic habitats. Others developed oxygen tolerance.

It must have been a short evolutionary step from having a capacity to neutralize oxygen to using it in metabolism. Why? Aerobic species soon emerged in nearly all bacterial groups.

The metabolic innovation was the start of rampant competition for resources. In the presence of so much oxygen, energy-rich organic compounds no longer could accumulate by geochemical processes. Organic compounds formed by living cells became the premier source of carbon and energy. Novel ways of acquiring and using those organic compounds developed. Many diverse bacteria engaged in new kinds of partnerships, predations, and parasitic interactions. As outlined in Section 21.4, some of those evolutionary experiments gave rise to eukaryotic cells.

Of all existing organisms, **protistans** are the ones most like those earliest, structurally simple eukaryotes. They differ from their bacterial ancestors in several respects. At the least, protistan cells have a nucleus, large ribosomes, mitochondria, endoplasmic reticulum, and Golgi bodies. Their chromosomes consist of DNA molecules closely associated with many histones and other proteins. They assemble microtubules for use in a cytoskeleton, in spindles that move chromosomes, and in a 9 + 2 core of flagella or cilia. Many species contain chloroplasts. And, depending on the species, protistans divide by way of mitosis, meiosis, or both.

Photosynthetic types range from microscopic single cells to giant seaweeds (Figure 23.1). Saprobes resemble certain bacteria and fungi. Some of the predators and

Figure 23.1 A tiny sampling of protistan diversity. (**a**) Along a rocky shoreline, a stand of *Postelsia palmaeformis*, one of the multicelled brown algae. (**b**) *Micrasterias*, a single-celled freshwater green alga, here dividing in two. (**c**) *Physarum*, a plasmodial slime mold. This mass of yellow-gold cells is migrating on a rotting log. (**d**) Mealtime for *Didinium*, a ciliated protozoan with a big mouth. *Paramecium*, a different ciliated protozoan, is poised at the mouth (*left*) and swallowed (*right*).

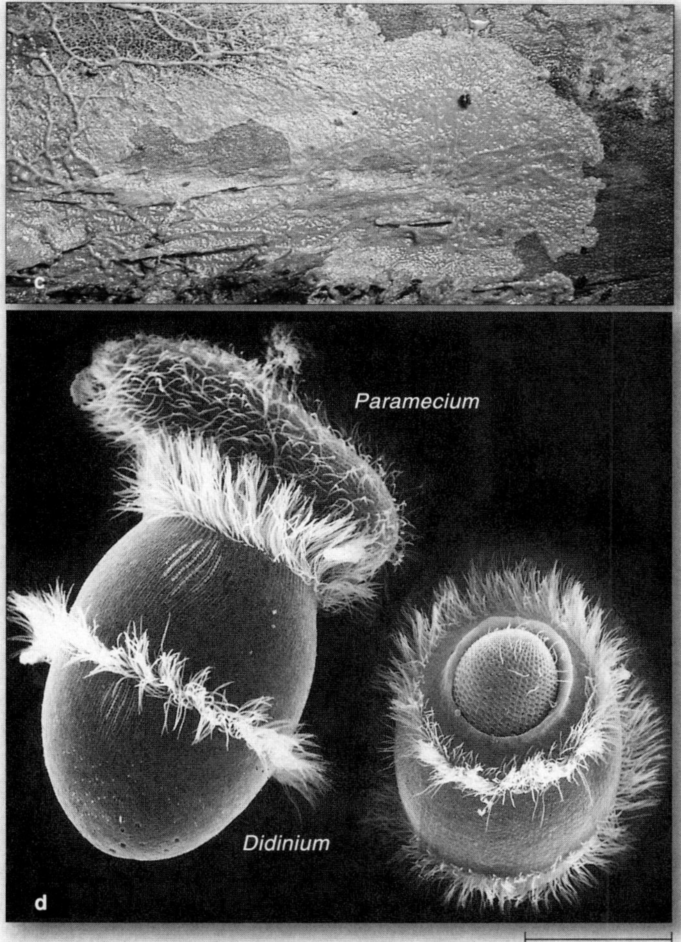
c
Paramecium
Didinium
d
50 µm

parasites even resemble animals. The vast majority of protistans are single-celled, but nearly every lineage also includes multicelled forms. Some groups have notably close evolutionary ties to other kingdoms.

We turn now to the major lineages. Chytrids, water molds, slime molds, protozoans, and sporozoans are heterotrophs. Among the euglenoids, chrysophytes, and dinoflagellates are species that are photosynthetic, heterotrophic, or both. Most red, brown, and green algae are evolutionarily committed to photosynthesis.

Opinions differ on how to classify many of these confoundingly diverse organisms. For instance, many biologists view the multicelled algae as protistans, and about as many view them as plants. Don't worry about memorizing the classification schemes. Simply become familiar with the various groups. Ongoing studies are bringing the evolutionary picture into sharper focus.

KEY CONCEPTS

1. Protistans are easily distinguishable from prokaryotes (the bacteria) but are difficult to classify with respect to other eukaryotes. They also differ enormously from one another in their morphology and life-styles.

2. Among the fungus-like protistans are parasitic and predatory molds that produce spores. The majority of the chytrids and water molds are single-celled decomposers in aquatic habitats. The phagocytic slime molds live as single amoeboid cells and as aggregations of cells that migrate together and form spore-producing structures.

3. The animal-like protistans include nonphotosynthetic flagellated protozoans, sporozoans, and ciliates. Among their ranks are single-celled predators, grazers, and parasites. In any given year, hundreds of millions of people are affected by infections caused by a few dozen species of flagellated protozoans and sporozoans.

4. Among the plant-like protistans are single-celled, photosynthetic flagellates. Astounding numbers of them live freely or as colonies in the open ocean and other bodies of water. These protistans are important members of phytoplankton, the food producers that are the start of nearly all food webs of aquatic habitats.

5. The protistans most like plants are the red, brown, and green algae; indeed, many botanists prefer placing them in the plant kingdom. Most species are photosynthetic. They range from single-celled to multicelled forms that show great diversity in size, morphology, life-styles, reproductive modes, and habitats.

23.1 PARASITIC AND PREDATORY MOLDS

Three groups of protistans, the **chytrids**, **water molds**, and **slime molds**, have members that resemble fungi in certain respects. Like fungi, all produce spore-bearing structures and are heterotrophs. Like fungi, many are saprobic decomposers or parasites. A **saprobe** secretes digestive enzymes that break down organic compounds made by other organisms, then absorbs the breakdown products. Unlike fungi, the slime molds are phagocytic predators. Also, members of all three groups differ from fungi in producing motile cells during the life cycle.

Chytrids

The chytrids (Chytridiomycota) are common in marine and freshwater habitats, where they absorb nutrients from plant debris or from necrotic plant tissues. Most of the 575 species are saprobic decomposers, and some are parasites.

In damp environments, chytrid population growth results in fuzzy, pale masses that resemble fungal molds. In some biochemical respects, also, chytrids are similar to fungi. Chitin reinforces the cell wall of some species, as it does in fungi.

The single-celled species produce flagellated asexual spores. After spores are released, they may settle onto a host cell, then germinate and develop into globular cells (Figure 23.2*a*). These cells have rootlike absorptive structures of a type called rhizoids. When the cells are mature, they become spore-producing structures.

Like many fungi, complex multicelled species have a **mycelium** (plural, mycelia), a mesh of fine absorptive filaments called **hyphae** (singular, hypha). Individual cell walls may or may not cut across hyphae. Either way, cytoplasmic streaming through a mycelium distributes enzymes and nutrients from digested food (Section 4.9).

Water Molds

Water molds (Oomycota) might be distantly related to yellow-green and brown algae. Most of the 580 known types are key saprobic decomposers in aquatic habitats. Of these, many are free-living species that get nutrients from plant debris in ponds, lakes, and streams. Others live inside necrotic tissues of living plants. Some water molds parasitize aquatic organisms. The pale, cottony growths you may have seen on some aquarium fish are mycelia of a parasitic type, *Saprolegnia* (Figure 23.2*b*).

Most water molds produce an extensive mycelium. Some of the hyphae differentiate into gamete-producing structures. (An *antheridium* produces male gametes; an *oogonium* produces female gametes, as in Figure 23.2*b*.) At fertilization, a male and female gamete fuse to form a diploid zygote, which develops into a thick-walled resting spore. A new mycelium develops from the spore

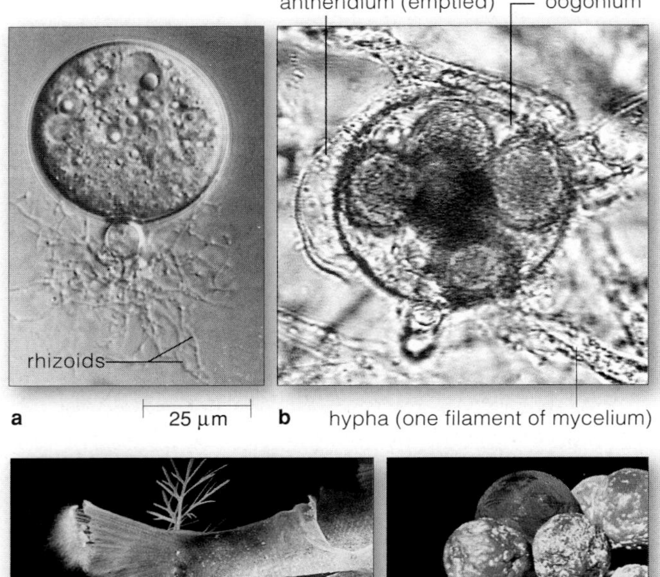

antheridium (emptied)
oogonium
rhizoids
a
25 µm
b
hypha (one filament of mycelium)
c
d

Figure 23.2 (**a**) *Chytridium confervale*, a common chytrid. This globe-shaped cell will develop into a spore-producing structure. (**b**) *Saprolegnia*, one of the parasitic water molds. (**c**) Observable outcome of a *Saprolegnia* attack on the tissues of an aquarium fish. (**d**) Grapes with downy mildew, an outcome of an attack by *Plasmopara viticola*.

after it germinates. Water molds reproduce asexually also, by way of flagellated, asexual spores.

Certain water molds are major plant pathogens. For example, grapes with *downy mildew* have been attacked by *Plasmopara viticola* (Figure 23.2*d*). Another pathogen seriously influenced human affairs. Over a century ago, Irish peasants grew potatoes as their main food crop. Between 1845 and 1860, the growing seasons were cool and damp, year after year, and conditions encouraged the rapid spread of *Phytophthora infestans*. This water mold causes *late blight*, a rotting of potato (and tomato) plants. Its abundant spores were dispersed unimpeded through the watery film on the plants. Destruction was rampant. During a fifteen-year period, one-third of the population in Ireland starved to death, died during the outbreak of typhoid fever that followed as a secondary effect, or fled to the United States and other countries.

Today, potato late blight is the single most costly biological constraint on global food production. Efforts to control it are also the cause of one of the largest uses of pesticides. Researchers at the University of Maryland and elsewhere are utilizing expressed sequence tagging and other methods of recombinant DNA technology to develop genetic databases for *P. infestans*. Their goal is

Further reading: Student Guide to InfoTrac on web site

Figure 23.3 A cellular slime mold, *Dictyostelium discoideum*. (**a**) Its life cycle includes a spore-producing stage. Spores give rise to free-living amoebas, which grow and divide until food (soil bacteria) dwindles. In response to cyclic AMP, a chemical signal that they secrete, amoebas stream toward one another. As the amoebas aggregate, their plasma membranes become sticky and they adhere to one another. A cellulose sheath forms around the aggregation, which starts crawling like a slug (**b–d**). Some slugs may incorporate 100,000 amoebas.

As a slug migrates, amoebas develop and differentiate into prestalk (*red*), prespore (*white*), and anteriorlike cells (*brown dots*). Prestalk cells secrete ammonia in amounts that vary with temperature and light intensity. The slug moves most rapidly in response to intermediate levels of ammonia (not too little, not too much), which correspond to warm, moist conditions (not too cold or hot, not soppy or dry). Where such conditions occur at the surface of a substrate, prestalk cells and prespore cells differentiate and form a stalked, spore-bearing structure (**e–g**). The anteriorlike cells, which sort into two groups, may function in elevating the nonmotile spores for dispersal from the top of the spore-bearing structure.

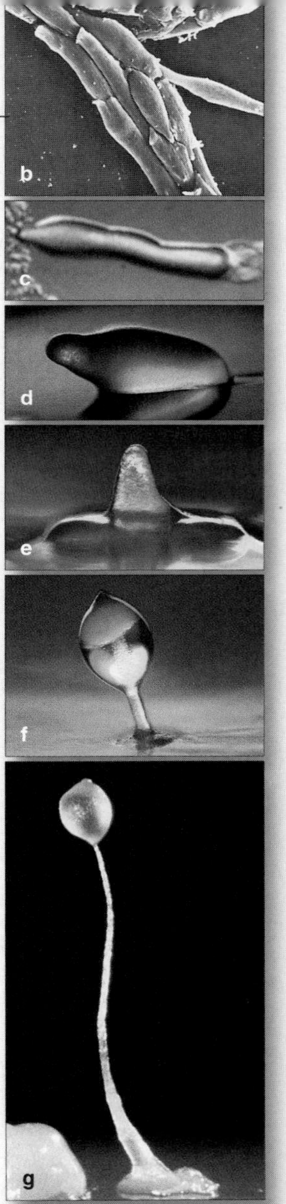

1 Stalked, spore-producing structure releases spores.
MITOTIC CELL DIVISION
2 Spores give rise to free-living amoebas that feed, grow, and reproduce by mitotic cell division.
AGGREGATION
3 When food gets scarce, the amoebas stream together to form an aggregate that crawls like a slug.
either
or
4 The slug may start developing at once into a spore-bearing structure, or it may migrate elsewhere first.
MIGRATING SLUG STAGE
CULMINATION
MATURE FRUITING BODY
a
b
c
d
e
f
g

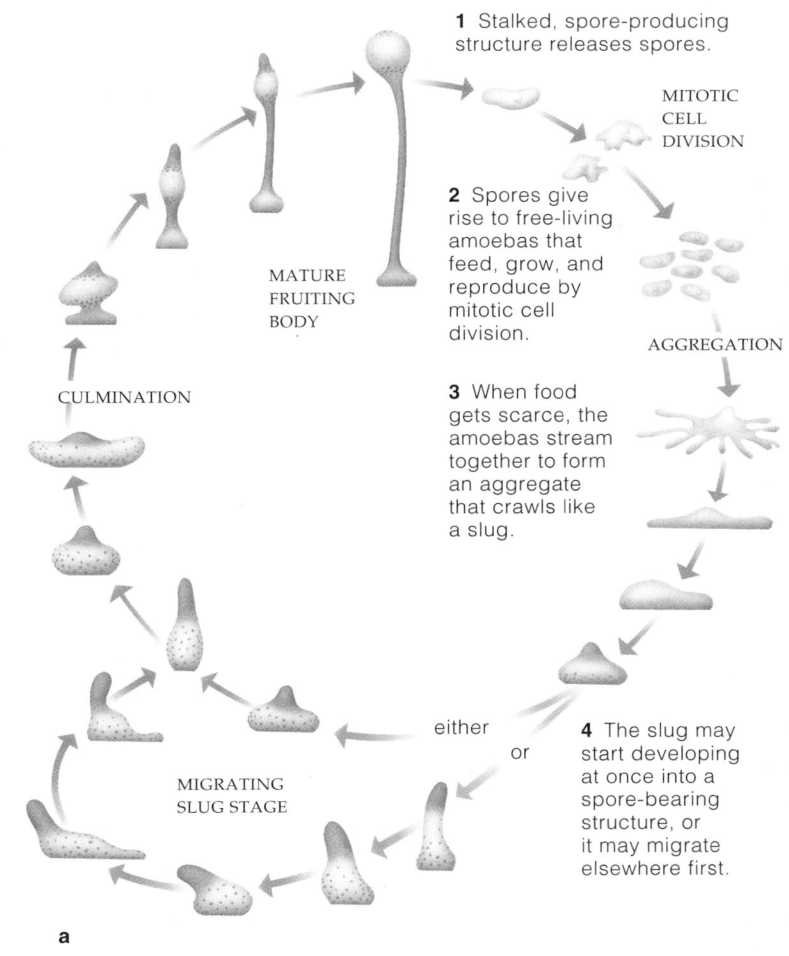

to help ensure an adequate global food supply as world population increases and resources dwindle.

Slime Molds

All slime molds produce free-living, amoebalike cells during part of the life cycle. The group includes *cellular* slime molds (Acrasiomycota) and *plasmodial* slime molds (Myxomycota). Their cells crawl on rotting plant parts, such as decaying leaves and bark. Like true amoebas, they are phagocytic predators. They engulf bacteria and yeasts, spores, and various organic compounds. When nutrients dwindle, many of the starving cells aggregate and form a slimy mass. The mass may migrate to a new location where conditions favor growth. Collectively, the contractions of single cells move the mass. Later in the life cycle, the amoebalike cells develop into a few cell types that differentiate and form a spore-bearing structure. After the spores are released, they germinate on warm, damp surfaces. Each spore gives rise to one amoebalike cell. Sexual reproduction (by gametes) also is common among slime molds.

Dictyostelium discoideum, one of 70 species of cellular slime molds, is often used for laboratory studies of development. Figure 23.3 shows its life cycle. Figure 23.1*c* shows one of the 500 species of plasmodial slime molds. When the amoebalike cells of members of this phylum aggregate, each cell's plasma membrane breaks down. Cytoplasm flows unimpeded and so distributes nutrients and oxygen through the mass.

Commonly, a streaming mass will occupy several square meters and migrate if food runs out. The mass is the plasmodium. You may have observed one crossing a lawn or a road, or even climbing a tree.

Most chytrids and water molds are saprobic decomposers of aquatic habitats. Some are single cells. Like slime molds, many are free-living predators some of the time and simple experiments in multicellularity at other times.

During a slime mold life cycle, amoeboid cells aggregate to form a migrating mass. Cells in the mass differentiate, then they form reproductive structures and spores or gametes.

23.2 THE ANIMAL-LIKE PROTISTANS

About 65,000 named species of protistans are known informally as **protozoans** ("first animals"), because they may resemble single-celled, heterotrophic protistans that gave rise to animals. They include amoeboid protozoans, ciliated protozoans, and animal-like flagellates. Table 23.1, which lists all of the groups, provides a preview of their key features.

Table 23.1 Major Groups of Animal-Like Protistans

SARCODINA
Amoeboid protozoans. Soft or shelled bodies; locomotion by pseudopods. Free-living or endosymbiotic heterotrophs.
RHIZOPODS. Naked amoebas, foraminiferans.
ACTINOPODS. Radiolarians, heliozoans
CILIOPHORA
Diverse free-living, sessile, and motile heterotrophs with numerous cilia. Predators or symbionts, some parasitic.
MASTIGOPHORA
Animal-like flagellates. Includes free-living and parasitic heterotrophs. Locomotion by one or more flagella.
APICOMPLEXA
Parasitic heterotrophs with a complex of structures, such as rings, tubules, and cones, at the head end. The most familiar members of a group commonly called sporozoans.

Different protozoans are predators, parasites, and grazers that actively move through their surroundings to secure a meal. All four groups include species that house endosymbiotic algae. Free-living species live in damp soil, in a variety of freshwater habitats such as ponds and lakes, and in marine habitats. Symbiotic and parasitic species live inside or on the moist tissues of other organisms. Some species are major pathogens.

Protozoans reproduce either asexually or sexually. Many species alternate between reproductive modes in response to environmental conditions. Most commonly, asexual reproduction is by **binary fission**, a process by which the body of the individual divides in two. The division plane is random for amoebas, longitudinal for flagellates, and transverse for ciliates. Budding from the parent organism also occurs. Some species undergo multiple fission. (More than two nuclei form, then each nucleus and a bit of cytoplasm separate as a daughter cell.) Many parasitic types go through an encysted stage. The **cyst**, a resistant body covering made of their own secretions, helps them wait out stressful conditions.

Like animals, protozoans include predatory, parasitic, and grazing species. They reproduce sexually or asexually, as by binary fission. Some species are major pathogens.

23.3 AMOEBOID PROTOZOANS

Let's first look at **amoeboid protozoans** (Sarcodina): the naked amoebas, foraminiferans, heliozoans, and radiolarians. None has permanent motile structures. Each moves by a combination of cytoplasmic streaming and the formation of **pseudopods** ("false feet"). As you read in Section 4.9, pseudopods are dynamic, reversible cytoplasmic extensions of a cell body. Figure 23.4 shows the rather thick pseudopods of a naked amoeba.

Rhizopods

Naked amoebas and foraminiferans are members of a group known as **rhizopods**. The constantly changing cytoskeleton lends structural support to the soft body of a naked amoeba, which never shows any symmetry. You find these protistans in damp soil, fresh water, and saltwater. Most species are free-living phagocytes that engulf other protozoans and bacteria. Some live in the gut of invertebrates or vertebrates and normally do no harm. A few are opportunistic parasites. So are certain free-living species that happen to enter the gut along with food or water. When conditions in the gut fan their population growth, they cause intestinal problems.

Entamoeba histolytica, a pathogenic type, completes its life cycle in human intestines. The infective stage becomes encysted and departs from the body in feces. The encysted stage is inactive in water or soil. But after it enters a new host, the trophozoite—the active, motile feeding stage—emerges from it. Trophozoites multiply rapidly by binary fission and release lysozymes, which destroy the cells that make up the lining of the host's large intestine (colon) and rectum. Painful abdominal cramps and diarrhea are symptoms of this infectious disease, which is called *amoebic dysentery*.

Amoebic dysentery is most prevalent in subtropical and tropical regions with contaminated water supplies

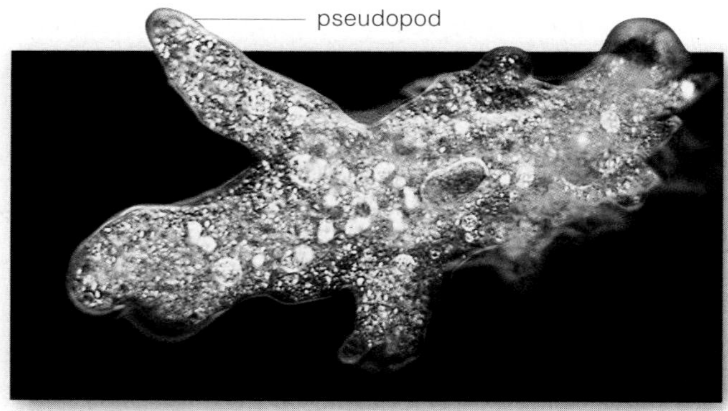
pseudopod

Figure 23.4 *Amoeba proteus*, one of the naked amoebas. Compared with most of the other amoeboid protozoans, it has stubbier pseudopods. This species is a favorite for laboratory experiments in biology classes.

Further reading: Student Guide to InfoTrac on web site —

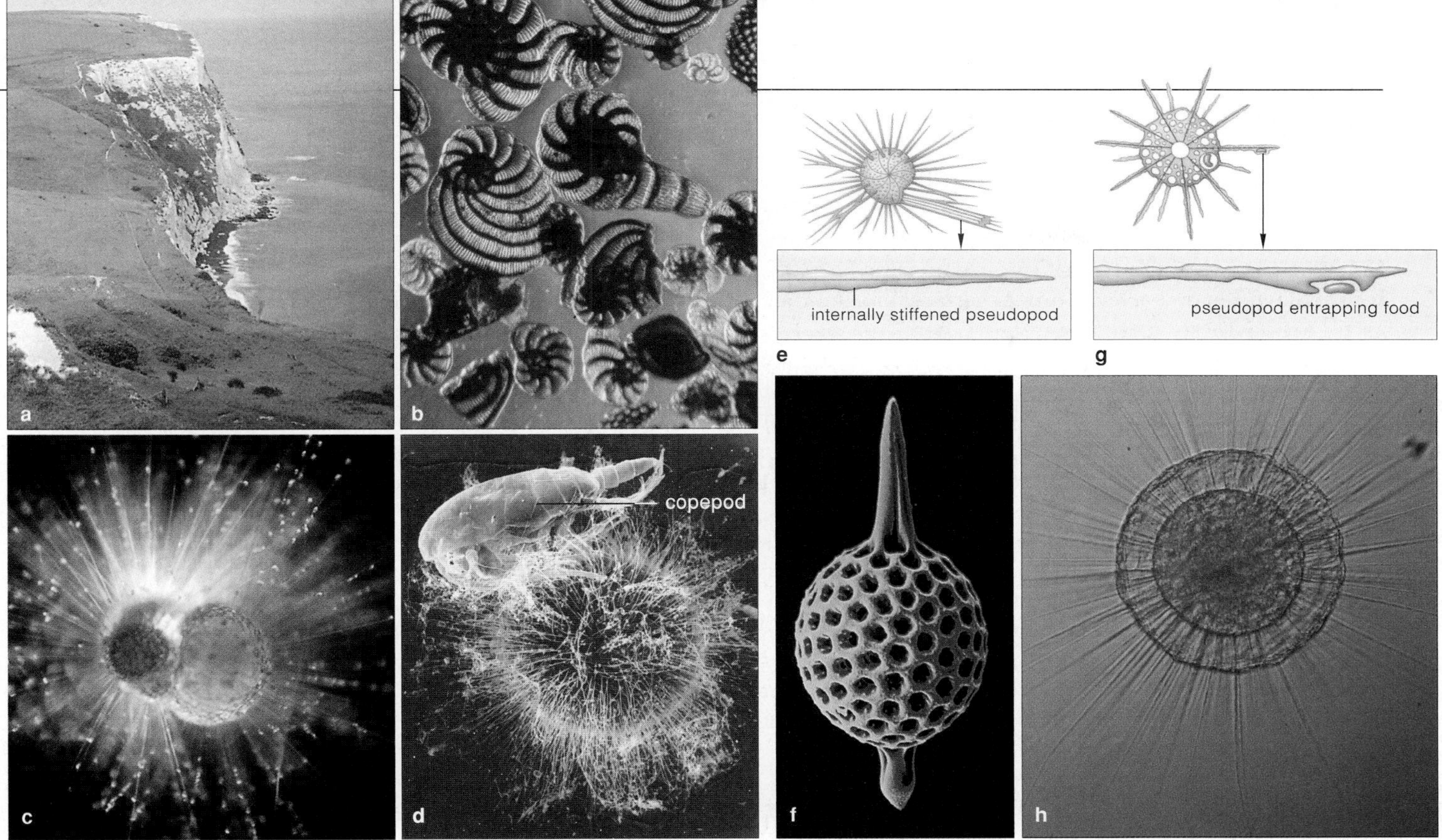
internally stiffened pseudopod
pseudopod entrapping food
copepod

and poor sanitation. Where cyst-harboring water, fruits, or vegetables abound, between 10 and 50 percent of the human population may be infected. The severity of an infection depends on the virulence of the *E. histolytica* strain and the state of an individual's immune system. People who recover show full or partial resistance to subsequent infection. Worldwide, amoebic dysentery is a leading cause of death among infants and toddlers, whose immune systems are not yet well developed.

Unlike the naked amoebas, foraminiferans have a highly perforated external "shell" of secreted organic substances, which are hardened with calcium carbonate (Figure 23.5*a–e*). Thin pseudopods extend through the perforations, and prey becomes trapped in mucus that covers them. Most foraminiferans live on the seafloor. All but 1 percent of the named species are extinct. The fossilized, compacted shells of foraminiferans make up great limestone beds, including the famed white cliffs of Dover, England (Figure 23.5*a*). We use their calcified remains in cement and blackboard chalk.

Actinopods

Actinopods ("ray feet") include the radiolarians and heliozoans. Their name refers to the numerous slender, reinforced pseudopods that radiate from the body. Like foraminiferans, radiolarians are well represented in the fossil record; their ornate, silica-hardened parts resist degradation (Figure 23.5*f*). Many skeletal parts are thin and project outward with stiffened pseudopods. Nearly all radiolarians drift with ocean currents as members of **plankton**. The word refers to aquatic communities of drifting or motile organisms that are mostly microscopic in size. A few species of radiolarians form colonies in which numerous cells are cemented together.

Figure 23.5 Foraminiferans: a look at how tiny cells often make grand structures. (**a**) The white cliffs of Dover, England, formed more than 200 million years ago by fossilization and compression of countless foraminiferan shells (**b**). Living species usually host symbionts, such as the golden algae visible in (**c**). Copepods and other prey become trapped in the sticky pseudopods (**d**). (**e**) Foraminiferan body plan. A core of microtubules reinforces the pseudopods. (**f**) Radiolarian skeleton. (**g**,**h**) Body plan and micrograph of a heliozoan, with its reinforced pseudopods.

Heliozoans ("sun animals") have many pseudopods radiating like the sun's rays (Figure 23.5*g*,*h*). Most are free-floating or bottom-dwelling species of freshwater habitats. As with radiolarians, a membrane or capsule divides the single-celled body into two biochemically distinct zones. The inner zone contains the nucleus. The outer zone contains so many digestive vacuoles that the cell appears almost frothy in micrographs. All those vacuoles impart buoyancy to the cell, which remains suspended in the water.

Amoeboid protozoans are soft-bodied single cells, some with hardened skeletal elements. All form pseudopods for use in motility and prey capture. Free-living and gut-dwelling species, some parasitic, belong to this group.

CILIATED PROTOZOANS

Ciliated protozoans (Ciliophora) typically have profuse arrays of cilia at their surface. You read about these fine motile structures in Section 4.9. Most ciliates use them for swimming through freshwater and marine habitats, where they prey on bacteria, tiny algae, and one another, as Figure 23.1*d* so aptly suggested. Their cilia beat in such a synchronized pattern over the body surface, they call to mind a soft wind through a field of tall grasses.

Paramecium is typical of the group (Figure 23.6). A fully grown cell is about 150 to 200 micrometers long. Outer membranes form a **pellicle**, a body covering that may be rigid or quite flexible, depending on how those membranes are organized. Inside the gullet, which starts at an oral depression at the body surface, some cilia sweep food-laden water into the cell body. There, the food becomes enclosed in enzyme-filled vesicles and is digested. As with amoeboid protozoans, *Paramecium*'s internal solute concentrations are greater than those of its surroundings. Therefore, it must continually counter

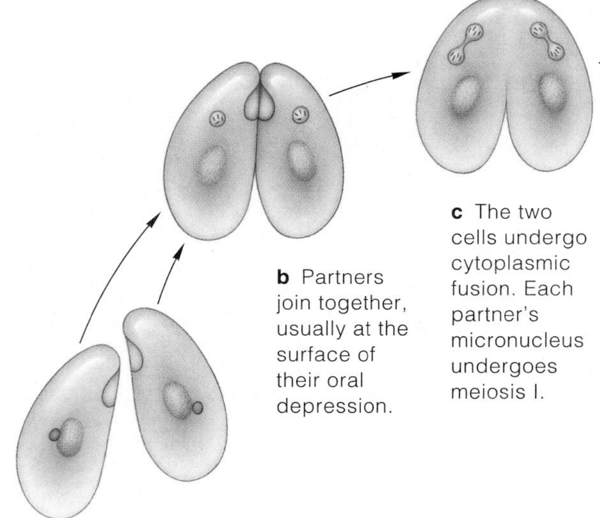

b Partners join together, usually at the surface of their oral depression.
c The two cells undergo cytoplasmic fusion. Each partner's micronucleus undergoes meiosis I.
a Prospective partners meet up.

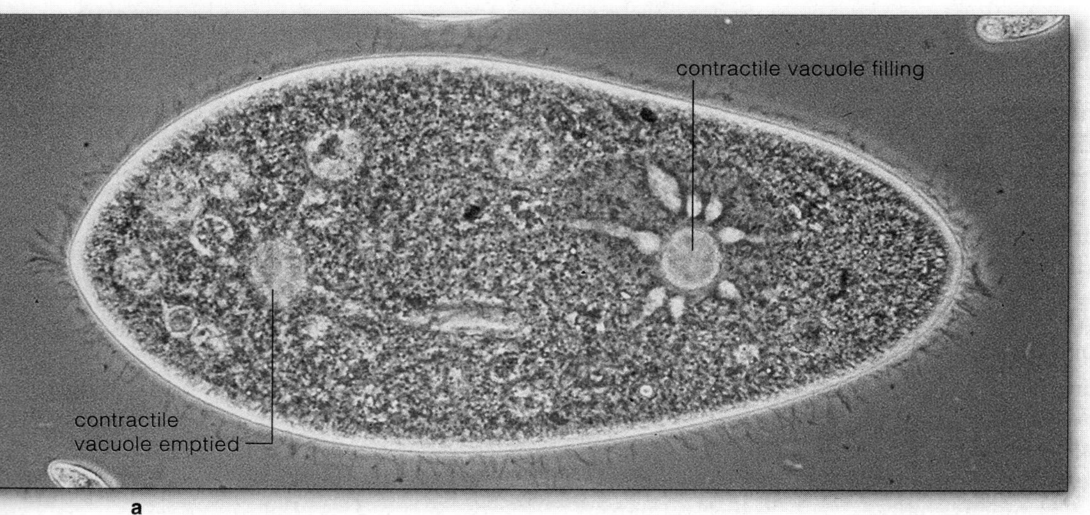

contractile vacuole filling
contractile vacuole emptied
a
20 µm

Figure 23.6 Representatives of two groups of ciliated protozoans.

Left and below: Paramecia. (**a**) Light micrograph of a living paramecium. (**b**) Generalized diagram of the body plan for the genus *Paramecium.* (**c**) Components of the pellicle. Trichocysts, which span the pellicle, are organelles that discharge threads when the cell is irritated. They might be a defense against predators.

Facing page: (**d**) Scanning electron micrograph of the arrays of cilia at the surface of these single-celled predators. (**e**) Hypotrichs. This free-living species lives in the Bahamas. The hypotrichs are the most animal-like of the ciliates. They run around on leglike tufts of cilia. Some have a "head" end with modified sensory cilia.

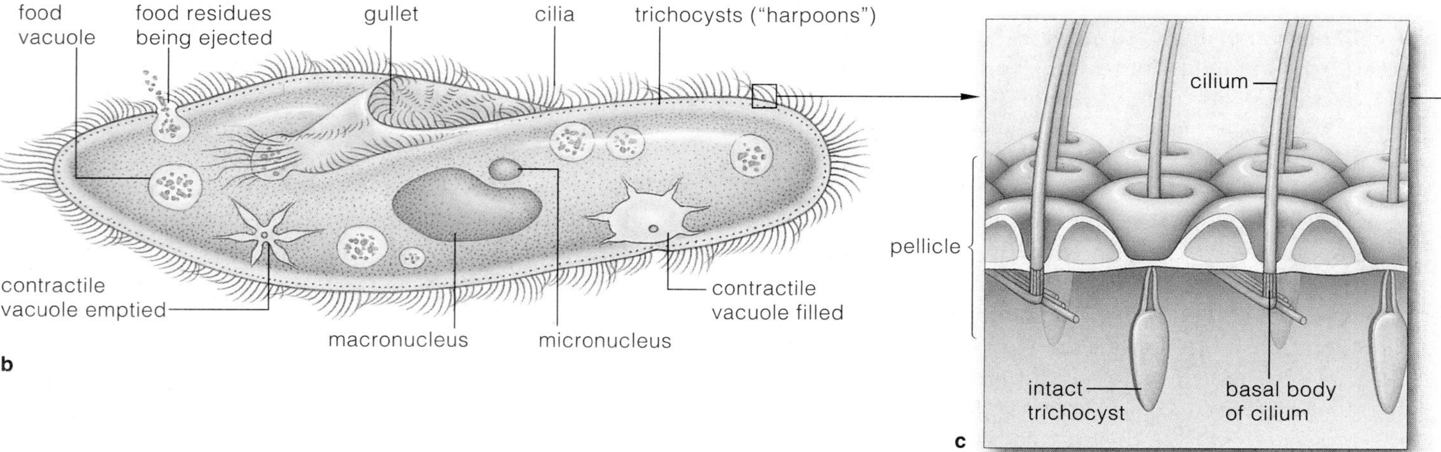

food vacuole
food residues being ejected
gullet
cilia
trichocysts ("harpoons")
contractile vacuole emptied
macronucleus
micronucleus
contractile vacuole filled
b
cilium
pellicle
intact trichocyst
basal body of cilium
c

Further reading: Student Guide to InfoTrac on web site

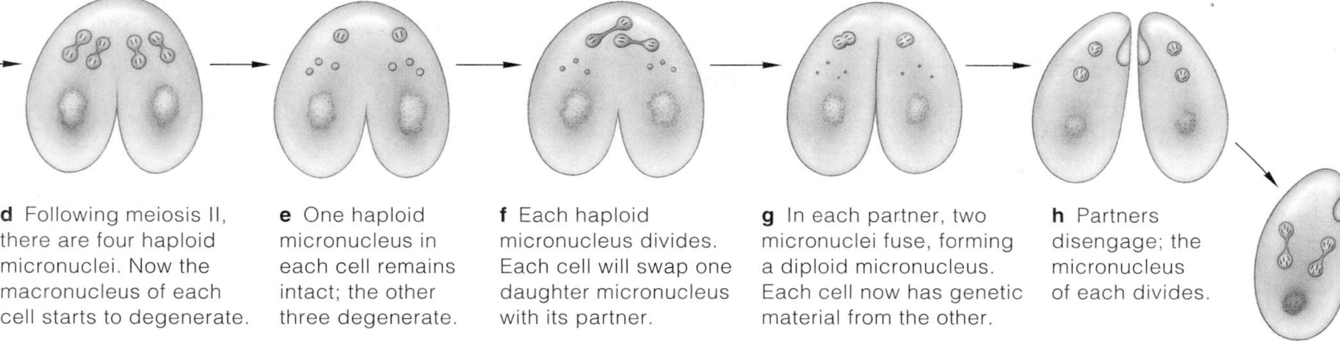
d Following meiosis II, there are four haploid micronuclei. Now the macronucleus of each cell starts to degenerate.
e One haploid micronucleus in each cell remains intact; the other three degenerate.
f Each haploid micronucleus divides. Each cell will swap one daughter micronucleus with its partner.
g In each partner, two micronuclei fuse, forming a diploid micronucleus. Each cell now has genetic material from the other.
h Partners disengage; the micronucleus of each divides.

Figure 23.7 Generalized diagram of protozoan conjugation, an unusual form of sexual reproduction, as demonstrated by the two ciliates that first encounter each other in (**a**).

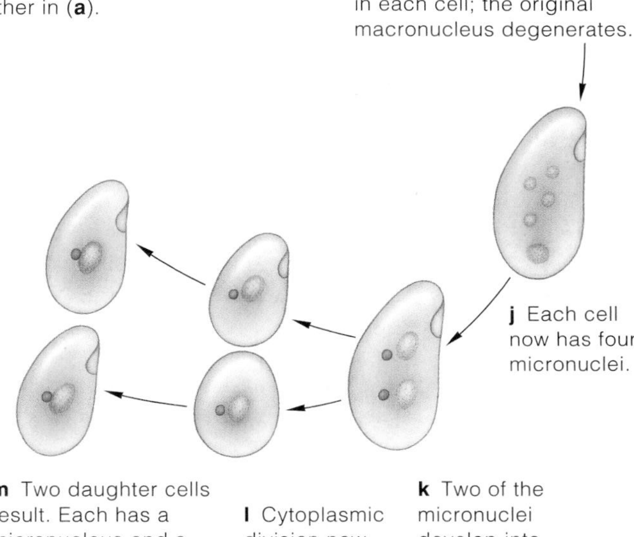
i Micronuclei divide again in each cell; the original macronucleus degenerates.
j Each cell now has four micronuclei.
k Two of the micronuclei develop into macronuclei.
l Cytoplasmic division now begins.
m Two daughter cells result. Each has a micronucleus and a macronucleus.

water's tendency to diffuse inward by osmosis. (Here you may wish to refer to Section 5.5.) Like amoebas, it uses **contractile vacuoles**. Tiny tubes extending from the center of these organelles collect excess water that moves osmotically into the cell body. Filled vacuoles contract and force water through a pore to the outside.

As is the case for other protistan groups, the ciliates show diversity in life-styles. Like *Paramecium*, about 65 percent of the known species are free-living and motile. Others attach themselves permanently or temporarily to substrates, often by means of a stalked component. Some form colonies. About 30 percent live as symbionts in or on other organisms. *Balantidium coli* is the largest protozoan parasite of humans, with effects like those of *Entamoeba histolytica* infections.

Ciliates can reproduce sexually and asexually, and things get interesting because each cell commonly has two types of nuclei. When a cell reproduces asexually by binary fission, a small, diploid *micro*nucleus divides by mitosis. And the large *macro*nucleus lengthens and splits in two a bit sloppily; some of the DNA may spill out. Most often, sexual reproduction occurs by a unique form of **conjugation**, as shown in Figure 23.7. In brief, the partner ciliates repeatedly divide their micronuclei, swap two daughter micronuclei, and allow two others to fuse and form a diploid macronucleus to replace the one that disappears. And you probably thought sex among the single-celled critters was simple!

e

d

Ciliated protozoans bear a profusion of cilia. These motile structures are used for swimming and for beating food into an oral cavity. Each ciliate usually has two types of nuclei, which are distributed to daughter cells in unique ways during asexual and sexual reproduction.

Like other protistans, the ciliates show great diversity in life-styles and diverse variations on the basic body plan.

23.5 ANIMAL-LIKE FLAGELLATES

Animal-like flagellates (Mastigophora) are free-living predators and parasites bearing one to several flagella. Members of one group are equipped with flagella *and* pseudopods, which is evidence of close evolutionary links with the amoebas. The free-living types abound in freshwater or marine habitats. Parasitic types live in the moist tissues of plants and animals.

Giardia lamblia, for example, is an internal parasite of humans, foraging cattle, and numerous wild animals such as beavers. Cells (trophozoites) survive outside the body in cysts, usually in feces-contaminated water. Ingesting the cysts is a prelude to infection. Stomach acid stimulates the cysts into releasing the trophozoites, which attach to the epithelial lining of the small intestine (Figure 23.8). After cells reproduce, they move into the large intestine (colon), then are expelled with feces. They infect new hosts who ingest contaminated water or food. The parasite also may be transmitted by way of anal intercourse with an infected partner.

Often the parasite causes only mild intestinal upsets. But it also causes a severe form of gastroenteritis called *Giardiasis*. Disease symptoms include bloating, nausea, intestinal cramps, and explosive, foul-smelling, watery diarrhea. The illness may recur over months or years.

Worldwide, 20 percent of the human population may be infected at a given time. Giardiasis is prevalent in developing countries, in overcrowded regions with poor sanitation and water quality control. In the United States, waterborne outbreaks have occurred mainly in mountains from coast to coast. For safety, hikers and campers should boil any water drawn from wilderness sources, including remote streams, before drinking it.

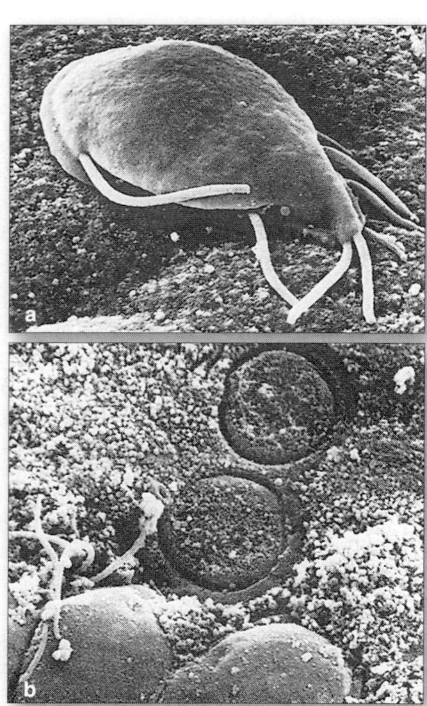

Figure 23.8 Scanning electron micrographs of *Giardia lamblia*, an animal-like flagellate that causes intestinal disturbances. (**a**) This cell is adhering to epithelium by means of its sucking disk. (**b**) When a cell detaches, its disk often leaves an impression on the epithelial surface that is distinctive when viewed with the aid of electron microscopes.

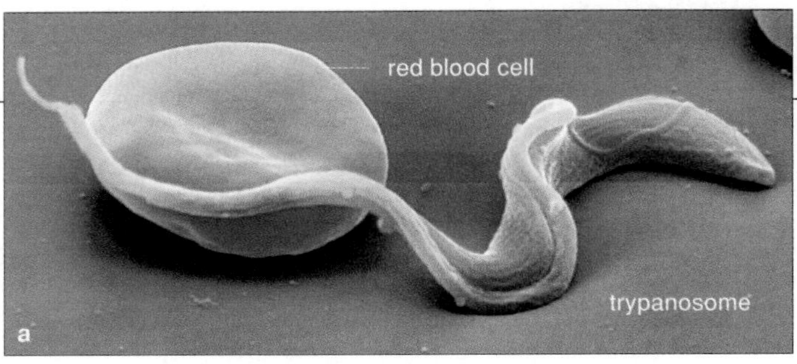
red blood cell
trypanosome
a

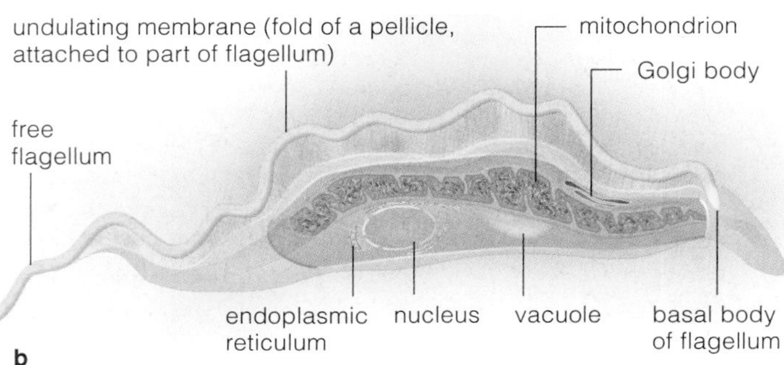
undulating membrane (fold of a pellicle, attached to part of flagellum)
mitochondrion
Golgi body
free flagellum
endoplasmic reticulum
nucleus
vacuole
basal body of flagellum
b

Figure 23.9 Light micrograph and diagram of *Trypanosoma brucei*. This animal-like flagellate causes African sleeping sickness.

Figure 23.10 Scanning electron micrograph of *Trichomonas vaginalis*. This pathogenic species causes trichomoniasis, a highly prevalent, sexually transmitted disease.

Trypanosomes also are serious parasites in many regions. *Trypanosoma brucei* is a type that causes *African sleeping sickness*, a severe disease of the nervous system (Figure 23.9). The tsetse fly is its vector between hosts. Another type, *T. cruzi*, is prevalent in South America and Mexico. It causes *Chagas disease*. Bugs pick up the parasite when they feed on infected humans and other animals. The parasite multiplies in the insect gut, then may be excreted onto a host. Any scratches in the skin invite infection. The liver and spleen enlarge, eyelids and the face swell up, then the brain and heart become severely damaged. There is no treatment or cure.

Many of the trichomonads are among the parasitic types. *Trichomonas vaginalis*, a worldwide nuisance, can infect human hosts during sexual intercourse (Figure 23.10). Unless trichomonad infections are treated, they can damage the urinary and reproductive tracts.

Animal-like flagellates include internal parasites that are responsible for serious disease throughout the world.

Further reading: Student Guide to InfoTrac on web site

23.6 SPOROZOANS

Sporozoan is an informal name for parasitic protistans that complete part of the life cycle *inside specific cells of host organisms*. These parasites form sporozoites, a type of motile infective stage. Some have encysted stages. At one end of the cell body is a complex of distinctive structures that function in penetrating host cells. The sporozoans are often grouped as phylum Apicomplexa.

Many sporozoans cause serious human diseases. For example, *cryptosporidiosis* is a waterborne disease caused by *Cryptosporidium*. The parasite invades epithelial cells of the small intestine or the respiratory system (Figure 23.11). Two to ten days after the infection, people suffer stomach cramps, watery diarrhea, and a slight fever. The symptoms may recur, often months or years later in people with weak immune systems. Infected people who do not show symptoms can still infect others.

The parasite can survive outside the body in lakes, rivers, streams, swimming pools, jacuzzis, chlorinated drinking water, and ice. (*Cryptosporidium*, unlike most pathogens, does not succumb to chlorine.) It may now contaminate most water supplies, and the only way to get rid of it is to pass water through a reverse osmosis filter or bring it to a rolling boil for a full minute. It should not be present in distilled water, and apparently it cannot survive in hot coffee or tea.

Another sporozoan, *Pneumocystis carinii,* lives in the lungs of humans and many domestic and wild animals. Malnutrition or a weakened immune system give it the opportunity to threaten its host. For example, it causes a deadly form of pneumonia in about two-thirds of all AIDS cases. A resistant cyst forms during its life cycle. Sporozoites released during an asexual phase develop into a stage that lives in interstitial tissues of the lungs (Figure 23.12). Tiny air sacs in the lungs fill with foamy material teeming with parasites. Fever, coughing, rapid breathing, and blue skin around the mouth and eyes follow. Untreated patients die from asphyxia; they stop breathing. Even with treatment, death rates are high.

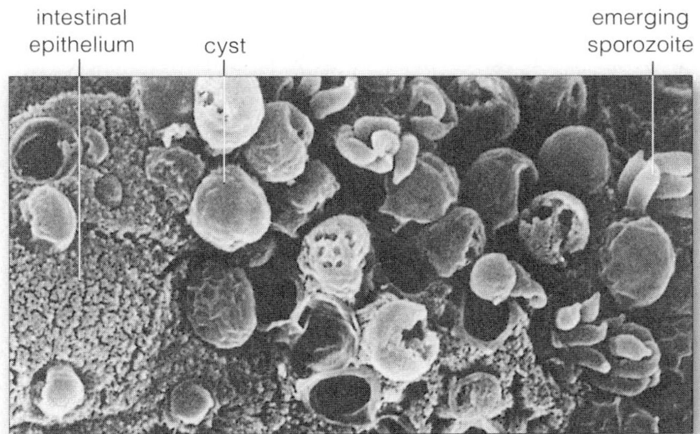
intestinal epithelium
cyst
emerging sporozoite

Figure 23.11 *Cryptosporidium* in intestinal epithelium.

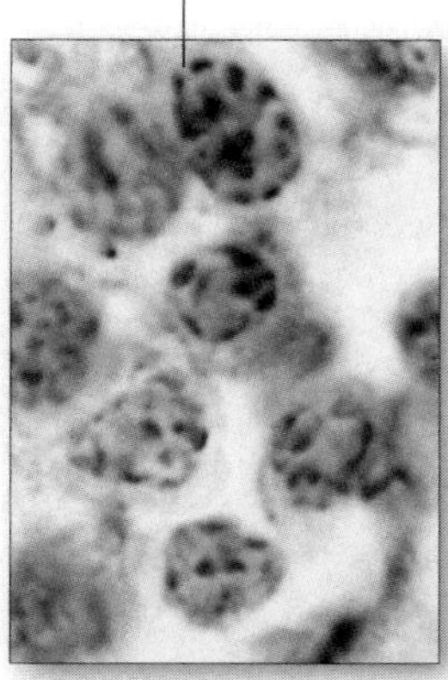
Encysted infective bodies in lung tissue

Figure 23.12 *Above:* An encysted stage of *Pneumocystis carinii*, the agent of a form of pneumonia that can kill people with compromised immune systems, as in AIDS.

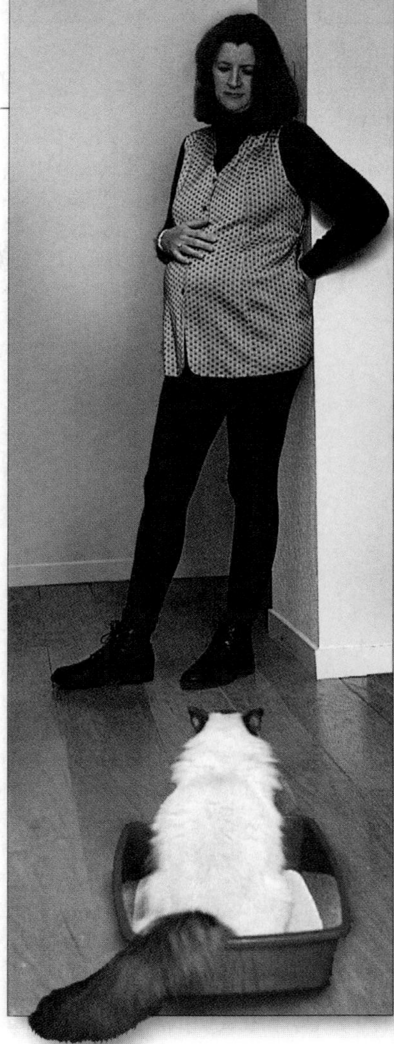

Figure 23.13 *Right:* Standoff between a well-informed mother-to-be and a possible reservoir for the sporozoan *Toxoplasma*.

The sporozoan *Toxoplasma* uses domestic and wild cats as definitive hosts. Its intermediate hosts are other domestic and wild animals—and humans. Its cysts may be present in raw or undercooked meat. Cockroaches and flies also can move cysts from cat feces onto food. The disease *toxoplasmosis* has flu-like symptoms. It is not common, but small epidemics do occur. The disease is dangerous for immune-compromised people and for embryos of pregnant women. It can cause miscarriages or birth defects if the parasite infects the unborn child. Any cat, no matter how coddled, can pass cysts. That is why all pregnant women should avoid stray cats and never empty litterboxes, clean up housecat "accidents," or clean out children's sandboxes (Figure 23.13).

The next section takes a close look at *Plasmodium*, the sporozoan agent of malaria. Why have we focused so far on the most notorious animal-like flagellates and sporozoans? After all, fewer than two dozen species cause serious diseases in humans. The reason is that they infect hundreds of millions of people every year. There are no effective vaccines against them.

Sporozoans are internal parasites that produce infective, motile stages. Many species form encysted stages.

23.7

MALARIA AND THE NIGHT-FEEDING MOSQUITOES

Each year in Africa alone, about a million people die of *malaria*, a long-lasting disease that four different parasitic species of *Plasmodium* can cause. This type of sporozoan currently has infected more than 100 million people.

Only female mosquitoes of the genus *Anopheles* can transmit the parasites to human hosts. Their bite delivers sporozoites (an infective, motile stage) to the host's blood. The bloodstream transports sporozoites to the liver, where they undergo multiple fission. Some of the resulting cells, called merozoites, penetrate and asexually reproduce in red blood cells, which they destroy. Others develop into male and female gametocytes, which eventually mature into gametes (Figure 23.14).

Symptoms of malaria begin after infected cells abruptly rupture and release merozoites, metabolic wastes, and cellular debris into the individual's bloodstream. Shaking, chills, a burning fever, and drenching sweats are classic symptoms. After one episode, symptoms subside for a few weeks or months. Infected individuals might even feel normal, but relapses inevitably recur. In time, anemia and gross enlargement of the liver and spleen may result.

Amazingly, the *Plasmodium* life cycle is sensitive to the body temperature and oxygen levels inside humans and mosquitoes. The gametocytes cannot mature in humans, who are warm-bodied and have little free oxygen in their blood (most is bound to hemoglobin). They are induced to mature inside the mosquito, which has a lower body temperature and which incidentally slurps in oxygen from the air along with gametocyte-containing blood.

Mature gametes fuse to form zygotes. These repeatedly divide and form many sporozoites, which migrate to the female mosquito's salivary glands and await the next bite.

Malaria has been around for a long time. People were describing its symptoms more than 2,000 years ago. It got its name in the seventeenth century, when Italians made a connection between the disease and noxious gases from swamps near Rome, where mosquitoes flourished (*mal*, bad; *aria*, air). By severely incapacitating so many people, malaria contributed to the decline of the ancient Greek and Roman empires. Much later in time, it incapacitated soldiers during the United States Civil War, World War II, and the Korean and Vietnam conflicts.

Throughout human history, malaria has been most prevalent in tropical and subtropical parts of Africa. Now, however, the numbers of cases reported in North America and elsewhere are increasing dramatically, owing to the hordes of globe-hopping travelers and unprecedented levels of human immigration.

Travelers who intend to visit countries with high rates of malaria are advised to use antimalarial drugs such as chloroquine. But certain strains of *Plasmodium* are now resistant to the drugs, and a vaccine has been difficult to develop. *Vaccines* are preparations that induce the body to build up resistance to a specific pathogen. Experimental vaccines for malaria are not equally effective against all the stages that develop during sporozoan life cycles. This is generally the case for vaccines that researchers hope to develop against most parasites with complex life cycles.

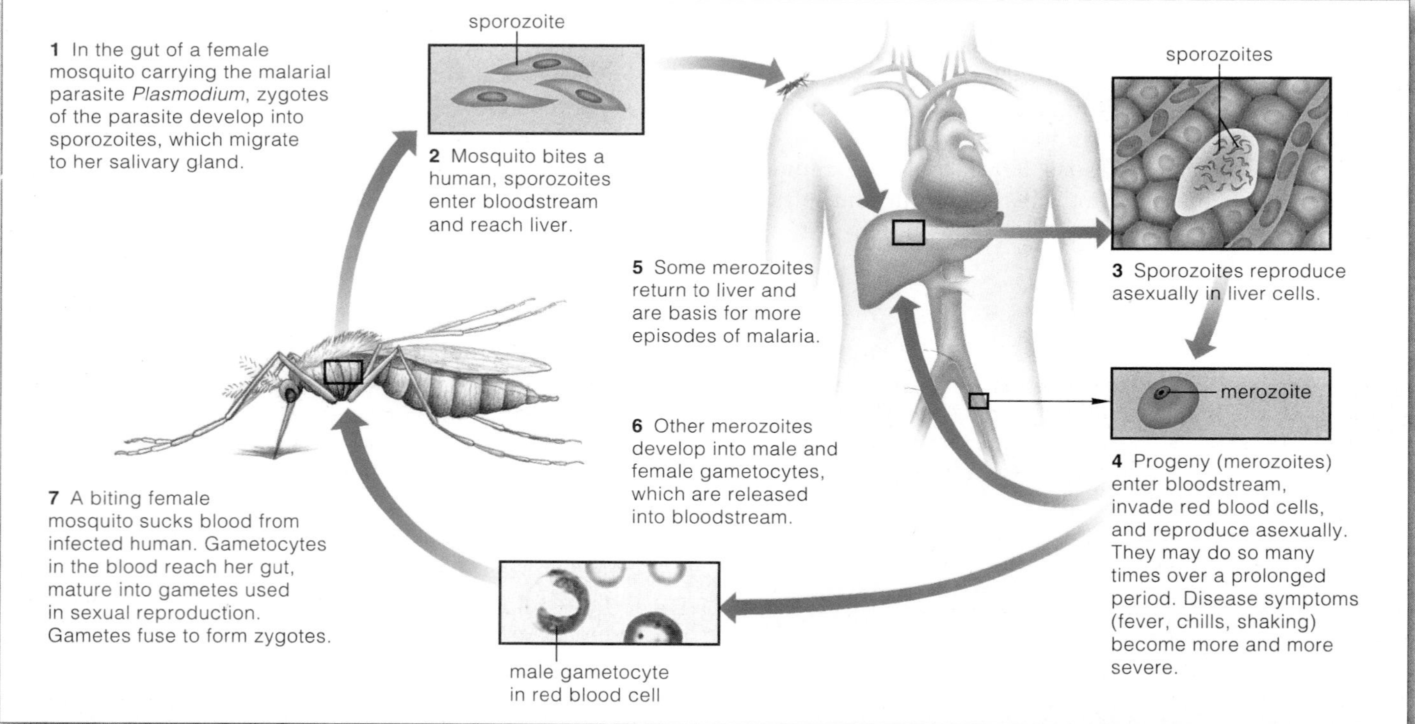

1 In the gut of a female mosquito carrying the malarial parasite *Plasmodium*, zygotes of the parasite develop into sporozoites, which migrate to her salivary gland.
sporozoite
2 Mosquito bites a human, sporozoites enter bloodstream and reach liver.
sporozoites
3 Sporozoites reproduce asexually in liver cells.
merozoite
4 Progeny (merozoites) enter bloodstream, invade red blood cells, and reproduce asexually. They may do so many times over a prolonged period. Disease symptoms (fever, chills, shaking) become more and more severe.
5 Some merozoites return to liver and are basis for more episodes of malaria.
6 Other merozoites develop into male and female gametocytes, which are released into bloodstream.
male gametocyte in red blood cell
7 A biting female mosquito sucks blood from infected human. Gametocytes in the blood reach her gut, mature into gametes used in sexual reproduction. Gametes fuse to form zygotes.

Figure 23.14 Life cycle of one of the sporozoans (*Plasmodium*) that causes malaria. 

Further reading: Student Guide to InfoTrac on web site —

23.8 REGARDING "THE ALGAE"

The remainder of this chapter surveys the protistans, largely photosynthetic, that are informally known as "the algae." Table 23.2 lists six groups, although no one yet knows how many there really are.

First we will consider the euglenoids, chrysophytes, and dinoflagellates. Single-celled species dominate all three groups, which is not the case for the red, brown, and green algae. The majority belong to **phytoplankton**: communities of aquatic photosynthetic species, mainly microscopic, that drift or swim weakly through water. As food producers, these algae are major contributors to nearly all food webs in aquatic habitats. They form impressively large populations in freshwater, brackish water, and seawater (Section 7.9).

Table 23.2 The Mostly Photosynthetic Protistans

Mostly free-living cells. Mostly photoautotrophs, the majority being components of phytoplankton	
EUGLENOPHYTA	Euglenoids. Single cells. Photoautotrophs, heterotrophs, or both
CHRYSOPHYTA	Chrysophytes. Mostly single cells, some colonial. Mostly photoautotrophs, some heterotrophs. Differences in pigment arrays, cell wall, and type of flagellated cell
	Golden algae
	Yellow-green algae
	Diatoms
	Coccolithophores
PYRRHOPHYTA	Dinoflagellates. Single cells; some symbionts. Photoautotrophs, heterotrophs
Mostly multicelled, aquatic, photoautotrophs with ecological roles comparable to those of land plants	
RHODOPHYTA	Red algae
PHAEOPHYTA	Brown algae
CHLOROPHYTA	Green algae

Most red, brown, and green algae are multicelled photoautotrophic species, some as tall as trees. Reflect on the variation and you might wonder how one kingdom can accommodate multicelled species in addition to all the one-celled forms. The answer is that considerable research has revealed strong resemblances in cellular structures, metabolism, chromosomal organization, life cycles, and genetics among these groups.

Many of the single-celled and multicelled protistans are photosynthetic species informally called "the algae."

Most algae are aquatic. Many single-celled species are key components of phytoplankton. Multicelled species play ecological roles comparable to those of land plants.

23.9 EUGLENOIDS

The **euglenoids** (Euglenophyta) are a classic example of evolutionary experimentation. Euglenoids are free-living, flagellated cells that abound in freshwater and in stagnant ponds and lakes. Most of the 1,000 known species are *photoautotrophs*. The rest are *heterotrophs* that subsist on organic compounds dissolved in the water.

A *Euglena* cell has a profusion of organelles (Figure 23.15). Among these are chloroplasts with chlorophylls *a* and *b*, and carotenoids—the same pigments that occur in plant chloroplasts. *Euglena* cells also sport flagella (one long, one short) and a contractile vacuole—just as animal-like protozoans do. Moreover, like some protozoans, *Euglena* has a pellicle, this one being a flexible cover that has many spiral strips of a translucent, protein-rich material. Some of the pigments form an "eyespot" that partly shields a light-sensitive receptor. By moving the long flagellum, this single cell can keep the receptor exposed to light and so stays where the light is most suitable for its photosynthetic activities.

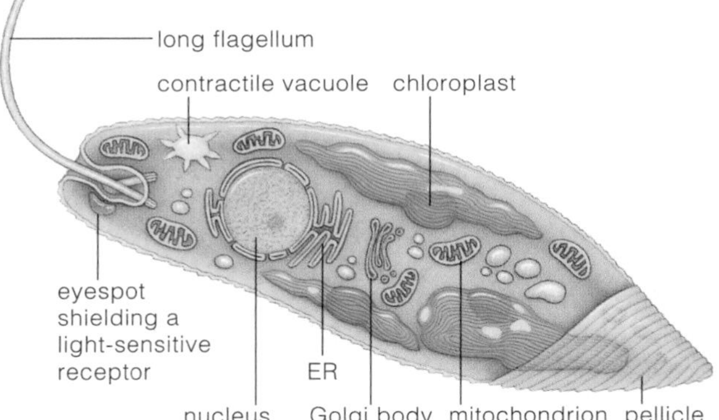
long flagellum
contractile vacuole
chloroplast
eyespot shielding a light-sensitive receptor
ER
nucleus
Golgi body
mitochondrion
pellicle

Figure 23.15 *Euglena* body plan. Compare this diagram with the electron micrograph in Section 21.4.

Generally, "self-feeders" make their own vitamins, which are required for growth. But all photoautotrophic euglenoids must get vitamin B_{12} from their surroundings (most cannot make vitamin B_1, either). Probably their chloroplasts originated from endosymbiotic green algae, in the manner described in Section 21.4.

Given their array of traits, could euglenoids be very close relatives of some existing protozoans? Maybe, but probably not. Researchers believe their similarities are more likely an outcome of convergent evolution.

The euglenoids are free-living, flagellated cells in fresh and stagnant bodies of water. Most are photoautotrophs; others are heterotrophs that feed on dissolved organic compounds.

23.10 CHRYSOPHYTES AND DINOFLAGELLATES

Chrysophytes

Most of the **chrysophytes** (Chrysophyta) are free-living photosynthetic cells that contain chlorophylls *a*, c_1, and c_2. The golden algae, yellow-green algae, diatoms, and coccolithophores belong to this large group. From time to time, some species undergo stupendous increases in population size, an event called an **algal bloom**.

GOLDEN AND YELLOW-GREEN ALGAE Silica scales or other hard parts cover cells of many of the 500 species of the chrysophytes called **golden algae** (Figure 23.16*a*). Chlorophylls of these cells are masked by fucoxanthin, a golden-brown carotenoid. Except for the chlorophylls, some amoeboid forms closely resemble true amoebas.

Yellow-green algae are common chrysophytes in many aquatic habitats. Fucoxanthin is absent from the 600 known species. Most are not motile, but they have flagellated gametes. Figure 23.16*b* shows an example.

DIATOMS The 5,600 existing species of **diatoms** have a silica "shell" of two perforated parts that overlap like a pillbox (Figure 23.16*c*). For 100 million years, finely crushed shells of at least 35,000 extinct species of these mostly photosynthetic cells accumulated at the bottom of lakes and seas. Many sediments contain the deposits, which we quarry for use in insulation, abrasives, and filters. Each year, for instance, more than 270,000 metric tons are quarried near Lompoc, California.

COCCOLITHOPHORES Calcium carbonate plates encase the **coccolithophores** (Figure 23.16*d*). Most of the 500 or so existing species are single-celled photosynthesizers of marine habitats, especially in the tropics. Accumulations of the plates in the past helped form ocean sediments, and chalk and limestone deposits (including Dover's white cliffs). During algal blooms, mucous secretions around these cells clog fish gills. Also, a by-product of their metabolism (dimethyl sulfide) is noxious enough to make migratory fishes deviate from normal routes.

Dinoflagellates

We know of more than 1,200 species of **dinoflagellates** (Pyrrhophyta). Most are single photosynthetic cells that are major producers in marine phytoplankton. A few are symbionts with corals. Some bioluminescent types make warm seawater shimmer at night.

Dinoflagellates bear flagella and plates of cellulose (Figure 23.17*a*). When they reproduce, the cell and its cellulose plates divide into two daughter cells. The cells are yellow-green, green, blue, brown, or red, depending on the pigments and endosymbiotic history.

Red tides form near coasts when certain species of dinoflagellates undergo blooms that turn the water rust-red or brown (Figure 23.17*b*). Then, the concentration of dinoflagellates may briefly reach 6 to 8 million cells per liter of water. A few species produce toxins that kill fish and other animals. In one year, for example, 150,000 grebes and 5,000 brown pelicans died at the Salton Sea in southern California. People who eat seafood tainted with some of the toxins can suffer brain damage.

Algal Blooms and the Cell From Hell

Since 1991, dinoflagellate blooms have destroyed more than a billion fish near the coasts of North Carolina, Virginia, and Maryland. In 1999, for example, close to a million fish died in tributaries of the Pocomoke River. Atlantic menhaden piled up on banks of the Bullbeggar Creek. The agent of their death was *Pfiesteria piscicida*, known for good reason as the "cell from hell."

The life cycle of *P. piscicida* proceeds through at least twenty-four encysted, flagellated, and amoeboid forms

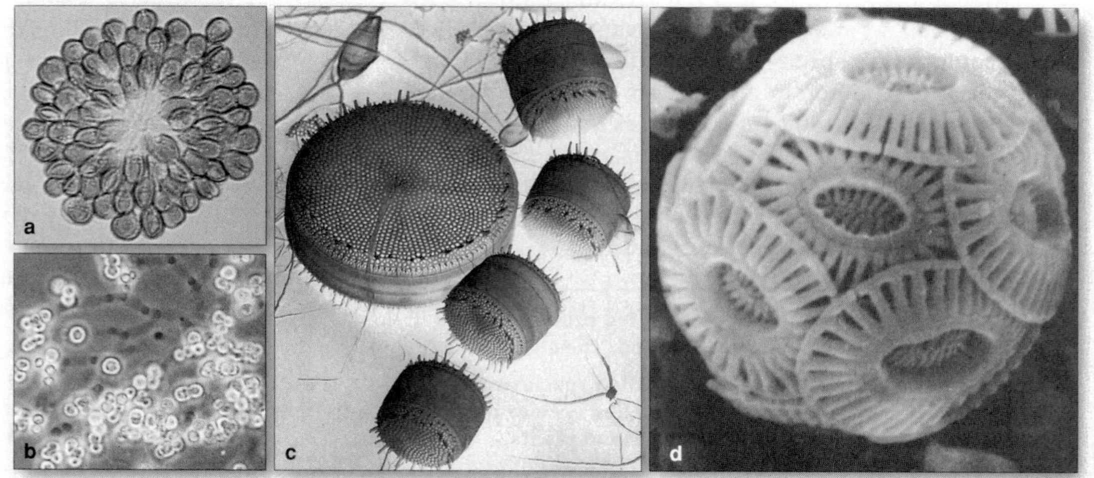

a
b
c
d

Figure 23.16 A sampling of chrysophytes. (**a**) *Synura*, a golden alga with a fishy odor, forms colonies in phytoplankton. (**b**) This colonial yellow-green alga, *Mischococcus*, occurs in phytoplankton. (**c**) Diatom shells, which fit together like a pillbox. The four small shells formed by cell division. New walls form inside a parent cell wall, so diatoms become smaller with each division. When too small, a spore forms inside the shell. It germinates and grows into a full-size diatom. (**d**) One of the coccolithophores, which have calcium carbonate plates.

Further reading: Student Guide to InfoTrac on web site –

a Perforated plates of a dinoflagellate

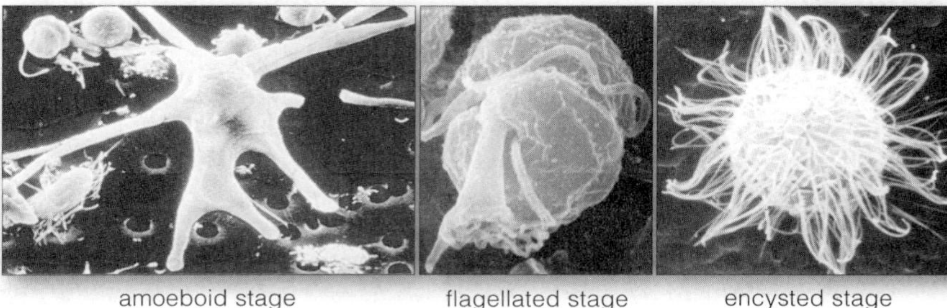
amoeboid stage
flagellated stage
encysted stage

c A few stages in the life cycle of *Pfiesteria*

b
d

Figure 23.17 (**a**) Scanning electron micrograph of a dinoflagellate. (**b**) *Left:* Red tide near the coast of central California. *Right:* Small portion of a fish kill resulting from a dinoflagellate bloom. (**c**) Stages in the life cycle of *Pfiesteria piscicida*. (**d**) The kind of damage inflicted by the cell from hell.

(Figure 23.17*c*). *P. piscicida* waits out adverse conditions as cysts in sediments of estuaries. Its flagellated forms (sexual and asexual) usually feed by attaching to prey and sucking out fluids. When a large school of fish (such as oily menhaden) linger to feed, their excretions incite encysted cells to emerge and release toxins, which slow down the fish and stop them from escaping. The toxins also eat away patches of fish skin. *P. piscicida* feeds on the sloughed epidermis, blood, and tissue juices oozing from the open sores (Figure 23.17*d*). Now the flagellates give rise to amoeboids that feast on dead fish. All of the changes may proceed within a matter of hours.

Nitrogen, phosphorus, and other nutrients in fertilizer runoff and raw sewage fan such algal blooms around the world. The blooms usually occur in shallow, warm water that has become enriched with nutrients. JoAnn Burkholder, a North Carolina State University botanist, correlated *Pfiesteria* blooms with hundreds of millions of gallons of raw sewage from hog and chicken farms. More than 16 million hogs are being raised in 3,500 industrial-scale hog farms in eastern North Carolina. Untreated wastes in holding lagoons often spill into rivers that drain into the poorly flushed Chesapeake Bay estuaries and into rivers in Alabama, Delaware, Florida, and Virginia. For instance, after one heavy rain, a hog farm spill exceeded by three times the volume of oil spilled from the *Valdez* (Section 51.8). In 1997 alone, *Pfiesteria* inflicted a loss of 60 million dollars on fishing and tourism industries. The huge loss finally prodded health, agricultural, and environmental agencies into starting coordinated research to address the problem.

Chrysophytes and dinoflagellates help form the food base for marine communities. Human activities that fan their explosive population growth can cause enormous damage.

23.11 RED ALGAE

Of 4,100 known species of **red algae** (Rhodophyta), nearly all are marine; only 200 live in freshwater. Red algae are most abundant in warm currents and tropical seas, often at surprising depths (265 meters below the surface) in clear water. A few occur in phytoplankton. Some encrusting types contribute to the formation of coral reefs and banks (Sections 28.3 and 50.12).

Cells of some red algae have walls hardened with calcium carbonate. Different species appear red, green, purple, or nearly black, depending on which accessory pigments (phycobilins, mostly) mask the chlorophyll *a*. Phycobilins are good at absorbing green and blue-green wavelengths, which are able to penetrate deep waters. (Chlorophylls are more efficient at absorbing red and blue wavelengths, which may not reach far below the water's surface.) The chloroplasts of red algae resemble cyanobacteria, which suggests endosymbiotic origins.

The life cycle of most species includes multicelled stages, but these lack tissues or organs (Figure 23.18). Reproductive modes are diverse, with complex asexual and sexual phases. Figure 23.19 shows an example.

Red algae have a flexible, slippery texture because of mucous material in their cell walls. Agar is made of extracts of wall material from several species. The inert, gelatinous substance is used as a moisture-preserving agent in baked goods and cosmetics, as a setting agent

a
b

Figure 23.18 (**a**) A red alga (*Bonnemaisonia hamifera*). Such a growth pattern (branching, filamentous) is the most common. (**b**) From a tropical reef, a red alga showing sheetlike growth.

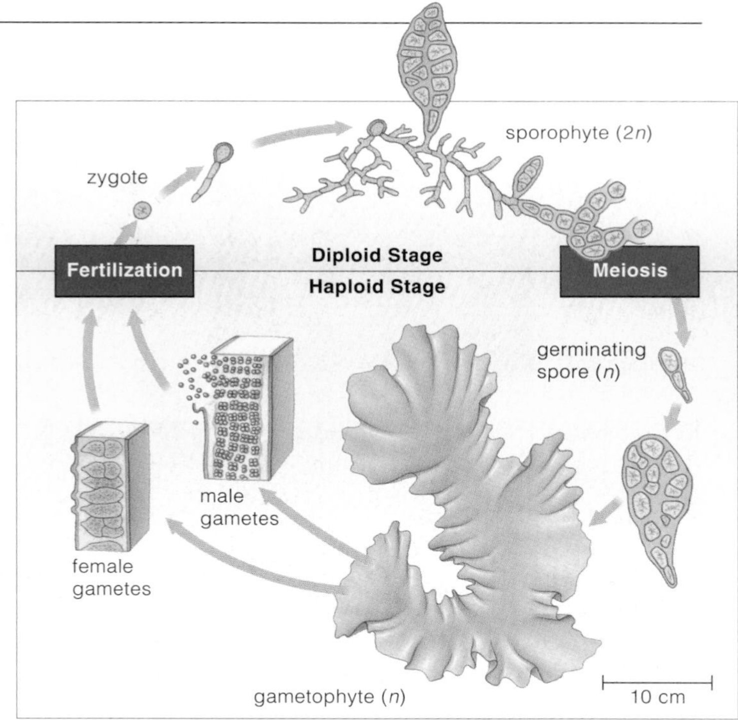
sporophyte (2n)
zygote
Fertilization
Diploid Stage
Haploid Stage
Meiosis
germinating spore (n)
male gametes
female gametes
gametophyte (n)
10 cm

Figure 23.19 Life cycle of *Porphyra*. From 1623 to the 1950s, Japanese fishermen cultivated and harvested a species of this red alga in early fall. The rest of the year, it seemed to vanish. Kathleen Drew-Baker studied the sheetlike form of *P. umbilicus* in the laboratory. She observed gametes forming in packets interspersed between vegetative cells near the sheet margins; the sheet was a gametophyte (gamete-producing body). She also observed gametes in a petri dish. Zygotes that formed after fertilization grew on bits of shell in the dish before developing into a tiny branching, filamentous form. This was how *Porphyra* spent most of the year! People already recognized the filamentous form as common pinkish growths on shells. But it was so different from the sheetlike form that it was viewed as a separate species. It is the alga's diploid sporophyte (spore-producing body).

Drew-Baker's discovery of alternating stages in the alga's life history, and the realization that the diploid stage could be grown on shells or other calcium-rich surfaces, revolutionized the nori industry. Within a few years, researchers worked out the life cycle of *P. tenera*, the harvested species. By 1960, *nori cultivation* was a billion-dollar industry.

for jellies, and as culture gels. We also shape it into soft capsules for delivery of drugs and food supplements. Carrageenan, extracted from *Eucheuma*, is a stabilizer in paints, dairy products, and many other emulsions.

Humans find different species of *Porphyra* tasty as well as nutritious. You may know it from sushi bars as *nori*, a wrapping for rice and fish (Figure 23.19).

Red algae are photosynthetic protistans with phycobilins and other accessory pigments that mask their chlorophyll *a*. Usually, multicelled stages develop during the life cycles. Most species are aquatic. They show great diversity in size, morphology, life-styles, reproductive modes, and habitats.

Further reading: Student Guide to InfoTrac on web site

23.12 BROWN ALGAE

Walk along a rocky shore at low tide and you may come across olive-green or brown seaweeds, as in Figure 23.20. They are among the 1,500 species called **brown algae** (Phaeophyta). Nearly all thrive in cool or temperate seawater, ranging from the intertidal zone to the open ocean. Masses of a floating brown alga, *Sargassum,* are the basis of a great floating ecosystem in the Sargasso Sea, which lies between the Azores and the Bahamas.

Different species appear olive-green, golden, or dark brown, depending on the pigments. Like chrysophytes, to which they may be related, the brown algae contain carotenoids such as fucoxanthin as well as chlorophylls *a*, c_1, and c_2. They range from microscopic, filamentous *Ectocarpus* to giant kelps, twenty to thirty meters long. Like red algae, they have diverse, complex life cycles that include sexual and asexual phases. In their life cycles, too, gametophytes alternate with sporophytes.

Macrocystis, *Laminaria*, and the other giant kelps are the largest, most complex protistans. Their multicelled sporophytes have stipes (stemlike parts), blades (leaflike parts), and holdfasts (anchoring structures). Hollow, gas-filled bladders impart buoyancy to the stipes and blades and keep them upright in the water. The stipes contain tubelike arrays of elongated cells. These tubes swiftly carry dissolved sugars and other products of photosynthesis to living cells through the kelp body. Flowering plants also transport sugars through similar kinds of tubes. This is a case of convergent evolution in two unrelated groups of large, multicelled organisms.

Giant kelp beds function as productive ecosystems. Think of them as underwater forests within which great numbers of diverse bacteria and protistans, as well as fishes and other animals, carry out their lives. The sizes of kelp forests are variable, depending on changes in ocean currents and in abundances of sea urchins (which feed on kelp debris) and sea otters (which feed on sea urchins). For example, in the late 1950s, a warm current displaced the cooler currents off the California coast. *Macrocystis* did not fare well with the temperature shift, and extensive beds off La Jolla and Palos Verdes almost disappeared. Without organic debris for the sea urchins to feed upon, these invertebrates fed directly on the kelps instead. Quantities of quicklime were dumped in the water to reduce the number of sea urchins, and the kelps made a comeback. So did fishes, lobster, abalone, and other species that make the kelp beds their home.

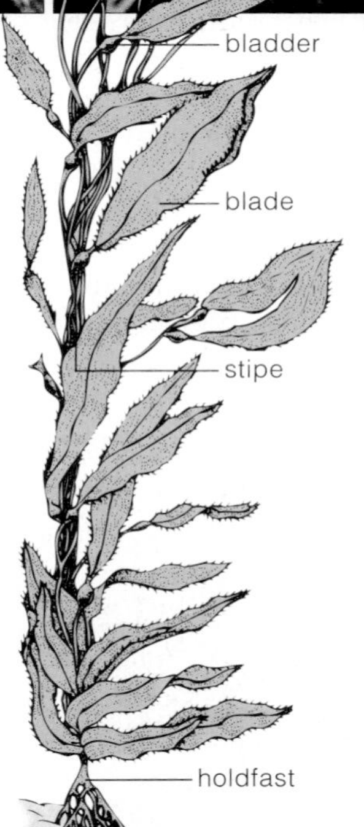

bladder
blade
stipe
holdfast

Figure 23.20 (**a**) Closer view of *Postelsia*, the brown alga shown in Figure 23.1*a*. You will find it thriving along coasts exposed to heavy surf from Vancouver Island on down to central California. More than a hundred deeply grooved blades top a highly resilient stipe, which is fastened to rocky substrates by a mound of anchoring structures. Its spores never disperse far from the parent sporophyte. At low tide, they simply drip onto rocks from the grooved blades. (**b**) Diagram of *Macrocystis* and an underwater view of a kelp forest. A few species live in the coastal waters of North and South America, New Zealand, Tasmania, and most of the islands in subantarctic waters.

Macrocystis and certain other brown algae are commercially harvested. Extracts from them are used in ice creams, puddings, jelly beans, salad dressings, beers, canned and frozen foods, cough syrups, toothpastes, cosmetics, floor polishes, and paper. Alginic acid from the cell walls of some species is used to make algins, which are added to various products as thickening, emulsifying, and suspension agents. In the Far East especially, people harvest kelps as sources of food and mineral salts, and as a fertilizer for crops.

The brown algae range in size from the microscopic to the largest of all of the multicelled protistans. Nearly all species live in marine habitats ranging from the intertidal zone to the surface waters of the open ocean.

23.13 GREEN ALGAE

Of all protistans, the **green algae** (Chlorophyta) are structurally and biochemically most like the plants and may be their closest relatives. All are photosynthetic. As in plants, their chlorophylls are the molecular types designated *a* and *b*, and they, too, store carbohydrates as starch grains inside their chloroplasts. The cell walls of some species are composed of cellulose, pectins, and other polysaccharides typical of plants.

Figure 23.21 is a sampling of the more than 7,000 known species. They include single-celled, sheetlike, tubular, filamentous, and colonial forms. You won't be able to see many without the aid of a microscope. Most, including the *Micrasterias* cell shown in Figure 23.1*b*, live in freshwater. *Micrasterias* is one of thousands of desmid species, which are important food producers in nutrient-poor ponds and peat bogs. Green algae also grow at the ocean surface, in marine sediments, just below the surface of soil, and on rocks, tree bark, other organisms, and snow. Some are symbionts with fungi, protozoans, and a few marine animals. A colonial form (*Volvox*) is a hollow, whirling sphere of 500 to 60,000 flagellated cells. Those white, powdery beaches in the tropics are largely the work of uncountable numbers of *Halimeda* cells, which formed calcified walls, then died and disintegrated (Figure 23.21*d*). Someday, green algae

Figure 23.21 Representative green algae. (**a**) One marine species (*Codium*) with a pronounced branching form. Although many green algae are microscopic, *C. magnum* is taller than you are. In 1956 still another species of *Codium*, from Washington, was accidentally introduced into the Connecticut River where it empties into the Atlantic. Puget Sound populations of this alga were kept in check by herbivores that evolved with them. In its new habitat, there were no native *Codium* species or herbivores that were adapted to grazing on it. The introduced alga spread along the coast as far north as Maine and as far south as North Carolina in a little more than two decades.

(**b**) Also from a marine habitat, *Acetabularia*, fancifully called the mermaid's wineglass. Each individual is a multinucleate cell mass with a rootlike structure, stalk, and cap in which gametes form.

(**c**) Sea lettuce (*Ulva*) grows in estuaries and attaches to kelps in the seas. Reproductive cells form within and are released from the margins of the sporophyte and gametophyte, both of which have the form shown in the diagram.

(**d**) *Halimeda incrassata*. Its large, branched cells form by repeated nuclear division. Cell walls form only during reproductive phases.

(**e**) *Udotea cyathiformis*, one of 100+ microscopic *Udotea* species in tropical and subtropical waters. Many are highly calcified.

(**f**) *Volvox*, a colony of interdependent cells that resemble free-living, flagellated cells of the genus *Chlamydomonas*.

(**g**,**h**) Above Utah's summer timberline, phycologist Ron Hoham's footprint in "red snow" reveals the presence of dormant cells of *Chlamydomonas nivalis*, a snow alga. Red accessory pigments protect the chlorophylls of these cells.

a
c
f
b
d
e
g
h Dormant cell

Further reading: Student Guide to InfoTrac on web site

Zygote (cross-section)
A thick-walled resistant zygote develops.
Nuclear Fusion
Diploid Stage
Haploid Stage
Meiosis and Germination
haploid cell (+ strain)
haploid cell (– strain)
Mitosis occurs. Whether the resulting cells develop into spores or gametes will depend on environmental conditions.
Cytoplasmic Fusion
SEXUAL REPRODUCTION:
Mainly when nitrogen levels are low and light is of a certain quality and intensity, the cells develop into gametes.
ASEXUAL REPRODUCTION:
More spores are produced.
ASEXUAL REPRODUCTION:
More spores are produced.
Gametes of different mating types meet.
–
+
+
–

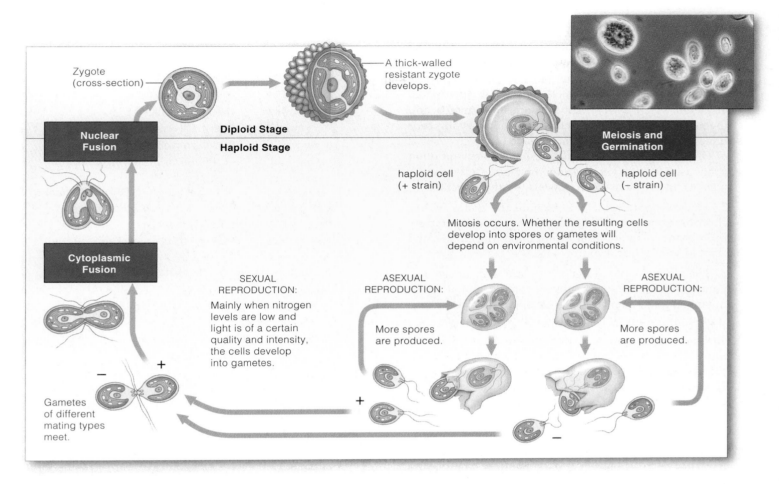

Figure 23.22 Life cycle of a species of *Chlamydomonas*, one of the most common green algae of freshwater habitats. This single-celled species reproduces asexually most of the time and sexually under certain environmental conditions.

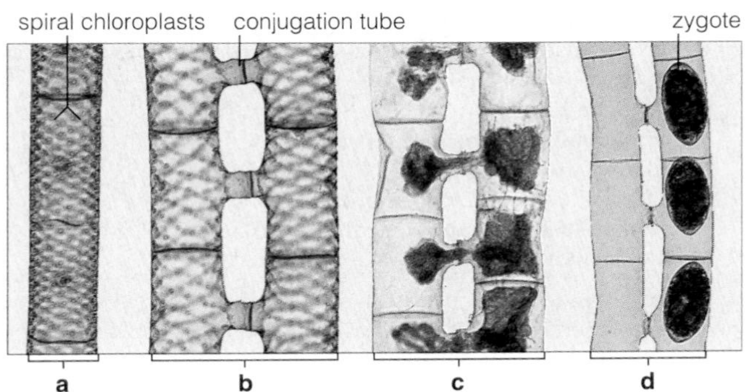
spiral chloroplasts
conjugation tube
zygote
a
b
c
d

Figure 23.23 One mode of sexual reproduction in *Spirogyra*, or watersilk. (**a**) The chloroplasts of this green alga are spiral and ribbonlike.

(**b**) Here, two different mating strains have made contact, and a conjugation tube has formed between cells of the adjacent haploid filaments. (**c**,**d**) The cellular contents of one strain pass through the tubes into cells of the other strain, where zygotes form. Zygotes develop thick walls. They undergo meiosis when they germinate and give rise to haploid filaments.

might even accompany astronauts on long missions in outer space. Green algae can grow in very small spaces, they release oxygen as a by-product of photosynthesis (the crew can't live without oxygen supplies), and they can take up carbon dioxide exhaled by the aerobically respiring crew.

Green algae employ diverse modes of reproduction. *Chlamydomonas* provides a classic example. This single-celled alga is twenty-five or so micrometers wide. It is able to reproduce sexually. Most of the time, however, it engages in asexual reproduction, with up to sixteen daughter cells forming by mitotic cell division within the confines of the parent cell wall. Daughter cells may live at home for a while, but sooner or later they leave by secreting enzymes that digest what's left of their parent. Figure 23.22 shows the life cycle of one species. To give a final example, Figure 23.23 shows *Spirogyra*, a filamentous green alga, reproducing sexually.

The green algae show great diversity in size, morphology, life-styles, and habitats. The structure and biochemistry of some groups indicate they have close evolutionary ties to the plant kingdom.

23.14 SUMMARY

1. Recall that protistans and other eukaryotes differ from bacteria in key respects. At the least, eukaryotic cells have a double-membraned nucleus, mitochondria, endoplasmic reticulum, and larger ribosomes. These cells have microtubules, which are used as cytoskeletal elements, as spindles for chromosome movements, and in the 9 + 2 core of flagella and cilia. Eukaryotic cells have two or more chromosomes, each of which consists of DNA complexed with numerous histones and other proteins. Eukaryotic cells alone engage in mitosis and meiosis. Many have chloroplasts and other plastids.

2. Heterotrophic protistans include the water molds, chytrids, slime molds (cellular and plasmodial types), protozoans (true amoebas, animal-like flagellates, and ciliated species), and sporozoans. See Table 23.3.

3. The chytrids and water molds are decomposers or parasites. Like fungi, they secrete enzymes that digest organic matter, then absorb breakdown products. Some form a mycelium (a mesh of absorptive filaments).

4. Like animals, slime molds are predatory. For part of the life cycle, they crawl on substrates as free-living, phagocytic, amoebalike cells. Like fungi, they produce spore-producing structures.

5. Like animals, protozoans are predators, grazers, or parasites. The amoebas, foraminiferans, heliozoans, and radiolarians are amoeboid protozoans with or without skeletal elements. Animal-like protozoans are internal parasites or live freely in aquatic habitats. Many cause human diseases. Sporozoans are internal parasites that may form cysts during their life cycle. An example is *Plasmodium*, which causes malaria. Ciliated protozoans such as *Paramecium* typically use their cilia for motility.

6. The majority of euglenoids, chrysophytes (diatoms, golden algae, and yellow-green algae), dinoflagellates, and red, brown, and green algae are all photoautotrophic. Many are members of phytoplankton (Table 23.3).

7. The red, brown, and green algae differ from one another in their body plan, size, reproductive modes, and habitats. Most are multicelled photosynthesizers. Some species of brown algae are the largest protistans. Plants may be descended from early green algae.

8. It has taken us more than one chapter to consider features of protistans, the simplest eukaryotes, and to speculate on their evolutionary ties to other kingdoms. Table 23.4 summarizes key similarities and differences among the prokaryotes and eukaryotes.

Table 23.3 Summary of Major Groups of Protistans

Group	Common Name
HETEROTROPHIC (decomposers, predators, grazers, parasites)	
Chytridiomycota	Chytrids
Oomycota	Water molds
Acrasiomycota	Cellular slime molds
Myxomycota	Plasmodial slime molds
Sarcodina	Amoeboid protozoans
	Rhizopods (naked amoebas, foraminiferans)
	Actinopods (radiolarians, heliozoans)
Mastigophora	Animal-like flagellated protozoans
Ciliophora	Ciliated protozoans
Apicomplexa	Major group of sporozoans
PHOTOAUTOTROPHIC AND HETEROTROPHIC	
Euglenophyta	Euglenoids
Pyrrhophyta	Dinoflagellates
MOSTLY PHOTOAUTOTROPHIC (photosynthesizers)	
Chrysophyta	Chrysophytes
	Golden algae
	Yellow-green algae
	Diatoms
	Coccolithophores
Rhodophyta	Red algae
Phaeophyta	Brown algae
Chlorophyta	Green algae

Review Questions

1. Outline some of the general characteristics of protistans. Then explain what is meant by this statement: Protistans are often classified by what they are *not*. *CI*

2. Review Table 23.3, then cover the right column with a sheet of paper. Now list the common names for the major categories of protistans. *Table 23.3*

3. Correlate some structural features of a protistan from each of the groups listed with conditions in their environments:
   a. chytrids, water molds *23.1*
   b. slime molds *23.1*
   c. amoeboid protozoans *23.3*
   d. ciliated protozoans *23.4*
   e. animal-like flagellates, sporozoans *23.5, 23.6*
   f. euglenoids, chrysophytes, dinoflagellates *23.9, 23.10*
   g. red, brown, and green algae *23.11–23.13*

4. Select three different protistan species, then briefly explain how they adversely affect our affairs, such as crop yields and human health. *23.1, 23.3, 23.5–23.7*

5. Select and briefly explain how activities of photosynthetic protistan species have positive benefits for some communities of organisms, including human communities. *23.10–23.13*

Self-Quiz (*Answers in Appendix III*)

1. Most chytrids and water molds are ______ decomposers in aquatic habitats.
   a. parasitic
   b. saprobic
   c. autotrophic
   d. chemosynthetic

Table 23.4 Comparison of Prokaryotes With Eukaryotes

	Prokaryotes	Eukaryotes
Organisms represented:	Bacteria only	Protistans, fungi, plants, and animals
Ancestry:	Two major, ancient lineages (archaebacteria, eubacteria) that evolved more than 3.5 billion years ago	Equally ancient prokaryotic ancestors gave rise to forerunners of eukaryotes which evolved more than 1.2 billion years ago.
Level of organization:	Single-celled	Protistans, single-celled or multicelled. Nearly all others are multicelled, with a division of labor among differentiated cells, tissues, and often organs.
Typical cell size:	Small (1–10 micrometers)	Large (10–100 micrometers)
Cell wall:	Most have distinctive walls.	Cellulose or chitin; none in animal cells
Membrane-bound organelles:	Very rarely	Typically profuse
Modes of metabolism:	Both anaerobic and aerobic	Aerobic modes predominate.
Genetic material:	Bacterial chromosome (and sometimes plasmids)	Complex chromosomes (DNA, many associated proteins) within a nucleus
Mode of cell division:	Prokaryotic fission, mostly; also budding	Nuclear division (mitosis, meiosis, or both), associated with one of various modes of cytoplasmic division

2. _______ are free-living, amoebalike cells that crawl on rotting plant parts and engulf bacteria, spores, and organic compounds.
   a. water molds   c. sporozoans
   b. amoeboid protozoans   d. slime molds

3. In a _______ life cycle, amoeboid cells aggregate and migrate as a mass. Cells in the mass differentiate, forming reproductive structures and spores or gametes.
   a. slime mold   b. water mold   c. protozoan   d. chytrid

4. Amoebas, foraminiferans, and radiolarians are _______.
   a. ciliated protozoans   c. amoeboid protozoans
   b. animal-like protozoans   d. sporozoans

5. Which disease is associated with trypanosomes, a type of parasitic, flagellated protozoan?
   a. Chagas disease   c. malaria
   b. toxoplasmosis   d. amoebic dysentery

6. Euglenoids and chrysophytes are mostly _______.
   a. photoautotrophic   c. heterotrophic
   b. chemoautotrophic   d. omnivorous

7. Single-celled photosynthetic protistans, including most of the euglenoids, chrysophytes, and dinoflagellates, are members of the _______, the "pastures" of most aquatic habitats.
   a. zooplankton   c. brown algae
   b. red algae   d. phytoplankton

8. Algin is used in ice cream, pudding, salad dressing, jelly beans, beer, cough syrup, toothpaste, cosmetics, and other products. Certain _______ are sources of algin.
   a. green algae   c. red algae
   b. brown algae   d. dinoflagellates

Critical Thinking

1. Suppose you decide to vacation in a developing country where sanitation practices and standards of personal hygiene are poor. Having read about some of the parasitic protozoans lurking in water and damp soil, what would you consider safe to drink once you arrive there? Which kinds of foods might be best to avoid and what kinds of food preparations might make them safe to eat?

2. As you read in this chapter, red tides are associated with "algal blooms." Such blooms may follow the enrichment of aquatic habitats with water that drains into them from heavily fertilized croplands or from concentrated sources of raw sewage. After thinking about it, do you accept the resulting destruction of aquatic species, birds, and other forms of wildlife as an unfortunate but necessary side effect of human activities? If you do not accept it, how would you stop the water pollution?

And if you could stop it, what sorts of measures would you take to feed the enormous human population, which is now extremely dependent on high-yield (and very heavily fertilized) crops? How would you propose to dispose of or prevent the accumulation of fecal matter and other wastes of 6 billion people?

Selected Key Terms

actinopod *23.3*
algal bloom *23.10*
amoeboid protozoan *23.3*
animal-like flagellate *23.5*
binary fission *23.2*
brown alga *23.12*
chrysophyte *23.10*
chytrid *23.1*
ciliated protozoan *23.4*
coccolithophore *23.10*
conjugation (protozoan) *23.4*
contractile vacuole *23.6*
cyst *23.2*
diatom *23.10*
dinoflagellate *23.10*
euglenoid *23.9*
golden alga *23.10*
green alga *23.13*
hypha *23.1*
mycelium *23.1*
pellicle *23.4*
phytoplankton *23.8*
plankton *23.3*
protistan *CI*
protozoan *23.2*
pseudopod *23.3*
red alga *23.11*
red tide *23.10*
rhizopod *23.3*
saprobe *23.1*
slime mold *23.1*
sporozoan *23.6*
water mold *23.1*
yellow-green alga *23.10*

Readings *See also www.infotrac-college.com*

Bold, H., and M. Wynne. 1985. *Introduction to the Algae*. Second edition. Englewood Cliffs, New Jersey: Prentice-Hall. Includes descriptions of the economic importance of major algal groups.

Margulis, L. 1993. *Symbiosis in Cell Evolution*. Second edition. New York: Freeman. Paperback.

Margulis, L., and K. Schwartz. 1992. *Five Kingdoms*. Second edition. New York: Freeman. Paperback.

Satchell, M. 28 July 1997. "The Cell From Hell." *U.S. News and World Report*, 26–28. Among other cases, the article correlates hundreds of millions of gallons of raw sewage from North Carolina's factory-like hog farms with an algal bloom (by the dinoflagellate *Pfiesteria piscicada*) that killed 14 million fish.

24 FUNGI

Ode to the Fungus Among Us

When push comes to shove, plants need certain fungi more than they need us. (In fact, they don't need us at all.) Fungi were there, as symbionts, when plants first invaded the land. **Symbiosis**, recall, refers to species that live closely together. In cases called **mutualism**, their interaction benefits both partners or does one of them no harm. Lichens and mycorrhizae are like this.

A **lichen** is a vegetative body in which a fungus has become intertwined with one or more photosynthetic organisms. From the Antarctic to the Arctic, you find lichens surviving in a variety of habitats that are just too hostile to support most organisms.

Lichens absorb mineral ions from substrates; some absorb nitrogen from the air. They make antibiotics against bacteria that can decompose lichens. They also make toxins against invertebrate larvae that graze on them. Coincidentally, their metabolic products help form new soils or enrich existing ones. This is what happens when lichens colonize new habitats, such as bare soil exposed by a retreating glacier. Conditions improve so much, other species move in and replace the pioneers. Actually, this may have happened when plants first invaded the land. Cyanobacteria-containing lichens even help maintain ecosystems. They put captured nitrogen into a form that plants use. *Lobaria* provides old-growth forests in the Pacific Northwest with up to 20 percent of their required nitrogen (Figure 24.1).

Lichens serve as early warnings of deteriorating environmental conditions. How? They absorb toxins but cannot get rid of them. From extensive studies in New York City and in England, we know that when lichens die around cities, air pollution is getting bad.

Figure 24.1 *Lobaria oregana*, one of the lichens.

a SULFUR SHELF FUNGUS *Polyporus*

Fungi, too, enter into mutualistic interactions with young tree roots. Together they form a **mycorrhiza** (plural, mycorrhizae), which means "fungus-root." The fungus absorbs sugars from the plant, which absorbs minerals from its partner. The underground parts of a fungus thread through soil and afford a huge surface area for absorption. The fungus rapidly takes up many ions of phosphorus and other minerals when they are abundant and releases them to the plant when the ions are scarce. Many plants do not grow well at all in the absence of mycorrhizae.

Fungi help plants in still another way, because many are premier **decomposers**. Figure 24.2 shows just a few of these. Like all other decomposers, fungi break down organic compounds in their surroundings. However, few organisms besides fungi digest their dinner out on the table, so to speak. As fungi grow in or on organic matter, they secrete enzymes that digest it into bits that individual cells can absorb. Such a mode of nutrition is called **extracellular digestion and absorption**. Plants —the primary producers of nearly all ecosystems—also benefit because they can take up some of the released carbon and other nutrients.

Keep this global perspective in mind. Why? As you will see, the activities of some fungi do cause diseases in humans, pets and farm animals, ornamental plants,

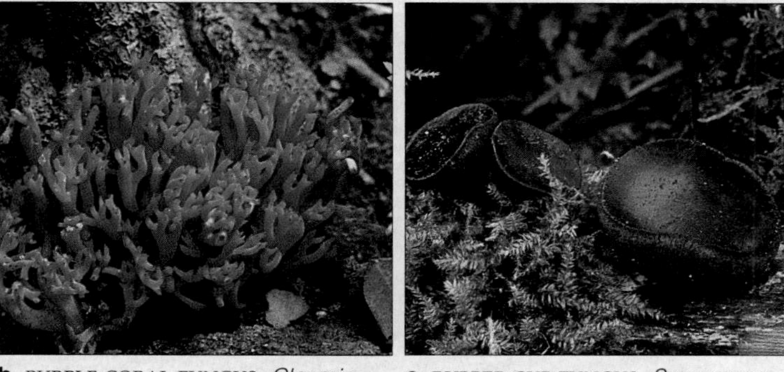
b PURPLE CORAL FUNGUS *Clavaria* **c** RUBBER CUP FUNGUS *Sarcosoma*

d BIG LAUGHING MUSHROOM *Gymnophilus*

e TRUMPET CHANTERELLE *Craterellus*

f SCARLET HOOD *Hygrophorus*

Figure 24.2 Fungal species from southeastern Virginia. This sampling hints at the rich diversity within the kingdom Fungi.

and important crop plants. Some species are notorious spoilers of food supplies. Others help us manufacture substances ranging from antibiotics to cheeses. We tend to assign "value" to fungi and other organisms in terms of their direct effect on our lives. There is nothing wrong with battling dangerous species and admiring beneficial ones, as long as we do not lose sight of the greater roles of fungi or any other kind of organism in nature.

KEY CONCEPTS

1. Fungi are heterotrophs. Together with heterotrophic bacteria, they are the biosphere's decomposers. Saprobic types obtain nutrients from nonliving organic matter. Parasitic types obtain them from tissues of living hosts.

2. Fungi secrete enzymes that digest food outside their body, then fungal cells absorb breakdown products. Their metabolic activities release carbon dioxide to the atmosphere and return many nutrients to the soil, where they become available to producer organisms.

3. Most fungi are multicelled. A mycelium, which is the food-absorbing portion of a fungal body, develops during the fungal life cycle. Each mycelium is a mesh of hyphae. The hyphae are elongated filaments that grow and develop by repeated mitotic cell divisions.

4. Commonly, a portion of the fungal hyphae becomes modified and weaves together to form a reproductive structure in or upon which fungal spores develop. A "mushroom" is such a structure. Germinating spores grow and develop into a new mycelium.

5. Many fungi are symbionts. Lichens consist of certain fungi that are partnered with algae and other organisms. Mycorrhizae consist of fungal species locked in mutually beneficial relationships with young roots of land plants. Metabolic activities of cells making up the fungal hyphae provide the plants with nutrients, and the plants provide the fungi with carbohydrates.

6. We tend to assign value to plants and fungi in terms of their direct effect on our lives. Our battles with the "bad" ones and reliance on the "good" ones should start from a solid understanding of their long-established roles in nature.

24.1 CHARACTERISTICS OF FUNGI

Mode of Nutrition

Fungi are heterotrophs, meaning they require organic compounds that other organisms synthesize. Most are **saprobes**; they obtain nutrients from nonliving organic matter and so cause its decay. Others are **parasites**; they extract nutrients from tissues of a living host. When the cells of any species grow in or on organic matter, they secrete digestive enzymes and then absorb breakdown products. Again, their mode of extracellular digestion helps plants, which readily absorb some of the released nutrients and carbon dioxide by-products. Without the fungi and heterotrophic bacteria, communities would slowly become buried in their own garbage, nutrients would not be cycled, and life could not go on.

Major Groups

For many of us, "fungi" are drab mushrooms sold in grocery stores. These are simply fungal body parts, and they are produced by a fungus with stunningly diverse relatives. The few species in Figure 24.2 don't do justice to the 56,000 fungal species we know about—and there may be at least a million more we don't know about!

We know, from the fossil record, that fungi evolved before 900 million years ago. Some accompanied plants onto the land 430 million years ago. About 100 million years later, three major lineages were well established. We call them the **zygomycetes** (Zygomycota), **sac fungi** (Ascomycota), and **club fungi** (Basidiomycota). Other, puzzling kinds known as "imperfect fungi" are lumped together but are not a formal taxonomic group. The vast majority of species in all these groups are multicelled.

Key Features of Fungal Life Cycles

Fungi reproduce asexually quite often, but given the opportunity, they also reproduce sexually. They form great numbers of nonmotile spores. As in plants, their **spores** are reproductive cells or multicelled structures, often walled, that germinate after dispersal from the parent. In multicelled species, spores give rise to a mesh of branched filaments. The mesh, a **mycelium** (plural, mycelia), rapidly grows over or into organic matter and has a good surface-to-volume ratio for food absorption. Each filament in a mycelium is called a **hypha** (plural, hyphae). Hyphal cells commonly have chitin-reinforced walls. Their cytoplasm interconnects, so that nutrients flow unimpeded throughout the mycelium.

Fungi are major decomposers that engage in extracellular digestion and absorption of organic matter. Multicelled types form absorptive mycelia and spore-producing structures.

24.2 CONSIDER THE CLUB FUNGI

A Sampling of Spectacular Diversity

Fungal life cycles and life-styles show dizzying variety. The most we can do here is to sample a few species, starting with the club fungi. The 25,000 or so club fungi include mushrooms, shelf fungi, coral fungi, puffballs, and stinkhorns. Figures 24.2 through 24.5 and 1.7*d* show examples. Some of the saprobic species are important decomposers of plant litter. As described later, other species are symbionts that live in close association with the young roots of trees in forests. The fungal rusts and smuts can destroy fields of wheat, corn, and other crop plants. Cultivation of the common mushroom (*Agaricus brunnescens*) is a multimillion-dollar business. It is the mushroom of grocery-store and pizza-topping fame. Yet some of its relatives produce toxins that can kill you or any other organism that nibbles on them.

Have you ever wondered which organisms are the oldest and the largest? *Armillaria bulbosa* is one of them. The mycelium of one individual, discovered in a forest in northern Michigan, extends through fifteen hectares of soil. (Each hectare is the equivalent of 10,000 square meters.) By some estimates, this fungus weighs more than 10,000 kilograms and has been spreading beneath the forest floor for more than 1,500 years!

a
b

Figure 24.3 Two club fungi. (**a**) The light-red coral fungus *Ramaria*. (**b**) The shelf fungus *Polyporus*. With the exception of the rubber cup fungus, all of the fungal species shown in Figure 24.2 are club fungi.

Further reading: Student Guide to InfoTrac on web site

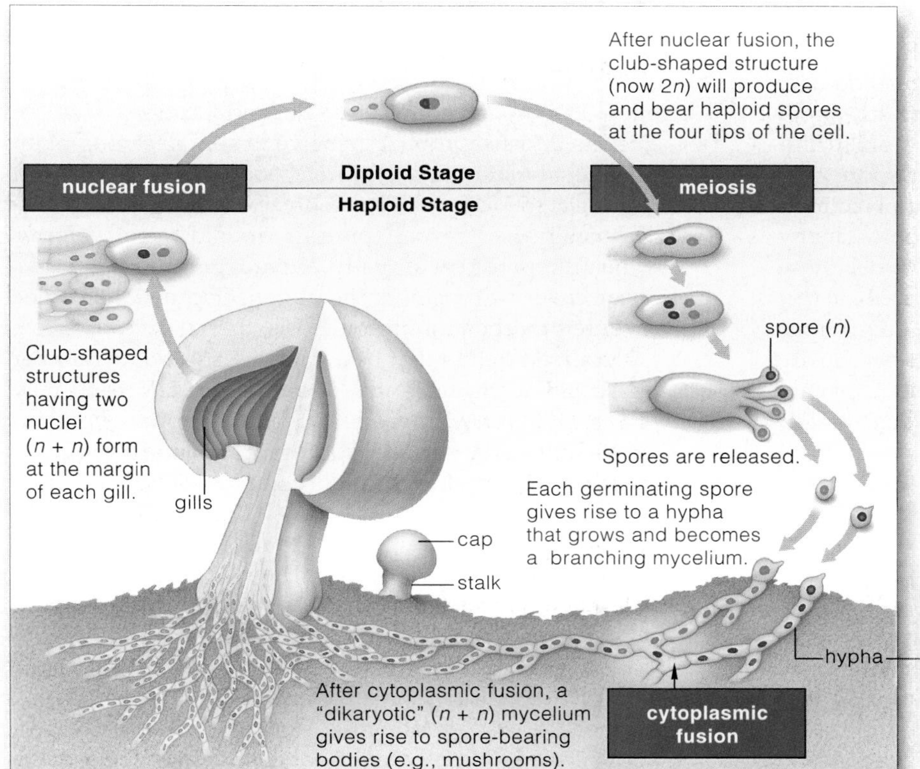
After nuclear fusion, the club-shaped structure (now 2n) will produce and bear haploid spores at the four tips of the cell.
nuclear fusion
Diploid Stage
Haploid Stage
meiosis
spore (n)
Club-shaped structures having two nuclei (n + n) form at the margin of each gill.
gills
Spores are released.
Each germinating spore gives rise to a hypha that grows and becomes a branching mycelium.
cap
stalk
hypha
After cytoplasmic fusion, a "dikaryotic" (n + n) mycelium gives rise to spore-bearing bodies (e.g., mushrooms).
cytoplasmic fusion

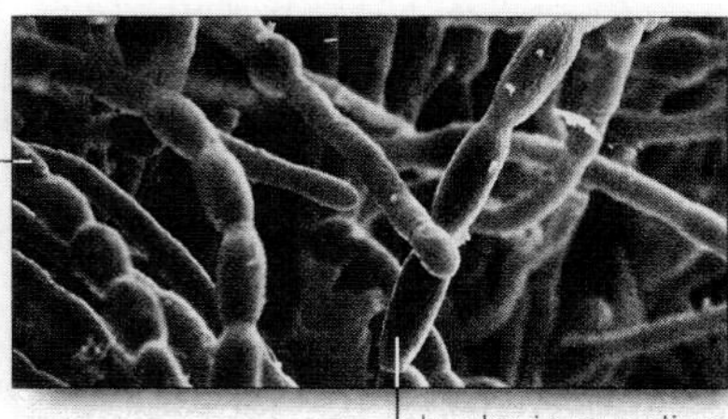
hypha in mycelium

Figure 24.4 Generalized life cycle for many club fungi. When hyphal cells of two compatible mating strains grow together, their cytoplasm (but not nuclei) fuse. Cell divisions result in a dikaryotic mycelium (its cells each have two nuclei). Mushrooms form, with club-shaped, spore-bearing structures on their gill surface. The two nuclei in each structure fuse to form a diploid zygote, which starts the cycle anew.

When you see mushrooms or any other fungus growing outdoors, think twice before devouring them. Unlike cultivated species, such as the common mushroom, many are toxic, including the two species shown in Figure 24.5. *No one should eat any fungi that were gathered in the wild until they have been accurately identified as edible.* As the saying goes, there are old mushroom hunters and bold mushroom hunters, but no old, bold mushroom hunters.

a
b

Figure 24.5 (**a**) Fly agaric mushroom (*Amanita muscaria*). It causes hallucinations when eaten. It was used to induce trances in ancient rituals in Central America, Russia, and India. (**b**) From California, *A. ocreata*. It can be fatal if eaten. A close relative, the death cap mushroom (*A. phalloides*), lives up to its name. Nibble as little as five milligrams of its toxin, and vomiting and diarrhea will begin eight to twenty-four hours later. Your liver and kidneys will degenerate; death may follow within a few days.

Example of a Fungal Life Cycle

Probably you already know what a common mushroom (*A. brunnescens*) looks like, so let's use it as our example of a fungal life cycle. Like most of the other club fungi, this species produces short-lived reproductive bodies—**mushrooms**—that are merely its aboveground parts; its living mycelium is buried in soil or decaying wood. A mushroom has a stalk and a cap. Fine tissue sheets, or gills, line the cap's inner surface. On these gills, club-shaped, spore-bearing structures form. The spores that form here are **basidiospores**. When a spore dispersed from a particular strain of mushroom lands on a suitable site, it germinates and gives rise to a haploid mycelium.

Suppose hyphae of two compatible mating strains make contact. They may undergo cytoplasmic fusion, but their nuclei do not fuse at once. The fused part may be the start of a *dikaryotic* mycelium, in which hyphal cells have one nucleus of each mating type (Figure 24.4). An extensive mycelium forms. And when conditions are favorable, mushrooms form. Each spore-producing structure of the mushroom is initially dikaryotic, but then its two nuclei fuse and form a short-lived zygote. The zygote undergoes meiosis, haploid spores develop outside on small stalks, then air currents disperse them.

Club fungi, the fungal group with the greatest diversity, have distinctive club-shaped, spore-bearing structures.

24.3 SPORES AND MORE SPORES

A fungus has a thing about spores. It produces sexual spores, asexual spores, or both, depending on contact with a suitable hypha, food availability, and how cool or damp conditions are. Its spores are usually small and dry, and air currents disperse them. Each spore that germinates can be the start of a hypha and a mycelium. Stalked reproductive structures may develop on many of the hyphae and produce asexual spores. After these spores germinate, each may be the start of still *another* extensive mycelium. In no time at all, that one fungus and staggering numbers of its descendants are busily decomposing organic stuff or pirating nutrients from a host! Look at what can happen to a slice of stale bread:

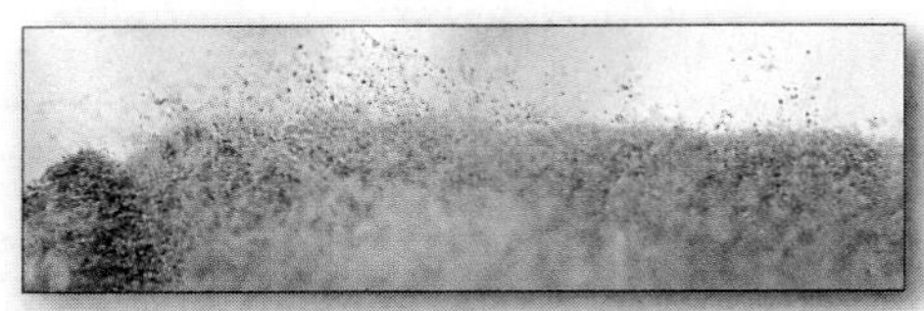

Each fungal class produces unique sexual spores. Club fungi form basidiospores, zygomycetes form spores by way of zygosporangia, and sac fungi form ascospores.

Producers of Zygosporangia

Consider the zygomycetes. Parasitic species feed on insects. Most saprobic types live in soil, decaying plant or animal material, and stored food. You just saw what *Rhizopus stolonifer*, the black bread mold, does to bread. A thick-walled sexual spore—a diploid zygote—forms when it reproduces sexually. A **zygosporangium**, a thin, clear covering, encloses the zygote (Figure 24.6*a*). The zygote proceeds through meiosis, then it gives rise to a specialized hypha that bears a spore sac. Some number of spores form inside the sporangium, and each may give rise to a new mycelium. Stalked hyphae grow out of such mycelia. Asexual spores form inside a spore sac perched on top of each stalk (Figure 24.6*b*).

Producers of Ascospores

We know of more than 30,000 kinds of sac fungi. The vast majority are multicelled. Most form sexual spores called **ascospores** within sac-shaped cells. They alone form these cells, which are called asci (singular, ascus). Reproductive structures that consist of tightly interwoven hyphae enclose the asci of multicelled species. They resemble flasks, globes, and shallow cups. Figures 24.2*c* and 24.7 show some examples.

Figure 24.6 Life cycle of the black bread mold *Rhizopus stolonifer.* Asexual phases are common. Different mating strains (+ and −) also reproduce sexually. Either way, haploid spores form and give rise to mycelia. Chemical attraction between a + hypha and a − hypha causes them to fuse. Two gametangia form, each with several haploid nuclei. Later their nuclei fuse to form a zygote. The zygote develops a thick wall, thus becoming a zygospore, and may remain dormant for several months. Meiosis occurs as the zygospore germinates, and asexual spores form.

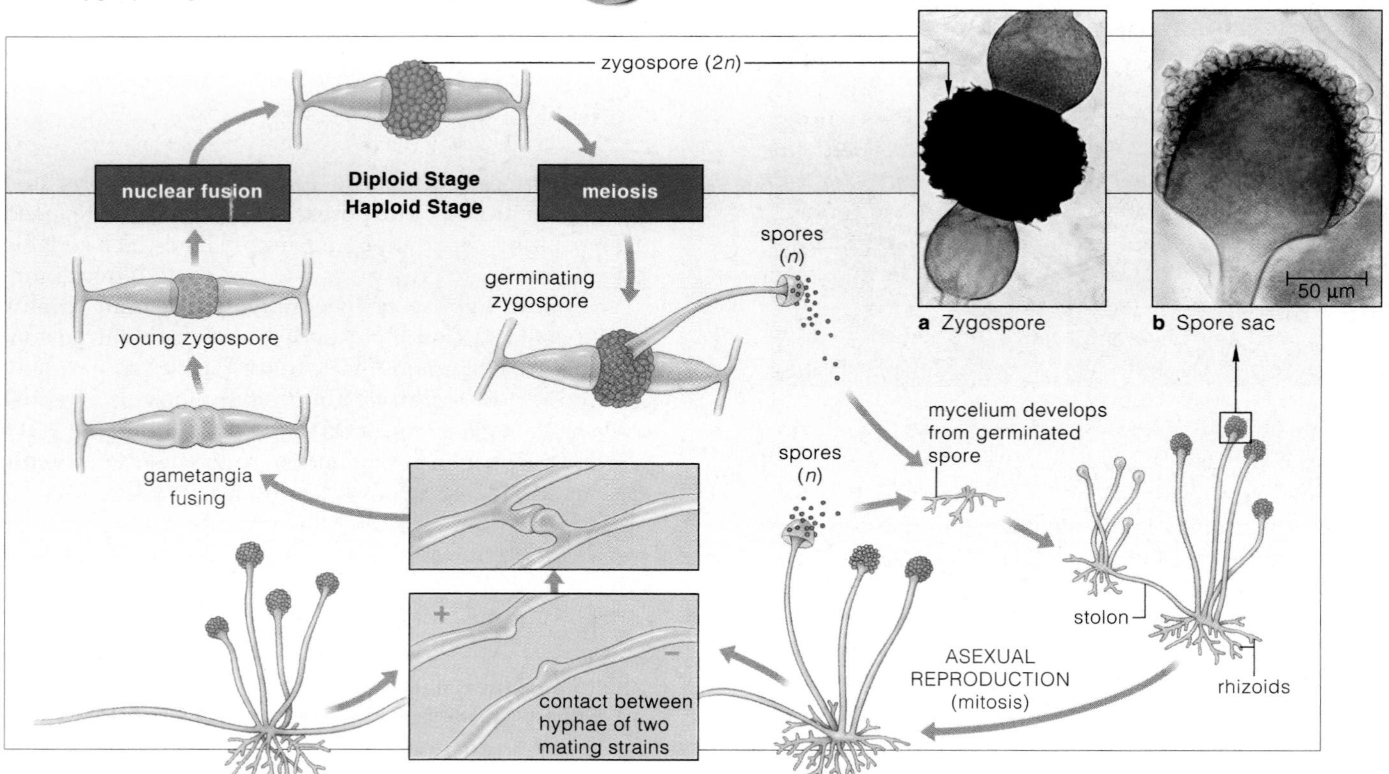
zygospore (2*n*)
nuclear fusion
Diploid Stage
Haploid Stage
meiosis
spores
(*n*)
germinating
zygospore
young zygospore
gametangia
fusing
spores
(*n*)
mycelium develops
from germinated
spore
+
−
contact between
hyphae of two
mating strains
stolon
ASEXUAL
REPRODUCTION
(mitosis)
rhizoids
50 µm
a Zygospore
b Spore sac

Further reading: Student Guide to InfoTrac on web site

ascospore (sexual spore)
spore sac
spore-bearing hypha of this ascocarp
a
b ascocarp
c ascocarp
d conidia (chains of asexual spores)
e budding yeast cell

Figure 24.7 Sac fungi. (**a**) Diagram and (**b**) photograph of *Sarcoscypha coccinia*, the scarlet cup fungus. Saclike structures on the cup's inner surface produce sexual spores (ascospores) by meiosis. (**c**) One of the morels (*Morchella esculenta*). This edible species has a poisonous relative. (**d**) From *Eupenicillium*, chains of asexual spores of a type called conidiospores. These drift away from the chains, like dust, after even the slightest jiggling. "Conidia" means dust. (**e**) Cells of *Candida albicans*, agent of "yeast" infections of the vagina, mouth, intestines, and skin.

The vast majority of sac fungi are multicelled. They include high-priced truffles and morels (Figure 24.7*c*). Truffles are underground symbionts with roots of hazelnut and oak trees. Pigs and dogs are trained to snuffle out truffles in the woods. In France, truffles are now being cultivated on the roots of inoculated seedlings.

Other multicelled sac fungi include certain species of *Penicillium* that "flavor" Camembert and Roquefort cheeses and species that make penicillins, widely used as antibiotics. We use *Aspergillus* to make citric acid for candies and soft drinks, and to ferment soybeans for soy sauce. Most red, bluish-green, and brown fungal molds that spoil stored food also are multicelled. One species, the salmon-colored *Neurospora sitophila*, can run amok in bakeries and in research laboratories. It produces so many spores that it is extremely difficult to eradicate. One of its relatives, *N. crassa*, is an important organism in genetic research.

Sac fungi also include about 500 species of single-celled yeasts (although still other yeasts are classified as club fungi). Yeasts reproduce sexually when two cells fuse and become a spore-producing sac. Some types live in the nectar of flowers and on fruits and leaves. Bakers and vintners put fermenting by-products of vast populations of yeasts to use. For example, the carbon dioxide by-products of *Saccharomyces cerevisiae* leaven bread. The commercial production of wine and beer depends on its ethanol end product. Many yeast strains with desirable properties have been developed through artificial selection and genetic engineering. Then again, *Candida albicans*, a notorious relative of "good" yeasts, causes vexing infections in humans (Figure 24.7*e*).

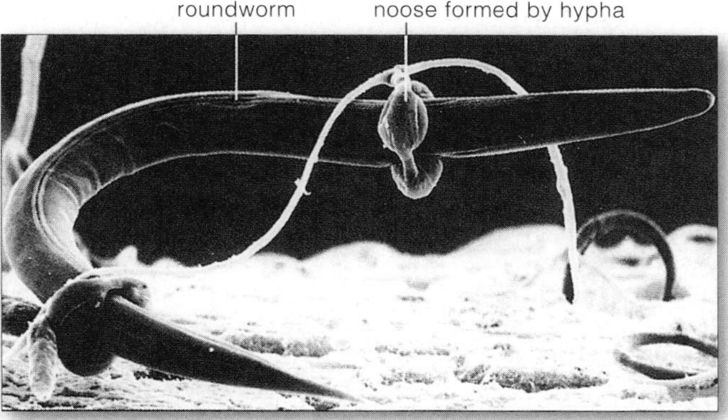
roundworm
noose formed by hypha

Figure 24.8 *Arthrobtrys dactyloides*, an imperfect fungus. This predatory species forms a nooselike ring that swells rapidly with turgor pressure when stimulated. The "hole" in the noose shrinks and captures a worm, into which hyphae will grow.

Elusive Spores of the Imperfect Fungi

Imperfect fungi are set aside in a taxonomic holding station not because they are somehow defective but mainly because no one has yet discovered what kind of sexual spores they produce (if any). Figure 24.8 shows one of the species, a puzzling predatory fungus, that is awaiting formal classification. Investigators recently reunited the previously orphaned *Aspergillus, Candida,* and *Penicillium* with their kin—other sac fungi.

Through their exuberant and rapid production of asexual and sexual spores, fungi take quick advantage of available organic matter, whether it has been discarded or is part of a living or dead organism. Their penchant for spore production is central to their success as decomposers and parasites.

24.4 THE SYMBIONTS REVISITED

Recall, from the introduction, that symbiosis refers to species that live together in close ecological association. Often one is a parasite's victim, not a partner. In cases of mutualism, interaction benefits both partners or does one of them no harm. Here are more detailed examples.

Lichens

In the single vegetative body called a lichen, a fungus is intertwined with one or more photosynthetic species. The fungal part is the *myco*biont. The photosynthetic part is the *photo*biont. Of about 13,500 known types of lichens, nearly half incorporate sac fungi. Only 100 or so species serve as photobionts, and most often these are green algae and cyanobacteria.

A lichen forms after the tip of a fungal hypha binds with a suitable host cell. Both lose their wall, and their cytoplasm fuses or the hypha induces the host cell to cup around it. The mycobiont and the photobiont grow and multiply together. The lichen commonly has distinct layers. The overall pattern of growth may be leaflike, flattened, pendulous, or erect (Figures 24.1 and 24.9).

Lichens typically colonize sites that are hostile for most organisms, including sunbaked or frozen rocks, fence posts, gravestones, and plants, even the tops of

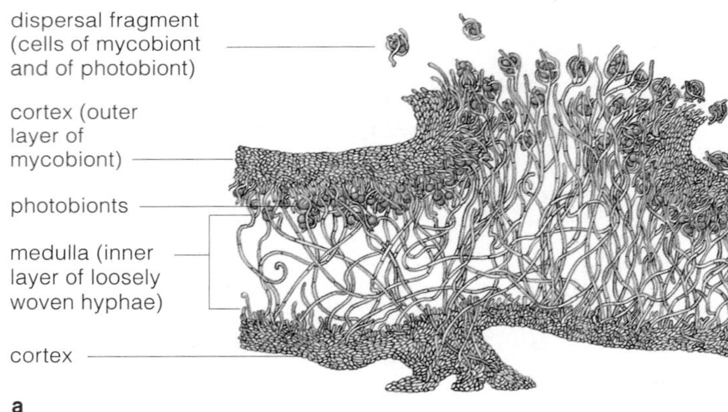
dispersal fragment (cells of mycobiont and of photobiont)
cortex (outer layer of mycobiont)
photobionts
medulla (inner layer of loosely woven hyphae)
cortex
a

b

c

d

e

Figure 24.9 (**a**) Sketch of a stratified lichen, cross-section. (**b**) Encrusting lichens on a rock. (**c**) Leaflike lichen on a birch tree. (**d**) *Usnea*, a pendant lichen commonly called old man's beard. (**e**) *Cladonia rangiferina*, an erect, branching lichen.

Further reading: Student Guide to InfoTrac on web site

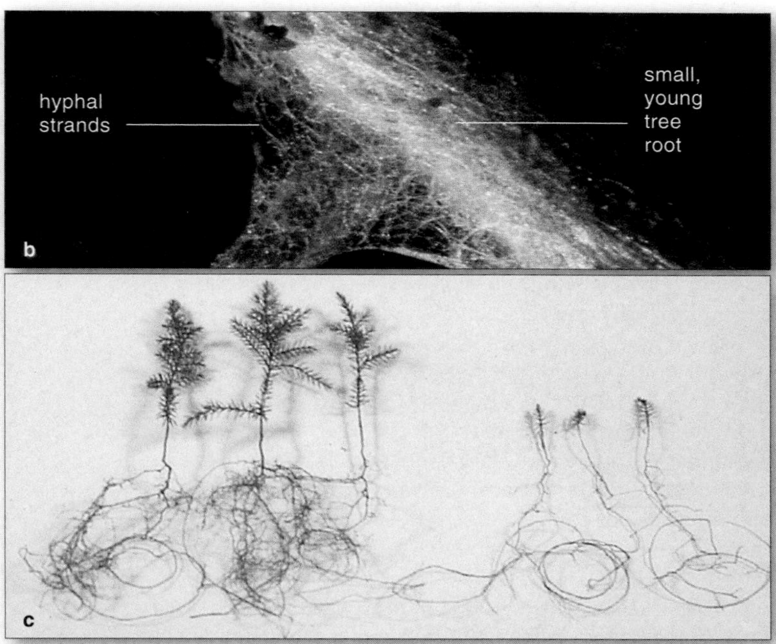
hyphal
strands
small,
young
tree
root

Figure 24.10 (**a**) Lodgepole pine (*Pinus contorta*) seedling, longitudinal section. Notice the extent of the mycorrhiza compared to the shoot system, which is only about four centimeters tall. (**b**) The mycorrhiza of a hemlock tree. (**c**) Effects of the presence or absence of mycorrhizae on plant growth. The juniper seedlings at the left are six months old. They were grown in sterilized, phosphorus-poor soil with a mycorrhizal fungus. The seedlings at the right were grown without the fungus.

giant Douglas firs. Almost always, the fungus is the largest component. Cyanobacteria typically reside in a separate structure inside or outside the main body. The fungus gets a long-term source of nutrients, which it absorbs from photobiont cells. Nutrient withdrawals affect the photobiont's growth a bit, but the lichen may help shelter it. If more than one fungus is present in the lichen, it might be a mycobiont, a parasite, or even an opportunist that is using the lichen as a substrate.

Mycorrhizae

Fungi, recall, also are mutualists with young tree roots, as mycorrhizae (Figure 24.10*a*). Without mycorrhizae, plants cannot grow as efficiently. In *ecto*mycorrhizae, hyphae form a dense net around living cells in roots but do not penetrate them (Figure 24.10*b*). Other hyphae form a velvety wrapping around roots as the mycelium radiates through soil. Ectomycorrhizae are common in temperate forests. They help the trees survive seasonal shifts in temperature and rainfall. About 5,000 fungal species, mostly club fungi, enter into such associations.

The more common *endo*mycorrhizae form in about 80 percent of all vascular plants. These fungal hyphae penetrate plant cells, as they do in lichens. Fewer than 200 species of zygomycetes serve as the fungal partner. Their hyphae branch extensively, forming tree-shaped absorptive structures in plant cells. Hyphae also extend for several centimeters into the soil. Chapter 30 offers a closer look at these beneficial species.

As Fungi Go, So Go the Forests

Since the early 1900s, collectors have recorded data on wild mushroom populations in European forests. As the records tell us, the number and kinds of fungi are declining at alarming rates. Mushroom gatherers can't be the cause, because inedible as well as edible species are vanishing. However, the decline does correlate with rising air pollution. Vehicle exhaust, smoke from coal burning, and emissions from nitrogen fertilizers pump ozone, nitrogen oxides, and sulfur oxides into the air. Normally as a tree ages, one species of mycorrhizal fungus gives way to another, in predictable patterns. When fungi die, trees lose this vital support system, and they become vulnerable to severe frost and drought. Are the North American forests at risk, also? Conditions there are deteriorating in comparable ways.

Lichens and mycorrhizae are both symbiotic associations between fungi and other organisms, with mutual benefits.

Focus on Health

24.5 A LOOK AT THE UNLOVED FEW

You know you are a serious student of biology when you view organisms objectively in terms of their place in nature, not in terms of their impact on humans in general and you in particular. As a student you salute saprobic fungi as vital decomposers and praise parasitic fungi that help keep populations of harmful insects and weeds in check. The true test is when you open the fridge to get a bowl of high-priced raspberries and discover a fungus beat you to them. The true test is when a fungus starts feeding on warm, damp tissues between your toes and turns skin scaly, reddened, and cracked (Figure 24.11*a*).

Which home gardeners wax poetic about black spot or powdery mildew on their roses? Which farmers happily give up millions of dollars a year to sac fungi that attack corn, wheat, peaches, and apples (Figure 24.11*b*)? Who cares that the sac fungus *Cryphonectria parasitica* blitzed most of the chestnut trees in eastern North America, leaving those once-magnificent trees to sprout as stubby versions of their former selves?

Who willingly inhales airborne spores of *Ajellomyces capsulatus*? After landing on soil, these dimorphic beasties form mycelia. When they alight in moist lung tissues, they form populations of yeastlike cells that cause a respiratory disease, *histoplasmosis*. Macrophages converge on infected tissues and engulf invading cells. Usually the defenders eliminate the threat, but their aggregations become calcified as the tissue heals. With heavy spore exposure, pneumonia develops. In rare cases, calcified masses form in the lymph nodes, liver, and other organs. Progressive histoplasmosis is usually lethal. The fungus thrives in regions where the nitrogen-rich droppings of many chickens, starlings, pigeons, and other birds pile up, as in spore-containing regions drained by the Ohio and Mississippi rivers. Kick up dust here and you put yourself at risk. Go spelunking in caves where roosting bats busily pile up droppings on the cave floor, and you invite the spores into your lungs.

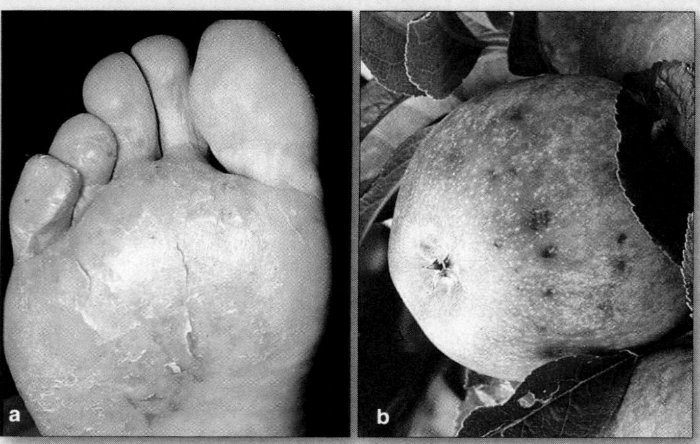

Figure 24.11 Love those fungi! (**a**) Athlete's foot, courtesy of *Epidermophyton floccosum*. (**b**) Apple scab, trademark of *Venturia inaequalis*.

Table 24.1 Some Pathogenic and Toxic Fungi*

ZYGOMYCETES	
Rhizopus	Food spoilage
ASCOMYCETES	
Ajellomyces capsulatus	Histoplasmosis
Aspergillus (some)	Aspergilloses (allergic reactions, sinus, ear, and lung infections; *A. flavus* toxin linked to cancers)
Candida albicans	Infection of mucous membranes
Claviceps purpurea	Ergot of rye, ergotism
Coccidioides immitis	Valley fever
Cryphonectria parasitica	Chestnut blight
Microsporum, Trichophyton, Epidermophyton	Various species cause ringworms, of scalp, body, nails, beard, athlete's foot (Figure 24.11*a*)
Monilinia fructicola	Brown rot of peaches, other stone fruits
Ophiostoma ulmi	Dutch elm disease
Venturia inaequalis	Apple scab (Figure 24.11*b*)
Verticillium	Plant wilt
BASIDIOMYCETES	
Amanita (some species)	Severe mushroom poisoning
Puccinia graminis	Black stem wheat rust
Tilletia indica	Smut of cereal grains
Ustilago maydis	Smut of corn

* After C. Alexopoulos, C. Mims, and M. Blackwell. 1996.

Household molds cause asthma as well as sinus, ear, and lung infections, hearing loss, memory loss, dizziness, and bleeding from the lungs. They have boosted asthma rates 300 percent in the past twenty years and have been linked to nearly 100 percent of the chronic sinus infections. *Stachybotrys atra*, *Memnoliella*, *Cladosporium*, and some *Aspergillus* and *Penicillium* species are in this nasty group.

Certain fungi even have had impact on human history. Consider one notorious species, *Claviceps purpurea*, which parasitizes rye and other cereal grains. Give it credit; we use some of its by-products (alkaloids) to treat migraine headaches and, following childbirth, to shrink the uterus to prevent hemorrhaging. However, the alkaloids are toxic when ingested in large amounts. Eat a lot of bread made with tainted rye flour and you end up with *ergotism*. The symptoms include vomiting, diarrhea, hallucinations, hysteria, and convulsions. Untreated, the disease can turn limbs gangrenous, and it can end in death.

Ergotism epidemics were common in Europe during the Middle Ages, when rye was a major crop. Ergotism also thwarted Peter the Great, the Russian czar who was obsessed with conquering ports along the Black Sea for his nearly landlocked empire. Soldiers laying siege to the ports ate mostly rye bread and fed rye to their horses. The former went into convulsions and the latter into "blind staggers." Possibly, outbreaks of ergotism were used as an excuse to launch witch-hunts in early American colonies.

Further reading: Student Guide to InfoTrac on web site

24.6 SUMMARY

1. Fungi are heterotrophs and major decomposers. The saprobes feed on nonliving organic matter; parasites feed on the tissues of living organisms. Some species of fungi are symbiotic partners with other organisms. The cells of all species secrete digestive enzymes that break down food to small molecules, which the cells absorb.

2. Nearly all fungi are multicelled. The food-absorbing portion, the mycelium, consists of a mesh of filaments (hyphae). Aboveground reproductive structures, such as mushrooms, form from tightly interwoven hyphae.

3. The major groups of fungi are the zygomycetes, the ascomycetes (sac fungi), and the basidiomycetes (club fungi). Each is characterized by distinctive sexual and asexual spores. When a sexual phase cannot be detected or is absent from the life cycle, a fungus is assigned to an informal category called the imperfect fungi.

4. Lichens are mutualistic associations of fungi with photosynthetic partners (green algae and cyanobacteria, mostly). Mycorrhizae are mutualistic associations of a fungus and the young roots of plants. Fungal hyphae provide nutrients for their symbiont, which provides the fungus with carbohydrates.

Review Questions

1. Describe the fungal mode of nutrition, and explain how the structure of mycelia facilitates this mode. *24.1*
2. How does a lichen differ from a mycorrhiza? *CI, 24.4*

Self-Quiz (*Answers in Appendix III*)

1. New mycelia form after ______ germinate.
   a. hyphae b. mycelia c. spores d. mushrooms
2. A "mushroom" is ______ .
   a. the food-absorbing part of a fungal body
   b. the part of the fungal body not constructed of hyphae
   c. a reproductive structure
   d. a nonessential part of the fungus
3. A mycorrhiza is a ______ .
   a. fungal disease of the foot c. parasitic water mold
   b. fungus-plant relationship d. fungus of barnyards
4. Parasitic fungi obtain nutrients from ______ .
   a. tissues of living hosts c. only living animals
   b. nonliving organic matter d. none of the above
5. Saprobic fungi derive nutrients from ______ .
   a. nonliving organic matter c. root hairs
   b. living organisms d. both b and c
6. Match the terms appropriately.

____ zygomycetes	a. mushrooms, shelf fungi
____ conidia	b. sac-shaped cells
____ hypha	c. chains of asexual spores in *Eupenicillium*
____ club fungi	d. each filament in a mycelium
____ asci	e. black bread mold
____ sac fungi	f. truffles, morels, some yeasts

Figure 24.12 Reproductive structures of *Pilobolus*, a name from a Greek word for "hat-thrower." The "hats" actually are spore sacs.

Critical Thinking

1. *Pilobolus* is a type of fungus that commonly dines on horse dung. Each morning, stalked reproductive hyphae emerge from irregularly spaced piles of dung. By early afternoon, they have dispersed spores to sunlit grasses where horses feed. The spores pass through the horse gut unharmed and exit with their own pile of dung. At the tip of each stalked hypha is a dark-walled, spore-containing sac (Figure 24.12). Just below the sac, the stalk is differentiated into a vesicle, swollen with a fluid-filled central vacuole. At the base of the vesicle is a ring of light-sensitive, pigmented cytoplasm. The stalk bends as it grows until its wall is parallel with the sun's rays and light strikes all of the ring. When that happens, turgor pressure builds up inside the central vacuole until the vesicle ruptures. The forceful blast can propel spore sacs two meters away—which is amazing, considering that the stalk is less than ten millimeters tall. Reflect on the examples of fungi discussed in this chapter. Would you say *Pilobolus* is a zygomycete, club fungus, or sac fungus?

2. Diana sees in the laboratory that the fungus *Trichoderma* grows well in distilled water. It continues to do so even after she rigorously treats the water and glassware to remove all traces of organic carbon. This fungus is not a photoautotroph. Suggest a metabolic life-style that lets it grow under these conditions.

3. *Trichoderma* is being tested as a natural pest control agent. Laboratory experiments demonstrated that some strains of this fungus combat other fungi that cause plant diseases. Some even promote seed germination and plant growth. During one set of twenty trials, workers increased lettuce yields by 54 percent. What concerns must be addressed before *Trichoderma* can be released into the environment for commercial applications?

Selected Key Terms

ascospore *24.3*
basidiospore *24.2*
club fungus (basidiomycetes) *24.1*
decomposer *CI*
extracellular digestion and absorption *CI*
fungus *24.1*
hypha *24.1*
lichen *CI*
mushroom *24.2*
mutualism *CI*
mycelium *24.1*
mycorrhiza *CI*
parasite *24.1*
sac fungus (ascomycetes) *24.1*
saprobe *24.1*
spore (fungal) *24.1*
symbiosis *CI*
zygomycetes *24.1*
zygosporangium *24.3*

Readings *See also www.infotrac-college.com*

Moore-Landecker, E. 1996. *Fundamentals of the Fungi*. Fourth edition. Englewood Cliffs, New Jersey: Prentice-Hall.

25 PLANTS

Pioneers In a New World

Seven hundred million years ago, no shorebirds stirred and noisily announced the dawn of a new day. There were no crabs to clack their claws together and skitter off to burrows. The only sounds were the rhythmic muffled thuds of waves in the distance, at the outer limits of another low tide.

More than 3 billion years before, life had its beginning somewhere in the waters of the Earth. And now, quietly, the invasion of the land was under way.

Why did it happen? Astronomical numbers of photosynthetic cells had come and gone, and the oxygen-producing types had slowly changed the atmosphere. High above the Earth, the sun's energy had converted much of the oxygen into a dense ozone layer. That layer became a shield against lethal doses of ultraviolet radiation, which had kept early organisms beneath the water's surface.

Were cyanobacteria the first to adapt to intertidal zones, where mud dried out with each retreating tide? Were they the first to spread into shallow, freshwater streams flowing down to the coasts? Probably so. From fossil evidence, we know that later in time, green algae and fungi made the same journey together.

Every plant around you today is a descendant of ancient species of green algae that lived near the water's edge or made it onto the land. Diverse fungi still associate with nearly all of them. Together, plants and fungi became the basis of communities in coastal lowlands, near the snow line of mountains, and in just about all places in between (Figure 25.1).

a

Figure 25.1 (**a**) Filaments of a green alga, massed in a shallow stream. More than 400 million years ago, green algal species that might have been ancestral to all plants, past and present, lived in similar streams that meandered down to the shores of early continents. (**b**) One land-dwelling descendant of those ancestral forms—a Ponderosa pine high above the floor of Yosemite Valley in the Sierra Nevada of California. (**c**) Flowers of one of the most highly prized flowering plants—orchids—growing on a branch of a living tree in a tropical rain forest. With this chapter, we turn to the beginning—and end of the line—of some ancient lineages.

We have a few tantalizing fossils of the first pioneers. We also are learning about them through comparative biochemistry and studies of existing species. Today, as in the late Precambrian, cyanobacteria and green algae grow in mats in nearshore waters and on the banks of freshwater streams (Figure 25.1*a*). After a volcano erupts or a glacier retreats, cyanobacteria are the first to colonize the barren rocks. Symbiotic associations between green algae and fungi follow. Gradually their organic products and remains accumulate and create pockets of soil. Then mosses and other species of plants can become established in the newly forming soil and further enrich it.

With this chapter we turn to the plant kingdom. Nearly all plants are multicelled photoautotrophs. They absorb energy from the sun, carbon dioxide from the air, and some minerals dissolved in water to synthesize organic compounds. These metabolic wizards also can split water molecules. In doing so, they get stupendous numbers of the electrons and hydrogen atoms required for growth into multicellular forms as tall as the giant redwoods, as extensive as an aspen forest that is one continuous clone.

We know of at least 295,000 kinds of existing plants. Be glad their ancient ancestors left the water. Without them, we humans and other land-dwelling animals never would have made it onto the evolutionary stage.

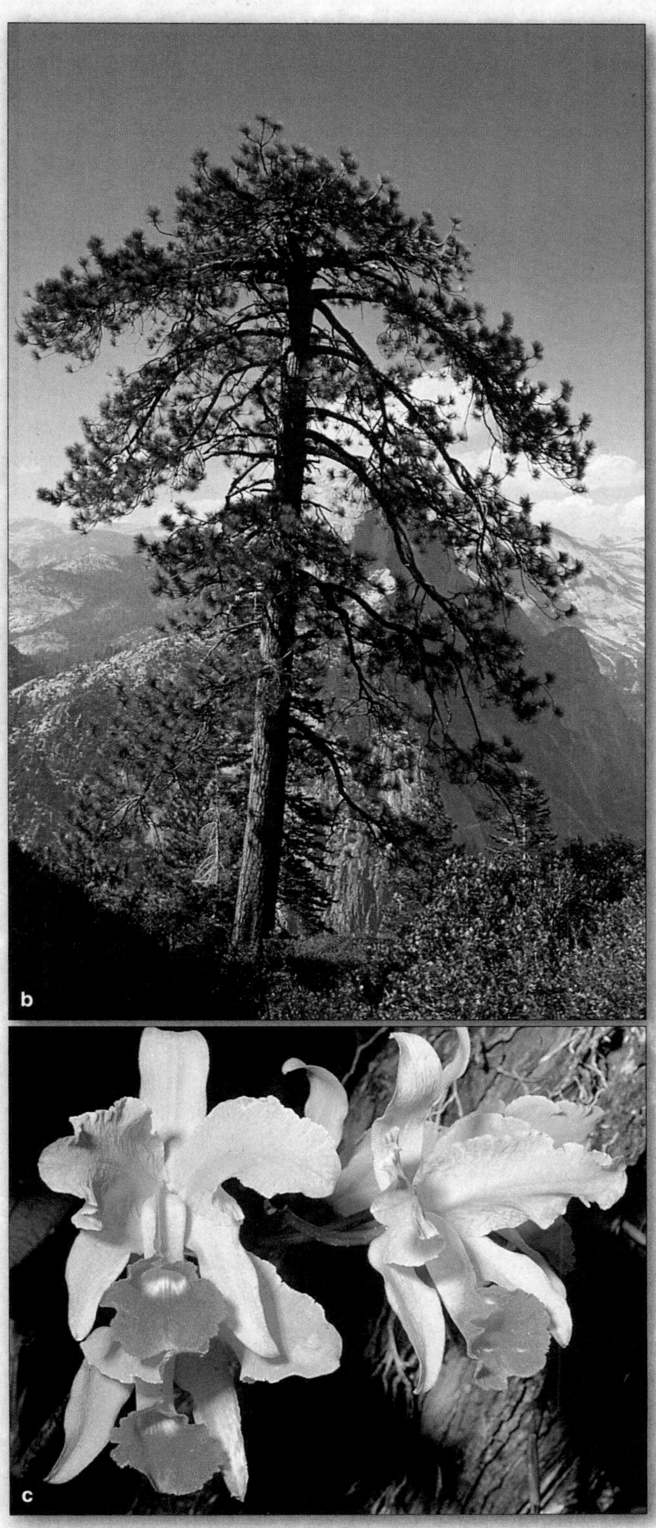

KEY CONCEPTS

1. With very few exceptions, the plant kingdom consists of multicelled photoautotrophs. From earlier chapters, you know that plants use chlorophylls *a* and *b* as their main photosynthetic pigments. In this respect they are like green algae, which are their closest relatives.

2. Unlike their algal ancestors, which were adapted to aquatic habitats, nearly all existing plants live on land.

3. In general, plants are structurally adapted to intercept sunlight, absorb water and mineral ions, and conserve water. Their lignin-reinforced tissues permit upright growth. Root systems mine the soil for water and ions, and internal tissues conduct water and solutes to all living cells in belowground and aboveground parts.

4. Land plants are reproductively adapted to withstand dry periods. During the life cycle, a sporophyte develops roots, stems, and leaves. It holds on to its developing gametes and supplies them with water and food resources. And it disperses the new generation in ways that are responsive to the conditions of specific habitats.

5. Early divergences gave rise to the bryophytes, then seedless vascular plants, and then seed-bearing vascular plants. Of these categories, the seed producers were the most successful in radiating into drier environments.

6. The seed-bearing vascular plants called gymnosperms include the cycads, ginkgos, gnetophytes, and conifers. The angiosperms, another group of vascular plants, bear flowers as well as seeds. There are two classes of flowering plants, informally called the dicots and monocots.

25.1 EVOLUTIONARY TRENDS AMONG PLANTS

Overview of the Plant Kingdom

The plant kingdom includes at least 295,000 species of photoautotrophs and a few heterotrophs. Most kinds are **vascular plants**, defined in part by internal tissues that conduct water and solutes through roots, stems, and leaves. Fewer than 19,000 species are *non*vascular plants called **bryophytes**. Plants, like photoautotrophic bacteria and protistans, are producers for ecosystems.

Liverworts, hornworts, and mosses are bryophytes. The whisk ferns, lycophytes, horsetails, and ferns are *seedless* vascular plants. Cycads, ginkgos, gnetophytes, and conifers belong to a group of *seed-bearing* vascular plants called **gymnosperms**. The **angiosperms**, another group of vascular plants, bear flowers and seeds. Dicots and monocots are two classes of flowering plants.

The ancestors of plants evolved in the seas by 700 million years ago. About 265 million more years passed before simple stalked plants appeared along coasts and streams. Evolutionarily, the pace picked up after that. Within 60 million years, plants radiated through much of the land. Long-term changes in their structure and reproduction explain how the diversity came about.

Evolution of Roots, Stems, and Leaves

Simple underground structures started to evolve when plants first colonized the land. In the lineages that led to vascular plants, they developed into root systems. Most **root systems** have many underground absorptive structures that collectively afford a large surface area. These rapidly take up soil water and dissolved mineral ions. In many species, root systems anchor the plant.

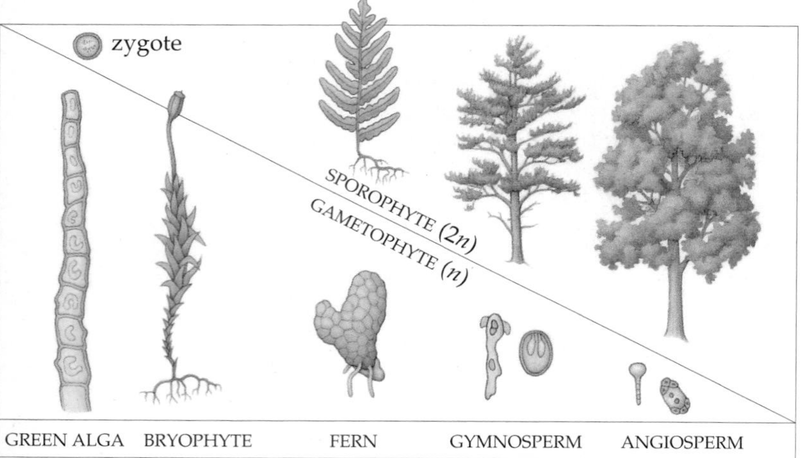
zygote
SPOROPHYTE (2n)
GAMETOPHYTE (n)
GREEN ALGA
BRYOPHYTE
FERN
GYMNOSPERM
ANGIOSPERM

Figure 25.2 Evolutionary trend among plants, from gametophyte (haploid) dominance to sporophyte (diploid) dominance in the life cycle. These representative existing species range from a green alga (*Ulothrix*) to a flowering plant. The trend occurred when early plants were colonizing habitats on land. See also Section 10.5.

Aboveground, **shoot systems** evolved. These have stems and leaves, which efficiently absorb energy from the sun and carbon dioxide from the air. Stems grew and branched extensively only after plants developed a biochemical capacity to synthesize and deposit **lignin**, an organic compound, in cell walls. Collectively, cells with lignified walls are strong enough to structurally support stems, which grow in patterns that increase the total light-intercepting surface of leaves.

Cellular pipelines for water and solutes evolved in many plants. The pipelines were major factors in the evolution of roots, stems, and leaves. They developed as components of xylem and phloem, which are two vascular tissues. **Xylem** distributes water and dissolved ions to all of the plant's living cells. **Phloem** distributes dissolved sugars and other photosynthetic products.

Life on land also depended on water conservation, which had not been a problem in most aquatic habitats. Shoots became protected by a **cuticle**, a waxy coat that helps conserve water on hot, dry days. Also, **stomata** (singular, stoma), tiny openings across the surfaces of leaves and some stems, helped control the absorption of carbon dioxide and restrict evaporative water loss. Later chapters describe these tissue specializations.

From Haploid to Diploid Dominance

As early plants radiated into higher, drier places, their life cycles changed. Think about the gametes of algae, which can get together only in the presence of liquid water. As earlier chapters showed, a *haploid* (n) phase in the form of **gametophytes** (gamete-producing bodies) dominates their life cycles. The diploid ($2n$) phase is the zygote, which forms when gametes fuse at fertilization.

Now look at Figure 25.2. *In most plant life cycles, the diploid phase dominates.* After a diploid zygote forms at fertilization, mitotic cell divisions and cell enlargements transform it into a multicelled diploid body, of a type called a **sporophyte**. Pine trees are an example. In time, some cells of the sporophyte undergo meiosis and give rise to haploid cells of a type called **spores** (sporophyte means spore-producing body). Later, the spores divide by way of mitosis and give rise to the gametophytes.

The shift to diploid dominance was an adaptation to land habitats, most of which show seasonal changes in the availability of free water and dissolved nutrients. Long ago in those challenging habitats, natural selection must have favored sporophytes with well-developed root systems. Young roots of such systems interact with fungal symbionts (Section 24.4). The association, called a mycorrhiza, enhances the plant's uptake of water and scarce minerals, even during dry seasons.

Unlike algae and bryophytes, vascular plants have a sporophyte that is larger and structurally far more

Further reading: Student Guide to InfoTrac on web site →

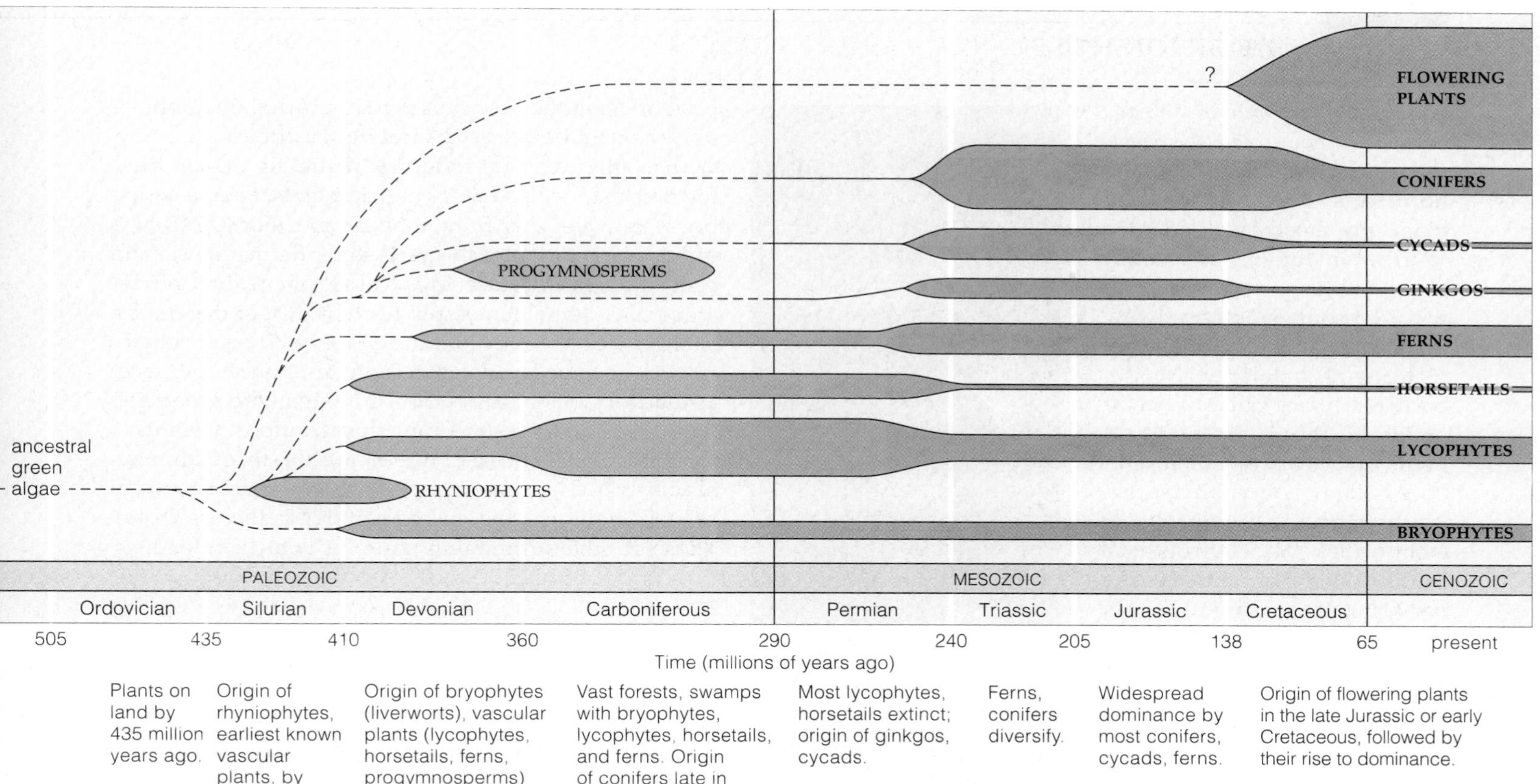

FLOWERING PLANTS
?
CONIFERS
CYCADS
PROGYMNOSPERMS
GINKGOS
FERNS
HORSETAILS
LYCOPHYTES
ancestral green algae
RHYNIOPHYTES
BRYOPHYTES
PALEOZOIC
MESOZOIC
CENOZOIC
Ordovician
Silurian
Devonian
Carboniferous
Permian
Triassic
Jurassic
Cretaceous
505
435
410
360
290
240
205
138
65
present
Time (millions of years ago)
Plants on land by 435 million years ago.
Origin of rhyniophytes, earliest known vascular plants, by mid-Silurian.
Origin of bryophytes (liverworts), vascular plants (lycophytes, horsetails, ferns, progymnosperms) by end of Devonian.
Vast forests, swamps with bryophytes, lycophytes, horsetails, and ferns. Origin of conifers late in Carboniferous.
Most lycophytes, horsetails extinct; origin of ginkgos, cycads.
Ferns, conifers diversify.
Widespread dominance by most conifers, cycads, ferns.
Origin of flowering plants in the late Jurassic or early Cretaceous, followed by their rise to dominance.

Figure 25.3 Milestones in plant evolution. Where a dashed line turns solid, this indicates when the lineage originated. The width of each lineage indicates variations in the range of diversity over time.

complex than the gametophyte. As you will see shortly, the gametophytes of seedless vascular plants develop independently of the sporophyte that produces them. But they protect their gametes, and after fertilization they nourish the embryo sporophytes. The sporophyte became most dominant among the gymnosperms and, later, the angiosperms. It retains, nourishes, and protects developing gametophytes as well as young sporophytes. *And it does so until environmental conditions are suitable for fertilization and for dispersal of the new generation.*

Evolution of Pollen and Seeds

Like some seedless species, seed-bearing plants produce not one but two types of spores. We call this condition *hetero*spory, as opposed to *homo*spory (only one type). In both gymnosperms and angiosperms, one type of spore develops into female gametophytes, where eggs form and become fertilized. The other spore type gives rise to **pollen grains**, which develop into the mature, sperm-bearing male gametophytes. Pollen grains hitch rides on air currents, insects, birds, and so on; they do not require free-standing water to reach the eggs. In this respect they differ greatly from algae. The evolution of pollen grains contributed to the successful radiation of seed-bearing plants into high and dry habitats.

Seed production also was adaptive in drier habitats. Female gametophytes (and eggs) of seed-bearing plants form inside nutritive tissues and a jacket of cell layers. Each **seed** consists of an embryo sporophyte, nutritive tissues, and a protective coat, which develops from the jacket. Seeds can withstand hostile conditions. It was no coincidence that seed plants rose to dominance during Permian times, when shifts in climate were extreme.

Before turning to the spectrum of diversity among plants, take a look at Figure 25.3. You may wish to use it as a map of the branching evolutionary roads.

The plant kingdom includes multicelled, photosynthetic species called bryophytes, seedless vascular plants, and seed-bearing vascular plants. Most of these live on land.

In most lineages, structural adaptations to life on land included root and shoot systems, waxy cuticles, stomata, vascular tissues, and lignin-reinforced tissues.

Sporophytes with well-developed roots, stems, and leaves came to dominate the life cycle of most land plants. Parts of these complex sporophytes nourish and protect fertilized eggs and embryos until conditions favor their growth.

Some plants started to produce two types of spores, not one. This led to the evolution of male gametes well adapted for dispersal without liquid water and to the evolution of seeds.

25.2 THE BRYOPHYTES

Today, the bryophyte lineage consists of about 18,600 species called **mosses**, **liverworts**, and **hornworts**. These nonvascular plants are mostly well adapted to grow in fully or seasonally moist habitats. However, you will find some mosses growing in deserts and on windswept plateaus of Antarctica. Mosses particularly are sensitive to air pollution. Where air quality is poor, mosses are few or absent.

All known bryophytes are less than twenty centimeters, or eight inches, tall. They do have leaflike, stemlike, and rootlike parts, but these do not contain xylem or phloem. Like lichens and some algae, bryophytes can dry out and then revive after absorbing moisture. Most have rhizoids. Rhizoids are elongated cells or threadlike structures that attach gametophytes to the soil and serve as absorptive structures.

Bryophytes are the simplest plants to display three features that evolved early in land plants. *First*, a cuticle prevents water loss from aboveground parts. *Second*, a cellular jacket around the parts that produce sperm and eggs holds in moisture. *Third*, of all plants, bryophytes alone have large gametophytes that do not depend on sporophytes for nutrition. Instead, embryo sporophytes start to develop inside gametophyte tissues—and even at maturity, they are *attached to* the gamete-producing body and still gain some nutritional support from it.

With 10,000 species, mosses are the most common bryophytes. The gametophytes of some species grow in clusters and form low, cushiony mounds (Figure 25.4*a*). Those of others commonly grow in branched, feathery

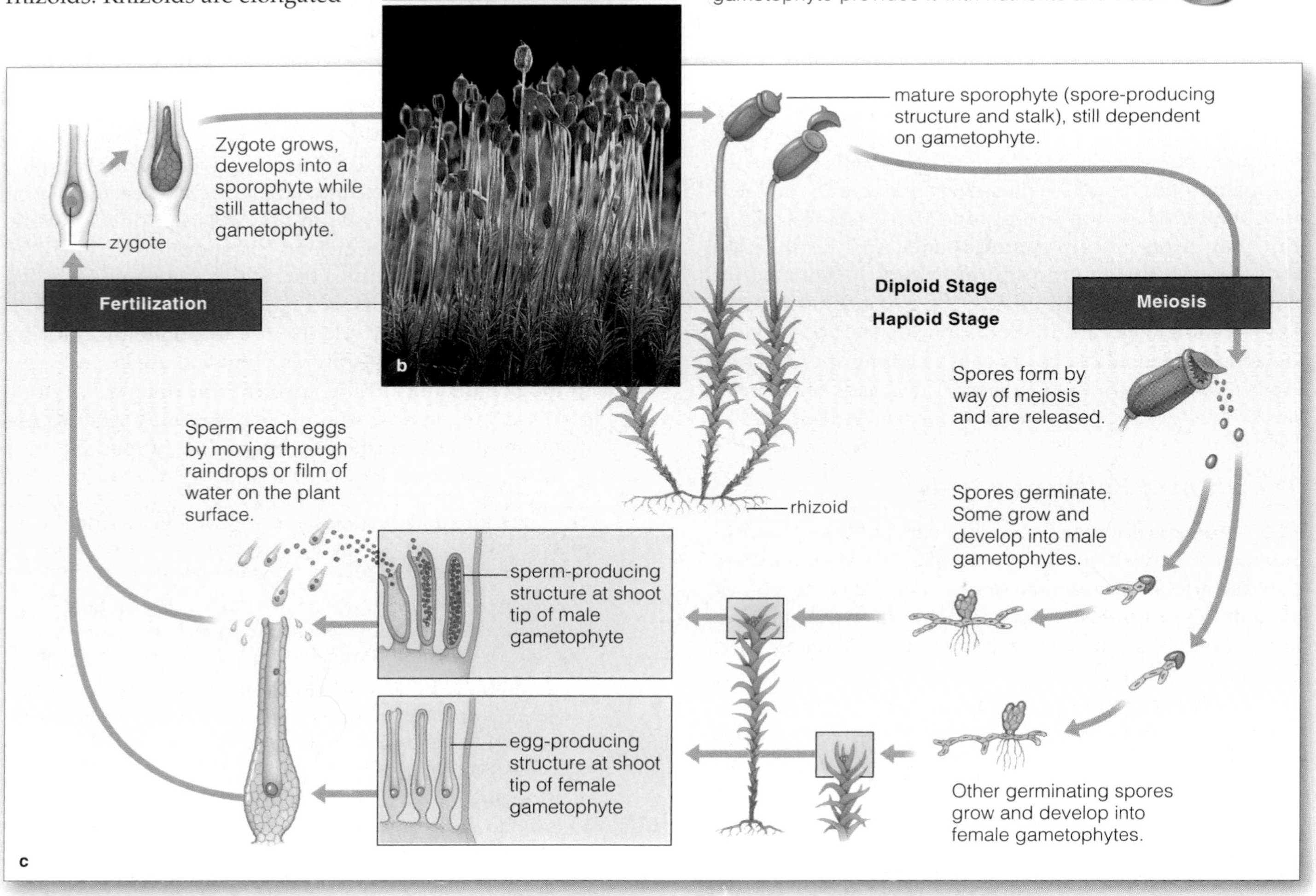

Figure 25.4 (**a**) Moss-covered rocks near a small cascading stream. (**b**,**c**) Photograph and life cycle of a moss (*Polytrichum*), one of the common bryophytes. The moss sporophyte remains attached to the gametophyte and is dependent upon it. The gametophyte provides it with nutrients and water.

a

b

mature sporophyte (spore-producing structure and stalk), still dependent on gametophyte.

Zygote grows, develops into a sporophyte while still attached to gametophyte.

zygote

Fertilization

Diploid Stage

Haploid Stage

Meiosis

Spores form by way of meiosis and are released.

Sperm reach eggs by moving through raindrops or film of water on the plant surface.

rhizoid

Spores germinate. Some grow and develop into male gametophytes.

sperm-producing structure at shoot tip of male gametophyte

egg-producing structure at shoot tip of female gametophyte

Other germinating spores grow and develop into female gametophytes.

c

Further reading: Student Guide to InfoTrac on web site

b
a

Figure 25.5 (**a**) Peat bog in Ireland. This family is cutting blocks of peat and stacking them to dry as a fuel source for their home. Peat also is harvested on a scale that is large enough to generate electricity in peat-burning power plants.

(**b**) Gametophyte of a peat moss (*Sphagnum*). A few sporophytes are attached to them. Brown, jacketed structures on the white stalks are the sporophytes. Because of its fine antiseptic properties and high absorbency, peat moss was used as an emergency poultice on wounds of soldiers during World War I.

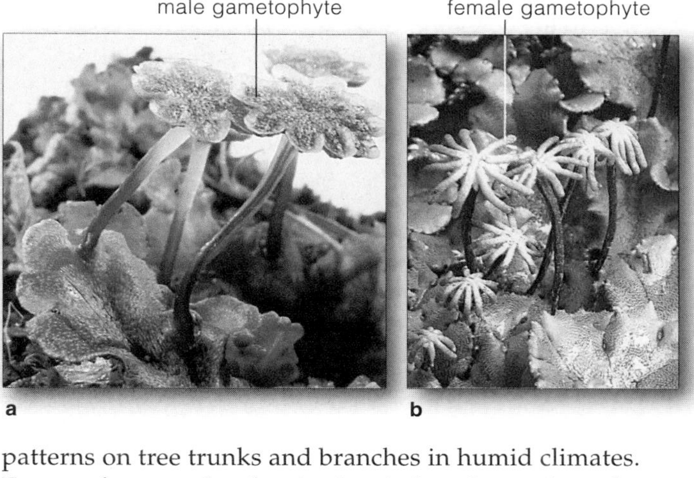
male gametophyte
female gametophyte
gemmae
a
b
c

Figure 25.6 *Marchantia*, one of the liverworts. Like other liverworts, this nonvascular plant reproduces sexually. Unlike the other types, it forms (**a**) male reproductive parts and (**b**) female reproductive parts on different plants. (**c**) *Marchantia* also reproduces asexually by way of gemmae, multicelled vegetative bodies that develop in tiny cups on the plant body. Gemmae grow and develop into individual plants after splashing raindrops transport them to suitable sites.

patterns on tree trunks and branches in humid climates. Eggs and sperm develop in tiny, jacketed vessels at the shoot tips of gametophytes. Sperm reach the eggs by swimming through a film of water on plant parts. After fertilization, the zygotes give rise to sporophytes, each composed of a stalk and a jacketed structure in which spores will develop.

Figure 25.5 shows one of 350 kinds of peat mosses (*Sphagnum*). Whereas most bryophytes grow slowly, the peat mosses can grow fast enough to yield twelve metric tons of organic matter per hectare, which is about twice as much as corn plants yield. They soak up five times as much water as cotton does, owing to large, dead cells in their leaflike parts. They also produce acids that inhibit the growth of bacterial and fungal decomposers. Their remains accumulate into compressed, excessively moist mats called **peat bogs**. In cold and temperate regions, peat bogs cover an area equal to one-half of the United States. Only the most acid-tolerant plants, such as larch, cranberries, blueberries, and Venus flytraps, can grow in the bogs, which can be as acidic as vinegar.

Nearly all the peat harvested and dried in Ireland and elsewhere is burned to generate electricity in power plants. Compared to coal burning, peat fires generate fewer pollutants. Every so often, peat harvesters come across exceptionally well-preserved bodies of humans who lived 2,000 to 3,000 years ago. Some ancient bogs apparently were sites of ceremonial human sacrifices.

So as not to dwell on the macabre, let us leave this section with the liverworts and their interesting ways of reproducing, as described in Figure 25.6.

Bryophytes are nonvascular plants with flagellated sperm that require liquid water to reach and fertilize the eggs.

A sporophyte of these plants develops within gametophyte tissues. It remains attached to the gametophyte and receives some nutritional support from it.

25.3 EXISTING SEEDLESS VASCULAR PLANTS

Figure 25.7*a* shows one of the early seedless vascular plants. Descendants of certain lineages are still with us; we call them **whisk ferns**, **lycophytes**, **horsetails**, and **ferns**. Like their ancestors, they differ from bryophytes in three key respects. The sporophyte does not remain attached to a gametophyte, it has true vascular tissues, and it is the larger, longer lived phase of the life cycle.

Most seedless vascular plants live in wet, humid places, and their gametophytes lack vascular tissues. Water droplets clinging to the plants are the only means by which flagellated sperm can reach the eggs. The few species in dry habitats reproduce sexually during brief, seasonal pulses of heavy rains. In a sense, whisk ferns, lycophytes, horsetails, and ferns are the "amphibians" of the plant kingdom. They have not fully escaped the aquatic habitats of their ancestors.

Whisk Ferns

Whisk ferns (Psilophyta), which are not ferns, resemble a whisk broom. Florist suppliers commonly cultivate them in Hawaii, Texas, Louisiana, Florida, Puerto Rico, and other tropical or subtropical regions. One genus, *Psilotum*, is a unique vascular plant, for its sporophytes have no roots or leaves. The photosynthetic, branched stems have scalelike projections and, internally, xylem and phloem (Figure 25.7*b*). Belowground are **rhizomes**, branching, short, mostly horizontal absorptive stems.

Lycophytes

About 350 million years ago, lycophytes (Lycophyta) included tree-sized members of swamp forests. About 1,100 far tinier species exist today. The most familiar are club mosses, members of communities in the Arctic, the tropics, and regions in between. Many form mats on forest floors. One type, called the resurrection plant, is common in Texas, New Mexico, and Mexico.

Sporophytes of most club mosses have leaves and a branching rhizome that gives rise to vascularized roots and stems. Some have nonphotosynthetic, cone-shaped leaf clusters with spore-producing structures (Figure 25.7*c*). Each cluster is a **strobilus** (plural, strobili). After spores disperse, they germinate and develop into small, free-living gametophytes. *Selaginella* is heterosporous; two kinds of spores develop in the same strobilus.

Horsetails

Tree-sized sphenophytes (Sphenophyta) flourished in ancient swamp forests. Twenty-five or so smaller species of one genus, *Equisetum*, made it to the present. These are the horsetails. Their body plan changed very little over the past 300 million years.

Horsetails thrive in streambank muds, vacant lots, roadsides, and other disrupted habitats. Figure 25.7*d–f* shows the vegetative, photosynthetic stems and fertile stems of one species. Its spores give rise to free-living

Figure 25.7 (**a**) *Cooksonia*, one of the earliest known vascular plants, no more than a few centimeters tall. It probably grew in mud flats. Its upright, branching stems had a cuticle. Its spores formed in structures at stem tips. Compare Figure 18.2*c*. (**b**) Sporophytes of a whisk fern (*Psilotum*), a seedless vascular plant. Pumpkin-shaped, spore-producing structures form at the ends of stubby branchlets. (**c**) Sporophyte of one of the lycophytes (*Lycopodium*). (**d**) Vegetative stem of *Equisetum*, which resembles a horsetail. (**e**) Fertile, nonphotosynthetic stems of *Equisetum*. (**f**) Closer look at the spore-producing structure of a fertile stem.

a

b

c

d

e

strobilus, an aggregation of spore-producing structures, at tip of a vegetative shoot of the horsetail sporophyte

f Each petal-shaped structure of a strobilus contains many spores that formed by way of meiotic cell divisions.

Further reading: Student Guide to InfoTrac on web site

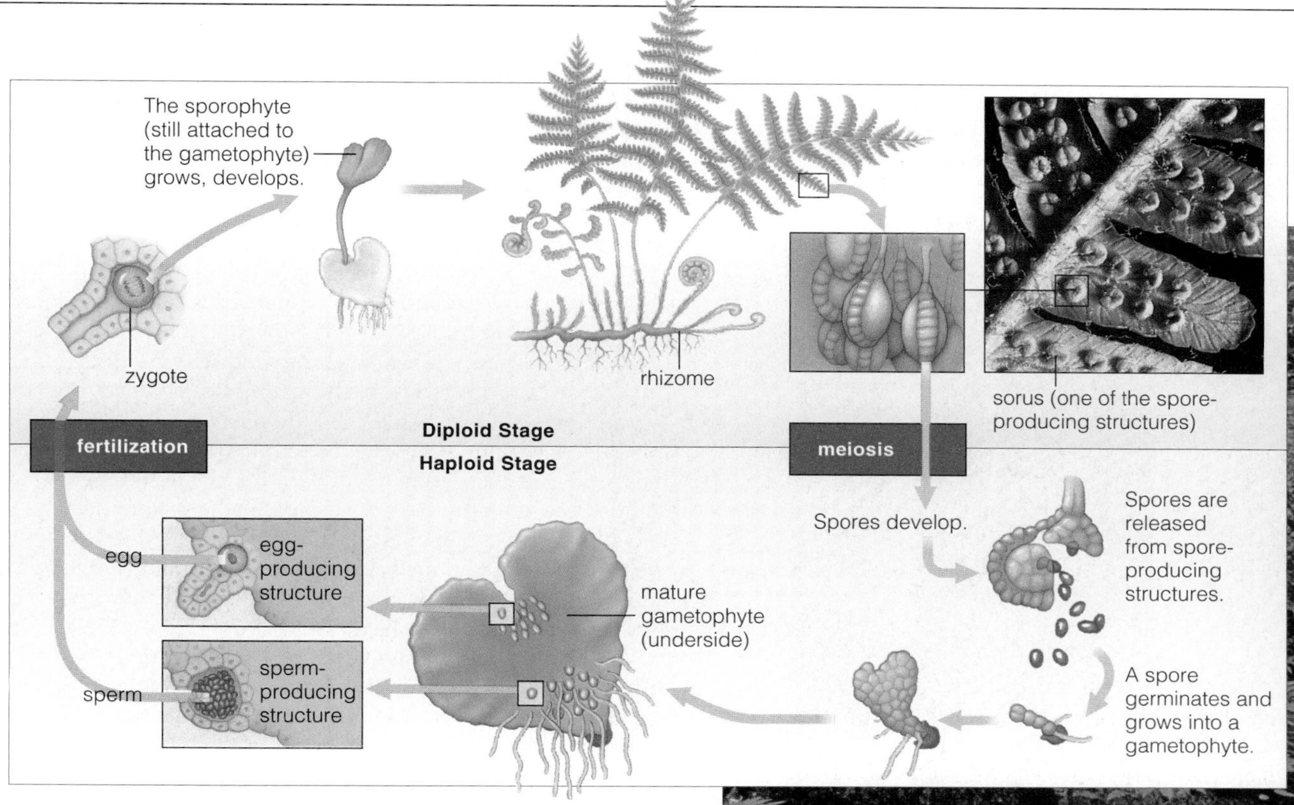

The sporophyte (still attached to the gametophyte) grows, develops.
zygote
rhizome
sorus (one of the spore-producing structures)
fertilization
Diploid Stage
Haploid Stage
meiosis
Spores develop.
Spores are released from spore-producing structures.
egg
egg-producing structure
sperm
sperm-producing structure
mature gametophyte (underside)
A spore germinates and grows into a gametophyte.

gametophytes 1 millimeter to 1 centimeter across. The sporophytes of most horsetails have rhizomes, hollow photosynthetic stems, and scale-shaped leaves. Strands of xylem and phloem are arrayed in a ring in the stems. Silica-reinforced ribs structurally support these stems and give them a texture like sandpaper. Pioneers of the American West, who did not have many places to store and wash towels, gathered horsetails on the westward journeys and used them as pot scrubbers.

Ferns

With 12,000 or so species, the ferns (Pterophyta) are the largest and most diverse group of seedless vascular plants. All but about 380 are native to the tropics, but they are popular houseplants all over the world. Their size range is stunning. Some floating species are less than 1 centimeter across. Some tropical tree ferns are 25 meters (82 feet) tall. One climbing fern has a modified leaf stalk about 30 meters long.

Most ferns have vascularized rhizomes that give rise to roots and leaves. Exceptions include tropical tree ferns and epiphytes. (*Epiphyte* refers to any aerial plant that grows attached to tree trunks or branches.) While they develop, young fern leaves are coiled, rather like a fiddlehead. At maturity, the leaves (fronds) commonly are divided into leaflets.

You may have noticed rust-colored patches on the lower surface of many fern fronds. Each patch, a cluster of spore-producing structures, is a sorus (plural, sori). At dispersal time, the spore-producing structures snap open with such force that the released spores catapult through the air. Each spore that germinates develops into a small gametophyte. You can see an example, a green, heart-shaped gametophyte, in Figure 25.8.

Figure 25.8 Life cycle of a fern. The photograph shows ferns growing in a moist habitat in Indiana. Ferns with finely divided fronds are in the foreground.

Seedless vascular plants (whisk ferns, lycophytes, horsetails, and ferns) have sporophytes adapted to conditions on land. Yet they have not entirely escaped their aquatic ancestry. When they reproduce sexually, their flagellated sperm cannot reach the eggs unless liquid water is clinging to the plant.

ANCIENT CARBON TREASURES

Three hundred million years ago, about halfway through the Carboniferous, mild climates prevailed and swamp forests carpeted the wet lowlands of the continents. The absence of pronounced seasonal swings in temperature favored plant growth through much of the year. The plants having lignin-reinforced tissues and well-developed root and shoot systems had the competitive edge under these growth conditions, and some of them evolved into giants. Massive-stemmed lycophyte trees—the giant club mosses—topped out at nearly forty meters (Figure 25.9). Each of their strobili produced as many as 8 billion microspores or hundreds of megaspores (Section 25.5). Being so high above the forest floor, dispersal of the new generations was a cinch. Giant horsetails, including species of *Calamites*, were close to twenty meters tall. Often their aboveground stems, which grew from rapidly spreading underground rhizomes, formed dense thickets.

As it happened, the sea level rose and fell fifty times during the Carboniferous. Each time the sea receded, the swamp forests flourished. When the sea moved back in, forest trees became submerged and buried in sediments that protected them from decay. Gradually the sediments compressed the saturated, undecayed remains into peat. Each time more sediments accumulated, they subjected the peat to increased heat and pressure that made it even more compact. In this way, compressed organic remains were transformed into great seams of **coal** (Figure 25.9).

With its high percentage of carbon, coal is energy rich and is one of our premier "fossil fuels." It took a fantastic amount of photosynthesis, burial, and compaction to form each major seam of coal in the Earth. It has taken us only a few centuries to deplete much of the world's known coal deposits. Often you will hear about annual "production rates" for coal or some other fossil fuel. But how much do we really produce each year? None. We simply *extract* it from the Earth. Coal is a nonrenewable source of energy.

Lepidodendron

stem of a giant lycophyte (*Lepidodendron*)
seed fern (*Medullosa*); probably related to the progymnosperms, which may have been among the earliest seed-bearing plants
stem of a giant horsetail (*Calamites*)

Figure 25.9 Reconstruction of a Carboniferous forest. The boxed inset shows part of a seam of coal.

Further reading: Student Guide to InfoTrac on web site

25.5 RISE OF THE SEED-BEARING PLANTS

About 360 million years ago, as the Devonian gave way to the Carboniferous, the first seed-bearing plants arose. In terms of diversity, numbers, and distribution, they would become the most successful groups of the plant kingdom. Seed ferns, gymnosperms, and (much later) angiosperms were the dominant groups. They differed from seedless vascular plants in three crucial respects.

First, seed-bearing plants produce pollen grains, the sperm-bearing male gametophytes. Remember, these plants produce two types of spores. Their **microspores** give rise to pollen grains. Unlike the spores of seedless vascular plants, they do not have a "tetrad scar," which marks the cleavage planes between four spores that form during meiotic cell division (Figure 25.10).

Like a suitcase, a pollen grain is a means of getting its contents (the sperm) to the eggs, even during times of prolonged drought. Seedless vascular plants do not have such an advantage; without predictable rains and moisture, their sperm simply cannot reach the eggs, and this has adverse effects on reproductive success. By contrast, pollen grains of gymnosperms simply drift with air currents. Those of angiosperms also are loaded onto insects, birds, bats, and other animals that truck them to the eggs. **Pollination** is the name for the arrival of pollen grains on the female reproductive structures. By this process, seed-bearing plants escaped dependence on free water for fertilization.

Second, besides microspores, seed-bearing plants also produce **megaspores**. These develop inside **ovules**, the female reproductive structures which, at maturity, are seeds (Figure 25.11). Each ovule consists of a female gametophyte (with its egg cell), nutrient-rich tissue, and a jacket of cell layers which, recall, develops into the seed coat. One zygote will form inside the ovule when a sperm reaches and fertilizes the egg. An embryo sporophyte will develop, and when the time comes to say good-bye to the parent plant, the coat around the seed will afford protection for the journey.

Third, compared to seedless vascular plants, the gymnosperms had water-conserving traits, including thicker cuticles and stomata recessed below the leaf surface. The traits gave gymnosperms competitive advantages in drier, cooler environments. Such environments were ushered in when Carboniferous gave way to the Permian. Before then, **seed ferns** of the type shown in Figure 25.9 rose to dominance, and they prevailed for about 70 million years. The seed ferns probably have evolutionary links with **progymnosperms**, which might have been one of the earliest plants to produce seeds. (At the least, they bore seedlike structures.) When the global climate did become cooler and drier, the swamplands disappeared. So did seed ferns. New kinds of seed-bearers—cycads, conifers, and other gymnosperms—would replace them.

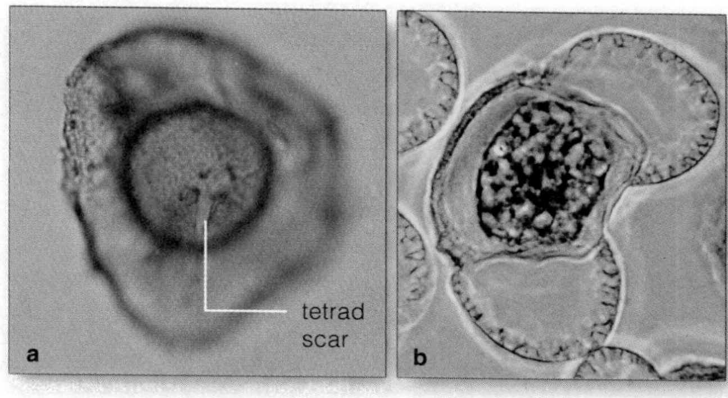
tetrad scar
a
b

Figure 25.10 (**a**) From the Devonian-Carboniferous boundary, a fossilized spore of a lycophyte. Its tetrad scar is typical of the spores of seedless plants. (**b**) A pine pollen grain (*Pinus*). Like the pollen of other seed-bearing plants, it has no tetrad scar.

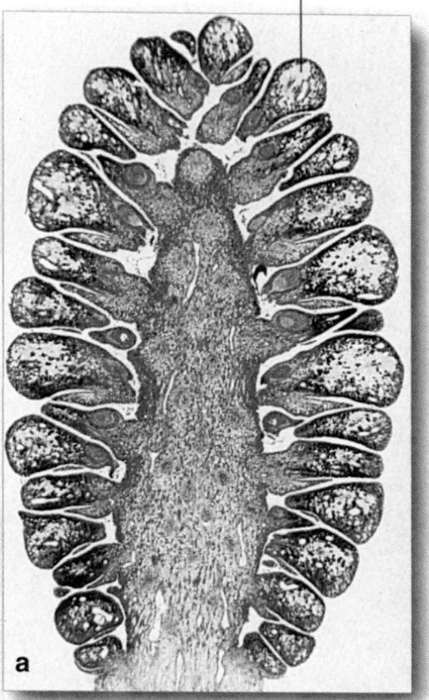
ovule, which may mature into a seed
a

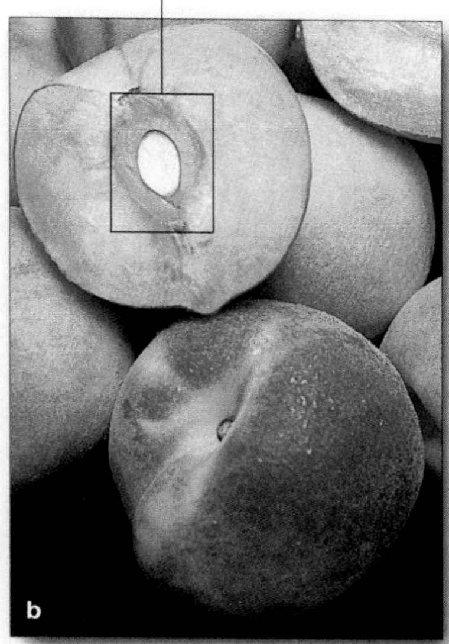
seed (mature ovule); the surrounding ovary matured to form the fleshy fruit
b

Figure 25.11 (**a**) Longitudinal section through a seed-bearing cone of a pine. These seeds are not protected by sporophyte tissues; they are exposed on scales of a cone. (**b**) A mature ovule, or seed, of a peach (*Prunus*), one of the angiosperms. It is protected by a seed coat, which is enclosed within the fleshy, edible tissue of the fruit.

In time, gymnosperms were the primary producers that sustained the dinosaurs. What some folks call the Age of Dinosaurs, botanists call the Age of Cycads.

Seed-bearing plants rely on pollen grains, ovules that mature into seeds, and tissue adaptations to dry conditions.

GYMNOSPERMS—PLANTS WITH "NAKED" SEEDS

With a bit of history behind us, we turn now to a survey of some of the existing gymnosperms. Unlike the seeds of flowering plants, which are enclosed in a chamber called an ovary, gymnosperm seeds are perched, in an exposed way, on a spore-producing structure. (*Gymnos* means naked; *sperma* is taken to mean seed.)

Conifers

Conifers (Coniferophyta) are woody trees and shrubs that have needlelike or scalelike leaves and that bear seeds exposed on cone scales. Conifer **cones** are clusters of modified leaves that surround the spore-producing structures. Figure 25.12 gives examples.

Most conifers shed some leaves all year long yet stay leafy, or *evergreen*. A few are *deciduous*, meaning that they shed all of their leaves in the fall. Among the conifers are the most abundant trees of the Northern Hemisphere (pines), the tallest (the coast redwoods), as well as the oldest (the bristlecone pine; one 4,725-year-old tree sprouted when Egyptians were building the Great Sphinx). Also in this group are firs, yews, spruces, junipers, larches, cypresses, the bald cypress, podocarps, and the dawn redwood.

Figure 25.12 (**a**) Bristlecone pine (*Pinus longaeva*) growing near the timberline in the Sierra Nevada. (**b**) Male pine cones releasing pollen. (**c**) Appearance of a female pine cone at the time of pollination. (**d**) From a juniper (*Juniperus*), cones with a berrylike appearance. These cones are made of fused-together, fleshy scales.

Lesser Known Gymnosperms

CYCADS About 100 species of **cycads** (Cycadophyta) made it to the present day. Figure 25.13 shows examples of their pollen-bearing and seed-bearing cones, which form on separate plants. Insects and air currents transfer pollen from "male" plants to "female" plants. At first glance, you might mistake cycad leaves for those of a palm tree, but palms are flowering plants.

These plants mainly inhabit tropical and subtropical areas. Two species (*Zamia*) grow wild in Florida and are planted as ornamentals. Elsewhere, cycad seeds and cycad trunks that are ground into a flour are made edible by removing their toxic alkaloids. Many cycad species are vulnerable to extinction.

Figure 25.13 (**a**) Pollen-bearing cone of a "male" cycad (*Zamia*). (**b**) Seed-bearing cone of a "female" cycad. Of all the existing species of gymnosperms, the cycads produce the largest seed-bearing cones. Some of these grow as long as one meter and weigh more than fifteen kilograms.

Further reading: Student Guide to InfoTrac on web site

Figure 25.14 (**a**) Ginkgo tree. (**b**) A fossilized ginkgo leaf compared with a leaf from its existing descendant. The fossil formed at the Cretaceous–Tertiary boundary. Even though 65 million years have passed since then, the leaf structure has not changed much, if at all. (**c**) Pollen-bearing cones and (**d**) fleshy-coated seeds of this type of gymnosperm.

Figure 25.15 (**a**) Sporophyte of *Ephedra* and (**b**) its pollen-bearing cones and (**c**) a seed-bearing cone. (**d**) Sporophyte of *Welwitschia mirabilis* and (**e**) its seed-bearing cones.

GINKGOS The **ginkgos** (Ginkgophyta) were a diverse group in dinosaur times. The only surviving species is the maidenhair tree, *Ginkgo biloba*. Like a few species of larch and some other gymnosperms, these plants are deciduous. Several thousand years ago, ginkgo trees were widely planted around temples in China. Then the natural populations nearly became extinct, even though ginkgos seem hardier than many other trees. Perhaps they became targets for firewood. Today, male ginkgo trees are again widely planted. They have attractive, fan-shaped leaves and are resistant to insects, disease, and air pollutants. Female trees are not favored. Their thick, fleshy seeds, which are the size of small plums, give off an awful stench (Figure 25.14).

GNETOPHYTES At present, there are three genera of woody plants known as the **gnetophytes** (Gnetophyta). Trees and leathery leafed vines of *Gnetum* thrive in the humid tropics. The shrubby *Ephedra* lives in California deserts and some other arid regions (Figure 25.15*a–c*). Photosynthesis proceeds in its green stems. *Welwitschia mirabilis* grows in hot deserts of south and west Africa. Its sporophyte is mainly a deep-reaching taproot. Its exposed part, a woody disk-shaped stem, has cones and one or two strap-shaped leaves that split lengthwise repeatedly as the plant ages (Figure 25.15*d, e*).

Conifers, cycads, ginkgos, and gnetophytes are groups of existing gymnosperms. Like their ancestors, they bear seeds on the exposed surfaces of cones and other spore-producing structures.

25.7 A CLOSER LOOK AT THE CONIFERS

Before we leave the gymnosperms, let's use conifers as an example of reproductive strategies. Depending on the species, a conifer's life cycle lasts a year or more.

Who among us hasn't noticed the woody, shelflike scales of "a pine cone"? The scales are actually parts of a mature female cone in which megaspores formed and developed into female gametophytes. Pine trees also produce male cones, in which microspores form and develop into pollen grains (Figure 25.16). Each spring, millions of pollen grains drift away from the male cones. Pollination is completed when some land on ovules of female cones. After each germinates, a tubular structure forms from it. This germinating pollen grain, the sperm-bearing male gametophyte, grows toward the egg in a female gametophyte. For species of pines, fertilization occurs months or a year after pollination.

As in other gymnosperms, seed formation begins at the ovule (Figure 25.16). An embryo sporophyte starts developing from the fertilized egg. The outer layers of the jacket around the female gametophyte and embryo mature into a hard coat. The seed coat will protect the embryo sporophyte after it is dispersed from the parent plant. The nutrients will help it through the critical time

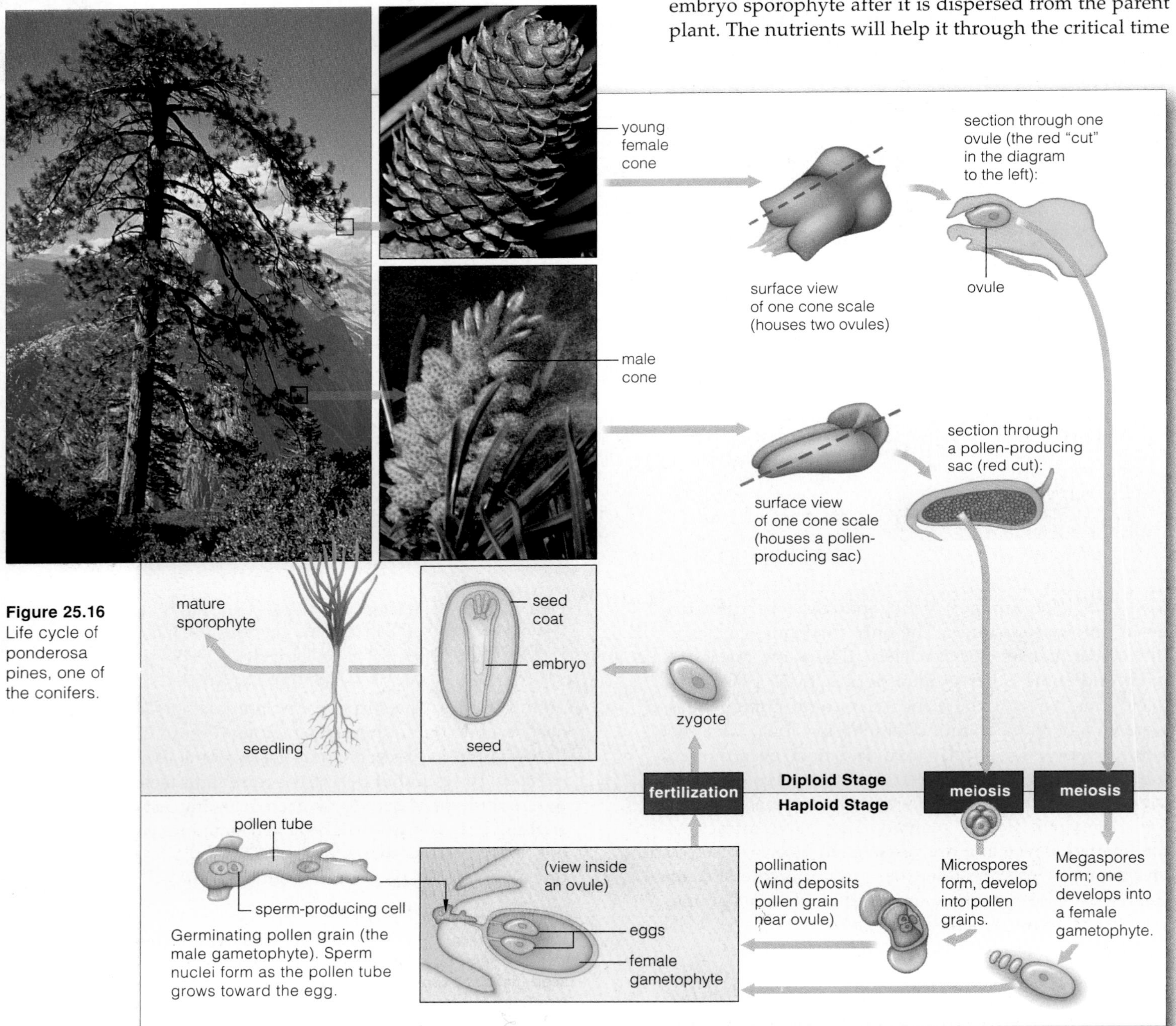
young female cone
section through one ovule (the red "cut" in the diagram to the left):
surface view of one cone scale (houses two ovules)
ovule
male cone
section through a pollen-producing sac (red cut):
surface view of one cone scale (houses a pollen-producing sac)
mature sporophyte
seed coat
embryo
zygote
seedling
seed
fertilization
Diploid Stage
Haploid Stage
meiosis
meiosis
pollen tube
sperm-producing cell
Germinating pollen grain (the male gametophyte). Sperm nuclei form as the pollen tube grows toward the egg.
(view inside an ovule)
eggs
female gametophyte
pollination (wind deposits pollen grain near ovule)
Microspores form, develop into pollen grains.
Megaspores form; one develops into a female gametophyte.

Figure 25.16 Life cycle of ponderosa pines, one of the conifers.

Further reading: Student Guide to InfoTrac on web site

a

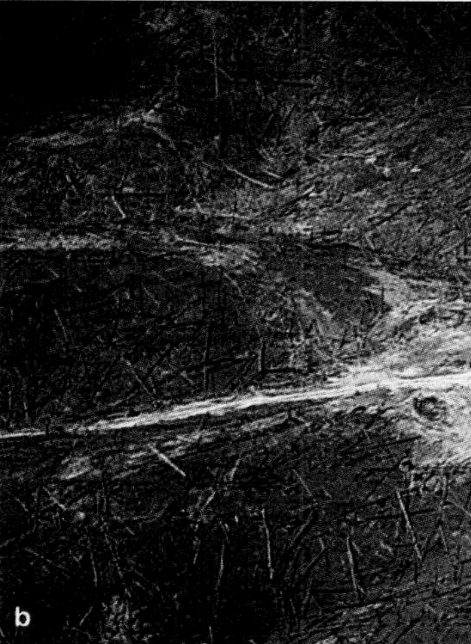
b

c

d

Figure 25.17 Sampling of the rampant deforestation under way around the world. Only deforested tracts in North America are shown; Section 51.4 focuses on the tropical rain forests.

From America's heartland, logged-over acreage in Arkansas (**a**). From the eastern seaboard, one of the denuded bits of North Carolina (**b**). Clear-cut peaks in Washington (**c**) and Alaska (**d**). These are not isolated examples. In the early 1980s, 400 million board feet of timber were being cut in Washington's Olympic Peninsula every year. In Arkansas, about one-third of the Ouachita National Forest was clear-cut. Its once-diverse forest communities have been replaced by "tree farms" of a single species of pine. Throughout the world, huge tracts of land that were deforested years ago still show no signs of recovery.

of germination, before its roots and shoots become fully functional.

Conifers dominated many land habitats during the Mesozoic, but their slow reproductive pace put them at a competitive disadvantage when the flowering plants began their stunning adaptive radiation (Section 21.6). Coniferous forests still predominate in the far north, at higher elevations, and in some parts of the Southern Hemisphere. However, existing conifers face more than competition with flowering plants for resources. Now they are vulnerable to **deforestation**: the removal of all trees from large tracts, as by clear-cutting (Figure 25.17). Conifers just have the bad luck to be premier sources of lumber, paper, and other wood products required in human societies. We return to this topic in Chapter 51.

Where flowering plants flourish, conifers are at a competitive disadvantage, partly because they take so long to reproduce. Rampant deforestation isn't helping them one bit, either.

25.8 ANGIOSPERMS—THE FLOWERING, SEED-BEARING PLANTS

Only angiosperms produce specialized reproductive structures called **flowers** (Figure 25.18). *Angeion,* which means vessel, refers to the female reproductive parts at the center of a flower. The enlarged base of the "vessel" is the floral ovary, where ovules and seeds develop.

Most flowering plants coevolved with **pollinators**—insects, bats, birds, and other animals that withdraw nectar or pollen from a flower and, in so doing, transfer pollen to its female reproductive parts. The recruitment of animals as assistants in reproduction probably has contributed to the success of flowering plants, which have dominated the land for 100 million years.

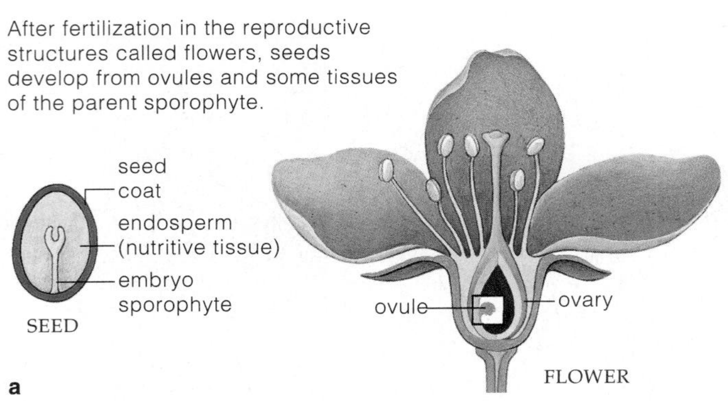
After fertilization in the reproductive structures called flowers, seeds develop from ovules and some tissues of the parent sporophyte.
seed coat
endosperm (nutritive tissue)
embryo sporophyte
SEED
ovule
ovary
FLOWER
a

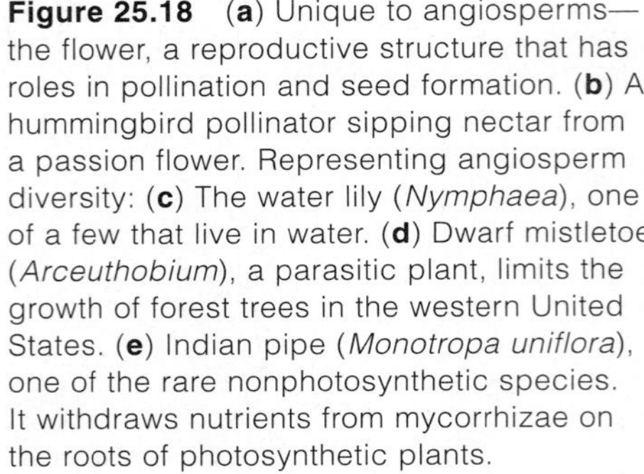

Figure 25.18 (**a**) Unique to angiosperms—the flower, a reproductive structure that has roles in pollination and seed formation. (**b**) A hummingbird pollinator sipping nectar from a passion flower. Representing angiosperm diversity: (**c**) The water lily (*Nymphaea*), one of a few that live in water. (**d**) Dwarf mistletoe (*Arceuthobium*), a parasitic plant, limits the growth of forest trees in the western United States. (**e**) Indian pipe (*Monotropa uniflora*), one of the rare nonphotosynthetic species. It withdraws nutrients from mycorrhizae on the roots of photosynthetic plants.

b
c
d
e

At least 260,000 species thrive in a variety of habitats. They range in size from tiny duckweeds (a millimeter or so long) to towering *Eucalyptus* trees (some are more than 100 meters tall). A few species, including mistletoes and Indian pipe, aren't even photosynthetic. They withdraw nutrients directly from other plants or from mycorrhizae.

There are two classes of flowering plants, called the **dicots** and **monocots** (more formally, the Dicotyledonae and Monocotyledonae). Among the 180,000 dicots are most herbaceous (nonwoody) plants, such as cabbages and daisies; most flowering shrubs and trees, such as oaks and apple trees; water lilies; and cacti. Among 80,000 or so species of monocots are the orchids, palms, lilies, and grasses, including rye, sugarcane, corn, rice, and wheat, as well as many other highly valued crop plants (Appendix I).

The next unit deals with the structure and function of flowering plants. For now, simply start thinking about how a large sporophyte dominates the life cycles. It retains and nourishes gametophytes; its sperm are dispersed within pollen grains. Endosperm, a nutritive tissue, surrounds embryo sporophytes inside the seeds of flowering plants. As the seeds develop, the ovaries (along with other structures) mature into fruits. Fruits protect and help disperse embryos. Figure 25.19, in the next section, is an overview of these life cycle events.

Angiosperms are the most successful plants in terms of their diversity, numbers, and distribution. They alone produce flowers. Most species coevolved with animal pollinators.

Further reading: Student Guide to InfoTrac on web site

VISUAL OVERVIEW OF FLOWERING PLANT LIFE CYCLES

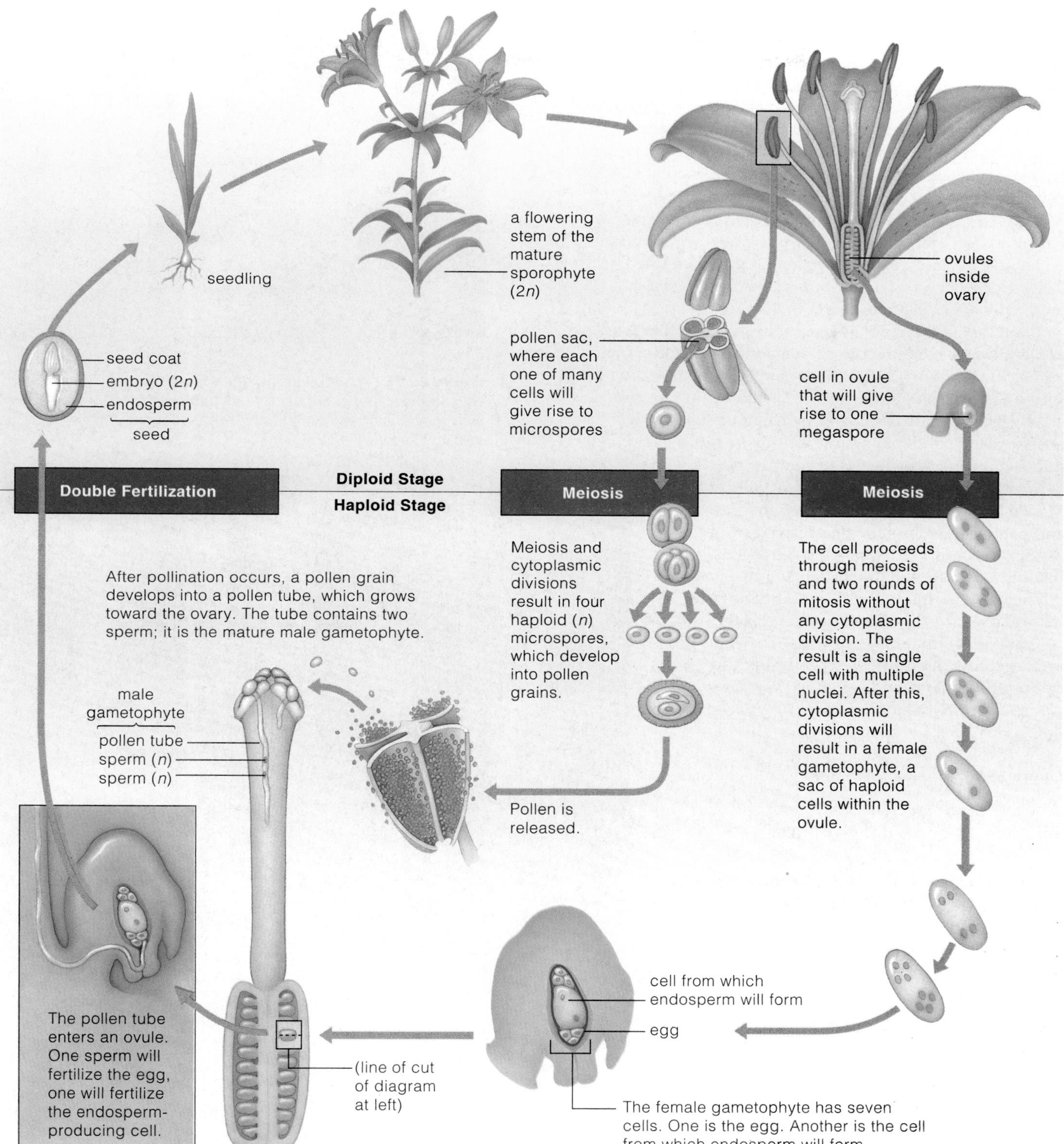

seedling
a flowering stem of the mature sporophyte (2n)
ovules inside ovary
seed coat
embryo (2n)
endosperm
seed
pollen sac, where each one of many cells will give rise to microspores
cell in ovule that will give rise to one megaspore
Double Fertilization
Diploid Stage
Haploid Stage
Meiosis
Meiosis
Meiosis and cytoplasmic divisions result in four haploid (n) microspores, which develop into pollen grains.
The cell proceeds through meiosis and two rounds of mitosis without any cytoplasmic division. The result is a single cell with multiple nuclei. After this, cytoplasmic divisions will result in a female gametophyte, a sac of haploid cells within the ovule.
After pollination occurs, a pollen grain develops into a pollen tube, which grows toward the ovary. The tube contains two sperm; it is the mature male gametophyte.
male gametophyte
pollen tube
sperm (n)
sperm (n)
Pollen is released.
cell from which endosperm will form
egg
The pollen tube enters an ovule. One sperm will fertilize the egg, one will fertilize the endosperm-producing cell.
(line of cut of diagram at left)
The female gametophyte has seven cells. One is the egg. Another is the cell from which endosperm will form.

Figure 25.19 Representative flowering plant life cycle. This example is for a lily (*Lilium*), one of the monocots. "Double" fertilization is a distinctive feature of flowering plant life cycles. A male gametophyte delivers two sperm to an ovule. One sperm fertilizes the egg, and the other fertilizes a cell that gives rise to endosperm, a tissue that will nourish the forthcoming embryo. Section 31.3 provides a closer look at flowering plant life cycles, using a dicot as the example.

25.10 SEED PLANTS AND PEOPLE

Which plants provide taste thrills and which kill? Starting with trials and errors of the earliest human species, we started collecting intimate knowledge of plants. Certainly by 300,000 years ago, *Homo erectus* clans in China were stashing pine nuts, walnuts, hazelnuts, and rose hips in caves and roasting seeds. At least by then, grown-ups were teaching children which plants are edible and which are toxic. Through language, youngsters learned about the plants that had evolved in their parts of the world.

About 11,000 years ago, people started to domesticate wheat, barley, and other plants, this being a way to count on having reliable quantities of food. Of an estimated 3,000 species that different populations recognized as food, only about 200 became *the* major crops (Figure 25.20).

Plant lore still threads through our lives in other ways. We learned to use flowers to grace our homes and enrich our customs (Figure 25.21). We learned to use fast-growing, soft-wooded conifers for lumber, fuel, and paper—and slow-growing, hard-wooded flowering plants, such as cherry, maple, and mahogany, for fine furniture. We make twine and rope from leaves of century plants (*Agave*), cords and textiles from leaf fibers of Manila hemp, and thatched roofs from grasses and palm fronds. Insecticides derived from Mexican cockroach plants kill cockroaches, fleas, lice, and flies. Extracts of neem tree leaves kill nematodes, insects, and mites but not the natural predators of those common pests.

And what about those flowers! Their oils impart scents to perfumes. Eucalyptus and camphor oils have medical uses. Digitalin extracted from foxglove (*Digitalis purpurea*)

Figure 25.21 A bride tossing her bouquet to single women who attended her wedding, a ritual that supposedly reveals who will be married next. (A sixty-eight-year-old optimist caught this one.)

a
c
b

Figure 25.20 A few of the prized flowering plants.

(**a**) Triticale, a popular hybrid grain; parental stocks are wheat and rye (*Secale*). It combines the high yield of wheat with rye's tolerance of truly harsh environmental conditions. (**b**) From the American Midwest, mechanized harvesting under way in a field of common bread wheat, *Triticum*.

(**c**) Fruits of *Theobroma cacao*. Each fruit holds as many as forty seeds, which are processed to make cocoa butter or chocolate essences. Americans who on average buy 8–10 pounds of chocolate per year might not be happy that a pathogenic fungus (*Monilia*) may be driving *T. cacao* to extinction.

(**d**) Indonesians gathering tender shoots of tea plants, evergreen shrubs related to camellias. Plants growing on hillsides in moist, cool regions yield leaves with the best flavors. Only the terminal bud and two or three of the youngest leaves are picked for the finest teas

(**e**) From Hawaii, a field of sugarcane (*Saccharum officinarum*). Wild stock of this cultivated species may have evolved in New Guinea. Sap extracted from its cut stems is boiled down to make sucrose crystals (table sugar) and syrups.

Further reading: Student Guide to InfoTrac on web site –

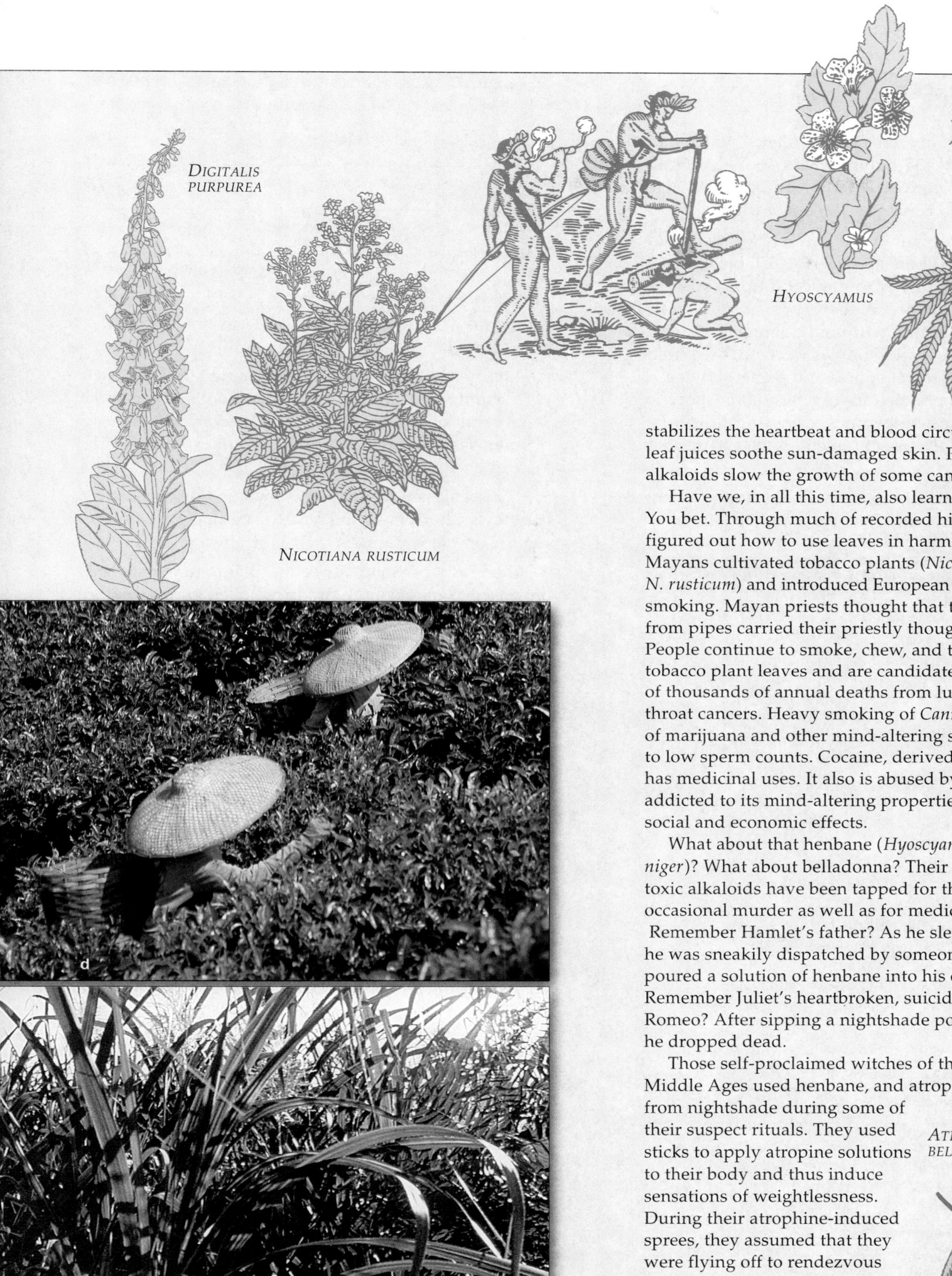
DIGITALIS PURPUREA
NICOTIANA RUSTICUM
HYOSCYAMUS
CANNABIS SATIVA
d
e
ATROPA BELLADONNA

stabilizes the heartbeat and blood circulation. *Aloe vera* leaf juices soothe sun-damaged skin. Periwinkle leaf alkaloids slow the growth of some cancer cells.

Have we, in all this time, also learned to abuse plants? You bet. Through much of recorded history, people also figured out how to use leaves in harmful ways. Ancient Mayans cultivated tobacco plants (*Nicotiana tabacum* and *N. rusticum*) and introduced European explorers to tobacco smoking. Mayan priests thought that the smoke rising from pipes carried their priestly thoughts to the gods. People continue to smoke, chew, and tuck in their mouth tobacco plant leaves and are candidates for the hundreds of thousands of annual deaths from lung, mouth, and throat cancers. Heavy smoking of *Cannabis sativa*, source of marijuana and other mind-altering substances, is linked to low sperm counts. Cocaine, derived from coca leaves, has medicinal uses. It also is abused by millions who are addicted to its mind-altering properties, with devastating social and economic effects.

What about that henbane (*Hyoscyamus niger*)? What about belladonna? Their toxic alkaloids have been tapped for the occasional murder as well as for medicine. Remember Hamlet's father? As he slept, he was sneakily dispatched by someone who poured a solution of henbane into his ear. Remember Juliet's heartbroken, suicidal Romeo? After sipping a nightshade potion, he dropped dead.

Those self-proclaimed witches of the Middle Ages used henbane, and atropine from nightshade during some of their suspect rituals. They used sticks to apply atropine solutions to their body and thus induce sensations of weightlessness. During their atrophine-induced sprees, they assumed that they were flying off to rendezvous with demons. Hence those Halloween cartoons of witches flying hither and yon on their broomsticks.

25.11 SUMMARY

1. Green algae probably gave rise to plants, which had invaded the land by 435 million years ago. Nearly all species of plants are multicelled photoautotrophs. Table 25.1 summarizes and compares the major phyla.

2. Several trends in plant evolution may be identified by comparing different lineages (see also Table 25.2):

a. Structural adaptations to dry conditions, such as vascular tissues (xylem and phloem).

b. A shift from haploid to diploid dominance during the life cycle. Complex sporophytes evolved; they hold on to, nourish, and protect spores and gametophytes.

c. A shift from one to two spore types (homospory to heterospory) that, among gymnosperms and flowering plants, led to the evolution of pollen grains and seeds.

3. Mosses, liverworts, and hornworts are bryophytes, nonvascular plants that have no well-developed xylem or phloem and that require free water for fertilization.

Table 25.1 Comparison of Major Existing Plant Groups

Nonvascular land plants. Fertilization requires free water. Haploid dominance. Cuticle, stomata present in some.	
BRYOPHYTES	18,600 species. Moist, humid habitats.
Seedless vascular plants. Fertilization requires free water. Diploid dominance. Cuticle, stomata present.	
WHISK FERNS	7 species, sporophytes with no obvious roots or leaves. *Psilotum.*
LYCOPHYTES	1,100 species with simple leaves. Mostly wet or shady habitats.
HORSETAILS	25 species of single genus. Swamps, disturbed habitats.
FERNS	12,000 species. Wet, humid habitats in mostly tropical, temperate regions.
Gymnosperms—vascular plants with "naked seeds." Free water not required for fertilization. Diploid dominance. Cuticle, stomata present.	
CONIFERS	550 species, mostly evergreen, woody trees and shrubs having pollen- and seed-bearing cones. Widespread distribution.
CYCADS	185 slow-growing tropical, subtropical species.
GINKGO	1 species, a tree with fleshy-coated seeds.
GNETOPHYTES	70 species. Limited to some deserts, tropics.
Angiosperms—vascular plants with flowers and protected seeds. Free water not required for fertilization. Diploid dominance. Cuticle, stomata present.	
FLOWERING PLANTS	
Monocots	80,000 species. Floral parts often arranged in threes or in multiples of three; one seed leaf; parallel leaf veins common.
Dicots	At least 180,000 species. Floral parts often arranged in fours, fives, or multiples of these; two seed leaves; net-veined leaves common.

Table 25.2 Evolutionary Trends Among Plants

Bryophytes	Ferns	Gymnosperms	Angiosperms
Nonvascular →	Vascular →	→	→
Haploid dominance →	Diploid dominance →	→	→
Spores of one type →	→	Spores of two types →	→
Motile gametes →	→	→	Nonmotile gametes* →
Seedless →	→	Seeds →	→

* Require pollination by wind, insects, animals, etc.

4. Nearly all vascular plants are adapted to land. Their cuticle and stomata conserve water. Root systems mine soil for nutrients. Upright and branched growth patterns of shoot systems intercept sunlight and carbon dioxide. Tissues enclose and protect spores and gametes.

5. The seedless vascular plants include the whisk ferns, lycophytes, horsetails, and ferns. Their flagellated sperm require ample water to swim to the eggs.

6. Gymnosperms and flowering plants (angiosperms) are vascular plants. Both produce pollen grains (mature microspores that develop into male gametophytes) and megaspores (which give rise to female gametophytes).

a. Megaspores develop within ovules: reproductive structures that contain (1) the egg-producing female gametophytes, (2) the precursor of nutritive tissue, and (3) a jacket of cell layers, the outer portion of which will develop into the seed coat. A mature ovule is a seed.

b. The evolution of pollen grains freed these plants from dependence on water for fertilization. Their seeds are efficient means of dispersing new generations, even during hostile conditions. Pollen grains and seeds were key adaptations in the move to high and dry habitats.

7. Only angiosperms produce flowers. Most coevolved with pollinators, such as insects, which enhance the transfer of pollen grains to female reproductive parts. Their seeds contain a nutritive tissue (endosperm) and are usually surrounded by fruit, which aids in dispersal.

Review Questions

1. Identify a few structural and reproductive modifications that helped plants invade and diversify in habitats on land. *25.1*

2. Does the haploid phase or diploid phase dominate the life cycles of most plants? *25.1*

3. Name representatives of the following groups of plants and then compare their main characteristics: (*also refer to Table 25.1*)
   a. Bryophytes and seedless vascular plants *25.2–25.4*
   b. Gymnosperms and angiosperms *25.6–25.8*

4. Distinguish between:
   a. Root system and shoot system *25.1*
   b. Xylem and phloem *25.1*
   c. Sporophyte and gametophyte *25.1*
   d. Ovule and seed *25.1, 25.5*
   e. Microspore and megaspore *25.5*

Figure 25.22 Where many conifers end up.

Self-Quiz (*Answers in Appendix III*)

1. Which of the following statements is *not* true?
   a. Monocots and dicots are two classes of angiosperms.
   b. Bryophytes are nonvascular plants.
   c. Lycophytes and angiosperms are both vascular plants.
   d. Gymnosperms are the simplest vascular plants.

2. Of all land plants, bryophytes alone have independent _______ and attached, dependent _______.
   a. sporophytes; gametophytes
   b. gametophytes; sporophytes
   c. rhizoids; zygotes
   d. rhizoids; stalked sporangia

3. Whisk ferns, lycophytes, horsetails, and ferns are classified as _______ plants.
   a. multicelled aquatic
   b. nonvascular seed
   c. seedless vascular
   d. seed-bearing vascular

4. Which does *not* apply to gymnosperms and angiosperms?
   a. vascular tissues
   b. diploid dominance
   c. single spore type
   d. all of the above

5. A seed is _______.
   a. a female gametophyte
   b. a mature ovule
   c. a mature pollen tube
   d. an immature embryo

6. Match the terms appropriately.

_____ gymnosperm	a. gamete-producing body
_____ sporophyte	b. help control water loss
_____ lycophyte	c. "naked" seeds
_____ ovary	d. protects, nourishes, disperses embryo sporophyte
_____ bryophyte	e. spore-producing body
_____ gametophyte	f. nonvascular land plant
_____ stomata	g. seedless vascular plant
_____ angiosperm seed	h. usually a fruit at maturity

Critical Thinking

1. Figure 25.22 shows a forest in the Nahmint Valley of British Columbia, before and after logging. It also shows wood frames of homes that are in the process of being built. Reflect on these photographs and Figure 25.17. To stop the loggers, would you chain yourself to a tree in an old-growth forest scheduled for clear-cutting? If your answer is yes, would you also give up the chance of owning a wood-frame home (as most homes are in developed countries)? What about forest products, including newspapers, toilet tissue, and fireplace wood?

2. With respect to question 1, multiply each of your answers by 6 billion (there are almost that many people in the world) and describe what might happen when, inevitably, we run out of trees. Also describe what you might consider to be some of the pros and cons of tree farms of, say, a single species of pine.

3. Elliot Meyerowitz of the California Institute of Technology has studied the genetic basis of flower formation in *Arabidopsis thaliana*. By inducing mutations in seeds of this small weed, he discovered three genes (call them *A*, *B*, and *C*) that interact in different parts of a developing flower. Gene interactions lead to the formation of different structures—sepals, petals, stamens, and carpels—from the same mass of undifferentiated tissue. From what you know of gene regulation, suggest ways in which the *A*, *B*, and *C* genes might be controlling flower development.

4. Genes nearly identical to the *A*, *B*, and *C* genes of *A. thaliana* also have been isolated from snapdragons and other flowering plants. These plants had evolved by 150 million years ago. They quickly rose to dominance in nearly all land habitats.

   Review the general introduction to adaptive radiation in Section 21.6. Then speculate on how the spectacular and rapid adaptive radiation of flowering plants came about.

Selected Key Terms

angiosperm *25.1*	hornwort *25.2*	pollinator *25.8*
bryophyte *25.1*	horsetail *25.3*	progymnosperm *25.5*
coal *25.4*	lignin *25.1*	rhizome *25.3*
conifer *25.6*	liverwort *25.2*	root system *25.1*
cuticle *25.1*	lycophyte *25.3*	seed *25.1*
cycad *25.6*	megaspore *25.5*	seed fern *25.5*
deforestation *25.7*	microspore *25.5*	shoot system *25.1*
dicot *25.8*	monocot *25.8*	spore *25.1*
fern *25.4*	moss *25.2*	sporophyte *25.1*
flower *25.8*	ovule *25.5*	stoma (stomata) *25.1*
gametophyte *25.1*	peat bog *25.2*	strobilus *25.3*
ginkgo *25.6*	phloem *25.1*	vascular plant *25.1*
gnetophyte *25.6*	pollen grain *25.1*	whisk fern *25.3*
gymnosperm *25.1*	pollination *25.5*	xylem *25.1*

Readings *See also www.infotrac-college.com*

Gray, J., and W. Shear. September–October 1992. "Early Life on Land." *American Scientist* 80:444–456.

Moore, R., W. D. Clark, and K. Stern. 1995. *Botany*. Dubuque, Iowa: W. C. Brown. Beautifully illustrated.

26

ANIMALS: THE INVERTEBRATES

Madeleine's Limbs

In August of 1994, about 900 million years after the first animals appeared on Earth, Madeleine made *her* entrance. As they are wont to do, her grandmothers and aunts made a quick count on the sly—arms, legs, ears, and eyes, two of each; fully formed mouth and nose—just to be sure these were present and accounted for.

One grandmother, having been too long in the company of biologists, experienced an epiphany as she witnessed Madeleine's birth. In that profound instant she sensed ancestral connections emerging from the distant past and through her, into the future.

Madeleine's body plan did not emerge out of thin air. Thirty-five thousand years ago, people just like us were having children just like Madeleine. And if we are interpreting the fossil record correctly, then five million years ago the offspring of individuals on the road to modern humans resembled her in some respects but not others. Sixty million years ago, primate ancestors of those individuals were giving birth precariously, up in the trees. Two hundred and fifty million years ago, mammalian ancestors of those primates were giving birth—and so on back in time to the very first animals, which had no limbs or eyes or noses at all.

We have very few clues to what those first animals looked like, but one thing is clear. By the dawn of the Cambrian, they had given rise to all major groups of invertebrates—animals without backbones—even to Madeleine's backboned but limbless ancestors.

And what stories those Cambrian animals tell! One bunch flourished 530 million years ago, in a submerged basin that had formed between a reef and the coast of an early continent. Protected from ocean currents, the sediments had piled up against the steep reef. About 500 feet below the surface, the water was oxygenated and clear. Small, well-developed animals lived in, on, and above the dimly lit, muddy sediments (Figure 26.1).

Like castles built from wet sand along a seashore, their living quarters were unstable. Part of the bank above the community slumped abruptly and obliterated it. The sediments from that underwater avalanche kept scavengers from reaching and removing all traces of the dead. Gradually through time, muddy silt rained down on the tomb. Increased pressure and chemical changes transformed the sediments into finely stratified shale, and the soft parts of the flattened animals became shimmering, mineralized films.

Figure 26.1 Reconstruction of a few Cambrian animals known from fossils of the Burgess Shale, British Columbia.

Sixty-five million years ago, part of the seafloor was plowing under the North American plate, and western Canada's mountain ranges were slowly rising. By 1909, the fossils had traveled high into the eastern mountains of British Columbia. In that year a fossil hunter tripped over a chunk of shale, which split apart into thin, fine layers—and so the Burgess Shale story came to light.

In this chapter and the next, you will be comparing body plans of different groups of animals. Such comparisons give insight into evolutionary relatedness and help us to construct family trees, such as the one in Figure 26.2. Don't assume that structurally simple animals of the most ancient lineages are somehow primitive or evolutionarily stunted. As you will see, they, too, are exquisitely adapted to their environment.

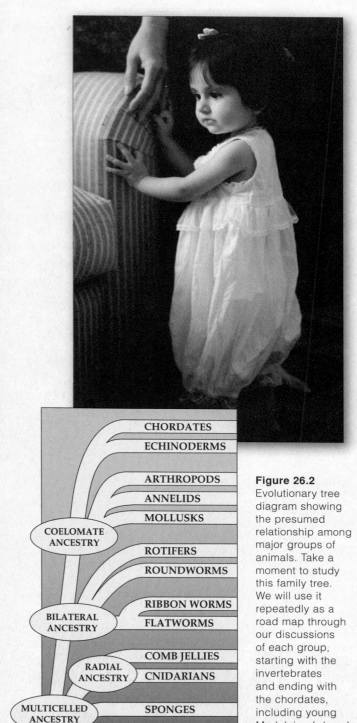

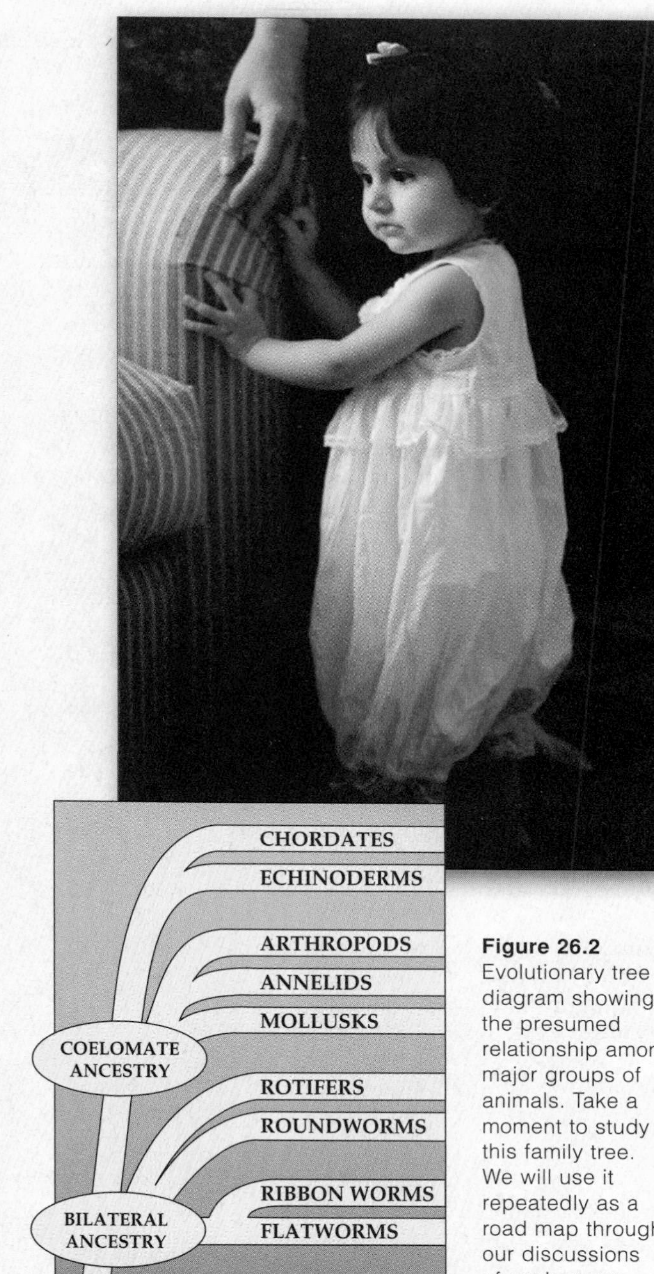
CHORDATES
ECHINODERMS
ARTHROPODS
ANNELIDS
MOLLUSKS
COELOMATE ANCESTRY
ROTIFERS
ROUNDWORMS
RIBBON WORMS
BILATERAL ANCESTRY
FLATWORMS
COMB JELLIES
RADIAL ANCESTRY
CNIDARIANS
MULTICELLED ANCESTRY
SPONGES
SINGLE-CELLED, PROTISTAN-LIKE ANCESTORS

Figure 26.2 Evolutionary tree diagram showing the presumed relationship among major groups of animals. Take a moment to study this family tree. We will use it repeatedly as a road map through our discussions of each group, starting with the invertebrates and ending with the chordates, including young Madeleine (*above*).

As you poke through the branches of the animal family tree, keep the greater evolutionary story in mind. At each branch point, microevolutionary processes gave rise to workable changes in body plans. Madeleine's uniquely human traits, and yours, emerged through modification of certain traits that had evolved earlier in countless generations of vertebrates and, before them, in ancient invertebrate forms.

KEY CONCEPTS

1. All animals are multicelled, aerobic heterotrophs that ingest or parasitize other organisms. Nearly all kinds have tissues, organs, and organ systems, and most are motile during at least part of their life cycle. Animals reproduce sexually and often asexually, and their embryos develop through a series of continuous stages.

2. Animals originated in the late Precambrian. More than 2 million existing animal species have been identified. Of these, more than 1,950,000 are invertebrates (animals with no backbone). Fewer than 50,000 species are vertebrates (animals with a backbone).

3. Comparisons of the body plans of existing animals, in conjunction with the fossil record, reveal that there were several trends in the evolution of certain lineages. The most revealing aspects of an animal's body plan are its type of symmetry, gut, and cavity (if any) between the gut and body wall; whether it has a distinct head end; and whether it is divided into a series of segments.

4. The placozoans and sponges are structurally simple animals with no body symmetry. Both are at the cellular level of body construction. Cnidarians and comb jellies show radial symmetry, and they are at the tissue level of body construction.

5. Flatworms, roundworms, rotifers, and nearly all other animals that are more complex than the cnidarians show bilateral symmetry. They consist of tissues, organs, and organ systems.

6. Not long after the flatworms evolved, divergences gave rise to two major lineages. One evolutionary branching gave rise to the mollusks, annelids, and arthropods. The other gave rise to the echinoderms and chordates.

7. By biological measures, including diversity, sheer numbers, and distribution, the arthropods—especially insects—have been the most successful animal group.

26.1 OVERVIEW OF THE ANIMAL KINGDOM

General Characteristics of Animals

What, exactly, are **animals**? We can only define them by a list of characteristics, not with a sentence or two. *First,* animals are multicelled. In most cases their body cells form tissues that become arranged as organs and organ systems. The body cells of nearly all species have a diploid chromosome number. *Second,* all animals are heterotrophs that get carbon and energy by ingesting other organisms or by absorbing nutrients from them. *Third,* animals require oxygen for aerobic respiration. *Fourth,* animals reproduce sexually and, in many cases, asexually. *Fifth,* most animals are motile during at least part of the life cycle. *Sixth,* the life cycle includes stages of embryonic development. Briefly, mitotic cell divisions transform the animal zygote into a multicelled embryo. The embryonic cells give rise to primary tissue layers, **ectoderm**, **endoderm**, and usually **mesoderm**. These layers in turn give rise to all tissues and organs of the adult, as described in Sections 33.6 and 44.2.

Diversity in Body Plans

Mammals, birds, reptiles, amphibians, and fishes are the most familiar animals. All are **vertebrates**, the only animals with a "backbone." And yet, of probably more than 2 million species of animals, fewer than 50,000 are vertebrates! What we call the **invertebrates** are animals with plenty of diverse features, but not a backbone.

We group the animals into more than thirty phyla. Table 26.1 lists the groups described in this book. The characteristics they share with one another arose early in time, before divergences from a common ancestor gave rise to separate lineages. Later, as morphological differences accumulated among them, the lineages took off in amazingly diverse directions. How might we get a conceptual handle on their modern-day descendants—on animals as different as flatworms, hummingbirds, spiders, toads, humans, and giraffes? We can compare their similarities and differences with respect to five basic features. These are body symmetry, cephalization, type of gut, type of body cavity, and segmentation.

Table 26.1 Animal Phyla Described in This Book

Phylum	Some Representatives	Existing Species
PLACOZOA (*Trichoplax*)	Simplest animal; like a tiny plate, but with layers of cells	1
PORIFERA (poriferans)	Sponges	8,000
CNIDARIA (cnidarians)	Hydrozoans, jellyfishes, corals, sea anemones	11,000
PLATYHELMINTHES (flatworms)	Turbellarians, flukes, tapeworms	15,000
CTENOPHORA (comb jellies)	Venus's girdle	100
NEMERTEA (ribbon worms)	Proboscis-equipped worms closely related to flatworms	800
NEMATODA (roundworms)	Pinworms, hookworms	20,000
ROTIFERA (rotifers)	Tiny body with crown of cilia, great internal complexity	2,000
MOLLUSCA (mollusks)	Snails, slugs, clams, squids, octopuses	110,000
ANNELIDA (segmented worms)	Leeches, earthworms, polychaetes	15,000
ARTHROPODA (arthropods)	Crustaceans, spiders, insects	1,000,000+
ECHINODERMATA (echinoderms)	Sea stars, sea urchins	6,000
CHORDATA (chordates)	Invertebrate chordates: Tunicates, lancelets	2,100
	Vertebrates:	
	Fishes	21,000
	Amphibians	4,900
	Reptiles	7,000
	Birds	8,600
	Mammals	4,500

BODY SYMMETRY AND CEPHALIZATION With very few exceptions, animals are radial or bilateral. Those with **radial symmetry** have body parts arranged regularly around a central axis, like spokes of a bike wheel. Thus a cut down the center of a hydra (Figure 26.3*a*) divides it into equal halves; another cut at right angles to the first divides it into equal quarters. Radial animals live in water. Their body plan is adapted to intercepting food that is coming toward them from any direction.

Animals with **bilateral symmetry** have a right half and left half that are mirror images of each other. Most

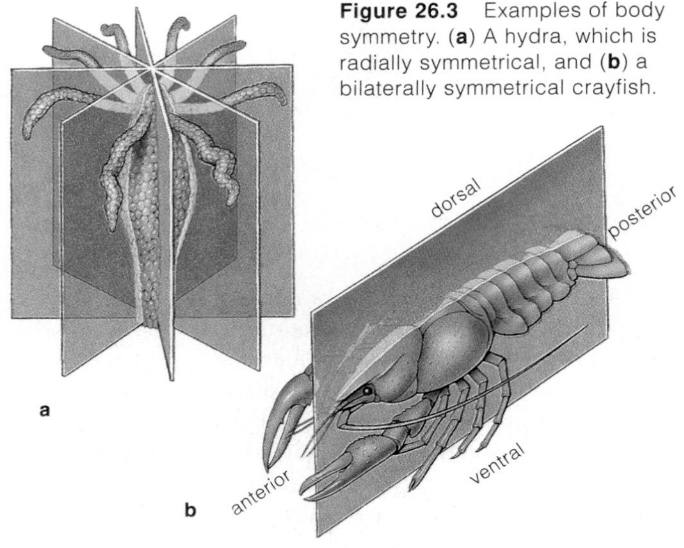
a
b
dorsal
posterior
anterior
ventral

Figure 26.3 Examples of body symmetry. (**a**) A hydra, which is radially symmetrical, and (**b**) a bilaterally symmetrical crayfish.

Further reading: Student Guide to InfoTrac on web site

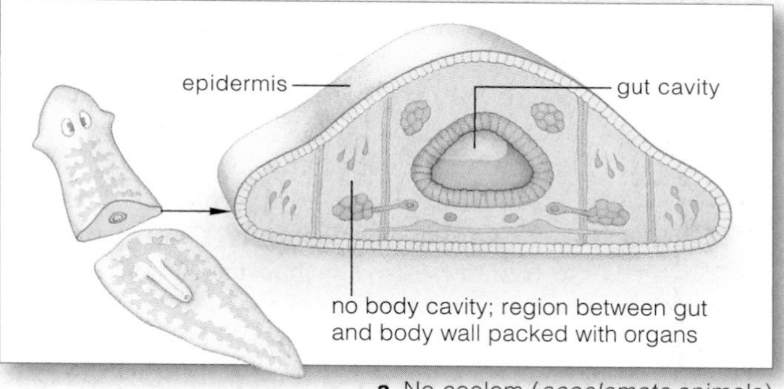
epidermis
gut cavity
no body cavity; region between gut
and body wall packed with organs

a No coelom (*acoelomate* animals)

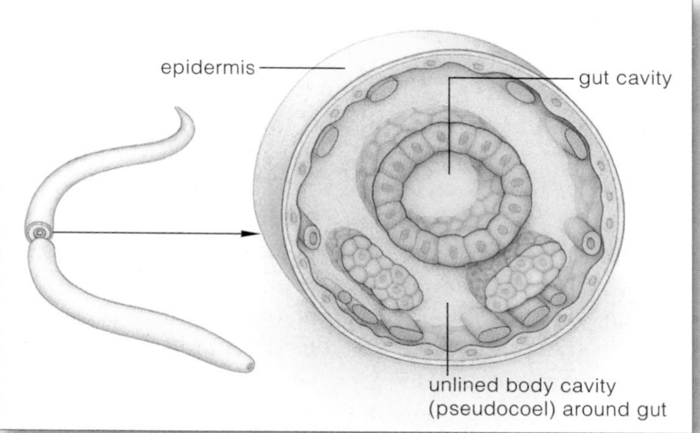
epidermis
gut cavity
unlined body cavity
(pseudocoel) around gut

b Pseudocoel (*pseudocoelomate* animals)

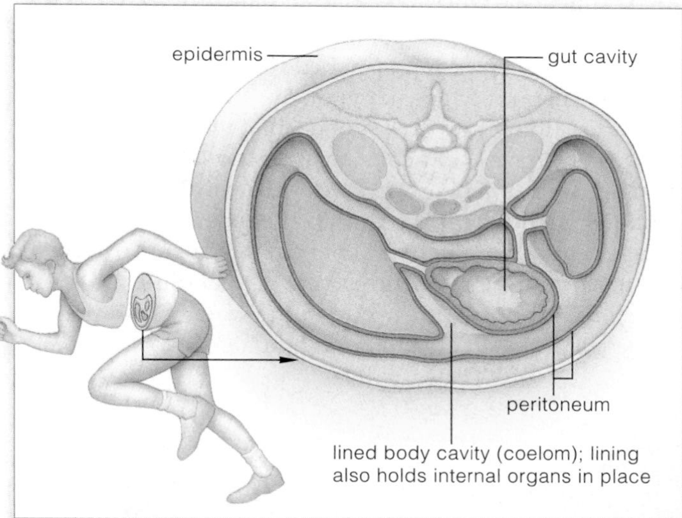
epidermis
gut cavity
peritoneum
lined body cavity (coelom); lining
also holds internal organs in place

c Coelom (*coelomate* animals)

Figure 26.4 Type of body cavity (if any) in animals. 

have an *anterior* end (head) and an opposite, *posterior* end. They have a *dorsal* surface (a back) and an opposite, *ventral* surface (Figure 26.3*b*). As fossils show, this body plan evolved among the first forward-creeping species. Their forward end would have encountered food and other stimuli first, so there must have been selection for **cephalization**. With this evolutionary process, sensory structures and nerve cells became concentrated in the head. The joint evolution of bilateral body plans and cephalization resulted in pairs of muscles and pairs of sensory structures, nerves, and brain regions.

TYPE OF GUT The **gut** is a tubular or saclike region in the body in which food is digested, then absorbed into the internal environment. Saclike guts have one opening (a mouth) for taking in food and expelling residues. Other guts are part of a tubelike, "complete" digestive system with an opening at two ends (mouth and anus). Different parts of the system have specialized functions, such as preparing, digesting, and storing material. As more efficient digestive systems evolved, this helped pave the way for increases in body size and activity.

BODY CAVITIES In between the gut and the body wall of most bilateral animals is a body cavity (Figure 26.4). One type of cavity, a **coelom**, has a unique tissue lining called a peritoneum. This lining also encloses organs in the coelom and helps hold them in place. For example, your body has a coelom, and a sheetlike muscle called a diaphragm divides it into two smaller cavities. Your heart and lungs are positioned in the upper (thoracic) cavity, and your stomach, intestines, and other organs occupy the lower (abdominal) cavity.

Some invertebrates don't have a body cavity; tissues fill the region between their gut and body wall. Others have a pseudocoel ("false coelom"), a body cavity with no peritoneum. The coelom was a key innovation by which larger and more complex animals evolved from ancestral forms. It favored increases in size and activity by cushioning and protecting internal organs.

SEGMENTATION Segmented animals have a repeating series of body units that may or may not be similar to one another. The many segments of earthworms have a similar outward appearance. Insect segments are fused into three units (head, thorax, and abdomen) and differ greatly from one another. Especially among the insects, diverse head parts, legs, wings, and other appendages evolved from less specialized segments.

Animals are multicelled, and most have tissues, organs, and organ systems. Most have a diploid chromosome number.

Animals are aerobically respiring heterotrophs that ingest other organisms or absorb nutrients from them.

Animals reproduce sexually and, in many cases, asexually. They go through a period of embryonic development, and most are motile during at least part of the life cycle.

Body plans of animals differ with respect to five features: body symmetry, cephalization, type of gut, type of body cavity, and segmentation.

26.2 PUZZLES ABOUT ORIGINS

Judging from recent genetic evidence and radiometric dating of fossilized tracks, burrows, and microscopic embryos, animals originated between 1.2 billion and 670 million years ago, during the Precambrian. *Where did they come from?* They probably evolved from protistan lineages, but we don't know which ones (Section 23.3).

By one hypothesis, the forerunners of animals were ciliates, like *Paramecium*, and had multiple nuclei in a one-celled body. As they evolved, each nucleus became compartmentalized in individual cells of a multicelled body. However, we don't know of any existing animal that develops by compartmentalization.

By another hypothesis, multicelled animals arose from spherical colonies of a number of flagellated cells, maybe like the *Volvox* colonies shown in Section 23.13. In time, as a result of mutation, some cells in the colony became modified in ways that enhanced reproduction and other specialized tasks. And so began the division of labor that characterizes multicellularity.

Suppose such colonies became flattened and started creeping around on the seafloor. A creeping life-style could have favored the evolution of layers of cells, such as those of *Trichoplax adhaerens*. This is the only known **placozoan** (Placozoa, after *plax*, meaning plate, and *zoon*, which means animal). *Trichoplax* is a soft-bodied marine animal, shaped a bit like a tiny pita bread. It has no symmetry and no mouth. Its several thousand cells are arranged into two distinctly different layers. It briefly humps up when its body glides over food, as in Figure 26.5. Gland cells in the lower layer secrete digestive enzymes onto the food, then individual cells absorb the breakdown products. Reproduction might be asexual (by budding or fission) or sexual, by mechanisms not yet understood. In sum, structurally and functionally, *Trichoplax* is as simple as animals get.

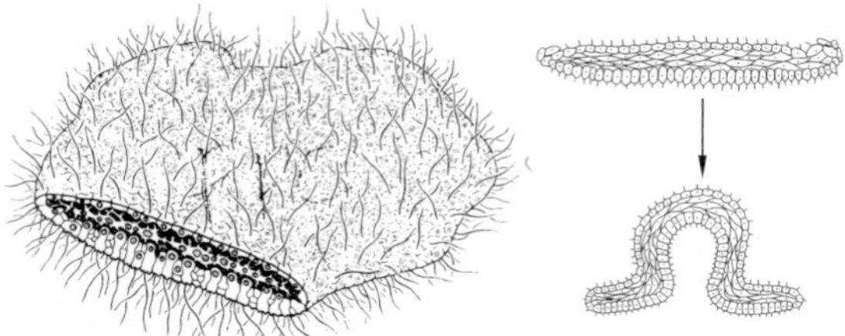

Figure 26.5 Cutaway views of *Trichoplax adhaerens*, an animal with a two-layer body measuring about three millimeters across.

Possibly the question of origins requires more than one answer. It may be that different lineages descended from more than one group of protistan-like ancestors.

Multicelled animals arose from protistan-like ancestors that may have resembled ciliates, colonial flagellates, or both.

26.3 SPONGES—SUCCESS IN SIMPLICITY

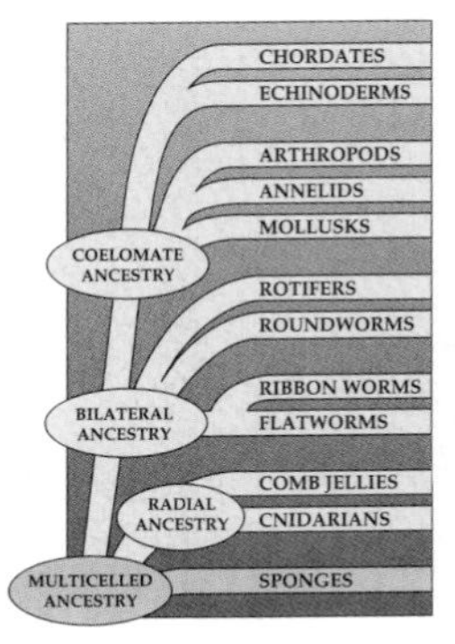
CHORDATES
ECHINODERMS
ARTHROPODS
ANNELIDS
MOLLUSKS
COELOMATE ANCESTRY
ROTIFERS
ROUNDWORMS
RIBBON WORMS
FLATWORMS
BILATERAL ANCESTRY
COMB JELLIES
CNIDARIANS
RADIAL ANCESTRY
SPONGES
MULTICELLED ANCESTRY

Sponges (Porifera) are animals with no symmetry, tissues, or organs. Yet they are one of nature's success stories. They have been abundant in the seas ever since the Precambrian, especially in the waters off coasts and along coral reefs. Of about 8,000 known species, only a hundred or so live in freshwater habitats. Many small worms, shrimps, and other animals make their home in or on a sponge body. Some sponges are large enough to sit in. Others are as small as a fingernail. Figures 26.6 and 26.7 show only a few of the flattened, sprawling, compact, lobed, tubular, cuplike, and vaselike shapes.

Regardless of its shape, the body of a sponge is not symmetrical. Flattened cells do line the outer surface and inner cavities. But these linings are not much more specialized than the cell layers of *Trichoplax*, and they differ from the tissues of other animals. Amoeboid cells live in a gelatin-like substance between the two linings (Figure 26.7*b*). Spicules, tough fibers, or both stiffen the sponge body. The fibers are made of spongin, a protein, and the sharp, glasslike spicules are made of calcium carbonate or silica.

The skeletal elements may be a reason why sponges as a group have endured so long. Cleveland Hickman put it this way: Most potential predators discover that sampling a sponge is about as pleasant as eating a mouthful of glass splinters embedded in fibrous gelatin. Besides, chemically speaking, many sponges stink.

Water flows into the sponge body through many microscopic pores and chambers, then out through one or more large openings. It does so when thousands or millions of **collar cells** beat their flagella. The cells are components of the body's inner lining. Their "collars" are food-trapping structures called microvilli (Figure 26.7*c*). Bacteria and other food in the water get trapped in the collars, then are engulfed by phagocytosis. Some

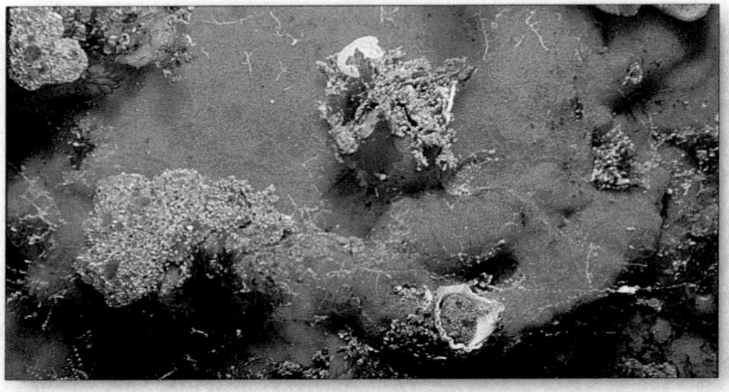

Figure 26.6 A sprawling, red-orange sponge, one of many types that encrust underwater ledges in temperate seas.

Further reading: Student Guide to InfoTrac on web site

Figure 26.7 (**a**,**b**) Body plan of a simple sponge. The outer lining consists of flattened cells. It also has some contractile cells, most of which are arranged around the large opening at the top. These cells contract slowly, independently of the other cell types, and so influence water flow through the body. Amoeboid cells inside the gelatin-like matrix between the inner and outer linings secrete materials from which the body's spicules and fibers are constructed. Other amoeboid cells digest and transport food. They do not lose the capacity to divide, and their cellular descendants can differentiate into any other type of sponge cell. They also have roles in asexual processes, such as gemmule formation.

(**c**) A great many flagellated cells line inner canals and chambers of the sponge body. Each of these phagocytic cells has a collar of food-trapping structures called microvilli. Fine filaments connect the microvilli to each other; they form a "sieve" that strains food particles from the water. At the base of the collar, the cell engulfs the trapped food.

(**d**) An example of skeletal elements—Venus's flower basket (*Euplectella*). This marine sponge has six-rayed spicules of silica fused in a rigid, elaborate network. A thin layer of cells stretches over all interconnecting spicules. At the base of the body is an anchoring tuft of spicules.

(**e**) Basket sponge releasing a cloud of sperm into the water.

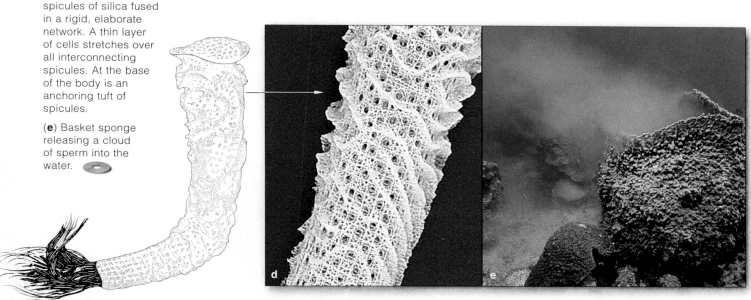

water out
central cavity
water in
a
glasslike structural elements
amoeboid cell
pore
semifluid matrix
flattened surface cells
b
flagellum
microvilli
nucleus
c Collar cell

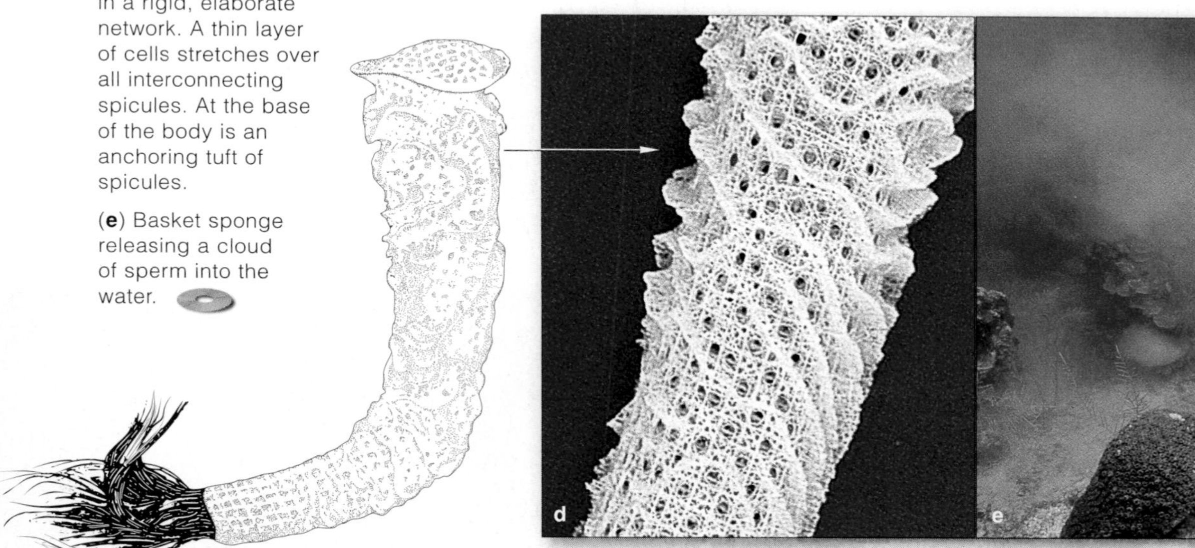
d
e

of the food also is transferred to the amoebalike cells for further breakdown, storage, and distribution.

Sponges reproduce sexually. Most release sperm into the water (Figure 26.7*e*). But sponges retain eggs until after fertilization and new embryos have started their development. A young sponge proceeds through a microscopic, swimming larval stage. A **larva** (plural, larvae) is a sexually immature stage that grows and develops into an adult, the sexually mature form of the species. As you will see, the life cycle of many animal species includes larval stages.

Some kinds of sponges also reproduce asexually by fragmentation; small fragments break away from the parent and grow into new sponges. Most freshwater species also reproduce asexually by way of gemmules. These are clusters of sponge cells, some of which form a hard covering around others. The clusters inside are protected from extreme cold or drying out. Later, when favorable conditions return, the gemmules germinate and establish a new colony of sponges.

Sponges have no symmetry, tissues, or organs; they are at the cellular level of construction. Yet they have successfully endured through time, possibly because most predators find their spicule-rich and often stinky bodies unappetizing.

26.4 CNIDARIANS—TISSUES EMERGE

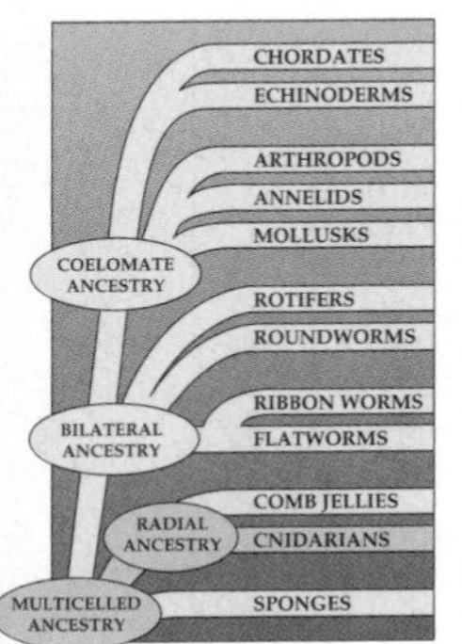
CHORDATES
ECHINODERMS
ARTHROPODS
ANNELIDS
MOLLUSKS
COELOMATE ANCESTRY
ROTIFERS
ROUNDWORMS
RIBBON WORMS
BILATERAL ANCESTRY
FLATWORMS
COMB JELLIES
RADIAL ANCESTRY
CNIDARIANS
MULTICELLED ANCESTRY
SPONGES

Look at the two jellyfishes in Figure 26.8. They are scyphozoans which, along with anthozoans (such as sea anemones) and hydrozoans (such as *Hydra*) are tentacled, radial animals known as **cnidarians**. Most live in the seas. Of 11,000 known species of the phylum Cnidaria, fewer than 50 are adapted to freshwater habitats.

Regarding the Nematocysts

Of all animals, cnidarians alone produce **nematocysts**: capsules that house dischargeable, tubular threads. The threads of some types have prey-piercing barbs and an open tip that delivers toxins (Figure 26.9). Other nematocysts discharge threads that ooze a sticky substance from their tip or long threads that entangle prey. Many swimmers have learned that the toxin-tipped threads of some species can sting. Hence the name of the phylum, Cnidaria, after the Greek word for nettle.

Cnidarian Body Plans

The **medusa** (plural, medusae) and **polyp** are the most common cnidarian body forms. Both have a saclike gut (Figure 26.8*a*,*b*). Medusae float. Some look like bells and others like upside-down saucers. The mouth, centered under the bell, may have extensions that assist in prey capture and feeding. Polyps have a tubelike body with a tentacle-fringed mouth at one end. Usually the other end is attached to a substrate. Unlike *Trichoplax* (which is rather like a gut on the run when draped over food), a cnidarian has a permanent food-processing chamber. Its gut has gastrodermis, a sheetlike lining that incorporates glandular cells. The cells secrete digestive enzymes. An epidermis lines the rest of the body's surfaces (Figure 26.10). Each lining is an **epithelium** (plural,

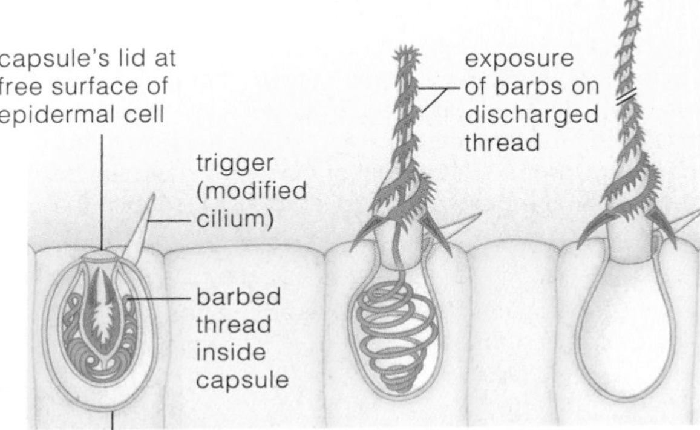
capsule's lid at free surface of epidermal cell
trigger (modified cilium)
barbed thread inside capsule
exposure of barbs on discharged thread
nematocyst (capsule at free surface of epidermal cell)

Figure 26.9 One type of nematocyst before and after a prey organism (not shown) touched its trigger. The contact made the capsule more "leaky" to water. Water diffused inward, turgor pressure built up within the capsule, and the thread was forced to turn inside out. The thread's tip pierced the prey's body.

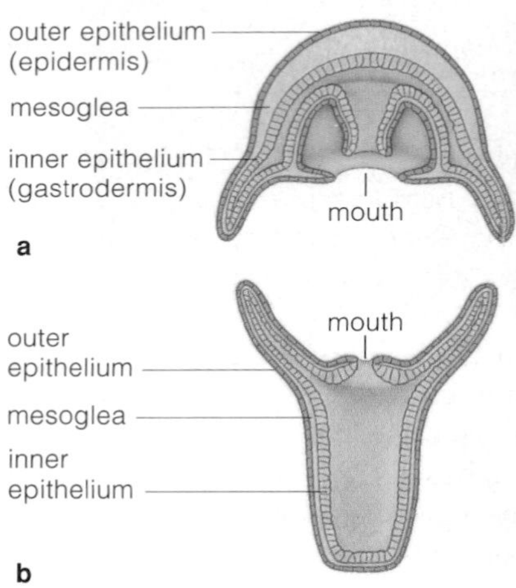
outer epithelium (epidermis)
mesoglea
inner epithelium (gastrodermis)
mouth
a
mouth
outer epithelium
mesoglea
inner epithelium
b

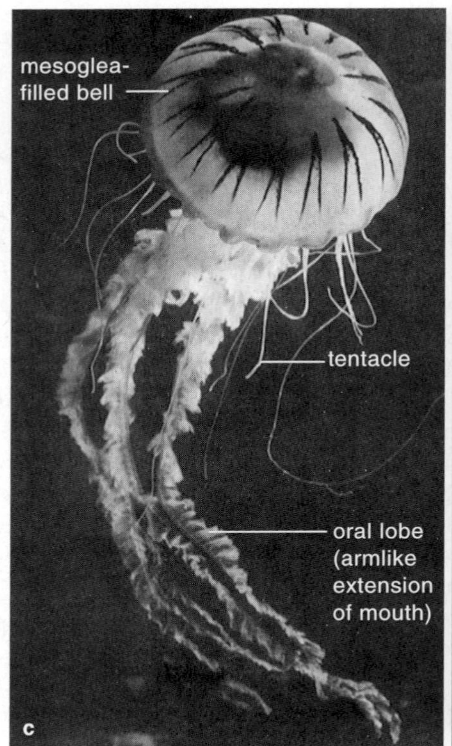
mesoglea-filled bell
tentacle
oral lobe (armlike extension of mouth)
c

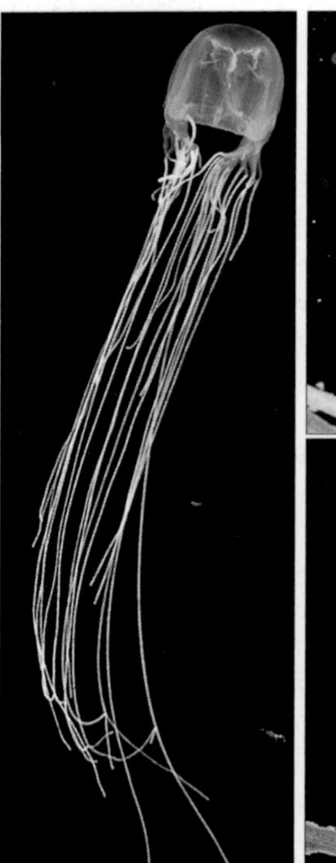
d

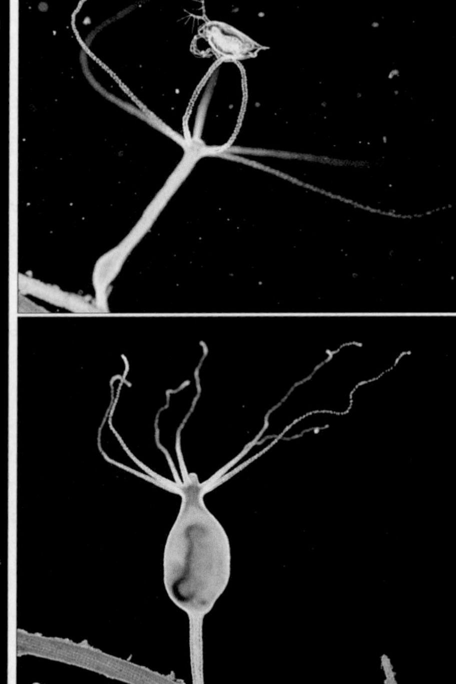
e

Figure 26.8 Cnidarian body plans. (**a**,**b**) Medusa and polyp, midsection. Two jellyfish: (**c**) sea nettle (*Chrysaora*) and (**d**) sea wasp (*Chironix*), which has tentacles up to fifteen meters long. Its toxin can kill a human in minutes. (**e**) A hydrozoan polyp (*Hydra*) attached to a substrate, capturing and then digesting its prey.

Further reading: Student Guide to InfoTrac on web site —

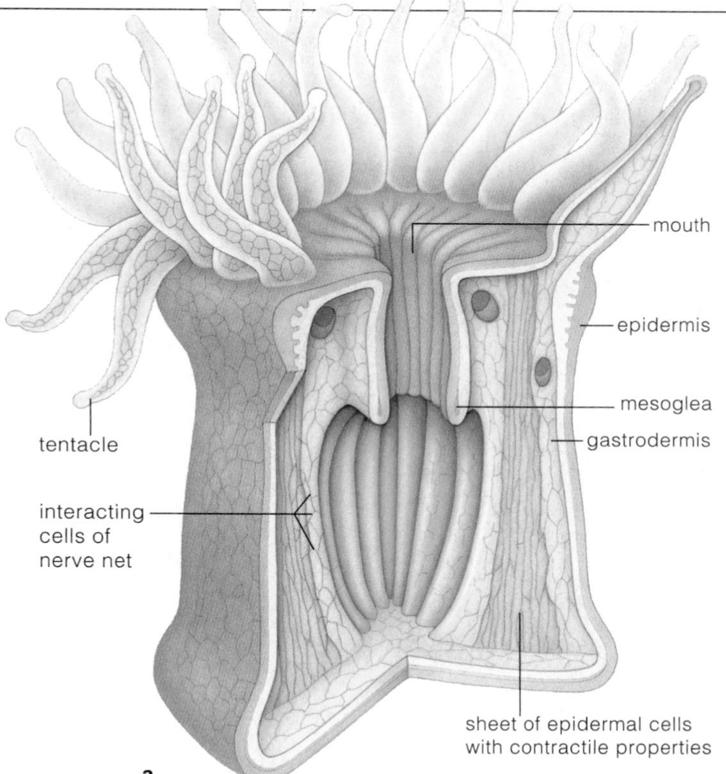
mouth
epidermis
mesoglea
gastrodermis
tentacle
interacting cells of nerve net
sheet of epidermal cells with contractile properties
a

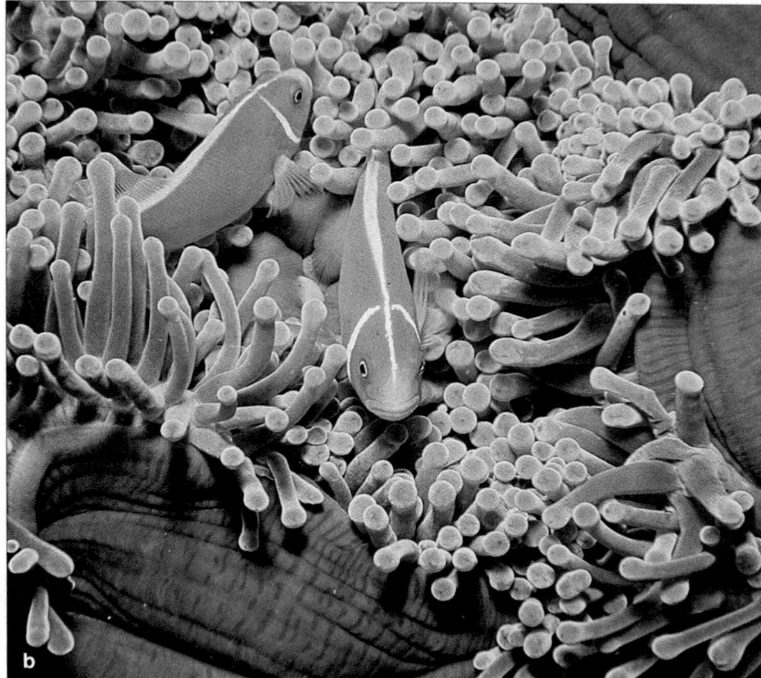
b

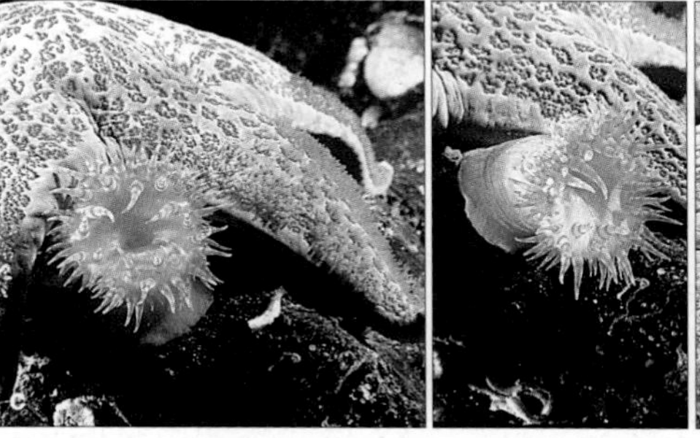

Figure 26.10 (**a**) Tissue organization of sea anemones. (**b**) Tentacles fringing the mouth of a sea anemone, which eats many fishes but not clownfishes. A clownfish swims out, captures food, and returns to the tentacles, which protect it from predators. The sea anemone eats food scraps falling from the fish mouth. This is a case of mutualism, a two-way flow of benefits between species. (**c**) A sea anemone using its hydrostatic skeleton to escape from a sea star. It closes its mouth, and the force generated by contractile cells in epithelial tissues acts against water in the gut. The body changes shape and the anemone thrashes about, generally in a direction away from the predator.

epithelia), a tissue having a free surface that faces the environment or some type of fluid inside the body. Any animal that is structurally more complex than sponges incorporates epithelia. Cnidarian epithelia house **nerve cells.** Such cells receive signals from receptors that can detect changes in the surroundings. They send signals to **contractile cells**, which carry out suitable responses. (When stimulated, contractile cells *shorten*, then return to their original length when stimulation stops.) The nerve cells interact as a "nerve net," a simple nervous tissue, to control movement and changes in shape.

Between the epidermis and gastrodermis is a layer of gelatinous secreted material, the mesoglea ("middle jelly"). Jellyfishes contain enough mesoglea to impart buoyancy and serve as a firm yet deformable skeleton against which contractile cells act. Imagine many cells contracting in coordinated ways in a jellyfish bell. Their action narrows the bell and forces water to jet out from underneath it, and the jet propels the jellyfish forward. The bell returns to its original position, cells contract again, and the animal is propelled forward. Although the swimming movements are not Olympian, they work well enough for an animal that secures food by dragging its tentacles through the water for small prey.

Any fluid-filled cavity or cell mass against which contractile cells can act is a **hydrostatic skeleton**. With coordinated contractions, the cavity's volume or mass does not change but rather is shunted about, so that the shape of the body changes. The contractile cells of most polyps, which have little mesoglea, act against water in their gut. As you will see, nearly all animals have some form of skeletal–muscular system of movement.

Cnidarians are radial animals with tentacles, a saclike gut, epithelia, a nerve net, and a hydrostatic skeleton. They alone produce nematocysts.

Cnidarians are at the tissue level of construction; compared with sponges, they have layers of cells interacting in more coordinated fashion in the performance of specific tasks.

26.5 VARIATIONS ON THE CNIDARIAN BODY PLAN

Various Stages in Cnidarian Life Cycles

From the preceding section, you may have concluded that the body form of a particular cnidarian is a medusa or a polyp. This is indeed the case for many species. However, the life cycle of *Obelia, Physalia,* and some other cnidarians includes both forms. By using the life cycle of *Obelia* as an example, you can get a general sense of how these forms grow and develop (Figure 26.11).

The medusa is the sexual stage of the cnidarian life cycle. It has simple **gonads**, which are primary (gamete-producing) reproductive organs. Either the epidermis or gastrodermis houses the gonads, which release gametes by rupturing. Most of the zygotes formed at fertilization develop into **planulas**—a kind of swimming or creeping larva, usually with ciliated epidermal cells. In time, a mouth opens at one end, the larva is transformed into a polyp or medusa, and the cycle begins anew.

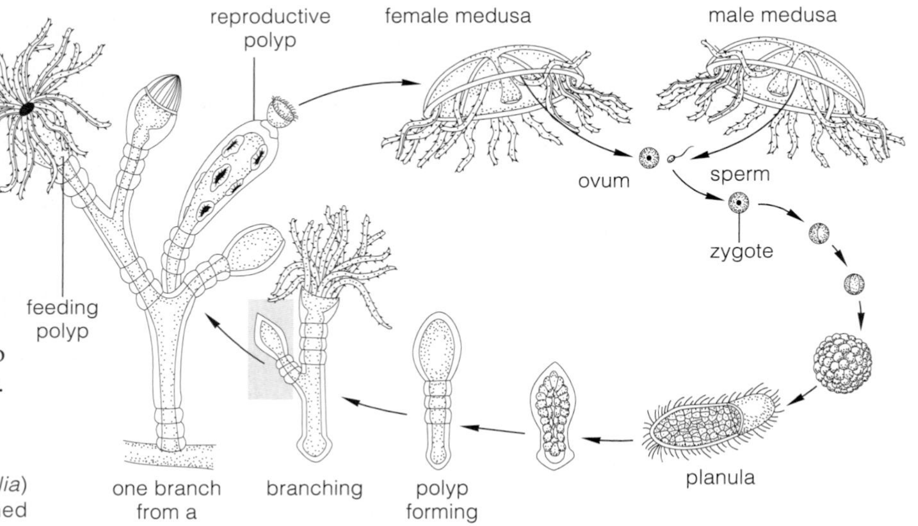
reproductive polyp
female medusa
male medusa
ovum
sperm
zygote
feeding polyp
planula
one branch from a mature colony
branching
polyp forming

Figure 26.11 *Right*: Life cycle of a hydrozoan (*Obelia*) that includes medusa and polyp stages. An established colony may contain thousands of feeding polyps.

A Sampling of Colonial Types

The reef-forming corals and other colonial anthozoans are a fine example of variations on the basic cnidarian body plan. The reef formers thrive in clear, warm water that is at least 20°C (68°F). As Figure 26.12 shows, the colonies consist of polyps that have secreted calcium-

a
c

land mass
barrier reef
open ocean
b

Figure 26.12 (**a**) One of the reef-building, colonial corals. External skeletons of their polyps interconnect with one another. Dinoflagellate mutualists live in the polyp tissues. (**b**) Aerial view of a barrier reef, mainly an accumulation of the compacted skeletons of certain coral species. (**c**) Not all corals are reef builders; cup corals such as *Tubastrea* are solitary forms. With tentacles extended, their polyps look like tiny sea anemones.

Further reading: Student Guide to InfoTrac on web site →

reinforced external skeletons, which interconnect with one another. Over time, the skeletons accumulate and so become the main building material for reefs. Today the most impressive accumulations parallel the eastern coast of Australia for about 1,600 kilometers. We call their remains the Great Barrier Reef.

Reef-building corals receive nutrient inputs from the changing tides and from dinoflagellate mutualists living in their tissues. Masses of these photosynthetic protistans supply corals with oxygen, recycle their mineral wastes, and adjust the pH of the surrounding water in a way that enhances the rate of calcium deposition for their host's skeletons. The host corals reciprocate by giving the dinoflagellates a relatively safe, sunlit habitat that has plenty of dissolved carbon dioxide and mineral ions. In the sunlit parts of a reef, nutrients are cycled quickly, directly, and efficiently.

As a final example of cnidarian diversity, consider *Physalia*, informally called the Portuguese man-of-war. The toxin in the nematocysts of this infamous colonial hydrozoan poses a danger to bathers and fishermen as

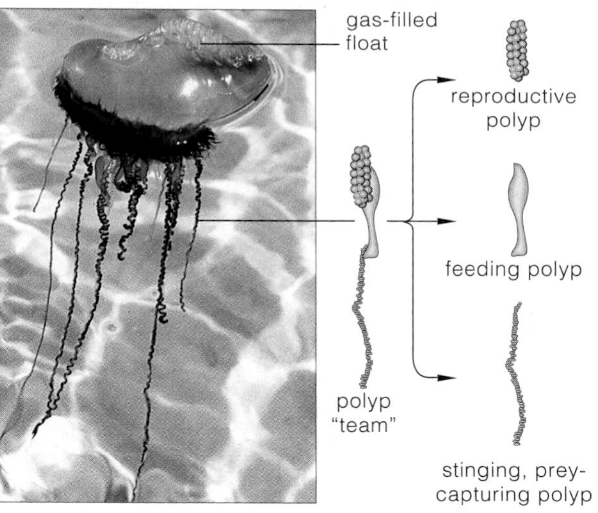
gas-filled float
reproductive polyp
feeding polyp
polyp "team"
stinging, prey-capturing polyp

Figure 26.13 Portuguese man-of-war (*Physalia*).

well as to prey organisms (fish). Although *Physalia* lives mainly in warm waters, currents sometimes move it up to the Atlantic coasts of North America and Europe. A blue, gas-filled float that develops from the planula keeps the colony near the water's surface, where winds move it about (Figure 26.13). Under the float, groups of polyps and medusae interact as "teams" in feeding, reproduction, defense, and other specialized tasks.

Cnidarians, including the colonial forms, show notable variation in their body plans and life-styles.

26.6 COMB JELLIES

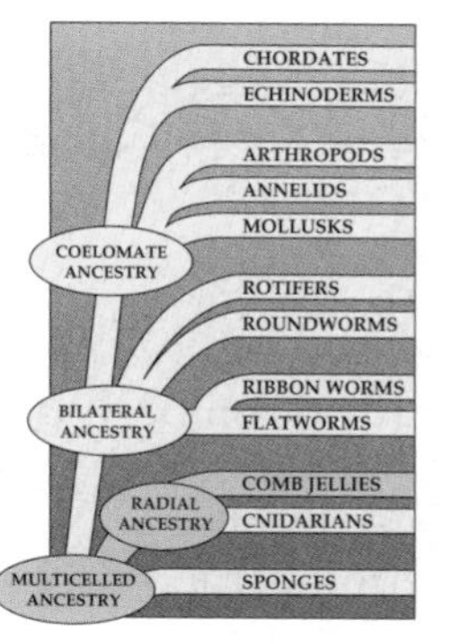
CHORDATES
ECHINODERMS
ARTHROPODS
ANNELIDS
MOLLUSKS
COELOMATE ANCESTRY
ROTIFERS
ROUNDWORMS
RIBBON WORMS
BILATERAL ANCESTRY
FLATWORMS
COMB JELLIES
RADIAL ANCESTRY
CNIDARIANS
MULTICELLED ANCESTRY
SPONGES

If you have ever been diving in coastal waters anywhere from the tropics to the poles, you may have observed cnidarians that look a bit like jellyfishes with combs running down the sides. These are **comb jellies**, of phylum Ctenophora (meaning "comb-bearing"). All are weak-swimming predators in planktonic communities, and all show modified radial symmetry. Slice them in two equal halves, then slice them into quarters, and two of those quarters will be mirror images of the other two.

A comb jelly has eight rows of comblike structures made of thick, fused cilia (Figure 26.14). All the combs in a row beat in waves and propel the animal forward, usually (and suitably) mouth first. Some species have two long, muscular tentacles with branches that are equipped with sticky cells. Comb jellies do not produce nematocysts, but sometimes they opportunistically save and use the ones from jellyfish they have eaten. Others use their sticky lips to capture prey.

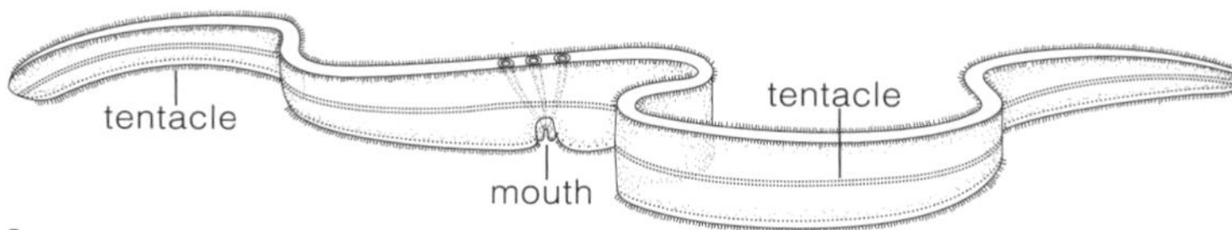
tentacle
tentacle
mouth
a

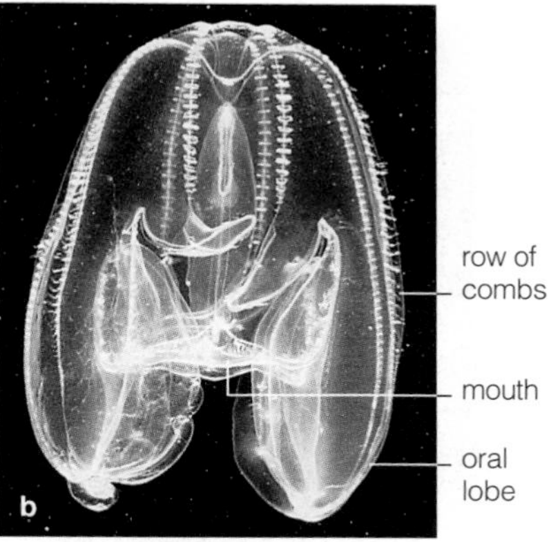
row of combs
mouth
oral lobe
b

Figure 26.14 Examples of comb jellies. (**a**) Sketch of Venus's girdle (*Cestum veneris*), a transparent animal about 1.5 meters (more than 4 feet) long.

(**b**) *Mnemiopsis*. Its oral lobes alternately expand and contract to waft prey into the mouth. It is about 15 centimeters (6 inches) long.

Evolutionarily, comb jellies are interesting because they have cells with multiple cilia. This trait evolved in many of the complex animals. In addition, comb jellies are the simplest animals having embryonic tissues that resemble mesoderm. Mesoderm is the embryonic source of muscles and organs of the circulatory, excretory, and reproductive systems of most complex animals.

Comb jellies have a modified radial body and comblike structures of modified cilia. They have an embryonic tissue that resembles the mesoderm of more complex animals.

26.7 ACOELOMATE ANIMALS—AND THE SIMPLEST ORGAN SYSTEMS

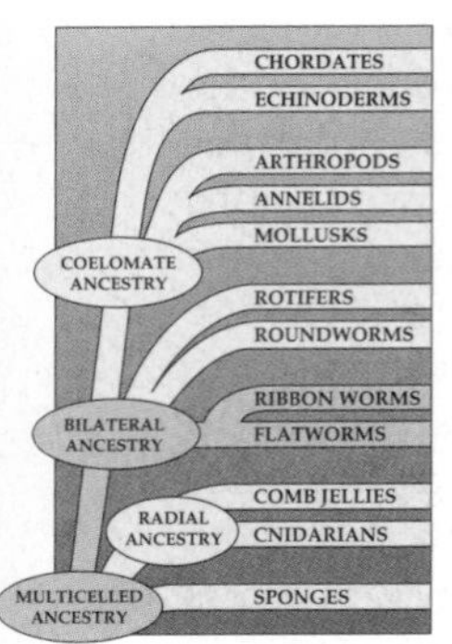
CHORDATES
ECHINODERMS
ARTHROPODS
ANNELIDS
MOLLUSKS
COELOMATE ANCESTRY
ROTIFERS
ROUNDWORMS
RIBBON WORMS
BILATERAL ANCESTRY
FLATWORMS
COMB JELLIES
RADIAL ANCESTRY
CNIDARIANS
MULTICELLED ANCESTRY
SPONGES

When we move beyond the cnidarians in our survey, we find animals that range from flatworms to humans. All of these animals have simple or complex organs. An **organ** is an association of one or more kinds of tissues, arranged in particular proportions and patterns. Most often, one organ interacts with others to carry out an activity that helps the body function. By definition, two or more organs that are interacting efficiently in the performance of some task represent an **organ system**. We turn now to the simplest animals at the *organ-system* level of construction. They are not what you would call breathtakingly complex but, as you will see shortly, a few parasitic types can make our lives miserable.

Parasites, recall, reside in or on living hosts and feed on their tissues. Most do not kill their hosts, at least not until after they reproduce. A *definitive* host harbors the mature stage of the parasite's life cycle. One or more *intermediate* hosts harbor immature stages.

Flatworms

Among the 15,000 or so known species of **flatworms** (phylum Platyhelminthes) are turbellarians, flukes, and tapeworms. Most of these bilateral, cephalized animals have simple organ systems in a flattened body, as in Figure 26.15. Their digestive system, for example, has a pharynx (a muscular tube, which flatworms use for feeding) as well as a saclike, often branching gut. They differ in their reproductive systems, but most species are **hermaphrodites**: individuals have female *and* male gonads. Each has a penis (sperm-delivery structure), so two flatworms can reproduce sexually by exchanging sperm. Secretions from certain flatworm glands form a protective capsule around the fertilized eggs.

TURBELLARIANS Most turbellarians (class Turbellaria) live in the seas; only planarians and a few others live in freshwater. Some eat tiny animals or suck tissues from dead or wounded ones. A planarian can divide in half at its midsection, then each half can regenerate missing parts. Thus it reproduces asexually by *transverse* fission. Like you, a planarian can adjust the composition and volume of its body fluids. Its water-regulating system has one or more branched tubes called protonephridia (singular, protonephridium). These tubes extend from pores at the body surface to bulb-shaped flame cells in tissues. When excess water diffuses into the flame cells, a tuft of cilia "flickering" in the bulb drives the water through the tubes to the outside (Figure 26.15*b*).

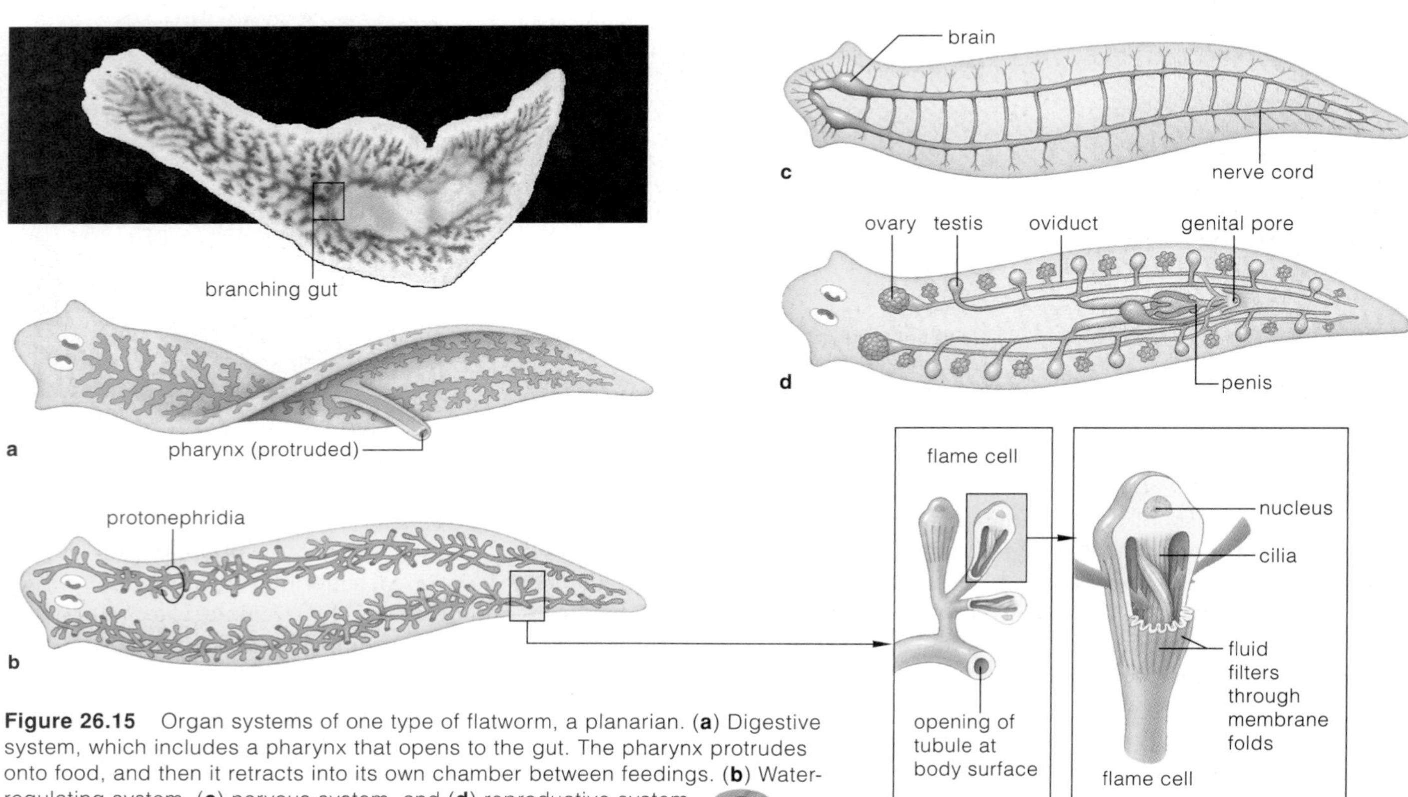
branching gut
a
pharynx (protruded)
protonephridia
b
brain
c
nerve cord
ovary
testis
oviduct
genital pore
d
penis
flame cell
opening of tubule at body surface
nucleus
cilia
fluid filters through membrane folds
flame cell

Figure 26.15 Organ systems of one type of flatworm, a planarian. (**a**) Digestive system, which includes a pharynx that opens to the gut. The pharynx protrudes onto food, and then it retracts into its own chamber between feedings. (**b**) Water-regulating system, (**c**) nervous system, and (**d**) reproductive system.

Further reading: Student Guide to InfoTrac on web site →

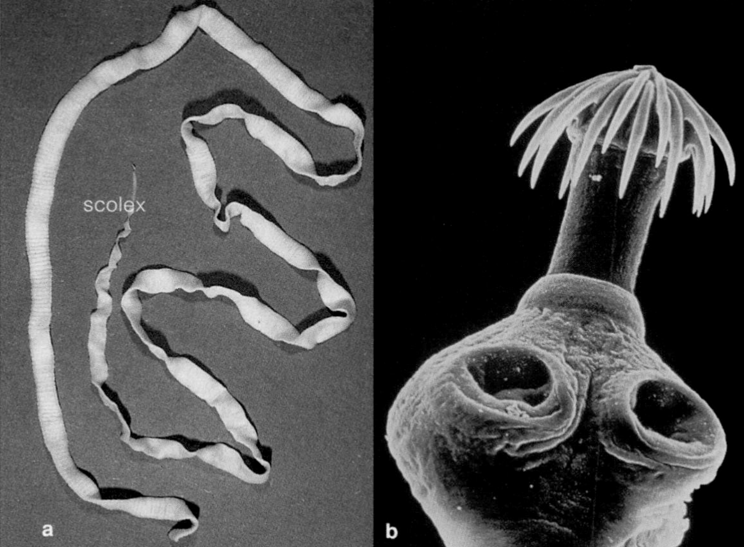
scolex
a
b

Figure 26.16 (**a**) Tapeworm scolex. This one attaches to the gut lining of a shorebird, its primary host. (**b**) Sheep tapeworm.

FLUKES Flukes (class Trematoda) are parasitic worms. Their complicated life cycle has sexual phases, many asexual phases, and one to four hosts. Almost always, a snail or a clam is the initial host for larval or juvenile stages, and a vertebrate is the definitive host. Section 29.5 takes a closer look a blood fluke (*Schistosoma*). Schistosomes parasitize as many as 200 million people.

TAPEWORMS Tapeworms (class Cestoda) parasitize intestines of vertebrates. It seems probable that the ancestral tapeworms had a gut but later lost it during their evolution in animal intestines, which happen to be habitats that are rich in predigested food. Their existing descendants attach to the intestinal wall with a scolex, a structure that has suckers, hooks, or both (Figure 26.16). **Proglottids** bud just behind the scolex; each is a new unit of the tapeworm body. These units are hermaphroditic; they mate and transfer sperm with one another. The older proglottids (those farthest from the scolex) store fertilized eggs. They break off from the younger ones, then they leave the body in feces. Later, the eggs may enter an intermediate host. Section 26.9 describes how proglottids form during the life cycle of a representative tapeworm.

In some respects, the simplest turbellarians, larval flukes, and larval tapeworms resemble the planulas of cnidarians. The resemblance inspires speculation that bilateral animals evolved from planula-like ancestors. They may have done so by increased cephalization and the development of tissues derived from mesoderm.

Ribbon Worms

Ribbon worms (phylum Nemertea) are bilateral, soft-bodied, elongated predators that swallow or suck tissue fluids from small worms, mollusks, and crustaceans. Most crawl, burrow, or lurk in shallow marine habitats, as in Figure 26.17, but some live in freshwater or humid tropical habitats. Like flatworms, which may be their close relatives, ribbon worms have a ciliated surface, and they secrete mucus and then move by beating their cilia through it. Ribbon worms also resemble flatworms in their tissue organization. However, they differ from flatworms in having a circulatory system, a complete gut, and separation of sexes. Ribbon worms also have a proboscis, which in this case is a tubular, prey-piercing, venom-delivering device tucked inside the head end. Muscle contractions force the proboscis inside out and into prey, which the venom paralyzes.

Figure 26.17 Ribbon worms, poised to ambush small prey that pass by their lair in an orange-colored gorgonian (horny coral) colony.

Flatworms are among the simplest bilateral, cephalized animals with organ systems. Ribbon worms may be related to them, but they have a complete gut, a circulatory system, and other traits that are notable departures from flatworms.

26.8 ROUNDWORMS

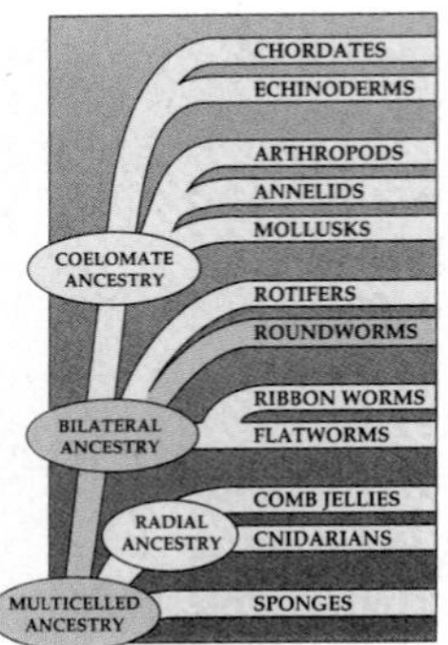
CHORDATES
ECHINODERMS
ARTHROPODS
ANNELIDS
MOLLUSKS
COELOMATE ANCESTRY
ROTIFERS
ROUNDWORMS
RIBBON WORMS
BILATERAL ANCESTRY
FLATWORMS
COMB JELLIES
RADIAL ANCESTRY
CNIDARIANS
MULTICELLED ANCESTRY
SPONGES

The **Roundworms** (phylum Nematoda) are pseudocoelomate worms. They thrive in nearly all environments, and they may be the most abundant animals alive. In most sediments beneath shallow fresh water or saltwater, there might be up to a million roundworms in each square meter. Many thousands may occupy a handful of rich topsoil in which scavenging types make fast work of dead earthworms or rotting plants. There are 20,000 known species, but there may be a hundred times more. Figure 26.18 shows a typical roundworm's body plan.

A roundworm is bilateral. But its body is cylindrical, typically tapered at both ends, and covered by a protective cuticle. Animal **cuticles** are tough, often flexible body coverings. A roundworm is the simplest animal equipped with a complete digestive system. Between the gut and body wall is a false coelom, typically jam-packed with reproductive organs. The cells in every tissue absorb nutrients from the coelomic fluid and give up wastes to it.

Parasitic species can severely damage their hosts, which include humans, cats, dogs, cows, sheep, soybeans, potatoes, and other crop plants. Although thin, they can grow long; the roundworms in female sperm whales may be nine meters long, stretched out. Elephantiasis, one of the world's most rapidly spreading diseases, is the work of a few species. (See also Sections 26.9 and 48.6.)

Most roundworms are free-living types that are harmless or beneficial, as when they cycle nutrients for communities. One species, *Caenorhabditis elegans*, is used in research into inheritance, development, and aging. The *C. elegans* genome was the first of any multicelled organism to be fully sequenced (Section 44.7).

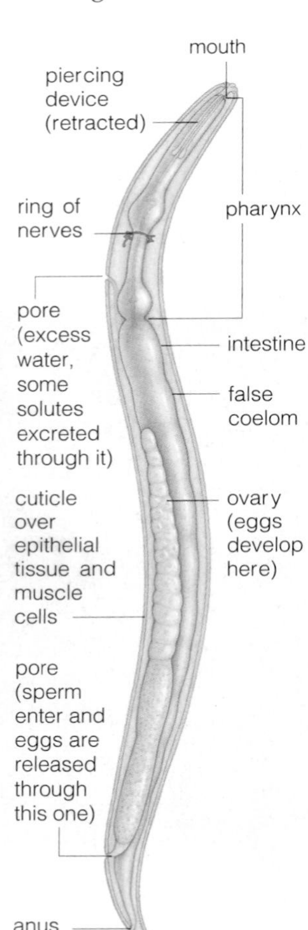
mouth
piercing device (retracted)
pharynx
ring of nerves
pore (excess water, some solutes excreted through it)
intestine
false coelom
cuticle over epithelial tissue and muscle cells
ovary (eggs develop here)
pore (sperm enter and eggs are released through this one)
anus

Figure 26.18 Body plan of a roundworm (*Paratylenchus*). This one parasitizes the roots of plants.

Roundworms are cylindrical, bilateral, cephalized animals with a false coelom and a complete digestive system. Most cycle nutrients in communities; the parasites are notorious.

Focus on Health

26.9 A ROGUE'S GALLERY OF WORMS

A number of parasitic flatworms and roundworms call the human body home. In any year, for instance, 200 million people house blood flukes responsible for *schistosomiasis*. One of these, the Southeast Asian blood fluke *Schistosoma japonicum*, requires a human definitive host, standing water in which its larvae can swim, and an aquatic snail as an intermediate host. The flukes grow, become sexually mature, and mate inside a human host (Figure 26.19*a*). After being fertilized, a female's eggs (*b*) leave the human body in feces and hatch into ciliated, swimming larvae (*c*) that burrow into a snail (*d*) and multiply asexually. In time, many fork-tailed larvae develop (*e*). These leave the snail and swim about until they contact human skin (*f*). They bore in and migrate to thin-walled intestinal veins, then the cycle begins anew. In infected humans, white blood cells that defend the body attack the masses of fluke eggs, and masses of grainy debris form in tissues. In time, the liver, spleen, bladder, and kidneys deteriorate.

Some tapeworms parasitize humans. Different species use pigs, freshwater fish, or cattle as intermediate hosts. Humans become infected when they eat pork, fish, or beef that is raw, improperly pickled, or insufficiently cooked—and contaminated with tapeworm larvae (Figure 26.20).

Or consider a parasitic roundworm that causes thin, serpentlike ridges in human skin. For several thousand years, healers have been extracting the "serpents" by winding them out slowly, painfully, around a stick. The roundworms called pinworms and hookworms cause other problems. *Enterobius vermicularis*, a pinworm of temperate regions, parasitizes humans. It lives in the large intestine, but at night the centimeter-long females migrate to the anal region of the host and lay eggs. Their presence causes itching, and scratchings made in response transfer

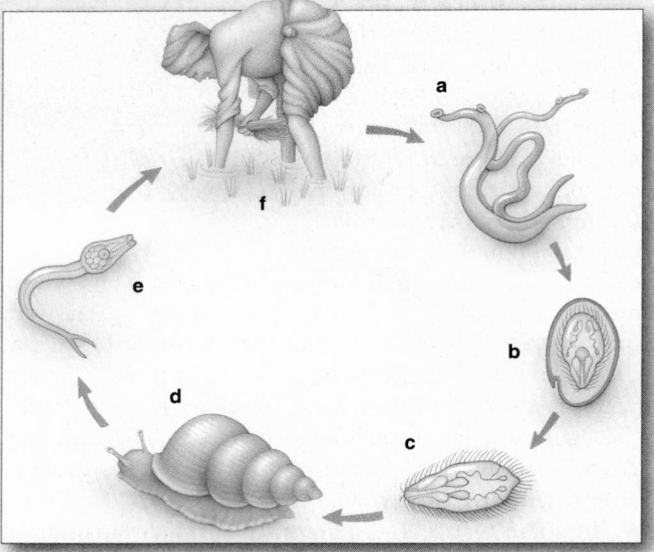
a
b
c
d
e
f

Figure 26.19 Life cycle of a dangerous blood fluke, *Schistosoma japonicum*.

Further reading: Student Guide to InfoTrac on web site —

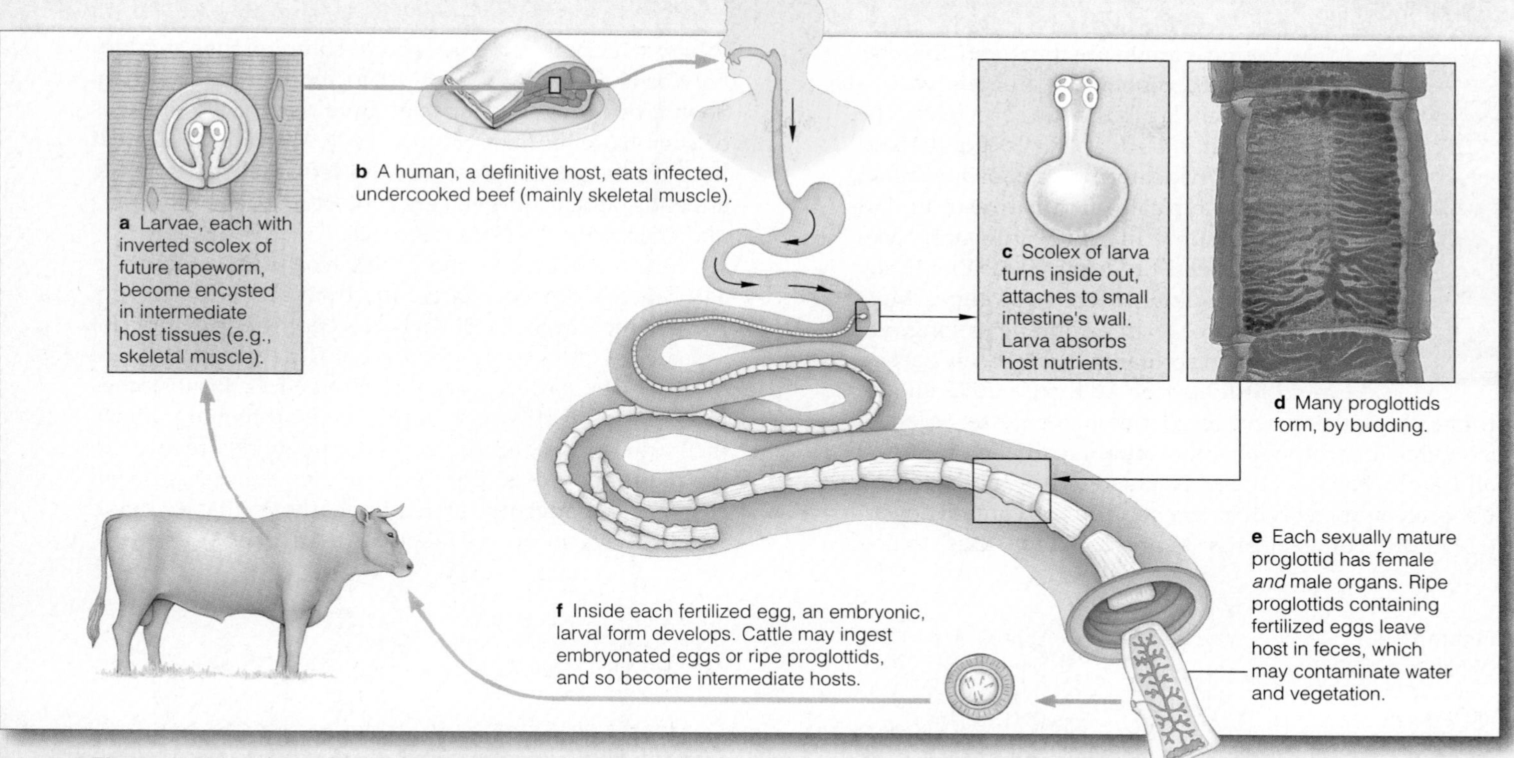
a Larvae, each with inverted scolex of future tapeworm, become encysted in intermediate host tissues (e.g., skeletal muscle).
b A human, a definitive host, eats infected, undercooked beef (mainly skeletal muscle).
c Scolex of larva turns inside out, attaches to small intestine's wall. Larva absorbs host nutrients.
d Many proglottids form, by budding.
e Each sexually mature proglottid has female and male organs. Ripe proglottids containing fertilized eggs leave host in feces, which may contaminate water and vegetation.
f Inside each fertilized egg, an embryonic, larval form develops. Cattle may ingest embryonated eggs or ripe proglottids, and so become intermediate hosts.

Figure 26.20 Life cycle of a beef tapeworm, *Taenia saginata*.

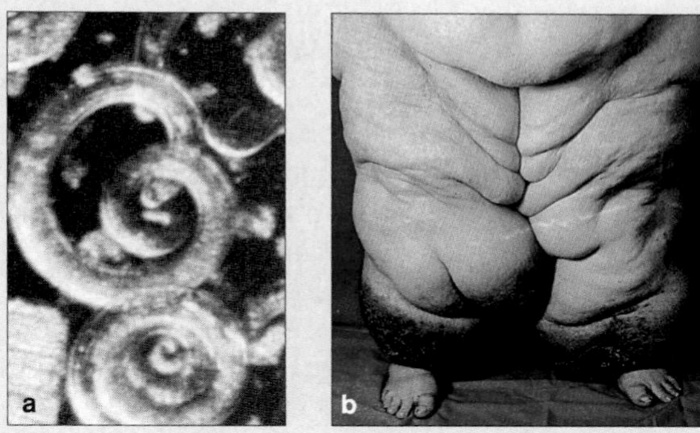
a
b

Figure 26.21 (**a**) Juveniles of a roundworm, *Trichinella spiralis*, inside the muscle tissue of a host animal. (**b**) Legs of a woman parasitized by the roundworm *Wuchereria bancrofti*.

some eggs to hands, then to other objects. Newly laid eggs contain embryonic pinworms. Within a few hours, they develop into juveniles and are ready to hatch if another human inadvertently ingests them.

The hookworms are especially vexing in impoverished areas of the tropics and subtropics. Adult hookworms live in a host's small intestine. Using their toothlike devices or sharp ridges around the mouth, they cut into the intestinal wall, feed on blood and other tissues, and so compete with their host for nutrients. Adult females, about a centimeter long, can release a thousand eggs daily. These leave the body in feces, then hatch into juveniles. When it meets up with bare skin of a host, a juvenile hookworms cuts its way inside. The parasite travels the bloodstream to the lungs. It works its way into the air spaces and moves up the windpipe. When the host swallows, the parasite is transported to the stomach and then the small intestine, where it may mature and live for several years.

Another roundworm, *Trichinella spiralis*, causes painful, sometimes fatal symptoms. Adults live in the lining of the small intestine. Females release juveniles (Figure 26.21*a*), which work their way into blood vessels and travel to muscles. There they become encysted; they secrete a covering around themselves and enter a resting stage. Humans become infected mainly by eating insufficiently cooked meat from pigs or some game animals. It is not easy to detect the encysted juveniles when fresh meat is being examined, even in a slaughterhouse.

Figure 26.21*b* shows the results of prolonged, repeated infections by *Wuchereria bancrofti*, another roundworm. Adult worms become lodged in lymph nodes, organs that filter lymph (excess tissue fluid) that normally flows into the bloodstream. The worms obstruct lymph flow. When an obstruction causes fluid to back up and accumulate in tissues, legs and other body regions may enlarge grossly. We call this condition *elephantiasis*.

A mosquito is *Wuchereria*'s intermediate host. Females of this parasite produce active young that travel about at night in the bloodstream. When a mosquito sucks blood from an infected person, the juveniles enter the insect's tissues. In time they move close to the insect's sucking device and enter a new host when it draws blood again.

26.10 ROTIFERS

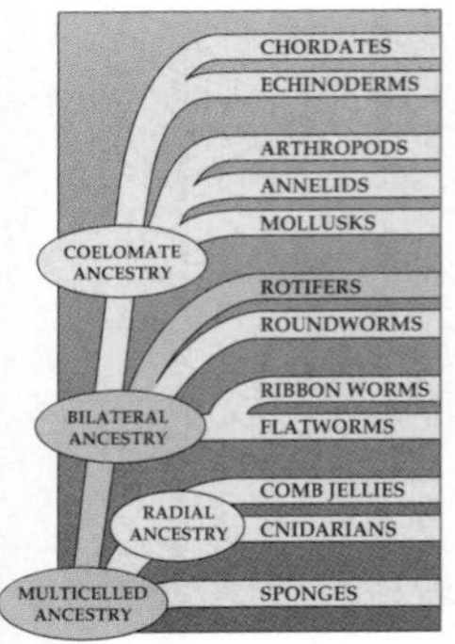

CHORDATES
ECHINODERMS
ARTHROPODS
ANNELIDS
MOLLUSKS
COELOMATE ANCESTRY
ROTIFERS
ROUNDWORMS
RIBBON WORMS
BILATERAL ANCESTRY
FLATWORMS
COMB JELLIES
RADIAL ANCESTRY
CNIDARIANS
MULTICELLED ANCESTRY
SPONGES

Like roundworms, the **rotifers** (Rotifera) are bilateral, cephalized animals with a false coelom. All but about 5 percent live in freshwater, such as lakes, ponds, and even films of water on mosses and other plants. We typically find between 40 and 500 rotifers in a liter of pondwater; 5,000 were recorded on a few occasions. They eat bacteria and microscopic algae. Most types are not even a millimeter long, yet rarely have so many organs been packed in so little space. As Figure 26.22 shows, rotifers have a pharynx, an esophagus, digestive glands, a stomach, protonephridia, and usually an intestine and anus. Nerve cell bodies clustered in the head end integrate body activities. Certain rotifers contain "eyes" (clusterings of absorptive pigments). Two "toes" exude substances that attach free-living species to substrates at feeding time. A feature unique to rotifers is a crown of cilia at the head end that assists in swimming and in wafting food toward the mouth. Its rhythmic motions reminded early microscopists of a turning wheel; hence the name of the phylum ("rotifer" means wheel-bearer).

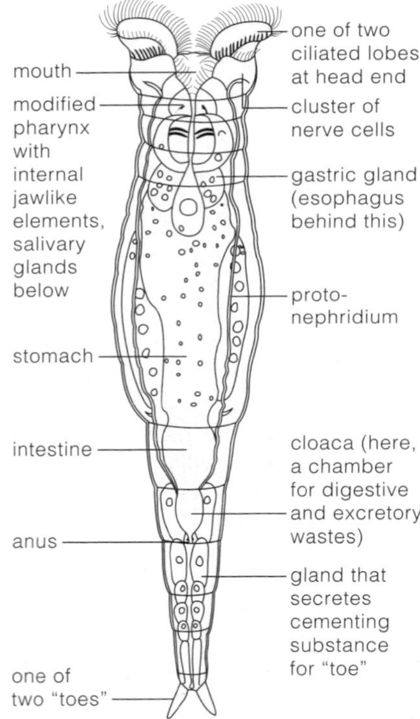

mouth
modified pharynx with internal jawlike elements, salivary glands below
stomach
intestine
anus
one of two "toes"
one of two ciliated lobes at head end
cluster of nerve cells
gastric gland (esophagus behind this)
proto-nephridium
cloaca (here, a chamber for digestive and excretory wastes)
gland that secretes cementing substance for "toe"

Figure 26.22 A rotifer (*Philodina roseola*). Males are unknown in this species and many others. Females produce diploid eggs that become diploid females. The females of other species do the same, but they also can produce haploid eggs that develop directly into haploid males. If a haploid egg happens to be fertilized by a male, it develops into a female. The males show up only occasionally. They are dwarfed, and they are short-lived.

The rotifers are bilateral, cephalized animals with ciliated lobes at their head end and a false coelom that is packed with diverse organs.

26.11 TWO MAJOR DIVERGENCES

Bilateral animals not much more complex than modern flatworms apparently evolved in the late precambrian. Soon afterward, some species gave rise to two lineages of coelomate animals (Figure 26.2). We call these great lineages the **protostomes** and **deuterostomes**. Mollusks, annelids, and arthropods are protostomes. Echinoderms and chordates are deuterostomes.

As a result of mutations, embryos of the animals in each lineage develop differently from the fertilized egg. For instance, mitotic cell divisions cut the egg cytoplasm repeatedly along prescribed planes to form a tiny ball of cells—the early embryo (Section 44.3). Protostomes undergo *spiral* cleavage. In this developmental pattern, the earliest cuts are made at oblique angles relative to the main body axis. But deuterostomes undergo *radial* cleavage, a pattern in which the earliest cuts are made parallel with and perpendicular to the axis:

Early protostome embryo. Its four cells are undergoing cleavages *oblique to* the original body axis:

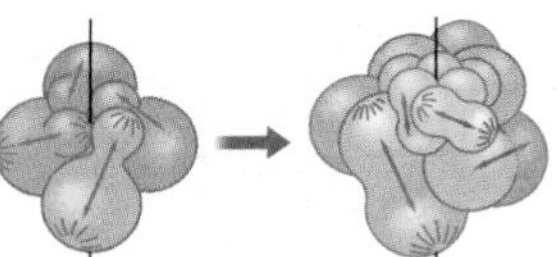

Early deuterostome embryo. Its four cells are undergoing cleavages *parallel with* and *perpendicular to* the original body axis:

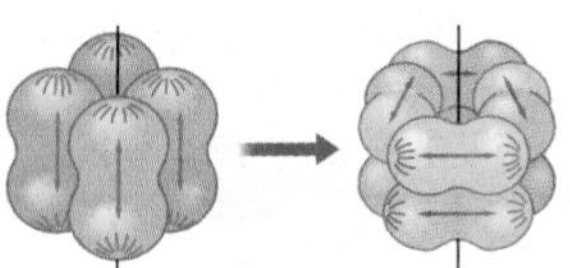

Also, the very first opening to form at the surface of a protostome embryo becomes the mouth; an anus forms elsewhere. In a deuterostome embryo, the first opening becomes the anus; the second becomes the mouth. As another example, the coelom of a protostome arises from spaces in the mesoderm, but a deuterostome coelom forms from outpouchings of the gut wall:

How a coelom forms in a protostome embryo: pouch will form mesoderm around coelom; developing gut

coelom

How a coelom forms in a deuterostome embryo: solid mass of mesoderm; developing gut

Such modifications to the embryonic stages of the two kinds of animals led to major differences in body plans.

Soon after the coelomate animals evolved in Cambrian times, two great lineages—the protostomes and deuterostomes—evolved through mutations that affected how their embryos develop, and this led to major differences in body plans.

Further reading: Student Guide to InfoTrac on web site

26.12 A SAMPLING OF MOLLUSCAN DIVERSITY

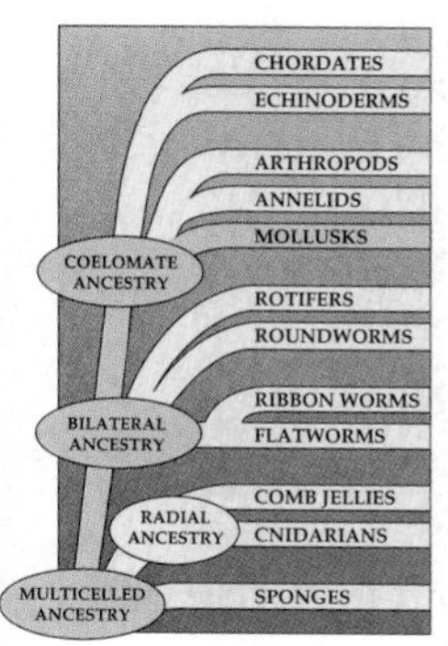

CHORDATES
ECHINODERMS
ARTHROPODS
ANNELIDS
MOLLUSKS
COELOMATE ANCESTRY
ROTIFERS
ROUNDWORMS
RIBBON WORMS
BILATERAL ANCESTRY
FLATWORMS
COMB JELLIES
RADIAL ANCESTRY
CNIDARIANS
MULTICELLED ANCESTRY
SPONGES

From children's books and explorations in gardens, few of us would have trouble recognizing a land snail when we see one (Figure 26.23*a*). Yet few of us know much about its 110,000 relatives in one of the largest of all animal groups, the phylum Mollusca. As the name implies, **mollusks** have fleshy soft bodies (*molluscus*, a Latin word, means soft). These bilateral animals have a small coelom. *Most* have a shell, or a reduced version of one, made of calcium carbonate and protein. These components are secreted from cells of a tissue that drapes like a skirt over the body mass. This tissue, the **mantle**, is unique to mollusks. The respiratory organs, gills of a type called ctenidia, contain thin-walled leaflets for gas exchange. *Most* mollusks have a fleshy foot. *Many* have a radula, a tonguelike, toothed organ that shreds food destined for the gut. Mollusks with a well-developed head have eyes and tentacles, but not all have a head. Beyond the generalizations, there are no "typical" mollusks. They range from tiny snails in treetops to huge predators of the seas. In this section and the next, we sample four classes: chitons, gastropods, bivalves, and cephalopods.

Gastropods ("belly foots") make up the largest class (90,000 species of snails and slugs). They are so named because their foot spreads out as they crawl. Many species have spirally coiled or conical shells. Coiling compacts the organs into a mass that can be balanced above the body, rather like a backpack. Other species have a reduced shell or none at all (Figure 26.23*a,b*).

Chitons are slow-moving or sedentary grazers with a dorsal shell divided into eight plates (Figure 26.23*c*). The bivalves—animals having a "two-valved shell"—include clams, scallops, oysters, and mussels (Figure 26.23*d*). Some bivalves are only a millimeter across. A few giant clams are over a meter across and weigh 225 kilograms (close to 500 pounds). Humans have eaten one type of bivalve or another since prehistoric times.

Cephalopods are highly active predators of the seas. This class includes the swiftest invertebrates (squids), the largest known invertebrate (giant squid), and the smartest (octopuses, Figure 26.23*e*). For instance, show an octopus an object with a distinctive shape and then give it a mild electric shock, and it will thereafter avoid that object. With respect to memory and the capacity to learn, octopuses and some of the squids are the world's most complex invertebrates.

Figure 26.23 A few mollusks. (**a**) Land snail, a gastropod. (**b**) Two sea slugs, a type of gastropod called nudibranchs. Different gastropods creep, swim, or float as they graze upon, prey upon, or parasitize other organisms. (**c**) Chiton from the intertidal zone of California's Monterey Bay. With its broad foot, it creeps over and clings to substrates. (**d**) Scallop, one of the bivalves. Light-sensitive "eyes" (small dark dots) fringe two halves of its shell. Many bivalves, including a few pearl producers, have shells lined with iridescent mother-of-pearl. (**e**) Octopus. Its eyes resemble yours, although they form in a different way. Like other cephalopods, it has a well-developed nervous system.

Mollusks are bilateral, soft-bodied, coelomate animals that differ tremendously in body details, size, and life-styles.

26.13 EVOLUTIONARY EXPERIMENTS WITH MOLLUSCAN BODY PLANS

Maybe it was their fleshy, soft bodies—so forgiving of chance evolutionary changes in morphology—that gave the ancestors of mollusks the potential to diversify in so many ways and to radiate into so many habitats. Let's explore this idea by using a few characteristics of our representative mollusks. As a point of departure, start by studying the body plan shown in Figure 26.24.

Twisting and Detwisting of Soft Bodies

Notice, in Figure 26.24*a,* how evolution put an unusual twist in the soft snail body. Its anus dumps wastes near the mouth! As a gastropod embryo develops, a cavity between its mantle and shell twists counterclockwise by 180°. So does nearly all of the visceral mass (the gut, heart, gills, and other internal organs). This process, called **torsion**, occurs only in gastropods:

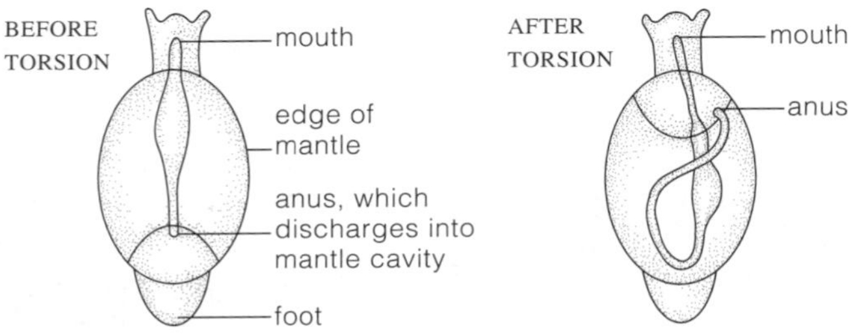
BEFORE TORSION
mouth
edge of mantle
anus, which discharges into mantle cavity
foot
AFTER TORSION
mouth
anus

Such a drastic rearrangement of body parts could have come about through mutations that affected retractor muscles, which attach a gastropod embryo or larva to its shell. The muscles on the body's right side develop before those on the left. As they grow toward the front of the body, they drag the mantle cavity and organs with them. It takes them a few hours at most to do this.

By one hypothesis, torsion proved to be adaptive, for the head could withdraw into the mantle cavity in times of danger. However, in some 1985 experiments by J. Pennington and F. Chia, predators ate just as many torsioned larvae as "pre-torsioned" ones. Was torsion a bad evolutionary experiment? By putting the gills, anus, and kidneys above the mouth, it certainly created a potentially awful sanitation problem. At the very least, the uptake of discharged wastes in ancestral torsioned species must have been distasteful.

In fact, we find evolutionary compensations for this state of affairs. Most gastropods now have enough cilia in this region to create currents that sweep the wastes away. Also, torsion is not as pronounced as it once was in some lineages. Nudibranchs have even undergone an apparent detorsion; the soft larval body twists, but then it untwists. At some point in their evolution, they also ended up losing most of their mantle cavity and all of their ctenidia. Most species have other outgrowths that function in gas exchange (Figures 26.23*b* and 26.24*c*).

Hiding Out, One Way or Another

If you were small, edible, and soft of body, an external shell would be a distinct advantage, as it is for chitons and clams. When a chiton is disturbed by predators or surf or exposed by a receding tide, it hunkers under its shell. Muscles in its foot pull the body mass down, and the mantle's edge around the shell's rim presses like a suction cup against a rock. That eight-plated shell is flexible. Pluck a chiton from a rock, and it rolls up in a ball until it can unroll and become reattached elsewhere.

Besides having a shell, protection also can be had by hiding in sediments and other substances. A bivalve's head is not much to speak of, but its foot is usually large and specialized for burrowing. Bivalves that dig

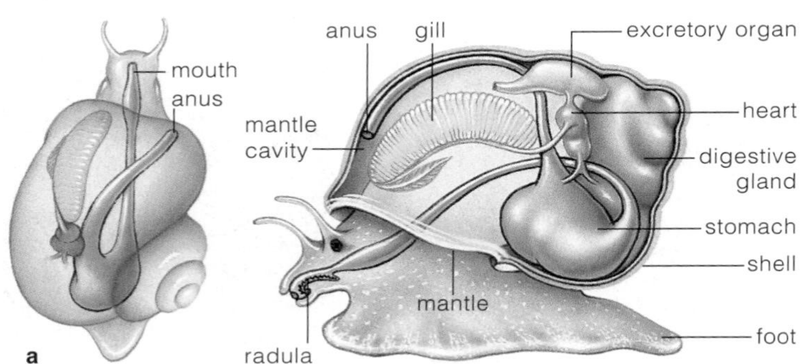
mouth
anus
a
anus
gill
excretory organ
mantle cavity
heart
digestive gland
stomach
shell
mantle
radula
foot

b
c

Figure 26.24 (**a**) Body plan of an aquatic snail, a gastropod. (**b**) Close-up of a radula, a feeding device of some mollusks, including snails. As a radula rhythmically protracts and retracts, it rasps food and draws it toward the gut on the retraction stroke. (**c**) Sea slug *Aplysia*, also called a sea hare. Two flaps above its dorsal surface are foot extensions that undulate and ventilate the mantle cavity. Like many other gastropods, *Aplysia* is hermaphroditic. It can function as a male, a female, or both.

Further reading: Student Guide to InfoTrac on web site

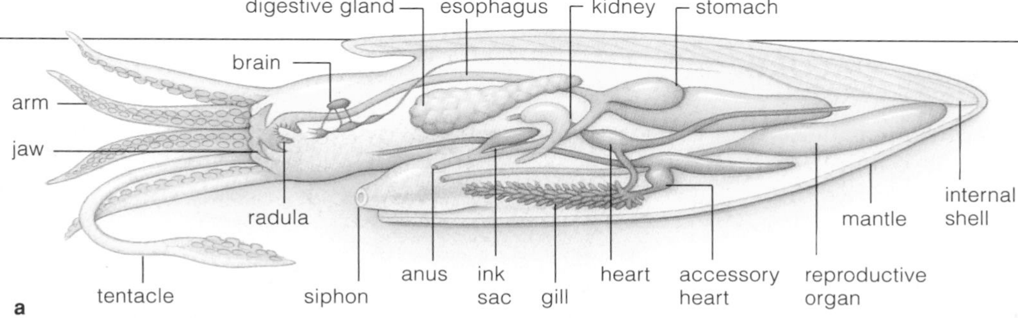
digestive gland
esophagus
kidney
stomach
brain
arm
jaw
radula
tentacle
siphon
anus
ink sac
gill
heart
accessory heart
reproductive organ
mantle
internal shell
a

in sand or mud have paired siphons: extensions of the mantle edges, fused into tubes (Figure 26.25*a*). Water is drawn into the mantle cavity through one siphon and leaves through the other, carrying wastes. One wonders what preyed on the ancestors of geoducks of the Pacific Northwest, which have siphons more than a meter long.

On the Cephalopod Need for Speed

Some 500 million years ago the cephalopods, with their buoyant, chambered shells, were the supreme predators in Ordovician seas (Section 21.5). And yet, of a lineage having more than 7,000 ancestral species, the shell of all but one of the existing descendant species is reduced or gone (Figure 26.26). What happened? This evolutionary trend coincided with an adaptive radiation of the bony fishes—which preyed on cephalopods or were strong competitors for the same prey. During what may have been a long-term race for speed and wits, cephalopods lost their thick external shell and became streamlined and highly active. Of all mollusks, they now have the largest brain relative to body size and display the most complex behavior. Nerves connect their brain to muscles that respond swiftly to food or danger. They have highly efficient blood circulation and respiration (Figure 26.26). Except for the chambered nautilus, they can discharge dark fluid from their ink sac, maybe to confuse predators.

Jet propulsion became the name of the game. Cephalopods force a jet of water from the mantle cavity and a funnel-shaped siphon. As mantle muscles relax, water is drawn into the cavity. As they contract, a water jet is squeezed out. When the mantle's free edge closes down on the head at the same time, a jet shoots through the siphon. The brain controls the siphon's activity and the direction of escape or pursuit.

b

c

Figure 26.26 (**a**) Body plan (generalized) of a cuttlefish, a cephalopod. Its tentacles, thinner than the arms, are specialized for capturing prey. (**b**) A squid (*Dosidiscus*) and a diver inspecting each other. (**c**) A chambered nautilus, the only existing cephalopod that has an external shell. Being so active, cephalopods have great demands for oxygen. They are the only mollusks having a closed circulatory system. Blood is pumped from a main heart to two gills, each with an accessory (booster) heart at its base that speeds blood flow, oxygen uptake (for muscle cells especially), and carbon dioxide removal.

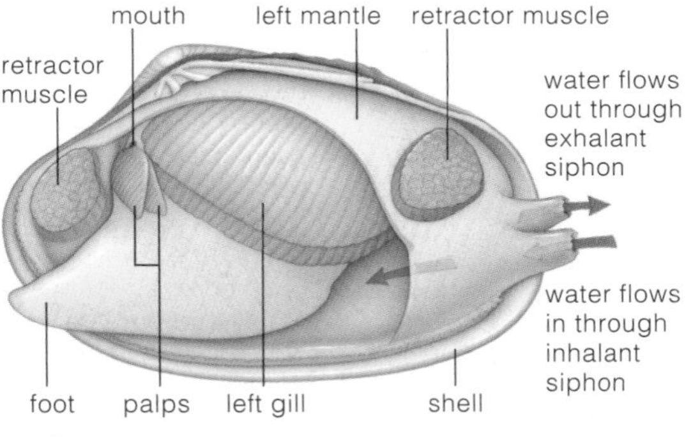
mouth
left mantle
retractor muscle
retractor muscle
water flows out through exhalant siphon
water flows in through inhalant siphon
foot
palps
left gill
shell
a

b

Figure 26.25 (**a**) Body plan of a clam, with half of its shell removed. In nearly all bivalves, gills serve in collecting food and in respiration. As water moves through the mantle cavity, mucus on the gills traps food. Cilia move the mucus and food to palps, where final sorting takes place before suitable bits are driven to the mouth. (**b**) A scallop escaping from a sea star by clapping its valves and producing a propulsive water jet.

Lively stories emerge when evolutionary theory is used to interpret the fossil record and the range of existing species diversity, as we have done for the mollusks.

26.14 ANNELIDS—SEGMENTS GALORE

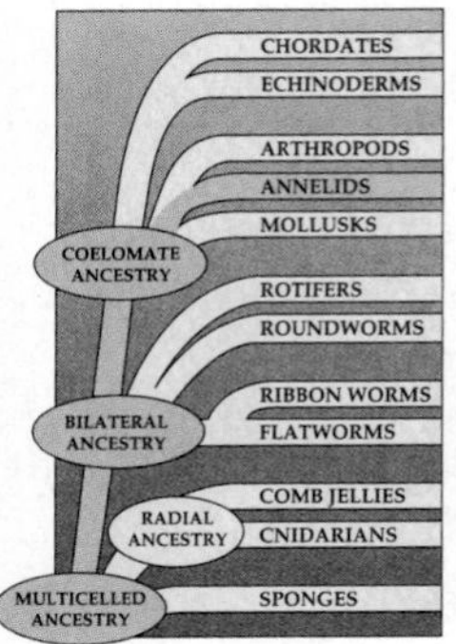

CHORDATES
ECHINODERMS
ARTHROPODS
ANNELIDS
MOLLUSKS
COELOMATE ANCESTRY
ROTIFERS
ROUNDWORMS
RIBBON WORMS
FLATWORMS
BILATERAL ANCESTRY
COMB JELLIES
CNIDARIANS
RADIAL ANCESTRY
SPONGES
MULTICELLED ANCESTRY

Maybe you've noticed earthworms after a downpour, when they wriggle out of their burrows to avoid drowning. Earthworms are among 15,000 or so species of bilateral, segmented animals known as the **annelids** (Annelida). Their relatives are a few kinds of leeches and the polychaetes, which are far more diverse but not nearly as well known (Figure 26.27). The phylum name means "ringed forms." But the rings are actually a series of repeating body units, and the segmentation is pronounced. Also, except for leeches, nearly all segments have pairs or clusters of chitin-reinforced bristles on each side of the body. The bristles are also called setae or chaetae, but these are just formal names for bristles. When pushed into soil, the bristles provide the traction required for crawling or burrowing. They have become broadened paddles in some swimming species. Earthworms, one of the oligochaetes, have few setae per body segment, and marine polychaete worms typically have many of them (*oligo*–, few; *poly*–, many).

Advantages of Segmentation

A segmented body has great evolutionary potential, for individual parts can undergo modification and become highly adapted for specialized tasks. Although most of an earthworm's segments are similar, the leeches have suckers at both ends, and polychaetes have an elaborate head and fleshy-lobed appendages known as parapods ("closely resembling feet"). By analyzing the existing species, we catch glimpses of developments that led to increases in size and to more complex internal organs.

Annelid Adaptations—A Case Study

The earthworms are examples of familiar annelids. As Figure 26.28 shows, partitions divide their body into a series of coelomic chambers. In most of the chambers we find repeats of muscles, blood vessels, branching nerves, and other

before feeding
after feeding

Figure 26.27 Leeches. Most leeches have sharp jaws and a blood-sucking device. This one is shown before and after gorging on human blood. For at least 2,000 years, *Hirudo medicinalis*, a freshwater leech, has been employed as a blood-letting tool to "cure" problems ranging from nosebleeds to obesity.

Leeches are still used, but more selectively, as when they draw off pooled blood after surgeons reattach a severed ear, lip, or fingertip. A patient's body cannot do this on its own until the severed blood circulation routes are reestablished.

a

b

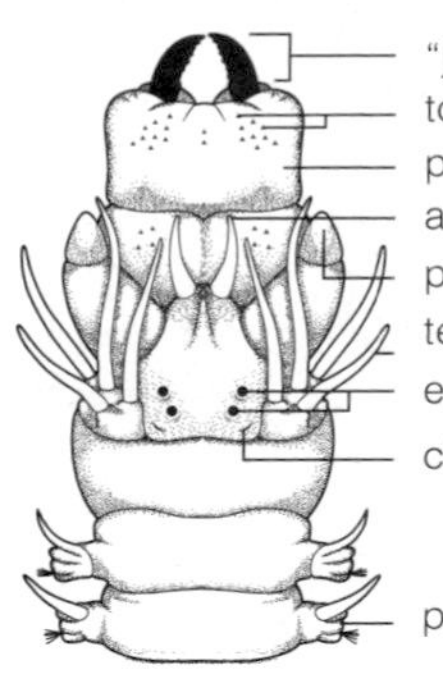

"jaws"
toothlike structures
pharynx (everted)
antenna
palp (food handling)
tentacle
eyes
chemical-sensing pit
parapod
c

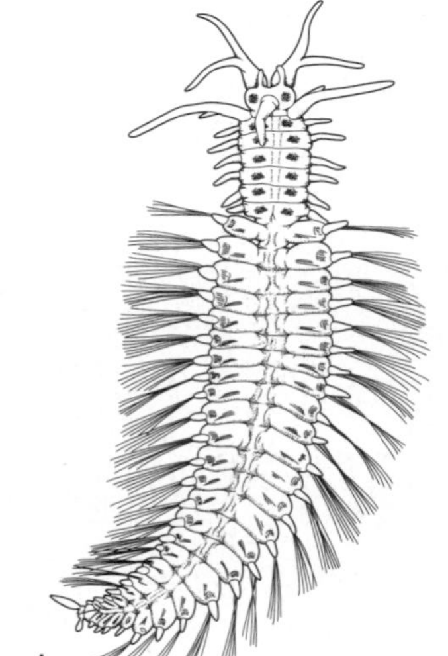

d

Figure 26.28 (**a**) A familiar annelid—one of the earthworms. (**b**–**d**) Polychaetes. These give a sense of the dizzying variety of modifications that have evolved in this group, starting with a segmented, coelomic body plan. The type shown in (**b**) is one of the tube-dwellers. Featherlike structures at its head end are coated with mucus. After the mucus has trapped bacteria and other bits of food, coordinated beating of cilia sweeps them to the mouth. Most polychaetes live in marine habitats. They actually are one of the most common types of animals along coasts. Many are predators or scavengers; others dine on algae.

Further reading: Student Guide to InfoTrac on web site

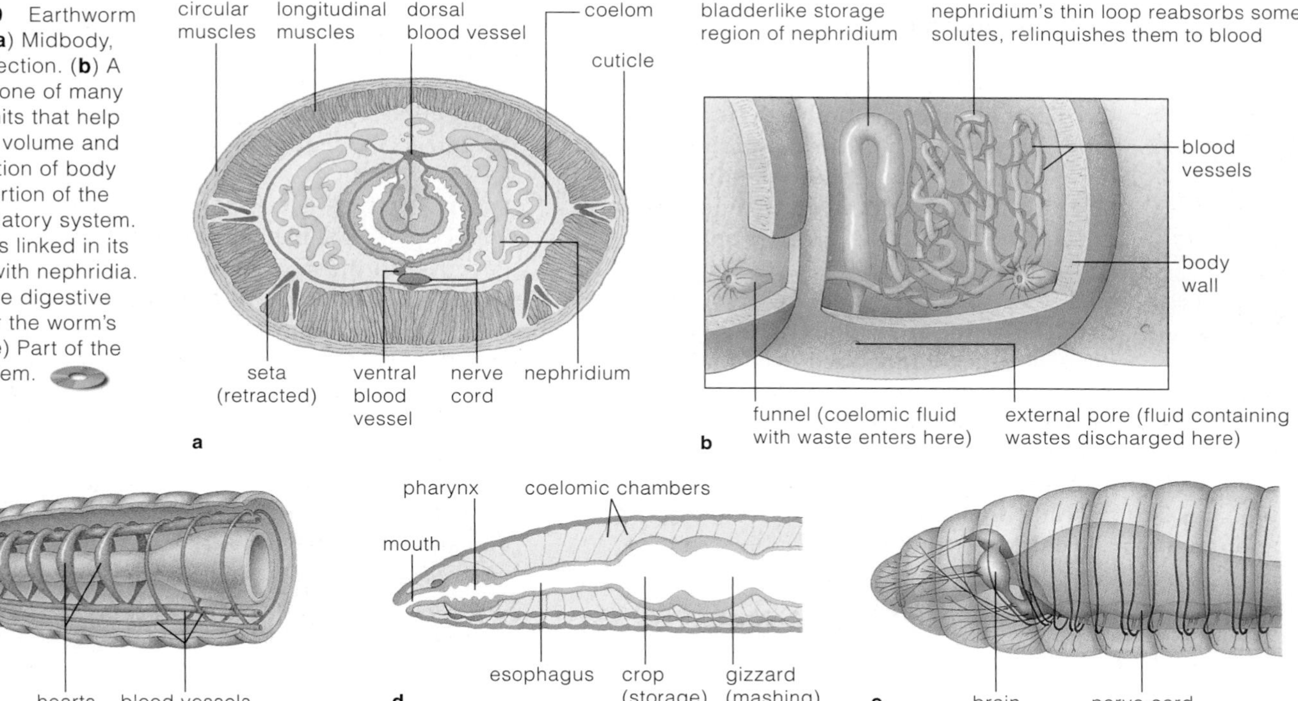

circular muscles
longitudinal muscles
dorsal blood vessel
coelom
cuticle
seta (retracted)
ventral blood vessel
nerve cord
nephridium
a
bladderlike storage region of nephridium
nephridium's thin loop reabsorbs some solutes, relinquishes them to blood
blood vessels
body wall
funnel (coelomic fluid with waste enters here)
external pore (fluid containing wastes discharged here)
b
c
hearts
blood vessels
pharynx
coelomic chambers
mouth
esophagus
crop (storage)
gizzard (mashing)
d
e
brain
nerve cord

Figure 26.29 Earthworm body plan. (**a**) Midbody, transverse section. (**b**) A nephridium, one of many functional units that help maintain the volume and the composition of body fluids. (**c**) Portion of the closed circulatory system. The system is linked in its functioning with nephridia. (**d**) Part of the digestive system, near the worm's head end. (**e**) Part of the nervous system.

organs. The gut extends through all the chambers, from mouth to anus. Like all annelids, an earthworm has a cuticle of secreted material that encapsulates the body surface. It bends easily, and it is permeable to water as well as to gases, which is one reason why annelids are restricted to aquatic habitats or moist habitats on land.

Earthworms are scavengers. They ingest moistened soil and mud that contains decomposing plant material and other organic matter. Each worm can eat its own weight every twenty-four hours. As numerous worms burrow and feed, they collectively aerate soil and lift nutrients to the surface, to the benefit of many plants.

As in other annelids, the fluid-cushioned coelomic chambers serve as a hydrostatic skeleton against which muscles act. Each segment's wall incorporates a layer of circular muscles (Figure 26.29*a*). When longitudinal muscles that bridge several segments contract, circular muscles relax, so the segments shorten and fatten. When the pattern reverses, the segments lengthen. While this is going on, bristles on different segments protract and retract. When the first few segments lengthen and their setae are not touching the ground, the front part of the body is extended. Bristles of the segments behind them plunge into the ground and anchor the worm. Next, the first few segments plunge *their* bristles into the ground, and the segments toward the posterior end of the body retract their bristles and are pulled forward. The whole worm moves forward when alternating contractions and elongations proceed along the body's length.

Figure 26.29*b* shows part of a system of **nephridia** (singular, nephridium), units that regulate the volume and composition of body fluids. In many annelids, cells of the units are similar to flame cells, which implies an evolutionary link between the flatworms and annelids. More often, a nephridium starts out as a funnel that collects excess fluid from one coelomic chamber. The funnel connects with a tubular part of the nephridium, which delivers fluid to a pore at the surface in the body wall of the next coelomic chamber.

The worm's head end has a rudimentary **brain**, an aggregation of nerve cell bodies that integrate sensory input as well as commands for muscle responses for the whole body. Paired **nerve cords**, each a bundle of long extensions of nerve cell bodies, lead away from the brain. They are pathways for rapid communication. In each body segment, the paired nerve cords broaden into a ganglion (plural, ganglia), a cluster of nerve cell bodies that controls local activity.

Finally, as is the case for most annelids, earthworms have a closed circulatory system, with blood confined in hearts and muscularized blood vessels. Contractions keep blood circulating in one direction. Smaller blood vessels service the gut, nerve cord, and body wall.

Annelids are bilateral, coelomate, segmented worms that have complex organ systems. Some species show the degree of specialization possible with a segmented body plan.

ARTHROPODS—THE MOST SUCCESSFUL ORGANISMS ON EARTH

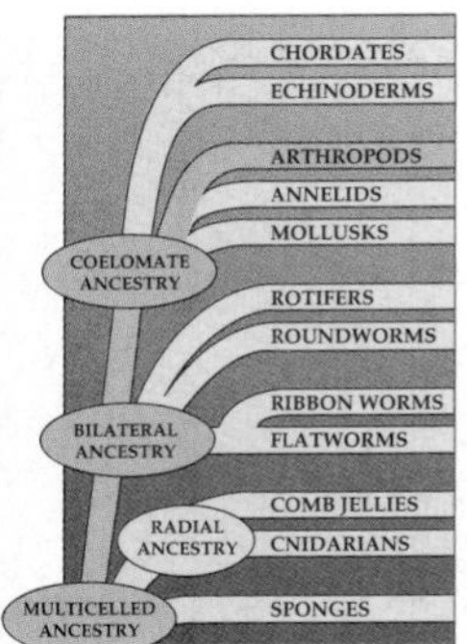
CHORDATES
ECHINODERMS
ARTHROPODS
ANNELIDS
MOLLUSKS
COELOMATE ANCESTRY
ROTIFERS
ROUNDWORMS
RIBBON WORMS
BILATERAL ANCESTRY
FLATWORMS
COMB JELLIES
RADIAL ANCESTRY
CNIDARIANS
MULTICELLED ANCESTRY
SPONGES

Arthropod Diversity

Evolutionarily speaking, "success" means having the greatest number of species, producing the most offspring, occupying the most habitats, effectively fending off threats and competition, and having the capacity to exploit the greatest amounts and kinds of food. These are the features that come to mind when we attempt to characterize the **arthropods** (Arthropoda). We have already identified over a million species—mostly insects—and researchers discover new ones weekly, mainly in tropical forests and the seas.

Of four major lineages, trilobites are extinct (Section 21.5). The other three are chelicerates (spiders and their relatives), crustaceans (including barnacles and crabs), and uniramians (centipedes, millipedes, and insects).

Adaptations of Insects and Other Arthropods

Six important adaptations contributed to the success of arthropods in general and the insects in particular: a hardened exoskeleton, jointed appendages, fused and modified segments, specialized respiratory structures, efficient nervous system and sensory organs, and often a division of labor in the life cycle.

HARDENED EXOSKELETONS Arthropods have a cuticle of chitin, proteins, and surface waxes that may even be impregnated with calcium carbonate deposits. It acts as a rigid, protective external skeleton—an **exoskeleton**. Such cuticles might have evolved as defenses against predation. They took on added functions when some arthropods first invaded the land. They support a body deprived of water's buoyancy, and their waxy surface restricts evaporative water loss. Hard cuticles stop size increases, but arthropods grow in spurts by **molting**. At certain stages of their life cycle, they secrete a soft new cuticle under the old one, which they shed (Figure 26.30). The body mass increases first by uptake of air or water, then by repeated, rapid cell division before the new cuticle can harden.

JOINTED APPENDAGES If an arthropod had a cuticle that was uniformly hardened, it wouldn't move much. But the arthropod cuticle is thinner at *joints,* where two body parts abut. Muscles associated with the joints make the thinner cuticle bend in specific directions and move body parts. This jointed exoskeleton was a key innovation that led to appendages as diverse as wings, antennae, and legs (*arthropod* means "jointed foot").

Figure 26.30 Molting, demonstrated by a red-orange centipede wriggling out of its old exoskeleton.

FUSED AND MODIFIED SEGMENTS The first arthropods were segmented, like the annelid stock that presumably gave rise to them. In most existing descendants, serial repeats of the body wall and organs are masked; many fused-together segments are modified to perform more specialized functions. For example, in the ancestors of insects, different segments fused to form three regions —head, thorax, and abdomen—which morphologically diverged from one another in astonishing ways.

RESPIRATORY STRUCTURES Many aquatic arthropods depend on gills for gas exchange. Air-conducting tubes evolved among insects and other land-dwellers. Insect tracheas begin as pores on the body surface and branch into tubes that deliver oxygen directly to tissues. They support energy-consuming activities, such as flight.

SPECIALIZED SENSORY STRUCTURES Intricate eyes and other sensory organs contributed to arthropod success. Numerous species have a wide angle of vision and can process visual information from many directions.

DIVISION OF LABOR Moths, butterflies, beetles, flies, and many other species divide the job of surviving and reproducing among different stages of development. For many species, the new individual is a *juvenile*—a miniaturized version of the adult that simply changes in size and proportion until reaching sexual maturity. Other species show **metamorphosis**, meaning the body form changes from embryo to adult. Under hormonal commands, their size increases, tissues reorganize, and body parts are remodeled. Sections 26.19, 37.8, and 44.8 have examples of this transitional time. Immature stages such as caterpillars typically specialize in feeding and growing in size. The adult stage specializes primarily in dispersal and reproduction. For such species, then, the life cycle turns on a *division of labor*: the different stages of development become specialized in ways that are adaptive to environmental conditions, such as seasonal variation in food resources and water supplies.

As a group, the arthropods are exceptionally abundant and widespread, and they have enormously different life-styles.

Their success arises largely from their hardened, jointed exoskeletons; fused, modified body segments; specialized appendages; specialized respiratory, nervous, and sensory organs; and often a division of labor in the life cycle.

Further reading: Student Guide to InfoTrac on web site

26.16 A LOOK AT SPIDERS AND THEIR KIN

The chelicerates originated in shallow seas early in the Paleozoic. The only surviving marine species are a few mites, sea spiders, and horseshoe crabs (Figure 26.31). The familiar chelicerates on land—spiders, scorpions, ticks, and chigger mites—are all classified as arachnids. We might say this about the whole group: Never have so many been loved by so few.

a
b

Figure 26.31 Horseshoe crab, (**a**) ventral and (**b**) dorsal views. The *five* pairs of legs of this chelicerate are one of its defining features. They are hidden beneath the hard, shieldlike cover.

a

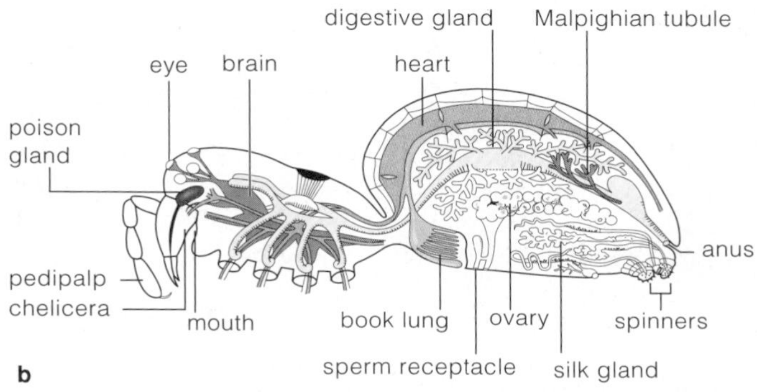

digestive gland
Malpighian tubule
eye
brain
heart
poison gland
anus
pedipalp
chelicera
mouth
book lung
ovary
spinners
sperm receptacle
silk gland
b

Figure 26.32 (**a**) Wolf spider. Like most spiders, it helps keep populations of insects in check. Its bite is harmless to humans. Section 26.20 describes two of its dangerous relatives. (**b**) A spider's internal organization.

Spiders and scorpions are efficient predators; they sting or bite and may subdue prey with venom. When most kinds sting or bite people, the reaction might be painful but only rarely does it lead to serious problems. Collectively, the spiders especially are beneficial in that they prey upon great numbers of pestiferous insects.

Bites of the blood-sucking ticks that parasitize deer, mice, and other vertebrates cause maddening itches and often serious diseases. For example, bites of some ticks transmit the bacterial agents of Rocky Mountain spotted fever or Lyme disease to humans (Sections 22.6 and 26.20). Most mites are free-living scavengers.

Arachnids have segments fused into a forebody and hindbody. The forebody's jointed appendages include four pairs of legs, a pair of pedipalps with primarily sensory functions, and a pair of chelicerae that inflict wounds and discharge venom. The appendages of the hindbody spin out silk threads for webs and for egg cases. Most webs are netlike. One type of spider spins a vertical thread with a ball of sticky material at the end. It uses a leg to swing the ball at insects passing by!

Inside the body is an *open* circulatory system, with a heart that pumps blood into body tissues, then receives blood through small openings in its wall. Some of the slowly circulating blood travels through moist folds of book lungs. These respiratory organs, which resemble loose pages of a book, greatly increase the surface area that is available for gas exchange with the air. Figure 26.32*b* shows the arrangement of book lungs and other major organs inside the spider body.

The spiders, scorpions, and their relatives have a variety of appendages specialized for predatory or parasitic life-styles.

A LOOK AT THE CRUSTACEANS

Shrimps, lobsters, crabs, barnacles, pillbugs, and other crustaceans got their name because they have a hard yet flexible "crust" (an external skeleton)—but so do nearly all arthropods. Only some of the 35,000 species live in fresh water or on land. The vast majority live in marine habitats, where they are so abundant they have been dubbed the insects of the seas. Lobsters and crabs are "giants" of this subphylum; most crustaceans are less than a few centimeters long. All have major roles in food webs, and humans harvest many edible types.

By having many pairs of similar appendages along most of their length, the simplest crustaceans might resemble their annelid ancestors. In the more complex lineages, unspecialized appendages evolved into diverse structures of the sort shown in Figure 26.33. The strong claws of lobsters and crabs, for example, are used to collect food, intimidate other animals, and sometimes dig burrows. Feathery appendages of barnacles comb microscopic bits of food from the water.

Many crustaceans have sixteen to twenty segments; some have more than sixty. In crabs, lobsters, and some other crustaceans, the dorsal cuticle extends back from the head as a shieldlike cover over some or all of the segments. The head has two pairs of antennae, one pair of jawlike appendages (mandibles), and two pairs of appendages for handling food. Crayfish, crabs, lobsters,

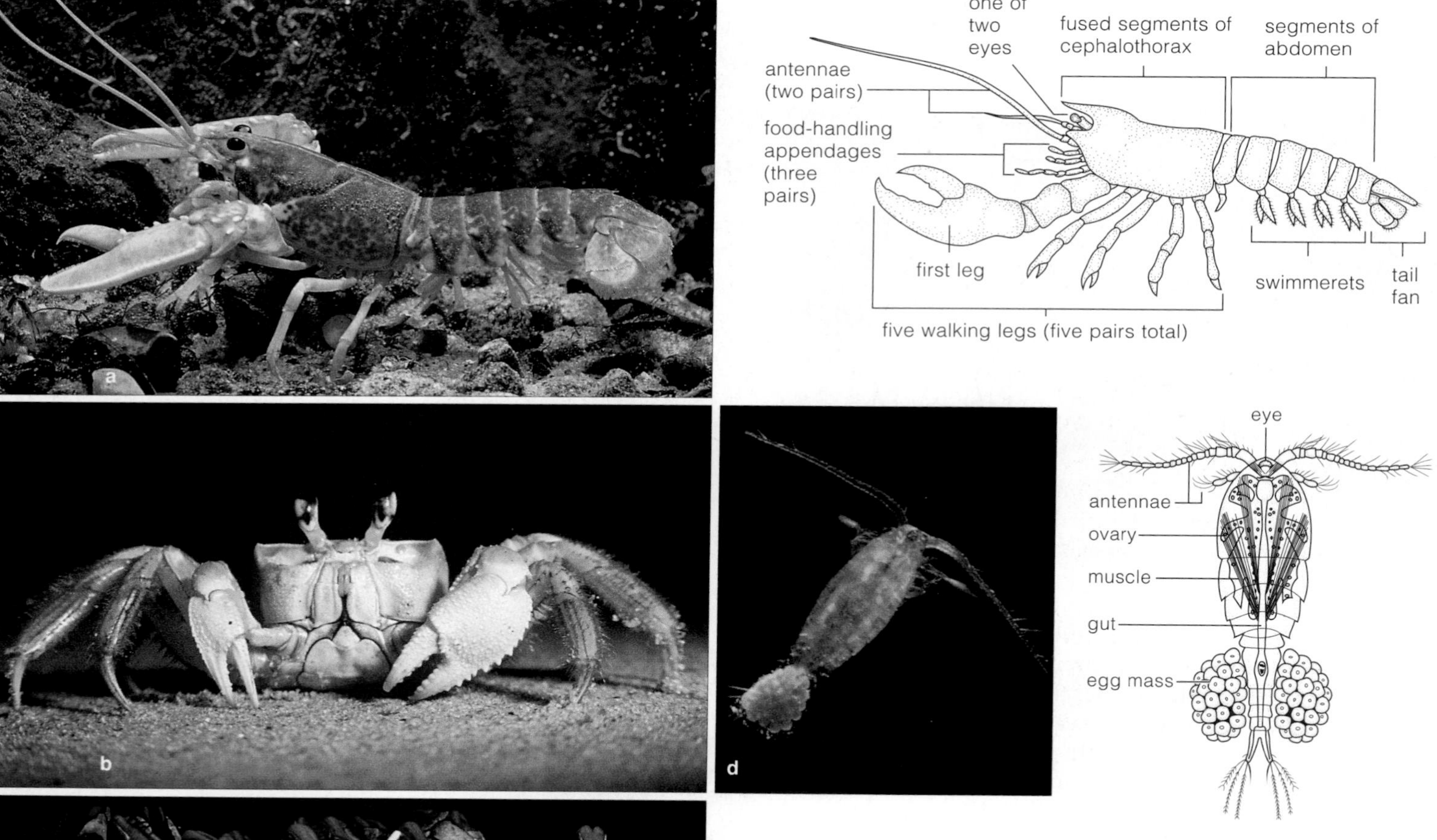
a
one of two eyes
fused segments of cephalothorax
segments of abdomen
antennae (two pairs)
food-handling appendages (three pairs)
first leg
swimmerets
tail fan
five walking legs (five pairs total)
b
d
eye
antennae
ovary
muscle
gut
egg mass
c

Figure 26.33 A sampling of crustaceans and their diverse life-styles. (**a**) Photograph and body plan of a lobster. For most of their lives, lobsters are secretive. (**b**) Crab. Guess why crabs skitter actively and openly across sand and rocks at night but not in the day. (**c**) Stalked barnacles. Adults cement themselves to one spot. You might mistake them for mollusks, but as soon as they open their hinged shell to filter-feed, you can observe their jointed appendages—the hallmark of arthropods. See also Figure 5.1. (**d**) Photograph and body plan of a copepod. Copepods are free-living filter feeders, predators, or parasites. This female has a pair of long antennae and is carrying her eggs with her.

Further reading: Student Guide to InfoTrac on web site –

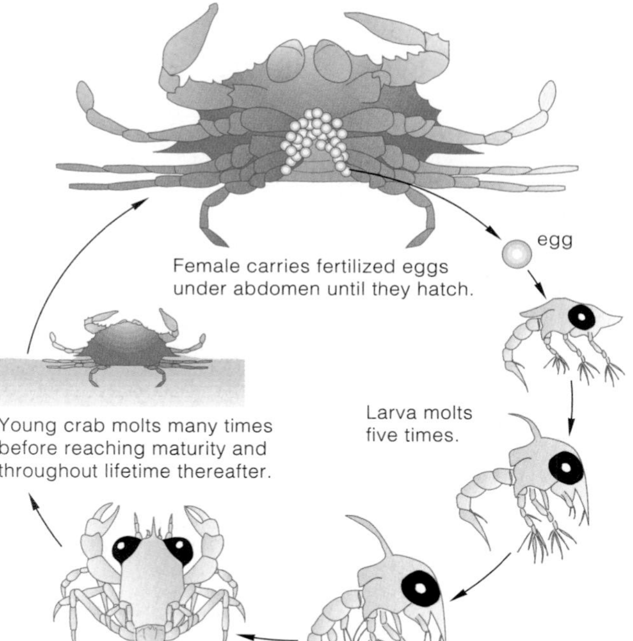
egg
Female carries fertilized eggs under abdomen until they hatch.
Larva molts five times.
Young crab molts many times before reaching maturity and throughout lifetime thereafter.

Figure 26.34 Life cycle of a crab. The larval and juvenile stages molt repeatedly and grow in size.

shrimps, and their various relatives are equipped with five pairs of walking legs.

Of all the arthropods, only barnacles have a calcified "shell," a modified external skeleton that protects them from predators, drying winds, battering surf, and strong currents. Adult barnacles cement themselves to rocks, wharf pilings, and similar surfaces (Figure 26.33*c*). A few kinds attach themselves only to the skin of whales.

Figure 26.33*d* shows a copepod. Copepods are less than two millimeters long and are the most numerous animals in aquatic habitats, maybe even in the world. About 1,500 kinds parasitize various invertebrates and fishes. The majority—8,000 species—are consumers of phytoplankton, the "pastures" of aquatic habitats. Some also eat larval or small adult invertebrates, fish eggs, and fish larvae, which they grab with their pair of food-handling appendages. In turn, the copepods are food for different invertebrates, fishes, and baleen whales.

Like other arthropods, crustaceans molt repeatedly to shed the exoskeleton during their life cycle. Figure 26.34 shows a crab's larval stages and increases in size.

Crustaceans differ greatly in the number and kind of their appendages. As is the case for arthropods in general, they repeatedly replace their external skeleton by molting.

26.18 *HOW* MANY LEGS?

The **millipedes** and **centipedes** have a long, segmented body with many legs. Of course, millipedes don't have "a thousand," as their name implies. Most have about 100 legs, although one exuberant individual grew 752. Centipedes have between 15 and 177 pairs of legs, not a nicely rounded number of "one hundred."

As the millipedes develop, pairs of segments fuse, and each segment in the cylindrical body ends up with two pairs of legs (Figure 26.35*a*). Millipedes scavenge decaying plant material in soil and forest litter.

a
b

Figure 26.35 (**a**) Millipede. (**b**) A Southeast Asian centipede.

Centipedes have a flattened body, and all but two segments have a pair of walking legs. All species are fast-moving, aggressive predators outfitted with fangs and venom glands. They prey on insects, earthworms, and snails. The one in Figure 26.35*b* can subdue small lizards, toads, and frogs. A house centipede (*Scutigera*) often hides out in buildings, where it hunts cockroaches, flies, and other pests. Athough helpful in this respect, its vaguely terrifying body keeps it from being welcomed.

The mild-mannered, scavenging millipedes and aggressive, predatory centipedes do not lend themselves to leg counts as they walk by.

26.19 A LOOK AT INSECT DIVERSITY

As a group, insects share the adaptations listed earlier in Section 26.15. Here we expand the list a bit. Insects have a head, a thorax, and an abdomen. The head has paired sensory antennae and paired mouthparts, which specialize in biting, chewing, sucking, or puncturing (Figure 26.36). Three pairs of legs and usually two pairs of wings project from the thorax. Most appendages of the abdomen are reproductive structures, such as egg-laying devices. An insect has a foregut, midgut (where most digestion proceeds), and hindgut (where water is reabsorbed). Insects get rid of waste material through **Malpighian tubules**, small tubes that connect with the midgut. When they break down proteins, the nitrogen-containing wastes diffuse from blood into the tubules and are converted into harmless crystals of uric acid. The crystals are eliminated with feces. This system lets land-dwelling insects get rid of potentially toxic wastes without losing precious water.

We've already catalogued more than 800,000 species of insects. If we use sheer numbers and distribution as the yardstick, the most successful insects are small in size and have a staggering reproductive capacity. For example, you may find some of these species growing and reproducing in great numbers on a single plant that might be only an appetizer for another animal. By one estimate, if all the progeny of a single female fly were to survive and reproduce through six more generations, that fly would have more than 5 trillion descendants!

Besides this, the most successful insect species are winged. In fact, they are the *only* winged invertebrates. They can move among food sources that are too widely scattered to be exploited by other kinds of animals. The capacity for flight contributed to their success on land.

Finally, insect life cycles commonly proceed through stages that allow exploitation of different resources at different times. As an insect embryo develops, organs required for feeding and other vital activities form and become functional. Before an insect becomes an adult (the sexually mature form of the species), it proceeds through immature, post-embryonic stages. **Nymphs** and **pupae**, as well as larvae, are examples of the stages.

Like human infants, nymphs of some insects are shaped like miniature adults. But unlike children, they undergo growth and molting (Figure 26.36*a*). Other insects go through post-embryonic stages of reactivated growth, tissue reorganization, and remodeling of body parts. Metamorphosis, remember, is the name for this resumption of growth and transformation into an adult form. Figure 26.37 indicates how the transformation is more drastic in some species than in others.

All the factors that contribute to insect success also make them our most aggressive competitors. Insects destroy crops, stored food, wool, paper, and timber. As some stealthily draw blood from us and from our pets,

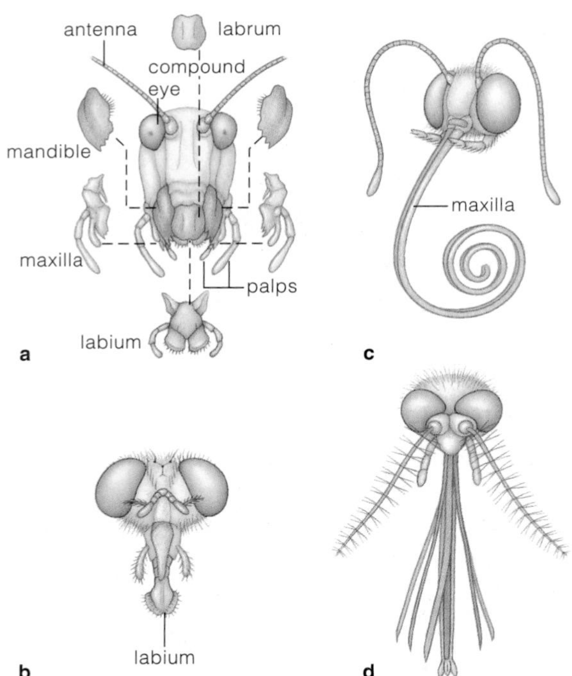
antenna
labrum
compound eye
mandible
maxilla
palps
labium
a
maxilla
c
labium
b
d

Figure 26.36 Examples of insect appendages. Headparts of (**a**) grasshoppers, a chewing insect; (**b**) flies, which sponge up nutrients with a specialized labium; (**c**) butterflies, which siphon up nectar with a specialized maxilla; and (**d**) mosquitoes, with piercing and sucking appendages.

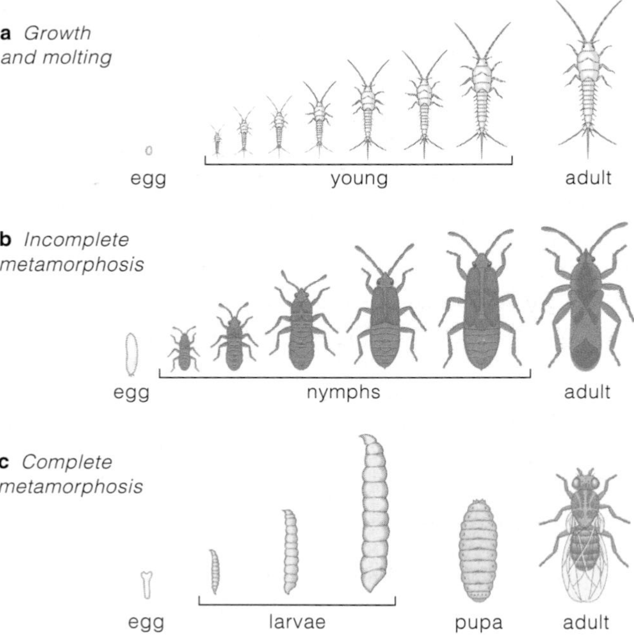
a Growth and molting
egg
young
adult
b Incomplete metamorphosis
egg
nymphs
adult
c Complete metamorphosis
egg
larvae
pupa
adult

Figure 26.37 Examples of post-embryonic development. (**a**) Young silverfish are adults in miniature, changing little except in size and proportion as they mature into adults. (**b**) True bugs show *incomplete* metamorphosis, which involves gradual, partial change from the first immature form until the last molt. (**c**) Fruit flies show *complete* metamorphosis. Tissues of immature forms are destroyed and replaced before emergence of the adult.

Further reading: Student Guide to InfoTrac on web site

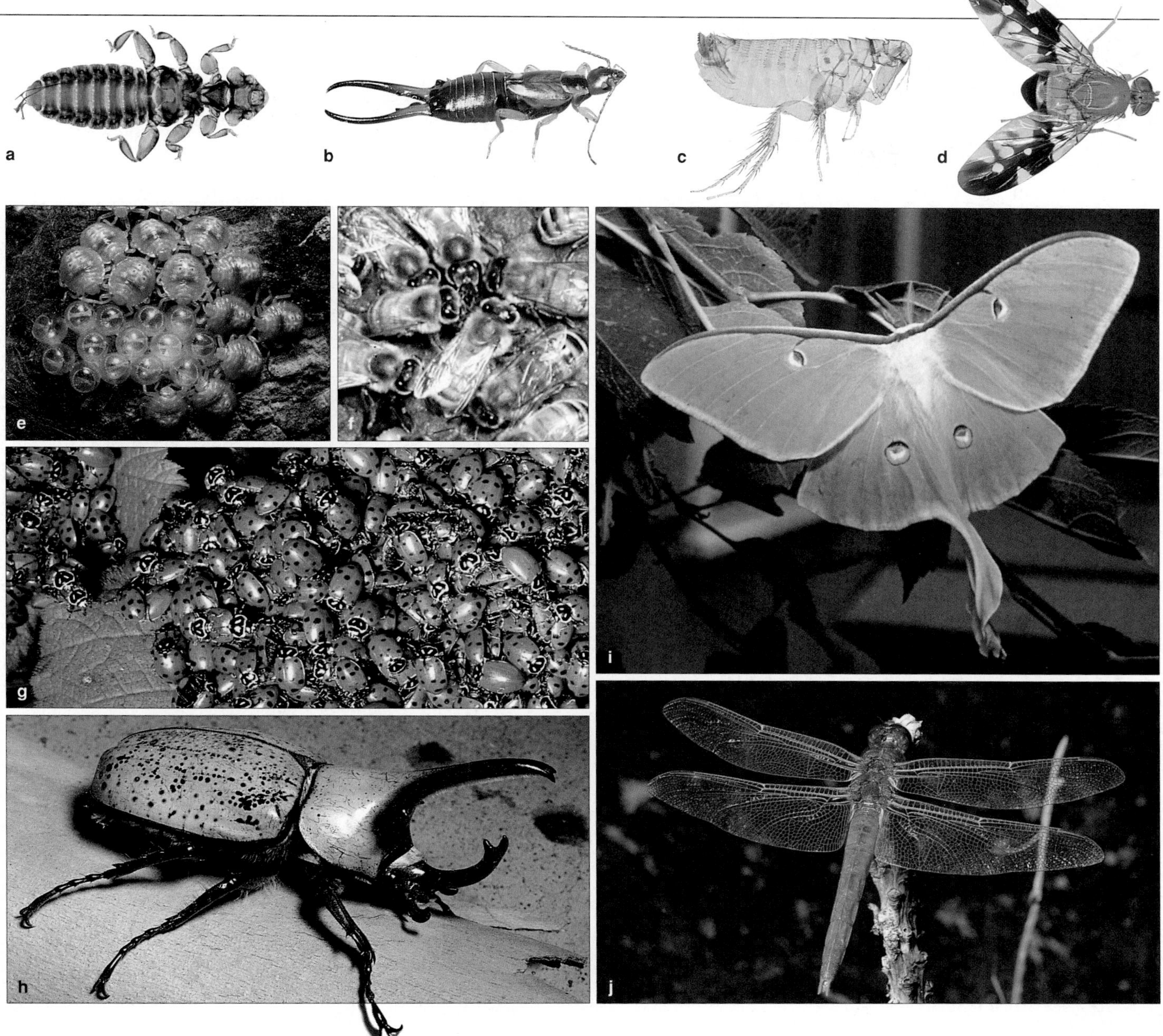

Figure 26.38 Representative insects. (**a**) Duck louse (order Mallophaga). It eats bits of feathers and skin. (**b**) European earwig (order Dermaptera), a common household pest. (**c**) Flea (order Siphonaptera), with strong legs for jumping onto and off animal hosts. (**d**) Mediterranean fruit fly (order Diptera). Its larvae destroy citrus fruit and other crops.

(**e**) Stinkbugs (order Hemiptera), newly hatched. (**f**) At center, a honeybee (order Hymenoptera) attracting its hive mates with a dance, as described in Section 47.4. (**g**) Ladybird beetles (order Coleoptera) swarming. These beetles are commercially raised and released as biological controls of aphids and other pests. Also in this order, the scarab beetle (**h**). With more than 300,000 species, Coleoptera is the largest order of the animal kingdom. (**i**) Luna moth (order Lepidoptera) of North America. Microscopic scales cover the wings and body of most butterflies and moths, including this one. (**j**) One of the dragonflies (order Odonata). It swiftly captures and eats insects in midflight.

they often transmit pathogenic microorganisms. Yet many pollinate flowering plants, which include highly valued crop plants. And many "good" insects attack or parasitize the ones we would rather do without. Figure 26.38 highlights representatives of some major orders of insects. In the next section, we conclude our arthropod survey with a selection of the notorious species.

As a group, insects show immense variation on the basic arthropod body plan. Many have wings, and their life cycles have stages that allow exploitation of different and often widely scattered food sources. Many species produce great numbers of small individuals that pass through immature, post-embryonic stages before the adult form emerges.

26.20 UNWELCOME ARTHROPODS

Given that there are now more than 6 billion of us on the planet, very, very few of us encounter the truly nasty arachnids—certain spiders, ticks, and scorpions that can, directly or indirectly, inflict memorable pain. All things considered, humans collectively do more harm to more species than, say, the spiders, most of which do great good by popping off uncountable numbers of insect pests. But the "good" ones have some "bad" relatives, and it doesn't hurt to know them when we see them.

HARMFUL SPIDERS All spiders of the genus *Loxosceles* are poisonous to humans and other mammals, but we seldom meet up with most of them. One exception, the brown recluse (*L. reclusa*), favors hiding in dry, warm places. These include houses, especially near refrigerator motors and clothes dryers, and in the folds of clothing or linens in closets. You can recognize it by the violin-shaped marking on its cephalothorax (Figure 26.39*a*). A bite ulcerates and heals poorly. Some bitten people have required skin grafts; a few died. *L. laeta*, a related species in Chile, Argentina, and Peru, is similarly dangerous.

About five people die each year in the United States as a result of black widow bites (Figure 26.39*b*)—orders of magnitude less than deaths attributed to, say, drug overdoses or accidents involving SUVs. Even so, besides being painful, this spider's neurotoxin causes muscles to stiffen, so medical attention should be prompt.

MIGHTY MITES Of all arachnids, mites are among the smallest, most diverse, and most widely distributed. Of an estimated 500,000 species, a few parasitic mites—ticks especially—are harmful. Figure 26.40 shows a deer tick (*Ixodes*), a vector for *Lyme disease* in the northeastern and north-central United States. (Different ticks spread the

a
b

Figure 26.39 (**a**) Brown recluse, with a violin-shaped mark on its forebody. Its bite can be severe to fatal. (**b**) Female black widow, with a red, hourglass-shaped marking on the underside of her shiny black abdomen. Males are smaller and do not bite.

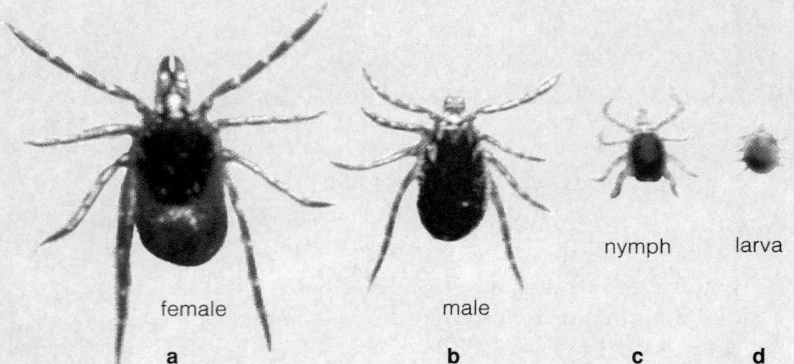
female
male
nymph
larva
a
b
c
d

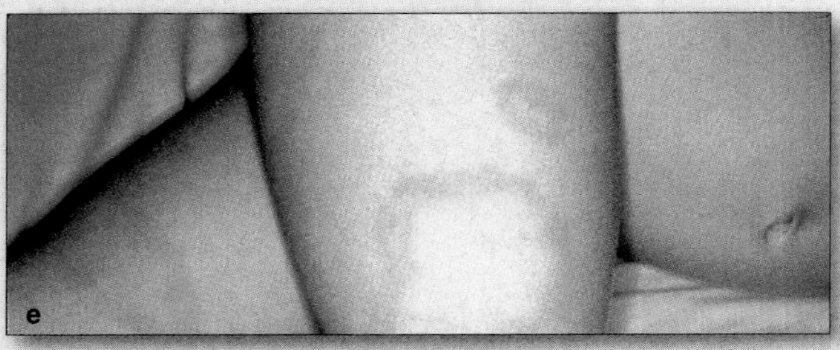
e

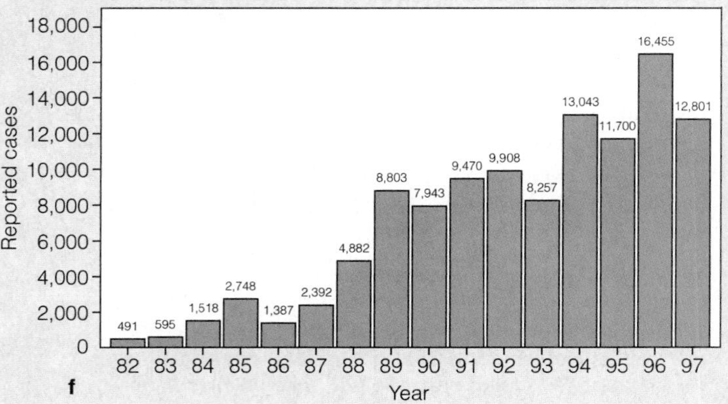
Reported cases
18,000
16,000
14,000
12,000
10,000
8,000
6,000
4,000
2,000
0
491
595
1,518
2,748
1,387
2,392
4,882
8,803
7,943
9,470
9,908
8,257
13,043
11,700
16,455
12,801
82 83 84 85 86 87 88 89 90 91 92 93 94 95 96 97
Year
f

Figure 26.40 (**a–d**) Deer ticks (*Ixodes dammini*), the most common vector for Lyme disease. The female is about three millimeters long. All stages of the life cycle are capable of transmitting the disease agent, *Borrelia burgdorferi*, to humans. (**e**) An extreme reaction to an infection by *Borrelia burgdorferi*, a spirochete. The rash is a symptom of what is now the most common tick-borne disease in the United States: Lyme disease. Tick bites deliver the spirochete to new hosts.

(**f**) Graph showing the increasing number of cases of Lyme disease in the United States between 1982 and 1997.

Further reading: Student Guide to InfoTrac on web site

Figure 26.41 *Centruroides sculpuratus*, a scorpion from Arizona that is one of the most dangerous species known.

a
b
c

Figure 26.42 (**a**) Adult western corn rootworm (*Diabrotica virgifera*). (**b**) Larvae on corn roots. (**c**) Damaged corn plants.

disease in the western and southeastern states.) Normally the tick feeds on white-tail deer, mice, other mammals, and birds. *Borrelia burgdorferi*, the agent of Lyme disease, is transmitted to humans mainly by nymphal stages, probably because these are no bigger than a pinhead and escape detection. Detection is critical, because ticks usually transmit the bacterium after they have been feeding two days or more. Adults are easier to spot, so they are more likely to be removed quickly.

Ticks don't fly or jump. They crawl onto grasses and shrubs, then onto animals that brush past, then typically into hairy, hidden body parts such as the scalp and groin. Look for them after walks in the wild. Different types can transmit pathogens responsible for many serious diseases, including Rocky Mountain spotted fever (Section 22.6), scrub typhus, tularemia, babesiasis, and encephalitis. A select few can even kill directly. Their toxin interferes with motor neurons in a way that causes progressive paralysis from the lower part of the body upward. The tick has to be removed before the bitten person stops breathing.

Ending on a less scary but still irritating note, think of the house dust mites. They don't kill you, but the allergies they provoke in people can be irritating to the extreme.

SCORPION COUNTRY Scorpions have that look about them. Large, prey-seizing pincers, a venom-dispensing stinger at the tip of a narrowed, jointed abdomen—what's not to love? Actually, scorpions seldom dispense venom unless their prey—other small arthropods, the occasional small lizard or mouse—puts up a mighty fight. In the wild they hunt at night and rest during the day in burrows or under logs, stones, and bark. They can stay there without starving for as long as a year.

We find scorpions in cold places, such as Canada, the southern Andes, and the southern Alps. Some live in moist forests. But most live in hot, dry places, and there they are active throughout the year. The ones that can hurt or kill people aren't the fiercest looking, and there aren't many of them. *Centruroides sculpturatus*, an Arizona scorpion, is the most dangerous species in the United States (Figure 26.41). Its close relatives in Mexico have a reputation for causing fatalities, especially amoung small children.

When in scorpion country, it is not a good idea to walk about barefoot at night or to put on shoes without first turning them upside-down and shaking them vigorously.

BEETLES, BEETLES EVERYWHERE First in America's Midwest and now in the Balkans, *Diabrotica virgifera* is making the rounds in cornfields as a representative of one of our major competitors for food. Figure 26.42 shows what it looks like and the kind of damage it causes. Its larvae feed on corn roots until the severely damaged plants simply keel over. Each year we lose about a billion dollar's worth of crops to these little critters alone.

Faced with such astronomical losses, farmers spread about 30 million pounds of pesticides a year on cornfields, which is about half the total dispensed for all row crops grown in the United States. So far rootworms are not impressed. Many have developed pesticide tolerance.

Inventive researchers are targeting the adults, which show an inordinate fondness for bitter plant juices—specifically, for curcurbitacin. This is one of the steroid terpenes that evolved as a natural pesticide in melons, squashes, cucumbers, and other members of the family Cucurbitaceae. Rootworms apparently appropriated the chemical as a defense against bird predators. (Birds learn to avoid rootworms after a taste trial and vomiting.) Robert Schroder and others are concocting to-die-for bait for the pests. For example, watermelon juice has been laced with red dye 28, a photoactive chemical. In taste trials, adult rootworms gorged themselves, turned deep red, and after five minutes in sunlight, dropped dead. Apparently sunlight triggered oxidation reactions that destroyed tissues throughout the body. Beneficial insects avoid the dyed juice. In this particular ongoing contest, chalk one up for the researchers.

26.21 THE PUZZLING ECHINODERMS

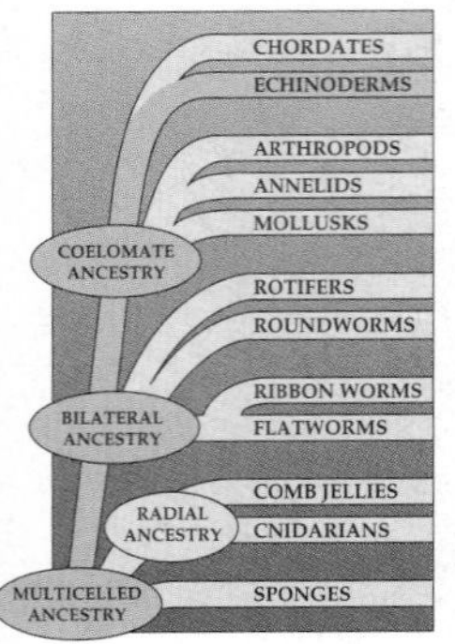
CHORDATES
ECHINODERMS
ARTHROPODS
ANNELIDS
MOLLUSKS
COELOMATE ANCESTRY
ROTIFERS
ROUNDWORMS
RIBBON WORMS
BILATERAL ANCESTRY
FLATWORMS
COMB JELLIES
RADIAL ANCESTRY
CNIDARIANS
MULTICELLED ANCESTRY
SPONGES

We turn, finally, to the second lineage of the coelomate animals, the deuterostomes. The major invertebrate members of this lineage are **echinoderms** (Echinodermata). The feather star, sea urchin, sea cucumber, and brittle stars shown in Figure 26.43 belong to this phylum. So do the sea lilies (crinoids), sand dollars, and sea biscuits. The sea lilies, which look a bit like stalked plants, flourished in Silurian times (Figure 21.14). About 13,000 echinoderm species are known from the fossil record, although most of these became extinct. Nearly all of the 6,000 or so existing species live in marine habitats.

An echinoderm body wall bears a number of spines, spicules, or plates made rigid with calcium carbonate. (*Echinodermata* means spiny-skinned.) These structures serve defensive functions, as you might suspect if you have ever stepped barefoot on a sea urchin. Its spines trigger painful swelling when they break off under the skin. Most echinoderms have a well-developed internal skeleton, which is composed of calcium carbonate and other substances secreted from specialized cells.

Oddly, the adult echinoderms are radial with some bilateral features. Most species even produce bilateral larvae. Did bilateral invertebrates give rise to ancestors of echinoderms, in which some radial features evolved at a later time? Maybe.

Adult echinoderms have no brain. However, their decentralized nervous system allows them to respond to information about food, predators, and so forth that is coming from different directions. For instance, any arm of a sea star that senses the shell of a tasty scallop

a

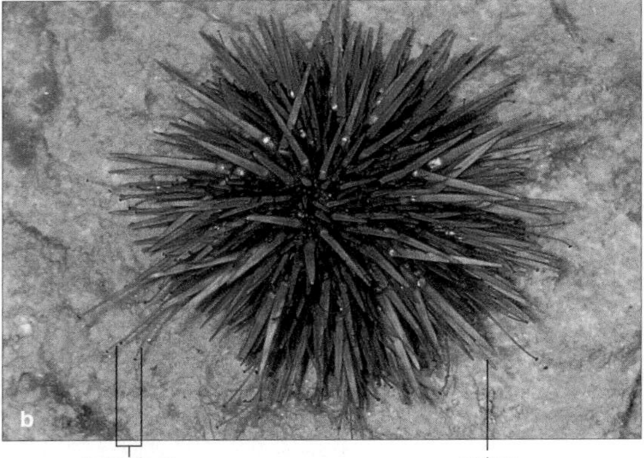
b
tube feet
spine

c

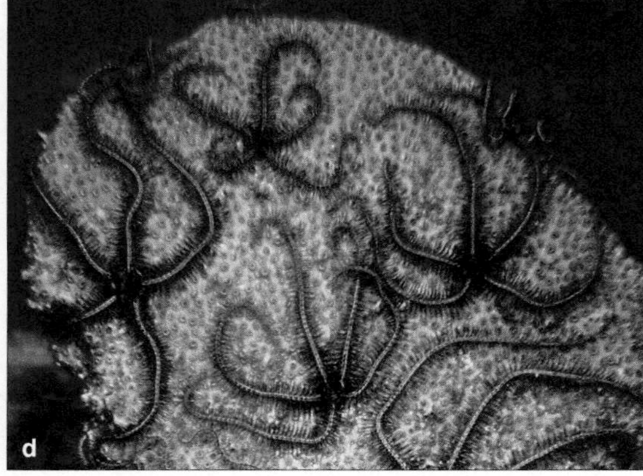
d

Figure 26.43 Representative echinoderms. (**a**) Feather star, with finely branched food-gathering appendages. (**b**) Sea urchin, which moves about on spines and tube feet. (**c**) Sea cucumber, with rows of tube feet along its body. (**d**) Brittle stars. Their arms (rays) make rapid, snakelike movements.

Further reading: Student Guide to InfoTrac on web site

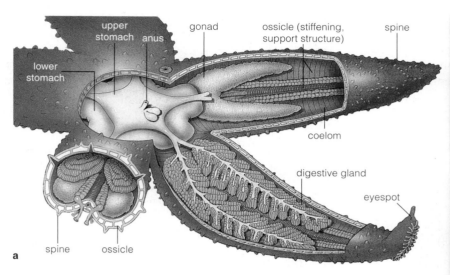

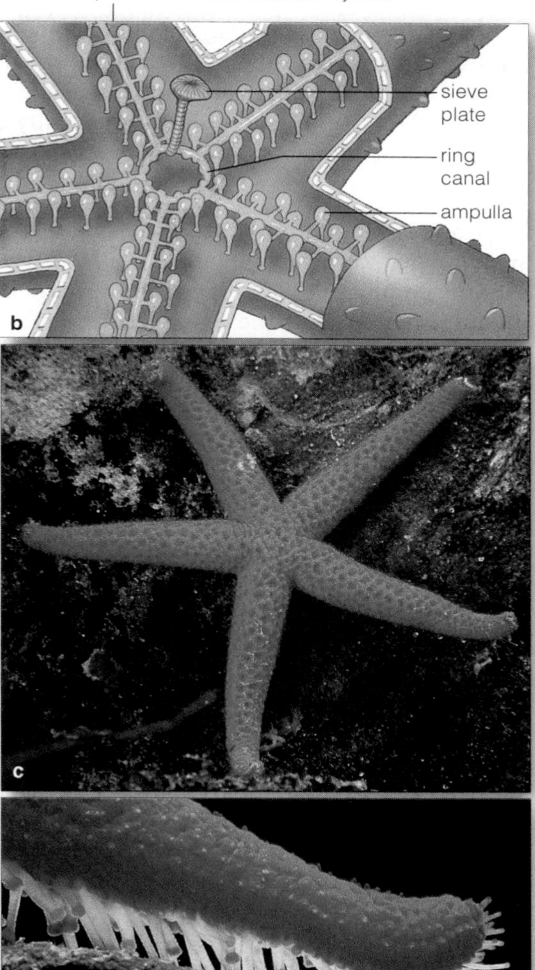
upper stomach
anus
gonad
ossicle (stiffening, support structure)
spine
lower stomach
coelom
digestive gland
eyespot
spine
ossicle
a
part of the water-vascular system
sieve plate
ring canal
ampulla
b
c
d
tube feet on the ventral surface of a sea star arm

Figure 26.44 (**a**) Key aspects of the radial body plan of a sea star. Its water-vascular system, combined with great numbers of tube feet, is the basis of locomotion. (**b–d**) Five-armed sea star, with closer views of tube feet.

can become the leader, directing the rest of the body to move in a direction suitable for capturing prey.

Figures 26.43 and 26.44 show examples of tube feet. These fluid-filled, muscular structures have suckerlike adhesive disks. Sea stars use their tube feet for walking, burrowing, clinging to rocks, or gripping a clam or snail about to become a meal. Tube feet are components of a **water-vascular system** unique to echinoderms. In sea stars, that system includes a main canal in each arm. Short side canals extend from them and deliver water to the tube feet. Each tube foot has an ampulla, a fluid-filled, muscular structure shaped rather like the rubber bulb on a medicine dropper. As an ampulla contracts, it forces fluid into the foot and causes it to lengthen.

Tube feet change shape constantly as muscle action redistributes fluid through the water-vascular system. Hundreds of tube feet may move at a time. After being released, each one swings forward, reattaches to the substrate, then swings backward and is released before swinging forward again. Their motions are splendidly coordinated, so sea stars glide rather than lurch along.

On their ventral surface, sea stars have a formidable feeding apparatus, such as the one shown here:

Some eager sea stars simply swallow their prey whole. Others push part of their stomach outside the mouth and around their prey, then start digesting their meal even before swallowing it. Sea stars get rid of coarse, undigested residues through the mouth. They do have a small anus, but this is of no help in getting rid of an empty clam shell or snail shell.

With their curious traits, echinoderms are a suitable point of departure for this chapter. Even though we can identify broad trends in animal evolution, we should keep in mind that there are confounding exceptions to the perceived macroevolutionary patterns.

Echinoderms are coelomate animals with spines, spicules, or plates in the body wall. From the evolutionary perspective, they are a puzzling mix of bilateral and radial features.

26.22 SUMMARY

1. Animals are multicelled, aerobic heterotrophs that ingest or parasitize other organisms. Nearly all have diploid body cells organized into tissues, organs, and organ systems. Animals reproduce sexually and often asexually. They undergo embryonic development. Most are motile during at least part of the life cycle.

2. Animals range from structurally simple placozoans and sponges to vertebrates.

a. By comparing major animal phyla and integrating the information with the fossil record, biologists have identified major evolutionary trends among them.

b. Revealing aspects of body plans are the type of symmetry, gut, and cavity (if any) between the gut and body wall; whether there is a head end; and whether the body is divided into segments (Figure 26.45).

3. *Trichoplax*, the only known placozoan, is the simplest animal. It is composed of little more than two layers of cells with a fluid matrix in between.

4. The sponge body has no symmetry and it is at the cellular level of construction. Although it consists of several kinds of cells, these are not organized like the epithelia and other tissues seen in complex animals.

5. Cnidarians include the jellyfishes, sea anemones, and hydras. They have radial symmetry, and they are at the tissue level of construction. Cnidarians alone produce nematocysts, capsules with dischargeable threads that are used mainly in prey capture.

6. Nearly all animals more complex than cnidarians show bilateral symmetry, and they form tissues, organs, and organ systems. Their gut may be saclike, as it is in flatworms, but it usually is complete, with an anus and a mouth. Most animals more complex than flatworms have a coelom or false coelom (cavities between the gut and body wall). A coelom has a lining, a peritoneum.

7. Two major lineages diverged shortly after flatworms evolved. One (protostomes) gave rise to the mollusks, annelids, and arthropods. The other (deuterostomes) gave rise to echinoderms and chordates.

8. All mollusks have a fleshy, soft body and a mantle. Most have either a shell or a remnant of one. Mollusks vary greatly in size, body details, and life-styles.

9. The annelids (earthworms, polychaetes, and leeches) have a segmented body, complex organs, and a series of coelomic chambers.

10. Collectively, arthropods are the most successful of all groups in terms of diversity, numbers, distribution, defenses, and capacity to exploit food resources.

a. All arthropods, including arachnids, crustaceans, and insects, have hardened, jointed exoskeletons. They also have modified segments, specialized appendages,

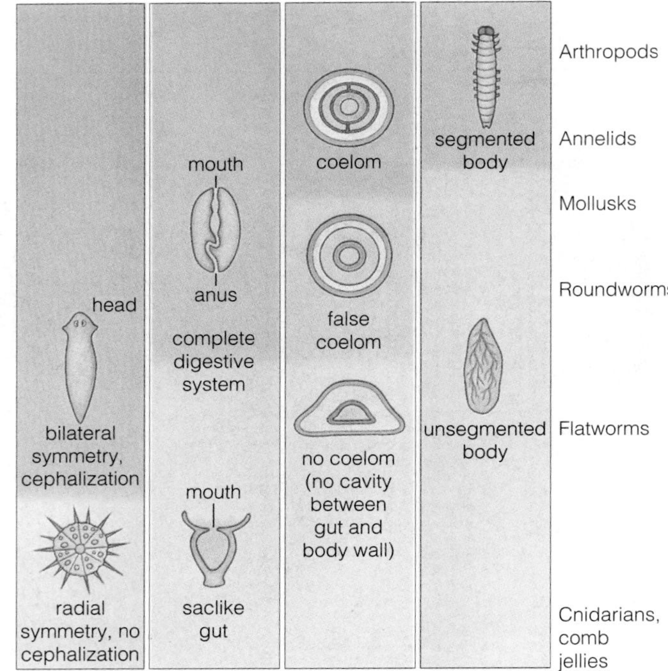

Arthropods
Annelids
Mollusks
Roundworms
Flatworms
Cnidarians, comb jellies
head
bilateral symmetry, cephalization
radial symmetry, no cephalization
mouth
anus
complete digestive system
mouth
saclike gut
coelom
false coelom
no coelom (no cavity between gut and body wall)
segmented body
unsegmented body

Figure 26.45 Summary of key trends in the evolution of animals, as identified by comparing body plans of major phyla. All of these features did not appear in every group.

notably specialized respiratory, nervous, and sensory organs, and (in insects only) wings.

b. Arthropods develop by growth in size and molting or develop through a series of immature stages, such as larvae and nymphs. Many metamorphose; the tissues of immature forms undergo major reorganization, and body parts are remodeled before the adult emerges.

11. Echinoderms have spines, spicules, or plates in their body wall. Evolutionarily, the phylum is puzzling. The larvae of most species form bilateral features, but they go on to develop into basically radial adults.

Review Questions

1. List the six main features that characterize animals. *26.1*

2. When attempting to discern evolutionary relationships among major groups of animals, which aspects of their body plans provide the most useful clues? *26.1*

3. What is a coelom? Why was it important in the evolution of certain animal lineages? *26.1*

4. Name some animals with a saclike gut. Evolutionarily, what advantages does a complete gut afford? *26.1, 26.4–26.6*

5. Choose a species of insect that lives in your neighborhood and describe some of the observable adaptations that underlie its success. *26.15, 26.19*

Self-Quiz (*Answers in Appendix III*)

1. Which is *not* a general characteristic of the animal kingdom?
   a. multicellularity; most have tissues, many form organs
   b. exclusive reliance on sexual reproduction
   c. motility at some stage of the life cycle
   d. embryonic development during the life cycle

Figure 26.46 *Telesto*, one of the soft branching corals.

2. Most animals that are more complex than cnidarians have ______ symmetry, and ______ forms in their embryos.
   a. radial; mesoderm c. bilateral; mesoderm
   b. bilateral; endoderm d. radial; endoderm

3. More complex animals have a ______ between their gut and body wall.
   a. pharynx c. coelom
   b. pseudocoelom d. archenteron

4. Jellyfishes, sea anemones, and their relatives have ______ symmetry, and their cells form ______.
   a. radial; mesoderm c. radial; tissues
   b. bilateral; tissues d. bilateral; mesoderm

5. Which phylum contains members that are notorious for causing serious diseases in humans?
   a. cnidarians c. segmented worms
   b. flatworms d. chordates

6. ______ have a coelom and pronounced segmentation, and a dizzying variety of modifications to the segments.
   a. arthropods c. sponges e. sea stars
   b. annelids d. snails and clams f. vertebrates

7. In sheer numbers and distribution, ______ are the most successful animals.
   a. arthropods c. sponges e. sea stars
   b. annelids d. snails and clams f. vertebrates

8. Match the terms with the appropriate groups.
   ___ sponges a. spiny-skinned
   ___ cnidarians b. vertebrates and kin
   ___ flatworms c. flukes and tapeworms
   ___ roundworms d. no tissue organization
   ___ rotifers e. no males for some
   ___ mollusks f. nematocysts, radial symmetry
   ___ annelids g. hookworms, elephantiasis
   ___ arthropods h. jointed exoskeleton
   ___ echinoderms i. "belly-foots" and kin
   ___ chordates j. segmented worms

Critical Thinking

1. A carnivorous sponge was discovered in an underwater cave in the Mediterranean Sea. Unlike other sponges, it has no pores or canals. Projecting from branching outgrowths of its body surface are hooklike spicules that act like Velcro to trap shrimp and other animals. The outgrowths envelop prey, which sponge cells then digest. Which components of the sponge body had to evolve for such a feeding strategy?

2. Tapeworms are hermaphroditic. What selective advantages might this feature offer, and in what kinds of environments?

3. People who eat raw oysters or clams harvested from sewage-polluted waters develop mild to severe gastrointestinal ailments. Think about the feeding modes of these mollusks and develop a hypothesis to explain why mollusk eaters can get sick.

4. You are diving in calm, warm waters behind a large tropical reef. You observe something that looks like a red and white plant, but it has a profusion of tiny tentacles (Figure 26.46). This is a branching soft coral with a great many tiny polyps. You will never see such a coral growing on reef surfaces exposed to the open sea. Formulate two hypotheses that might explain their distribution.

5. The animal in Figure 26.47*a* is packed with organs and sports a crown of cilia at its head end. The marine worm in Figure 26.47*b* has a highly segmented body. Most of its segments are similar to one another; each has bristles on the sides that were employed to dig a burrow in sediments. To which groups do the two animals belong?

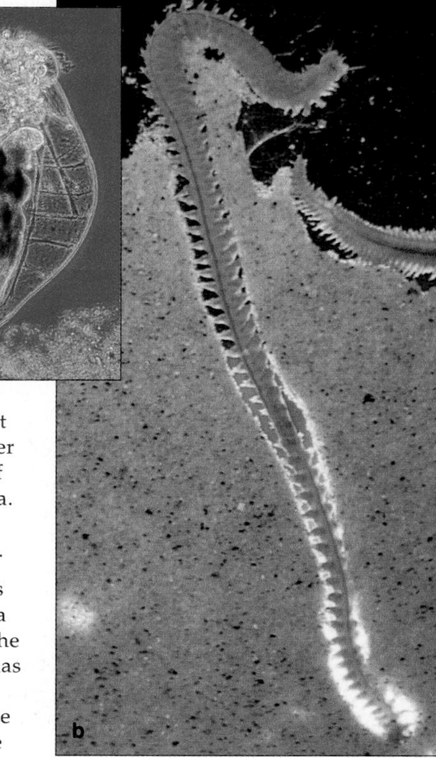

a
b

Figure 26.47 Go ahead, name the mystery animals.

Selected Key Terms

animal *26.1*
annelid *26.14*
arthropod *26.15*
bilateral symmetry *26.1*
brain *26.14*
centipede *26.18*
cephalization *26.1*
cnidarian *26.4*
coelom *26.1*
collar cell *26.3*
comb jelly *26.6*
contractile cell *26.4*
cuticle *26.8*
deuterostome *26.11*
echinoderm *26.21*
ectoderm *26.1*
endoderm *26.1*
epithelium *26.4*
exoskeleton *26.15*
flatworm *26.7*
gonad *26.5*
gut *26.1*
hermaphrodite *26.7*
hydrostatic skeleton *26.4*
invertebrate *26.1*
larva *26.3*
Malpighian tubule *26.19*
mantle *26.12*
medusa *26.4*
mesoderm *26.1*
metamorphosis *26.15*
millipede *26.18*
mollusk *26.12*
molting *26.15*
nematocyst *26.4*
nephridium *26.14*
nerve cell *26.4*
nerve cord *26.14*
nymph *26.19*
organ *26.7*
organ system *26.7*
placozoan *26.2*
planula *26.5*
polyp *26.4*
proglottid *26.7*
protostome *26.11*
pupa *26.19*
radial symmetry *26.1*
ribbon worm *26.7*
rotifer *26.10*
roundworm *26.8*
sponge *26.3*
torsion *26.13*
vertebrate *26.1*
water-vascular system *26.21*

Readings *See also www.infotrac-college.com*

Kerr, E. 2 October 1998. "Tracks of Billion-Year-Old Animals?" *Science.* 282:19–21.

Pearse, V., et al. 1987. *Living Invertebrates.* Palo Alto, California: Blackwell.

Pechenik, J. 1995. *Biology of Invertebrates.* Third edition. Dubuque: W. C. Brown.

Raloff, J. July 10, 1999. "The Bitter End: Enticing Agricultural Pests to Their Last Repast." *Science News* 156:24–26.

27

ANIMALS: THE VERTEBRATES

Making Do (Rather Well) With What You've Got

In 1798, a few naturalists were skeptically probing a specimen delivered to the British Museum in London, looking for signs that a prankster had slyly stitched the bill of an oversized duck onto the pelt of a small furry mammal. They didn't know it, but they were examining the remains of a platypus, a web-footed mammal about half the size of a housecat. Like the other mammals, the duck-billed platypus (*Ornithorhynchus anatinus*) has mammary glands and hair. However, like birds and reptiles, it has a cloaca, an enlarged duct through which gametes, feces, and excretions from the kidneys pass. It lays shelled eggs, as birds and most reptiles do. Its young hatch pink and unfinished, as embryonic stages too helpless to fend for themselves (Figure 27.1*a*). Its fleshy bill does appear ducklike and its broad, flat, furry tail looks like the one on a beaver (Figure 27.1*b*).

With its unusual traits, the platypus invites us to challenge preconceived notions of what constitutes "an animal." This particular combination of traits started coming together when the supercontinent Pangea was breaking up. As you will see, the platypus's ancestors happened to be stuck on a huge fragment that remained isolated from the other continents for 150 million years; eventually it became Australia. The geographic separation was a springboard for genetic divergence, and unique mutations in combination with selection pressures gave rise to the platypus body. Even though that body plan is much ridiculed, we scarcely can call it a failure. It has endured far longer than the body plan we humans inherited.

The platypus inhabits streams and lagoons in remote regions of Australia and Tasmania. During the day it hides in underground burrows. At night it slips into the water, where it hunts for small invertebrates. Even when it stays in water all night, its dense fur

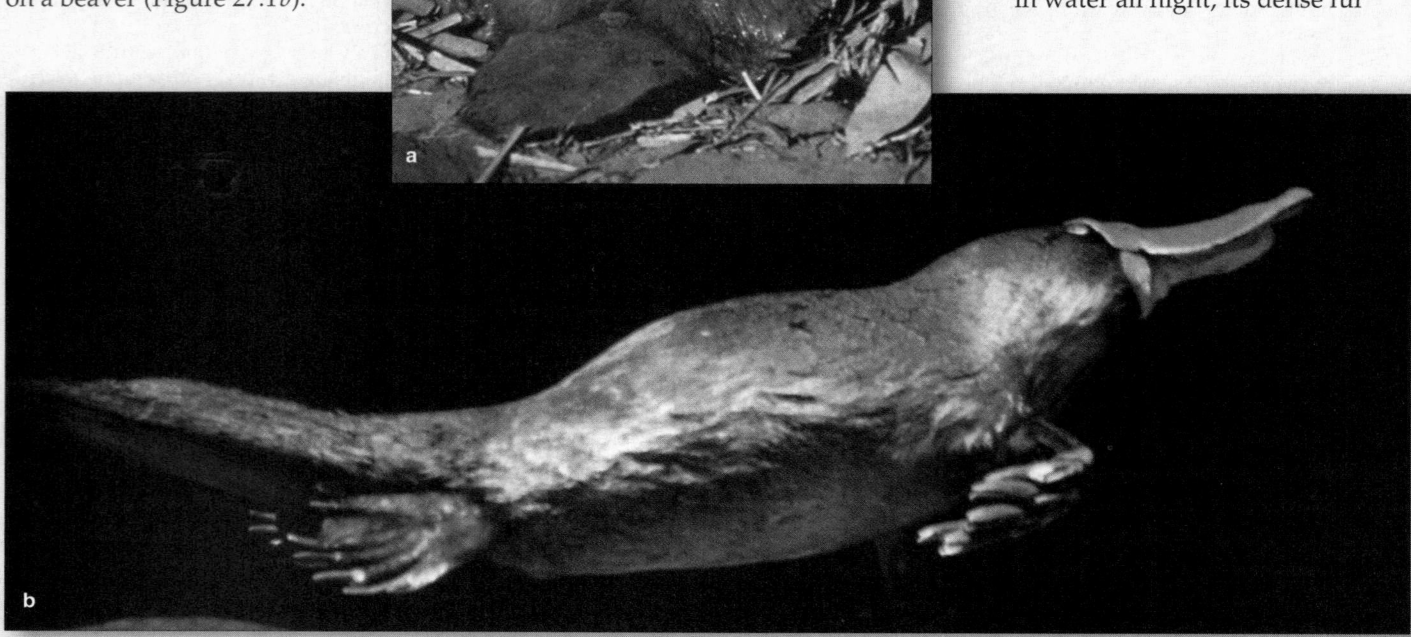

Figure 27.1 One of evolution's success stories—the platypus, (**a**) raising its incompletely formed offspring in a burrow and (**b**) underwater, eyes and ears shut, yet homing in on prey.

helps maintain body temperature, just as it does all day long when the platypus is resting in its cool burrow. The broad, thick tail is a most excellent rudder during underwater maneuvers and a fine storehouse for energy-rich reserves of fat. The broad, flattened platypus bill is well shaped for scooping up aquatic snails, shrimps, worms, mussels, and insect larvae. The horny pads that line the jaws grind up and make short work of the shelled and hard-cuticled meals. Back at its burrow, the platypus uses the strong claws that project from its hind feet for digging out underground chambers.

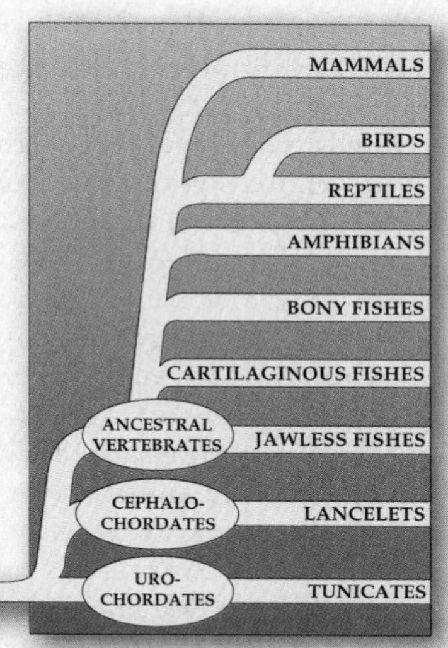
MAMMALS
BIRDS
REPTILES
AMPHIBIANS
BONY FISHES
CARTILAGINOUS FISHES
ANCESTRAL VERTEBRATES
JAWLESS FISHES
CEPHALO-CHORDATES
LANCELETS
URO-CHORDATES
TUNICATES

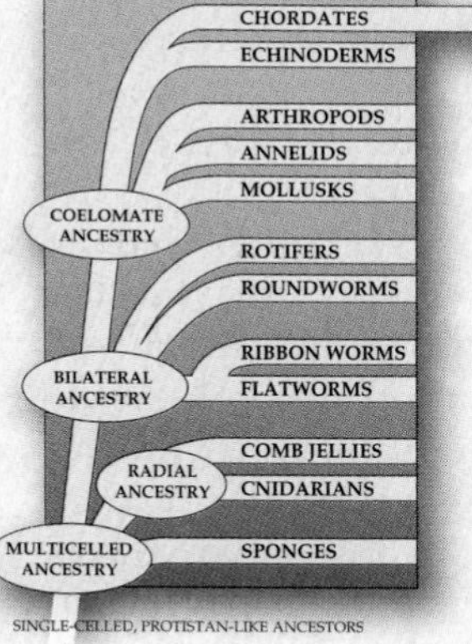
CHORDATES
ECHINODERMS
ARTHROPODS
ANNELIDS
MOLLUSKS
COELOMATE ANCESTRY
ROTIFERS
ROUNDWORMS
RIBBON WORMS
BILATERAL ANCESTRY
FLATWORMS
COMB JELLIES
RADIAL ANCESTRY
CNIDARIANS
MULTICELLED ANCESTRY
SPONGES
SINGLE-CELLED, PROTISTAN-LIKE ANCESTORS

Figure 27.2 Family tree for the vertebrates and other chordates.

Like a submarine, a platypus closes the hatches, so to speak, when it dives into the water. Its nostrils as well as a fleshy groove around the eyes and ears snap shut. No need for eyes or ears to zero in on prey; many of the 800,000 sensory receptors in its bill detect tiny oscillations in water pressure as prey swim past. Other receptors detect changes in electric fields as weak as those initiated by the flicks of a shrimp tail.

The platypus is one of 4,500 kinds of mammals that now occupy the vertebrate branch of the animal family tree (Figure 27.2). Its vertebrate relatives include fishes, amphibians, reptiles, and birds, which are described in this chapter. This chapter surveys all vertebrate groups. It concludes with a look at the evolutionary history of the human species, starting with its mammalian and primate stocks. As you consider each group, keep this key concept in mind: *Each animal is a mosaic of traits, many conserved from remote ancestors and others unique to its branch on the animal family tree.*

KEY CONCEPTS

1. The chordate branch of the family tree for animals includes invertebrate and vertebrate species. All are bilateral animals. While they are embryos or larvae, each typically develops a supporting rod for the body (notochord), a dorsal nerve cord, a pharynx, gill slits in the pharynx wall, and a tail that extends past the anus. Some or all of these features persist in adults.

2. Existing invertebrate chordates include the tunicates and lancelets.

3. Of eight classes of vertebrates, seven have living representatives. These are jawless fishes, cartilaginous fishes, bony fishes, amphibians, reptiles, birds, and mammals. The other class, the placoderms, became extinct early in vertebrate history.

4. Four major trends occurred when certain vertebrate lineages evolved. Support and movement of the body came to depend less on the notochord and more on a backbone. After jaws evolved, the nerve cord evolved into a spinal cord and brain. When some lineages moved onto land, gills became less important than lungs for gas exchange, a trend enhanced by the evolution of more efficient circulatory systems. Among pioneers on land, fleshy fins with skeletal supports evolved into limbs, which evolved in diverse ways among the amphibians, reptiles, birds, and mammals.

5. In the primate branch of the mammalian lineage are prosimians, tarsioids, and anthropoids (monkeys, apes, and humans). Apes and humans are hominoids. Humans and some extinct species with a mosaic of apelike and humanlike traits are further classified as hominids.

6. A long-term cooling trend started during the Miocene. It led to seasonal changes in habitats and food sources. As food sources became scarcer, early hominoids spread out through Africa and, later, entered Europe and southern Asia. One lineage gave rise to the hominids.

7. Unlike earlier hominids, *Homo erectus* and *H. sapiens* displayed remarkable behavioral flexibility and creative experimentation with their environment, as when they started using fire. This characteristic allowed them to survive the challenges of dispersing into novel and often harsh environments around the world.

27.1 THE CHORDATE HERITAGE

Characteristics of Chordates

The preceding chapter left off with echinoderms, one of the most ancient lineages of the deuterostome branch of the animal family tree. Dominating this branch are their more recently evolved relatives, the **chordates** (phylum Chordata). Of these bilateral animals, about 2,100 are, like the echinoderms, invertebrates. The vast majority—about 48,000 species—are vertebrates.

Vertebrates are chordates with a backbone, of either cartilage or bone, and a brain located inside a protective chamber of skull bones. The "invertebrate chordates" share certain features with these dominant members of the phylum, but a backbone isn't one of them.

Four features are evident in chordate embryos, and in many species these features persist into adulthood. *First*, a **notochord**, a long rod of stiffened tissue (not cartilage or bone), helps support the body. *Second*, the nervous system of chordate embryos develops from a tubular, dorsal **nerve cord**, which runs parallel to the notochord and gut. The cord's anterior end increases in mass and undergoes modifications to form the brain. *Third*, the embryos have distinctive slits in the wall of their **pharynx**, which is a muscularized tube. A pharynx functions in feeding, respiration, or both. *Fourth*, a tail forms in embryos and extends past the anus.

Chordate Classification

Biologists group nearly all of the chordates into three subphyla. These are named Urochordata (tunicates and their kin), Cephalochordata (lancelets), and Vertebrata (vertebrates). There are eight classes of vertebrates:

Agnatha	*Jawless fishes*
Placodermi	*Jawed, armored fishes (extinct)*
Chondrichthyes	*Cartilaginous fishes*
Osteichthyes	*Bony fishes*
Amphibia	*Amphibians*
Reptilia	*Reptiles*
Aves	*Birds*
Mammalia	*Mammals*

Appendix I has an expanded classification scheme for the vertebrates. Unit VI provides details of their body plans and functions. Here we become acquainted with major trends in their evolution. We find clues to those trends among the invertebrate chordates.

The embryos of chordates alone have this combination of features: a notochord, a tubular dorsal nerve cord, a pharynx with slits in its wall, and a tail extending past the anus.

27.2 INVERTEBRATE CHORDATES

Tunicates

The 2,000 species of existing urochordates are baglike animals one to several centimeters long. More often they are called the **tunicates**, after the gelatinous or leathery "tunic" that adults secrete around themselves. Adults of the most common species, the "sea squirts," squirt water through a siphon when something irritates them.

Tunicates live in marine habitats from the intertidal zone to surprising depths. Most adults remain attached to rocks and other suitably hard substrates. Some live solitary lives; others are colonial. Figure 27.3*a* shows an adult sea squirt. It develops from a bilateral, swimming larva that resembles a tadpole (Figure 27.3*b*,*c*). A larva, recall, is an immature stage between the embryonic and adult stages of an animal life cycle. The firm, flexible notochord of a tunicate larva is a series of cells that acts like a torsion bar. As muscles on one side of the tail or the other contract, the notochord bends, then it springs back as muscles relax. The strong, side-to-side motion propels the larva forward. Most fishes use muscles and their backbone for the same kind of motion.

Like some other animals, tunicates are **filter feeders**: they filter food from a current of water that is directed through part of their body. Water flows in through one siphon and passes through **gill slits**, or openings in the thin pharynx wall. Water flows out through a different siphon. Their pharynx also acts as a respiratory organ. Dissolved oxygen is less concentrated in blood vessels adjoining the pharynx than in the surrounding water. Therefore, it diffuses from water into the blood. Carbon dioxide's concentration gradient is such that it diffuses from blood into the water leaving the pharynx.

Sea squirts undergo metamorphosis. By this process of development, recall, remodeling and reorganization of body tissues transform an immature stage into the adult (Section 26.19). As a sea squirt larva develops, its tail and notochord disappear, and a tunic forms. The pharynx enlarges, and perforations in its wall become subdivided into many slits. The nerve cord regresses, so all that remains is a very simplified nervous system.

Lancelets

About twenty-five species of fish-shaped, translucent animals called cephalochordates live in nearshore marine sediments around the world. Most are shorter than your little finger. Much of the time they are buried, almost up to their mouth, in sand and sediments. Their common name, **lancelets**, refers to the sharp tapering of their body at both ends. The lancelet body plan sketched in Figure 27.4*a* clearly shows the four chordate features. Notice the segmented pattern of muscles on both sides of the notochord. Like tunicate larvae, lancelets use the

Further reading: Student Guide to InfoTrac on web site —

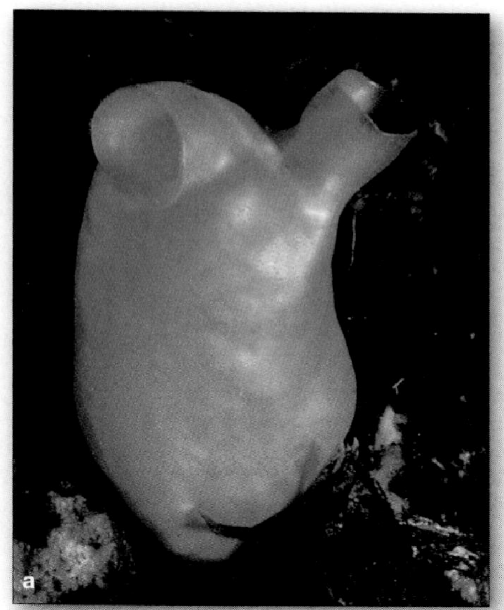

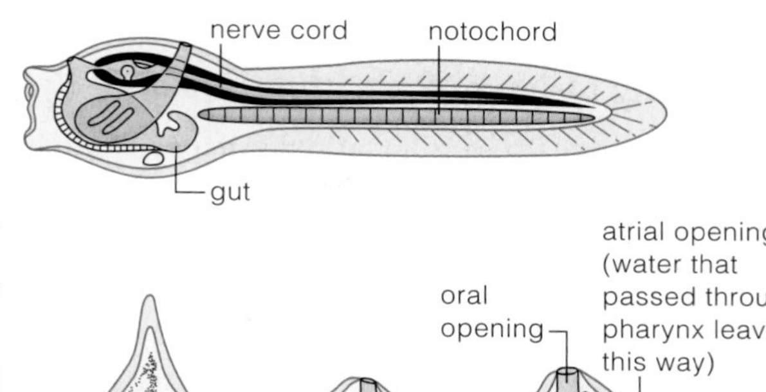

nerve cord
notochord
gut

b Body plan of the tadpole-like larva, shown midsagittal section.

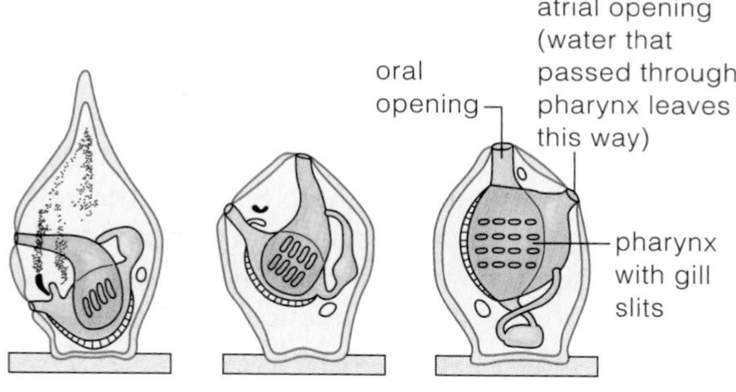

oral opening
atrial opening (water that passed through pharynx leaves this way)
pharynx with gill slits

c A new larva swims about for a brief period. Metamorphosis begins when its head attaches to a substrate. The notochord, tail, and most of the nervous system are resorbed (recycled to form new tissues). Slits in the pharynx wall multiply. Organs rotate until the openings through which water enters and leaves the pharynx are directed away from the substrate.

Figure 27.3 Example of a tunicate. (**a**) Adult form of a sea squirt. (**b**) Diagram of a sea squirt larva. (**c**) Metamorphosis of the larva into the adult.

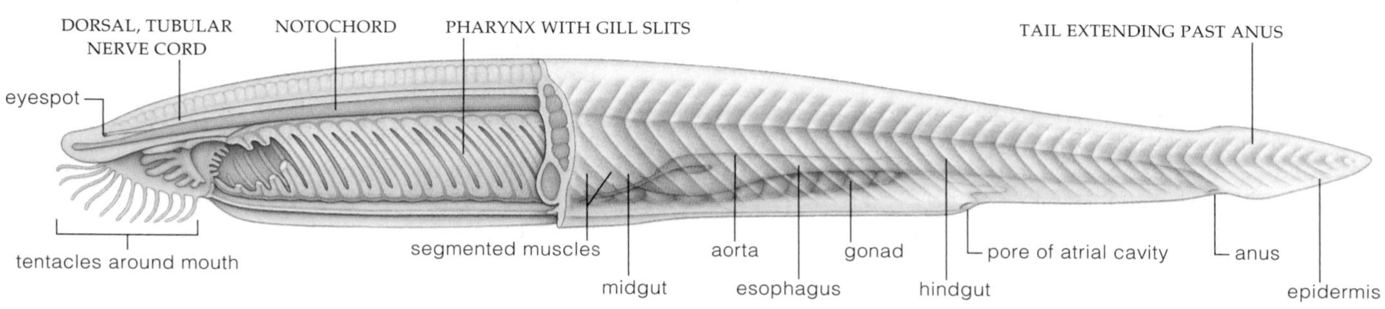

DORSAL, TUBULAR NERVE CORD
NOTOCHORD
PHARYNX WITH GILL SLITS
TAIL EXTENDING PAST ANUS
eyespot
tentacles around mouth
segmented muscles
midgut
aorta
esophagus
gonad
hindgut
pore of atrial cavity
anus
epidermis
a

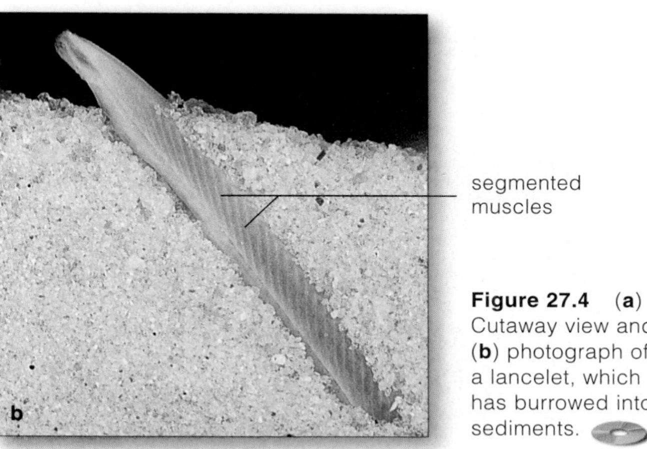

segmented muscles

Figure 27.4 (**a**) Cutaway view and (**b**) photograph of a lancelet, which has burrowed into sediments.

Like tunicates, lancelets are filter-feeding animals. Cilia line their mouth cavity and create a current that draws water through it. The water then moves through the pharynx, where food becomes trapped in mucus. The trapped food is delivered to the rest of the gut. By itself, the collective beating of tiny cilia cannot deliver sufficient food to a filter-feeding animal. Delivery also depends on having a large food-trapping surface area. In lancelets, the pharynx provides the area. It is large relative to the overall body length (Figure 27.4). Besides this, as many as 200 ciliated, food-trapping gill slits perforate its wall. As you will read in sections to follow, the pharynx turned out to be an organ with interesting evolutionary possibilities for the vertebrates.

notochord and muscles to produce swimming motions. Unlike tunicates, they have a closed circulatory system (but no red blood cells). A complex brain is nowhere in sight, but the anterior end of the dorsal nerve cord is expanded, and pairs of nerves extend into each muscle segment. The mode of respiration is simple yet effective for such a small body. Dissolved oxygen and carbon dioxide diffuse across the body's thin skin.

Tunicate larvae and the lancelets use their notochord and muscles for fishlike swimming movements. Their pharynx, a muscularized tube, has finely divided openings across its thin wall.

The tunicates and lancelets are filter feeders. They use their pharynx to filter microscopic food from a current of water, which they draw in through the body. Tunicate larvae also use the pharynx as a respiratory organ, for gas exchange.

EVOLUTIONARY TRENDS AMONG THE VERTEBRATES

Puzzling Origins, Portentous Trends

Did vertebrates arise from tadpole-shaped chordates? The hemichordates, an obscure phylum, suggest this may well have happened (*hemi-* meaning half, as in "halfway to chordates"). In morphology, these marine invertebrates are midway between echinoderms and chordates. They don't have a notochord, but they have a gill-slitted pharynx and a dorsal, tubular nerve cord. Their larvae resemble echinoderm *and* tunicate larvae. What if gene mutations in an echinoderm lineage accelerated the rates at which sex organs developed? If those organs started functioning earlier, in tadpole-shaped *larvae*, then metamorphosis—and the original adult form—could be dispensed with. This idea is not far-fetched. A few modern tunicates resemble larvae but have sex organs. Some amphibians become sexually mature even though they are larvae in some respects, and they can reproduce generation after generation.

Assuming tadpole-shaped chordates emerged, how did they evolve into vertebrates? The key trends started in fishes (Figures 27.5 and 27.6). One involved a shift from the notochord to reliance on a column of separate, hardened segments: **vertebrae** (singular, vertebra). This was the start of an endoskeleton—an internal skeleton —that muscles could work against and initiate motion.

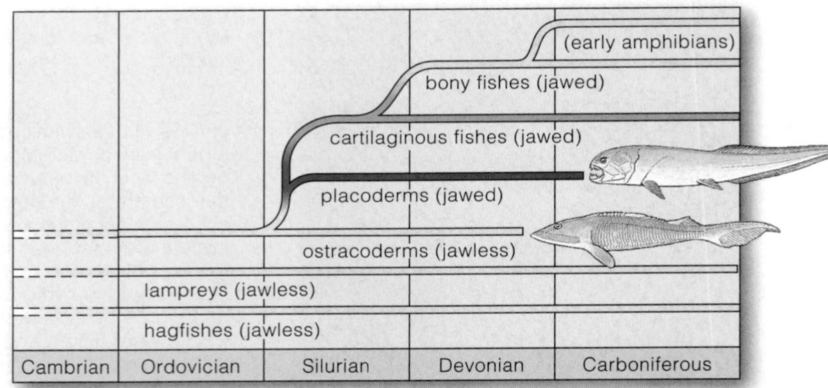

(early amphibians)
bony fishes (jawed)
cartilaginous fishes (jawed)
placoderms (jawed)
ostracoderms (jawless)
lampreys (jawless)
hagfishes (jawless)
Cambrian
Ordovician
Silurian
Devonian
Carboniferous

Figure 27.5 Evolutionary history of the fishes.

A vertebral column evolved in fast-moving predators, some of which were ancestral to all other vertebrates.

In a related trend, part of the nerve cord expanded into a brain. It did so after **jaws** evolved from the first of a series of structures that lent support to the gill slits. Those structural elements were a key innovation; they meant new feeding possibilities and greater competition among predators. Fishes able to smell or see predators or food farther away now had the advantage—provided that the brain also had evolved enough to process and respond to the new sensory information. *A trend toward complex sensory organs and nervous systems did indeed begin among fishes, and it continued in land vertebrates.*

Another trend began when paired fins evolved. **Fins** are appendages that help propel, stabilize, and guide the body through water (compare Figure 27.9). In some lineages, ventral fins became fleshy and equipped with skeletal supports that turned out to be forerunners of limbs. *Paired, fleshy fins were the starting point for all of the legs, arms, and wings that evolved among amphibians, reptiles, birds, and mammals.*

A major change in respiration was another trend. Think of the lancelet. Except when it is buried in sediments, oxygen and carbon dioxide merely diffuse across its body surface. But **gills** evolved in most vertebrate lineages. Gills are respiratory organs having a moist, thin, greatly folded surface. They are richly endowed with blood vessels, and they offer a large surface area for gas exchange. For example, five to seven pairs of a shark's gill slits are continuous with gills extending from the pharynx to the body's surface. When a shark opens its mouth and closes its external gill openings, the pharynx expands. Oxygen diffuses

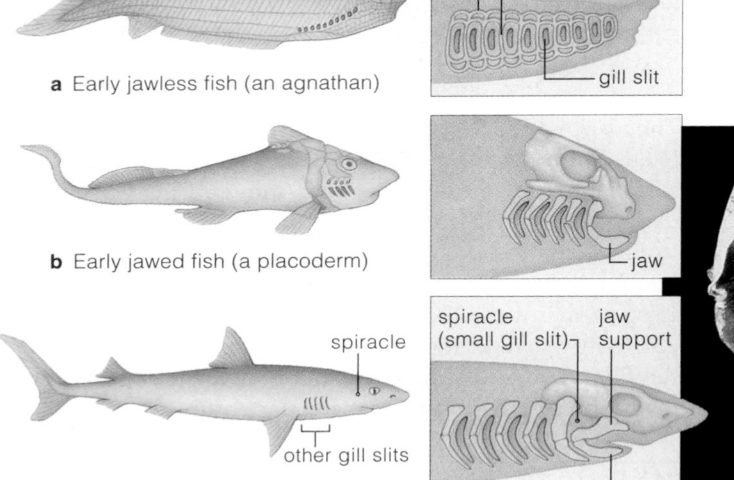

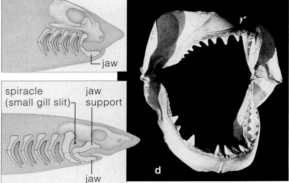

supporting structures
gill slit
a Early jawless fish (an agnathan)
jaw
b Early jawed fish (a placoderm)
spiracle
other gill slits
spiracle (small gill slit)
jaw support
jaw
c Modern jawed fish (a shark)
d

Figure 27.6 Comparison of gill-supporting structures in jawless fishes and jawed fishes. In the placoderms and other early jawed vertebrates, cartilage supported the rim of the mouth. In modern jawed fishes, a gill slit between the jaws and a nearby supporting element is a spiracle, an opening through which water is drawn. In (**a**), gill supports are just under the skin. In (**b**) and (**c**), they are internal to the gill surface. (**d**) Jaws of a modern shark.

Further reading: Student Guide to InfoTrac on web site

from water in the mouth into the gills, while carbon dioxide diffuses *into* that water. Muscles constrict the pharynx and force the oxygen-depleted, carbon dioxide-enriched water out through the gill slits.

As some fishes became larger and more active, gills became more efficient. But gills can't work out of water; they stick together unless water flows through them and keeps them moist. In fishes ancestral to the land vertebrates, pouches developed from the gut wall. The pouches evolved into **lungs**—internally moistened sacs for gas exchange. In a related trend, modifications to the heart enhanced the pumping of oxygen and carbon dioxide through the body. *Ancestors of land vertebrates relied less on gills and more on lungs. And more efficient circulatory systems accompanied the evolution of lungs.*

The First Vertebrates

Figure 27.5 has a time frame for vertebrate evolution. Free-swimming species originated in Cambrian times, and they gave rise to two categories of fishes—those without and those with jaws. Among them were the ancestors of all vertebrate lineages that followed. The earliest jawless fishes (Agnatha) included **ostracoderms**. Figure 27.6*a* shows one of those bottom-dwelling filter feeders. Probably the skeleton was a notochord plus a protective case for the enlarged brain. Armorlike plates on the body consisted of bony tissue and dentin, a hardened tissue still present in vertebrate teeth. The armor might have been useful against the pincers of giant sea scorpions but apparently wasn't much good against jaws. Ostracoderms disappeared when jawed fishes began their first adaptive radiation.

Placoderms were among the earliest fishes with jaws and paired fins (Figure 27.6*b*). Those bottom-dwelling scavengers or predators had a notochord reinforced with bony elements. The first gill-supporting structures were enlarged and had bony projections something like teeth, and they functioned as jaws. Before placoderms, feeding strategies had been limited to filtering, sucking, or rasping away at food material. When placoderms started biting and tearing off large chunks of prey, they set off an evolutionary race of offensive and defensive adaptations that has continued to the present.

Placoderms diversified, but they all became extinct in the Carboniferous period. New kinds of predators, the cartilaginous and bony fishes, replaced them in the seas. We will consider the living descendants of those new predators after a look at existing jawless fishes.

The vertebral column, jaws, paired fins, and lungs were pivotal developments in the evolution of lineages that gave rise to fishes, amphibians, reptiles, birds, and mammals.

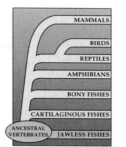

27.4 EXISTING JAWLESS FISHES

The **hagfishes** and **lampreys** are living descendants of early jawless fishes. All seventy-five species have a cylindrical body and a skeleton of cartilage (Figure 27.7). None has the paired fins seen in jawed fishes (Figure 27.6). Most species are less than 1 meter (3.3 feet) long.

Hagfishes live together in groups on sediments of continental shelves. They prey on polychaete worms or scavenge for weakened or dead organisms. Even though they lack jaws, hagfishes eat quite well. They have sensory tentacles around the mouth and a tongue that rasps soft tissues from their prey. They defend themselves by secreting copious amounts of sticky, smelly, slimy mucus from a series of glands along their body.

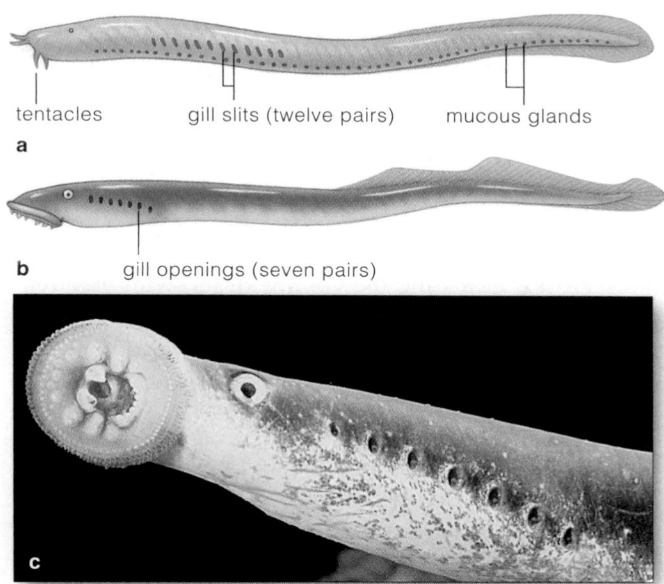

tentacles
gill slits (twelve pairs)
mucous glands
a
b
gill openings (seven pairs)
c

Figure 27.7 Body plan of (**a**) a hagfish and (**b**) a lamprey. (**c**) A lamprey's oral disk, pressed against aquarium glass.

Lampreys are specialized predators that are almost parasitic. Their suckerlike oral disk has horny, toothlike parts that rasp flesh from prey. Some types latch onto salmon, trout, and other commercially valuable fishes, then suck out juices and tissues. Just before the turn of the century, lampreys began invading the Great Lakes of North America. Their introduction resulted in the collapse of populations of lake trout and other large, valued fishes, as it has done in other aquatic habitats.

Hagfishes and lampreys get along without jaws; they latch onto prey with their efficient, specialized mouthparts.

27.5 EXISTING JAWED FISHES

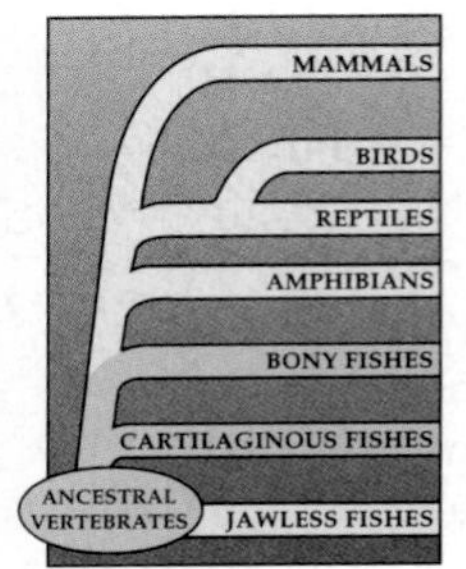
MAMMALS
BIRDS
REPTILES
AMPHIBIANS
BONY FISHES
CARTILAGINOUS FISHES
ANCESTRAL VERTEBRATES
JAWLESS FISHES

Few of us make a career of studying life beneath the surface of the seas and other bodies of water, so fishes are not widely recognized as being the world's dominant vertebrates. Their numbers exceed those of all other vertebrate groups combined. Fishes also show far more diversity than other vertebrates; we know of more than 21,000 species of bony fishes alone.

The form and behavior of a fish tell us something about the challenges it faces in water. For example, being about 800 times more dense than air, water resists fast motion through it. As an adaptation to this constraint, many marine fishes are streamlined for pursuit or for escape (Figure 27.8*a*). Sharks are an example. Their long, trim body reduces friction, and their tail muscles are organized for great propulsive force and forward motion. By contrast, some bottom-dwelling fishes, such as the rays, have a flattened body (Figure 27.8*b*). Theirs is not a high-speed body plan; it is good for hiding out from predators or prey.

A motionless trout, suspended in shallow water, is another example of adaptation to water's density. Like many fishes, it maintains neutral buoyancy with a **swim bladder**, an adjustable flotation device that exchanges gases with blood. When a fish gulps air at the water's surface, it is adjusting the volume of its swim bladder.

Cartilaginous Fishes

Cartilaginous fishes (Chondrichthyes) include about 850 species of skates, sharks, and chimaeras, as shown in Figure 27.8. These marine predators are equipped with prominent fins, a skeleton of cartilage, and five to seven gill slits on both sides of their pharynx. Most species have a few scales or many rows of them. **Scales** are small, bony plates at the body surface. Fish scales typically protect the body without weighing it down.

Skates and rays are mainly bottom dwellers with flattened teeth suitable for crushing hard-shelled prey. Both have enlarged fins that extend onto the side of the head. The largest species, the manta ray, measures up to six meters from fin tip to fin tip. A venom gland in the tail of stingrays probably helps to deter predators. Other rays have electric organs in the tail or fins that can stun prey with as much as 200 volts of electricity.

At fifteen meters from head to tail, some sharks are among the largest living vertebrates. As Figure 27.6*d* shows, sharks have formidable jaws. They continually shed and replace sharp-edged teeth (modified scales), which they use to grab prey and rip off chunks of flesh. The relatively few shark attacks on humans have given the group a bad reputation. Surfboards with human legs dangling over the sides are a new temptation for some sharks, which have been hunting invertebrates, fishes, and marine mammals for many millions of years.

Figure 27.8 Cartilaginous fishes: (**a**) shark, (**b**) blue-spotted reef ray, and (**c**) chimaera, sometimes called a ratfish.

The thirty or so species of chimaeras feed mostly on mollusks. With their bulky body and slender tail, they look a bit like a rat; hence the common name, ratfishes. They have a venom gland in front of the dorsal fin.

Bony Fishes

Bony fishes (Osteichthyes) are the most numerous and diverse vertebrates. They make up all but 4 percent of modern fish species. Their ancestors arose in Silurian times and gave rise to three lineages: ray-finned fishes, lobe-finned fishes, and lungfishes. Descendants of the early forms radiated into nearly all aquatic habitats.

Body plans vary greatly. Marine predators typically have a torpedo shape, a flexible body, and strong tail fins used in swift pursuit. Many reef dwellers are small finned and box shaped; they move easily in narrow spaces. Those with an elongated, flexible body, such as the moray eel, wriggle through mud and into concealing crevices. The cryptic body shape of sea horses and many bottom-dwelling species helps conceal them from prey, predators, or both. Figure 27.9 shows examples.

In ray-finned fishes, outwardly projecting rays that are derived from dermis (skin's inner layer) support paired fins. Most species have highly maneuverable fins

Further reading: Student Guide to InfoTrac on web site —

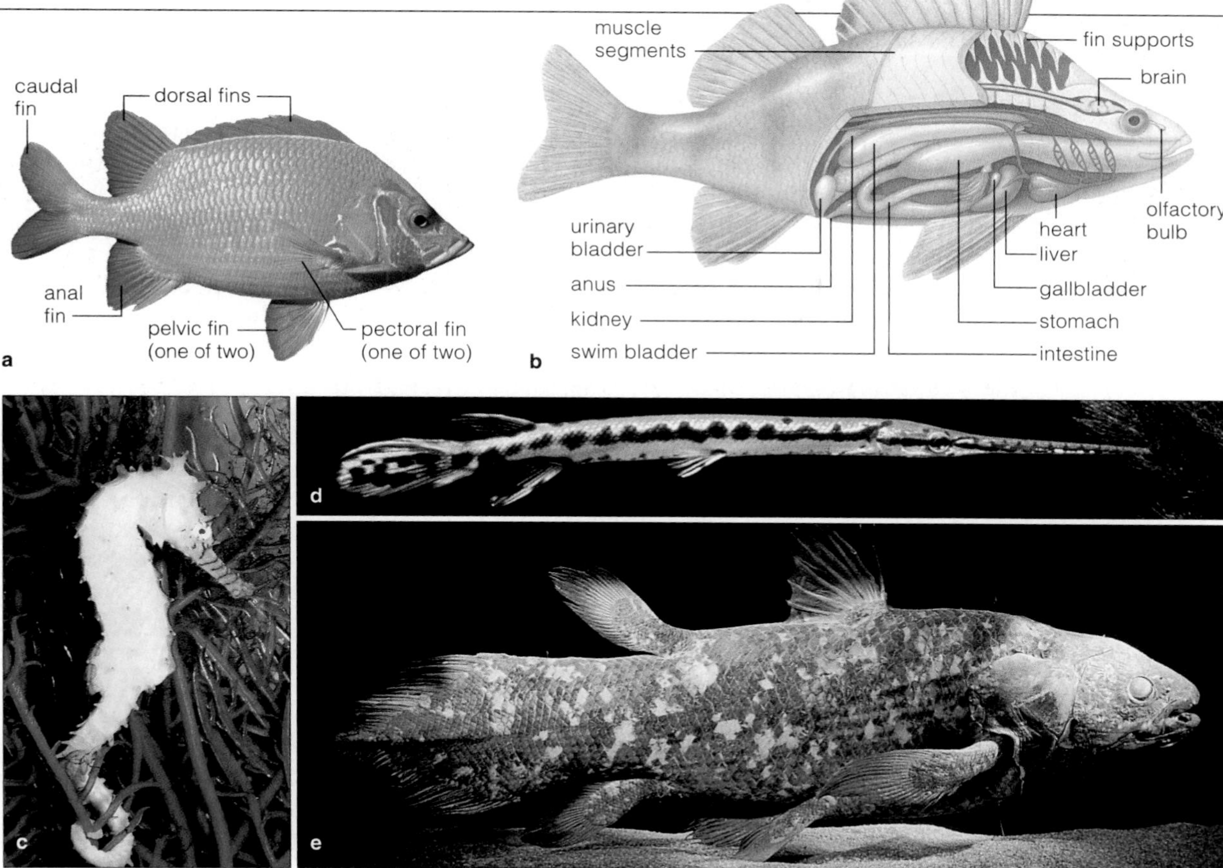
caudal fin
dorsal fins
anal fin
pelvic fin (one of two)
pectoral fin (one of two)
a
muscle segments
fin supports
brain
olfactory bulb
urinary bladder
heart
liver
anus
gallbladder
kidney
stomach
swim bladder
intestine
b
c
d
e

Figure 27.9 Some common features of the teleosts, the most diverse bony fishes: (**a**) Fins of a soldierfish. (**b**) Internal organs of a perch.

Just a few of the variations on the basic body plan: (**c**) Sea horse, which uses its tail to attach to substrates. (**d**) Long-nose gar. (**e**) Coelacanth (*Latimeria*), a "living fossil." It shares traits with early lobe-finned fishes. (**f**) Sketch of a moray eel. See Section 28.3.

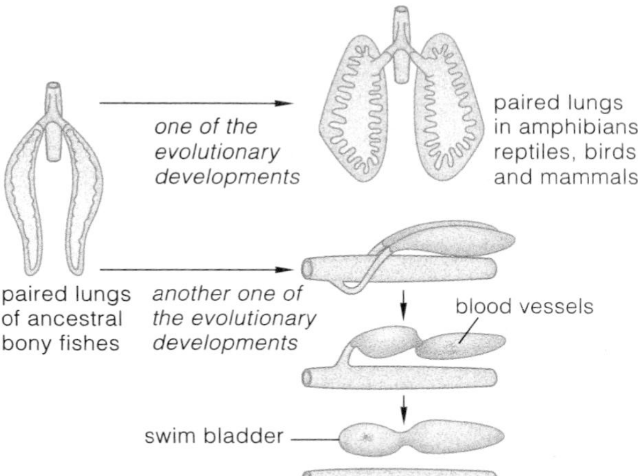
one of the evolutionary developments
paired lungs in amphibians, reptiles, birds, and mammals
paired lungs of ancestral bony fishes
another one of the evolutionary developments
blood vessels
swim bladder

Figure 27.10 Evolution of swim bladders and lungs. Respiratory surfaces are coded *pink* and the esophagus (a tube leading to the stomach), *gold*. Lungs evolved as outpouchings of the esophagus. Being in close contact with blood vessels, they increased the surface area for gas exchange in oxygen-poor habitats. The lung sacs became swim bladders (buoyancy devices) in some species.

Trout and other less specialized fishes surface and gulp air, which enters the swim bladder through a duct from the esophagus. Most bony fishes have no such duct. Gases dissolved in blood diffuse into a swim bladder that has a dense mesh of arteries and veins. Blood flow through the vessels increases the concentrations of gases in the swim bladder. Another area of the swim bladder enhances reabsorption of gases by cells in body tissues.

and light, flexible scales that don't hamper complex motions. In the ancestors of these fishes, lungs evolved as sac-shaped outpouchings of the wall of the esophagus, a tube to the stomach. The sacs supplemented gills for gas exchange (Figure 27.10). In certain species, those sacs evolved into swim bladders, which help a fish hold its position at different depths.

One group of ray-finned fishes resembles their early ancestors. It includes the sturgeons and paddlefishes of the Mississippi River basin. Teleosts, the most abundant fishes, include salmon, tuna, rockfish, catfish, perch, minnows, moray eels, flying fish, sculpins, blennies, scorpionfish, and pikes. All are thin scaled or scaleless.

By contrast, only one species of lobe-finned fishes and three genera of lungfishes made it to the present. As their name suggests, a **lobe-finned fish** is unique in having paired fins that incorporate fleshy extensions from the body. As you will see next, neither it nor the lungfishes have changed much from their ancient stock.

Of all existing vertebrates, the bony fishes are the most spectacularly diverse and the most abundant.

27.6 AMPHIBIANS

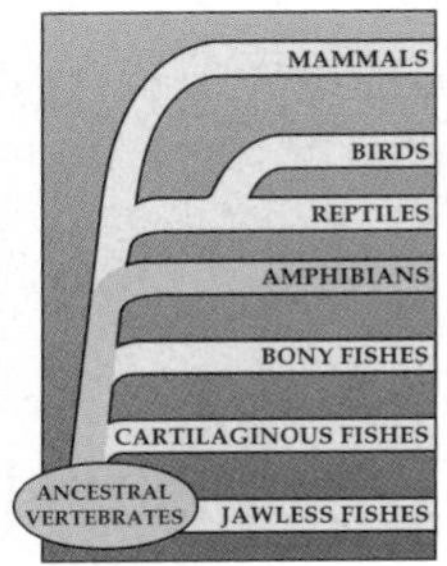
MAMMALS
BIRDS
REPTILES
AMPHIBIANS
BONY FISHES
CARTILAGINOUS FISHES
ANCESTRAL VERTEBRATES
JAWLESS FISHES

Origin of Amphibians

Take another look at the coelacanth, the lobe-finned fish shown in Figure 27.9*e*. It is a "living fossil," a relic of an early time when some vertebrates first moved onto land. Remember the earlier description of living conditions in the Devonian? Sea levels rose and fell repeatedly, and the swamps fringing the coasts flooded and drained many times. Ancestors of the lobe-finned fishes evolved during those trying times. And they probably used their lobed fins to pull themselves from dried-up ponds to ones that still had water and were habitable. What else made their pond-to-pond lurchings possible? They gulped air, and they had lungs.

Existing lungfishes provide more clues to how the ancestral forms might have made it through stressful times. They live in stagnant water but surface to gulp air. In dry seasons, when streams dwindle to mud, they encase themselves in a slathering of mud and slime that keeps them from drying out until the next rainy season.

Devonian lobe-finned fishes lurched over land only as a way of reaching more hospitable ponds. Yet the very act of traveling out of water favored the evolution of stronger fins and more efficient lungs. Among the evolving forms were the ancestors of amphibians. An **amphibian** (Amphibia) is a vertebrate with a body plan and reproductive mode somewhere between fishes and reptiles. Most kinds have a largely bony endoskeleton, and they have four legs (or four-legged ancestors).

Early amphibians were spending time on land by the close of the Devonian (Figure 27.11*a*). For them, life in those drier habitats was dangerous—and promising. Temperatures shifted more on land than in the water, air didn't support the body as well as water did, and water was not always plentiful. However, air is richer in oxygen. Amphibian lungs continued to be modified in ways that enhanced oxygen uptake. Also, circulatory systems became better at rapidly distributing oxygen to living cells throughout the body. Both modifications increased the energy base for more active life-styles.

New sensory information also challenged the early amphibians. Swamp forests supported vast numbers of edible insects and other invertebrate prey. Animals with good vision, hearing, and balance—the senses that are most advantageous on land—were favored. And brain regions concerned with those senses expanded.

All of the salamanders, frogs, toads, and caecilians alive today are descended from those first amphibians. None has escaped the water entirely. Even when gills or lungs are present, amphibians can use their thin skin as a respiratory surface (that is, for gas exchange). But respiratory surfaces must be kept moist, and skin dries easily in air. Some species still spend their entire lives in water; others lay their eggs in water or produce aquatic larvae. Even species adapted to land lay eggs in moist places or their embryos develop inside the moist body.

a

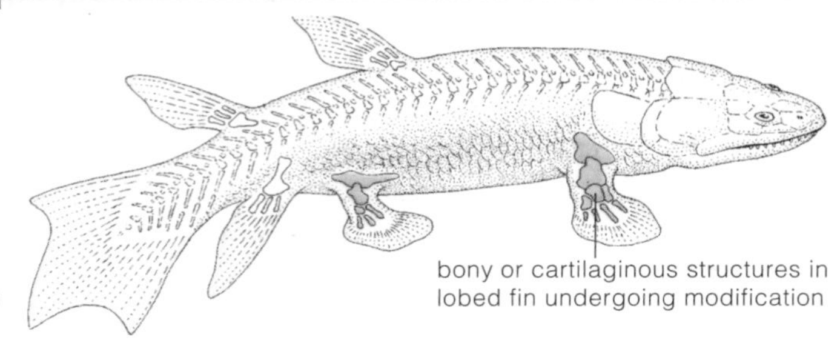
b
bony or cartilaginous structures in lobed fin undergoing modification

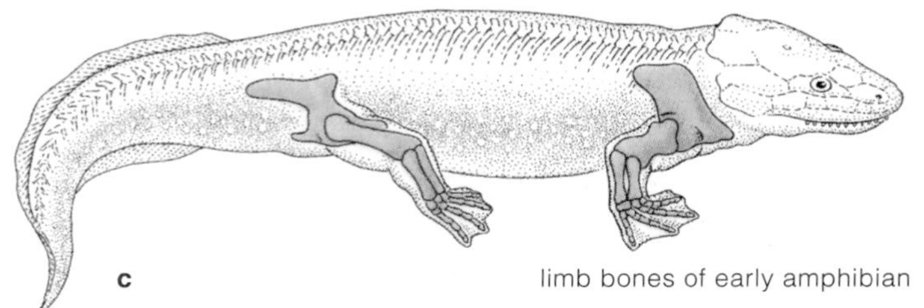
c
limb bones of early amphibian

Figure 27.11 (**a**) *Ichthyostega*, one of the early Devonian amphibians. Fossils of this species have been recovered in Greenland. The skull, deep tail, and fins were decidedly fishlike. Unlike fishes, this species had four limbs adapted for moving on land, and a short neck intervened between its head and the rest of the body. Its vertebral column and rib cage were adapted to support the body's weight out of water.

(**b**,**c**) Proposed evolution of skeletal elements inside the lobed fins of certain fishes into the limb bones of early amphibians.

Further reading: Student Guide to InfoTrac on web site

Figure 27.12 Amphibians. (**a**) Terrestrial stage in the life cycle of a red-spotted salamander. (**b**) A frog, splendidly jumping. (**c**) An American toad. (**d**) A caecilian.

Salamanders

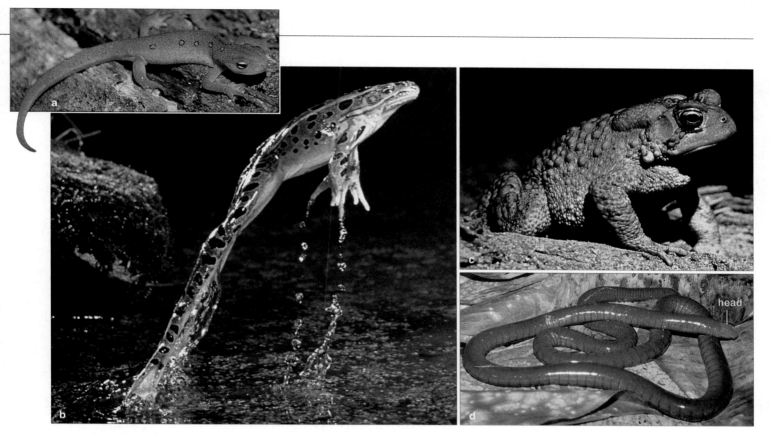

About 380 species of salamanders and their kin, the newts, live in north temperate zones and tropical areas of Central and South America. Most are not even fifteen centimeters long. Most have legs at right angles to the body, with forelimbs and hindlimbs of about the same size (Figure 27.12*a*). Like fishes and early amphibians, salamanders bend from side to side when they walk:

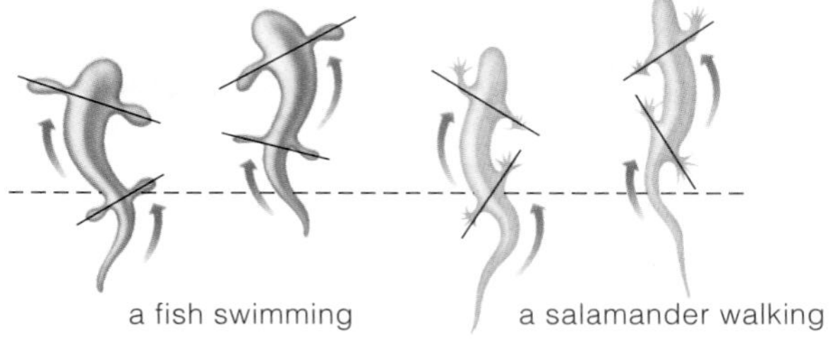

a fish swimming
a salamander walking

Probably the first four-legged vertebrates also walked this way. Larval and adult salamanders are carnivores. Adults of some species retain several larval features. For example, the Mexican axolotl retains the external gills of the larval form, and the development of its teeth and bones is arrested at an early stage. As in some other species, its larvae are sexually precocious; they breed.

Frogs and Toads

With close to 4,000 species, the frogs and toads are the most successful amphibians (Figure 27.12*b,c*). Their long hindlimbs and powerful muscles allow them to catapult into the air or barrel through the water. Most frogs flip a sticky-tipped, prey-capturing tongue from their mouth. An adult eats just about any animal it can catch; only its head size limits prey sizes. Like all amphibians, frogs have mucous and poison glands in their skin. Notably poisonous types often have bright warning coloration. Frog skin contains antibiotics that provides protection against many pathogens in aquatic habitats. In Unit VI, you will have opportunities to take a look at how frogs are put together and how they function.

Caecilians

The ancestors of caecilians lost their limbs and most of their scales. They gave rise to decidedly worm-shaped amphibians (Figure 27.12*d*). Nearly all of the 160 or so existing species burrow through soft, moist soil as they pursue insects and earthworms. A few live in shallow freshwater habitats. Adults of most species are blind.

In body form and behavior, amphibians show resemblances to fishes and certain reptiles. Regardless of how far they venture onto land, most have not fully escaped dependency on aquatic or moist habitats to complete the life cycle.

27.7 THE RISE OF REPTILES

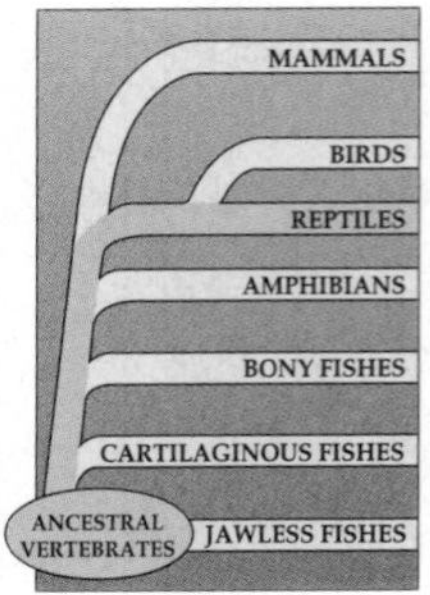
MAMMALS
BIRDS
REPTILES
AMPHIBIANS
BONY FISHES
CARTILAGINOUS FISHES
ANCESTRAL VERTEBRATES
JAWLESS FISHES

Late in the Carboniferous, a divergence from the amphibian lineage gave rise to the **Reptiles** (Reptilia). Of all vertebrates, reptiles were first to escape dependency on standing water. They did so mainly by four adaptations, as listed below.

First, reptiles have tough, dry, scaly skin that restricts loss of water from the body (Figure 27.13*a*). Second, fertilization is internal (as it is in some amphibians). A copulatory organ deposits sperm inside a female's body; sperm do not require free water to reach eggs. Third, reptilian kidneys are good at conserving water. Fourth, **amniote eggs** form during the life cycle of most species. Embryos in such eggs develop to an advanced stage before being hatched or born into dry habitats. Amniote eggs have membranes that conserve water and protect or provide metabolic support to the embryo. Most have a leathery or calcified shell (Figure 27.13*b*,*c* and Section 27.9). We will consider the structure and development of amniote eggs in Chapter 45.

Compared to most amphibians, early reptiles chased prey with far greater cunning and speed. With their well-muscled jawbones and formidable teeth, reptiles could seize and apply sustained, crushing force on prey. Their prey included other vertebrates as well as insects. Also, reptilian limbs generally were more efficient at supporting the trunk of the body on land. For many species, the nervous system increased in complexity. A reptile's brain is small compared to the rest of the body mass, but it governs complex forms of behavior that are unknown among amphibians. For example, the cerebral cortex is the most complex part of the forebrain. Here, information from diverse sensory organs is integrated and stored, and commands for responses are issued. The cerebral cortex evolved first among the reptiles.

b
c

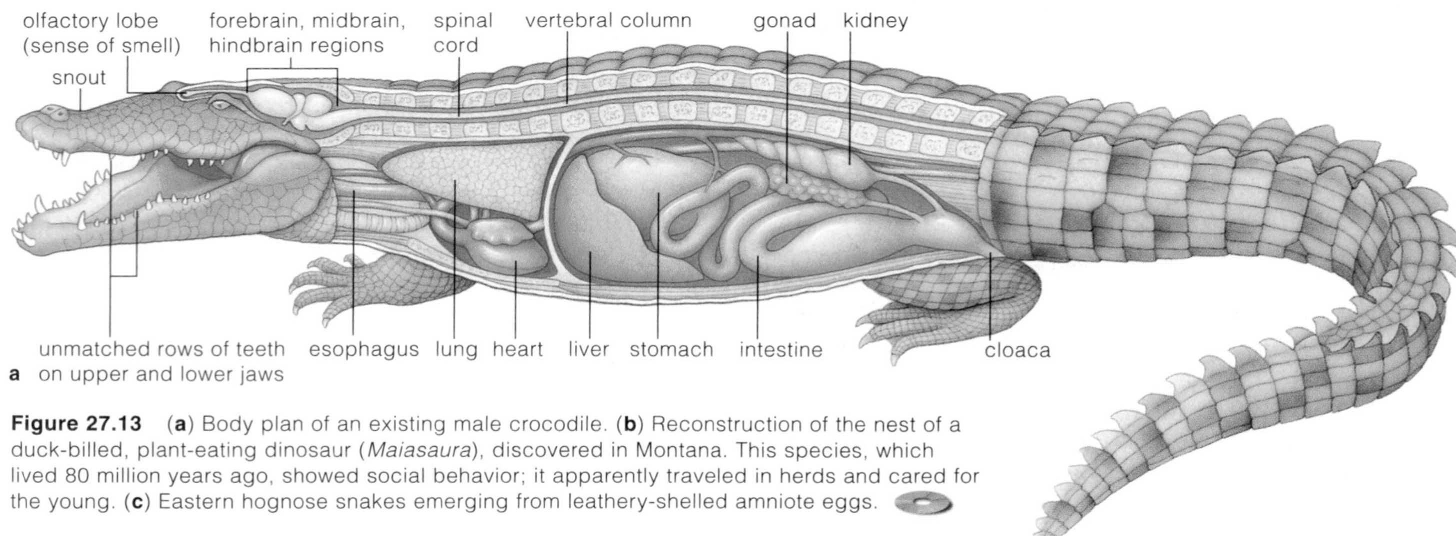
olfactory lobe (sense of smell)
forebrain, midbrain, hindbrain regions
spinal cord
vertebral column
gonad
kidney
snout
unmatched rows of teeth on upper and lower jaws
esophagus
lung
heart
liver
stomach
intestine
cloaca
a

Figure 27.13 (**a**) Body plan of an existing male crocodile. (**b**) Reconstruction of the nest of a duck-billed, plant-eating dinosaur (*Maiasaura*), discovered in Montana. This species, which lived 80 million years ago, showed social behavior; it apparently traveled in herds and cared for the young. (**c**) Eastern hognose snakes emerging from leathery-shelled amniote eggs.

Further reading: Student Guide to InfoTrac on web site

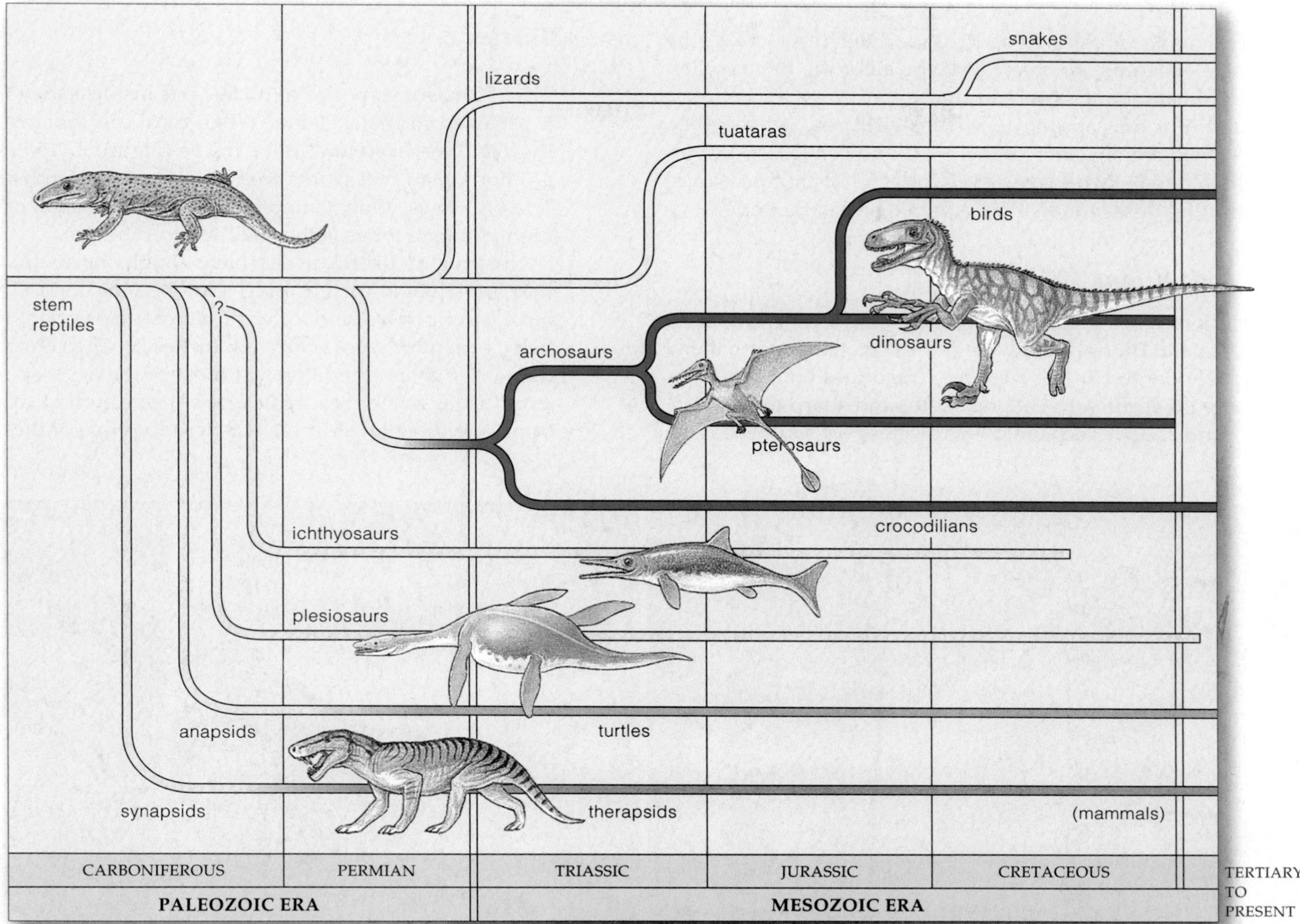
snakes
lizards
tuataras
birds
stem
reptiles
?
archosaurs
dinosaurs
pterosaurs
crocodilians
ichthyosaurs
plesiosaurs
anapsids
turtles
synapsids
therapsids
(mammals)
CARBONIFEROUS
PERMIAN
TRIASSIC
JURASSIC
CRETACEOUS
TERTIARY TO PRESENT
PALEOZOIC ERA
MESOZOIC ERA

Figure 27.14 Evolutionary history of the reptiles.

Crocodilians, which diverged from ancestral reptiles, were the first animals with a muscular, four-chambered heart fully separated into two halves. (Blood enters the first chamber of each half; the second chamber pumps it out.) As Chapter 39 describes, such separation permits oxygen-rich blood to travel from the lungs to the rest of the body, and oxygen-poor blood from the body to the lungs, in two independent circuits. Also, gas exchange across the skin, so vital for amphibians, was abandoned by reptiles. Nearly all reptiles depend on lungs.

Early reptiles underwent adaptive radiations that led to fabulously diverse forms. In the seas, some of the marine reptiles called masosaurs evolved into the largest lizards ever known; some were ten meters long. Monitor lizards, such as the Komodo dragon, are their living relatives (compare Figure 28.3). The group called dinosaurs evolved during the Triassic. For the next 125 million years they were the dominant land vertebrates. The fossilized nests of one kind (*Maiasaura*) contain not only eggs but also juveniles a few months old, at most (Figure 27.13*b*). Such evidence indicates that at least some dinosaurs showed parental behavior; that is, they took care of their young during a period of dependency. (*Maiasaura* means "good mother lizard.")

When the Cretaceous ended abruptly, so did the last of the dinosaurs (Sections 21.6 and 21.7). As you can see from Figure 27.14 and the next section, the crocodilians, turtles, tuataras, snakes, and lizards are reptilian groups that made it to the present.

Reptiles, with their tough, scaly skin, reliance on internal fertilization, water-conserving kidneys, and amniote eggs, were the first vertebrates to escape dependency on free-standing water in their habitats.

The move onto land also required major modifications in the nervous, circulatory, and respiratory systems.

27.8 MAJOR GROUPS OF EXISTING REPTILES

The name of class Reptilia is derived from the Latin *repto*, meaning "to creep." Maybe all of the first reptiles did creep about slowly in muddy swamps and on dry land. But the capacity to move swiftly and with agility evolved among some of their early descendants, such as the bipedal (two-legged) *Velociraptors* of the Mesozoic. Existing descendants race, lumber, and slither about.

Crocodilians

Modern crocodiles and alligators, the closest relatives of birds and dinosaurs, live in or near water. Among them are the largest living reptiles. Crocodiles and alligators have powerful jaws, a long snout, and sharp teeth, as in Figure 27.15*a*. The feared "man-eaters" of southern Asia and Nile crocodiles weigh as much as 1,000 kilograms. They drag a mammal or bird into water, tear it apart by violently spinning about, then gulp down the chunks.

Like the other reptiles and birds, crocodilians adjust body temperature with behavioral and physiological mechanisms. They are like birds in displaying complex social behaviors, as when parents guard nests and assist hatchlings into water. This trait and others suggest that crocodilians and birds share a common ancestor.

Turtles

The 250 existing species of turtles live inside a shell that is attached to the skeleton. When most are threatened, they pull the head and limbs inside (Figure 27.15*b*,*c*). It is a body plan that works well; it has been around since Triassic times. Only among the sea turtles and other highly mobile types is the shell reduced in size.

Instead of teeth, turtles have tough, horny plates that are suitable for gripping and chewing food. They have strong jaws and often a fierce disposition that helps keep predators at bay. All turtles lay eggs on land, then leave them. Predators eat most of the eggs, so few new turtles hatch. Sea turtles have been hunted to the brink of extinction. Section 41.8 describes one of these.

a

b

Figure 27.15 Representative reptiles. (**a**) Spectacled caiman. As is true of other crocodilians, its peglike upper teeth do not match up with peglike lower ones. The body plan and life-styles of crocodilians have not changed much for nearly 200 million years. Now their future is not rosy. Housing developments are encroaching on many of their habitats, and their belly skin is in demand for wallets, shoes, and handbags. (**b**) A heavy-shelled Galápagos tortoise. (**c**) Section through the skeleton and shell of a turtle, with head withdrawn and extended. *Facing page:* (**d**) Frilled lizard, flaring a ruff of neck skin as defensive behavior. (**e**) Rattlesnake in mid-strike. (**f**) Tuatara (*Sphenodon*).

ribs of endoskeleton
hard shell
c

Lizards and Snakes

Lizards and snakes, which make up 95 percent or so of living reptiles, are distant relatives of dinosaurs. Most are small, but the Komodo monitor lizard is big enough to capture a young water buffalo. The longest modern snake would stretch across ten yards of a football field.

Most of the 3,750 kinds of lizards are insect eaters of deserts and tropical forests. They include aggressive,

Further reading: Student Guide to InfoTrac on web site →

adhesive-toed geckos and iguanas; the photograph of the Unit VI introduction shows one of the more colorful types. Most lizards grab prey with small, peglike teeth. Chameleons capture prey with accurate flicks of their tongue, which is longer than their body (Section 33.4).

Being small themselves, most lizards are prey for many other animals. Some attempt to startle predators and intimidate rivals by flaring their throat fan (Figure 27.15*d*). Many give up their tail when a predator grabs them. The detached tail wriggles for a bit and might be distracting enough to permit a getaway.

During the early Cretaceous, certain short-legged, long-bodied lizards gave rise to the elongated, limbless snakes. Most of the 2,300 existing species slither in S-shaped waves, much like salamanders. "Sidewinders" make J-shaped movements and leave distinctive trails across loose sand and sediments. Some species have bony remnants of ancestral hindlimbs (Section 17.1).

f

d

venom gland
hollow fang
e

All snakes are carnivores. They have flexible skull bones and highly movable jaws; some swallow animals wider than they are. Pythons and boas coil around prey and suffocate it into submission before gulping it down. Fanged types, including coral snakes and rattlesnakes (a pit viper) bite and subdue prey with venom (Figure 27.15*e*). Snakes usually do not act aggressively toward people. Even so, during an average year in the United States, the rattlesnakes and other poisonous types bite 8,000 or so people and kill about 12 of them. In India, king cobras and other snakes bite 200,000 or so people and kill about 9,000 of them.

Even the most feared snakes are vulnerable during their life cycle; birds and other predators relish snake eggs. The female snakes store sperm and lay several clutches of fertilized eggs at intervals after they mate, which improves the odds that at least some will hatch.

Tuataras

Besides having reptilian traits, tuataras are like modern amphibians in some aspects of their brain and in their way of walking. The two existing species live on small, windswept islands near New Zealand (Figure 27.15*f*). Their body plan has not changed much for the past 140 million years. Like lizards, tuataras have a third "eye," with retina, lens, and nerves to the brain. Being covered with skin, it can only register changes in daylength and light intensity. Does it have roles in hormonal controls over reproduction? Maybe (compare Section 37.8). The tuataras, like turtles, may live longer than sixty years. They engage in sex only after they are twenty years old.

Existing crocodilians are the closest relatives of dinosaurs and birds. Turtles, like tuataras, have changed little in body plan for millions of years. Lizards and snakes represent about 95 percent of all living reptiles.

27.9 BIRDS

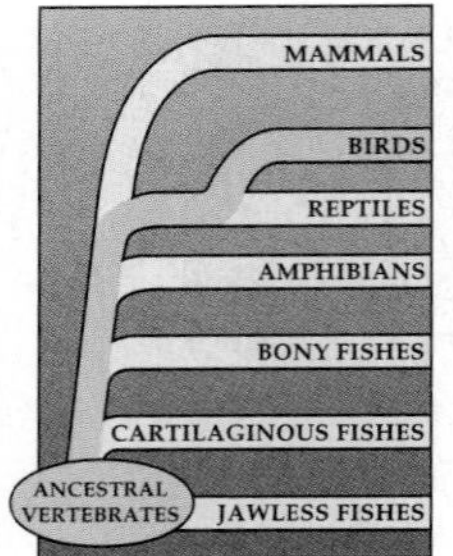
MAMMALS
BIRDS
REPTILES
AMPHIBIANS
BONY FISHES
CARTILAGINOUS FISHES
ANCESTRAL VERTEBRATES
JAWLESS FISHES

Of all organisms, only **birds** (Aves) grow feathers. **Feathers**, lightweight structures derived from skin, are used for flight, body insulation, or both. Judging from the fossil record, birds descended from tiny reptiles that ran around on two legs 160 million years ago. *Archaeopteryx* was in or near the lineage leading to modern birds. It had reptilian *and* avian traits—including feathers (Section 17.4). In fact, feathers evolved as highly modified reptilian scales.

In many respects, birds still resemble reptiles. For example, they have scales on their legs and a number of the same internal structures. They, too, lay eggs and commonly engage in parenting behavior, as when they guard the nest (Figure 27.16). Thomas Huxley, one of Darwin's supporters, was the first to argue that birds are glorified reptiles. Today, most biologists do indeed classify birds as a branching of the reptilian lineage.

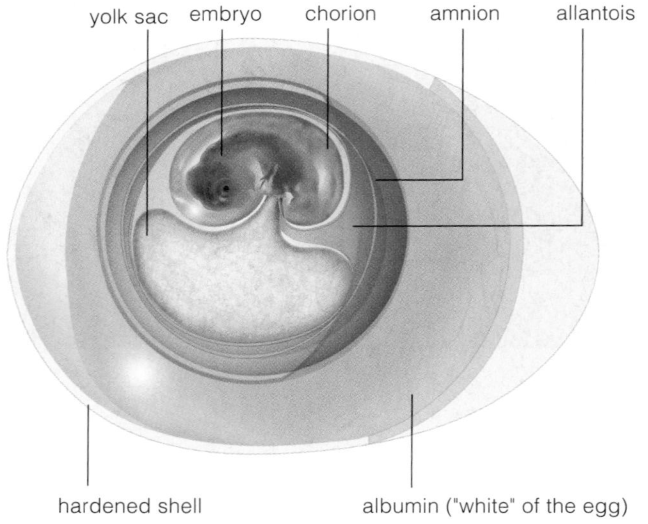
yolk sac
embryo
chorion
amnion
allantois
hardened shell
albumin ("white" of the egg)
a

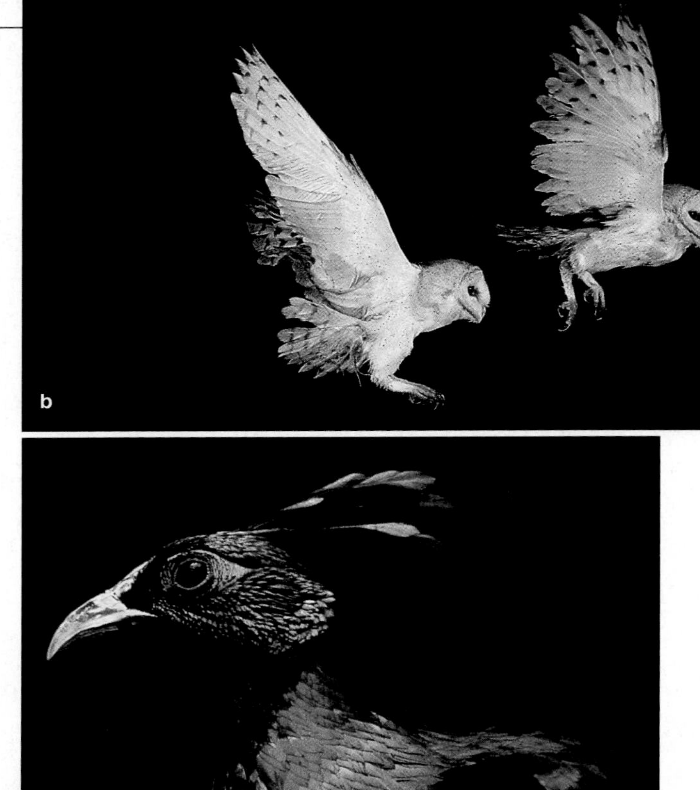
b

c

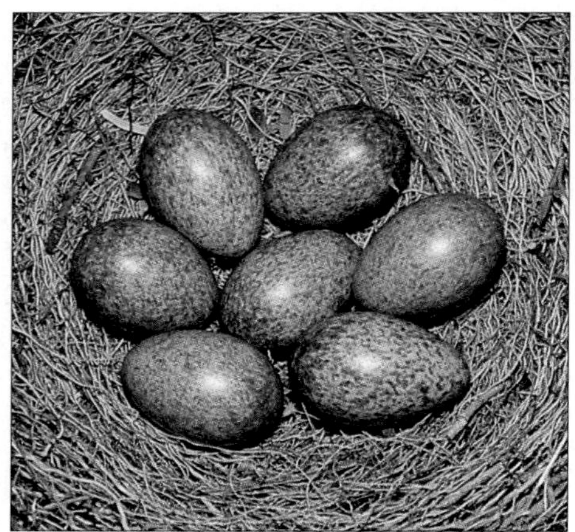
e

Figure 27.16 Characteristics of birds. (**a**) Generalized sketch of an amniote egg. All birds lay these eggs. (**b**) Flight. Of all living vertebrates, only birds and bats fly by flapping their wings. (**c**) Feathers, the key defining characteristic of birds. The flamboyant plumage of this male pheasant is an outcome of sexual selection. This bird is a native of the Himalaya Mountains of India, and it is an endangered species. As is the case for many other kinds of birds, its bright, jewel-colored feathers end up adorning humans—in this case, on the caps of native tribespeople.

(**d**) Many birds, including these Canada geese, display migratory behavior. *Animal migration* is a recurring pattern of movement between two or more places in response to environmental rhythms. For instance, seasonal change in daylength is a cue that acts on internal timing mechanisms (biological clocks), which trigger physiological and behavioral changes in individuals. Such changes induce migratory birds to make round trips between distant regions that differ in climate. Canada geese spend the summer nesting in marshes and lakes in the northern United States and Canada. Their wintering grounds are in New Mexico and other parts of the southern United States.

(**e**) Speckled, hard-shelled eggs of a magpie, unhatched in the nest. 

Further reading: Student Guide to InfoTrac on web site

There are almost 9,000 named species of birds. They show stunning variation in their body size, proportions, coloration, and capacity for flight. One of the smallest hummingbirds barely tips the scales at 2.25 grams (0.08 ounce). The largest existing bird, the ostrich, weighs about 150 kilograms (330 pounds). Ostriches cannot fly, but they are impressively long-legged sprinters (Figure 17.2). Many birds, such as warblers and other perching types, differ markedly in feather coloration and in their territorial songs. Bird songs and other complex social behaviors are topics of later chapters.

Flight demands high metabolic rates, which require a good flow of oxygen through the body. Like mammals, birds have a large, durable, four-chambered heart. The heart pumps oxygen-enriched blood to the lungs and to the rest of the body along separate routes, as it probably did in reptilian ancestors of birds. Also, a bird's respiratory system incorporates a unique ventilating apparatus that enormously enhances oxygen uptake. Bird respiration is described in Section 41.3.

Flight also demands an airstream, low weight, and a powerful downstroke that can provide lift (a force at right angles to the airstream). The bird wing, a modified forelimb, is composed of feathers and lightweight bones attached to powerful muscles. The bones do not weigh much, owing to profuse air cavities in the bone tissue. For example, the skeleton of a frigate bird, with its seven-foot wingspan, weighs only four ounces. That is less than the feathers weigh! A bird's flight muscles attach to an enlarged breastbone (sternum) and to the upper limb bones adjacent to it (Figure 27.17). Muscle contraction creates the powerful downstroke required for flight. Wings, with their long flight feathers, serve as airfoils. Usually, a bird spreads out long feathers on a downstroke and thus increases the size of the surface pushing against air (Figure 27.16*b*). On the upstroke, it folds the feathers somewhat, so each wing presents the least possible resistance to air.

Of all animals, birds alone have feathers, which they use in flight, in heat conservation, and in socially significant communication displays.

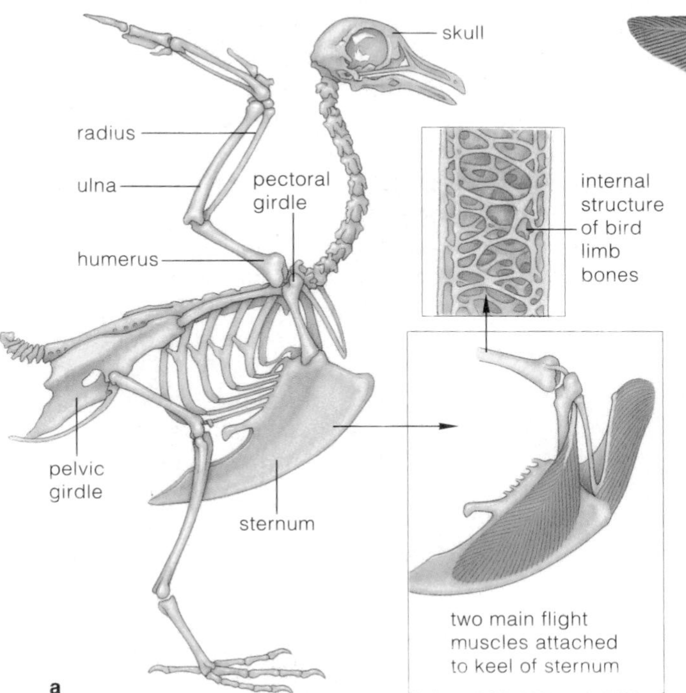
skull
radius
ulna
pectoral girdle
humerus
internal structure of bird limb bones
pelvic girdle
sternum
two main flight muscles attached to keel of sternum
a
barbule
barb
shaft
b

Figure 27.17 (**a**) Body plan of a typical bird. Flight muscles attach to the large, keeled breastbone (sternum). A bird wing is a complicated system of lightweight bones and feathers. (**b**) Feathers gain strength from a hollow central shaft and from tiny barbules that are interlocked in a latticelike array.

27.10 THE RISE OF MAMMALS

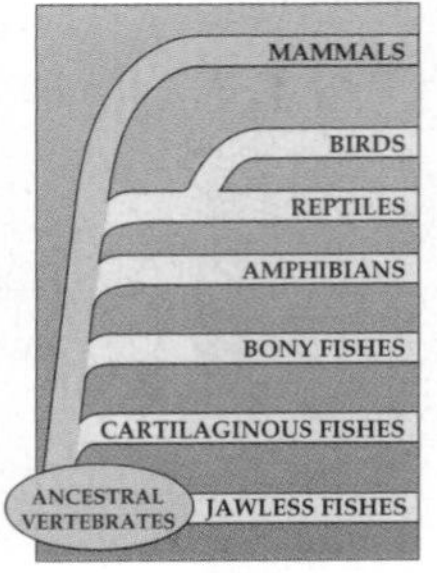

MAMMALS
BIRDS
REPTILES
AMPHIBIANS
BONY FISHES
CARTILAGINOUS FISHES
ANCESTRAL VERTEBRATES
JAWLESS FISHES

Distinctly Mammalian Traits

Mammals are vertebrates with hair and mammary glands (hence the name of the class Mammalia, from the Latin *mamma*, which means breast). Of all organisms, mammals are the only ones having these traits. Females feed the young with milk, a nutritious fluid that mammary glands secrete into ducts that lead to openings at the body's ventral or anterior surface (Figure 27.18*a*). Mammals also are highly diverse. For example, in size alone, existing species range from the 1.5-gram Kitti's hognosed bat to 100-ton great whales. They also differ greatly in the amount, distribution, and type of hair.

A few aquatic mammals, including the whales, lost most of their hair after their land-dwelling ancestors returned to the seas. But look closely and you see that even the whale snout has some "whiskers"—modified hairs that serve sensory functions, just as they do in dogs and cats. More typically, mammals have a furry coat of underhair (a dense, soft, insulative layer that traps heat) and coarser, longer guard hairs that protect the insulative layer from wear and tear (Figure 27.18b). When wet, the guard hairs of platypuses, otters, and other aquatic mammals become matted down like a protective blanket, and the underhair stays dry.

Mammals also are distinctive in that most care for the young for an extended period, and adults serve as models for their behavior. Young mammals are born with a capacity to learn and to repeat behaviors that have survival value. It is among mammals that we find the most stunning shows of **behavioral flexibility**—a capacity to expand on basic activities with novel forms of behavior—although the trait is far more pronounced in some species than in others. Behavioral flexibility coevolved with expansion of the brain, especially the cerebral cortex. Remember, this outermost layer of the forebrain receives, processes, and stores information from sensory structures, and it issues commands for complex responses. We find the most highly developed cerebral cortex among the mammals called primates. We will soon turn to their story.

Unlike reptiles, which generally swallow their prey whole, most mammals secure, cut, and sometimes chew food before swallowing. They also differ from reptiles in **dentition** (that is, in the type, number, and size of teeth). Mammals have four distinctive types of upper and lower teeth that match up and work together to crush, grind, or cut food (Figure 27.18*c*). Their incisors (flat chisels or cones) nip or cut food. Horses and other grazing mammals have large incisors. Canines, with piercing points, are longest in meat-eating mammals. Premolars and molars (cheek teeth) are a platform with surface bumps (cusps); they crush, grind, and shear food. If a mammal has large, flat-surfaced cheek teeth, you can safely bet its ancestors evolved in places where tough, fibrous plants were abundant foods. As you will see later on in the chapter, fossilized jaws and teeth from the early primates as well as from species that apparently were on the road to modern humans offer clues to their life-styles.

a

b

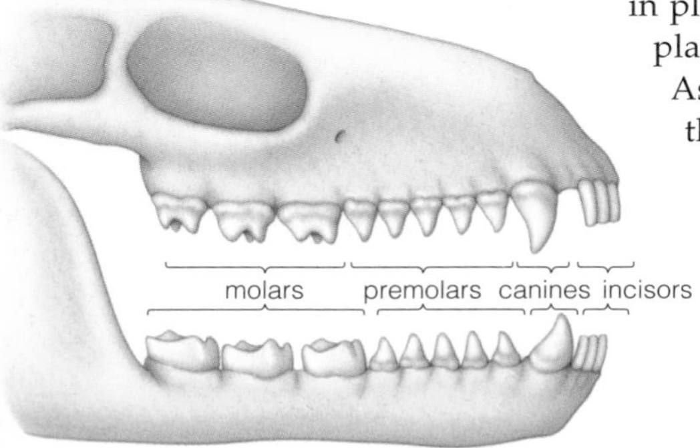

molars
premolars
canines
incisors
c

Figure 27.18 Three distinctly mammalian traits. (**a**) A human baby busily demonstrating a key defining feature: It derives nourishment from mammary glands. (**b**) A pair of juvenile raccoons displaying their fur coat. (**c**) Unlike the teeth of their reptilian ancestors, the upper and lower rows of mammalian teeth match up. (Compare Figure 27.15*a*.)

Further reading: Student Guide to InfoTrac on web site

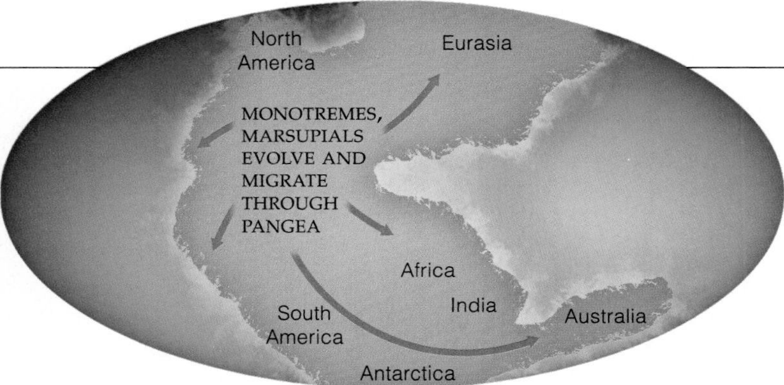
North America
Eurasia
MONOTREMES, MARSUPIALS EVOLVE AND MIGRATE THROUGH PANGEA
Africa
India
South America
Australia
Antarctica

a About 150 million years ago, during the Jurassic

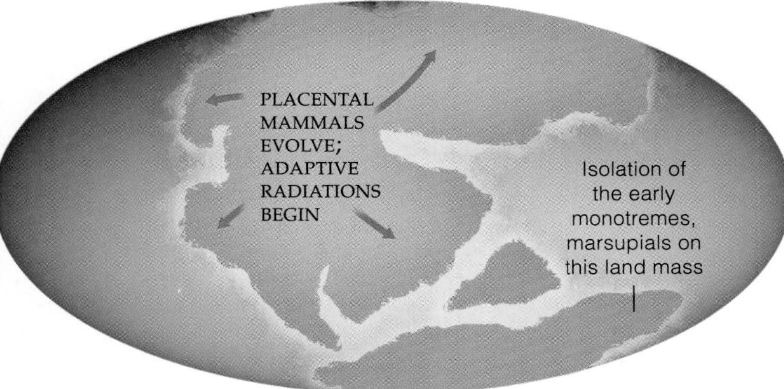
PLACENTAL MAMMALS EVOLVE; ADAPTIVE RADIATIONS BEGIN
Isolation of the early monotremes, marsupials on this land mass

b Between 100 and 85 million years ago, during the Cretaceous

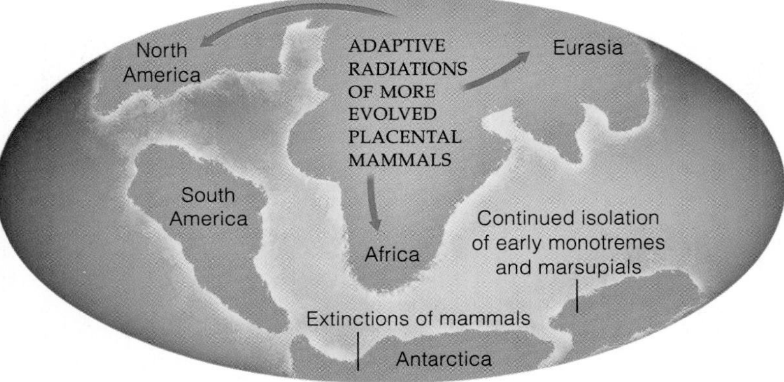
North America
ADAPTIVE RADIATIONS OF MORE EVOLVED PLACENTAL MAMMALS
Eurasia
South America
Africa
Continued isolation of early monotremes and marsupials
Extinctions of mammals
Antarctica

c About 20 million years ago, during the Miocene

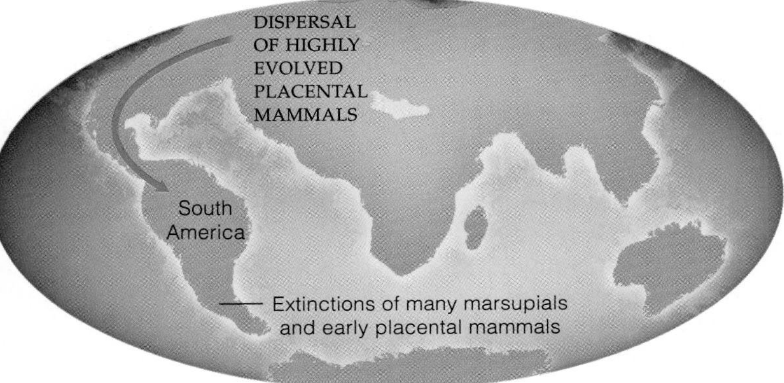
DISPERSAL OF HIGHLY EVOLVED PLACENTAL MAMMALS
South America
Extinctions of many marsupials and early placental mammals

d About 5 million years ago, during the Pliocene

Figure 27.19 Adaptive radiations of mammals. 

Mammalian Origins and Radiations

More than 200 million years ago, during the Triassic, a genetic divergence from small, hairless reptiles called synapsids gave rise to the **therapsids**, the ancestors of mammals (Figure 27.14). And by Jurassic times, diverse plant-eating and meat-eating mammals called **therians** had evolved. Most were the size of a mouse, and they were endowed with hair and major changes in the jaws, teeth, and body form. For instance, their four limbs were positioned upright under the body's trunk. This skeletal arrangement made it easier to walk erect, but a trunk higher from the ground was not as stable. At that time the cerebellum, a brain region dealing with the body's balance and spatial positioning, started expanding.

The therians coexisted with dinosaurs through the Cretaceous. When the last of the dinosaurs vanished, diverse adaptive zones opened up for three lineages of those previously inconspicuous mammals (Sections 19.5 and 21.8). New opportunities, differences in traits, and key geologic events put those lineages—the **monotremes** (egg-laying mammals), **marsupials** (pouched mammals), and **eutherians** (placental mammals)—on very different paths to the present. Compared with monotremes and marsupials, both of which retained many archaic traits, placental mammals had the competitive edge. They had higher metabolic rates, more precise ways of regulating body temperature, and a new way of nourishing their developing embryos. You will read about those traits in later chapters. For now, it is enough to know that the placental mammals radiated into many new adaptive zones at the expense of their less competitive relatives.

As Figure 27.19*a* shows, ancestors of monotremes and marsupials had entered the southern part of the supercontinent Pangea by the late Jurassic. Following the breakup of Pangea, those on the land mass that would become Australia were already isolated from the ancestors of placental mammals, which were evolving elsewhere (Figure 27.19*b*). In the future South America, monotremes were replaced by marsupials and early placental mammals that had evolved independently of their relatives on other continents. When a land bridge rejoined South and North America during the Pliocene, highly evolved placental mammals radiated southward and rapidly drove many of those previously isolated mammals to extinction. Only opossums and a few other species successfully invaded land in the other direction.

Mammals alone have hair and mammary glands. They have distinctive dentition, a highly developed nervous system, and a notable capacity for behavioral flexibility.

Much of mammalian history was a matter of luck, of species with particular traits being in the right or wrong places on a changing geologic stage at particular times.

27.11 PORTFOLIO OF EXISTING MAMMALS

Evolutionarily distant, geographically isolated lineages have sometimes evolved in similar ways in similar habitats, so that they come to resemble each other in form and function. We call this **convergent evolution**, as defined in Section 20.4. With their interesting history, the three lineages of mammals offer classic examples, as the lists in Table 27.1 suggest.

The only living monotremes are two species of spiny anteaters and the duck-billed platypus described at the start of the chapter. One spiny anteater (*Tachyglossus*, in Figure 27.20*a*) is common in Australia. The other lives in mountains of New Guinea. These small, burrowing mammals feed mostly on ants. Like porcupines, they bristle with protective spines that are modified hairs. As is the case for platypuses, females lay eggs. Unlike platypuses, they do not dig out nests. They incubate a single egg, and the hatchling suckles and completes its development in a skin pouch that forms *temporarily* (by muscle contractions) on the mother's ventral surface.

Of the 260 existing species of marsupials, most are native to Australia and nearby islands; only a few live in the Americas (Figure 27.20*b*,*c*). The tiny, blind, and hairless newborns suckle and finish their development in a *permanent* pouch on the mother's ventral surface.

Table 27.1 Convergences Among Mammalian Groups

Life-Style	Home	Mammalian Family
Aquatic invertebrate eater	North America Central America Australia	Water shrew (Soricidae) Water mouse (Cricedidae) Platypus (Ornithorhynchidae)
Land-dwelling carnivore	North America Australia	Wolf (Canidae) Tasmanian wolf (Thylacinedae)
Land-dwelling anteater	South America Africa Australia	Giant anteater (Myrmecophagidae) Aardvark (Orycteropodidae) Spiny anteater (Tachyglossidae)
Ground-dwelling leaf, tuber eater	North America South America Eurasia	Pocket gopher (Geomyidae) Tuco-tuco (Ctenomyidae) Mole rat (Spalacidae)
Tree-dwelling leaf eater	South America Africa Madagascar Australia	Howler monkey (Cebidae) Colobus monkey (Cercopithecidae) Woolly lemur (Indriidae) Koala (Phascolarctidae)
Tree-dwelling nut, seed eater	Southeast Asia Africa Australia	Flying squirrel (Sciuridae) Flying squirrel (Anomaluridae) Flying squirrel (Phalangeridae)

Figure 27.20 A representative monotreme: (**a**) Spiny anteater (*Tachyglossus*). Two Australian marsupials: (**b**) A female koala and her albino offspring living in relative safety in the San Diego Zoo. Their ancestors evolved millions of years ago, when the climate turned drier. Drought-resistant plants evolved, and koalas came to depend on eucalyptus (gum trees) for food and, often, for shelter. Koalas were abundant until Europeans arrived. Eucalyptus forests were cleared for farmland; millions of the slow-moving mammals were shot for their pelts. Koalas became a protected species in the 1930s. They still declined in numbers because nothing protected the eucalyptus trees. This still is the case through much of their home range. (**c**) Tasmanian devil, the largest carnivorous marsupial. It is about the size of a small dog. Its fierce show of fangs (a threat display used as a communication signal) as well as its hair-raising screeches, coughs, and snarls have given it an undeserved bad reputation. The Tasmanian devil is a nocturnal scavenger, a bit famous for its rowdy communal feeds at carcasses.

Every other existing descendant of therians is a placental mammal. A **placenta** is a spongy tissue of maternal and fetal membranes (Figure 27.21). It forms inside a pregnant female's uterus, the chamber in which an embryo develops in relative freedom from harsh conditions in the outer world. The female sends nutrients and oxygen to her embryo across the placenta and removes its

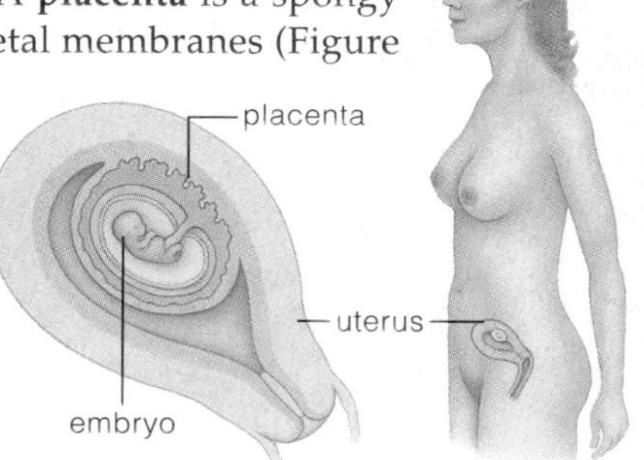
placenta
uterus
embryo

Figure 27.21 Location of the placenta in a human female.

Further reading: Student Guide to InfoTrac on web site —

Figure 27.22 A few placental mammals. (**a**) Herd of camels traversing an extremely hot desert with ease. (**b**) One of the bats, the only truly flying mammals, which dominate the night sky vacated by birds. (**c**) Manatee. It lives in the sea and eats submerged seaweed. (**d**) Arctic fox, with thick, insulative fur that camouflages it from prey. In summer, its light-brown fur blends with golden-brown grasses. In winter, the fur turns white and blends with the snow-covered ground. (**e**) Walruses swim in frigid waters after their prey and sunbathe on ice.

metabolic wastes. Section 45.9 provides a look at the structure and function of the placenta. For now, it is enough simply to remember that placental mammals generally have a developmental advantage. They grow faster in the uterus than marsupials do in their pouch, and many species are fully formed (and less vulnerable to predators) at birth. Figure 27.22 shows a few species; Appendix I lists the major groups.

The body form and function, behavior, and ecology of the mammals will occupy our attention later in the book. Chapter 28, for example, will provide insight into why many existing mammalian species are threatened by a mass extinction that humans are bringing about.

Convergent evolution occurred among the families of all three lineages of mammals that now occupy similar habitats in different regions of the world.

27.12 EVOLUTIONARY TRENDS AMONG PRIMATES

Having traversed the mammalian branch of the animal family tree, we are now ready to follow it along the branchings leading to primates, then to humans. As we explore the branchings, keep this key point in mind: *"Uniquely" human traits emerged through modification of traits that evolved earlier, in ancestral forms.*

Primate Classification

In the order **Primates** are prosimians, tarsioids, and anthropoids (Appendix I). Figure 27.23 shows a few. The first prosimians were arboreal (tree-dwelling). They dominated northern forests for millions of years before monkeys and apes evolved and almost displaced them. Small tarsiers of Southeast Asia, the only living tarsioids, are somewhere between prosimians and anthropoids in traits. Monkeys, apes, and humans are all anthropoids. In biochemistry and body form, the apes are closer to humans than to monkeys. Apes, humans, and extinct species of their lineages are classified as hominoids. **Hominids** include all humanlike and human species of a line of descent that started with its divergence from the last shared ancestor of apes and humans.

Table 27.2 Primate Classification

Taxon	Representatives
PROSIMIANS	
Lemuroids	Lemur, loris
TARSIOIDS	
Tarsioids	Tarsier
ANTHROPOIDS	
Ceboids (New World monkeys)	Spider monkey, howler monkey
Cercopithecoids (Old World monkeys)	Baboon, langur, macaque
Hominoids:	
Hylobatids	Gibbon, siamang
Pongids	Bonobo, orangutan, chimpanzee, gorilla
Hominids	Humans, extinct humanlike forms

Most primates live in tropical or subtropical forests, woodlands, or savannas (open grasslands with sparse stands of trees). Like their ancestors, most are *arboreal*, or tree dwellers. No one feature sets them apart from other mammals, and each lineage has its own defining traits. Five trends that help define the lineage leading to humans were set in motion when certain primates started adapting to life in the trees. *First*, there was less reliance on their sense of smell and more on daytime vision. *Second*, skeletal changes led to upright walking, which freed hands for novel tasks. *Third*, changes in bones and muscles led to refined hand movements. *Fourth*, teeth became less specialized. *Fifth*, evolution of the brain, behavior, and culture became interlocked.

Overview of Key Trends

ENHANCED DAYTIME VISION Early primates observed the world through one eye on each side of their head. Later ones had forward-directed eyes, an arrangement that is far better for sampling shapes and movements in three dimensions. Other modifications allowed eyes to respond to variations in color and light intensity (dim

Figure 27.23 Representative primates. (**a**) Bonobos, our closest primate relatives. Like us, they walk upright. (**b**) Gibbon, with body and limbs well adapted for swinging arm over arm through trees. (**c**) Spider monkey, a four-legged climber, leaper, and runner. (**d**) Tarsiers, vertical climbers and leapers.

Further reading: Student Guide to InfoTrac on web site

Figure 27.24 Comparison of the skeletal organization and stance of a monkey, ape, and human. Monkeys climb and leap. Gorillas and chimps use their forelimbs to climb and help support their weight, but most of the time they walk on all fours on the ground. Humans are two-legged walkers. Differences in modes of locomotion arose through modifications of the basic skeletal plan of mammals.

to bright light). Being able to interpret and respond swiftly to the novel stimuli proved advantageous for life in the trees.

UPRIGHT WALKING How a primate walks depends on its arm length and the shape and positioning of its shoulder blades, pelvic girdle, and backbone. With arm and leg bones of about the same length, a monkey can climb, leap, and run on four legs, but not two (Figures 27.23*c* and 27.24). A gorilla walks on two legs and on the knuckles of its two long arms. Only humans and bonobos show **bipedalism**: they can routinely stride and run about on two legs. Compared with an ape or monkey backbone, the human backbone is shorter, S-shaped, and flexible. Bipedalism was a key innovation that evolved in the ancestors of hominids.

POWER GRIP AND PRECISION GRIP How did we get our versatile hands? The first mammals spread their toes apart to support the body as they walked or ran on four legs. Primates still spread fingers or toes. Many also cup their fingers, as when monkeys lift food to the mouth. Among ancient tree-dwelling primates, modifications to handbones allowed them to wrap fingers around objects (*prehensile* movements) and to touch a thumb to the tip of a finger (*opposable* movements). In time, hands became freed from load-bearing functions. Later, when hominids evolved, so did power and precision grips:

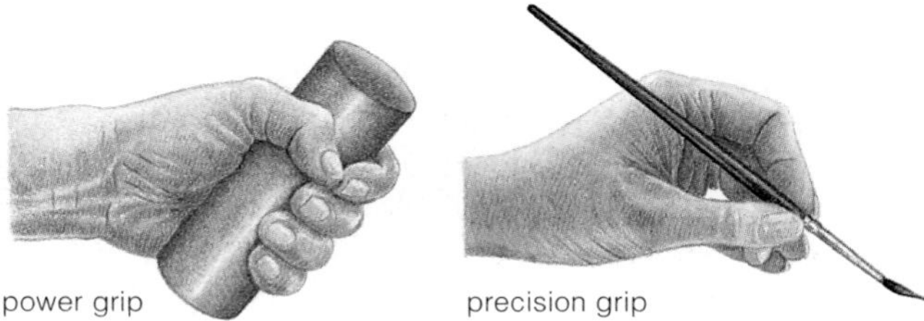
power grip
precision grip

These hand positions gave early humans a capacity to make and use tools. They were a foundation for unique technologies and cultural development.

TEETH FOR ALL OCCASIONS Before hominids evolved, modifications in primate jaws and teeth accompanied a shift from eating insects to fruits and leaves, and on to a mixed diet. Later, rectangular jaws and long canines came to be more defining features of monkeys and apes. Along the road that led to modern humans, a notably bow-shaped jaw and smaller teeth of about the same length evolved.

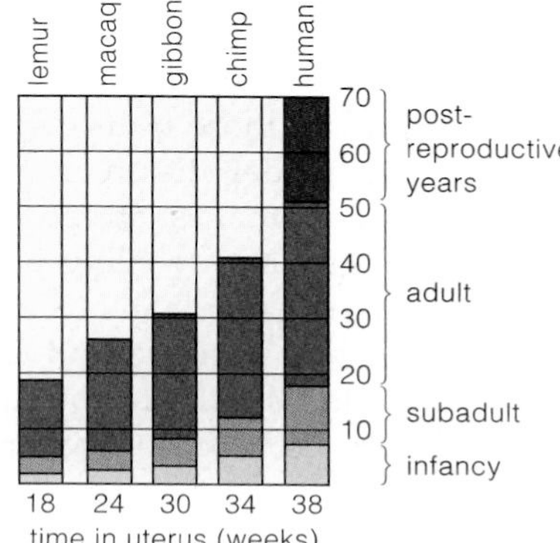
lemur
macaque
gibbon
chimp
human
70
60
50
40
30
20
10
post-reproductive years
adult
subadult
infancy
18
24
30
34
38
time in uterus (weeks)

Figure 27.25 Trend toward longer life spans and greater dependency of offspring on adults among primates.

THE BRAIN, BEHAVIOR, AND CULTURE Among some of the early primates, an arboreal life-style required shifts in reproductive and social behavior. In many lineages, parents put more effort in fewer offspring. They formed stronger bonds with offspring, maternal care grew more intense, and the time for dependency and learning grew longer (Figure 27.25).

Expansion of brain regions became interlocked with selection for more complex behavior. Novel behaviors promoted new neural connections in brain regions that dealt with sensory inputs, a brain with more intricate wiring favored more novel behavior, and so on. We find evidence of such interlocking in the parallel evolution of the human brain and culture. We define **culture** as the sum of behavior patterns of a social group, handed down among successive generations through learning and symbolic behavior, especially language. A capacity for language arose in ancestral human species mainly as an outcome of changes in skull bones and the brain.

These key adaptations evolved along the pathway from arboreal primates to modern humans: complex, forward-directed vision; bipedalism; refined hand movements; generalized dentition; and an interlocked elaboration of brain regions, behavior, and culture.

27.13 FROM EARLY PRIMATES TO HOMINIDS

Origins and Early Divergences

Primates evolved from mammals more than 60 million years ago, in tropical forests of the Paleocene. The first ones resembled small rodents or tree shrews (Figures 27.26 and 27.27). Like tree shrews, they probably had huge appetites and foraged at night for insects, seeds, buds, and eggs beneath the trees. They had a long snout and a refined sense of smell, suitable for snuffling food and predators. They clawed their way up stems, although not with much speed or grace.

During the Eocene (between 54 and 38 million years ago), certain primates were staying up in the trees. They had a shorter snout, better daytime vision, a larger brain, and a capacity for refined grasping motions. How did those traits evolve?

Consider the trees. Arboreal life-styles had advantages: abundant food and safety from ground-dwelling predators. However, they also were habitats of uncompromising selection. Visualize an Eocene morning—dappled sunlight, boughs swaying in the breeze, colorful, tasty fruit hidden among the leaves, perhaps predatory birds. A long, odor-sensitive snout would not have been much good in a dense forest canopy, where air currents disperse odors. But a brain that could assess movement, depth, shape, and color would have been a definite plus. So would a brain that could work fast when its owner was running and leaping (especially!) from branch to branch. Body weight, wind speed, and the distance and suitability of a destination had to be estimated swiftly, all at the same time. Adjustments for miscalculations had to be quick.

Figure 27.26 A tree shrew of Indonesia.

By 36 million years ago, before the dawn of the Oligocene epoch, the tree-dwelling anthropoids had evolved in the forests. One form, the squirrel-size *Catopithecus*, was on or near the evolutionary road leading to monkeys and apes. It had forward-directed eyes. Given its snoutless, flattened face and upper jaw with its shovel-shaped front teeth, it must have grabbed fruit and insects with its hands.

Some of the anthropoids lived in swamplands that were infested with fearsome, predatory reptiles. They rarely ventured down to the ground. Maybe that's why it was imperative to think fast and grip strongly. Many primates still fall out of the trees, so we can assume early anthropoids also slipped up.

Between 25 and 5 million years ago, in the Miocene, apelike forms—*the first hominoids*—evolved. They underwent an adaptive radiation into Africa, Europe, and southern Asia. While these early hominoids were evolving, continents were on the move, and ocean circulation patterns were shifting in a major way. The shift triggered a long-term cooling trend that would continue into the Pliocene (Figure 27.28).

The climate in eastern and southern Africa became cooler, drier, and more seasonal. Tropical forests—with their bounty of edible soft fruits, leaves, and abundant insects—started to give way to dense woodlands, then to grasslands. Food was drier and harder (and harder to find). Hominids that had evolved in the forests had two options: move into new adaptive zones or die out.

Not all of the early hominoids made the transition; most were extinct by the time the Miocene ended. But

a b c

Figure 27.27 Comparison of the skull shape and teeth of three early primates. (**a**) From the Paleocene, *Plesiadapis* was as tiny as a tree shrew and had rodentlike teeth. (**b**) The monkey-sized *Aegyptopithecus*, an Oligocene anthropoid, probably predates the divergence that gave rise to Old World monkeys and apes. (**c**) One of the apelike dryopiths that lived during the Miocene. Some dryopiths were as large as a chimpanzee.

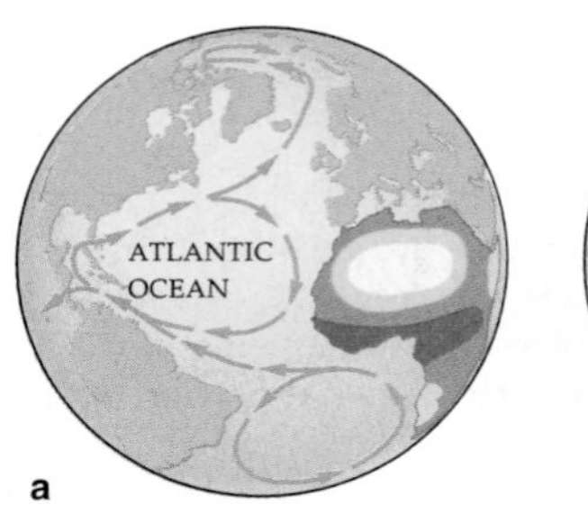
ATLANTIC
OCEAN
a

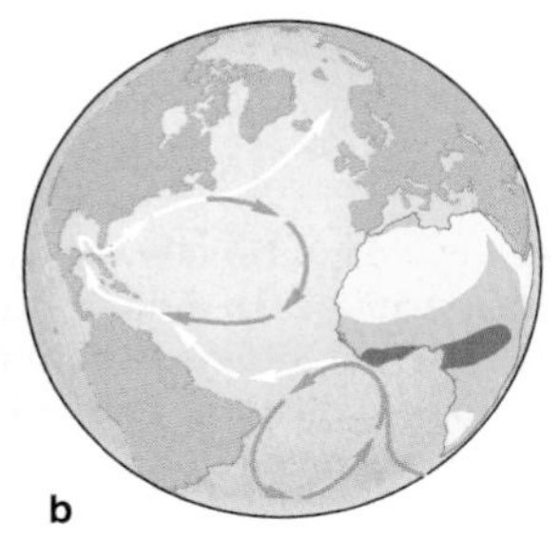
b

Figure 27.28 Model of a long-term shift in global climate. (**a**) Before the Isthmus of Panama formed, salinity levels of currents near the ocean surface were much the same around the world. Circulation patterns kept Arctic waters warm. (**b**) After the isthmus formed, North Atlantic surface currents became saltier and heavier, so they sank before they reached the Arctic region. With colder water in polar regions, the Arctic ice cap formed and ushered in a long-term trend toward a cooler, drier climate in Africa.

Further reading: Student Guide to InfoTrac on web site

Figure 27.29 (**a**) Remains of "Lucy" (*Australopithecus afarensis*), who lived 3.2 million years ago. (**b**) At Laetoli in Tanzania, Mary Leakey found these footprints, made in soft damp volcanic ash 3.7 million years ago. (**c**,**d**) The arch, big toe, and heel marks of such footprints are signs of bipedal hominids. Unlike apes, early hominids didn't have a splayed big toe, as the chimpanzee in (**c**) obligingly demonstrates.

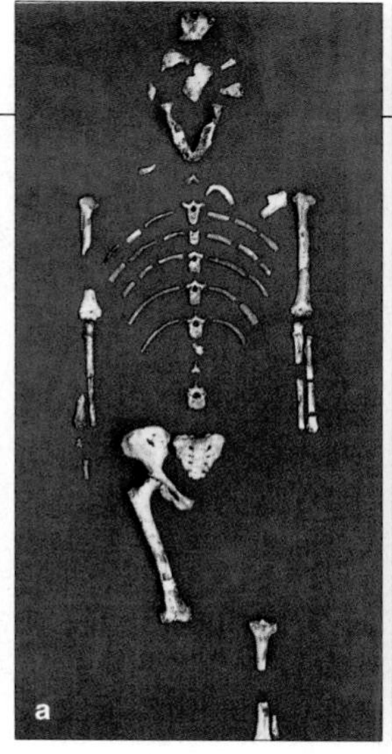

Ardipithecus ramidus
4 million years old

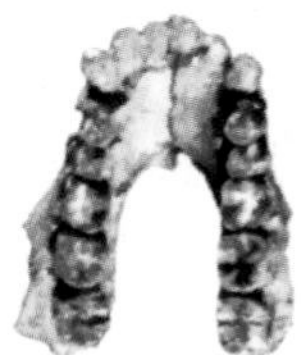

Australopithecus anamensis
4.2–3.9 million years
(walked upright)

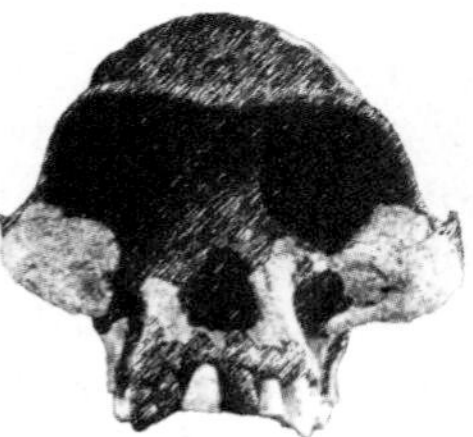

A. afarensis
3.6–2.9 million years

A. africanus
3.2–2.3 million years

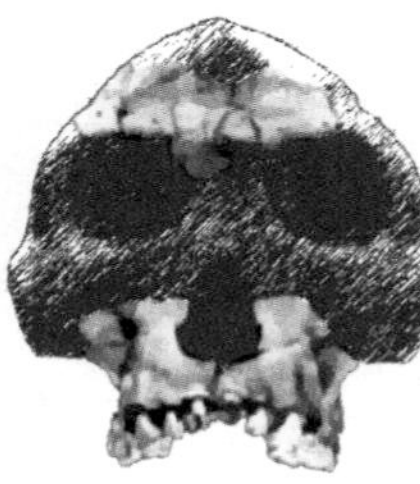

A. garhi
2.5 million years
(first tool user?)

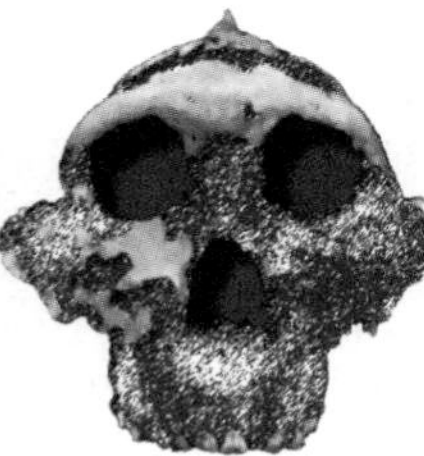

A. boisei
2.3–1.4 million years
(huge molars)

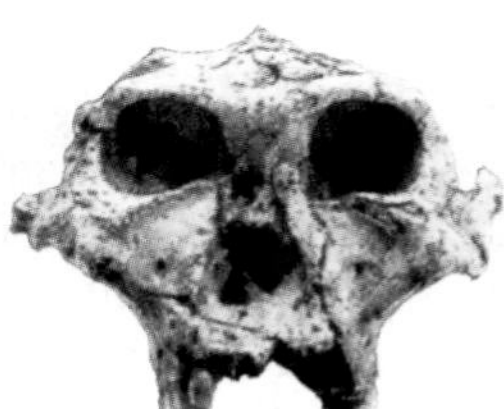

A. robustus
1.9–1.5 million years

Figure 27.30 Representative fossils of African hominids that lived between 4.4 and 1.4 million years ago.

genetic analyses suggests that two lineages branched from a common ancestor some time between 6 million and 4 million years ago. One lineage gave rise to the great apes, and the other gave rise to the first *hominids*.

The First Hominids

Fossil fragments of early hominids, some older than 5 million years, have been found in eastern and southern Africa (Figures 27.29 and 27.30). One species, *Ardipithecus ramidus*, lived 4.4 million years ago. It was more apelike than humanlike. Its small molars, large canines, and thin tooth enamel suggest that its diet consisted of easily chewed fruits and vegetables. Paleontologist Tim White puts it in the minor leagues of brain development.

Diverse hominids evolved over the next 2 million years. We still don't know how they are related, so we informally call them **australopiths**, or "southern apes." At this writing, the oldest known australopith species is *Australopithecus anamensis*. Like *A. afarensis* and *A. africanus,* it was gracile (slightly built). The ones called *A. boisei* and *A. robustus* were robust (muscular and heavily built). Like apes, australopiths had a large face, protruding jaws, and a small skull (and brain) size. Yet they differed in key respects from the hominoids. For example, their molars had thicker enamel and could grind harder foods. They could walk upright. We know this from fossilized hip bones and limb bones and from footprints. Around 3.7 million years ago, *A. afarensis* walked on newly fallen volcanic ash, which a light rain turned into quick-drying cement (Figure 27.29*b*).

Bipedal hominids had started to evolve in the late Miocene. In the trees, their hands had become adapted to grip objects strongly and with precision. When their descendants left the trees for life on the ground, rather than becoming specialized in running fast on all fours, they used manipulative skills to advantage. They kept their hands free to hold offspring and probably to carry precious food during their foraging expeditions.

Primates evolved from small, rodentlike mammals that had moved into arboreal habitats more than 60 million years ago. The earliest hominoids evolved between 25 million and 5 million years ago in Africa, Europe, and southern Asia.

Between 6 million and 4 million years ago, lineages that would give rise to the great apes and to the first hominids branched from a common hominoid ancestor.

Early hominids called australopiths were apelike in many skeletal details but humanlike in walking upright.

27.14 EMERGENCE OF EARLY HUMANS

What can the fossilized fragments of early hominids tell us about our own origins? The record is still too sketchy for us to know how the diverse australopiths were related to one another, let alone which ones may have been ancestral to humans. Besides, which traits do characterize **humans**—members of the genus *Homo*?

Well, what about the brain? The brain of modern humans is the basis of great analytical and verbal skills, complex social behavior, and technological innovation. It clearly sets us apart from apes, which have a much smaller skull volume and brain size (Figure 27.31). Yet this feature alone cannot tell us when certain hominids made the leap to being human. Why? Their brain size almost certainly fell within the ape range. Even though they made simple tools, so do chimps and some parrots. We have no clues to their social behavior.

We are left to speculate on a continuum of physical traits among a number of fossils—a skeleton adapted for bipedalism; manual dexterity; larger brain volume; a smaller face; and smaller, more thickly enameled teeth. Those traits, which originated in the late Miocene, were evident in what may have been the earliest humans—*Homo habilis,* a name that means "handy man."

Figure 27.31 Comparison of brain size for (**a**) a chimpanzee and (**b**) a modern human. (**c**) Artist's view of *Homo habilis* in an East African woodland. Australopiths are in the distance.

Figure 27.32 Olduvai Gorge stone tools: (**a**) a crude chopper, (**b**) a more refined chopper, (**c**) a hand ax, and (**d**) a cleaver,

Between 2.4 and 1.6 million years ago, early forms of *Homo* lived in dry woodlands adjoining the savannas of eastern and southern Africa. Their dentition implies they could handle hard-shelled nuts and seeds as well as soft fruits, leaves, and insects. Those food supplies were seasonal. Most likely, *H. habilis* had to think ahead, plan when to venture about to gather and maybe store foods that would help it survive the cold dry seasons.

H. habilis shared its habitat with saber-tooth cats and other predators. The cats' teeth could impale prey and shear off flesh but couldn't crush open marrow bones. So carcasses of their kills, with shreds of meat clinging to marrow bones, were concentrated stores of nutrients in nutrient-stingy places. *H. habilis* was a forager, not a full-time carnivore. But it opportunistically enriched its diet by scavenging carcasses (Figure 27.31*c*).

Fossil hunters have found a great many stone tools that date to the time of *H. habilis*. They cannot say with certainty that *H. habilis* was the only species that made them. Possibly australopiths as well as *H. habilis* used sticks and other perishable materials for tools before then, as modern apes do, but we have no way of knowing.

Maybe ancestors of modern humans started down a toolmaking road by picking up rocks to crack marrow bones. Maybe they scraped flesh from bones with sharp flakes split naturally from rocks. However it happened, at some point they started *shaping* stone implements. Paleontologist Mary Leakey was first to find evidence of stone toolmaking at Africa's Olduvai Gorge, which cuts through many sedimentary layers. The oldest tools there are crudely chipped pebbles in the deepest layers (Figure 27.32). Early humans might have used them to smash marrow bones, dig for roots, and poke insects from bark. More recent layers have more complex tools. Where shorelines of ancient lakes prevailed, we observe numerous bones and tools. Why? In arid habitats, lakes would have beckoned plenty of thirsty prey animals.

At such sites we find fossils of an early form of *Homo* that was twice as brainy as australopiths and obviously

Further reading: Student Guide to InfoTrac on web site

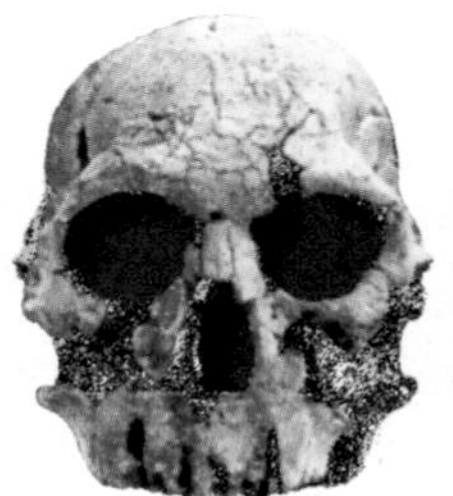

Homo rudolfensis
2.4–1.8 million years

H. habilis
1.9–1.6 million years

H. erectus
2 million–53,000? years

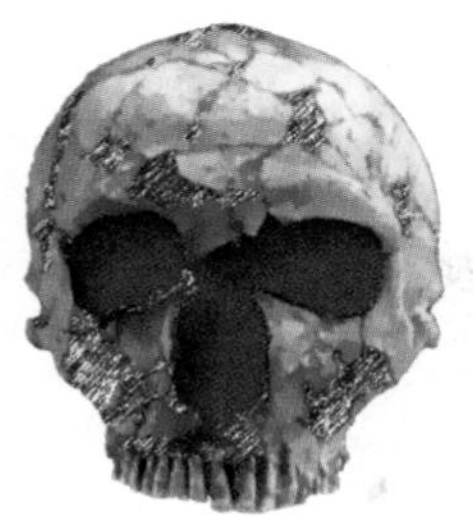

H. neanderthalensis
200,000–30,000 years

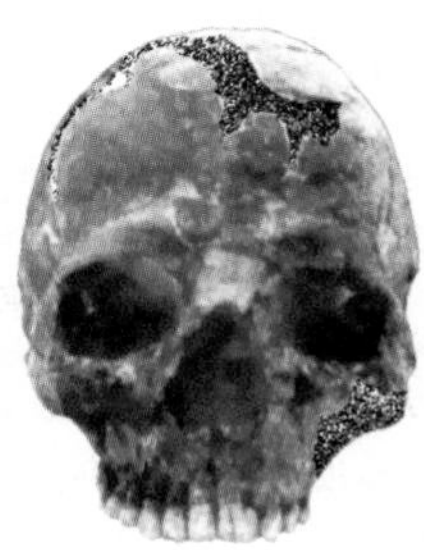

H. sapiens
100,000–?

Figure 27.33 Representative fossils of early humans (*Homo*). 

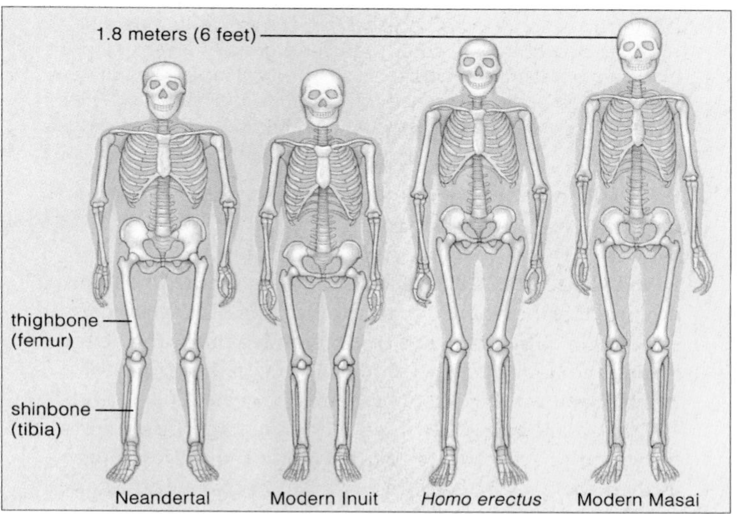
1.8 meters (6 feet)
thighbone (femur)
shinbone (tibia)
Neandertal
Modern Inuit
Homo erectus
Modern Masai

Figure 27.34 Body build correlated with climate. Humans adapted to cold climates have a heat-conserving body (stockier, shorter legs, compared with humans adapted to hot climates).

ate well. There apparently was no selection pressure for more creativity in securing food resources; stone tools did not change much for the next 500,000 years.

Apparently, the ancestors of modern humans stayed put in Africa until about 2 million years ago. Then, a genetic divergence gave rise to *Homo erectus*, a species clearly related to fully modern humans (Figure 27.33). The name means upright man. Its forerunners also were upright, two-legged walkers, but *H. erectus* populations did the name justice. They trekked out of Africa, turned left into Europe, and right into Asia. Some traversed the 14,000 kilometers to China; fossils from Southeast Asia and the former Soviet republic of Georgia are 1.8 million and 1.6 million years old. More than once, *H. erectus* survived when ice sheets advanced down into northern Europe, southern Asia, and North America.

Whatever pressures triggered the far-flung travels, this was a time of physical changes, as in skull size and leg length. It also was a time of cultural lift-off for the human lineage. *H. erectus* had a larger brain and was a more creative toolmaker. Its social organization and communication skills must have been well developed. How else can we explain its successful dispersal? From southern Africa to England, different populations used the same kinds of hand axes and other tools to pound, scrape, shred, chop, cut, and whittle material. *H. erectus* withstood environmental challenges by building fires and using furs for clothing. Clear evidence of fire use dates from an ice age in the early Pleistocene.

As Middle Eastern fossils show, *Homo sapiens* evolved by 100,000 years ago. Early *H. sapiens* had smaller teeth and jaws than *H. erectus* (Figure 27.33). Many had a novel feature: a chin. Facial bones were smaller, the skull higher and rounder, and the brain larger. They might have developed complex language. But their origin and geographic dispersal are hotly debated (Section 27.15).

One group of early humans, the Neandertals, lived in Europe and in the Near East from 200,000 to 30,000 years ago. Massively built and large brained, some were the first to adapt to the coldest regions (Figure 27.34). Their disappearance coincided with the appearance of anatomically modern humans in the same regions about 40,000 to 30,000 years ago. We have no evidence that Neandertals warred or interbred with the later arrivals. We still do not know what happened to them. We do know that Neandertal DNA has unique sequences, so they might not have contributed to the gene pools of modern European populations.

From 40,000 years ago to today, human evolution has been almost entirely cultural, not biological. And so we leave the story with these conclusions: Humans spread rapidly through the world by devising *cultural* means to deal with a broader range of environments. Compared to their predecessors, they developed rich, varied cultures. Although hunters and gatherers persist in parts of the world, others moved from "stone-age" technology to the age of "high tech," attesting to the great plasticity and depth of human adaptations.

Cultural evolution has outpaced the biological evolution of the only remaining human species, *H. sapiens*. Today, humans everywhere rely on cultural innovation to adapt rapidly to a broad range of environmental challenges.

27.15 OUT OF AFRICA—ONCE, TWICE, OR . . .

If researchers are interpreting the fossil record of human evolution correctly, then it would seem Africa was the cradle for us all. At this writing, at least, no one has any fossils of humans older than 1.8 million years *except* in Africa. *H. erectus* coexisted for a time with earlier humans (*H. habilis*) before dispersing from Africa to the cooler grasslands, forests, and mountains of Europe and Asia. This form apparently left Africa in waves between about 2 million and 500,000 years ago. Judging from *H. erectus* fossils from Java, isolated populations may have endured until some time between 53,000 and 37,000 years ago.

Where, on the larger geologic stage, do we place the origin of *H. sapiens*? *Here we find a good example of how the same body of evidence can be interpreted in different ways.* The interpretations are called the **multiregional model** and **African emergence model** for modern human origins. Both attempt to explain the world distribution of fossils of particular ages and measured genetic distances among existing human populations. For example, biochemical and immunological studies imply that the greatest genetic distance separates *H. sapiens* populations native to Africa from populations everywhere else; the next greatest distance separates Southeast Asia and Australia from everywhere else (Figure 27.35). As still another example, Figure 27.36 correlates *H. sapiens* fossils to specific times.

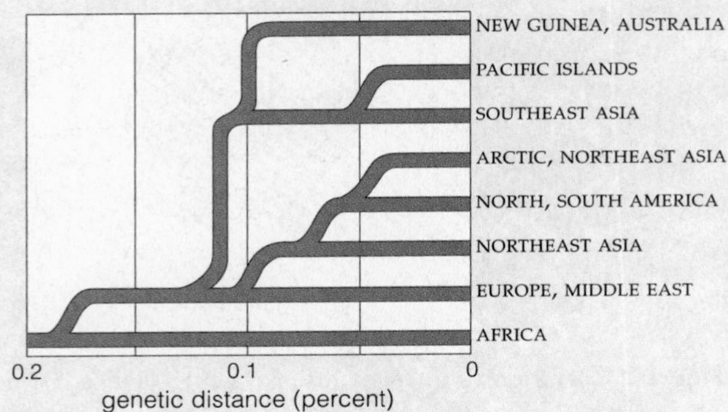
NEW GUINEA, AUSTRALIA
PACIFIC ISLANDS
SOUTHEAST ASIA
ARCTIC, NORTHEAST ASIA
NORTH, SOUTH AMERICA
NORTHEAST ASIA
EUROPE, MIDDLE EAST
AFRICA
0.2
0.1
0
genetic distance (percent)

Figure 27.35 One proposed family tree for populations of modern humans (*Homo sapiens*) that are native to different regions. Branch points show presumed genetic divergences. The tree is based on nucleic acid hybridization studies of many genes (including those for mitochondrial DNA and the ABO blood group) and immunological comparisons.

MULTIREGIONAL MODEL By this model, *H. erectus* populations spread through much of the world by about 1 million years ago. The geographically separated groups faced different selection pressures, so their traits evolved in regionally distinctive ways. Then the subpopulations ("races") of *H. sapiens* evolved from them in different places. They differed phenotypically but did not evolve into separate species because gene flow continued among them, even to the present. Thus, for example, while the armies of Alexander the Great were sweeping eastward, they also contributed "blue-eye genes" from Greeks to the allele pool of generally brown-eyed subpopulations in Africa, the Near East, and Asia.

AFRICAN EMERGENCE MODEL This model does not dispute fossil evidence that populations of *H. erectus* evolved in different ways in different regions. However, it holds that *H. sapiens*—modern humans—originated in sub-Saharan Africa somewhere between 200,000 and 100,000 years ago. Only later did *H. sapiens* populations move out of Africa, into regions along the routes shown in the Figure 27.36 map. In each region that populations of *H. sapiens* settled, they replaced the archaic *H. erectus* populations that had preceded them. Only then did regional phenotypic differences become superimposed on the original *H. sapiens* body plan.

In support of this model, the oldest known *H. sapiens* fossils are from Africa. Also, in Zaire, a finely wrought barbed-bone harpoon and other exquisite tools suggest that African populations were as skilled at making tools as *Homo* populations known earlier from Europe. Also, in 1998, researchers at the University of Texas and in China announced the findings from the Chinese Human Genome Diversity Project. Detailed analysis of gene patterns from forty-three ethnic groups in Asia suggest that modern humans moved from central Asia, along the coast of India, then on into Southeast Asia and southern China. Populations later moved north and northwest into China, then into Siberia, and then on down into the Americas.

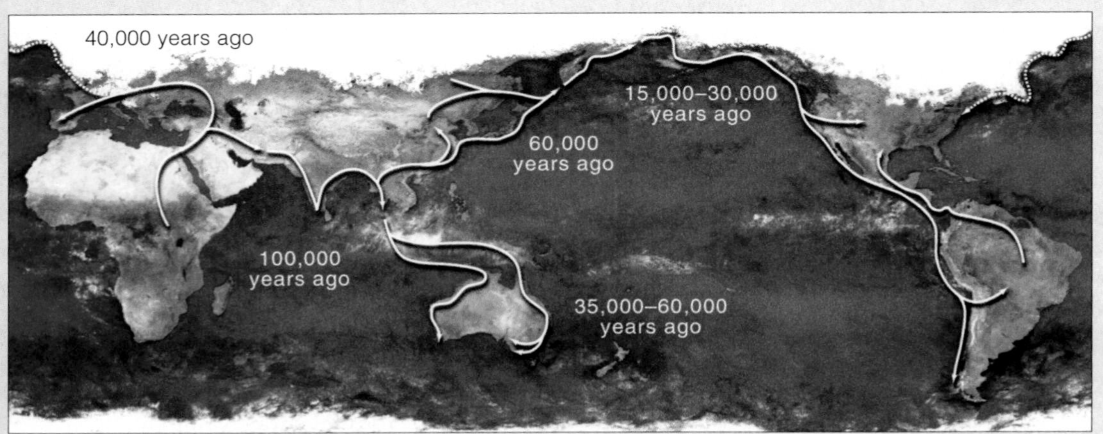
40,000 years ago
15,000–30,000 years ago
60,000 years ago
100,000 years ago
35,000–60,000 years ago

Figure 27.36 Estimated times when populations of early *H. sapiens* were colonizing different regions of the world, based on radiometric dating of fossils. The presumed dispersal routes (white lines) seem to support the African emergence model.

Further reading: Student Guide to InfoTrac on web site

27.16 SUMMARY

1. Almost all chordate embryos have a notochord, a dorsal hollow nerve cord, a pharynx with gill slits (or hints of these), and a tail that extends past the anus. Some or all of these traits persist in adults. Chordates with a backbone are vertebrates. Tunicates (including sea squirts) and lancelets are invertebrate chordates.

2. The eight classes of vertebrates are jawless fishes, jawed armored fishes (now extinct), cartilaginous fishes, bony fishes, amphibians, reptiles, birds, and mammals.

3. The earliest vertebrates, the jawless fishes that arose in the Cambrian, included ostracoderms. Their modern descendants are lampreys and hagfishes. Jawed fishes also arose in Cambrian times. A once-dominant lineage, the placoderms, became extinct. Other lineages gave rise to the cartilaginous fishes and bony fishes.

4. Four evolutionary trends are associated with certain vertebrate lineages:

a. A vertebral column supplanted the notochord as a structural element against which muscles can act. This development led to fast-moving predatory animals.

b. Jaws evolved from gill-supporting elements. This led to increased predator–prey competition; it favored more efficient nervous systems and sensory organs.

c. In one lineage of bony fishes, paired fins evolved into fleshy lobes having internal structural elements. The lobes were the forerunners of paired limbs.

d. In some fish lineages, respiration by gills came to be supplemented by paired lungs, which proved to be adaptive during the invasion of the land. In a related development, the circulatory system became far more efficient at distributing oxygen through the body.

5. Amphibians were the first vertebrates to invade the land, but they never fully escaped the water. Their skin dries out, and aquatic stages persist in most life cycles.

6. Unlike amphibians, reptiles escaped dependence on standing free water. The key adaptations that allowed them to do so include toughened, scaly skin that can restrict loss of water from the body; a reliance upon internal fertilization; more efficient, water-conserving kidneys; and amniote eggs, often leathery or shelled, which protect and metabolically support the embryos.

7. Reptiles, and birds and mammals (which descended from certain reptilian lineages), have highly efficient circulatory and respiratory systems. They also have well-developed nervous system and sensory organs.

8. Of all vertebrates, birds alone have feathers, which they use in flight, heat conservation, and social displays.

9. Of all animals, mammals alone have milk-producing mammary glands, and they have hair or thick skin that functions in insulation. They have distinctive dentition and a highly developed cerebral cortex, the brain's most complex region. Most adults nurture offspring through an extended period of dependency and learning.

a. Therians, the first mammals, evolved from small, hairless reptiles (therapsids) during the Triassic. At the end of the Cretaceous, new adaptive zones opened up and, combined with geologic events, set the stage for three previously inconspicuous mammalian lineages.

b. Those lineages still persist. They are monotremes (egg-laying mammals), marsupials (pouched mammals), and eutherians (placental mammals). In similar habitats in different parts of the world, some of their members converged evolutionarily.

c. In many places, placental mammals outcompeted the monotremes and marsupials, which retained many archaic traits. Placental mammals had higher metabolic rates, more precise control of body temperature, and an efficient way to nourish and protect embryos.

10. Primates include the prosimians (lemurs and related forms), the tarsioids, and the anthropoids (which include monkeys, apes, and humans). Apes and humans alone are hominoids. Only the anatomically modern humans (*H. sapiens*) and others of the lineage (from *Ardipithecus* and australopiths to *H. erectus*) are hominids.

11. The first primates were small, rodentlike mammals that evolved by 60 million years ago in tropical forests.

12. Apelike forms (the first hominoids) were evolving in Miocene times, 25 to 5 million years ago. In eastern and southern Africa, some gave rise to hominids by at least 5 million years ago.

a. The origin and early evolution of hominids might coincide with selection pressures that emerged during a long-term change in climate, brought about by shifts in land masses and ocean circulation patterns. Tropical forest gave way to dense woodlands, then grasslands.

b. Some hominids became adapted to the changes. They had physical modifications that allowed upright walking (bipedalism), a more diverse diet (changes in dentition), and a capacity to find scarcer and seasonally available food (possible with a more complex brain).

13. *H. habilis*, the earliest known species of the genus *Homo*, evolved by 2.5 million years ago. It might have been the first stone toolmaker. *Homo erectus*, possibly ancestral to modern humans, evolved by 2 million years ago. *H. erectus* populations radiated out of Africa, into Asia and Europe. The oldest known fossils of the early modern humans (*H. sapiens*) date from about 100,000 years ago. About 40,000 years ago, cultural evolution outstripped biological evolution of the human form.

14. Modern humans are adapted to a broad range of environments. This capacity resulted from evolutionary modifications in certain primate lineages. Starting with arboreal primate ancestors, there was less reliance on the sense of smell and more on enhanced daytime vision.

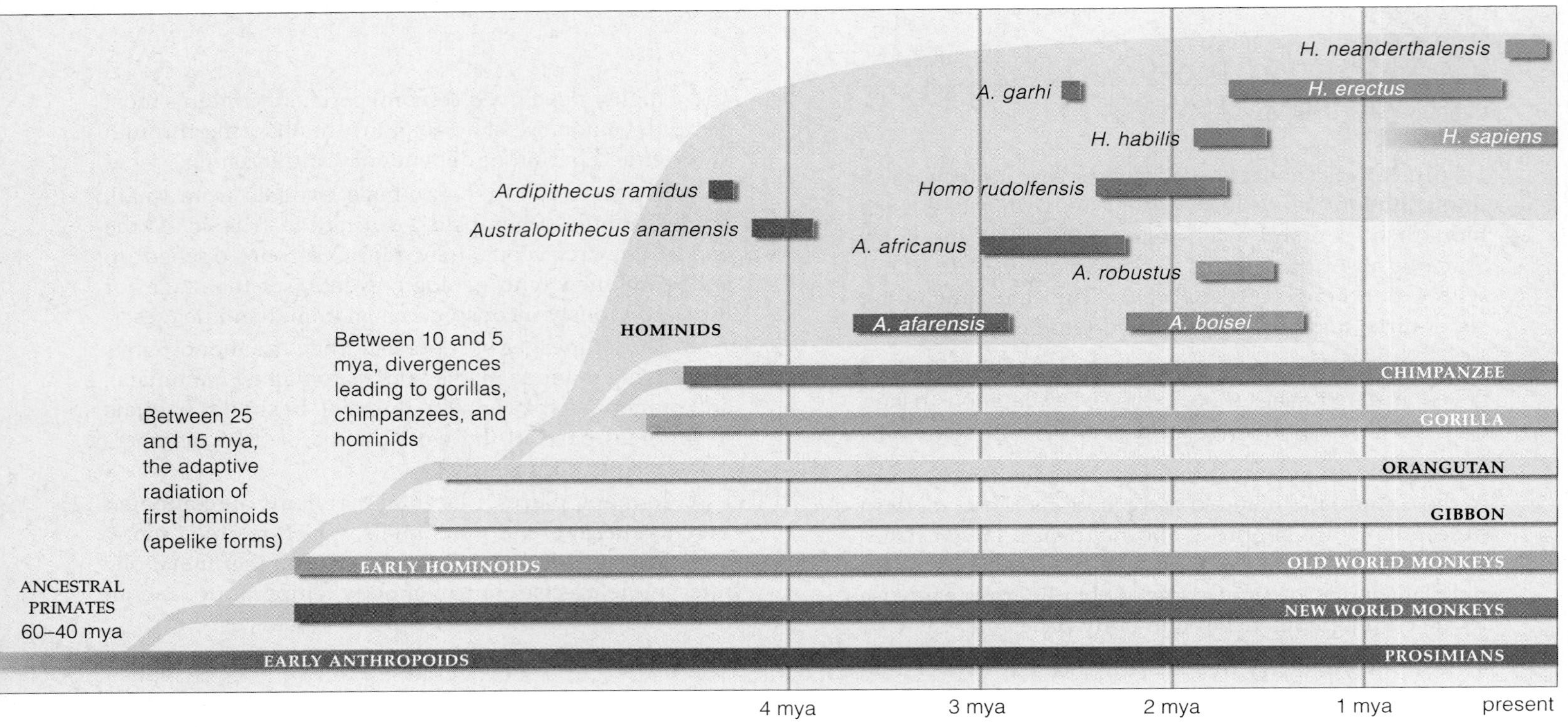

H. neanderthalensis
A. garhi
H. erectus
H. habilis
H. sapiens
Ardipithecus ramidus
Homo rudolfensis
Australopithecus anamensis
A. africanus
A. robustus
HOMINIDS
A. afarensis
A. boisei
Between 10 and 5 mya, divergences leading to gorillas, chimpanzees, and hominids
CHIMPANZEE
Between 25 and 15 mya, the adaptive radiation of first hominoids (apelike forms)
GORILLA
ORANGUTAN
GIBBON
EARLY HOMINOIDS
OLD WORLD MONKEYS
ANCESTRAL PRIMATES 60–40 mya
NEW WORLD MONKEYS
EARLY ANTHROPOIDS
PROSIMIANS
4 mya
3 mya
2 mya
1 mya
present

Figure 27.37 Summary of presumed evolutionary branchings leading to modern humans in the primate family tree.

Also among the arboreal primates, manipulative skills increased as hands began to be freed from load-bearing functions. Starting with Miocene apes, there was a shift from four-legged climbing to bipedalism, a shift from specialized to omnivorous eating habits, and increases in brain complexity and behavior.

15. Figure 27.37 summarizes what we currently know or suspect about primate origins and evolution.

Review Questions

1. Which features distinguish chordates from other groups of animals? *27.1*
2. List four major trends that occurred during the evolution of at least some vertebrate lineages. *27.3*
3. Describe some features of jawless fishes, cartilaginous fishes, and bony fishes. Which group is the most abundant? *27.3*
4. Which traits distinguish reptiles from amphibians? *27.6, 27.7*
5. Birds, crocodilians, and dinosaurs share what traits? *27.7–27.9*
6. List some characteristics that distinguish each of the three mammalian lineages from the reptiles. *27.10, 27.11*
7. What is the difference between a "hominoid" and a "hominid"? *27.12*
8. Briefly describe some of the conserved physical traits that connect anatomically modern humans with their mammalian ancestors, then their primate ancestors. *27.12, 27.13*

Self-Quiz (*Answers in Appendix III*)

1. Only ______ have a notochord, a tubular dorsal nerve cord, a pharynx with slits in the wall, and a tail extending past the anus.
   a. echinoderms
   b. tunicates and lancelets
   c. vertebrates
   d. both b and c
   e. all of the above

2. Gills function in ______ .
   a. respiration
   b. circulation
   c. food trapping
   d. water regulation
   e. both a and c

3. A shift from a reliance on ______ to reliance on ______ was pivotal in the evolution of all vertebrates.
   a. the notochord; a backbone
   b. filter feeding; jaws
   c. gills; lungs
   d. all of the above

4. The first vertebrates probably were ______ .
   a. bony fishes
   b. jawless fishes
   c. lobe-finned fishes
   d. both a and b

5. Of all existing vertebrates, ______ are the most diverse.
   a. cartilaginous fishes
   b. bony fishes
   c. amphibians
   d. reptiles
   e. birds
   f. mammals

6. Generally, the only amphibian groups that entirely escaped dependency on free-standing water are ______ .
   a. salamanders
   b. toads
   c. caecilians
   d. none of the above

7. Reptiles moved fully onto land owing to ______ .
   a. tough skin
   b. internal fertilization
   c. good kidneys
   d. amniote eggs
   e. none of the above
   f. all of the above

8. A four-chambered heart evolved first in ______ .
   a. bony fishes
   b. amphibians
   c. birds
   d. mammals
   e. crocodilians
   f. both c and d

9. ______ have highly efficient circulatory and respiratory systems, and a complex nervous system and sensory organs.
   a. Reptiles
   b. Birds
   c. Mammals
   d. all of the above

10. Birds use feathers in ______ .
   a. flight
   b. heat conservation
   c. social functions
   d. all of the above

11. Various mammals ________ .
   a. hatch
   b. complete their embryonic development in pouches
   c. complete their embryonic development in the uterus
   d. both b and c
   e. all of the above

12. Early humans ________ .
   a. adapted to a wide range of environments
   b. adapted to a narrow range of environments
   c. had flexible bones that cracked easily
   d. were limber enough to swing through the trees

13. Match the organisms with the appropriate features.

____ jawless fishes	a. complex cerebral cortex, thick skin or hair
____ cartilaginous fishes	b. respiration by skin and lungs
____ bony fishes	c. include coelacanths
____ amphibians	d. include hagfishes
____ reptiles	e. include sharks and rays
____ birds	f. complex social behavior, feathers
____ mammals	g. first with amniote eggs

Critical Thinking

1. Describe the factors that might contribute to the collapse of native fish populations in a lake following the introduction of a novel predator, such as the lamprey.

2. Think about the flight muscles of birds and their demands for oxygen and ATP energy. What type of organelle would you expect to be profuse in these muscles? Explain your reasoning.

3. Kathie and Gary, two amateur fossil hunters, have unearthed the complete fossilized remains of a mammal. How can they determine whether that mammal was a herbivore, carnivore, or omnivore?

4. In Australia, many species of marsupials are competing with recently introduced placental mammals (such as rabbits) for resources—but they are not winning. Explain how it is that placental mammals that did not even evolve in the Australian habitats show greater fitness than the native mammals.

5. About seventy-five species of coral snakes live in the rain forests, grasslands, mountains, and desertlike regions of North, Central, and South America. Coral snakes of family Colubridae are harmless; those of family Elapidae are poisonous. In both families, the most poisonous types bite mostly to kill prey, yet moderately poisonous types can be extremely vicious. All coral snakes, including the deadliest (*Micrurus*, Figure 27.38), have distinctive color banding. Even snake experts (herpetologists) may have trouble knowing which is which. However, relatives of the two families in other parts of the world have no such banding; neither do bigger snakes.

Thus, color banding in the two families can't be explained in terms of environmental differences, phylogeny, or lethality. What brought it about? According to a hypothesis proposed by herpetologist R. Mertens, the ancestors of *Micrurus* migrated across a land bridge that connected Asia with North America more than 60 million years ago. Because predators learned to avoid them by associating the color banding with "taste trials," the ancestral snakes must *not* have been as deadly as some of their descendants. (If the snakes were *too* deadly, all trials might have ended in death.) Therefore, the protective function of the color banding must depend on unpleasant trials with only *moderately* poisonous snakes.

Among the harmless snakes (Colubridae), identical warning coloration also evolved. This was a case of *mimicry*, in which one group (mimics) came to bear deceptive resemblance to another group (models) that enjoy a survival advantage. For this to work, of course, mimics can't outnumber models. That said, explain the following numbers of trapped snakes that were brought to the Butantan Institut in São Paulo for the years shown (trappers did not distinguish among poisonous and nonpoisonous types):

Figure 27.38 One of the extremely poisonous coral snakes (*Micrurus*).

	1950	*1951*	*1952*	*1953*
Micrurus:	59	41	58	56
Others:	218	246	265	284

Selected Key Terms

<u>*Vertebrates*</u>
amniote egg *27.7*
amphibian *27.6*
behavioral flexibility *27.10*
bird *27.9*
bony fish *27.5*
cartilaginous fish *27.5*
chordate *27.1*
convergent evolution *27.11*
dentition *27.10*
eutherian *27.10*
feather *27.9*
filter feeder *27.2*
fin *27.3*
gill *27.3*
gill slit *27.2*
hagfish *27.4*
jaw *27.3*
lamprey *27.4*
lancelet *27.2*
lobe-finned fish *27.5*
lung *27.3*
mammal *27.10*
marsupial *27.10*
monotreme *27.10*
nerve cord *27.1*
notochord *27.1*
ostracoderm *27.3*
pharynx *27.1*
placenta *27.11*
placoderm *27.3*
Primates (order) *27.12*
reptile *27.7*
scale (fish) *27.5*
swim bladder *27.5*
therapsid *27.10*
therian *27.10*
tunicate *27.2*
vertebra *27.3*
vertebrate *27.1*

<u>*Human Evolution*</u>
African emergence model *27.15*
australopith *27.13*
bipedalism *27.12*
culture *27.12*
hominid *27.12*
human *27.14*
multiregional model *27.15*

Readings *See also www.infotrac-college.com*

Gould, S. J. (general editor). 1993. *The Book of Life*. New York: Norton. Splendid, easy-to-read essays, gorgeous illustrations.

Romer, A. S., and T. S. Parsons. 1986. *The Vertebrate Body*. Sixth edition. Philadelphia: Saunders.

Strickberger, M. 1996. *Evolution*. Second edition. Boston: Jones and Bartlett.

Tattersall, I. 1998. *Becoming Human: Human Evolution and Human Uniqueness*. Pennsylvania: Harvest Books.

Wickler, W. 1968. *Mimicry in Plants and Animals*. New York: McGraw-Hill. Paperback.

28

BIODIVERSITY IN PERSPECTIVE

The Human Touch

In 1722, on Easter morning, a European explorer landed on a small volcanic island and found a few hundred skittish, hungry Polynesians living in caves. He also found more than 200 massive stone statues near the coast and 700 unfinished, abandoned ones in inland quarries (Figure 28.1). There were no trees whatsoever on the island, only dry, withered grasses and scorched, shrubby plants. Without trees for timber and fibrous plants to make ropes, how did the islanders erect the statues? Some statues weighed fifty tons. And there were no wheeled carts or strong animals. How did the statues get from the quarries to the coast?

Two years later James Cook visited the island. When islanders paddled out to his ships, he noted that they had neither the knowledge nor the materials to keep their canoes from leaking; they spent half the time bailing water. He saw only four canoes on the whole island. Nearly all the statues had been tipped over, often onto spikes that shattered the faces on impact.

Later, researchers solved the mystery of the statues. Easter Island, as it came to be called, has a mere 165 square kilometers (64 square miles) of surface land. Voyagers from the Marquesas discovered this eastern outpost of Polynesia around A.D. 350, possibly after being blown off course by a series of storms.

The place was a paradise. Its fertile volcanic soil supported dense palm forests, hauhau trees, toromino shrubs, and lush grasses. The new arrivals built large, buoyant, ocean-going canoes by using long, straight palms strengthened with rope made of fibers from the hauhau trees. They used toromino wood as fuel to cook fishes and dolphins; they cleared forests to plant crops of taro, bananas, sugarcane, and sweet potatoes.

By 1400, between 10,000 and 15,000 descendants were living on the island. Society had become highly structured, given the need to cultivate, harvest, and distribute food for so many. Crop yields had declined by then, for every crop withdrew precious nutrients from the soil, as did ongoing, neglected erosion. Edible marine species vanished from the safe waters around the island, so fishermen had to build larger canoes to sail out ever farther, into the open sea.

Figure 28.1 On Easter Island, a few of the massive stone tributes to the gods. Apparently islanders erected the statues as a plea for divine intervention after their once-large population wiped out the biodiversity of their tropical paradise. The plea didn't have any effect whatsoever on reversing the losses on land and in the surrounding sea. The human population didn't recover, either.

Survival was at stake. Those in power appealed to the gods. They redirected community resources into carving divine images of unprecedented size and power, and to moving the newly carved statues over miles of rough terrain to the coast. As Jo Anne Van Tilburg recently demonstrated, they probably lashed the statues to canoe-shaped rigs, then simply rolled them along a horizontal "ladder" made of greased logs. (She got the idea after observing how existing islanders haul canoes from water onto land.)

By now, islanders had all of the arable land under cultivation. They had eaten all the native birds, and no new birds came to nest. They were raising and eating rats, the descendants of hitchhikers on the first canoes. They coveted palm seeds, now a scarce delicacy.

Wars broke out over dwindling food and space. By about 1550 no one was venturing offshore to harpoon fishes or dolphins. They couldn't build canoes because the once-rich forests were gone. They had cut down all of the palms. Hauhau trees were extinct; islanders had used every last one as firewood for cooking.

The islanders turned to the only remaining source of animal protein. They started to hunt and eat one another.

Central authority crumbled, and gangs replaced it. As gang wars raged, those on the rampage burned the remaining grasses to destroy hideouts. The rapidly dwindling population retreated to caves and launched raids against perceived enemies. Winners ate the losers and tipped over the statues. Even if the survivors had wanted to leave the island, they no longer had a way to do so. What possibly could they have been thinking when they chopped down the last palm?

Easter Island initially sustained a society that rose to greatness amid abundant resources, then abruptly fell apart. In life's greater evolutionary story, its loss of biodiversity might not even warrant a footnote. After all, compare the loss to the uncountable numbers of species that appeared and then became extinct in the distant past!

Yet there is a lesson here. The near-total destruction that 15,000 Polynesians wrought on biodiversity was confined to a small bit of land in the vastness of the Pacific Ocean. What are the requirements and whims of *6 billion* people doing to global biodiversity today? Is the loss of species on Easter Island an isolated case? Or should we take what happened there as a warning of a worldwide extinction crisis of our own making, one that is even now under way? Let's take a look.

KEY CONCEPTS

1. The current range of global biodiversity is largely an outcome of an overall pattern of abrupt extinction crises and slow recoveries. It takes tens of millions of years to reach the level of biodiversity that prevailed before such mass extinctions.

2. The loss of individual species as well as extinctions of major groups may have complicated causes, the understanding of which requires scientific analysis.

3. Species found only in a restricted geographic region and extremely vulnerable to extinction are classified as endangered species. Over the past four decades, rates of extinction have been rising through a combination of habitat loss and fragmentation, species introductions, overharvesting, and illegal trading in wildlife.

4. The rapidly growing human population is expected to reach 9 billion by 2050. Its demands for food, materials, and living space are threatening biodiversity around the world and are the basis of a new extinction crisis.

5. Conservation biology takes a three-pronged approach to preserving biodiversity. It includes the systematic survey of the full range of biodiversity, analysis of its evolutionary and ecological origins, and identification of methods to maintain and use biodiversity for the benefit of the human population.

6. Systematic surveys focus initially on identifying hot spots, which are limited areas where many species are in danger of extinction owing to human activities. At the next research level, broader or multiple hot spots are researched. Data are combined into a global picture of ecoregions where biodiversity is most threatened.

28.1 ON MASS EXTINCTIONS AND SLOW RECOVERIES

Based on many lines of evidence accumulated over the past few centuries, an estimated 99 percent of all species that have ever lived are extinct. Even so, the full range of biodiversity is greater now than it has ever been at any time in the past.

Reflect on the evolutionary stories of the preceding chapters in this unit. For at least the first 2 billion years of life's history, single-celled prokaryotes dominated the evolutionary stage. They did so until the Cambrian period, when atmospheric oxygen started to approach current levels. At that time, some of the single-celled eukaryotes began their genetic divergences and intricate species interactions, which led to the origin of diverse protistans, plants, fungi, and animals.

You saw how mass extinctions reduced biodiversity on land and in the seas. Each global episode spurred evolutionary change and radiations into newly vacated adaptive zones. However, recovery to the same level of biodiversity was exceedingly slow, requiring 20 to 100 million years. Figure 28.2*a* is a review of the pattern of five major extinctions and slow recoveries.

That pattern is only a composite of what happened to the major taxa. Lineages, remember, differ in their time of origin, the extent to which the member species evolved, and how long they persisted. If you consider ongoing survival and reproduction to be the measures of success, then each became a loser or a winner when environmental conditions changed in drastic or novel ways. To appreciate this point, reflect on Figure 28.2*b*, which shows the evolutionary history of representative lineages. Many such histories were combined to give us the overall pattern of Figure 28.2*a*.

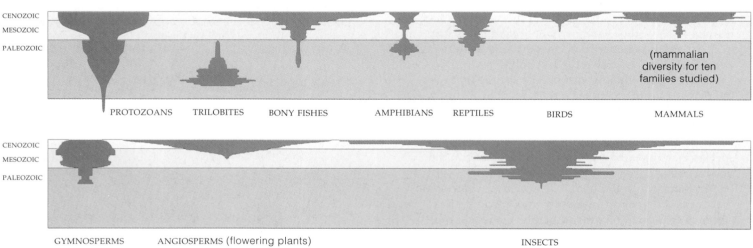

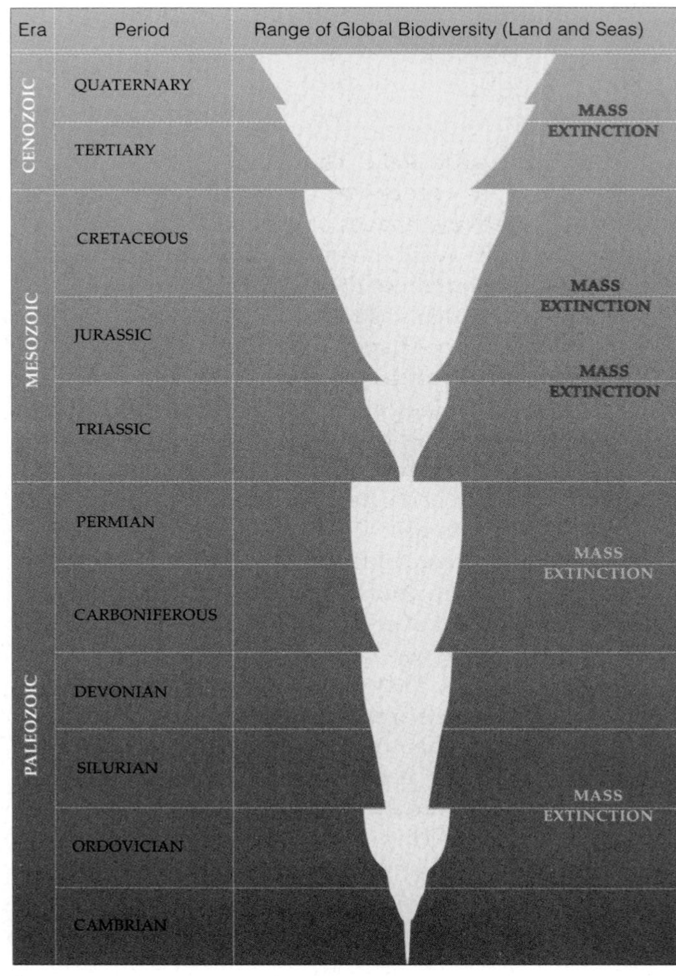

Era
Period
Range of Global Biodiversity (Land and Seas)
CENOZOIC
MESOZOIC
PALEOZOIC
QUATERNARY
TERTIARY
CRETACEOUS
JURASSIC
TRIASSIC
PERMIAN
CARBONIFEROUS
DEVONIAN
SILURIAN
ORDOVICIAN
CAMBRIAN
MASS EXTINCTION
MASS EXTINCTION
MASS EXTINCTION
MASS EXTINCTION
MASS EXTINCTION

a

CENOZOIC
MESOZOIC
PALEOZOIC

(mammalian diversity for ten families studied)

PROTOZOANS TRILOBITES BONY FISHES AMPHIBIANS REPTILES BIRDS MAMMALS

CENOZOIC
MESOZOIC
PALEOZOIC

GYMNOSPERMS ANGIOSPERMS (flowering plants) INSECTS

b

Figure 28.2 (**a**) Review of the range of global diversity through geologic time, on land and in the seas, taking into account five of the greatest mass extinctions and subsequent slow recoveries. (**b**) Within the framework of this generalized graph, patterns of extinction and recoveries differed significantly for different groups of organisms, as these selected examples indicate.

Further reading: Student Guide to InfoTrac on web site

a

Figure 28.3 Two extinct and one threatened species. (**a**) Charles Knight's magnificent painting of *Tylosaurus*, one of the mosasaurs. This powerful marine lizard flourished in shallow, nearshore waters of the Cretaceous seaways. Indonesia's Komodo dragon, which grows 10 feet (3 meters) long, is a living relative. We think the Komodo dragon is a giant, yet imagine meeting up with a mosasaur. Some were 40 feet (12 meters) long.

(**b**) The dodo (*Raphus cucullatus*), a large, flightless bird that evolved on Mauritius, then vanished more than 300 years ago. Certain trees (*Calvaria major*) also evolved on that island, as did tortoises. Did the tree coevolve with dodos, tortoises, or both? Either may have fed on the tree's large fruits. For example, the large dodo gizzard could have partially digested the thick-walled coat of seeds inside fruits, just enough to help the seeds germinate after leaving the gut. Either way, after the dodo became extinct, the seeds stopped germinating. Only thirteen trees were still standing by the mid-1970s, and by some estimates they are more than 300 years old. Each year they still produce seeds, but these apparently cannot break out of the seed coats without help. Today, botanists employ turkeys or gem polishers as substitute grinders.

b

What causes mass extinctions? The answer may not be obvious. For example, considerable evidence suggests that many lineages never made it past the K–T boundary (Section 21.1). Many believe "the asteroid did it." It did deliver the coup de grace for some lineages, including the dinosaurs and mosasaurs (Figure 28.3*a*). But other factors were at work. Biodiversity had been declining for 10 million years. Pterosaurs were gone before the K–T event. The event had little effect on insects. Was tectonic change a factor? When a land mass destined to become Australia was being rafted away from Antarctica, deep, cold currents from the south were able to move into the warmer, equatorial seas—where many groups were hit. Seawater composition, sea levels, and the climate itself shifted. The consequences affected life on land as well as in the seas. The K–T asteroid impact did indeed end some lineages abruptly. But for others, it may simply have been the final blow in a long streak of bad luck.

Now extend this thought about hidden causes of extinctions to individual species. Long ago, Dutch sailors clubbed to death every last dodo, a flightless bird that lived only on Mauritius. After that, a tree species also native to the island simply stopped reproducing. It was not until the 1970s that a hypothesis emerged: If the tree depended intimately on a coevolved species that became extinct, then the tree would be vulnerable to extinction, also. Was its partner the dodo? Maybe (Figure 28.3*b*).

Biodiversity is greater now than it has ever been in the past.

The current range of global biodiversity is an outcome of an overall pattern of mass extinctions and slow recoveries in the history of life. Within that pattern, lineages differ in which member species persisted and which became extinct.

The loss of individual species, as well as mass extinctions, may have obvious or complicated causes.

28.2 THE NEWLY ENDANGERED SPECIES

No biodiversity-shattering asteroids have hit the Earth for 65 million years. *Yet the sixth major extinction event is under way.* Throughout the world, human activities are swiftly driving many species to extinction. Think of the mammalian lineage, which outlasted the dinosaurs. About 2 million years ago, early humans started to hunt mammals in earnest. About 11,000 years ago, they started to encroach on mammalian habitats in a big way; with agriculture, domesticated species became favored at the expense of wild stock. Think of the once-stupendous herds of bison, which nearly vanished in the 1800s. As adventurers moved westward, they shot too many bison for sport, sometimes from the platforms of trains.

Only 4,500 or so species of mammals made it to the present. Of those, 300 (including most whales, wild cats, otters, and primates other than humans) have the bad luck to be enrolled in the endangered species club. An **endangered species** is an endemic species extremely vulnerable to extinction. *Endemic* means it originated in only one geographic region and lives nowhere else.

We are just one species among millions. Yet because of our rapid population growth, we are threatening others by way of habitat losses, species introductions, overharvesting, and illegal wildlife trading. For the past forty years, combinations of these factors have raised extinction rates. Figure 28.4 shows the most threatened regions. In time, nature probably will again heal itself. But recoveries after mass extinctions are known to take millions of years. We certainly will not be around to congratulate ourselves on not having caused irrevocable harm, if and when healing is complete.

Habitat Loss and Fragmentation

Edward O. Wilson defines **habitat loss** as the physical reduction in suitable places to live, as well as the loss of suitable habitat as an outcome of chemical pollution. Habitat loss is one of the major threats to more than 90 percent of endemic species facing extinction.

Biodiversity is greatest in the tropics. There, habitats on land are assaulted daily by deforestation, a topic you will read about in Sections 51.4 and 51.5. Grasslands, wetlands, and freshwater and marine habitats are also under attack. Section 28.3 gives one example. As other examples, 98 percent of the tallgrass prairie, 50 percent of the wetlands, and 85–95 percent of old-growth forests of the United States are gone.

Habitats also may become chopped up into isolated patches. Such **habitat fragmentation** has three effects on biodiversity. First, it increases a habitat's boundaries (edges), so species are more vulnerable to predators, temperature changes, winds, fires, and disease. Second, the patches might not be large enough to support the population sizes required for breeding. (Generally, each tenfold decrease in habitat area leads to a 50 percent reduction in the number of species.) Third, patches may not have enough food and other resources to sustain a population of the species.

ISLAND BIOGEOGRAPHY AND HABITAT ISLANDS Section 48.10 gives insight into factors that shape biodiversity on islands, which represent but a fraction of the Earth's surface. Nevertheless, about half of the plant and animal

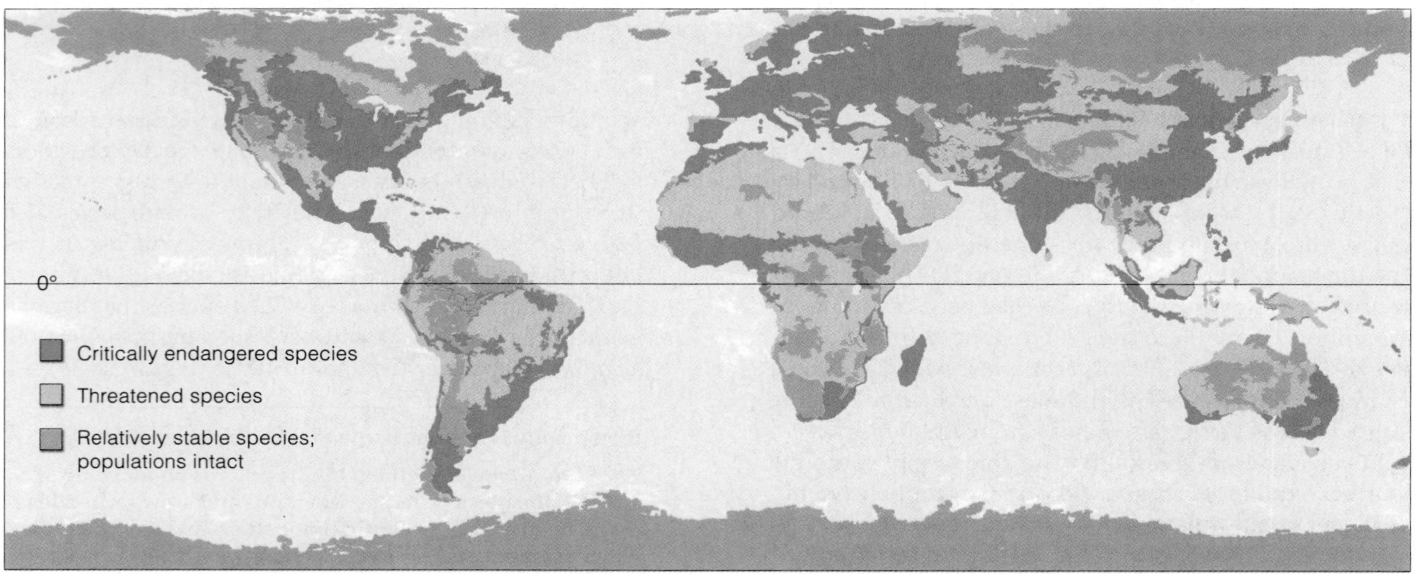
0°
Critically endangered species
Threatened species
Relatively stable species; populations intact

Figure 28.4 Threats to biodiversity: major regions of the world in which species are critically endangered, threatened, or relatively stable, as projected from 1998 through 2018.

Further reading: Student Guide to InfoTrac on web site

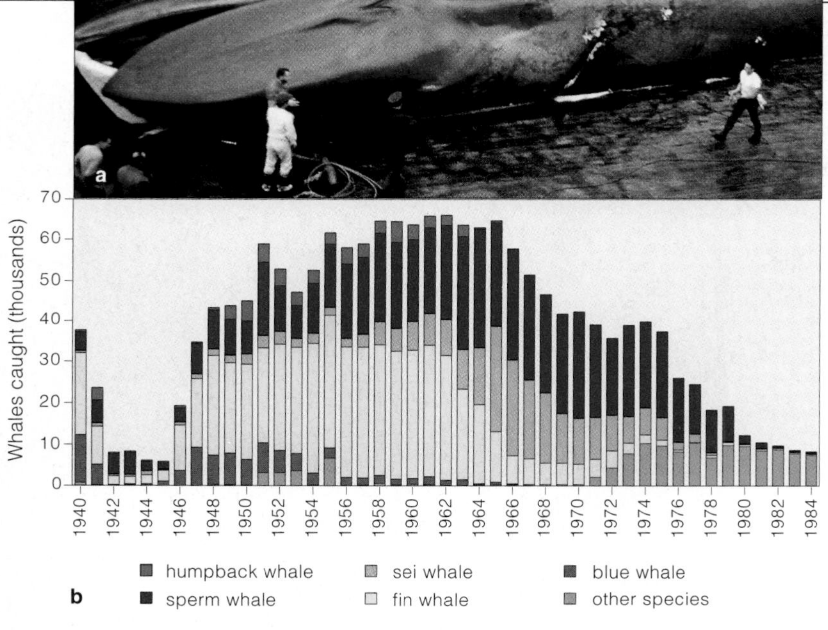

a
Whales caught (thousands)
70
60
50
40
30
20
10
0
1940
1942
1944
1946
1948
1950
1952
1954
1956
1958
1960
1962
1964
1966
1968
1970
1972
1974
1976
1978
1980
1982
1984
b
humpback whale
sperm whale
sei whale
fin whale
blue whale
other species

c

Figure 28.5 (**a**) Blue whale being butchered in British Columbia. (**b**) Graph of whale harvesting that reflects the declining numbers of great whales, now at the brink of extinction. (**c**) Confiscated products made of endangered species. All but 10 percent of the illegal wildlife trade, which threatens more than 600 species of animals and plants, goes undetected. North American examples and their black market prices: live large saguaro cactus ($5,000–15,000), bighorn sheep head ($10,000–60,000), polar bear ($6,000), grizzly ($5,000), bald eagle ($2,500), and peregrine falcon ($10,000). Other examples: live chimpanzee ($50,000), live mountain gorilla ($150,000), live imperial Amazon macaw ($30,000), Bengal tiger hide ($100,000), snow leopard hide ($14,000), and rhinoceros horn ($28,600 per kilogram).

species that have become extinct since 1600 were native to islands, mainly because they had nowhere else to go.

R. MacArthur and Wilson have developed a model of island biogeography as a way to estimate the number of current and future extinctions in regions surrounded by logging, mining, urbanization, and other destructive activities. National parks, tropical forests, reserves, and lakes are examples. Think of them as **habitat islands** in a sea of unsuitable habitat. By the MacArthur–Wilson model, destruction of 50 percent of an island habitat or a habitat island will drive about 10 percent of endemic species to extinction. A 90 percent loss will drive about 50 percent of endemic species to extinction.

INDICATOR SPECIES Conservation biologists often study **indicator species**, which provide warning of changes in habitat and impending widespread loss of biodiversity. Birds are an example. Different types live in all major land regions and climate zones, they respond quickly to changes in habitats, and they are relatively easy to track.

Think of the 700 migratory species that make annual trips to and from their summer breeding grounds in North America. Many live most of the year in tropical forests. Biologists surveyed sixty-four species between 1978 and 1987. The population sizes of forty-four insect-eating, migratory songbird species declined. For twenty species, it plummeted 25 to 45 percent. Factors in the decline were deforestation of winter habitats and fragmentation of summer habitats. Intrusion of farms, freeways, and suburbs also cut forests into patches. In North America, this made it easier for skunks, opossums, squirrels, raccoons, and blue jays to feast on eggs and juveniles of migratory songbirds. About 69 percent of the 9,600 known species of birds are now facing habitat loss.

Other Contributing Factors

Whether by accident or deliberate importation, a species sometimes moves from its home range and successfully insinuates itself into a different habitat. Each habitat has only so many resources, and its occupants compete for their share. Introduced species have adaptations that make them highly competitive, and they displace one or more endemic species. You will read about these *exotic* species in Sections 48.7 and 48.8. For now, it is enough to know that species introductions are a major factor in almost 70 percent of the cases where endemic species are being driven to extinction.

Overharvesting also is reducing biodiversity. Whales are an example (Figure 28.5*a*). Humans have killed them for food, lubricating oil, fuel, cosmetics, fertilizer, even corset stays. Substitutes exist for all whale products. Yet fishermen of several nations still ignore a moratorium on slaughtering large whales, and they routinely cross boundaries of sanctuaries set aside for their recovery.

Finally, in a sad commentary on human nature, the more rare a wild animal becomes, the more its value soars in the black market. Figure 28.5*b* gives examples.

An endangered species is any endemic species extremely vulnerable to extinction.

For the past forty years, human activities have raised rates of extinction through habitat losses, species introductions, overharvesting, and illegal wildlife trading.

Habitat loss is the physical reduction in suitable living spaces and habitat closure by chemical pollution.

CASE STUDY: THE ONCE AND FUTURE REEFS

Coral reefs are wave-resistant formations that develop mainly in clear, warm waters between latitudes 25° north and south. All show spectacular biodiversity. All suffer losses from hurricanes, warming of ocean currents, and other natural calamities, but usually they recover within a few decades. Reefs also suffer from human assaults and are less likely to recover from them.

Each coral reef began with the accumulated remains of countless organisms. Hard parts of corals became the structural foundation. Coralline algae contributed to it; deposits of calcium and magnesium carbonate hardened the cell walls of these red algae, including *Corallina* (Figure 28.6). Secretions from other organisms helped cement things together.

The massive, pocketed spine of a present-day reef is home to hundreds of species of corals, and a staggering variety of red algae and other organisms. Figure 28.6 merely hints at the wealth of warning colors, spines, tentacles, and stealthy behavior—clues to dangers and fierce competition for resources among species that are packed together in limited space.

Dinoflagellates often live as symbionts in tissues of reef-building corals (Section 26.5). In return for protection, they provide a coral polyp with oxygen and recycle its mineral wastes. When stressed, the polyps expel their protistan symbionts. When stressed for more than a few months, they die, and only bleached hard parts remain.

Abnormal, widespread bleaching in the Caribbean and tropical Pacific began in the 1980s. So did increases in sea surface temperature, which may be a major stress factor. Is the damage one outcome of human activities that are contributing to global warming (Section 49.9)? If so, as marine biologists Lucy Bunkley-Williams and Ernest Williams suggest, the future looks grim for reefs.

Human activities are destroying reefs in more direct ways than this. For example, pollutants such as raw sewage are being discharged into the nearshore waters around populated islands. Massive oil spills, as occurred during the Persian Gulf war, have calamitous effects. So do dredging operations and mining for coral rock.

CORALLINE ALGA

Figure 28.6 A small sampling of the stunning biodiversity of tropical reefs, such as the one shown at lower left.

coral reef
island
lagoon
open ocean

Where commercial fishermen from Japan, Indonesia, and Kenya move in, reef life is being decimated. No simple nets for these fellows. They drop dynamite in the water, so fish hiding among the coral are blasted out and float dead to the surface. They also squirt sodium cyanide into the water to stun fish, which float to the surface. Some survivors end up as tropical fish in pet stores. The endangered species are transported to Asian restaurants, where they are dispatched and served up as exorbitantly expensive status symbols. On small, native-owned islands, fishing rights are traded for paltry sums. Then fishermen destroy the reefs, which can no longer sustain the small human populations that once depended on them for survival.

Reef biodiversity is in great danger around the world, in places extending from Australia and Southeast Asia to such distant places as the Hawaiian Islands, Galápagos Islands, the Gulf of Panama, Florida, and Kenya. Reef formations off Florida's Key Largo have already been diminished by one-third, mostly since 1970.

Further reading: Student Guide to InfoTrac on web site

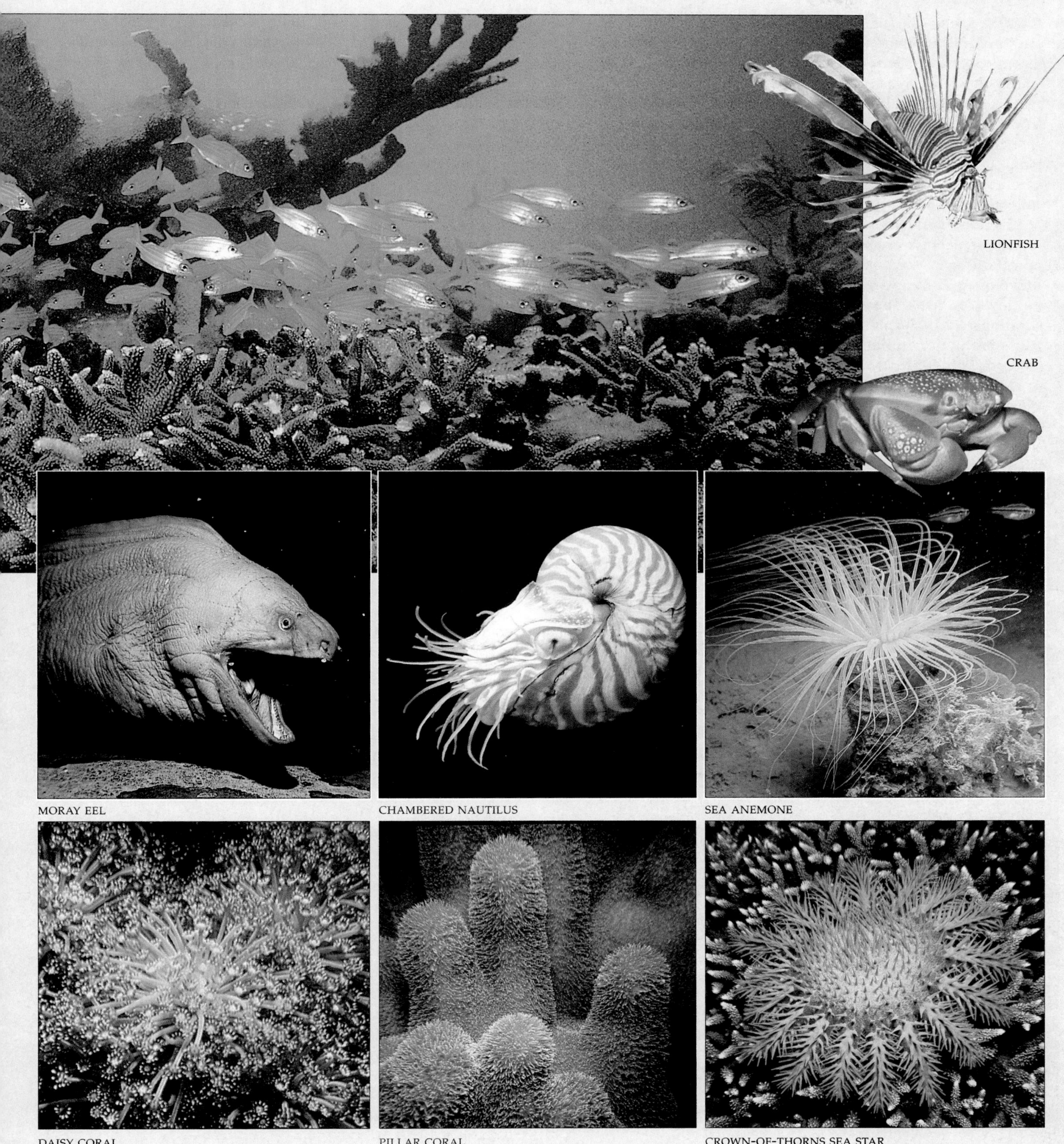

LIONFISH

CRAB

MORAY EEL

CHAMBERED NAUTILUS

SEA ANEMONE

DAISY CORAL

PILLAR CORAL

CROWN-OF-THORNS SEA STAR

Focus on Science

28.4 RACHEL'S WARNING

In 1951 Rachel Carson (Figure 28.7), an oceanographer and marine biologist, wrote a book about human impact on the ocean. *The Sea Around Us* was translated into thirty-two languages and sold more than 2 million copies. Seven years later, officials happened to spray mosquito-controlling DDT near the home and private bird sanctuary of one of her friends, who became agitated over the subsequent agonizing deaths of several birds. That friend begged Carson to find someone who might study the effects of pesticides on birds and other wildlife.

Carson discovered that independent, critical research on the subject was almost nonexistent. She conducted an extensive survey of the literature and then methodically developed information about the harmful effects of the widespread use of pesticides. In 1962, she published her findings. *Silent Spring*, the book's title, was an allusion to the silencing of "robins, catbirds, doves, jays, wrens, and scores of other bird voices" following pesticide exposure.

Many scientists, politicians, and policymakers read *Silent Spring*, and the public embraced it. Manufacturers of chemicals viewed Carson as a threat to their booming pesticide sales, and they swiftly mounted a campaign to discredit her. Critics and industry scientists claimed the book was full of inaccuracies, made selective and biased use of research findings, and failed to balance the picture with an account of the benefits of pesticides. Some even claimed that, as a woman, Carson simply was incapable of understanding such a highly technical subject. Others charged she was a hysterical woman and a radical nature lover trying to scare the public in order to sell books.

During those intense attacks, Carson knew she had terminal cancer. Yet she strongly defended her research and successfully countered her critics. She died in 1964—eighteen months after publication of *Silent Spring*—without knowing that her efforts would be the start of what is now known as the environmental movement in the United States. The new field of conservation biology is one outgrowth of that movement.

Figure 28.7 Rachel Carson, who helped awaken the public's interest in human impact on nature. She died without knowing she helped start the environmental movement.

28.5 CONSERVATION BIOLOGY

All countries can count their wealth in three forms: material, cultural, and biological. Until recently, many undervalued their biological wealth—*their biodiversity*, which is a source of food, medicine, and other goods. No one knows how much countries have already lost.

For instance, in the 1970s, a Mexican college student poking around in Jalisco discovered *Zea diploperennis*, a wild species of maize. Unlike domesticated corn, it resists disease and lives for more than one growing season. Gene transfers from the wild species to *Z. mays* have the potential to enormously boost production of corn for hungry people in Mexico and elsewhere. That species was growing in a mountain habitat no larger than ten hectares (twenty-five acres). The student saved it in the nick of time. A week later, humans started to clear the habitat with machetes and controlled burns. They would have pushed *Z. diploperennis* into oblivion.

There is a monumental problem. In the next fifty years, human population size may reach 9 billion, with most growth proceeding in the developing countries. Like all organisms, each of those 9 billion people must have raw materials, energy, and living space. How many species will they crowd out? In Wilson's words, "An awful symmetry . . . binds the rise of humanity to the fall of biodiversity: the richest nations preside over the smallest and least interesting biotas, while the poorest nations, burdened by explosive population growth and little scientific knowledge, are stewards of the largest."

Awareness of the impending extinction crisis gave rise to **conservation biology**. We define this field of pure and applied research as (1) a systematic survey of the full range of biological diversity, (2) an attempt to decipher the evolutionary and ecological origins of diversity, and (3) an attempt to identify methods that might maintain and use biodiversity for the good of the human population. Its goal is to conserve and use, in sustainable fashion, as much biodiversity as possible.

The Role of Systematics

Realistically, we simply don't have enough time, money, or people to complete a global survey. Possibly 99 percent of existing species have not even been discovered, and we don't know where most of them live. Thus initial work has focused on identifying **hot spots**. These are habitats with many species found nowhere else and in greatest danger of extinction because of human activities.

At the first survey level, researchers target a limited area, such as an isolated valley. A complete survey is out of the question, so they inventory birds, mammals, fishes, butterflies, and other indicator species for the habitat. At the next level, broader areas that are major or multiple hot spots are systematically explored. The widely separated forests of Mexico are an example of a

Further reading: Student Guide to InfoTrac on web site

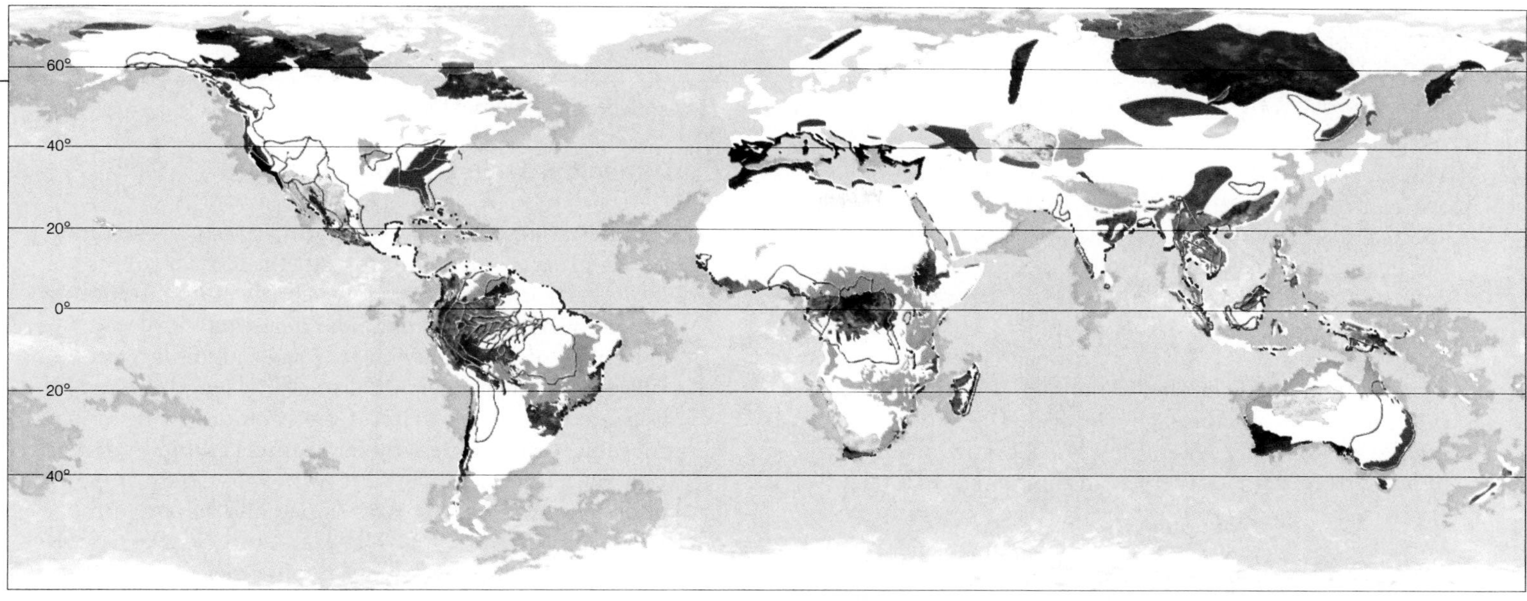
60°
40°
20°
0°
20°
40°

Figure 28.8 One map of the most vulnerable areas of land and seas, compiled by the World Wildlife Fund.

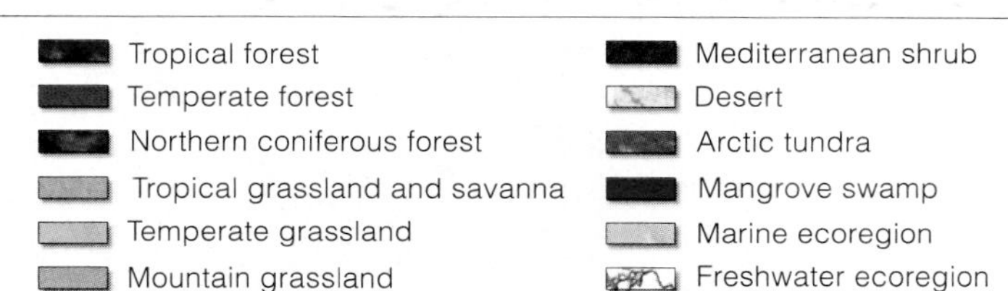
Tropical forest
Temperate forest
Northern coniferous forest
Tropical grassland and savanna
Temperate grassland
Mountain grassland
Mediterranean shrub
Desert
Arctic tundra
Mangrove swamp
Marine ecoregion
Freshwater ecoregion

major hot spot. Research stations are set up at different latitutes and elevations across the region.

At the highest survey level, hot spot inventories are combined with existing information on biodiversity of ecoregions. An **ecoregion** is a broad land or ocean region defined by climate, geography, and producer species. For example, Figure 28.8 is a work in progress by the World Wildlife Fund. Currently this organization has identified 233 regions crucial to global biodiversity (136 land, 36 fresh water, and 61 marine). Of these, it selected 25 for immediate conservation efforts. Such efforts on behalf of biodiversity will be refined over the coming decades.

Bioeconomic Analysis

The systematic expansion of species inventories serves as a baseline for economic analysis—that is, for assigning future value to ecoregions. This concept is something new in human history. For example, when the ancestors of Polynesians journeyed eastward from Southeast Asia, they literally ate their way across the Pacific, through New Zealand, Tonga, and Easter Island, all the way to the Hawaiian Islands. Flightless birds, eagles, and other endemic species that had no evolutionary experience with humans didn't stand a chance. Another wave of extinction accompanied the expansion of Paleo-Indians throughout the Americas. What counted in the past, as counts today, is food on the table for self and family. It counted for the Mexican truckdriver who said, after shooting one of the two remaining imperial woodpeckers on Earth, "It was a great piece of meat."

Probably, short-sighted thinking can be countered only by showing economically pressed individuals that sustaining biodiversity has more economic value than destroying it. Should governments—and individuals—protect a habitat, withdraw resources in a sustainable way, or obliterate it? Answers will require analysis of what a threatened species may offer. For example, some may be sources of new medicines and other chemical products. Developing countries do not have advanced laboratories for chemical analysis. Large pharmaceutical companies do. One of the largest has been paying Costa Rica's National Institute of Biodiversity to collect and identify promising species and send in chemical samples extracted from them. If natural chemicals end up being marketed, Costa Rica will get a share of the royalties, which are earmarked for conservation programs.

Sustainable Development

Ultimately, biodiversity will be best protected when its species can be used over the long term for the good of local economies. Identifying and implementing such uses is easier said than done. The technical problems and social barriers are huge. However, there are a few success stories, as the next section describes.

Conservation biology entails a systematic survey of the full range of biodiversity, analysis of its evolutionary and ecological origins, and identification of methods to maintain and use biodiversity for the benefit of humans.

Researchers identify local hot spots (limited areas with many endemic species in greatest danger of extinction by human activities), then broader or multiple hot spots, which they combine into a global picture of biodiversity.

To counter economic demands of the human population, the future economic value of biodiversity must be determined.

RECONCILING BIODIVERSITY WITH HUMAN DEMANDS

The fact remains that, when people locked in poverty are confronted with the choice of saving endangered species or themselves, the species will lose. Deprived of the education and resources that people in developed countries take for granted, they destroy tropical forests. They hunt animals and dig up plants in "protected" parks and reserves. They try to raise crops and herds on marginal land. They will take the last tree, whale, or primate, the last imperial woodpecker. We can expect people to stop only when they are shown ways to earn a living from the biodiversity in threatened habitats.

uncut forest
cut 1 year ago
dirt road
cut 3–5 years ago
cut 6–10 years ago
uncut forest
stream in watershed

Figure 28.9 Strip logging, a practice that might sustain biodiversity even as it enhances the regeneration of economically valued forest trees.

A Case for Strip Logging

Section 51.4 describes the staggering pace of destruction of once-great tropical rain forests, the richest sources of biodiversity on land. For now, simply think about one concept of how to minimize the destruction.

Besides providing wood for local economies, trees of tropical rain forests yield diverse, exotic woods prized by developed countries. What if they could be logged in a profitable, sustainable way that preserved biodiversity? Gary Hartshorn was the first to propose **strip logging** in portions of forests that are sloped and have a number of streams (Figure 28.9). The idea is to clear a narrow corridor that parallels contours of the land, and use the upper part of it as a road to haul away logs. After a few years, saplings start to grow in the corridor. Another corridor is cleared above the road. Precious nutrients that runoff leaches from the exposed soil trickle into the first corridor. There they are taken up by saplings, which benefit from all the nutrient input by growing more rapidly. Later, a third corridor is cut above the second one —and so on in a profitable cycle of logging, which the habitat sustains over time.

Ranching and the Riparian Zones

Look again at the Figure 28.8 map, and you sense at once that the biodiversity crisis is not confined to Third World countries. Also take a look at Figure 28.10, which shows a riparian zone in the western United States. Each **riparian**

Figure 28.10 Riparian zone along the San Pedro River in Arizona, shown before and after restoration.

Further reading: Student Guide to InfoTrac on web site

zone is a relatively narrow corridor of vegetation along a stream or river. Its plants afford a major line of defense against flood damage by sponging up water during spring runoffs and summer storms. Shade from the canopy of taller shrubs and trees helps conserve water during droughts. Riparian zones also offer food, shelter, and shade for wildlife, especially in arid and semiarid regions. For example, in the western United States, 67 to 75 percent of the endemic species must spend all or part of their life cycle in riparian zones. They include more than 136 species of songbirds, some of which nest only in riparian vegetation.

Cattle raised in the American West provide beef for much of the human population. Compared with wild ungulates, cattle drink a lot more water, so they tend to congregate at rivers and streams. There they trample and feed until grasses and herbaceous shrubs are gone. It takes only a few head of cattle to destroy riparian vegetation. All but 10 percent of the riparian vegetation of Arizona and New Mexico has already disappeared, mainly into the stomachs of grazing cattle.

Restricted access and development of watering sites for cattle away from streams can help conserve riparian zones. So can the rotation of livestock and provision of supplemental feed at different grazing areas. Cattle ranchers resist implementing such measures. Putting in more fencing, for example, is costly.

Jack Turnell is one rancher who knows almost as much about riparian biodiversity as he knows about cattle. Turnell runs cattle on his 32,000-hectare (80,000-acre) ranch south of Cody, Wyoming, and on 16,000 hectares (40,000 acres) of Forest Service land for which he has grazing rights. Ten years after he took over the ranch, he decided to raise cattle in an unconventional way. He began systematically rotating cattle away from the riparian areas. He also crossed Hereford and Angus cattle with a breed from France that does not consume as much water. He made most of his ranching decisions in consultation with specialists in range and wildlife management.

Willows and other plants are sustaining diversity in the restored zones. And Turnell's ranching approach is profitable; grasses maintained in the riparian zones help put more meat on his cattle. His may be only a small step toward sustainable ranching, but it appears to be a step in the right direction.

Throughout much of the world, the unprecedented rate of human population growth is the elephant in the room that no one wants to talk about.

Protecting biodiversity depends on finding ways for people to make a living from it without destroying it. Can such a balance be struck among so many billions of people?

28.7 SUMMARY

1. Global biodiversity is greater now than ever, but the current rate of species losses is high enough to suggest that an extinction crisis is under way.

2. The history of life indicates that, after global mass extinctions, 20 million to 100 million years pass before biodiversity recovers to the preceding level.

a. Different taxa have different evolutionary histories. Some were eliminated by major extinction events. Others passed through them relatively unscathed.

b. Mass extinctions, and extinction of single species, may have obvious or hidden, complicated causes.

3. Growth of the human population, which is projected to reach 9 billion within the next fifty years, is causing the sixth extinction crisis. That growth is most rapid in regions with the richest, most vulnerable biodiversity.

4. Habitat losses, habitat fragmentation, introduction of species into novel habitats, overharvesting, and illegal wildlife trading threaten endemic species. "Endemic" means it originated in and is presently restricted to one geographic region. An endangered species is defined as an endemic species extremely vulnerable to extinction.

5. Habitat loss refers to physical reduction or chemical pollution of suitable places for species to live. Habitat fragmentation is carving a habitat into isolated patches. It puts species at risk, as by chopping up populations to sizes that cannot promote successful breeding.

6. Island biogeography models help predict the number of current and future extinctions. A habitat island is a habitat in a sea of possibly destructive human activities, such as logging. Generally, destruction of 50 percent of an island habitat (or habitat island) will drive one-tenth of its species to extinction. Destruction of 90 percent will drive one-half to extinction.

7. Indicator species are birds and other easily tracked species that can provide warning of changes in habitats and impending widespread loss of biodiversity.

8. Humans are directly destroying coral reefs and other regions, as when commercial fishermen dynamite reefs to harvest fish. They may be destroying them indirectly, as by contributing to global warming (with concurrent rises in sea surface temperatures and sea levels).

9. Conservation biology is a field of pure and applied research. Its goal is to conserve and use biodiversity in sustainable ways. The field encompasses:

a. A systematic survey at three ever more inclusive levels of the full range of biodiversity.

b. Analysis of biodiversity's origins in evolutionary and ecological terms.

c. Identification of methods that might maintain and use biodiversity for the good of the human population, which may otherwise destroy it.

10. Today the systematic survey of conservation biology is proceeding at three levels:

a. Local hot spots (for example, an isolated valley) are identified and indicator species inventoried. The hot spots are habitats with many species in greatest danger of extinction because of human activities.

b. Major hot spots (or multiple ones) are inventoried. Research stations are set up across these broader areas to gather data by latitude and elevation.

c. Data gathered at the first two levels are combined with data on ecoregions; these are the most vulnerable, broad regions of land and seas throughout the world.

d. Data gathered for such maps will be refined into an increasingly detailed picture of global biodiversity.

11. Protecting biodiversity depends on finding ways for people to make a living from it without destroying it. To counter growing economic demands, biodiversity's future economic value must be determined. Methods by which local economies can exploit that biodiversity in sustainable ways must be developed.

12. When people are given a choice between saving a species or themselves, that species will lose. Currently, millions of people are making such choices because of their demands for food, shelter, and material goods.

Review Questions

1. How many major mass extinctions have occurred, including the one that is now under way? *28.1*
2. List four human activities that are major contributors to the current extinction crisis. Which threaten coral reefs? *28.2, 28.3*
3. Define endangered, endemic, and indicator species. *28.2*
4. Distinguish between island habitat and habitat island. *28.2*
5. State the goal of conservation biology and briefly describe its three-pronged approach to achieving that goal. *28.5*
6. Define hot spot. Why are conservation biologists focusing on hot spots rather than quickly completing a global survey? *28.5*

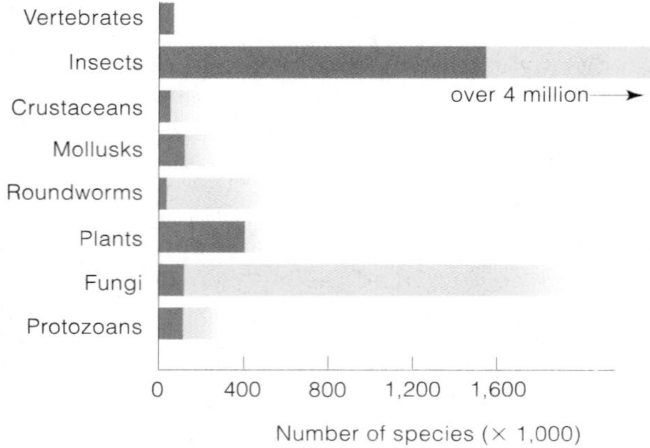

Vertebrates
Insects
Crustaceans
Mollusks
Roundworms
Plants
Fungi
Protozoans
over 4 million
0
400
800
1,200
1,600
Number of species (× 1,000)

Figure 28.11 Current species diversity for a few major taxa. *Red* indicates the number of named species; *gold* indicates the estimated number of species not yet discovered.

Self-Quiz (Answers in Appendix III)

1. Following mass extinctions in the past, recovery to the same level of biodiversity has taken many ______ of years.
   a. hundreds b. millions c. billions
2. Species may be driven to extinction by ______.
   a. obvious factors c. human activities
   b. complicated, hidden factors d. all of the above
3. The goal(s) of conservation biology is (are) to ______.
   a. conduct a three-level, systematic survey of all biodiversity
   b. analyze biodiversity's evolutionary and ecological origins
   c. identify ways to maintain and use biodiversity for people
   d. all of the above
4. Strip logging ______.
   a. destroys wild habitats c. sustains forests
   b. is profitable d. both b and c

Critical Thinking

1. Figure 28.11 compares the current estimated number of species for some major taxa. Given what you learned about taxa from chapters in this unit, which are most vulnerable in the current extinction crisis? Which are likely to pass through it with much of their biodiversity intact? List some reasons why (for example, compare the distribution of member species).

2. Visit or study a riparian zone near where you live. Imagine visiting it five years from now, then ten years after that. Given its location, what kinds of changes do you predict for it?

3. Mentally transport yourself to a tropical rain forest of South America. Imagine you don't have a job. There are no jobs in sight, not even in overcrowded cities some distance away. You have no money or contacts to get you anywhere else. Yet you are the sole supporter of your large family. Today a stranger approaches you. He tells you he will pay good money if you can capture alive a certain brilliantly feathered parrot in the forest. You know the parrot is rare; it is seldom seen. However, you have an idea of where it lives. What will you do?

4. In his review of this chapter, ecologist Robin Tyser offered these comments: Students might find the current biodiversity crisis too overwhelming to contemplate. But people are making a difference. For example, they helped reestablish viable bald eagle populations in the continental United States and wolves in northern Wisconsin. Daniel Janzen is working to recreate a dry forest ecosystem in Costa Rica. Do a literature search or computer research and report on one or two success stories that you personally might consider inspiring.

Selected Key Terms

conservation biology *28.5*
coral reef *28.3*
ecoregion *28.5*
endangered species *28.2*
habitat fragmentation *28.2*
habitat island *28.2*
habitat loss *28.2*
hot spot *28.5*
indicator species *28.2*
riparian zone *28.6*
strip logging *28.6*

Readings

See also www.infotrac-college.com

Cox, G. 1999. *Alien Species in North America and Hawaii: Impacts on Natural Ecosystems.* Washington, D.C.: Island Press.

Wilson, Edward O. 1992. *The Diversity of Life.* Cambridge, Massachusetts: Belknap Press. This slender book proves that Wilson is one of the most eloquent defenders of life's diversity.

FACING PAGE: *A flowering plant (Prunus) busily doing what it does best: producing flowers for the fine art of reproduction.*

29

PLANT TISSUES

Plants Versus the Volcano

On a clear spring day in 1980, in a richly forested part of the Cascade Range of southwestern Washington, Mount Saint Helens exploded and 500 million metric tons of ash were blown skyward. Within minutes, shock waves from the blast flattened or incinerated hundreds of thousands of mature trees growing near the northern flanks of the mountain.

Rivers of hot volcanic ash surged down the slopes at rates exceeding forty-four meters per second. They rapidly turned into rivers of cementlike mud when the intense heat from the blast melted and released more than 75 billion liters of water that had been locked up in the mountain's snowfields and glacial ice.

In one mind-numbing moment, about 40,500 hectares—100,000 acres—of magnificent forests dominated by hemlock and Douglas fir were transformed into barren sweeps of land (Figure 29.1*a*,*b*). In the aftermath of the violent eruption, more than a few observers gained stunning insight into what the world must have looked like long ago, before the first plants started to colonize habitats on land.

Figure 29.1 (**a**,**b**) Grim reminder of what the world would be like without plants: the aftermath of the violent eruption of Mount Saint Helens in 1980. Nothing remained of the extensive forest that had surrounded this Cascade volcano. (**c**) In less than a decade, however, seed-bearing vascular plants were making a comeback. (**d**) Twelve years after the volcanic eruption, seedlings of a dominant species (Douglas fir) were starting to reclaim the land.

Yet it did not take long for existing plants to move back into habitats that their ancestors had claimed. In less than a year's time, seeds of a variety of flowering plants, including fireweed and blackberry, sprouted near the grayed trunks of fallen trees around Mount Saint Helens.

Even before ten years had passed, young willows and alders took hold near riverbanks, and low shrubs cloaked the land (Figure 29.1*c*). Their presence afforded pockets of shade, which favored germination of the seeds of slower growing but ultimately dominant species, the hemlocks and Douglas fir (Figure 29.1*d*). In less than a century, the forest will be as it once was.

With this example, we open a unit dedicated to seed-bearing vascular plants, with emphasis on the flowering types. In terms of distribution and diversity, they are the most successful plants on Earth.

This first chapter provides an overview of plant tissues and body plans. Chapter 30 explains how the seed-bearing plants absorb water and mineral ions, conserve water, and distribute organic substances among roots, stems, and leaves. Chapter 31 takes a closer look at their patterns of growth, development, and reproduction. As you will see, their structure and physiology (that is, the functioning of the plant body) help them survive sometimes hostile conditions on land—even momentary takeovers by volcanoes.

d

KEY CONCEPTS

1. Angiosperms (flowering plants) and, to a lesser extent, gymnosperms are groups that now dominate the plant kingdom. These seed-bearing vascular plants all have complex aboveground shoot systems, which consist of stems, leaves, and some other structures. They also have complex root systems that typically grow downward and outward through soil.

2. There are three major categories of tissue systems in these plants. A ground tissue system makes up the bulk of the plant body. A vascular tissue system distributes water, dissolved minerals, and photosynthetic products through roots and shoots. A dermal tissue system covers and protects plant surfaces exposed to the surroundings.

3. The simple plant tissues—parenchyma, collenchyma, and sclerenchyma—are each composed of no more than one type of cell.

4. Complex plant tissues incorporate two or more types of cells. Xylem and phloem, which are vascular tissues, are like this. So are the dermal tissues called epidermis and periderm.

5. Plants grow by way of mitotic cell divisions and cell enlargements at meristems, which are localized regions of dividing cells.

6. Each growing season, shoots and roots lengthen. The lengthening, called primary growth, originates only at apical meristems, in the tips of shoots and roots.

7. Each growing season, the shoots and roots of many plants also thicken. Typically, lateral meristems inside shoots and roots give rise to an increase in diameter, which is called secondary growth. Wood is one outcome of secondary growth.

29.1 OVERVIEW OF THE PLANT BODY

Earlier, in Chapter 25, we surveyed representatives of the 295,000 known species of plants. Even that sprint through diversity revealed why no one species can be used as a typical example of plant body plans. Even so, when we hear the word "plant," we usually think of well-known species of seed-bearing vascular plants—gymnosperms (including pine trees) and angiosperms (flower-producing plants, such as roses, corn, cactuses, and elms). With 260,000 species, angiosperms dominate the plant kingdom, and they will be our main focus.

Flowering plant life cycles extend from germination to seed formation, then death. "Annuals" complete the life cycle in a single growing season, and they generally are *nonwoody*, or herbaceous, plants. The marigolds and alfalfa are like this. "Biennials" such as carrots live for two consecutive growing seasons. Their roots, stems, and leaves form the first season; flowers form, seeds form, and the plant dies the next season. "Perennials" continue vegetative growth and seed formation year after year. Roots and stems thicken in many of them.

Shoots and Roots

Many flowering plants have a body plan similar to that shown in Figure 29.2. Its aboveground parts, or **shoots**, are stems, leaves, flowers (reproductive shoots), and other structures. Its stems afford structural support for upright growth, and some parts of the stems conduct water and solutes. Upright growth gives photosynthetic cells in young stems and leaves favorable exposure to sunlight. **Roots** are specialized structures that typically penetrate the soil and spread downward and outward through it. A root system absorbs water and dissolved minerals, and it commonly anchors aboveground parts. It also stores food, then releases it as required for root cells and for distribution to aboveground parts.

Three Plant Tissue Systems

A flowering plant's stems, branches, leaves, and roots are alike in a key respect. Each has three major tissue systems. The **ground tissue system**, the most extensive, makes up the bulk of the plant body. The **vascular tissue system** has two kinds of tissues that distribute water and solutes through the plant body. The **dermal tissue system** covers and protects the plant's surfaces. Figure 29.2 shows the general locations of these systems.

Some tissues in each system are simple, in that they have only one type of cell. Parenchyma, collenchyma, and sclerenchyma fall in this category. Other tissues are complex, with organized arrays of two or more types of cells. Xylem, phloem, and epidermis are like this.

The next sections describe the tissue organization of shoots and roots. You may find it easier to interpret the

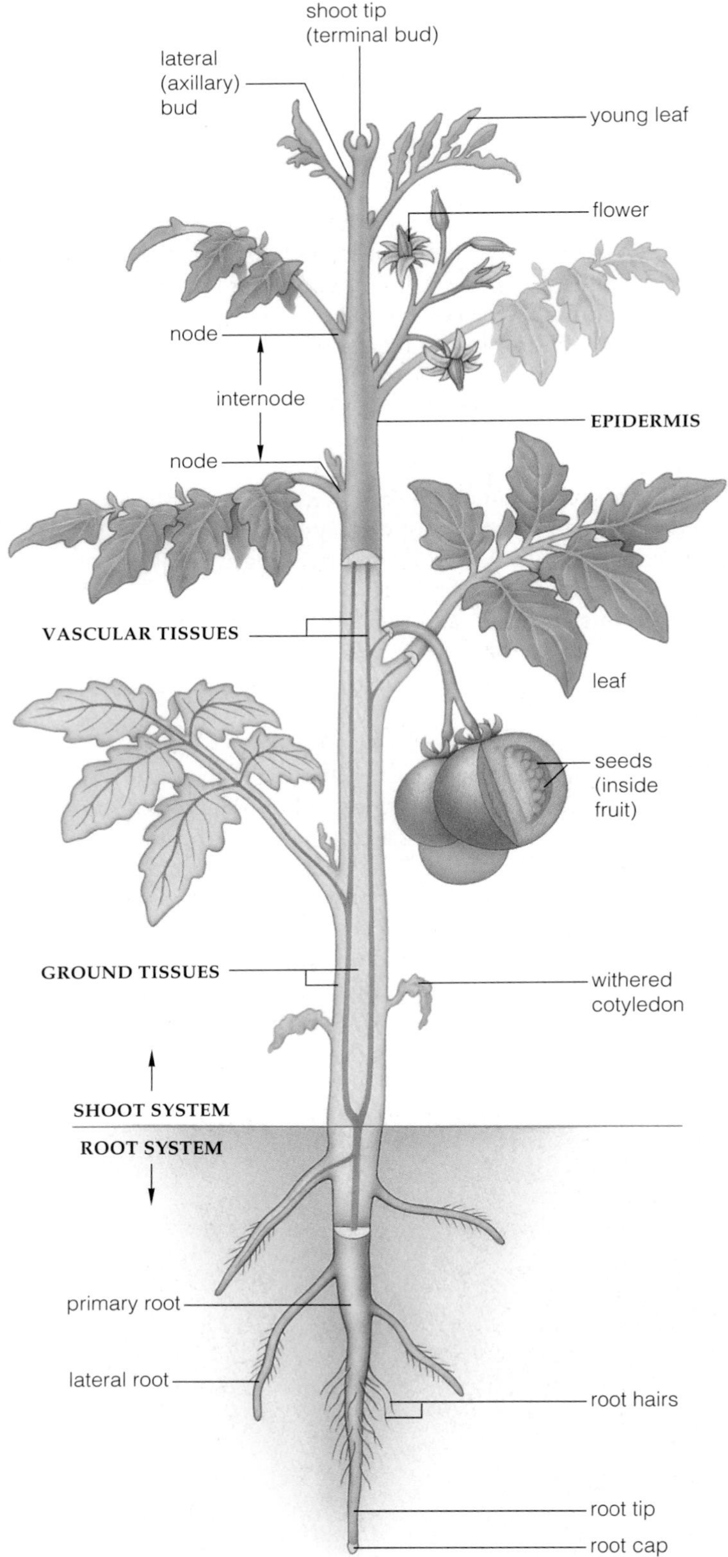

shoot tip
(terminal bud)
lateral
(axillary)
bud
young leaf
flower
node
internode
node
EPIDERMIS
VASCULAR TISSUES
leaf
seeds
(inside
fruit)
GROUND TISSUES
withered
cotyledon
SHOOT SYSTEM
ROOT SYSTEM
primary root
lateral root
root hairs
root tip
root cap

Figure 29.2 Body plan for a tomato plant (*Lycopersicon*), a typical angiosperm. Vascular tissues (*purple*) conduct water, dissolved minerals, and organic substances. They thread through ground tissues, which make up most of the plant body. A dermal tissue (epidermis in this case) covers the surfaces of both the root system and shoot system.

Further reading: Student Guide to InfoTrac on web site

Figure 29.3 Terms used to identify how tissue specimens are cut from a plant. Cuts perpendicular to a stem or root's long axis give *transverse* sections (cross-sections). Along the radius of a stem or root, longitudinal cuts give *radial* sections. Cuts at right angles to a root or a stem radius give *tangential* sections.

transverse

radial

tangential

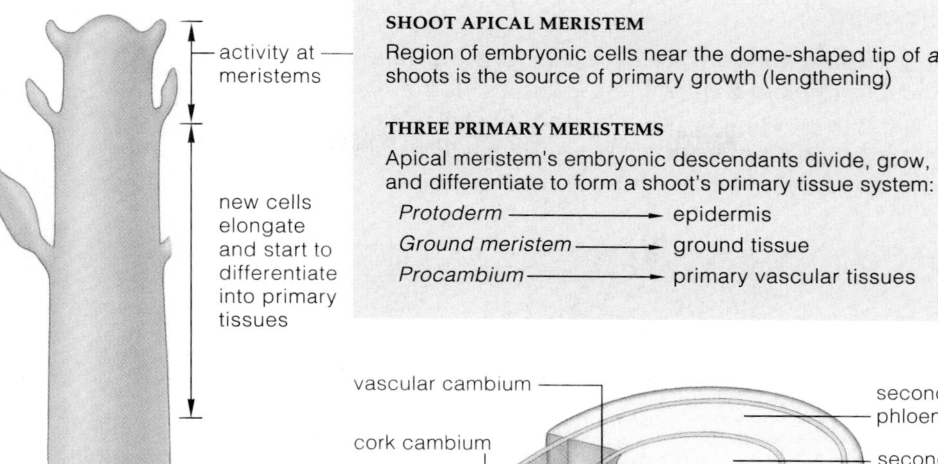

activity at meristems
new cells elongate and start to differentiate into primary tissues
SHOOT APICAL MERISTEM
Region of embryonic cells near the dome-shaped tip of *all* shoots is the source of primary growth (lengthening)
THREE PRIMARY MERISTEMS
Apical meristem's embryonic descendants divide, grow, and differentiate to form a shoot's primary tissue system:
Protoderm → epidermis
Ground meristem → ground tissue
Procambium → primary vascular tissues
vascular cambium
cork cambium
secondary phloem
secondary xylem
thickening
LATERAL MERISTEMS
Two lateral meristems in older stems and roots of woody plants produce secondary growth (increases in diameter):
Vascular cambium → secondary vascular tissues
Cork cambium → periderm (replaces epidermis)

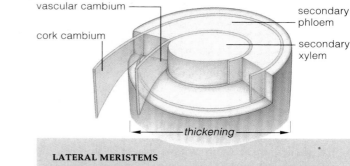

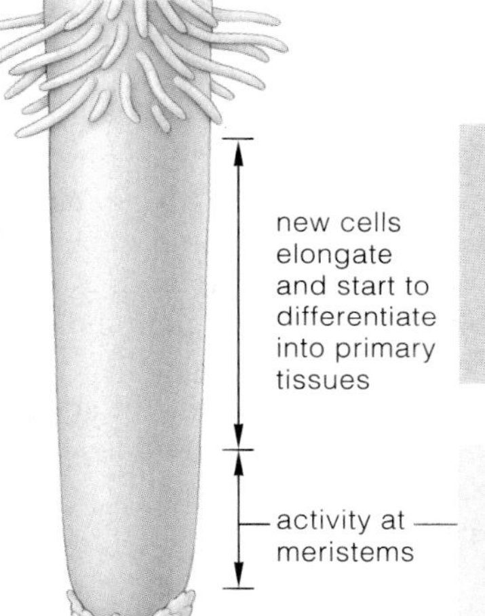

new cells elongate and start to differentiate into primary tissues
activity at meristems
ROOT APICAL MERISTEM
Apical meristem near all root tips gives rise to protoderm, ground meristem, and procambium. These transitional meristems give rise to the root's primary tissue systems—epidermis, ground tissues, and vascular tissues.

Figure 29.4 *Right:* Types of meristems that are responsible for increases in the length and diameter of the shoots and roots of a vascular plant.

micrographs taken of those tissues by studying Figure 29.3. It can help you understand the different ways in which tissue specimens are cut from plants.

Meristems—Where Tissues Originate

During a growing season, vascular plants do not grow everywhere at the same time. Most growth is confined to **meristems**, which are local regions of dividing cells. In other regions, cellular descendants of the meristems are maturing or have already reached maturity. Figure 29.4 shows the locations of meristems.

The lengthening of every shoot and root originates at **apical meristem**, located in their dome-shaped tip. Some populations that form here become the *primary* meristems known as protoderm, ground meristem, and procambium. These are immature forms of the primary tissues: epidermis, ground tissue, and vascular tissues respectively. Taken as a whole, the *lengthening* of stems and roots represents the plant's primary growth.

Also during the growing season, the older stems and roots of many plants thicken. The increases in girth start with lateral meristems, each a cylindrical array of cells that forms inside stems and roots. One lateral meristem, the **vascular cambium**, produces secondary vascular tissues. The other, **cork cambium**, gives rise to a sturdier covering (periderm) that replaces epidermis. Taken as a whole, the *thickening* of a plant's stems and roots represents secondary growth.

Vascular plants have stems that support upright growth and conduct substances, leaves that function in photosynthesis, shoots specialized for reproduction, and other structures. They also have roots that absorb water and solutes, anchor aboveground parts, and often store food.

A ground tissue system makes up most of the plant body. A vascular tissue system distributes water, dissolved ions, and photosynthetic products through it. A dermal tissue system covers and protects plant surfaces.

Shoots and roots lengthen (put on primary growth) when their apical and primary meristems are active. In many plants, older stems and roots also thicken (add secondary growth) when lateral meristems called vascular cambium and cork cambium are active.

29.2 TYPES OF PLANT TISSUES

We turn now to an overview of the organization and functions of plant tissues. Simple tissues are composed of one type of cell. The vascular and dermal tissues are complex, with a variety of cell types. Figures 29.5 through 29.9 show examples from these tissue categories.

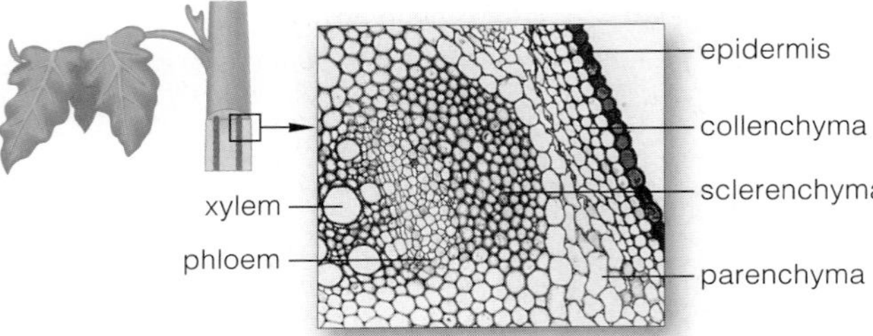

epidermis
collenchyma
sclerenchyma
parenchyma
xylem
phloem

Figure 29.5 Locations of simple tissues and complex tissues in one kind of plant stem, transverse section.

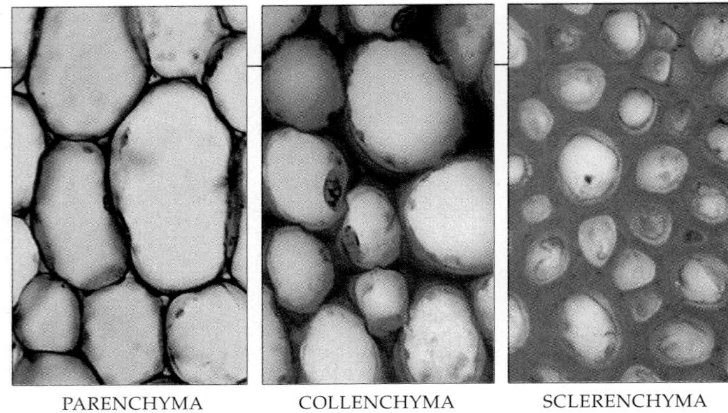

PARENCHYMA
COLLENCHYMA
SCLERENCHYMA

Figure 29.6 From the stem of a sunflower plant (*Helianthus*), examples of simple tissue, transverse section.

Simple Tissues

Tissues of **parenchyma** make up most of the soft, moist, primary growth of roots, stems, leaves, flowers, and fruits. Most parenchyma cells are pliable, thin-walled, and many-sided (Figure 29.6). When mature, they are alive and retain the capacity to divide. Often their cell divisions heal wounds in plant parts. Mesophyll, a type of parenchyma in leaves, is photosynthetic. Air spaces between photosynthetic parenchyma cells enhance gas exchange. Other types of parenchyma serve in storage, secretion, and other specialized tasks. Parenchyma cells also are components of vascular tissue systems.

Collenchyma provides flexible support for primary tissues (Figure 29.6). Its living cells often form patches or a cylindrical array near a lengthening stem's surface. They also form pliable ribs in many leaf stalks. Most of the cells are elongated, with unevenly thickened walls layered with pectin. This polysaccharide glues cellulose fibrils together and makes the tissue pliable.

Sclerenchyma supports mature plant parts and also protects many seeds. Most of its cells have thick walls impregnated with lignin (Figures 29.6 and 29.7). Lignin, recall, strengthens and waterproofs walls. Without it, plants would not have evolved on land (Section 25.1).

Sclerenchyma consists of fibers or sclereids. *Fibers* are long, tapered cells in the vascular tissue systems of some stems and leaves. They can flex and twist without stretching. We use certain fibers to make cloth, rope, paper, and other valued commodities (Figure 29.7*a*,*b*). *Sclereids* are stubbier cells. Think of a hard seed coat, coconut shell, and peach pit or a pear's gritty texture; sclereids are the source of such features (Figure 29.7*c*).

Complex Tissues

VASCULAR TISSUES Two vascular tissues, called xylem and phloem, function in the distribution of substances throughout the plant. Often their conducting cells are associated with a sheath of fibers and parenchyma cells.

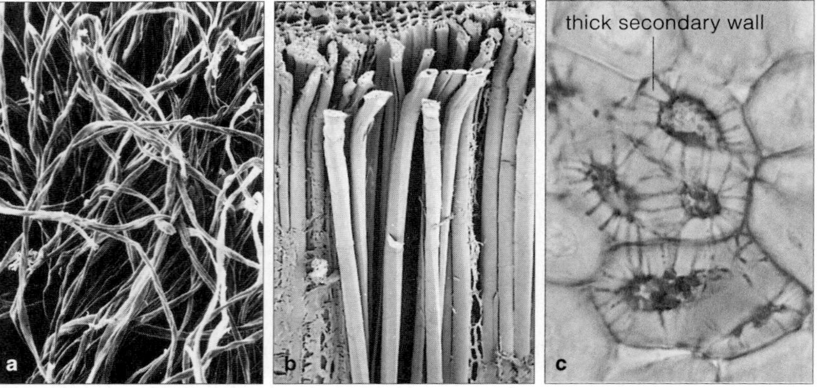

thick secondary wall
a
b
c

Figure 29.7 Examples of sclerenchyma. (**a**) From cotton plant seeds, loosely twisted fibers that can interlock and form threads when spun. (**b**) Strong fibers from flax stems. (**c**) From a pear, one type of sclereid: stone cells, each with a thick, lignified wall.

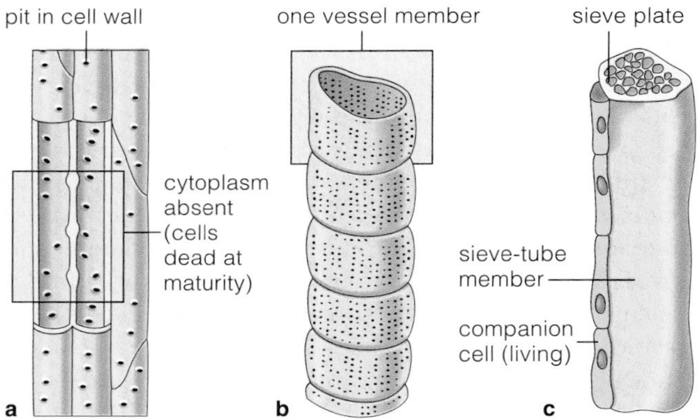

pit in cell wall
one vessel member
sieve plate
cytoplasm absent (cells dead at maturity)
sieve-tube member
companion cell (living)
a
b
c

Figure 29.8 From xylem, portions of (**a**) tracheids and (**b**) a vessel. Pipelines made of such cells conduct water and dissolved ions. (**c**) One type of phloem cell. Long tubes of many such cells conduct sugars and other organic compounds.

Further reading: Student Guide to InfoTrac on web site

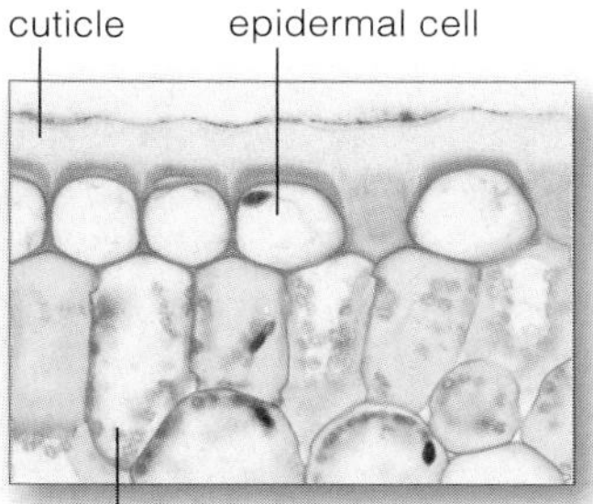
cuticle
epidermal cell
parenchyma cell inside leaf

Figure 29.9 Light micrograph of a section through the upper surface of a kaffir lily leaf. The plant cuticle is made of secretions from epidermal cells. Inside the leaf are many photosynthetic parenchyma cells.

Xylem conducts water and dissolved mineral ions. It also helps mechanically support a plant. Figure 29.8*a,b* shows examples of its conducting cells. The cells, called *vessel members* and *tracheids*, are dead at maturity, and their lignified walls interconnect. Collectively, the cell walls form water-conducting pipelines and strengthen plant parts. Water flows into and out of the adjoining cells through numerous pits in the cell walls.

Phloem conducts sugars and other solutes. Its main conducting cells, called *sieve-tube members*, are alive at maturity (Figure 29.8*c*). Adjacent cells interconnect at openings in their side walls and at open or perforated end walls. Sugars synthesized inside leaves are loaded into sieve-tube members. *Companion cells* (specialized, living parenchyma cells) commonly assist the loading process. Sugars moving through phloem's pipelines are unloaded in regions where cells are growing or storing food. The chapter to follow describes this process.

DERMAL TISSUES All surfaces of primary plant parts are covered and protected by a dermal tissue system called **epidermis**. In most plants, epidermis is mainly a single layer of unspecialized cells. Waxes and the fatty substance cutin coat the outermost cell walls. We call the surface coating a **cuticle**. A plant cuticle functions in restricting water loss and often in resisting attacks by certain microorganisms (Figure 29.9).

Stem and leaf epidermis contains many specialized cells. For example, pairs of guard cells change shape in response to changing conditions. As they do, a gap—or **stoma** (plural, stomata)—closes or opens between them. The next chapter looks at how stomata serve as control points for the movement of water vapor, oxygen, and carbon dioxide across the epidermis. **Periderm** replaces epidermis in stems and roots with secondary growth. Dead cork cells in this protective covering have a wall heavily impregnated with suberin, a fatty substance.

Dicots and Monocots—Same Tissues, Different Features

The **dicots** and **monocots**, recall, are the two classes of flowering plants (Section 25.8). Most trees and shrubs other than conifers—such as maples, elms, roses, cacti, peas, beans, lettuces, cotton, and carrots—are dicots. Palm trees, lilies, orchids, ryegrass, bamboos, wheat, corn, sugarcane, and pineapples are familiar monocots.

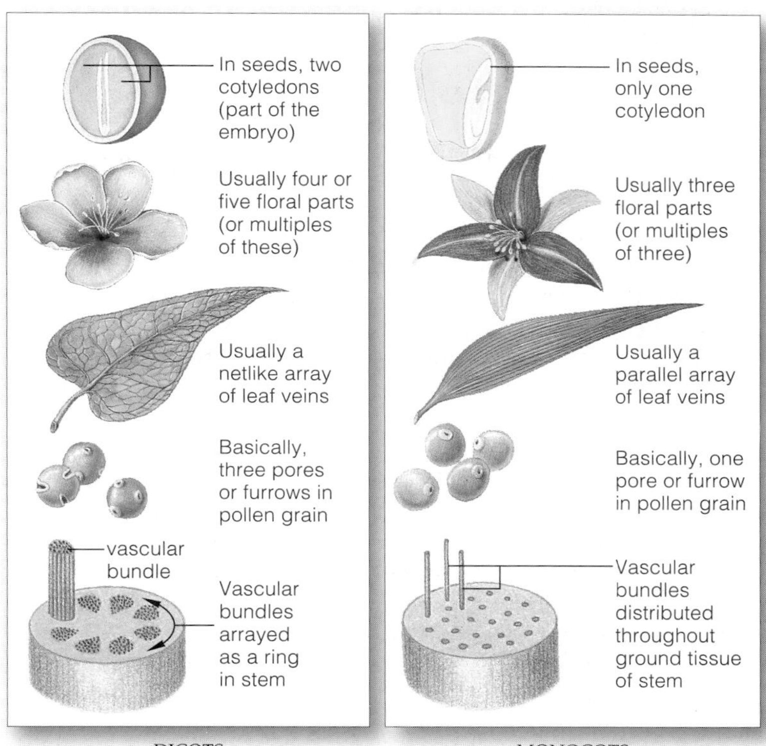
In seeds, two cotyledons (part of the embryo)
Usually four or five floral parts (or multiples of these)
Usually a netlike array of leaf veins
Basically, three pores or furrows in pollen grain
vascular bundle
Vascular bundles arrayed as a ring in stem
DICOTS
In seeds, only one cotyledon
Usually three floral parts (or multiples of three)
Usually a parallel array of leaf veins
Basically, one pore or furrow in pollen grain
Vascular bundles distributed throughout ground tissue of stem
MONOCOTS

Figure 29.10 Comparison of the defining features of dicots and monocots. Both classes of flowering plants consist of the same simple and complex tissues, but their body parts show some differences in structural organization.

Dicots and monocots are similar in structure and function, but they differ in some distinctive ways. For instance, dicot seeds have two cotyledons and monocot seeds have only one. Cotyledons are leaflike structures commonly known as seed leaves. They form in seeds as part of a plant embryo, and they store or absorb food for it. After the seed germinates, cotyledons wither and new leaves grow and start to make food. Figure 29.10 shows other differences between dicots and monocots.

Most of the plant body (ground tissue system) consists of parenchyma, collenchyma, and sclerenchyma. Each of these simple tissues is composed of only one type of cell.

Xylem and phloem are vascular tissues. In xylem, pipelines made of tracheids and vessel members conduct water and dissolved ions. In phloem, sieve-tube members interact with companion cells to distribute organic compounds.

Of two dermal tissues, epidermis covers the surfaces of the primary plant body. Periderm replaces it on plant parts that have extensive secondary growth.

Monocots and dicots consist of the same tissues, but each has some of the tissues organized in distinctive ways.

29.3 PRIMARY STRUCTURE OF SHOOTS

How Stems and Leaves Form

Next time you or a friend eats a bundle of bean sprouts or alfalfa sprouts, pull one aside to look at its structure. That seedling started forming while it was still inside a seed coat. It already has a primary root and shoot. In the primary shoot's tip, apical meristem and its derivative meristematic tissues are laying out orderly frameworks for the stem's primary structure (Figure 29.11). Beneath the apical meristem, most tissues become progressively specialized as cells divide at different rates, in different directions, and as they differentiate in size, shape, and function. The meristematic activities result in distinct stem regions, leaves, and lateral (axillary) buds from which lateral shoots develop. In turn, the lateral shoots give rise to side branches and reproductive structures.

Briefly, as a typical shoot lengthens, bulges of tissue develop along the sides of apical meristem. Each bulge is an immature leaf (Figures 29.11 and 29.12). While growth continues, the stem lengthens between tier after tier of new leaves. Each part of the stem where one or more leaves are attached is a node. As shown in Figure 29.2, the stem region between two successive nodes is an internode. Buds develop in the upper angle where leaves attach to the stem. Each **bud** is an undeveloped shoot of mostly meristematic tissue, often protected by bud scales (modified leaves). As Chapter 32 describes, buds give rise to new stems, leaves, and flowers.

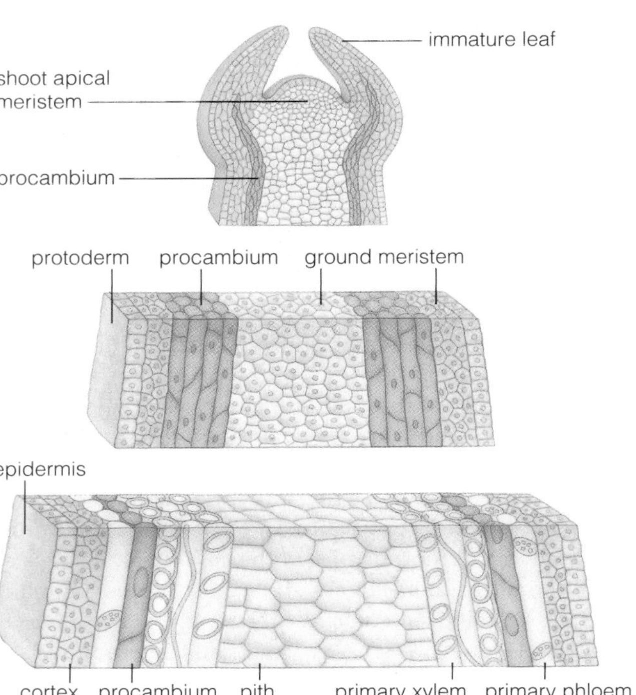

immature leaf
shoot apical meristem
procambium
protoderm
procambium
ground meristem
epidermis
cortex
procambium
pith
primary xylem
primary phloem

Figure 29.11 Successive stages in primary growth that started with activity at the shoot apical meristem of a typical dicot, then continued at the primary meristems derived from it. Notice the progressive differentiation of most of the tissue regions.

Internal Structure of Stems

While the primary plant body of a monocot or dicot is forming, ground, vascular, and dermal tissues of the stem become organized in distinctive ways. Most often, primary xylem and phloem develop inside the same sheath of cells, as **vascular bundles**. Such bundles are multistranded cords threading lengthwise through the ground tissue system of the primary and lateral shoots. They commonly develop according to two genetically dictated patterns. In most dicot stems, the long bundles are arranged as a ring that divides the ground tissue into a cortex and pith (Figure 29.13*a*). The stem's **cortex** is the region in between the vascular bundles and the epidermis. Its **pith** is the stem's center, inside the ring of vascular bundles. Ground tissue of the plant's roots becomes similarly divided, into root cortex and pith. A

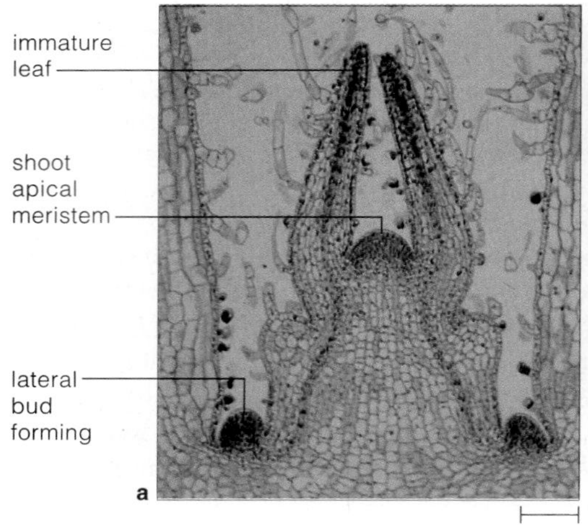

immature leaf
shoot apical meristem
lateral bud forming
a
100 µm

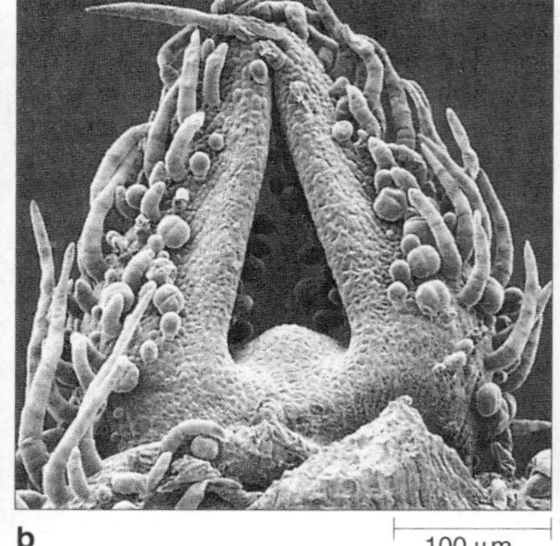

b
100 µm

c

Figure 29.12 (**a**) Light micrograph of a *Coleus* shoot tip, cut longitudinally through its center. (**b**) Scanning electron micrograph of its surface. (**c**) New *Coleus* leaves.

Further reading: Student Guide to InfoTrac on web site

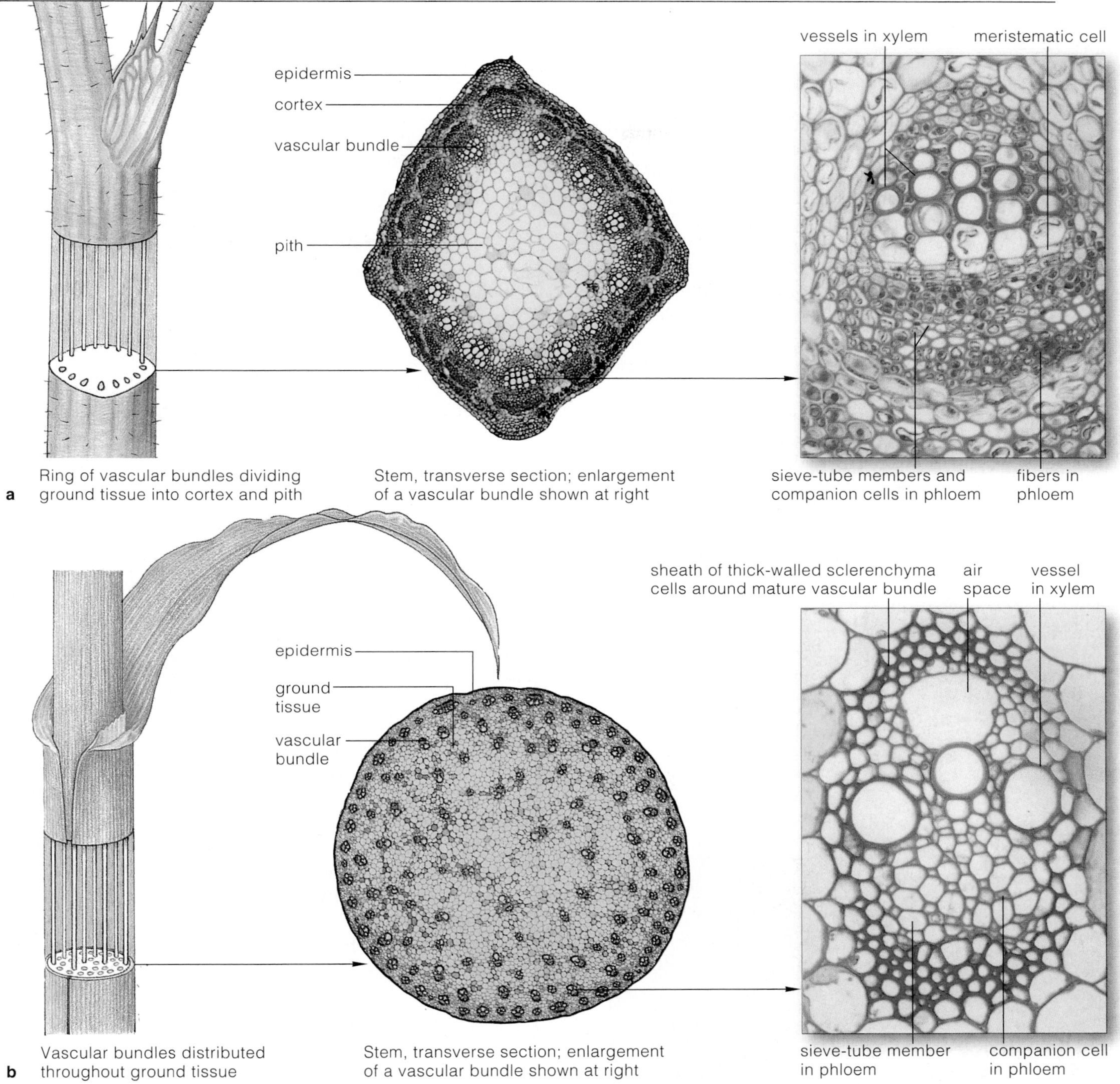

epidermis
cortex
vascular bundle
pith
vessels in xylem
meristematic cell
sieve-tube members and companion cells in phloem
fibers in phloem
a
Ring of vascular bundles dividing ground tissue into cortex and pith
Stem, transverse section; enlargement of a vascular bundle shown at right
epidermis
ground tissue
vascular bundle
sheath of thick-walled sclerenchyma cells around mature vascular bundle
air space
vessel in xylem
sieve-tube member in phloem
companion cell in phloem
b
Vascular bundles distributed throughout ground tissue
Stem, transverse section; enlargement of a vascular bundle shown at right

Figure 29.13 Internal organization of cells and tissues inside the stems from a dicot and a monocot.

(**a**) Part of a stem from alfalfa (*Medicago*), a dicot. In many species of dicots and conifers, the vascular bundles develop in a more or less ringlike array in the ground tissue system, as shown here. The portion of the ground tissue between the ring and the surface of the stem is the cortex. The portion enclosed within the ring is the pith.

(**b**) Part of a stem from corn (*Zea mays*), a monocot. In most monocots and some nonwoody dicots, vascular bundles are scattered through the ground tissue, as shown here.

different pattern is common inside the stems of most monocots and some dicots. The long vascular bundles are scattered through the ground tissue (Figure 29.13*b*). How different substances are conducted through such vascular systems is a topic of the next chapter.

Shoot apical meristem gives rise to the primary plant body, which develops a distinctive internal structure (as in the pattern in which its vascular bundles are arranged).

29.4 A CLOSER LOOK AT LEAVES

Similarities and Differences Among Leaves

Every **leaf** that forms is a metabolic factory, equipped with many photosynthetic cells. Yet leaves vary greatly in size, shape, surface details, and internal structure. A duckweed leaf is no more than 1 millimeter (0.04 inch) across; the leaves of some water lilies are 2 meters (6.5 feet) wide. Various leaves resemble blades, spikes, cups, needles, feathers, tubes, and other structures. They also differ greatly in coloration, odor, and edibility (many produce toxins). Leaves of birches and other species of *deciduous* plants wither and drop away from stems as winter approaches. The leaves of camellias and other *evergreen* plants also drop, but not all at once.

A typical leaf has a flat blade, as in Figures 29.14*a* and 29.15. It has a petiole, or stalk, that attaches it to a stem. *Simple* leaves are undivided, although many are lobed as in Figure 29.14*c*. *Compound* leaves have blades divided into leaflets oriented in the same plane (Figure 29.14*d*). Most monocot leaves, such as those of ryegrass and corn, are flat surfaced, a bit like a knife blade. The blade's base encircles a stem and thus sheathes it (Figure 29.14*b*).

The leaves of most species are thin, with a high surface-to-volume ratio. Their flat surface grows and moves itself so it becomes oriented perpendicular to rays of the sun. A plant's individual leaves commonly project from stems in patterns that minimize shading of neighboring leaves. For example, leaves of *Coleus* form in pairs on opposite sides of the stem, and the petioles of each pair attach to the stem at right angles to those of the pair above it (Figure 29.12*c*).

a
b

Figure 29.15 (**a**) Decaying dicot leaf, with its netlike veins. (**b**) Parallel veins of a monocot leaf (in this case, *Agapanthus*).

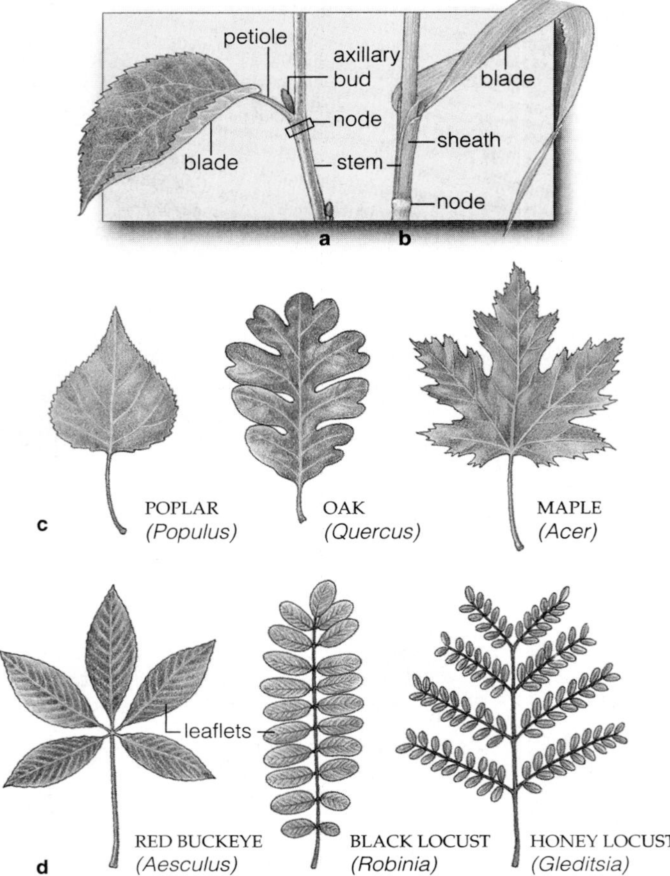

petiole
axillary bud
blade
node
sheath
blade
stem
node
a
b
c
POPLAR
(Populus)
OAK
(Quercus)
MAPLE
(Acer)
leaflets
d
RED BUCKEYE
(Aesculus)
BLACK LOCUST
(Robinia)
HONEY LOCUST
(Gleditsia)

Figure 29.14 Common leaf forms of (**a**) dicots and (**b**) monocots. Examples of (**c**) simple leaves and (**d**) compound leaves.

Such leaf adaptations afford maximum interception of sunlight, which is the plant's energy source. They also promote an inward diffusion of carbon dioxide and an outward diffusion of oxygen. When a leaf is thick, you can assume that it belongs to a succulent or some other plant of arid habitats and functions in water storage as well as in photosynthesis. Leaves of many desert plants orient themselves parallel with the sun's rays; by doing so, they reduce heat absorption to tolerable levels.

Leaf Fine Structure

In fine structure, also, each leaf is adapted to intercept sunlight energy and promote gas exchange. In addition, many leaves have distinctive surface specializations.

Further reading: Student Guide to InfoTrac on web site —

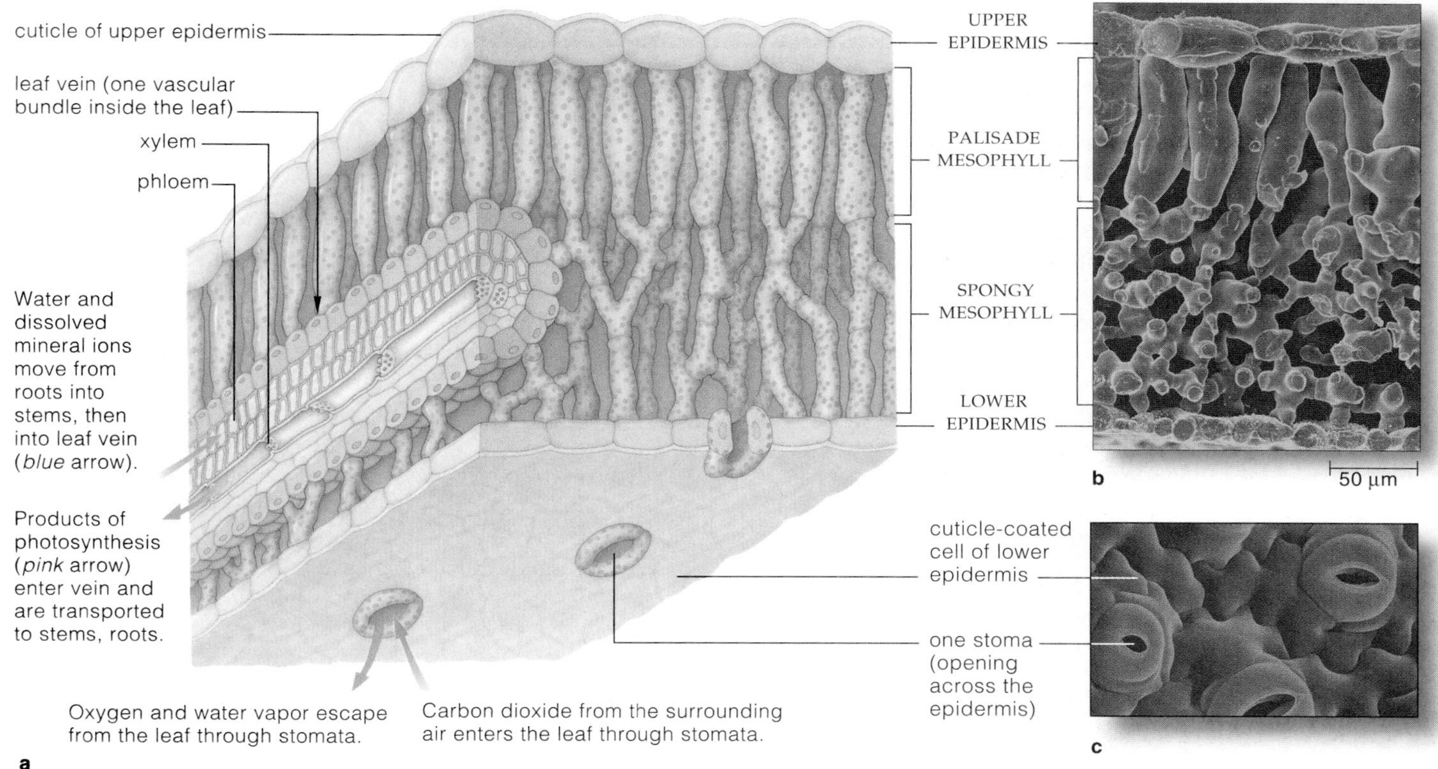
cuticle of upper epidermis
leaf vein (one vascular bundle inside the leaf)
xylem
phloem
Water and dissolved mineral ions move from roots into stems, then into leaf vein (blue arrow).
Products of photosynthesis (pink arrow) enter vein and are transported to stems, roots.
Oxygen and water vapor escape from the leaf through stomata.
Carbon dioxide from the surrounding air enters the leaf through stomata.
a
UPPER EPIDERMIS
PALISADE MESOPHYLL
SPONGY MESOPHYLL
LOWER EPIDERMIS
b
50 µm
cuticle-coated cell of lower epidermis
one stoma (opening across the epidermis)
c

Figure 29.16 (**a**) Diagram of leaf structure for many kinds of flowering plants. (**b**) Scanning electron micrograph of the tissue organization of a leaf from the kidney bean plant (*Phaseolus*). Notice the compact organization of epidermal cells. (**c**) Stomata, each a tiny opening across the epidermis, appear when paired guard cells are in their plumped configuration.

LEAF EPIDERMIS Epidermis covers every leaf surface exposed to the air. It may be smooth surfaced, sticky, or slimy, with hairs, scales, spikes, hooks, glands, and other surface specialties. The Chapter 30 introduction has two fine examples. Covering the sheetlike, compact array of epidermal cells is a cuticle, which helps minimize the loss of precious water (Figures 29.9 and 29.16). Most leaves have far more stomata on the lower surface than on their upper surface. Water lily leaves are one of the exceptions. Stomata of plants growing in arid habitats commonly are located inside depressions in the leaf's surface, along with thickly coated epidermal "hairs." Their location and the hairs help restrict water loss.

MESOPHYLL—PHOTOSYNTHETIC GROUND TISSUE As you read earlier, **mesophyll**, a type of parenchyma with photosynthetic cells and a large volume of air spaces, extends throughout the interior of a leaf. Figure 29.16 shows an example. The air spaces, which connect with outside air through stomata, promote rapid diffusion of carbon dioxide to cells and oxygen away from them. Leaves that are oriented perpendicular to the sun have two layers of mesophyll. Columnar parenchymal cells are attached to the upper epidermis. They seem densely packed, but most of their cell walls are exposed to the air. These are cells of *palisade* mesophyll; they have far more chloroplasts and engage in more photosynthesis, compared to the cells making up the layer of *spongy* mesophyll below them (Figure 29.16). Monocot leaves grow vertically and intercept light from all directions; their mesophyll is not organized as two layers.

VEINS—THE LEAF'S VASCULAR BUNDLES Leaf **veins** are vascular bundles, often strengthened with fibers. Their continuous strands of xylem rapidly move water and dissolved nutrients to all of the mesophyll cells, and the continuous strands of phloem carry photosynthetic products—especially sugars—away from them. In most dicots, the veins branch lacily into a number of minor veins that are embedded inside the mesophyll. In most monocots, veins are more or less similar in length and run parallel with the leaf's long axis (Figure 29.15).

A leaf's structure is adapted for sunlight interception, gas exchange, and distribution of water, dissolved nutrients, and photosynthetic products. Leaves of each species have a characteristic size, shape, and often surface specializations.

29.5 PRIMARY STRUCTURE OF ROOTS

Taproot and Fibrous Root Systems

When you think you can't bear one more gratuitously violent action-hero movie, give it a rest and watch a seed germinate. The first part to emerge from the seed coat is a **primary root**. In most dicot seedlings, this root increases in diameter while it grows downward. Later, **lateral roots** form in internal tissue, perpendicular to the primary root's axis, then erupt through epidermis. The youngest lateral roots are closest to the root tip. A **taproot system** is a primary root and lateral branchings. Dandelions, carrots, oak trees, and poppies are among the plants with a taproot system (Figure 29.17*a*).

By comparison, the primary root of most monocots, such as rye and other grasses, is short-lived. However, in its place, adventitious roots arise from the stem, then lateral roots branch from these. (*Adventitious* structures form at an unusual location.) All of the lateral roots are more or less alike in diameter and length. Collectively, roots that form this way are a **fibrous root system** (Figure 29.17*b*).

Internal Structure of Roots

Figure 29.18 shows the meristems in the tip of one root of such systems. Many of the cellular descendants of these meristems divide, enlarge, elongate, and become cells of primary tissue systems. Notice the root cap, a dome-shaped mass of cells at the tip. Apical meristem produces the cap, which in turn protects the meristem.

Protoderm gives rise to the root epidermis, a plant's absorptive interface with the soil. Some epidermal cells send out extensions called **root hairs** (Figure 29.18*a*). Collectively, the root hairs greatly increase the surface area available for taking up water and nutrients. Only

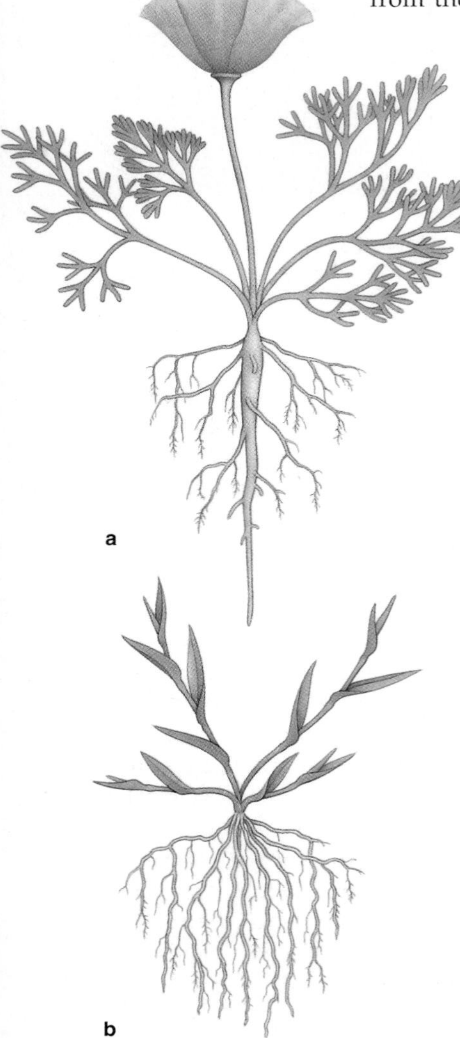
a
b

Figure 29.17 (**a**) Taproot system of a California poppy (*Eschscholzia*). (**b**) Fibrous root system of a grass plant.

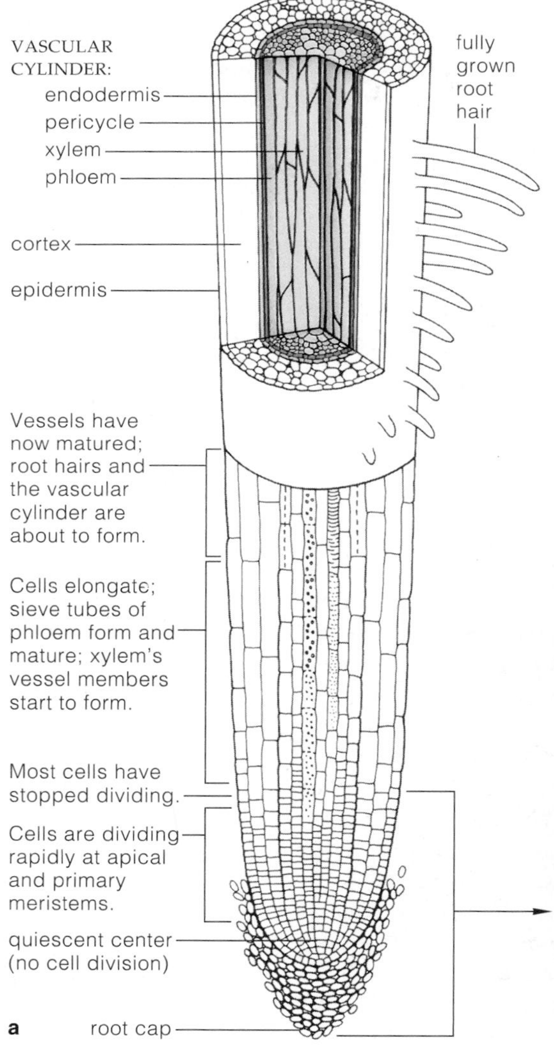
VASCULAR CYLINDER:
endodermis
pericycle
xylem
phloem
cortex
epidermis
fully grown root hair
Vessels have now matured; root hairs and the vascular cylinder are about to form.
Cells elongate; sieve tubes of phloem form and mature; xylem's vessel members start to form.
Most cells have stopped dividing.
Cells are dividing rapidly at apical and primary meristems.
quiescent center (no cell division)
a
root cap

b
100 µm

Figure 29.18 (**a**) Generalized sketch of a primary root, showing the location of the zones where cells divide, then elongate, then differentiate. (**b**) This is a light micrograph of a root tip of corn (*Z. mays*), longitudinal section. The oldest cells are farthest from apical meristem, which a root cap protects. Root cap cells secrete mucigel, a polysaccharide-rich slime that lubricates the root as cell divisions push it through the soil. Mucigel ends up covering all of the root's epidermis. It may enhance the uptake of dissolved mineral ions and the formation of mycorrhizae.

Further reading: Student Guide to InfoTrac on web site

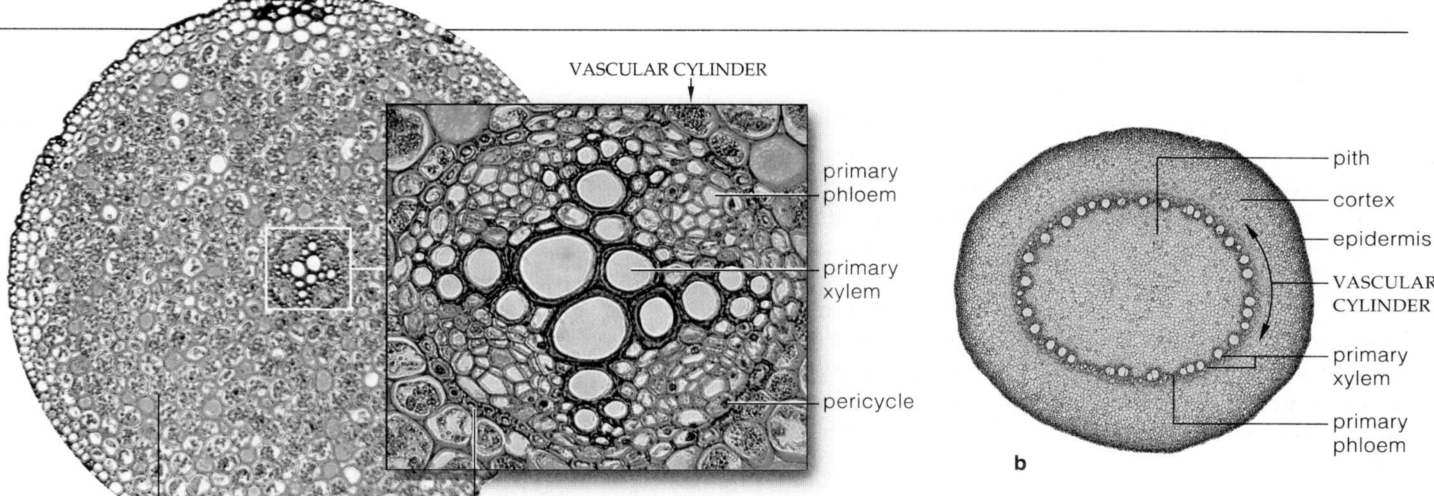
VASCULAR CYLINDER
primary phloem
primary xylem
pericycle
endodermis
epidermis
root cortex
a
pith
cortex
epidermis
VASCULAR CYLINDER
primary xylem
primary phloem
b

Figure 29.19 (**a**) Young root of a buttercup (*Ranunculus*), transverse section. The inset is a detail of its vascular cylinder. (**b**) Root of corn (*Z. mays*), transverse section. Its vascular cylinder divides the ground tissue into cortex and pith.

first-time or foolish gardeners would yank a plant from the ground when transplanting it. They would tear off too much of the highly fragile absorptive surface.

Apical meristem also gives rise to the ground tissue system and to a **vascular cylinder**. A vascular cylinder consists of primary xylem and phloem and one or more layers of cells called the pericycle. Figure 29.19*a* shows an example of a vascular cylinder at the center of the cortex inside a dicot root. Figure 29.19*b* shows how one monocot's vascular cylinder divides the ground tissue system into cortex and pith regions. Either way, there are plenty of air spaces in between cells of the ground tissue system, and oxygen can easily diffuse through them. Like other cells in the plant, the living root cells depend on oxygen for aerobic respiration.

When water enters a root, it moves from cell to cell until it reaches the endodermis, a layer of cells around the vascular cylinder. Where endodermal cells abut, their walls are waterproofed, so incoming water is forced to pass through their cytoplasm. As described in Chapter 30, this arrangement controls the movement of water and dissolved substances into the vascular cylinder.

The pericycle lies just inside the endodermis. Its cells divide repeatedly and form lateral roots, which then erupt through the cortex and epidermis (Figure 29.20).

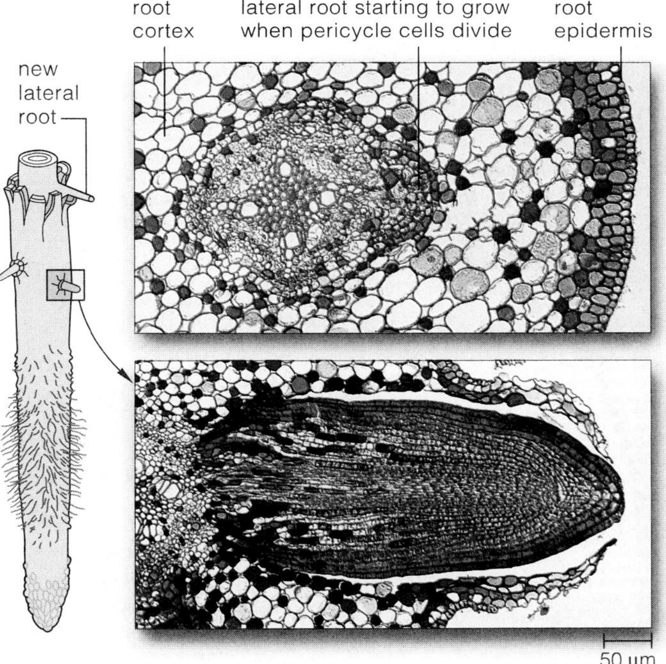
root cortex
lateral root starting to grow when pericycle cells divide
root epidermis
new lateral root
50 µm

Figure 29.20 Lateral root formation in a primary root from a willow tree (*Salix*), transverse section.

Regarding the Sidewalk-Buckling, Record-Breaking Root Systems

Unless tree roots start to buckle a sidewalk or choke off a sewer line, most of us don't pay much attention to the root systems of flowering plants. Roots mine the soil for water and minerals, and most reach a depth of 2 to 5 meters. In hot deserts, where free water is scarce, one hardy mesquite shrub is known to have sent its roots down 53.4 meters (175 feet) near a streambed. Some "simple" cacti have shallow roots radiating outward for 15 meters. Someone once measured the roots of a young rye plant that had been growing for four months in only 6 liters of soil water. If the surface area of that root system were laid out as one sheet, it would occupy more than 600 square meters!

Primary roots provide a plant with a tremendous surface area for absorbing water and solutes.

ACCUMULATED SECONDARY GROWTH—THE WOODY PLANTS

Woody and Nonwoody Plants Compared

Flowering plant life cycles extend from germination to seed formation, then death. **Annuals** complete the life cycle in a single growing season, and they generally are *nonwoody*, or herbaceous, plants. Alfalfa and marigolds are like this. **Biennials** such as carrots can live for two growing seasons. Their roots, stems, and leaves form the first season; flowers form, seeds form, and the plant dies in the next season. **Perennials** continue vegetative growth and seed formation year after year. Secondary tissues form in a number of them.

Like all gymnosperms, some of the monocots and many dicots add secondary growth during two or more growing seasons; they are *woody* plants. Early in life, their stems and roots are similar to those of nonwoody plants. Differences emerge after their lateral meristems become active and start producing large amounts of secondary vascular tissues, especially secondary xylem. The differences are especially pronounced in some of the perennial plants in which the vascular cambium has been reactivated every growing season for hundreds or thousands of years. Such ongoing meristematic activity has produced giants. For example, at last measure, the massive trunk of a coast redwood of the sort shown in Figure 29.21*a* was towering more than 110 meters above the forest floor. Its accumulated secondary growth has been estimated to weigh close to 100 metric tons. The tree with greatest girth is a chestnut (*Castanea*) growing in Sicily. To walk completely around the base of it, you would have to pace off 58 meters.

Activity at the Vascular Cambium

Massive, woody stems and roots originate at the lateral meristem called vascular cambium. Take a look at the stem in Figure 29.21*b*. Every spring, primary growth resumes at its buds, and secondary growth proceeds inside it. When vascular cambium in the stem is fully developed, it is like a cylinder one cell or a few cells thick. Some cells (*fusiform initials*) give rise to secondary xylem and phloem, which extend longitudinally through

a

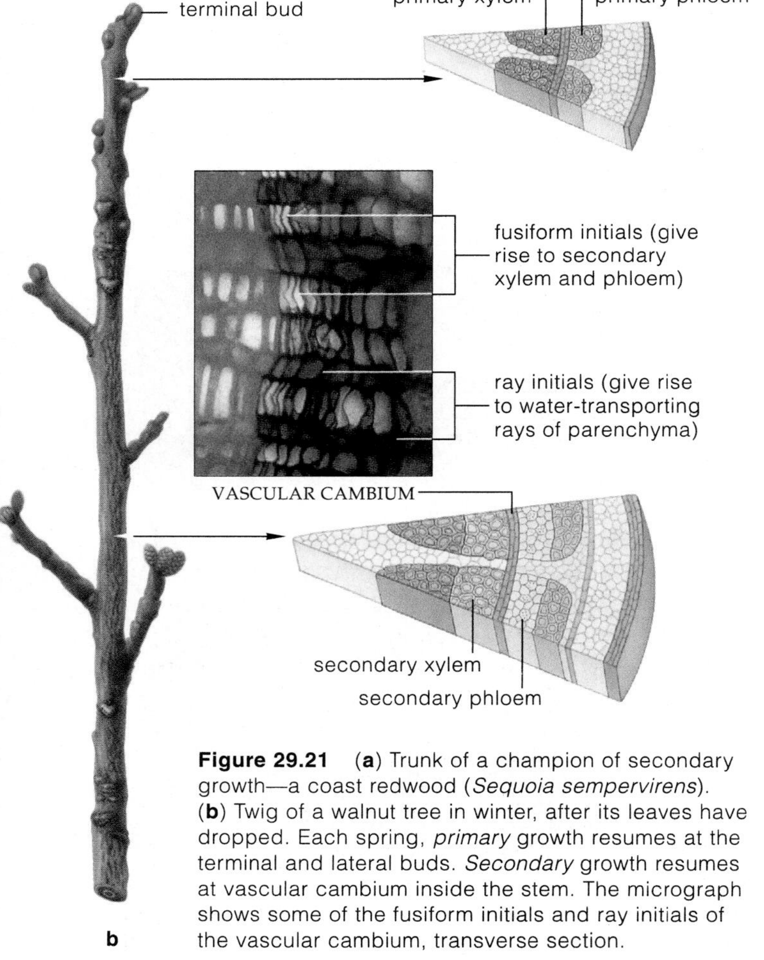
terminal bud
primary xylem
primary phloem
fusiform initials (give rise to secondary xylem and phloem)
ray initials (give rise to water-transporting rays of parenchyma)
VASCULAR CAMBIUM
secondary xylem
secondary phloem
b

Figure 29.21 (**a**) Trunk of a champion of secondary growth—a coast redwood (*Sequoia sempervirens*). (**b**) Twig of a walnut tree in winter, after its leaves have dropped. Each spring, *primary* growth resumes at the terminal and lateral buds. *Secondary* growth resumes at vascular cambium inside the stem. The micrograph shows some of the fusiform initials and ray initials of the vascular cambium, transverse section.

Further reading: Student Guide to InfoTrac on web site

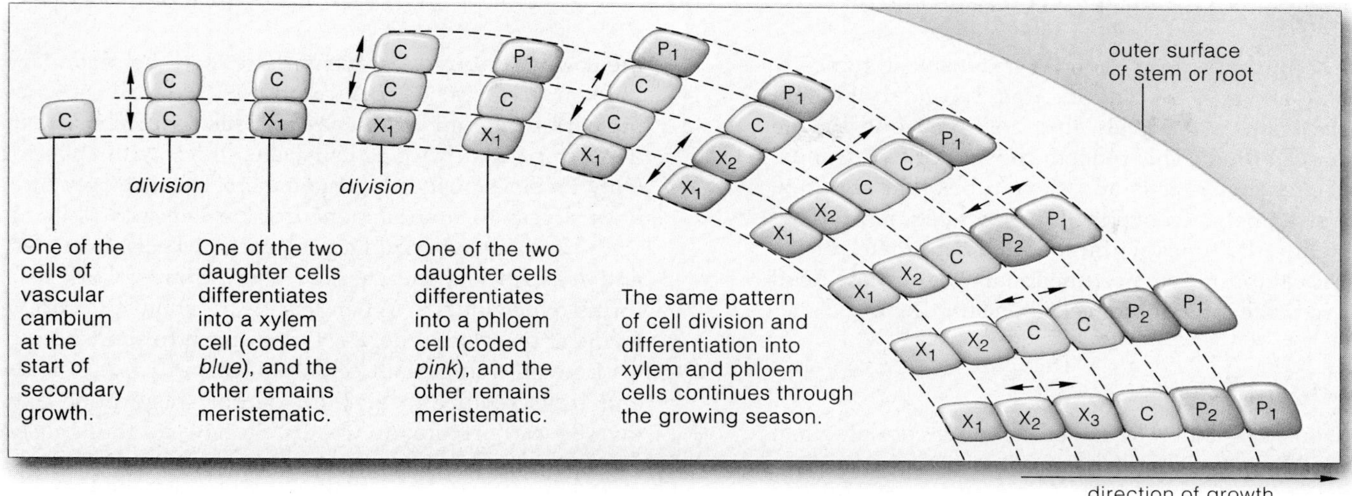
outer surface of stem or root
C
C
C
C
X1
C
C
X1
P1
C
X1
P1
C
X1
P1
C
X2
X1
P1
C
C
X2
X1
P1
P2
C
X2
X1
P1
P2
C
C
X2
X1
P1
P2
C
X3
X2
X1
division
division
One of the cells of vascular cambium at the start of secondary growth.
One of the two daughter cells differentiates into a xylem cell (coded blue), and the other remains meristematic.
One of the two daughter cells differentiates into a phloem cell (coded pink), and the other remains meristematic.
The same pattern of cell division and differentiation into xylem and phloem cells continues through the growing season.
direction of growth

Figure 29.22 Composite drawing of activity at the vascular cambium inside a woody stem. Ongoing cell divisions enlarge the inner core of secondary xylem and displace the vascular cambium toward the surface of the stem or root.

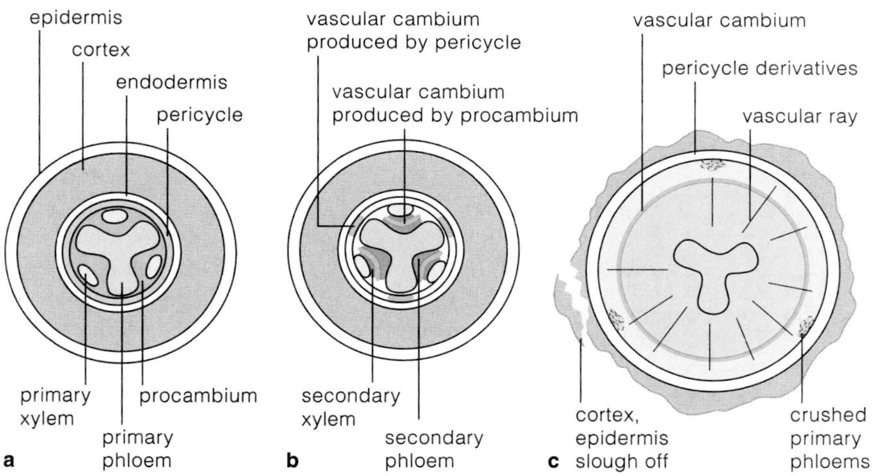
epidermis
cortex
endodermis
pericycle
primary xylem
procambium
primary phloem
a
vascular cambium produced by pericycle
vascular cambium produced by procambium
secondary xylem
secondary phloem
b
vascular cambium
pericycle derivatives
vascular ray
cortex, epidermis slough off
crushed primary phloems
c

Figure 29.23 Secondary growth in the root of one type of woody plant. (**a**) This is how tissues are organized as primary growth ends. (**b**,**c**) A thin cylinder of vascular cambium forms and gives rise to secondary xylem and phloem. Cell divisions are parallel with the vascular cambium. The cortex ruptures as the root thickens.

the stem. Other cells of vascular cambium (*ray initials*) produce many rays of parenchyma cells. The rays deliver water "sideways" through the stem, in a radial pattern that is like a sliced pie. Without all of these longitudinal and radial vascular tissues, substances could not travel up, down, and sideways in enlarging woody stems.

Figure 29.22 shows a simple way to think about the pattern of secondary growth at vascular cambium. As you can see, secondary xylem forms on the *inner* face of this meristematic tissue, and secondary phloem forms on the *outer* face. Does the accumulation of secondary growth eventually squash the vascular cambium deep inside the stem? Not at all. As the inner core of xylem gets thicker and thicker, it automatically and steadily displaces cells of the vascular cambium toward the stem surface. The meristematic cells also maintain the ring of vascular cambium by dividing sideways in an ever widening ring never far from the surface.

We have been focusing on how stems thicken, but bear in mind that secondary xylem and phloem also form at vascular cambium in the plant's roots. Figure 29.23 shows one of the patterns of secondary growth at vascular cambium in the root of a typical plant.

What are the selective advantages of woody stems and roots? Remember, plants as well as other organisms compete for resources. In this case, plants with taller stems or broader canopies that can defy the pull of gravity intercept more energy streaming in from the sun. With a greater supply of energy for photosynthesis, they have the metabolic means to develop large root and shoot systems, hence to be more competitive in acquiring resources—and ultimately to be reproductively successful in particular habitats.

In woody plants, secondary vascular tissues form at a ring of vascular cambium inside older stems and roots. Wood is an accumulation of secondary xylem especially.

With their sturdier tissues, woody plants defy gravity and grow taller and broader. Where competition for sunlight is intense, the ones that intercept the most sunlight win. Other factors being equal, having more energy to drive photosynthesis provides advantages in terms of metabolic capabilities, growth, and reproductive success.

29.7 A LOOK AT WOOD AND BARK

Where secondary xylem (wood) is extensive, it typically makes up about 90 percent of a tree. Secondary phloem is restricted to a relatively thin zone just outside the vascular cambium. This phloem consists of thin-walled, living parenchyma cells and sieve tubes that are often interspersed between bands of thick-walled, reinforcing fibers. Only the tubes within about a centimeter of the vascular cambium remain functional. The rest are dead and help protect the living cells beneath them.

Formation of Bark

As seasons pass and a tree ages, its inner core of xylem continues its outward expansion. The resulting pressure is directed toward the stem or root surface. Eventually it ruptures the cortex and the outer part of secondary phloem. Some cortex and epidermis split away. A new surface cover, the periderm, forms from cork cambium. Together, the periderm and secondary phloem constitute **bark**. In other words, bark is composed of all tissues outside the vascular cambium (Figure 29.24).

Periderm consists of cork, secondary cortex, and the cork cambium that produces these tissues. Soon after vascular cambium forms, cork cambium forms from the outermost parenchyma cells of the stem or root cortex. Such cells, recall, retain the capacity to divide. (When the cortex ruptures, parenchyma cells in the secondary phloem give rise to cork cambium.) Cell divisions at the cork cambium produce **cork**. This tissue consists of densely packed rows of cell walls, thick with suberin. Only its innermost cells remain alive, because only they have access to nourishment from xylem and phloem. With its many suberized layers, cork protects, insulates, and waterproofs the stem or root surface. Cork also forms over wounds. When leaves are about to drop, it forms at the place where petioles attach to stems.

Like all living plant cells, the cells in woody stems and roots require oxygen for aerobic respiration and give off carbon dioxide wastes. So how do these gases get across the suberized, corky surface of bark? They do so through lenticels, which are localized areas where the packing of cork cells is loosened up a bit. Those dark spots you might have noticed on a wine bottle's cork are all that is left of lenticels.

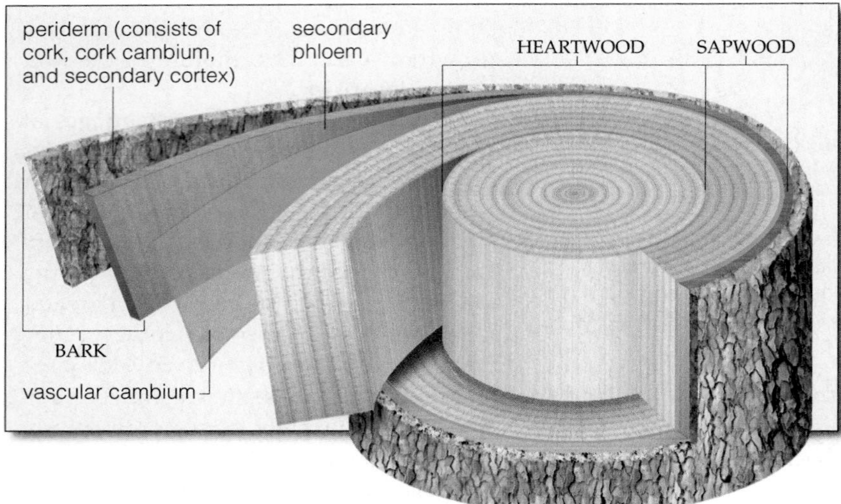
periderm (consists of cork, cork cambium, and secondary cortex)
secondary phloem
HEARTWOOD
SAPWOOD
BARK
vascular cambium

Figure 29.24 Structure of a stem with extensive secondary growth. Heartwood, the mature tree's core, has no living cells. Sapwood, a cylindrical zone of xylem between heartwood and vascular cambium, contains some living parenchyma cells among nonliving conducting cells of xylem. Everything outside the vascular cambium is bark.

Girdling is a deliberate stripping away of a band of secondary phloem around a trunk's circumference. With cuts through all of its vertical phloem, photosynthetically derived food cannot reach roots, which die—and so, in time, does the tree.

Heartwood and Sapwood

As a tree ages, changes also unfold in the appearance and function of the wood itself. In the center of its older stems and roots is **heartwood**, which is dry tissue that no longer transports water and solutes and is a dumping ground for some metabolic wastes. Resins, oils, gums, and tannins are among these metabolites. In time, they clog and fill in the oldest xylem pipelines. Typically, they darken heartwood, strengthen it, and make it more aromatic.

Heartwood helps the tree defy gravity, but the big trees can get along without it. In the early 1900s, when lumbermen were active in California's groves of old-growth redwoods, someone with knowledge of heartwood cut a tunnel through a few of the biggest trees, the better to drive an automobile through them. Was this a cute idea, or was it analogous to cutting a tunnel through Grandpa?

By contrast, **sapwood** is secondary growth located between heartwood and the vascular cambium. Compared to heartwood, it is wet, usually pale, and not as strong. Maple trees provide a famous example. Each spring, from early March through early April, farmers in New England insert metal tubes into the sapwood of sugar maples (*Acer saccharum*). Sap, the sugar-rich fluid in secondary phloem, drips out into buckets.

Early Wood, Late Wood, and Tree Rings

Vascular cambium becomes inactive during parts of the year in regions having cool winters or prolonged dry spells. The first xylem cells produced at the start of the

Further reading: Student Guide to InfoTrac on web site →

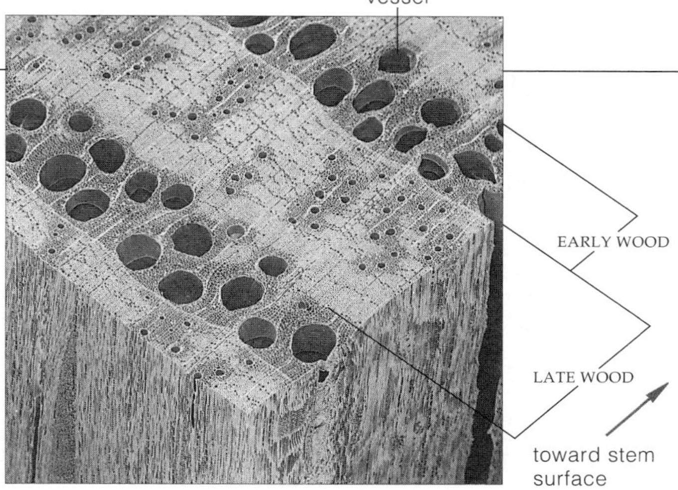
vessel
EARLY WOOD
LATE WOOD
toward stem surface

Figure 29.25 Scanning electron micrograph of early and late wood in a block cut from a red oak (*Quercus rubra*).

spruces, redwoods, and other conifers have tracheids and parenchyma rays but no vessels or fibers. Conifers are **softwood** trees. Without fibers, they are weaker and less dense than hardwoods (Figure 29.26*b*,*c*).

Limits to Secondary Growth

Unlike people, trees cannot run away from impending attacks. And if a pathogen penetrates their tissues, they cannot count on an immune system because they have none. Some trees live in habitats that are too harsh and too remote for most invaders, and their meristematic capacity for renewal has kept them going for many thousands of years. (Remember the bristlecone pines?)

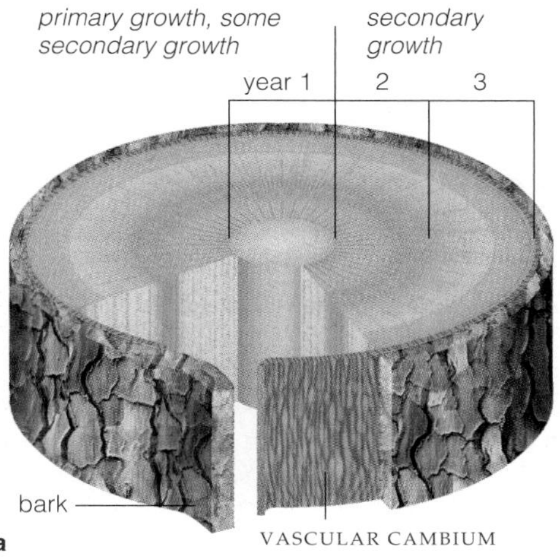
primary growth, some secondary growth
secondary growth
year 1
2
3
bark
VASCULAR CAMBIUM
a

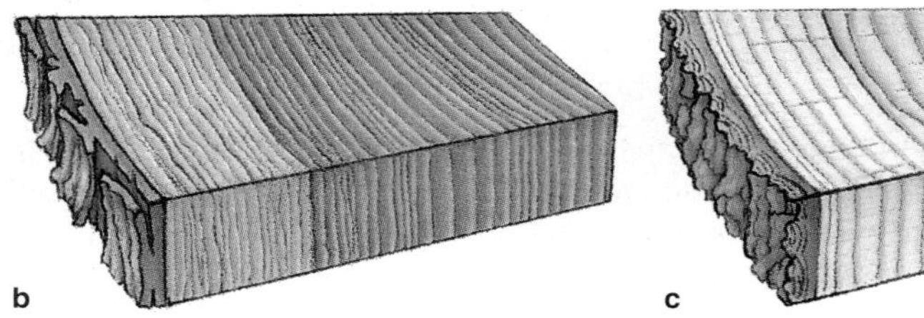
b

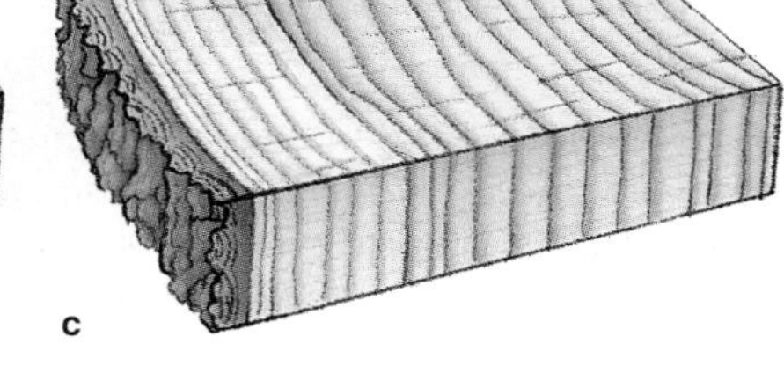
c

Figure 29.26 (**a**) Radial cut through a woody stem that has three annual rings. (**b**) Hickory, one of the hardwood dicots, is durable, strong, and yet resilient. It is used to make handles of hammers and other high-impact tools. (**c**) Pine, a softwood, is lightweight yet resists warping. Pine grows faster than hardwoods and is commercially "farmed" as a source of relatively cheap lumber.

growing season tend to have large diameters and thin walls. They form *early* wood. During summer's drier days, vascular cambium gives rise to cells with smaller diameters and thicker walls. Such cells form *late* wood. When you see "tree rings" in a transverse section from a trunk, these are alternating bands of early and late wood, which reflect light differently (Figures 29.25 and 29.26). The differences are called **growth rings**.

Seasonal change is predictable in temperate regions, and trees growing there usually add one growth ring per year. In deserts, thunderstorms rumble through at different times of year, and trees respond by adding more than one growth ring. Seasonal change is almost nonexistent in the tropics; as you might suspect, growth rings are not a feature of tropical trees.

Notice the large vessels in the block of oak shown in Figure 29.25. Oak is a **hardwood**. Like hickory, maple, and other dicot trees that evolved in temperate and tropical regions, it has vessels, tracheids, and fibers in its xylem. Depending on the species, the vessels may occur only in early wood or they may occur in both early and late wood. By contrast, the xylem of pines,

Most trees growing elsewhere limit their vulnerability by walling off invaders, building a fortress of thickened cell walls around wounds, and deploying phenols and other toxic compounds. Taken as a whole, the responses to attack are called **compartmentalization**.

So potent are some tree toxins that they also kill cells of the tree itself. As compartments form around injured or infected or poisoned sites, the tree lays down new tissues over them. This works well when the tree is not under massive attack or when it can respond fast enough. Some don't make it. They die when too many compartments form and interfere with the flow of water and solutes through the vascular system.

Bark consists of all living and nonliving tissues outside the vascular cambium—that is, secondary phloem and periderm.

Periderm consists of cork (the outermost covering of woody stems and roots), cork cambium, and secondary cortex.

Wood may be classified by its location and functions (as in heartwood versus sapwood) and by the type of plant (many dicots produce hardwood, and conifers produce softwood).

29.8 SUMMARY

1. Seed-bearing vascular plants include gymnosperms and the angiosperms, which are flowering plants. Their shoots (stems, leaves, and other structures) and roots consist of dermal, ground, and vascular tissue systems.

2. All plant growth originates at meristems, which are localized regions of dividing cells.

a. Primary growth (lengthening of stems and roots) originates at apical meristems in root and shoot tips.

b. In many plants, secondary growth (increases in diameter) originates inside stems and roots, at lateral meristems called vascular cambium and cork cambium.

3. Parenchyma, sclerenchyma, and collenchyma are the simple tissues, each with only one cell type (Table 29.1).

a. Parenchyma cells, alive and metabolically active at maturity, make up the bulk of ground tissue systems. They function in a variety of tasks. The ones making up mesophyll, for example, are photosynthetic.

b. Collenchyma helps to strengthen certain growing plant parts. Sclerenchyma supports mature plant parts.

4. Complex tissues include vascular tissues (xylem and phloem) and dermal tissues (epidermis and periderm). Each consists of two or more cell types (Table 29.1).

a. Vascular tissues distribute water and dissolved substances throughout a plant. Vascular bundles (each having xylem and phloem clustered together inside a cellular sheath) thread through the ground tissue.

b. Water-conducting cells of xylem are not alive at maturity. Their lignified, pitted walls interconnect and serve as pipelines for water and dissolved minerals.

c. Phloem's conducting cells are alive at maturity. The cytoplasm of adjoining cells interconnects across perforated end walls and side walls. In leaves, sugars and other photosynthetic products are loaded into the cells, often with the help of companion cells. They are unloaded where cells are growing or storing food.

d. Epidermis covers and protects the outer surfaces of primary plant parts. Periderm replaces epidermis on plants showing extensive secondary growth.

5. Stems function in support of upright growth and in conducting substances through a plant body by way of vascular bundles. Most monocot stems contain vascular bundles distributed through ground tissue. Most dicot stems contain a ring of bundles that divides the ground tissue into cortex and pith.

6. Leaves have veins and mesophyll (photosynthetic parenchyma) between the upper and lower epidermis. Air spaces around the photosynthetic cells enhance gas exchange. Water vapor and gases cross the epidermis through numerous tiny openings called stomata.

7. Roots absorb water and mineral ions for distribution to aboveground parts. They also anchor the plant. Most store food, and some help support the shoot.

Table 29.1 Summary of Flowering Plant Tissues and Their Components

SIMPLE TISSUES	
Parenchyma	Parenchyma cells
Collenchyma	Collenchyma cells
Sclerenchyma	Fibers or sclereids
COMPLEX TISSUES	
Xylem	Conducting cells (tracheids, vessel members); parenchyma cells; sclerenchyma cells
Phloem	Conducting cells (sieve-tube members); parenchyma cells; sclerenchyma cells
Epidermis	Undifferentiated cells; also guard cells and other specialized cells
Periderm	Cork; cork cambium; secondary cortex

8. Wood (secondary xylem) is classified by location and function (as in heartwood or sapwood) and plant type (as in hardwood of many dicots, softwood of conifers). Bark consists of secondary phloem and periderm.

Review Questions

1. Choose a flowering plant and list some functions of its roots and shoots. *29.1*

2. Name and define the basic functions of a flowering plant's three main tissue systems. *29.1*

3. Describe the differences between:
   a. apical, transitional, and lateral meristems *29.1*
   b. parenchyma and sclerenchyma *29.2*
   c. xylem and phloem *29.2*
   d. epidermis and periderm *29.2, 29.6*

4. Is the plant with the yellow flower in Figure 29.27 a dicot or monocot? What about the plant with the purple flower? *29.2*

a
b

Figure 29.27 Flower of (**a**) St. John's wort (*Hypericum*) and (**b**) a lily (*Lilium*).

5. Which of the two stem sections below is typical of most dicots? Which is typical of most monocots? Label the main tissue regions of both sections. *29.2, 29.3*

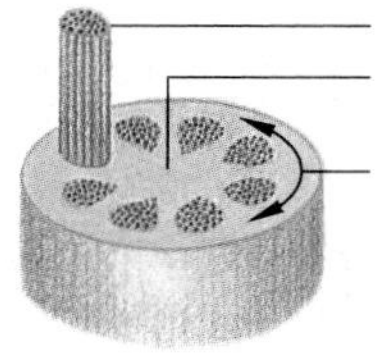

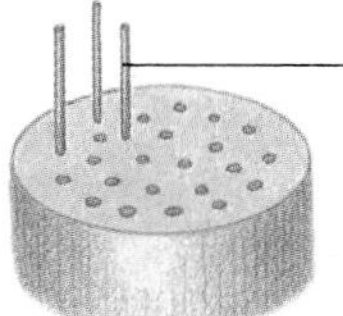

6. Label the components of the three-year-old tree section to the right. Then make a rough count of the number of growth rings in the photograph of a stem section below it. *29.7*

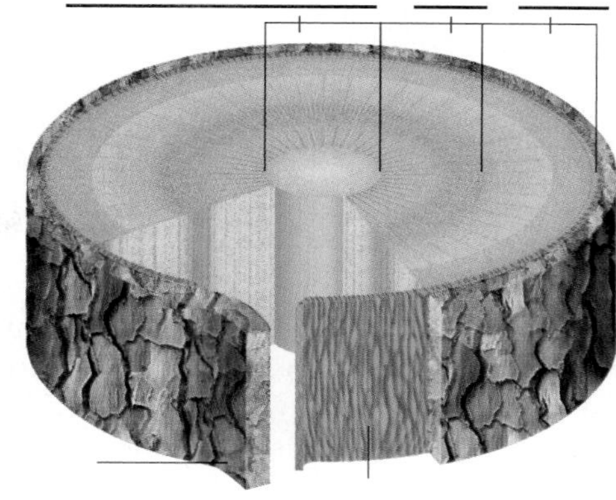

Self-Quiz (*Answers in Appendix III*)

1. Roots and shoots lengthen through activity at ______.
   a. apical meristems c. vascular cambium
   b. lateral meristems d. cork cambium
2. Older roots and stems thicken through activity at ______.
   a. apical meristems c. vascular cambium
   b. lateral meristems d. both b and c
3. Soft, moist plant parts consist mostly of ______ cells.
   a. parenchyma c. collenchyma
   b. sclerenchyma d. epidermal
4. Xylem and phloem are ______ tissues.
   a. ground b. vascular c. dermal d. both b and c
5. ______ conducts water and ions; ______ conducts food.
   a. Phloem; xylem c. Xylem; phloem
   b. Cambium; phloem d. Xylem; cambium
6. Buds give rise to ______.
   a. leaves c. stems
   b. flowers d. all of the above
7. Mesophyll consists of ______.
   a. waxes and cutin c. photosynthetic cells
   b. lignified cell walls d. cork but not bark
8. In early wood, cells have ______ diameters, ______ walls.
   a. small; thick c. large; thick
   b. small; thin d. large; thin
9. Match the plant parts with the most suitable description.
   ___ apical meristem a. masses of xylem
   ___ lateral meristem b. source of primary growth
   ___ xylem, phloem c. corky surface covering
   ___ periderm d. source of secondary growth
   ___ vascular cylinder e. conduct water, food
   ___ wood f. central column in roots

Critical Thinking

1. Think about the kinds of conditions that might prevail in a hot desert in New Mexico or Arizona; on the floor of a shady, moist forest in Georgia or Oregon; and in the arctic tundra in Alaska. Also consider the kinds of animals living in each type of environment. Then "design" a type of flowering plant that would be suitably adapted to each place.

2. Fitzgerald, known for his highly underdeveloped sense of nature, sneaks into a forest preserve late at night and then maliciously girdles an old-growth redwood. He settles down and watches the tree, waiting for it to die. But the tree will not die that night or any time soon. Explain why.

3. Sylvia lives in Santa Barbara, where droughts are common and a long-term abundance of water is not. She replaced most of her garden with drought-tolerant plants and cut back the size of the lawn. The lawn does not get a light sprinkling every day. Sylvia only waters it twice a week in the evening, after the sun goes down. Then the lawn gets a good soak to a depth of several inches. Why is her strategy good for lawn grasses?

4. Oscar and Lucinda meet on a trip through a tropical rain forest and fall in love. In the exuberance of the moment, he carves their initials into the bark of a small tree. They never do get together, though. Ten years later, the still-heartbroken Oscar searches for the tree. Given what you know about primary and secondary growth, will he find the carved initials higher relative to ground level? If he goes berserk and cuts down the tree, what kind of growth rings will he see?

Selected Key Terms

annual *29.6*
apical meristem *29.1*
bark *29.7*
biennial *29.6*
bud *29.3*
collenchyma *29.2*
compartmentalization *29.7*
cork *29.7*
cork cambium *29.1*
cortex *29.3*
cuticle (plant) *29.2*
dermal tissue system *29.1*
dicot *29.2*
epidermis *29.2*
fibrous root system *29.5*
ground tissue system *29.1*
growth ring *29.7*
hardwood *29.7*
heartwood *29.7*
lateral root *29.5*
leaf *29.4*
meristem *29.1*
mesophyll *29.4*
monocot *29.2*
parenchyma *29.2*
perennial *29.6*
periderm *29.2*
phloem *29.2*
pith *29.3*
primary root *29.5*
root *29.1*
root hair *29.5*
sapwood *29.7*
sclerenchyma *29.2*
shoot *29.1*
softwood *29.7*
stoma (stomata) *29.2*
taproot system *29.5*
vascular bundle *29.3*
vascular cambium *29.1*
vascular cylinder *29.5*
vascular tissue system *29.1*
vein (leaf) *29.4*
xylem *29.2*

Readings *See also www.infotrac-college.com*

Esau, K. 1977. *Anatomy of Seed Plants.* Second edition. New York: Wiley. Well, sometimes the oldies are still goodies.

Rost, T., M. Barbour, C. R. Stocking, and T. Murphy. 1998. *Plant Biology.* Belmont, California: Wadsworth.

Simpson, B., and M. Conner-Ogorzaly. 1995. *Economic Botany.* Second edition. New York: McGraw-Hill. Fascinating book on uses and abuses of plants.

30

PLANT NUTRITION AND TRANSPORT

Flies for Dinner

How often do we think that plants actually do anything impressive? Being mobile, intelligent, and emotional, we tend to be fascinated more with ourselves than with immobile, expressionless plants. Yet plants don't just stand around soaking up sunlight. Consider the Venus flytrap (*Dionaea muscipula*), a native flowering plant of bogs in North and South Carolina. Its two-lobed, spine-fringed leaves open and close much like a steel trap (Figure 30.1*a–d*). Like all other plants, it cannot grow properly without nitrogen and other nutrients, which happen to be scarce in the soil of bogs. But plenty of insects fly in from places around the bogs.

Sticky sugars ooze from epidermal glands onto the surface of the flytrap's leaf. The sugars entice insects to land. As they do, they brush against hairlike structures that project from the leaf surface. These are triggers for the trap. When an insect touches two hairs at the same time or the same hair twice in rapid succession, the two lobes of the leaf snap shut. Now digestive juices pour out from cells of the leaf. They pool around the insect, dissolve it, and so release nutrients from it. In other words, the Venus flytrap makes its own nutrient-rich water, which it proceeds to absorb!

The Venus flytrap is only one of several species of **carnivorous plants**. We call them this even though it takes a flying leap of the imagination to put their mode of nutrient acquisition (a form of extracellular digestion and absorption) in the same category as the chompings of lions, dogs, and similar meat eaters. Besides, not all carnivorous plants have active traps. Certain species, including the one shown in Figure 30.2, lure prey into fluid-filled traps and then simply let them drown.

All carnivorous plants evolved in habitats where nitrogen and other nutrients are hard to come by. For example, plants with bizarre nutrient-acquiring habits also live in the shallow waters of many freshwater lakes and streams, which contain only dilute concentrations of dissolved minerals.

Given the variety and numbers of insects and other animals that attack plants, you can just imagine how endearing the carnivorous plants are to botanists. With their plucky modes of nutrition, these plants also are a fine way to start thinking about **plant physiology**—the study of adaptations by which plants function in their environment. As you already know, nearly all plants are photoautotrophs that use energy from sunlight to drive the synthesis of organic compounds from water, carbon dioxide, and some minerals. Like people, they do not have unlimited supplies of the resources required to nourish themselves. Of every 1 million molecules of air, only 350 are carbon dioxide. Unlike the soggy habitats

VENUS FLYTRAP, OPEN FOR DINNER

a

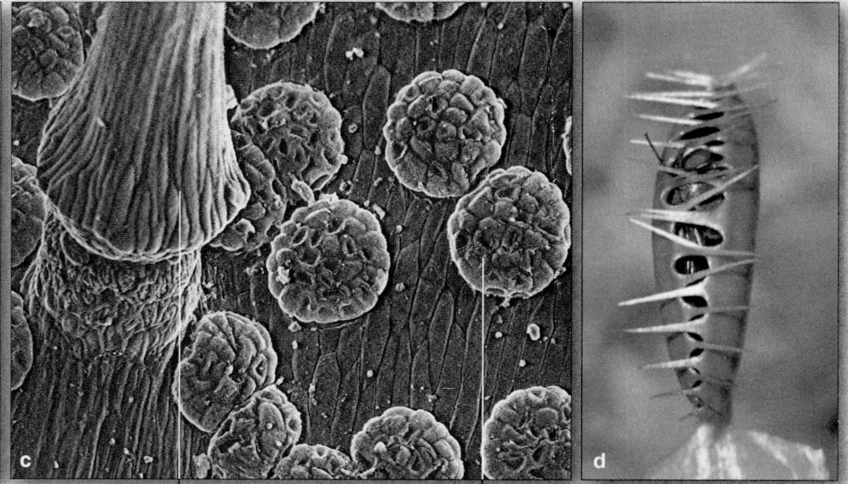
b
c
d
base of epidermal hair
epidermal gland

Figure 30.1 Do plants take nutrition seriously? You bet. (**a**) A Venus flytrap (*Dionaea muscipula*). This carnivorous plant makes up for scarce nutrients by turning the dinner table on animals that alight upon its leaves. (**b**) A fly stuck in the sugary goo on a lobed leaf. (**c**) It brushes against hairlike triggers on the leaf surface; the base of one is shown here. (**d**) Activated, the leaf snaps shut.

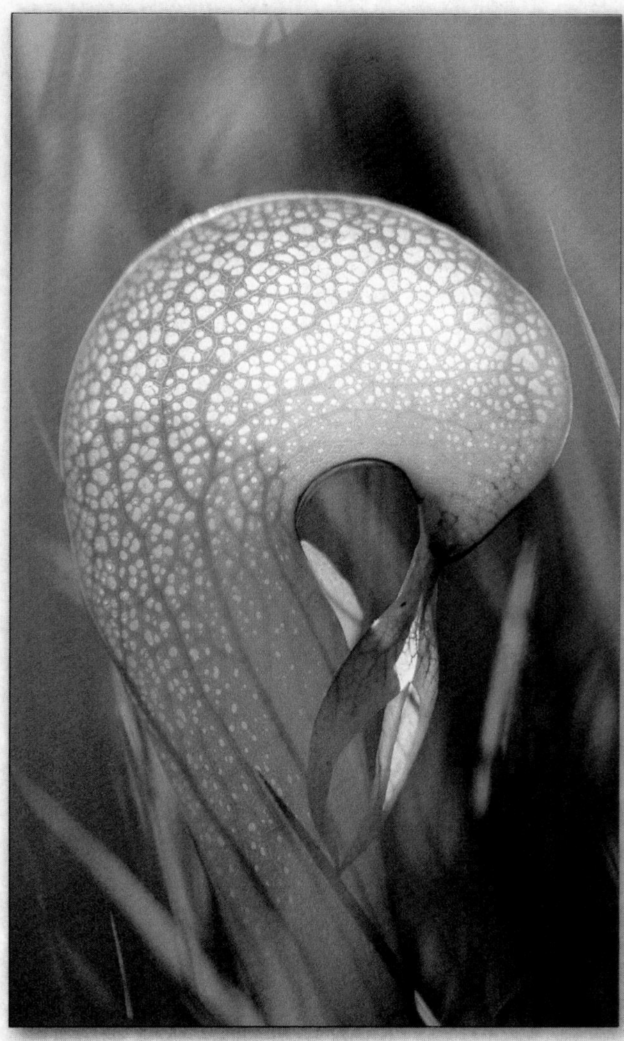

Figure 30.2 Cobra lily (*Darlingtonia californica*). Its leaves form a "pitcher" that is partly filled with digestive juices. Insects lured in by irresistible odors often cannot find the way back out; light shining through the pitcher's patterned dome confuses them. They just wander around and down, adhering to downward-pointing leaf hairs—which become slickened with wax above the potent vat.

of Venus flytraps, most soils are frequently dry. And nowhere except in overfertilized gardens does soil water hold lavish amounts of dissolved minerals. As you continue with these chapters on vascular plants, keep this point in mind: *Many aspects of plant structure and function are responses to low concentrations of vital resources in the environment.*

KEY CONCEPTS

1. Many aspects of the structure and function of plants are adaptive responses to low concentrations of water, minerals, and other environmental resources.

2. A plant's root system takes up water and mines the soil for nutrients. For many land plants, mycorrhizae and bacterial symbionts assist in nutrient uptake. The properties of soil in a given habitat influence the availability of nutrients and water.

3. A plant's cuticle and its many stomata function in the conservation of water, a scarce resource in most habitats on land. Stomata are passageways across the epidermis of leaves and, to a lesser extent, stems. They help control water loss, yet they permit gas exchange.

4. During the day, stomata remain open so that carbon dioxide, required for photosynthesis, is able to diffuse into leaves. Water loss is rapid when stomata are open. Most plants conserve water by closing stomata at night.

5. In flowering plants and other vascular plants, the flow of water and solutes through xylem and phloem functionally connects all living cells of the roots, stems, and leaves. Xylem serves in the uptake and distribution of water and dissolved nutrients. Phloem serves in the distribution of photosynthetically produced sugars and other organic compounds through the plant body.

6. Water absorbed from soil moves on up through xylem and into leaves. By the process of transpiration, the dry air around leaves promotes evaporation through stomata. The force of evaporation is enough to pull up continuous columns of water molecules, which are hydrogen-bonded to one another, from roots to aboveground parts.

7. By the energy-requiring process of translocation, sucrose and other organic compounds are distributed throughout the plant. Organic compounds produced by photosynthetic cells in leaves are loaded into conducting cells of phloem. They are unloaded at actively growing regions or storage regions of the plant body.

30.1 SOIL AND ITS NUTRIENTS

Properties of Soil

At the surface of most habitats on land is a thin cloak of soil. **Soil** consists of particles of minerals mixed with variable amounts of decomposing organic material, or **humus**. Its minerals come from the weathering of hard rocks. Dead organisms and organic litter (fallen leaves, feces, and so on) make up humus. Water and oxygen occupy spaces between the particles and organic bits.

In different regions, even in different parts of the same habitat, soils differ in the proportions of mineral particles and in the extent to which these have become compacted together. Generally, these particles come in three sizes, known as sand, silt, and clay. If you have ever let beach sand dribble between your fingers, you already have an idea that sand particles are large (0.05 to 2 millimeters across). Rub some silt from pond mud between your fingers. You won't be able to distinguish among individual particles; they are only 0.002 to 0.05 millimeter across. Clay particles are the finest of all.

How suitable is a given soil for plant growth? Is it gummy when wet because it does not have enough air spaces? Does it form hard clods when dry? The answer depends partly on the relative proportions of sand, silt, and clay. The more clay, the finer the soil's texture.

Each clay particle consists of thin, stacked layers of aluminosilicates with negatively charged ions at their surfaces. Clay attracts and holds (adsorbs) positively charged mineral ions dissolved in the water trickling through soil, as well as water molecules themselves. Ions and water cling reversibly to clay, and this chemical behavior is vital for plants. With its high adsorption capacity, clay holds on to many nutrients for plants even as the water percolates on past and drains away.

Too much clay, however, is bad for plants. Clay particles pack together so tightly, they don't leave enough space for all the oxygen that root cells require for aerobic respiration. The tight packing also retards penetration of water into the soil. In heavy clay soils, runoff is pronounced, so water and nutrients dissolved in it are not available for a plant's growth. Plants do best in **loams**, which are soils that have roughly the same proportions of sand, silt, and clay.

The amount of humus in a given soil also affects plant growth. Generally, humus has an abundance of negatively charged organic acids, so that it can weakly bind with and retain dissolved mineral ions of opposite charge. Humus also has a high capacity to absorb and swell with water, then shrink as the water is gradually released. Its alternating swelling and shrinking aerate the soil. When decomposers work it over, humus slowly releases nutrients, which become available for plants.

In general, soils that are 10 to 20 percent humus are most favorable for plant growth. The worst have less than 10 percent humus or more than 90 percent humus, which is characteristic of swamps and bogs.

Soils can be classified by *profile* properties. The word refers to the layered characteristics of soils, which are in different stages of development in different places. Figure 30.3 has an example. **Topsoil**, the uppermost part of soil, is called the A horizon. This is the most essential layer for plant growth, and its depth is highly variable.

Nutrients Essential for Plant Growth

We have mentioned nutrients in passing, but what does the word mean? **Nutrients** are elements that are essential for a given organism because, directly or indirectly, they have roles in metabolism (hence growth and survival) that no other element can fulfill. For plants, the essential elements include oxygen, hydrogen, and carbon secured from water and carbon dioxide during photosynthesis.

O HORIZON
Fallen leaves and other organic material littering the surface of mineral soil
A HORIZON
Topsoil, which contains some percentage of decomposed organic material and which is variably deep (only a few centimeters deep in deserts, but elsewhere extending as far as thirty centimeters below the soil surface)
B HORIZON
Compared with the A horizon, larger soil particles, not much organic material, but greater accumulation of minerals; extends thirty to sixty centimeters below soil surface
C HORIZON
No organic material, but partially weathered fragments and grains of rock from which soil forms; extends to underlying bedrock
BEDROCK

Figure 30.3 Some of the soil horizons that have developed in one habitat in Africa.

Further reading: Student Guide to InfoTrac on web site

Table 30.1 Essential Elements and Plant Function

MACRONUTRIENT	Some Functions	Some Deficiency Symptoms
Carbon, Hydrogen, Oxygen	Basic ingredients for photosynthesis	Available in abundance from water and from carbon dioxide in the air
Nitrogen	Component of proteins, nucleic acids, coenzymes, chlorophylls	Stunted growth; light-green older leaves; older leaves yellow and die (these symptoms define a condition called chlorosis)
Potassium	Activation of enzymes; key role in maintaining water-solute balance and so influences osmosis*	Reduced growth; curled, mottled, or spotted older leaves; burned leaf edges; weakened plant
Calcium	Regulation of many cell functions; cementing of cell walls	Leaves deformed; terminal buds die; poor root growth
Magnesium	Component of chlorophyll; activation of enzymes	Chlorosis; drooped leaves
Phosphorus	Component of nucleic acids, phospholipids, ATP	Purplish veins; stunted growth; fewer seeds, fruits
Sulfur	Component of most proteins, two vitamins	Light-green or yellowed leaves; reduced growth
MICRONUTRIENT	**Some Functions**	**Some Deficiency Symptoms**
Chlorine	Role in root and shoot growth; role in photolysis	Wilting; chlorosis; some leaves die
Iron	Roles in chlorophyll synthesis and in electron transport	Chlorosis; yellow and green striping in grasses
Boron	Roles in germination, flowering, fruiting, cell division, nitrogen metabolism	Terminal buds, lateral branches die; leaves thicken, curl, and become brittle
Manganese	Chlorophyll synthesis; coenzyme action	Dark veins, but leaves whiten and fall off
Zinc	Role in formation of auxin, chloroplasts, and starch; enzyme component	Chlorosis; mottled or bronzed leaves; abnormal roots
Copper	Component of several enzymes	Chlorosis; dead spots in leaves; stunted growth
Molybdenum	Part of enzyme used in nitrogen metabolism	Pale green, rolled or cupped leaves

* All mineral elements contribute to the water-solute balance, but potassium is notable because there is so much of it.

Besides carbon, hydrogen, and oxygen, plants depend on the uptake of at least thirteen other elements (Table 30.1). These typically are dissolved in soil water in ionic forms that reversibly bind with clay. Ions of calcium (Ca^{++}) and potassium (K^{+}) are examples. Plants exchange hydrogen ions for these weakly bound elements in clay.

Nine essential elements are *macro*nutrients. Normally they are required in amounts above 0.5 percent of the plant's dry weight (weighed after all the water has been removed from the plant). The other elements listed are *micro*nutrients; they make up traces (usually a few parts per million) of the dry weight. Even the trace amounts are essential for normal growth.

Figure 30.4 In agricultural fields, the erosive force of water can form gullies that channel runoff from the land. When gullies grow deeper and wider, erosion becomes more and more rapid. When topsoil is depleted, productivity declines and fertilizers usually must be trucked in to replace the lost nutrients.

Leaching and Erosion

Leaching refers to the removal of some of the nutrients in soil as water percolates through it. Leaching is most pronounced in sandy soils, which are not as good as clay at binding nutrients. It is not the same as **erosion**, which is the movement of land under the force of wind, running water, and ice (Figure 30.4). For example, each year, erosion from farmlands in the Mississippi River watershed puts about 25 billion metric tons of topsoil into the Gulf of Mexico. Whether by leaching or erosion, the loss of nutrients from soil is bad for plants and for all the organisms that depend on plants for survival.

The mineral component of soil includes particles ranging from large-grained sand to silt and fine-grained clay. These particles, clay especially, reversibly bind water molecules and dissolved mineral ions and thereby make them more accessible for uptake by plant roots.

Soil also contains humus, which is a reservoir of organic material, rich in organic acids, in different stages of decay.

Most plants grow best in soils having equal proportions of sand, silt, and clay, as well as 10 to 20 percent humus.

Nutrients are essential elements. No other element can substitute for their direct or indirect roles in the metabolic activities that sustain growth and keep organisms alive.

30.2

HOW DO ROOTS ABSORB WATER AND MINERAL IONS?

Energetically speaking, mining the soil for mineral ions and water molecules that are clinging to clay particles is an expensive task. Plants spend considerable energy on building extensive root systems. Wherever the soil's texture and composition change, new roots must form to replace old ones and snake out in different regions. It isn't that roots "explore" soil for resources. Rather, the patches of soil where concentrations of water and mineral ions are greater stimulate the outward growth.

Absorption Routes

Think back on the preceding chapter's discussion of a typical root's structure (Section 29.5). Water molecules in soil are only weakly bound to clay particles, so they readily move across root epidermis and continue into the **vascular cylinder**, a column of vascular tissue at the root's center. There, a cylindrical layer of cells, an **endodermis**, wraps around the column. A waxy band, a **Casparian strip**, is deposited in abutting parts of endodermal cell walls (Figure 30.5). Water molecules are not able to penetrate the wax, so they are directed to wall regions that are not waxed, then into the cells proper, and out the other side. This is the only way water and solutes can get into the vascular cylinder. Like all cells, endodermal cells have many transport proteins embedded in the plasma membrane. The proteins let some solutes but not others cross it (Section 5.3). *Transport proteins of endodermal cells are control points where the plant adjusts the quantity and types of solutes that it absorbs from soil water.*

Many roots have an **exodermis**, a cell layer beneath their surface (Figure 30.5*a*). Walls of exodermal cells often have a Casparian strip that functions just like the one next to the root vascular cylinder.

Specialized Absorptive Structures

ROOT HAIRS Vascular plants require huge amounts of water. Roots of a mature corn plant absorb more than three liters of water daily. They could not do so without **root hairs**. Recall, from the preceding chapter, that root hairs are slender extensions of specialized epidermal cells. They greatly increase the surface area available for absorption (Section 29.5 and Figure 30.6). When a plant is putting on primary growth, its system of roots may develop millions or billions of root hairs.

ROOT NODULES Bacteria and fungi help many plants absorb nutrients and receive something in return. A two-way flow of benefits between species, remember, is a symbiotic interaction known as **mutualism** (Section 24.4).

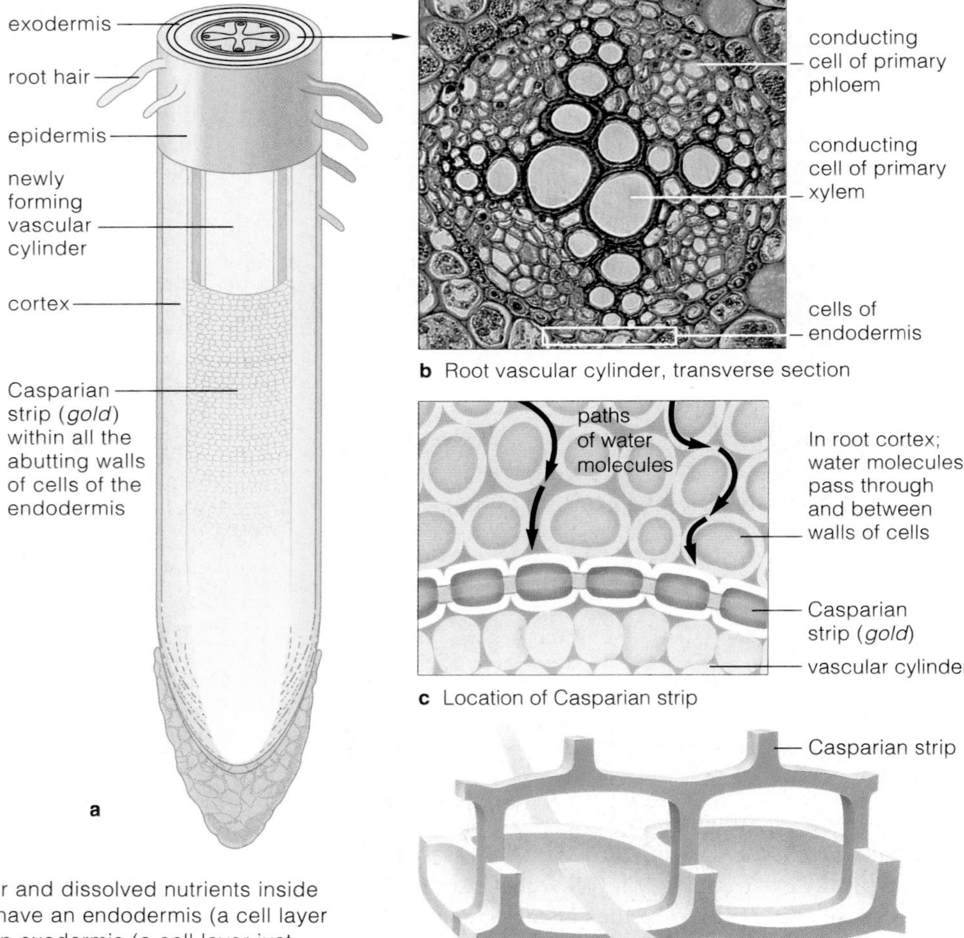

exodermis
root hair
epidermis
newly forming vascular cylinder
cortex
Casparian strip (*gold*) within all the abutting walls of cells of the endodermis
a
conducting cell of primary phloem
conducting cell of primary xylem
cells of endodermis
paths of water molecules
In root cortex; water molecules pass through and between walls of cells
Casparian strip (*gold*)
vascular cylinder
Casparian strip

b Root vascular cylinder, transverse section

c Location of Casparian strip

d Two endodermal cells. For clarity, the artist removed the cytoplasm and most of their wall. Water and solutes can only move into the vascular cylinder by passing through the cells (and the unwaxed part of their walls), not in between cells.

Figure 30.5 Control of the uptake of water and dissolved nutrients inside roots. (**a**,**b**) Roots of most flowering plants have an endodermis (a cell layer that surrounds the vascular cylinder) and an exodermis (a cell layer just beneath the epidermis). (**c**) Abutting walls of cells of both layers contain a waxy Casparian strip. The strip keeps water from moving indiscriminately *around* the cells and into the vascular column. It makes water move *through* the cells (**d**). In this way, transport proteins that span the plasma membrane of these cells selectively control the uptake of water and nutrients.

Further reading: Student Guide to InfoTrac on web site

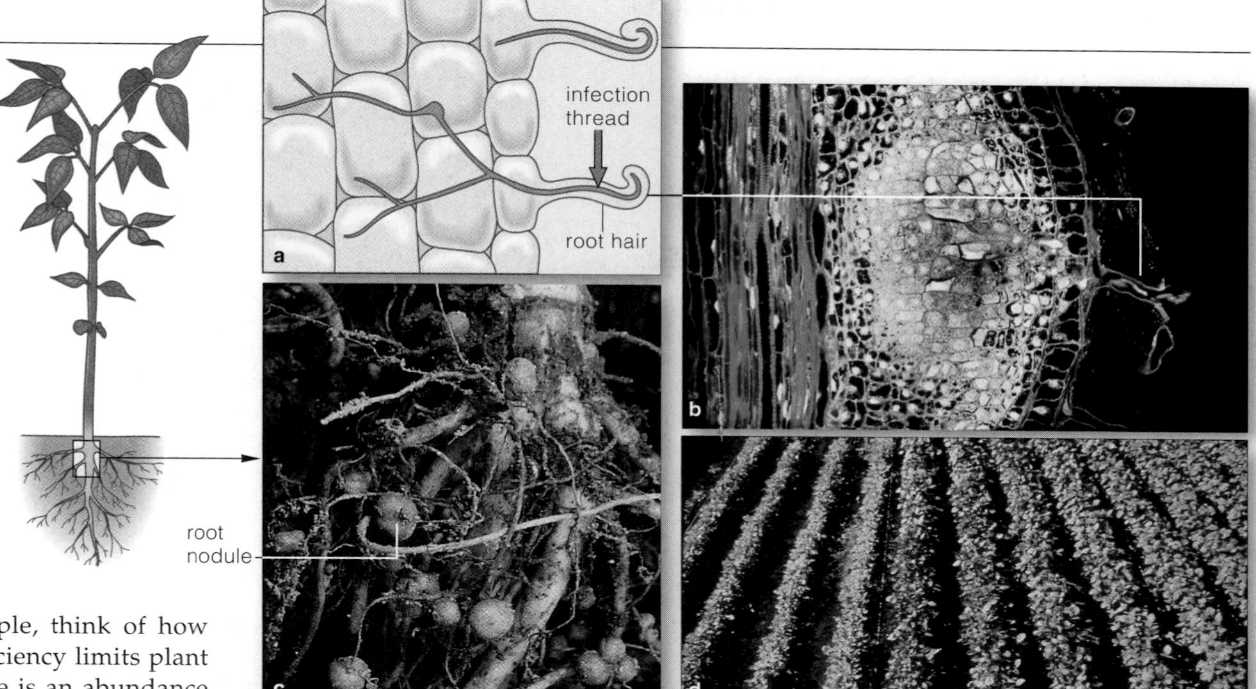
infection thread
root hair
a
b
root nodule
c
d

Figure 30.7 (**a**) Nutrient uptake at root nodules of legumes that are mutualists with nitrogen-fixing bacteria (*Rhizobium* and *Bradyrhizobium*). When infected by such bacteria, root hair cells are induced to form an "infection thread" of cellulose deposits. The bacteria use the thread like a highway to invade plant cells inside the root cortex. (**b**,**c**) Infected plant cells and the bacterial cells within them divide repeatedly, forming a swollen mass that becomes a root nodule. The bacteria start fixing nitrogen when plant cell membranes surround them. The plant takes up some of the nitrogen; the bacteria take up some photosynthetic compounds. (**d**) To the *left* in this photograph, rows of soybean plants growing in nitrogen-poor soil. The plants in the rows to the *right* were inoculated with *Rhizobium* cells and developed root nodules.

For example, think of how nitrogen deficiency limits plant growth. There is an abundance of gaseous nitrogen (N≡N) in the air. But plants do not have the metabolic means to engage in **nitrogen fixation**. With this process, certain enzymes break all three of the covalent bonds in gaseous nitrogen, and then they attach the nitrogen atoms to organic compounds. To get high crop yields, farmers apply nitrogen-rich fertilizers or they encourage growth of nitrogen-fixing bacteria that naturally inhabit soil. Such bacteria convert gaseous nitrogen to forms that they—and plants—can use. String beans, peas, alfalfa, clover, and other legumes have an advantage in this respect. Nitrogen-fixing bacteria live in their roots, inside local swellings called **root nodules**. Figure 30.7*b* shows an example. The symbiotic bacteria withdraw some photosynthetically produced organic compounds that were distributed from leaves to the roots. But in return they provide the plants with some of the nitrogen that they secured.

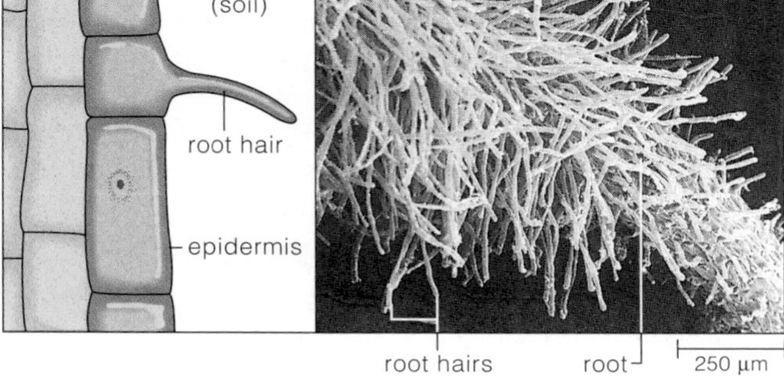
(soil)
root hair
epidermis
root hairs
root
250 μm

Figure 30.6 Sketch of a root hair (a slender extension of a specialized root epidermal cell), and a scanning electron micrograph of the root hairs of a portion of a young root.

MYCORRHIZAE Also think back on the **mycorrhizae** (singular, mycorrhiza). As described in Section 24.4, a mycorrhiza is a symbiotic interaction between a young root and a fungus. Hence the name, meaning "fungus-root." Fungal filaments (hyphae) form a velvety cover around roots or they penetrate root cells. Collectively, hyphae have a large surface area that absorbs mineral ions from a larger volume of soil than the roots can do. The fungus can absorb sugars and nitrogen-containing compounds from root cells. The root cells obtain some scarce minerals that the fungus is better able to absorb.

Gymnosperms and flowering plant roots control the uptake of water and dissolved nutrients at the vascular cylinder's endodermis and at a similar layer near the root surface.

Root hairs, root nodules, and mycorrhizae enhance the uptake of water and scarce nutrients at the root.

HOW IS WATER TRANSPORTED THROUGH PLANTS?

Transpiration Defined

By now, you have a sense of how the distribution of water and dissolved mineral ions to all living cells is central to plant growth and functioning. Let's turn now to the prevailing model of how water actually moves from a plant's roots to its stems, then into leaves.

Recall that plants use only a fraction of the water they absorb for growth and metabolism. Most of that water is lost, mainly through the numerous stomata in leaves. The evaporation of water from leaves, stems, and other plant parts is a process called **transpiration**.

Cohesion–Tension Theory of Water Transport

This brings up an interesting question. Assuming plants lose most of the absorbed water from their leaves, how does water actually get *to* the leaves? What gets it to the uppermost leaves of plants, including redwoods and other trees that may be more than 100 meters tall?

In a plant's vascular tissues, water moves through a complex tissue called **xylem**. Recall, from Section 29.2, that the water-conducting cells of xylem are **tracheids** and **vessel members**. Figure 30.8 provides examples. These cells are dead at maturity, and only their lignin-reinforced walls remain. So this means the conducting cells in xylem cannot be actively pulling the water "uphill."

Some time ago, the botanist Henry Dixon came up with a useful way to explain water transport in plants. By his **cohesion–tension theory**, the water inside xylem is pulled upward by air's drying power, which creates continuous negative pressures—that is, tensions. These tensions extend downward from leaves to roots.

Figure 30.9 illustrates the cohesion–tension theory of water transport in plants. Think about these three points as you review the illustration:

1. Air has drying power that causes transpiration: the evaporation of water from all plant parts exposed to air, but most notably at stomata. Transpiration puts water that is confined inside the conducting tubes of xylem into a state of tension. That tension extends from veins inside leaves, down through the stems, and on into young roots where water is being absorbed.

2. Unbroken, fluid columns of water show *cohesion*; they resist rupturing as they are pulled up under *tension*. (Here you may wish to refer to Section 2.5.) The collective strength of hydrogen bonds between water molecules that are confined within the narrow, tubular xylem cells imparts this cohesion.

3. For as long as molecules of water continue to escape from a plant, the continuous tension inside the xylem permits more molecules to be pulled upward from the roots, and therefore to replace them.

Figure 30.8 Scanning electron micrographs and sketches of some types of tracheids and of vessel members from xylem. These water-conducting tubes are made of the walls of cells that are dead at maturity. The cell walls still remain interconnected, and they form the tubes. Tracheids probably evolved before vessel members. Both occur in nearly all vascular plants.

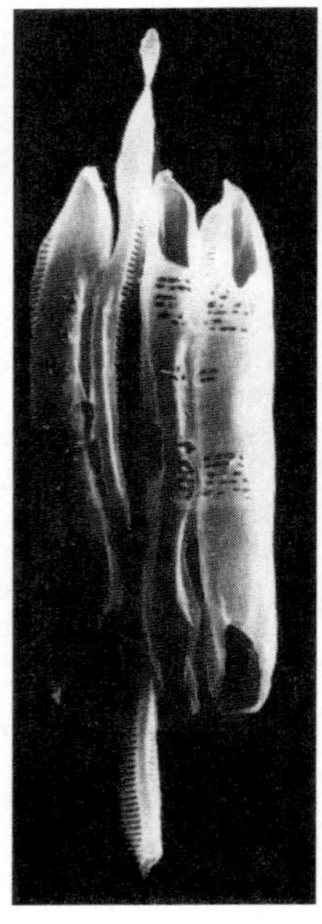

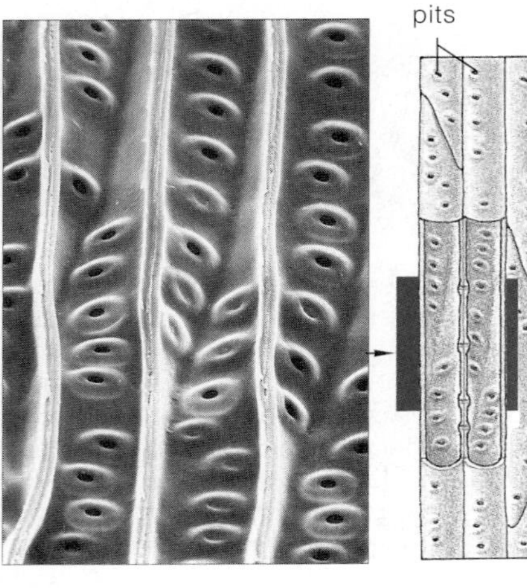
pits

a *Left:* Tracheids have tapered, unperforated end walls. *Right:* In intact xylem, small pits in walls of adjoining tracheids match up. The pits are too small to permit rapid flow, but they do confine air bubbles to individual tracheids in the system. Bubbles can obstruct water transport.

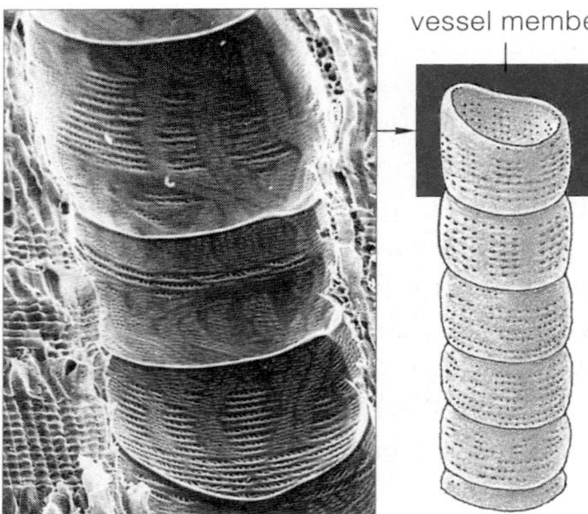
vessel member

b The micrograph is a close-up of three individual vessel members. The thick, finely perforated walls of these dead cells interconnect one after another to form vessels, another type of water-conducting tube in xylem.

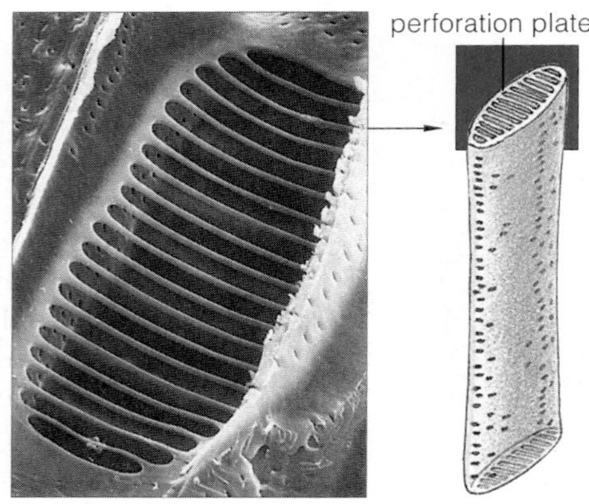
perforation plate

c Perforation plate at the end wall of one type of vessel member. The perforated ends allow water and air bubbles to flow unimpeded through the conducting tube. This may be why natural selection favored conserving tracheids and vessel members in the same plants.

Further reading: Student Guide to InfoTrac on web site –

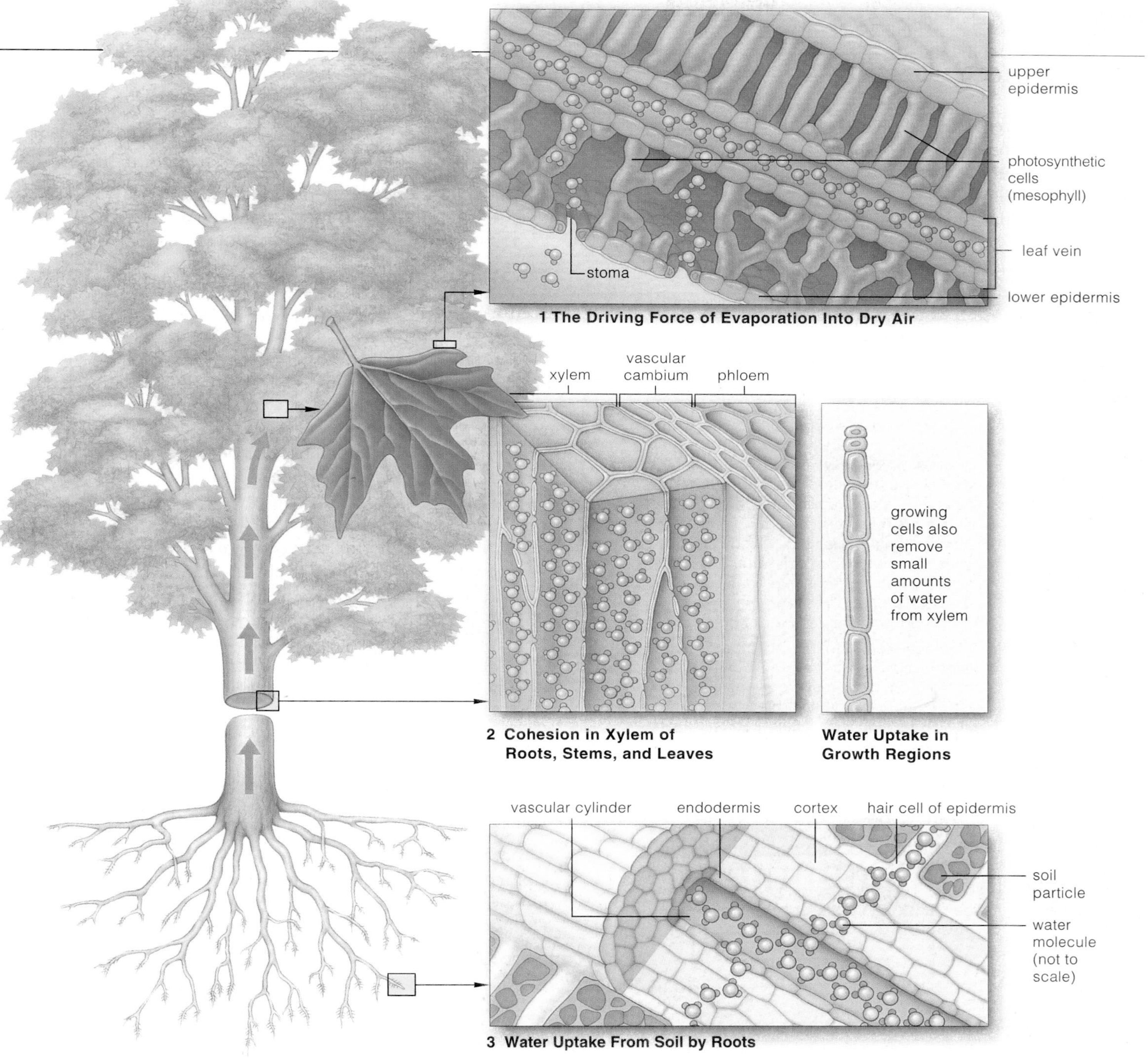
upper epidermis
photosynthetic cells (mesophyll)
leaf vein
lower epidermis
stoma
1 The Driving Force of Evaporation Into Dry Air
xylem
vascular cambium
phloem
growing cells also remove small amounts of water from xylem
2 Cohesion in Xylem of Roots, Stems, and Leaves
Water Uptake in Growth Regions
vascular cylinder
endodermis
cortex
hair cell of epidermis
soil particle
water molecule (not to scale)
3 Water Uptake From Soil by Roots

Figure 30.9 Diagram illustrating the cohesion–tension theory of water transport.

Point 1: Transpiration is the evaporation of water molecules from aboveground plant parts, especially at stomata. The process puts the water in xylem in a state of tension that extends from roots to leaves.

Point 2: The collective strength of hydrogen bonds among water molecules, which are confined within the narrow water-conducting tubes in xylem, imparts cohesion to the water. Hence the narrow columns of water in xylem resist rupturing under the continuous tension.

Point 3: As long as water molecules continue to escape by transpiration, that tension drives the uptake of replacement water molecules from soil water.

Hydrogen bonds are strong enough to hold water molecules together inside the xylem. However, they are not strong enough to prevent the water molecules from breaking away from one another during transpiration and then escaping from leaves, through stomata.

Transpiration is the process of evaporation from plant parts.

By the cohesion–tension theory of water transport, this process is the key source of tensions in water in xylem. The tensions extend from leaves to roots, and they allow columns of water molecules that are hydrogen-bonded to one another to be pulled upward through the plant body.

30.4 HOW DO STEMS AND LEAVES CONSERVE WATER?

At least 90 percent of the water that enters a leaf goes right on through it and evaporates into the surrounding air. Cells use only 2 percent of the water that stays in the leaf for photosynthesis, membrane functions, and other events. That tiny amount is vital. Plants wilt and water-dependent events are severely disrupted when water loss exceeds water uptake at roots for extended periods (Figure 30.10). Yet land plants are not entirely at the mercy of changes in the availability of soil water. They have a cuticle, and they have stomata.

Figure 30.10 Osmosis and the wilting of leafy plants. When a plant with soft green leaves is growing well, you can safely bet that the soil water is dilute, or hypotonic, compared to fluids in the plant's living cells.

Recall, from Section 5.5, that when water responds to solute concentration gradients and moves osmotically into a plant cell, internal fluid pressure builds up against the cell wall. Botanists call this turgor pressure. Water also is squeezed out when the turgor pressure is enough to counter the attractive force of cytoplasmic fluid, which usually has more solutes than soil water has. Both forces have the potential to cause water to move directionally. The sum of these two opposing forces is the "water potential."

When soft plant parts are erect, as much water is moving into cells as is moving out, and the constant pressure keeps cell walls plump. Suppose the soil dries or gets too salty. Water's concentration gradient reverses and cells lose water. Osmotically induced shrinkage of cytoplasm in all of the young cells of soft plant parts results in wilting; the parts droop with the loss of turgor pressure.

The experiment at *left* demonstrates the wilting effect. Place 10 grams of table salt (NaCl) in 60 milliliters of water. Pour the salty solution into the soil around a tomato plant. The plant starts to collapse after 5 minutes. In less than 30 minutes, wilting is severe.

The Water-Conserving Cuticle

Even mildly water-stressed plants would rapidly wilt and die without their **cuticle** (Figure 30.11). Epidermal cells secrete this translucent, water-impermeable layer, which coats their wall regions exposed to the air. At the cuticle surface are deposits of waxes, which are water-insoluble lipids with long fatty-acid tails. The cuticle proper consists of waxes embedded in **cutin**, which is an insoluble lipid polymer. Beneath the waxes, cellulose threads through the cutin. A layer of polysaccharides (pectins) often helps bind the cuticle to cell walls.

A cuticle does not bar the passage of light rays into photosynthetic parts of the plant. It does restrict water loss. It also restricts the *inward* diffusion of the carbon dioxide required for photosynthesis and the *outward* diffusion of oxygen by-products. Recall, from Section 7.7, that a buildup of oxygen in the air spaces inside a leaf has bad effects on the rate of photosynthesis.

Controlled Water Loss at Stomata

How do carbon dioxide and oxygen get past the cuticle-covered, water-conserving epidermis? They do this at **stomata** (singular, stoma). At these openings, water evaporates from a plant and carbon dioxide moves in. A pair of specialized cells, the **guard cells**, define each opening (Figure 30.12). When each swells with water,

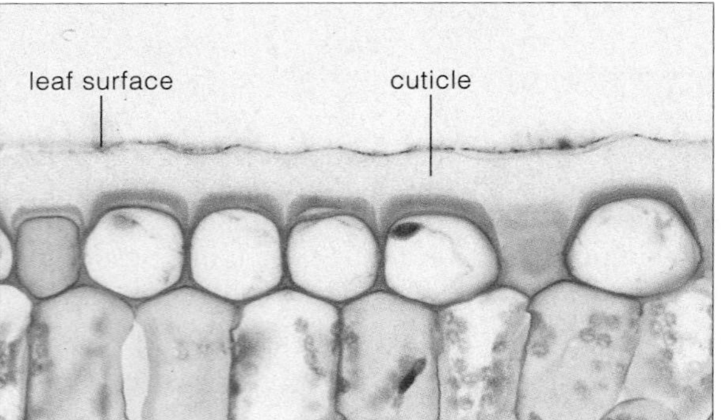
leaf surface
cuticle

Figure 30.11 Waxy cuticle on upper leaf epidermis. Water, carbon dioxide, and oxygen cannot cross the cuticle proper. They enter or escape from the leaf primarily at stomata. In general, plants in deserts and other seasonally dry habitats have thick cuticles, which provide greater protection against water loss. Aquatic plants have thin cuticles or none at all.

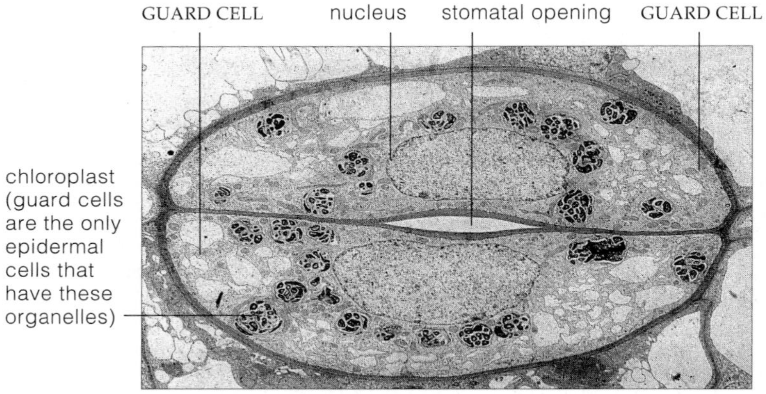

GUARD CELL nucleus stomatal opening GUARD CELL

chloroplast (guard cells are the only epidermal cells that have these organelles)

Figure 30.12 Guard cells. This pair spans stem epidermis from a prickly pear (*Opuntia*), also called the beavertail cactus. Cacti have their stomata on stems and on tiny leaves (if any). They also bear spines. These modified nonphotosynthetic shoots protect succulent stems from browsing animals while reducing the surface area from which water can evaporate.

Further reading: Student Guide to InfoTrac on web site

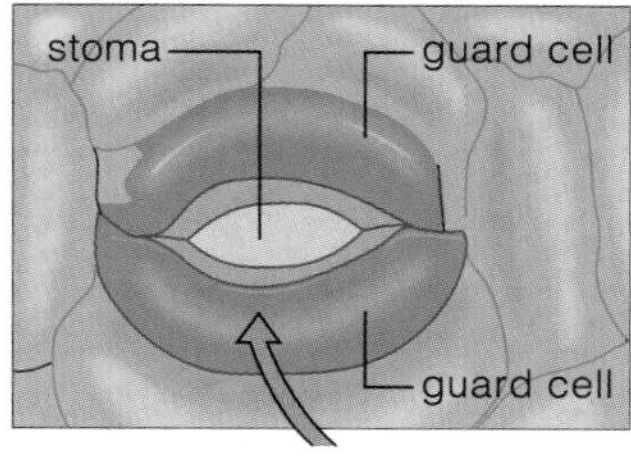
stoma
guard cell
guard cell

a Stomatal opening. Water enters collapsed guard cells, which swell under turgor pressure and move apart, forming the stoma.

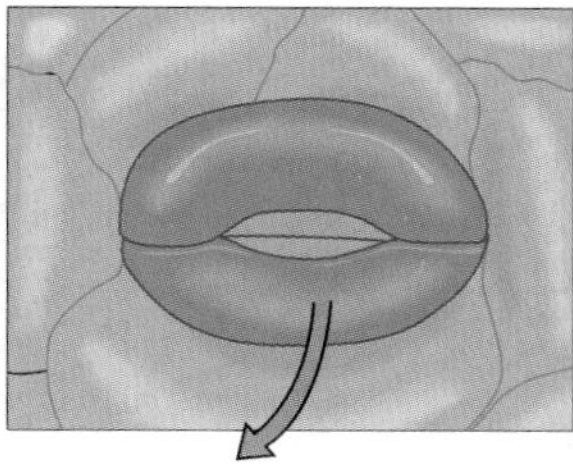

b Stomatal closing. Water leaves swollen guard cells, which collapse against each other and close the stoma.

Figure 30.13 Stomatal action. Whether a stoma is open or closed at any time of day or night depends on changes in the shape of its paired guard cells, which together define one opening across the cuticle-covered epidermis.

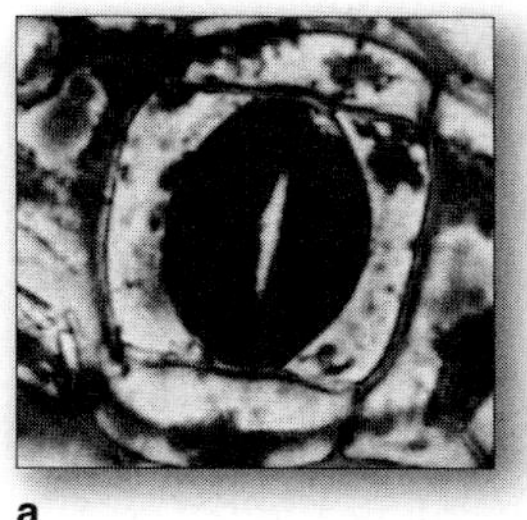

a

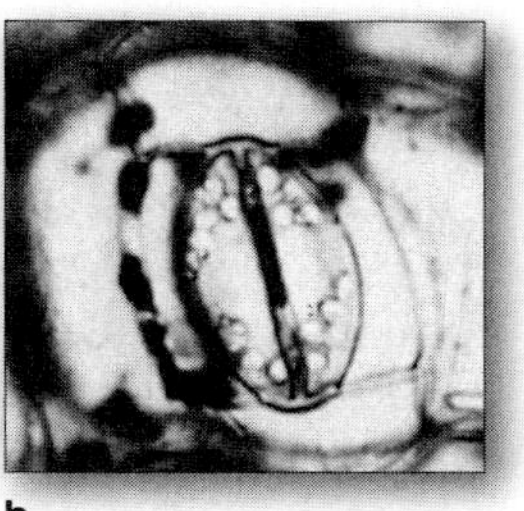

b

Figure 30.14 Experimental evidence of an accumulation of potassium ions in guard cells that are expanding with incoming water. Strips from leaf epidermis of a dayflower (*Commelina communis*) were immersed in solutions containing dark-staining substances that bind preferentially with potassium ions. (**a**) In leaf samples with opened stomata, most of the potassium ions were concentrated in the guard cells. (**b**) In leaf samples with closed stomata, very little of the potassium was in guard cells; most of it was present in other epidermal cells.

a
b

Figure 30.15 (**a**) Stomata at holly leaf surface. (**b**) What they look like when a holly plant grows in industrialized regions. Gritty airborne pollutants clog its stomata and prevent much of the sun's rays from reaching photosynthetic cells in the leaf.

the force of internal fluid pressure distorts its wall. An osmotically induced, internal fluid pressure that builds up against cell walls is called **turgor pressure**. It makes guard cells bend in such a way that a gap (one stoma) forms between them. When two guard cells lose water and their internal fluid pressure drops, they collapse against each other and so close the gap (Figure 30.13).

In most plants, stomata stay open during the day, when photosynthesis takes place. They do lose water—but they gain carbon dioxide for the reactions. Stomata stay closed at night. Then, plants conserve water, and precious carbon dioxide accumulates inside the leaf as the cells busily engage in aerobic respiration.

Guard cells are the only epidermal cells that contain chloroplasts. Whether a stoma opens or closes depends on how much water and carbon dioxide its guard cells hold. Photosynthesis starts after the sun comes up. As the morning progresses, carbon dioxide levels drop in cells—including guard cells. The decrease helps trigger active transport of potassium ions into the guard cells (Figure 30.14). So do blue wavelengths of light, which penetrate the atmosphere better as the sun arcs higher in the sky. Water follows potassium ions into the cells, by osmosis. That inward movement of water provides the fluid pressure to open the stoma.

Carbon dioxide levels in the cells rise when the sun sets and photosynthesis is shut down. Potassium, then water, move out of the guard cells, and stoma close.

CAM plants, including most cacti, conserve water differently. They open stomata at night, when they fix carbon dioxide by the metabolic C4 pathway described in Section 7.7. The next day, when their stomata close, CAM plants use carbon dioxide in photosynthesis.

As this section makes clear, plant survival depends on stomata. Think about this when you are out and about on smog-shrouded days (Figure 30.15).

Water-dependent events in plants are severely disrupted when water loss exceeds uptake at roots for extended periods. Wilting is one observable outcome.

Transpiration and gas exchange occur mainly at stomata. These numerous small openings span the waxy cuticle, which covers all plant epidermal surfaces exposed to air.

Plants open and close stomata at different times to control water loss, carbon dioxide uptake, and oxygen disposal, all of which affect rates of photosynthesis and plant growth.

30.5 HOW ARE ORGANIC COMPOUNDS DISTRIBUTED THROUGH PLANTS?

Whereas xylem distributes water and minerals through the plant, the vascular tissue called **phloem** distributes organic products of photosynthesis. Like xylem, phloem consists of many conducting tubes, fibers, and strands of parenchyma cells, all of which extend through the plant body. Unlike xylem, phloem has conducting tubes of a type called **sieve tubes**. These tubes consist of *living* cells, positioned side by side and end to end in the vascular tissues. Figure 30.16 shows an example of the cells themselves, which are called sieve-tube members. Dissolved sugars and other organic compounds flow rapidly through large pores in the abutting end walls of sieve-tube members.

Adjoining the long sieve tubes are **companion cells**. Like the sieve-tube members themselves, companion cells are differentiated parenchyma cells. As you will see, companion cells help load organic compounds into sieve tubes.

To understand what goes on inside phloem, think about what happens to sucrose and other organic products of photosynthesis. Leaf cells use some of these products for their own activities. The remainder travels to roots, stems, buds, flowers, and fruits. In most cells, carbohydrates become stored as starch, in plastids. The proteins and fats, which the cells synthesize from carbohydrates and the amino acids distributed to them, are stored in many seeds. Avocados and some other fruits also accumulate fats.

Starch molecules are too large for transport across the plasma membrane of cells that produce them. Also, they are too insoluble for transport to other plant parts. Overall, proteins are too large and fats are too insoluble to be transported from the storage sites. *But cells convert the storage forms of organic compounds to solutes of smaller size that are more easily transported through the phloem.* For instance, the cells degrade starch to glucose monomers. When one of these monomers then combines with fructose, the result is sucrose—which is an easily transportable sugar.

Experiments with aphids revealed that sucrose is the main carbohydrate being transported inside phloem. These insects were anesthetized by exposure to high levels of carbon dioxide as they were feeding on juices in conducting tubes of phloem. Then their body was detached from the mouthparts, which were still embedded in the sieve tubes. For most of the plant species investigated, sucrose was the most abundant carbohydrate in the fluid that was being forced out of the tubes.

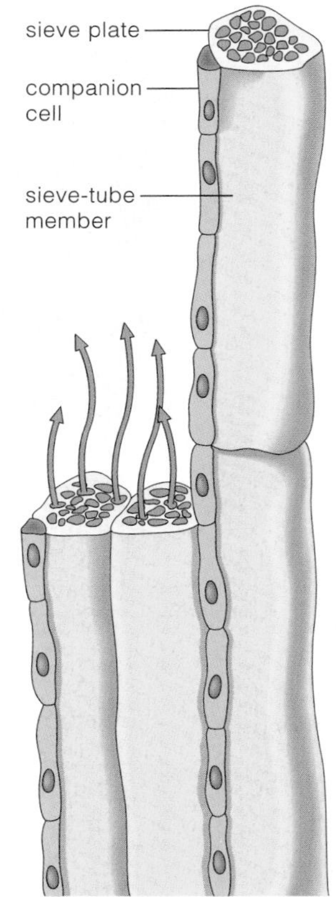
sieve plate
companion cell
sieve-tube member

Figure 30.16 Examples of sieve-tube members, living cells that interconnect to form a conducting tube in phloem. Living companion cells expend ATP energy to load organic compounds into adjacent sieve-tube members.

Translocation

Translocation is the technical name for the transport of sucrose and all other organic compounds through the phloem of a vascular plant. High pressure drives this process. Often that pressure is five times as high as it is in automobile tires. Aphids demonstrate the magnitude of it when they force their mouthparts into sieve tubes and feed upon the dissolved sugars. The high pressure can force fluid through an aphid gut and out the other end, as "honeydew" (Figure 30.17). Park a car under some trees being attacked by aphids, and it might get spattered by sticky honeydew droplets thanks to the high fluid pressure in phloem.

Pressure Flow Theory

Phloem translocates organic compounds along gradients of decreasing pressure and solute concentrations. The *source* of the flow is any region of the plant where the organic compounds are being loaded into sieve tubes. Common sources are mesophylls, the photosynthetic tissues in leaves. The flow ends at a *sink*, which is any plant region where the organic compounds are being unloaded, used, or stored. As they are developing, flowers and fruits are common sink regions.

Why do the organic compounds flow from a source to a sink? According to the **pressure flow theory**, internal pressure builds up at the source end of the sieve-tube system and *pushes* the solute-rich

Figure 30.17 A honeydew droplet exuding from the end of an aphid gut. The insect's tubular mouthpart penetrated a conducting tube in phloem. The sugary fluid was under high pressure in phloem and was forced out through the gut's terminal opening.

Further reading: Student Guide to InfoTrac on web site

upper epidermis
photosynthetic cell
sieve tube
companion cell
section from a leaf
lower epidermis

a *Loading at a source.* Photosynthetic cells in leaves are a common source of organic compounds that must be distributed through a plant. Small, soluble forms of these compounds move from the cells into phloem (a leaf vein).

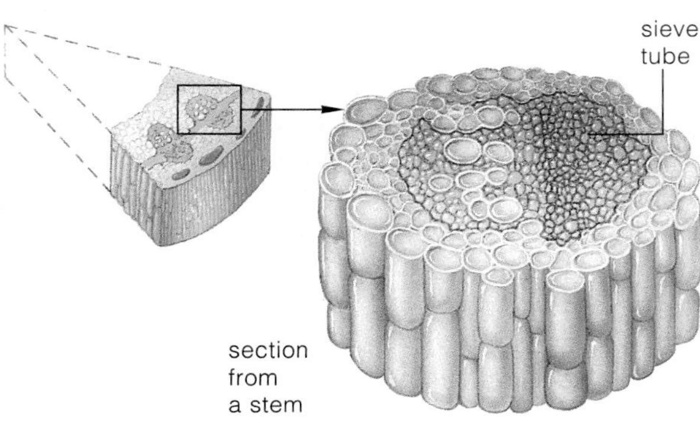
sieve tube
section from a stem

b *Translocation along a distribution path.* Fluid pressure is greatest inside sieve tubes at the source. It pushes the solute-rich fluid to a sink, which is any region where cells are growing or storing food. There, the pressure is lower because cells are withdrawing solutes from the tubes.

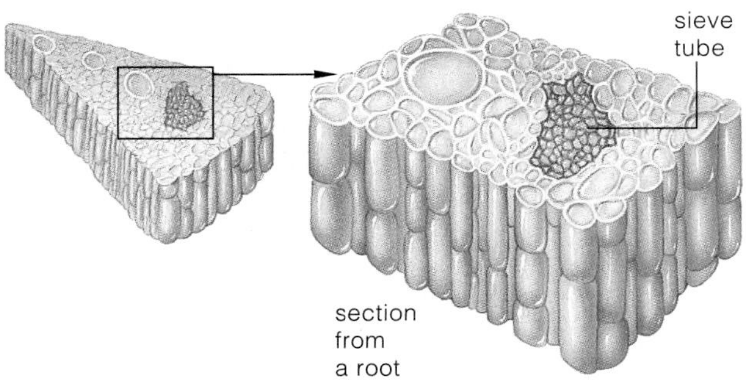
sieve tube
section from a root

c *Unloading at the sink.* Solutes are unloaded from sieve tubes into cells at the sink; water follows. Translocation continues as long as solute concentration gradients and a pressure gradient exist between the source and the sink.

Figure 30.18 Translocation of organic compounds in sow thistle (*Sonchus*). Review Figure 7.1 to get an idea of how translocation relates to photosynthesis in vascular plants.

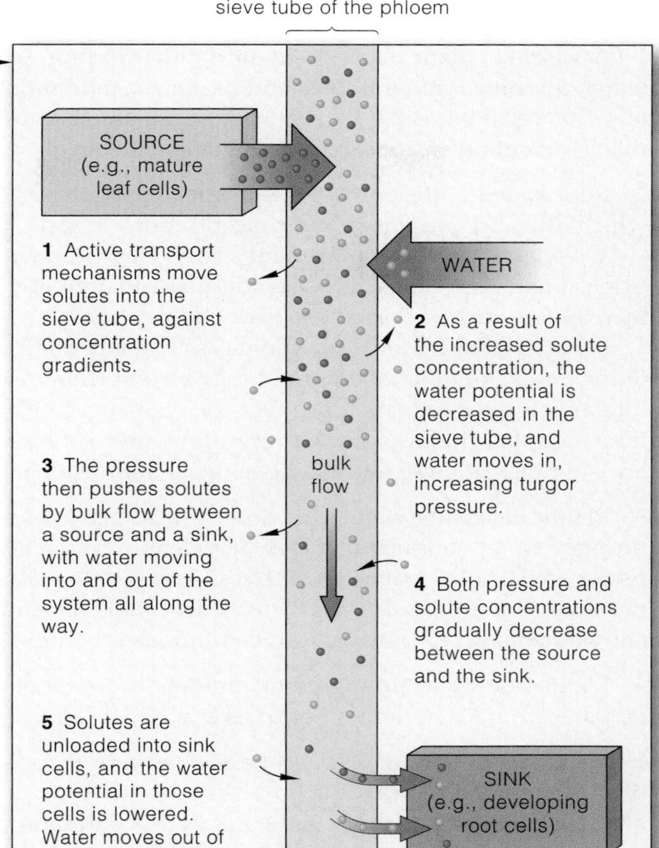
sieve tube of the phloem
SOURCE (e.g., mature leaf cells)
WATER
1 Active transport mechanisms move solutes into the sieve tube, against concentration gradients.
2 As a result of the increased solute concentration, the water potential is decreased in the sieve tube, and water moves in, increasing turgor pressure.
3 The pressure then pushes solutes by bulk flow between a source and a sink, with water moving into and out of the system all along the way.
bulk flow
4 Both pressure and solute concentrations gradually decrease between the source and the sink.
5 Solutes are unloaded into sink cells, and the water potential in those cells is lowered. Water moves out of the sieve tube and into sink cells.
SINK (e.g., developing root cells)

solution on toward any sink, where solutes are being removed. Experimental evidence for the pressure flow theory was obtained from investigations of sow thistle (*Sonchus*). Use Figure 30.18 to track what happens after sucrose moves from photosynthetic cells into small veins in the leaves of this plant. Companion cells in the veins spend energy to load sucrose into adjacent sieve-tube members. As the sucrose concentration increases in the tubes, water also moves in, by osmosis. More internal fluid pressure is exerted against the tube walls. As the pressure increases, it pushes the sucrose-laden fluid out of the leaf, into the stem, and on to the sink.

Plants store organic compounds in the form of starch, fats, and proteins. They convert the storage forms to sucrose and other small units that are soluble and easily translocated.

Translocation is the distribution of organic compounds to different plant regions. It depends on concentration and pressure gradients in the sieve-tube system of phloem.

Gradients exist as long as companion cells load compounds into the sieve tubes at sources, such as mature leaves, and as long as compounds are removed at sinks, such as roots.

30.6 SUMMARY

1. A vascular plant depends upon the distribution of water, dissolved mineral ions, and organic compounds to all of its living cells. Figure 30.19 summarizes how that distribution sustains plant growth and survival.

2. Root systems efficiently take up water and nutrients, which often are present in low concentrations in soil.

a. Collectively, a vascular plant's root hairs (slender extensions of specialized root epidermal cells) greatly increase the surface area available for absorption.

b. Bacterial and fungal symbionts assist many plants in the uptake of mineral ions, and they benefit from the interaction by using some products of photosynthesis. Root nodules (induced by bacteria) and mycorrhizae are examples of these mutually beneficial interactions.

3. Plants distribute water and dissolved mineral ions through water-conducting tubes of xylem, a vascular tissue. Cells called tracheids and vessel members form the tubes but are dead at maturity. Their walls remain and interconnect to form narrow, continuous pipelines.

4. Plants lose water through transpiration: evaporation of water from leaves and other parts exposed to air.

5. Here are the points of the cohesion–tension theory of water transport in plants:

a. Inside xylem's water-conducting cells, continuous negative pressures (tensions) extend from leaves to the roots. Transpiration causes the tension.

b. When water molecules escape from the leaves, replacements are pulled into the leaf under tension.

c. The collective strength of numerous hydrogen bonds between water molecules imparts cohesion that allows water molecules to be pulled up as continuous fluid columns. The "cohesion" part of the theory refers to the capacity of the hydrogen bonds between water molecules to resist rupturing when put under tension.

6. Most plants have a waxy, water-impermeable cuticle covering their aboveground parts. They lose water and take up carbon dioxide at stomata. A pair of guard cells (specialized parenchyma cells) defines each of these small openings across leaf (and stem) epidermis.

7. Stomata open and close at different times. Control of their action balances water conservation with carbon dioxide uptake and the release of oxygen.

a. In many kinds of plants, stomata open during the day. Such plants lose water but take in carbon dioxide for photosynthesis. Stomata close at night; thus, water and the carbon dioxide released by aerobically respiring cells are conserved for the next day.

b. CAM plants open stomata and fix carbon dioxide by the C4 pathway at night. During the day, they close their stomata and use the carbon fixed the night before for photosynthesis.

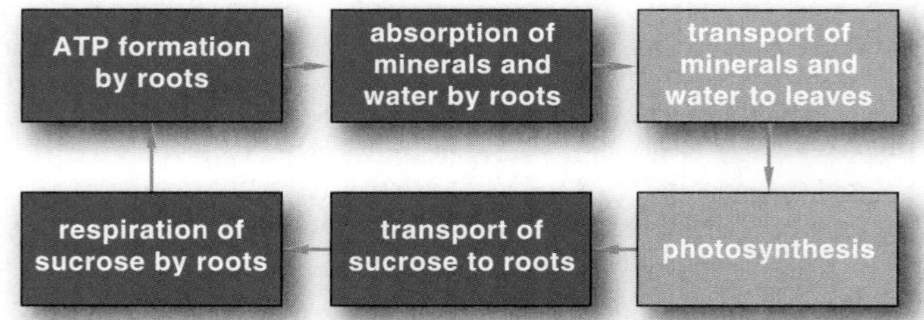
ATP formation by roots
absorption of minerals and water by roots
transport of minerals and water to leaves
photosynthesis
transport of sucrose to roots
respiration of sucrose by roots

Figure 30.19 Summary of the interdependent processes that sustain the growth of vascular plants. All living cells in plants require oxygen, carbon, hydrogen, and at least thirteen mineral ions. They all produce ATP to drive metabolic activities.

8. Plants distribute organic compounds through sieve tubes of phloem, a vascular tissue. The distribution of organic compounds in phloem is called translocation.

9. According to the pressure flow theory, translocation is driven by differences in solute concentrations and pressure between source and sink regions. A source is any site where organic compounds are being loaded into sieve tubes, such as mature leaves. A sink is any site where compounds are being unloaded from sieve tubes. Roots are examples. Solute concentration gradients and pressure gradients exist for as long as companion cells expend energy to load solutes into the sieve tubes and for as long as solutes are removed at the sink.

Review Questions

1. Define soil, then distinguish between: *30.1*
   a. humus and loam
   b. leaching and erosion
   c. macronutrient and micronutrient (for plants)

2. Define nutrient. What are some signs that a plant is deficient in one of the essential nutrients listed in Table 30.1? *30.1*

3. What is the function of the Casparian strip in roots? *30.2*

4. Using Dixon's model, explain how water moves from soil upward through tall plants. *30.3*

5. Describe the structure and function of a plant cuticle. *30.4*

6. Which type of ion influences stomatal action? *30.4*

7. Explain translocation according to the pressure flow theory described in this chapter. *30.5*

Self-Quiz (*Answers in Appendix III*)

1. Water can be pulled up through a plant by the cumulative strength of ______ between water molecules.

2. In plants, mineral ions ______.
   a. have roles in metabolism
   b. help establish gradients across membranes
   c. influence water movement into cells
   d. maintain cell shape and growth
   e. all of the above

3. The nutrition of some plants depends on a root–fungus association known as a ______.
   a. root nodule c. root hair
   b. mycorrhiza d. root hypha

4. The nutrition of some plants depends on a root–bacterium association known as a ______ .
   a. root nodule c. root hair
   b. mycorrhiza d. root hypha

5. Water evaporation from plant parts is called ______ .
   a. translocation c. transpiration
   b. expiration d. tension

6. Water transport from roots to leaves is explained by ______ .
   a. the pressure flow theory
   b. differences in source and sink solute concentrations
   c. the pumping force of xylem vessels
   d. the cohesion–tension theory

7. A ______ in endodermal cell walls forces water and solutes that entered roots to move through the cells, not around them.
   a. cutin strip c. Casparian strip
   b. lignin strip d. cellulose strip

8. During the day, most plants lose ______ and take up ______ .
   a. carbon dioxide; water c. oxygen; water
   b. water; oxygen d. water; carbon dioxide

9. At night, most plants conserve ______ , and ______ accumulates.
   a. carbon dioxide; oxygen c. oxygen; water
   b. water; oxygen d. water; carbon dioxide

10. In phloem, organic compounds flow through ______ .
   a. collenchyma cells c. vessels
   b. sieve tubes d. tracheids

11. Match the concepts of plant nutrition and transport.

____ stomata	a. evaporation from plant parts
____ nutrient	b. response to scarce soil nutrients
____ sink	c. balancing water loss with carbon dioxide requirements
____ root system	d. cohesion in water transport
____ hydrogen bonds	e. sugars unloaded from sieve tubes
____ transpiration	f. organic compounds distributed through the plant body
____ translocation	g. element with roles in metabolism that no other element can fulfill for an organism

Critical Thinking

1. Home gardeners, like farmers, must ensure that their plants have access to nitrogen from either nitrogen-fixing bacteria or fertilizer. Insufficient nitrogen stunts plant growth; leaves turn yellow and die. Which major classes of biological molecules incorporate nitrogen? How would a low nitrogen level in plants affect biosynthesis and cause symptoms of nitrogen deficiency?

2. When moving a plant from one place to another, it helps to include some native soil around the roots. Explain why, given what you know about mycorrhizae and root hairs.

3. Henry discovered a way to keep all of a plant's stomata open at all times. He also figured out how to keep all the stomata of another plant closed all the time. Both plants died. Explain why.

4. Allen is studying the rate of transpiration from tomato plant leaves. He notices that several environmental factors, including wind and relative humidity, affect the rate. Explain why.

5. You have just returned home from a three-day vacation. Your plants tell you, by their severe wilting, that you forgot to water them before you departed. Being aware of the cohesion–tension theory of water transport, explain what happened to them.

6. Not having a green thumb, your friend Stephanie decides to grow zucchini plants, which are most forgiving of poor soils and amateur gardeners. She plants too many seeds and ends up with far too many plants. Among them you happen to notice a stunted plant and decide to find out what happened to it. After many experiments, you decide that its leaves are producing a mutated, malfunctioning form of an enzyme that is necessary for the formation of sucrose. Knowing what you do about the pressure flow theory, explain why plant growth was hindered.

Figure 30.20 Middle fork of the Salmon River, Idaho.

7. At the middle fork of the Salmon River in Idaho, clear water from a wilderness area (visible at *right* in Figure 30.20) converges on brown-colored water (visible at *left* and *center*). The brown water is enriched with silt from a wilderness area that was disturbed by cattle ranching and some other human activities. Knowing what you do about the nature of topsoil, what is probably happening to plant growth in the disturbed habitat? Knowing how fishes acquire oxygen for aerobic respiration, how might the silt be affecting their survival?

Selected Key Terms

CAM plant *30.4*
carnivorous plant *CI*
Casparian strip *30.2*
cohesion–tension theory *30.3*
companion cell *30.5*
cuticle *30.4*
cutin *30.4*
endodermis *30.2*
erosion *30.1*
exodermis *30.2*
guard cell *30.4*
humus *30.1*
leaching *30.1*
loam *30.1*
mutualism *30.2*
mycorrhiza *30.2*
nitrogen fixation *30.2*
nutrient *30.1*
phloem *30.5*
plant physiology *CI*
pressure flow theory *30.5*
root hair *30.2*
root nodule *30.2*
sieve tube *30.5*
soil *30.1*
stoma (stomata) *30.4*
topsoil *30.1*
tracheid *30.3*
translocation *30.5*
transpiration *30.3*
turgor pressure *30.4*
vascular cylinder *30.2*
vessel member *30.3*
xylem *30.3*

Readings *See also www.infotrac-college.com*

Galston, A., P. Davies, and R. Satter. 1980. *The Life of a Green Plant.* Englewood Cliffs, New Jersey: Prentice-Hall.

Hausenbuiller, R. 1985. *Soil Science: Principles and Practices.* Third edition. Dubuque, Iowa: W. C. Brown.

Hopkins. W. G. 1995. *Introduction to Plant Physiology.* New York: Wiley.

Salisbury, F., and C. Ross. 1992. *Plant Physiology.* Fourth edition. Belmont, California: Wadsworth.

31 PLANT REPRODUCTION

A Coevolutionary Tale

Flowering plants thrive almost everywhere, from icy tundra to deserts to oceanic islands. What accounts for their distribution and diversity? Consider the **flower**, which is a specialized reproductive shoot. About 435 million years ago, when plants first invaded the land, insects that fed on decaying plants and spores probably were not far behind them. The smorgasbord of aboveground plant parts seems to have favored natural selection of winged insects with an amazing array of sucking, piercing, and chewing mouthparts.

By 390 million years ago, in humid coastal forests, seed-bearing plants were producing pollen grains. Those small, sperm-bearing suitcases can travel to eggs, which develop inside ovules in separate plant parts. Pollen is rich in nutrients. At first, it simply may have drifted on air currents to the ovules. Then insects made the connection between "plant parts with pollen" and "food." Plants lost some pollen to hungry insects but gained a reproductive advantage. How? Compared to air currents, pollen-dusted insects clambering over the plants could make more accurate deliveries of pollen to ovules. The tastier the pollen, the more home deliveries, and the more seeds. *And the greater the number of seeds, the greater the chances of reproductive success.*

What we are describing here is a case of **coevolution**. The word refers to two or more species jointly evolving as an outcome of close ecological interactions. When one of the species evolves, the change affects selection pressures operating between the two, and so the other species evolves, also.

In our coevolutionary tale, new or modified plant structures that were more enticing to pollen-delivering insects were favored. Individual insects that were quicker to recognize and locate particular plants also enjoyed a competitive edge. Hence the evolution of plants with distinctive flowers, fragrances, and sugar-rich nectar—and of pollinators specifically adapted to take advantage of them. A **pollinator** is any agent that transfers pollen from male to female reproductive parts of flowers of the same plant species. Besides insects, pollinating agents include air currents, water currents, bats, birds, and other animals (Figures 31.1 and 31.2).

Today you can correlate many floral features with specific mutualists. For example, the positioning of a flower's reproductive parts ensures that its pollinating agent will brush past them. Many parts are strategically located above nectar-filled floral tubes that are as long as the feeding devices of their preferred pollinators. Red and yellow flowers attract birds, which have excellent daytime vision but a poor sense of smell. As you might suspect, plants that birds visit do not divert metabolic resources to producing fragrances. By contrast, the red

Figure 31.1 Example of adaptations uniting a flowering plant with its pollinator. The giant saguaro of Arizona's Sonoran Desert has large, showy white flowers at the tips of its spiny arms. Insects and birds visit the flowers by day, and bats visit by night. The plant offers these animals nectar, and the animals transport pollen grains that stick to their body from one cactus plant to another.

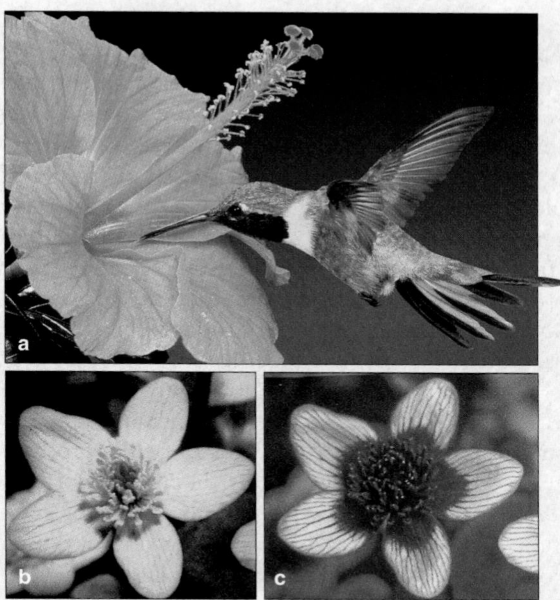
a
b
c
How we see it
How bees see it

Figure 31.2 (**a**) Bahama woodstar sipping nectar from a hibiscus blossom. Like other hummingbirds, it forages for nectar in midflight. Its long, narrow bill coevolved with long, narrow floral tubes. (**b**,**c**) Shine ultraviolet light on a gold-petaled marsh marigold to reveal its bee-attracting pattern.

flowers do not attract beetles, and neither do flowers with large, deep nectar cups in which the beetles might drown. Flowers that beetles (and flies) pollinate smell like rotten meat, moist dung, or the decaying litter on a forest floor, where beetles first evolved.

Daisies and other fragrant flowers with distinctive patterns, shapes, and red or orange components attract butterflies, which forage by day. Nectar-sipping bats and most moths forage by night. They pollinate intensely sweet-smelling flowers with white or pale petals, which are more visible in the dark than colored petals. Long, thin mouthparts of moths and butterflies reach nectar in narrow floral tubes or floral spurs. The Madagascar hawkmoth uncoils a mouthpart the same length as a narrow floral spur of an orchid, *Angraecum sesquipedale*. It is 22 centimeters (more than 8-1/2 inches) long!

Flowers with sweet odors and yellow, blue, or purple parts attract bees. Their ultraviolet-light-absorbing pigments form patterns that say, "Nectar here!" Unlike us, bees can see ultraviolet light, and they find nectar guides alluring (Figure 31.2*b*,*c*). Unlike beetles, bees have long mouthparts that extend into floral tubes.

In short, far from being merely decorative, flowers contribute to reproductive success. This chapter focuses on modes of reproduction of the plants that bear them.

KEY CONCEPTS

1. Sexual reproduction is the premier reproductive mode of flowering plant life cycles. It involves the formation of spores and gametes, both of which develop inside specialized reproductive shoots called flowers.

2. Microspores form in a flower's male reproductive parts and develop into pollen grains, the sperm-bearing male gametophytes. Animals, air currents, and other pollinating agents transfer pollen grains to the female reproductive parts of flowers.

3. Megaspores develop inside ovules, which form on the inner ovary wall of a flower's female reproductive parts. Within each ovule, a mature female gametophyte forms from a megaspore. One of its cells is the egg.

4. After sperm fertilize the eggs, ovules mature into seeds. Each seed consists of an embryo sporophyte and tissues that function in its nutrition and protection.

5. While seeds are developing, tissues of the ovary and sometimes neighboring tissues mature into fruits, which function in seed dispersal. Air currents, water currents, and animals function as dispersal agents.

6. Many species of flowering plants also can reproduce asexually by various mechanisms of vegetative growth, parthenogenesis, and tissue culture propagation.

31.1 REPRODUCTIVE STRUCTURES OF FLOWERING PLANTS

Think Sporophyte and Gametophyte

You probably don't think about this very often, but flowering plants engage in sex. Like humans, they have splendid reproductive systems that produce, nourish, and protect sperm and eggs. Like human females, they house embryos during early development. They issue floral invitations to third parties—pollinators that help sperm and egg get together. Long before we thought of it, flowering plants were using tantalizing fragrances and colors to improve the odds for sexual success.

When you hear the word "plant," you may think of something like a cherry tree (Figure 31.3*a*). The tree is an example of a typical **sporophyte**, a vegetative body that grows, by mitotic cell divisions, from a fertilized egg. During the life cycle, a flowering plant sporophyte bears floral shoots—flowers—that are specialized for reproduction. In those flowers, haploid spores form and develop into the haploid bodies called **gametophytes**. Sperm form inside the male gametophytes; eggs form inside the female gametophytes. As you know, the first cell that forms at fertilization has two sets of genetic instructions, from two individual gametes.

Components of Flowers

Flowers form at floral shoots of the primary plant body. During their formation, they differentiate into nonfertile parts (sepals and petals) and fertile parts (stamens and carpels). Figures 31.3*b* and 31.4 show how components of flowers are arranged. Directly or indirectly, all parts connect to a receptacle, the modified base of the floral shoot. Peel open a rosebud, and you see that sepals—the outermost whorl of leaflike parts—enclose leaflike petals. In turn, petals enclose the fertile components of the flower. This is a common arrangement. The sepals are a flower's calyx. Petals ringing the male and female parts are a flower's corolla.

Like leaves, sepals and petals have ground tissues, vascular tissues, and epidermis. Why are many flowers sweet smelling? Some epidermal cells of petals produce fragrant oils. What gives the petals their shimmer and color? Cells of the ground tissue contain pigments, such as carotenoids (yellow to red-orange) and anthocyanins (red to blue), and small, light-refracting crystals. What does a flower's fragrance or its coloration, patterning, or arrangement of petals do? It attracts pollinators.

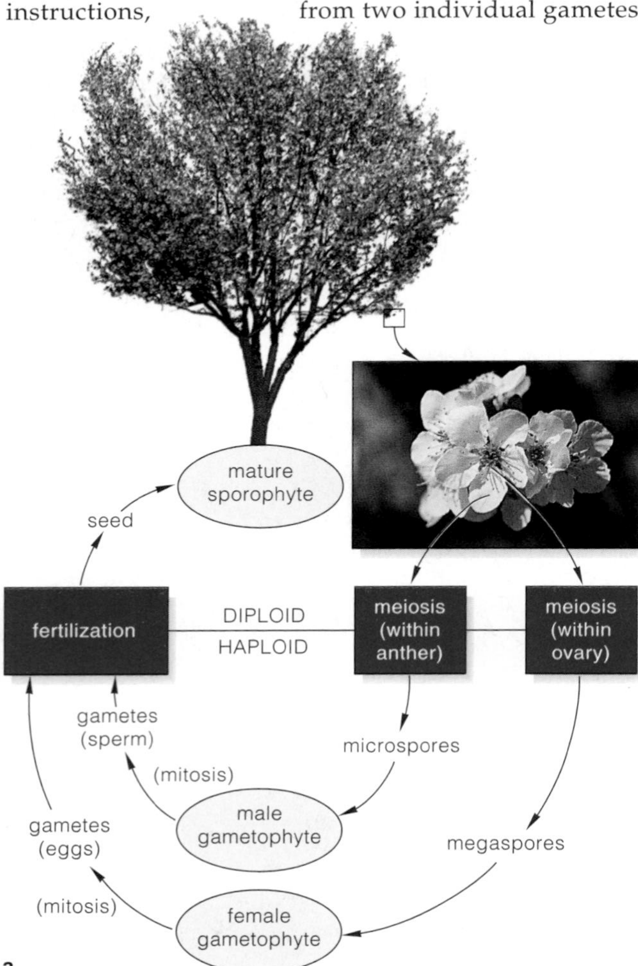
mature sporophyte
seed
fertilization
DIPLOID
HAPLOID
meiosis (within anther)
meiosis (within ovary)
gametes (sperm)
microspores
(mitosis)
male gametophyte
gametes (eggs)
megaspores
(mitosis)
female gametophyte
a

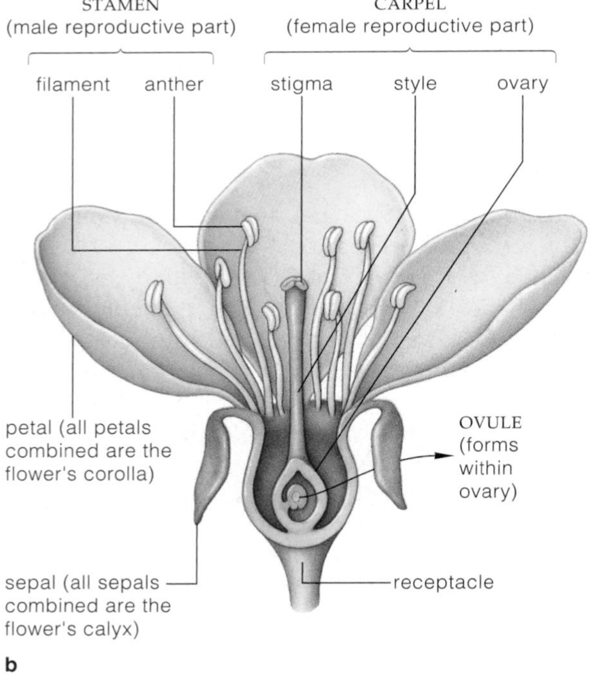
STAMEN
(male reproductive part)
CARPEL
(female reproductive part)
filament
anther
stigma
style
ovary
petal (all petals combined are the flower's corolla)
OVULE (forms within ovary)
sepal (all sepals combined are the flower's calyx)
receptacle
b

Figure 31.3 (**a**) Overview of flowering plant life cycles, using the cherry (*Prunus*) tree as the example. (**b**) Structure of a flower, a cherry blossom. As is the case for flowers of many plants, this blossom has a single carpel (a female reproductive part) and stamens (the male reproductive parts). Flowers of other plants have two or more carpels, often fused into a single structure. Whether single or fused, carpels consist of an ovary and a stigma. Often an ovary extends upward as a slender column, the style.

Further reading: Student Guide to InfoTrac on web site

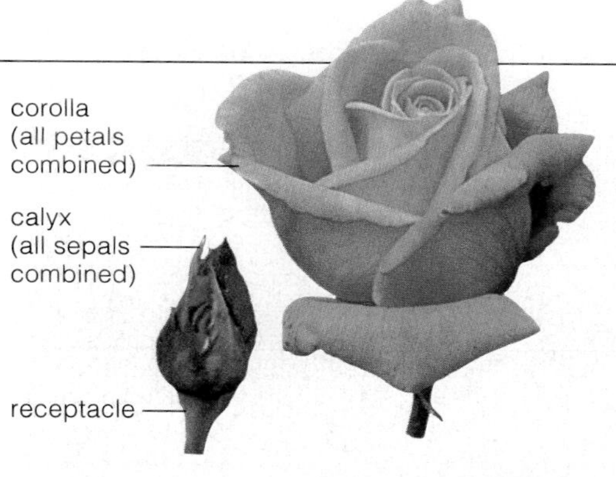
corolla
(all petals
combined)
calyx
(all sepals
combined)
receptacle

Figure 31.4 Location of floral parts in a prized cultivated plant, the rose (*Rosa*).

Where Pollen and Eggs Develop

Just inside the corolla of a flower are **stamens**, the male reproductive parts. Nearly all stamens consist of one anther and a single-veined stalk (filament). An anther is internally divided into pollen sacs, which are chambers in which walled, haploid spores form and develop into male gametophytes called **pollen grains** (Figure 31.5).

A flower's female reproductive parts are located at its center. You may have heard someone call these parts pistils, but their more recent name is **carpels**.

Many flowers have one carpel. Others have more, often fused together to form a compound structure. The lower portion of either single or fused carpels is the **ovary**, where the eggs develop, fertilization takes place, and seeds mature. "Angiosperm," a more formal name for flowering plants, refers to the carpel (from the Greek *angeion*, meaning vessel; and *sperma*, meaning seed). The upper portion of a carpel, the stigma, is a sticky or hairy surface tissue that captures the pollen grains and favors their germination. Commonly the stigma is elevated on a style, a slender, upward extension of the ovary's wall.

Not every kind of flower has stamens and carpels. Some plant species produce *perfect* flowers, with both male and female parts. Other species produce *imperfect* flowers, which contain male or female parts. In certain species, such as oaks, the same individual plant bears male and female flowers. In willows and other species, they are on separate, individual plants.

Sexual reproduction is the dominant reproductive mode of flowering plant life cycles. Such cycles alternate between the production of sporophytes (spore-producing bodies) and gametophytes (gamete-producing bodies).

Flowering plant sporophytes grow by mitotic cell divisions from a fertilized egg. A sporophyte is all of the plant body except for its male and female gametophytes.

Gametophytes arise from haploid spores, which form inside stamens and carpels—the male and female reproductive parts of the flower.

Each male gametophyte develops into a sperm-producing pollen grain. Each female gametophyte develops inside a carpel's ovary; one of its cells is an egg. Also inside the ovary, fertilization takes place and seeds mature.

31.2 *Focus on Health*

POLLEN SETS ME SNEEZING

In the year 1835 Sidney Smith wrote, "I am suffering from my old complaint, the hay-fever . . . that sets me sneezing; and if I begin sneezing at twelve, I don't leave off till two o'clock, and am heard distinctly in Taunton [six miles distant] when the wind sets that way."

Sidney suffered what is now called *allergic rhinitis*. This is the name for hypersensitivity to a normally harmless substance, and such hypersensitivity is common.

The pollen of ragweed, shown in Figure 31.5, brought on Sidney's dreaded hay fever, but it wasn't out to get him. Every spring and summer, flowering plants release pollen grains. In many millions of people, white blood cells respond by mounting an immune response against some of the proteins that project from the surface of pollen grain walls. The pollen grains from different plants have differences in their wall proteins. The immune system of a person who is hypersensitive might chemically respond to some of these but not others. The misdirected response results in a profusely runny nose, reddened and itchy eyelids, congestion, and bouts of sneezing.

Hay fever is a genetic abnormality; it runs in families. Some individuals simply are genetically predisposed to overreact to certain kinds of pollen. In addition, infections, emotional stress, and changes in temperature may trigger immune reactions to pollen.

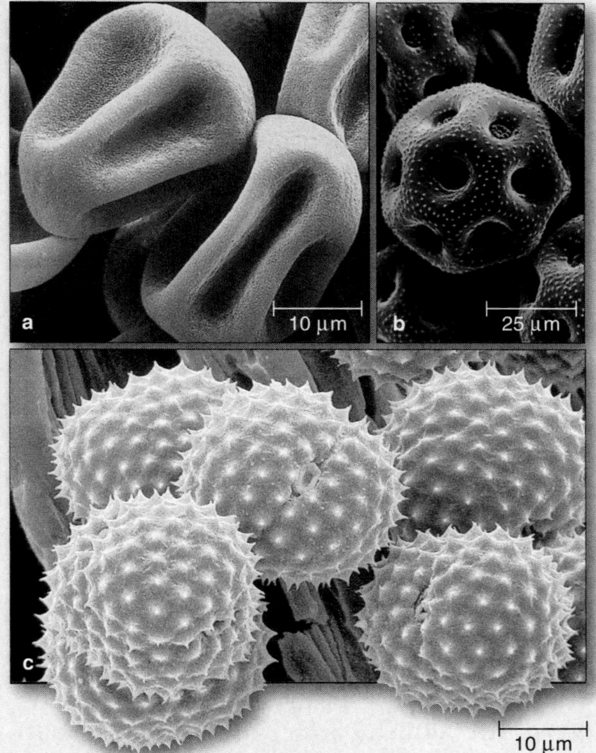
a
10 μm
b
25 μm
c
10 μm

Figure 31.5 Pollen grains from (**a**) grass, (**b**) rose, and (**c**) ragweed plants. Pollen grains of most families of plants differ in size, wall sculpturing, and number of wall pores.

31.3 A NEW GENERATION BEGINS

From Microspores to Pollen Grains

We turn now to pollen grain formation. While anthers are growing, four masses of spore-producing cells form by mitotic cell divisions. Walls develop around them. Each anther now has four chambers, called pollen sacs (Figure 31.6*a*). Cells in the sacs undergo meiosis and cytoplasmic division. This results in **microspores**. The haploid spores go on to develop an elaborately sculpted wall. The walled microspores divide once or twice by way of mitosis and become pollen grains, which now enter a period of arrested growth. In time, they will be released from the anther. Their wall components will protect them from decomposers in their surroundings.

As soon as many types of pollen grains form, they produce sperm nuclei, which are the male gametes of flowering plants. Other types do not do this until after they arrive at a carpel and start growing toward its ovule. In short, a pollen grain is a mature *or* immature male gametophyte, depending on the plant species.

From Megaspores to Eggs

Meanwhile, one or more masses of cells are forming on the inner wall of a flower's ovary. Each is the start of an **ovule**, a structure that houses a female gametophyte and that may become a seed. As each cell mass grows, a tissue forms within it, and one or two protective layers called integuments form around it. Inside the mass, a cell divides by meiosis, and four haploid spores form. Spores formed in flowering plant ovaries are generally larger than microspores and are called **megaspores**.

Commonly, all megaspores but one disintegrate. The one remaining undergoes mitosis three times without cytoplasmic division. At first it is a cell with eight nuclei (Figure 31.6*b*). Its cytoplasm divides after each nucleus migrates to a specific location. In this case, the result is a seven-celled embryo sac (a female gametophyte). One cell, the "endosperm mother cell," has two nuclei and will help form the **endosperm**, a nutritive tissue for the forthcoming embryo. Another cell is the egg.

From Pollination to Fertilization

Every spring, flowering plants release pollen. You are acutely aware of this reproductive event if you are one of the millions of people who suffer from hay fever, the allergic reaction to wall proteins of pollen grains from ragweed and many other plants (Section 31.2).

Specifically, **pollination** means a transfer of pollen grains to a receptive stigma. Air or water currents, birds, and other agents of pollination make the transfer. You already looked at the relationship between flowering plants and their pollinators, in the chapter introduction.

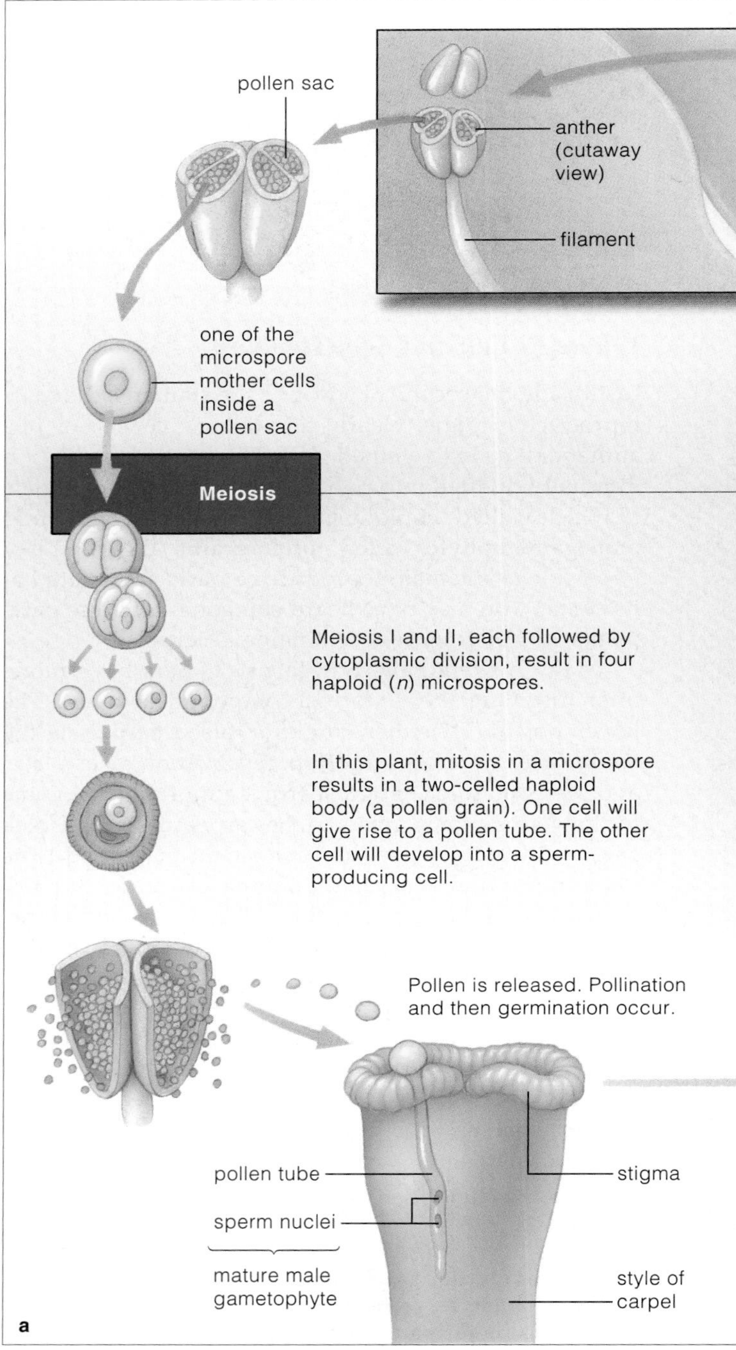
pollen sac
anther (cutaway view)
filament
one of the microspore mother cells inside a pollen sac
Meiosis
Meiosis I and II, each followed by cytoplasmic division, result in four haploid (*n*) microspores.
In this plant, mitosis in a microspore results in a two-celled haploid body (a pollen grain). One cell will give rise to a pollen tube. The other cell will develop into a sperm-producing cell.
Pollen is released. Pollination and then germination occur.
pollen tube
sperm nuclei
mature male gametophyte
stigma
style of carpel
a

Figure 31.6 Life cycle of cherry (*Prunus*), one of the flowering plants classified as a dicot. (**a**) How pollen grains (male gametophytes) develop and germinate. (**b**) Events inside the ovule in this ovary.

Once a pollen grain lands on a receptive stigma, it germinates. For this event, germination means a pollen grain resumes growth and then develops into a tubular structure. This "pollen tube" grows down through the tissues of the ovary and carries the sperm nuclei with it

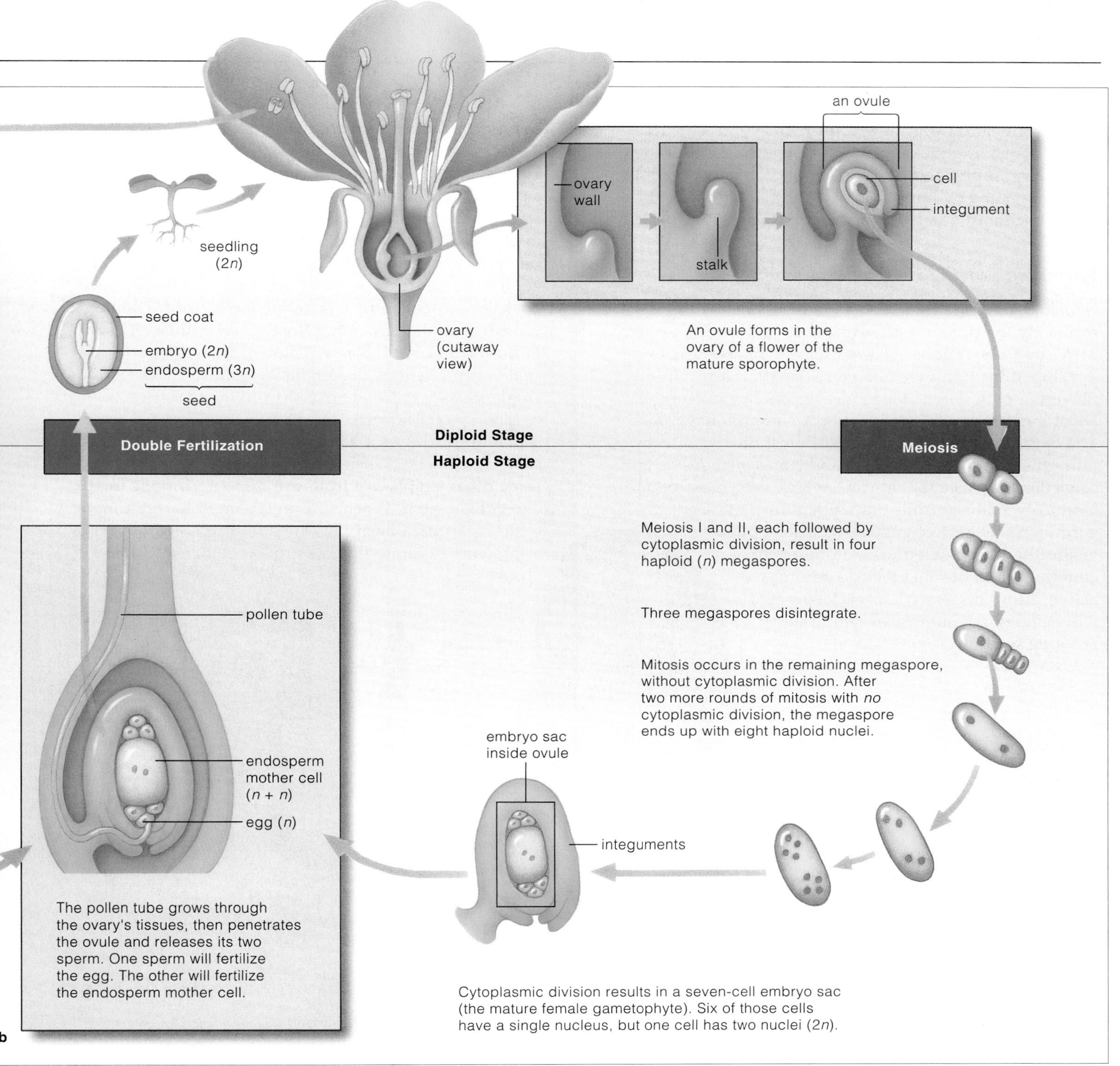
an ovule
ovary wall
stalk
cell
integument
seedling (2n)
seed coat
embryo (2n)
endosperm (3n)
seed
ovary (cutaway view)
An ovule forms in the ovary of a flower of the mature sporophyte.
Double Fertilization
Diploid Stage
Haploid Stage
Meiosis
Meiosis I and II, each followed by cytoplasmic division, result in four haploid (n) megaspores.
Three megaspores disintegrate.
Mitosis occurs in the remaining megaspore, without cytoplasmic division. After two more rounds of mitosis with no cytoplasmic division, the megaspore ends up with eight haploid nuclei.
pollen tube
endosperm mother cell (n + n)
egg (n)
embryo sac inside ovule
integuments
The pollen tube grows through the ovary's tissues, then penetrates the ovule and releases its two sperm. One sperm will fertilize the egg. The other will fertilize the endosperm mother cell.
Cytoplasmic division results in a seven-cell embryo sac (the mature female gametophyte). Six of those cells have a single nucleus, but one cell has two nuclei (2n).
b

(Figure 31.6*b*). Chemical and molecular cues guide the growth of a pollen tube through the ovary's tissues, toward the egg chamber and sexual destiny. When the pollen tube reaches an ovule, it penetrates the embryo sac and its tip ruptures to release two sperm.

"Fertilization" generally means fusion of a sperm nucleus with an egg nucleus. But **double fertilization** occurs in flowering plants. In species having a diploid chromosome number, one sperm nucleus fuses with an egg nucleus to form a diploid ($2n$) zygote. Meanwhile, the other sperm nucleus fuses with both nuclei of the endosperm mother cell. The resulting cell has a triploid ($3n$) nucleus and will give rise to the nutritive tissue.

In flowering plants, sperm cells form within pollen grains, the male gametophytes. Ovules form inside ovaries, and female gametophytes (with egg cells) develop within them.

After pollination and double fertilization, an embryo and nutritive tissue form in the ovule, which becomes a seed.

FROM ZYGOTE TO SEEDS AND FRUITS

After fertilization, the newly formed zygote embarks on a course of mitotic cell divisions that lead to a mature embryo sporophyte. It develops as part of an ovule and is accompanied by formation of a **fruit**: a mature ovary, with or without other, neighboring tissues (Table 31.1).

Formation of the Embryo Sporophyte

Consider how the shepherd's purse (*Capsella*) embryo forms. By the time it reaches the stage shown in Figure 31.7*e*, two **cotyledons**, or seed leaves, have started to develop from two lobes of meristematic tissue. In all flowering plants, cotyledons form as part of an embryo. Dicot embryos have two; monocot embryos have one. The *Capsella* embryo, like those of most dicots, absorbs nutrients from the endosperm and stores them inside its cotyledons. By contrast, in corn, wheat, and most other monocots, endosperm is not tapped until the seed germinates. Digestive enzymes are stockpiled inside the thin cotyledons of a monocot embryo. When the enzymes do become active, nutrients stored in endosperm will be released and transferred to the growing seedling.

Seeds and Fruit Formation

From the time a zygote forms until an embryo matures, a parent plant transfers nutrients to ovule tissues. Food reserves accumulate in endosperm or cotyledons. The ovule eventually separates from the ovary wall, and its integuments thicken and harden into a seed coat. The embryo, food reserves, and coat are a self-contained package—a **seed**, which is defined as a mature ovule.

When seeds form, other floral parts change and start to form fruits. As Table 31.1 shows, three categories are *simple* fruits (either dry or fleshy, and derived from a single ovary), *aggregate* fruits (from numerous separate ovaries of one flower), and *multiple* fruits (from many separate ovaries of a single flower, all attached to the same receptacle). When they have matured, acorns and some other simple dry fruits are intact; pea pods unzip to release seeds. Peaches, as in Figure 31.8*a*, are simple fleshy fruits. Pineapples are a multiple fruit; ovaries of many separate flowers grow together into a fleshy

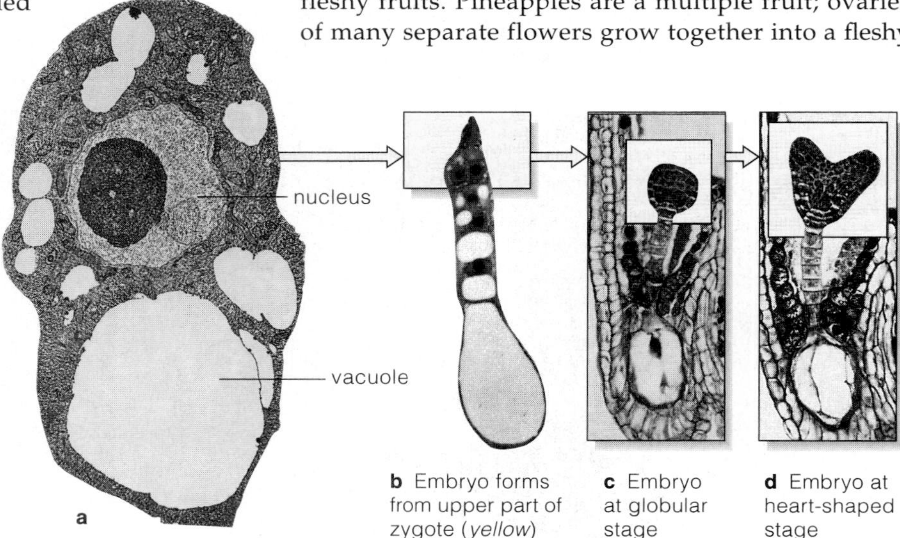
nucleus
vacuole
a
b Embryo forms from upper part of zygote (*yellow*)
c Embryo at globular stage
d Embryo at heart-shaped stage

Figure 31.7 Stages in a dicot seed's development. The micrographs of this *Capsella* seed are not to the same scale. (**a**) Single-celled zygote. (**b–d**) The early embryo is identified by the *yellow* boxes. The row of cells below each box transfers nutrients to the embryo sporophyte from the parent plant. The embryo in (**e**) is well developed, and the one in (**f**) is mature. (**g**) *Capsella* fruit.

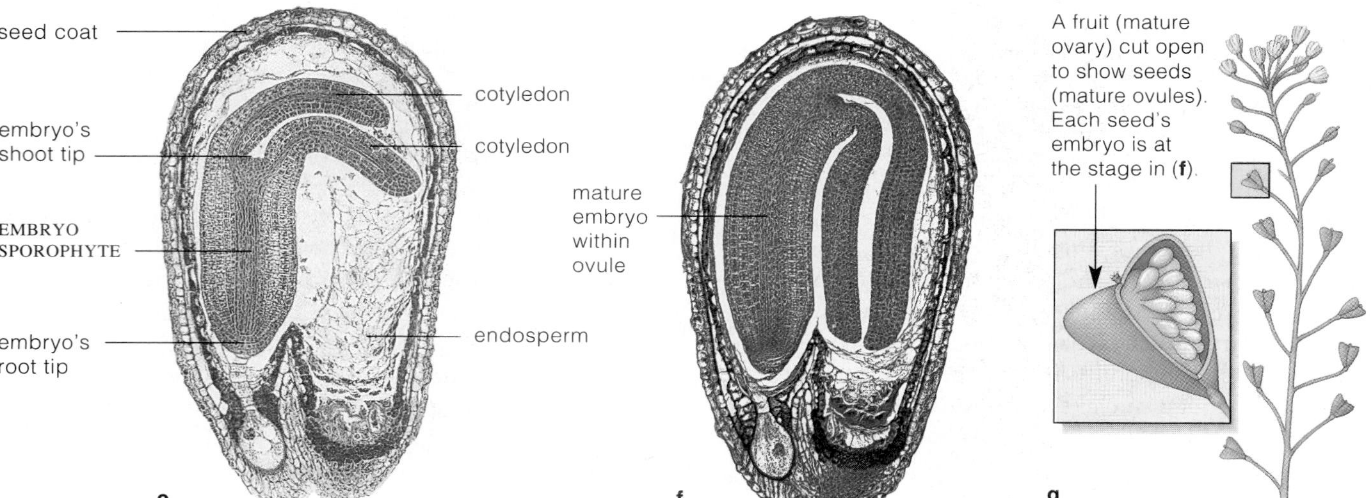
seed coat
cotyledon
embryo's shoot tip
cotyledon
EMBRYO SPOROPHYTE
embryo's root tip
endosperm
e
mature embryo within ovule
f
A fruit (mature ovary) cut open to show seeds (mature ovules). Each seed's embryo is at the stage in (**f**).
g

Further reading: Student Guide to InfoTrac on web site

fleshy fruit (from ovary tissues)
seed
a

b
one of many individual fruits

c

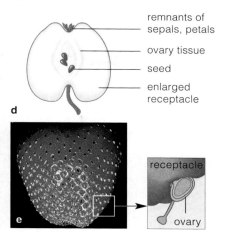

remnants of sepals, petals
ovary tissue
seed
enlarged receptacle
d

receptacle
ovary
e

Figure 31.8 Fruits. (**a**) Peach, a drupe. (**b**) Pineapple (*Ananas*), a multiple fruit. Besides running into the New World, Christopher Columbus ran into cultivated pineapples, a symbol of hospitality to Native Americans. To Columbus, they resembled pine cones, hence the name. Natives used pineapple juice as the base for an alcoholic beverage and for arrow tip poison. (**c**) Fruit formation on an apple (*Malus*) tree. By the time petals drop, the eggs usually are fertilized. (**d**) An apple is an accessory fruit; its flesh consists mainly of an enlarged receptacle and calyx around the core. (**e**) Strawberry (*Fragaria*), an enticingly fragrant accessory fruit.

Table 31.1 Main Categories and Examples of Fruits

SIMPLE FRUITS From one ovary of one flower.

Dry fruit

- *Dehiscent.* Fruit wall splits along definite seams to release seeds. Legume (e.g., pea, bean), poppy, larkspur, mustard
- *Indehiscent.* Fruit wall does not split on seams to release seeds. Acorn, grains (e.g., corn), sunflower, carrot, maple

Fleshy fruit

- *Berry.* Compound ovary, many seeds. Tomato, grape, banana
  - Pepo. Ovary wall has hard rind. Cucumber, watermelon
  - Hesperidium. Ovary wall has leathery rind. Orange, lemon
- *Drupe.* One or two seeds. Thin skin, part of flesh around a usually hard seed. Peach, cherry, apricot, almond, olive

AGGREGATE FRUITS
Many ovaries of one flower, all attached to the same receptacle. Many seeds. Raspberry, blackberry (not really "berries")

MULTIPLE FRUITS
Combined from ovaries of many flowers. Pineapple, fig

ACCESSORY FRUITS
Most tissues of the flesh are not derived from ovary; e.g., mainly from the receptacle. Pome (apple, pear), strawberry

mass as they enlarge (Figure 31.8*b*). Apples, an accessory fruit, are mainly an enlarged receptacle and calyx (Figure 31.8*c,d*). Strawberries are an aggregate fruit with many seeds on a fleshy, expanded receptacle (Figure 31.8*e*).

Botanists view fleshy fruits as having three divisions, although sometimes the three are not clearly visible. The *endo*carp is the innermost portion around the seed or seeds. *Meso*carp is the fleshy portion. *Exo*carp is the skin. Together, the three fruit regions are called a **pericarp**. In walnuts and other dry fruits, the pericarp is typically thin. In true berries, such as tomatoes and grapes, the pericarp is thin skinned and relatively soft at maturity.

A mature ovule, which encases an embryo sporophyte and food reserves inside a protective coat, is a seed.

A mature ovary, with or without additional floral parts that have become incorporated into it, is a fruit.

Fruits are classified according to whether they are dry or fleshy, whether they are derived from one or more ovaries, and whether they incorporate other tissues besides those of the ovary.

Fleshy fruits have three regions. Innermost is the endocarp, which surrounds the seeds. The mesocarp is the fleshy part, and the exocarp is the skin. Together, the three regions are known as the pericarp.

31.5 DISPERSAL OF FRUITS AND SEEDS

The fruits you read about in the preceding section have a common function: *seed dispersal*. To gain insight into their structure, think about how they coevolved with particular dispersing agents—currents of air or water, or animals passing by.

Consider the wind-dispersed fruit of maples (*Acer*), as shown in Figure 31.9*a*. The pericarp of a maple fruit extends out like wings, and when the fruit drops from a tree, it interacts with air currents and spins sideways. The wings can whirl the fruit far enough away that the embryo sporophytes inside the seeds will not have to compete with the parent plant for water, minerals, and sunlight. Brisk winds have been known to transport fruits as far as ten kilometers from a tree. Similarly, air currents easily lift the "parachute" pluming outward from each tiny fruit of dandelions and easily disperse orchid seeds, which are as fine as dust particles.

Other fruits taxi to new locations on or in animals, including many birds, mammals, and insects. Some can adhere to feathers, feet, or fur with hooks, spines, hairs, and sticky surfaces. Fruits of cocklebur, bur clover, and bedstraw are like this. Embryo sporophytes tucked in the seed coats of fleshy fruits survive being eaten and assaulted by the digestive enzymes in an animal's gut. Besides digesting the fruit's flesh, these enzymes digest some of the seed coat. The digested portions make it easier for an embryo to break through a hard coat after the seeds are expelled from the animal body and start to germinate. We turn to this topic in the next chapter.

Some water-dispersed fruits have heavy wax coats, and others, including sedges, have sacs of air that help them float. With its thick, water-repelling pericarp, the coconut palm fruit is adapted to travel across the open ocean. If it doesn't wash onto a beach soon enough, saltwater might penetrate the pericarp and kill the embryo. Coconuts can drift hundreds of kilometers before this happens.

a
seed
wing
b
c

Figure 31.9 (**a**) Winged seeds of maple (*Acer*). (**b**) Students taking a chocolate ice cream break and indirectly contributing to the reproductive success of cacao (*Theobroma cacao*). (**c**) Cacao fruit, or pod. Each contains as many as forty seeds ("beans") that are processed into cocoa butter and essences of chocolate products. Interactions among at least a thousand compounds in chocolate exert compelling (some say addictive) effects on the human brain. As you read in Section 25.10, the average American buys 8 to 10 pounds of chocolate annually. The average Swiss citizen craves a whopping 22 pounds. With that kind of demand, *T. cacao* seeds are cherished and "dispersed" in orderly arrays in plantations, where the trees that grow from them are carefully tended.

Humans are grand agents of dispersal. In the past, explorers transported seeds all over the world, although most places now control imports. Cacao (Figure 31.9*c*), oranges, corn, and other plants were domesticated and encouraged to reproduce in new habitats, in far greater numbers than they did on their own. And in evolutionary terms, reproductive success is what life is all about.

Seeds and fruits are structurally adapted for dispersal by air currents, water currents, and many kinds of animals.

Further reading: Student Guide to InfoTrac on web site

Focus on Science

31.6 WHY SO MANY FLOWERS AND SO FEW FRUITS?

In the Sonoran Desert of Arizona (Figure 31.10*a*), the giant saguaro produces as many as a hundred flowers at the tips of its huge, spiny arms. Remember its large, showy flowers as shown in Figure 31.1? Each flower only blooms for twenty-four hours. Again, insects visit the flowers by day, and bats by night. The animals benefit by getting nectar, and the giant saguaro benefits because the animals deliver their dusting of pollen grains to other giant saguaros.

The flowers that do get pollinated, courtesy of insects or bats, now begin the task of producing seeds and fruits. Petals wilt, then they wither as the many egg cells inside the ovary become fertilized. Each egg-containing ovule expands and matures into a seed. During the same interval, the ovary develops into a fruit (Figure 31.10). Weeks later, the plum-sized fruit splits open, exposing its bright red interior and dark seeds.

White-winged doves feast on the ripe fruits. When they fly away, each might carry hundreds of seeds in its gut. A few seeds may elude the gut's digestive enzymes and later end up on the ground. There, each seed will have a tiny chance of growing into a giant saguaro.

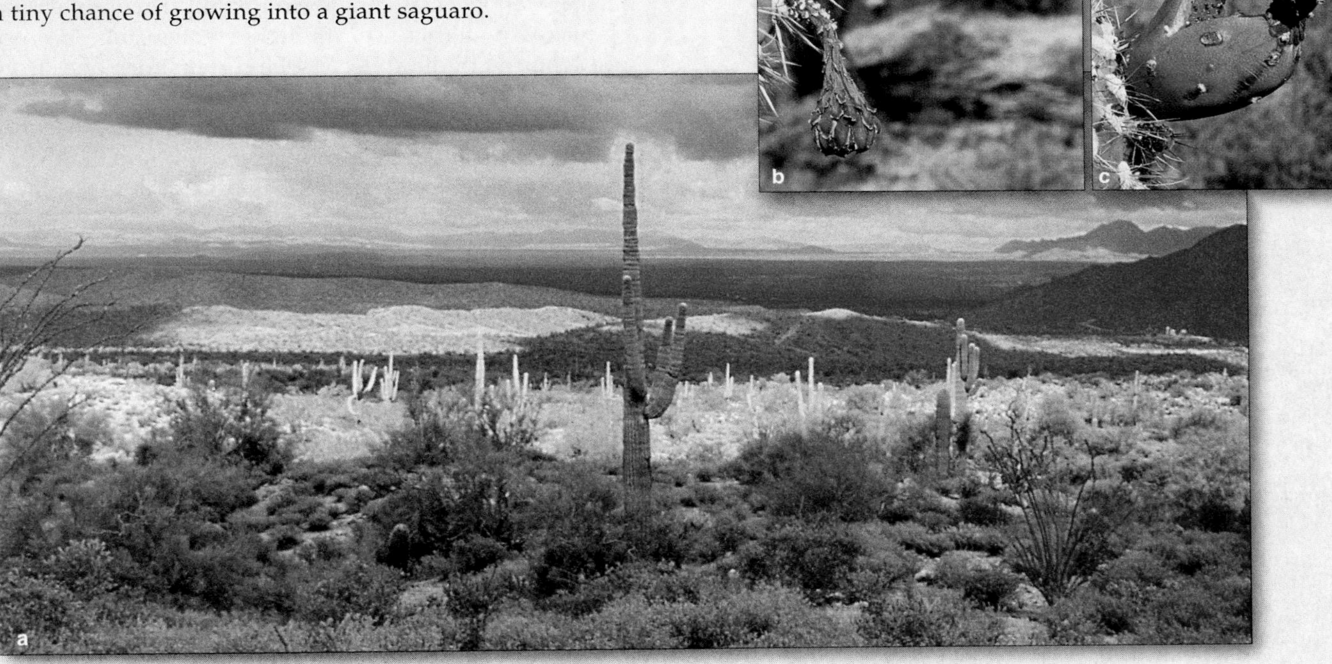

Figure 31.10 (**a**) Giant saguaro growing in Arizona's Sonoran Desert. (**b**) Two fruits. The one above set, and the one below it failed to set. (**c**) A red, maturing fruit.

One of the puzzling aspects of this annual event is the frequency with which saguaro flowers fail to give rise to fruit (Figure 31.10*b*, for example). Why would a cactus invest so much energy constructing a hundred flowers if only thirty or so of them will set fruit? Compare one of these "inefficient" plants with a species that produces a hundred flowers, all of which develop into fruit. Which do you think will leave more descendants?

Common sense might lead you to conclude, "The more fruits, the better." Yet saguaros—like many other plants—commonly produce far more flowers than they do mature fruits. They seem to be producing flowers excessively and giving up chances to make seeds and leave descendants. Doing so is not a reproductive adaptation that you would expect, based on Darwinian evolutionary theory.

Well, *suppose that some of the plants simply did not receive enough pollen to have all their egg cells fertilized*. After all, unfertilized flowers cannot produce seeds and fruits. This hypothesis has been experimentally tested for some plant species. Researchers carefully brushed pollen on every single flower. In many instances, the plants still failed to produce fruit for every flower!

Alternatively, *suppose the presumed "excess" flowers are formed strictly to produce pollen for export to other plants.* Pollen grains are small. Energetically speaking, they are inexpensive to produce compared to large, calorie-rich, seed-containing fruits. Thus, for a fairly small investment, a plant might reap a large reward in offspring that carry its genes.

The idea that some kinds of plants actually set aside flowers exclusively for pollen export is only now being tested for saguaros. Perhaps you might like to design and carry out experiments that will help clear up the mystery of the "excess" flowers of saguaros and similar plants.

31.7 ASEXUAL REPRODUCTION OF FLOWERING PLANTS

Asexual Reproduction in Nature

The features of sexual reproduction that we considered in the preceding sections dominate flowering plant life cycles. Bear in mind, many species also can reproduce asexually by various modes of **vegetative growth**. Table 31.2 lists some of these modes. In essence, new roots and shoots grow right out from extensions or fragments of parent plants. As you know, such asexual reproduction proceeds by way of mitosis, so offspring are genetically identical to the parent; they are a clone.

One "forest" of quaking aspen (*Populus tremuloides*) provides us with an impressive example of vegetative reproduction. The leaves of this flowering plant species tremble even in the slightest breeze, hence the name. Figure 31.11 is a panoramic view of the shoot systems of a single individual. The root system of the parent plant gave rise to adventitious shoots, which became separate shoot systems. Barring rare mutations, individuals at the north end of this vast clone are genetically identical to individuals at the south end. Water travels from roots near a lake all the way to shoot systems in much drier soil. Dissolved ions can travel in the opposite direction.

As long as environmental conditions favor growth and regeneration, such clones are about as close as one can get to being immortal. No one knows how old the aspen clones are. The oldest known clone is a ring of creosote bushes (*Larrea divaricata*) growing in the Mojave Desert. It has been around for the past 11,700 years.

The reproductive possibilities are amazing. Watch strawberry plants send out horizontal aboveground stems (runners), then watch the new roots and shoots develop at every other node. The oranges you eat may be descended from a single tree in southern California that reproduced by **parthenogenesis**. By this process, an embryo develops from an unfertilized egg.

Parthenogenesis can be stimulated when pollen has contacted a stigma even though a pollen tube has not grown through the style. Maybe certain hormones from the stigma or pollen grains diffuse to the unfertilized egg and trigger formation of an embryo, which is $2n$ as a result of fusion of the products of mitotic cell division in the egg. A $2n$ cell outside the gametophyte also may be stimulated to develop into an embryo.

Table 31.2 Asexual Reproductive Modes of Flowering Plants

Mechanism	Examples	Characteristics
VEGETATIVE REPRODUCTION ON MODIFIED STEMS		
1. Runner	Strawberry	New plants arise at nodes along aboveground, horizontal stems.
2. Rhizome	Bermuda grass	New plants arise at nodes of underground horizontal stems.
3. Corm	Gladiolus	New plants arise from axillary buds on short, thick, vertical underground stems.
4. Tuber	Potato	New shoots arise from axillary buds on tubers, which are the enlarged tips of slender underground rhizomes.
5. Bulb	Onion, lily	New bulbs arise from axillary buds on short underground stems.
PARTHENOGENESIS		
	Orange, rose	An embryo develops without nuclear or cellular fusion (for example, from an unfertilized haploid egg or by developing adventitiously, from tissue surrounding the embryo sac).
VEGETATIVE PROPAGATION		
	Jade plant, African violet	A new plant develops from tissue or structure (a leaf, for instance) that drops from the parent plant or is separated from it.
TISSUE CULTURE PROPAGATION		
	Orchid, lily, wheat, rice, corn, tulip	A new plant is induced to arise from a parent plant cell that has not become irreversibly differentiated.

Induced Propagation

Most houseplants, woody ornamentals, and orchard trees are clones. People typically propagate them from cuttings or fragments of shoot systems. For example, with suitable encouragement, a severed African violet leaf may form a callus from which adventitious roots develop. A callus is one type of meristem. Remember, a meristem is a localized region of plant cells that retain the potential for mitotic cell division.

Or consider a twig or bud from one plant, grafted onto a different variety of some closely related species. Vintners in France, for instance, graft prized grapevines onto disease-resistant root stock from America.

Frederick Steward and his coworkers pioneered in **tissue culture propagation**. They cultured small bits of phloem from the differentiated roots of carrot plants (*Daucus carota*) in rotating flasks. They used a liquid growth medium that contained sucrose, minerals, and vitamins. The liquid also contained coconut milk, which Steward knew was rich in then-unidentified growth-inducing substances. As the flasks rotated, individual cells that were torn away from the tissue bits divided and formed multicelled clumps, which sometimes gave rise to roots (Figure 31.12*a*). Steward's experiments were among the first to demonstrate that some cells of

Further reading: Student Guide to InfoTrac on web site —

Figure 31.11 A mere portion of Pando the Trembling Giant, so named by Michael Grant and his coworkers at the University of Colorado, who studied its genetic makeup. This "forest" in Utah is actually a single asexually reproducing male organism, of a type called quaking aspen (*Populus tremuloides*). Its root system extends through about 106 hectares (about 262 acres) of soil and functionally supports its 47,000 genetically identical shoots; that is, the trees. By one estimate, this clone weighs more than 5,915,000 metric tons.

Figure 31.12 (**a**) Cultured cells from a carrot plant. At the time this photograph was taken, roots and shoots of embryonic plants were already forming. (**b**) Young orchid plants, developed from cultured meristem and young leaf primordia. Orchids are one of the most highly prized cultivated plants. Before the meristems were cloned, they were difficult to hybridize. From the time seeds form, it can take seven years or more until a new plant flowers. In natural habitats, the seeds normally will not germinate unless they interact with a specific fungus.

specialized tissues still house the genetic instructions required to produce an entire individual.

Researchers now use shoot tips and other parts of individual plants for tissue culture propagations. The techniques are useful when an advantageous mutant arises. Such a mutant may show resistance to a disease that is crippling to wild-type plants of the same species. Tissue culture propagation can result in hundreds, even thousands of identical plants from merely one mutant specimen. The techniques are already being employed in efforts to improve major food crops, including corn, wheat, rice, and soybeans. They also are being used to increase production of hybrid orchids, lilies, and other prized ornamental plants (Figure 31.12*b*).

Besides reproducing sexually, flowering plants engage in asexual reproduction, as by vegetative growth.

31.8 SUMMARY

1. Sexual reproduction is the main reproductive mode of flowering plant life cycles. Diploid sporophytes, or spore-producing plant bodies, develop, as do haploid gametophytes (gamete-producing bodies).

a. The sporophyte is a multicelled vegetative body with roots, stems, leaves, and, at some point, flowers. Flowers of many species coevolved with animals that feed on nectar or pollen and also serve as pollinators. Gametophytes form in male and female floral parts.

b. Many flowering plants also reproduce asexually. They can do this naturally (as by runners, rhizomes, and bulbs) and artificially (as by cuttings and grafting).

2. Flowers typically have sepals, petals, and one or more stamens, which are male reproductive structures. They also have carpels, the female reproductive structures. Most or all of the reproductive structures are attached to a receptacle, the modified end of a floral shoot.

a. Anthers of stamens contain pollen sacs in which cells divide by meiosis. A wall develops around each resulting haploid cell (microspore), which develops into a sperm-bearing pollen grain (male gametophyte).

b. A carpel (or two or more carpels fused together) has an ovary where the eggs develop, fertilization takes place, and seeds mature. A stigma is a sticky or hairy surface tissue above the lower portion of the ovary. It captures pollen grains and promotes their germination.

3. Inside carpels, ovules form on the inner ovary wall. Each consists of a female gametophyte with an egg cell, an endosperm mother cell, a surrounding tissue, and one or two protective layers called integuments.

a. At maturity, each ovule is a seed; its integuments form the seed coat.

b. The ovary, and sometimes other tissues, matures into a fruit, which is a seed-containing structure.

4. Commonly, female gametophytes form as follows:

a. Four haploid megaspores form after meiosis. All but one usually disintegrate.

b. The one remaining megaspore undergoes mitosis three times but not cytoplasmic division. Its multiple nuclei move to prescribed positions in the cytoplasm.

c. With cytoplasmic division, a female gametophyte (in this case, a seven-celled, eight-nuclei body) results. One cell is an egg. The cell with two nuclei (endosperm mother cell) will help give rise to endosperm, which is a nutritive tissue for the forthcoming embryo.

5. Pollination refers to the arrival of pollen grains on a suitable stigma. Once it has been transferred, a pollen grain germinates. It becomes a pollen tube that grows down through tissues of the ovary, carrying the two sperm nuclei with it.

6. At double fertilization, one sperm nucleus fuses with one egg nucleus, thereby forming a diploid ($2n$) zygote. The other sperm nucleus fuses with both nuclei of the endosperm mother cell to form a cell that will give rise to nutritive tissue in the seed.

7. After double fertilization, the endosperm forms, the ovule expands, and the seed and fruit mature.

8. Fruits function to protect and disperse seeds, which germinate after dispersal from the parent plant. As an example, fruits can attract animals and other dispersal agents; and lightweight fruits can be dispersed by wind. Some have hooks and such that attach to animals.

9. Many flowering plants can reproduce asexually by vegetative growth. For example, new shoot systems may arise by mitotic divisions at nodes or buds along modified stems of the parent plant. New plants also may arise by vegetative propagation, from fragments or parts severed from a parent plant.

Review Questions

1. Define flower and define pollinator. What kinds of animals are attracted to flowers with red and orange components? What kinds respond to the flowers that smell like decaying organic material? What are some of the ways in which night-foraging animals find flowers in the dark? *CI*

2. Label the floral parts. Explain the role each part plays in the reproduction of flowering plants. *31.1*

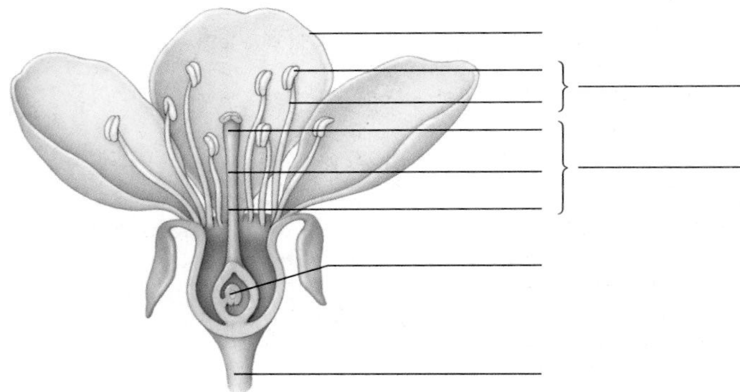

3. Distinguish between these terms:
   a. Sporophyte and gametophyte *31.1*
   b. Stamen and carpel *31.1*
   c. Ovule and ovary *31.1, 31.3*
   d. Microspore and megaspore *31.3*
   e. Pollination and fertilization *CI, 31.3*
   f. Pollen grain and pollen tube *31.1, 31.3*

4. Describe the steps by which an embryo sac, a type of female gametophyte, forms in a dicot such as cherry (*Prunus*). *31.3*

5. Define the difference between a seed and a fruit. *31.4*

6. Do food reserves accumulate in endosperm, cotyledons, or both? If both, in what ways do these structures differ? *31.4*

7. Name an example of a simple dry fruit, simple fleshy fruit, accessory fruit, aggregate fruit, and multiple fruit. *31.4*

8. Name the three regions of a fleshy fruit. What is the name for all three regions combined? *31.4*

9. Define and give an example of an asexual reproductive mode used by a flowering plant. *31.7*

Self-Quiz (*Answers in Appendix III*)

1. The flowers of many species coevolved with insects, birds, and other agents that function as ______.

2. The ______, which bears flowers, roots, stems, and leaves, dominates the life cycle of flowering plants.
   a. sporophyte
   b. gametophyte
   c. sporangium and its derivatives
   d. gametangium and its derivatives

3. Male gametophytes of flowering plants produce ______, and the female gametophytes produce ______.
   a. megaspores; eggs c. eggs; sperm
   b. sperm; microspores d. sperm; eggs

4. A ______ is a closed vessel that contains an ovary in which eggs develop, fertilization occurs, and seeds mature.
   a. pollen sac c. receptacle
   b. carpel d. sepal

5. Seeds are mature ______; fruits are mature ______.
   a. ovaries; ovules c. ovules; ovaries
   b. ovules; stamens d. stamens; ovaries

6. After meiosis within pollen sacs, haploid ______ form.
   a. megaspores c. stamens
   b. microspores d. sporophytes

7. Following meiosis in ovules, ______ megaspores form.
   a. two c. six
   b. four d. eight

8. The seed coat forms from which structure(s)?
   a. ovule integuments c. endosperm
   b. ovary d. residues of sepals

9. Cotyledons develop as part of all flowering plant ______.
   a. seeds c. fruits
   b. embryos d. ovaries

10. The outermost region of a fleshy fruit is the ______.
   a. pericarp c. mesocarp
   b. endocarp d. exocarp

11. Development of a new plant from a tissue or structure that drops or is separated from the parent plant is called ______.
   a. parthenogenesis
   b. exocytosis
   c. vegetative propagation
   d. nodal growth

12. Match the terms with the most suitable description.

___ double fertilization	a. formation of zygote and first cell of endosperm
___ ovule	b. outcome of two species interacting in close ecological fashion over geologic time
___ mature female gametophyte	c. contains a female gametophyte and has potential to be a seed
___ asexual reproduction	d. an embryo sac, commonly with seven cells (one has two nuclei)
___ coevolution	e. mitotic cell division at bud or node produces a new plant

Critical Thinking

1. Wanting to impress her many friends with her sophisticated botanical knowledge, Dixie Bee is preparing a plate of tropical fruits for a party and cuts open a papaya (*Carica papaya*) for the first time. Inside are great numbers of slime-covered seeds, surrounded by soft flesh and soft skin (Figure 31.13). Knowing that her friends will ask what sort of fruit this is and which parts are the endocarp, mesocarp, and exocarp, she panics, runs to her biology book, and opens it to Section 31.4. What does she find out? What will she tell them when they ask what kinds of agents disperse the seeds of such a fruit from the parent plant?

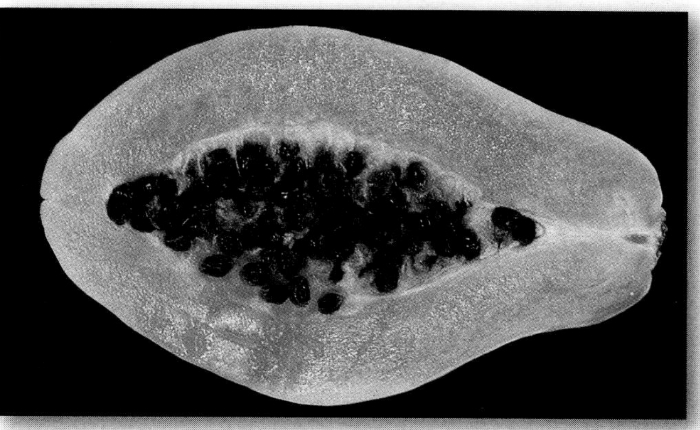

Figure 31.13 Seeds in the fruit of a papaya (*Carica papaya*).

2. Observe several kinds of flowers growing in the area where you live. Given the coevolutionary links between flowering plants and their pollinators, describe what sorts of pollination agents your floral neighbors might depend upon.

3. Elaine, a plant physiologist, succeeded in cloning genes for pest resistance into petunia cells. How can she use tissue culture propagation to produce many petunia plants with those genes?

4. Before cherries, apples, peaches, and many other fruits ripen and the seeds inside them mature, their flesh is bitter or sour. Only later does it become tasty to animals that assist in seed dispersal. Develop a hypothesis of how this feature improves the odds for the plant's reproductive success.

Selected Key Terms

carpel *31.1*
coevolution *CI*
cotyledon *31.4*
double fertilization *31.3*
endosperm *31.3*
flower *CI*
fruit *31.4*
gametophyte *31.1*
megaspore *31.3*
microspore *31.3*
ovary *31.1*
ovule *31.3*
parthenogenesis *31.7*
pericarp *31.4*
pollen grain *31.1*
pollination *31.3*
pollinator *CI*
seed *31.4*
sporophyte *31.1*
stamen *31.1*
tissue culture propagation *31.7*
vegetative growth *31.7*

Readings *See also www.infotrac-college.com*

Grant, M. October 1993. "The Trembling Giant." *Discover* 14(10): 82–89.

Proctor, M., and P. Yeo. 1973. *The Pollination of Flowers*. New York: Taplinger.

Raven, P., R. Evert, and S. Eichhorn. 1992. *Biology of Plants*. Fifth edition. New York: Worth.

Salisbury, F., and C. Ross. 1991. *Plant Physiology*. Fourth edition. Belmont, California: Wadsworth.

Stern, K. 2000. *Introductory Plant Biology*. Eighth edition. Dubuque, Iowa: W. C. Brown. The photographs in this slender, accessible book are exceptional.

32 PLANT GROWTH AND DEVELOPMENT

Foolish Seedlings, Gorgeous Grapes

A few years before the American stock market grew feverishly and then collapsed catastrophically in 1929, a researcher in Japan came across a substance that caused runaway growth and subsequent collapse of rice plants. Ewiti Kurosawa was studying what the Japanese call *bakane*, or the "foolish seedling" effect on rice plants. Stems of rice seedlings that had become infected with *Gibberella fujikuroi*, a fungus, grew twice as long as the stems of uninfected plants. The elongated stems were weak and spindly. Eventually they fell over, and the infected plants died. Kurosawa soon discovered that he could trigger the disease by applying extracts of the fungus to plants. Many years later, other researchers purified the disease-causing substance from fungal extracts. The substance was named gibberellin.

Gibberellin, as we now know, is one of the premier plant hormones. **Hormones**, remember, are signaling molecules secreted by some cells that travel to target cells, where they stimulate or inhibit gene activity (Section 15.5). Any cell that bears molecular receptors for a given hormone is its target. Plant hormones never do travel far from the cells that secrete them.

Researchers have isolated more than eighty different forms of gibberellin from the seeds of flowering plants as well as from fungi. Possibly this same hormone is present in other groups of plants as well. The particular changes that gibberellins trigger make stems lengthen (Figure 32.1). In nature, gibberellins help seeds and buds break dormancy and resume growth in spring. They also influence the flowering process.

Expose a cabbage plant to a suitable concentration of a gibberellin and its growth may well astound you (Figure 32.2). Applications of gibberellins can make celery stalks longer and crispier, and they can prevent

a
b
c

Figure 32.1 Demonstrations of hormonal effects on plant growth and development. (**a**) Seedless grapes radiating market appeal. Gibberellin made the stems lengthen, which improved air circulation around grapes and gave them more room to grow. Grapes got larger and weighed more, making growers happy (grapes are sold by weight). (**b**) This young California poppy (*Eschscholzia californica*) was left alone. (**c**) Gibberellin was applied to this young poppy plant.

two untreated cabbages (controls)
cabbages treated with gibberellins

Figure 32.2 Foolish cabbages!

the skin of navel oranges in orchard groves from ripening too quickly. Walk past those plump seedless grapes in the produce bins of grocery stores and observe how the fleshy fruits of the grape plant (*Vitis*) grow in bunches along stems. Gibberellin applications made the stems lengthen between the internodes. This opened up more space between individual grapes, which thereby grew larger. And air was able to circulate more freely between grapes, which made it harder for pathogenic, grape-loving fungi to take hold and do damage.

Gibberellin and a number of other plant hormones orchestrate plant growth and development by responding to signals from one another. They also respond to cues afforded by the rhythmic changing of the seasons, as when days grow longer and warmer in spring, after the short, cold nights of winter. Those grapes, cabbage leaves, or celery stalks that find their way into your mouth are the culmination of a plant's own exquisitely controlled programs of growth and development.

Here we continue with a story that we started in the preceding chapter. That chapter traced events by which a zygote of a flowering plant forms and then develops into a mature embryo, housed inside a protective seed coat. At some point following its dispersal from the parent plant, the tiny embryo becomes transformed into a seedling, which in turn grows and develops into the mature sporophyte. In due time, the sporophyte forms flowers, then new seeds and fruits. Depending on the species, it drops old leaves throughout the year or all at once, in autumn.

This part of the story surveys the heritable, internal mechanisms that govern plant development, and the environmental cues that turn such mechanisms on or off at different times, in different seasons.

KEY CONCEPTS

1. From the time a plant seed germinates, a number of hormones influence growth and development.

2. Hormones are signaling molecules between cells. One cell type produces and then secretes a particular kind of hormone, which stimulates or inhibits gene activity in other cell types that take up molecules of the hormone. The changes in gene activity have predictable effects, as when they trigger the mitotic cell divisions and other processes that make stems grow longer.

3. The known plant hormones are auxins, gibberellins, cytokinins, abscisic acid, and ethylene. We have indirect evidence of other kinds of hormones, which have not yet been identified.

4. Plant hormones help bring about predictable patterns of development, including the extent and direction of cell growth and differentiation in particular plant parts.

5. Plant hormones help adjust patterns of growth and development in response to environmental rhythms, including seasonal changes in daylength and temperature. Additionally, they help adjust the patterns of response to environmental circumstances in which an individual plant finds itself, such as the amount of sunlight or shade, moisture, and so on at a given site.

6. Commonly, two or more kinds of plant hormones must interact with one another to bring about specific effects on growth and development.

PATTERNS OF EARLY GROWTH AND DEVELOPMENT—AN OVERVIEW

Seed Germination

Let's first consider the overall patterns of growth and development for dicots and monocots, starting with the events that take place inside a seed. Figure 32.3 shows the embryo sporophyte inside a grain of corn. (A grain, remember, is a seed-containing dry fruit.) The growth of the embryo idles before or after the seed is dispersed from the parent plant. Later, if all goes well, the seed germinates. **Germination** is the process by which some immature stage in the life cycle of a species resumes growth after a period of arrested development.

Germination depends on environmental factors, such as the temperature, moisture, and oxygen level of soil and the number of daylight hours. The factors can vary seasonally. For example, mature seeds do not contain enough water for cell expansion or metabolism. In many habitats on land, ample water only becomes available on a seasonal basis, so that seed germination coincides with the return of spring rains. By a process known as **imbibition**, water molecules move into the seed, being attracted mainly to the hydrophilic groups of proteins stored inside endosperm or cotyledons. As more water moves in, the seed swells and its coat ruptures.

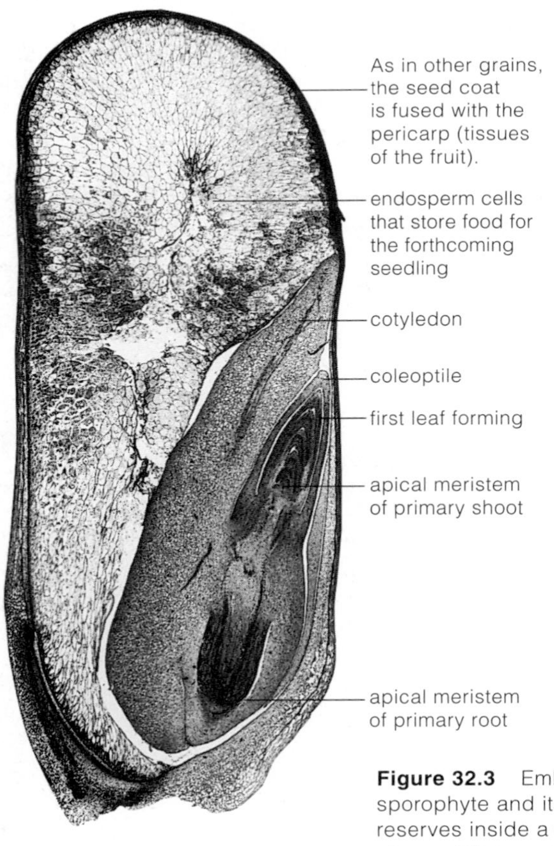
As in other grains, the seed coat is fused with the pericarp (tissues of the fruit).
endosperm cells that store food for the forthcoming seedling
cotyledon
coleoptile
first leaf forming
apical meristem of primary shoot
apical meristem of primary root

Figure 32.3 Embryo sporophyte and its food reserves inside a grain of corn (*Zea mays*).

Figure 32.4 (**a**,**b**) Pattern of growth and development of a dicot, the common bean plant (*Phaseolus vulgaris*). When a bean seed germinates, the embryo inside resumes growth. Then a hypocotyl (a hook-shaped portion of the shoot beneath the cotyledons) forces a channel through the soil. The food-storing cotyledons can be pulled up through the channel without being torn apart. At the surface, sunlight exposure makes the hook straighten. For several days, photosynthetic cells produce food in the seedling's cotyledons, which then wither and fall off. In time, foliage leaves (each divided into three leaflets) take over photosynthesis. Flowers will form in buds at the nodes. (**c**) Cotyledons breaking through a seed coat.

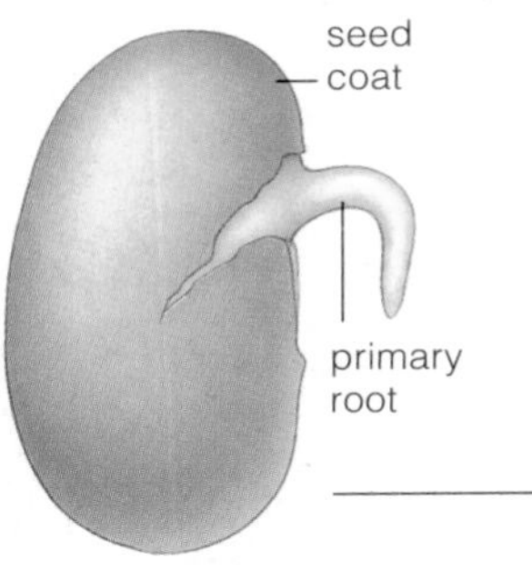
seed coat
primary root

a Bean seedling at the close of germination.

Figure 32.5 Pattern of growth and development of a monocot, using corn (*Zea mays*) as the example. (**a**,**b**) After germination of a corn grain, the coleoptile protects new leaves as the seedling grows through the soil. In corn seedlings, adventitious roots develop at the coleoptile's base. (**c**) Coleoptile and primary root of a corn seedling. (**d**) Coleoptile and first foliage leaf of two seedlings poking above the soil surface.

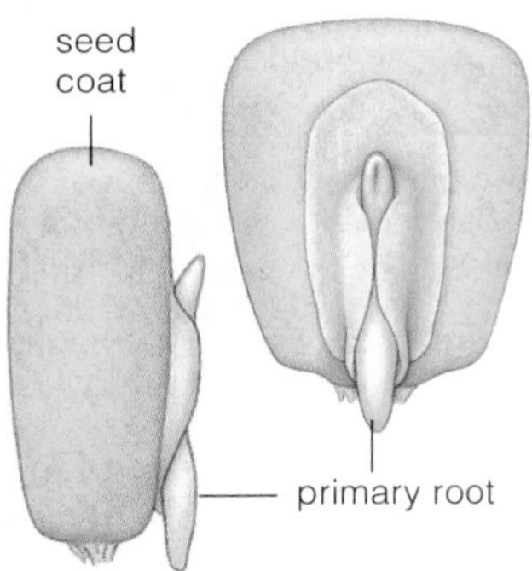
seed coat
primary root

a Corn grain at the close of germination.

Once the seed coat splits, more oxygen reaches the embryo, and aerobic respiration moves into high gear. The embryo's meristematic cells start to divide rapidly. In general, the root meristem is the first to be activated. Its meristematic descendants divide, elongate, and give rise to the primary root, the first root of the seedling sporophyte. Germination is over when the seedling's primary root breaks through the seed coat.

Genetic Programs, Environmental Cues

Figures 32.4 and 32.5 show patterns of germination, growth, and development for dicots and monocots. The patterns have a heritable basis, being dictated by genes. All cells in a plant arise from the same cell (a zygote), so they generally inherit the same genes. But unequal divisions between daughter cells and their positions in

Further reading: Student Guide to InfoTrac on web site

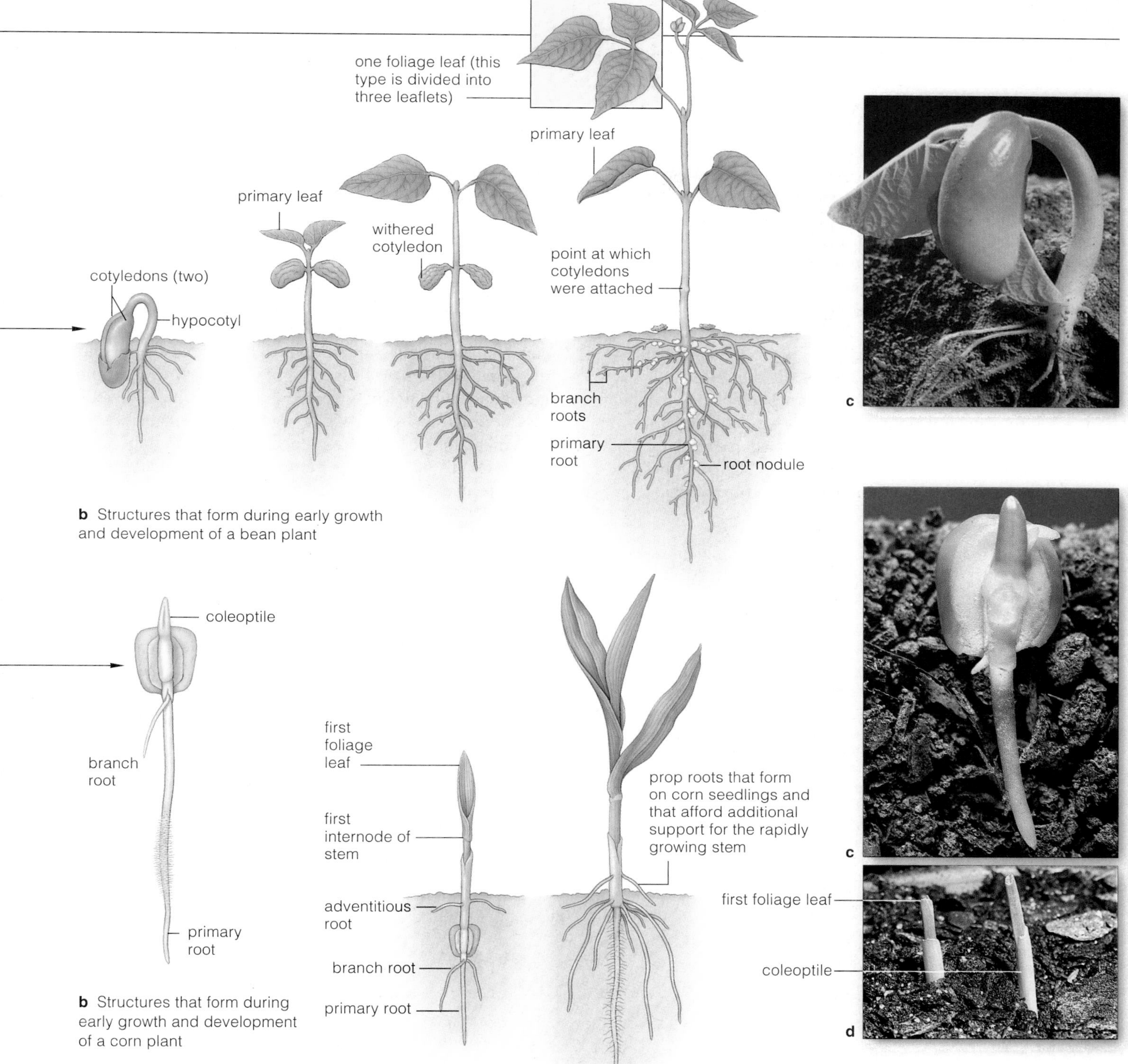
one foliage leaf (this type is divided into three leaflets)
primary leaf
primary leaf
withered cotyledon
cotyledons (two)
hypocotyl
point at which cotyledons were attached
branch roots
primary root
root nodule
c
coleoptile
branch root
primary root
first foliage leaf
first internode of stem
adventitious root
branch root
primary root
prop roots that form on corn seedlings and that afford additional support for the rapidly growing stem
c
first foliage leaf
coleoptile
d

b Structures that form during early growth and development of a bean plant

b Structures that form during early growth and development of a corn plant

the developing plant lead to differences in their metabolic equipment and output. Activities in those daughter cells start to vary as a result of selective gene expression, as described in Chapter 15. For instance, genes governing the synthesis of growth-stimulating hormones are activated in some cells but not others. As you will see shortly, such events seal the developmental fate of different cell lineages.

Bear in mind, plants commonly adjust prescribed growth patterns in response to unusual environmental pressures. Suppose a seed germinates in a vacant lot and a heavy paper bag blows on top of it. The primary shoot will quickly bend and grow out from under the bag, in the direction of sunlight impinging on it. Interactions among enzymes, hormones, and other gene products in its cells result in this growth response.

Plant growth involves cell divisions and cell enlargements. Plant development requires cell differentiation, as brought about by selective gene expression.

Interactions among genes, hormones, and the environment govern how an individual plant grows and develops.

HORMONAL EFFECTS ON PLANT GROWTH AND DEVELOPMENT

Growth Versus Development

Before getting into the effects of hormones on growth and development, make sure you understand the basic difference between these processes. In general, **growth** of a multicelled organism means the number, size, and volume of cells increase. **Development** is the emergence of specialized, morphologically different body parts. In other words, we measure growth in *quantitative* terms and development in *qualitative* terms.

As you read in Section 29.1, the cell divisions and enlargements underlying plant growth occur at apical and laternal meristems. About half of the daughter cells don't grow in size but they retain the capacity to divide (Figure 32.6). The rest enlarge, often by twenty times. Cellular descendants divide in prescribed planes and expand in certain directions (Figure 32.7). The outcome will be plant parts with specific shapes and functions.

When young cells of an embryo sporophyte or a seedling take up water, turgor pressure (internal fluid pressure) increases against primary cell walls and drives their enlargement. Imagine blowing up a balloon. While it is soft, the balloon inflates easily. Similarly, cells with a soft wall expand rapidly under turgor pressure. The wall thickens when polysaccharides are added to it, and more cytoplasm forms between it and a central vacuole.

Hormones help control the selective gene expression that underlies cell differentiation and the emergence of basic patterns of development. At least five different classes of hormones are known to have predictable and important effects on flowering plants, starting with the germination of the new seed. As with animals, genetic instructions in a plant's DNA govern the synthesis of enzymes and other proteins necessary for metabolism, hence for all of the cell's activities. *And how and when each type of enzyme functions depend on hormonal action.*

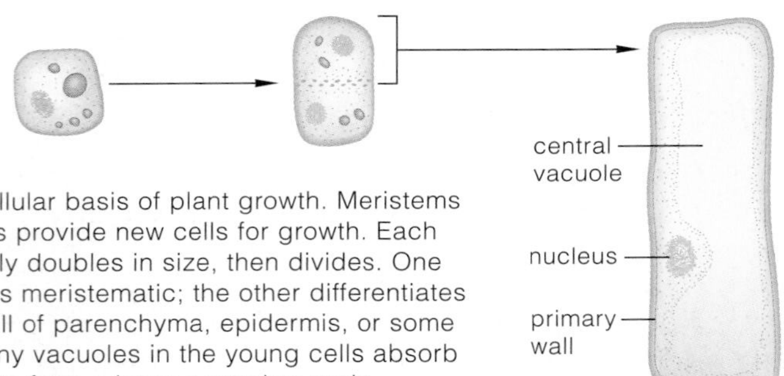
central vacuole
nucleus
primary wall

Figure 32.6 The cellular basis of plant growth. Meristems of root and shoot tips provide new cells for growth. Each small new cell roughly doubles in size, then divides. One daughter cell remains meristematic; the other differentiates into a specialized cell of parenchyma, epidermis, or some other plant tissue. Tiny vacuoles in the young cells absorb water and then fuse to form a large central vacuole.

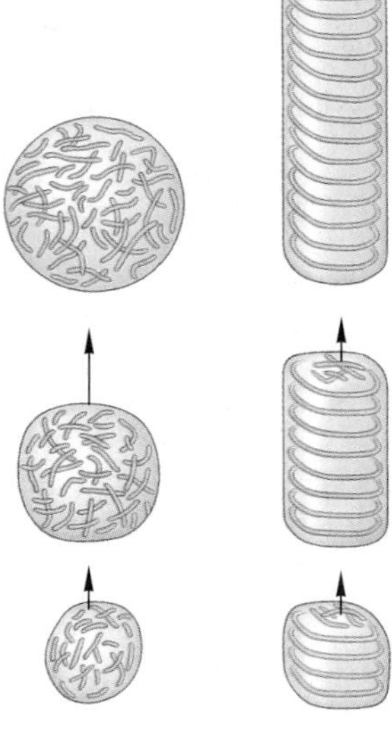

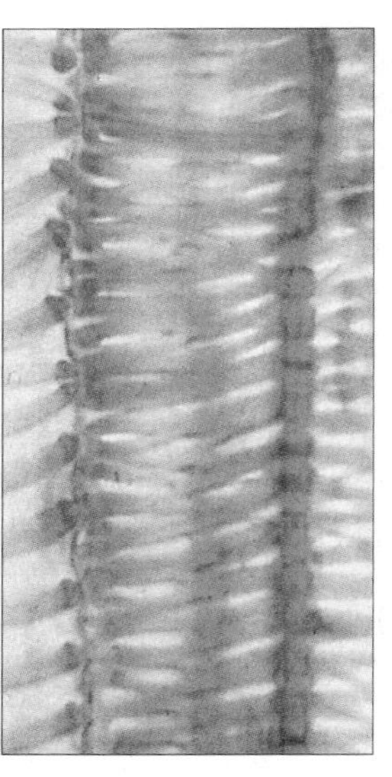

Figure 32.7 How different cell shapes can arise. In each cell, microtubules in the cytoplasm that it inherited from the parent cell are already oriented in prescribed patterns that govern how cellulose microfibrils will be oriented in the cell wall. With random orientation (**a**), the primary wall is elastic all over, so the cell can expand in all directions and assume a spherical shape. (**b**,**c**) When the cellulose strands become transversely oriented, the cell can only lengthen.

a Growth proceeds in all directions.

b Growth in longitudinal direction only.

c Cellulose rings thicken the secondary wall of this tracheid, a conducting tube of xylem.

Categories of Plant Hormones

Table 32.1 lists the five main categories of plant hormones. Besides gibberellins, they are **auxins**, **cytokinins**, **ethylene**, and **abscisic acid**. Here we define their known functions. The remainder of the chapter gives examples of their effects during different stages of the life cycle.

GIBBERELLINS As you saw earlier, the gibberellins promote stem lengthening (Figures 32.1 and 32.2). They also help buds as well as seeds break dormancy and resume growth in the spring. In at least some plant species, they influence the flowering process.

AUXINS Auxins affect the lengthening of stems and responses to gravity and light; they also make coleoptiles grow longer (Figure 32.8). A **coleoptile** is a sheath around a primary shoot of grass seedlings, such as corn plants. It helps keep a shoot from shredding as growth pushes it up through soil. Indoleacetic acid (IAA) is the most pervasive auxin in nature. Orchardists use auxins to thin flowers in spring so that the trees yield fewer but larger fruits. They also use auxins to prevent premature fruit drop, which cuts labor costs; all fruit in the orchard can be picked at the same time.

Certain synthetic auxins are used as **herbicides**. At suitable concentrations,

Further reading: Student Guide to InfoTrac on web site

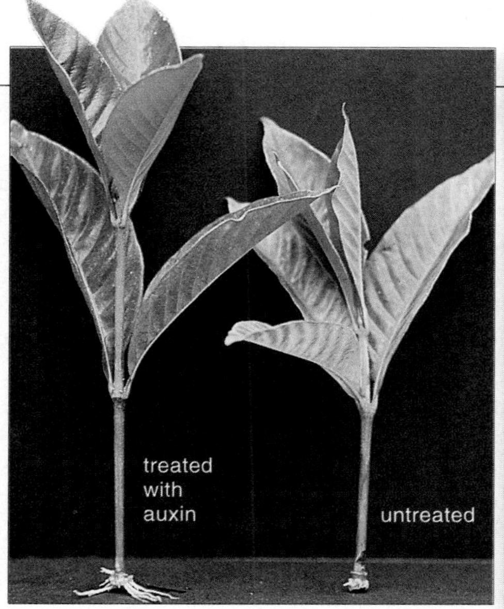
treated
with
auxin
untreated

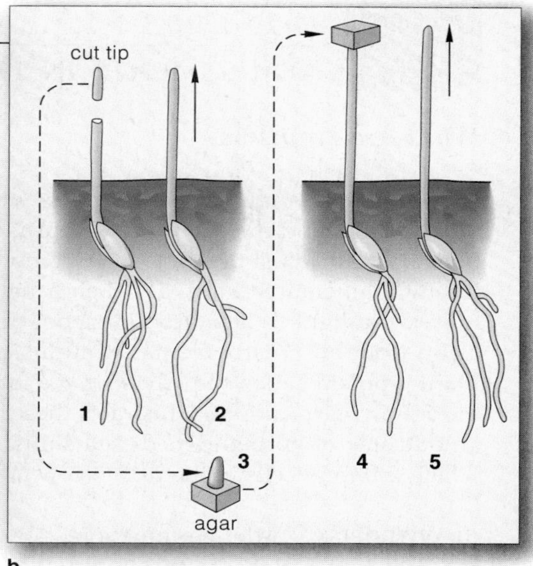
cut tip
1
2
3
agar
4
5

Figure 32.8 Examples of the effects of auxin. (**a**) A cutting from a gardenia plant (*left*), four weeks after an auxin had been applied to its base. Another cutting, used as a control (*right*), was untreated.

(**b**) Experiments demonstrating how IAA (an auxin) in a coleoptile tip promotes elongation of cells below it. (1) Cut off the tip of an oat coleoptile. The cut stump does not elongate as much as a normal oat coleoptile (2) used as the control. (3) Next, place a tiny block of agar under the cut tip and leave it for several hours. During that interval, IAA diffuses into the agar. (4) Now place the agar on another de-tipped coleoptile. Its elongation will proceed about as rapidly as in an intact coleoptile growing next to it (5).

Table 32.1 Main Plant Hormones and Some of Their Known or Suspected Effects

GIBBERELLINS. Promote stem elongation; help end dormancy of seeds and buds; contribute to flowering. Notable amounts in apical meristems of buds, roots, and leaves and in embryos.

AUXINS. Promote cell elongation in coleoptiles and stems; roles in phototropism and gravitropism. Notable amounts in bud and leaf apical meristems and in embryos in seeds.

CYTOKININS. Promote cell division and leaf expansion; retard leaf aging. Synthesized in roots and travel elsewhere.

ABSCISIC ACID. Promotes stomatal closure; promotes bud and seed dormancy. Notable amounts in leaves, stems, and unripened fruit.

ETHYLENE. Promotes ripening of fruit and abscission of leaves, flowers, and fruits. Notable amounts in seeds, fruits, stems, leaves, and roots.

these compounds can kill some plant species but not others. 2,4-Dichlorophenoxyacetic acid (2,4-D) is able to control broadleaf weeds that compete vigorously with cereal plants for nutrients. Apparently 2,4-D does not harm humans when properly handled. This is not the case for 2,4,5-T, a related compound. When the two are mixed in equal proportions, *Agent Orange* results. During the Vietnam conflict, the United States Armed Forces used this herbicide to defoliate thickly forested, nearly impenetrable war zones. Later, results from laboratory experiments suggested that traces of dioxin, which is a contaminant of 2,4,5-T, may cause miscarriages, birth defects, and liver disorders. Dioxin also is a carcinogen that may cause leukemias. The manufacture and use of 2,4,5-T is now banned in the United States.

CYTOKININS Cytokinins stimulate cell division (hence the name, which refers to cytoplasmic division). They are most abundant in root and shoot meristems, and in the tissues of maturing fruits. Natural and synthetic versions are used in basic research, and to prolong the shelf life of cut flowers and other horticultural goods.

ABSCISIC ACID Abscisic acid (ABA) helps plants adapt to seasonal changes, as by inducing bud dormancy and inhibiting cell growth or premature seed germination. ABA also contributes to stomatal closure when a plant is water stressed. Growers often apply ABA to nursery stock before shipping it; plants are not as vulnerable to injury when they are dormant.

ETHYLENE Ethylene induces various aging responses, including fruit ripening and leaf drop. Ancient Chinese burned incense to make fruit ripen faster, and growers in the 1900s piled citrus fruits in sheds with kerosene stoves to ripen them; the smoke contained ethylene. Today, food distributors use ethylene to ripen green tomatoes and other fruit after they ship it. Fruit picked green doesn't bruise or deteriorate as fast. Oranges and other citrus fruits are exposed to ethylene to brighten their rind before being displayed in the market.

Besides the hormones just described, root and leaf cells make their own hormones. One or more unknown hormones might trigger flowering; and another kind, associated with shoot tips, blocks the growth of lateral buds. This form of growth inhibition is called **apical dominance**. By pinching off shoot tips, gardeners can prevent the hormone from diffusing through a plant's stems and exerting its inhibitory effect. Lateral buds are free to branch out, and gardeners get bushier plants.

Growth is a quantitative increase in the number, size, and volume of cells. Development, the emergence of qualitative differences in body parts, is basically a result of selective gene expression, hormone action, and cell differentiation.

The main categories of plant hormones are gibberellins, auxins, cytokinins, abscisic acid, and ethylene.

32.3 ADJUSTMENTS IN THE RATE AND DIRECTION OF GROWTH

What Are Tropisms?

Generally, the young roots of land plants grow down through soil, and the shoots grow upright through air. Both also can adjust the direction of growth in response to environmental stimuli, as when a new shoot turns toward sunlight. When a root or shoot turns toward or away from an environmental stimulus, we call this a **plant tropism** (after the Greek *trope*, for turning). As the following examples illustrate, these responses are an outcome of hormone-mediated shifts in the rates at which different cells in the plant grow and elongate.

GRAVITROPISM After germination, the first root to emerge through a seed coat always curves downward and coleoptiles and stems curve up. Any growth response to the Earth's gravitational force is called a form of **gravitropism**.

Auxin, together with a growth-inhibiting hormone in roots, might trigger the response. Turn a young root on its side and remove its root cap, and it will *not* curve down. Put the cap back on, and the root will do so. Elongating root cells will not stop growing if you remove the root cap; if anything, the cells will grow faster. Suppose a growth inhibitor that is present in root cap cells becomes *redistributed* in a root turned on its side. If gravity somehow causes the inhibitor to move out of the cap and accumulate in cells on the lower side of the root, those cells will not elongate as much as cells on the upper side—and the root will curve downward.

The gravity-sensing mechanisms of plants are based on **statoliths**, which generally are clusters of particles in a number of different cells. In plants, they are clusters of unbound starch grains in modified plastids. Figure 32.9*a,b* provides examples. These plastids, made more dense by the statoliths, collect near the bottom of root cells in response to the tug of gravity. They settle downward until coming to rest in the lowest region of the cytoplasm. Possibly, redistribution of the statoliths triggers a redistribution of auxin within the cells, and thereby triggers a gravitropic response.

Similarly, Figure 32.9*c* shows how you may track what can happen after you turn a potted seedling on its side in a dark room. The stem curves up, even in the absence of light. Why? In the horizontally positioned stem, cell elongations on the stem side facing upward slow down—markedly so—and the elongations on the lower side rapidly increase. The different rates of elongation on opposing sides of the stem are enough to make it bend in an upward direction. It appears that something makes cells on the bottom of a "sideways" stem *more* sensitive to auxin or some other hormone, and those on top *less* so.

a
statoliths

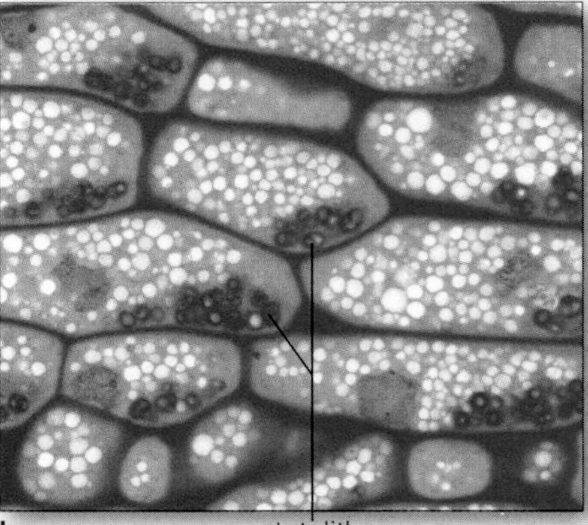
b
statoliths

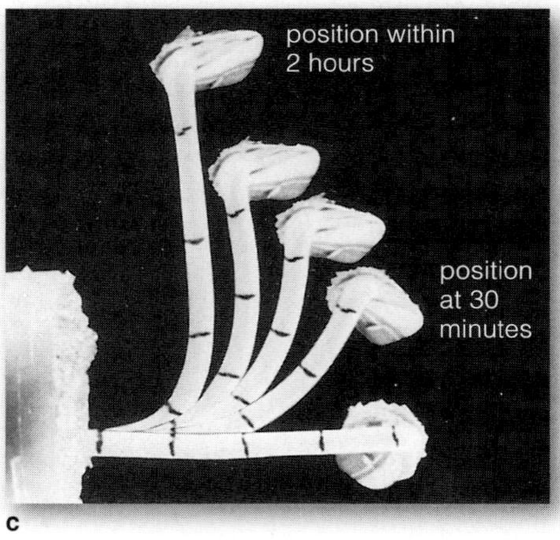
position within
2 hours
position
at 30
minutes
c

Figure 32.9 Two observational tests to demonstrate gravitropic responses by young shoots and roots. (**a**) For one test, examine the root cap on a corn root. Normally, plastids of the root cap that contain statoliths settle downward. This light micrograph clearly shows their normal orientation. (**b**) Now turn the root sideways. Five to ten minutes later, the plastids will settle to the bottom of the root cap cells. A gravity-sensing mechanism based on statoliths may be sensitive to such redistribution of auxin through a root tip. A difference in auxin concentrations might cause cells on the "top" of a root turned sideways to elongate faster than cells on the bottom. Different elongation rates will make the root tip curve downward.

(**c**) For a different test, force a sunflower seedling to grow in the dark for five days. Then turn it on its side and mark the shoot at 0.5-centimeter intervals. A gravity-sensing mechanism will cause the stem to turn upright.

PHOTOTROPISM Whenever stems or leaves adjust the rate and direction of growth in response to light, they are showing a form of **phototropism**. The adaptive advantage is clear. Without sunlight, the light-dependent reactions of photosynthesis stop, and so plants move to maximize light interception.

Charles Darwin wondered about phototropism after it dawned on him that a coleoptile was growing toward light that was striking one side of its tip. But it wasn't until the 1920s that Fritz Went, a graduate student in Holland, linked phototropism with a growth-promoting substance. He was the person who named the substance

Further reading: Student Guide to InfoTrac on web site

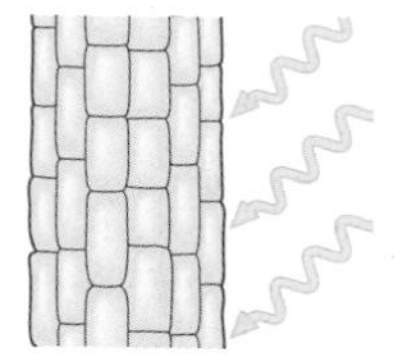

a Rays from sun strike one side of a coleoptile.

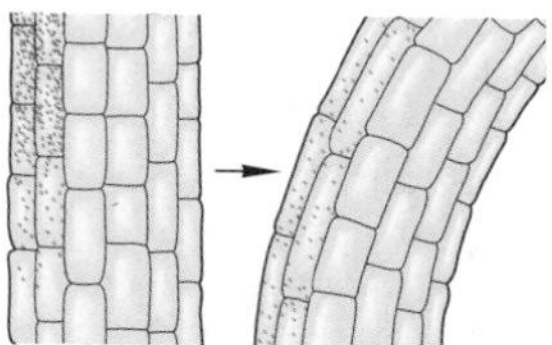

b The coleoptile bends after auxin diffuses down from tip to cells on its shaded side.

bean
pea
oat

c

Figure 32.10 (**a**,**b**) Hormone-mediated differences in cell elongation rates that cause coleoptiles and stems to bend toward light. (**c**) Phototropism in seedlings. The plant physiologist Frank Salisbury grew these seedlings in darkness, then allowed light to strike their right side for a few hours before photographing them.

auxin (after the Greek *auxein*, meaning "to increase"). Went demonstrated that auxin moves from the tip of a coleoptile into cells less exposed to light and makes them elongate faster than cells on the illuminated side. The difference in their growth rates brings about the bending toward light (Figure 32.10*a*,*b*).

You can observe a phototropic response by putting bean sprouts or seedlings of another sun-loving plant in shade, next to a window through which sunlight is streaming in. They will start curving in the direction where the most light is available (Figure 32.10*c*).

As we now know, blue wavelengths of light induce plants to make their strongest phototropic response. Therefore, **flavoprotein**, a yellow pigment molecule that absorbs blue wavelengths, may be a central component of the phototropic bending mechanism.

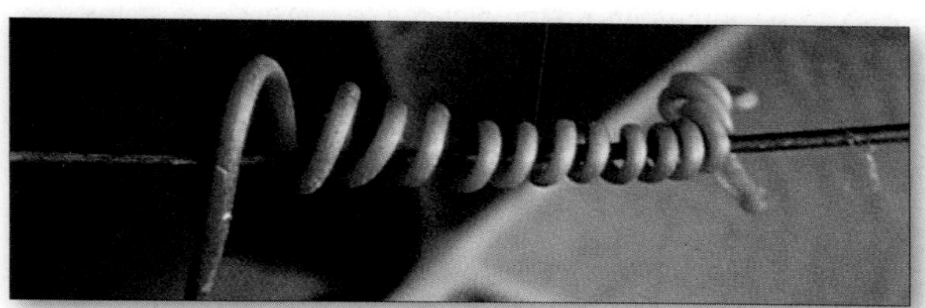

Figure 32.11 Passion flower (*Passiflora*) tendril twisting thigmotropically.

a
b
c

Figure 32.12 Effect of mechanical stress on tomato plants. (**a**) This plant, the control, grew in a greenhouse. (**b**) Each day for twenty-eight days, this plant was mechanically shaken for thirty seconds. (**c**) This one had two shakings each day.

THIGMOTROPISM Plants also shift their direction of growth when they contact solid objects. This response is called **thigmotropism** (after the Greek *thigma*, which means "touch"). Auxin and ethylene may have roles in the response. Vines, which are stems too slender or soft to grow upright without support, make this contact response. So do tendrils, which are modified leaves or stems that wrap around objects and thus help support the plant (Figure 32.11). You can observe what happens when they grow against a stem of another plant. Within minutes, cells on the contact side stop elongating and the vine or tendril starts curling around the stem, often more than once. Afterward, the cells on both sides will resume growth at the same rate.

Responses to Mechanical Stress

Mechanical stress, as inflicted by prevailing winds and grazing animals, can inhibit stem elongation and plant growth. You can see such effects on trees growing near the snowline of windswept mountains; they are stubby compared to trees of the same species that are growing at lower elevations. Similarly, plants grown outdoors commonly have shorter stems than plants grown inside a greenhouse. You can observe this response to stress by shaking a plant daily for a brief period. Doing so will inhibit growth of the entire plant (Figure 32.12).

Plants adjust their direction and rate of growth in response to environmental stimuli.

32.4 BIOLOGICAL CLOCKS AND THEIR EFFECTS

Like other organisms, flowering plants have internal mechanisms that preset the time for recurring changes in biochemical events. These internal timing mechanisms —**biological clocks**—trigger shifts in daily activities. They also help induce seasonal adjustments in basic patterns of growth, development, and reproduction.

The alarm button for some biological clocks in plants is the blue-green pigment molecule **phytochrome**. This pigment can absorb both red and far-red wavelengths of light, with different results (Figure 32.13). At sunrise, red wavelengths dominate the sky, and phytochrome converts to Pfr, its active molecular form. Its activation may induce cells to take up free calcium ions (Ca^{++}) or induce certain plant organelles to release them. Either way, when the ions combine with calcium-binding proteins in cells, they initiate rhythmic movements of leaves and some other responses to light.

Phytochrome reverts to its inactive form (Pr) at sunset, at night, or in shaded areas. Then, far-red wavelengths predominate.

Rhythmic Leaf Movements

Each day, some plants position their leaves horizontally and then, at night, fold them closer to stems (Figure 32.14). Keep such a plant in full sunlight or darkness for a few days and it continues to move its leaves into and out of the "sleep" position!

Rhythmic leaf movements are one type of **circadian rhythm**, a biological activity that recurs in cycles, each of which lasts for about twenty-four hours. (*Circadian* means about a day.) Some experiments by Ruth Satter, Richard Crain, and their University of Connecticut colleagues demonstrated that the phytochrome molecule is part of controls over movements of leaves. Satter was a pioneer in studies of internal timing mechanisms, one of the first to correlate sleep movements of plants with "hands of a biological clock."

1 A.M.

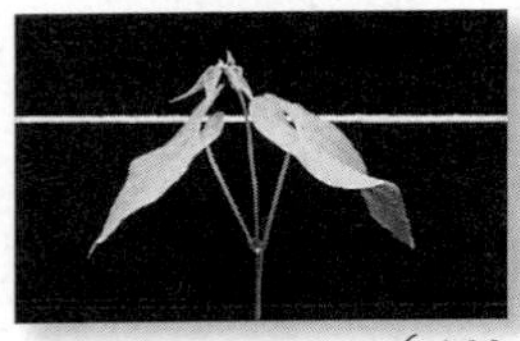
6 A.M.

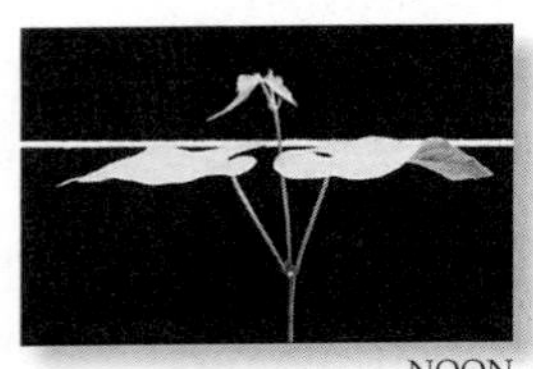
NOON

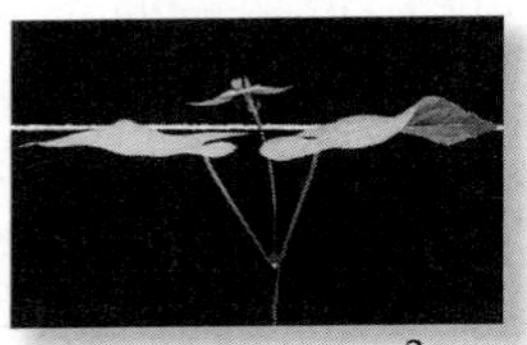
3 P.M.

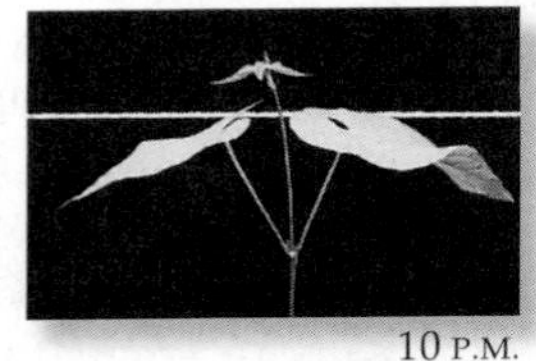
10 P.M.

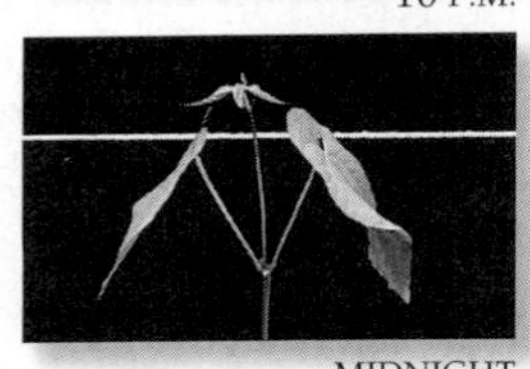
MIDNIGHT

Figure 32.14 Rhythmic movements of bean plant leaves. The investigator kept a bean plant in total darkness for twenty-four hours. Its leaves continued to move independently of sunrise (6 A.M.) and sunset (6 P.M.). Leaves folded close to stems might keep moonlight from activating phytochrome and interrupting the dark period that triggers flowering. Or folding might reduce heat loss from leaves exposed to cold night air.

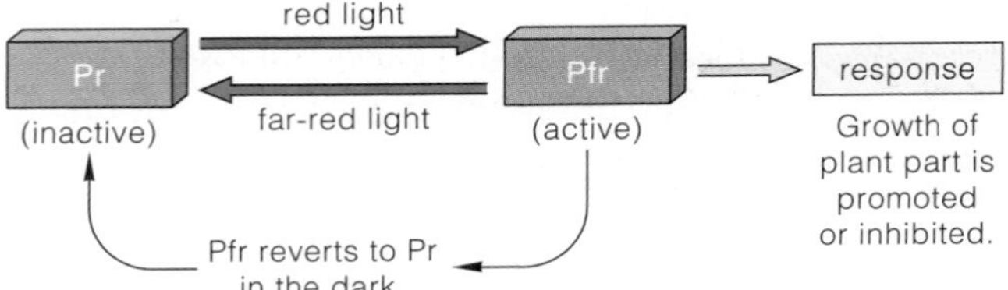
red light
Pr
(inactive)
far-red light
Pfr
(active)
response
Growth of plant part is promoted or inhibited.
Pfr reverts to Pr in the dark.

Figure 32.13 Interconversion of the phytochrome molecule from active form (Pfr) to inactive form (Pr). This blue-green pigment is part of a switching mechanism that promotes or inhibits the growth of a variety of plant parts.

Flowering—A Case of Photoperiodism

As another example of internal timekeeping, let's look at one way in which a flowering plant resets biological clocks as seasons change. At a certain time of year, the plant starts diverting more energy to the formation of flowers. Figure 32.15 can give you an idea of how this activity and others are predictable responses to rhythmic environmental cues, such as more hours of light and warmer temperatures in summer.

Photoperiodism is any biological response to change in the relative length of daylight and darkness in the cycle of twenty-four hours. Pfr, phytochrome's active form, may be a switching mechanism for the process; it might trigger the synthesis of specific enzymes in specific types of plant cells. We know that different enzymes take part in seed germination, stem branching and lengthening, leaf expansion, and formation of flowers, fruits, and seeds.

Plants do not all flower in response to the same environmental cues. For example, think of poinsettias, chrysanthemums, and cockleburs. These **short-day plants** produce flowers during late summer or early fall, when daylength is shorter than a critical value (Figures 32.16 and 32.17). By contrast, spinach, irises, and other **long-day plants** form flowers during the spring, when daylength exceeds some critical value (Figure 32.16). **Day-neutral plants** simply flower when they are mature enough to do so.

Spinach plants will not flower and produce seeds unless they are exposed to ten hours of darkness (that is, fourteen hours of daylight) for two weeks. That is why you would not want to start, say, a spinach seed farm in the tropics, where this set of cues doesn't occur. By contrast, flower growers stall the flowering process by exposing chrysanthemum plants to a flash of light each night, thus breaking up one long night into two

Further reading: Student Guide to InfoTrac on web site

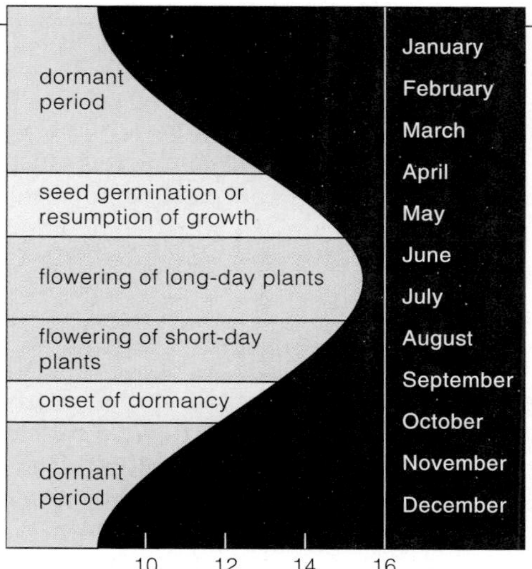
dormant period
seed germination or resumption of growth
flowering of long-day plants
flowering of short-day plants
onset of dormancy
dormant period
January
February
March
April
May
June
July
August
September
October
November
December
10
12
14
16
daylength (hours)

Figure 32.15 Plant growth and development as correlated with the number of hours of light available each day. The number changes with the passing of the seasons. The data shown reflect photoperiodic responses of plants that grow in temperate regions of North America. In such regions, rainfall and temperature shift with the seasons.

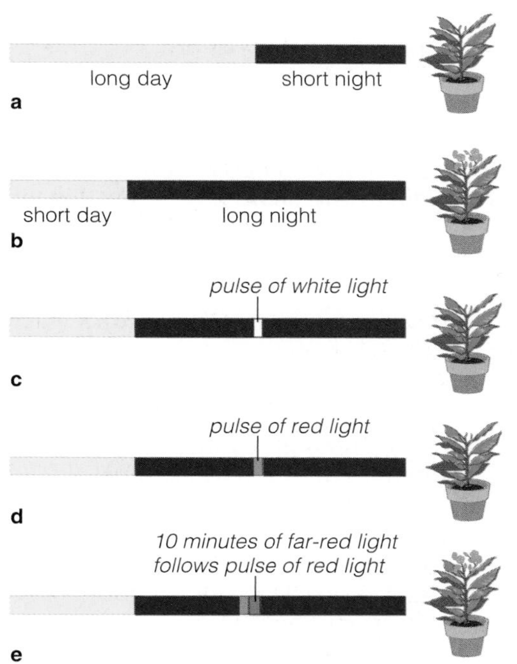
a
long day
short night
b
short day
long night
c
pulse of white light
d
pulse of red light
e
10 minutes of far-red light follows pulse of red light

Figure 32.16 (**a–c**) Results of experiments demonstrating that short-day plants flower by measuring the length of night, not daylength. Each horizontal bar signifies twenty-four hours. Daylight hours are *yellow*; hours of night are *dark blue*. (**a**) Plants grown inside a room in which simulated nights were shorter than required for flowering did not flower. (**b**) Those grown when nights were longer flowered. (**c**) Plants exposed to a pulse of light in the middle of the night also did not flower. Experimental results showed that phytochrome is the receptor that responds to light in short-day plants. (**d**) A short pulse of red light at night stops flowering. (**e**) Flowering occurs when 10 minutes of far-red light follows a pulse of red light.

flowers
a
b

Figure 32.17 Experimental results showing flowering responses of (**a**) spinach, a long-day plant, and (**b**) chrysanthemum, a short-day plant. In each photograph, the plant positioned at the left grew under short-day conditions, and the one at the right grew under long-day conditions.

"short" nights. Thus you can buy chrysanthemums in spring even though they naturally bloom in autumn. Or think of cockleburs, which normally flower after a single night longer than eight and one-half hours. If artificial light interrupts their dark period for even a minute, they refuse to flower.

As an example of a poor landscaping decision, why didn't poinsettias that were planted along California's interstate highways put out flowers? Headlights from cars and trucks zipping by during the night blocked the flowering response.

Besides phytochrome, other kinds of hormones that are not yet identified probably influence flowering and other growth responses. Some of the hormones may be produced in leaves and transported to new buds. For example, if you trim all but one leaf from a cocklebur, and then cover the remaining leaf with black paper for eight and one-half hours, the plant will flower. If you cut off that one leaf right after the dark period is over, you will not see any cocklebur flowers.

Like other organisms, flowering plants have internal time-keeping mechanisms called biological clocks.

Phytochrome, a blue-green pigment, is part of a switching mechanism for phototrophic responses to light of red and far-red wavelengths. Its active form, Pfr, might interact with other hormones to control which kinds of enzymes are being produced in particular cells.

Different enzymes are required to complete flowering and other growth responses, which are influenced by daylength and other environmental cues.

32.5 LIFE CYCLES END, AND TURN AGAIN

Senescence

While leaves and fruits are growing, cells inside them produce auxin (IAA), which moves into stems. There it interacts with cytokinins and gibberellins to maintain growth. When autumn approaches and the length of daylight decreases, plants start to withdraw nutrients from leaves, stems, and roots, then distribute them to flowers, fruits, and seeds. Deciduous plants, which shed leaves when a growing season ends, channel nutrients to storage sites in twigs, stems, and roots before the leaves die and drop. The dropping of flowers, leaves, fruits, and other plant parts is called **abscission**. The abscission zone is composed of thin-walled parenchyma cells at the base of a petiole or some other part about to drop from the plant. Figure 32.18 shows one example.

Senescence is the sum total of processes that lead to death of a plant or some of its parts. The recurring cue for senescence is a decrease in daylength, but other environmental factors, such as drought, wounds, and nutrient deficiencies, also bring it about. Either way, the cue triggers a decline in IAA production in leaves and fruits. A different signal, possibly produced by the plant itself, stimulates abscission zone cells to produce ethylene. The cells enlarge, deposit suberin in their walls, and produce enzymes that digest cellulose and pectin in middle lamellae. A middle lamella is the cementing layer between plant cell walls (Section 4.10). As cells continue to enlarge and enzymes destroy their walls, they separate from one another. Then the leaf or other plant part above the abscission zone is free to fall away.

Interrupt the diversion of nutrients into flowers, seeds, or fruits, and you can prevent senescence in a plant's leaves, stems, and roots. For example, if you remove each new flower or seed pod from a plant, its leaves and stems will remain vigorous and green much longer (Figure 32.19). Gardeners routinely remove the flower buds from many species of plants to maintain vegetative growth.

Entering and Breaking Dormancy

As autumn approaches and days grow shorter, many perennials and biennials start to shut down growth. They do so even when temperatures are still mild, the sky is bright, and water is plentiful. When a plant stops growing under conditions that seem (to us) suitable for growth, it has entered a state of **dormancy** in which its metabolic activities idle. Ordinarily, the plant's buds will not resume growth until there is a convergence of precise environmental cues in early spring.

Short days and long, cold nights are strong cues for dormancy. (So is dry, nitrogen-deficient soil.) Researchers tested this by interrupting the long dark period of some Douglas firs with a short period of red light. The plants responded as if nights were shorter and days longer. They continued to grow taller (Figure 32.20). In this experiment, conversion of Pr to Pfr by red light during the dark period prevented dormancy. In nature, buds might enter dormancy because less Pfr can form when daylength shortens in late summer.

The requirement for multiple cues for dormancy has adaptive value. For example, if temperature were the only cue, warm autumn weather might make plants flower and seeds germinate—and winter frost would kill them. By contrast, with artificial selection, growers have developed seeds that will germinate promptly in greenhouses any time of year.

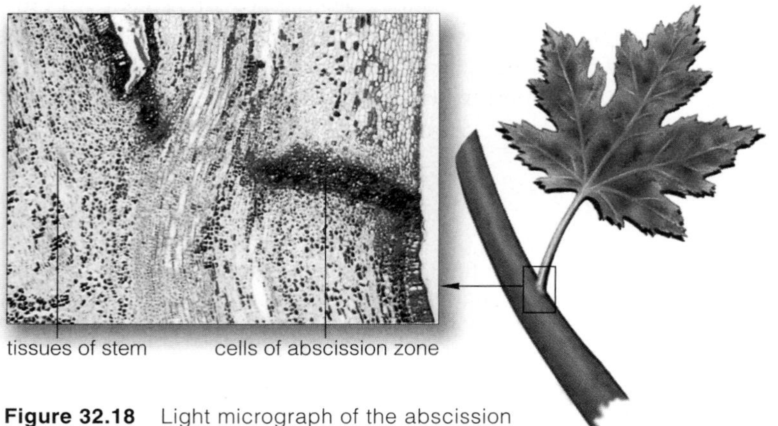
tissues of stem
cells of abscission zone

Figure 32.18 Light micrograph of the abscission zone in maple (*Acer*). This is a longitudinal section through the base of a leaf petiole.

control (pods not removed)
experimental plant (pods removed)

Figure 32.19 Observable results from an experiment in which seed pods were removed from a soybean plant. Removal delayed its senescence.

Further reading: Student Guide to InfoTrac on web site

Three Douglas fir plants exposed to different day-night cycles during three year-long experiments

Figure 32.20 Experiment to test the effect of the relative length of day and night on Douglas firs. The fir at left was exposed to 12-hour light/12-hour dark cycles for one year. Its buds stayed dormant; "daylength" was too short. The fir at right was exposed to 20-hour light/4-hour dark cycles. It continued growing. The fir in the middle was exposed to 12-hour light/11-hour dark cycles with an episode of 1 hour of light in the middle of the dark period. The interruption with light prevented bud dormancy; it caused Pfr formation at a sensitive time in the normal day–night cycle.

Potted plant grown inside a greenhouse did not flower.
Branch exposed to cold outside air flowered.

Figure 32.21 Localized effect of cold temperature on dormant buds of a lilac plant. For this experiment, a branch of the plant was positioned so that it protruded from a greenhouse during a cold winter. The rest of the plant was kept inside and exposed only to warmer temperatures. Only buds on the branch exposed to low outside temperatures resumed growth in spring.

Vernalization

Temperatures as well as daylength change seasonally in most regions. Changes in temperature influence many plant responses, such as flowering. For example, unless buds of some biennials and perennials are exposed to low winter temperatures, flowers will not form on their stems in spring. The low-temperature stimulation of flowering is called **vernalization** (from *vernalis*, which means "to make springlike"). Figure 32.21 shows how you can gather experimental evidence of this effect.

As long ago as 1915, the plant physiologist Gustav Gassner studied flowering of some cereal plants after exposing their seeds to controlled temperatures. As an example, he germinated some seeds of winter rye (*Secale cereale*) at near-freezing temperatures. Plants flowered the next summer even when planted in late spring. Vernalization is common in agriculture.

Figure 32.22 Profusion of new leaves harvesting sunlight, a sure sign of spring and the resurgence of plant growth.

Breaking Dormancy

A dormancy-breaking process is at work between fall and spring. Temperatures often become milder, and rains and nutrients are available again. With the return of favorable conditions, life cycles turn once more. Seeds germinate; buds resume growth and give rise to new leaves, then to flowers (Figure 32.22).

Depending on the plant species, breaking dormancy probably involves gibberellins and abscisic acid, and it requires exposure to low winter temperatures at specific times. The temperature required to break dormancy varies quite a bit among plants. For example, Delicious apple trees grown in Utah require 1,230 hours near 43°F (6°C), whereas apricot trees grown there need only 720 hours. Generally, trees growing in the southern United States require less cold exposure than the trees growing in northern states and Canada. If you live in Colorado and order a young peach tree from a Georgia nursery, the tree might start spring growth too soon and be killed by late frost or heavy snow.

Multiple cues from the environment influence hormonal secretions that stimulate or inhibit processes of growth and development during the life cycle of plants. These cues include changes in daylength, temperature, moisture, and nutrient availability.

Commentary

32.6

THE RISE AND FALL OF A GIANT

In this unit, you surveyed the spectrum of vascular plant tissues and the mechanisms and outcomes of plant growth, development, and reproduction. To pull all of this into a coherent picture, imagine walking in the steep, rounded, sandstone hills near the central California coast. Those hills started forming 65 million years ago as a jagged new coastal range. Over time, erosive rains and winds softened their stark contours and sent mineral-laden sediments into the canyons. Grasses took hold, and their remains slowly accumulated. More than 10 million years ago, the coast live oak (*Quercus agrifolia*) evolved in that enriched soil.

As you walk, you pass under trees thirty meters tall, with evergreen branches that spread even wider. One of those giants is 300 years old. It's early spring, so you see golden catkins—male flowers—among its leaves. A light breeze is dispersing their pollen to female flowers near branch tips of the same tree or neighboring ones. Long after you pass by, new zygotes will undergo the first of millions of cell divisions. Seed coats will form, and the ovarian walls will become hard shells. By early fall, the trees will shed seeds in fruits called acorns.

Three centuries ago, long before Gaspar de Portola sent landing parties ashore to found colonies through upper California, oaks were shedding seeds of a new generation. By chance, one particular acorn escaped the attention of foraging squirrels and blue jays. It germinated the next spring and embarked on a journey of continued growth. Meristems in the primary root gave rise to the root cortex, epidermis, and a vascular cylinder through which water and ions flowed. Lateral roots emerged. As new roots grew longer, their absorptive surfaces increased. In the primary shoot, vascular bundles began forming.

The oak seedling's developing roots, stems, and leaves demanded more water and dissolved nutrients. As fungi formed velvety, mycorrhizal partnerships with the roots, the uptake of water and mineral ions increased. Stomatal action in leaves conserved precious water. As the seasons passed, woody stems developed and supported upright growth. By transpiration, water and dissolved nutrients flowed from roots, through secondary xylem, to higher stems and leaves. Secondary phloem shuttled sugars to cells growing or storing food throughout the tree.

By chance, the seed had sprouted in a well-drained, sunlit basin at the foot of a canyon. Each winter, rainwater accumulated and kept the soil moist enough to encourage growth through spring and the dry summers. The sun's red wavelengths activated phytochrome and so triggered hormone-induced stem branching and leaf expansion. Other intricate, hormone-mediated responses were made to winds, the sun, gravity's tug, and the changing seasons.

And so the oak flourished every season. Century after century, roots snaked through a huge volume of the moist soil. Branches continued to spread beneath the sun. Leaves proliferated, and mesophyll cells put together food with sunlight energy, water, carbon dioxide, and the few simple minerals being mined from the soil.

In 1849, prospectors on their way to California's gold fields rested in the shade of the oak's canopy (Figure 32.23). The great earthquake of 1906 scarcely disturbed the giant, anchored as it was by a root system extending twenty-four meters through the soil. The few small brush fires that swept through the canyon did not seriously damage the giant tree. Parasitic fungi that could have rotted its roots never took hold in the well-drained soil. Birds and the tree's own defensive chemicals kept insects in check.

In the 1960s, suburban housing started to encroach on the once-wild hills. A developer turned his tractors into the canyon but spared the giant oak. Death came later.

The new homeowners were ignorant of the ancient, fragile relationships between the giant tree and the land that sustained it. They graded the soil between the trunk and drip line of the tree's canopy, mounded flower beds against the trunk, and planted lawns beneath the branches, then kept sprinklers busy. Overwatering put standing water next to the trunk. Before then, the tree had resisted the oak root fungus (*Armillaria*). In the changed environment, the fungus took hold. With its roots rotting away, the oak suffered massive disruption to the feedback relations among its roots, stems, and leaves. Eventually it had to be cut down. In their fifth winter, in their small brick fireplace, the homeowners began burning three centuries of firewood.

Figure 32.23 Looking up through part of the canopy of a coast live oak (*Quercus agrifolia*).

32.7 SUMMARY

1. Plant growth involves increases in the number, size, and volume of cells, by way of mitotic cell divisions and cell enlargements. Plant development refers to the emergence of different body parts, as brought about by selective gene expression and cell differentiation. Thus, plant growth is measurable in quantitative terms, and plant development in qualitative terms.

2. After seed germination, intricate control mechanisms govern the growth and development of the new plant. (Before then, the embryo sporophyte inside the seed was dormant. By the process of imbibition, it absorbed water, resumed growth, and broke through the seed coat.) Now the seedling increases in volume and mass. Its tissues and organs develop. Later, fruits and new seeds form and older leaves drop.

3. How individual plants grow and develop depends on interactions among their genes, their hormones, and cues from the environment.

a. A plant's genes govern the synthesis of enzymes and other proteins necessary for metabolism, hence for all cell activities. How and when each type of enzyme functions depend on hormonal action.

b. Hormones are a category of signaling molecules. After being produced and secreted by some cells, they travel to target cells in different parts of the plant body and stimulate or inhibit gene activity. As is the case for other kinds of organisms, any cell that bears receptors for a particular hormone is its target.

c. The prescribed growth patterns are influenced by predictable environmental cues. Often they are adjusted in response to unusual environmental pressures.

4. Five major categories of plant hormones have been identified, and the existence of others is suspected.

a. Gibberellins promote stem elongation, they help seeds and buds break dormancy in spring, and they may help induce the flowering process.

b. Auxins promote coleoptile and stem elongation. They have roles in phototropism and gravitropism.

c. Cytokinins stimulate cell division, promote leaf expansion, and retard leaf aging.

d. Abscisic acid promotes bud and seed dormancy, and it limits water loss by promoting stomatal closure.

e. Ethylene promotes fruit ripening and abscission.

5. These plant hormones trigger predictable patterns of growth and development. They also trigger responses to environmental rhythms (such as the seasonal changes in daylength and temperature) and to variations in the amount of sunlight, shade, and other circumstances.

6. Plant parts make tropic responses to light, gravity, and other environmental conditions. Hormones induce a difference in the rate and direction of growth on two sides of the part, which causes it to turn or move.

a. By gravitropism, roots grow downward and stems grow upright in response to the Earth's gravitational force. Some gravity-sensing mechanisms in plants are based on statoliths (clusters of particles in cells).

b. With phototropism, stems and leaves adjust rates and directions of growth in response to light. A yellow pigment (flavoprotein) might be involved; it absorbs blue wavelengths that trigger the strongest response.

c. With thigmotropism, plants adjust their direction of growth in response to contact with solid objects.

7. Plants respond to mechanical stress, as when strong winds inhibit stem elongation and plant growth.

8. A biological clock is any internal, time-measuring mechanism that has a biochemical basis.

a. Circadian rhythms are biological activities recurring in about twenty-four hour cycles. Rhythmic movements of leaves are an example.

b. Photoperiodism is a biological response to change in the relative length of daylight and darkness in the twenty-four-hour cycle. Seasonal photoperiodism occurs in plants. Phytochrome, a blue-green pigment, serves as a switching mechanism of a clock that promotes or inhibits germination, stem elongation, leaf expansion, stem branching, and flower, fruit, and seed formation.

c. Long-day plants flower during spring or summer, when there are more hours of daylight than darkness. Short-day plants flower when daylength is less. Day-neutral plants flower regardless of daylength.

9. Senescence is the sum of processes leading to the death of a plant or plant structure. Dormancy is a state in which a biennial or perennial stops growing even when conditions appear to be suitable for continued growth. A decrease in Pfr levels may trigger dormancy.

10. Dormancy is a state in which a perennial or biennial stops growing even when conditions appear suitable for continued growth. A decrease in Pfr levels might trigger dormancy. Breaking dormancy might involve exposure to certain temperatures and hormonal action, including gibberellins and abscisic acid.

Review Questions

1. Distinguish between growth and development. *32.2*
2. Name the three key factors that interact to dictate patterns of plant growth and development. *32.2*
3. List the five known types of plant hormones and describe the known functions of each. *32.2*
4. Define plant tropism. What is the difference between phototropism and photoperiodism? *32.3, 32.5*
5. What is phytochrome, and what role does it play in the flowering process? *32.4*
6. Explain the differences between long-day, short-day, and day-neutral plants. *32.4*
7. Define dormancy and senescence. Give examples. *32.5*

Figure 32.24 Field of sunflowers (*Helianthus*) that are busily demonstrating solar tracking.

Self-Quiz (*Answers in Appendix III*)

1. Seed germination is over when the ______.
   a. embryo sporophyte absorbs water
   b. embryo sporophyte resumes growth
   c. primary root pokes out of the seed coat
   d. cotyledons unfurl

2. Which of the following statements is false?
   a. Auxins and gibberellins promote stem elongation.
   b. Cytokinins promote cell division but retard leaf aging.
   c. Abscisic acid promotes water loss and dormancy.
   d. Ethylene promotes fruit ripening and abscission.

3. Plant hormones ______.
   a. interact with one another
   b. are influenced by environmental cues
   c. are active in plant embryos within seeds
   d. are active in adult plants
   e. all of the above

4. Plant growth depends on ______.
   a. cell division c. hormones
   b. cell enlargement d. all of the above

5. Light of ______ is the strongest stimulus for phototropism.
   a. red wavelengths c. green wavelengths
   b. far-red wavelengths d. blue wavelengths

6. Light of ______ wavelengths causes phytochrome to switch from inactive to active form; light of ______ wavelengths has the opposite effect.
   a. red; far-red c. far-red; red
   b. red; blue d. far-red; blue

7. The flowering process is a ______ response.
   a. phototropic c. photoperiodic
   b. gravitropic d. thigmotropic

8. Abscission occurs during ______.
   a. seed germination c. senescence
   b. flowering d. dormancy

9. Senescence involves a decrease in ______ in leaves and fruits and an increase in ______ at abscission zones.
   a. IAA; ethylene c. Pfr; gibberellin
   b. ethylene; IAA d. gibberellin; abscisic acid

10. Match the plant reproduction and development terms.
   ___ vernalization a. water moves into seeds
   ___ senescence b. unequal growth following contact with solid objects
   ___ imbibition c. inhibition of lateral bud formation
   ___ thigmotropism d. low-temperature stimulation of the flowering process
   ___ apical dominance e. all processes leading to death of plant or plant part

Critical Thinking

1. Given what you know about the growth of plants (Chapter 29), would you expect hormones to influence primary growth only? What about secondary growth in, say, a redwood tree?

2. Plant growth depends on photosynthesis, which depends on inputs of energy from the sun. How, then, can seedlings that were germinated in a dark room grow taller than seedlings that germinated in the sun?

3. *Solar tracking* refers to the observation that many plants are able to maintain the flat blades of their leaves at right angles to the sun throughout the day. Figure 32.24 gives an example. This tropic response maximizes the harvesting of the sun's rays by leaves. Suggest the name of one type of molecule that might be involved in the response.

4. Belgian scientists isolated a mutated gene in wall cress (*Arabidopsis thaliana*) that produces excess amounts of auxin. Predict what some of the resulting phenotypic traits might be.

5. All flowers are variations on a basic pattern of growth and development. Two groups led by Elliot Meyerowitze and Detlef Weigel recently identified the genes responsible for that plan in *Arabidopsis*. The master gene (*leafy*) activates other genes that contribute to the formation of sepals (gene *A*), petals (gene *B*), and reproductive structures (gene *C*). Discovering these genes and their interactions has been called the botanical equivalent of isolating master genes of *Drosophila* development. Speculate on what kind of internal and external signals switch on the leafy gene in the first place.

6. Cattle typically are given somatotropin, an animal hormone that makes them grow bigger (the added weight means greater profits). There is growing concern that such hormones may have unforeseen side effects on beef-eating humans. Would you think plant hormones applied to crop plants can affect humans also? Why or why not?

Selected Key Terms

abscisic acid *32.2*
abscission *32.5*
apical dominance *32.2*
auxin *32.2*
biological clock *32.4*
circadian rhythm *32.4*
coleoptile *32.2*
cytokinin *32.2*
day-neutral plant *32.4*
development *32.2*
dormancy *32.5*
ethylene *32.2*
flavoprotein *32.3*
germination *32.1*
gibberellin *CI*
gravitropism *32.3*
growth *32.2*
herbicide *32.2*
hormone *CI*
imbibition *32.1*
long-day plant *32.4*
photoperiodism *32.4*
phototropism *32.3*
phytochrome *32.4*
plant tropism *32.3*
senescence *32.5*
short-day plant *32.4*
statolith *32.3*
thigmotropism *32.3*
vernalization *32.5*

Readings *See also www.infotrac-college.com*

Bewley, J. D., and M. Black. 1985. *Seeds: Physiology of Development and Germination.* New York: Plenum.

Meyerowitz, E. M. November 1994. "Genetics of Flower Development." *Scientific American* 271: 56–65.

Rost, T., M. Barbour, C. Stocking, and T. Murphy. 1998. *Plant Biology.* Chapter 15. Belmont, California: Wadsworth. Paperback.

Salisbury, F., and C. Ross. 1992. *Plant Physiology.* Fourth edition. Belmont, California: Wadsworth.

FACING PAGE: *How many and what kinds of body parts does it take to function as a lizard in a tropical forest? Make a list of what comes to mind as you start reading Unit VI, then see how resplendent the list can become at the unit's end.*

ANIMAL STRUCTURE AND FUNCTION 

33 TISSUES, ORGAN SYSTEMS, AND HOMEOSTASIS

Meerkats, Humans, It's All the Same

After a cold night in Africa's Kalahari Desert, animals small enough to fit inside a coat pocket emerge stiffly from their burrows. These "meerkats" are a type of mongoose. They stand on their hind legs and face east, exposing their chilled bodies to the warm rays of the morning sun (Figure 33.1). Meerkats don't know it, but sunning behavior helps their enzymes. If their body's internal temperature were to fall below a tolerable range, the action of countless enzyme molecules in their cells would falter and metabolism would suffer.

Once meerkats warm up, they fan out from their burrows and look for food. Into the meerkat gut go insects and an occasional lizard. These are pummeled, dissolved, and digested into glucose and other nutritious bits small enough to move across the gut wall, into the blood, and on to the body's cells. Aerobic machinery in the cells cracks apart molecules of glucose and other organic compounds and releases energy. A respiratory system contributes to supplying the machinery with oxygen and taking away its carbon dioxide leftovers.

All of this activity changes the composition and volume of the **internal environment**, which consists of blood and interstitial fluid (tissue fluid) that bathes the living cells of any complex animal. Drastic changes in these fluids would kill the cells, but a urinary system works to keep this from happening. Governing that system and all others are the nervous and endocrine systems. The two interact as a central command post to mobilize the body as a whole for everything from simple housekeeping tasks to heart-thumping flights from predators.

And so meerkats start us thinking about this unit's topics: how the animal body is structurally put together (its *anatomy*) and how it functions (its *physiology*). This chapter is an overview of the animal tissues and organ systems we will be considering. It also introduces the key concept of **homeostasis**. With respect to animals, the word refers to stable operating conditions in the internal environment, as brought about by coordinated activities of cells, tissues, organs, and organ systems.

Amazingly, all complex animals consist of only four basic types of tissues. These are epithelial, connective, muscle, and nervous tissues. A **tissue** is a group of cells and intercellular substances, all interacting in one or more tasks. As one example, muscle tissue takes part in contraction. An **organ** consists of different tissues that are organized in specific proportions and patterns. Thus every vertebrate heart has predictable proportions and arrangements of epithelial, connective, muscle, and nervous tissues. An **organ system** consists of two or

more organs that are interacting physically, chemically, or both in a common task, as when interconnected arteries and other vessels transport blood through the body under the driving force of a beating heart.

Cells, tissues, organs, and organ systems split up the work, so to speak, in ways that contribute to survival of the body as a whole. This is sometimes referred to as a **division of labor**. By the end of this unit, you may have an abiding appreciation of the sheer magnitude of the division of labor among the separate parts and the extent to which their activities are integrated.

As you will see, whether you look at a flatworm or salmon, a meerkat or human being, the animal body is structurally and physiologically adapted to perform four overriding tasks:

1. *Maintain conditions in the internal environment within ranges that are most favorable for cell activities.*
2. *Acquire nutrients and other raw materials, distribute them through the body, and dispose of wastes.*
3. *Afford protection against injury or attack from viruses, bacteria, and other agents of disease.*
4. *Reproduce, then often help nourish and protect the new individuals during their early growth and development.*

Figure 33.1 In the Kalahari Desert, gray meerkats (*Suricata suricatta*) face the sun's warming rays, as they do every morning. This simple behavior helps them maintain internal body temperature. How animals function in their environment is the focus of this unit.

KEY CONCEPTS

1. The cells of most animals interact at three levels of organization—in tissues, many of which are combined in organs, which are components of organ systems.

2. Most animals are constructed of four types of tissues, which are called epithelial, connective, muscle, and nervous tissues.

3. Each animal cell engages in basic metabolic activities that ensure its own survival. At the same time, animal cells of a given tissue perform one or more activities that contribute to the survival of the animal as a whole.

4. The internal environment consists of all fluids that are not inside the body's cells—that is, blood and interstitial fluid.

5. The combined contributions of cells, tissues, organs, and organ systems help maintain stability in the internal environment, which is required for the survival of each individual cell. This concept helps us understand the functions of any organ or organ system.

6. Homeostasis is the formal name for stable operating conditions in the internal environment.

33.1 EPITHELIAL TISSUE

General Characteristics

We commonly refer to an epithelial tissue as **epithelium** (plural, epithelia). This tissue has a free surface, which faces either a body fluid or the outside environment. *Simple* epithelium, with a single layer of cells, functions as a lining for body cavities, ducts, and tubes. *Stratified* epithelium, which has two or more cell layers, typically functions in protection, as it does in your skin. Figure 33.2 shows examples of this type of animal tissue.

All cells in epithelium are close together, with little intervening material. As is the case for nearly all animal tissues, specialized junctions afford both structural and functional links between its individual cells, which absorb, make, and secrete a variety of substances.

Cell-to-Cell Contacts

In epithelium and other tissues, we observe three classes of cell junctions. **Tight junctions** help stop substances from leaking across a tissue. **Adhering junctions** are spot welds; they cement neighbor cells together. **Gap junctions** are channels connecting the cytoplasm of abutting cells. By promoting rapid transfers of ions and small molecules, they help cells communicate with each other (Figure 33.3).

Think of gastric fluid inside the stomach to get an idea of how important such junctions are. If this highly acidic fluid were to leak past the stomach's epithelial lining, it would digest proteins of your own body instead of those brought in with meals. Actually, that happens in patients with peptic ulcers (Section 42.4). Many tight junctions in such linings form a leakproof barrier between all the cells near their free surface.

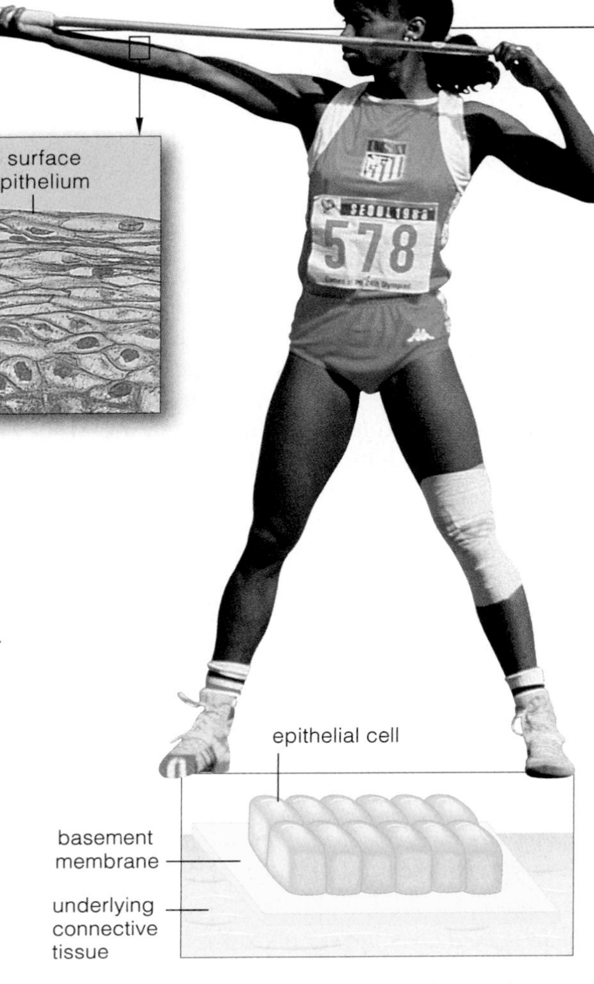

free surface
of epithelium
epithelial cell
basement
membrane
underlying
connective
tissue

a Some characteristics of epithelium

Figure 33.2 (**a**) Some basic characteristics of epithelium. Epithelia have a free surface exposed to a body fluid or to the environment. Between the epithelium's other surface and the connective tissue on which it rests is a basement membrane. The diagram below the athlete shows this arrangement for a simple epithelium, which has a single layer of cells. The light micrograph shows the upper part of stratified epithelium. This type of epithelium has more than one layer of cells, which are flattened out near the surface.

(**b**) Light micrographs and sketches of three simple epithelia, showing the three shapes of cells that are common in this type of tissue.

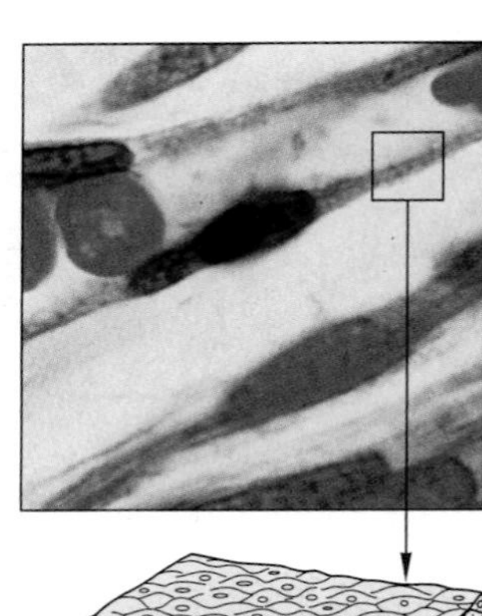

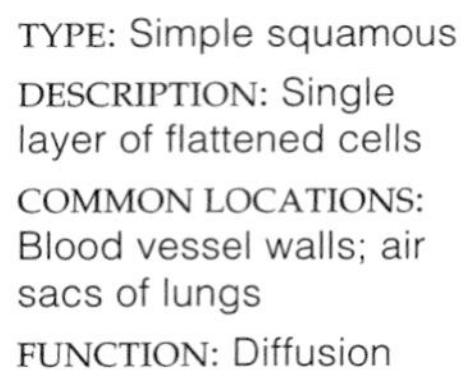

TYPE: Simple squamous

DESCRIPTION: Single layer of flattened cells

COMMON LOCATIONS: Blood vessel walls; air sacs of lungs

FUNCTION: Diffusion

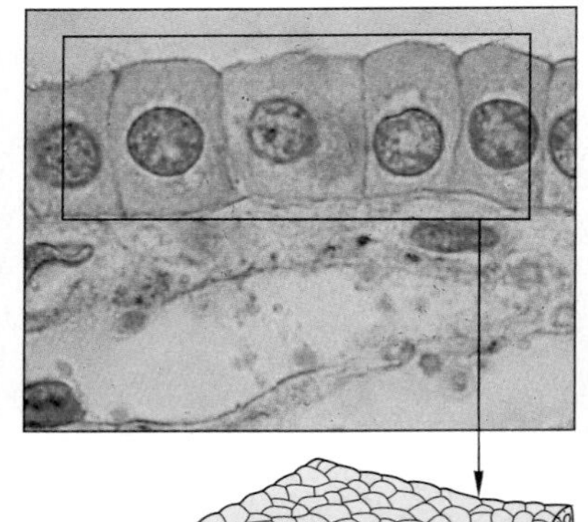

TYPE: Simple cuboidal

DESCRIPTION: Single layer of cubelike cells; free surface may have microvilli (absorptive structures)

COMMON LOCATIONS: Glands and tubular parts of nephrons in kidneys

FUNCTION: Secretion, absorption

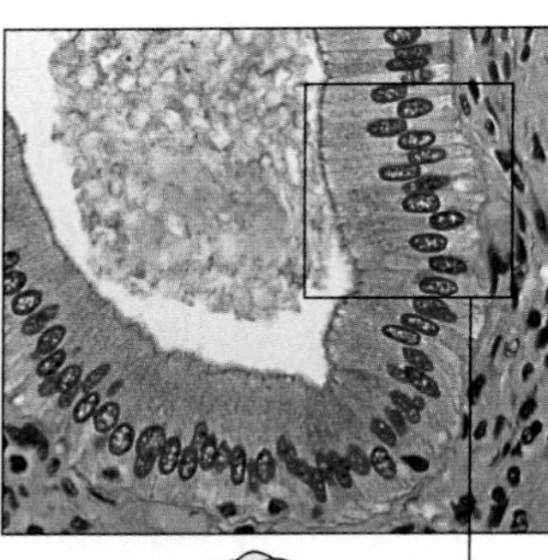

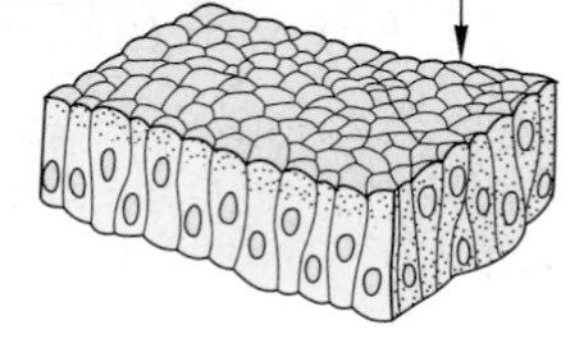

TYPE: Simple columnar

DESCRIPTION: Single layer of tall, slender cells; free surface may have microvilli

COMMON LOCATIONS: Part of lining of gut and respiratory tract

FUNCTION: Secretion, absorption

b Examples of cell shape in simple epithelium

Further reading: Student Guide to InfoTrac on web site

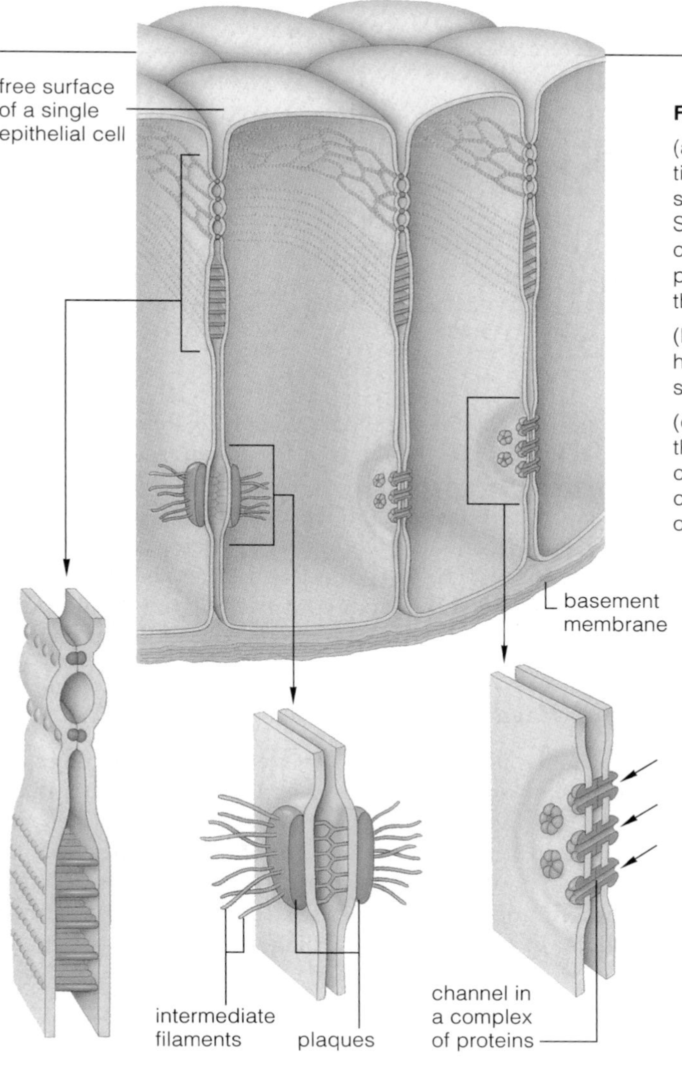
free surface
of a single
epithelial cell
basement
membrane
intermediate
filaments
plaques
channel in
a complex
of proteins
a
TIGHT JUNCTION
Strands (rows of proteins) running parallel with the free surface of the tissue; they stop leaks between adjoining cells.
b
ADHERING JUNCTION
Adjoining cells adhere at a mass of proteins (a plaque) anchored beneath their plasma membrane by many intermediate filaments of the cytoskeleton.
c
GAP JUNCTION
Cylindrical arrays of proteins that span the plasma membrane of adjoining cells pair up, forming open channels between the cytoplasm of both cells.

Figure 33.3 *Left:* Examples of cell junctions.

(**a**) In some epithelia, cells have parallel rows of proteins that form tight seals and fuse each cell with its neighbors. These tight junctions stop most substances from leaking across the tissue's free surface. Substances reach tissues below only by passing *through* epithelial cells. Built-in controls make the cells' plasma membrane selectively permeable. They let some substances but not others move across, through the interior of transport proteins (Sections 5.5 and 5.6).

(**b**) Adhering junctions (spot welds or collars) function as a unit to hold cells of all animal tissues together. They are profuse in skin's surface layer and other tissues subjected to abrasion.

(**c**) Gap junctions, channels across the plasma membrane, connect the cytoplasm of adjoining cells. Gap junctions promote chemical communication (the diffusion of ions and small molecules from cell to cell). They are especially abundant in the heart, stomach, and other organs in which cell activities must be rapidly coordinated.

pore that opens at skin surface

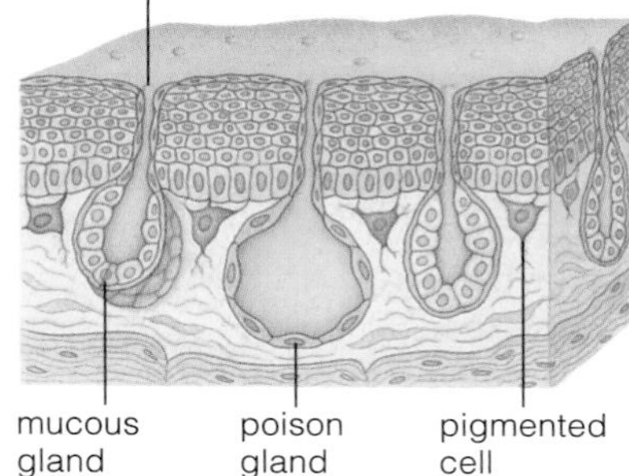

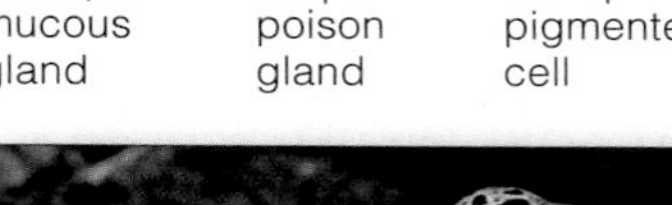
mucous
gland
poison
gland
pigmented
cell

Figure 33.4 Section through glandular epithelium of a frog. The photograph shows a frog of the genus *Dendrobates*, which produces one of the most lethal glandular secretions known. (Some natives of a tribe in Colombia use the exocrine gland secretion to poison dart tips, which they shoot through blowguns.) Pigment-rich cells that branch through the epithelium impart color to the skin. The striking coloration of all poisonous frogs serves as a strong warning signal to potential predators.

Glandular Epithelium and Glands

Epithelia of structurally simple invertebrates contain **gland cells**, which secrete (release) products, unrelated to their own metabolism, that are to be used elsewhere. In more complex animals, such cells occur in glandular epithelium and in glands, which are secretory organs derived from epithelia. ("Secretion" is not the same as "excretion," a concentration and removal of metabolic wastes or excess substances of no use to the body.)

Exocrine glands secrete mucus, saliva, earwax, oil, milk, digestive enzymes, and other cell products. Most exocrine products are released onto the free epithelial surface through ducts or tubes, as in Figure 33.4. By contrast, **endocrine glands** do not have ducts. Their products are hormones, which they secrete directly into fluid bathing the gland. Hormone molecules typically enter the bloodstream, which distributes them to target cells elsewhere in the body.

Epithelia are sheetlike tissues lining the body's surface and its cavities, ducts, and tubes. Epithelia have one free surface facing a body fluid or the outside environment. Their cells are structurally and functionally connected at junctions.

Glands are secretory organs derived from epithelium.

33.2 CONNECTIVE TISSUE

Of all tissues in complex animals, connective tissues are the most abundant and widely distributed. They range from soft connective tissues to the specialized types called cartilage, bone tissue, adipose tissue, and blood (Table 33.1). In all connective tissues except blood, cells secrete fibers of structural proteins: collagen or elastin. (This is the collagen that plastic surgeons use to plump wrinkled skin and lips.) The cells also secrete modified polysaccharides, which accumulate between cells and fibers as the connective tissue's "ground substance."

Table 33.1 Categories of Connective Tissues

SOFT	SPECIALIZED
Loose connective tissue	Cartilage
Dense, irregular connective tissue	Bone tissue
Dense, regular connective tissue (ligaments, tendons)	Adipose tissue
	Blood

Soft Connective Tissues

These tissues generally have the same components but different proportions of them. **Loose connective tissue** contains fibers and fibroblasts (cells that produce and secrete the fibers), all loosely arranged in a semifluid ground substance (Figure 33.5*a*). This tissue often acts as a support framework for epithelium. It also houses white blood cells that can mount early counterattacks against foreign invaders—as when bacterial cells enter skin through small cuts and abrasions or breach the digestive, respiratory, or urinary tract lining.

Dense, irregular connective tissue has fibroblasts and many fibers (mostly collagen-containing ones) that are oriented every which way. This tissue is present in skin and forms protective capsules around organs that do not stretch much (Figure 33.5*b*). In **dense, regular connective tissue**, the fibroblasts occur in rows between many parallel bundles of fibers. Tendons, which attach skeletal muscle to bone, have this tissue. The bundles of collagen fibers help tendons resist being torn (Figure 33.5*c*). Dense, regular connective tissue also is present in elastic ligaments, which attach one bone to another. The elastic fibers allow movement around joints.

Specialized Connective Tissues

Like rubber, the intercellular material called **cartilage** is pliable yet solid, and it resists compression. Cells that made and released the material became imprisoned in small cavities in their own secretions (Figure 33.5*d*). Cartilage deposited in vertebrate embryos serves as the structural models for bones. Bone replaces most of the models. In adults, some cartilage maintains the shape

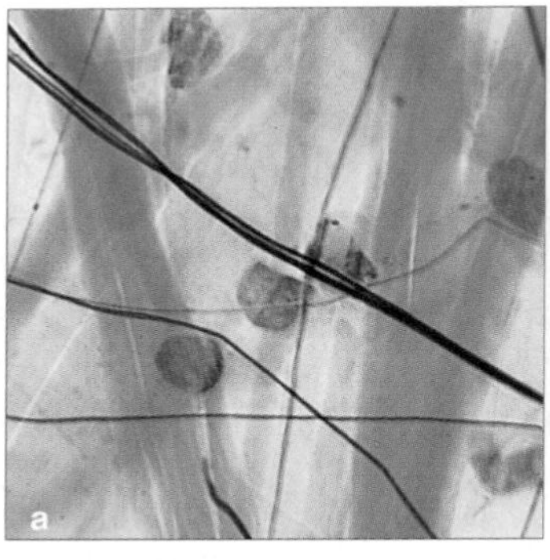
a

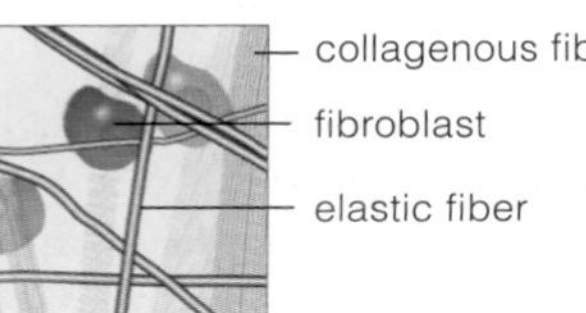
collagenous fiber
fibroblast
elastic fiber

TYPE: Loose connective tissue
DESCRIPTION: Fibroblasts, other cells, plus fibers loosely arranged in semifluid ground substance
COMMON LOCATIONS: Under the skin and most epithelia
FUNCTION: Elasticity, diffusion

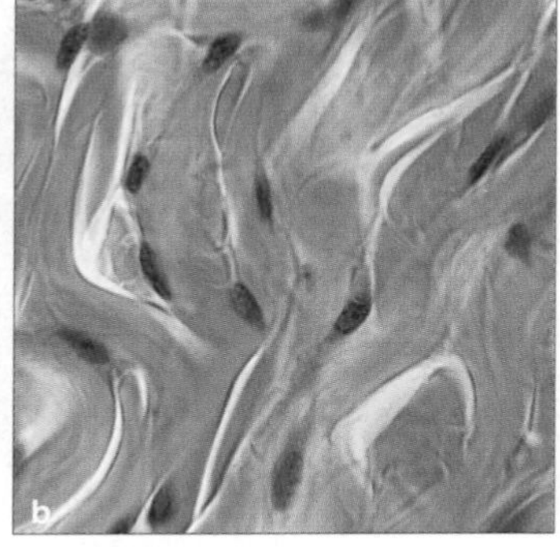
b

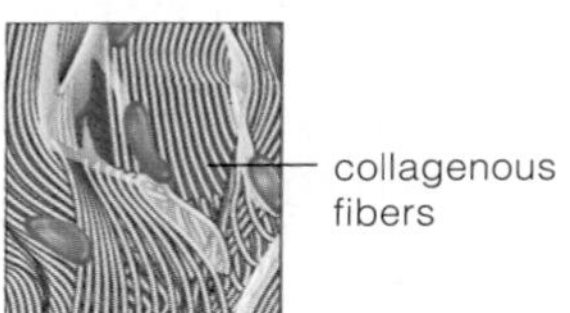
collagenous fibers

TYPE: Dense, irregular connective tissue
DESCRIPTION: Collagenous fibers, fibroblasts, less ground substance
COMMON LOCATIONS: In skin and capsules around some organs
FUNCTION: Support

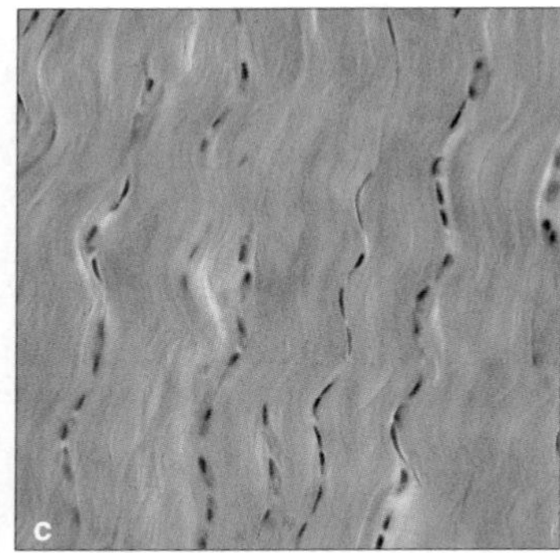
c

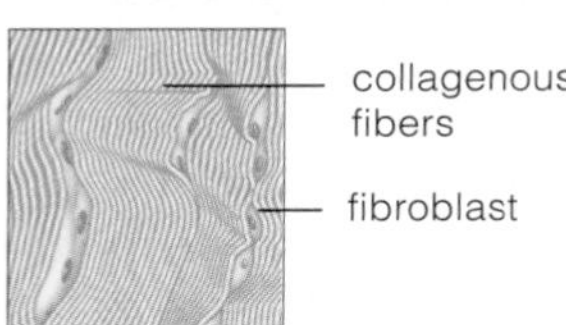
collagenous fibers
fibroblast

TYPE: Dense, regular connective tissue
DESCRIPTION: Collagen fibers in parallel bundles, long rows of fibroblasts, little ground substance
COMMON LOCATIONS: Tendons, ligaments
FUNCTION: Strength, elasticity

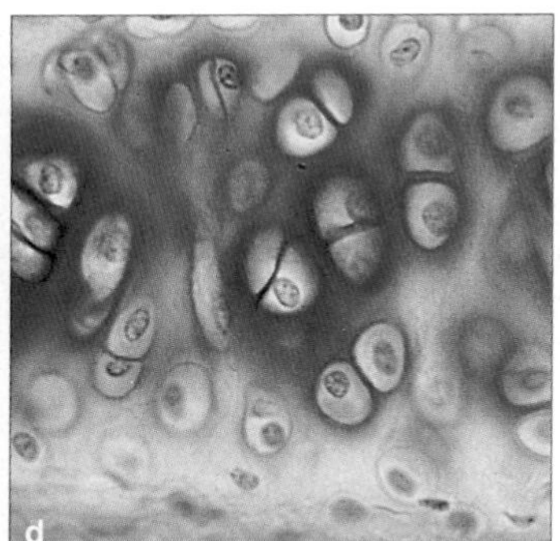
d

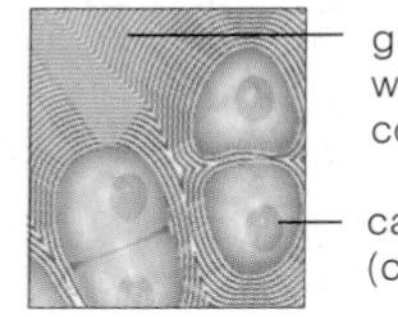
ground substance with very fine collagen fibers
cartilage cell (chondrocyte)

TYPE: Cartilage
DESCRIPTION: Cells embedded in pliable, solid ground substance
COMMON LOCATIONS: Ends of long bones, nose, parts of airways, skeleton of vertebrate embryos
FUNCTION: Support, flexibility, low-friction surface for joint movement

Figure 33.5 Examples of soft connective tissues and specialized connective tissues. 

Further reading: Student Guide to InfoTrac on web site

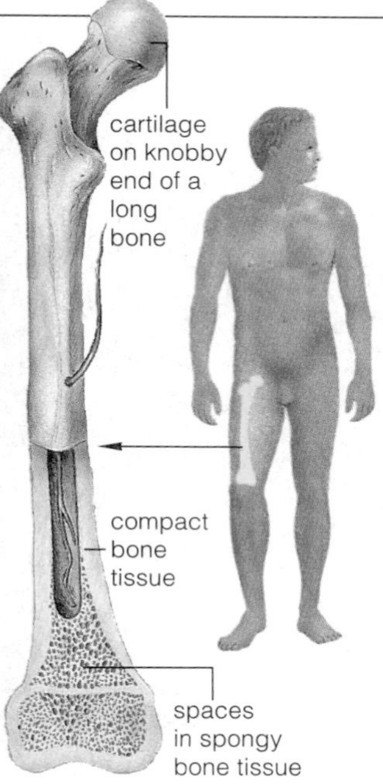

cartilage on knobby end of a long bone
compact bone tissue
spaces in spongy bone tissue

Figure 33.6 Cartilage and bone tissue. Spongy bone tissue has tiny, needlelike hard parts with spaces in between. Compact bone tissue is more dense. Bone is a load-bearing tissue that resists compression. Over time, it served as a basis for increases in the body size of many land vertebrates, such as giraffes. It gives large animals selective advantages. Among other things, they can ignore most predators with impunity, roam farther for food and water, and gain or lose heat more slowly than small animals. They can do so because of their lower surface-to-volume ratio and greater heat production.

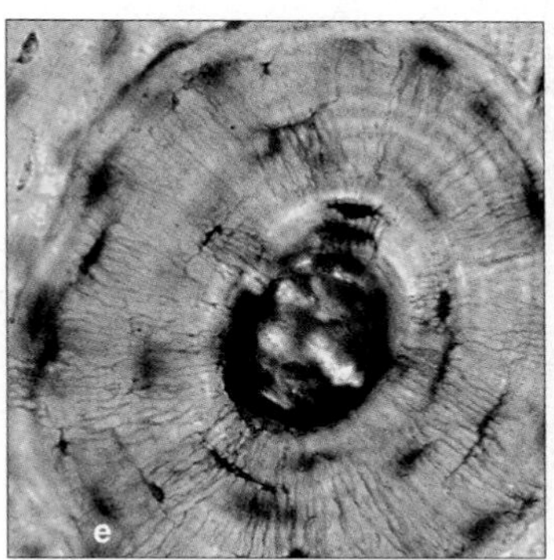

e

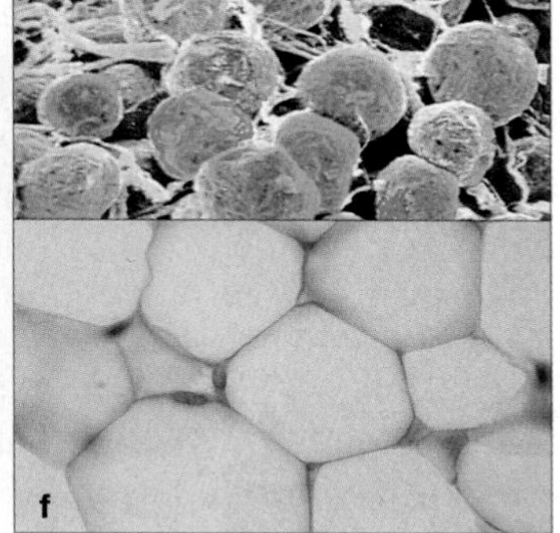

f

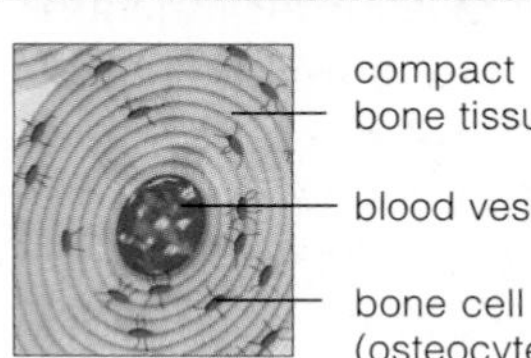

compact bone tissue
blood vessel
bone cell (osteocyte)

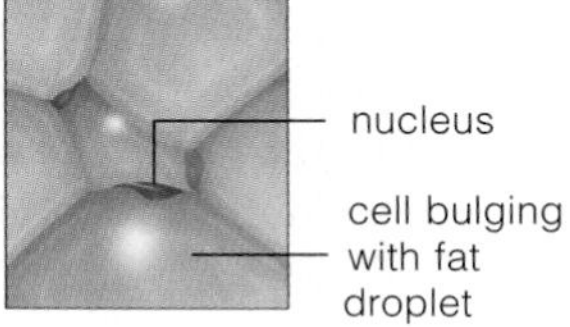

nucleus
cell bulging with fat droplet

TYPE: Bone tissue
DESCRIPTION: Collagen fibers, ground substance hardened with calcium
COMMON LOCATIONS: Bones of vertebrate skeleton
FUNCTION: Movement, support, protection

TYPE: Adipose tissue
DESCRIPTION: Large, tightly packed fat cells occupying most of ground tissue
COMMON LOCATIONS: Under skin, around heart, kidneys
FUNCTION: Energy reserves, insulation, padding

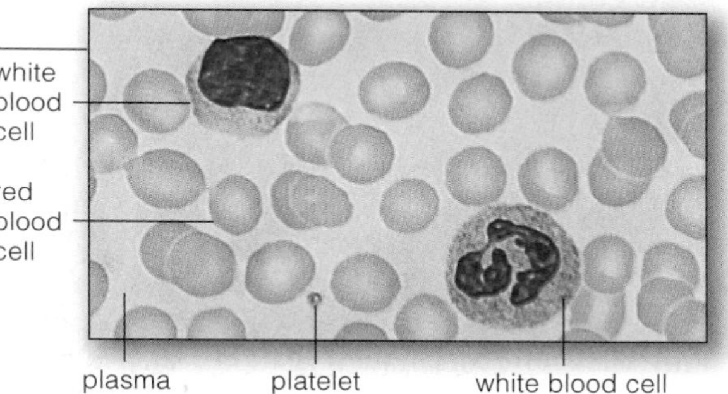

white blood cell
red blood cell
plasma
platelet
white blood cell

Figure 33.7 Some components of human blood. This tissue's straw-colored, liquid matrix (plasma) is mostly water in which many nutrients, diverse proteins, oxygen, carbon dioxide, ions, and other substances are dissolved.

of the nose, outer ear, and other body parts. It cushions joints between adjacent bones of the vertebral column, limbs, hands, and elsewhere.

Bone tissue is mineral hardened; its collagen fibers and ground substance are rich in calcium salts (Figures 33.5*e* and 33.6). It is the main tissue of bones. In vertebrate skeletons, different bones support and protect softer tissues and organs. Limb bones, such as the long bones of your own legs, serve weight-bearing functions. They also interact with skeletal muscles attached to them to bring about movements. Parts of some bones also are production sites for blood cells.

Excess carbohydrates and proteins that cells do not use at once are converted to storage forms, especially fats. A few fat droplets collect in the cytoplasm of various cells. But cells in **adipose tissue**, located primarily beneath skin, specialize in fat storage; fat droplets almost fill their cytoplasm (Figure 33.5*f*). Fats readily move to and from cells by way of fine blood vessels that thread through the tissue.

Because **blood** is derived primarily from connective tissue, many biologists classify it as a connective tissue. Blood serves transport functions. Circulating within its *plasma*, a fluid medium, are great numbers of red blood cells, white blood cells, and platelets (Figure 33.7). Red blood cells efficiently deliver oxygen to metabolically active tissues and then carry carbon dioxide and other wastes away from them. Plasma is largely water, but it has a great number of different kinds of proteins, ions, and other substances dissolved in it. We will be taking a closer look at this complex tissue in Section 39.2.

Diverse types of connective tissues bind together, support, strengthen, protect, and insulate other tissues in the body.

Soft connective tissues consist of protein fibers as well as a variety of cells arranged in a ground substance.

Cartilage, bone, blood, and adipose tissue are specialized connective tissues. Cartilage and bone are both structural materials. Blood is a fluid tissue with transport functions. Adipose tissue is a reservoir of stored energy.

33.3 MUSCLE TISSUE

In all muscle tissues, cells *contract* (that is, shorten) in response to stimulation from the outside, then lengthen and so return to their uncontracted state. These tissues have many long, cylindrical cells arranged in parallel arrays. Layers of muscles and muscular organs contract and relax in coordinated fashion. Their action moves the body through the surroundings, and it maintains and changes the positions of the body's various parts. The three types of muscle tissue are skeletal, smooth, and cardiac muscle tissues.

a

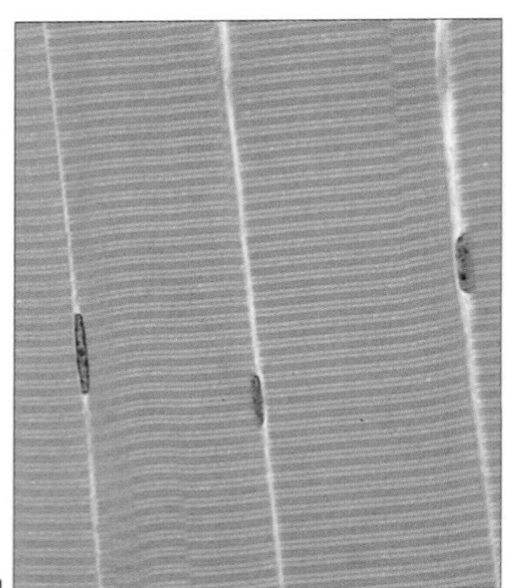

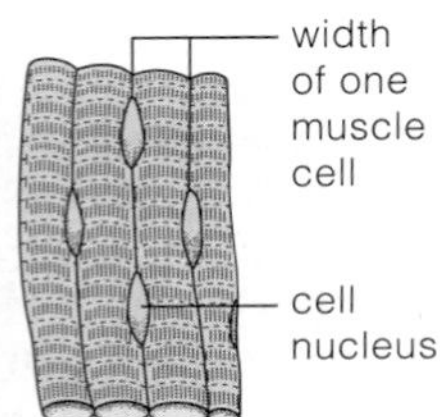
width of one muscle cell
cell nucleus

TYPE: Skeletal muscle

DESCRIPTION: Bundles of long, cylindrical, striated, contractile cells

LOCATION: Associated with skeleton

FUNCTION: Locomotion, movement of body parts

b

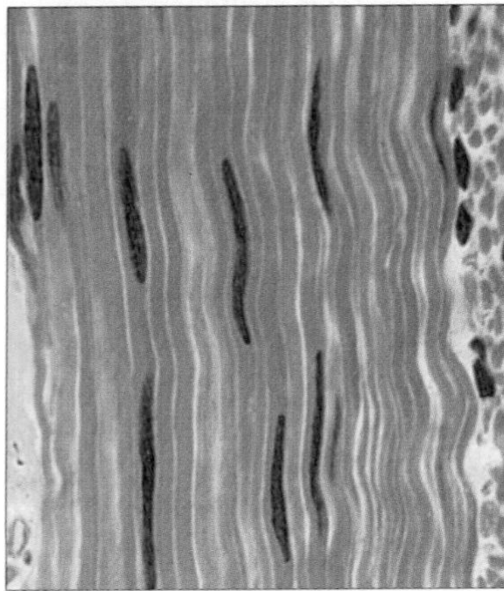

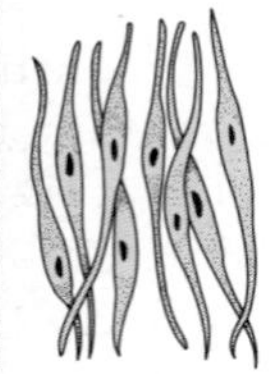
cells teased apart for clarity

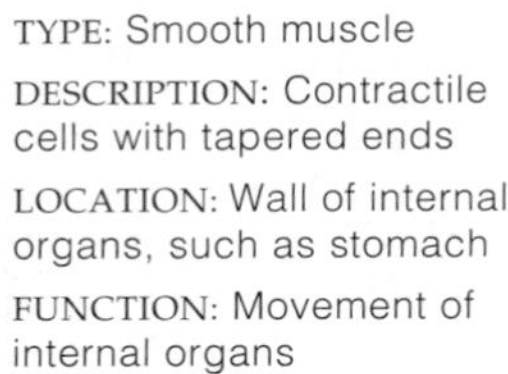

TYPE: Smooth muscle

DESCRIPTION: Contractile cells with tapered ends

LOCATION: Wall of internal organs, such as stomach

FUNCTION: Movement of internal organs

c

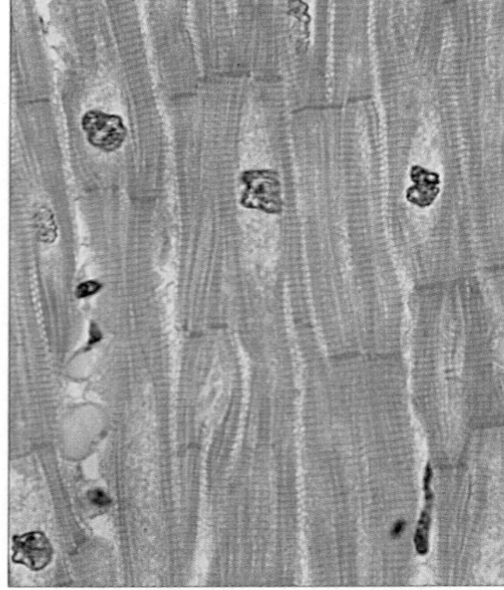

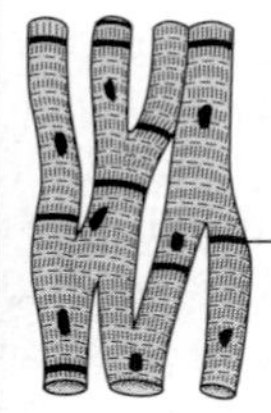
junction between adjacent cells

TYPE: Cardiac muscle

DESCRIPTION: Cylindrical, striated cells that have specialized end junctions

LOCATION: Wall of heart

FUNCTION: Pump blood within circulatory system

Figure 33.8 Characteristics and examples of skeletal muscle, smooth muscle, and cardiac muscle tissues.

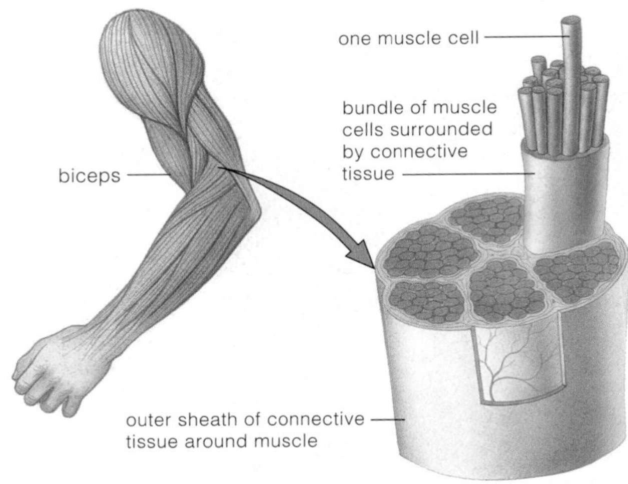
one muscle cell
bundle of muscle cells surrounded by connective tissue
biceps
outer sheath of connective tissue around muscle

Figure 33.9 Location and general arrangement of the muscle cells in a typical skeletal muscle.

Skeletal muscle tissue is located in muscles that are securely fastened to skeletal bones (Figure 33.8*a*). In a typical muscle, such as the biceps, striated skeletal muscle cells are bundled closely together, in parallel. (*Striated* means striped.) A sheath of tough connective tissue encloses several bundles of muscle cells, as you can see from Figure 33.9. The structure and function of skeletal muscle tissue are topics of Chapter 38.

The contractile cells of **smooth muscle tissue** taper at both ends (Figure 33.8*b*). Cell junctions hold them together, and they are bundled together in a connective tissue sheath. The wall of internal organs—including blood vessels, the stomach, and the intestines—contains this type of muscle tissue. Contraction of smooth muscle action is sometimes said to be "involuntary," because we usually are not able to make it contract merely by thinking about it. We can do this with skeletal muscle.

Cardiac muscle tissue is a contractile tissue that is present only in the heart (Figure 33.8*c*). Cell junctions fuse the plasma membranes of cardiac muscle cells and make them stick together. Communication junctions at some fusion points allow the cells to contract as a unit. That is, when one cell receives a signal to contract, its neighbors are stimulated to contract, also.

Muscle tissue, which can contract (shorten) in response to stimulation, helps move the body and specific body parts.

Skeletal muscle is the only muscle tissue attached to bones. Smooth muscle is a component of internal organs. Cardiac muscle alone makes up the contractile walls of the heart. Connective tissue sheathes all three types of tissues.

Further reading: Student Guide to InfoTrac on web site

33.4 NERVOUS TISSUE

Of all tissues, **nervous tissue** exerts the greatest control over the body's responsiveness to changing conditions. In your own body, **neuroglia** makes up more than one-half the volume of nervous tissue. That word refers to a variety of cells that protect, structurally support, and metabolically support neurons, which make up the rest of your nervous tissue. **Neurons** are excitable cells, the communication units of most nervous systems.

When a neuron is suitably stimulated, an electrical disturbance swiftly travels along its plasma membrane. Arrival of the disturbance at the neuron's endings, or output zone, triggers events that may cause stimulation or inhibition of adjacent neurons and other cells.

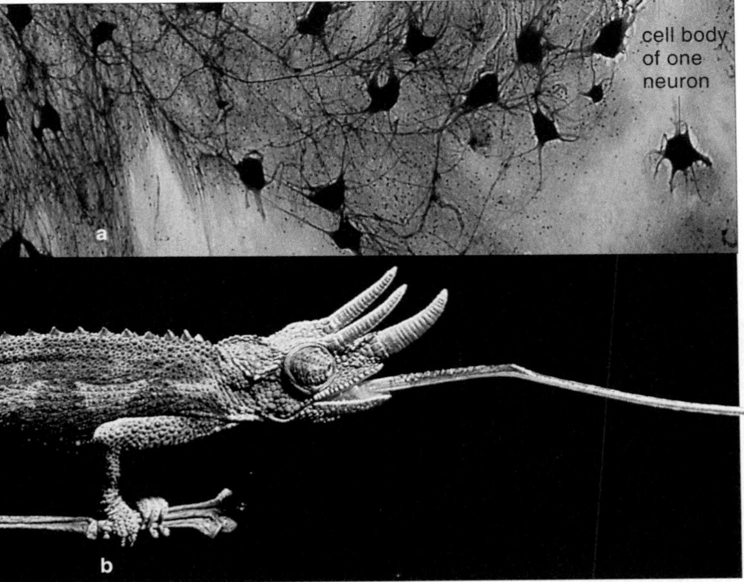
cell body of one neuron
a
b

Figure 33.10 (**a**) A few motor neurons from the human nervous system. This type relays signals from the brain or spinal cord to muscles and glands. Collectively, different neurons detect and process a great number and great variety of signals about environmental change and initiate suitable responses. Without neurons, for instance, a chameleon (**b**) could not detect an edible insect, calculate its distance, and command a long, sticky, prey-capturing tongue to uncoil with stunning speed.

For example, more than a hundred billion neurons are organized as communication lines throughout your body. Some detect specific changes in environmental conditions. Others coordinate immediate and long-term responses to change. The type shown in Figure 33.10*a* relays signals from the brain to muscles and glands. How these cells function is a topic of later chapters.

Neurons are the basic units of communication in nervous tissue. Different kinds detect specific stimuli, integrate information, and issue or relay commands for response.

Focus on Science

33.5 FRONTIERS IN TISSUE RESEARCH

As you know by now, a tissue is more than the sum of its cells. When each new animal grows and develops, its cells interact and become organized in prescribed ways to give rise to the body's diverse tissues. Also, cells of each new tissue synthesize specific gene products that are vital for normal body functioning.

For many decades, medical researchers attempted to grow artificial tissues in quantity in the laboratory. Such *lab-grown epidermis* could be used to speed the healing of extensive burns, ulcers, cancerous lesions, and certain blistering disorders. Now lab-grown epidermis is a reality. One company produces it by mixing cells of foreskins discarded from circumscribed male infants and proteins from cattle tendons. The mixture is added to rows of small, shallow dishes holding a culture medium that is enriched with nutrients and growth factors. The cells multiply to form round, paper-thin grafts. A bit of foreskin no larger than a postage stamp contains enough undifferentiated cells to make 200,000 grafts, each with a five-day shelf life. Surgeons apply such grafts to a wound and bind it with a gauze dressing. Over the next few weeks, the graft's cells interact biochemically and structurally with a patient's cells to regenerate damaged or missing tissue. Use of such lab-grown skin is less costly and less risky than surgery to obtain a patch of the patient's own skin.

On the horizon are *designer organs*. These encapsulated, preselected groups of living cells might synthesize specific hormones, enzymes, growth factors, and other substances. The idea is to surgically snip a bit of epithelium from a patient and then use it to enclose the cells, like a capsule. Because the capsule is derived from epithelial cells of the patient's body, it will not be chemically recognized as "foreign" and attacked by the patient's immune system. As you will see in Chapter 40, this is what happens when the body rejects organ or tissue implants. Such rejection can have serious medical consequences.

Currently, biotechnologists are close to understanding how to synthesize molecular cues that will allow designer organs to stick to appropriate sites in the body. Once the synthetic organs stick, they may become integral parts of normal body functioning.

Ultimately, the goal of this research is to put together packages of cells capable of producing specific life-saving substances that are absent in patients who suffer from severe genetic disorders or chronic diseases. For example, imagine the potential for people who have type 1 *diabetes mellitus*. This metabolic disorder results in an elevated concentration of glucose in the blood. Affected people produce little if any insulin, which is the hormone that signals cells to take up glucose from the blood. Unless they receive insulin injections on a regular basis, they will die. However, if a customized, insulin-secreting organ can be successfully installed inside the body of such individuals, their daily injections of insulin might be a thing of the past.

ORGAN SYSTEMS

Overview of Major Organ Systems

Figure 33.11 gives an overview of the organ systems of a typical vertebrate, the adult human. Figure 33.12 lists some terms that are used when describing positions of various organs. It also shows the major body cavities in which a number of important organs are located.

Each organ system contributes to the survival of all living cells in the animal body. You might think this is stretching things a bit. For example, how could muscles and bones be helping each microscopically small cell stay alive? Yet interactions between the skeletal system and muscular system allow us to move about—toward sources of nutrients and water, for example. Some parts of the two organ systems help keep blood circulating to cells, as when leg muscle contractions help move blood in veins back to the heart. Blood inside the circulatory system rapidly transports oxygen, nutrients, and other substances to cells, and transports products and wastes away from them. The respiratory system swiftly delivers oxygen from the outside air to the circulatory system and takes up carbon dioxide wastes from it, skeletal muscles assist the respiratory system—and so it goes, throughout the entire body.

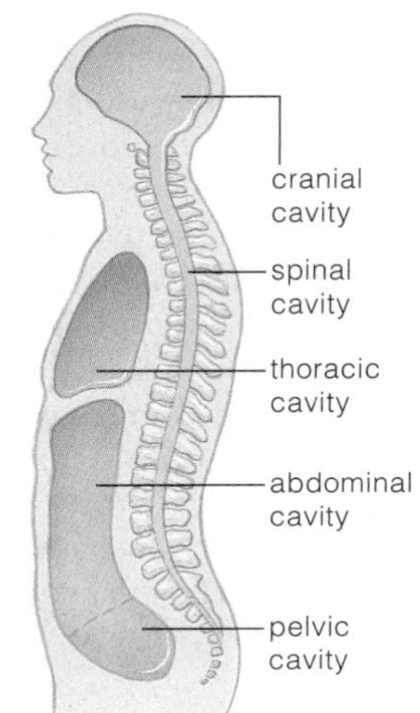
cranial cavity
spinal cavity
thoracic cavity
abdominal cavity
pelvic cavity
a

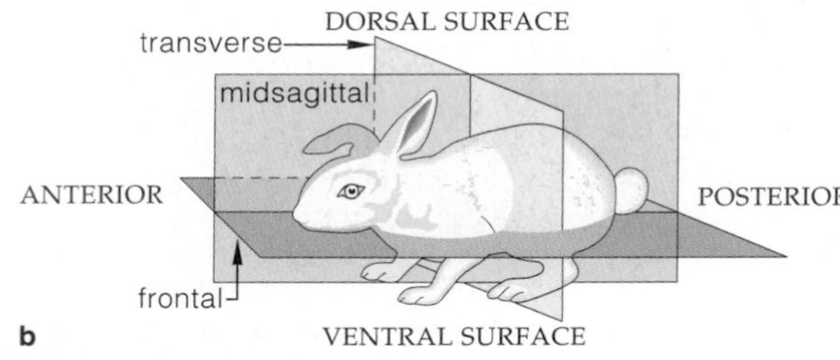
DORSAL SURFACE
transverse
midsagittal
ANTERIOR
POSTERIOR
frontal
b
VENTRAL SURFACE

Figure 33.12 (**a**) Major cavities in the human body. (**b**,**c**) Directional terms and planes of symmetry for the vertebrate body. Notice how the *midsagittal* plane divides the body into right and left halves. Most vertebrates, such as fishes and rabbits, move with the main body axis parallel with the Earth's surface. For them, *dorsal* pertains to their back or upper surface, and *ventral* pertains to the opposite, lower surface.

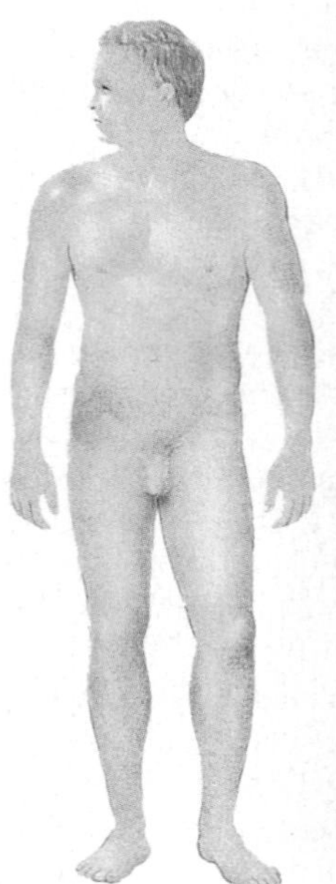

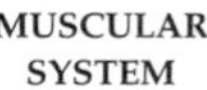

INTEGUMENTARY SYSTEM
Protect body from injury, dehydration, and some pathogens; control its temperature; excrete some wastes; receive some external stimuli.

MUSCULAR SYSTEM
Move body and its internal parts; maintain posture; generate heat by increases in metabolic activity.

SKELETAL SYSTEM
Support and protect body parts; provide muscle attachment sites; produce red blood cells; store calcium, phosphorus.

NERVOUS SYSTEM
Detect both external and internal stimuli; control and coordinate responses to stimuli; integrate all organ system activities.

ENDOCRINE SYSTEM
Hormonally control body function; work with nervous system to integrate short-term and long-term activities.

CIRCULATORY SYSTEM
Rapidly transport many materials to and from cells; help stabilize internal pH and temperature.

Figure 33.11 Overview of human organ systems and their functions. 

Further reading: Student Guide to InfoTrac on web site

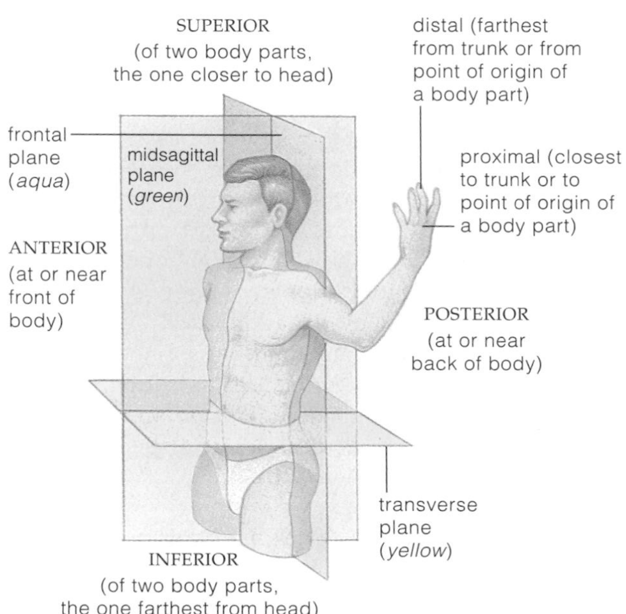
SUPERIOR
(of two body parts, the one closer to head)
distal (farthest from trunk or from point of origin of a body part)
frontal plane (aqua)
midsagittal plane (green)
proximal (closest to trunk or to point of origin of a body part)
ANTERIOR
(at or near front of body)
POSTERIOR
(at or near back of body)
transverse plane (yellow)
INFERIOR
(of two body parts, the one farthest from head)

c Unlike quadrupedal animals, humans walk upright, with the main body axis perpendicular to the ground. *Anterior* refers to the front of an upright walker; it corresponds to ventral, as in (**b**). *Posterior* refers to the back; it corresponds to dorsal.

Tissue and Organ Formation

Where do tissues of organ systems come from? To get a sense of how they originate, start with sperm and eggs. (These form from *germ* cells, or immature reproductive cells. All other cells in the body are *somatic*, after a Greek word for body.) A zygote forms after a sperm fertilizes an egg, then mitotic cell divisions form an early embryo. In vertebrates, cells become arranged as three primary tissues—ectoderm, mesoderm, and endoderm. These three are embryonic forerunners of all tissues in an adult. **Ectoderm** gives rise to skin's outer layer and to tissues of a nervous system. **Mesoderm** gives rise to tissues of muscle, bone, and most of the circulatory, reproductive, and urinary systems. **Endoderm** gives rise to the lining of the digestive tract and to organs derived from it.

In general, vertebrates have the same kinds of organ systems. Each organ system serves specialized functions, such as gas exchange, blood circulation, and locomotion.

Vertebrate tissues, organs, and organ systems arise from three primary tissues that form in the early embryo. The primary tissues are ectoderm, mesoderm, and endoderm.

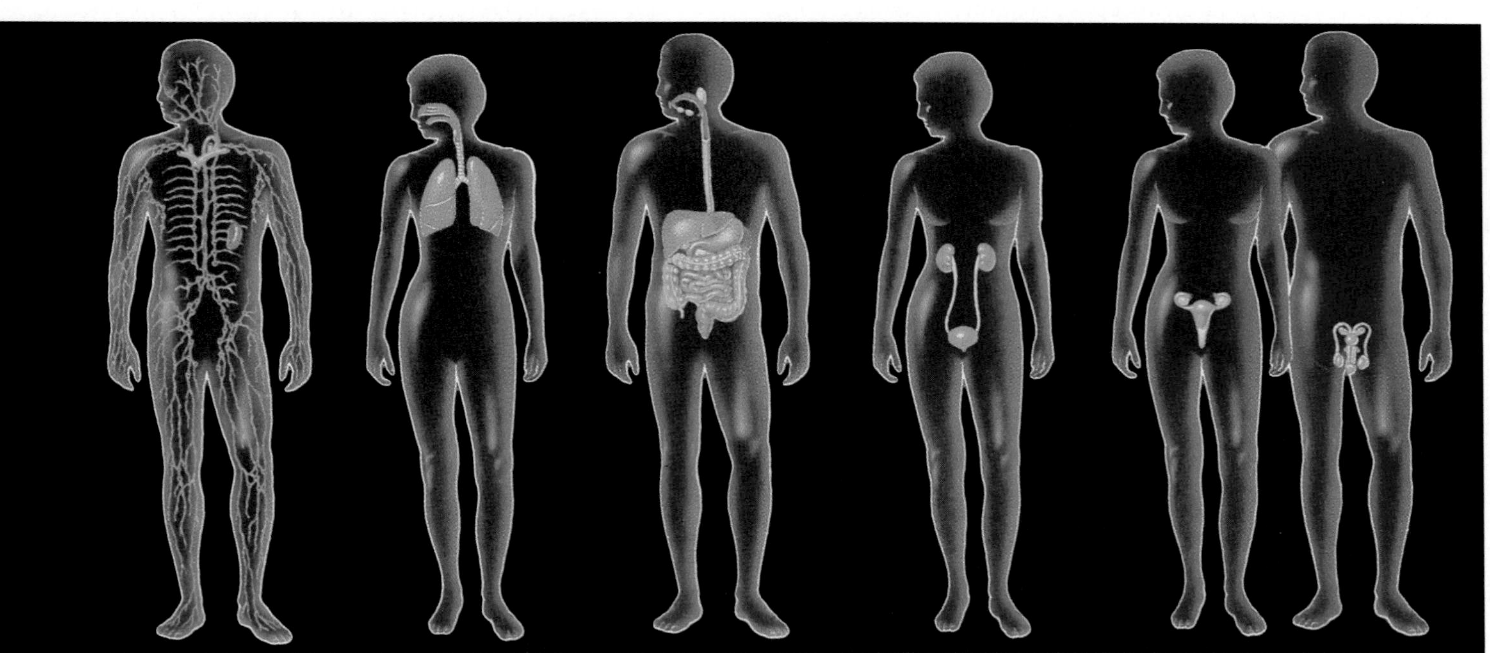

LYMPHATIC SYSTEM
Collect and return some tissue fluid to the bloodstream; defend the body against infection and tissue damage.

RESPIRATORY SYSTEM
Rapidly deliver oxygen to the tissue fluid that bathes all living cells; remove carbon dioxide wastes of cells; help regulate pH.

DIGESTIVE SYSTEM
Ingest food and water; mechanically, chemically break down food, and absorb small molecules into internal environment; eliminate food residues.

URINARY SYSTEM
Maintain the volume and composition of internal environment; excrete excess fluid and blood-borne wastes.

REPRODUCTIVE SYSTEM
Female: Produce eggs; after fertilization, afford a protected, nutritive environment for the development of new individual. *Male:* Produce and transfer sperm to the female. Hormones of both systems also influence other organ systems.

33.7 HOMEOSTASIS AND SYSTEMS CONTROL

Concerning the Internal Environment

To stay alive, your cells must remain bathed in a fluid that offers nutrients and carries away metabolic wastes. In this they are no different from an amoeba or some other free-living, single-celled organism. The difference is, trillions of cells coexist in your body. They all must draw nutrients from and dump wastes into the same fifteen liters of fluid, which is less than sixteen quarts.

The fluid *not* inside cells is **extracellular fluid**. Much of it is *interstitial*, meaning it occupies spaces between cells and tissues. The remainder is *plasma*, which is the fluid portion of blood. The interstitial fluid exchanges substances with the cells it bathes and with blood.

In functional terms, changes in extracellular fluid cause changes in cells. That is why drastic changes in the composition and volume of extracellular fluid have drastic effects on cell activities. The type and number of ions are especially crucial, for they must be maintained at concentrations that are compatible with metabolism. Otherwise, the animal itself cannot survive.

It makes no difference whether an animal is simple or complex. *The component parts of every animal work together to maintain the stable fluid environment that all of its living cells require*. This concept is absolutely central to understanding the structure and function of animals, and its key points may be summarized as follows. First, each cell of the animal body engages in basic metabolic activities that ensure its own survival. Second, the cells of a given tissue perform one or more activities that contribute to survival of the whole organism. Third, the combined contributions of individual cells, tissues, organs, and organ systems that are engaged in a division of labor help maintain a stable internal environment—extracellular fluid—required for individual cell survival.

Mechanisms of Homeostasis

Homeostasis, recall, means stable operating conditions in the internal environment. Three components interact to maintain this state. They are called sensory receptors, integrators, and effectors. **Sensory receptors** are cells or parts of cells that detect forms of energy, such as pressure and light. We call any specific form of energy that a receptor detects a **stimulus**. When someone kisses you, for example, there is a change in pressure on your lips. Receptors in the skin of your lips translate the stimulus into signals that are sent to the brain. Your brain is an **integrator**, a central command post that pulls together different bits of information to select a response. Integrators send out signals to muscles or glands (or both). Muscles and glands are the body's **effectors**—they carry out the response. In this case, the response may be flushing with pleasure and kissing the person back. Of course, you can't engage in a kiss indefinitely, for this would prevent you from eating and carrying out other activities necessary to maintain operating conditions inside your body.

So how does your brain reverse the physiological changes induced by the kiss? Receptors only provide it with information about how things *are* operating. The brain also receives information about how things *should be* operating—that is, information from "set points." When physical or chemical conditions deviate sharply from a set point, the brain functions to bring them back to an effective operating range. It does this by way of signals that cause specific muscles and specific glands to increase or decrease their activity.

NEGATIVE FEEDBACK Feedback mechanisms are among the controls that operate to keep physical and chemical aspects of the body within tolerable ranges. To give one example, with a **negative feedback mechanism**, some activity alters a condition in the internal environment and thereby triggers a response that reverses the altered condition (Figure 33.13).

Think of a furnace with a thermostat. A thermostat can sense the temperature of the surrounding air and "compare" it against a preset point on a thermometer built into the furnace's control system. Suppose the temperature falls below the preset point. The thermostat sends signals to a switching mechanism that turns on the furnace. When the air becomes heated enough to match the prescribed level, the thermostat again signals the switching mechanism, which shuts off the furnace.

Similarly, feedback mechanisms help keep the body temperature of meerkats, humans, huskies, and many other animals near 37°C (98.6°F) even during hot or cold weather. Visualize a young husky running around on a hot summer day. Soon its body gets hot, and receptors trigger events that slow down the whole dog *and* its cells. The husky searches for shade and rests under a tree. Moisture from its respiratory system evaporates from its tongue and carries away some body heat with

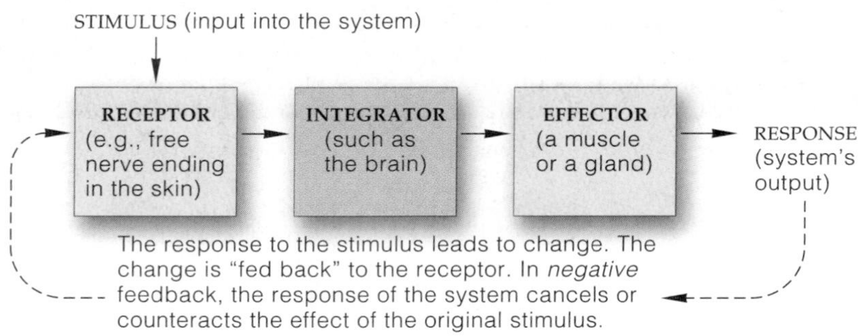
STIMULUS (input into the system)
RECEPTOR (e.g., free nerve ending in the skin)
INTEGRATOR (such as the brain)
EFFECTOR (a muscle or a gland)
RESPONSE (system's output)
The response to the stimulus leads to change. The change is "fed back" to the receptor. In *negative* feedback, the response of the system cancels or counteracts the effect of the original stimulus.

Figure 33.13 Components required for negative feedback at the organ level.

Further reading: Student Guide to InfoTrac on web site

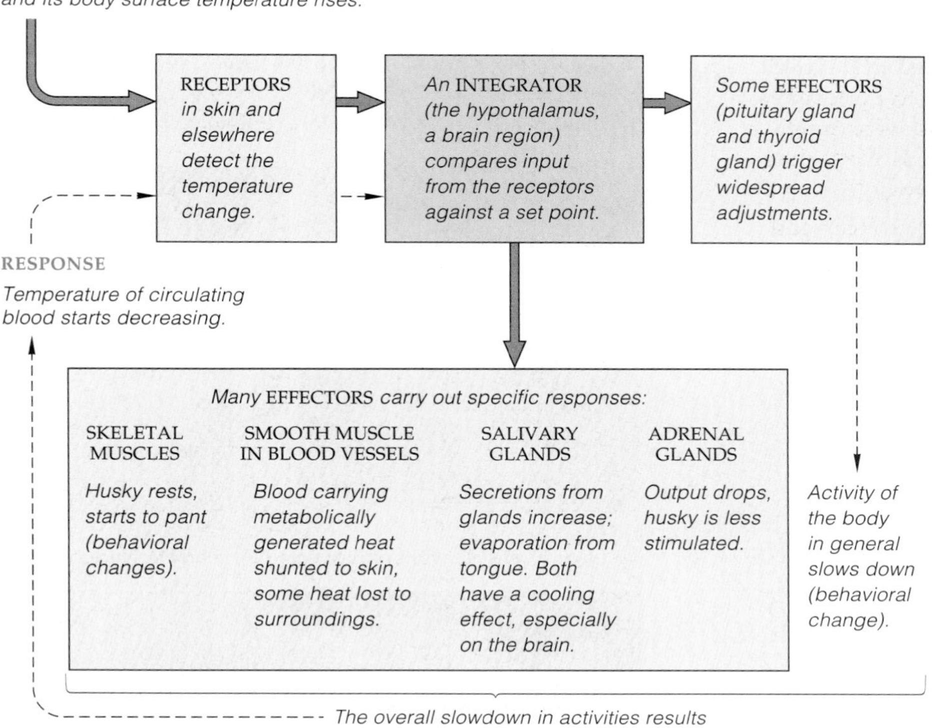
STIMULUS
The husky is overactive on a hot, dry day and its body surface temperature rises.
RECEPTORS in skin and elsewhere detect the temperature change.
An INTEGRATOR (the hypothalamus, a brain region) compares input from the receptors against a set point.
Some EFFECTORS (pituitary gland and thyroid gland) trigger widespread adjustments.
RESPONSE
Temperature of circulating blood starts decreasing.
Many EFFECTORS carry out specific responses:
SKELETAL MUSCLES
Husky rests, starts to pant (behavioral changes).
SMOOTH MUSCLE IN BLOOD VESSELS
Blood carrying metabolically generated heat shunted to skin, some heat lost to surroundings.
SALIVARY GLANDS
Secretions from glands increase; evaporation from tongue. Both have a cooling effect, especially on the brain.
ADRENAL GLANDS
Output drops, husky is less stimulated.
Activity of the body in general slows down (behavioral change).
The overall slowdown in activities results in less metabolically generated heat.

Figure 33.14 Homeostatic controls over the internal temperature of a husky's body. *Blue* arrows indicate the main control pathways. The dashed line shows how a feedback loop is completed.

it, as Figure 33.14 indicates. These control mechanisms and others counter overheating by curbing the activities that naturally generate metabolic heat and by giving up the body's excess heat to the surrounding air.

POSITIVE FEEDBACK In some cases, **positive feedback mechanisms** operate. These controls set in motion a chain of events that *intensify* a change from an original condition—and after a limited time, the intensification reverses the change. Such positive feedback is associated with instability in a system. For example, during sexual intercourse, chemical signals from the nervous system of a human female might induce her to make intense physiological responses to stimulation from her partner. Those responses may stimulate changes in her partner that stimulate the female even more, and so on until she reaches an explosive, climax level of excitation. Normal conditions now return, and homeostasis prevails.

As another example, at childbirth, the fetus exerts pressure on the wall of its mother's uterus. Pressure stimulates the production and secretion of oxytocin, a hormone. Oxytocin causes wall muscles to contract and exert pressure on the fetus, which exerts more pressure on the wall, and so on until the fetus is expelled.

What we have been describing is a general pattern of detecting, evaluating, and responding to a continual flow of information about the animal's internal and external environments. During all of this activity, organ systems operate together in astoundingly coordinated fashion. Throughout this unit of the book, you will be asking the following questions about their operation:

1. *What physical or chemical aspects of the internal environment are organ systems working to maintain as conditions change?*
2. *By what means are organ systems kept informed of the various changes?*
3. *By what means do they process incoming information?*
4. *What mechanisms are set in motion in response?*

As you will see in later chapters, operation of all organ systems is under neural and endocrine control.

Each living cell of an animal body engages in basic metabolic activities that ensure its own survival. Concurrently, the cells of any given tissue are performing one or more activities that contribute to the survival of the whole animal.

The combined contributions of cells, tissues, organs, and organ systems maintain the stable internal environment (extracellular fluid) required for individual cell survival.

Homeostatic control mechanisms help maintain physical and chemical aspects of the body's internal environment within ranges that are most favorable for cell activities.

33.8 SUMMARY

1. A tissue is an aggregation of cells and intercellular substances that perform a common task. An organ is a structural unit of different tissues combined in definite proportions and patterns that allow them to perform a common task. An organ system has two or more organs interacting chemically, physically, or both in ways that contribute to the survival of the body as a whole.

2. Epithelial tissues cover external body surfaces and line internal cavities and tubes. These tissues have one free surface exposed to body fluids or the environment.

3. Different connective tissues bind together, support, strengthen, protect, and insulate the other tissues. Most have fibers of structural proteins (especially collagen), fibroblasts, and other cells within a ground substance.

a. Loose connective tissue, with a semifluid ground substance, is present under skin and most epithelia.

b. Dense, irregular connective tissue contains mostly collagen fibers and fibroblasts. It is present in skin and forms protective capsules around a number of organs.

c. Dense, regular connective tissue, such as that in tendons, contains parallel bundles of collagen fibers. It protects and structurally supports organs.

d. Cartilage, with its solid yet pliable intercellular material, has structural and cushioning roles. Bone, the weight-bearing tissue of a vertebrate skeleton, interacts with skeletal muscle to bring about movement.

e. Blood, a specialized connective tissue, consists of plasma, cellular components, and dissolved substances. Adipose tissue, another specialized connective tissue, is a reservoir of energy; it consists mainly of fat cells.

4. Muscle tissues contract (shorten), then return to the resting position. They help move the body or parts of it. The three types of muscle tissue are skeletal muscle, smooth muscle, and cardiac muscle tissue.

5. Nervous tissue intercepts and integrates information about internal and external conditions, and governs the body's responses to change. Neurons of this tissue are the basic units of communication in nervous systems.

6. Tissues, organs, and organ systems work together to maintain the stable internal environment (that is, the extracellular fluid) required for individual cell survival. At homeostasis, conditions in the internal environment are balanced at levels most favorable for cell activities.

7. Feedback controls help maintain internal operating conditions for the body's cells. With negative feedback, for example, a change in a particular condition triggers a response that results in reversal of the change.

8. Homeostasis depends on receptors, integrators, and effectors. Stimuli are specific forms of energy detected by receptors. Integrating centers such as a brain receive and process information from the receptors and direct effectors (muscles and glands) to carry out responses.

Review Questions

1. Describe the characteristics of epithelial tissue in general. Then describe the various types of epithelial tissues in terms of specific characteristics and functions. *33.1*

2. List the major types of connective tissues; add the names and characteristics of their specific types. *33.2*

3. Identify and describe the following tissues: *33.1–33.3*

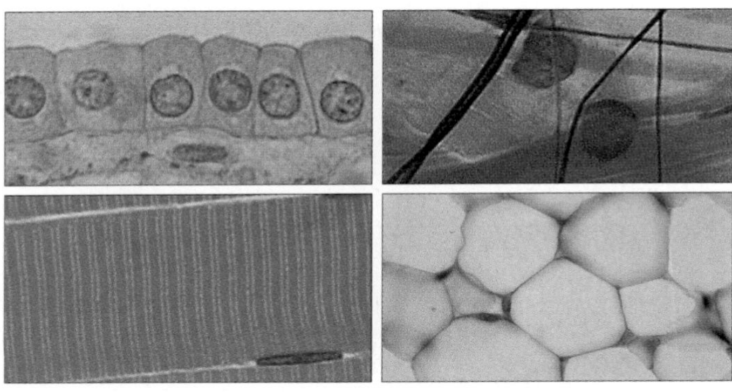

4. Identify this category of tissue and its characteristics: *33.4*

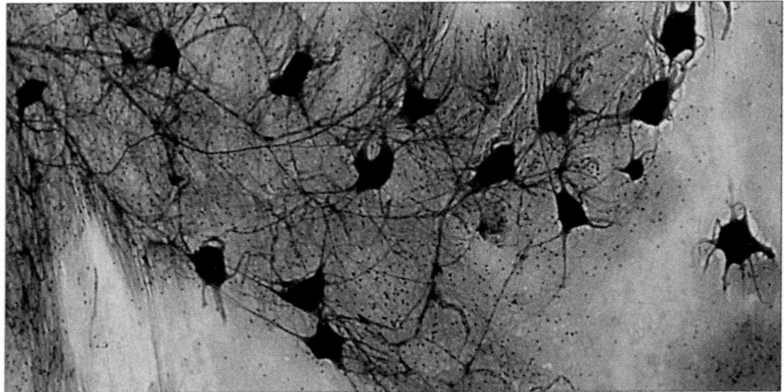

5. What type of cell serves as the basic unit of communication in nervous systems? *33.4*

6. Define animal tissue, organ, and organ system. List and define the functions of the eleven major organ systems of the human body. *CI, 33.6*

7. Define extracellular fluid, interstitial fluid, and plasma. *33.7*

8. Define homeostasis. *CI, 33.7*

9. Briefly describe two major categories of the homeostatic mechanisms operating in the human body. *33.7*

Self-Quiz (*Answers in Appendix III*)

1. ______ tissues have closely linked cells and one free surface.
   a. Epithelial c. Nervous
   b. Connective d. Muscle

2. In most ______, cells secrete fibers of collagen and elastin.
   a. epithelial tissue c. muscle tissue
   b. connective tissue d. nervous tissue

3. ______ has a semifluid ground substance and occurs under most epithelia.
   a. Dense, irregular connective tissue c. Dense, regular connective tissue
   b. Loose connective tissue d. Cartilage

4. ________, a specialized connective tissue, is mostly plasma with cellular components and various dissolved substances.
   a. Irregular connective tissue   c. Cartilage
   b. Blood   d. Bone

5. After you eat too many carbohydrates and proteins, your body converts the excess to storage fats, which accumulate in ________.
   a. connective tissue proper   c. adipose tissue
   b. dense connective tissue   d. both b and c

6. Components of ________ detect and coordinate information about changes and control responses to those changes.
   a. epithelial tissue   c. muscle tissue
   b. connective tissue   d. nervous tissue

7. In your own body, ________ can shorten (contract).
   a. epithelial tissue   c. muscle tissue
   b. connective tissue   d. nervous tissue

8. Cells of complex animals ________.
   a. survive by their own metabolic activities
   b. contribute to the survival of the whole animal
   c. help maintain extracellular fluid
   d. all of the above

9. At ________, physical and chemical conditions in the internal environment are within tolerable ranges.
   a. the time of positive feedback
   b. the time of negative feedback
   c. homeostasis
   d. metastasis

10. With negative feedback mechanisms, ________.
   a. a stimulus brings about a response that tends to return internal operating conditions to the original state
   b. a stimulus suppresses internal operating conditions to levels below a set point for the body
   c. a stimulus raises internal operating conditions to levels above a set point for the body
   d. fewer solutes are fed back to the affected cells

11. Of the components that exert feedback control of organ activity, ________ detect specific changes in the environment, an ________ pulls together different bits of information and selects a suitable response, and then ________ carry out the response.

12. Match the terms with the suitable description.

____ epithelium	a. strong, pliable; like rubber
____ cartilage	b. covers or lines body surfaces
____ homeostasis	c. stable internal environment
____ muscles and glands	d. integrating center
____ positive feedback	e. most common homeostatic control mechanism
____ negative feedback	f. effectors
____ brain	g. chain of events intensifies original condition in body

Critical Thinking

1. *Anhidrotic ectodermal dysplasia* is a disorder associated with a recessive allele on the mammalian X chromosome (Section 15.4). As one symptom of this genetic disorder, affected males and females have no sweat glands in tissues where the recessive allele is expressed. What type of tissue are we talking about?

2. Adipose tissue and blood are often said to be "atypical" connective tissues. Compared with other connective tissues, which of their features are *not* typical?

3. *Porphyria*, a genetic disorder, shows up in about 1 in 25,000 individuals. Affected people lack certain enzymes that are part of a metabolic pathway leading to formation of heme, which is the iron-containing group of hemoglobin. An accumulation of porphyrins (intermediates of the pathway) results in visually jarring symptoms, especially after exposure to sunlight. Lesions and scars form on the skin. Hair grows thickly on the face and hands. Gums retreat from the teeth, and this makes the canines take on a fanglike appearance.

Symptoms worsen upon exposure to a variety of substances, including garlic and alcoholic beverages. Affected individuals avoid sunlight and aggravating substances. They also may receive injections of heme from normal red blood cells.

If you are familiar with vampire stories, which began before the Middle Ages, speculate on how they may have evolved among superstitious folk who lacked medical knowledge of porphyria.

4. After graduating from high school, Jeff and Ryan set out on a trip along the desert roads of California and Arizona. One hot, dry morning in Joshua Tree National Monument, they saw an unusual rock formation that didn't appear to be too far from the road. They left the car and started hiking toward it.

Their destination turned out to be farther away than they thought, and the sun's rays were more relentless than they had anticipated. They reached the shade of the rocks in the early afternoon. Their canteen was nearly empty, however, and the physiological meaning of "thirst" made itself known in a scary way. They knew they had to locate and drink water (or some other fluid), which is what the brain usually tells us to do when dehydration sets in.

From what you read in this chapter, would you suspect that Ryan and Jeff's *thirst behavior* is part of a positive or negative feedback control mechanism?

Selected Key Terms

adhering junction *33.1*
adipose tissue *33.2*
blood *33.2*
bone tissue *33.2*
cardiac muscle tissue *33.3*
cartilage *33.2*
dense, irregular connective tissue *33.2*
dense, regular connective tissue *33.2*
division of labor *CI*
ectoderm *33.6*
effector *33.7*
endocrine gland *33.1*
endoderm *33.6*
epithelium *33.1*
exocrine gland *33.1*
extracellular fluid *33.7*
gap junction *33.1*
gland cell *33.1*
homeostasis *CI*
integrator *33.7*
internal environment *CI*
loose connective tissue *33.2*
mesoderm *33.6*
negative feedback mechanism *33.7*
nervous tissue *33.4*
neuroglia *33.4*
neuron *33.4*
organ *CI*
organ system *CI*
positive feedback mechanism *33.7*
sensory receptor *33.7*
skeletal muscle tissue *33.3*
smooth muscle tissue *33.3*
stimulus *33.7*
tight junction *33.1*
tissue *CI*

Readings *See also www.infotrac-college.com*

Bloom, W., and D. W. Fawcett. 1995. *A Textbook of Histology*. Twelfth edition. Philadelphia: Saunders.

Leeson, C. R., T. Leeson, and A. Paparo. 1988. *Textbook of Histology*. Philadelphia: Saunders.

Telford, I., and C. Bridgman. 1995. *Introduction to Functional Histology*. Second edition. New York: HarperCollins.

Wright, C. November 1999. "Ready-to-Wear Flesh." *Discover*.

34

INFORMATION FLOW AND THE NEURON

Figure 34.1 A terrifying view across the Kansas prairie. Imagine yourself alone and in the open when you first see a tornado, knowing you have only minutes to remove yourself from harm's way. By what means do you conceive of plans of action? Can the plans be put together and evaluated quickly enough? The answers begin with the functioning of neurons in your nervous system.

TORNADO!

It is spring in the American Midwest, and you happen to be engrossed in photographing wildflowers in an expanse of shortgrass prairie. So intent are you on capturing all the species on film that you fail to notice the rapidly darkening sky.

What's that rumbling you hear in the distance? It sounds something like a freight train. You wonder how can that be, when there is no train track, anywhere, in this part of the prairie. You turn to identify the source of the sound. Then you see it, but you don't want to believe your eyes. A funnel cloud is advancing across the prairie and heading right for you! *TORNADO!* The image rivets your attention as nothing else has ever done (Figure 34.1). You know that you cannot remain where you are and survive. Suddenly you recall that hiding in a low area is better than standing out in the open. Commands rush from your brain to your four limbs: *GET MOVING OUT OF HERE!* With heart thumping, you run along a path, looking frantically for safety. You're in luck! Just ahead is a steep-banked creek and you reach it in less than a minute. You find a small ledge in the creek's muddy bank and wedge yourself under it, wishing wildly to be inconspicuous, to be overlooked by a force of nature on the rampage.

It takes a few minutes before you realize the tornado has roared past the creek. You remain motionless, fingers clutching mud. Finally you sit up and look around. Not too far away, you see a swath of twisted grass that marks the tornado's path. Your heart no longer feels as if it is slamming against your chest wall, but your legs feel like rubber when you stand up.

Thank your nervous system for every perception, every memory,

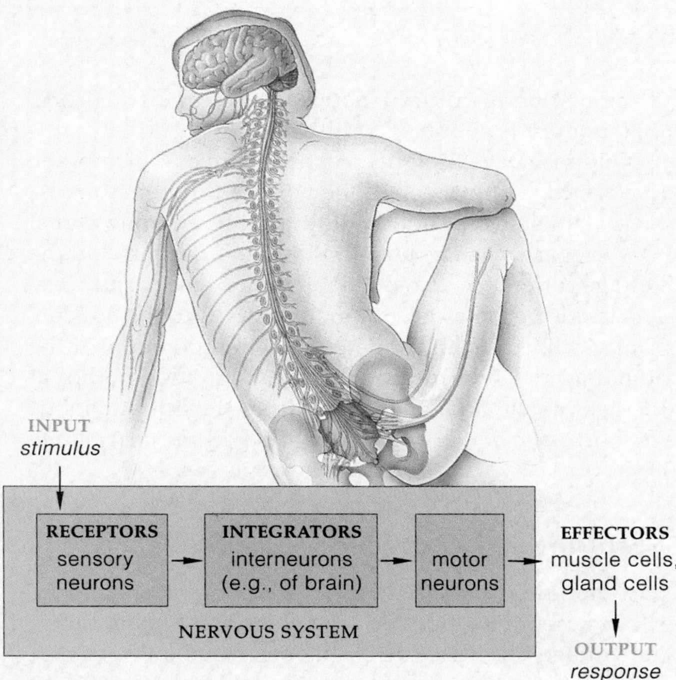

INPUT
stimulus
RECEPTORS
sensory neurons
INTEGRATORS
interneurons (e.g., of brain)
motor neurons
NERVOUS SYSTEM
EFFECTORS
muscle cells, gland cells
OUTPUT
response

Figure 34.2 Overview of the communication lines of vertebrate nervous systems.

and nearly every action that collectively helped you escape the tornado. Thank the information that traveled swiftly, in suitable directions, among interacting cells called **neurons**. Neurons work together to monitor the conditions in and around the body. They issue commands for responsive actions that help the body as a whole. They interact as the communication lines of your brain, spinal cord, and nerves.

This chapter starts with the structure and function of neurons. In chapters to follow, you will consider how these cells interact with one another in nervous systems and sensory systems. Through their interactions, you can detect tornadoes and other events that have bearing on whether you survive from one day to the next.

Three classes of neurons interact in all vertebrate nervous systems. **Sensory neurons** respond to specific stimuli and relay information about them to the spinal cord and brain. A **stimulus** is a specific form of energy, such as light. In the spinal cord and brain you find **interneurons**. These receive sensory input, integrate it with other incoming or stored information, and then influence the activity of other neurons. **Motor neurons** relay information from the brain and spinal cord to the body's effectors—muscles or glands—which carry out the specified responses (Figure 34.2).

Neurons are not the only cells in vertebrate nervous systems. As you will see in this chapter and the next, a variety of cells, collectively called **neuroglia**, make up more than half the volume of the vertebrate systems. Different kinds of neuroglial cells metabolically assist, protect, and structurally support the neurons.

KEY CONCEPTS

1. Neurons are the basic units of communication in nearly all nervous systems. Collectively, many neurons interact to detect and integrate information about external and internal conditions, then select or control muscles and glands in ways that produce suitable responses.

2. Neurons are a type of excitable cell, which means a suitable stimulus can disturb the distribution of electric charge across their plasma membrane. Such disturbances are the basis of messages that travel from the input zone of a neuron to its output zone near a neighboring cell.

3. In an undisturbed neuron, the cytoplasm is negatively charged with respect to the fluid just outside the plasma membrane. With suitable stimulation, the polarity of charge across the membrane may undergo an abrupt, brief reversal. Such reversals are action potentials.

4. Information flow through the nervous system starts with action potentials, which are self-propagating along the plasma membrane of individual neurons.

5. Action potentials start and end at small gaps between neurons. Chemical signals released from one neuron bridge the gap and stimulate or inhibit the adjoining neuron, muscle cell, or gland cell.

6. Information flow through nervous systems depends on moment-by-moment integration of excitatory and inhibitory signals that act upon the neurons of given pathways.

NEURONS—THE COMMUNICATION SPECIALISTS

Functional Zones of a Neuron

To understand how your own nervous system works, start with how its neurons function. Neurons consist of a nucleated cell body and cytoplasmic extensions that differ in number and length (Figure 34.3).

Typically, the cell body and the slender extensions known as **dendrites** are *input* zones, where the neuron receives information. Another slender but often longer extension, an **axon**, is a *conducting* zone. Axons rapidly propagate signals that arise at a neuron's trigger zone, where the incoming information is added up. Except in sensory neurons, the *trigger* zone is a patch of plasma membrane at the junction between the cell body and an axon. An axon has branched endings. These are *output* zones, where messages are sent to other cells.

A Neuron at Rest, Then Moved to Action

Different forms of signals arise in the nervous system. Let's start with one that begins and ends on the same neuron and is not transferred to another cell.

When a neuron is not being bothered, a difference in electric charge is being maintained across its plasma membrane. The cytoplasmic fluid next to the membrane remains negatively charged, compared to the interstitial fluid right outside. We measure these charges in units called millivolts. The amount of energy inherent in the steady voltage difference across the plasma membrane is the **resting membrane potential**. For many neurons, that amount is about −70 millivolts.

Suppose a weak signal reaches a patch of membrane in a neuron's input zone. The voltage difference across the patch changes slightly, if at all. But a strong signal may trigger an **action potential**, a fleeting reversal in the voltage difference across a neuron's plasma membrane. The inside becomes positive with respect to the outside. That localized reversal invites an action potential at the adjoining patch of membrane, which invites another at the next patch, and so on away from the initiation site. In short, *stimulation of a neuron disturbs the distribution of electric charge across its plasma membrane.*

Restoring and Maintaining Readiness

Each neuron is part of the body's communication lines. It can accept and pass on signals only by maintaining the voltage difference across its plasma membrane and restoring it between action potentials. How does it do this? The answer lies with two membrane properties.

First, every cell membrane has a lipid bilayer, which bars the passage of potassium ions (K^+), sodium ions (Na^+), and other charged substances. Because of this, a neuron can build up differences in ion concentrations across its plasma membrane. Second, ions can still flow from one side of the membrane to the other, through the interior of transport proteins that span the bilayer (Figure 34.4). And that flow is controlled.

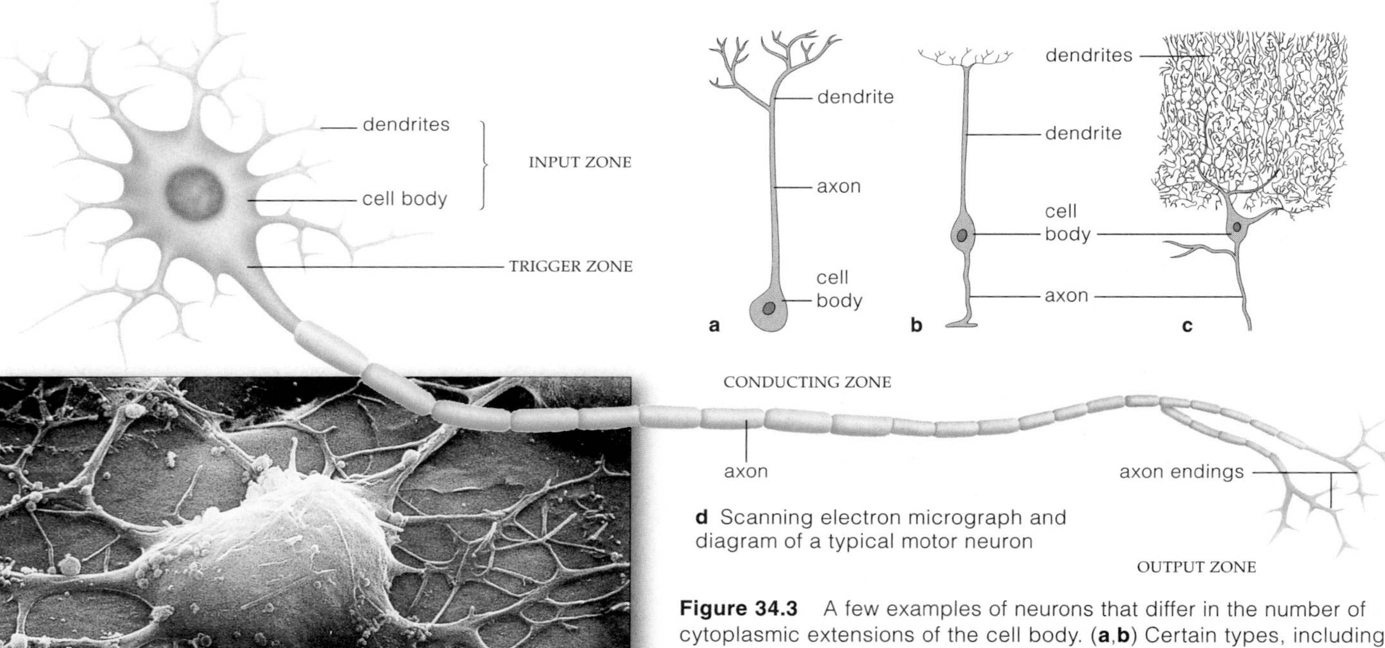
dendrites
cell body
INPUT ZONE
TRIGGER ZONE
dendrite
axon
cell body
a
dendrites
dendrite
cell body
axon
b
c
CONDUCTING ZONE
axon
axon endings
d Scanning electron micrograph and diagram of a typical motor neuron
OUTPUT ZONE
10 µm

Figure 34.3 A few examples of neurons that differ in the number of cytoplasmic extensions of the cell body. (**a**,**b**) Certain types, including many sensory neurons, have one axon with dendritic branchings or one axon and one dendrite. Many sensory neurons are like this. (**c**) Other types have one axon and a profusion of dendrites; this kind occurs in the mammalian brain. (**d**) Micrograph and sketch of the functional zones of a motor neuron.

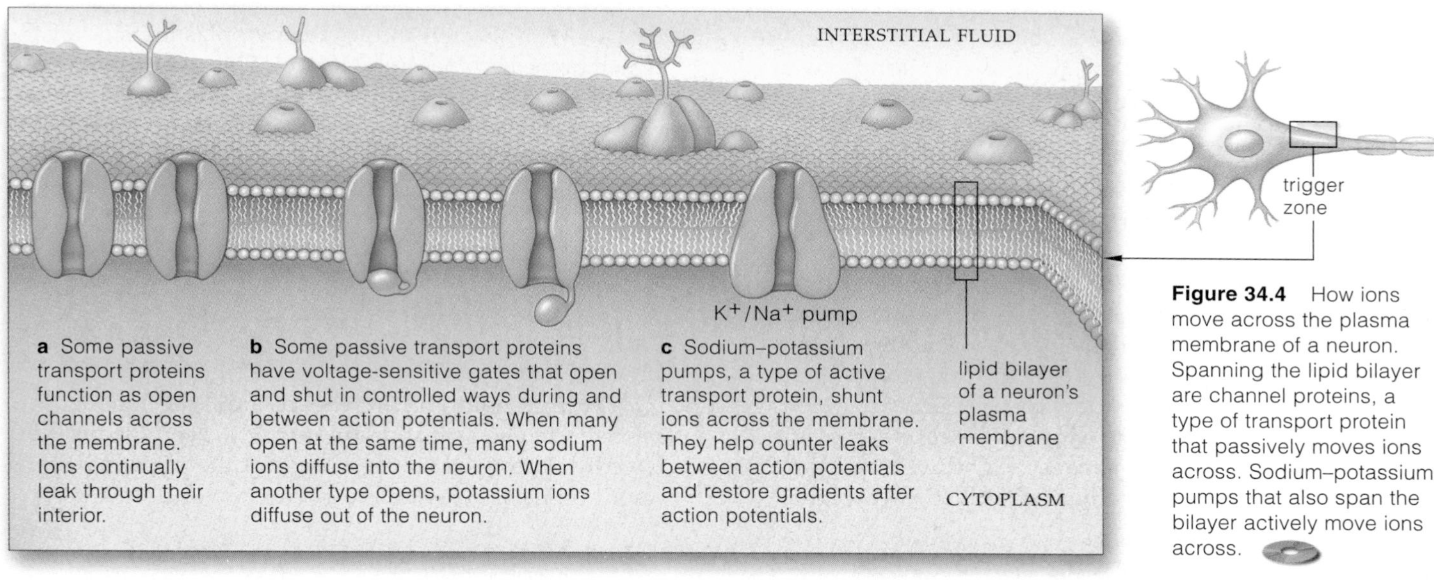
INTERSTITIAL FLUID
trigger zone
K+/Na+ pump
a Some passive transport proteins function as open channels across the membrane. Ions continually leak through their interior.
b Some passive transport proteins have voltage-sensitive gates that open and shut in controlled ways during and between action potentials. When many open at the same time, many sodium ions diffuse into the neuron. When another type opens, potassium ions diffuse out of the neuron.
c Sodium–potassium pumps, a type of active transport protein, shunt ions across the membrane. They help counter leaks between action potentials and restore gradients after action potentials.
lipid bilayer of a neuron's plasma membrane
CYTOPLASM

Figure 34.4 How ions move across the plasma membrane of a neuron. Spanning the lipid bilayer are channel proteins, a type of transport protein that passively moves ions across. Sodium–potassium pumps that also span the bilayer actively move ions across.

Suppose a motor neuron has 15 sodium ions inside the membrane for every 150 outside. Also suppose that it has 150 potassium ions inside for every 5 outside. You can depict each ion's concentration gradient in this way (from the large to the small letters):

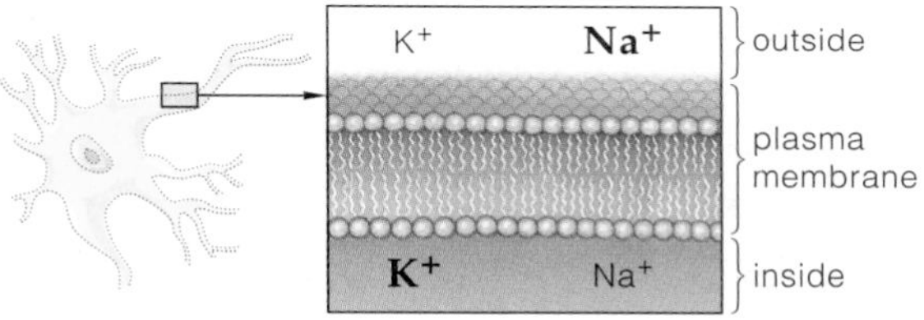
K+
Na+
outside
plasma membrane
K+
Na+
inside

Gradients such as these determine the net direction in which sodium and potassium ions diffuse. The ions can diffuse through the interior of channel proteins, a type of transport protein in the membrane. Some of the channels never shut, so ions leak (diffuse) through them all of the time. Others have molecular gates, which can open only after the neuron is adequately stimulated.

Suppose that motor neuron is not being stimulated. Its sodium channels are shut, so sodium ions can't rush inside. Some potassium is leaking out through a few open channels and making the cytoplasm a bit more negative, so some potassium is attracted back in. When the inward pull of electric charge balances the outward force of diffusion, there is no more net movement of potassium ions. The concentration and electric gradients now existing across the plasma membrane will permit the neuron to respond to stimulation.

The gradients reverse during an action potential, and then **sodium–potassium pumps** restore them. These are transport proteins that span the plasma membrane (Section 5.4). When they get an energy boost from ATP, the pumps actively transport potassium into the neuron and sodium out at the same time. Sodium–potassium pumps must maintain as well as restore the gradients. Why? Even in a resting neuron, a tiny fraction of the outward-leaking potassium isn't attracted back in. Also, a tiny fraction of sodium leaks in through a few open channels (Figure 34.5). If the leaks went unattended, the crucial gradients would gradually disappear.

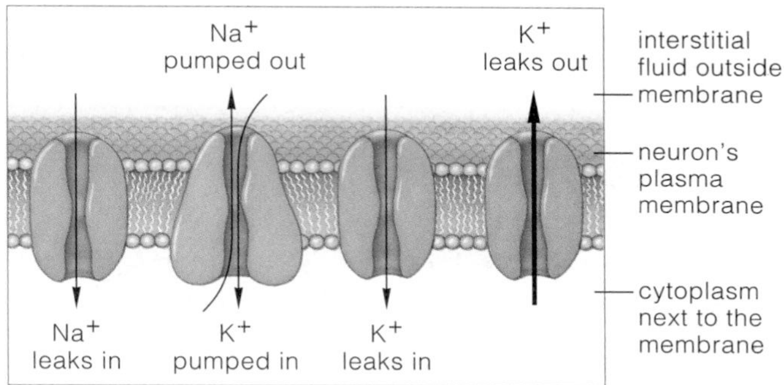
Na+ pumped out
K+ leaks out
interstitial fluid outside membrane
neuron's plasma membrane
cytoplasm next to the membrane
Na+ leaks in
K+ pumped in
K+ leaks in

Figure 34.5 Pumping and leaking processes that dictate the distribution of sodium and potassium ions across the plasma membrane of a neuron at rest. The arrow widths indicate the magnitude of the movements. Notice how the total inward and outward movements for each kind of ion are balanced.

An undisturbed neuron is maintaining a resting membrane potential—a voltage difference across its plasma membrane. An action potential is an abrupt, short-lived reversal in that voltage difference in response to adequate stimulation.

After an action potential, sodium–potassium pumps restore and maintain the resting membrane potential.

34.2 A CLOSER LOOK AT ACTION POTENTIALS

The preceding section gave an overview of a neuron's structure and function. Let's now take a closer look at how signals arise and are propagated at its surface.

Approaching Threshold

Weakly stimulate a neuron at its input zone and you disturb the ion balance across the membrane, but not much. Suppose your toes gently tap a cat snoozing at your feet and put a bit of pressure on its skin. Tissues beneath the skin surface have receptor endings—input zones of sensory neurons. Patches of plasma membrane at the endings deform under the pressure. Some ions now flow across, so the voltage difference across the membrane changes slightly. The pressure has produced a graded, local signal.

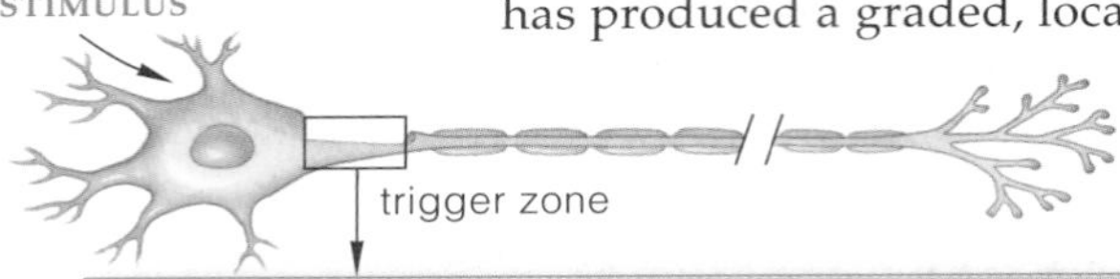

STIMULUS
trigger zone

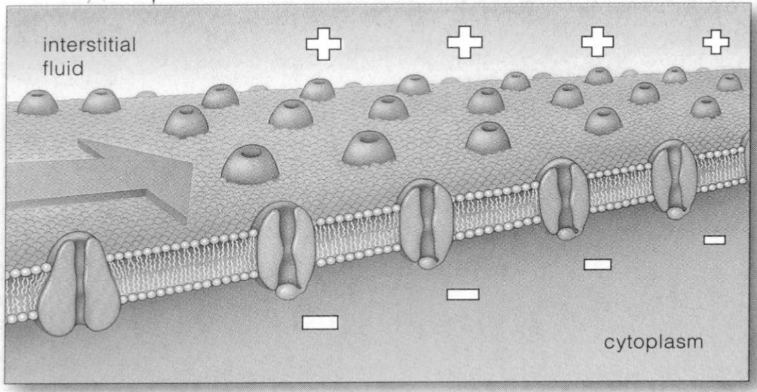

interstitial fluid
cytoplasm

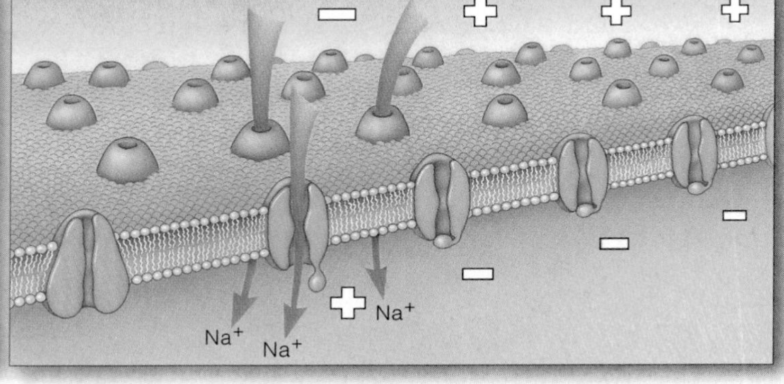

Na+
Na+
Na+

a Membrane at rest (inside negative with respect to the outside). An electrical disturbance (*red* arrow) spreads from an input zone to an adjacent trigger region of the membrane, which has a great number of gated sodium channels.

b A strong disturbance initiates an action potential. Sodium gates open. The sodium inflow decreases the negativity inside the neuron. The change causes more gates to open, and so on until threshold is reached and the voltage difference across the membrane reverses.

Figure 34.6 Propagation of an action potential along the axon of a motor neuron.

Graded means that signals arising at an input zone vary in magnitude; they are small to large, depending on the stimulus intensity or duration. *Local* means that signals do not spread far from the site of stimulation. Why? It takes certain kinds of ion channels to propagate a signal, and input zones simply don't have them.

When a stimulus is intense or long lasting, graded signals spread from an input zone into an adjoining trigger zone. This patch of membrane is richly endowed with voltage-sensitive gated channels for sodium ions. *And this is where a certain amount of change in the voltage difference across the plasma membrane triggers an action potential*. The amount is the **threshold level**.

The stimulus causes positively charged sodium ions to flow across the membrane, into the neuron (Figure 34.6). With the ion influx, the cytoplasmic side of the membrane becomes less negative. This causes more gates to open and more sodium to enter. The ever increasing inward flow of sodium is a case of **positive feedback**; the event intensifies as a result of its own occurrence:

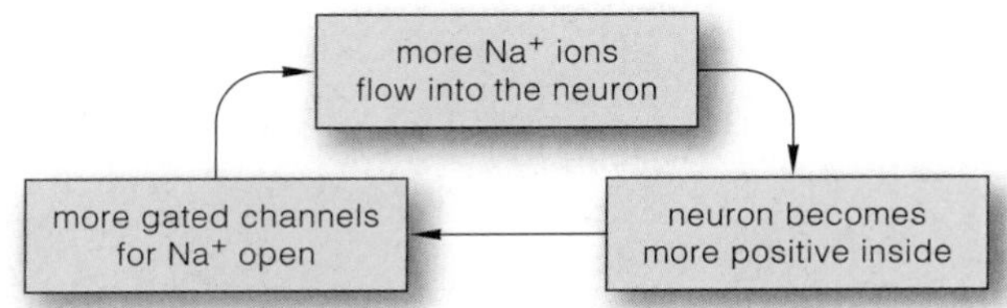

more Na+ ions flow into the neuron
neuron becomes more positive inside
more gated channels for Na+ open

At threshold, the opening of more sodium gates no longer depends on stimulus strength. Now the positive-feedback cycle is under way, and the inward-rushing sodium itself is enough to open more sodium gates.

An All-or-Nothing Spike

Figure 34.7 shows a recording of the voltage difference across the plasma membrane before, during, and after an action potential. Notice how the membrane potential spikes once threshold is reached. Every single action potential in the neuron spikes to the same level above threshold as an *all-or-nothing* event. That is, once the positive-feedback cycle starts, nothing will stop the full spiking. If threshold is not reached, the disturbance to the plasma membrane will subside just as soon as the stimulus is removed.

Each spike lasts only for a millisecond or so. Why? At the membrane site of the charge reversal, the gated sodium channels closed and shut off the sodium inflow.

Further reading: Student Guide to InfoTrac on web site

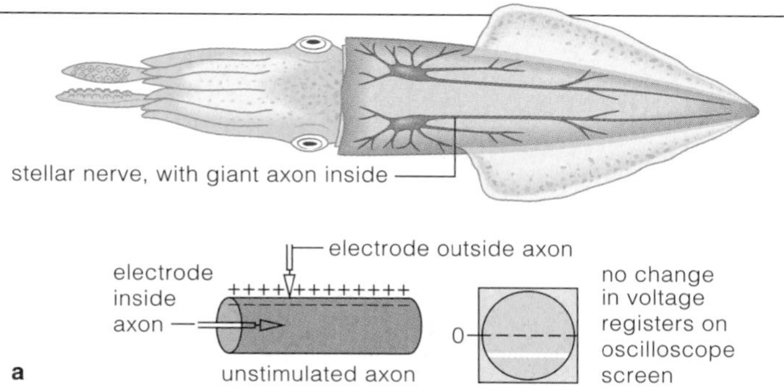
stellar nerve, with giant axon inside
electrode outside axon
electrode inside axon
unstimulated axon
no change in voltage registers on oscilloscope screen
a

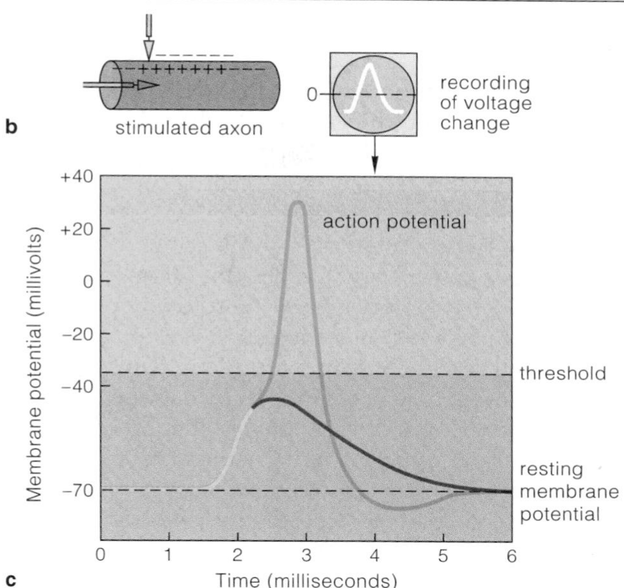
b
stimulated axon
recording of voltage change
action potential
threshold
resting membrane potential
Membrane potential (millivolts)
Time (milliseconds)
+40
+20
0
-20
-40
-70
0
1
2
3
4
5
6
c

Figure 34.7 Action potentials. (**a**) When researchers first started to investigate neural function, the squid *Loligo* provided them with evidence of action potential spiking. This squid's "giant" axons are large enough to slip electrodes inside.

(**b**) The researchers placed electrodes inside and outside of the axon, then stimulated it. The electrodes detected the change in voltage, which showed up as deflections in a beam of light across the screen of an oscilloscope connected to the electrodes. (**c**) This is a typical waveform (*yellow* line) for an action potential on an oscilloscope screen.

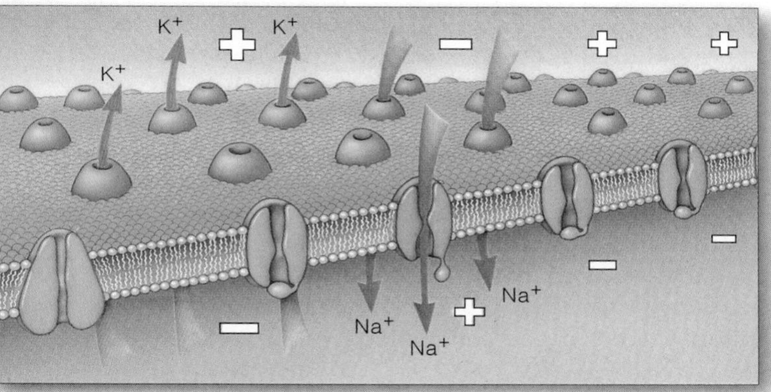
K^+
K^+
K^+
Na^+
Na^+
Na^+

c With the reversal, sodium gates shut and potassium gates open (*pink* arrows). Potassium follows its gradient out of the neuron. Voltage is restored. The disturbance triggers an action potential at the adjacent site, and so on, away from the point of stimulation.

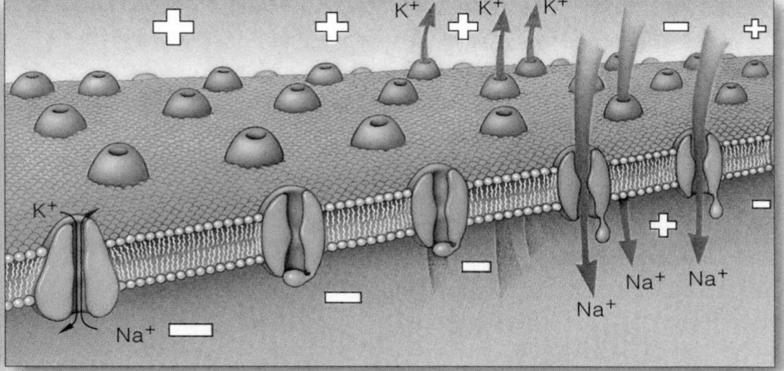
K^+
K^+
K^+
K^+
Na^+
Na^+
Na^+
Na^+

d Following each action potential, the inside of the plasma membrane becomes negative once again. However, the sodium and potassium concentration gradients are not yet fully restored. Active transport at sodium–potassium pumps restores them.

Also, about halfway through the reversal, potassium channels opened, so many more potassium ions flowed out and restored the original voltage difference across the membrane. And sodium–potassium pumps restored the ion gradients. Later on, after the resting membrane potential has been restored, most potassium gates close and sodium gates are in their initial state, ready to be opened with the arrival of a suitable disturbance.

Propagation of Action Potentials

The membrane disturbances leading up to an action potential are self-propagating, and they do not diminish in magnitude. As they spread to an adjacent membrane patch, an equivalent number of gated channels open. With this new disturbance, gated channels open in the *next* adjacent patch, then the next, and so on. For a brief period after each membrane patch has been excited, it is insensitive to stimulation. Sodium gates in the patch are inactivated and ions cannot move through them. This is the reason why action potentials do not spread back to the trigger zone (the site where they were initiated) but rather are self-propagating away from it.

The cytoplasm next to the plasma membrane of a neuron at rest is more negative than the interstitial fluid just outside the membrane.

During an action potential, the inside of a disturbed patch of membrane becomes more positive than the outside.

After an action potential, resting conditions are restored at the membrane patch.

34.3 CHEMICAL SYNAPSES

When action potentials reach a neuron's output zone, they usually do not proceed farther than this. But their arrival may induce the neuron to release one or more **neurotransmitters**, which are signaling molecules that diffuse across chemical synapses. A **chemical synapse** is the narrow cleft between the output zone of a neuron and an input zone of an adjacent cell (Figures 34.8 and 34.9). Some clefts intervene between two neurons, and others between a neuron and a muscle cell or gland cell.

At each chemical synapse, one of the two cells stores neurotransmitter molecules inside synaptic vesicles in its cytoplasm. Think of it as the *pre*synaptic cell. Gated channels for calcium ions span its membrane, and they open when an action potential arrives. There are more calcium ions outside the cell than inside, so they flow into the cell, down the gradient. The flow induces the synaptic vesicles to fuse with the plasma membrane and thereby release neurotransmitter into the synaptic cleft.

These neurotransmitter molecules diffuse across the cleft. On the *post*synaptic cell's membrane are protein receptors that bind specific neurotransmitters. Binding changes the receptor shape and creates a passageway through its interior. Ions cross the plasma membrane by diffusing through the passageway (Figure 34.9*c*).

A postsynaptic cell's response depends on the type and concentration of neurotransmitter in the cleft, what kinds of receptors the cell bears, and the number and responsiveness of gated channels in its membrane. Such factors influence whether a neurotransmitter will have an *excitatory* effect and help drive the postsynaptic cell membrane toward the threshold of an action potential.

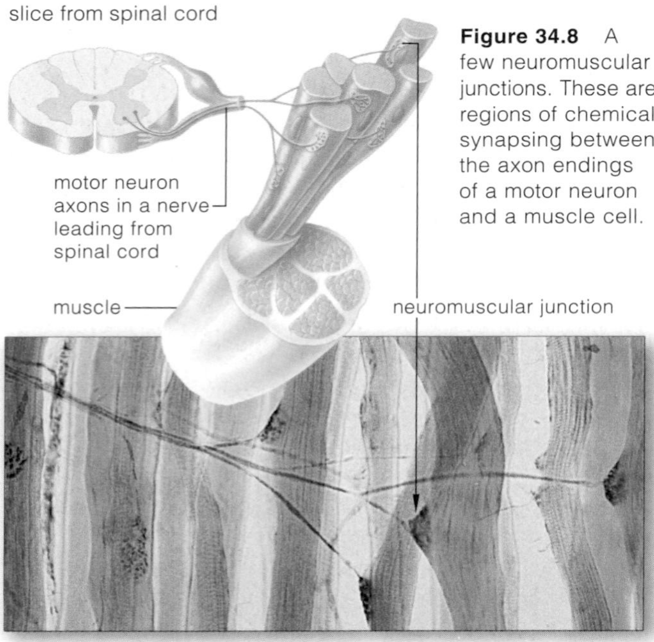
slice from spinal cord
motor neuron
axons in a nerve
leading from
spinal cord
muscle
neuromuscular junction

Figure 34.8 A few neuromuscular junctions. These are regions of chemical synapsing between the axon endings of a motor neuron and a muscle cell.

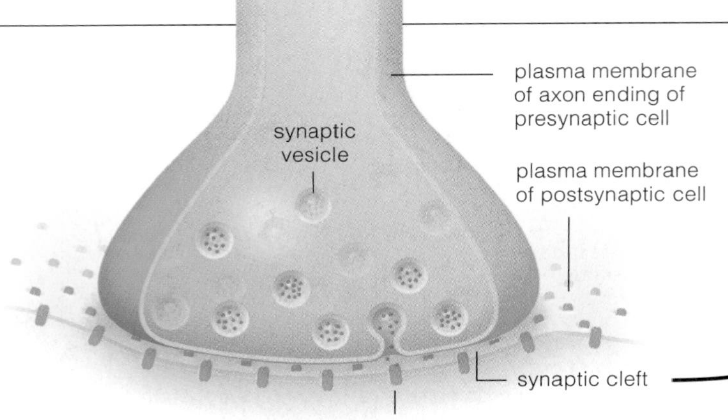
plasma membrane
of axon ending of
presynaptic cell
synaptic
vesicle
plasma membrane
of postsynaptic cell
synaptic cleft
a
membrane receptor for neurotransmitter

Figure 34.9 Closer look at a chemical synapse. (**a**,**b**) The plasma membranes of a motor neuron and a muscle cell face each other across a cleft between them. The arrival of an action potential at an axon ending triggers the release of molecules of neurotransmitter from the neuron. (**c**) These molecules diffuse across the cleft and bind to receptors on gated channel proteins of the muscle cell membrane. The gates open; ions flow in and trigger a graded potential at the membrane site.

They also influence whether it will have an *inhibitory* effect and drive the membrane away from threshold.

Consider the neurotransmitter **acetylcholine** (ACh). It has both excitatory and inhibitory effects on the brain, spinal cord, glands, and muscles. For example, it acts at chemical synapses between a motor neuron and muscle cell, as in Figure 34.9. ACh is released from the motor neuron, diffuses across the cleft, and binds to receptors on the muscle cell membrane. In this kind of cell it has excitatory effects; it can trigger action potentials, which in turn initiate muscle contraction (Section 38.10).

A Smorgasbord of Signals

Acetylcholine is only one of a veritable smorgasbord of signals that neurons deliver to target cells. For example, serotonin acts on neurons in brain regions that govern sleeping, sensory perception, temperature control, and emotions. Norepinephrine works in brain regions that control emotions, dreaming, and waking up. Dopamine also works in brain regions that deal with emotions. GABA (gamma aminobutyric acid) is the most common inhibitory signal in the brain. Antianxiety drugs, such as Valium, may exert their effects by enhancing GABA's effects. Except for ACh, these neurotransmitters and the other types are amino acids or are derived from them.

Signaling molecules known as **neuromodulators** can magnify or reduce the effects of a neurotransmitter on neighboring or distant neurons. They include substance P, which induces pain perception, and endorphins—the natural painkillers that inhibit the release of substance P from sensory nerves. Neuromodulators might also influence memory and learning, sexual activity, control of body temperature, and emotional states.

Further reading: Student Guide to InfoTrac on web site

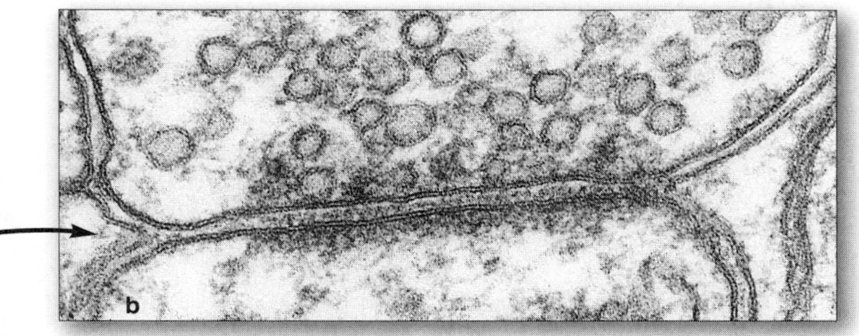
b

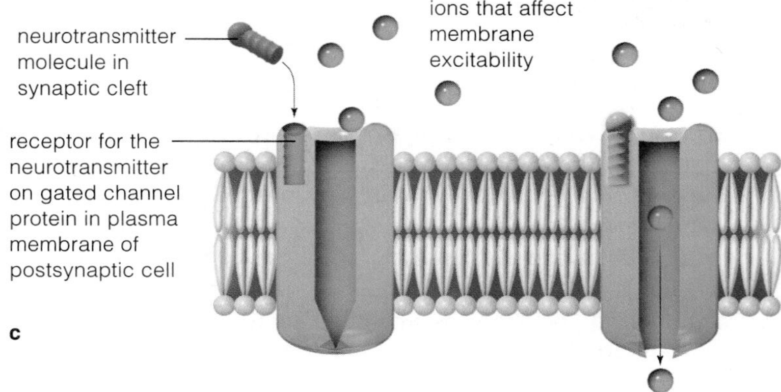
neurotransmitter molecule in synaptic cleft
ions that affect membrane excitability
receptor for the neurotransmitter on gated channel protein in plasma membrane of postsynaptic cell
c

Synaptic Integration

Anywhere from 1,000 to 10,000 communication lines form synapses with a typical neuron in your brain, which has at least 100 billion neurons. As long as you are alive, those neurons hum with messages about doing what it takes to be a human. At any moment, a great number of excitatory and inhibitory signals may wash over input zones of a postsynaptic cell. Some drive its membrane closer to threshold; others maintain the resting level or drive it away from threshold. Said another way, signals compete for control of the neuron's membrane.

All synaptic signals are graded potentials. The type we call an **EPSP** (for excitatory postsynaptic potential) has a *depolarizing* effect. This simply means it drives the membrane closer to threshold. An **IPSP** (for inhibitory postsynaptic potential) may have a *hyperpolarizing* effect and thus drive the membrane away from threshold, or it may help maintain the membrane at its resting level.

With **synaptic integration**, competing signals that reach an input zone of a neuron at the same time are summed. By this process, two or more signals arriving at a neuron may be dampened, suppressed, reinforced, or sent onward to other cells in the body.

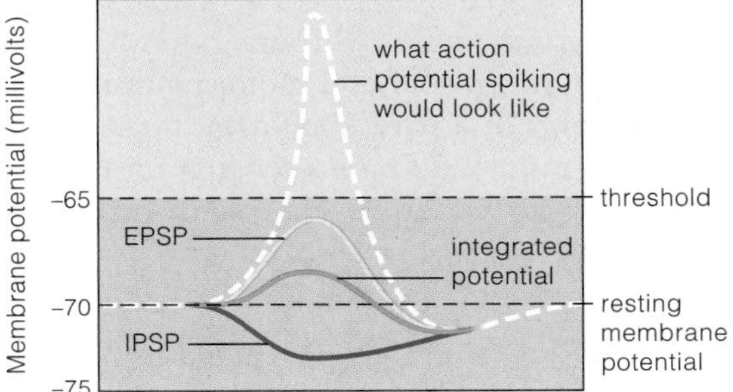
Membrane potential (millivolts)
what action potential spiking would look like
-65
threshold
EPSP
integrated potential
-70
resting membrane potential
IPSP
-75

Figure 34.10 Example of synaptic integration. The *yellow* line shows how an EPSP of a certain magnitude would register on an oscilloscope screen if it were acting alone. The *purple* line shows the effect of an IPSP if *it* were acting alone. Suppose both signals arrive at a postsynaptic cell membrane at the same time. The *red* line shows their effect. In this case, when the two signals are integrated, threshold is not reached. Thus an action potential cannot be initiated in the target cell.

Figure 34.10 shows what two separate recordings of an EPSP and an IPSP might look like, and it includes a recording of their summation. Such integration occurs when neurotransmitter from several presynaptic cells reaches the input zone of a neuron at the same time. It also occurs after neurotransmitter is released swiftly and repeatedly from a single presynaptic cell that has been whipped into a frenzy of excitability by a rapid series of action potentials.

How Is Neurotransmitter Removed From the Synaptic Cleft?

The flow of information through the nervous system depends on the prompt, precisely controlled removal of neurotransmitter molecules from synaptic clefts. Some number of the molecules simply diffuse away from the cleft. Enzymes present inside the cleft cleave others, as when acetylcholinesterase breaks apart ACh. Besides this, transport proteins in the membrane of presynaptic cells actively pump the molecules back inside or into neighboring neuroglial cells.

What happens if neurotransmitter accumulates in the cleft? As one example, cocaine blocks the uptake of dopamine. Molecules of this neurotransmitter linger in synaptic clefts and just keep on stimulating target cells. At first the abnormal stimulation produces euphoria (intense pleasure). Later on, it has disastrous effects, as you will read in Section 35.9.

Neurotransmitters are signaling molecules that bridge a synaptic cleft, which is a tiny gap between two neurons or between a neuron and a muscle cell or gland cell.

Neurotransmitters have excitatory or inhibitory effects on different kinds of receiving cells. Synaptic integration is the moment-by-moment combining of excitatory and inhibitory signals acting on a postsynaptic cell.

With the summation process, messages traveling through the nervous system can be reinforced or downplayed, sent onward or suppressed. The process is essential for normal body functioning.

34.4 PATHS OF INFORMATION FLOW

Blocks and Cables of Neurons

Through synaptic integration, messages arriving at any neuron in the body might be reinforced and sent on to neighboring neurons. What determines the direction in which a particular message will travel? It depends on the organization of neurons in different body regions.

For example, your brain deals with its staggering numbers of neurons in a manner analogous to block parties. Regional blocks of hundreds or thousands of neurons receive excitatory and inhibitory signals. They integrate signals entering the block, then send out new ones in response. Some regions have neurons organized as *divergent* circuits, with processes fanning out from one block to form connections with others. Some have neurons arranged in *convergent* circuits, with signals from many sent on to just a few. In still other regions, neurons synapse back on themselves and repeat signals like gossip that just won't go away. Such *reverberating* circuits include the ones that make your eye muscles rhythmically twitch while you sleep.

In the cablelike **nerves**, long axons of many sensory neurons, motor neurons, or both permit long-distance communication between the brain or spinal cord and the rest of the body. Connective tissue bundles most of the axons in parallel array (Figure 34.11). Each axon has a **myelin sheath** that enhances the rate of action potential propagation. The sheath is a series of **Schwann cells**, a type of neuroglial cell, wrapped like jellyrolls around the long axon. An exposed node, or gap, separates each cell from adjacent ones. There, voltage-sensitive, gated sodium channels pepper the plasma membrane (Figure 34.12). The sheathed regions in between nodes hamper ion movements across the membrane. Ion disturbances tend to flow along the membrane until the next node in line. At each node, ion flow can produce a new action potential. In large sheathed axons, action potentials are propagated at a remarkable 120 meters per second.

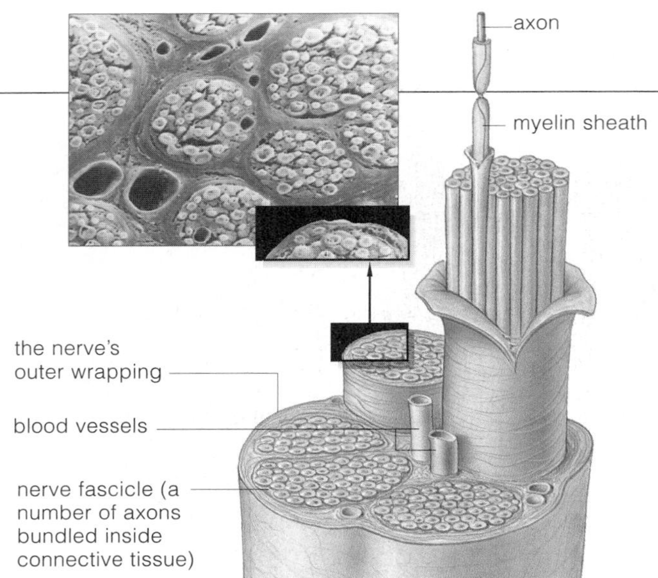
axon
myelin sheath
the nerve's outer wrapping
blood vessels
nerve fascicle (a number of axons bundled inside connective tissue)

Figure 34.11 Structure of a nerve. Axons in the nerve are bundled together inside wrappings of connective tissue.

In *multiple sclerosis*, myelin sheaths around axons in the spinal cord degenerate slowly. A mutant gene may predispose a person to the disease, but viral infection might trigger it. Symptoms include serious progressive weakening of the muscles, fatigue, and numbness.

Reflex Arcs

Figure 34.13 is a specific example of the direction of information flow through nervous systems. It shows how sensory and motor neurons of certain nerves take part in a path known as the stretch reflex. **Reflexes** are automatic movements made in response to stimuli. Sensory neurons synapse directly on motor neurons in the simplest reflex arcs.

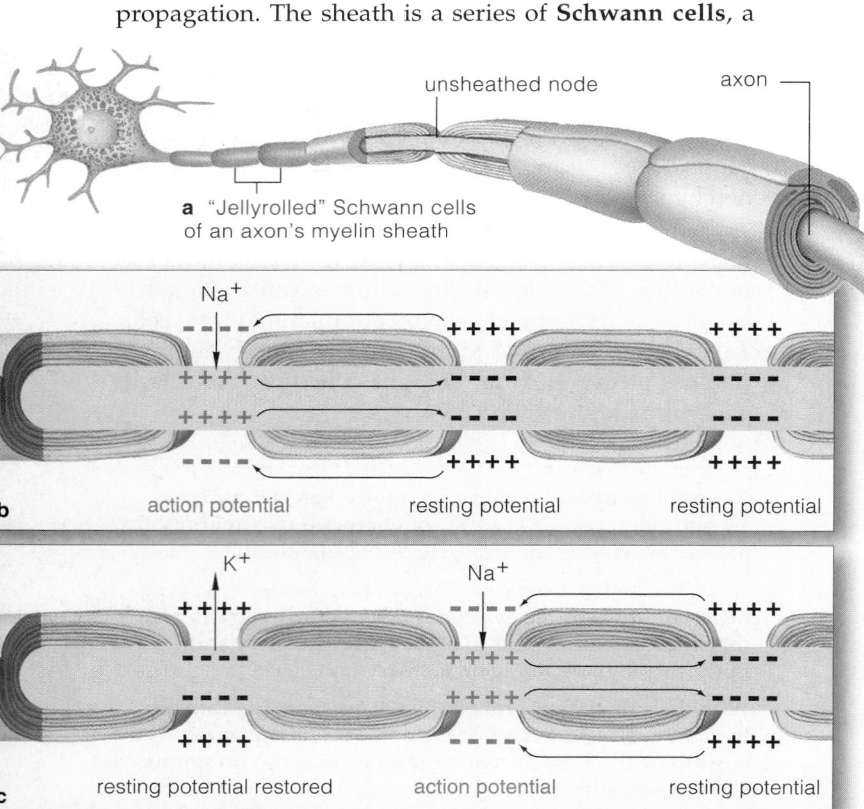
unsheathed node
axon
a "Jellyrolled" Schwann cells of an axon's myelin sheath
Na+
b
action potential
resting potential
resting potential
K+
Na+
c
resting potential restored
action potential
resting potential

Figure 34.12 Action potential propagation along a sheathed neuron. (**a**) A myelin sheath is a series of Schwann cells, each wrapped like a jellyroll around an axon. It blocks ion movements across the membrane. But ions can cross at unsheathed nodes between the jellyrolls. The nodes have dense arrays of gated sodium channels. (**b**) A disturbance caused by an action potential spreads along the axon. When it reaches a node, sodium gates open, sodium ions rush inward, and another action potential results. (**c**) The new disturbance spreads swiftly to the next node and triggers a new action potential, and so on down the line.

Further reading: Student Guide to InfoTrac on web site

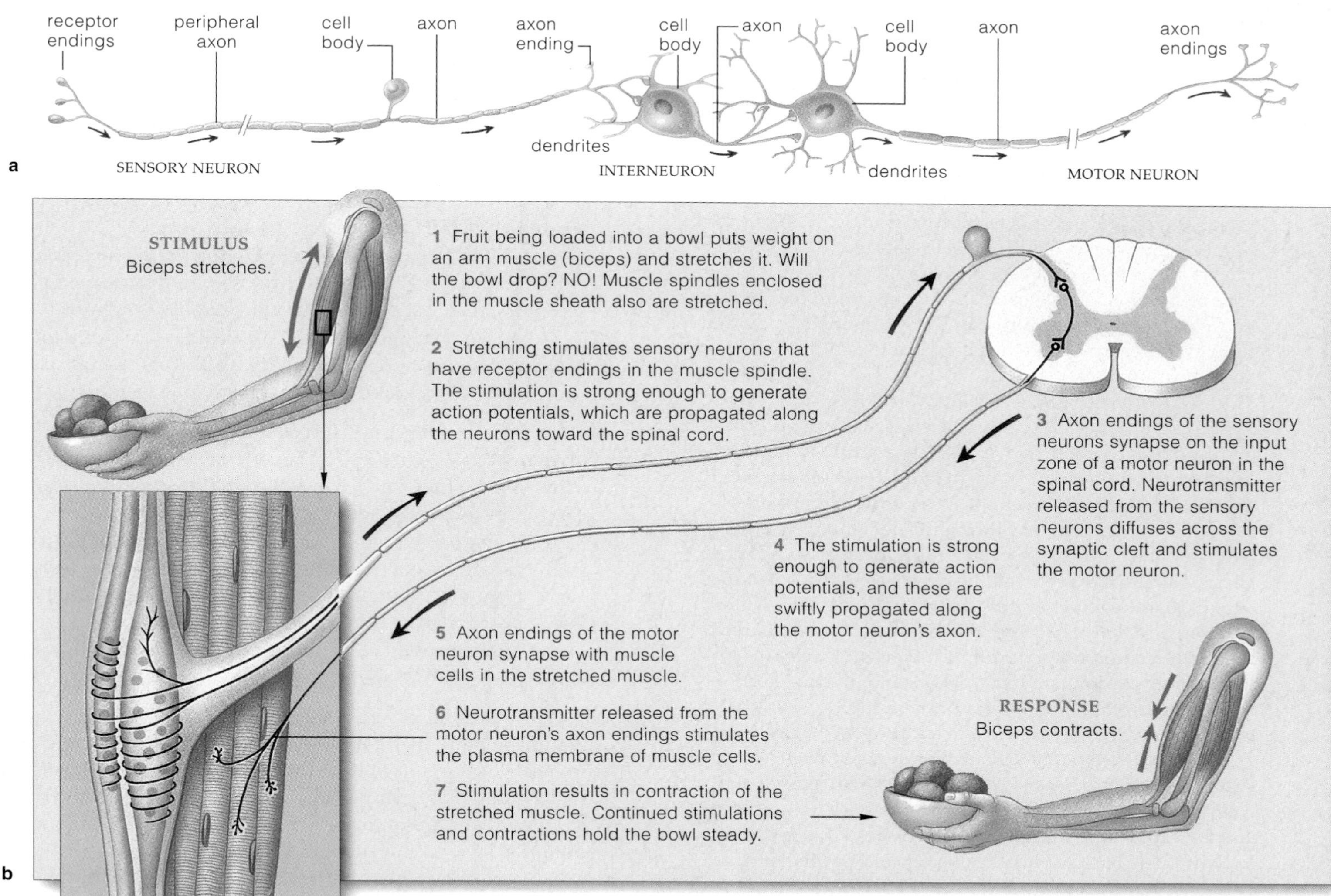

receptor endings
peripheral axon
cell body
axon
axon ending
cell body
axon
cell body
axon
axon endings
dendrites
dendrites
a
SENSORY NEURON
INTERNEURON
MOTOR NEURON
STIMULUS
Biceps stretches.
1 Fruit being loaded into a bowl puts weight on an arm muscle (biceps) and stretches it. Will the bowl drop? NO! Muscle spindles enclosed in the muscle sheath also are stretched.
2 Stretching stimulates sensory neurons that have receptor endings in the muscle spindle. The stimulation is strong enough to generate action potentials, which are propagated along the neurons toward the spinal cord.
3 Axon endings of the sensory neurons synapse on the input zone of a motor neuron in the spinal cord. Neurotransmitter released from the sensory neurons diffuses across the synaptic cleft and stimulates the motor neuron.
4 The stimulation is strong enough to generate action potentials, and these are swiftly propagated along the motor neuron's axon.
5 Axon endings of the motor neuron synapse with muscle cells in the stretched muscle.
6 Neurotransmitter released from the motor neuron's axon endings stimulates the plasma membrane of muscle cells.
7 Stimulation results in contraction of the stretched muscle. Continued stimulations and contractions hold the bowl steady.
RESPONSE
Biceps contracts.
b
muscle spindle
muscle cell

Figure 34.13 (**a**) General direction of information flow in nervous systems. Sensory neurons relay information *into* the spinal cord and brain, where they synapse with interneurons. Interneurons *within* the spinal cord and brain integrate signals. Many synapse with motor neurons, which carry signals *away* from the spinal cord and brain. (**b**) Organization of nerves in a reflex arc that deals with muscle stretching. In a skeletal muscle, stretch-sensitive receptors of a sensory neuron are located in muscle spindles. The stretching generates action potentials, which reach axon endings in the spinal cord. These synapse with a motor neuron that carries signals to contract, from the spinal cord back to the stretched muscle.

The stretch reflex works to contract a muscle after gravity or some other load has caused the muscle to stretch. Suppose you hold out a large bowl and keep it stationary as someone puts several peaches into it. The peaches add weight to the bowl, and when your hand starts to drop, a muscle in your arm (the biceps) is stretched.

In the muscle, stretching activates receptor endings that are a part of muscle spindles—sensory organs in which specialized cells are enclosed in a sheath that runs parallel with the muscle. The receptor endings are the input zones of sensory neurons, the axons of which synapse with motor neurons within the spinal cord (Figure 34.13). Axons of the motor neurons lead back to the stretched muscle. Action potentials that reach the axon endings trigger the release of ACh, which initiates contraction. As long as receptor activity continues, the motor neurons are excited even more, and this allows them to maintain your hand's position.

In the vast majority of reflex pathways, the sensory neurons also interact with a number of interneurons, which then activate or suppress all the motor neurons necessary for a coordinated response.

Vertebrates have interneurons organized in information-processing blocks. Cablelike nerves that have long axons of sensory neurons, motor neurons, or both connect the brain and spinal cord with the rest of the body.

Reflex arcs, in which sensory neurons synapse directly on motor neurons, are the simplest paths of information flow.

34.5 SKEWED INFORMATION FLOW

What happens if something disrupts information flow in a nervous system? Ask *Clostridium botulinum*, an anaerobic, endospore-forming bacterium that normally lives in soil. It can cause the disease *botulism*. Improperly preserved, canned, or stored food may contain its endospores. These can germinate in someone who ingests the food. One metabolic product of this bacterium is toxic. It enters neurons that synapse with muscle cells and blocks the release of acetylcholine (ACh). Without ACh, muscles cannot contract. They become more and more flaccid and paralyzed. Unless antitoxins are administered within ten days, breathing becomes impossible and the heart stops beating.

A related bacterium, *C. tetani*, lives in the gut of horses, cattle, and other grazing animals, even many people. Its endospores survive in soil, especially manure-rich ones. They persist for years if sunlight and oxygen do not reach them. They resist disinfectants, heat, and boiling water. When they enter the body through a deep puncture or a deep cut, they germinate in anaerobic, dead tissues. These bacteria do not spread from the dead tissue. They produce a toxin that the blood or nerves deliver to the spinal cord and brain. In the spinal cord, the toxin affects interneurons that help control motor neurons. It blocks the release of inhibitory neurotransmitters (GABA and glycine) and frees the motor neurons from normal inhibitory control. Symptoms of the disease *tetanus* are about to begin.

After four to ten days, the overstimulated muscles stiffen and go into spasm. They cannot be released from contraction, and prolonged, spastic paralysis follows. Fists and jaws may remain clenched (the disease is also called lockjaw). The back may arch severely and permanently. Once respiratory and cardiac muscles become paralyzed, death nearly always follows.

Vaccines were not available for soldiers of early wars, when dead cavalry horses and manure littered battlefields (Figure 34.14). Today, vaccines have all but eradicated tetanus in the United States.

Figure 34.14 Painting of a young victim of a contaminated battle wound, as he lay dying in a military hospital.

34.6 SUMMARY

1. Nervous systems sense and interpret specific aspects of the environment and issue commands for responses to them. In nearly all animals, communication lines are made of sensory neurons, interneurons, and motor neurons, which activate muscle and gland cells.

a. Sensory neurons are receptors that detect specific stimuli (specific forms of energy, such as light).

b. Interneurons are integrators in the brain and spinal cord. They receive and interpret signals from sensory receptors, then issue commands for suitable responses.

c. Motor neurons carry commands away from the brain and spinal cord to the body's effectors, which are muscle cells and gland cells that carry out responses.

2. A neuron's dendrites and cell body are input zones. If arriving signals spread to a trigger zone (such as the start of an axon), they may trigger an action potential that can be propagated to an output zone (axon endings).

a. Transport proteins pepper the plasma membrane from the trigger zone to axon endings of a neuron. They serve as gated or open channels for the passage of ions, Na^+ and K^+ especially, across the membrane.

b. The controlled flow of ions across the membrane is the basis of message propagation along a neuron.

3. For a neuron at rest, the ion distribution between the cytoplasm and the interstitial fluid results in a difference in electric charge (voltage difference) across the plasma membrane. The cytoplasmic fluid is more negatively charged by an amount known as the resting membrane potential. An action potential is an abrupt, short-lived reversal in the voltage difference across the membrane in response to adequate stimulation.

a. Stimuli give rise to local, graded potentials, which vary in magnitude and do not spread far from the point of stimulation. A disturbance from a number of graded potentials may spread to a trigger zone and drive the membrane to the threshold of an action potential.

b. A strong disturbance opens many gated channels for sodium ions in an ever accelerating, self-propagating way. Thus, the voltage difference across the membrane reverses abruptly. Accelerated openings proceed in patch after patch of membrane, to the neuron's output zone.

c. In between action potentials, sodium–potassium pumps counter small ion leaks across the membrane. By doing so, they help maintain the gradients. After an action potential, they restore the gradients.

4. Action potentials arriving at an output zone trigger the release of neurotransmitters. The neurotransmitters diffuse across chemical synapses, which are small clefts between one neuron and another neuron, a muscle cell, or a gland cell. They may excite the plasma membrane of the postsynaptic cell (drive it closer to threshold) or inhibit it (drive it away from threshold).

Further reading: Student Guide to InfoTrac on web site —

5. Integration is the moment-by-moment summation of excitatory and inhibitory signals reaching all of the synapses on a neuron. It is a means of playing down, suppressing, reinforcing, or sending on information to other neurons of the nervous system.

Review Questions

1. Describe sensory neuron, interneuron, and motor neuron in terms of their structure and functions. *CI, 34.1, 34.4*
2. Label the functional zones of this motor neuron. *34.1*

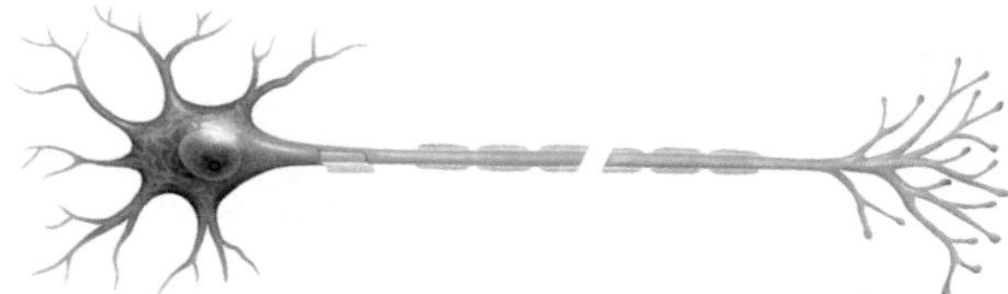

3. Define resting membrane potential, graded potential, and action potential. *34.1, 34.2*
4. A neuron at rest is controlling the ion distribution across its plasma membrane. Identify the two major kinds of ions. Do they leak across the membrane, are they pumped across, or both? As part of your answer, describe channel proteins and transport proteins embedded in the neural membrane. *34.1*
5. With respect to action potentials, explain threshold level, all-or-nothing spikes, and self-propagation. *34.2*
6. Define chemical synapse and neurotransmitter. Choose an example of a neurotransmitter and state where it acts. *34.3*
7. Define synaptic integration. Include definitions of EPSPs and IPSPs as part of your answer. *34.3*
8. What is a myelin sheath? Do all neurons have one? *34.4*
9. How do divergent, convergent, and reverberating circuits among blocks of neurons differ from one another? *34.4*
10. Define reflex, then give an example of a reflex arc. *34.4*

Self-Quiz (*Answers in Appendix III*)

1. Action potentials occur when ______.
   a. a neuron receives adequate stimulation
   b. sodium gates open in an ever accelerating way
   c. sodium–potassium pumps kick into action
   d. both a and b

2. Compared with interstitial fluid near the plasma membrane of a neuron at rest, the cytoplasm near the membrane's inner surface carries a slight ______ charge.
   a. positive
   b. negative
   c. graded and local
   d. both b and c

3. The resting membrane potential is maintained by ______.
   a. ion leaks
   b. ion pumps
   c. neurotransmitters
   d. both a and b

4. Neurotransmitters diffuse across a ______.
   a. chemical synapse
   b. membrane pump
   c. myelin sheath
   d. both a and b

5. An action potential lasts only briefly because ______ at the membrane region where it occurred.
   a. gates for sodium open, gates for potassium close
   b. gates for sodium close, gates for potassium open
   c. sodium–potassium pumps restore gradients
   d. both b and c

6. A nerve may consist of bundled-together axons of ______.
   a. sensory neurons
   b. motor neurons
   c. sensory and motor neurons
   d. all of the above

7. Integration is the ______.
   a. self-propagation of ion flow along a neuron
   b. trigger that releases neurotransmitters
   c. summation of signals acting at all synapses on a neuron
   d. collection of all information about several stimuli

8. Match the terms with their most suitable description.

___ synaptic integration	a. occurs at input zone of the neuron
___ muscle spindle	b. spatial, temporal summation of all signals arriving at a neuron at the same time
___ graded, local potential	c. arises at trigger zone
___ action potential	d. stretch-sensitive receptor

Critical Thinking

1. With *multiple sclerosis*, recall, myelin sheaths around axons in the spinal cord slowly degenerate. Affected individuals progressively lose the ability to control movements of their skeletal muscles. Reflect on the role of myelin sheaths in the nervous system, then formulate a hypothesis of how their degeneration results in a loss of control over skeletal muscles.

2. *Epilepsy* is a neurological disorder characterized by short, recurring episodes of sensory and motor malfunctioning. The episodes, or epileptic seizures, might arise when reverberating circuits involving millions of interneurons in the brain become abnormally activated. Affected individuals typically experience involuntary muscle contractions and often sense lights, sounds, and odors even if receptors in the eyes, ears, and nose have not been stimulated. The drug valproic acid can eliminate or lessen the severity of the seizures by stimulating the body's synthesis of GABA. Why would stepping up GABA production help?

3. A mud-covered nail punctured Evita's foot. To protect Evita from tetanus, her doctor administered an antitoxin. This one had enough protein molecules that could specifically bind with and neutralize the *C. tetanus* toxin molecules. Will Evita now have long-lasting protection against future attacks by this bacterium?

Selected Key Terms

acetylcholine (ACh) *34.3*
action potential *34.1*
axon *34.1*
chemical synapse *34.3*
dendrite *34.1*
EPSP *34.3*
interneuron *CI*
IPSP *34.3*
motor neuron *CI*
myelin sheath *34.4*
nerve *34.4*
neuroglia *CI*
neuromodulator *34.3*
neuron *CI*
neurotransmitter *34.3*
positive feedback *34.2*
reflex *34.4*
resting membrane potential *34.1*
Schwann cell *34.4*
sensory neuron *CI*
sodium–potassium pump *34.1*
stimulus *CI*
synaptic integration *34.3*
threshold level *34.2*

Readings See also *www.infotrac-college.com*

Kandell, E., J. Schwartz, and T. Jessell. 1995. *Essentials of Neural Science and Behavior*. Norwalk, Connecticut: Appleton & Lange.

Nicholls, J., A. Martin, and B. Wallace. 1992. *From Neuron to Brain*. Third edition. Sunderland, Massachusetts: Sinauer.

35 INTEGRATION AND CONTROL: NERVOUS SYSTEMS

Why Crack the System?

Suppose your biology instructor asks you to volunteer for an experiment. You will get a microchip implanted in your brain. It will make you feel really good. But it may mess up your health, lop ten years off your life, and destroy a good part of your brain. Your behavior will change for the worse, so you might have trouble completing school, getting or keeping a job, or even having a normal family life.

The longer the chip is implanted, the less you will want to give it up. You won't get paid. You will pay the experimenter—first at bargain rates, then a little more each week. The chip is illegal. If you get caught using it, you and the experimenter will go to jail.

Sometimes Jim Kalat, a professor at North Carolina State University, proposes this experiment, which of course is hypothetical. Hardly any students volunteer. Then he substitutes *drug* for microchip and *dealer* for experimenter, and an amazing number of students come forward! Like 30 million other Americans, the "volunteers" seem ready to engage in self-destructive uses of drugs that alter emotional and behavioral states.

The destruction shows up in unexpected places. Each year, for instance, about 300,000 newborns are already addicted to crack, thanks to their addicted mothers. *Crack* is a cheap form of cocaine. It causes relentless stimulation of brain regions that govern the sense of pleasure. It dampens normal urges to eat and to sleep, and blood pressure rises. Elation and sexual desire intensify. In time, however, the brain cells that produce the stimulatory chemicals cannot keep up with the abnormal demand. The chemical vacuum makes crack users frantic and then profoundly depressed. Only crack makes them feel good again.

Addicted babies cannot know all of this. They can only quiver with "the shakes" and respond to the world with chronic irritation. And they are abnormally small. While they were developing inside their mother, their body tissues simply were not provided with enough

Figure 35.1 Owners of an evolutionary treasure—a complex brain that is the foundation for our memory and reasoning, and our future.

oxygen and nutrients. As one of its side effects, crack causes blood vessels to constrict, and maternal blood vessels are the only supply lines that the developing individual has.

Paradoxically, crack babies are exceptionally fussy, yet they cannot respond to rocking and other forms of stimulation that normally have soothing effects. It may be a year or more before they recognize even their own mother. Without treatment, they are likely to grow up as emotionally unstable children, prone to aggressive outbursts and stony silences. Why? Their mother's drug habit crippled their nervous system.

Think about it. The nervous system evolved as a way to sense and respond, with exquisite precision, to changing conditions inside and outside the body. Awareness of sounds and sights, of odors, of hunger and passion, fear and rage—all such things begin with the flow of information along the communication lines of the nervous system. Do those lines remain silent until they receive outside signals, much as telephone lines wait to carry calls from all over the country? Absolutely not. Even before you were born, excitable cells called neurons became organized into extensive gridworks in your newly forming tissues and started chattering among themselves. All through your life, in moments of danger or reflection, supreme excitement or sleep, their chattering has never ceased and will not cease until the time you die.

In the preceding chapter, our primary focus was on the nature of messages that travel through any nervous system. We turn now to how the systems themselves are organized. As you will see, they differ in vital ways among animals. And the differences correspond to differences in life-styles.

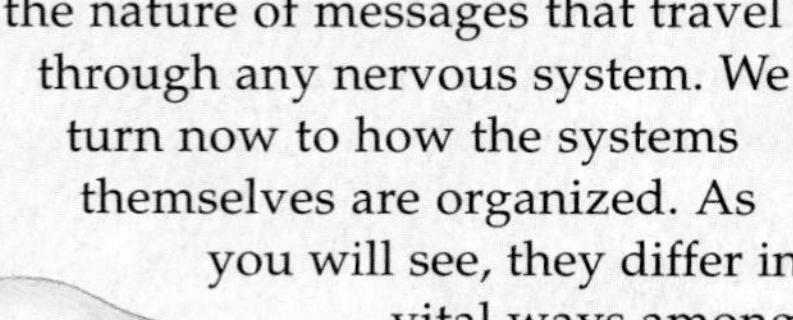

KEY CONCEPTS

1. Nervous systems are composed of neurons and a variety of cells called neuroglia, which structurally and functionally support the neurons. The neurons interact in the detection and integration of information about external and internal conditions, and in commanding muscles and glands to carry out suitable responses.

2. The simplest nervous systems are the nerve nets of radial animals, such as hydras and sea anemones. The nervous systems of most animals show pronounced cephalization and bilateral symmetry.

3. Vertebrate nervous systems are functionally divided into central and peripheral regions. The brain and spinal cord make up the central nervous system. Paired nerves that thread through the remainder of the body are the key components of the peripheral nervous system.

4. The somatic nerves of the peripheral nervous system deal with skeletal muscles. The autonomic nerves deal with the heart, lungs, and other internal organs that are generally enclosed in body cavities.

5. The vertebrate brain has three functional divisions called the hindbrain, midbrain, and forebrain. Its most ancient parts deal with reflex control of breathing, blood circulation, and other basic functions that are essential for staying alive.

6. Long ago, in certain vertebrate lineages, the brain expanded in volume and complexity. Its newer regions appropriated more and more control over the ancient reflex functions.

7. In birds and mammals especially, the portion of the forebrain called the cerebrum contains the most complex centers for receiving, integrating, storing, and responding to sensory information.

8. The hypothalamus, another part of the forebrain, is the main homeostatic control center over the internal environment and the functioning of internal organs. Being part of the limbic system, the hypothalamus also influences emotional states and memory.

35.1 INVERTEBRATE NERVOUS SYSTEMS

Regarding the Nerve Net

At this point in the book, you know that all animals except sponges have some type of a **nervous system** in which nerve cells, such as neurons, are oriented relative to one another in signal-conducting and information-processing highways. At the minimum, the cells making up the communication lines receive information about changing conditions outside and inside the body, then elicit suitable responses from muscle and gland cells.

To appreciate the diversity of nervous systems, start by recalling that animals first evolved in the seas. It is in the seas that we still find animals with the simplest nervous systems. They are the cnidarians, such as sea anemones and jellyfishes. These invertebrates display *radial* symmetry. That is, their body parts are arranged about a central axix much like spokes of a bike wheel, as shown in Section 26.1.

Such animals rely on a **nerve net**, a loose mesh of nerve cells intimately associated with epithelial tissue (Figure 35.2). The nerve cells interact with sensory cells and contractile cells along reflex pathways in the same epithelial tissue. In **reflex pathways**, remember, sensory stimulation triggers simple, stereotyped movements.

In all cnidarians, one pathway dealing with feeding behavior extends from sensory receptors in the tentacles, along nerve cells, to contractile cells organized around the mouth. In jellyfishes, other reflexes are the basis of movements necessary to swim (slowly) and to keep the soft body right-side up.

The nerve net itself extends through the animal's body, but the flow of information flow through it is not focused. It simply commands the body wall to slowly contract and expand or move tentacles through water. Its owners will never dazzle you with bursts of speed or precision acrobatics. Either they are weak swimmers or they are sedentary types that spend most of their life attached to substrates. In their watery world, the nerve net is adaptive; it is equally responsive to bits of food or danger that might come from any direction.

On the Importance of Having a Head

Flatworms are the simplest animals having a bilateral nervous system (Figure 35.3). *Bilateral* symmetry means having equivalent body parts on the left and right sides of the body's midsagittal plane. (Imagine sliding down a staircase banister that turns into a razor and you may never forget the location of the midsagittal plane.) Both sides have the same array of muscles that function in moving the body forward. Both have the same array of nerves to control the muscles, and so on.

A flatworm has a ladderlike nervous system with two cordlike nerves. As Section 34.4 describes, a **nerve** is like a cable in which sensory axons, motor axons, or both are bundled together inside a sheath of connective tissue. The flatworm's nerve cords run longitudinally through the body and have many side branches.

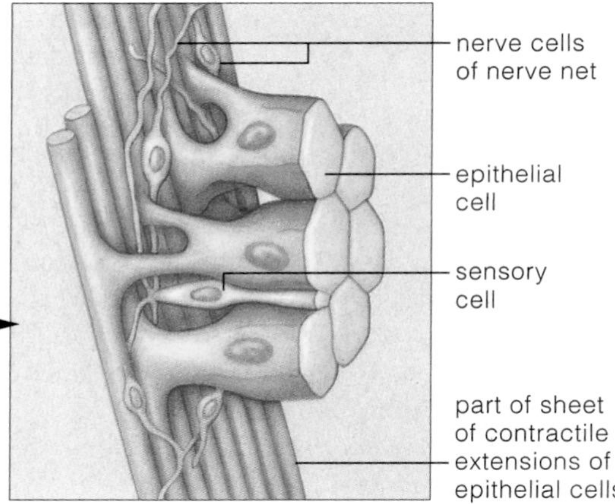

nerve cells of nerve net
epithelial cell
sensory cell
part of sheet of contractile extensions of epithelial cells

Figure 35.2 Nerve net of a sea anemone, one of the cnidarians. Its nerve cells interact with sensory and contractile cells. The two kinds of cells are embedded in an epithelial tissue, between the outer epidermis and a jellylike middle layer of the body wall. The middle layer is called mesoglea.

Further reading: Student Guide to InfoTrac on web site

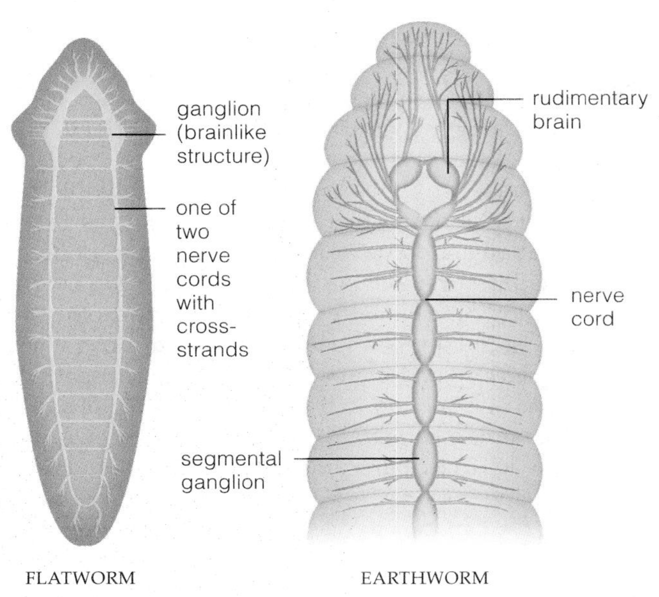
ganglion (brainlike structure)
one of two nerve cords with cross-strands
rudimentary brain
nerve cord
segmental ganglion
FLATWORM
EARTHWORM

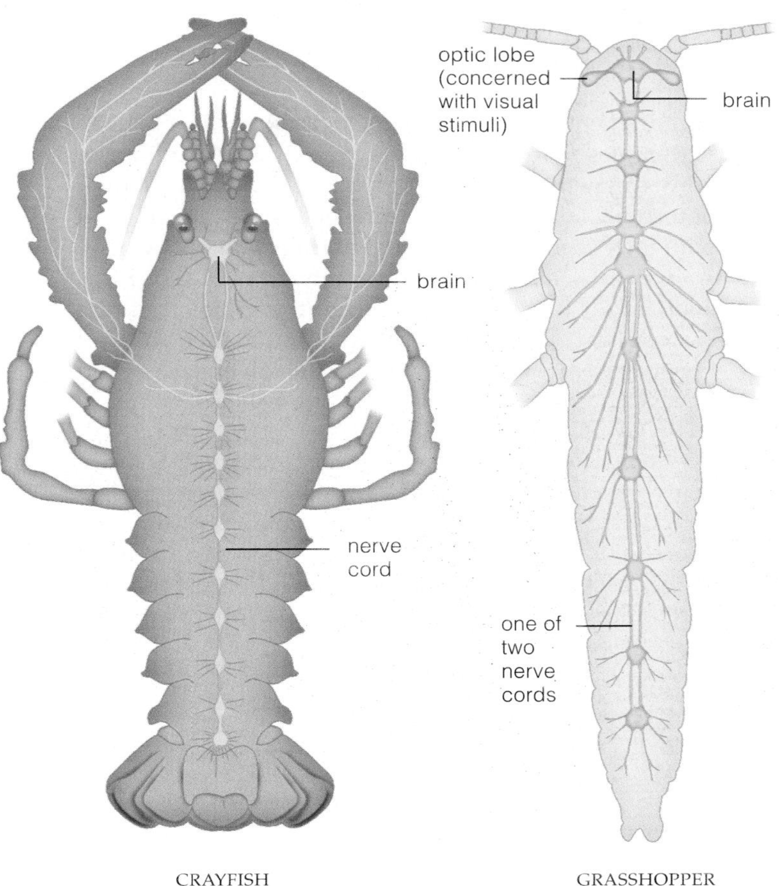
optic lobe (concerned with visual stimuli)
brain
brain
nerve cord
one of two nerve cords
CRAYFISH
GRASSHOPPER

Figure 35.3 Bilateral nervous systems of a few invertebrates. The sketches are not to scale relative to one another.

Also, in the head end of certain flatworms are two **ganglia** (singular, ganglion). Every ganglion is a local cluster of nerve cell bodies, and it is a local integrating center. In this case, the ganglia coordinate signals from paired sensory organs (two eyespots, for example), and they provide a bit of control over the nerves.

Did bilateral nervous systems evolve from nerve nets? Maybe. In nearly all animals more complex than flatworms, we find local nerve nets—plexuses—such as the one in your intestinal wall. More intriguing, a self-feeding larval stage called a **planula** develops in some cnidarian life cycles. Like the flatworms, a planula has a flattened body and uses cilia to swim or crawl about:

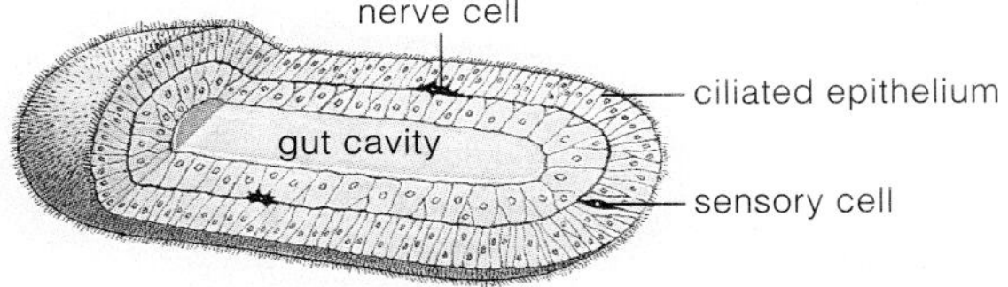
nerve cell
ciliated epithelium
gut cavity
sensory cell

Imagine an ancient planula crawling on the seafloor in Cambrian times. By chance, a mutation of a regulatory gene blocked the scheduled transformation into adults but did not block the maturing of reproductive organs. (This actually happens in the larvae of some animals; refer to Sections 20.3 and 26.5) The planula kept right on crawling, but now it had the capacity to reproduce. Its offspring inherited the mutant gene responsible for forward mobility. They crawled longer and entered new parts of their habitat—food-rich, food-poor, and maybe dangerous. The mutation proved to have survival value in the new places. Therefore, a concentration of sensory cells in the body's leading end—not the trailing end—was favored, because it permitted more rapid, effective responses to diverse stimuli.

It seems probable that natural selection favored a concentration of sensory cells at the leading end of the body in some animal lineages. Regardless of how this happened, cephalization (the formation of a head) and bilateral symmetry did develop in most invertebrate lineages (Chapter 26). Also, as you will see shortly, the patterns of both cephalization and bilateral symmetry are evident in the paired sensory structures, paired brain centers, paired nerves, and paired skeletal muscles of yourself and all other vertebrates.

Radial animals have a nerve net, a diffuse mesh of nerve cells that take part in simple reflex pathways involving sensory cells and contractile cells of an epithelial tissue.

Nervous systems of cephalized, bilateral animals include a brain (or ganglia at the head end) as well as paired nerves and paired sensory structures.

Nerves are cablelike communication lines of sensory axons, motor axons, or both bundled in a connective tissue sheath.

35.2 VERTEBRATE NERVOUS SYSTEMS—AN OVERVIEW

Evolutionary High Points

Hundreds of millions of years ago, the earliest fishlike vertebrates were evolving (Section 27.3). A column of bony segments was taking over the functions of their notochord, a long rod of stiffened tissue that worked with their segmented muscles to bring about movement. Above the notochord, a hollow, tubular nerve cord was evolving. It was the forerunner of the spinal cord and brain. These developments were a foundation for new life-styles. Not long afterward, genetic divergences from lineages of filter-feeding, scavenging fishes gave rise to swift, jawed predators of the seas.

At first, simple reflex pathways prevailed. Sensory neurons synapsed directly with motor neurons, which directly signaled muscles to contract. No other neurons altered the flow of signals from reception of a stimulus to the response. In the changing world of fast-moving vertebrates, however, the predators and prey that were better equipped to sense the presence of food or danger had the competitive edge.

The senses of smell, hearing, and balance became keener among the vertebrates that invaded land. Bones and muscles evolved in ways that allowed specialized movements. Also, the brain became variably thickened with nervous tissue that could integrate rich sensory information and issue orders for complex responses.

The oldest regions of the vertebrate brain still deal with reflex coordination of respiration and other vital functions. But a great many interneurons now synapse on the sensory and motor neurons of ancient pathways and with one another in the newer brain regions. In the most complex vertebrates, interneurons receive, store, retrieve, and compare information about experiences. They weigh possible responses. And they give our own species the capacity to reason, remember, and learn.

The nerve cord persists in all vertebrate embryos. We call it the **neural tube**. While the embryo grows and develops, the tube expands and becomes regionally modified into the brain and spinal cord (Figure 35.4*a*). A vertebral column encloses the spinal cord. Adjacent tissues give rise to nerves that thread through all body regions and connect with the spinal cord and brain.

FOREBRAIN. Receives, integrates sensory information from nose, eyes, and ears; in land-dwelling vertebrates, contains the highest integrating centers

MIDBRAIN. Coordinates reflex responses to sight, sounds

HINDBRAIN. Reflex control of respiration, blood circulation, other basic tasks; in complex vertebrates, coordination of sensory input, motor dexterity, and possibly mental dexterity

(start of spinal cord)

a Diagram of how the anterior end of the dorsal, hollow nerve cord of vertebrates expanded into functionally distinct regions, which also increased in complexity in certain lineages

Functional Divisions of the Vertebrate Nervous System

Figure 35.5 gives you a general sense of the expansions of nervous tissue in the human nervous system. This diagram shows the major paired nerves of this bilateral system. What it cannot possibly show are the 100 billion interneurons in the brain alone. To be sure, humans do have the most intricately wired nervous system in the animal world. Even so, you find similar patterns among other vertebrates.

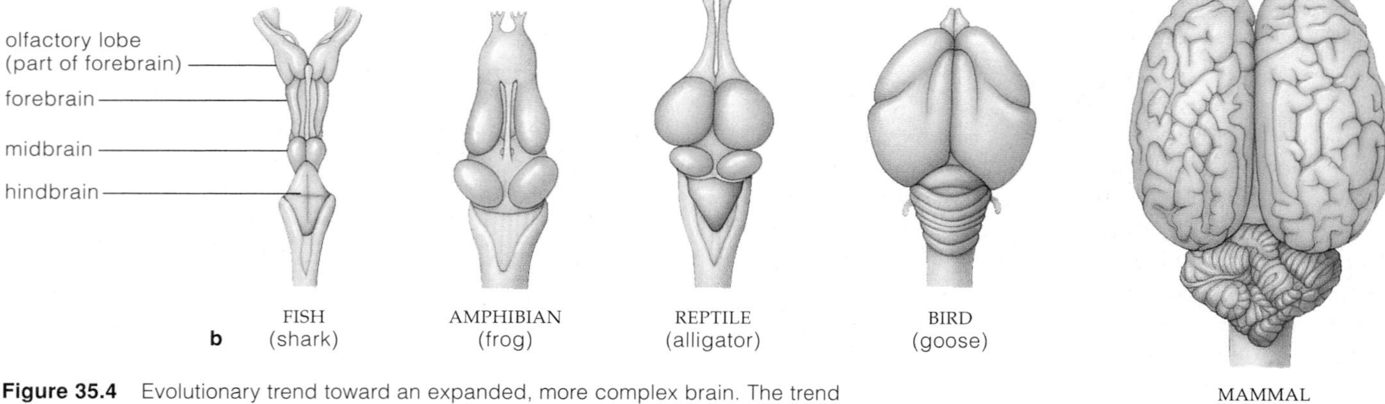

olfactory lobe (part of forebrain)
forebrain
midbrain
hindbrain
b
FISH (shark)
AMPHIBIAN (frog)
REPTILE (alligator)
BIRD (goose)
MAMMAL (horse)

Figure 35.4 Evolutionary trend toward an expanded, more complex brain. The trend became apparent after morphological comparisons were made of the brains of some vertebrates. These dorsal views are not to the same scale.

Further reading: Student Guide to InfoTrac on web site

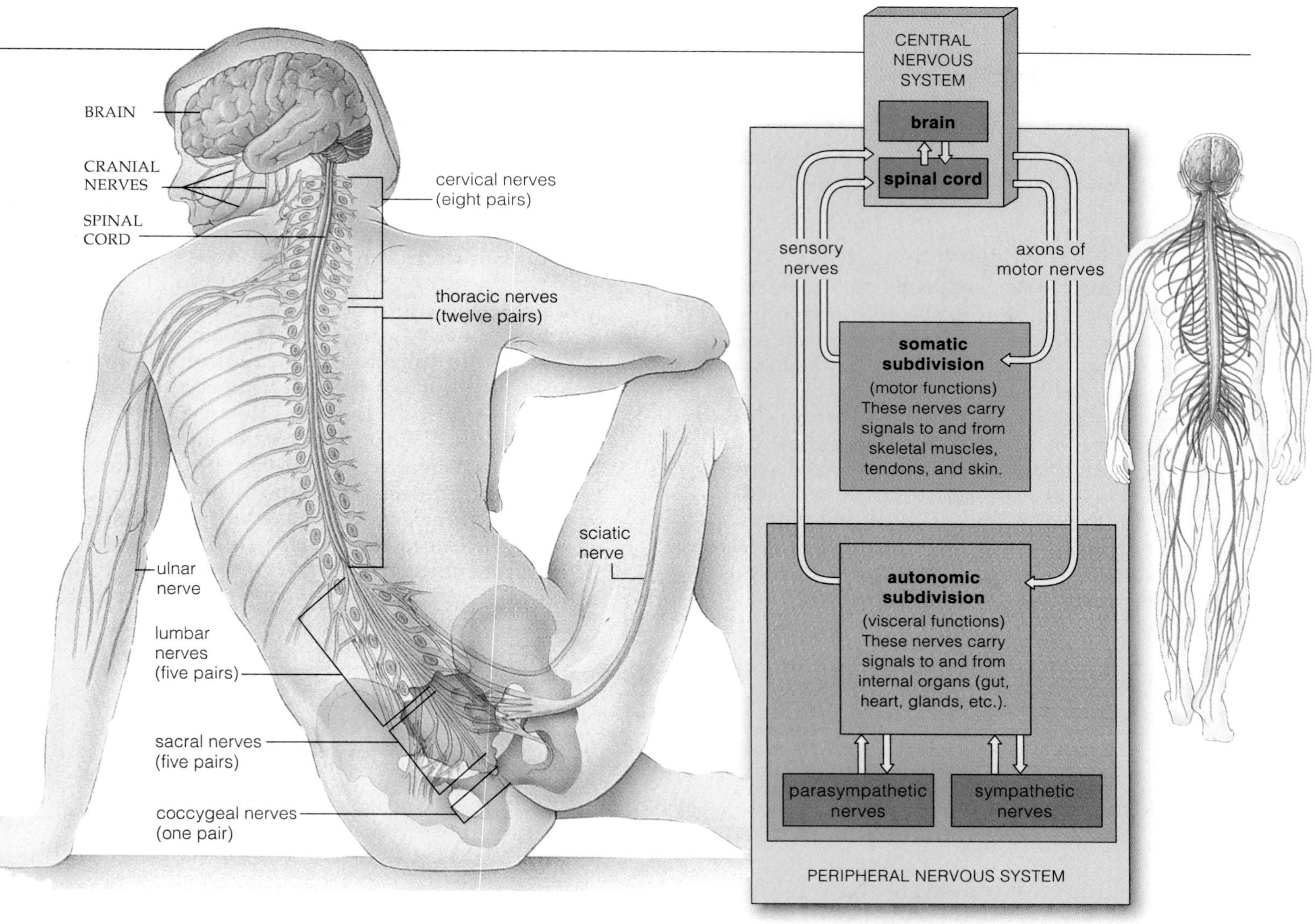
BRAIN
CRANIAL NERVES
SPINAL CORD
cervical nerves (eight pairs)
thoracic nerves (twelve pairs)
ulnar nerve
sciatic nerve
lumbar nerves (five pairs)
sacral nerves (five pairs)
coccygeal nerves (one pair)
CENTRAL NERVOUS SYSTEM
brain
spinal cord
sensory nerves
axons of motor nerves
somatic subdivision
(motor functions)
These nerves carry signals to and from skeletal muscles, tendons, and skin.
autonomic subdivision
(visceral functions)
These nerves carry signals to and from internal organs (gut, heart, glands, etc.).
parasympathetic nerves
sympathetic nerves
PERIPHERAL NERVOUS SYSTEM

Figure 35.5 Sketch of the brain, spinal cord, and some major peripheral nerves of the human nervous system. The system also incorporates twelve pairs of cranial nerves that originate from the brain. Other vertebrates have a similar system.

Figure 35.6 Functional divisions of the human nervous system. The central nervous system is color-coded *blue*, somatic nerves *green*, and autonomic nerves *red*. Sometimes nerves that carry sensory input to the central nervous system are called *afferent* (a word meaning "to bring to"). Nerves that carry motor output away from the central nervous system to muscles and glands are *efferent* ("to carry outward").

Investigators typically approach the complexity of the vertebrate nervous system by functionally dividing it into central and peripheral regions (Figure 35.6). All of the interneurons are confined to the **central nervous system**, which consists of the spinal cord and brain. The **peripheral nervous system** consists mainly of nerves, which thread through the rest of the body and carry signals into and out of the central nervous system.

Inside the brain and spinal cord, the communication lines are called tracts, not nerves. The tracts making up **white matter** have axons with glistening white myelin sheaths and specialize in rapid signal transmission. By contrast, **gray matter** consists of unmyelinated axons, dendrites, and cell bodies of neurons, plus neuroglial cells. **Neuroglia**, remember, protects or structurally and functionally supports neurons. It makes up more than half the volume of vertebrate nervous systems.

The coevolution of nervous, sensory, and motor systems made more complex life-styles possible among vertebrates.

The vertebrate nervous system has become so intricately wired that the structure and functions of its central and peripheral regions are described separately.

The central nervous system consists of the brain and spinal cord. The peripheral nervous system consists of nerves that thread through the rest of the body and carry signals into and out of the central region.

Myelinated axons make up the white matter of tracts inside the spinal cord and the brain. Neuroglia, unmyelinated axons, dendrites, and cell bodies of neurons make up the gray matter of those tracts.

35.3 THE MAJOR EXPRESSWAYS

Let's now take a look at the peripheral nervous system and the spinal cord. The two interconnect as the major expressways for information flow through the body.

Peripheral Nervous System

SOMATIC AND AUTONOMIC SUBDIVISIONS In humans, the peripheral nervous system includes thirty-one pairs of *spinal* nerves, which make connections with the spinal cord. The system also includes twelve pairs of *cranial* nerves, which connect directly with the brain.

The cranial and spinal nerves are further classified according to function. The ones that carry signals about moving the head, trunk, and limbs are **somatic nerves**. Sensory axons inside these nerves deliver information from receptors in skin, skeletal muscles, and tendons to the central nervous system. Their motor axons deliver commands issued by the brain and spinal cord to the body's skeletal muscles.

By contrast, spinal and cranial nerves dealing with smooth muscle, cardiac (heart) muscle, and glands are the **autonomic nerves**. They deal with visceral parts of the body—in other words, with its internal organs and structures. Autonomic nerves carry signals to and from these organs and structures.

PARASYMPATHETIC AND SYMPATHETIC NERVES Figure 35.7 shows the two categories of autonomic nerves. We call them parasympathetic and sympathetic. Normally they work antagonistically, with signals from one kind opposing signals from the other. However, both carry excitatory and inhibitory signals to the internal organs. Often their signals arrive at the same time at muscle or gland cells and compete for control over them. In such cases, synaptic integration at the cellular level leads to minor adjustments in the organ's level of activity.

Parasympathetic nerves dominate when the body is not receiving much outside stimulation. They tend to

SYMPATHETIC OUTFLOW FROM THE SPINAL CORD

Examples of Responses
- Heart rate increases
- Pupils of eyes dilate (widen, let in more light)
- Glandular secretions in airways to lungs decrease
- Salivary gland secretions thicken
- Stomach and intestinal movements slow down
- Sphincters (rings of muscle) contract

PARASYMPATHETIC OUTFLOW FROM THE SPINAL CORD AND BRAIN

Examples of Responses
- Heart rate decreases
- Pupils of eyes constrict (keep more light out)
- Glandular secretions in airways to lungs increase
- Salivary gland secretions become dilute
- Stomach and intestinal movements increase
- Sphincters (rings of muscle) relax

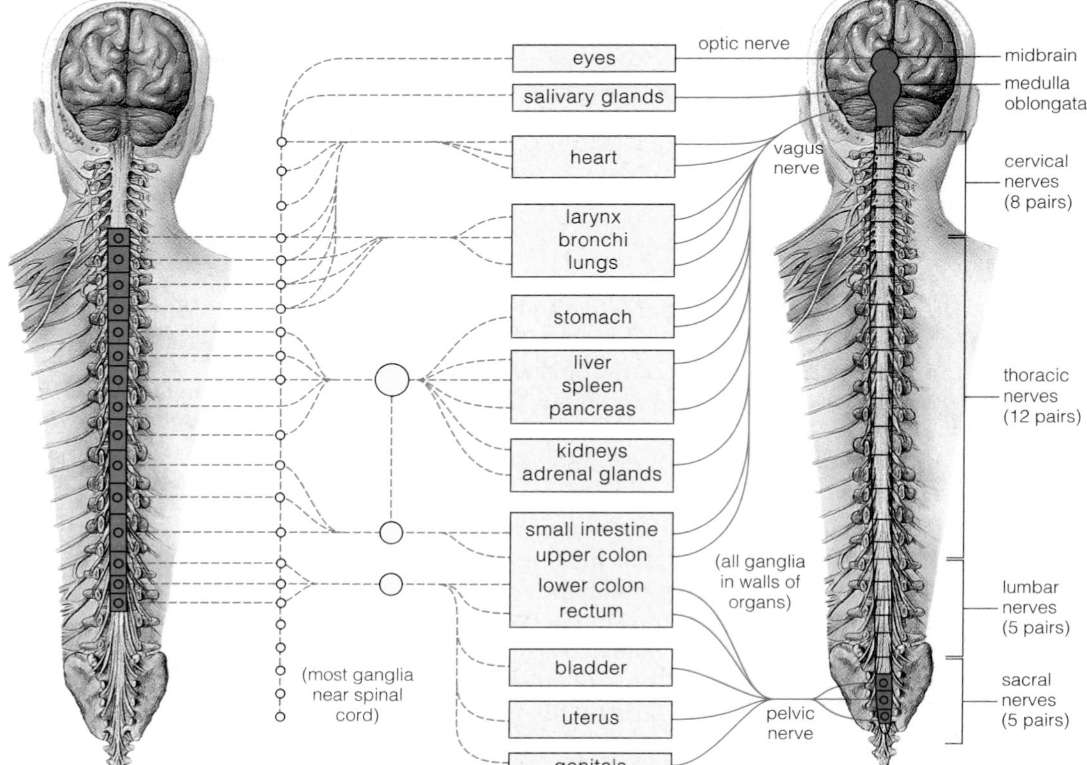

eyes
salivary glands
heart
larynx
bronchi
lungs
stomach
liver
spleen
pancreas
kidneys
adrenal glands
small intestine
upper colon
lower colon
rectum
bladder
uterus
genitals
(most ganglia near spinal cord)
optic nerve
vagus nerve
(all ganglia in walls of organs)
pelvic nerve
midbrain
medulla oblongata
cervical nerves (8 pairs)
thoracic nerves (12 pairs)
lumbar nerves (5 pairs)
sacral nerves (5 pairs)

Figure 35.7 Autonomic nervous system. This diagram shows the major sympathetic nerves and parasympathetic nerves leading from the central nervous system to some major organs. Remember, there are *pairs* of both kinds of nerves, servicing the right and left halves of the body. The ganglia are clusters of the cell bodies of neurons that have their axons bundled together in nerves.

Further reading: Student Guide to InfoTrac on web site

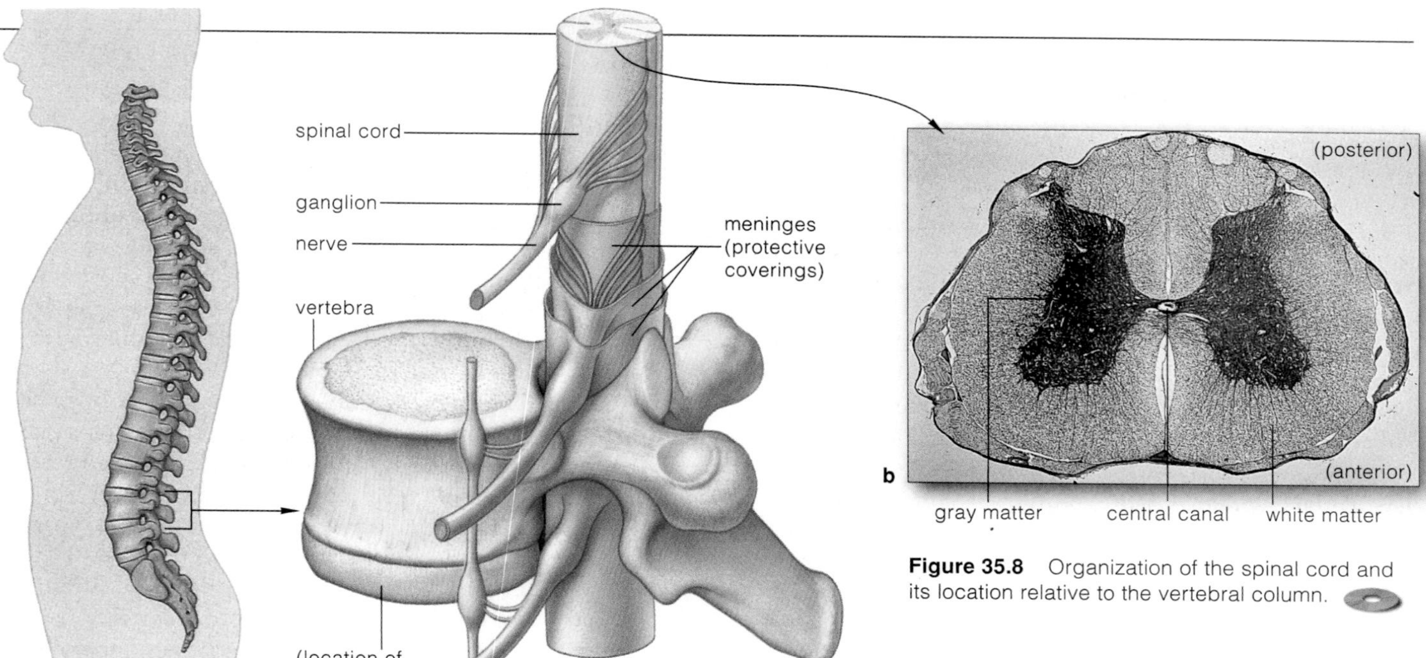
spinal cord
ganglion
nerve
vertebra
meninges (protective coverings)
(location of intervertebral disk)
a
(posterior)
(anterior)
b
gray matter
central canal
white matter

Figure 35.8 Organization of the spinal cord and its location relative to the vertebral column.

slow down the body overall and divert energy to basic "housekeeping" tasks, such as digestion.

Sympathetic nerves dominate in times of sharpened awareness, excitement, or danger. They tend to shelve housekeeping tasks. At the same time, they prepare the animal to fight or escape when threatened or to frolic, as in play or sexual behavior. Right now, sympathetic nerves are commanding your heart to beat a bit faster, and parasympathetic nerves are commanding it to beat a bit slower. Integration of the opposing signals adjusts the heart rate. If something scares or excites you, the parasympathetic input to the heart drops. Sympathetic signals cause the release of epinephrine, which makes your heart beat faster. They also incite you to breathe faster and start to sweat. In this state of intense arousal, you are primed to either fight (or play) hard or get away fast. Hence the name **fight–flight response**.

Suppose that the stimulus for a fight–flight response ends. Sympathetic activity may decrease abruptly, and parasympathetic activity may rise suddenly. You might observe this "rebound effect" after someone has been instantly mobilized to rush onto a road to save a child from an oncoming car. The person might well faint as soon as the child has been swept out of danger.

The Spinal Cord

By definition, the **spinal cord** is a vital expressway for signals between the peripheral nervous system and the brain. Also, sensory and motor neurons make direct reflex connections in the spinal cord. The stretch reflex described earlier is an example of such a connection. The spinal cord threads through a canal made of bones of the vertebral column (Figure 35.8*a*). The bones, and the ligaments attached to them, protect the cord. So do the **meninges**, three tough, tubelike coverings around the spinal cord and brain. The coverings are not impervious to all attacks. *Meningitis*, an often-fatal disease, results from certain viral or bacterial infections. Disease symptoms include severe headaches, fever, a stiff neck, and nausea.

Signals swiftly travel up and down the spinal cord in bundles of axons that have glistening myelin sheaths. These sheaths distinguish the cord's white matter from gray matter (Figure 35.8*b*). Gray matter, again, consists of dendrites and cell bodies of neurons, and neuroglial cells. It plays an important role in controlling reflexes for limb movements, as when you walk or wave your arms, and organ activity, such as bladder emptying.

Experiments with frogs provide evidence for these pathways. Between the frog's spinal cord and brain are neural circuits that deal with straightening legs after they have been bent. If these circuits are severed at the base of the brain, the legs become paralyzed—but only for about a minute. So-called extensor reflex pathways in the spinal cord recover quickly and have the frog hopping about in no time. The recovery time is minimal after similar damage in humans and other primates—the vertebrates with the greatest cephalization.

Nerves of the peripheral nervous system connect the brain and spinal cord with the rest of the body.

Somatic nerves of the peripheral nervous system deal with skeletal muscle movements. Autonomic nerves govern the functions of internal organs, such as the heart and glands.

The spinal cord is a vital expressway for signals between the brain and peripheral nerves. Some of its interneurons also exert direct control over certain reflex pathways.

35.4 FUNCTIONAL DIVISIONS OF THE VERTEBRATE BRAIN

The anterior end of the spinal cord is continuous with the **brain**, which is the body's master control center. The brain receives, integrates, stores, retrieves, and issues information. It coordinates responses to information by adjusting activities throughout the body. Like the spinal cord, the brain is protected by bones and membranes.

Reflect on Figures 35.4*a* and 35.9. In the vertebrate embryo, the forebrain, midbrain, and hindbrain form from three successive regions of the neural tube. The nervous tissue that evolved first in all three regions is called the **brain stem**. The brain stem is still identifiable in an adult's brain, and it still contains many simple, basic reflex centers. Over evolutionary time, expanded layers of gray matter developed from the brain stem. Biologists have correlated those more recent additions with increasing reliance on three major sensory organs —the nose, ears, and eyes. The forebrain and possibly parts of the hindbrain region called the cerebellum have the newest additions of gray matter.

Hindbrain

The medulla oblongata, cerebellum, and pons are all components of the hindbrain. The **medulla oblongata** contains reflex centers for a number of vital tasks, such as respiration and blood circulation. It also coordinates motor responses with certain complex reflexes, such as coughing. In addition, it influences other brain regions that help you sleep or wake up.

The **cerebellum** integrates sensory signals from the eyes, ears, and muscle spindles with motor signals from the forebrain. It helps control motor skills. More recent expansions of the cerebellum in humans may contribute to language and some other forms of mental dexterity.

Bands of many axons extend from both sides of the cerebellum to the **pons** (meaning "bridge"). Although lodged inside the brain stem, the pons is a major traffic center for information passing between the cerebellum and the higher integrating centers of the forebrain.

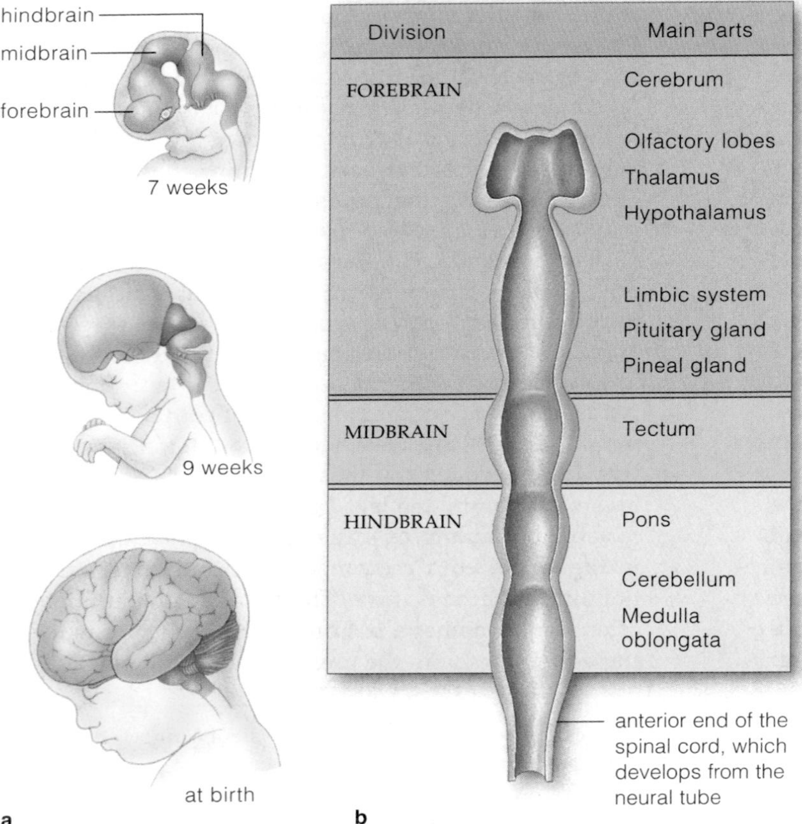

hindbrain
midbrain
forebrain
7 weeks
9 weeks
at birth
a
Division
Main Parts
FOREBRAIN
Cerebrum
Olfactory lobes
Thalamus
Hypothalamus
Limbic system
Pituitary gland
Pineal gland
MIDBRAIN
Tectum
HINDBRAIN
Pons
Cerebellum
Medulla oblongata
anterior end of the spinal cord, which develops from the neural tube
b

Figure 35.9 (**a**) Sketches showing how the anterior end of the neural tube of a human embryo develops into a brain with three major subdivisions: the hindbrain, midbrain, and forebrain. (**b**) Overview of the components of the brain's three subdivisions, correlated with the neural tube at an early stage of development.

Midbrain

The midbrain coordinates reflex responses to sights and sounds. Its roof is called the **tectum**, which is a more recent layering of gray matter. In fishes and amphibians, the tectum is a center for coordinating nearly all sensory input and for initiating motor responses. To demonstrate its importance, someone might surgically deprive a frog of its highest brain center and yet leave the tectum intact. That frog will still be able to do almost everything frogs normally do.

In most vertebrates (but not mammals), the midbrain also includes a pair of brain centers called optic lobes, which deal with sensory input from eyes. Mammalian eyes deserted the tectum, so to speak. The eyes formed important functional connections with integrating centers in the forebrain. The tectum of mammals became a reflex center that quickly relays sensory signals to those higher integrating centers.

Forebrain

For much of vertebrate history, chemical odors that slowly diffused through aquatic habitats were the most important clues to survival. An olfactory lobe that primarily dealt with odors from predators, prey, and potential mates was a significant forebrain structure (Figure 35.9*b*). So were a pair of nervous tissue outgrowths from the brain

Further reading: Student Guide to InfoTrac on web site –

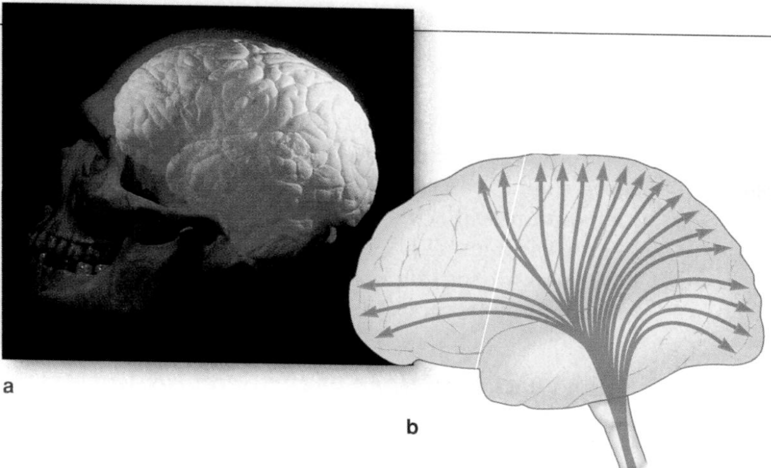
a
b

Figure 35.10 (**a**) Model showing the surface of the cerebral cortex of the human brain. (**b**) Communication pathways of the reticular formation, an evolutionarily ancient, diffuse network of neurons. The formation extends from the anterior end of the spinal cord on up to the highest integrative centers of the cerebral cortex.

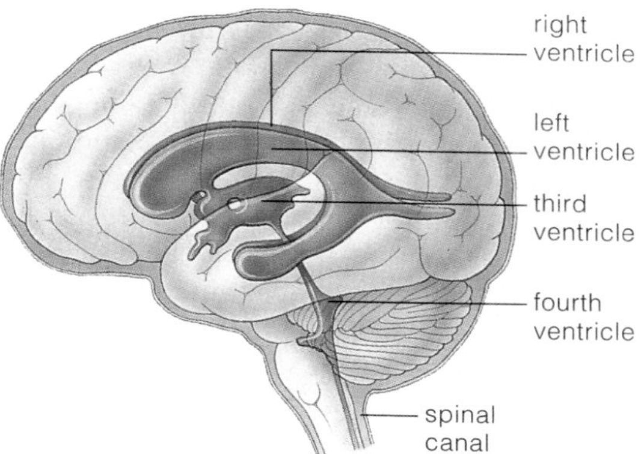
right ventricle
left ventricle
third ventricle
fourth ventricle
spinal canal

Figure 35.11 Cerebrospinal fluid (*blue*) in the human brain. This extracellular fluid surrounds and cushions the spinal cord and the brain. It also fills four interconnected cavities (cerebral ventricles) within the brain and the spinal cord's central canal.

stem, where olfactory input and responses to it became integrated. In time the outgrowths expanded greatly, especially after certain vertebrates invaded the land. We now call them the two hemispheres of the **cerebrum**.

Another forebrain region, the **thalamus**, evolved as a coordinating center for sensory input and as a relay station for input to the cerebrum. Below the thalamus, the **hypothalamus** evolved into the premier center for homeostatic control over the internal environment. The hypothalamus became central to behaviors related to internal organ activities, such as thirst, hunger, and sex, and to emotional expression, such as sweating with fear.

The Reticular Formation

By now, you might be thinking that the brain is tidily subdivided into three main regions. This is not the case. An evolutionarily ancient network of interneurons still extends from the uppermost part of the spinal cord, on through the brain stem, and on into higher integrative centers of the cerebral cortex (Figure 35.10). This major network of interneurons is the **reticular formation**. It persists as a low-level pathway to motor centers of the medulla oblongata and spinal cord. It also can activate centers in the cerebral cortex and thereby help govern activities of the nervous system as a whole.

Brain Cavities and Canals

The hollow neural tube that first develops in vertebrate embryos persists in adults, as a continuous system of fluid-filled cavities and canals. The clear extracellular fluid within the system is called **cerebrospinal fluid**. It cushions and protects tissues of the brain and spinal cord from sudden, jarring movements (Figure 35.11).

A mechanism called the **blood–brain barrier** protects the brain and spinal cord by exerting some control over which solutes enter the cerebrospinal fluid. No other portion of extracellular fluid has solute concentrations maintained within such narrow limits. Even normal changes in the composition of extracellular fluid that accompany, say, eating or exercising are opposed here. Why? Some of the hormones and amino acids that are dissolved in blood can disrupt the function of neurons. Also, changes in the levels of certain ions, such as K^+, can skew the threshold for action potentials.

The barrier operates at the plasma membrane of certain cells. The cells, which are interconnected, make up the wall of blood capillaries that service the brain. In most brain regions, continuous tight junctions fuse the abutting walls of cells, so water-soluble substances must move *through* cells to reach the brain. Membrane transport proteins let glucose and other vital nutrients as well as some ions move into and out of the cells but bar some toxins and metabolic wastes, such as urea. The barrier does not keep out fat-soluble substances, such as oxygen, carbon dioxide, anesthetics, alcohol, caffeine, and nicotine. It is nonexistent around the brain stem's vomiting center and the hypothalamus (guess why).

The vertebrate brain develops from a hollow neural tube, which persists in adults as a system of cavities and canals filled with cerebrospinal fluid. The fluid cushions nervous tissue from sudden, jarring movements. The nervous tissue is subdivided into the hindbrain, forebrain, and midbrain.

The brain stem, the nervous tissue first to evolve in all three regions, affords reflex control over basic functions necessary for survival. The highest integrative centers are in the forebrain, especially the cerebral hemispheres.

35.5 A CLOSER LOOK AT THE HUMAN CEREBRUM

How Are Cerebral Hemispheres Organized?

Study any nervous system or even the best computers and you realize the human brain is the most complex integration center of all. It only weighs 1,300 grams (3 pounds) in a person of average size, and more than half of that mass consists of neuroglial cells. But remember, a human brain also has at least *100 billion* interneurons!

The human cerebrum, housed in a chamber of hard skull bones, resembles the much-folded nut in a walnut shell (Figure 35.12). Interneurons here contribute most to your "humanness." A longitudinal fissure divides the cerebrum into left and right **cerebral hemispheres**. Each has a thin outer layer of gray matter, the **cerebral cortex**. Below this is the white matter (axons), within which are patches of gray matter called basal nuclei.

Each cerebral hemisphere receives, processes, and coordinates responses to sensory input mainly from the opposite side of the body. For instance, pressure signals from the right arm travel to the left hemisphere. The left hemisphere deals primarily with speech, analytical skills, and mathematics. In most people it dominates the right hemisphere, which deals more with visual–spatial relationships, music, and other creative enterprises. A transverse band of nerve tracts—the corpus callosum—carries signals back and forth between the hemispheres and coordinates their functioning.

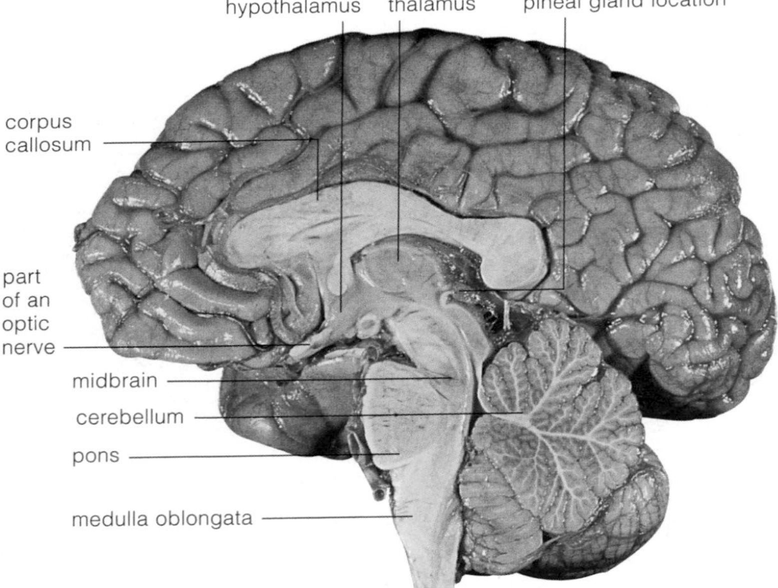
hypothalamus
thalamus
pineal gland location
corpus callosum
part of an optic nerve
midbrain
cerebellum
pons
medulla oblongata

Figure 35.12 Right cerebral hemisphere of a human brain, sagittal view. Not visible is the reticular formation, which extends between the upper spinal cord and cerebrum. Each hemisphere has a layer of gray matter, the cerebral cortex, about 2–4 millimeters (1/8 inch) thick. Interneuron cell bodies and dendrites, unmyelinated axons, neuroglia, and blood vessels make up the gray matter.

Each hemisphere is divided into four tissue regions: the frontal, occipital, temporal, and parietal lobes, which receive and process different signals. EEGs and PET scans can reveal activity in each lobe. Figure 35.13 gives a few examples. (*EEG*, short for electroencephalogram, is simply a recording of summed electrical activity in a brain region. Section 2.2 describes PET scans.)

Functional Divisions of the Cerebral Cortex

What humans comprehend, communicate, remember, and voluntarily act upon arises in the cerebral cortex; it governs our conscious behavior. We functionally divide it into *motor* areas (control of voluntary motor activity), *sensory* areas (perception of the meaning of sensations), and *association* areas (integration of information which precedes conscious action). These areas do not function alone. Consciousness arises by interactions throughout the cortex. As examples, read Section 35.6. Also read "Music of the Hemispheres," an article in the October 1996 issue of *Discover* magazine that describes circuits dealing with melodies and language.

MOTOR AREAS In the frontal lobe of each hemisphere, the entire body is spatially mapped out in the primary motor cortex. This area controls coordinated movements of skeletal muscles. Thumb, finger, and tongue muscles get much of the area's attention. This gives you an idea of how much control is required for voluntary hand movements and verbal expression (Figure 35.14).

The frontal lobe also contains the premotor cortex, Broca's area, and the frontal eye field. Learned patterns of motor skills are the domain of the premotor cortex. Play a piano concerto, use a computer keyboard, dribble a basketball—such repetitive movements are evidence that this region is coordinating the simultaneous and sequential movements of many muscle groups. Broca's area (usually in the left hemisphere) and a corresponding area in the right hemisphere control the tongue, throat, and lip muscles used in speech. It becomes active when we are about to speak, even when we plan voluntary motor activities other than speech. The frontal eye field, above Broca's area, controls voluntary eye movements.

SENSORY AREAS Sensory areas occur in different parts of the cortex. In the parietal lobe, the body is spatially mapped out in the primary somatosensory cortex. This area is the main receiving center for sensory input from the skin and joints (Section 36.2). The parietal lobe also has a primary cortical area that deals with perception of taste. The primary visual cortex, at the back of the occipital lobe, receives sensory signals from the eyes (Sections 35.6 and 36.9). Perception of sounds and odors arises in primary cortical areas in each temporal lobe.

Further reading: Student Guide to InfoTrac on web site

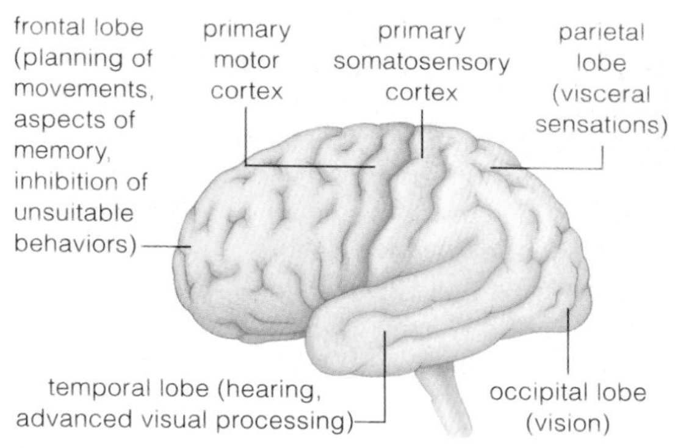
frontal lobe (planning of movements, aspects of memory, inhibition of unsuitable behaviors)
primary motor cortex
primary somatosensory cortex
parietal lobe (visceral sensations)
temporal lobe (hearing, advanced visual processing)
occipital lobe (vision)
a

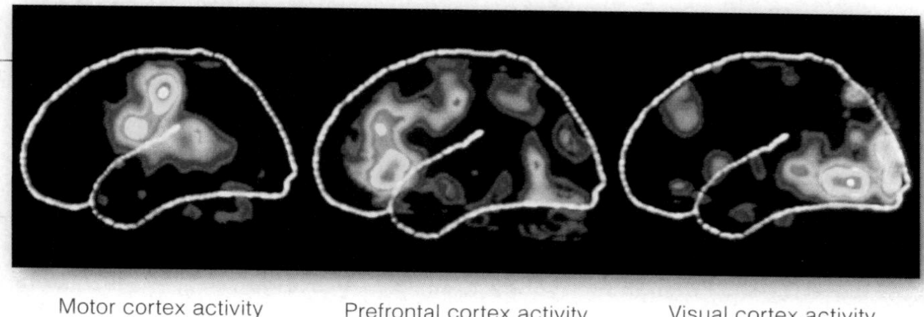
b
Motor cortex activity when speaking
Prefrontal cortex activity when generating words
Visual cortex activity when observing words

Figure 35.13 (**a**) Primary receiving and integrating centers of the human cerebral cortex. Primary cortical areas receive signals from receptors on the body's periphery. Association areas coordinate and process sensory input from different receptors. (**b**) Three PET scans identifying which brain regions were active when a person performed three specific tasks: speaking, generating words, and observing words.

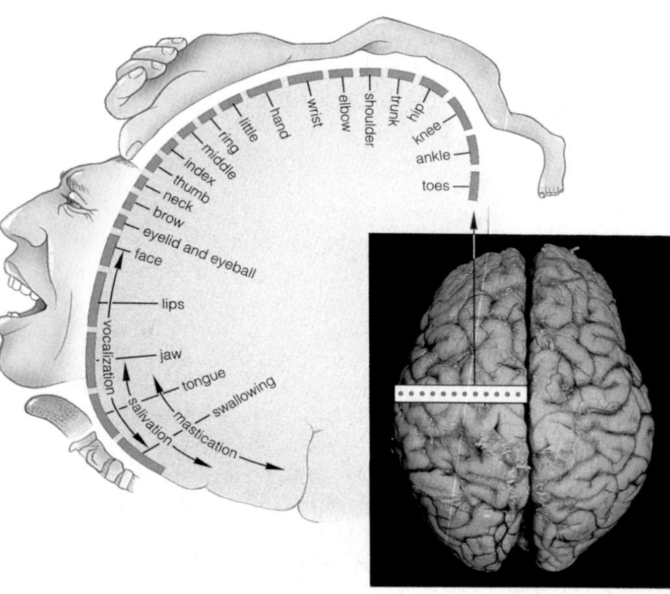
hip
trunk
shoulder
elbow
wrist
hand
little
ring
middle
index
thumb
neck
brow
eyelid and eyeball
face
knee
ankle
toes
lips
jaw
tongue
swallowing
vocalization
salivation
mastication

Figure 35.14 Diagram of a slice through the primary motor cortex of the left cerebral hemisphere of the human brain. Sizes of different body parts draped over the diagram are distorted to show which receive the most precise control. The photograph is a dorsal view of both cerebral hemispheres.

ASSOCIATION AREAS Association areas occupy all parts of the cortex except primary motor and sensory regions. Each integrates, analyzes, and responds to many inputs. For instance, the visual association area surrounds the primary visual cortex. It helps us recognize something we see by comparing it with visual memories. The most complex association area, the prefrontal cortex, is the basis of complex learning, intellect, and personality. Without it, we would be incapable of abstract thought, judgment, planning, and concern for others.

Connections With the Limbic System

Encircling the upper brain stem is the **limbic system**, which controls emotions and has roles in memory. It includes the hypothalamus, amygdala, cingulate gyrus, hippocampus, and parts of the thalamus (Figure 35.15). The hypothalamus serves as a clearinghouse for emotions and visceral activity. The amygdala is important for emotional stability, interpretation of social cues and, with the hippocampus, conversion of stimuli to long-term memory. The cingulate ("belt-shaped") gyrus is central to our will to act.

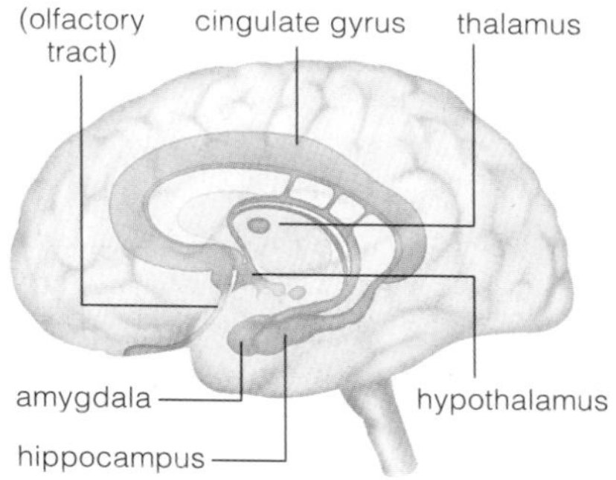
(olfactory tract)
cingulate gyrus
thalamus
amygdala
hypothalamus
hippocampus

Figure 35.15 Key structures of the limbic system.

The limbic system is distantly related to olfactory lobes and still deals with the sense of smell. That is one reason why you may feel warm and fuzzy when your brain recalls the cologne of a special person.

By its connections with the prefrontal cortex and other brain centers, the limbic system correlates organ activities with self-gratifying behavior, such as eating and sex. For example, it can make you feel your stomach or heart is on fire with passion or indigestion. Signals based on reasoning in the cerebral cortex often override or dampen rage, hatred, or other "gut reactions." Hence the limbic system is called our emotional–visceral brain.

The cerebrum of humans and other complex vertebrates is divided into two hemispheres. Each half receives, processes, and coordinates responses to sensory input from primarily the opposite side of the body. The corpus callosum carries signals between the two halves.

The left hemisphere deals mainly with speech, analytical skills, and mathematics. It usually dominates the right hemisphere, which deals more with creative activity, such as visual–spatial relationships and music.

The cerebral cortex (each hemisphere's outermost layer of gray matter) contains motor, sensory, and association areas. Interactions among the areas govern conscious behavior. The cerebral cortex also interacts with the limbic system, which governs emotions and contributes to memory.

35.6 SPERRY'S SPLIT-BRAIN EXPERIMENTS

Some time ago, the neurosurgeon Roger Sperry and his colleagues demonstrated some intriguing differences in perception between the two cerebral hemispheres of epileptics. Severe *epilepsy* is characterized by seizures, sometimes as often as every half hour. The seizures are analogous to an electrical storm in the brain. Sperry asked: Would *cutting* the corpus callosum of epileptics confine the electrical storm to one hemisphere, leaving at least the other hemisphere to function normally? Earlier studies of laboratory animals and humans whose corpus callosum had been damaged suggested this might be so.

Sperry performed the surgery on some patients. The electrical storms did subside in frequency and intensity. Cutting the neural bridge ended what must have been a positive feedback loop of ever intensifying electrical disturbances between the two hemispheres. The "split-brain" patients were able to lead what seemed, on the surface, entirely normal lives.

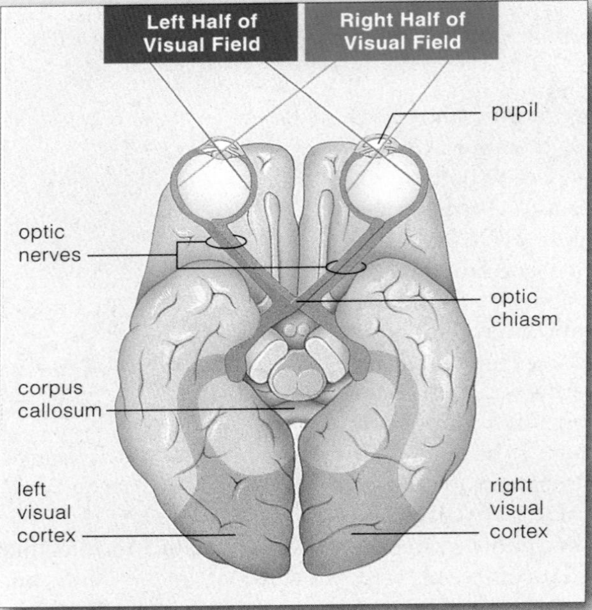
Left Half of Visual Field
Right Half of Visual Field
pupil
optic nerves
optic chiasm
corpus callosum
left visual cortex
right visual cortex

Figure 35.16 Pathway by which sensory input about visual stimuli reaches the visual cortex of the human brain.

Each eye gathers visual information at the retina, a layer of densely packed photoreceptors at the back of the eyeball (Section 36.9). Light from the *left* half of the visual field strikes receptors on the right side of both retinas. Parts of two optic nerves carry signals from the receptors to the right cerebral hemisphere. Light from the *right* half of the visual field strikes receptors on the left side of both retinas. Parts of the optic nerves carry signals from them to the left hemisphere.

But then Sperry devised some elegant experiments to test whether their conscious experience was indeed "normal." Given that the corpus callosum contains 200 million axons, surely *something* was different. Something was. "The surgery," Sperry later reported, "left these people with two separate minds, that is, two spheres of consciousness. What is experienced in the right hemisphere seems to be entirely outside the realm of awareness of the left."

Sperry presented the two hemispheres of split-brain patients with two different portions of the same visual stimulus. Researchers knew at the time that the visual connections to and from one hemisphere are mainly concerned with the opposite half of the visual field, as described in Figure 35.16. Sperry projected words—say, COWBOY—onto a screen so that COW fell in the left half of the visual field, and BOY fell in the right (Figure 35.17).

The subjects of this experiment reported *seeing* the word BOY. The left hemisphere, which controls language, perceived only the letters BOY. However, when asked to write the perceived word with the left hand—a hand that was deliberately blocked from a subject's view—the subject wrote COW. The right hemisphere "knew" the other half of the word (COW) and had directed the left hand's motor response. But it could not tell the left hemisphere what was going on because of the severed corpus callosum. The subject knew that a word was being written, but could not say what it was!

Thus Sperry showed that signals across the corpus callosum coordinate the functioning of the two cerebral hemispheres, each of which had responded to visual signals from the opposite side of the body.

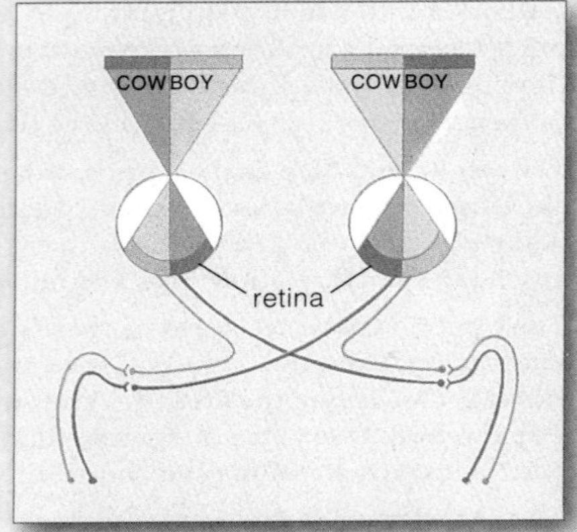
COWBOY
COWBOY
retina

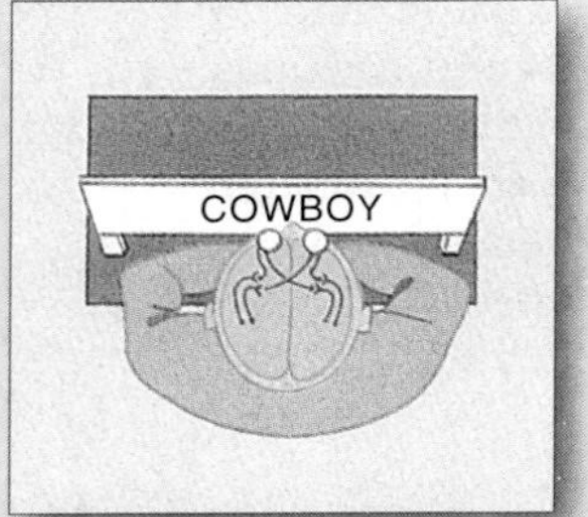
COWBOY

Cow
COWBOY

Figure 35.17 Response of a split-brain patient to different portions of a visual stimulus.

Further reading: Student Guide to InfoTrac on web site →

35.7 MEMORY

Memory refers to a brain's capacity to store and retrieve information about past sensory experience. Without it, learning and adaptive modifications of behavior would be impossible. Information is stored in stages. In *short-term* storage, neural excitation lasts a few seconds to a few hours. This stage is limited to a few bits of sensory information—numbers, words of a sentence, and so on. In *long-term* storage, seemingly unlimited amounts of information get tucked away more or less permanently, as shown in Figure 35.18.

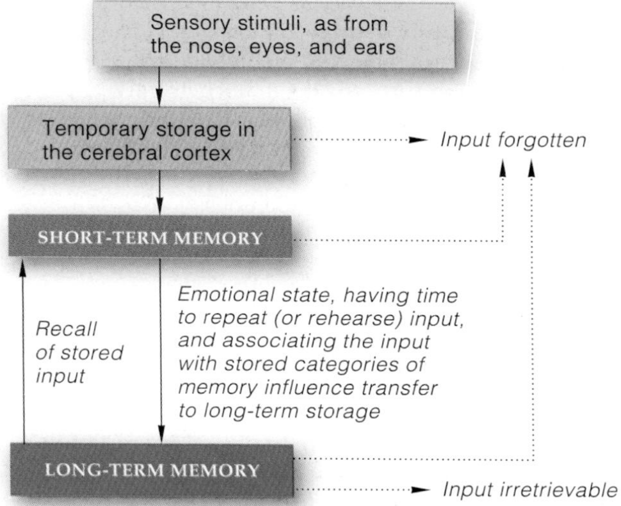

Sensory stimuli, as from the nose, eyes, and ears
Temporary storage in the cerebral cortex
Input forgotten
SHORT-TERM MEMORY
Recall of stored input
Emotional state, having time to repeat (or rehearse) input, and associating the input with stored categories of memory influence transfer to long-term storage
LONG-TERM MEMORY
Input irretrievable

Figure 35.18 Stages of memory processing, starting with the temporary storage of sensory inputs in the cerebral cortex.

Not all of the sensory input bombarding a cerebral cortex ends up in memory storage. Only some is chosen for transfer to brain structures involved in short-term memory. Information in these temporary holding bins is processed for relevance, so to speak. If irrelevant, it is forgotten. If not, it is consolidated with the banks of information in long-term storage structures.

The human brain processes facts separately from skills. Dates, names, faces, words, odors, and other bits of explicit information are *facts*, soon forgotten or filed away in long-term storage along with the circumstances in which the facts were learned. That is why you might associate, say, the smell of sun-warmed watermelon with a special picnic held long ago at the beach. By contrast, *skills* are gained by practicing specific motor activities. A skill such as slam-dunking a basketball or playing a violin concerto is best recalled by actually performing it, rather than by recalling the circumstances in which the skill was initially learned.

Separate memory circuits handle different kinds of input. A circuit leading to fact memory (Figure 35.19*a*) starts with inputs at the sensory cortex that flow to the two structures of the limbic system, the amygdala and hippocampus. The amygdala acts as the gatekeeper; it connects the sensory cortex with parts of the thalamus and hypothalamus that govern emotional states. The hippocampus mediates learning and spatial relations. Information flows on to the prefrontal cortex. There, multiple banks of fact memories are retrieved and used to stimulate or inhibit other parts of the brain. The new input also flows to basal ganglia, structures that send it back to the cortex in a feedback loop. The loop reinforces the input until it is consolidated in long-term storage.

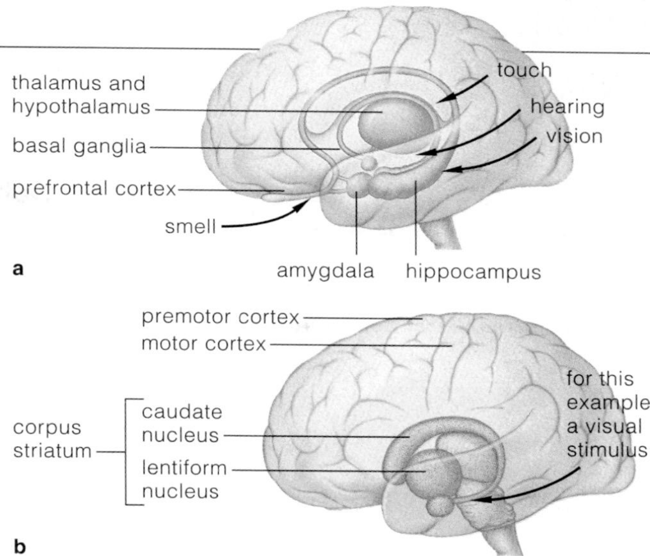

thalamus and hypothalamus
basal ganglia
prefrontal cortex
smell
touch
hearing
vision
amygdala
hippocampus
a
premotor cortex
motor cortex
corpus striatum
caudate nucleus
lentiform nucleus
for this example, a visual stimulus
b

Figure 35.19 Diagrams of possible circuits involved in (**a**) fact memory and (**b**) skill memory.

Skill memory also starts at the sensory cortex, but this circuit routes sensory input to the corpus striatum, which promotes motor responses (Figure 35.19*b*). Motor skills involve muscle conditioning. Thus, as you might suspect, the circuit extends to the cerebellum, the brain region that coordinates motor activity.

Amnesia is a loss of memory. Its severity depends on whether the hippocampus, the amygdala, or both are damaged, as by a severe head blow. Amnesia does not affect the capacity to learn new skills. Basal ganglia are destroyed and learning ability is lost during *Parkinson's disease*, yet skill memory is retained. *Alzheimer's disease* usually starts late in life and involves structural changes in the cerebral cortex and hippocampus. Often, affected people are able to remember long-standing facts, such as their Social Security number. But they have trouble remembering what has just happened to them. In time they become confused, depressed, and incoherent.

Memory, the storage and retrieval of sensory information, results from circuits between the cerebral cortex and parts of the limbic system, thalamus, and hypothalamus. Sensory input is processed through short-term and long-term storage.

35.8 STATES OF CONSCIOUSNESS

The spectrum of **consciousness** includes sleeping and aroused states, during which neural chattering shows up as wavelike patterns in EEGs. Specifically, EEGs are electrical recordings of the frequency and strength of membrane potentials at the surface of the brain. Figure 35.20 gives examples. Also, PET scans show the precise location of brain activity while it is proceeding.

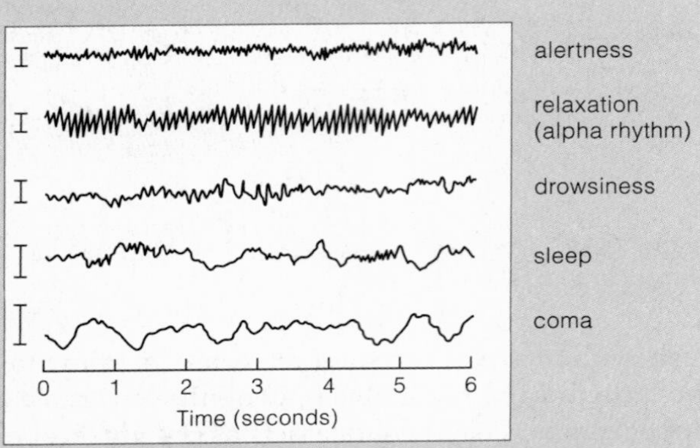
alertness
relaxation (alpha rhythm)
drowsiness
sleep
coma
0
1
2
3
4
5
6
Time (seconds)

Figure 35.20 EEG patterns. Vertical bars mark a fifty-microvolt range of electrical responses (EEG waves). Irregular horizontal graph lines show waves recorded in "trains" of one after the other, about ten per second.

The *alpha rhythm* is the prominent wave pattern for someone meditating (relaxed, with eyes closed). Wave trains become larger, slower, and more erratic during the transition to sleep. Low sensory input and a mind that is more or less idling invite this *slow-wave sleep* pattern. Awaken subjects from slow-wave sleep, and they usually say they weren't dreaming; they often seemed to be mulling over recent, ordinary events.

REM sleep punctuates slow-wave sleep. *R*apid *E*ye *M*ovements accompany this pattern (eyes jerk under closed lids), as do irregular breathing, faster heartbeat, and twitching fingers. Most people awakened from REM sleep say they were experiencing vivid dreams. A shift to low-amplitude, higher frequency wave trains marks the transition from sleep or from deep relaxation to alert wakefulness, *EEG arousal*. Then, someone consciously focuses on external stimuli or one's own thoughts.

Part of the reticular formation promotes chemical changes that influence whether you stay awake or fall asleep. Serotonin, a neurotransmitter released from one of its sleep centers, inhibits other neurons that arouse the brain and maintain wakefulness. At high levels, serotonin triggers drowsiness and sleep. Substances released from another brain center inhibit serotonin's effects and bring about wakefulness.

The spectrum of consciousness, which includes sleeping and states of arousal, is influenced by the reticular formation.

Focus on Health

35.9 DRUGGING THE BRAIN

Broadly speaking, a **drug** is a substance introduced into the body to provoke a specific physiological response. Some drugs help a person cope with illness or emotional stress. Others artificially fan pleasure associated with sex and other self-gratifying behaviors.

Many drugs are habit-forming. That is, even if the body is able to function well without them, a person continues to use drugs for the real or imagined relief they afford. Often the body develops tolerance of such drugs; it takes larger or more frequent doses to produce the same effect. Habituation and tolerance are signs of **drug addiction**, a chemical dependence on a drug (Table 35.1). *In cases of addiction, a drug has assumed an "essential" biochemical role in the body.* Abruptly deprive addicts of the drug, and they go through physical pain and mental anguish. The entire body goes through an episode of biochemical upheaval. Stimulants, depressants, hypnotics, narcotic analgesics, hallucinogens, and psychedelics have such effects.

STIMULANTS Stimulants increase alertness and body activity—then cause depression. *Caffeine* in coffee, tea, chocolate, and many soft drinks is a common stimulant. Low doses arouse the cerebral cortex first and increase alertness. Higher doses act at the medulla oblongata to disrupt motor coordination and mental coherence. Another stimulant, the *nicotine* in tobacco, has potent effects on the nervous system. It mimics acetylcholine and directly stimulates a variety of sensory receptors. In the short term, nicotine results in water retention, irritability, high blood pressure, and gastric upsets.

Possibly 2 million Americans are *cocaine* abusers. This stimulant gives a rush of pleasure by blocking reabsorption of norepinephrine, dopamine, and other neurotransmitters. Because the molecules accumulate in synaptic clefts, they incessantly stimulate postsynaptic cells for an extended period. Heart rate, blood pressure, and sexual appetite increase. In time the molecules diffuse away. However, neurons cannot synthesize replacements fast enough to

Table 35.1 Warning Signs of Drug Addiction*

1. Tolerance—it takes increasing amounts of the drug to produce the same effect.
2. Habituation—it takes continued drug use over time to maintain self-perception of functioning normally.
3. Inability to stop or curtail use of the drug, even if there is persistent desire to do so.
4. Concealment—not wanting others to know of the drug use.
5. Extreme or dangerous behavior to get and use a drug, as by stealing, asking more than one doctor for prescriptions, or jeopardizing employment by drug use at work.
6. Deterioration of professional and personal relationships.
7. Anger and defensive behavior when someone suggests there may be a problem.
8. Preference of drug use over previously customary activities.

* Three or more of these signs may be cause for concern.

Further reading: Student Guide to InfoTrac on web site —

counter the loss, and the sense of pleasure evaporates as hypersensitized postsynaptic cells demand stimulation. After prolonged, heavy use of cocaine, "pleasure" is no longer possible. Addicts become anxious and depressed. They lose weight and cannot sleep properly. The immune system weakens, and heart abnormalities begin.

Granular cocaine is inhaled (snorted). Abusers burn crack cocaine and inhale the smoke. As suggested at the start of this chapter, crack is extremely addictive. Its highs are higher, its crashes are more devastating, and the social and economic tolls are extreme.

Amphetamines induce massive release of dopamine and norepinephrine. Addicts often smoke, snort, inject, or swallow a form called *crank*. The effects last two to twelve hours and range from euphoria and sexual arousal to a pounding heart, dry mouth, tremors, agitation, inability to sleep, and paranoia. Crank also kills the appetite; initially it attracted women who want to lose weight. In time, the production of dopamine and norepinephrine dwindles as the brain depends more on artificial stimulation. Long-term abuse leads to malnutrition, psychosis, depression, memory loss, and damage to the brain, heart, lungs, and liver. Crank is made cheaply in home kitchens (dubbed Beavis and Butthead labs) from such caustic ingredients as drain cleaners. Abuse of this "poor man's cocaine" is epidemic, especially in the American West and Midwest.

DEPRESSANTS, HYPNOTICS These drugs lower activity in nerves and parts of the brain. Some act at synapses in the reticular formation and thalamus. Depending on the dosage, physiological state, and emotional state, responses range from emotional relief through drowsiness, sleep, anesthesia, coma, and on to death. Low doses have the most effect on inhibitory synapses. A person feels excited or euphoric at first. Increased doses suppress excitatory synapses as well, and they lead to depression. Depressants and hypnotics are addictive. One amplifies another, as when alcohol plus barbiturates heightens the depression.

Alcohol, or ethyl alcohol, acts directly on the plasma membrane to alter cell function. Like nicotine and cocaine, it is lipid soluble and easily crosses the blood–brain barrier to exert rapid effects. Some people mistakenly think of it as a harmless stimulant because of the initial "high" that it produces. However, alcohol is one of the most powerful psychoactive drugs and a factor in a great many deaths. In the short term, small amounts cause disorientation, uncoordinated motor functions, and diminished judgment. Long-term addiction can lead to *cirrhosis*. Then, connective tissue permanently replaces damaged liver cells, and the self-regenerating capacity of liver tissue slowly diminishes. In time, the outcome is chronic liver failure and death.

ANALGESICS When severe stress leads to physical or emotional pain, the brain produces natural *analgesics*, or pain relievers. Endorphins and enkephalins are two examples. They act on many parts of the nervous system, including brain centers for pain and emotions. *Codeine*, *heroin*, and other narcotic analgesics sedate the body and relieve pain. They also are among the most addictive substances known. Deprivation after massive doses of heroin results in hyperactivity and anxiety, fever, chills, violent vomiting, cramping, and diarrhea.

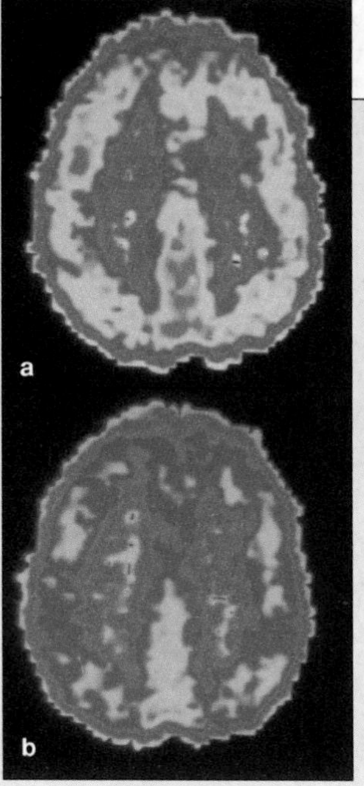

a
b

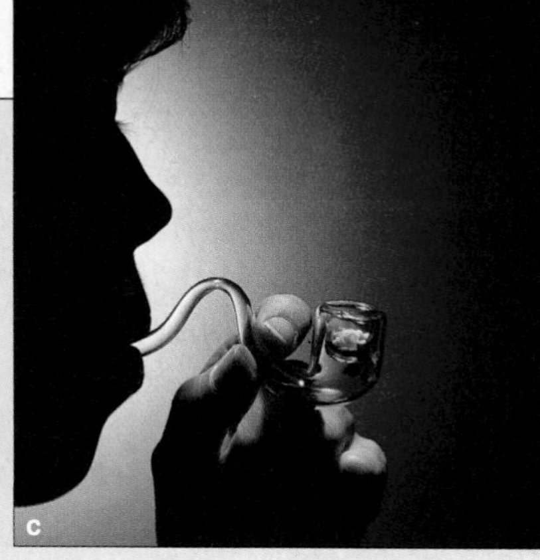

c

Figure 35.21 (**a**) Normal brain activity revealed by a PET scan. (**b**) PET scan of a comparable section that reveals cocaine's effect. *Red* indicates greatest activity; and *yellow*, *green*, and *blue* indicate increasingly inhibited activity. (**c**) In less than eight seconds, smoking crack puts cocaine in the brain.

PSYCHEDELICS, HALLUCINOGENS These drugs skew sensory perception by interfering with the action of acetylcholine, norepinephrine, or serotonin. *LSD* (lysergic acid diethylamide) alters serotonin's roles in inducing sleep, controlling body temperature, and mediating sensory perception. Even in small doses, LSD warps perceptions. For example, some users "perceived" they could fly and "flew" off buildings.

Marijuana, another hallucinogen, is made from crushed leaves, flowers, and stems of the plant *Cannabis*. In low doses it is like a depressant. It slows down but does not impair motor activity; it relaxes the body and elicits mild euphoria. It also causes disorientation, anxiety bordering on panic, delusions, and hallucinations. Like alcohol, it affects the performance of complex tasks, such as driving a car. In one study, pilots showed a marked deterioration in instrument-flying ability for more than two hours after smoking marijuana. In time, marijuana smoking impairs the immune system and mental functions.

Each of us possesses a body of great complexity. Its architecture, its functioning are legacies of millions of years of evolution. Its nervous system is unparalleled in the living world. One of its most astonishing products is language, the encoding of shared experiences of groups of individuals in time and space. Through the evolution of our nervous system, the sense of history was born, and the sense of destiny. Through this system we can ask how we have come to be what we are and where we are headed from here. Perhaps the sorriest consequence of drug abuse is its implicit denial of this legacy—the denial of self when we cease to ask, and cease to care.

Focus on Science

35.10 THE NOT-QUITE-COMPLETE TEEN BRAIN

All these years, teenagers have had a bad rap. As a group, they are known for wild mood swings, insolence, rages, narcissism, a breathtaking lack of concentration in school, defiance of authority in general and parents in particular, body-piercing and other acts of self-mutilation, and self-endangering impulsiveness.

Are they being swept away by a predictable surge of hormones, by a molecular call for sexual awakening and pimples? Are they experimenting with how best to assert their independence? As the saying goes, Why can't they just act like adults?

MY LIMBIC SYSTEM MADE ME DO IT Conventional wisdom was that the brain is fully developed before puberty ends, with all 100 billion neurons hardwired. As it turns out, all of the necessary connections aren't completed before the early twenties!

Like arms, legs, and other body parts, different parts of the brain are on different developmental schedules. For example, the prefrontal cortex is one of the last parts to be completed. As you know, this major coordinating center keeps tabs on other brain regions, including the limbic system. The prefrontal cortex mediates decision making, sifts through and makes sense of ambiguous information, and reinforces or dampens raw emotions generated in the limbic system. However, even while its circuitry is still under construction, the limbic system is rapidly developing! In effect, a teenager is like a car with a revved-up engine and no brakepads.

How would you go about testing the hypothesis that teenagers do not yet have the neural wiring necessary to exercise sound judgment and control emotions? Deborah Yurgelum–Todd, a neuropsychologist, and Abigail Baird, a graduate student, started by showing standardized photographs of human faces that were expressing fear. (Fear is a primal emotion, a "gut reaction.") They asked fifteen adults and fifteen teenagers to state which emotion the photographs expressed. All of the adults identified the expression correctly. All but four teenagers gave one or more wrong answers; they said the facial expressions were conveying anger or discomfort.

Then the researchers used *MRI* (magnetic resonance imaging) to compare adult and teenage brains. MRI is a method of forming images of organ structure or activity every few seconds. Yurgelum-Todd and Baird wanted to see which parts of the brain were being activated in response to the photographs. In the adult brain, an image of fear elicited activity in the limbic system *and* in the prefrontal cortex. MRIs of the teenage brain also revealed a strong limbic system response to the same image—but very little activity in the prefrontal cortex.

On the basis of such results, Yurgelum-Todd put forth a hypothesis: Teenagers are still learning to interpret the content of the social world, and their prefrontal cortex is not helping enough with the interpretations.

Since 1991, Jay Giedd and coworkers at the National Institutes of Health have been using MRI to map the brain structure of nearly a thousand children who range from three to eighteen years old. They discovered that interneurons of the prefrontal cortex begin sprouting rapidly to form new connections around ages nine or ten. By the time the individual is twelve or so, most of the connections are starting to wither. It is as if the brain undergoes a pruning process to remove synapses that turn out to be useless over the long haul.

In the meantime, teenagers might not be able to work their way through the unpruned forest of synapses with much proficiency. If this is so, then it might explain why they have trouble organizing several tasks or tracking several thoughts at once. It may be that they do not yet have the brainpower to quickly call up memories and emotions necessary for making such decisions.

A CHEMICAL BREW MADE ME DO IT Other factors are at work here. When the prefrontal cortex is still far from finished, neurotransmitters and the sex hormones are busy at work in the teenage brain. For example, as Giedd's team discovered, the amygdala swells at puberty as an outcome of a surge of testosterone. The swelling is more notable in boys, but it also happens in girls. (The female body uses estrogen as a testosterone precursor.) The amygdala, recall, helps control emotional states—especially anger and fear. Its sudden growth spurt may

Figure 35.22 (**a–c**) Teenagers busily demonstrating the primacy of a limbic system revved up by predictable surges of neurotransmitters and sex hormones.

Further reading: Student Guide to InfoTrac on web site

be why both sexes become more aggressive and irritable during the teen years.

As neurobiologist Sara Leibowitz points out, starting at puberty, most girls also find themselves gaining weight. This is expected; the female body must have a certain percentage of fat to mature sexually. Why does this happen? The hypothalamus steps up production and secretion of an appetite-stimulating hormone.

And what about teen zombies in the classroom? Mary Carskadon argues that they just need more sleep, 9–1/4 hours per night, to be specific. Why? Perhaps because most of the somatotropin and other hormones required for growth and sexual maturation are released while the body sleeps. The brain has a biological clock, in the form of melatonin secretions, that governs circadian rhythms; it dictates when you awaken and fall asleep. In one study, Carskadon's team asked students to fall asleep during the day. Many students must have been monumentally sleep deprived; they did so within three or four minutes.

Also, memory and learning are boosted during REM sleep. Then, chemical levels in brain centers are adjusted and short-term memory banks are emptied for the new day. Students deprived of REM sleep are depressed and irritable. They cannot retrieve memories rapidly, their judgment is compromised, and they do not perform well on tests designed to measure their reaction time. Sleep-deprived teens get the most C's and D's on tests. Well-rested teens get the most A's and B's.

As other examples, most adolescents experience a brief decline in serotonin levels, and impulsive behavior is associated with the decline. Also, without a finished prefrontal cortex to act as traffic cop, teenagers are far more open than adults to invading the brain's pleasure center. Interneurons in that center release dopamine, a neurotransmitter governing arousal and motivation. Novel behaviors, especially those with an element of risk or danger, stimulate the center. So do cocaine and some other drugs. Sneak out late for a rock concert? Sure. Snort cocaine? Why not? Screw around? Hah!

If you are not yet in your twenties, don't jump to the conclusion that, since your brain isn't ready, you don't have to challenge it (as by reading) or make choices (as in behaving responsibly or impulsively). How you decide to exercise your brain will shape your forming neural circuits, which in turn will profoundly influence your behavior and your success later in life.

35.10 www.brookscole.com/biology

35.11 SUMMARY

1. Nervous systems sense and interpret specific aspects of the environment and issue commands for responses to them. In nearly all animals, communication lines are composed of sensory neurons, interneurons, and motor neurons, which activate muscle and gland cells.

a. The simplest nervous systems are the nerve nets, such as those of cnidarians and other radial animals. Their meshwork of nerve cells forms reflex connections with contractile and sensory cells of the epithelium.

b. Most animals have a bilateral, cephalized nervous system, with a brain or ganglia at their anterior end. Such animals have cordlike nerves and tracts (bundled axons of sensory neurons, motor neurons, or both).

2. The vertebrate central nervous system consists of a brain and a spinal cord. The peripheral nervous system consists mainly of nerves that carry signals between all body regions and the spinal cord and brain.

a. Somatic nerves of the peripheral nervous system deal with skeletal muscles. Autonomic nerves deal with the heart, lungs, glands, and other soft internal organs.

b. Autonomic nerves are either parasympathetic or sympathetic. Parasympathetic nerves dominate when outside stimulation is low and tend to divert energy to basic housekeeping tasks. Overall, sympathetic nerves step up body activities during heightened awareness or danger. They govern the fight–flight response.

3. Reflex connections for limb movements and internal organ activity are made in the spinal cord. Many tracts in the spinal cord also carry signals between the brain and the peripheral nervous system.

4. During embryonic development, the brain develops by regional expansions of a hollow neural tube. The tube's anterior end becomes the brain stem, which is still discernible in an adult brain. The brain's most recent expansions of gray matter receive, store, compare, and use experiences to initiate novel action.

a. A neural tube's interior develops into a system of canals and cavities, filled with cerebrospinal fluid. A protective mechanism, the blood–brain barrier, exerts some control over which water-soluble substances enter cerebrospinal fluid, which must be maintained within very narrow limits for neurons to function properly.

b. The reticular formation, a mesh of interneurons extending from the top of the spinal cord to the cerebral cortex, is a low-level route by which signals travel to centers of motor activity.

5. The vertebrate brain has three main divisions:

a. The hindbrain includes the medulla oblongata, pons, and cerebellum. It contains reflex centers for vital functions and muscle coordination.

b. The midbrain coordinates and relays visual and auditory information to other brain regions.

c. The forebrain includes the cerebrum, thalamus, and hypothalamus. The cerebral hemispheres contain the highest centers for receiving, processing, storing, and responding to sensory information. The thalamus relays signals and helps coordinate motor responses. The hypothalamus is the main control center governing the internal environment and the viscera (soft internal organs). It is a clearinghouse for thirst, sexual activity, and other forms of behavior related to organ functions. It is part of the limbic system, which governs emotions and memory.

6. Table 35.2 summarizes the main components of the vertebrate nervous system, and it also lists their most important functions.

Table 35.2 Summary of Components and Functions of the Vertebrate Brain and Spinal Cord*

FOREBRAIN	Cerebrum	Localizes, processes sensory inputs; initiates, controls skeletal muscle activity. Governs memory, emotions, and abstract thought in the most complex vertebrates
	Olfactory lobe	Relays sensory input from the nose to olfactory centers of cerebrum
	Thalamus	Has relay stations for conducting sensory signals to and from cerebral cortex; also has role in memory
	Hypothalamus	With the pituitary gland, a homeostatic control center that adjusts volume, composition, and temperature of the internal environment. Governs behaviors affecting organ functions (e.g., hunger and thirst), and physical expression of emotion (e.g., sweating)
	Limbic system	A complex of structures that governs emotions; has roles in memory
	Pituitary gland (Chapter 37)	With hypothalamus, provides endocrine control over metabolism and over growth and development
	Pineal gland (Chapter 37)	Serves in the control of some circadian rhythms; also has role in mammalian reproductive physiology
MIDBRAIN	Tectum	In fishes and amphibians, its centers coordinate sensory input (as from optic lobes) and motor responses. In mammals, its (mainly) reflex centers rapidly relay sensory input to forebrain
HINDBRAIN	Pons	Serves as a "bridge" of tracts between cerebrum and cerebellum; its other tracts connect spinal cord with forebrain. Works with medulla oblongata to control the rate and depth of respiration
	Cerebellum	Coordinates motor activity for moving limbs and maintaining posture, and for spatial orientation
	Medulla oblongata	Its tracts relay signals between pons and spinal cord; its reflex centers help control heart rate, adjustments in blood vessel diameter, respiratory rate, vomiting, coughing, and other vital functions
SPINAL CORD		Has reflex connections for movements of limbs. Many of its tracts carry signals between the brain and the peripheral nervous system

* The reticular formation extends from the spinal cord to the cerebral cortex.

Review Questions

1. Generally describe the type of nervous systems found in radially symmetrical animals and in bilaterally symmetrical animals. *35.1*

2. What constitutes the central nervous system? The peripheral nervous system? *35.2*

3. Distinguish between the following:
   a. central and peripheral nervous system *35.2*
   b. cranial and spinal nerves *35.3*
   c. somatic and autonomic nerves *35.3*
   d. parasympathetic and sympathetic nerves *35.3*

4. Distinguish between:
   a. white matter and gray matter *35.2*
   b. nerve and tract *35.2*

5. Describe the structure and function of the spinal cord. *35.3*

6. Label the major parts of the human brain in the following photograph: *35.4, 35.5*

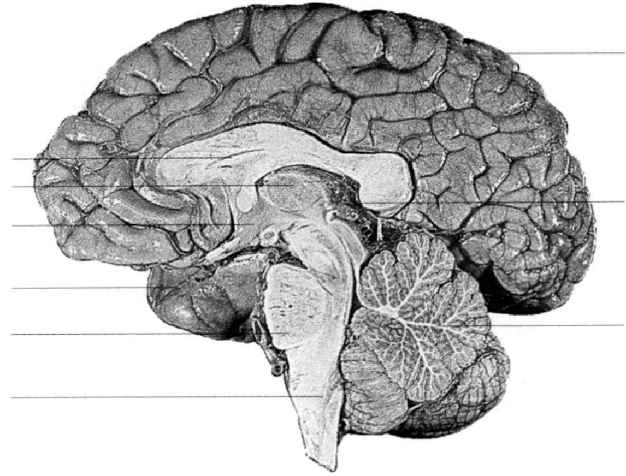

7. Explain the difference between the reticular formation and the limbic system in terms of their component parts and main functions. *35.4, 35.5*

8. Define cerebrospinal fluid and the blood–brain barrier. *35.4*

9. Briefly describe one of the brain centers of the cerebral cortex; mention the tissue lobe in which it is located. *35.5*

10. Distinguish between: *35.7*
    a. short-term memory and long-term memory
    b. fact memory and skill memory

11. Which part of the brain has major influence over states of consciousness, which include sleeping and arousal? *35.8*

12. What is a habit-forming drug? Briefly describe the effects of one such drug on the central nervous system. *35.9*

Self-Quiz (*Answers in Appendix III*)

1. In sea anemones and jellyfishes, the ______ is a loose mesh of nerve cells that interact with sensory and contractile cells of an epithelium to bring about reflex responses to stimuli.

2. The oldest parts of the vertebrate brain provide ______ .
   a. reflex control of breathing, circulation, and other basic activities
   b. coordinating and relaying visual and auditory input
   c. storing, comparing, and retrieving memories
   d. both a and c

3. Is this statement true or false: White matter and gray matter are components of the spinal cord alone.

4. Is this statement true or false: The peripheral nervous system has nerves; the brain and spinal cord have tracts.

5. Overall, the ______ nerves slow down the body and divert energy to digestion and other basic housekeeping tasks; and the ______ nerves slow down housekeeping tasks and increase overall activity in times of heightened awareness or excitement.
   a. autonomic; somatic
   b. sympathetic; parasympathetic
   c. parasympathetic; sympathetic
   d. peripheral; central

6. Parasympathetic and sympathetic nerves ______ , to bring about minor adjustments in internal organ activity.
   a. service entirely different internal organs
   b. work antagonistically, most often
   c. come into play only during a fight–flight response
   d. none of the above

7. In tracts of the brain and spinal cord, ______ is associated with myelinated axons; ______ consists of neuroglia as well as unmyelinated axons, dendrites, and cell bodies of neurons.
   a. gray matter; white matter
   b. white matter; gray matter

8. In an adult brain, the brain stem is present in ______ .
   a. the hindbrain only c. the forebrain only
   b. the midbrain only d. all three divisions of brain

9. The hindbrain, which includes the medulla oblongata, pons, and cerebellum, contains ______ .
   a. the tectum c. part of the reticular formation
   b. the limbic system d. all of the above

10. The ______ are located in the cerebrum.
   a. medulla oblongata, pons, cerebellum
   b. thalamus, hypothalamus, limbic system
   c. medulla oblongata, pons, cerebral cortex
   d. cerebellum, medulla oblongata, pons, limbic system
   e. hypothalamus, limbic system, pons, cerebral cortex

11. The ______ is central to planning movements and blocking unsuitable behavior, and it also has roles in memory.
   a. visual cortex c. somatosensory cortex
   b. motor cortex d. prefrontal cortex

12. Match the component with its main functions.

___ spinal cord	a. its motor, sensory, association areas deal with conscious behavior, emotions, memory
___ medulla oblongata	b. homeostatic control of internal environment, internal organs
___ hypothalamus	c. deals with emotions, memory
___ limbic system	d. reflex control of respiration, blood circulation, other basic activities
___ cerebral cortex	e. expressway between brain and peripheral nervous system; also, direct reflex connections

Critical Thinking

1. In human newborns and premature babies, the blood–brain barrier is not fully developed. Explain why this might be reason enough to pay careful attention to their diet.

2. When Jennifer was only six years old, a man lost control of his car and hit a tree in front of her house. She ran over to the car and screamed when she saw blood from a deep head wound dripping onto a huge bouquet of red roses. Thirty-five years later, someone gave Jennifer a bottle of *Tea Rose* perfume for a birthday present. When she sniffed the perfume, she became frightened and extremely anxious. A few minutes later she also had a vivid recollection of the accident. Explain this incident in terms of what you have learned about memory.

3. Eric typically drinks one cup of coffee nearly every hour, all day long. Today, by midafternoon, he is having trouble concentrating on his studies, and he feels tired and more than a little clumsy. Drinking still another cup of coffee didn't make Eric more alert. Explain how the caffeine in coffee might produce such symptoms.

4. *Epilepsy* is a neurological disorder characterized by short, recurring episodes of sensory and motor malfunctioning. The episodes, or epileptic seizures, might arise when reverberating circuits involving millions of interneurons in the brain become abnormally activated. Affected individuals typically experience involuntary muscle contractions and often sense lights, sounds, and odors even if receptors in the eyes, ears, and nose have not been stimulated.

The drug valproic acid can eliminate or lessen the severity of the seizures by stimulating the body's synthesis of GABA. Why would stepping up GABA production help?

Selected Key Terms

autonomic nerve *35.3*
blood–brain barrier *35.4*
brain *35.4*
brain stem *35.4*
central nervous system *35.2*
cerebellum *35.4*
cerebral cortex *35.5*
cerebral hemisphere *35.5*
cerebrospinal fluid *35.4*
cerebrum *35.4*
consciousness *35.8*
drug *35.9*
drug addiction *35.9*
fight–flight response *35.3*
ganglion (ganglia) *35.1*
gray matter *35.2*
hypothalamus *35.4*
limbic system *35.5*
medulla oblongata *35.4*
memory *35.7*
meninges (singular, meninx) *35.3*
nerve *35.1*
nerve net *35.1*
nervous system *35.1*
neural tube *35.2*
neuroglia *35.2*
parasympathetic nerve *35.3*
peripheral nervous system *35.2*
planula *35.1*
pons *35.4*
reflex pathway *35.1*
reticular formation *35.4*
somatic nerve *35.3*
spinal cord *35.3*
sympathetic nerve *35.3*
tectum *35.4*
thalamus *35.4*
tract *35.2*
white matter *35.2*

Readings *See also www.infotrac-college.com*

Romer, A., and T. Parsons. 1986. *The Vertebrate Body*. Sixth edition. Philadelphia: Saunders. Fine insights into the evolution of vertebrate nervous systems.

Shepherd, G. 1994. *Neurobiology*. Third edition. New York: Oxford. Paperback.

36 SENSORY RECEPTION

Different Strokes for Different Folks

Even though you might be reluctant to pet a snake or scratch a bat behind the ears, you have to give them credit for being two of your vertebrate relatives with some special sensory traits.

Think of a python. Figure 36.1*a* shows one of these reptiles, which have thermoreceptors in rows of pits above and below their mouth. Thermoreceptors detect infrared energy—in this case, the body heat of small, night-foraging mammals that are the snake's prey of choice. When stimulated, the receptors send messages to the brain, where several centers process the signals and issue commands to muscle cells. In no time at all, a strike is aimed and executed with stunning accuracy.

A motionless, edible frog would not have the same stimulatory effect on the receptors, so the snake would slither past it. The skin of a frog is not warm, and it blends with the colors of the frog's habitat. A python does not have the types of receptors that can detect frogs or a neural program for responding to them.

Or consider the mammals called bats. Nearly all species of bats sleep during the day and spread their webbed wings at dusk. Different kinds take to the air in search of nectar, fruit, frogs, or insects. Many sensory receptors located inside their eyes, nose, ears, mouth, and skin are not all that different from yours. However, other receptors provide the bats with a sense of hearing that you cannot even begin to match. Even tiny-eyed, nearly blind species navigate and capture flying insects with precision in the dark!

Bats are masters of **echolocation**. They emit calls, as the bat in Figure 36.1*b* is doing. When sound waves of the calls bounce off insects, trees, and other objects, acoustical receptors inside the bat's ears detect the echoes and send signals about them to the bat brain.

As an echolocating bat flies, it emits a steady stream of about ten clicking sounds per second—sounds you cannot hear. The clicks are intense "ultrasounds," above the range of sound waves that receptors in human ears

a

Figure 36.1 Examples of sensory receptors. (**a**) Thermoreceptors in pits above and below this python's mouth detect body heat, or infrared energy, of nearby prey. (**b**) Some bats listen to echoes of their own high-frequency sounds. Their brain constructs a sound map based on echoes bouncing back from objects in the surroundings. Such maps help this night-flying bat capture insects in midair without the help of eyes.

are able to detect. When a bat hears a pattern of distant echoes from, say, an airborne mosquito or moth, it increases the rate of ultrasonic clicks to as many as 200 per second, which is faster than a machine gun fires bullets. In the few milliseconds of silence between the clicks, sensory receptors detect the echoes and deliver messages about them to the bat brain. The brain rapidly constructs a "map" of the sounds that the bat uses in its maneuvers through the night world.

b

With this chapter, we turn to the means by which animals receive signals from the external and internal environments, then decode the signals in ways that give rise to awareness of sounds, sights, odors, pain, and other sensations. These tasks require sensory neurons, nerve pathways, and specific brain regions. Together, they represent the parts of the nervous system called sensory systems.

KEY CONCEPTS

1. Sensory systems are portions of the nervous system. Each consists of specific types of sensory receptors, nerve pathways from receptors to the brain, and brain regions that receive and process the sensory information.

2. A stimulus is a form of energy that activates a specific type of sensory receptor, which is either a sensory neuron or a specialized cell adjacent to it. Photoreceptors detect light energy, thermoreceptors detect infrared energy, and so on.

3. We define a sensation as conscious awareness of change in some aspect of the external or internal environment. It begins when sensory receptors detect a specific stimulus. The stimulus energy becomes converted to a graded, local signal that may contribute to initiating an action potential.

4. Information concerning the stimulus becomes encoded in the number and frequency of action potentials sent to the brain along particular nerve pathways. Specific brain regions translate the information into a sensation.

5. The somatic sensations include touch, pressure, pain, temperature, and muscle sense.

6. Taste, smell, hearing, and vision are special senses.

36.1 SENSORY RECEPTORS AND PATHWAYS—AN OVERVIEW

Sensory systems, the front doors of the nervous system, receive information about specific changes inside and outside the body and notify the spinal cord and brain of what is going on. They consist of sensory receptors, nerve pathways leading from receptors to the brain, and brain regions where sensory information is translated into sensations. A **sensation** is conscious awareness of a stimulus. It is not the same as **perception** (understanding of what the sensation means). *Compound* sensations arise when information about different stimuli is integrated at the same time. For example, "wetness" is not a single stimulus; our perception of it arises from simultaneous inputs concerning pressure, touch, and temperature.

Types of Sensory Receptors

Sensory receptors are receptionists at the front door of nervous systems. Table 36.1 lists the six major categories. We define each category in terms of a **stimulus**, which is a specific form of energy that a receptor is able to detect:

Mechanoreceptors detect forms of mechanical energy, or changes in pressure, position, or acceleration.

Thermoreceptors detect infrared energy, or heat.

Pain receptors (nociceptors) detect tissue damage.

Chemoreceptors detect chemical energy of specific substances dissolved in the fluid surrounding them.

Osmoreceptors detect changes in the water volume or solute concentration of the surrounding fluid.

Photoreceptors detect visible and ultraviolet light.

Depending upon the kinds and numbers of their sensory receptors, animals sample the environment in different ways and differ in their awareness of it. Unlike bees, for example, you have no receptors for ultraviolet light and do not see many flowers the way they do. The introduction to Chapter 31 provides an example of this. Unlike many bats, you and bees have no receptors for ultrasound. Unlike pythons, you, bees, and bats have no receptors for detecting warm-blooded prey in the dark.

Sensory Pathways

Recall, from Chapter 34, that sensory axons carry signals from receptors to the brain. Before they can do so, the stimulus energy must be converted to action potentials, the basis of neural messages. Briefly, when a stimulus disturbs the plasma membrane of a receptor's ending, certain ions flow across a local patch of the membrane. In other words, the disturbance triggers a local, graded potential. This type of signal does not spread far from the point of stimulation, and it can vary in magnitude.

When a stimulus is intense or when it is repeated fast enough for a summation of local signals, action potentials may be the result. Action potentials propagate themselves from the receptor to the axon endings of sensory neurons (Figure 36.2). There, neurotransmitter is released from the presynaptic cell, and it influences the electrical activity of the cell adjacent to it (another interneuron or a motor neuron). The disturbance may trigger action potentials in the postsynaptic cell, which is part of an information pathway leading to the brain.

Table 36.1 Major Categories of Sensory Receptors

Category	Examples	Stimulus
MECHANORECEPTORS		
Touch, pressure	Certain free nerve endings and Pacinian corpuscles in skin	Mechanical pressure against body surface
Baroreceptor	Carotid sinus (Section 39.7)	Pressure changes in the fluid that bathes them
Stretch	Muscle spindle in skeletal muscle	Stretching
Auditory	Hair cells in organ inside ear	Vibrations (sound or ultrasound waves)
Balance	Hair cells in organ inside ear	Fluid movement
THERMORECEPTORS	Certain free nerve endings	Change in temperature (heating, cooling)
PAIN RECEPTORS (NOCICEPTORS)*	Certain free nerve endings	Tissue damage (e.g., distortions, burns)
CHEMORECEPTORS		
Internal chemical sense	Carotid bodies in blood vessel wall	Substances (O_2, CO_2, etc.) dissolved in extracellular fluid
Taste	Taste receptors of tongue	Substances dissolved in saliva, etc.
Smell	Olfactory receptors of nose	Odors in air, water
OSMORECEPTORS	Hypothalamic osmoreceptors (Section 43.2)	Change in water volume (solute concentration) of the fluid that bathes them
PHOTORECEPTORS		
Visual	Rods, cones of eye	Wavelengths of light

* Extremely intense stimulation of any sensory receptor also may be perceived as pain.

Further reading: Student Guide to InfoTrac on web site

Action potentials traveling along a sensory neuron are not like a wailing ambulance siren. *They do not vary in amplitude.* How, then, does the brain assess the nature of a particular stimulus? The answer lies with (1) *which* nerve pathways happen to be carrying action potentials, (2) the *frequency* of action potentials traveling along each axon that is part of the pathway, and (3) the *number* of axons that the stimulus recruited.

First, the wiring in each animal's brain is genetically programmed to interpret the action potentials only in certain ways. That's why you "see stars" when one of your eyes gets poked, even in a darkened room. A number of photoreceptors in that eye were mechanically disturbed enough to trigger messages that traveled along one of two optic nerves into the brain. Your brain always interprets any signals arriving from an optic nerve as "light."

Second, when a stimulus is strong, receptors fire action potentials more frequently than they do with a weak stimulus. The same receptor can detect the sounds of a throaty whisper or a wild screech. And the brain can sense the differences through frequency variations in signals that the receptor sends to it.

Third, a stronger stimulus can recruit more sensory receptors than a weaker stimulus can do. Gently tap a spot of skin on one of your arms and you activate a few receptors. Press hard on the spot and you activate more receptors in a larger area. The increased disturbance translates into action potentials in many sensory axons at the same time. The brain interprets the combined activity as an increase in stimulus intensity. Figure 36.3 is an example of this effect.

In some cases, the frequency of action potentials decreases or stops even when the stimulus is being maintained at constant strength. Any decrease in a response to a stimulus is a **sensory adaptation**. For example, after you put on clothing, your awareness of the pressure it exerts against your skin ceases. Inside your skin, some of the mechanoreceptors adapt rapidly to the sustained stimulation; they are of the type that can only signal a change in the stimulus (its onset and its removal). By contrast, other receptors adapt slowly or not at all; they help the brain monitor particular stimuli all the time. The stretch receptors you read about earlier, in Section 34.4, are like this. Many stretch receptors continually inform the brain about changes in the length of muscles that help maintain balance and posture.

In sections to follow, we will turn to some specific examples of sensory receptors. The types present at more than one body location contribute to what we call **somatic sensations**. Other types of sensory receptors are restricted to particular locations, such as inside the eyes or ears. They contribute to the **special senses**.

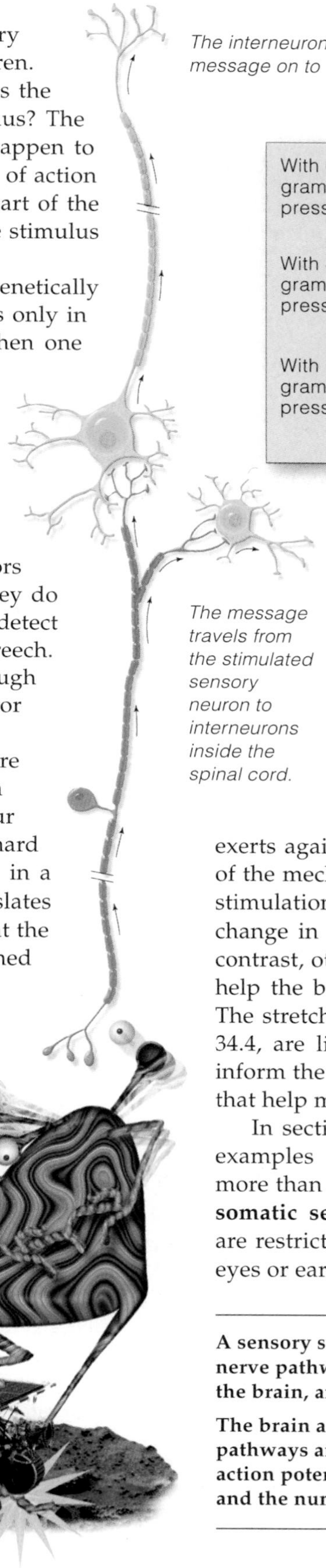

A sensory system has sensory receptors for specific stimuli, nerve pathways that conduct information from receptors to the brain, and brain regions that receive the information.

The brain assesses a given stimulus based on which nerve pathways are carrying action potentials, the frequency of action potentials traveling along each axon of that pathway, and the number of axons that have been recruited to action.

The interneuron sends the message on to the brain.

The message travels from the stimulated sensory neuron to interneurons inside the spinal cord.

Figure 36.2 NASA Rover running over a Martian's foot and stimulating the receptor endings of a sensory neuron. This is an example of a sensory nerve pathway leading away from a sensory receptor to a brain. The sensory neuron is color-coded *red*; interneurons are *yellow*.

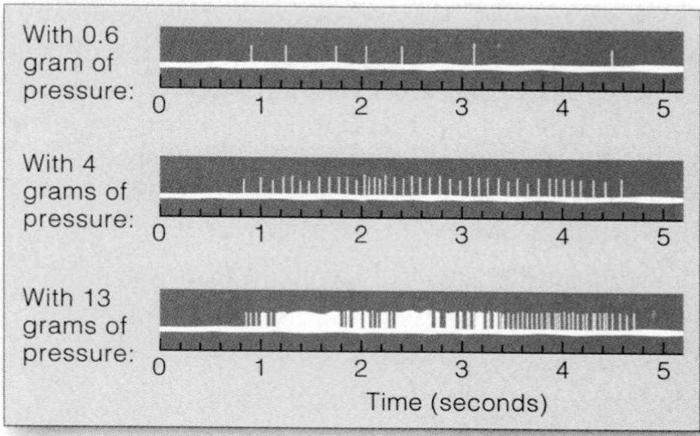

With 0.6 gram of pressure:
With 4 grams of pressure:
With 13 grams of pressure:
0 1 2 3 4 5
Time (seconds)

Figure 36.3 *Above:* Recordings of action potentials from a single pressure receptor with endings in the skin of a human hand. They correspond to variations in stimulus strength. Here, an investigator pressed a thin rod against skin with the amount of pressure indicated on the left side of each diagram. Vertical bars above each thick horizontal line represent individual action potentials. In each case, the increases in frequency correspond to increases in the strength of the stimulus.

36.2 SOMATIC SENSATIONS

Somatic sensations begin with receptors in the body's surface tissues, skeletal muscles, and walls of internal organs. The receptors are most highly developed in birds and mammals. Amphibians have few, and apparently fishes have none. Receptor inputs travel into the spinal cord and on into the **somatosensory cortex**, part of the gray matter of the cerebral hemispheres (Section 35.5). The interneurons in this brain region are organized like maps corresponding to particular parts of the body's surface. Map regions with the largest areas correspond to body parts that have the greatest sensory acuity and that require the most intricate control. Such body parts include the fingers, thumbs, and lips (Figure 36.4).

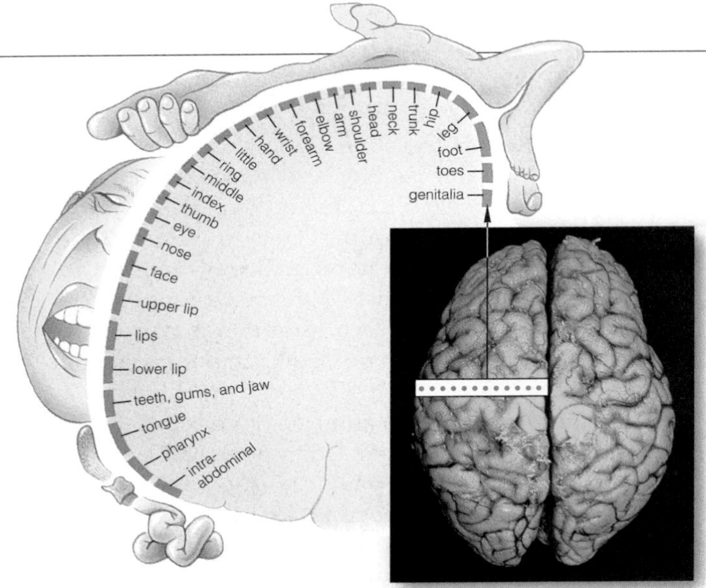

little
ring
middle
index
thumb
eye
nose
face
upper lip
lips
lower lip
teeth, gums, and jaw
tongue
pharynx
intra-abdominal
hand
wrist
forearm
elbow
arm
shoulder
head
neck
trunk
hip
leg
foot
toes
genitalia

Figure 36.4 Differences in the representation of different body parts in the primary somatosensory cortex. This region is a strip of cerebral cortex, a little wider than 2.5 centimeters (about an inch), from the top of the human head to above the ear.

Receptors Near the Body Surface

You and other mammals discern sensations of touch, pressure, cold, warmth, and pain near the body surface. Regions with the greatest number of sensory receptors, such as the fingertips and the tip of the tongue, are the most sensitive to stimulation. Other regions, such as the back of the hand and neck, do not have nearly as many receptors and are far less sensitive.

Free nerve endings are the simplest receptors. These thinly myelinated or unmyelinated (naked) branched endings of sensory neurons are distributed in skin and internal tissues. Different types are mechanoreceptors, thermoreceptors, and pain receptors. All adapt slowly to stimulation. One subpopulation of free nerve endings gives rise to a sense of prickling pain, as when you jab a finger with a pin. Another subpopulation contributes to itching or warming sensations that are provoked by chemicals, including histamine. Two thermoreceptive types have peak sensitivities that are higher and lower than the body's normal core temperature, respectively. One mechanoreceptive type coils around hair follicles and detects hair movements (Figure 36.5).

Encapsulated receptors are common near the body surface. The Meissner's corpuscle adapts very slowly to low-frequency vibrations. It is abundant in fingertips, lips, eyelids, nipples, and genitals. The bulb of Krause, an encapsulated thermoreceptor, is activated at 20°C (68°F) or lower. Below 10°C, it contributes to painful freezing sensations. Ruffini endings are slowly adapting encapsulated types. These respond to steady touching and pressure, and to temperatures above 45°C (113°F).

The Pacinian corpuscle, an encapsulated receptor, is widely distributed deep in the skin, where it responds to fine textures. It also is located near freely movable joints and inside some internal organs. Onionlike layers of membrane alternating with fluid-filled spaces enclose this sensory ending (see Figure 36.5). The construction enhances the receptor's ability to detect rapid pressure changes associated with touch and vibrations.

Muscle Sense

Sensing limb motions and the body's position in space requires mechanoreceptors in skeletal muscle, joints, tendons, ligaments, and skin. Examples include stretch receptors of muscle spindles. As explained in Section 34.4, these sensory organs run parallel with skeletal muscle cells. Their responses to stimulation depend on how much and how fast the muscle stretches.

The Nature of Pain

Pain is a perception of injury to some body region. In humans, the major pain receptors are subpopulations of free nerve endings. Several million thread through all tissues and organs except the brain. Sensations of *somatic* pain start with free nerve endings in skin, joints, tendons, and skeletal muscles. Sensations of *visceral* pain, which is associated with the internal organs, are related to excessive chemical stimulation, muscle spasm, muscle fatigue, inadequate blood flow to organs, and other abnormal conditions.

Superficial somatic pain arises at or near the skin surface. It is usually sharp or prickling and doesn't last long. Deep somatic pain, which arises farther down in skin, muscles, or joints, is more diffuse and lasts longer. This is also the case for visceral pain. We characterize both as burning or gnawing pain, or a dull ache.

When cells are damaged, they release chemicals that activate neighboring pain receptors. The most potent, the bradykinins, open the floodgates for an outpouring of histamine, prostaglandins, and other participants in the inflammatory response (Section 40.10). They might also bind with and activate pain receptor endings.

Further reading: Student Guide to InfoTrac on web site

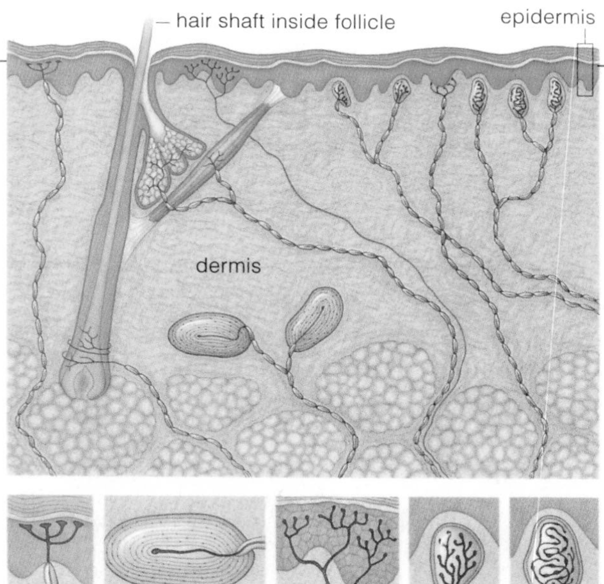
hair shaft inside follicle
epidermis
dermis
free nerve endings
Pacinian corpuscle
Ruffini endings
bulb of Krause
Meissner's corpuscle

Figure 36.5 Sensory receptors in human skin. Subpopulations of free nerve endings function as mechanoreceptors (such as those around hair follicles, which detect hair movements), thermoreceptors, and pain receptors. Pacinian corpuscles detect rapid pressure changes associated with vibrations and touch. Ruffini endings detect steady touching and pressure. The bulb of Krause is most sensitive to cold. Meissner's corpuscles detect vibrations of lower frequency than those detected by Pacinian corpuscles.

Signals from pain receptors reach interneurons in the spinal cord and induce them to release substance *P*. This neurotransmitter causes signals to be relayed along two pathways. Some signals reach the thalamus and the sensory cortex, which rapidly analyzes the type and intensity of pain. Other signals alert the brain stem and limbic system, which mediates arousal and emotional responses to pain. Now signals from the hypothalamus and midbrain stimulate the release of endorphins and enkephalins, natural opiates that reduce pain perception. Morphine, hypnosis, and natural childbirth techniques may also stimulate release of these natural opiates.

We all tolerate pain differently even though we all perceive tissue damage at the same level of stimulation. For instance, "heat" becomes "pain" between 44°C and 46°C, which is when it starts to damage tissues. Pain tolerance depends on emotional state, cultural factors, and possibly how old we are (tolerance might increase with age). Extremely intense or long-lasting stimulation can lead to *hyperalgesia*, or pain amplification. Someone with a severe sunburn experiences this amplified effect when they step into a hot shower, for example.

Referred Pain

Responses to pain depend on the ability of the brain to identify the affected tissue and to project the sensation back to it. For example, get smacked in the face with a snowball and you "feel" the contact on facial skin. However, sensations of pain from some internal organs may be wrongly projected to part of the skin. We all have this capacity for *referred pain*, whereby visceral sensations are perceived as somatic sensations. As one example, people suffering a heart attack may perceive pain radiating from the chest wall, from the neck, and along the left shoulder and arm (Figure 36.6).

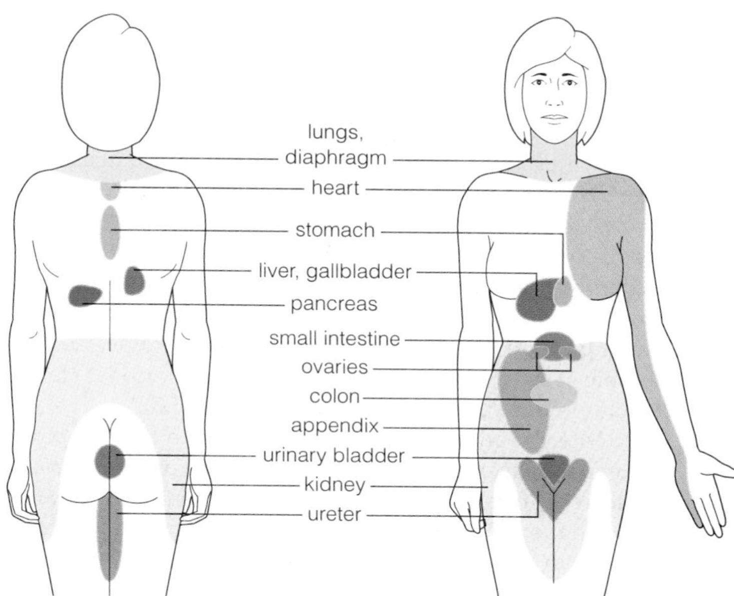
lungs, diaphragm
heart
stomach
liver, gallbladder
pancreas
small intestine
ovaries
colon
appendix
urinary bladder
kidney
ureter

Figure 36.6 Referred pain. Receptors in some internal organs detect painful stimuli. Instead of localizing the pain at the organs, the brain projects the sensation to the skin areas indicated.

How can the brain be so mistaken? The answer lies in the way the nervous system is constructed. Sensory inputs from the skin and from certain internal organs enter the spinal cord in the same segments of the cord. Compared with internal organs, the skin is hit more often with painful stimuli, so signals travel more often along the somatic pathway to the brain. The brain may be interpreting most of the sensory inputs as arriving from the more common source.

Referred pain is not the same as the *phantom pain* reported by amputees. They often sense the presence of a missing body part, as if it were still there. In some undetermined way, severed sensory nerves continue to respond to the amputation. The brain projects the pain back to the missing part, past the healed region.

Diverse sensory receptors near the body surface and in internal organs detect touch, pressure, temperatures, pain, motion, and positional changes of body parts. Their input travels through the spinal cord to the somatosensory cortex and other brain regions where somatic sensations arise.

36.3 SENSES OF TASTE AND SMELL

We turn now to two examples of the special senses, taste and smell. Both are *chemical* senses; their sensory pathways start at chemoreceptors, which are activated when they bind with certain chemical substances that are dissolved in the fluid bathing them. These receptors wear out and new ones replace them on an ongoing basis. In both pathways, sensory input travels from the receptors through the thalamus and on to the cerebral cortex, where perceptions of the stimulus take shape and undergo fine-tuning. The input also travels to the limbic system, which integrates it with emotional states and stored memories (Sections 35.5 and 35.7).

Different animals taste substances with their mouth, antennae, legs, tentacles, or fins depending on where their **taste receptors** are. In the throat and mouth, especially the tongue's upper surface, these chemoreceptors are located in about 10,000 sensory organs called taste buds (Figure 36.7). Taste buds have a pore through which fluids in the mouth contact receptor cells. All of the thousands of perceived tastes are a combination of five primary sensations: sweet (elicited by glucose and other simple sugars), sour (acids), salty (NaCl and other salts), bitter (alkaloids and other plant toxins), and *umami* (glutamate, a brothy or savory taste typical of aged cheese and meat).

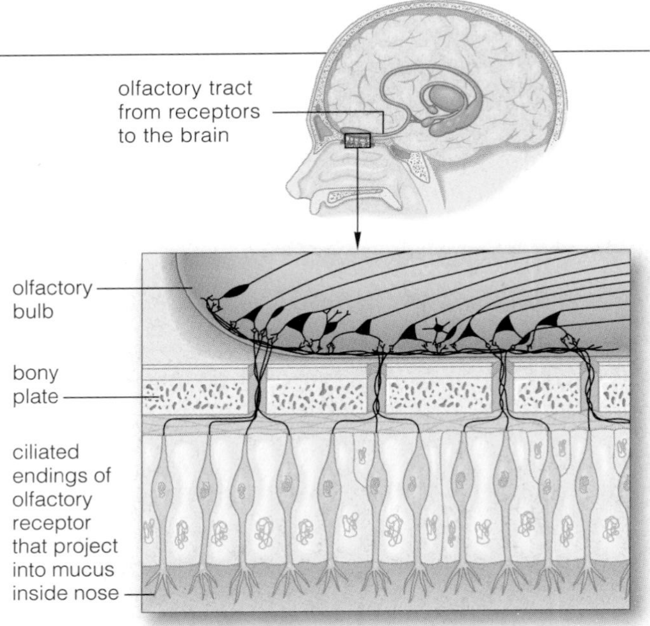
olfactory tract
from receptors
to the brain
olfactory bulb
bony plate
ciliated endings of olfactory receptor that project into mucus inside nose

Figure 36.8 Sensory pathway from the sensory endings of olfactory receptors in the nose to the cerebral cortex and limbic system. Receptor axons pass through holes in a bony plate between the nasal lining and the brain. In the earliest vertebrates, an olfactory bulb and olfactory lobe dominated the forebrain. In certain lineages, the olfactory bulb became less important when the cerebrum expanded greatly in size and functions, as Sections 35.2 and 35.4 indicate.

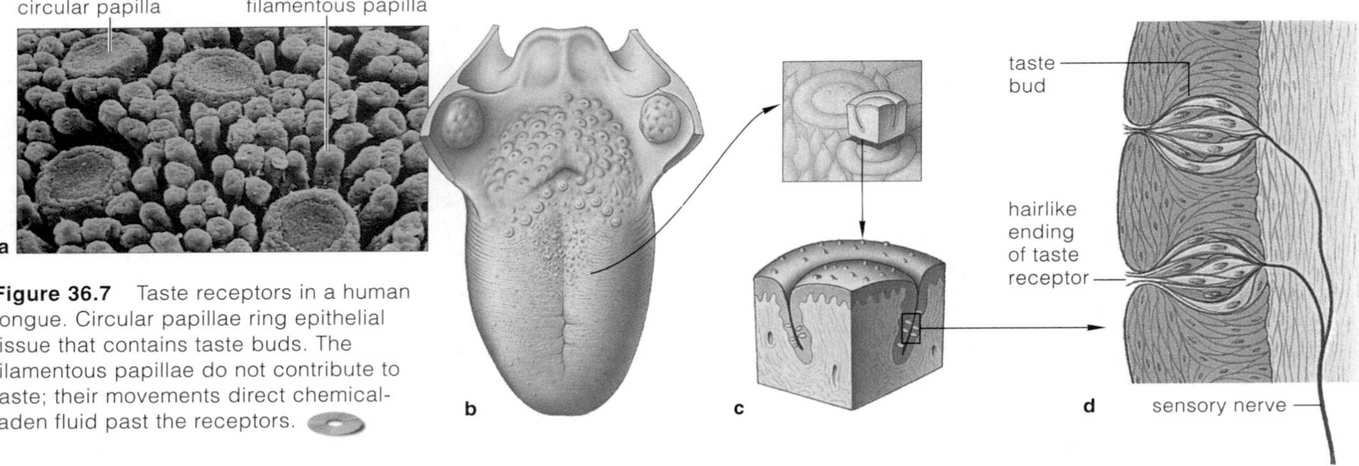
circular papilla
filamentous papilla
a
b
c
taste bud
hairlike ending of taste receptor
d
sensory nerve

Figure 36.7 Taste receptors in a human tongue. Circular papillae ring epithelial tissue that contains taste buds. The filamentous papillae do not contribute to taste; their movements direct chemical-laden fluid past the receptors.

Olfactory receptors detect water-soluble or volatile (easily vaporized) substances. A bloodhound nose has more than 200 million; a human nose has about 5 million. The receptor axons lead into one of two olfactory bulbs. Inside these small brain structures, they synapse with groups of cells that sort out the components of a given scent and relay the information along an olfactory tract for further processing in the cerebrum (Figure 36.8).

A vomeronasal organ, or "sexual nose," is common among animals, including humans. Its receptors detect **pheromones**, a type of signaling molecule with roles in the social aspects of reproduction. These exocrine gland secretions affect the behavior of other individuals of the same species, especially potential mates. Think about the effect of bombykol on olfactory receptors of a male silk moth. Contact with merely one bombykol molecule per second sends action potentials to the moth brain. They help a male locate a female in the dark, even if she is more than a kilometer upwind from him.

The senses of taste and smell both start at chemoreceptors. Both involve sensory pathways that lead to processing regions in the cerebral cortex and in the limbic system.

Further reading: Student Guide to InfoTrac on web site

36.4 SENSE OF BALANCE

Most animals assess and respond to displacements from their natural, *equilibrium* position, in which the body is balanced in relation to gravity, velocity, acceleration, and other forces that influence its positions and movement. For example, animals right themselves after tilting or turning upside-down. A variety of fishes, amphibians, and reptiles have paired **inner ears**, which evolved first as organs of equilibrium and had little, if anything, to do with hearing. These are systems of fluid-filled sacs and canals on both sides of the brain. Some parts detect rotational, accelerated motions of the head, as when we ride a looping roller coaster (Figure 36.9). Other parts detect linear motion of the head. In amphibians, birds, and mammals, the brain integrates sensory input from internal ears with input from sensory receptors in the eyes, skin, and joints to arrive at sensations of balance.

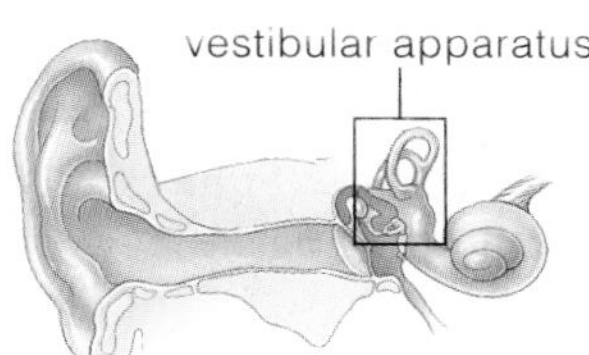
vestibular apparatus

Figure 36.9 Location of the internal ear in humans. Inside the vestibular apparatus (a system of fluid-filled sacs and canals), organs detect the head's linear, rotational, and accelerated motions.

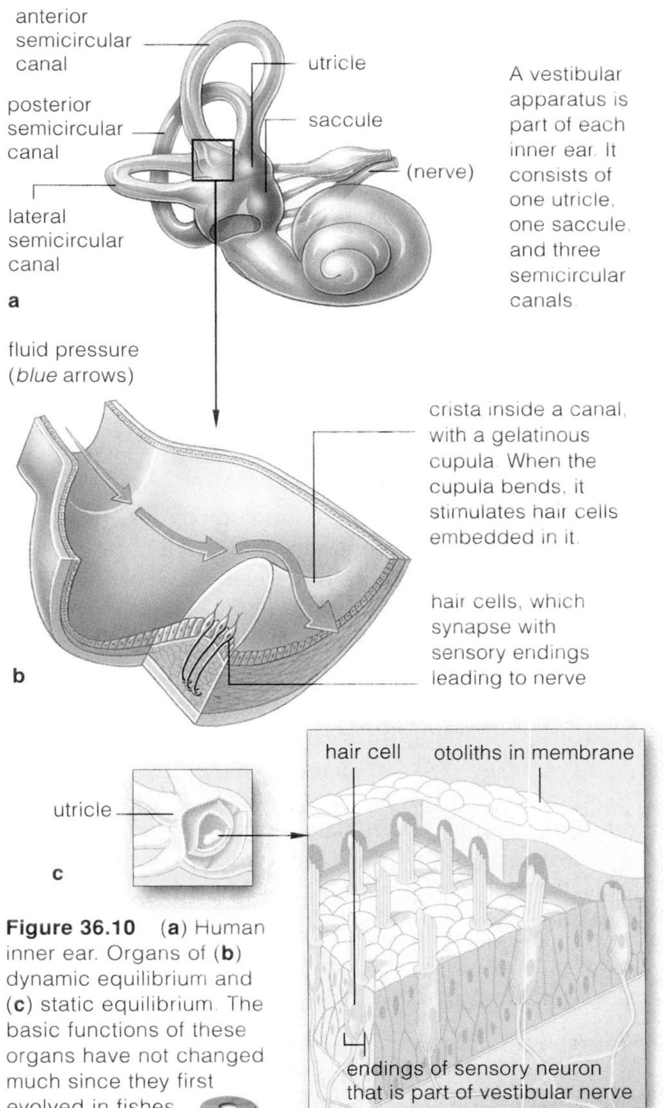
anterior semicircular canal
utricle
A vestibular apparatus is part of each inner ear. It consists of one utricle, one saccule, and three semicircular canals.
posterior semicircular canal
saccule
(nerve)
lateral semicircular canal
a
fluid pressure (*blue* arrows)
crista inside a canal, with a gelatinous cupula. When the cupula bends, it stimulates hair cells embedded in it.
hair cells, which synapse with sensory endings leading to nerve
b
hair cell
otoliths in membrane
utricle
c
endings of sensory neuron that is part of vestibular nerve

Figure 36.10 (**a**) Human inner ear. Organs of (**b**) dynamic equilibrium and (**c**) static equilibrium. The basic functions of these organs have not changed much since they first evolved in fishes.

In humans, organs of equilibrium are located in the parts of the inner ear called the **vestibular apparatus** (Figures 36.9 and 36.10*a*). Cristae, the organs of *dynamic* equilibrium, detect accelerated, rotational motions of the head. A crista is present at the swollen base of three semicircular canals, oriented in three directions. Rotate your head horizontally, vertically, or diagonally and you displace fluid within the canal corresponding to that direction. The fluid presses against a gelatinous mass, a cupola, into which hair cells project (Figure 36.10*b*). **Hair cells** are mechanoreceptors. Pressure can bend them and deform their plasma membrane, thus sparking action potentials. Hair cells associated with a cupola synapse with the endings of neighboring sensory neurons, the axons of which help form a vestibular nerve.

The vestibular apparatus also has organs of *static* equilibrium, which are activated when an animal starts and stops moving in a straight line. Two structures, a utricle and saccule, have one of these organs on their floor. Positioned above the floor is a thick membrane into which hair cells project (Figure 36.10*c*). Crystals of calcium carbonate are deposited in the membrane. We call the crystalline masses **otoliths**, or "ear stones." Thus weighted, the membrane slides in response to linear acceleration. Sliding bends and activates the hair cells.

Motion sickness may result when monotonous linear, angular, or vertical motion overstimulates hair cells in the canals of the organs of balance. It also may result from conflicting signals from the ears and eyes about motion or the head's position. People who are carsick, airsick, or seasick suffer nausea and may vomit. Why? Action potentials triggered by the sensory input reach a brain center that governs the vomiting reflex.

Organs of equilibrium help keep the body balanced in relation to gravity, velocity, acceleration, and other forces that influence its position and movement. The vestibular apparatus of the human inner ear includes this type of organ.

36.5 SENSE OF HEARING

Properties of Sound

Many arthropods and most vertebrates have a sense of **hearing**, which is the perception of sounds. Sounds are traveling vibrations, or wavelike forms of mechanical energy. Think about what happens when you clap your hands and produce waves of compressed air. Each time hands clap together, molecules are forced outward, so a low-pressure state is created in the region they vacated. Such pressure variations are depicted as wave forms in which *amplitude* corresponds to loudness (Figure 36.11). Typically, amplitude is measured in decibels. Every ten of these units signifies a tenfold increase in intensity above the faintest sound that humans are able to hear. The *frequency* of a sound is the number of wave cycles per second. Every cycle extends from the start of one wave peak to the start of the next wave peak. The more cycles per second, the higher the frequency and higher the perceived pitch.

Evolution of the Vertebrate Ear

After vertebrates invaded the land, the sense of hearing became far more important than it had been in aquatic habitats. Although the parts of the *internal* ear dealing with the sense of balance did not change much, other parts slowly expanded into structures that could receive and process faint sounds traveling through air.

Thus the **middle ear** evolved, and structures inside it amplified and transmitted air waves to the inner ear.

Sounds differ in *amplitude* (they vary in pressure, depicted here as wave forms) and in *frequency* (number of wave cycles/unit time).

amplitude
one cycle
frequency per unit time

A sound's *intensity* (loudness) depends on its amplitude.

soft:
loud:
same frequency, different amplitude

A sound's *pitch* (tone) depends on its frequency.

low note:
high note:
same loudness, different pitch

Figure 36.11 Some properties and examples of sound waves. A tuning fork gives a pure tone. By comparison, most sounds are combinations of sound waves of different frequencies. Combinations of overtones give the sounds their timbre, or quality, which is one of the properties that help you recognize, say, voices of people you know over the phone.

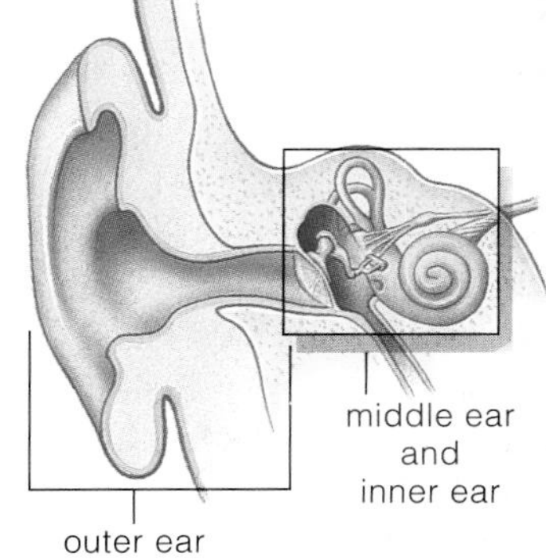
middle ear
and
inner ear
outer ear

a The three functional divisions of the human ear. They are the outer ear (the pinna and inner channel), middle ear (which includes the eardrum and ear bones), and inner ear (cochlea).

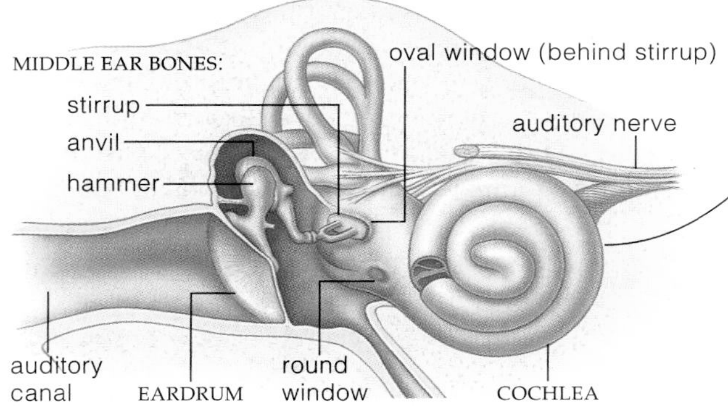
MIDDLE EAR BONES:
stirrup
anvil
hammer
oval window (behind stirrup)
auditory nerve
auditory canal
EARDRUM
round window
COCHLEA

b Components of the human ear. The outer ear's external flaps collect sound waves, which move through an auditory canal to the eardrum (also called the tympanic membrane).

Figure 36.12 Sensory reception at the human ear, a sensory organ that collects, amplifies, and sorts sound waves (acoustical stimuli).

In reptiles, a shallow depression formed on each side of the head and later evolved into an eardrum. This thin membrane rapidly vibrates in response to air waves. Behind the eardrum of all existing crocodiles, birds, and mammals is an air-filled cavity and small bones, which transmit vibrations to the inner ear. The forerunners of the bones were present in early fishes. They structurally supported gill pouches and then became part of the jaw joint. In other words, the bones that once supported gas exchange became specialized for feeding functions, then for hearing among reptiles, birds, and mammals.

An **external ear** that can collect sound waves also became well developed among mammals. The external ear of nearly all species has a pinna: a sound-collecting, skin-covered flaps of cartilage, typically with elaborate folds, that project from both sides of the head. They also contain a deep channel leading to the middle ear.

Sensory Reception in the Human Ear

Figure 36.12 illustrates the outer, middle, and inner ear of humans. Humans (like other mammals and also like birds) have a pea-sized but highly developed **cochlea**,

Further reading: Student Guide to InfoTrac on web site

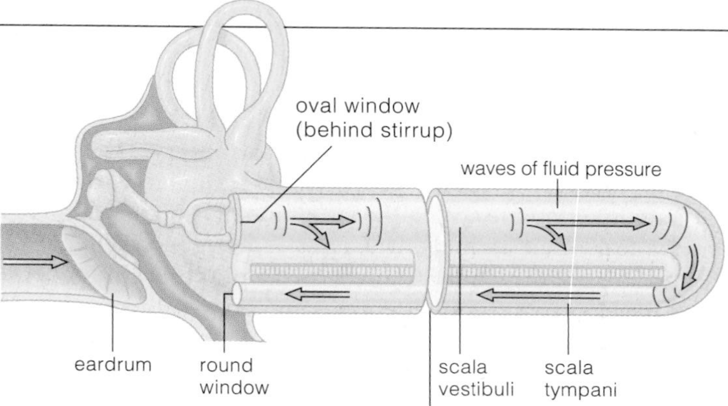
oval window (behind stirrup)
waves of fluid pressure
eardrum
round window
scala vestibuli
scala tympani

c Sound reception in the inner ear. Sound waves that arrive at an eardrum make it vibrate. Bones of the middle ear pick up the vibrations, and they amplify the stimulus by transmitting the force of the pressure waves to a smaller surface, the oval window. The "window" is an elastic membrane over the entrance to the part of the inner ear called the cochlea (shown uncoiled in this diagram for clarity).

When the oval window bows in and out, it produces waves of fluid pressure in two ducts that are called the scala vestibuli and scala tympani. The waves reach another membrane, the round window. When this membrane bulges under incoming pressure, fluid moves back and forth in the inner ear.

The pressure waves are sorted out at the cochlear duct, the third duct inside the coiled inner ear. One of its membranes (the basilar membrane) starts out narrow and stiff. It is broader and flexible deeper in the coil. At different points, the basilar membrane vibrates more strongly to sounds of different frequencies.

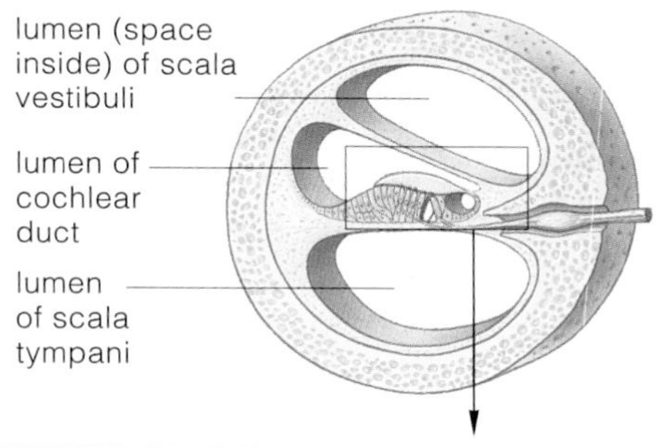
lumen (space inside) of scala vestibuli
lumen of cochlear duct
lumen of scala tympani

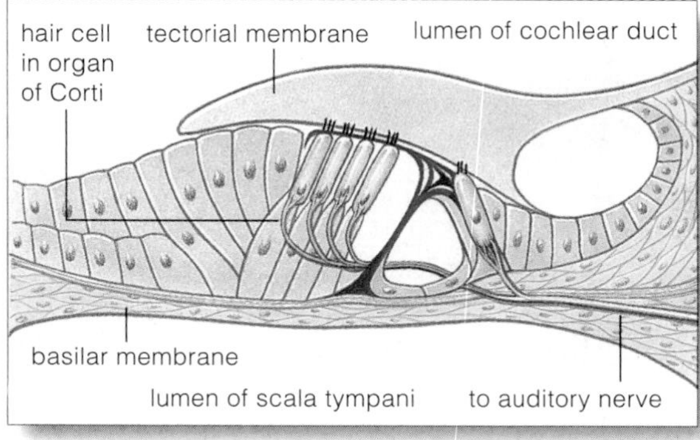
hair cell in organ of Corti
tectorial membrane
lumen of cochlear duct
basilar membrane
lumen of scala tympani
to auditory nerve

ORGAN OF CORTI
(organ of hearing that rests on top of basilar membrane)

d At the organ of Corti, 16,000 hair cells project into a jellylike flap (tectorial membrane) over the basilar membrane. When the flap vibrates suitably, hair cells bend, action potentials arise, and messages travel along an auditory nerve to the brain.

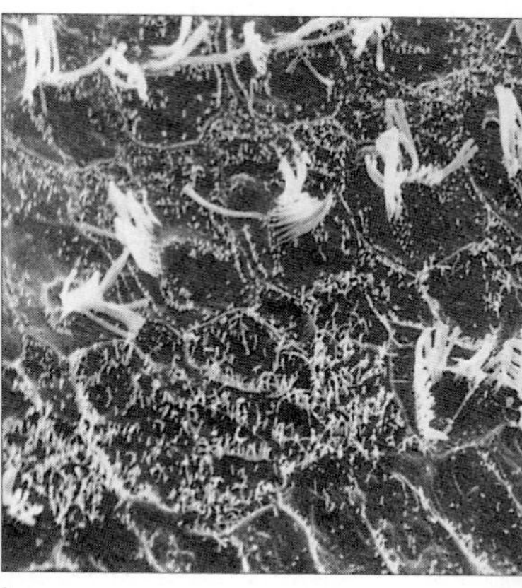

a b

Figure 36.13 Results of an experiment on the effect of intense sound on the inner ear. (**a**) From a guinea pig ear, three rows of hair cells that project into the tectorial membrane in the organ of Corti. (**b**) Hair cells in the same organ after twenty-four hours of exposure to noise levels comparable to extremely loud music.

To give you a sense of what "loud" is, a ticking watch measures 10 decibels (100 times louder than the threshold of hearing for humans). A normal conversation is about 60 decibels (about a million times louder), a food blender operating at high speed is about 90 decibels (a billion times louder), and a raging rock concert, about 120 decibels (a trillion times louder).

an inner ear structure that accepts sound waves from the outer ear and sorts them out. The cochlea has vibration-sensitive mechanoreceptors, or **acoustical receptors**. The ones in the human ear are hair cells that bend back and forth in response to pressure waves. They start a flow of information along an auditory nerve that leads from the receptors into the brain. Prolonged exposure to intense sounds permanently damages hair cells. These receptors are not adapted to amplified music, jet planes taking off, and other recent developments (Figure 36.13).

Echolocation

Earlier, you read that the bat brain responds to signals generated by echolocation, or the use of echoes from ultrasounds that the bat itself produces. Ultrasounds are extremely high-frequency waves, in the 100-decibel range, which humans cannot even hear. Dolphins and whales also emit ultrasounds that travel through water. By perceiving frequency variations in the echoes, these mammals also can pinpoint the distance and direction of movement of one another and of predators or prey.

Hearing among land vertebrates involves structures that collect, amplify, and sort out sound waves traveling through air. In the inner ear, sound waves produce fluid pressure variations, which hair cells transduce into action potentials.

36.6 SENSE OF VISION

Requirements for Vision

All organisms are sensitive to light. Although they have no photoreceptor cells, sunflowers track the sun as it arcs across the sky. Even single-celled amoebas abruptly stop moving when you shine a light on them. We do not find photoreceptor cells until we poke about among the invertebrates. Even then, not all species that have photoreceptors "see" as you do. Many might detect a change in light intensity, as a mollusk on the seafloor might do when a fish swims above it. However, they cannot discern the size or shape of such objects.

At the minimum, the sense of **vision** requires two components: eyes, and a capacity for image formation in brain centers that receive and interpret patterns of visual stimulation being sent to them. Such images are pieced together from information about the shape, brightness, position, and movement of visual stimuli. **Eyes** are sensory organs that incorporate a tissue with a dense array of photoreceptors. All photoreceptors have this in common: *They incorporate pigment molecules that can absorb photon energy, which can be converted into excitation energy in sensory neurons.*

Nearly all photoreceptor cells have some portion of their surface infolded to a small or large extent, which increases the surface area available for photochemical reactions. Their structure can be traced to the kind of epidermal cells that gave rise to them. The *rhabdomeric* photoreceptors are derived from cells with microvilli. They predominate in flatworms, mollusks, annelids, arthropods, and echinoderms. By contrast, the *ciliary* photoreceptors have a photoreceptive surface derived from the membrane of a cilium. We find these among the cnidarians, some flatworms, and all vertebrates.

A Sampling of Invertebrate Eyes

SIMPLE EYES Earthworms and some other animals have photoreceptors dispersed through their integument, so they can respond to light even though they have no eyes. The simplest eye is an **ocellus** (plural, ocelli), a patch or cuplike depression of the integument in which photoreceptors and pigmented cells are concentrated. Figure 36.14*a* shows one of these. Such eyes allow the animal to use light for orienting the body, detecting a predator's shadow, or influencing biological clocks.

COMPLEX EYES Vision evolved among fast-moving predators that had to discriminate quickly among prey and other objects that crossed their rapidly changing visual field. Remember, a **visual field** is simply the part of the outside world that the animal sees. At the least, different groups of photoreceptors must sample the light intensity of different parts of the visual field. The brain interprets the intensities as contrasting parts of a visual image. The quality of the resulting image varies among invertebrates. With the exception of cephalopods and a few other animals, it usually is not good. Why? Good image formation requires many photoreceptors, and the eyes of most invertebrates are too small to have a lot of them. For example, a planarian's pigment cup only has about 200 photoreceptors, compared with the 70,000 per square millimeter in an octopus eye.

Image formation benefits from a **lens**, a transparent body that bends all light rays from a given point in the visual field so that they converge onto photoreceptors (compare Section 4.2). Abalones have a spherical lens that bends incoming rays but not to the same points, so images that form are blurred (Figure 36.14*b*). Images improve when some space intervenes between the lens and photoreceptors, as in a snail eye. They get better when the eye also has a **cornea**, a transparent cover that directs light rays onto the lens, as happens in a snail or conch (Figure 36.14*c*,*d*).

Invertebrates on land are better adapted for visually discriminating objects. This is the case for spiders and

epidermis
photoreceptor
pigmented cell
a

epidermis
transparent body (lens)
photoreceptor
sensory cell
b

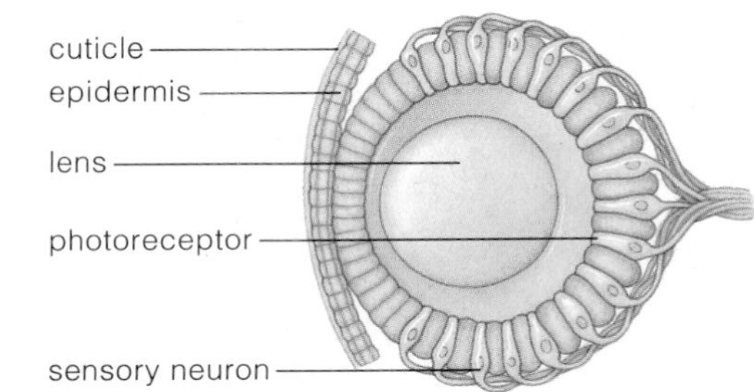
cuticle
epidermis
lens
photoreceptor
sensory neuron
c

d

Figure 36.14 Organization of invertebrate eyes, longitudinal section. There are far more photoreceptors than can be shown in these diagrams. (**a**) Limpet ocellus, a shallow depression in the epidermis that incorporates light-sensitive receptors. (**b**) Abalone eye, with its spherical, transparent lens. (**c**) Eye of a land snail. (**d**) Well-developed eyes of a conch, which is peering into the waters of the Great Barrier Reef along the east coast of Australia.

Further reading: Student Guide to InfoTrac on web site

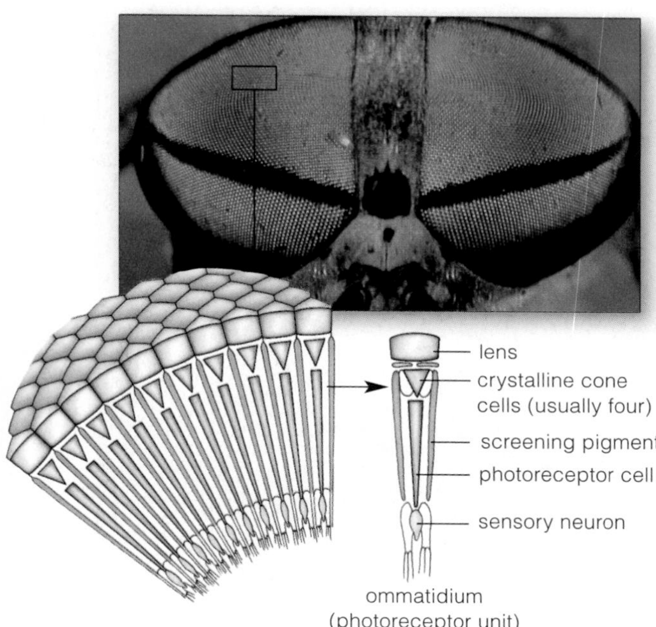
lens
crystalline cone
cells (usually four)
screening pigment
photoreceptor cell
sensory neuron
ommatidium
(photoreceptor unit)

Figure 36.15 Compound eyes of a deerfly. The lens of each photosensitive unit (ommatidium) directs light onto a crystalline cone, which focuses light on photoreceptor cells below it.

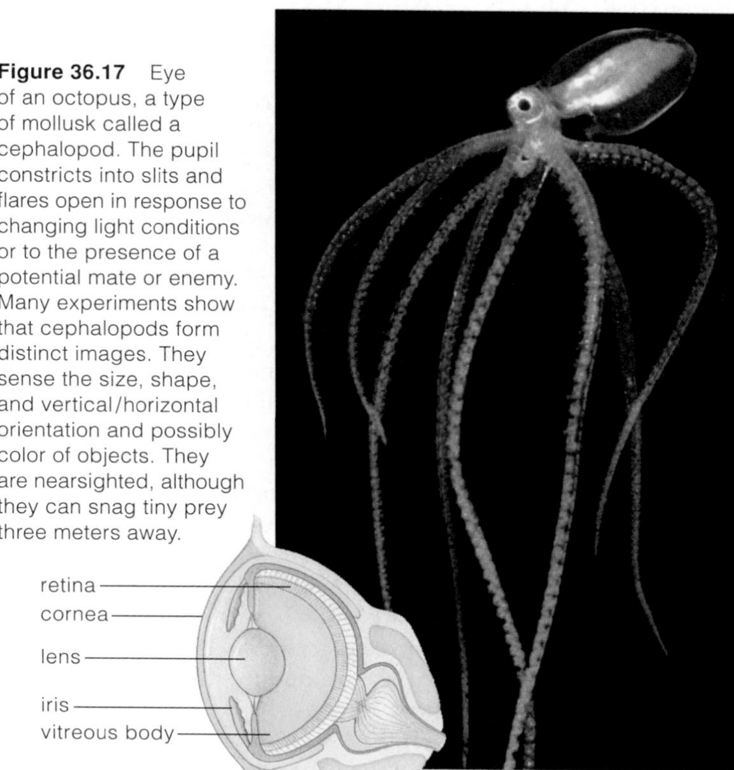
retina
cornea
lens
iris
vitreous body

Figure 36.17 Eye of an octopus, a type of mollusk called a cephalopod. The pupil constricts into slits and flares open in response to changing light conditions or to the presence of a potential mate or enemy. Many experiments show that cephalopods form distinct images. They sense the size, shape, and vertical/horizontal orientation and possibly color of objects. They are nearsighted, although they can snag tiny prey three meters away.

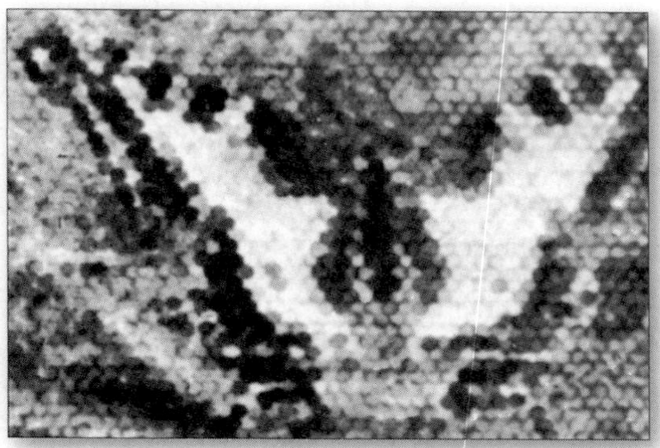

Figure 36.16 An approximation of light reception in an insect eye. This image of a butterfly formed after a researcher took a photograph through the outer surface of a compound eye that had been detached from an insect. It might not be what the insect "sees," because integration of signals sent to the brain from photoreceptors might produce a sharper image. But this representation is useful insofar as it suggests how the separate ommatidia *sample* the overall visual field.

most other arthropods having *simple* eyes, in which a single lens services all the photoreceptors. It also is the case with the complex, *compound* eyes of crustaceans and insects. A **compound eye** has many closely packed photosensitive units, each with a bundle of rhabdomeric photoreceptors (Figure 36.15). Some have thousands of units called ommatidia (singular, ommatidium).

According to the mosaic theory of image formation, each unit samples only a small part of the visual field. The brain builds images based on signals about differences in light intensities, with each unit contributing a small bit to the formation of a visual mosaic (Figure 36.16).

Of all invertebrates, octopuses and squids have the most complex eyes. Like vertebrates, they have **camera eyes**, so named because the eyeball is structured along the lines of a camera. The interior is a darkened chamber. Light enters it only through the pupil, an opening in a ring of contractile tissue called the iris (equivalent to a camera's diaphragm). Behind the pupil, a lens focuses light onto an array of photoreceptors (a camera's light-sensitive film). Axons of sensory nerves converge as an optic tract that leads to the brain (Figure 36.17).

The cephalopods and vertebrates are only remotely related, so similarities in the structure and functions of their eyes might be a result of convergent evolution. By one theory, a group of genes that originally had roles in the development of the nervous system was coopted for the task of eye formation. Whether this might have happened more than once in different animal lineages is not yet known.

Vision requires eyes (sensory organs with a dense array of photoreceptors) and a capacity for image formation in the brain, based on patterns of visual stimulation.

STRUCTURE AND FUNCTION OF VERTEBRATE EYES

The Eye's Structure

Reflect on some points made in the preceding section. *Vision* requires eyes (sensory organs that incorporate a tissue of densely packed photoreceptors) as well as image formation in brain centers that not only receive but also interpret patterns of stimulation from different parts of the eyeballs. Such images, again, require information about the shape, brightness, position, and movement of visual stimuli. With this in mind, take a look at Figure 36.18, which shows a human eye. Like most vertebrate eyes, this eyeball has a three-layered wall (Table 36.2).

The eyeball's outermost layer consists of the sclera and cornea. The sclera, the dense, fibrous "white" of the eye, protects most of the eyeball. The cornea, made of transparent collagen fibers, covers the rest of it.

The middle layer has a choroid, ciliary body, iris, and pupil. The choroid is a dark-pigmented, vascularized tissue. It absorbs the light that photoreceptors have not absorbed and prevents it from scattering inside the eye. Suspended in back of the cornea is a doughnut-shaped, pigmented iris (the Latin *iris* refers to the rings of a rainbow). The dark "hole" in the center of an iris is the pupil, the entrance for light. When rays of bright light strike the eye, circular muscles embedded within the iris contract, so the pupil's diameter shrinks. In dim light, radial muscles in the iris contract, so the pupil enlarges.

The inner layer, which includes the retina, is at the back of your eyeball. In birds of prey, such as owls and hawks, it is closer to the eyeball's roof (Figure 36.19).

The eye's interior has a lens. A clear fluid called the aqueous humor bathes the lens, which consists of layers of transparent proteins. The vitreous body, a jellylike substance, fills the chamber behind the lens.

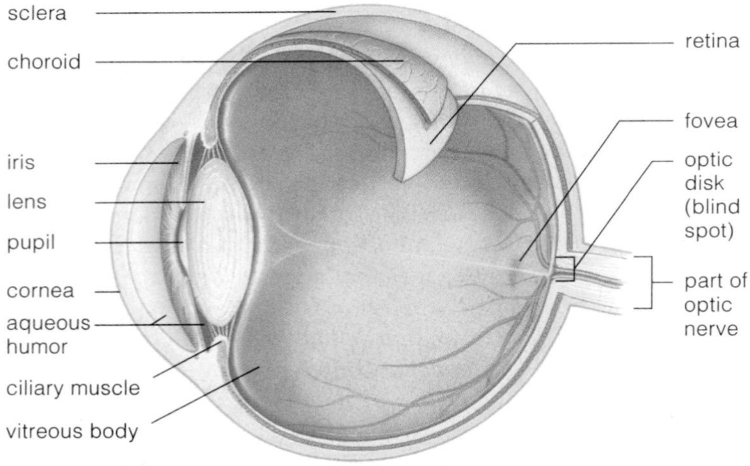

sclera
choroid
iris
lens
pupil
cornea
aqueous humor
ciliary muscle
vitreous body
retina
fovea
optic disk (blind spot)
part of optic nerve

Figure 36.18 Structure of the human eye. 

Figure 36.19 Here's looking at you! In owls and some other birds of prey, photoreceptors are concentrated more on top of the inner eyeball, not at back. Such birds look down more than up when they fly and scan the ground for a meal. When they are on the ground, they cannot see something overhead very easily unless they turn their head almost upside-down.

Eyes with a large pupil and iris tell us something about the life-style of their owners. Animals that move about actively at night or in dimly lit habitats are less likely to stumble, bump into objects, or fall off cliffs when they intercept as much of the available light as they can in a given interval. The greater the amount of incoming light, the better is the visual discrimination.

Table 36.2 Components of the Vertebrate Eye

THREE LAYERS FORMING THE WALL OF EYEBALL	
Fibrous Tunic	*Sclera.* Protects eyeball *Cornea.* Focuses light
Vascular Tunic	*Choroid.* Blood vessels nutritionally support wall cells; pigments prevent light scattering *Ciliary body.* Its muscles control lens shape; its fine fibers hold lens in upright position *Iris.* Adjusting iris controls incoming light *Pupil.* Serves as entrance for light
Sensory Tunic	*Retina.* Absorbs and transduces light energy *Fovea.* Increases visual acuity *Start of optic nerve.* Carries signals to brain
INTERIOR OF EYEBALL	
Lens	Focuses light on photoreceptors
Aqueous humor	Transmits light, maintains pressure
Vitreous body	Transmits light, supports lens and eyeball

Further reading: Student Guide to InfoTrac on web site →

Figure 36.20 Pattern of retinal stimulation in the human eye. A transparent cornea positioned in front of the pupil is curved, so it changes the trajectories of light rays as they enter the eye. The pattern is upside-down and reversed left to right, compared with the stimulus in the visual field.

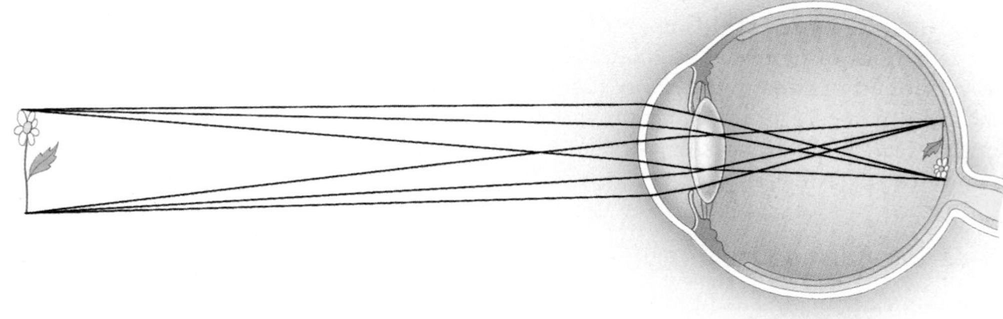

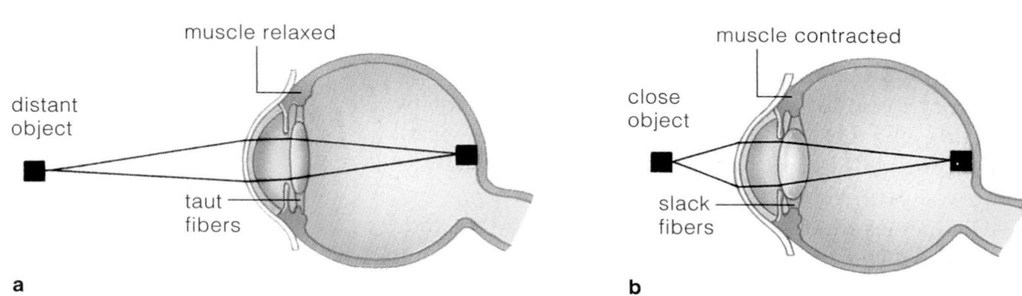
muscle relaxed
distant object
taut fibers
a
muscle contracted
close object
slack fibers
b

Figure 36.21 Two focusing mechanisms. A ciliary muscle encircles the lens and attaches to it. (**a**) When it relaxes, the lens flattens; the focal point moves farther back and brings distant objects into focus. (**b**) When the muscle contracts, the lens bulges; so the focal point moves closer and brings close objects into focus.

A large pupil lets in large quantities of light. A pupil ringed by a large iris may be dilated or constricted to admit more or less light, a useful option for animals that are active at night as well as during the day.

When light rays do converge at the back of the eye, they stimulate the retina in a distinct pattern. Because the cornea has a curved surface, light rays coming from a particular point in the visual field strike it at different angles, so that their trajectories change. (As described in Section 4.2, rays of light bend at boundaries between substances of different densities. Bending sends them off in new directions.) Because of their newly angled trajectories, light rays that converge at the back of the eyeball stimulate the retina in a pattern that is upside-down and reversed left to right, relative to the original source of the light rays. Figure 36.20 gives a simplified diagram of this outcome.

Visual Accommodation

Light rays emanating from sources at varying distances from the eye strike the cornea at different angles, which means that they could end up being focused at different distances behind it. This would do the brain no good. However, adjustments in the position or shape of the lens normally focus all incoming stimuli onto the retina. We call these lens adjustments **visual accommodation**. Without them, light rays from distant objects would be improperly focused in front of the retina, and rays from close objects would be focused behind it.

In the fish and reptilian eye, muscles move the lens forward or back, like a camera's focusing apparatus. Extending the distance between the lens and the retina moves the focal point forward; shrinking the distance moves it back. By contrast, in birds and mammals, the lens shape becomes adjusted. A ciliary muscle encircles the lens and attaches to it by fiberlike ligaments (Figure 36.18). When this muscle relaxes, the lens flattens and thereby moves the focal point farther back, as in Figure 36.21*a*. When the muscle contracts, the lens bulges and moves the focal point forward, as in Figure 36.21*b*.

In some cases, the lens can't be adjusted enough to make the focal point match up precisely with the retina. For example, in some people, the eyeball is not shaped quite right, and the position of the lens is either too close or too far away from the retina. When this is the case, accommodation alone cannot bring about an exact match. Eyeglasses often correct both visual disorders, which you undoubtedly know by their familiar names: nearsightedness and farsightedness (Section 38.9).

For most vertebrate eyes, the eyeball has three layers, and it encloses a lens, an aqueous humor, and a vitreous body.

A protective sclera and a light-focusing cornea make up the outer layer. The middle layer has a vascularized, pigmented choroid and other parts that admit and control incoming light. Photoreception occurs at the retina of the inner layer.

Adjustments in the positioning or shape of the lens focus incoming visual stimuli onto the retina.

36.8 CASE STUDY: FROM SIGNALING TO VISUAL PERCEPTION

We conclude the basics of this chapter with one of the classic examples of neuronal architecture—the sensory pathway from the retina to the brain. It shows how raw visual input is received, transmitted, and combined in ways that lead to awareness of light and shadows, of colors, of near and distant objects in the outside world.

How Is the Retina Organized?

Information flow begins as light strikes the retina. The retina's basement layer, a pigmented epithelium, covers the choroid. Resting on this layer are densely packed arrays of **rod cells** and **cone cells**, which are two classes of ciliary photoreceptors (Figure 36.22). Rod cells detect very dim light. During the night or in darkened places, they contribute to coarse perception of movement by detecting changes in light intensity across the visual field. Cone cells detect bright light. They contribute to sharp vision and color perception during the day.

Distinct layers of neurons are located above the rods and cones. The first to accept the visual information from photoreceptors are bipolar sensory neurons, then ganglion cells. Axons of these ganglion cells form the two optic nerves to the brain (Figure 36.23).

Before signals depart from the retina, they converge dramatically. The input from 125 million photoreceptors converges on only 1 million ganglion cells. Signals also flow laterally among horizontal cells and amacrine cells. Both kinds of neurons interact to dampen or enhance signals before ganglion cells get them. Thus, *a great deal of synaptic integration and processing goes on even before visual information is sent to the brain.*

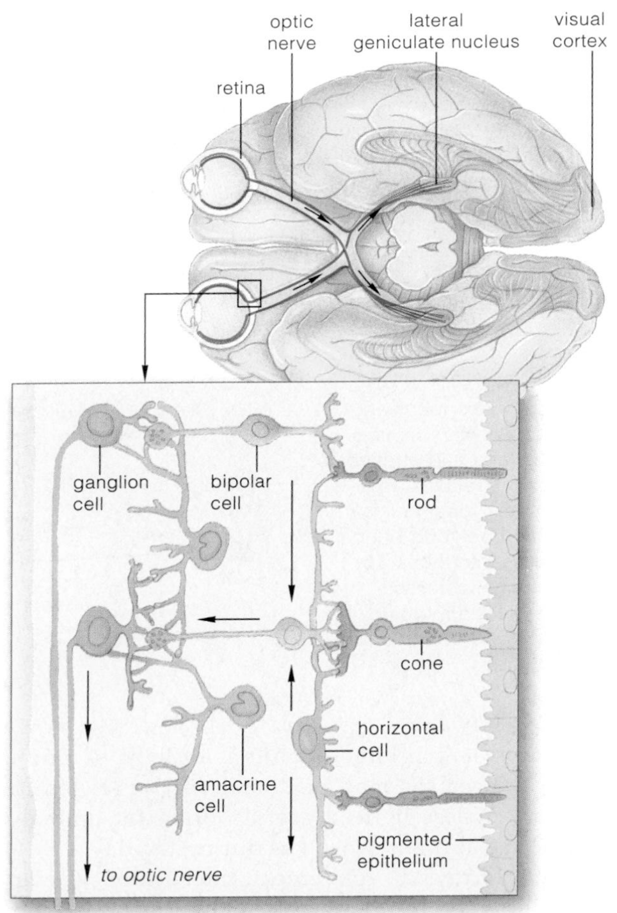
optic nerve
lateral geniculate nucleus
visual cortex
retina
ganglion cell
bipolar cell
rod
cone
horizontal cell
amacrine cell
pigmented epithelium
to optic nerve

Figure 36.23 Diagram of the sensory pathway from the retina to the brain.

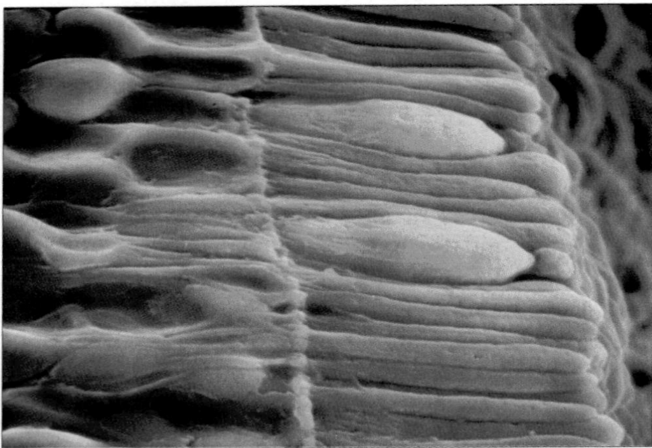

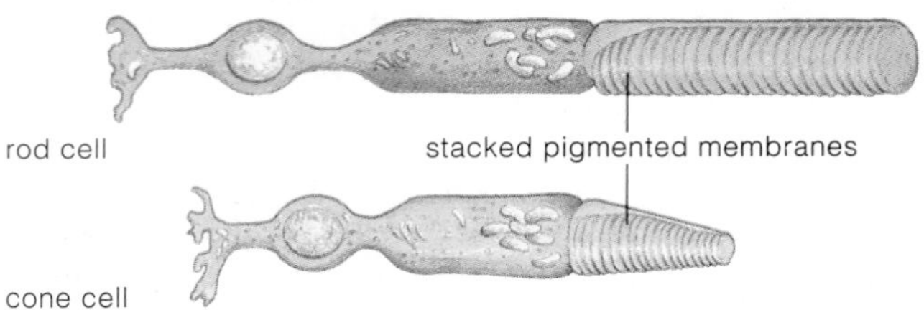
rod cell
stacked pigmented membranes
cone cell

Figure 36.22 Mammalian photoreceptors: rods and cones. 

Neuronal Responses to Light

ROD CELLS A rod cell's outer segment contains several hundred membrane disks, each peppered with about 10^8 molecules of a visual pigment called rhodopsin. The membrane stacking and the extremely high density of pigments greatly increase the likelihood of intercepting packets of light energy—that is, photons of particular wavelengths. The action potentials that result from the absorption of even a few photons can lead to conscious awareness of objects in dimly lit surroundings, as in dark rooms or late at night.

Rhodopsin effectively absorbs wavelengths in the blue-to-green portion of the visible spectrum (Section 7.2). Absorption changes the shape of the pigment. This triggers a cascade of reactions that alter activity at ion

Further reading: Student Guide to InfoTrac on web site

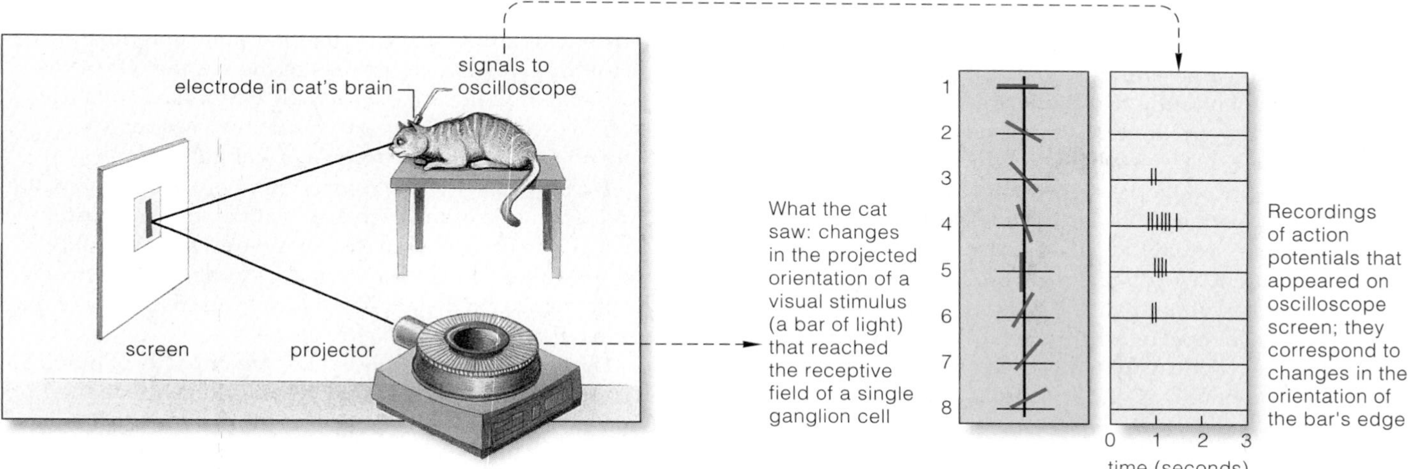

signals to oscilloscope
electrode in cat's brain
screen
projector
What the cat saw: changes in the projected orientation of a visual stimulus (a bar of light) that reached the receptive field of a single ganglion cell
1
2
3
4
5
6
7
8
Recordings of action potentials that appeared on oscilloscope screen; they correspond to changes in the orientation of the bar's edge
0
1
2
3
time (seconds)

Figure 36.24 Example of experiments into the nature of receptive fields for visual stimuli. David Hubel and Torsten Wiesel implanted an electrode in an anesthetized cat's brain. They placed the cat in front of a small screen upon which different patterns of light were projected—in this case, a hard-edged bar. Light or shadow falling on part of the screen excited or inhibited signals sent to a single neuron in the visual cortex.

Tilting the bar at different angles produced changes in the neuron's activity. A vertical bar image produced the strongest signal (*numbered 5 in the sketch*). When the bar image tilted slightly, signals were less frequent. When it tilted past a certain angle, signals stopped.

channels and ion pumps across the rod cell's plasma membrane. Gated sodium channels close, the voltage difference changes across the membrane, and a graded potential results. This potential reduces the ongoing release of a neurotransmitter with inhibitory effects on adjacent sensory neurons. Released from inhibition, the sensory neurons start sending signals about the visual stimulus on toward the brain.

CONE CELLS The sense of color and of daytime vision starts with photon absorption by red, green, and blue cone cells, each with a different kind of visual pigment. Here again, photon absorption reduces the release of a neurotransmitter that otherwise inhibits the neurons that are adjacent to the photoreceptors. Near the center of the retina is a funnel-shaped depression called the fovea. This is the retinal area with the greatest density of photoreceptors and the greatest visual acuity. Cone cells of the fovea are the ones that discriminate the most precisely between adjacent points in space.

RESPONSES AT RECEPTIVE FIELDS Neurons in the eye collectively respond to information about light in highly organized ways. The retinal surface is organized into receptive fields, or restricted areas that influence the activity of individual sensory neurons. For example, the field is a tiny circle for each ganglion cell. Some cells respond best to a spot of light ringed by dark in the field's center. Others respond to rapidly changing light intensity, to a spot of one color, or to motion. In one experiment, sensory cells generated signals after detecting signals about a suitably oriented bar (Figure 36.24). Such cells do not respond to diffuse, uniform illumination, which is just as well. Doing so would send a confusing array of signals to the brain.

ON TO THE VISUAL CORTEX The part of the outside world that an individual actually sees is its visual field. The right side of both retinas intercepts light from the *left* half of the visual field; the left side intercepts light from the *right* half. The optic nerve leading away from each eye delivers signals about a stimulus from the *left* visual field to the right cerebral hemisphere. It delivers signals from the *right* visual field to the left hemisphere. (Here you may wish to refer to Sections 35.5 and 35.6.)

Axons of optic nerves end in the lateral geniculate nucleus, an island of gray matter inside the cerebrum. Each layer has a map corresponding to receptive fields of the retina. Its interneurons deal with one aspect of the stimulus—form, movement, depth, color, texture, and so on. After initial processing, all the visual signals travel rapidly, at the same time, to different parts of the visual cortex. There, final integration results in highly organized electrical activity and the sensation of sight.

Each eye is an outpost of the brain, collecting and analyzing information about the distance, shape, brightness, position, and movement of visual stimuli. Its sensory pathway starts at the retina and proceeds along an optic nerve to the brain.

Organization of visual signals begins at receptive fields in the retina, it is further processed in the lateral geniculate nucleus, and it is finally integrated in the visual cortex.

36.9 DISORDERS OF THE HUMAN EYE

Of all the sensory receptors that the human brain requires to help keep you alive and functioning independently in the environment, fully two-thirds are located in your pair of eyes. They are photoreceptors, and they do more than detect light. They also allow you to see the world in a rainbow of colors. The eyes are the single most important source of information about the outside world.

Given their essential role, it is no surprise that so much attention is paid to eye disorders, as brought about by injuries, inherited abnormalities, diseases, and advancing age. The consequences range from relatively harmless conditions, such as nearsightedness, to total blindness. Each year, many millions of people must deal with such consequences.

FOCUSING PROBLEMS Other heritable abnormalities arise from misshapen features of the eye that affect the focusing of light. *Astigmatism*, for example, results from corneas with an uneven curvature; they cannot bend incoming light rays to the same focal point.

Nearsightedness, or myopia, commonly occurs when the horizontal axis of the eyeball is longer than the vertical axis. It also occurs when the ciliary muscle responsible for adjusting the lens contracts too strongly. The outcome is that images of distant objects are focused in front of the retina instead of on it (Figure 36.25).

Farsightedness, or hyperopia, is the opposite of myopia. Here, the vertical axis of the eyeball is longer than the horizontal axis (or the lens is "lazy"). As a result, close images are focused behind the retina (Figure 36.26).

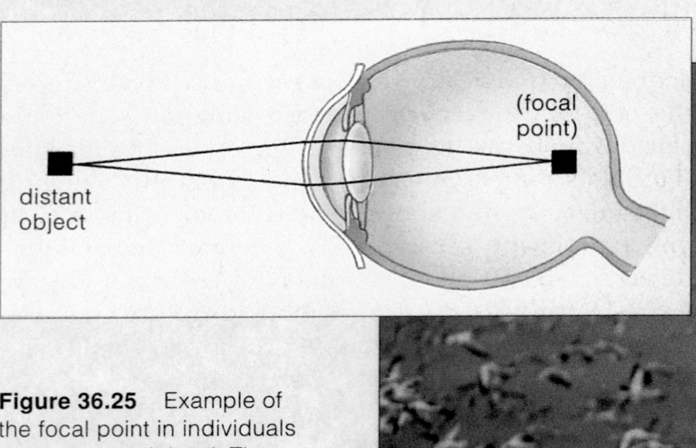
(focal point)
distant object

Figure 36.25 Example of the focal point in individuals who are nearsighted. The birds are flamingos at a lake in Tanzania, East Africa. To a nearsighted person, the birds in flight in the distance would be out of focus, as shown here.

COLOR BLINDNESS Occasionally, some or all of the cone cells that selectively respond to light of red, green, or blue wavelengths fail to develop in individuals. The rare individuals who have only one of three kinds of cones are totally *color-blind*. They can perceive the world only in shades of gray.

Consider a common genetic abnormality, *red–green color blindness*. It is an X-linked, recessive trait that shows up most often in males. The retina of affected individuals lacks some or all of the cone cells with pigments that normally respond to light of red or green wavelengths. Most of the time, people who are affected by red–green color blindness merely have trouble distinguishing red from green in dim light. Some cannot distinguish between the two even in bright light.

Even a normal lens loses some of its natural flexibility as a person grows older. That is why people over forty years old often start wearing eyeglasses.

EYE DISEASES The structure and functioning of the eyes are vulnerable to infectious diseases. For example, *histoplasmosis*, a fungal infection of the lungs that is described in Section 24.5, can lead to retinal damage, which can cause partial or total loss of vision. As another example, *Herpes simplex*, a virus that causes skin sores, also can infect the cornea and cause it to ulcerate.

Trachoma is an extremely contagious disease that has blinded millions, mostly in North Africa and the Middle East. The culprit is a bacterium that also is responsible for the sexually transmitted disease chlamydia (Section 45.14).

Further reading: Student Guide to InfoTrac on web site —

The eyeball and the lining of the eyelids (the conjunctiva) become damaged. The damaged tissues are entry points for bacteria that can cause secondary infections. In time the cornea can become so scarred that blindness follows.

AGE-RELATED PROBLEMS You have probably heard of *cataracts*, a gradual clouding of the lens. It is a problem associated with aging, although it also may arise through eye injury or diabetes. Possibly cataracts form when the transparent proteins that make up the eye's lens undergo structural changes. The clouding may skew the trajectory of incoming light rays. If the lens becomes opaque, light cannot enter the eye at all.

Glaucoma, a different age-related problem, results when excess aqueous humor accumulates inside the eyeball. The blood vessels that service the retina collapse under the increased fluid pressure. Vision deteriorates as sensory neurons of the retina and optic nerve die off.

Although we often associate chronic glaucoma with advanced age, the conditions that give rise to the disorder actually start to develop in middle age. If doctors detect the fluid pressure that tends to build up in the eye early enough, they can relieve the fluid pressure with drugs or surgery before damage becomes severe.

EYE INJURIES *Retinal detachment* is the eye injury that we read about most often. It may follow a physical blow to the head or illness that causes a tear in the retina. As the jellylike vitreous body oozes through the torn region, the retina is lifted away from the underlying choroid. In time the retina may peel away entirely, leaving its blood supply behind.

Early symptoms of a detached retina include blurred vision, flashes of light that occur even in the absence of outside visual stimulation, and loss of peripheral vision. Without medical intervention, the person may become totally blind in the damaged eye.

NEW TECHNOLOGIES Today a variety of tools can be used to correct some eye disorders. In *corneal transplant surgery*, for example, a defective cornea can be removed, and then an artificial cornea (made of clear plastic) or a natural cornea from a donor can be stitched in its place. Within a year, the patient can be fitted with eyeglasses or contact lenses. Similarly, cataracts can be surgically corrected by removing the lens and replacing it with an artificial one, although the operation is difficult and not always successful.

Severely nearsighted people sometimes opt for *radial keratotomy*, a still-controversial surgical procedure in which tiny, spokelike incisions are made around the edge of the cornea to flatten it more. When all goes well, the adjustment eliminates the need for corrective lenses. But vision sometimes is overcorrected or undercorrected, in which case the patient must undergo further surgery.

As a final example, retinal detachment may be treated with *laser coagulation*, a painless technique in which a laser beam seals off leaky blood vessels and "spot welds" the retina to the underlying choroid.

(focal point)
close object

Figure 36.26 Example of the focal point in farsighted vision. To a farsighted person, the birds standing in water in the foreground would be out of focus, as shown here.

36.10 SUMMARY

1. A stimulus is a specific form of energy that the body detects by means of sensory receptors. A sensation is an awareness that stimulation has occurred. Perception is understanding what the sensation means.

2. Sensory receptors are endings of sensory neurons or specialized cells next to them. They respond to stimuli, which are specific forms of energy, such as mechanical pressure and light. Animals respond to aspects of the internal or external environment only when they have receptors that are sensitive to the energy of stimuli.

a. Mechanoreceptors, including free nerve endings and hair cells, detect mechanical energy related to touch, pressure, motion, and changes in position.

b. Thermoreceptors detect radiant energy (heat).

c. Pain receptors (nociceptors) detect tissue damage.

d. Chemoreceptors, such as olfactory receptors and taste receptors, detect chemical substances dissolved in the fluid bathing them.

e. Osmoreceptors detect changes in water volume (hence solute concentrations) in the surrounding fluid.

f. Photoreceptors detect light. Rods and cones of the retina in the human eye are examples.

3. A sensory system has sensory receptors for specific stimuli and nerve pathways that lead from receptors to receiving and processing centers in the brain. The brain assesses a particular stimulus based upon which nerve pathway is delivering signals, the frequency of signals traveling along individual axons of that pathway, and the number of axons that were recruited into action.

4. Somatic sensations include touch, pressure, pain, temperature, and muscle sense. The receptors associated with these sensations are not localized in a single organ or tissue. Special senses include taste, smell, hearing, balance, and vision. Receptors associated with these senses typically reside in sensory organs, such as eyes, or some other particular body region.

5. The senses of taste and smell both involve sensory pathways from chemoreceptors to processing regions in the cerebral cortex and limbic system.

6. Organs of equilibrium, as in the vertebrate inner ear, detect gravity, velocity, acceleration, and other forces that affect the body's positions and movements. The vertebrate sense of hearing (sound perception) requires components of the outer, middle, and inner ear that respectively collect, amplify, and sort out sound waves.

7. Vision requires eyes: sensory organs having a dense array of photoreceptors, as in a retina. It requires a capacity for image formation in the brain, based upon incoming patterns of visual stimulation. The sense of vision and discrimination among objects evolved first among the earliest fast-moving, predatory animals.

Review Questions

1. Define stimulus. *CI (see also Chapter 35 CI)*
2. Name six categories of sensory receptors and the type of stimulus energy that each kind detects. *36.1*
3. What are the basic components of a sensory system? How does the brain assess the nature of a given stimulus? *36.1*
4. How do somatic sensations differ from special senses? *36.1*
5. What is pain? Describe one type of pain receptor. *36.2*
6. Name the five primary taste sensations. Which substances elicit each type of sensation? *36.3*
7. Define pheromone and give an example. *36.3*
8. Which evolved first, a sense of balance or sense of hearing? Do both involve the outer, middle, and inner ear? *36.4, 36.5*
9. How does vision differ from light sensitivity? What sensory organs and structures does vision require? *36.6*
10. Label the component parts of the human eye shown in the diagram below: *36.7*

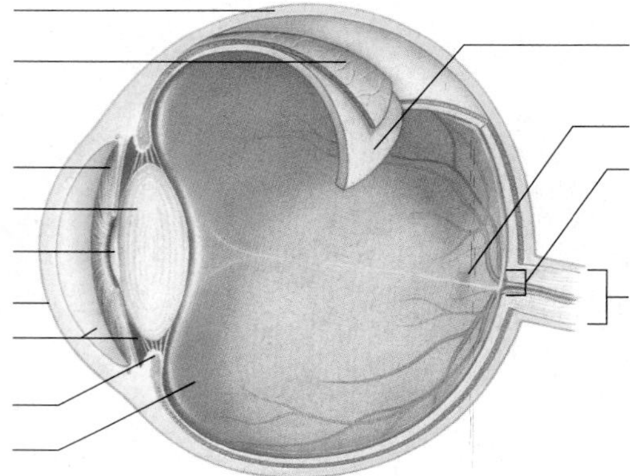

Self-Quiz (*Answers in Appendix III*)

1. A ______ is a specific form of energy that is detected by a sensory receptor.

2. Conscious awareness of a stimulus is called a ______.

3. ______ is actually understanding what particular sensations mean.

4. Each sensory system consists of ______.
   a. nerve pathways from specific receptors to the brain
   b. sensory receptors
   c. brain regions that deal with sensory information
   d. all of the above

5. ______ may be defined as a decrease in the response to an ongoing stimulus.
   a. Perception
   b. Sensory adaptation
   c. Visual accommodation
   d. both b and c

6. ______ detect mechanical energy associated with changes in pressure, position, or acceleration.
   a. Chemoreceptors
   b. Mechanoreceptors
   c. Photoreceptors
   d. Thermoreceptors

7. Detecting light energy is the function of ______.
   a. chemoreceptors
   b. mechanoreceptors
   c. photoreceptors
   d. thermoreceptors

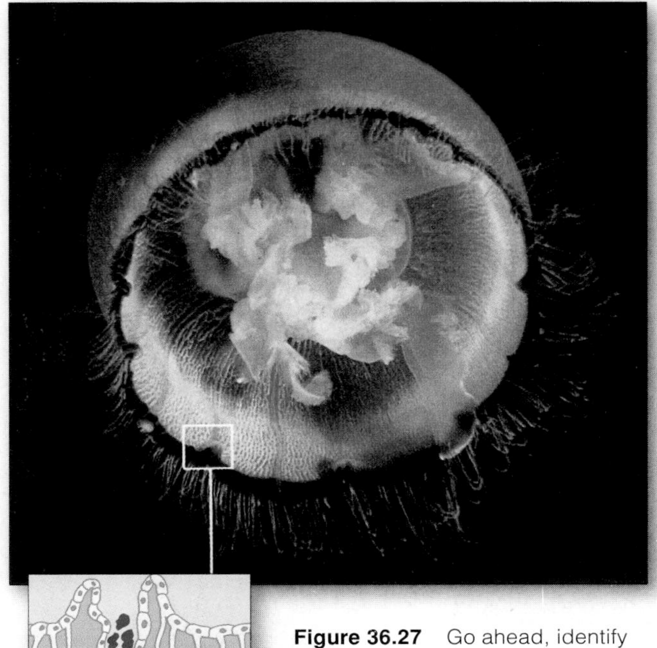
particles

Figure 36.27 Go ahead, identify the function of the mystery particles located on the bell-shaped margin of this jellyfish medusa.

8. Free nerve endings are ______.
   a. mechanoreceptors c. pain receptors
   b. thermoreceptors d. all of the above

9. Is this statement true or false: Visceral pain sensations are associated with internal organs.

10. We now recognize ______ primary sensations of taste.
   a. two b. three c. four d. five

11. In humans, the ______ ear contains organs that can detect accelerated, rotational motions of the head.
   a. outer b. middle c. inner e. all of the above

12. The inner layer of the human eyeball includes the ______.
   a. retina c. lens and choroid
   b. sclera and cornea d. both a and c

13. ______ detect bright light; ______ detect very dim light.
   a. rods; cones b. cones; rods

14. Match each term with the most suitable description.

____ somatic sensations	a. the part of the outside world that an animal sees
____ somatosensory cortex	b. e.g., vision (by way of eyes) and hearing (by way of ears)
____ special senses	c. form of energy that a specific sensory receptor can detect
____ variations in stimulus intensity	d. encoded in the frequency and number of action potentials
____ visual field	e. touch, pressure, temperature, pain, and muscle sense
____ stimulus	f. has maps of body surface

Critical Thinking

1. On the bell-shaped rim of the jellyfish shown in Figure 36.27 are tiny saclike structures that hold calcium particles next to a sensory cilium. When the bell tilts, the particles slide over the cilium. Would you assume that these structures contribute to the sense of taste, smell, balance, hearing, or vision?

2. Wayne, on standby for the last flight from San Francisco to New York, was assigned the last available seat on the plane. It was in the last row, where vibrations and noise from the jet engines are most pronounced. When Wayne got off the plane in New York, he was speaking loudly and having trouble hearing what other people were saying. The next day, things were back to normal. What happened to his hearing during the flight?

3. Juanita made an appointment with her doctor because she was experiencing recurring episodes of dizziness. Her doctor immediately asked her whether "dizziness" meant she had sensations of lightheadedness, as if she were going to faint, or whether it meant she had sensations of vertigo—that is, a feeling that she herself or objects near her were spinning around. Why did her doctor consider this clarification important early in his evaluation of her condition?

4. Michael, now three years old, experiences chronic *middle-ear infection*, which is now common among youngsters enrolled in crowded day-care centers. This year, despite antibiotic treatment, an infection became so advanced that he had some trouble hearing. Then Michael's left eardrum ruptured and a thickened, jellylike substance dribbled out. His mother was startled and upset, but the pediatrician told her not to worry, that if the eardrum had not ruptured on its own, he would have had to insert a very small drainage tube into it. Speculate on why the pediatrician concluded that this would have been a necessary surgical intervention.

Selected Key Terms

acoustical receptor *36.5*
camera eye *36.6*
chemoreceptor *36.1*
cochlea *36.5*
compound eye *36.6*
cone cell *36.8*
cornea *36.6*
echolocation *CI*
encapsulated receptor *36.2*
external ear *36.5*
eye *36.6*
free nerve ending *36.2*
hair cell *36.4*
hearing *36.5*
inner ear *36.4*
lens *36.6*
mechanoreceptor *36.1*
middle ear *36.5*
ocellus *36.6*
olfactory receptor *36.3*
osmoreceptor *36.1*
otolith *36.4*
pain *36.2*
pain receptor *36.1*
perception *36.1*
pheromone *36.3*
photoreceptor *36.1*
retina *36.6*
rod cell *36.8*
sensation *36.1*
sensory adaptation *36.1*
sensory system *36.1*
somatic sensation *36.1*
somatosensory cortex *36.2*
special sense *36.1*
taste receptor *36.3*
thermoreceptor *36.1*
vestibular apparatus *36.4*
vision *36.6*
visual accommodation *36.7*
visual field *36.6*

Readings See also *www.infotrac-college.com*

Delcomyn, F. 1998. *Foundations of Neurobiology*. New York: Freeman.

Romer, A., and T. Parsons. 1986. *The Vertebrate Body*. Sixth edition. New York: Saunders.

Sherwood, L. 1997. *Human Physiology*. Third edition. Belmont, California: Wadsworth.

Travis, J. 10 April 1999. "Making Sense of Scents." *Science News* 155:236–238.

Wright, K. April 1994. "The Sniff of Legend." *Discover* 15(4): 60–67. Speculation on the possibility of a sensory pathway that might be activated by human pheromones.

Zeki, S. September 1992. "The Visual Image in Mind and Brain." *Scientific American* 267(3): 68–76.

37 ENDOCRINE CONTROL

Hormone Jamboree

In the 1960s, at her camp in a forest by the shore of Lake Tanganyika in Tanzania, the primatologist Jane Goodall let it be known that bananas were available. One of the first chimpanzees attracted to the delicious food was a female—Flo, as she came to be called (Figure 37.1). Flo brought along her two offspring, an infant female and a juvenile male, and exhibited commendable parental behavior toward them.

Three years passed, and Goodall observed that Flo's preoccupation with motherhood gave way to a preoccupation with sex. She also observed that male chimpanzees followed Flo to the camp by the lake and stayed for more than the bananas.

Sex, Goodall discovered, is *the* premier binding force in the social life of chimpanzees.

These primates do not mate for life as eagles do, or wolves. Before the rainy season begins, mature females that are entering their fertile cycle become sexually active. As one of the more dramatic responses to changing concentrations of hormones in their bloodstream, the external sexual

b
a

Figure 37.1 (**a**) Primatologist Jane Goodall in Gombe National Park, near the shores of Lake Tanganyika, scouting for chimpanzees. (**b**) Flo and three of her offspring, all subjects of long-term field observations. Flo helped Goodall clarify the central role of sex—and of the hormones that orchestrate it—in the social life of these primate relatives of humans.

organs of the females become swollen and vivid pink. The swellings are strong visual signals to males. They are the flags of sexual jamborees, of great gatherings of highly stimulated chimps in which any males present may have a turn at copulating with the same female.

The gathering of many flag-waving females draws together individuals that forage alone or in small family groups for most of the year. It reestablishes bonds that unite a rather fluid community.

As infants and juveniles play with one another and with the adults, their aggressive and submissive jostlings help map out future dominance hierarchies. Flo happened to be a high-ranking individual in the social hierarchy. Through her sexual attractiveness and direct solicitations, she built alliances with many male chimps. Through her status and aggressive behavior, she helped her male offspring win confrontations with other young male chimps.

The hormone-induced swelling during the fertile period of Flo and other female chimps lasts somewhere between ten and sixteen days. Yet females are fertile for only one to five days. Sex hormones induce swelling even after a female gets pregnant. We can hypothesize that sexual selection has favored prolonged swelling. Males groom a sexually attractive female more often, protect her, give her more food, and let her tag along to new foraging sites. The more that males accept a female, the higher she rises in the social hierarchy and the more her individual offspring benefit.

Through their effects, hormones help orchestrate the growth, development, and reproductive cycles of nearly all animals, from the invertebrate worms to chimpanzees and humans. They influence minute-by-minute and day-to-day metabolic functions. Through their interplays with one another and with the nervous system, hormones have profound influence over the physical appearance, the well-being, and the behavior of individuals. Even the behavior of individuals affects whether they will survive, either on their own or as part of social groups.

This chapter focuses mainly on hormones—on their sources, targets, and interactions as well as the mechanisms involved in their secretion. If the details start to seem remote, remember that this is the stuff of life. Hormones underwrote Flo's appearance, behavior, and rise through the chimpanzee social hierarchy. Just imagine what they have been doing for you.

KEY CONCEPTS

1. For nearly all animals, hormones and other signaling molecules have central roles in integrating the activities of individual cells in ways that benefit the whole body.

2. Only the cells with molecular receptors for a specific hormone are its targets. Hormones operate by serving as signals for change in the activities of their targets.

3. Many types of hormones influence gene transcription and protein synthesis in target cells. Other types call for alterations in existing molecules and structures in cells. Some exert their effects by binding with and altering membrane characteristics, such as the permeability of the plasma membrane to a particular solute.

4. Some hormones help the body adjust to short-term shifts in diet and levels of activity. Others help induce long-term adjustments in cell activities that bring about the body's growth, development, and reproduction.

5. Among vertebrates, the hypothalamus and pituitary gland interact in ways that coordinate the activities of a number of endocrine glands. Together, they exert control over many aspects of the body's functioning.

6. Hormones, neural signals, shifts in local chemical conditions, and environmental cues function as triggers for the secretion of particular hormones.

37.1 THE ENDOCRINE SYSTEM

Hormones and Other Signaling Molecules

Throughout their lives, cells must respond to changing conditions by taking up and releasing various chemical substances. In vertebrates, the responses of millions to many billions of cells must be integrated in ways that help keep the whole body alive and functioning.

Signaling molecules help integrate activities within cells, between cells, and between tissues. We call them hormones, neurotransmitters, local signaling molecules, and pheromones. Each acts on target cells. A "target" is any cell that has receptors for a signaling molecule and that may change its activities in response to it. Targets may or may not be adjacent to cells that send signals.

By definition, **hormones** are secretory products of endocrine glands, endocrine cells, and certain neurons. Animal hormones travel the bloodstream to nonadjacent target cells. They are this chapter's focus.

By contrast, **neurotransmitters** are released from the axon endings of neurons, then act swiftly on target cells by diffusing across a gap (synaptic cleft) between them. You read about their sources and effects in Section 34.3. **Local signaling molecules** released by many types of body cells alter conditions in local tissues. For instance, some prostaglandins decrease or increase blood flow to certain tissues. They do this by causing smooth muscle cells in blood vessel walls to constrict or dilate.

Pheromones, nearly odorless secretions of certain exocrine glands, diffuse through water or air to targets outside the animal body. These hormone-like secretions act on cells of other individuals of the same species. They help integrate social behavior, such as behaviors related to sexual reproduction. Pheromones are common signals among animals. As examples, female silk moths secrete bombykol as a sex attractant (Section 47.4), and soldier termites secrete alarm signals when ants attack their colony (Section 47.9). The vomeronasal organ, a common pheromone detector in animals, was recently discovered in humans. Its presence invites speculation: Do human pheromones act at the subconscious level to trigger those inexplicable impressions, such as instant good or bad "feelings" about somebody you just met? We do not know the answer, but studies are under way.

Discovery of Hormones and Their Sources

The word "hormone" dates to the early 1900s. W. Bayliss and E. Starling were trying to find out what triggers the secretion of pancreatic juices when food is traveling through the canine gut. As they knew, acids are mixed with food in the stomach, and the pancreas secretes an alkaline solution after the acidic mixture is propelled forward into the small intestine. Was the nervous system or something else stimulating the pancreatic response?

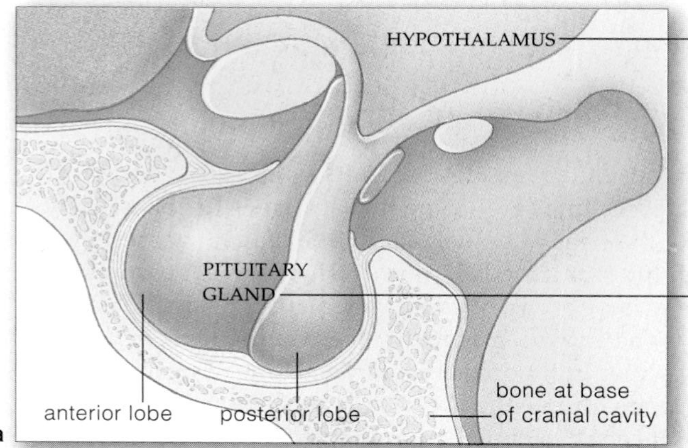

HYPOTHALAMUS
PITUITARY GLAND
anterior lobe
posterior lobe
bone at base of cranial cavity
a

Figure 37.2 (**a**) A major neural–endocrine control center. The pituitary gland interacts intimately with the hypothalamus, a brain region that secretes some hormones. (**b**) *Facing page:* Overview of the key components of the human endocrine system and of the primary effects of their main hormonal secretions. The system also includes endocrine cells of many organs, such as the liver, kidneys, heart, small intestine, and skin.

To find an answer, Bayliss and Starling blocked the nerves—but not blood vessels—leading to a laboratory animal's upper small intestine. Later, when acidic food entered the small intestine, the pancreas still secreted the alkaline solution. Even more telling, extracts of cells from the intestinal lining—a glandular epithelium—also induced the response. Glandular cells had to be the source of the pancreas-stimulating substance.

The substance came to be called secretin. Proof of its existence and mode of action confirmed a centuries-old idea: *The bloodstream picks up internal secretions that can influence the activities of organs inside the body.* Starling coined the word "hormone" for such internal glandular secretions (after *hormon*, meaning to set in motion).

Later on, researchers identified many other kinds of hormones and their sources. Figure 37.2 is a simplified picture of the locations of the following major sources of hormones for the human body. Bear in mind, most vertebrates have the same hormone sources:

- Pituitary gland
- Adrenal glands (*two*)
- Pancreatic islets (*numerous cell clusters*)
- Thyroid gland
- Parathyroid glands (*in humans, four*)
- Pineal gland
- Thymus gland
- Gonads (*two*)
- Endocrine cells of the hypothalamus, stomach, small intestine, liver, kidneys, heart, placenta, skin, adipose tissue, and other organs

Further reading: Student Guide to InfoTrac on web site →

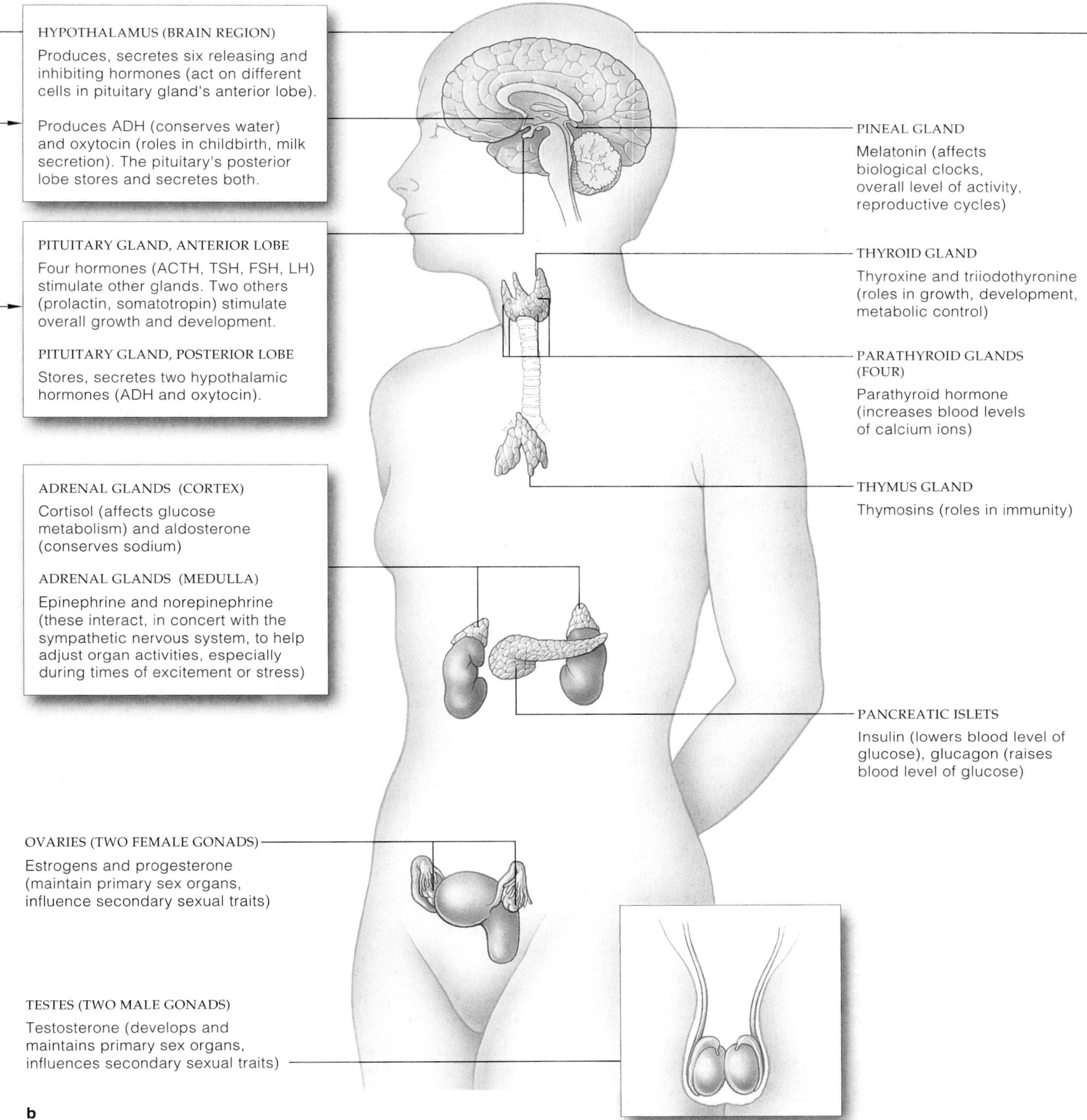

HYPOTHALAMUS (BRAIN REGION)
Produces, secretes six releasing and inhibiting hormones (act on different cells in pituitary gland's anterior lobe).
Produces ADH (conserves water) and oxytocin (roles in childbirth, milk secretion). The pituitary's posterior lobe stores and secretes both.
PITUITARY GLAND, ANTERIOR LOBE
Four hormones (ACTH, TSH, FSH, LH) stimulate other glands. Two others (prolactin, somatotropin) stimulate overall growth and development.
PITUITARY GLAND, POSTERIOR LOBE
Stores, secretes two hypothalamic hormones (ADH and oxytocin).
ADRENAL GLANDS (CORTEX)
Cortisol (affects glucose metabolism) and aldosterone (conserves sodium)
ADRENAL GLANDS (MEDULLA)
Epinephrine and norepinephrine (these interact, in concert with the sympathetic nervous system, to help adjust organ activities, especially during times of excitement or stress)
OVARIES (TWO FEMALE GONADS)
Estrogens and progesterone (maintain primary sex organs, influence secondary sexual traits)
TESTES (TWO MALE GONADS)
Testosterone (develops and maintains primary sex organs, influences secondary sexual traits)
PINEAL GLAND
Melatonin (affects biological clocks, overall level of activity, reproductive cycles)
THYROID GLAND
Thyroxine and triiodothyronine (roles in growth, development, metabolic control)
PARATHYROID GLANDS (FOUR)
Parathyroid hormone (increases blood levels of calcium ions)
THYMUS GLAND
Thymosins (roles in immunity)
PANCREATIC ISLETS
Insulin (lowers blood level of glucose), glucagon (raises blood level of glucose)
b

Collectively, the body's sources of hormones came to be called the **endocrine system**. The name implies there is a separate control system for the body, apart from the nervous system. (*Endon* means "within" and *krinein* is taken to mean "secrete.") Later, biochemical research and electron microscopy studies revealed that endocrine sources and the nervous system function in intricately connected ways, as you will see shortly.

Integration of cell activities depends on hormones and other signaling molecules. Each type of signaling molecule acts on target cells, which are any cells that have receptors for it and that may alter their activities in response to it.

Components of the endocrine system and certain neurons produce and secrete hormones, which the bloodstream takes up and distributes to nonadjacent target cells.

37.2 SIGNALING MECHANISMS

The Nature of Hormonal Action

Hormones and other signaling molecules interact with protein receptors of target cells, and their interactions have diverse effects on physiological processes. Some kinds of hormones induce a target cell to increase its uptake of glucose, calcium, or some other substance from the surroundings. Others stimulate or inhibit the target in ways that alter its rates of protein synthesis, modify the structure of proteins or other elements in the cytoplasm, or change cell shape.

Two factors greatly influence responses to hormonal signals. *First,* different hormones activate different kinds of mechanisms in target cells. *Second,* not all types of cells are equipped to respond to a particular signal. For example, many cell types have receptors for cortisol, so this hormone has widespread effects through the body. By contrast, only a few cell types have the receptors for hormones that stimulate a highly directed response.

Let us now briefly consider the effects of two main categories of these signaling molecules: the steroid and peptide hormones (Table 37.1).

Characteristics of Steroid Hormones

Cholesterol is the precursor for lipid-soluble molecules called **steroid hormones**. Steroid-secreting cells in the adrenal glands and in the primary reproductive organs (gonads) remodel the cholesterol at multiple-enzyme systems in endoplasmic reticulum and mitochondria. Testosterone, one of the sex hormones, an example of the derivatives.

Think of testosterone's effects on the development of secondary sexual traits associated with "maleness." The developmental steps will proceed normally only if target cells have working receptors for testosterone. In a genetic disorder called *testicular feminization syndrome,* these receptors are defective. Genetically, the affected person is male (XY), because he has functional testes that are able to secrete testosterone. Target cells cannot respond to the hormone, however, so secondary sexual traits that do develop are like those of females.

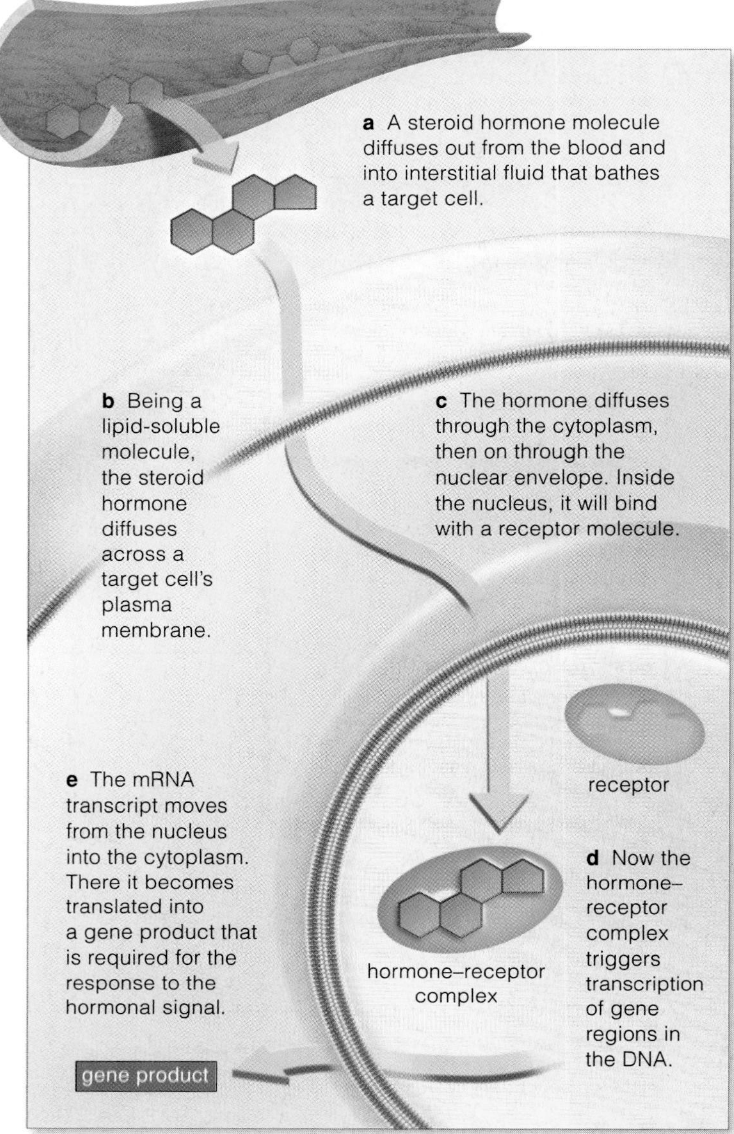

a A steroid hormone molecule diffuses out from the blood and into interstitial fluid that bathes a target cell.
b Being a lipid-soluble molecule, the steroid hormone diffuses across a target cell's plasma membrane.
c The hormone diffuses through the cytoplasm, then on through the nuclear envelope. Inside the nucleus, it will bind with a receptor molecule.
receptor
d Now the hormone–receptor complex triggers transcription of gene regions in the DNA.
hormone–receptor complex
e The mRNA transcript moves from the nucleus into the cytoplasm. There it becomes translated into a gene product that is required for the response to the hormonal signal.
gene product

Figure 37.3 Example of a mechanism by which a steroid hormone initiates changes in a target cell's activities.

How does a steroid hormone exert its effects on a target cell? Being lipid soluble, it can diffuse directly across the lipid bilayer of the cell's plasma membrane (Figure 37.3). Once inside the cytoplasm, the hormone molecule usually moves into the nucleus and binds to some type of protein receptor. In some cases, however, it binds to a receptor molecule in the cytoplasm, then the hormone–receptor complex enters the nucleus. The specific shape of the complex allows it to interact with a specific region of the cell's DNA. Different complexes inhibit or stimulate transcription of target gene regions

Table 37.1 Two Main Categories of Hormones

Type	Examples
Steroid and steroid-like hormones	Estrogens (feminizing effects), progestins (related to pregnancy), androgens (such as testosterone; masculinizing effects), cortisol, aldosterone. Thyroid hormones and vitamin D act like steroid hormones.
Peptide hormones	
Peptides	Glucagon, ADH, oxytocin, TRH
Proteins	Insulin, somatotropin, prolactin
Glycoproteins	FSH, LH, TSH

Further reading: Student Guide to InfoTrac on web site —

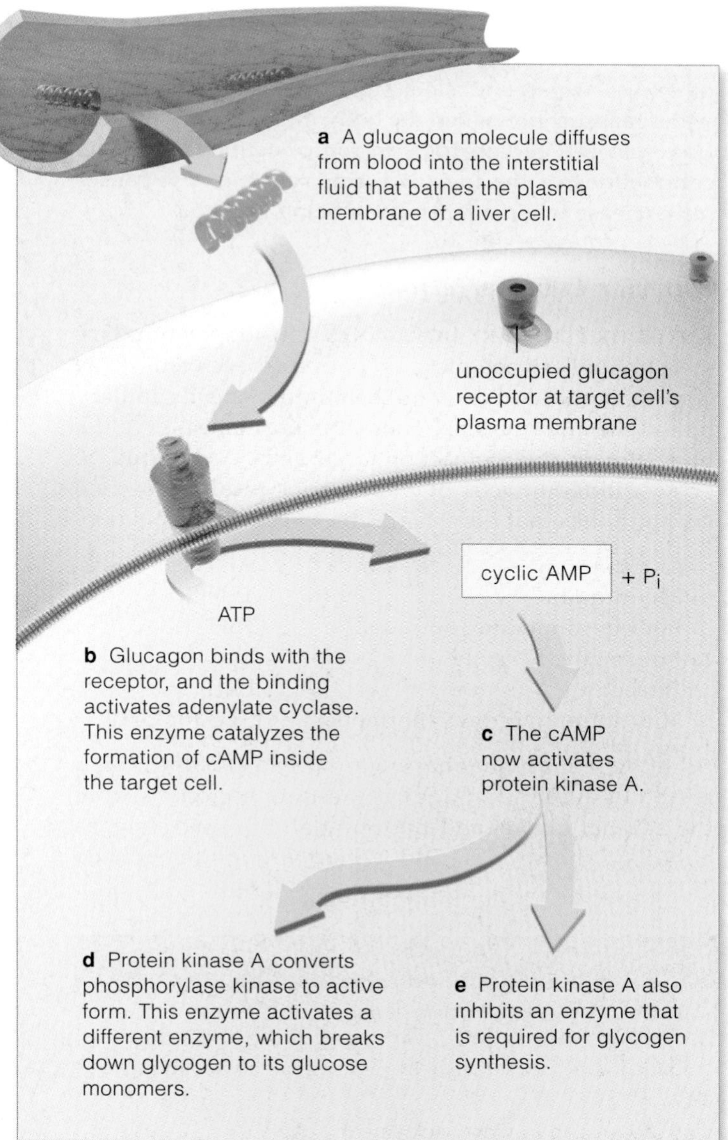
a A glucagon molecule diffuses from blood into the interstitial fluid that bathes the plasma membrane of a liver cell.
unoccupied glucagon receptor at target cell's plasma membrane
cyclic AMP + Pi
ATP
b Glucagon binds with the receptor, and the binding activates adenylate cyclase. This enzyme catalyzes the formation of cAMP inside the target cell.
c The cAMP now activates protein kinase A.
d Protein kinase A converts phosphorylase kinase to active form. This enzyme activates a different enzyme, which breaks down glycogen to its glucose monomers.
e Protein kinase A also inhibits an enzyme that is required for glycogen synthesis.

Figure 37.4 Example of a mechanism by which a peptide hormone initiates changes in a target cell's activities. When the hormone glucagon binds at a receptor, it initiates reactions in the cell. In this case cyclic AMP, which is one type of second messenger, relays the signal into the cell interior.

into mRNA. The translation of such mRNA transcripts results in enzymes and other proteins that can carry out a response to the hormonal signal.

Steroid hormones may exert effects in another way. It now appears that some types with receptors on cell membranes may alter the properties of the membrane itself to modify the functions of the target cell.

Part of a group of genes that specifies receptors for steroid hormones also specifies thyroid hormone and vitamin D receptors. Thyroid hormones and vitamin D are not steroid hormones, but they do behave like them.

Characteristics of Peptide Hormones

Traditionally, **peptide hormones** have been defined as water-soluble signaling molecules having anywhere from 3 to 180 amino acids. Many peptides, polypeptides, and glycoproteins fall in this category. Each peptide hormone binds to a receptor at the plasma membrane of a cell, and the receptor activates specific membrane-bound enzyme systems. The systems initiate reactions leading to the cellular response.

Let's focus on a liver cell that has receptors for the peptide hormone glucagon. This type of receptor spans the plasma membrane and extends into the cytoplasm. When glucagon binds to its receptor, the action triggers formation of a **second messenger**—a small molecule in the cytoplasm that relay signals from hormone–receptor complexes at the plasma membrane into a cell. In this case, cAMP (cyclic adenosine monophosphate) is the second messenger.

The response to glucagon binding takes place at a membrane-bound enzyme system (Figure 37.4). First an activated form of the enzyme adenylate cyclase starts a cascade of reactions by converting ATP to cyclic AMP. Many cyclic AMP molecules form. They act as signals to convert many molecules of a protein kinase, another type of enzyme, to its active form. These act on other enzymes, and so on until the final reaction converts stored glycogen in the cell to glucose. In short order, the number of molecules involved in the final response to the glucagon–receptor complex is enormous.

Or consider a muscle cell with receptors for insulin, a protein hormone. Among other things, when insulin binds to the receptor, the complex induces molecules of proteins called glucose transporters to move through the cytoplasm and insert themselves into the plasma membrane. They help cells take up glucose faster. The signal also activates enzymes of glucose storage.

Bear in mind, there are other hormone categories, including the catecholamines such as epinephrine. Like glucagon, epinephrine combines with specific surface receptors and triggers the release of cyclic AMP. This second messenger assists in the cellular response.

Hormones interact with receptors at the plasma membrane or inside the cytoplasm of target cells. Hormone binding directly influences protein synthesis in target cells.

Steroid hormones interact with a cell's DNA after entering the nucleus directly or after binding with a receptor in the cytoplasm. Some may act by altering membrane properites.

Peptide hormones do not exert their effects by entering a cell. When they bind to a membrane receptor, the binding itself is a signal for enzyme-mediated, intracellular events. Often a second messenger inside the cytoplasm relays the signal into the cell interior.

37.3 THE HYPOTHALAMUS AND PITUITARY GLAND

Deep in the forebrain is the **hypothalamus**. This brain region monitors internal organs and activities related to their functioning, such as eating. It influences certain forms of behavior, such as sexual behavior. It secretes some hormones. Suspended from its base by a slender stalk of tissue is a lobed, pea-size gland. This **pituitary gland** and the hypothalamus interact closely as a major neural–endocrine control center.

The pituitary's *posterior* lobe secretes two hormones that are synthesized by the hypothalamus. Its *anterior* lobe produces and secretes its own hormones, most of which help control the release of hormones from other endocrine glands (Table 37.2). The pituitary of many vertebrates—*not* humans—has an intermediate lobe as well. In many cases, the third lobe secretes a hormone that governs reversible changes in skin or fur color.

Posterior Lobe Secretions

Figure 37.5*a* shows the cell bodies of some neurons in the hypothalamus. Their axons extend down into the posterior lobe, then terminate next to a capillary bed. The neurons produce antidiuretic hormone (ADH) and oxytocin, then store them in the axon endings. After either hormone is released into the interstitial fluid, it diffuses into capillaries, then travels the bloodstream to targets. ADH acts on cells of nephrons and collecting ducts in the kidneys, which filter blood and rid the body of excess water and salts (in urine). ADH promotes water reabsorption when the body must conserve water. Oxytocin has certain roles in reproduction. It triggers contractions of the uterus during labor, and it causes milk release when offspring are being nursed.

Anterior Lobe Secretions

ANTERIOR PITUITARY HORMONES Inside the pituitary stalk, a capillary bed picks up hormones secreted by the hypothalamus and delivers them into a *second* capillary bed in the anterior lobe. There, the hormones leave the bloodstream and then act on target cells. As Figure 37.6 shows, different cells of the anterior pituitary secrete six hormones that they themselves produce:

Corticotropin	ACTH
Thyrotropin	TSH
Follicle-stimulating hormone	FSH
Luteinizing hormone	LH
Prolactin	PRL
Somatotropin (growth hormone)	STH (or GH)

All of these hormones have widespread effects through the body. ACTH and TSH orchestrate secretions from the adrenal glands and thyroid gland, respectively, as described shortly. FSH and LH act through the gonads

Table 37.2 Hormones Released From the Mammalian Pituitary Gland

Pituitary Lobe	Secretions	Designation	Main Targets	Primary Actions
POSTERIOR Nervous tissue (extension of hypothalamus)	Antidiuretic hormone	ADH	Kidneys	Induces water conservation as required during control of extracellular fluid volume and solute concentrations
	Oxytocin	OCT	Mammary glands	Induces milk movement into secretory ducts
			Uterus	Induces uterine contractions
ANTERIOR Glandular tissue, mostly	Corticotropin	ACTH	Adrenal cortex	Stimulates release of cortisol, an adrenal steroid hormone
	Thyrotropin	TSH	Thyroid gland	Stimulates release of thyroid hormones
	Follicle-stimulating hormone	FSH	Ovaries, testes	In females, stimulates estrogen secretion, egg maturation; in males, helps stimulate sperm formation
	Luteinizing hormone	LH	Ovaries, testes	In females, stimulates progesterone secretion, ovulation, corpus luteum formation; in males, stimulates testosterone secretion; sperm release
	Prolactin	PRL	Mammary glands	Stimulates and sustains milk production
	Somatotropin (or growth hormone)	STH (GH)	Most cells	Promotes growth in young; induces protein synthesis, cell division; roles in glucose, protein metabolism in adults
INTERMEDIATE* Glandular tissue, mostly	Melanocyte-stimulating hormone	MSH	Pigmented cells in skin and other integuments	Induces color changes in response to external stimuli; affects some behaviors

* Present in most vertebrates (not humans). MSH is associated with the anterior lobe in humans.

Further reading: Student Guide to InfoTrac on web site —

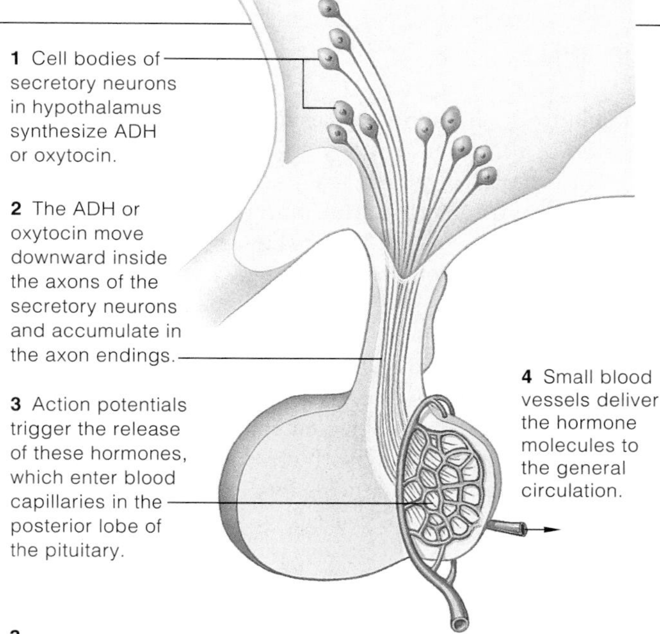

1 Cell bodies of secretory neurons in hypothalamus synthesize ADH or oxytocin.
2 The ADH or oxytocin move downward inside the axons of the secretory neurons and accumulate in the axon endings.
3 Action potentials trigger the release of these hormones, which enter blood capillaries in the posterior lobe of the pituitary.
4 Small blood vessels deliver the hormone molecules to the general circulation.
a

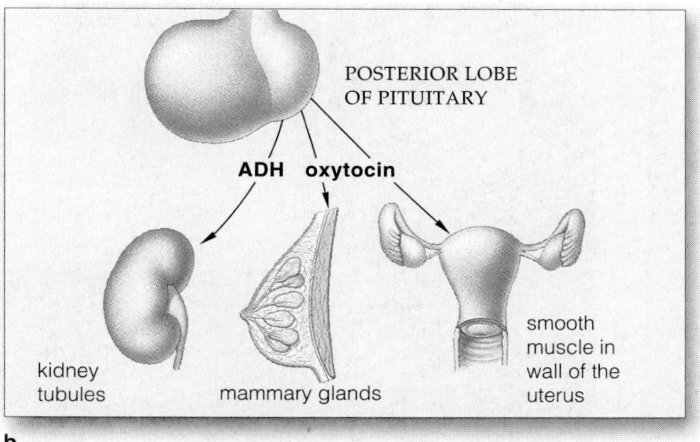

POSTERIOR LOBE OF PITUITARY
ADH
oxytocin
kidney tubules
mammary glands
smooth muscle in wall of the uterus
b

Figure 37.5 (**a**) Functional links between the hypothalamus and the posterior lobe of the pituitary gland. (**b**) Main targets of the posterior lobe secretions.

1 Cell bodies of different secretory neurons in the hypothalamus secrete releasing and inhibiting hormones.

2 The hormones are picked up by a capillary bed at the base of the hypothalamus,

3 Bloodstream delivers hormones to a second capillary bed in anterior lobe of pituitary.

4 Molecules of the releasing or inhibiting hormone diffuse out of the capillaries and act on endocrine cells in the anterior lobe.

5 Hormones secreted from anterior lobe cells enter small blood vessels that lead to the general circulation.

a

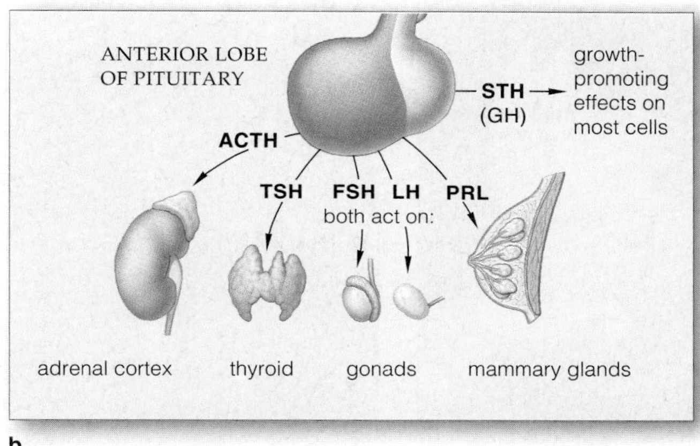

ANTERIOR LOBE OF PITUITARY
STH (GH)
growth-promoting effects on most cells
ACTH
TSH
FSH
LH
both act on:
PRL
adrenal cortex
thyroid
gonads
mammary glands
b

Figure 37.6 (**a**) Functional links between the hypothalamus and the anterior lobe of the pituitary gland. (**b**) Main targets of the anterior lobe secretions.

to influence gamete formation and secretion of the sex hormones required in sexual reproduction, the central topic of Chapter 45. Somatotropin affects metabolism in many tissues and triggers liver secretions that affect growth of bone and soft tissues (Table 37.2 and Figure 37.6*b*). Prolactin acts on different cell types. However, it is best known for stimulating and then sustaining milk production in mammary glands, after other hormones prime the tissues. In some species, prolactin also affects hormone production in the ovaries.

REGARDING THE HYPOTHALAMIC TRIGGERS Most of the hypothalamic hormones acting in the anterior lobe of the pituitary are **releasers**, meaning they stimulate secretion of hormones from target cells. For example, GnRH (gonadotropin-releasing hormone) brings about secretion of FSH and LH, which are gonadotropins. As another example, TRH (thyrotropin-releasing hormone) stimulates the secretion of thyrotropin.

Other hypothalamic hormones are **inhibitors** of the secretions from their targets in the anterior pituitary. For instance, the one called somatostatin brings about a decrease in somatotropin and thyrotropin secretion.

The hypothalamus and pituitary gland produce eight kinds of hormones and interact to control their secretion.

The posterior lobe of the pituitary stores and secretes two hypothalamic hormones, ADH and oxytocin, both of which target specific cell types.

The anterior lobe of the pituitary produces and secretes six hormones, ACTH, TSH, FSH, LH, PRL, and STH. These trigger release of other hormones from other glands, with a variety of effects throughout the body.

37.4 EXAMPLES OF ABNORMAL PITUITARY OUTPUT

The body does not churn out enormous quantities of hormone molecules. Two researchers, Roger Guilleman and Andrew Schally, realized this when they isolated the first known releasing hormone. In their four-year attempt to secure TRH, they dissected 500 metric tons of brains and 7 metric tons of hypothalamic tissue from sheep and ended up with only a single milligram of it.

Yet normal body function depends on the tiny but significant amounts of endocrine gland secretions, which commonly are released in short bursts. Elegant controls over the frequency of secretion work to block the underproduction and the overproduction of a hormone. If something interferes with the controls, the body's form and functioning may be altered in abnormal ways.

For instance, *gigantism* results from overproduction of somatotropin. Affected adults are proportionally like an average-size person but larger (Figure 37.7*a,b*). *Pituitary dwarfism* results from underproduction of somatotropin. Affected adults are proportionally similar to an average person but much smaller (Figure 37.7*b*).

What if somatotropin output becomes excessive in adulthood, when long bones no longer are lengthening? Bone, cartilage, and other connective tissues inside the hands, feet, and jaws will thicken abnormally. So will the epithelia of the skin, nose, eyelids, lips, and tongue. We call this outcome *acromegaly* (Figure 37.7*c*).

Bullets
10
a
b

age 9
age 16
age 33
age 52
c

Figure 37.7 Examples of the outcome of abnormalities in the secretion of a hormone.

(**a**) Manute Bol, an NBA center, is 7 feet 6–3/4 inches tall owing to excessive secretion of somatotropin (STH) during childhood.

(**b**) More examples of the effect of STH on body growth. The male at the center of this photograph is affected by gigantism, which resulted from excessive STH production in childhood. The person at the right displays pituitary dwarfism, which resulted from underproduction of STH in childhood. The person at the left is of average height.

(**c**) Acromegaly, which resulted from excessive production of STH during adulthood. Before this female reached maturity, she was symptom-free.

As another example, ADH secretion may dwindle or stop if the pituitary's posterior lobe is damaged, as by a blow to the head. This is one cause of *diabetes insipidus*. Symptoms include excessive excretion of dilute urine and life-threatening dehydration. Patients may respond to hormone replacement therapy based on injections or nasal spray applications of synthetic ADH.

Generally, endocrine glands release very small amounts of hormones. The frequency of those releases depends on control mechanisms. When controls fail, the resulting oversecretion or undersecretion may cause disorders.

Further reading: Student Guide to InfoTrac on web site

37.5 SOURCES AND EFFECTS OF OTHER HORMONES

Table 37.3 lists hormones from endocrine sources other than the pituitary. The remainder of this chapter will provide you with a few examples of their effects and of the controls over their output. The examples will make more sense if you keep the following points in mind.

First, hormones often interact with one another. In other words, one or more hormones may oppose, add to, or prime target cells for another hormone's effects. *Second,* negative feedback mechanisms often control the secretions. When a hormone's concentration increases or decreases in some body region, the change triggers events that respectively dampen or stimulate further secretion. *Third,* a target cell may react differently to a hormone at different times. Its response depends on the hormone's concentration as well as on the functional state of the cell's receptors. *Fourth,* environmental cues may be important mediators of hormonal secretion.

The secretion of a hormone and its effects are influenced by hormone interactions, feedback mechanisms, variations in the state of target cells, and sometimes environmental cues.

Table 37.3 Hormone Sources Other Than the Mammalian Hypothalamus and Pituitary

Source	Secretion(s)	Main Targets	Primary Actions
ADRENAL CORTEX	Glucocorticoids (including cortisol)	Most cells	Promote protein breakdown and conversion to glucose
	Mineralocorticoids (including aldosterone)	Kidney	Promote sodium reabsorption (sodium conservation); help control the body's salt–water balance
ADRENAL MEDULLA	Epinephrine (adrenaline)	Liver, muscle, adipose tissue	Raises blood level of sugar, fatty acids; increases heart rate and force of contraction
	Norepinephrine	Smooth muscle of blood vessels	Promotes constriction or dilation of certain blood vessels; thus helps control the flow of blood volume to different body regions
THYROID	Triiodothyronine, thyroxine	Most cells	Regulate metabolism; have roles in growth, development
	Calcitonin	Bone	Lowers calcium level in blood
PARATHYROIDS	Parathyroid hormone	Bone, kidney	Elevates calcium level in blood
GONADS			
Testes (in males)	Androgens (including testosterone)	General	Required in sperm formation, development of genitals, maintenance of sexual traits; growth, development
Ovaries (in females)	Estrogens	General	Required for egg maturation and release; preparation of uterine lining for pregnancy and its maintenance in pregnancy; genital development; maintenance of sexual traits; growth, development
	Progesterone	Uterus, breasts	Prepares, maintains uterine lining for pregnancy; stimulates development of breast tissues
PANCREATIC ISLETS	Insulin	Liver, muscle, adipose tissue	Lowers sugar level in blood
	Glucagon	Liver	Raises sugar level in blood
	Somatostatin	Insulin-secreting cells	Inhibits digestion of nutrients, hence their absorption from gut
THYMUS	Thymosins, etc.	Lymphocytes	Have roles in immune responses
PINEAL	Melatonin	Gonads (indirectly)	Influences daily biorhythms, seasonal sexual activity
STOMACH, SMALL INTESTINE	Gastrin, secretin, etc.	Stomach, pancreas, gallbladder	Stimulate activities of stomach, pancreas, liver, gallbladder required for food digestion, absorption
LIVER	Somatomedins	Most cells	Stimulate cell growth and development
KIDNEYS	Erythropoietin	Bone marrow	Stimulates red blood cell production
	Angiotensin*	Adrenal cortex, arterioles	Helps control secretion of aldosterone (hence sodium reabsorption, and blood pressure)
	1,25-hydroxyvitamin D_6* (calcitriol)	Bone, gut	Enhances calcium reabsorption from bone and calcium absorption from gut
HEART	Atrial natriuretic hormone	Kidney, blood vessels	Increases sodium excretion; lowers blood pressure

* Kidneys produce *enzymes* that modify precursors of this substance, which enters the general circulation as an activated hormone.

37.6 FEEDBACK CONTROL OF HORMONAL SECRETIONS

By considering just a few of the endocrine glands listed in Table 37.3, you can sense how feedback mechanisms control hormonal secretions. Briefly, the hypothalamus, pituitary, or both signal these glands to alter secretory activity. The outcome is a change in the concentration of a secreted hormone in blood or elsewhere. With the shift in chemical signals, a feedback mechanism trips into action and blocks or promotes further change.

In cases of **negative feedback**, an increase in the concentration of a secreted hormone triggers activities that *inhibit* further secretion. With **positive feedback**, an increase in the concentration of a secreted hormone triggers events that *stimulate* further secretion.

Negative Feedback From the Adrenal Cortex

Humans have a pair of adrenal glands, one above each kidney (Figure 37.2*b*). Some cells of the **adrenal cortex**, the outer part of an adrenal gland, secrete hormones such as glucocorticoids. Glucocorticoids help increase the level of glucose in blood. Cortisol is one of these hormones. It acts when the body is under stress, as when the glucose level declines below a set point. Glucose itself provides the information for a negative feedback mechanism that works to counter the decline.

Take a look at Figure 37.8. When the hypothalamus detects the decline, it secretes CRH in response. This releasing hormone stimulates the anterior pituitary to secrete corticotropin (ACTH). In turn, ACTH stimulates cells of the adrenal cortex to secrete cortisol. In response to cortisol, liver cells break down glycogen to glucose, skeletal muscle cells especially break down proteins, and adipose cells break down fats. Now glucose, amino acids, and fatty acids enter the blood and are mobilized as energy sources or building blocks to build or repair damaged cell structures. Later, the hypothalamus and pituitary detect the increase in the blood glucose level and respond by inhibiting cortisol secretion.

During chronic stress, injury, or illness, the nervous system initiates a stress response in which cortisol also helps to suppress inflammation. Unchecked, prolonged inflammation can damage tissues. That is why doctors may prescribe cortisol-like drugs for asthma and other chronic inflammatory disorders.

Local Feedback in the Adrenal Medulla

The **adrenal medulla** is the inner part of the adrenal gland (Figure 37.8). It incorporates neurons that secrete epinephrine and norepinephrine. These substances are neurotransmitters in some contexts and hormones in others. Sympathetic nerves service the medulla. Suppose they deliver hypothalamic signals calling for the release of norepinephrine. Molecules of norepinephrine collect in the synaptic cleft between the axon endings of these nerves and target cells. In this case, a localized negative feedback mechanism operates at receptors on the axon endings. Excess norepinephrine binds to the receptors and causes a shutdown of its further release.

During times of excitement or stress, epinephrine and norepinephrine help adjust blood circulation and fat and carbohydrate metabolism. They increase heart rate, trigger vasoconstriction and vasodilation of arterioles in different regions, and dilate airways to the lungs. The controlled activity directs more of the total volume of blood to heart and muscle cells, and more oxygen flows to energy-demanding cells through the body. These are features of the *fight–flight response* (Section 35.3).

Cases of Skewed Feedback From the Thyroid

The human **thyroid gland** is located at the base of the neck in front of the trachea, or windpipe (Figures 37.2*b* and 37.9*a*,*b*). Thyroxine and triiodothyronine, its main hormones, have widespread effects. In their absence, many tissues cannot develop normally. Also, the overall metabolic rates of warm-blooded animals, including

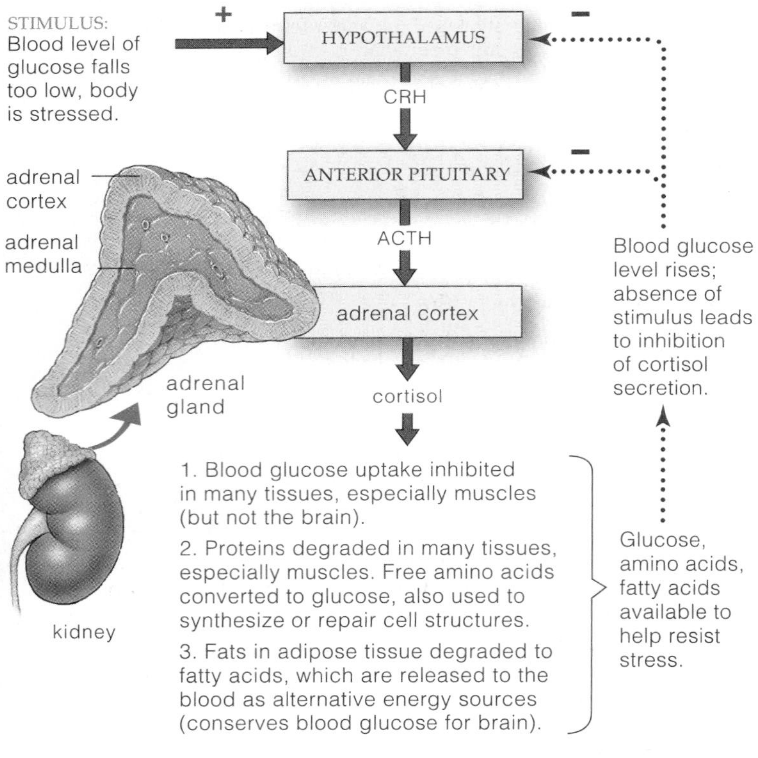

STIMULUS:
Blood level of glucose falls too low, body is stressed.
+
HYPOTHALAMUS
CRH
ANTERIOR PITUITARY
ACTH
adrenal cortex
cortisol
−
−
Blood glucose level rises; absence of stimulus leads to inhibition of cortisol secretion.
1. Blood glucose uptake inhibited in many tissues, especially muscles (but not the brain).
2. Proteins degraded in many tissues, especially muscles. Free amino acids converted to glucose, also used to synthesize or repair cell structures.
3. Fats in adipose tissue degraded to fatty acids, which are released to the blood as alternative energy sources (conserves blood glucose for brain).
Glucose, amino acids, fatty acids available to help resist stress.
adrenal cortex
adrenal medulla
adrenal gland
kidney

Figure 37.8 Location of the two human adrenal glands. One gland rests on top of each kidney. The diagram shows a negative feedback loop that governs cortisol secretion from the adrenal cortex.

Further reading: Student Guide to InfoTrac on web site →

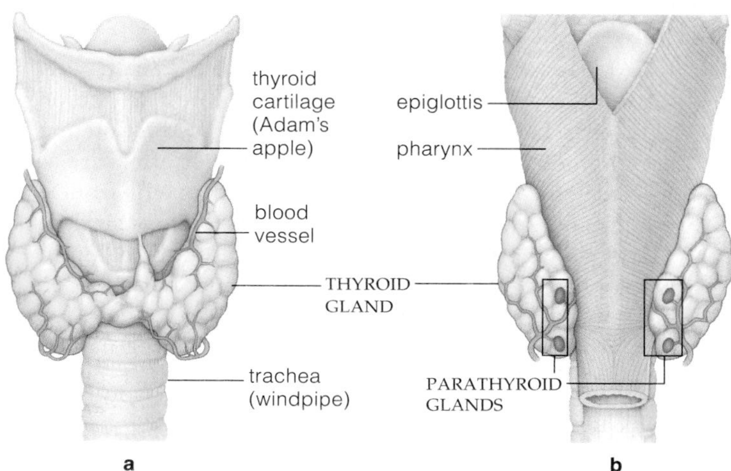
thyroid cartilage (Adam's apple)
blood vessel
THYROID GLAND
trachea (windpipe)
epiglottis
pharynx
PARATHYROID GLANDS
a
b

c

Figure 37.9 Human thyroid gland. (**a**) Anterior and (**b**) posterior views showing the location of four parathyroid glands. (**c**) A mild case of goiter, displayed by Maria de Medici in the year 1625. A rounded neck was considered to be a sign of great beauty during the late Renaissance. It occurred regularly in parts of the world where iodine supplies were insufficient for normal thyroid function.

humans, depend upon them. The importance of feedback control of the secretions of these hormones comes into sharp focus with cases of abnormal output of thyroid hormones.

As a case in point, synthesis of a thyroid hormone requires iodine, which we obtain from food. Iodine is converted to an iodized form, iodide, when absorbed from the gut. Without iodide, the blood levels of thyroid hormones decrease. The anterior pituitary responds by secreting its thyroid-stimulating hormone (TSH). But without iodine, thyroid hormones cannot be made. The feedback signal continues, and so does TSH secretion. Over time, a sustained response is made to the low blood level of TSH level and the thyroid gland enlarges. The enlargement is one form of *goiter*. Goiter resulting from iodine deficiency is no longer common in countries where people use iodized salt (Figure 37.9*c*).

When blood levels of thyroid hormones are too low, *hypothyroidism* results. Hypothyroid adults commonly are overweight, sluggish, dry-skinned, intolerant of cold, confused, and depressed. Affected women commonly have menstrual disturbances.

When blood levels of the thyroid hormones are too high, *hyperthyroidism* results. Increased heart rate, heat intolerance, elevated blood pressure, profuse sweating, and weight loss even when caloric intake increases are symptoms of the disorder. Affected adults typically are nervous and agitated, and have trouble sleeping.

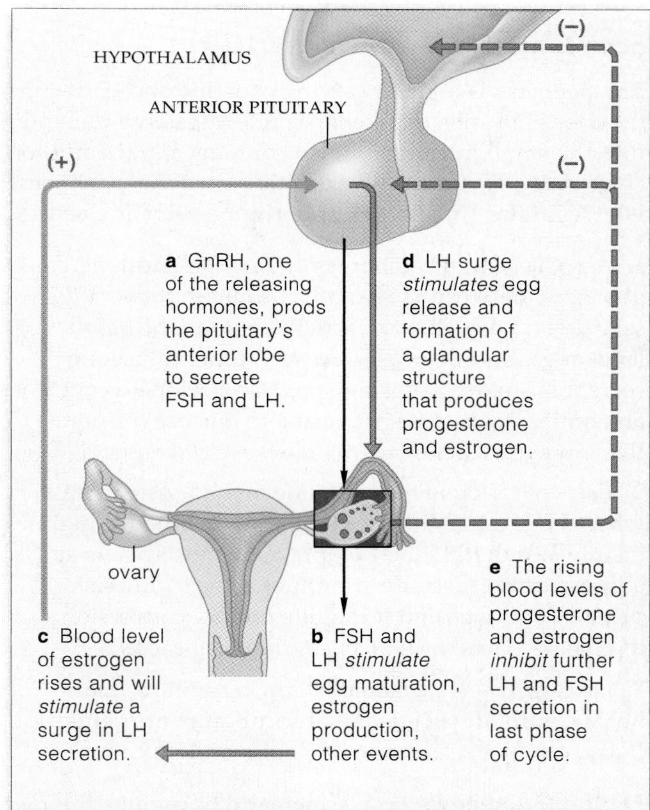
HYPOTHALAMUS
ANTERIOR PITUITARY
(–)
(+)
(–)
a GnRH, one of the releasing hormones, prods the pituitary's anterior lobe to secrete FSH and LH.
d LH surge *stimulates* egg release and formation of a glandular structure that produces progesterone and estrogen.
ovary
e The rising blood levels of progesterone and estrogen *inhibit* further LH and FSH secretion in last phase of cycle.
c Blood level of estrogen rises and will *stimulate* a surge in LH secretion.
b FSH and LH *stimulate* egg maturation, estrogen production, other events.

Figure 37.10 Feedback loops to the hypothalamus and the pituitary gland from the ovaries during the menstrual cycle, a recurring reproductive event. Positive feedback triggers egg release from an ovary. Negative feedback after its release prevents release of another egg until the cycle is completed.

Feedback Control of the Gonads

Gonads are *primary* reproductive organs, which make and secrete gametes and certain sex hormones. Testes (singular, testis) in males and ovaries in females are examples. Testes secrete testosterone; ovaries secrete estrogens and progesterone. These hormones influence secondary sexual traits (as they did for those chimps described earlier), and feedback controls govern their secretion. Figure 37.10 is a preview of the feedback loops from ovaries to the hypothalamus and pituitary during the menstrual cycle, a key topic of Chapter 45.

Feedback mechanisms control secretions from endocrine glands. In many cases, feedback loops to the hypothalamus, pituitary, or both govern the secretory activity.

Negative feedback slows down the further secretion of a hormone. Positive feedback enhances the further secretion of a hormone.

RESPONSES TO LOCAL CHEMICAL CHANGES

Some endocrine glands or cells don't respond primarily to signals from other hormones or nerves. They respond homeostatically to chemical changes in the immediate surroundings, as the following examples illustrate.

Secretions From Parathyroid Glands

Humans have four **parathyroid glands** located on the posterior surface of the thyroid gland (Figure 37.9*b*). The glands secrete parathyroid hormone, or PTH, the main regulator of the calcium level in blood. Calcium ions, recall, have roles in muscle contraction, enzyme action, blood clot formation, and other tasks. The parathyroids secrete PTH in response to a low calcium level in blood. Their secretory activity slows when the calcium level rises. PTH acts on cells of the skeleton and kidneys.

PTH induces living bone cells to secrete enzymes that digest bone tissue. Digestion releases calcium and other minerals to the interstitial fluid, then to blood. It enhances calcium reabsorption from the filtrate flowing through the nephrons of kidneys. PTH also prods some kidney cells to secrete enzymes that act on blood-borne precursors of an active form of vitamin D_3, a hormone (Table 37.3). The activated form stimulates intestinal cells to absorb much more calcium from the lumen of the gut. In children who have vitamin D deficiency, not enough calcium and phosphorus are absorbed, so rapidly growing bones develop improperly. The resulting bone abnormality, *rickets*, is characterized by bowed legs, a malformed pelvis, and in many cases a malformed skull and rib cage (Figure 37.11).

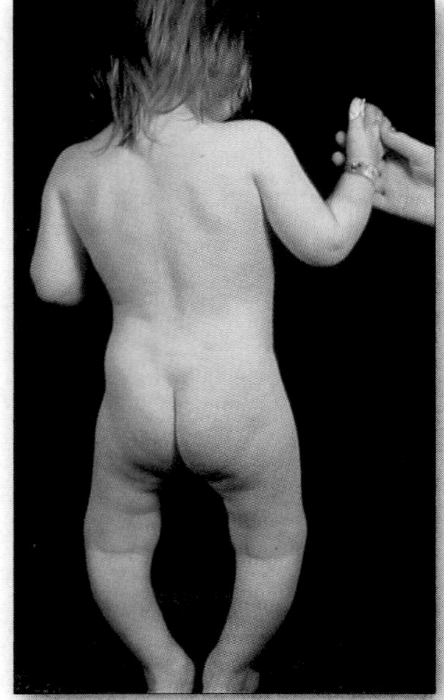

Figure 37.11 A child who is affected by rickets.

Effects of Local Signaling Molecules

Many cells detect changes in the surrounding chemical environment and alter their activity, often in ways that counteract or amplify those changes. The cells secrete various local signaling molecules, the effects of which are confined to the immediate vicinity of change. Target cells take up most signaling molecules so rapidly that few enter the general circulation.

At least sixteen types of prostaglandins, each having a twenty-carbon fatty acid structure with a five-carbon ring, have been found in many tissues. Production and secretion of these local signaling molecules often rise as local chemical conditions change, with potent results. Some make smooth muscle cells of arterioles constrict or dilate; they help redirect blood flow through the body. Others have roles in inflammation, control of stomach acid and intestinal motility, and childbirth. They also are important players in overall endocrine regulation. Researchers are looking into their possible therapeutic roles in asthma, ulcers, hypertension, and heart attacks.

Other local signaling molecules are growth factors that affect cell division rates in tissues. An epidermal growth factor (EGF) discovered by Stanley Cohen acts on many cell types. Rita Levi-Montalcini discovered a nerve growth factor (NGF) that helps neurons survive and guides their direction of growth in the embryo. Experimenters showed that some immature neurons can survive indefinitely in tissue culture when NGF is present but die within a few days when it is absent.

Secretions From Pancreatic Islets

The pancreas is a gland serving exocrine and endocrine functions. Its *exocrine* cells secrete digestive enzymes into the small intestine. It also contains about 2 million clusters of *endocrine* cells. Each tiny cluster, a **pancreatic islet**, contains three types of hormone-secreting cells:

1. *Alpha* cells in the pancreas secrete the hormone glucagon. Between meals, cells throughout the body take up and use glucose from the blood. The blood level of glucose decreases. At such times, glucagon secretion causes the storage polysaccharide glycogen and amino acids to be converted to glucose inside the liver. In such ways, *glucagon raises the glucose level.*

2. *Beta* cells secrete the hormone insulin. After meals, when the level of glucose circulating in blood is high, insulin stimulates glucose uptake by muscle cells and adipose cells especially. It promotes the synthesis of proteins and fats, and it inhibits protein conversion to glucose. Thus, *insulin lowers the glucose level.*

3. *Delta* cells secrete somatostatin, a hormone that helps control digestion and absorption of nutrients. It also can block secretion of insulin and glucagon.

Figure 37.12 shows how pancreatic hormones interact to maintain the level of glucose in blood even though the times and amounts of food intake vary. Bear in mind, insulin is the only hormone that prods cells to take up and store glucose in forms that can be rapidly tapped when required. Its central role in carbohydrate, protein, and fat metabolism becomes clear when we

Further reading: Student Guide to InfoTrac on web site

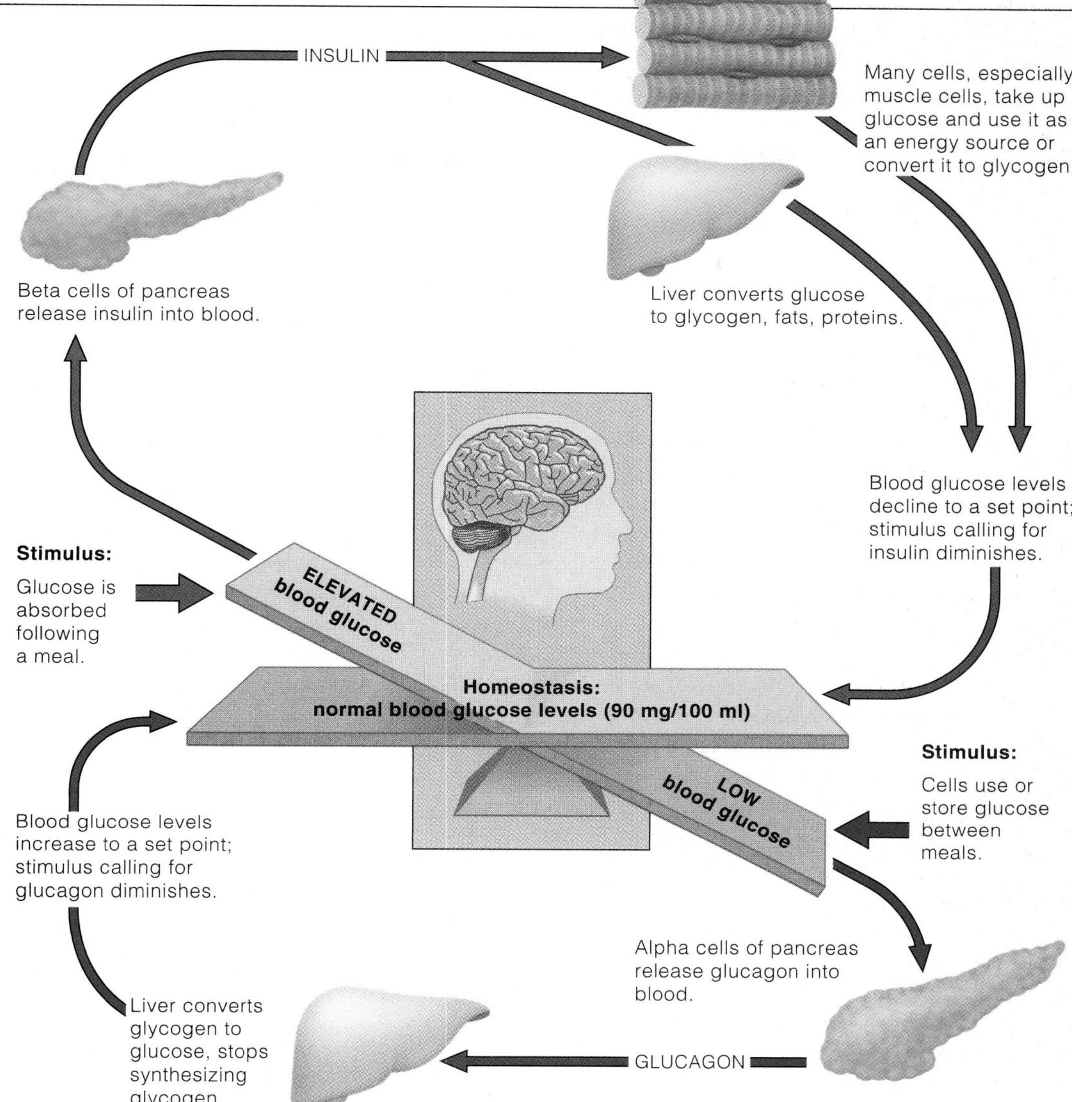
INSULIN
Many cells, especially muscle cells, take up glucose and use it as an energy source or convert it to glycogen.
Beta cells of pancreas release insulin into blood.
Liver converts glucose to glycogen, fats, proteins.
Blood glucose levels decline to a set point; stimulus calling for insulin diminishes.
Stimulus:
Glucose is absorbed following a meal.
ELEVATED blood glucose
Homeostasis:
normal blood glucose levels (90 mg/100 ml)
Stimulus:
Cells use or store glucose between meals.
LOW blood glucose
Blood glucose levels increase to a set point; stimulus calling for glucagon diminishes.
Alpha cells of pancreas release glucagon into blood.
Liver converts glycogen to glucose, stops synthesizing glycogen.
GLUCAGON

Figure 37.12 Some of the homeostatic controls over glucose metabolism.

Following a meal, glucose enters the bloodstream faster than cells can use it. The blood glucose level rises, and pancreatic beta cells are stimulated to secrete insulin. Insulin's main targets—liver and muscle cells—not only use glucose but store excess amounts of it in the form of glycogen.

Between meals, the blood glucose level decreases. Pancreatic alpha cells are stimulated to secrete glucagon. The target cells with receptors for this hormone convert glycogen back to glucose, which enters the blood.

Glucose metabolism also is subjected to indirect control. For example, the hypothalamus commands the adrenal medulla to secrete certain hormones. The hormones speed the conversion of glycogen to glucose in the liver and slow the reverse process, especially in the liver and muscle tissue.

consider people who cannot produce enough insulin or who lack body cells that can respond to it.

For example, insulin deficiency may lead to *diabetes mellitus*, a disorder in which excess glucose accumulates in blood, then in urine. Urination becomes excessive, so the body's water–solute balance is disrupted. Affected people become dehydrated and thirsty—abnormally so. Without a steady supply of glucose, their body cells start depleting their own fats and proteins as sources of energy. Weight loss is one outcome. Another is that ketones accumulate in the blood and urine. Ketones are normal acidic products of fat breakdown. When they accumulate, they contribute to water losses and alter the body's acid–base balance. Such imbalances disrupt brain function. In extreme cases, death may follow.

In *type 1 diabetes*, the body mistakenly mounts an autoimmune response against its insulin-secreting beta cells. Certain lymphocytes identify beta cells as foreign and work to destroy them. A combination of genetic susceptibility and environmental triggers produces the disorder, which is less common but more immediately dangerous than the other forms of diabetes. Usually the symptoms first appear in childhood and adolescence (the disorder is also known as juvenile-onset diabetes). Type 1 diabetic patients survive with insulin injections.

In *type 2 diabetes*, insulin levels are close to or above normal, but target cells are not equipped to respond to insulin. As affected persons grow older, their beta cells produce less and less insulin. The symptoms of type 2 diabetes usually emerge during middle age. Affected persons lead normal lives by controlling their diet and weight, and sometimes by taking prescription drugs to enhance insulin action or secretion.

Secretions from some endocrine glands and endocrine cells are direct homeostatic responses to a change in the localized chemical environment..

37.8 HORMONES AND THE ENVIRONMENT

This last section of the chapter invites you to reflect on a key point. An individual's growth, development, and reproduction begin with genes and hormones, and so does behavior. *But certain environmental factors commonly influence gene expression and hormonal secretion, and they do this in predictable ways.* Later chapters invite analysis of specific environmental effects on animals. For now, it is enough to consider the following examples.

Figure 37.13 (**a**) A male white-throated sparrow, belting out a song that began, indirectly, with an environmentally induced decrease in the level of melatonin secretion. (**b**) Annie blanketing her winter blues.

Daylength and the Pineal Gland

Embedded in the brain is a photosensitive organ, the **pineal gland** (Section 35.5). In the absence of light, the gland secretes the hormone melatonin. Thus the level of melatonin in the blood varies from day to night and over the seasons. The variations influence growth and development of gonads (primary reproductive organs). For many species, they are components of a **biological clock**, an internal timing mechanism, that exerts control over reproductive cycles and reproductive behavior.

Think of a hamster in winter, when there are more hours of darkness than in summer. The melatonin level in its blood is high, and this suppresses sexual activity. When daylength is longest in summer, the blood level is low and hamster sex peaks. Or think of a male white-throated sparrow (Figure 37.13*a*). In the fall and winter, melatonin indirectly suppresses growth of its gonads by inhibiting gonadotropin secretion. It does so until daylight increases in spring. Now stepped-up gonadal activity leads to production of hormones that influence singing behavior, as described in Section 47.1. With his distinctive song, the male sparrow defines his territory and may hold the interest of a mate.

Does melatonin influence human behavior as well? Perhaps. Clinical observations and studies suggest that decreased melatonin levels may trigger **puberty**, the age at which reproductive organs and structures start to maturing. To give an example, puberty was premature in individuals who were deprived of a functional pineal gland.

Melatonin is known to influence the neurons that call for a decrease in core temperature and that make you drowsy after sunset, when the light is waning. At sunrise, when melatonin secretion fall, the core temperature of your body increases. In response, you wake up and you become active. A biological clock governs this cycle of sleep and arousal. This one ticks in synchrony with daylength. Think of night workers who simply cannot get to sleep in the morning. Think of the travelers from the United States to Paris who go through four days of "jet lag." Two or three hours past midnight they sit up in bed, but two hours past noon they are ready for bed. They may shift to a new routine when melatonin signals arrive at their target neurons on Paris time.

Seasonal affective disorder (SAD) hits some people in winter. They become abnormally depressed, binge on carbohydrates, and have an almost overwhelming need to sleep (Figure 37.13*b*). Such "winter blues" might develop when a biological clock is out of sync with the seasonally shorter daylengths. Clinically administered melatonin doses make the seasonal symptoms worse. Exposure to intense light, which can shut down pineal activity, may lead to dramatic improvement.

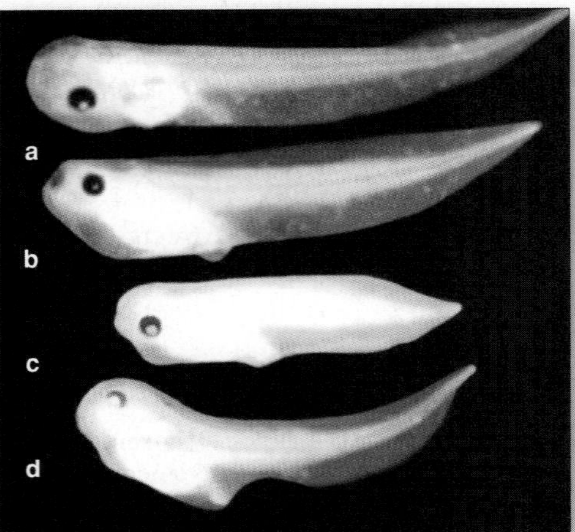

Figure 37.14 *Xenopus laevis* tadpoles showing possible environmental effects on the thyroid, hence on development. (**a**) Tadpole raised in water taken from a lake where there were few deformed frogs. (**b–c**) Three tadpoles raised in water taken from three "hot spot" lakes that had increasingly higher concentrations of dissolved chemical compounds, both natural and synthetic.

Water Pollutants and Deformed Frogs

Since 1994, in natural habitats around the world, the number of deformed frogs has been skyrocketing. The reasons are not fully understood. In 1999, researchers decided to expose *Xenopus laevis* embryos to Minnesota and Vermont lakewater. Half the samples came from lakes where deformity rates were low. The other half came from "hot spots" with as many as twenty kinds of dissolved pesticides—and with high deformity rates.

When the frog embryos developed into tadpoles, the ones that had been raised in the hot-spot water had bent spines, malformed eyes, and a malformed mouth. Some did not metamorphose into adults (Figure 37.14). The other frog embryos of the sampling were normal.

Further reading: Student Guide to InfoTrac on web site

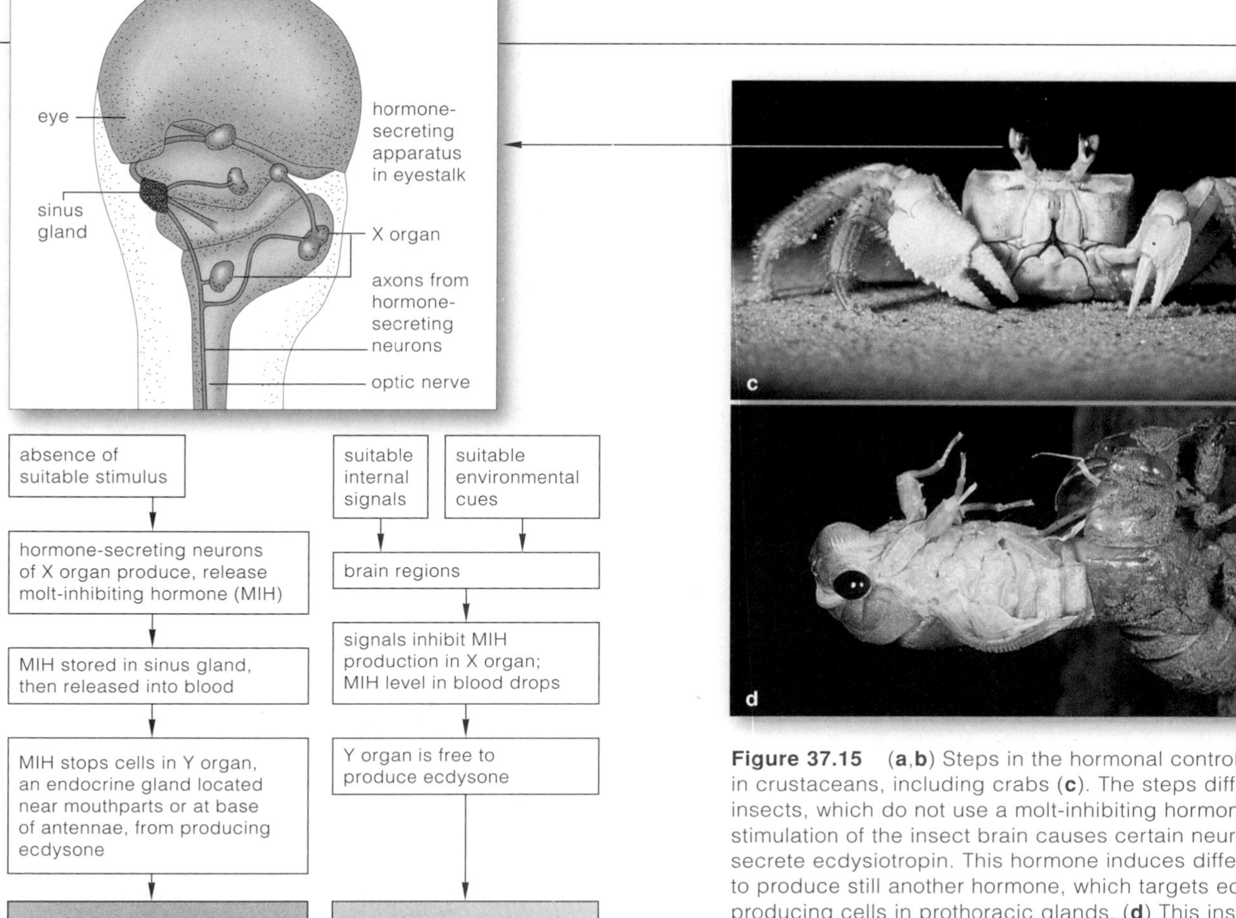

eye
sinus gland
hormone-secreting apparatus in eyestalk
X organ
axons from hormone-secreting neurons
optic nerve
absence of suitable stimulus
hormone-secreting neurons of X organ produce, release molt-inhibiting hormone (MIH)
MIH stored in sinus gland, then released into blood
MIH stops cells in Y organ, an endocrine gland located near mouthparts or at base of antennae, from producing ecdysone
MOLTING INHIBITED
a
suitable internal signals
suitable environmental cues
brain regions
signals inhibit MIH production in X organ; MIH level in blood drops
Y organ is free to produce ecdysone
MOLTING STIMULATED
b
c
d

Figure 37.15 (**a**,**b**) Steps in the hormonal control of molting in crustaceans, including crabs (**c**). The steps differ a bit in insects, which do not use a molt-inhibiting hormone. Rather, stimulation of the insect brain causes certain neurons to secrete ecdysiotropin. This hormone induces different neurons to produce still another hormone, which targets ecdysone-producing cells in prothoracic glands. (**d**) This insect, a cicada, is emerging from its old cuticle.

Thyroid hormones orchestrate much of vertebrate development. Were some chemicals in the hot-spot water interfering with thyroid functioning? Maybe. Embryos that developed in that water showed few symptoms (or none) if they also were given extra thyroid hormones.

Other studies correlated some frog deformities with ultraviolet radiation and with infections by a parasitic protistan. Chemical soup habitats that disrupt hormone function may be added to the list.

Comparative Look at a Few Invertebrates

Although this chapter's focus has been on vertebrates, do not lose sight of the fact that all organisms produce signaling molecules of one sort or another. Let's look at hormonal control of **molting**, a periodic discarding and replacement of a hardened cuticle that otherwise would limit increases in body mass. As described in Section 26.15, molting occurs during the life cycle of all insects, crustaceans, and other invertebrates with thick cuticles.

Although details vary from group to group, molting is largely under the control of **ecdysone**. This steroid hormone is derived from cholesterol and is chemically related to many important vertebrate hormones. In the insects and crustaceans, molting glands produce and store ecdysone, then release it for distribution through the body at molting time. Hormone-secreting neurons in the brain seem to regulate its release. The hormone-secreting neurons apparently respond to a combination of environmental cues, including light and temperature, as well as internal signals.

Figure 37.15 provides examples of the control steps, which differ in crustaceans and insects. During premolt and molting periods, coordinated interactions among ecdysone and other hormones bring about structural and physiological changes. The interactions make the old cuticle separate from the epidermis and muscles. They induce events that dissolve the inner layers of the cuticle and recycle the remnants. The interactions also trigger changes in metabolism and in the composition and volume of the internal environment. They promote cell divisions, secretions, and pigment formation, all of which go into producing a new cuticle. Simultaneously, hormonal interactions control heart rate, muscle action, color changes, and other physiological processes.

Environmental cues, such as changes in light intensity from day to night and seasonal changes in daylength, influence certain hormonal secretions.

37.9 SUMMARY

1. The cells of complex animals continually exchange substances with the body's internal environment. Their myriad withdrawals and secretions are integrated in ways that ensure cell survival through the whole body.

2. Integration of cell activities requires the stimulatory or inhibitory effects of signaling molecules.

a. Signaling molecules are chemical secretions from a cell that adjust the behavior of other, target cells.

b. Any cell with molecular receptors for a signaling molecule is a target. Target cells might or might not be next to the cell that sends the signal.

c. There are different kinds of signaling molecules. Hormones as well as neurotransmitters, local signaling molecules, and pheromones are the main kinds.

d. Certain steroids, steroid-like molecules, amines, peptides, proteins, and glycoproteins are hormones.

3. In target cells, hormones influence gene activation, protein synthesis, and alterations in existing enzymes, membranes, and other cellular components. Hormones exert their physiological effects through interactions with specific protein receptors at the plasma membrane or in the cytoplasm of target cells.

a. Steroid hormones enter the target cell's nucleus directly, after binding with an intracellular receptor, or possibly by binding with plasma membrane receptors.

b. Being water soluble, the protein hormones cannot enter target cells. They bind to membrane receptors at the cell surface. Their effect is exerted with the help of transport proteins as well as second messengers in the cytoplasm, some of which trigger the actual response.

4. The posterior lobe of the pituitary stores and secretes two hypothalamic hormones, ADH and oxytocin. ADH targets cells in kidneys and affects extracellular fluid volume. Oxytocin acts on cells in mammary glands and the uterus to influence reproductive events.

5. The hypothalamic hormones called releasing and inhibiting hormones control secretions from different cells of the anterior lobe of the pituitary gland.

6. The anterior lobe makes and secretes six hormones: ACTH, TSH, FSH, LH, PRL, and STH. These trigger secretions from the adrenal cortex, thyroid, gonads, and mammary glands. They exert a variety of responses throughout the body.

7. The vertebrate body has other sources of hormones, including the adrenal medulla, parathyroid, thymus, and pineal glands; pancreatic islets; and endocrine cells in the stomach, small intestine, liver, and heart.

8. Interactions among hormones, feedback mechanisms, the number and kind of target cell receptors, variations in the state of target cells, and often environmental cues influence secretion of a hormone and its effects.

9. Many cellular responses to hormones help the body adjust to short-term shifts in diet and levels of activity. Other responses help bring about long-term adjustments for growth, development, and reproduction.

a. In general, secretion of hormones such as insulin and parathyroid hormone can change rapidly when the extracellular concentration of some substance must be homeostatically controlled.

b. Hormones such as somatotropin have prolonged, gradual, often irreversible effects, as on development.

Review Questions

1. Name the endocrine glands typical of most vertebrates and state where each is located in the human body. *37.1*

2. Distinguish among hormones, neurotransmitters, local signaling molecules, and pheromones. *37.1*

3. A hormone molecule binds to a receptor on a cell membrane. It does not enter the cell. Binding activates a second messenger in the cell, which triggers an amplified response to the hormonal signal. State whether the molecule is a steroid hormone or a peptide hormone. *37.2*

4. Which secretions of the posterior lobe of the pituitary gland have the targets indicated? (Fill in the blanks.) *37.3*

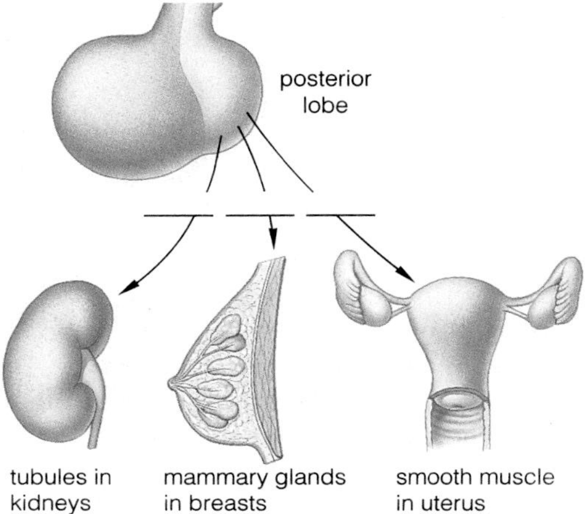
posterior lobe
tubules in kidneys
mammary glands in breasts
smooth muscle in uterus

5. Which secretions of the anterior lobe of the pituitary gland have the targets indicated? (Fill in the blanks.) *37.3*

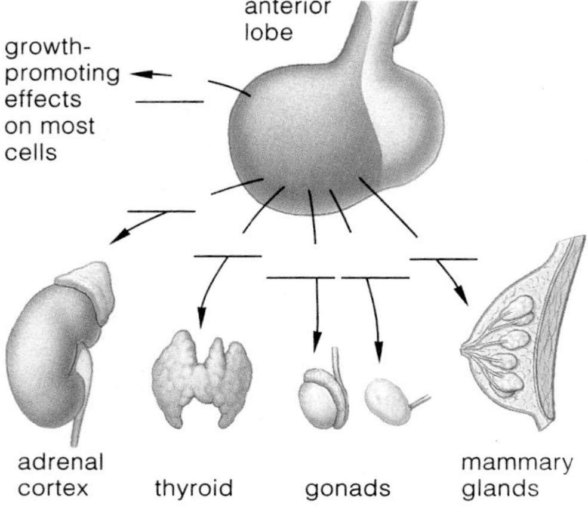
anterior lobe
growth-promoting effects on most cells
adrenal cortex
thyroid
gonads
mammary glands

Self-Quiz (*Answers in Appendix III*)

1. ______ are molecules released from a signaling cell that have effects on target cells.
   a. Hormones
   b. Neurotransmitters
   c. Pheromones
   d. Local signaling molecules
   e. both a and b
   f. a through d

2. Hormones are products of ______.
   a. endocrine glands
   b. some neurons
   c. exocrine cells
   d. a and b
   e. a and c
   f. a, b, and c

3. Second messengers include ______.
   a. steroid hormones
   b. protein hormones
   c. cyclic AMP
   d. both a and b

4. ADH and oxytocin are hypothalamic hormones secreted from the ______ lobe of the pituitary gland.
   a. anterior
   b. posterior
   c. intermediate
   d. secondary

5. GnRH is a ______ secreted by hypothalamic neurons.
   a. releasing hormone
   b. inhibiting hormone
   c. corticotropin
   d. somatotropin

6. Which do *not* stimulate hormone secretions?
   a. neural signals
   b. local chemical changes
   c. hormonal signals
   d. environmental cues
   e. All of the above can stimulate secretion.

7. ______ lowers blood sugar levels; ______ raises it.
   a. Glucagon; insulin
   b. Insulin; glucagon
   c. Gastrin; insulin
   d. Gastrin; glucagon

8. The pituitary detects a rising hormone concentration in blood and inhibits the gland secreting the hormone. This is a ______ feedback loop.
   a. positive
   b. negative
   c. long-term
   d. b and c

9. Match the hormone source with the closest description.
   ____ adrenal medulla
   ____ thyroid gland
   ____ parathyroids
   ____ pancreatic islets
   ____ pineal gland
   ____ prostaglandin
   a. affected by daylength
   b. potent local effects
   c. raise blood calcium level
   d. epinephrine source
   e. insulin, glucagon
   f. hormones require iodine

Critical Thinking

1. The zebra offspring being nursed in Figure 37.16 is too young to nourish itself by eating grasses. Its source of nutrients is its mother's milk. Explain how secretions from the hypothalamus and both lobes of the pituitary gland influence the production and secretion of milk.

2. In winter, with its far fewer daylight hours compared to the summer, Maxine became very depressed, craved carbohydrate-rich foods, and stopped exercising regularly. And she put on a great deal of weight. Her doctor diagnosed her condition as *seasonal affective disorder* (SAD), or the winter blues. Maxine was advised to purchase a cluster of intense, broad-spectrum lights and to sit near them for at least an hour every day. The cloud of depression started to lift quickly. Use your understanding of the secretory activity of the pineal gland to explain why Maxine's symptoms appeared and why the prescribed therapy worked.

3. Marianne is affected by *type 1 insulin-dependent diabetes*. One day, after injecting herself with too much insulin, she starts to shake and feels confused. Her doctor recommends a glucagon injection. What caused her symptoms? How would an injection of glucagon help?

Figure 37.16 Female zebra nursing her offspring.

4. By application of recombinant DNA technology, somatotropin (growth hormone) is now commercially available for treating pituitary dwarfism. Although it is illegal to do so, some athletes use somatotropin instead of anabolic steroids, which Section 38.12 describes. Why? Somatotropin cannot be detected by the drug test procedures currently employed in sports medicine. Explain how athletes might believe this hormone can improve their performance.

5. *Osteoporosis* is a condition in which loss of calcium results in thin, brittle bones. Combined with other treatments, vitamin D_3 injections are sometimes recommended. Explain why.

Selected Key Terms

adrenal cortex *37.6*
adrenal medulla *37.6*
biological clock *37.8*
ecdysone *37.8*
endocrine system *37.1*
gonad *37.6*
hormone *37.1*
hypothalamus *37.3*
inhibitor (hypothalamic) *37.3*
local signaling molecule *37.1*
molting *37.8*
negative feedback *37.6*
neurotransmitter *37.1*
pancreatic islet *37.7*
parathyroid gland *37.7*
peptide hormone *37.2*
pheromone *37.1*
pineal gland *37.8*
pituitary gland *37.3*
positive feedback *37.6*
puberty *37.8*
releaser (hypothalamic) *37.3*
second messenger *37.2*
steroid hormone *37.2*
thyroid gland *37.6*

Readings *See also www.infotrac-college.com*

Goodall, J. 1986. *The Chimpanzees of Gombe*. Cambridge, Massachusetts: Belknap Press of Harvard University Press.

Goodman, H. 1994. *Basic Medical Endocrinology*. Second edition. New York: Raven Press.

Hadley, M. 1995. *Endocrinology*. Fourth edition. Englewood Cliffs, New Jersey: Prentice-Hall.

Raloff, J. 2 October 1999. "Thyroid Linked to Some Frog Defects." *Science News* 156:212.

Sherwood, L. 1997. *Human Physiology*. Third edition. Belmont, California: Wadsworth.

38

PROTECTION, SUPPORT, AND MOVEMENT

Of Men, Women, and Polar Huskies

In 1989 Will Steger and his dog-sled team walked on ice for seven months, endured temperatures of −113°F, and lived through a blizzard that lasted for more than seven weeks. They crossed Antarctica—all 6,023 kilometers (3,741 miles) of it. In 1995 that legendary polar explorer set out with four men, two women, and thirty-three sled dogs to cross 3,220 kilometers of the Arctic Ocean in one season. Ice blankets this northernmost ocean in winter, but the ice becomes treacherously thin during the spring thaw. The sled dogs were with the team for two-thirds of the journey. They were flown out only when the team encountered too much melting ice and had to switch to using canoes.

To Steger's mind, the polar huskies were the heroes of the polar crossings, the members of the team that worked hardest and pulled all the weight (Figure 38.1). They are a mixed breed, the traits of which have been modified through years of artificial selection among Canadian and Greenland huskies (bred for size and strength), Siberian huskies (bred for intelligence), and Alaskan racing dogs (bred for spirit and endurance). Steger's polar huskies show a combination of these traits as well as the loyalty of cared-for pets.

A husky's leg bones are sturdy yet lightweight. Its forelegs move freely, thanks to a rib cage that is deep but not too broad. Its hind legs have massive muscles. These are not the muscles of sprinting greyhounds or cheetahs. They are the muscles of a load-pulling, long-distance runner. The husky also has tough, calloused foot pads—cushions against sharp ice and frozen rock. Like many other mammals, it has a fur coat. The coat's underhair, a dense, soft, insulative layer, traps heat. Its coarser, longer, and slightly oily guard hairs protect the insulative layer from wear and tear. On winter nights, the husky settles into a comfortable position and covers its nose with its furry tail, oblivious of drifting snow.

Steger and his teammates, Victor Boyarsky, Julie Hanson, Martin Hignell, Paul Pregont, and Takako Takano, could not even approach the polar husky's stamina and built-in protection against the elements. Long before the polar crossings, they were adhering to a regimen of diet and exercise to put their arm and

Figure 38.1 In Ely, Minnesota, Will Steger and his polar huskies warming up for an Arctic crossing.

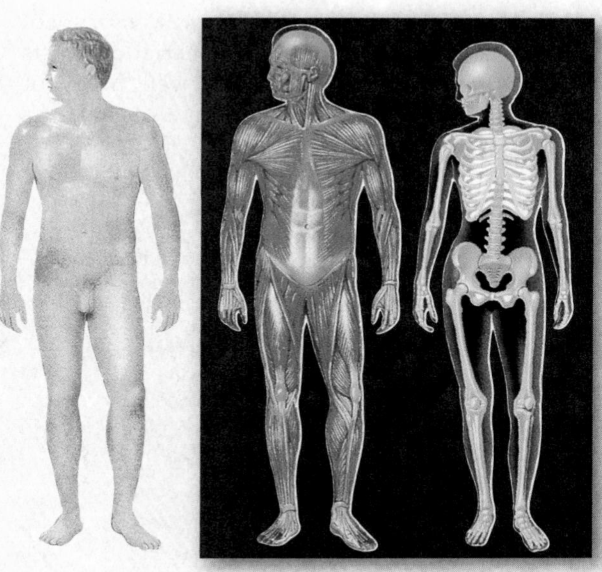

Figure 38.2 From left to right, overview of the human integumentary, muscular, and skeletal systems.

leg muscles in peak condition for the extraordinary effort that lay ahead. Human legs are not adapted for load-pulling motion, but rather for long-distance walking. Also, human skin cannot withstand bitter cold. Lacking the fur coat of mammals that evolved in polar climates, the team had to depend on special clothing that could insulate and protect them from cold without restricting body movements. From this perspective, it was human ingenuity that allowed humans to keep company with the huskies, which are supremely adapted for the challenges of life on ice.

With this chapter, we turn to three systems that together are responsible for the superficial features, shape, and movements of most animals. Figure 38.2 serves as the starting point for our consideration of the structural organization and functions of these systems. Traveling from the outside in, it depicts the integumentary system, muscle system, and skeletal system of one of the more familiar vertebrates.

As you will see, the body of nearly all animals has an outer covering, or integument. Regardless of the species, movement of the body or parts of it requires contractile cells and some medium or structure against which contractile force may be applied. Hydrostatic skeletons, exoskeletons, and endoskeletons receive the applied force among different animal groups.

KEY CONCEPTS

1. Nearly all animals have an integument, some type of skeleton, and muscles. An integument is the outer covering of the animal body, and a prime example is vertebrate skin.

2. Skin protects the body from abrasion, ultraviolet radiation, bacterial attack, and other environmental insults. It also contributes to overall body functioning, as when it helps control moisture loss.

3. Three categories of skeletal systems are common in the animal kingdom. We call them hydrostatic skeletons, exoskeletons, and endoskeletons. Each has body fluids or structural elements, such as bones, against which a contractile force can be applied.

4. Bones are collagen-rich, mineralized organs. They function in movement, protection and support of soft organs, and mineral storage. Blood cells form in some. Ligaments or cartilage bridges joints between bones, and tendons attach them to skeletal muscles.

5. Many responses to changes in external and internal conditions involve muscles that move the animal body or parts of it. In response to suitable stimulation, the cells of muscle tissue contract, or shorten.

6. Smooth muscle and cardiac muscle are responsible for the motions of internal organs. Skeletal muscle helps move the body's limbs and other structural elements and maintain their spatial positions.

7. In each muscle cell, many threadlike structures called myofibrils are divided into sarcomeres. The sarcomere is the basic unit of contraction. It has parallel arrays of actin and myosin filaments. ATP-driven interactions between the actin and myosin shorten the sarcomeres of a muscle and collectively account for its contraction.

INTEGUMENTARY SYSTEM

Animals ranging from invertebrate worms to humans have an outer covering, or **integument** (after the Latin *integere*, meaning to cover). Most coverings are tough, pliable barriers against many environmental insults.

The integument of roundworms and insects, crabs, and other arthropods is a protective **cuticle**, hardened with chitin. Chitin, remember, is a polysaccharide that incorporates nitrogen atoms. Sections 3.2 and 26.15 describe this type of covering.

Vertebrate Skin and Its Derivatives

For vertebrates, the integument consists of a covering called **skin** as well as a variety of structures derived from epidermal cells of the skin's outer tissue layers (Figures 38.3 and 38.4). Beyond the typical assortment of epithelial tissues and glands, great variation exists within vertebrate groups as well as between them. For example, birds have the unique epidermal derivatives called feathers as well as bills and claws. Many herbivorous mammals have hooves and horns, porcupines bristle with quills, you and the rest of the primates have nails, and so on. Hagfishes have bare skin, which they coat with a lathering of very slimy secretions (Section 27.4). Epidermal cells of many fishes divide and differentiate to form hardened scales of a variety of thicknesses, shapes, and colors (Section 27.5).

The skin itself has two regions: an outermost **epidermis** and underlying **dermis**. Below this, a tissue region called the hypodermis anchors the skin to underlying structures and yet still allows it to move a bit (Figure 38.3*a*). Fats that become stored in the hypodermis help insulate the body and cushion some of its parts.

Your skin weighs about four kilograms (nine pounds). Stretched out, its surface area would be fifteen to twenty square feet. Most parts of human skin are as thin as a paper towel. It thickens only on the soles of the feet and in other regions that encounter recurring abrasions or pounding.

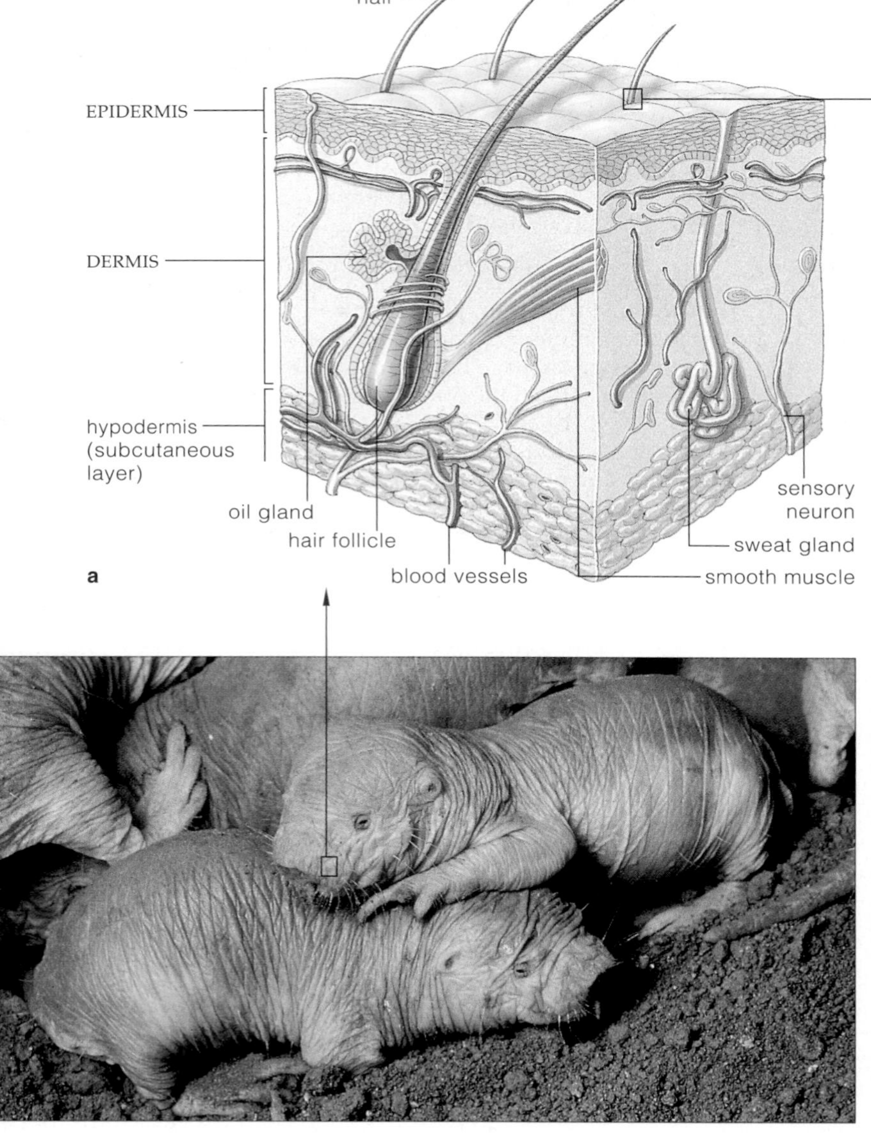
hair
EPIDERMIS
DERMIS
hypodermis (subcutaneous layer)
oil gland
hair follicle
blood vessels
sensory neuron
sweat gland
smooth muscle
a
c

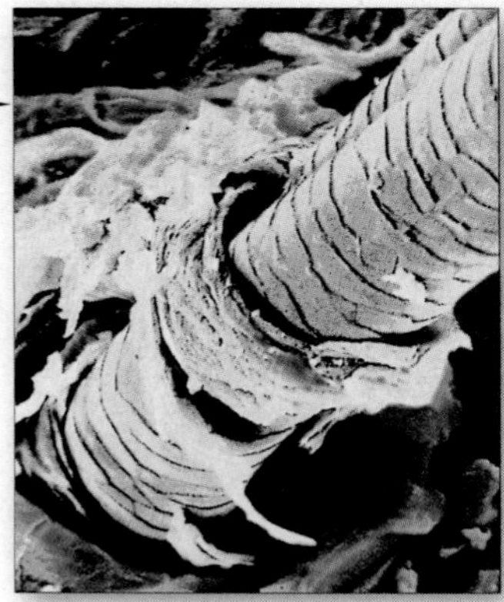
b

Dead, flattened epidermal cells around a shaft of hair that is projecting from the skin surface. Refer to Section 3.5, which includes a description of the molecular structure of hair.

Figure 38.3 (**a**) Structure of vertebrate skin. The uppermost portion is the epidermis; the lower portion is the dermis. (**b**) Scanning electron micrograph of a hair. (**c**) This display of skin is provided by naked mole-rats (*Heterocephalus glaber*), which live underground in burrows. Most of their hairs are located on the snout and are modified for sensory functions.

Functions of Skin

No garment ever made comes close to skin in its qualities. What besides skin holds its shape after

Further reading: Student Guide to InfoTrac on web site

a

b

c

Figure 38.4 A few examples of skin and of structures that are derived from it.

(**a**) Richly pigmented skin of a tree frog. Its pigment-producing cells, called chromatophores, reside primarily in the dermis. The kind deepest in the dermis produces a silvery pigment. Wavelengths reflected and scattered from such pigment molecules appear blue. Other kinds of pigment cells higher in the dermis produce a yellow pigment, which filters the blue wavelengths to produce a rich green coloration. Section 33.1 has a detailed diagram of the skin of a poisonous frog.

(**b**) Richly pigmented skin and hairs of the mountain gorilla. Hair follicles have a profusion of melanin-producing cells that supply the pigments. Each hair consists of a cuticle around a central shaft. Dead, flattened cells packed with the protein keratin make up the cuticle.

Very little melanin is distributed among the epidermal layers of a naked mole-rat (Figure 38.3*c*). Blood inside vessels that thread through the dermis of this small mammal contributes to the skin's reddish hue.

(**c**) Peacock feathers, one of the more spectacular examples of structures derived from skin.

(**d–f**) This series of diagrams shows how the feather of a typical bird grows from its base, at a region of actively dividing epidermal cells.

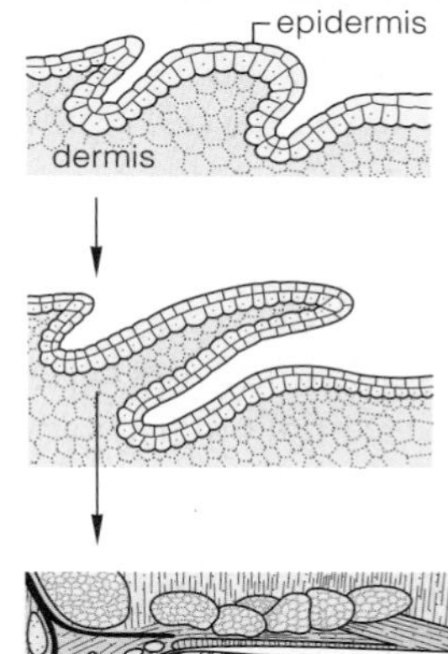
epidermis
dermis

d One of many small buds that appear on a chick embryo's surface (in this case, on the sixth day of incubation).

e One of the buds growing into a cone-shaped structure. Its base sinks into the skin. The depression around it will become one feather follicle.

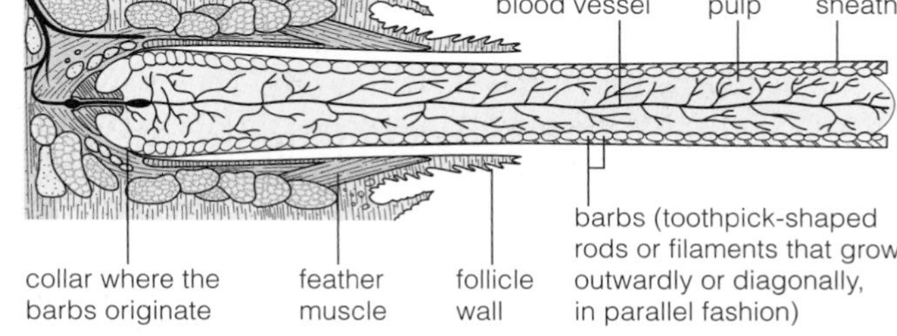
blood vessel
pulp
sheath
collar where the barbs originate
feather muscle
follicle wall
barbs (toothpick-shaped rods or filaments that grow outwardly or diagonally, in parallel fashion)

f A layer of horny cells at the cone's surface differentiates into a sheath. An epidermal layer just beneath the sheath will give rise to a feather. The dermis beneath this region of epidermis is richly supplied with blood vessels. It becomes the pulp, which nourishes the growing feather but does not contribute to its structure.

repeated stretchings and washings, blocks dangerous rays from the sun, kills many bacteria on contact, holds in moisture, fixes small cuts and burns, *and* can last as long as you do? Skin also produces vitamin D, required for calcium metabolism.

Finally, skin plays a passive role in mechanisms by which the internal body temperature of humans and other mammals is adjusted. The nervous system rapidly adjusts the flow of blood (which transports metabolic heat) to and from skin's great numbers of small blood vessels. And signals from sensory receptor endings in skin help the brain assess the surrounding world.

The body of most animals has an integument, a protective covering that usually is tough yet pliable. A number of invertebrate species have a cuticle, and vertebrates have skin, which consists of epidermis and dermis.

38.2 A LOOK AT HUMAN SKIN

Structure of Epidermis and Dermis

Like puff pastry, epidermis consists of sheetlike layers; it is a *stratified* epithelium. Its cells are structurally and functionally knit together by an abundance of junctions of the sort described in Section 33.1. Its inner sheets are composed of many living, rapidly dividing cells. The most abundant are **keratinocytes**, which synthesize the tough, water-insoluble protein keratin. One reason why skin is such a strong, cohesive integument is that the adhesion junctions between keratinocytes are anchored to numerous, crosslinked keratin fibers inside them.

Other cells, the **melanocytes**, produce and donate the brownish black pigment melanin to keratinocytes. Melanin screens out harmful ultraviolet radiation from the sun. Humans in general have the same number of melanocytes, but skin color varies owing to differences in the distribution and metabolic activity of these cells. For example, melanocytes in albinos cannot produce all of the enzymes required for melanin production. Pale skin contains little melanin, so the pigment hemoglobin inside red blood cells is not masked. The skin appears pinkish because hemoglobin's red color shows through thin-walled blood vessels and the epidermis itself, both of which are transparent. Carotene, which is a yellow-orange pigment, also contributes to skin color.

Section 38.3 describes two less abundant cell types in epidermis, the Langerhans and Granstein cells.

Rapid, ongoing mitotic divisions push epidermal cells from deeper layers toward the skin's free surface. Pressure from the continually growing cell mass, and from normal wear and tear at the skin surface, kills and flattens the older cells before they reach the outer layers (Figure 38.5). There, cells are abraded off or flake away on an ongoing basis. Rapid divisions not only replace the outermost, keratinized layer. They help skin mend quickly after cuts or burns.

Beneath the epidermis is the dermis. This is mostly dense connective tissue with many elastin fibers (which counter stretching) and collagen fibers (which provide strength). Blood vessels, lymph vessels, and receptor endings of sensory nerves thread through the dermis. Nutrients from the bloodstream reach living epidermal cells by diffusing through the dermal ground tissue.

Sweat Glands, Oil Glands, and Hairs

In general, human skin generally has sweat glands, oil glands, and husklike cavities (follicles) for hairs, mainly in the dermis. These structures arise from epidermal cells.

Fluid secreted by **sweat glands** is 99 percent water, with dissolved salts, traces of ammonia, vitamin C, and other substances. You have 2.5 million sweat glands controlled by sympathetic nerves. One type abounds in your palms, soles, forehead, and armpits. They have roles in adjusting body temperature and in *cold sweats*, a response to frightening, unsettling situations (Section 35.3). Secretions from a different type of sweat gland increase during stress, pain, and sexual foreplay. These secretions also increase prior to menstruation.

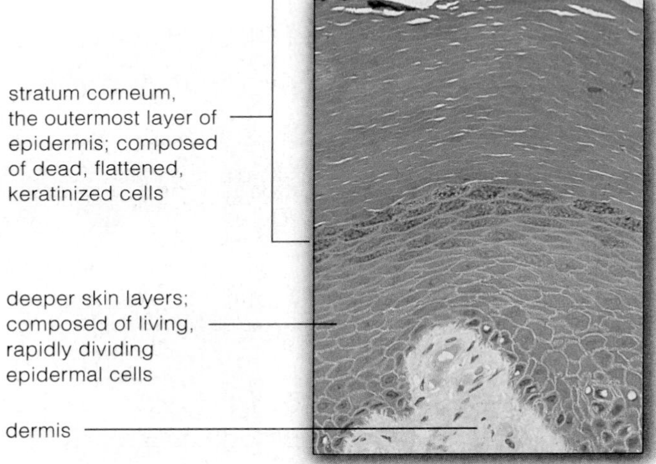

stratum corneum, the outermost layer of epidermis; composed of dead, flattened, keratinized cells
deeper skin layers; composed of living, rapidly dividing epidermal cells
dermis

Figure 38.5 Micrograph of a section through human skin.

Except on palms of the hands and soles of the feet, skin contains **oil glands** (also called sebaceous glands). Oil glands lubricate and soften hair and the skin, and their secretions kill many potentially harmful surface bacteria. *Acne*, an inflammation of skin, develops after bacteria successfully infect oil gland ducts.

Each **hair**, a flexible structure of mostly keratinized cells, has a root embedded in skin and a shaft above its surface (Figure 38.3). Cells divide near the root's base, are pushed upward, then flatten and die. Flattened cells of the shaft's outer layer overlap like roof shingles. When mechanically abused, these frizz as "split ends." The average human scalp includes about 100,000 hairs, although genes, nutrition, and hormones influence hair growth and density. Protein deficiency causes hair to thin (amino acids are required for keratin synthesis). So do high fever, emotional stress, and excess vitamin A intake. When the body produces abnormal amounts of testosterone, *hirsutism*, or excessive hairiness, might be one result. This hormone influences patterns of hair growth and other secondary sexual traits.

Skin's multiple layers of keratinized, melanin-shielded epidermal cells help the body conserve water, avoid damage by ultraviolet radiation, and resist mechanical stress.

Sweat glands, oil glands, hairs, and other structures derived from epidermal cells are embedded largely in the dermis. Blood vessels, lymph vessels, and the receptor endings of sensory neurons also reside in the dermis.

Further reading: Student Guide to InfoTrac on web site

38.3 SUNLIGHT AND SKIN

THE VITAMIN D CONNECTION Even when you do little more than sit outside in the sun, you are giving some of the epidermal cells in your skin the opportunity to make cholecalciferol, a form of vitamin D. **Vitamin D** actually is a generic name for steroid-like compounds that help the body absorb calcium from food. When some type of cell in skin is exposed to sunlight, it produces vitamin D from a precursor molecule related to cholesterol. Such cells then release vitamin D to the bloodstream, which transports it to absorptive cells of the intestinal lining. This is a hormone-like action, which means skin acts like an endocrine gland when exposed to sunlight.

Humans, remember, evolved beneath the intense sun of the African savanna, so skin alone would have supplied our early ancestors with enough vitamin D. Only after some human populations moved out of the tropics and into caves, animal skins, and clothing did they start to depend more on dietary sources of the essential vitamin D.

SUNTANS AND SHOE-LEATHER SKIN Do you like to tan your body by rotating under the sun's rays or a tanning lamp, like a chicken in an oven broiler? If so, think about what a broiler does to the chicken. The sun's ultraviolet wavelengths stimulate melanin production in skin cells. Continued exposure increases melanin concentrations in light skin and visibly darkens it, thereby producing the "tan" that so many people covet (Figure 38.6). Tanning does protect the body against ultraviolet radiation. Even in naturally dark skin, however, prolonged exposure to sunlight causes the elastin fibers in connective tissue of the dermis to clump together, so skin loses its resiliency. In time it starts to look like old shoe leather.

In this respect, tanning accelerates the *aging* of skin. As any person grows older, epidermal cells divide less often. Skin gets thinner and more vulnerable to injury. Glandular secretions that once kept it soft and moistened dwindle. Collagen and elastin fibers in the dermis break down and become sparser, so skin loses elasticity and its wrinkles deepen. Either by itself or in combination with excessive tanning, prolonged exposure to dry wind and tobacco smoke also accelerates the aging process.

SUNLIGHT AND THE FRONT LINE OF DEFENSE Besides having cells that produce melanin and keratin, skin also contains two other cell types that help protect the body against invasion by pathogenic cells and against cancer. The defenders of the epidermis are called Langerhans cells and Granstein cells.

Langerhans cells are phagocytes that develop in bone marrow, then take up stations in skin. After they engulf virus particles or bacterial cells, they display molecular alarm signals at the surface of their plasma membrane. These signals can mobilize the body's immune system.

Ultraviolet radiation can damage Langerhans cells. This might be why sunburns often trigger *cold sores*, the small, painful blisters that announce the recurrence of a *Herpes simplex* infection. Nearly everyone harbors the *H. simplex* virus. It remains hidden within the face, inside a ganglion. (A ganglion, remember, is a cluster of neuron cell bodies.) Sunburns and other stress factors can activate the virus. Virus particles move down the neurons to the axonal endings in skin. There they infect epithelial cells and cause skin eruptions.

When ultraviolet radiation damages Langerhans cells, it weakens one of the body's first lines of defense. It also might activate proto-oncogenes and trigger cancerous transformation of skin cells. As described in the Chapter 15 introduction, skin cancers grow rapidly, and they may spread to adjacent lymph nodes unless surgically removed.

In some as-yet-undetermined way, **Granstein cells** apparently interact with the white blood cells that can put the brakes on immune responses in the skin. By issuing suppressor signals, they help keep the responses from spiraling out of control. Even though the functions of Granstein cells are not completely understood, researchers have learned that these cells are less vulnerable than the Langerhans cells to the damaging effects of ultraviolet radiation.

Figure 38.6 Demonstration of how shoe-leather skin forms.

TYPES OF SKELETONS

Operating Principles for Skeletons

Many of the responses that an animal makes to external and internal conditions require movements of the whole body or parts of it. The movements come about through activation, contraction, and relaxation of muscle cells. But muscle cells alone do not produce them. *All muscles require the presence of some medium or structural element against which the force of contraction may be applied.* A skeletal system fulfills this requirement.

Three types of skeletons predominate in the animal world. With a **hydrostatic skeleton**, the muscles work against an internal body fluid and redistribute it within a confined space. Like a filled waterbed, the enclosed fluid resists compression. By contrast, an **exoskeleton** has rigid, *external* body parts, such as a hinged shell, that receives the applied force of muscle contraction. An **endoskeleton** has rigid, *internal* body parts, such as bones, that receive the applied force of contraction.

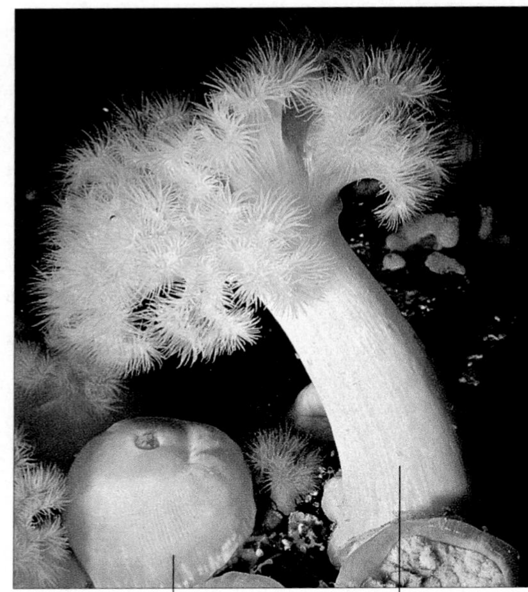
a Resting position
b Feeding position

Figure 38.7 Outcomes of contractile force applied against the sea anemone's hydrostatic skeleton. This invertebrate's body wall has contractile cells running longitudinally (parallel with the body axis) and radially around the gut cavity. In (**a**), radial cells are relaxed and longitudinal ones are contracted. Anemones typically assume this resting position at low tide, when currents cannot bring food morsels to them. In (**b**), the radial cells are contracted, longitudinal ones are relaxed, and the body is extended to its upright feeding position.

Examples From the Invertebrates

Many invertebrates with a soft body have a hydrostatic skeleton. Reflect on a sea anemone, with its soft, vase-shaped body and saclike gut (Figure 38.7). Its body wall incorporates longitudinal and radial muscles. Between meals, its longitudinal muscles are contracted (that is, shortened) and radial ones are relaxed (lengthened), so the sea anemone looks short and squat. When its body lengthens into an upright feeding position, its radial muscles are contracting and forcing some fluid out of the gut cavity, and longitudinal muscles are relaxing.

Or think about an earthworm. This annelid, recall, has a series of coelomic compartments, each with its own muscles, nerves, and bristles (setae). Its body moves forward by contracting and relaxing its fluid-filled segments, one after the other as described in Section 26.14. By coordinating contractions on one side or the other side of these segments, the earthworm also can thrash sideways or move forward and back.

As another example, a jumping spider has a hinged exoskeleton, as other arthropods do. In addition, it uses body fluids to transmit force when leaping at prey. Its heart pumps blood into body tissues. By contracting muscles, the spider makes blood surge into its hind leg spines. It's like quickly squeezing a water-filled rubber glove so its skinny fingers become rigidly erect. Figure 38.8 illustrates this resourceful use of hydraulic pressure. (*Hydraulic* means fluid pressure inside tubes.)

The hinged arthropod exoskeleton has advantages. Some of its hard parts can be moved like levers by sets of muscles attached to them. Thus, small contractions can bring about large movements of wings or some other body parts. This is especially true of the cuticle of a winged insect. It extends over all body segments and over gaps between segments (Figure 38.9). At the gaps, the cuticle remains pliable. It acts like a hinge when muscles alternately raise and lower either the wing or the body parts to which the wings are attached.

Figure 38.8 Time-lapse photographs of the leap of the jumping spider *Sitticus pubescens*, an impressive predator. Its leaping depends partly on the hydraulic extension of its hind legs when blood surges into them under high pressure.

Further reading: Student Guide to InfoTrac on web site

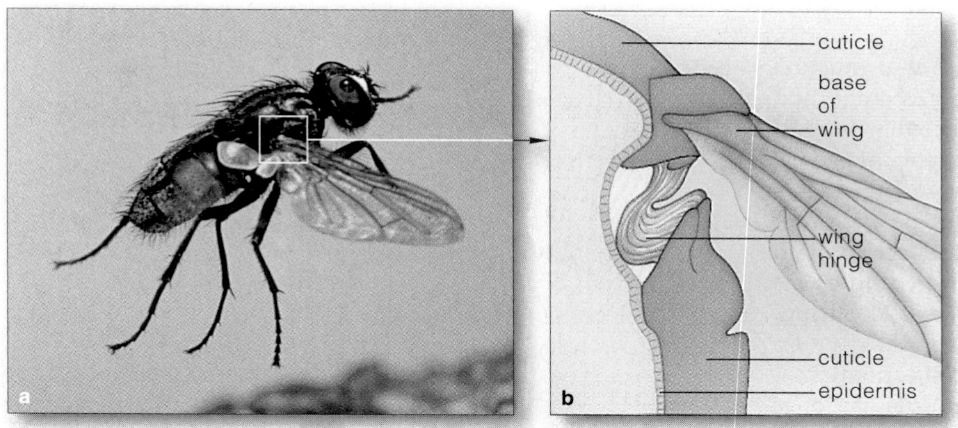
a
cuticle
base
of
wing
wing
hinge
cuticle
epidermis
b

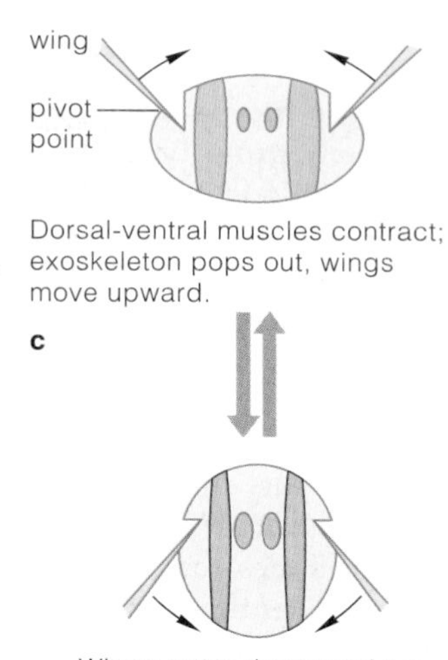
wing
pivot
point
Dorsal-ventral muscles contract; exoskeleton pops out, wings move upward.
c
Wings move downward as exoskeleton pops back and swiftly stretch the muscles.
d

Figure 38.9 Housefly wing movement, an outcome of the contraction of sets of muscles that extend from dorsal to ventral regions of the exoskeleton. The contractile force works against the exoskeleton, near hinge points where wings are attached. When the muscles contract, the exoskeleton pops inward and wings move up. When the exoskeleton pops outward by elastic force, the wings move down. The quick-stretched muscles invite a fast repeat of the events.

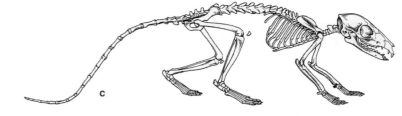

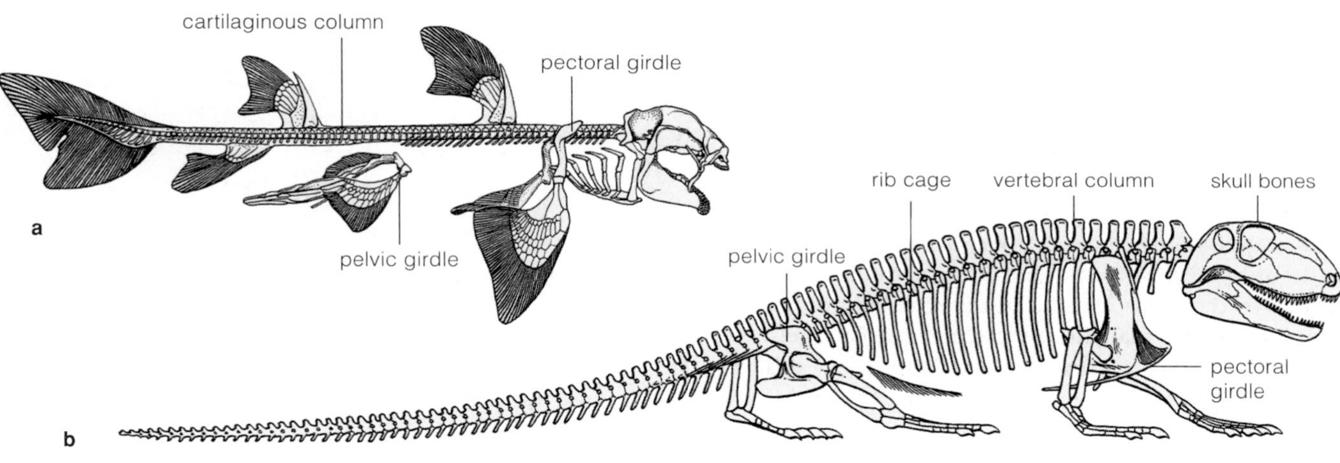
cartilaginous column
pectoral girdle
a
pelvic girdle
rib cage
vertebral column
skull bones
pelvic girdle
pectoral
girdle
b

Figure 38.10 Comparison of skeletons of a shark (**a**), generalized early reptile (**b**), and generalized mammal (**c**). In sharks, cartilage hardened with calcium deposits superficially resembles bone. As in bony fishes, the girdles function as a stable base for fin motions. In four-legged vertebrates, they function in transferring body weight to limbs. These girdles have more surface area for attaching limb muscles, and they are tied more closely to the body, thus offering better support and assisting in the thrust of locomotion.

c

Examples From the Vertebrates

Vertebrates have endoskeletons of one sort or another. For example, sharks have a skeleton of an opaque form of cartilage, hardened with calcium deposits (Figure 38.10*a*). Some other fishes have a flexible skeleton of an elastic, translucent form of cartilage that almost looks like glass. Most vertebrates have an endoskeleton that is largely bone (Figure 38.10*b,c*). We turn next to the functions and characteristics of bones. Afterward, we will consider how different types of bones are arranged in the human skeletal system.

Animal skeletons have structural elements or body fluids against which the force of contraction may be applied.

38.5 CHARACTERISTICS OF BONE

Bone Structure and Function

By definition, **bones** are complex organs that function in movement, protection, support, mineral storage, and formation of blood cells (Table 38.1). Bones that support and anchor skeletal muscles help maintain or change the positions of body parts. Some form hard compartments that enclose and protect the brain, the lungs, and other internal organs. Bones are reservoirs for calcium and phosphorus ions. Continual deposits and withdrawals of ions from bone help maintain blood levels of calcium and phosphorus, and thus support metabolic activities. Some bones (not all) are sites of blood cell formation.

Human bones range in size from middle earbones smaller than lentils to clublike femurs, or thighbones (Figure 38.11). Different bones are long, flat, short or cubelike, and irregular. All consist of connective tissues (bone tissue especially) and epithelia. Bone tissue has mature bone cells, or **osteocytes**, and collagen fibers in a calcium-hardened ground substance.

For the thighbone in Figure 38.11, *compact* bone tissue in the shaft and at both ends resists mechanical shock. This bone tissue is deposited as many thin, cylindrical, dense layers around interconnected canals for blood vessels and nerves that service living bone cells. Each array is a Haversian system. *Spongy* bone tissue in the bone ends and shaft imparts strength, yet it does not weigh much. Abundant spaces make the tissue appear spongy, but its flattened parts are firm. **Red marrow**, a major site of blood cell formation, fills spaces in some bones, such as the breastbone. Cavities in most mature bones contain **yellow marrow**. Yellow marrow is largely fat. It converts to red marrow and produces more red blood cells to counter severe blood loss from the body.

Bone Formation and Remodeling

Bone tissue starts to form in the embryo, where many bones are constructed on cartilage models. **Osteoblasts**, or bone-forming cells, emerge and start to secrete organic

Table 38.1 Functions of Bone

1. *Movement.* Bones interact with skeletal muscle to maintain or change the position of body parts.
2. *Support.* Bones support and anchor muscles.
3. *Protection.* Many bones form hard compartments that enclose and protect soft internal organs.
4. *Mineral storage.* Bones are a reservoir for calcium and phosphorus, the deposits and withdrawals of which help maintain ion concentrations in body fluids.
5. *Blood cell formation.* Some bones contain regions where blood cells are produced.

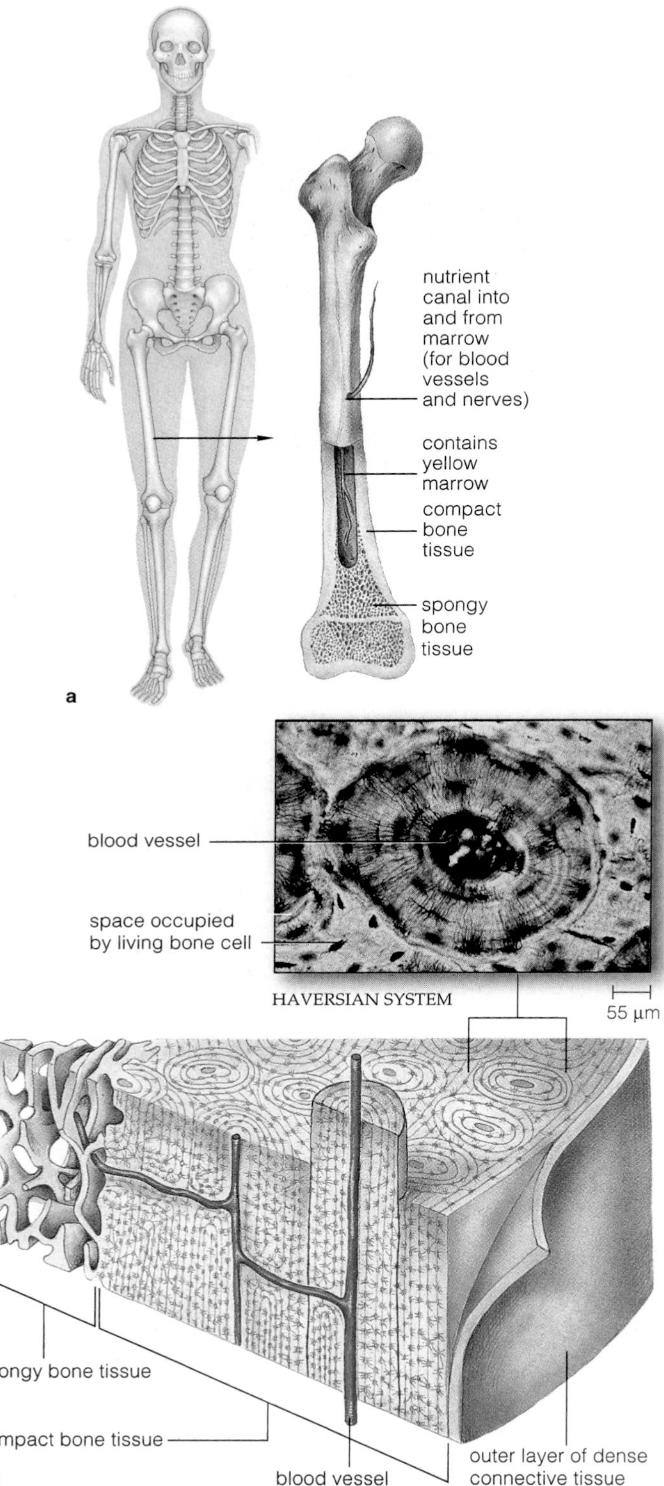
nutrient canal into and from marrow (for blood vessels and nerves)
contains yellow marrow
compact bone tissue
spongy bone tissue
a
blood vessel
space occupied by living bone cell
HAVERSIAN SYSTEM
55 µm
spongy bone tissue
compact bone tissue
b
blood vessel
outer layer of dense connective tissue

Figure 38.11 (**a**) Structure of a thighbone. (**b**) Appearance of its spongy and compact bone tissue. Thin, dense layers of compact bone tissue form cylindrical, interconnected arrays around canals for blood vessels and nerves. Each array is a Haversian system. The blood vessel inside services osteocytes, living bone cells in small tissue spaces. Tunnels connect adjacent spaces.

Further reading: Student Guide to InfoTrac on web site

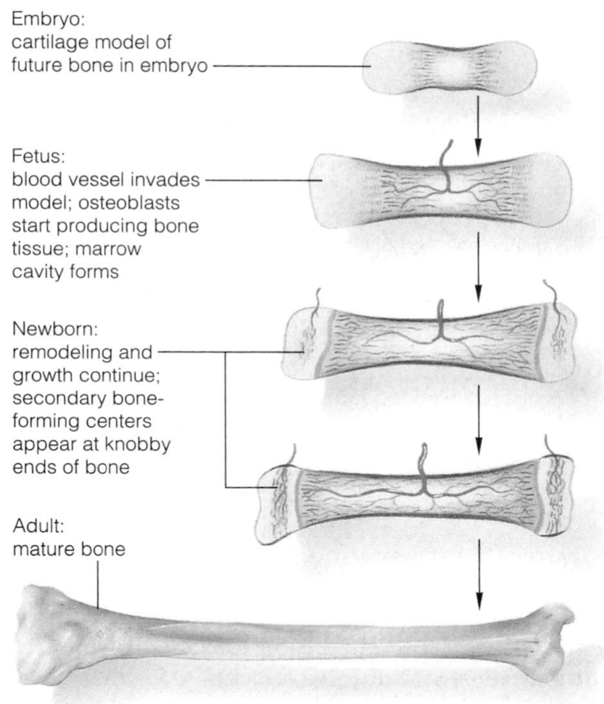
Embryo:
cartilage model of
future bone in embryo
Fetus:
blood vessel invades
model; osteoblasts
start producing bone
tissue; marrow
cavity forms
Newborn:
remodeling and
growth continue;
secondary bone-
forming centers
appear at knobby
ends of bone
Adult:
mature bone

Figure 38.12 Long bone formation, starting with osteoblast activity in a cartilage model (here, already formed in the embryo). These bone-forming cells are active first in the shaft region, then at the knobby ends. In time, the only cartilage left is at the ends.

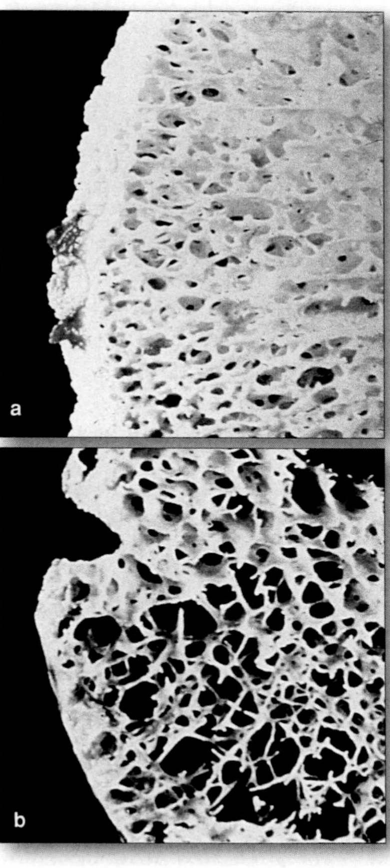
a
b

Figure 38.13 An example of bone tissue affected by osteoporosis.

(**a**) Section through normal bone tissue. In such tissue, the mineral deposits continually replace the withdrawals.

(**b**) After the onset of osteoporosis, replacements of mineral ions lag behind withdrawals. In time the tissue erodes, and bones become hollow and brittle.

substances on the model. The secretions then become mineralized, and cartilage inside the shaft breaks down to open a marrow cavity (Figure 38.12). Once osteoblasts have become imprisoned by their own secretions, we call them **osteocytes**, or mature, living bone cells.

The total bone mass in healthy, young adults does not change much, but large amounts of mineral deposits and thousands of osteocytes are removed and replaced on an ongoing basis. Usually with **bone remodeling**, minerals are deposited and removed at the same time. The process adjusts bone strength and helps maintain blood levels of calcium and phosphorus. Two cell types are partners in the process. Osteoblasts deposit bone, and **osteoclasts** secrete enzymes that digest the organic matrix of bone. Thanks to these bone-degrading cells, the liberated ions enter the interstitial fluid, from which they can be reabsorbed by the bloodstream.

Maintaining Blood Levels of Calcium

Given the roles of calcium in neural function, muscle contraction, and other vital activities, its blood level is one of the most tightly controlled aspects of metabolism. Bones and teeth store all but 1 percent or so of a body's calcium. Negative feedback loops between two glands and the blood govern the rates of calcium release and uptake. When the calcium concentration in blood rises, the thyroid gland releases calcitonin in response. This hormone suppresses osteoclast activity and slows down the release of calcium into blood.

When calcium's blood level falls, parathyroid glands release parathyroid hormone (PTH), which promotes calcium release from bone and from filtrate inside the kidneys. PTH also enhances osteoclast activity and helps activate vitamin D, which stimulates calcium absorption from the small intestine (Sections 37.7 and 38.3).

When a bone is stressed, as when it is subjected to compression, mineral deposition by osteoblasts exceeds withdrawals by osteoclasts. That is why bones of people who exercise rigorously are more dense and stronger than bones of couch potatoes. But mineral withdrawals predominate when bones are injured. Also, as a person ages, the backbone and other bones decline in mass, a condition is called *osteoporosis* (Figure 38.13). Decreased physical activity, declining activity of the bone-forming cells, loss of calcium, excessive protein intake, and sex hormone deficiencies contribute to the disorder.

Bones are collagen-rich, mineralized organs that function in movement, protection, support, storage of calcium and other minerals, and blood cell formation.

38.6 HUMAN SKELETAL SYSTEM

Appendicular and Axial Portions

The human skeleton has 206 bones, which anatomists subdivide into appendicular and axial portions (Table 38.2). The *appendicular* portion has a pair of pectoral girdles (at the shoulders), a pelvic girdle (at the hips), and pairs of arms, hands, legs, and feet. Each pectoral girdle has a slender collarbone and flat shoulder blade. Fall on an outstretched arm and you might dislocate a shoulder or fracture a collarbone. The flimsily arranged collarbone is the most frequently broken bone.

The human skeleton's *axial* portion includes skull bones, twelve pairs of ribs, a breastbone, and twenty-six **vertebrae** (singular, vertebra). These are bony segments of a curved backbone (vertebral column). The backbone extends from the base of the skull to the pelvic girdle. There the backbone transmits the torso's weight to the lower limbs. A spinal cord threads through a series of canals in segments at the back of the column. Between these segments are **intervertebral disks**, cartilaginous shock absorbers and flex points that permit movement.

Sometimes a severe or rapid shock forces a disk to slip out of place or rupture. Such painful, *herniated disks* are a less-than-advantageous outcome of bipedalism. Remember, the primate ancestors of the human lineage were quadrupedal. The first hominids started to walk upright about 4 million years ago, and a pronounced S-shaped curve in the backbone was a result. Today, the older we get, the longer we have been fighting gravity in a compromised way, and the more back pain we get.

Table 38.2 Components of the Human Skeleton

APPENDICULAR PORTION	
	Pectoral girdles: clavicle (collarbone) and scapula (shoulder blade)
	Arm bones: humerus, radius, ulna
	Hand bones: carpals, metacarpals, phalanges (of fingers)
	Pelvic girdle (six fused bones at the hip)
	Leg bones: femur (thighbone), patella, tibia, fibula
	Foot bones: tarsals, metatarsals, phalanges (of toes)
AXIAL PORTION	
	Skull: cranial bones and facial bones
	Rib cage: sternum (breastbone) and ribs (12 pairs)
	Vertebral column: vertebrae (26) and intervertebral disks

Cartilaginous joints bridging the vertebrae, ribs, and the breastbone allow slight movements. Synovial joints, such as knee joints, move freely, and ligaments stabilize them (Figures 38.14 and 38.15). Cartilage cushions the abutting bones and absorbs shocks. A flexible capsule of dense connective tissue surrounds the contact area. Cells of a membrane lining the interior of the capsule secrete a fluid that lubricates the joint.

Like many other joints, the knee joint is vulnerable to stress. This is the joint that lets you swing, bend, and turn the long bones below it. When you run, it absorbs the force of your weight each time the foot below it hits the ground. Stretch or twist a knee joint suddenly and too far, and you *strain* it. Tear its ligaments or tendons and you *sprain* it. Move the wrong way and you may well dislocate the attached bones. During collision sports such as football, a blow to a knee often severs a ligament. The severed part must be reattached within ten days. Why? Phagocytes in a lubricating fluid present in the joint normally clean up after everyday wear and tear. When presented with torn ligaments, they indiscriminately turn the tissue to mush.

Joint inflammation as well as degenerative disorders are collectively called "arthritis." In *osteoarthritis*, cartilage at the knees and other freely movable joints wears off as a person ages. Joints in the fingers, knees, hips, and vertebral column are affected most often.

In *rheumatoid arthritis*, synovial membranes in joints become inflamed and thickened, cartilage degenerates, and bone deposits build up. This degenerative disorder

Skeletal Joints

Joints are areas of contact or near-contact between bones, and each has a distinctive bridge of connective tissue. Very short connecting fibers join bones at *fibrous* joints. Straps of cartilage join them at *cartilaginous* joints. Long straps of dense connective tissue, or **ligaments,** bridge the gap between the bones at *synovial* joints.

Fibrous joints hold teeth in their sockets. They link together the flat skull bones of a fetus. At childbirth, the loose connections allow the bones to slide over each other a bit and prevent skull fractures. The skull of a newborn still has fibrous joints as well as membranous areas called "soft spots," or fontanels. By childhood, the fibrous tissue hardens and the skull bones are fused into a single unit.

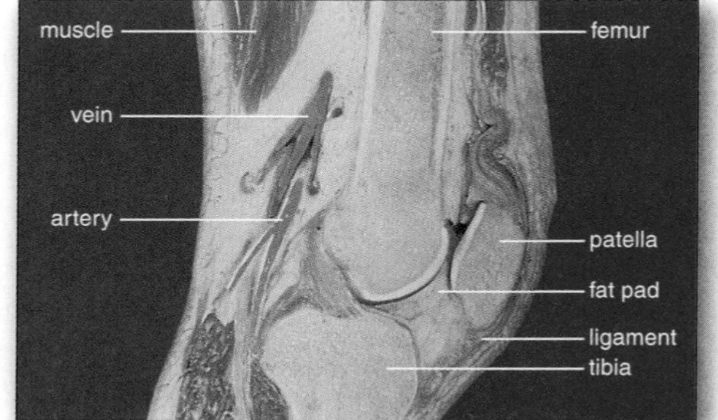
muscle
vein
artery
femur
patella
fat pad
ligament
tibia

Figure 38.14 Human knee joint, longitudinal section.

Further reading: Student Guide to InfoTrac on web site

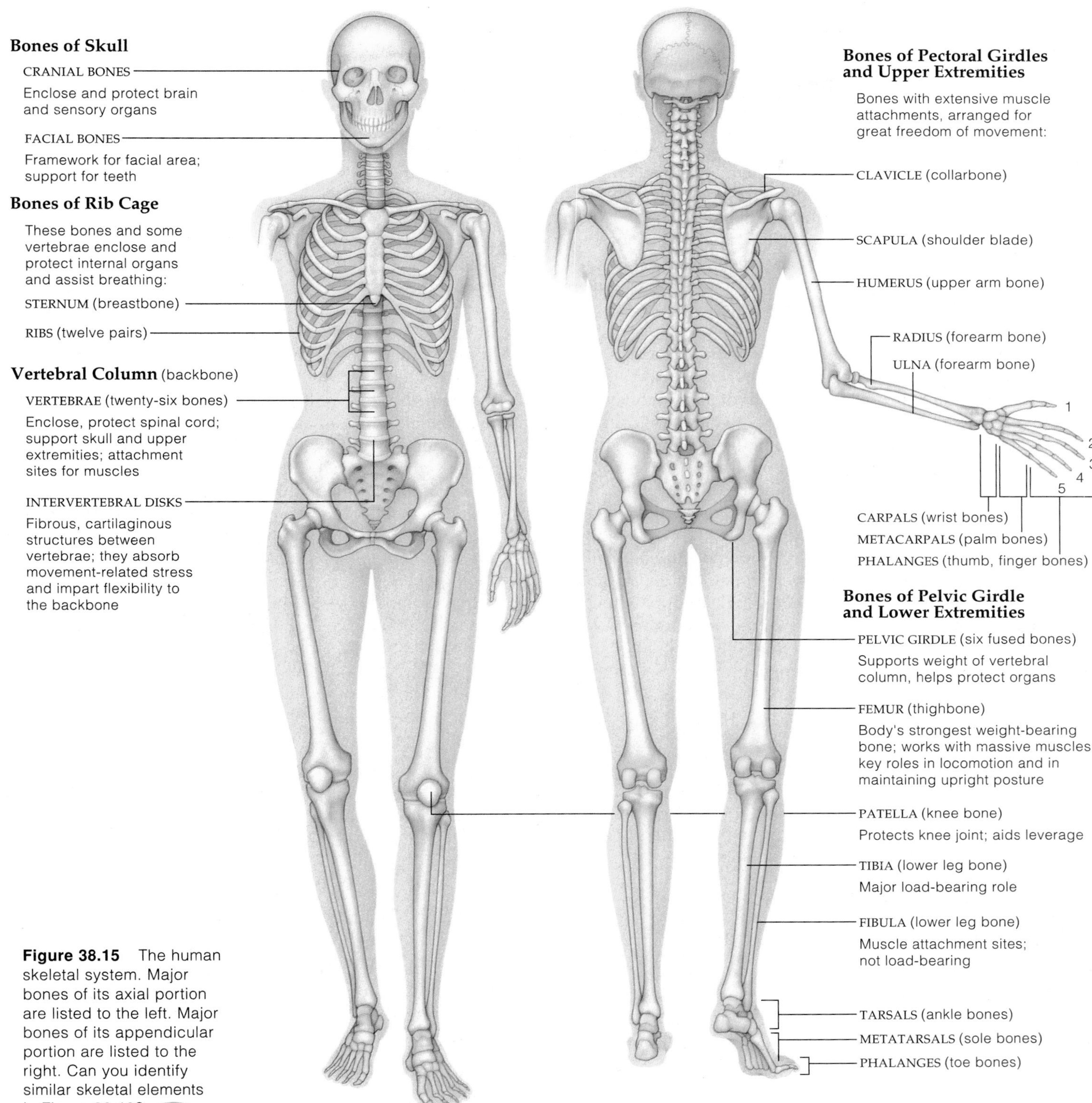

Bones of Skull
CRANIAL BONES
Enclose and protect brain and sensory organs
FACIAL BONES
Framework for facial area; support for teeth
Bones of Rib Cage
These bones and some vertebrae enclose and protect internal organs and assist breathing:
STERNUM (breastbone)
RIBS (twelve pairs)
Vertebral Column (backbone)
VERTEBRAE (twenty-six bones)
Enclose, protect spinal cord; support skull and upper extremities; attachment sites for muscles
INTERVERTEBRAL DISKS
Fibrous, cartilaginous structures between vertebrae; they absorb movement-related stress and impart flexibility to the backbone
Bones of Pectoral Girdles and Upper Extremities
Bones with extensive muscle attachments, arranged for great freedom of movement:
CLAVICLE (collarbone)
SCAPULA (shoulder blade)
HUMERUS (upper arm bone)
RADIUS (forearm bone)
ULNA (forearm bone)
1
2
3
4
5
CARPALS (wrist bones)
METACARPALS (palm bones)
PHALANGES (thumb, finger bones)
Bones of Pelvic Girdle and Lower Extremities
PELVIC GIRDLE (six fused bones)
Supports weight of vertebral column, helps protect organs
FEMUR (thighbone)
Body's strongest weight-bearing bone; works with massive muscles; key roles in locomotion and in maintaining upright posture
PATELLA (knee bone)
Protects knee joint; aids leverage
TIBIA (lower leg bone)
Major load-bearing role
FIBULA (lower leg bone)
Muscle attachment sites; not load-bearing
TARSALS (ankle bones)
METATARSALS (sole bones)
PHALANGES (toe bones)

Figure 38.15 The human skeletal system. Major bones of its axial portion are listed to the left. Major bones of its appendicular portion are listed to the right. Can you identify similar skeletal elements in Figure 38.10?

appears to be triggered by a bacterial or viral infection. It also seems to have a genetic component. Rheumatoid arthritis can begin at any age, but symptoms usually emerge before age fifty.

A human skeleton has an axial portion (a backbone, skull bones, and rib cage) and an appendicular portion (pelvic girdle, pectoral girdles, and arm, hand, leg, and foot bones).

38.7 SKELETAL–MUSCULAR SYSTEMS

How Muscles and Bones Interact

Skeletal muscles are the functional partners of bones. Each skeletal muscle contains bundles of hundreds to many thousands of muscle cells, which look like long, striped fibers. In muscle tissue, remember, muscle cells contract (shorten) in response to adequate stimulation. They lengthen in response to gravity and other loads. Whenever you dance, breathe, scribble notes, or tilt your head, contracting muscle cells are helping to move your body or change the positions of some of its parts.

Connective tissue bundles the muscle cells together and extends past them to form **tendons**. Each tendon, a cord or strap of dense connective tissue, attaches some muscle to bone (Figure 38.16). Most of the attachment sites are like a car's gearshift. *They form a lever system, in which a rigid rod is attached to a fixed point yet moves about at it.* Muscles connect to bones (rigid rods) near a joint (fixed point). As they contract, they transmit force to the bones and make them move. Tendons often rub against bones. But they slide inside fluid-filled sheaths that help reduce friction. Your knees, wrists, and finger joints have such sheaths (Figure 38.16).

Skeletal muscles interact with one another as well as with bones. Some are arranged in pairs or groups to promote some movement. Others work in opposition, with the action of one opposing or reversing the action of another. Figure 38.17 shows how opposing muscle groups work to move frog legs.

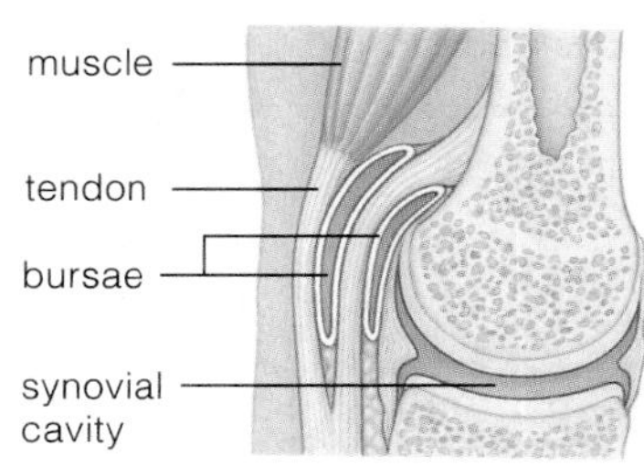

muscle
tendon
bursae
synovial cavity

Figure 38.16 Tendon. A bursa is a flattened sac filled with synovial fluid. Bursae are located between tendons and bones (or some other structure). They help reduce friction during movements.

Also look at Figure 38.18. Extend your right arm forward, then place your left hand over the biceps in the upper right arm and slowly "bend the elbow." Feel the biceps contract? Even when a biceps contracts only a bit, it causes a large motion of the bone connected to it. This is the case for most leverlike arrangements.

Bear in mind, only *skeletal* muscle is the functional partner of bone. As mentioned earlier, smooth muscle is mainly a component of the walls of internal organs, such as the stomach (Section 33.3). Cardiac muscle is present only in the wall of the heart. We will consider the structure and functioning of smooth muscle as well as cardiac muscle in later chapters in this unit.

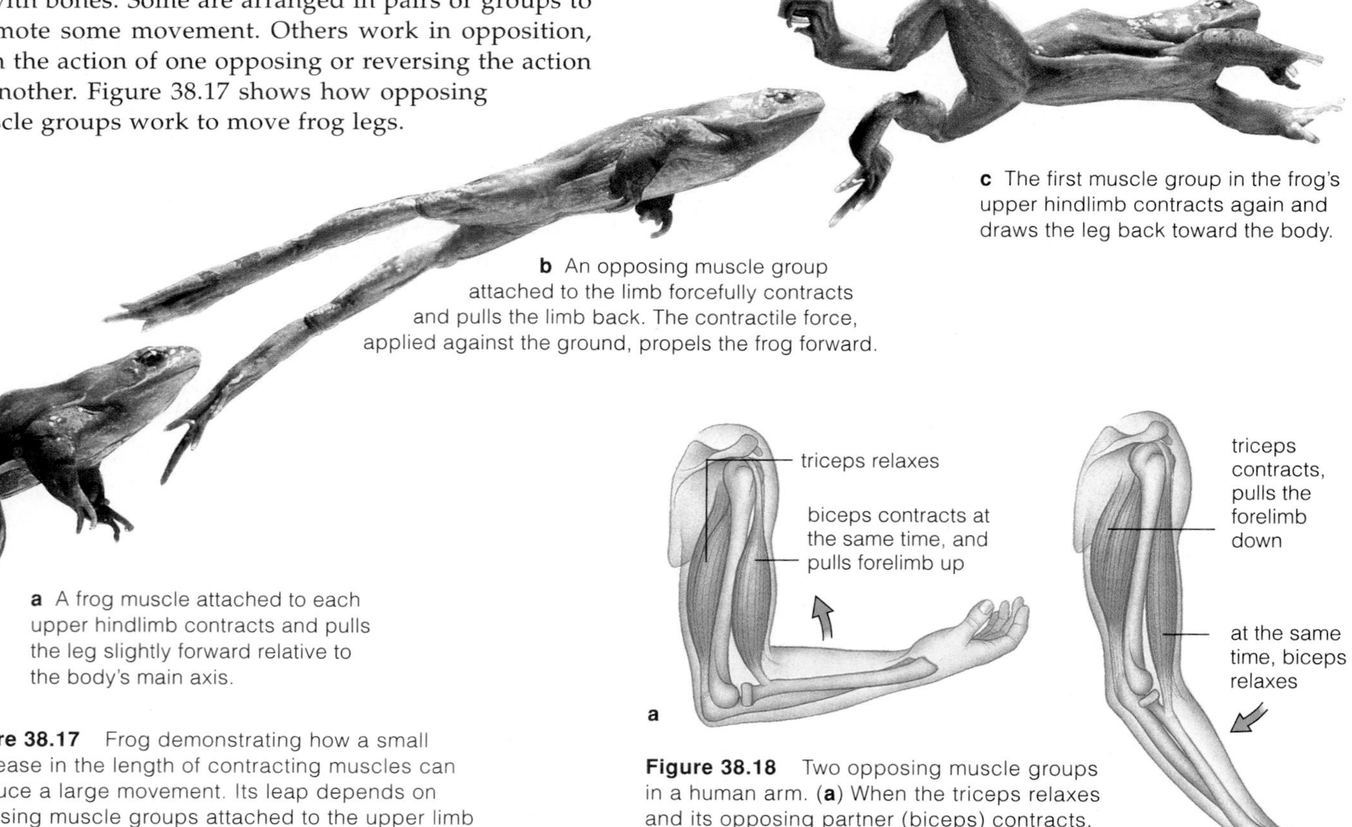

triceps relaxes
biceps contracts at the same time, and pulls forelimb up
a
triceps contracts, pulls the forelimb down
at the same time, biceps relaxes
b

c The first muscle group in the frog's upper hindlimb contracts again and draws the leg back toward the body.

b An opposing muscle group attached to the limb forcefully contracts and pulls the limb back. The contractile force, applied against the ground, propels the frog forward.

a A frog muscle attached to each upper hindlimb contracts and pulls the leg slightly forward relative to the body's main axis.

Figure 38.17 Frog demonstrating how a small decrease in the length of contracting muscles can produce a large movement. Its leap depends on opposing muscle groups attached to the upper limb bone of each hind leg. (**a**) One muscle group pulls the limb forward a bit, toward the body's midline. (**b**) Another pulls it back and away from the body.

Figure 38.18 Two opposing muscle groups in a human arm. (**a**) When the triceps relaxes and its opposing partner (biceps) contracts, the elbow joint flexes and the forearm bends upward. (**b**) When the triceps contracts, the forearm straightens out.

Further reading: Student Guide to InfoTrac on web site

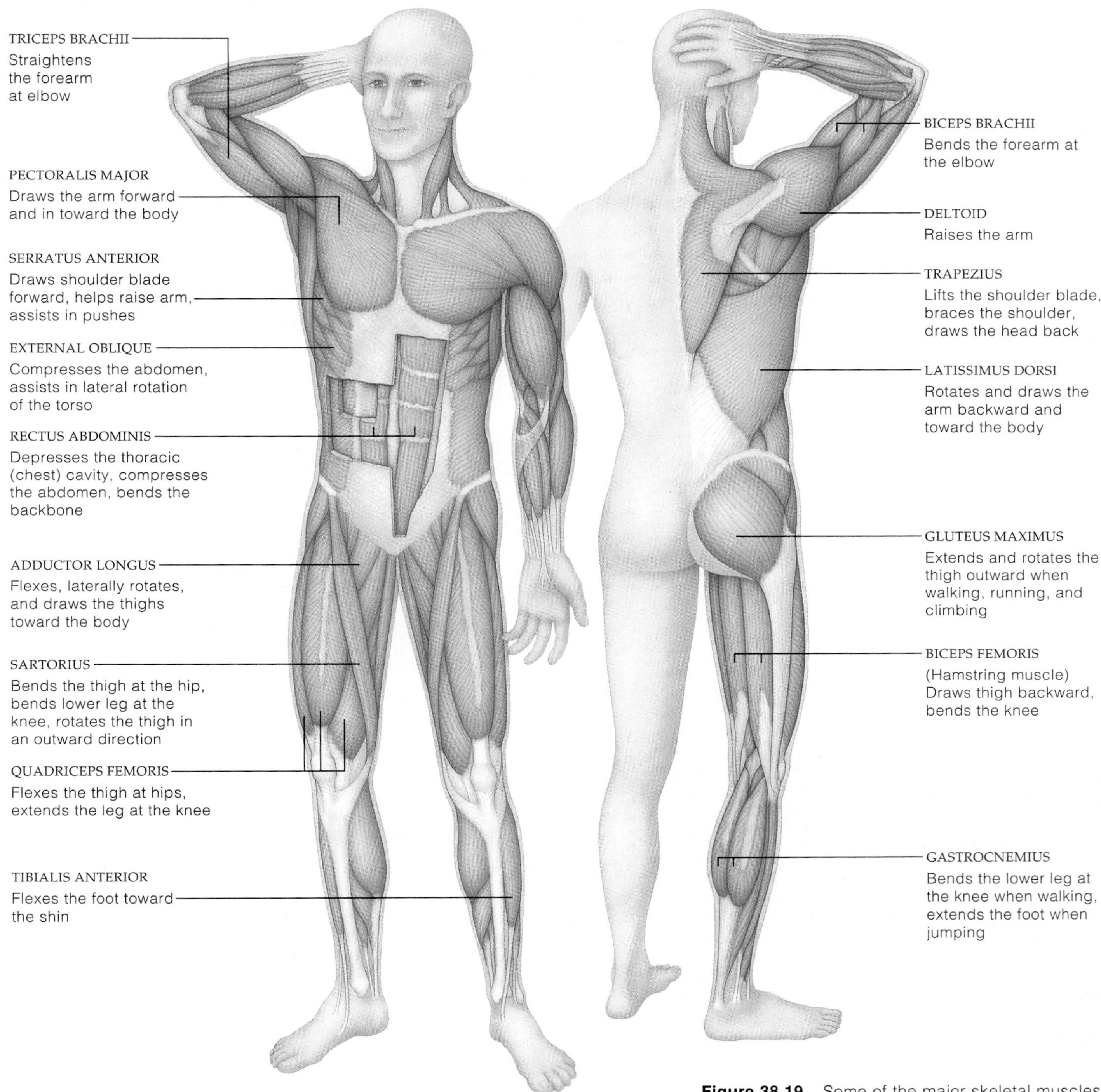
TRICEPS BRACHII
Straightens the forearm at elbow
PECTORALIS MAJOR
Draws the arm forward and in toward the body
SERRATUS ANTERIOR
Draws shoulder blade forward, helps raise arm, assists in pushes
EXTERNAL OBLIQUE
Compresses the abdomen, assists in lateral rotation of the torso
RECTUS ABDOMINIS
Depresses the thoracic (chest) cavity, compresses the abdomen, bends the backbone
ADDUCTOR LONGUS
Flexes, laterally rotates, and draws the thighs toward the body
SARTORIUS
Bends the thigh at the hip, bends lower leg at the knee, rotates the thigh in an outward direction
QUADRICEPS FEMORIS
Flexes the thigh at hips, extends the leg at the knee
TIBIALIS ANTERIOR
Flexes the foot toward the shin
BICEPS BRACHII
Bends the forearm at the elbow
DELTOID
Raises the arm
TRAPEZIUS
Lifts the shoulder blade, braces the shoulder, draws the head back
LATISSIMUS DORSI
Rotates and draws the arm backward and toward the body
GLUTEUS MAXIMUS
Extends and rotates the thigh outward when walking, running, and climbing
BICEPS FEMORIS
(Hamstring muscle) Draws thigh backward, bends the knee
GASTROCNEMIUS
Bends the lower leg at the knee when walking, extends the foot when jumping

Figure 38.19 Some of the major skeletal muscles of the human skeletal–muscular system.

Human Skeletal–Muscular System

The human body has more than 600 skeletal muscles, some superficial, others deep in the body wall. Some, such as facial muscles, attach to the skin. The trunk has muscles of the thorax, backbone, abdominal wall, and pelvic cavity. Other groups of muscles attach to upper and lower limb bones. Figure 38.19 shows a few of the main skeletal muscles and lists their functions. We turn next to the mechanisms underlying their contraction.

Skeletal muscles transmit contractile force to bones and make them move. Tendons strap skeletal muscles to bones.

38.8 A CLOSER LOOK AT MUSCLES

Skeletal Muscle Structure and Function

Take time to study Figure 38.20. The ballerina's bones move—they are pulled in a certain direction—when the skeletal muscles that are attached to them shorten. When a skeletal muscle shortens, its component muscle cells are shortening. When each muscle cell shortens, many units of contraction within the cell are shortening. The basic units of contraction are called **sarcomeres**.

Figure 38.20 shows how bundles of cells in a skeletal muscle run parallel with the muscle. In each muscle cell are myofibrils, threadlike structures bundled together in parallel. Each myofibril is divided transversely into bands. When stained, these show up as an alternating light–dark pattern. The bands of all the myofibrils in the cell are in close register, which gives each muscle cell a cross-striated appearance (Figure 38.20*d*). That is why skeletal (and cardiac) muscle have cross-stripes as well.

The dark bands, or Z bands, define each sarcomere. Inside a sarcomere are many thin and thick filaments. Each *thin* filament is about six nanometers in diameter, with one end attached to a Z band. It is like two strands of pearls twisted together. The "pearls" are molecules of **actin**, a globular protein with contractile properties

muscle's outer sheath (connective tissue)
two bundles of muscle cells (each has its own connective tissue sheath)
one muscle cell
one myofibril

b Sketch of skeletal muscle structure. The muscle cells are bundled together in parallel inside a sheath.

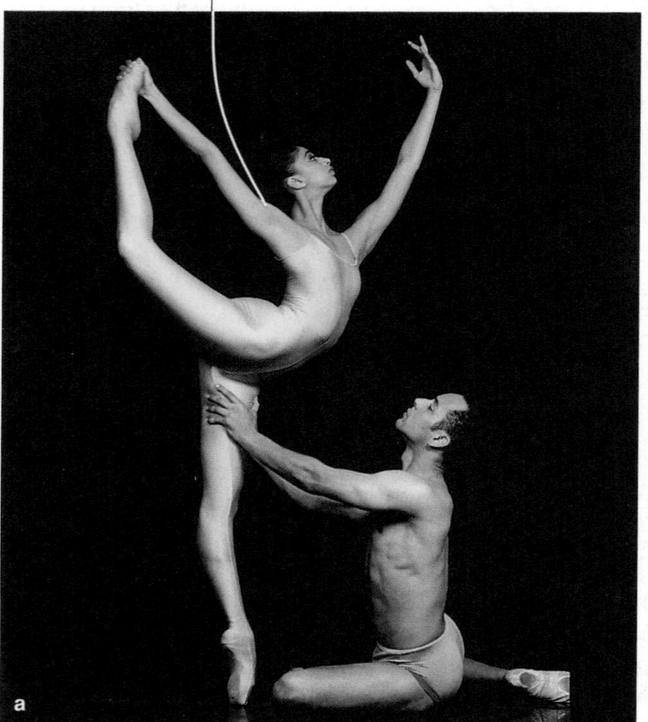

a

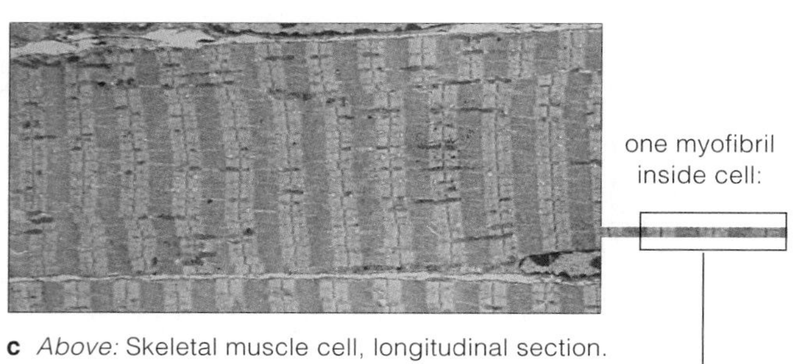

one myofibril inside cell:

c *Above:* Skeletal muscle cell, longitudinal section. Many myofibrils run longitudinally inside it. Their bands are in register (except as an artififact of specimen preparation) and give the cell a striped appearance.

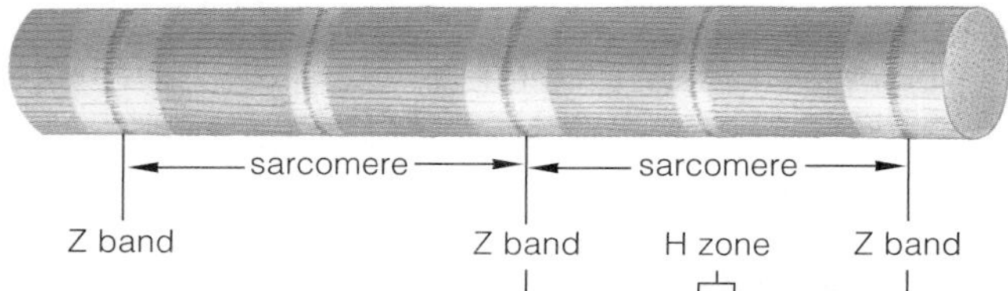

sarcomere
sarcomere
Z band
Z band
H zone
Z band

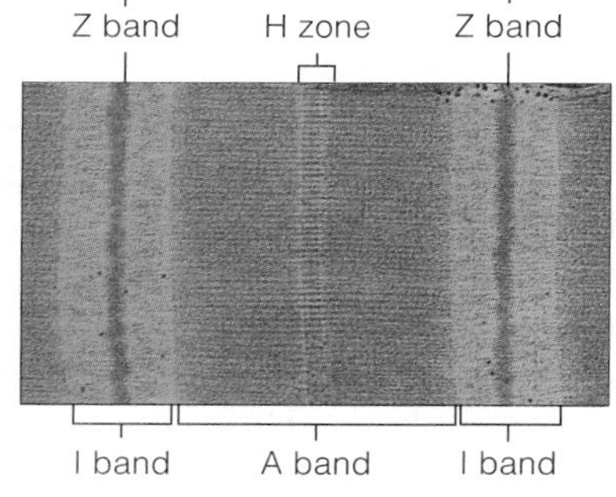

I band
A band
I band

d Sarcomeres (units of contraction), each extending between two Z bands. Many thick and thin filaments overlap in the A band, but only thick filaments extend across the H zone. Only thin filaments extend across I bands to the Z bands. Different proteins organize and stabilize the array. For example, they connect all thick filaments at the H zone and anchor actin filaments at the Z bands.

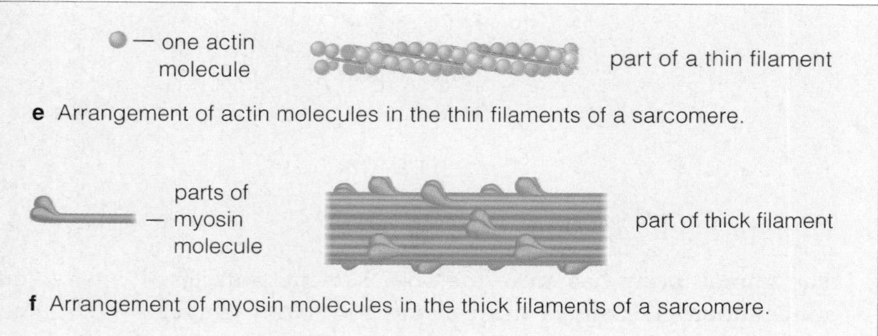

one actin molecule
part of a thin filament
parts of myosin molecule
part of thick filament

e Arrangement of actin molecules in the thin filaments of a sarcomere.

f Arrangement of myosin molecules in the thick filaments of a sarcomere.

Figure 38.20 (**a**) From the Dance Theatre of Harlem, an example of exquisite control of skeletal muscle movements. (**b–f**) This sequence shows skeletal muscle components from a biceps down to molecules with contractile properties.

Further reading: Student Guide to InfoTrac on web site →

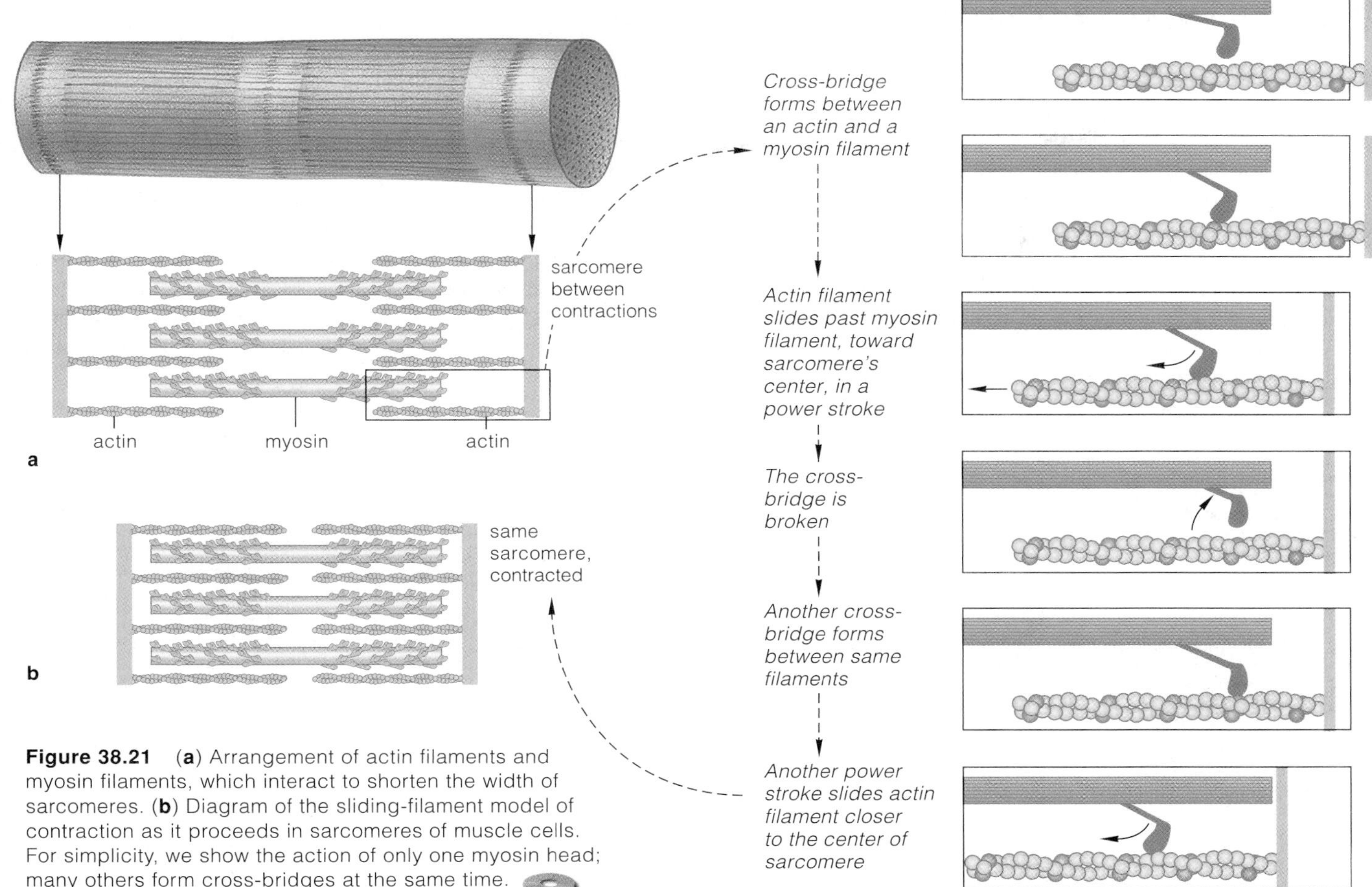

sarcomere between contractions
actin
myosin
actin
a
same sarcomere, contracted
b
Cross-bridge forms between an actin and a myosin filament
Actin filament slides past myosin filament, toward sarcomere's center, in a power stroke
The cross-bridge is broken
Another cross-bridge forms between same filaments
Another power stroke slides actin filament closer to the center of sarcomere

Figure 38.21 (**a**) Arrangement of actin filaments and myosin filaments, which interact to shorten the width of sarcomeres. (**b**) Diagram of the sliding-filament model of contraction as it proceeds in sarcomeres of muscle cells. For simplicity, we show the action of only one myosin head; many others form cross-bridges at the same time.

(Sections 4.8, 4.9, and Figure 38.20*e*). Other proteins are positioned near surface grooves in the actin filament.

Each *thick* filament (about twelve nanometers across) is assembled from a number of molecules of **myosin**, a motor protein with roles in contraction (Section 4.8). A myosin molecule has a tail and a double head. Many are bundled as a thick filament in which all the heads project outward (Figure 38.20*f*). Their precisely angled heads point away from the center of the sarcomere.

Thus muscle bundles, muscle cells, and myofibrils and their filaments run in the same direction. Of what use is this parallel orientation? It focuses the force of muscle contraction onto a bone in a particular direction.

Sliding-Filament Model of Contraction

How do sarcomeres shorten to contract a muscle? The answer lies with coordinated sliding and pulling motions that occur between Z bands. Two sets of actin filaments, one anchored at each Z band, extend toward the center of the sarcomere. A parallel set of myosin filaments overlaps them but is not attached to Z bands (Figure 38.21). During contraction, both sets of actin filaments slide over their stationary myosin neighbors. (Imagine two pocket doors closing and pulling both walls along with them to understand how the sarcomere shortens.)

By a **sliding-filament model** of contraction, myosin and actin interact by **cross-bridge formation**, whereby activated myosin heads briefly attach to binding sites on an adjacent actin filament. Driven by ATP energy, they tilt in a short stroke toward the sarcomere's center. The heads pull the actin filament along with them. Another energy input makes the heads let go, attach to another binding site, tilt in another stroke, and so on. A single contraction of each sarcomere involves many myosin heads performing a series of short power strokes down the length of their neighbors, the actin filaments.

A skeletal muscle shortens through combined decreases in the length of its numerous sarcomeres. Sarcomeres are the basic units of muscle contraction.

The parallel orientation of a skeletal muscle's component parts directs the force of contraction toward a bone that must be pulled in some direction.

By energy-driven interactions between myosin and actin filaments, the many sarcomeres of a muscle cell shorten and collectively account for its contraction.

38.9 ENERGY FOR CONTRACTION

All cells require ATP, but only in muscle cells does the demand skyrocket in so short a time. When a muscle cell at rest is called upon to contract, phosphate donations from ATP must occur twenty to a hundred times faster. But a cell has only a small supply of ATP at the start of contractile activity. At such times, it forms ATP by a quick reaction. An enzyme simply transfers phosphate from **creatine phosphate**, an organic compound, to ADP. A cell has about five times as much creatine phosphate as ATP, so this reaction is good for a few contractions. And that is enough to buy time for a relatively slower ATP-forming pathway to kick in (Figure 38.22).

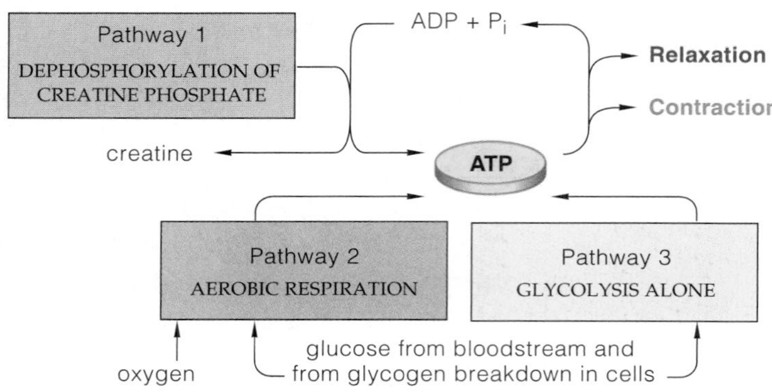

Pathway 1
DEPHOSPHORYLATION OF CREATINE PHOSPHATE
ADP + Pi
Relaxation
Contraction
creatine
ATP
Pathway 2
AEROBIC RESPIRATION
Pathway 3
GLYCOLYSIS ALONE
oxygen
glucose from bloodstream and from glycogen breakdown in cells

Figure 38.22 Three metabolic routes by which ATP forms in muscle cells in response to the demands of physical exercise.

During prolonged, moderate exercise, the oxygen-requiring reactions of aerobic respiration typically can supply most of the ATP needed for contraction. Suppose a muscle cell taps its store of glycogen for glucose (the starting substrate) in the first five to ten minutes. For the next half hour or so of sustained activity, that muscle cell depends on glucose and fatty acid deliveries from the blood. For contractile activity longer than this, fatty acids are the main fuel source (Section 8.6).

What happens when exercise is so intensive that it exceeds the capacity of the respiratory and circulatory systems to deliver oxygen for the aerobic pathway? At such times, glycolysis alone will contribute more of the total ATP that is being produced. Remember, by this set of anaerobic reactions, a glucose molecule is not fully broken down, so the net ATP yield is small. But muscle cells use this metabolic route as long as glycogen stores continue to provide glucose before fatigue sets in.

After intense exercise, deep, rapid breathing helps repay the body's **oxygen debt**, incurred when ATP use by muscles exceeded the aerobic pathway's deliveries.

During exercise, the availability of ATP inside muscle cells affects whether contraction will proceed, and for how long.

38.10 CONTROL OF CONTRACTION

The Control Pathway

When skeletal muscles contract, they help move the body and its assorted parts at certain times, in certain ways—and they do so in response to commands from the nervous system. The commands, which reach the muscles by way of motor neurons, stimulate or inhibit contraction of the sarcomeres in muscle cells.

Like all other cells, a muscle cell shows a difference in electric charge across its plasma membrane. That is,

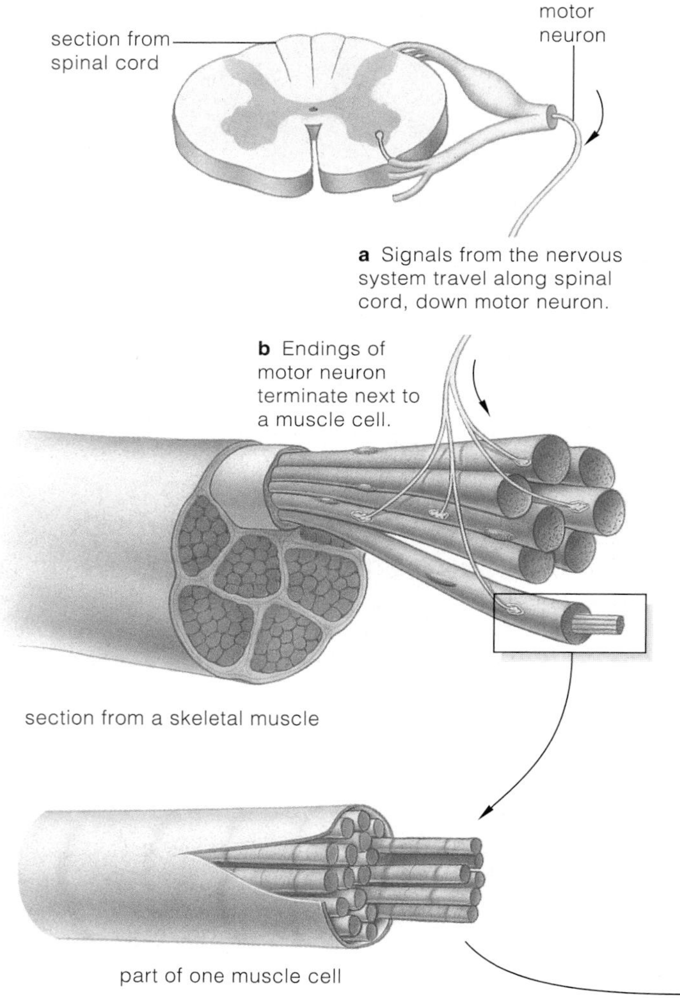

section from spinal cord
motor neuron
a Signals from the nervous system travel along spinal cord, down motor neuron.
b Endings of motor neuron terminate next to a muscle cell.
section from a skeletal muscle
part of one muscle cell
c Signals travel along muscle cell's plasma membrane to sarcoplasmic reticulum around cell's myofibrils.

Figure 38.23 Pathway for signals from the nervous system that stimulate or inhibit contraction of skeletal muscle. The plasma membrane of each muscle cell surrounds the cell's myofibrils and connects with inward-threading T tubules. These membranous tubes are close to the sarcoplasmic reticulum, a calcium-storing system that functions in the control of contraction.

Further reading: Student Guide to InfoTrac on web site —

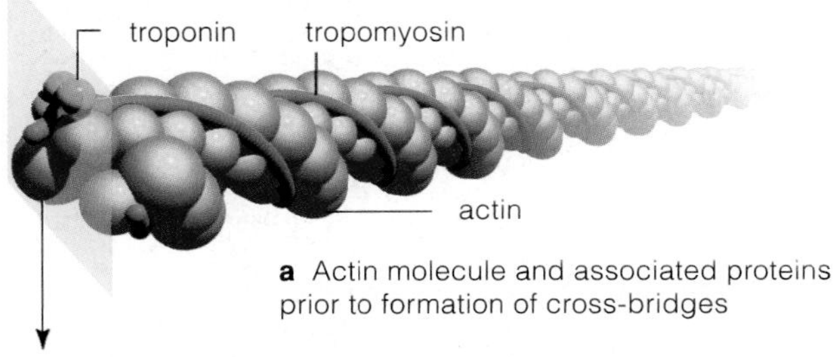
troponin
tropomyosin
actin

a Actin molecule and associated proteins prior to formation of cross-bridges

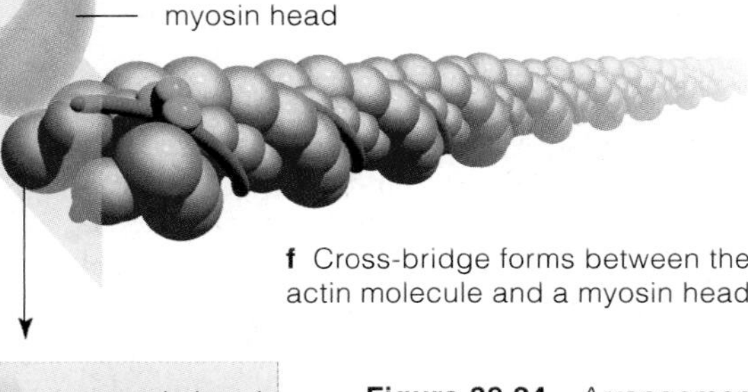
myosin head

f Cross-bridge forms between the actin molecule and a myosin head

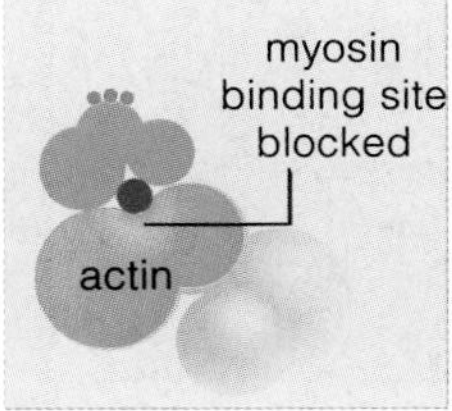
myosin binding site blocked
actin

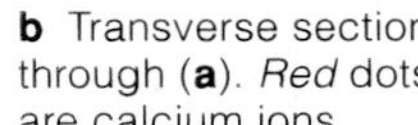

b Transverse section through (**a**). *Red* dots are calcium ions.

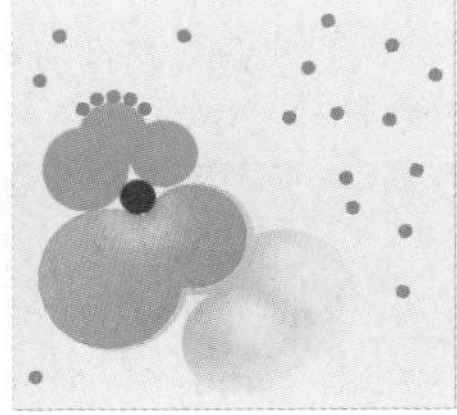

c Influx of calcium ions; troponin binds more of them.

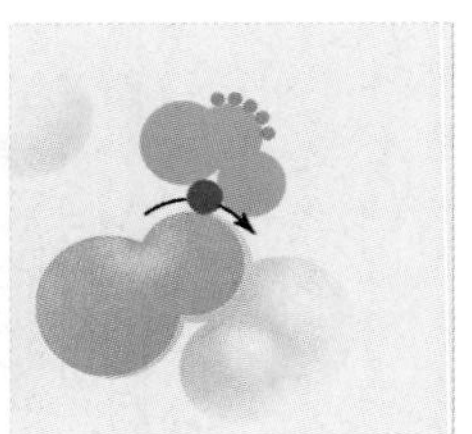

d The troponin changes shape and moves away from myosin binding site

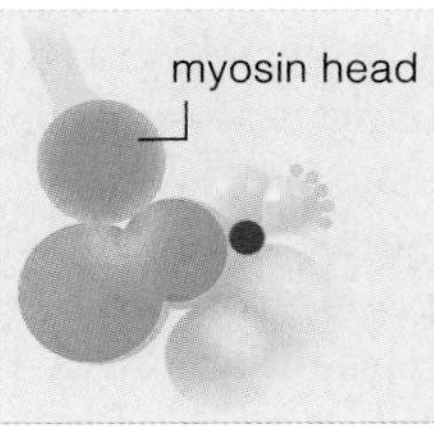
myosin head

e Binding site is now exposed, actin is free to bind myosin head.

Figure 38.24 Arrangement of actin and tropomyosin filaments, and troponins, in a skeletal muscle cell. When troponin binds more calcium ions, it changes shape and makes the tropomyosin move to expose a binding site for a myosin head. This action occurs in series at many binding sites along the actin molecule.

the cytoplasm just beneath the membrane is a bit more negative than the interstitial fluid outside it. But only in muscle cells, neurons, and other *excitable* cells does the difference in charge reverse abruptly, briefly, and in a predictable way in response to adequate stimulation.

The abrupt reversal in charge, an **action potential,** occurs as charged ions flow across the membrane in an accelerating way. The electrical commotion, remember, spreads without diminishing along the membrane, away from the point of stimulation (Section 34.2).

Suppose that action potentials arise in a muscle cell. They spread rapidly away from the stimulation point, then along the small, tubelike extensions of the plasma membrane shown in Figure 38.23. The tubes connect with a system of membranous chambers, which thread lacily around the muscle cell's myofibrils. That system, called the **sarcoplasmic reticulum,** takes up, stores, and releases calcium ions in controlled ways.

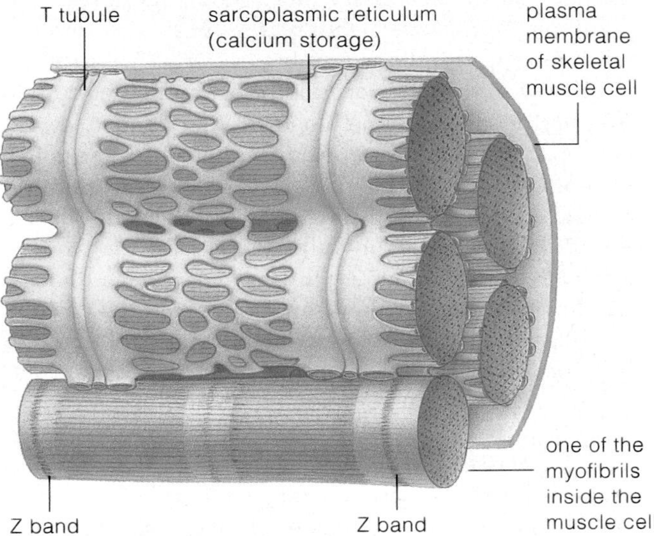
T tubule
sarcoplasmic reticulum (calcium storage)
plasma membrane of skeletal muscle cell
one of the myofibrils inside the muscle cell
Z band
Z band

d Signals trigger the release of calcium ions from sarcoplasmic reticulum threading among the myofibrils. Arrival of calcium allows actin and myosin filaments in the myofibrils to interact and bring about contraction, as in Figure 38.24.

The Control Mechanism

The arrival of action potentials causes an outward flow of calcium ions from the sarcoplasmic reticulum. The released ions diffuse into the myofibrils and reach actin filaments. Before this happened, the muscle was at rest (it was not contracting). Its actin binding sites were blocked and myosin could not form cross-bridges with them. However, the arrival of calcium ions clears the binding sites, so that contraction can proceed. After the contraction, the calcium ions are actively transported back into the membrane storage system.

What blocks cross-bridge binding sites in a muscle at rest? A tropomyosin filament as well as troponins are located in or near grooves at the surface of the actin filaments (Figure 38.24). When the calcium level is low, these proteins are joined so tightly that the tropomyosin is forced slightly outside the groove. In that position, it blocks the cross-bridge binding site. However, when commands from the nervous system trigger an inflow of calcium into the sarcomere, enough calcium ions bind with the troponin to alter its shape. When that happens, the troponin has a different molecular grip on the tropomyosin filament, which is now free to move into the groove and expose the binding site.

It takes commands from the nervous system to initiate action potentials in muscle cells. The action potentials are signals for cross-bridge formation, hence for contraction.

38.11 PROPERTIES OF WHOLE MUSCLES

Muscle Tension and Muscle Fatigue

Whether a muscle actually shortens during cross-bridge formation depends on the external forces acting on it. Collectively, the cross-bridges exert **muscle tension**. By definition, this is a mechanical force that a contracting muscle exerts on an object, such as a bone. Opposing it is a load, either the weight of an object or gravity's pull on the muscle. Only when muscle tension exceeds the load does a stimulated muscle shorten.

An *isometrically* contracting muscle develops tension but it does not shorten. It supports a load in a constant position, as when you hold a glass of lemonade in front of you. An *isotonically* contracting muscle shortens and moves a load. With *lengthening* contraction, though, an external load is greater than the muscle tension, so the muscle lengthens during the period of contraction. This happens to leg muscles when you walk down stairs.

A muscle's tension relates to the formation of cross-bridges in its cells and to the number of cells recruited into action. Consider a **motor unit**: a motor neuron and all muscle cells that form junctions with its endings. By stimulating that motor unit with an electrical impulse, we can induce an action potential and make a recording of an isometric contraction. It takes a few milliseconds for tension to increase, then it peaks and declines. This response is a **muscle twitch** (Figure 38.25*a*). Its duration depends on the load and on cell type. For example, fast-acting muscle cells rely on glycolysis—not efficient, but fast—and use up ATP faster than slow-acting cells do.

If another stimulus is applied before the response is finished, the muscle twitches again. **Tetanus** is a large contraction resulting from the repeated stimulation of a motor unit, so that twitches mechanically run together. (In *tetanus*, a disease by the same name, toxins disrupt muscle relaxation, as described in Section 34.5.) Figure 38.25*d* shows a recording of tetanic contraction.

Continuous, high-frequency stimulation that keeps a muscle in a state of tetanic contraction leads to *muscle fatigue*, or a decline in tension. After a few minutes of rest, a fatigued muscle will contract again in response to stimulation. The extent of recovery depends largely on how long and how frequently it was stimulated before. Muscles associated with brief, intense exercise (such as weightlifting) fatigue fast but recover fast. The muscles associated with prolonged, moderate exercise fatigue slowly but take longer to recover, often up to twenty-four hours. The molecular mechanisms causing muscle fatigue are unknown, but glycogen depletion is a factor.

Effects of Exercise and Aging

A muscle's properties depend on how often, how long, and how intensely it is put to use. With regular **exercise**

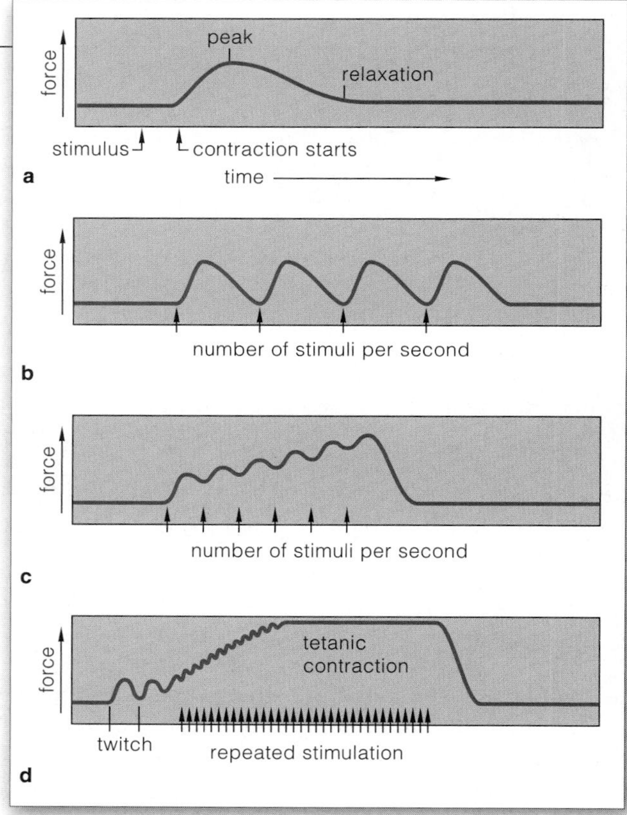

force
peak
relaxation
stimulus
contraction starts
time
a
force
number of stimuli per second
b
force
number of stimuli per second
c
force
tetanic
contraction
twitch
repeated stimulation
d

Figure 38.25 Recordings of twitches in muscles artificially stimulated in different ways. (**a**) A single twitch. (**b**) Four stimulations per second cause a series of twitches. (**c**) Six per second cause a summation of twitches, and (**d**) about twenty per second cause tetanic contraction.

(that is, increased levels of contractile activity), muscle cells do not increase in number. Rather, they increase in size and metabolic activity, and become more resistant to fatigue. Think of *aerobic exercise*, which is not intense but is long in duration. Aerobic exercise increases the number of mitochondria in both fast and slow muscle cells, and it increases the blood capillaries that service them. Such physiological changes improve endurance. By contrast, *strength training* (intensive, short-duration exercise such as weightlifting) affects fast-acting muscle cells. These form more myofibrils and more enzymes of glycolysis. Strong, bulging muscles may result, although they don't have much endurance. They fatigue rapidly.

Muscle tension decreases in adult humans after age thirty or forty. Older people might exercise just as long and intensely as younger ones, but their muscles cannot adapt (change) in response to the same extent. Even so, some adaptation can be beneficial. Aerobic exercise can improve blood circulation. And, as it turns out, modest strength training slows the loss of muscle tissue that is an inevitable part of the aging process.

Properties of muscles vary with age and levels of activity.

Further reading: Student Guide to InfoTrac on web site —

PEBBLES, FEATHERS, AND MUSCLE MANIA

A male penguin has a bit of a problem when scouting for a female penguin, because superficially they both look too much alike. The only way he can find one is to drop a pebble at the feet of a likely prospect. If the feet belong to a male, the pebble may be perceived as an insult, and it may start a fight. If they are female feet, his stony overture to courtship might be ignored or fancied, depending on her receptivity at that moment.

Peacocks and the males of many other species have no such problem. For them, sexual dimorphism is visible and quite pronounced (Figure 38.4*c*). Maybe the peacock's eye-stopping feathers evolved through sexual selection, with peahens serving as the deciders of their reproductive success. Charles Darwin certainly thought so. He viewed cases of extreme sexual dimorphism as the outcome of male competition for females and of females choosing among males.

What about extreme sexual dimorphism in body size, which involves increases in muscle mass? This might be one measure of how much mammalian males invest in fighting capacity. Possibly the cost of securing and using resources for growth and maintenance of a massive body is offset by great reproductive rewards.

All of this might make you wonder: Exactly what are the "rewards" for astoundingly muscled human athletes (Figure 38.26)? Do such muscles speak of what it takes to win in modern athletic competition? Consider this: Each year in the United States alone, about a million athletes use anabolic steroids. The vast majority of professional football players have used them to increase their "brute power." Even adolescent boys use them as a way to gain the winning edge in wrestling, football, and weightlifting tournaments.

Anabolic steroids are synthetic hormones. They mimic testosterone, a sex hormone that (among other things) governs secondary sexual traits. Testosterone makes boys get a deeper voice; more hair on their face, underarms, and pubic skin; and greater muscle mass in shoulders, arms, legs, and elsewhere. Besides this, testosterone stimulates heightened aggressive behavior, which is often associated with maleness.

Figure 38.26 A human male with pumped-up biceps.

Anabolic steroids stimulate the synthesis of protein molecules, including proteins in muscle cells. Supposedly they induce very rapid gains in muscle mass and muscle strength when taken during weight-training and exercise programs. This claim is disputed; results of most studies are based on too few subjects. Even so, not everyone believes that these drugs do enough damage to outweigh the edge they presumably give in athletic competition or the wealth and hero status they accord the "winners."

Yet steroid-using athletes do suffer minor and major side effects. In men, acne, baldness, shrinking testes, and infertility are early signs of toxicity. The symptoms begin when a high level of anabolic steroids in blood triggers a sharp decline in the body's production of testosterone. Anabolic steroids also may trigger early heart disease. Even brief or occasional use may damage the kidneys or set the stage for cancer of the liver, testes, and prostate gland. Among women, anabolic steroids deepen the voice, and they trigger the formation of excessive facial hair. Menstrual cycles become irregular. Breasts may shrink, and the clitoris may become grossly enlarged.

Not every steroid user develops severe physical side effects. Far more common are mental difficulties, called *'roid rage* or *body-builder's psychosis*. In such cases, users become irritable and increasingly aggressive. Some men become wildly aggressive, uncontrollably manic, and delusional. For example, one steroid user accelerated his car to high speed and deliberately drove it into a tree; and doesn't that make you wonder just how superior some of us are to a pebble-toting penguin.

38.13 SUMMARY

1. Most animals have an integumentary system, which covers the body's surface. Examples include the cuticle of roundworms and arthropods, as well as vertebrate skin and the structures derived from it.

2. Skin protects against abrasion, ultraviolet radiation, dehydration, and many pathogenic bacteria. It assists in controlling internal temperature (blood flow to skin can dissipate heat). Its sensory receptors detect external stimuli. Skin exposed to sunlight has an endocrine role; it helps produce a hormone-like substance (vitamin D) required for absorption of calcium from food.

3. Skin consists of two regions: an outer epidermis and underlying dermis. Keratinocytes (keratin producers) are its most abundant cells. Other cells are melanocytes (melanin producers), as well as Langerhans cells and Granstein cells (which help defend the body against pathogens and cancer cells).

a. Epidermis consists mainly of multiple layers of dead, keratinized, and melanin-shielded epithelial cells.

b. The dermis is where rapid, ongoing cell divisions produce the replacements for cells that are continually shed or abraded away. Hair, oil glands, sweat glands, and other structures derived from epidermal cells are embedded mainly in the dermis.

4. Movement of the animal body or parts of it requires contractile cells and some medium or structure against which contractile force can be applied.

a. With hydrostatic skeletons, such as that of a sea anemone, body fluids accept the contractile force and, as a result, are redistributed within a confined space.

b. With exoskeletons, such as that of an insect, rigid external body parts accept the force of contraction.

c. With endoskeletons, rigid internal structures such as bones receive the applied force of contraction.

5. Bones are complex organs with osteocytes—living bone cells—embedded in a mineralized, collagen-rich ground substance. Bones have roles in the movement, protection, and support of body parts; in the storage of minerals; and, in bones with red and yellow marrow, blood cell formation.

6. The human skeleton is divided into two portions.

a. The appendicular portion has a pelvic girdle, two pectoral girdles (each with a collarbone and shoulder blade), arm and hand bones, and leg and foot bones.

b. The axial portion includes skull bones, a rib cage, and a vertebral column, or backbone. The backbone is composed of vertebrae (bony segments) cushioned by cartilaginous, intervertebral disks.

7. Skeletal joints are areas of contact or near-contact at which connective tissue bridges neighboring bones. The gap between bones at different joints may be bridged by a fibrous joint (short fibers), a cartilaginous joint (of cartilage), or a synovial joint (with straplike ligaments).

8. The cells of smooth, cardiac, and skeletal muscle contract (shorten) in response to adequate stimulation.

9. Tendons attach skeletal muscles to bones. Skeletal muscles and bones interact as a system of levers, with rigid rods (bones) moving at fixed points (joints). Many muscles work together or in opposition to bring about movement or positional changes in body parts.

10. Inside each skeletal muscle cell are many myofibrils arranged parallel with the long axis. These threadlike structures contain filaments of actin and myosin, also organized in parallel. Every myofibril is transversely divided into sarcomeres, the basic units of contraction. The parallel orientation of these parts directs the force of contraction at a bone to be pulled in some direction.

11. In response to stimulation, skeletal muscles shorten by combined decreases in the length of all sarcomeres. The nervous system controls muscle cells. Its commands are delivered by motor neuron endings that terminate on muscle cells, and they can trigger action potentials at the muscle cell's plasma membrane.

12. Here are the main points about the sliding-filament model of muscle contraction:

a. In a muscle cell at rest, tropomyosin, a threadlike protein, wraps around each actin filament in a way that blocks binding sites for myosin. Other, small proteins (troponins) are tightly bound to the tropomyosin. Action potentials trigger the release of calcium ions from a membrane system (sarcoplasmic reticulum) around the cell's myofibrils. Calcium diffuses into sarcomeres. The influx changes the shape of the troponins, which makes the tropomyosin filament slide away from the blocked binding sites. Heads of myosin filaments are now free to form cross-bridges with actin filaments.

b. Each cross-bridge is a brief, reversible attachment between a myosin head projecting from a thick filament and a binding site on an adjacent actin filament. Cross-bridges form by repeated, ATP-driven power strokes. The repeated strokes cause actin filaments to slide past myosin filaments. Together they shorten the sarcomere.

13. Muscle cells obtain the ATP required for contraction by three metabolic pathways: *The dephosphorylation of creatine phosphate*, a direct, fast pathway good for a few seconds of contraction. *Aerobic respiration*, a pathway that predominates during prolonged, moderate exercise. *Glycolysis*, which takes over if intense exercise exceeds the body's capacity to deliver oxygen to muscle cells.

14. Muscle tension refers to a mechanical force created by cross-bridge formation. The load (gravity or weight of objects) is an opposing force. A stimulated muscle shortens when tension exceeds the load and lengthens when tension is less than the load. Levels of exercise and aging affect the properties of muscles.

15. A motor unit is a motor neuron and all muscle cells that form junctions with its endings. A muscle twitch is a brief, weak contraction in response to a single action potential at a motor unit. Tetanus is a large contraction resulting from repeated stimulation at a motor unit.

Review Questions

1. List the functions of skin, then describe the distinguishing features of its regions. Is the hypodermis part of skin? *38.1*

2. Identify four cell types in vertebrate skin and briefly describe their functions. *38.2, 38.3*

3. Distinguish between:
   a. hydrostatic skeleton, exoskeleton, and endoskeleton *38.4*
   b. vertebra and intervertebral disk *38.6*
   c. ligament and tendon *38.6, 38.7*

4. What are the functions of bones? What is a joint? *38.5*

5. Which hormones help control calcium ion concentrations in blood? Name their general effects on bone tissue turnover. *38.5*

6. In what respect are tendons like a car's gearshift? *38.7*

7. Name the three categories of muscle tissue. *38.7*

8. Review Figure 38.20. On your own, sketch and label the fine structure of a muscle, down to one of its individual myofibrils. Identify the basic unit of contraction in the myofibrils. *38.8*

9. What role does calcium play in control of contraction? What role does ATP play, and by what routes does it form? *38.9, 38.10*

Self-Quiz (*Answers in Appendix III*)

1. Which is *not* a function of skin?
   a. resist abrasion
   b. restrict dehydration
   c. initiate movement
   d. help control temperature

2. ________ are shock pads and flex points.
   a. Vertebrae
   b. Femurs
   c. Marrow cavities
   d. Intervertebral disks

3. Blood cells form in ________.
   a. red marrow
   b. all bones
   c. certain bones only
   d. a and c

4. The ________ is the basic unit of contraction.
   a. myofibril
   b. sarcomere
   c. muscle fiber
   d. myosin filament

5. Muscle contraction requires ________.
   a. calcium ions
   b. ATP
   c. action potential arrival
   d. all of the above

6. ATP for muscle contraction can be formed by ________.
   a. aerobic respiration
   b. glycolysis
   c. creatine phosphate breakdown
   d. all of the above

7. Match the M words with their defining feature.

______ muscle	a. actin's partner
______ muscle twitch	b. all in the hands
______ muscle tension	c. blood cell production
______ melanin	d. decline in tension
______ myosin	e. brownish black pigment
______ marrow	f. motor unit response
______ metacarpals	g. force exerted by cross-bridges
______ myofibrils	h. muscle cells bundled in connective tissue
______ muscle fatigue	i. threadlike parts in muscle cell

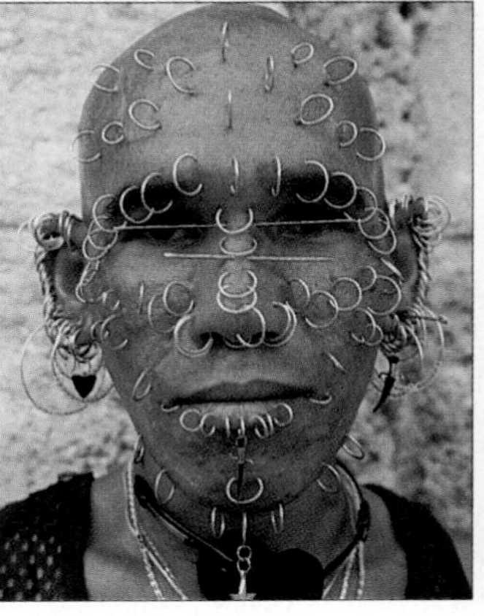

Figure 38.27 Demonstration of curious invasions of the integument with no obvious adaptive value.

Critical Thinking

1. The nose, lips, tongue, navel, and private parts are common targets for *body piercing*, cutting holes into the body so jewelry and other objects can be threaded through them. *Tattooing*, using permanent dyes to make patterns in skin, is another common fad (Figure 38. 27). Besides being painful, both invasions of the skin invite bacterial infections, chronic viral hepatitis, AIDS, and other nasty diseases if piercers and tattooers reuse unsterilized needles, razors, dye, gloves, swabs, or trays. Months may pass before disease symptoms arise, so the cause-and-effect connection is not always obvious. If you dismiss such risks and decide that unregulated tissue invasions are okay, how could you make sure the equipment used is sterile?

2. For young women, the recommended daily allowance (RDA) of calcium is 800 milligrams/day. During Hilde's pregnancy, the RDA is 1,200 milligrams/day. Why, and what might happen to a pregnant woman's bones without the larger amount?

3. Compared to most people, Joe and other long-distance runners have a greater number of muscle fibers with more mitochondria. Sprinters have muscle fibers that contain more of the enzymes necessary for glycolysis but fewer mitochondria. Think about how these two forms of exercise differ. Then explain why the muscle fibers differ between the two kinds of runners.

Selected Key Terms

actin *38.8*
action potential *38.10*
anabolic steroid *38.12*
bone *38.5*
bone remodeling *38.5*
creatine phosphate *38.9*
cross-bridge formation *38.8*
cuticle *38.1*
dermis *38.1*
endoskeleton *38.4*
epidermis *38.1*
exercise *38.11*
exoskeleton *38.4*
Granstein cell *38.3*
hair *38.2*
hydrostatic skeleton *38.4*
integument *38.1*
intervertebral disk *38.6*
joint *38.6*
keratinocyte *38.2*
Langerhans cell *38.3*
ligament *38.6*
melanocyte *38.2*
motor unit *38.11*
muscle tension *38.11*
muscle twitch *38.11*
myosin *38.8*
oil gland *38.2*
osteoblast *38.5*
osteoclast *38.5*
osteocyte *38.5*
oxygen debt *38.9*
red marrow *38.5*
sarcomere *38.8*
sarcoplasmic reticulum *38.10*
skeletal muscle *38.7*
skin *38.1*
sliding-filament model *38.8*
sweat gland *38.2*
tendon *38.7*
tetanus *38.11*
vertebra (vertebrae) *38.6*
vitamin D *38.3*
yellow marrow *38.5*

Readings *See also www.infotrac-college.com*

Brusca, R., and G. Brusca. 1990. *Invertebrates*. Sunderland, Massachusetts: Sinauer.

Sherwood, L. 1997. *Human Physiology*. Third edition. Belmont, California: Wadsworth.

Travis, J. 15 January 2000. "Boning Up." *Science News* 157: 41–43.

39 CIRCULATION

Heartworks

For Dr. Augustus Waller, Jimmie the bulldog was no ordinary pooch. Connected to wires and soaked to his ankles in buckets of salty water, Jimmie was a four-footed explorer of the workings of the heart (Figure 39.1*a*). Press your fingers to your chest a few inches to the left of center, between the fifth and sixth ribs, and feel the repetitive thumpings of your heart. The same rhythms intrigued Waller and other nineteenth-century physiologists. They wondered: Does each beat of the heart produce a pattern of electrical currents? Could those early researchers find out by devising a painless way to record such currents at the body surface?

That's where Jimmie and the buckets of salty water came in. Saltwater happens to be an efficient conductor of electricity. In Waller's experiment, it picked up faint signals from Jimmie's beating heart through the skin of his legs and conducted them to a crude monitoring device. With that device, Waller made one of the first recordings of heart activity (Figure 39.1*c*). Today we call such recordings ECGs, an abbreviation for electrocardiograms.

A graph of the normal electrical activity of your own heart would look much the same. The pattern emerged a few weeks after you started to grow, by way of mitotic cell divisions, from a single fertilized egg inside your mother. Early on, some of the embryonic cells differentiated into cardiac muscle cells, and these started to contract spontaneously. One small patch of the cells took the lead and has functioned as your heart's pacemaker ever since.

If all goes well, that patch of cardiac muscle cells will continue to contract as it should until the day you die. It is a natural pacemaker; it sets the baseline rate at which blood is pumped out of the heart, through the blood vessels, then back to the heart. The rate is moderate, about seventy beats every minute. But commands from the nervous

a
BUCKET

b
JIMMIE
DR. WALLER

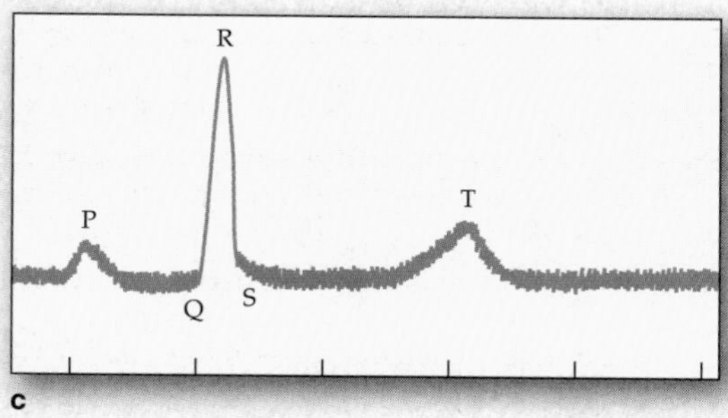
R
P
T
Q
S
c

Figure 39.1 A bit of history in the making. (**a**) Jimmie the bulldog taking part in a painless experiment. (**b**) Augustus Waller and his beloved pet bulldog sharing a quiet moment in Waller's study after the experiment, which yielded one of the world's first electrocardiograms (**c**).

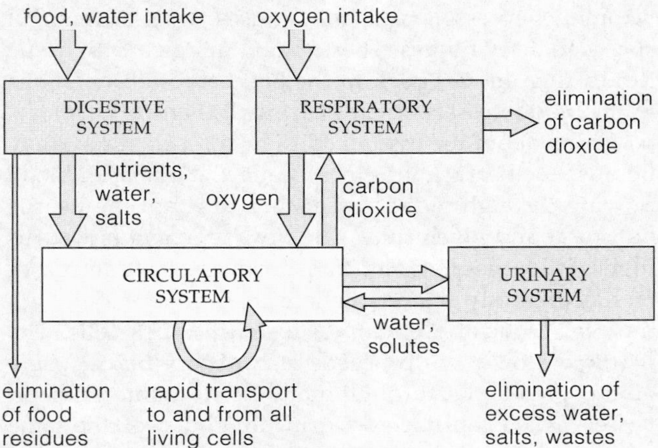

food, water intake
oxygen intake
DIGESTIVE SYSTEM
RESPIRATORY SYSTEM
elimination of carbon dioxide
nutrients, water, salts
oxygen
carbon dioxide
CIRCULATORY SYSTEM
URINARY SYSTEM
water, solutes
elimination of food residues
rapid transport to and from all living cells
elimination of excess water, salts, wastes

Figure 39.2 Diagram of the functional connections between the circulatory, respiratory, and digestive systems, which interact in transporting substances to and from all living cells in the animal body. Their integrated activities help maintain favorable operating conditions in the internal environment.

system and endocrine system continually adjust it. When you jog, for example, your skeletal muscle cells demand much more blood-borne oxygen and glucose than they do when you sleep. At such times, your heart starts pounding more than twice as fast, and this helps deliver sufficient blood to cells.

We have come a long way from Waller and Jimmie in our monitorings of the heart. Sensors can now detect the faint signals characteristic of an impending heart attack. Internists now use computers to analyze a patient's beating heart, and they use ultrasound probes to build images of it on a video screen. Cardiologists routinely substitute battery-powered pacemakers for faltering natural ones.

With this chapter, we turn to the circulatory system, the means by which substances move rapidly to and from the interstitial fluid that bathes all living cells in almost all animals. The circulatory system continually accepts the oxygen, nutrients, and other substances that an animal secures, by way of respiratory and digestive systems, from the outside environment (Figure 39.2). Simultaneously it picks up carbon dioxide and other wastes from cells and delivers them to the respiratory and urinary systems for disposal. Its smooth operation is absolutely central to maintaining operating conditions in the internal environment within a tolerable range—a state we call homeostasis.

KEY CONCEPTS

1. All cells survive by exchanging substances with their surroundings. In most animals, substances rapidly move to and from cells by way of a closed circulatory system.

2. Blood, a fluid connective tissue, is the transport medium of circulatory systems. It transports oxygen, carbon dioxide, plasma proteins, vitamins, hormones, lipids, and many other solutes. Blood also transports metabolically generated heat.

3. Birds and mammals have a four-chambered, muscular heart that pumps blood through two separate circuits of blood vessels, both of which lead back to the heart. One circuit flows through the lungs, where it picks up oxygen. The other circuit delivers oxygenated blood to all body regions, from which it picks up carbon dioxide wastes.

4. Arteries are large-diameter, low-resistance vessels that rapidly transport oxygenated blood from the heart. Veins are large-diameter, low-resistance vessels that transport oxygen-poor blood to the heart and serve as blood volume reservoirs.

5. The flow volume through each organ is regulated at arterioles. In response to control signals, the diameter of these blood vessels widens in some parts of the body and narrows in others. The coordinated responses of arterioles throughout the body divert more of the flow volume to organs that are most active at a given time.

6. Numerous small-diameter, thin-walled capillaries interconnect in capillary beds, which are diffusion zones. As a volume of blood spreads out through a bed, it slows owing to the total cross-sectional area of the capillaries, which is greater than that of arterioles. The slowdown allows time for exchanges between blood and interstitial fluid.

7. Pressure forces some fluid out of capillaries, into interstitial fluid. Drainage vessels of the lymphatic system return it to the blood. Organs of that system work to cleanse blood of infectious agents and other threats to the body.

39.1 CIRCULATORY SYSTEMS—AN OVERVIEW

General Characteristics

Imagine that an earthquake has closed off the highways around your neighborhood. Grocery trucks can't enter and waste-disposal trucks can't leave, so food supplies dwindle and garbage piles up. Cells would face similar predicaments if your body's highways were disrupted. The highways are part of a **circulatory system**, which functions in the rapid internal transport of substances to and from cells. The system helps maintain favorable neighborhood conditions, so to speak, and this is vital work. All of your differentiated cells perform specialized tasks and cannot fend for themselves. Different types interact in coordinated ways to maintain the volume, composition, and temperature of **interstitial fluid**, the tissue fluid that bathes them. A circulating connective tissue—**blood**—interacts with interstitial fluid. Blood makes continual deliveries and pickups that help keep conditions tolerable for enzymes and other molecules that carry out cell activities. Together, interstitial fluid and blood are the body's "internal environment."

Blood flows inside blood vessels, or tubes that differ in wall thickness and diameter. A muscular pump, the **heart**, generates pressure that keeps blood flowing. Like many animals, you have a *closed* circulatory system in which blood is confined within continuously connected walls of the heart and blood vessels. This is not true of *open* circulatory systems of arthropods, such as insects, and most mollusks. In an open system, blood flows out through the vessels and into sinuses, which are small spaces in body tissues. There, blood mingles with tissue fluids, then moves back to the heart through openings in the blood vessels or the heart wall (Figure 39.3*a*).

Think about the overall "design" of a closed system. Because the heart pumps incessantly, the *volume* of blood flowing through blood vessels has to equal the heart's output at any given time. The flow's *velocity* (speed) is highest in large-diameter transport vessels. It decreases in specific body regions, where there must be enough time for blood to exchange substances with cells. The required slowdown proceeds at **capillary beds**, where blood spreads out through many small-diameter blood vessels called **capillaries**. During any interval, the same volume of blood is moving forward through the beds as elsewhere in the system. But it is doing so at a more leisurely pace because of the great total cross-sectional area of the blood capillaries. The simple analogy given in Figure 39.3*e* may help you grasp this concept.

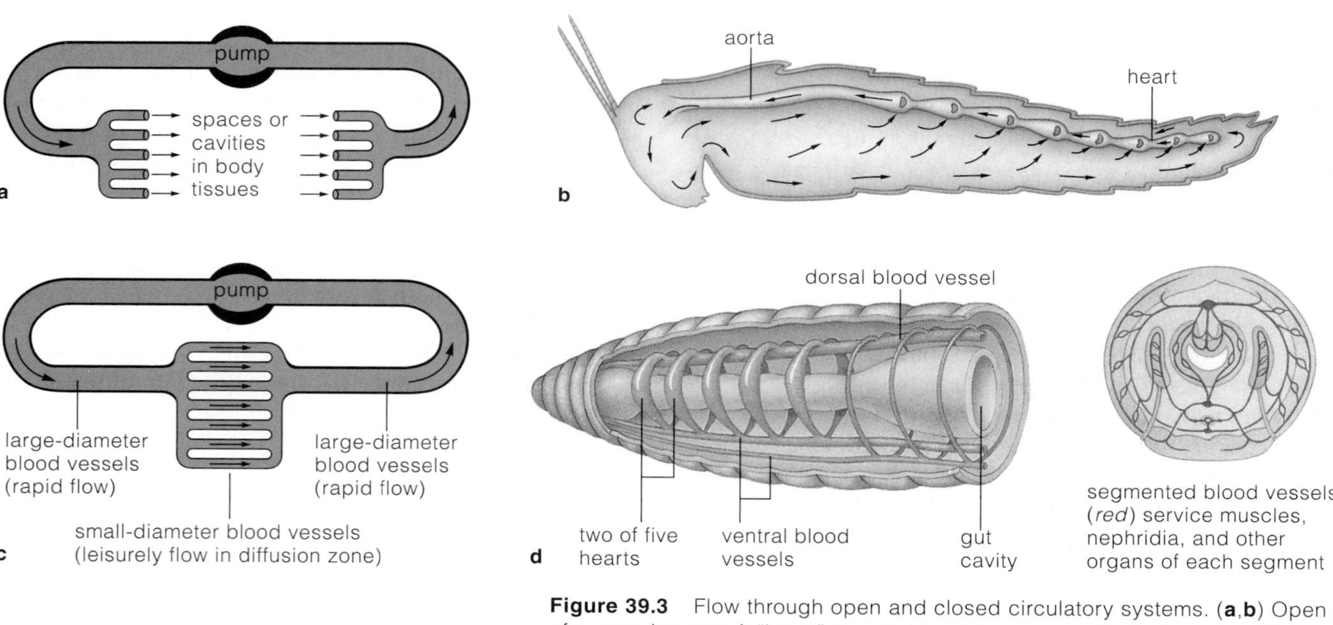
a
pump
spaces or cavities in body tissues
b
aorta
heart
c
pump
large-diameter blood vessels (rapid flow)
large-diameter blood vessels (rapid flow)
small-diameter blood vessels (leisurely flow in diffusion zone)
d
dorsal blood vessel
two of five hearts
ventral blood vessels
gut cavity
segmented blood vessels (*red*) service muscles, nephridia, and other organs of each segment
e
river in
lake
river out
1 2 3
1 2 3
1 2 3

Figure 39.3 Flow through open and closed circulatory systems. (**a**,**b**) Open system of a grasshopper. A "heart" pumps blood through a vessel, the aorta. Blood moves into tissue spaces and mingles with fluid that bathes cells. It reenters the heart at openings in the heart wall. (**c**,**d**) Closed system of an earthworm. Blood is confined in several pairs of muscular "hearts" near the head end and within blood vessels.

(**e**) Relationship between flow velocity and total cross-sectional area in any closed circulatory system. Visualize two fast rivers flowing into and out from a lake. The flow *rate* is the same in all three places; an identical volume of water moves from point *1* to point *3* during the same interval. Flow *velocity* decreases within the lake. Why? The volume spreads out through a larger cross-sectional area and moves forward a shorter distance during the same time.

Further reading: Student Guide to InfoTrac on web site

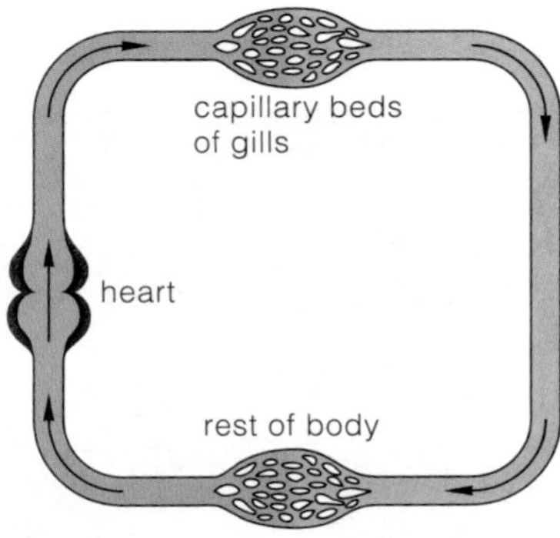
capillary beds
of gills
heart
rest of body

a In fishes, a two-chambered heart (atrium, ventricle) pumps blood in one circuit. Blood picks up oxygen in gills, delivers it to rest of body. Oxygen-poor blood flows back to heart.

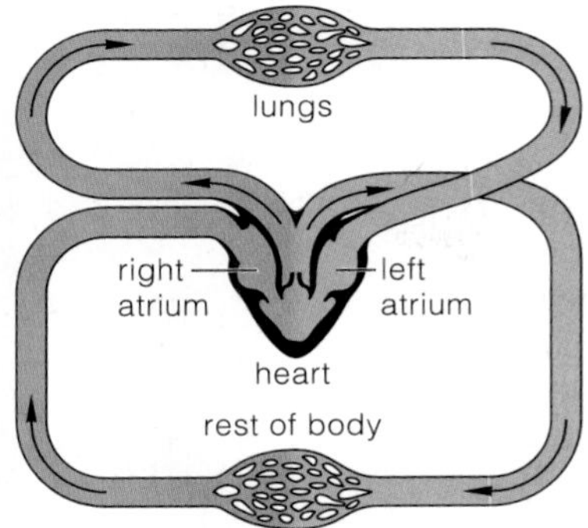
lungs
right atrium
left atrium
heart
rest of body

b In amphibians, a heart pumps blood through two partially separate circuits. Blood flows to lungs, picks up oxygen, returns to heart. It mixes with oxygen-poor blood still in heart, flows to rest of body, returns to heart.

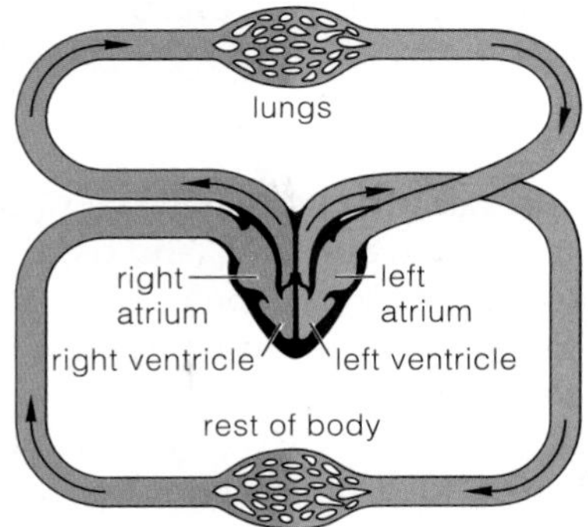
lungs
right atrium
left atrium
right ventricle
left ventricle
rest of body

c In birds and mammals, the heart is fully partitioned into two halves. Blood circulates in two circuits: from the heart's right half to lungs and back, then from the heart's left half to oxygen-requiring tissues and back.

Figure 39.4 Comparison of the closed circulatory system of vertebrate groups. 

Evolution of Vertebrate Circulatory Systems

Humans and other existing vertebrates have a closed circulatory system, although the pump and plumbing for fishes, amphibians, birds, and mammals differ in their details. The differences evolved over hundreds of millions of years and corresponded to the move of some vertebrate lineages onto land. Recall from Section 27.3 that fishes, the first vertebrates, had gills. Gills, like all respiratory structures, have a thin, moist surface that oxygen and carbon dioxide can diffuse across. Later, in the ancestors of all land vertebrates, lungs evolved that supplemented gas exchange. Being *internally moistened* sacs, lungs had advantages for the move onto dry land. Also advantageous were concurrent modifications in circulatory systems, which pick up oxygen from lungs and deliver carbon dioxide wastes to them.

Consider this: In fishes, blood flows in *one* circuit (Figure 39.4*a*). Pressure generated by a two-chambered heart forces it through gill capillary beds, into the largest artery, through capillary beds of organs, and back to the heart. Extensive gill capillaries offer so much resistance to flow that pressure drops considerably before blood enters the main artery. Blood delivery is sufficent for the activity level of most fishes. It would not be good enough for the more active life-styles of most land vertebrates.

When amphibians were evolving, their heart became partitioned into right and left halves—only partly so, but enough to pump blood through *two* partially separated circuits (Figure 39.4*b*). The separation of flow continued in crocodilians, which are reptiles. It became complete in birds and mammals. Their heart acts like two side-by-side pumps, as Figure 39.4*c* shows. The *right* half of their heart pumps oxygen-poor blood to the lungs, where the blood picks up oxygen and gives up carbon dioxide. Then the freshly oxygenated blood flows to the heart's left half. This route is the **pulmonary circuit**.

In the **systemic circuit**, the heart's *left* half pumps the freshly oxygenated blood to every tissue and organ where oxygen is used and carbon dioxide forms. Then the oxygen-poor blood flows to the heart's right half.

The double circuit is a rapid and efficient mode of blood delivery. It supports the high levels of activity typical of vertebrates whose ancestors evolved on land.

Links With the Lymphatic System

The heart's pumping action puts pressure on blood flowing through the circulatory system. Partly because of the pressure, small amounts of water and a few of the proteins dissolved in blood are forced out of capillaries and become part of interstitial fluid. However, a rather elaborate network of drainage vessels picks up excess interstitial fluid and reclaimable solutes, then returns them to the circulatory system. This network is part of the **lymphatic system**. Later, you will see how other parts of the lymphatic system help cleanse bacteria and other threats from fluid being returned to the blood.

Closed circulatory systems confine blood within one or more hearts and a network of blood vessels. In the open systems, blood also intermingles with tissue fluids.

The closed system of vertebrates transports substances to and from interstitial fluid bathing the body's cells. It is functionally connected with the lymphatic system.

Blood flows rapidly in large-diameter vessels between the heart and capillary beds. There, the flow velocity slows and exchanges are made between blood and interstitial fluid.

In fishes, blood flows in one circuit from and back to the heart. In birds and mammals, blood flows in two circuits, through a heart partitioned as two side-by-side pumps. The double circuit supports the high levels of activity typical of vertebrates that evolved on land.

39.2 CHARACTERISTICS OF BLOOD

Functions of Blood

Blood is a connective tissue with multiple functions. It transports oxygen, nutrients, and other solutes to cells. It carries away their metabolic wastes and secretions, including hormones. Blood helps stabilize internal pH. Blood also serves as a highway for phagocytic cells that fight infections and scavenge debris in tissues. In birds and mammals, blood helps equalize body temperature. It does so by carrying excess heat from skeletal muscles and other regions of high metabolic activity to the skin, where heat can be dissipated.

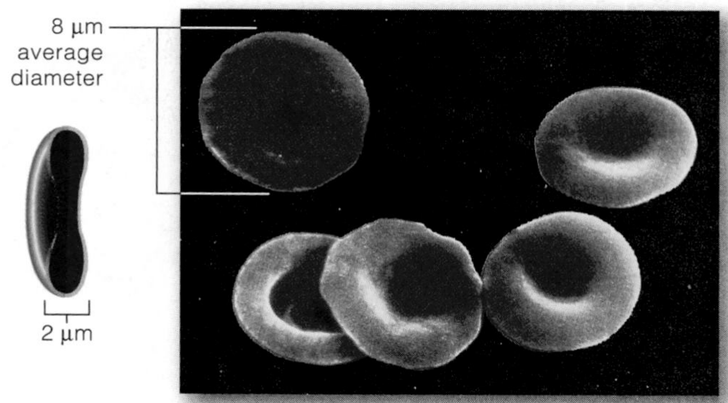

8 µm
average
diameter
2 µm

Figure 39.6 Size and shape of red blood cells.

Blood Volume and Composition

The volume of blood depends on body size and on the concentrations of water and solutes. Blood volume for average-size adult humans is about 6 to 8 percent of the total body weight. That amounts to about four or five quarts. As is the case for all vertebrates, human blood is a sticky fluid, thicker than water and slower flowing. Its components are the **plasma**, **red blood cells**, **white blood cells**, and **platelets**. Plasma normally accounts for 50 to 60 percent of the total blood volume.

PLASMA Prevent a blood sample in a test tube from clotting, and it separates into a red, cellular portion and the plasma, a straw-colored liquid, which floats on the cellular portion (Figure 39.5). Plasma is mostly water, and it serves as a transport medium for blood cells and platelets. Plasma also functions as a solvent for ions and molecules, including hundreds of different plasma proteins. Some of the plasma proteins transport lipids and fat-soluble vitamins through the body. Others have roles in blood clotting or in defense against pathogens. Collectively, the concentration of the plasma proteins influences the fluid volume of blood, for it affects the movement of water between the blood and interstitial fluid. Glucose and other simple sugars as well as lipids, amino acids, vitamins, and hormones are dissolved in plasma. So are oxygen, carbon dioxide, and nitrogen.

RED BLOOD CELLS Erythrocytes, or red blood cells, are biconcave disks, like doughnuts with a squashed-in center instead of a hole (Figure 39.6). They transport the oxygen used in aerobic respiration, and they carry away some carbon dioxide wastes. When oxygen first diffuses

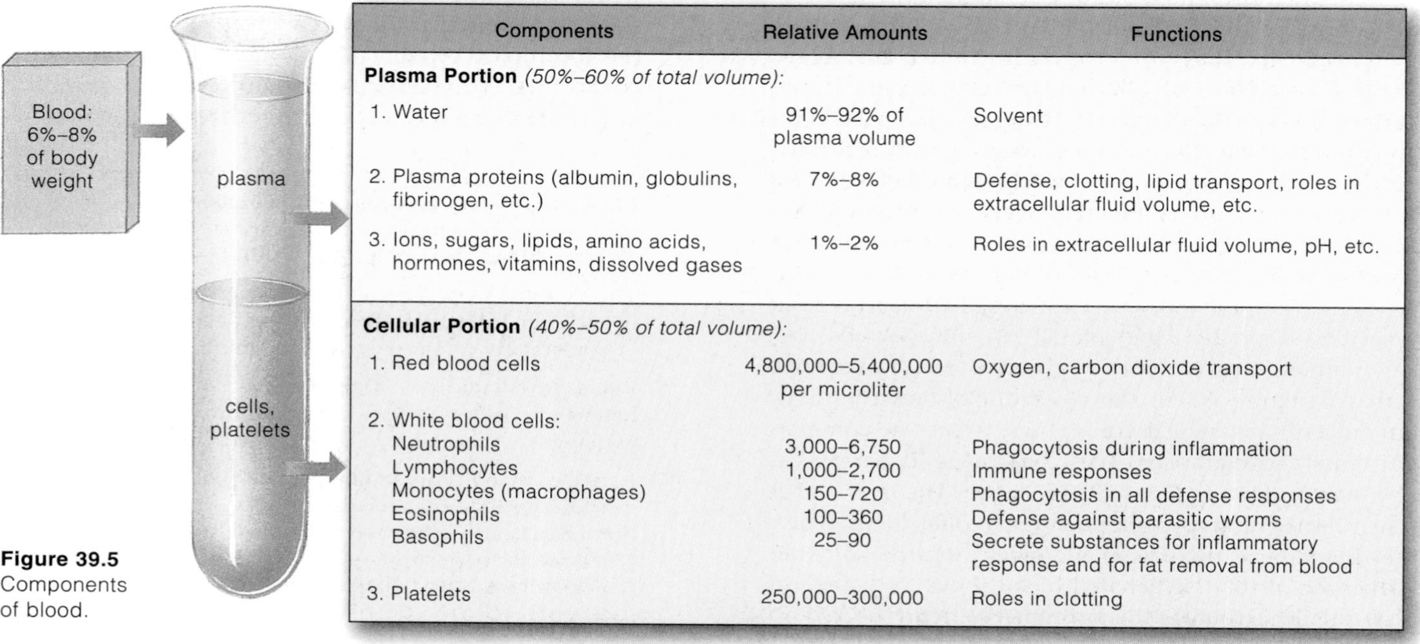

Blood:
6%–8%
of body
weight
plasma
cells,
platelets

Components	Relative Amounts	Functions
Plasma Portion *(50%–60% of total volume):*		
1. Water	91%–92% of plasma volume	Solvent
2. Plasma proteins (albumin, globulins, fibrinogen, etc.)	7%–8%	Defense, clotting, lipid transport, roles in extracellular fluid volume, etc.
3. Ions, sugars, lipids, amino acids, hormones, vitamins, dissolved gases	1%–2%	Roles in extracellular fluid volume, pH, etc.
Cellular Portion *(40%–50% of total volume):*		
1. Red blood cells	4,800,000–5,400,000 per microliter	Oxygen, carbon dioxide transport
2. White blood cells:		
Neutrophils	3,000–6,750	Phagocytosis during inflammation
Lymphocytes	1,000–2,700	Immune responses
Monocytes (macrophages)	150–720	Phagocytosis in all defense responses
Eosinophils	100–360	Defense against parasitic worms
Basophils	25–90	Secrete substances for inflammatory response and for fat removal from blood
3. Platelets	250,000–300,000	Roles in clotting

Figure 39.5 Components of blood.

Further reading: Student Guide to InfoTrac on web site

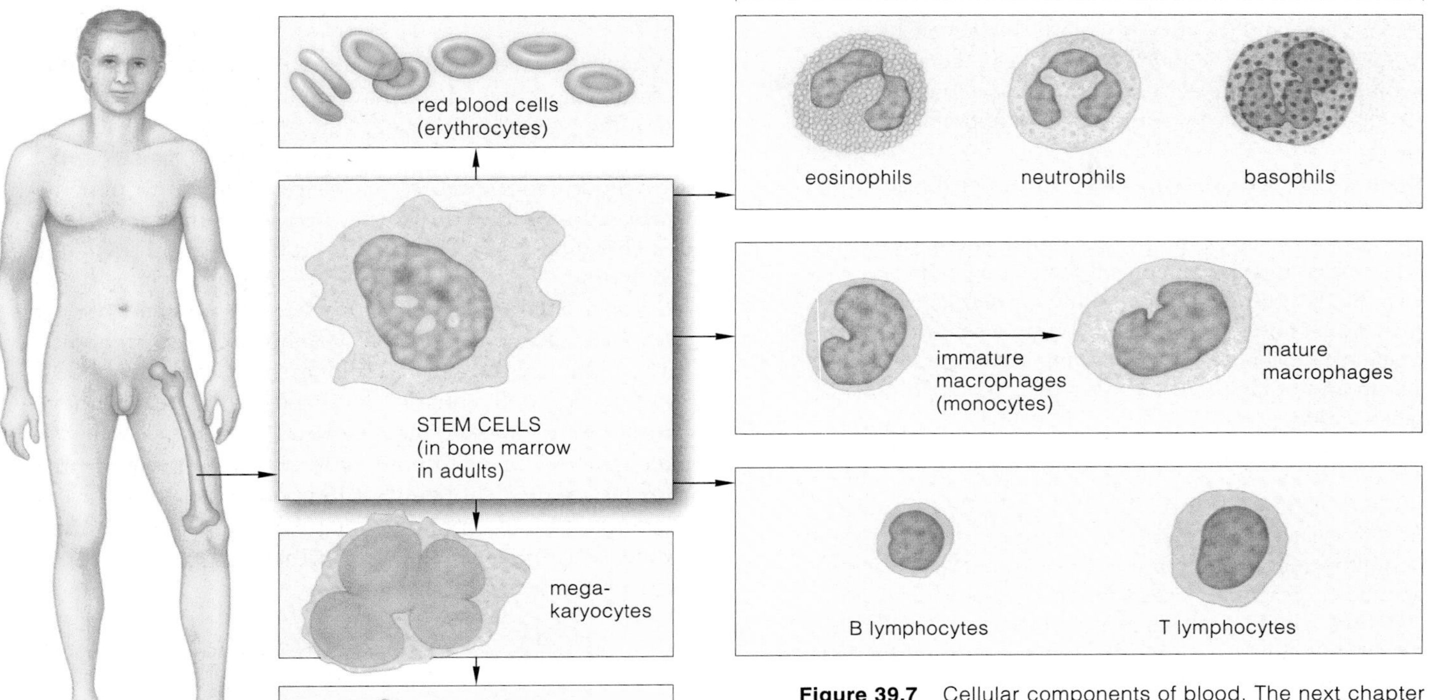

white blood cells (leukocytes)
red blood cells
(erythrocytes)
eosinophils
neutrophils
basophils
immature
macrophages
(monocytes)
mature
macrophages
STEM CELLS
(in bone marrow
in adults)
mega-
karyocytes
B lymphocytes
T lymphocytes
platelets

Figure 39.7 Cellular components of blood. The next chapter takes a close look at white blood cell functions. Section 39.10 describes the role of platelets in blood clotting.

into blood, it binds to hemoglobin, an iron-containing pigment that gives red blood cells their color. (Here you may wish to review hemoglobin's molecular structure, as shown in Section 3.5.) Oxygenated blood is bright red. Poorly oxygenated blood is darker red but appears blue inside blood vessel walls near the body surface.

Stem cells in bone marrow give rise to red blood cells (Figure 39.7). Generally, **stem cells** remain unspecialized and retain the capacity for mitotic cell division. Their daughter cells also divide, but only a portion go on to differentiate into specialized types.

Mature red blood cells no longer have their nucleus, nor do they require it. They have enough hemoglobin, enzymes, and other proteins to function for about 120 days. Phagocytes continually engulf the oldest red blood cells or those already dead. But ongoing replacements keep the cell count fairly stable. A **cell count** is a measure of the number of cells of a given type in a microliter of blood. For example, the average number of red blood cells is 5.4 million in males and 4.8 million in females.

WHITE BLOOD CELLS Leukocytes, or white blood cells, arise from stem cells in bone marrow. They function in daily housekeeping and defense. Some types patrol tissues. They target or engulf damaged or dead cells and anything chemically recognized as foreign to the body. Many others are massed together in lymph nodes and the spleen, which are organs of the lymphatic system. There they divide to produce armies of cells that battle specific viruses, bacteria, and other threats to health.

White blood cells differ in size, nuclear shape, and staining traits. There are five categories: the neutrophils, eosinophils, basophils, monocytes, and lymphocytes (Figure 39.7). Their numbers can change, depending on whether an individual is active, healthy, or under siege, as explained in the next chapter. The neutrophils and monocytes are search-and-destroy cells. The monocytes follow chemical trails to inflamed tissues. There they develop into macrophages ("big eaters") that can engulf invaders and debris. Two classes of lymphocytes, B cells and T cells, make highly specific defense responses.

PLATELETS Some stem cells in bone marrow give rise to megakaryocytes. These "giant" cells shed cytoplasmic fragments wrapped in a bit of plasma membrane. The membrane-bound fragments are what we call platelets. Each platelet only lasts five to nine days, but hundreds of thousands are always circulating in blood. They can release substances that initiate blood clotting.

Vertebrate blood has major roles in transporting substances, defense, clotting, and maintaining the volume, composition, and temperature of the internal environment.

Focus on Health

39.3 BLOOD DISORDERS

The body continually replaces blood cells, for good reason. Besides aging and dying off regularly, blood cells typically encounter a variety of pathogens that use them as places to complete their life cycle. Besides this, sometimes blood cells malfunction as a result of gene mutations.

RED BLOOD CELL DISORDERS Consider the **anemias**, disorders that result from too few red blood cells or deformed ones. Oxygen levels in blood cannot be kept high enough to support normal metabolism. Shortness of breath, fatigue, and chills follow. *Hemorrhagic* anemias result from sudden blood loss, as from a severe wound; *chronic* anemias result from ongoing but slight blood loss, as from an undiagnosed bleeding ulcer, hemorrhoids, or menstruation.

Some bacteria and protozoans replicate in blood cells, then escape by lysis and kill the cells. They cause some *hemolytic* anemias. Insufficient iron in the diet results in *iron deficiency* anemia, because red blood cells can't make enough normal hemoglobin without iron. *B_{12} deficiency* anemia is a potential hazard for strict vegetarians and alcoholics. Red blood cells form but they cannot divide without vitamin B_{12}. Meats, poultry, and fish in the diet usually provide enough of the vitamin.

As described elsewhere in the book, a gene mutation that gives rise to an abnormal form of hemoglobin can result in *sickle-cell* anemia. Another mutation partially or fully inhibits synthesis of the globin chains that make up hemoglobin, which causes *thalassemias*. Too few red blood cells can form, and those that do are thin and fragile.

Polycythemias—symptoms of far too many red blood cells—make blood flow sluggish. Some bone marrow cancers can result in this condition. So can *blood doping* by some athletes who compete in strenuous events. They withdraw and store their own red blood cells, then reinject them a few days prior to competition. The withdrawal triggers red blood cell formation as the body attempts to replace "lost" cells, so the cell count is bumped up when withdrawn cells are put back in the body. The idea is to increase oxygen-carrying capacity and endurance. Blood doping leads to temporary high blood pressure and lowers blood's viscosity. Some believe the practice works, but others call it unethical. It is banned from the Olympics.

WHITE BLOOD CELL DISORDERS Possibly you have heard of *infectious mononucleosis*. An Epstein–Barr virus causes this highly contagious disease, which results from the formation of too many monocytes and lymphocytes. Following a few weeks of fatigue, aches, low fever, and a chronic sore throat, patients usually recover.

Recovery is chancy for *leukemias*. As you read in Chapter 12, these cancers suppress or impair the formation of white blood cells in bone marrow. The cancer cells can be killed by radiation therapy and chemotherapy, but side effects are severe. Remissions may last for months or years. A "remission" is a symptom-free period of a chronic illness.

39.4 BLOOD TRANSFUSION AND TYPING

Concerning Agglutination

Whenever blood volume or blood cell counts decline, countermeasures kick in automatically. However, if the volume were to decrease by more than 30 percent, then circulatory shock would follow and could cause death.

Blood from donors can be transfused into patients affected by a blood disorder or substantial blood loss. Such **blood transfusions** cannot be hit-or-miss events. Why not? *A potential donor and the recipient may not have the same kinds of recognition proteins at the surface of their red blood cells.* Some of the proteins are "self" markers. They identify the cells as belonging to one's own body. During a blood transfusion, if cells from a donor have the "wrong" marker, the recipient's immune system will recognize them as foreign, with serious consequences.

Figure 39.8 shows what happens when blood from incompatible donors and recipients intermingles. In a defense response called **agglutination**, proteins called

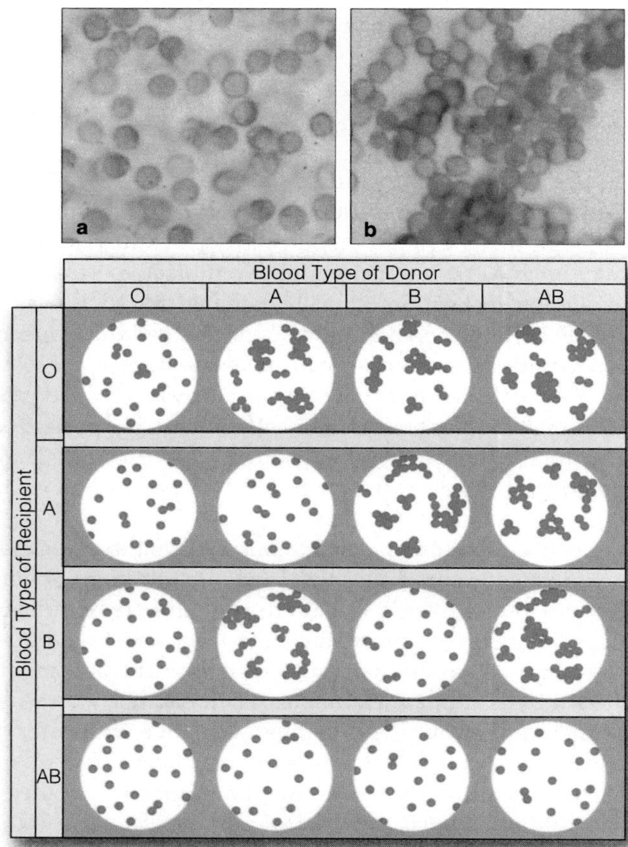

a
b
Blood Type of Donor
O
A
B
AB
Blood Type of Recipient
O
A
B
AB
c

Figure 39.8 Light micrographs showing (**a**) the absence of agglutination in a mixture of two different yet compatible blood types and (**b**) agglutination in a mixture of incompatible types. (**c**) Responses in blood types A, B, AB, and O when mixed with samples of the same and different types.

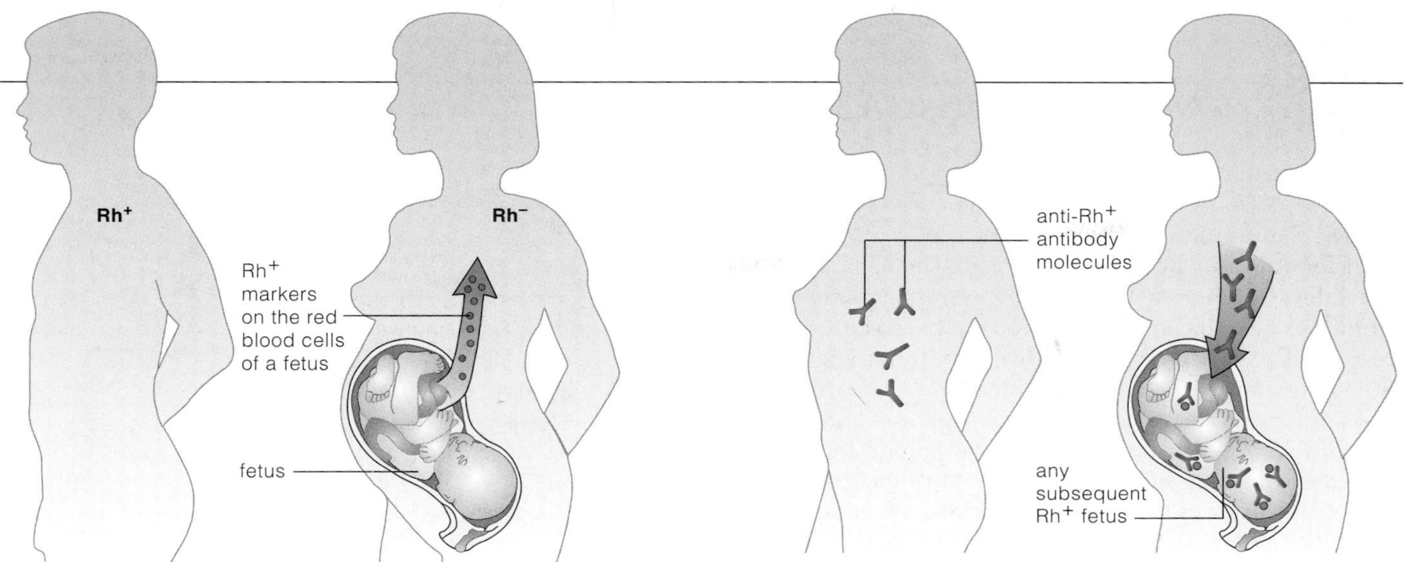

Rh+
Rh−
Rh+
markers
on the red
blood cells
of a fetus
fetus
anti-Rh+
antibody
molecules
any
subsequent
Rh+ fetus

a A forthcoming child of an Rh^- woman and Rh^+ man inherits the gene for the Rh^+ marker. During childbirth, some of its cells bearing the marker may leak into the maternal bloodstream.

b The foreign marker stimulates antibody formation. If the same woman becomes pregnant again and if her second fetus (or any other) inherits the gene for the marker, her circulating anti-Rh^+ antibodies will act against it.

Figure 39.9 Antibody production in response to Rh^+ markers on red blood cells of a fetus.

antibodies that are circulating in the plasma act against the foreign cells and cause them to clump together. As described in the next chapter, antibodies bind to specific foreign markers and target any cell or particle bearing them for destruction by the immune system. Also, the sheer number of foreign cells transfused into a patient end up in numerous clumps that can clog small blood vessels and damage tissues. Without treatment, death may follow. The same thing can happen during certain pregnancies when antibodies diffuse from the mother's circulatory system into that of her unborn child.

Based on an understanding of cell surface markers and antibodies, scientists have devised ways to analyze the forms of self markers on a person's red blood cells. Blood typing is based on these analyses.

ABO Blood Typing

Molecular variations in one kind of self marker on red blood cells are analyzed with **ABO blood typing**. The genetic basis of such variation is described in Section 11.4. People with one form of the marker are said to have type A blood; those with another form have type B blood. Those with both forms of the marker on their red blood cells have type AB blood. Others do not have either form of the marker; they have type O blood.

If you are type A, your antibodies ignore A markers but will act against B markers. If you are type B, your antibodies will ignore B markers but will act against A markers. If you are type AB, your antibodies ignore both forms of the marker, so you can tolerate donations of type A, B, AB, or O blood. If you are type O, you have antibodies against both forms of the marker, so your options are limited to type O donations.

Rh Blood Typing

Rh blood typing is based on the presence or absence of an Rh marker, so named because it was first identified in blood samples of *Rh*esus monkeys. If you are type Rh^+, your blood cells bear this marker at their surface. If you are type Rh^-, they don't. People ordinarily do not have antibodies against Rh markers. But a recipient of transfused Rh^+ blood produces antibodies against them, and the antibodies remain in the blood.

If an Rh^- woman becomes impregnated by an Rh^+ man, there is a chance that the fetus will be Rh^+. At childbirth, some fetal red blood cells may leak into the woman's bloodstream. If they do so, then her body will produce antibodies against Rh (Figure 39.9). And if she becomes pregnant again, Rh antibodies will enter the bloodstream of this new fetus. If its blood is type Rh^+, then her antibodies will cause fetal red blood cells to swell, rupture, and release hemoglobin.

Erythroblastosis fetalis is a severe outcome of mixing Rh^+ and Rh^- types. The fetus dies; too many cells are destroyed. If the condition is diagnosed before birth, the newborn can survive if its blood is slowly replaced with transfusions that are free of Rh antibodies. Today, a known Rh^- woman can be treated right after her first pregnancy with an anti-Rh gamma globulin (RhoGam) that will protect her next fetus. The drug will inactivate any Rh^+ fetal blood cells that are circulating through her bloodstream before she can become sensitized and begin producing potentially dangerous antibodies.

To avoid the symptoms of blood incompatibilities, red blood cells should be typed before transfusions or pregnancies.

39.5 HUMAN CARDIOVASCULAR SYSTEM

"Cardiovascular" comes from the Greek *kardia* (heart) and Latin *vasculum* (vessel). In a human cardiovascular system, a muscular heart pumps the blood into large-diameter **arteries**. Then blood flows into small, muscular **arterioles**, which branch into the even smaller diameter capillaries introduced earlier. Blood flows continuously from capillaries into small **venules**, and from there into large-diameter **veins** that return blood to the heart.

As in most vertebrates, a partition separates a human heart into a double pump, which drives blood through two cardiovascular circuits (Figure 39.10). Each circuit has its own arteries, arterioles, capillaries, venules, and veins. The *pulmonary* circuit—a short loop—oxygenates blood. It leads from the heart's right half to capillary beds in both lungs, then returns to the heart's left half. The *systemic* circuit is a longer loop starting at the left half of the heart. Its main artery, the **aorta**, accepts the oxygenated blood, which flows on through arterioles, capillary beds in all body regions, and then veins that deliver the oxygen-poor blood to the heart's right half. Figure 39.11 identifies the location of the major blood vessels of both circuits and defines their functions.

For most of the systemic branchings, a given volume of blood flows through one capillary bed. The branch to the intestines is one of the exceptions. First blood picks up glucose and other absorbed substances from one bed, then moves through another capillary bed, in the liver—an organ with a key role in nutrition. The second bed gives the liver time to process absorbed substances.

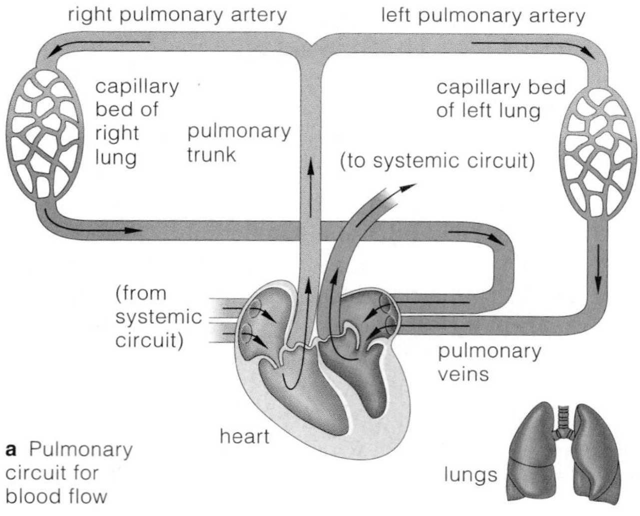
right pulmonary artery
left pulmonary artery
capillary bed of right lung
pulmonary trunk
capillary bed of left lung
(to systemic circuit)
(from systemic circuit)
pulmonary veins
heart
lungs
a Pulmonary circuit for blood flow

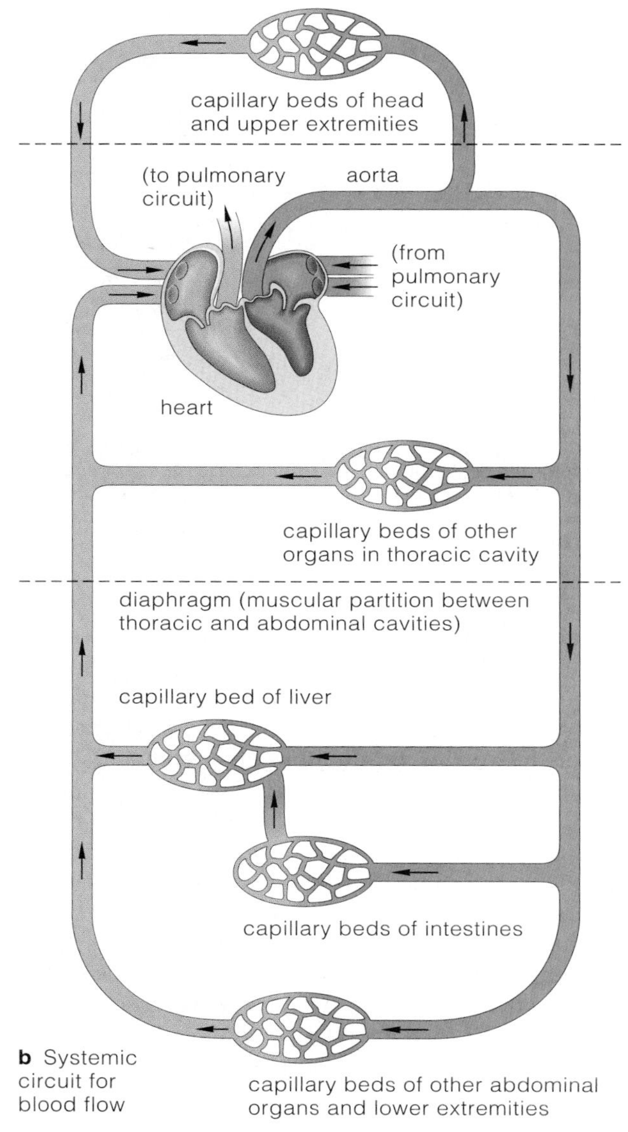
capillary beds of head and upper extremities
(to pulmonary circuit)
aorta
(from pulmonary circuit)
heart
capillary beds of other organs in thoracic cavity
diaphragm (muscular partition between thoracic and abdominal cavities)
capillary bed of liver
capillary beds of intestines
capillary beds of other abdominal organs and lower extremities
b Systemic circuit for blood flow

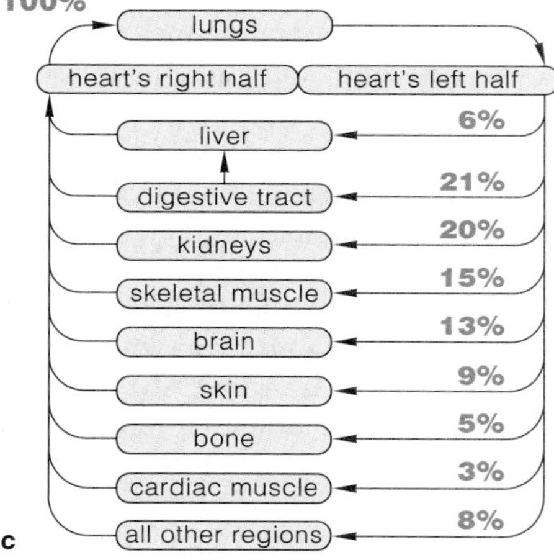
100%
lungs
heart's right half
heart's left half
liver
6%
digestive tract
21%
kidneys
20%
skeletal muscle
15%
brain
13%
skin
9%
bone
5%
cardiac muscle
3%
all other regions
8%
c

Figure 39.10 (**a**,**b**) Diagrams of the pulmonary and systemic circuits for blood flow through the human cardiovascular system. The blood vessels that are carrying oxygenated blood are color-coded *red*. Those carrying oxygen-poor blood are color-coded *blue*. (**c**) Distribution of the heart's output in a person at rest. The lungs receive all blood pumped out of the heart's right half. Organs serviced by the systemic circuit receive the portions indicated from the heart's left half. Control mechanisms adjust the flow distribution when necessary.

Further reading: Student Guide to InfoTrac on web site

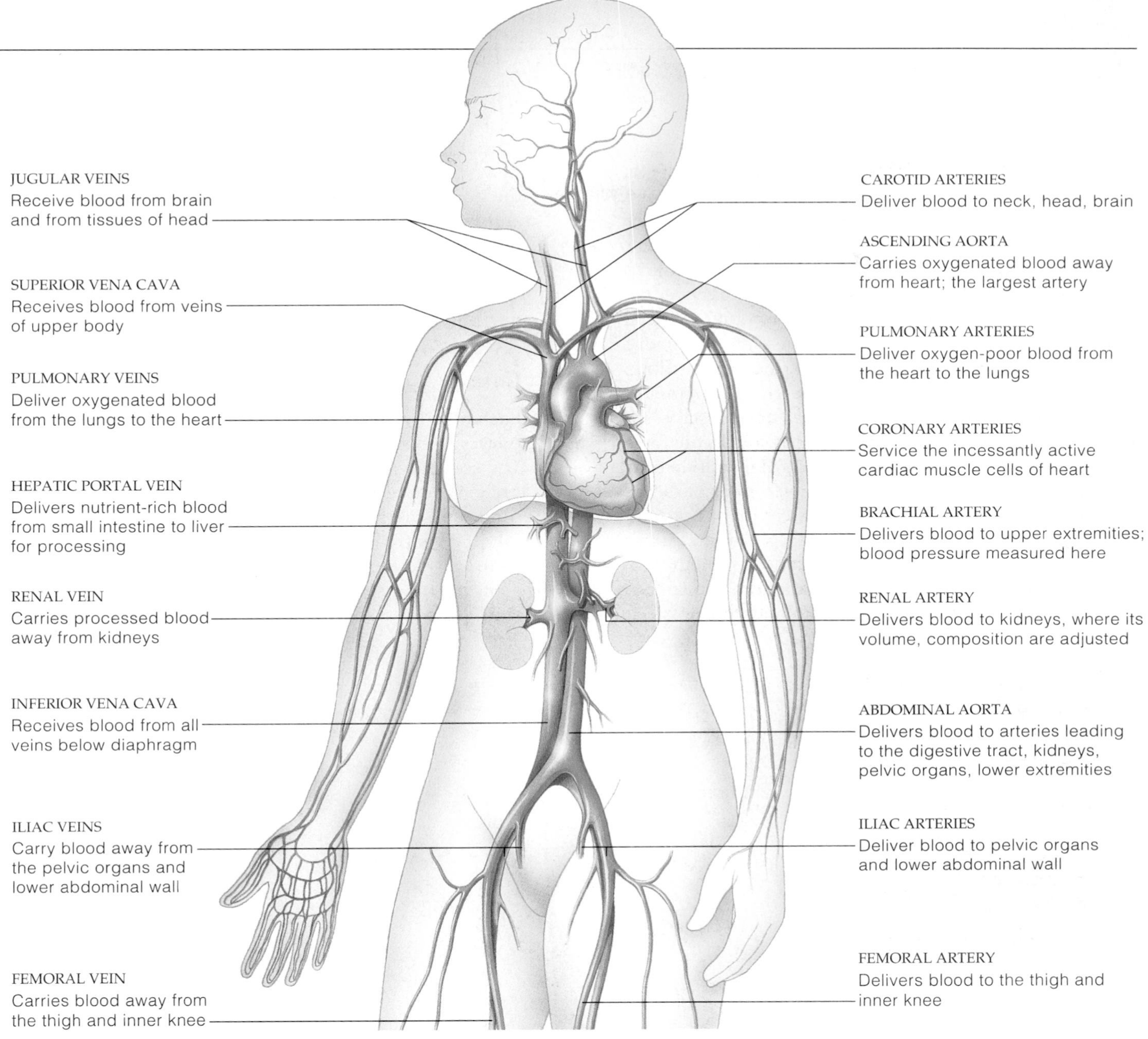
JUGULAR VEINS
Receive blood from brain and from tissues of head
SUPERIOR VENA CAVA
Receives blood from veins of upper body
PULMONARY VEINS
Deliver oxygenated blood from the lungs to the heart
HEPATIC PORTAL VEIN
Delivers nutrient-rich blood from small intestine to liver for processing
RENAL VEIN
Carries processed blood away from kidneys
INFERIOR VENA CAVA
Receives blood from all veins below diaphragm
ILIAC VEINS
Carry blood away from the pelvic organs and lower abdominal wall
FEMORAL VEIN
Carries blood away from the thigh and inner knee
CAROTID ARTERIES
Deliver blood to neck, head, brain
ASCENDING AORTA
Carries oxygenated blood away from heart; the largest artery
PULMONARY ARTERIES
Deliver oxygen-poor blood from the heart to the lungs
CORONARY ARTERIES
Service the incessantly active cardiac muscle cells of heart
BRACHIAL ARTERY
Delivers blood to upper extremities; blood pressure measured here
RENAL ARTERY
Delivers blood to kidneys, where its volume, composition are adjusted
ABDOMINAL AORTA
Delivers blood to arteries leading to the digestive tract, kidneys, pelvic organs, lower extremities
ILIAC ARTERIES
Deliver blood to pelvic organs and lower abdominal wall
FEMORAL ARTERY
Delivers blood to the thigh and inner knee

Figure 39.11 Location and functions of major blood vessels of the human cardiovascular system.

Figure 39.10*c* shows how the pulmonary and systemic circuits distribute the heart's output to different organs. The percentages listed are for a person at rest. As you will see, change in levels of physical activity and shifts in internal and external conditions trigger adjustments in the flow distribution along the systemic route. For example, the blood flow to and from a skeletal muscle varies greatly, depending on what it's being called upon to do at a given time. The same is true of skin. When the body becomes too cold, the blood flow—and metabolic heat—is diverted away from skin's vast capillary beds. When the body is hot, flow to the skin's beds increases and heat radiates away from the skin's surface. Only the brain is exempt; it cannot tolerate flow variations.

The human cardiovascular system consists of two separate circuits, pulmonary and systemic, for blood flow.

The pulmonary circuit is a short loop from the heart's right half, through both lungs, to the heart's left half. Oxygen-poor blood flowing into the circuit rapidly picks up oxygen and gives up carbon dioxide at capillary beds in the lungs.

The longer systemic circuit starts at the heart's left half and aorta. It delivers oxygen and accepts carbon dioxide at capillary beds of all metabolically active regions. Then its veins deliver oxygen-poor blood to the heart's right half.

39.6 THE HEART IS A LONELY PUMPER

Heart Structure

Think about the 2.5 billion times a human heart beats during a seventy-year life span, and you know it must be a durable pump. Its structure speaks of its durability (Figure 39.12). The pericardium, a double sac of tough connective tissue, protects the heart and anchors it to nearby structures (*peri*, around). A fluid in between its layers lubricates the heart during its perpetual twisting motions. The inner layer is actually the outer portion of the heart wall. The bulk of that wall, the myocardium, consists of cardiac muscle cells tethered to elastin and collagen fibers. These fibers are so densely crisscrossed, they act as the "skeleton" against which contractile force is applied. The oxygen-demanding cardiac muscle cells have their own, *coronary* circulation. Coronary arteries branch off the aorta and lead into a capillary bed that services only them. The heart wall's glistening, inner layer consists of connective tissue and endothelium, a one-cell-thick epithelial sheet. Endothelium is a special tissue located only in the heart and blood vessels.

Each half of the heart has two chambers: an atrium (plural, atria) and a ventricle. Blood flows into the atria, down into the ventricles, then out through great arteries: the aorta and pulmonary trunk. Between each atrium and ventricle is an AV valve (short for atrioventricular). Between each ventricle and the artery leading out from it is a semilunar valve. Both are *one-way* valves that are alternately forced open and shut to help keep the blood moving in the forward direction only.

Cardiac Cycle

Each time the heart beats, its four chambers go through phases of systole (contraction) and diastole (relaxation). The sequence of contraction and relaxation is a **cardiac cycle**. When relaxed, the atria fill with blood. Increasing fluid pressure forces both AV valves open. Blood flows into the ventricles, which completely fill when the atria contract (Figure 39.13*a*). As the filled ventricles start to contract, the rising fluid pressure forces the AV valves shut. It rises so sharply above the pressure in the great

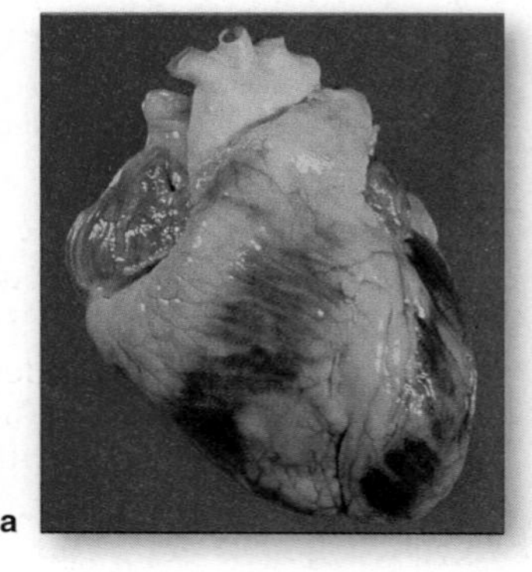
a

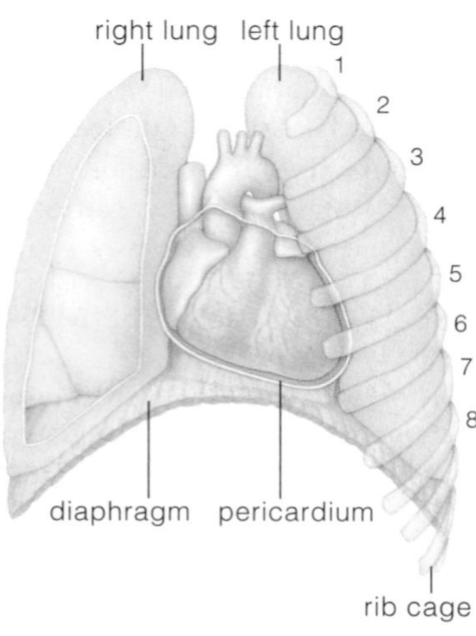
right lung
left lung
1
2
3
4
5
6
7
8
diaphragm
pericardium
rib cage
b

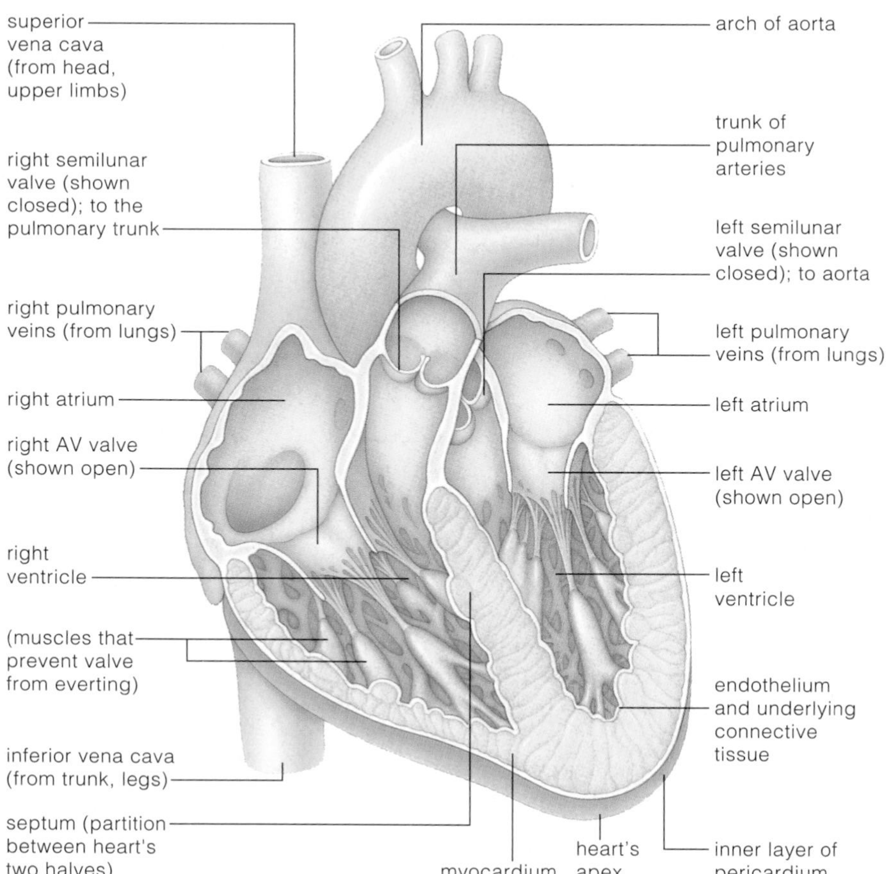
superior vena cava (from head, upper limbs)
right semilunar valve (shown closed); to the pulmonary trunk
right pulmonary veins (from lungs)
right atrium
right AV valve (shown open)
right ventricle
(muscles that prevent valve from everting)
inferior vena cava (from trunk, legs)
septum (partition between heart's two halves)
myocardium
heart's apex
arch of aorta
trunk of pulmonary arteries
left semilunar valve (shown closed); to aorta
left pulmonary veins (from lungs)
left atrium
left AV valve (shown open)
left ventricle
endothelium and underlying connective tissue
inner layer of pericardium
c

Figure 39.12 (**a**) Photograph of the human heart and (**b**) its location in the thoracic cavity. (**c**) Cutaway view of the heart, showing its wall and internal organization.

Further reading: Student Guide to InfoTrac on web site

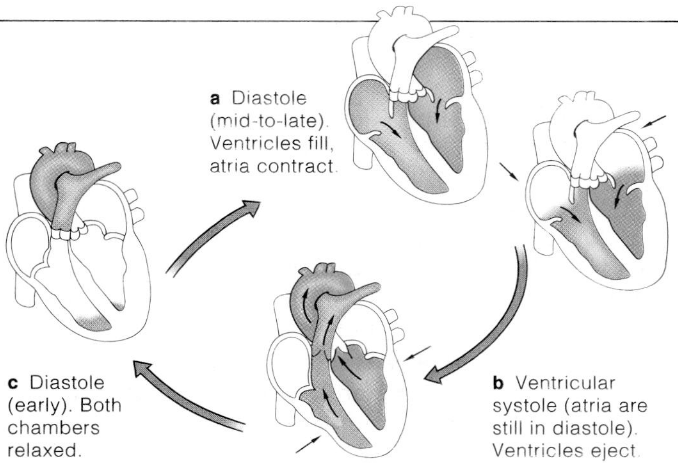
a Diastole (mid-to-late). Ventricles fill, atria contract.
b Ventricular systole (atria are still in diastole). Ventricles eject.
c Diastole (early). Both chambers relaxed.

Figure 39.13 Blood flow during part of a cardiac cycle. Blood and heart movements generate a "lub-dup" sound at the chest wall. At each "lub," AV valves are closing as ventricles contract. At each "dup," semilunar valves are closing as ventricles relax.

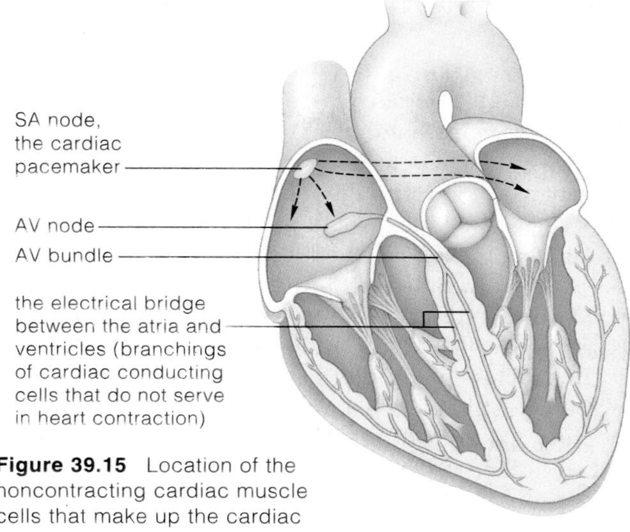
SA node, the cardiac pacemaker
AV node
AV bundle
the electrical bridge between the atria and ventricles (branchings of cardiac conducting cells that do not serve in heart contraction)

Figure 39.15 Location of the noncontracting cardiac muscle cells that make up the cardiac conduction system.

arteries, it forces the semilunar valves open—and blood leaves the heart (Figure 39.13*b*). The ventricles relax, the semilunar valves close, and the already filling atria are ready to repeat the cycle. Thus, in a cardiac cycle, atrial contraction simply helps fill the ventricles. *Contraction of the ventricles is the driving force for blood circulation.*

Cardiac Muscle Contraction

Like skeletal muscle, cardiac muscle is striated (striped). In response to action potentials, the sarcomeres of its cells contract in the manner predicted by the sliding-filament model (Section 38.8). This tissue, too, requires energy for contraction; large numbers of mitochondria in the myocardium provide the ATP. But cardiac muscle cells are structurally unique. They are branching, short, and joined at the end regions. Communication junctions span the plasma membranes of abutting regions and let action potentials spread swiftly among cells, in waves of excitation that wash over the heart (Figure 39.14).

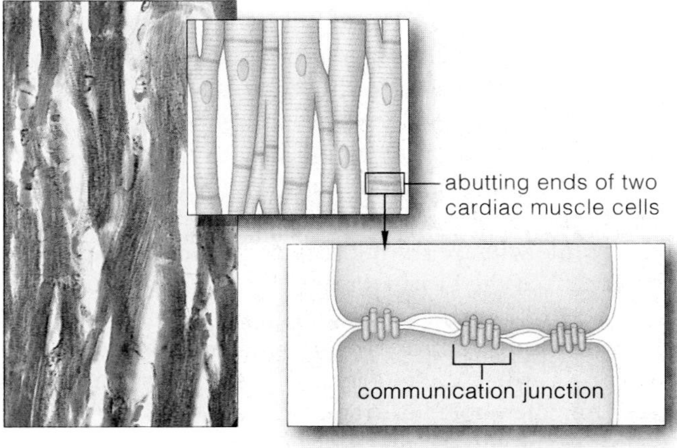
abutting ends of two cardiac muscle cells
communication junction

Figure 39.14 Light micrograph of cardiac muscle cells. The sketches show communication junctions between cells.

In addition, about 1 percent of the cardiac muscle cells *do not* contract. They function instead as a **cardiac conduction system**. About seventy times each minute, these specialized cells initiate and propagate waves of excitation that travel in a rhythmic, orderly sequence from the atria to ventricles. The synchronized excitation underlies the heart's efficient pumping. Each wave starts at the SA node, a cluster of cell bodies in the wall of the right atrium (Figure 39.15). It passes through the wall to another cell body cluster, the AV node. This is the only electrical bridge between the atria and ventricles, which connective tissue insulates everywhere else. After the AV node, conducting cells are arranged as a bundle in the partition between the heart's two halves. These cells branch, and their branches deliver the excitatory wave up the ventricle walls. The ventricles contract in response with a twisting movement, upward from the heart's apex, that ejects blood into the great arteries.

The SA node fires action potentials faster than the rest of the system and serves as the **cardiac pacemaker.** Its spontaneous, rhythmic signals are the foundation for the normal rate of heartbeat. The nervous system can only *adjust* the rate and strength of contractions dictated by the pacemaker. Even if all nerves leading to a heart are severed, the heart will keep on beating!

The heart's construction reflects its role as a durable pump. Under the spontaneous, rhythmic signals from the cardiac pacemaker, its branching, abutting cardiac muscle cells contract in synchrony, almost as if they were a single unit.

Although the heart has four chambers (each half has one atrium and one ventricle), contraction of the ventricles is the driving force for blood circulation away from the heart.

39.7 BLOOD PRESSURE IN THE CARDIOVASCULAR SYSTEM

As you have seen, blood flows to and from the human heart through arteries, arterioles, capillaries, venules, and finally veins. Figure 39.16 shows the structure of these vessels. The arteries are the major transporters of oxygenated blood through the body. Arterioles in each body region are sites where the volume of blood flow through organs is controlled. Blood capillaries and, to a lesser extent, venules are diffusion zones, and veins are blood volume reservoirs and the major transporters of oxygen-poor blood back to the heart.

Two factors greatly influence the flow rate through each type of blood vessel. First, the flow rate is directly proportional to the pressure gradient between the start and end of the vessel. Second, the flow rate is inversely proportional to the vessel's resistance to flow.

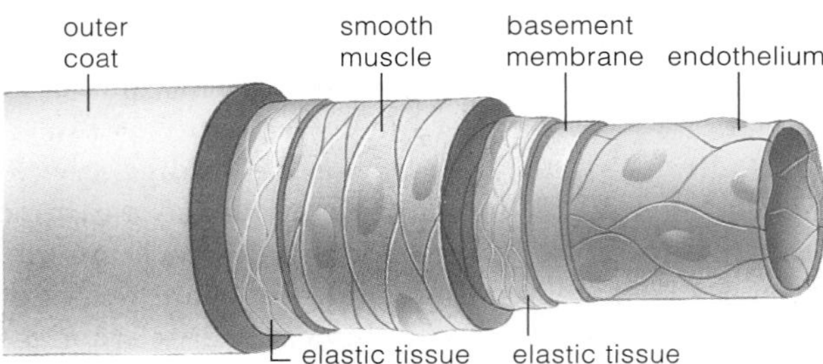
outer coat
smooth muscle
basement membrane
endothelium
elastic tissue
elastic tissue

a ARTERY

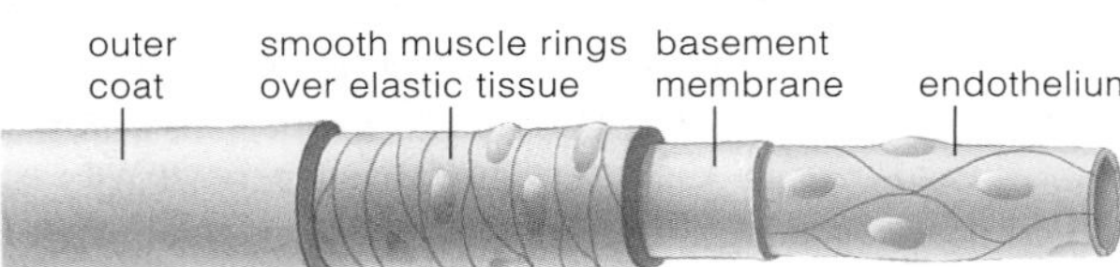
outer coat
smooth muscle rings over elastic tissue
basement membrane
endothelium

b ARTERIOLE

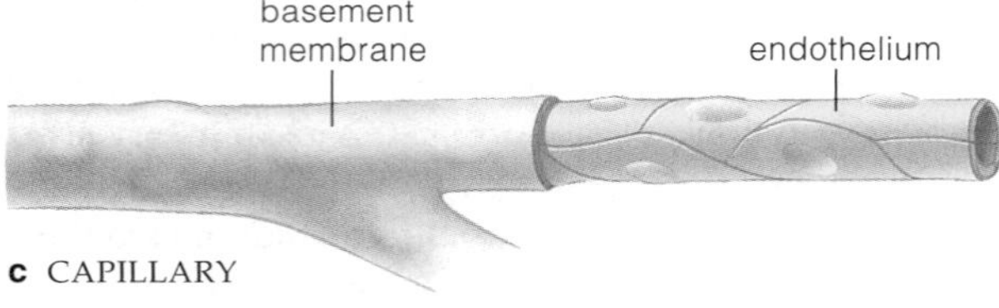
basement membrane
endothelium

c CAPILLARY

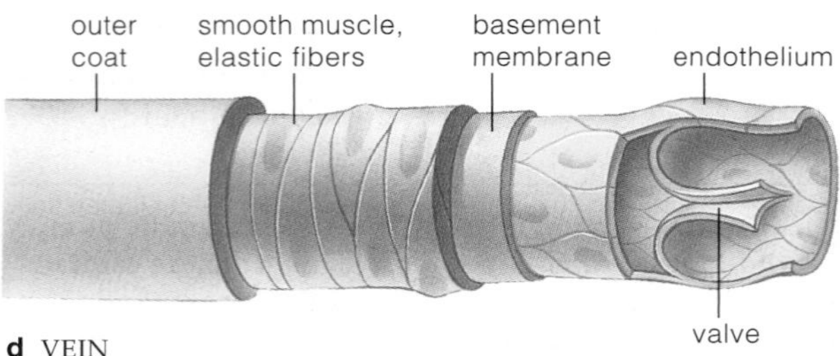
outer coat
smooth muscle, elastic fibers
basement membrane
endothelium
valve

d VEIN

Figure 39.16 Structure of blood vessels. The basement membrane around each vessel's endothelium is a noncellular layer, rich with proteins and polysaccharides.

Blood pressure is fluid pressure imparted to blood by heart contractions. The beating heart establishes the higher blood pressure at the beginning of a vessel. As flowing blood rubs against the vessel's inner wall, the friction causes some energy (in the form of pressure) to be lost. The lower pressure at the end of the vessel is due to the frictional losses. Even small changes in blood vessel diameter have major influence over the flow rate. In the pulmonary and systemic circuits, the diameters decrease and present more resistance to flow. And so, because of heart contractions and subsequent events (mainly frictional losses), blood pressure is highest in the contracting ventricles, still high at the beginning of arteries, then continues to drop until the relaxed atria. There it reaches its lowest value (Figure 39.17).

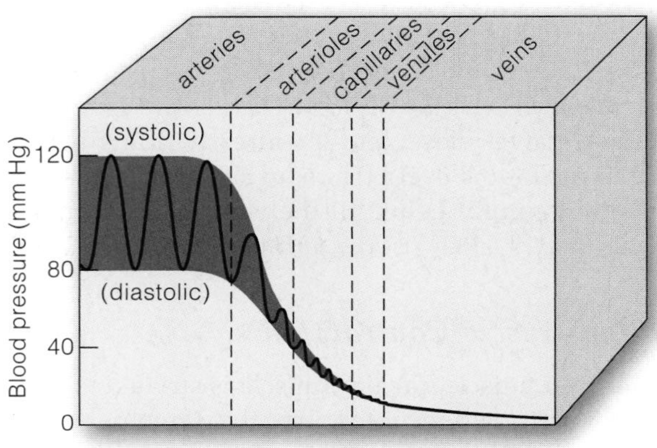
arteries
arterioles
capillaries
venules
veins
(systolic)
(diastolic)
Blood pressure (mm Hg)
120
80
40
0

Figure 39.17 Plot of the decline in fluid pressure for a volume of blood moving through the systemic circuit.

Arterial Blood Pressure

From Figure 39.16*a* and the preceding discussion, you see that arteries have a large diameter and present low resistance to flow. Thus they serve as rapid transporters of oxygenated blood. They also are pressure reservoirs that smooth out pulsations in pressure caused by each cardiac cycle. Their thick, muscular, elastic wall bulges as ventricular contraction forces a large volume of blood into them. Then the wall recoils and so forces the blood forward through the circuit while the heart is relaxing.

Suppose you decide to measure your blood pressure daily while you are resting. Typically, you would take a reading from the brachial artery of an upper arm, as in Figure 39.18. *Systolic* pressure is the peak pressure that the contracting ventricles exert against the artery's wall during a cardiac cycle. *Diastolic* pressure is the lowest pressure when blood is draining into the vessels after it. You can estimate the mean arterial pressure as *diastolic pressure* + *1/3 pulse pressure*. (The difference between the highest and lowest readings is the pulse pressure.)

Further reading: Student Guide to InfoTrac on web site

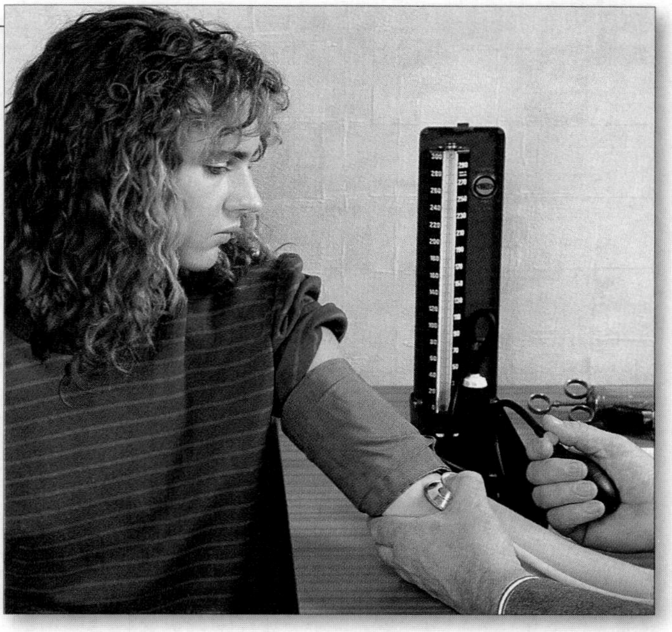

Figure 39.18 Measuring blood pressure.

A hollow cuff attached to a pressure gauge is wrapped around a person's upper arm. The cuff is inflated with air to a pressure above the highest pressure of the cardiac cycle—at systole, when the ventricles contract. Above this pressure, no sounds can be heard through a stethoscope positioned below the cuff and above the artery, because no blood is flowing through the vessel. Air inside the cuff is slowly released, so that some blood flows into the artery. The turbulent flow causes soft tapping sounds. When the tapping first occurs, the value on the gauge is the systolic pressure, or about 120 mm mercury (Hg) in young adults at rest. This particular value means that the measured pressure would force mercury to move upward 120 millimeters in a narrow glass column.

More air is released from the cuff. Just after the sounds become dull and muffled, blood is flowing continuously, so the turbulence and tapping end. The silence corresponds to diastolic pressure at the end of a cardiac cycle, just before the heart pumps out blood. The reading is usually about 80 mm Hg. In this example, pulse pressure (the difference between the highest and the lowest pressure readings) is 120 – 80, or 40 mm Hg. Assuming you are an adult in good health, this average resting value stays fairly constant over a few weeks or even months.

How Can Arterioles Resist Flow?

Track a volume of blood through the systemic circuit and you find the greatest pressure drop at arterioles (Figure 39.17), for these offer the greatest resistance to blood flow. The slowdown at arterioles allows time for control mechanisms to divert greater or lesser portions of the total flow volume to different regions. Control is exerted at arteriole walls, which incorporate rings of smooth muscle. Some control signals cause the smooth muscle cells to relax, which results in **vasodilation**. The word means an enlargement of blood vessel diameter (that is, dilation). Other signals cause contraction of the smooth muscle cells, which brings about a decrease in the blood vessel diameter, or **vasoconstriction**.

The nervous system and endocrine system provide control over changes in arteriole diameter. For example, sympathetic nerve endings terminate on smooth muscle cells of many arterioles. Increased activity along these nerves initiates action potentials in the cells and brings about contraction, hence widespread vasoconstriction. Epinephrine and angiotensin are two of the signaling molecules that trigger changes in arteriole diameter.

Arteriole diameter also is adjusted when changes in metabolic activity shift the localized concentrations of substances in a tissue. Such local chemical changes are "selfish" in that they invite or divert the blood flow to meet the tissue's own metabolic needs. Think of what happens when you run. The oxygen level falls in skeletal muscle tissue, and levels of carbon dioxide, hydrogen and potassium ions, and other substances increase. The changed conditions cause vasodilation of arterioles in the vicinity. Now more blood can flow through active muscles to deliver more raw materials and carry away cell products and metabolic wastes. When muscles relax and demand fewer blood deliveries, the oxygen level increases and cause vasoconstriction of the arterioles.

Controlling Mean Arterial Blood Pressure

Mean arterial pressure depends on cardiac output and on total resistance through the vascular system. Cardiac output is adjusted by controls over the rate and strength of heartbeats, and total resistance is adjusted mainly by vasoconstriction at the arterioles. Given that organs and tissues have variable demands for blood, maintaining blood pressure is not easy. The balance between cardiac output and total resistance must be juggled continually.

A **baroreceptor reflex** is the main short-term control over arterial pressure. It starts at baroreceptors, which detect fluctuations in arterial mean pressure and pulse pressure (Section 36.1). The most vital of these sensory receptors occur in carotid arteries (which deliver blood to the brain) and the aortic arch (which deliver blood to the rest of the body). They continually generate action potentials, and their signals reach a control center in a hindbrain region, the medulla oblongata (Section 35.4). In response, the control center adjusts its commands flowing along sympathetic and parasympathetic nerves to the heart and blood vessels. For example, when the mean arterial pressure rises, the center commands the heart to beat more slowly and contract less forcefully.

Long-term controls over blood pressure are exerted at kidneys, which adjust the volume and composition of the blood. And that is a topic of another chapter.

Mean arterial pressure is an outcome of controls over the heart's output and over the total resistance to blood flow through the vascular system, as exerted mainly at arterioles.

39.8 FROM CAPILLARY BEDS BACK TO THE HEART

Capillary Function

THE NATURE OF THE EXCHANGES Capillary beds, again, serve as diffusion zones for exchanges between blood and interstitial fluid. Living cells in any body tissue die quickly when deprived of these exchanges. Brain cells die within four minutes.

By some estimates, between 10 billion and 40 billion capillaries service the human body. Collectively, they afford an astonishing surface area for gas exchanges. These components of the circulatory system extend into nearly all tissues, and at least one is as close as 0.001 centimeter to every living cell. Their proximity to cells is vital because shorter distances mean faster diffusion rates. Think about this: It would take years for oxygen to diffuse on its own from your lungs all the way down your legs. By that time your toes, and the rest of you, would long be dead.

Red blood cells (8 micrometers wide) squeeze single file through capillaries, which are 3 to 7 micrometers wide and deform only a bit. The squeeze puts oxygen-transporting red blood cells, and substances dissolved in plasma, in direct contact with the exchange area—the capillary wall—or only a short distance away from it.

And remember, taken as a whole, capillaries present a greater cross-sectional area than the arterioles leading into them, so the flow velocity cannot be as great in the capillary beds. The slowdown means there is plenty of time for interstitial fluid to exchange substances with the 5 percent or so of the total volume of blood moving forward through the beds during a given interval. Here you may wish to reflect on Figure 39.3*e*.

Each capillary is a single sheet of endothelial cells in the form of a tube (Figure 39.16*c*). The cells abut one another, but the "fit" differs in different body regions. In most cases, narrow, water-filled clefts occur between endothelial cells. In capillaries that service the brain, tight junctions join the cells, and clefts are nonexistent; substances cannot "leak" between cells but must move through them. Thus the junctions are a functional part of the blood–brain barrier described in Section 35.4.

Oxygen, carbon dioxide, and most other small, lipid-soluble substances cross the capillary wall by diffusing through the lipid portion of an endothelial cell's plasma membrane and through its cytoplasm. Certain proteins enter and leave the cells by endocytosis or exocytosis. Small, water-soluble substances such as ions enter and leave at the clefts between endothelial cells. So do white blood cells, as you will see in the chapter to follow.

MECHANISMS OF EXCHANGE Diffusion is not the only means by which substances are exchanged across the wall of capillaries. At any time, fluid pressures acting on the wall may not be balanced, and there may be bulk flow one way or the other across it. Bulk flow, recall, is a movement of water and solutes in the same direction in response to fluid pressure.

As Figure 39.19 shows, the concentrations of water and solutes in blood and interstitial fluid influence the direction of flow. At the beginning of a capillary bed, the outward-directed force of blood pressure is greater than the inward-directed osmotic force because of the plasma proteins. A small amount of protein-free plasma is pushed out in bulk through the clefts in the capillary wall. This process is called **ultrafiltration**. Farther on in the bed, the balance shifts. Whereas blood pressure has been steadily declining, the osmotic force has stayed the same. When the inward-directed osmotic force exceeds the outward force of blood pressure, tissue fluid moves through the clefts between endothelial cells and into the capillary. This process is called **reabsorption**.

Normally, the outcome is a very small *net* outward movement of fluid from a capillary bed. The lymphatic system returns fluid to the blood. Such bulk flow helps maintain the fluid balance between the interstitial fluid and blood. This is important, because blood pressure is maintained only when there is adequate blood volume. When blood volume plummets during hemorrhage and certain other abnormal events, interstitial fluid may be tapped by way of reabsorption to help counter the loss.

Sometimes blood pressure increases so much that it causes excessive ultrafiltration. When too much fluid accumulates in interstitial spaces, we call this condition *edema*. Edema occurs to some extent during physical exercise, when arterioles dilate in many tissue regions. It also can result from an obstructed vein or from heart failure. Edema becomes extreme during *elephantiasis*. As described in Section 26.9, elephantiasis is a disease brought on by roundworm infection and a subsequent obstruction of lymphatic vessels.

Venous Pressure

What happens to the flow velocity after it continues on past the vast cross-sectional area of the capillary beds? Remember, the capillaries merge into venules, or "little veins." These in turn merge into large-diameter veins. Once again the total cross-sectional area is reduced, and flow velocity increases as blood returns to the heart.

In functional terms, venules are a bit like capillaries. Some solutes diffuse across their wall, which isn't much thicker than the thin capillary wall. Some control over capillary pressure also is exerted at these vessels.

Veins are large-diameter, low-resistance transport tubes to the heart (Figure 39.16*d*). They have valves that prevent backflow. When gravity beckons, venous flow reverses direction and pushes the valves shut. The vein wall can bulge considerably under pressure, more so

Further reading: Student Guide to InfoTrac on web site

ARTERIOLE END OF CAPILLARY BED	
Outward-Directed Pressure:	
Hydrostatic pressure of blood in capillary:	35 mm Hg
Osmotic force due to plasma proteins:	28 mm Hg
Inward-Directed Pressure:	
Hydrostatic pressure of interstitial fluid:	0
Osmotic force due to interstitial proteins:	3 mm Hg
Net Ultrafiltration Pressure:	
(35 – 0) – (28 – 3) =	10 mm Hg
ULTRAFILTRATION FAVORED	

VENULE END OF CAPILLARY BED	
Outward-Directed Pressure:	
Hydrostatic pressure of blood in capillary:	15 mm Hg
Osmotic force due to plasma proteins:	28 mm Hg
Inward-Directed Pressure:	
Hydrostatic pressure of interstitial fluid:	0
Osmotic force due to interstitial proteins:	3 mm Hg
Net Reabsorption Pressure:	
(15 – 0) – (28 – 3) =	– 10 mm Hg
REABSORPTION FAVORED	

blood from arteriole
outward-directed bulk flow
cells of tissue
inward-directed osmotic movement
blood to venule

Figure 39.19 Bulk flow in an idealized capillary bed. The fluid movement plays no significant role in diffusion of solutes. However, it is important in maintaining the distribution of extracellular fluid between blood and interstitial fluid.

The fluid movements across a capillary wall result from the two opposing forces of ultrafiltration and reabsorption.

At a capillary's arteriole end, the difference between blood pressure and interstitial fluid pressure forces some plasma, but very few plasma proteins, to leave the capillary. Plasma moves by bulk flow through clefts between endothelial cells of the capillary wall. Ultrafiltration is the bulk flow of fluid *out* of the capillary.

Reabsorption is an osmotic movement of some interstitial fluid *into* the capillary as a result of a difference in the water concentration between interstitial fluid and plasma. Plasma, with its dissolved protein components, has a greater solute concentration and therefore a lower water concentration.

Reabsorption near the end of a capillary bed tends to balance ultrafiltration at the beginning. Normally there is only a small *net* filtration of fluid, which the lymphatic system returns to the blood.

than an arterial wall. Thus veins can serve as reservoirs for variable volumes of blood. Collectively, veins of an adult human can hold up to 50 to 60 percent of the total blood volume.

The vein wall contains some smooth muscle. When blood must circulate faster, as during physical exercise, the smooth muscle contracts. The wall stiffens and the vein bulges less, so that venous pressure increases and drives more blood back to the heart. Also, when limbs move, skeletal muscles bulge against the veins in their vicinity. They help raise the venous pressure and drive blood back to the heart (Figure 39.20). Rapid breathing also contributes to increased venous pressure. Inhaled air pushes down on internal organs and thereby alters the pressure gradient between the heart and veins.

Capillary beds are diffusion zones for exchanges between blood and interstitial fluid. Here also, a slight amount of bulk flow helps maintain the fluid balance between blood and interstitial fluid.

Venules overlap somewhat with capillaries in function.

Veins are highly distensible blood volume reservoirs and help adjust flow volume back to the heart.

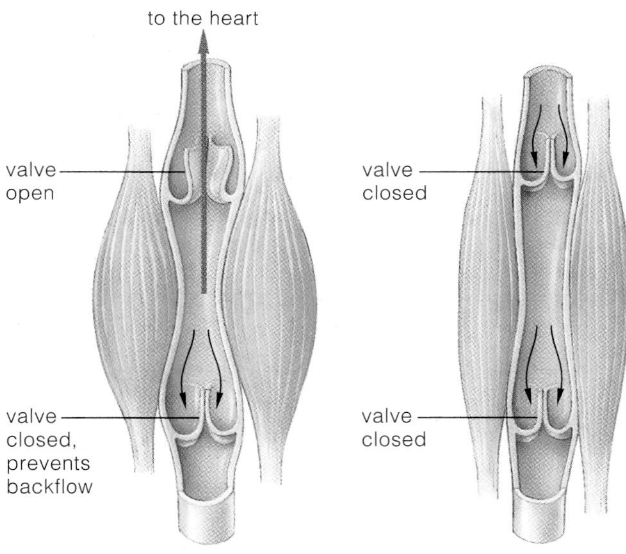
to the heart
valve open
valve closed, prevents backflow
valve closed
valve closed
a Contracted skeletal muscles against nearby vein assist blood flow to heart.
b Relaxed skeletal muscles; venous valves shut, no backflow.

Figure 39.20 How the bulging of contracting skeletal muscles helps increase fluid pressure inside a vein. Notice the structure of each valve (membrane flaps) in the vein's lumen.

39.9 CARDIOVASCULAR DISORDERS

Cardiovascular disorders are the leading cause of death in the United States. They affect at least 40 million people and kill about 1 million each year. The most common are *hypertension*, which is sustained high blood pressure; and *atherosclerosis*, a progressive thickening of the arterial wall and narrowing of the arterial lumen. Both disorders affect blood circulation and cause most *heart attacks* (damaged or destroyed heart muscle) and *strokes* (brain damage).

Warning signs of a heart attack are pain or a sensation of squeezing behind the breastbone, numbness or pain in the left arm, sweating, and nausea. In women more often than in men, other signs are neck and back pain, tiredness, shortness of breath, anxiety, and a sense of indigestion along with a fast heartbeat and low blood pressure.

RISK FACTORS Researchers correlate these factors with an increased risk of developing cardiovascular disorders:

1. Smoking (Section 41.7)
2. Genetic predisposition to heart failure
3. High level of cholesterol in the blood
4. High blood pressure
5. Obesity (Section 42.10)
6. Lack of regular exercise
7. Diabetes mellitus (Section 37.7)
8. Age (the older you get, the greater the risk)
9. Gender (until age fifty, males are at much greater risk than females)

Exercising regularly, eating properly, and not smoking help lower the risk. With too little exercise and too much food, adipose tissue masses increase in the body, more blood capillaries develop to service them, and the heart has to work harder to pump blood through increasingly divided vascular circuits. In people who smoke, nicotine in tobacco makes the adrenal glands secrete epinephrine. This hormone causes blood vessel diameters to constrict, and thereby boosts heart rate and blood pressure. Also, carbon monoxide present in cigarette smoke competes with oxygen for binding sites on hemoglobin, so the heart has to pump harder to deliver enough oxygen to cells. Smoking also predisposes people to atherosclerosis even when their blood level of cholesterol is normal.

The next paragraphs describe some of the damage that results from cardiovascular disorders.

HYPERTENSION Hypertension results from gradual increases in the flow resistance through small arteries. In time, blood pressure remains above 140/90 even when a person is resting. Heredity may be a factor; the disorder tends to run in families. Diet is also a factor. For instance, high salt intake raises blood pressure in some susceptible people and increases the heart's workload. Eventually the heart may enlarge and fail to pump blood effectively. High blood pressure also may contribute to a "hardening" of arterial walls, which hampers the delivery of oxygen to the brain, heart, and other vital organs.

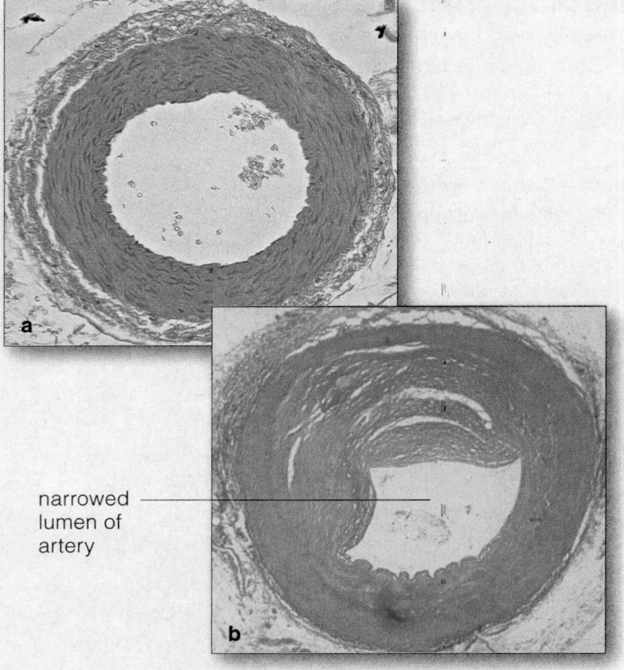
a
b
narrowed lumen of artery

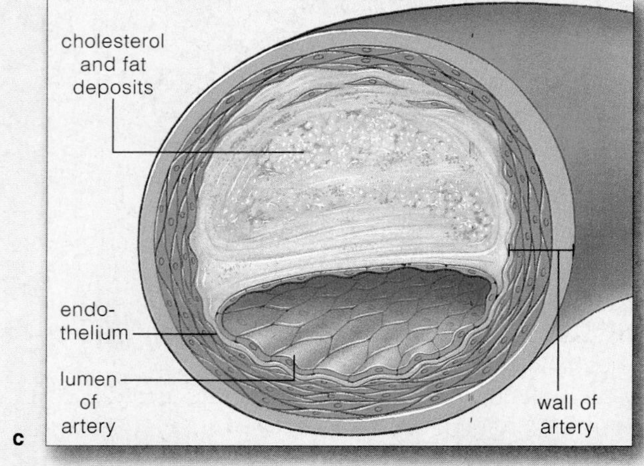
cholesterol and fat deposits
endo-thelium
lumen of artery
wall of artery
c

Figure 39.21 Sections from (**a**) a normal artery and (**b**) one with a narrowed lumen. (**c**) Atherosclerotic plaque.

Hypertension is called a "silent killer" because affected individuals may show no outward symptoms. Even when they know their blood pressure is high, some people tend to resist helpful medication, changes in diet, and regular exercise. Of 23 million hypertensive Americans, most do not seek treatment. About 180,000 die each year.

ATHEROSCLEROSIS With *arteriosclerosis*, arteries thicken and lose their elasticity. With atherosclerosis, the condition worsens as cholesterol and other lipids build up in the arterial wall and cause the lumen to narrow (Figure 39.21).

Recall, from the introduction to Chapter 16, that the liver produces enough cholesterol to satisfy the body's

Further reading: Student Guide to InfoTrac on web site –

needs. Habitually eating cholesterol-rich foods typically increases the blood cholesterol level. When circulating in blood, cholesterol is bound to proteins as *low-density* lipoproteins, or **LDLs**. Cells throughout the body have LDL receptors (Section 5.6). They take up LDL—with its cholesterol cargo—for use in their activities. They don't take up all of it. Some excess cholesterol becomes attached to proteins as *high-density* lipoproteins, or **HDLs**, and is transported back to the liver, where it is metabolized.

For a variety of reasons, not enough LDL is removed from the blood of some people. As the blood level of LDL increases, so does the risk of atherosclerosis. With their bound cholesterol, LDLs infiltrate the walls of arteries. In these walls, abnormal smooth muscle cells multiply, connective tissue components increase in mass, and cholesterol accumulates in endothelial cells and the clefts between them. Calcium deposits actually form microscopic slivers of bone on top of the lipids! Then a fibrous net forms over the entire mass. This *atherosclerotic plaque* sticks out into the arterial lumen (Figure 39.21*c*).

The bony slivers of the plaque shred the endothelium. Platelets gather at the damaged site and, in the manner described in Section 39.10, they secrete chemicals that initiate clot formation. The condition worsens as fatty globules in the plaque become oxidized. Many globules take on a form that resembles the surface components of common bacteria—including a type that instigates the formation of bonelike calcium deposits in the lungs. As an awful consequence, the call goes out to bacteria-fighting monocytes. Soon an inflammatory response is under way, and certain chemicals released during the fray activate the genes for bone formation. Normally, this is a good thing; it helps wall off invaders and prevents infection from spreading. It is bad news for arteries.

As plaques and clots grow, they can narrow or block an artery. Blood flow to tissues that the artery services may dwindle to a trickle or stop. A clot that stays in place is a *thrombus*. When it becomes dislodged and travels the bloodstream, it is an *embolus*. Think of coronary arteries, which have narrow diameters. They are highly vulnerable to clogging by a plaque or clot. When they narrow to one-quarter of their former diameter, the outcome ranges from *angina pectoris*, or mild chest pains, to a heart attack.

Physicians use *stress electrocardiograms* to diagnose atherosclerosis in coronary arteries. These are recordings of electrical activity of the cardiac cycle made as a person exercises on a treadmill. They also diagnose the condition by *angiography*. By this procedure, an opaque dye injected into blood reveals blockages of blood flow in x-ray film.

Surgery may alleviate severe blockage. With *coronary bypass surgery*, a section of an artery from the chest is stitched to the aorta and to the coronary artery below the narrowed or blocked region, as shown in Figure 39.22. With *laser angioplasty*, laser beams directed at the plaques vaporize them. With *balloon angioplasty*, a small balloon inflated within a blocked artery breaks up the plaques.

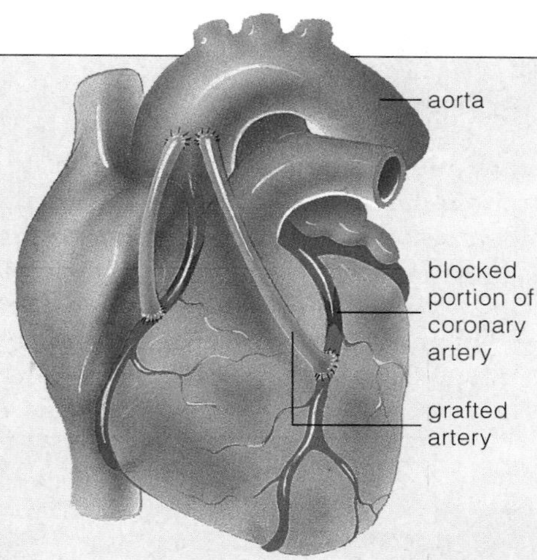
aorta
blocked portion of coronary artery
grafted artery

Figure 39.22 Two coronary bypasses (color-coded *green*), which extend from the aorta around two clogged portions of coronary arteries.

ARRHYTHMIAS ECGs can be used to detect *arrhythmias*, or irregular heart rhythms (Figure 39.23). Arrhythmias are not always a sign of abnormal conditions. Endurance athletes, for example, may have a below-average resting cardiac rate, or *bradycardia*. As an adaptation to ongoing strenuous exercise, their nervous system has adjusted the cardiac pacemaker's rate of contraction downward. Exercise or stress often causes 100+ heartbeats a minute, or *tachycardia*.

Atrial fibrillation, an irregular heartbeat, affects more than 10 percent of the elderly as well as young people with heart diseases. Coronary occlusions or some other disorder may cause irregular rhythms that rapidly lead to a dangerous condition called ventricular fibrillation. In portions of the ventricles, cardiac muscle haphazardly contracts and blood pumping falters. Within seconds, the individual loses consciousness, which might signify impending death. A strong electric shock to the chest may restore normal cardiac function.

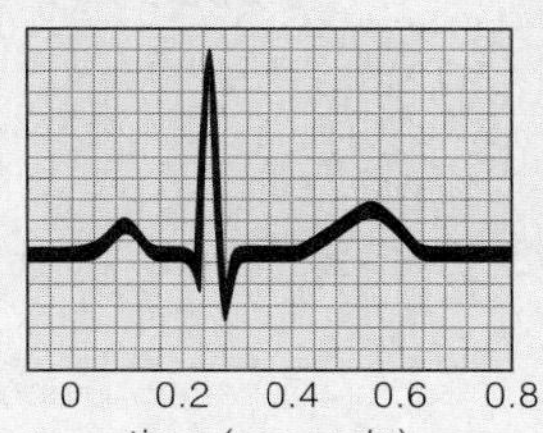
0
0.2
0.4
0.6
0.8
a
time (seconds)

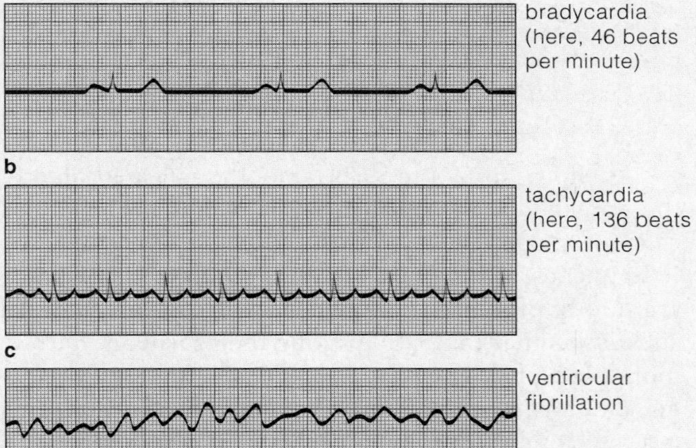
bradycardia (here, 46 beats per minute)
b
tachycardia (here, 136 beats per minute)
c
ventricular fibrillation
d

Figure 39.23 (**a**) ECG of a single, normal human heartbeat. (**b–d**) Three examples of recordings of arrhythmias.

39.10 HEMOSTASIS

Small blood vessels are vulnerable to ruptures, cuts, and similar injuries. **Hemostasis**, a process involving blood vessel spasm, platelet plug formation, and blood coagulation, may repair the damage and stop blood loss.

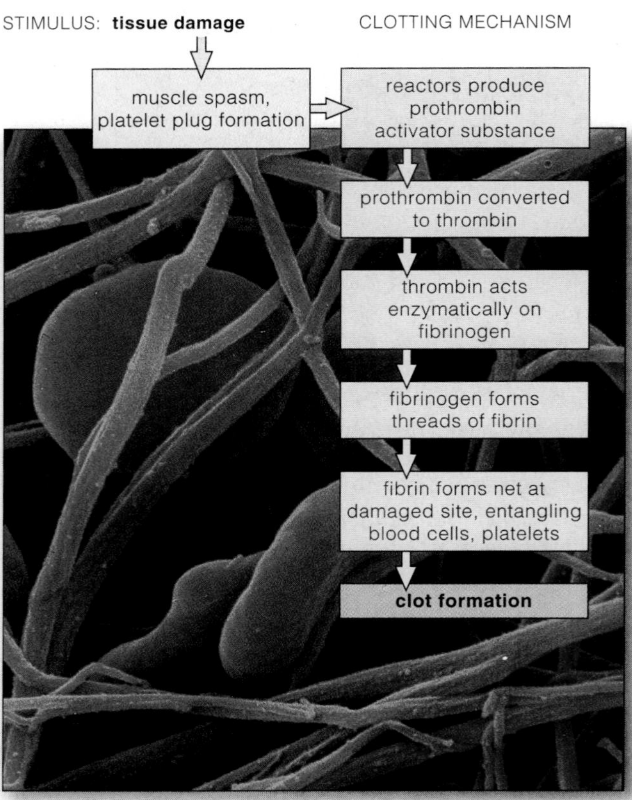

STIMULUS: tissue damage
CLOTTING MECHANISM
muscle spasm, platelet plug formation
reactors produce prothrombin activator substance
prothrombin converted to thrombin
thrombin acts enzymatically on fibrinogen
fibrinogen forms threads of fibrin
fibrin forms net at damaged site, entangling blood cells, platelets
clot formation

Figure 39.24 Plasma proteins involved in clot formation. Blood coagulates when damage exposes the collagen fibers in blood vessel walls. Exposure initiates reactions that cause rod-shaped plasma proteins (fibrinogens) to stick together as long, insoluble threads. These adhere to the exposed collagen and form a net that traps blood cells and platelets. The entire mass is a clot.

As shown in Figure 39.24, smooth muscle located in the damaged wall contracts in an automatic response called a spasm. The vasoconstriction is a temporary fix; it staunches blood flow. Platelets clump together as a temporary plug in the damaged wall. They also release substances that can help prolong the spasm and attract more platelets. Then blood coagulates, or converts to a gel, and forms a clot. Finally the clot retracts and forms a compact mass, and the breach in the wall is sealed.

The body routinely repairs damage to small blood vessels and prevents blood loss. The repair process involves blood vessel spasm, platelet plug formation, and coagulation.

39.11 LYMPHATIC SYSTEM

We conclude this chapter with a brief section on the manner in which the lymphatic system supplements blood circulation. Think of this section as a bridge to the next chapter, on immunity, for the lymphatic system also helps defend the body against injury and attack. The system is composed of drainage vessels, lymphoid organs, and lymphoid tissues. When tissue fluid moves into the vessels, we call this lymph.

Lymph Vascular System

The portion of the lymphatic system called the **lymph vascular system** consists of many tubes that collect and transport water and solutes from interstitial fluid to ducts of the circulatory system. Its main components are lymph capillaries and lymph vessels (Figure 39.25).

The lymph vascular system serves three functions. First, its vessels are drainage channels for water and plasma proteins that have leaked out from the blood at capillary beds and that must be delivered back to the blood circulation. Second, the system also takes up fats that the body has absorbed from the small intestine and delivers them to the general circulation, in the manner described in Section 39.5. Third, it delivers pathogens, foreign cells and material, and cellular debris from the

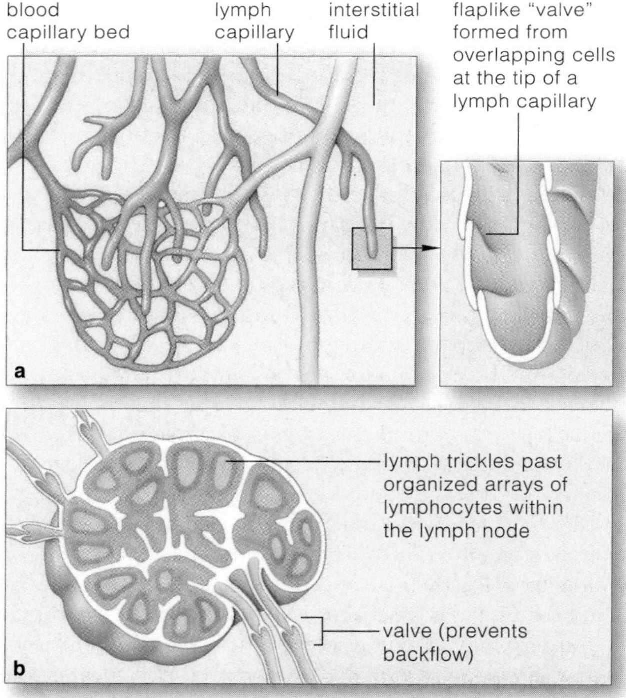

blood capillary bed
lymph capillary
interstitial fluid
flaplike "valve" formed from overlapping cells at the tip of a lymph capillary
a
lymph trickles past organized arrays of lymphocytes within the lymph node
valve (prevents backflow)
b

Figure 39.25 (**a**) Diagram of lymph capillaries at the start of a drainage network, the lymph vascular system. (**b**) Cutaway view of a lymph node. Its inner compartments are packed with organized arrays of infection-fighting white blood cells.

Further reading: Student Guide to InfoTrac on web site

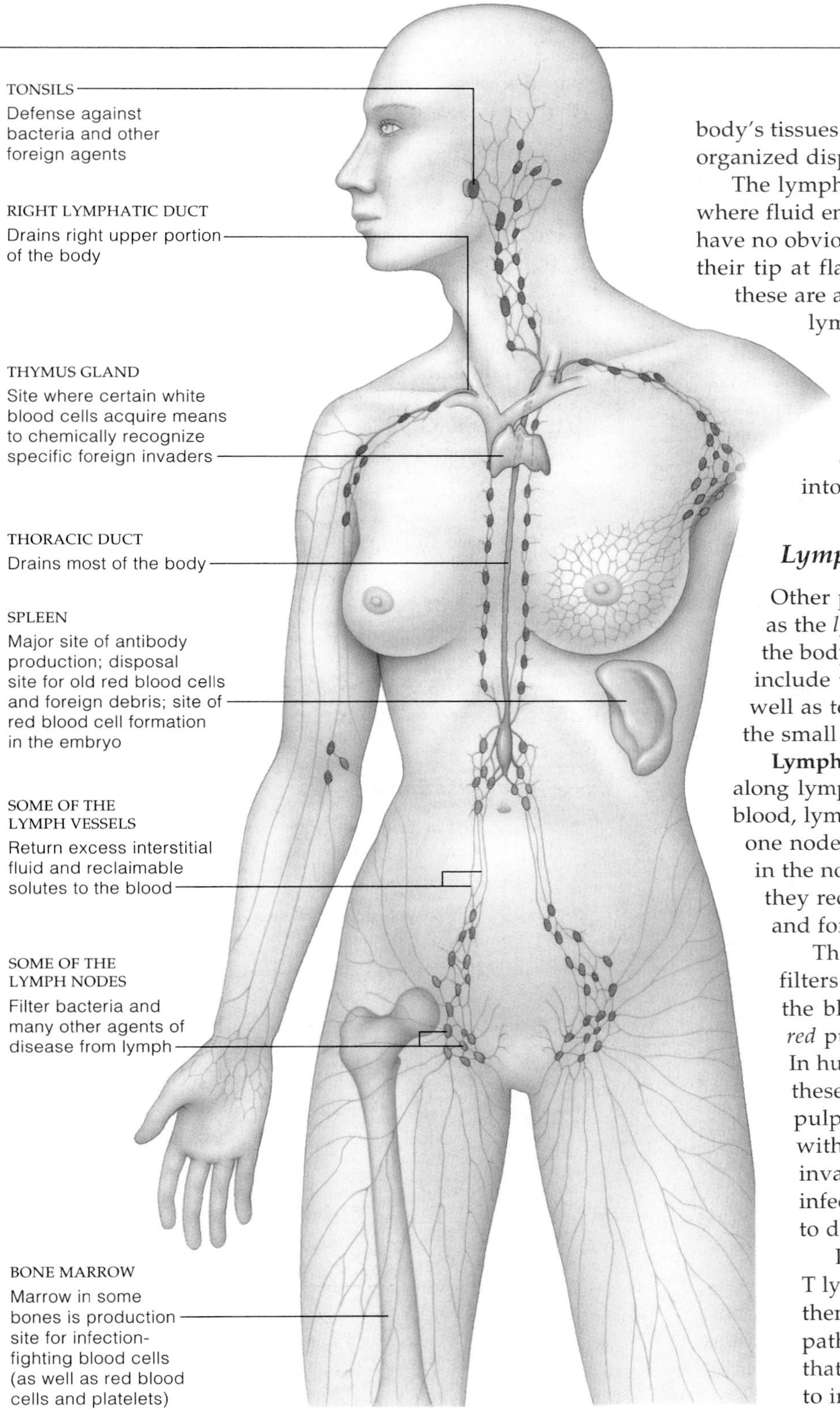

TONSILS
Defense against bacteria and other foreign agents
RIGHT LYMPHATIC DUCT
Drains right upper portion of the body
THYMUS GLAND
Site where certain white blood cells acquire means to chemically recognize specific foreign invaders
THORACIC DUCT
Drains most of the body
SPLEEN
Major site of antibody production; disposal site for old red blood cells and foreign debris; site of red blood cell formation in the embryo
SOME OF THE LYMPH VESSELS
Return excess interstitial fluid and reclaimable solutes to the blood
SOME OF THE LYMPH NODES
Filter bacteria and many other agents of disease from lymph
BONE MARROW
Marrow in some bones is production site for infection-fighting blood cells (as well as red blood cells and platelets)

Figure 39.26 Components of the human lymphatic system and their functions.

The small *green* ovals represent some of the major lymph nodes. Not shown are the patches of lymphoid tissue in the small intestine and in the appendix, which also are part of the lymphatic system.

body's tissues to the lymph vascular system's efficiently organized disposal centers, the lymph nodes.

The lymph vascular system starts at capillary beds, where fluid enters the lymph capillaries. The capillaries have no obvious entrance; water and solutes move into their tip at flaplike "valves." As Figure 39.25*a* shows, these are areas where endothelial cells overlap. The lymph capillaries merge into lymph vessels, which have a larger diameter. The lymph vessels contain some smooth muscle in their wall as well as valves in their lumen that prevent backflow. They converge into collecting ducts that drain into veins in the lower neck (Figure 39.26).

Lymphoid Organs and Tissues

Other portions of the lymphatic system, known as the *lymphoid* organs and tissues, are central to the body's defenses against injury or attack. They include the lymph nodes, spleen, and thymus, as well as tonsils and patches of tissue in the wall of the small intestine and appendix.

Lymph nodes are strategically located at intervals along lymph vessels (Figure 39.26). Before entering blood, lymph is filtered as it trickles through at least one node. Masses of lymphocytes take up stations in the nodes after forming in bone marrow. When they recognize an invader, they multiply rapidly and form large armies to destroy it.

The **spleen** is the largest lymphoid organ. It filters pathogens and used-up blood cells from the blood itself. One of its compartments, the *red* pulp, is a huge reservoir of red blood cells. In human embryos, the red pulp also produces these cells. The other compartment, the *white* pulp, has masses of lymphocytes associated with blood vessels. If and when a particular invader reaches the spleen during a severe infection, the lymphocytes become mobilized to destroy it, just as they are in lymph nodes.

It is in the **thymus gland** that immature T lymphocytes differentiate in ways that allow them to recognize and respond to particular pathogens. The thymus produces hormones that influence these vital actions. It is central to immunity, the focus of the next chapter.

The lymph vascular portion of the lymphatic system returns water and solutes from tissue fluid to blood, and delivers fats and foreign material to lymph nodes. The system's lymph nodes and other lymphoid organs help defend the body against tissue damage and infectious diseases.

39.12 SUMMARY

1. The closed circulatory system of humans and other vertebrates consists of a heart (muscular pump), many blood vessels (arteries, arterioles, capillaries, venules, and veins), and blood. The system functions in the rapid internal transport of substances to and from cells.

2. Blood helps maintain favorable conditions for cells. This fluid connective tissue consists of plasma, red and white blood cells, and platelets. It delivers oxygen and other substances to the interstitial fluid around cells. It also picks up cell products and wastes from that fluid.

a. Plasma, the liquid portion of blood, is a transport medium for blood cells and platelets. Plasma is a solvent for plasma proteins, simple sugars, lipids, amino acids, mineral ions, vitamins, hormones, and several gases.

b. Red blood cells transport oxygen from the lungs to all body regions. They are packed with hemoglobin, an iron-containing pigment that binds reversibly with oxygen. In addition, red blood cells also transport some carbon dioxide from interstitial fluid to the lungs.

c. Phagocytic white blood cells cleanse tissues by engulfing dead cells, cellular debris, and anything else recognized as not belonging to the body. White blood cells called lymphocytes form great armies that destroy specific bacteria, viruses, and other disease agents.

3. The heart of birds and mammals is an incessantly beating double pump. Each half of the heart has two chambers, called an atrium and a ventricle. Blood flows into atria, then ventricles, then into the great arteries. One-way valves enforce the forward-directed flow.

4. In birds and mammals, blood flows in two separate circuits, one pulmonary and the other systemic.

a. The pulmonary circuit loops between the heart and lungs. Oxygen-poor blood from the systemic veins enters the heart's *right* atrium, is pumped through pulmonary arteries to the two lungs, picks up oxygen, then flows through pulmonary veins to the heart's left atrium.

b. The systemic circuit loops between the heart and all body tissues. Oxygenated blood in the *left* atrium flows into the left ventricle, is pumped into the aorta, then is distributed to capillary beds. There, the blood gives up oxygen and picks up carbon dioxide. Systemic veins return the blood to the heart's right atrium.

5. Ventricular contraction drives blood through both circuits. Blood pressure is high in contracting ventricles. It drops successively in arteries, arterioles, capillaries, venules, and veins. It is lowest in relaxed atria.

a. Arteries are rapid-transport vessels and a pressure reservoir; they smooth out pressure changes resulting from heartbeats and thereby smooth out blood flow.

b. Arterioles are vessels with roles in controlling the flow volume through different body regions.

c. Capillary beds are diffusion zones where blood and interstitial fluid exchange substances.

d. Venules overlap capillaries in function. Veins are rapid-transport vessels and a blood volume reservoir that is tapped to adjust flow volume back to the heart.

6. The cardiac conduction system serves as the basis for the heart's rhythmic, spontaneous contractions.

a. One percent of the cardiac muscle cells are not contractile. They are specialized to initiate and distribute action potentials, independently of the nervous system. The nervous system only adjusts the rate and strength of the basic contractions; it does not initiate them.

b. The system's SA node fires action potentials the fastest and is the cardiac pacemaker. Waves of excitation starting here wash over the atria, then down the heart's partition, then up the ventricles. The ventricles contract in a wringing motion that ejects blood from the heart.

7. The lymphatic system has these functions:

a. Its vascular portion takes up water and plasma proteins that seep out of blood capillaries, then returns them to circulating blood. It transports absorbed fats and delivers pathogens and foreign material to disposal centers. Lymph capillaries and vessels are components.

b. Its lymphoid organs and tissues have production centers for lymphocytes. Some are battlegrounds where organized arrays of lymphocytes fight disease agents.

Review Questions

1. Define the functions of the circulatory system and lymphatic system. Distinguish between blood and interstitial fluid. *39.1*

2. Distinguish between systemic and pulmonary circuits. *39.1*

3. Describe the cellular components of blood. Describe the plasma portion of blood. *39.2*

4. State the functions of arteries, arterioles, capillaries, veins, venules, and lymph vessels. *39.1, 39.5, 39.7, 39.8*

5. Distinguish between the functions of the human heart's atria and ventricles. Then label the heart's components: *39.6*

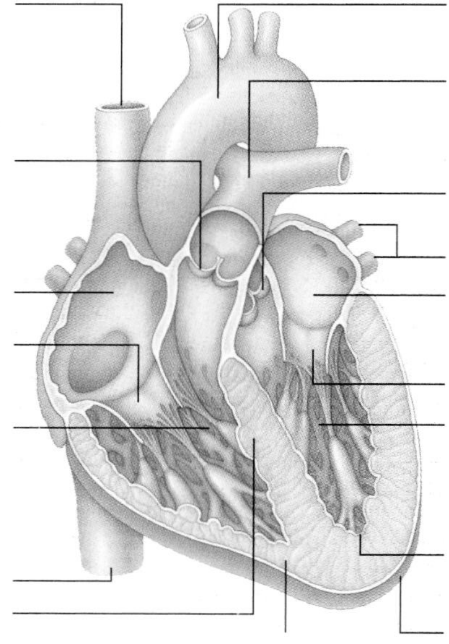

Self-Quiz (*Answers in Appendix III*)

1. Cells directly exchange substances with ______.
   a. blood vessels   c. interstitial fluid
   b. lymph vessels   d. both a and b

2. Which are *not* components of blood?
   a. plasma
   b. blood cells and platelets
   c. gases and other dissolved substances
   d. All of the above are components of blood.

3. The ______ produces red blood cells, which transport ______ and some ______.
   a. liver; oxygen; mineral ions
   b. liver; oxygen; carbon dioxide
   c. bone marrow; oxygen; hormones
   d. bone marrow; oxygen; carbon dioxide

4. The ______ produces white blood cells, which function in ______ and ______.
   a. liver; oxygen transport; defense
   b. lymph nodes; oxygen transport; pH stabilization
   c. bone marrow; housekeeping; defense
   d. bone marrow; pH stabilization; defense

5. In the pulmonary circuit, the heart's ______ half pumps blood to lungs, then ______ blood flows to the heart.
   a. right; oxygen-poor   c. right; oxygen-rich
   b. left; oxygen-poor   d. left; oxygen-rich

6. In the systemic circuit, the heart's ______ half pumps ______ blood to all body regions.
   a. right; oxygen-poor   c. right; oxygen-rich
   b. left; oxygen-poor   d. left; oxygen-rich

7. Blood pressure is high in ______ and lowest in ______.
   a. arteries; veins   c. arteries; ventricles
   b. arteries; relaxed atria   d. arterioles; veins

8. ______ contraction drives blood through the pulmonary circuit and the systemic circuit; blood pressure is highest in contracting ______.
   a. Atrial; ventricles   c. Ventricular; arteries
   b. Atrial; atria   d. Ventricular; ventricles

9. Which is *not* a function of the lymphatic system?
   a. delivers disease agents to disposal centers
   b. produces lymphocytes
   c. delivers oxygen to cells
   d. returns water and plasma proteins to blood

10. Match the type of blood vessel with its major function.
    ____ arteries   a. diffusion
    ____ arterioles   b. control of blood volume distribution
    ____ capillaries   c. transport, blood volume reservoirs
    ____ venules   d. overlap capillary function
    ____ veins   e. transport, pressure reservoirs

11. Match the components with their most suitable description.
    ____ capillary beds   a. two atria, two ventricles
    ____ lymph vascular system   b. bathes body's living cells
    ____ human heart chambers   c. driving force for blood
    ____ blood   d. zones of diffusion
    ____ heart contractions   e. starts at capillary beds
    ____ interstitial fluid   f. fluid connective tissue

Critical Thinking

1. Shirelle, who is using a light microscope to examine a human tissue specimen, sees red blood cells moving single file through thin-walled tubes. She makes a photomicrograph (Figure 39.27). What type of blood vessel has she captured on film?

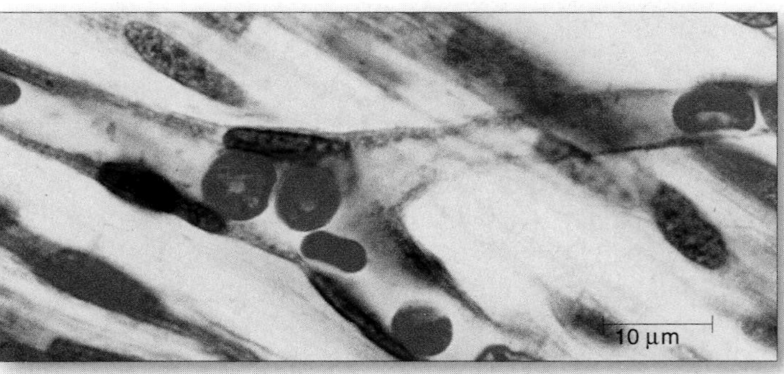
10 µm

Figure 39.27 Light micrograph of a branching blood vessel.

2. In individuals who have weak or leaky valves in their veins, fluid pressure associated with the backflow of blood causes venous walls below the valves to bulge outward. In time, the walls become stretched and flabbily distorted, a condition called *varicose veins*. Some people are genetically predisposed to develop the condition, but the cumulative mechanical stress associated with prolonged standing, pregnancy, and aging can contribute to it. With chronic varicosity, the legs themselves become swollen. Explain why this swelling might happen. Also explain why veins close to the leg surface are more susceptible to varicosity than those deeper in the leg tissues.

3. Infection by a hemolytic bacterium (*Streptococcus pyrogenes*) may trigger an inflammation that ultimately damages valves in the heart. The disease symptoms of *rheumatic fever* follow. Explain how this disease must affect the heart's functioning and what kinds of symptoms would arise as a consequence.

Selected Key Terms

ABO blood typing *39.4*
agglutination *39.4*
anemia *39.3*
aorta *39.5*
arteriole *39.5*
artery *39.5*
baroreceptor reflex *39.7*
blood *39.1*
blood pressure *39.7*
capillary (blood) *39.1*
capillary bed *39.1*
cardiac conduction system *39.6*
cardiac cycle *39.6*
cardiac pacemaker *39.6*
cell count *39.2*
circulatory system *39.1*
HDL *39.9*
heart *39.1*
hemostasis *39.10*
interstitial fluid *39.1*
LDL *39.9*
lymph node *39.11*
lymphatic system *39.1*
lymph vascular system *39.11*
plasma *39.2*
platelet *39.2*
pulmonary circuit *39.1*
reabsorption *39.8*
red blood cell *39.2*
Rh blood typing *39.4*
spleen *39.11*
stem cell *39.2*
systemic circuit *39.1*
thymus gland *39.11*
transfusion (blood) *39.4*
ultrafiltration *39.8*
vasoconstriction *39.7*
vasodilation *39.7*
vein *39.5*
venule *39.5*
white blood cell *39.2*

Readings See also *www.infotrac-college.com*

Baraga, M. 3 May 1996. "Finding New Drugs to Treat Stroke." *Science*, 274.

Little, R., and W. Little. 1989. *Physiology of the Heart and Circulation*. Fourth edition. Chicago: Year Book Medical.

Zamir, M. September–October 1996. "Secrets of the Heart." *The Sciences*. Effects of exercise on heart function.

40 IMMUNITY

Russian Roulette, Immunological Style

Until about a century ago, smallpox epidemics swept repeatedly through the world's cities. Some outbreaks were so severe that only half of those who were stricken survived. The survivors were left with permanent scars on the face, neck, shoulders, and arms. But they seldom contracted the same disease again; they were said to be "immune" to smallpox.

No one knew what caused smallpox, but the idea of acquiring immunity was dreadfully appealing. In twelfth-century China, people in good health gambled with deliberate infections. They sought out individuals with mild cases of smallpox (who displayed only mild scarring) and then removed bits of crusts from the scars, ground them up, and inhaled the powder.

By the seventeenth century Mary Montagu, wife of the English ambassador to Turkey, was championing inoculation. She went so far as to poke bits of smallpox scabs into the skin of her own children. Other people soaked threads in fluid from smallpox sores, then poked the threads into their own skin.

Some individuals who survived the chancy practices acquired immunity to smallpox, but many developed raging infections. As if the odds were not dangerous enough, the crude inoculation procedures also invited the acquisition of several other infectious diseases.

While this immunological version of Russian roulette was going on, Edward Jenner was growing up in the countryside of England. At the time, it was common knowledge that people who contracted cowpox never got smallpox. (Cowpox is a mild disease that can be transmitted from cattle to humans.) No one thought much about this until 1796, when Jenner, by now a physician, injected material from a cowpox sore into the arm of a healthy boy. Six weeks later, after the reaction subsided, Jenner injected some material from *smallpox* sores into the boy (Figure 40.1*a*). He had hypothesized that the earlier injection might provoke immunity to

Figure 40.1 (**a**) Statue in honor of Edward Jenner's work to develop an immunization procedure against smallpox, one of the most dreaded diseases in human history. (**b**) False-color scanning electron micrograph of a white blood cell being attacked by HIV (*blue particles*), the virus that causes AIDS. Immunologists are working to develop effective weapons against this modern-day scourge.

smallpox. Fortunately he was right; the boy remained free of smallpox. Jenner had developed an effective immunization procedure against a specific pathogen.

The French mocked Jenner's procedure by calling it **vaccination**. The word literally means "encowment." Much later Louis Pasteur, an influential French chemist, developed similar immunization procedures for other diseases. In acknowledgment of Jenner's pioneering work, Pasteur called his procedures vaccinations also, and only then did the word become respectable.

By Pasteur's time, improved light microscopes were revealing diverse bacteria, fungal spores, and other previously invisible forms of life. As Pasteur himself discovered, microorganisms abound in ordinary air. Did some cause diseases? Probably. Could they drift onto food or drink and make it spoil? He demonstrated that they could and did.

Pasteur found a way to kill most of the suspected agents of disease (pathogens) in food or beverages. As he and others knew, boiling killed the suspects. He also knew that people cannot boil wine (or beer or milk, for that matter) and end up with the same beverage. He devised a way to heat beverages at a temperature low enough not to ruin them but high enough to kill most of the pathogens lurking within them. We still use his antimicrobial method, which was named pasteurization in his honor.

In the late 1870s Robert Koch, a German physician, linked a specific pathogen to a specific disease; namely, anthrax. In one experiment, Koch injected blood from animals with anthrax into healthy animals. Recipients of the injection ended up with blood that teemed with cells of the bacterium *Bacillus anthracis*—and they developed anthrax. Even more convincing, injections of bacterial cells cultured outside of the animal body also caused the disease!

Thus, by the beginning of the twentieth century, the promise of understanding the basis of infectious disease and immunity loomed large—and the battles against those diseases were about to begin in earnest. Since that time, spectacular advances in microscopy, biochemistry, and molecular biology have increased our understanding of the body's defenses. Today we have greater insight into responses to tissue damage in general and immune responses to specific pathogens or tumor cells. The responses are the focus of this chapter.

KEY CONCEPTS

1. The vertebrate body has physical, chemical, and cellular defenses against pathogenic microorganisms, malignant tumor cells, and other agents that can destroy tissues and the individual itself.

2. In the early stages of tissue invasion and damage, white blood cells and certain proteins in the plasma portion of blood escape from capillaries. They execute a rapid counterattack in response to a general alarm, not to the presence of a particular pathogen, so we call this a nonspecific inflammatory response. Phagocytic white blood cells ingest invading agents and clean up tissue debris. Plasma proteins promote phagocytosis, and some also destroy invaders directly.

3. If the invasion persists, certain white blood cells make immune responses. Those cells can chemically recognize distinct configurations on molecules that are abnormal or foreign to the body, such as those on bacterial cells and viruses. If the foreign or abnormal molecule triggers an immune response, it is called an antigen.

4. In one type of immune response, some of the white blood cells produce enormous quantities of antibodies. Antibodies are molecules that bind to a specific antigen and tag it for destruction.

5. In another type of immune response, executioner cells directly destroy body cells that have become abnormal, as by infection or by a tumor-producing process.

40.1 THREE LINES OF DEFENSE

You continually cross paths with astoundingly diverse **pathogens**—the viruses, bacteria, fungi, protozoans, parasitic worms, and other agents that cause diseases. Having coevolved with most of them, you and other vertebrates have three lines of defense, so you need not lose sleep over this. Most pathogens cannot breach the body surface. If they do, they face chemical weapons and white blood cells, or leukocytes, that attack anything perceived as foreign. Other white blood cells zero in on specific targets. Table 40.1 lists the three lines of defense.

Surface Barriers to Invasion

Most often, pathogens cannot get past skin or the other linings of body surfaces. Think of skin as a habitat of low moisture, low pH, and thick layers of dead cells. Normally harmless bacteria tolerate these conditions. Few pathogens can compete with dense populations of established types unless conditions change. Repeatedly subject your toes to warm, damp shoes, for example, and you may be inviting certain fungi to penetrate the sodden, weakened tissues and cause *athlete's foot.*

Similarly, resident bacteria on the mucous lining of the gut or vagina help protect you. For instance, lactate, a fermentation product of *Lactobacillus* populations in the vagina, helps maintain a low pH that most bacteria and fungi cannot tolerate. Barriers in branching, tubular airways leading to the lungs stop airborne pathogens. Air rushing down the tubes flings bacteria against their sticky, mucus-coated walls. In the mucus are protective substances, especially **lysozyme**. This type of enzyme digests cell walls of bacteria and invites their death. As a final touch, broomlike cilia lining the airways sweep away trapped and enzymatically whapped pathogens.

The body has additional defenses. Lysozyme and other substances in tears, saliva, and gastric fluid offer protection, as when tears give eyes a sterile washing. Urine's low pH and flushing action help bar pathogens from moving up into the urinary tract. Diarrhea swiftly flushes irritating pathogens from the gut, and blocking its action in adults may prolong infection. (However, in children especially, severe diarrhea causes dehydration. A glucose–salt solution used in oral hydration therapy can reverse the dangerous loss of water and ions.)

Table 40.1 The Vertebrate Body's Three Lines of Defense Against Pathogens

BARRIERS AT BODY SURFACES (*nonspecific* targets)
Intact skin; mucous membranes at other body surfaces
Infection-fighting chemicals (e.g., lysozyme in tears, saliva)
Normally harmless bacterial inhabitants of skin and other body surfaces that can outcompete pathogenic visitors
Flushing effect of tears, saliva, urination, and diarrhea
NONSPECIFIC RESPONSES (*nonspecific* targets)
Inflammation:
1. Fast-acting white blood cells (neutrophils, eosinophils, and basophils)
2. Macrophages (also take part in immune responses)
3. Complement proteins, blood-clotting proteins, and other infection-fighting substances
Organs with pathogen-killing functions (such as lymph nodes)
Some cytotoxic cells (e.g., NK cells) with a range of targets
IMMUNE RESPONSES (*specific* targets only)
T cells and B cells; macrophages interact with them
Communication signals (e.g., interleukins) and chemical weapons (e.g., antibodies, perforins)

Nonspecific and Specific Responses

All animals react to tissue damage. Even small aquatic invertebrates have phagocytic cells and antimicrobial substances, including lysozymes. But the more complex the animal, the more complex are the systems to defend it. Think back on the evolution of the circulatory and lymphatic systems of vertebrates that invaded the land (Section 39.1). As the circulation of body fluids became more efficient, so did the means to defend the body. Phagocytic cells and plasma proteins could be swiftly transported to tissues under attack and could intercept pathogens trickling along the vascular highways.

Vertebrates came to be equipped with sets of plasma proteins. Some proteins promote rapid clot formation after tissue damage; others destroy invading pathogens or target them for phagocytosis. In addition, exquisitely focused responses to specific dangers evolved.

In short, *internal defenses against a tremendous variety of pathogens are in place even before damage occurs*. Ready and waiting are specialized white blood cells as well as plasma proteins. They take part in *nonspecific* responses to damaged tissue in general, not to one pathogen or another. Other white blood cells may recognize unique molecular configurations of a *specific* pathogen. If they do, the resulting "immune" response will run its course whether tissues are damaged or not.

Intact skin, mucous membranes, antimicrobial secretions, and other barriers at the body surface constitute the first line of defense against invasion and tissue damage.

Inflammation and other internal, nonspecific responses to invasion are the second line of defense.

Immune responses against specific invaders, as executed by armies of white blood cells and their chemical weapons, are the third line of defense.

Further reading: Student Guide to InfoTrac on web site →

40.2 COMPLEMENT PROTEINS

A set of plasma proteins has roles in nonspecific *and* specific defenses. Collectively, they are the **complement system**. About twenty kinds circulate in the blood in inactive form. If even a few molecules of one kind are activated, they trigger cascading reactions that activate many molecules of another complement protein. Each of these activates many molecules of another kind of protein at the next reaction step, and so on. Deployment of so many molecules has the following effects.

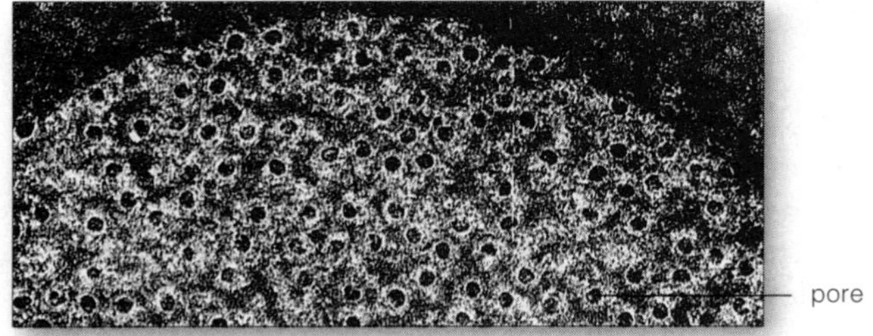

pore

Figure 40.2 Micrograph of a cell surface, showing pores formed by membrane attack complexes.

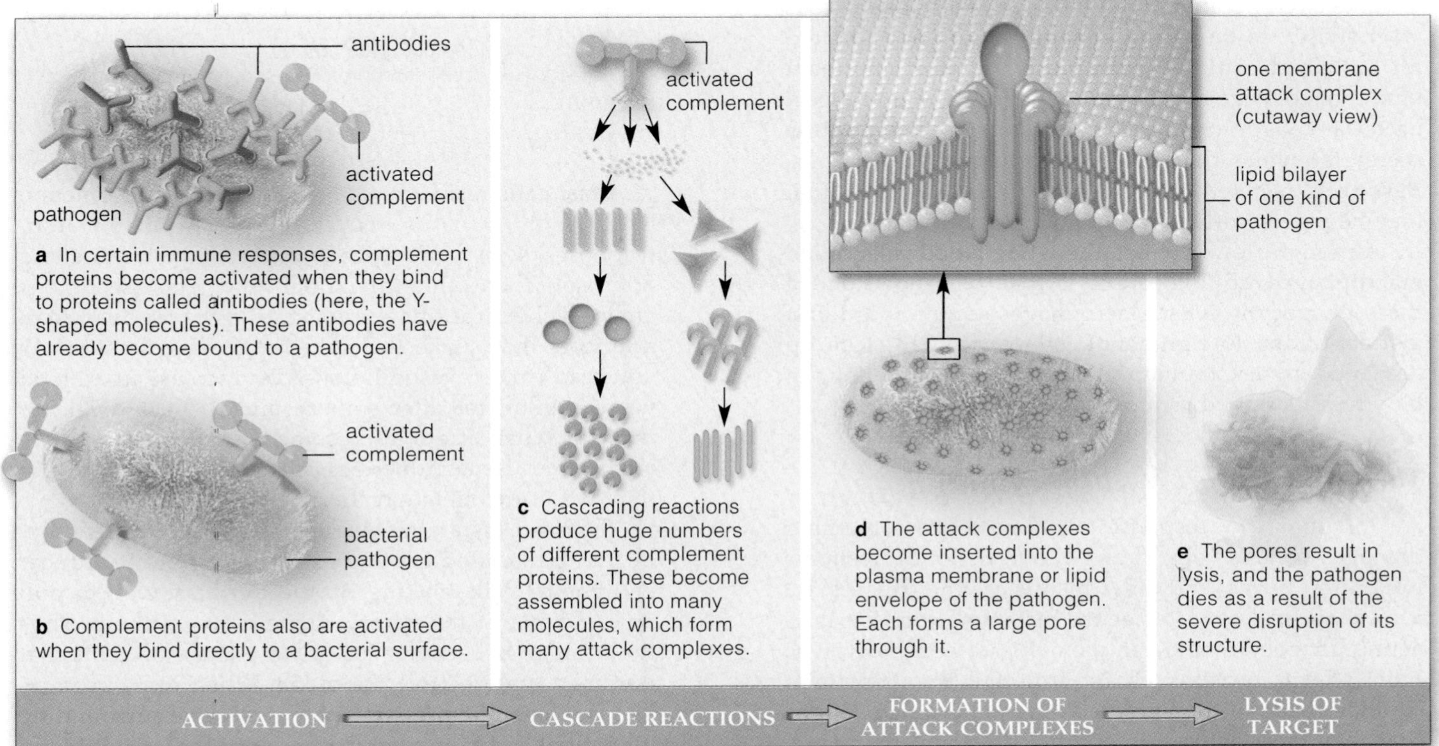

antibodies
activated complement
pathogen
a In certain immune responses, complement proteins become activated when they bind to proteins called antibodies (here, the Y-shaped molecules). These antibodies have already become bound to a pathogen.
activated complement
bacterial pathogen
b Complement proteins also are activated when they bind directly to a bacterial surface.
activated complement
c Cascading reactions produce huge numbers of different complement proteins. These become assembled into many molecules, which form many attack complexes.
one membrane attack complex (cutaway view)
lipid bilayer of one kind of pathogen
d The attack complexes become inserted into the plasma membrane or lipid envelope of the pathogen. Each forms a large pore through it.
e The pores result in lysis, and the pathogen dies as a result of the severe disruption of its structure.
ACTIVATION
CASCADE REACTIONS
FORMATION OF ATTACK COMPLEXES
LYSIS OF TARGET

Figure 40.3 Formation of membrane attack complexes. One reaction pathway starts when complement proteins bind to bacterial surfaces. Another operates in immune responses to specific invaders. Both result in membrane pore complexes that induce lysis in pathogens.

Some complement proteins join together and form pore complexes. These molecular structures have an inner channel (Figures 40.2 and 40.3). When inserted in the plasma membrane of a variety of pathogens, they form pores that lead to lysis. **Lysis**, remember, is a gross structural disruption of a plasma membrane or cell wall that results in cell death. Pore complexes also become inserted into the wall of Gram-negative bacteria. Such cell walls consist of a lipid-rich, outer surface above a peptidoglycan layer. Molecules of lysozyme can diffuse through the pores and digest the peptidoglycan.

Certain activated proteins promote inflammation, a nonspecific defense response described next. Through their cascades of reactions, they create concentration gradients that attract phagocytic white blood cells to an irritated or damaged tissue. The complement proteins also encourage phagocytes to dine. They can bind to the surface of many types of invaders. When they do so, an invader ends up having a complement "coat." The coat adheres to phagocytes, and this is rather like putting a basted turkey on the dinner table.

In such ways, the complement proteins target many bacteria, parasitic protistans, and enveloped viruses.

The complement system, a set of about twenty kinds of plasma proteins circulating in blood, takes part in cascades of reactions that help defend against many bacteria, some parasitic protistans, and enveloped viruses.

The complement system takes part in both nonspecific and specific defenses.

40.3 INFLAMMATION

The Roles of Phagocytes and Their Kin

Certain types of white blood cells take part in an initial response to tissue damage. White blood cells, recall, arise from stem cells in bone marrow. Many of the cells circulate in blood and lymph. A great many take up stations in lymph nodes as well as in the spleen, liver, kidneys, lungs, and brain. Here you may wish to scan Sections 39.2 and 39.11 once more; they introduce the components of blood and the lymphatic system.

Like SWAT teams, three kinds of white blood cells react swiftly to danger in general but are not adapted for sustained battles. **Neutrophils**, the most abundant kind, phagocytize bacteria. They ingest, kill, and digest bacterial cells into simple molecular bits. **Eosinophils** secrete enzymes that make holes in parasitic worms. **Basophils** secrete histamine and other substances that may help keep inflammation going after it starts.

Although slower to act, the white blood cells called **macrophages** are "big eaters." Figure 40.4 shows one of these phagocytic cells. Macrophages engulf and digest just about any foreign agent. They also help clean up damaged tissues. Immature macrophages circulating in blood are classified as monocytes.

The Inflammatory Response

An inflammatory response develops when something damages or kills cells of any tissue region. Infections, punctures, burns, and other insults are the triggers. By a mechanism known as **acute inflammation**, the fast-acting phagocytes and complement proteins, as well as other plasma proteins, escape from the bloodstream at capillary beds in the vulnerable tissue. Localized signs that they have entered interstitial fluid and that acute inflammation is under way include redness, swelling, heat, and pain, as listed in Table 40.2.

Table 40.2 Localized Signs of Inflammation and Their Causes

Redness	Arteriolar vasodilation increases blood flow to the affected tissue.
Warmth	Arteriolar vasodilation delivers more blood, carrying more metabolic heat, to the tissue.
Swelling	Chemical signals increase capillary permeability; plasma proteins escape and disrupt fluid balance across capillary walls; localized edema results.
Pain	Nociceptors (pain receptors) are stimulated by increased fluid pressure, local chemical signals.

Mast cells reside in connective tissues and function like basophils; that is, they take part in inflammatory responses. They synthesize and release **histamine** and other substances into interstitial fluid. Their secretions are local chemical signals, which trigger vasodilation of arterioles that snake through the damaged tissue. As you read earlier, vasodilation is an increase in a blood vessel's diameter after smooth muscle in its wall has relaxed. When arterioles become engorged with blood, an affected tissue reddens and becomes warmer owing to blood-borne metabolic heat.

Released histamine also increases the permeability of the thin-walled capillaries in the tissue. It induces endothelial cells making up the capillary wall to pull apart farther at the narrow clefts between them. Thus the capillaries become "leaky" to plasma proteins that normally do not leave the blood. When some proteins leak out, osmotic pressure increases in the surrounding interstitial fluid. In combination with the higher blood pressure brought about by the increased blood flow to the tissue, ultrafiltration increases and reabsorption

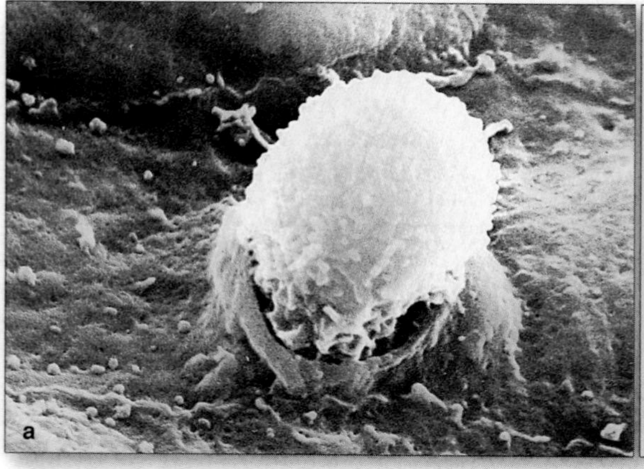
a

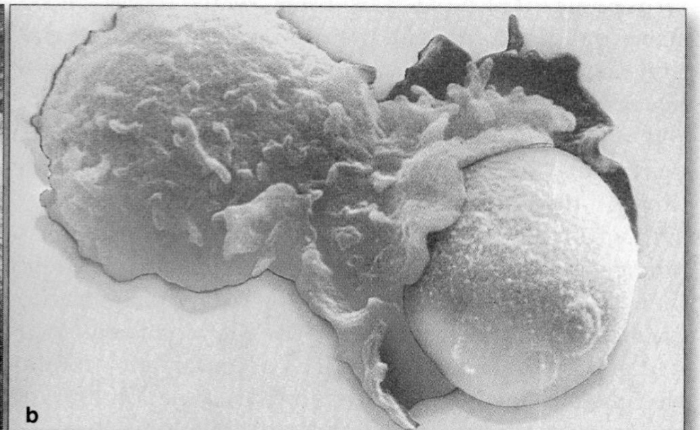
b

Figure 40.4 (**a**) White blood cell squeezing out of a blood capillary at a cleft between endothelial cells. (**b**) Macrophage about to engulf a yeast cell.

Further reading: Student Guide to InfoTrac on web site —

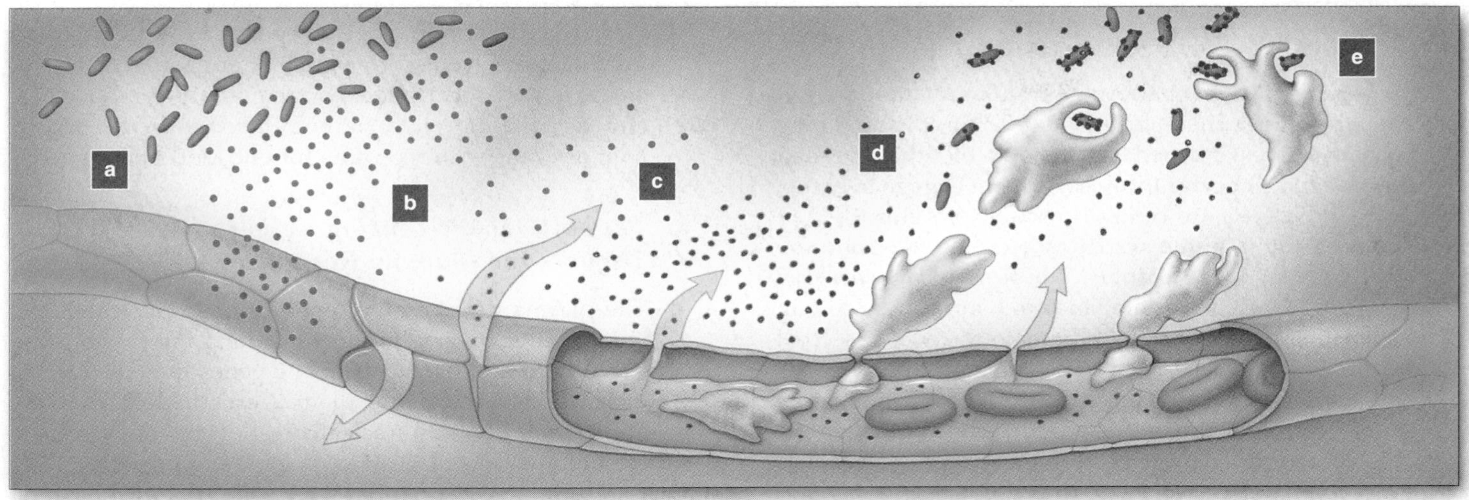
a
b
c
d
e

a Bacteria invade a tissue and directly kill cells or release metabolic products that damage tissue.

b Mast cells in tissue release histamine, which then triggers arteriolar vasodilation (hence redness and warmth) as well as increased capillary permeability.

c Fluid and plasma proteins leak out of capillaries; localized edema (tissue swelling) and pain result.

d Complement proteins attack bacteria. Clotting factors wall off inflamed area.

e Neutrophils, macrophages, and other phagocytes engulf invaders and debris. Macrophage secretions attract even more phagocytes, directly kill invaders, and call for fever and for T and B cell proliferation.

Figure 40.5 Acute inflammation in response to a bacterial invasion. The response involves delivering phagocytes and plasma proteins to the tissue. Together, these components of blood inactivate, destroy, or isolate the invaders, remove chemicals and cellular debris, and prepare the tissue for subsequent repair. These are their functions in all inflammatory responses.

decreases across the capillary wall. Localized edema is the outcome of the shift in the fluid balance across the capillary wall. (Here you may wish to review Section 39.8.) The tissue swells with fluid, and its nociceptors give rise to sensations of pain. Typically an individual avoids voluntary movements that might aggravate the pain. This behavior promotes tissue repair.

Within hours of the first physiological responses to the damage, neutrophils are squeezing across capillary walls. They swiftly go to work. Monocytes arrive later, differentiate into macrophages, and engage in sustained action (Figure 40.5). While macrophages are engulfing the invaders and debris, they secrete a number of local signaling molecules that serve as chemical mediators.

The chemical mediators known as *chemotaxins* attract more phagocytes to the tissue. One of the *interleukins* stimulates formation of B and T cell armies, as described shortly. *Lactoferrin* directly kills bacteria. *Endogenous pyrogen* may trigger the release of prostaglandins, which in turn promotes an increase in the "set point" on the hypothalamic thermostat that regulates the body's core temperature. What we call **fever** is a core temperature that has reached the higher set point.

A fever of about 39°C (100°F) is not a bad thing. It increases body temperature to a level that is "too hot" for the functioning of most pathogens. It also promotes an increase in a host's defense activities. *Interleukin-1* induces drowsiness, which reduces the body's demands for energy. Thus, more energy may be diverted to the tasks of defense and tissue repair. Macrophages take part in the cleanup and repair operations.

Among the plasma proteins that leak into the tissue are complement proteins and clotting factors of the sort described in Section 39.10. Upon exposure to chemicals that phagocytes secreted and to tissue thromboplastin, fibrin forms and clots develop in the spaces around the inflamed tissue. These clots wall off the inflamed area and typically prevent or delay the spread of invaders and toxic chemicals into the surrounding tissues. After the inflammation subsides, anticlotting factors that had also escaped from the capillaries dissolve the clots.

An inflammatory response develops in a local tissue when cells are damaged or killed, as by infection. It proceeds during both nonspecific and specific defenses of tissues.

Mast cells in damaged or invaded tissues secrete histamine, which causes arterioles to vasodilate and increases capillary permeability to fluid and plasma proteins. The localized vasodilation reddens and warms the tissue. Edema results from the fluid imbalance across the capillary wall. The tissue swelling causes pain.

The response involves phagocytes such as macrophages, which engulf invaders and debris and secrete chemical mediators. It requires plasma proteins, such as complement proteins that target invaders for destruction, as well as clotting factors that wall off the inflamed tissue.

40.4 THE IMMUNE SYSTEM

Defining Features

Sometimes physical barriers and inflammation are not enough to overwhelm an invader, so an infection may become well established. Then, white blood cells called **B** and **T lymphocytes** form armies that engage in battle.

B and T cells are central to the body's third line of defense—the immune system. We define the **immune system** by two key features. The first is *immunological specificity*, whereby certain kinds of lymphocytes zero in on specific pathogens and eliminate them. The second feature is *immunological memory*, whereby a portion of T and B cells formed during a first-time confrontation is set aside for a future battle with the same pathogen.

The operating principle for the immune system is this: *Each kind of cell, virus, and substance bears unique molecular configurations that give it a unique identity*. An individual's own cells bear certain configurations that function as *self* markers. Lymphocytes of that individual recognize the self markers and normally ignore them. They also recognize unique molecular configurations of specific foreign agents. When they identify these *nonself* markers, they are stimulated to divide again and again, by way of mitosis, and form huge populations.

As the divisions proceed, subpopulations of the new cells become specialized to respond to the foreign agent in different ways. Some consist of *effector* cells—fully differentiated cells that engage and destroy the enemy. Other subpopulations are *memory* cells. These enter a resting phase. Instead of taking part in the attack on the agent that triggered the initial response, memory cells "remember" it. If the same kind of agent shows up again, they will join a larger, more rapid response to it.

Any molecular configuration that triggers formation of lymphocyte armies and is their target is an **antigen**. The most important antigens are certain proteins at the surface of pathogens and tumor cells. As you will see, lymphocytes synthesize molecules of receptors that are able to bind to the configurations. Binding is the means by which lymphocytes recognize nonself.

In short, immunological specificity and memory involve three events: *recognition* of antigen, *repeated cell divisions* that form huge populations of lymphocytes, and *differentiation* into subpopulations of effector and memory cells with receptors for one kind of antigen.

The Antigen-Presenting Cells—Triggers for Immune Responses

The plasma membrane of every human nucleated cell incorporates diverse proteins. Among the proteins are **MHC markers**, named after the genes that encode the instructions for making them. Certain MHC markers are common to each nucleated cell in the body. Others are unique to its macrophages and lymphocytes.

Think of what happens after a cut allows bacteria to enter a tissue inside your finger. Lymph vessels pick up some of the inflamed tissue's interstitial fluid and deliver it, along with some invading cells, to the lymph nodes in the vicinity. There, macrophages join the fray. They engulf foreign cells and enclose them in vesicles with digestive enzymes that cleave antigen molecules into fragments. The fragments bind to MHC molecules and form **antigen–MHC complexes**. Later, vesicles fuse with the plasma membrane. This means the complexes are automatically displayed at the macrophage surface.

Any cell displaying processed antigen that is bound with a suitable MHC molecule is an **antigen-presenting cell**. When lymphocytes do encounter such a cell, they take notice (Figure 40.6). *This is the antigen recognition that promotes the divisions that form lymphocyte armies.*

Key Players in Immune Responses

The same categories of white blood cells are called into action during each immune response. Figure 40.7 is an overview of how the cells interact. In brief, recognition of antigen–MHC complexes activates **helper T cells**. These produce and secrete substances that induce any responsive T or B lymphocyte to divide and give rise to large populations of effector cells and memory cells. Recognition also activates **cytotoxic T cells**, which can eliminate infected body cells or tumor cells by "touch killing." When they contact a target, they deliver

MHC marker designating "self" (present only at the surface of the body's own cells)

T cells and B cells ignore this.

Processed antigen bound to MHC marker at surface of an antigen-presenting cell

T cell recognition initiates immune response.

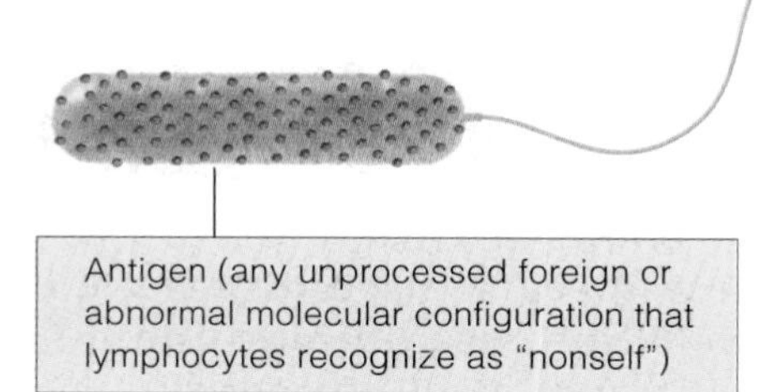

Antigen (any unprocessed foreign or abnormal molecular configuration that lymphocytes recognize as "nonself")

B cell recognition initiates immune response.

Figure 40.6 Molecular cues that T and B cells either ignore or recognize as a signal to initiate immune responses.

Further reading: Student Guide to InfoTrac on web site

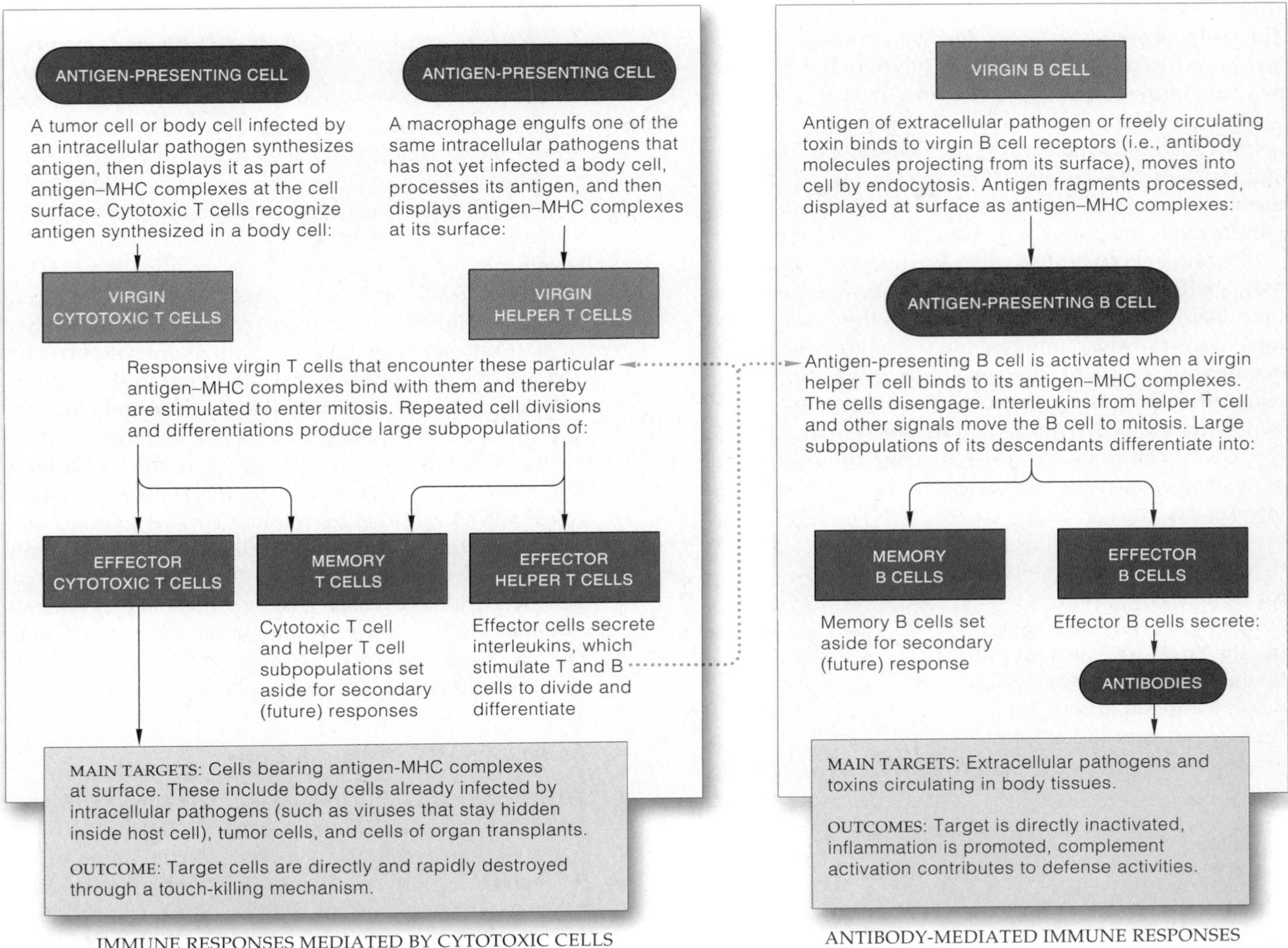

ANTIGEN-PRESENTING CELL
A tumor cell or body cell infected by an intracellular pathogen synthesizes antigen, then displays it as part of antigen–MHC complexes at the cell surface. Cytotoxic T cells recognize antigen synthesized in a body cell:
ANTIGEN-PRESENTING CELL
A macrophage engulfs one of the same intracellular pathogens that has not yet infected a body cell, processes its antigen, and then displays antigen–MHC complexes at its surface:
VIRGIN CYTOTOXIC T CELLS
VIRGIN HELPER T CELLS
Responsive virgin T cells that encounter these particular antigen–MHC complexes bind with them and thereby are stimulated to enter mitosis. Repeated cell divisions and differentiations produce large subpopulations of:
EFFECTOR CYTOTOXIC T CELLS
MEMORY T CELLS
EFFECTOR HELPER T CELLS
Cytotoxic T cell and helper T cell subpopulations set aside for secondary (future) responses
Effector cells secrete interleukins, which stimulate T and B cells to divide and differentiate
MAIN TARGETS: Cells bearing antigen-MHC complexes at surface. These include body cells already infected by intracellular pathogens (such as viruses that stay hidden inside host cell), tumor cells, and cells of organ transplants.
OUTCOME: Target cells are directly and rapidly destroyed through a touch-killing mechanism.
IMMUNE RESPONSES MEDIATED BY CYTOTOXIC CELLS
VIRGIN B CELL
Antigen of extracellular pathogen or freely circulating toxin binds to virgin B cell receptors (i.e., antibody molecules projecting from its surface), moves into cell by endocytosis. Antigen fragments processed, displayed at surface as antigen–MHC complexes:
ANTIGEN-PRESENTING B CELL
Antigen-presenting B cell is activated when a virgin helper T cell binds to its antigen–MHC complexes. The cells disengage. Interleukins from helper T cell and other signals move the B cell to mitosis. Large subpopulations of its descendants differentiate into:
MEMORY B CELLS
EFFECTOR B CELLS
Memory B cells set aside for secondary (future) response
Effector B cells secrete:
ANTIBODIES
MAIN TARGETS: Extracellular pathogens and toxins circulating in body tissues.
OUTCOMES: Target is directly inactivated, inflammation is promoted, complement activation contributes to defense activities.
ANTIBODY-MEDIATED IMMUNE RESPONSES

Figure 40.7 Overview of key interactions among B and T lymphocytes during an immune response. Most often, both types of white blood cells are activated when antigen has been detected. An antigen is any large molecule that lymphocytes recognize as not being "self" (normal body molecules). A first-time encounter with antigen elicits a *primary* response. A subsequent encounter with the same type of antigen elicits a *secondary* immune response. A secondary response is larger and more rapid. Memory cells that formed but that were not used during the first battle immediately engage in the second one.

cell-killing chemicals into it. By contrast, B cells make antigen-binding receptor molecules called **antibodies**. When a response is under way, effector B cells secrete staggering numbers of antibody molecules. Only B cells are the basis of *antibody-mediated* responses.

Control of Immune Responses

Antigen provokes an immune response—and removal of antigen stops it. For example, by the time the tide of battle turns, effector cells and their chemical secretions have already killed most of the antigen-bearing agents inside the body. With fewer antigen molecules around to stimulate the cells, the response declines, then stops. As another example, inhibitory signals from cells with suppressor roles help shut down immune responses.

Antigens are any nonself molecular configurations which, when recognized by certain lymphocytes, trigger immune responses. Helper T cells, cytotoxic T cells, B cells, and their secretions execute these responses.

Following antigen recognition, large T and B cell armies form by repeated mitotic cell divisions. These differentiate into subpopulations of effector cells and memory cells, all of which are sensitized to that one kind of antigen.

40.5 LYMPHOCYTE BATTLEGROUNDS

The antigen-presenting cells and lymphocytes we have just introduced interact inside of lymphoid organs that promote immune responses (Figures 39.11 and 40.8).

Think about the tonsils or a lymph nodule located below mucous membranes of the respiratory, digestive, and reproductive systems. Just after invaders penetrate the body's surface barriers, antigen-presenting cells and lymphocytes inside these nodules intercept them.

Or think about antigen dissolved in interstitial fluid that is entering the lymph vascular system. Because the fluid inside lymph vessels drains into the expressways for blood transport, antigen could become distributed throughout the body. However, before the antigen can reach the blood, it must trickle through lymph nodes, which are packed with defending cells. Even in those few cases where antigen does manage to enter blood, defending cells inside the spleen intercept it.

Inside the lymph nodes, the defending cells are organized for utmost effectiveness. The antigen-presenting cells make up the front line and engulf the invaders. They process and display antigen, thus calling lymphocyte comrades into action.

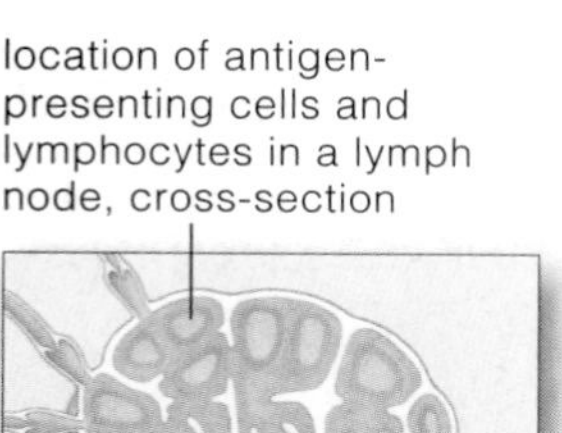

location of antigen-presenting cells and lymphocytes in a lymph node, cross-section

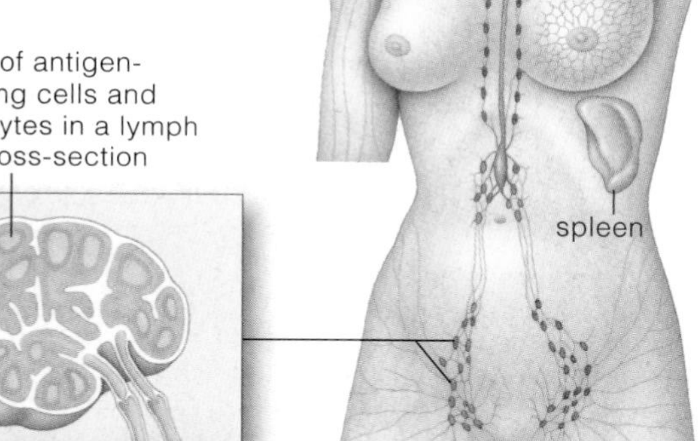

tonsils
thymus gland
spleen

Figure 40.8 Organized arrays of antigen-presenting cells and lymphocytes in lymph nodes.

The cell divisions that produce subpopulations of effector and memory cells proceed in lymph nodes. As lymph drains through, it moves the effector activities to the back of the organ and beyond. And all the while, virgin and memory cells circulate through the lymph node, reconnoitering at the front line.

Antigen-presenting cells and lymphocytes intercept and battle pathogens in organized ways within lymphoid organs and tissues, especially the lymph nodes.

40.6 CELL-MEDIATED RESPONSES

T Cell Formation and Activation

Consider first the functions of T lymphocytes during an immune response. As you read in Section 39.2, T cells arise from stem cells in bone marrow. However, they do not fully develop in bone marrow. Rather, they travel to an organ called the thymus, where they become fully differentiated into helper T cells and cytotoxic T cells. Specifically, these immature cells acquire their **TCRs** (short for *T*-Cell *R*eceptors) inside the thymus. Bristling with receptors, the cells depart from the thymus. They circulate in blood or take up stations in lymph nodes and the spleen as virgin T cells. In this context, "virgin" means the cells are as yet undisturbed by antigen.

The TCRs of virgin T cells ignore unadorned MHC markers on the body's cells. They ignore free antigen. *But they recognize and bind with antigen–MHC complexes at the surface of antigen-presenting cells.* As Figure 40.9 shows, binding induces the T cells to divide repeatedly and give rise to large clones. (A clone is a population of genetically identical cells.) Then the clonal descendants differentiate into subpopulations of effector cells and memory cells. *And every one of those descendants has the same TCR for one kind of antigen–MHC complex.*

Functions of Effector T Cells

What actions do subpopulations of effector T cells take? Effector helper T cells secrete interleukins, which fan repeated mitotic cell divisions and differentiation of any responsive T and B cell. Figure 40.9 (and Figure 40.10 in the next section) show this action. Effector cytotoxic T cells are killers turned on by antigen–MHC complexes at the surface of tumor cells and body cells already infected with intracellular pathogens or viruses. Such complexes are a "double signal" that tells the killers to attack the cell bearing it (Figure 40.9*e*).

Effector cytotoxic T cells destroy infected cells with a touch-kill mechanism. They secrete *perforins*, protein molecules that form doughnut-shaped pores in a target cell's plasma membrane. The pores look similar to the ones shown in Figure 40.2. These cells also secrete the chemicals that induce cell death by way of **apoptosis**. As Section 15.6 explains, the target "commits suicide." Its cytoplasm dribbles out, its organelles are disrupted, and its DNA becomes fragmented. The cytotoxic T cell, having made its lethal hit, quickly disengages from the doomed target and resumes its reconnoitering.

Cytotoxic T cells also contribute to the rejection of tissue and organ transplants. Parts of MHC markers on donor cells are different enough from the recipient's to be recognized as antigens. But other parts are similar enough to complete the double signal. MHC typing and matching donors to recipients minimize the risk.

Further reading: Student Guide to InfoTrac on web site

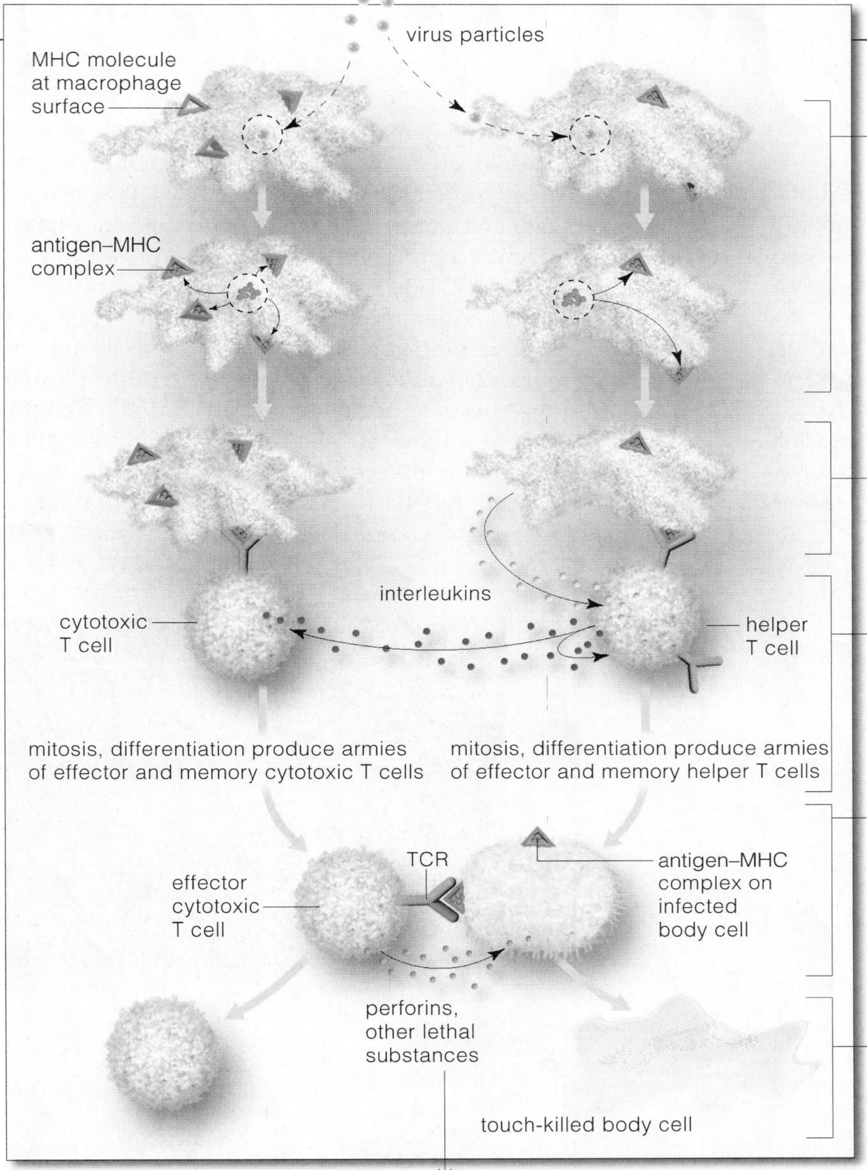
virus particles
MHC molecule at macrophage surface
antigen–MHC complex
interleukins
cytotoxic T cell
helper T cell
mitosis, differentiation produce armies of effector and memory cytotoxic T cells
mitosis, differentiation produce armies of effector and memory helper T cells
TCR
effector cytotoxic T cell
antigen–MHC complex on infected body cell
perforins, other lethal substances
touch-killed body cell

a A virus inserts its genetic material into a macrophage. The host cell's pirated metabolic machinery synthesizes viral proteins, which are antigenic. These proteins are processed into fragments, bound to MHC molecules, and displayed at the cell surface. TCRs of cytotoxic T cells can recognize antigen synthesized inside a host cell.

Another macrophage engulfs a particle of the same virus and encloses it in an endocytic vesicle. Digestive enzymes cleave the particle's antigen into fragments. These bind to MHC molecules and are presented at the macrophage surface. *TCRs of helper T cells also recognize engulfed and processed antigen.*

b A responsive T cell binds with the antigen–MHC complexes. Binding stimulates the macrophage to secrete interleukins (*yellow* dots).

c These communication signals stimulate the helper T cell to secrete different kinds of interleukins (*blue* dots). The new signals stimulate cell divisions and differentiations by which large populations of effector T cells and memory T cells form.

d The same virus penetrated a cell in the lining of the respiratory tract, and antigen synthesized in the cell's cytoplasm is processed. Antigen becomes bound to MHC molecules, and the complexes are displayed at the cell's surface.

An effector cytotoxic T cell encounters the target. This type of effector specializes in touch-killing. It releases perforins and toxic substances (*green* dots) onto its target and so programs it for death.

e The effector disengages from the doomed cell and reconnoiters for more targets. Meanwhile, perforins make holes in the cell's plasma membrane. Toxins move into the cell, disrupt its organelles, and make the DNA disassemble. The infected cell dies.

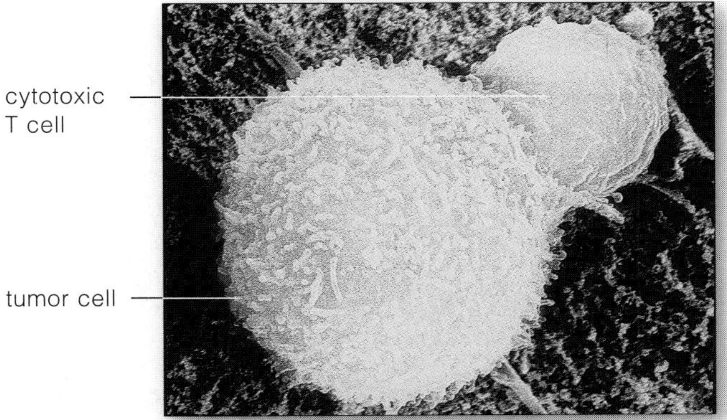
cytotoxic T cell
tumor cell

Figure 40.9 *Above*: Diagram of a T cell–mediated immune response. In this example, the response involves an antigen–MHC complex that activates T cells. *Left*: Scanning electron micrograph of a cytotoxic T cell, caught in the act of touch-killing a tumor cell.

Regarding the Natural Killer Cells

Other cytotoxic cells, including **natural killer cells** (NK cells), also arise from stem cells in bone marrow. They appear to be a type of lymphocyte, but not a T or B cell. Their arousal does not depend on the double signal (an antigen–MHC complex). NK cells reconnoiter for tumor cells and virus-infected cells, then proceed to touch-kill their finds. Possibly these killer cells can recognize odd molecular configurations at the surface of their targets.

T cells arise in bone marrow, then acquire TCRs (receptors for self markers and bound antigen) in the thymus gland.

Effector helper T cells secrete interleukins that trigger cell divisions and differentiation into immense armies against specific antigens. Effector cytotoxic cells touch-kill infected cells or tumor cells, even foreign cells of transplants.

40.7 ANTIBODY-MEDIATED RESPONSES

B Cells and the Targets of Antibodies

Like T cells, B cells also arise from stem cells in bone marrow and start down a pathway that will culminate in full differentiation. Along *their* pathway, however, the B cells start synthesizing a great many copies of a single kind of antibody molecule.

All antibodies are proteins, but the antigen binding site of each kind matches only one kind of antigen. Antibody molecules are more or less Y-shaped, with a tail and two arms that bear identical antigen receptors. Section 40.9 provides a closer look at these molecules. For now, simply think of them as Y-shaped structures of the sort shown in Figure 40.10.

When a B cell is maturing, each freshly synthesized antibody molecule moves to the plasma membrane. Its tail becomes embedded in the membrane's lipid bilayer and its two arms project above it. Soon the cell bristles with antigen receptors (the bound antibodies), and it is ready to join the body's defenses as a virgin B cell.

When its antigen receptors lock on to a target, the B cell does something you might not expect. *It becomes an antigen-presenting cell.* First, an endocytic vesicle forms around the bound antigen and moves it into the cell for digestion into fragments. The fragments bind to MHC molecules and are presented at the B cell surface. Now suppose the TCRs of a responsive helper T cell bind to the antigen–MHC complex, and signals are transferred

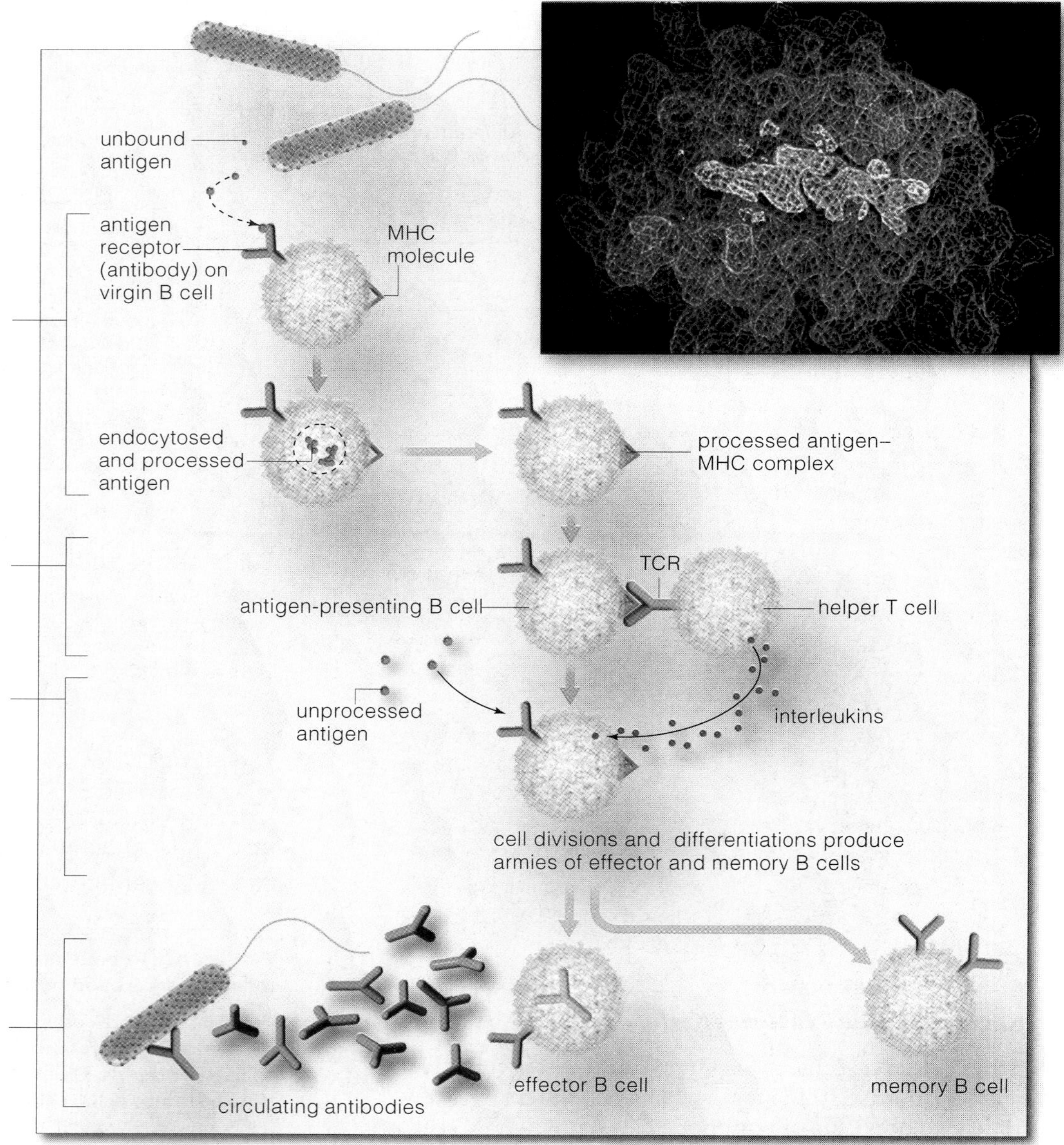

unbound antigen
antigen receptor (antibody) on virgin B cell
MHC molecule
endocytosed and processed antigen
processed antigen–MHC complex
TCR
antigen-presenting B cell
helper T cell
unprocessed antigen
interleukins
cell divisions and differentiations produce armies of effector and memory B cells
circulating antibodies
effector B cell
memory B cell

Figure 40.10 Antibody-mediated immune response. This example is a response to a bacterial invasion. The inset is a computer model of an antigen fragment (*pink*) bound to the cleft of an MHC protein, color-coded *blue*.

a A virgin B cell encounters unbound antigen in tissue fluid. Antigen receptors (in this case, membrane-bound antibodies) bind the antigen. An endocytic vesicle moves bound antigen into the cell for processing. Antigen–MHC complexes are displayed at the cell surface. Thus the B cell has become an antigen-presenting cell.

b TCRs of a helper T cell bind to antigen–MHC complexes on the B cell. Binding activates the T cell and stimulates the B cell to prepare for mitosis. Then the cells disengage.

c Unprocessed antigen binds to the B cell. Meanwhile, the helper T cell secretes interleukins (*blue* dots). Both events trigger repeated cell divisions and differentiations that yield large armies of antibody-secreting effector B and memory B cells.

d Antibody molecules released from the effector B cells enter extracellular fluid. When they contact a bacterial cell that is the target, they bind to antigen on its surface. Binding tags the cell for destruction. (Compare Figures 40.3 and 40.7.)

Further reading: Student Guide to InfoTrac on web site

STOP!
Don't throw this away!

See the other side for your

FREE

InfoTrac® College Edition pass code!

Here's how InfoTrac® College Edition can help you . . .

Time saver—why travel to the library when you just need to look up a few articles?

Topic finder—you can easily determine topics for presentations and group projects.

Research assistant—helps you locate resources for papers and projects.

Study aid—many Thomson Learning™ texts tie into *InfoTrac® College Edition,* providing study and review questions that make use of its research capabilities.

pass code on reverse

between the B cell and a T cell. The cells soon disengage. Then the B cell encounters unprocessed antigen, and its surface antibodies bind to it. The binding, along with interleukins secreted from nearby helper T cells, drives the B cell to mitosis. Its clonal descendants differentiate into effector B and memory B cells. The effectors (also called plasma cells) make and secrete huge numbers of antibody molecules. When freely circulating antibody binds antigen, it tags an invader for destruction, as by phagocytes and complement activation.

The main targets of antibody-mediated responses are extracellular pathogens and toxins, which are freely circulating in tissues or body fluids. *Antibodies cannot bind to pathogens or toxins that are hidden in a host cell.*

The Immunoglobulins

During immune responses, B cells produce four classes of antibodies in abundance and lesser quantities of a different class. Collectively, the five classes of antibodies are known as the **immunoglobulins**, or **Igs**. They are the protein products of gene shufflings that proceed while B cells mature and while an immune response is under way. The molecules in each class have antigen-binding sites *and* other sites with specialized functions.

IgM antibodies are the first to be secreted during immune responses. They trigger the cascade reactions that produce complement and bind targets together in clumps that are more handily destroyed by phagocytes. (Remember the agglutination responses described in Section 39.4?) The *IgD* antibodies associate with IgM on virgin B cells, but their function is not yet understood.

IgG antibodies activate complement proteins and neutralize many toxins. These long-lasting antibodies are the only ones that can cross the placenta. They help protect both the fetus and newborn with the mother's acquired immunities. IgGs also are secreted into early milk produced by the mammary glands. Later, they are absorbed into the suckling newborn's bloodstream.

IgA antibodies enter mucus-coated surfaces of the respiratory, digestive, and reproductive tracts. There, they may neutralize infectious agents. Mother's milk delivers them to the mucous lining of a newborn's gut.

IgE triggers inflammation after attacks by parasitic worms and other pathogens. As described later, it also figures in allergies. The tails of IgE antibodies bind to basophils and mast cells, and the antigen receptors face outward. Antigen binding induces basophils and mast cells to release substances that promote inflammation.

Antibodies that are secreted by B cells bind to antigens of extracellular pathogens or toxins and tag them for disposal, as by phagocytes and complement activation.

Focus on Health

40.8 CANCER AND IMMUNOTHERAPY

Carcinomas, sarcomas, leukemia—these chilling words refer to malignant tumors in skin, bone, and other tissues. Such tumors arise when viral attack, irradiation, or chemicals alter genes and cells turn cancerous (Section 15.6). The transformed cells divide repeatedly. Unless they can be destroyed or surgically removed, they kill the individual.

Transformed cells often bear abnormal proteins and protein fragments bound to MHC. Immune responses to them may make a tumor regress but may not be enough to kill it. Also, some tumors "hide" by releasing copies of the abnormal proteins, which saturate antigen receptors. Hidden tumors that reach a critical mass may overwhelm the immune system. The idea behind *immunotherapy* is to enlist white blood cells to destroy this threat and others.

MONOCLONAL ANTIBODIES Two decades ago, Cesar Milstein and Georges Kohler made limited amounts of antibodies that home in on tumor-specific antigens. They injected an antigen into mice, which made antibodies against it. Then they fused antibody-producing B cells from the mice with cells extracted from the B cell tumors. Clonal descendants of such hybrid cells made *monoclonal antibodies* (identical copies of antibodies). However, mice cannot produce useful amounts. Genetic engineers have now designed cows that can secrete antibodies into milk, but it may be difficult to isolate these molecules from bacteria or other pathogens that may be in the milk.

Today, genetically engineered plants can manufacture monoclonal antibodies. Cornfields might mass-produce them. Besides being cost-effective, *plantibodies* are safer than antibodies from cattle. (Few plant pathogens infect people.) The first plantibody used on human volunteers prevented infection by a bacterial agent of tooth decay.

MULTIPLYING THE TUMOR KILLERS Lymphocytes often infiltrate tumors. Researchers have removed them from a tumor and exposed them to lymphokine, an interleukin. Large populations of tumor-infiltrating lymphocytes with enhanced killing abilities have formed. These *LAK* cells (short for Lymphokine-Activated Killers) appear to be somewhat effective when injected back into a patient.

THERAPEUTIC VACCINES On the horizon are *therapeutic vaccines*—wake-up calls against specific tumor cells. One idea is to genetically engineer antigen so that it becomes obvious to killer lymphocytes. For example, remember Langerhans cells (Section 38.3)? Like other dendritic cells, these phagocytes prowl on many pseudopods. They also are antigen-presenting cells that swiftly migrate to lymph nodes, where they sound the alarm. Dirk Schadendorf harvested dendritic cells from melanoma patients. He cultured them with ground-up tumor cells, then injected them at intervals into his patients' lymph nodes or skin. A year later, two patients had no trace of melanoma. In three others, tumors shrank more than half their size. Such vaccines may be available to the public within five years.

IMMUNE SPECIFICITY AND MEMORY

How Do Antigen-Specific Receptors Form?

The variety of antigens in your surroundings is mind-boggling. Collectively, however, antigen receptors of all T and B cell populations in your body show staggering diversity—enough to recognize about a billion different antigens. How does antibody diversity arise?

For the answer, start with the knowledge that all antigen receptors of a single T cell or B cell are identical, and all are proteins. For example, a Y-shaped antibody molecule consists of four polypeptide chains bonded together (Figure 40.11). Certain parts of each chain, the *variable* regions, fold in ways that produce grooves and bumps with a certain charge distribution. Only antigen that has complementary grooves, bumps, and charge distribution will be able to bind with them.

Receptor diversity begins with rearrangements in DNA sequences that code for such variable regions. Many of the sequences are V segments and, some distance away, J segments (Figure 40.12). As a B cell matures, one of its V segments randomly contacts and binds with one of its J segments. The stretch of DNA intervening between the two joined segments loops outward, and enzymes snip it off. The B cell ends up with a unique sequence randomly put together from a choice of segments.

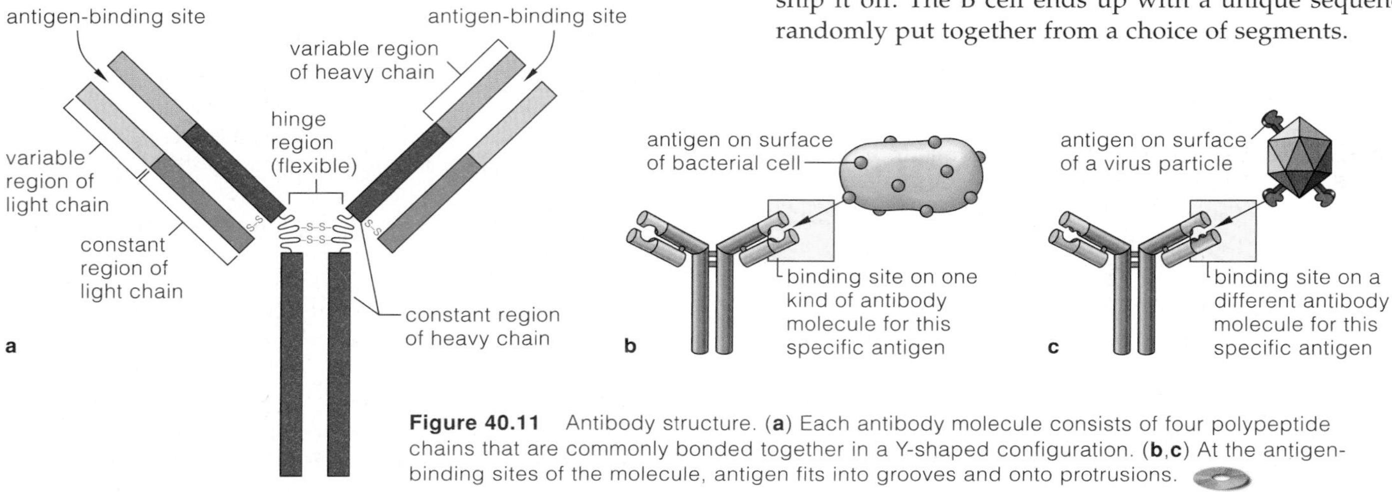

antigen-binding site
antigen-binding site
variable region of heavy chain
hinge region (flexible)
variable region of light chain
constant region of light chain
constant region of heavy chain
a
antigen on surface of bacterial cell
binding site on one kind of antibody molecule for this specific antigen
b
antigen on surface of a virus particle
binding site on a different antibody molecule for this specific antigen
c

Figure 40.11 Antibody structure. (**a**) Each antibody molecule consists of four polypeptide chains that are commonly bonded together in a Y-shaped configuration. (**b**,**c**) At the antigen-binding sites of the molecule, antigen fits into grooves and onto protrusions.

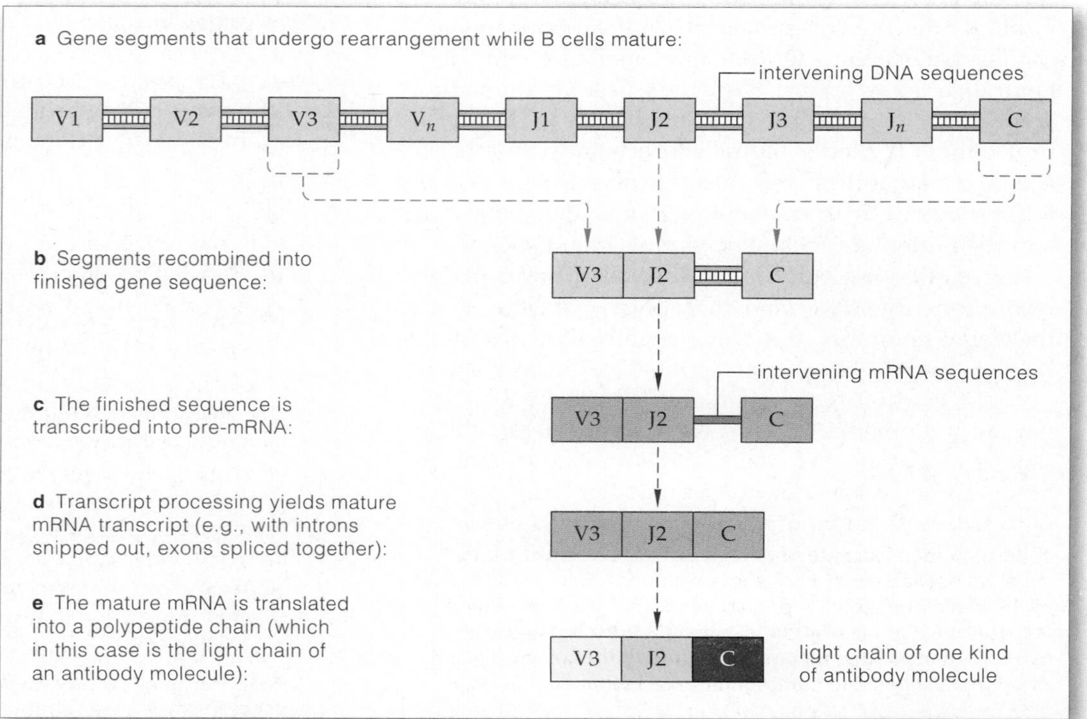

a Gene segments that undergo rearrangement while B cells mature:
intervening DNA sequences
V1
V2
V3
Vn
J1
J2
J3
Jn
C
b Segments recombined into finished gene sequence:
V3
J2
C
c The finished sequence is transcribed into pre-mRNA:
intervening mRNA sequences
V3
J2
C
d Transcript processing yields mature mRNA transcript (e.g., with introns snipped out, exons spliced together):
V3
J2
C
e The mature mRNA is translated into a polypeptide chain (which in this case is the light chain of an antibody molecule):
V3
J2
C
light chain of one kind of antibody molecule

Figure 40.12 Generation of antibody diversity. Antibodies are proteins, and instructions for building proteins are encoded in genes. In the chromosomes having genes for antibodies, extensive DNA regions contain different versions of segments that code for variable regions of an antibody molecule.

The different V and J segments shown are examples. They code for the variable region of a light chain (compare Figure 40.11*a*). A recombination event occurs in this region as each B cell matures. In this example, any V segment may be joined to any J segment. Afterward, the DNA intervening between them is excised. The new sequence gets attached to a C (*C*onstant) segment. This completes a rearranged antibody gene—which also will be present in every one of the descendants of the cell.

Further reading: Student Guide to InfoTrac on web site

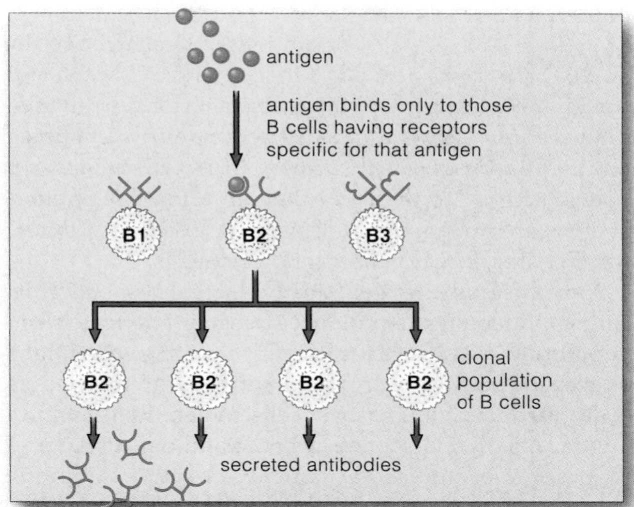
antigen
antigen binds only to those B cells having receptors specific for that antigen
B1
B2
B3
B2
B2
B2
B2
clonal population of B cells
secreted antibodies

Figure 40.13 Clonal selection of a B cell that produced the specific antibody that can combine with a specific antigen. Only antigen-selected B cells (and T cells) are activated and give rise to a clonal population of immunologically identical cells.

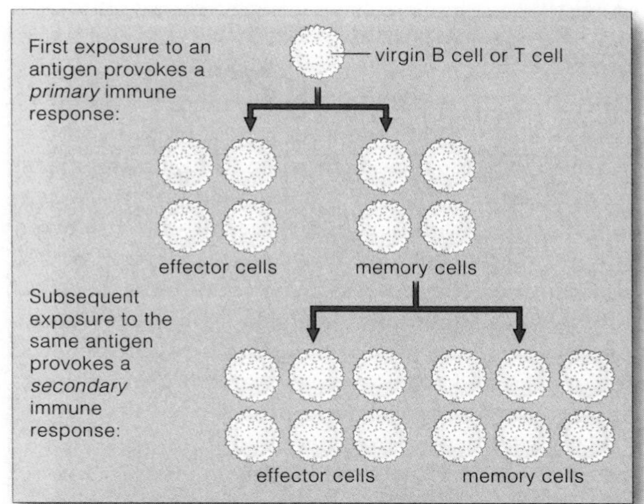
First exposure to an antigen provokes a *primary* immune response:
virgin B cell or T cell
effector cells
memory cells
Subsequent exposure to the same antigen provokes a *secondary* immune response:
effector cells
memory cells

Figure 40.14 Immunological memory. Not all B and T cells are used in a primary immune response to an antigen. A large number continue to circulate as memory cells, which become activated during a secondary immune response.

The same kinds of DNA rearrangements have roles in producing the variable regions of the TCR molecules of maturing T cells.

Some time ago, Macfarlane Burnet devised a *clonal selection* hypothesis that helped point the way to our current view of receptor diversity. He proposed that antigen "chooses" (binds to) one lymphocyte from all the various types in the body, because that lymphocyte has the receptor specific for it. Repeated mitotic cell divisions then give rise to a clone of cells that carry out the response (Figure 40.13).

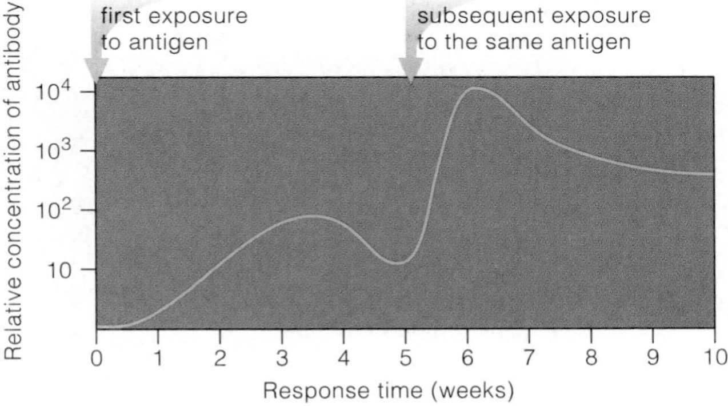
first exposure to antigen
subsequent exposure to the same antigen
Relative concentration of antibody
10^4
10^3
10^2
10
0
1
2
3
4
5
6
7
8
9
10
Response time (weeks)

Figure 40.15 Example of the differences in magnitude and duration between a primary and a secondary immune response to the same antigen. The primary response peaked twenty-four days after it started. The secondary response peaked after only seven days (the span shown between weeks five and six). Antibody concentration during the secondary response was 100 times greater (10^4 compared to 10^2).

Immunological Memory

The clonal selection theory explains how an individual can have "immunological memory" of a first encounter with antigen. The term refers to the body's capacity to make a *secondary* immune response to any subsequent encounter with the same type of antigen that provoked the primary response (Figure 40.14).

Memory cells that form during a primary immune response do not engage in that battle. They circulate for years or for decades. Compared to the virgin cells that initiate a primary response, these patrolling battalions consist of far more cells and intercept antigen far sooner. Effector cells form sooner, in greater numbers, so the infection is terminated before the host gets sick. Even greater numbers of memory T and B cells form during a secondary response. Figure 40.15 has an example of this. In evolutionary terms, the advance preparations against subsequent encounters with a pathogen bestow a survival advantage on the individual.

By recombination of segments drawn at random from the receptor-encoding regions of DNA, each T cell or B cell receives a gene sequence for one of a billion possible kinds of antigen receptors.

Immunological specificity means the clonal descendants of an antigen-selected cell will react only with the selecting antigen.

Immunological memory refers to the capacity to make a secondary (faster, greater) immune response to a pathogen that caused a primary response in an individual.

40.10 DEFENSES ENHANCED, MISDIRECTED, OR COMPROMISED

Immunization

After reflecting on the chapter introduction, you might already sense that **immunization** refers to a variety of processes which promote increased immunity against specific diseases. With *active* immunization, an antigen-containing preparation called a **vaccine** is either taken orally or injected into the body, as in Figure 40.16. An initial injection triggers a primary immune response to an antigen. A subsequent injection—a booster—elicits a secondary response, with rapid formation of more effector cells and memory cells that can provide long-lasting protection against the disease.

Many vaccines are manufactured from weakened or killed pathogens. Others use inactivated natural toxins, such as a bacterial toxin that causes tetanus. Others are made of harmless viruses that are genetically engineered so genes from three or more pathogens are inserted into their DNA or RNA. After vaccination, an individual's cells use the new genes to produce the antigens, and immunity is established.

Passive immunization is used for individuals who are already infected with pathogens that cause diphtheria, tetanus, measles, hepatitis B, and some other diseases. A person at risk receives injections of purified antibody, the best source of which is some other individual who already has made a large amount of the required antibody. The effects don't last long; the patient's B cells have not made any antibodies. But antibody injections may counter the initial attack.

Vaccines may fail or have adverse effects. In a few cases, they result in chronic immunological or neurological problems. It is important to assess the risks and benefits before agreeing to a procedure.

RECOMMENDED VACCINES	RECOMMENDED AGES
Hepatitis B	Birth–2 months
Hepatitis B booster	1–4 months
Hepatitis B booster	6–18 months
Hepatitis B assessment	11–12 years
DTP (Diphtheria; Tetanus; and Pertussin, or whooping cough)	2, 4, and 6 months
DTP booster	15–18 months
DTP booster	4–6 years
DT	11–16 years
HiB (*Haemophilus influenzae*)	4 and 6 months
HiB booster	12–15 months
Polio	2 and 4 months
Polio booster	6–18 months
Polio booster	4–6 years
MMR (Measles, Mumps, Rubella)	12–15 months
MMR booster	4–6 years
MMR assessment	11–12 years
Varicella	12–18 months
Varicella assessment	11–12 years

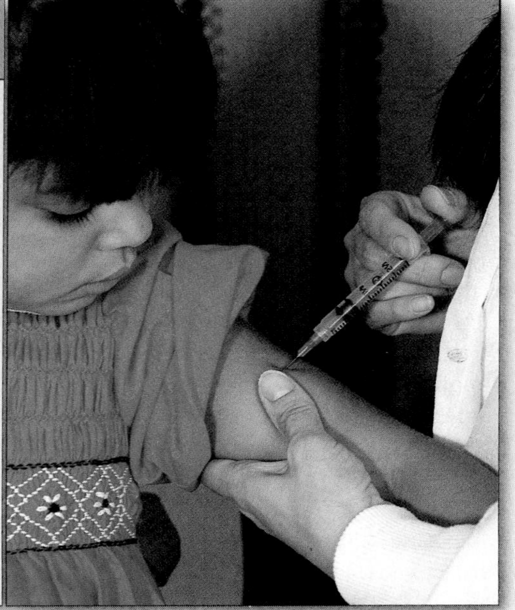

Figure 40.16 From the Centers for Disease Control and Prevention, the 1998 immunization guidelines for children in the United States. Doctors routinely immunize infants and children. Low-cost or free vaccinations are available at many community clinics and health departments.

Allergies

In many people, normally harmless substances provoke inflammation, excess mucus secretion, and sometimes immune responses. Such substances are **allergens**, and hypersensitivity to them is called an **allergy**. Common allergens are pollen, many drugs and foods, dust mites, fungal spores, insect venom, and cosmetics.

Some people are genetically inclined to develop allergies. Infections, emotional stress, or changes in air temperature also trigger reactions that otherwise might not occur. Upon exposure to an antigen, IgE antibodies are secreted and bind to mast cells. When the IgE binds antigen, mast cells secrete prostaglandins, histamine, and other substances that fan inflammation. Copious amounts of mucus are secreted, and airways constrict. Stuffed-up sinuses, labored breathing, a drippy nose, and sneezing are symptoms of the allergic response in *asthma* and *hay fever* (Figure 40.17).

In a few cases, the inflammatory reactions wash through the body in a life-threatening event called *anaphylactic shock*. For example, someone allergic to wasp or bee venom may die within minutes of one sting. Airways constrict massively, and fluid escapes rapidly from grossly permeable capillaries. Blood pressure plummets, and circulation may collapse.

Antihistamines (anti-inflammatory drugs) often relieve the mild, short-term symptoms of allergies. Over time, a patient can follow a desensitization program. First, skin tests may identify offending allergens. Inflammatory responses to some can be blocked if the patient can be stimulated to make

RAGWEED POLLEN . . .

. . . AND SOMETHING IT CAN PROVOKE

Figure 40.17 Example of the effects of pollen and other allergens in sensitive people. Allergy sufferers who moved to deserts to escape pollen brought it with them. Half the human population in Tucson, Arizona, is now sensitized to pollen from olive and mulberry trees, planted far and wide in cities and the suburbs.

Figure 40.18 A case of severe combined immunodeficiency, or SCID. Ashanthi DeSilva was born without an immune system. She has a mutated gene for ADA (adenosine deaminase), an enzyme. Without ADA, her cells cannot break down adenosine. As a result, a reaction product accumulates that is toxic to lymphocytes. The disorder's symptoms are caused by infections that cannot be controlled. They include high fever, severe ear and lung infections, diarrhea, and an inability to gain weight.

Ashanthi's parents consented to the first federally approved gene therapy for a human patient. Genetic engineers spliced the ADA gene into the genetic material of a harmless virus. Then they used the modified virus rather like a hypodermic needle: they allowed it to deliver copies of the "good" gene into Ashanthi's bone marrow cells. Some of those cells incorporated the gene in their DNA and started to synthesize the missing enzyme. At this writing, Ashanthi is in her teens. Like other ADA-deficient patients who have undergone this gene therapy, she is doing well.

In other cases, researchers take bone marrow stem cells from blood in the umbilical cord of affected newborns. (The cord, which connects the fetus and placenta during pregnancy, is discarded after birth.) They expose the cells to viruses that deliver copies of the ADA gene into them and to factors that stimulate mitotic cell division and growth. The cells are reinserted into newborns.

IgG instead of IgE. Larger doses of specific allergens are administered slowly. Each time, the body produces more circulating IgG and memory cells. IgG binds with the allergen to block its attachment to IgE and thereby blocks inflammation.

Autoimmune Disorders

In an **autoimmune response**, the immune system acts against self antigens. For example, in *Grave's disorder*, the body makes too many thyroid hormone molecules, which control metabolic rates and the growth of many tissues. Affected individuals produce antibodies that bind to receptors on cells that make the hormones. The antibodies do not respond to the feedback mechanisms that normally control production of thyroid hormones, so too many molecules are produced. Typical symptoms are elevated rates of metabolism, heart fibrillations, excessive sweating, nervousness, and weight loss.

As other examples, *myasthenia gravis*, a progressive weakening of muscles, arises when antibodies bind to receptors on skeletal muscle cells and interfere with the normal action of acetylcholine. *Rheumatoid arthritis* is a chronic inflammation of the skeletal joints. Patients are genetically predisposed to this disorder. Macrophages, T cells, and B cells are activated when antigens act at skeletal joints. Immune responses are made against the body's own collagen molecules and against antibody that has become bound to as-yet-unidentified antigen. Complement activation and inflammation cause more damage in the tissues of skeletal joints. So do skewed repair mechanisms. Eventually, joints fill with synovial membrane cells and become immobilized.

Deficient Immune Responses

When the body has inadequate numbers of functioning lymphocytes, immune responses are not effective. Such *severe combined immunodeficiencies* (SCIDs) result from heritable disorders or from assaults by outside agents. Deficient or nonexistent immune responses make the person highly vulnerable to infections that are not life threatening to the general population.

ADA deficiency is one of the heritable SCIDs (Figure 40.18). Acquired immunodeficiency syndrome (AIDS) is one of the SCIDs resulting from viral attack. The next section describes how this virus, HIV, replicates inside lymphocytes and destroys the body's capacity to fight infections. We also return to this topic in Section 45.14.

Immunization programs boost immunity to specific diseases.

Certain heritable disorders or attacks by certain pathogens and other agents can result in misdirected, compromised, or nonexistent immunity.

40.11 AIDS—THE IMMUNE SYSTEM COMPROMISED

CHARACTERISTICS OF AIDS AIDS is a constellation of disorders that follow infection by HIV (short for *H*uman *I*mmunodeficiency *V*irus). The virus cripples the immune system, which makes the body extremely susceptible to usually harmless infections and rare forms of cancer. Worldwide, over 33.6 million are infected (Table 40.3). More than 95 percent of all cases and all deaths occur in developing countries, mostly among young adults and increasingly among women. At present there is no way to rid the body of known forms of the virus (HIV-I and HIV-II). *There is no cure for those already infected.*

At first an infected person might appear to be in good health, suffering no more than a bout of "the flu." Then symptoms that foreshadow AIDS emerge. They typically include chronic weight loss, fever, many enlarged lymph nodes, fatigue, and bed-drenching night sweats. In time, diseases result from certain opportunistic infections. Rare in the population at large, the diseases are signs of AIDS. Among these are yeast infections of the mouth, esophagus, vagina, and elsewhere, and a form of pneumonia caused by *Pneumocystis carinii* (Section 23.6). Bruises often develop on legs and feet especially (Figure 40.19). They are signs of Kaposi's sarcoma, a cancer of blood vessel endothelium. The infections or cancers end up killing the person.

HOW HIV REPLICATES HIV infects antigen-presenting macrophages and helper T cells, which are also called CD4 lymphocytes. HIV is a retrovirus. Each virus particle has an outermost lipid envelope: a bit of plasma membrane that surrounded the particle when it departed from an infected cell. Various proteins spike outward from the envelope, span it, and line its inner surface. Sequestered inside the lipid envelope, viral coat proteins enclose two RNA strands and several copies of reverse transcriptase, a retroviral enzyme (Sections 16.1 and 22.8).

Once inside an effector or memory host cell, the viral enzyme uses the RNA as a template for making DNA, which gets inserted into a host chromosome (Figure 40.20*a*). In some cells, the inserted viral genes remain silent, only to become activated during a later round of infection. Whenever it occurs, transcription yields copies of the viral RNA. Some transcripts are translated into viral proteins. Others become enclosed by viral coat proteins when new particles are put together; they will function as the viral hereditary material. As Figure 40.20*b* shows, the particles are released by budding from the host cell's plasma membrane, after which they start a new round of infection in other cells. With each round, more of the body's macrophages, antigen-presenting cells, and helper T cells are impaired or killed.

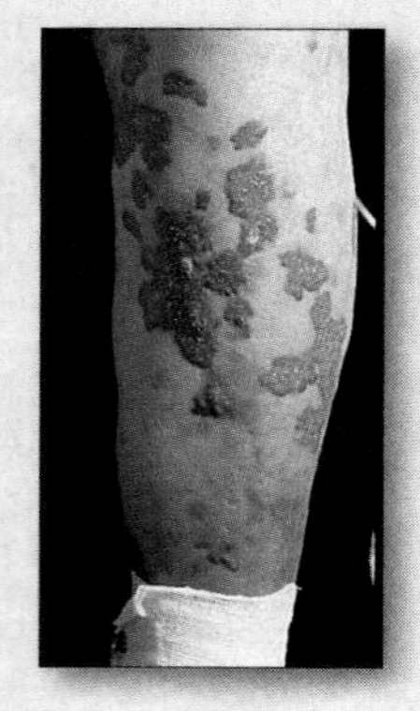

Figure 40.19 Lesions that are signs of Kaposi's sarcoma.

A TITANIC STRUGGLE BEGINS Infection marks the onset of a titanic battle between the enemy and the immune system. B cells synthesize antibodies in response to HIV antigenic proteins. These are antibodies that are the basis of diagnostic tests to identify HIV infection. Armies of helper T cells and cytotoxic T cells also form. However, the virus infects an estimated 2 billion helper T cells and produces 100 million to 1 billion virus particles per day during some phases of the infection. Every two days, the immune system destroys about half the virus particles and replaces half the helper T cells lost in the battle. Immense reservoirs of HIV and masses of infected T cells accumulate in lymph nodes. As the battle proceeds, the number of virus particles in the general circulation rises. Gradually the numbers tilt. The body produces fewer and fewer helper T cells to replace the ones it lost. It may take a decade or more, but erosion of the helper T cell count inevitably causes the body to lose its capacity to mount effective immune responses.

Table 40.3 Global Estimates of AIDS Cases*

Sub-Saharan Africa:	23,800,000
South/Southeast Asia:	6,000,000
Latin America:	1,300,000
North America:	920,000
East Asia and Pacific:	530,000
Western Europe:	520,000
Caribbean Islands:	360,000
Eastern Europe/Central Asia:	360,000
Middle East/North Africa:	220,000
Australia/New Zealand:	12,000

* Estimates as of December 1999.

Some viruses, including the measles virus, produce far more virus particles in a given day, but the immune system usually wins out. Other viruses, including herpes viruses, can lurk in the body for a lifetime, but the immune system keeps them in check. With HIV, the immune system loses, and infections and tumors always kill the individual.

HOW HIV IS TRANSMITTED Like any other virus that infects humans, HIV requires a medium that allows it to leave one host, survive in the environment into which it is released, then enter another host. HIV is transmitted when some body fluid of an infected person enters tissues of another person. In the United States, transmission initially occurred most often between homosexual males, as by anal intercourse, and then among drug abusers who share blood-contaminated syringes and needles. It also has spread in the heterosexual population, increasingly by vaginal intercourse.

HIV travels from infected mothers to offspring during pregnancy, birth, and breast-feeding. Before 1985, contaminated blood supplies accounted for some AIDS cases. Health care providers have since implemented careful screening. Tissue transplants caused a few infections. In some of the developing countries, health care providers accidentally spread HIV by way of contaminated transfusions and reuse of unsterile needles and syringes.

The molecular structure of HIV does not remain stable outside the human body, which is why it must be directly transmitted from one host to another. At

Further reading: Student Guide to InfoTrac on web site –

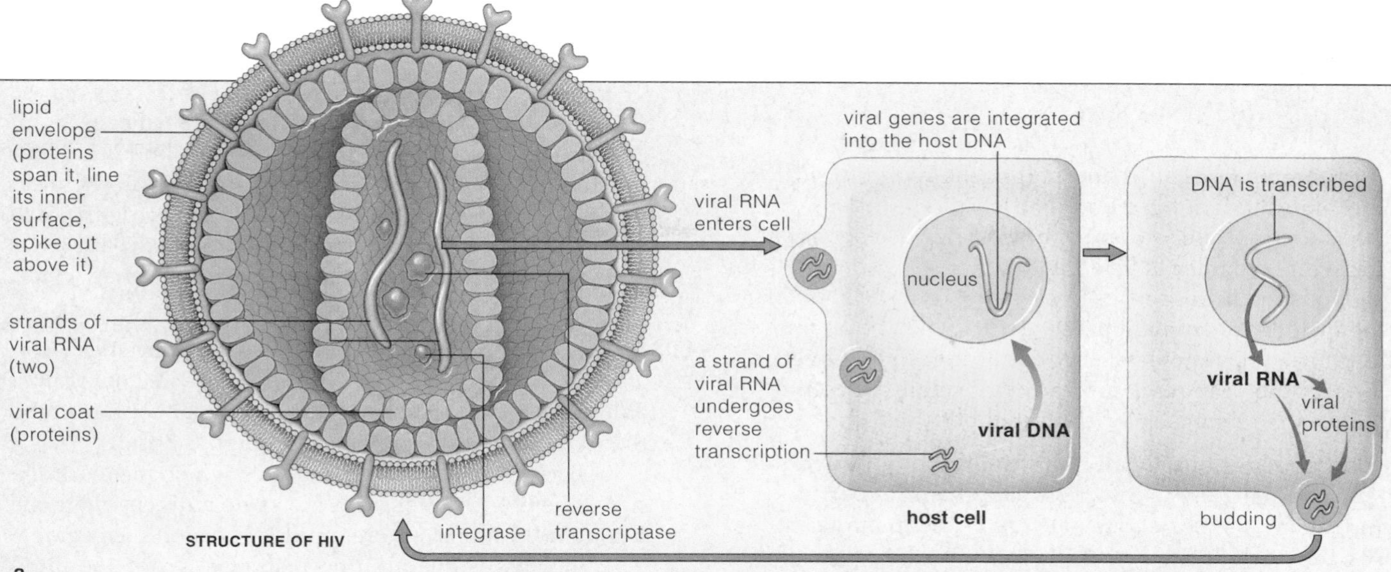
lipid envelope (proteins span it, line its inner surface, spike out above it)
strands of viral RNA (two)
viral coat (proteins)
integrase
reverse transcriptase
STRUCTURE OF HIV
a
viral RNA enters cell
viral genes are integrated into the host DNA
nucleus
a strand of viral RNA undergoes reverse transcription
viral DNA
host cell
DNA is transcribed
viral RNA
viral proteins
budding

Figure 40.20 (**a**) Replication cycle of HIV, one of the retroviruses. (**b**) Electron micrographs of an HIV particle budding from a host cell.

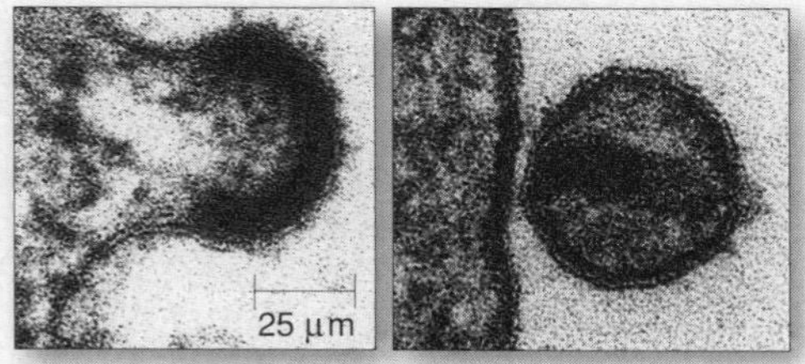
b
25 μm

this writing, HIV has been isolated from human blood, semen, vaginal secretions, saliva, tears, urine, breast milk, amniotic fluid, urine, and cerebrospinal fluid. Probably it is present in other fluids, secretions, and excretions. Until recently, only infected blood, semen, vaginal secretions, and breast milk were thought to contain HIV particles in concentrations high enough for successful transmission. Transmission by oral sex is now known to have resulted in some cases of AIDS. HIV is not effectively transmitted by way of food, air, water, casual contact, or insect bites.

REGARDING PREVENTION OR TREATMENT Developing effective drugs or vaccines against HIV is a formidable challenge. High mutation rates characterize the viral genome owing to HIV's replication mechanisms and the staggering number of replications in an infected person. Natural selection inevitably operates in patients who are undergoing drug therapies; it favors drug resistance and fans the evolution of drug-resistant HIV populations. Chemical "cocktails" help slow replication. Among the current drugs of choice are protease inhibitors, AZT (azidothymidine), and ddI (dideoxyinosine). However, they cannot *cure* infected people; drugs cannot eliminate HIV genes already incorporated in someone's DNA.

High mutation rates have another worrisome outcome: they give rise to variations in HIV antigens. The variation makes it hard for researchers to select effective antigens for vaccines. Another problem is the scarcity of HIV strains with poor cell-killing abilities. Such strains are central to producing antigen that can stimulate the formation of cytotoxic T cell armies, the best protection against HIV.

Over a decade ago, many people in Australia received blood transfusions from a donor whose infection had not been diagnosed. At this writing, neither they nor the donor show any immunodeficiency. And they all carry HIV with a similar defect in the same gene (the *nef* gene).

Recently, x-ray diffraction methods revealed the three-dimensional structure of gp120, the viral protein that binds to helper T cell receptors (Figure 40.21). Portions of gp120 must remain stable; if they were to change, HIV would lose its binding capacity. They hide behind an oligosaccharide "forest" and in crevices that are too small for antibody molecules to reach. Genetic engineers might be able to design molecules small enough to infiltrate the crevices and inactivate the binding site. Such an approach might or might not work, but research is under way.

In short, *until researchers develop effective vaccines and treatments, checking the spread of HIV depends absolutely on persuading people to avoid or modify social behaviors that put them at risk*. We return to this topic in Section 45.14.

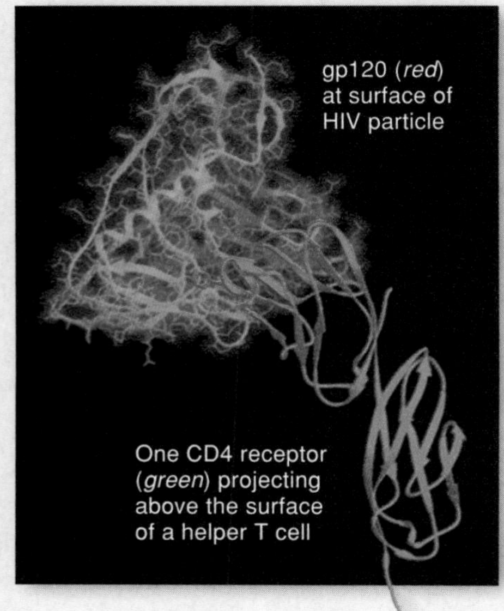
gp120 (*red*) at surface of HIV particle
One CD4 receptor (*green*) projecting above the surface of a helper T cell

Figure 40.21 Computer model of gp120, the HIV protein that can bind to two receptors of host cells (chemokine and CD4 receptors). The most stable parts of gp120 are hidden by a thick "forest" of sugar molecules or in the bottom of a crevice. There they elude the rather bulky antibody molecules. One part becomes exposed, but only briefly, after it hooks on to a CD4 receptor and before it hooks on to a nearby chemokine receptor.

40.12 SUMMARY

1. Vertebrates fend off many pathogens with physical and chemical barriers at body surfaces. They also are protected by nonspecific and specific defense responses of the white blood cells listed in Table 40.4.

a. Nonspecific responses to irritation or damage of a tissue include inflammation and participation of organs with phagocytic functions, such as the spleen and liver.

b. Immune responses are mounted against specific pathogens, foreign cells, or abnormal body cells.

2. Skin and mucous membranes lining body surfaces are physical barriers to infection. The chemical barriers include glandular secretions (such as lysozyme in tears, saliva, and gastric fluid) and some metabolic products of bacteria that normally reside on body surfaces.

3. An inflammatory response develops in body tissues that have become damaged, as by infection.

a. An inflammatory response starts with arteriolar vasodilation that increases the blood flow to the tissue, which reddens and becomes warmer as a result. Blood capillary permeability increases and the resulting local edema causes swelling and pain.

Table 40.4 Summary of White Blood Cells and Their Roles in Defense

Cell Type	Main Characteristics
MACROPHAGE	Phagocyte; acts in nonspecific and specific responses; presents antigen to T cells, and cleans up and helps repair tissue damage
NEUTROPHIL	Fast-acting phagocyte; takes part in inflammation, not in sustained responses, and is most effective against bacteria
EOSINOPHIL	Secretes enzymes that attack certain parasitic worms
BASOPHIL AND MAST CELL	Secrete histamine, other substances that act on small blood vessels to produce inflammation; also contribute to allergies
LYMPHOCYTES:	*All take part in most immune responses; following antigen recognition, form clonal populations of effector and memory cells.*
B cell	Effectors secrete five types of antibodies (IgM, IgG, IgA, and IgD known to protect a host in specialized ways; also IgD)
Helper T cell	Effectors secrete interleukins that stimulate rapid divisions and differentiation of both B cells and T cells
Cytotoxic T cell	Effectors kill infected cells, tumor cells, and foreign cells by a touch-kill mechanism
NATURAL KILLER (NK) CELL	Cytotoxic cell of undetermined affiliation; kills virus-infected cells and tumor cells by a touch-kill mechanism

b. Pathogens and dead or damaged body cells release the substances that trigger increased permeability of capillaries. White blood cells leave the blood and enter the tissue, where they release a number of chemical mediators and engulf invaders. Plasma proteins also enter the tissue. Complement proteins bind pathogens and induce their lysis, and they also attract phagocytes. Blood-clotting proteins wall off the damaged tissue.

4. An immune response has these characteristics:

a. It shows specificity, meaning it is directed against antigen. An antigen is a specific molecular configuration that lymphocytes recognize as foreign (nonself).

b. Each response shows memory, which means that a subsequent encounter with the same antigen triggers a secondary response (more rapid, of greater magnitude).

c. Normally, an immune response is not mounted against the body's own self-marker proteins.

5. Antigen-presenting cells process and bind fragments of antigen to their own MHC markers. Lymphocyte receptors can bind to displayed antigen–MHC complexes. Binding is a start signal for immune responses.

6. After recognition of antigen, an immune response proceeds by repeated cell divisions that form clones of B and T cells. These differentiate into subpopulations of effector and memory cells. Chemical mediators such as interleukins secreted by white blood cells drive the responses. The effector helper T cells, cytotoxic T cells, effector B cells, and antibodies act at once. The memory cells are set aside for secondary responses.

7. T cells arise in bone marrow but continue to develop in the thymus, where they acquire T-cell receptors that recognize and bind antigen–MHC complexes on antigen-presenting cells. B cells arise in bone marrow. While they mature, they synthesize antibodies (a type of antigen receptor), which become positioned at their surface.

8. Effector cytotoxic T cells directly kill virus-infected cells, tumor cells, and cells making up tissue or organ transplants. Effector B cells produce and secrete great numbers of antibodies that freely circulate.

9. Antibodies are protein molecules, often Y-shaped, each with binding sites for one kind of antigen. Only B cells make them. When antibody binds antigen, toxins are neutralized, pathogens are tagged for destruction, or attachment of pathogens to body cells is prevented.

10. By active immunization, vaccines provoke immune responses involving production of effector and memory cells. By passive immunization, injections of purified antibodies help the individual through an infection.

11. Allergic reactions are immune responses to some generally harmless substance. Autoimmune responses are misguided attacks triggered by configurations on the body's own cells. Immunodeficiency is a weakened or nonexistent capacity to mount an immune response.

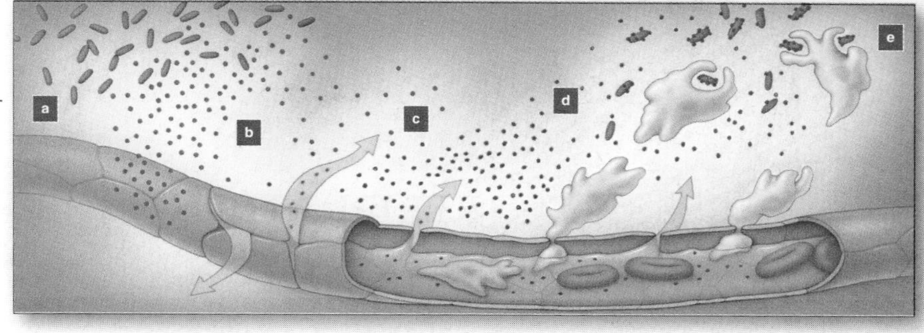
a
b
c
d
e

Review Questions

1. While jogging barefoot along a seashore, some of your toes accidentally land on a jellyfish. Soon the toes are swollen, red, and warm to the touch. Using the diagram at right as a guide, describe events that result in these signs of inflammation. *40.3*

2. Distinguish between:
   a. neutrophil and macrophage *40.3*
   b. cytotoxic T cell and natural killer cell *40.4, 40.6*
   c. effector cell and memory cell *40.4*
   d. antigen and antibody *40.4*

3. Describe the events by which a macrophage turns into an antigen-presenting cell. *40.4*

4. Why is a vaccine to control AIDS so elusive? *40.11*

Self-Quiz (*Answers in Appendix III*)

1. ______ are barriers to pathogens at body surfaces.
   a. Intact skin and mucous membranes
   b. Tears, saliva, gastric fluid
   c. resident bacteria
   d. urine flow
   e. all of the above

2. Macrophages are derived from ______.
   a. basophils b. neutrophils c. monocytes d. eosinophils

3. Complement functions in defense. They ______.
   a. neutralize toxins
   b. enhance resident bacteria
   c. promote inflammation
   d. form pore complexes that cause lysis of pathogens
   e. both a and b
   f. both c and d

4. ______ are certain molecules that lymphocytes recognize as foreign and that elicit an immune response.
   a. Interleukins
   b. Antibodies
   c. Immunoglobulins
   d. Antigens
   e. Histamines

5. Immunoglobulins ______ increase antimicrobial activity in mucus-coated surfaces of some organ systems.
   a. IgA b. IgE c. IgG d. IgM e. IgD

6. Antibody-mediated responses work best against ______.
   a. intracellular pathogens
   b. extracellular pathogens
   c. extracellular toxins
   d. both b and c
   e. all of the above

7. The most important antigens are ______.
   a. nucleotides
   b. triglycerides
   c. steroids
   d. proteins

8. ______ would be a target of an effector cytotoxic T cell.
   a. Extracellular virus particles in blood
   b. A virus-infected body cell or tumor cell
   c. Parasitic flukes in the liver
   d. Bacterial cells in pus
   e. Pollen grains in nasal mucus

9. Development of a secondary immune response is based on populations of ______.
   a. memory cells
   b. circulating antibodies
   c. effector B cells
   d. effector cytotoxic T cells
   e. mast cells

10. Match the immunity concepts.
    ____ inflammation
    ____ antibody secretion
    ____ a phagocyte
    ____ immunological memory
    ____ vaccination
    ____ allergy
    a. neutrophil
    b. effector B cell
    c. nonspecific response
    d. deliberately provoking an immune response
    e. basis of secondary response
    f. nonprotective immune response

Critical Thinking

1. As described in the chapter introduction, Edward Jenner lucked out. He performed a potentially harmful experiment on a boy who managed to survive it. What would happen if a would-be Jenner tried to do the same thing today?

2. Rob's bumper sticker reads, "Have you thanked your resident bacteria today?" Explain why he appreciates the bacteria that normally reside on the body's skin and mucous membranes.

3. Researchers have been attempting to develop a way to get the immune system to accept foreign tissue as "self." Speculate on some of the clinical applications of such a development.

4. Before each flu season, you get an influenza vaccination. This year you come down with "the flu" anyway. What do you suppose happened? (There are at least three explanations.)

5. Infection by *Ebola* virus results in a hemorrhagic fever with a 90 percent mortality rate (Section 22.10). A patient received a blood serum transfusion from another who survived the disease. Explain why the transfusion might increase chances of survival.

6. Ellen developed *chicken pox* when she was in kindergarten. Later in life, when her children developed chicken pox, she remained healthy even though she was exposed to countless virus particles daily. Explain why.

7. Quickly review Section 33.7 on homeostasis. Then write a short essay on how immune responses contribute to stability in the internal environment.

Selected Key Terms

allergen *40.10*
allergy *40.10*
antibody *40.4*
antigen *40.4*
antigen–MHC complex *40.4*
antigen-presenting cell *40.4*
apoptosis *40.6*
autoimmune response *40.10*
B lymphocyte (B cell) *40.4*
basophil *40.3*
complement system *40.2*
cytotoxic T cell *40.4*
eosinophil *40.3*
fever *40.3*
helper T cell (CD4 lymphocyte) *40.4*
histamine *40.3*
immune system *40.4*
immunization *40.10*
immunoglobulin (Ig) *40.7*
inflammation, acute *40.3*
lysis *40.2*
lysozyme *40.1*
macrophage *40.3*
mast cell *40.3*
MHC marker *40.4*
natural killer (NK) cell *40.6*
neutrophil *40.3*
pathogen *40.1*
T lymphocyte (T cell) *40.4*
TCR *40.6*
vaccination *CI*
vaccine *40.10*

Readings *See also www.infotrac-college.com*

Bartlett, J., and R. Moore. July 1998. "Improving HIV Therapy." *Scientific American*.

Nowak, M., and A. McMichael. August 1995. "How HIV Defeats the Immune System." *Scientific American* 273(2): 58–65.

Tizard, I. 1995. *Immunology: An Introduction*. Fourth edition. Philadelphia: Saunders.

41 RESPIRATION

Conquering Chomolungma

To experienced climbers, Chomolungma may be the ultimate challenge (Figure 41.1). The summit of this Himalayan mountain, also known as Everest, is 9,700 meters (29,128 feet) above sea level. It is the highest place on Earth. Iced-over vertical rock, driving winds, blinding blizzards, and heart-stopping avalanches await the challengers. So does the extreme danger that oxygen-poor air poses to the brain.

Most of us live at low elevations. Of the air that we breathe, one molecule in five is oxygen. When we travel 2,400 meters (about 8,000 feet) or more above sea level, the Earth's gravitational pull is weaker, gas molecules spread out more, and the breathing game changes. We face *hypoxia*, or cellular oxygen deficiency. Sensing the deficiency, the brain makes us breathe faster and deeper than usual, or hyperventilate. Past 3,300 meters (10,000 feet), hyperventilating causes significant ion imbalances in cerebrospinal fluid. It may lead to heart palpitations, shortness of breath, headaches, nausea, and vomiting. These are strong clues that the body's cells are, in a manner of speaking, screaming for oxygen.

Living for months at high elevations helps climbers adapt physiologically to the thinner air. For example, mechanisms kick in that raise the red blood cell count, hence the body's oxygen-carrying capacity. Professional climbers know that, above 7,000 meters, oxygen scarcity and low air pressure combine to make blood capillaries leaky. More plasma escapes through the expanded gaps

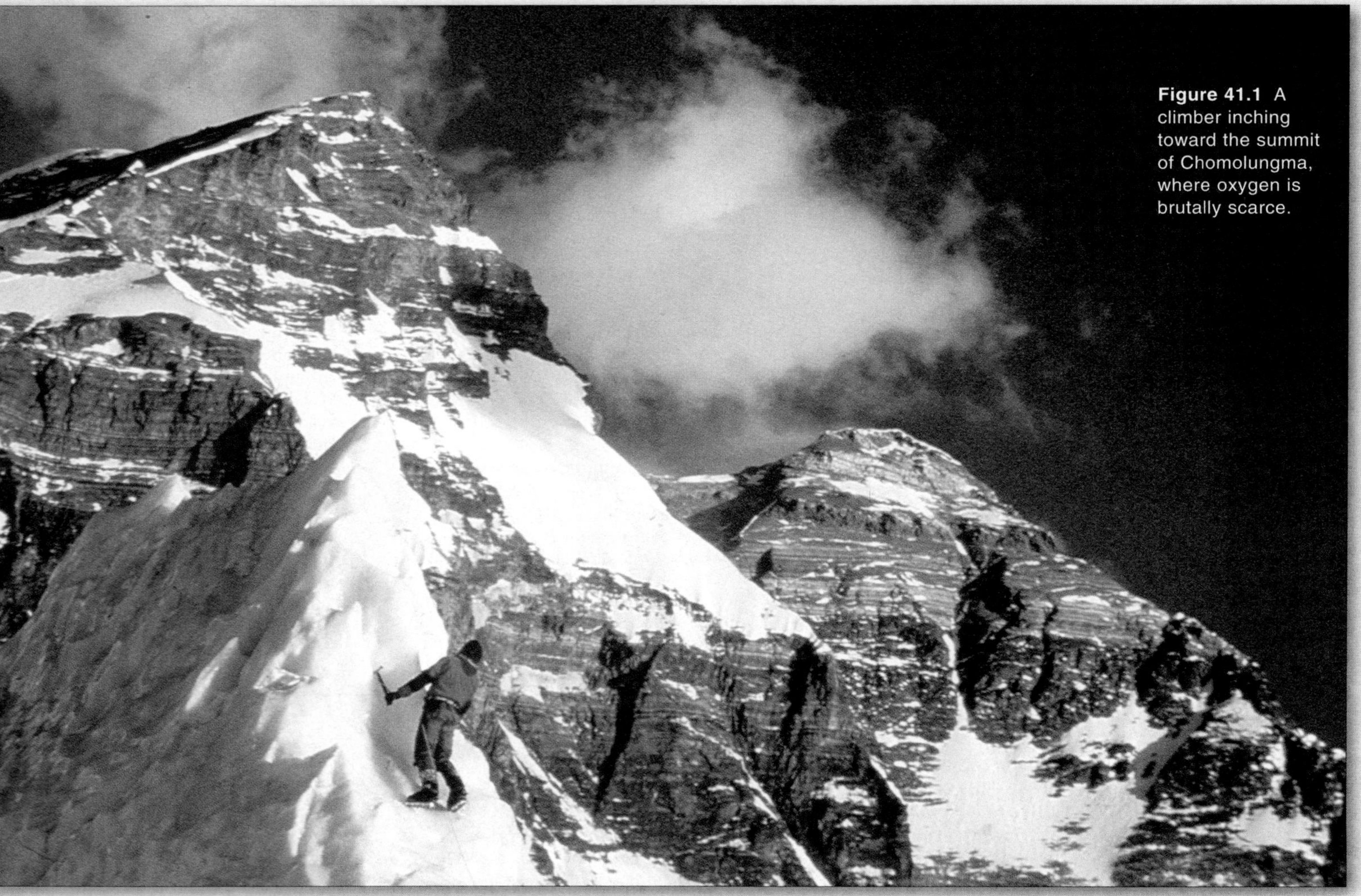

Figure 41.1 A climber inching toward the summit of Chomolungma, where oxygen is brutally scarce.

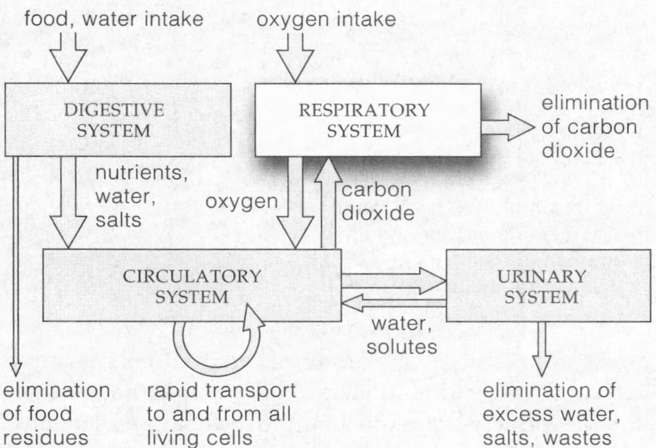
food, water intake
oxygen intake
DIGESTIVE SYSTEM
RESPIRATORY SYSTEM
elimination of carbon dioxide
nutrients, water, salts
oxygen
carbon dioxide
CIRCULATORY SYSTEM
URINARY SYSTEM
water, solutes
elimination of food residues
rapid transport to and from all living cells
elimination of excess water, salts, wastes

Figure 41.2 Interactions between the respiratory system and other organ systems in complex animals.

between endothelial cells that make up the capillary wall. Tissues inside the brain and lungs swell with excess fluid. Climbers will become comatose and die if this dangerous form of edema is not reversed. When prepared rescuers reach stricken climbers in time, they often zip them up inside airtight bags. Then they use a device that pumps bottled oxygen into the rescue bag and removes carbon dioxide from it until the "air" in the bag more closely approximates air at 2,400 meters.

Few of us will ever find ourselves near the peak of Chomolungma, pushing our reliance on oxygen to the limits. Here in the lowlands, disease, smoking, and other environmental insults push it in more ordinary ways, although the risks can be just as great.

The point is this: *Each animal has a body plan that is adapted to the oxygen levels of a particular habitat*. One way or another, by a physiological process known as **respiration**, the body plan allows oxygen to move into the internal environment and carbon dioxide to move out. Why is this important? Animals, remember, have great energy demands. Their cells require a great deal of oxygen—especially for aerobic respiration. This metabolic pathway, remember, produces carbon dioxide wastes as it releases energy from organic compounds.

This chapter samples a few **respiratory systems**, which function in the exchange of gases between the body and the environment. Together with other organ systems, they also contribute to homeostasis—that is, to maintaining internal operating conditions for all of the body's living cells (Figure 41.2).

KEY CONCEPTS

1. Of all organisms, multicelled animals require the most energy to drive their metabolic activities. At the cellular level, the energy comes mainly from aerobic respiration, an ATP-producing metabolic pathway that requires oxygen and produces carbon dioxide wastes.

2. By a physiological process called respiration, animals move oxygen into their internal environment and give up carbon dioxide to the external environment.

3. Oxygen diffuses into the animal body as a result of a pressure gradient. The pressure of this gas is higher in air than it is in metabolically active tissues, where cells rapidly use oxygen. Carbon dioxide follows its own gradient, in the opposite direction. Its pressure is higher in tissues, where it is a by-product of metabolism, than it is in the air.

4. In most respiratory systems, oxygen and carbon dioxide diffuse across a respiratory surface, such as the thin, moist respiratory membrane inside human lungs. Blood flowing through the body's circulatory system picks up oxygen and gives up carbon dioxide at this respiratory membrane.

5. Gas exchange is most efficient when the rate of air flow matches the rate of blood flow. The nervous system brings the rates into balance by controlling the rhythmic pattern and magnitude of breathing.

41.1 THE NATURE OF RESPIRATION

The Basis of Gas Exchange

A concentration gradient, recall, is a difference in the number of molecules of a substance between two regions. Like all substances, oxygen or carbon dioxide tends to diffuse down its concentration gradient, or show a net outward movement from a region where its molecules are colliding more frequently (Section 5.6). The process of respiration is based on the tendency of both gases to follow their respective concentration gradients—or, as we say for gases, to diffuse down **pressure gradients**—that form between the animal and its surroundings.

The gases do not exert the *same* pressure. Pump air into a flat tire near a beach in San Diego or Miami, and you fill it with about 78 percent nitrogen, 21 percent oxygen, 0.04 percent carbon dioxide, and 0.96 percent other gases. This is true of dry air anywhere at sea level. Atmospheric pressure is about 760 mm Hg at sea level, as measured by a mercury barometer of the sort shown in Figure 41.3. Oxygen exerts only part of that total pressure on the tire wall, and its "partial" pressure is greater than that of carbon dioxide. Said another way, the **partial pressure** of oxygen—its contribution to the total atmospheric pressure—is $760 \times 21/100$, or about 160 mm Hg. When similarly measured, carbon dioxide's partial pressure is about 0.3 mm Hg.

Gases enter and leave the animal body by crossing a respiratory surface. A **respiratory surface** is a thin layer of epithelium or some other body tissue. It must be kept moist at all times, for gaseous molecules cannot diffuse across it without being dissolved in fluid. What dictates the *number* of gas molecules moving across a respiratory surface in a given interval? According to **Fick's law**, the more extensive the surface area and the larger the partial pressure gradient, the faster will be the diffusion rate.

Which Factors Influence Gas Exchange?

SURFACE-TO-VOLUME RATIO All animal body plans promote favorable rates of inward diffusion of oxygen and outward diffusion of carbon dioxide. For example, animals with no respiratory organs are tiny, tubelike, or flattened, and gases diffuse directly across the body surface. These body plans meet a constraint imposed by the surface-to-volume ratio, as Section 4.1 describes. To get a sense of the ratio's effects, imagine a flatworm growing in all directions, like an inflating balloon. The worm's surface area does not increase at the same rate as its volume. Once its girth exceeds a single millimeter, the diffusion distance between the body's surface and internal cells will be so great that the worm will die.

VENTILATION Large-bodied, highly active animals have great demands for gas exchange, more than diffusion alone could satisfy. A variety of adaptations make the exchange rate more efficient. For example, above their gills (respiratory organs), many fishes have tissue flaps that stir the surrounding water by moving back and forth. Stirred water puts more dissolved oxygen closer to the gills and carries more carbon dioxide away from them. Among vertebrates, a circulatory system rapidly transports oxygen to cells and carbon dioxide to gills or lungs for disposal. As another example, you breathe in and out to ventilate your lungs.

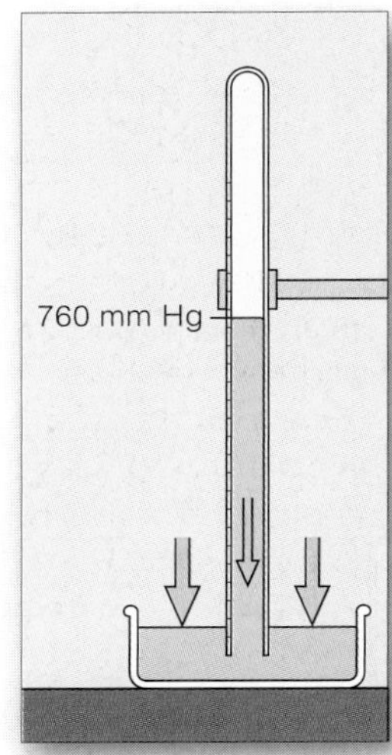
760 mm Hg

Figure 41.3 Atmospheric pressure as measured with a device called a mercury barometer. Part of the device is a glass tube in which the height of a column of mercury (Hg) can increase or decrease, depending upon air pressure outside the device. At sea level, the mercury rises to about 760 millimeters (29.91 inches) from the base of the tube. At this level, the pressure that the column of mercury exerts inside the tube is equal to the atmospheric pressure on the outside.

TRANSPORT PIGMENTS Rates of gas exchange get a boost with transport pigments—mainly **hemoglobin**—that help maintain the steep pressure gradients across the respiratory surface. For example, at the respiratory surfaces in human lungs, the oxygen concentration is high, and each hemoglobin molecule in blood weakly binds as many as four oxygen molecules. (Here you may wish to refer to Section 3.5.) The circulatory system swiftly transports hemoglobin away from the respiratory surface. In oxygen-poor tissues, the oxygen follows its gradient and diffuses out of hemoglobin. By its oxygen-transporting activity, then, hemoglobin helps maintain a pressure gradient that entices oxygen into the blood.

As the chapter introduction states, respiration is a process by which the animal body takes in oxygen for aerobic respiration (an ATP-producing pathway), then removes the pathway's carbon dioxide wastes.

Oxygen and carbon dioxide enter and leave the body by diffusing across a moist respiratory surface. Like other atmospheric gases, they tend to move down their respective pressure gradients. Each type of gas exerts only part of the total pressure across a respiratory surface.

Gas exchange depends on steep partial pressure gradients between the outside and inside of the animal body. The greater the area of the respiratory surface and the larger the partial pressure gradient, the faster diffusion will proceed.

Further reading: Student Guide to InfoTrac on web site

41.2 INVERTEBRATE RESPIRATION

Flatworms, earthworms, and many other invertebrates are not massive, and their life-styles do not depend on high metabolic rates (Figure 41.4*a*). Demands for gas exchange are simply met by **integumentary exchange**, in which gases diffuse directly across the body's surface covering (integument). This mode of respiration only works when the surface is moist, and invertebrates that rely exclusively on it are restricted to aquatic or damp habitats. (By contrast, amphibians and some other large animals use integumentary exchange also, but only as a supplement to other modes of respiration.)

Many invertebrates of aquatic habitats have moist, thin-walled respiratory organs called **gills**. Extensively folded gill walls have an increased respiratory surface area that enhances the rates of exchange between blood or some other body fluid and the surroundings. Figure 41.4*b* shows the much-folded gill of a sea hare (*Aplysia*). By supplementing integumentary exchange, the gill helps provide adequate oxygen for this large mollusk. Some sea hares are 40 centimeters (nearly 16 inches) long.

Spiders and other invertebrates of dry habitats have a small, thick, or hardened integument that is not well endowed with blood vessels. Their integument holds in precious water, but it is not a good respiratory surface. Such animals have an *internal* respiratory surface. Most spiders, for instance, have book lungs: respiratory organs with thin, folded walls that resemble book pages (Section 26.16). But some species have a system of internal tubes for **tracheal respiration**. Most insects, millipedes, and centipedes also rely on tracheal respiration.

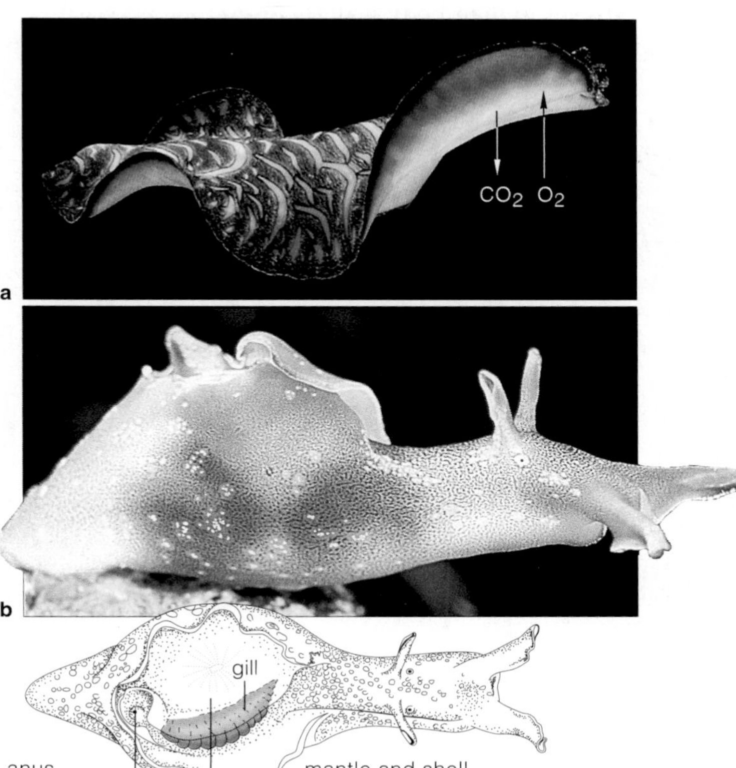
CO_2 O_2
a
b
gill
anus
mantle and shell

Figure 41.4 Invertebrates of aquatic habitats. (**a**) A flatworm, small enough to get along well without an oxygen-transporting circulatory system. Dissolved oxygen in such habitats reaches individual cells by diffusing across the body surface. (**b**) Gill of a sea hare (*Aplysia*), one of the gastropods.

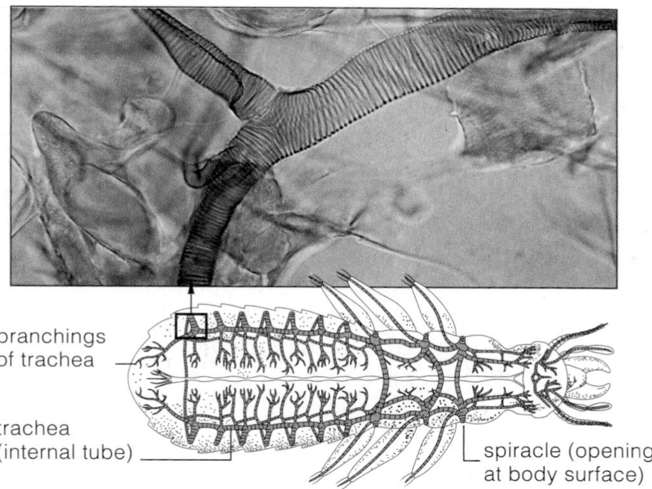
branchings of trachea
trachea (internal tube)
spiracle (opening at body surface)

Figure 41.5 General plan of insect tracheal systems. Chitin rings reinforce many of the branching tubes of the system.

Figure 41.5 shows the tracheal system of an insect. Small openings perforate the insect's integument. Each opening, a spiracle, is the start of a tube that branches within the body. Each of the last branchings dead-ends at its fluid-filled tip, where gases diffuse directly into tissues. The tube tips are especially profuse in muscle tissue and other tissues with high demands for oxygen.

We find hemoglobin or other respiratory pigments in many invertebrates, although they are rare in insects. Respiratory pigments increase the capacity of body fluids to transport oxygen, as they do in vertebrates. In species having a well-developed head, oxygenated blood tends to circulate first through the head end and then through the rest of the body.

Flatworms and some other invertebrates of aquatic or moist habitats do not have a massive body and use integumentary exchange. By this mode of respiration, oxygen and carbon dioxide diffuse directly across the body surface.

Most marine invertebrates and many freshwater types have gills of one sort or another. These respiratory organs have moist, thin, and often highly folded walls.

Most insects, millipedes, centipedes, and some spiders use tracheal respiration. Gases flow through open-ended tubes that start at the body surface and end directly in tissues.

Gills of Fishes and Amphibians

Gills are the respiratory organs of many vertebrates. A few kinds of fish larvae and a few amphibians have *external* gills projecting into water. Adult fishes have a pair of *internal* gills. These are rows of slits or pockets at the back of the mouth that extend to the surface of the body, as in Figure 41.6*a*. Whatever its form, each gill has a wall of moist, thin, vascularized epithelium.

In fishes, water flows into the mouth and pharynx, then over arrays of filaments in the gills (Figure 41.6*b*). Blood vessels thread through respiratory surfaces in each filament. First the water flows past a vessel that leads to the rest of the body. The blood inside has less oxygen than the water does, so oxygen diffuses into the blood. The same volume of water flows over a vessel leading into the gills. The water already gave up some oxygen, but it still has more than the blood inside the vessel does, so more oxygen diffuses into the filament. The movement of two fluids in opposing directions is called **countercurrent flow**. By this mechanism, a fish extracts about 80 to 90 percent of the dissolved oxygen flowing past. That is more than the fish would get from a one-way flow mechanism, at far less energy cost.

Lungs

Some fishes and all amphibians, birds, and mammals have one pair of **lungs**, internal respiratory surfaces in the shape of a cavity or sac. Lungs originated in some lineages of fishes more than 450 million years ago as pouches off the anterior part of the gut wall (Section 27.5). They evolved rapidly by way of natural selection, probably because they afforded survival advantages. Paired lungs increased the surface area available for gas exchange in oxygen-poor habitats. They worked better than gills could when some vertebrates moved onto dry land. Gills stick together and cannot function unless water flows through them and keeps them moist.

Lungfishes of oxygen-poor habitats still have gills. They also use tiny lungs as a backup for gas exchange.

Amphibians never completed the transition to land. Salamanders especially use their skin as a respiratory surface for integumentary exchange. Frogs and toads rely more on small lungs for oxygen uptake, but most of their carbon dioxide wastes diffuse outward across the skin. Frogs also are heavy-duty breathers; they *force* air into their lungs, which they empty by contracting muscles in their body wall (Figure 41.7).

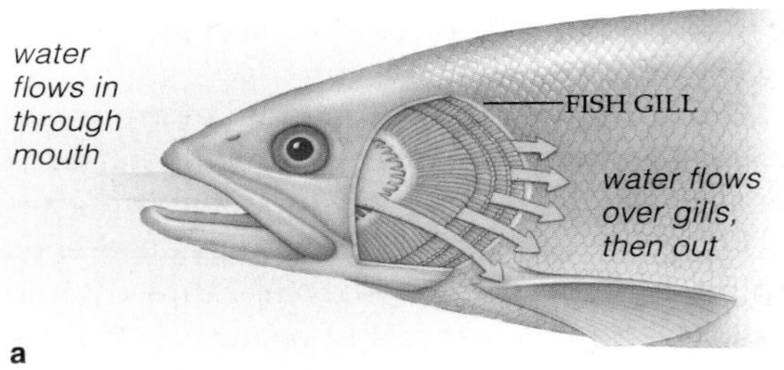
water flows in through mouth
FISH GILL
water flows over gills, then out
a

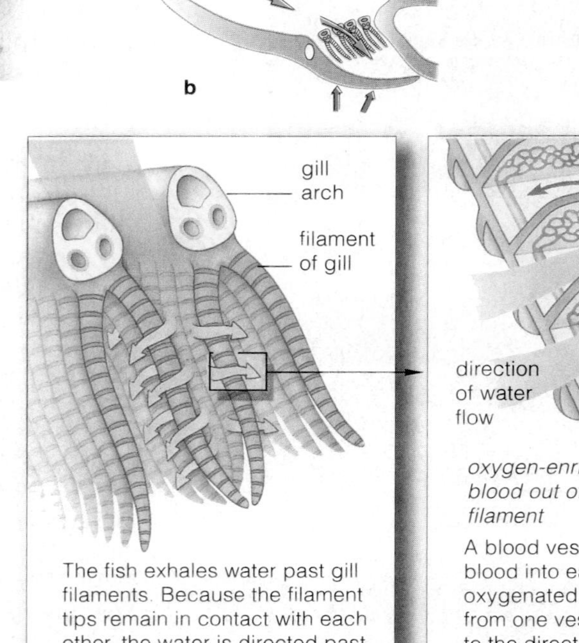
mouth open
lid closed
b
gill arch
filament of gill
d

The fish exhales water past gill filaments. Because the filament tips remain in contact with each other, the water is directed past a gas exchange surface.

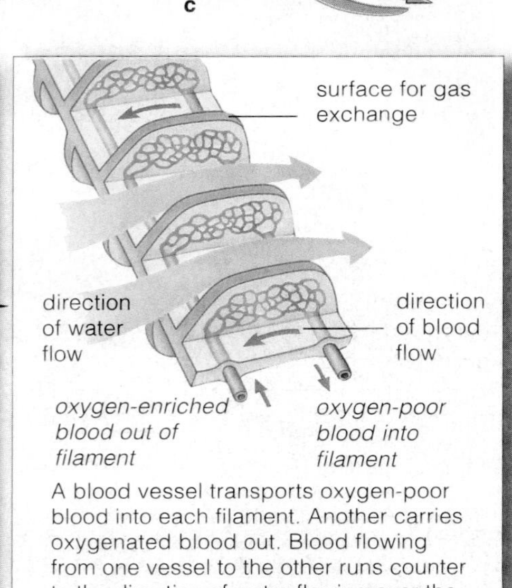
mouth closed
lid open
c
surface for gas exchange
direction of water flow
direction of blood flow
oxygen-enriched blood out of filament
oxygen-poor blood into filament
e

A blood vessel transports oxygen-poor blood into each filament. Another carries oxygenated blood out. Blood flowing from one vessel to the other runs counter to the direction of water flowing over the gas exchange surfaces.

Figure 41.6 Example of fish gills. (**a**) One of a pair of gills. Each gill is located under a bony lid, removed for this sketch.

(**b**) Gill ventilation. Water is pulled across a fish gill when the mouth opens and the lid closes. (**c**) Water is forced back out when the mouth closes and the lid opens.

(**d**) Gas exchange at a gill. Filaments in the gill have vascularized respiratory surfaces. Tips of adjacent filaments touch, so water passing over them is directed past the gas exchange surfaces before leaving the body. (**e**) One blood vessel from tissues deep in the body delivers oxygen-poor blood into a filament. Another carries oxygenated blood away from it. Blood flowing from one vessel into the other runs counter to the direction of water flow over the gas exchange surfaces. Countercurrent flow favors the movement of oxygen (down its partial pressure gradient) from water to the blood.

Further reading: Student Guide to InfoTrac on web site

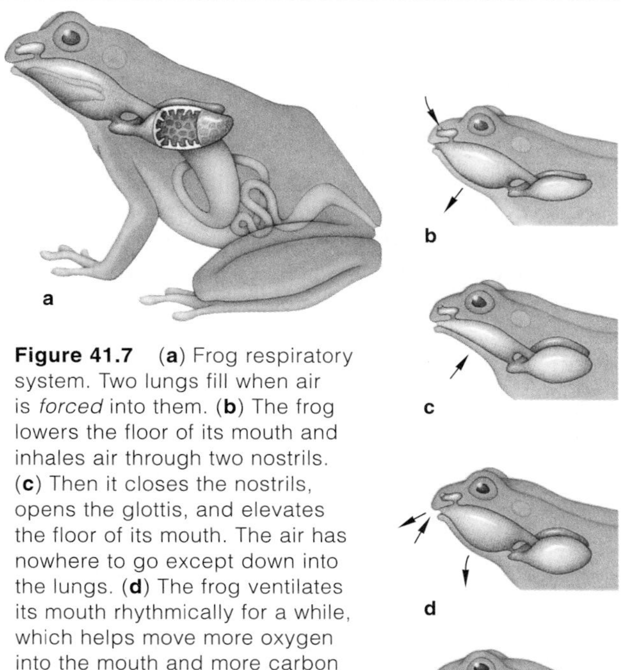

a
b
c
d
e

Figure 41.7 (**a**) Frog respiratory system. Two lungs fill when air is *forced* into them. (**b**) The frog lowers the floor of its mouth and inhales air through two nostrils. (**c**) Then it closes the nostrils, opens the glottis, and elevates the floor of its mouth. The air has nowhere to go except down into the lungs. (**d**) The frog ventilates its mouth rhythmically for a while, which helps move more oxygen into the mouth and more carbon dioxide out from it. (**e**) Muscles in the body wall above the lungs contract, the lungs elastically recoil, and air is forced out.

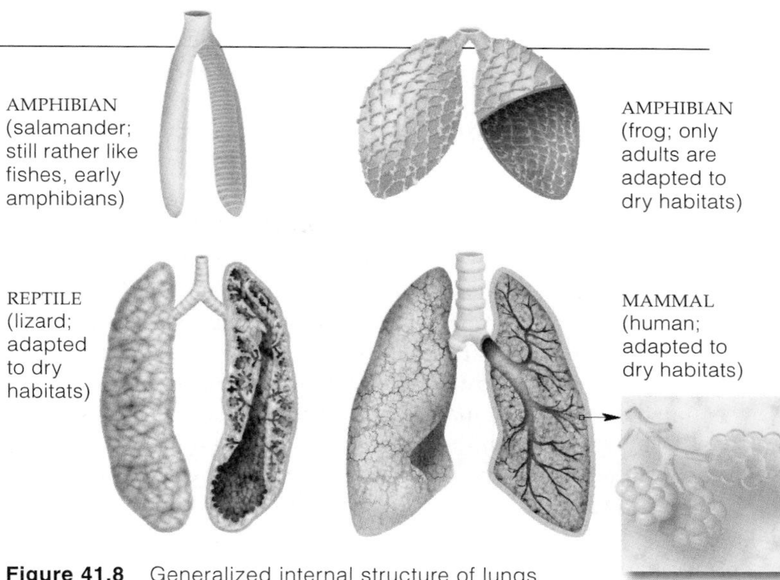

AMPHIBIAN (salamander; still rather like fishes, early amphibians)
AMPHIBIAN (frog; only adults are adapted to dry habitats)
REPTILE (lizard; adapted to dry habitats)
MAMMAL (human; adapted to dry habitats)

Figure 41.8 Generalized internal structure of lungs from four vertebrates, suggestive of an evolutionary trend from simple gas exchange sacs to complex lungs with a greater respiratory surface area.

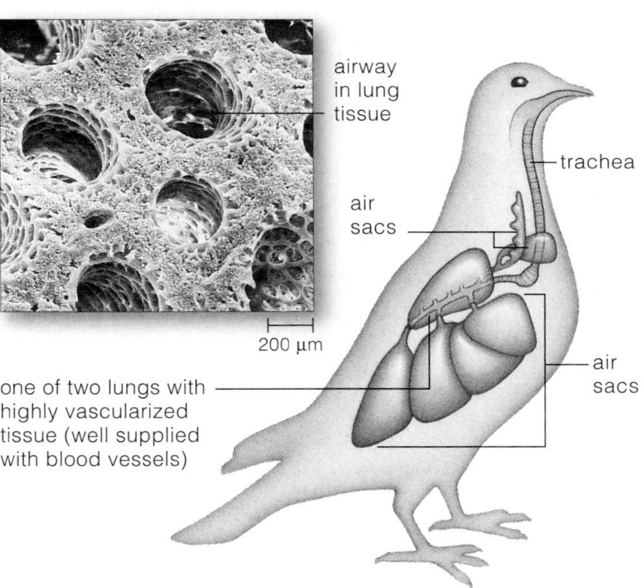

airway in lung tissue
trachea
air sacs
200 μm
air sacs
one of two lungs with highly vascularized tissue (well supplied with blood vessels)

Figure 41.9 Bird respiratory system. Two small, inelastic lungs have a number of attached air sacs. When the bird inhales, it draws air into the air sacs through tubes, open at both ends, that thread through the vascularized lung tissue. This tissue is the respiratory surface, where gases are exchanged.

When the bird exhales, it forces air out of the sacs, through the small tubes, and out of the trachea. Thus, air is not merely drawn into the bird lungs. Air is continuously drawn through them and across respiratory surfaces. This unique ventilating system supports the high metabolic rates that birds require for flight and other energy-intensive activities.

Frogs also use their lungs for sound production. So do all mammals except whales. Sound originates near the entrance to a larynx, an airway to the lungs. Here, part of a mucous membrane folds perpendicularly to the airway. The folds are **vocal cords**, and the gap between them is a **glottis**. Frogs produce sounds by forcing air back and forth through the glottis, between the lungs and paired pouches on the floor of the mouth. Air flow causes the cords to vibrate. Like other vertebrates, frogs produce different sounds by controlling the vibrations.

Paired lungs are the dominant respiratory organs in reptiles, birds, and mammals (Figure 41.8). Breathing moves air by bulk flow into and out of the lungs, where blood capillaries wrap lacily around the respiratory surface. In the lungs, oxygen and carbon dioxide have steep concentration gradients and diffuse rapidly across the respiratory surface. Oxygen enters the capillaries and is circulated quickly through the body. In regions with low oxygen levels, oxygen diffuses into interstitial fluid, then into cells. Carbon dioxide moves rapidly in the opposite direction and is expelled from the lungs.

This mode of gas exchange is embellished a bit only in birds. As described in Figure 41.9, birds are unique in that air not only flows into and out from their lungs; it also flows *through* their lungs. With that exception in mind, we turn next to the human respiratory system, for its operating principles apply to most vertebrates.

A countercurrent flow mechanism in fish gills compensates for low oxygen levels in aquatic habitats. Internal air sacs—lungs—are more efficient in dry habitats on land.

Amphibians use integumentary exchange and force air into and out of small lungs. Ventilation of paired lungs is the major mode of respiration in reptiles, birds, and mammals.

HUMAN RESPIRATORY SYSTEM

Functions of the Respiratory System

Getting oxygen from air and expelling carbon dioxide are the key functions of the human respiratory system. Breathing, a form of ventilation, alternately moves air into and out of a pair of lungs, each of which has about 300 million outpouchings. Each outpouching is a tiny air sac: an **alveolus** (plural, alveoli). Controls adjust the rate of breathing so that the inflow and outflow of air match metabolic demands for gas exchange.

The respiratory system's role in respiration ends at alveoli, where the circulatory system takes over the task of moving gases. Oxygen and carbon dioxide move by diffusion between the alveoli and pulmonary capillaries adjacent to them. (The Latin *pulmo* means lung.)

The respiratory system has other functions. Breathing is used for vocalizations, such as speech. It enhances the return of venous blood to the heart, and it helps rid the body of excess body heat and water. Controls over breathing adjust the body's acid–base balance. Carbon dioxide, remember, is used in the formation of carbonic acid (H_2CO_3), which works with bicarbonate (HCO_3^-) as a buffer system. HCO_3^- accepts or releases hydrogen ions (H^+), depending on the blood's pH (Section 2.6). Rapid, deep breathing expels more carbon dioxide, so less carbonic acid forms—and so the body loses acid.

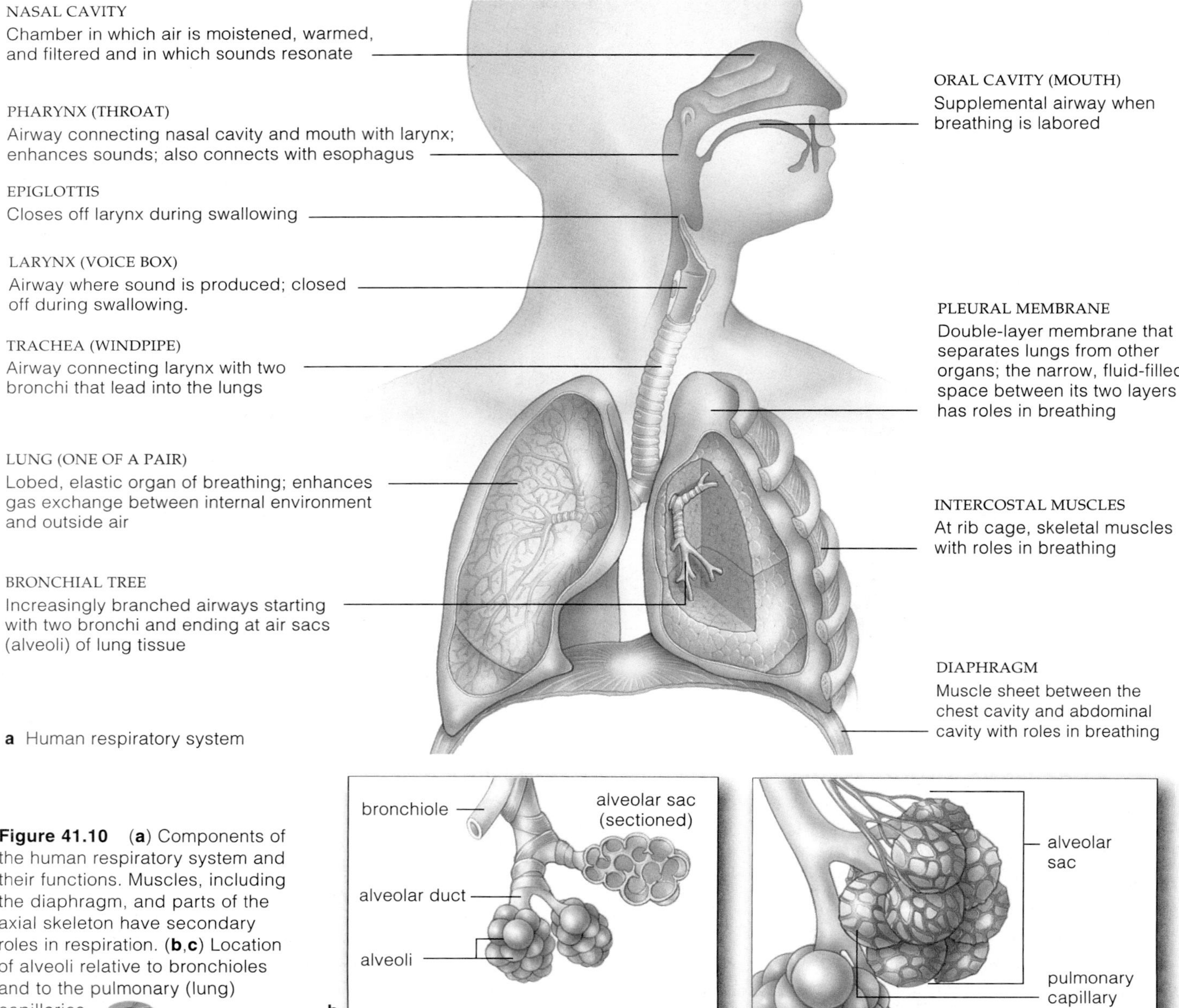
NASAL CAVITY
Chamber in which air is moistened, warmed, and filtered and in which sounds resonate
PHARYNX (THROAT)
Airway connecting nasal cavity and mouth with larynx; enhances sounds; also connects with esophagus
EPIGLOTTIS
Closes off larynx during swallowing
LARYNX (VOICE BOX)
Airway where sound is produced; closed off during swallowing.
TRACHEA (WINDPIPE)
Airway connecting larynx with two bronchi that lead into the lungs
LUNG (ONE OF A PAIR)
Lobed, elastic organ of breathing; enhances gas exchange between internal environment and outside air
BRONCHIAL TREE
Increasingly branched airways starting with two bronchi and ending at air sacs (alveoli) of lung tissue
a Human respiratory system
ORAL CAVITY (MOUTH)
Supplemental airway when breathing is labored
PLEURAL MEMBRANE
Double-layer membrane that separates lungs from other organs; the narrow, fluid-filled space between its two layers has roles in breathing
INTERCOSTAL MUSCLES
At rib cage, skeletal muscles with roles in breathing
DIAPHRAGM
Muscle sheet between the chest cavity and abdominal cavity with roles in breathing
bronchiole
alveolar sac (sectioned)
alveolar duct
alveoli
b
alveolar sac
pulmonary capillary
c

Figure 41.10 (**a**) Components of the human respiratory system and their functions. Muscles, including the diaphragm, and parts of the axial skeleton have secondary roles in respiration. (**b**,**c**) Location of alveoli relative to bronchioles and to the pulmonary (lung) capillaries.

Slow, shallow breathing has the opposite effect. Carbon dioxide builds up, more H_2CO_3 forms, and the body gains acid.

The respiratory system also has built-in mechanisms for dealing with airborne foreign agents and substances inhaled with air. Finally, it removes and inactivates or otherwise modifies a number of blood-borne substances before they can circulate through the rest of the body.

From Airways Into the Lungs

It will take at least 300 million breaths to get you to age seventy-five. You might find yourself going without food for a few hours or days. But stop breathing even for five minutes and normal brain function is over.

Take a deep breath. Now look at Figure 41.10 to get an idea of where the air will travel in your respiratory system. Unless you are out of breath and panting, air just entered two nasal cavities, not your mouth. There, mucous secretions warm and moisten it. In the cavities, ciliated epithelium and hairs filter dust and particles from air. Their olfactory receptors function in the sense of smell (Section 36.3). Now air is poised at the **pharynx**, or throat. The pharynx is the entrance to the **larynx**, an airway with two paired folds of mucous membrane. The lower pair are vocal cords, as shown in Figure 41.11. When you breathe, air is forced in and out through the glottis, which is the gap between the cords. As in frogs (Section 41.3), air flow makes the vocal cords vibrate in ways that can be controlled to make different sounds.

Within the folds are thick bands of elastic ligaments connected to various cartilage tissues. When muscles of the larynx contract and relax, the ligaments tighten or slacken, which changes the extent to which the folds are stretched. Under commands from the nervous system, the coordinated action of muscles narrows or widens the glottis. For example, by increasing muscle tension in the vocal cords, you can decrease the gap between them and make high-pitched sounds or squeaks. The lips, teeth, tongue, and the soft roof over the tongue are enlisted to modify the different sounds into patterns of vocalization, such as speech and song.

When vocal cords are inflamed as an outcome of an infection or irritation, swelling of their mucous lining interferes with their capacity to vibrate. If the swelling causes hoarseness, this condition is called *laryngitis*.

The **epiglottis** is a tissue flap at the entrance to the larynx. When it points upward, air can move into the **trachea**, or windpipe. The epiglottis points downward and shuts off the trachea's entrance when you swallow food or fluid, which enters a different tube (esophagus) that connects the pharynx with the stomach.

The trachea branches into two airways, one leading into the tissue of each lung. Each airway is a **bronchus**

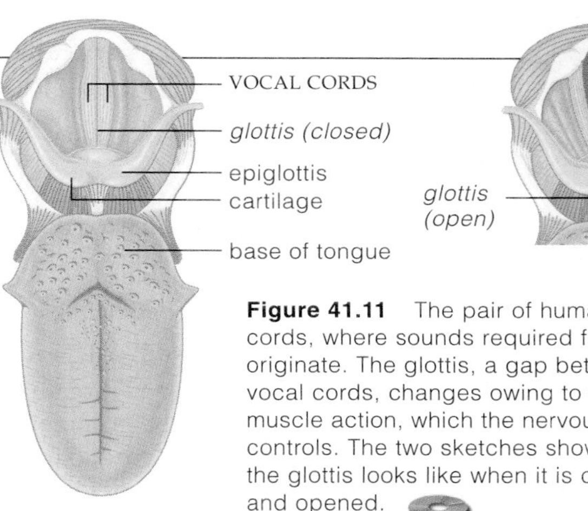
VOCAL CORDS
glottis (closed)
epiglottis
cartilage
base of tongue
glottis (open)

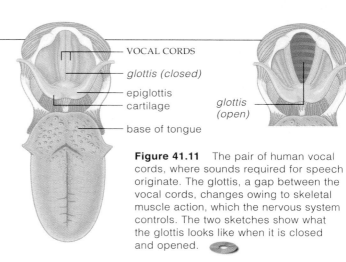

Figure 41.11 The pair of human vocal cords, where sounds required for speech originate. The glottis, a gap between the vocal cords, changes owing to skeletal muscle action, which the nervous system controls. The two sketches show what the glottis looks like when it is closed and opened.

(plural, bronchi). Its epithelial lining has a profusion of cilia and mucus-secreting cells. The lining is a barrier to infection. Bacteria and airborne particles stick to the mucus. Then cilia sweep the debris-laden mucus toward the mouth. The mucus is expelled or swallowed.

Human lungs are elastic, cone-shaped organs of gas exchange. They are positioned inside the rib cage to the left and right of the heart and above the diaphragm. The **diaphragm** is a muscular partition between the thoracic and abdominal cavities. A thin, saclike pleural membrane lines the outer surface of the lungs and inner surface of the thoracic cavity wall. The thoracic wall and the lungs press its opposing surfaces together. By analogy, think of each lung as a baseball pushed so far into one side of a softly inflated balloon that the opposing sides of the balloon press together as they wrap around the balls.

A thin film of lubricating fluid separates the pleural membrane surfaces and reduces friction between them. In a respiratory ailment called *pleurisy*, the membrane becomes inflamed and swollen, friction occurs between its surfaces, and breathing can be painful.

Inside each lung, air moves through finer and finer branchings of a "bronchial tree" (Figure 41.10*a*). These airways are **bronchioles**. Their finest branchings, the *respiratory* bronchioles, end in the cup-shaped alveoli. Alveoli usually are clustered as larger pouches called alveolar sacs, as in Figure 41.10*b*,*c*. Collectively, alveolar sacs offer a tremendous surface area for gas exchange with blood. If all the alveolar sacs were stretched out in one layer, they would cover a racquetball court floor!

Oxygen uptake and carbon dioxide removal are the major functions of the human respiratory system. Inside its pair of lungs, the circulatory system takes over the remaining tasks of respiration.

The respiratory system also has roles in moving venous blood to the heart, in vocalization, in adjusting the body's acid–base balance, in defense against harmful airborne agents or substances, in removing or modifying a number of blood-borne substances, and in the sense of smell.

BREATHING—CYCLIC REVERSALS IN AIR PRESSURE GRADIENTS

The Respiratory Cycle

There is a cyclic pattern to breathing, which ventilates the lungs. Each **respiratory cycle** consists of two actions: *inhalation* (a single breath of air drawn into the airways) and *exhalation* (a single breath out).

Look at Figure 41.12. Inhalation always is an active, energy-requiring action. When someone is breathing quietly, inhalation is brought about by the contraction of the diaphragm and, to a lesser extent, the external intercostal muscles. The outcome is an increase in the thoracic cavity volume. Breathe hard, and the volume increases further because neck muscles contract and so elevate the sternum and first two ribs attached to them.

During each respiratory cycle, the thoracic cavity's volume increases, then decreases. *And pressure gradients between air inside and outside the respiratory tract change.* Think about the different pressures exerted during one respiratory cycle. The *atmospheric* pressure, 760 mm Hg at sea level, is exerted by the combined weight of all atmospheric gases on all airways. Before inhalation, *intrapulmonary* pressure (the pressure inside all alveoli) is also 760 mm Hg (Figure 41.12*d*).

A different pressure gradient helps keep the lungs close to the thoracic cavity wall during a respiratory cycle even through exhalation, when the lungs have a smaller volume than the thoracic cavity (Figure 41.12*e,f*). When that cavity expands, so do the lungs, as a result of a pressure gradient across the lung wall.

In a person who is resting, the *intrapleural* pressure (inside the pleural sac) averages 756 mm Hg, which is less than atmospheric pressure. Intrapleural

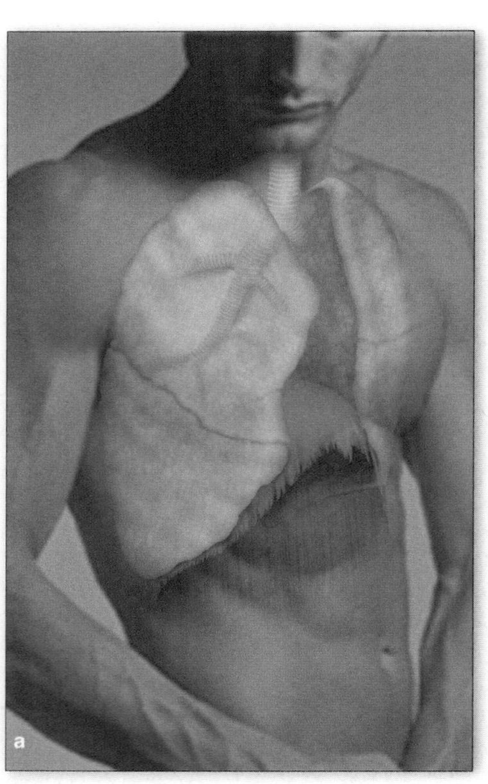
a

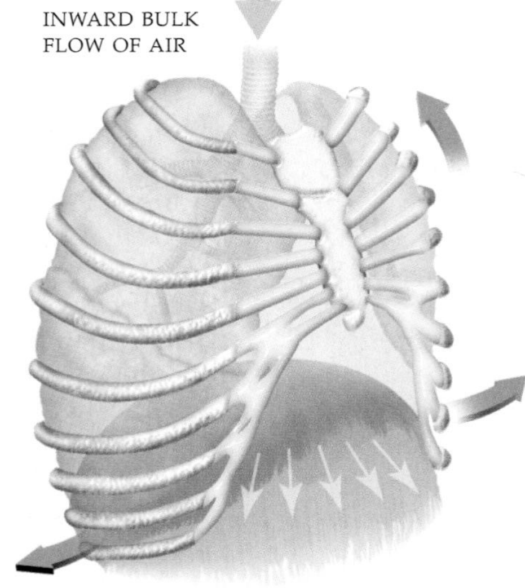
INWARD BULK
FLOW OF AIR

b Inhalation. Diaphragm contracts and moves down. The external intercostal muscles contract and lift rib cage upward and outward. The lung volume expands.

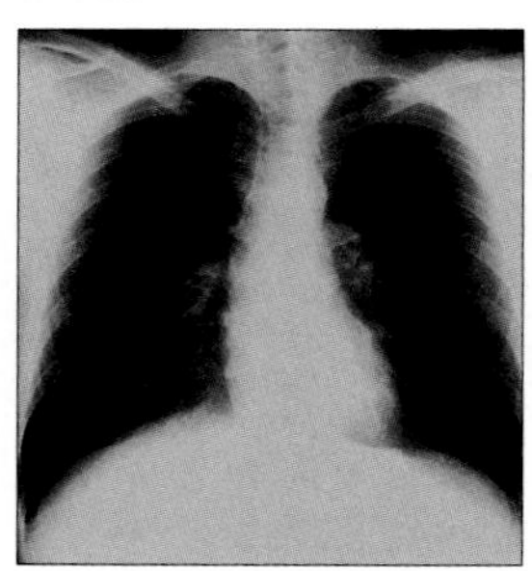

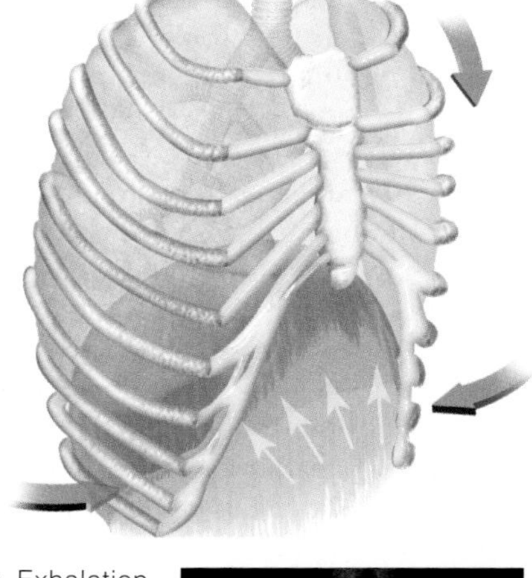
OUTWARD BULK
FLOW OF AIR

c Exhalation. Diaphragm and external intercostal muscles return to the resting positions. Rib cage moves down. Lungs recoil passively.

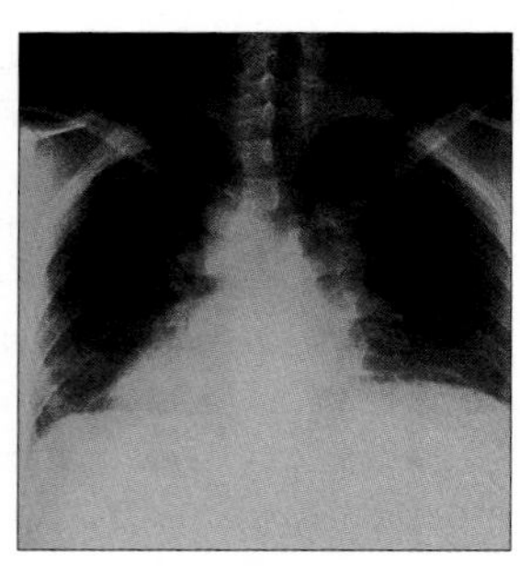

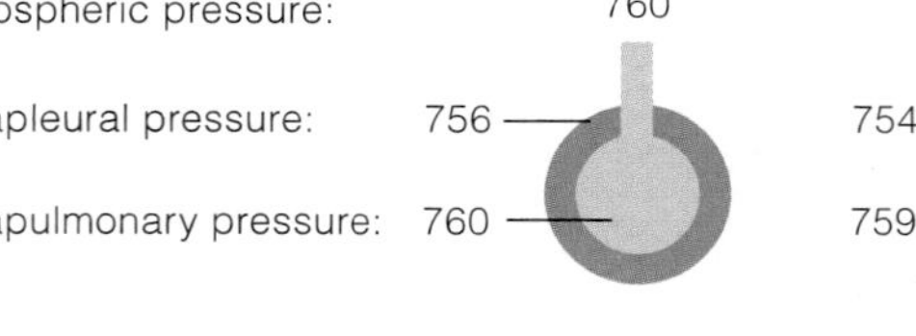
Atmospheric pressure:
Intrapleural pressure:
Intrapulmonary pressure:
760
756
760

d BEFORE INHALATION

760
754
759

e DURING INHALATION (LUNGS EXPANDED)

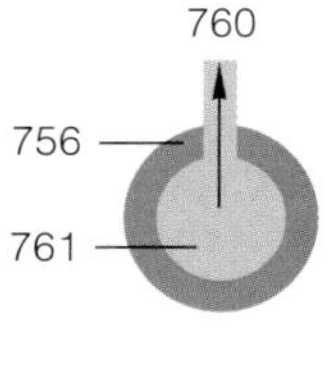
760
756
761

f DURING EXHALATION

Figure 41.12 (**a**) Location of the lungs relative to the diaphragm. (**b**,**c**) Changes in the size of the thoracic cavity during one respiratory cycle. The x-ray images indicate how the maximum inhalation possible changes the volume of the thoracic cavity. (**d–f**) Changes in lung volume and intrapulmonary pressure during a respiratory cycle.

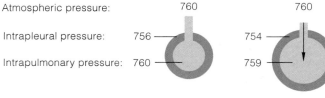

Further reading: Student Guide to InfoTrac on web site

pressure is exerted outside the lungs—that is, inside the thoracic cavity. As intrapleural pressure is pushing inward on a lung wall, the intrapulmonary pressure is pushing outward. The difference in pressure between them—4 mm Hg—is large enough to make the lungs stretch and fill the thoracic cavity.

The cohesiveness of water molecules present in the intrapleural fluid (the fluid inside the pleural sac) also helps keep lungs close to the thoracic wall. By analogy, wet two small panes of glass and press them together. The panes can easily slide back and forth but strongly resist being pulled apart. Similarly, intrapleural fluid "glues" lungs to the wall. What is the outcome? *When the thoracic cavity is expanding during inhalation, the lungs must expand, also.*

Figure 41.12*b* shows what happens as you start to inhale. The dome-shaped diaphragm flattens and moves downward, and the rib cage is lifted both upward and outward. When the thoracic cavity expands, the lungs expand along with it. At this time, the air pressure in all alveolar sacs combined is lower than the atmospheric pressure. Fresh air follows the pressure gradient and flows down the airways, then into the alveoli.

When someone breathes quietly, the second action of the respiratory cycle is passive. The muscles that brought about inhalation relax and the lungs passively recoil, without further energy outlays. The resultant decrease in lung volume compresses air in the alveolar sacs. At this time, the pressure inside the sacs is greater than atmospheric pressure. Air follows the gradient and flows out from the lungs, as in Figure 41.12*c*. Exhalation only becomes active and energy-requiring when you exercise vigorously and more air must be expelled.

With active exhalation, muscles in the abdominal wall contract, so pressure inside the abdomen increases and exerts upward-directed force on the diaphragm. As the diaphragm is pushed upward, the thoracic cavity volume decreases. How? Internal intercostal muscles contract and thus pull the thoracic wall downward and inward. The chest wall flattens, so the dimensions of the thoracic cavity decrease even more. Lung volume decreases as well when the elastic tissue of the lungs passively recoils.

Lung Volumes

The lungs can hold up to 5.7 liters or so of air in adult males and 4.2 liters in adult females who are young and in good health. These are only average values; age, build, and respiratory health influence the total lung capacity. During quiet breathing, the lungs are far from being fully inflated. In general, they hold 2.7 liters at the end of one inhalation and 2.2 liters at the end of one exhalation. The lungs never do deflate completely.

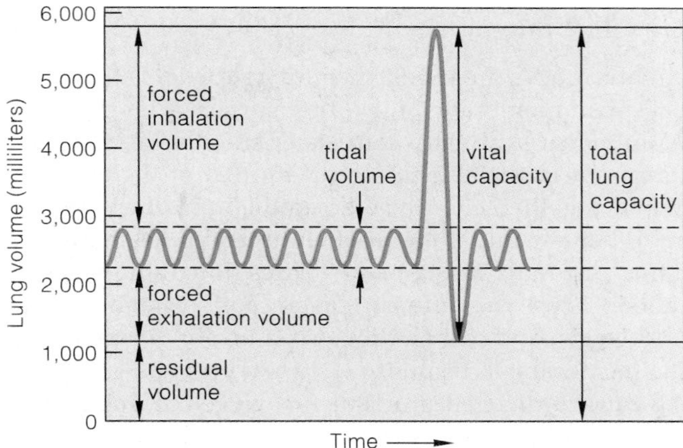

6,000
5,000
4,000
3,000
2,000
1,000
0
Lung volume (milliliters)
forced inhalation volume
tidal volume
vital capacity
total lung capacity
forced exhalation volume
residual volume
Time

Figure 41.13 Lung volume. During quiet breathing, a tidal volume of air enters and leaves the lungs. Forced inhalation delivers more air to them; forced exhalation releases some air that normally stays in them. A residual volume is trapped in partially filled alveoli even during the strongest exhalation.

When air flows out and lung volume is low, walls of the smallest airways collapse and prevent further air loss.

The volume of air that can move out of the lungs in one breath after maximal inhalation is called the **vital capacity**. Humans rarely use more than half of the total vital capacity, even when taking in very deep breaths during strenuous exercise (Figure 41.13). Because the lungs are never empty, gas exchange between alveolar air and the blood proceeds even at the end of maximum exhalation. *Hence the concentrations of gases in the blood remain fairly constant throughout the respiratory cycle.*

The volume of air flowing into or out of the lungs in the respiratory cycle, the **tidal volume**, averages 0.5 liter. How much of that volume is available for gas exchange? In between breaths, about 0.15 liter stays in the airways; only 0.35 liter of fresh air reaches alveoli. When you breathe, say, ten times a minute, you are supplying your alveoli with (0.35×10) or 3.5 liters of fresh air per minute.

Breathing, which ventilates the lungs, has a cyclic pattern. The respiratory cycle consists of inhalation (one breath of air in) and exhalation (one breath of air out).

Inhalation is always an active, energy-requiring process involving contractions mainly of the diaphragm and the external intercostal muscles.

During quiet breathing, exhalation is a passive process. Muscles relax, the thoracic cavity volume decreases, and the lungs recoil elastically. Forceful exhalation is an active process that requires abdominal muscle contraction.

Breathing reverses pressure gradients between the lungs and the air outside the body.

41.6 GAS EXCHANGE AND TRANSPORT

Gas Exchange

An alveolus is a cupped sheet of epithelial cells with a basement membrane at its outer surface (Figure 41.14). A pulmonary capillary consists of endothelial cells, and its basement membrane fuses with that of the alveolus. Taken together, the alveolar epithelium, the capillary endothelium, and their basement membranes represent a thin respiratory membrane. Gases flow rapidly across them because the diffusion distance is so small.

Oxygen and carbon dioxide passively diffuse across the membrane in response to partial pressure gradients. An inward-directed gradient for oxygen is maintained because inhalations replenish oxygen and cells use it on an ongoing basis. An outward-directed gradient for carbon dioxide is maintained because cells continually produce carbon dioxide and exhalation removes it.

Oxygen Transport

Oxygen, like carbon dioxide, does not dissolve well in blood, so it cannot be efficiently transported on its own. In all large animals, demands for oxygen are met with the assistance of hemoglobin molecules. These are the respiratory pigments that are packed inside red blood cells. A hemoglobin molecule, recall, is a compact array of four polypeptide chains and four **heme groups**, each of which contains an iron atom that binds reversibly with oxygen. *Of all oxygen inhaled into the human body, 98.5 percent of it is bound to heme groups of hemoglobin.*

Normally, inhaled air that reaches alveoli has plenty of oxygen. The opposite is true of blood flowing past in the pulmonary capillaries. Thus, in the lungs, oxygen tends to diffuse into the plasma portion of blood. Then it diffuses into red blood cells and rapidly binds with hemoglobin. Hemoglobin that has oxygen bound to it is known as **oxyhemoglobin**, or HbO_2.

In a given interval, the amount of HbO_2 that forms depends on oxygen's partial pressure. The higher the pressure, the greater will be the oxygen concentration. When plenty of oxygen molecules are available, they tend to randomly collide with the heme binding sites at a faster rate. The encounters continue until all four of the binding sites in hemoglobin are saturated.

HbO_2 molecules bind oxygen weakly and give it up where oxygen's partial pressure is lower than in the lungs. They give it up faster in tissues where blood is warmer, the pH is lower, and carbon dioxide's partial pressure is high. Such conditions prevail in contracting muscle and other metabolically whipped-up tissues.

Carbon Dioxide Transport

Carbon dioxide diffuses into blood capillaries in any tissue where its partial pressure is higher than it is in the blood flowing past. From there, three mechanisms transport it to the lungs. About 10 percent of the gas stays dissolved in blood. Another 30 percent binds with hemoglobin to form **carbamino hemoglobin** ($HbCO_2$). Most of it—60 percent—is transported in the form of bicarbonate (HCO_3^-). How do these HCO_3^- molecules form? Carbon dioxide combines with water in blood to form carbonic acid, which then dissociates (separates) into bicarbonate and hydrogen ions (H^+):

$$CO_2 + H_2O \rightleftharpoons \underset{\text{CARBONIC ACID}}{H_2CO_3} \rightleftharpoons \underset{\text{BICARBONATE}}{HCO_3^-} + H^+$$

In blood plasma, that reaction converts only 1 of every 1,000 carbon dioxide molecules—which doesn't amount to much. It is a different story in red blood cells, which have the enzyme **carbonic anhydrase**. Enzyme action enhances the reaction rate 250 times! Most of the carbon dioxide not bound to hemoglobin becomes converted to carbonic acid this way. Conversion makes the blood level of carbon dioxide decline rapidly. Thus it helps maintain the gradient that promotes diffusion of carbon dioxide from interstitial fluid into the bloodstream.

What about the HCO_3^- that forms in the reactions? It tends to move out of red blood cells and into blood

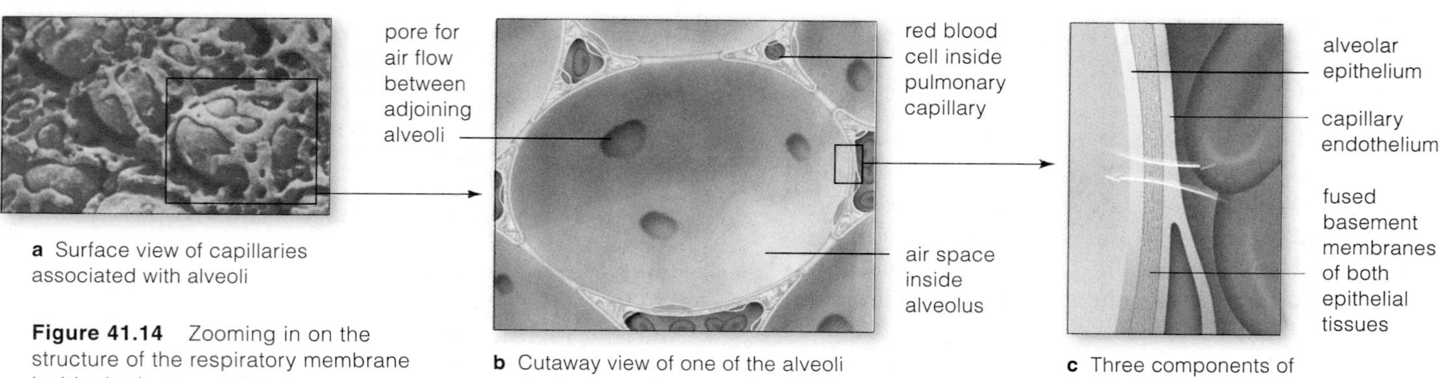

pore for air flow between adjoining alveoli
red blood cell inside pulmonary capillary
air space inside alveolus
alveolar epithelium
capillary endothelium
fused basement membranes of both epithelial tissues

a Surface view of capillaries associated with alveoli

b Cutaway view of one of the alveoli and adjacent pulmonary capillaries

c Three components of the respiratory membrane.

Figure 41.14 Zooming in on the structure of the respiratory membrane inside the lungs.

Further reading: Student Guide to InfoTrac on web site

plasma. What about the H^+ ions? Hemoglobin acts as a buffer for them and keeps blood from getting too acidic. A *buffer* is any molecule that combines with or releases H^+ in response to a shift in cellular pH (Section 2.6).

The reactions are reversed in alveoli, where carbon dioxide's partial pressure is lower than it is in the lung capillaries. Carbon dioxide and water form here. The gas diffuses into the alveolar sacs, then it is exhaled.

At this point, reflect on Figure 41.15. It summarizes the partial pressure gradients for oxygen and carbon dioxide through the human respiratory system.

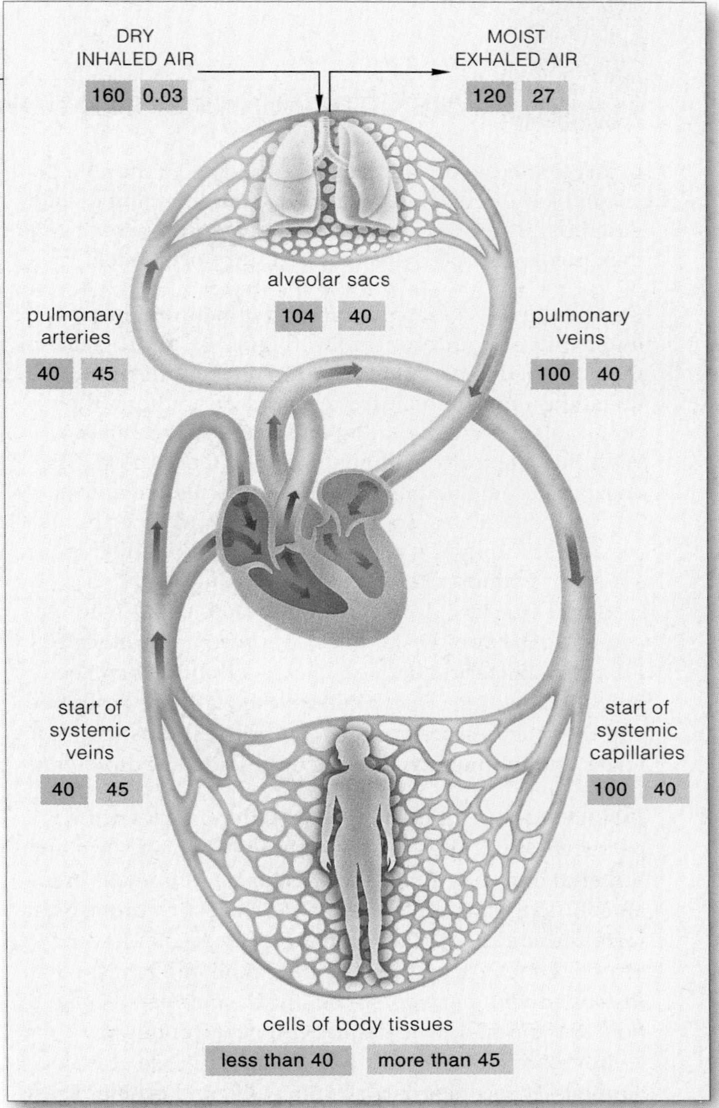
DRY
INHALED AIR
160
0.03
MOIST
EXHALED AIR
120
27
alveolar sacs
104
40
pulmonary
arteries
40
45
pulmonary
veins
100
40
start of
systemic
veins
40
45
start of
systemic
capillaries
100
40
cells of body tissues
less than 40
more than 45

Figure 41.15 Partial pressure gradients for oxygen (*blue* boxes) and carbon dioxide (*pink* boxes) in the respiratory tract.

Matching Air Flow With Blood Flow

Gas exchange is most efficient when the rate of air flow matches the rate of blood flow. The nervous system acts to balance them by controlling the *rhythm* of breathing and its *magnitude* (that is, its rate and depth).

A respiratory center in the medulla oblongata, which is part of the brain stem, controls the rhythmic pattern of breathing. A group of its neurons shows pacemaker activity; it sets the pace for neurons that send messages to muscles involved in inhalation. The messages trigger contraction. In their absence, the muscles relax and air is exhaled. When exhalation must be active and strong, different neurons in the respiratory center issue signals for activating the muscles involved in exhalation.

Higher in the brain stem, in the pons, other centers smooth out the breathing rhythm set by the respiratory pacemakers. The apneustic center prolongs inhalation; the pneumotaxic center curtails it.

Controlling the magnitude of breathing depends on the partial pressures of oxygen and carbon dioxide as well as the H^+ concentration in arterial blood. Carbon dioxide levels have greatest influence over the ongoing adjustments. Chemoreceptors in the brain are sensitive to a rise in H^+ in the cerebrospinal fluid bathing them. (Remember, H^+ is also a by-product of reactions that kick in when blood has too much carbon dioxide.) The brain responds by commanding the diaphragm and other muscles to alter their activity. In this way it adjusts the rate and depth of breathing.

Aortic bodies (in the aortic arch) and carotid bodies (at a branching of the carotid artery) will notify the brain if oxygen's partial pressure plummets below 60 mm Hg in arterial blood. Such life-threatening decreases occur at very high altitudes and during severe lung diseases.

In some situations, a person can "forget" to breathe. Breathing that is briefly interrupted and then resumes spontaneously is known as *apnea*. It may stop for one or two seconds or minutes—in a few cases as often as 500 times a night. With *sudden infant death syndrome* (SIDS), a sleeping infant cannot awaken from an apneic episode. SIDS may have a heritable basis. It has been linked to gene mutations that give rise to an irregular heartbeat. Also, the risk is three times as high among infants of women who smoked cigarettes or had been exposed to secondhand smoke during pregnancy. Infants who sleep on their back or sides are less vulnerable to SIDS than those positioned on their abdomen.

Driven by its partial pressure gradient, oxygen diffuses from alveolar air spaces, through interstitial fluid, and into lung capillaries. Carbon dioxide, driven by its partial pressure gradient, diffuses in the opposite direction.

Hemoglobin in red blood cells enormously enhances the oxygen-carrying capacity of blood. Most carbon dioxide is transported in blood in the form of bicarbonate, nearly all of which forms by an enzyme action in red blood cells.

Respiratory centers in the brain stem control the rhythmic pattern of breathing and the magnitude of breathing, in ways that maintain appropriate levels of carbon dioxide, oxygen, and hydrogen ions in arterial blood.

WHEN THE LUNGS BREAK DOWN

In large cities, in certain workplaces, even in the cloud around a cigarette smoker, airborne particles and certain gases are present in abnormally high concentrations. And they put extra workloads on the respiratory system.

BRONCHITIS Ciliated, mucous epithelium lines the inner wall of your bronchioles (Figure 41.16). It is one of the built-in defenses that protect you from respiratory infections. Toxins in cigarette smoke and other airborne pollutants irritate the lining and may lead to *bronchitis*. With this respiratory ailment, epithelial cells along the airways become irritated and secrete excessive mucus. As mucus accumulates, so do the bacteria and other particles stuck in it. Coughing brings up some of the gunk. If the source of irritation persists, so does coughing.

Initial attacks of bronchitis are treatable. But when the aggravation continues, bronchioles become chronically inflamed. Bacteria, chemical agents, or both attack the bronchiole walls. Ciliated cells in the walls are destroyed, and the mucus-secreting cells multiply. Fibrous scar tissue forms and in time may narrow or obstruct the airways.

free surface of a mucus-secreting cell
free surface of a cluster of ciliated cells

Figure 41.16 False-color scanning electron micrograph of ciliated and mucus-secreting cells that line a bronchus.

EMPHYSEMA Thick mucus clogs the airways during persistent bronchitis. Inside the lungs, tissue-destroying bacterial enzymes attack the stretchable, thin walls of alveoli. The walls break down, and inelastic fibrous tissue forms around them. Gas exchange proceeds at fewer alveoli, which become enlarged. In time, the lungs remain distended and inelastic, so the fine balance between air flow and blood flow is permanently compromised.

Compare Figure 41.17*a* with *b* to get a sense of what happens. It becomes hard to run, walk, and exhale. These are symptoms of *emphysema*, a respiratory ailment that affects about 1.3 million people in the United States alone.

A few people are genetically predisposed to developing emphysema. They do not have a workable gene for the enzyme antitrypsin, which can inhibit bacterial attack on alveoli. Poor diet and persistent or recurring colds and other respiratory infections also invite emphysema later in life. However, *smoking is the major cause of the disease*. Emphysema may develop slowly, over twenty or thirty years. When severe damage is not detected in time, the lung tissues cannot be repaired.

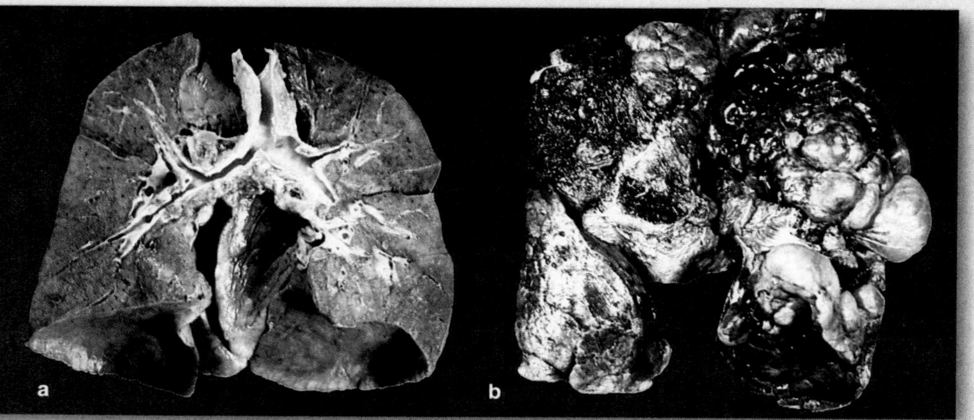
a
b

Figure 41.17 (**a**) Normal appearance of tissues of human lungs. (**b**) Lungs from someone affected by emphysema.

EFFECTS OF SMOKING Worldwide, 3 million people die each year as a result of complications arising from tobacco smoking. The World Health Organization, the Harvard School of Health, and the World Bank estimate that deaths from smoking may surpass even the AIDS epidemic by the year 2020. In the United States, every thirteen seconds one of 50 million smokers dies from emphysema, chronic bronchitis, or heart disease, and one nonsmoker dies of ailments brought on by *secondhand smoke*—of prolonged exposure to tobacco smoke in the surrounding air. And how much thought is given to children who breathe secondhand smoke? Parents and others who otherwise care about children indirectly heighten their vulnerability to allergies and lung ailments. Smoking is the major cause of lung cancer. Yet every day, 3,000 to 5,000 Americans light a cigarette for the first time. Even children spend a billion dollars a year on cigarettes. Each year, direct medical costs of treating smoke-induced respiratory disorders drain 22 billion dollars from the economy.

How does cigarette smoke damage lungs? Noxious particles in the smoke from just one cigarette immobilize cilia in the bronchioles for several hours. The particles also trigger mucus secretions, which in time clog the airways. They can kill infection-fighting

Further reading: Student Guide to InfoTrac on web site

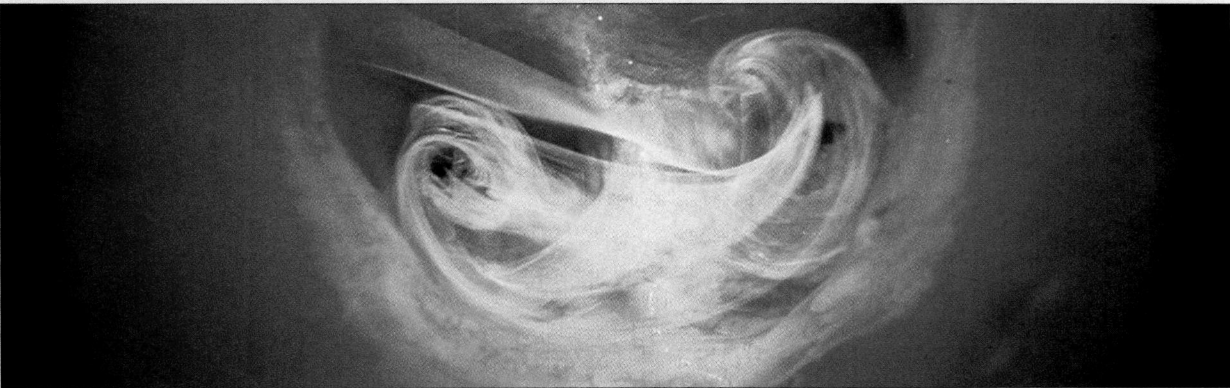

RISKS ASSOCIATED WITH SMOKING	REDUCTION IN RISKS BY QUITTING
SHORTENED LIFE EXPECTANCY: Nonsmokers live 8.3 years longer on average than those who smoke two packs daily from the midtwenties on.	Cumulative risk reduction; after 10 to 15 years, life expectancy of ex-smokers approaches that of nonsmokers.
CHRONIC BRONCHITIS, EMPHYSEMA: Smokers have 4–25 times more risk of dying from these diseases than do nonsmokers.	Greater chance of improving lung function and slowing down rate of deterioration.
LUNG CANCER: Cigarette smoking is the major cause of lung cancer.	After 10 to 15 years, risk approaches that of nonsmokers.
CANCER OF MOUTH: 3–10 times greater risk among smokers.	After 10 to 15 years, risk is reduced to that of nonsmokers.
CANCER OF LARYNX: 2.9–17.7 times more frequent among smokers.	After 10 years, risk is reduced to that of nonsmokers.
CANCER OF ESOPHAGUS: 2–9 times greater risk of dying from this.	Risk proportional to amount smoked; quitting should reduce it.
CANCER OF PANCREAS: 2–5 times greater risk of dying from this.	Risk proportional to amount smoked; quitting should reduce it.
CANCER OF BLADDER: 7–10 times greater risk for smokers.	Risk decreases gradually over 7 years to that of nonsmokers.
CORONARY HEART DISEASE: Cigarette smoking is a major contributing factor.	Risk drops sharply after a year; after 10 years, risk reduced to that of nonsmokers.
EFFECTS ON OFFSPRING: Women who smoke during pregnancy have more stillbirths, and weight of liveborns averages less (hence, babies are more vulnerable to disease, death).	When smoking stops before fourth month of pregnancy, risk of stillbirth and lower birthweight eliminated.
IMPAIRED IMMUNE SYSTEM FUNCTION: Increase in allergic responses, destruction of defensive cells (macrophages) in respiratory tract.	Avoidable by not smoking.
BONE HEALING: Evidence suggests that surgically cut or broken bones require up to 30 percent longer to heal in smokers, possibly because smoking depletes the body of vitamin C and reduces the amount of oxygen reaching body tissues. Reduced vitamin C and reduced oxygen interfere with production of collagen fibers, a key component of bone. Research in this area is continuing.	Avoidable by not smoking.

Figure 41.18 From the American Cancer Society, a list of the risks incurred by smoking and the benefits of quitting. The photograph shows swirls of cigarette smoke poised at the entrance to the two bronchi that lead into the lungs.

macrophages in the respiratory tract. What starts out as "smoker's cough" can end in bronchitis and emphysema.

Or consider how cigarette smoke contributes to lung cancer. Inside the body, certain compounds in coal tar and in cigarette smoke become converted to highly reactive intermediates. These are the carcinogens; they provoke uncontrolled cell divisions in lung tissues. On the average, 90 of every 100 smokers who develop lung cancer will die from it. If you now smoke, or are thinking about starting or quitting, you may wish to give serious thought to the information in Figure 41.18.

EFFECTS OF MARIJUANA SMOKE In 1994, 17 million Americans smoked *pot*—marijuana (*Cannabis*)—at least once to induce light-headed euphoria. The number is rising, especially in junior high and high schools. About 1.5 million are chronic users. They become enamored of the euphoria. However, the sense of well-being does not last long; it fades to apathy, depression, and fatigue. Users keep smoking to avoid the negative effects.

Besides causing psychological dependency, long-term use can result in chronic throat irritations, persistent coughing, bronchitis, and emphysema.

41.8 RESPIRATION IN UNUSUAL ENVIRONMENTS

Breathing at High Altitudes

From the chapter introduction, you already know that breathing becomes an agonizing challenge for climbers of Chomolungma and other cloud-piercing peaks. Even so, like llamas and other species that evolved at high elevations, millions of people live quite comfortably 4,800 meters (16,000 feet) or more above sea level.

Compared with people living at lower elevations, the permanent residents of high mountains have lungs with far more alveoli and blood vessels, which formed as they were growing up. Also, their heart developed larger ventricles, so it pumps larger volumes of blood. And more mitochondria formed in their muscle tissue.

Llamas have another advantage. Compared to human hemoglobin, llama hemoglobin has greater affinity for oxygen. It picks up oxygen more efficiently at the lower pressures characteristic of high altitudes (Figure 41.19).

Does this mean a healthy person who grew up by the seashore cannot ever pick up roots and move to the mountains? No. By mechanisms of **acclimatization**, an individual often can make long-lasting physiological and behavioral adaptations to a new environment that is markedly different from the one left behind. At high altitudes, gradual changes in the pattern and magnitude of breathing and in the heart's output can supplant the initially acute compensations that an individual makes to cellular oxygen deficiency (hypoxia).

Within a few days, the reduction in oxygen delivery stimulates kidney cells to secrete more **erythropoietin**, a hormone that induces stem cells in bone marrow to divide repeatedly and give rise to red blood cells. Each second in adult humans, 2 million to 3 million red blood cells are churned out as replacements for the ones that continually die off. The stepped-up erythropoietin secretion can increase that astounding pace by six times during episodes of extreme stress. Increased numbers of circulating red blood cells increase the blood's oxygen-delivery capacity. When the concentration of oxygen in the blood increases sufficiently, the kidneys slow down erythropoietin secretion.

Erythropoietin is the major hormonal mediator of red blood cell production. However, human males have a larger muscle mass and greater demands for oxygen. They depend as well upon testosterone. Among other things, this male sex hormone stimulates increases in the basic rate of red blood cell production. That is why a sample of blood taken from males normally has a greater percentage of red blood cells than an equivalent sample from females. But the compensatory increase in red blood cells comes at a cost. Having many more cells in the bloodstream increases resistance to blood flow, because the blood is more viscous, or "thicker." The heart must work harder to pump blood.

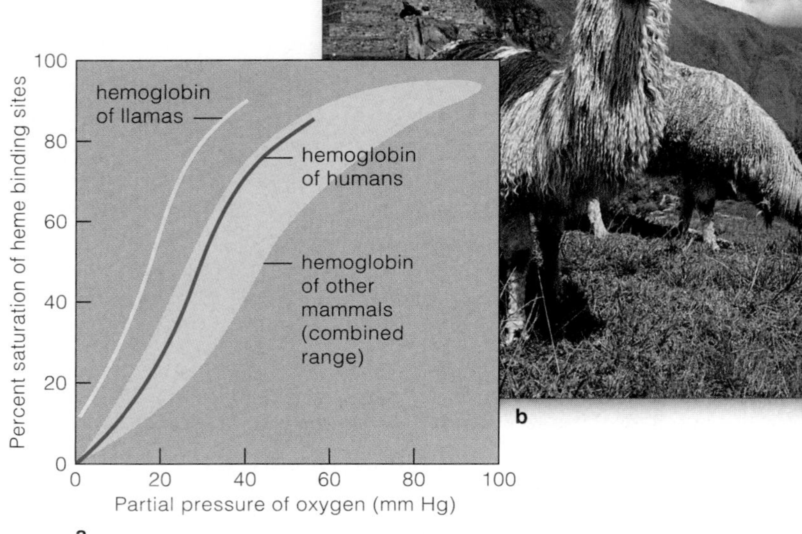

Percent saturation of heme binding sites
100
80
60
40
20
0
hemoglobin of llamas
hemoglobin of humans
hemoglobin of other mammals (combined range)
0
20
40
60
80
100
Partial pressure of oxygen (mm Hg)
a
b

Figure 41.19 (**a**) Comparison of the binding capacity and releasing capacity of hemoglobin of humans, llamas, and other animals. (**b**) Llamas high up in the Peruvian Andes.

Carbon Monoxide Poisoning

High elevations are not the exclusive culprits behind cellular oxygen deficiency. Hypoxia also occurs when the partial pressure of oxygen in arterial blood falls owing to *carbon monoxide poisoning*. Carbon monoxide (CO), a colorless, odorless gas, occurs in exhaust fumes from gas-driven vehicles. The smoke from burning coal and wood and from tobacco contains it. CO competes with oxygen for hemoglobin binding sites. Its binding capacity is at least 200 times greater than oxygen's! Even small amounts of this gas can tie up 50 percent of the hemoglobin in the body. That is how it can impair oxygen delivery to tissues to dangerous degrees.

Respiration in Diving Mammals

THE PHYSIOLOGY OF DIVING How do whales, seals, and other air-breathing mammals that routinely dive survive underwater, where they cannot inhale oxygen and exhale carbon dioxide? Apparently such mammals (as well as diving birds) share a common, specialized pattern of metabolism during their underwater forays.

We are not talking about a quick bob under the water's surface. Weddell seals (*Leptonychotes weddelli*) of the Antarctic waters typically remain submerged for twenty to twenty-five minutes. Some do so for more than an hour. They usually limit their dives to 400 meters (1,312 feet) or less, although a few have been observed at depths that approach 600 meters—more than a third of a mile down. One sperm whale (Figure 41.20) submerged itself for eighty-two minutes; another, tracked by sonar, reached a depth of 2,250 meters.

Further reading: Student Guide to InfoTrac on web site —

By comparison, humans trained to dive without oxygen tanks can fully submerge themselves for about three minutes at most. Career divers, including the ama of Korea and Japan, do only a bit better.

When diving mammals leave the air behind, they take oxygen with them. Hemoglobin binds much of it. So does **myoglobin**, an oxygen-binding protein that is abundant in skeletal muscles. And the lungs serve as a suitcase filled with oxygen. During short dives, aerobic respiration proceeds as usual. During prolonged dives, however, the oxygen is preferentially distributed to the heart and central nervous system, both of which cannot function without the high energy yields of the aerobic pathway. Basically, blood is directed away from most organs. Blood pumped from the heart travels mainly to the lungs, brain, and back to the heart. Skeletal muscles use their own myoglobin-bound oxygen as well as the hemoglobin-bound oxygen in the dense capillary beds that service them. Later on, they switch to anaerobic pathways, and lactate accumulates in tissues. When the animal surfaces, circulation to the muscles is restored.

Diving mammals also have specialized respiratory adaptations. For example, a sperm whale's respiratory system is adapted to collect and store oxygen. When the whale surfaces, it rapidly blows stale air from its lungs, through a tubelike epiglottis and then a blowhole at the body surface. Before the whale dives again, it breathes rapidly, so that its elastic, extensible lungs fill quickly with large volumes of air.

During a dive, special valves close its nostrils; and rings of muscle and of cartilage clamp its bronchioles. The whale's respiratory surface is not that large. But strategically located valves and local networks of blood vessels (plexuses) store and distribute the blood volume and gases in economical fashion. The heart rate and metabolism slow, and so does the rate of oxygen use and carbon dioxide formation. In addition, compared to land mammals, the whale's respiratory center is less sensitive to carbon dioxide levels.

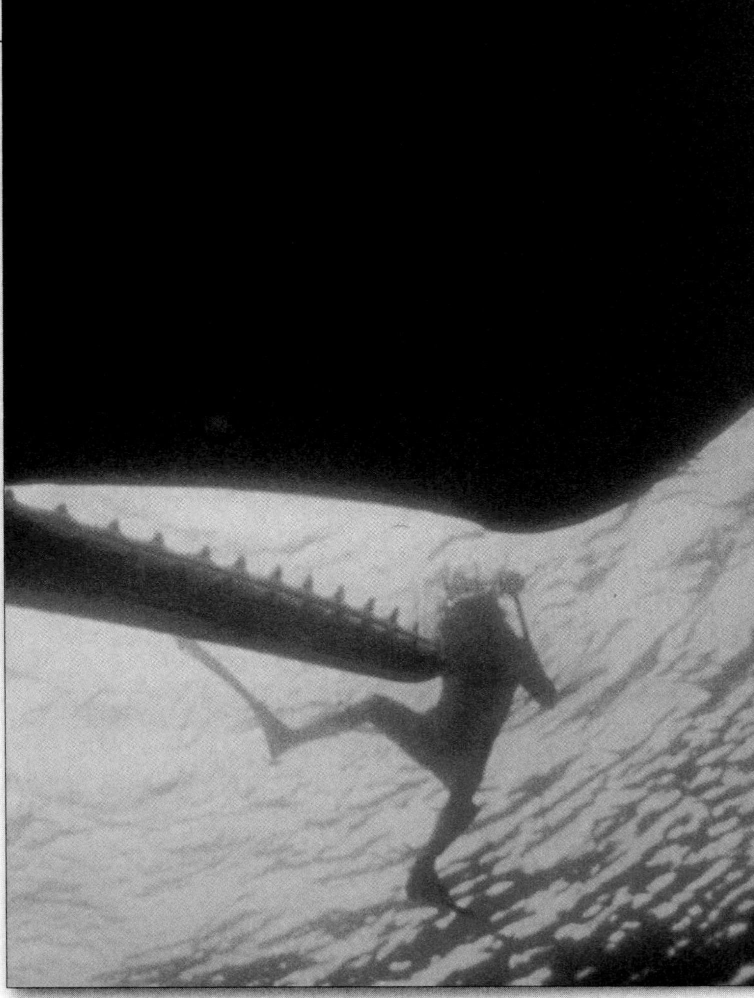

Figure 41.20 Underwater view of a sperm whale and a human diver inspecting each other.

HUMAN DEEP-SEA DIVERS Professional divers know that water pressure increases greatly with depth. They do not dive in deep water without tanks of compressed air (air under pressure). The increased pressure at depths causes much more gaseous nitrogen (N_2) than usual to dissolve in their body tissues. The outcome is nitrogen narcosis, or "raptures of the deep." At depths of 45 meters (150 feet), divers become euphoric and drowsy, as if they were tipsy with alcohol. Some have been known to offer the mouthpiece of their airtank to fishes. At lower depths, the divers become clumsy and weak. Below 105 meters, they can slip into a coma.

Gaseous nitrogen is a lipid-soluble gas, so it readily dissolves in the lipid bilayer of cell membranes. By one hypothesis, when it does this to neurons, it suppresses their capacity to respond to action potentials.

Deep-sea divers also are cautious when they return to the surface. When the pressure decreases, N_2 moves back out of the tissues and into the bloodstream. If an ascent is too rapid, N_2 enters the blood faster than the lungs can dispose of it. Then, bubbles of nitrogen may form in the blood and tissues. Too many bubbles cause pain, especially at joints. Hence the common name, "the bends," for what is otherwise known as decompression sickness. If bubbles obstruct the blood flow to the brain, deafness, impaired vision, and paralysis may result.

During brief stays at high elevations, the human body can make acute compensatory responses to the thinner air. Over time, the body makes adaptive adjustments in breathing and cardiac output.

Diving mammals bind oxygen to hemoglobin and to the myoglobin of skeletal muscles. They also use respiratory adaptations that store and distribute oxygen when diving.

41.9 RESPIRATION IN LEATHERBACK SEA TURTLES

Late at night, near the shoreline of a Caribbean island, biologist Molly Lutcavage and several colleagues move out on a turtle patrol. Finally a large, dark shape, with flippers flailing, emerges from the pale surf. A female Atlantic leatherback sea turtle (*Dermochelys coriacea*) is returning from the sea to nest in the sand (Figure 41.21).

Sea turtles are members of the reptilian lineage, which extends 300 million years back in time. Green turtles, ridleys, loggerheads, leatherbacks—all are endangered or threatened species. The race is on to gather information about their life histories and physiology that may help pull them back from the brink of extinction.

Leatherbacks are the largest and least understood sea turtles. Adult female leatherbacks may weigh more than 400 kilograms (880 pounds). Adult males may weigh nearly twice that amount. Leatherbacks normally leave the water only to breed and lay eggs. During the rest of their lives, they migrate across vast stretches of the open ocean.

Leatherbacks do something no other reptile on Earth can do. They have the capacity to dive 1,000 meters (3,000 feet) below sea level! Before biologists conducted tracking experiments during the mid-1980s, Weddell seals, fin whales, and some other marine mammals were the only nonhuman deep-sea divers known.

To reach such depths, a turtle would have to swim for nearly forty minutes without taking a breath. How does it dive so deeply, for so long, and still have enough oxygen for aerobic metabolism?

Oxygen reserves in the lungs are not enough for the leatherback's prolonged underwater excursions. Even at depths between 80 and 160 meters, pressure exerted by the surrounding water causes air-filled lungs to collapse. However, like the diving marine mammals described in Section 41.8, the leatherback's skeletal muscle cells have an abundance of the oxygen-binding protein myoglobin. Also, compared to other diving reptiles, they have more blood per unit of body weight, and their blood has more red blood cells—hence more hemoglobin (Table 41.1).

Imagine yourself as an observer on Lutcavage's team. Their plan is to draw blood samples from a leatherback so that they can study the oxygen-binding capacity of its hemoglobin. Samples of respiratory gases dissolved in the blood may provide insight into how leatherbacks use oxygen. As the researchers know, a nesting female enters a trancelike state during the twenty or so minutes it takes her to deposit eggs in the sand. During that time, she will not resist being gently handled.

The team gets to work. They fit the head of a nesting female with a helmet that is equipped with an expandable plastic sleeve. The fit is snug but not restrictive, so the turtle is able to breathe normally. With each exhalation, the expired gases flow through a one-way valve into a collecting sac. At the same time, the researchers count the number of breaths required to completely fill the sac. (As a baseline, they also tallied the breathing frequency before placing the helmet over her head.) They store the gas samples according to established procedures and set them aside for laboratory analysis.

Working quickly, a member of the team draws blood from a superficial vein in the turtle's neck, and then packs the sample in ice to preserve it. Once they are back at the laboratory, the sample will be subjected to hemoglobin analysis and a red blood cell count. The oxygen level, carbon dioxide level, and pH also will be measured.

Lutcavage and her coworkers did gather samples of blood and respiratory gases from several turtles. They also analyzed skeletal muscle tissue, taken earlier from a drowned turtle. The data provided insight into how leatherbacks manage their spectacularly deep dives.

As expected, calculations of the oxygen uptake and total tidal volume during breathing revealed that the leatherback cannot inhale and hold enough air in the lungs to sustain aerobic metabolism during a prolonged dive. Other oxygen-supplying mechanisms had to be at work. As further analysis revealed, myoglobin levels are so high in leatherbacks, oxygen remains available for

Table 41.1 Physiological Comparison of a Few Diving Reptiles and Mammals

Characteristic	LEATHERBACK TURTLE	GREEN TURTLE	CROCODILE	KILLER WHALE	WEDDELL SEAL
Maximum diving depth (meters)	Greater than 1,000	Less than 100	Less than 30	260	600
Red blood cell count (percentage of total volume)	39	30	28	44	58
Hemoglobin level (grams/deciliter of blood)	15.6	8.8	8.7	16.0	17–22
Myoglobin level (milligram/gram of muscle tissue)	4.9	—*	—*	—*	44.6
Oxygen-carrying capacity (volume percent)	21	7.5–11.9	12.4	23.7	31.6

* Not known.

Further reading: Student Guide to InfoTrac on web site —

skeletal muscle activity during a dive after the air in the lungs gives out (or when their lungs collapse).

Also, leatherbacks have a notable abundance of red blood cells and a type of hemoglobin with a high affinity for binding oxygen. A leatherback's blood may contain a remarkable 21 percent oxygen by volume. That amount is typical of a human, not a "plodding" turtle.

The oxygen-carrying capacity turns out to be the highest ever recorded for a reptile—and it is close to the capacity of deep-diving mammals. Add to this a streamlined body and massive front flippers, and you have a reptile uniquely adapted for diving and maneuvering at great depths in the sea.

Studies of leatherbacks are only now under way. Researchers are starting to investigate newer tracking methods to help them monitor a leatherback's metabolism at sea and during dives. So far, it has proved extremely difficult and expensive to track individual turtles in the open ocean.

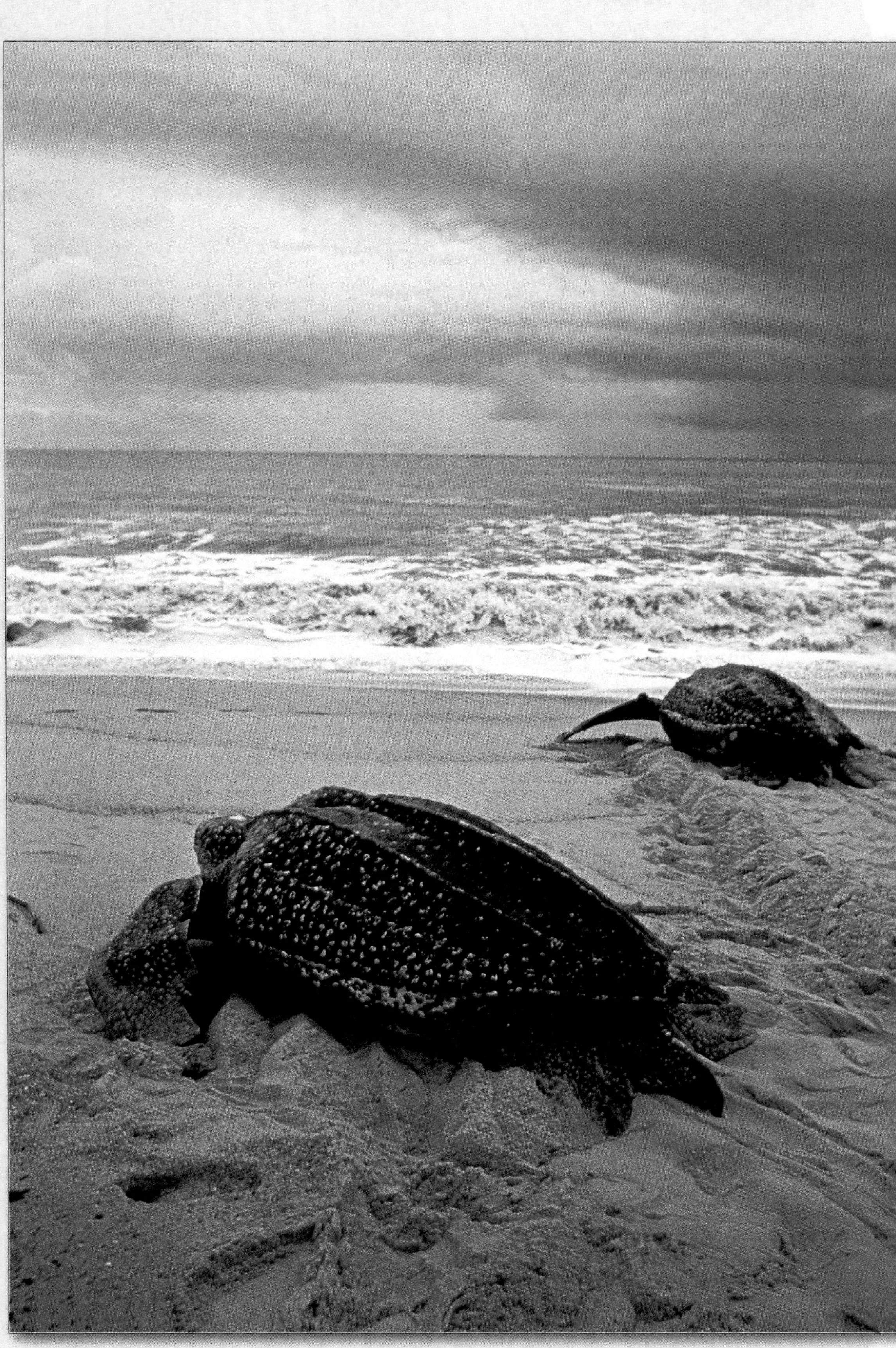

Figure 41.21 Two female leatherback sea turtles (*Dermochelys coriacea*) returning to the water after laying their eggs in the sands of an island in the Caribbean Sea.

41.10 SUMMARY

1. Aerobic respiration is the main metabolic pathway that yields enough energy for active life-styles. It uses oxygen and has carbon dioxide by-products. Respiration is the process by which the animal body as a whole acquires oxygen and disposes of carbon dioxide wastes.

2. Air is a mixture of oxygen, carbon dioxide, and other gases, each exerting a partial pressure. Each gas tends to move from areas of higher to lower partial pressure. Respiratory systems make use of this tendency.

3. In respiratory systems, oxygen and carbon dioxide diffuse across a respiratory surface. In humans, this is a thin membrane of alveolar epithelium, lung capillary epithelium, and their fused basement membranes.

4. Modes of respiration differ among animal groups.

a. Invertebrates with a small body mass depend only on integumentary exchange; oxygen and carbon dioxide simply diffuse across the body surface. This mode also persists in some large animals, including amphibians.

b. Many marine and some freshwater invertebrates have gills: respiratory organs with moist, thin, and often highly folded walls. Most insects and some spiders use tracheal respiration. Gases flow through open-ended tubes leading from the body surface directly to tissues. Most spiders have book lungs, with leaflike folds.

c. Fishes have greater energy demands than small invertebrates. At their two gills, a countercurrent flow mechanism compensates for the low oxygen levels of aquatic habitats. Paired lungs are the dominant means of respiration in reptiles, birds, and mammals.

5. The airways of the human respiratory system are the nasal cavities, pharynx, larynx, trachea, bronchi, and bronchioles. Many millions of alveoli at the end of the terminal bronchioles are the main sites of gas exchange.

6. Breathing ventilates the lungs. Each respiratory cycle consists of inhalation (one breath in) and exhalation (one breath out). During inhalation, the chest cavity expands, lung pressure decreases below atmospheric pressure, and air flows into the lungs. The events are reversed during normal exhalation.

7. Driven by its partial pressure gradient, oxygen in lungs diffuses from alveolar air spaces into pulmonary capillaries. It diffuses into red blood cells and binds weakly with hemoglobin. Hemoglobin gives up oxygen at capillary beds in metabolically active tissues. Oxygen diffuses across the interstitial fluid, then into cells.

8. Driven by its partial pressure gradient, carbon dioxide in tissues diffuses from cells, across interstitial fluid, and into the blood. Most of it reacts with water to form bicarbonate. The reactions are reversed in the lungs; carbon dioxide diffuses from the pulmonary capillaries into air spaces of alveoli, then is exhaled.

Figure 41.22 Body surfing.

Review Questions

1. Distinguish between respiration and aerobic respiration. *CI*
2. Define respiratory surface. Why must the oxygen and carbon dioxide partial pressure gradients across it be steep? *41.1*
3. What is the name of the main respiratory pigment? *41.1*
4. A few of your friends who have not taken a biology course ask you what insect lungs look like. How do you answer them (assuming your instructor is listening)? *41.2*
5. Briefly describe how countercurrent flow through a fish gill is so efficient at taking up dissolved oxygen from water. *41.3*
6. Does the respiratory system of fishes, amphibians, reptiles, birds, or mammals have air sacs that ventilate the lungs? *41.3*
7. Distinguish between:
   a. tracheal respiration and trachea (windpipe) *41.2, 41.4*
   b. pharynx and larynx *41.4*
   c. bronchiole and bronchus *41.4*
   d. pleural sac and alveolar sac *41.4, 41.5*
8. Define the functions of the human respiratory system. In the diagram below, label its components and the major bones and muscles with which it interacts during breathing. *41.4*

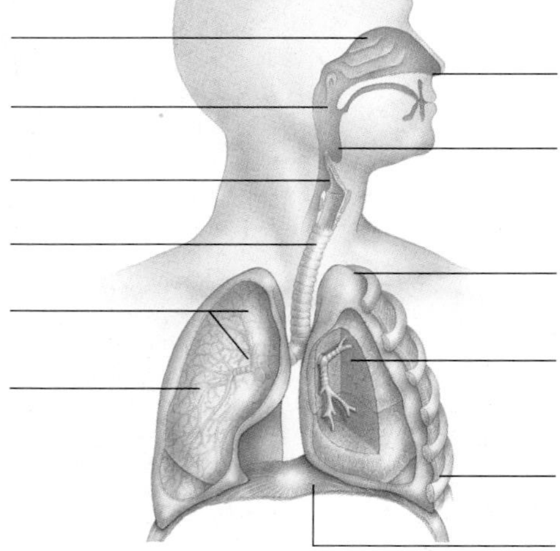

9. Explain why the human male in Figure 41.22 cannot survive on his own for very long under water.

Self-Quiz (*Answers in Appendix III*)

1. Most spiders and insects depend on ______ .
   a. tracheal respiration   c. integumentary exchange
   b. lungs   d. both a and c

2. ______ have an abundance of respiratory pigments.
   a. invertebrates   c. insects
   b. vertebrates   d. both a and b

3. In birds, air flows ______ .
   a. into and out of lungs   c. into air sacs
   b. through lungs   d. all of the above

4. Each human lung encloses a ______ .
   a. diaphragm   c. pleural sac
   b. bronchial tree   d. both b and c

5. In human lungs, gas exchange occurs at the ______ .
   a. two bronchi   c. alveolar sacs
   b. pleural sacs   d. both b and c

6. When you breathe quietly, inhalation is ______ and exhalation is ______ .
   a. passive; passive   c. passive; active
   b. active; active   d. active; passive

7. After oxygen diffuses into pulmonary capillaries, it diffuses into ______ and binds with ______ .
   a. interstitial fluid; red blood cells
   b. interstitial fluid; carbon dioxide
   c. red blood cells; hemoglobin
   d. red blood cells; carbon dioxide

8. Oxyhemoglobin (HbO_2) gives up oxygen faster where the ______ than it is in the lungs.
   a. pH is lower   c. O_2 partial pressure is higher
   b. blood is cooler   d. CO_2 partial pressure is lower

9. Sixty percent of the carbon dioxide in human blood is in the form of ______ and thirty percent in the form of ______ .
   a. carbon dioxide; carbonic acid
   b. carbonic anhydrase; bicarbonate
   c. bicarbonate; carbamino hemoglobin
   d. carbamino hemoglobin; bicarbonate

10. Hypoxia triggers increased ______ secretion.
    a. carbonic anydrase   c. erythropoeitin
    b. carbon monoxide   d. myoglobin

11. Match the components with their descriptions.
    ____ trachea   a. airway leading into a lung
    ____ pharynx   b. gap between vocal cords
    ____ alveolus   c. fine branch of bronchial tree
    ____ hemoglobin   d. windpipe
    ____ bronchus   e. respiratory pigment
    ____ bronchiole   f. site of gas exchange
    ____ glottis   g. throat

Critical Thinking

1. Some cigarette manufacturers are urging people to "smoke responsibly." What social and biological issues are involved here? For example, what about risks to a nonsmoking spouse or children of a smoker? To nonsmoking patrons in a restaurant? To the unborn child of a pregnant smoker? In your opinion, what behavior would constitute "responsible smoking"?

2. People sometimes poison themselves with carbon monoxide by building a charcoal fire in an enclosed area. Assuming help arrives in time, what would be the best treatment: (1) placing the victim outdoors in fresh air or (2) rapidly administering pure oxygen? Explain how you arrived at your answer.

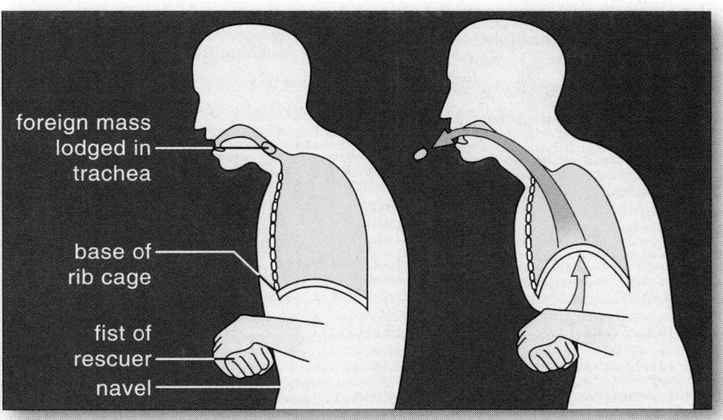
foreign mass lodged in trachea
base of rib cage
fist of rescuer
navel

Figure 41.23 Heimlich maneuver to dislodge food stuck in the trachea. Stand behind the victim, make a fist with one hand, then position the fist, thumb-side in, against the victim's abdomen. The fist must be slightly above the navel and well below the rib cage. Now press the fist into the abdomen with a sudden upward thrust. Repeat the thrust several times if needed. The maneuver can be performed on someone who is standing, sitting, or lying down.

3. When you swallow food, muscle contractions force the epiglottis down, to its closed position. This prevents food from going down the trachea and blocking gas flow. Yet each year, several thousand people choke to death after food enters the trachea and blocks air flow for as little as four or five minutes. The Heimlich maneuver—which is an emergency procedure only—often dislodges food from the trachea (Figure 41.23).

When correctly performed, the Heimlich maneuver elevates the diaphragm forcibly, causing a sharp decrease in the volume of the thoracic cavity and an abrupt increase in the alveolar pressure. Air forced up the trachea by the increased pressure may be enough to dislodge the obstruction. Once the obstacle is dislodged, a doctor must see the person at once; inexperienced rescuers can inadvertently cause internal injuries or crack a rib.

Reflect now on the current social climate in which lawsuits abound. Would you risk performing the Heimlich maneuver to save a relative's life? A stranger's life? Why or why not?

Selected Key Terms

acclimatization *41.8*
alveolus (alveoli) *41.4*
bronchiole *41.4*
bronchus *41.4*
carbamino hemoglobin *41.6*
carbonic anhydrase *41.6*
countercurrent flow *41.3*
diaphragm *41.4*
epiglottis *41.4*
erythropoietin *41.8*
Fick's law *41.1*
gill (fish) *41.2*
glottis *41.3*
heme group *41.6*
hemoglobin *41.1*
integumentary exchange *41.2*
larynx *41.4*
lung *41.3*
myoglobin *41.8*
oxyhemoglobin *41.6*
partial pressure *41.1*
pharynx *41.4*
pressure gradient *41.1*
respiration *CI*
respiratory cycle *41.5*
respiratory surface *41.1*
respiratory system *CI*
tidal volume *41.5*
trachea *41.4*
tracheal respiration *41.2*
vital capacity *41.5*
vocal cord *41.3*

Readings *See also www.infotrac-college.com*

Hill, R., and G. Wyse. 1992. *Animal Physiology*. Second edition. New York: HarperCollins.

Sherwood, L. 1997. *Human Physiology*. Second edition. Belmont, California: Wadsworth.

42

DIGESTION AND HUMAN NUTRITION

Lose It—And It Finds Its Way Back

America's fixation on Beautiful People who have nary an ounce of extra fat on their utterly perfect selves is bad enough. After all, we can always rationalize our own extra ounce with the truism that beauty is only skin deep. But the American Medical Association's dire warnings of the Fat Connection to atherosclerosis, heart attacks, strokes, colon cancer, and other deadly ailments is beyond rationalization.

By current standards, the proportion of body fat relative to total tissue mass should be 18 to 24 percent for human females who are less than thirty years old. For males, it should be no more than 12 to 18 percent. An estimated 34 million Americans do not even come close to the standards.

No matter how much we diet, the lost weight seems to find its way back. This physiological dilemma is an outcome of our evolutionary heritage. Like most other mammals, we have an abundance of fat-storing cells in adipose tissue. Collectively, the cells are an adaptation for survival, an energy warehouse that opens up when food is scarce. Once these fat-storing cells have formed, they are in the body to stay. Variations in food intake only influence how empty or full each one gets.

When we diet and the fat warehouse opens, the brain interprets this as "starvation," and it issues commands for a metabolic slowdown. The body uses energy far more efficiently even for basic tasks, such as breathing and digesting food. It now takes less food to do the same things!

As you have probably heard, dieting does no good without a long-term commitment to physical exercise. Why? Your skeletal muscles also adapt to "starvation" and burn less energy than before. Jog for four hours or play tennis for eight hours straight, and you might lose a pound of fat. Meanwhile, your appetite surges. If and when you do stop dieting, the "starved" fat cells quickly refill. Unless you eat less *and* exercise moderately throughout your life, you can't keep off extra weight (Figure 42.1).

What about gaining weight? A weight gain triggers an *increase* in metabolism. Now the body uses 15 to 20 percent more energy than before until weight drops back to what it was.

As if this were not enough, your emotions influence weight gains and losses. As an extreme case, *anorexia nervosa* is a potentially fatal eating disorder based on a seriously flawed assessment of body weight. Typically, anorexics have an overwhelming dread of being fat *and* hungry. They starve themselves and often overexercise. They commonly have fears about growing up and of being sexually mature. They also have irrationally high expectations about their personal performance.

We find another extreme case in *bulimia*, an out-of-control "oxlike appetite." During an hour-long eating binge, a bulimic might take in about 50,000 kilocalories' worth of food, then vomit or use laxatives to purge the body. Some follow a binge–purge routine as an "easy" way to lose weight. Others suffer severe emotional

Figure 42.1 Organisms positively engaged in countering imbalances between caloric intake and energy output.

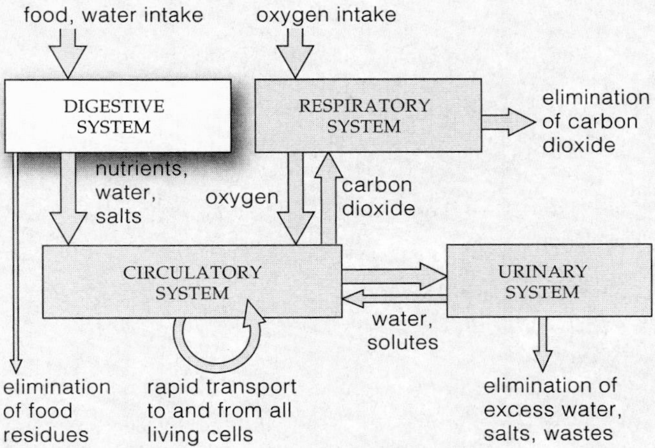
food, water intake
oxygen intake
DIGESTIVE SYSTEM
RESPIRATORY SYSTEM
elimination of carbon dioxide
nutrients, water, salts
oxygen
carbon dioxide
CIRCULATORY SYSTEM
URINARY SYSTEM
water, solutes
elimination of food residues
rapid transport to and from all living cells
elimination of excess water, salts, wastes

Figure 42.2 Functional links between the digestive, respiratory, circulatory, and urinary systems. These organ systems and others work together to supply cells with raw materials and eliminate wastes.

stress. They may not even like to eat, but at some level the purging relieves them of anger and frustration.

Some bulimics binge–purge once a month. Others do so several times a day. Do they know that repeated purgings can damage the gut? That chronic vomiting brings gastric fluid into the mouth and erodes teeth to stubs? That, at its extreme, bulimia can rupture the stomach and cause heart or kidney failure?

Are severe eating disorders rare? No. In the United States, an estimated 7 million females and 1 million males are anorexic or bulimic. Most are in their teens and early twenties, but the number also is increasing among preadolescents. Each year, 5 to 6 percent of the weight-obsessed individuals die from complications arising from their disorders.

And with these sobering thoughts in mind, we start a tour of **nutrition**. The word encompasses all those processes by which an animal ingests and digests food, then absorbs the released nutrients for later conversion to the body's own carbohydrates, lipids, proteins, and nucleic acids.

The nutritional processes proceed at the **digestive system**. This body tube or cavity mechanically and chemically reduces food to particles, then to molecules that are small enough to be absorbed into the internal environment. It also eliminates unabsorbed residues. Other organ systems, especially those shown in Figure 42.2, contribute to the nutritional processes.

KEY CONCEPTS

1. In complex animals, interactions among digestive, circulatory, respiratory, and urinary systems supply all living cells of the body with raw materials, dispose of wastes, and maintain the volume and composition of extracellular fluid.

2. Most digestive systems include specialized regions for food transport, processing, and storage. Different regions mechanically break apart and chemically break down food, absorb breakdown products, and eliminate the unabsorbed residues.

3. "Nutrition" refers to all processes by which an animal ingests food, digests food, and then absorbs the released nutrients, which become converted to the body's own carbohydrates, lipids, proteins, and nucleic acids.

4. A subcategory of nutritional processes is the intake of foods that are good sources of vitamins, minerals, and a number of amino acids and fatty acids that the body itself cannot produce.

5. To maintain an acceptable body weight and overall health, energy intake must balance energy output by way of metabolic activity, physical exertion, and so on. Complex carbohydrates are the main source of dietary glucose, which typically is the body's main source of immediately usable energy.

42.1 THE NATURE OF DIGESTIVE SYSTEMS

Incomplete and Complete Systems

Recall, from Chapter 26, that we do not see many organ systems until we get to the flatworms. Flatworms are among the invertebrates with an **incomplete digestive system**, a saclike, branching gut cavity with a single opening at the start of a tubular pharynx (Figure 42.3*a*). Food enters the sac, where it is partly digested, then is circulated to cells even as residues are being sent out.

The saclike gut of flatworms, cnidarians, and a few other small invertebrates has a single opening for food intake *and* for waste disposal. The two-way traffic does not favor regional specialization. Other animals have a **complete digestive system**, a tube with an opening at one end for taking in food and an opening at the other end for eliminating unabsorbed residues (for example, mouth and anus). In between, the tube is subdivided into specialized regions, which function in the one-way transport, processing, and storage of raw materials.

Think about the complete digestive system of a frog (Figure 42.3*b*). Between that frog's mouth and the anus are a pharynx, stomach, and small and large intestines. Functionally connected with the small intestine are the liver, gallbladder, and pancreas, which are organs with accessory roles in digestion. Birds, too, have a complete digestive system, which includes a few unique regional specializations (Figure 42.3*c*).

Regardless of its complexity, a complete digestive system carries out five overall tasks:

1. **Mechanical processing and motility**. Movements that break up, mix, and propel food material.

2. **Secretion**. Release of digestive enzymes and other substances into the space inside the tube.

3. **Digestion**. Breakdown of food into particles, then into nutrient molecules small enough to be absorbed.

4. **Absorption**. Passage of digested nutrients and fluid across the tube wall and into body fluids.

5. **Elimination**. Expulsion of the undigested and unabsorbed residues from the end of the gut.

Correlations With Feeding Behavior

Let's now correlate the specializations of any digestive system with feeding behavior. Look at the pigeon in Figure 42.3*c*. Its *food-gathering* region is a bill adapted to pecking seeds off the ground. Seeds enter a mouth, then a long tube (esophagus), and then a *food-processing* region that is compactly centered in the body mass. The compact arrangement helps a pigeon balance its body during flight. As in other seed-eating birds, part of the esophagus balloons out as a *crop* (a stretchable storage organ). A crop lets the bird gulp down a lot of food and

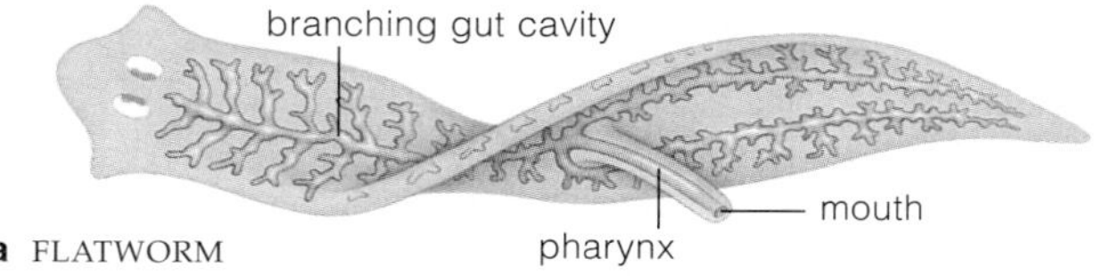
branching gut cavity
mouth
pharynx
a FLATWORM

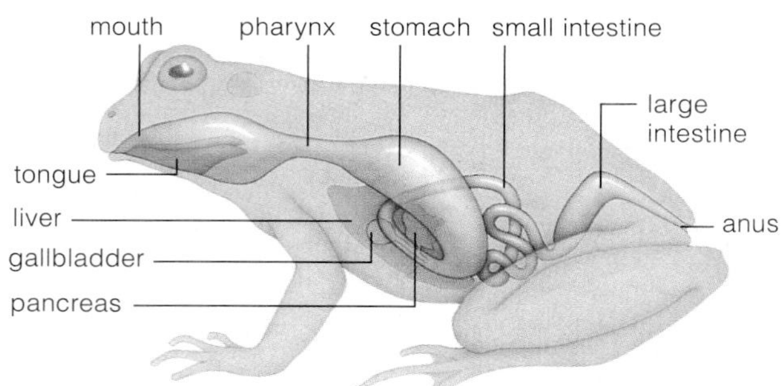
mouth
pharynx
stomach
small intestine
large intestine
tongue
liver
gallbladder
pancreas
anus
b FROG

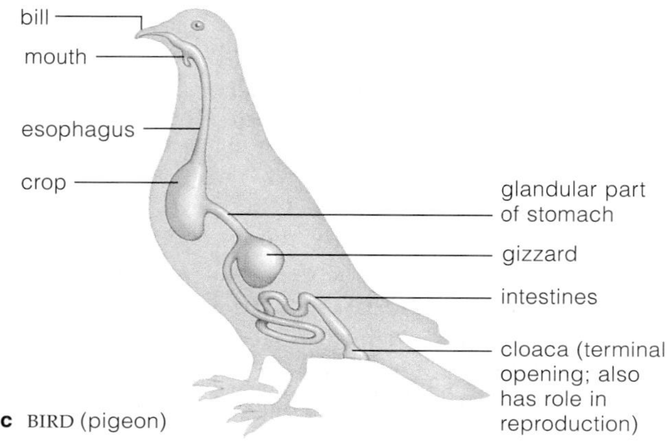
bill
mouth
esophagus
crop
glandular part of stomach
gizzard
intestines
cloaca (terminal opening; also has role in reproduction)
c BIRD (pigeon)

Figure 42.3 (**a**) Incomplete digestive system of a flatworm, with two-way traffic of food and undigested material through one opening. (**b**,**c**) Examples of a complete digestive system, a tube with specialized regions and an opening at each end.

make quick getaways from predators. Like all birds, this one eats during the day. Its crop fills before the sun goes down, then it releases food in the first few hours after dark and so lessens the time of overnight fasting.

The first part of the pigeon stomach has a glandular lining, which secretes enzymes and other substances that function in digestion. The second part, the gizzard, is a muscular organ that grinds up food much as teeth and jaws do. A pigeon's intestines are proportionally shorter than they are in ducks, ostriches, and other birds that eat plant parts rich in tough, fibrous cellulose—which requires a much longer processing time than seeds do.

We also can correlate feeding behavior with the digestive system of a pronghorn antelope (*Antilocapra americana*). Fall through winter, on mountain ridges from central Canada down into northern Mexico, pronghorn

Further reading: Student Guide to InfoTrac on web site

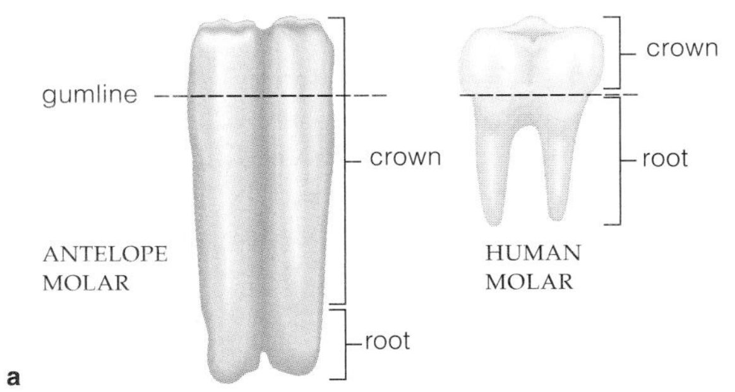
gumline
crown
root
ANTELOPE MOLAR
crown
root
HUMAN MOLAR
a

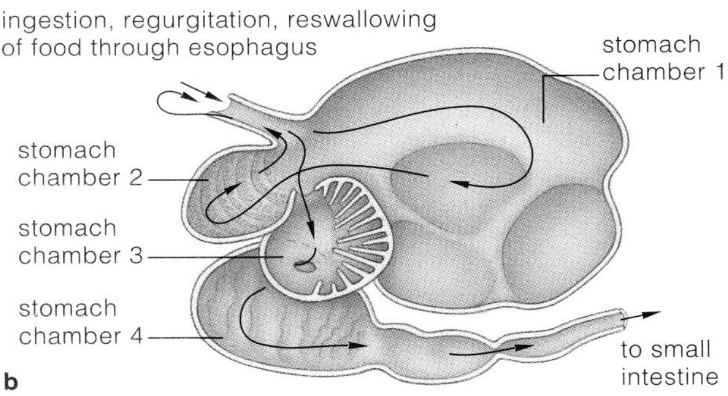
ingestion, regurgitation, reswallowing of food through esophagus
stomach chamber 1
stomach chamber 2
stomach chamber 3
stomach chamber 4
to small intestine
b

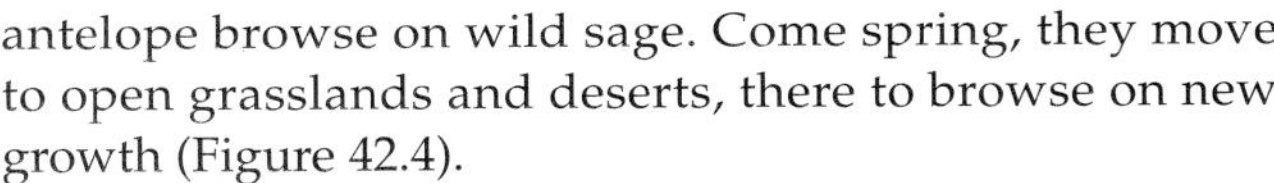

Figure 42.4 Some regional specializations of the complete digestive system of pronghorn antelope (*Antilocapra americana*).

(**a**) Comparison of antelope and human molars. (**b**) Multiple-chambered antelope stomach. The first chamber is a large pouch. The second is smaller, with a honeycombed inner surface (what butchers call "tripe"). In both chambers, food is mixed with fluid, kneaded, and exposed to fermentation activities of bacterial and protozoan symbionts. Some of the symbionts degrade cellulose; others synthesize organic compounds, fatty acids, and vitamins. The host uses a portion of these substances. Kneaded food is regurgitated into the mouth, rechewed, then swallowed again. It enters the third chamber, where it is pummeled once more before entering the last stomach chamber.

antelope browse on wild sage. Come spring, they move to open grasslands and deserts, there to browse on new growth (Figure 42.4).

Now think of your own cheek teeth, or molars, each with a flattened crown that acts as a grinding platform. The crown of an antelope's molars dwarfs yours (Figure 42.4*a*). Why the difference? You probably do not brush your mouth against dirt while eating. An antelope does. Abrasive bits of soil enter its mouth along with tough plant material, so its teeth wear down rapidly. Natural selection has favored more crown to wear down.

Antelopes are **ruminants**, a type of hoofed mammal having multiple stomach chambers in which cellulose is slowly broken down. The breakdown gets under way inside the first two of four stomach sacs (Figure 42.4*b*). There, bacterial symbionts produce cellulose-digesting enzymes, to the antelope's benefit as well as their own. As enzymes act, the antelope regurgitates and rechews the contents of the first two sacs, then swallows again. (That is what "chewing cud" means.) Pummeling plant material more than once exposes more surface area to enzymes and gives them more time to act. Thus, the stomach system of ruminants accepts a steady flow of plant material during extended feeding times, and then it slowly liberates nutrients when the animal rests.

Predatory and scavenging mammals have different specializations. They gorge on food when they can get it, then may not eat again for some time (Figure 42.5). Part of their digestive system stores the food, which is digested and absorbed at a more leisurely pace. As you will see, other organs with accessory roles in digestion help ensure that there will be an adequate distribution of nutrients between meals.

Figure 42.5 Exceptionally Big Mac of the snake world.

An incomplete digestive system is a saclike body cavity. Both food and residues enter and leave through the same opening. A complete digestive system is a tube with two openings and regional specializations in between.

Complete digestive systems carry out the following tasks in controlled ways: movements that break up, mix, and propel food material; secretion of digestive enzymes and other substances; breakdown of food; absorption of nutrients; and expulsion of undigested and unabsorbed residues.

42.2 OVERVIEW OF THE HUMAN DIGESTIVE SYSTEM

Let's look at your own body as an example of the kinds of organs in a complete digestive system. The human digestive system is a tube with two openings and many specialized organs (Figure 42.6). If the tube of an adult were fully stretched out, it would extend 6.5 to 9 meters (21 to 30 feet). From beginning to end, mucus-coated epithelium lines all surfaces facing the lumen. (A *lumen* is the space inside a tube.) The thick, moist mucus protects the wall of the tube and promotes diffusion across its inner lining. Substances advance in one direction, from the mouth through the pharynx, esophagus, and gastrointestinal tract, or gut. The **gut** starts at the stomach and extends through the small intestine, the large intestine (colon) and rectum, to the anus. Accessory organs called the salivary glands, gallbladder, liver, and pancreas secrete a variety of substances into different regions of the tube.

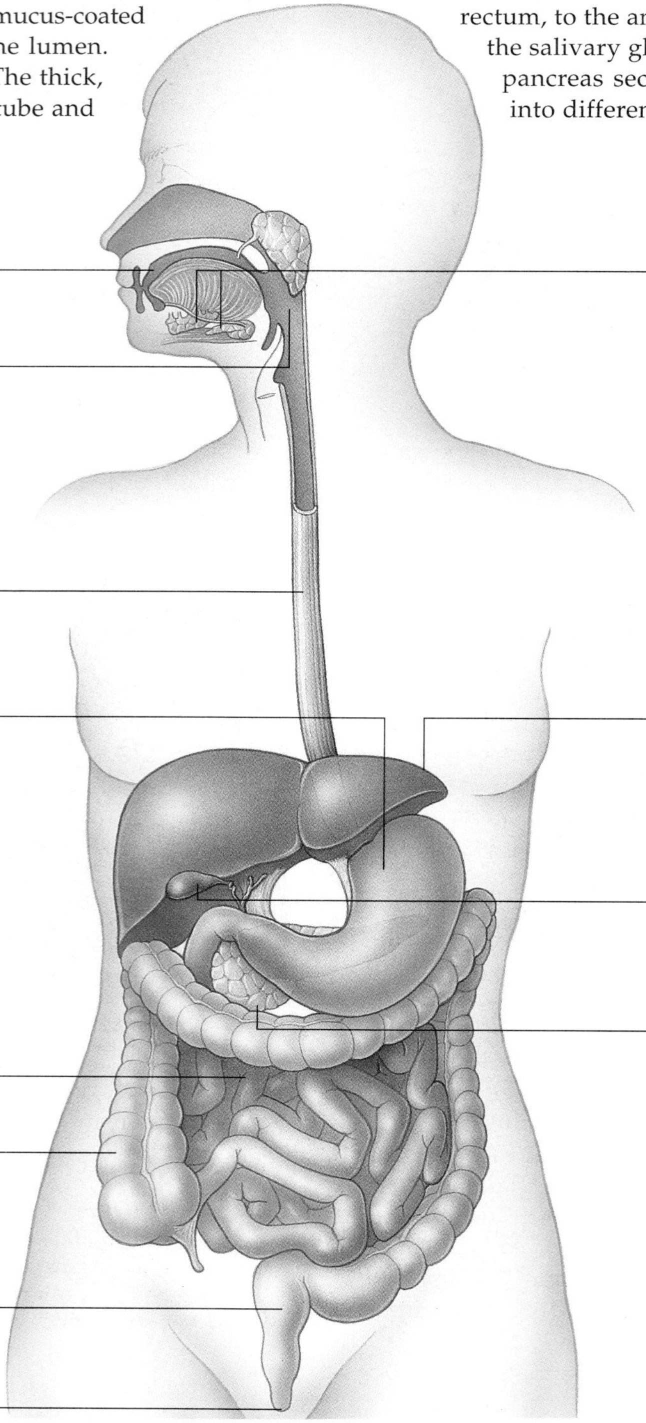

Major Components:
MOUTH (ORAL CAVITY)
Entrance to system; food is moistened and chewed; polysaccharide digestion starts.
PHARYNX
Entrance to tubular part of system (and to respiratory system); moves food forward by contracting sequentially.
ESOPHAGUS
Muscular, saliva-moistened tube that moves food from pharynx to stomach.
STOMACH
Muscular sac; stretches to store food taken in faster than can be processed; gastric fluid mixes with food and kills many pathogens; protein digestion starts.
SMALL INTESTINE
First part (duodenum, C-shaped, about 10 inches long) receives secretions from liver, gallbladder, and pancreas.
In second part (jejunum, about 3 feet long), most nutrients are digested and absorbed.
Third part (ileum, 6–7 feet long) absorbs some nutrients; delivers unabsorbed material to large intestine.
LARGE INTESTINE (COLON)
Concentrates and stores undigested matter by absorbing mineral ions, water; about 5 feet long: divided into ascending, transverse, and descending portions.
RECTUM
Distension stimulates expulsion of feces.
ANUS
End of system; terminal opening through which feces are expelled.
Accessory Organs:
SALIVARY GLANDS
Glands (three main pairs, many minor ones) that secrete saliva, a fluid with polysaccharide-digesting enzymes, buffers, and mucus (which moistens and lubricates food).
LIVER
Secretes bile (for emulsifying fat); roles in carbohydrate, fat, and protein metabolism.
GALLBLADDER
Stores and concentrates bile that the liver secretes.
PANCREAS
Secretes enzymes that break down all major food molecules; secretes buffers against HCl from the stomach.

Figure 42.6 Overview of the component parts of the human digestive system and their specialized functions. Organs with accessory roles in digestion are also listed.

Further reading: Student Guide to InfoTrac on web site

INTO THE MOUTH, DOWN THE TUBE

Food is chewed and polysaccharide breakdown begins in the oral cavity, or mouth. Thirty-two teeth typically project into an adult's mouth (Figure 42.7). Each **tooth** has an enamel coat (hardened calcium deposits), dentin (a thick, bonelike layer), and an inner pulp with nerves and blood vessels. It is an engineering marvel, able to withstand years of chemical insults and mechanical stress. Recall, from Section 27.12, that the chisel-shaped incisors shear off chunks of food. Cone-shaped canines tear it. Premolars and molars, with broad crowns and rounded cusps, are good at grinding and crushing food.

Also in the mouth is a **tongue**, an organ consisting of membrane-covered skeletal muscles that function in positioning food in the mouth, swallowing, and speech. On that tongue's surface are many circular structures with taste buds embedded in their tissues (Figure 42.8). A taste bud contains sensory receptors that respond to chemical differences in dissolved substances. The brain uses this information to give rise to our sense of taste.

Chewing mixes food with **saliva**. This fluid contains an enzyme (salivary amylase), a buffer (bicarbonate, or HCO_3^-), mucins, and water. Salivary glands beneath and in back of the tongue produce and secrete saliva through ducts to the free surface of the mouth's lining. Salivary amylase breaks down starch. The HCO_3^- helps maintain the mouth's pH when you eat acidic foods. Modified proteins called mucins help form the mucus that binds food into a softened, lubricated ball (bolus).

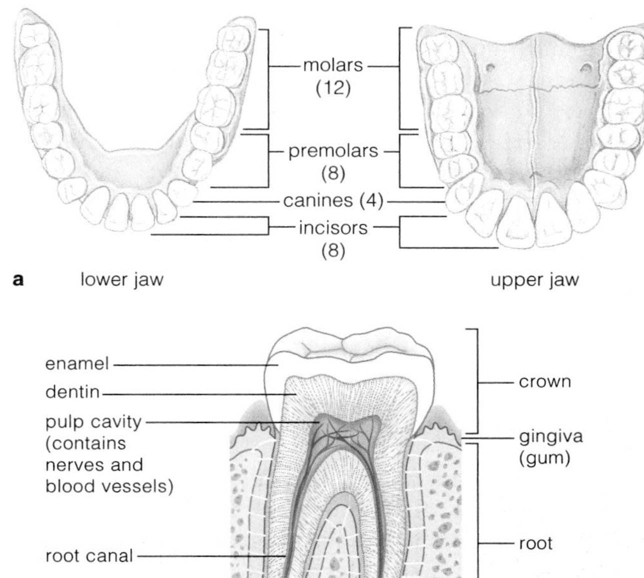
molars
(12)
premolars
(8)
canines (4)
incisors
(8)
a
lower jaw
upper jaw
enamel
dentin
pulp cavity
(contains
nerves and
blood vessels)
root canal
periodontal
membrane
bone
b
crown
gingiva
(gum)
root

Figure 42.7 (**a**) Number and arrangement of human teeth in the upper and lower jaw. (**b**) A molar's main regions: the crown and root. Enamel caps the crown. It consists of calcium deposits and is the hardest substance in the body.

Normally harmless bacteria live on and between teeth. Daily flossing, gentle brushing, and avoidance of too many sweets help keep their populations in check. Without such preventive measures, conditions favor bacterial infection. These may cause *caries* (tooth decay), *gingivitis* (inflamed gums), or both. Infections also can spread to the periodontal membrane, which anchors teeth to the jawbone. In *periodontal disease*, infection slowly destroys the bone tissue around a tooth.

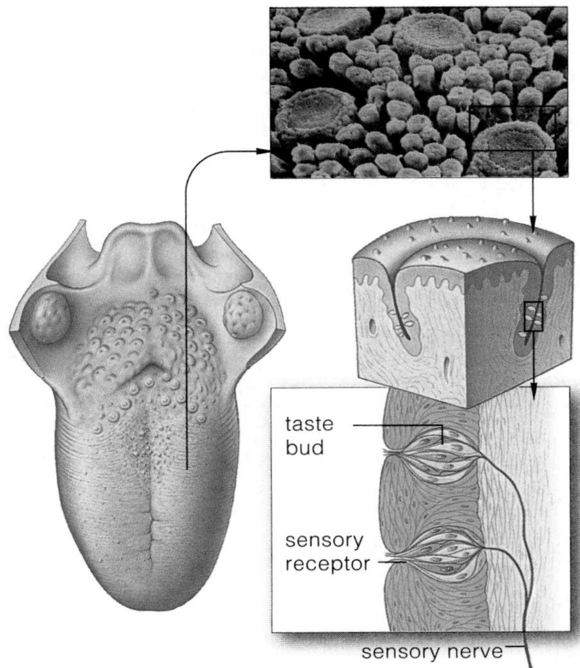
taste
bud
sensory
receptor
sensory nerve

Figure 42.8 Human tongue. Circular structures at its surface house many taste buds. Filamentous structures adjacent to them help move food in the mouth.

When you swallow, contractions of tongue muscles force boluses into the **pharynx**, the tubular entrance to the esophagus *and* trachea, an airway to the lungs. As food leaves the pharynx, a flaplike valve (epiglottis) and the vocal cords close off the trachea and so prevent breathing. The controlled closure is why you normally don't choke on food (Sections 41.4 and *Critical Thinking* question 3 in Section 41.10).

The tubular **esophagus** connects the pharynx with the stomach. Contractions of its muscular wall propel food past a sphincter, into the stomach. A **sphincter** is a ring of smooth muscles, the contractions of which close off a passageway or an opening to the body surface.

By the action of the mouth's teeth and tongue, food gets chewed, mixed with saliva, and bound into soft, lubricated balls that will be propelled through tubes to the stomach. Enzymes in saliva start the digestion of polysaccharides.

42.4 DIGESTION IN THE STOMACH AND SMALL INTESTINE

We arrive now at the premier food-processing organs, the stomach and small intestine. Both have layers of smooth muscles, the contractions of which break apart, mix, and directionally move food. The lumen of both organs receives digestive enzymes and other secretions that help break up nutrients into fragments, then into molecules small enough to be absorbed. Table 42.1 lists the names and sources of the major digestive enzymes. Carbohydrate breakdown *starts* in the mouth. Protein breakdown *starts* in the stomach. However, digestion of nearly all of the carbohydrates, lipids, proteins, and nucleic acids in food is *completed* in the small intestine.

The Stomach

The **stomach**, a muscular, stretchable sac (Figure 42.9*a*), serves three major functions. First, it mixes and stores ingested food. Second, its secretions help dissolve and degrade the food, particularly proteins. Third, it helps control passage of food into the small intestine.

The stomach wall surface exposed to the lumen is lined with glandular epithelium. Each day, the lining's glandular cells secrete two liters or so of hydrochloric acid (HCl), mucus, pepsinogens, and other substances that make up **gastric fluid**, or stomach fluid. The potent stomach acidity, in combination with strong stomach contractions, converts food into a thick, liquid mixture called **chyme**. Chyme's high acidity kills many pathogens that are ingested with food. It also can cause *heartburn* when gastric fluid backs up into the esophagus.

Protein digestion starts when the high acidity alters the structure of proteins and exposes the peptide bonds. It also converts pepsinogens to pepsins, active enzymes that cleave the bonds. Protein fragments accumulate in the stomach's lumen. Meanwhile, glandular cells of the stomach lining secrete gastrin. This hormone stimulates the cells that secrete HCl and pepsinogen.

A *peptic ulcer*, an eroded wall region in the stomach or small intestine, results from insufficient secretion of mucus and buffers or excess pepsin. Heredity, chronic emotional stress, smoking, and excessive use of alcohol and aspirin are contributing factors. In addition, nearly everyone who has intestinal ulcers and 70 percent of those having stomach ulcers are infected by *Helicobacter pylori*. Section 22.1 has a micrograph of this bacterium. A full course of antibiotic therapy cures peptic ulcers in patients who test positive for *H. pylori.*

The stomach empties by waves of contraction and relaxation, of a type called peristalsis. These waves mix chyme and gain force as they approach a sphincter at the base of the stomach (Figure 42.9*b*). With the arrival of a strong contraction, the sphincter closes, so most of the chyme is squeezed backward. Only a small amount moves into the small intestine at a given time. In due course, however, it all moves out of the stomach.

The Small Intestine

Figure 42.6 shows three regions of the small intestine, called the duodenum, jejunum, and ileum. The stomach and ducts from three organs, the **pancreas**, **liver**, and **gallbladder**, deliver an average of 9 liters of fluid per day to the duodenum. At least 95 percent of the fluid is absorbed across the small intestine's epithelial lining.

Cells of the intestinal lining and the pancreas secrete enzymes that digest food to monosaccharides (such as

Table 42.1 Major Digestive Enzymes and Their Breakdown Products

Enzyme	Source	Where Active	Substrate	Main Breakdown Products
CARBOHYDRATE DIGESTION				
Salivary amylase	Salivary glands	Mouth, stomach	Polysaccharides	Disaccharides
Pancreatic amylase	Pancreas	Small intestine	Polysaccharides	Disaccharides
Disaccharidases	Intestinal lining	Small intestine	Disaccharides	MONOSACCHARIDES* (e.g., glucose)
PROTEIN DIGESTION				
Pepsins	Stomach lining	Stomach	Proteins	Protein fragments
Trypsin and chymotrypsin	Pancreas	Small intestine	Proteins	Protein fragments
Carboxypeptidase	Pancreas	Small intestine	Protein fragments	AMINO ACIDS*
Aminopeptidase	Intestinal lining	Small intestine	Protein fragments	AMINO ACIDS*
FAT DIGESTION				
Lipase	Pancreas	Small intestine	Triglycerides	FREE FATTY ACIDS, MONOGLYCERIDES*
NUCLEIC ACID DIGESTION				
Pancreatic nucleases	Pancreas	Small intestine	DNA, RNA	NUCLEOTIDES*
Intestinal nucleases	Intestinal lining	Small intestine	Nucleotides	NUCLEOTIDE BASES, MONOSACCHARIDES*

* Breakdown products small enough to be absorbed into the internal environment.

Further reading: Student Guide to InfoTrac on web site

glucose), monoglycerides (each consists of glycerol with one fatty acid tail), free fatty acids, amino acids, and nucleotide bases and nucleotides. For instance, like pepsin secreted from cells in the stomach lining, two pancreatic enzymes (trypsin and chymotrypsin) break down proteins to peptides. Another cleaves the peptides to free amino acids. The pancreas also secretes bicarbonate, a buffer that helps neutralize HCl from the stomach.

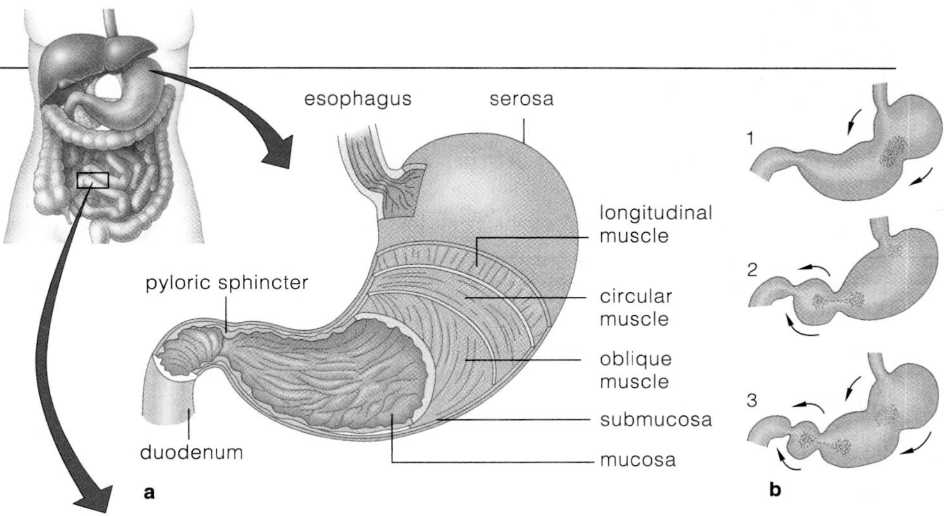
esophagus
serosa
longitudinal muscle
circular muscle
oblique muscle
submucosa
mucosa
pyloric sphincter
duodenum
a
1
2
3
b

The Role of Bile in Fat Digestion

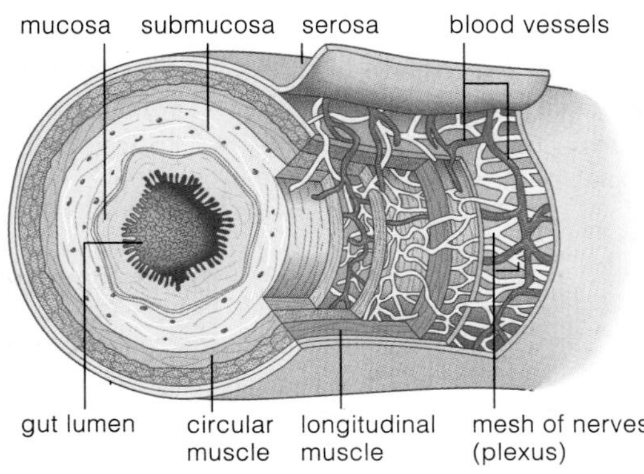
mucosa
submucosa
serosa
blood vessels
gut lumen
circular muscle
longitudinal muscle
mesh of nerves (plexus)
c

Figure 42.9 (**a**) Stomach structure. (**b**) Peristaltic wave down the stomach. (**c**) Structure of the small intestine. The gut wall usually has an innermost mucosa (epithelium and an underlying layer of connective tissue). Next is the submucosa, a connective tissue layer with blood and lymph vessels and a mesh of nerves that locally control digestion. Next in the wall are smooth muscle layers that differ in orientation, hence in the direction of contraction. The outermost layer of connective tissue is the serosa.

Fat digestion requires enzyme action. It also requires **bile**, a fluid that the liver continually secretes. The fluid contains bile salts and bile pigments (including bilirubin, an iron pigment derived from degraded red blood cells). It contains cholesterol and a phospholipid, lecithin. When the stomach is empty, a sphincter closes off the main bile duct from the liver. Bile backs up into the gallbladder, which stores and concentrates it.

By a process of **emulsification**, bile salts accelerate fat digestion. Most fats in the diet are triglycerides. Being insoluble in water, triglycerides tend to cluster into large fat globules in chyme. As muscles in the intestinal wall move and agitate chyme (Figure 42.9*c*), the globules break apart into small droplets that become coated with bile salts. Bile salts carry negative charges, so the coated droplets repel each other and stay separated. This suspension of fat droplets, formed by mechanical and chemical action, is the "emulsion."

Compared with fat globules, emulsion droplets give fat-digesting enzymes a much greater surface area to act upon. Thus triglycerides can be broken down much more rapidly to fatty acids and monoglycerides.

Controls Over Digestion

Homeostatic controls counter changes in the internal environment. By contrast, controls over digestion act *before* food is absorbed into the internal environment. The nervous system, endocrine system, and meshes of nerves (plexuses) inside the gut wall exert control.

For example, incoming food distends the stomach and stimulates mechanoreceptors in the stomach wall. The resulting signals travel along short reflex pathways to smooth muscles and glands; they also travel longer reflex pathways to the brain. Either way, they stimulate muscles in the gut wall to contract or glandular cells to secrete enzyme-rich fluids into the lumen or hormones into blood. Stomach emptying depends on the chyme's volume and composition. For instance, when a large meal activates more receptors in the stomach wall, contractions become more forceful and the stomach empties faster. High acidity or a high fat content in the small intestine calls for secretion of hormones that trigger a slowdown in stomach emptying, so food is not moved faster than it can be processed. Such slowdowns also result from fear, depression, and other emotional upsets.

Gastrointestinal hormones are part of the controls. If chyme contains amino acids and peptides, cells of the stomach lining will secrete the gastrin that stimulates secretion of acid into the stomach. Secretin induces the pancreas to secrete bicarbonate. CCK (cholecystokinin) promotes pancreatic enzyme secretion and gallbladder contraction. When the small intestine contains glucose and fat, GIP (glucose insulinotropic peptide) calls for insulin secretion, which prods cells to absorb glucose.

Carbohydrate breakdown starts in the mouth, and protein breakdown starts in the stomach. In the small intestine, most large organic compounds are digested to molecules small enough to be absorbed into the internal environment.

Signals from the nervous system and a nerve plexus in the gut wall, along with endocrine secretions, control digestion.

42.5 ABSORPTION IN THE SMALL INTESTINE

Structure Speaks Volumes About Function

Unlike the stomach, which absorbs very few substances (such as alcohol), the small intestine is where the vast majority of nutrients are absorbed. Figure 42.10 shows the structure of its wall. Focus first on the profuse folds of the mucosa that project into the lumen. Look closer and you see astounding numbers of tinier projections from each large fold. Look closer still and you see that epithelial cells at the surface of those tiny projections have a brushlike crown of great numbers of even tinier projections, all exposed to the intestinal lumen.

What is the significance of so much convolution? It has a highly favorable surface-to-volume ratio (Section 4.1). Collectively, all the projections from the intestinal mucosa ENORMOUSLY increase the surface area available to interact with the chyme and absorb nutrients from it. Without that immense surface area, absorption would proceed hundreds of times more slowly, which of course would not be enough to sustain human life.

Figure 42.10*c,d* shows **villi** (singular, villus), which are absorptive structures on mucosal folds. Each villus is about a millimeter long. Millions project outward from the mucosa. Their density gives the mucosa a velvety appearance. Inside each villus is an arteriole, a venule, and a lymph vessel that function in moving substances to and from the general circulation (Figure 42.10*e*).

Most cells in the epithelial lining of each villus bear **microvilli** (singular, microvillus), which are ultrafine, threadlike projections from their free surface. Each of these cells has about 1,700 microvilli. Hence its name, "brush border" cell. Glandular cells in the lining make and secrete digestive enzymes (Table 42.1), Also, some phagocytic cells patrol and help protect the lining.

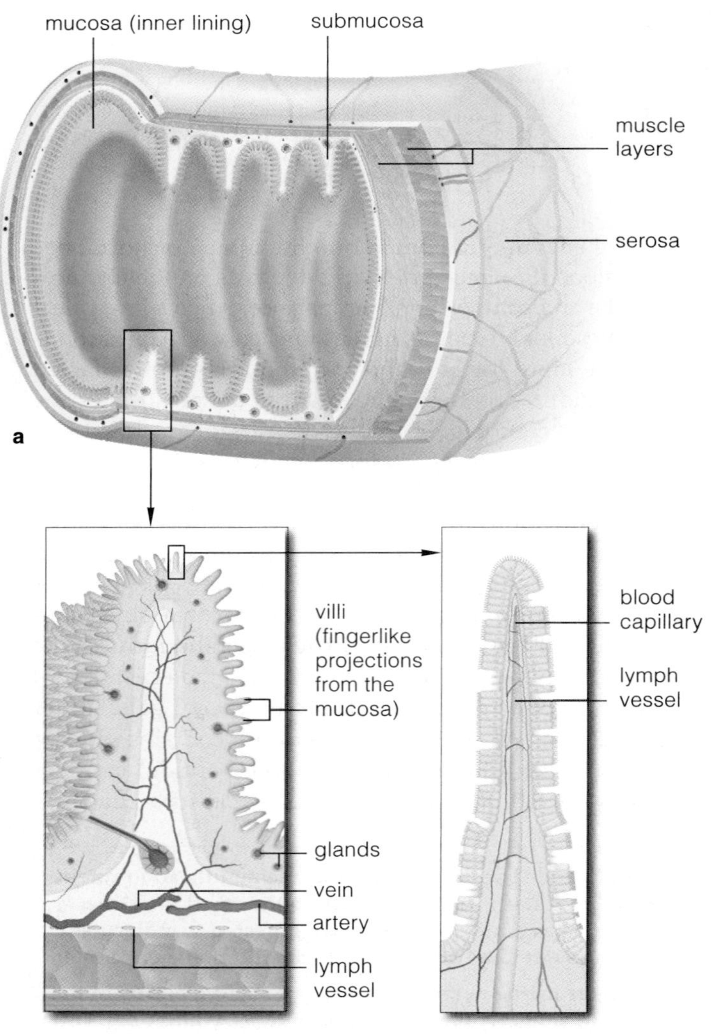

mucosa (inner lining)
submucosa
muscle layers
serosa
a
villi (fingerlike projections from the mucosa)
glands
vein
artery
lymph vessel
blood capillary
lymph vessel

c Villi of the intestinal mucosa

d One villus

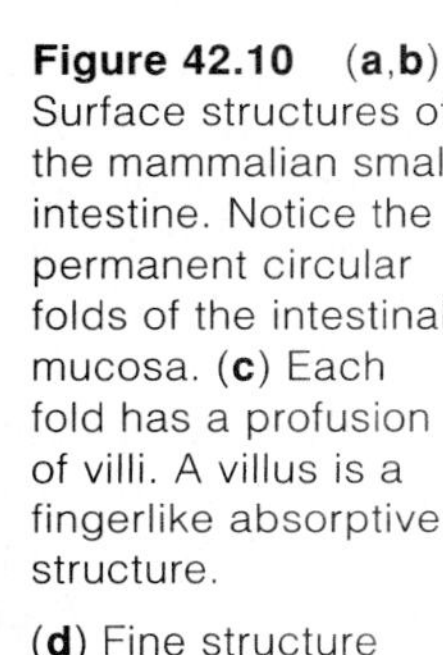

b Circular folds of the mucosa

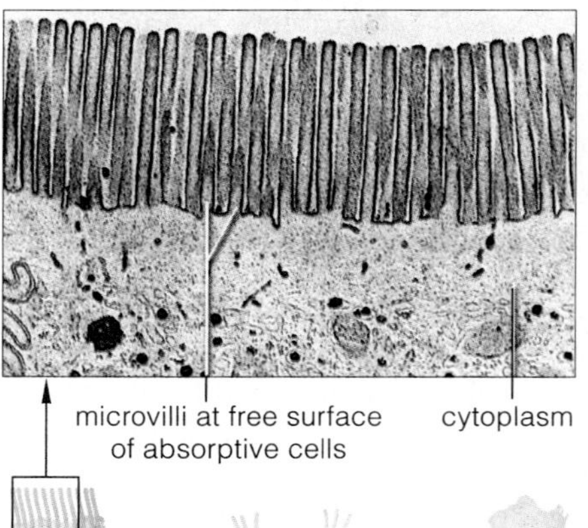

microvilli at free surface of absorptive cells
cytoplasm

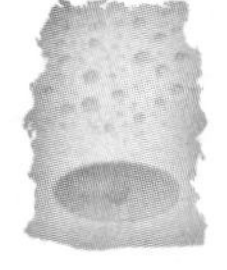

absorption
mucus secretion
hormone secretion
phagocytosis, lysozyme secretion

e Functions of epithelial cells of villus

Figure 42.10 (**a**,**b**) Surface structures of the mammalian small intestine. Notice the permanent circular folds of the intestinal mucosa. (**c**) Each fold has a profusion of villi. A villus is a fingerlike absorptive structure.

(**d**) Fine structure of a single villus. Monosaccharides and most amino acids that cross the intestinal lining enter blood capillaries in the villus. Fats enter the lymph vessels.

(**e**) Epithelial cell types at the free surface of a villus. Absorptive cells have a crown of microvilli facing the intestinal lumen.

Further reading: Student Guide to InfoTrac on web site

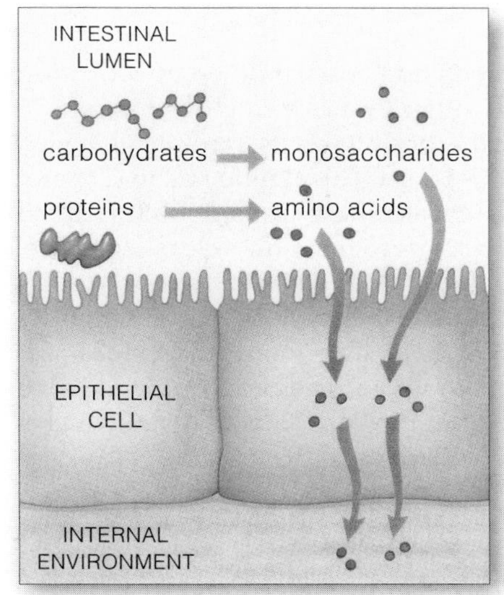
INTESTINAL LUMEN
carbohydrates
monosaccharides
proteins
amino acids
EPITHELIAL CELL
INTERNAL ENVIRONMENT

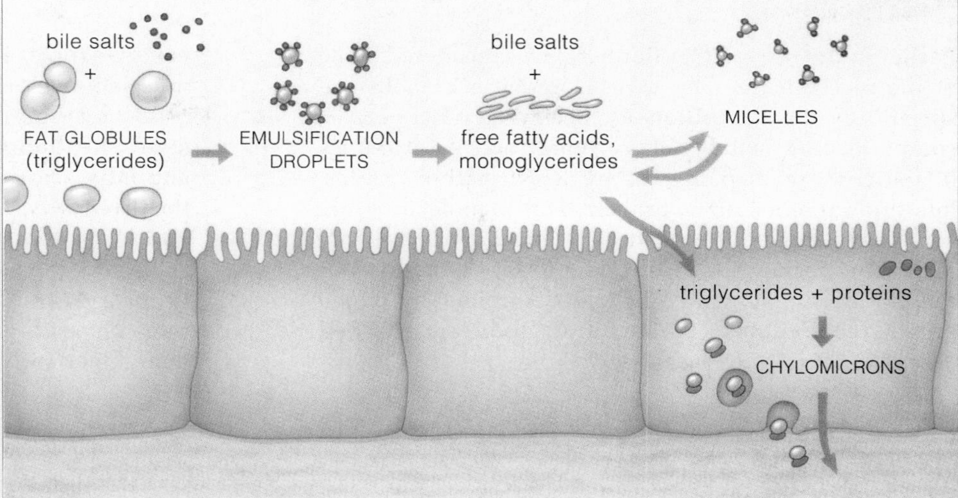
bile salts
+
FAT GLOBULES (triglycerides)
EMULSIFICATION DROPLETS
bile salts
+
free fatty acids, monoglycerides
MICELLES
triglycerides + proteins
CHYLOMICRONS

a Digestion of carbohydrates to monosaccharides, and proteins to amino acids, is completed with enzymes secreted by the pancreas and by cells of the epithelial lining.

b Monosaccharides and amino acids are actively transported across the plasma membrane of the epithelial cells, then out of the same cells and into the internal environment.

c Emulsification: The constant movement of the intestinal wall breaks up fat globules into small emulsification droplets. Bile salts prevent the globules from re-forming. Pancreatic enzymes digest the droplets to fatty acids and monoglycerides.

d Micelles form as bile salts combine with digestion products and phospholipids. Products can readily slip into and out of the micelles.

e Concentration of monoglycerides and fatty acids in micelles enhances gradients that lead to diffusion of both substances across the lipid bilayer of the plasma membrane of cells making up the lining.

f Products of fat digestion reassemble into triglycerides in cells of the intestinal lining. These become coated with proteins, then are expelled (by exocytosis from the cells) into the internal environment.

Figure 42.11 Digestion and absorption in the small intestine. 

What Are the Absorption Mechanisms?

Absorption, recall, is the passage of nutrients, water, salts, and vitamins into the internal environment. The intestinal wall's vast absorptive surface facilitates the process, but so does the action of its smooth muscle. In **segmentation**, rings of circular muscle contract and relax repeatedly. This creates an oscillating (back and forth) movement that constantly mixes and forces the lumen contents against the wall's absorptive surface:

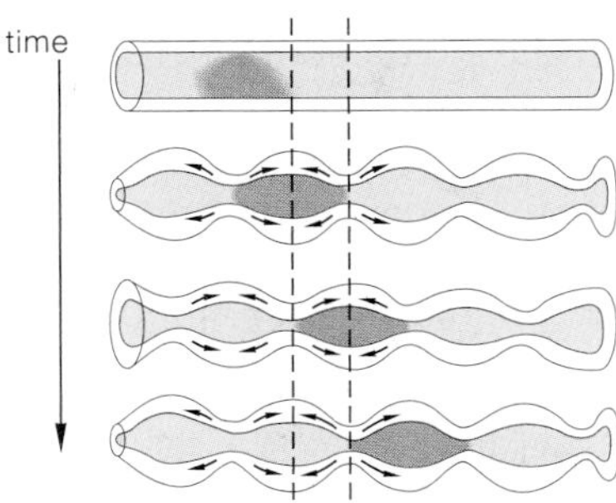
time

By the time a meal is halfway through the small intestine, most is broken apart and digested. Molecules of water cross the intestinal lining by osmosis. Mineral ions are selectively absorbed. Transport proteins in the plasma membrane of brush border cells actively shunt some breakdown products, including monosaccharides and amino acids, across the lining. By contrast, fatty acids and monoglycerides diffuse across the lipid bilayer of the plasma membrane. And bile salts help them do this (Figure 42.11).

By a process called **micelle formation**, the bile salts combine with the products of fat digestion and with phospholipids to form tiny droplets (micelles). Product molecules in the micelles continuously exchange places with those dissolved in chyme. However, the micelles concentrate them next to the intestinal lining, and when they are concentrated enough, the gradients favor their diffusion out of micelles and into epithelial cells. Fatty acids and monoglycerides recombine in the cells to form triglycerides. Then triglycerides combine with proteins into particles (chylomicrons) that leave the cells by way of exocytosis and enter the internal environment.

Once absorbed, the glucose and amino acids enter blood vessels directly. Triglycerides enter lymph vessels that eventually drain into blood vessels.

With its richly folded intestinal mucosa, millions of villi, and hundreds of millions of microvilli, the small intestine offers a vast surface area for absorbing nutrients.

Substances pass through the brush border cells that line the free surface of each villus by active transport, osmosis, and diffusion across the lipid bilayer of plasma membranes.

42.6 DISPOSITION OF ABSORBED ORGANIC COMPOUNDS

Earlier in the book, in Section 8.6, you considered some of the mechanisms that govern organic metabolism—specifically, the disposition of glucose and other organic compounds in the body as a whole. You saw examples of the conversion pathways by which carbohydrates, fats, and proteins can be broken apart to molecules that serve as intermediates in the ATP-producing pathway of aerobic respiration. Here, Figure 42.12 rounds out the picture by showing all the major routes by which organic compounds obtained from food are shuffled and reshuffled in the body as a whole.

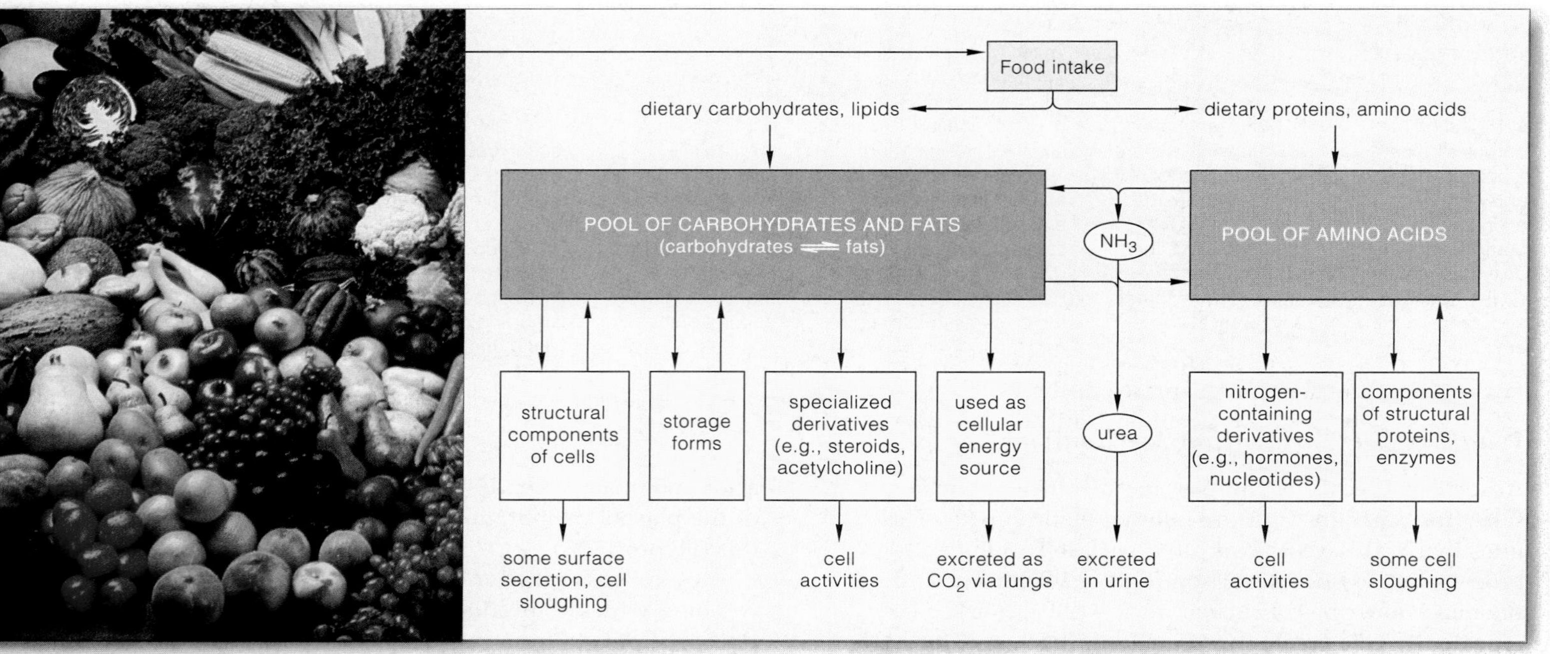

Food intake
dietary carbohydrates, lipids
dietary proteins, amino acids
POOL OF CARBOHYDRATES AND FATS
(carbohydrates ⇌ fats)
NH_3
POOL OF AMINO ACIDS
structural components of cells
storage forms
specialized derivatives (e.g., steroids, acetylcholine)
used as cellular energy source
urea
nitrogen-containing derivatives (e.g., hormones, nucleotides)
components of structural proteins, enzymes
some surface secretion, cell sloughing
cell activities
excreted as CO_2 via lungs
excreted in urine
cell activities
some cell sloughing

Figure 42.12 Summary of the major pathways of organic metabolism. Cells continually synthesize and tear down carbohydrates, fats, and proteins. Urea forms mainly in the liver. See also Sections 8.6 and 37.7.

All living cells in the body continually break apart most of their carbohydrates, lipids, and proteins, then use the breakdown products as energy sources or as building blocks. The nervous and endocrine systems interact to integrate this massive molecular turnover, but a treatment of how they do this would be beyond the scope of this book.

For now, simply reflect on a few key points about organic metabolism. When you eat, your body adds to its pools of materials. Excess amounts of carbohydrates and other organic compounds absorbed from the gut are transformed mostly into fats, which become stored in adipose tissue. Some are converted to glycogen in the liver and muscle tissue. *While* compounds are being absorbed and stored, most cells use glucose as the main energy source. There is no net breakdown of protein in muscle tissue or any other tissue during this period.

Between meals, the body taps its fat and glycogen stores. In adipose tissue, cells dismantle fats to glycerol and fatty acids and release these to the blood. Cells in the liver break apart glycogen and release glucose, which also enters blood. Body cells take up the fatty acids and glucose and use them for ATP production.

Bear in mind, the liver does more than store and interconvert organic compounds. For example, it helps maintain their concentrations in blood. It inactivates most hormone molecules, which are sent to the kidneys for excretion, in urine. The liver also removes worn-out blood cells from the circulation and inactivates many potentially toxic compounds. For example, ammonia (NH_3), produced during amino acid breakdown, can be toxic in high concentrations. The liver converts NH_3 to urea, a less toxic waste product. Urea is excreted from the kidneys, as a component of urine.

Just after a meal, the body's cells take up the glucose being absorbed from the gut and use it as a quick energy source. Excess amounts of glucose and other organic compounds become converted mainly to fats that are stored in adipose tissue. Some of the excess is converted to glycogen, which gets stored mainly in the liver and in muscle tissue.

Between meals, the body taps the fat reservoirs as the main energy source. Fats get converted to glycerol and fatty acids, both of which can enter ATP-producing pathways.

42.7 THE LARGE INTESTINE

What happens to the material *not* absorbed in the small intestine? It moves into the large intestine, or **colon**. The colon concentrates and stores feces: a mixture of water, undigested and unabsorbed matter, and bacteria. The colon starts at a cup-shaped pouch, the cecum. It ascends on the right side of the abdominal cavity and continues across to the other side. Then it descends and connects with a short tube, the rectum (Figures 42.6 and 42.13).

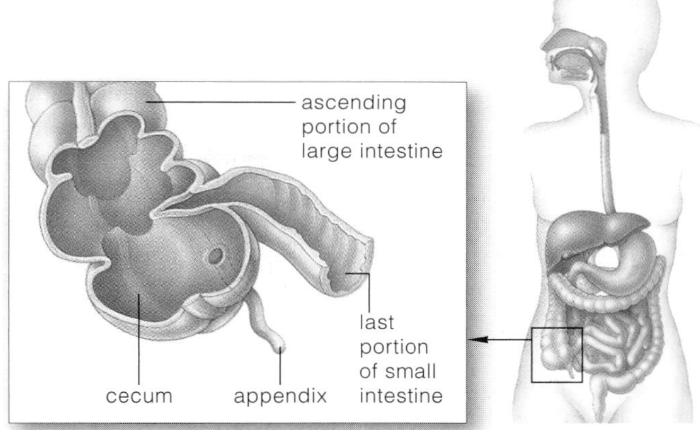
ascending portion of large intestine
last portion of small intestine
cecum
appendix

Figure 42.13 Cecum and appendix of the large intestine (colon).

Colon Functioning

Material becomes concentrated as water moves across the colon's lining. Cells of the lining actively transport sodium ions out of the lumen. As ion concentrations in the lumen fall, the water concentration increases. Thus water also moves out of the lumen, by osmosis. Cells of the lining secrete mucus, which lubricates the feces and helps keep them from mechanically damaging the wall of the colon. Additionally, the cells secrete bicarbonate, which helps buffer the acidic fermentation products of ingested bacteria that managed to colonize the colon.

Short, longitudinal bands of smooth muscle in the colon wall are gathered at their ends, like a series of full skirts nipped in at elastic waistbands. As they contract and relax, the contents of the lumen move back and forth against the absorptive surface of the wall. Nerve plexuses largely control the motion, which is similar to segmentation in the small intestine but much slower.

Whereas the quick transit time through the small intestine does not favor the rapid population growth of ingested bacteria, the colon's slower motion favors it. Generally speaking, the colonizers do no harm unless they breach the wall and enter the abdominal cavity.

With each new meal, hormonal signals (gastrin) and commands from autonomic nerves induce large portions of the ascending and transverse colon to contract at the same time. Within a few seconds, the lumen's existing contents move as much as three-fourths of the colon's length and make way for incoming food.

The contents are stored in the last part of the colon until they trigger defecation—they distend the rectal wall enough to trigger a reflex action that leads to their expulsion. The nervous system controls the expulsion by stimulating or inhibiting the contraction of a muscle sphincter at the anus, the terminal opening of the gut.

The volume of cellulose fiber and other undigested material that cannot be decreased by absorption in the colon is called **bulk**. Bulk contributes to the volume and normal transit time of material through the colon.

Colon Malfunctioning

The normal frequency of defecation ranges from three times a day to once a week. Aging, emotional stress, a low-bulk diet, injury, or disease can result in delayed defecation, or *constipation*. The longer the delay, the more water is absorbed, so feces become hard and dry. The abdominal discomfort is accompanied by loss of appetite, headaches, and often nausea and depression.

Hard feces may become lodged inside the **appendix**, a narrow projection from the cecum (Figure 42.13). The appendix has no known digestive functions. But it is colonized by lymphocytes that defend the body against ingested bacteria. Feces that obstruct the normal blood flow and mucus secretion to this projection can cause *appendicitis*, or an inflamed appendix. Unless surgically removed, an inflamed appendix may rupture. Bacteria that enter the abdominal cavity through the breach can cause life-threatening infections.

The colon also is vulnerable to cancer, which occurs most often among the world's wealthiest and best-fed populations. Too many of these people skip meals, eat too much and too fast when they do sit down at the table, and generally give their gut erratic workouts. Their diet tends to be rich in sugar, cholesterol, and salt, and low in bulk. Too little bulk extends the transit time of feces through the colon. The longer that irritating, potentially carcinogenic material is in contact with the colon wall, the more damage it may cause. Symptoms of *colon cancer* include a change in bowel functioning, rectal bleeding, and blood in feces. Some people appear to be genetically predisposed to develop colon cancer, but a low-fiber diet might also be a factor.

Colon cancer is almost nonexistent in rural India and Africa, where people can't afford to eat much more than fiber-rich whole grains. When those same people move to cities in the affluent nations and change their eating habits, the incidence of colon cancer increases.

The large intestine, or colon, functions in the absorption of water and mineral ions from the gut lumen. It also functions in the compaction of undigested residues into feces, for expulsion at the terminal opening of the gut.

42.8 HUMAN NUTRITIONAL REQUIREMENTS

You grow and maintain yourself by eating certain foods that are suitable sources of energy and raw materials. Nutritionists measure the energy in **kilocalories**. Each kilocalorie is 1,000 calories of heat energy, the amount needed to raise the temperature of 1 kilogram of water by 1°C. The raw materials will now be described.

New, Improved Food Pyramids

A few million years ago, our hominid ancestors dined mostly on fresh fruits and other fibrous plant material. From that nutritional beginning, humans in many parts of the world came to prefer low-fiber, high-fat foods. They still do even when they study **food pyramids**, or charts of well-balanced diets, from elementary school onward. Nutritionists revise the charts on an ongoing basis to reflect additional research. Figure 42.14 shows a recent version. It gives daily portions of foods that will supply an adult male of average size with 55–60 percent complex carbohydrates, 15–20 percent proteins (less for females), and 20–25 percent fats. Not everyone agrees on proportions. For example, according to *the zone diet*, daily intake should be 40 percent complex carbohydrates, 30 percent proteins, and 30 percent fats to keep the body at peak performance levels.

MILK, YOGURT, CHEESE GROUP
Choose 2–3 servings from these alternatives:
1 cup nonfat or lowfat milk
1 cup nonfat or lowfat yogurt
1-1/2 ounces lowfat, unprocessed cheese
2 ounces lowfat, processed cheese

FRUIT GROUP
Choose 2–4 servings from these alternatives:
1 medium-size fruit, such as apple, mango
1 wedge cantaloupe or pineapple
1 cup fresh berries
3/4 cup unsweetened fruit juice
1/2 cup canned unsweetened fruit
1/4 cup dried fruit

ADDED FATS AND SIMPLE SUGARS
Restrict intake (other foods provide ample amounts)
Limit salad dressings, butter, cream, margarine, syrups, table sugar, soft drinks, candies, sweet desserts

LEGUME, NUT, POULTRY, FISH, MEAT GROUP
Choose 2–3 servings from these alternatives:
1-1/4 to 1-1/2 cups cooked legumes
2-1/2 to 3 ounces cooked poultry (no skin)
2-1/2 to 3 ounces cooked fish
2-1/2 to 3 ounces trimmed, cooked, lean red meat
1 egg (or two egg whites)
5 tablespoons peanut butter

VEGETABLE GROUP
Choose 3–5 servings from these alternatives:
1/2 cup chopped raw or cooked vegetables
1 cup raw, leafy vegetables

BREAD, CEREAL, RICE, PASTA GROUP
Choose 6–11 servings from these alternatives:
1 slice whole-grain bread
1/2 cup cooked rice (brown rice has more nutrients)
1/2 cup cooked pasta
1/2 cup cooked unsweetened cereal, such as oatmeal

1 ounce dry unsweetened cereal, such as shredded wheat

Figure 42.14 1992 food pyramid diagram. Daily portions of foods give you an idea of what mainstream nutritionists call a well-balanced diet. The alternatives listed are a general guide to the kinds of foods in each group. Good health requires daily selections from all groups *except* the pyramid's "red-hot" tip. Added fats and simple sugars pile on the calories but provide few vitamins and minerals. Where a range of serving sizes is shown, choosing the smallest helps keep total caloric intake to 1,600 kilocalories or so. Choosing the largest will raise the total to about 2,800 kilocalories.

Carbohydrates

Starch and, to a lesser extent, glycogen should be the main carbohydrates in the human diet. These complex carbohydrates are easily degraded to glucose, the body's main energy source. Starch is abundant in fleshy fruits, cereal grains, and legumes, including beans and peas.

Foods rich in complex carbohydrates have another advantage: they typically are high in fiber. Section 42.7 highlighted the sobering consequences of a low-fiber diet. Epidemiologist Denis Burkitt may have been only half-joking when he put it this way: If you pass a small volume of feces, you have large hospitals.

Simple sugars do not have the fiber component of complex carbohydrates. Nor do they have the vitamins and minerals of whole foods. Yet each week, an American typically consumes as much as two pounds of refined sugar (sucrose). If you dismiss this as a far-fetched statistic, start taking a critical look at all the ingredients

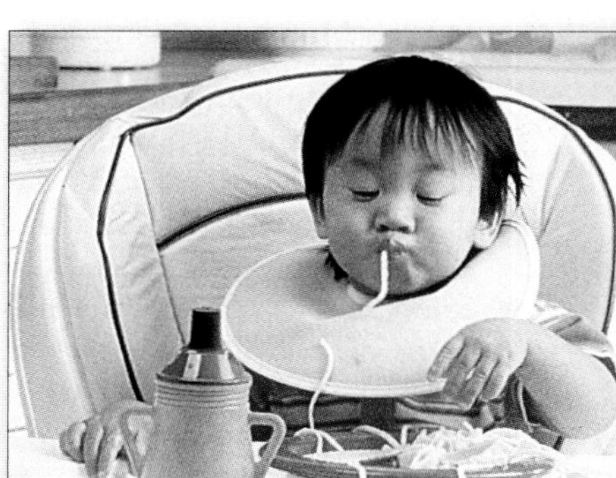

Further reading: Student Guide to InfoTrac on web site

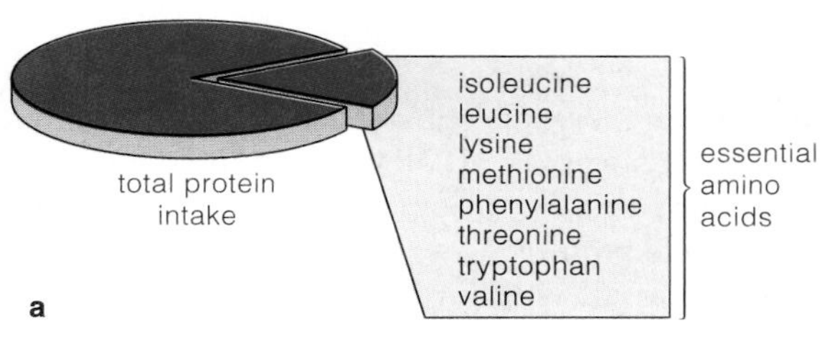

total protein intake
isoleucine
leucine
lysine
methionine
phenylalanine
threonine
tryptophan
valine
essential amino acids
a

No Limiting Amino Acid	Low in Lysine	Low in Methionine, Other Sulfur-Containing Amino Acids	Low in Tryptophan
legumes: soybean tofu soy milk cereal grains: wheat germ milk cheeses (except cream cheese) yogurt eggs meats	legumes: peanuts cereal grains: barley buckwheat cornmeal oats rice rye wheat nuts, seeds: almonds cashews coconut English walnuts hazelnuts pecans pumpkin seeds sunflower seeds	legumes: beans (dried) black-eyed peas garbanzos lentils lima beans mung beans peanuts nuts: hazelnuts fresh vegetables: asparagus broccoli green peas mushrooms parsley potatoes soybeans Swiss chard	legumes: beans (dried) garbanzos lima beans mung beans peanuts cereal grains: cornmeal nuts: almonds English walnuts fresh vegetables: corn green peas mushrooms Swiss chard

b

Figure 42.15 (**a**) The eight essential amino acids, a small portion of the total protein intake. All must be available at the same time, in certain amounts, if cells are to build their own proteins. Milk and eggs have high amounts of all eight in proportions that humans require, meaning they are among the *complete* proteins.

Nearly all plant proteins are *incomplete*. Vegetarians must be very careful to avoid protein deficiency. For instance, they can combine different foods from the three columns of incomplete proteins shown in (**b**). Also, strict vegetarians (who also avoid dairy products and eggs) should be sure to take vitamins B_{12} and B_2 (riboflavin) supplements. Animal protein is a luxury in most traditional societies. Their cuisines have good combinations of plant proteins, including rice/beans, chili/cornbread, tofu/rice, and lentils/wheat bread.

listed on the containers for cereals, frozen foods, soft drinks, and numerous other prepared foods. Commonly, sugars are "hidden" in the labels as corn syrup, corn sweeteners, dextrose, and so on.

Lipids

The body cannot function without fats and other lipids. For example, the phospholipid lecithin is a necessary component of cell membranes; and fats serve as energy reserves, cushion many internal organs, and serve as insulation beneath the skin. Dietary fats also help the body store fat-soluble vitamins. However, the body can synthesize most of its own fats, including cholesterol, from proteins and carbohydrates. The ones it cannot make are called **essential fatty acids**, but whole foods provide plenty of them. Linoleic acid is one example. One teaspoon a day of corn oil, olive oil, or some other polyunsaturated fat in food provides enough of it.

On average, butter and other fats now make up 40 percent of the American diet. It should be 30 percent or less, according to most of the medical community.

Like other animal fats, butter is a saturated fat that tends to raise the level of cholesterol in the blood. As described in earlier chapters, cholesterol is used in the synthesis of bile acids and steroid hormones, and it is a required component of animal cell membranes. But too much cholesterol may interfere seriously with blood circulation (Section 39.9). Certain potato chips and other snack foods are made with edible but nondigestible oils such as *sucrose polyester*. They are aimed at people who crave fats enough to put up with the digestive upsets.

Proteins

Amino acid components of dietary proteins are needed for the body's own protein-building programs. Of the twenty common types, eight are **essential amino acids**. That is, our body cells cannot synthesize them, so we must get enough of them from food. The eight amino acids are either methionine or cysteine (its metabolic equivalent), isoleucine, leucine, lysine, phenylalanine (or tyrosine), threonine, tryptophan, and valine.

Most animal proteins are *complete*: their amino acid ratios match human nutritional needs. Plant proteins are *incomplete*; they lack one or more essential amino acids. To get enough amino acids, vegetarians must eat certain combinations of different plants (Figure 42.15).

A measure called the net protein utilization (NPU) is used to compare proteins from different sources. NPU values range from 100 (all essential amino acids occur in ideal proportions) to 0 (one or more amino acids are absent; when eaten alone, the protein is not complete).

Because enzymes and other proteins are crucial for the body's structure and functioning, protein-deficient diets may have severe consequences. Protein deficiency is most damaging to fetuses and to infants. Why? The brain grows and develops rapidly early in life. Unless a mother-to-be takes in enough protein, especially just before and just after birth, her newborn will be affected by irreversible mental retardation. Even seemingly mild protein starvation retards growth and can affect mental and physical performance.

Individuals grow and maintain themselves in good health by eating certain amounts of complex carbohydrates, lipids, and proteins that provide them with adequate raw materials and energy, as measured in kilocalories.

42.9 VITAMINS AND MINERALS

Vitamins are *organic* substances that are essential for growth and survival, for no other substances can play their metabolic roles. Most plants are able to synthesize vitamins on their own. Most animals lost the ability to do so and must obtain vitamins from food.

At the minimum, human cells require the thirteen vitamins listed in Table 42.2. Each vitamin has specific metabolic functions. Many reactions use several types, and the absence of one affects the functions of others.

Minerals are *inorganic* substances that are essential for growth and survival; no other substances can play their metabolic roles. For instance, all of your cells need iron for their electron transport chains. Red blood cells don't function at all without iron in hemoglobin. And neurons stop functioning in the absence of sodium and potassium (Table 42.3).

People who are in good health get all the vitamins and minerals they need from a balanced diet of whole

Table 42.2 Vitamins: Sources, Functions, and Effects of Deficiencies or Excesses*

Vitamin	Common Sources	Main Functions	Effects of Chronic Deficiency	Effects of Extreme Excess
FAT-SOLUBLE VITAMINS				
A	Its precursor comes from beta-carotene in yellow fruits, yellow or green leafy vegetables; also in fortified milk, egg yolk, fish liver	Used in synthesis of visual pigments, bone, teeth; maintains epithelia	Dry, scaly skin; lowered resistance to infections; night blindness; permanent blindness	Malformed fetuses; hair loss; changes in skin; liver and bone damage; bone pain
D	D_3 formed in skin and in fish liver oils, egg yolk, fortified milk; converted to active form elsewhere	Promotes bone growth and mineralization; enhances calcium absorption	Bone deformities (rickets) in children; bone softening in adults	Retarded growth; kidney damage; calcium deposits in soft tissues
E	Whole grains, dark green vegetables, vegetable oils	Counters effects of free radicals; helps maintain cell membranes; blocks breakdown of vitamins A and C in gut	Lysis of red blood cells; nerve damage	Muscle weakness, fatigue, headaches, nausea
K	Enterobacteria form most of it; also in green leafy vegetables, cabbage	Blood clotting; ATP formation via electron transport	Abnormal blood clotting; severe bleeding (hemorrhaging)	Anemia; liver damage and jaundice
WATER-SOLUBLE VITAMINS				
B_1 (thiamin)	Whole grains, green leafy vegetables, legumes, lean meats, eggs	Connective tissue formation; folate utilization; coenzyme action	Water retention in tissues; tingling sensations; heart changes; poor coordination	None reported from food; possible shock reaction from repeated injections
B_2 (riboflavin)	Whole grains, poultry, fish, egg white, milk	Coenzyme action	Skin lesions	None reported
Niacin	Green leafy vegetables, potatoes, peanuts, poultry, fish, pork, beef	Coenzyme action	Contributes to pellagra (damage to skin, gut, nervous system, etc.)	Skin flushing; possible liver damage
B_6	Spinach, tomatoes, potatoes, meats	Coenzyme in amino acid metabolism	Skin, muscle, and nerve damage; anemia	Impaired coordination; numbness in feet
Pantothenic acid	In many foods (meats, yeast, egg yolk especially)	Coenzyme in glucose metabolism, fatty acid and steroid synthesis	Fatigue, tingling in hands, headaches, nausea	None reported; may cause diarrhea occasionally
Folate (folic acid)	Dark green vegetables, whole grains, yeast, lean meats; enterobacteria produce some folate	Coenzyme in nucleic acid and amino acid metabolism	A type of anemia; inflamed tongue; diarrhea; impaired growth; mental disorders	Masks vitamin B_{12} deficiency
B_{12}	Poultry, fish, red meat, dairy foods (not butter)	Coenzyme in nucleic acid metabolism	A type of anemia; impaired nerve function	None reported
Biotin	Legumes, egg yolk; colon bacteria produce some	Coenzyme in fat, glycogen formation and in amino acid metabolism	Scaly skin (dermatitis), sore tongue, depression, anemia	None reported
C (ascorbic acid)	Fruits and vegetables, especially citrus, berries, cantaloupe, cabbage, broccoli, green pepper	Collagen synthesis; possibly inhibits effects of free radicals; structural role in bone, cartilage, and teeth; role in carbohydrate metabolism	Scurvy, poor wound healing, impaired immunity	Diarrhea, other digestive upsets; may alter results of some diagnostic tests

* Guidelines for appropriate daily intakes are being worked out by the Food and Drug Administration.

Further reading: Student Guide to InfoTrac on web site

Table 42.3 Major Minerals: Sources, Functions, and Effects of Deficiencies or Excesses*

Mineral	Common Sources	Main Functions	Signs of Severe Long-Term Deficiency	Signs of Extreme Excess
Calcium	Dairy products, dark green vegetables, dried legumes	Bone, tooth formation; blood clotting; neural and muscle action	Stunted growth; possibly diminished bone mass (osteoporosis)	Impaired absorption of other minerals; kidney stones in susceptible people
Chloride	Table salt (usually too much in diet)	HCl formation in stomach; contributes to body's acid–base balance; neural action	Muscle cramps; impaired growth; poor appetite	Contributes to high blood pressure in susceptible people
Copper	Nuts, legumes, seafood, drinking water	Used in synthesis of melanin, hemoglobin, and some transport chain components	Anemia, changes in bone and blood vessels	Nausea, liver damage
Fluorine	Fluoridated water, tea, seafood	Bone, tooth maintenance	Tooth decay	Digestive upsets; mottled teeth and deformed skeleton in chronic cases
Iodine	Marine fish, shellfish, iodized salt, dairy products	Thyroid hormone formation	Enlarged thyroid (goiter), with metabolic disorders	Goiter
Iron	Whole grains, green leafy vegetables, legumes, nuts, eggs, lean meat, molasses, dried fruit, shellfish	Formation of hemoglobin and cytochrome (transport chain component)	Iron-deficiency anemia, impaired immune function	Liver damage, shock, heart failure
Magnesium	Whole grains, legumes, nuts, dairy products	Coenzyme role in ATP–ADP cycle; roles in muscle, nerve function	Weak, sore muscles; impaired neural function	Impaired neural function
Phosphorus	Whole grains, poultry, red meat	Component of bone, teeth, nucleic acids, ATP, phospholipids	Muscular weakness; loss of minerals from bone	Impaired absorption of minerals into bone
Potassium	Diet provides ample amounts	Muscle and neural function; roles in protein synthesis and body's acid–base balance	Muscular weakness	Muscular weakness, paralysis, heart failure
Sodium	Table salt; diet provides ample to excessive amounts	Key role in body's salt–water balance; roles in muscle and neural function	Muscle cramps	High blood pressure in susceptible people
Sulfur	Proteins in diet	Component of body proteins	None reported	None likely
Zinc	Whole grains, legumes, nuts, meats, seafood	Component of digestive enzymes; roles in normal growth, wound healing, sperm formation, and taste and smell	Impaired growth, scaly skin, impaired immune function	Nausea, vomiting, diarrhea; impaired immune function and anemia

* The guidelines for appropriate daily intakes are being worked out by the Food and Drug Administration.

foods. Generally, vitamin and mineral supplements are necessary only for strict vegetarians, the elderly, and people suffering a chronic illness or taking medication that interferes with utilization of specific nutrients. As examples, vitamin K supplements aid calcium retention and lessen osteoporosis in elderly women. Vitamin C, vitamin E, and the beta-carotene precursor for vitamin A may lessen some aging effects and improve immune functioning by inactivating free radicals. A free radical, remember, is an atom or group of atoms that is highly reactive owing to an unpaired electron.

However, metabolism varies in its details from one person to the next, so no one should take massive doses of any vitamin or mineral supplement except under medical supervision. Also, excessive amounts of many vitamins and minerals can harm everyone (Tables 42.2 and 42.3). Large doses of vitamins A and D are good examples. As is the case for all fat-soluble vitamins, excess amounts of these vitamins accumulate in tissues and interfere with normal metabolic activity. Similarly, sodium is present in plant and animal tissues and is a component of table salt. Sodium has roles in the body's salt–water balance, muscle activity, and nerve function. However, prolonged, excessive intake of sodium may contribute to high blood pressure in some people.

Severe shortages or self-prescribed, massive excesses of vitamins and minerals can disturb the delicate balances in body function that promote health.

42.10 *Focus on Science*

WEIGHTY QUESTIONS, TANTALIZING ANSWERS

Have you ever asked yourself *why* you want to weigh a given amount? Are you merely fearful of "being fat"—that is, obese? By definition, **obesity** is an excess of fat in adipose tissues, most often caused by imbalances between caloric intake and energy output. To be sure, one standard of what constitutes obesity is simply cultural, and it varies from one culture to the next. Think of the female student who despaired over her plumpness until she took part in a graduate studies program in Africa. There, great numbers of males considered her to be one of the most desirable females on the planet. However, a lot depends on how healthy you want to be and how long you want to live. Thinner people generally live longer.

In the United States, obesity is now the second leading cause of death; 300,000 or so people die each year from preventable, weight-related conditions. Each year, the weight-related cases of type 2 diabetes, heart disease, hypertension, breast cancer, colon cancer, gout, gallstones, and osteoarthritis add up to a 100-billion-dollar drain on the economy. The future doesn't look any rosier: children today are 42 percent fatter than they were in 1980.

IN PURSUIT OF THE "IDEAL" WEIGHT Figure 42.16 shows charts for estimating the "ideal" weight for adults. Another indicator of obesity-related health risk is the *body mass index* (BMI), as determined by this formula:

$$\text{BMI} = \frac{\text{weight (pounds)} \times 700}{\text{height (inches)}^2}$$

Shape Up America, a national initiative for promoting health, will quickly do the calculation for you. Their web site is http://www.shapeup.org. If your BMI value is 27 or higher, the health risk increases dramatically. Other factors that influence the risk are smoking habits, family history of heart disorders, intake of sex hormones after menopause, and fat distribution. Fat stored above the belt, as with "beer bellies," definitely is not good.

Dieting alone cannot lower a BMI value. Eat less, and the body slows its metabolic rate to conserve energy. So how do you keep functioning normally over the long term while maintaining acceptable weight? *You must balance caloric intake with energy output.* For most of us, this means eating controlled portions of low-calorie, nutritious foods *and* exercising regularly.

To figure out how many kilocalories you should take in daily to maintain a desired weight, multiply that weight (in pounds) by 10 if you are not active physically, by 15 if moderately active, and by 20 if highly active. From the value you get this way, subtract the following amount:

Age:	*Subtract:*
25–34	0
35–44	100
45–54	200
55–64	300
Over 65	400

For instance, if you wish to weigh 120 pounds and are very active, 120 × 20 = 2,400 kilocalories. If you are thirty-five years old, then (2,400 − 100) or 2,300 kilocalories. Such calculations give a rough estimate of caloric intake. Other factors, including height, must also be taken into consideration. An active person 5 feet, 2 inches tall does not require as much energy as an active person who weighs the same but is 6 feet tall.

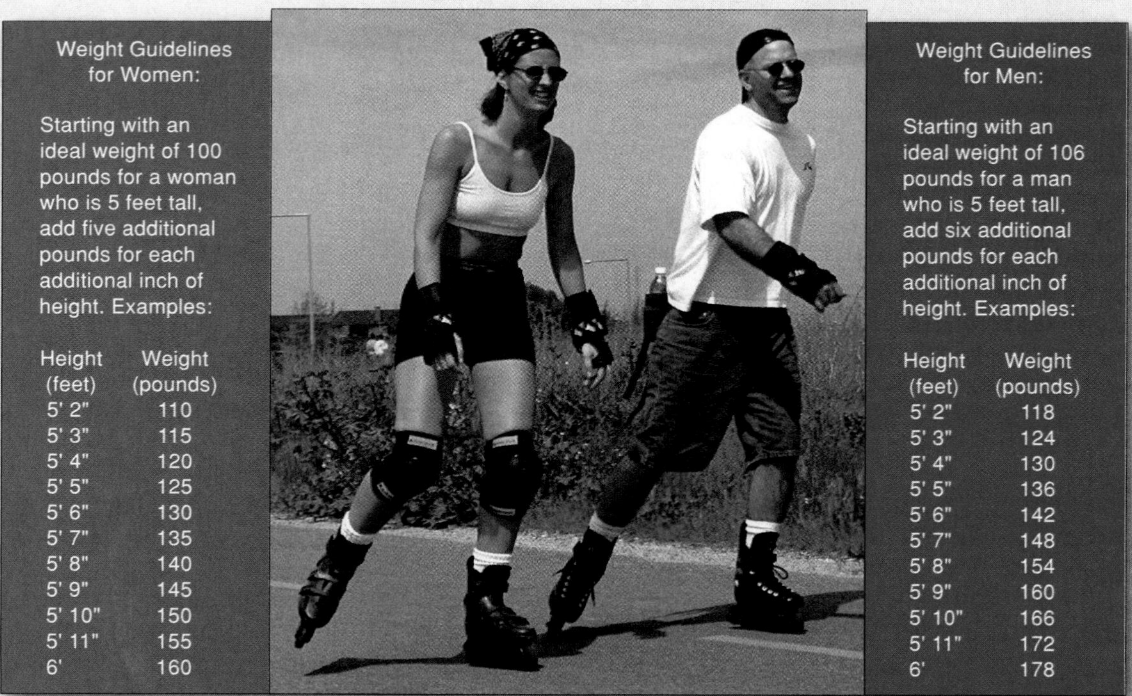

Weight Guidelines for Women:
Starting with an ideal weight of 100 pounds for a woman who is 5 feet tall, add five additional pounds for each additional inch of height. Examples:

Height (feet)	Weight (pounds)
5' 2"	110
5' 3"	115
5' 4"	120
5' 5"	125
5' 6"	130
5' 7"	135
5' 8"	140
5' 9"	145
5' 10"	150
5' 11"	155
6'	160

Weight Guidelines for Men:
Starting with an ideal weight of 106 pounds for a man who is 5 feet tall, add six additional pounds for each additional inch of height. Examples:

Height (feet)	Weight (pounds)
5' 2"	118
5' 3"	124
5' 4"	130
5' 5"	136
5' 6"	142
5' 7"	148
5' 8"	154
5' 9"	160
5' 10"	166
5' 11"	172
6'	178

Figure 42.16 How to estimate the "ideal" weight for an adult. The values shown are consistent with a long-term Harvard study into the link between excessive weight and increased risk of cardiovascular disorders. Depending on certain factors (such as having a small, medium, or large skeletal frame), the "ideal" might vary by plus or minus 10 percent.

Further reading: Student Guide to InfoTrac on web site

MY GENES MADE ME DO IT Some people have far more trouble keeping off excess weight than others do. On average, one of every three Americans is like this, and genes have a lot to do with it. Researchers suspected as much for a long time. Studies of identical twins provided clues. (*Identical twins* are born with identical genes.) As an outcome of family problems, certain twins who were studied had been separated at birth and had been raised apart, in different households. Yet, at adulthood, body weights of those separated twins were similar! People, it seemed, were born with a "set point" for body fat. Could it be that whatever set point an individual inherits is the one he or she will be stuck with for life?

Many experiments yielded evidence that confirmed this suspicion. The first experiments started in 1950 with a seriously obese mouse (Figure 42.17). In 1995, molecular geneticists identified one of its genes that influences the set point for body fat. They named it the ***ob* gene** (guess why). Which clues led to its discovery? The nucleotide sequence in the gene region of obese mice differs from that in normal mice. The gene is active in adipose tissue, nowhere else. Its biochemical message translates into a type of protein released from cells and picked up by the bloodstream. The gene, apparently, specifies a hormone.

Jeffrey Friedman discovered the hormone in 1994 and named it **leptin**. A nearly identical hormone also was isolated in humans. Leptin is only one of a number of factors that influence the brain's commands to suppress or whip up appetite, but it's a crucial one. By assessing the blood concentrations of incoming hormonal signals, an appetite control center in the hypothalamus "decides" whether the body has taken in enough food to provide enough fat for the day. If so, hypothalamic commands go out to increase metabolic rates—and to stop eating.

If the *ob* gene is mutated, then its expression alone may be enough to disrupt the control center so a mouse's appetite skyrockets and metabolic furnaces burn less. By extension, the same disruptions might occur in humans with a similar mutation. During one experiment that supports the hypothesis, researchers injected obese mice with leptin. The mice quickly shed their excess weight.

Will genetic researchers go on to develop therapies for obesity in humans? Maybe. As you saw in Chapter 37, however, hormonal control in humans is tricky business. It now appears that, besides its role in controlling body weight, leptin also acts as a selective inhibitor of bone formation, as mediated by the central nervous system. Thus the danger is that anti-obesity therapies will bring on the osteoporosis.

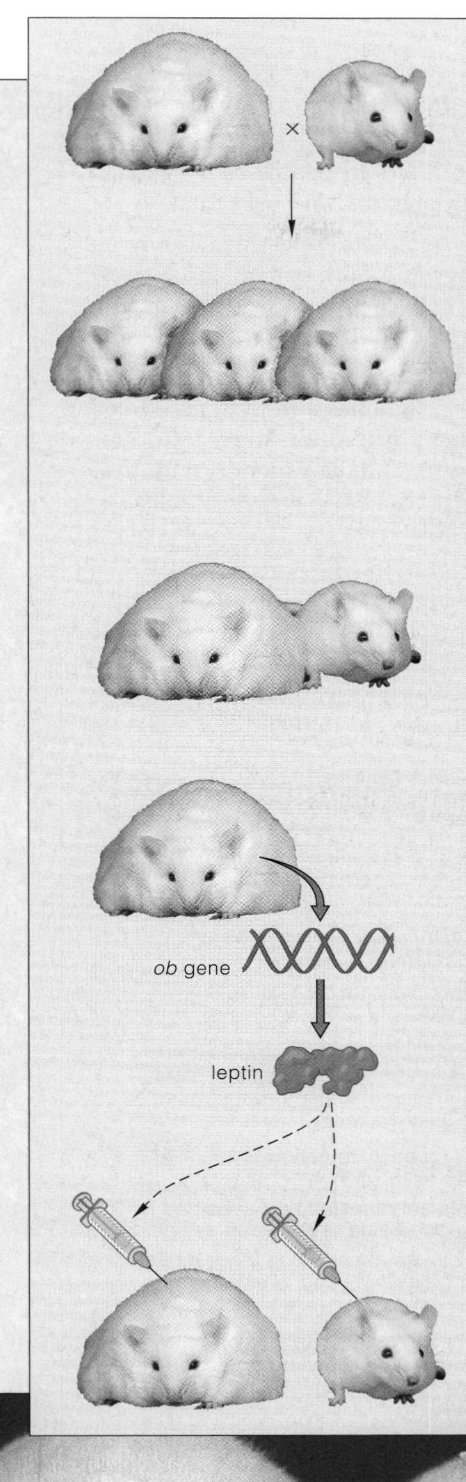

×
ob gene
leptin

a *1950*. Researchers at the Jackson Laboratories in Maine notice that one of their laboratory mice is extremely obese, with an uncontrollable appetite. Through crossbreeding of this apparent mutant individual with a normal mouse, they produce a strain of obese mice.

b *Late 1960s*. Douglas Coleman of the Jackson Laboratories surgically joins the bloodstreams of an obese mouse and a normal one. The obese mouse now loses weight. Coleman hypothesizes that a factor circulating in blood may be influencing its appetite, but he is not able to isolate it.

c *1994*. Late in the year, Jeffrey Friedman of Rockefeller University discovers a mutated form of what is now called the *ob* gene in obese mice. Through DNA cloning and gene sequencing, he defines the protein that the mutated gene encodes. The protein, now called leptin, is a hormone that influences the brain's commands to suppress appetite and increase metabolic rates.

d *1995*. Three different research teams develop and use genetically engineered bacteria to produce leptin, which, when injected in obese and normal mice, triggers significant weight loss, apparently without harmful side effects.

Figure 42.17 Chronology of research developments that revealed the identity of a key factor in the genetic basis of body weight.

42.11 SUMMARY

1. For animals, *nutrition* refers to processes by which the body takes in, digests, absorbs, and uses food.

2. Most animals have a complete digestive system, a tube that has two openings (a mouth and an anus), and regional specializations between them. Mucus-coated epithelium lines and protects all exposed surfaces of the tube and facilitates diffusion across the tube wall.

3. The human digestive system has a mouth, pharynx, esophagus, stomach, small intestine, large intestine (colon), rectum, and anus. Salivary glands and the liver, gallbladder, and pancreas have accessory roles in the system's functions (Table 42.4).

4. These activities proceed in the digestive system:

a. Mechanical processing and motility: Movements that break up, mix, and propel food material.

b. Secretion: Release of digestive enzymes and other substances from the pancreas and liver, as well as from glandular epithelium, into the gut lumen.

Table 42.4 Summary of the Digestive System

MOUTH (oral cavity)	Start of digestive system, where food is chewed and moistened; polysaccharide digestion begins here
PHARYNX	Entrance to tubular part of digestive and respiratory systems
ESOPHAGUS	Muscular tube, moistened by saliva, that moves food from pharynx to stomach
STOMACH	Sac where food mixes with gastric fluid and protein digestion begins; stretches to store food taken in faster than can be processed; gastric fluid destroys many microbes
SMALL INTESTINE	The first part (duodenum) receives secretions from the liver, gallbladder, and pancreas Most nutrients are digested and absorbed in the second part (jejunum) Some nutrients are absorbed in the last part (ileum), which delivers unabsorbed material to the colon
COLON (large intestine)	Concentrates and stores undigested matter (by absorbing mineral ions and water)
RECTUM	Distension triggers expulsion of feces
ANUS	Terminal opening of digestive system
Accessory Organs:	
SALIVARY GLANDS	Glands (three main pairs, many minor ones) that secrete saliva, a fluid with polysaccharide-digesting enzymes, buffers, and mucus (which moistens and lubricates ingested food)
PANCREAS	Secretes enzymes that digest all major food molecules; secretes buffers against HCl from the stomach
LIVER	Secretes bile (used in fat emulsification); roles in carbohydrate, fat, and protein metabolism
GALLBLADDER	Stores and concentrates bile from the liver

c. Digestion: Breakdown of food into particles, then into nutrient molecules small enough to be absorbed.

d. Absorption: The diffusion or transport of digested organic compounds, fluid, and ions from the gut lumen into the internal environment.

e. Elimination: The expulsion of undigested as well as unabsorbed residues at the end of the system.

5. Starch digestion starts in the mouth, and protein digestion starts in the stomach. Digestion is completed and most nutrients are absorbed in the small intestine. The pancreas secretes the main digestive enzymes. Bile from the liver assists in fat digestion.

6. In absorption, cells of the intestinal lining actively transport glucose and most amino acids out of the gut lumen. Fatty acids and monoglycerides diffuse across the lipid bilayer of these cells, then are recombined in the cytoplasm as triglycerides. The cells release these triglycerides, by exocytosis, into interstitial fluid.

7. The nervous and endocrine systems, and meshes of neurons (plexuses) in the gut wall, interact to govern the digestive system. Many controls operate in response to the volume and composition of food in the stomach and intestines. They cause changes in muscle activity and in the rate at which hormones and enzymes are secreted.

8. Nutritionists advise a daily food intake in certain proportions. For example, for an adult male of average body weight: 55–60 percent complex carbohydrates, 15–20 percent proteins (less for females), 20–25 percent fats and other lipids. A well-balanced diet of whole foods normally provides all required vitamins and minerals.

9. To maintain a particular body weight and overall health, caloric intake must balance energy output.

Review Questions

1. Define the five key tasks carried out by a complete digestive system. Then correlate some organs of such a system with the feeding behavior of a particular kind of animal. *42.1*

2. Using Figure 42.18, list the organs and accessory organs shown and the main functions of each. *Table 42.4*

3. Using the *black* lines shown in Figure 42.19 as a guide, name the breakdown products small enough to be absorbed across the intestinal lining, into the internal environment. *42.5*

4. Define segmentation. Does it proceed in the stomach? Does it proceed in the small intestine, colon, or both? *42.5, 42.7*

Self-Quiz (*Answers in Appendix III*)

1. The ______ maintains the internal environment, supplies cells with raw materials, and disposes of metabolic wastes.
   a. digestive system
   b. circulatory system
   c. respiratory system
   d. urinary system
   e. interaction of all of the systems listed

2. Most digestive systems have regions for ______ food.
   a. transporting
   b. processing
   c. storing
   d. all of the above

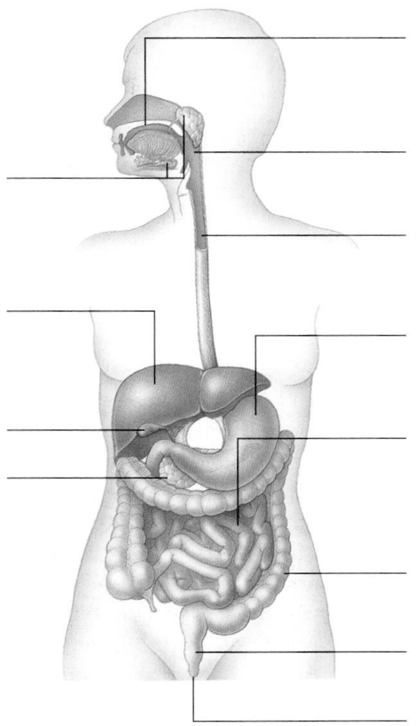

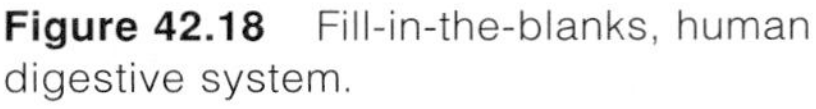

Figure 42.18 Fill-in-the-blanks, human digestive system.

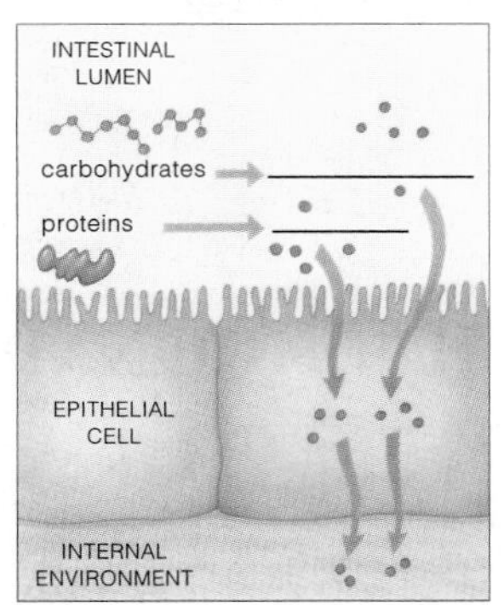
INTESTINAL LUMEN
carbohydrates
proteins
EPITHELIAL CELL
INTERNAL ENVIRONMENT

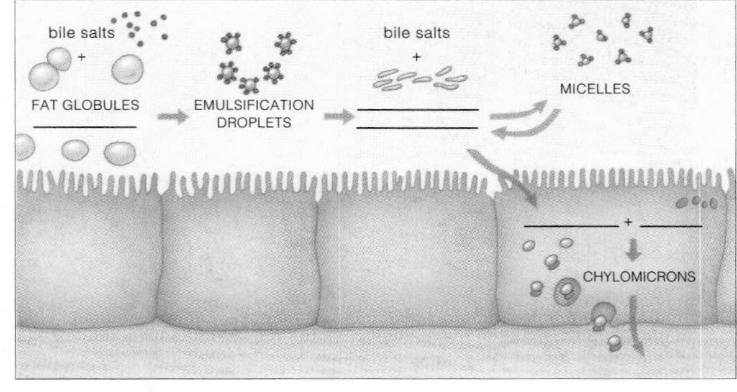
bile salts
FAT GLOBULES
EMULSIFICATION DROPLETS
bile salts
MICELLES
CHYLOMICRONS

Figure 42.19 Fill in the blanks for substances that cross the lining of the small intestine.

3. Maintaining good health and normal body weight requires that ______ intake be balanced by ______ output.

4. Most of our caloric intake should come from ______.
   a. complex carbohydrates   c. proteins
   b. simple carbohydrates   d. lipids

5. On its own, the human body cannot produce all of the ______ it requires.
   a. vitamins and minerals   d. a through c
   b. fatty acids   e. a and c
   c. amino acids

6. ______ secretions do not assist in digestion and absorption.
   a. Salivary gland   c. Liver
   b. Thymus gland   d. Pancreas

7. Digestion is completed and most nutrients are absorbed in the ______.
   a. mouth   c. small intestine
   b. stomach   d. colon

8. Glucose and most amino acids are absorbed across the gut lining ______.
   a. by active transport   c. at lymph vessels
   b. by diffusion   d. as fat droplets

9. Bile has roles in ______ digestion and absorption.
   a. carbohydrate   c. protein
   b. fat   d. amino acid

10. Match the organ with its key digestive function(s).

______ gallbladder	a. secrete bile; many metabolic roles
______ stomach	b. digest, absorb most nutrients
______ colon	c. store, mix, dissolve food; start protein breakdown
______ pancreas	d. store, concentrate bile
______ salivary gland	e. concentrate undigested matter
______ small intestine	f. secrete substances that moisten food, start polysaccharide breakdown
______ liver	g. secrete digestive enzymes and bicarbonate

Critical Thinking

1. A glassful of whole milk contains lactose, proteins, butterfat (mostly triglycerides), vitamins, and minerals. Explain what will happen to each component in your digestive tract.

2. When people age, the number of their body cells steadily decreases and energy needs decline. If you were planning an older person's diet, which foods would you emphasize, and why? Which ones would you deemphasize?

3. Using Section 42.10 as a reference, determine your ideal weight and design a well-balanced program of diet and exercise that will help you achieve or maintain that weight.

4. Holiday meals often are larger than everyday ones and have a high fat content. After stuffing themselves at an early dinner on Thanksgiving Day, Richard and other members of his family feel uncomfortably full for the rest of the afternoon. Based on what you have learned about controls over digestion, propose a biochemical explanation for their discomfort.

5. Your digestive system helps protect against many pathogenic bacteria that may contaminate the kinds of food you eat. Explain some ways in which it destroys or washes away pathogens. (You may wish to refer to Section 40.1, also.)

Selected Key Terms

appendix *42.7*
bile *42.4*
bulk *42.7*
chyme *42.4*
colon *42.7*
complete digestive system *42.1*
digestive system *CI*
emulsification *42.4*
esophagus *42.3*
essential amino acid *42.8*
essential fatty acid *42.8*
food pyramid *42.8*
gallbladder *42.4*
gastric fluid *42.4*
gut *42.2*
incomplete digestive system *42.1*
kilocalorie *42.8*
leptin *42.10*
liver *42.4*
micelle formation *42.5*
microvillus (microvilli) *42.5*
mineral *42.9*
nutrition *CI*
ob gene *42.10*
obesity *42.10*
pancreas *42.4*
pharynx *42.3*
ruminant *42.1*
saliva *42.3*
segmentation *42.5*
sphincter *42.3*
stomach *42.4*
tongue *42.3*
tooth *42.3*
villus (villi) *42.5*
vitamin *42.9*

Readings *See also www.infotrac-college.com*

Blaser, M. J. February 1996. "The Bacteria Behind Ulcers." *Scientific American*, 274.

Kassirer, J. 17 September 1998. *New England Journal of Medicine.* Editorial introduces reports on unproven and harmful uses of dietary supplements, some of which were contaminated.

Sherwood, L. 1997. *Human Physiology.* Third edition. Monterey, California: Brooks-Cole.

43 THE INTERNAL ENVIRONMENT

Tale of the Desert Rat

Look closely at a fish or some other marine animal, and you will find that the cells of its body are exquisitely adapted to life in a salty fluid. And yet, some lineages of animals that evolved in the seas moved onto dry land about 375 million years ago. They were able to do so partly because they brought some salty fluid along with them, as an *internal* environment for their cells. Even so, it was not a simple transition. Those pioneers on land and their descendants encountered intense sunlight, dry winds, more pronounced swings in temperature, water of dubious salt content, and sometimes no water at all.

How did the pioneers conserve or replace the water and the salts that they lost as a result of their everyday activities? How did they manage to stay comfortably warm when their surroundings became too cold or too hot? They must have done these things. Otherwise the volume, composition, and temperature of their internal environment would have spun out of control. In short, the question is this: *How did land-dwelling descendants of marine animals maintain internal operating conditions and thereby prevent cellular anarchy?*

Observe any of their existing descendants and you get a sense of what some answers might be. Suppose you focus on a tiny mammal, a kangaroo rat living in an isolated desert of New Mexico (Figure 43.1). After a brief rainy season, the sun bakes the desert sand for months. The only obvious water is imported, sloshing about in the canteens of the occasional researcher and tourist. Yet with nary a sip of free water, a kangaroo rat routinely counters threats to its internal environment.

It waits out the daytime heat inside a burrow, then forages in the cool of night for dry seeds and maybe a succulent. It is not sluggish about this. It hops rapidly and far, searching for seeds and fleeing from coyotes and snakes. All that hopping requires ATP energy and water. Seeds, chockful of energy-rich carbohydrates, provide both. Metabolic reactions that release energy from carbohydrates and other organic compounds also yield water. Every day, such "metabolic water" represents a whopping 90 percent of a kangaroo rat's total water intake. By comparison, it is only 12 percent or so of the total water intake for your body.

When resting inside the cool burrow, a kangaroo rat conserves and recycles water. As it inhales the cool air into its warm lungs, water vapor condenses onto the

	KANGAROO RAT	HUMAN
WATER GAIN (milliliters)		
by ingesting solids	6.0	850
by ingesting liquids	0.0	1,400
by metabolism	54.0	350
	60.0	2,600
WATER LOSS (milliliters)		
in urine	13.5	1,500
in feces	2.6	200
by evaporation	43.9	900
	60.0	2,600

Figure 43.1 Kangaroo rat, master of water conservation in a New Mexico desert. The chart will give you an idea of how kangaroo rats and humans gain and lose water. They do so in different ways, Even so, in both cases, *water losses balance out the gains* —just as they do in all animals. How this happens, and why it absolutely must happen, is the initial focus of this chapter.

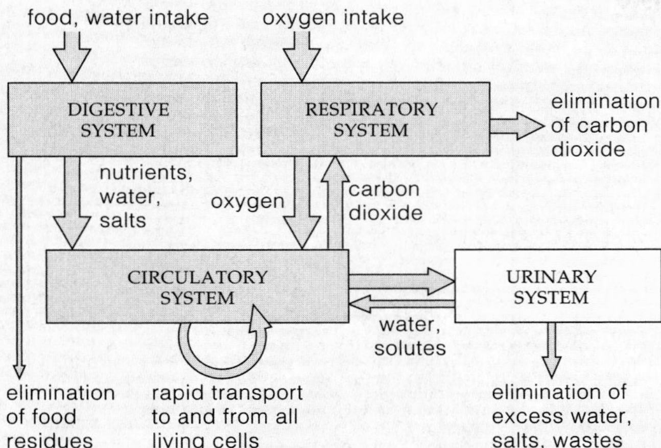
food, water intake
oxygen intake
DIGESTIVE SYSTEM
RESPIRATORY SYSTEM
elimination of carbon dioxide
nutrients, water, salts
oxygen
carbon dioxide
CIRCULATORY SYSTEM
URINARY SYSTEM
water, solutes
elimination of food residues
rapid transport to and from all living cells
elimination of excess water, salts, wastes

Figure 43.2 Links between the urinary system and other organ systems that contribute to homeostasis, or stability in favorable operating conditions in the animal body.

epithelial lining inside its nose—and some of that water diffuses back into the body. Also, after a busy night of foraging, it empties its cheek pouches of seeds. As the seeds leave its mouth, they soak up the small amount of water dripping from its nose. When the kangaroo rat eats the dripped-upon seeds, it reclaims the water.

The kangaroo rat cannot lose water by perspiring, because it has no sweat glands. It loses water when it urinates, but its two specialized kidneys do not let it piddle away much. The kidneys filter the blood's water and solutes, including dissolved salts. They continually adjust *how much water* and *which solutes* will return to the blood or leave the body as urine.

Overall, kangaroo rats and all other animals take in enough water and solutes to replace their daily losses (Figure 43.1). How they accomplish their balancing acts will be our initial focus in the chapter. Later, we will consider some of the means by which mammals withstand hot, cold, and sometimes unpredictable changes in environmental temperatures on land.

As a starting point, remind yourself of the kinds of fluids inside most animals. **Interstitial fluid** fills the spaces between living cells and other components of tissues. Another fluid, **blood**, transports substances to and from all tissue regions by way of a circulatory system. Taken together, the interstitial fluid and blood are the **extracellular fluid**. In many animals, a well-developed urinary system helps keep the volume and composition of extracellular fluid within tolerable ranges. As you will see, other organ systems, especially those indicated in Figure 43.2, interact with the urinary system in the performance of this homeostatic task.

KEY CONCEPTS

1. Animals are continually gaining and losing water and dissolved substances (solutes). They continually produce metabolic wastes. Even with all the inputs and outputs, the overall volume and composition of the extracellular fluid in the body remain relatively constant.

2. In humans as in other vertebrates, a urinary system is crucial to balancing the intake and output of water and solutes. This system continually filters water and solutes from the blood. It reclaims some amount of both and eliminates the rest. Different amounts are reclaimed at different times, depending on what is necessary to maintain extracellular fluid.

3. Kidneys are blood-filtering organs, and the urinary system of vertebrates has a pair of them. Packed inside each kidney are a great number of nephrons.

4. At its beginning, each nephron cups around a set of blood capillaries and receives water and solutes from them. The nephron returns most of the filtrate to the blood by giving it up to a second set of capillaries, which intertwine around the nephron's tubular parts.

5. Water and solutes not returned to the blood leave the body as a fluid called urine. During any interval, control mechanisms influence whether the urine is concentrated or dilute. Two hormones, ADH and aldosterone, have key roles in the adjustments.

6. The internal body temperature of animals depends on the balance between heat produced through metabolism, heat absorbed from the environment, and heat lost to the environment.

7. The internal body temperature is maintained within a favorable range through controls over metabolic activity and adaptations in body form and behavior.

43.1 URINARY SYSTEM OF MAMMALS

The Challenge—Shifts in Extracellular Fluid

Different solid foods and fluids intermittently enter a mammal's gut. Afterward, variable amounts of absorbed water, nutrients, and other substances move into the blood, then into interstitial fluid and on into cells. Such events could easily shift the volume and composition of extracellular fluid beyond tolerable limits. But the body makes compensatory adjustments that balance out the gains and losses. Within a given time frame, it takes in as much water and solutes as it gives up.

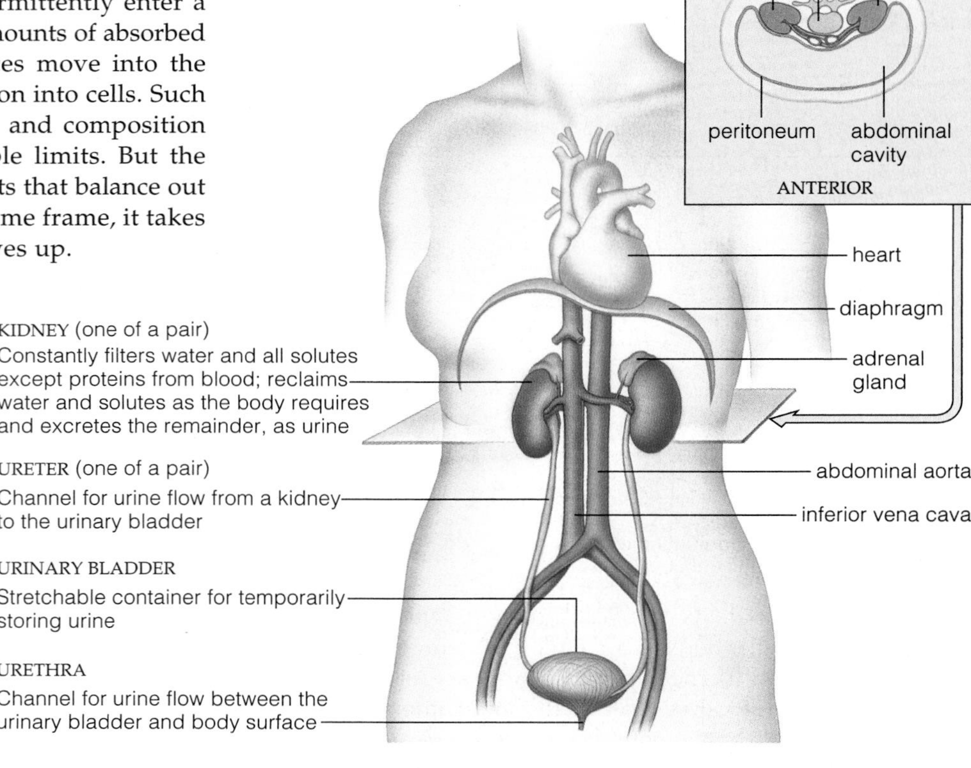

POSTERIOR
right kidney
vertebral column
left kidney
peritoneum
abdominal cavity
ANTERIOR
heart
diaphragm
adrenal gland
abdominal aorta
inferior vena cava
KIDNEY (one of a pair)
Constantly filters water and all solutes except proteins from blood; reclaims water and solutes as the body requires and excretes the remainder, as urine
URETER (one of a pair)
Channel for urine flow from a kidney to the urinary bladder
URINARY BLADDER
Stretchable container for temporarily storing urine
URETHRA
Channel for urine flow between the urinary bladder and body surface

Figure 43.3 Organs of the human urinary system and their functions. The two kidneys, two ureters, and urinary bladder are located *outside* the peritoneum, the membranous lining of the abdominal cavity. Compare Section 26.1.

WATER GAINS AND LOSSES To start, think about how humans and other mammals *gain* water, mainly by two processes:

Absorption from the gut
Metabolism

Considerable water is absorbed from solids and liquids inside the gut. As you know, water also forms as a normal by-product of many metabolic reactions. In land mammals, how much water enters the gut in the first place depends on a thirst mechanism. When the body loses too much water, land mammals seek out streams, waterholes, and so on. We will be looking at the thirst mechanism later in the chapter.

Normally, the mammalian body *loses* water mainly by way of four physiological processes, as listed here:

Urinary excretion
Evaporation from lungs and the skin
Sweating, by mammals that sweat
Elimination, in feces

Urinary excretion affords the most control over water loss. The process eliminates excess water and solutes as urine, the fluid that forms in a urinary system such as that shown in Figure 43.3. Some water also evaporates from respiratory surfaces. Some animal species lose it in sweat. A mammal in good health loses very little water from the gut (most is absorbed, not eliminated in feces).

SOLUTE GAINS AND LOSSES There are several ways in which mammals *gain* solutes. But they do so mainly by four processes:

Absorption from the gut
Secretion from cells
Respiration
Metabolism

For example, nutrients and mineral ions are absorbed from the gut. So are drugs and food additives. Cellular secretions and wastes, including carbon dioxide, enter interstitial fluid, then the blood. The respiratory system puts oxygen into blood, and aerobically respiring cells put carbon dioxide into it.

Mammals typically *lose* solutes by these processes:

Urinary excretion
Respiration
Sweating, by some species

The urine of mammals includes various wastes formed by the breakdown of organic compounds. For example, ammonia forms as amino groups are split from amino acids. Then **urea**, a major waste, forms in the liver when two ammonia molecules join with carbon dioxide. Also in urine are uric acid (from nucleic acid breakdown), hemoglobin breakdown products (which give the urine much of its color), drugs, and food additives. Besides these solute losses, some mammals also lose mineral ions in sweat. And all mammals lose carbon dioxide, the most abundant waste, by way of respiration.

Further reading: Student Guide to InfoTrac on web site –

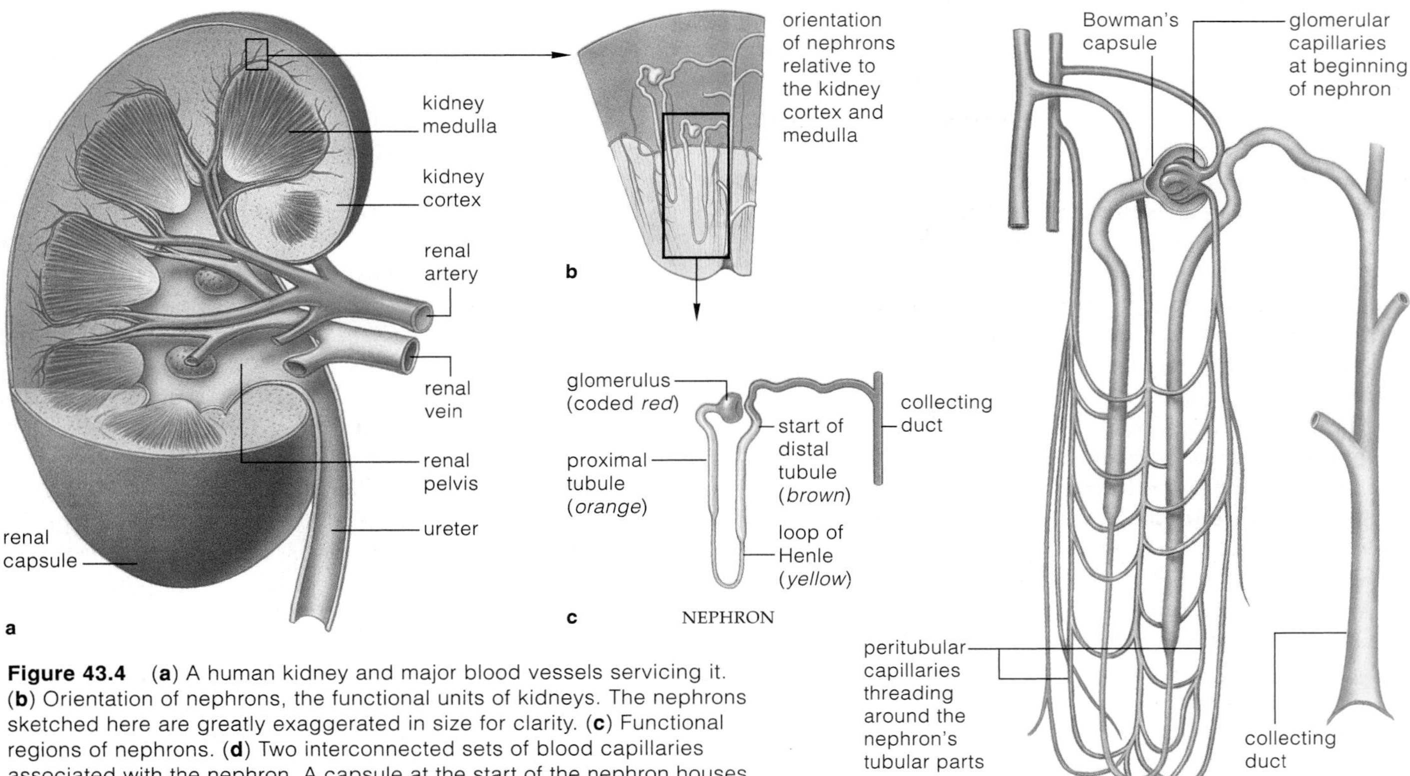
kidney medulla
kidney cortex
renal artery
renal vein
renal pelvis
ureter
renal capsule
a
orientation of nephrons relative to the kidney cortex and medulla
b
glomerulus (coded red)
proximal tubule (orange)
start of distal tubule (brown)
loop of Henle (yellow)
collecting duct
c
NEPHRON
Bowman's capsule
glomerular capillaries at beginning of nephron
peritubular capillaries threading around the nephron's tubular parts
collecting duct
d

Figure 43.4 (**a**) A human kidney and major blood vessels servicing it. (**b**) Orientation of nephrons, the functional units of kidneys. The nephrons sketched here are greatly exaggerated in size for clarity. (**c**) Functional regions of nephrons. (**d**) Two interconnected sets of blood capillaries associated with the nephron. A capsule at the start of the nephron houses the first set, the glomerular capillaries. The second set, the peritubular capillaries, threads around all tubular parts of the nephron.

Components of the Urinary System

Mammals routinely counter shifts in the composition and volume of extracellular fluid, mainly with a **urinary system**. This consists of two kidneys, two ureters, one urinary bladder, and one urethra. **Kidneys** are a pair of bean-shaped organs, about as large as a fist in an adult human (Figures 43.3 and 43.4). Each has an outer capsule of connective tissue. Blood capillaries thread through its two inner regions, the cortex and medulla.

Kidneys filter water, mineral ions, organic wastes, and other substances from blood. Then they adjust the filtrate's composition and return all but about 1 percent to the blood. The small portion of unreclaimed water and solutes is urine. By definition, **urine** is a fluid that rids the body of water and solutes that are in excess of the amounts required to maintain extracellular fluid.

Urine flows from each kidney into a **ureter**, a tubular channel between it and the **urinary bladder**, a muscular sac. Urine is briefly stored here before flowing into the **urethra**, a muscular tube leading out to body surface.

Flow from the bladder (urination) is a reflex action. When the bladder is full, the sphincter around its neck opens, smooth muscle in its balloonlike wall contracts, and so urine is forced out through the urethra. Skeletal muscle surrounds the urethra. Its contraction, which is under voluntary control, prevents urination.

Nephrons—Functional Units of Kidneys

Each human kidney has more than a million **nephrons**, slender tubules packed inside lobes that extend from the kidney cortex down through the medulla. *It is at the nephrons that water and solutes are filtered from blood and adustments are made in the amounts to be reclaimed.*

Each nephron starts as a **Bowman's capsule**. Here its wall cups around *glomerular* capillaries. The cupped wall region and blood vessels are a blood-filtering unit called a **glomerulus** (Figure 43.4*c*). Next, the nephron has a **proximal tubule** (closest to the capsule), then a hairpin-shaped **loop of Henle** and **distal tubule** (most distant from the capsule). It ends as a **collecting duct**, which is part of a duct system leading into the kidney's central cavity (renal pelvis) and then into a ureter.

Blood does not give up all of its water and solutes. The unfiltered part flows into a second set of capillaries around the nephron's tubular parts. In these *peritubular* capillaries, the blood reclaims water and solutes. Then it flows into veins and back to the general circulation.

A urinary system counters unwanted shifts in the volume and composition of extracellular fluid. In its paired kidneys, water and solutes are filtered from blood. The body reclaims most of this, but the excess leaves the kidneys as urine.

43.2 URINE FORMATION

Urine forms in nephrons by three processes: filtration, tubular reabsorption, and tubular secretion. All three depend on properties of cells of the nephron wall. The cells differ in membrane transport mechanisms and in permeability from one part of the nephron to the next.

Blood pressure generated by the heart's contractions drives **filtration**, which takes place at the glomerulus (Figure 43.5). Pressure "filters" blood by forcing water and all solutes except proteins out from the glomerular capillaries. The protein-free filtrate moves out from the cupped part of the nephron, into the proximal tubule.

Tubular reabsorption proceeds along the nephron's tubular regions. Most of the filtrate's water and solutes move out from the space enclosed by the nephron wall (the lumen) and into peritubular capillaries (Figure 43.6).

Tubular secretion proceeds at the tubule wall but in the opposite direction of reabsorption. Cells making up the wall accept solutes from the peritubular capillaries, then secrete them into the nephron's lumen. The main solutes are hydrogen and potassium ions (H^+ and K^+). The process also prevents certain metabolites (such as uric acid) and foreign substances (such as drugs) from accumulating in blood.

Factors Influencing Filtration

Each minute, about 1.5 liters (1–1/2 quarts) of blood flow through an adult's kidneys, and 120 milliliters of water and small solutes are filtered into the nephrons. That's 180 liters of filtrate per day! That high filtration rate is possible mainly because glomerular capillaries are 10 to 100 times more permeable to water and small solutes than other capillaries are. Also, blood pressure is high in glomerular capillaries. Why? Arterioles delivering blood to the glomerulus have a wider diameter and less resistance to flow than those carrying blood away. As a result, blood dams up inside the glomerulus, which increases the glomerular capillary blood pressure.

Filtration rates also depend on flow volume to the kidneys. Neural, endocrine, and local controls maintain the flow even when blood pressure changes. When you run in a race or dance until dawn, the nervous system diverts an above-normal volume of blood away from kidneys, toward the heart and skeletal muscles. It also directs coordinated vasoconstriction and vasodilation in different regions of the body, so less of the blood flow reaches the kidneys.

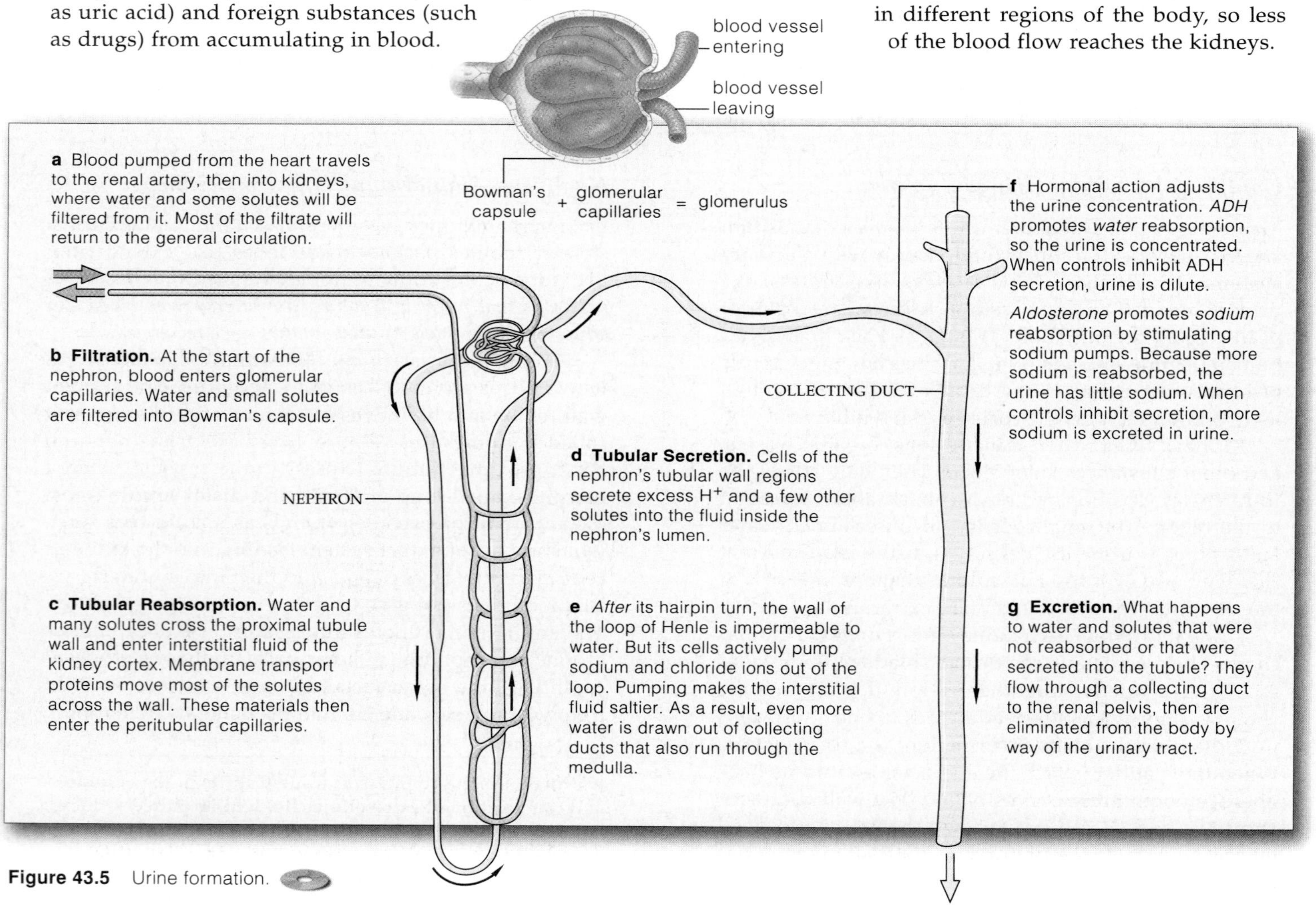

blood vessel entering
blood vessel leaving
a Blood pumped from the heart travels to the renal artery, then into kidneys, where water and some solutes will be filtered from it. Most of the filtrate will return to the general circulation.
Bowman's capsule + glomerular capillaries = glomerulus
f Hormonal action adjusts the urine concentration. *ADH* promotes *water* reabsorption, so the urine is concentrated. When controls inhibit ADH secretion, urine is dilute.
Aldosterone promotes *sodium* reabsorption by stimulating sodium pumps. Because more sodium is reabsorbed, the urine has little sodium. When controls inhibit secretion, more sodium is excreted in urine.
b Filtration. At the start of the nephron, blood enters glomerular capillaries. Water and small solutes are filtered into Bowman's capsule.
COLLECTING DUCT
d Tubular Secretion. Cells of the nephron's tubular wall regions secrete excess H^+ and a few other solutes into the fluid inside the nephron's lumen.
NEPHRON
c Tubular Reabsorption. Water and many solutes cross the proximal tubule wall and enter interstitial fluid of the kidney cortex. Membrane transport proteins move most of the solutes across the wall. These materials then enter the peritubular capillaries.
e *After* its hairpin turn, the wall of the loop of Henle is impermeable to water. But its cells actively pump sodium and chloride ions out of the loop. Pumping makes the interstitial fluid saltier. As a result, even more water is drawn out of collecting ducts that also run through the medulla.
g Excretion. What happens to water and solutes that were not reabsorbed or that were secreted into the tubule? They flow through a collecting duct to the renal pelvis, then are eliminated from the body by way of the urinary tract.

Figure 43.5 Urine formation.

Further reading: Student Guide to InfoTrac on web site

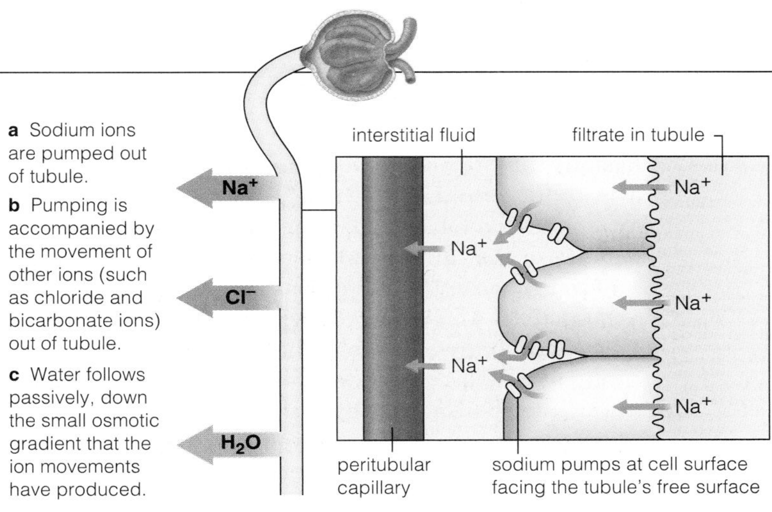
a Sodium ions are pumped out of tubule.
b Pumping is accompanied by the movement of other ions (such as chloride and bicarbonate ions) out of tubule.
c Water follows passively, down the small osmotic gradient that the ion movements have produced.
Na+
Cl−
H2O
interstitial fluid
filtrate in tubule
Na+
Na+
Na+
Na+
Na+
peritubular capillary
sodium pumps at cell surface facing the tubule's free surface

Figure 43.6 Membrane pumps in the nephron wall. Sodium and water are reabsorbed here in response to, say, gulping too much water or ingesting too much salt.

Also, cells in the walls of arterioles leading to the glomeruli respond to changes in pressure. When blood pressure decreases, arterioles vasoconstrict, so kidneys receive less of the total blood volume. But when blood pressure rises, they vasodilate, so more blood flows in.

Reabsorption of Water and Sodium

REABSORPTION MECHANISM Kidneys precisely adjust how much water and sodium ions the body excretes or conserves. Suppose you drink too much or too little water or wolf down salty potato chips or lose too much sodium in sweat. Responses start promptly as filtrate enters proximal tubules of nephrons. Cells in the tubule wall actively transport some sodium out of the filtrate. Other ions follow sodium into interstitial fluid. Water also leaves the filtrate, by osmosis. The nephron wall is highly permeable in this region, so about two-thirds of the filtrate's water is reabsorbed here (Figure 43.5*c*,*d*).

Interstitial fluid is saltiest around the hairpin turn of the loop of Henle. Water moves out of the filtrate, by osmosis, *before* the turn. The fluid left behind becomes saltier until it matches the interstitial fluid. The loop wall *after* the turn is impermeable to water. But sodium is pumped out by active transport mechanisms (Figure 43.5*e*). The interstitial fluid becomes saltier and attracts more water out of the filtrate just entering the loop.

Fluid arriving at the distal tubule is dilute. The stage is set for adjustments, which can lead to urine that is highly dilute, concentrated, or anywhere in between.

HORMONE-INDUCED ADJUSTMENTS The cells of distal tubules and collecting ducts have receptors for ADH and aldosterone. If the hypothalamus detects a drop in extracellular fluid volume, it calls for secretion of **ADH** (antidiuretic hormone). ADH makes the tubule walls more permeable to water, so more water is reabsorbed and urine gets more concentrated (Figure 43.5*f*). When the body holds too much water, the secretion of ADH decreases. The walls become less permeable, less water is reabsorbed, and urine remains dilute.

Aldosterone promotes reabsorption of sodium. The extracellular fluid volume falls when too much sodium is lost. Sensory receptors in the heart and blood vessels detect the decreases and signal gland cells in the wall of the arteriole at the glomerulus. The cells secrete renin. This enzyme splits a plasma protein, and a fragment is converted to **angiotensin II**. This hormone's targets are aldosterone-secreting cells of the adrenal cortex, part of a gland atop each kidney (Figure 43.3). In response, the cells step up secretion of aldosterone, which stimulates cells of distal tubules and collecting ducts to reabsorb sodium faster and excrete less sodium. Conversely, when there is too much sodium, aldosterone secretion slows. Less sodium is reabsorbed, so more is excreted.

THIRST BEHAVIOR A **thirst center** in the hypothalamus induces water-seeking behavior. Osmoreceptors located nearby warn of decreases in blood volume and rising blood solute levels. The signals activate the thirst center *and* neighboring ADH-secreting cells. Thus, while water intake is being encouraged, urinary output is reduced.

In addition to stimulating secretion of aldosterone, angiotensin II acts on the brain to promote thirst and ADH secretion. Thirst behavior also is initiated when free nerve endings detect dryness in the mouth, which is an early sign of dehydration.

Concentrated or dilute urine forms in kidneys by filtration, tubular reabsorption, and tubular secretion.

Filtration rates depend mainly on heart contractions, which generate high hydrostatic pressure at the glomerulus of each nephron. They also depend on neural, endocrine, and local controls over blood flow being directed to the kidneys.

Reabsorption, which can be adjusted by hormonal controls, helps maintain extracellular fluid. The adjustments rid the body of suitable amounts of water and solutes, in urine.

ADH promotes water conservation and concentrated urine. When ADH secretion is inhibited, urine is dilute.

Aldosterone promotes sodium conservation. When its secretion is inhibited, more sodium is excreted in urine.

Focus on Health

43.3 WHEN KIDNEYS BREAK DOWN

By this point in the chapter, you probably have sensed that the functioning of nephrons is central to good health. Whether by illness or accident, when the nephrons of both kidneys become damaged and no longer perform their regulatory and excretory functions, we call this **renal failure**. Chronic renal failure is irreversible.

Infectious agents that reach the kidneys by way of the bloodstream or through the urethra can cause renal failure. So can ingestion of lead, arsenic, pesticides, and other toxins. Continued high doses of aspirin and some other drugs can do the same thing. Abnormal retention of metabolic wastes, such as the by-products of protein breakdown, results in *uremic toxicity*. Atherosclerosis, heart failure, hemorrhage, and shock diminish blood flow and skew filtration pressure in the kidneys.

In a rare type of disease called *glomerulonephritis*, the kidneys become inflamed, but not as a result of being infected. Antibody–antigen complexes become trapped in glomeruli. Unless phagocytes remove them, they go on activating complement and other agents that bring about widespread inflammation and tissue damage.

Kidney stones form when uric acid, calcium salts, and other wastes settle out of urine and collect in the renal pelvis. These hard deposits are usually passed in urine but can become lodged in the ureter or urethra. If they disrupt urine flow, they must be medically or surgically removed to prevent renal failure.

About 13 million people in the United States alone suffer from renal failure. A *kidney dialysis machine* often can restore the proper solute balances. Like the kidney, it helps maintain extracellular fluid by selectively adjusting the solutes in blood. "Dialysis" refers to an exchange of substances across an artificial membrane interposed between two solutions that differ in composition.

In *hemodialysis*, the machine is connected to an artery or a vein. Then the patient's blood is pumped on through tubes made of a material that is similar to sausage casing or cellophane. The tubes are submerged in a warm saline bath. The mix of salts, glucose, and other substances of the bath sets up the correct concentration gradients with blood. The blood then returns to the body. For kidney dialysis to have optimum effect, it must be performed three times a week. Each time, the procedure takes about four hours, because the patient's blood must circulate repeatedly to improve the solute concentrations in her or his body. In *peritoneal dialysis*, fluid of an appropriate composition is introduced into the patient's abdominal cavity, left in place for a specific length of time, then drained out. In this case, the cavity's lining itself, the peritoneum, serves as the membrane for dialysis.

Bear in mind, kidney dialysis is used as a temporary measure in reversible kidney disorders. In chronic cases, the procedure must be used for the rest of the patient's life or until a transplant operation provides her or him with a functional kidney. With treatment and controlled diets, many patients can resume fairly normal activity.

43.4 THE ACID–BASE BALANCE

Besides maintaining the volume and composition of extracellular fluid, kidneys help keep it from becoming too acidic or too basic (alkaline). The overall **acid–base balance** of that fluid is an outcome of controls over its concentrations of H^+ and other dissolved ions. *Metabolic acidosis* hints at the importance of the balancing acts. This condition results when the kidneys cannot secrete enough H^+ to keep pace with all the H^+ that is forming during metabolism. It is life-threatening.

Buffer systems, respiration, and urinary excretion all work in concert to provide control over the acid–base balance. A buffer system, remember, consists of weak acids or bases that can reversibly latch onto and release ions, and so help minimize shifts in pH (Section 2.6).

Normally, the extracellular pH of the human body should be maintained between 7.37 and 7.43. As you know, acids lower the pH and bases raise it. A variety of acidic and basic substances enter blood after they are absorbed from the gut and as an outcome of normal metabolism. Cell activities typically produce an excess of acids. These dissociate into H^+ and other fragments, and pH decreases. The effect is minimized when excess hydrogen ions react with buffer molecules. An example is the *bicarbonate–carbon dioxide* buffer system:

$$H^+ + \underset{\text{BICARBONATE}}{HCO_3^-} \rightleftharpoons \underset{\text{CARBONIC ACID}}{H_2CO_3} \rightleftharpoons CO_2 + H_2O$$

In this case, the buffer system neutralizes excess H^+, and the carbon dioxide that forms during the reactions is exhaled from the lungs. Like other buffer systems in the body, however, this one has only temporary effects; it does not *eliminate* excess H^+. Only the urinary system can do so and thereby restore the buffers.

The same reactions proceed in reverse in cells of the nephron's tubular walls. HCO_3^- formed by the reverse reactions moves into interstitial fluid, then peritubular capillaries. Afterward, it enters the general circulation and buffers excess acid. The H^+ formed in the cells is secreted into the nephron and may join with HCO_3^-. The CO_2 that formed may be returned to blood, then exhaled. H^+ also may combine with phosphate ions or ammonia (NH_3), then leave the body in urine.

The kidneys work in concert with buffering systems, which neutralize acids, and with the respiratory system to help keep the extracellular fluid from becoming too acidic or too basic (alkaline).

A bicarbonate–carbon dioxide buffer system temporarily neutralizes excess hydrogen ions. The urinary system alone eliminates the excess ions and thus restores these buffers.

The bicarbonate–carbon dioxide buffer system is one of the key mechanisms that help maintain the acid–base balance.

Further reading: Student Guide to InfoTrac on web site →

ON FISH, FROGS, AND KANGAROO RATS

Now that you have a general sense of how your body maintains water and solute levels, consider what goes on in some other vertebrates, including that kangaroo rat hopping about at the start of the chapter.

Bony fishes and amphibians of freshwater habitats gain water and lose solutes (Figure 43.7*a*). Water moves into their internal environment by osmosis; they don't gain water by drinking it. The water diffuses across thin gill membranes in fishes and across the skin in adult amphibians. Excess water leaves as dilute urine, formed inside a pair of kidneys. In both groups of vertebrates, solute losses are balanced when more solutes come in with food and when gill or skin cells pump in sodium.

Body fluids of herring, snapper, and other marine fishes are about three times less salty than seawater. These fishes lose water by osmosis and replace it by drinking more. They excrete ingested solutes against concentration gradients (Figure 43.7*b*). Fish kidneys do not have loops of Henle, so urine cannot ever become saltier than body fluids. Cells in the fish gills actively pump out most of the excess solutes in blood.

Figure 43.7*c* describes the water–solute balancing act in a salmon. This type of fish spends part of its life cycle in fresh water and another part in seawater.

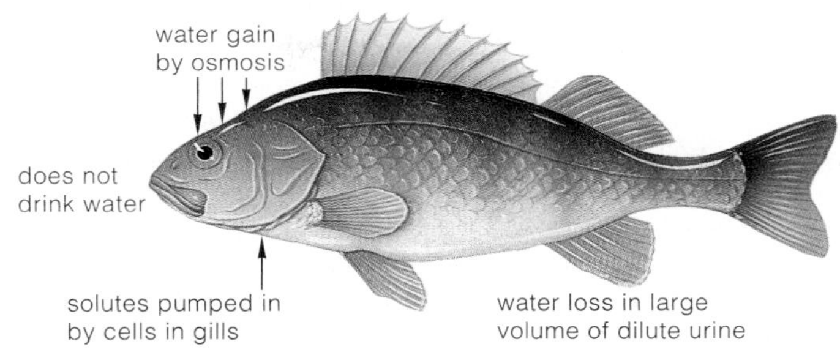
water gain
by osmosis
does not
drink water
solutes pumped in
by cells in gills
water loss in large
volume of dilute urine

a Freshwater bony fish (body fluids far saltier than surroundings)

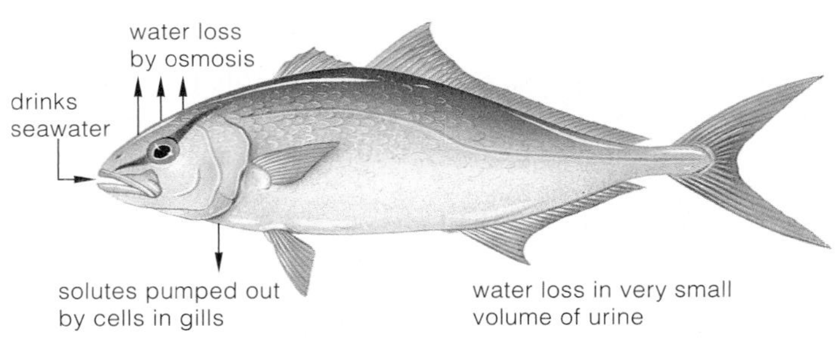
water loss
by osmosis
drinks
seawater
solutes pumped out
by cells in gills
water loss in very small
volume of urine

b Marine bony fish (body fluids less salty than surroundings)

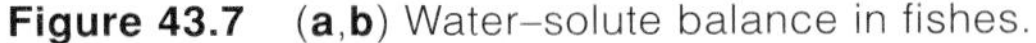

Figure 43.7 (**a**,**b**) Water–solute balance in fishes.

(**c**) Water–solute balance by salmon, a type of fish that lives in saltwater and fresh water. Salmon hatch in streams and later move downstream to the open sea, where they feed and mature. They return to home streams to spawn.

For most salmon, salt tolerance is one outcome of changing concentrations of certain hormones. The changes appear to be triggered by increasing daylength in spring. Prolactin, a pituitary hormone, plays a key role in sodium retention in freshwater habitats. We know this because a freshwater fish that has its pituitary gland removed will die from sodium loss—but that fish will live if prolactin is administered to it.

Cortisol, a steroid hormone secreted by the adrenal cortex, is crucial to the development of salt tolerance in salmon. Cortisol secretions correlate with an increase in sodium excretion, in sodium–potassium pumping by cells in the salmon's gills, and in absorption of ions and water in the gut. In young salmon, cortisol secretion increases prior to the seaward movement—and so does salt tolerance.

salmon avoiding
grizzly while
maintaining
water–solute
balance

c

And about that kangaroo rat! Proportionally, the loops of Henle of its nephrons are astonishingly long, compared to yours. This means a great deal of sodium is pumped out from the nephron. Therefore, the solute concentration in the interstitial fluid around the loops becomes very high. The osmotic gradient between the fluid in the loops of Henle and the urine is *so* steep that nearly all of the water that does reach the equally long collecting ducts gets reabsorbed. In fact, kangaroo rats give up only a tiny volume of urine. And that urine is three to five times more concentrated than the most concentrated urine of humans.

The urinary systems of vertebrates differ in their details, such as the length of the nephron's loop of Henle. They are adapted to balance the body's gains in water and solutes with its losses of water and solutes in particular habitats.

43.6 MAINTAINING THE BODY'S CORE TEMPERATURE

We turn now to one other major aspect of the internal environment—its temperature. Start by thinking about the temperature around your body. Is the air too hot? Too cold? If you have been sitting for some time, your cells have not been producing much metabolic heat. If you just finished exercising strenuously, metabolic rates have soared, and so has metabolic heat production.

Despite such differences, and assuming you are in good health, the internal temperature of your body remains much the same. A variety of physiological and behavioral responses to change work to maintain that constancy.

Temperatures Suitable for Life

Enzyme-mediated reactions proceed simultaneously in the millions to trillions of cells of a large-bodied animal. Enzymes of most animals typically function within the range of 0°–40°C (32°–104°F) as Table 43.1 shows. Above 41°C, chemical bonds holding an enzyme molecule in its required shape are disrupted. So is enzyme function. When the temperature decreases by 10 degrees, the rate of enzyme activity commonly plummets by 50 percent or more. Therefore metabolism, and life itself, depends on maintaining the **core temperature** within the range of tolerance of the body's enzymes. "Core" refers to the internal temperature of the animal body, as opposed to temperatures of the tissues near its surface.

Heat Gains and Heat Losses

Each metabolic reaction generates heat. If heat were to accumulate internally, core temperature would steadily increase. However, a warm body tends to lose heat to a cooler environment. The core temperature holds steady when the rate of heat loss balances the rate of metabolic heat production. In general, the heat content of large, complex animals depends on a balance between heat gains and heat losses, which we can express in the following way:

Table 43.1 Temperatures Favorable for Metabolism, Compared With Environmental Temperatures

General range of internal temperatures favorable for metabolism:	0°C to 40°C (32°F to 104°F)
Range of air temperatures above land surfaces:	−70°C to +85°C (−94°F to +185°F)
Range of surface temperatures of open ocean:	−2°C to +30°C (+28.4°F to +86°F)

$$\text{CHANGE IN BODY HEAT} = \text{HEAT PRODUCED} + \text{HEAT GAINED} - \text{HEAT LOST}$$

For this summary equation, the heat gains and losses occur by exchanges at body surfaces, such as the skin. Four processes, called radiation, conduction, convection, and evaporation, drive the exchanges.

With **radiation**, an animal gains heat after exposure to radiant energy, as from sunlight, or to any surface warmer than the surface temperature of its body.

With **conduction**, an animal contacting a solid object directly gains heat from it or gives up heat to it as a response to a thermal gradient between them. Animals lose heat by resting on objects that are cooler than their own temperature, such as a frozen deck chair (Figure 43.8*a*). They gain heat when they are in direct contact with warmer objects, such as hot sand (Figure 43.8*b*).

With **convection**, moving air or water transfers heat. Conduction plays a part in this process, for heat moves down a thermal gradient between the body and air or water next to it. Mass transfer also plays a part, with currents carrying heat away from or toward the body.

Figure 43.8 (**a**) How might this intrepid tourist sitting on a deck chair of a ship off the coast of Antarctica be gaining and losing heat? (**b**) A sidewinder making its signature J-shaped track across hot desert sand at dusk. As near as you can tell, how might this rattlesnake be gaining and losing heat?

Further reading: Student Guide to InfoTrac on web site

When heated, air becomes less dense and it moves away from the body. Even when there is no breeze, the body loses heat, for its movements create convection.

With **evaporation**, a liquid converts to gaseous form and heat is lost in the process. As described earlier, in Section 2.5, the liquid's heat content provides energy for the conversion. Evaporation from the body surface has a cooling effect, because the water molecules that are escaping carry away some energy with them.

Evaporation rates depend on humidity and on the rate of air movement. If air next to the body is already saturated with water—that is, when the local relative humidity is 100 percent—water won't evaporate. If air next to the body is hot and dry, evaporation may be the only means of countering the metabolic production of heat and the heat gains from radiation and convection.

Figure 43.9 Meerkats keeping warm on a cold winter night in a zoo in Germany.

Ectotherms, Endotherms, and In-Betweens

Animals can adjust the amount of heat lost or gained through changes in behavior and physiology, although some are far better equipped than others to do so. Like most animals, snakes, lizards, and other reptiles have low metabolic rates and poor insulation (Figure 43.8*b*). They rapidly absorb and gain heat, especially the small species. They protect their core temperature primarily by gaining heat from their environment, not by their metabolic activities. Hence we classify these animals as **ectotherms**, which means "heat from outside."

When outside temperatures change, an ectotherm must alter its behavior. This is *behavioral* temperature regulation. For example, that iguana shown at the start of this unit basks on warm rocks, thus gaining heat by conduction. It keeps reorienting its body to expose the most surface area to the sun's infrared radiation. It loses heat after sunset. Before its metabolic rates decrease, it crawls into a crevice or under a rock, where heat loss isn't as great and it is not as vulnerable to predators. As another example, meerkats bask in morning sunlight when they emerge from burrows (Figure 33.1). They also learn to bask beneath heat lamps after being shipped to zoos in cold climates (Figure 43.9).

Most birds and mammals are **endotherms** ("heat from within"). With high metabolic rates, they can stay active under a wide temperature range. (Compared to a foraging lizard of the same weight, a foraging mouse uses up to thirty times *more* energy.) Core temperatures of these animals depend on a balancing of metabolism, controlled heat loss and conservation, and complex behavior. Adaptations in morphology help conserve or dissipate heat associated with the high metabolic rates. Feathers, fur, fat layers, even clothing reduce heat loss (Figure 43.8*a*). Some mammals in cold habitats have more massive bodies than close relatives in warmer ones. For example, the snowshoe hare of Canada has a more compact body, far shorter legs, and far shorter ears than the jackrabbit of the American Southwest. Its body has a greater volume of cells for generating metabolic heat relative to the surface area available for heat loss. And heat dissipation from its legs and ears does not begin to match the losses from a jackrabbit's lengthy extremities.

Certain birds and mammals are **heterotherms**. At some times, their core temperature shifts (as it does in ectotherms). At other times, heat exchange is controlled (as in endotherms). For example, given their small size, hummingbirds have high metabolic rates. They locate and sip nectar only in the day. At night, they may shut down almost entirely. Their metabolic rates plummet, and they may get almost as cool as their surroundings. This way, they conserve precious energy.

In general, ectotherms have an advantage in warm, humid regions, especially in the tropics. They need not spend much energy to maintain core temperatures, and more energy can be devoted to reproduction and other tasks. In terms of numbers and diversity, reptiles far exceed mammals in tropical regions. Endotherms have the edge in moderate to cold regions. For example, with their high metabolic rates, snowshoe hares, arctic foxes, and some other endotherms occupy polar habitats —where you would never find a lizard.

The internal, core temperature of an animal's body is being maintained when heat gains and heat losses are in balance.

Metabolic reactions generate heat inside the body. Radiation, conduction, and convection can move heat down thermal gradients that exist between the body and its surroundings. Evaporative heat loss carries heat away from the body.

Besides being morphologically adapted to their habitats, animals make behavioral and physiological adjustments to environmental temperatures.

43.7 TEMPERATURE REGULATION IN MAMMALS

Control centers that maintain the core temperature of the mammalian body reside in the hypothalamus. They receive ongoing input from peripheral thermoreceptors in skin and from thermoreceptors deeper in the body (Figure 43.10). When the temperature deviates from a set point, the centers integrate complex responses that involve skeletal muscles, smooth muscle in arterioles that service skin, and, often, sweat glands. (Here you may wish to review Section 33.7.) Negative feedback loops back to the hypothalamus shut off the responses when a suitable temperature is reinstated.

Responses to Cold Stress

Table 43.2 lists the major responses that mammals make to cold stress. (Birds make the same responses.) They are called peripheral vasoconstriction, the pilomotor response, shivering, and nonshivering heat production.

Suppose peripheral thermoreceptors detect a decline in outside temperature. They notify the hypothalamus, which commands smooth muscle in arterioles in the skin to contract. The response is **peripheral vasoconstriction**. The diameters of these arterioles constrict, which limits the blood's convective delivery of heat to body surfaces. When your fingers or toes are chilled, all but 1 percent of blood that otherwise would flow to the skin is diverted to other regions of the body.

Table 43.2 Responses to Cold Stress

Core Temperature	Responses
36°–34°C	Shivering response, increase in respiration. Increase in metabolic heat output (about 95°F). Peripheral vasoconstriction puts more blood deeper in body. Dizziness and nausea set in.
33°–32°C	Shivering response stops. Metabolic heat (about 91°F) output drops.
31°–30°C (about 86°F)	Capacity for voluntary motion is lost. Eye and tendon reflexes inhibited. Consciousness is lost. Cardiac muscle action becomes irregular.
26°–24°C	Ventricular fibrillation sets in (Section 39.9). Death follows (about 77°F).

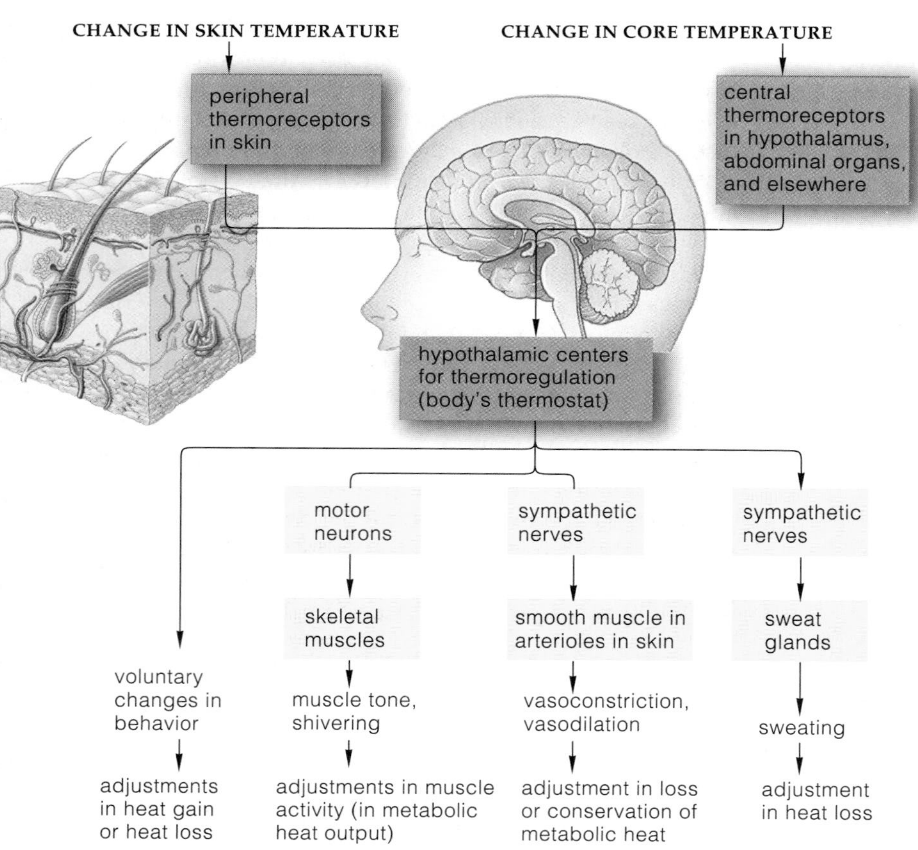

CHANGE IN SKIN TEMPERATURE
CHANGE IN CORE TEMPERATURE
peripheral thermoreceptors in skin
central thermoreceptors in hypothalamus, abdominal organs, and elsewhere
hypothalamic centers for thermoregulation (body's thermostat)
motor neurons
sympathetic nerves
sympathetic nerves
skeletal muscles
smooth muscle in arterioles in skin
sweat glands
voluntary changes in behavior
muscle tone, shivering
vasoconstriction, vasodilation
sweating
adjustments in heat gain or heat loss
adjustments in muscle activity (in metabolic heat output)
adjustment in loss or conservation of metabolic heat
adjustment in heat loss

Figure 43.10 Physiological and behavioral adjustments of a typical mammal to changes in outside temperature. This example shows the main pathways for thermoregulation in humans.

Also, muscle contractions make hairs (or feathers) "stand up." This **pilomotor response** creates a layer of still air next to the skin and reduces convective and radiative heat loss. Behavioral changes also minimize exposed surface areas and reduce heat loss, as when cats curl up or when you hold both of your arms tightly against the body.

A **shivering response** might be made to prolonged cold. Rhythmic tremors begin as skeletal muscles contract, ten to twenty times a second, in response to commands from the hypothalamus. The shivering increases heat production by several times. But it has a high energy cost and is not effective for long.

Prolonged or severe cold exposure also leads to a hormonal response that elevates the rate of metabolism. This **nonshivering heat production** is most characteristic of *brown* adipose tissue. Animals that hibernate or are acclimatized to cold environments have this connective tissue, as do human infants. Adults have little unless they have become adapted to cold. The tissue is present in ama, Japanese and Korean divers who spend six hours per day collecting shellfish in very cold water.

Failure to defend against cold results in *hypothermia*, a condition in which the core temperature falls below normal. In humans, a decrease by even a few degrees adversely affects brain function and leads to confusion;

Further reading: Student Guide to InfoTrac on web site —

Table 43.3 Summary of Mammalian Responses to Changes in Core Temperature

Stimulus	Main Responses	Outcome
Cold stress	Widespread vasoconstriction in skin; behavioral adjustments (e.g., minimizing surface parts exposed)	Conservation of body heat
	Increased muscle action; shivering; nonshivering heat production	Heat production increases
Heat stress	Widespread vasodilation in skin; behavioral adjustments; in some species, sweating, panting	Dissipation of heat from body
	Decreased muscle action	Heat production decreases

cooling leads to coma and death. Many mammals can recover from profound hypothermia. However, frozen cells may die unless tissues thaw under close medical supervision. Tissue destruction as a result of localized freezing is called *frostbite*.

Responses to Heat Stress

Peripheral vasodilation and evaporative heat loss are the main responses to heat stress (Table 43.3). With **peripheral vasodilation**, hypothalamic signals cause blood vessels in skin to dilate. More blood flows from deeper body regions to skin, where the excess heat it carries is dissipated. **Evaporative heat loss** occurs at moist respiratory surfaces and across skin. Animals that sweat lose more water by this process (Figure 43.11).

For example, humans and some other mammals have sweat glands, which release water and certain solutes through pores at the skin surface. An average-size human has 2–1/2 million or more sweat glands that can produce 1 to 2 liters of sweat in an hour. For every liter of sweat that evaporates, 600 kilocalories of heat energy are lost. Bear in mind, sweat dripping from skin does not dissipate heat; the water in sweat must evaporate for that to happen. On hot, humid days, the evaporation rates typically cannot match the rate of sweat secretion; the air's high water content slows it down.

During strenuous exercise, sweating can balance the elevated rates of heat production in skeletal muscle. With extreme sweating, as might occur in a marathon race, the body loses an important salt—sodium chloride—as well as water. Such losses disrupt the composition and the volume of extracellular fluid. When losses are great enough, the runner collapses and faints.

What about mammals that sweat little or not at all? Some kinds make behavioral responses, such as licking fur, panting, or resting in shade. "Panting" is shallow, very rapid breathing that increases evaporative water loss from the respiratory tract (compare Figure 33.14). Cooling results when water evaporates from the nasal cavity, mouth, and tongue.

Figure 43.11 Evaporative water loss. Besides losing water at the moist respiratory surface of their lungs, horses, humans, and some other mammals have a large number of sweat glands that move water and solutes through pores to the skin surface.

Sometimes peripheral blood flow and evaporative heat loss are not enough to counter heat stress in the body, and *hyperthermia* results. With this condition, the core temperature increases above normal. For humans and other endotherms, increases of only a few degrees above normal can be dangerous.

Fever

A *fever*, remember, is part of the inflammatory response to tissue damage. The hypothalamus resets the body's thermostat, which dictates what the core temperature is supposed to be (Sections 33.7 and 35.4). Mechanisms that increase metabolic heat production and decrease heat loss operate, but their operation maintains a higher temperature. A person feels chilled at a fever's onset. When the fever "breaks," peripheral vasodilation and sweating increase as the body attempts to restore the normal core temperature. Then, the person feels warm.

By bringing down a fever, anti-inflammatory drugs such as aspirin may actually prolong the healing time. Only when fevers approach dangerous levels should drugs be prescribed, under medical supervision.

Mammals counter cold stress by widespread vasoconstriction in skin, behavioral adjustments, increased muscle activity, and shivering and nonshivering heat production. They counter heat stress by widespread peripheral vasodilation in skin and by evaporative heat loss.

43.8 SUMMARY

Control of Extracellular Fluid

1. For animal cells, the environment consists of certain types and amounts of substances that are dissolved in water. Extracellular fluid fills tissue spaces and blood vessels. Its volume and composition are maintained when the animal's daily intake and output of water as well as solutes are in balance. The following processes maintain the balance in mammals:

a. Water is gained by absorption from the gut and by metabolism. It is lost by urinary excretion, evaporation from lungs and skin, sweating, and elimination of feces.

b. Solutes are gained by absorption from the gut, secretion, respiration, and metabolism. They are lost by excretion, respiration, and sweating.

c. Losses of water and solutes are controlled mainly by adjusting the volume and composition of urine.

2. The vertebrate urinary system has a pair of kidneys, a pair of ureters, a urinary bladder, and a urethra.

3. Kidneys have many nephrons that filter blood and form urine. Each nephron interacts intimately with two sets of blood capillaries: glomerular and peritubular.

a. The start of a nephron is cup-shaped (Bowman's capsule). It continues as three tubular regions (proximal tubule, loop of Henle, and distal tubule, which empties into a collecting duct).

b. Together, Bowman's capsule and the set of highly permeable glomerular capillaries within it are a blood-filtering unit (glomerulus). Blood pressure forces water and small solutes out of the capillaries, into the cup. Most of the filtrate is reabsorbed at tubular regions and is returned to blood. A portion is excreted as urine.

4. Urine forms in the nephron by three processes:

a. Filtration of blood at the glomerulus, which puts water and small solutes into the nephron.

b. Reabsorption. Water and solutes to be retained leave the nephron's tubular parts and enter capillaries that thread around them. A small volume of water and solutes remains in the nephron.

c. Secretion. A few substances can leave peritubular capillaries and enter the nephron, for disposal in urine.

5. Urine is made more concentrated or less so by two hormones that act on cells of the wall of distal tubules and collecting ducts. ADH conserves water by enhancing reabsorption across the wall. In its absence, more water is excreted. Aldosterone enhances sodium reabsorption. In its absence, sodium is excreted. Angiotensin II induces aldosterone secretion (sodium conservation) and thirst; it also promotes ADH secretion (water conservation).

6. The urinary system acts in concert with buffers and with the respiratory system to maintain the acid–base balance of extracellular fluid.

Control of Body Temperature

1. Maintaining an animal's core (internal) temperature depends on balancing metabolically produced heat and the heat absorbed from and lost to the environment.

2. Animals exchange heat with their environment by four processes:

a. Radiation. Emission from the body of infrared and other wavelengths. Radiant energy can be absorbed at the body surface, then converted to heat energy.

b. Conduction. Direct transfer of heat energy from one object to another object in contact with it.

c. Convection. Heat transfer by air or water currents; involves conduction and mass transfer of heat-bearing currents away from or toward the animal body.

d. Evaporation. Conversion of liquid to a gas, driven by energy inherent in the heat content of the liquid. Some animals lose heat by evaporative water loss.

3. Core temperatures depend on metabolic rates and anatomical, behavioral, and physiological adaptations.

a. For ectotherms, core temperature depends more on heat exchange with the environment than on heat generated by metabolism.

b. For endotherms, core temperature depends more on high metabolic rates and precise controls over heat produced and heat lost.

c. For heterotherms, the core temperature fluctuates some of the time, and controls over heat balance come into play at other times.

Review Questions

1. State the function of the urinary system in terms of gains and losses for the internal environment. Name the components of the mammalian urinary system and state their functions. *43.1*

2. Label the component parts of this kidney and nephron. *43.1*

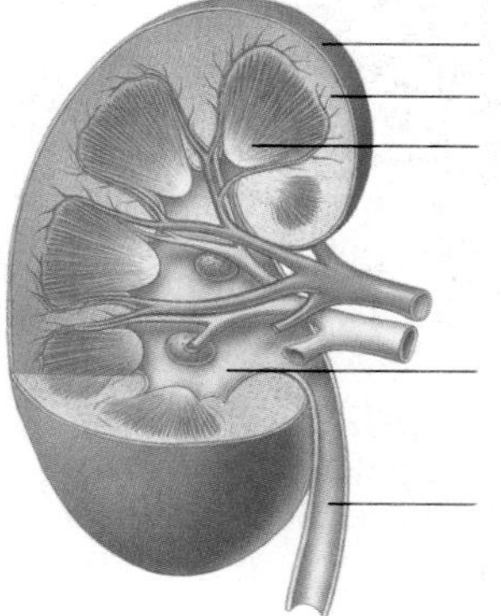

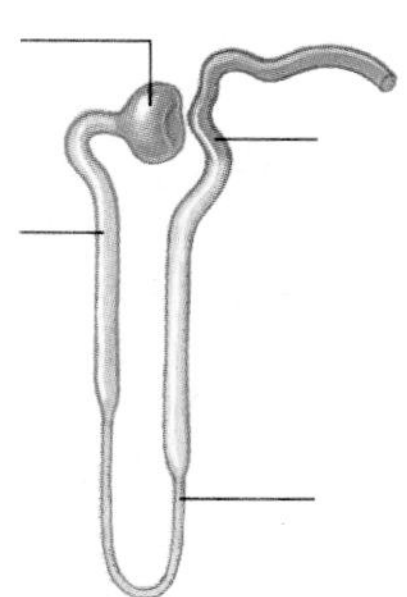

3. Define filtration, tubular reabsorption, and secretion. Explain how urine formation helps maintain the internal environment. *43.2*

4. Which hormone or hormones promote (a) water conservation, (b) sodium conservation, and (c) thirst behavior? *43.2*

5. Name and define the physical processes by which animals gain and lose heat. What are the main physiological responses to cold stress and to heat stress in mammals? *43.6, 43.7*

Self-Quiz (*Answers in Appendix III*)

1. In mammals, water intake depends on ______.
   a. absorption from gut c. a thirst mechanism
   b. metabolism d. all of the above

2. In mammals, water is lost by way of the ______.
   a. skin d. urinary system
   b. respiratory system e. c and d
   c. digestive system f. a through d

3. Water and small solutes enter nephrons during ______.
   a. filtration c. tubular secretion
   b. tubular reabsorption d. both a and c

4. Kidneys return water and small solutes to blood by ______.
   a. filtration c. tubular secretion
   b. tubular reabsorption d. both a and b

5. A few substances move out of the peritubular capillaries that thread around tubular parts of the nephron. The substances are moved into the nephron during ______.
   a. filtration c. tubular secretion
   b. tubular reabsorption d. both a and c

6. A nephron's reabsorption mechanism depends on ______.
   a. osmosis across nephron wall
   b. active transport of sodium across nephron wall
   c. a steep solute concentration gradient
   d. all of the above

7. ______ promotes water conservation.
   a. ADH c. Low extracellular fluid volume
   b. Aldosterone d. Both a and c

8. ______ enhances sodium reabsorption.
   a. ADH c. Low extracellular fluid volume
   b. Aldosterone d. Both b and c

9. Match the term with the most suitable description.
   ___ glomerulus a. surrounded by saltiest fluid
   ___ distal tubule b. extra-long loops of Henle
   ___ loop of Henle c. involves buffer systems
   ___ acid–base balance d. blood-filtering unit
   ___ kangaroo rat e. ADH, aldosterone act here

10. Match the term with the most suitable description.
   ___ ectotherm a. heat transfer by air or water currents
   ___ endotherm b. body temperature fluctuates some of the time, is controlled at other times
   ___ evaporation c. emission of radiant energy
   ___ heterotherm d. direct heat transfer between an object and another object in contact with it
   ___ radiation e. metabolism dictates core temperature
   ___ conduction f. environment dictates core temperature
   ___ convection g. conversion of liquid to gas

Critical Thinking

1. Fatty tissue holds kidneys in place. Rarely, extremely rapid weight loss may cause the tissue to shrink and kidneys to slip from their normal position. If slippage puts a kink in one or both ureters and blocks urine flow, what may happen to the kidneys?

2. Drink one quart of water in one hour. What changes might you expect in your kidney function and in urine composition?

3. In 1912, the ocean liner *Titanic* left Europe on her maiden voyage to America—and a chunk of the leading edge of a glacier in Greenland broke away and floated out to sea. Late at night, off the Newfoundland coast, the iceberg and the *Titanic* made an ill-fated rendezvous (Figure 43.12). The *Titanic* was said to be unsinkable. Survival drills were neglected. There weren't enough lifeboats to hold even half the 2,200 passengers. The *Titanic* sank in about 2–1/2 hours. Rescue ships were on the scene in less than two hours, but 1,517 bodies were recovered from a calm sea. All the dead had on life jackets; none had drowned. They probably died from ______. If so, how did their blood flow, metabolism, and skeletal muscle action change prior to death?

Figure 43.12 Sinking of the *Titanic*, based on eyewitness accounts.

4. When iguanas have an infection, they rest for a prolonged period in the sun. Propose a hypothesis to explain why.

5. Out on a first date, Jon takes Geraldine's hand in a darkened theater. *"Aha!"* he thinks. *"Cold hands, warm heart!"* What does this tell us about the regulation of core temperature, let alone Jon?

Selected Key Terms

<u>*Salt–Water Balance*</u>

acid–base balance *43.4*
ADH *43.2*
aldosterone *43.2*
angiotensin II *43.2*
blood *CI*
Bowman's capsule *43.1*
collecting duct *43.1*
distal tubule *43.1*
extracellular fluid *CI*
filtration *43.2*
glomerulus *43.1*
interstitial fluid *CI*
kidney *43.1*
loop of Henle *43.1*
nephron *43.1*
proximal tubule *43.1*
renal failure *43.3*
thirst center *43.2*
tubular reabsorption *43.2*
tubular secretion *43.2*
urea *43.1*
ureter *43.1*
urethra *43.1*
urinary bladder *43.1*
urinary excretion *43.1*
urinary system *43.1*
urine *43.1*

<u>*Body Temperature*</u>

conduction *43.6*
convection *43.6*
core temperature *43.6*
ectotherm *43.6*
endotherm *43.6*
evaporation *43.6*
evaporative heat loss *43.7*
heterotherm *43.6*
nonshivering heat production *43.7*
peripheral vasoconstriction *43.7*
peripheral vasodilation *43.7*
pilomotor response *43.7*
radiation *43.6*
shivering response *43.7*

Readings *See also www.infotrac-college.om*

Flieger, K. March 1990. "Kidney Disease: When Those Fabulous Filters Are Foiled." *FDA Consumer* 24: 26–29.

Sherwood, L. 1997. *Human Physiology*. Second edition. Monterey, California: Brooks-Cole.

Smith, H. 1961. *From Fish to Philosopher*. New York: Doubleday.

44

PRINCIPLES OF REPRODUCTION AND DEVELOPMENT

From Frog to Frog and Other Mysteries

With a quavering, low-pitched call that only a female of its kind could find seductive, a male frog proclaims the onset of warm spring rains, of ponds, of sex in the night. By August the summer sun will have parched the earth, and his pond dominion will be gone. But tonight is the hour of the frog!

Through the dark, a hormone-primed female moves toward the vocal male. They meet; they dally in the behaviorally prescribed ways of their species. He clamps his forelegs above her swollen abdomen and gives her a prolonged squeeze (Figure 44.1*a*). Out into the water streams a ribbon of hundreds of eggs, which the male blankets with a milky cloud of sperm. Soon afterward, fertilized eggs—**zygotes**—are suspended in the water.

For the leopard frog, *Rana pipiens*, a drama begins to unfold that has been reenacted each spring, with only minor variations, for many millions of years. Within a few hours after fertilization, each zygote divides into two cells, the two divide into four, then the four into eight. In less than twenty hours after fertilization, the mitotic cell divisions have produced a ball of cells no larger than the zygote. It is an early embryonic stage, of a type known as a blastocyst.

The cells continue to divide, but now they start to interact by way of their surface structures and chemical secretions. At prescribed times, many change shape and migrate to prescribed positions. They all inherited the same genetic instructions from the zygote, yet now they start to differ in appearance and function!

Through their associations, the cells form layers of embryonic tissues and then embryonic organs. A pair of tissue regions at the embryo's surface interact with the tissues beneath them. Together they give rise to a pair of eyes. Within the embryo a heart is forming, and soon it starts an incessant, rhythmic beating. In less than a week, events have transformed the frog embryo into a swimming, algae-eating larva called a tadpole.

Several months pass. Legs form; the tail shortens and disappears. The mouth develops jaws that snap shut on insects and worms. Eventually the transformations lead to an adult frog. With luck the frog will avoid predators, disease, and other threats in the months ahead. In time

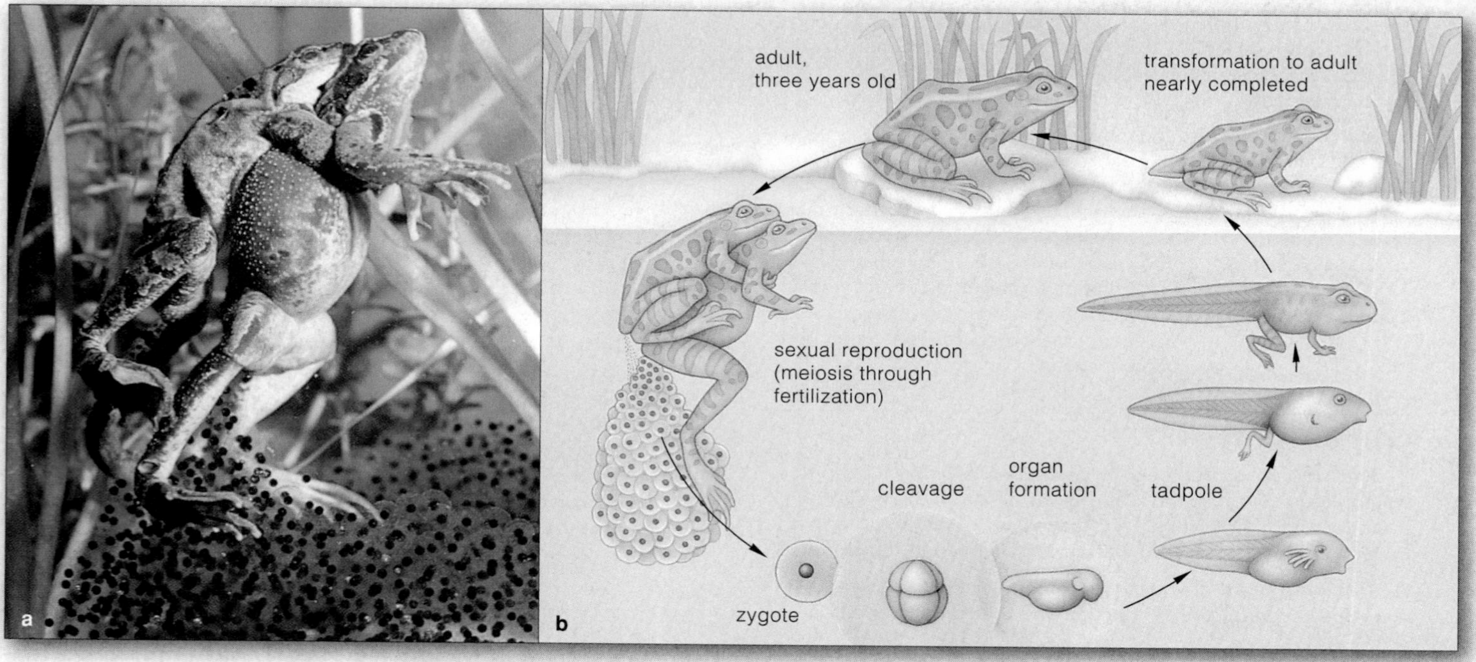
a
b
adult,
three years old
transformation to adult
nearly completed
sexual reproduction
(meiosis through
fertilization)
cleavage
organ
formation
tadpole
zygote

Figure 44.1 Reproduction and development of *Rana pipiens*, the leopard frog. (**a**,**b**) We zoom in on the life cycle as a male clasps a female in a reproductive behavior called amplexus. The female releases eggs into the water. The male releases sperm over the eggs. A zygote forms when an egg nucleus and a sperm nucleus fuse at fertilization. (**c**) Frog embryos suspended in the water. (**d**) A tadpole. (**e**) A transitional form between a tadpole and the young adult frog (**f**).

it may even call out quaveringly across a moonlit pond, and the life cycle will turn again.

Many years ago you, too, started a developmental journey when a zygote carved itself up. Three weeks into the journey, your embryonic body had the stamp of "vertebrate" on it. A mere five weeks after that, it was a recognizable human in the making!

With this chapter we turn to one of life's greatest dramas—the development of offspring in the image of sexually reproducing parents. The guiding question is this: *How does a single-celled zygote of a frog, human, or any other complex animal become transformed into all the specialized cells and structures of the adult form?* Some answers will become apparent as we move through a survey of basic principles in this chapter, then through a case study of human reproduction and development in the chapter to follow.

DEVELOPING EMBRYO
c
d
e
f

KEY CONCEPTS

1. Sexual reproduction dominates the life cycle of nearly all animals, but separation into sexes has biological costs. It requires the construction and maintenance of specialized reproductive structures. It also requires hormonal control mechanisms and complex forms of behavior attuned to the environment and to potential mates and rivals.

2. Separation into sexes affords a selective advantage. Offspring show variation in traits, which improves the odds that at least some will survive and reproduce despite unexpected challenges from the environment. This reproductive advantage offsets the biological cost of the separation.

3. The life cycles of humans and many other animals proceed through six stages of embryonic development—gamete formation, fertilization, cleavage, gastrulation, organ formation, and growth and tissue specialization.

4. Each stage of embryonic development builds on the tissues and structures that formed during the stage that preceded it.

5. In a developing embryo, the fate of each type of cell depends partly on cleavage, which distributes different regions of the fertilized egg's cytoplasm to different daughter cells. It also depends on interactions among cells of the embryo. These activities are the foundation for cell differentiation and morphogenesis.

6. With cell differentiation, each cell type selectively uses certain genes and synthesizes proteins not found in other types, and so becomes unique in structure and function. With morphogenesis, tissues and organs change in size, shape, and proportion. They become organized relative to one another in prescribed patterns.

7. All complex, multicelled animals that show extensive cell differentiation undergo aging. Their cells gradually break down in structure and function, which leads to the decline of tissues, organs, and eventually the body.

44.1 THE BEGINNING: REPRODUCTIVE MODES

Sexual Versus Asexual Reproduction

In earlier chapters, you learned about the cellular basis of **sexual reproduction**. Briefly, by this reproductive mode, meiosis and gamete formation typically proceed in two prospective parents. At fertilization, a gamete from one parent fuses with a gamete from the other to form the zygote, the first cell of the new individual. You also read about **asexual reproduction**, by which a single parent organism produces offspring by various means (but not by gamete formation). We now turn to a few structural, behavioral, and ecological aspects of the two reproductive modes.

Picture a scuba diver accidentally kicking a sponge. A tissue fragment breaks away from the sponge body, then grows and develops by mitotic cell divisions and cell differentiation into a new sponge. Or picture one of the flatworms undergoing transverse fission while it glides along through the water. First its body constricts below the midsection. The part behind the constriction grips a substrate and starts a tug-of-war with the part in front. After a few hours, it splits off. Both parts go their separate ways, regenerate the missing part, and become a whole worm. Only some species do this.

In such cases of *asexual* reproduction, all offspring are genetically the same as their individual parent, or nearly so. Phenotypically they are much the same, also. We can speculate that phenotypic uniformity is useful when each individual's gene-encoded traits are highly adapted to a limited and more or less consistent set of environmental conditions. Most variations introduced into the finely tuned gene package would not do much good, and often they would do harm.

However, most animals live where opportunities, resources, and danger are highly variable. For the most part, such animals reproduce sexually, with female and male parents bestowing different mixes of alleles on their offspring (Section 10.1). The resulting variation in traits improves the odds that some of those offspring, at least, will survive and reproduce even if conditions change in the environment.

Costs and Benefits of Sexual Reproduction

Among animals, separation into sexes is not without cost. Some cells that can serve as gametes must be set aside and nurtured. Housing and delivering gametes near or within a prospective mate require specialized reproductive structures. Mating often requires special forms of behavior, such as courtship, that can promote fertilization. Mating also requires built-in controls that can synchronize the timing of gamete formation, sexual readiness, even parental behavior in two individuals.

a
b

Figure 44.2 Examples of where invertebrate and vertebrate embryos develop, how they are nourished, and how (if at all) parents protect them.

(**a**) Snails are *oviparous*, meaning they release eggs from which the young later hatch (*ovi-*, egg; *parous*, produce). Snails also are hermaphroditic parents, and they leave eggs unprotected.

(**b**) Birds also are oviparous. Their fertilized eggs have large yolk reserves, and they, too, develop and hatch outside the mother's body. Unlike snails, one or both parent birds expend considerable energy feeding and caring for the young.

Facing page: Some fishes, all lizards, and many snakes are *ovoviviparous*. Their fertilized eggs develop inside the mother, then offspring are born live (*vivi-*, alive). Yolk reserves, not the mother's own tissues, sustain the eggs. Example: (**c**) These liveborn copperhead offspring are still in their egg sacs.

Most mammals are *viviparous*; their young are born live. (**d**,**e**) In kangaroos and some other species, embryos are born in unfinished form. The young of this marsupial complete development in a pouch on the mother's ventral surface. The juvenile stages (joeys) continue to draw nourishment from mammary glands in the mother's pouch. (**f**) By contrast, a human female retains the fertilized egg inside her body. Maternal tissues nourish the developing individual until the time of birth, in the manner described in the next chapter.

Just look at the question of *reproductive timing*. How do mature sperm in one individual become available exactly when the eggs mature in a different individual? Timing depends upon energy outlays for constructing, maintaining, and operating neural as well as hormonal control mechanisms in each parent. Also, parents must produce mature gametes in response to the same cues, such as a seasonal change in daylength, that mark the onset of the most suitable time of reproduction for their species. For example, male and female moose become sexually active only in late summer and early fall. The coordinated timing ensures that their offspring will be born next spring—when the weather will improve and food will be plentiful for many months.

Further reading: Student Guide to InfoTrac on web site

liveborn snake inside egg sac
c

Finding and actually recognizing a potential mate of the same species is another challenge. Different species invest energy when they synthesize pheromones. They construct visual mating signals such as feathers of certain colors and patterns, and complex sensory receptors to detect the specific signals being sent. In addition, males often expend astonishing amounts of energy to execute courtship routines, as you will read in Chapter 47.

Ensuring the survival of offspring is also costly. For example, many invertebrates, bony fishes, and frogs simply release eggs and motile sperm into the watery surroundings, as in Figure 44.1*a*. If each adult produced only *one* sperm or *one* egg each season, the odds for fertilization would not be good. Such species invest energy in producing numerous gametes, often thousands of them. As another example, nearly all land animals rely on **internal fertilization**, a union of sperm and egg *inside* the female's body. They invest energy to construct elaborate reproductive organs, such as a penis (by which a male deposits its sperm inside a female) and a uterus (a chamber inside the female where the embryo grows and develops).

Finally, animals set aside energy in forms that can *nourish the developing individual* until it is developed enough to feed itself. For instance, nearly all animal eggs contain **yolk**, which is a protein-rich, lipid-rich substance that nourishes embryonic stages. The eggs of some species have much more yolk than others. Sea urchins make tiny eggs with very little yolk, release large numbers of them, and thus limit the biochemical investment in each one. Within twenty-four hours, each fertilized egg has developed into a self-feeding, free-moving larva. Sea stars and other predators consume most of the eggs. So for sea urchins, bestowing as little as possible on as many gametes as possible does pay off in terms of reproductive success.

By contrast, mother birds lay truly yolky eggs. Yolk nourishes the bird embryo through an extended period of development inside an eggshell that forms after the egg is fertilized. Your mother put tremendous demands on her body to protect and nourish you through nine months of development inside her, starting from the time you were an egg with almost no yolk. After you implanted yourself into her uterus, physical exchanges with her tissues supported you through the extended pregnancy (Figure 44.2*f*).

d
e
f

As these few examples suggest, animals show great diversity in reproduction and development. However, as you will see in the sections to follow, some patterns are widespread throughout the animal kingdom, and they can serve as a framework for our reading.

Separation into male and female sexes requires special reproductive cells and structures, neural and hormonal control mechanisms, and forms of behavior. A selective advantage—variation in traits among offspring—offsets biological costs associated with the separation into sexes.

44.2 STAGES OF DEVELOPMENT—AN OVERVIEW

Embryos are a class of transitional forms on the road from a fertilized egg to an adult. Although they all start out as a single cell, embryos of different species often look different as they grow and develop. For example, you do not look like a frog now, and you did not look like one when you and the frog were early embryos, either. However, despite the differences in appearance, it is possible to identify certain patterns in the way the embryos of nearly all animal species develop.

Figure 44.3 is an overview of the stages of animal development. During **gamete formation**, the first stage, eggs or sperm develop inside the reproductive organs of one parent's body. **Fertilization**, the second stage, starts when the plasma membrane of a sperm fuses with the plasma membrane of an egg. It is over when the egg nucleus and sperm nucleus fuse and form a zygote.

The third stage, **cleavage**, is a program of mitotic cell divisions that divides the volume of egg cytoplasm into a number of **blastomeres**—smaller cells, each with its own nucleus. There is no growth during this stage. Cleavage only increases the number of cells; it does not change the original volume of the egg cytoplasm.

As cleavage draws to a close, the pace of mitotic cell division slackens. The embryo enters **gastrulation**. This fourth stage of animal development is a time of major cellular reorganization. The newly formed cells become arranged into two or three primary tissues, often called germ layers. The cellular descendants of the primary tissues will form all the tissues and organs of the adult:

1. **Ectoderm**. This is the *outermost* primary tissue layer, the one that forms first in the embryos of every animal. Ectoderm is the embryonic forerunner of the cell lineages that give rise to tissues of the nervous system and to the integument's outer layer.

2. **Endoderm**. Endoderm is the *innermost* primary tissue layer. It is the embryonic forerunner of the gut's inner lining and organs derived from the gut.

3. **Mesoderm**. This *intermediate* primary tissue layer is the forerunner of muscle and most of the skeleton; of circulatory, reproductive, and excretory organs; and of connective tissue layers of the gut and integument. Mesoderm originated hundreds of millions of years ago, and it was a pivotal step in the evolution of nearly all large, complex animals.

After they have formed, the primary tissue layers give rise to subpopulations of cells. This marks the onset of **organ formation**. The subpopulations become unique in structure and function, and their descendants give rise to different kinds of tissues and organs.

During the final stage of animal development, **growth and tissue specialization**, organs increase in size, and they gradually assume specialized functions. This stage continues into adulthood.

Now take a look at Figure 44.4. Its photographs and sketches show several stages in the embryonic development of a typical animal, a frog. Take a moment to study this

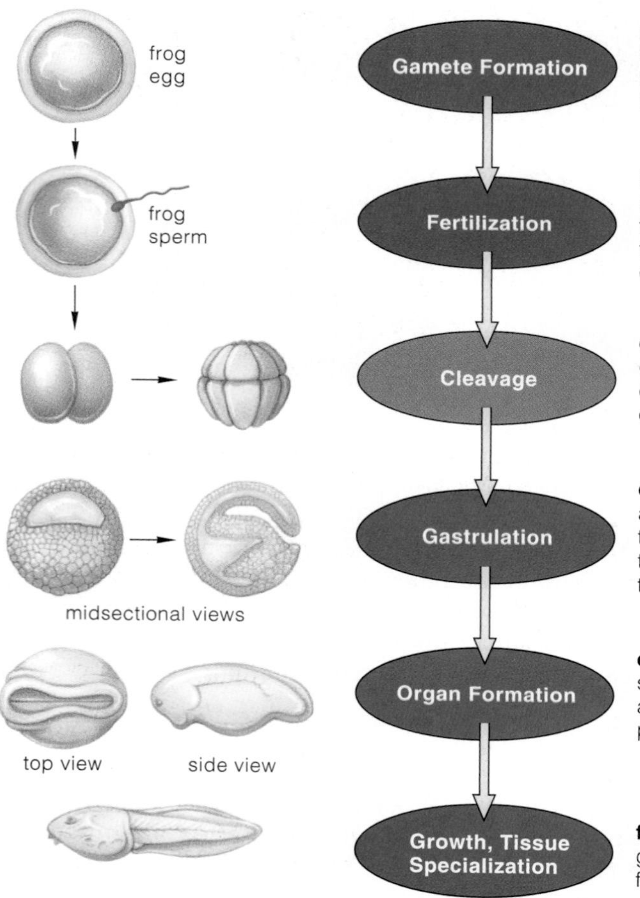

frog egg
Gamete Formation
frog sperm
Fertilization
Cleavage
Gastrulation
midsectional views
Organ Formation
top view
side view
Growth, Tissue Specialization

a Eggs form and mature in female reproductive organs, and sperm form and mature in male reproductive organs.

b A sperm and an egg fuse at their plasma membrane, then the nucleus of one fuses with the nucleus of the other to form the zygote.

c By a series of mitotic cell divisions, different daughter cells receive different regions of the egg cytoplasm.

d Cell divisions, migrations, and rearrangements produce two or three primary tissues, the forerunners of specialized tissues and organs.

e Subpopulations of cells are sculpted into specialized organs and tissues in prescribed spatial patterns at prescribed times.

f Organs increase in size and gradually assume specialized functions.

Figure 44.3 Overview of the stages of animal development. We use a few forms that appear during the frog life cycle as examples.

Further reading: Student Guide to InfoTrac on web site

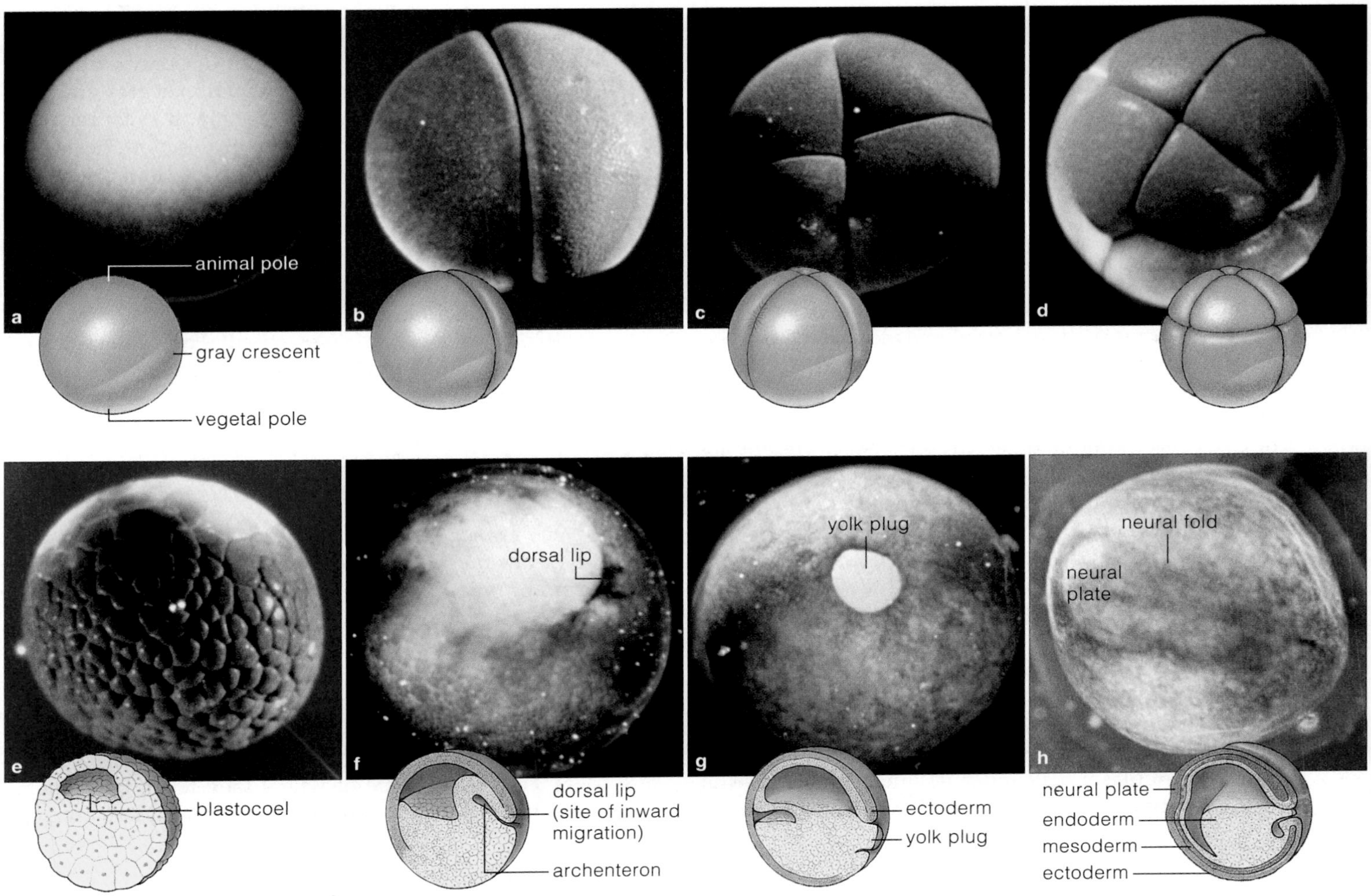

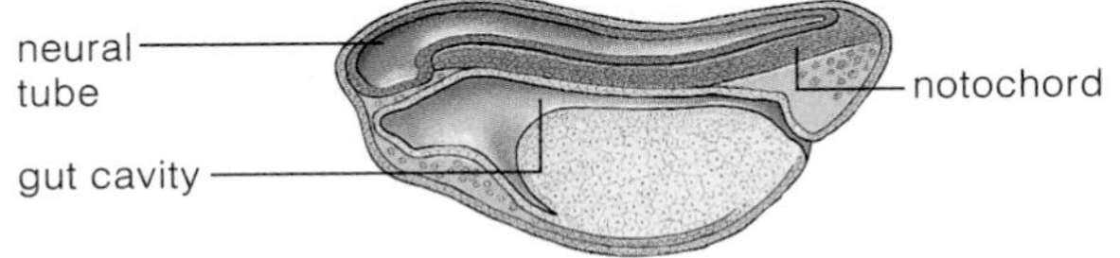
animal pole
gray crescent
vegetal pole
a
b
c
d
dorsal lip
yolk plug
neural fold
neural plate
e
f
g
h
blastocoel
dorsal lip (site of inward migration)
archenteron
ectoderm
yolk plug
neural plate
endoderm
mesoderm
ectoderm
i
neural tube
gut cavity
notochord

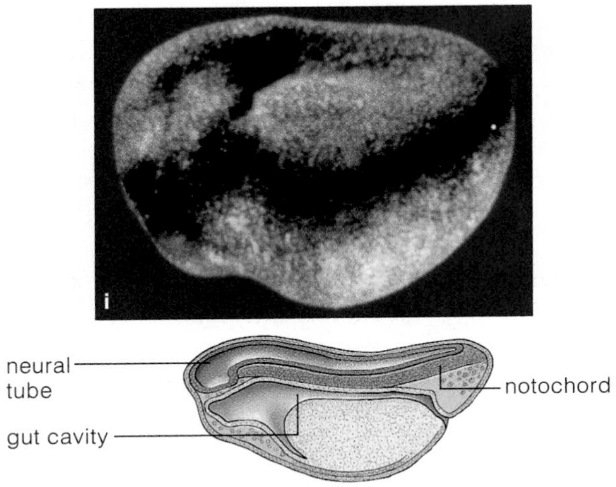

Figure 44.4 Examples of stages of animal development. The sketches and micrographs show the changing appearance of a frog embryo over time. For all micrographs except (**h**), a jellylike layer that surrounds the frog egg has been removed.

(**a**) About one hour after fertilization, a feature called the gray crescent appears on the surface of this type of embryo and establishes the body's anterior–posterior axis. In frogs, this is the head-to-tail axis. Gastrulation will start here. (**b–e**) Cleavage produces a blastula, a ball of cells with a fluid-filled cavity called a blastocoel.

(**f, g**) Cells actively migrate to new locations and are rearranged during gastrulation. (**h,i**) Primary tissue layers form, then a primitive gut cavity—in this case, a blastocoel (fluid-filled body cavity) in which internal organs will be suspended. Cell differentiation proceeds, moving the embryo on its way to becoming a tadpole, as shown in Figure 44.1.

illustration, for it reinforces an important concept. The structures that form during one stage of development serve as the foundation for the stage that comes after it. Successful development depends on the formation of all of those structures according to normal patterns, in prescribed sequence.

Animal development proceeds from gamete formation and fertilization through cleavage, gastrulation, then organ formation, and finally growth and tissue specialization.

Development cannot proceed properly unless each stage is successfully completed before the next begins.

EARLY MARCHING ORDERS

Information in the Egg Cytoplasm

Why don't you have an arm attached to your nose or toes growing from your navel? The patterning of body parts for any complex animal, including yourself, starts with messages built into the cytoplasm of an immature egg—an **oocyte**—even before a sperm enters the picture. A **sperm**, remember, consists of paternal DNA and a bit of equipment, including a tail, that helps the sperm reach and penetrate an egg. Compared to a sperm, an oocyte is much larger and more complex (Section 10.5).

As an oocyte matures, its volume increases. Enzymes, mRNA transcripts, and other factors become stockpiled in different parts of the cytoplasm and will be activated after fertilization. Typically they take part in early rounds of DNA replication and cell division. Also present are tubulin molecules, plus factors that will govern the angle and timing of tubulin assembly into the microtubules of a mitotic spindle. Such factors influence the pattern of cleavage. Also, the cytoplasm has yolk. The amount and distribution of yolk dictate where cleavage cuts will proceed and how large the blastomeres will be.

Such regionally localized aspects of the oocyte are "maternal messages." We find evidence of their effects as early as fertilization. As an example, a frog egg has pigment granules concentrated near one pole and yolk near the other. The egg's cortex (the plasma membrane and the cytoplasm just under it) undergoes structural reorganization when a sperm fertilizes the egg. Part of the cortex shifts toward the point of sperm entry, and it exposes a crescent-shaped area of yolky cytoplasm. This area of intermediate pigmentation, a **gray crescent**, forms near the frog egg's midsection. It establishes the anterior–posterior axis for the body. Section 33.6 shows the directional planes for this body axis and others.

In itself, a gray crescent is not evidence of regional differences in maternal messages. Such evidence comes from experiments of the sort shown in Figure 44.5. It also comes from observing embryos in which localized cytoplasmic differences are pronounced enough to be tracked easily during development. For example, if you were to continue tracking the development of fertilized frog eggs, you would see that a gray crescent is the site where gastrulation normally begins.

Cleavage—The Start of Multicellularity

Once an egg is fertilized, the zygote enters cleavage. Beneath the plasma membrane, its midsection bears a ring of microfilaments made of the contractile protein actin. The ring tightens as microfilaments slide past one another and pinch in the cytoplasm. The cell surface above the tightening ring is drawn inward as a cleavage furrow (Section 9.4). It is this force of contraction that splits the cytoplasm into two blastomeres. A new ring forms from actin subunits in each daughter cell.

Simply by virtue of where they form, blastomeres end up with different maternal messages. This outcome

Figure 44.5 Examples of experiments that illustrate how the cytoplasm of a fertilized egg has localized differences that help determine the fate of cells in a developing embryo. The cortex of frog eggs contains granules of dark pigment concentrated near one pole. At fertilization, a portion of the granule-containing cortex shifts toward the point of sperm entry. The shift exposes lighter colored, yolky cytoplasm in a crescent-shaped gray area:

pigmented cortex
yolk-rich cytoplasm
sperm penetrating frog egg
gray crescent

The first cleavage normally puts part of the gray crescent in both of the resulting blastomeres.

(**a**) For one experiment, the first two blastomeres that formed were physically separated from each other. Each blastomere still gave rise to a complete tadpole.

(**b**) For another experiment, a fertilized egg was manipulated so the cut through the first cleavage plane missed the gray crescent. Of two resulting blastomeres, only one received the gray crescent. It alone developed into a normal tadpole. The other, deprived of maternal messages present in the gray crescent, gave rise to a ball of undifferentiated cells.

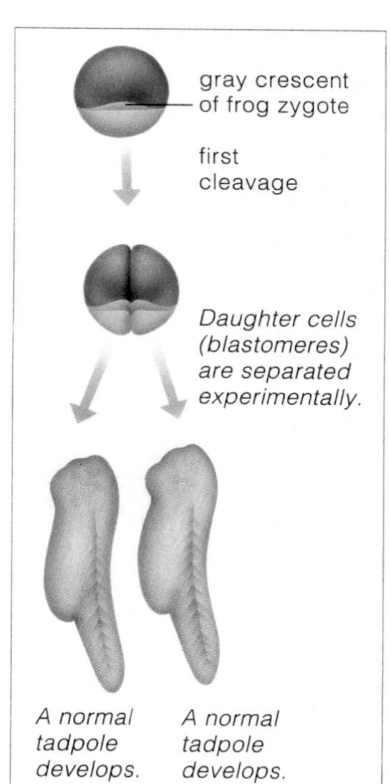

gray crescent of frog zygote
first cleavage
Daughter cells (blastomeres) are separated experimentally.
A normal tadpole develops.
A normal tadpole develops.

a EXPERIMENT 1

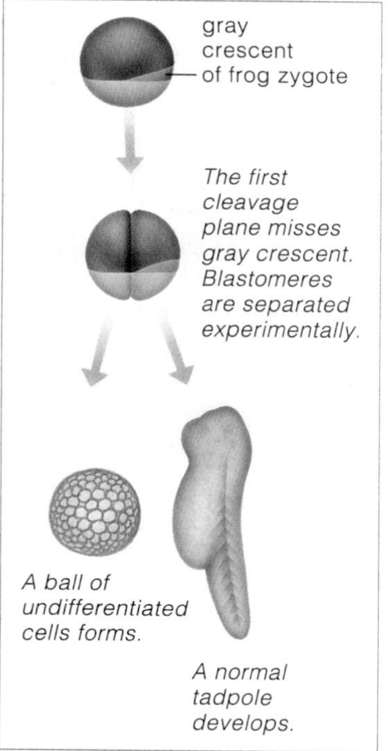

gray crescent of frog zygote
The first cleavage plane misses gray crescent. Blastomeres are separated experimentally.
A ball of undifferentiated cells forms.
A normal tadpole develops.

b EXPERIMENT 2

Further reading: Student Guide to InfoTrac on web site –

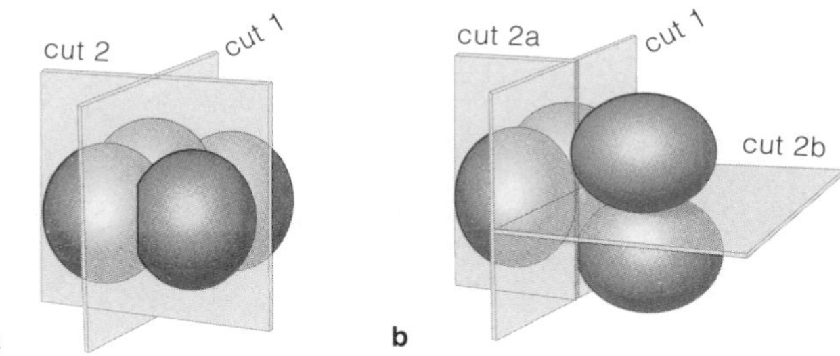
cut 2
cut 1
a
cut 2a
cut 1
cut 2b
b

Figure 44.6 Comparison of early cleavage planes for (**a**) sea urchins and (**b**) mammals. The early cuts of a fertilized egg are radial in sea urchins and rotational in mammals.

Figure 44.7 Identical twins, who started out from the same zygote. The first two blastomeres (formed at cleavage), the inner cell mass, or some other early embryonic stage split. The split was the beginning of two genetically identical, look-alike individuals.

of cleavage, called **cytoplasmic localization**, helps seal the developmental fate of each cell's descendants. Its cytoplasm alone may have molecules of a protein that can activate, say, a gene coding for a certain hormone. And its descendants alone will make that hormone.

Cleavage Patterns

In the simplest cleavage pattern, complete cuts occur after each nuclear division, and each cleavage plane is perpendicular to the mitotic spindle. The cuts parcel out a copy of the nucleus to each blastomere, so all the blastomeres are genetically equivalent. Blastomeres do not grow in size before they divide. Cleavage divides the volume of cytoplasm into increasingly smaller cells.

In most animals, the zygote's genes are silent during early cleavage. How fast the cuts proceed and how the blastomeres become arranged are under the control of the proteins and mRNAs stockpiled in the cytoplasm. In mammals, however, certain genes must be activated first. The proteins specified by those genes take part in cleavage, which cannot be completed without them.

We can correlate cleavage patterns of major animal groups with the amount and distribution of yolk, which typically inhibits cleavage. If an egg has little yolk, the cuts may go right through it. If yolk is concentrated at one end, the early cuts will only go partway through the cytoplasm. Such eggs have polarity. Their yolk-rich end is the *vegetal* pole; the *animal* pole is the end closest to the nucleus. The frog egg in Figure 44.5 is like this.

Cleavage differs among animals with little yolk in their eggs, so heritable factors must influence the cuts. For instance, frog eggs and sea urchin eggs undergo *radial* cleavage; cleavage furrows run horizontally and vertically with respect to the animal–vegetal axis (Figure 44.6*a*). Successive cuts produce a **blastula**, an embryonic stage in which blastomeres enclose a fluid-filled cavity (a blastocoel). A sea urchin egg is nearly yolkless, so the cuts produce horizontal row after row of blastomeres.

Now take a look at Figure 44.4*e*. The yolk of a frog egg impedes the cuts near the vegetal pole. Cleavage is faster and yields more, smaller blastomeres near the animal pole and an offset blastocoel. Between the time that 16 to 64 blastomeres form, any amphibian blastula resembles a mulberry and is called a morula (after the Latin word for mulberry).

What about eggs of reptiles, birds, and most fishes? They undergo *incomplete* cleavage. The large volume of yolk restricts the early cuts to a small, caplike region near the animal pole. The result is two flattened layers of cells with a narrow cavity in between.

What about a mammal's egg? It undergoes *rotational* cleavage, with blastomeres dividing slowly at different times. The first cut is vertical through both poles. For the second cleavage, one blastomere is cut vertically and one horizontally (Figure 44.6*b*). The third cleavage yields a loose arrangement with plenty of space between eight cells. Next, the cells undergo compaction; they abruptly huddle into a compact ball. Tight junctions stabilize the outer cells, which seal off the inner cells. Gap junctions form between the inner cells and enhance rapid chemical communication between them. Descendants of the outer cells form a thin surface layer and secrete fluid into the ball, a fluid-filled cavity forms, and the cells inside mass together against one side of the surface layer. This type of blastula is called a **blastocyst**. In humans and other mammals, the inner cell mass gives rise to the embryo proper. Section 45.7 has a detailed example of this.

Once in awhile, the first two blastomeres, the inner cell mass, or even a later stage split. Any such split may result in *identical twins*, which have the same genetic makeup and share a common placenta (Figure 44.7). By contrast, *fraternal twins* arise from two oocytes that matured and were fertilized during the same menstrual cycle. Each is serviced by its own placenta.

The egg cytoplasm contains maternal instructions in the form of regionally distributed enzymes and other proteins, mRNAs, cytoskeletal elements, yolk, and other factors.

Cleavage divides a zygote into blastomeres, each with a localized part of maternal messages inherent in the egg cytoplasm. This outcome is called cytoplasmic localization.

Differences in the amount and distribution of yolk and other, heritable factors give rise to different patterns of cleavage that affect body plans of different animal groups.

44.4 HOW DO SPECIALIZED TISSUES AND ORGANS FORM?

Nearly all animals have a gut, with tissues and organs that function in digestion and absorption of nutrients. They have surface parts that protect internal parts and detect what is going on outside. In between, most have organs, such as those dealing with structural support, movement, and blood circulation.

This type of three-layer body plan starts to emerge as cleavage ends and gastrulation begins. The embryo's size increases little, if any. But cells start migrating to new positions to form primary tissue layers—ectoderm, endoderm, and mesoderm. Figure 44.8 shows some outer cells of a sea urchin embryo migrating inward to form a lining for a cavity that will become a gut. Figure 44.9 shows the formation of a bird embryo's anterior–posterior axis, the outcome of cell migrations and other rearrangements. In all vertebrates, the anterior–posterior axis defines where the **neural tube**, the forerunner of a brain and spinal cord, will form. All such organs now start to form by cell differentiation and morphogenesis.

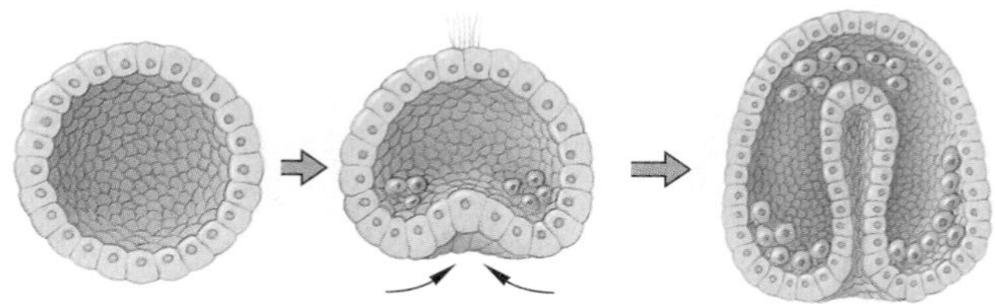
a Blastula
b Cell migrations in early gastrula

endoderm cells
c migrating inward

Figure 44.8 (**a**,**b**) A few early stages of development of a sea urchin (*Lytechinus*), cross-section. After cleavage, the resulting blastula becomes transformed into a gastrula. (**c**) Scanning electron micrograph of cells at the gastrula surface cells. Some cells have begun their inward migration.

Cell Differentiation

All cells of an embryo descend from the same zygote, hence they have the same number and kinds of genes. They all activate genes that specify histones, enzymes of glucose metabolism, and other proteins that are basic to cell survival. However, from gastrulation onward, certain groups of genes are activated in some cells but not others. When a cell selectively activates genes and synthesizes proteins not found in other cell types, we call this process **cell differentiation**. You read about the molecular basis of selective gene expression in Section 15.3. Basically, it results in proteins that are required for distinctive cell structures, products, and functions.

For instance, when your eye lenses developed, some cells started synthesizing crystallin, a family of proteins that became incorporated in transparent fibers of each lens. Only those cells could activate the required genes. Long crystallin fibers formed and forced the cells to lengthen and flatten. Collectively, those differentiated cells provided each lens with unique optical properties. And those crystallin-producing cells are only 1 of 150 or so differentiated cell types now present in your body.

As many experiments tell us, nearly all cells become fully differentiated without loss of genetic information. For example, John Gurdon removed the nucleus from unfertilized eggs of the African clawed frog (*Xenopus laevis*). He also isolated intestinal cells from tadpoles of the same species and ruptured their plasma membrane. He left the nucleus (and most of the cytoplasm) intact,

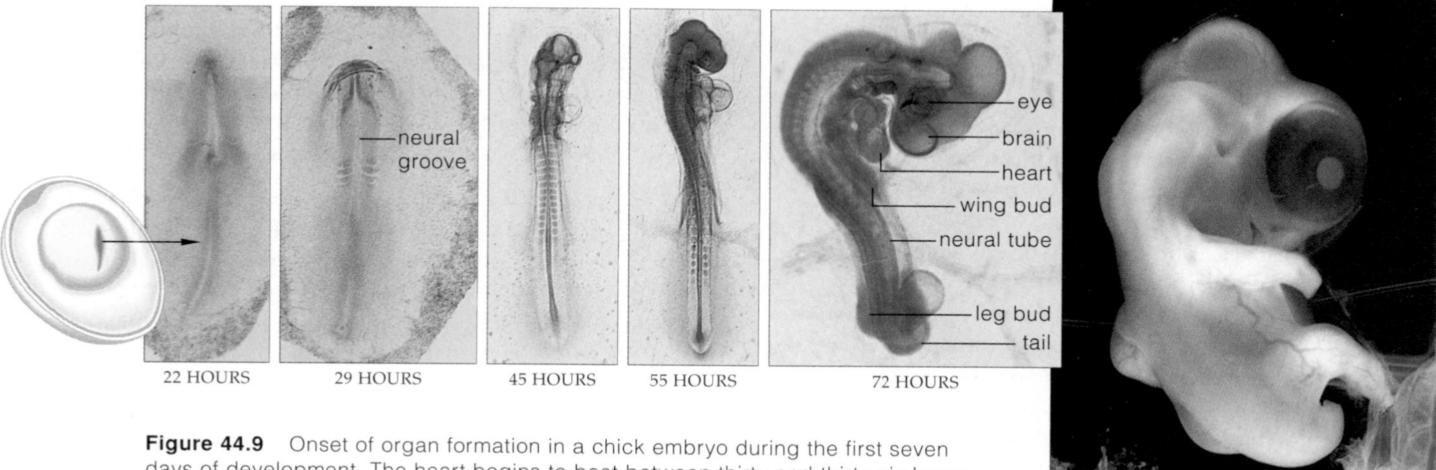
neural groove
eye
brain
heart
wing bud
neural tube
leg bud
tail
22 HOURS
29 HOURS
45 HOURS
55 HOURS
72 HOURS
168 HOURS (SEVEN DAYS OLD)

Figure 44.9 Onset of organ formation in a chick embryo during the first seven days of development. The heart begins to beat between thirty and thirty-six hours. You may have observed such embryos at the yolk surface of raw, fertilized eggs.

Further reading: Student Guide to InfoTrac on web site

then inserted it into the enucleated egg. In some cases the egg gave rise to a complete frog! The intestinal cell had the same number and kinds of genes as the zygote; its nucleus had all the genes required to make all cell types that make up a frog. When it had differentiated into an intestinal cell, those genes had not been lost.

Embryonic human cells also retain the capacity to give rise to a whole individual. This is what happens after spontaneous separation of human blastomeres at the two-cell stage. The split does not result in two half-embryos, but rather in identical twins (Figure 44.7).

Morphogenesis

Morphogenesis refers to a program of orderly changes in an embryo's size, shape, and proportions, the result being specialized tissues and early organs. As part of the program, cells divide, grow, migrate, and change in size. Tissues expand and fold, and the cells in some of them die in controlled ways at prescribed locations.

Consider active cell migration. *Cells send out and use pseudopods that move them along prescribed routes.* When they reach their destination, they establish contact with cells already there. For example, forerunners of neurons interconnect this way as a nervous system is forming.

How do cells "know" where to move? They respond to adhesive cues, as when migrating Schwann cells stick to adhesion proteins on the surface of axons but not blood vessels. And they respond to chemical gradients. Their migrations may be coordinated by the synthesis, release, deposition, and removal of specific chemicals in the extracellular matrix. Adhesive cues also tell cells when to stop. Cells will migrate to regions of strongest adhesion, but once there, further migration is impeded. Section 23.1 describes the chemical-induced migrations among cells of a slime mold, *Dictyostelium discoideum*.

Also, *whole sheets of cells expand and fold inward and outward.* Microtubules lengthen and microfilament rings constrict within the cells. The assembly and disassembly of these components of the cytoskeleton are aspects of a controlled program of changes in cell shape.

Through such controlled, localized events, the size, shape, and proportions of body parts emerge. As one example, take a look at Figure 44.10*a*, which shows what happens after three primary tissues form in the embryos of amphibians, reptiles, birds, and mammals. At the embryo's midline, ectodermal cells elongate to form a neural plate, the first sign that a region of ectoderm is on its way to becoming nervous tissue. Microtubules lengthen in some cells, and other cells become wedge shaped as a microfilament ring contracts and constricts each cell at one end. Collectively, the changes in shape cause tissue flaps to fold over and meet at the midline, thus forming the neural tube (compare Section 45.8).

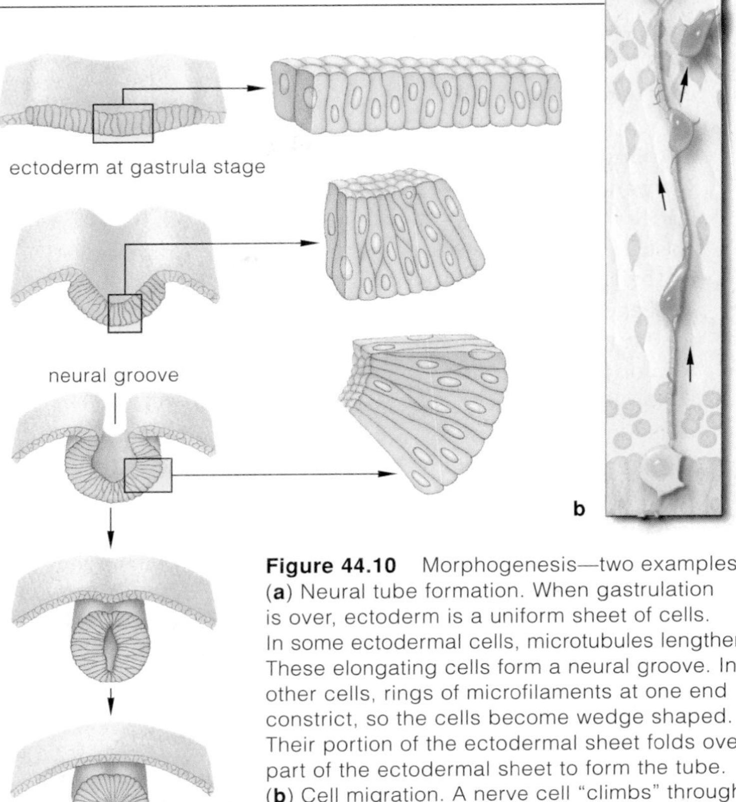
ectoderm at gastrula stage
neural groove
neural tube
a
b

Figure 44.10 Morphogenesis—two examples. (**a**) Neural tube formation. When gastrulation is over, ectoderm is a uniform sheet of cells. In some ectodermal cells, microtubules lengthen. These elongating cells form a neural groove. In other cells, rings of microfilaments at one end constrict, so the cells become wedge shaped. Their portion of the ectodermal sheet folds over part of the ectodermal sheet to form the tube. (**b**) Cell migration. A nerve cell "climbs" through the developing brain to its final position, using a glial cell as its highway.

As a last example, *programmed cell death helps sculpt body parts.* Cells that function for only limited periods in embryonic tissues execute themselves. By this form of cell death, called **apoptosis**, molecular signals from some cells activate weapons of self-destruction already stockpiled in other, target cells. Section 15.6 describes what happens to the targets at the molecular level. For now, simply think about a hand of a human embryo, which first looks like a paddle (Section 9.4). Cartilage models of the digits are forming inside it. Cells in the tissue zones between digits have receptors for certain proteins, which are produced when certain genes are expressed in the embryo. When the receptors bind those proteins, the cells commit mass suicide and the digits separate from one another. On rare occasions, a gene mutation blocks apoptosis and digits stay webbed.

In cell differentiation, a cell selectively uses certain genes and makes certain proteins, not found in other cell types, for distinctive cell structures, products, and functions.

Morphogenesis is a program of orderly changes in body size, shape, and proportions. It results in specialized tissues and organs at prescribed locations, at prescribed times.

Morphogenesis involves cell division, active cell migration, tissue growth and foldings, changes in cell size and shape, and programmed cell death by way of apoptosis.

44.5 EXPERIMENTAL EVIDENCE OF CELL INTERACTIONS

Pattern Formation

In a developing embryo, differences between cells arise in spatially organized ways. Cells divide, differentiate, and live or die; they migrate and stick to like cells in tissues. All of these cells must have a chemical memory of their spatial position in the embryo, for their cellular descendants fill in body details in an orderly pattern. Certain tissues of chick embryos give rise to a pair of wing buds and, at locations lower in the body, to a pair of leg buds. Cells of an amphibian embryo give rise to a head at one end and a tail at the other, and so on.

The developmental fate of each cell in the embryo depends partly on which bit of cytoplasm it acquired during cleavage. This effect of cytoplasmic localization is especially predominant in *Drosophila*, for it stamps each cell with its final identity (Section 44.6). Sidney Brenner refers to this as the "European style" of cell determination, in which ancestry is everything.

By contrast, in echinoderm and vertebrate embryos, an "American style" of cell determination predominates in which the emphasis is on the neighborhood. That is, the fate of each cell depends more on whom it meets up with in the embryo. Such an embryonic cell retains the capacity to alter its fate, for it contains all of the instructions required to give rise to a whole animal. (Remember how two blastomeres at the two-cell stage of a human embryo can give rise to identical twins?)

Either way, once a cell is locked in at a given place in the embryo, it has chemical memory; it selectively reads some genes but not others. Now it produces and secretes molecular signals that change the activities of its neighbors, and it responds to neighborhood signals.

A cell's memory is passed on to all its descendants, and its descendants go on to form tissues in the places where we expect them to form. Now we see evidence of **embryonic induction**: the developmental fates of the embryonic cell lineages change upon exposure to gene products from adjacent tissues. Through these changes, specialized tissues and organs emerge from clumps of embryonic cells, and they do so in an ordered, spatial pattern. We call their emergence **pattern formation**.

Cases of Embryonic Induction

Researchers have evidence of cell memory. In one set of experiments, they modified the wing buds of chick embryos. Normally, mitotic cell divisions in mesoderm and the overlying ectoderm produce buds at a site in the embryo where a wing is supposed to grow. When they grow and lengthen, ectodermal cells form an **AER**, a ridge at the bud apex. An AER is a population of self-perpetuating cells that do not mingle with cells around them. It forms on newly developing limbs of mammals as well as birds. Rapid divisions of mesodermal cells beneath an AER generate the tissues that elongate the developing limb. The newest cells push out others from this location and give rise to tissues of the limb's outermost parts. Mesodermal cells pushed out early give rise to bones closest to the body's trunk. Cells pushed out much later give rise to the outermost bones and to other limb structures.

When researchers removed an AER from a bud, they arrested the development of that body structure (Figure 44.11*a*). When they grafted a bit of extra AER into a bud, extra wing parts formed (Figure 44.11*b*).

In other experiments, researchers excised mesoderm from a leg bud and grafted it below ectoderm of a

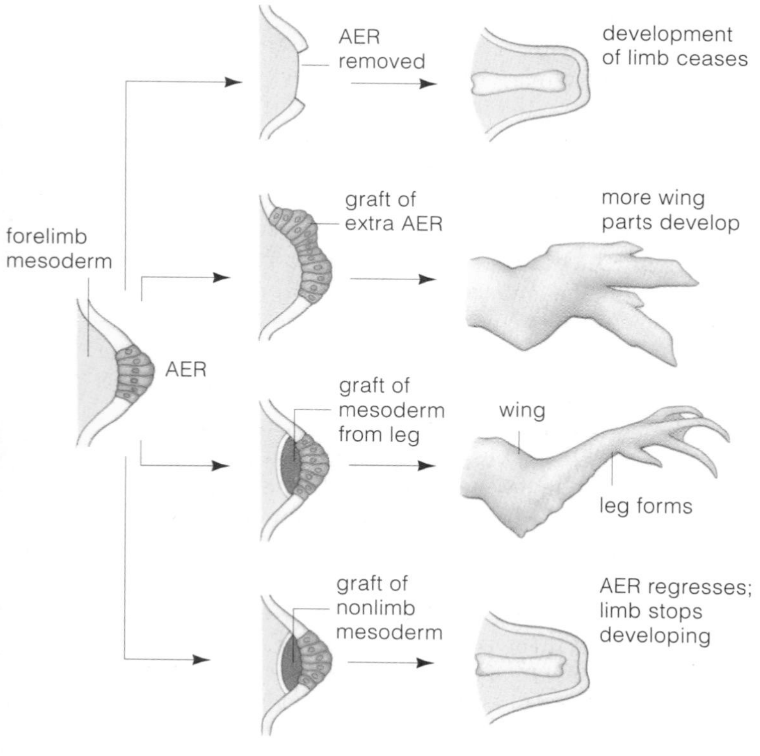

a Remove a wing bud's AER and its development stops.

b Graft extra AER onto a bud and the wing that forms has extra parts.

c Graft a bit of leg mesoderm beneath the AER and part of a leg (even some toes) develops.

d Graft nonlimb mesoderm below the AER, and the AER regresses and wing development stops.

Figure 44.11 Experimental evidence of an interaction between mesoderm and ectoderm in the development of a chick wing from a wing bud. AER is a narrow ridge of self-perpetuating cells at the bud's apex.

Further reading: Student Guide to InfoTrac on web site

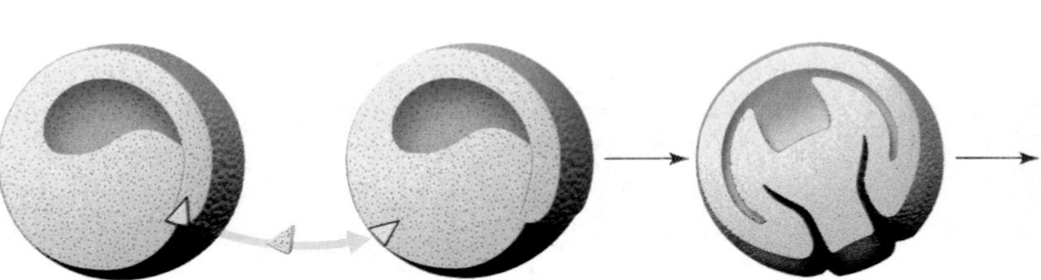
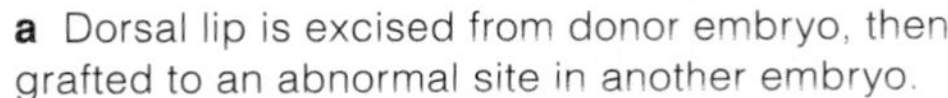
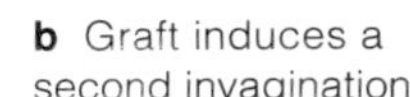
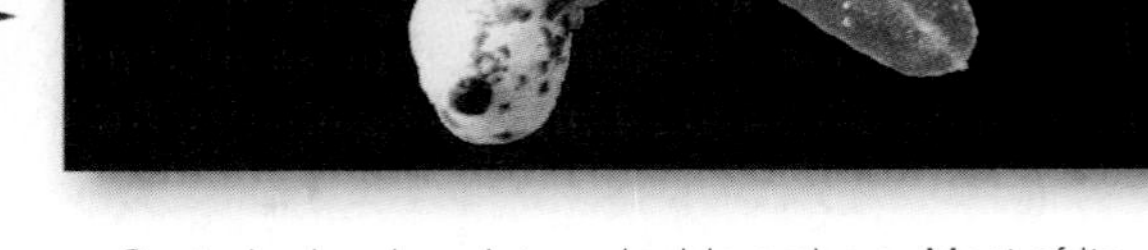

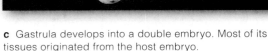

a Dorsal lip is excised from donor embryo, then grafted to an abnormal site in another embryo.

b Graft induces a second invagination.

c Gastrula develops into a double embryo. Most of its tissues originated from the host embryo.

Figure 44.12 Experimental evidence that the dorsal lip of an axolotl embryo triggers and controls gastrulation. (An axolotl is an amphibian.) When transplanted to a different location in another amphibian embryo, it organized the formation of a second set of body structures.

wing bud. As Figure 44.11*c* shows, a leg—not a wing—formed at that location. They also grafted mesoderm from part of an embryo where legs normally would not develop. The nonlimb mesoderm instructed ectoderm located above it *not* to develop a wing; development of the wing stopped (Figure 44.11*d*).

As another example, researchers excised the dorsal lip of a normal amphibian embryo (an axolotl). For this experiment, they grafted its dorsal lip at a different site in another amphibian embryo. Gastrulation proceeded at the recipient's own dorsal lip *and* at the graft site. Siamese twins—a double embryo with two sets of body parts—developed. Figure 44.12 shows this outcome.

Hans Spemann, a pioneer in developmental biology, gave us our first experimental evidence of embryonic induction. Through his studies of amphibian embryos, he suspected that chemical signals arising in the dorsal lip were coordinating cell movements at gastrulation and maybe the pattern by which the neighboring tissues become specialized. Experiments that produced double embryos strongly indicated that the dorsal lip is central to organizing the body's main axis. (How else would the experiment shown in Figure 44.12 produce a two-headed, two-tailed tadpole?)

In time, the dorsal lip came to be called Spemann's Organizer. It was the first embryonic signaling center to be discovered.

Embryonic Signals in Pattern Formation

During cleavage, cytoplasmic localization ensures that different chemical signals will operate within different cells. While the embryo develops, cells that have been neighbors since the beginning interact through chemical signals. So do cells that meet during gastrulation and other morphogenetic movements.

For example, Spemann's Organizer is the source of a morphogen. **Morphogens** are degradable molecules that diffuse from an embryonic signaling center into adjoining tissues. The concentration gradient helps cells assess their position in the body, and it influences their differentiation. The signal is strongest at the start of the gradient and weakens with distance. As a result, cells at different positions along the gradients are exposed to different chemical information that guides the selective expression of different genes in different body regions.

Morphogens are long-range signals; some diffuse from one end of the embryo to another. Other signals are short range, involving cell-to-cell contacts. To give an example, changes in the expression of the genes for cadherins, integrins, and other glycoproteins at the cell surface lead to changing patterns of how cells associate with one another during gastrulation and early organ formation. When a cell starts producing a cytoskeletal protein, it might lengthen or move in some direction. When adhesion proteins at its surface change, the cell may cohere with its neighbors, break free from them, or become segregated from cells of an adjoining tissue. Such chemical signals guide cells that migrate through the embryo. Most vertebrate tissues are a mosaic of cell types as a result of such invasions.

Also, cells respond differently to signals depending on when they receive them. Graft a bit of animal pole epithelium from an early gastrula over a rudimentary eye of a later embryo, and it gives rise to something like neural tube tissue. Leave it alone in vitro for a few hours before the graft, and it differentiates into a lens. Leave it even longer in vitro and it will not respond at all to the inductive signals from the eye tissue.

Pattern formation is the orderly, sequential sculpting of embryonic cells into specialized tissues and organs. Cells differentiate partly through cytoplasmic localization and, later, according to their position in the developing embryo.

The pattern starts with different chemical signals inside blastomeres with different maternal messages. Cell-to-cell inductive interactions fill in the details of the pattern.

Morphogens and other positional signals are long-range and short-range beacons that help cells assess their position in the embryo. Cells respond differently to the signals at different times of development.

44.6 TO KNOW A FLY

"Although small children have taboos against stepping on ants because such actions are said to bring on rain, there has never seemed to be a taboo against pulling off the legs or wings of flies. Most children eventually outgrow this behavior. Those who do not either come to a bad end or become biologists."

So wrote Vincent Dethier in *To Know a Fly*, a whimsical yet scientifically sound tribute to flies and their suitability for controlled laboratory experiments. Even before Dethier published his booklet in 1962, plenty of geneticists and developmental biologists had already come to the same conclusion. *Drosophila melanogaster* was and still is the fly of choice. It costs almost nothing to feed or house this tiny insect. It reproduces rapidly in bottles, it has a short life cycle, and disposing of spent bodies is a snap (Section 12.4).

For much of this century, the embryonic development of *Drosophila* has been studied in detail at the anatomical, cytological, biochemical, and genetic levels. The findings tell us much about the embryonic development and even the evolutionary history of animals in general. They tell us something about our own development, even though fruit fly and mammalian embryos proceed through the stages of development according to different patterns.

SUPERFICIAL CLEAVAGE IN *DROSOPHILA* For instance, mammalian eggs contain relatively little yolk, so cleavage furrows can cut through the entire egg. But the eggs of *Drosophila* (and those of most insects, fishes, reptiles, and birds) contain a large mass of yolk that restricts cleavage to a small portion of egg cytoplasm. The yolk mass of a *Drosophila* egg is centrally located and confines cleavage to the cytoplasm's periphery. This is one example of a *superficial* cleavage pattern.

Compared to a mammalian egg, which undergoes slow cuts every twelve to twenty-four hours, the superficial cleavage of a fertilized *Drosophila* egg is rapid. The egg nucleus repeatedly divides but the cytoplasm does not, forming what is called a syncytial blastoderm (Figure 44.13). Nuclear division proceeds every eight minutes until 256 nuclei crowd together in the yolky cytoplasm. This embryonic stage is a type of cellular blastoderm. Each nucleus in it has its own tiny array of microtubules and microfilaments, which will influence the forthcoming patterns of cell formation and elongation.

Most of the nuclei migrate toward the egg periphery. First, the plasma membrane folds inward around each nucleus that reaches the egg's posterior pole axis. The

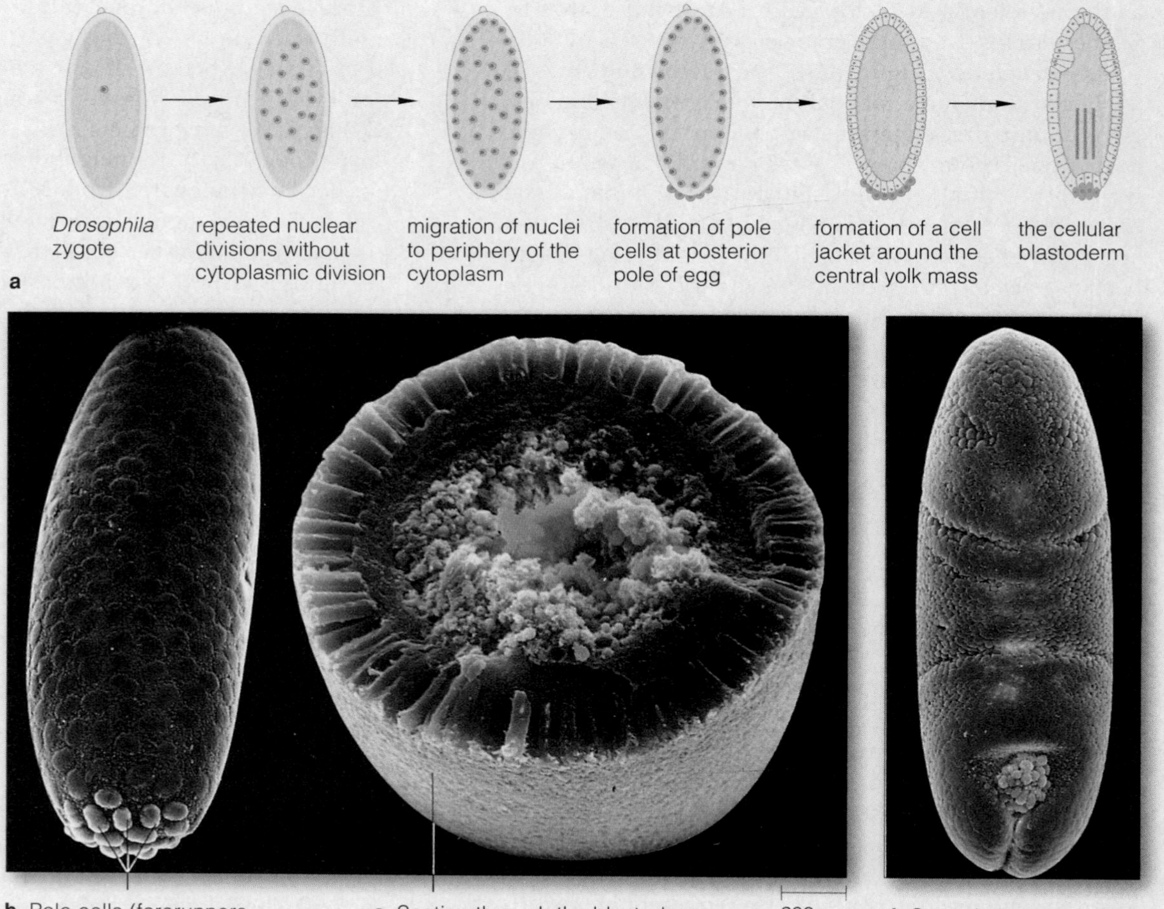

a
Drosophila zygote
repeated nuclear divisions without cytoplasmic division
migration of nuclei to periphery of the cytoplasm
formation of pole cells at posterior pole of egg
formation of a cell jacket around the central yolk mass
the cellular blastoderm
b Pole cells (forerunners of germ cells of the adult)
c Section through the blastoderm showing jacket of cells around yolk
200 μm
d Gastrulation, well under way

Figure 44.13 Superficial cleavage in the yolk-rich egg of *Drosophila*.

(**a**) A fertilized egg's nucleus repeatedly divides. Cytoplasmic division is delayed until after the nuclei move to the periphery of the cytoplasm.

(**b**) Pole cells form. (**c**,**d**) Then other cells form; they are arrayed like a jacket around the central mass of yolk. This early stage of development is a type of blastoderm.

As the blastoderm forms, polar granules in the cytoplasm migrate and become isolated in pole cells that form at one end of the blastoderm. Pole cells are forerunners of germ cells that give rise to eggs or sperm in the adult fly.

Further reading: Student Guide to InfoTrac on web site —

resulting pole cells are forerunners of germ cells of the adult fly. Physical separation of the future reproductive cells from the rest of the blastoderm is one of the first events of insect development. After separation occurs, the plasma membrane folds inward and forms a partition around each remaining nucleus and a bit of cytoplasm. Within four hours after fertilization, about 6,000 cells are arranged like a jacket around the yolky mass.

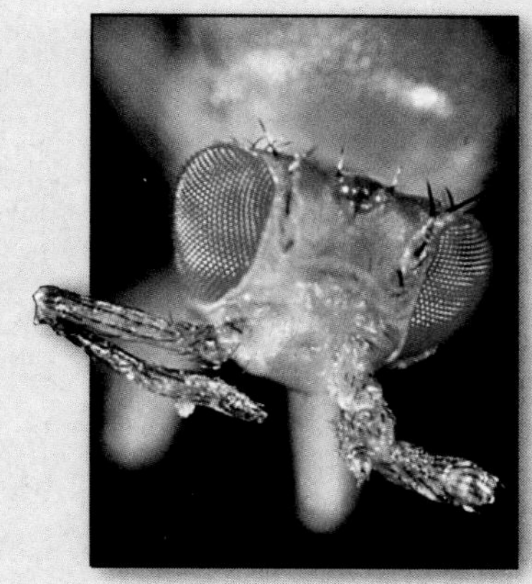

Figure 44.14 Experimental evidence that certain genes control the development of body parts. In a normal *Drosophila* larva, a cluster of cells is the embryonic source of antennae on the head. If the larva carries a certain mutant form of the *antennapedia* gene, the adult fly will have legs instead of antennae on its head.

MASTER ORGANIZER GENES During the late 1940s, Edward Lewis of the California Institute of Technology discovered that the products of certain mutated genes introduce bizarre blips in pattern formation in *Drosophila*. To give an example, as a result of a single gene mutation, a leg grew out from the head region where an antenna should have grown, as the photograph in Figure 44.14 shows

Look at the *Drosophila* zygote in Figure 44.15*a*. A **fate map** of its surface is a map that shows where each kind of differentiated cell in the forthcoming adult will originate. The embryo develops a head and a tail. A series of body segments forms in between, each with a gene-specified identity. In an adult fly, for example, the first segment (part of the thorax) has only legs, the next has legs and wings, the next has legs and balancing devices, and so on.

Researchers have put together a model to explain how polarity in the *Drosophila* egg gives rise to segmented polarity of the adult body. First, *maternal effect* genes that specify regulatory proteins are transcribed, translated, or both (Figure 44.15*b*). The resulting mRNAs and proteins get localized in different portions of the egg cytoplasm. These gene products are switched on in the zygote. They activate or suppress *gap* genes, which map out broad regions of the body. When the product of one gap gene (hunchback) diffuses through the syncytial blastoderm, it creates a chemical gradient that influences the extent to which different gap genes are expressed. The resulting differences in concentrations of gap gene products switch on *pair-rule* genes. The products of these genes accumulate in bands (stripes) corresponding to two body segments. They activate *segment polarity* genes, the products of which divide the embryo into segment-sized units.

Interactions among the products of the gap, pair-rule, and segment polarity genes control expression of another class of genes—*homeotic* genes, which collectively govern the developmental fate of each segment. When mutated, homeotic genes can transform one body segment into the likeness of another. Thus mutations in the *antennapedia* gene result in abnormal responses to regulatory proteins. In the example cited earlier, a mutated gene product activated the wrong set of homeotic genes in cells that were supposed to produce two antennae on the head. Their expression gave rise to a pair of legs instead.

Researchers do not yet have a complete understanding of the multiple levels of gene interactions that specify body parts in *Drosophila*. But their studies of this tiny organism give us fascinating glimpses into the kinds of regulatory mechanisms that must come into play in other organisms as well. And think about *that* next time you swat at fruit flies, each not much bigger than a rounded pinhead, as they hover over a bowl of pungently ripe fruit.

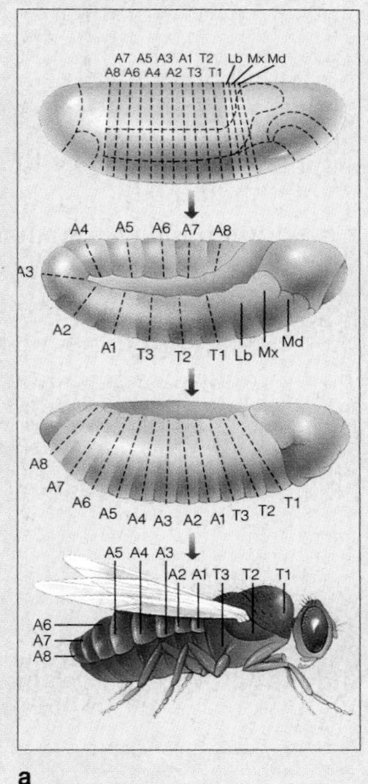

A7 A5 A3 A1 T2 Lb Mx Md
A8 A6 A4 A2 T3 T1
A4 A5 A6 A7 A8
A3
A2
A1 T3 T2 T1 Lb Mx Md
A8
A7
A6 A5 A4 A3 A2 A1 T3 T2 T1
A5 A4 A3
A2 A1 T3 T2 T1
A6
A7
A8
a

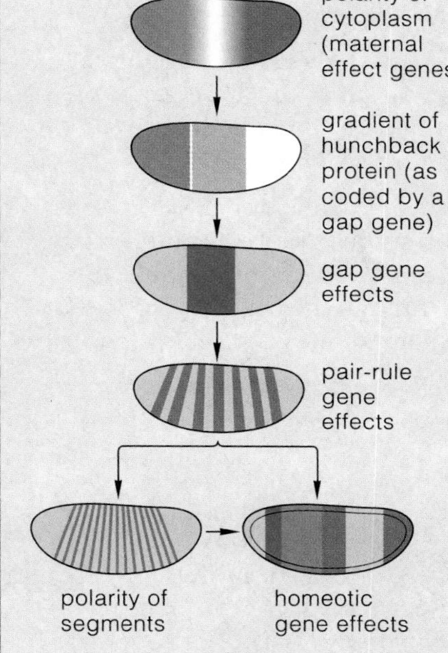

polarity of cytoplasm (maternal effect genes)
gradient of hunchback protein (as coded by a gap gene)
gap gene effects
pair-rule gene effects
polarity of segments
homeotic gene effects
b

Figure 44.15 (**a**) Fate map of a *Drosophila* zygote. The dashed lines indicate the regions that normally will become different segments of the body, each with specialized parts. For this organism, each segment's fate is sealed at the time of cytoplasmic localization. (**b**) A generalized model of pattern formation in *Drosophila*, as explained in the text.

AN UNDERLYING UNITY IN ANIMAL DEVELOPMENT

A Theory of Pattern Formation

Reflect now on the preceding sections, which should have left you with a general sense of how the animal body develops according to plan. As you have seen, different cells emerge in a regular spatial pattern. So at this point in your reading, you might well ask: Do the same classes of genes govern that spatial patterning in all animals? It now appears so. After studies of many diverse animals, researchers have put together a **theory of pattern formation**. Here are its key points:

1. The formation of embryonic cells in ordered, spatial patterns usually starts with cytoplasmic localization and proceeds by gene-directed cell-to-cell interactions.

2. As the embryo develops, classes of master genes are activated in orderly sequence, at prescribed times.

3. Interactions among the master genes are guided by regulatory proteins. They result in the appearance of different gene products that are spatially organized relative to one another in the embryo.

4. Different genes are activated and suppressed in cells along the embryo's anterior–posterior axis and dorsal–ventral axis. Certain protein products of this selective gene expression diffuse through the embryo and create chemical gradients that help define each cell's identity.

5. **Homeotic genes**, a class of master genes, specify the development of specific body parts.

As you read in the preceding sections, morphogens and other inducers switch on whole blocks of genes in sequence. Their targets include homeotic genes, which have now been identified in animals as evolutionarily distant as roundworms and vertebrates. When organs first form, these gene products interact with regulatory elements. Together, they activate and inhibit blocks of genes—and they do so in similar ways among major animal groups. They direct inductions that map out the overall body plan, including its major axes. If master genes fail to do so when organs are forming, disaster follows, as when a heart forms at the wrong location. Once expression of master genes has locked in the basic body plan, the many inductions remaining have only localized effects. As one example, if a lens fails to form, then only the eye will be affected.

Evolutionary Constraints on Development

How long have the master genes been operating? Take a look at the velvet worm in Figure 44.16. Its basic body plan has not changed much for about 500 million years. Neither has the plan for sponges, flatworms, mollusks, flies, vertebrates, and all other major animal groups. Many millions of animal species have appeared on the evolutionary stage—but they are only variations on a few dozen basic plans that were set by Cambrian times.

Why is this so? Apparently there are very few new master genes, so variations in body plans may be more

Figure 44.16 Onychophoran, of a small phylum commonly known as velvet worms (for their velvety integument) or walking worms (they have walking legs and a wormlike body). Seventy or so species live in Central America, the Caribbean, Central Africa, Malaysia, parts of Chile, South Africa, Australia, New Guinea, and New Zealand. Their wide distribution suggests their common ancestor evolved before Gondwana broke up. A spine-bearing velvet worm is even represented in the Burgess shale fossils from Cambrian times. The body plan of some existing relatives has not changed much for millions of years.

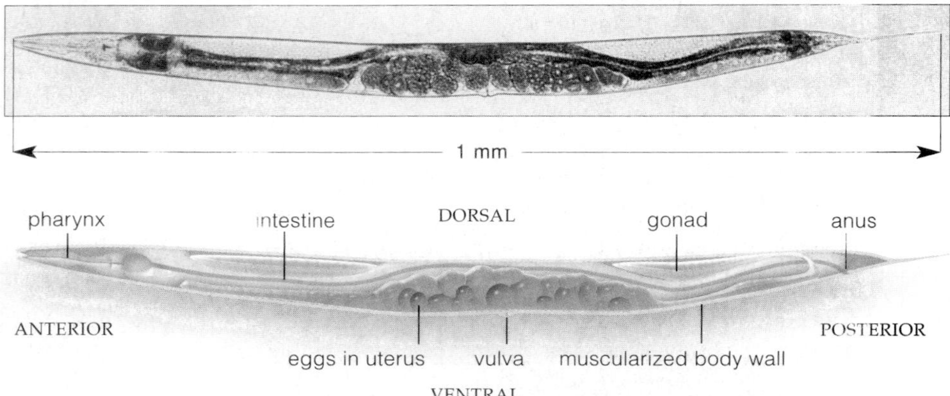

1 mm
pharynx
intestine
DORSAL
gonad
anus
ANTERIOR
POSTERIOR
eggs in uterus
vulva
muscularized body wall
VENTRAL

Figure 44.17 Micrograph and body plan of a roundworm, *Caenorhabditis elegans*. More complex animals have the same basic body plan (cephalized, bilateral, with a complete digestive system and the same basic tissues). The individual shown is a hermaphrodite that can cross-fertilize or fertilize itself. The ability to produce clones of oneself has proved to be wonderfully convenient for genetic studies. So have the worm's transparency, small size, short generation time, and small genome (19,900 genes, only 3,000 of which are vital for normal structure and functioning). An adult body only has about 1,000 somatic cells and 1,000 to 2,000 germ cells. Researchers deciphered each cell's lineage, from egg to adult, by using electron microscopy and serial sections at different times of development.

Further reading: Student Guide to InfoTrac on web site →

of an outcome of how control genes control each other. For example, the *eyeless* gene regulates eye formation in fruit flies. Humans have a nearly identical gene which, in mutated form, results in eyeless babies. As another example, the *distalless* gene in fruit flies causes legs to grow from leg buds. Sean Carroll and other biologists found that the same gene determines the fate of cells that give rise to lobster legs, beetle legs, butterfly wings, sea star arms, fish fins, and mouse feet. As you know, legs, fins, and wings are very different structures with different functions. For example, beetle legs have their skeletal structures on the outside, your legs have them on the inside. Yet the structures all start out as buds from the main body axis. Similarly, insect, squid, and vertebrate eyes differ greatly in their structure, yet the genes that govern their formation are nearly identical.

As a final example, all 19,900 genes of *Caenorhabditis elegans* are decoded (Figure 44.17). About 70 percent of the genes of this roundworm are identical or similar to many of your genes. Biologists know how all 957 of its cells develop in sequence and interact. And its organ systems and yours operate in the same basic ways.

And so, rather than starting from scratch with new genes, diverse animals are using identical or similar master genes to control development. By analogy, they have old software programs in more recent computers.

For a long time, we have known about the *physical* constraints (such as the surface-to-volume ratio) and *architectural* constraints (as imposed by body axes). The kinds of events described in this chapter indicate there also are *phyletic* constraints on change. These are the constraints imposed on each lineage by interactions of the master organizer genes, which operate when organs form and control induction of the basic body plan.

That might be why we have so many species and so few body plans. Once master genes had evolved and were engaged in intricate interactions, it would have been difficult to change the body's basic characteristics without destroying the embryo. Mutations have indeed added marvelous variations to animal lineages. But the basic body plans have prevailed through time; maybe it simply has proved unworkable to start all over again.

Inductive interactions among classes of master genes map out the basic body plan and specify where and how body parts develop. Products of their selective expression create chemical gradients in the embryo that help seal each cell's developmental fate.

There are physical, architectural, and phyletic constraints on the evolution of animal body plans. There are millions of species, but only a few basic body plans.

The master genes that guide development are similar or identical in all major animal groups.

44.8 FROM THE EMBRYO ONWARD

Now that you have a sense of how embryos develop, think about what goes on after they are released or hatched from parents. In many species, a new individual is a **juvenile**, a miniaturized version of the adult that changes in size and proportion until it reaches sexual maturity. In other species, **metamorphosis** occurs: the body form changes during the transition to an adult. Metamorphosis involves size increases, reorganization of tissues, and remodeling of body parts. It is under hormonal control. Section 37.8 gives examples of this transitional time for crustaceans and an insect.

Section 26.19, on insect metamorphosis, gives some examples of transitional stages called larvae, nymphs, and pupae. Some insect nymphs are miniature adults that go through **molting** episodes in which they shed a hardened cuticle and grow rapidly before a new cuticle hardens. Some insects show *incomplete* metamorphosis; they undergo gradual, partial change between the first and last molt. Bugs are like this. *Drosophila* is one of the insects that show *complete* metamorphosis. Tissues of immature stages are destroyed and replaced before the adult emerges (Section 44.6).

Some insect predators metamorphose too soon or not at all after exposure to plant toxins. Either way, they die before adulthood. Harvard researchers found this out when they tried raising a bug (*Pyrrhocoris apterus*) on paper in petri dishes. The larvae metamorphosed normally when raised on European paper (including pages torn from the journal *Nature*) but could not when raised on American paper (such as the journal *Science*). Later on, the researchers discovered that the American paper was manufactured from balsam fir, which is a North American species. And balsam fir synthesizes a toxin that resembles juvenile hormone. Together with the active form of ecdysone, juvenile hormone is one of the major signaling molecules in insect development. (Here you may wish to compare Section 37.8.)

Amphibians undergo striking metamorphic change that affects nearly all organs (Figure 44.1). The thyroid hormones thyroxine and triiodothyronine bring about transformation of a tadpole into the adult. Increasing hormone levels cause some immature structures, such as the tail, to degenerate. In addition, they trigger further development of other organs, including the eyes.

As you probably have noticed, metamorphosis is not a characteristic of the human life cycle. We undergo gradual proportional changes in the body, and that is a topic of Section 45.11.

After embryonic development, different animals show variation in patterns of further development—as when amphibians metamorphose and humans simply undergo changes in size and proportion to reach adulthood.

44.9 WHY DO ANIMALS AGE?

Every multicelled animal that undergoes extensive cell differentiation gradually deteriorates. Through processes collectively called **aging**, its cells gradually break down in structure and function, which leads to the decline of tissues, organs, and eventually the whole body.

The Programmed Life Span Hypothesis

Does an internal, biological clock control aging? After all, each species has a maximum life span. For example, we know the maximum is 20 years for dogs, 12 weeks for butterflies, and 35 days for fruit flies. No human has lived past 122 years. The consistency within species indicates that genes have something to do with aging. Indeed, researchers have doubled the life span of the fruit fly by manipulating its genes.

Suppose the animal body is like a clock shop, with each type of cell, tissue, and organ ticking at its own genetically set pace. Many years ago, Paul Moorhead and Leonard Hayflick investigated such a possibility. They cultured normal human embryonic cells, all of which divided about fifty times, then died off. Hayflick also withdrew cultured cells that were partway through the series of divisions and froze them for several years. After he thawed the cells and placed them in a culture medium, they proceeded to complete the in vitro cycle of fifty doublings and died on schedule.

No cell in the human body divides more than eighty or ninety times. You might well be wondering: If an internal clock does tick off their life span, then how can *cancer* cells go on dividing? The answer gives us a clue to why *normal* cells can't beat the clock.

As you know, cells duplicate their chromosomes before they divide. Capping the ends of chromosomes are repetitive DNA sequences called **telomeres**. A bit of each telomere is lost during each nuclear division, and when only a nub remains, cells stop dividing and die. The exceptions are cancer cells and reproductive cells. Both make telomerase—an enzyme that causes telomeres to lengthen. In recent experiments, somatic cells exposed to the enzyme go on dividing well past their normal life span, with no apparent bad effects.

The Cumulative Assaults Hypothesis

A different hypothesis holds that aging is an outcome of ongoing damage at the molecular and cellular levels. According to this hypothesis, damage accumulates in DNA because of ongoing environmental assaults and a decline in its self-repair mechanisms.

For example, free radicals, those rogue, short-lived molecular fragments you read about in Section 6.8, bombard DNA and all other biological molecules. This includes the DNA of mitochondria, the power plants of eukaryotic cells. Structural changes in DNA endanger the synthesis of functional enzymes and other proteins required for normal life processes. And free radicals are implicated in many age-related problems, such as atherosclerosis, cataracts, and Alzheimer's disease.

Problems in DNA replication and repair operations also have been implicated in aging. Researchers have correlated *Werner's syndrome*, an aging disorder, with a mutation in the gene coding for one of the helicases. These enzymes have roles in unwinding the nucleotide strands from each other. The mutated helicase probably does not compromise the replication of DNA; affected people don't die immediately. (Starting in their thirties, they age abnormally fast and die before reaching fifty.) But its nonmutated form might be essential for repair. The DNA of people with Werner's syndrome seems to accumulate mutations at above-normal rates. Sooner or later, the damage may interfere with cell division. Like skin cells from very old folks, skin cells from Werner's patients do not divide many times at all.

It may be that both hypotheses have merit. Aging may involve many interconnected processes in which genes, hormones, environmental insults, and a decline in DNA repair mechanisms come into play. Consider how living cells of all tissues require exchanges of materials between cytoplasm and extracellular fluid. Collagen is a structural component of many connective tissues. If something shuts down or mutates collagen-encoding genes, the missing or altered gene product may disrupt the flow of oxygen, nutrients, hormones, and so forth to and from cells through all those connective tissues. And so the repercussions would ripple through the body.

Or consider mutations in the genes for membrane proteins that serve as self markers. If markers change, do T cells of the immune system perceive the body's cells as foreign and attack them? If such autoimmune responses were to increase over time, they certainly would promote the greater vulnerability to disease and stress associated with old age.

In evolutionary terms, reproductive success means surviving long enough to produce offspring. Humans can reach sexual maturity *and* help their own children reach adulthood in thirty years. We don't *need* to live longer. But we among all animals have the capacity to think about it, and we generally have decided that we like life better than the alternative. Eventually, though, even the most tenacious clingers to life must confront their own mortality; and perhaps in time we all might learn to accept this with wisdom and grace.

Aging may result from many factors, including an internal clock that ticks out the life spans of cells, tissues, and organs, and an accumulation of errors and damage to DNA.

Further reading: Student Guide to InfoTrac on web site

44.10

Commentary

DEATH IN THE OPEN

Everything in the world dies, but we only know about it as a kind of abstraction. If you stand in a meadow, at the edge of a hillside, and look around carefully, almost everything you can catch sight of is in the process of dying, and most things will be dead long before you are. If it were not for the constant renewal and replacement going on before your eyes, the whole place would turn to stone and sand under your feet.

There are some creatures that do not seem to die at all; they simply vanish totally into their own progeny. Single cells do this. The cell becomes two, then four, and so on, and after a while the last trace is gone. It cannot be seen as death; barring mutation, the descendants are simply the first cell, living all over again. . . .

There are said to be a billion billion insects on the Earth at any moment, most of them with very short life expectancies by our standards. Someone estimated that there are 25 million assorted insects hanging in the air over every temperate square mile, in a column extending upward for thousands of feet, drifting through the layers of atmosphere like plankton. They are dying steadily, some by being eaten, some just dropping in their tracks, tons of them around the Earth, disintegrating as they die, invisibly.

Who ever sees dead birds, in anything like the huge numbers stipulated by the certainty of the death of all birds? A dead bird is an incongruity, more startling than an unexpected live bird, sure evidence to the human mind that something has gone wrong. Birds do their dying off somewhere, behind things, under things, never on the wing.

Animals seem to have an instinct for performing death alone, hidden. Even the largest, most conspicuous ones find ways to conceal themselves in time. If an elephant missteps and dies in an open place, the herd will not leave him there; the others will pick him up and carry the body from place to place, finally putting it down in some inexplicably suitable location. When elephants encounter the skeleton of an elephant in the open, they methodically take up each of the bones and distribute them, in a ponderous ceremony, over neighboring acres.

It is a natural marvel. All of the life on earth dies, all of the time, in the same volume as the new life that dazzles us each morning, each spring. All we see of this is the odd stump, the fly struggling on the porch floor of the summer house in October, the fragment on the highway. I have lived all my life with an embarrassment of squirrels in my backyard, they are all over the place, all year long, and I have never seen, anywhere, a dead squirrel.

I suppose that is just as well. If the Earth were otherwise, and all the dying were done in the open, with the dead there to be looked at, we would never have it out of our minds. We can forget about it much of the time, or think of it as an accident to be avoided, somehow. But it does make the process of dying seem more exceptional than it really is, and harder to engage in at the times when we must ourselves engage.

In our way, we conform as best we can to the rest of nature. The obituary pages tell us of the news that we are dying away, while birth announcements in finer print, off at the side of the page, inform us of our replacements, but we get no grasp from this of the enormity of the scale. There are now 4 billion of us on the Earth, and all 4 billion must be dead, on a schedule, within this lifetime. The vast mortality, involving something over 50 million each year, takes place in relative secrecy. We can only really know of the deaths in our households, among our friends. These, detached in our minds from all the rest, we take to be unnatural events, anomalies, outrages. We speak of our own dead in low voices; struck down, we say, as though visible death can occur only for cause, by disease or violence, avoidably. We send off for flowers, grieve, make ceremonies, scatter bones, unaware of the rest of the 4 billion on the same schedule. All of that immense mass of flesh and bone and consciousness will disappear by absorption into the earth, without recognition by the transient survivors.

Less than half a century from now, our replacements will have more than doubled in numbers. It is hard to see how we can continue to keep the secret, with such multitudes doing the dying. We will have to give up the notion that death is a catastrophe, or detestable, or avoidable, or even strange. We will need to learn more about the cycling of life in the rest of the system, and about our connection in the process. Everything that comes alive seems to be in trade for everything that dies, cell for cell. There might be some comfort in the recognition of synchrony, in the information that we all go down together, in the best of company.

— LEWIS THOMAS, 1973

44.11 SUMMARY

1. For animals, sexual reproduction is the dominant reproductive mode. It requires specialized reproductive structures, control mechanisms, and forms of behavior that assist fertilization and support the offspring.

2. Among animals, embryonic development commonly proceeds through six stages:

a. Gamete formation, when oocytes (immature eggs) and sperm form and develop in reproductive organs. Molecular and structural components are stockpiled and become localized in different parts of the oocyte.

b. Fertilization, from the time a sperm penetrates an egg to the fusion of sperm and egg nuclei that results in a zygote (fertilized egg).

c. Cleavage, when mitotic cell divisions transform a zygote into smaller cells called blastomeres. Cleavage doesn't increase the egg's original volume of cytoplasm; it simply increases the number of embryonic cells. It regionally divides yolk, mRNAs, proteins, cytoskeletal elements, and other "maternal messages" among new blastomeres, an outcome called cytoplasmic localization.

d. Gastrulation, when primary tissue layers (or germ layers) form. Cell divisions, cell migrations, and other events lead to the formation of endoderm, ectoderm, and (in most species) mesoderm. All tissues making up the adult body develop from primary tissue layers.

e. Onset of organ formation, when different organs start developing by a tightly orchestrated program of cell differentiation and morphogenesis.

f. Growth and tissue specialization, when organs enlarge overall and acquire their specialized chemical and physical properties. The maturation of tissues and organs continues into post-embryonic stages.

3. A new embryo cannot develop properly unless each stage of development is successfully completed before the next stage begins.

4. In cell differentiation, a cell selectively uses certain genes and synthesizes proteins not found in other cell types. The outcome is subpopulations of specialized lineages of cells that differ from one another in their structure, biochemistry, and functioning.

5. Starting at the time of gastrulation, morphogenesis is a program by which the embryo changes in size, shape, and proportions. Cell divisions, cell migrations, changes in cell size and shape, the growth and folding of tissues, and controlled cell death (by apoptosis) are involved.

6. Pattern formation is the emergence of the basic body plan and specific body parts in specific regions, in an orderly sequence. Cytoplasmic localization helps seal the fate of cell lineages in this patterning. Among echinoderms and vertebrates, cell determination relies more on interactions among classes of master genes that specify certain products. These gene products are spatially organized and create chemical gradients that help seal the identity of each cell lineage in the embryo.

7. Change in the developmental fate of an embryonic cell lineage, as brought about by exposure to products released from an adjacent tissue, is called embryonic induction.

8. Following embryonic development, some animals grow directly into the adult form. Other animals show metamorphosis. By this process, reactivated growth and tissue reorganization produce larvae or some other immature forms before the emergence of the sexually mature adult.

9. All animals that show extensive cell differentiation also age. They undergo gradual changes in structure and a decline in efficiency that typically ends in death.

Review Questions

1. What is the main benefit of sexual reproduction? What are some of its biological costs? *44.1*

2. Define the key events during gamete formation, fertilization, cleavage, gastrulation, and organ formation. At what stage is the frog embryo in the photograph at right? *44.2*

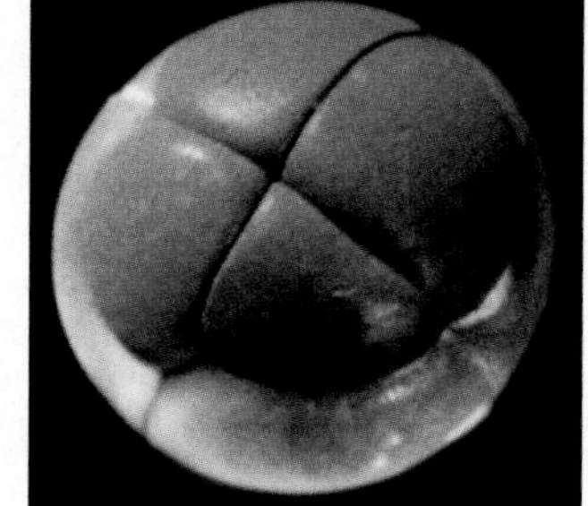

3. Compared to the zygote, does cleavage increase the volume of cytoplasm, the number of cells, or both? *44.3*

4. Define blastula. What is a blastocyst? *44.3*

5. Define cell differentiation and morphogenesis. *44.4*

6. Define cytoplasmic localization. At which stage does it occur? Does it play a larger role in cell determination in insects such as *Drosophila* than in vertebrates? *44.3, 44.5, 44.6*

7. Define pattern formation. Then summarize the key points of the theory of pattern formation. *44.5, 44.7*

Self-Quiz (*Answers in Appendix III*)

1. Sexual reproduction among animals is ______.
   a. biologically costly
   b. diverse in its details
   c. evolutionarily advantageous
   d. all of the above

2. Development cannot proceed properly unless each stage is successfully completed before the next begins, starting with ______.
   a. gamete formation
   b. fertilization
   c. cleavage
   d. gastrulation
   e. organ formation
   f. growth, tissue specialization

3. During cleavage, the new blastomeres are allocated different regions of the fertilized egg's cytoplasm. This is called ______.
   a. cytoplasmic localization
   b. embryonic induction
   c. cell differentiation
   d. morphogenesis

4. Primary tissue layers first appear ______.
   a. in the egg cortex
   b. during cleavage
   c. in the gastrula
   d. in primary organs

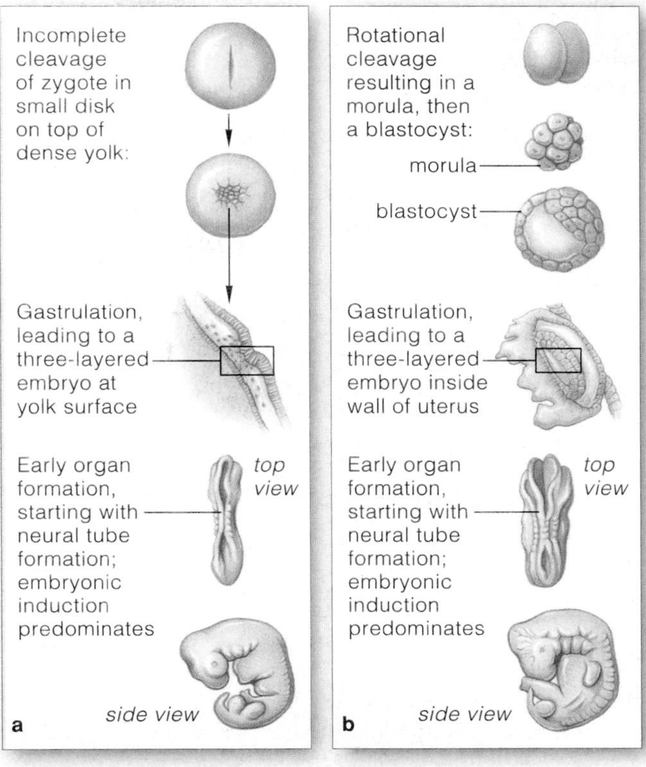
Incomplete cleavage of zygote in small disk on top of dense yolk:
Gastrulation, leading to a three-layered embryo at yolk surface
Early organ formation, starting with neural tube formation; embryonic induction predominates
top view
side view
a
Rotational cleavage resulting in a morula, then a blastocyst:
morula
blastocyst
Gastrulation, leading to a three-layered embryo inside wall of uterus
Early organ formation, starting with neural tube formation; embryonic induction predominates
top view
side view
b

Figure 44.18 Comparison of the pattern of embryonic development for (**a**) a chick and (**b**) a human.

5. The astonishing internal complexity characteristic of most animals became possible following the evolution of ______ .
   a. ectoderm b. mesoderm c. myoderm d. endoderm

6. During development, the formation of subpopulations of different cell types is the outcome of ______ .
   a. selective gene expression c. morphogenesis
   b. cell differentiation d. a and b

7. A cell formed during cleavage is a ______ .
   a. blastula b. morula c. blastomere d. gastrula

8. ______ distributes different parts of the fertilized egg's cytoplasm to different blastomeres.
   a. Gametogenesis c. Morphogenesis
   b. Cleavage d. Pattern formation

9. Pattern formation requires ______ .
   a. master genes c. regulatory proteins
   b. morphogens d. all of the above

10. Metamorphosis does *not* occur in ______ .
   a. crustaceans c amphibians
   b. insects d. humans

11. A juvenile is a ______ .
   a. totally immature dweeb
   b. miniature version of adult that changes in size
   c. immature stage in which tissues are destroyed and replaced

12. Match the development stage with its description.
   ____ cleavage
   ____ gametogenesis
   ____ organ formation
   ____ growth, tissue specialization
   ____ gastrulation
   ____ fertilization

   a. egg and sperm mature in parents
   b. sperm nucleus, egg nucleus fuse
   c. formation of primary tissue layers
   d. cytoplasmic localization dominates in most animals (not mammals) rather than embryonic induction
   e. organs, tissues increase in size, acquire specialized properties
   f. starts when primary tissue layers split into subpopulations of cells

Critical Thinking

1. Experimentally, it is possible to divide an amphibian egg so that the gray crescent is wholly within one of the two cells formed. If the two cells are separated from each other, only the cell with the gray crescent will form an embryo with a long axis, notochord, nerve cord, and back musculature. The other cell will form a shapeless mass of immature gut and blood cells. Reflect on Section 44.3. Does cytoplasmic localization or embryonic induction play a greater role in these outcomes?

2. Differences in the amount and distribution of yolk and other factors in the egg cytoplasm give rise to different cleavage patterns among animal groups. Think about the stages shown in Figure 44.18. Then speculate on how some of the differences may contribute to a pattern of incomplete cleavage of fertilized bird eggs (in this example, a chick egg) and to the pattern of complete, rotational cleavage of fertilized mammalian eggs (a human egg).

Also notice this: Despite obvious differences in the early stages of development, chick and human embryos resemble one another once early organ formation is under way. Using the theory of pattern formation, speculate on the role of embryonic induction in bringing about the similarities between these vertebrate lineages.

Selected Key Terms

AER *44.5*
aging *44.9*
apoptosis *44.4*
asexual reproduction *44.1*
blastocyst *44.3*
blastomere *44.2*
blastula *44.3*
cell differentiation *44.4*
cleavage *44.2*
cytoplasmic localization *44.3*
ectoderm *44.2*
embryo *44.2*
embryonic induction *44.5*
endoderm *44.2*
fate map *44.6*
fertilization *44.2*
gamete formation *44.2*
gastrulation *44.2*
gray crescent *44.3*
growth, tissue specialization *44.2*
homeotic gene *44.7*
internal fertilization *44.1*
juvenile *44.8*
mesoderm *44.2*
metamorphosis *44.8*
molting *44.8*
morphogen *44.5*
morphogenesis *44.4*
neural tube *44.4*
oocyte *44.3*
organ formation *44.2*
pattern formation *44.5*
pattern formation theory *44.6*
sexual reproduction *44.1*
telomere *44.9*
sperm *44.3*
yolk *44.1*
zygote *CI*

Readings See also *www.infotrac-college.com*

Caldwell, M. November 1992. "How Does a Single Cell Become a Whole Body?" *Discover* 13(11): 86–93.

Gilbert, S. 1994. *Developmental Biology*. Fourth edition. Sunderland, Massachusetts: Sinauer.

McGinnis, W., and M. Kuziora. February 1994. "The Molecular Architects of Body Design." *Scientific American* 270(2): 58–66.

Nusslein-Volhard, C. August 1996. "Gradients That Organize Embryonic Development." *Scientific American*, 54–61.

Raff, R., and T. Kaufman. 1983. *Embryos, Genes, and Evolution*. New York: Macmillan.

45 HUMAN REPRODUCTION AND DEVELOPMENT

The Journey Begins

At first nothing appears to be happening to the egg, so recently fertilized in the billowing folds of an oviduct (Figure 45.1*a*). Before the clock ticks off twenty-four hours, however, a spectacular journey is under way.

Cleavage furrows herald the onset of a programmed series of cuts through the egg cytoplasm. Every twelve to twenty-four hours, a new cut carves up the cytoplasm until, by the fifth day, a fluid-filled ball of thirty-two tiny cells is tumbling along the moist, soft lining of the uterus. It is a blastocyst, an embryonic stage you read about in the preceding chapter. Its surface cells are organized as a thin layer that is rather like the shell of a Ping-Pong ball. Inside, some cells huddle against part of the surface layer's wall. This is the inner cell mass, the forerunner of an adult body that will someday consist of trillions upon trillions of cells. Yet now the blastocyst is smaller than the tip of a pin.

By week's end, molecular signals urge the blastocyst to anchor itself to the uterine lining and start burrowing into it. Soon, some surface cells of the blastula send out fingerlike cytoplasmic projections. These grow into the domain of maternal blood vessels threading through the lining. The connections that they establish with the mother's tissues will metabolically support the new individual through the months ahead.

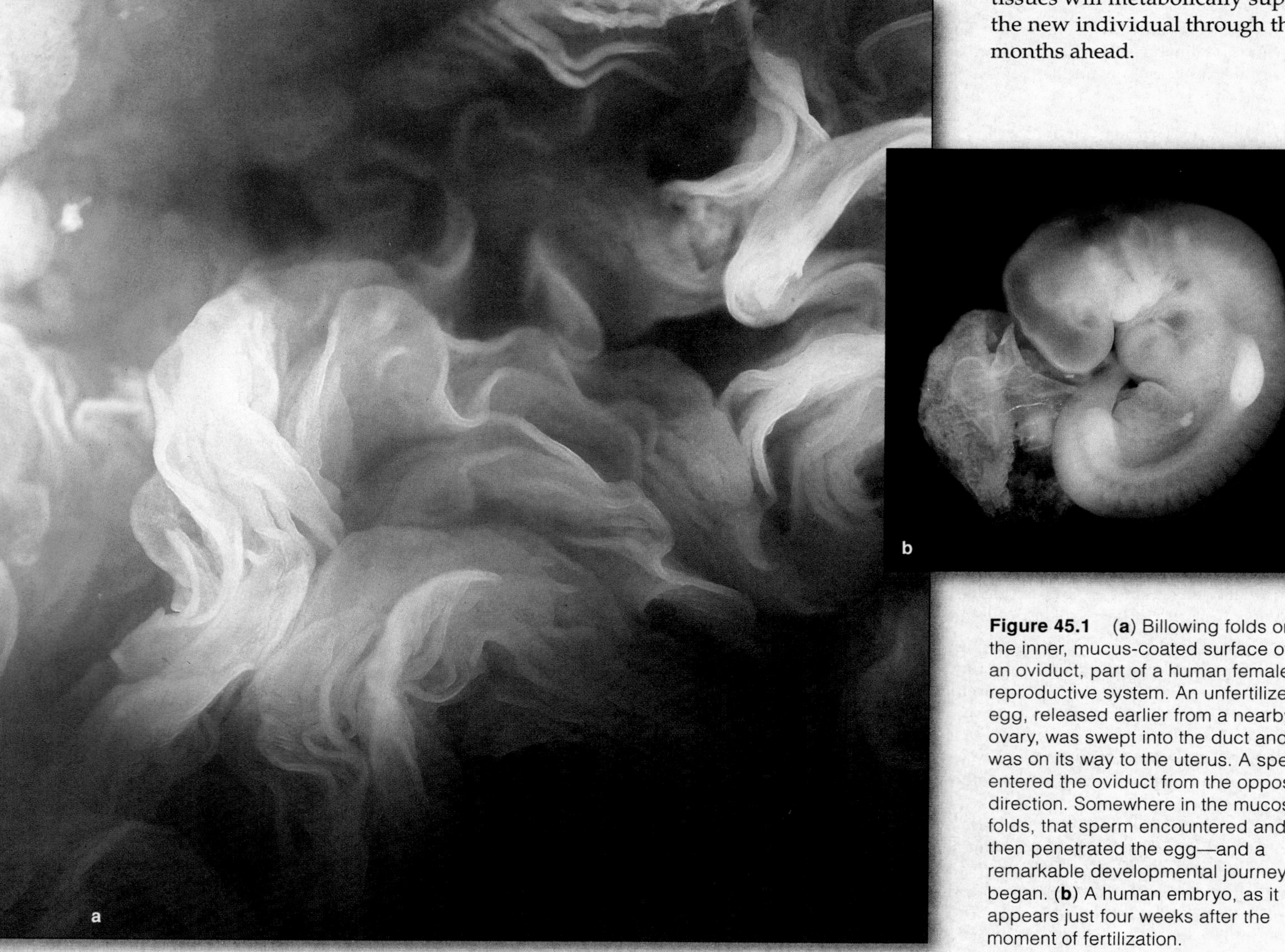

Figure 45.1 (**a**) Billowing folds on the inner, mucus-coated surface of an oviduct, part of a human female's reproductive system. An unfertilized egg, released earlier from a nearby ovary, was swept into the duct and was on its way to the uterus. A sperm entered the oviduct from the opposite direction. Somewhere in the mucosal folds, that sperm encountered and then penetrated the egg—and a remarkable developmental journey began. (**b**) A human embryo, as it appears just four weeks after the moment of fertilization.

Now, in an astonishing feat of self-organization, the inner cell mass transforms itself into a three-layered, oval disk. It does so as embryonic cells change in size and shape and migrate to new locations. Gradually, cell differentiation and morphogenesis transform the disk into a pale, crescent-shaped embryo.

Day after day the sculpting continues. Three weeks after fertilization a neural tube, the forerunner of the central nervous system, starts to form. Paired limb buds appear by the fourth week. A patch of cells in a tubular, primordial heart starts its rhythmic beating. From now on, through birth and adulthood until death, they will beat incessantly as the cardiac pacemaker.

Cephalization, bilateral symmetry, and other features tell us that this early embryo is one of the vertebrates (Figure 45.1*b*). Is it a human? A minnow? A duckling? We don't know the answer until eight weeks into the journey, when we recognize a human in the making. Later on, in a body not much bigger than a peanut, cell interactions will fill in details of the body plan.

These events parallel your own origin inside your mother. The story you are about to read is *your* story.

The preceding chapter acquainted you with some of the principles that govern animal reproduction and development. Now you will see how these principles apply to humans, starting with gamete formation and the mechanisms of sexual reproduction.

In both men and women, the reproductive system is a pair of primary reproductive organs, or gonads, with accessory glands and ducts. The male gonads, **testes** (singular, testis), produce sperm. The female gonads, **ovaries**, produce eggs. Both also secrete sex hormones that influence reproductive functions as well as the development of **secondary sexual traits**. Such traits are features we associate with maleness and femaleness, although they do not play a direct role in reproduction. Prime examples are the amount and distribution of body fat, hair, and skeletal muscles.

You already know that early human embryos have neither male nor female traits (Section 12.3). Seven weeks after fertilization, however, ovaries start to develop in embryos that do not have a Y chromosome, which carries the master gene for sex determination; testes develop only XY embryos. Those ovaries and testes will be fully formed at the time of birth. More than a decade will pass before they grow to their full size and become reproductively functional, and that is where our story picks up.

KEY CONCEPTS

1. The human reproductive system consists of a pair of primary reproductive organs, or gonads, as well as a number of accessory glands and ducts. Testes are sperm-producing male gonads, and ovaries are egg-producing female gonads.

2. In response to signals from the hypothalamus and the pituitary gland, gonads also release sex hormones that orchestrate reproductive functions and the development of secondary sexual traits.

3. Human males continually produce sperm from puberty onward. The hormones testosterone, LH, and FSH are central to the control of male reproductive functions.

4. Human females are fertile on a cyclic basis. Each month during their reproductive years, one of their two ovaries releases an egg, and the lining of their uterus is primed for pregnancy. The hormones estrogen, progesterone, FSH, and LH dominate this cyclic activity.

5. As is the case for nearly all animals, human embryonic development starts with gamete formation and proceeds through fertilization, then cleavage, gastrulation, organ formation, and growth and tissue specialization.

45.1 REPRODUCTIVE SYSTEM OF HUMAN MALES

Figure 45.2 shows the organs of the male reproductive system, and Table 45.1 lists their functions. Again, an adult male has a pair of gonads, of a type called testes. These primary reproductive organs, the equivalent of a female's ovaries, produce sperm and the sex hormones that govern male reproductive function and secondary sexual traits. Enlarging testes are often the first sign that boys have entered **puberty**, the time when gonads start producing mature gametes and secondary sexual traits emerge. Usually between ages twelve and sixteen, boys also undergo a growth spurt, more hair sprouts on their face and elsewhere, and their voice deepens.

Where Sperm Form

In an embryo that is destined to become male, a pair of testes form on the abdominal cavity wall. Before birth, the testes descend from the abdominal cavity into the scrotum, an outpouching of skin that hangs below the pelvic region. At the time of birth, testes are fully formed miniatures of the adult organs. They start to produce sperm at puberty.

Figure 45.2*a* shows the position of the scrotum in an adult male. If the sperm cells are to develop properly, the temperature within the scrotum must remain a few degrees cooler than the core temperature of the rest of the body. A control mechanism works by stimulating and inhibiting the contraction of smooth muscles that are positioned in the scrotum's wall. It helps ensure that the scrotum's internal temperature does not stray far from 95°F. When the air just outside the body becomes too cold, contractions draw the pouch closer to the body mass, which is warmer. When it is warmer outside, muscles relax and thereby lower the pouch.

Packed inside each testis are a large number of small, highly coiled tubes, the **seminiferous tubules**. Sperm formation begins in these tubules, in the manner described in Section 45.2.

Where Semen Forms

Mammalian sperm travel from each testis through a series of ducts that lead to the urethra. They are not quite mature when they enter the epididymis, the first duct, which is lengthy and coiled. The secretions from glandular cells located in the duct wall trigger events that put the finishing touches on the maturing sperm cells. When fully mature, the sperm are stored in the last stretch of each epididymis. There they will remain until they are ejaculated from the body.

When a male is sexually aroused, muscle contractions in the walls of the reproductive organs propel mature sperm into and through a pair of thick-walled tubes, the vasa deferentia (singular, vas deferens). From there, contractions propel sperm through a pair of ejaculatory ducts, then on through the urethra. That last tube in the series threads through the interior of the penis, the male sex organ, and opens at its tip. The urethra is a duct that also functions in urinary excretion.

Glandular secretions become mixed with sperm as they travel to the urethra. The result is **semen**, a thick fluid that eventually is expelled from the penis during sexual activity. Early in the formation of semen, a pair

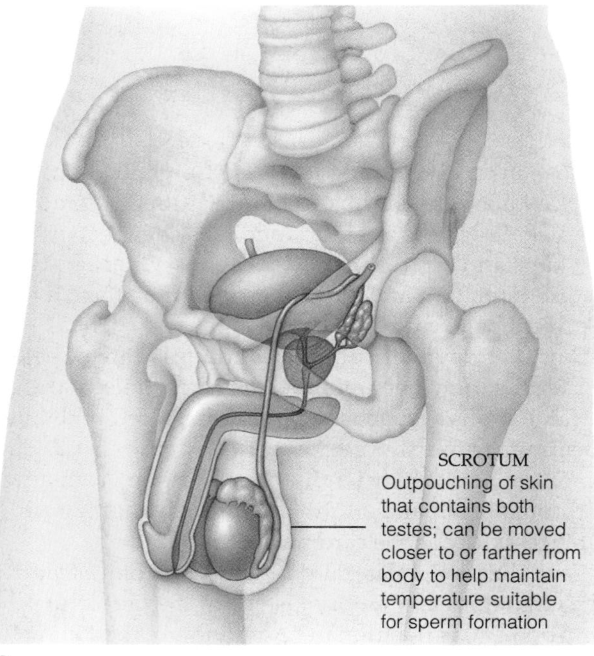
SCROTUM
Outpouching of skin that contains both testes; can be moved closer to or farther from body to help maintain temperature suitable for sperm formation
a

Figure 45.2 *Above and facing page:* (**a**) Position of the human male reproductive system relative to the pelvic girdle and urinary bladder. The midsagittal section in (**b**) shows the components of the system and lists their functions.

Table 45.1 Organs and Accessory Glands of the Male Reproductive Tract

REPRODUCTIVE ORGANS	
Testis (2)	Production of sperm, sex hormones
Epididymis (2)	Sperm maturation site and sperm storage
Vas deferens (2)	Rapid transport of sperm
Ejaculatory duct (2)	Conduction of sperm to penis
Penis	Organ of sexual intercourse
ACCESSORY GLANDS	
Seminal vesicle (2)	Secretion of large part of semen
Prostate gland	Secretion of part of semen
Bulbourethral gland (2)	Production of lubricating mucus

Further reading: Student Guide to InfoTrac on web site

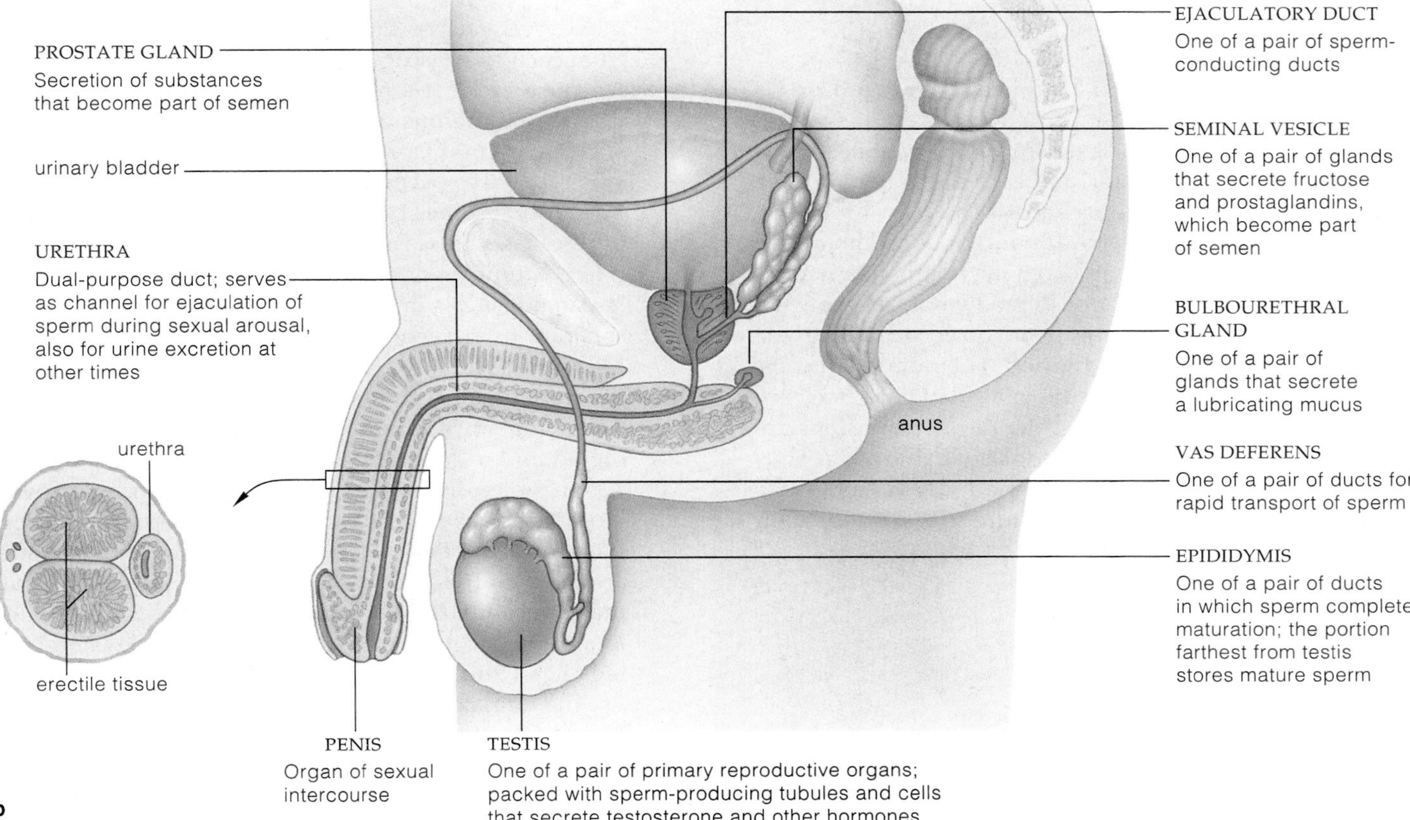

PROSTATE GLAND
Secretion of substances that become part of semen
urinary bladder
URETHRA
Dual-purpose duct; serves as channel for ejaculation of sperm during sexual arousal, also for urine excretion at other times
urethra
erectile tissue
PENIS
Organ of sexual intercourse
TESTIS
One of a pair of primary reproductive organs; packed with sperm-producing tubules and cells that secrete testosterone and other hormones
EJACULATORY DUCT
One of a pair of sperm-conducting ducts
SEMINAL VESICLE
One of a pair of glands that secrete fructose and prostaglandins, which become part of semen
BULBOURETHRAL GLAND
One of a pair of glands that secrete a lubricating mucus
anus
VAS DEFERENS
One of a pair of ducts for rapid transport of sperm
EPIDIDYMIS
One of a pair of ducts in which sperm complete maturation; the portion farthest from testis stores mature sperm

b

of seminal vesicles secrete fructose. The sperm use this sugar as an energy source. Seminal vesicles also secrete certain kinds of prostaglandins that can induce muscle contractions. Possibly, these signaling molecules exert their effects during sexual activity. At that time, they might induce contractions in the female's reproductive tract and thereby assist sperm movement through it.

Secretions from a prostate gland may help buffer the acidic conditions in the female reproductive tract. (The pH of vaginal fluid is about 3.5–4.0, but sperm are able to swim more efficiently at pH 6.) Two bulbourethral glands secrete some mucus-rich fluid into the urethra when a male is sexually aroused.

Cancers of the Prostate and Testis

Until recently, cancers of the male reproductive tract did not get much media coverage in the United States. Yet *prostate cancer* is the second leading cause of death in men, exceeded only by cancers of the respiratory tract. At this writing, estimates are that 180,400 new cases will be diagnosed in 2000, and 31,900 will die. The death rate for breast cancer, which is highly publicized, is not much higher. Even so, this is a dramatic decline from prostate cancer rates for 1989–1992, possibly as a result of earlier diagnostic screenings. Also, *testicular cancer* is a frequent cause of death among young men. About 5,000 cases are diagnosed each year.

Both cancers are painless in early stages. In cases where they are not detected in time, the cancers spread silently into lymph nodes of the abdomen, chest, neck, then the lungs. Once a cancer metastasizes, prospects are not good. Testicular cancer, for example, now kills as many as half of those stricken.

Doctors can detect prostate cancer through physical examinations and blood tests for an increase in prostate-specific antigen (PSA). Once a month from high school onward, men should examine each testis after a warm bath or shower, when scrotal muscles are relaxed. The testis should be rolled gently between the thumb and the forefinger to check for any enlargement, hardening, or lump. Such changes may or may not cause noticeable discomfort. But they must be reported so a physician can order a full examination. The treatment of testicular cancer has one of the highest success rates, provided that the cancer is caught before it starts to spread.

Human males have a pair of testes—primary reproductive organs that produce sperm and sex hormones—as well as accessory glands and ducts. The hormones influence sperm formation and the development of secondary sexual traits.

45.2 MALE REPRODUCTIVE FUNCTION

Sperm Formation

Each testis is only about 5 centimeters long, and that is smaller than a golfball. Yet packed inside are 125 meters of seminiferous tubules. As many as 300 wedge-shaped lobes partition the interior, and each holds two or three coiled tubules (Figures 45.3 and 45.4).

Inside the wall of each tubule are undifferentiated cells called spermatogonia (singular, spermatogonium). Ongoing cell divisions force them away from the wall, toward the interior. During their forced departure, the cells undergo mitotic cell division. Their daughter cells, primary spermatocytes, are the ones that enter meiosis.

As Figure 45.4*a* indicates, the nuclear divisions are accompanied by *incomplete* cytoplasmic divisions. Thus the developing cells are interconnected by cytoplasmic bridges. Ions and molecules required for development move freely across the bridges, so successive divisions from each spermatogonium produce clones of cells that mature together. **Sertoli cells**, the only other type of cell in the seminiferous tubules, provide the forerunners of sperm with nourishment and molecular signals.

Secondary spermatocytes form by way of meiosis I. The chromosomes of these haploid cells are still in the duplicated state; each consists of two sister chromatids. (Here you may wish to review the general overview of spermatogenesis in Section 10.5.) The sister chromatids separate from each other at meiosis II, then daughter cells form. These are haploid spermatids that gradually develop into sperm, the male gametes.

Each mature sperm is a flagellated cell with a head region and a long tail that has a core of microtubules. An enzyme-containing cap called an acrosome covers most of the head, which contains a DNA-packed nucleus (Figure 45.4*b*). Before fertilization can occurs, enzymes are released from the cap. They help a sperm penetrate the extracellular material around an egg. Mitochondria in a midpiece just behind the head provide energy for the tail's whiplike movements.

The formation of each sperm takes nine to ten weeks. From puberty onward, a human male produces sperm on a continual basis, so that many millions of cells are in different stages of development on any given day.

Hormonal Controls

Coordinated secretions of LH, FSH, testosterone, and other hormones govern reproductive function in males. Take a look at Figure 45.3*b*, and notice the **Leydig cells** between lobes in the testes. They secrete **testosterone**, a steroid hormone that is essential for the growth, form, and functioning of the male reproductive tract.

Besides having the primary role in sperm formation, testosterone is the hormone that promotes development

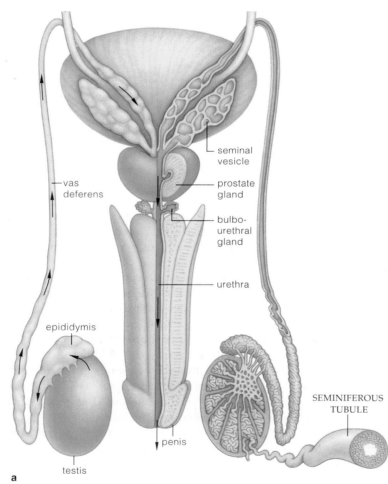

seminal vesicle
vas deferens
prostate gland
bulbo-urethral gland
urethra
epididymis
SEMINIFEROUS TUBULE
penis
testis
a

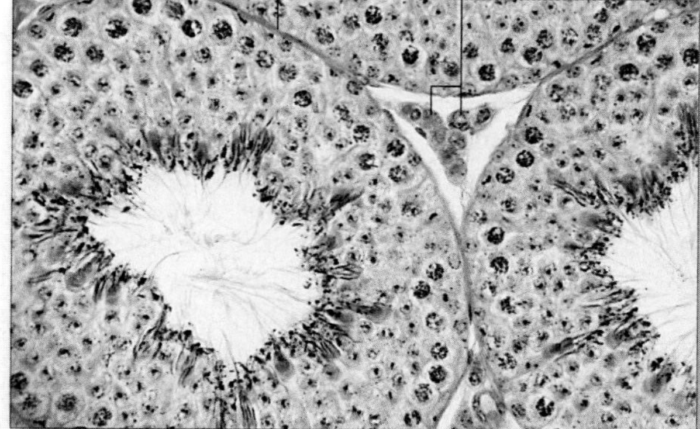

wall of seminiferous tubule
Leydig cells between tubules
b

Figure 45.3 (**a**) Male reproductive tract, posterior view. The arrows show the route that sperm take before ejaculation from a sexually aroused male. (**b**) Light micrograph of cells inside three adjacent seminiferous tubules, cross-section. Leydig cells occupy tissue spaces between the tubules.

Further reading: Student Guide to InfoTrac on web site

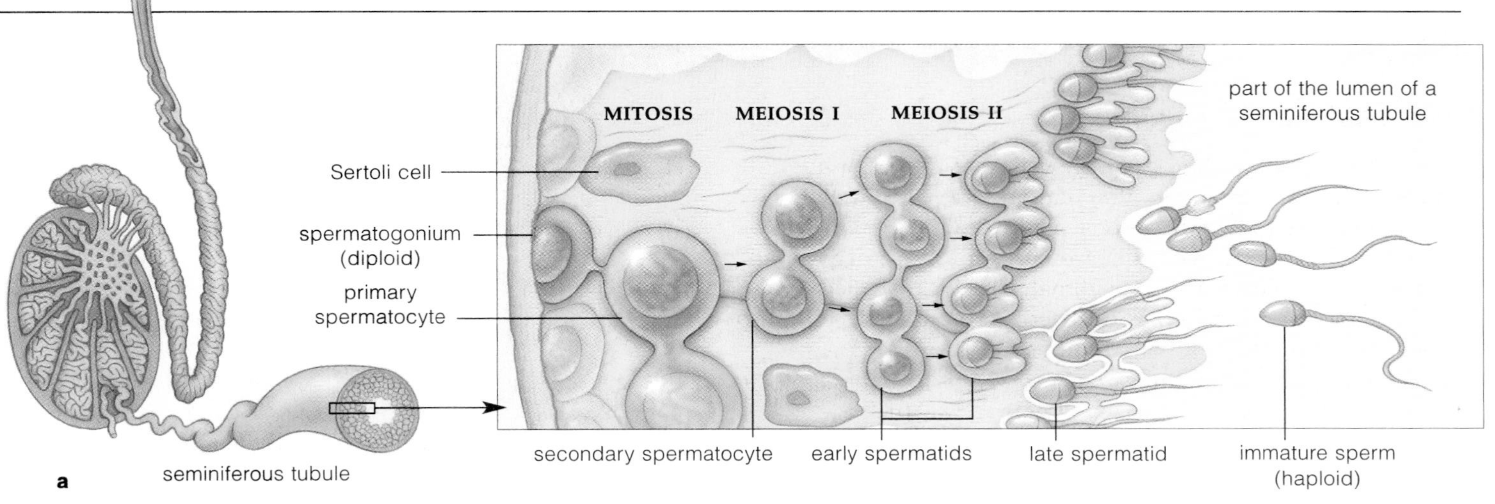
MITOSIS
MEIOSIS I
MEIOSIS II
part of the lumen of a seminiferous tubule
Sertoli cell
spermatogonium (diploid)
primary spermatocyte
secondary spermatocyte
early spermatids
late spermatid
immature sperm (haploid)
a
seminiferous tubule

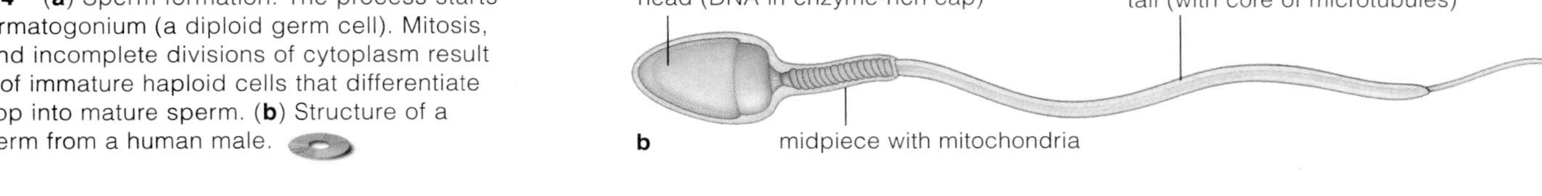
head (DNA in enzyme-rich cap)
tail (with core of microtubules)
b
midpiece with mitochondria

Figure 45.4 (**a**) Sperm formation. The process starts with a spermatogonium (a diploid germ cell). Mitosis, meiosis, and incomplete divisions of cytoplasm result in a clone of immature haploid cells that differentiate and develop into mature sperm. (**b**) Structure of a mature sperm from a human male.

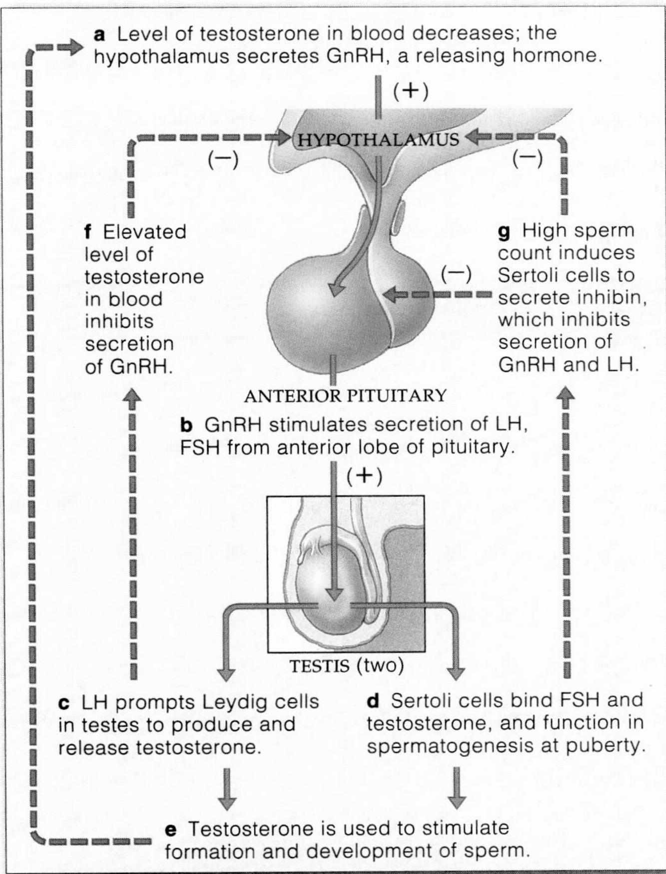
a Level of testosterone in blood decreases; the hypothalamus secretes GnRH, a releasing hormone.
(+)
HYPOTHALAMUS
(−)
(−)
f Elevated level of testosterone in blood inhibits secretion of GnRH.
g High sperm count induces Sertoli cells to secrete inhibin, which inhibits secretion of GnRH and LH.
(−)
ANTERIOR PITUITARY
b GnRH stimulates secretion of LH, FSH from anterior lobe of pituitary.
(+)
TESTIS (two)
c LH prompts Leydig cells in testes to produce and release testosterone.
d Sertoli cells bind FSH and testosterone, and function in spermatogenesis at puberty.
e Testosterone is used to stimulate formation and development of sperm.

Figure 45.5 Negative feedback loops to the hypothalamus and to the anterior lobe of the pituitary gland from the testes. Through these loops, excess testosterone production shuts off the mechanisms leading to its production. This helps maintain the testosterone level in amounts required for sperm formation.

of secondary sexual traits in the male. It also stimulates sexual behavior and aggressive behavior.

LH and **FSH**, remember, are secreted by the anterior lobe of the pituitary gland (Section 37.6). Initially, both hormones were named for their effects in females. (One abbreviation stands for *L*uteinizing *H*ormone, the other for *F*ollicle-*S*timulating *H*ormone.) Later on, researchers discovered that the molecular structure of LH and FSH is identical in both males and females.

The hypothalamus, a part of the forebrain, controls sperm formation by controlling secretion of LH, FSH, and testosterone. As Figure 45.5 shows, in response to low blood levels of testosterone and other factors, the hypothalamus secretes **GnRH**. This releasing hormone stimulates the anterior lobe to step up the release of LH and FSH, which have targets in the testes.

LH prods Leydig cells to secrete testosterone, which assists in stimulating the formation and development of sperm. Sertoli cells have receptors for FSH, which is required to start up spermatogenesis at puberty. We don't know whether FSH is also necessary for normal functioning of mature human testes.

Figure 45.5 also shows how feedback loops to the hypothalamus lead to decreased testosterone secretion and sperm formation. An elevated testosterone level in blood slows down the release of GnRH. Also, when the sperm count is high, Sertoli cells release **inhibin**. This protein hormone acts on the hypothalamus and pituitary to cut back on the release of GnRH and FSH.

Sperm formation depends on the hormones LH, FSH, and testosterone. Negative feedback loops from the testes to the hypothalamus and pituitary gland control their secretion.

45.3 REPRODUCTIVE SYSTEM OF HUMAN FEMALES

The Reproductive Organs

We turn now to the reproductive system of the human female. Figure 45.6 shows its components, and Table 45.2 summarizes their functions. The female's *primary* reproductive organs, her pair of ovaries, produce eggs and secrete sex hormones. Her immature eggs are called **oocytes**. After each oocyte is released from an ovary, it passes through the entrance of an adjacent oviduct. A female has a pair of oviducts. Each serves as a channel to her **uterus**, a hollow, pear-shaped organ in which the embryo can grow and develop.

A thick layer of smooth muscle, the myometrium, makes up most of the uterine wall. As you will see, the **endometrium**, the inner lining of the wall, is vital for embryonic development. It is composed of connective tissues, glands, and blood vessels. The narrowed-down portion of the uterus is the cervix. A muscular tube, the vagina, extends from the cervix to the surface of the body. The vagina receives sperm and functions as part of the birth canal.

At the body surface are external genitals (vulva), which include organs for sexual stimulation. Outermost are a pair of fat-padded skin folds (the labia majora). They enclose a smaller pair of skin folds (labia minora) that are highly vascularized but have no fatty tissue. The smaller folds partly enclose the clitoris, a sex organ that is sensitive to stimulation. At the body's surface, the opening of the urethra is positioned about midway between the clitoris and the vaginal opening.

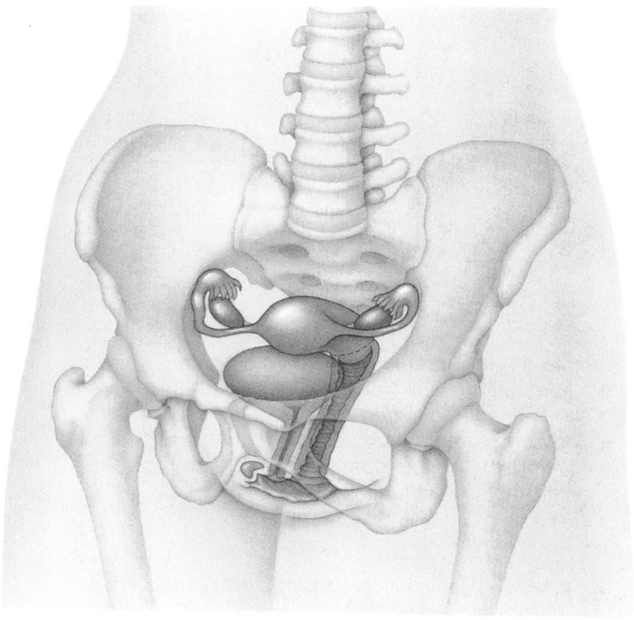

a

Figure 45.6 *Above and facing page:* (**a**) Position of the human female reproductive system relative to the pelvic girdle and urinary bladder. The midsagittal section in (**b**) shows the components of the system and lists their functions.

Table 45.2	Female Reproductive Organs
Ovaries	Oocyte production and maturation, sex hormone production
Oviducts	Ducts for conducting oocyte from ovary to uterus; fertilization normally occurs here
Uterus	Chamber in which new individual develops
Cervix	Secretion of mucus that enhances sperm movement into uterus and (after fertilization) reduces embryo's risk of bacterial infection
Vagina	Organ of sexual intercourse; birth canal

Overview of the Menstrual Cycle

Most mammalian females follow an *estrous* cycle. They are fertile and in heat (sexually receptive to males) only at certain times of year. By contrast, female primates, including humans, follow a **menstrual cycle**. They are fertile intermittently, on a cyclic basis, and heat is not synchronized with their fertile periods. Said another way, female primates of reproductive age can become pregnant only at certain times of year, but they may be receptive to sex at any time.

Briefly, an oocyte matures and escapes from an ovary during each menstrual cycle. Also during the cycle, the endometrium becomes primed to receive and nourish a forthcoming embryo *if* a sperm penetrates that oocyte and fertilization occurs. However, if the oocyte does not become fertilized, then about four to six tablespoons of blood-rich fluid from the uterus will start flowing out through the vaginal canal. Such a recurring blood flow is called menstruation. It means "there is no embryo at this time," and it marks the first day of a new cycle. The uterine nest is being sloughed off and is about to be constructed once again.

The events just sketched out proceed through three phases. The cycle starts with a **follicular phase**. This is the time of menstruation, endometrial breakdown and rebuilding, and oocyte maturation. The next phase is restricted to the release of an oocyte from the ovary. We call it **ovulation**. During the **luteal phase** of the cycle, an endocrine structure (corpus luteum) forms, and the endometrium is primed for pregnancy (Table 45.3).

All three phases are governed by feedback loops to the hypothalamus and pituitary gland from the ovaries. FSH and LH promote cyclic changes in the ovaries. As

Further reading: Student Guide to InfoTrac on web site →

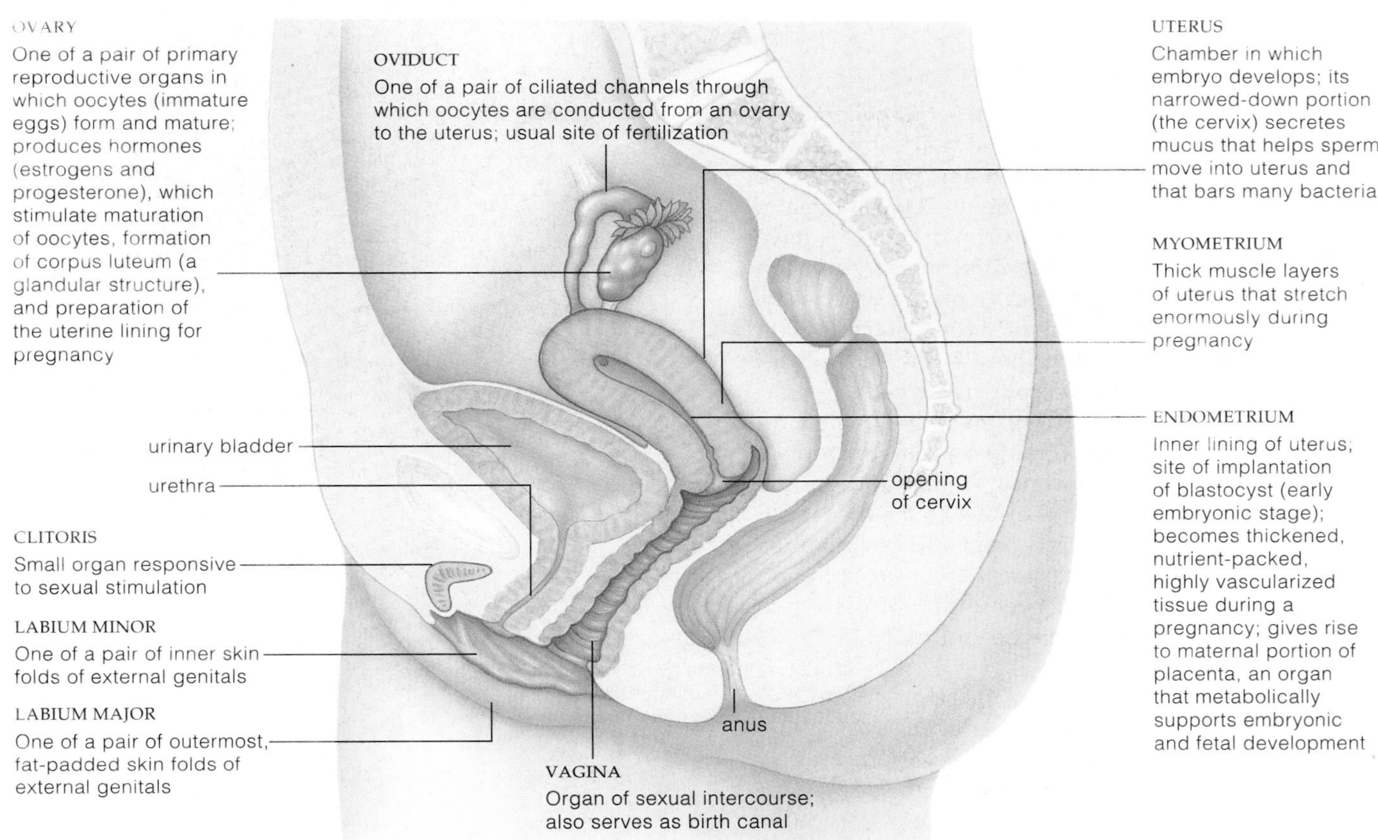

OVARY
One of a pair of primary reproductive organs in which oocytes (immature eggs) form and mature; produces hormones (estrogens and progesterone), which stimulate maturation of oocytes, formation of corpus luteum (a glandular structure), and preparation of the uterine lining for pregnancy
OVIDUCT
One of a pair of ciliated channels through which oocytes are conducted from an ovary to the uterus; usual site of fertilization
UTERUS
Chamber in which embryo develops; its narrowed-down portion (the cervix) secretes mucus that helps sperm move into uterus and that bars many bacteria
MYOMETRIUM
Thick muscle layers of uterus that stretch enormously during pregnancy
ENDOMETRIUM
Inner lining of uterus; site of implantation of blastocyst (early embryonic stage); becomes thickened, nutrient-packed, highly vascularized tissue during a pregnancy; gives rise to maternal portion of placenta, an organ that metabolically supports embryonic and fetal development
urinary bladder
urethra
opening of cervix
CLITORIS
Small organ responsive to sexual stimulation
LABIUM MINOR
One of a pair of inner skin folds of external genitals
LABIUM MAJOR
One of a pair of outermost, fat-padded skin folds of external genitals
anus
VAGINA
Organ of sexual intercourse; also serves as birth canal

b

Table 45.3 Events of the Menstrual Cycle

Phase	Events	Days of Cycle*
Follicular phase	Menstruation; endometrium breaks down	1–5
	Follicle matures in ovary; endometrium rebuilds	6–13
Ovulation	Oocyte released from ovary	14
Luteal phase	Corpus luteum forms, secretes progesterone; the endometrium thickens and develops	15–28

* Assuming a 28-day cycle.

you will see, FSH and LH also stimulate the ovaries to secrete sex hormones—**estrogens** and **progesterone**—to promote the cyclic changes in the endometrium.

A human female's menstrual cycles start between ages ten and sixteen. Each cycle lasts for about twenty-eight days, but this is merely the average. It runs longer for some women and shorter for others. The menstrual cycles continue until a woman is in her late forties or early fifties, when her supply of eggs is dwindling and hormonal secretions slow down. This is the onset of *menopause*, the twilight of reproductive capacity.

You may have heard about *endometriosis*, a condition that arises when endometrial tissue abnormally spreads and grows outside the uterus. Endometrial scar tissue may form on ovaries or oviducts and lead to infertility. In the United States, 10 million women may be affected annually. Possibly the condition arises when menstrual flow backs up through the oviducts and spills into the pelvic cavity. Or perhaps some embryonic cells became positioned in the wrong place before birth and were stimulated to grow during puberty, when sex hormones became active. Whatever the case, estrogen still acts on cells in the mislocated tissue. The resulting symptoms include pain during menstruation, sex, or urination.

Ovaries, the female primary reproductive organs, produce oocytes (immature eggs) and sex hormones. Endometrium lines the uterus, a chamber in which embryos develop.

Secretion of sex hormones (estrogens and progesterone) is coordinated on a cyclic basis through the reproductive years.

A menstrual cycle starts with menstruation, breakdown and rebuilding of the endometrium, and maturation of an oocyte. After an oocyte escapes from an ovary (ovulation), the cycle ends with the formation of a corpus luteum and an endometrium that has become primed for pregnancy.

45.4 FEMALE REPRODUCTIVE FUNCTION

Cyclic Changes in the Ovary

Take a moment to review Section 10.5, the generalized picture of meiosis in an oocyte. A normal baby girl has about 2 million primary oocytes in her ovaries. By the time she is seven years old, only about 300,000 remain; her body resorbed the rest. Her primary oocytes have already entered meiosis I, but this nuclear division process was arrested in a genetically programmed way. Meiosis resumes in one oocyte at a time, starting with the first menstrual cycle. Only about 400 to 500 oocytes will be released during the reproductive years.

In Figure 45.7, sketch *a* is a primary oocyte near an ovary's surface. A cell layer (granulosa cells) surrounds and nourishes it. We call a primary oocyte and the cell layer around it a **follicle**. At the start of the menstrual cycle, the hypothalamus is secreting GnRH in amounts that cause the anterior pituitary to step up *its* secretion of FSH and LH. The blood concentration of these two hormones increases. That increase causes the follicle to grow (Figure 45.8).

The oocyte starts to increase in size, and more layers of cells form around it. Glycoprotein deposits accumulate between the oocyte and the layers. As they do, they widen the space between them. In time, all the deposits form the **zona pellucida**, a noncellular coating around the oocyte.

FSH and LH stimulate cells outside the zona pellucida to secrete estrogens. An estrogen-containing fluid accumulates in the follicle, and estrogen levels in blood begin to increase. About eight to ten hours before being released from the ovary, the oocyte completes meiosis I. And then its cytoplasm divides, forming two cells.

One cell, the **secondary oocyte**, ends up with nearly all of the cytoplasm. The other cell is the first of three **polar bodies**. The meiotic parceling of the chromosomes among all four cells gives the secondary oocyte a haploid chromosome number—which is the precise number required for gametes and for sexual reproduction.

About halfway through the menstrual cycle, the pituitary gland detects the rise in the blood level of estrogens. It responds with a brief outpouring of LH. The LH surge causes rapid vascular changes that make the follicle swell quickly. The surge also induces enzymes to digest the bulging follicle wall. The weakened wall ruptures. Fluid escapes and carries the secondary oocyte with it (Figures 45.7 and 45.8).

Thus, the midcycle surge of LH triggers ovulation—the release of a secondary oocyte from the ovary.

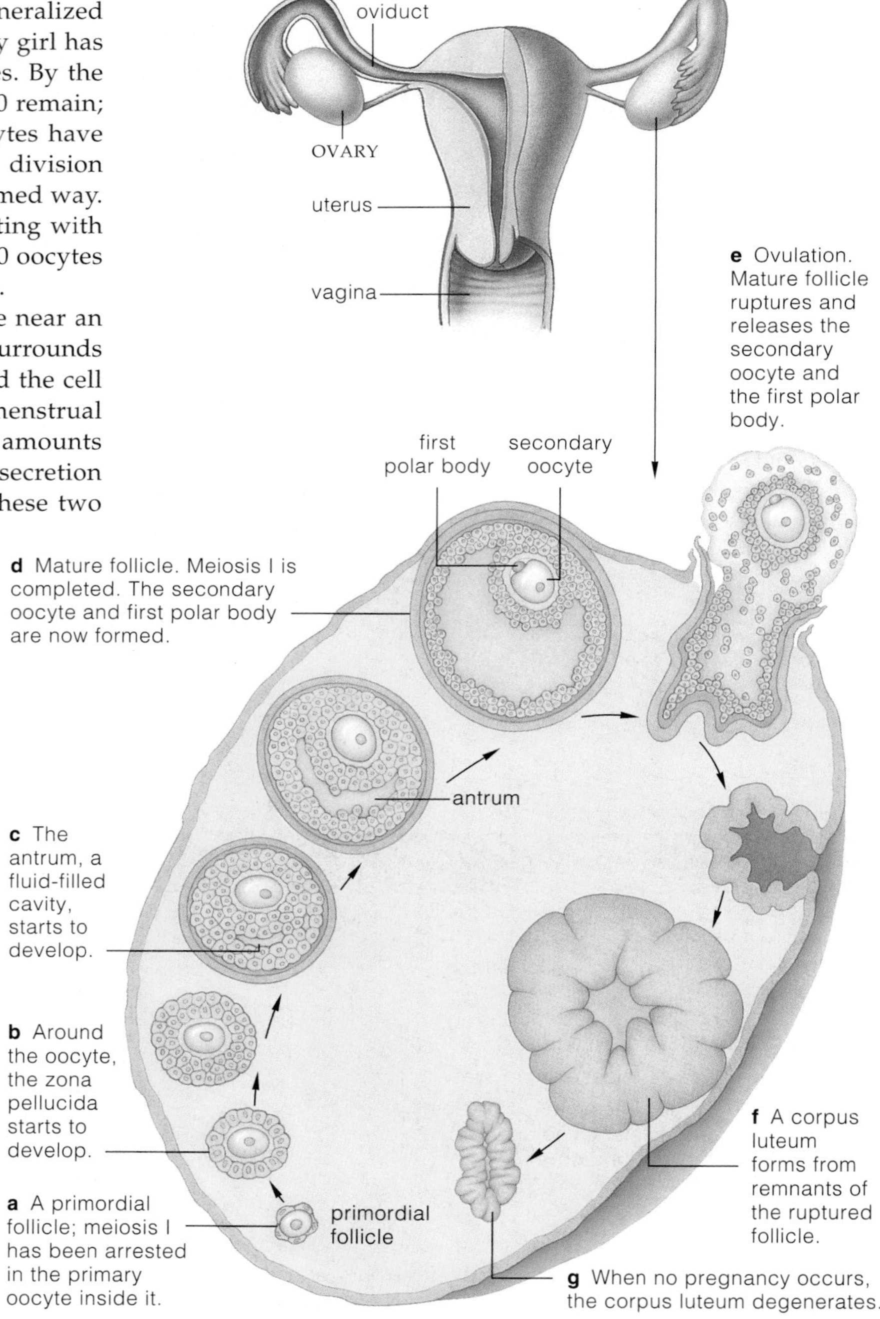
oviduct
OVARY
uterus
vagina
e Ovulation. Mature follicle ruptures and releases the secondary oocyte and the first polar body.
first polar body
secondary oocyte
d Mature follicle. Meiosis I is completed. The secondary oocyte and first polar body are now formed.
antrum
c The antrum, a fluid-filled cavity, starts to develop.
b Around the oocyte, the zona pellucida starts to develop.
a A primordial follicle; meiosis I has been arrested in the primary oocyte inside it.
primordial follicle
f A corpus luteum forms from remnants of the ruptured follicle.
g When no pregnancy occurs, the corpus luteum degenerates.

Figure 45.7 Cyclic events in a human ovary, cross-section.

A follicle remains in the same place in an ovary all through one menstrual cycle. It does not move around as in this diagram, which only shows the *sequence* in which events occur. In the cycle's first phase, a follicle grows and matures. At ovulation, the second phase, the mature follicle ruptures and releases a secondary oocyte. In the third phase, a corpus luteum forms from the follicle's remnants. If the woman does not get pregnant during the cycle, the corpus luteum self-destructs.

Further reading: Student Guide to InfoTrac on web site →

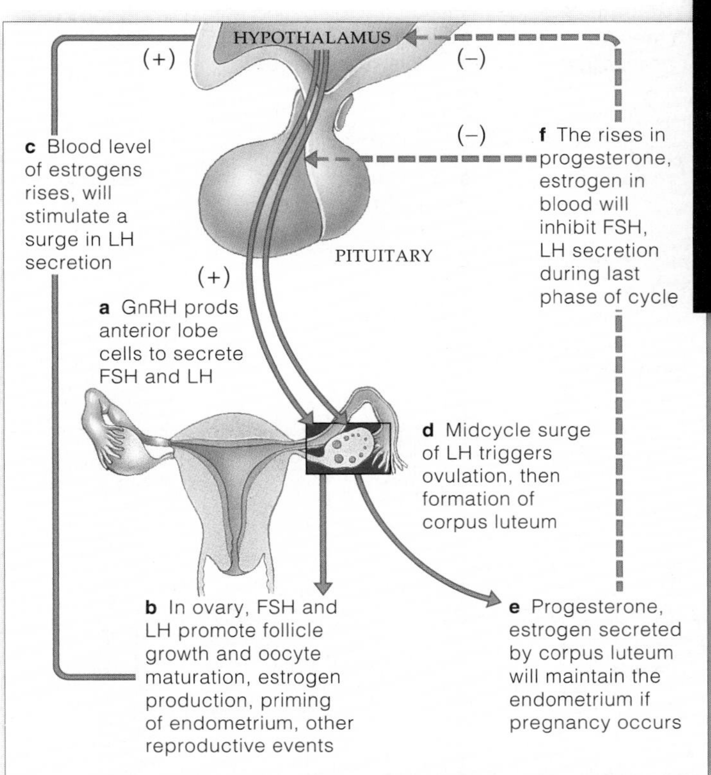

HYPOTHALAMUS
(+)
(–)
(–)
c Blood level of estrogens rises, will stimulate a surge in LH secretion
f The rises in progesterone, estrogen in blood will inhibit FSH, LH secretion during last phase of cycle
PITUITARY
(+)
a GnRH prods anterior lobe cells to secrete FSH and LH
d Midcycle surge of LH triggers ovulation, then formation of corpus luteum
b In ovary, FSH and LH promote follicle growth and oocyte maturation, estrogen production, priming of endometrium, other reproductive events
e Progesterone, estrogen secreted by corpus luteum will maintain the endometrium if pregnancy occurs

secondary oocyte surrounded by follicle cells
surface of ovary

Figure 45.8 Feedback control of hormonal secretion during a menstrual cycle. A positive feedback loop from an ovary to the hypothalamus causes a surge in LH secretion. The surge triggers ovulation. The light micrograph shows a secondary oocyte being released from an ovary at this time. Afterward, negative feedback loops to the hypothalamus and pituitary inhibit FSH secretion. They prevent another follicle from maturing until the cycle is over.

Cyclic Changes in the Uterus

The estrogens released early in the menstrual cycle also help pave the way for a possible pregnancy. Estrogens stimulate growth of the endometrium and its glands. Just before the midcycle surge of LH, follicle cells start to secrete some progesterone as well as estrogens. Blood vessels grow rapidly in the thickened endometrium. At ovulation, the estrogens act on tissue around the cervical canal, the narrowed portion of the uterus that leads to the vagina. The cervix starts to secrete large quantities of a thin, clear mucus that is an ideal medium for sperm to swim through.

The midcycle surge of LH that triggers ovulation also induces the formation of a **corpus luteum** in the ovary. This yellowish glandular structure forms by cell differentiation among granulosa cells left behind in the follicle. (Its name means yellow body.) Secretions from the corpus luteum influence the rest of the cycle.

The corpus luteum secretes progesterone and some estrogen. Progesterone prepares the reproductive tract for the arrival of a **blastocyst**, the type of blastula that forms from a fertilized mammalian egg (Section 44.3). For example, this hormone makes cervical mucus turn thick and sticky. The mucus may keep normal bacterial inhabitants of the vagina out of the uterus. Progesterone also will maintain the endometrium during pregnancy.

A corpus luteum persists for about twelve days. All the while, the hypothalamus is calling for minimal FSH secretion, which stops other follicles from developing. If a blastocyst does not burrow into the endometrium, the corpus luteum will self-destruct during the last days of the cycle. It will secrete certain prostaglandins that apparently can disrupt its own functioning.

After this, progesterone and estrogen levels in blood decline rapidly, and so the endometrium starts to break down. Deprived of oxygen and nutrients, the lining's blood vessels constrict and tissues die. Blood escapes as the walls of weakened capillaries start rupturing. Blood as well as the sloughed endometrial tissues make up a menstrual flow, which continues for three to six days. Then the cycle begins anew, with rising estrogen levels stimulating the repair and growth of the endometrium.

By menopause, the supply of oocytes is dwindling, hormone secretions slow down, and in time menstrual cycles—and fertility—will be over. Eggs that are still to be released late in a woman's life are at some risk of alterations in chromosome number or structure when meiosis resumes. A newborn with Down syndrome, as described in Section 12.10, is one possible outcome.

During a menstrual cycle, FSH and LH stimulate growth of an ovarian follicle (a primary oocyte and the surrounding layer of cells). The first meiotic cell division results in a secondary oocyte and the first polar body.

A midcycle surge of LH triggers ovulation, the release of the secondary oocyte and the polar body from the ovary.

Early on, estrogens call for endometrial repair and growth. Then estrogens and progesterone prepare the endometrium and other parts of the reproductive tract for pregnancy.

45.5 VISUAL SUMMARY OF THE MENSTRUAL CYCLE

By now, you probably have come to the conclusion that the menstrual cycle is not a simple tune on a biological banjo. It is a full-blown hormonal symphony!

Before continuing with your reading, take a moment to review Figure 45.9. It correlates the cyclic events in the ovary and uterus with all the coordinated changes in hormone levels that bring about those events. The illustration may leave you with a better understanding of what goes on.

Figure 45.9 Changes in the ovary and uterus, as correlated with changing hormone levels during each turn of the menstrual cycle. *Green* arrows indicate which hormones dominate the cycle's first phase (the time when the follicle matures), then the second phase (when the corpus luteum forms).

(**a–c**) FSH and LH secretions bring about the changes in ovarian structure and function. (**d**,**e**) Estrogen and progesterone secretions from the ovary stimulate the changes in the endometrium.

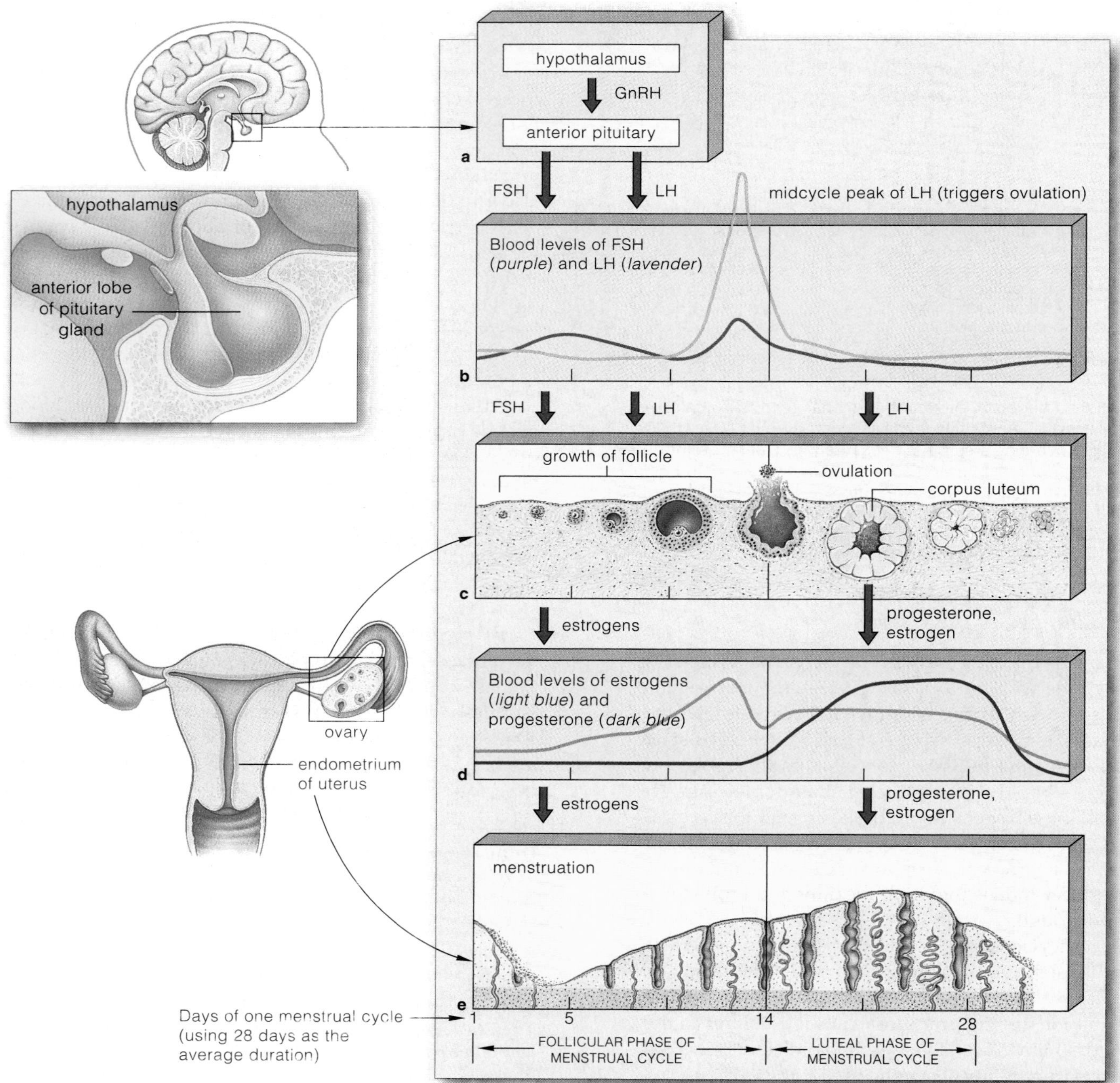
hypothalamus
GnRH
anterior pituitary
a
FSH
LH
midcycle peak of LH (triggers ovulation)
hypothalamus
anterior lobe of pituitary gland
Blood levels of FSH (purple) and LH (lavender)
b
FSH
LH
LH
growth of follicle
ovulation
corpus luteum
c
estrogens
progesterone, estrogen
Blood levels of estrogens (light blue) and progesterone (dark blue)
ovary
endometrium of uterus
d
estrogens
progesterone, estrogen
menstruation
e
Days of one menstrual cycle (using 28 days as the average duration)
1
5
14
28
FOLLICULAR PHASE OF MENSTRUAL CYCLE
LUTEAL PHASE OF MENSTRUAL CYCLE

Further reading: Student Guide to InfoTrac on web site

45.6 PREGNANCY HAPPENS

Sexual Intercourse

Suppose the secondary oocyte is on its way down an oviduct when a female and male are engaged in sexual intercourse, or **coitus**. The male sex act requires *erection*, whereby a normally limp penis stiffens and lengthens; and *ejaculation*, a forceful expulsion of semen into the urethra and out from the penis. As Figure 45.2 shows, the penis incorporates cylinders of spongy tissue. Many friction-activated sensory receptors are on its mushroom-shaped tip, the glans penis. In the sexually unaroused male, large blood vessels that lead into the cylinders are vasoconstricted. In aroused males, they vasodilate and blood flows into the penis faster than it flows out, so blood collects in the spongy tissue. The resulting engorgement stiffens and lengthens the organ, this being helpful for penetration into the vaginal canal.

During coitus, pelvic thrusts stimulate the penis, the female's clitoris, and the vaginal wall. The mechanical stimulation induces involuntary contractions in the male's reproductive tract. These rapidly force sperm out from each epididymis. They force the contents of the seminal vesicles and the prostate gland into the urethra. The substances mix together and form semen. When semen is ejaculated, a sphincter closes the neck of the bladder and prevents urination. During coitus, the semen is released into the vagina.

Emotional intensity, hard breathing, heart pounding, and contractions of skeletal muscles in general accompany rhythmic throbbing of the pelvic muscles. During *orgasm*, the end of the sex act, strong sensations of physical release, warmth, and relaxation dominate. Similar sensations typify female orgasm. It is a common misconception that a female cannot become pregnant if she doesn't reach orgasm. Don't believe it.

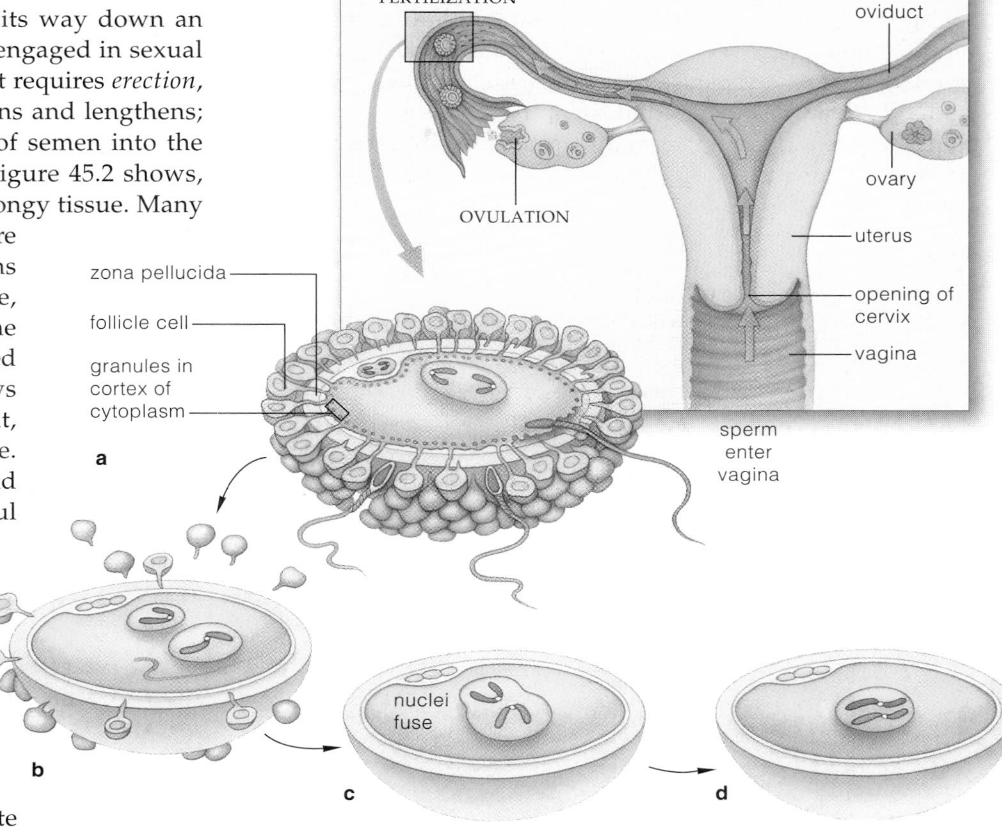

FERTILIZATION
oviduct
ovary
OVULATION
uterus
opening of cervix
vagina
zona pellucida
follicle cell
granules in cortex of cytoplasm
sperm enter vagina
a
b
nuclei fuse
c
d

Figure 45.10 Fertilization. (**a**) Many sperm surround a secondary oocyte. Acrosomal enzymes clear a path through the zona pellucida. (**b**) When a sperm does penetrate the secondary oocyte, granules in the egg cortex release substances that make the zona pellucida impenetrable to other sperm. Penetration also stimulates meiosis II of the oocyte's nucleus. (**c**) The sperm's tail degenerates and its nucleus enlarges and fuses with the oocyte nucleus. (**d**) With fusion, fertilization is over. The zygote has formed.

Fertilization

Now sperm are in the vagina. A single ejaculation can put 150 million to 350 million there. If they arrive a few days before or after ovulation or anytime in between, fertilization may be the outcome. Less than thirty minutes after ejaculation, muscle contractions move the sperm deeper into the female's reproductive tract. Only a few hundred sperm actually reach the upper portion of the oviduct, where fertilization usually takes place.

The stunning micrograph that opens Unit II shows living sperm around a secondary oocyte. When sperm contact an oocyte, they release enzymes that clear a path through the zona pellucida (Figure 45.10). Although many sperm might get this far, usually only one fuses with the oocyte. It degenerates in the oocyte's cytoplasm until only its nucleus and centrioles remain. Penetration induces the secondary oocyte and the first polar body to finish meiosis II. There are now three polar bodies and a mature egg, or **ovum** (plural, ova). When the sperm nucleus and egg nucleus fuse, their chromosomes restore the diploid number for a brand new zygote.

The intense physiological events that accompany coitus have one function: to put sperm on a collision course with an egg. Fertilization is over with the fusion of a sperm nucleus and egg nucleus, which results in a diploid zygote.

45.7 FORMATION OF THE EARLY EMBRYO

Pregnancy lasts an average of thirty-eight weeks from the time of fertilization. It takes about two weeks for a blastocyst to form. The time span from the third to the end of the eighth week is the *embryonic* period, when the major organ systems form. When it ends, the new individual has distinctly human features and is called a **fetus**. In the *fetal* period, from the start of the ninth week until birth, organs enlarge and become specialized.

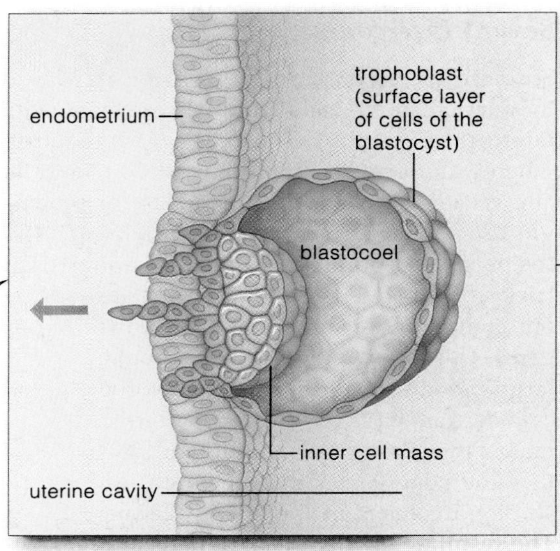

endometrium
trophoblast (surface layer of cells of the blastocyst)
blastocoel
inner cell mass
uterine cavity

d DAY 5. A blastocoel (a fluid-filled cavity) forms in the morula as a result of secretions from the surface cells. By the thirty-two-cell stage, cells of an inner cell mass are already differentiating. They will give rise to the embryo proper. This embryonic stage is called a blastocyst.

inner cell mass

e DAYS 6–7. Surface cells of the blastocyst attach to the endometrium and start to burrow into it. Implantation is under way.

actual size

c DAY 4. By 96 hours, there is a ball of sixteen to thirty-two cells that is shaped like a mulberry. It is a morula (after *morum*, Latin for mulberry). Cells of the surface layer will function in implantation and will give rise to a membrane, the chorion.

b DAY 3. After the third cleavage, the cells suddenly huddle together into a compacted ball, which becomes stabilized by numerous tight junctions among the outer cells. Gap junctions form among the interior cells and enhance intercellular communication.

a DAYS 1–2. Cleavage begins within 24 hours after fertilization. The first cleavage furrow extends between the two polar bodies. Subsequent cuts are rotational, so the resulting cells are not symmetrically arranged (compare Section 44.3). Until the eight-cell stage forms, the cells are loosely arranged, with considerable space between them.

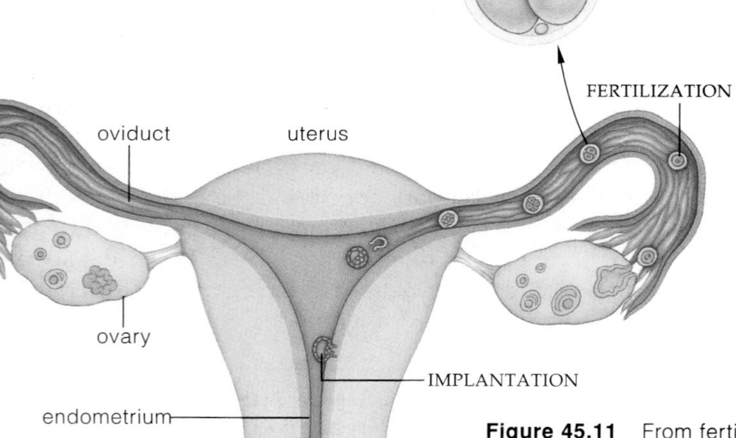

FERTILIZATION
oviduct
uterus
ovary
IMPLANTATION
endometrium

Figure 45.11 From fertilization through implantation. A blastocyst forms during cleavage. Within the blastocyst, the inner cell mass gives rise to the disk-shaped early embryo. Three of the extraembryonic membranes (amnion, chorion, and yolk sac) start forming. The fourth extraembryonic membrane (allantois) forms after the blastocyst is implanted.

We typically call the first three months of pregnancy the *first* trimester. The *second* trimester extends from the start of the fourth month to the end of the sixth. The *third* trimester extends from the seventh month until birth. Beginning with Figure 45.11, the next series of illustrations shows the characteristic features of the new individual at progressive stages of development.

Cleavage and Implantation

Three to four days after fertilization, the zygote is already undergoing cleavage as it tumbles through the oviduct. Genes are already being expressed; the early divisions depend on their products. At the eight-cell stage, the cells huddle into a compact ball. By the fifth day, there is a surface layer of cells (a trophoblast), a cavity filled with their secretions (a blastocoel), and a tiny cluster of interior cells (an inner cell mass). These are the defining features of a human blastocyst (Figure 45.11*d*).

Six or seven days after fertilization, **implantation** is under way. By this process, the blastocyst adheres to the uterine lining, some of its cells send out projections that invade the mother's tissues, and connections start forming that will metabolically support the developing embryo through the months ahead. While the invasion is proceeding, the inner cell mass develops into two

Further reading: Student Guide to InfoTrac on web site —

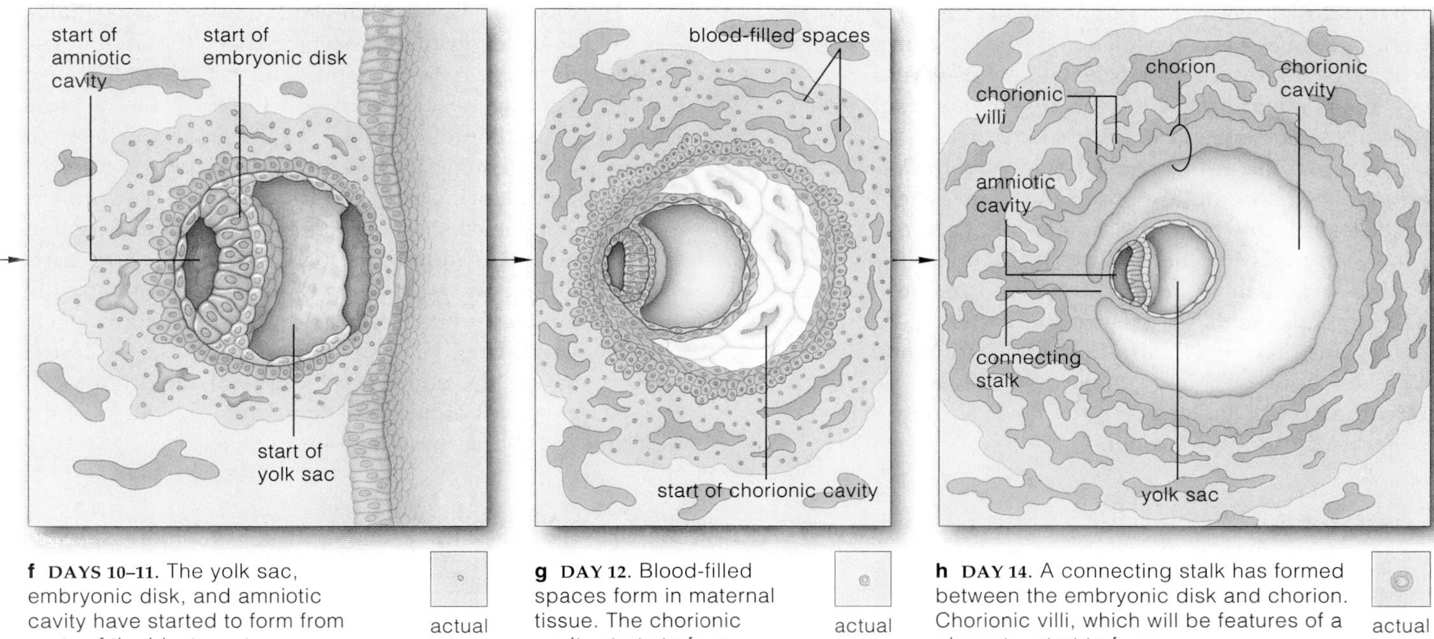
start of
amniotic
cavity
start of
embryonic disk
start of
yolk sac
blood-filled spaces
start of chorionic cavity
chorion
chorionic
cavity
chorionic
villi
amniotic
cavity
connecting
stalk
yolk sac

f DAYS 10–11. The yolk sac, embryonic disk, and amniotic cavity have started to form from parts of the blastocyst. actual size

g DAY 12. Blood-filled spaces form in maternal tissue. The chorionic cavity starts to form. actual size

h DAY 14. A connecting stalk has formed between the embryonic disk and chorion. Chorionic villi, which will be features of a placenta, start to form. actual size

cell layers of a flattened and somewhat circular shape. The two layers make up the embryonic disk—and in short order they will give rise to the embryo proper.

Extraembryonic Membranes

As implantation progresses, membranes start to form outside the embryo. First a fluid-filled *amniotic* cavity opens up between the embryonic disk and part of the blastocyst's surface (Figure 45.11*f*). Then cells migrate around the wall of the cavity and form the **amnion**, a membrane that will enclose the embryo. Fluid inside the cavity will function as a buoyant cradle where the embryo can grow, move freely, and be protected from abrupt temperature changes and mechanical impacts.

While the amnion forms, other cells migrate around the inner wall of the blastocyst's first cavity. They form a lining that becomes the **yolk sac**. This extraembryonic membrane speaks of the evolutionary heritage of land vertebrates (Sections 27.7 and 27.9). For most animals that produce shelled eggs, the sac holds nutritive yolk. In humans, part of the yolk sac becomes a site of blood cell formation, and part will give rise to germ cells, the forerunners of gametes.

Before the blastocyst is fully implanted, spaces open in maternal tissues and fill with blood seeping in from ruptured capillaries. Inside the blastocyst, another cavity opens around the amnion and yolk sac. Now fingerlike projections start to form on the cavity's lining, which is the **chorion**. This new membrane will become part of a spongy, blood-engorged tissue called the placenta.

After the blastocyst is finally implanted, another extraembryonic membrane will form as an outpouching of the yolk sac. This third membrane will become the **allantois**. An allantois functions differently in different animal groups. Among reptiles, birds, and some of the mammals, it has roles in respiration and in the storage of metabolic wastes. In humans, the urinary bladder as well as blood vessels for the placenta form from it.

One more point should be made here. Cells of the blastocyst secrete the hormone **HCG** (*H*uman *C*horionic *G*onadotropin), which stimulates the corpus luteum to keep on secreting progesterone and estrogen. Thus the blastocyst itself prevents menstrual flow and works to avoid being sloughed off until the placenta takes over the task, some eleven weeks later. By the start of the third week, HCG can be detected in the mother's blood or urine. At-home *pregnancy tests* use a treated "dip-stick" that changes color when HCG is present in urine.

A human blastocyst is composed of a surface layer of cells around a fluid-filled cavity (blastocoel) and an inner cell mass, which will give rise to the embryo proper.

Six or seven days after fertilization, the blastocyst implants itself in the endometrium. Now projections from its surface invade maternal tissues, and connections start to form that in time will metabolically support the developing embryo.

Some parts of the blastocyst give rise to an amnion, yolk sac, chorion, and allantois. These extraembryonic membranes serve different functions. Together they are vital for the structural and functional development of the embryo.

45.8 EMERGENCE OF THE VERTEBRATE BODY PLAN

By the time a woman has missed her first menstrual period, cleavage is completed and gastrulation is under way. **Gastrulation**, recall, is a stage when extensive cell divisions, migrations, and rearrangements result in the primary tissue layers (Sections 44.2 and 44.3).

By now, the embryonic disk is surrounded by the amnion and chorion, except at the point where a stalk connects it to the chorion's inner wall. Before then, the inner cell mass had behaved *as if* it were still perched on the large ball of yolk that evolved in the reptilian ancestors of mammals. (Hence the flattened, two-layer embryonic disk, which also forms in reptiles and birds.) Cell divisions and migrations of one layer produced the lining of the yolk sac. The other layer now starts to generate the embryo proper.

For example, by the eighteenth day, two neural folds have appeared on the embryonic disk (Figure 45.12*b*). They will merge to form a neural tube. Early in the third week, a sausage-shaped outpouching forms on the yolk sac. It is the allantois (Greek *allas*, meaning sausage). In humans, remember, the allantois takes part only in the formation of a urinary bladder and blood vessels for the placenta. Some mesoderm folds into a tube that becomes the notochord. A human notochord is only a structural framework; bony segments of the vertebral column soon form around it. Toward the end of the third week, some mesoderm gives rise to the **somites**. These paired segments are the embryonic source of most bones, of skeletal muscles of the head and trunk, as well as of the dermis overlying these regions.

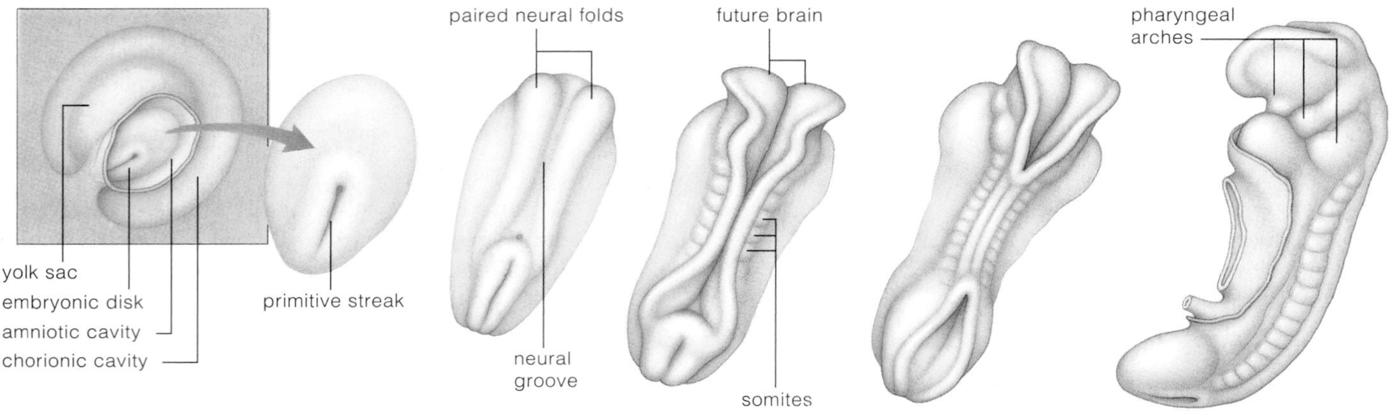
yolk sac
embryonic disk
amniotic cavity
chorionic cavity
primitive streak
paired neural folds
neural groove
future brain
somites
pharyngeal arches

a DAY 15. A faint band appears around a depression along the axis of the embryonic disk. This is the primitive streak, and it marks the onset of gastrulation in vertebrate embryos.

b DAYS 18–23. Organs start to form through cell divisions, cell migrations, tissue folding, and other events of morphogenesis. Neural folds will merge to form the neural tube. Somites (bumps of mesoderm) appear near the embryo's dorsal surface. They will give rise to most of the skeleton's axial portion, skeletal muscles, and much of the dermis.

c DAYS 24–25. By now, some embryonic cells have given rise to pharyngeal arches. These will contribute to the formation of the face, neck, mouth, nasal cavities, larynx, and pharynx.

Figure 45.12 Hallmarks of the embryonic period of humans and other vertebrates. A primitive streak, then a notochord form. Neural folds, somites, and pharyngeal arches form later. (**a**,**b**) Dorsal views, of the embryo's back. (**c**) Side view. Compare the diagrams with the Figure 45.14 photographs.

Through cell divisions and migrations, the midline of that layer thickens faintly around a depression at its surface. This "primitive streak" lengthens and thickens further the next day. It marks the onset of gastrulation (Figure 45.12*a*). It also defines the anterior–posterior axis of the embryo and, in time, its bilateral symmetry. Here, cells migrating inward give rise to endoderm and mesoderm. Now pattern formation begins, as described in Section 44.5. Interactions among classes of master genes map out the basic body plan characteristic of all vertebrates. Through embryonic inductions, specialized tissues and organs start forming in orderly sequence in prescribed parts of the embryo.

Pharyngeal arches start to form. They will contribute to the face, neck, mouth, nose, larynx, and pharynx. Small spaces open up in parts of the mesoderm. In time, all the spaces will interconnect as the coelomic cavity.

During the third week after fertilization, a time when the woman has missed her first menstrual period, the basic vertebrate body plan emerges in the new individual.

A primitive streak, neural tube, somites, and pharyngeal arches form during the embryonic period of all vertebrates. Formation of the primitive streak establishes the body's anterior–posterior axis and its bilateral symmetry.

Further reading: Student Guide to InfoTrac on web site

45.9 WHY IS THE PLACENTA SO IMPORTANT?

Even before the onset of the embryonic period, the extraembryonic membranes have been collaborating with the uterus to sustain the embryo's rapid growth. By the third week, tiny fingerlike projections from the chorion have grown profusely into the maternal blood that has pooled in endometrial spaces. The projections, the chorionic villi, enhance the exchange of substances between the mother and the new individual. They are functional components of the placenta.

The **placenta**, a blood-engorged organ, consists of endometrial tissue and extraembryonic membranes. At full term, it will make up a fourth of the inner surface of the uterus (Figure 45.13).

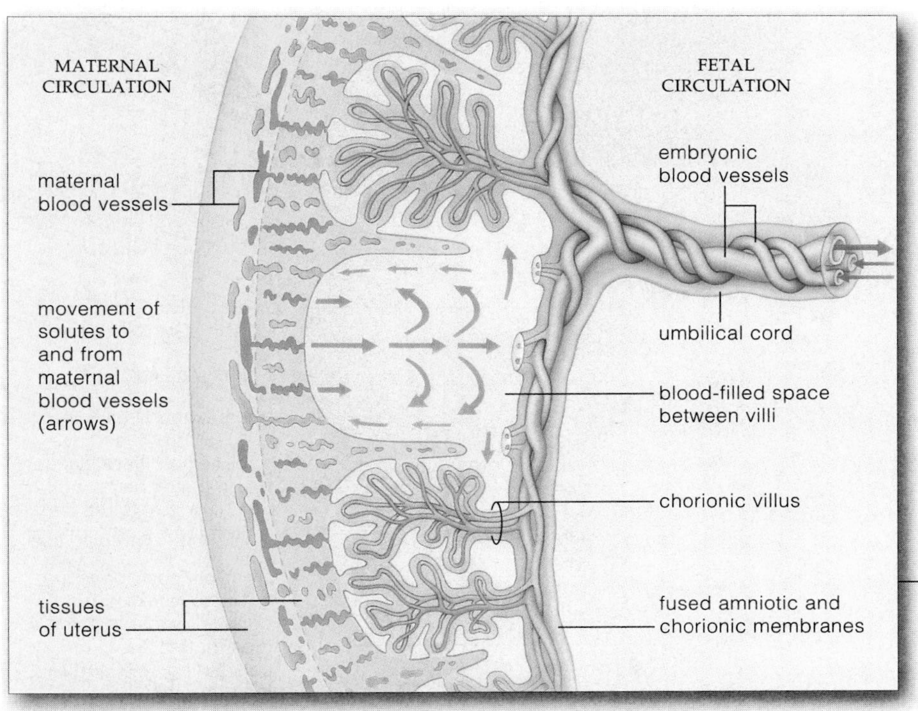

MATERNAL CIRCULATION
FETAL CIRCULATION
maternal blood vessels
embryonic blood vessels
movement of solutes to and from maternal blood vessels (arrows)
umbilical cord
blood-filled space between villi
chorionic villus
tissues of uterus
fused amniotic and chorionic membranes
a

4 weeks

8 weeks

12 weeks

appearance of the placenta at full term

b

Figure 45.13 (**a**) Relationship between fetal and maternal blood circulation in a full-term placenta (**b**). Blood vessels extend from the fetus, through the umbilical cord, and into chorionic villi. Maternal blood spurts into spaces between the villi, but the two bloodstreams do not intermingle. Oxygen, carbon dioxide, and other small solutes diffuse across the placental membrane surface.

A placenta is the body's way of sustaining the new individual while allowing its blood vessels to develop apart from the mother's. Oxygen and nutrients diffuse out of the maternal blood vessels, across the placenta's blood-filled spaces, then into embryonic blood vessels. (The vessels converge in the umbilical cord, the lifeline between the placenta and the new individual.) Carbon dioxide and other wastes diffuse in the other direction. The mother's lungs and kidneys quickly dispose of them.

After the third month, the placenta secretes progesterone and estrogens—and so maintains the uterine lining.

The placenta is a blood-engorged organ of endometrial and extraembryonic membranes. It permits exchanges between the mother and the new individual without intermingling their bloodstreams. Thus it sustains the individual and allows its blood vessels to develop apart from the mother's.

45.10 EMERGENCE OF DISTINCTLY HUMAN FEATURES

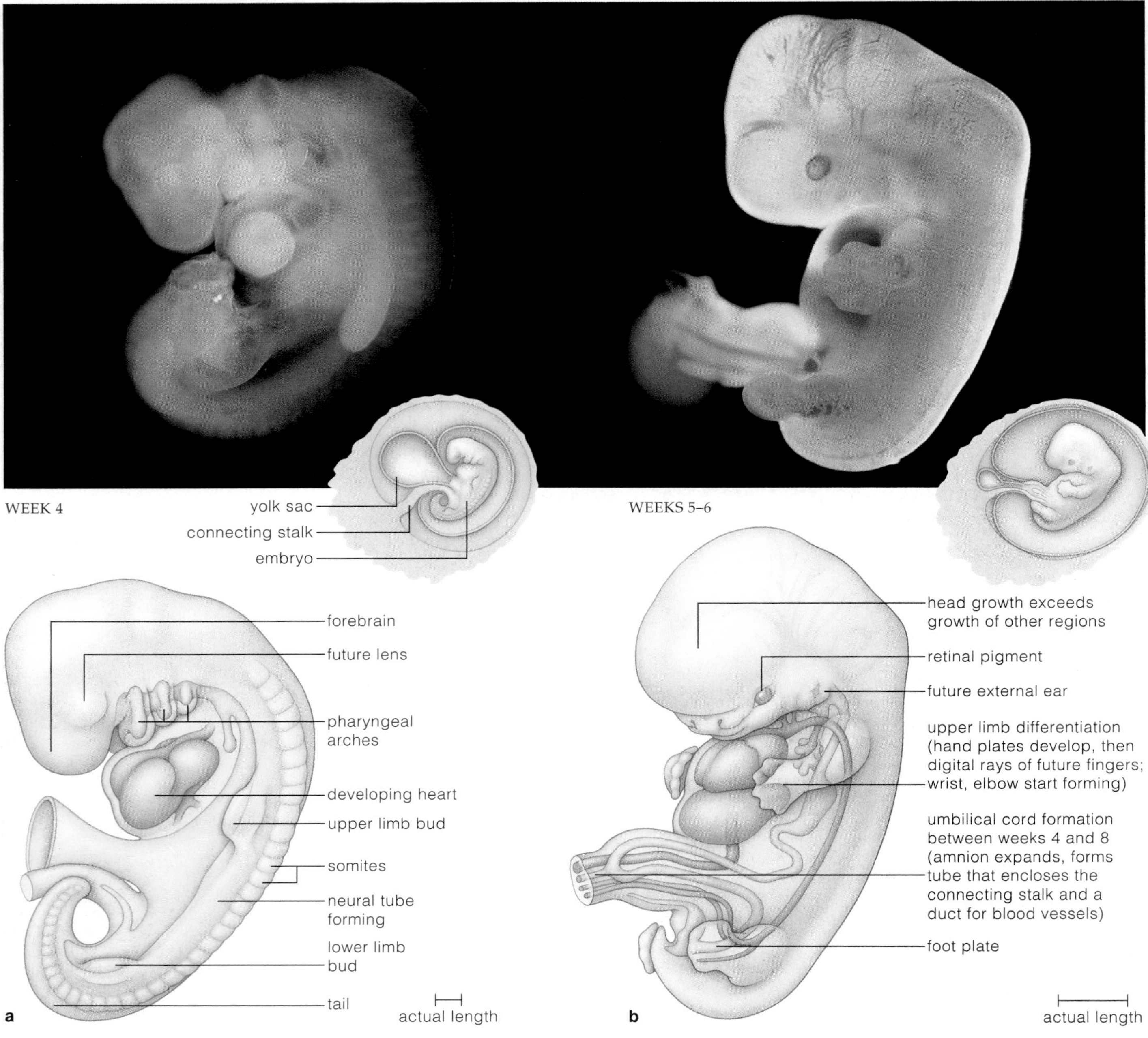

WEEK 4
yolk sac
connecting stalk
embryo
forebrain
future lens
pharyngeal arches
developing heart
upper limb bud
somites
neural tube forming
lower limb bud
tail
actual length
a
WEEKS 5–6
head growth exceeds growth of other regions
retinal pigment
future external ear
upper limb differentiation (hand plates develop, then digital rays of future fingers; wrist, elbow start forming)
umbilical cord formation between weeks 4 and 8 (amnion expands, forms tube that encloses the connecting stalk and a duct for blood vessels)
foot plate
b
actual length

Figure 45.14 (**a**) Human embryo four weeks after fertilization. Like all vertebrates, it has a tail and pharyngeal arches. (**b**) Embryo at five to six weeks. (**c**) Embryo at the boundary between the embryonic and fetal periods. It now has human features. It floats in amniotic fluid. The chorion covers the amniotic sac but has been pulled aside here. (**d**) Fetus at sixteen weeks. Movements begin as nerves make functional connections with forming muscles. Legs kick, arms wave, fingers grasp, the mouth puckers. These reflex actions will be vital skills in the world outside the uterus.

As the fourth week of the embryonic period draws to a close, the embryo has grown to 500 times its original size. The placenta has been sustaining the growth spurt, but now the pace slows as details of organs fill in. Limbs form. Fingers and toes develop from embryonic paddles. The umbilical cord also develops, and the circulatory system becomes intricate. Growth of the all-important head now surpasses that of any other body region (Figure 45.14). The embryonic period ends after the eighth week is over. No longer is the embryo merely "a vertebrate." By now its features clearly define it as a human fetus.

During the second trimester, the fetus is moving its facial muscles. It frowns; it squints. It busily practices

Further reading: Student Guide to InfoTrac on web site

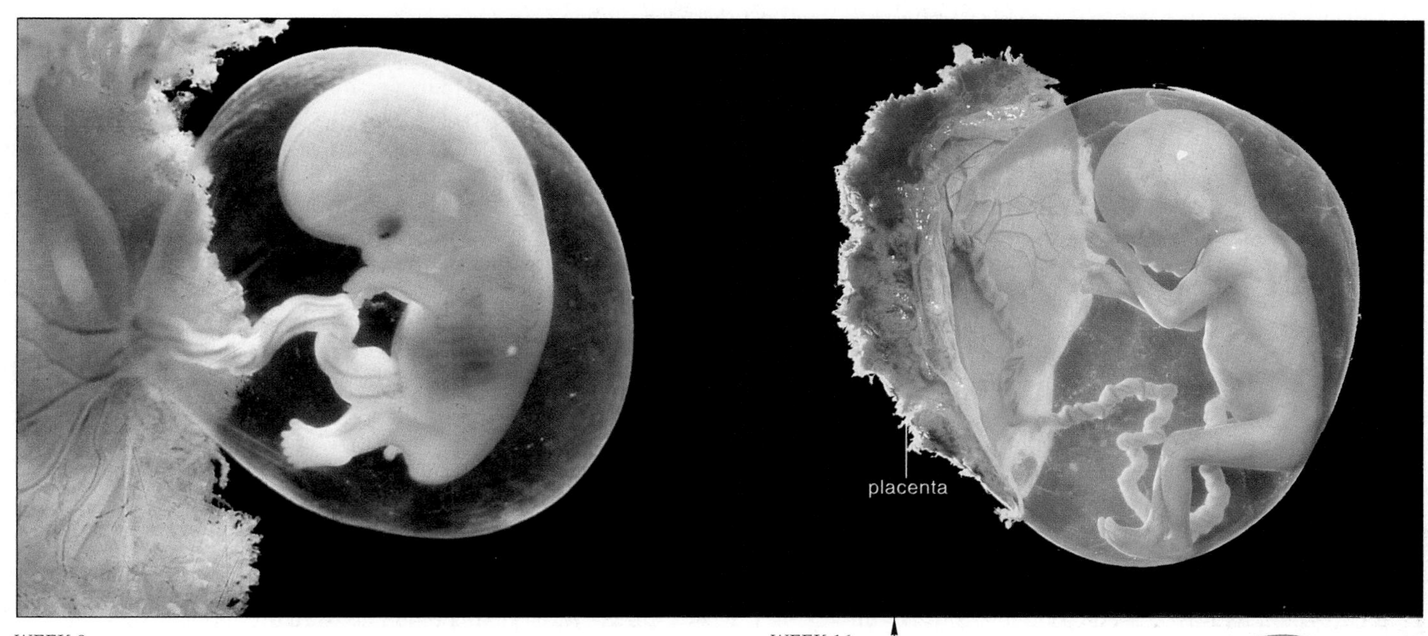
placenta

WEEK 8

final week of embryonic period; embryo looks distinctly human compared to other vertebrate embryos

upper and lower limbs well formed; fingers and then toes have separated

primordial tissues of all internal, external structures now developed

tail has become stubby

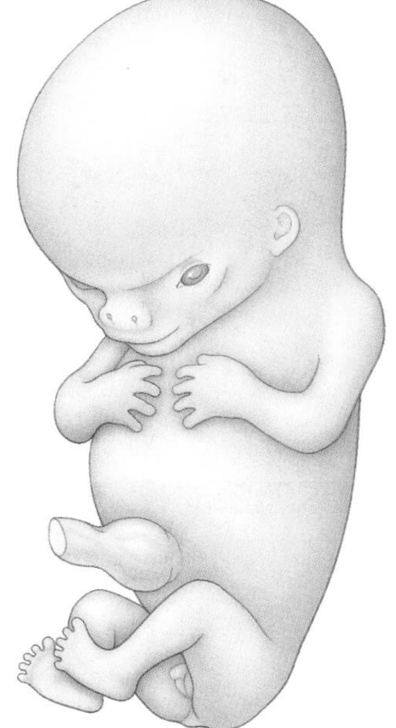

c actual length

WEEK 16
Length: 16 centimeters (6.4 inches)
Weight: 200 grams (7 ounces)

WEEK 29
Length: 27.5 centimeters (11 inches)
Weight: 1,300 grams (46 ounces)

WEEK 38 (full term)
Length: 50 centimeters (20 inches)
Weight: 3,400 grams (7.5 pounds)

During fetal period, length measurement extends from crown to heel (for embryos, it is the longest measurable dimension, as from crown to rump).

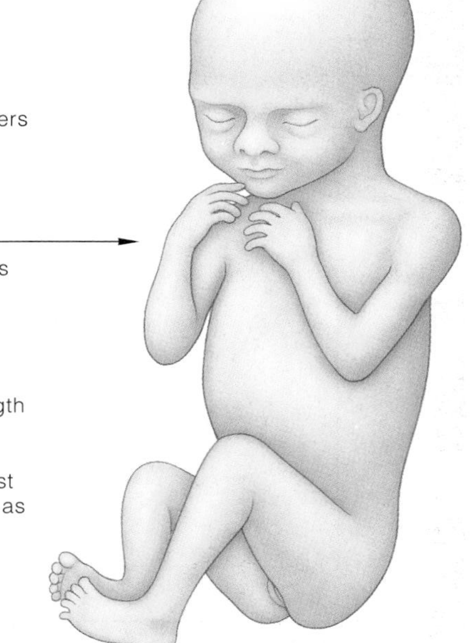

d

the sucking reflex, as shown in the next section. Now the mother can easily sense movements of the fetal arms and legs. When the fetus is five months old, she can hear its heart through a stethoscope positioned on her abdomen. Soft, fuzzy hair (the lanugo) covers the fetal body. The skin is wrinkled, reddish, and protected from abrasion by a thick, cheesy coating. In the sixth month, delicate eyelids and eyelashes form. During the seventh month, the eyes open.

A fetus born prematurely (before 22 weeks) cannot survive. The situation also is grave for births before 28 weeks, mainly because the lungs have not developed sufficiently. The risks start to drop after this. But a fetus born before the optimal birthing time (about 38 weeks after the estimated time of fertilization) still has trouble breathing and maintaining a normal core temperature even with the best medical care. By 36 weeks, however, the survival rate is 95 percent.

In the fetal period, primary tissues that formed in the early embryo become sculpted into distinctly human features.

45.11 MOTHER AS PROVIDER, PROTECTOR, POTENTIAL THREAT

A woman who decides to become pregnant is committing a large part of her body's resources and functions to the development of a new individual. From fertilization until birth, her future child is absolutely at the mercy of her diet, health habits, and life-style (Figure 45.15).

SOME NUTRITIONAL CONSIDERATIONS How does a pregnant woman best provide nutrients for the embryo, then the fetus? The same balanced diet that is good for her should provide her future child with all of the required carbohydrates, lipids, and proteins. (Chapter 42 is one starting point for an understanding of human nutritional requirements.) The woman's own demands for vitamins and minerals increases during pregnancy. The placenta absorbs enough for her embryo from the bloodstream, except for folate (folic acid). She can reduce the risk that her embryo will develop severe neural tube defects by taking more B-complex vitamins (under supervision by her doctor) before conception and during early pregnancy. Besides this, one study showed that women who smoked while pregnant had depressed blood levels of vitamin C even when their vitamin C intake was identical to that of a control group. *And so did their fetuses.* Smoking may affect utilization of other nutrients as well.

A pregnant woman also must eat enough so that her body weight increases by between 20 and 25 pounds, on the average. If her weight gain is a great deal lower than that, she is stacking the deck against the fetus. Compared to newborns of normal weight, significantly underweight newborns have more postdelivery complications. They are also at greater risk of having impaired mental functions later in life.

As birth approaches, a fetus makes greater nutritional demands of the mother. Clearly her diet will profoundly influence the remaining developmental events. The brain, like most of the other fetal organs, is especially vulnerable in the weeks just before and after birth, when it undergoes its greatest expansion. All of the fetal body's neurons have formed, so poor nutrition now will have repercussions on intelligence and other neural functions later in life.

RISK OF INFECTIONS The antibodies circulating in a pregnant woman's blood continually move across the placenta. They protect the new individual from all but

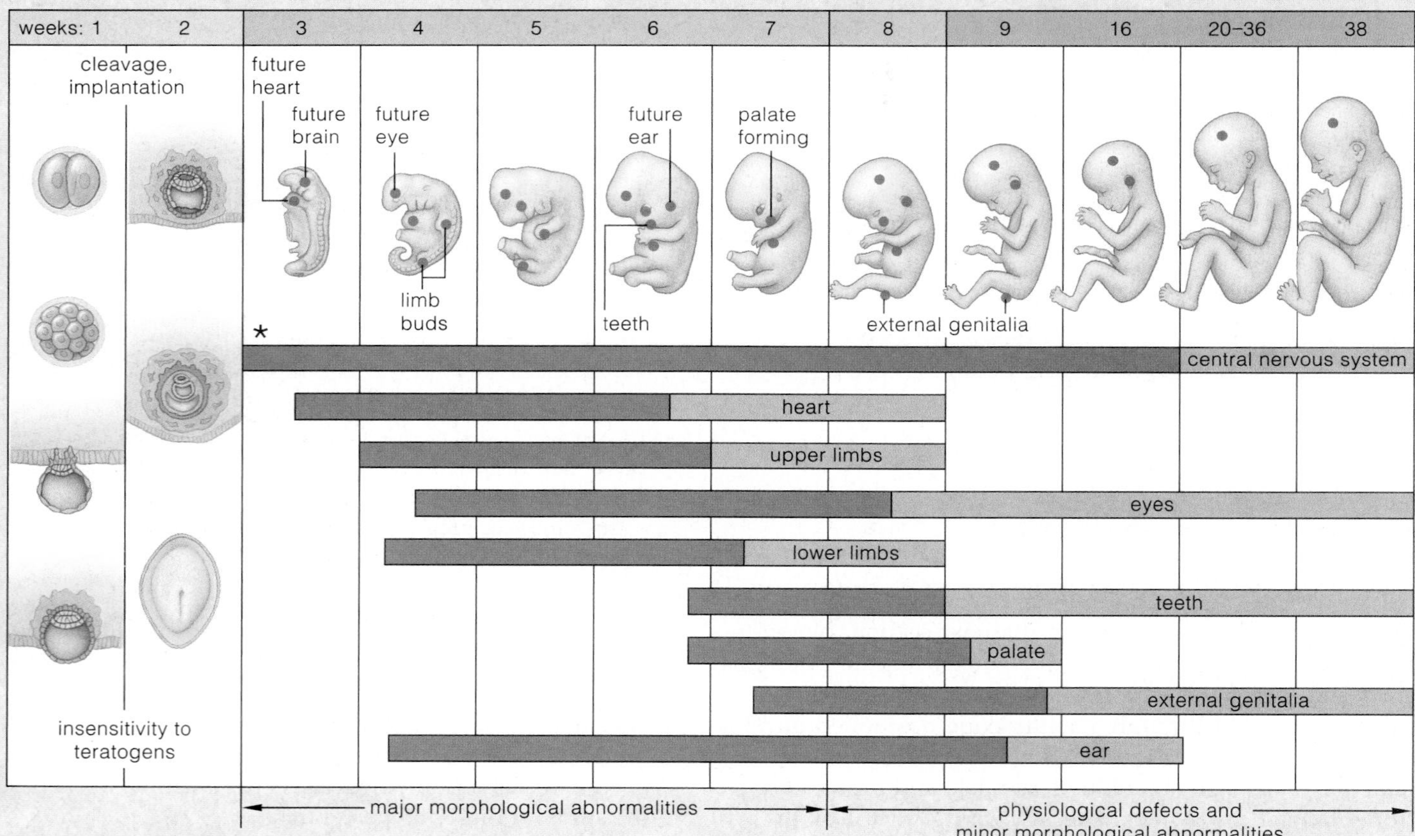

weeks: 1
2
3
4
5
6
7
8
9
16
20–36
38
cleavage, implantation
future heart
future brain
future eye
future ear
palate forming
limb buds
teeth
external genitalia
*
central nervous system
heart
upper limbs
eyes
lower limbs
teeth
palate
external genitalia
ear
insensitivity to teratogens
major morphological abnormalities
physiological defects and minor morphological abnormalities

Figure 45.15 Sensitivity to teratogens during pregnancy. *Teratogens* are drugs and any other environmental factors that may induce deformities in the embryo or fetus. They usually have no effect before the onset of organ formation. After that, they can block or abnormally stimulate growth, tissue remodeling, and tissue resorption.

* *dark blue* denotes highly sensitive period; *light blue* denotes less severe sensitivity to teratogens

Further reading: Student Guide to InfoTrac on web site —

the most serious bacterial infections. Some virus-induced diseases can be dangerous during the first six weeks after fertilization, which is a critical time of organ formation. Suppose a pregnant woman contracts *rubella* (German measles) during this critical period. There is a 50 percent chance that some organs of her embryo will not form properly. For example, if she becomes infected when embryonic ears are forming, her newborn may be deaf. If she becomes infected during or after the fourth month of pregnancy, the disease will have no notable effect. However, a woman can avoid this risk entirely, because vaccination before pregnancy can prevent rubella.

EFFECTS OF PRESCRIPTION DRUGS A pregnant woman absolutely should not take any drugs unless she is under close medical supervision. Think about what happened when the tranquilizer *thalidomide* was being routinely prescribed in Europe. Women who had used thalidomide during the first trimester gave birth to infants who were either missing arms and legs or had grossly deformed ones. As soon as its connection with the deformities was apparent, thalidomide was withdrawn from the market.

However, other tranquilizers, sedatives, and barbiturates are still being prescribed. There is a risk that they may cause similar, although less severe, damage. Even certain *anti-acne drugs* increase the risk of facial and cranial deformities. Or consider two overprescribed antibiotics. One of these, tetracycline, yellows teeth. Streptomycin causes hearing problems and may adversely affect the nervous system.

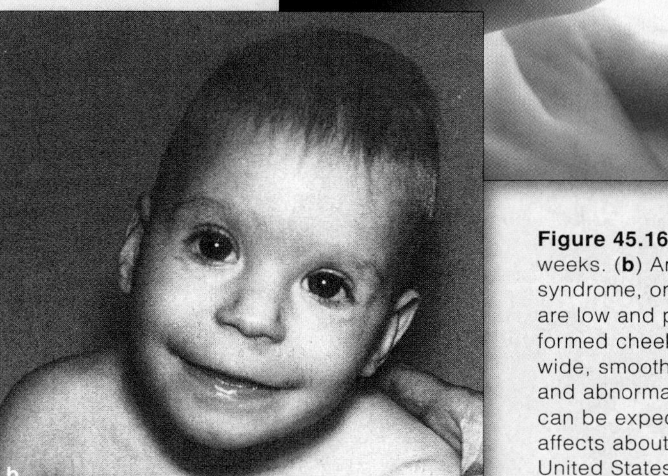

Figure 45.16 (**a**) The fetus at eighteen weeks. (**b**) An infant with fetal alcohol syndrome, or FAS. Obvious symptoms are low and prominent ears, improperly formed cheekbones, and an abnormally wide, smooth upper lip. Growth problems and abnormalities of the nervous system can be expected. This disorder currently affects about 1 in 750 newborns in the United States.

EFFECTS OF ALCOHOL As a fetus grows, its physiology becomes increasingly like its mother's. Because the fetus is so small and grows so rapidly, alcohol is more harmful to it—and alcohol passes freely across the placenta.

Drinking alcohol while pregnant invites *fetal alcohol syndrome*, or FAS. Symptoms include reduced brain and head size, mental retardation, facial deformities, poor growth and coordination, and often heart defects (Figure 45.16*b*). One in every 750 newborns shows the abnormal features. The symptoms are irreversible; children affected by FAS never "catch up," physically or mentally.

Many researchers suspect there is no "safe" drinking level; any alcohol may harm the fetus. More doctors are now urging near- or total abstinence during pregnancy.

EFFECTS OF COCAINE A pregnant woman who uses cocaine, crack especially, disrupts the nervous system of her future child as well as her own. Here you may wish to read again the start of Chapter 35, which describes the kind of future that a crack addict's offspring will face.

EFFECTS OF CIGARETTE SMOKE Again, cigarette smoke impairs fetal growth and development. Besides this, a long-term study at Toronto's Hospital for Sick Children showed that toxic elements in tobacco accumulate even in the fetuses of pregnant nonsmokers who are exposed to *secondhand smoke* at home or work.

Daily smoking during pregnancy leads to underweight newborns even if the woman's weight, nutritional status, and all other relevant variables match those of pregnant nonsmokers. Smoking has other effects. In Great Britain, all infants born during the same week were tracked for seven years. Infants of smokers were smaller, had twice as many heart abnormalities, and died of more postdelivery complications. At age seven, they were nearly half a year behind children of nonsmokers in average "reading age."

The mechanisms by which smoking affects a fetus are not known. Its demonstrated effects are evidence that the placenta, marvelous structure that it is, cannot prevent all assaults on the fetus that the human mind can dream up.

FROM BIRTH ONWARD

Giving Birth

On average, pregnancy ends thirty-eight weeks after fertilization, give or take a few weeks. We call the birth process **labor** (or delivery). Labor involves dilation of the cervical canal, so the fetus can move out from the uterus, through the vagina, and out into the world. It also requires strong uterine contractions as the driving force behind the expulsion.

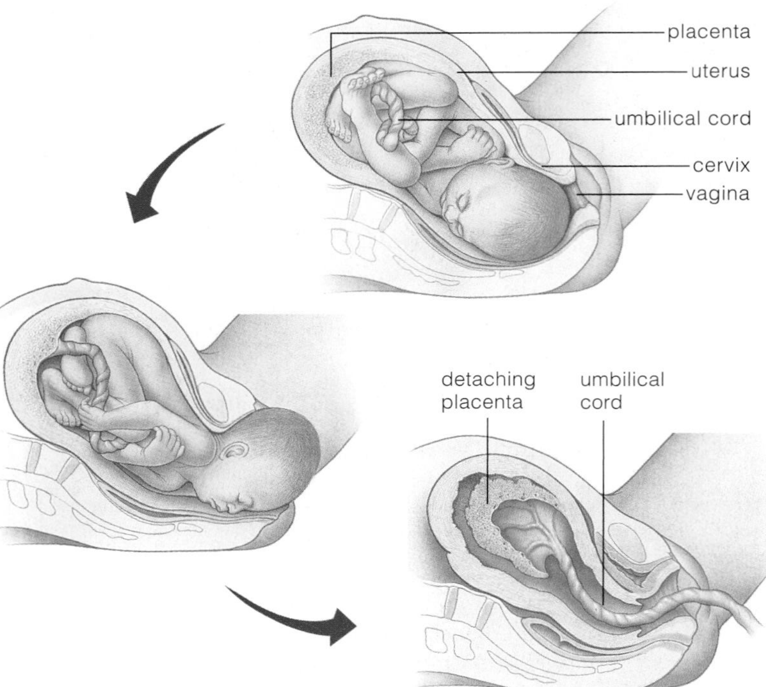
placenta
uterus
umbilical cord
cervix
vagina
detaching placenta
umbilical cord

Figure 45.17 Expulsion of a human fetus and the afterbirth (the placenta, tissue fluid, and blood) during the process of birth.

During the last trimester, mild uterine contractions begin. The cervix softens as its connective tissues loosen. Relaxin, a peptide hormone secreted from the corpus luteum and placenta, brings about the softening. Also, relaxin induces connections between the pelvic bones to loosen up. Meanwhile the fetus "drops," or is shifted downward, usually with its head touching the cervix (Figure 45.17). A *breech birth* is the likely outcome when any part of the fetus other than its head gets positioned near the birth canal first.

Rhythmic, usually painless contractions herald the onset of labor. They increase in frequency and intensity over the next two to eighteen hours.

Just before birth, the amnion typically ruptures, and "water" (amniotic fluid) rushes from the vagina. Usually contractions expel the fetus less than an hour after the cervix has fully dilated. Fifteen to thirty minutes later, contractions make the placenta detach from the uterus; it is expelled as the "afterbirth." The contractions also constrict blood vessels at the placental attachment site and prevent hemorrhaging. Someone ties and cuts the umbilical cord. A few days after it shrivels, the stump is the newborn's navel. Once the lifeline to the mother is severed, the newborn embarks on its course of post-embryonic development, starting with an extended time of dependency and learning that is typical of primates.

New mothers typically sink into the *afterbaby blues*, or postpartum depression. **CRH** (corticotropin releasing hormone) produced by the placenta might have a role in this. Its level in blood rises by as much as three times during pregnancy, and it might influence the timing of labor. Possibly cortisol, a stress hormone, is secreted in response to elevated CRH levels to help the mother cope with extraordinary stresses that pregnancy and labor place on her body. Once the placenta is expelled, her CRH levels crash to levels typical of some clinical depressions. Afterbaby blues continue until secretion of CRH from the hypothalamus returns to normal.

Nourishing the Newborn

Survival of the newborn requires an ongoing supply of milk or its nutritional equivalent. Milk production, or **lactation**, occurs in mammary glands in the mother's breasts (Figure 45.18). Before pregnancy, breasts consist mainly of adipose tissue and a system of undeveloped

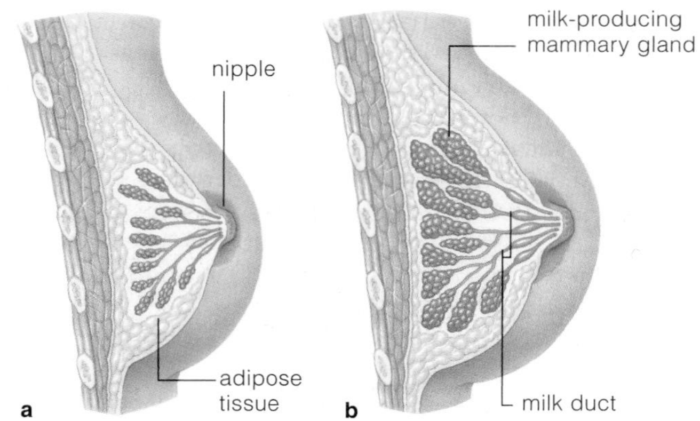
nipple
adipose tissue
a
milk-producing mammary gland
milk duct
b

Figure 45.18 (**a**) Breast of a woman who is not pregnant. (**b**) Breast of a lactating woman.

Further reading: Student Guide to InfoTrac on web site

ducts. Their size depends on how much fat they contain, not on their milk-producing ability. During pregnancy, breasts respond to estrogen and progesterone. A glandular system of milk production develops as the outcome (Figure 45.18*b*).

For the first few days following birth, both mammary glands produce a fluid that is rich in proteins and lactose. The anterior pituitary secretes **prolactin**, a hormone that induces synthesis of the enzymes required for milk production (Section 37.3). When a newborn suckles, the pituitary gland also releases **oxytocin**. This hormone triggers contractions that force the milk into breast tissue ducts and induce the uterus to shrink back to its pre-pregnancy size.

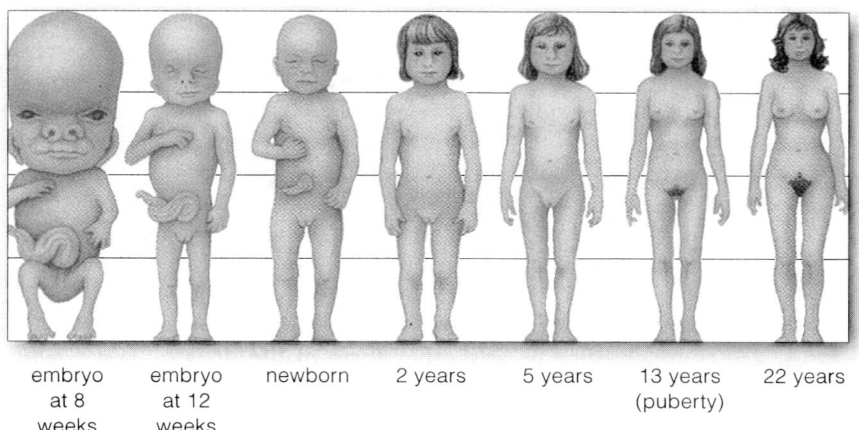
embryo at 8 weeks
embryo at 12 weeks
newborn
2 years
5 years
13 years (puberty)
22 years

Figure 45.19 Observable, proportional changes in the human body during prenatal and postnatal growth. Changes in overall physical appearance are slow but noticeable until the teens. For example, the head gets proportionally smaller, compared to the embryonic period. Legs get longer; the trunk gets shorter.

Regarding Breast Cancer

On average, 100,000 women develop *breast cancer* each year. Obesity, high cholesterol, and high estrogen levels contribute to the cancerous transformation. The chances of cure are excellent with early detection and treatment. Once a month, about a week after she has menstruated, a woman should examine her breasts. For recommended examination procedures, she can contact her physician or the American Cancer Society. The Society has listings in local telephone directories. It also posts information on its web site, http://www.cancer.org.

Table 45.4 Stages of Human Development

PRENATAL PERIOD	
Zygote	Single cell resulting from fusion of sperm nucleus and egg nucleus at fertilization
Morula	Solid ball of cells produced by cleavages
Blastocyst	Ball of cells with surface layer, fluid-filled cavity, and inner cell mass (the mammalian blastula)
Embryo	All developmental stages from two weeks after fertilization until end of eighth week
Fetus	All developmental stages from the ninth week to birth (about thirty-eight weeks after fertilization)
POSTNATAL PERIOD	
Newborn	Individual during the first two weeks after birth
Infant	Individual from two weeks to about fifteen months after birth
Child	Individual from infancy to about ten or twelve years
Pubescent	Individual at puberty, when secondary sexual traits develop; girls between ten and fifteen years, boys between twelve and sixteen years
Adolescent	Individual from puberty until about three or four years later; physical, mental, emotional maturation
Adult	Early adulthood (between eighteen and twenty-five years); bone formation and growth finished. Changes proceed very slowly afterward
Old age	Aging follows late in life

Postnatal Development and Aging

After birth, a new individual follows a course of further growth and development that leads to the adult, the mature form of the species. Table 45.4 summarizes the *prenatal* (before birth) stages and *postnatal* (after birth) stages. Figure 45.19 shows examples of the proportional changes that take place in the body as the human life cycle unfolds.

Postnatal growth is most rapid between thirteen and nineteen years. Then, secretions of sex hormones step up and bring about the emergence of secondary sexual traits as well as sexual maturity. Not until adulthood are the bones fully mature. Body tissues normally are maintained in top condition during early adulthood. As the years pass, it becomes more and more difficult to maintain and repair existing tissues, and so the body gradually deteriorates. By processes collectively known as aging, the body's cells gradually start to break down. The processes are not fully understood, but they bring about structural changes and gradual loss of functions. As Section 44.9 explains, aging and death are the fate of all animals that show extensive cell differentiation.

The human life cycle flows naturally from the time of birth, growth, and development, to production of the individual's own offspring, and on through aging to the time of death.

45.13 CONTROL OF HUMAN FERTILITY

Some Ethical Considerations

The transformation of a zygote into an adult of intricate detail raises profound questions. *When does development begin?* As you have seen, major developmental events unfold even before fertilization. *When does life begin?* During her lifetime, a woman can produce as many as 500 eggs, all of which are alive. During one ejaculation, a man can release a quarter of a billion sperm, which are alive. Even before sperm and egg merge by chance and establish the genetic makeup of a new individual, they are as much alive as any other form of life. It is scarcely tenable, then, to say life begins at fertilization. *Life began billions of years ago—and every gamete, every zygote, every sexually mature individual is but a fleeting stage in the continuation of that beginning.*

This greater perspective on life cannot diminish the meaning of conception. It is no small thing to entrust a new individual with the gift of life, wrapped in the unique evolutionary threads of our species and handed down through an immense sweep of time.

Yet how can we reconcile the marvel of individual birth with growing awareness of the astounding birth rate for the human species? Almost 15,000 newborns are entering the world every hour. By the time you go to bed tonight, there will be 358,300 more newborns on Earth than there were last night at that hour. In three months there will be 32,790,000 more—about as many people as there are now in the entire state of California. *Within three months.*

Worldwide, human population growth is rapidly outstripping resources. Each year, many millions face the horrors of starvation. Living as we do on one of the most productive continents on Earth, few of us know what it means to give birth to a child, to give it the gift of life, and have no food to keep it alive.

And how can we reconcile the marvel of birth with the confusion that surrounds unwanted pregnancies? Even highly developed countries do not have adequate educational programs concerning fertility. And many people are not inclined to exercise control. Each year, there are 800,000 unintended teenage pregnancies in the United States. Many parents actually encourage early boy–girl relationships without thinking through the risks of premarital intercourse and unplanned pregnancy. Advice is often condensed to a terse "Don't do it. But if you do it, be careful!" And each year, there are between 1.2 and 1.3 million abortions among all age groups.

The motivation to engage in sex has been evolving for more than 500 million years. And a few centuries of moral and ecological arguments for its suppression have not stopped all that many unwanted pregnancies. Besides this, complex social factors have contributed to a population growth rate that is out of control.

How will we reconcile our biological past and the need for a stabilized cultural present? Whether and how human fertility is to be controlled is one of the most volatile issues of our time. We will return to this issue in the next chapter, in the context of principles that govern the growth and stability of populations. Here, we briefly consider some control options.

Birth Control Options

The most effective method of birth control is complete *abstinence,* no sexual intercourse whatsoever. Data show that it is unrealistic to expect many people to practice it.

A less reliable variation of abstinence is the *rhythm method*. The idea is to avoid intercourse in the woman's fertile period, starting a few days before ovulation and ending a few days after. A woman identifies and tracks her fertile period. She also keeps records of the length of her menstrual cycles, takes her temperature each morning when she wakes up, or both. (The body's core temperature rises by one-half to one degree just before a fertile period.) But ovulation may not be regular, and miscalculations are frequent. Also, sperm deposited in the vagina a few days before ovulation may survive until ovulation. The rhythm method *is* inexpensive; it costs nothing after you buy a thermometer. It does not require fittings and periodic medical checkups. But its practitioners run a risk of pregnancy (Figure 45.20).

Withdrawal, or removing the penis from the vagina before ejaculation, dates back at least to biblical times. But withdrawal requires very strong willpower, and the practice may fail anyway. Fluid released from the penis just before ejaculation may contain some sperm.

Douching (rinsing the vagina with a chemical right after intercourse) is next to useless. Sperm move past the cervix and out of reach of the douche within ninety seconds after ejaculation.

Controlling fertility by surgical intervention is less chancy. In *vasectomy*, a physician makes a tiny incision in a man's scrotum, then severs and ties off each vas deferens. The operation takes only twenty minutes and requires only a local anesthetic. After the operation, sperm cannot leave the testes and cannot be present in semen. So far, there is no firm evidence that vasectomy disrupts hormonal functions or adversely affects sexual activity. Vasectomies can be reversed. But half of those who submit to surgery later develop antibodies against sperm and may not be able to regain fertility.

Females may opt for a *tubal ligation*. By this surgical intervention, oviducts are cauterized or cut and tied off, usually in a hospital. When the operation is performed correctly, tubal ligation is the most effective means of birth control. A few women have recurring pain in the pelvic area. The operation sometimes can be reversed.

Further reading: Student Guide to InfoTrac on web site

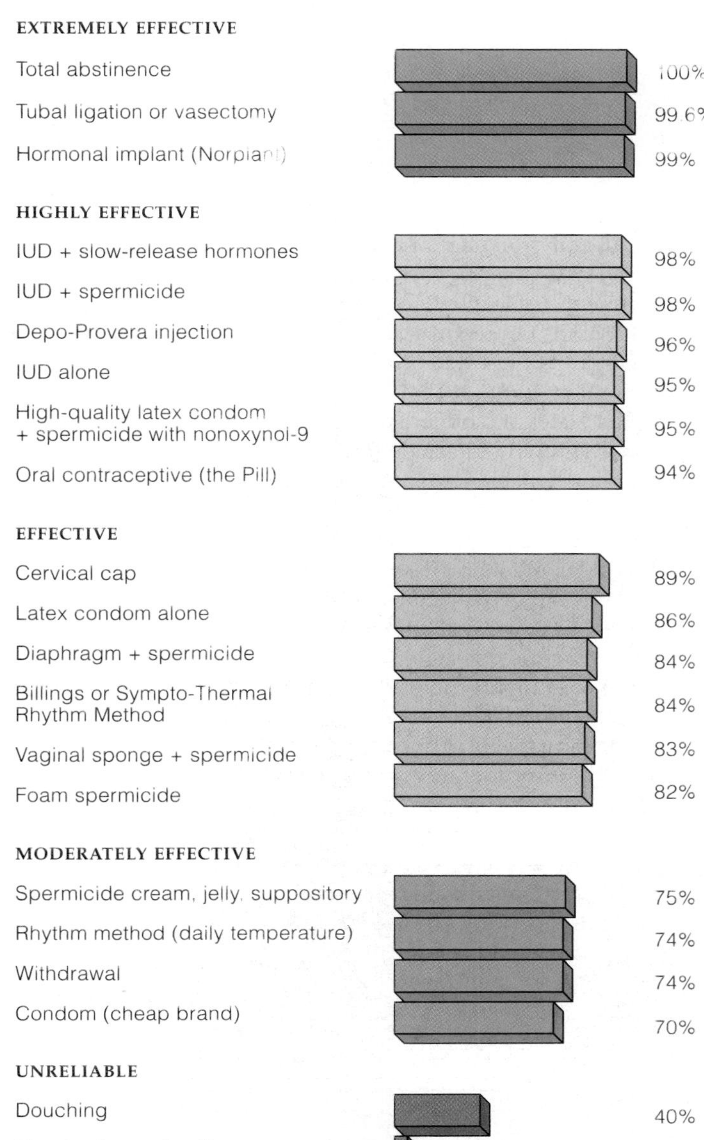
EXTREMELY EFFECTIVE
Total abstinence
100%
Tubal ligation or vasectomy
99.6%
Hormonal implant (Norplant)
99%
HIGHLY EFFECTIVE
IUD + slow-release hormones
98%
IUD + spermicide
98%
Depo-Provera injection
96%
IUD alone
95%
High-quality latex condom + spermicide with nonoxynol-9
95%
Oral contraceptive (the Pill)
94%
EFFECTIVE
Cervical cap
89%
Latex condom alone
86%
Diaphragm + spermicide
84%
Billings or Sympto-Thermal Rhythm Method
84%
Vaginal sponge + spermicide
83%
Foam spermicide
82%
MODERATELY EFFECTIVE
Spermicide cream, jelly, suppository
75%
Rhythm method (daily temperature)
74%
Withdrawal
74%
Condom (cheap brand)
70%
UNRELIABLE
Douching
40%
Chance (no method)
10%

Figure 45.20 Comparison of the effectiveness of methods of contraception in the United States. The percentages also indicate the number of unplanned pregnancies per 100 couples who used only that method of birth control for a year. For example, "94% effectiveness" for oral contraceptives (the Pill) means that 6 of every 100 women will still become pregnant, on the average.

Less drastic methods of controlling fertility involve physical or chemical barriers that prevent sperm from entering the uterus and oviducts. *Spermicidal foam* and *spermicidal jelly* are toxic to sperm. They are transferred from an applicator into the vagina before intercourse. These products aren't always reliable unless used with another device, such as a diaphragm or condom. *IUDs* are coils inserted into the uterus by a physician. They sometimes invite pelvic inflammatory disease.

A *diaphragm* is a flexible, dome-shaped device. It is inserted into the vagina and positioned over the cervix before intercourse. It is relatively effective when fitted initially by a doctor, used with foam or jelly before each sexual contact, inserted correctly each time, and left in place only for a prescribed length of time.

Good brands of *condoms*—thin, tight-fitting sheaths worn over the penis during intercourse—are up to 95 percent effective if used with a spermicide. Only latex condoms offer protection against sexually transmitted diseases (Section 45.14). However, condoms often tear and leak, at which time they become absolutely useless.

The *birth control pill* is an oral contraceptive made of synthetic estrogens and progestins (progesterone-like hormones). Taken daily except for the last five days of a menstrual cycle, it suppresses oocyte maturation and ovulation. Often it corrects erratic menstrual cycles and reduces cramps, but in some women it causes nausea, weight gain, tissue swelling, and headaches. With more than 50 million women taking it, "the Pill" is the most-used fertility control method. It is 94 percent effective. Earlier formulations were linked to blood clots, high blood pressure, and maybe breast cancer. Lower doses of newer formulations might lessen the risk of breast and endometrial cancer. The possibility of a correlation between the Pill and breast cancer is still under study.

Progestin injections or implants inhibit ovulation. A *Depo-Provera* injection works for three months and is 96 percent effective. *Norplant* (six rods implanted under the skin) works for five years and is 99 percent effective. Both may cause sporadic, heavy bleeding, and doctors have some trouble surgically removing Norplant rods.

A pregnancy test doesn't register positive until after implantation. According to one view, a woman is not pregnant until that time, so *morning-after pills* intercept pregnancy. Actually, such pills work up to seventy-two hours after unprotected intercourse by interfering with the hormones that control events between ovulation and implantation. One is Preven, a kit with high doses of birth control pills that suppress ovulation and block corpus luteum function. RU-486, more commonly used as an abortion pill, also can be used as a morning-after pill. It can block fertilization and progesterone-dependent implantation. It also may induce abortion when taken within seven weeks of implantation (progesterone also is essential to maintain pregnancy). As is the case with oral contraceptives, RU-486 may trigger elevated blood pressure, blood clots, or breast cancer in some women. At this writing, RU-486 is available in Europe, but its use in the United States is controversial. By contrast, Preven had been used previously as a contraceptive, and its repackaging as a morning-after pill has not been a subject of controversy. It gained FDA approval in 1998.

Whether and how human fertility is to be controlled is a volatile issue. It centers on reconciling our biological past with seriously divergent views about our cultural present.

45.14 SEXUALLY TRANSMITTED DISEASES

At some point in his or her life, at least one of every four people in the United States who engages in sexual intercourse will become infected by the pathogens that cause **sexually transmitted diseases**, or **STDs**. Tens of millions are already infected; STDs have reached epidemic proportions. Two-thirds are under age twenty-five; one-fourth are teenagers. Women and children are hardest hit. Worse yet, antibiotic-resistant strains of bacteria are on the rise, and some viral diseases simply cannot be cured.

Urban poverty, prostitution, intravenous drug abuse, and sex-for-drugs are fanning the epidemic. Yet STDs also are rampant in high schools and colleges, where students too often think, "It can't happen to me." They reject the idea that no sex—abstinence—is the only safe sex. In one poll of high school students, two-thirds of the respondents said they don't use condoms. More than 40 percent of them have two or more sex partners.

The economics of this health problem are staggering. In 1993, the Centers for Disease Control tallied the annual cost of treating the most prevalent STDs: *Herpes*, $759 million; gonorrhea, $1 billion; chlamydial infection, $2.4 billion; and pelvic inflammatory disease, $4.2 billion. This does not include the accelerating cost of treating patients with AIDS. In many developing countries, AIDS alone threatens to overwhelm health-care delivery systems and to unravel decades of economic progress.

The social consequences are sobering. Mothers bestow a chlamydial infection on 1 of every 20 newborns in the United States alone. They bestow type II *Herpes* virus on 1 in 10,000 newborns; one-half of the infected babies die, and one-fourth have severe neural defects. Every year, 1 million women develop pelvic inflammatory disease, and 100,000 to 150,000 become infertile.

AIDS Someone can become infected by HIV, the human immunodeficiency virus, and not even know it. However, the infection marks the start of a titanic battle that the immune system almost certainly will lose (Section 40.11). At first there may be no outward symptoms. Five to ten years later, a set of chronic disorders develops. They are evidence of *AIDS* (*A*cquired *I*mmune *D*eficiency *S*yndrome). When the immune system finally is weakened, the stage is set for opportunistic infections. Normally harmless, resident bacteria are the first to take advantage of the lowered resistance. Dangerous pathogens also take their toll. In time, the immune-compromised person simply is overwhelmed.

Most commonly, HIV spreads through anal, vaginal, and oral intercourse and by IV drug users. Most infections occur by the transfer of blood, semen, urine, or vaginal secretions between people. Cuts or abrasions on the penis, vagina, rectum, and oral membranes invite HIV into the internal environment of a new host.

Today there is no vaccine against HIV and no effective treatment for AIDS. If you get it, you die. *There is no cure.*

HIV was not identified until 1981, but it was present in some parts of Central Africa for at least several decades before that. In the 1970s and early 1980s, it spread to the United States and other developed countries, where most of those initially infected were male homosexuals. Today, many in the heterosexual population are infected or at risk.

Free or low-cost, confidential testing for HIV exposure is available at public health facilities and at many doctor's offices. People who are worried should know there may be a time lag between their first exposure and the first test to come out positive. It takes a few weeks to six months or more before detectable amounts of antibodies will form in response to infection. (The presence of antibodies in the blood indicates exposure to the virus.) Anyone who tests positive is capable of spreading the virus.

Public education programs attempt to stop the spread of HIV. Most health-care workers advocate safe sex, yet there is great confusion about what "safe" means. Many advocate the use of high-quality latex condoms *and* a spermicide that contains nonoxynol-9 to help prevent the transmission of the virus. Even then, there is a slight risk. As an unfortunate couple learned in 1997, no one should participate in open-mouthed kissing with someone who tests positive for HIV. Caressing is not risky *if* there are no lesions or cuts where body fluids that harbor the virus can enter the body. Pronounced lesions caused by other sexually transmitted diseases may increase susceptibility to HIV infection.

In sum, AIDS reached pandemic proportions mainly for three reasons. First, it took a while to discover that the virus can travel in semen, blood, and vaginal fluid and that *behavioral* controls can limit its spread. Second, it took time to develop ways to test symptom-free carriers, who infect others. Third, many still don't realize the medical, economic, and social consequences affect everyone.

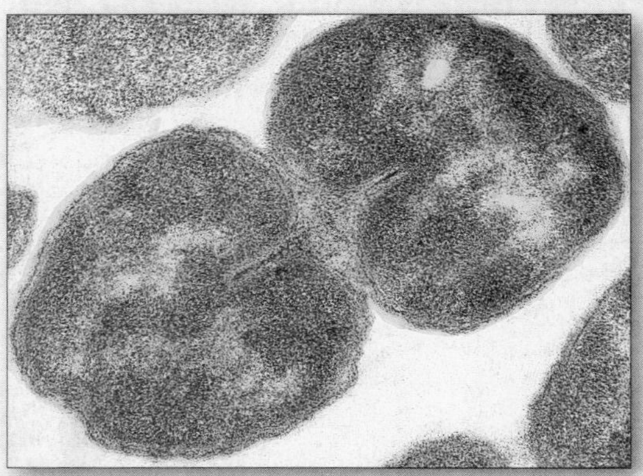

Figure 45.21 Bacterial agents of gonorrhea.

GONORRHEA During intercourse, *Neisseria gonorrhoeae* (Figure 45.21) can enter the body at mucous membranes of the urethra, cervix, and anal canal. This bacterium causes *gonorrhea*. An infected female may notice a slight vaginal discharge or burning sensation while she is urinating. If the infection spreads to her oviducts, it may induce severe cramps, fever, vomiting, and scarring, which may cause sterility. Males have more obvious symptoms. Within a week of infection, the penis discharges yellow pus. Urinating is more frequent and may be painful.

Further reading: Student Guide to InfoTrac on web site

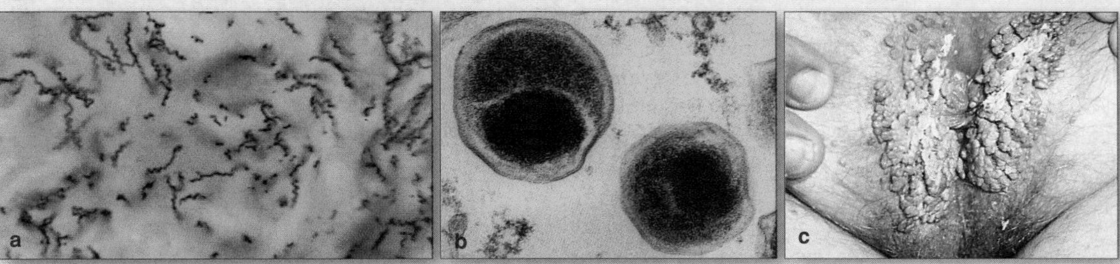
a
b
c

Figure 45.22 The bacterial agents of (**a**) syphilis and (**b**) chlamydia. (**c**) Genital warts on the external genitalia of a young woman.

This STD is rampant even though prompt treatment quickly cures it. Why? Women experience no troubling symptoms during early stages. Also, infection does not confer immunity to the bacterium, perhaps because there are sixteen or more different strains of it. Contrary to common belief, someone can get gonorrhea over and over again. Also, oral contraceptives, used so widely, promote infection by altering vaginal pH. Populations of resident bacteria decline—and *N. gonorrhoeae* is free to move in.

SYPHILIS *Syphilis*, a dangerous STD, is caused by the spirochete *Treponema pallidum* (Figure 45.22*a*). Sex with an infected partner puts the motile, spiral bacterium on the surface of genitals or into the cervix, vagina, or oral cavity. It can enter the body through tiny epidermal cuts. One to eight weeks later, new treponemes are twisting about in a flattened, painless chancre (local ulcer). This first chancre is a symptom of the primary stage of syphilis. By then, treponemes are in the blood. Treponemes can cross the placenta of an infected, pregnant woman. The result will be a miscarriage, stillbirth, or syphilitic newborn.

The chancres usually heal, but treponemes multiply in the spinal cord, brain, eyes, mucous membranes, joints and bones. More chancres and a skin rash develop during this infectious, secondary stage of syphilis. Immune responses counter the disease in about 25 percent of the cases, so the symptoms subside. Another 25 percent remain infected but symptom free. In the rest, minor to significant lesions and scars appear in the skin and liver, bones, and other internal organs. Few treponemes form during this tertiary stage. But the host's immune system is hypersensitive to them. Chronic immune reactions may severely damage the brain and spinal cord and lead to general paralysis.

Probably because the symptoms are so alarming, more people seek early treatment for syphilis than for gonorrhea. Later stages require prolonged treatment.

PELVIC INFLAMMATORY DISEASE *Pelvic inflammatory disease* (PID) is one of the most serious complications of gonorrhea, chlamydial infections, and other STDs. It also can arise when normal bacterial residents of the vagina ascend to the pelvic region. Most commonly, the uterus, oviducts, and ovaries are affected. There is bleeding and vaginal discharge. Pain in the lower abdomen may be as severe as an acute appendicitis attack. The oviducts may become scarred. Pelvic inflammatory disease invites abnormal pregnancies as well as sterility.

GENITAL HERPES About 25 million Americans have *genital herpes*, caused by the type II *Herpes simplex* virus. Infection requires direct contact with active *Herpes* viruses or sores that contain them. Mucous membranes of the mouth and the genitals are highly susceptible to invasion. The symptoms are often mild or absent. Among infected women, small, painful blisters may appear on the vulva, cervix, urethra, or anal tissues. Among men, blisters form on the penis and anal tissues. Within three weeks, the virus enters latency, and the sores crust over and heal.

The virus is reactivated sporadically. Each time, it causes painful sores at or near the original infection site. Sexual intercourse, menstruation, emotional stress, or other infections trigger *Herpes* flare-ups. Acyclovir, an antiviral drug, decreases healing time and often decreases the pain.

CHLAMYDIAL INFECTION *Chlamydia trachomatis* (Figure 45.22*b*), a parasitic bacterium, spends part of its life cycle in genital and urinary tract cells. It causes several diseases, including *NGU* (chlamydial nongonococcal urethritis). Chlamydial infections are now the most common reported STDs in the United States; 4 million are infected annually. Rates are highest for individuals between ages fifteen and nineteen. *C. trachomatis* and *N. gonorrhoeae* can enter new hosts at the same time by vaginal, oral, or anal intercourse. Prompt doses of penicillin cure the gonorrhea—but not NGU, which requires both tetracycline and sulfonamide.

NGU leads to inflammation of the cervix and, in both sexes, the urethra. Infected people may notice a burning sensation while urinating. When they are symptom-free, they do not seek treatment and complications develop. In males, the prostate becomes swollen and inflamed. In females, the infection may spread into the uterus and the oviducts to cause pelvic inflammatory disease.

Symptom-free infected individuals are unwittingly spreading destructive chlamydial infections through all ethnic groups, among poor and affluent alike.

GENITAL WARTS More than sixty types of the human papillomaviruses (HPV) are known. A few cause *genital warts*, or benign, bumplike growths (Figure 45.22*c*). HPV infection of the genitals and the anus is one of the most prevalent STDs in this country. Usually type 16 HPV does not cause obvious warts, but it may cause precancerous sores and cancers of the cervix, vagina, vulva, penis, and anus. In one Seattle study, 22 percent of female college students who were examined tested positive for the virus.

Focus on Bioethics

45.15 TO SEEK OR END PREGNANCY

Because of sterility or infertility, about 15 percent of all couples in the United States cannot conceive. For example, hormonal imbalances in the female body may prevent ovulation. Or the male's sperm count may be so low that fertilization would be next to impossible without medical intervention.

In vitro fertilization means conception outside the body (literally "in glass" petri dishes or test tubes). It may occur if the couple can produce normal sperm and oocytes. First the woman receives injections of a hormone that prepares her ovaries for ovulation. Afterward, the preovulatory oocyte is removed from her with a suction device. In the meantime, sperm from the male are placed in a solution that simulates fluid in the oviducts. A few hours after the sperm and suctioned oocyte make contact in a petri dish, fertilization may occur. Twelve hours later, the zygote is transferred to a solution that can sustain it through the initial cleavages. Two to four days later, the resulting ball of cells is transferred to the female's uterus.

Each attempt at in vitro fertilization costs about 8,000 dollars, and most attempts aren't successful. In 1994, each "test-tube" baby cost the nation's health-care system about 60,000 to 100,000 dollars, on average. The childless couple may believe no cost is too great. But in an era of increased population growth and shrinking medical coverage, is the cost too great for society to bear? Court battles are being waged over this issue.

At the other extreme is **abortion**, the dislodging and removal of the blastocyst, embryo, or fetus from the uterus. At one time in the United States, abortions were forbidden by law unless the pregnancy endangered the mother's life. Later the Supreme Court ruled that the government does not have the right to forbid abortions during the early stages of pregnancy (typically up to five months). Before this ruling, there were dangerous, traumatic, and often fatal attempts to abort embryos, either by pregnant women themselves or by quacks.

From a clinical standpoint, vacuum suctioning and other methods make abortion painless for the woman, rapid, and free of complications when performed during the first trimester. Abortions performed in the second and third trimesters are extremely controversial unless the mother's life is threatened. Even so, for both medical and humanitarian reasons, the majority of people in this country generally agree that the preferred route to birth control is not through abortion. Rather, it is through sexually responsible behavior that prevents unwanted pregnancy from happening in the first place.

This biology textbook cannot offer you the "right" answer to a moral question because of the reasons given in Section 1.7. It *can* provide you with a serious, detailed description of how a new human individual develops. Your choice of the "right" answer to the question of the morality of abortion will be just that—your choice—and one that can be based on objective insights into the nature of life.

45.16 SUMMARY

1. Humans have a pair of primary reproductive organs (sperm-producing testes in males, and egg-producing ovaries in females), accessory ducts, and glands. Testes and ovaries also produce sex hormones that influence reproductive functions and secondary traits.

a. In men, the hormones LH, FSH, and testosterone control sperm formation. They are part of feedback loops from the testes to the hypothalamus and anterior lobe of the pituitary gland that involve GnRH and inhibin.

b. In women, estrogens, progesterone, FSH, and LH, and GnRH control the maturing and release of oocytes from an ovary, and cyclic changes in the endometrium (the inner lining of the uterus). Their secretion is part of feedback loops from ovaries to the hypothalamus and to the anterior lobe of the pituitary gland.

2. A menstrual cycle is a recurring cycle of intermittent fertility in the reproductive years of female humans and other primates. These events occur during a cycle:

a. Follicular phase: One of many follicles matures inside an ovary. Each follicle is an oocyte and the cell layer surrounding it. Meanwhile, the endometrium is prepared for a possible pregnancy. It breaks down each cycle if pregnancy does not occur.

b. Ovulation: A midcycle surge of the blood level of LH triggers ovulation, the release of a secondary oocyte from the ovary.

c. Luteal phase: After ovulation the corpus luteum, a glandular structure, develops from the remnants of the follicle. It secretes progesterone and some estrogen, which prime the endometrium for fertilization. If fertilization occurs, the corpus luteum will be maintained.

3. After fertilization, a blastocyst forms by cleavage and implants itself in the endometrium. Three primary tissue layers—ectoderm, endoderm, and mesoderm—form. They give rise to all organs.

4. Four extraembryonic membranes form as embryos of humans (and other vertebrates) develop:

a. The amnion becomes a fluid-filled sac around the embryo, which it protects from drying out, mechanical shock, and abrupt temperature changes.

b. The yolk sac stores nutritive yolk in most shelled eggs. In humans, part of the sac becomes a major site of blood formation and some of its cells give rise to germ cells. (Germ cells later give rise to sperm and eggs.)

c. The chorion is protective membrane surrounding the embryo and the other membranes. It also becomes a major component of the placenta.

d. In humans, the blood vessels for the placenta arise from the allantois, as does the urinary bladder.

5. The placenta, a blood-engorged organ, is composed of endometrium and extraembryonic membranes. The placenta allows embryonic blood vessels to develop

Further reading: Student Guide to InfoTrac on web site –

independently of the mother's, but it allows oxygen, nutrients, and wastes to diffuse between them.

6. To some extent, the placenta serves as a protective barrier for the fetus. But it cannot protect the fetus from harmful effects of the mother's nutritional deficiencies, infections, intake of prescription drugs, illegal drugs, alcohol, and cigarette smoking.

7. During pregnancy, estrogens as well as progesterone stimulate the growth and development of mammary glands. During labor, strong uterine contractions expel the fetus and the afterbirth. After delivery, nursing stimulates the release of prolactin and oxytocin, which in turn stimulate milk production and release.

8. Whether and how to control human sexuality and fertility are major ethical issues of our time. There is widespread agreement that sexually responsible behavior rather than abortion is the preferred route to birth control.

Review Questions

1. Name the two primary reproductive organs of the human male and where sperm formation starts inside them. Does semen consist of sperm, glandular secretions, or both? *45.1*

2. Does sperm formation require mitosis, meiosis, or both? *45.2*

3. Study Figure 45.5. Then, on your own, sketch the feedback loops (to the hypothalamus and anterior pituitary from the testes) that govern sperm formation. Include the names of the releasing hormone, hormones, and cells in the testes involved in these loops. *45.2*

4. Name the two primary reproductive organs of the human female. What is the endometrium? *45.3*

5. Distinguish between:
   a. oocyte and ovum *45.3, 45.6*
   b. follicular phase and luteal phase of menstrual cycle *45.3*
   c. follicle and corpus luteum *45.4*
   d. ovulation and implantation *45.3, 45.7*

6. What is the menstrual cycle? Name four of the hormones that influence this cycle. Which one triggers ovulation? *45.3, 45.4*

7. Study Figure 45.8. Then, on your own, sketch the feedback loops (to the hypothalamus and anterior pituitary from the ovaries) that govern the menstrual cycle. Include the names of the releasing hormone, hormones, and ovarian structures involved in these loops. *45.4*

8. Describe the cyclic changes that occur in the endometrium during the menstrual cycle. What role does the corpus luteum play in the changes? *45.4*

9. Distinguish between the embryonic period and fetal period of human development. Then describe the general organization of a human blastocyst. *45.7*

10. State the embryonic source of the amnion, yolk sac, chorion, and allantois. Also state the functions of each extraembryonic membrane with respect to the structure or functioning of the developing individual. *45.7*

11. Name some of the early organs that are the hallmark of the embryonic period of humans and other vertebrates. *45.8*

12. Describe the placenta's structure and function. Do maternal and fetal blood vessels interconnect in this organ? *45.9*

13. Name the releasing hormone with a key role in labor. Then state the role of estrogen, progesterone, prolactin, and oxytocin in ensuring that the newborn will have an ongoing supply of milk. *45.12*

14. Label the components of the human male reproductive system and state their functions: *45.1, 45.2*

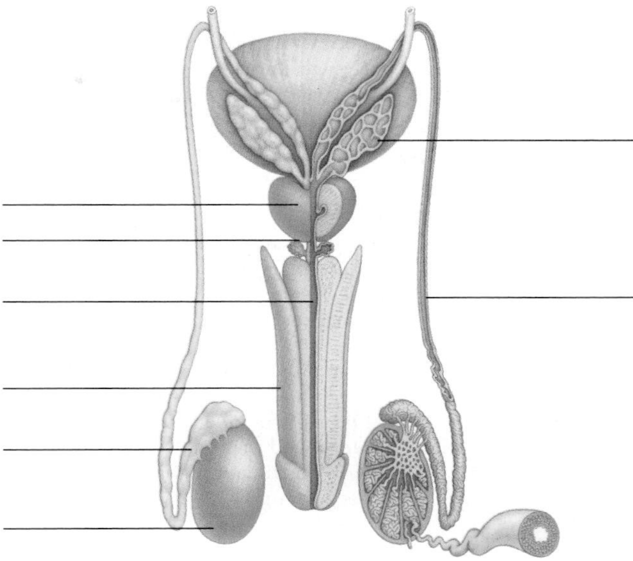

15. Label the components of the human female reproductive system and state their functions: *45.3, 45.4*

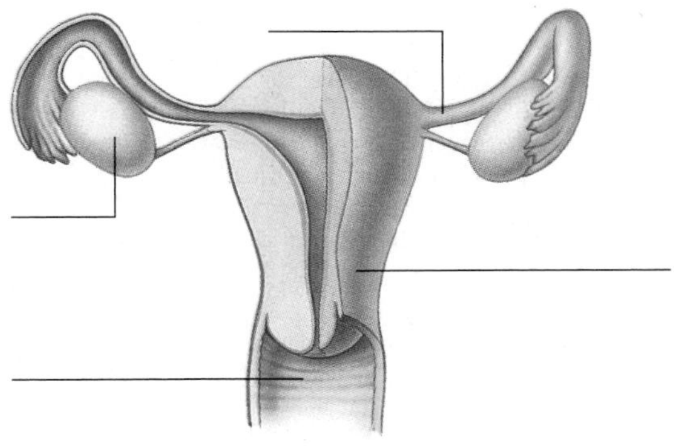

Self-Quiz (*Answers in Appendix III*)

1. The ______ and the pituitary gland control secretion of sex hormones from the testes and the ovaries.

2. Reproductive function in males requires ______.

a. FSH	d. GnRH
b. LH	e. inhibin
c. Testosterone	f. all of the above

3. During a menstrual cycle, a midcycle surge of ______ triggers ovulation.
   a. estrogen b. progesterone c. LH d. FSH

4. During a menstrual cycle, ______ and ______ secreted by the corpus luteum prime the uterus for pregnancy.

a. FSH; LH	c. estrogen; progesterone
b. FSH; testosterone	d. Only estrogens prime it.

5. In implantation, a ______ burrows into the endometrium.

a. zygote	c. blastocyst
b. gastrula	d. morula

6. The ______, a fluid-filled sac, surrounds and protects the embryo from mechanical shocks and keeps it from drying out.
   a. yolk sac b. allantois c. amnion d. chorion

7. At full term, a placenta ______.
   a. is composed of extraembryonic membranes
   b. directly connects maternal and fetal blood vessels
   c. keeps maternal and fetal blood vessels separated

8. (A) ______ form(s) in all vertebrate embryos.
   a. neural plate c. pharyngeal arches e. a through c
   b. somites d. primitive streak f. a through d

9. Distinctly human features emerge in the embryo by the end of the ______ week after fertilization.
   a. second c. fourth e. eighth
   b. third d. fifth f. sixteenth

10. Match each term with the most suitable description.
   ____ blastocyst a. secondary sexual traits develop
   ____ seminiferous tubule b. secretions influence labor
   ____ corpus luteum c. sperm start to form here
   ____ secondary oocyte d. secretes estrogens, progesterone to prepare for blastocyst and maintain endometrium during pregnancy
   ____ CRH, cortisol e. type of blastula formed from a fertilized mammalian egg
   ____ puberty f. after fertilization, becomes mature egg (ovum)

Critical Thinking

1. Suppose that, at the time a human zygote is undergoing cleavage, the first two blastomeres that form separate from each other completely. Both of the blastomeres and their cellular descendants continue to divide on schedule, on the prescribed developmental program. The result is *identical twins*, or two normal, genetically identical individuals (compare Section 44.3). *Nonidentical twins* form when two different secondary oocytes are fertilized at the same time by two different sperm. On the basis of this information, explain why nonidentical twins show considerable genetic variability and identical twins do not.

2. Infection by the rubella virus apparently has an inhibitory effect on mitosis. Serious birth defects result when a woman is infected during the first trimester of pregnancy, but not later. Review the developmental events that unfold during pregnancy and explain why this might be so.

3. Imagine you are an obstetrician advising a woman who has just learned she is pregnant. What instructions would you provide concerning her diet and behavior during pregnancy?

4. In the United States, teenage pregnancies and STD infections are rampant. Suppose the office of the Surgeons General asks you to take part in a task force that will recommend practices that might reduce the incidence of both. What practices might meet with the greatest success? What kind of enthusiasm or resistance might they provoke among the teenagers in your community? Among adults? Explain why.

Selected Key Terms

abortion *45.15*
allantois *45.7*
amnion *45.7*
blastocyst *45.7*
chorion *45.7*
coitus *45.6*
corpus luteum *45.4*
CRH *45.12*
endometrium *45.3*
estrogen *45.3*
fetus *45.7*
follicle *45.4*
follicular phase, menstrual cycle *45.3*
FSH *45.2*
gastrulation *45.8*
GnRH *45.2*
HCG *45.7*
implantation *45.7*
inhibin 45.2
in vitro fertilization *45.15*
labor (birth process) *45.12*
lactation *45.12*
Leydig cell *45.2*
LH *45.2*
luteal phase, menstrual cycle *45.3*
menstrual cycle *45.3*
oocyte *45.3*
ovary *CI*
ovulation *45.3*
ovum *45.6*
oxytocin *45.12*
placenta *45.9*
polar body *45.4*
progesterone *45.3*
prolactin *45.12*
secondary oocyte *45.4*
secondary sexual trait *CI*
semen *45.1*
seminiferous tubule *45.1*
Sertoli cell *45.2*
sexually transmitted disease (STD) *45.14*
somite *45.8*
testis (testes) *CI*
testosterone *45.2*
uterus *45.3*
yolk sac *45.7*
zona pellucida *45.4*

Readings *See also www.infotrac-college.com*

Caldwell, M. November 1992. "How Does a Single Cell Become a Whole Body?" *Discover* 13(11): 86–93.

Cohen, J., and I. Stewart. April 1994. "Our Genes Aren't Us." *Discover* 15(4): 78–84. Argument that DNA means nothing in the absence of a developmental context.

Gilbert, S. 1994. *Developmental Biology*. Fourth edition. Sunderland, Massachusetts: Sinauer.

Larsen, W. 1993. *Human Embryology*. New York: Churchill Livingston. Paperback. Recommended for the serious student, but maybe not for the fainthearted.

McGinnis, W., and M. Kuziora. February 1994. "The Molecular Architects of Body Design." *Scientific American* 270(2): 58–66.

Nilsson, L., et al. 1986. *A Child Is Born*. New York: Delacorte Press/Seymour Lawrence.

Nusslein-Volhard, C. August 1996. "Gradients That Organize Embryonic Development." *Scientific American*, 54–61.

Rathus, S. and S. Boughn. 1993. "AIDS: What Every Student Needs to Know." Harcourt Brace. Dallas: Texas. Paperback; also has sections on other sexually tranmitted diseases.

Sherwood, L. 1997. *Human Physiology*. Fourth edition. Belmont, California: Brooks/Cole.

Zack, B. July 1981. "Abortion and the Limitations of Science." *Science* 213:291.

FACING PAGE: *Two organisms—a fox in the shadows cast by a snow-dusted spruce tree. What are the nature and consequences of their interactions with each other, with other organisms, and with their environment? By the end of this last unit, you possibly will see worlds within worlds in such photographs.*

46 POPULATION ECOLOGY

Tales of Nightmare Numbers

Across from Sausalito, California, the steep flanks of Angel Island rise from the waters of San Francisco Bay (Figure 46.1*a*). The island, set aside as a game reserve, escaped urban development. It did not escape from the descendants of a few deer that well-meaning nature lovers shipped over in the early 1900s. With no natural predators to keep them in check, the few deer became many—far too many for the limited food supply of their isolated habitat. Yet the island attracted a steady stream of picnickers from the mainland. They felt sorry for the malnourished animals and made sure to load the picnic baskets with extra food for them.

The visitors imported so much food that scrawny deer kept on living and reproducing. In time, the herd nibbled away the native grasses, the roots of which had helped slow soil erosion on the steep hillsides. Hungry deer chewed off all the new leaves of seedlings; they killed small trees by stripping the bark and its phloem. The herd was destroying the environment.

In desperation, game managers proposed using a few skilled hunters to thin the herd. They were strongly denounced as being cruel. They proposed importing a few coyotes to the island to thin the herd naturally. Animal rights advocates opposed that solution, also.

As a compromise, about 200 of the 300+ deer were captured, loaded onto a boat, and shipped to suitable mainland habitats. A number of them received collars with radio transmitters so that game managers could track them after the release. In less than sixty days, dogs, coyotes, bobcats, hunters, and speeding cars and trucks had killed off most of them. In the end, relocating each surviving deer had cost taxpayers almost 3,000 dollars. The State of California refused to do it again. And no one else, anywhere, volunteered to pick up future tabs.

It is not difficult to define the boundaries of Angel Island or track its inhabitants, so it is easy to draw a lesson from this tale: *A population's growth depends on the resources of its environment. And attempts to "beat nature" by altering the sometimes cruel outcome of limited resources only postpone the inevitable.* Does the same lesson apply to other populations, in other places? Yes, it does, as the next tale makes clear.

When 1999 drew to a close, there were over 6 billion people on Earth. About 2 billion already live in poverty. Each year 40 million more join the ranks of the starving. Next to China, India is the most populous country, with more than a billion inhabitants (Figure 46.1*b*). By 2010

Figure 46.1 (**a**) Angel Island, which turned out to be a laboratory for studying population growth. (**b**) Bathers crowded along a bank of the Ganges River in India—just a small sampling of a human population that has exceeded 6 billion. In this chapter, we turn to principles that govern the growth and sustainability of all populations.

there may be 182 million more. Forty percent of those people live in rat-infested shantytowns, without enough food or fresh water. They are forced to wash clothes and dishes in open sewers. Land available to raise their food shrinks by 365 acres a day, on average. Why? Irrigated soil becomes too salty when it drains poorly, and there is not enough water to flush away the salts.

Can wealthier, less densely populated nations help? After all, they use most of the world's resources. Maybe they should learn to get by more efficiently, on less. For example, people might limit their meals to cereal grains and water; give up their private cars, living quarters, air conditioners, televisions, and dishwashers; stop taking vacations and stop laundering so much; close all the malls, restaurants, and theaters at night; and so on.

Maybe wealthier nations also should donate more surplus food than they already donate to less fortunate ones. Then again, would huge donations help, or would they encourage dependency and spur more increases in population size? And what if surpluses run out?

It is a monumental dilemma. At one extreme, the redistribution of resources on a global scale would allow the greatest number of people to survive, but at the lowest comfort level. At the other extreme, foreign aid rationed only to nations that restrict population growth would allow fewer individuals to be born, but the quality of life would be greater.

Currently, the foreign aid program of the United States is based on two premises: (1) that individuals of every nation have an irrevocable right to bear children, even if unrestricted reproduction ruins the environment that must sustain them; and (2) that because human life is precious above all else, the wealthiest nations have an absolute moral obligation to save lives everywhere.

Regardless of the positions that nations take on this issue, ultimately they must come to terms with this fact: *Certain principles govern the growth and sustainability of populations over time.* These principles are the bedrock of **ecology**—the systematic study of how organisms interact with one another and with their physical and chemical environment. Ecological interactions start within and between populations, and they extend on through communities, ecosystems, and the biosphere. They are the focus of this last unit of the book.

In this chapter we look first at the relationships that influence the size, structure, and distribution of populations. Later, we will apply the basic principles of population growth to the past, present, and future of the human species.

KEY CONCEPTS

1. Certain ecological principles govern the growth and sustainability of all populations, including our own.

2. In addition to genetic factors, a population's size, density, distribution, and the number of individuals in various age categories influence its patterns of growth.

3. When the birth rate exceeds the death rate, and when immigration and emigration are in balance, populations may show a pattern of exponential growth.

4. With exponential growth, a population expands by ever increasing increments during successive intervals, because the number of individuals in its reproductive age category (or about to enter it) becomes ever larger.

5. A shortage of any resource that individuals require is a limiting factor on population growth.

6. For all populations, carrying capacity is the maximum number of individuals that can be sustained indefinitely by the resources of a given environment. That number may rise or fall with changes in resource availability.

7. A population may show a pattern of logistic growth. By this pattern, population density—that is, the number of individuals in a specified area at a given point in time—is initially low. Then the population size rapidly increases. It finally levels off as resource scarcity limits its further increase or triggers a decline in numbers.

8. Some populations show fluctuations in numbers that cannot be explained by a single growth model.

9. All populations face limits to growth, because no environment can indefinitely sustain a successively increasing number of individuals. Competition, disease, predation, and other factors control population growth. The controls vary in their relative effects on populations of different species, and they vary over time.

46.1 CHARACTERISTICS OF POPULATIONS

By this point in the book you know that a population is a group of individuals of the same species occupying a given area. As described in Section 18.1, its individuals share a gene pool, which is the basis for characteristic ranges of morphological, physiological, and behavioral traits. When studying a population, ecologists consider the genetic make-up as well as the reproductive modes and reproductive behavior of individuals. In addition, they consider the **demographics**, or vital statistics of a population, and that will be our focus here. Among the vital statistics are population size, density, distribution, and age structure.

Population size is the number of individuals that contribute to a population's gene pool. A population's **age structure** is the number of individuals in each of several to many age categories. For instance, all members may be grouped into *pre-reproductive*, *reproductive*, and *post-reproductive* ages. Individuals in the first category will have the capacity to produce offspring at maturity. Together with the actually and potentially reproducing individuals in the second category, they help make up the population's **reproductive base**.

Population density is the number of individuals in some specified area or volume of a habitat, such as the number of guppies per liter of water in a stream. (A *habitat*, remember, is the type of place where a species normally lives. We characterize a habitat by its physical features, chemical features, and the presence of other species.) **Population distribution** is the general pattern in which individuals of the population are dispersed through a specified area.

Crude density is a measured number of individuals in a specified area. Section 46.2 outlines how ecologists make their counts, which help them track changes in population density over time. The counts do not reveal how much of a habitat is used as living space. Even areas that appear to be uniform, such as a long, sandy beach, are more like fine tapestries of light, moisture, temperature, mineral composition, and other variables. Also, the tapestry of environmental conditions often changes with the seasons. So a population might find one part of a habitat more suitable for occupancy than other parts, and it may do so some or all of the time.

Also, different species typically share the same area, and they compete for energy, nutrients, living space, and other resources. Most interact as predators, prey, or parasites. Later chapters describe these interactions. For now, simply note that species interactions influence population density and dispersion through a habitat.

Theoretically, populations show three distribution patterns. By these patterns, individuals are dispersed in clumps, nearly uniformly, or randomly (Figure 46.2). The individuals of most populations form aggregations at specific sites in a habitat. Why is clumping the most common dispersion pattern? Three possible reasons come to mind.

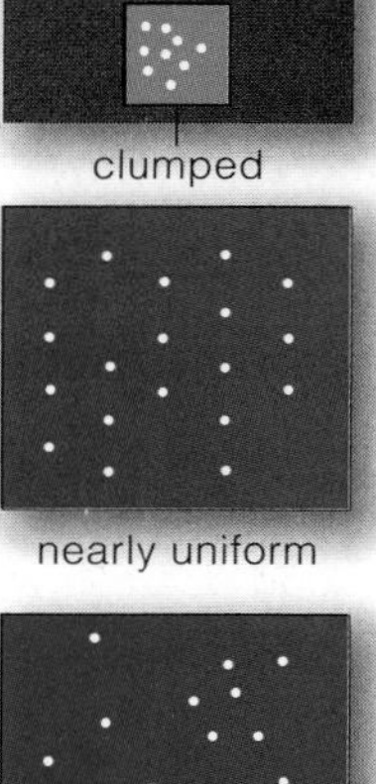
clumped
nearly uniform
random

Figure 46.2 Three generalized patterns of population distribution.

First, each species is adapted to a limited set of ecological conditions, which usually are patchy through a habitat. For example, some parts of a habitat offer more or less shade, moisture, hiding spots, and hunting spots. Second, many animal species gather in social groups that afford advantages in terms of survival and reproduction. Such groups offer more opportunities for, say, mating or mutual defense against predators. Third, adults of many species cannot disperse their seeds, larvae, or other immature forms of the new generation over large distances. For example, sponge larvae can swim, but not very far from a parent sponge. They simply settle down on substrates near parents.

Where the individuals are more evenly spaced than they would be by chance alone, we call this a nearly uniform disperal pattern. Regularly spaced fruit trees in orchards are often cited as examples. Few populations live this way in nature. You might see the pattern where habitat conditions are fairly uniform and where there is strong competition for resources or territorial behavior.

Creosote bushes (*Larrea*) that live in arid habitats of the American Southwest may show this pattern. Large, mature plants deplete soil water all around them. Seed-eating ants and rodents forage near the plants, which offer cover from predators and from the hot sun. Most seeds are devoured, which puts seeds and seedlings at a competitive disadvantage. So the seeds typically take hold only at patches where established, mature plants are weakened or have died. As a result, creosote bushes are not clumped together (Figure 46.3*a*). Neither are they dispersed at random; seeds that rodents and ants do overlook and that await opportunity for life in the sun cannot travel far from the mature parent plants.

We observe random dispersion only when habitat conditions are nearly uniform, resource availability is fairly steady, and individuals of the population neither attract nor avoid one another. For instance, wolf spiders are solitary hunters on forest floors. Each generation might well be randomly spaced. However, a dispersion pattern such as this is extremely rare in nature.

Each population has its own gene pool and range of traits. It also has a characteristic size, density, distribution pattern, and age structure. Environmental conditions and species interactions influence these characteristics.

Further reading: Student Guide to InfoTrac on web site

Focus on Science

46.2 ELUSIVE HEADS TO COUNT

Does it seem like Bambis are just about everywhere, munching through gardens and smacking into more and more cars on the roads? The estimated number of deer in the United States has risen from fewer than 1 million at the turn of the century to more than 18 million today. Suppose you live in a northern California county and wonder how many deer live there with you. How would you go about determining their population density?

To get some idea of what your study will entail, you start with a literature search. Ecologists already have baselines of approximate population densities for many organisms in their natural habitats. For example, in some habitats, baselines indicate that you can expect to find as many as 5 million diatoms per cubic meter, 500 trees per hectare (a hectare is 2.47 acres), 250 field mice per hectare, and so on. Your search turns up one estimate of 4 deer per square kilometer. And this should tell you that deer will be easier to count than, say, diatoms.

Total counts are the most straightforward way to measure the absolute density of a population. Census takers do this every ten years for the human population in the United States. Ecologists do the same for large animals in small areas, such as songbirds in a forest, northern fur seals at their breeding grounds, sea stars in a tidepool, and creosote bushes in a specified part of a desert (Figure 46.3*a*). More often, ecologists must content themselves with sampling a small part of the population and then estimating the total density.

You could get a map of your county and divide it into a number of small plots, or quadrats. Quadrats are sampling areas of the same size and shape (such as rectangles, squares, and hexagons). Then you could count all the deer in several plots and extrapolate the average for the entire county. Ecologists conduct such counts for migrating herds and flocks. For example, each year, the North American Breeding Bird Survey makes counts along more than 2,000 flyways for migratory waterfowl (Figure 46.3*b*).

Deer are among the animals that do not stay put in their habitat. How can you be absolutely sure that the individuals you count in one plot aren't the same individuals you counted earlier in a different plot?

For mobile animals, ecologists sample population density by using a **capture–recapture method**. The idea is to capture individual animals and mark them in some way. (Squirrels get tattoos, salmon get tags, migratory birds get leg rings, deer get bright collars, and so on.) Marked animals are released at time 1, then animals are captured and checked for marks at time 2. In later samples taken from the population, the proportion of marked individuals should represent the proportion marked in the whole population:

$$\frac{\text{Marked individuals in sampling 2}}{\text{Total capture in sampling 2}} = \frac{\text{Marked individuals in sampling 1}}{\text{Total population size}}$$

Ideally, marked and unmarked individuals of the population are captured at random, no marked animal dies during the study interval, and none of the marked animals leaves the population or is overlooked. In the real world, however, recapturing of marked individuals might *not* be random. Squirrels that were marked after being attracted to yummy bait in boxes might now be trap-happy or trap-shy, so they might overrepresent or underrepresent the population. Instead of mailing the tags of marked fish back to ecologists, some fishermen keep them as good-luck charms. Birds may lose their leg rings. Undocumented immigrants may not answer the door when the census taker comes knocking.

Your estimate also depends on the time of year when you do a sampling. Population distribution varies over time, as in migratory responses to environmental rhythms. Few places yield abundant resources all year long, and many animals move from one habitat to another with changes in the seasons (Figure 46.3*b*). Deer are like this, so you would probably use capture–recapture methods more than once a year, for several years.

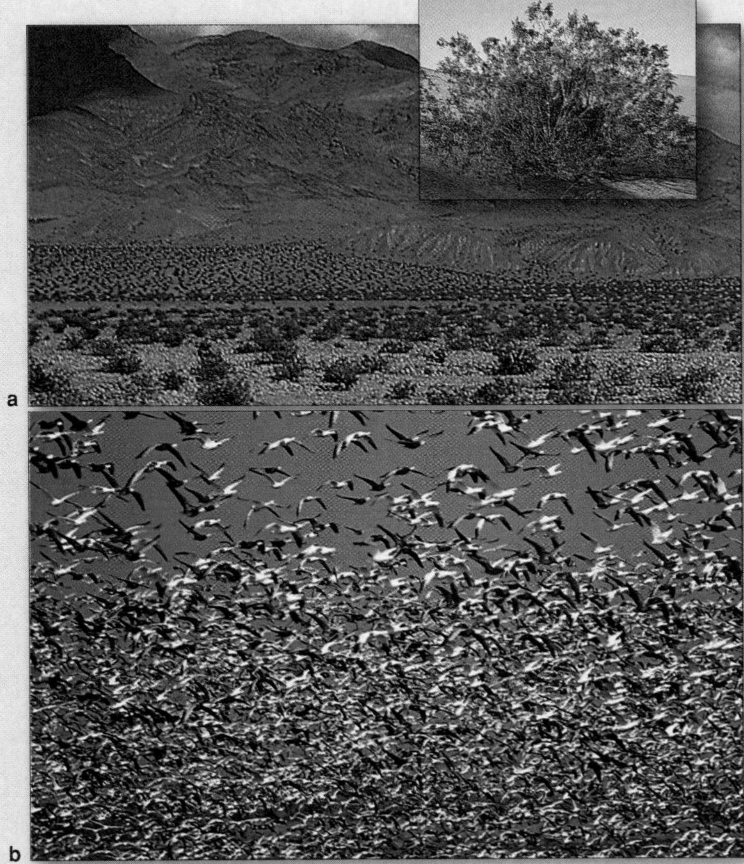

a
b

Figure 46.3 (**a**) Example of nearly uniform spacing, represented by creosote bushes near Death Valley, California. (**b**) Snow geese, one of many species of waterfowl that migrate through 3.4 million square kilometers of air space above Canada and the United States.

46.3 POPULATION SIZE AND EXPONENTIAL GROWTH

Gains and Losses in Population Size

Populations are dynamic units of nature. Depending on the species, they might add or lose individuals every half hour or every day, season, or year. Sometimes they glut portions of the habitat with individuals. At other times individuals may be scarce. Populations even can drive themselves or be driven to extinction. During a specified interval, such changes in population size can be measured in terms of birth rates, death rates, and the number of individuals entering and leaving.

Population size increases as a result of (1) births and (2) **immigration**—the arrival of new residents from other populations of the species. It decreases as a result of (1) deaths and (2) **emigration**, whereby individuals permanently move out of the population.

For many species, population size also changes on a predictable basis as a result of daily or seasonal events called **migrations**. However, migration is a recurring round trip between two areas, so we need not consider its transient effects during this initial consideration of population size.

From Ground Zero to Exponential Growth

For our purposes, assume that immigration is balancing emigration over time so that we can ignore the effects of both on population size. Doing so allows us to define **zero population growth** as an interval during which the number of births is balanced out by the number of deaths. The population size is stabilized during such an interval, with no overall increase or decrease.

Births, deaths, and other variables that might affect population size can be measured in terms of **per capita** rates, or rates per individual. (*Capita* means heads, as in head counts.) Visualize 2,000 mice living in a cornfield. Twenty or so days after their eggs are fertilized, the females produce a litter, then nurse their offspring for a month or so, then get pregnant again. If the female mice collectively give birth to 1,000 mice per month, the birth rate would be 1,000/2,000 = 0.5 per mouse per month. If 200 of the 2,000 die during that interval, the death rate would be 200/2,000 = 0.1 per mouse per month.

If we assume the birth rate and death rate remain constant, we can combine both into a single variable—the **net reproduction per individual per unit time**, or ***r*** for short. For that mouse population in the cornfield, r is 0.5 − 0.1 = 0.4 per mouse per month. This example gives us a way to represent population growth:

$$\left(\text{population growth per unit time}\right) = \left(\text{net population growth rate per individual per unit time}\right) \times \left(\text{number of individuals}\right)$$

or, more simply, $G = rN$.

			Net Monthly Increase:		New Population Size:
$G = r \times$	3,920	=	1,568	=	5,488
$r \times$	5,488	=	2,195	=	7,683
$r \times$	7,683	=	3,073	=	10,756
$r \times$	10,756	=	4,302	=	15,058
$r \times$	15,058	=	6,023	=	21,081
$r \times$	21,081	=	8,432	=	29,513
$r \times$	29,513	=	11,805	=	41,318
$r \times$	41,318	=	16,527	=	57,845
$r \times$	57,845	=	23,138	=	80,983
$r \times$	80,983	=	32,393	=	113,376
$r \times$	113,376	=	45,350	=	158,726
$r \times$	158,726	=	63,490	=	222,216
$r \times$	222,216	=	88,887	=	311,103
$r \times$	311,103	=	124,441	=	435,544
$r \times$	435,544	=	174,218	=	609,762
$r \times$	609,762	=	243,905	=	853,667
$r \times$	853,677	=	341,467	=	1,195,134

a

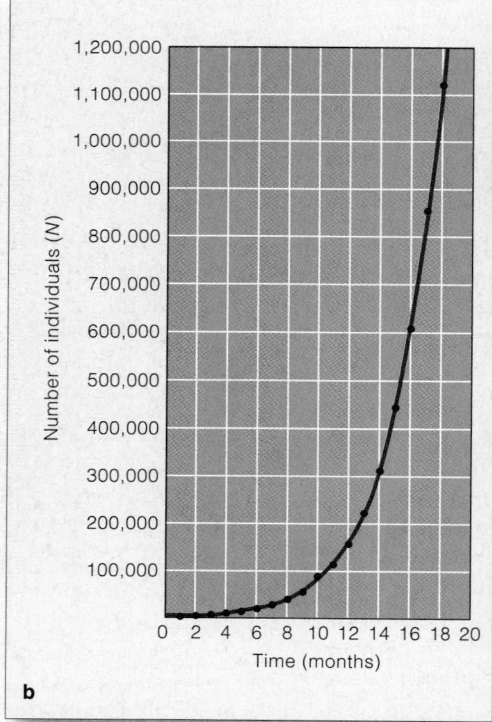
1,200,000
1,100,000
1,000,000
900,000
800,000
700,000
600,000
500,000
400,000
300,000
200,000
100,000
Number of individuals (N)
0 2 4 6 8 10 12 14 16 18 20
Time (months)
b

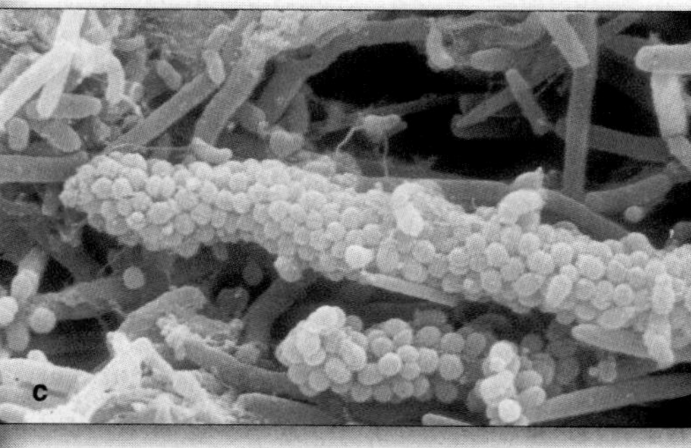
c

Figure 46.4 (**a**) Data showing the net monthly increases in a population of field mice living in a cornfield. From start to finish, the numbers listed show a pattern that is typical of exponential growth. (**b**) Graphing the data yields a J-shaped growth curve.

(**c**) Do such growth curves seem far removed from your own experience? Think about the growth of populations of, say, the resident bacteria in your mouth after you present them with a buffet of nutrients in candy or some other sugar-laden food.

Further reading: Student Guide to InfoTrac on web site →

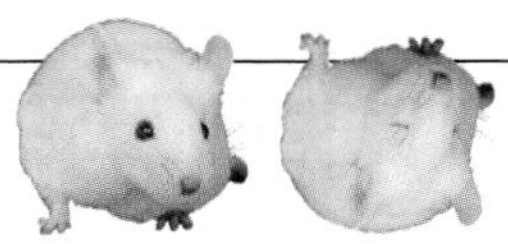

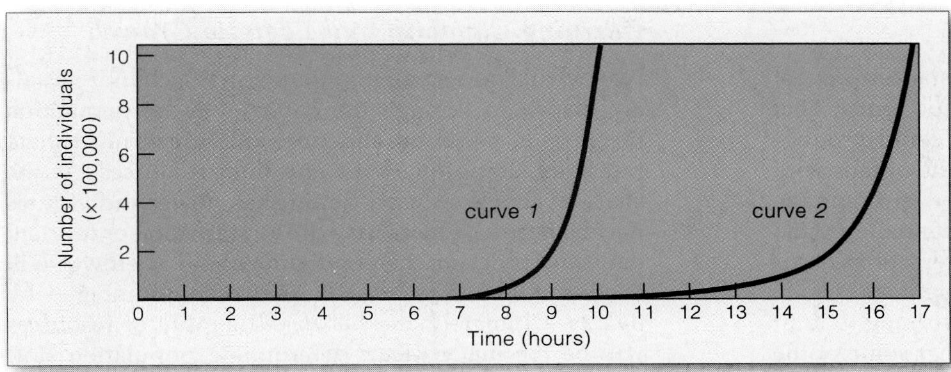

Number of individuals (× 100,000)
10
8
6
4
2
curve 1
curve 2
0 1 2 3 4 5 6 7 8 9 10 11 12 13 14 15 16 17
Time (hours)

Figure 46.5 Effect of deaths on the rate of increase in two bacterial populations. Growth curve *1* represents a population of bacterial cells that reproduced every half hour. Growth curve *2* represents a different population in which cells divided every half hour but 25 percent died between divisions.

The graph shows that deaths can slow the rate of increase but cannot in themselves stop exponential growth. This is so for populations of bacteria, mice, and all other organisms.

As the next month begins, 2,800 mice are scurrying about. With the net increase of 800 fertile critters, the reproductive base has expanded. If r does not change, the population size will expand this month also, by a net increase of $0.4 \times 2,800 = 1,120$ mice. The population is now 3,920. In the months ahead, it just so happens that r remains constant, and a growth pattern emerges. As Figure 46.4*a* shows, *less than two years from the time we started counting, the number of mice running around in the cornfield increased from 2,000 to more than a million!*

Plot the monthly increases against time, as in Figure 46.4*b*, and you end up with a graph line in the shape of a "J." When growth of a population over increments of time plots out as a J-shaped curve, you know that you are tracking **exponential growth**.

"Exponential" refers to a relationship in which one variable increases much faster than another in a specific mathematical way. In this case, a population's size is a variable that depends on how many individuals make up the reproductive base in successive increments of time. *The larger the reproductive base, the greater will be the expansion in population size during each specified interval.*

Now look at other aspects of exponential growth by supplying a single bacterium in a culture flask with all the nutrients required for growth. Thirty minutes later, the one cell divides in two. Thirty minutes pass, the two cells divide, and so on every thirty minutes. Assuming no cells die between divisions, the population size will double in each interval—from 1 to 2, then 4, 8, 16, 32, and so on. The length of time it takes for a population to double in size is its **doubling time**.

The larger the population gets, the more cells there are to divide. After 9–1/2 hours (nineteen doublings), it has more than 500,000 bacterial cells. After 10 hours (twenty doublings), it has more than 1 million. Plot the doublings in size against time, and you end up with the J-shaped curve typical of unrestricted, exponential growth. Figure 46.5, curve *1*, shows this.

To examine the effect of deaths on the growth rates, start over with one bacterium. Assume that 25 percent of the cells die in each thirty-minute interval. Because of deaths, it takes about seventeen hours (not ten) for population size to reach a million. *But deaths changed only the time scale for population growth.* You still end up with a J-shaped curve, which is curve *2* in Figure 46.5.

What Is the Biotic Potential?

Finally, visualize a population occupying a place where conditions are ideal. Every one of its individuals has adequate shelter, food, and other vital resources. No predators, pathogens, or pollutants lurk anywhere in the habitat. That population might well display its **biotic potential**, which is the maximum rate of increase per individual under ideal conditions.

Each species has a characteristic maximum rate of increase. For many bacteria, it is 100 percent every half hour or so. For humans and other large mammals, it is between 2 and 5 percent per year. But the *actual* rate depends on the age at which each generation starts to reproduce, how often each individual reproduces, and how many offspring are produced. Now think about this. A human female is biologically capable of bearing twenty or more children, yet in each generation, many females do not reproduce at all. The human population has not been displaying its biotic potential. Even so, since the mid-eighteenth century, its growth has been exponential, for reasons that will soon be apparent.

During a specified interval, population size is generally an outcome of births, deaths, immigration, and emigration.

With exponential growth, population size expands by ever increasing increments during successive time intervals, because the reproductive base becomes ever larger.

Population size against time plots out as a characteristic J-shaped curve if the population is growing exponentially.

As long as the per capita birth rate remains even slightly above the per capita death rate, a population can grow exponentially.

46.4 LIMITS ON THE GROWTH OF POPULATIONS

What Are the Limiting Factors?

Most of the time, environmental circumstances prevent any population from fulfilling its biotic potential. That is why sea stars, the females of which could produce 2,500,000 eggs each year, do not fill up the oceans with sea stars. That also is why humans will never fill up the planet, even with our capacity for exponential growth.

In natural environments, complex interactions occur within and between populations of different species, so it is not easy to identify all the factors working to limit population growth. To get a sense of what some of the factors might be, start again with a lone bacterium in a culture flask, where you can control the variables. First you enrich the culture medium with glucose and other nutrients required for bacterial growth. Then you allow bacterial cells to reproduce for many generations. At first the growth pattern appears to be exponential. Then growth slows, and after that, population size remains rather stable. But after the stable period, the population size declines rapidly and all of the bacteria die.

What happened? When the population expanded by ever increasing amounts, cells tapped more and more nutrients. The dwindling nutrients signaled the cells to stop dividing (Section 9.2). And when the existing cells exhausted the nutrient supply, they starved to death.

Any essential resource that is in short supply is a **limiting factor** on population growth. Food, minerals of certain types, refuge from predators, living quarters, and a pollution-free environment are examples. The number of such factors can be huge, and their effects can vary. Even so, one factor alone is often enough to put the brakes on population growth at any given time.

Suppose you keep on freshening the supply of all required nutrients for the growing bacterial culture. After an episode of exponential growth, the population still crashes. Like all other organisms, bacteria produce metabolic wastes. The wastes of that huge population of bacterial cells were so great, they drastically altered living conditions in the culture. Therefore, by its own metabolic activities, the population polluted the experimentally designed habitat and put a stop to exponential growth.

Carrying Capacity and Logistic Growth

Now visualize a small population in which individuals are dispersed through the habitat. As the population increases in size, more and more individuals must share nutrients, living quarters, and other resources. As the share available to each diminishes, fewer individuals may be born and more may die by starvation or nutrient deficiencies. Then, the population's rate of growth will decline until births are balanced or outnumbered by deaths. Ultimately, the *sustainable* supply of resources will be the major factor determining population size. The name **carrying capacity** refers to the maximum number of individuals of a population (or species) that a given environment can sustain indefinitely.

The pattern of **logistic growth** is a fine example of how carrying capacity can affect a population. By this pattern, a low-density population starts growing slowly in size, then it grows rapidly, and finally its size levels off once the carrying capacity is reached. How can you represent the pattern? Starting with the exponential growth equation given earlier, in Section 46.3, add the term $(K - N)/K$, where K designates carrying capacity:

$$G = r_{max}\, N\, \frac{K - N}{K}$$

The term indicates how many individuals can be added to the population, based on the proportion of resources still available. When population size is small, $(K - N)$ is close to 1. It approaches 0 when the size is close to the carrying capacity. Representing this another way,

$$\begin{pmatrix}\text{population}\\ \text{growth per}\\ \text{unit time}\end{pmatrix} = \begin{pmatrix}\text{maximum}\\ \text{net population}\\ \text{growth rate}\\ \text{per individual}\\ \text{per unit time}\end{pmatrix} \times \begin{pmatrix}\text{number}\\ \text{of}\\ \text{individuals}\end{pmatrix} \times \begin{pmatrix}\text{proportion}\\ \text{of resources}\\ \text{not yet used}\end{pmatrix}$$

A plot of logistic growth gives an S-shaped curve (Figure 46.6). Such curves are only an approximation of

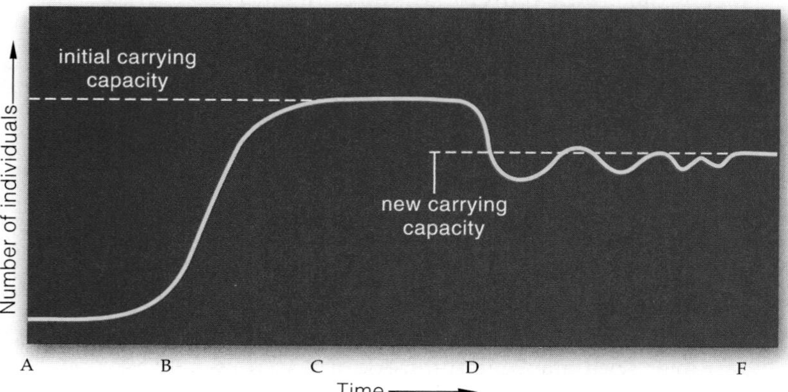
initial carrying capacity
new carrying capacity
Number of individuals
A
B
C
D
F
Time

Figure 46.6 Idealized S-shaped curve characteristic of logistic growth. After a rapid growth phase (time B to C), growth slows and the curve flattens out as the carrying capacity is reached (time C to D). S-shaped growth curves can show variations, as when changes in the environment lower the carrying capacity (time D to F). This happened to the human population of Ireland before 1900, when late blight, a disease caused by a water mold, destroyed potato crops that were the mainstay of the diet (Section 23.1).

Further reading: Student Guide to InfoTrac on web site

Figure 46.7 Carrying capacity and a reindeer herd. In 1910, four male and twenty-two female reindeer were introduced on St. Matthew Island in the Bering Sea. In less than thirty years, the size of the herd increased to two thousand. Individuals had to compete for dwindling vegetation, and overgrazing destroyed most of it. In 1950, herd size plummeted to eight. The growth curve reflects how the population size overshot the carrying capacity, then crashed.

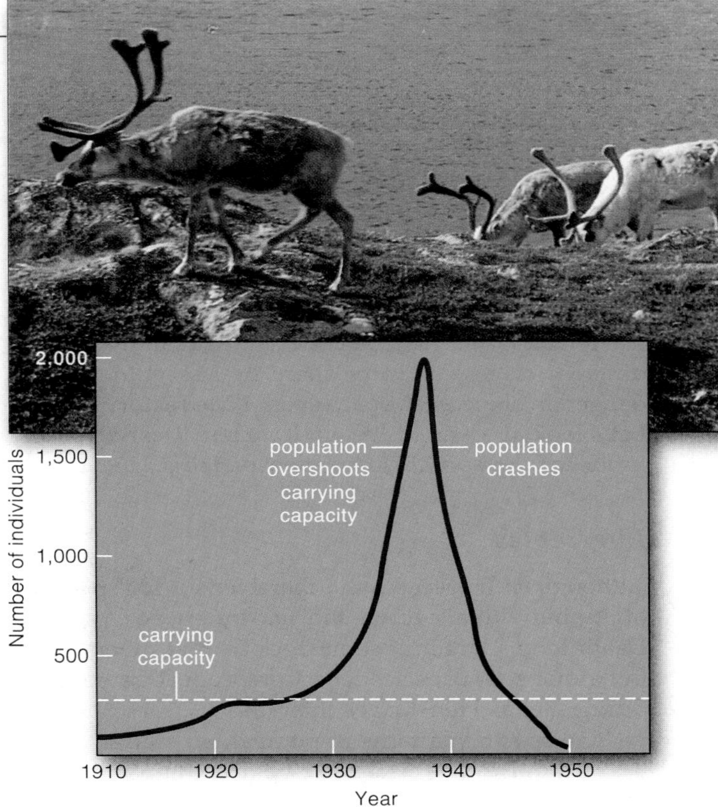

2,000
1,500
1,000
500
Number of individuals
population overshoots carrying capacity
population crashes
carrying capacity
1910
1920
1930
1940
1950
Year

what goes on in nature. For example, a population that grows too rapidly can overshoot the carrying capacity. The death rate skyrockets and the birth rate plummets. These two responses drive the number of individuals down to the carrying capacity—or lower (Figure 46.7).

Density-Dependent Controls

The logistic growth equation just described deals with **density-dependent control** of populations. That is, a small population may grow rapidly, but when it finally bumps up against its carrying capacity, so to speak, its size will stabilize or even decline.

In addition to carrying capacity, other natural factors help control population density. When they come into play, they, too, put the number of individuals below the maximum sustainable level. For example, individuals are more likely to die under crowded conditions. Then, predators, parasites, and pathogens are able to interact more intensely with their host population and trigger a decline in density. Once population size declines, the density-dependent interactions relax, and population size may increase once again.

Bubonic plague and *pneumonic plague*, two dangerous diseases, are examples of density-dependent controls. Both are caused by the bacterium *Yersinia pestis*. A large reservoir of *Y. pestis* persists in rabbits, rats, and certain other small mammals. Fleas transmit it by biting new hosts. The bacterial cells reproduce so much in the flea gut that their numbers and metabolic activities disrupt digestion. Hunger sensations drive the host flea to feed more often on more hosts, and so the disease spreads.

A devastating episode of bubonic plague occurred in the fourteenth century. The plague swept through European cities where humans were crowded together, sanitary conditions were poor, and rats were abundant. By the time the epidemic subsided, urban populations in Europe had declined by 25 million.

Both diseases are still threats. For example, in 1994 in India, they raced through rat-infested cities where garbage and animal carcasses had piled up for months in the streets. Terrified residents fled by the thousands. Some carried the pathogen with them as far away as London. Global panic ensued before concerted efforts to burn the garbage, poison the rats and the fleas, and rapidly dispense antibiotics averted a pandemic.

Density-Independent Factors

Sometimes events result in more deaths or fewer births regardless of population density. For example, every year, monarch butterflies migrate from Canada to spend the winter in forested mountains of Mexico. But logging opened up stands of the forest trees—which normally buffer temperatures. In 1995 a sudden freeze combined with the deforestation to kill millions of butterflies. Both were **density-independent factors**, meaning their effects came into play independently of population density.

Similarly, heavy applications of pesticides in your backyard may also kill most insects, mice, cats, birds, and other animals. They will do this regardless of how uncrowded or dense the populations are.

Resources in short supply put limits on population growth. Together, all of the limiting factors acting on a population dictate how many individuals can be sustained.

Carrying capacity is the maximum number of individuals of a population that can be sustained indefinitely by the resources in a given environment. The number may rise or fall with changes in resource availability.

The size of a low-density population may increase slowly, go through a rapid growth phase, then level off once the carrying capacity for the population is reached. This is a logistic growth pattern.

Density-dependent controls and density-independent factors can bring about decreases in population size.

46.5 LIFE HISTORY PATTERNS

So far, we have looked at populations as if all of their members are identical throughout a given interval. For most species, however, individuals are at different stages of development. They are interacting in different ways with other organisms and with their environment. At different times in the life cycle, they may be adapted to exploiting different resources, as when caterpillars eat leaves and butterflies prefer nectar. Also, individuals at different stages might be more or less vulnerable to danger. In short, each species has a **life history pattern**. Let's look at a few of the environmental variables that influence age-specific life history patterns.

Life Tables

Although each species has a characteristic life span, few of its individuals reach the maximum age possible. Death looms larger at some ages than at others. Also, individuals of a species tend to reproduce or emigrate during a characteristic age interval.

Age-specific patterns in populations first intrigued life insurance and health insurance companies, then ecologists. Such investigators typically track a **cohort**, a group of individuals recorded from the time of birth until the last one dies. They also track the number of offspring born to individuals in each age interval. Life tables list data gathered on a population's age-specific death schedule. They are often converted to more cheery "survivorship" schedules, or the number of individuals that reach some specified age (x). Table 46.1 is a typical example. Table 46.2 is another; it was constructed for the 1989 human population of the United States.

Dividing a given population into age classes and assigning birth rates and mortality risks to each one has practical applications. Unlike a crude census (or head count), the resulting data might be a basis for informed policy decisions in pest management, endangered species protections, social planning for human populations, and other current issues. For instance, birth and death schedules for the northern spotted owl figured in federal court rulings that put a stop to mechanized logging in the owl's habitat (old-growth forests).

Patterns of Survival and Reproduction

Evolutionarily speaking, we measure the reproductive success of individuals in terms of the number of their surviving

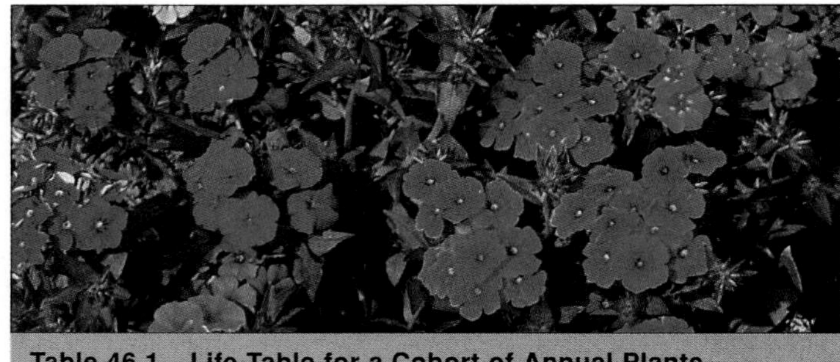

Table 46.1 Life Table for a Cohort of Annual Plants (*Phlox drummondii*)*

Age Interval (days)	Survivorship (number surviving at start of interval)	Number Dying During Interval	Death Rate per Individual During Interval	"Birth" Rate (number of seeds produced per individual) During Interval
0–63	996	328	0.329	0
63–124	668	373	0.558	0
124–184	295	105	0.356	0
184–215	190	14	0.074	0
215–264	176	4	0.023	0
264–278	172	5	0.029	0
278–292	167	8	0.048	0
292–306	159	5	0.031	0.33
306–320	154	7	0.045	3.13
320–334	147	42	0.286	5.42
334–348	105	83	0.790	9.26
348–362	22	22	1.000	4.31
362–	0	0	0	0
		996		

* Data from W. J. Leverich and D. A. Levin. 1979. *American Naturalist*, 113:881–903.

Table 46.2 Life Table for the United States Human Population, 1989*

Age Interval (category for individuals between the two ages listed)	Survivorship (number alive at start of age interval, per 100,000 individuals)	Mortality (number dying during the age interval)	Life Expectancy (average lifetime remaining at start of age interval)	Reported Live Births for Total Population
0–1	100,000	896	75.3	
1–5	99,104	192	75.0	
5–10	98,912	117	71.1	
10–15	98,795	132	66.2	11,486
15–20	98,663	429	61.3	506,503
20–25	98,234	551	56.6	1,077,598
25–30	97,683	606	51.9	1,263,098
30–35	97,077	737	47.2	842,395
35–40	96,340	936	42.5	293,878
40–45	95,404	1,220	37.9	44,401
45–50	94,184	1,766	33.4	1,599
50–55	92,418	2,727	28.9	
55–60	89,691	4,334	24.7	
60–65	85,357	6,211	20.8	
65–70	79,146	8,477	17.2	
70–75	70,669	11,470	13.9	
75–80	59,199	14,598	10.9	
80–85	44,601	17,448	8.3	
85+	27,153	27,153	6.2	Total: 4,040,958

* Compiled by Marion Hansen, based on data from U.S. Bureau of the Census, *Statistical Abstract of the United States*, 1992 (edition 112).

Further reading: Student Guide to InfoTrac on web site →

Number of survivors
TYPE
I
Age

a

TYPE
II
TYPE
III

b

c

Figure 46.8 Three generalized types of survivorship curves. Type I populations show high survivorship until some age and then show high mortality. Type II populations show a fairly constant death rate. Type III populations show low survivorship early in life.

offspring. However, the number differs among species, which differ in how much energy and time are allocated to producing gametes, securing mates, and parenting, and in the size of offspring. Apparently, many tradeoffs have been made in response to selection pressures, such as habitat conditions and species interactions.

A **survivorship curve** is a graph line that emerges when ecologists plot a cohort's age-specific survival in a habitat. Each species has a characteristic survivorship curve, and three types are common in nature.

Type I curves reflect high survivorship until fairly late in life, then a large increase in deaths. Such curves are typical of elephants and other large mammals that bear only one or a few large offspring at a time, then engage in extended parental care (Figure 46.8*a*). For example, a female elephant gives birth to four or five calves and devotes several years to parenting each one.

Type I curves also are typical of human populations in which individuals have access to good health care services. However, today as in the past, infant deaths cause a sharp drop at the start of the curve in regions where health care is poor. Following the drop, it levels off from childhood to early adulthood.

Type II curves reflect a fairly constant death rate at all ages. They are typical of organisms just as likely to be killed or to die of disease at any age, such as lizards, small mammals, and some songbirds (Figure 46.8*b*).

Type 1 curves signify a death rate that is highest early in life. They are typical of species that produce many small offspring, then show little, if any, parental behavior. Figure 46.8*c* shows how the curve plummets for sea stars. Sea stars produce mind-boggling numbers of tiny larvae, which must rapidly eat, grow, and finish developing on their own without support, protection, or guidance from the parents. Corals and other animals swiftly eat most of them. The plummeting survivorship curve is typical of many other marine invertebrates, most insects, and many fishes, plants, and fungi.

At one time, ecologists thought selection processes favored *either* the early, rapid production of many small offspring *or* the late production of a few large ones. They now know that the two patterns are extremes at opposite ends of a range of possible life histories. Also, both life history patterns—as well as intermediate ones—are sometimes evident in different populations of the same species, as the next section makes clear.

Tracking a cohort (a group of individuals from birth until the last one dies) reveals patterns of reproduction, death, and migration that typify the populations of a species.

Survivorship curves can reveal differences in age-specific survival among species. In some cases, such differences exist even between populations of the same species.

46.6 NATURAL SELECTION AND THE GUPPIES OF TRINIDAD

Several years ago, drenched with sweat and with fish nets in hand, two evolutionary biologists were busily engaged in fieldwork in the mountains of Trinidad, an island in the southern Caribbean Sea. They were after some of the small, live-bearing fishes that were darting about in the shallow freshwater streams (Figure 46.9). The fishes were guppies (*Poecilia reticulata*). For eleven years the biologists, David Reznick and John Endler, identified environmental variables that affect guppy life history patterns.

The researchers easily distinguished the male guppies from the females. The males are smaller. They also have bright-colored scales in a patterning that functions as a visual signal during mating behavior. The males engage in intricate courtship maneuvers. And once they reach sexual maturity, they stop growing. By contrast, the females are drab colored. They also continue to grow in size during the reproductive phase of the life cycle.

In Trinidad's mountains, different guppy populations occupy different streams. In some cases, different guppy populations even occupy different parts of the same stream. And they encounter different kinds of predators.

A type of killifish (*Rivulus hartii*) shares some streams with guppies. It is not an especially large fish. It preys efficiently on smaller, immature guppies but not on larger adults. A larger fish, a type of pike-cichlid (*Crenicichla alta*), shares different streams with guppies (Figure 46.10). It preys on larger, sexually mature guppies, and it tends not to bother with the small ones.

Using their understanding of natural selection as the starting point, Reznick and Endler hypothesized that predation has profound effects on the life history patterns of guppies. As they had noticed, in streams where the pike-cichlids reign supreme, individual guppies mature faster and their body size is smaller at maturity, compared to guppies that live and reproduce in killifish-dominated streams. In addition, the guppies targeted by pike-cichlids reproduce earlier in life, they produce far more offspring, and they do so more frequently.

It was possible that other variables influenced guppy life history patterns in killifish and pike-cichlid streams. To test this possibility, the researchers carefully shipped two groups of guppies from the two kinds of streams back to a laboratory in the United States. There they allowed them to reproduce in separate aquariums for two generations in the absence of predation. They also made sure that the physical and chemical conditions in the aquariums were identical for the two experimental populations. As it turned out, offspring of the experimental guppy populations showed the same differences as the natural populations. The conclusion? The differences between

Figure 46.9 Evolutionary biologist David Reznick contemplating interactions among guppies (*Poecilia reticulata*) and their predators in a freshwater stream in Trinidad.

GUPPY FROM A KILLIFISH STREAM
GUPPY FROM A PIKE-CICHLID STREAM
PIKE-CICHLID

Figure 46.10 Two guppies and one of the eaters of guppies.

Further reading: Student Guide to InfoTrac on web site →

a

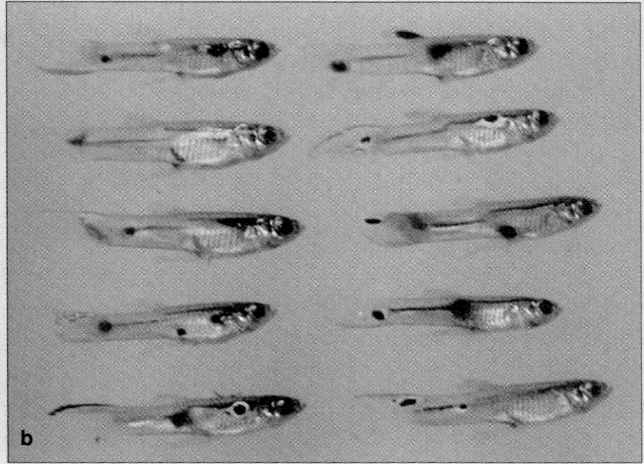
b

Figure 46.11 Representative guppies that live in streams with killifish (**a**) and streams with pike-cichlids (**b**). Guppies of populations targeted by killifish tend to be larger, less streamlined, and more brightly colored. Guppies of populations targeted by pike-cichlids tend to be smaller, more streamlined, and duller in color patterning. The differences between the groups are accompanied by clear differences in life history patterns.

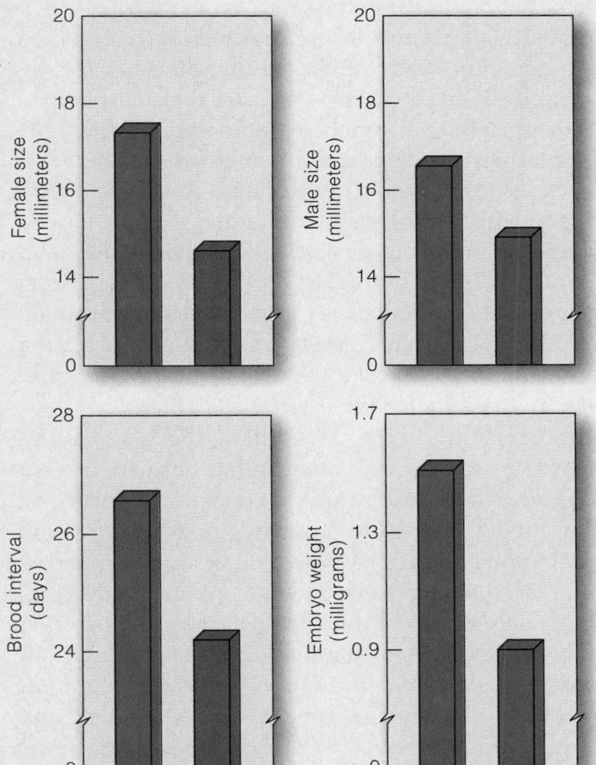
20
18
16
14
0
Female size (millimeters)
20
18
16
14
0
Male size (millimeters)
28
26
24
0
Brood interval (days)
1.7
1.3
0.9
0
Embryo weight (milligrams)

Figure 46.12 Graphs of some of the experimental evidence of natural selection among populations of guppies subjected to different predation pressures. Differences arise in body size and in the intervals between broods for guppies raised with killifish (*red* bars) or pike-cichlids (*green* bars). Killifish are small and prey on smaller guppies, pike-cichlids are large fish and prey on larger guppies, and so the two predators select for the differences.

guppies preyed upon by different predators have a genetic basis. This finding is consistent with the predictions of a theory of life history evolution that is based on such differences in mortality schedules. Figure 46.11 shows representatives of both kinds of guppy populations.

The researchers looked into the role of predation in the evolution of differences in body size, the length of time between generations, and other aspects of the guppy life history pattern. They raised many guppy generations in the laboratory. As predicted, the body sizes of the guppy lineage subjected to predation over time by killifish were larger at maturity (Figure 46.12). The lineage raised with pike-cichlids showed a trend toward earlier maturity.

When they made their first visit to Trinidad, Reznick and Endler instituted a field experiment by introducing guppies to a site upstream from a small waterfall. Before the experiment, the waterfall had been an effective barrier to dispersal; it had prevented guppies and all predators except the killifish from moving into the upstream site. The researchers selected the introduced guppies from a population that evolved with pike-cichlids downstream from the waterfall. They designated that part of the stream as the control site.

Eleven years later, researchers revisited the stream. As they discovered, the guppies had evolved. They made comparisons between guppies from the experimental site and the control site. Just as Reznick and Endler had predicted, the body size, frequency of reproduction, and other aspects of guppy life history patterns correlated with the preferences of the neighborhood predator. Later, a laboratory experiment involving two generations of guppies confirmed that the differences were indeed genetic. The researchers concluded that the differences are the product of natural selection.

46.7 HUMAN POPULATION GROWTH

The human population surpassed 6 billion in 1999. By midyear 2000, rates of increase for different countries ranged from more than 3 percent to below 1 percent, for an annual average of 1.26 percent. Visualize enough people to fill another New York City every month and you get the picture. Now think about this: The annual additions to that base population will result in a *larger* absolute increase each year into the foreseeable future.

Our staggering population growth continues even though as many as 2 billion are already malnourished or starving, without clean drinking water and adequate shelter. It continues when 1.5 billion are already going without the benefits of health care delivery and sewage treatment facilities. And it continues mainly in regions that are already overcrowded. Nearly all of the 6 billion live on 10 percent of the land; 4 billion crowd together within 480 kilometers (about 300 miles) of the seas.

Suppose it were possible, by monumental efforts, to double the food supply to keep pace with growth. We would do little more than maintain marginal living conditions for most. The annual deaths from starvation could still be 20 million to 40 million. Even this would come at great cost, for we are drastically modifying the environment that must sustain us. Salted-out cropland, desertification, deforestation, pollution—these are some of the consequences you will read about in Chapter 51, and they do not bode well for our future.

For a while, it will be like the Red Queen's garden in Lewis Carroll's *Through the Looking Glass*, where one is forced to run as fast as one can to remain in the same place. But what happens when our population doubles again? Can you brush the doubling aside as being too far in the future to warrant concern? *It is no further removed from you than the sons and daughters of the next generation.*

How We Began Sidestepping Controls

How did we get into this predicament? For most of its history, the human population grew slowly. During the past two centuries, increases in growth rates became astounding. There are three possible reasons for this:

1. Humans steadily developed the capacity to expand into new habitats and new climate zones.

2. Humans increased the carrying capacity in their existing habitats.

3. Human populations sidestepped limiting factors.

The human population did all three of these things.

Reflect on the first point. The first humans evolved in woodlands and savannas. They were vegetarians, for the most part, but they also scavenged bits of meat. Small bands of hunter–gatherers started moving out of Africa about 2 million years ago. By 40,000 years ago, different populations of their descendants had become established in much of the world (Section 27.15).

Most other species could not have expanded into such a broad range of habitats. Humans, with their highly complex brains, could use learning and memory to figure out how to build fires, assemble shelters, make clothes and tools, and plan community hunts. Learned experiences did not die with individuals. They spread quickly from one band to another through language—the capacity for extraordinary cultural communication. *Therefore, the human population expanded into diverse new environments in an exceptionally short time, compared to the long-term geographic dispersal of other kinds of organisms.*

What about the second possibility? Starting about 11,000 years ago, many hunter–gatherer bands shifted to agriculture. Instead of following migratory game herds, they settled in fertile valleys and other places where the seasonal harvesting of fruits and grains was favored. By doing so, they developed a more dependable basis for life. A pivotal factor was the domestication of wild grasses, including species ancestral to modern wheats and rice. People harvested, stored, *and planted* seeds in one place. They domesticated animals and kept them close to home for food and for pulling plows. They dug ditches and diverted water to irrigate their croplands.

Their agricultural practices increased productivity. And the larger, more dependable food supplies favored increases in population growth rates. Towns and cities developed. A social hierarchy emerged that provided a labor base for more intensive agriculture. Much later, food supplies increased again by the use of fertilizers, herbicides, and pesticides. As transportation improved, so did food distribution. *Therefore, even at its simplest, management of food supplies through agriculture increased the carrying capacity for the human population.*

What about the third possibility—the sidestepping of limiting factors? Think about what happened when medical practices and sanitary conditions improved. Until about 300 years ago, poor hygiene, malnutrition, and contagious disease kept death rates high enough to more or less balance birth rates. Contagious diseases, which are density-dependent controls, swept unchecked through overcrowded settlements and cities that were infested with fleas and rodents. Then came plumbing and methods of sewage treatment. Over time, vaccines, antitoxins, and antibiotics were developed as weapons against many of the pathogens. Death rates dropped sharply. Births began to exceed deaths—and extremely rapid population growth was under way.

In the mid-eighteenth century, people discovered how to harness the energy stored in fossil fuels, starting with coal. Within a few decades, large industrialized societies started forming in western Europe and North America. Then efficient technologies developed after

Further reading: Student Guide to InfoTrac on web site →

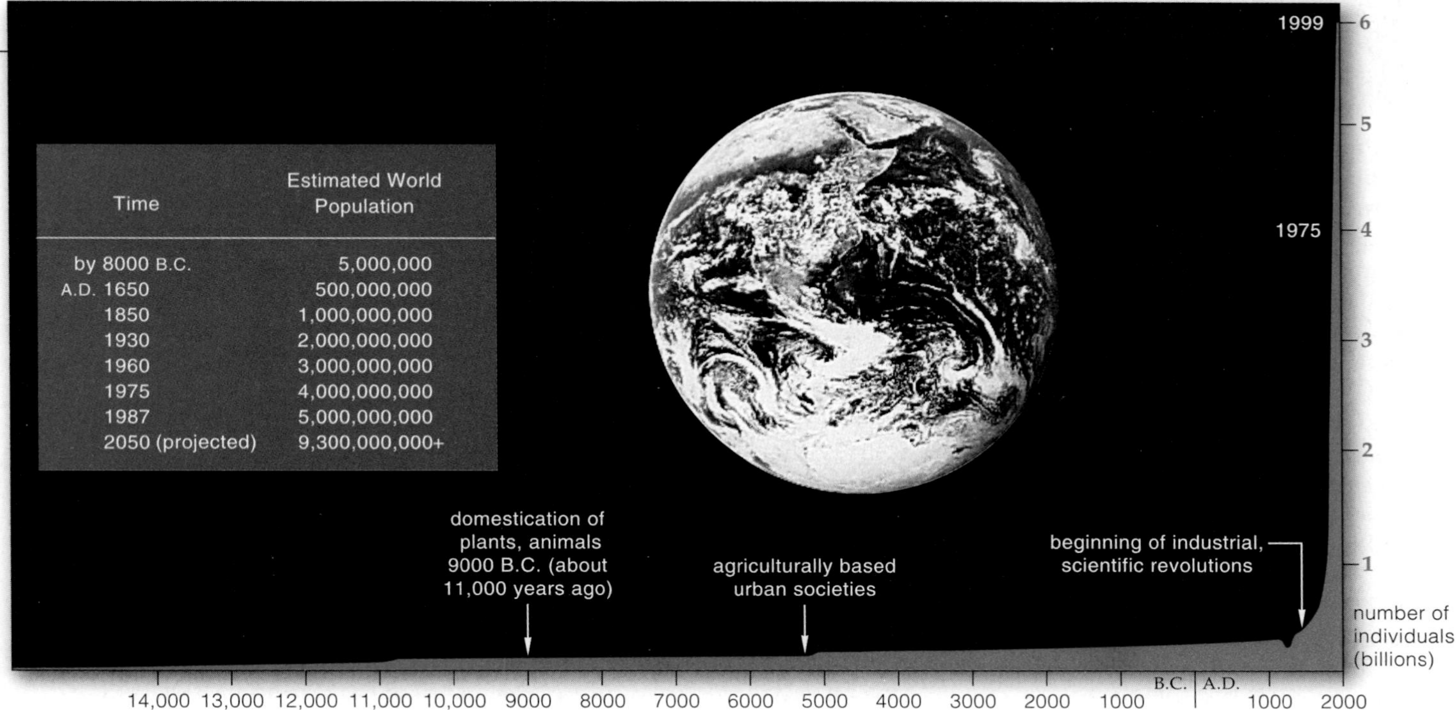
1999
1975
6
5
4
3
2
1
number of individuals (billions)
domestication of plants, animals 9000 B.C. (about 11,000 years ago)
agriculturally based urban societies
beginning of industrial, scientific revolutions
14,000 13,000 12,000 11,000 10,000 9000 8000 7000 6000 5000 4000 3000 2000 1000
B.C. A.D.
1000 2000

Time	Estimated World Population
by 8000 B.C.	5,000,000
A.D. 1650	500,000,000
1850	1,000,000,000
1930	2,000,000,000
1960	3,000,000,000
1975	4,000,000,000
1987	5,000,000,000
2050 (projected)	9,300,000,000+

Figure 46.13 Growth curve (*red*) for the human population. The graph's vertical axis represents world population, in billions. (The dip between the years 1347 and 1351 is the time when 25 million people died in Europe as a result of bubonic plague.) Agricultural revolutions, industrialization, and improvements in health care have sustained the accelerated growth pattern for the past two centuries. The data in the *blue* box show how long it took for the human population to increase from 5 million to 5 billion.

World War I. Large factories mass-produced tractors, cars, and other affordable goods. Machines replaced many of the farmers that had produced the food, and fewer farmers were able to support a larger population.

Therefore, by controlling many disease agents and by tapping into concentrated, existing forms of energy (fossil fuels), humans managed to sidestep major factors that had previously limited their population growth.

Present and Future Growth

Where have all the far-flung dispersals and spectacular advances in agriculture, industrialization, and health care taken us? Starting with *Homo habilis* (Section 27.14), it took about 2.5 million years for human population size to reach 1 billion. As Figure 46.13 shows, it took only 80 years to reach 2 billion, then 30 to reach 3 billion, 15 to reach 4 billion, and 12 to reach 5 billion. And it took merely 11 more years to reach 6 billion!

From what we know about the principles governing population growth—and unless more breakthroughs in technology can increase the carrying capacity—we can expect a dramatic increase in death rates. *Although the stupendously accelerated growth of the human population continues, it cannot be sustained indefinitely.*

Besides having adverse effects on resource supplies, these skyrocketing numbers invite density-dependent controls. For example, infection by *Vibrio cholerae* results in a disease called cholera. The bacterium enters hosts who drink water or eat food that is contaminated with raw sewage. It multiplies in the gut, where it produces a toxin that triggers severe diarrhea and massive fluid loss. Within two to seven days, infected people who are not treated can die from extreme dehydration.

For centuries, *V. cholerae* has thrived and mutated into new strains in India's sewage-enriched rivers and estuaries (Figure 50.15). In the slums of Calcutta alone, millions are forced to bathe in polluted waterways and ponds. In 1992, cholera struck tens of thousands in that city. Since then, the largest cholera epidemic in 100 years has been sweeping through southern Asia. Like the six previous epidemics, it began in India and then spread through Africa, the Middle East, and the Mediterranean countries. It has reached Central America and the United States. And this latest round of cholera resurgence may claim 5 million people. Existing vaccines do not work against the mutant bacterial strain causing it.

At this writing, a new era of human migration has begun. By some estimates, economic hardship and civil strife have put 50 million people on the move within and between many countries. Will the relocations prove peaceable? Where will they find sustainable supplies of food, clean water, and other basic resources?

Through expansion into new habitats, cultural intervention, and technological innovation, the human population has temporarily skirted environmental resistance to growth. Its accelerated growth cannot be sustained indefinitely.

46.8 CONTROL THROUGH FAMILY PLANNING

Figure 46.14 presents the annual rates of increase for populations in different parts of the world in 1997. The current average rate, 1.26 percent, is expected to decline. Even so, the world population is still projected to reach more than 9.1 billion by 2050.

It is mind numbing to think about all the natural resources required to sustain that many people. We will have to increase food production, supplies of drinkable water, energy reserves, and supplies of wood products and other materials to meet everyone's basic needs—something we are not even doing now. Most likely, the gross manipulation of resources will intensify pollution, which will adversely affect water supplies, air quality, and food productivity on land and in the seas.

Population growth, resource depletion, pollution, and the quality of life are all interconnected. Currently, most governments are attempting to lower birth rates, as through family planning programs. The programs educate individuals about choosing how many children they will have, and when. Details vary from country to country, but all provide information on the available methods of fertility control of the sort listed in Section 45.13. Carefully developed and administered programs might bring about a long-term decline in birth rates.

The **total fertility rate** (TFR) is the average number of children being born to women of a given population during their reproductive years. TFRs are estimated on the basis of current age-specific rates. The TFR for the world's population has declined dramatically since 1950, when the average was 6.5. That number was far above the replacement level of 2.1, which is considered to be the level necessary to arrive at zero population growth. The average replacement level is slightly higher than two children per couple because some female children die before reaching reproductive age.

At this writing the world population's TFR is 2.9. In many of the developed countries, TFRs are at or below the replacement level, but in developing countries it is about 3.2.

Even if every couple on the planet decided that they will bear no more than two children, the world population will keep growing for another sixty years! Why? A huge number of existing children will soon be of reproductive age.

Figure 46.15 shows some age structure diagrams for populations growing at different rates. The central part of each diagram includes individuals of reproductive age. The lower part includes children who will move into the reproductive age category over the next fifteen years, with 15–44 the average range for childbearing years. The age structure diagrams for rapidly growing populations, including Mexico, have a broad base. The human population of the United States has a relatively narrow base and is an example of slow growth. Figure 46.16 tracks its 78 million *baby-boomers*. This is a cohort that started to form in 1946, when American soldiers returned home after World War II and started families.

Today, *more than one-third of the world population falls in the broad pre-reproductive base*. The numbers give us an idea of the magnitude of the effort it will take to control the world population growth.

The rate of birth slows when women bear children in their early thirties rather than in mid-teens or early twenties. Delayed reproduction slows the growth rate

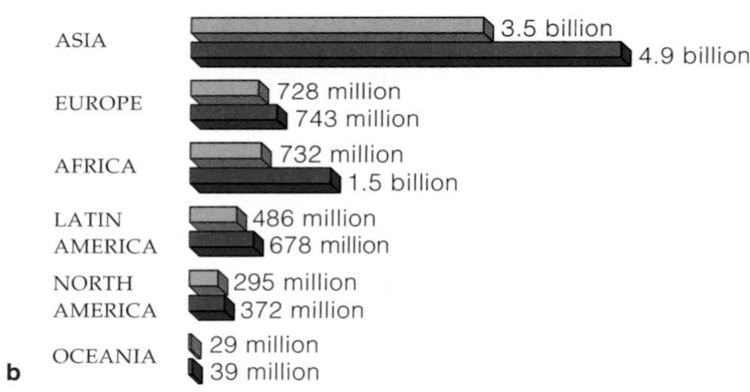

b
ASIA
3.5 billion
4.9 billion
EUROPE
728 million
743 million
AFRICA
732 million
1.5 billion
LATIN AMERICA
486 million
678 million
NORTH AMERICA
295 million
372 million
OCEANIA
29 million
39 million

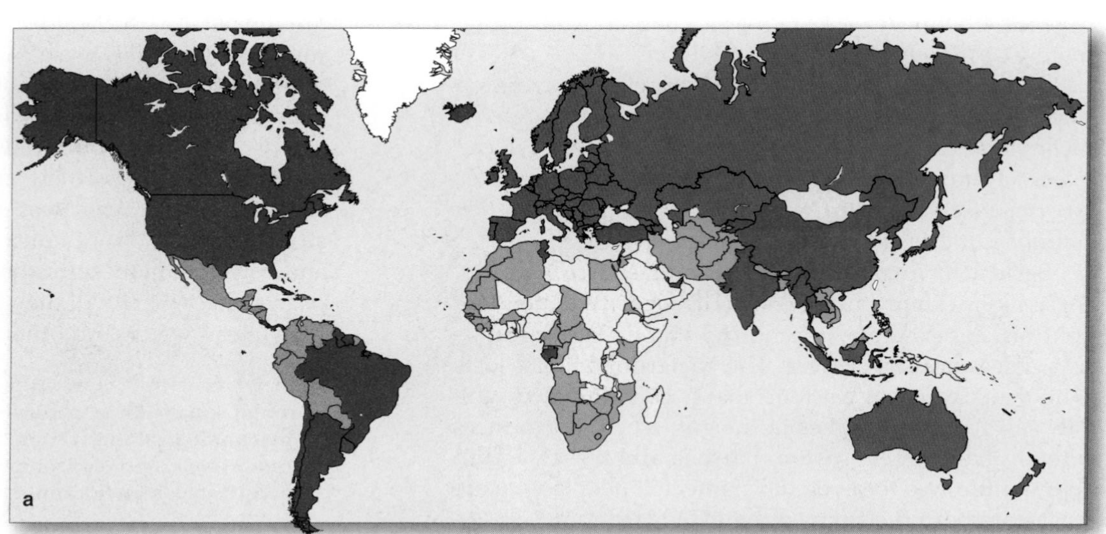

a

Figure 46.14 (**a**) Average population growth rate in major regions for 1997. *Purple* signifies less than 1%, *green* 1–9%, *orange*, 2–2.9 percent, and *gold*, over 3%. (**b**) Population sizes in 1997 (*orange* bars) and projected for the year 2025 (*blue* bars).

Further reading: Student Guide to InfoTrac on web site →

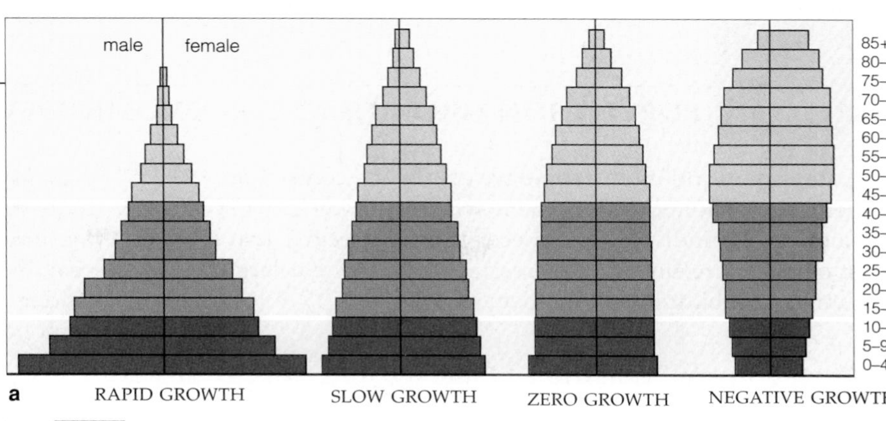
male
female
85+
80–84
75–79
70–74
65–69
60–64
55–59
50–54
45–49
40–44
35–39
30–34
25–29
20–24
15–19
10–14
5–9
0–4
a
RAPID GROWTH
SLOW GROWTH
ZERO GROWTH
NEGATIVE GROWTH

Figure 46.15 (**a**) General age structure diagrams for countries with rapid, slow, zero, and negative population growth rates. Pre-reproductive years are *green* bars; reproductive years, *purple*; and post-reproductive years, *light blue*. The vertical axis divides each graph into males (*left*) and females (*right*). Bar widths correspond to the proportion of individuals in each age group. (**b**) Age structure diagrams for representative countries in 1997. The population sizes are measured in millions. 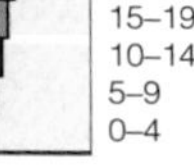

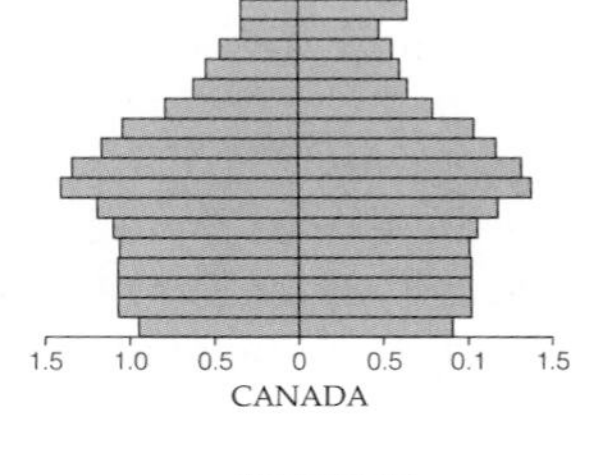
1.5 1.0 0.5 0 0.5 0.1 1.5
CANADA

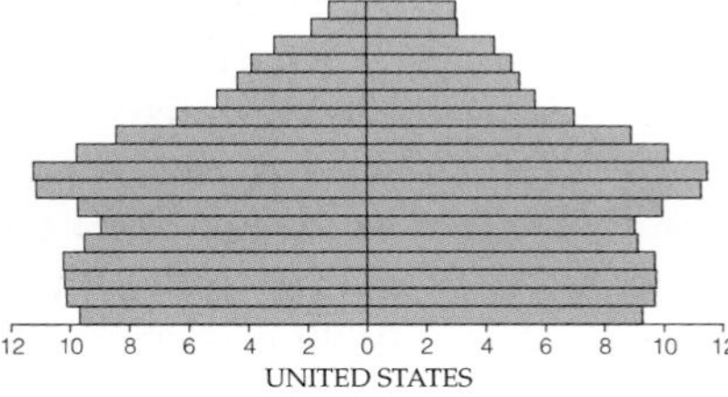
12 10 8 6 4 2 0 2 4 6 8 10 12
UNITED STATES

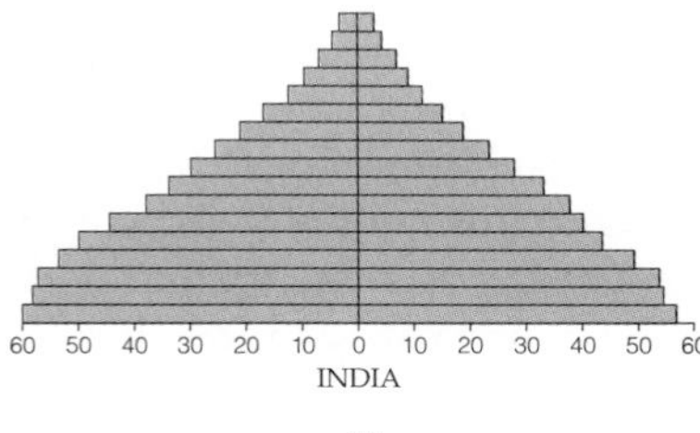
60 50 40 30 20 10 0 10 20 30 40 50 60
INDIA

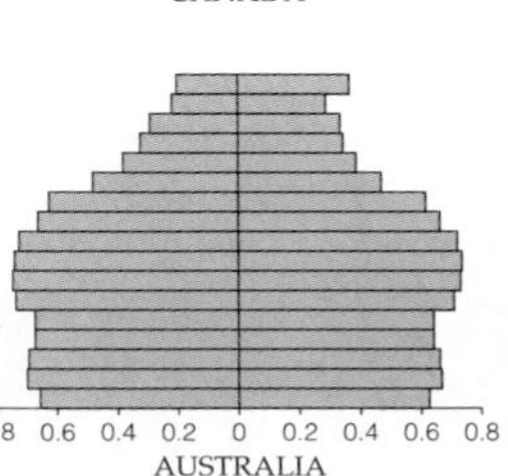
0.8 0.6 0.4 0.2 0 0.2 0.4 0.6 0.8
AUSTRALIA

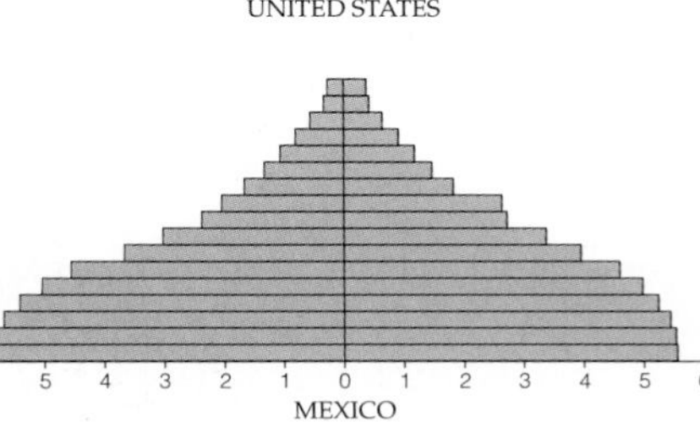
6 5 4 3 2 1 0 1 2 3 4 5 6
MEXICO

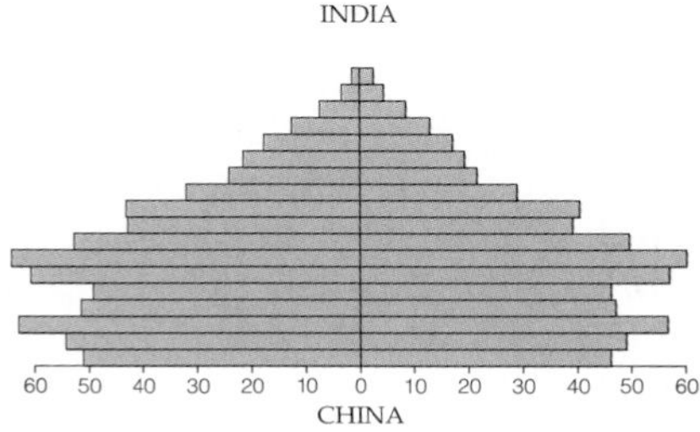
60 50 40 30 20 10 0 10 20 30 40 50 60
CHINA

b

and lowers the average number of children in families. China, for instance, supports the world's most extensive family planning program. The government discourages premarital sex, and asks people to postpone marriage and limit family size to one child. It makes available free abortions, contraceptives, and sterilization to married couples; paramedics and mobile units ensure access to these measures even in remote rural areas.

The couples who pledge to have no more than one child receive more food, free medical care, much better housing, and salary bonuses. Their child will be granted free tuition and preferential treatment when he or she is old enough to enter the job market. Breaking the pledge means losing benefits and paying more taxes. Are the measures inhumane? Then what alternatives would you suggest? Between 1958 and 1962 alone, 30 million Chinese may have died of starvation because of widespread famines.

Billboard in China promoting family planning

1955 1985 2015 2035

Figure 46.16 Age structure diagrams for the United States population. *Gold* bars track the baby-boomer generation.

Since 1972, China's TFR has declined sharply, from 5.7 to 1.8. Even so, the population time bomb has not stopped ticking. Currently its population is almost 1.27 billion—and 150 million of its young females are in the pre-reproductive age category. By 2010, the population in China is projected to surpass 1.3 billion.

Family planning programs on a global scale are designed to help stabilize the size of the human population.

Even if it reaches a level of zero population growth, the human population will continue to grow for sixty years, for its reproductive base already consists of a staggering number of individuals.

46.9 POPULATION GROWTH AND ECONOMIC DEVELOPMENT

Today, more and more people are aware of a connection between economic development and population growth rates. When individuals are economically secure, they apparently are under less pressure to produce a large number of children to help them survive.

Demographic Transition Model

We can correlate changes in population growth with changes that typically unfold in four stages of economic development. The four stages are at the heart of the **demographic transition model** (Figure 46.17).

According to this model, living conditions are harsh during a *preindustrial* stage, before widespread use of technology and medical advances. Birth and death rates are both high, so the rate of population growth is low.

Next, during the *transitional* stage, industrialization begins; food production as well as health care improves. Death rates drop, but birth rates stay fairly high. Thus the population grows rapidly. And its growth continues at high rates for an extended length of time. Annual growth rates tend to be 2.5 to 3 percent, on the average. When living conditions improve and birth rates begin to decline, the growth starts to level off.

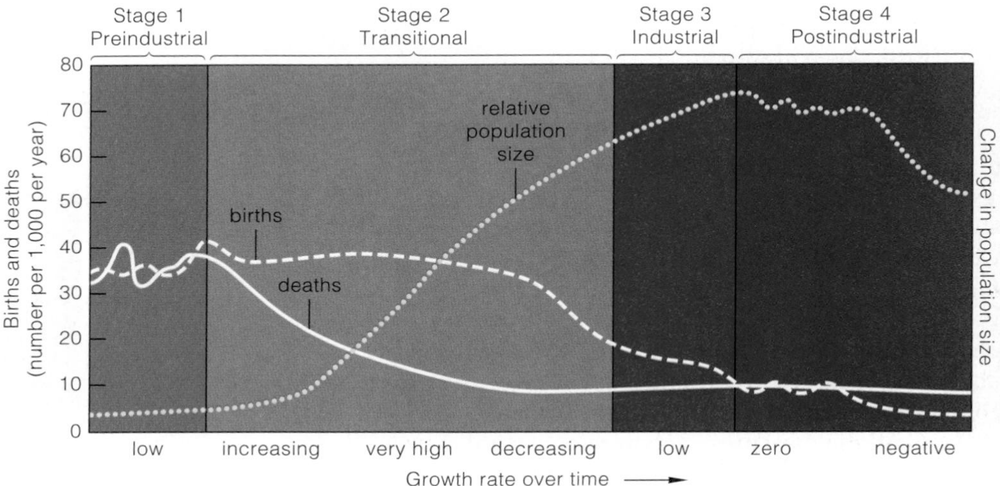

Stage 1
Preindustrial
Stage 2
Transitional
Stage 3
Industrial
Stage 4
Postindustrial
80
70
60
50
40
30
20
10
0
Births and deaths
(number per 1,000 per year)
relative population size
births
deaths
Change in population size
low
increasing
very high
decreasing
low
zero
negative
Growth rate over time

Figure 46.17 Diagram of the demographic transition model of changes in the growth characteristics and size of populations, correlated with changing economic development. The model explains past changes in western Europe and other industrialized regions.

During the *industrial* stage, when industrialization is in full swing, the population's growth slows dramatically. A slowdown emerges mostly because people move from the country to cities —and urban couples tend to control the size of their families. Many couples get caught up in the accumulation of goods, and often decide that the time and cost of raising more than a few children conflict with that goal. As you will see, this is a sticky issue with respect to population size.

In the *postindustrial* stage, zero population growth is reached. The birth rate falls below the death rate, and population size slowly decreases.

The United States, Canada, most of western Europe, Australia, Japan, and the nations of the former Soviet Union are in the industrial stage. Their growth rate is slowly decreasing. In Germany, Bulgaria, Hungary, and some other countries, the death rates exceed the birth rates, and the populations are getting smaller.

Mexico and other less developed countries are in the transitional stage. But they do not have enough skilled workers to complete the transition to a fully industrial economy. Fossil fuels and other resources that drive industrialization are being used up there as well as in the industrialized countries. Fuel costs might become prohibitive for countries at the bottom of the economic ladder even before they enter the industrial stage.

If population growth continues to outpace economic growth, death rates will increase. Thus many countries may be stuck in the transitional stage. And some may return to the harsh conditions of the preceding stage.

Enormous disparities in economic development are driving forces for immigration and emigration. In 1996 close to 916,000 legal immigrants and 75,000 political refugees crossed the borders of the United States. So did 275,000 undocumented immigrants, mainly from Latin America and Asia. They brought the estimated total of undocumented immigrants residing in this developed country to 5 million.

Or think of how California's population increased by 12 *million* between 1970 and 1996. With its 2.5 percent growth rate, mainly among legal and illegal newcomers, it may jump by another 16 million by the year 2020. For some time, California has been a symbol of the good life, drawing people from the world over. However, its schools are overcrowded, funds for welfare and other social services are shrinking, sewage treatment plants are nearing capacity, and water shortages are severe and frequent. Also, both salinization of croplands and air pollution are eroding California's agricultural base, which is a major factor in the state's economic health.

Claiming that population growth affects economic health, many governments restrict immigration. Only the United States, Canada, Australia, and a few other

Further reading: Student Guide to InfoTrac on web site →

countries allow large annual increases in size. Elsewhere, appalling living conditions, harsh governmental policies, and civil strife combine to promote the exodus of more individuals than the better-off nations can easily absorb.

A Question of Resource Consumption

This chapter opened with a brief look at the conditions that confront most people in India, with its whopping 16 percent of the world human population. By comparison, the United States has merely 4.7 percent.

Yet which country is the most "overpopulated"—not in terms of actual numbers, but rather in terms of resource consumption and instigation of environmental damage?

The highly industrialized United States is producing 21 percent of the world's goods and services. And its people consume fifty times as many goods and services than the average person in India. Its people use 25 percent of the world's processed minerals and available, nonrenewable sources of energy. Also, they generate at least 25 percent of the world's pollution and trash. By contrast, India produces about 1 percent of all goods and services. It uses 3 percent of the available minerals and nonrenewable energy resources. And it generates about 3 percent of the pollution and trash.

Extrapolating from these numbers, Tyler Miller, Jr., estimated that it would take 12.9 billion impoverished individuals living in India to have as much impact on the environment as 258 million Americans.

Differences in population growth among countries correlate with levels of economic development, hence with economic security (or lack of it) of individuals. Growth rates are low in preindustrial, industrial, and post-industrial stages. They are greatest during the transition to industrialization.

46.10 SOCIAL IMPACT OF NO GROWTH

For us, as for all species, the biological implications of extremely rapid growth are staggering. Yet so are the social implications of what will happen when (and if) the human population decreases to the point of zero population growth—and stays there.

For instance, in a growing population, most people fall in younger age brackets. If living conditions ensure constant growth over time, the age distribution should guarantee availability of a future work force. This has social implications. Why? *It takes a large work force to support the older age brackets.* In the United States, older, nonproductive people expect the federal government to provide them with subsidized medical care, low-cost housing, and many other social programs. However, as a result of improved medicine and hygiene, people in these brackets are living far longer than the elderly did when the nation's social security program was set up. The current cash benefits exceed the contributions that they made to the program when *they* were younger.

If the population ever does reach and maintain zero growth, a larger proportion of individuals will end up in the older age brackets. Even slower growth will pose problems. What will happen when baby-boomers retire? Will all those nonproductive members continue to get goods and services if productive ones must carry more and more of the economic burden? This is not an abstract question. Put it to yourself. Exactly how much economic hardship are you willing to bear for the sake of your parents? For your grandparents? And how much will your children be willing or able to bear for you?

We have arrived at a major turning point, not only in our biological evolution but in our cultural evolution as well. The decisions awaiting us are among the most difficult we will ever have to make—yet it is clear that they must be made, and soon.

All species face limits to growth. We might think we are different from the rest, for our special ability to undergo rapid cultural evolution has enabled us to postpone the action of most of the factors that limit growth. But the crucial word is *postpone*. No amount of cultural intervention can hold back the ultimate check of limited resources and a damaged environment.

We have sidestepped a number of the smaller laws of nature. In doing so, we have become more vulnerable to those laws which cannot be repealed. Today, there may be only two options available. Either we make a global effort to limit population growth in accordance with environmental carrying capacity, or we wait until the environment does it for us.

In the final analysis, no amount of cultural intervention can repeal the ultimate laws governing population growth, as imposed by the carrying capacity of the environment.

46.11 SUMMARY

1. A population is a group of individuals of the same species occupying a given area. It has a characteristic size, density, distribution, and age structure as well as characteristic ranges of heritable traits.

2. The growth rate for a population during a specified interval may be determined by calculating the rates of birth, death, immigration, and emigration. To simplify calculations, we may put aside effects of immigration and emigration, and combine the birth and death rates into a variable r (net reproduction per individual per unit time). Then we may represent population growth (G) as $G = rN$, where N is the number of individuals during the interval specified.

a. In cases of exponential growth, the reproductive base of a population increases and its size expands by ever increasing increments during successive intervals. This trend plots out as a J-shaped growth curve.

b. Any population will show exponential growth as long as its per capita birth rate stays even slightly above its per capita death rate.

c. In logistic growth, a low-density population slowly increases in size, goes through a rapid growth phase, then levels off in size once carrying capacity is reached.

3. Carrying capacity is the name ecologists give to the maximum number of individuals in a population that can be sustained indefinitely by the resources available in their environment.

4. The availability of sustainable resources as well as other factors that limit growth dictates population size during a specified interval. The limiting factors vary in their relative effects and vary over time, so population size also changes over time.

5. Limiting factors such as competition for resources, disease, and predation are density-dependent. Density-independent factors, such as weather on the rampage, tend to increase the death rate or decrease the birth rate more or less independently of population density.

6. Patterns of reproduction, death, and migration vary over the life span of a species. Environmental variables also help shape the life history (age-specific) patterns.

7. The human population surpassed 6 billion in 1999. Its annual growth rate now varies from below zero in a few developed countries to above 3 percent in some less developed countries. In 2000 the annual growth rate for the entire human population was 1.26 percent.

8. The human population's rapid growth in the past two centuries occurred through its capacity to expand into new habitats, and because of agricultural, medical, and technological developments that increased the carrying capacity. Ultimately, we must confront the reality of the carrying capacity and limits to our population growth.

Review Questions

1. Define population size, population density, and population distribution. Describe a typical population in terms of several categories for its age structure. *46.1*

2. Define exponential growth. Be sure to state what goes on in the age category that is a foundation for its occurrence. *46.3*

3. Define carrying capacity and describe its effect as evidenced by a logistic growth pattern. *46.4*

4. Give examples of the limiting factors that come into play when a population of mammals (for example, rabbits or humans) reaches very high density. *46.4, 46.5*

5. Define doubling time. In what year is the human population expected to surpass 9 billion? *46.3, 46.7*

6. How did earlier human populations expand steadily into new environments? How did they increase the carrying capacity in their habitats? Have we now avoided some limiting factors on population growth? Or is the avoidance an illusion? *46.7*

Self-Quiz (*Answers in Appendix III*)

1. ______ is the study of how organisms interact with one another and with their physical and chemical environment.

2. A ______ is a group of individuals of the same species that occupy a certain area.

3. The rate at which a population grows or declines depends upon the rate of ______.
 a. births c. immigration e. all of the above
 b. deaths d. emigration

4. Populations grow exponentially when ______.
 a. birth rate exceeds death rate, immigration and emigration rates stay the same
 b. birth rate and death rate stabilize
 c. emigration rates exceed immigration rates
 d. both b and c

5. For a given species, the maximum rate of increase per individual under ideal conditions is the ______.
 a. biotic potential c. environmental resistance
 b. carrying capacity d. density control

6. Resource competition, disease, and predation are ______ controls on population growth rates.
 a. density-independent c. age-specific
 b. population-sustaining d. density-dependent

7. Which of the following factors does *not* affect sustainable population size?
 a. predation c. resources e. All of the above can
 b. competition d. pollution affect population size.

8. In 2000, the average annual growth rate for the human population at midyear was ______ percent.
 a. 0 b. 1.05 c. 1.26 d. 1.55 e. 2.7 f. 4.0

9. Match each term with its most suitable description.

____ carrying capacity	a. depends on rates of birth, death, emigration, immigration
____ exponential growth	b. slow growth then rapid growth phase, then leveling off
____ population growth rate	c. the maximum number of individuals sustainable by an environment's resources
____ limiting factor	d. population growth plots out as a J-shaped curve
____ logistic growth	e. short supply of any resource essential to population growth

Critical Thinking

1. If house cats that have not been neutered or spayed live up to their biotic potential, two can be the start of many kittens—12 the first year, 72 the second year, 429 the third, 2,574 the fourth, 15,416 the fifth, 92,332 the sixth, 553,019 the seventh, 3,312,280 the eighth, and 19,838,741 kittens in the ninth year. Is this a case of logistic growth? Exponential growth? Irresponsible cat owners?

2. A third of the world population is below age fifteen. Describe the effect of this age distribution on the future growth rate of the human population. If you conclude that it will have severe impact, what sorts of humane recommendations would you make to encourage individuals of this age group to limit family size? What are some social, economic, and environmental factors that might keep them from following the recommendations?

3. Write a short essay about a population having one of the age structures shown below. Describe what may happen to younger and older groups when individuals move into new categories.

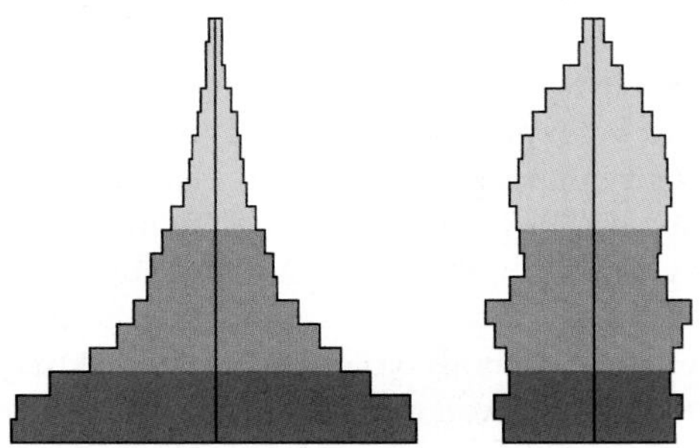

4. Figure 46.18 charts the legal immigration to the United States between 1820 and 1997. (The Immigration Reform and Control Act of 1986 accounted for the most recent dramatic increase; it granted legal status to illegal immigrants who could prove they had lived in the country for years.) During the 1980s and 1990s, an economic downturn fanned resentment against newcomers. Many people now say legal immigration should be restricted to 300,000–450,000 annually and we should crack down on illegal immigrants. Others argue that such a policy would diminish our reputation as a land of opportunity. They also say it would discriminate against legal immigrants during crackdowns on others of the same ethnic background. Do some research, then write an essay on the pros and cons of both positions. (You may wish to start with web site http://www.census.gov.)

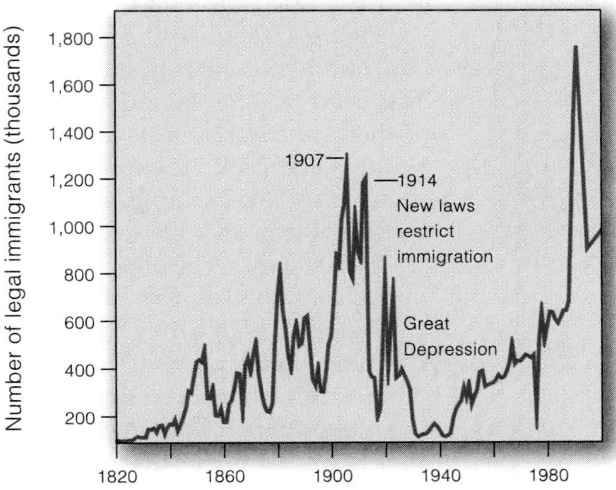

Number of legal immigrants (thousands)
1,800
1,600
1,400
1,200
1,000
800
600
400
200
1907
1914
New laws restrict immigration
Great Depression
1820
1860
1900
1940
1980

Figure 46.18 Chart of legal immigration to the United States between 1820 and 1997.

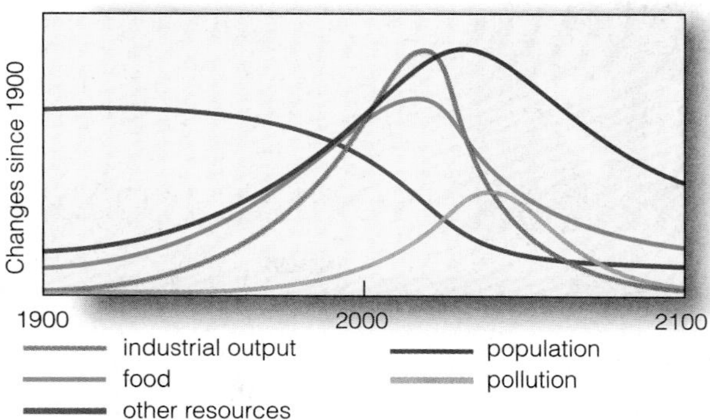

Changes since 1900
1900
2000
2100
industrial output
food
other resources
population
pollution

Figure 46.19 Computer-based projection of what may happen if the size of the human population continues to skyrocket without dramatic policy changes and technological innovation. The assumptions are that the population has already overshot the carrying capacity and current trends continue unchanged.

5. In his book *Environmental Science*, Miller points out that the unprecedented projected increase in the human population to 9+ billion by the year 2050 raises serious questions. Will there be enough food, energy, water, and other resources to sustain that many people? Will governments be able to provide adequate education, housing, medical care, and other social services for all of them? Computer models suggest that the answers are no (Figure 46.19). Yet some people claim we can adapt socially and politically to a far more crowded world, assuming harvests improve through technological innovation, every inch of arable land is cultivated, and everyone eats only grain.

There are no easy answers to these questions. If you have not yet been doing so, start following the arguments in your local newspapers, in magazines, on television, and on the Web. This will allow you to become an informed participant in a global debate that surely will have impact on your future.

Selected Key Terms

age structure *46.1*
biotic potential *46.3*
capture–recapture method *46.2*
carrying capacity *46.4*
cohort *46.5*
demographic transition model *46.9*
demographics *46.1*
density-dependent control *46.4*
density-independent factor *46.4*
doubling time *46.3*
ecology *CI*
emigration *46.3*
exponential growth *46.3*
immigration *46.3*
life history pattern *46.5*
limiting factor *46.4*
logistic growth *46.4*
migration *46.3*
per capita *46.3*
population density *46.1*
population distribution *46.1*
population size *46.1*
r (net reproduction per individual per unit time) *46.3*
reproductive base *46.1*
survivorship curve *46.5*
total fertility rate *46.8*
zero population growth *46.3*

Readings See also *www.infotrac-college.com*

Bloom, D. and D. Canning. 18 February 2000. "The Health and Wealth of Nations." *Science* 287:1207–1209. Argues that poor health is more than a consequence of low income; that it is also one of its fundamental causes.

Cohen, J. E. 1995. *How Many People Can the Earth Support?* New York: Norton. No pat answer to title's question.

Miller, G. T. 2000. *Living in the Environment*. Eleventh edition. Belmont, California: Brooks/Cole.

47

SOCIAL INTERACTIONS

Deck the Nest With Sprigs of Green Stuff

In 1890, bird fanciers imported more than a hundred starlings (*Sturnus vulgaris*) from Europe and released them in New York City's Central Park. In less than a century, the descendants of the introduced species had expanded their geographic range from coast to coast. They are so good at gleaning food from the agricultural fields of North America that they have outmultiplied native birds. They also have evicted great numbers of them from scarce nesting sites in the cavities of trees.

Once a male and female commandeer a tree cavity, they gather dry grass and twigs and use them to build or rebuild a nest (Figure 47.1). Theirs is no ordinary nest. They *decorate* the nest bowl with sprigs, freshly plucked.

Why do starlings do this? Do the sprigs of greenery camouflage the nest from predators? Not likely. The nests are already concealed inside the cavities of trees. Do the sprigs function as insulative material that might help keep the forthcoming eggs warm? Actually, the still-moist, green plant parts would promote heat loss, not heat conservation. Well, then, do the green sprigs combat parasites? Mites, for example, parasitize birds and infest their nest cavities. In short order, even a few tiny mites can produce thousands of descendants. When present in large numbers, mites can suck enough blood from a nestling to weaken it. They interfere with the nestling's growth, development, and survival.

Larry Clark and Russell Mason decided to test the third hypothesis. As both biologists knew, starlings do not weave just any green plant material into the nests. They choosily favor the leaves of certain plants, such as wild carrot (Figure 47.1). Thus Clark and Mason built a set of experimental nests, some with freshly cut wild carrot leaves and some without. They removed the natural nests that pairs of starlings had constructed and had already started using. Fifty percent of the nesting pairs received replacement nests decorated with sprigs of wild carrot. No sprigs decorated the replacement nests for the other pairs of starlings.

Figure 47.1 A most excellent fumigator in nature—a European starling (*Sturnus vulgaris*). It combats infestations of mites by decorating previously owned nests with fresh sprigs of wild carrot (*Daucus carota*), shown at the left.

Figure 47.2 shows the results. The number of mites in greenery-free nests was consistently greater than the number in nests decorated with greenery. At the end of one experiment, for example, sprig-free nests teemed with an average of 750,000 mites. The ones with sprigs contained a mere 8,000.

Shoots of wild carrot happen to contain a highly aromatic steroid compound. Almost certainly, the compound repels herbivores and thus helps the plant survive. By coincidence, the compound also prevents mites from becoming sexually mature—and thereby

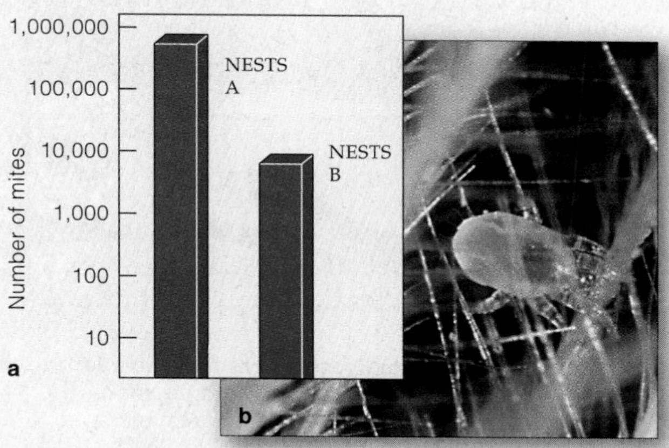
1,000,000
100,000
10,000
1,000
100
10
Number of mites
NESTS
A
NESTS
B
a
b

Figure 47.2 (**a**) Experimental results that point to the adaptive value of starling nest-decorating behavior. Nests designated *A* were kept free of fresh sprigs of wild carrot and other plants that contain aromatic compounds. The compounds prevent baby mites from developing into (**b**) adult mites. Other nests, designated *B*, had fresh sprigs added every seven days. Head counts of mites (*Ornithonyssus sylviarum*) infesting the nests were made at the time the starling chicks left the nest, twenty-one days after this experiment was under way.

prevents mite population explosions inside the nests festooned with wild carrot.

Far from being a trivial behavior, then, "decorating" a nest with aromatic greenery has adaptive value. It fumigates the nest and thereby increases an individual's chance of producing healthy, surviving offspring.

And so starlings lead us into the fascinating world of research into behavior. As you will see, some studies focus on the adaptive value of specific behavioral trait to an individual's reproductive success. Others focus on the internal mechanisms that enable individuals to behave as they do.

As a starting point, recall that information encoded in genes directs the formation of tissues and organs that make up the animal body, including those of its nervous system. That system detects, processes, and integrates information about stimuli in the internal and external environments, then commands muscles and glands to make suitable responses. Other gene products called hormones contribute to the responses. Because genes specify all of the substances required for the structure and function of nervous and endocrine systems, *they are the heritable foundation for responses that animals make to stimuli*. We call the observable, coordinated responses to stimuli **animal behavior**.

KEY CONCEPTS

1. Animals make coordinated responses to stimuli. We call these responses animal behavior. The cooperative, interdependent relationships among individuals of a species are forms of social behavior.

2. Genes underlie the behavioral ability of individuals, as when instructions encoded in certain genes govern the development of the nervous and endocrine systems. These organ systems allow an animal to detect, process, and issue commands for behavioral responses to stimuli.

3. Most newly born or hatched animals have the neural wiring and motor systems necessary to perform some behaviors without having to learn them through actual experience. Sign stimuli, which are simple, well-defined environmental cues, trigger such instinctive behaviors.

4. For nearly all animal species, the nervous system has a capacity to process and retain information about specific experiences in the environment, then to use the information to vary or change behavioral responses.

5. Like other traits having a genetic basis, behavior has evolved by way of natural selection, which occurs when genetically different individuals in a population have different numbers of surviving offspring. Behavioral mechanisms that enhance the ability of individuals to pass on their genes to offspring have been favored.

6. Evolved modes of communication underlie social behavior. Communication signals, which are encoded in stimuli such as species-specific odors, body coloration and patterning, postures, and movements, hold clear meaning for both the sender and the receiver of signals.

7. Living in a social group has costs and benefits, as measured by an individual's ability to pass on its genes to offspring. Not every environment favors the evolution of such groups. In some cases, a solitary life-style allows an individual to leave more descendants than belonging to a large group would allow.

8. In altruism, an individual helps others in ways that require it to sacrifice its own reproductive success. The evolution of altruism requires special circumstances in which individuals can propagate their genes indirectly, by helping their relatives reproduce successfully.

BEHAVIOR'S HERITABLE BASIS

Genes and Behavior

An animal's nervous system, recall, is wired to detect, interpret, and issue commands for response to stimuli—that is, to specific aspects of the external and internal environments. One way or another, each of the steps required to assemble and operate the system's receptors, nerve pathways, and brain involves the animal's genes. If we define animal behavior as coordinated responses to stimuli, then behavior starts with genes.

Stevan Arnold found evidence of the genetic basis of behavior by studying the feeding preferences of coastal and inland populations of a snake species in California. Garter snakes near the coast prefer banana slugs (Figure 47.3*a*). Snakes living inland prefer tadpoles and small fishes. Offer them a banana slug and they ignore it.

In one set of experiments, Arnold offered captive newborn garter snakes a chunk of slug as the first meal. Offspring of coastal snakes usually ate the chunk. They even flicked their tongue at cotton swabs drenched in essence of slug. (A snake "smells" by tongue-flicking, which draws chemical odors into the mouth.) But the offspring of snakes living inland ignored those swabs, and only rarely did they eat slug meat. Here was a clear difference between captive baby snakes that had no prior, direct experience with slugs. Arnold concluded that the snakes had been programmed before birth to accept or reject slugs; they certainly didn't learn their feeding preferences through taste trials. Maybe the coastal and inland populations had allelic differences for the genes that affect the formation of odor-detecting mechanisms when a garter snake embryo is developing.

To test his hypothesis, Arnold crossbred coastal and inland snakes. If the different food preferences between snake populations has a genetic basis, then hybrid offspring might make an *intermediate* response to slug chunks and odors. Results of the crosses matched the prediction. Compared with typical newborn inland snakes, many baby snakes of mixed parentage tongue-flicked more often at the slug-scented cotton swabs—but less often than newborn coastal snakes did.

a
b
c

Figure 47.3 (**a**) Banana slug, food for (**b**) an adult garter snake of coastal California. (**c**) A newborn garter snake from a coastal population, tongue-flicking at a cotton swab drenched with tissue fluids from a banana slug.

Hormones and Behavior

The gene products called hormones also contribute to bird song and other behaviors. Consider the male zebra finch in Figure 47.4. Each spring he repeats a song with splendid consistency and clarity. In part, his singing is caused by seasonal differences in how much melatonin, a hormone, is secreted from his pineal gland (Section 37.7). A high level of melatonin in blood suppresses the growth and functions of songbird gonads. Its effect is reversed when photoreceptors in the pineal gland absorb sunlight energy and issue signals that lead to a decrease in melatonin secretion.

Winter has fewer hours of daylight than spring; there is not enough light to inhibit secretion of melatonin. In

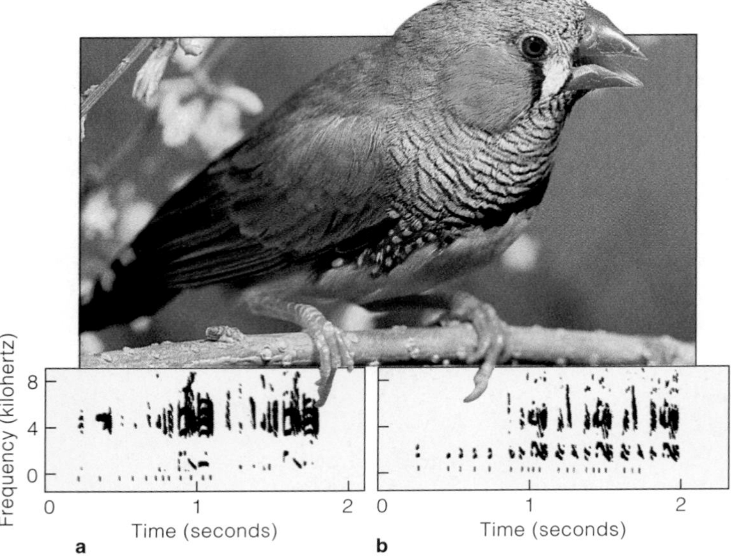
Frequency (kilohertz)
8
4
0
0
1
2
Time (seconds)
a
b

Figure 47.4 Hormones and the territorial song of zebra finches. (**a**) Sound spectrogram of the male's song. The spectrogram is a visual record of each note's pitch (that is, its frequency, measured in kilohertz). (**b**) Sound spectrogram of a female zebra finch that was experimentally converted to a singer. When she was a nestling, she received extra estrogen. At adulthood, she received an implant of testosterone—and started singing. Estrogen normally organizes the song system (certain parts of the brain) in developing songbird embryos. Testosterone activates the system in adult males.

Further reading: Student Guide to InfoTrac on web site

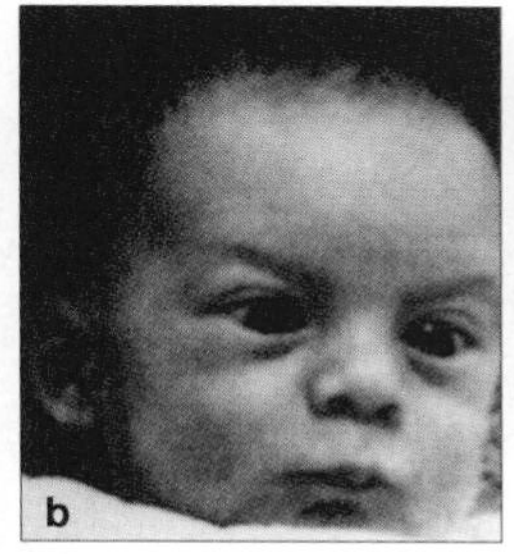

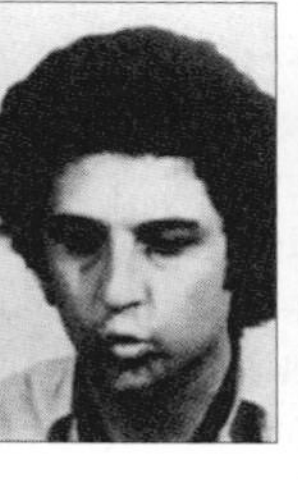

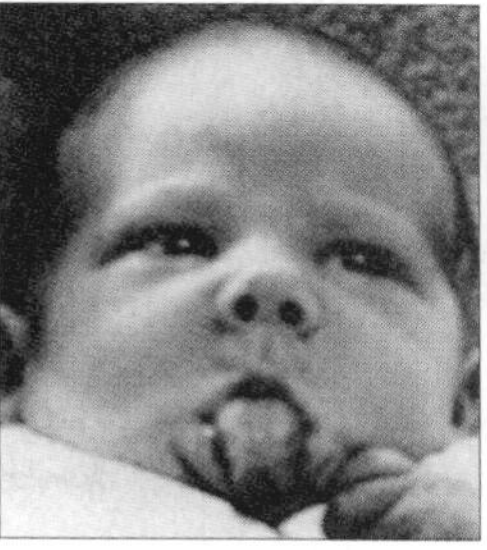

Figure 47.5 Examples of instinctive responses that human offspring make to sign stimuli. (**a**) A close-up face of an adult triggers smiling behavior in very young infants. (**b**) Somewhat older infants instinctively attempt to imitate facial expressions of adults.

spring, when daylength increases, gonads are released from hormonal suppression. They increase in size and step up their secretions of estrogen and testosterone. These sex hormones trigger gender-related differences in bird singing behavior. How? While embryos of most songbirds are developing, estrogen controls formation of a **song system**. The song system consists of several brain structures that will govern the activity of a vocal organ's muscles. Male and female birds differ in the size and structural components of their sound system.

Unlike humans, females of songbird species are XY and males are XX. Certain genes on the Y chromosome specify products that block estrogen production. Before a *male* bird hatches, his gonads secrete estrogen which, at high levels, stimulates development of a masculinized brain. (Estrogen gets converted to testosterone in his brain.) Later on, his gonads enlarge. At the start of the breeding season, his gonads secrete more testosterone. When the hormone binds to receptors on cells in the song system, it induces metabolic changes that will prepare the male to sing when he is suitably stimulated.

Instinctive Behavior Defined

With their tongue-flicking, body orientation, and strikes at prey, newborn garter snakes offer a fine example of **instinctive behavior**. This means a particular behavior is performed without having been learned by actual experience in the environment. Instead, the nervous system of a newly born or hatched animal is already wired to recognize one or two simple, well-defined cues in the environment, or **sign stimuli**, that can trigger a suitable response. The snake's response is a stereotyped motor program. When the snake recognizes certain sign stimuli, a **fixed action pattern** follows. It is a program of coordinated muscle activity that runs to completion independently of feedback from the environment.

As another example, when human infants are two or three weeks old, they tend to smile instinctively when an adult's face comes close to their own (Figure 47.5*a*). Infants make the same response to an overly simplified

Figure 47.6 Example of a social parasite. European cuckoos lay eggs in nests of other species. In response to an environmental cue, this cuckoo hatchling is executing a behavior that is innate. It has inherited the knowledge of what to do without having to learn it. Even before its eyes open, it responds to the spherical shape of the host's eggs and shoves them out of the nest. The foster parents keep on feeding the usurper.

stimulus—a flat, face-sized mask with two dark spots where eyes would be on a human face. A mask with one "eye" won't do the trick. As Figure 47.5*b* suggests, older infants continue to show instinctive behavior.

And remember the cuckoo, a social parasite? Adult females lay their eggs in nests of other species. If newly hatched cuckoos eliminate the natural-born offspring, they will receive the undivided attention of their foster parents (Section 48.6). They are blind at birth. But if they contact an egg or any other round object, they will carry out a fixed action pattern. They maneuver the egg onto their back, then push it from the nest (Figure 47.6).

Animals clearly have a genetically based capacity to respond automatically to certain environmental cues. But they also have the means to process information about specific experiences, then use the information to vary or change their responses. As you will see next, the outcome is what we call learned behavior.

Genes underlie animal behavior—coordinated responses to stimuli. Certain gene products are essential in constructing and operating the nervous system, which governs behavior. Other gene products called hormones also influence the mechanisms required for particular forms of behavior.

Animals start out life neurally wired to recognize simple but important cues in the environment. These sign stimuli trigger suitable responses, such as fixed action patterns.

47.2 LEARNED BEHAVIOR

Animals process and integrate information gained from experiences, then use it to vary or change responses to stimuli. We call this **learned behavior**. For example, a young toad's nervous system commands it to flip its sticky tongue instinctively at any dark object crossing its field of vision. In the toad world, such objects usually are edible insects. What if a dark object is a bumblebee that stings the tongue? After this experience, the toad will avoid "bumblebee-sized black-and-yellow-banded objects that sting." **Imprinting**, another example, is a time-dependent form of learning triggered by exposure to sign stimuli, most often at a sensitive period when an animal is young. Imprinting of baby geese is a classic example (Table 47.1 and Figure 47.7).

Figure 47.7 No one can tell these imprinted baby geese that Konrad Lorenz is not Mother Goose. (Refer to Table 47.1.)

Table 47.1 A Few Categories of Learned Behavior

IMPRINTING This time-dependent form of learning involves exposure to sign stimuli, most often early in development. For instance, in response to a moving object and probably to certain sounds, baby geese imprint on the mother and follow her during a short, sensitive period after hatching. They are neurally wired to learn crucial information—the identity of the individual that will protect them in the months ahead. Normally that individual is the mother or father. Imprinting occurs among many animals.

Konrad Lorenz, an early ethologist, must have presented the baby geese in Figure 47.7 with sign stimuli that made them form an attachment to him. As another outcome of imprinting, birds also direct sexual attention to members of the species they had been sexually imprinted upon when young.

CLASSICAL CONDITIONING Ivan Pavlov's early experiments with dogs are an example of classical conditioning. Dogs will salivate just before eating. Pavlov's dogs were conditioned to salivate even in the absence of food. They did so in response to the sound of a bell or a flash of light that was initially associated with the presentation of food. In this case, the animals learned to associate an automatic, unconditioned response with a novel stimulus that does not normally trigger the response.

OPERANT CONDITIONING An animal learns to associate a voluntary activity with its consequences, as when a toad learns to avoid stinging insects after its first attempt to eat them.

HABITUATION An animal learns by experience *not* to respond to a situation if the response has neither positive nor negative consequences. Thus pigeons and other birds living in cities learn not to flee from people who pose no threat to them.

SPATIAL (LATENT) LEARNING By inspecting its environment, an animal acquires a mental map of a particular region, often by learning the position of local landmarks. For example, blue jays can store information about the position of dozens, if not hundreds, of places where they have stashed food.

INSIGHT LEARNING An animal abruptly solves some problem without trial-and-error attempts at the solution. Chimpanzees often exhibit insight learning in captivity when they suddenly solve a novel problem that their captors devise for them. Some chimps abruptly stacked and stood on several boxes *and* used a stick to reach bananas suspended out of their reach.

Don't fall into the trap of thinking that instinctive behavior is governed only by genes and learned behavior only by the environment. Behavior arises through gene expression *and* experiences. A bird does have a nervous system prewired to recognize sounds (acoustical cues), but what it *actually hears* affects its response to them. Male white-crown sparrows have an internal capacity to sing—but birds in different habitats use variations, or dialects, of the species song. As Peter Marler found out, male songbirds acquire the full song ten to fifty days after hatching by listening to other birds sing it. During this sensitive period, their primed learning mechanism responds to select information from the environment.

Also, male birds learn parts of a song by picking up cues when other males sing. Marler raised nestlings to maturity in soundproof chambers so they wouldn't hear adult males singing. At maturity, the song of captive males had none of the detailed structure of a typical adult's song. Marler also exposed isolated captives to recordings of white-crown sparrows *and* song sparrows. As adults, the captives sang only the white-crown song. They even mimicked the dialects. Evidently, birdsong partly requires *a genetically based capacity to learn* from specific sounds, or acoustical cues.

This isn't the whole story. In another experiment, Marler let young, hand-reared white-crowns interact with a "social tutor" of a different species, as opposed to listening to taped songs. The males tended to learn the tutor's song. Social experience as well as acoustical cues must influence their primed learning mechanisms.

In instinctive behavior, animals make complex, stereotyped responses to specific, often simple environmental cues. In learned behavior, responses may vary or change as a result of individual experiences in the environment.

Whether instinctive or learned, behavior develops through interactions between genes and environmental experiences.

Further reading: Student Guide to InfoTrac on web site —

47.3 THE ADAPTIVE VALUE OF BEHAVIOR

If you accept that genes directly or indirectly govern forms of behavior, then it follows that forms of behavior are subject to evolution by natural selection. **Natural selection** is a result of differences in reproductive success among individuals of a population that differ from one another in their heritable traits. Some versions of a trait may be better than others at helping individuals leave offspring. Thus alleles for those versions increase in a population, while other alleles do not. (Alleles, recall, are different molecular forms of the same gene.) In time, genetic changes that yield greater reproductive success for individuals spread through the population.

By using the theory of evolution by natural selection as our point of departure, we should be able to develop and test possible explanations of why a behavior has endured. We should be able to discern how some actions bestow reproductive benefits that offset reproductive costs (disadvantages) associated with them. If a behavior is adaptive, it must promote the *individual's* production of offspring. Keep this thought in mind while you read through the following list of terms, which you will use repeatedly in the remainder of this chapter:

1. **Reproductive success**: The number of surviving offspring that the individual produces.
2. **Adaptive behavior**: Any behavior that promotes the propagation of an individual's genes; its frequency is maintained or increases in successive generations.
3. **Social behavior**: Cooperative, interdependent relationships among individuals of the species.
4. **Selfish behavior**: Within a population, any form of behavior that increases an individual's chance to produce or protect offspring of its own regardless of the consequences for the group as a whole.
5. **Altruistic behavior**: Within a population, a self-sacrificing behavior. The individual behaves in a way that helps others but diminishes or precludes its own chance of producing offspring.

When biologists speak of selfish or altruistic behavior, they don't mean an individual is consciously aware of what it is doing or of the behavior's reproductive goal. A lion doesn't have to know that eating zebras is good for its reproductive success. Its nervous system simply calls for hunting behavior when the lion is hungry and sees a zebra. The behavior persists in the population because the genes responsible for it are persisting also.

As a case in point, Norwegian lemmings disperse from a population when density skyrockets and food is scarce. Many die during the exodus. Are they helping the species by committing suicide to get rid of "excess" individuals? Or do some just happen to die as a result of starvation, predation, or accidental drowning while scurrying to new locations where they may reproduce?

An instructive cartoon by Gary Larson depicts a population of lemmings plunging over a cliff above water, presumably in the act of suicide. But one has an inflated inner tube about its waist! If many lemmings were suicidally altruistic, then only "selfish" lemmings would reproduce. Over time, then, genes underlying altruism would disappear from the population. More likely, individual lemmings disperse to less crowded places where each will have a better chance to survive and reproduce.

Another case: In northern forests, ravens scavenge carcasses of deer, elk, or moose, which are few and far between. When one of these large birds comes across a carcass—even in winter, when food is scarce—it often calls loudly and attracts a crowd of similarly hungry ravens. The calling behavior puzzled Bernd Heinrich, for it might seem to go against the caller's interests. Wouldn't a quiet raven eat more, increase its chances of surviving, and also leave more descendants than an "unselfish" vocalizer? If the behavior is an outcome of natural selection, then the cost in terms of lost calories and nutrients must be offset by a reproductive benefit for the individual caller. What could be the benefit?

Maybe a lone bird picking at a carcass is vulnerable to predators that lie in wait for it. If that were so, then other ravens attracted by the calls could help keep an eye out for danger. However, ravens are big, agile birds that have very few known enemies. Heinrich stealthily watched ravens feeding alone and feeding in pairs at a carcass. He never saw a predator attack any of them.

Then he realized that territory may have something to do with it. A **territory** is an area that one or more individuals defend against competitors. After hauling a cow carcass into a Maine forest, Heinrich observed that solitary or paired ravens do not always vocally advertise food. Maybe those silent ravens were adults that had already staked out a large territory, which happened to include the spot where he put the carcass.

A pair of ravens would gain nothing by attracting others to their territory. But what if that Maine forest had been *subdivided* into territories and was defended by powerful adults? A wandering young bird would be lucky to eat at all in an aggressive pair's territory. But recruiting a gang of other, nonterritorial ravens might overwhelm the resident pair's defensive behavior.

As it turned out, only wandering ravens advertise carcasses. Their calling behavior is basically selfish and adaptive. It gives them a shot at otherwise off-limits food and therefore promotes their reproductive success.

Behavioral biologists generally find it profitable to look for evidence of natural selection of the individual's traits rather than something that benefits the species as a whole.

47.4 COMMUNICATION SIGNALS

Competing for food, defending territory, alerting others to danger, advertising sexual readiness, forming bonds with a potential mate, caring for offspring—these are the kinds of intraspecific interactions that depend on communication among animals, as in the case of those vocalizing ravens.

The Nature of Communication Signals

Intraspecific interactions depend on intricate mixes of instinctive and learned behaviors by which individuals send and respond to **communication signals**. These are information-laden cues, encoded in stimuli that hold unambiguous meaning for members of the species. The cues include specific odors, colors, patterning, sounds, postures, and movements.

Communication signals sent by one individual, the **signaler**, induce behavioral changes in others of the species. Responding individuals are **signal receivers**. A signal will evolve or persist when it tends to increase the reproductive success of the sender as well as the receiver. When a signal proves disadvantageous to either party, then natural selection will favor individuals that either don't send the signal or don't respond to it.

Remember **pheromones**? They are chemical signals between individuals of the same species. As you know, chemical odors from food and danger were the most important stimuli that early animals had to deal with. As animals evolved, most started to rely on *signaling* pheromones to bring about rapid responses from a receiver. Some, including chemical alarm signals, call for aggressive or defensive behaviors. Others, such as the bombykol molecules released by female silk moths, serve as sex attractants (Section 37.1). The *priming* pheromones call for generalized physiological responses. One, a volatile odor in the urine of certain male mice, triggers and enhances estrus in female mice.

Figure 47.8 A dog soliciting play behavior with a play bow.

Acoustical signals also abound in nature, as when male songbirds sing to secure territory and attract a female. Similarly, tungara frogs issue nighttime calls to females and rival males. This frog call is a distinctive "whine," followed by an equally distinctive "chuck."

Some signals never vary. For instance, ears laid back against the head of a zebra convey hostility, but ears pointing up convey its absence. Other signals convey the intensity of the signaler's message. A zebra with laid-back ears isn't too riled up when its mouth is just a bit open. But when its mouth is gaping, watch out. The combination is a **composite signal**: a communication signal with information encoded in two cues (or more).

Signals can take on different meaning in different contexts. A lion may emit a spine-tingling roar to keep in touch with others of its pride or to threaten a rival. Also, one signal often conveys information about other signals to follow. Dogs and wolves solicit play behavior by means of a play bow, as in Figure 47.8. Because of the bow, subsequent behavioral patterns that a signal receiver construes as aggressive, sexual, or exploratory in other contexts will be construed as playful only.

Examples of Communication Displays

The play bow is a **communication display**—a pattern of behavior that is a social signal. Another common pattern, the **threat display**, is an unambiguous message that a signaler is prepared to attack a signal receiver. When a rival for a receptive female confronts him, a dominant male baboon will role his eyes upward and "yawn" to expose his formidable canines (Figure 47.9*a*).

Figure 47.9 (**a**) Exposed canines, part of a male baboon's threat display. (**b**) Part of a courtship display with visual, tactile, and acoustical signals that commonly precedes copulation. Here a male albatross spreads his wings, an information-laden cue for the female. See also Section 19.2.

Further reading: Student Guide to InfoTrac on web site

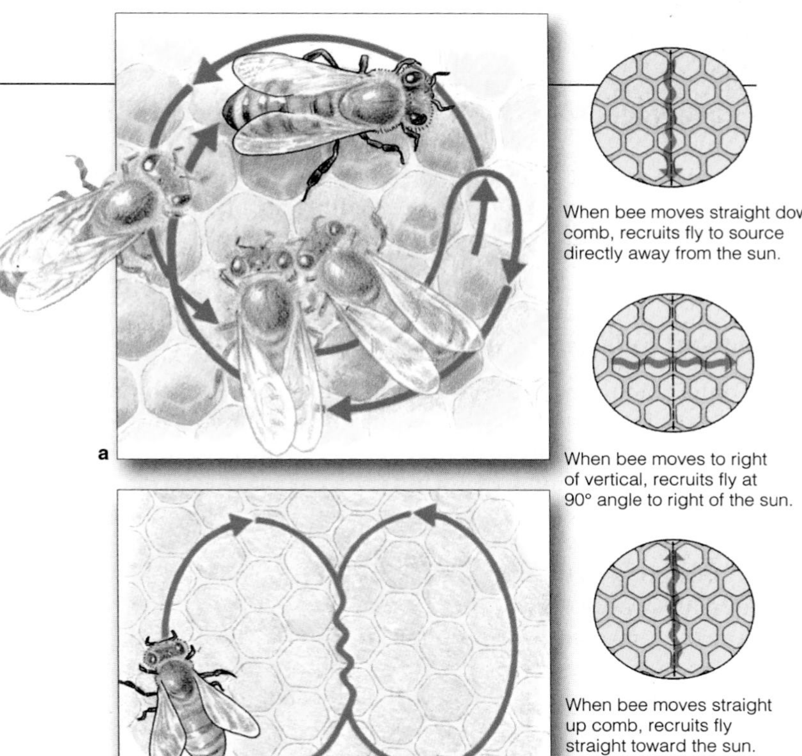
When bee moves straight down comb, recruits fly to source directly away from the sun.
When bee moves to right of vertical, recruits fly at 90° angle to right of the sun.
When bee moves straight up comb, recruits fly straight toward the sun.
a
b
c

Figure 47.10 Dances of the honeybees, examples of tactile displays. (**a**) Honeybees that visit food sources close to their hive perform a *round* dance on the honeycomb. Worker bees that maintain contact with the forager through the dance will then go out and search for food close to the hive.

(**b**) Bees trained to visit feeding stations more than 100 meters from the hive perform a *waggle* dance. During the dance, a bee makes a straight run and waggles its abdomen. (**c**) As Karl von Frisch discovered, the orientation of a straight run can vary, depending on the direction in which food is located. When he put a dish of honey on a direct line between the hive and the sun, foragers that located it returned to the hive and oriented their straight runs right up the honeycomb. When he put the honey at right angles to a line between the hive and the sun, the foragers made their straight runs at 90 degrees to vertical. Thus, a honeybee "recruited" into foraging for food can orient its flight *with respect to the sun and the hive.* By doing so, it will waste less time and energy during its food-gathering expedition.

Waggle dancers also vary the dance speed to convey even more information about the distance of a food source. When a site is 150 meters away, the dance is executed much faster, with more waggles per straight run, compared with a dance regarding a food source that is, say, 500 meters away.

Often the rival backs down. If so, the signaler benefits, for he retains access to the female without putting up a fight. The signal receiver benefits, because he avoids a serious beating, infected wounds, and possibly death.

Such displays are *ritualized*, with intended changes in the function of common behavior patterns. Normal movements may be exaggerated yet simplified; postures may be frozen. Body parts such as feathers, manes, and claws are often conspicuously enlarged, patterned, and colored. Ritualization is often developed to an amazing degree in **courtship displays** between potential mates. For example, food-enticing behavior is common among many birds. A male bird might emit calls while bowing low, as if to peck the ground for food. When a female comes running, he may spread his wings, as if to focus attention on the ground in front of him (Figure 47.9*b*). Here, one behavior (searching for food) has changed to attract a female, the intent being copulation. Another example: Courtship displays of fireflies and some other nocturnal animals incorporate bioluminescent flashes (Chapter 6). A male firefly uses a light-generating organ to emit a bright, flashing signal. A few seconds later, a receptive female of his species may answer him with a flash. Both may flash back and forth until they meet.

Similarly, in **tactile displays**, a signaler touches the receiver in ritualized ways. After discovering a source of pollen or nectar, a foraging honeybee returns to its colony (a hive) and performs a complex dance. It moves about in a circle, jostling in the dark with a crowd of workers. Other bees may follow and maintain physical contact with the dancer. Honeybees that do this acquire information about the general location, distance, and direction of pollen or nectar (Figure 47.10).

Illegitimate Signalers and Receivers

Sometimes the wrong parties intercept communication signals. Termites will respond with defensive behavior if they catch a whiff of a scent from an invading ant. That scent is meant to identify the ant as a member of an ant colony, and it elicits cooperative behavior from other ants. Thus the scent has evolved functions, but announcing an invasion of a termite colony is not one of them. A termite also might detect the scent and kill the ant. In that case, it is an **illegitimate receiver** of a signal *meant for individuals of a different species.*

We even see **illegitimate signalers**. Certain assassin bugs hook dead and drained bodies of termite prey on their dorsal surface to acquire the odor of their victims. By deceptively signaling that they "belong" to a termite colony, they hunt termite victims more easily. Another illegitimate signaler is the female of certain predatory fireflies. If one observes a flash from a male firefly, she flashes back. If she can lure him into attack range, he becomes her meal—an evolutionary cost of having an otherwise useful response to a come-hither signal.

A communication signal between individuals of the same species is an action that has a net beneficial effect on both the signaler and the receiver. Natural selection tends to favor communication signals that promote reproductive success.

MATES, PARENTS, AND INDIVIDUAL REPRODUCTIVE SUCCESS

For reasons that we need not explore here, most people find the mating and parenting behaviors of different animals fascinating. How useful is selection theory in helping us interpret such behavior? Let's take a look.

Sexual Selection Theory and Mating Behavior

Competition among members of one sex for access to mates is common. So is choosiness in selecting a mate. Recall, from Section 18.6, that such activities are forms of a microevolutionary process called **sexual selection**. Sexual selection favors traits that give the individual a competitive edge in attracting and holding on to mates.

Sexual selection typically occurs because the male animals produce great numbers of tiny sperm, but the females produce considerably larger and fewer eggs. Reproductive success for a male generally depends on how many eggs he fertilizes. For the female, success depends largely on how many eggs she produces or how many offspring she can care for. In most cases, *the key factor influencing her sexual preference is the quality of a mate, not the quantity of partners*. Hangingflies, sage grouse, and bison (Figures 47.11 through 47.13) provide wonderful examples of how such females dictate the rules of male competition. They also illustrate how the males employ tactics that might help them fertilize as many eggs as possible.

Female hangingflies (*Harpobittacus apicalis*) select males that offer them superior material goods. And so you might observe males capturing and killing a moth or some other insect. After males do this, they release a sex pheromone that might attract females to the "nuptial

Figure 47.11 A male hangingfly dangling a moth as a nuptial present for a future mate. The females of certain hangingfly species choose sexual partners on the basis of the size of prey that males offer to them.

Figure 47.12 Ornamental feathers of a male sage grouse. These visual signals function in a courtship display—a dance performed at a lek. Males vigorously defend a small patch of this communal mating area. Females (the smaller brown birds) observe the prancing males before choosing the one they will mate with.

Further reading: Student Guide to InfoTrac on web site —

gift" (Figure 47.11). A female chooses males that display the larger, calorie-rich offering. She permits a male to mate, but only *after* she's been eating a gift for five minutes or so. Then she accepts sperm and holds them in a special storage sac, but only as long as the food holds out. Before twenty minutes are up, she can break off the mating at any point. If she does so, she might well mate with another male and accept his sperm. And doing so will dilute the reproductive success of her first partner.

In western regions of North and South Dakota, we find sage grouse (*Centrocercus urophasianus*) dispersed among the stands of sagebrush. We never find them far from the protective cover of sage during spring courtship and summer nesting. Male sage grouse make no attempt to hold and keep a large territory. In the breeding season they congregate in a type of communal display ground called a **lek**. Every male stakes out only a few square meters as his territory. Females are attracted to the lek—not to feed or nest, but rather to observe males. With tail feathers erect and large neck pouches puffed, each male emits booming calls and stamps about, like a wind-up toy, on his display ground (Figure 47.12). Female sage grouse tend to select and mate with one male only. Then they go off to nest alone in sagebrush, unassisted by a sexual partner. Many females often choose the same male, so most of the males never mate.

As another example, females of some species cluster in defendable groups when they are sexually receptive. Where you find such a group, most likely you will observe male competition for access to the clusters. Competition for ready-made harems favors extremely combative male lions, sheep, elk, elephant seals, and bison, to name a few animals (Figures 1.7*g* and 47.13).

Figure 47.13 Sexual competition between male bison, which are fighting for access to a cluster of females.

Figure 47.14 Male and female Caspian terns protecting their chicks. Parental care has costs as well as benefits.

Costs and Benefits of Parenting

What happens after mating, when the offspring arrive? Until the offspring have developed enough to survive on their own, the parents of some species will care for them. For example, adult Caspian terns (Figure 47.14) incubate the eggs, shelter the nestlings and feed them, then accompany and protect them after they start to fly. Such parental behavior comes at a reproductive cost. It drains time and energy that might otherwise be given to improving their own chances of living to reproduce another time. Yet for many species, parenting improves the likelihood that the current generation of offspring will survive. The benefit of devoting time and energy to immediate reproductive success outweighs the cost of reduced reproductive success at some later time.

Selection theory helps explain some aspects of mating behavior, for sexual selection favors behavioral traits that give the individual a competitive edge in reproductive success. Selection theory applies also to parental behavior that contributes to reproductive success.

47.6 BENEFITS OF LIVING IN SOCIAL GROUPS

Survey the animal kingdom and you observe a range of social groupings. For some species, individuals spend most of their lives alone or in small family groups. For some other species, individuals live in huge groups of thousands of related individuals. Termite and honeybee societies are like this. Individuals of still other species live in social units composed primarily of nonrelatives. Populations of the human species are like this.

Given such differences, how might we gain insight into the basis for any social group? To find answers, evolutionary biologists commonly take a cost–benefit approach. That is, they attempt to identify the costs and benefits of sociality in terms of the reproductive success of an individual, as measured by its contribution to the gene pool of the next generation.

Cooperative Predator Avoidance

First, look at how a group of animals that are acting cooperatively against a predator can reduce the net risk to any one individual. In a flock or herd vulnerable to predation, simply having more pairs of eyes to scan the surrounding area helps individuals detect a predator sooner. Individuals of the group also may join forces to make a counterattack or perform a defensive behavior, as the musk-oxen in Figure 47.15 are doing.

The biologist Birgitta Sillén-Tullberg found tangible evidence of such benefits of sociality. She was studying Australian sawfly caterpillars, which live together in clumps on tree branches. Figure 47.16 shows one of the clumps. When something disturbs the caterpillars, they collectively rear up from the branch and writhe about, all the while regurgitating partially digested food. The food of choice is eucalyptus leaves. These leaves are heavily impregnated with chemical compounds that are toxic to most kinds of predatory animals—including the songbirds that prey on the caterpillars.

Sillén-Tullberg hypothesized that individual sawfly caterpillars benefit from the coordinated act of repelling bird predators. She used her hypothesis to predict that the birds are more likely to eat a solitary individual, not a cluster of individuals.

To test her prediction, Sillén-Tullberg offered young, hand-reared Great Tits (*Parus major*) a chance to feed on sawfly caterpillars, which she offered either one by one or in a group of twenty per offering. She did this for a standard number of presentations. Ten birds that were offered one individual at a time consumed an average of 5.6 caterpillars. But ten birds that were each offered a clump of caterpillars only ate an average of 4.1. As she had predicted for this experiment, the individuals were somewhat safer in a group than on their own.

The Selfish Herd

Simply by their physical position within a group, some individuals form a living shield against predation on others in the group. They all belong to a **selfish herd**, a

Figure 47.16 Social defensive behavior of Australian sawfly caterpillars, which clump together in tree branches. The larvae collectively regurgitate and hold fluid in their mouth (the yellow blobs), where it helps repel predators that try to grab them. After danger passes, they swallow the fluid, which is toxic to most animals.

Figure 47.15 Social defensive behavior of musk-oxen (*Ovibos moschatus*). In the presence of a perceived threat—usually wolf predators—adults form a circle around the young. They face outward, and their "ring of horns" successfully deters the wolves.

Further reading: Student Guide to InfoTrac on web site

Figure 47.17 Appeasement behavior between baboons. Notice the assured position of the dominant animal (*left*) and the abject stare and groveling posture of the subordinate one, who is making little conciliatory smacking noises with its lips.

relatively simple society held together by reproductive self-interest, although not consciously so. Investigators have tested the selfish-herd hypothesis for male bluegill sunfishes, which build adjacent nests on the bottom of lakes. Males use their fins to hollow out a depression in lake mud, where females deposit eggs.

If a colony of bluegill males is a selfish herd, then we can predict competition for the "safe" sites—at the center of the colony. Compared to the periphery, eggs laid in nests at the center are less likely to be attacked by snails and largemouth bass. The competition does indeed exist. The largest, most powerful males tend to claim central locations. Other, smaller males assemble around them and bear the brunt of predatory attacks. Even so, they are better off in the group than on their own, fending off a bass singlehandedly, so to speak.

Dominance Hierarchies

As you have seen, individuals of a selfish herd make no personal sacrifice for the others; the personal benefits of living with others simply seem to outweigh the costs. By contrast, some individuals may *help* others survive and reproduce but at their own expense. Their helpful behavior may be a cost of belonging to the social group. And it may be that they do not entirely give up their chance at reproductive success. We find some evidence of both possibilities in **dominance hierarchies**, a type of social group in which some of the individuals have adopted subordinate status to others.

In baboon troops, for example, individuals help one another, but reproductive opportunity is unequal. With a threat signal from a dominant member, subordinates relinquish safe sleeping places, choice bits of food, and sexually receptive females. Each individual recognizes its social status with respect to the other individuals, as the abject baboon at the left, in Figure 47.17, suggests. In her studies of chimpanzees at Gombe, Jane Goodall investigated dominance hierarchies that develop among them. You read about their social behaviors in Chapter 37.

Why do subordinate adults remain in a group where they have such low social status? As is the case for the bluegill sunfish, they probably reap long-term benefits that offset their low status. Consider that it simply may not be possible to survive alone, outside the protection afforded by the group. A solitary baboon out in the open surely quickens the pulse of the first leopard that spies it, as shown by Figure 48.7 in the next chapter. Besides, challenging a stronger member may result in injuries that could shorten a life. And self-sacrificing behavior may give a subordinate individual a chance to reproduce, if it lives long enough and if predation or weakness in old age removes dominant peers. Some subordinate wolves and baboons do eventually move up the social ladder if dominant members slip down a rung or fall off. In short, accepting subordinate status can pay off in the long term for the patient individual.

We may evaluate the costs and benefits of social life in terms of reproductive success, as measured by the genes that each individual contributes to the next generation.

47.7 COSTS OF LIVING IN SOCIAL GROUPS

So far, we have used individual selection theory to help illuminate the advantages of communication signals, territoriality and dominance hierarchies, mating tactics, parenting, and other aspects of living in social groups. Now the question becomes this: If social behavior is so advantageous, *then why are there so few social species among most groups of animals?* One plausible answer is that, in certain environments, costs to the individual outweigh the benefits of life in a social group.

Most obviously, when more individuals of a species live together in their habitat, the competition for food increases. For example, royal penguins, herring gulls, cliff swallows, and prairie dogs are among the animals that live in huge colonies (Figure 47.18). Great numbers of individuals must compete for their fair share of the same ecological pie.

Figure 47.18 Colony of royal penguins on Macquarie Island, between New Zealand and Antarctica.

In addition, living in social groups can encourage the spread of contagious diseases and parasites. Under crowded living conditions, the individual as well as its offspring is more likely to be weakened by pathogens and parasites that are readily transmitted from host to host in crowded groups. Similarly, plagues spread like wildfire through densely crowded human populations. As described in Section 46.7, this is especially the case for settlements and cities with recurring infestations of rats and fleas, and with very poor or nonexistent sewage treatment and medical care.

Another cost is the risk of being killed or exploited by others in the group. For example, when presented with the opportunity to do so, breeding pairs of herring gulls will cannibalize the eggs or young chicks of their neighbors in an instant. Or consider lions. These long-lived cats compete for permanent hunting territories, even though they can live for an extended time between kills. When three or more lions hunt cooperatively, they are better at capturing large prey. Yet George Schaller's data revealed that individuals in prides of three or more actually eat *less* well than one or a pair of lions. Also, males intent on taking over a pride show infanticidal behavior; they kill the cubs. For lionesses, then, group living costs, in terms of food intake and reproductive success. They still stick together. Maybe doing so helps them defend their territories against smaller groups of rivals. Besides, aggressive males almost always kill the cubs of a single lioness, but occasionally a group of two or more lionesses can save some of the cubs.

Individuals that belong to a social group pay costs in terms of increased competition for food, living quarters, mates, and other limited resources. They also are more vulnerable to contagious diseases and parasitic infections.

Individuals also may risk being exploited or having their offspring killed by others in the social group.

Further reading: Student Guide to InfoTrac on web site

47.8 EVOLUTION OF ALTRUISM

A subordinate animal that gives way to a dominant one acts in its own interest. What about truly altruistic animals? We find them among many vertebrate groups. Consider the wolf pack. Although the females of most mammalian groups make the largest investment in raising offspring, male wolves also make contributions by providing their pups with prey, defending a feeding territory, and driving off infanticidal intruders. Unlike herbivores, which forage as individuals, wolves are carnivores that share captured prey with others of the social group. Therefore, a female's reproductive success increases when she monopolizes the benefits that males offer. Usually a wolf pack contains only one dominant breeding female and male (Figure 47.19). The others are nonbreeding sisters, aunts, brothers, and uncles. They altruistically hunt and bring back food to members that stay inside the den and guard the young. Nonbreeding females may ovulate and court males, but they fail to reproduce when a dominant female is present.

Altruistic behavior is extreme in some insect societies, as demonstrated when a worker bee plunges its stinger into an invader of the hive. With this act of defending the colony, the bee is commiting suicide. Section 47.9 focuses on termite and honeybee altruists.

Yet if the altruistic individuals of a social group do not contribute their genes to the next generation, then how are genes for their altruistic behavior perpetuated over evolutionary time? According to William Hamilton's **theory of indirect selection**, genes associated with caring for one's *relatives*, not one's direct descendants, may be favored in some situations. Remember, when a sexually reproducing parent cares for offspring, it is not helping exact genetic copies of itself. Such parents commonly are diploid; they have pairs of genes. Each gamete they produce, hence each offspring, has one-half of its genes. If other individuals of the social group have the same ancestors, they share the same genes with parents. Two siblings (brothers, sisters) are as genetically similar as a parent is to one of its offspring. Nephews and nieces have inherited about one-fourth of their uncle's genes.

We might therefore think of selective altruism as an extension of parenting. For example, suppose an uncle helps his niece survive long enough to reproduce. He has made an *indirect* genetic contribution to the next generation, as measured in terms of the genes that he and his niece share. Altruism costs him; he might lose his own opportunities to reproduce. But if the cost is less than the benefit, altruistic actions will propagate the uncle's genes and favor the spread of his kind of altruism in the species. If an uncle saves two nieces, this is equivalent to saving his own daughter.

Similarly, nonbreeding workers in insect societies indirectly promote their "self-sacrifice" genes through altruistic behavior directed toward relatives. Colonies of honeybees, ants, and termites actually are extended families. The family's worker force labors on behalf of their siblings, some of which are the future kings and queens. Thus, when a guard bee drives her stinger into a raccoon, she inevitably dies—but her siblings in the hive might perpetuate some of her genes. Sterility and extreme self-sacrifice are rare among social groups of vertebrates. The known exceptions include naked mole-rats. Section 47.10 describes one study of the genetic relationships in a clan of these nearly hairless mammals.

Figure 47.19 Dominant male of a wolf pack, surrounded by subordinate members of the pack engaged in a ritual display involving lowered heads and drooped tails.

Altruistic behavior occurs among many vertebrate groups, such as wolf packs. It reaches its most extreme among some insect societies, such as honeybee and termite colonies.

Altruistic behavior may persist when individuals pass on genes indirectly, by helping relatives survive and reproduce.

By the theory of indirect selection, genes associated with altruistic behavior that is directed toward relatives may spread through a population in certain situations.

47.9 WHY SACRIFICE YOURSELF?

CONSIDER THE TERMITE Picture yourself in a eucalyptus forest in Queensland, Australia. You see a narrow, brittle tube running along the trunk of a dead tree and decide to chip a few fragments from it. Sunlight pours in through the breached tube wall. Some small, nearly white insects bang their head against the wall, then scurry away.

Head banging, an acoustical signal, makes vibrations that alert golden-brown soldier termites. Soldiers run to the breach and make a defensive stand (Figure 47.20). Each one has a swollen, eyeless, tapering head with a long, pointed "nose." Disturb a soldier termite and it shoots thin jets of silvery goo out of its nose! The silvery strands release volatile compounds that attract still more soldier termites. Together the soldiers battle the danger, which more typically is an invasion by ants.

Insects in that ruptured tunnel belong to a complex termite colony. Whereas the soldier termites protect the colony, the pale ones are workers, which build tunnels that lead to places where they can safely gather fibers of wood. They carry fibers to an underground nest where termites live and cultivate an edible fungus. Fungal hyphae grow into the wood fibers and absorb nutrients from them. The termites eat some portions of the fungus. They chew the wood into small particles that they can swallow. Unlike most termite species, they don't produce enzymes that can digest cellulose in wood. Symbiotic microorganisms in their gut do this, and the symbionts and termites both absorb the released nutrients.

Soldier termites and worker termites are sterile; neither can reproduce. They engage in self-sacrificing behavior that contributes to the survival of others. A single queen and one or more kings serve as "parents" for the entire society.

COOPERATIVE HONEYBEES The only fertile female in a honeybee colony, or hive, is the queen bee. Figure 47.21*a* shows one, with her court of sterile worker daughters.

Figure 47.20 Soldier termites defending their social colony by guarding a break in a foraging tunnel, which worker termites constructed on a dead tree trunk. Members of this caste shoot out strands of glue from their pointed "nose." The glue entangles invading ants and other intruders.

Further reading: Student Guide to InfoTrac on web site —

a
e
b
c
d
early larval stage
nearly mature pupa
f

Figure 47.21 Life in a honeybee colony. (**a**) A court of sterile worker daughters surrounds a queen bee, the only female of the colony that reproduces. (**b**) A stingless drone. (**c**) One of the 30,000 to 50,000 worker bees in the hive. (**d**) Worker bees store honey or pollen in the honeycomb, which also houses new bee generations. Young workers feed the larvae. (**e**) Worker bees transferring food to one another. (**f**) Worker females at the hive entrance serve as guards.

The daughters feed the queen and relay her pheromones throughout the hive, and these influence the activities of all members. The queen is much larger than the worker bees, partly because of the relatively enormous egg-producing ovaries in her abdomen. Unlike her sterile daughters, she has fully developed ovaries.

Only at certain times of year do the stingless drones develop and mature (Figure 47.21*b*). Drones do not work for the colony. Instead, they leave the colony and attempt to mate with queens of other hives. If successful, they may perpetuate part of their family's genes.

The hive contains 30,000 to 50,000 workers (Figure 47.21*c*). They feed bee larvae, clean and maintain the hive, and construct honeycomb from wax secretions. Workers store honey or pollen in honeycomb, which houses bee generations from eggs, through a series of larval stages, to pupae, to the adults (Figure 47.21*d*).

Adult workers live for about six weeks in spring and summer. Workers also engage in scent-fanning. Fanned air passes over a bee's exposed scent gland. As pheromones waft away from the scent gland, they help bees become oriented relative to the hive's entrance on their way to or from foraging expeditions. When foragers return to the hive after locating a rich source of nectar or pollen, they start a dance—a tactile display that recruits more workers into taking off for the source (Figure 47.10).

In other cooperative behaviors, workers transfer food to one another, and worker females at the hive entrance act as guards (Figure 47.21 *e,f*). Guard bees readily sacrifice themselves to repel intruders.

After these glimpses into the lives of self-sacrificing insects, see if you can answer a few basic questions. First, *by what means do the members of a termite or honeybee colony cooperate and benefit the group and themselves?* For instance, identify some of their communication signals and the manner in which individuals respond to them. Second, *what is the adaptive value of some of their social behaviors?* How, for instance, might sterility increase an individual's chances for genetic success?

47.10 INDIRECT SELECTION AMONG THE NAKED MOLE-RATS

Think about a fat sausage with a pale, wrinkled casing. Make it bucktoothed and put a few sprouts of hairs on it. What you have visualized comes close to the appearance of *Heterocephalus glaber*, the highly social mammals called naked mole-rats (Figure 47.22). Naked mole-rats live in arid regions of eastern Africa, in cooperatively excavated burrows. They always live in clans (small social units) of anywhere from 25 to 300 individuals.

A single reproducing female dominates each clan, and she mates with one to three males. As is the case for honeybees and termites—and not for any other known vertebrate—other members of the clan are nonbreeding. They live out their lives protecting and caring for the "queen" and "king" (or kings) and their offspring.

Nonreproducing "diggers" busily excavate extensive subterranean tunnels and special chambers, which serve as living rooms or waste-disposal centers. They locate large tubers growing underground. These they chop up into edible bits, which they then deliver to the queen, her retinue of males, and her offspring.

Also, digger mole-rats deliver food to certain other helpers that loaf about, shoulder to shoulder and belly to back, with the reproductive royals. Usually the "loafers" are larger than the diggers. And they aren't really loafers. They spring to action when a snake or some other enemy threatens the clan. Collectively, at great personal risk, they chase away or attack and kill the predator.

Using indirect selection theory as our guide, we can formulate a hypothesis to account for the altruism of the mole-rats that never do breed.

If their helpful behavior is genetically advantageous to some individuals in a naked mole-rat clan (*the hypothesis*), then it follows that helpers will be related to reproductive members of the clan that benefit from their altruistic behavior (*the prediction*).

To *test* the prediction, we require information about genetic relationships among the members of the naked mole-rat colony. One way to obtain it would be to know which animals were offspring of which others. However, constructing a family tree for an entire colony would be most laborious and difficult. Why? There are a number of breeding males. In addition, queen mole-rats die and are replaced.

H. Kern Reeve and his colleagues decided on a more practical approach. They relied upon *DNA fingerprinting*, a method of establishing degrees of genetic relatedness among individuals. As described in Section 16.3, this laboratory method starts with the formation of restriction fragments of DNA molecules. Then technicians construct a visual record of different sets of fragments of DNA from different individuals. Essentially, a set of identical twins would have identical DNA fingerprints (their DNA would have the same base sequence). Individuals that have the same father and mother would have similar DNA; so they would have similar although not identical DNA fingerprints. On average, the DNA fingerprints of genetically unrelated individuals should show far more differences than the DNA fingerprints of siblings or some other relatives.

When Reeve constructed DNA fingerprints for naked mole-rats, he discovered that all individuals from the same clan are *very* close relatives. He also found out that they are very different genetically from members of other clans. The findings suggest that each naked mole-rat clan is highly inbred as a result of generations of brother-sister, mother-son, and father-daughter matings. As you know from Section 18.8, inbreeding among individuals of a population results in extremely reduced genetic variability.

Therefore, a self-sacrificing naked mole-rat is helping to perpetuate a high proportion of the alleles that it carries. As it turns out, the genotypes of helpers and the helped might be as much as 90 percent identical!

Figure 47.22 A peek at a few naked mole-rats (*Heterocephalus glaber*).

Further reading: Student Guide to InfoTrac on web site

47.11 AN EVOLUTIONARY VIEW OF HUMAN SOCIAL BEHAVIOR

If we can analyze the evolutionary basis of the behavior of termites, naked mole-rats, and other animals, would it not be rewarding to analyze such a basis of human behavior also? Many people resist the idea. It seems they believe that attempts to identify the adaptive value of a particular human trait are attempts to define its moral or social advantage. Clearly, however, there is a difference between trying to explain something in terms of its evolutionary history and attempting to justify it. "Adaptive" does not mean "morally right." It means valuable with respect to the transmission of an individual's genes.

Consider the case of altruistic behavior that we call **adoption**, or the acceptance of offspring of other individuals as one's own. If we wish to develop understanding of adoptive behavior, we might use evolutionary theory and formulate a hypothesis about it. But this does not mean that we are passing judgment on whether adoption is moral or even socially desirable. These are two separate issues—about which biologists really have no more to say than anybody else.

Figure 47.23 Two adult emperor penguins competing to adopt an orphan, which penguins accept as substitute offspring.

Start with a premise that all adaptations have costs and benefits. One cost is that certain adaptive behaviors may be *redirected* under rare or unusual circumstances. In many species, adults that lost offspring will adopt a substitute. Certain cardinals have fed goldfish. A whale tried to lift a log out of water as if it were a distressed infant that needed help in reaching the water's surface. Emperor penguins fight to adopt orphans (Figure 47.23).

As John Alcock suggests, such examples constitute a test of a hypothesis about adoption by humans: Human adoptive behavior occurs "by mistake" when otherwise adoptive parental behavior is directed toward unrelated offspring. Suppose it arises through a frustrated desire to bear children (the hypothesis). If so, then we can expect that couples who lost an only child or are sterile should be especially prone to adopt strangers (the prediction).

Now consider this. For most of their evolutionary history, humans have lived in small groups and, later, small villages. They probably had few opportunities or showed little inclination to adopt strangers. They were likely to accept a child of a deceased relative. But how would adoptive parents gain genetic representation in the next generation? According to theories of natural selection and indirect selection, individuals act in ways that promote genetic self-interest. So we might predict that parents with dependent, care-requiring children of their own are much less likely to adopt substitutes, compared with adults who have no children or who already raised children and are living by themselves. We could test the prediction by piecing together a large enough sampling of information, ideally from diverse existing human cultures, on possible connections between adoption and a childless condition.

Indirect selection also favors adults who focus their parenting behavior on relatives and thereby perpetuate their shared genes in an indirect way. We might even predict that such adults are more likely to be related to the adopted child than we would expect by chance alone. A researcher, Joan Silk, tested the prediction in traditional societies. Her results showed that people will adopt related children far more often than nonrelated children.

Particularly in large, industrialized societies with welfare agencies and other means of adoption assistance, individuals become parents of nonrelated children. Such societies create evolutionary environments. Parenting mechanisms evolved in the past, and it may be that their redirection toward nonrelatives says more about our evolutionary history than it does about the transmission of one's genes.

The point is this: Evolutionary hypotheses about the adaptive value of behavior lend themselves to testing. And through such testing, we can gain understanding about the evolution of human behavior.

It is possible to test evolutionary hypotheses regarding the adaptive value of human behaviors.

There may be greater acceptance of such testing when more people come to understand that adaptive behavior and socially desirable behavior are separate issues.

In biology, "adaptive" means only that some specified trait has proved beneficial in the transmission of the genes of an individual that are responsible for that trait.

47.12 SUMMARY

1. Animal behavior (coordinated responses to stimuli) originates with genes that directly or indirectly specify products required for the development and operation of the nervous, endocrine, and skeletal–muscular systems.

2. A behavior performed without having been learned by actual experience is instinctive, a prewired response to one or two simple, well-defined environmental cues (sign stimuli) that trigger a suitable response, such as a fixed action pattern. Individual experiences can lead to variations or changes in responses (learned behavior). Both are outcomes of genetic and environmental inputs.

3. Behavior with a genetic basis is subject to evolution by natural selection. It evolved as a result of individual differences in reproductive success in past generations, when the reproductive benefits of a particular behavior exceeded its reproductive costs (or disadvantages).

4. Members of the same species often create obstacles to one another's reproductive success, as by selective mate choice and by competition for access to mates.

5. Social groups require cooperative interdependency among individuals of a species. Sociality is promoted by communication signals, cues sent by one individual (a signaler) that can change the behavior of another of the same species (a signal receiver). For a signal to be adaptive, it must benefit the signaler and the receiver.

6. Chemical, visual, acoustical, and tactile signals are components of communication displays.

a. Pheromones function in chemical communication. Signaling pheromones, such as sex attractants and alarm signals, can induce immediate change in the behavior of a receiver. Priming pheromones elicit a generalized physiological response by the receiver.

b. Visual signals (observable actions or cues) are key components of courtship displays and threat displays.

c. Acoustical signals (sounds with precise, species-specific information) include the mating calls of frogs.

d. Tactile signals are specific forms of physical contact between a signaler and a receiver.

7. Costs and benefits of social life are reflected in the individual's reproductive success, as measured by its genetic contribution to the next generation.

8. Social groups have costs, such as competition for limited resources and greater exposure to contagious diseases and parasites. Benefits outweigh the costs in some cases, as when predation pressure is severe.

9. In self-sacrificing (altruistic) behavior, individuals give up chances to reproduce while helping others of their social group. In most social groups, individuals do not sacrifice reproductive chances to help others.

a. Dominance hierarchies are social rankings in which some of the individuals give way to others of their group. Often, the dominant individuals force their subordinates to relinquish food or some other resource.

b. When subordinates give in to dominant members of their group, they may receive compensatory benefits of group living, such as safety from predators. In some species, subordinates may reproduce if they live long enough and dominant ones slip in the hierarchy or die.

10. By one theory of indirect selection, genes associated with caring for relatives (not one's direct descendants but individuals that bear a fraction of the same genes) are favored in some cases. Indirect selection can lead to extreme altruism, as among workers of some species of social insects and naked mole-rats. The workers usually do not reproduce. Altruism helps reproducing relatives survive, so altruistic individuals pass on "by proxy" the genes underlying the development of this behavior.

Review Questions

1. Explain how genes and their products, including hormones, influence the mechanisms required for forms of behavior. *47.1*

2. Define these terms: instinctive behavior, sign stimulus, fixed action pattern, and learned behavior. *47.1, 47.2*

3. Contrast altruistic behavior with selfish behavior. Why does either form of behavior persist in a population? *47.3*

4. Describe some characteristics of communication signals. Then give an example of a communication display. *47.4*

5. Describe some feeding behavior or mating behavior in terms of natural (individual) selection. *47.3, 47.5*

6. List some of the benefits and costs of sociality. *47.6, 47.7*

Self-Quiz (*Answers in Appendix III*)

1. "Starlings minimize nest mites by festooning nests with wild carrot," said Clark and Mason. Their statement was ______.
   a. an untested hypothesis
   b. a prediction
   c. a test of a hypothesis
   d. a proximate conclusion

2. Genes affect the behavior of individuals by ______.
   a. influencing the development of nervous systems
   b. affecting the kinds of hormones in individuals
   c. governing the development of muscles and skeletons
   d. all of the above

3. Many kinds of female mammals live in herds. Which of these mating systems might you see in such herds?
   a. lek mating system
   b. males competing for clusters of receptive females
   c. territorial defense of feeding sites by males
   d. males capturing food to lure females

4. A statement that overcrowding causes lemmings to disperse to areas that are more favorable for reproduction ______.
   a. is consistent with Darwinian evolutionary theory
   b. is based on a theory of evolution by group selection
   c. is supported by the finding that most animals behave altruistically during their lives

5. The genetic similarity between an uncle and his nephew is ______.
   a. the same as between a parent and his offspring
   b. greater than between two full siblings
   c. dependent on how many other nephews the uncle has
   d. less than that between a mother and her daughter

Figure 47.24 Caterpillar under siege.

Figure 47.25 A black heron, possibly on a fish-enticing expedition.

6. Match the terms with their most suitable description.

____ fixed action pattern	a.	time-dependent form of learning requiring exposure to key stimulus
____ altruism	b.	genes and environmental experience
____ basis of instinctive *and* learned behavior	c.	stereotyped motor program that runs to completion independently of feedback from environment
____ imprinting	d.	assisting another individual at one's own expense

Critical Thinking

1. A large caterpillar is crawling along a branch in a tropical forest. You poke it, and it partly responds by letting go of the branch and puffing up the anterior end of its body (Figure 47.24). Propose a mechanism that may underlie this defensive behavior. Also propose how the behavior has adaptive value. How would you test your two hypotheses?

2. A cheetah scent-marks plants in its territory by releasing specific chemicals from certain glands. What evidence would you require to demonstrate that the cheetah's action is an evolved communication signal?

3. Suppose you are traveling through Africa and you see a black heron standing in the water with its wings held over its head like an umbrella, as shown in Figure 47.25. You might suppose that, like other herons, it is on the lookout for fish. But most other herons keep their wings down when they are hunting. What is the adaptive value of this behavior? Possibly the black heron is creating shade on the water, and perhaps minnows are being attracted to the shade because it appears to offer shelter. How would you test this hypothesis?

4. You observe mallard ducks in a small pond, then you notice a rooster wading out to the ducks with amorous intent (Figure 47.26). What probably happened to the rooster early in life?

5. Develop an adaptive value hypothesis for the observation that male lions kill offspring of females that they acquire after chasing away previous pride holders that had mated with those females. How would you test your infanticide hypothesis?

6. What are the likely evolutionary costs and benefits to a male hangingfly of offering a nuptial gift to its potential mate? How might a female's choosiness depend on those costs and benefits?

Figure 47.26 Behaviorally confused rooster.

Selected Key Terms

adaptive behavior *47.3*
adoption *47.11*
altruistic behavior *47.3*
animal behavior *CI*
communication display *47.4*
communication signal *47.4*
composite signal *47.4*
courtship display *47.4*
dominance hierarchy *47.6*
fixed action pattern *47.1*
illegitimate receiver *47.4*
illegitimate signaler *47.4*
imprinting *47.2*
indirect selection *47.8*
instinctive behavior *47.1*
learned behavior *47.2*
lek *47.5*
natural selection *47.3*
pheromone *47.4*
reproductive success *47.3*
selfish behavior *47.3*
selfish herd *47.6*
sexual selection *47.5*
sign stimulus *47.1*
signal receiver *47.4*
signaler *47.4*
social behavior *47.3*
song system *47.1*
tactile display *47.4*
territory *47.3*
threat display *47.4*

Readings

See also www.infotrac-college.com

Alcock, J. 1998. *Animal Behavior: An Evolutionary Approach*. Sixth edition. Sunderland, Massachusetts: Sinauer.

Dawkins, R. 1989. *The Selfish Gene*. New York: Oxford University Press.

Frisch, K. von. 1961. *The Dancing Bees*. New York: Harcourt Brace Jovanovich. A classic on the behavior of honeybees.

Wilson, E. O. 1975. *Sociobiology: The New Synthesis*. Cambridge, Massachusetts: Harvard University Press.

48 COMMUNITY INTERACTIONS

No Pigeon Is An Island

Flying through the rain forests of New Guinea is an extraordinary pigeon with cobalt blue feathers and lacy plumes on its head (Figure 48.1). It is about as big as a turkey, and it flaps so slowly and noisily that its flight sounds like an idling truck. As is true of eight species of smaller pigeons living in the same forest, it perches on branches to eat fruit. How is it possible that *nine* species of large and small fruit-eating pigeons live in the space of the same forest? Wouldn't you think that competition for food would leave one the winner? In fact, in that rain forest, every species lives, grows, and reproduces in a characteristic way, as defined

Figure 48.1 A cobalt blue, turkey-sized Victoria crowned pigeon, one of nine species of pigeons living in the same tropical rain forest of New Guinea. Within this habitat, each species has its own niche.

by its relationships with other organisms and with the surroundings.

Big pigeons perch on the sturdiest branches when they feed, and they eat big fruit. Smaller pigeons, with their smaller bills, cannot open big fruit. They eat small fruit hanging from slender branches that are not sturdy enough to support the weight of a turkey-sized pigeon. The species of trees in the forest differ with respect to the diameter of their fruit-bearing branches and the size of their fruit. So they attract different pigeons with different characteristics. In such ways, the nine species of pigeons *partition* the fruit supply.

And how do individual trees benefit from enticing the pigeons to dine? The seeds inside their fruits have tough coats, which can resist the action of digestive enzymes inside the pigeon gut. During the time it takes for ingested seeds to travel through the gut, the pigeons fly about, so they dispense seed-containing droppings in more than one place. In this way, the pigeons tend to disperse seeds some distance from the parent plants. Later, when seedlings grow, the odds are better that at least some will not have to compete with their parents for sunlight, water, and nutrients. Seeds that drop close to home cannot compete in any significant way with the resource-gathering capacity of the mature trees, which already have extensive, well-developed roots and leafy crowns.

Within the same forest, leaf-eating, fruit-munching, and bud-nipping insects interact with other organisms and their surroundings in certain ways. So do nectar-drinking, flower-pollinating bats, birds, and insects. And so do great numbers of beetles, worms, and other invertebrates that busily extract energy from remains and wastes of other organisms on the forest floor. By their activities, they cycle nutrients back to the trees.

Like humans, then, no pigeon is an island, isolated from the rest of the living world. The nine species of New Guinea pigeons eat fruit of different sizes. They disperse seeds from different sorts of trees. Dispersal influences where new trees will grow and where the decomposers will flourish. Ultimately, tree distribution and decomposition activities influence how the entire forest community is organized.

Directly or indirectly, interactions among coexisting populations organize the community to which they belong. With this chapter, we turn to community interactions that influence all populations over time and in the space of their environment.

KEY CONCEPTS

1. A habitat is the type of place where individuals of a species normally live. A community is an association of all the populations of species that occupy the same habitat.

2. Every species in the community has its own niche, defined as the sum of all activities and relationships in which its individuals engage as they secure and use the resources required for their survival and reproduction.

3. The structure of a community starts with adaptive traits that give individuals of each species the capacity to respond to the physical and chemical features of their habitat, and to levels and patterns of resource availability over time.

4. Interactions among species influence the structure of a community. They include mutually beneficial interactions, competition, predation, and parasitism.

5. Community structure also depends on the geographic location and size of the habitat, the rates at which the member species arrive and disappear, and the history of physical disturbances to the habitat.

6. The first species to occupy a particular type of habitat are replaced by others, which are replaced by still others, and so on in sequence. This process, known as primary succession, produces a climax community. A climax community is a stable, self-perpetuating array of species in balance with one another and with the environment.

7. Different stages of succession often prevail within the same habitat owing to local differences in the soil and other habitat conditions, recurring disturbances such as seasonal fires, and chance events.

48.1 WHICH FACTORS SHAPE COMMUNITY STRUCTURE?

Think of a clownfish darting above a coral reef, a maple tree on a Vermont hillside, or a mole making a burrow in soft earth. The type of place where you will normally observe a clownfish, maple, or mole is its **habitat**. The habitat of an organism is characterized by physical and chemical features, such as its temperature and salinity, and by the array of other species living in it. Directly or indirectly, the populations of all species in a habitat associate with one another as a **community**.

Five factors shape the structure of a community. *First*, interactions between climate and topography help dictate the habitat's temperatures, rainfall, soil types, and other conditions. *Second*, the kinds and amounts of food and other resources that become available through the year influence which species can live there. *Third*, individuals of each species have adaptive traits that allow them to survive and exploit specific resources in the habitat. *Fourth*, species in the habitat interact, as by competition, predation, and mutually helpful activities. *Fifth*, community structure is influenced by the overall pattern of population size (and the actual history of changes in its size), by the arrival and disappearances of species, and by physical disturbances to the habitat.

Together, the five factors help dictate the number of species at different "feeding levels," starting with the producers and continuing through levels of consumers. They influence population sizes. They also help dictate the overall number of species. For example, high solar radiation, warm temperatures, and high humidity in tropical habitats favor growth of many kinds of plants, which support many kinds of animals. Conditions in arctic habitats do not favor great numbers of species.

The chapters to follow deal with energy flow through feeding levels and with geographic factors influencing community structure. Here we begin with interactions among species, using the niche concept as our guide.

The Niche

If the organisms within a community all share the same habitat—that is, if they all have the same "address"—in what respects do they differ? Each kind is distinct in terms of its "profession" within the community—that is, in the sum of activities and relationships in which it engages to secure and use the resources necessary for its survival and reproduction. This is its **niche**.

For each species, the *fundamental* niche is the one that might prevail in the absence of competition and other factors that could constrain its acquisition and use of resources. However, as you will see, such constraining factors do come into play in all communities. They tend to bring about a more constrained, *realized* niche that shifts in large and small ways over time, as individuals of the species respond to a mosaic of changes.

Categories of Species Interactions

Dozens to hundreds of species interact in diverse ways, even in simple communities. In spite of this diversity, we can identify six categories of interactions that have different effects on population growth (Table 48.1).

Table 48.1 Categories of Two-Species Interactions*

Type of Interaction	Direct Effect on Species 1	Direct Effect on Species 2
Neutral relationship	0	0
Commensalism	+	0
Mutualism	+	+
Interspecific competition	−	−
Predation	+	−
Parasitism	+	−

* 0 means no direct effect on population growth; + means positive effect; − means negative effect.

Each species has a neutral relationship with most species in its habitat. For example, Canadian lynx and grasses do not affect each other directly. Interactions with other species link them only indirectly. The lynx preys on and decreases the number of snowshoe hares that eat plants; and plants fatten the hares, the prey of lynx.

Commensalism directly helps one species but does not affect the other much, if at all. For instance, some birds use tree branches for roosting sites. The trees get nothing but are not harmed. With **mutualism**, benefits flow both ways between the interacting species. Don't think of this as cozy cooperation. Benefits flow from a two-way exploitation. With **interspecific competition**, disadvantages flow both ways between species. Finally, **predation** and **parasitism** are interactions that directly benefit one species (either the predator or the parasite) and directly hurt the other (the prey or host).

Commensalism, mutualism, and parasitism are cases of **symbiosis**, or "living together." For at least part of the life cycle, individuals of two (or more) species interact with neutral, positive, or negative effects on one another.

A habitat is the type of place where individuals of a species normally live. A community consists of all populations that live in the habitat.

Community structure arises from the habitat's physical and chemical features, resource availability over time, adaptive traits of its members, how the members interact, and the history of the habitat and its occupants. Species may have neutral, positive, or negative effects on one another.

A niche is the sum of all activities and relationships in which individuals of a species engage as they secure and use the resources necessary to survive and reproduce.

Further reading: Student Guide to InfoTrac on web site →

48.2 MUTUALISM

Mutualistic interactions, in which positive benefits flow both ways, abound in nature. When trees provide New Guinea pigeons with food and when the pigeons help disperse seeds from the trees to new germination sites, both function as mutualists. Many other kinds of plants and animals enter into such interactions. For example, most flowering plants and the insects and birds, bats, and other animals that pollinate them are mutualists. The introduction to Chapter 31 gives vivid examples.

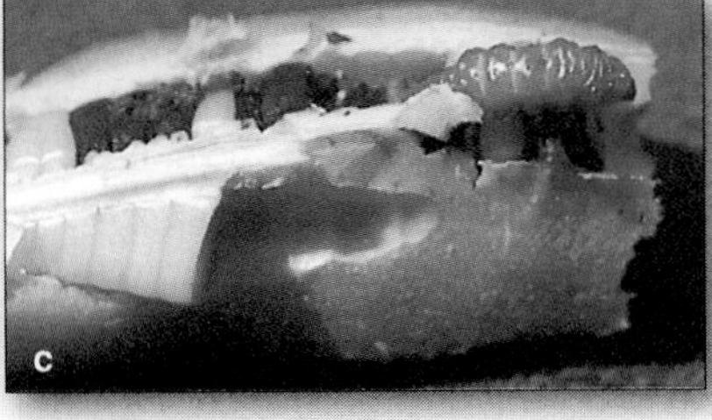

Figure 48.2 One mutualistic interaction on a rocky slope of Colorado's high desert.

(**a**) Different kinds of flowering plants of the genus *Yucca* are each pollinated exclusively by one species of yucca moth (**b**). This insect cannot complete its life cycle with any other plant.

The adult stage of the moth life cycle coincides with blossoming of yucca flowers. By using her specialized mouthparts, a female moth gathers sticky pollen and rolls it into a ball. Then she wings her way to another flower. She pierces the wall of the flower's ovary, where seeds form and develop, and lays her eggs inside. As she crawls out of the flower, she pushes a ball of pollen onto a pollen-receiving surface.

Pollen grains germinate and grow down through tissues of the ovary. They deliver sperm to the flower's eggs. The seeds develop after fertilization. Meanwhile, the moth eggs develop to the larval stage. (**c**) When larvae emerge, they eat a few seeds, then gnaw their way out of the ovary. The seeds that moth larvae do not eat give rise to new yucca plants.

Some forms of mutualism are *obligatory*. That is, the individuals of one species cannot grow and reproduce in the absence of intimate dependency with individuals of another species during the life cycle. This is the case for an interaction between yucca plants and yucca moths. Each plant of this genus (*Yucca*) can be pollinated only by one species of the yucca moth genus (*Tegeticula*). In addition, the larval stages of the moth grow only in the yucca plant; they eat only yucca seeds (Figure 48.2).

We also find cases of obligatory mutualism between many fungi and plants. Mycorrhizae, remember, are intimate ecological interactions between fungal hyphae and young roots (Chapters 24 and 30). The two kinds of absorptive structures interact in two ways. Either the fungal hyphae penetrate root cells or they grow as a dense, velvety mat around them. The fungus is good at absorbing mineral ions from soil, and the plant comes to depend on tapping into some of these. In turn, the fungus withdraws some sugar molecules that the plant makes by photosynthesis. It also depends on the plant for its own reproductive success. When photosynthesis ceases, the fungus stops producing spores.

And what about the apparent endosymbiotic origins of eukaryotes? Long ago, phagocytic bacteria may have engulfed aerobic bacterial cells that resisted digestion, tapped host nutrients, and reproduced independently. In time, the hosts came to depend on ATP produced by the guests, which evolved into mitochondria and chloroplasts. If those prokaryotic cells had not become so mutually interdependent, you and all other eukaryotic organisms would not even be around today (Section 21.4).

In cases of mutualism, each of the participating species reaps benefits from the interaction.

48.3 COMPETITIVE INTERACTIONS

Organisms do not come into the world with guarantees that they will secure enough energy, nutrients, living space, and other necessities to survive and reproduce. They typically compete for a share of limited resources. *Intraspecific* competition means that individuals of the same population or species compete with one another. As you probably deduced from the preceding chapter, their interactions can be fierce. *Interspecific* competition, which occurs between populations of different species, usually is not as intense. Why? *The requirements of two species might be similar, but they never can be as close as they are for individuals of the same species.*

Think about two forms of competitive interactions. Sometimes individuals have equal access to a required resource, but some are better than others at exploiting it. In such cases, competition tends to reduce the supply of a shared, limited resource. (When you and a friend both use straws to share a small milkshake, you might not get nearly as much if your friend uses a jumbo straw.) In other cases, some individuals control access to a resource and partially or completely prevent others from using it regardless of its scarcity or abundance. (For instance, even if you shared a ten-gallon milkshake, you still would get less if your friend pinched your straw.)

Competition abounds in nature. Chipmunks exclude other species of chipmunks from their habitats (Figure 48.3). A strangler fig tree wraps around other trees as a framework for its own growth and finally kills them. From early spring until late summer, a male *broadtailed* hummingbird busily chases other males and females of its species from his blossoming territory in the Rockies. In August, however, *rufous* hummingbirds of the Pacific Northwest migrate through the Rockies on their way to wintering grounds in Mexico. The rufous males prove to be more aggressive and stronger competitors for the available food. And they evict the male broadtails from broadtail territories all along their migratory route.

ALPINE CHIPMUNK

LODGEPOLE CHIPMUNK

YELLOW PINE CHIPMUNK

LEAST CHIPMUNK

Figure 48.3 Example of competition in nature. On the eastern slopes of the Sierra Nevada, different chipmunk species occupy different habitats. The alpine habitat is at the highest elevation. Below it are the lodgepole pine, piñon pine, then the sagebrush habitats. The least chipmunk lives at the base of the mountains in sagebrush. Its adaptations would allow it to move up into the piñon pine habitat, but the aggressively competitive behavior of yellow pine chipmunks in that habitat won't let it. Food preferences keep the yellow pine chipmunk out of the sagebrush habitat.

Competitive Exclusion

To a greater or lesser extent, any two species differ in their adaptations for getting food and avoiding enemies. Thus one usually competes more effectively for scarce resources. The two are less likely to coexist in the same habitat when they use resources in very similar ways. G. Gause demonstrated this by growing two species of *Paramecium* separately, then together (Figure 48.4*a–c*). Both species exploited the same food (bacterial cells) and competed intensely for it. As results from Gause's experimental tests implied, any two species that utilize identical resources cannot coexist indefinitely. Many other experiments have since supported Gause's concept, which ecologists have named **competitive exclusion**.

Gause also studied two other species of *Paramecium*, which did not overlap as much in their requirements. When grown together, one

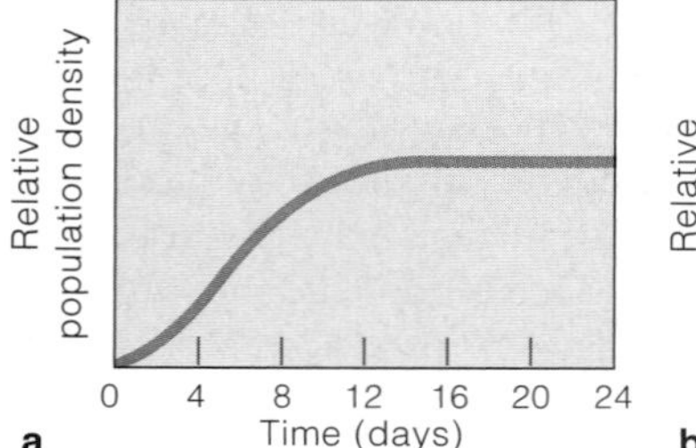

Relative population density
0
4
8
12
16
20
24
Time (days)
a

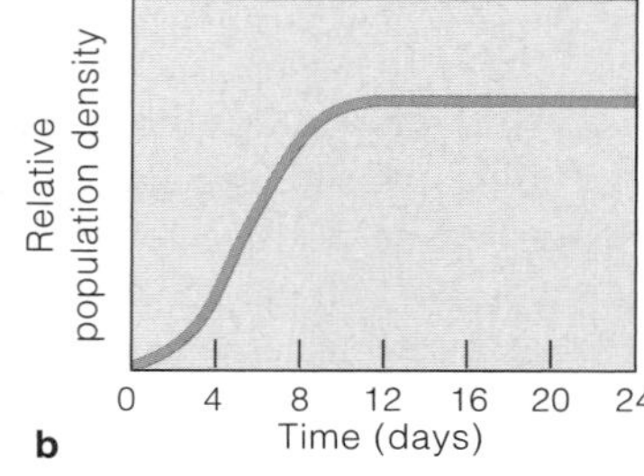

Relative population density
0
4
8
12
16
20
24
Time (days)
b

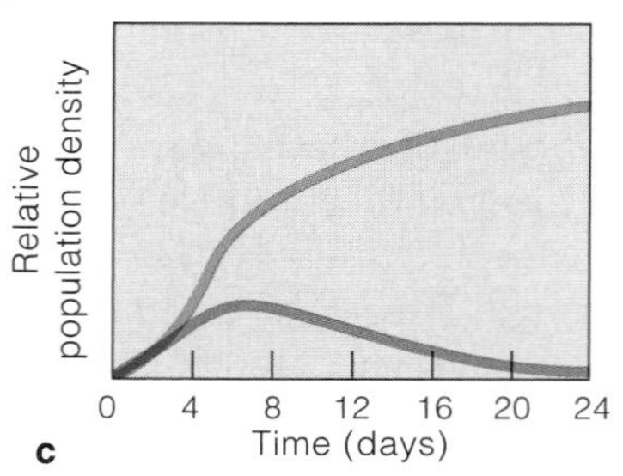

Relative population density
0
4
8
12
16
20
24
Time (days)
c

Figure 48.4 Results of competitive exclusion between two protistan species that compete for the same food. (**a**) *Paramecium caudatum* and (**b**) *P. aurelia* were grown in separate culture tubes and established stable populations. S-shaped growth curves in these graphs indicate stability. (**c**) Then the populations were grown together. *P. aurelia* (*red* curve) drove the other species toward extinction (*blue* curve in **c**). This experiment and others suggest that two species cannot coexist indefinitely in the same habitat *when they require identical resources*. If their requirements do not overlap much, one might influence the population growth rate of the other, but they may still coexist.

Further reading: Student Guide to InfoTrac on web site →

a
b

Figure 48.5 Two species of salamanders that coexist in the same habitat: (**a**) *Plethodon glutinosus* complex and (**b**) *P. jordani.*

tended to feed on bacterial cells suspended in liquid inside a culture tube. The other fed on yeast cells at the bottom of the test tube. The population growth rate slowed down for both species, but the overlap in resource use was not enough for one species to completely exclude the other. In other words, these two species continued to coexist.

Field experiments reveal certain effects of competition, also. For example, N. Hairston studied salamanders in their native habitat in the Great Smoky Mountains and the Balsam Mountains (Figure 48.5). One kind, *Plethodon glutinosus,* lives at lower elevations than its relative *P. jordani,* but their home ranges do overlap in some areas.

For one experiment, Hairston removed one species or the other species from different test plots in the overlap areas. He also left some plots untouched; these were the control plots. After five years, nothing had changed in the control plots; the two species simply were coexisting. By contrast, salamander populations in the test plots were increasing in size. Plots cleared of *P. jordani* had a greater proportion of *P. glutinosus*. And plots cleared of *P. glutinosus* had a greater proportion of *P. jordani.*

Hairston concluded that, where populations of the two salamander species coexist in nature, competitive interactions suppress the growth rate of both of them.

Resource Partitioning

Think back on the nine species of fruit-eating pigeons in the same New Guinea forest. They require the same resource: fruit. Yet they overlap only slightly in their use of the resource, because each specializes in fruits of a particular size. Together, they are a good example of **resource partitioning**—the *subdividing* of some category of similar resources that lets competing species coexist.

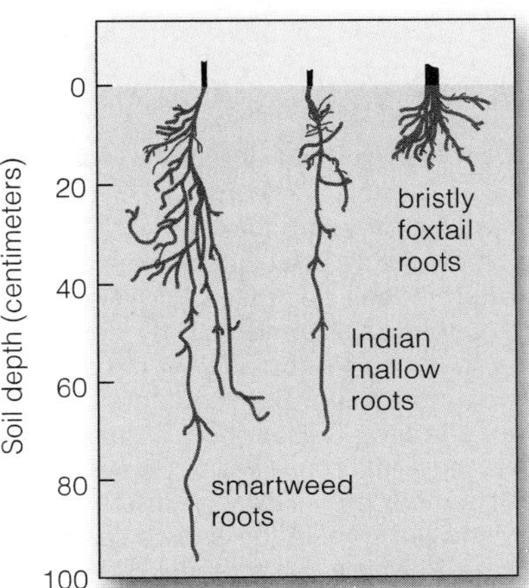
0
20
40
60
80
100
Soil depth (centimeters)
bristly foxtail roots
Indian mallow roots
smartweed roots

Figure 48.6 An example of resource partitioning (nutrients and water in soil) among three annual plants in a plowed, abandoned field. All three species require water and the same mineral ions. But they differ in the adaptations by which they secure these resources.

A similar competitive situation arises among three species of annual plants in certain plowed, abandoned fields. Like other plants, all three require sunlight, water, and dissolved mineral ions. Yet each species is adapted to exploiting a different portion of the habitat (Figure 48.6). Drought-tolerant foxtail grasses have a shallow, fibrous root system that quickly absorbs rainwater. They grow where moisture in the soil varies from day to day. Mallow plants, with a taproot system, grow in deeper soil that is moist early in the growing season but drier later. Smartweed's taproot system branches in topsoil and soil below the roots of the other species. It grows where soil is continuously moist.

In sum, species may coexist in the same habitat even if their niches do overlap. Many species compete when experimentally grown together but not in their natural habitats, where they tend to use different foods or other resources. Hairston's salamanders compete yet coexist at suppressed population sizes. New Guinea pigeons coexist through resource partitioning. Birds and people overlap in their need for oxygen, but atmospheric oxygen is so abundant that there is no need to compete for it.

In some competitive situations, all individuals have equal access to a resource that they all require, but some are better than others at exploiting it. In other competitive situations, some individuals control access to a resource.

The more two species in the same habitat differ in their use of resources, the more likely they can coexist.

Two competing species also may coexist through resource partitioning—by sharing the same resource in different ways or at different times.

48.4 PREDATION

Predation Versus Parasitism

Of all community interactions, predation is so dynamic that it often rivets our attention, not to mention the prey's. But what exactly is a predator–prey interaction?

Consider the leopard in Figure 48.7 as it closes in on a baboon, one of its preferred foods. Although far less dramatic than a baboon–leopard confrontation, a goat pulling up a thistle plant for breakfast is a predator, also. Its prey is a living organism, killed for food. Can the same be said about horses grazing on plants? What about a mosquito or vampire bat withdrawing blood from a mammal before they fly away? What about ticks or fleas that withdraw blood leisurely before they jump off one host and lay their eggs elsewhere? What about tapeworms, mistletoe, and other organisms that live in or on other species? Definitions can get fuzzy.

For simplicity, let's use two fairly broad definitions for interactions between various consumers and their victims. **Predators** are animals that feed on other living organisms but do not take up residence in or on them. Most often, the targets of predators—their **prey**—are killed outright or mutilated. In this respect, predators are unlike **parasites**, most of which feed on tissues of living organisms—**hosts**—and reside in or on them for at least part of their life cycle. A host might or might not die as a result of the interaction.

Many adaptations of predators (or parasites) and their victims arose by **coevolution**, the joint evolution of two or more species that exert selection pressure on each other as an outcome of close ecological interaction. Suppose, as an outcome of mutation, a prey organism displays a new, heritable means of defense, which later spreads through the prey population. Some individual predators are more effective than others at countering the defense. They tend to eat more and have a better chance to survive and reproduce. In time the forms of traits that are most effective at overcoming the defense increase in frequency in the population. Now the more effective predators exert selection pressure that favors better defenses among prey—and so on through time.

Dynamics of Predator–Prey Interactions

In any specified interval, the outcome of predator–prey interactions depends partly upon the carrying capacity for the prey population. **Carrying capacity**, remember, is defined as the maximum number of individuals that

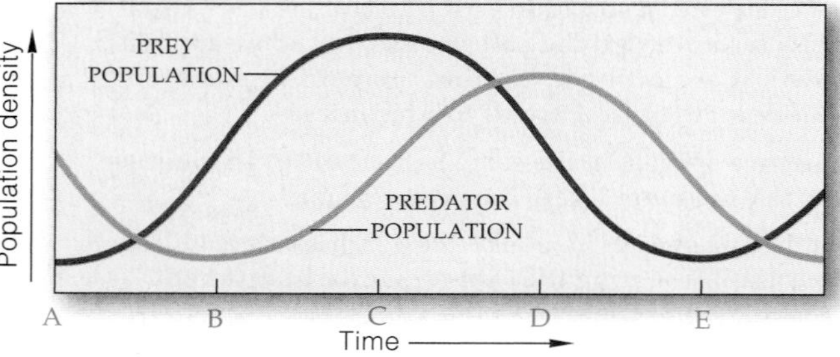
PREY
POPULATION
PREDATOR
POPULATION
Population density
A
B
C
D
E
Time

Figure 48.7 Idealized cycling of predator and prey abundances. (For clarity, the diagram exaggerates predator density; predators are usually less common than their prey during the cycle.) This pattern arises because of time lags in predator responses to changes in prey abundance. At time A, prey density is low; predators have a harder time getting food; their population is declining. Prey respond to the decline in predators by increasing. But predators do not start increasing until they start to reproduce at time B. Both populations grow until the predator increase causes the prey population to decline (time C to E). Predators continue to increase and take out more prey. However, the lower prey density leads to starving predators, and the population growth rate for predators slows (starting at time D). At time E, a new cycle starts.

Further reading: Student Guide to InfoTrac on web site

160
140
120
100
80
60
40
20
0
Number of pelts taken(x 1,000)
1845
1865
1885
1905
1925

Figure 48.8 Study of the predator–prey interactions between Canadian lynx and snowshoe hare populations. The graph's correspondence between abundances of both populations is based on counts of pelts sold by trappers to Hudson's Bay Company over ninety years. The *dashed* line tracks abundances of lynx, and the *solid* line tracks hare abundances. The photograph underscores Charles Krebs' observation that predation causes mass paranoia among the hares, which continually look over their shoulders during the declining phase of each cycle.

The graph is a good test of whether you will readily accept someone else's conclusions without questioning their scientific basis. (Remember the sections in Chapter 1 on scientific methods?) What other factors may have affected the cycle? Did weather vary, with more severe winters imposing greater demand for hares (to keep lynx warmer) and higher death rates? Did lynx compete with other predators, such as owls? Did predators turn to alternative prey during low points of the hare cycle? When fur prices rose in Europe, did the trapping increase? When the supply of pelts outstripped the demand, did the trapping decline?

the resources of an environment are capable of maintaining indefinitely. Rates of reproduction for predator and prey populations also influence the outcome in a given interval. So do responses of the population of predators to increased prey density.

During times when predation is keeping the prey population from exceeding the carrying capacity, both populations tend to coexist at fairly steady levels. More prey organisms are available, and predators reproduce promptly and eat more prey. Population densities fluctuate if predators don't reproduce as fast as the prey, if they can eat only so many prey organisms in a given interval, and when the carrying capacity for the prey population is high.

Figure 48.7 shows an idealized pattern of the cyclic changes corresponding to time lags in the predator's response to changes in prey abundance. In nature, we sometimes observe such a correspondence, but other factors also influence the pattern. Consider Figure 48.8. It demonstrates a ten-year cycling in populations of the Canadian lynx and the showshoe hare. Charles Krebs led a study in Alaska's Yukon to identify the sources of this cycling. For ten years, he tracked hare densities in 1-square-kilometer control plots and experimental plots that kept out mammalian predators (by electric fences), had additional food, or were fertilized to increase plant growth. The team attached radio collars to 1,000+ hares, squirrels, lynx, and coyotes. In predator-free plots, hare density doubled. Extra food tripled the density. Predator reduction and extra food increased density elevenfold. Yet the manipulations only delayed the declines in the cycle; they did not prevent them. It turns out that the fence couldn't keep out owls and other raptors; predators got 83 percent of the collared hares, only 9 percent of which starved to death. Thus, a simple predator–prey or plant–herbivore model can't explain the results. More variables are operating in a *three-level* interaction among plants, herbivores, and carnivores.

Some predator and prey populations coexist at more or less steady levels. Others undergo recurring cycles of abundance and crashes, erratic cycles, or prey extinction.

When predation keeps a prey population from overshooting the carrying capacity, the predator and prey populations may coexist at stable levels. Delays in a predator's response to changes in the abundance of prey contribute to cyclic or irregular fluctuations in population levels.

Other factors, such as additional predators and variations in food supplies, contribute to changes in prey abundance.

Focus on the Environment

48.5 A COEVOLUTIONARY ARMS RACE

Populations of other species are part of any organism's environment, and the ones that interact as predators and prey exert continual selection pressure on each other. One must defend itself and the other must overcome the defenses. This is the basis of a coevolutionary arms race that has resulted in some truly amazing adaptations.

CAMOUFLAGE Consider prey species that **camouflage** themselves; they can hide in the open. Such organisms have adaptations in form, patterning, color, and behavior that help them blend with their surroundings and escape detection. Figure 48.9 shows classic examples, including a desert plant (*Lithops*) that resembles a small rock. Only during a brief rainy season does *Lithops* flower. That is when other plants grow profusely and divert herbivores from *Lithops*—and when free water instead of juicy plant tissues is available to quench an animal's thirst.

WARNING COLORATION Many prey species taste bad, are highly toxic, or inflict pain on attackers. The toxic types often have **warning coloration**, or conspicuous patterns and colors that predators learn to recognize as "avoid me" signals. For example, maybe a young, inexperienced bird will spear a yellow-banded wasp or an orange-patterned monarch butterfly—once. It will quickly learn to associate the distinctive colors and patterning with a painful sting or with vomiting foul-tasting butterfly toxins.

Truly dangerous or repugnant species make little or no attempt to conceal themselves. Skunks are like this. So are frogs of the genus *Dendrobates*; they are among the most vivid and most poisonous organisms (Section 33.1).

MIMICRY Many prey organisms bear close resemblance to dangerous, unpalatable, or hard-to-catch species. Any close resemblance in form, behavior, or both between one species that serves as a *model* for deception and another that is its *mimic* is called **mimicry**. Figure 48.11 shows how some tasty but weaponless mimics physically resemble species that predators have learned to ignore. In "speed" mimicry, sluggish prey species look like swift-moving species that predators have given up trying to catch.

MOMENT-OF-TRUTH DEFENSES When luck runs out, survival of prey organisms that are cornered or under attack may turn on a last-ditch trick. Suppose a leopard manages to run down a tasty baboon, as in Figure 48.7. By turning abruptly and displaying formidable canines, the baboon may startle and confuse this predator long enough for a getaway. Other cornered animals spew chemicals as disgusting repellents or toxins. Earwigs, skunks, and stink beetles produce awful odors. Several beetle species take aim and let loose with noxious sprays.

Similarly, many plants synthesize predator repellents. Tannins in the foliage and seeds of certain plants taste

Figure 48.9 A few prey organisms demonstrating the fine art of camouflage. (**a**) What bird??? When a predator approaches its nest, the least bittern stretches its neck (which is colored like the surrounding withered reeds), thrusts its beak upward, and sways gently like reeds in the wind. (**b**) An unappetizing bird dropping? No. This edible caterpillar's body coloration and the stiff body positions it assumes help camouflage it (hide it in the open) from predatory birds. (**c**) Find the plants (*Lithops*) hiding in the open from herbivores by their stonelike form, pattern, and coloring.

Further reading: Student Guide to InfoTrac on web site

Figure 48.10 A few examples of mimicry among the insects.

Many predators avoid prey that taste awful, secrete toxins, or inflict painful bites or stings. Commonly, such prey display warning coloration (bright colors, bold markings, or both), and many do not even bother to hide. Many species of prey unrelated to the dangerous or unpalatable ones have evolved striking behavioral and morphological resemblances to them. The inedible butterfly in (**a**) is a model for the edible mimic *Dismorphia* (**b**). The yellowjacket in (**c**) stings aggressively and probably is the model for nonstinging, edible wasps (**d**) and beetles (**e**) that have a similar appearance.

c The dangerous model . . .

d . . . one of its edible mimics . . .

e . . . and another edible mimic

bitter and make the plant tissues hard to digest. Make the mistake of nibbling on the seemingly luscious yellow petals of a buttercup (*Ranunculus*), and you will inflict a chemical burn upon the lining of your mouth.

PREDATOR RESPONSES TO PREY As part of the coevolutionary arms race, predators counter prey defenses with their own marvelous adaptations. Among the countering measures are stealth, camouflage, and clever ways of avoiding repellents.

Consider the edible beetles that direct sprays of noxious chemicals at their attackers. Grasshopper mice grab such beetles and plunge the "sprayer" end into the ground, then feast on the unprotected head (Figure 48.11*a,b*). Chameleons typically hold themselves motionless for extended intervals. Prey might not even "see" them until the amazingly swift chameleon tongue zaps them (Section 33.4). And it is no accident that stealthy predators blend with backgrounds. Think of snow-white polar bears camouflaged against snow, golden tigers crouched in tall-stalked, golden grasses, and pastel predatory insects lurking in pastel flowers. And hope you never step barefoot on a scorpionfish concealed on the seafloor.

Figure 48.11 Examples of adaptive responses of predators to prey defenses. (**a**) Certain beetles spray noxious chemicals at attackers, which works as a deterrent some of the time. (**b**) At other times, however, grasshopper mice plunge the chemical-spraying tail end of their beetle prey into the ground and feast on the head end. (**c**) Find the scorpionfish, a venomous predator with camouflaging fleshy flaps, multiple colors, and profuse spines. (**d**) Where do pink parts of the flower end and pink parts of a praying mantis begin?

48.6 PARASITIC INTERACTIONS

Evolution of Parasitism

Think of parasitism as an ecological interaction in which one organism drains nutrients from another, typically while living in or on it. The parasite benefits; its host does not. This also is true of predation, but you would not expect to come across a predator that characteristically exploits another organism as food *and* habitat.

Parasites have pervasive influences in populations. By draining host nutrients, they alter how much energy and how many nutrients the population is withdrawing from the habitat. Weakened hosts are more vulnerable to predation and less attractive to potential mates. Some infections result in sterility. Some may alter the ratio of host males to females. In such ways, parasitic infections lower the birth rate, raise the death rate, and influence intraspecific and interspecific competition.

Sometimes the gradual drain of nutrients during a parasitic infection indirectly causes death, for the host becomes so weakened that it succumbs to secondary infections. In evolutionary terms, however, killing a host is not very good for a parasite's reproductive success. An infection of longer duration obviously will give the parasite more time to produce more offspring. We may therefore speculate that natural selection tends to favor parasite and host adaptations that promote some level of mutual tolerance and less-than-fatal effects.

Usually, death results only when a parasite attacks a novel host (which has no coevolved defenses against it) or when too many individual parasites are active at the same time during an infection.

Kinds of Parasites

Previous chapters introduced you to diverse parasitic species. They include *ecto*parasites (which live on a host body surface) and *endo*parasites (which live within the host's body). Some parasites live on or in one host for their entire life cycle, and others are free-living some of the time and residents of different hosts at different stages of their life cycle. Many enlist insects and other arthropods as taxis from one host organism to another.

The **microparasites** are microscopic in size and rapid reproducers. They include many viruses, bacteria, and protistans, especially some protozoans and sporozoans. The trematode in Figure 48.12*b* is an example. Diverse invertebrates are among the **macroparasites**. The ones with worldwide notoriety include numerous flatworms and roundworms (also called nematodes), such as the type in Figure 48.12*b*. Also in this category are many arthropods, including fleas, ticks, and mites, and lice that attack birds and mammals. They lay eggs either directly on the host or in its nests or bedding. As you know from Section 24.5, the fungal kingdom also has its share of macroparasites.

The plant kingdom also includes a few. *Holo*parasitic plants are nonphotosynthetic; they withdraw nutrients and water from young roots of host plants. Members of the broomrape family do this to the roots of oaks and beech trees. *Hemi*parasitic plants retain the capacity for photosynthesis but still withdraw nutrients and water from the host plant. Mistletoe is like this; it puts out long extensions that invade the sapwood of host trees.

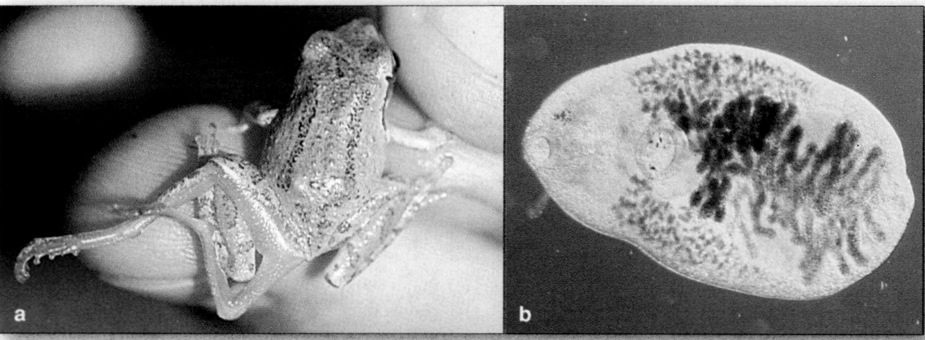
a
b

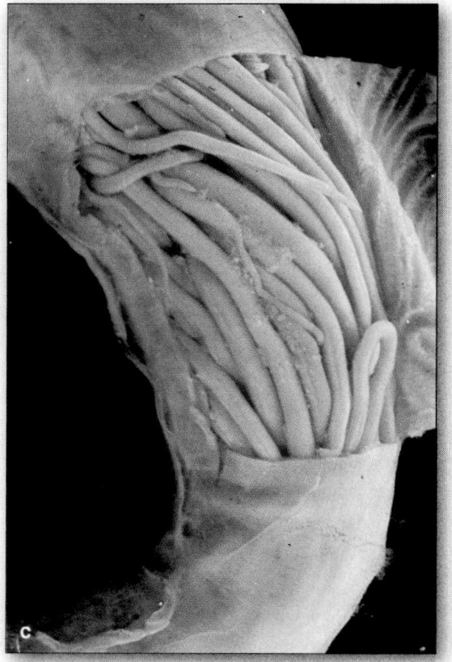
c

Figure 48.12 (**a**) Deformed frog with extra legs, an outcome of a parasitic infection. Frog population sizes started to plummet in the 1990s. Increased parasitism, predation, ultraviolet radiation, habitat losses, and chemical pollution are contributing to the decline; even mud-caked hiking boots import some parasites and pathogens into frog habitats. As an example, Stanley Sessions of Hartwick College and Stanford graduate Pieter Johnson identified a parasitic infection that affects frog limbs. (**b**) Trematodes (*Ribeiroia*) are the culprits. They burrow into tadpole limb buds and physically or chemically alter cells. Infected tadpoles grow extra legs or none at all. With high exposure to *Ribeiroia*, the number of tadpoles that successfully complete metamorphosis declines. Trematode cysts have been found in weirdly legged frogs and salamanders in California, Oregon, Arizona, and New York. (**c**) Packed inside the small intestine of a host pig, adult roundworms (*Ascaris*), a type of endoparasite.

Further reading: Student Guide to InfoTrac on web site

Figure 48.13 Example of a social parasite. European cuckoos lay eggs in nests of other species. In response to an environmental cue, a hatchling executes an innate behavior. As Section 47.1 describes, even before its eyes open, it responds to the spherical shape of the host's eggs by shoving them out of the nest. The foster parents raise the usurper even after it is old enough to show its true feathers, so to speak.

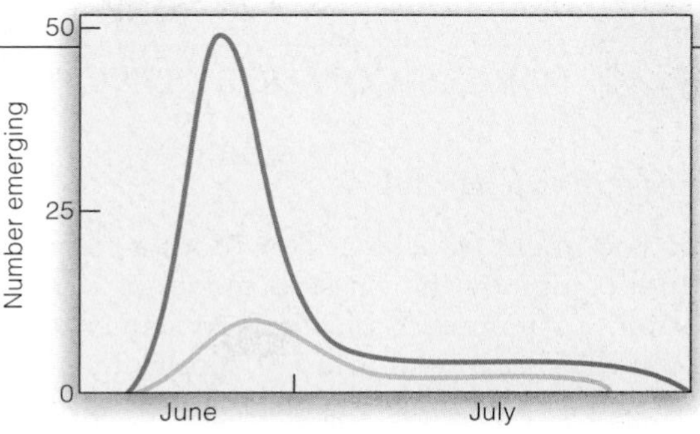

50
25
0
Number emerging
June
July

Figure 48.14 Graph of the effect of the activity of a parasitoid wasp on the emergence of adult sawflies from cocoons on the forest floor. The *brown* line represents the number emerging when wasp attacks were experimentally prevented. The *pink* line represents the number emerging after the attacks. Notice that the wasp made most of its attacks on the would-be *early* emergers from cocoons not buried as deeply in forest litter.

True parasites feed on host tissues. A different kind, the **social parasites**, manipulates the social behavior of another species to complete their life cycle. Cuckoos and North American cowbirds are examples. The adult females lay their eggs in the nests of other species. Their hatchlings either get rid of the rightful occupants or demand and get most of the food (Section 47.1).

Regarding the Parasitoids

In their own category between predators and parasites are **parasitoids**, insect larvae that always kill what they eat. As parasitoids are growing up, they consume all the soft tissues of their hosts, which are the larvae or pupae of other insect species.

The biologist Peter Price studied the evolutionary effects of a parasitoid wasp that lays eggs on cocoons of a sawfly species. Sawflies reproduce once a year and lay their eggs in trees. In time, fertilized eggs develop into fly larvae, which feed on foliage and grow about as large as a pencil in diameter. Then larvae drop to the forest floor and spin cocoons after they burrow into leaf litter. Some larvae burrow more deeply than others.

As Price noted, the first adult sawflies to emerge did so from cocoons that were the least deeply buried. Later in the season, more sawflies emerged from the more deeply buried cocoons (Figure 48.14). Parasitoid wasps tend to lay eggs on the cocoons closest to the surface of leaf litter. They exert strong selection pressure on the sawfly population, because the deep-burrowing sawfly individuals are more likely to escape detection. There are only so many cocoons near the surface, so the wasp that is able to locate cocoons deeper in the litter will be more competitive in securing food for her larvae.

In this coevolutionary contest, the host organism is staying ahead. Every time the sawfly larvae burrow deeper, female wasps must spend more time searching for them. Fewer wasp eggs are laid, and so the wasp population is held in check.

Parasites as Biological Control Agents

You might well conclude that parasites and parasitoids exert control over the population growth of other insect species. Indeed, many kinds are commercially raised and selectively released as *biological controls.* They are often touted as a fine alternative to chemical pesticides. But less than 20 percent of existing selections qualify as effective defenses against pests.

As C. Huffaker and C. Kennett point out, effective biological control agents display five attributes. They are well adapted to a host species and to their habitat. They are exceptionally good at searching for a host. Their population growth rate is high relative to that of the host species. Their offspring are mobile enough to ensure adequate dispersal. And the lag time between their responses to changes in the numbers of the host population is minimal.

Releasing more than one kind of biological control agent in an area may trigger competition among them and eventually lessen their overall effectiveness. Also, a shotgun approach to biological control is risky. There is a possibility that the parasites may attack nontargeted species. In 1983, for example, F. Howarth reported that the populations of butterflies and moths native to the Hawaiian Islands are declining. Introductions of wasps that were supposed to serve as biological controls over something else are partly to blame. Later in the chapter, we will return to the effects of introducing new species into a habitat on native populations.

Like predators and their prey victims, parasites and their hosts are locked into long-term, coevolutionary contests.

Natural selection favors parasitic species that temper their demands in ways that ensure an adequate supply of hosts.

Too great a demand quickly kills the hosts and limits the duration of an attack, which sets a limit on the number of parasitic offspring that can be produced.

48.7 FORCES CONTRIBUTING TO COMMUNITY STABILITY

A Successional Model

By now you might be asking: How does a community come into being? By the classical model of **ecological succession**, a community develops in sequence, from pioneer species to an end array of species that remain in equilibrium over some region. **Pioneer species** are opportunistic colonizers of vacant or vacated habitats. They enjoy high dispersal rates and rapid growth. In time, more competitive species replace the pioneers, then are themselves replaced until the array of species stabilizes under the prevailing habitat conditions. This persistent array of species is the **climax community**.

Primary succession is a process that begins as pioneer species colonize a barren habitat such as a new volcanic island or land exposed when a glacier retreats (Figure 48.15). Pioneer species include lichens and small plants that are small, have brief life cycles, and are adapted to exposed sites having intense sunlight, large temperature changes, and nutrient-deficient soil. In the early years, hardy flowering plants produce great numbers of small seeds, which are quickly dispersed.

Once established, the pioneers improve conditions for other species and often set the stage for their own replacement. Many types are mutualists with nitrogen-fixing bacteria, and they outcompete the early arrivals in nitrogen-poor habitats. Also, the pioneers form low-growing mats that shelter the seeds of later species and cannot shade out seedlings. In time, accumulated wastes and remains add volume and nutrients to the soil that help other species take hold. The successional species eventually crowd out the pioneers, whose spores and seeds travel as fugitives on wind and water—destined, perhaps, for a new but temporary habitat.

In **secondary succession**, a disturbed area within a community recovers and moves again toward a climax state. The pattern is typical of abandoned fields, burned forests, and storm-battered intertidal zones. It emerges after falling trees open part of an established forest's canopy. Sunlight reaches seeds and seedlings that are already on the forest floor and spurs their growth.

By one hypothesis, the colonizers *facilitate* their own replacement. By another, the earliest colonizers compete against the species that could replace them, so that the sequence of succession depends on who gets there first.

The Climax-Pattern Model

At one time, some ecologists thought the same general type of community always develops in a given region because of constraints imposed by climate. However, stable communities other than "the climax community" commonly persist in a region, as when tallgrass prairie extends from the west into Indiana's deciduous forests.

a

Figure 48.15 (**a**) Primary succession in Alaska's Glacier Bay area. The glacier in this photograph has been retreating since 1794. *Facing page*: (**b**) As a glacier retreats, meltwater leaches nitrogen and other minerals from the newly exposed soil. Less than ten years ago, ice buried this nutrient-poor soil.The first invaders are lichens, horsetails, and seeds of fireweed and mountain avens (*Dryas*). (**c**) *Dryas*, a pioneer species that benefits from nitrogen-fixing activities of mutualistic microbes, grows and spreads rapidly over the glacial till.

(**d**) Within twenty years, alder, cottonwood, and willow seedlings have taken hold in drainage channels. They, too, are symbionts with nitrogen-fixing microbes. (**e**) Within fifty years, alders have formed mature, dense thickets in which cottonwood, hemlock, and a few evergreen spruce can grow quickly. (**f**) After eighty years, cottonwood and spruce crowd out the mature alders. (**g**) In areas deglaciated for more than a century, dense forests of Sitka spruce and western hemlock dominate.

By the **climax-pattern model**, a community is adapted to many environmental factors—topography, climate, soil, wind, species interactions, recurring disturbances, chance events, and so on—that vary in their influence over a region. As a result, even in the same region you may see one climax stage extending into another along gradients of influential environmental conditions.

Cyclic, Nondirectional Changes

Many small-scale changes occur over and over again in patches of a habitat. Such recurring changes contribute to the internal dynamics of the community as a whole. Thus, observe a disturbed patch of habitat, and you might conclude that great shifts in species composition are afoot. Observe the community on a larger scale, and you may find that the overall composition includes all of the pioneer species that are colonizing the patch as well as the dominant climax species.

Think of a tropical forest. It slowly develops through phases of colonization by pioneer species, increases in

small fires are prevented. The underbrush gradually thickens with fire-susceptible species. The dense underbrush prevents the sequoia seeds from germinating and also fuels hotter fires. And hotter fires damage the giants. Park rangers now set controlled fires. By periodically removing underbrush, the controlled fires can promote conditions that are favorable for the community's cyclic replacements.

species diversity, and then reaches maturity. Whipping sporadically through the successional pattern, however, are heavy winds that cause treefalls. Where trees fall, gaps open in the forest canopy and more light reaches the forest floor. In that local patch, conditions favor the growth of previously suppressed small trees as well as germination of pioneers or shade-intolerant species.

Or think of the groves of sequoia trees in the Sierra Nevada of California. Some trees in this type of climax community are giants, more than 4,000 years old. Their persistence depends partly on recurring brush fires that sweep through parts of the forests. Sequoia seeds only germinate in the absence of smaller, shade-tolerant plant species. Too much litter on the forest floor inhibits germination. Modest fires eliminate trees and shrubs that compete with young sequoias but do not damage the sequoias themselves. Mature sequoias have thick bark that burns poorly and insulates the living phloem cells of the giant trees against modest heat damage.

At one time, fires were prevented in many sequoia groves in national and state parks—not just accidental fires from campsites and discarded cigarettes, but also natural, lightning-sparked fires. Litter builds up when

Restoration Ecology

Reflect on the introduction to Chapter 29, which described the regrowth after the Mount Saint Helens eruption during 1980, and you already know that secondary succession can heal disturbances on a massive scale. Such *natural* restoration of communities often takes a long time. Today we also see *active* restoration: efforts to reestablish biodiversity in key areas adversely affected or lost by agriculture and other human activities. Chapter 28 describes the urgency associated with this work. For now, simply reflect on the following examples. Ecologists and volunteers are building artificial reefs along coasts, and repairing wetlands and grasslands. In 1972, "lost" species of natural prairie communities were identified in old cemeteries and other neglected patches of land in Illinois. Those species were transplanted by hand to a common, shared site, which now measures 180 hectares (445 acres). Volunteers maintain this small restored patch of prairie, as with controlled burns and hand weeding. Their goal is to find all of the 150–200 known species of the original community.

Community structure is an outcome of a balance of forces, including predation and competition, operating over time.

A climax community is a stable, self-perpetuating array of species in equilibrium with one another and their habitat.

Similar climax stages can persist along gradients dictated by environmental factors and by species interactions. Also, recurring, small-scale changes help shape many communities.

Natural or deliberate ecological restoration often can repair a damaged climax community, provided that suitable species are available to reinstate the original biodiversity.

48.8 COMMUNITY INSTABILITY

The preceding sections might lead you to believe that all communities become stabilized in predictable ways. But this is not always the case. *Community stability is an outcome of forces that have come into uneasy balance.* Resources are sustained, as long as populations do not flirt dangerously with the carrying capacity. Predators and prey coexist, as long as neither wins. Competitors have no sense of fair play. Mutualists are stingy, as when plants produce as little nectar as necessary to attract pollinators, and pollinators take as much nectar as they can for the least effort.

In the short term, disturbances can hurt the growth of some populations. You saw one example of this in Section 48.3. Also, long-term changes in climate or some other environmental variable often have destabilizing effects. If the instability is great enough, a community might change in ways that will persist even when the disturbance ends or is reversed. If some of its member species happen to be rare or don't compete well with others, they might become extinct.

How Keystone Species Tip the Balance

The uneasy balancing of forces in a community comes into sharp focus with studies of a **keystone species**, a dominant species that can dictate community structure. Robert Paine identified a keystone species by removal experiments in intertidal zones along North America's west coast. Pounding surf, tides, and storms continually disturb these zones, and living space is scarce. In his control plots, Paine included a sea star (*Pisaster*) and its invertebrate prey. The sea star proved to be a keystone species; it controls the abundances of resident mussels (*Mytilus*), limpets, chitons, and assorted barnacles.

After Paine removed all sea stars from experimental plots, mussels took over and crowded out seven other invertebrate species. Mussels are the main prey of sea stars but are the strongest competitors when sea stars are absent. Predation (by sea stars) normally maintains the diversity of prey species by preventing competitive exclusion (by mussels). Remove the sea stars, and the community shrinks from fifteen species to eight.

Or consider periwinkles (*Littorina littorea*), an alga-eating snail of intertidal zones. Jane Lubchenco found that periwinkles can increase *or* decrease the diversity of algal species in different settings. In tidepools, they eat a dominant algal species (*Enteromorpha*). By doing so, they help many other, less competitive algal species survive. By contrast, on rocks exposed only at high tide, *Chondrus* and other red algae predominate. Periwinkles leave these tough, unpalatable species alone—and eat the competitively weak algal species that they pass up

a
b
c

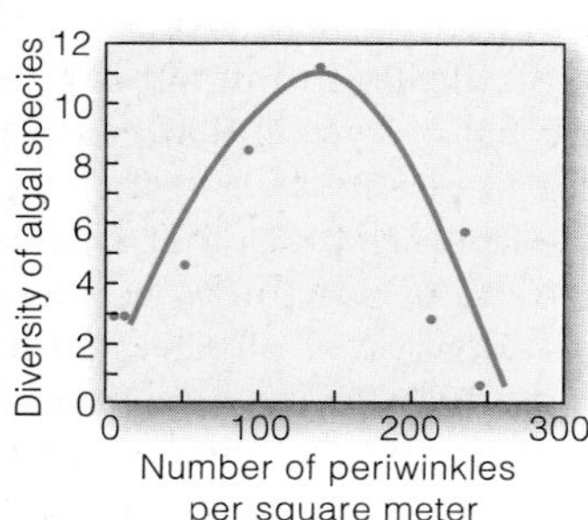
Diversity of algal species
12
10
8
6
4
2
0
0
100
200
300
Number of periwinkles per square meter

d Algal diversity in tidepools

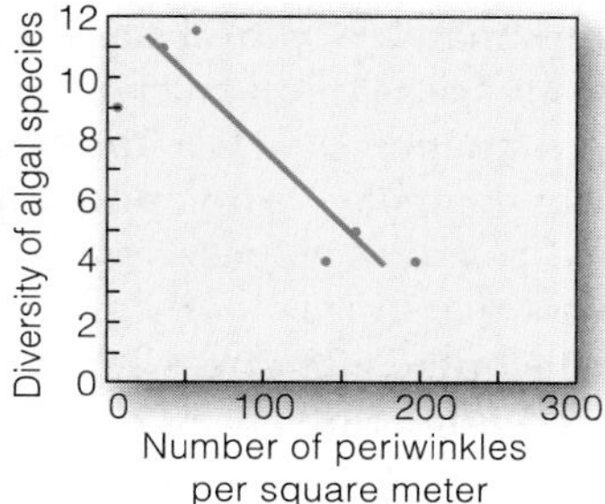
Diversity of algal species
12
10
8
6
4
2
0
0
100
200
300
Number of periwinkles per square meter

e Algal diversity on rocks that are alternately exposed and submerged

Figure 48.16 Effect of competition and predation on community structure. (**a**) Periwinkles (*Littorina littorea*) influence the number of algal species in different ways in different marine habitats. (**b**) *Chondrus* and (**c**) *Enteromorpha*, two algae in their natural habitat. (**d**) In tidepools, periwinkles graze on a dominant alga (*Enteromorpha*). Thus less competitive algae that might otherwise be overwhelmed have an advantage. (**e**) Algal diversity is lower on rocks exposed at low tide; periwinkles do not eat the dominant species (*Chondrus* and other red algae).

Further reading: Student Guide to InfoTrac on web site →

Species Introduced	Origin	Mode of Introduction	Outcome
Table 48.2 Detrimental Effects of Some Species Introduced Into the United States			
Water hyacinth	South America	Intentionally introduced (1884)	Clogged waterways; shading out of other vegetation
Dutch elm disease: *Ophiostoma ulmi* (the fungal pathogen)	Europe	Accidentally imported on infected elm timber (1930)	Destruction of millions of elms; great disruption of forest ecology
Bark beetle (carrier of the pathogen)		Accidentally imported on unbarked elm timber (1909)	
Chestnut blight fungus	Asia	Accidentally imported on nursery plants (1900)	Destruction of nearly all eastern American chestnuts; disruption of forest ecology
Argentine fire ant	Argentina	In coffee shipments from Brazil? (1891)	Crop damage; destruction of native ant communities; death of ground-nesting birds
Japanese beetle	Japan	Accidentally imported on irises or azaleas (1911)	Defoliation of more than 250 plant species, including commercially important species such as citrus
Sea lamprey	North Atlantic Ocean	Through Erie Canal (1860s), then through Welland Canal (1921)	Destruction of lake trout and lake whitefish in the Great Lakes
European starling	Europe	Released intentionally in New York City (1890)	Competition with native songbirds; crop damage; transmission of swine diseases; airport runway interference; very noisy and messy in large flocks
House sparrow	England	Released intentionally (1853)	Crop damage; displacement of native songbirds; transmission of some diseases

in tidepools. Thus, periwinkles increase the diversity of algae in tidepools and yet reduce it on rocks that are recurringly exposed (Figure 48.16).

How Species Introductions Tip the Balance

Finally, consider what can happen when some residents of established communities move out from their home range and successfully take up residence elsewhere. We call such a directional movement **geographic dispersal**. It can proceed in three ways.

First, over a number of generations, a population might expand its home range by slowly moving into outlying regions that prove hospitable. Second, some individuals might be rapidly transported across great distances. This is called **jump dispersal**. It often takes individuals across regions where they could not survive on their own, as when insects travel from the mainland to Maui in a ship's cargo hold. Third, with imperceptible slowness, a population might be moved away from its home range over geologic time, as by continental drift.

The dispersal and colonization of vacant places can be amazingly rapid as well as successful. Consider one of Amy Schoener's experiments in the Bahamas. She set out plastic sponges on the barren, sandy floor of Bimini Lagoon. How fast did aquatic species take up residence in chambers of the artificial hotels? Schoener recorded occupancy by 220 species within thirty days.

Or think of the 4,500 or so nonindigenous species that successfully established themselves in the United States following jump dispersal. These are just the ones we know about. Some, including soybeans, rice, wheat, corn, and potatoes, have been put to good use as food resources. Most adversely impact natural communities and farmlands (Table 48.2 and Section 48.9).

If you live in the northeastern United States, you know about imported gypsy moths. They escaped from a research facility near Boston in 1869, and today their descendants defoliate whole forests. You might know about zebra mussels, which probably entered the Great Lakes on a cargo ship's hull. They attach to and block the water intake pipes for cities. The estimated damage exceeds 5 billion dollars. And *Cryphonectria parasitica*, a fungus introduced to North America almost a century ago, killed nearly all American chestnut trees.

If you live in the southeastern United States, you know about the hot sting of fire ants (*Solenopsis invicta*). Nearly a half century ago, the ants were accidentally imported from South America, possibly in the hold of a ship that docked in Mobile, Alabama. Each year, fire ants inflict enough multiple bites to make 70,000 or so people seek medical attention.

And remember those African bees released in South America? As you read in Chapter 8, their Africanized, "killer bee" descendants traveled northward, through Mexico and on into the United States. Swarms have already killed several hundred people in Latin America. At this writing, these aggressive bees have attacked 140 Americans, two of whom died from multiple stings.

Long-term shifts in climate, the rapid introduction and successful establishment of a new species, and many other disturbances can permanently alter community structure.

48.9 EXOTIC AND ENDANGERED SPECIES

When you hear someone bubbling enthusiastically about an **exotic species,** you can safely bet the speaker is not an ecologist. This is a name for a resident of an established community that has moved from its home range and successfully taken up residence elsewhere. It makes no difference whether the importation was deliberate or accidental. Unlike most imports, which cannot take hold outside their home range, an exotic species insinuates itself into the new community.

Sometimes the additions are harmless and even have beneficial effects. More often, they make native species **endangered species**, which by definition are extremely vulnerable to extinction (Section 28.2). Of all species that are now on rare or endangered lists or have already become extinct, *close to 70 percent owe their precarious existence or demise to displacement by exotic species*.

HELLO VICTORIA, GOOD-BYE CICHLIDS Finding better ways to manage our food supplies is essential, given the astounding growth rate of the human population. Such efforts are well intentioned, but they can have disastrous consequences when ecological principles are not taken into account. For example, several years ago, someone thought it would be a great idea to introduce the Nile perch into Lake Victoria in East Africa. People had been using simple, traditional methods of fishing there for thousands of years (Figure 48.17). Now they were taking too many fish. Soon there would be too few fish to feed the local populations and no excess catches to sell for profit. But Lake Victoria is a very big lake, and the Nile perch is a very big fish (more than two meters long). A big fish in a big lake seemed like an ideal combination to attract commercial fishermen with big, elaborate nets from the outside world—right? Wrong.

Native fishermen had been harvesting native fishes called cichlids, which eat mostly detritus and aquatic plants. The Nile perch eats other fish—including cichlids. Having had no prior evolutionary experience with the new predator, the 200 coexisting species of cichlids that were native to Lake Victoria had no defenses against it.

And so the Nile perch ate its way through the cichlid populations and destroyed the lake's biodiversity. Dozens of cichlid species found nowhere else are extinct. Without the cichlids to clean up the lake bottom, levels of dissolved oxygen plummeted and contributed to frequent fish kills. By 1990, fishermen were catching mostly Nile perch. Now there are signs that the Nile perch population is about to crash. By destroying its food source, the Nile perch has undercut its own population growth. It has ceased to be a potentially large, exploitable food source for the people who live around the lake.

As if that weren't enough, the Nile perch is an oily fish. Unlike cichlids, which can be sun dried, the Nile perch must be preserved by smoking—and smoking requires firewood. Local people started cutting down more trees in the local forests, and trees are not rapidly renewable resources. To add insult to injury, the people living near Lake Victoria never liked to eat Nile perch anyway. They prefer the flavor and texture of cichlids.

What is the lesson? A little knowledge and some simple experiments in a contained setting could have prevented the whole mess at Lake Victoria.

THE RABBITS THAT ATE AUSTRALIA During the 1800s, British settlers in Australia just couldn't bond with koalas and kangaroos, so they started to import familiar animals from their homeland. In 1859, in what would be the start of a wholesale disaster, a landowner in northern Australia imported and released two dozen wild European rabbits (*Oryctolagus cuniculus*). Good food and good sport hunting, that was the idea. An ideal rabbit habitat with no natural predators—that was the reality.

Six years later, the landowner had killed 20,000 rabbits and was besieged by 20,000 more. The rabbits displaced livestock, even kangaroos. Now Australia has 200 to 300 million hippityhopping through the southern half of the country. They overgraze perennial grasses in good times and strip bark from shrubs and trees during droughts. You know where they've been; they transform grasslands and shrublands into eroded deserts (Figure 48.18). They have been shot and poisoned. Their warrens have been plowed under, fumigated, and dynamited. Even when all-out assaults reduced their population size by 70 percent, the rapidly reproducing imports made a comeback in less than a year. Did the construction of a 2,000-mile-long fence protect western Australia? No. Rabbits made it to the other side before workers completed the fence.

Figure 48.17 Native fishermen at the shore of Lake Victoria, East Africa.

Further reading: Student Guide to InfoTrac on web site –

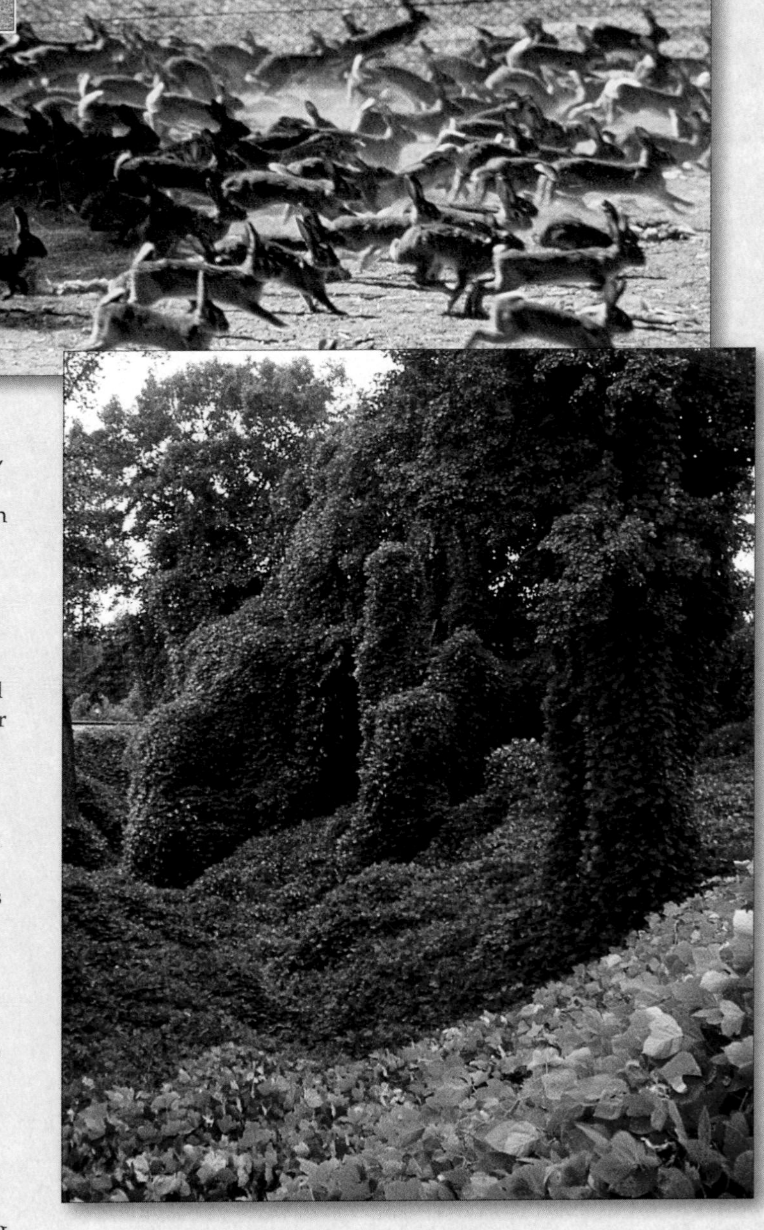

Figure 48.18 Part of the fence erected to keep out 200–300 million rabbits that are munching their way through Australia's vegetation.

In 1951, government researchers introduced myxoma virus by way of mildly infected South American rabbits, its normal hosts. This virus causes *myxomatosis*. The disease has mild effects on the South American rabbits that coevolved with the virus but nearly always had lethal effects on *O. cuniculus*. Biting insects, mainly mosquitoes and fleas, quickly transmit the virus from host to host. Having no coevolved defenses against the novel virus, the European rabbits died in droves. As you might expect, natural selection has since favored rapid growth of populations of *O. cuniculus* that are resistant to the virus.

In 1991, on an uninhabited island in Spencer Gulf, Australian researchers released a population of rabbits that they had injected with a calicivirus. The rabbits died quickly and relatively painlessly from blood clots in their lungs, heart, and kidneys. In 1995, the test virus escaped from the island, possibly on insect vectors. It has been killing 80 to 95 percent of the adult rabbits in Australian regions. At this writing, researchers are now questioning whether the calicivirus should be used on a widespread scale, whether it can jump boundaries and infect animals other than rabbits (such as humans), and what the long-term consequences will be.

THE PLANTS THAT ATE GEORGIA A vine called kudzu (*Pueraria lobata*) was deliberately imported from Japan to the United States, where it faces no serious threats from herbivores, pathogens, or competitor plants. In temperate parts of Asia, kudzu is a well-behaved legume with a well-developed root system. It *seemed* like a good idea to import it for erosion control on hills and near highways in the southeastern United States. However, with nothing to stop it, kudzu's shoots can grow one-third of a meter per day. Vines now blanket streambanks, trees, telephone poles, houses, hills, and almost everything else in their path (Figure 48.19). Attempts to dig them up or burn them are futile. Grazing goats and herbicides help, but goats are indiscriminate eaters and herbicides contaminate water supplies. If the global temperature continues to rise, kudzu could reach the Great Lakes by the year 2040.

On the bright side, a Japanese firm is constructing a kudzu farm and processing plant in Alabama. Asians use a starch extract from kudzu in beverages, candy, and herbal medicines. The idea is to export the starch to Asia, where the demand currently exceeds the supply. (And perhaps this gives new meaning to the expression, What comes around goes around.) Also, kudzu might eventually help reduce the extent of logging operations. At the Georgia Institute of Technology, researchers have reported that kudzu may be useful as an alternative source of paper.

Figure 48.19 Kudzu (*Pueraria lobata*) taking over part of Lyman, South Carolina. You can now find kudzu vines from East Texas to Florida and as far north as Pennsylvania.

48.10 PATTERNS OF BIODIVERSITY

As you might well conclude from Chapter 28 and the preceding discussions, biodiversity differs greatly from one habitat to another. Also, as ecologists have found, through many field investigations, biodiversity differs in significant ways between geographic regions.

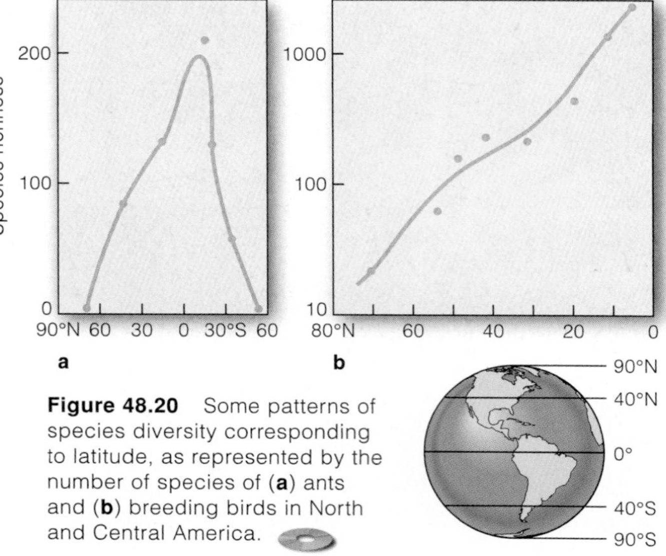

Species richness
200
100
0
90°N 60 30 0 30°S 60
a
1000
100
10
80°N 60 40 20 0
b
90°N
40°N
0°
40°S
90°S

Figure 48.20 Some patterns of species diversity corresponding to latitude, as represented by the number of species of (**a**) ants and (**b**) breeding birds in North and Central America.

Number of recorded colonizing species
70
60
50
40
30
20
10
1965 1967 1969 1971 1973
MOSSES
VASCULAR PLANTS

Figure 48.21 Aerial photograph of Surtsey, a volcanic island, at the time of its formation in the sea. Newly formed, isolated islands are natural laboratories for ecologists. The chart gives the number of colonizing species of mosses and vascular plants recorded on Surtsey between 1965 and 1973.

What Causes Mainland and Marine Patterns?

The most striking pattern of biodiversity corresponds to distance from the equator. For most groups of plants and animals, *the number of coexisting species on land and in the seas is greatest in the tropics, and it systematically declines from the equator to the poles.* Figure 48.20 shows two examples of this biodiversity pattern. What factors underlie it? Three are paramount.

First, as described in Chapter 50, tropical latitudes intercept more sunlight of consistently greater intensity, the rainfall is heavier, and the growing season is longer. As one result, *resource availability tends to be higher and more reliable in the tropics than elsewhere.* All year long, different tree species in humid tropical forests put out new leaves, flowers, and fruit. Year in and year out, they support diverse herbivores, nectar foragers, and fruit eaters. Such diverse specializations cannot evolve to a comparable degree in temperate and arctic regions.

Second, *species diversity might be self-reinforcing.* The diversity of tree species in tropical forests is far greater than in comparable forests at higher latitudes. When more plant species compete and coexist, more herbivore species also evolve and coexist, partly because no single herbivore can overcome the chemical defenses of all of the different plants. Typically, then, more predators and parasites evolve in response to the diversity of prey and hosts. The same applies to diversity on tropical reefs.

Third, evolutionary history tells us this: *The rates of speciation in the tropics have exceeded those of background extinction.* Decreases in biodiversity have occurred, but mainly during mass extinctions. Global temperatures drop following such episodes. Species adapted to cool climates tend to survive, but the ones adapted to the temperatures of the tropics do not. Also, millions of species in tropical forests may disappear within the next decade, as described in Chapters 28 and 51.

What Causes Island Patterns?

Islands often serve as terrific laboratories for studying biodiversity. For example, in 1965, a volcanic eruption formed a new island southwest of Iceland. Within six months, bacteria, fungi, seeds, flies, and some seabirds were established on it. One vascular plant appeared after two years, and a moss two years after that (Figure 48.21). As soils improved, the number of plant species increased. All the colonists originated in Iceland. None

Further reading: Student Guide to InfoTrac on web site —

a

c

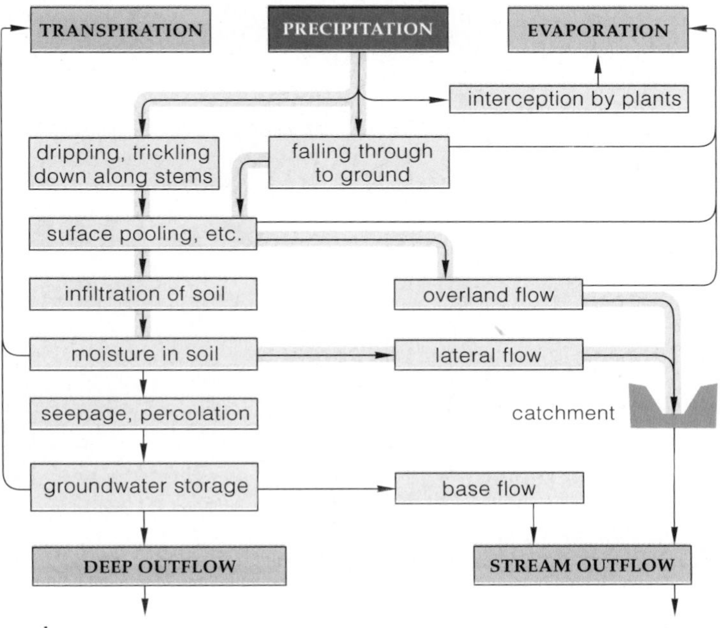
TRANSPIRATION
PRECIPITATION
EVAPORATION
interception by plants
dripping, trickling down along stems
falling through to ground
suface pooling, etc.
infiltration of soil
overland flow
moisture in soil
lateral flow
seepage, percolation
catchment
groundwater storage
base flow
DEEP OUTFLOW
STREAM OUTFLOW
b

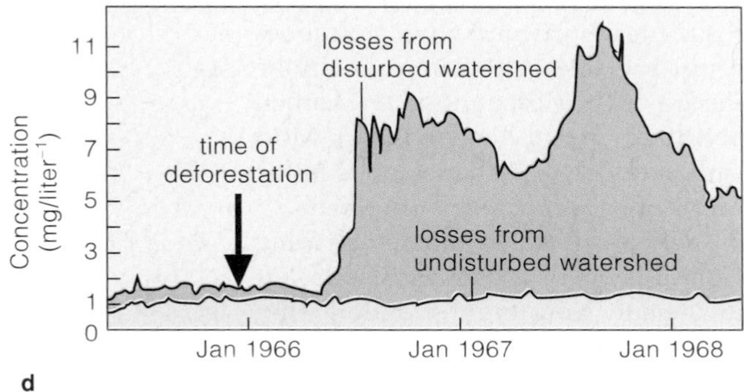
losses from disturbed watershed
time of deforestation
losses from undisturbed watershed
Concentration (mg/liter⁻¹)
0
1
3
5
7
9
11
Jan 1966
Jan 1967
Jan 1968
d

Figure 49.16 Results from experiments involving deliberate disturbances to forests of experimental watersheds in Hubbard Brook Valley (**a**). (**b**) In this model of the movement of water through watersheds, the *yellow* lines indicate the flow of water that researchers monitored during these experiments.

In this model, *dark blue* indicates inputs to the watershed; *medium blue*, distribution within the watershed; and *light blue*, outputs from it. Transpiration, remember, is the evaporation of water from leaves and other plant parts that are exposed to air. The plants tap water in the soil and in groundwater stores.

Researchers stripped the vegetation cover in forested areas but did not disturb the soil. They applied herbicides to the soil for three years to prevent regrowth. All of the surface water draining from each watershed was measured as it flowed over V-shaped concrete catchments, as in (**c**). The arrow in (**d**) points out the time of deforestation. The concentrations of calcium and other minerals were compared against those in water passing over a control catchment in an undisturbed area. Calcium losses were *six times* greater from the deforested area.

Most water entering a watershed seeps into soil or becomes part of surface runoff, which enters streams. Plants withdraw water and its dissolved minerals from the soil, then lose it by transpiration (Figure 49.16*b*).

Measurements of watershed inputs and outputs have many practical applications. For instance, cities that draw upon surface water supplies of watersheds adjust their usage in compliance with seasonal shifts in water volume. Measurements also reveal the extent to which vegetation cover influences the movement of nutrients through the ecosystem phase of biogeochemical cycles.

For example, you might think that surface runoff in a watershed would swiftly leach calcium ions and other minerals. But in young, undisturbed forests in Hubbard Brook watersheds, each hectare lost only 8 kilograms or so of calcium. Also, rainfall and weathering of rocks were bringing in calcium replacements. And tree roots were "mining" the soil, so calcium was being stored in a growing biomass of tree tissues.

In experimental watersheds in the Hubbard Brook Valley, deforestation caused a shift in nutrient outputs. The results are shown in Figure 49.16*d*. The results are sobering. Calcium and other nutrients cycle so slowly that deforestation may disrupt nutrient availability for entire ecosystems. This is especially so for forests that cannot regenerate themselves in the short term. Tropical rain forests and northern coniferous forests are like this.

In the hydrologic cycle, water slowly moves on a global scale from the oceans (the main reservoir), through the atmosphere, onto land, then back to the oceans.

In ecosystems on land, plants stabilize the soil and absorb dissolved minerals. By doing so, they minimize the loss of soil nutrients in runoff from land.

49.8 CARBON CYCLE

In a vital atmospheric cycle, carbon moves through the atmosphere and food webs on its way to and from the ocean, sediments, and rocks (Figure 49.17). Its global movement is called the **carbon cycle**. Sediments and rocks hold most of the carbon, followed by the ocean, soil, atmosphere, and land biomass. It enters the atmosphere as cells engage in aerobic respiration, as fossil fuels burn, and when volcanoes erupt and release it from rocks in the Earth's crust. Most atmospheric carbon occurs as carbon dioxide (CO_2). Most carbon dissolved in the ocean is in the forms of bicarbonate and carbonate.

When you watch bubbles escaping from a glass of carbonated soda left in the sun, you might wonder: Why doesn't all the CO_2 dissolved in warm ocean surface waters escape to the atmosphere? Driven by winds and regional differences in water density, water makes a gigantic loop from the surface of the Pacific and Atlantic oceans to the Atlantic and Antarctic seafloors. There its CO_2 moves into deep storage reservoirs before water loops upward (Figure 49.18). This is a factor in carbon's distribution and global budget, as listed below:

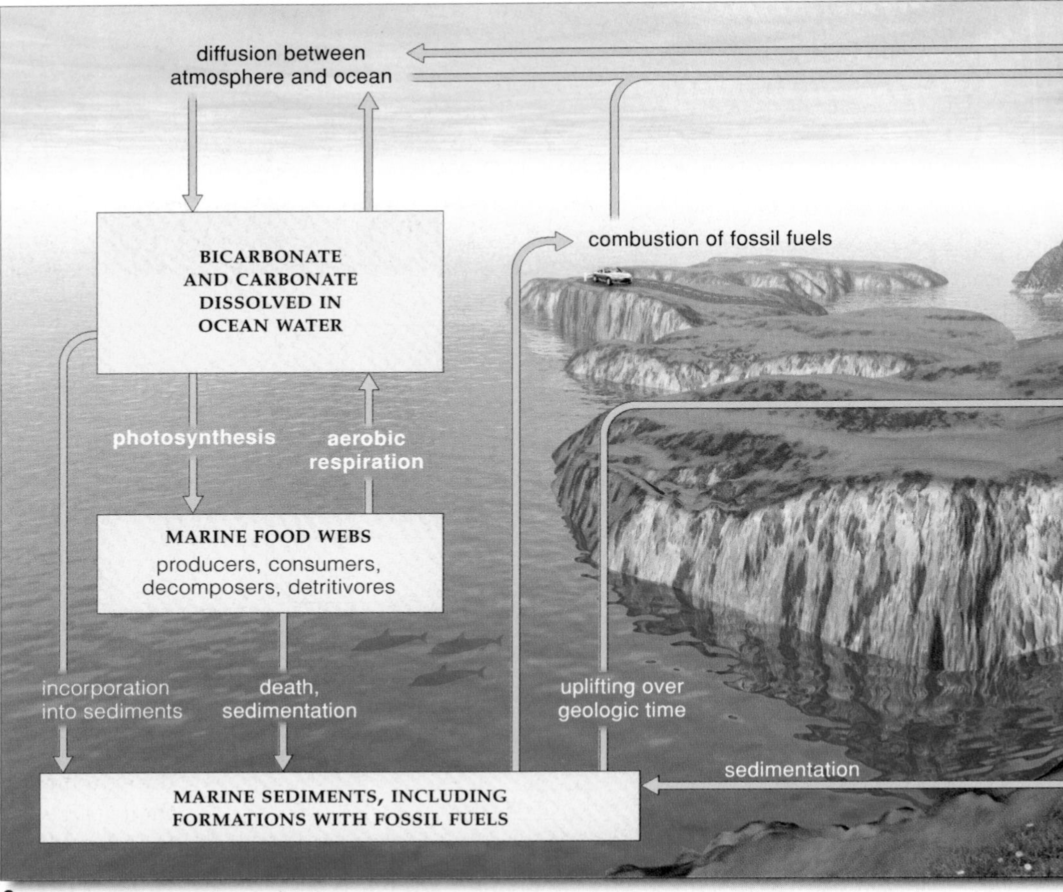

diffusion between atmosphere and ocean
BICARBONATE AND CARBONATE DISSOLVED IN OCEAN WATER
combustion of fossil fuels
photosynthesis
aerobic respiration
MARINE FOOD WEBS
producers, consumers, decomposers, detritivores
incorporation into sediments
death, sedimentation
uplifting over geologic time
sedimentation
MARINE SEDIMENTS, INCLUDING FORMATIONS WITH FOSSIL FUELS
a

MAIN CARBON RESERVOIRS AND HOLDING STATIONS:

Sediments and rocks	77,000,000 (10^{15} grams)
Ocean (dissolved forms)	39,700
Soil	1,500
Atmosphere	750
Biomass on land	715

ANNUAL FLUXES IN GLOBAL DISTRIBUTION OF CARBON:

From atmosphere to plants (carbon fixation)	120
From atmosphere to ocean	107
To atmosphere from ocean	105
To atmosphere from plants	60
To atmosphere from soil	60
To atmosphere from fossil fuel burning	5
To atmosphere from net destruction of plants	2
To ocean from runoff	0.4
Burial in ocean sediments	0.1

b

Figure 49.17 (**a**) Global carbon cycle. The part of the diagram on this page shows carbon's movement through typical marine ecosystems. The *facing page* shows carbon's movement through ecosystems on land. *Gold* boxes indicate key reservoirs.

(**b**) Los Angeles freeway under its self-generated blanket of smog at twilight. Every day, fossil fuel burning by vehicles, industries, and homes puts carbon and other substances into the atmosphere.

When photosynthetic autotrophs engage in carbon dioxide fixation, they lock up billions of metric tons of carbon atoms in organic compounds each year (Sections 7.9 and 49.4). Yet the average time that an ecosystem holds a given carbon atom varies greatly. For example, organic wastes and remains decompose so rapidly in tropical rain forests that not much carbon accumulates

Further reading: Student Guide to InfoTrac on web site →

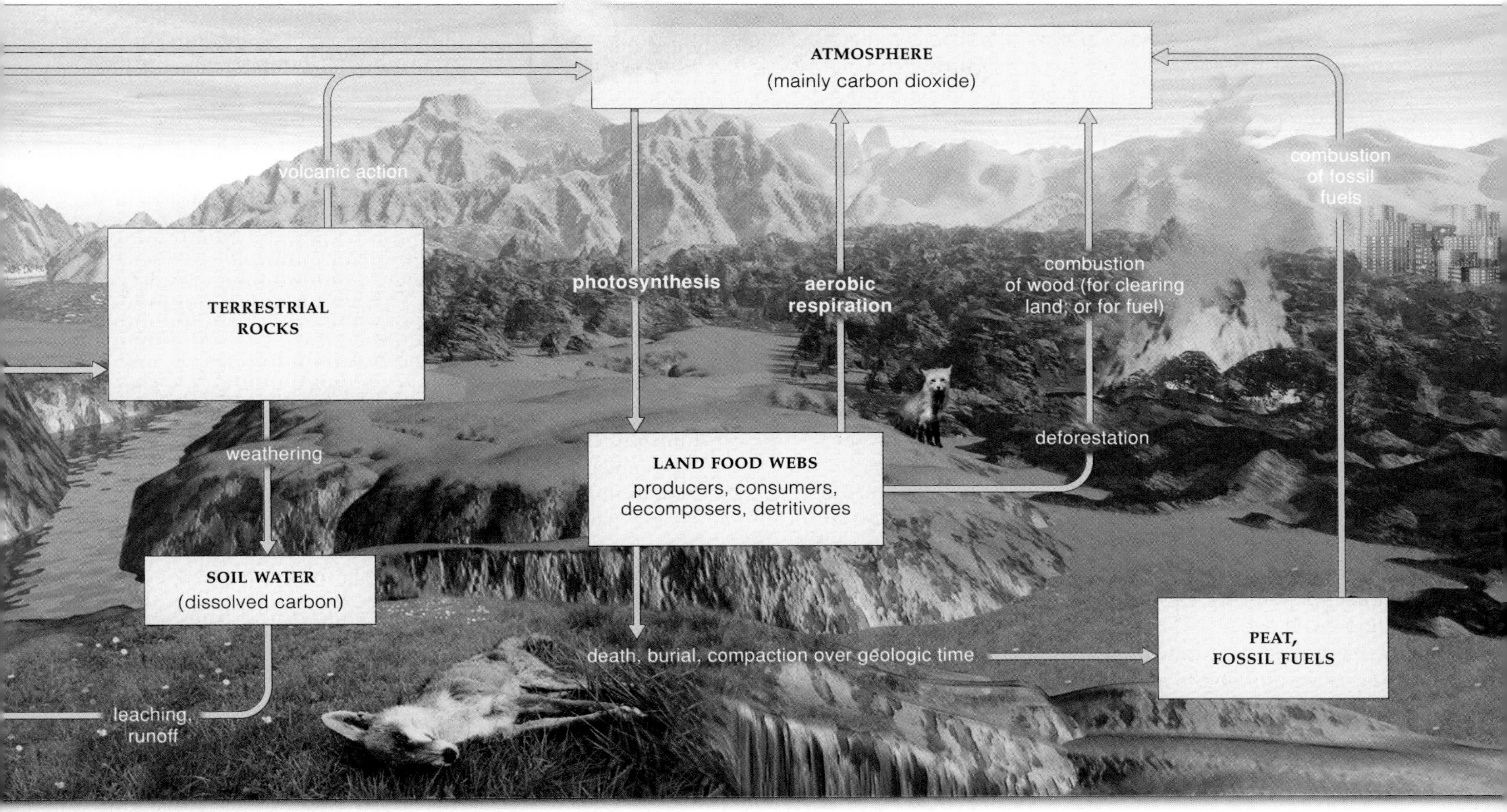

ATMOSPHERE
(mainly carbon dioxide)
volcanic action
combustion of fossil fuels
TERRESTRIAL ROCKS
photosynthesis
aerobic respiration
combustion of wood (for clearing land; or for fuel)
deforestation
weathering
LAND FOOD WEBS
producers, consumers, decomposers, detritivores
SOIL WATER
(dissolved carbon)
death, burial, compaction over geologic time
PEAT, FOSSIL FUELS
leaching, runoff

Warm, less salty, shallow current
Cold, salty, deep current

Figure 49.18 Loop of ocean water that delivers carbon dioxide to its deep ocean reservoir. It sinks in the cold, salty North Atlantic and rises in the warmer Pacific.

at soil surfaces. In bogs, marshes, and other anaerobic sites, decomposers cannot degrade organic compounds to smaller bits, so carbon gradually accumulates in peat and other forms of compressed organic matter. Also, in food webs of ancient aquatic ecosystems, carbon was incorporated in staggering numbers of shells and other hard parts. Shelled organisms died, sank through water, then were buried in sediments. The carbon remained buried for many millions of years in deep sediments until part of the seafloor was uplifted above the ocean surface by geologic forces (Sections 20.10 and 23.3).

As a final example, plants of the vast swamp forests of the Carboniferous also incorporated carbon atoms in their organic compounds (Section 25.4). In time, those compounds became converted to petroleum, coal, and gas reserves—which we now tap as fossil fuels. Fossil fuel burning and other human activities are currently putting more carbon into the atmosphere than can be cycled naturally to the ocean reservoirs. As you might deduce from the preceding page, the ocean removes only about 2 percent of the excess carbon that enters the air. The excess is amplifying the greenhouse effect, which means it may be contributing to global warming. The next section describes this effect and some possible outcomes of modifications to it.

The ocean and atmosphere interact in the global cycling of carbon. Fossil fuel burning and other human activities may be contributing to imbalances in the global carbon budget.

49.9 FROM GREENHOUSE GASES TO A WARMER PLANET?

GREENHOUSE EFFECT Atmospheric concentrations of gaseous molecules play a profound role in shaping the average temperature near the surface of the Earth. That temperature has enormous effects on the global climate.

Countless molecules of carbon dioxide, water, ozone, methane, nitrous oxide, and chlorofluorocarbons are key players in interactions that dictate the global temperature. Collectively, the gases act somewhat like a pane of glass in a greenhouse—hence their name, "greenhouse gases." Wavelengths of visible light can pass around them and reach the Earth's surface. But greenhouse gases impede the escape of longer, infrared wavelengths—heat—from the Earth into space. How? Gaseous molecules can absorb these wavelengths and radiate much of the absorbed energy back toward the Earth, as shown in Figure 49.19.

The constant radiation of heat energy from greenhouse gases toward Earth proceeds lockstep with the constant bombardment and absorption of wavelengths from the sun, and so heat builds up in the lower atmosphere. The **greenhouse effect** is the name for this warming action.

GLOBAL WARMING DEFINED Without the action of greenhouse gases, the Earth's surface would be cold and lifeless. However, there can be too much of a good thing. Largely as an outcome of human activities, greenhouse gases are building to atmospheric levels that are higher than they were in the past. Figure 49.20 documents the increase. As many researchers suspect, greenhouse gases may be contributing to long-term higher temperatures at the Earth's surface, an effect called **global warming**.

What is so alarming about a warmer planet? Suppose the temperature of the lower atmosphere were to rise by only 4°C (7°F). The increase might cause sea levels to rise by about 0.6 meter (2 feet). Why? Temperatures near the ocean surface would increase—and water expands when heated. Also, global warming could make glaciers and the polar ice sheets melt faster (see *Critical Thinking* question 5 in Section 49.12). The volume of water released in this way alone would flood low coastal regions.

Imagine a long-term rise in sea level combined with high tides and storm waves. The waterfronts of Vancouver, Seattle, San Diego, New York, Boston, Galveston, Hilo, and all other cities perched along the rim of the world's ocean would be submerged. So would the agricultural lowlands and deltas in India, China, and Bangladesh—where much of the world's rice is grown. Huge tracts of Florida and Louisiana would face saltwater intrusions.

And what if global warming disturbed the regional patterns of precipitation and temperature? We might predict deserts will expand and the interiors of the great continents will become much drier than they already are. Will the nations that control the world's great grain belts be affected? Will other nations be better or worse off? Imagine the economic and political consequences.

Some predicted effects of climate change are emerging. For example, in 1996, we found that the water 200 meters below the Arctic ice cap is 1°C warmer than it was just five years ago. If the rapid warming continues, the ice cap may disappear in the next century. The rise in temperature may already be disrupting the loop of ocean water that extends like a gigantic conveyor belt from the surface to the bottom of the great oceans (Section 49.8). In addition, during the 1990s, record-breaking hurricanes, flooding, and prolonged droughts had devastating effects on crop production around the world. Sandy beaches have been diminishing by 0.6 meter or more per year.

EVIDENCE OF AN INTENSIFIED GREENHOUSE EFFECT In the 1950s, researchers on Hawaii's highest volcano started measuring the atmospheric concentrations of different greenhouse gases. They chose that remote site because it is almost free of local airborne contamination and is representative of conditions for the Northern Hemisphere. Monitoring activities continue.

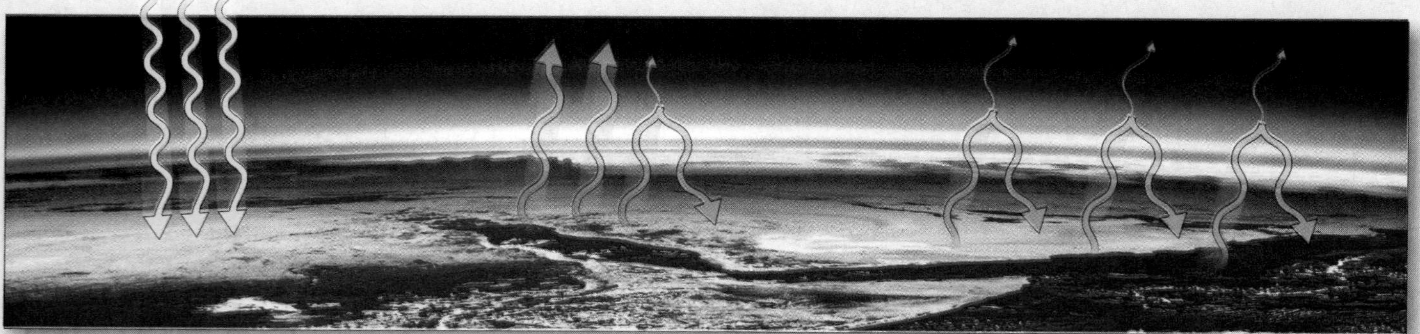

a Rays of sunlight penetrate the lower atmosphere and warm the Earth's surface.

b The surface radiates heat (infrared wavelengths) to the lower atmosphere. Some heat escapes into space. But greenhouse gases and water vapor absorb some infrared energy and radiate a portion of it back toward Earth.

c Increased concentrations of greenhouse gases trap more heat near Earth's surface. Sea surface temperature rises, more water evaporates into atmosphere, and Earth's surface temperature rises.

Figure 49.19 The greenhouse effect, as executed mainly by carbon dioxide, water, ozone, methane, nitrous oxide, and chlorofluorocarbons in the atmosphere.

Further reading: Student Guide to InfoTrac on web site

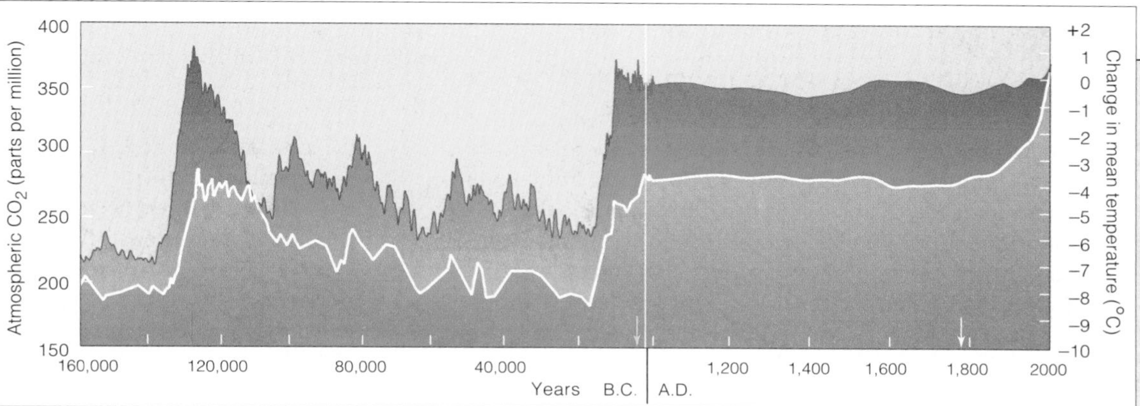
Atmospheric CO_2 (parts per million)
400
350
300
250
200
150
160,000
120,000
80,000
40,000
1,200
1,400
1,600
1,800
2000
Years B.C. | A.D.
Change in mean temperature (°C)
+2
1
0
−1
−2
−3
−4
−5
−6
−7
−8
−9
−10

Figure 49.20 Changes in atmospheric carbon dioxide levels (the *white* graph line) correlated with past ice ages and interglacials. The 1985 mean temperature was used as a baseline for tracking the global temperature changes (*red* graph line). Arrows: *red*, start of last ice age; *yellow*: start of last interglacial; and *white*, start of industrial revolution.

The Chapter 3 introduction outlined what researchers found out. Briefly, in the Northern Hemisphere, carbon dioxide levels follow annual cycles of plant growth. They decline in summer, when photosynthesis rates are highest. Atmospheric levels rise in winter, when photosynthesis declines and aerobic respiration still proceeds.

The troughs and peaks around the graph line in Figure 49.21*a* represent annual lows and highs. For the first time, scientists glimpsed the integrated effects of carbon balances for land and water ecosystems of an entire hemisphere. Notice the midline of the troughs and peaks in the cycle. *It is steadily increasing*. Many take this as evidence of a buildup in carbon dioxide levels. They predict that it will intensify the greenhouse effect over the next century.

Probably global burning of fossil fuels is contributing most to the rise in carbon dioxide levels. Deforestation is adding to it; when wood burns, its carbon is released. In the past four decades especially, vast forested regions have been cleared by deliberate burns, highly mechanized logging, and other practices, at astounding rates (Section 51.4). Also, as plant biomass plummets, global absorption of carbon dioxide in photosynthesis may decline.

Will atmospheric levels of greenhouse gases continue to increase until the middle of the twenty-first century? Will global temperature rise by several degrees? If the warming trend is already in motion, we will not be able to reverse it simply by waiting until the last minute to stop fossil fuel burning and deforestation.

There is widespread agreement among scientists that nations must begin preparing for the consequences. For example, we might increase funding levels for genetic engineering studies to develop drought-resistant and salt-resistant plants. Such plants may prove crucial in regions of saltwater intrusions and climatic change.

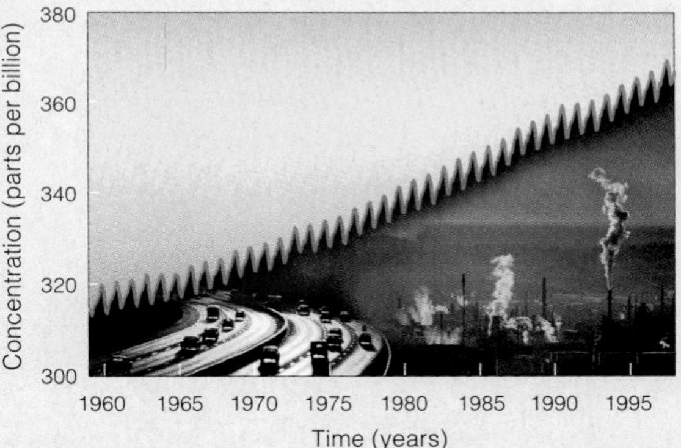
Concentration (parts per billion)
380
360
340
320
300
1960
1965
1970
1975
1980
1985
1990
1995
Time (years)

a CARBON DIOXIDE (CO_2). Of all human activities, the burning of fossil fuels and deforestation contribute the most to increasing atmospheric levels.

Concentration (parts per million)
1.80
1.70
1.60
1.50
1.40
1976
1980
1985
1990
1995
1998
Time (years)

c METHANE (CH_4). Termite activity and anaerobic bacteria living in swamps, landfills, and the stomachs of cattle and other ruminants produce great amounts of methane as a by-product of metabolism.

Concentration (parts per trillion)
1200
1000
800
600
400
200
1976
1980
1985
1990
1995
1998
Time (years)

b CFCS. Until restrictions were in place, CFCs were widely used in plastic foams, refrigerators, air conditioners, and industrial solvents.

Concentration (parts per billion)
320
310
300
290
280
270
260
1976
1980
1985
1990
1995
1998
Time (years)

d NITROUS OXIDE (N_2O). Denitrifying bacteria produce N_2O in metabolism. Also, fertilizers and animal wastes release enormous amounts; this is especially the case for vast livestock feedlots.

Figure 49.21 Recently documented increases in atmospheric concentrations of four greenhouse gases: carbon dioxide, CFCs, methane, and nitrous oxide.

49.10 NITROGEN CYCLE

Since the time of life's origin, the atmosphere and ocean have contained nitrogen. This component of all proteins and nucleic acids moves in an atmospheric cycle called the **nitrogen cycle**. Gaseous nitrogen (N_2) makes up 80 percent or so of the atmosphere, the largest reservoir. Successively smaller reservoirs are seafloor sediments, ocean water, soil, land biomass, atmospheric nitrous oxide (N_2O), and the smallest, marine biomass.

The atoms of N_2 are joined by triple covalent bonds ($N{\equiv}N$). Few organisms have the metabolic means to break them. Only certain bacteria, volcanic action, and lightning convert N_2 into forms that enter food webs.

Of all nutrients required for plant growth, nitrogen often is the scarcest. Nearly all of the nitrogen in soils was put there by nitrogen-fixing organisms. Ecosystems lose it when other bacteria "unfix" the fixed nitrogen. Land ecosystems lose more by leaching, although this is the basis of nitrogen inputs to aquatic ecosystems such as streams, lakes, and the seas (Figure 49.22).

Cycling Processes

Let's follow nitrogen atoms through the portion of the nitrogen cycle that proceeds in an ecosystem on land. They are objects of processes called nitrogen fixation, assimilation and biosynthesis, decomposition, and then ammonification.

In **nitrogen fixation**, a few kinds of bacteria convert N_2 to ammonia (NH_3), which quickly dissolves in the cytoplasm, thus forming ammonium (NH_4^+). In aquatic ecosystems, *Anabaena, Nostoc*, and other cyanobacteria are nitrogen fixers. In many land ecosystems, *Rhizobium* and *Azotobacter* fix nitrogen. Collectively, these bacteria fix about 200 million metric tons of nitrogen each year!

Decomposition and **ammonification** are processes by which bacteria and fungi degrade nitrogenous wastes and remains of organisms. Decomposers use part of the released proteins and amino acids during metabolism. But most of the nitrogen remains in the decay products, in the form of ammonia or ammonium, which plants take up. Nitrifying bacteria also act on the ammonia or ammonium. In **nitrification**, they strip the compounds of electrons, and nitrite (NO_2^-) is the result. Other kinds of bacteria use the nitrite during their metabolism and produce nitrate (NO_3^-), which plants then take up.

Most plants growing in nitrogen-poor soil interact as mutualists with fungi. Their roots form mycorrhizae with the fungal hyphae, which have a very large surface area for absorbing mineral ions from soil. Clover, beans, peas, and other legumes are mutualists with nitrogen-

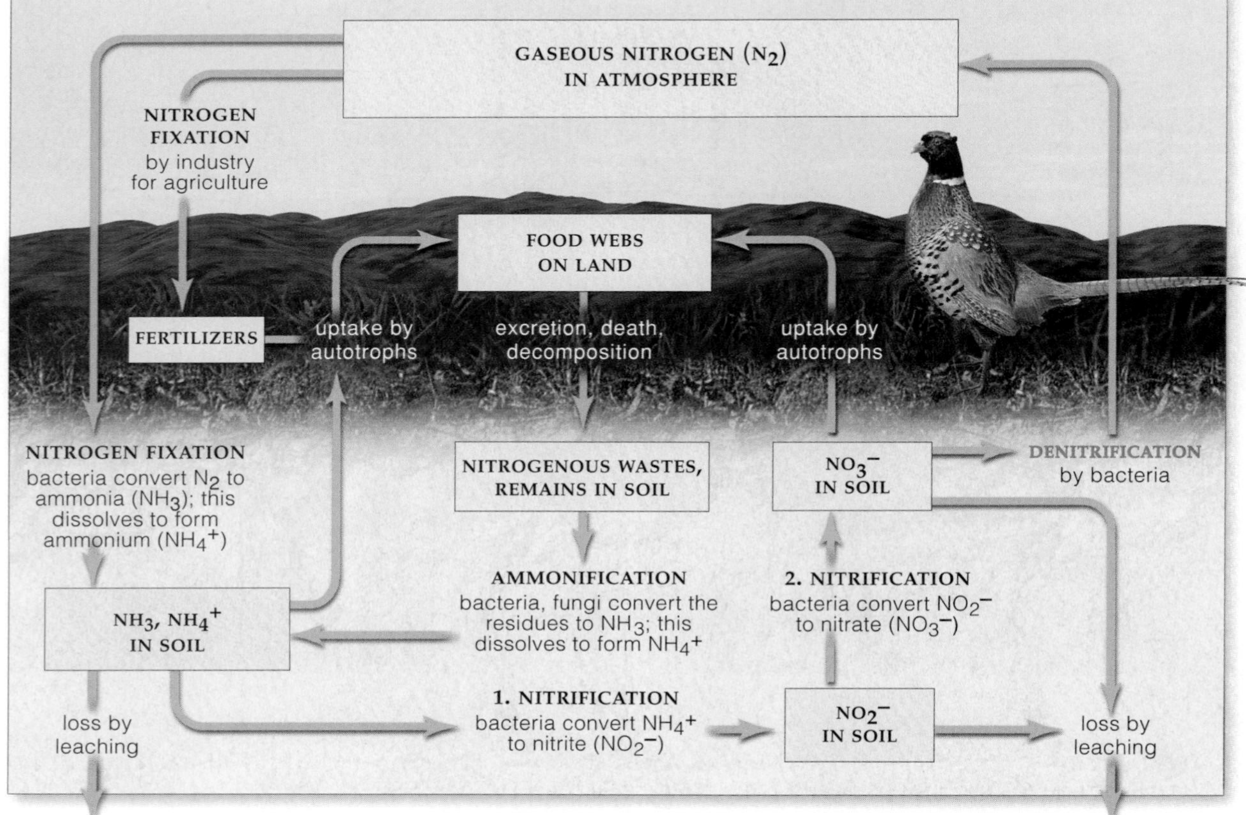

GASEOUS NITROGEN (N_2) IN ATMOSPHERE
NITROGEN FIXATION by industry for agriculture
FERTILIZERS
FOOD WEBS ON LAND
uptake by autotrophs
excretion, death, decomposition
uptake by autotrophs
NITROGEN FIXATION bacteria convert N_2 to ammonia (NH_3); this dissolves to form ammonium (NH_4^+)
NITROGENOUS WASTES, REMAINS IN SOIL
NO_3^- IN SOIL
DENITRIFICATION by bacteria
AMMONIFICATION bacteria, fungi convert the residues to NH_3; this dissolves to form NH_4^+
2. NITRIFICATION bacteria convert NO_2^- to nitrate (NO_3^-)
NH_3, NH_4^+ IN SOIL
1. NITRIFICATION bacteria convert NH_4^+ to nitrite (NO_2^-)
NO_2^- IN SOIL
loss by leaching
loss by leaching

Figure 49.22 The nitrogen cycle in an ecosystem on land. Activities of nitrogen-fixing bacterial species make nitrogen available to plants. Other bacterial species cycle nitrogen to plants when they convert organic wastes to ammonium and nitrates. By far, the atmosphere is the largest nitrogen reservoir (primarily N_2), followed by ocean sediments, ocean water, and soil. Lesser amounts of nitrogen also are present in the atmosphere as nitrous oxide (N_2O) and in the biomass of the seas.

Further reading: Student Guide to InfoTrac on web site

fixing bacteria that form root nodules. Chapter 24 and Section 30.2 describe these interactions. How do plants use the nitrogen provided by bacterial activities? Their cells assimilate nitrogen and channel much of it into synthesizing amino acids, proteins, and nucleic acids. Plant tissues, remember, are the sole nitrogen source for animals, which directly or indirectly feed on plants.

Nitrogen Scarcity

The ammonium, nitrite, and nitrate that form during the nitrogen cycle are vulnerable to leaching and runoff. Leaching, recall, is the removal of some soil nutrients as water percolates through it (Section 30.1). Besides losses from leaching, some nitrogen is lost to the air by **denitrification**. By this process, certain bacteria convert nitrate or nitrite to N_2 and N_2O (nitrous oxide). Usually, most denitrifying bacteria rely on aerobic respiration. In waterlogged or poorly aerated soil, they use anaerobic pathways in which nitrate, nitrite, or N_2O (not oxygen) is the final electron acceptor. Section 8.5 describes this metabolic strategy. During the reactions, fixed nitrogen is converted to N_2, much of which escapes to the air.

Also, nitrogen fixation comes at high metabolic cost to plants that are mutualists with the nitrogen fixers. In exchange for nitrogen, plants give up sugars and other photosynthetic products that depend on investments of ATP and NADPH. Such plants have a competitive edge in nitrogen-poor soil. In nitrogen-rich soil, species that don't pay the metabolic price often displace them.

Human Impact on the Nitrogen Cycle

Humans are having disproportionately harmful effects on the nitrogen cycle, compared to other species. With rampant deforestation and conversion of grasslands for agriculture, losses of nitrogen from soil are huge. Each forest clearing and harvest removes nitrogen (stored in plant tissues). Soil erosion and leaching remove more.

Many farmers counter nitrogen losses by rotating crops; for example, by alternating wheat with legumes. Crop rotation can help keep soil stable and productive. In developed countries, farmers also depend heavily on nitrogen-rich fertilizers. They even select new strains of crop plants with a greater capacity to take up fertilizers. Through such practices, the crop yields per hectare have doubled and even quadrupled over the past forty years.

But heavy fertilizer applications alter soil water and can harm plants. Remember, the surface charges of soil particles attract the ions that weathering, irrigation, and fertilizers introduce to soils. By a pH-dependent process called **ion exchange**, ions dissociate from soil particles, and other ions dissolved in soil water replace them. Any change in the composition of soil water disrupts the number and kinds of ions in exchangeable form—which are the kinds that plants take up. Calcium and magnesium are the most abundant exchangeable ions. When nitrogen fertilizers enter soil, ammonium ions also assume importance. This is especially the case for ammonium sulfate $(NH_4)_2SO_4$, for all of its nitrogen is in the form of ammonium. Heavy applications of such a fertilizer greatly increase the acidity of soil water. Its many hydrogen ions compete for binding sites on soil particles. As one outcome, the calcium and magnesium ions necessary for plant growth decline in number.

Figure 49.23 Dead and dying spruce trees in a forest in Germany. They are cited as casualties of extremely high concentrations of nitrogen oxides and other forms of air pollution.

Humans introduce excess nitrogen into ecosystems in other ways. For example, nitrogen-rich sewage enters rivers, lakes, and estuaries. Section 51.7 describes the effects. As another example, fossil fuel burning by power plants and vehicles having combustion engines releases NO_2 and other oxides of nitrogen into the atmosphere (Section 51.1). Winds carry these air pollutants to forests at high elevations and high northern latitudes, which have nitrogen-poor soils. Nitrogen is a limiting factor on tree growth, and trees of these forests are adapted to storing and cycling it. With more nitrogen, they grow more rapidly, well into the summer. But the late growth is vulnerable to autumn frost and dies during winter. In addition, their nitrogen-stimulated growth outstrips the availability of phosphorus and other nutrients, so the trees suffer the effects of nutrient deficiencies. Their needles yellow and drop, so the rates of photosynthesis decline. The decline has repercussions for growth of all plant parts. The weakened trees are more susceptible to disease and other environmental stresses (Figure 49.23).

The nitrogen cycle turns on processes of nitrogen fixation, assimilation and biosynthesis by all organisms, and on decomposition. Its natural cycling depends on the activity of nitrogen-fixing bacteria and on mycorrhizae.

Ecosystems lose nitrogen by the activities of denitrifying bacteria, which convert nitrite or nitrate to forms that escape into the atmosphere.

Ecosystems lose or gain too much nitrogen through human activities, in ways that compromise plant growth.

SEDIMENTARY CYCLES

A Look at the Phosphorus Cycle

We continue our survey of biogeochemical cycling with one of the key sedimentary cycles. In the **phosphorus cycle**, phosphorus passes quickly through food webs as it moves from land, to ocean sediments, and then back to land. The Earth's crust is the largest reservoir, just as it is for other minerals (Figure 49.24).

Rock formations incorporate phosphorus mainly as phosphate ions (PO_4^{3-}). By weathering and soil erosion, phosphates enter streams and rivers, then enter ocean sediments. There they slowly accumulate together with other minerals. They form insoluble deposits mainly on submerged shelves of continents. Millions of years go by. In regions where the movements of crustal plates uplift part of the seafloor, phosphates become exposed on drained land surfaces. Over time, weathering and erosion releases phosphates from exposed rocks, so the cycle's geochemical phase begins again.

The ecosystem phase of the cycle is more rapid than the long-term geochemical phase. All organisms require phosphorus for synthesizing phospholipids, NADPH, ATP, nucleic acids, and other compounds. Plants take up dissolved phosphates rapidly and efficiently from soilwater. Herbivores get phosphates by eating plants; carnivores get them by eating herbivores. All animals excrete phosphates as a waste in urine and feces.

Bacterial and fungal decomposers in the soil release phosphates, then plants take up dissolved forms of this mineral. By this action the plants help cycle phosphorus rapidly through the ecosystem.

Eutrophication

Sedimentary cycles, in combination with the hydrologic cycle, take most mineral elements through ecosystems on land and in water. Rainfall that evaporates from the ocean falls over land, and on its way back to the ocean, it transports silt and dissolved minerals that primary producers use as nutrients.

Of all minerals, phosphorus is the most prevalent limiting factor in natural ecosystems around the world. Soils hold little of it, and sediments tie up most of it in aquatic ecosystems. It helps that phosphorus does not have a gaseous phase, as nitrogen does, so ecosystems lose little of their scarce supply to the atmosphere. It also helps that phosphorus resists leaching. Also, as you might predict, producers are adapted to respond rather efficiently to low phosphorus concentrations.

Some human activities are lowering the phosphorus levels of natural ecosystems. This is especially the case for tropical and subtropical regions of the developing countries. There, highly weathered soils have very little phosphorus. Enough is stored in biomass and is slowly released from decomposing organic matter to sustain undisturbed forests or grasslands. Even so, when people harvest trees or clear grassland for agriculture, they deplete the phosphorus. Crop yields start out low and soon become nonexistent. Fields are quickly abandoned, but regrowth of natural vegetation is sparse. About 1 to 2 billion hectares are already depleted of phosphorus, mostly where poor farmers cannot afford fertilizers and do not engage in soil management practices.

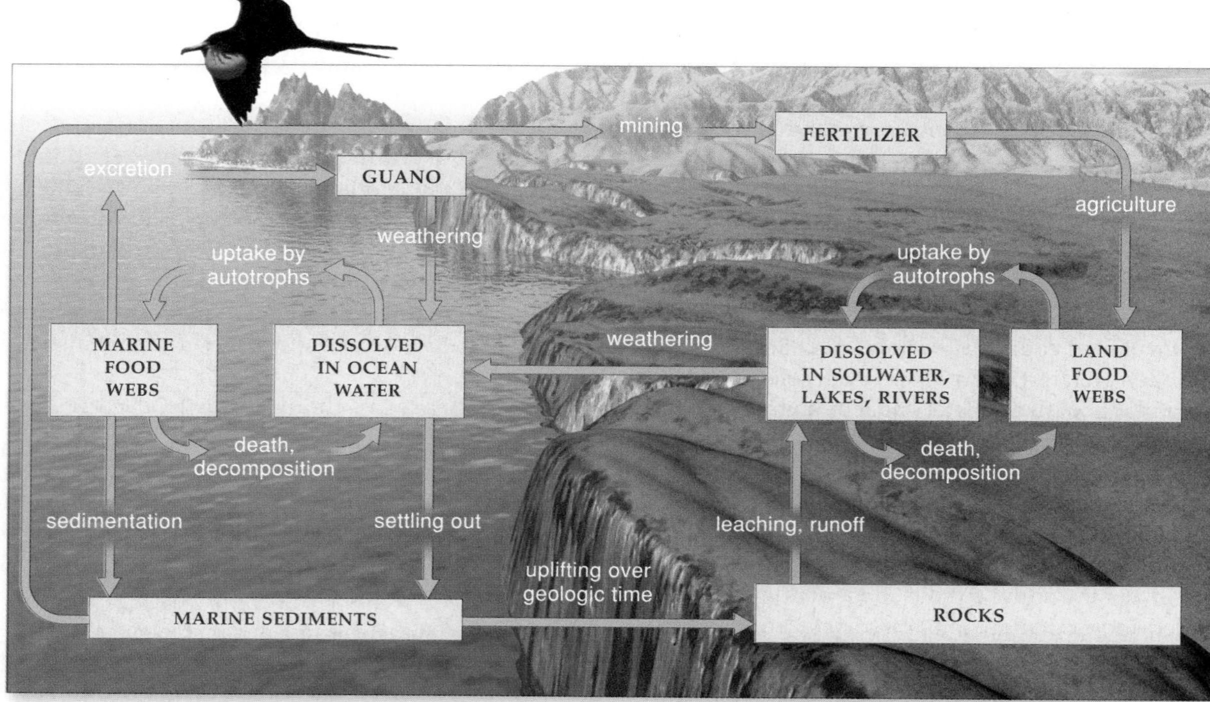

mining
FERTILIZER
excretion
GUANO
agriculture
weathering
uptake by autotrophs
uptake by autotrophs
weathering
MARINE FOOD WEBS
DISSOLVED IN OCEAN WATER
DISSOLVED IN SOILWATER, LAKES, RIVERS
LAND FOOD WEBS
death, decomposition
death, decomposition
sedimentation
settling out
leaching, runoff
uplifting over geologic time
MARINE SEDIMENTS
ROCKS

Figure 49.24 Phosphorus cycle. In this sedimentary cycle, most phosphorus moves in the form of phosphate ions (PO_4^{3-}).

Further reading: Student Guide to InfoTrac on web site

Figure 49.25 Experiment demonstrating lake eutrophication in Ontario, Canada. Researchers stretched a plastic curtain across a channel between two basins of the same lake. They added phosphorus, carbon, and nitrogen to one basin (the basin in the *background*) and carbon and nitrogen to the other (the basin in the *foreground*). Within two months, the phosphorus-enriched basin showed signs of accelerated eutrophication; a dense algal bloom turned the water green.

Developed countries have a different problem. Years of heavy phosphorus applications have raised the level of phosphorus in soils. Without stringent monitoring, phosphorus becomes concentrated in eroded sediments and in runoff from agricultural fields. Phosphorus also is present in outflows from sewage treatment plants and from industries. It is present in runoff from cleared land, even from lawns.

Dissolved phosphorus that enters streams, rivers, lakes, and estuaries can promote dense algal blooms. Like plants, photosynthetic species of algae cannot grow without phosphorus, nitrogen, and other minerals. In many freshwater ecosystems, nitrogen-fixing bacteria are abundant, so nitrogen is not a limiting factor on growth of algae. Phosphorus is. Decomposition of the remains from algal blooms depletes water of oxygen, which kills off the fish and other organisms.

Eutrophication is the name for nutrient enrichment of an ecosystem that is naturally low in nutrients. As you will be reading in Section 50.10, eutrophication is a natural process, but phosphorus inputs accelerate it. Field experiments, including the one shown in Figure 49.25, nicely demonstrate this outcome.

Sedimentary cycles, in combination with the hydrologic cycle, take most mineral elements, such as phosphorus, through terrestrial and aquatic ecosystems.

Agriculture, deforestation, and other human activities upset the nutrient balance of an ecosystem when they accelerate the amount of minerals entering and leaving it.

49.12 SUMMARY

1. An ecosystem consists of an array of producers, consumers, detritivores, and decomposers along with their environment. It is an open system, with inputs, internal transfers, and outputs of energy and nutrients. There is a one-way flow of energy into and out from it and a cycling of materials among its organisms.

2. Sunlight is the initial energy source for nearly all ecosystems. Photoautotrophs, the primary producers, convert that energy to other forms, such as chemical bond energy of ATP. They also assimilate many of the nutrients required by the ecosystem's heterotrophs.

a. Many heterotrophs are consumers. The herbivores feed upon algae and plants. Carnivores ingest animals. Parasites withdraw nutrients from the tissues of living hosts. Omnivores use a variety of food sources.

b. Decomposers also are heterotrophs. Most (certain fungi and bacteria) obtain energy and nutrients from organic remains and wastes.

c. Detritivores, such as crabs and earthworms, are heterotrophs that ingest small particles of decomposing or dead organisms or their organic products.

3. Feeding relationships are structured as trophic levels, a hierarchy of energy transfers in an ecosystem. (This is sometimes referred to as "Who eats whom.")

a. Primary producers are at the first trophic level, and consumers are at successively higher trophic levels.

b. Herbivores are second-level consumers. Carnivores and parasites are at higher levels.

c. Decomposers and detritivores are assigned to all trophic levels above the first. So are parasites.

d. Because humans and other omnivores get energy from more than one source, they cannot be assigned to a single trophic level.

4. Each food chain is a linear series of steps by which energy stored in tissues of autotrophs passes on through higher trophic levels. Isolated food chains are rare in nature. They cross-connect, as food webs.

a. In grazing food webs, energy is transferred in one direction: from primary producers to herbivores, then through carnivores.

b. In detrital food webs, energy is transferred in one direction: from primary producers through detritivores and decomposers.

5. The amount of useful energy that flows through successive levels of consumers declines at each energy transfer in a food web because of metabolic heat loss and shunting of food energy into organic wastes. This inefficiency typically limits the number of trophic levels in a food web to no more than four or five.

a. In environments that show variations in salinity, temperature, and other environmental conditions, food webs tend to have short chains. In stable environments, such as parts of the deep ocean, food chains are longer.

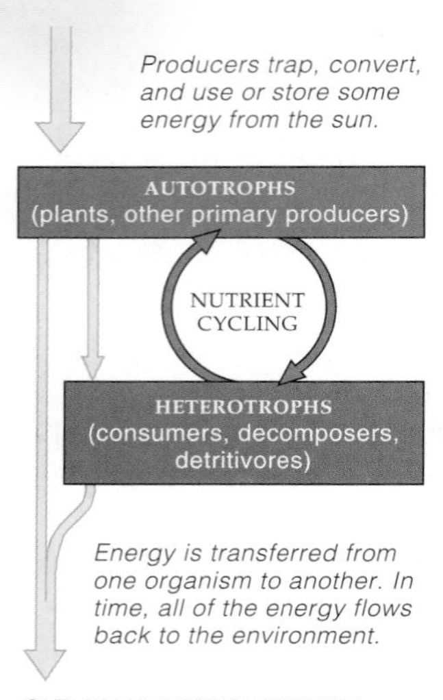

Producers trap, convert, and use or store some energy from the sun.
AUTOTROPHS
(plants, other primary producers)
NUTRIENT CYCLING
HETEROTROPHS
(consumers, decomposers, detritivores)
Energy is transferred from one organism to another. In time, all of the energy flows back to the environment.
ONE-WAY FLOW OF ENERGY

Figure 49.26 Summary of energy flow and nutrient cycling in ecosystems.

b. The most complex food webs have short chains; the simplest have the most top carnivores.

6. Disturbance to one aspect of an ecosystem usually has unexpected effects on other, seemingly unrelated parts. To predict consequences of a disruption to an ecosystem, as by computer modeling, researchers must identify all relationships and incorporate them into the model.

7. Primary productivity is the rate at which primary producers capture and store a given amount of energy in a specified interval. The net primary productivity is the rate of energy storage in producers in excess of their rate of aerobic metabolism.

8. In addition to energy inputs, primary productivity and ecosystem structure require a cycling of nutrients (Figure 49.26). Water and minerals move slowly through the physical environment, rapidly through organisms, and back to the environment, in biogeochemical cycles.

a. Water moves through a hydrologic cycle. In land ecosystems, plants stabilize soil and minimize nutrient loss during the cycle, as by runoff.

b. In atmospheric cycles, a nutrient prevails mainly in gaseous form (such as carbon, in carbon dioxide).

c. In a sedimentary cycle, a nutrient is in solid form and the main reservoir is the Earth's crust.

9. In the carbon cycle, carbon dioxide is the main gas in the atmosphere. The ocean is carbon's main reservoir. Burning of fossil fuels, logging, and conversion of natural ecosystems for farming and grazing disrupt the global carbon budget and may be factors in global warming.

10. Nitrogen is a limiting factor in the total net primary productivity of ecosystems on land. Gaseous nitrogen is abundant in the atmosphere. Nitrogen-fixing bacteria convert N_2 to ammonia and nitrates, which producers take up. Mycorrhizae and root nodules, two symbiotic interactions, enhance the nitrogen uptake.

11. Sedimentary cycles interact with the hydrologic cycle to move mineral nutrients to and from ecosystems.

a. Certain human activities are depleting minerals from ecosystems, as when weathered soils of tropical forests are cleared for agriculture.

b. Some human activities are accelerating the process of eutrophication. They are adding nutrients to aquatic ecosystems that are naturally low in those nutrients, and so promote destructive algal blooms.

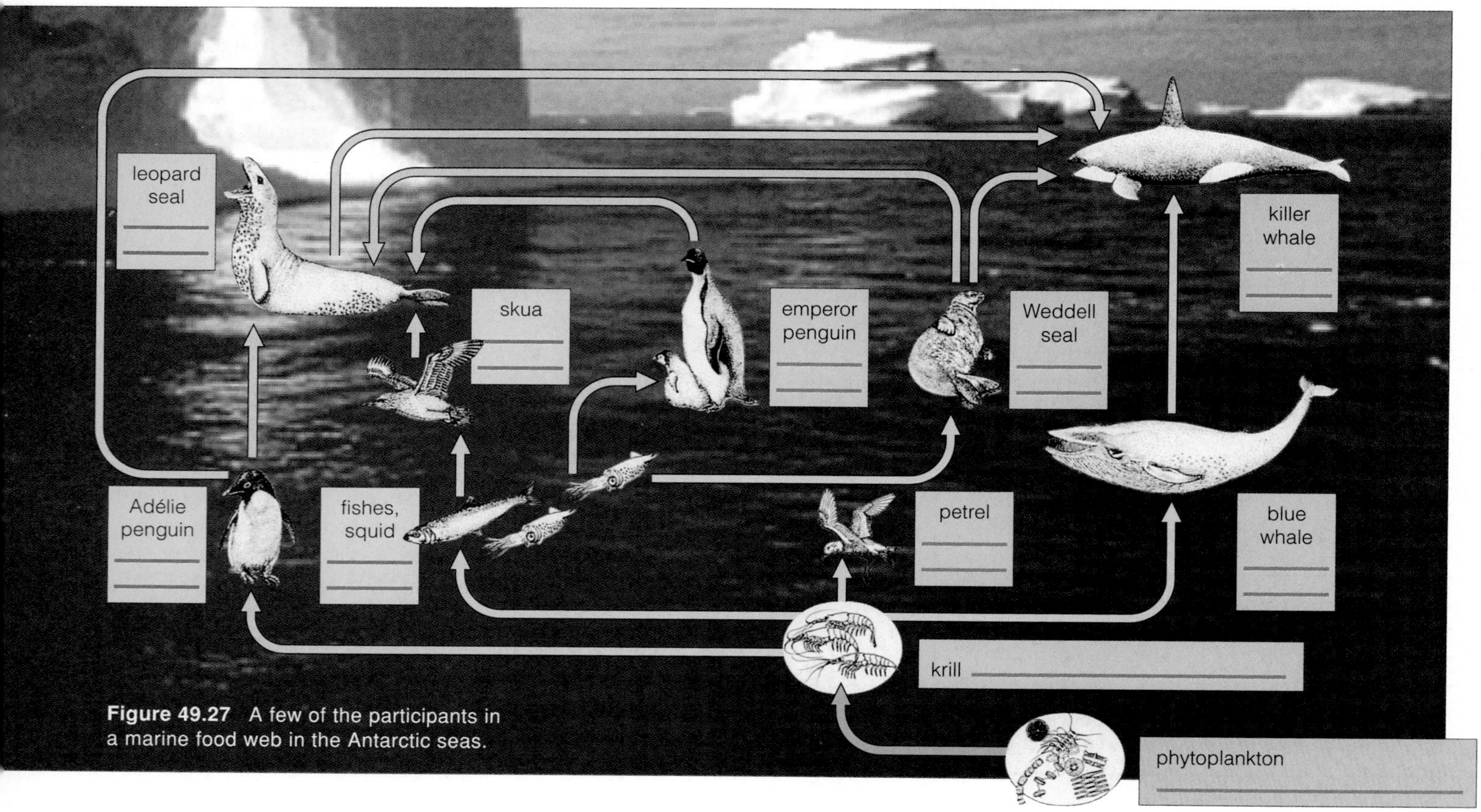

Figure 49.27 A few of the participants in a marine food web in the Antarctic seas.

Review Questions

1. Define an ecosystem in terms of inputs and outputs. *49.1*

2. Distinguish among primary producer, consumer, decomposer, and detritivore. Give an example of each. *49.1*

3. Define and describe trophic levels. Which class of organisms is farthest from the energy input to an ecosystem? *49.1, 49.2*

4. Compare Figure 49.27 with Figure 49.4, then identify and label the trophic level to which each organism belongs. Also indicate whether the organism is a primary producer, first-level (primary consumer), or second-, third-, or fourth-level consumer. *49.2*

5. Define food chain and food web. How do grazing food webs differ from detrital food webs? *491, 49.2*

6. Distinguish between biomass pyramid and energy pyramid. What did the pyramids for Silver Springs reveal? *49.4*

7. Define three types of biogeochemical cycles and give an example of each. *49.6–49.11*

8. Define and describe the connections among nitrogen fixation, nitrification, ammonification, and denitrification. *49.10*

Self-Quiz (*Answers in Appendix III*)

1. Ecosystems have ______ .
   a. energy inputs and outputs
   b. one trophic level
   c. nutrient cycling but not outputs
   d. a and b

2. Trophic levels are ______ .
   a. structured feeding relationships
   b. who eats whom in an ecosystem
   c. a hierarchy of energy transfers
   d. all of the above

3. Primary productivity on land is affected by ______ .
   a. photosynthesis and respiration by plants
   b. how many plants are neither eaten nor decomposed
   c. rainfall and temperature
   d. all of the above

4. Match the ecosystem terms with the suitable description.
   ____ producers   a. herbivores, carnivores, omnivores
   ____ consumers   b. feed on partly decomposed matter
   ____ decomposers   c. degrade organic remains, wastes
   ____ detritivores   d. photoautotrophs

Critical Thinking

1. Imagine and describe an extreme situation in which you would be a participant in a food chain rather than a food web.

2. Marguerite is growing a vegetable garden in Maine. What are the variables that can affect its net primary production?

3. List as many of the agricultural products and manufactured goods as you can identify that you depend upon. Are any of them implicated in the amplified greenhouse effect?

4. Of all crops grown in the United States, less than 10 percent are free of pesticide or fungicide applications. Such *organically grown produce* costs more. It also spoils faster and is nonuniform in size and appearance. Do you prefer to buy such produce? Why or why not?

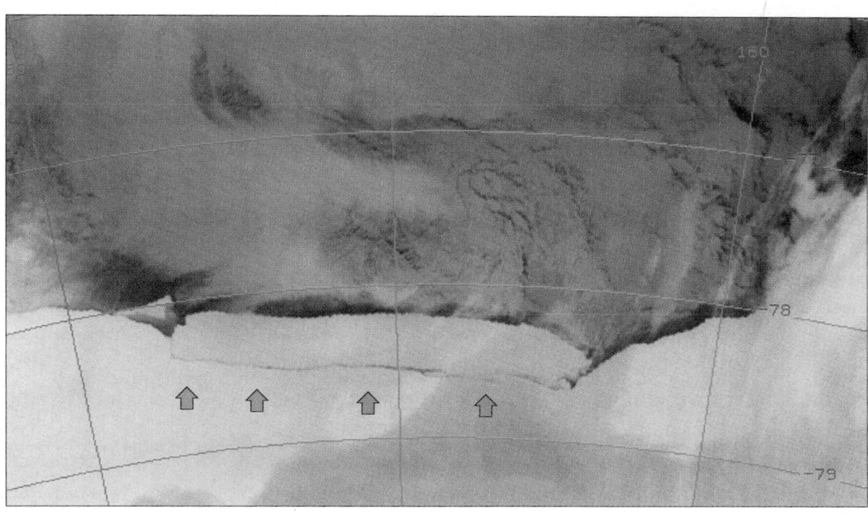

160
-78
-79

Figure 49.28 Satellite image of an iceberg roughly the same size as Connecticut. This one started to break away from Antarctica in 2000 and may be the longest ever recorded.

5. *Polar ice shelves* are vast, enormously thickened sheets of ice floating on seawater. Measurements taken between 1978 and the present suggest that Antarctic ice shelves are retreating. A chunk of the Larsen Ice Shelf the size of Rhode Island broke away from Antarctica in 1995 and drifted off into the open ocean. Figure 49.28 shows a gargantuan chunk, 37 kilometers wide and 295 kilometers long, that started to split away from the Ross Ice Shelf in 2000.

   Do you suppose that the retreating and fragmentation of the polar ice sheets are evidence of a long-term trend in global warming? If not, what other factors might be at work that could produce such major events?

Selected Key Terms

ammonification *49.10*
biogeochemical cycle *49.6*
biological magnification *49.3*
biomass pyramid *49.4*
carbon cycle *49.8*
consumer *49.1*
decomposer *49.1*
decomposition *49.10*
denitrification *49.10*
detrital food web *49.2*
detritivore *49.1*
ecosystem *49.1*
ecosystem modeling *49.3*
energy pyramid *49.4*
eutrophication *49.11*
food chain *49.1*
food web *49.1*
global warming *49.9*
grazing food web *49.2*
greenhouse effect *49.9*
hydrologic cycle *49.7*
ion exchange *49.10*
nitrification *49.10*
nitrogen cycle *49.10*
nitrogen fixation *49.10*
phosphorus cycle *49.11*
primary producer *49.1*
primary productivity *49.4*
trophic level *49.1*
watershed *49.7*

Readings *See also www.infotrac-college.com*

Brady, N., and R. Weil. 1996. *The Nature and Properties of Soils*. Eleventh edition. New Jersey: Prentice-Hall.

Krebs, C. 1994. *Ecology*. Fourth edition. New York: HarperCollins.

Smith, R. 1992. *Elements of Ecology*. Third edition. New York: HarperCollins.

50 THE BIOSPHERE

Does a Cactus Grow in Brooklyn?

Suppose you live in the American Southwest but find yourself touring one of the deserts of Africa. There you come across a flowering plant with spines, tiny leaves, and columnlike, fleshy stems—just like some cactus plants back home (Figure 50.1). Or suppose you live in the coastal hills of California and decide to tour the Mediterranean coast, the southern tip of Africa, or even central Chile. There you come across many-branched, tough-leafed woody plants—very much like the many-branched, tough-leafed chaparral plants back home.

In both cases, the plants are separated by enormous geographic and evolutionary distances. Why, then, are they so much alike? The question intrigues you, so you decide to compare their locations on a global map. As you quickly discover, the American and African desert plants grow approximately the same distance from the equator. Chaparral plants and their distant look-alikes grow along the western or southern coasts of continents between latitudes 30° and 40°. As Charles Darwin and other naturalists did long ago, you have just stumbled onto one of many predictable patterns in nature.

What causes such patterns? With this question, we turn to **biogeography**—the study of the distribution of organisms, past and present, and of diverse processes that underlie the distribution patterns. For example, "accidents of history" put many species in particular places. Think back on the colossal breakup of Pangea, more than 100 million years ago. When chunks of that supercontinent began to drift apart, many species had no choice but to go along for the ride. Over evolutionary time, the travelers were dispersed to different isolated locations. And there they proceeded to undergo genetic

Figure 50.1 Life in the biosphere—a case of morphological convergence. (**a**) *Echinocerus*, of the cactus family (Cactaceae), grows in deserts of the American Southwest. (**b**) In deserts of southwestern Africa, we find *Euphorbia*, of the spurge family (Euphorbiaceae). Although the lineages appear to be closely related, they are geographically and evolutionarily distant. Long ago, plants in both lineages put similar structures (including leaves) to similar uses in similar habitats. Their descendants ended up resembling each other.

divergence from the parent populations. Among those changing populations on the fragment that became Australia were the ancestors of eucalyptus trees and platypuses, of wombats and kangaroos.

Species also owe their distribution to topography, climate, and species interactions. With diligence, you might grow a cactus under artificial lights in a heated room in Brooklyn or some other New York City borough. Plant that cactus outside, and it won't last one winter.

This last example reminds us that we humans tinker with the distribution of species. Not all of our tinkering is as harmless as growing a cactus in Brooklyn. Think of our predatory effects on the world's fisheries or the effects of our pesticide battles with insect competitors for food. Earlier chapters provided you with a general picture of predation, competition, and other species interactions. This chapter considers the physical forces shaping the biosphere itself. It serves as a foundation for addressing the impact of the human species on the biosphere—the topic of the chapter to follow.

Start with the definition of the **biosphere**—the sum total of all places in which organisms live. They live in the waters of the Earth, including the ocean, polar ice caps, and other forms of liquid and frozen water (the hydrosphere). They live in the soils and sediments of the Earth's outer, rocky layer (the lithosphere). And they live in the lower **atmosphere**, which is made up of gases and airborne particles that envelop the Earth. Air in the upper atmosphere, seventeen kilometers or so above the Earth's surface, is too thin to support life.

In this vast biosphere you will find ecosystems that range from continent-straddling forests to tiny pools in cup-shaped leaves of plants. As you will see, climate profoundly influences all ecosystems except for a few at hydrothermal vents on the ocean floor.

Climate refers to the average weather conditions, such as temperature, humidity, wind speed, cloud cover, and rainfall, over time. Many factors contribute to climate. The main ones are variations in the amount of incoming solar radiation, the Earth's daily rotation and its path around the sun, the world distribution of continents and oceans, and land mass elevations. All of these factors interact to produce prevailing winds and ocean currents that influence the global patterns of climate. Climate affects the physical and chemical development of sediments and soils.

Climate, together with the composition of sediments and soils, influences the growth of primary producers—and, through them, the distribution of ecosystems.

KEY CONCEPTS

1. Energy from the sun is the initial energy source for nearly all ecosystems on Earth. In addition, solar energy influences the global distribution of ecosystems. It does so by continually providing heat energy that warms the atmosphere and drives the Earth's weather systems.

2. Interactions among global air circulation patterns, ocean currents, and diverse topographic features result in regional variations in patterns of temperature and rainfall. The variations influence the composition of soils and sediments, the growth and distribution of primary producers, and, through them, the distribution of ecosystems.

3. A biome is a large, regional unit of land that can be characterized by the climax vegetation of the ecosystems within its boundaries. Deserts and broadleaf forests are examples. Their distribution corresponds roughly with regional variations in climate, topography, and soil type.

4. The water provinces cover more than 71 percent of the Earth's surface. The oceans hold all but 3 percent of the free-flowing water. The remainder is in inland seas, estuaries, and bodies of fresh water, including lakes, streams, and rivers.

5. Each freshwater and marine ecosystem has gradients in light availability, temperature, and dissolved gases. The gradients vary daily and seasonally. They affect primary productivity and the composition of species.

50.1 AIR CIRCULATION PATTERNS AND REGIONAL CLIMATES

Each winter Pacific Grove, California, is host to great gatherings of tourists and monarch butterflies (Figure 50.2). Similarly, caribou, Canada geese, sea turtles, and whales undertake vast migrations, moving to and from overwintering grounds. Other kinds of animals do stay put. They have adaptations in body form, physiology, and behavior that help them endure seasonal change.

Figure 50.2 Monarch butterflies, migratory insects that each winter gather in the trees of California coastal regions and of central Mexico. Monarchs travel hundreds of kilometers south to those places, which are cool and humid in winter. If they were to remain in their northern breeding grounds, they would risk being killed by more severe climatic conditions.

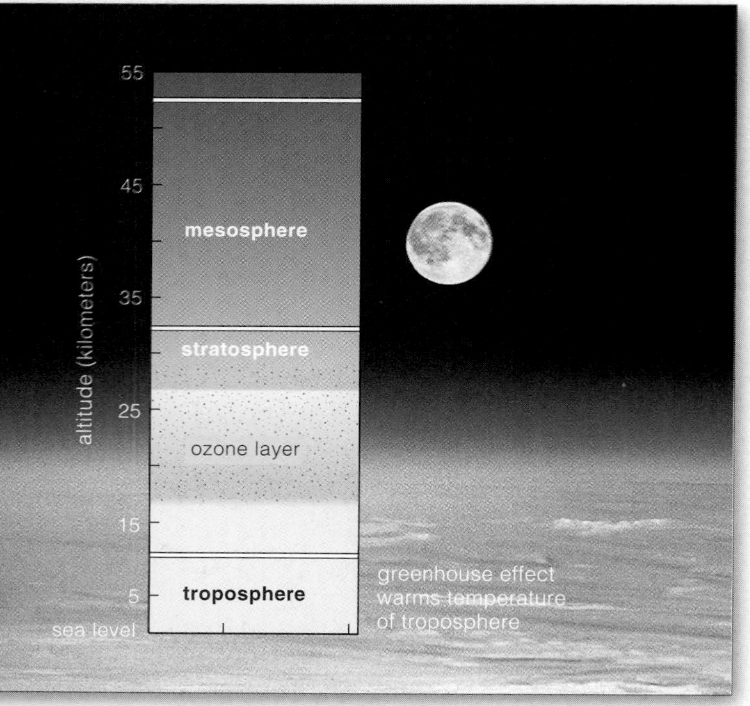
55
45
35
25
15
5
altitude (kilometers)
mesosphere
stratosphere
ozone layer
troposphere
sea level
greenhouse effect
warms temperature
of troposphere

Figure 50.3 Earth's atmosphere. Most of the global air circulation proceeds in the troposphere, where temperature decreases rapidly with altitude. Most of the ultraviolet wavelengths in the sun's rays are absorbed in the upper atmosphere, mainly at the ozone layer. The ozone layer typically ranges between seventeen and twenty-seven kilometers above sea level.

Throughout Canada and the United States, flowering plants form leaves, flower, bear fruit, and drop leaves. In the ocean, uncountable numbers of photosynthetic microorganisms undergo seasonal bursts of primary productivity. Clearly, organisms are exquisitely attuned to climatic conditions, which differ from one region to the next and as one season gives way to the next.

Climate begins with incoming rays from the sun. Of the total amount of solar radiation arriving at the outer atmosphere, only about half gets through to the Earth's surface. Molecules of ozone (O_3) and of oxygen (O_2) in the atmosphere absorb most wavelengths of ultraviolet radiation. Such wavelengths, recall, are lethal for most organisms. Absorption is greatest in a region between seventeen and twenty-seven kilometers above sea level, where ozone molecules are most concentrated (Figure 50.3). Hence the name "ozone layer." Clouds, dust, and water vapor suspended in the atmosphere absorb some other wavelengths or reflect them back into space.

Radiation that penetrates the atmosphere warms the surface of the Earth, which gives up heat by way of radiation or evaporation. The molecules making up the lower atmosphere absorb some heat and reradiate part of it toward the Earth. The effect is somewhat like heat retention in a greenhouse, which lets in the sun's rays while retaining heat that is being lost from the plants and soil inside (Section 49.9). Why is the greenhouse effect important? *Heat energy derived from the sun warms the atmosphere, and it ultimately drives the Earth's great weather systems.*

The sun's effect differs from one latitude to the next. Its rays are more concentrated at the equator than at the poles, so air is heated more at the equator (Figure 50.4). The global pattern of air circulation starts as warm air at the equator rises and spreads north and south. Under the moving air, the Earth rotates faster at the equator than it does at the poles, and this creates worldwide belts of prevailing east and west winds. The differences in solar heating at different latitudes and the modified air circulation patterns help define the world's major **temperature zones**, which are shown in Figure 50.5*a*.

Global air circulation patterns give rise to differences in rainfall at different latitudes as well. Think about this: Warm air can hold more moisture than cool air. At the equator, air picks up moisture from the seas and rises to cooler altitudes. There it gives up moisture as rain, which supports the luxuriant growth of tropical

Further reading: Student Guide to InfoTrac on web site

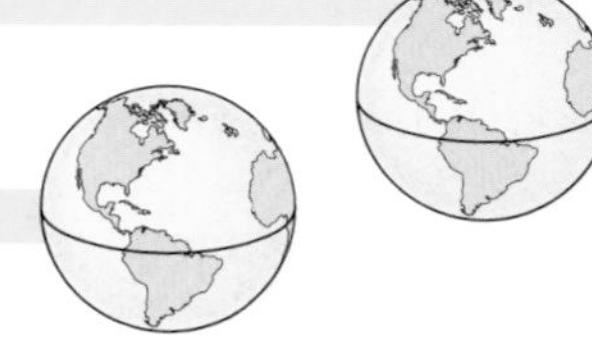
rays of sunlight
a

Figure 50.4 Global air circulation patterns, as brought about by three interrelated factors. *First*, the sun's rays are less spread out in equatorial regions than in polar regions (**a**). Warm equatorial air rises and spreads northward and southward to produce the initial pattern of air circulation (**b**).

Second, the nonuniform distribution of land masses creates variations in air pressure. Land absorbs and gives up heat faster than the ocean. Thus, depending on what's below them, some parcels of air rise or sink faster than others. Air pressure decreases where air rises and increases where air sinks. The differences create winds that disrupt the overall movement from the equator to the poles.

Third, the rotation and overall shape of the Earth introduce easterly and westerly deflections in wind directions. Each time the ball-shaped Earth makes a full rotation, its surface turns faster beneath air masses at the equator and slower beneath those at the poles. Therefore, a rising air mass cannot really move "straight north" or "straight south," but rather is deflected to the east or west. This is the source of prevailing east and west winds (**c**).

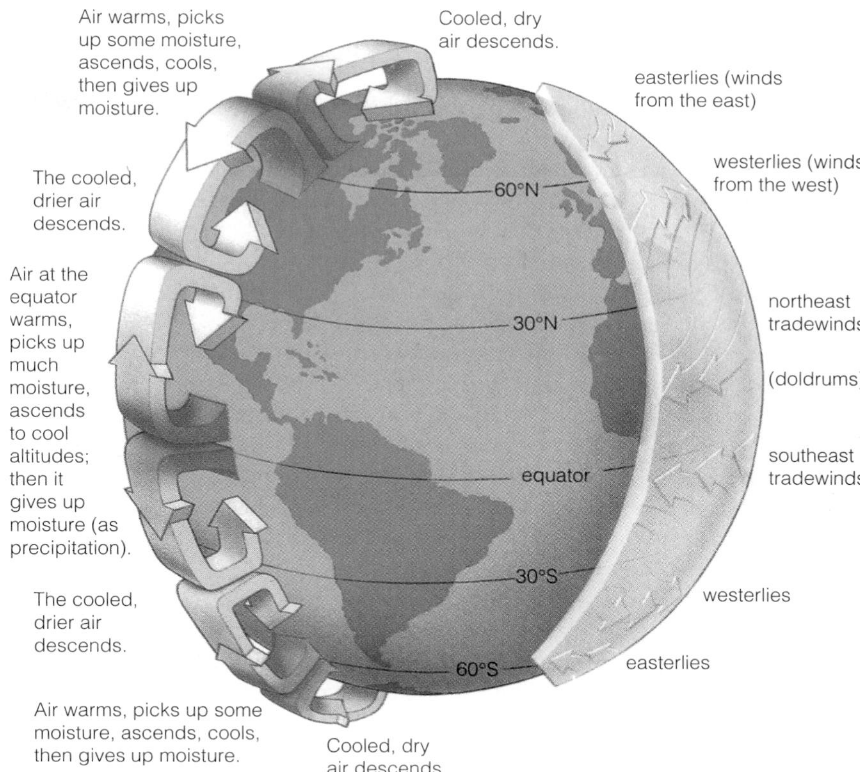
Air warms, picks up some moisture, ascends, cools, then gives up moisture.
Cooled, dry air descends.
easterlies (winds from the east)
The cooled, drier air descends.
westerlies (winds from the west)
60°N
Air at the equator warms, picks up much moisture, ascends to cool altitudes; then it gives up moisture (as precipitation).
30°N
northeast tradewinds
(doldrums)
equator
southeast tradewinds
30°S
westerlies
The cooled, drier air descends.
60°S
easterlies
Air warms, picks up some moisture, ascends, cools, then gives up moisture.
Cooled, dry air descends.

b Initial pattern of air circulation

c Deflections in the paths of air flow near the Earth's surface

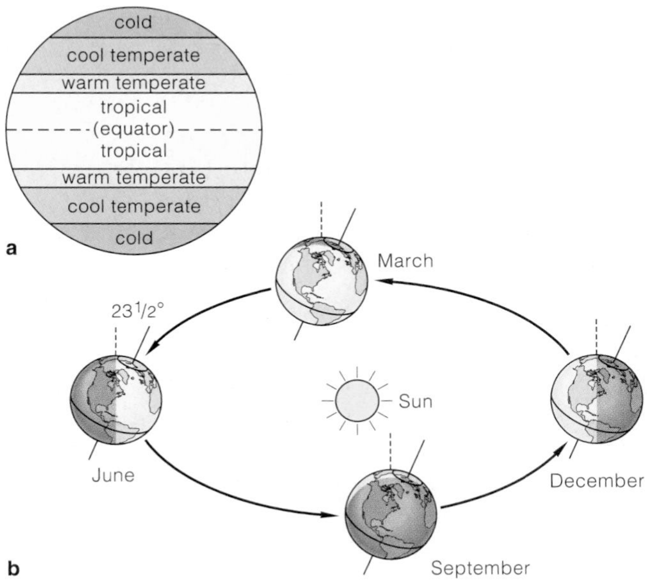
cold
cool temperate
warm temperate
tropical
(equator)
tropical
warm temperate
cool temperate
cold
a
March
23 1/2°
Sun
June
December
September
b

Figure 50.5 (**a**) World temperature zones. (**b**) Annual variation in incoming solar radiation. The northern end of the Earth's fixed axis tilts toward the sun in June and away from it in December. As a result, the equator's position relative to the boundary of illumination between day and night varies over the year. Such variations in sunlight intensity and daylength cause seasonal variations in temperature in the two hemispheres. The seasonal change becomes more pronounced with distance from the equator. Change is greatest in the central regions of continents, where the moderating effects of the seas are minimal.

forests. The air, which is drier now, moves away from the equator. Later, it descends at latitudes of about 30°, and it becomes warmer and drier during this descent. Deserts tend to form at these latitudes. Farther from the equator, air again picks up moisture, ascends to high altitudes, and creates another moist belt at latitudes of about 60°. Then it descends in polar regions, where the low temperatures and almost nonexistent precipitation give rise to the cold, dry polar deserts.

Finally, the amount of solar radiation reaching the surface varies seasonally because of the Earth's rotation around the sun (Figure 50.5*b*). This leads to seasonal changes in daylength, prevailing wind directions, and temperature. And so primary productivity alternately rises and falls on land and in the seas—and migrations shift many kinds of animals to new locations.

Latitudinal differences in the amount of solar radiation reaching the Earth produce global air circulation patterns. The Earth's rotation and its overall shape affect the patterns to produce latitudinal belts of temperature and rainfall.

Also, the Earth's annual rotation around the sun introduces seasonal changes in winds, temperatures, and rainfall.

These factors influence the locations of different ecosystems.

50.2 THE OCEAN, LANDFORMS, AND REGIONAL CLIMATES

Ocean Currents and Their Effects

The **ocean** is a continuous body of water cloaking more than 71 percent of the Earth. The uppermost 10 percent of the water moves in currents that distribute nutrients in marine ecosystems and affect regional climates. Solar heating and wind friction drive the surface currents.

Latitudinal and seasonal variations in solar heating warm and cool ocean water on a vast scale. The volume of water increases if heated and decreases if cooled. So the sea level is about 8 centimeters (3 inches) higher at the (hot) equator than it is at the (cold) poles. The sheer volume of water in this "slope" is enough to get the surface waters moving in response to gravity, generally from the equator to the poles. Along the way, surface waters warm the air parcels above them. Every second at midlatitudes, warm water transfers about *10 million billion* calories of heat energy to the atmosphere!

The "tug" of mainly the trade winds and westerlies causes currents—rapid mass flows of water. The Earth's rotation, land mass distribution, and shapes of oceanic basins affect the direction and properties of the currents. Water circulates clockwise in the Northern Hemisphere and counterclockwise in the Southern Hemisphere, as shown in Figure 50.6.

Swift, deep, and narrow currents of nutrient-poor waters move parallel with the east coast of continents. For instance, 55 million cubic meters per second of warm water move northward as the Gulf Stream along the east coast of North America. The slower, shallow, and broad currents paralleling the west coast of continents move cold water toward the equator. As you will see, these currents move deep, nutrient-rich waters to the surface.

Why do cool, mild, foggy summers prevail along the Pacific Northwest coastline? The California Current moves cold water near the coast as it flows toward the equator, and winds approaching the coast cool off as they give up heat to the cold water. Why are Baltimore and Boston so muggy in summer? Air above the Gulf Stream gains heat and moisture, which southerly and easterly winds carry to those cities. Compared to Ontario in central Canada, why are winters milder in London and Edinburgh, which are at the same latitude? The North Atlantic Current picks up warm water from the Gulf Stream, and as it flows past northwestern Europe, it gives up heat energy to the prevailing winds.

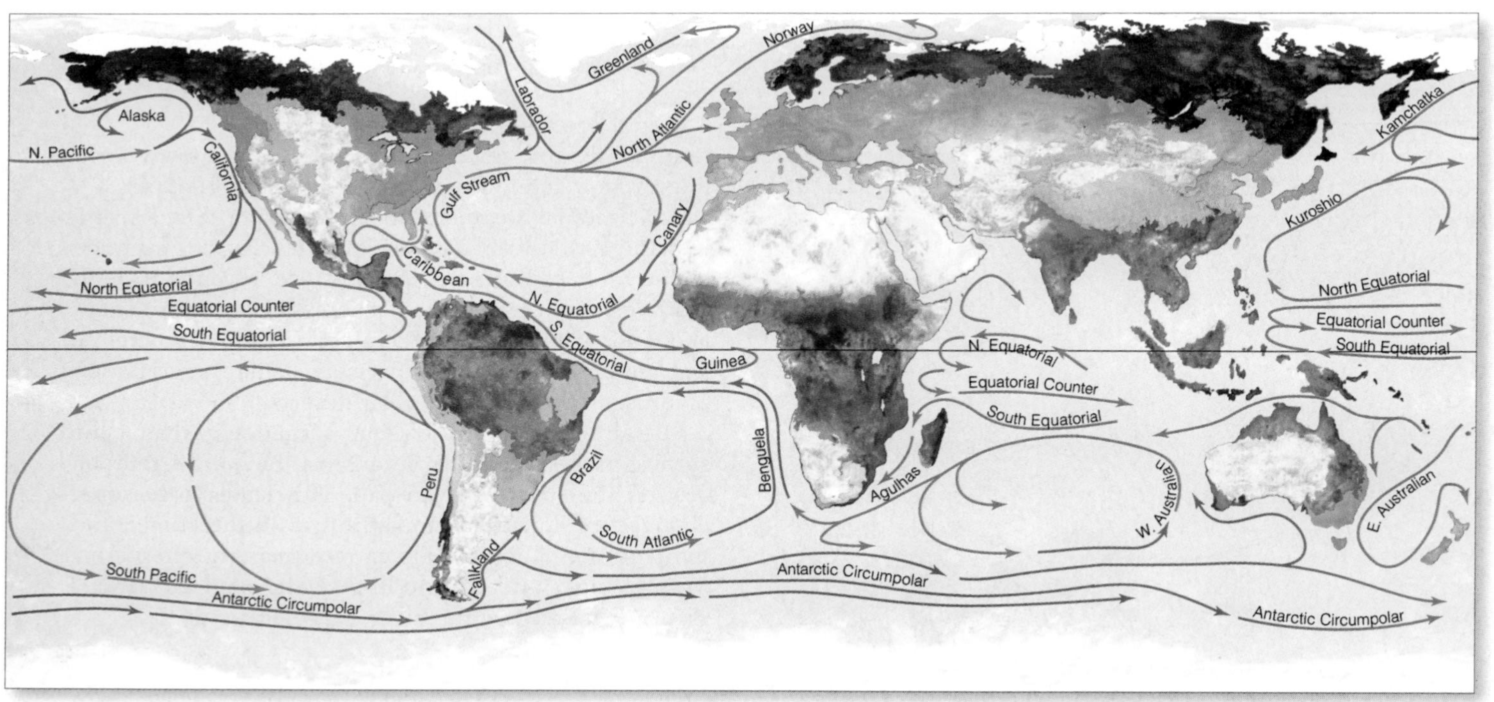

Alaska
N. Pacific
California
Labrador
Greenland
Norway
North Atlantic
Kamchatka
Gulf Stream
Canary
Kuroshio
Caribbean
North Equatorial
Equatorial Counter
South Equatorial
N. Equatorial
S. Equatorial
Guinea
N. Equatorial
Equatorial Counter
South Equatorial
North Equatorial
Equatorial Counter
South Equatorial
Peru
Brazil
Benguela
Agulhas
W. Australian
E. Australian
South Atlantic
Falkland
South Pacific
Antarctic Circumpolar
Antarctic Circumpolar
Antarctic Circumpolar

Figure 50.6 Generalized map of major climate zones correlated with surface currents and drifts of the world ocean. Warm surface currents move from the equator toward the poles. Water temperatures differ with latitude and depth. The differences contribute to regional differences in temperature and rainfall. In which direction does a given current flow? It depends on prevailing winds, the Earth's rotation, gravity, the shapes of ocean basins, and the positions of land masses.

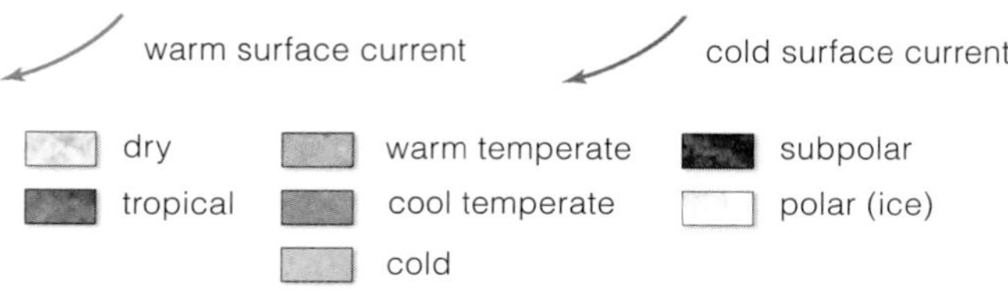

warm surface current
cold surface current
dry
tropical
warm temperate
cool temperate
cold
subpolar
polar (ice)

Further reading: Student Guide to InfoTrac on web site —

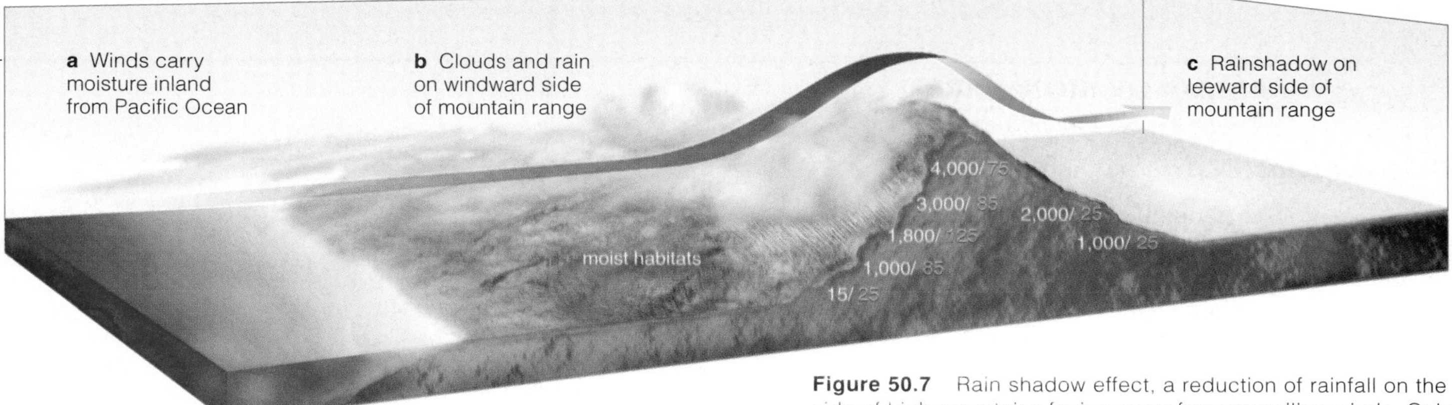
a Winds carry moisture inland from Pacific Ocean
b Clouds and rain on windward side of mountain range
c Rainshadow on leeward side of mountain range
moist habitats
4,000/75
3,000/85
1,800/125
1,000/85
15/25
2,000/25
1,000/25

Figure 50.7 Rain shadow effect, a reduction of rainfall on the side of high mountains facing away from prevailing winds. Only plants adapted to arid or semiarid conditions grow in rain shadows. *Blue* numbers show average yearly precipitation in centimeters, as measured at different locations on both sides of the mountain range. *White* numbers signify elevation in meters.

Regarding Rain Shadows and Monsoons

Mountains, valleys, and other aspects of topography influence regional climates. "Topography" means the physical features of a region, such as its elevation. For example, imagine a warm air mass picking up moisture off California's western coast. Moving inland, it reaches the Sierra Nevada, a mountain range that parallels the coast. The air cools as it ascends to higher altitudes, then loses moisture as rain (Figure 50.7). The vegetation belts at different elevations correspond to changes in air temperature and moisture. Grasslands with species adapted to semiarid conditions prevail at the western base of the range. At higher elevations, more moisture and cooler air support deciduous and evergreen forests. Higher still, a subalpine belt supports a few species of evergreen trees that can withstand the rigors of a cold habitat. Above the subalpine belt, only low, nonwoody plants and dwarfed shrubs can grow.

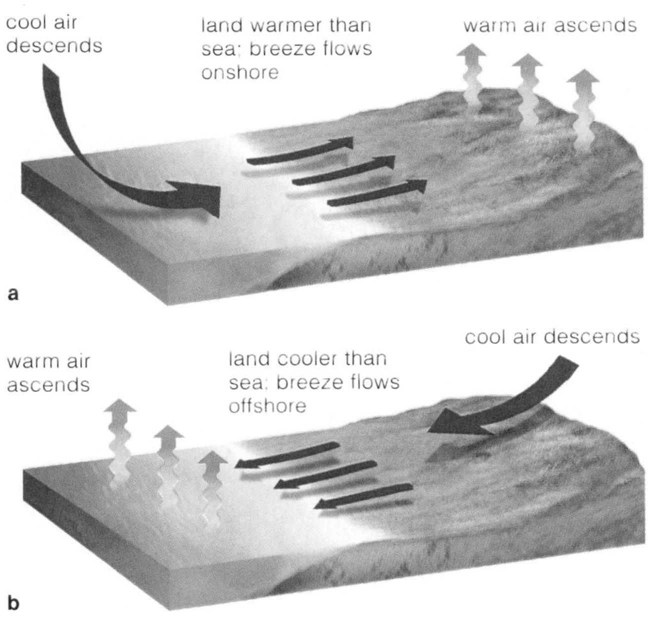
cool air descends
land warmer than sea: breeze flows onshore
warm air ascends
a
warm air ascends
land cooler than sea: breeze flows offshore
cool air descends
b

Figure 50.8 Coastal breeze in the afternoon (**a**) and at night (**b**).

After air flows over mountain crests and starts its descent, it warms up. Warm air can hold more water, so now it retains moisture and draws more out of plants and soil. The outcome is a **rain shadow**, a semiarid or arid region of sparse rainfall on the leeward side of high mountains. *Leeward* is the direction not facing a wind; *windward* is the direction from which the wind blows. Europe's mountain ranges, Asia's Himalayas, and South America's Andes create rain shadows. Tropical forests thrive on the windward side of Hawaii's high volcanic peaks, and arid conditions prevail on the leeward side.

Monsoons are air circulation patterns that influence continents north or south of warm ocean water. The land heats so intensely, low-pressure air parcels form above it. The low pressure causes moisture-laden air above the neighboring ocean to move inland, as from the Bay of Bengal into India and Bangladesh or from the Gulf of Mexico into the hot deserts of the American Southwest. Fantastic summer thunderstorms are one outcome. In a zone called the doldrums, trade winds converge and the equatorial sun causes intense heating and heavy rain. The air circulates north and south seasonally. It creates patterns of wet and dry seasons, as in East Africa.

The recurring sea breezes along coastlines might be viewed as "mini-monsoons." Here again, the water and the adjacent land have different heat capacities. During the morning, the water's temperature does not rise as rapidly as the land's temperature. When the warmed air above the land rises, the cooler marine air moves in. After sunset, the land loses heat faster than the water; land breezes flow in the reverse direction (Figure 50.8).

Surface ocean currents, in combination with global air circulation patterns, influence regional climates and help distribute nutrients in marine ecosystems.

Air circulation patterns, ocean currents, and landforms interact in ways that influence regional temperatures and moisture levels. Thus they also influence the distribution and dominant features of ecosystems.

50.3 REALMS OF BIODIVERSITY

Biogeographic Distribution

The circulation patterns in the atmosphere and at the ocean surface, in concert with topography, give rise to regional differences in temperature and moisture. Their influence helps explain why tundra, forests, grasslands, and deserts form in some places but not others. It helps us understand the distribution of ecosystems and their species, each adapted to regional conditions.

For example, the information provides clues to why many evolutionarily distant species look alike. Often such species are results of convergent evolution, which Section 18.2 describes. Briefly, their ancestors were not closely related but lived in similar environments and faced similar selection pressures. By natural selection, their body plans underwent similar modifications and ended up looking alike. Ancestors of those cactus and spurge plants in Figure 50.1 both evolved in hot, dry deserts, where water is scarce. Fleshy plant stems that have thick cuticles conserve precious water. And rows of sharp spines help deter hungry, thirsty herbivores, which might otherwise chew on juicy plant parts.

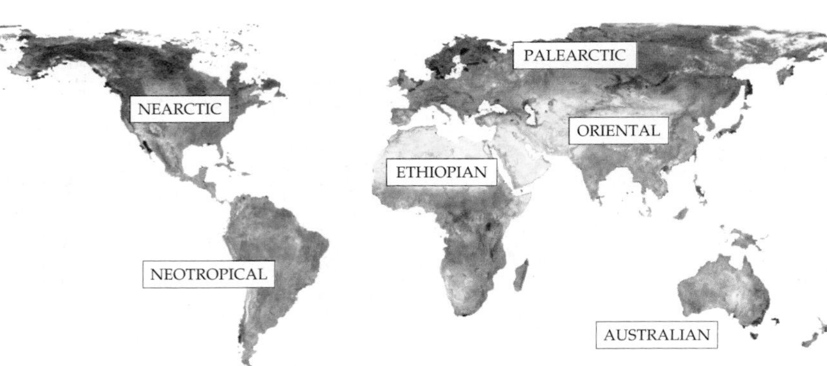

NEARCTIC
PALEARCTIC
ORIENTAL
ETHIOPIAN
NEOTROPICAL
AUSTRALIAN

Figure 50.9 Biogeographic realms.

60°
40°
20°
0°
20°
40°

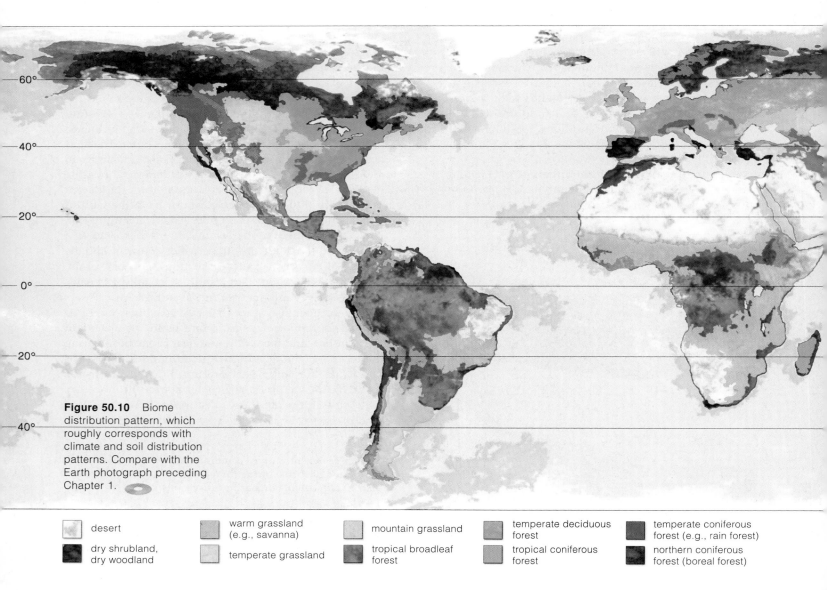

Figure 50.10 Biome distribution pattern, which roughly corresponds with climate and soil distribution patterns. Compare with the Earth photograph preceding Chapter 1.

desert
dry shrubland, dry woodland
warm grassland (e.g., savanna)
temperate grassland
mountain grassland
tropical broadleaf forest
temperate deciduous forest
tropical coniferous forest
temperate coniferous forest (e.g., rain forest)
northern coniferous forest (boreal forest)

Further reading: Student Guide to InfoTrac on web site

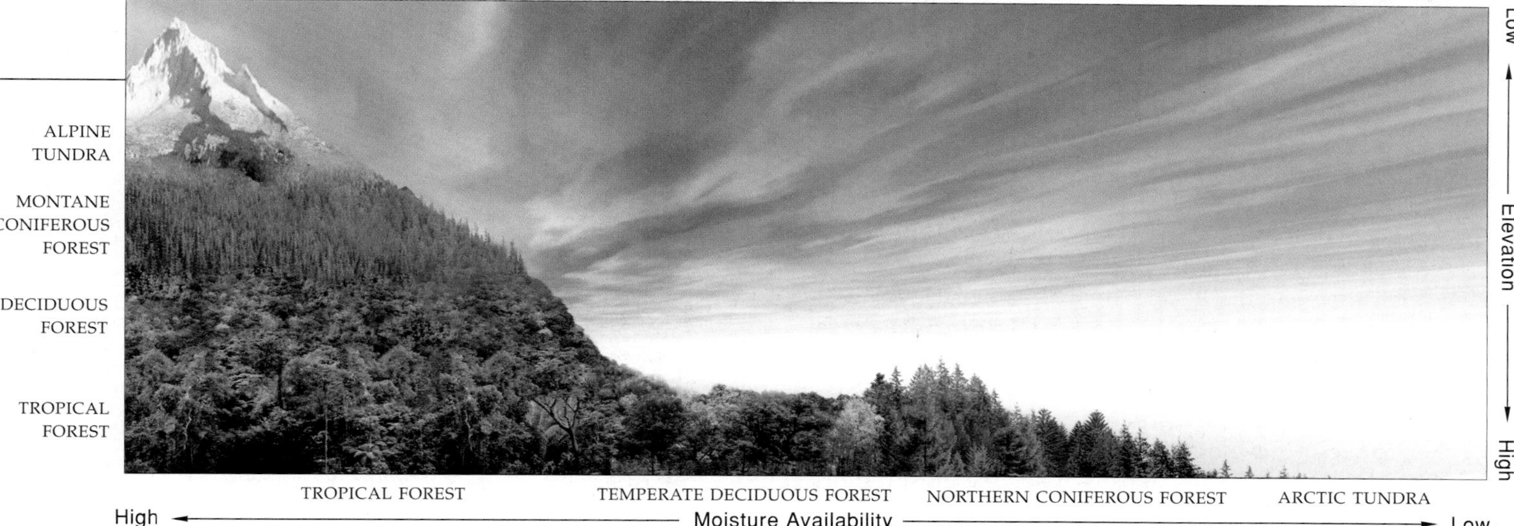
ALPINE TUNDRA
MONTANE CONIFEROUS FOREST
DECIDUOUS FOREST
TROPICAL FOREST
TROPICAL FOREST
TEMPERATE DECIDUOUS FOREST
NORTHERN CONIFEROUS FOREST
ARCTIC TUNDRA
High
Moisture Availability
Low
Low
Elevation
High

Figure 50.11 For North America, general changes in plant form along environmental gradients. Gradients in elevation and availability of water greatly influence a biome's primary productivity. Other factors, such as mean annual air temperature and soil drainage characteristics, also influence plant form.

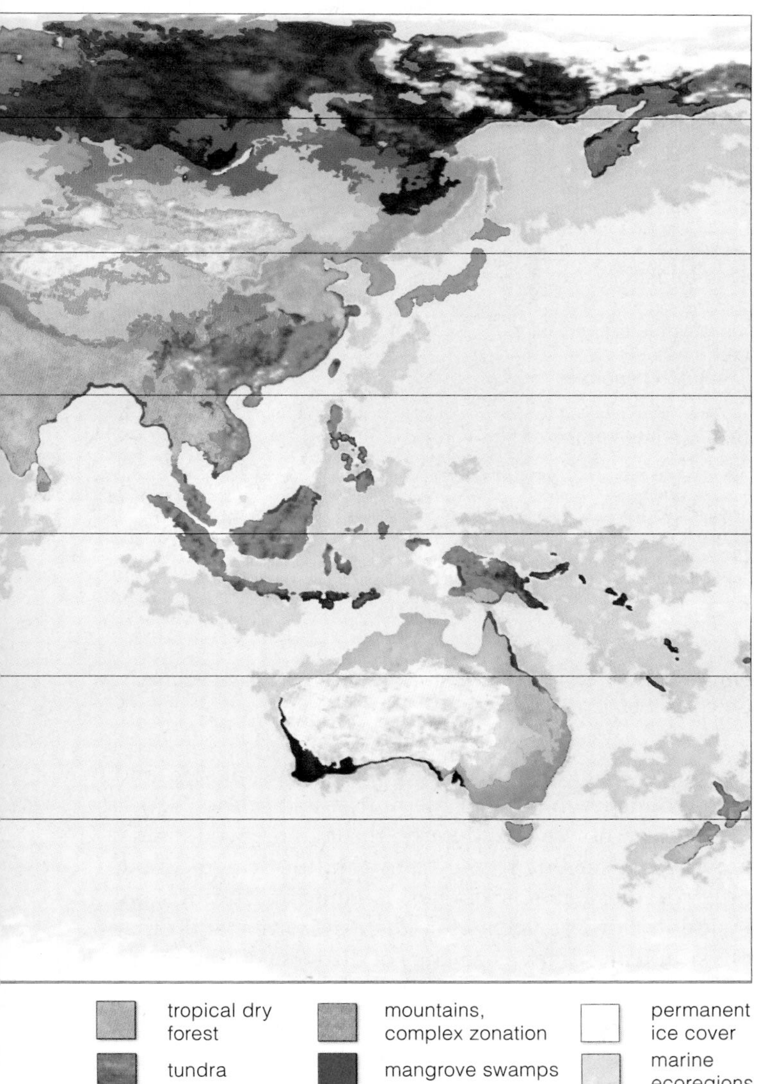
tropical dry forest
tundra
mountains, complex zonation
mangrove swamps
permanent ice cover
marine ecoregions

Long ago W. Sclater, then Alfred Wallace, attached the name **biogeographic realms** to six vast land areas, each with distinctive kinds and numbers of plants and animals (Figure 50.9). Biogeographic realms maintain their identity partly because of climate. They also tend to maintain a distinct identity when mountain ranges, oceans, and other physical barriers restrict gene flow and thus keep their component species isolated.

Biomes and Ecoregions

Each biogeographic realm may be further divided into biomes. A **biome** is a large region of land characterized by habitat conditions and by its community structure, including endemic species, which evolved nowhere else. Take a look at Figure 50.10, then see how it correlates with the environmental factors shown in Figure 50.11.

As you might deduce, distinctive biomes prevail at certain latitudes and elevations. For example, biomes that are dominated by species of short plants prevail in dry regions, at high elevations, and at high latitudes. Biomes dominated by tall, leafy plant species prevail at tropical and temperate latitudes, and at low elevations with warm temperatures and high rainfall. As you will see shortly, soils also affect biome distribution.

Conservationists are working to locate, inventory, and protect "hot spots." These are portions of certain biomes that are the richest in biodiversity and the most vulnerable to species loss (Section 28.5). Twenty-four of the hot spots hold more than half of all land species.

In its Global 200 program, the World Wildlife Fund is targeting ecoregions as well as hot spots. **Ecoregions** are large areas representative of the globally important biomes *and* water provinces that are now vulnerable to extinction. Figure 50.10 identifies the marine ecoregions.

The land and the seas contain realms of biodiversity. The unique identity of each biogeographic realm, biome, and ecoregion is an outcome of habitat conditions, the kinds and numbers of species, and their evolutionary history.

50.4 SOILS OF MAJOR BIOMES

Most regions of land have **soils**, which are mixtures of mineral particles and variable amounts of decomposing organic material (humus). As described in Section 30.1, the weathering of hard rocks produces coarse-grained gravel, then sand, silt, and finely grained clay. Water and air infiltrate spaces between particles. How much the particles compact together and the proportions of each kind vary within and between regions.

Soils have a layered structure, or *profile*, that reflects their stage of development. Figure 50.12 shows a few examples. Uppermost is topsoil. It has the most humus and is the most vulnerable to erosion. Topsoil may be less than a centimeter thick on steep slopes and more than a meter thick in grasslands. The lowest layer of soil consists of rocks in varying degrees of weathering.

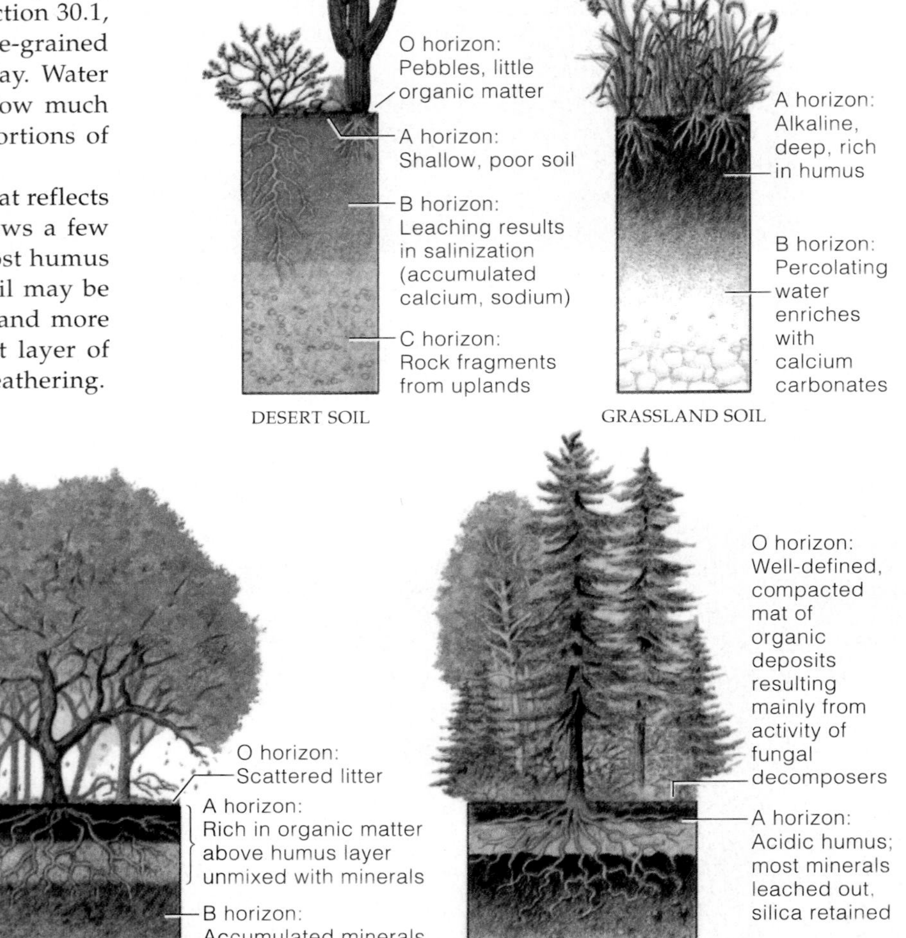
O horizon: Pebbles, little organic matter
A horizon: Shallow, poor soil
B horizon: Leaching results in salinization (accumulated calcium, sodium)
C horizon: Rock fragments from uplands
DESERT SOIL
A horizon: Alkaline, deep, rich in humus
B horizon: Percolating water enriches with calcium carbonates
GRASSLAND SOIL
O horizon: Sparse litter
A–E horizons: Continually leached; iron, aluminum left behind impart red color to acidic soil
B horizon: Clays with silicates, other residues of chemical weathering
TROPICAL RAIN FOREST SOIL
O horizon: Scattered litter
A horizon: Rich in organic matter above humus layer unmixed with minerals
B horizon: Accumulated minerals leached from above
C horizon: Poorly weathered rocks
DECIDUOUS FOREST SOIL
O horizon: Well-defined, compacted mat of organic deposits resulting mainly from activity of fungal decomposers
A horizon: Acidic humus; most minerals leached out, silica retained
B horizon: Accumulated clays with oxides of iron and aluminum
CONIFEROUS FOREST SOIL

Figure 50.12 Soil profiles from a few representative biomes. Compare Section 29.1.

The growth of most plants suffers in poorly aerated, poorly draining soils. Remember, loam topsoils have the best mix of sand, silt, and clay for agriculture. They have enough coarse particles to promote drainage and enough fine particles to retain water-soluble mineral ions that serve as nutrients for plant growth. Gravelly or sandy soils promote rapid leaching, which depletes them of water and vital minerals. Clay soils with fine, closely packed particles are poorly aerated and do not drain well. Few plants grow in waterlogged clay soils.

As farmers know, soil affects primary productivity. They grow most crops in cleared, former grasslands. Many burn or clear-cut tropical forests for agriculture. But these biomes have very little topsoil above poorly draining sublayers. Clearing exposes the topsoil, and heavy rains leach most of the nutrients (Section 28.6).

We turn now to the deserts, shrublands, woodlands, grasslands, various forests, and tundra. Bear in mind, none of these major biomes is uniform throughout. Inside its borders, local climates, landforms, and other physical features favor patches of distinct communities.

Primary productivity depends on the soil profile and its proportions of sand, silt, clay, gravel, and humus.

Further reading: Student Guide to InfoTrac on web site

50.5 DESERTS

Deserts form in lands with less than ten centimeters or so of annual rainfall and high potential for evaporation. Such conditions prevail at latitudes of about 30° north and south. There we find great deserts of the American Southwest, northern Chile, Australia, northern and southern Africa, and Arabia. Farther north are the high deserts of eastern Oregon, and Asia's vast Gobi and the Kyzyl-Kum east of the Caspian Sea. Rain shadows are the main reason why these northern deserts are so arid.

Deserts do not have lush vegetation. Rain falls in heavy, brief, infrequent pulses that swiftly erode the exposed topsoil. Humidity is so low that the sun's rays easily penetrate the air. They quickly heat the ground surface, which radiates heat and cools quickly at night.

Although arid or semiarid conditions do not favor large, leafy plants, deserts show plenty of biodiversity. In one patch of Arizona's desert, you may see creosote bushes and other deep-rooted, evergreen, woody shrubs as well as fleshy-stemmed, shallow-rooted cacti (Figure 50.13). You might see tall saguaros, short prickly pears, and ocotillos, which drop leaves more than once a year and grow new ones after a rain. Desert perennials and annuals flower profusely, spectacularly, but briefly after the rains. Deep-rooted plants, including mesquite and cottonwood, commonly grow near the few streambeds that have a permanent underground water supply.

Of all the land surfaces, more than one-third is arid or semiarid, without enough rainfall to support crops. Crops growing in California's Imperial Valley and some other deserts require unflagging soil management and extensive irrigation. Without drainage programs, such croplands become unproductive from waterlogging and salt buildup. Alarmingly, many parts of the world are now undergoing **desertification**. The term refers to the conversion of vast grasslands and similarly productive biomes to desertlike wastelands with low biodiversity. We take a closer look at this trend in Section 51.6.

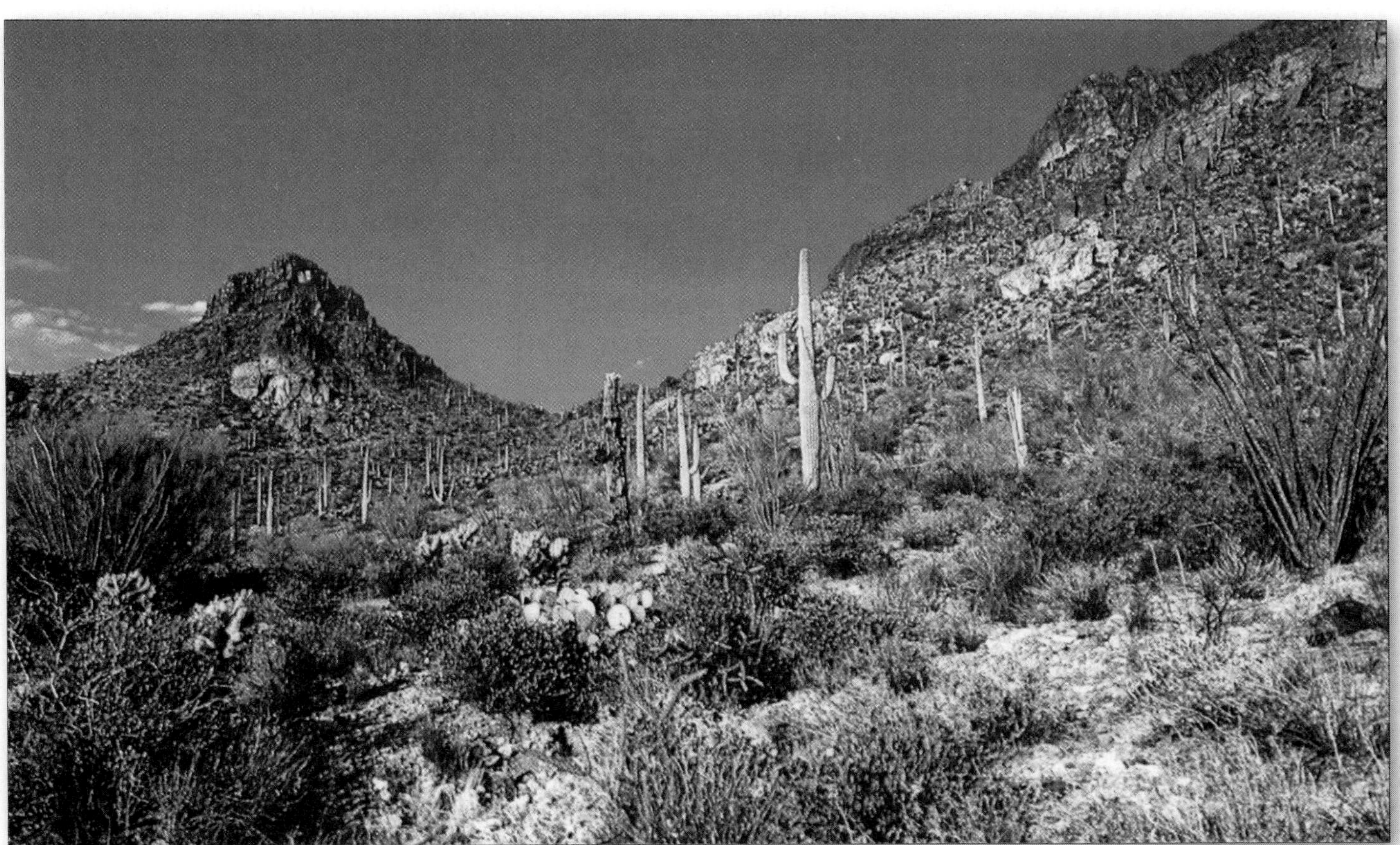

Figure 50.13 Warm desert near Tucson, Arizona. Primary producers include creosote bushes, multistemmed ocotillos, tall saguaro cacti, and prickly pear cacti with rounded pads.

Where the potential for evaporation greatly exceeds sparse rainfall, deserts form. This condition prevails at latitudes 30° north and south and in rain shadows.

50.6 DRY SHRUBLANDS, DRY WOODLANDS, AND GRASSLANDS

Dry shrublands and **dry woodlands** prevail in western or southern coastal regions of the continents between latitudes 30° and 40°. These semiarid regions get more rain than deserts, but not much more. Most falls in mild winters; the summers are long, hot, and dry. Dominant plants often have hardened, tough, evergreen leaves.

We find examples of dry shrublands, which get less than 25 to 60 centimeters of rain per year, in California, South Africa, and Mediterranean regions. Their exotic local names include fynbos and chaparral. California has 2.4 million hectares (6 million acres) of chaparral. In summer, lightning-sparked, wind-driven firestorms can sweep through these biomes (Figure 50.14). Shrubs that have highly flammable leaves quickly burn to the ground. Yet they are highly adapted to episodes of fire and soon resprout from their root crowns. Trees do not fare as well during firestorms. Shrubs, which "feed" the fires, have the competitive edge.

Dry woodlands dominate where annual rainfall is about 40 to 100 centimeters. The dominant trees can be tall, but they do not form a dense, continuous canopy. Eucalyptus woodlands of southwestern Australia and oak woodlands of California and Oregon are like this.

Grasslands cloak much of the interior of continents in zones between deserts and temperate forests. Warm temperatures prevail in the summer, and winters are extremely cold. The 25 to 100 centimeters of annual rainfall keep deserts from forming but aren't enough to support forests. Drought-tolerant primary producers survive strong winds, light and infrequent rainfall, and rapid evaporation. Dominant animals are grazing and burrowing species. Their grazing activities, combined with periodic fires, keep shrublands and forests from encroaching on the fringes of many grasslands.

The main grasslands of North America are *shortgrass* prairie and *tallgrass* prairie. Usually they form where the land is flat or rolling, as in Figure 50.15*a*. Roots of perennial plants extend profusely through the topsoil. In the 1930s, shortgrass prairie of the American Great Plains was overgrazed and also plowed under to grow wheat, which requires more water than this region sometimes receives. Strong winds, prolonged droughts, and unsuitable farming practices turned much of the prairie into a Dust Bowl (Section 51.6). John Steinbeck's *The Grapes of Wrath* and James Michener's *Centennial*, which are two historical novels, speak eloquently of the consequences.

Also in North America's interior, tallgrass prairie once extended west from temperate deciduous forests (Figures 49.4 and 50.15*b*). Diverse legumes and composites, including daisies, thrived in the interior, with its richer topsoil and slightly more frequent rainfall. Farmers converted nearly all of the original tallgrass prairie for agriculture, but certain areas are now being restored (Section 48.7).

Between the tropical forests and deserts of Africa, South America, and Australia are broad belts of grasslands with a smattering of shrubs or trees. These are the **savannas** (Figure 50.15*c*). Rainfall averages 90 to 150 centimeters a year, and prolonged seasonal droughts are common. Where rainfall is low, fast-growing grasses dominate. Acacia and other shrubs grow in regions that get a bit more moisture. Where the rainfall is higher,

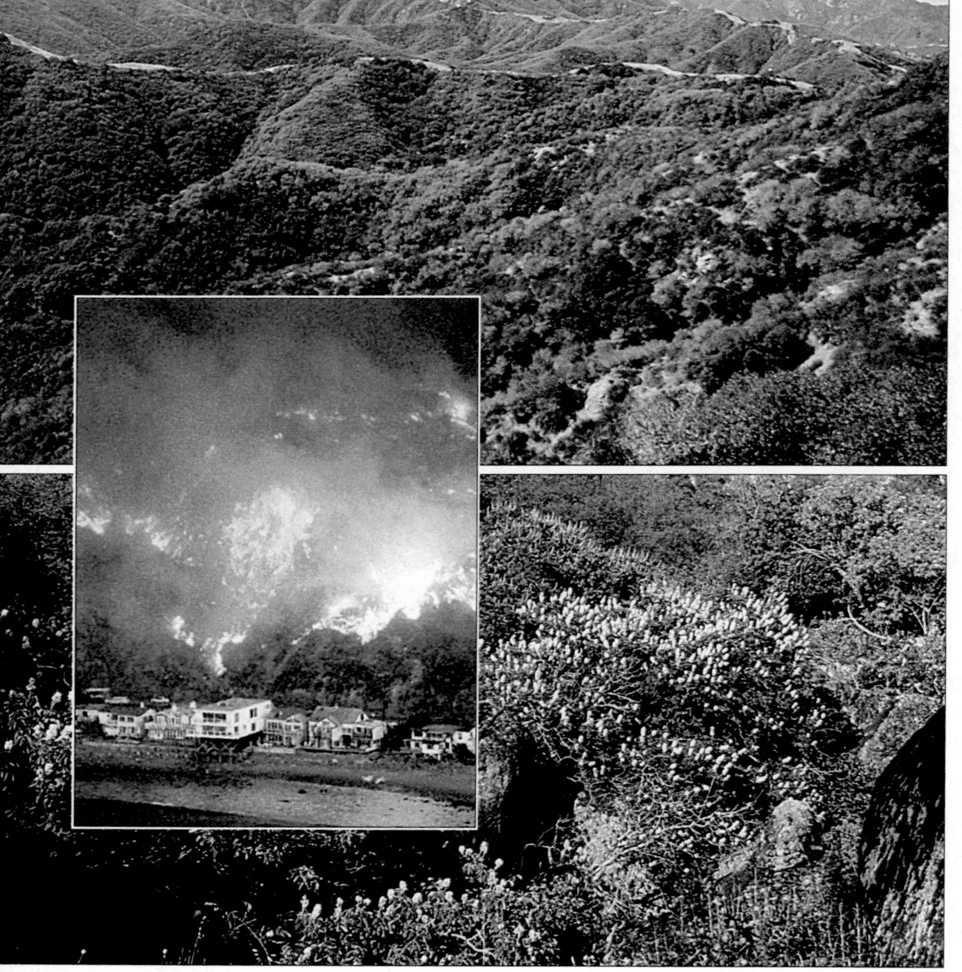

Figure 50.14 California chaparral. The dominant plants are multibranched, woody, and typically only a few meters tall. Without periodic fires, they form a nearly impenetrable vegetation cover. The boxed inset shows a firestorm racing through a chaparral-choked canyon above Malibu.

Further reading: Student Guide to InfoTrac on web site —

Figure 50.15 (**a**) Rolling shortgrass prairie to the east of the Rocky Mountains. Once, bison were the dominant large herbivore of North American grasslands. At one time, the astoundingly abundant grasses supported 60 million of these hefty ungulates. Ungulates are hooved, plant-eating mammals.

(**b**) A rare patch of natural tallgrass prairie in eastern Kansas.

(**c**) The African savanna—warm grasslands with scattered stands of shrubs and trees. More varieties and greater numbers of large ungulates live here than anywhere else. They include migratory wildebeests (shown in this photograph), giraffes, Cape buffalos, zebras, and impalas.

savannas grade into tropical woodlands with tall, coarse grasses, hardy shrubs, and low trees. *Monsoon* grasslands form in southern Asia where heavy rains alternate with a dry season. Dense stands of tall, coarse grasses form, then die back and often burn during the dry season.

Plants of dry shrublands, dry woodlands, and grasslands are supremely adapted to surviving strong winds, grazing animals, and recurring episodes of drought and fire.

50.7 TROPICAL RAIN FORESTS AND OTHER BROADLEAF FORESTS

In forest biomes, tall trees grow close together and form a fairly continuous canopy over a broad stretch of land. In general, there are three types of trees. Which type prevails in a region depends partly on distance from the equator. Evergreen broadleafs dominate between latitudes 20° north and south. Deciduous broadleafs are common at moist, temperate latitudes where winters are not severe. Evergreen conifers are common at high, colder latitudes and in mountains of temperate zones.

Evergreen broadleaf forests sweep across tropical zones of Africa, the East Indies and Malay Archipelago, Southeast Asia, South America, and Central America. Annual rainfall can exceed 200 centimeters and is never less than 130 centimeters. One forest biome, the **tropical rain forest**, depends on regular and heavy rainfall, an annual mean temperature of 25°C, and humidity of 80 percent or more (Figure 50.16). In this highly productive forest, evergreen trees produce leaves and shed old ones throughout the year. Even so, litter doesn't accumulate; decomposition and mineral cycling are extremely rapid in the hot, humid climate. Soils are weathered, humus-deficient, and poor nutrient reservoirs. We will take a closer look at these forests in the next chapter.

Leaving tropical rain forests, we enter regions where temperatures remain mild but rainfall dwindles during part of the year. This is the start of **deciduous broadleaf**

Figure 50.16 Tropical rain forests, biomes of great biodiversity. Among millions of known species are jaguars and *Rafflesia*, a leafless, foul-smelling plant with a fly-pollinated flower that grows three meters across. Bromeliads, orchids, and other epiphytes grow on trees, obtaining minerals from organic remains of leaves, insects, and other litter that have dissolved in tiny pockets of water.

Further reading: Student Guide to InfoTrac on web site

SPRING

SUMMER

WINTER

AUTUMN

Figure 50.17 Part of a temperate deciduous forest south of Nashville, Tennessee, in spring, summer, autumn, and winter.

forests. Trees of *tropical* deciduous forests drop some or all of their leaves during a pronounced dry season. The *monsoon* forests of India and southeastern Asia have such trees, also. Farther north, in the temperate zone, rainfall is even lower. The winters are cold, and water is locked away as snow and ice. *Temperate* deciduous forests, such as those of the southeastern United States, prevail here (Figure 50.17). Decomposition is not as rapid as in the humid tropics, and many nutrients are conserved in accumulated litter on the forest floor.

Complex forests of ash, beech, birch, chestnut, elm, and deciduous oaks once stretched across northeastern North America, Europe, and East Asia. They declined drastically when farmers cleared the land. Pathogenic species introduced to North America destroyed nearly all of the chestnuts and many elms (Section 48.8). Now, maple and beech predominate in the Northeast. Farther west, oak–hickory forests prevail, then oak woodlands that eventually grade into tallgrass prairie.

In forest biomes, conditions favor dense stands of tall trees that form a continuous canopy over a broad region.

50.8 CONIFEROUS FORESTS

Conifers (cone-bearing trees) are primary producers of **coniferous forests**. Most have needle-shaped leaves with a thick cuticle and recessed stomata—adaptations that help the trees conserve water through dry times. They dominate the boreal forests, montane coniferous forests, temperate rain forests, and pine barrens.

Boreal forests stretch across northern Europe, Asia, and North America. Such forests also are called *taigas*, meaning "swamp forests." Most are in glaciated regions with cold lakes and streams (Figure 50.18*a*). It rains mostly in summer, and evaporation is low in the cool summer air. The cold, dry winters are more severe in eastern parts of these biomes than in the west, where ocean winds moderate the climate. Spruce and balsam fir dominate the boreal forests of North America. Pines, birches, and aspens take hold in the burned or logged areas. Where soil is poorly drained, highly acidic bogs dominated by peat mosses, shrubs, and stunted trees prevail. Boreal forests become much less dense to the north, where they grade into arctic tundra.

Figure 50.18 (**a**) Spruce-dominated boreal forest. (**b**) Montane coniferous forest of Yosemite Valley in California's Sierra Nevada. (**c**) A firefighter retreats from a wall of flames in a pine–oak forest near Daytona Beach, Florida.

In the Northern Hemisphere, montane coniferous forests extend southward through the great mountain ranges. Spruces and firs dominate in the north and at higher elevations. They give way to firs and pines in the south and at lower elevations (Figures 3.1 and 50.18*b*).

Some temperate lowlands support coniferous forests. One temperate rain forest that parallel the coast from Alaska into northern California contains some of the world's tallest trees—Sitka spruce to the far north and redwoods to the south. Logging destroyed much of this biome. One bright note: In 1994, a timber corporation surrendered its rights to log one of the largest intact temperate rain forests outside of the tropics. The trees, 800 years old, cloak British Columbia's Kitlope Valley.

Southern pine forests dominate the coastal plains of the south Atlantic and the Gulf states. Pine species are adapted to the dry, sandy, nutrient-poor soil and to natural fires or controlled burns. These fires do not get hot enough to damage the trees, and they open up the understory. Forests of pine, scrub oak, and wiregrass grow in New Jersey. Palmettos grow beneath pines and loblolly in the Deep South. During a heat wave in 1998, fires swept through more than 320,000 acres in Florida. Smaller burns had been suppressed in counties where housing and other developments encroached on the forests. Dense, tinder-dry understory accumulated and was the basis for the destruction (Figure 50.18*c*).

Coniferous forests prevail in regions in which a cold, dry season alternates with a cool, rainy season.

Further reading: Student Guide to InfoTrac on web site —

50.9 ARCTIC AND ALPINE TUNDRA

Tundra is derived from *tuntura*, a Finnish word for the great treeless plain between the polar ice cap and belts of boreal forests in Europe, Asia, and North America. Much of this *arctic* tundra is flat, windswept, and wet (Figure 50.19*a*). Temperatures are cool in summer and below freezing in winter, so not much water evaporates. Little rain or snow falls. Sunlight is nearly continuous during the summer months. Then, short plants grow and flower profusely, and seeds ripen fast.

Snow does not cloak arctic tundra all year long, but summers are too short to thaw much more than surface soil. Just beneath the surface is a perpetually frozen layer, the **permafrost**. It is more than 500 meters thick in some places. Because permafrost prevents drainage, the soil above it remains waterlogged even in summer. The anaerobic conditions and low temperatures limit nutrient cycling. Organic matter decomposes slowly, so it accumulates in soggy masses of organic matter. All but about 5 percent of the carbon in the arctic tundra is locked up in peat bogs.

Figure 50.19 (**a**) Extensive ponding in the arctic tundra. The rain and snowmelt cannot percolate downward because of the permafrost. (**b**) Short, hardy plants typical of alpine tundra.

A similar type of biome prevails at high elevations in mountains throughout the world, although there is no permafrost beneath the soil. Figure 50.19*b* shows an example of this *alpine* tundra. Dominant plants often form low cushions and mats that are able to withstand the buffeting of strong winds. Winter temperatures fall below freezing. Even in the summer, shaded patches of snow persist. The thin, fast-draining soil of an alpine tundra is nutrient-poor, so primary productivity is low.

At high latitudes with short, cool summers and long, very cold winters, we find arctic tundra. In high mountains where seasonal changes vary with latitude but where the climate is too cold to support forests, we find alpine tundra.

50.10 FRESHWATER PROVINCES

Freshwater and saltwater provinces cover more of the Earth's surface than all biomes combined. They include lakes, ponds, estuaries, wetlands, coral reefs, oceans, and hydrothermal vents. "Typical" examples are not that easy to find. Some ponds can be waded across; Lake Baikal in Siberia is more than 1.7 kilometers deep. All aquatic ecosystems have gradients in light penetration, temperature, and dissolved gases, but the values differ greatly. All we can do here is sample the diversity.

Lake Ecosystems

A **lake** is a body of standing fresh water produced by geologic processes, as when an advancing glacier carves out the land beneath it. When the glacier retreats, water collects in the exposed basin (Figure 50.20). Over time, erosion and sedimentation alter the dimensions of the lake, which usually ends up filled or drained completely.

A lake has littoral, limnetic, and profundal zones (Figure 50.21). The littoral extends all around the shore to a depth at which rooted aquatic plants stop growing. The water is shallow and usually well lit. Diversity is greatest here. The limnetic is open, sunlit water past the littoral; it extends to a depth where photosynthesis is insignificant. Communities of tiny organisms (plankton) abound here. Cyanobacteria, diatoms, and green algae dominate the *phyto*plankton. The *zoo*plankton include rotifers and copepods. The profundal extends through all open water below the depth at which wavelengths suitable for photosynthesis can penetrate. Detritus sinks through this zone to bottom sediments. Communities of diverse bacterial decomposers live in and on bottom sediments, and they enrich the water with nutrients.

SEASONAL CHANGES IN LAKES In temperate regions where summers are warm and winters cold, lakes show seasonal changes in density and temperature from the surface to the bottom. A layer of ice forms over many of them in midwinter. Water near the freezing point is the least dense and accumulates beneath the ice. Water at 4°C is the most dense. It collects in deeper layers that are a bit warmer than the surface layer in midwinter.

Figure 50.20 In the Canadian Rockies, the basin of Moraine Lake, formed by glacial action during the most recent ice age.

In spring, daylength increases and the air is not as cold. Lake ice melts, the surface water slowly warms to 4°C, and temperatures turn uniform throughout. Winds blowing over the surface now cause a **spring overturn**: strong vertical movements carry dissolved oxygen from the surface layer to its depths, and nutrients released by decomposition move from sediments to the surface.

By midsummer, the surface layer is above 4°C. Now the lake has a *thermocline*—a midlayer of water where the temperature changes abruptly and blocks vertical mixing (Figure 50.22). Being warmer and less dense, the surface water floats on the thermocline. In cooler water beneath it, decomposers typically deplete the dissolved oxygen. Come autumn, the upper layer cools, becomes denser, then sinks, so the thermocline vanishes. During this **fall overturn**, water mixes vertically, so dissolved oxygen moves down and nutrients move up.

Primary productivity corresponds with the seasons. After a spring overturn, the longer daylengths and the cycled nutrients support higher rates of photosynthesis. Phytoplankton and rooted aquatic plants quickly take up phosphorus, nitrogen, and other nutrients. During a growing season, the thermocline cuts off vertical mixing. Nutrients locked in the remains of organisms sink to deeper water, so photosynthesis slows. By late summer,

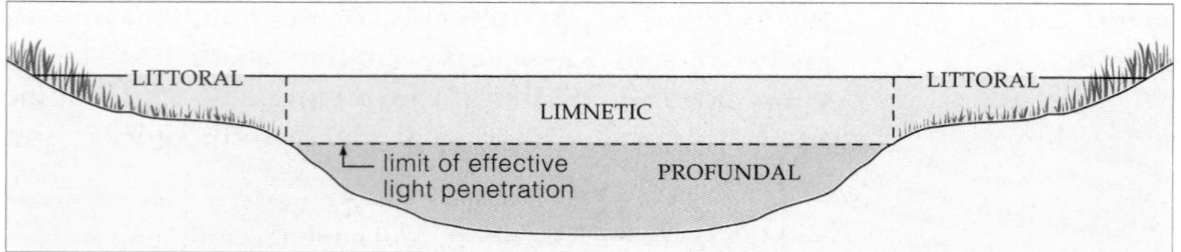
LITTORAL
LIMNETIC
LITTORAL
limit of effective light penetration
PROFUNDAL

Figure 50.21 Lake zonation. The littoral extends around the lake's edge from the shore to the depth where aquatic plants stop growing. The profundal is all water below the depth of light penetration. Above the profundal are open, sunlit waters of the limnetic zone.

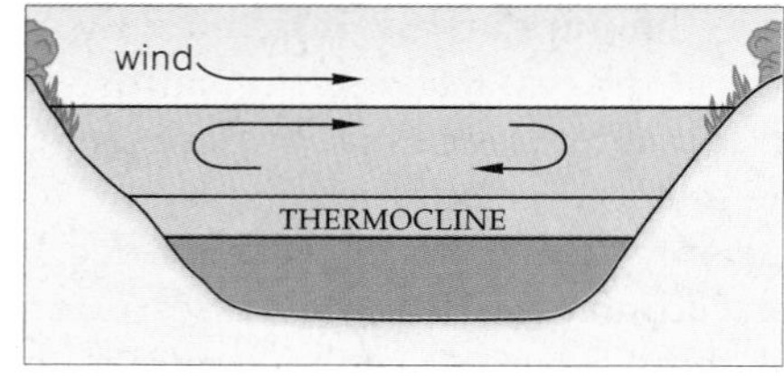
wind
THERMOCLINE

Figure 50.22 Thermal layering in a temperate lake in summer.

Further reading: Student Guide to InfoTrac on web site

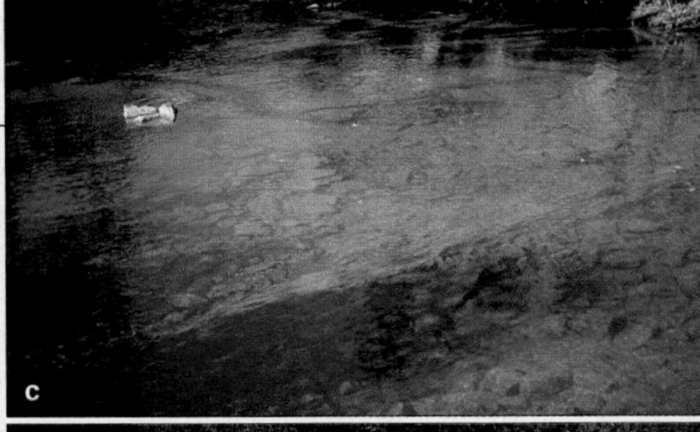
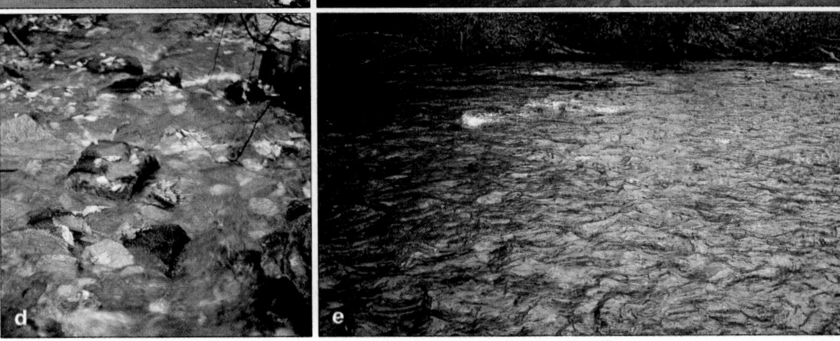

Figure 50.23 Stream habitats in North Carolina and Virginia. (**a**) Pool. (**b**) Pool leading into a riffle. (**c**) A run, Sinking Creek. (**d**) Leaf detritus in a riffle. (**e**) Closer look at a riffle.

shortages are limiting photosynthesis. A fall overturn cycles nutrients to the surface, and this drives a burst of primary productivity. But the shorter daylight hours do not sustain a long-lasting burst. Not until spring will primary productivity increase again.

TROPHIC NATURE OF LAKES The topography, climate, and geologic history of a lake dictate the numbers and kinds of residents, their dispersal, and the cycling of nutrients among them. Soils of the lake basin and the surrounding regions contribute to the type and amount of nutrients available to support organisms.

Interplays among the climate, soil, basin shape, and metabolic activities of a lake's residents contribute to conditions ranging from oligotrophy to eutrophy. Water in *oligotrophic* lakes often is deep, clear, and nutrient poor. Its primary productivity is not great. The water in *eutrophic* lakes typically is shallow, nutrient rich, and rather high in primary productivity. These conditions develop naturally as sediments gradually accumulate in the lake basin. The water becomes less deep and less transparent, compared with water in oligotrophic lakes, and phytoplankton dominate aquatic communities. More sediments usually accumulate, and the final successional stage is a filled-in basin—hence no more lake.

Eutrophication is a name for processes that enrich a body of water with nutrients. As Section 49.11 described, human activities can bring it about. Here, think about Val Smith's report on the way people brought about eutrophication of Lake Washington in Seattle—and then reversed it. From 1941 to 1963, this lake received too much phosphorus-rich sewage, which promoted huge blooms of cyanobacteria. These formed thickened, slimy mats over the lake each summer and made it useless for recreation. With phosphorus enrichment, nitrogen became the lake's limiting resource. So cyanobacteria—which are superior competitors for nitrogen by their ability to fix N_2—became dominant. From 1963 to 1968, sewage discharges slowed, then stopped. By 1975, the lake neared full recovery. Lower density populations of diatoms and green algae became dominant.

Stream Ecosystems

The flowing-water ecosystems called **streams** start out as freshwater springs or seeps. They grow and merge as they flow downslope, then they often combine to form a river. Between the headwaters and a river's end, we find three kinds of habitats—riffles, pools, and runs. The *riffles* are shallow, turbulent stretches where water flows swiftly over a rough bottom of sand and rock. Figure 50.23 shows two examples. The *pools* have deep water flowing slowly over a smooth, sandy, or muddy bottom. The *runs* are smooth-surfaced but fast-flowing stretches over bedrock or rock and sand.

A stream's average flow volume and temperature depend on rainfall, snowmelt, geography, altitude, and even the shade cast by plants. Its solute concentrations are influenced by the streambed's composition as well as by agricultural, industrial, and urban wastes.

Especially in forests, streams import most of the organic matter that supports food webs. Where trees cast shade, photosynthesis is diminished. Most of the organic matter is litter that enters detrital food webs. The aquatic organisms continually take up and release nutrients as water flows downstream. Nutrients move upstream only as components of migrating fishes and other animals. Think of nutrients as spiraling between water and aquatic organisms as the stream flows on its one-way course, usually to a river that flows to the sea.

Ever since cities formed, streams have been sewers for industrial and municipal wastes. The wastes, as well as other pollutants from poorly managed farmlands, choked many streams with sediments and chemically poisoned them. However, streams are resilient. They can recover impressively when pollution is controlled.

Freshwater and saltwater provinces are far more extensive than the Earth's biomes. All of the aquatic ecosystems in these water provinces have characteristic gradients in light penetration, temperature, and dissolved gases.

50.11 THE OCEAN PROVINCES

Beyond land's end are two vast provinces of the world ocean, which covers almost three-fourths of the Earth's surface. The *benthic* province includes all sediments and rocks of the ocean bottom (Figure 50.24). It starts at continental shelves and extends to deep-sea trenches. The *pelagic* province is the full volume of ocean water. Its neritic zone is all water above continental shelves, and its oceanic zone is the water of the ocean basins.

Walk along the ocean and you might be humbled by its vastness (Figure 50.25). An immense volume of water extends to the distant horizon, with no mountains or plains or valleys to break it up visually into something less overwhelming. What you do not see are *submerged* mountains and valleys and plains—which are as varied as the ones configuring the dry continents.

Primary Productivity in the Ocean

Within the ocean's upper surface waters, photosynthesis proceeds on a stupendous scale, just as it does on land. Primary productivity there varies seasonally, just as it does on land (Section 7.9 and Figure 50.25). Its huge "pastures" of phytoplankton are the start of food webs that include copepods, shrimplike krill, whales, squids, and a great variety of fishes. The organic remains and wastes from all of its marine communities slowly sink through the water and support detrital food webs for most of the benthic communities.

Close to the ocean surface, 70 percent of primary productivity might be the work of **ultraplankton**. These are all photosynthetic bacterial cells about 2 micrometers wide (40 millionths of an inch). In subtropical and tropical waters once thought to be nearly devoid of producers, 0.035 gram (1 ounce) of water holds up to 3 million of the cells. In deep ocean water too dark to support photosynthesis, food webs start with **marine snow**. These bits of organic matter collectively drift down to become the food base for staggering biodiversity in the midoceanic waters—possibly up to 10 million species. In what may be the greatest of all circadian migrations, some species rise thousands of feet to feed at night in shallower water, then move down the next morning. Top carnivores include siphonophores, a type of soft-bodied cnidarian. At thirty-nine meters, *Praya dubia*—a newly discovered siphonophore—is one of the world's longest animals!

Hydrothermal Vents

In 1977, researchers studying the ocean floor near the Galápagos Islands discovered a new ecosystem. In the Galápagos Rift, a volcanically active boundary between two crustal plates, they found communities thriving near **hydrothermal vents**. At such vents, near-freezing water seeps down into fissures and becomes heated to extremely high temperatures. As the pressure forces up heated water, minerals are leached from rocks before the water spews out through vents in the seafloor. Iron, zinc, copper sulfides, and sulfates of magnesium and calcium are dissolved in hydrothermal outpourings. These settle out and form rich mineral deposits. Sulfides

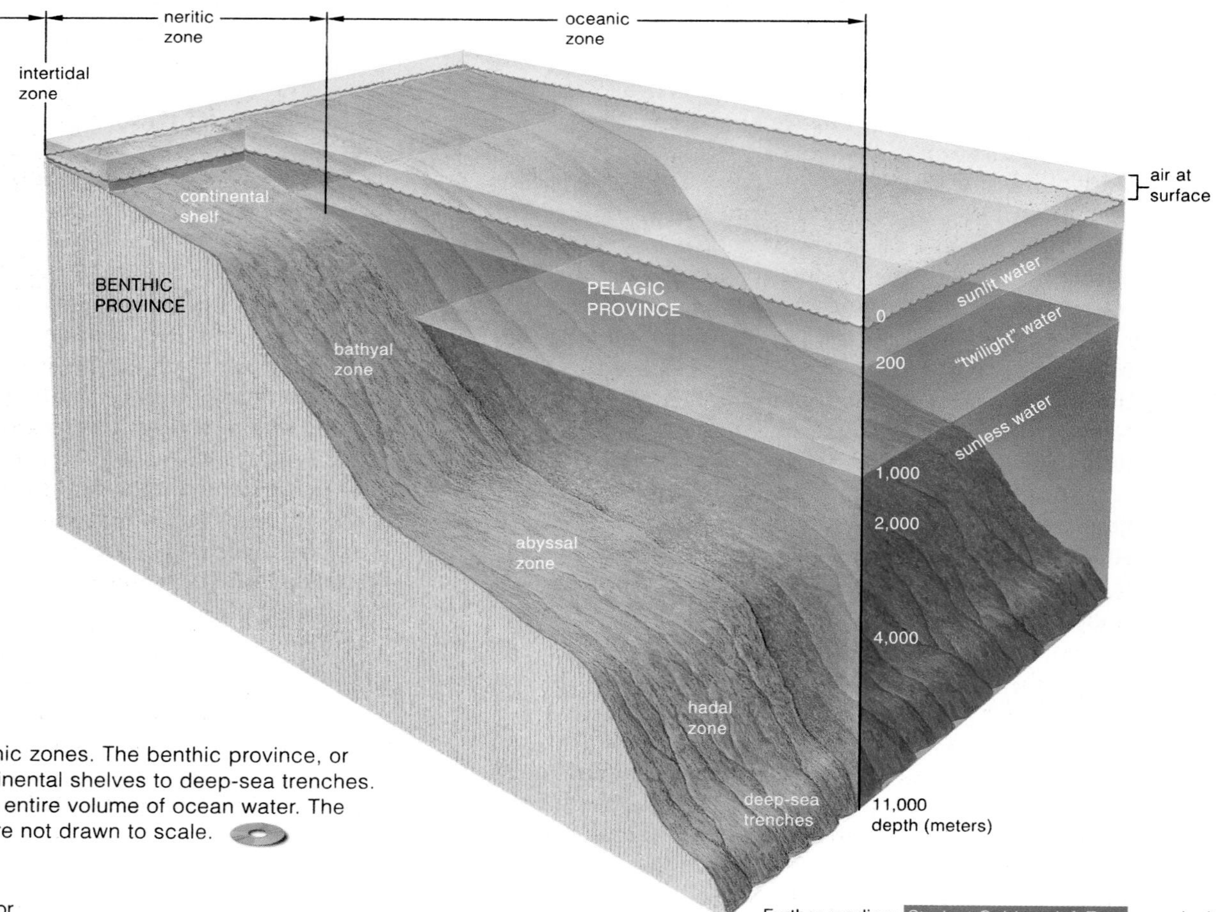

neritic zone
oceanic zone
intertidal zone
air at surface
continental shelf
BENTHIC PROVINCE
PELAGIC PROVINCE
sunlit water
"twilight" water
sunless water
bathyal zone
abyssal zone
hadal zone
deep-sea trenches
0
200
1,000
2,000
4,000
11,000
depth (meters)

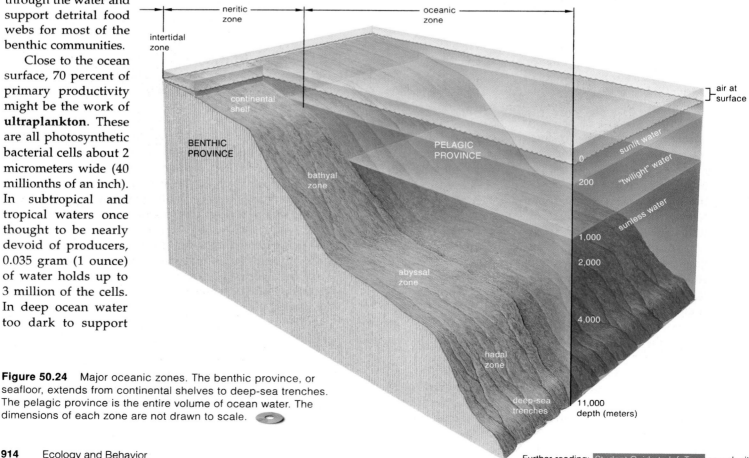

Figure 50.24 Major oceanic zones. The benthic province, or seafloor, extends from continental shelves to deep-sea trenches. The pelagic province is the entire volume of ocean water. The dimensions of each zone are not drawn to scale.

Further reading: Student Guide to InfoTrac on web site

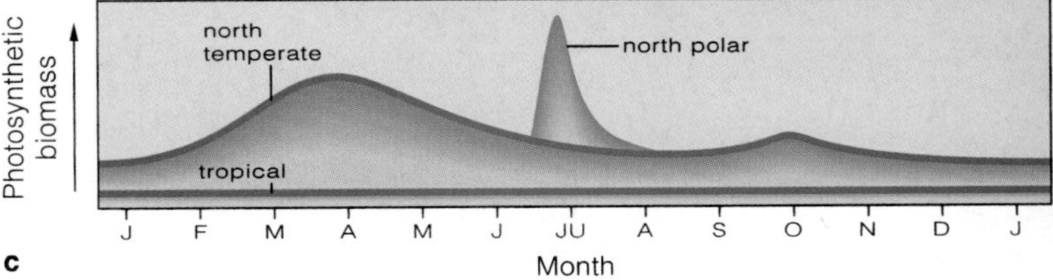
north temperate
north polar
tropical
Photosynthetic biomass
J
F
M
A
M
J
JU
A
S
O
N
D
J
Month

Figure 50.25 (**a**) A wave breaking on the ocean shore. (**b**) Near the British Columbia coastline, a whale breaching (leaping out of the water).

(**c**) Seasonal variations in primary production in the ocean, which correspond to latitude. The strongest peak in the *dark-green* graph lines for the north polar and temperate seas corresponds to phytoplankton blooms brought about by a seasonal increase in daylength. The lower peak corresponds to an increase in the availability of nutrients, something like the fall overturn in lakes. In most tropical seas, daylength and nutrient availability do not vary much; neither does primary productivity.

Tube worms (**d**) and crustaceans (**e**) at a hydrothermal vent ecosystem on the ocean floor.

in the deposits are energy sources for chemoautotrophs, the starting point for hydrothermal vent communities. Chemoautotrophic bacteria are primary producers for food webs that include red tube worms, crustaceans, clams, and fishes (Figure 50.25*d*,*e*).

Still more hydrothermal vent ecosystems have been found in the South Pacific near Easter Island; the Gulf of California, about 150 miles south of the tip of Baja California, Mexico; and the Atlantic. In 1990, a team of United States and Russian scientists found one in Lake Baikal, the world's deepest lake. This lake basin seems to be splitting apart (hence the vents) and may mark the beginning of a new ocean.

Did life originate in such nutrient-rich places on the seafloor? Conditions at the surface of the early Earth were about as inhospitable as you might imagine. At the least, cells at the seafloor would have had protection from the destructive radiation bombarding the Earth before the oxygen-rich atmosphere formed. Were some descendants of those cells carted closer to the surface as the seafloor was uplifted during episodes of crustal crunchings? These are some unanswered questions that evolutionary detectives are now asking. Some of their interesting hypotheses are described in Section 7.8.

Although we are most familiar with features of the land, a continuous ocean dominates the Earth's surface. Its primary productivity and biodiversity are staggering.

Communities of organisms thrive near hydrothermal vents on the ocean floor, which suggests to some that life itself may not have originated in the shallow waters of the Earth.

50.12 CORAL REEFS AND CORAL BANKS

Types of Coral Reefs

Coral reefs are wave-resistant, nutrient-rich formations in the nutrient-poor tropical seas. They consist mainly of accumulated remains of corals and coralline algae (Section 28.3). The reefs themselves are the foundation for complex ecosystems where a variety of habitats and rapid nutrient cycling sustain staggering biodiversity.

Long ago, Charles Darwin recognized three types of formations: fringing reefs, barrier reefs, and atolls. He understood that the formation of many reefs is related to the emergence of submarine volcanoes above the sea surface and their eventual subsidence (Figure 50.26). However, he couldn't explain why volcanoes grew and disappeared below the waves because he did not know about hot spots and plate tectonics (Section 20.10).

Fringing reefs form near the land in regions where rainfall and runoff are light. You commonly find them on the leeward (downwind) side of the most recently formed volcanic islands, such as the Hawaiian Islands. Figure 50.26*a* shows an example from Moorea, one of the popular tourist destinations in the South Pacific.

Barrier reefs form parallel with the shore of volcanic islands or continents. Figure 50.26*b* shows the barrier reef along the island of Bora Bora in the South Pacific. Calm lagoons form behind these reefs, and they may be a few meters to sixty meters deep. The largest example of biological "architecture" is Australia's Great Barrier Reef. This is not one continuous reef. It is a string of thousands of them, some of which are 150 kilometers (95 miles) across. They parallel Queensland's coast for 2,500 kilometers (1,560 miles). The biodiversity here is incredible. You might observe 50 coral species on other reefs, but about 500 live on the Great Barrier Reef. So (for example) do 3,000 species of fishes, 1,000 species of mollusks, and 40 species of sea snakes.

Ring-shaped islands called *atolls* are actually coral reefs and coral debris that wholly or partly enclose a shallow lagoon (Figure 50.26*c*). Channels typically link the lagoon to the open sea, but biodiversity is not great here. Shallow water can become too hot for corals. Also, during heavy rainfall, the water becomes so freshened that corals go into osmotic shock. That's one reason why you will not find coral reefs where a river drains into the sea or near rainfall-drenched land.

Coral Banks

Beyond latitudes 25° north and south, solitary corals and small colonies construct **coral banks** in temperate seas. We observe coral banks even in cold seas above the continental shelves. Examples are the smooth, vertical banks in cold, deep waters near the coasts of England, California, Japan, and New Zealand. One colonial branched coral, *Lophelia*, has built great banks in the cold waters of Norway's fjords.

a
b

c

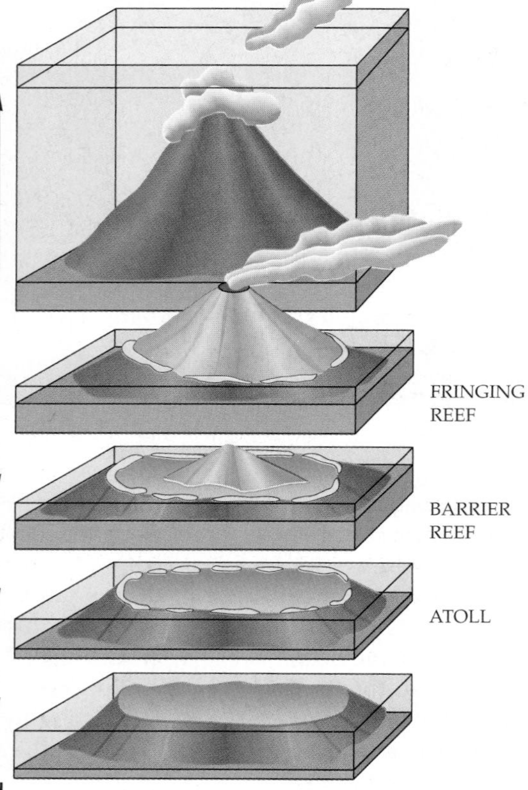
FRINGING REEF
BARRIER REEF
ATOLL
d

Figure 50.26 Three types of coral reef formations: (**a**) fringing reef, (**b**) barrier reef, and (**c**) atoll. (**d**) Correlation between reef formation and the origin and later subsidence of volcanic islands. Fringing reefs start to form when the top of a submerged and actively growing volcano has risen above the sea surface. When volcanic activity ceases, the island gradually subsides even as corals continue their reef-building activity. An atoll is the last formation before full subsidence.

Further reading: Student Guide to InfoTrac on web site

Figure 50.27 Distribution map for coral reefs (color-coded *orange*), coral banks (*green*), and solitary corals (*red*). Nearly all of the reef-building corals are confined to areas with warm sea surface temperatures (inside the dark lines). The photograph shows a tiny sampling of coral reef biodiversity in the Red Sea.

Distribution of Coral Formations

As you can see from Figure 50.27, the most substantial coral reefs form in warm waters between latitudes 25° north and south. The restriction is an outcome of their dependence on mutualistic dinoflagellates. Corals offer safe, stable habitats for their captive partners, which provide corals with oxygen, carbohydrates, and a local pH that promotes deposition of the calcium carbonate used to make coral skeletons (Section 26.5).

Dinoflagellates cannot survive in the cold waters of higher latitudes. As you might imagine, coral species there cannot grow as rapidly, and the structures found at coral banks are smooth, not lacily exuberant as they are in the tropics. Rare, solitary corals live in the cold, deep waters of Antarctica.

Reef biodiversity is endangered. Coral bleaching is one threat (Section 28.3). So are fast-growing seaweeds that are now smothering the reefs. Humans have killed nearly all the turtles and fishes that fed on the seaweed populations and kept them in check. Also, nearly all of the sea urchins that fed on these seaweeds were wiped out during recent, unusually destructive hurricanes.

Coral reefs generally form in clear, warm waters between latitudes 25° north and south. Vertical coral banks form in temperate seas and in cold waters above continental shelves.

50.13 MANGROVE WETLANDS

At tropical latitudes we find a different kind of nutrient-rich, saltwater ecosystem. It forms in tidal flats near the sea, and it is called a **mangrove wetland**. "Mangrove" refers to forests in sheltered regions along tropical coasts (Figure 50.28*a*). Ocean waves cannot reach these regions, so anaerobic sediments and mud accumulate. The term also refers to a particular group of salt-tolerant plants (halophytes), such as red mangrove (*Rhizophora*). Plants have shallow, spreading roots or branching prop roots that extend from the trunk (Figure 50.28*b*). Many have pneumatophores, root extensions that take up oxygen.

Figure 50.28 (**a**) Aerial view of a mangrove wetland, Florida Everglades. (**b**) Red mangrove prop roots.

Florida's mangrove wetlands contain few mangrove species, whereas those in Southeast Asia contain as many as thirty or forty. Seeds of all of these species germinate while still attached to the parent tree. Seedlings drop to the water and float upright until currents deposit them in shallow water, where their roots take hold in mud.

The net primary productivity of mangrove wetlands varies, depending partly on the volume and flow rate of water moving in and out with the tides. It depends also on salinity and nutrient availability, which is affected by litter accumulation and turnover. But these are all rich ecosystems. Tidal circulation and nutrient inputs from land, by way of streams and rivers, combine to support considerable biomass. Particularly along the Malaysian peninsula and the Gulf of Mexico, nutrient-rich detritus from mangrove wetlands enriches nearby estuaries, a topic of the section that follows.

In tidal flats of tropical regions, mangrove wetlands are rich ecosystems dominated by salt-tolerant species.

50.14 ESTUARIES AND THE INTERTIDAL ZONE

Mangrove wetlands are not the only rich ecosystems that develop near the coasts of continents and islands. Here also, we find estuaries and intertidal zones. Like freshwater ecosystems, these marine ecosystems differ in their physical and chemical properties, such as light penetration and water temperature, depth, and salinity.

Life in Estuaries

An **estuary** is a partially enclosed coastal region where seawater slowly mixes with nutrient-rich freshwater from rivers, streams, and runoff from the surrounding land. The confined conditions, slow mixing of water, and tidal action combine to trap dissolved nutrients. The continually freshened water replenishes nutrients and allows estuaries to support productive ecosystems.

Primary producers are phytoplankton, salt-tolerant plants that withstand submergence at high tide, and algae living in mud and on plants. Much of the primary production enters detrital food webs in which bacterial and fungal decomposers are the first to feed. Detritus (and bacteria on and in it) feeds nematodes, snails, crabs, and fish. Filter feeders such as clams eat food particles suspended in the slowly moving water. So many larval and juvenile stages of invertebrates and some fishes are present that estuaries are often called marine nurseries. Also, many migratory birds use estuaries as rest stops.

Chesapeake Bay, Mobile Bay, and San Francisco Bay are broad, shallow estuaries. Estuaries in Alaska and British Columbia are narrow and deep; so are Norway's fjords. In Texas and Florida, estuaries lie behind long spits of sand and mud. The New England coast has many estuarine salt marshes (Figure 50.29). Mangrove swamps, described in the preceding section, function much like salt marshes in estuarine ecology.

Many estuarine ecosystems are under serious assault from raw sewage; agricultural runoff; industrial, urban, and suburban wastes; and upstream diversion of fresh water for human use (Section 49.3). Normal conditions, such as suitable salinity levels, cannot be maintained without inflows of unpolluted fresh water.

Life Along Rocky and Sandy Coastlines

Along rocky and sandy coastlines we find ecosystems of the **intertidal zone**, which is not renowned for its creature comforts. Waves batter its resident organisms. Tides alternately submerge and expose them. The higher they are, the more they dry out, freeze in winter, or bake in summer, and the less food comes their way. The lower they are, the more they must compete in limited spaces. At low tides, birds, rats, and raccoons move in to feed on them. High tides bring the predatory fishes.

Generalizing about coastlines isn't easy, for waves and tides constantly resculpt them. But one feature that rocky and sandy shores do share is vertical zonation.

Rocky shores often have three zones (Figure 50.30*a*). The highest zone, the upper littoral, is submerged only during the highest tide of the lunar cycle; it is sparsely populated. The midlittoral (middle zone) is submerged during the highest regular tide and exposed during the lowest. In its tidepools we typically find red, brown, and green algae, hermit crabs, nudibranchs, sea stars, and small fishes (Figure 50.30*b*). Diversity is greatest in the lowest zone, the lower littoral, which is exposed only during the lowest tide of the lunar cycle. In all three shore zones, rapid erosion prevents detritus from accumulating, so grazing food webs prevail.

Waves and currents continually rearrange stretches of loose sediments, called *sandy* and *muddy* shores. Few

SALT MARSH (estuary)
open ocean
sound
shallow bay
creek
tidal river

Figure 50.29 Salt marsh of a New England estuary. The view above shows the open ocean in the distance. The closer view at right shows *Spartina*, a marsh grass that is the main producer. Its microbe-enriched litter feeds consumers in the shallow estuarine water and muddy sediments.

Further reading: Student Guide to InfoTrac on web site →

UPPER LITTORAL
MIDLITTORAL
LOWER LITTORAL
a

b

Figure 50.30 (**a**) Example of vertical zonation in the intertidal zone of a rocky shore in the Pacific Northwest. The range between high and low tides is about three meters. It varies from a low of a few centimeters in the Mediterranean Sea to more than fifteen meters in the Bay of Fundy, near Nova Scotia. (**b**) A tidepool that is typical of the Pacific Northwest.

large plants grow in these unstable places, so you will not find many grazing food webs. Detrital food webs start with organic debris imported from offshore or nearby landforms. The vertical zonation is less obvious than along rocky shores. Below the low tide mark in temperate regions are sea cucumbers and blue crabs. Marine worms, crabs, and other invertebrates thrive between high and low tide marks. At night, at the high tide mark, beach hoppers and ghost crabs pop out of their burrows, seeking food.

Upwelling Along the Coasts

In the Northern Hemisphere, prevailing winds from the north parallel the west coasts of continents and tug on the ocean surface. Wind friction causes the surface waters to begin moving. Under the force of the Earth's rotation, the slow-moving water is deflected westward,

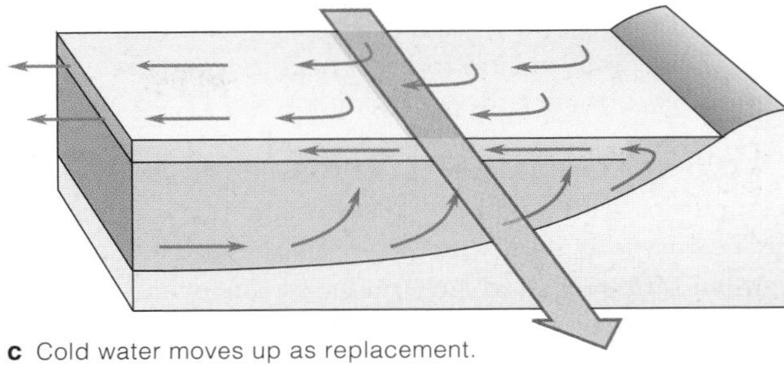
a Wind from the north starts surface ocean water moving.
b Force of Earth's rotation deflects the moving water westward.
c Cold water moves up as replacement.

Figure 50.31 Upwelling along the west coasts in the Northern Hemisphere. Prevailing winds from the north start surface water moving. The force of Earth's rotation deflects the moving water westward. Cold, deeper water moves up to replace it.

away from coasts. And cold, deep, often nutrient-rich water moves in vertically to replace it (Figure 50.31). Whenever cold, deep water moves up this way, we call it **upwelling**. It happens in equatorial currents as well as along the coasts of continents in both hemispheres, and it cools air masses above it. For example, fogbanks along the California coast are an outcome of warm air encountering upwellings of cold water.

In the Southern Hemisphere, the commercial fishing industries of Peru and Chile depend on wind-induced upwelling. When prevailing coastal winds blow in from the south and southeast, they tug surface water away from shore; and cold, deeper water that the Humboldt Current delivers to the continental shelf moves to the surface. Tremendous amounts of nitrate and phosphate are pulled up, then carried northward by the cold Peru Current. Phytoplankton that depend on these nutrients are the basis of one of the world's richest fisheries.

Every three to seven years, warm surface waters of the western equatorial Pacific Ocean move eastward. This massive displacement of warm water affects the prevailing wind direction. The eastward flow speeds up so much that it influences the movement of water along the coastlines of Central and South America.

Waters driven toward any coastline will be forced downward and will flow seaward along the continental shelf. Near Peru's coast, a prolonged "downwelling" displaces the cooler waters of the Humboldt Current—and prevents upwelling. The phenomenon, which local fishermen named **El Niño**, has catastrophic effects on productivity, on seabirds that feed on anchovetas and other fishes, and on fishing industries. The next section provides a closer look at the El Niño phenomenon.

Major shifts in the circulation patterns of the ocean and atmosphere can have repercussions on the functioning of ecosystems that are global in scope.

Focus on Science

50.15 RITA IN THE TIME OF CHOLERA

We turn now to an application that reinforces a unifying concept of this chapter—that events in the atmosphere, in the ocean, and on land interconnect in ways that can profoundly influence the world of life.

EL NIÑO SOUTHERN OSCILLATION Let's set the stage for our story with a glimpse at a climatic event called the *El Niño Southern Oscillation* (ENSO). Massive dislocations in global rainfall patterns characterize this event, which corresponds to changes in sea surface temperatures and air circulation patterns. "Southern Oscillation" refers to a recurring seesaw in atmospheric pressure in the western equatorial Pacific—the world's largest reservoir of warm water. More warm air rises here than anywhere else. It is the source of heavy rainfall, which releases heat energy that drives the world's air circulation system.

Between ENSOs, the warm reservoir and heavy rainfall associated with it move to the west (Figure 50.32). In an ENSO, surface winds prevailing in the western equatorial Pacific pick up speed and "drag" the surface waters east (Figure 50.33). As eastward water transport increases, westward transport slows. Even more warm water in the vast reservoir moves eastward—and so on in a feedback loop between the ocean and the atmosphere. Sea surface temperatures rise, evaporation from the ocean accelerates, and air pressure falls. The humid updrafts trigger violent storms and flooding along coasts and often heavier rain inland. Elsewhere, extended droughts prevail.

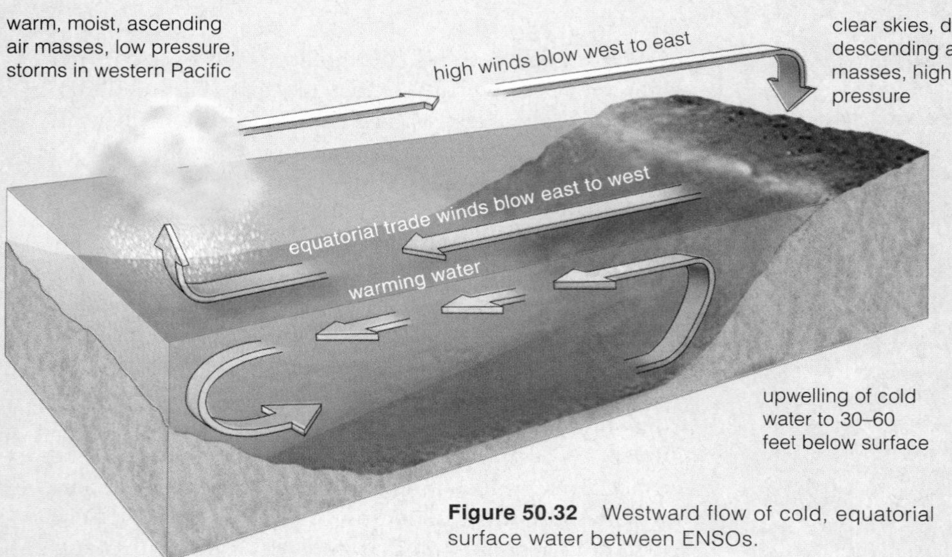
warm, moist, ascending air masses, low pressure, storms in western Pacific
high winds blow west to east
clear skies, dry descending air masses, high pressure
equatorial trade winds blow east to west
warming water
upwelling of cold water to 30–60 feet below surface

Figure 50.32 Westward flow of cold, equatorial surface water between ENSOs.

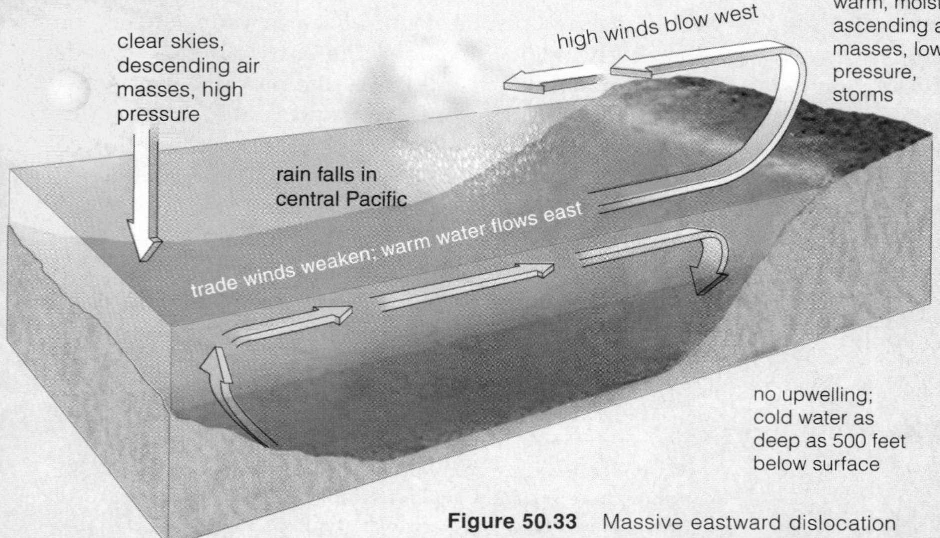
clear skies, descending air masses, high pressure
high winds blow west
warm, moist ascending air masses, low pressure, storms
rain falls in central Pacific
trade winds weaken; warm water flows east
no upwelling; cold water as deep as 500 feet below surface

Figure 50.33 Massive eastward dislocation of warm ocean water during an ENSO.

As you read earlier, reversal in a westward flow of both air and water displaces the cold, deep Humboldt Current and stops upwelling along the western coast of South America. Usually a warm, nutrient-poor current from the east reaches the coast around Christmas. Peru's fishermen named the current "El Niño" ("the little one," a reference to the baby Jesus).

Today, satellites and ocean-bobbing buoys gather data on winds, currents, and sea surface temperatures, which are loaded into supercomputers that simulate global weather patterns. The simulations are getting good: In 1997, scientists were able to predict an ENSO a season in advance. It turned out to be the most powerful ENSO of the century. Average sea surface temperatures rose nine degrees in the eastern Pacific. Warm water extended 9,660 kilometers west from Peru's coastline and 320 kilometers north (Figure 50.34).

Storms slamming into California, Mexico, Ecuador, and Peru caused heavy flooding and mudslides. An ice storm hit New England and knocked out power for 4 million people in one of Canada's worst natural disasters. Tornadoes devastated parts of the southeastern United States. A numbing heat wave caused hundreds of deaths and triggered wildfires in Mexico and the United States. Australia reeled under a prolonged drought. Heavy rains and floods struck Kenya and Somalia. Monsoons were delayed in Southeast Asia—and fires set to clear

Further reading: Student Guide to InfoTrac on web site →

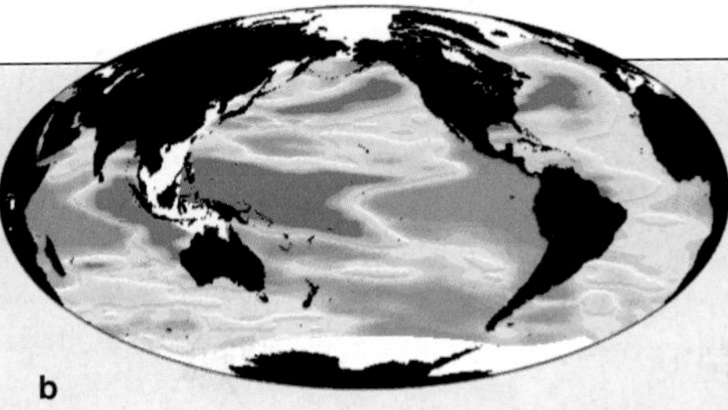

Figure 50.34 (**a**) Sea levels for the 1993–1994 winter between ENSOs. (**b**) Sea levels for the El Niño winter of 1997–1998. Values are based on measured increases in sea surface temperatures.

−8 −6 −4 −2 0 2 4 6 8
height (centimeters)

tracts of tropical rain forests swept out of control through 400,000 hectares (1 million acres). The strongest hurricane ever recorded hit the eastern Pacific; eight cyclones hit the central Pacific, compared to two the year before.

A CHOLERA CONNECTION During the 1997–1998 El Niño episode, 30,000 cases of cholera were reported in Peru, compared to only 60 from January to August 1997. For some time, people knew that cholera was linked to water contaminated by *Vibrio cholerae* (Figure 50.35*a*). Epidemics develop after this bacterium hitches rides in the gut of human travelers, then enters water and food supplies by way of feces expelled during the severe diarrhea that characterizes the disease.

What people did *not* know was where the bacterium lurked between cholera outbreaks. Year after year, no one could find it in humans or in water supplies. Then cholera would emerge simultaneously in places some distance apart—most often in coastal cities where the urban poor drew water from rivers that entered the sea.

Rita Colwell, a microbiologist who later became the director of the National Science Foundation, suspected that humans did not harbor the pathogen between the outbreaks. Was there an environmental reservoir? Maybe, but nobody had detected any cells of *V. cholerae* in water samples subjected to standard culturing. Then Colwell had a flash of insight: What if no one could find the pathogen because it changes form and enters a dormant, sporelike stage between outbreaks?

During a cholera outbreak in Louisiana, she decided to employ labeled antibodies that would bind to a certain protein at the bacterium's surface. Later, antibody tests in Bangladesh pinpointed the bacterium in fifty-one of fifty-two water samples—but culture methods missed it in all but seven of the samples.

V. cholerae thrives in brackish water, rivers, estuaries, and the seas. Colwell knew that plankton (communities of mostly microscopic aquatic species) also thrive in these aquatic environments. She focused her search on the waters near Bangladesh, where outbreaks of cholera are endemic and seasonal. In time she discovered the dormant stage of *V. cholerae* in copepods, which graze on algae and other phytoplankton (Figure 50.35*b*). And the abundances of copepods—hence of *V. cholerae*—rise and fall with changes in the abundance of phytoplankton, the "pastures of the seas."

As it happened, Colwell already knew about seasonal variations in sea surface temperatures. Remember the saying, Chance favors the prepared mind? She chanced upon a correlation between seasonal temperature peaks in the Bay of Bengal with cholera admissions to hospitals in the area. Her correlation held for the 1990–1991 and 1997–1998 El Niño episodes (Figure 50.35*c*,*d*). Four to six weeks after sea surface temperatures go up, so do the cases of cholera!

Together with her colleague Anwarul Huq, Colwell is working to identify other factors, such as shifts in salinity and nutrient content, that trigger algal blooms, hence rises in copepod population size. Their goal is to integrate the data into a model that can be used to predict precisely where cholera outbreaks will occur and to give advance warning to filter drinking water. Meanwhile, Colwell is advising women in Bangladesh to use four layers of old sari cloth as filters, which remove 99 percent of *V. cholerae* cells from the water (Figure 50.35*e*). They can be rinsed in clean water, sun-dried, and used again and again.

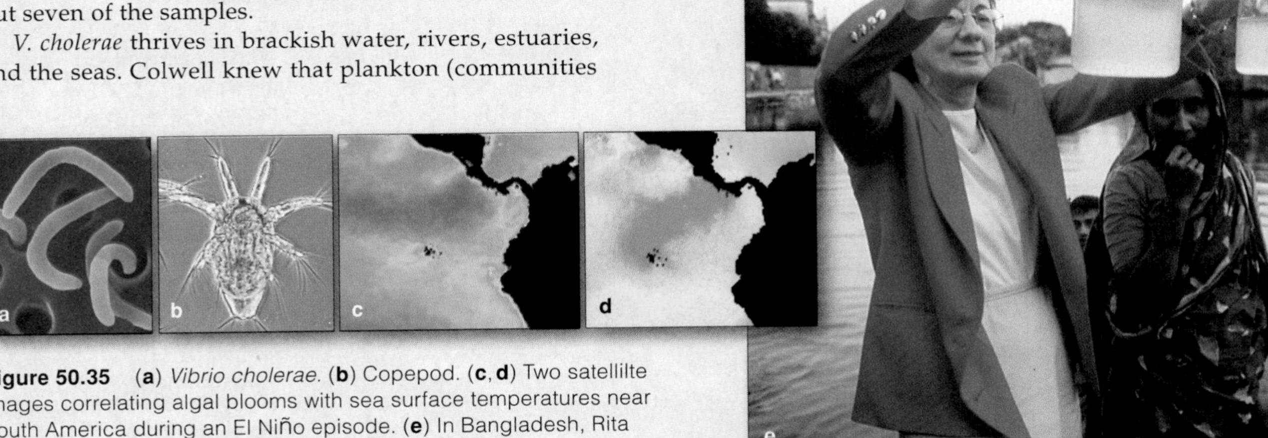

Figure 50.35 (**a**) *Vibrio cholerae.* (**b**) Copepod. (**c**, **d**) Two satellilte images correlating algal blooms with sea surface temperatures near South America during an El Niño episode. (**e**) In Bangladesh, Rita Colwell with unfiltered and filtered drinking water.

50.16 SUMMARY

1. The biosphere encompasses the Earth's waters, the lower atmosphere, and the uppermost portions of its crust in which organisms live. Energy flows one way through the biosphere, and materials move through it on a grand scale to influence ecosystems everywhere.

2. The distribution of species through the biosphere is an outcome of the Earth's history, topography, climate, and interactions among species.

a. "Climate" (average weather conditions, including temperature, humidity, wind velocity, cloud cover, and rainfall over time) results from differences in the amount of solar radiation reaching equatorial and polar regions, the Earth's daily rotation and annual path around the sun, the distribution of continents and oceans, and the elevation of land masses.

b. Interacting climatic factors produce the prevailing winds and ocean currents, which shape global weather patterns. The weather affects soil composition and water availability, which affects the growth and distribution of primary producers in ecosystems.

3. The world's land masses are classified as six major biogeographic realms. Each is more or less isolated by oceans, mountain ranges, or desert barriers, which tend to restrict gene flow between realms. As a result, each tends to maintain a characteristic array of species.

4. A biome (a category of major ecosystems on land) is shaped by regional variations in climate, landforms, and soils. Dominant plant species are adapted to the set of conditions prevailing in the major biomes: deserts, dry shrublands, dry woodlands, grasslands, broadleaf forests (such as tropical forests), coniferous forests, and tundra.

5. Water provinces cover more than 71 percent of the Earth's surface. They include standing fresh water (such as lakes), running fresh water (such as streams), as well as the world oceans and seas. All aquatic ecosystems show gradients in light penetration, water temperature, salinity, and dissolved gases. These factors vary daily and seasonally. They influence primary productivity.

6. Estuaries, intertidal zones, rocky and sandy shores, tropical reefs, and regions of the open ocean are major marine ecosystems. Photosynthetic activity is greatest in shallow coastal waters and in regions of upwelling. Upwelling is an upward movement of deep, cool ocean water that often carries nutrients to the surface.

Review Questions

1. Define biosphere. As part of your answer, include definitions of the atmosphere, lithosphere, and hydrosphere. *CI*

2. List the major interacting factors that influence climate. *CI*

3. List some of the ways in which air currents, ocean currents, or both may influence the region where you live. *50.1, 50.2*

4. Indicate where these biomes tend to be located in the world, and describe some of their defining features. *50.3, 50.5–50.9*

a. deserts
b. dry shrublands
c. dry woodlands
d. grasslands
e. evergreen broadleaf forests
f. deciduous forests
g. arctic tundra
h. alpine tundra
i. boreal forest
j. montane coniferous forests

5. Define soils, then explain how the composition of regional soils affects ecosystem distribution. *50.4*

6. Describe the littoral, limnetic, and profundal zones of a large temperate lake in terms of seasonal primary productivity. *50.10*

7. Define the two major provinces of the world ocean. Is the open ocean devoid of life? *50.11*

8. Define and characterize the following ecosystems: coral reef, coral bank, mangrove wetland, and estuary. *50.12–50.14*

Self-Quiz (*Answers in Appendix III*)

1. Solar radiation drives the distribution of weather systems and so influences ______.
a. temperature zones
b. rainfall distribution
c. seasonal variations
d. all of the above

2. The ______ is a shield against ultraviolet wavelengths.
a. upper atmosphere
b. lower atmosphere
c. ozone layer
d. greenhouse effect

3. Regional variations in the global patterns of rainfall and temperature depend on ______.
a. global air circulation
b. ocean currents
c. topography
d. all of the above

4. A rain shadow is a reduction in rainfall on the ______ of a mountain range.
a. windward side
b. leeward side
c. highest elevation
d. lowest elevation

5. Biogeographic realms are ______.
a. land and water provinces
b. six major land provinces
c. divided into biomes
d. b and c

6. Biome distribution corresponds roughly with regional variations in ______.
a. climate b. soils c. topography d. all of the above

7. Dominant plants of ______ are highly adapted to recurring episodes of lightning-sparked fires.
a. dry shrublands
b. grasslands
c. southern pine forests
d. all of the above

8. During ______, deeper, often nutrient-rich water moves to the surface of a body of water.
a. spring overturns
b. fall overturns
c. upwellings
d. all of the above

9. Match the terms with the most suitable description.

____ boreal forest	a. high productivity; rapid cycling of nutrients (poor reservoirs)
____ permafrost	b. "swamp forest"
____ chaparral	c. a type of dry shrubland
____ tropical rain forest	d. feature of arctic tundra

10. Match the terms with the most suitable description.

____ marine snow	a. deep, cool, often nutrient-rich ocean water moves upward
____ upwelling	b. sediments, rocks of ocean bottom
____ eutrophication	c. partially enclosed mix of seawater and fresh water
____ estuary	d. nutrient enrichment of body of water; reduced transparency, phytoplankton blooms
____ benthic province	e. basis of midocean food webs

OLIGOTROPHIC LAKE	EUTROPHIC LAKE
Deep, steeply banked	Shallow with broad littoral
Large deep-water volume relative to surface-water volume	Small deep-water volume relative to surface-water volume
Highly transparent	Limited transparency
Water blue or green	Water green to yellow- or brownish-green
Low nutrient content	High nutrient content
Oxygen abundant through all levels throughout year	Oxygen depleted in deep water during summer
Not much phytoplankton; green algae and diatoms dominant	Abundant, thick masses of phytoplankton; and cyanobacteria dominant
Abundant aerobic decomposers favored in profundal zone	Anaerobic decomposers
Low biomass in profundal	High biomass in profundal

b

Figure 50.36 (**a**) Crater Lake, Oregon, a collapsed volcanic cone filled with water from rains and melted snow. Like other volcanoes of the Cascade range, it started forming as a result of tectonic forces that prevailed at the dawn of the Cenozoic. (**b**) Major characteristics of oligotrophic and eutrophic lakes.

Critical Thinking

1. Kangaroos are furry, pouched mammals. Raccoons are furry placental mammals. Speculate on why Australia but not North America is the original home of kangaroos and why the reverse is true of raccoons. Use your knowledge of geologic history when devising an explanation. (Compare Section 27.10.)

2. Reflect on the world distribution of land masses and ocean water, then develop an explanation of why grassland biomes tend to form in the interior of continents, not along their coasts.

3. *Wetlands,* remember, are transitional zones between aquatic and terrestrial ecosystems in which plants are adapted to grow in periodically or permanently saturated soil. As mentioned in Section 28.6, they include riparian zones and mangrove swamps. Wetlands everywhere are being converted for agriculture, home building, and other human activities. Does the great ecological value of wetlands for the nation as a whole outweigh the rights of the private citizens who own many of the remaining wetlands in the United States? Should the private owners be forced to transfer title to the government? If so, who should decide the value of a parcel of land *and* pay for it? Or would such seizure of private property violate laws of the United States?

4. Observe conditions in a lake near your home or a place where you vacation. If lakes aren't part of your life, think about Oregon's Crater Lake instead (Figure 50.36*a*). Using the data in Figure 50.36*b* as a guide, would you conclude Crater Lake is oligotrophic or eutrophic? Will it remain so over time?

5. The United States Forest Service and Geophysical Dynamics Laboratory ran supercomputer programs to predict possible outcomes of a *global warming* trend. (Here you may wish to review Section 49.9.) According to the results, by the year 2030, the United States will experience recurring, devastating forest fires. Severe storms and more frequent rainfall in the western states will accelerate erosion, especially along the coasts and in steep foothills and mountain ranges.

Increases in the deposition of sediments will have adverse effects on rivers, streams, and estuaries. Dry shrublands and woodlands may replace hardwood forests in Minnesota, Iowa, and Wisconsin, and in parts of Missouri, Illinois, Indiana, and Michigan. The breadbasket of the American Midwest will shift north into Canada. Think about where you live now and where you plan to live in the future. How would such changes affect you? Should nations get serious about finding ways to counter global warming? Or do you believe that the scientists who are concerned about this are merely alarmist and that nothing bad will happen? Explain how you have reached your conclusion.

6. Think about the ocean, which covers the majority of the Earth's surface. Its deepest regions are remote from human populations, to say the least. Now think about a modern-day problem—where to dispose of nuclear wastes and other truly hazardous materials. You will be reading about this serious problem in the next chapter. For now, simply be aware that the United States government has stockpiles of very dangerous wastes and has nowhere to put them. Would it be feasible, let alone ethical, to use the deep ocean as a "burying ground" for them? Why or why not?

Selected Key Terms

atmosphere *CI*
biogeographic realm *50.3*
biogeography *CI*
biome *50.3*
biosphere *CI*
boreal forest *50.8*
climate *CI*
coniferous forest *50.8*
coral bank *50.12*
coral reef *50.12*
deciduous broadleaf forest *50.7*
desert *50.5*
desertification *50.5*
dry shrubland *50.6*
dry woodland *50.6*
ecoregion *50.3*
El Niño *50.14*
estuary *50.14*
eutrophication *50.10*
evergreen broadleaf forest *50.7*
fall overturn *50.10*
grassland *50.6*
hydrothermal vent *50.11*
intertidal zone *50.14*
lake *50.10*
mangrove wetland *50.13*
marine snow *50.11*
monsoon *50.2*
ocean *50.2*
permafrost *50.9*
rain shadow *50.2*
savanna *50.6*
soil *50.4*
southern pine forest *50.8*
spring overturn *50.10*
stream *50.10*
temperature zone *50.1*
tropical rain forest *50.7*
tundra *50.9*
ultraplankton *50.11*
upwelling *50.14*

Readings *See also www.infotrac-college.com*

Brown, J., and A. Gibson. 1983. *Biogeography*. St. Louis: Mosby.

Dold, C. February 1999. "The Cholera Lesson." *Discover*, 71–76.

Garrison, T. 1996. *Oceanography*. Second edition. Belmont, California: Wadsworth. Accessible; read its lyrical epilogue.

Olson, D. M., and E. Dinerstein, 1998. "The Global 200: A Representation Approach to Conserving the Earth's Most Biologically Valuable Ecoregions. *Conservation Biology* 12(3).

Smith, R. 1996. *Ecology and Field Biology*. Fifth edition. New York: HarperCollins.

51

HUMAN IMPACT ON THE BIOSPHERE

An Indifference of Mythic Proportions

Of all the concepts introduced in the preceding chapter, the one that should be foremost in your mind is this: *The atmosphere, ocean, and land interact in ways that help dictate living conditions throughout the biosphere.* Driven by energy streaming in continually from the sun, these stupendous interactions give rise to globe-spanning temperatures and circulation patterns upon which life ultimately depends. With this chapter, we turn to a related concept of equal importance. Simply put, *we have become major players in these interactions even before we fully comprehend how they work.*

To gain perspective on what is happening, think about something we take for granted—the air around us. The composition of the present-day atmosphere is a result of geologic and metabolic events—most notably, photosynthesis—that began billions of years ago. The first humans, recall, had evolved by 2.4 million years ago. Like us, they breathed oxygen from an atmosphere of ancient origins. Like us, they were protected from the sun's ultraviolet radiation by an ozone shield in the stratosphere. The size of their population was not much to speak of, and their effect on the biosphere was trivial. About 11,000 years ago, however, agriculture began in earnest, and it laid the foundation for huge increases in population size. A few centuries ago, medical and industrial revolutions expanded that foundation—and human population growth skyrocketed in a mere blip of evolutionary time.

Today we are extracting huge amounts of energy and resources from the environment and giving back monumental amounts of wastes. As we do so, we are destabilizing ecosystems everywhere, even though the magnitude of change might not even be recognized when measured on the scale of a lifetime (Figure 51.1).

In a few developed countries, population growth has more or less stabilized, and the resource use per individual has dropped a bit. But resource utilization levels in those places are already high. In developing countries in Central America, Africa, and elsewhere, population sizes and resource consumption are rapidly growing, even though hundreds of millions of people are already malnourished or starving to death.

1900
1940
1954
1962

Many of the problems sketched out in this chapter are not going to disappear tomorrow. It will take many decades, even centuries, to reverse some trends that are already in motion—and not everyone is ready to make the effort. A few enlightened individuals in Michigan or Alberta or New South Wales can commit themselves to resource conservation and to minimizing pollution. But scattered attempts will not be enough. Individuals of all nations will unite to reverse global trends only when they perceive that the dangers of *not* doing so outweigh the personal benefits of ignoring them.

Does this seem pessimistic? Think of the exhaust fumes released to the air each time you drive a car or truck. Think of oil refineries, food-processing plants, and paper mills that supply you with goods and also release chemical wastes into the nation's waterways. Think of Mexico and other developing countries that produce cheap food by using an unskilled labor force and toxic pesticides. Unregulated pesticide applications poison the people who work the land, the land itself, and sometimes people who buy the exported produce.

Who changes behavior first? We have no answer to the question. We can suggest, however, that a strained biosphere can rapidly impose an answer upon us. As an individual, you might choose to cherish or brood about or ignore any aspect of the world of life. Whatever your choice may be, the bottom line is that you and all other organisms are in this together. Our lives interconnect to degrees that we are only now starting to comprehend.

Figure 51.1 Tracking human population growth in one small part of the world, based on historical data and satellite imaging. *Red* denotes areas of dense human settlement in and around the San Francisco Bay Area and Sacramento since 1900.

1974

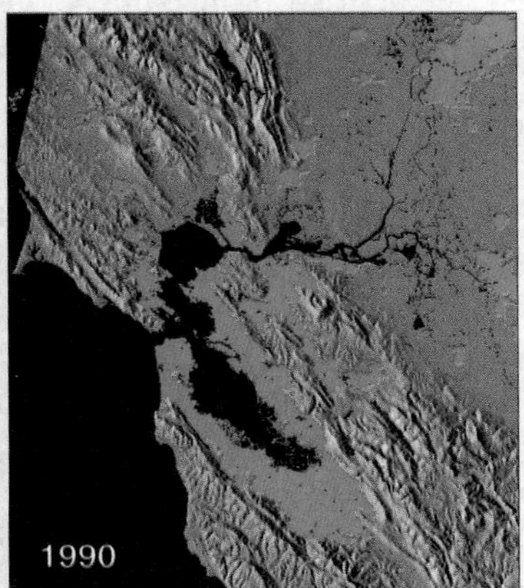
1990

KEY CONCEPTS

1. Human population growth has been skyrocketing ever since the mid-eighteenth century. At present, humans have the population size, the technology, and the cultural inclination to use energy and alter the environment at astonishing rates.

2. Pollutants are substances with which ecosystems have had no prior evolutionary experience, in terms of kinds or amounts, so adaptive mechanisms for dealing with them are not in place. In a more restrictive sense, pollutants are substances that accumulate in amounts that have adverse effects on human health, activities, or survival.

3. Many pollutants, including the kinds that contribute to the formation of smog and acid rain, exert regionally harmful effects. The effects of other pollutants, including chlorofluorocarbons that attack the ozone layer in the stratosphere, are global in scale.

4. Conversion of marginally fertile lands for agriculture, rampant deforestation, and other practices required to meet the demands of the growing human population are leading to loss of soil fertility and desertification. They are also resulting in a decline in the quality and quantity of one of the most crucial of all resources—fresh water.

5. Ultimately, the world of life depends on energy inputs from the sun. That energy drives the complex interactions among the atmosphere, ocean, and land. Our activities are disrupting the interactions in ways that may have severe consequences in the near future.

6. We as a species must come to terms with principles of energy flow and principles of resource utilization that govern all systems of life on Earth.

51.1 AIR POLLUTION—PRIME EXAMPLES

Let's start this survey of the impact of human activities by defining pollution, which is central to our problems. **Pollutants** are substances with which ecosystems have had no prior evolutionary experience, in terms of kinds or amounts, so adaptive mechanisms that might deal with them are not in place. From the human perspective, they are substances that accumulate to levels that have adverse effects on our health, activities, or survival.

Air pollutants are prime examples. As listed in Table 51.1, they include carbon dioxide, oxides of nitrogen and sulfur, and chlorofluorocarbons. Also among them are photochemical oxidants formed when the sun's rays interact with certain chemicals. The United States alone releases 700,000 metric tons of air pollutants each day. Whether these remain concentrated at the source or are dispersed during a given interval depends on the local climate and topography, as you will now see.

Table 51.1 Major Classes of Air Pollutants

Carbon oxides	Carbon monoxide (CO), carbon dioxide (CO_2)
Sulfur oxides	Sulfur dioxide (SO_2), sulfur trioxide (SO_3)
Nitrogen oxides	Nitric oxide (NO), nitrogen dioxide (NO_2), nitrous oxide (N_2O)
Volatile organic compounds	Methane (CH_4), benzene (C_6H_6), chlorofluorocarbons (CFCs)
Photochemical oxidants	Ozone (O_3), peroxyacyl nitrates (PANs), hydrogen peroxide (H_2O_2)
Suspended particles	Solids (dust, soot, asbestos, lead, etc.) and droplets (sulfuric acid, oils, pesticides, etc.)

Smog

When weather conditions trap a layer of cool, dense air under a layer of warm air, this is a **thermal inversion**. If the trapped air holds pollutants, winds can't disperse them, and they might accumulate to dangerous levels. Thermal inversions have been key factors in some of the worst air pollution disasters, because they intensify an atmospheric condition called smog (Figure 51.2).

Two types of smog—gray air and brown air—form in major cities. Where winters are cold and wet, **industrial smog** develops as a gray haze over industrialized cities that burn coal and other fossil fuels for manufacturing, heating, and generating electric power. Burning releases airborne pollutants, such as dust, smoke, soot, ashes, asbestos, oil, bits of lead and other heavy metals, and sulfur oxides. If not dispersed by winds and rain, the emissions may reach lethal concentrations. Industrial smog caused 4,000 deaths during the 1952 air pollution disaster in London. Before coal burning was restricted, New York, Pittsburgh, and Chicago also were gray-air cities. Today, most industrial smog forms in the cities of China, India, and other developing countries, as well as cities of coal-dependent countries of eastern Europe.

In warm climates, **photochemical smog** develops as a brown haze over large cities. It becomes concentrated when the surrounding land forms a natural basin, as it does around Los Angeles and Mexico City (Figure 51.2). The key culprit is nitric oxide. After it is released from vehicles, it reacts with oxygen in air to form nitrogen dioxide. When exposed to sunlight, nitrogen dioxide reacts with hydrocarbons; photochemical oxidants result. Most of the hydrocarbons are from spilled or partly burned gasoline. Key oxidants are ozone and **PANs** (for peroxyacyl nitrates). Even traces of PANs sting eyes, irritate lungs, and damage crops.

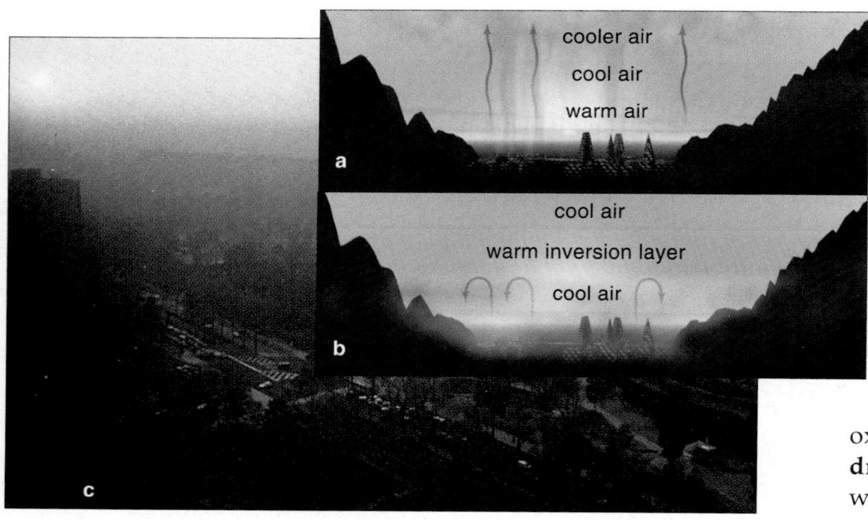

cooler air
cool air
warm air
a
cool air
warm inversion layer
cool air
b
c

Figure 51.2 (**a**) Normal pattern of air circulation in smog-forming regions. (**b**) Air pollutants trapped under a thermal inversion layer. (**c**) Mexico City on an otherwise bright, sunny morning. Topography, staggering numbers of people and vehicles, and industry combine to make its air among the world's smoggiest. Breathing this city's air is like smoking two packs of cigarettes a day.

Acid Deposition

Oxides of sulfur and nitrogen are among the worst air pollutants. Coal-burning power plants, metal smelters, and factories emit most sulfur dioxides. Motor vehicles, power plants that burn gas and oil, and nitrogen-rich fertilizers produce nitrogen oxides. In dry weather, fine particles of oxides may briefly stay airborne, then fall to Earth as **dry acid deposition**. When dissolved in atmospheric water, they form weak solutions of sulfuric and nitric acids. Winds may disperse them over great distances. If they fall to Earth in rain and snow, we call this wet acid deposition, or **acid rain**. The pH of normal rainwater is about 5. Acid rain can be 10 to 100 times more acidic, even as potent as lemon juice! The deposited acids eat away at marble buildings, metals, rubber, plastic, nylon

Further reading: Student Guide to InfoTrac on web site

Very high acidity (pH 4.2–4.4)
Sensitivity to acid deposition
Moderate to high acidity (pH 4.5–5.0)

Figure 51.3 Map of average acidities of precipitation and soil sensitivities to acid deposition in 1984 for North American regions. The photograph was taken in Germany. It is an example of how prolonged exposure to air pollutants is contributing to the rapid destruction of forests in much of the world. Weakened trees are more vulnerable to drought, disease, and insect pests.

stockings, and many other materials. They significantly disrupt the physiology of organisms, the chemistry of ecosystems, and biodiversity.

Depending on their soil type and vegetation cover, some regions are far more sensitive than others to acid deposition (Figure 51.3). Highly alkaline soil neutralizes acids before they enter streams and lakes of watersheds. Water having a high carbonate content also neutralizes acids. However, in many watersheds of northern Europe and southeastern Canada, and in regions throughout the United States, thin soils overlie solid granite. These soils cannot buffer much of the acidic inputs.

Rain in much of eastern North America is thirty to forty times more acidic than it was several decades ago. Crop yields are diminishing. Fish populations have already vanished from more than 200 lakes in New York's Adirondack Mountains. By some predictions, fish will disappear from 48,000 lakes in Ontario, Canada, in the next two decades. Pollution from industrial regions

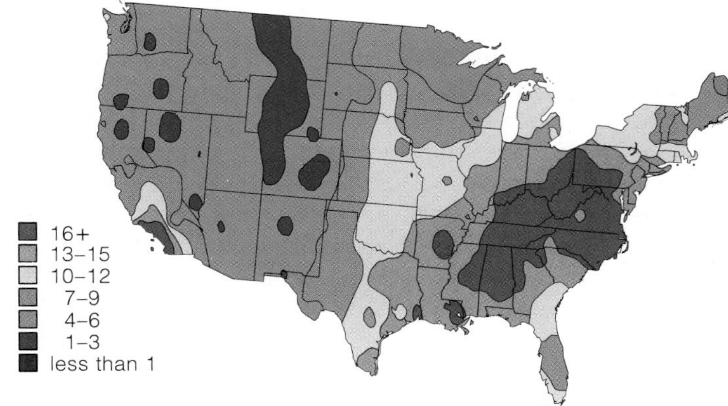
16+
13–15
10–12
7–9
4–6
1–3
less than 1

Figure 51.4 Regional differences in the concentrations of fine particles (2.5 μm or less) in micrograms per cubic meter of air, measured in 1992. Data are averages; they understate peak values in individual cities.

also is changing rainfall acidity enough to contribute to the decline of forest trees and mycorrhizae that support new growth (Section 24.4).

Researchers confirmed long ago that emissions from power plants, factories, and vehicles are the key sources of air pollutants. In 1995, researchers from the Harvard School of Public Health and Brigham Young University reported this finding from a comprehensive air quality study: Whack a year or so off your life span if you live in cities with the dirtiest air—especially air with fine particles of dust, soot, smoke, or acid droplets. Smaller particles are more easily inhaled and can damage lung tissue. Figure 51.4 summarizes air-quality data gathered in 1992 for the continental United States.

At one time the world's tallest smokestack, in the Canadian province of Ontario, accounted for 1 percent by weight of the annual worldwide emissions of sulfur dioxide. But Canada receives more acid deposition from industrialized regions of the midwestern United States than it sends across its southern border. Most of the air pollutants in Scandinavian countries, the Netherlands, Austria, and Switzerland arise in industrialized regions of western and eastern Europe. *Prevailing winds—hence air pollutants—do not stop at national boundaries.*

Pollutants are substances with which ecosystems have had no prior evolutionary experience, in terms of kinds or amounts, so adaptive mechanisms are not in place to deal with them.

Accumulated pollutants can harm organisms, as when they reach levels that adversely affect human health.

Smog formation and acid deposition are two examples of air pollution in specific regions, although prevailing winds often distribute pollutants beyond regional boundaries.

Smog forms mainly as a result of fossil fuel burning in urban and industrialized regions. Airborne acidic pollutants drift down as dry particles or as components of acid rain.

OZONE THINNING—GLOBAL LEGACY OF AIR POLLUTION

Now think about the ozone layer, almost twice as high above sea level as the top of Mount Everest, the highest place on Earth. Each September through mid-October, the layer becomes thinner at high latitudes. The seasonal **ozone thinning** is so pronounced, it was once called an "ozone hole." In 1998, ozone thinning over Antarctica was the most extensive ever recorded; it covered an area larger than North America. In both 1997 and 1999, the seasonal loss at high northern latitudes was 60 percent—the lowest ever recorded.

Why is ozone reduction alarming? It lets more ultraviolet radiation reach the Earth, which is triggering far more skin cancers, cataracts, and weakened immune systems. Ultraviolet radiation also affects photosynthesis. Substantial declines in the oxygen-releasing activity of phytoplankton alone could alter the composition of the atmosphere, given their staggering numbers (Section 7.9).

Chlorofluorocarbons (**CFCs**) are the major factors in the ozone reduction. These compounds of chlorine, fluorine, and carbon are odorless and invisible. You find them in refrigerators and air conditioners (as the coolants), solvents, and plastic foams. CFCs slowly escape into air and resist breakdown. When a free CFC molecule absorbs ultraviolet light, it gives up one chlorine atom. If chlorine reacts with ozone, this yields oxygen and chlorine monoxide—which can react with free oxygen to release another chlorine atom. Each released chlorine atom can convert 10,000 ozone molecules or more to oxygen!

The chlorine monoxide levels above Antarctica are 100 to 500 times higher than at midlatitudes. Why? During the winter, high-altitude ice clouds form there. Winds rotate around the South Pole for most of the winter, like a moving fence that keeps the ice clouds from spreading to other latitudes (Figure 51.5). This also happens on a lesser scale in the Arctic. The ice crystals serve as surfaces on which chlorine compounds can be swiftly degraded. When air warms in the spring, chlorine is free to destroy ozone—hence the ozone thinning.

CFCs aren't the only ozone eaters. Methyl bromide, a fungicide, is better at it. It persists only briefly in the atmosphere, but it will account for about 15 percent of the thinning in future years unless production stops.

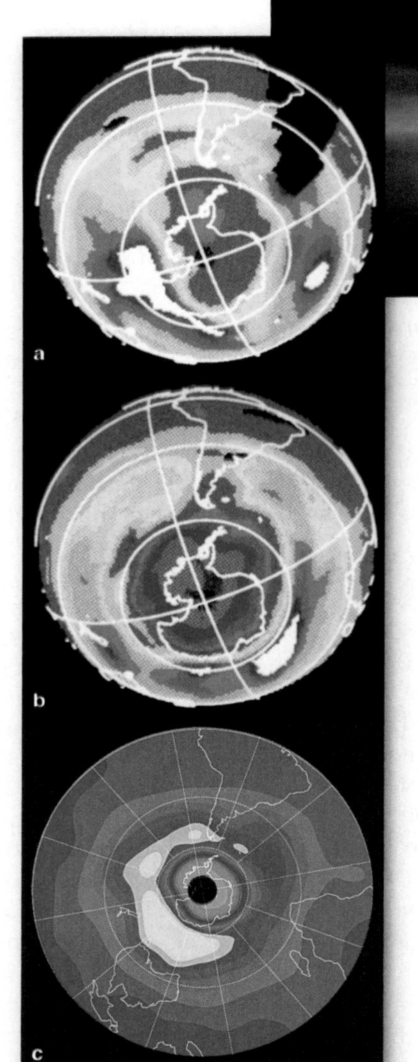
a
b
c
ice clouds
d

Figure 51.5 Examples of seasonal ozone thinning above Antarctica in (**a**) 1979, (**b**) 1991, and (**c**) 1996. Lowest ozone values are coded *magenta* and *purple*. (**d**) Ice clouds above Antarctica, which have a role in the ozone thinning each spring.

Some scientists dismiss the threat of the chemicals that contain chlorine and bromine. But the vast majority of those who have studied ecosystem modeling programs conclude that these chemicals do pose long-term threats—not only to human health and certain crops but also to animal life in general. Substitutes are now available for most applications of CFCs, and others are being developed.

Under international agreement, CFC production has been phased out in the developed countries. By 2010, developing countries will have phased out CFCs. Methyl bromide production will stop by 2010. By recent computer models, the area of thinning should not become any larger. Assuming international goals are met, it still will be about 50 years before the ozone layer is restored to 1985 levels and another 100 to 200 years to total recovery, to pre-1950 levels. Meanwhile, you and all the children and grandchildren of future generations will be living with the destructive effects of air pollutants. And if levels of greenhouse gases continue to rise, the stratosphere could cool enough to increase the size and duration of seasonal ozone depletions over the poles.

Air pollution can have global repercussions, as when CFCs and other compounds contribute to a thinning of the ozone layer that shields life from the sun's ultraviolet radiation.

Further reading: Student Guide to InfoTrac on web site →

51.3 WHERE TO PUT SOLID WASTES? WHERE TO PRODUCE FOOD?

Oh Bury Me Not, In Our Own Garbage

In natural ecosystems, one organism's wastes serve as resources for others, so the by-products of existence are cycled through the system. In the developing countries, many resources are scarce. People conserve what they can and discard little. In the United States and some other developed countries, most people use something once, discard it, then buy another. Each year, millions of metric tons of solid wastes are dumped, burned, and buried. Paper products make up half the total volume, which also includes 50 billion nonreturnable cans and bottles. Each week, paper manufactured from 500,000 trees ends up in Sunday newspapers for Americans. If every reader recycled merely one of ten newspapers, 25 million trees a year could be left standing. Recycling paper would reduce the airborne pollutants released during paper manufacturing by 95 percent and require 30 to 50 percent less energy than making new paper.

Our throwaway mentality is unique in the world of life. We are condoning the monumental accumulation of solid wastes in human ecosystems and natural ones. Bury the garbage in landfills? Then what happens when the space around cities runs out? Besides this, landfills "leak" and in time threaten supplies of groundwater. Burn wastes in inefficient incinerators? These spew great volumes of pollutants and ashes into the atmosphere.

Today, recycling is affordable and feasible. At the personal level, individuals can help bring about change by refusing to buy goods that are excessively wrapped, packaged in nondegradable containers, and designed for one-time use. They can engage in curbside recycling, by which they presort recyclable wastes. Such action is more efficient than reliance on huge resource recovery centers. These require so much trash to turn a profit that owners may end up encouraging the throwaway mentality. The centers also produce a toxic ash that must be disposed of in landfills, which eventually leak.

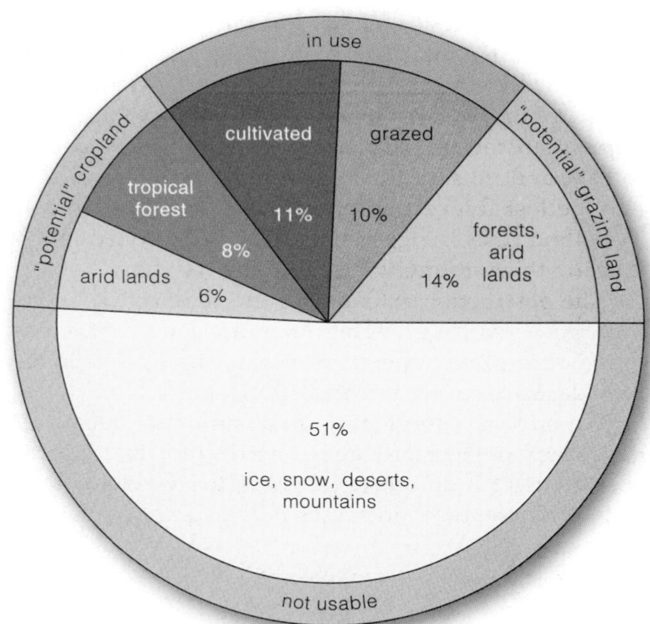

in use
"potential" cropland
cultivated
grazed
"potential" grazing land
tropical forest
11%
10%
forests, arid lands
8%
arid lands
6%
14%
51%
ice, snow, deserts, mountains
not usable

Figure 51.6 Classification of land with respect to its suitability for agriculture. Theoretically, clearing vast tracts of tropical forests and irrigating marginal lands could more than double the world's cropland. Doing so would destroy valuable forest resources, damage the environment, cause severe losses in biodiversity, and possibly cost more than it is worth.

Converting Marginal Lands for Agriculture

The human population already uses nearly 21 percent of the Earth's land surfaces for cropland or for grazing. Another 28 percent is said to be potentially suitable for agriculture, but productivity would be so low that the conversion may not be worth the cost (Figure 51.6).

Asia and several other heavily populated regions now experience recurring, severe food shortages. Yet more than 80 percent of their productive land is already under cultivation. Scientists have made valiant efforts to improve crop production on existing land. Under the banner of the **green revolution**, their research has been directed toward (1) improving the genetic character of crop plants for higher yields and (2) exporting modern agricultural practices and equipment to the developing countries. Many of these countries rely on *subsistence* agriculture, which runs on energy inputs from sunlight and human labor. They also rely very heavily on *animal-assisted* agriculture, with energy inputs from oxen and other draft animals. By contrast, *mechanized* agriculture requires massive inputs of fertilizers, pesticides, and ample irrigation to sustain high-yield crops. It requires fossil fuel energy to drive farm machines. Crop yields are four times as high. But the modern practices use up 100 times more energy and minerals. Also, there are signs that limiting factors are coming into play to slow down further increases in crop yields.

Pressures for increased food production are greatest in parts of Central and South America, Asia, the Middle East, and Africa where human populations are rapidly expanding into marginal lands. Repercussions extend beyond national boundaries, as you will see next.

Our astounding population growth has impact on the Earth's land. We generate huge amounts of solid wastes but reuse or recycle very little. We also are making the energetically and environmentally costly move of expanding into marginal lands for food production.

DEFORESTATION—AN ASSAULT ON FINITE RESOURCES

At one time, tropical forests cloaked regions that were, collectively, twice the size of Europe. For ten thousand years or more, those forests endured in rich complexity, as the homes of an estimated 50 to 90 percent of all land-dwelling species. In less than four decades, human populations destroyed more than half of those ancient forests, and most of their spectacular arrays of species may be lost forever. With every passing year, another 38 million acres is logged over. That is the equivalent of leveling thirty-four city blocks every minute.

The destruction extends beyond the tropics. Today, highly mechanized logging operations are proceeding in the once-vast temperate forests of the United States, Canada, Europe, Siberia, and elsewhere.

We have a name for the removal of all trees from large tracts of land for logging, agriculture, and grazing operations. It is **deforestation**. Why are we doing this? Paralleling the rapid increases in the size of the human population are rapidly increasing demands for lumber, fuel, and other forest products, as well as for cropland and grazing land. More and more people compete for diminishing resources. They do so for economic profit, but also because alternative ways of life simply are not available to most individuals and families.

The world's great forests have profound influences on ecosystems. Like enormous sponges, the watersheds of forested regions absorb, hold, and then release water gradually. By intervening in the downstream flow of water, they help control soil erosion, flooding, and the accumulation of sediments that can clog rivers, lakes, and reservoirs. When the vegetation cover gets stripped away, the exposed soil becomes vulnerable to leaching of nutrients and to erosion, especially on steep slopes.

Today, deforestation is greatest in Brazil, Indonesia, Colombia, and Mexico. If the clearing and destruction continue at present rates, only Brazil and Zaire will still have large tropical forests in the year 2010. By 2035, most of their forested regions will be cleared, also.

Figures 51.7 and 51.8 are close-up and panoramic views of what is now happening in the Amazon basin of South America. Section 25.7 includes examples of the outcome of rampant deforestation in North America.

In tropical regions, the clearing of forests for agriculture sets the stage for long-term losses in productivity. The irony is that tropical forests are one of the worst places to grow crops or to raise pasture animals. In intact forests, litter cannot accumulate, for the high temperatures and the heavy, frequent rainfall promote the rapid decomposition of organic remains and wastes. As fast as the decomposers release nutrients, the trees and other plants take them up. Deep, nutrient-rich topsoils simply cannot form.

Long before the advent of highly mechanized logging practices, people were practicing **shifting cultivation** (once referred to as "slash-and-burn

Figure 51.7 Countries that are allowing the greatest destruction of tropical forests. *Red* denotes regions where 2,000 to 14,800 square kilometers are deforested each year. *Orange* denotes "moderate" deforestation, which encompasses areas of 100 to nearly 2,000 square kilometers.

Further reading: Student Guide to InfoTrac on web site

ANDES
SMOKE
FROM
FIRES

Figure 51.8 The vast Amazon River basin of South America in September 1988. Smoke from fires that had been deliberately set during the dry season to clear tropical forests, pasturelands, and croplands completely obscured its features. The smoke extended to the Andes Mountains near the western horizon, about 650 miles (1,046 kilometers) away. That smoke cover was the largest that astronauts had ever observed. It extended almost 175 million square kilometers (1,044,000 square miles).

The smoke plume near the center of the photograph covered an area comparable to the immense smoke cover arising from the forest fire in Yellowstone National Park in that year. During the El Niño–induced drought of 1997, fires that had been set deliberately burned out of control. A smoke cover formed again.

Massive deforestation is not confined to the equatorial regions. For instance, during the past century, 2 million acres of redwood forests along the coast of California were logged over. Most of the destruction of such temperate forests has been proceeding since 1950, owing to the widespread use of chainsaws and tractors and the practice of exporting many logs to overseas lumber mills, where wages are low.

agriculture"). They cut and burn trees, then till ashes into the soil. The nutrient-rich ashes can sustain crops for one to several seasons. Afterward, cleared plots are abandoned, for heavy leaching leaves the soil infertile. When shifting cultivation is practiced on small, widely scattered plots, a forest ecosystem does not necessarily suffer extensive damage. But soil fertility plummets with increases in population size. Then, larger areas are cleared, and plots are cleared again at shorter intervals.

Figure 51.9 Wangari Maathai, a Kenyan who organized the internationally acclaimed Green Belt Movement in 1977. The 50,000 members of this women's group are committed to establishing nurseries, raising seedlings, and planting a tree for each of the 27 million Kenyans. Together with half a million schoolchildren, they had planted more than 10 million trees by 1990. Their success inspired similar programs in more than a dozen countries in Africa. Dr. Maathai's efforts are not appreciated by her own government. Kenyan police have jailed her twice, and in 1992 they severely beat her because of her efforts on behalf of the environment.

In the larger picture, deforestation alters rates of evaporation, transpiration, runoff, and perhaps regional patterns of rainfall. For example, trees release between 50 and 80 percent of the water vapor above tropical forests. In the logged-over regions, annual precipitation declines, and rain swiftly drains away from the exposed, nutrient-poor soil. The regions are now hotter and drier, and soil fertility and moisture have declined. In time, sparse grassland or desertlike conditions might prevail instead of the formerly rich, forested biomes.

Also, tropical forests absorb much of the sunlight reaching the equatorial regions of the Earth's surface. Deforested land is shinier, so to speak, and it reflects more incoming energy back into space. And because of the combined photosynthetic activity of so many trees, the forests help sustain the global cycling of carbon and oxygen. With extensive tree harvesting and burning, the carbon stored in the tree biomass is released to the atmosphere, as carbon dioxide. Thus deforestation may be a factor in the amplification of the greenhouse effect.

Conservation biologists are attempting to reverse the trend. As three examples, a coalition of 500 groups is dedicated to preserving Brazil's remaining tropical forests. In India, women have already built and installed 300,000 inexpensive, smokeless wood stoves. Over the past decade, the stoves saved more than 182,000 metric tons' worth of trees by reducing demands for fuelwood. In Kenya, women have planted millions of trees to hold soil in place and to provide fuelwood (Figure 51.9).

Once-vast forests helped sustain rapid increases in human population growth. Recent and highly mechanized modes of deforestation are rapidly depleting these finite resources.

Focus on Bioethics

51.5 YOU AND THE TROPICAL RAIN FOREST

Developing countries in Latin America, Southeast Asia, and Africa have the fastest-growing populations but limited food, fuel, and lumber. Thanks to the relatively recent inventions of chainsaws, tractors, and logging trucks, most of their forests will probably disappear within your lifetime.

Why does it matter? For purely ethical reasons, many condemn the destruction of so much biodiversity. Tropical rain forests have the greatest variety and numbers of insects, and the world's largest ones. They are home to the most bird species and to plants with the largest flowers. Within the forest canopy and understory are monkeys, tapirs, and jaguars in South America and apes, okapi, and leopards in Africa. Massive vines twist around trees. Orchids, mosses, lichens, and other organisms grow on branches, absorbing minerals that rains deliver to them. Entire communities of microbes, insects, spiders, and amphibians live, breed, and die in small pools of water that collect in furled leaves.

For practical reasons, the destruction affects your life. Only a few strains of crop plants and livestock, vulnerable to evolving pathogens, sustain most human populations. Tissue-culture specialists and genetic engineers use genes of forest species to develop new or hybrid strains that can make our food base less vulnerable. Geneticists use them to develop more effective antibiotics and vaccines. Aspirin, the most widely used painkiller, is based on a chemical blueprint of an extract from tropical willow leaves. Many ornamental plants and spices and foods, including cocoa, cinnamon, and coffee, originated in the tropics. So did latex, gums, resins, dyes, waxes, and oils for tires, shoes, toothpaste, ice cream, shampoo, compact discs, condoms, and perfumes. And think about this: Rampant burning of forests is releasing enough air pollutants to change the air you breathe and help heat the planet during your lifetime.

Thus conservation biologists rightly decry the mass extinction, the assaults on species diversity, the depletion of much of the world's genetic reservoir. Yet something else is going on here. Too many of us become uneasy when we hike or drive through destroyed forests in our own country. Is it because we are losing the comfort of our heritage—a connection with our evolutionary past? Many millions of years ago, our earliest primate ancestors moved into the trees of tropical forests. Through countless generations, their nervous and sensory systems evolved and became highly responsive to information-rich, arboreal worlds.

Does our neural wiring still resonate with rustling leaves, with shafts of light and mosaic shadows? Are we innately attuned to the forests of Eden—or have time and change buried recognition of home?

Figure 51.10 Tropical rain forest in Southeast Asia.

Further reading: Student Guide to InfoTrac on web site →

51.6 WHO TRADES GRASSLANDS FOR DESERTS?

Long-term shifts in climate can convert grasslands to deserts. So can human populations. **Desertification** is the name for the conversion of large tracts of natural grasslands to a more desertlike condition. It applies also when conversions of rain-fed or irrigated croplands result in a 10 percent or greater decline in agricultural productivity. Over the past fifty years, 9 million square kilometers worldwide have become desertified. At least 200,000 square kilometers are still being converted annually. Prolonged droughts accelerate the process, as one did in the Great Plains years ago (Figure 51.11). At present, overgrazing of livestock on marginal lands is the main cause of large-scale desertification.

Colorado
Kansas
Dust Bowl
New Mexico
Oklahoma
Texas

Figure 51.11 An awe-inspiring dust storm approaching Prowers County, Colorado, in 1934.

The Great Plains of the American Midwest are dry, windy grasslands that are subjected to pronounced, recurring droughts. Extensive conversion of these grasslands to agriculture began in the 1870s. Overgrazing left the ground bare across vast tracts. In May 1934, a cloud of topsoil that blew off the land blanketed the entire eastern portion of the United States, giving the Great Plains a dubious new name—the Dust Bowl. About 3.6 million hectares (9 million acres) of cropland were destroyed. Today, without large-scale irrigation and intensive conservation farming, desertlike conditions could prevail.

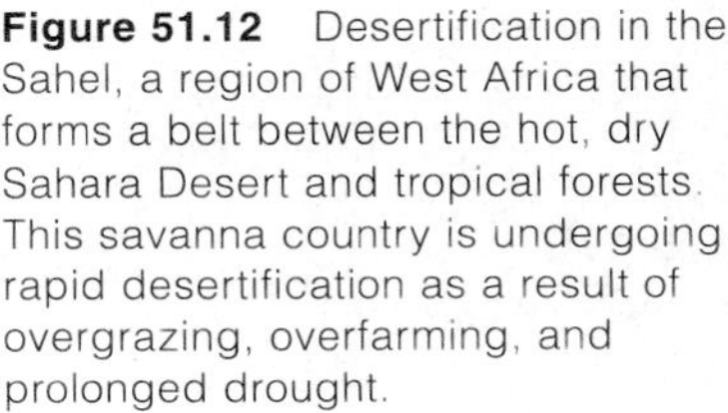

Figure 51.12 Desertification in the Sahel, a region of West Africa that forms a belt between the hot, dry Sahara Desert and tropical forests. This savanna country is undergoing rapid desertification as a result of overgrazing, overfarming, and prolonged drought.

In Africa, for example, there are too many cattle in the wrong places. Cattle require more water than the region's native wild herbivores, so they move back and forth between grazing areas and watering holes more often. As they do, they trample grasses and compact the soil surface (Figure 51.12). By contrast, gazelles and other endemic herbivores get most (if not all) of their water from plants. They also are better at conserving water; they lose little in feces, compared to cattle.

In 1978 a biologist, David Holpcraft, began ranching antelopes, zebras, giraffes, ostriches, and other native herbivores. He also raised cattle as control groups to compare costs and meat yields on the same land. The native herds increased and yielded tasty meat. Range conditions did not deteriorate; they improved. Vexing problems remained. Certain tribes in Africa have their own idea of what constitutes "good" meat, and some view cattle as the symbol of wealth in their society.

Without irrigation and conservation practices, grasslands that were converted for agriculture often end up as deserts.

51.7 A GLOBAL WATER CRISIS

The Earth has a tremendous supply of water, but most is too salty for human consumption or for agriculture. Imagine all that water in a bathtub. Withdraw all of the fresh, renewable portion (from lakes, rivers, reservoirs, groundwater, and other sources of surface water) and it would barely fill a teaspoon.

Why not consider **desalinization**—removal of salt from seawater? Basically, we have unlimited supplies of seawater. Desalinization processes exist. They either distill seawater or force it through membranes (a method called reverse osmosis). Expensive fuel energy drives these processes, so they may be feasible only in Saudi Arabia and a few other countries with limited population sizes, large reserves of energy, and lots of cash. Actually, in some situations they might be the only alternative to running out of drinkable water, as Santa Barbara and some other California cities nearly did during a recent prolonged drought. Yet desalinization can't solve the core problem. It may never be cost-effective for large-scale agriculture, and it makes mountains of salts.

Figure 51.13 Irrigated crops in the Sahara Desert, in Algeria.

Consequences of Heavy Irrigation

Large-scale agriculture accounts for nearly two-thirds of the human population's use of freshwater. In many cases, irrigation water from surface sources is piped into vast fields where water-demanding crops will not grow on their own. At its most extreme, irrigation has turned some hot deserts into lush gardens, although believing these can be maintained over the long term is a bit delusional (Figure 51.13).

Irrigation itself can change the land's suitability for agriculture. Concentrations of mineral salts commonly are high in piped-in water. In regions where soil drains poorly, evaporation may cause **salinization**: a buildup of salt in soil. Salinization can stunt the growth of crop plants, eventually kill them, and decrease yields.

Land that drains poorly also becomes waterlogged. When water accumulates underground, it slowly raises the **water table**, the upper limit at which the ground is fully saturated. When the water table rises close to the ground's surface, soil gets saturated with saline water, which can damage plant roots. Properly managing the water–soil system can reverse the salinization and the waterlogging. The economic cost of doing so is high.

Worse, groundwater overdrafts (the amount nature doesn't replenish) are high (Figure 51.14). For example, overdrafts have depleted half of the Ogallala aquifer, which supplies irrigation water for 20 percent of the croplands in the United States. So much groundwater has been withdrawn for irrigation in some parts of the San Joaquin Valley in California that the surface of the water table has subsided by as much as six meters.

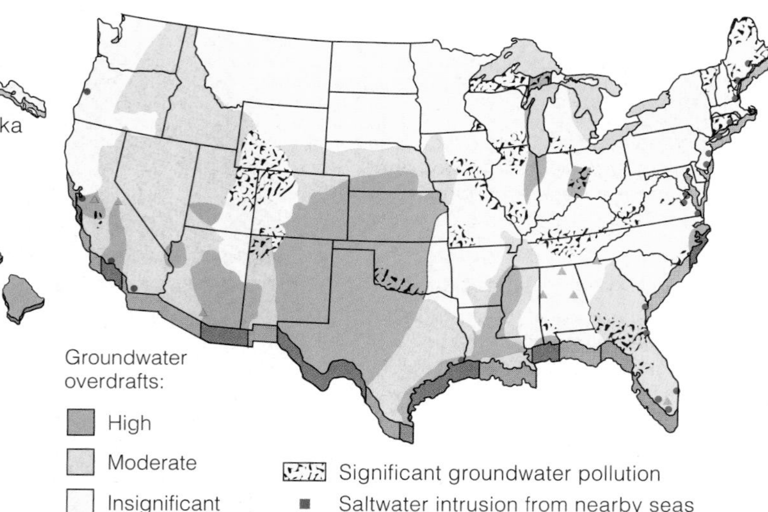

Alaska
Hawaiian Islands
Groundwater overdrafts:
High
Moderate
Insignificant
Significant groundwater pollution
Saltwater intrusion from nearby seas

Figure 51.14 Areas of aquifer depletion, saltwater intrusion, and groundwater contamination in the United States.

Water Pollution

Water pollution amplifies the problem of water scarcity. Inputs of sewage, animal wastes, and toxic chemicals make water unfit to drink, even to swim in. Pollutants encourage contamination by pathogens. Agricultural runoff pollutes water with sediments, pesticides, and plant nutrients. Power-generating plants and factories pollute the water with chemicals, radioactive materials, and excess heat (thermal pollution).

Pollutants collect in lakes, rivers, and bays before reaching the oceans. Many cities throughout the world dump untreated sewage into coastal waters. The cities

Further reading: Student Guide to InfoTrac on web site →

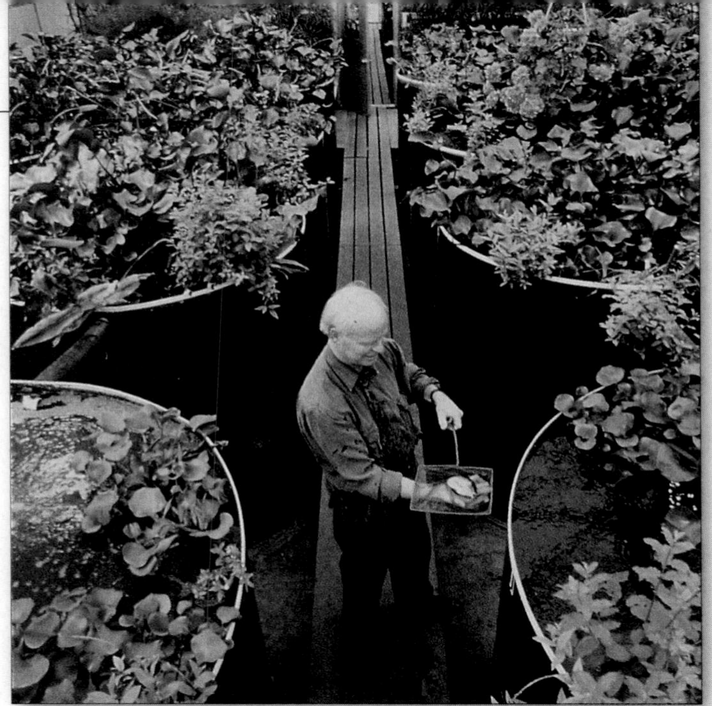

Figure 51.15 An experimental wastewater treatment facility in Rhode Island. Treatment begins when sewage flows into rows of large water tanks in which water hyacinths, cattails, and other aquatic plants are growing. Decomposers in the tank degrade wastes—which contain nutrients that promote plant growth. Heat from incoming sunlight speeds the decomposition. From these tanks, water flows through an artificial marsh of sand, gravel, and bulrushes that filter out algae and organic wastes. Then it flows into aquarium tanks, where zooplankton and snails consume microorganisms suspended in the water—and where zooplankton become food for crayfishes, tilapia, and other fishes that can be sold as bait. After ten days, the now-clear water flows into a second artificial marsh for final filtering and cleansing.

along rivers and harbors maintain shipping channels by dredging the polluted muck and barging it out to sea. They also barge sewage sludge—coarse, settled solids that contain bacteria, viruses, and toxic metals.

In the United States, about 15,000 facilities partially treat liquid wastes from 70 percent of the population and 87,000 industries. The remaining wastes are mostly from suburban and rural populations. These are treated in lagoons or septic tanks or are directly discharged—untreated—into waterways.

There are three levels of **wastewater treatment**. In *primary* treatment, screens and settling tanks remove sludge, which is dried, burned, dumped in landfills, or treated further. Chlorine often is used to kill pathogens. It does not kill them all, and it produces carcinogens whenever it reacts with certain industrial chemicals.

In *secondary* treatment, microbial populations break down organic matter after primary treatment but before chlorination. Wastewater trickles past microorganisms in gravel beds or is aerated in tanks and seeded with microorganisms. Toxic solutes can poison the microbes. At such times, the facilities shut down until populations of microorganisms become reestablished. Primary and secondary treatments remove most of the suspended solids and oxygen-demanding wastes, but not all of the nitrogen, phosphorus, and toxic substances, including heavy metals and pesticides. Then the water is usually chlorinated before being released into the waterways.

Tertiary treatment adequately reduces pollution but is largely experimental and expensive. It is applied to only 5 percent of our nation's wastewater.

In short, most wastewater is not treated adequately. A pattern gets repeated thousands of times along our waterways. Water for drinking is drawn upstream from a city, and wastes from industry and sewage treatment are discharged downstream. It takes no great leap of the imagination to see that water pollution intensifies as rivers flow to the oceans. In Louisiana, waters drained from the central states flow toward the Gulf of Mexico. Its high pollution levels threaten public health as well as ecosystems. Water destined for drinking gets treated to remove pathogens, but treatment does not remove toxic wastes dumped by numerous factories upstream.

This rather bleak picture might be numbing to most of us, but not to biologist John Todd. He constructed experimental wastewater treatment facilities in several greenhouses and artificial lagoons (Figure 51.15). When it works properly, the solar–aquatic treatment system produces water fit to drink. Such natural alternatives cannot work for very large urban areas. But they are an attractive alternative for small towns and rural areas.

The Coming Water Wars

If the current rates of population growth and water depletion hold, the amount of freshwater available for everyone on the planet will soon be 55 to 66 percent less than what it was in 1976. Already in the past decade, thirty-three nations have been engaged in conflicts over reductions in water flow, pollution, and silt buildup in major aquifers, rivers, and lakes. The United States and Mexico, Pakistan, India, and Israel and the occupied territories are among the squabblers.

Remember the Persian Gulf War, mainly about oil? Unless we pull off a blue revolution equivalent to the green one, we may be in for upheavals and wars over water rights. Does this sound farfetched? By building dams and irrigation systems at the headwaters of the Tigris and Euphrates rivers, Turkey can, in the view of one of its dam-site managers, stop the water flow into Syria and Iraq for as long as eight months "to regulate their political behavior." Regional, national, and global planning for the future is long overdue.

Water, not oil, may become the most important fluid of the twenty-first century. National, regional, and global policies for water usage and water rights have yet to be developed.

51.8 A QUESTION OF ENERGY INPUTS

Paralleling the J-shaped curve of human population growth is a dramatic rise in total and per capita energy consumption. It is due to increased numbers of energy users and to extravagant consumption and waste. For example, in one of the most pleasant of all climates, a major university constructed seven- and eight-story buildings with narrow, sealed windows. The windows can't be opened to catch prevailing ocean breezes. The buildings and windows were not designed or aligned to use sunlight for passive solar heating and breezes for passive cooling. Massive energy-demanding cooling and heating systems were installed.

When you hear talk of abundant energy supplies, bear in mind that there is a large difference between the *total* and net amounts available. *Net* refers to the amount left over after subtracting the energy required to locate, extract, transport, store, and deliver energy to consumers. Some sources, such as direct solar energy, are renewable. And others, such as coal, are not (Figure 51.16).

Fossil Fuels

Fossil fuels are the legacy of forests that disappeared many hundreds of millions of years ago, so they are nonrenewable resources (Section 25.4). Over time, the carbon-containing remains of the plants were buried in sediments, compacted, and chemically transformed into coal, petroleum (oil), and natural gas.

Even if conservation becomes strict, we may use up known petroleum and natural gas reserves in the next century. As known reserves run out in accessible areas, we explore wilderness areas in Alaska and other fragile environments, such as continental shelves. Net energy declines when the cost of extraction and transportation to and from remote areas increases. The environmental costs of extraction and transportation escalate. The widespread damage and clean-up costs following the 11-million-gallon oil spill from the supertanker *Valdez*, off the coast of Alaska, are a classic example.

What about coal? In theory, known world reserves may meet the energy needs of the human population for at least several centuries. However, coal burning has been the primary source of air pollution. Most of the known coal reserves contain low-quality material with a high sulfur content. Unless sulfur is removed before or after fuel is burned, sulfur dioxides enter the air and contribute to global acid deposition. Fossil fuel burning also releases carbon dioxide and adds to the greenhouse effect.

Extensive strip mining of coal reserves close to the surface carries its own problems. It reduces the land available for agriculture, grazing, and wildlife. Most strip mines are located in arid and semiarid regions where the absence of sufficient water supplies and poor soils make restoration efforts difficult.

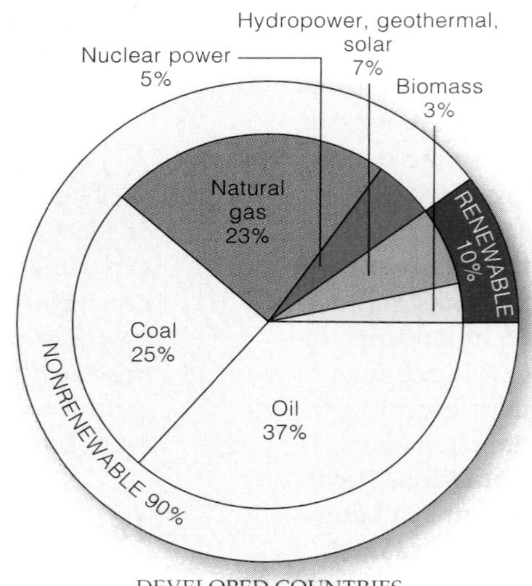
Hydropower, geothermal, solar 7%
Nuclear power 5%
Biomass 3%
Natural gas 23%
RENEWABLE 10%
Coal 25%
Oil 37%
NONRENEWABLE 90%
DEVELOPED COUNTRIES

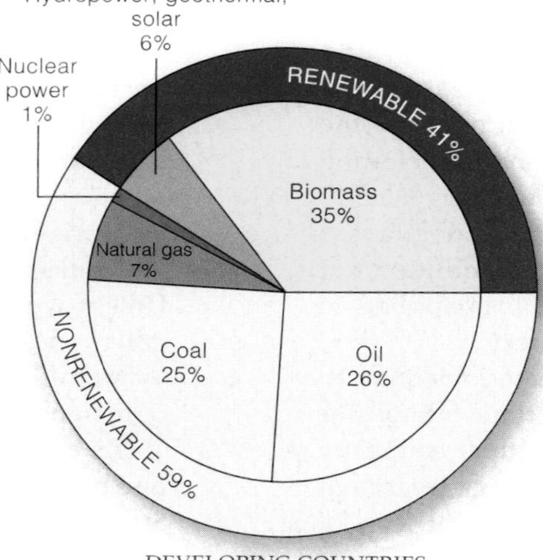
Hydropower, geothermal, solar 6%
Nuclear power 1%
RENEWABLE 41%
Biomass 35%
Natural gas 7%
Coal 25%
Oil 26%
NONRENEWABLE 59%
DEVELOPING COUNTRIES

Figure 51.16 Energy consumption in the developed and developing countries, which differ greatly in sources of energy and average per capita energy use. The values indicated do not take into account energy from the sun, which is the foundation for agriculture.

Nuclear Energy

NUCLEAR REACTORS In 1945, as Hiroshima burned, people recoiled in horror from the destructive force of nuclear energy. During the 1950s, however, many were championing nuclear energy as an instrument of progress. A number of energy-poor industrialized countries, including France, now depend heavily upon nuclear power. Yet, construction of more nuclear plants has been delayed or even cancelled in most countries. After 1970 in the United States alone, the plans to build 117 nuclear power plants were shelved. Plants already being constructed were abandoned before completion. A few are being converted, at great cost, to fossil fuel burning. What happened? We started to question the operating cost, efficiency, safety record, and environmental impact of reliance on nuclear power.

By 1990, nuclear power was generating electricity at a cost only slightly above that of coal-burning plants. Putting aside other factors, its electricity-generating costs are now lower. However, solar energy with natural gas backup should cost less in the near future.

What about safety? Compared to coal-burning plants of the same capacity, a nuclear plant emits less radioactivity and carbon dioxide, and no sulfur dioxide. However, there is greater danger in their potential for **meltdown**. As nuclear fuel undergoes radioactive decay, it releases considerable heat. Water

Further reading: Student Guide to InfoTrac on web site

a

Figure 51.17 Incident at Chernobyl. (**a**) On April 26, 1986, errors in judgment during a routine test procedure resulted in runaway reactions, explosions, and a full core meltdown at the Chernobyl power plant in Ukraine. Helicopter pilots who were supposed to drop 5,000 tons of lead, sand, clay, and other materials on the blazing core to suppress further release of radioactivity missed the target. Unimpeded and uncovered, nuclear fuel burned for nearly ten days, right on through a six-foot-thick steel and gravel barrier beneath it.

Between 185 and 250 million curies of radioactive matter may have escaped in those ten days alone. Inhaling as little as ten-millionths of a curie of plutonium can cause cancer. Thirty-one people died at once; others died of radiation sickness in the following weeks. Inhabitants of entire villages were relocated; their former homes were bulldozed under. In time, concrete entombed the 180 tons of partially burned nuclear fuel.

(**b**) Afterward, the number of people opposed to nuclear plants rose sharply, even in France. We get a sense of why opposition increased dramatically by studying maps of the global distribution of radioactive fallout within two weeks of the meltdown. The fallout put 300 to 400 million people at risk for leukemia and other radiation-induced disorders. Throughout Europe, hundreds of millions of dollars were lost when the fallout made crops and livestock unfit for consumption. In 1994, rainwater and air were still moving through 11,000 square meters of holes in the power plant's concrete. By 1998, the rate of thyroid abnormalities in children living immediately downwind from Chernobyl was nearly seven times as high as for those living upwind; their thyroid gland collected iodine radioisotopes from the fallout.

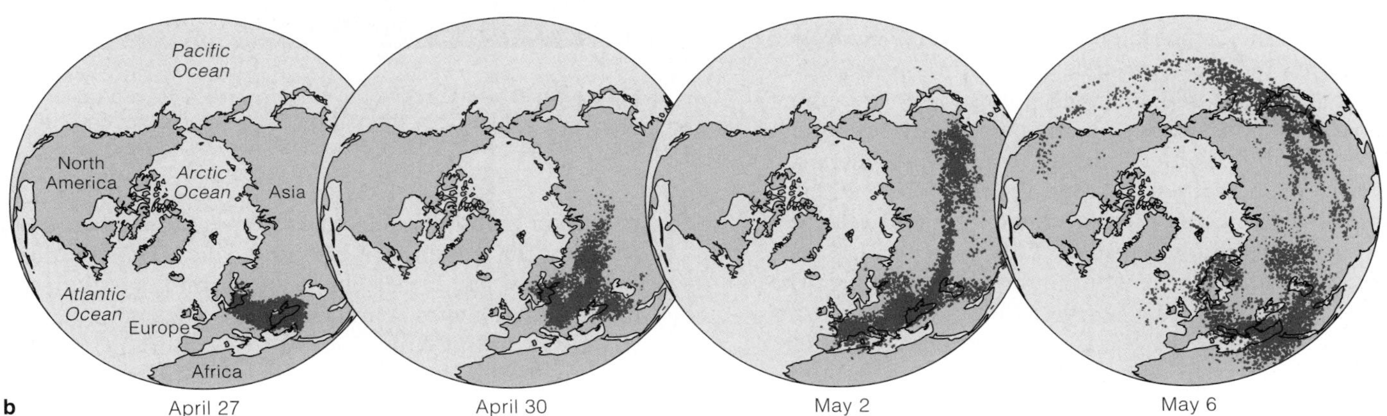
Pacific Ocean
North America
Arctic Ocean
Asia
Atlantic Ocean
Europe
Africa
b
April 27
April 30
May 2
May 6

typically circulates over the nuclear fuel, absorbs heat, and so produces steam to drive electricity-generating turbines. If the circulating water system develops a leak, water levels may plummet around the fuel, which may then heat past its melting point. Melting fuel on a generator floor would instantly convert the remaining water to steam. Combined with other reactions, steam formation could blow the system apart and release radioactive material. An overheated core could melt through its greatly thickened concrete containment slab. Figure 51.17 describes one incident that yielded compelling evidence of the consequences.

NUCLEAR WASTE DISPOSAL Unlike coal, nuclear fuel cannot be burned to harmless ashes. After three years or so, the fuel elements are spent, but they still contain uranium fuel as well as hundreds of new radioisotopes that formed in the reactions. The wastes are extremely radioactive and dangerous. They get extremely hot as they undergo radioactive decay, so they are plunged at once into water-filled pools. The water cools them and keeps radioactive material from escaping. Even after being stored for several months, the isotopes remaining are lethal. Some must be kept isolated for at least 10,000 years. If a certain isotope of plutonium (^{239}Pu) is not removed, the wastes must be kept isolated for a quarter of a million years! After nearly fifty years of research, scientists still cannot agree on what is the best way to store high-level radioactive wastes. Even if they could, there is no politically acceptable solution. No one wants radioactive wastes anywhere near where they live.

Finally, as if we don't have enough to worry about, following the Soviet Union's breakup, some underpaid workers of a Russian nuclear power plant have been selling fuel elements on the black market. The buyers? Some developing nations that want to produce nuclear weapons—and possibly deliver them into the hands of terrorist organizations.

The nuclear genie is out of the bottle, exploitable by the best and worst elements of the human population.

51.9 ALTERNATIVE ENERGY SOURCES

Less than thirty years from now, the projected size of the human population will be such that the demands for fuel will increase by 30 percent—and for electricity by 265 percent. More efficient use and conservation of our existing energy sources alone will not do the trick. We must make the transition to reliance on alternative sources of energy, of the sort described next.

Solar–Hydrogen Energy

Each year, incoming sunlight contains about ten times more energy than that in all of the known fossil fuel reserves. That is 15,000 times more energy than our population uses now. Isn't it about time that we start to collect it in earnest, in something besides crops?

For example, when exposed to sunlight, electrodes in "photovoltaic cells" produce an electric current that splits water molecules into oxygen and hydrogen gas (H_2)—which can be used directly as fuel or to generate electricity. The technology to tap such **solar–hydrogen energy** has been around since the 1940s (Figure 51.18*a*). Such energy is stored efficiently for as long as required. It costs less to distribute H_2 than electricity, and water is the only by-product of using it. Space satellites run on it. Here on Earth, fossil fuels are still "cheaper."

Unlike fossil fuels, however, sunlight and seawater are virtually unlimited resources. And the technology's potential to protect the environment is staggering. The environmental scientist G. Tyler Miller, Jr., puts it this way: "If we make the transition to an energy-efficient solar–hydrogen age, we can say good-bye to smog, oil spills, acid rain, and nuclear energy, and perhaps to global warming. The reason is simple. As hydrogen burns in air, it reacts with oxygen gas to produce water vapor—not a bad thing to have coming out of tailpipes, chimneys, and smokestacks." Also, if this technology becomes cost-effective for the developing countries, the great forests now being destroyed for timber and fuel might still be around for future generations.

In 1995, photovoltaic cells generated almost 4 million kilowatthours of net electricity in the United States (of the total electricity generation of 988.8 million). Thirty states now promote development of solar energy. In 1996, Hawaii, Iowa, and Washingon offered incentives to promote clean, renewable energy technologies.

Wind Energy

As you know, solar energy also is converted into the mechanical energy of winds. Where prevailing winds travel faster than 7.5 meters per second, we find **wind farms**. These arrays of cost-effective turbines exploit wind patterns that arise from latitudinal variations in the intensity of incoming sunlight (Figure 51.18*b*). One

Figure 51.18 Harnessing solar energy. (**a**) A large array of electricity-producing photovoltaic cells in panels that collect sunlight energy. (**b**) A field of turbines harvesting wind energy.

percent of California's electricity comes from wind farms. Possibly the winds of North and South Dakota alone can meet all but 20 percent of the current energy needs of the United States. Wind energy also has potential for islands and other areas remote from utility grids. One drawback: winds do not blow on a regular schedule, so they cannot be an exclusive or major energy source.

What About Fusion Power?

The sun's gravitational force is enough to compress atomic nuclei to high densities, and its temperatures are high enough to force atomic nuclei to fuse. We call this **fusion power**. Similar conditions do not exist on Earth, but maybe we can mimic them. Researchers confine a certain fuel (a heated gas of two isotopes of hydrogen) in magnetic fields, then bombard it with lasers. The fuel implodes, it is compressed to extremely high densities —and energy is released. The more energetic the lasers, the greater the compression, and the more the fuel will burn. The bad news is, although the amount of energy released has been steadily increasing, it will be at least fifty years before fusion reactors could be operating, and the costs will probably be high. The good news is, that's about the time fossil fuels may start running out.

Sunlight may end up sustaining the energy needs of the human population in more ways than one.

Further reading: Student Guide to InfoTrac on web site

51.10 BIOLOGICAL PRINCIPLES AND THE HUMAN IMPERATIVE

Molecules, single cells, tissues, organs, organ systems, multicelled organisms, populations, communities, then ecosystems and the biosphere. These are architectural systems of life, assembled in increasingly complex ways during the past 3.8 billion years. We are latecomers to this immense biological building program. Yet within the relatively short span of 10,000 years, our activities have been changing the very character of the land, ocean, and atmosphere, even the genetic character of species.

It would be presumptuous to think we alone have had profound effects on the world of life. As long ago as the Proterozoic era, photosynthetic organisms irrevocably changed the course of biological evolution by gradually enriching the atmosphere with oxygen. During the past as well as the present, competitive adaptations ensured the rise of some groups, whose dominance assured the decline of others. Change is nothing new to this biological building program. What *is* new is the capacity of a species —our own—to comprehend what might be going on.

We now have the population size, the technology, and the cultural inclination to use up energy and modify the environment at frightening rates. Where will rampant, accelerated change lead us? Will feedback controls begin to operate as they do, for example, when population growth exceeds the carrying capacity of the environment? In other words, will negative feedback controls come into play and keep things from getting too far out of hand?

Feedback control will not be enough, for it operates only when deviation already exists. Patterns of resource consumption and growth rates for the human population are founded on an illusion of unlimited resources and a forgiving environment. A prolonged, global shortage of food or the passing of a critical threshold for the global climate can come too fast to be corrected. At some point, deviations may have too great an impact to be reversed.

What about feedforward mechanisms that might serve as early warning systems? For example, when sensory receptors near the surface of skin sense a drop in outside air temperature, each sends messages to the nervous system. That system responds by triggering mechanisms that raise the body's core temperature before the body itself becomes dangerously chilled. If we develop feedforward control mechanisms, maybe we can begin corrective measures before we have altered the environment too significantly.

By themselves, feedforward controls won't work, for they start operating when change is under way. Think of the DEW line—the Distant Early Warning system. It is like a vast sensory receptor, one that can detect the launching of intercontinental ballistic missiles against North America. By the time the system actually detects what it is designed to detect, it may be too late to stop widespread destruction.

It would be naive to assume we can ever reverse who we are at this point in evolutionary time, to de-evolve ourselves culturally and biologically into becoming less complex in the hope of averting disaster. However, there is reason to believe we can avert disaster by using a third kind of control mechanism—a capacity to anticipate events even before they happen. We are not locked into responding only after irreversible change has begun. We have the capacity to anticipate the future—it is the essence of our visions of utopia or of nightmarish hell. *We all have the capacity to adapt to a future that we can partly shape.*

We can, for example, stop trying to "beat nature" and learn to work with it. Individually and collectively, we can work to develop long-term policies at the local, regional, and global levels—policies that take into account biotic and abiotic limits on population growth. Far from being a surrender, this would be one of the most complex and intelligent behaviors of which we are capable.

Having a capacity to adapt and actually using it are not the same thing. We have already put the world of life on dangerous ground because we have not yet mobilized ourselves as a species to work toward self-control.

Our survival depends on predicting possible futures. It depends on preserving, restoring, and even designing and constructing ecosystems that are in harmony with our definition of basic human values and with the biological models available to us. Human values can change; our expectations can and must be adapted to biological reality. *For the principles of energy flow and resource utilization, which govern the survival of all systems of life, do not change.* It is our biological and cultural imperative that we come to terms with these principles, and ask ourselves what our long-term contribution will be to the world of life.

51.11 SUMMARY

1. Accompanying the extraordinarily rapid growth of the human population are increases in energy demands and in environmental pollution.

2. Pollutants are substances with which ecosystems have had no prior evolutionary experience (in terms of kinds and amounts) and therefore have no mechanisms for absorbing or cycling them. Many pollutants result from certain human activities, and they adversely affect the health, activities, or survival of all organisms.

3. Smog, a form of air pollution, arises in industrialized and urban regions that rely on fossil fuels. It becomes most concentrated in land basins that promote thermal inversion (a layer of cool, dense air trapped below a warm air layer). Industrial smog forms in industrial coal-burning regions with cold, wet winters. In warm climates, large cities with many fuel-burning vehicles have photochemical smog. Mainly, sunlight makes nitric oxide emitted from vehicles react with hydrocarbons to form photochemical oxidants such as PANs.

4. During dry weather, acidic air pollutants, especially oxides of nitrogen and sulfur, fall to Earth as dry acid deposition. They also dissolve in atmospheric water, then fall to Earth as wet acid deposition, or acid rain.

5. Seasonal thinning of the ozone layer at high latitudes has become pronounced as CFCs (chlorofluorocarbons) and other air pollutants rise to the stratosphere, where they deplete ozone and allow more harmful ultraviolet radiation from the sun to reach the Earth's surface.

6. Human population growth presently depends upon the expansion of agriculture, made possible by large-scale irrigation and extensive applications of fertilizers and pesticides. Global freshwater supplies are limited, and yet they are being polluted by agricultural runoff (which includes sediments, pesticides, and fertilizers), industrial wastes, and human sewage.

7. Human populations are damaging land surfaces by:

a. Passively accepting the dumping, burning, and burial of solid wastes rather than making a concerted effort to recycle or reuse materials and to reduce waste.

b. Engaging in rampant deforestation (destruction of vast tracts of tropical and temperate forest biomes).

c. Contributing to desertification (the conversion of natural grasslands, croplands, and grazing lands, on a large scale, to desertlike conditions).

8. Energy in the form of fossil fuels is nonrenewable, dwindling, and environmentally costly to extract and use. Nuclear energy in itself is less polluting, but the costs and risks associated with fuel containment and storing radioactive wastes are enormous. The challenge is to develop affordable alternatives that are based on renewable energy resources, such as solar energy.

Review Questions

1. Define pollution and list some specific examples of water pollutants. *51.1, 51.7*

2. Distinguish among the following conditions: *51.1*
   a. industrial smog c. dry acid deposition
   b. photochemical smog d. wet acid deposition

3. Define CFCs and describe how they apparently contribute to seasonal thinning of the ozone layer in the stratosphere. *51.2*

4. What percent of the Earth's land is under cultivation? What percent is available for new cultivation? *51.3*

5. Define and describe possible consequences of deforestation and of desertification. *51.4, 51.5, 51.6*

6. Which human activity uses the most freshwater? *51.7*

Self-Quiz (*Answers in Appendix III*)

1. Since the mid-eighteenth century, human population growth has been ______.
   a. leveling off c. accelerating
   b. growing slowly d. not much to speak of

2. Pollutants disrupt ecosystems because ______.
   a. their components differ from those of natural substances
   b. only humans have uses for them
   c. there are no evolved mechanisms to deal with them
   d. their only effect is on ecosystems, not humans

3. During a thermal inversion, weather conditions trap a layer of ______ air under a layer of ______ air.
   a. warm; cool c. warm; sooty
   b. cool; warm d. cool; sooty

4. ______ is a case of regional air pollution.
   a. Smog c. Ozone layer thinning
   b. Acid rain d. a and b

5. ______ is a case of air pollution with global effects.
   a. Smog c. Ozone layer thinning
   b. Acid rain d. b and c

6. Worldwide, two-thirds of the freshwater used annually goes to ______.
   a. urban centers c. treatment facilities
   b. agriculture d. a and c

7. The upper limit at which the ground is fully saturated with water is called ______.
   a. groundwater c. the water table
   b. an aquifer d. the salinization limit

8. Energy from fossil fuels is ______; their extraction and use come at ______ cost to the environment.
   a. renewable; low c. renewable; high
   b. nonrenewable; low d. nonrenewable; high

9. Nuclear energy normally pollutes ______ than fossil fuels; it poses ______ dangers than fossil fuels.
   a. less; lesser b. more; greater c. more; lesser d. less; greater

10. Match each term with the most suitable description.

____ desertification	a. possibly one of our best options
____ deforestation	b. soil loss, watershed damage, altered rainfall patterns follow
____ green revolution	c. attempt to improve crop production on existing land
____ solar–hydrogen power	d. converting large tracts of natural grasslands to desertlike state
____ CFCs	e. invisible, odorless compounds that contribute to ozone thinning

Critical Thinking

1. Kristen, a recent college graduate, is finding her idealism on a collision course with reality. She strongly believes people who live in the United States are obliged to make the world a level playing field for all human beings, with equality in resources, health care, education, economic security, and a pristine environment for all. Yet she also understands that the sheer size of the human population makes this impossible.

Kristen recently said she cannot be party to hard choices and actions that go against her ideals, and she just wants nature to solve the problem for us. Comment on this true story.

2. Populations of every species use resources and produce wastes. Using Figure 51.19 as your starting point, write a brief essay on the accumulation and uses of energy and materials, including wastes, in a human population that is concentrated in a large city. Contrast your description with the flow of energy and the materials cycling that are characteristic of a natural population of some other organism.

3. In 1995, biologists Reed Noss, J. Michael Scott, and Edward LaRoe issued a report on *endangered ecosystems* in the United States. They categorize thirty natural ecosystems as being in danger of disappearing; the once-vast domains have diminished in size by more than 98 percent. Figure 51.20 is a map of this deforestation. Agricultural conversion, urban development, and other human activities account for the losses.

In the preceding chapter, you read about the type and extent of the ecosystems in question. The once-largest among them include tallgrass prairie, oak-studded savannas, and eastern deciduous forests, as well as pine forests that once covered much of the coastal plains of the southeastern states. As Figure 51.20 shows, the most imperiled regions are largely in the eastern part of the country. As the map indicates, more biodiversity was lost at the ecosystem level than is generally recognized. The finding may impact how (and whether) the government amends its conservation laws, such as the Endangered Species Act. Do we as a nation protect entire ecosystems, not just an endangered species? That is a goal of many conservationists. But property-rights advocates criticize the goal; they view it as a threat to the long-standing tradition of private ownership of land. They also worry that merely identifying ecosystems in need of protection will endanger property values.

Do some research on a natural habitat that is part of an imperiled ecosystem. Get a sense of its decline in biodiversity, of the kinds of organisms that once flourished there. Then ask yourself: Would you participate in efforts to set aside land for ecological restoration? Or would you consider such efforts too intrusive on individual rights of property ownership?

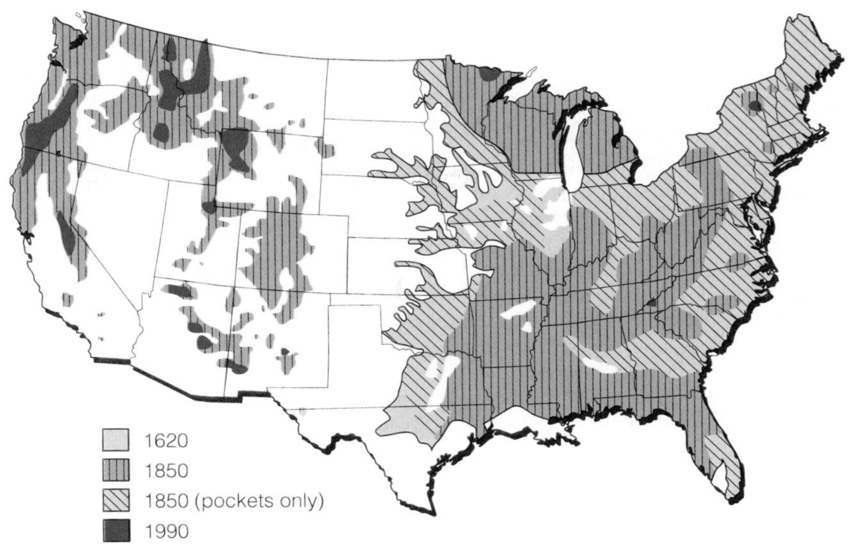
1620
1850
1850 (pockets only)
1990

Figure 51.20 Extent of deforestation in the continental United States, starting from the year 1620 through 1990.

Figure 51.19 City as ecosystem.

4. Investigate where the water for your own city comes from and where it has been. You may find the answer illuminating.

5. Make a list of advantages you personally enjoy as a member of an affluent, industrialized society. List some drawbacks. Do the benefits outweigh the costs? This is not a trick question.

6. It has been said that economic wars, more than military wars, will determine the winners and losers among nations in the near future. The economic growth of certain nations, including some in the former Soviet Union and the Far East, has had devastating impact on the environment. Elsewhere, efforts of conservation biologists to maintain standards of environmental protection adds to the cost of goods produced and puts practicing nations at serious disadvantage in this global competition.

Should the United States loosen some of its environmental laws as a way to help ensure its economic survival? Why or why not? Can you think of some examples of pressures that might be imposed on indifferent nations to encourage them to change harmful practices?

7. Many anti-environmental coalitions flourish in the United States. One group, the *Wise-Use Movement*, has these goals:

a. Cut all old-growth forests in national forests and replace them with tree plantations.

b. Modify the Endangered Species Act so economic factors override preservation of endangered or threatened species.

c. Eliminate restrictions on wetland development.

d. Open all national parks, national wildlife refuges, and wilderness areas to oil drilling, mining, off-road vehicles, and commercial development.

e. Launch a twenty-year program to build new concessions in national parks and run them by private firms instead of the National Park Service.

f. Continue mining on public lands under the 1872 Mining Law. Under certain provisions, taxpayers receive no recompense for minerals extracted on public lands, and miners are entitled to purchase public lands for negligible sums.

g. Recognize private property rights to mining claims, water, grazing permits, and timber contracts on public lands and do not raise fees for these rights. Better to sell resource-rich public lands to private enterprises.

h. Impose civil penalties against any individual who legally challenges economic action or development on federal lands.

i. Allow pro-industry groups or individuals to sue as harmed parties on behalf of industries that are threatened by environmentalists.

Focus on one of these Wise-Use Movement goals. Do you agree or disagree with it? Why or why not?

8. To reduce the backlog of unnamed new species and secure some funding, a group of German taxonomists is currently offering shoppers the opportunity, for $2,500 and up, to name a species (Biopat web site). For example, as a birthday present, investment banker Stan Lai's wife got a yellow-striped brown toad that lives in Bolivia named in her husband's honor (*Bufo stanlaii*, or "Stan Lai's toad").

Usually, a species name reflects its appearance or honors its discoverer. But not always; for example, one midge has the name *Dicrotendipes thanatogratus* (the "Grateful Dead" midge). The International Commission on Zoological Nomenclature in London denounces offering taxonomy to the public, which it believes will obscure science and hinder conservation efforts. What does Mrs. Lai think? "Now, every time I hear something about Bolivia, I'll pay attention. After all, I don't want to see our toad die."

Does inviting the public to pay their way into taxonomy undermine the science? Is it a good or bad way to raise public awareness of conservation biology?

9. Many psychologists believe that we spend much of our lives consciously or subconsciously searching for roots that might anchor us in bewildering or frightening times of change. Some philosophers have even argued that each of us must find a mountain, river, backyard, or any other place that elicits a sense of sanctuary, of rooted connections with nature. Others believe that forming an emotional connection with nature is based on a romanticized or mystical view of it—and that time would be better spent developing a scientifically sound understanding of nature and the technological means to protect and sustain its resources.

Do you agree with one or the other point of view? Or do you suspect that the two are not mutually exclusive?

Selected Key Terms

acid rain *51.1*
chlorofluorocarbon (CFC) *51.2*
deforestation *51.4*
desalinization *51.7*
desertification *51.6*
dry acid deposition *51.1*
fossil fuel *51.8*
fusion power *51.9*
green revolution *51.3*
industrial smog *51.1*
meltdown *51.8*
ozone thinning *51.2*
PAN (peroxyacyl nitrate) *51.1*
photochemical smog *51.1*
pollutant *51.1*
salinization *51.7*
shifting cultivation *51.4*
solar–hydrogen energy *51.9*
thermal inversion *51.1*
wastewater treatment *51.7*
water table *51.7*
wind farm *51.9*

Readings *See also www.infotrac-college.com*

Collins, M. 1990. *The Last Rain Forests*. New York: Oxford University Press.

Frosh, R. September 1995. "The Industrial Ecology of the Twenty-First Century." *Scientific American* 273(3):178–181.

Gruber, D. 1989. "Biological Monitoring and Water Resources." *Endeavour* 13(3):135–140.

Kaiser, J. 18 February 2000. "Ecologists on a Mission to Save the World." *Science* 287:1188–1192. Addresses a growing debate over how far environmental scientists should go when they are interpreting their findings for policy makers. Efforts to make advocacy over environmental issues an accepted practice among ecologists has provoked a backlash; some leading ecologists are warning colleagues that activism might erode credibility of research in general. Do read this brief, balanced article.

Miller, G. T., Jr. 1999. *Living in the Environmental*. Seventh edition. Belmont, California: Brooks/Cole. This author consistently puts information from numerous sources into a current, accessible survey of the present and future state of the environment.

Plucknett, D., and D. Winkelmann. September 1995. "Technology for a Sustainable Agriculture." *Scientific American* 273(3):182–186.

Wilson, E. 1988. *Biodiversity*. Washington, D.C.: National Academy of Sciences.

APPENDIX I. CLASSIFICATION SCHEME

The classification scheme that follows is a composite of several that microbiologists, botanists, and zoologists use. Major groupings are agreed upon, more or less. There is not always agreement on what to name a particular grouping or where it may fit within the overall hierarchy. There are several reasons for the lack of total consensus.

First, the fossil record varies in its quality and in its completeness. Therefore, the phylogenetic relationship of one group to others is sometimes open to interpretation. Comparative studies at the molecular level are firming up the picture, but this work is still under way.

Second, ever since the time of Linnaeus, classification schemes have been based on the perceived morphological similarities and differences among organisms. Although some original interpretations are now open to question, we are so used to thinking about organisms in certain ways that reclassification often proceeds slowly.

For example, birds and reptiles traditionally have been placed in separate classes (Reptilia and Aves), yet there are many compelling arguments for grouping the lizards and snakes in one class and the crocodilians, dinosaurs, and birds in a separate class. Some favor six kingdoms but others favor three domains: archaebacteria, eubacteria, and eukaryota (alternatively, archaea, bacteria, and eukarya).

Third, researchers in microbiology, mycology, botany, zoology, and the other fields of biological inquiry have inherited a wealth of literature, based on classification schemes that were developed over time in each of those fields. Many see no good reason to give up the established terminology and thereby disrupt access to the past.

For instance, many microbiologists and botanists use *division*, and zoologists *phylum*, for taxa that are equivalent in the hierarchy of classification. Also, opinions are still polarized with respect to the kingdom Protista, certain members of which could just as easily be grouped with plants, fungi, or animals. Indeed, the term protozoan is a holdover from an earlier scheme in which some single-celled organisms were ranked as simple animals.

Given the problems, why do we even bother imposing artificial frameworks on the history of life? We do this for the same reason that a writer might decide to break up the history of civilization into several volumes, a number of chapters, and many paragraphs. Both efforts are attempts to impart obvious structure to what might otherwise be an overwhelming body of knowledge and to enhance the retrieval of information from it.

Finally, bear in mind that we include this classification scheme primarily for your reference purposes. Besides being open to revision, it also is by no means complete. It does not include the most recently discovered species, as from the mid-ocean. Many existing and extinct organisms of the so-called lesser phyla are not represented here. Our strategy is to focus mainly on organisms mentioned in the text. A few examples of organisms also are listed.

SUPERKINGDOM PROKARYOTA. Prokaryotes. Almost all microscopic species. DNA organized at nucleoid (a region of cytoplasm), not inside a membrane-bound nucleus. All are bacteria, either single cells or simple associations of cells. Autotrophs and heterotrophs. Table A on the following page lists representative types. Reproduce by prokaryotic fission, sometimes by budding and by bacterial conjugation.

The authoritative reference in bacteriology, *Bergey's Manual of Systematic Bacteriology*, calls this "a time of taxonomic transition." It groups bacteria mostly by numerical taxonomy (Section 22.6), not on phylogeny. The scheme presented here reflects strong evidence of evolutionary relationships for at least some bacterial groupings.

KINGDOM EUBACTERIA. Gram-negative, gram-positive forms. Peptidoglycan present in cell wall. Photosynthetic autotrophs, chemosynthetic autotrophs, and heterotrophs.

PHYLUM GRACILICUTES. Typical Gram-negative, thin wall. Autotrophs (photosynthetic and chemosynthetic) and heterotrophs. *Anabaena* and other cyanobacteria. *Escherichia, Pseudomonas, Neisseria, Myxococcus.*

PHYLUM FIRMICUTES. Typical Gram-positive, thick wall. Heterotrophs. *Bacillus, Staphylococcus, Streptococcus, Clostridium, Actinomycetes.*

PHYLUM TENERICUTES. Gram-negative, wall absent. Heterotrophs (saprobes, pathogens). *Mycoplasma.*

KINGDOM ARCHAEBACTERIA. Methanogens, extreme halophiles, extreme thermophiles. Evolutionarily closer to eukaryotic cells than to eubacteria. Strict anaerobes. Distinctive cell wall, membrane lipids, ribosomes, RNA sequences. *Methanobacterium, Halobacterium, Sulfolobus.*

SUPERKINGDOM EUKARYOTA. Eukaryotes. Both single-celled and multicelled species. Cells start out life with a nucleus (encloses the DNA) and usually other membrane-bound organelles. Chromosomes have many histones and other proteins attached.

KINGDOM PROTISTA. Diverse single-celled, colonial, and multicelled eukaryotic species. Existing species are unlike bacteria in characteristics and are most like the earliest, structurally simple eukaryotes. Autotrophs, heterotrophs, or both (Table 23.3). Reproduce sexually and asexually (by meiosis, mitosis, or both). Many related evolutionarily to plants, fungi, and possibly animals.

PHYLUM CHYTRIDIOMYCOTA. Chytrids. Heterotrophs; saprobic decomposers or parasites. *Chytridium.*

PHYLUM OOMYCOTA. Water molds. Heterotrophs. Decomposers, some parasites. *Saprolegnia, Phytophthora, Plasmopara.*

PHYLUM ACRASIOMYCOTA. Cellular slime molds. Heterotrophs with free-living, phagocytic amoeboid cells and spore-bearing stages. *Dictyostelium.*

PHYLUM MYXOMYCOTA. Plasmodial slime molds. Heterotrophs with free-living, phagocytic amoeboid cells and spore-bearing stages. Aggregate into streaming mass of cells that discard plasma membranes. *Physarum.*

Table A Representative Eubacteria and Archaebacteria Grouped on the Basis of Numerical Taxonomy

Some Major Groups	Main Habitats	Characteristics	Representatives
EUBACTERIA			
Photoautotrophs:			
Cyanobacteria, green sulfur bacteria, and purple sulfur bacteria	Mostly lakes, ponds; some marine, terrestrial habitats	Photosynthetic; use sunlight energy, carbon dioxide; cyanobacteria use oxygen-producing noncyclic pathway; some also use cyclic route	*Anabaena, Nostoc, Rhodopseudomonas, Chloroflexus*
Photoheterotrophs:			
Purple nonsulfur and green nonsulfur bacteria	Anaerobic, organically rich muddy soils, and sediments of aquatic habitats	Use sunlight energy; organic compounds as electron donors; some purple nonsulfur may also grow chemotrophically	*Rhodospirillum, Chlorobium*
Chemoautotrophs:			
Nitrifying, sulfur-oxidizing, and iron-oxidizing bacteria	Soil; freshwater, marine habitats	Use carbon dioxide, inorganic compounds as electron donors; influence crop yields, cycling of nutrients in ecosystems	*Nitrosomonas, Nitrobacter, Thiobacillus*
Chemoheterotrophs:			
Spirochetes	Aquatic habitats; parasites of animals	Helically coiled, motile; free-living and parasitic species; some major pathogens	*Spirochaeta, Treponema*
Gram-negative aerobic rods and cocci	Soil, aquatic habitats; parasites of animals, plants	Some major pathogens; some fix nitrogen (e.g., *Rhizobium*)	*Pseudomonas, Neisseria, Rhizobium, Agrobacterium*
Gram-negative facultative anaerobic rods	Soil, plants, animal gut	Many major pathogens; one bioluminescent (*Photobacterium*)	*Salmonella, Escherichia, Proteus, Photobacterium*
Rickettsias and chlamydias	Host cells of animals	Intracellular parasites; many pathogens	*Rickettsia, Chlamydia*
Myxobacteria	Decaying organic material; bark of living trees	Gliding, rod-shaped; aggregation and collective migration of cells	*Myxococcus*
Gram-positive cocci	Soil; skin and mucous membranes of animals	Some major pathogens	*Staphylococcus, Streptococcus*
Endospore-forming rods and cocci	Soil; animal gut	Some major pathogens	*Bacillus, Clostridium*
Gram-positive nonsporulating rods	Fermenting plant, animal material; gut, vaginal tract	Some important in dairy industry, others major contaminators of milk, cheese	*Lactobacillus, Listeria*
Actinomycetes	Soil; some aquatic habitats	Include anaerobes and strict aerobes; major producers of antibiotics	*Actinomyces, Streptomyces*
ARCHAEBACTERIA (ARCHAEA)			
Methanogens	Anaerobic sediments of lakes, swamps; animal gut	Chemosynthetic; methane producers; used in sewage treatment facilities	*Methanobacterium*
Extreme halophiles	Brines (extremely salty water)	Heterotrophic; also, unique photosynthetic pigments (bacteriorhodopsin) form in some	*Halobacterium*
Extreme thermophiles	Acidic soil, hot springs, hydrothermal vents	Heterotrophic or chemosynthetic; use inorganic substances as electron donors	*Sulfolobus, Thermoplasma*

PHYLUM SARCODINA. Amoeboid protozoans. Heterotrophs, free-living or endosymbiotic, some pathogens. Soft-or shelled bodies, locomotion by pseudopods. The rhizopods (naked amoebas, foraminiferans), *Amoeba proteus, Entomoeba*. Also actinopods (radiolarians, heliozoans).

PHYLUM CILIOPHORA. Ciliated protozoans. Heterotrophs, predators or symbionts, some parasitic. All have cilia. Free-living, sessile, or motile. *Paramecium, Didinium,* hypotrichs.

PHYLUM MASTIGOPHORA. Animal-like flagellated protozoans. Heterotrophs, free-living, many internal parasites. All with one to several flagella. *Trypanosoma, Trichomonas, Giardia.*

APICOMPLEXA. Heterotrophs, sporozoite-forming parasites. Complex structures at head end. Most familiar members called sporozoans. *Cryptosporidium, Plasmodium, Toxoplasma.*

PHYLUM EUGLENOPHYTA. Euglenoids. Mostly heterotrophs, some autotrophs (photosynthetic), some switch depending on environmental conditions. Most with one short, one long flagellum; red, green, or colorless. *Euglena, Peranema.*

PHYLUM PYRRHOPHYTA. Dinoflagellates. Photosynthetic, mostly, but some heterotrophs. *Fiesteria, Gymnodinium breve.*

PHYLUM CHRYSOPHYTA. Golden algae, yellow-green algae, diatoms. Photosynthetic. Some flagellated, others not. *Mischococcus, Synura, Vaucheria.*

PHYLUM RHODOPHYTA. Red algae. Mostly photosynthetic, some parasitic. Nearly all marine, some in freshwater habitats. *Porphyra. Bonnemaisonia, Euchema.*

PHYLUM PHAEOPHYTA. Brown algae. Photosynthetic, nearly all in temperate or marine waters. *Macrocystis, Fucus, Sargassum, Ectocarpus, Postelsia.*

PHYLUM CHLOROPHYTA. Green algae. Mostly photosynthetic, some parasitic. Most freshwater, some marine or terrestrial. *Chlamydomonas, Spirogyra, Ulva, Volvox, Codium, Halimeda.*

KINGDOM FUNGI. Nearly all multicelled eukaryotic species. Heterotrophs, mostly saprobic decomposers, some parasites. Nutrition based upon extracellular digestion of organic matter and absorption of nutrients by individual cells. Multicelled species form absorptive mycelia within substrates and structures that produce asexual spores (and sometimes sexual spores).

PHYLUM ZYGOMYCOTA. Zygomycetes. Zygosporangia (zygote inside thick wall) formed by sexual reproduction. Bread molds, related forms. *Rhizopus, Philobolus.*

PHYLUM ASCOMYCOTA. Ascomycetes. Sac fungi. Sac-shaped cells form sexual spores (ascospores). Most yeasts and molds, morels, truffles. *Saccharomycetes, Morchella, Neurospora, Sarcoscypha, Claviceps, Ophiostoma, Candida, Aspergillus, Penicillium.*

PHYLUM BASIDIOMYCOTA. Basidiomycetes. Club fungi. Most diverse group. Produce basidiospores inside club-shaped structures. Mushrooms, shelf fungi, stinkhorns. *Agaricus, Amanita, Craterellus, Gymnophilus, Puccinia, Ustilago.*

IMPERFECT FUNGI. Sexual spores absent or undetected. The group has no formal taxonomic status. If better understood, a given species might be grouped with sac fungi or club fungi. *Arthobotrys, Histoplasma, Microsporum, Verticillium.*

LICHENS. Mutualistic interactions between fungal species and a cyanobacterium, green alga, or both. *Lobaria, Usnea, Cladonia.*

KINGDOM PLANTAE. Multicelled eukaryotes. Nearly all photosynthetic autotrophs with chlorophylls *a* and *b*. Some parasitic. Nonvascular and vascular species, generally with well-developed root and shoot systems. Nearly all adapted in form and function to survive dry conditions on land; a few in aquatic habitats. Sexual reproduction predominant; also asexual reproduction by vegetative propagation and other mechanisms.

PHYLUM RHYNIOPHYTA. Earliest known vascular plants; muddy habitats. Extinct. *Cooksonia, Rhynia.*

PHYLUM PROGYMNOSPERMOPHYTA. Progymnosperms. Ancestral to early seed-bearing plants; extinct. *Archaeopteris.*

PHYLUM PTERIDOSPERMOPHYTA. Seed ferns. Fernlike gymnosperms; extinct. *Medullosa.*

PHYLUM CHAROPHYTA. Stoneworts.

PHYLUM BRYOPHYTA. Bryophytes: mosses, liverworts, hornworts. Seedless, nonvascular, haploid dominance. *Marchantia, Polytrichum, Sphagnum.*

PHYLUM PSILOPHYTA. Whisk ferns. Seedless, vascular. No obvious roots, leaves on sporophyte. *Psilotum.*

PHYLUM LYCOPHYTA. Lycophytes, club mosses. Seedless, vascular. Leaves, branching rhizomes, vascularized roots and stems. *Lepidodendron* (extinct), *Lycopodium, Selaginella.*

PHYLUM SPHENOPHYTA. Horsetails. Seedless, vascular. Some sporophyte stems photosynthetic, others nonphotosynthetic, spore-producing. *Calamites* (extinct), *Equisetum.*

PHYLUM PTEROPHYTA. Ferns. Largest group of seedless vascular plants (12,000 species), mainly tropical, temperate habitats. *Pteris, Trichomanes, Cyathea* (tree ferns), *Polystichum.*

PHYLUM CYCADOPHYTA. Cycads. Gymnosperm group (vascular, bears "naked" seeds). Tropical, subtropical. Palm-shaped leaves, simple cones on male and female plants. *Zamia.*

PHYLUM GINKGOPHYTA. Ginkgo (maidenhair tree). Type of gymnosperm. Seeds with fleshy outer layer. *Ginkgo.*

PHYLUM GNETOPHYTA. Gnetophytes. Only gymnosperms with vessels in xylem and double fertilization (but endosperm does not form). *Ephedra, Welwitchia.*

PHYLUM CONIFEROPHYTA. Conifers. Most common and familiar gymnosperms. Generally cone-bearing species with needle-like or scale-like leaves.

Family Pinaceae. Pines, firs, spruces, hemlock, larches, Douglas firs, true cedars. *Abies, Cedrus, Pinus, Pseudotsuga.*
Family Cupressaceae. Junipers, cypresses. *Cupressus, Juniperus.*
Family Taxodiaceae. Bald cypress, redwoods, bigtree, dawn redwood. *Metasequoia, Sequoia, Sequoiadendron, Taxodium.*
Family Taxaceae. Yews. *Taxus.*

PHYLUM ANTHOPHYTA. Angiosperms (the flowering plants). Largest, most diverse group of vascular seed-bearing plants. Only organisms that produce flowers, fruits.

Class Dicotyledonae. Dicotyledons (dicots). Some families of some representative orders are listed:

Family Nymphaeaceae. Water lilies.
Family Papaveraceae. Poppies.
Family Brassicaceae. Mustards, cabbages, radishes.
Family Malvaceae. Mallows, cotton, okra, hibiscus.
Family Solanaceae. Potatoes, eggplant, petunias.
Family Salicaceae. Willows, poplars.
Family Rosaceae. Roses, apples, almonds, strawberries.
Family Fabaceae. Peas, beans, lupines, mesquite.
Family Cactaceae. Cacti.
Family Euphorbiaceae. Spurges, poinsettia.
Family Cucurbitaceae. Gourds, melons, cucumbers, squashes.
Family Apiaceae. Parsleys, carrots, poison hemlock.
Family Aceraceae. Maples.
Family Asteraceae. Composites. Chrysanthemums, sunflowers, lettuces, dandelions.

Class Monocotyledonae. Monocotyledons (monocots). Some families of several different orders are listed:

Family Liliaceae. Lilies, hyacinths, tulips, onions, garlic.
Family Iridaceae. Irises, gladioli, crocuses.
Family Orchidaceae. Orchids.
Family Arecaceae. Date palms, coconut palms.
Family Cyperaceae. Sedges.
Family Poaceae. Grasses, bamboos, corn, wheat, sugarcane.
Family Bromeliaceae. Bromeliads, pineapples, Spanish moss.

KINGDOM ANIMALIA. Multicelled eukaryotes, nearly all with tissues, organs, and organ systems and motility during at least part of their life cycle. Heterotrophs, predators (herbivores, carnivores, omnivores), parasites, detritivores. Reproduce sexually and, in many species, asexually. Embryonic development proceeds through a series of continuous stages.

PHYLUM PLACOZOA. Marine. Simplest known animal. Two cell layers, no mouth, no organs. *Trichoplax.*

PHYLUM MESOZOA. Ciliated, wormlike parasites, about the same level of complexity as *Trichoplax.*

PHYLUM PORIFERA. Sponges. No symmetry, tissues. *Euplectella.*

PHYLUM CNIDARIA. Radial symmetry, tissues, nematocysts.

Class Hydrozoa. Hydrozoans. *Hydra, Obelia, Physalia, Prya.*
Class Scyphozoa. Jellyfishes. *Aurelia.*
Class Anthozoa. Sea anemones, corals. *Telesto.*

PHYLUM CTENOPHORA. Comb jellies. Modified radial symmetry.

PHYLUM PLATYHELMINTHES. Flatworms. Bilateral, cephalized; simplest animals with organ systems. Saclike gut.

Class Turbellaria. Triclads (planarians), polyclads. *Dugesia.*
Class Trematoda. Flukes. *Clonorchis, Schistosoma.*
Class Cestoda. Tapeworms. *Diphyllobothrium, Taenia.*

PHYLUM NEMERTEA. Ribbon worms. *Tubulanus.*

PHYLUM NEMATODA. Roundworms. *Ascaris, Caenorhabditis elegans, Necator* (hookworms), *Trichinella.*

PHYLUM ROTIFERA. Rotifers. *Asplancha, Philodina.*

PHYLUM MOLLUSCA. Mollusks.

Class Polyplacophora. Chitons. *Cryptochiton, Tonicella.*

Class Gastropoda. Snails (periwinkles, whelks, limpets, abalones, cowries, conches, nudibranchs, tree snails, garden snails), sea slugs, land slugs. *Aplysia, Ariolimax, Cypraea, Haliotis, Helix, Liguus, Limax, Littorina, Patella.*

Class Bivalvia. Clams, mussels, scallops, cockles, oysters, shipworms. *Ensis, Chlamys, Mytelus, Patinopectin.*

Class Cephalopoda. Squids, octopuses, cuttlefish, nautiluses. *Dosidiscus, Loligo, Nautilus, Octopus, Sepia.*

PHYLUM BRYOZOA. Bryozoans (moss animals).

PHYLUM BRACHIOPODA. Lampshells.

PHYLUM ANNELIDA. Segmented worms.

Class Polychaeta. Mostly marine worms. *Eunice, Neanthes.*

Class Oligochaeta. Mostly freshwater and terrestrial worms, but many marine. *Lumbricus* (earthworms), *Tubifex*.

Class Hirudinea. Leeches. *Hirudo, Placobdella.*

PHYLUM TARDIGRADA. Water bears.

PHYLUM ONYCHOPHORA. Onychophorans. *Peripatus.*

PHYLUM ARTHROPODA.

Subphylum Trilobita. Trilobites; extinct.

Subphylum Chelicerata. Chelicerates. Horseshoe crabs, spiders, scorpions, ticks, mites.

Subphylum Crustacea. Shrimps, crayfishes, lobsters, crabs, barnacles, copepods, isopods (sowbugs).

Subphylum Uniramia.
Superclass Myriapoda. Centipedes, millipedes.
Superclass Insecta.
Order Ephemeroptera. Mayflies.
Order Odonata. Dragonflies, damselflies.
Order Orthoptera. Grasshoppers, crickets, katydids.
Order Dermaptera. Earwigs.
Order Blattodea. Cockroaches.
Order Mantodea. Mantids.
Order Isoptera. Termites.
Order Mallophaga. Biting lice.
Order Anoplura. Sucking lice.
Order Homoptera. Cicadas, aphids, leafhoppers, spittlebugs.
Order Hemiptera. Bugs.
Order Coleoptera. Beetles.
Order Diptera. Flies.
Order Mecoptera. Scorpion flies. *Harpobittacus.*
Order Siphonaptera. Fleas.
Order Lepidoptera. Butterflies, moths.
Order Hymenoptera. Wasps, bees, ants.

PHYLUM ECHINODERMATA. Echinoderms.
Class Asteroidea. Sea stars. *Asterias.*
Class Ophiuroidea. Brittle stars.
Class Echinoidea. Sea urchins, heart urchins, sand dollars.
Class Holothuroidea. Sea cucumbers.
Class Crinoidea. Feather stars, sea lilies.
Class Concentricycloidea. Sea daisies.

PHYLUM HEMICHORDATA. Acorn worms.

PHYLUM CHORDATA. Chordates.

Subphylum Urochordata. Tunicates, related forms.

Subphylum Cephalochordata. Lancelets.

Subphylum Vertebrata. Vertebrates.
Class Agnatha. Jawless vertebrates (lampreys, hagfishes).
Class Placodermi. Jawed, heavily armored fishes; extinct.
Class Chondrichthyes. Cartilaginous fishes (sharks, rays, skates, chimaeras).
Class Osteichthyes. Bony fishes.
Subclass Dipnoi. Lungfishes.
Subclass Crossopterygii. Coelacanths, related forms.
Subclass Actinopterygii. Ray-finned fishes.
Order Acipenseriformes. Sturgeons, paddlefishes.
Order Salmoniformes. Salmon, trout.
Order Atheriniformes. Killifishes, guppies.
Order Gasterosteiformes. Seahorses.
Order Perciformes. Perches, wrasses, barracudas, tunas, freshwater bass, mackerels.
Order Lophiiformes. Angler fishes.
Class Amphibia. Mostly tetrapods; embryo in amnion.
Order Caudata. Salamanders.
Order Anura. Frogs, toads.
Order Apoda. Apodans (caecilians).
Class Reptilia. Skin with scales, embryo enclosed in amnion.
Subclass Anapsida. Turtles, tortoises.
Subclass Lepidosaura. *Sphenodon,* lizards, snakes.
Subclass Archosaura. Dinosaurs (extinct), crocodiles, alligators.
Class Aves. Birds. (In some of the more recent schemes, dinosaurs, crocodilians, and birds are grouped in the same category.)
Order Struthioniformes. Ostriches.
Order Sphenisciformes. Penguins.
Order Procellariiformes. Albatrosses, petrels.
Order Ciconiiformes. Herons, bitterns, storks, flamingoes.
Order Anseriformes. Swans, geese, ducks.
Order Falconiformes. Eagles, hawks, vultures, falcons.
Order Galliformes. Ptarmigan, turkeys, domestic fowl.
Order Columbiformes. Pigeons, doves.
Order Strigiformes. Owls.
Order Apodiformes. Swifts, hummingbirds.
Order Passeriformes. Sparrows, jays, finches, crows, robins, starlings, wrens.
Class Mammalia. Skin with hair; young nourished by milk-secreting glands of adult.
Subclass Prototheria. Egg-laying mammals (duckbilled platypus, spiny anteaters).
Subclass Metatheria. Pouched mammals or marsupials (opossums, kangaroos, wombats, Tasmanian devil).
Subclass Eutheria. Placental mammals.
Order Insectivora. Tree shrews, moles, hedgehogs.
Order Scandentia. Insectivorous tree shrews.
Order Chiroptera. Bats.
Order Primates.
Suborder Strepsirhini (prosimians). Lemurs, lorises.
Suborder Haplorhini (tarsioids and anthropoids).
Infraorder Tarsiiformes. Tarsiers.
Infraorder Platyrrhini (New World monkeys).
Family Cebidae. Spider monkeys, howler monkeys, capuchin.
Infraorder Catarrhini (Old World monkeys and hominoids).
Superfamily Cercopithecoidea. Baboons, macaques, langurs.
Superfamily Hominoidea. Apes and humans.
Family Hylobatidae. Gibbon.
Family Pongidae. Chimpanzees, gorillas, orangutans.
Family Hominidae. Existing and extinct human species (*Homo*) and australopiths.
Order Lagomorpha. Rabbits, hares, pikas.
Order Rodentia. Most gnawing animals (squirrels, rats, mice, guinea pigs, porcupines, beavers, etc.).
Order Cetacea. Whales, porpoises.
Order Carnivora. Carnivores.
Suborder Feloidea. Cats, mongooses, hyenas.
Suborder Canoidea. Dogs, weasels, skunks, otters, raccoons, pandas, bears.
Order Pinnipedia. Seals, walruses, sea lions.
Order Proboscidea. Elephants; mammoths (extinct).
Order Sirenia. Sea cows (manatees, dugongs).
Order Perissodactyla. Odd-toed ungulates (horses, tapirs, rhinos).
Order Artiodactyla. Even-toed ungulates (camels, deer, bison, sheep, goats, antelopes, giraffes, etc.).
Order Edentata. Anteaters, tree sloths, armadillos.
Order Tubulidentata. African aardvarks.

Metric-English Conversions

Length

English		Metric
inch	=	2.54 centimeters
foot	=	0.30 meter
yard	=	0.91 meter
mile (5,280 feet)	=	1.61 kilometer
To convert	*multiply by*	*to obtain*
inches	2.54	centimeters
feet	30.00	centimeters
centimeters	0.39	inches
millimeters	0.039	inches

Weight

English		Metric
grain	=	64.80 milligrams
ounce	=	28.35 grams
pound	=	453.60 grams
ton (short) (2,000 pounds)	=	0.91 metric ton
To convert	*multiply by*	*to obtain*
ounces	28.3	grams
pounds	453.6	grams
pounds	0.45	kilograms
grams	0.035	ounces
kilograms	2.2	pounds

Volume

English		Metric
cubic inch	=	16.39 cubic centimeters
cubic foot	=	0.03 cubic meter
cubic yard	=	0.765 cubic meters
ounce	=	0.03 liter
pint	=	0.47 liter
quart	=	0.95 liter
gallon	=	3.79 liters
To convert	*multiply by*	*to obtain*
fluid ounces	30.00	milliliters
quart	0.95	liters
milliliters	0.03	fluid ounces
liters	1.06	quarts

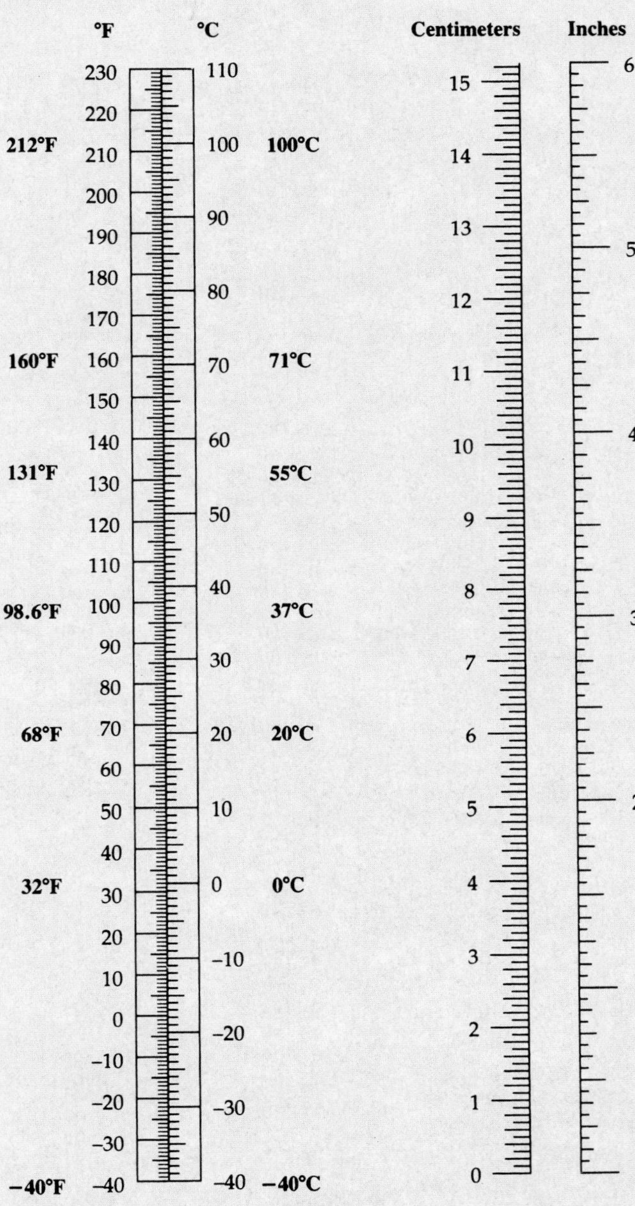
°F
°C
Centimeters
Inches
212°F
100°C
160°F
71°C
131°F
55°C
98.6°F
37°C
68°F
20°C
32°F
0°C
−40°F
−40°C

APPENDIX III. ANSWERS TO SELF-QUIZZES *Italicized numbers refer to relevant text pages*

1
1. **Metabolism** *5*
2. **Homeostasis** *5*
3. **Cell** *6*
4. **Adaptive** *10*
5. **Mutations** *10*
6. d *4*
7. d *4, 10*
8. d *10–11*
9. d *13*
10. c *13, 16*
11. a *15*
12. c *10*
e *11*
d *13*
b *12*
a *12*

2
1. a *24*
2. c *26–27*
3. f *28–29*
4. e *29*
5. f *30–31*
6. **acid, base** *30*
7. c *20*
a *31*
b *24*
d *29*

3
1. d *36*
2. e *37*
3. f *38*
4. b *40*
5. d *42, 47*
6. d *45*
7. d *47–48*
8. c *42–43*
e *47*
b *41*
d *48–49*
a *38*

4
1. c *54–55*
2. d *58*
3. d *58, 72*
4. **False (many cells have a wall)** *58, 72, 73*
5. c *74*
6. e *66*
d *67*
a *54, 58, 64*
b *64*
c *64*

5
1. c *80–81*
2. b *80*
3. a *81*
4. a **(e.g., receptors, adhesion proteins not part of internal cell membranes)** *79, 81*
5. a *89*
6. **centrifuge** *82*
7. d *84*
8. b *87*

6
1. c *97*
2. d *98–99*
3. d *100*
4. d *101*
5. b *103*
6. a *102*
7. e *104, 106*
8. a *104–105*
9. a *107*
10. c *107*
g *103*
a *103*
d *103*
e *103*
b *104*
f *103*

7
1. **Carbon dioxide; sunlight** *112*
2. d *114, 115i*
3. b *114, 115i*
4. e *114, 115i*
5. d *120*
6. c *114, 115i*
7. c *123*
8. c *123*
9. c *123*
d *123*
e *121*
b *120–121*
f *120*
a *118*

8
1. d *132*
2. c *134*
3. b *132*
4. c *132–133*
5. d *140*
6. c *140*
7. b *140–141*
8. d *142–143*
9. b *134*
c *140*
a *136–137*
d *138*

9
1. a *150*
2. b *150*
3. c *150*
4. c *151*
5. a *150, 153*
6. d *150*
7. b *151*
8. d *152–153*
b *152*
c *153*
a *153*

10
1. d *162*
2. d *163, 165*
3. a *162*
4. b *163*
5. d *162*
6. c *164*
7. c *164*
8. d *165*
9. d *162*
a *162*
c *164*
b *163*

11
1. a *177*
2. b *177*
3. a *177*
4. c *178–179*
5. b *177*
6. a *179*
7. d *180*
8. b *180–181*
d *178–179*
a *177*
c *177*

12
1. c *194*
2. c *199*
3. e *206–207*
4. c *208*
5. d *208*
6. d *203*
7. **True** *194*
8. c *199*
e *207*
d *208*
b *206*
a *192*

13
1. c *218*
2. d *218*
3. d *219*
4. c *219*
5. a *220*
6. d *220–221*
7. d *221*
b *220*
c *220*
a *221*

14
1. d *226–227*
2. c *228*
3. b *228*
4. c *228–229*
5. c *230*
6. a *230*
7. a *230–231*
8. e *235*
c *232–233*
a *229*
f *230*
d *230–231*
g *229*
b *230*

15
1. d *242*
2. d *242*
3. d *242*
4. a *242*
5. a *242*
6. d *244*
7. d *244*
8. c *247*
9. d *248*
10. b *251*
11. b *251*
12. d *249*
e *247*
b *250*
a *242*
c *246*

16
1. d *254*
2. c *256*
3. **Plasmids** *256*
4. a *257*
5. b *257*
6. a *258*
7. b *259*
8. d *260*
9. d *259*
c *263*
f *255*
e *261*
b *264*
a *265*

17
1. a *270–271*
2. b *273*
3. d *276–277*
4. c *277*
5. c *274*
d *274*
e *275*
b *275*
a *276–277*

18
1. **Populations** *282*
2. d *282*
3. c *291*
4. c *285*
5. b *286*
6. c *289*
7. c *291*
d *285*
a *283*
b *292*

19
1. d *298–299*
2. d *299*
3. c *300*
4. c *306*
5. b *306*
6. c *306*
e *306*
a *306–307*
d *307*
b *307*

20
1. c *317*
2. a *315*
3. c *321*
4. c *320–321*
5. d *320–321*
6. e *321*
b *312*
f *313*
c *314–315*
a *318–319*
d *315*

21
1. c *340*
2. e *334–349*
3. b *337*
4. d *336*
5. d *341*
6. d *341*
7. b *340–341*
d *341*
e *344–345*
a *346*
c *349*

22
1. c *358–359*
2. c *360*
3. b *362*
4. c *358*
5. d *357*
6. d *362*
7. e *364–365*
8. d *364*
9. d *360*
e *362*
b *364–365*
f *359*
g *358*
c *360*
a *367*

23
1. b *374*
2. d *375*
3. a *375*
4. c *376–377*
5. a *380*
6. a *383, 384*
7. d *383*
8. b *387*

24
1. c *394*
2. c *395*
3. b *CI, 399*
4. a *394*
5. a *394*
6. e *396*
c *397*
d *394*
a *394–395*
b *396*
f *396–397*

25
1. d *404*
2. b *406*
3. c *404*
4. c *405*
5. b *411*
6. c *412*
e *404–405*
g *408*
h *416*
f *404*
a *404*
b *416–417*
d *416–417*

26
1. b *424*
2. c *424*
3. c *425*
4. c *428–429*
5. b *432–435*
6. a *442*
7. a *442*
8. d *426*
f *428*
c *432*
g *434–435*
e *436*
i *437*
j *440*
h *442*
a *450*
b *424* (*Table 26.1*)

27
1. d *456–457*
2. a *458*
3. d *457–458*
4. b *459*
5. b *460*
6. d *462*
7. f *464*
8. e *465*
9. d *465, 469*
10. d *468, 469*
11. e *472*
12. a *478–479*
13. d *459*
e *460*
c *461*
b *462*
g *464*
f *468–469*
a *470*

28
1. b *486*
2. d *487*
3. d *492–493*
4. d *494*

29
1. a *501*
2. d *501*
3. a *502*
4. b *502–503*
5. c *503*
6. d *504*
7. c *507*
8. d *512–513*
9. b *501*
   d *501*
   e *502–503*
   c *501, 512*
   f *509*
   a *512*

30
1. **hydrogen bonds** *522*
2. e *519*
3. b *521*
4. a *520–521*
5. c *522*
6. d *522–523*
7. c *520*
8. d *525*
9. d *525*
10. b *526*
11. c *524–525*
    g *518*
    e *526*
    b *520*
    d *522*
    a *522*
    f *526*

31
1. **pollinators** *530–531*
2. a *532*
3. d *532*
4. b *533*
5. c *536*
6. b *534*
7. b *534*
8. a *536*
9. b *536*
10. d *537*
11. c *540*
12. a *534*
    c *535*
    d *534*
    e *540*
    b *530*

32
1. c *546*
2. c *549*
3. e *544–545*
4. d *548*
5. d *551*
6. a *552*
7. c *552*
8. c *554*
9. a *554*
10. d *555*
    e *554*
    a *546*
    b *551*
    c *549*

33
1. a *562*
2. b *564*
3. b *564*
4. b *565*
5. c *565*
6. d *567*
7. c *566*
8. d *570*
9. c *570*
10. a *570*
11. **receptors, integrator, effectors** *570*
12. b *562*
    a *564*
    c *560, 570*
    f *570*
    g *571*
    e *570*
    d *570*

34
1. d *576, 578*
2. b *576*
3. d *577, 579*
4. a *580*
5. d *578–579*
6. d *582*
7. c *581*
8. b *581*
   d *583*
   a *578*
   c *576, 578*

35
1. **nerve net** *588*
2. a *590*
3. **False** *591*
4. **True** *591*
5. c *592–593*
6. b *592*
7. b *591, 593*
8. d *594*
9. c *594–595*
10. b *594* (*Figure 35.9*)
11. d *597, 599, 602–603*
12. e *593*
    d *594*
    b *595*
    c *597*
    a *596–597*

36
1. **stimulus** *608*
2. **sensation** *608*
3. **Perception** *608*
4. d *608*
5. b *609*
6. b *608*
7. c *608*
8. d *610*
9. **True** *610*
10. d *612*
11. c *613*
12. a *618*
13. b *620*
14. e *610–611*
    f *610*
    b *609*
    d *609*
    a *616*
    c *608*

37
1. f *628*
2. d *628*
3. b *631*
4. b *633*
5. a *633, 637*
6. a *635, 638*
7. b *638*
8. b *636*
9. d *636*
   f *637*
   c *638*
   e *638*
   a *640*
   b *638*

38
1. c *646–647*
2. d *654*
3. d *652*
4. b *658*
5. d *659–661*
6. d *660*
7. h *656*
   f *662*
   g *662*
   e *648*
   a *659*
   c *652*
   b *654–655* (*Table 38.2*)
   i *658*
   d *662*

39
1. c *668*
2. d *670*
3. d *670–671*
4. c *671*
5. c *669*
6. d *669*
7. b *678*
8. d *677, 678*
9. c *684–685*
10. e *678*
    b *679*
    a *680*
    d *680*
    c *680–681*
11. d *668, 680*
    e *684*
    a *676*
    f *668*
    c *668*
    b *668*

40
1. e *690*
2. c *692*
3. f *691*
4. d *694*
5. a *699*
6. d *699*
7. d *694*
8. b *696*
9. a *701*
10. c *690*
    b *695, 699*
    a *692*
    e *701*
    d *702*
    f *702*

41
1. a *711*
2. d *711*
3. d *713*
4. b *714*i, *715*
5. c *714*i, *715*
6. d *717*
7. c *718*
8. a *718*
9. c *718*
10. c *722*
11. d *714*i, *715*
    g *715*
    f *714*
    e *710*
    a *715*
    c *715*
    b *713, 715*

42
1. e *745*
2. d *730*
3. **caloric (or energy); energy** *728*i, *744*
4. a *740*
5. d *741, 742*
6. b *732, 732*i
7. c *734*
8. a *737*
9. b *735*
10. d *735*
    c *734*
    e *739*
    g *734–735*
    f *733*
    b *734*
    a *732*i, *735, 738*

43
1. d *750*
2. f *750*
3. a *752*
4. b *752*
5. c *754*
6. d *753*
7. a *752*i, *753*
8. b *752*i, *753*
9. d *751*
   e *757*
   a *752*i, *753*
   c *754*
   b *755*
10. f *757*
    e *757*
    g *757*
    b *756*
    c *756*
    d *756–757*
    a *756*

44
1. d *764–765*
2. a *766–767*
3. a *768–769*
4. c *766*
5. b *766*
6. d *770*
7. c *766*
8. b *766, 768*
9. d *773, 775, 776*
10. d *777*
11. b *777*
12. d *768–769*
    a *766*
    f *766*
    e *766*
    c *766*
    b *766*

45
1. **hypothalamus** *787, 790*
2. f *787*
3. c *790*
4. c *791*
5. c *794*
6. c *795*
7. c *797*
8. f *796*
9. e *798*
10. e *791, 794*
    c *784*
    d *791*
    f *790*
    b *802*
    a *784*

46
1. **ecology** *813*
2. **population** *814*
3. e *816*
4. a *816–817*
5. a *817*
6. d *819*
7. e *818–819*
8. c *824*
9. c *818*
   d *817*
   a *816*
   e *818*
   b *818*

47
1. a *833*
2. d *834–835*
3. b *841*
4. a *837*
5. d *845*
6. c *835*
   d *837*
   b *834–836*
   a *836*

48
1. e *854*
2. b *854*
3. e *854*
4. d *857*
5. b *858–859*
6. d *858, 862*
7. d *864*
8. e *870–871*
9. c *867*
   d *871*
   a *864*
   e *864*
   b *866*

49
1. a *876*
2. d *877*
3. d *881*
4. d *876*
   a *876*
   c *876*
   b *876*

50
1. d *898–899*
2. c *898*
3. d *898–901*
4. b *901*
5. d *903*
6. d *903*
7. d *906, 910*
8. d *912, 919*
9. b *910*
   d *911*
   c *908*
   a *906*
10. e *914*
    a *919*
    d *913*
    c *918*
    b *914*

51
1. c *924–925*
2. c *926*
3. b *926*
4. d *926*
5. c *927–928*
6. b *934*
7. c *934*
8. d *936*
9. d *936–937*
10. d *933*
    b *930–931*
    c *929*
    a *938*
    e *926* (*Table 51.1*)

APPENDIX IV. ANSWERS TO GENETICS PROBLEMS

CHAPTER 11

1. a. *AB*

   b. *AB*, *aB*

   c. *Ab*, *ab*

   d. *AB*, *Ab*, *aB*, *ab*

2. a. All of the offspring will be *AaBB*.

   b. 1/4 *AABB* (25% each genotype)
   1/4 *AABb*
   1/4 *AaBB*
   1/4 *AaBb*

   c. 1/4 *AaBb* (25% each genotype)
   1/4 *Aabb*
   1/4 *aaBb*
   1/4 *aabb*

   d. 1/16 *AABB* (6.25%)
   1/8 *AaBB* (12.5%)
   1/16 *aaBB* (6.25%)
   1/8 *AABb* (12.5%)
   1/4 *AaBb* (25%)
   1/8 *aaBb* (12.5%)
   1/16 *AAbb* (6.25%)
   1/8 *Aabb* (12.5%)
   1/16 *aabb* (6.25%)

3. Yellow is recessive. Because F_1 plants have a green phenotype and must be heterozygous, green must be dominant over the recessive yellow.

4. a. *ABC*

   b. *ABc*, *aBc*

   c. *ABC*, *aBc*, *ABc*, *aBc*

   d. *ABC*, *aBC*, *AbC*, *abC*,
   ABc, *aBc*, *Abc*, *abc*

5. Because all F_1 plants of this dihybrid cross had to be heterozygous for both genes, then 1/4 (25%) of the F_2 plants will be heterozygous for both genes.

6. a. The mother must be heterozygous for both genes. Both the father and their first child are homozygous recessive for both genes.

   b. The probability that their second child will not be a tongue roller and will have detached earlobes is 1/4 (25%).

7. a. Bill *AaEeSs*, and Marie *AAEESS*

   b. There is a 100% probability that all their children will have Bill's phenotype (flat feet, long eyelashes, and tendency to sneeze).

   c. Zero probability; no child of theirs will have high arches or short eyelashes, and none will be sneezy.

8. a. The mother must be heterozygous I^Ai. The male with type B blood could have fathered the child if he were heterozygous I^Bi .

   b. Genotype alone cannot prove the accused male is the father. Even if he happens to be heterozygous, *any* male who carries the *i* allele could be the father, including those heterozygous for type A blood (I^Ai) or type B blood (I^Bi) and those with type O blood (*ii*).

9. A mating between a mouse from a true-breeding, white-furred strain and a mouse from a true-breeding, brown-furred strain would provide you with the most direct evidence. Because true-breeding strains typically are homozygous for a trait being studied, all F_1 offspring from this mating should be heterozygous. Record the phenotype of each F_1 mouse, then let them mate with one another. Assuming only one gene locus is involved, these are possible outcomes for the F_2 offspring:

   a. All F_1 mice are brown, and their F_2 offspring segregate 3 brown : 1 white. *Conclusion*: Brown is dominant to white.

   b. All F_1 mice are white, and their F_2 offspring segregate 3 white : 1 brown. *Conclusion*: White is dominant to brown.

   c. All F_1 mice are tan, and the F_2 offspring segregate 1 brown : 2 tan : 1 white. *Conclusion*: The alleles at this locus show incomplete dominance.

10. You cannot guarantee that the puppies won't develop the disorder without more information about Dandelion's genotype. You could do so only if she is a heterozygous carrier, if the male is free of the alleles, and if the alleles are recessive.

11. Fred could use a testcross to find out if his pet's genotype is *WW* or *Ww*. He can let his black guinea pig mate with a white guinea pig having the genotype *ww*.
    If any F_1 offspring are white, then his pet's genotype is *Ww*. If the two parents are allowed to mate repeatedly and all the offspring of the matings are black, there is a high probability that his pet is *WW*. (If, say, ten offspring are all black, then the probability that the male is *WW* is about 99.9 percent. The greater the number of offspring, the more confident Fred can be of his conclusion.)

12. a. 1/2 red, 1/2 pink

    b. All pink

    c. 1/4 red, 1/2 pink, 1/4 white

    d. 1/2 pink, 1/2 white

13. 9/16 walnut comb
3/16 rose comb
3/16 pea comb
1/16 single comb

14. Because both parents are heterozygotes (Hb^AHb^S), the following are the probabilities for each child:

a. 1/4 Hb^SHb^S

b. 1/4 Hb^AHb^A

c. 1/2 Hb^AHb^S

15. A mating of two M^LM cats yields:

1/4 homozygous dominant (MM)

1/2 heterozygous (M^LM)

1/4 homozygous recessive (M^LM^L)

Because M^LM^L is lethal, the probability that any one kitten among the survivors will be heterozygous is 2/3.

16. a. Both parents must be heterozygotes (*Aa*). Their children may be albino (*aa*) or unaffected (*AA* or *Aa*).

b. All are *aa*.

c. The albino father must be *aa*. They have an albino child, so the mother must be *Aa*. (If she were *AA*, they could not have an albino child.) The albino child is *aa*. The three unaffected children are *Aa*. There is a 50% chance that any child of theirs will be albino. The observed 3:1 ratio is not surprising, given the small number of offspring.

17. a. All of the offspring will have medium-red color corresponding to the genotype $A^1A^2B^1B^2$.

b. All possible genotypes could appear in the following proportions:

1/16	$A^1A^2B^1B^1$	dark red
1/8	$A^1A^1B^1B^2$	medium-dark red
1/16	$A^1A^1B^2B^2$	medium red
1/8	$A^1A^2B^1B^1$	medium-dark red
1/4	$A^1A^2B^1B^2$	medium red
1/8	$A^1A^2B^2B^2$	light red
1/16	$A^2A^2B^1B^1$	medium red
1/8	$A^2A^2B^1B^2$	light red
1/16	$A^2A^2B^2B^2$	white

CHAPTER 12

1. a. Human males (XY) inherit their X chromosome only from their mother.

b. In males, an X-linked allele will only be found on his one X chromosome. Males can produce two kinds of gametes: one kind with a Y chromosome free of the gene, and the other kind with an X chromosome bearing the X-linked allele.

c. One. Each gamete of a woman who is homozygous for an X-linked allele will have an X chromosome that carries the allele.

d. Two. If a female is heterozygous for an X linked allele, half of the gametes that she produces will contain one of the alleles and the other half will contain the other allele.

2. All of the offspring should be heterozygous for the gene and all should have long wings. However, because some have vestigial wings, the dominant allele might have mutated because of radiation.

3. a. Females normally do not have a Y chromosome, hence they would not have a Y-linked gene.

b. All sons inherit their Y chromosome from their father, but a daughter can only inherit his X chromosome. So the allele for hairy pinnae can be transmitted only to his sons, not to his daughters.

4. Because Marfan syndrome is a case of autosomal dominant inheritance and because one parent bears the allele, the probability of any child inheriting the mutant allele is 50%.

5. Because the phenotype appeared in every generation shown in the diagram, this must be a pattern of autosomal dominant inheritance.

6. A daughter could develop this type of muscular dystrophy only if she were to inherit two X-linked recessive alleles—one from her father and one from her mother. However, if a male bears the allele on his X chromosome, it will be expressed, then he will develop the disorder, and most likely he will not father children because of his early death.

7. If no crossover occurs between the two genes, then half the chromosomes will carry alleles *AB* and half will carry alleles *ab*.

8. If the alleles are close together, it is highly unlikely that a crossover will separate them from each other during meiosis. The greater the distance between the two loci, the greater the probability that a crossover will separate them from each other.

9. a. Nondisjunction could occur in anaphase I or anaphase II of meiosis.

b. As a result of a translocation, chromosome 21 (which is small) may become attached to the end of chromosome 14. Even though the chromosome number of the new individual would be 46, its somatic cells would contain the translocated chromosome 21, in addition to two normal chromosomes 21.

10. In the mother, a crossover between the two genes at meiosis generates an X chromosome that carries neither mutant allele.

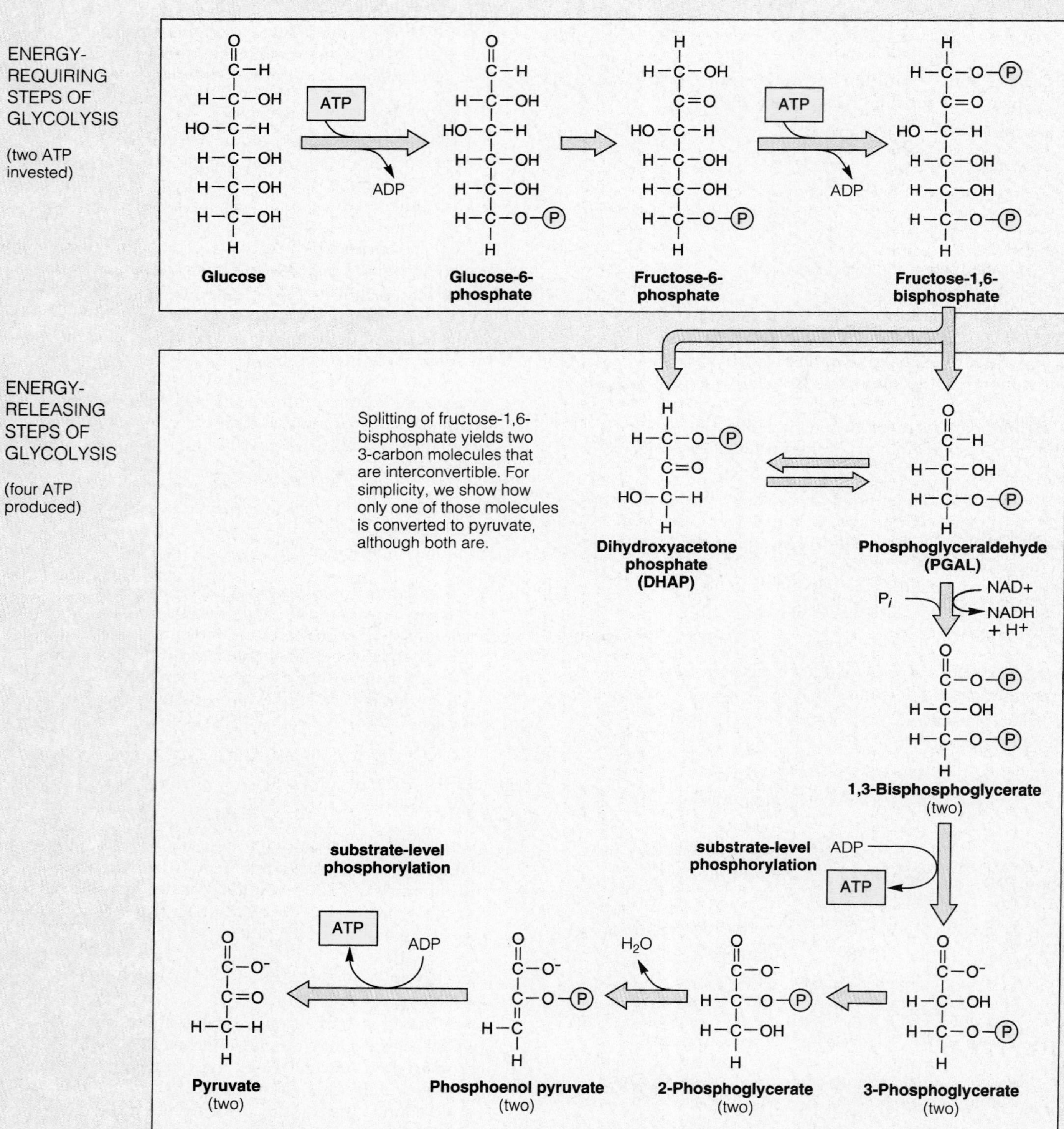
ENERGY-REQUIRING STEPS OF GLYCOLYSIS
(two ATP invested)
ATP
ADP
ATP
ADP
Glucose
Glucose-6-phosphate
Fructose-6-phosphate
Fructose-1,6-bisphosphate
ENERGY-RELEASING STEPS OF GLYCOLYSIS
(four ATP produced)
Splitting of fructose-1,6-bisphosphate yields two 3-carbon molecules that are interconvertible. For simplicity, we show how only one of those molecules is converted to pyruvate, although both are.
Dihydroxyacetone phosphate (DHAP)
Phosphoglyceraldehyde (PGAL)
P_i
NAD+
NADH + H+
1,3-Bisphosphoglycerate (two)
substrate-level phosphorylation
ADP
ATP
3-Phosphoglycerate (two)
H_2O
2-Phosphoglycerate (two)
Phosphoenol pyruvate (two)
substrate-level phosphorylation
ATP
ADP
Pyruvate (two)

Figure A Glycolysis, ending with two 3-carbon pyruvate molecules for each 6-carbon glucose entering the reactions. The *net* energy yield is two ATP molecules (two invested, four produced).

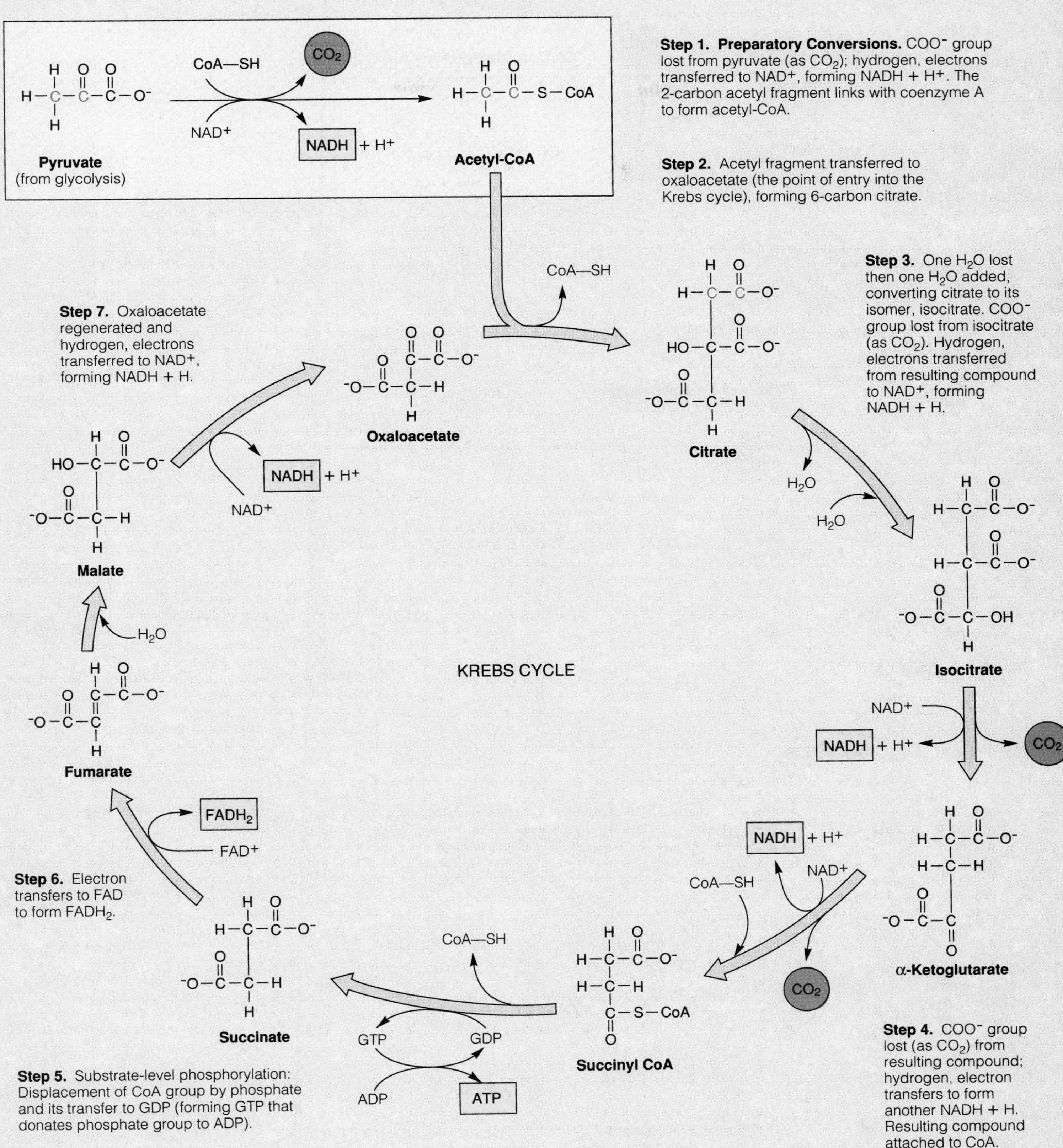

Step 1. Preparatory Conversions. COO⁻ group lost from pyruvate (as CO_2); hydrogen, electrons transferred to NAD^+, forming NADH + H^+. The 2-carbon acetyl fragment links with coenzyme A to form acetyl-CoA.
Step 2. Acetyl fragment transferred to oxaloacetate (the point of entry into the Krebs cycle), forming 6-carbon citrate.
Step 3. One H_2O lost then one H_2O added, converting citrate to its isomer, isocitrate. COO^- group lost from isocitrate (as CO_2). Hydrogen, electrons transferred from resulting compound to NAD^+, forming NADH + H.
Step 4. COO^- group lost (as CO_2) from resulting compound; hydrogen, electron transfers to form another NADH + H. Resulting compound attached to CoA.
Step 5. Substrate-level phosphorylation: Displacement of CoA group by phosphate and its transfer to GDP (forming GTP that donates phosphate group to ADP).
Step 6. Electron transfers to FAD to form $FADH_2$.
Step 7. Oxaloacetate regenerated and hydrogen, electrons transferred to NAD^+, forming NADH + H.
Pyruvate
(from glycolysis)
CoA—SH
CO_2
NAD^+
NADH + H^+
Acetyl-CoA
Citrate
Isocitrate
α-Ketoglutarate
Succinyl CoA
Succinate
Fumarate
Malate
Oxaloacetate
KREBS CYCLE
H_2O
$FADH_2$
FAD^+
GTP
GDP
ADP
ATP

Figure B Krebs cycle (also known as the citric acid cycle). *Red* identifies the carbon atoms entering the cyclic pathway by way of acetyl-CoA.

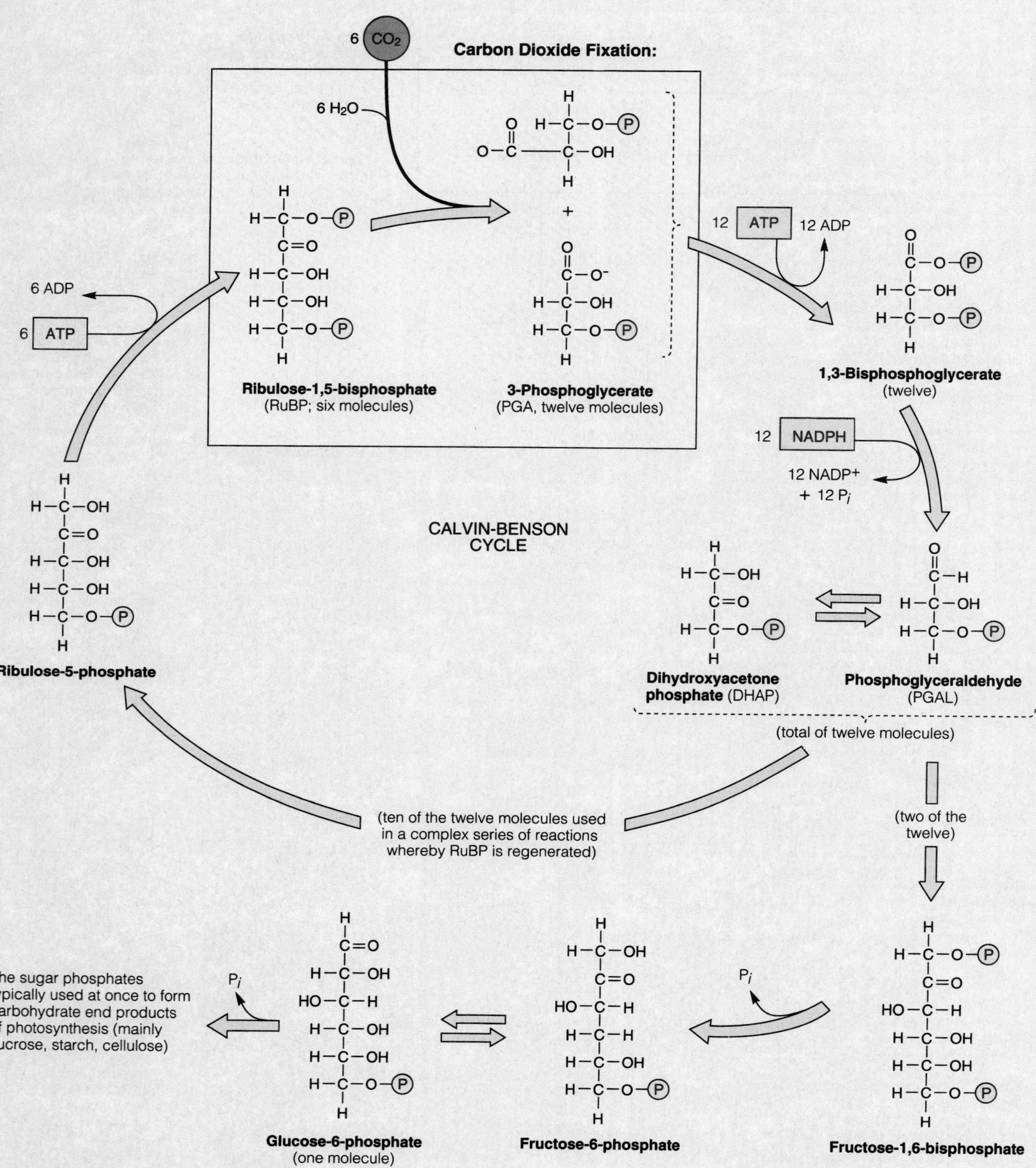

6 CO_2
Carbon Dioxide Fixation:
6 H_2O
Ribulose-1,5-bisphosphate
(RuBP; six molecules)
3-Phosphoglycerate
(PGA, twelve molecules)
12 ATP 12 ADP
1,3-Bisphosphoglycerate
(twelve)
12 NADPH
12 $NADP^+$
+ 12 P_i
6 ADP
6 ATP
CALVIN-BENSON
CYCLE
Ribulose-5-phosphate
Dihydroxyacetone
phosphate (DHAP)
Phosphoglyceraldehyde
(PGAL)
(total of twelve molecules)
(ten of the twelve molecules used
in a complex series of reactions
whereby RuBP is regenerated)
(two of the
twelve)
The sugar phosphates
typically used at once to form
carbohydrate end products
of photosynthesis (mainly
sucrose, starch, cellulose)
P_i
Glucose-6-phosphate
(one molecule)
Fructose-6-phosphate
P_i
Fructose-1,6-bisphosphate

Figure C Calvin–Benson cycle of the light-independent reactions of photosynthesis.

APPENDIX VI. PERIODIC TABLE OF THE ELEMENTS

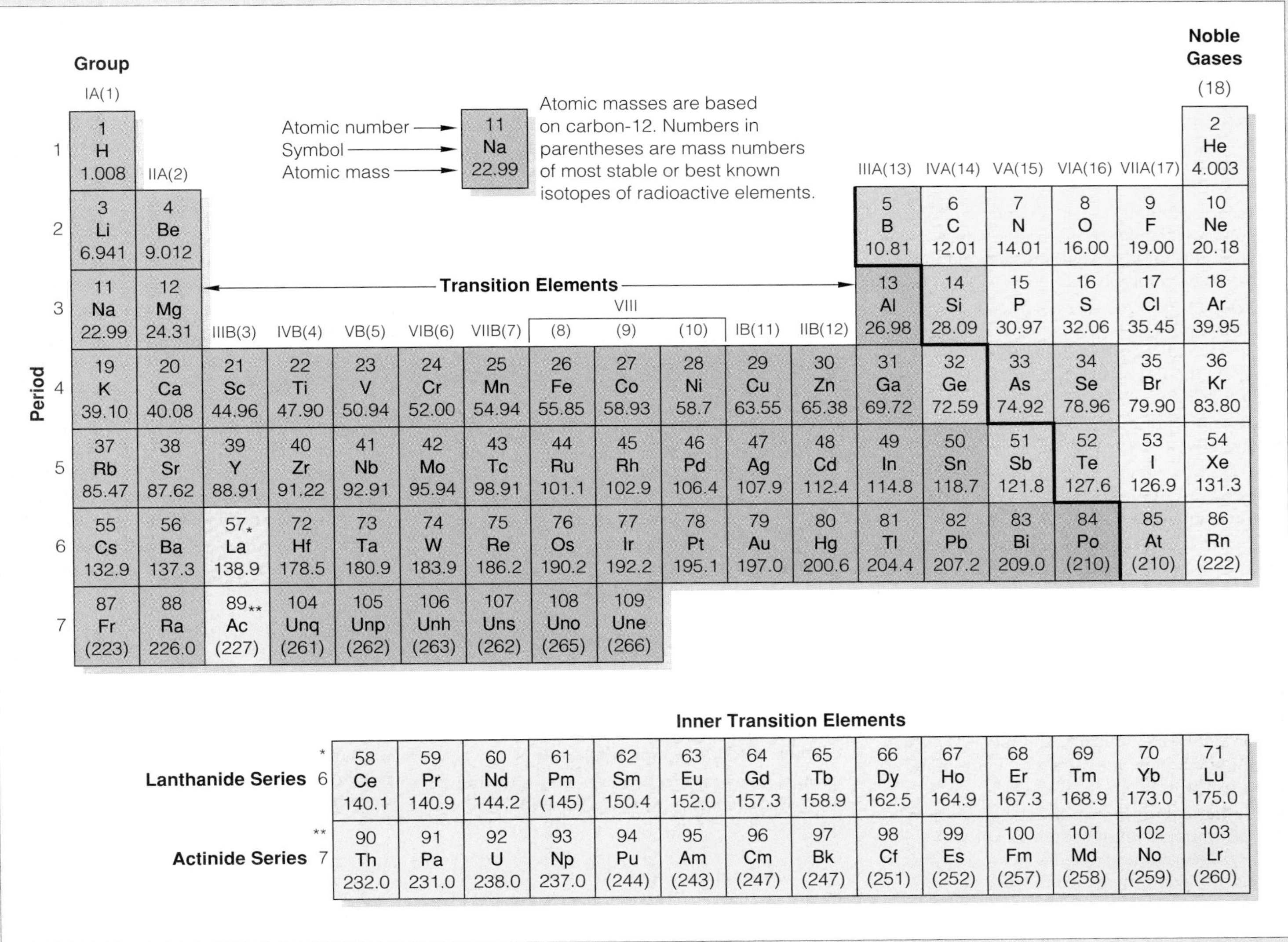
Atomic number → 11
Symbol → Na
Atomic mass → 22.99
Atomic masses are based on carbon-12. Numbers in parentheses are mass numbers of most stable or best known isotopes of radioactive elements.
Transition Elements
Inner Transition Elements

Period	IA(1)	IIA(2)	IIIB(3)	IVB(4)	VB(5)	VIB(6)	VIIB(7)	VIII (8)	VIII (9)	VIII (10)	IB(11)	IIB(12)	IIIA(13)	IVA(14)	VA(15)	VIA(16)	VIIA(17)	Noble Gases (18)
1	1 H 1.008																	2 He 4.003
2	3 Li 6.941	4 Be 9.012											5 B 10.81	6 C 12.01	7 N 14.01	8 O 16.00	9 F 19.00	10 Ne 20.18
3	11 Na 22.99	12 Mg 24.31											13 Al 26.98	14 Si 28.09	15 P 30.97	16 S 32.06	17 Cl 35.45	18 Ar 39.95
4	19 K 39.10	20 Ca 40.08	21 Sc 44.96	22 Ti 47.90	23 V 50.94	24 Cr 52.00	25 Mn 54.94	26 Fe 55.85	27 Co 58.93	28 Ni 58.7	29 Cu 63.55	30 Zn 65.38	31 Ga 69.72	32 Ge 72.59	33 As 74.92	34 Se 78.96	35 Br 79.90	36 Kr 83.80
5	37 Rb 85.47	38 Sr 87.62	39 Y 88.91	40 Zr 91.22	41 Nb 92.91	42 Mo 95.94	43 Tc 98.91	44 Ru 101.1	45 Rh 102.9	46 Pd 106.4	47 Ag 107.9	48 Cd 112.4	49 In 114.8	50 Sn 118.7	51 Sb 121.8	52 Te 127.6	53 I 126.9	54 Xe 131.3
6	55 Cs 132.9	56 Ba 137.3	57* La 138.9	72 Hf 178.5	73 Ta 180.9	74 W 183.9	75 Re 186.2	76 Os 190.2	77 Ir 192.2	78 Pt 195.1	79 Au 197.0	80 Hg 200.6	81 Tl 204.4	82 Pb 207.2	83 Bi 209.0	84 Po (210)	85 At (210)	86 Rn (222)
7	87 Fr (223)	88 Ra 226.0	89** Ac (227)	104 Unq (261)	105 Unp (262)	106 Unh (263)	107 Uns (262)	108 Uno (265)	109 Une (266)									

Inner Transition Elements

	Period														
* Lanthanide Series	6	58 Ce 140.1	59 Pr 140.9	60 Nd 144.2	61 Pm (145)	62 Sm 150.4	63 Eu 152.0	64 Gd 157.3	65 Tb 158.9	66 Dy 162.5	67 Ho 164.9	68 Er 167.3	69 Tm 168.9	70 Yb 173.0	71 Lu 175.0
** Actinide Series	7	90 Th 232.0	91 Pa 231.0	92 U 238.0	93 Np 237.0	94 Pu (244)	95 Am (243)	96 Cm (247)	97 Bk (247)	98 Cf (251)	99 Es (252)	100 Fm (257)	101 Md (258)	102 No (259)	103 Lr (260)

GLOSSARY

ABO blood typing Method of using the presence of proteins A, B, or both at surface of red blood cells to characterize an individual's blood. O signifies absence of both proteins.

abortion Premature, spontaneous or induced expulsion of the embryo or fetus from uterus. Spontaneous abortion also called miscarriage.

abscisic acid (ab-SISS-ik) Plant hormone that promotes stomatal closure, bud dormancy, and seed dormancy.

abscission (ab-SIH-zhun) [L. *abscindere*, to cut off] Hormone-induced dropping of leaves, flowers, fruits, or other parts from a plant.

absorption Uptake of water and solutes from the environment by cell or multicelled organism; e.g., movement of nutrients, fluid, and ions across gut lining and into internal environment.

accessory pigment Any of a variety of light-trapping pigments that extend the range of wavelengths for photosynthesis beyond those absorbed by chlorophylls.

acetyl-CoA (uh-SEED-ul) Coenzyme A with a two-carbon fragment from pyruvate attached. In the second stage of aerobic respiration, it transfers the fragment to oxaloacetate for the Krebs cycle.

acidity Of a solution, an excess of hydrogen ions relative to hydroxyl ions.

acid [L. *acidus*, sour] Any dissolved substance that donates hydrogen ions to other solutes or to water molecules.

acid rain Wet acid deposition; falling of rain (or snow) rich in sulfur and nitrogen oxides.

acoelomate (ay-SEE-luh-mate) Absence of a fluid-filled cavity between gut and body wall.

acoustical signal Sounds used as a form of intraspecific communication.

actin (AK-tin) Cytoskeletal protein; subunit of microfilaments.

action potential Abrupt, brief reversal in the resting membrane potential of a neuron and other excitable cells.

activation energy For each type of reaction, the minimum amount of collision energy that will drive reactant molecules to an activated state, from which the reaction will proceed spontaneously.

active site Crevice in the surface of an enzyme molecule where a specific reaction is catalyzed (made to proceed more rapidly than it would spontaneously).

active transport Pumping of a specific solute across a cell membrane, through the interior of a transport protein, against its concentration gradient. Requires an energy boost, as from ATP.

adaptation [L. *adaptare*, to fit] Of evolution, being adapted (or becoming more adapted) to a set of environmental conditions. Of a sensory neuron, a decrease or cessation in the frequency of action potentials when a stimulus is maintained at constant strength.

adaptive radiation Macroevolutionary pattern; a burst of genetic divergences from a lineage that gives rise to many new species, each adapted to using a novel resource or a new (or recently vacated) habitat.

adaptive trait Any aspect of form, function, or behavior that helps the individual survive and reproduce under prevailing conditions.

adaptive zone A way of life available for organisms that are physically, ecologically, and evolutionarily equipped to live it, such as "catching insects in the air at night."

adenine (AH-de-neen) A purine; a nitrogen-containing base in certain nucleotides.

ADH Antidiuretic hormone. Hypothalamic hormone that induces water conservation; helps control the solute concentrations and volume of extracellular fluid.

adipose tissue A connective tissue having an abundance of fat-storing cells and blood vessels for transporting fats.

ADP Adenosine diphosphate (ah-DEN-uh-seen die-FOSS-fate). Nucleotide coenzyme; typically accepts inorganic (unbound) phosphate or a phosphate group, thus becoming ATP.

aerobic respiration (air-OH-bik) [Gk. *aer*, air, + *bios*, life] Main ATP-forming pathway; proceeds from glycolysis through Krebs cycle and then electron transport phosphorylation. Final electron acceptor is oxygen. Typical net energy yield: 36 ATP per glucose molecule.

age structure Number of individuals in each age category for a population.

agglutination (ah-glue-tin-AY-shun) Forced clumping together of nonself markers when antibodies circulating in blood chemically recognize them. Clumping makes bearers of those markers more easily destroyed by phagocytes. Potential problem in recipients of transfused blood of a different type.

aging Of any multicelled organism showing extensive cell differentiation, a gradual and expected deterioration of the body over time.

AIDS Acquired immunodeficiency syndrome. A set of chronic disorders following infection by the human immunodeficiency virus (HIV), which destroys cells of the immune system.

alcohol Organic compound that includes one or more hydroxyl groups (—OH); it dissolves readily in water. Sugars are examples.

alcoholic fermentation One of the anaerobic ATP-forming pathways. The pyruvate from glycolysis is degraded to acetaldehyde, which accepts electrons from NADH to form ethanol; NAD^+ needed for the reactions is regenerated. Net yield: two ATP.

aldosterone (al-DOSS-tuh-rohn) Hormone of the adrenal cortex; helps control the body's reabsorption of sodium from nephrons.

allantois (ah-LAN-twahz) [Gk. *allas*, sausage] An extraembryonic membrane. Functions in respiration and in storing metabolic wastes of embryos of reptiles, birds, and certain mammals. In humans, the urinary bladder and placental blood vessels form from it.

allele (uh-LEEL) For a given gene locus, one of two or more slightly different molecular forms of a gene that arise through mutation and that code for different versions of the same trait.

allele frequency For a given locus, the relative abundance of each kind of allele among all the individuals of a population.

allergen Any normally harmless substance that provokes inflammation, excessive mucus secretion, and often immune responses.

allergy Hypersensitivity to an allergen.

allopatric speciation [Gk. *allos*, different, + L. *patria*, native land]. Model of what may be the most common speciation route. A physical barrier arises and separates populations or subpopulations of a species, stops gene flow, and favors divergences that end in speciation.

altruism (AL-true-IZ-um) Behavior that helps other members of a species but diminishes an individual's own chance of reproductive success.

alveolus (ahl-VEE-uh-lus), plural **alveoli** [L. *alveus*, small cavity] Cupped, thin-walled outpouching of respiratory bronchiole where oxygen diffuses from lungs into blood, and carbon dioxide diffuses from blood to lungs.

amino acid (uh-MEE-no) An organic molecule with a hydrogen atom, an amino group, an acid group, and an R group, all covalently bonded to a carbon atom. Twenty kinds are the subunits of polypeptide chains.

ammonification (uh-moan-ih-fih-KAY-shun) Process by which some soil bacteria and fungi break down nitrogenous wastes and organic remains; part of nitrogen cycle.

amnion (AM-nee-on) An extraembryonic membrane; the boundary layer of a fluid-filled sac (amniotic cavity) in which the embryos of some vertebrate embryos grow and develop, move freely, and are protected from sudden impacts and temperature shifts.

amniote egg Egg that has extraembryonic membranes and often a shell. Contributed to successful invasion of land by vertebrates.

amphibian Only type of vertebrate making the transition from water to land (evolutionarily and in their embryonic development). Existing gropus are salamanders, frogs, toads, and caecilians.

anaerobic electron transport ATP-forming pathway in which electrons stripped from an organic compound move through transport systems in the bacterial plasma membrane; an inorganic compound in the environment often serves as the final electron acceptor. Variable but always small net energy yield.

anaerobic pathway (an-uh-ROW-bik) [Gk. *an*, without, + *aer*, air] Metabolic pathway in which a substance other than oxygen is final acceptor of electrons stripped from substrates.

analogous structures (ann-AL-uh-gus) [Gk. *analogos*, similar to one another] Body parts that once differed in evolutionarily distant lineages but converged in their structure and function in response to similar environmental pressures.

anaphase (AN-uh-faze) Of mitosis, stage when sister chromatids of each chromosome move apart to opposite spindle poles. In anaphase I of meiosis, each duplicated chromosome and its homologue move to opposite spindle poles. In anaphase II of meiosis, sister chromatids of each chromosome move to opposite poles.

aneuploidy (AN-yoo-ploy-dee) Having one extra or one less chromosome relative to the parental chromosome number.

angiosperm (AN-gee-oh-spurm) [Gk. *angeion*, vessel, and *spermia*, seed] Flowering plant.

animal Multicelled heterotroph that feeds on other organisms, is motile for at least part of life cycle, develops through embryonic stages, has tissues (except for *Trichoplax* and sponges), and most often organs and organ systems.

Animalia Kingdom of animals.

annelid Type of invertebrate; a segmented worm (e.g., oligochaete, leech, or polychaete).

annual Flowering plant that completes its life cycle in one growing season.

anther [Gk. *anthos*, flower] Part of a stamen; pollen forms in it and is dispersed from it.

antibiotic Metabolic product of soil microbes that kills bacterial competitors for nutrients.

antibody [Gk. *anti*, against] Antigen-binding receptor. Only B cells make antibodies, then position them at their surface or secrete them.

anticodon Sequence of three nucleotide bases in a tRNA molecule that can base-pair with a codon in an mRNA molecule.

antigen (AN-tih-jen) [Gk. *anti*, against, + *genos*, race, kind] Any molecular configuration that certain lymphocytes recognize as nonself and that triggers an immune response.

antigen-presenting cell Cell that processes and bears antigen fragments, bound to MHC molecules, at its surface. The antigen–MHC complexes promote immune responses.

aorta (ay-OR-tah) Of vertebrates, main artery of systemic circulation; transports the volume of blood that the heart pumps into it.

apical dominance Growth-inhibiting effect of a terminal bud on growth of lateral buds.

apical meristem (AY-pih-kul MARE-ih-stem) [L. *apex*, top, + Gk. *meristos*, divisible] Mass of self-perpetuating cells underlying primary growth at root tip and shoot tip.

apoptosis (APP-oh-TOE-sis) Programmed cell death. Molecular signals activate weapons of self-destruction in body cells that finished their prescribed functions or became altered, as by infection or cancerous transformation.

appendicular skeleton (ap-en-DIK-yoo-lahr) Bones of limbs, hips, and shoulders.

Archaebacteria Kingdom of prokaryotes more like eukaryotic cells than eubacteria; includes methanogens, halophiles, and thermophiles.

archipelago An island chain some distance away from a continent.

area effect Idea that larger islands support more species than smaller ones at equivalent distances from sources of colonizing species.

arteriole (ar-TEER-ee-ole) Type of blood vessel between arteries and capillaries. Controls over arteriolar dilation and constriction selectively distribute blood volume throughout body.

artery Large-diameter rapid-transport vessel with thick, muscular wall; smooths out blood pressure pulses caused by heart contractions.

arthropod Invertebrate having a hardened exoskeleton, specialized body segments, and jointed appendages. Insects are examples.

artificial selection Selection of traits among individuals of a population in an artificial environment, under contrived conditions.

asexual reproduction Any of a number of modes of reproduction by which offspring arise from a single parent and inherit the genes of that parent only.

atmosphere The volume of gases, airborne particles, and water vapor enveloping Earth.

atmospheric cycle Any biogeochemical cycle in which the element occurs in gaseous phase (e.g., carbon dioxide of the carbon cycle).

atom Smallest particle unique to an element; has one or more positively charged protons, electrons, and (except for hydrogen), neutrons.

atomic number Number of protons in the nucleus of each atom of an element.

ATP Adenosine triphosphate (ah-DEN-uh-seen try-FOSS-fate). Energy-carrying nucleotide with adenine, ribose, and three phosphate groups. Phosphate-group transfers from ATP drive most energy-requiring metabolic reactions.

australopith (OHSS-trah-low-pith) [L. *australis*, southern, + Gk. *pithekos*, ape] One of earliest known hominids; a primate that may be on or near evolutionary road to modern humans.

autoimmune response Misdirected immune response in which lymphocytes mount an attack against normal body cells.

autonomic nervous system (auto-NOM-ik) All nerves from central nervous system to the smooth muscle, cardiac muscle, and glands of viscera (soft internal organs and structures).

autosome Type of chromosome that is the same in males and in females of the species.

autotroph (AH-toe-trofe) [Gk. *autos*, self, + *trophos*, feeder] Organism that makes its own organic compounds using an environmental energy source (e.g., sunlight) along with carbon dioxide as its carbon source.

auxin (AWK-sin) Plant hormone; promotes stem lengthening and responses to gravity and light.

axial skeleton (AX-ee-uhl) Skull, backbone, ribs, and breastbone (sternum).

axon Cylindrical extension of neuron cell body, specialized for the rapid propagation of action potentials.

B cell B lymphocyte; only cell that produces antibodies. Key player in immune responses.

bacterial conjugation The transfer of plasmid DNA from one bacterial cell to another.

bacteriophage (bak-TEER-ee-oh-fahj) [Gk. *baktērion*, small staff, rod, + *phagein*, to eat] Category of viruses that infect bacterial cells.

Barr body In body cells of female mammals, one of either of the two X chromosomes that was condensed to inactivate its genes.

basal body A centriole which, after giving rise to microtubules of a flagellum or cilium, remains attached to its base in the cytoplasm.

base Any substance that accepts hydrogen ions when dissolved in water.

base sequence Particular order in which one nucleotide base follows the next in a strand of DNA or RNA. The order is unique in at least some regions for each species.

basophil Fast-acting white blood cell; its secretions (e.g., histamine) cause vasodilation during an inflammatory response.

behavior Response to external and internal stimuli based on sensory, neural, endocrine, and effector components. Has a genetic basis, can evolve, and can be modified by learning.

benthic province All sediments and rocky formations of the ocean bottom.

biennial (bi-EN-yul) Flowering plant that completes life cycle in two growing seasons.

bilateral symmetry Body plan in which left and right halves generally are mirror images.

binary fission Asexual reproductive mode of protozoans and some other animals. The body divides into two parts of the same or different sizes. *Compare* prokaryotic fission.

binomial system Of taxonomy, assigning a generic and a specific name to each species.

biogeochemical cycle Slow movement of an element from the environment, through food webs, then back to the environment.

biogeographic realm [Gk. *bios*, life, + *geographein*, to describe the Earth's surface] One of six vast land areas, each with distinctive kinds and numbers of plants and animals.

biological clock Internal time-measuring mechanism that helps adjust an organism's daily activities, seasonal activities, or both in response to environmental cues.

biological magnification The increasing concentration of a nondegradable or slowly degradable substance in body tissues as it is passed along food chains.

biological species concept Defines a species as one or more populations of individuals that are interbreeding under natural conditions, that are producing fertile offspring, and that are isolated reproductively from other such populations. Applies to sexually reproducing species only.

bioluminescence Any organism flashing with fluorescent light by way of an ATP-driven reaction involving enzymes (luciferases).

biomass Combined weight of all organisms at a given trophic level in an ecosystem.

biome Large subdivision of a biogeographic realm, distinctive in its habitat conditions, community structure, and endemic species.

biosphere [Gk. *bios*, life, + *sphaira*, globe] All regions of the Earth's waters, crust, and atmosphere in which organisms live.

biosynthetic pathway Metabolic pathway by which organic compounds are synthesized.

biotic potential Of population growth for a given species, the maximum rate of increase per individual under ideal conditions.

bipedalism Habitually walking upright on two feet, as by ostriches and hominids.

bird Only vertebrate that produces feathers; evolutionarily linked with dinosaurs.

blastocyst (BLASS-tuh-sist) [Gk. *blastos*, sprout, + *kystis*, pouch] Type of blastula. Blastomeres form a surface layer, a cavity filled with their own secretions, and an inner cell mass.

blastomere One of the small, nucleated cells that form during cleavage of animal zygote.

blastula (BLASS-chew-lah) An early outcome of cleavage; a number of blastomeres enclosing a fluid-filled cavity.

blood Fluid connective tissue of water, solutes, and formed elements (blood cells and platelets). Blood transports substances to and from cells, and helps maintain internal environment.

blood pressure Fluid pressure, generated by heart contractions, that circulates blood.

blood-brain barrier Mechanism that controls which solutes enter cerebrospinal fluid; helps protect the brain and spinal cord.

bone Vertebrate organ with mineral-hardened connective tissue (bone tissue); helps move body, protects other organs, stores minerals. Some (e.g., breastbone) produce blood cells.

bottleneck Severe reduction in population size brought about by intense selection pressure or by a natural calamity.

Bowman's capsule Cup-shaped portion of a nephron that receives water and solutes being filtered from blood.

brain Of most nervous systems, integrating center that receives and processes sensory input and issues coordinated commands for responses by muscles and glands.

brain stem Of vertebrates, nervous tissue that evolved first and still persists in the hindbrain, midbrain, and forebrain.

bronchiole Finely branched airway; part of the bronchial tree inside the lung.

bronchus, plural **bronchi** (BRONG-cuss, BRONG-kee) [Gk. *bronchos*, windpipe] Tubular airway that branches from trachea and leads into lungs.

brown alga Photoautotrophic protistan with chlorophylls *a*, c_1, and c_2, and carotenoids (e.g., fucoxanthin). Mostly marine. Range in size from microscopic to giant multicelled kelps.

bryophyte Nonvascular land plant requiring free water to complete fertilization. Haploid dominance in life cycle. Cuticle and stomata present in some species. A moss (e.g., peat moss), liverwort, or hornwort.

bud Undeveloped shoot, mainly meristematic tissue. Small, protective scales often cover it.

buffer system A weak acid and the base that forms when it dissolves in water. The two work as a pair to counter slight shifts in pH.

bulk flow In response to a pressure gradient, a movement of more than one kind of molecule in the same direction in the same medium, as in blood, sap, or air.

C3 plant Plant that uses three-carbon PGA as the first intermediate for carbon fixation during the second stage of photosynthesis.

C4 plant Plant that uses oxaloacetate (a four-carbon compound) as the first intermediate for carbon fixation during the second stage of photosynthesis. CO_2 is fixed twice, in two cell types. Carbon dioxide accumulates in leaf and helps counter photorespiration.

Calvin–Benson cycle Cyclic, light-independent reactions; "synthesis" part of photosynthesis. Uses ATP and NADPH from light-dependent reactions. RuBP or some compound to which carbon has been affixed is rearranged and regenerated, and a sugar phosphate forms.

CAM plant Type of plant that conserves water by opening stomata only at night, when it fixes carbon dioxide by means of a C4 pathway.

camouflage Adaptation in coloration, form, patterning, or behavior that helps predators or prey hide in the open (blend with their surroundings) and escape detection.

cancer Malignant tumor; mass of cells that have grossly altered plasma membrane and cytoplasm, grow and divide abnormally, and adhere weakly to home tissue (which leads to metastasis). Lethal unless eradicated.

capillary, blood [L. *capillus*, hair] Blood vessel with thin endothelial wall and small diameter; functions in the exchange of carbon dioxide, oxygen, and other solutes between blood and interstitial fluid.

capillary bed Diffusion zone, consisting of great numbers of capillaries, where blood and interstitial fluid exchange substances.

carbohydrate [L. *carbo*, charcoal, + *hydro*, water] Molecule of carbon, hydrogen, and oxygen mostly in a 1:2:1 ratio. Carbohydrates are structural materials, energy stores, and transportable energy forms. Monosaccharide, oligosaccharide, or polysaccharide.

carbon cycle An atmospheric cycle. Carbon moves from its largest reservoirs (sediments, rocks, and the ocean), through the atmosphere (mostly as CO_2), through food webs, and back to the reservoirs.

carbon dioxide fixation First step of light-independent reactions. Enzyme action affixes carbon (from CO_2) to RuBP or to some other compound for entry into the Calvin–Benson cycle.

carcinogen (kar-SIN-uh-jen) Any substance or agent that can trigger cancer.

cardiac cycle (KAR-dee-ak) [Gk. *kardia*, heart, + *kyklos*, circle] Sequence of muscle contraction and relaxation in one heartbeat.

cardiac pacemaker Sinoatrial (SA) node; basis of normal rate of heartbeat. Self-excitatory cardiac muscle cells spontaneously generate rhythmic waves of excitation over heart.

cardiovascular system Organ system that has blood, one or more hearts, and blood vessels.

carnivore [L. *caro*, *carnis*, flesh, + *vovare*, to devour] Animal that eats other animals.

carotenoid (kare-OTT-en-oyd) An accessory pigment. Different kinds absorb blue-violet and blue-green wavelengths, the energy of which is transferred to chlorophylls. They reflect yellow, orange, and red wavelengths.

carpel (KAR-pul) Female reproductive part of a flower. One or more carpels make up the ovary and a stigma, and often a style.

carrying capacity The maximum number of individuals in a population (or species) that a given environment can sustain indefinitely.

cartilage Connective tissue with solid, pliable intercellular material that resists compression.

Casparian strip Narrow, waxy, impermeable band between walls of abutting cells making up root endodermis and exodermis.

cDNA DNA molecule copied from a mature mRNA transcript by reverse transcription.

cell [L. *cella*, small room] Smallest living unit; organized unit with a capacity to survive and reproduce on its own, given DNA instructions, energy sources, and raw materials.

cell count The number of cells of a given type in one microliter of blood.

cell cycle Events by which a cell increases in mass, roughly doubles its cytoplasmic components, duplicates its DNA, then divides its nucleus and cytoplasm. Extends from the time a cell forms until it completes division.

cell differentiation Key development process. Different cell lineages become specialized in their composition, structure, and function by activating and suppressing some fraction of the genome in different ways.

cell junction Site that joins cells physically, functionally, or both (e.g., tight junction in animals; plasmodesma in plants).

cell plate Disklike structure that forms in a plant cell after nuclear division; becomes a crosswall, with new plasma membrane on both surfaces, that divides the cytoplasm.

cell theory Theory stating that all organisms consist of one or more cells, the cell is the smallest unit with a capacity for independent life, and all cells arise from preexisting cells.

cell wall Of most bacteria, many protistan and fungal cells, and plant cells, the outermost, semirigid, permeable structure that helps the cell retain its shape and resist rupturing when the internal fluid pressure increases.

central nervous system Brain and spinal cord.

central vacuole Large, fluid-filled organelle of living, mature plant cell. Stores amino acids, sugars, ions, and toxic wastes. As it enlarges during growth, it forces the primary cell wall to expand and cell surface area to increase.

centriole (SEN-tree-ohl) Structure that gives rise to microtubules of cilia and flagella.

centromere (SEN-troh-meer) [Gk. *kentron*, center, + *meros*, a part] Constricted portion of each chromosome; location of a kinetochore to which spindle microtubules become attached.

cephalization (sef-ah-lah-ZAY-shun) [Gk. *kephalikos*, head] The concentration of sensory structures and nerve cells in a head; occurred during evolution of bilateral animals.

cerebellum (ser-ah-BELL-um) [L. diminutive of *cerebrum*, brain] Hindbrain region with reflex centers for maintaining posture and smoothing out limb movements.

cerebral cortex Thin surface layer of cerebral hemispheres; receives, integrates, and stores sensory information; coordinates responses.

cerebrospinal fluid Clear extracellular fluid, enclosed in a continuous system of canals and chambers, that bathes and protects the brain and spinal cord.

cerebrum (suh-REE-bruhm) Part of forebrain that integrates olfactory input, selects motor responses. In mammals, it evolved into the most complex integrating center.

channel protein Transport protein that serves as an open or gated channel where ions and other solutes move across a cell membrane.

chemical bond A union between the electron structures of two or more atoms or ions.

chemical energy Potential energy of molecules.

chemical synapse (SIN-aps) [Gk. *synapsis*, union] Narrow cleft between a presynaptic neuron and a postsynaptic cell. Molecules of neurotransmitter diffuse across it.

chemiosmotic theory (KIM-ee-oz-MOT-ik) Idea that an electrochemical gradient drives ATP formation. H^+ accumulates in a membranous compartment, then flows out through proteins (ATP synthases) in response to concentration and electric gradients. The flow drives the attachment of inorganic phosphate to ADP.

chemoautotroph (KEE-moe-AH-toe-trofe) Type of bacterium that can synthesize its own organic compounds using only carbon dioxide as the carbon source and an inorganic substance as the energy source.

chemoreceptor Sensory receptor that detects chemical energy (ions or molecules dissolved in the fluid bathing it).

chlorofluorocarbon (KLORE-oh-FLOOR-oh-car-bun) Compound of chlorine, fluorine, and carbon that has been contributing to ozone thinning.

chlorophyll (KLOR-uh-fill) [Gk. *chloros*, green, + *phyllon*, leaf] Main photosynthetic pigment; absorbs violet-to-blue and red wavelengths but transmits green.

chloroplast (KLOR-uh-plast) The organelle of photosynthesis in plants and many protistans.

chordate Animal with a notochord, dorsal hollow nerve cord, pharynx, and gill slits in pharynx wall during at least part of life cycle.

chorion (CORE-ee-on) Type of extraembryonic membrane that becomes part of placenta. Villi (absorptive structures) form at its surface and facilitate exchanges of substances between the embryo and mother.

chromatid (CROW-mah-tid) Of a duplicated eukaryotic chromosome, one of two DNA molecules (with associated proteins) attached at centromere. Mitosis or meiosis separates them; each becomes a separate chromosome.

chromatin A cell's collection of DNA and all of the proteins associated with it.

chromosome (CROW-moe-some) [Gk. *chroma*, color, + *soma*, body] Of eukaryotic cells, a DNA molecule, duplicated or unduplicated, with a profusion of associated proteins. Of prokaryotic cells (bacteria), a circular DNA molecule with few, if any, proteins attached.

chromosome number All chromosomes in a given type of cell. *See* haploidy; diploidy.

cilium (SILL-ee-um), plural **cilia** Short motile or sensory structure projecting from surface of certain eukaryotic cells; its core is a 9 + 2 array of microtubules.

circadian rhythm (ser-KAYD-ee-un) [L. *circa*, about, + *dies*, day] Cycle of physiological events completed every twenty-four hours or so independently of environmental change.

circulatory system Organ system that moves substances to and from cells, and often helps stabilize body temperature and pH. Typically consists of a heart, blood vessels, and blood.

cladogram [Gk. *clad-*, branch] Evolutionary tree diagram with groups arranged by branch points to show relative relationships. Groups closer together share a more recent common ancestor than those farther apart.

classification scheme A way of organizing and retrieving information about species.

cleavage Early stage of animal development. Mitotic cell divisions divide a fertilized egg into many smaller, nucleated cells; original volume of egg cytoplasm does not increase.

cleavage furrow Ringlike depression defining cleavage plane for dividing animal cells.

cleavage reaction Enzyme action that splits a molecule in two or more parts; hydrolysis is an example.

climate For a specified region, the prevailing weather conditions (e.g., temperature, cloud cover, wind speed, rainfall, and humidity).

climax community Array of species that has stabilized under prevailing habitat conditions.

climax pattern model Idea that physical conditions and other environmental factors often vary in their influence over a large region, so that stable communities other than the climax stage may also persist in that region.

cloaca (kloe-AY-kuh) Chamber or duct in last portion of gut of some animals that also serves in excretion, reproduction, and sometimes respiration.

cloning vector Plasmid that has been modified in the laboratory to accept foreign DNA.

club fungus Fungus with club-shaped cells that produce and bear spores.

cnidarian Radial invertebrate at tissue level of organization; the only nematocyst producer. Medusae and polyps are typical body forms.

coal Nonrenewable energy source that formed over 280 million years ago from submerged, undecayed, and compacted plant remains.

codominance In heterozygotes, simultaneous expression of a pair of nonidentical alleles that specify different phenotypes.

codon mRNA base triplet; its linear sequence corresponds to a linear sequence of amino acids in a polypeptide chain. Of 64 codons, 61 code for amino acids; 3 of these are also START signals for translation, 1 is a STOP signal for translation.

coelom (SEE-lum) [Gk. *koilos*, hollow] Cavity lined with peritoneum between the gut and body wall of most animals.

coenzyme Nucleotide that acts as an enzyme helper; it accepts electrons and hydrogen atoms stripped from substrates at a reaction site and transfers them to another reaction site.

coevolution Joint evolution of two species that interacting closely. When one evolves, the change affects selection pressures that are operating between the two, so that the other also evolves.

cofactor Metal ion or coenzyme; it helps an enzyme catalyze a reaction or it transfers electrons, atoms, or functional groups to a different substrate.

cohesion Capacity to resist rupturing when placed under tension (stretched).

cohesion theory of water transport Theory that the collective cohesive strength of their hydrogen bonds allows water molecules to be pulled up through a plant's xylem in response to transpiration from leaves.

collenchyma (coll-ENG-kih-mah) A simple plant tissue (one cell type only). Gives flexible support for primary growth, as in stems that are lengthening.

colon (CO-lun) Large intestine.

commensalism [L. *com*, together, + *mensa*, table] Ecological interaction between two (or more) species in which one benefits directly and the other is affected little, if at all.

communication display Behavior pattern that is a social signal. Often ritualized (has intended changes in function of common patterns).

communication signal Social cue encoded in stimuli (e.g., specific body coloration, body patterning, odors, sounds, and postures).

community All populations in a habitat. Also, a group of organisms with similar life-styles in a habitat, such as a bird community.

companion cell Living parenchyma cell that is specialized to help load organic compounds into adjacent conducting cells of phloem.

comparative morphology [Gk. *morph*, form] Scientific study of comparable body parts of adults or embryonic stages of major lineages.

competitive exclusion Theory that two or more species that require identical resources cannot coexist indefinitely.

complement system Set of proteins circulating in inactive form in vertebrate blood. Different kinds promote inflammation, induce lysis of pathogens, and stimulate phagocytes during nonspecific defenses and immune responses.

compound Substance consisting of two or more elements in unvarying proportions.

concentration gradient Between two adjoining regions, a difference in the number of molecules (or ions) of a substance. All molecules are in constant motion. As they collide, they careen outward to an adjoining region where they are less concentrated. Barring other forces, all substances diffuse down such a gradient.

condensation reaction Two molecules become covalently bonded into a larger molecule, and water often forms as a by-product.

cone cell In vertebrate eye, a photoreceptor that responds to intense light and contributes to sharp daytime vision and color perception.

conifer Type of gymnosperm; generally, an evergreen woody tree or shrub with needle-like or scale-like leaves.

connective tissue proper Animal tissue with a characteristic proportion of fibroblasts and other cells, fibers (e.g., collagen, elastin), and a ground substance consisting of modified polysaccharides.

conservation biology International efforts to identify all species (particularly in marine and land ecosystems most vulnerable to habitat destruction and high extinction rates) and to design long-term management programs based on ecological and evolutionary principles.

consumer [L. *consumere*, to take completely] A heterotroph that feeds on cells or tissues of other organisms for carbon and energy (e.g., herbivores, carnivores, and parasites).

continuous variation Of a population, a more or less continuous range of small differences in a given trait among its individuals.

contractile vacuole (kun-TRAK-till VAK-you-ohl) [L. *contractus*, to draw together] Organelle that takes up excess water in cell body and contracts to expel water through a pore.

control group A group used as a standard for comparison with an experimental group. Ideally, a control group is identical with the experimental group in all respects except for the one variable being studied.

cork cambium Lateral meristem that replaces epidermis with cork on woody plant parts.

corpus callosum (CORE-pus ka-LOW-sum) A band of numerous axons linking two cerebral hemispheres.

corpus luteum (CORE-pus LOO-tee-um) Of female mammals, a glandular structure that forms from cells of a ruptured ovarian follicle; it secretes progesterone and estrogen.

cortex [L. *cortex*, bark] Generally, a rindlike layer such as kidney cortex or cell cortex. In vascular plants, a ground tissue that supports plant parts and stores food.

cotyledon (KOT-uhl-EE-dun) Seed leaf. One develops in monocot seed and has digestive roles; two develop in dicot seed and store food for germination and early growth.

courtship display Pattern of ritualized social behavior between potential mates. Commonly incorporates frozen postures, exaggerated yet simplified movements, and visual signals.

covalent bond (koe-VAY-lunt) [L. *con*, together, + *valere*, to be strong] Sharing of one or more electrons between atoms or groups of atoms. If electrons shared equally, bond is nonpolar. If shared unequally, it is polar (slightly positive at one end, slightly negative at the other).

cross-bridge formation Reversible, ATP-driven interaction between a sarcomere's actin and myosin filaments; results in a short power stroke that is the basis of muscle contraction.

crossing over At prophase I of meiosis, an interaction in which nonsister chromatids (of a pair of homologous chromosomes) break at corresponding sites and exchange segments; genetic recombination is the result.

culture Sum of behavior patterns of a social group, passed between generations by way of learning and symbolic behavior, especially language.

cuticle (KEW-tih-kull) Body covering. Of plants, a transparent covering of waxes and cutin on outer epidermal cell walls. Of annelids, a thin, flexible coat. Of arthropods, a lightweight exoskeleton hardened with protein and chitin.

cyclic AMP (SIK-lik) Short for cyclic adenosine monophosphate. A nucleotide having roles in intercellular communication.

cyclic pathway of ATP formation Most ancient photosynthetic pathway, at plasma membrane of some bacteria and at chloroplast's thylakoid membrane. Photosystems in the membrane give up electrons to transport systems, which return them to photosystems. Electron flow across the membrane sets up H^+ gradients that drive ATP formation at nearby sites.

cyst Of many microorganisms, a resting stage with thick outer layers that typically forms under adverse conditions. Of skin, abnormal, fluid-filled sac without an external opening.

cytochrome (SIGH-toe-krome) [Gk. *kytos*, hollow vessel, + *chrōma*, color] Iron-containing protein molecule of electron transport systems, as used in photosynthesis, aerobic respiration.

cytokinesis (SIGH-toe-kih-NEE-sis) [Gk. *kinesis*, motion] Cytoplasmic division; splitting of a parent cell into daughter cells.

cytokinin (SIGH-toe-KYE-nin) Type of plant hormone; stimulates cell division, promotes leaf expansion, and retards leaf aging.

cytology Study of cell structure and function.

cytomembrane system System of organelles that modify, package, and distribute newly formed proteins and lipids. Endoplasmic reticulum, Golgi bodies, lysosomes, and various vesicles are its components.

cytoplasm (SIGH-toe-plaz-um) All cell parts, particles, and semifluid substances between the plasma membrane and the nucleus (or nucleoid, in bacteria).

cytoplasmic localization Parceling of a portion of maternal messages in the egg cytoplasm to each blastomere that forms during cleavage.

cytosine (SIGH-toe-seen) Pyrimidine; one of the nitrogen-containing bases in nucleotides.

cytoskeleton Dynamic internal framework of eukaryotic cells. Its microtubules and other components structurally support the cell and organize and move its internal parts. It helps free-living cells move in their environment.

cytotoxic T cell T lymphocyte that touch-kills body cells that are infected or altered, as by cancerous transformation.

decomposer [partly L. *dis-*, to pieces, + *companere*, arrange] Fungal or bacterial heterotroph. Obtains carbon and energy from remains, products, or wastes of organisms. Collectively, decomposers help cycle nutrients to producers in ecosystems.

deductive logic Pattern of thinking by which an individual makes inferences about specific consequences or specific predictions that must follow from a hypothesis.

deforestation Removal of all the trees from a large tract of land (e.g., Amazon River Basin).

degradative pathway A stepwise series of metabolic reactions that break down organic compounds to products of lower energy.

deletion At the cytological level, loss of a segment from a chromosome. At the molecular level, loss of one to several base pairs from a DNA molecule.

denaturation (deh-NAY-chur-AY-shun) Loss of a molecule's three-dimensional shape as weak bonds (e.g., hydrogen bonds) are disrupted.

dendrite (DEN-drite) [Gk. *dendron*, tree] Short, slender extension from cell body of a neuron; commonly a signal input zone.

denitrification (DEE-nite-rih-fih-KAY-shun) Conversion of nitrate or nitrite by certain soil bacteria to gaseous nitrogen (N_2) and a small amount of nitrous oxide (N_2O).

density-dependent control Factor that limits population growth by reducing the birth rate, increasing death and dispersal rates, or all of these. Predation, parasitism, disease, and competition for resources are examples.

density-independent factor Factor that causes a population's death rate to rise independently of population density. Storms and floods are examples.

dentition (den-TIH-shun) Collectively, the type, size, and number of an animal's teeth.

dermal tissue system Tissues that cover and protect all exposed surfaces of a plant.

dermis Skin layer beneath the epidermis that consists primarily of dense connective tissue.

desert Biome that forms where the potential for evaporation greatly exceeds rainfall, and where soil is thin and vegetation sparse.

desertification (dez-urt-ih-fih-KAY-shun) The conversion of grassland or irrigated or rain-fed cropland to a desertlike condition, with a drop in productivity of 10 percent or more.

detrital food web Network of food chains in which energy flows mainly from plants through arrays of detritivores and decomposers.

detritivore (dih-TRY-tih-vore) [L. *detritus*; after *deterere*, to wear down]. Heterotroph that ingests decomposing particles of organic matter (e.g., crab, earthworm, roundworm).

deuterostome (DUE-ter-oh-stome) [Gk. *deuteros*, second, + *stoma*, mouth] Category of bilateral animals in which the first indentation to form in the early embryo becomes an anus (e.g., an echinoderm or chordate).

development Of multicelled species, emergence of specialized, morphologically distinct body parts according to a genetic program.

diaphragm (DIE-uh-fram) [Gk. *diaphragma*, to partition] Muscular partition between thoracic and abdominal cavities with role in breathing. Also a contraceptive device inserted into the vagina to prevent sperm from entering uterus.

dicot (DIE-kot) [Gk. *di*, two, + *kotylēdōn*, cup-shaped vessel] Dicotyledon. Flowering plant generally characterized by embryos with two cotyledons; net-veined leaves; and floral parts arranged in fours, fives, or multiples of these.

diffusion Net movement of like molecules or ions down their concentration gradient. In the absence of other forces, they collide constantly and randomly owing to their inherent energy. It occurs most frequently where they are most crowded, so the net movement is outward from regions of higher to lower concentrations.

digestive system Body sac or tube having one or two openings and often specialized regions for ingesting, digesting, and absorbing food, then eliminating undigested residues.

dihybrid cross Experimental cross between true-breeding parents that differ in two traits. The hybrid F_1 offspring inherit two gene pairs, each with two nonidentical alleles.

diploidy (DIP-loyd-ee) Presence of two of each type of chromosome (pairs of homologous chromosomes) in a cell nucleus at interphase. *Compare* haploidy.

directional selection Mode of natural selection by which allele frequencies underlying a range of phenotypic variation shift in a consistent direction, in response to directional change or to new conditions in the environment.

disaccharide (die-SAK-uh-ride) [Gk. *di*, two, + *sakcharon*, sugar] A common oligosaccharide; two covalently bonded sugar monomers.

disease Outcome of infection when defenses against a pathogen cannot be mobilized fast enough, and the pathogen's activities interfere with normal body functions.

disruptive selection Mode of natural selection by which forms of a trait at both ends of range of variation are favored and the intermediate forms are selected against.

distal tubule Tubular portion of nephron that selectively reabsorbs water and sodium.

distance effect Idea that only species adapted for long-distance dispersal can be potential colonists of islands far from their home range.

diversity of life Sum total of all variations in form, function, and behavior that accumulated in different lineages. Such variations generally are adaptive to prevailing conditions or were adaptive to conditions in the past.

DNA Deoxyribonucleic acid (dee-OX-ee-RYE-bow-new-CLAY-ik). For cells and many viruses, a nucleic acid that is the molecule of inheritance. Hydrogen bonds hold its two helically twisted nucleotide strands together. DNA's nucleotide sequence encodes instructions for synthesizing proteins, hence new individuals of a species.

DNA clone Many identical copies of foreign DNA that was inserted into plasmids and later replicated repeatedly after being taken up by population of host cells (typically, bacteria).

DNA–DNA hybridization *See* nucleic acid hybridization.

DNA fingerprint DNA fragments, inherited in a Mendelian pattern from each parent, that give each individual a unique identity. For humans, the most informative fragments are from tandem repeats (short regions of repeated DNA) that differ greatly among individuals.

DNA ligase (LYE-gaze) Enzyme that seals new base-pairings during DNA replication; also used in recombinant DNA technology.

DNA polymerase (poe-LIM-uh-raze) Enzyme of replication and repair that assembles a new strand of DNA on a parent DNA strand; it uses exposed nucleotide bases as the template.

DNA repair Process that restores the original base sequence when part of a DNA molecule gets altered. DNA polymerases, DNA ligases, and other enzymes execute the repair.

DNA replication Process by which molecules of DNA are duplicated for later distribution to daughter nuclei. Completed before mitosis and meiosis in eukaryotic cells and during prokaryotic fission in bacterial cells.

dominance hierarchy Social organization in which some individuals of the group have adopted a subordinate status to others.

dominant allele Of diploid cells, an allele that masks phenotypic effect of any recessive allele paired with it.

dormancy [L. *dormire*, to sleep] A predictable time of metabolic inactivity for many spores, cysts, seeds, perennials, and some animals.

double fertilization Of flowering plants only, fusion of a sperm nucleus with an egg nucleus (a zygote forms), and fusion of another sperm nucleus with nuclei of endosperm mother cell, which gives rise to a nutritive tissue.

doubling time Time it takes for a population to double in size.

drug addiction Chemical dependence on a drug, which assumes an "essential" biochemical role in the body following habituation and tolerance.

dry shrubland Biome that forms when annual rainfall is less than 25 to 60 centimeters; short, multibranched woody shrubs dominate.

dry woodland Biome that forms when annual rainfall is about 40 to 100 centimeters; it may have tall trees, but no dense canopy.

duplication Gene sequence repeated several to many hundreds or thousands of times. Even normal chromosomes have such sequences.

ecdysone Hormone that has major influence over early development of many insects.

echinoderm Type of invertebrate with calcified spines, needles, or plates on body wall. Radially symmetrical, but with some bilateral features. Sea stars and sea urchins are examples.

ecology [Gk. *oikos*, home, + *logos*, reason] Scientific study of how organisms interact with one another and with their physical and chemical environment.

ecoregion Large area that is representative of a globally important biome or water province.

ecosystem Array of organisms and their physical environment, all interacting by a flow of energy and a cycling of materials.

ecosystem modeling An analytical method, based on computer programs and models, of predicting unforeseen effects of specific disturbances to an ecosystem.

ectoderm [Gk. *ecto*, outside, + *derma*, skin] The first-formed, outermost primary tissue layer of animal embryos; gives rise to nervous system tissues and integument's outer layer.

effector Muscle (or gland) that responds to signals from an integrator (e.g., a brain) by producing movement (or chemical change) to adjust the body to changing conditions.

effector cell Differentiated cell of a lymphocyte subpopulation that immediately engages and destroys an antigen-bearing agent during an immune response.

egg Mature female gamete; an ovum.

El Niño Massive eastward movement of warm surface waters of the western equatorial Pacific that displaces cooler water off coast of South America. A recurring event that disrupts the climate and ecosystems throughout the world.

electromagnetic spectrum All wavelengths from radiant energy less than 10^{-5} nm long to radio waves more than 10 km long.

electron Negatively charged unit of matter, with particulate and wavelike properties, that occupies one of the orbitals around the atomic nucleus. Atoms gain, lose, or share electrons.

electron transfer A molecule donates one or more electrons to another molecule.

electron transport phosphorylation (FOSS-for-ih-LAY-shun) Last stage of aerobic respiration, when electrons from reaction intermediates flow through a membrane transport system that gives them up to oxygen. The flow sets up an electrochemical gradient that drives ATP formation at other sites in the membrane.

electron transport system Organized array of membrane-bound enzymes and cofactors that accept and donate electrons in series. It sets up an electrochemical gradient that makes H^+ flow across the membrane. The flow energy drives ATP formation at other reaction sites.

element Substance that cannot be degraded by ordinary means into a substance having different properties.

embryo (EM-bree-oh) [Gk. *en*, in, + *bryein*, to swell] Of animals, a multicelled body formed by way of cleavage, gastrulation, and other early developmental events. Of plants, a young sporophyte, from the time of the earliest cell divisions after fertilization until germination.

embryo sac Female gametophyte of flowering plants.

embryonic induction In a growing embryo, release of a gene product from one tissue that affects the developmental fate of an adjacent tissue.

emerging pathogen Deadly pathogen, either a newly mutated strain of an existing type or one that evolved long ago and is now exploiting an increased presence of human hosts.

emulsification In chyme, a suspension of fat droplets coated with bile salts.

end product Substance present at the end of a metabolic pathway.

endangered species A species that is highly vulnerable to extinction by natural events (e.g., genetic drift after a severe bottleneck) or by human activities (e.g., accidental introduction of competitive, exotic species).

endergonic reaction (en-dur-GONE-ik) Chemical reaction having a net gain in energy.

endocrine gland Ductless gland that secretes hormones, which later enter the bloodstream.

endocrine system Integrative system of cells, tissues, and organs, functionally linked to the nervous system, that exerts control by way of its hormones and other chemical secretions.

endocytosis (EN-doe-sigh-TOE-sis) Cell uptake of substances when part of plasma membrane forms a vesicle around them. Three routes are receptor-mediated endocytosis, bulk transport of extracellular fluid, and phagocytosis.

endoderm [Gk. *endon*, within, + *derma*, skin] Inner primary tissue layer of animal embryos; source of inner gut lining and derived organs.

endodermis Sheetlike wrapping of single cells around root vascular cylinder; helps control uptake of water and dissolved nutrients.

endometrium (EN-doe-MEET-ree-um) [Gk. *metrios*, of the womb] Inner lining of uterus.

endoplasmic reticulum or **ER** (EN-doe-PLAZ-mik reh-TIK-yoo-lum) Organelle that starts at nucleus and curves through cytoplasm. New polypeptide chains acquire side chains inside rough ER (with ribosomes on its cytoplasmic side); smooth ER (with no ribosomes) is a site of lipid synthesis.

endoskeleton [Gk. *endon*, within, + *sklēros*, hard] Of chordates, an internal framework of cartilage, bone, or both that works with skeletal muscle to support and move body, and to maintain posture.

endosperm (EN-doe-sperm) Nutritive tissue that surrounds an embryo sporophyte inside the seed of a flowering plant.

endospore Resting structure formed by some bacteria; encloses a duplicate of the bacterial chromosome and a portion of cytoplasm.

endosymbiosis Continuing physical contact between one species and another species that lives in its body.

energy Capacity to do work.

energy carrier Molecule that delivers energy from one metabolic reaction site to another. ATP is the most common energy carrier.

energy flow pyramid Pyramidal diagram of an ecosystem's trophic structure. It depicts the energy losses at each transfer to another trophic level.

enhancer A short DNA base sequence that is a binding site for an activator protein.

entropy (EN-trow-pee) Measure of the degree of disorder in a system (how much energy has become so disorganized, usually as dissipating heat, that it is no longer available to do work). A system must receive and use energy to stay organized; it tends toward entropy when its energy outputs exceed energy inputs.

enzyme (EN-zime) A protein or one of a few RNAs that greatly speed (catalyze) reactions between substances, most often at functional groups.

eosinophil Fast-acting white blood cell; its enzyme secretions digest holes in parasitic worms during an inflammatory response.

epidermis Outermost tissue layer of plants and all animals above sponge level of organization.

epiglottis Flaplike structure between pharynx and larynx; its controlled positional changes direct air into trachea or food into esophagus.

epinephrine (ep-ih-NEF-rin) Adrenal hormone that raises blood levels of sugar and fatty acids; increases heart's rate and force of contraction.

epistasis (eh-PISS-tah-sis) Interaction among the products of two or more gene pairs.

epithelium (EP-ih-THEE-lee-um) Tissue that covers the animal body's external surfaces and lines its internal cavities and tubes. It has one free surface and one resting on a basement membrane that is next to a connective tissue.

equilibrium, dynamic [Gk. *aequus*, equal, + *libra*, balance] Point at which a reaction runs forward as fast as in reverse, so no net change in reactant or product concentrations.

erosion Movement of land under the force of wind, running water, and ice.

erythrocyte (eh-RITH-row-site) [Gk. *erythros*, red, + *kytos*, vessel] Red blood cell.

esophagus (ee-SOF-uh-gus) A muscular tube just after the pharynx in vertebrate and many invertebrate digestive tracts.

essential amino acid Any amino acid that an organism cannot synthesize for itself and must obtain from a food source.

essential fatty acid Any fatty acid that an organism cannot synthesize for itself and must obtain from a food source.

estrogen (ESS-trow-jen) Female sex hormone that helps oocytes mature, induces changes in the uterine lining during menstrual cycle and pregnancy, helps maintain secondary sexual traits, and affects growth and development.

estrus (ESS-truss) [Gk. *oistrus*, frenzy] Of most mammals, a cyclic period during which the female is sexually receptive to the male.

estuary (EST-you-ehr-ee) A partially enclosed coastal region where seawater mixes with fresh water and runoff from the land, as by streams and rivers.

ethylene (ETH-il-een) Plant hormone that stimulates fruit ripening and abscission.

Eubacteria Kingdom of all prokaryotic cells except archaebacteria.

eukaryotic cell (yoo-CARE-EE-oh-tic) [Gk. *eu*, good, + *karyon*, kernel] Cell having a nucleus and other membrane-bound organelles.

eutrophication The enrichment of a body of water with nutrients; typically leads to reduced transparency and a phytoplankton-dominated community.

evaporation [L. *e-*, out, + *vapor*, steam] The conversion of a substance from liquid state to gaseous state under input of heat energy.

evolution, biological [L. *evolutio*, unrolling] Genetic change in a line of descent over time, brought about by microevolutionary events (gene mutation, natural selection, genetic drift, and gene flow).

evolutionary tree Treelike diagram in which each branch signifies a separate line of descent from a common ancestor and each branch point signifies a time of divergence.

excitatory postsynaptic potential (EPSP) A graded potential that arises at an input zone of an excitable cell and that drives the cell membrane closer to threshold.

excretion Removal of excess water, solutes, and wastes, and some harmful substances from body by way of a urinary system or glands.

exergonic reaction (EX-ur-GONE-ik) Chemical reaction that shows a net loss in energy.

exocrine gland (EK-suh-krin) [Gk. *es*, out of, + *krinein*, to separate] Glandular structure that secretes products, usually through ducts or tubes, to a free epithelial surface.

exocytosis (EK-so-sigh-TOE-sis) Release of the contents of a vesicle at the cell surface, where the vesicle's membrane fuses with and becomes part of the plasma membrane.

exodermis Cylindrical sheet of cells just inside the root epidermis of most flowering plants; helps control uptake of water and solutes.

exon One of the base sequences of an mRNA transcript that will become translated.

exoskeleton [Gk. *sklēros*, hard, stiff] External skeleton, as in arthropods.

exotic species Species that left its established home range, deliberately or accidentally, and successfully took up residence elsewhere.

experiment Test that simplifies observation in nature or in the laboratory by manipulating and controlling conditions under which the observations are made.

exponential growth (EX-po-NEN-shul) Pattern of population growth in which population size expands by ever increasing increments during successive intervals, because its reproductive base becomes ever larger. Increases in size against time plot out as a J-shaped curve.

extinction Irrevocable loss of a species.

extracellular fluid Of most animals, all fluid not inside cells; plasma (the liquid portion of blood) plus interstitial fluid (the liquid that occupies spaces between cells and tissues).

extracellular matrix The ground substance, fibrous proteins, and other materials between cells of animal tissues (e.g., cartilage).

FAD Flavin adenine dinucleotide; a type of nucleotide coenzyme that transfers electrons and unbound protons (H^+) from one reaction site to another. At such times it is abbreviated $FADH_2$.

fall overturn Vertical mixing of a body of water in autumn as its upper layer cools. Cool water is more dense, and sinks. The oxygen dissolved in surface water moves down and nutrients from bottom sediments move up.

family pedigree Chart of genetic relationship of family individuals through the generations.

fat Lipid with a glycerol head and one, two, or three fatty acid tails. Tryglycerides have three. The carbon backbone of unsaturated tails has single covalent bonds; that of saturated tails has one or more double bonds.

fate map Surface diagram of certain early embryos (e.g., of *Drosophila*) showing where the differentiated cells of the adult originate.

fatty acid Molecule with a backbone of as many as thirty-six carbon atoms, a carboxyl group ($-COO^-$) at one end, and hydrogen atoms at most or all of the other bonding sites.

feedback inhibition Mechanism by which a cellular change resulting from some activity shuts down the activity that brought it about.

fermentation [L. *fermentum*, yeast] Anaerobic pathway of ATP formation that starts with glycolysis and ends with transfer of electrons to a breakdown product or an intermediate. NAD^+ is regenerated. Net energy yield: two ATP per glucose molecule.

fertilization [L. *fertilis*, to carry, to bear] The Fusion of a sperm nucleus with the nucleus of an egg, which thus becomes a zygote.

fever Any core temperature higher than the set point in the hypothalamic region that functions as the body's thermostat.

fibrous root system All lateral branchings of adventitious roots that arose from a young stem.

filter feeder Animal that filters food from a current of water directed through a body part (e.g, through a sea squirt's pharynx).

filtration First step in urine formation; the pressure of heart contractions filters blood by forcing water and all solutes except proteins from glomerular capillaries into Bowman's capsule of nephrons.

fin Of fishes generally, appendage that helps stabilize, propel, and guide body in water.

first law of thermodynamics [Gk. *therme*, heat, + *dynamikos*, powerful] Law of nature that the total amount of energy in the universe remains constant. Energy cannot be created from nothing and existing energy cannot be destroyed.

fish Aquatic animal of the most ancient, diverse vertebrate lineage; a jawless, cartilaginous, or bony fish.

fitness Increase in adaptation to environment brought about by genetic change.

fixation Loss of all but one kind of allele at a gene locus; all individuals in a population are homozygous for it.

fixed action pattern Program of coordinated, stereotyped muscle activity that is completed independently of feedback from environment.

flagellum (fluh-JELL-um), plural **flagella** Motile structure of many free-living eukaryotic cells. Its core has a 9 + 2 array of microtubules.

flower Of angiosperms only, a reproductive structure with nonfertile parts (sepals, petals) and fertile parts (stamens, carpels) attached to a receptacle (modified base of a floral shoot).

fluid mosaic model Idea that cell membranes consist of a lipid bilayer and proteins. The lipids impart basic structure, impermeability to water-soluble molecules, and (by packing variations and movements) fluidity. Diverse proteins span the bilayer or associate with one of its surfaces and perform most membrane functions (e.g., transport, signal reception).

follicle (FOLL-ih-kul) Small sac, pit, or cavity, as around a hair; also a mammalian oocyte with its surrounding layer of cells.

food chain Straight-line sequence of who eats whom in an ecosystem.

food web Cross-connecting, interlinked food chains consisting of producers, consumers, and decomposers, detritivores, or both.

forebrain Most complex portion of vertebrate brain; includes cerebrum (and cerebral cortex), olfactory lobes, and hypothalamus.

forest Biome where tall trees grow close enough to form a fairly continuous canopy.

fossil Recognizable, physical evidence of an organism that lived in the distant past.

fossil fuel Coal, petroleum, or natural gas; a nonrenewable energy source that formed long ago from remains of swamp forests.

fossilization How fossils form. An organism or traces of it become buried in sediments or volcanic ash. Water and dissolved inorganic compounds infiltrate remains. Accumulating sediments exert pressure above the burial site. Over time, the pressure and chemical changes transform the remains to stony hardness.

founder effect A form of bottlenecking. By chance, allele frequencies of a few individuals that establish a new population are not the same as those of original population. With no further gene flow, natural selection affects allele frequencies in drastically different ways owing to its interactions with genetic drift.

free radical Any highly reactive molecular fragment having an unpaired electron.

fruit [L. after *frui*, to enjoy] Flowering plant's mature ovary, often with accessory structures.

FSH Follicle-stimulating hormone produced and secreted by the anterior lobe of pituitary gland; has reproductive roles in both sexes.

functional group An atom or a group of atoms that is covalently bonded to the carbon backbone of an organic compound and that influences its chemical behavior.

functional-group transfer Enzyme-mediated event in which a molecule donates one or more functional groups to another molecule.

Fungi Kingdom of fungi which, as a group, are major decomposers. Also includes diverse pathogens and parasites.

fungus Eukaryotic heterotroph that secretes digestive enzymes to break down food outside the body into molecules that its cells absorb (e.g., extracellular digestion and absorption). Saprobes feed on nonliving organic matter, and parasites feed on cells or tissues of living organisms.

gall bladder Organ that stores bile secreted from liver; its duct connects to small intestine.

gamete (GAM-eet) [Gk. *gametēs*, husband, and *gametē*, wife] Haploid cell, formed by meiotic cell division of a germ cell; required for sexual reproduction. Eggs and sperm are examples.

gamete formation Formation of sex cells (e.g., sperm and eggs); occurs in reproductive tissues or organs in most eukaryotic species.

gametophyte (gam-EET-oh-fite) [Gk. *phyton*, plant] Haploid gamete-producing body that forms during plant life cycles.

ganglion (GANG-lee-on), plural **ganglia** Distinct cluster of cell bodies of neurons.

gastrulation (gas-tru-LAY-shun) Stage of animal development; major reorganization of new cells into two or three primary tissue layers.

gel electrophoresis Laboratory technique used to distinguish different molecules in a given sample. Application of an electric field forces molecules to migrate through a viscous gel and distance themselves from one another on the basis of length, size, or electric charge.

gene [short for German *pangan*, after Gk. *pan*, all, + *genes*, to be born] Unit of information about a heritable trait, passed from parents to offspring. Each gene has a specific location on a chromosome (e.g., its locus).

gene flow Microevolutionary process; alleles enter and leave a population as an outcome of immigration and emigration, respectively.

gene frequency Abundance of a given allele relative to other alleles at same locus in a population.

gene library Mixed collection of bacteria that house many different cloned DNA fragments.

gene locus A gene's chromosomal location.

gene pair Two alleles at the same gene locus on a pair of homologous chromosomes.

gene pool All genotypes in a population.

gene therapy Generally, the transfer of one or more normal genes into an organism to correct or lessen adverse effects of a genetic disorder.

genetic code [After L. *genesis*, to be born] The correspondence between nucleotide triplets in DNA (then mRNA) and specific sequences of amino acids in resulting polypeptide chains; the basic language of protein synthesis in cells.

genetic disease Illness in which expression of one or more genes increases susceptibility to an infection or weakens immune response to it.

genetic disorder Inherited condition that causes mild to severe medical problems.

genetic divergence Gradual accumulation of differences in gene pools of populations or subpopulations of a species after a geographic barrier arises and separates them; mutation, natural selection, and genetic drift thereafter are operating independently in each one.

genetic drift Change in allele frequencies over the generations, as brought about by chance alone. Population size influences its effect on genetic and phenotypic diversity, because small populations are more vulnerable to losing alleles entirely.

genetic engineering Altering the information content of DNa molecules with recombinant DNA technology.

genetic equilibrium In theory, a state in which a population is not evolving. These conditions must be met: no mutation, the population very large in size and isolated from others of same species, and no natural selection (all members reproduce equally by random mating).

genetic recombination Result of any process that can incorporate new genetic information into a chromosome or DNA fragment. As one example, allelic combinations in chromosomes emerging from meiosis usually differ from the parental combinations (say, *Ab* and *aB* compared to parental types *AB* and *ab*). Also, nonreciprocal gene transfers in nature or the laboratory result in genetic recombination.

genome All of the DNA in a haploid number of chromosomes for a given species.

genotype (JEEN-oh-type) Genetic constitution of an individual; a single gene pair or the sum total of an individual's genes.

genus, plural **genera** (JEEN-US, JEN-er-ah) [L. *genus*, race or origin] A grouping of all species perceived to be more closely related to one another in their morphology, ecology, and evolutionary history than to other species at the same taxonomic level.

geographic dispersal Directional movement in which residents of an established community leave their home range and successfully settle in a new location, where they are exotic species.

geologic time scale Time scale for the Earth's history. Its major subdivisions correspond to mass extinctions. Its absolute dates have been refined by radiometric dating.

germ cell Animal cell of a lineage set aside for sexual reproduction; gives rise to gametes.

germination (jur-min-AY-shun) Of seeds and spores, resumption of growth after dispersal, dormancy, or both.

gibberellin (JIB-er-ELL-un) Type of plant hormone that promotes elongation of stems, helps seeds and buds break dormancy, and contributes to flowering.

gill Organ of respiration. Most have a thin, moist, vascularized layer for gas exchange.

gland Secretory cell or structure derived from epithelium and often connected to it.

glomerular capillary One of a set of blood capillaries in Bowman's capsule of a nephron.

glomerulus (glow-MARE-you-luss) [L. *glomus*, ball] First portion of the nephron, where water and solutes are filtered from blood.

glucagon (GLUE-kuh-gone) Hormone secreted by pancreas; stimulates cells to convert their stores of glycogen and amino acids to glucose.

glyceride (GLISS-er-eyed) Molecule with one, two, or three fatty acid tails attached to a glycerol backbone; one of the fats or oils.

glycerol (GLISS-er-ohl) [Gk. *glykys*, sweet, + L. *oleum*, oil] Three-carbon compound with three hydroxyl groups; component of fats and oils.

glycocalyx Sticky mesh of polysaccharides, polypeptides, or both around the cell wall of many bacteria.

glycogen (GLY-kuh-jen) A highly branched polysaccharide made of glucose monomers; the main storage carbohydrate in animals.

glycolysis (gly-CALL-ih-sis) [Gk. *glykys*, sweet, + *lysis*, breaking apart] Breakdown of glucose or another organic compound to two pyruvate molecules. First stage of aerobic respiration, fermentation, and anaerobic electron transport. Oxygen has no role in glycolysis, which occurs in the cytoplasm of all cells. Two NADH form. Net yield: two ATP per glucose molecule.

glycoprotein Protein with linear or branched oligosaccharides covalently bonded to it. Nearly all surface proteins of animal cells and many proteins circulating in blood are examples.

gnetophyte Woody gymnosperm, unusual in having vessels in xylem; vine or shrub.

Golgi body (GOHL-gee) Organelle of lipid assembly, polypeptide chain modification, and packaging of both in vesicles for export or for transport to locations in cytoplasm.

gonad (GO-nad) Primary reproductive organ in which animal gametes are produced.

graded potential Of a neuron and other excitable cells, a local signal that can slightly alter the resting membrane potential and can vary in magnitude, depending on the stimulus. With prolonged or intense stimulation, such signals may spread to a trigger zone of the membrane and initiate an action potential.

granum, plural **grana** In many chloroplasts, a stack of flattened, membranous compartments that have chlorophyll and other light-trapping pigments and reaction sites for ATP formation.

grassland Biome that has flat or rolling land, 25–100 centimeters of annual rainfall, warm summers, distinct array of grazing animals, and recurring fires that regenerate dominant plant species.

gravitropism (GRAV-ih-TROPE-izm) [L. *gravis*, heavy, + Gk. *trepein*, to turn] Tendency of a plant to grow directionally in response to the Earth's gravitational force.

gray matter Unmyelinated axons, dendrites, and cell bodies of neurons, plus neuroglial cells, in the brain and spinal cord.

grazing food web Network of food chains in which energy flows from plants to an array of herbivores, then to carnivores.

green alga Type of protistan evolutionarily, structurally, and biochemically most similar to plants (e.g., nearly all are photoautotrophs with starch grains and chlorophylls *a* and *b* in chloroplasts; and some have cell walls of cellulose, pectin, and other polysaccharides typical of plants).

green revolution In developing countries, the use of improved crop varieties, modern agricultural equipment and practices (e.g., heavy fertilizer and pesticide application) to increase crop yields.

greenhouse effect Overall warming of lower atmosphere. Gaseous molecules (e.g., carbon dioxide and methane) impede the escape of infrared wavelengths (heat) from the Earth's sunlight-warmed surface. The gases continually absorb those wavelengths and radiate much of their energy back toward Earth.

ground meristem (MARE-ih-stem) [Gk. *meristos*, divisible] Primary meristem that gives rise to the ground tissue system).

ground substance Intercellular material in some animal tissues; made of cell secretions and other noncellular components.

ground tissue system Tissues (parenchyma, especially) making up most of the plant body.

growth Of multicelled species, increases in the number, size, and volume of cells. Of bacteria, increase in the number of cells in a population.

guanine Nitrogen-containing base in one of the four nucleotide monomers in DNA or RNA.

guard cell Either of two adjoining cells that influence movement of carbon dioxide, oxygen, and water vapor across leaf or stem epidermis. When both guard cells swell with water and move apart, an opening (stoma) forms; when they lose water and collapse into each another, the stoma closes.

gut Generally, a sac or tube from which food is absorbed into internal environment. Also, a gastrointestinal tract (from stomach onward).

gymnosperm (JIM-noe-sperm) [Gk. *gymnos*, naked, + *sperma*, seed] Vascular plant that bears seeds at exposed surfaces of its reproductive structures (e.g., on cone scales).

habitat [L. *habitare*, to live in] Type of place where an organism or species normally lives; characterized by physical and chemical features and by its array of species.

hair cell Mechanoreceptor that may give rise to action potentials when bent or tilted.

half-life The time it takes for half of a given quantity of any radioisotope to decay into a different, and less unstable, daughter isotope.

halophile Archaebacterium of saline habitats.

haploidy (HAP-loyd-ee) Presence of only half of the parental number of chromosomes in a spore or gamete, as brought about by meiosis.

Hardy–Weinberg rule Statement that allele frequencies stay the same over the generations when there is no mutation, the population is infinitely large and is isolated from other populations of the same species, mating is random, and all individuals are reproducing equally and randomly (no natural selection).

HCG Short for human chorionic gonadotropin. Hormone that helps maintains endometrium during the menstrual cycle and first trimester of pregnancy.

heart Muscular pump; its contractions keep blood circulating through the animal body.

heat Thermal energy; a form of kinetic energy.

helper T cell T lymphocyte with central roles in both antibody-mediated and cell-mediated immune responses. When activated, it makes and secretes chemicals that induce responsive T and B cells to undergo rapid divisions into populations of effector and memory cells.

hemoglobin (HEEM-oh-glow-bin) [Gk. *haima*, blood, + L. *globus*, ball] Iron-containing, oxygen-transporting protein of red blood cells.

hemostasis (HEE-mow-STAY-sis) [Gk. *haima*, blood, + *stasis*, standing] Process that stops blood loss from damaged blood vessel; involves coagulation, blood vessel spasm, platelet plug formation, and other mechanisms.

herbivore [L. *herba*, grass, + *vovare*, to devour] Plant-eating animal (e.g., snail, deer, manatee).

hermaphrodite Individual having both male and female gonads.

heterocyst (HET-er-oh-sist) Cyanobacterial cell that modifies itself and makes a nitrogen-fixing enzyme when nitrogen supplies dwindle.

heterotroph (HET-er-oh-trofe) [Gk. *heteros*, other, + *trophos*, feeder] Organism unable to make its own organic compounds; feeds on autotrophs, other heterotrophs, organic wastes.

heterozygous condition (HET-er-oh-ZYE-guss) [Gk. *zygoun*, join together] For a given trait, having a pair of nonidentical alleles at a gene locus (that is, on a pair of homologous chromosomes).

higher taxon (plural, **taxa**) One of ever more inclusive groupings that reflect relationships among species. Family, order, class, phylum, and kingdom are examples.

hindbrain Medulla oblongata, cerebellum, and pons of vertebrate brain. Includes reflex centers for respiration, blood circulation, and other basic functions; also helps coordinate motor responses and many complex reflexes.

histone Type of protein intimately associated with eukaryotic DNA and largely responsible for organization of eukaryotic chromosomes.

homeostasis (HOE-me-oh-STAY-sis) [Gk. *homo*, same, + *stasis*, standing] State in which physical and chemical aspects of internal environment (blood, interstitial fluid) are being maintained within ranges suitable for cell activities.

homeotic gene A master gene governing the development of specific body parts.

hominid [L. *homo*, man] All species on or near evolutionary road leading to modern humans.

hominoid Apes, humans, and recent ancestors.

homologous chromosome (huh-MOLL-uh-gus) [Gk. *homologia*, correspondence] Of cells with a diploid chromosome number, one of a pair of chromosomes that are identical in size, shape, and gene sequence, and that interact at meiosis. Nonidentical sex chromosomes (e.g., X and Y) also interact as homologues during meiosis.

homologous structures The same body parts that became modified differently, in different lines of descent from a common ancestor.

homology Similarity in one or more body parts in different species; attributable to descent from a common ancestor.

homozygous condition (HOE-moe-ZYE-guss) For a specified trait, having a pair of identical alleles at a gene locus (on a pair of homologous chromosomes).

homozygous dominant condition Having a pair of dominant alleles at a gene locus (on a pair of homologous chromosomes).

homozygous recessive condition Having a pair of recessive alleles at a gene locus (on a pair of homologous chromosomes).

hormone [Gk. *hormon*, stir up, set in motion] Signaling molecule secreted by one cell that stimulates or inhibits activities of any other cell having receptors for it. Animal hormones are picked up by bloodstream, which delivers them to cells some distance away. In plants, hormones do not travel far from source cells.

horsetail Seedless vascular plant with ancient, tree-size ancestors. The sporophytes of the one existing genus have rhizomes, scale-shaped leaves, and hollow photosynthetic stems with silica-reinforced ribs.

human genome project Worldwide research project to sequence all 3.2 billion nucleotides in the DNA of human chromosomes. Now scheduled for completion within a few years.

humus Decomposing organic matter in soil. Amount varies in soils of different types.

hybrid offspring Of a genetic cross, offspring with a pair of nonidentical alleles for a trait.

hydrogen bond Weak interaction between a small, highly electronegative atom of a molecule and a neighboring hydrogen atom already taking part in a polar covalent bond.

hydrogen ion Free (unbound) proton; that is, a hydrogen atom that lost its electron and now bears a positive charge (H^+).

hydrologic cycle A biogeochemical cycle, driven by solar energy, in which water moves through the atmosphere, on or through land, to the ocean, and back to the atmosphere.

hydrolysis (high-DRAWL-ih-sis) [L. *hydro*, water, + Gk. *lysis*, loosening] Cleavage reaction that breaks covalent bonds and splits a molecule into two or more parts. Commonly, H^+ and OH^- (derived from a water molecule) become attached to the exposed bonding sites.

hydrophilic substance [Gk. *philos*, loving] A polar substance that dissolves easily in water. Sugars are examples.

hydrophobic substance [Gk. *phobos*, dreading] A nonpolar substance; it strongly resists being dissolved in water. Oil is an example.

hydrosphere Collectively, all of the Earth's liquid or frozen water.

hydrostatic pressure Pressure exerted by a volume of fluid against a wall, membrane, or some other structure that encloses the fluid.

hydrothermal vent ecosystem Ecosystem near a steaming fissure in the ocean floor. Chemoautotrophic bacteria are the basis of its food webs.

hypertonic solution A fluid having a greater concentration of solutes relative to another fluid.

hypha (HIGH-fuh), plural **hyphae** [Gk. *hyphe*, web] Fungal filament with chitin-reinforced walls; component of a mycelium.

hypodermis Subcutaneous layer with stored fat that helps insulate the body; anchors skin but still allows it to move somewhat.

hypothalamus [Gk. *hypo*, under, + *thalamos*, inner chamber, or *tholos*, rotunda] Forebrain region that controls visceral activities (e.g., salt–water balance, core temperature, and reproduction); influences related behaviors (e.g., hunger, thirst, and sex) and emotional states (e.g., sweating with fear).

hypothesis In science, a possible explanation of a phenomenon, one that has the potential to be proved false by experimental tests.

hypotonic solution A fluid that has a lower concentration of solutes relative to another fluid.

immune response Events by which B cells and T cells recognize antigen andgive rise to antigen-sensitized populations of effector cells and memory cells.

immunoglobulin (Ig) One of five classes of antibodies, each with antigen-binding sites as well as other sites with specialized functions.

implantation Event in pregnancy. A blastocyst burrows into endometrium and establishes connections by which a mother will exchange substances with the embryo (and fetus) that develops from the blastocyst's inner cell mass.

imprinting Time-dependent form of learning, usually during a sensitive period for a young animal, triggered by exposure to sign stimuli.

in vitro fertilization Conception outside the body ("in glass" petri dishes or test tubes).

inbreeding Nonrandom mating among close relatives that share many identical alleles; a form of genetic drift in a small group of relatives that are preferentially interbreeding.

incomplete dominance Condition in which one allele of a pair is not fully dominant over the other; a heterozygous phenotype in between both homozygous phenotypes emerges.

independent assortment theory Mendelian theory that by the end of meiosis, each pair of homologous chromosomes (and linked genes on each one) are sorted out for shipment to gametes independently of how all the other pairs were sorted. Later modified to account for the disruptive effect of crossing over on linkages.

indirect selection theory Idea that altruistic individuals can pass on their genes indirectly by helping relatives survive and reproduce.

induced-fit model Idea that a substrate alters the shape of an enzyme's active site when bound to it, causing a more precise molecular fit between the two that promotes reactivity.

inductive logic Pattern of thinking by which an individual derives a general statement from specific observations.

infection Invasion and multiplication of a pathogen in a host. Disease follows if defenses are not mobilized fast enough; the pathogen's activities interfere with normal body functions.

inflammation, acute Rapid response to tissue injury by phagocytes and diverse proteins (e.g., histamine, complement, clotting factors). Signs include localized redness, heat, swelling, pain.

inheritance The transmission, from parents to offspring, of genes that specify structures and functions characteristic of the species.

inhibiting hormone Hypothalmic signaling molecule that suppresses a particular secretion by the anterior lobe of the pituitary gland.

inhibitor Substance able to bind with a specific molecule and interfere with its functioning.

inhibitory postsynaptic potential (IPSP) Graded potential at an input zone of an excitable cell that drives its membrane away from threshold.

instinctive behavior A behavior performed without having been learned by experience.

insulin Pancreatic hormone that lowers level of glucose in blood by causing cells to take up glucose; also promotes protein and fat synthesis and inhibits protein conversion to glucose.

integration, neural [L. *integrare*, coordinate] Moment-by-moment summation of excitatory and inhibitory synapses acting on a neuron.

integrator A control center (e.g., brain) that receives, processes, and stores sensory input, then puts together and issues commands for coordinated responses to it.

integument Of animals, protective body cover (e.g., skin). Of seed-bearing plants, one or more layers around an ovule; becomes a seed coat.

integumentary exchange (in-teg-you-MEN-tuh-ree) Respiration across a thin, moist, and often vascularized surface layer of animal body.

interleukin Type of protein that mediates the signaling between a variety of cell types (e.g., cells of immune system and of hypothalamus).

intermediate Substance that forms between the start and end of a metabolic pathway.

intermediate filament One of the ropelike cytoskeletal elements that impart mechanical strength to animal cells and tissues.

interneuron Neuron of brain or spinal cord.

internode In vascular plants, the stem region between two successive nodes.

interphase Of a cell cycle, interval between nuclear divisions when a cell increases in mass and roughly doubles the number of its cytoplasmic components. It also duplicates its chromosomes (replicates its DNA) during interphase, but *not* between meiosis I and II.

interspecific competition Ecological interaction in which individuals that belong to different species compete for a share of resources in the same habitat.

interstitial fluid (IN-ter-STISH-ul) [L. *interstitus*, to stand in the middle of something] The portion of extracellular fluid that occupies the spaces between animal cells and tissues.

intertidal zone The region above the low water mark and below the high water mark of a rocky or sandy shore; tides alternately submerge and expose its inhabitants.

intervertebral disk Disk-shaped, cartilaginous structure that is a shock absorber and flex point between a backbone's bony segments.

intraspecific competition Ecological interaction in which individuals that belong to the same population compete for a share of resources in their habitat.

intron One of the noncoding portions of a pre-mRNA transcript. All introns are excised before translation.

inversion A linear stretch of DNA within a chromosome that has become oriented in the reverse direction, with no molecular loss.

invertebrate Any animal without a backbone. Of the 2 million named species in the animal kingdom, all but 50,000 are invertebrates.

ion, negatively charged (EYE-on) An atom or a molecule that acquired an overall negative charge by gaining one or more electrons.

ion, positively charged An atom or a molecule that acquired an overall positive charge by losing one or more electrons.

ionic bond Two ions being held together by the attraction of their opposite charge.

isotonic solution A fluid having the same solute concentration as a fluid against which it is being compared.

isotope (EYE-so-tope) Of an element, an atom with more or fewer neutrons than the atoms having the most common number.

joint Area of contact or near-contact between bones.

juvenile Of some animals, a post-embryonic stage that changes only in size and proportion to become the adult (no metamorphosis).

J-shaped curve Type of diagrammatic curve that emerges when unrestricted exponential growth of a population is plotted against time.

karyotype (CARE-ee-oh-type) For an individual or a species, a preparation of metaphase chromosomes sorted by length, centromere location, and other defining features.

keratin Tough, water-insoluble protein made by vertebrate epidermal cells, which die and accumulate as keratinized bags at the surface of skin to form a barrier against dehydration, bacteria, and many toxins.

key innovation A structural or functional modification to the body that, by chance, gives a lineage the opportunity to exploit the environment in more efficient or novel ways.

keystone species A species that dominates a community and dictates its structure.

kidney One of a pair of vertebrate organs that filter ions and other substances from blood; it controls the amounts returned and so helps maintain the solute concentrations and volume of extracellular fluid.

kilocalorie 1,000 calories of heat energy (the amount required to raise the temperature of 1 kilogram of water by 1°C). Used as the unit of measure for the caloric content of foods.

kinase Enzyme that catalyzes phosphate-group transfers (e.g., a protein kinase).

kinetic energy Energy of motion.

kinetochore Cluster of proteins and DNA at the centromere region of a chromosome; spindle microtubules become attached to it at mitosis or meiosis. One is present on each chromatid of a duplicated chromosome.

Krebs cycle Cyclic pathway in mitochondria only; together with a few preparatory steps, the stage of aerobic respiration in which pyruvate is broken down to carbon dioxide and water. Coenzymes accept electrons and unbound protons (H^+) from intermediates and deliver them to next stage; two ATP form.

lactate fermentation An anaerobic pathway of ATP formation. Pyruvate from glycolysis is converted to three-carbon lactate, and NAD^+ is regenerated. Net energy yield: two ATP.

lactation Production and secr etion of milk by hormone-primed mammary glands.

large intestine Colon; gut region that receives unabsorbed residues from small intestine, and concentrates and stores feces until expulsion.

larva, plural **larvae** An immature stage that develops between the embryo and adult in many animal life cycles.

larynx (LARE-inks) Tubular airway leading to lungs. Contains vocal cords in some animals.

lateral meristem Meristem in plants that show secondary growth; vascular or cork cambium.

lateral root Outward branching from the first (primary) root of a taproot system.

leaching Removal of some nutrients from soil as water percolates through it.

leaf Chlorophyll-rich plant part adapted for sunlight interception and photosynthesis.

learned behavior Lasting modification in behavior as a result of experience or practice.

lek Communal display ground for courtship behavior by some animals, including birds.

lethal mutation Mutation with drastic effects on phenotype; usually causes death.

LH Luteinizing hormone. Anterior pituitary hormone that has reproductive roles in both males and females.

lichen (LY-kun) Symbiotic interaction between a fungus and a photoautotroph.

life cycle Recurring pattern of genetically programmed events from the time individuals are produced until they themselves reproduce.

life table Tabulation of age-specific patterns of birth and death for a population.

ligament A strap of dense connective tissue that bridges a joint.

light-dependent reactions The first stage of photosynthesis. Sunlight energy is trapped and converted to chemical energy of ATP, NADPH, or both, depending on the pathway.

light-independent reactions Second stage of photosynthesis. ATP makes phosphate-group transfers required to build sugar phosphates. NADPH delivers electrons and hydrogen atoms for the synthesis reactions, which also require carbon from carbon dioxide. Sugar phosphates enter other reactions by which starch, cellulose, and other end products are assembled.

lignification Deposition of lignin in secondary wall of many plant cells. Lignin imparts strength and rigidity, stabilizes and protects other wall components, and forms a waterproof barrier. Important in vascular plant evolution.

limbic system A system of centers in cerebral hemisphere that governs emotions and affects memory. Distantly related to olfactory lobes; it still deals with the sense of smell.

limiting factor Any essential resource that, in short supply, limits population growth.

lineage (LIN-ee-age) Line of descent.

linkage group All the genes on a chromosome.

lipid Mainly a greasy or oily hydrocarbon. Lipid molecules strongly resist dissolving in water but quickly dissolve in nonpolar substances. Some types serve as the main reservoirs of stored energy in all cells; others are structural materials (as in membranes) and cell products (e.g., surface coatings).

lipid bilayer Structural basis of cell membranes. Two layers of mostly phospholipid molecules. Hydrophobic tails of the lipids are sandwiched between the hydrophilic heads, and heads are dissolved in intracellular or extracellular fluid.

lipoprotein Molecule that forms when proteins circulating in blood combine with cholesterol, triglycerides, and phospholipids absorbed from the small intestine.

liver In vertebrates and many invertebrates, a large gland that stores, converts, and helps maintain blood levels of organic compounds; inactivates most hormone molecules that have completed their tasks; inactivates compounds that can be toxic at high concentrations.

local signaling molecule One of the secretions by many cell types that alter chemical conditions only in localized tissue regions. Prostaglandin is an example.

logic Thought patterns by which an individual draws a conclusion that does not contradict evidence used to support that conclusion.

logistic population growth (low-JISS-tik) Pattern of population growth. A low-density population slowly increases in size, enters a phase of rapid growth, then levels off in size once the carrying capacity has been reached.

loop of Henle Hairpin-shaped, tubular part of a nephron that reabsorbs water and solutes.

lung Internal sac-shaped respiratory surface that evolved in oxygen-poor habitats as an adaptation that increases the surface area for gas exchange. A pair occur in a few fishes and in amphibians, birds, reptiles, and mammals.

lycophyte Seedless vascular plant with tree-size ancestors of ancient swamp forests. They require free water to complete life cycle. Most have leaves, branching rhizome, vascularized roots and stems (e.g., club mosses).

lymph (LIMF) [L. *lympha*, water] Tissue fluid that drained into vessels of lymphatic system.

lymph capillary Small-diameter vessel where tissue fluid moves into lymph vascular system.

lymph node Lymphoid organ that serves as a battleground for immune responses. Organized arrays of lymphocytes packed inside it cleanse lymph before it can reach the bloodstream.

lymph vascular system [L. *lympha*, water, + *vasculum*, a small vessel] Parts of lymphatic system that take up excess tissue fluid, absorbed fats, and reclaimable solutes for delivery to the bloodstream.

lymphatic system Supplement to vertebrate circulatory system. Its vessels deliver fluid and solutes from interstitial fluid to blood; and its lymphoid organs have roles in body defenses.

lymphocyte A T cell or B cell.

lysis [Gk. *lysis*, a loosening] Gross damage to a plasma membrane, cell wall, or both that lets the cytoplasm to leak out; causes cell death.

lysogenic pathway Latent period that extends many viral replication cycles. Viral genes get integrated into host chromosome and may stay inactivated through many host cell divisions but eventually are replicated in host progeny.

lysosome (LYE-so-sohm) Important organelle of intracellular digestion.

lysozyme Infection-fighting enzyme present in mucous membranes (e.g., of mouth, vagina).

lytic pathway Of viruses, a rapid replication pathway that ends with lysis of host cell.

macroevolution Large-scale patterns, trends, and rates of change among higher taxa.

macrophage Phagocytic white blood cell; roles in nonspecific defenses and immune responses. One of the key antigen-presenting cells.

mammal Only vertebrate whose females nourish offspring with milk from mammary glands.

mass extinction Catastrophic event or phase in geologic time when entire families or other major groups are irrevocably lost.

mass number Sum of all protons and neutrons in an atom's nucleus.

mast cell A basophil-like cell that releases histamine during tissue inflammation.

mechanoreceptor Sensory cell or nearby cell that detects mechanical energy (changes in pressure, position, or acceleration).

medulla oblongata Hindbrain region with reflex centers for basic tasks (e.g., respiration); coordinates motor responses with complex reflexes (e.g., coughing); also influences brain centers concerned with sleep and arousal.

medusa (meh-DOO-sah) [Gk. *Medousa*, one of three sisters in Greek mythology having snake-entwined hair] Of cnidarian life cycles, a free-swimming, bell-shaped stage, often with oral lobes and tentacles extending below the bell.

megaspore Haploid spore that forms by way of meiosis in the ovary of seed-bearing plants; one of its cellular descendants develops into an egg.

meiosis (my-OH-sis) [Gk. *meioun*, to diminish] Two-stage nuclear division process that halves the chromosome number of a parental germ cell (to haploid number). Each daughter nucleus receives one of each type of chromosome. Basis of gamete formation. Also, basis of formation of spores that give rise to gamete-producing bodies (gametophytes).

memory The capacity to store and retrieve information about past sensory experience.

memory cell B or T cell that forms during an immune response but that does not act at once; it enters a resting phase, from which it is released for a secondary immune response.

menopause (MEN-uh-pozz) [L. *mensis*, month, + *pausa*, stop] The time when the reproductive potential of human females draws to a close.

menstrual cycle A recurring cycle, lasting twenty-eight days on average in adult human females. A secondary oocyte is released from one of a pair of ovaries, and the lining of the uterus is primed for pregnancy. The hormones estrogen, progesterone, FSH, and LH control the cyclic activity.

menstruation Sloughing of a blood-enriched endometrium when pregnancy does not occur.

mesoderm (MEH-zoe-derm) [Gk. *mesos*, middle, + *derm*, skin] Primary tissue layer important in evolution of all large, complex animals; gives rise to many internal organs and part of the integument.

mesophyll (MEH-zoe-fill) A photosynthetic parenchyma with abundant air spaces for gas exchange between its cells.

messenger RNA (mRNA) A single strand of ribonucleotides transcribed from DNA, then translated into a polypeptide chain. The only RNA encoding protein-building instructions.

metabolic pathway (MEH-tuh-BALL-ik) Orderly sequence of enzyme-mediated reactions by which cells maintain, increase, or decrease the concentrations of particular substances.

metabolism (meh-TAB-oh-lizm) [Gk. *meta*, change] All the controlled, enzyme-mediated chemical reactions by which cells acquire and use energy to synthesize, store, degrade, and eliminate substances in ways that contribute to growth, survival, and reproduction.

metamorphosis (me-tuh-MOR-foe-sis) [Gk. *meta*, change, + *morphe*, form] Major changes in body form during the transition from the embryo to the adult; involves hormonally controlled size increases, reorganization of tissues, and remodeling of body parts.

metaphase Of meiosis I, a stage when all pairs of homologous chromosomes have become positioned at the spindle equator. Of mitosis or meiosis II, a stage when all the duplicated chromosomes have become positioned at the equator of the microtubular spindle.

metazoan Any multicelled animal.

methanogen Anaerobic archaebacterium that produces methane gas as by-product.

MHC marker Self-marker protein. Some are on all body cells of the individual; others are unique to macrophages and lymphocytes.

micelle (my-SELL) Of fat digestion, tiny droplet of bile salts, fatty acids, and monoglycerides; role in fat absorption from small intestine.

microevolution Of a population, any change in allele frequencies resulting from mutation, genetic drift, gene flow, natural selection, or some combination of these.

microfilament [Gk. *mikros*, small, + L. *filum*, thread] Cytoskeletal element. Each consists of two thin, twisted polypeptide chains; it has roles in cell movement and in producing and maintaining cell shapes.

micrograph Photograph of an image that came into view with the aid of a microscope.

microorganism Organism, usually single celled, too small to be observed without a microscope.

microspore Walled haploid spore; becomes a pollen grain in gymnosperms and angiosperms.

microtubular spindle Bipolar array of many microtubules; forms during nuclear division and moves chromosomes apart in controlled ways.

microtubule (my-crow-TUBE-yool) Cylindrical, hollow cytoskeletal element that consists of tubulin subunits; roles in cell shape, growth, and motion (e.g., key skeletal element of cilia, flagella, spindle apparatus).

microtubule organizing center MTOC; mass of substances in cytoplasm of eukaryotic cells; number, type, and location dictate orientation and organization of cell's microtubules.

microvillus (MY-crow-VILL-us) [L. *villus*, shaggy hair] Slender extension from free surface of certain cells; arrays of many microvilli greatly increase absorptive or secretory surface area.

midbrain Part of vertebrate brain with centers for coordinating reflex responses to visual and auditory input; also relays signals to forebrain.

migration Recurring pattern of movement between two or more locations in response to environmental rhythms; e. g., circadian rhythms and seasonal changes in daylength. It requires activation or suppression of internal timing mechanisms that govern physiological and behavioral functions of migratory animals.

mimicry (MIM-ik-ree) Close resemblance in form, behavior, or both between one species (the mimic) and another (its model). Serves in deception, as when an orchid mimics a female insect and so attracts males that pollinate it.

mineral Any element or inorganic compound that formed by natural geologic processes and is required for normal cell functioning.

mitochondrion (MY-toe-KON-dree-on) Double-membrane organelle of ATP formation. Only site of aerobic respiration's second and third stages. May have endosymbiotic origins.

mitosis (my-TOE-sis) [Gk. *mitos*, thread] Type of nuclear division that maintains the parental chromosome number for daughter cells. The basis of growth in size, tissue repair, and often asexual reproduction for eukaryotes.

mixture Two or more elements intermingled in proportions that can and usually do vary.

model Theoretical, detailed description or analogy that helps people visualize something that has not yet been directly observed.

molar Tooth with a platform having cusps (surface bumps) that help crush, grind, and shear food; one of the cheek teeth.

molecular clock Model used to calculate the time of origin of one lineage or species relative to others. The underlying assumption is that neutral mutations accumulate in a lineage at predictable rates that can be measured as a series of ticks back through time.

molecule A unit of matter in which chemical bonds hold together two or more atoms of the same or different elements.

mollusk Only invertebrate with a tissue fold (mantle) draped over a soft, fleshy body; most have an external or internal shell. Enormous diversity in body plans and sizes, as in chitons, gastropods (e.g, snails), bivalves (e.g., clams), and cephalopods (e.g., octopuses, squids).

molting Periodic shedding of body structures that are too small, worn out, or both. Permits certain animals to grow in size or renew some parts (e.g., exoskeletons, shells, hairs, feathers, and horns). Especially characteristic of insects and other arthropods.

Monera In earlier classification schemes, a prokaryotic kingdom that encompasses both archaebacteria and eubacteria.

monocot (MON-oh-kot) Monocotyledon; a flowering plant with one cotyledon in seeds, floral parts generally in threes (or multiples of three), and often parallel-veined leaves.

monohybrid cross Experimental cross between two parents that are homozygous for different versions of the same trait (e.g., *AA* and *aa*). F_1 offspring are heterozygous; each inherits a pair of nonidentical alleles (*Aa*) for the trait.

monomer Small molecule used as a subunit of polymers, such as sugar monomers of starch.

monosaccharide (MON-oh-SAK-ah-ride) [Gk. *monos*, alone, single, + *sakcharon*, sugar] One of the simple carbohydrates; a single sugar monomer. Glucose is an example.

monosomy Presence of a chromosome that has no homologue in a diploid cell.

morphogenesis (MORE-foe-JEN-ih-sis) [Gk. *morphe*, form, + *genesis*, origin] Inherited program of orderly changes in size, shape, and proportions of an animal embryo, leading to specialized tissues and early organs.

morphological convergence Macroevolutionary pattern. In response to similar environmental pressures over time, evolutionarily distant lineages evolve in similar ways and end up being alike in appearance, functions, or both.

morphological divergence Macroevolutionary pattern. Genetically diverging lineages slowly undergo change from the body form of their common ancestor.

motor neuron Type of neuron specialized to swiftly relay commands from the brain or spinal cord to muscle cells, gland cells, or both.

multicelled organism Organism composed of many cells with coordinated metabolic activity; most show extensive cell differentiation into tissues, organs, and organ systems.

multiple allele system Three or more slightly different molecular forms of a gene that occur among individuals of a population.

muscle fatigue Decline in tension of a muscle kept in a state of tetanic contraction as a result of continuous, high-frequency stimulation.

muscle tension Mechanical force exerted by a contracting muscle; resists opposing forces (e.g., gravity or weight of an object being lifted).

muscle tissue Tissue with arrays of cells able to contract under stimulation, then passively lengthen and return to their resting position.

mutagen (MEW-tuh-jen) Any environmental agent, such as a virus or ultraviolet radiation, that can alter DNA's molecular structure.

mutation [L. *mutatus*, a change, + *-ion*, an act, a result, or a process] Heritable change in the molecular structure of DNA. Original source of all new alleles and, ultimately, the diversity of life.

mutation frequency Of a population, the number of times that a mutation at a particular locus has arisen.

mutation rate Of a gene locus, the probability that a spontaneous mutation will occur during or between DNA replication cycles.

mutualism [L. *mutuus*, reciprocal] Symbiotic interaction that benefits both participants.

mycelium (my-SEE-lee-um), plural **mycelia** [Gk. *mykes*, fungus, mushroom, + *helos*, callus] Mesh of tiny, branching filaments (hyphae); the food-absorbing portion of most fungi.

mycorrhiza (MY-coe-RIZE-uh) "Fungus-root." A form of mutualism between fungal hyphae and young plant roots. The plant gives up some carbohydrates and the fungus gives up some of its absorbed mineral ions.

myelin sheath Axonal sheath around many sensory and motor neurons; enhances the long-distance propagation of action potentials.

myofibril (MY-oh-FY-brill) One of the many internal threadlike structures of a muscle cell, each divided into sarcomeres.

myosin (MY-uh-sin) Motor protein, often bound to microtubules; key roles in cell movements.

NAD^+ Nicotinamide adenine dinucleotide. A nucleotide coenzyme; abbreviated NADH when carrying electrons and H^+ to a reaction site.

$NADP^+$ Nicotinamide adenine dinucleotide phosphate. A phosphorylated nucleotide coenzyme; abbreviated $NADPH_2$ when it is carrying electrons and H^+ to a reaction site.

natural killer cell Cytotoxic lymphocyte that reconnoiters for tumor cells and virus-infected cells, then touch-kills them.

natural selection Microevolutionary process; the outcome of differences in survival and reproduction among individuals that vary in details of heritable traits. Over generations, it typically leads to increased fitness.

necrosis (neh-CROW-sis) Passive death of many cells that results from severe tissue damage.

negative feedback mechanism A homeostatic mechanism by which a condition that has changed as a result of some activity triggers a response that reverses the change.

nematocyst (NEM-at-uh-sist) [Gk. *nema*, thread, + *kystis*, pouch] Cnidarian capsule that has a dischargeable, tube-shaped thread, sometimes barbed; releases a toxin or sticky substance.

nephridium (neh-FRID-ee-um), plural **nephridia** Unit that controls composition and volume of fluid in some invertebrates (e.g., earthworms).

nephron (NEFF-ron) [Gk. *nephros*, kidney] One of the urine-forming tubules in a kidney; it filters water and solutes from blood, then selectively reabsorbs adjusted amounts of both in ways that help maintain the volume and composition of extracellular fluid.

nerve Cordlike bundle of the axons of sensory neurons, motor neurons, or both sheathed in connective tissue.

nerve cord A prominent longitudinal nerve. Most animals have one, two, or three. The nervous system of chordate embryos develops from a tubular, dorsal nerve cord.

nerve impulse *See* action potential.

nerve net Simple nervous system in epidermis of cnidarians and some other invertebrates; a diffuse mesh of simple, branching nerve cells interacts with contractile cells, sensory cells, or both.

nervous system Integrative organ system with nerve cells interacting in signal-conducting and information-processing pathways. Detects and processes stimuli, and elicits responses from effectors (e.g., muscles and glands).

nervous tissue Connective tissue composed of neurons and often neuroglia.

net population growth rate per individual (r) For population growth equations, a single variable combining birth and death rates; the assumption is that both remain constant.

neural tube Embryonic and evolutionary forerunner of brain and spinal cord.

neuroglia (NUR-oh-GLEE-uh) Collectively, cells that structurally and metabolically support neurons. They make up about half the volume of nervous tissue in vertebrates.

neuromodulator Any of a variety of signaling molecules that magnify or reduce the effects of a neurotransmitter.

neuromuscular junction A chemical synapse between axonal endings of a motor neuron and a muscle cell.

neuron (NUR-on) Type of nerve cell; basic communication unit in most nervous systems.

neurotransmitter Any of a class of signaling molecules secreted by neurons. It acts on cell next to it, then is rapidly degraded or recycled.

neutral mutation Mutation that has little or no effect on phenotype. Natural selection cannot change its frequency in a population because it does not affect survival or reproduction.

neutron Unit of matter, one or more of which occupies the atomic nucleus and has mass but no electric charge.

neutrophil The most abundant, fastest acting white blood cell; it phagocytizes bacteria.

niche (NITCH) [L. *nidas*, nest] Sum total of all activities and relationships in which individuals of a species engage as they secure and use the resources required to survive and reproduce.

nitrification (nye-trih-fih-KAY-shun) Process by which certain bacteria in soil break down ammonia or ammonium to nitrite, then other bacteria break down nitrite to nitrate (which is a form that plants can take up). Key part of the nitrogen cycle.

nitrogen cycle Atmospheric cycle. Nitrogen moves from its largest reservoir (atmosphere), through the ocean, ocean sediments, soils, and food webs, then back to the atmosphere.

nitrogen fixation Process by which certain bacterial species convert gaseous nitrogen to ammonia, which swiftly dissolves in their cytoplasm to form ammonium. Ammonium can be used in biosynthesis.

nociceptor (NO-SEE-sep-tur) Pain receptor that detects tissue damage, as by burns.

node Stem site where one or more leaves form.

noncyclic pathway of ATP formation (non-SIK-lik) [L. *non*, not, + Gk. *kylos*, circle] Light-dependent reactions of photosynthesis that requires photolysis, two photosystems, and two transport chains. Water molecules are split. They release electrons and hydrogen that are used in ATP and NADPH formation, plus oxygen as a by-product (the basis of Earth's oxygen-rich atmosphere).

nondisjunction Failure of sister chromatids or failure of a pair of homologous chromosomes to separate at meiosis or mitosis. As a result, daughter cells end up with too many or too few chromosomes.

notochord (KNOW-toe-kord) Of chordates, a rod of stiffened tissue (not cartilage or bone) that is a supporting structure for the body.

nuclear envelope Outermost portion of a cell nucleus; composed of a double membrane (two lipid bilayers and associated proteins).

nucleic acid (new-CLAY-ik) Single- or double-stranded chain of four kinds of nucleotides joined at phosphate groups. Nucleic acids differ in their base sequences. DNA and RNA are examples.

nucleic acid hybridization Any base-pairing between sequences of DNA or RNA from different sources.

nucleoid (NEW-KLEE-oid) Portion of bacterial cell interior in which the DNA is physically organized but not enclosed by a membrane.

nucleolus (new-KLEE-oh-lus) [L. *nucleolus*, little kernel] In the nucleus of a nondividing cell, an assembly site for the protein and RNA subunits that will later form ribosomes in the cytoplasm.

nucleosome (NEW-klee-oh-sohm) A stretch of eukaryotic DNA looped twice around a spool of histone molecules; one of many units that give condensed chromosomes their structure.

nucleotide (NEW-klee-oh-tide) Small organic compound consisting of a five-carbon sugar (deoxyribose), a nitrogen-containing base, and a phosphate group. The structural unit of adenosine phosphates, nucleotide coenzymes, and nucleic acids.

nucleotide coenzyme Protein that assists an enzyme by delivering electrons and hydrogen atoms released at a reaction site to another reaction site.

nucleus (NEW-klee-us) [L. *nucleus*, a kernel] Of atoms, a central core of one or more protons and (in all but hydrogen atoms) neutrons. In a eukaryotic cell, the organelle that physically separates DNA from cytoplasmic machinery.

numerical taxonomy Study of the degree of relatedness between an unidentified organism and a known group through comparisons of traits. Used to classify prokaryotes, which are poorly represented in the fossil record.

nutrient Element with a direct or indirect role in metabolism that no other element fulfills.

nutrition Processes by which food is selectively ingested, digested, absorbed, and converted to the body's own organic compounds.

obesity Excess of fat in adipose tissue; caloric intake has exceeded the body's energy output.

oligosaccharide (oh-LIG-oh-SAC-uh-RID) Short-chain carbohydrate of two or more covalently bonded sugar monomers. Disaccharides (two monomers) are examples.

omnivore [L. *omnis*, all, + *vovare*, to devour] An animal that feeds at more than one trophic level.

oncogene (ON-koe-jeen) Any gene having the potential to induce cancerous transformation.

oocyte Immature egg of all animals and some protistans.

oogenesis (oo-oh-JEN-uh-sis) Process by which a germ cell develops into a mature oocyte.

operator Very short base sequence between a promoter and bacterial genes; a binding site for a repressor that can block transcription.

operon Promoter–operator sequence that services more than one bacterial gene; part of a control mechanism that adjusts transcription rates upward or downward.

orbital One of the volumes of space around the atomic nucleus in which one or at most two electrons are likely to be at any instant.

organ Body structure having definite form and function that consists of more than one tissue.

organ formation Developmental stage in which primary tissue layers give rise to cell lineages unique in structure and function. Descendants of those lineages give rise to all the different tissues and organs of the adult.

organ system Two or more organs that are interacting chemically, physically, or both in a common task.

organelle (or-GUN-ell) Membrane-bound sac or compartment in the cytoplasm having one or more specialized metabolic functions. Most eukaryotic cells have a profusion of them.

organic compound Molecule of one or more elements covalently bonded to some number of carbon atoms.

osmoreceptor Sensory receptor that detects changes in water volume (solute concentration) in the fluid bathing it.

osmosis (oss-MOE-sis) [Gk. *osmos*, pushing] The diffusion of water in response to water concentration gradient between two regions that are separated by a selectively permeable membrane. The greater the number of ions and molecules dissolved in a solution, the lower its water concentration.

osmotic pressure Force that operates after hydrostatic pressure develops in a cell or in an enclosed body region; the amount of force that stops further increases in fluid volume by countering the inward diffusion of water.

ovary (OH-vuh-ree) In most animals, a female gonad. In flowering plants, the enlarged base of one or more carpels. A fruit is a mature ovary often combined with other floral parts.

oviduct (OH-vih-dukt) One of a pair of ducts through which eggs travel from an ovary to the uterus. Formerly called a fallopian tube.

ovulation (OHV-you-LAY-shun) Release of a secondary oocyte from an ovary during one menstrual cycle.

ovule (OHV-youl) [L. *ovum*, egg] Tissue mass in a plant ovary that develops into a seed. Consists of female gametophyte with egg cell, nutrient-rich tissue, and a jacket (cell layers) that will become a seed coat.

ovum (OH-vum) Mature secondary oocyte.

oxaloacetate (OX-AL-oh-ASS-ih-tate) A four-carbon compound with roles in metabolism (e.g., the point of entry into the Krebs cycle).

oxidation–reduction reaction An electron transfer between atoms or molecules. Often an unbound proton (H^+) is transferred at the same time.

ozone thinning Pronounced seasonal thinning of the atmosphere's ozone layer, especially above the Earth's polar regions.

pancreas (PAN-cree-us) Gland with roles in digestion and organic metabolism. Secretes enzymes and bicarbonate into small intestine; also secretes insulin and glucagon, hormones that travel the bloodstream to target cells.

pancreatic islet Any of the 2 million or so clusters of endocrine cells in the pancreas.

parapatric speciation Idea that neighboring populations can become distinct species while maintaining contact along a common border.

parasite [Gk. *para*, alongside, + *sitos*, food] Organism that lives in or on a host organism for at least part of its life cycle. It feeds on specific tissues and usually does not kill its host outright.

parasitism Symbiotic interaction in which one species (a parasite) benefits and the other (its host) is harmed. The parasite lives inside or on a host and feeds on its cells or tissues.

parasitoid Type of insect larva that grows and develops in a host organism (usually another insect), consumes its soft tissues, and kills it.

parasympathetic nerve (PARE-uh-SIM-pu-THET-ik) An autonomic nerve. Signals carried by such nerves tend to slow overall body activities and divert energy to basic tasks, and to help make small adustments in internal organ activity by acting continually in opposition to sympathetic nerve signals.

parenchyma (par-ENG-kih-mah) Simple tissue that makes up the bulk of a plant; has roles in photosynthesis, storage, secretion, other tasks.

parthenogenesis (par-THEN-oh-GEN-uh-sis) An unfertilized egg giving rise to an embryo.

passive transport Process by which a transport protein that spans a cell membrane passively permits a solute to diffuse through its interior. Also called facilitated diffusion.

pathogen (PATH-oh-jen) [Gk. *pathos*, suffering, + *genēs*, origin] Any virus, bacterium, fungus, protistan, or parasitic worm that can infect an organism, multiply in it, and cause disease.

pattern formation theory Explanation of the orderly, sequential sculpting of embryonic cells into specialized animal tissues and organs. Cytoplasmic localization and, later, inductive interactions among master genes are responsible. Gene products map out the basic body plan and create chemical gradients that dictate how specific body parts develop.

PCR Polymerase chain reaction. A method of enormously amplifying the quantity of DNA fragments cut by restriction enzymes.

peat bog Compressed, soggy, highly acidic mat of accumulated remains of peat mosses.

pedigree Diagram of the genetic connections among related individuals through successive generations; uses standardized symbols.

pelagic province (peh-LAD-jik) Total volume of water in the world ocean.

penis Male copulatory organ by which sperm is deposited in a female reproductive tract.

peptide hormone Amino acid hormone that, when bound to a membrane receptor, activates enzyme systems that trigger a response. Often a second messenger in the cell relays its signal.

perennial [L. *per-*, throughout, + *annus*, year] Flowering plant that lives for three or more growing seasons.

pericycle (PARE-ih-sigh-kul) [Gk. *peri-*, around, + *kyklos*, circle] One or more cell layers just inside the endodermis that give rise to lateral roots and contribute to secondary growth.

periderm Protective cover that replaces plant epidermis during extensive secondary growth.

peripheral nervous system (per-IF-ur-uhl) [Gk. *peripherein*, to carry around] All nerves leading into and out from the spinal cord and brain. Includes ganglia of those nerves.

peristalsis (pare-ih-STAL-sis) Recurring waves of contraction and relaxation of muscles in the wall of a tubular or saclike organ.

peritoneum (pare-ih-tuh-NEE-um) Membrane that lines the coelom and helps maintain the positions of soft organs inside it.

peritubular capillary One of the set of blood capillaries around tubular parts of a nephron. Reabsorbs water and solutes; secretes excess H^+ and other substances to be excreted later.

permafrost Permanently frozen layer beneath the soil surface in arctic tundra. Water cannot penetrate it even in summer.

peroxisome Vesicle in which fatty acids and amino acids are first digested to hydrogen peroxide, then converted to harmless products.

pest resurgence Directional selection for an insecticide-resistant strain of a pest species.

PGA Phosphoglycerate (FOSS-foe-GLISS-er-ate) Intermediate of glycolysis and of the Calvin–Benson cycle.

PGAL Phosphoglyceraldehyde. Intermediate of glycolysis and of the Calvin-Benson cycle.

pH scale Measure of the concentration of free hydrogen ions (H^+) in blood, water, and other solutions. pH 0 is the most acidic, 14 the most basic, and 7, neutral.

phagocyte (FAG-uh-sight) [Gk. *phagein*, to eat, + *kytos*, hollow vessel] Cell that captures prey by phagocytosis (e.g., amoebas); also cells that use same process for defense and day-to-day tissue housekeeping (e.g., macrophages).

phagocytosis (FAG-uh-sigh-TOE-sis) [Gk. *phagein*, to eat, + *kytos*, hollow vessel] Engulfment of foreign cells or particles by way of pseudopod formation and endocytosis.

pharynx (FARE-inks) Among invertebrates, a muscular tube to the gut. In some chordates, a gas exchange organ. In land vertebrates, a dual entrance to the esophagus and trachea.

phenotype (FEE-no-type) [Gk. *phainein*, to show, + *typos*, image] Observable trait or traits of an individual that arise from gene interactions and gene–environment interactions.

pheromone (FARE-oh-moan) [Gk. *phero*, to carry, + *-mone*, as in hormone] Hormone-like, nearly odorless exocrine gland secretion. A signaling molecule between individuals of the same species that integrates social behavior.

phloem (FLOW-um) Plant vascular tissue that conducts sugars and other solutes. Includes living cells (sieve tubes) that connect to form conducting tubes, and adjoining companion cells that help load solutes into the tubes.

phospholipid Organic compound that has a glycerol backbone, two fatty acid tails, and a hydrophilic head of two polar groups (one being phosphate). Phospholipids are the main structural material of cell membranes.

phosphorus cycle Movement of phosphorus (mainly phosphate ions) from land, through food webs, to ocean sediments, then back to land. As for other minerals, Earth's crust is the largest reservoir in this sedimentary cycle.

phosphorylation (FOSS-for-ih-LAY-shun) A common means of activating molecules for a reaction. An enzyme either attaches inorganic phosphate to a molecule or mediates a transfer of a phosphate group from one molecule to another (as when ATP phosphorylates glucose).

photoautotroph Photosynthetic autotroph; any organism that synthesizes its own organic compounds using carbon dioxide as the source of carbon atoms and sunlight as the energy source. Nearly all plants, some protistans, and a few bacteria are photoautotrophs.

photolysis (foe-TALL-ih-sis) [Gk. *photos*, light, +*-lysis*, breaking apart] Reaction sequence in which photon energy splits water molecules. The released electrons and hydrogen take part in the noncyclic pathway of photosynthesis, and the oxygen is released as a by-product.

photoperiodism Biological response to change in relative lengths of daylight and darkness.

photoreceptor Light-sensitive sensory cell.

photosynthesis Trapping of sunlight energy, followed by its conversion to chemical energy (ATP, NADPH, or both) and then synthesis of sugar phosphates, which become converted into sucrose, cellulose, starch, and other end products. The main pathway by which energy and carbon enter the web of life.

photosystem One of many clusters of light-trapping pigment molecules embedded in photosynthetic membranes. A chlorophyll of the system gives up electrons necessary for the light-dependent reactions of photosynthesis.

phototropism [Gk. *photos*, light, + *trope*, a turning, direction] Change in the direction of cell movement or growth in response to light (e.g., as when differences in cell elongation cause a stem to bend toward light).

photovoltaic cell Device that converts energy of sunlight into electricity.

phycobilin (FIE-koe-BY-lin) Type of accessory pigment that extends the functional range of chlorophyll in photosynthesis. Abundant in red algae and in cyanobacteria especially.

phylogeny Evolutionary relationships among species, starting with an ancestral form and including branches leading to descendants.

phytochrome A light-sensitive pigment. Its controlled activation and inactivation take part in plant hormone activities that govern leaf expansion, stem branching, stem lengthening and often seed germination and flowering.

phytoplankton (FIE-toe-PLANK-tun) [Gk. *phyton*, plant, + *planktos*, wandering] Aquatic community of floating or weakly swimming photoautotrophs (e.g., "pastures of the seas").

pigment Any light-absorbing molecule.

pioneer species Any opportunistic colonizer of barren or disturbed habitats. Adapted for rapid growth and dispersal.

pituitary gland Endocrine gland that, with the hypothalamus, controls many physiological functions, including activity of many other endocrine glands. Its posterior lobe stores and secretes hypothalamic hormones. Its anterior lobe produces and secretes its own hormones.

placenta (play-SEN-tuh) Blood-engorged organ of pregnant female placental mammals; made of some endometrial tissue and extraembryonic membranes. Allows exchanges between the mother and fetus without an intermingling of their bloodstreams, thus sustaining the new individual and allowing its blood vessels to develop apart from the mother's.

plankton [Gk. *planktos*, wandering] Of aquatic habitats, a community of suspended or weakly swimming organisms, mostly microscopic.

plant Generally, a multicelled photoautotroph with well-developed root and shoot systems; photosynthetic cells that include starch grains as well as chlorophylls *a* and *b* ; and cellulose, pectin, and other polysaccharides in cell walls.

Plantae Kingdom of plants.

plasma (PLAZ-muh) Liquid portion of blood; mainly water in which ions, proteins, sugars, gases, and other substances are dissolved.

plasma membrane Outermost cell membrane; structural and functional boundary between cytoplasm and the fluid outside the cell.

plasmid Of many bacteria, a small, circular molecule of extra DNA that carries only a few genes and that is replicated independently of the bacterial chromosome.

plasmodesma, plural **plasmodesmata** (PLAZ-moe-DEZ-muh) Plant cell junction; membrane-lined channel that crosses both walls of two adjacent cells and connects their cytoplasm.

plasmolysis Osmotically induced shrinkage of a cell's cytoplasm.

plate tectonics Theory that great slabs (plates) of the Earth's outer layer (lithosphere) float on the hot, plastic, underlying mantle. All plates are in motion and have rafted continents to new positions over time. The geologic changes have profoundly affected the evolution of life.

platelet (PLAYT-let) Megakaryocyte fragment that releases substances used in clot formation.

pleiotropy (PLEE-oh-troe-pee) [Gk. *pleon*, more, + *trope*, direction] Positive or negative effects on two or more traits owing to expression of alleles at a single gene locus. Effects may or may not emerge at the same time.

polar body One of four cells that form by the meiotic cell division of an oocyte but that does not become the ovum.

pollen grain [L. *pollen*, fine dust] Immature or mature, sperm-bearing male gametophyte of gymnosperms and angiosperms.

pollen sac Chamber inside an anther in which pollen grains develop.

pollen tube Sperm-carrying tube that grows from a germinated pollen grain, through carpel tissues to the egg inside an ovule.

pollination Arrival of a pollen grain on the landing platform (stigma) of a carpel.

pollutant Natural or synthetic substance with which an ecosystem has no prior evolutionary experience, in terms of kinds or amounts; it accumulates to disruptive or harmful levels.

polymer (POH-lih-mur) [Gk. *polus*, many, + *meris*, part] Large molecule consisting of three to millions of monomers of the same or different kinds.

polymerase (puh-LIM-ur-aze) Enzyme that catalyzes a polymerization reaction (e.g., the DNA polymerase of DNA replication/repair).

polymorphism (poly-MORE-fizz-um) [Gk. *polus*, many, + *morphe*, form] The persistence of two or more qualitatively different forms of a trait (morphs) in a population.

polyp (POH-lip) Vase-shaped, sedentary stage of cnidarian life cycles.

polypeptide chain Organic compound with a sequence of three or more amino acids. Peptide bonds between them result in a regular pattern of nitrogen atoms in the carbon backbone: —N—C—C—N—C—C— . Every protein consists of one or more polypeptide chains.

polyploidy (POL-ee-PLOYD-ee) Having three or more of each type of chromosome in the nucleus of cells at interphase.

polysaccharide [Gk. *polus*, many, + *sakcharon*, sugar] Straight or branched chain of many covalently linked sugar units of the same or different kinds. In nature, the most common polysaccharides are cellulose, starch, and glycogen.

polysome A number of ribosomes translating the same mRNA molecule at the same time, one after the other, during protein synthesis.

pons Hindbrain region; traffic center for signals between cerebellum and forebrain centers.

population All individuals of the same species occupying the same area.

population density Count of individuals of a population occupying a specified area or specified volume of a habitat.

population distribution Dispersal pattern for individuals of a population through a habitat.

population size The number of individuals that make up the gene pool of a population.

positive feedback mechanism A homeostatic control mechanism. It sets in motion a chain of events that intensifies change from an original condition; after a limited time, intensification reverses the change.

potential energy Capacity of any stationary object to do work owing to its position in space or to the arrangement of its parts (e.g., a cat in a frozen posture, about to spring at a mouse).

predation Ecological interaction in which a predator feeds on a prey organism.

predator [L. *prehendere*, to grasp, seize] A heterotroph that feeds on other living organisms (its prey), that lives neither in or nor on them (as parasites do), and that may or may not end up killing them.

prediction Statement about what you should observe in nature if you were to go looking for a particular phenomenon; the if–then process.

pressure flow theory Explanation of how organic compounds move through phloem of vascular plant. The compounds follow solute concentration and pressure gradients between sources (e.g., photosynthetically active leaves where they form) and sinks (e.g., growing parts where they are being used or stored).

primary growth Plant growth originating at root tips and shoot tips.

primary immune response Defensive actions by white blood cells and their secretions, as elicited by first-time recognition of antigen. Includes antibody- and cell-mediated responses.

primary productivity, gross Of ecosystems, the rate at which primary producers capture and store a given amount of energy in their cells and tissues during a specified interval.

primary productivity, net Of ecosystems, the rate of energy storage in primary producer cells and tissues in excess of rate of aerobic respiration during a specified interval.

primary wall A wall of polysaccharides, glycoproteins, and cellulose that is flexible and thin enough to allow new plant cells to divide or change shape during growth and development.

primate Mammalian lineage dating from the Eocene; includes prosimians, tarsioids, and anthropoids (monkeys, apes, and humans).

primer Short nucleotide sequence designed to base-pair with any complementary DNA sequence; later, DNA polymerases recognize it as a START tag for replication.

prion Small infectious protein that causes rare, fatal degenerative diseases of nervous system.

probability The chance that each outcome of a given event will occur is proportional to the number of ways the outcome can be reached.

probe Very short stretch of DNA designed to base-pair with part of a gene being studied and labeled with an isotope to distinguish it from DNA in the sample being investigated.

procambium (pro-KAM-bee-um) A meristem that gives rise to primary vascular tissues.

producer Autotroph (self-feeder); it nourishes itself using sources of energy and carbon from its physical environment. Photoautotrophs and chemoautotrophs are examples.

progesterone (pro-JESS-tuh-rown) A sex hormone secreted by ovaries and the corpus luteum of female mammals.

prokaryotic cell (pro-CARE-EE-oh-tic) [L. *pro*, before, + Gk. *karyon*, kernel] Archaebacterium or eubacterium; single-celled organism, most often walled; lacks the profusion of membrane-bound organelles observed in eukaryotic cells.

prokaryotic fission Cell division mechanism by which a bacterial cell reproduces.

promoter Short stretch of DNA to which RNA polymerase can bind and start transcription.

prophase Of mitosis, a stage when duplicated chromosomes start to condense, microtubules form a spindle, and the nuclear envelope starts to break up. Duplicated pairs of centrioles (if present) are moved to opposite spindle poles.

prophase I The first stage of meoisis I. Each duplicated chromosome starts to condense. It pairs with its homologue; nonsister chromatids usually undergo crossing over. Each becomes attached to microtubular spindle. One of the duplicated pairs of centrioles (if present) is moved to opposite spindle pole.

prophase II First stage of meiosis II. In each daughter cell, spindle microtubules attach to kinetochores of each chromosome and move them toward spindle's equator. One centriole pair (if present) is already at each spindle pole.

protein Organic compound composed of one or more polypeptide chains.

Protista Kingdom of protistans. Chytrids; water molds; slime molds; protozoans; sporozoans; euglenoids; chrysophytes; dinoflagellates; and red, brown, and green algae are major groups.

protistan (pro-TISS-tun) [Gk. *prōtistos*, primal, very first] Diverse species, ranging from single cells to giant kelps, that are photoautotrophs, heterotrophs, or both. Some are thought to be most like the earliest eukaryotic cells. All are unlike bacteria in having a nucleus, large ribosomes, mitochondria, ER, Golgi bodies, chromosomes with many proteins attached, and cytoskeletal microtubules.

proton Positively charged particle; one or more reside in nucleus of each atom. An unbound (free) proton is called a hydrogen ion (H^+).

proto-oncogene Gene sequence similar to an oncogene but coding for a protein that is used in normal cell functions. When mutated, it may trigger cancerous transformation.

protostome (PRO-toe-stome) [Gk. *proto*, first, + *stoma*, mouth] Lineage of coelomate, bilateral animals that includes mollusks, annelids, and arthropods. The first indentation to form in protostome embryos becomes the mouth.

protozoan Type of protistan that may resemble the single-celled heterotrophs that gave rise to animals. Amoeboid, animal-like, and ciliated protozoans are major categories.

proximal tubule Nephron's tubular portion into which water and solutes enter after being filtered from blood at Bowman's capsule.

pulmonary circuit Vertebrate cardiovascular route in which oxygen-poor blood flows from the heart to the lungs, where it is oxygenated before flowing back to the heart.

Punnett-square method Construction of a diagram of a genetic cross that is a simple way to predict the probable outcomes.

purine Nucleotide base having a double ring structure (e.g., adenine or guanine).

pyrimidine (pih-RIM-ih-deen) Nucleotide base having a single ring structure (e.g., cytosine or thymine).

pyruvate (PIE-roo-vate) A small organic compound with a backbone of three carbon atoms. Two molecules form as end products of glycolysis.

r Variable in population growth equations that signifies net population growth rate. Birth and death rates are assumed to remain constant and are combined into this one variable.

radial symmetry Animal body plan having four or more roughly equivalent parts around a central axis (e.g., sea anemone).

radioisotope Unstable atom (uneven number of protons and neutrons). It spontaneously emits particles and energy; over a predictable time span, it decays into a different atom.

radiometric dating Method of measuring the proportions of (1) a radioisotope in a mineral trapped long ago in newly formed rock and (2) a daughter isotope that formed from it by radioactive decay in the same rock. Used to assign absolute dates to fossil-containing rocks and to the geologic time scale.

rain shadow A reduction in rainfall on the leeward side of a high mountain range that results in arid or semiarid conditions.

reabsorption In a kidney, diffusion or active transport of water and reclaimable solutes from a nephron into peritubular capillaries; under control of ADH and aldosterone. At a capillary bed, osmotic movement of interstitial fluid into a capillary when water concentration between plasma and interstitial fluid differ.

rearrangement, molecular Conversion of one organic compound to another through changes in its internal bonds.

receptor, molecular Type of membrane protein that binds an extracellular substance (e.g., hormone).

receptor, sensory Sensory cell or specialized cell adjacent to it that can detect a stimulus.

recessive allele [L. *recedere*, to recede] In heterozygotes, an allele whose expression is fully or partially masked by expression of its partner. Fully expressed only in homozygous recessives.

recombinant chromosome Of eukaryotes, a chromosome that emerges from meiosis with a combination of alleles that differs from a parental combination of alleles.

recombinant DNA Any molecule of DNA that incorporates one or more nonparental nucleotide sequences. Outcome of microbial gene transfer in nature or recombinant DNA technology.

recombinant DNA technology Procedures by which DNA molecules from different species are isolated, cut up, and spliced together. A population of rapidly dividing bacterial cells or PCR is then used to amplify the recombinant molecules to useful quantities.

recombination Any enzyme-mediated reaction that inserts one DNA sequence into another. "Generalized" recombination uses any pair of homologous sequences between chromosomes as substrates, as during crossing over. Site-specific recombination uses only a short stretch of homology between viral and bacterial DNA. A different reaction can insert transposable elements at new, random sites in bacterial or eukaryotic genomes; no homology is required.

red alga Type of protistan. Most are multicelled, aquatic photoautotrophs with an abundance of phycobilins that mask chlorophyll *a*.

red blood cell Erythrocyte. Cell that serves in rapid transport of oxygen in blood.

red marrow Site of blood cell formation in the spongy tissue of many bones.

reflex [L. *reflectere*, to bend back] Stereotyped, simple movement in response to stimuli. In simple reflex arcs, sensory neurons synapse directly on motor neurons.

refractory period Brief interval following an action potential when a small patch of neural membrane is insensitive to stimulation.

regulatory protein Component of mechanisms that control transcription, translation, and gene products by interacting with DNA, RNA, new polypeptide chains, or proteins (e.g., enzymes).

releasing hormone Hypothalamic signaling molecule that enhances or slows secretions from target cells in anterior lobe of pituitary gland.

repressor Protein that binds with an operator on bacterial DNA to block transcription.

reproduction Any process by which a parental cell or organism produces offspring. Among eukaryotes, asexual modes (e.g., binary fission, budding, vegetative propagation) and sexual modes. Bacteria employ prokaryotic fission. Viruses do not reproduce themselves; host organisms execute their replication cycle.

reproductive isolating mechanism A heritable feature of body form, functioning, or behavior that prevents interbreeding between two or more genetically divergent populations.

reproductive success Production of viable offspring by the individual.

reptile Carnivorous species belonging to the first vertebrate lineage to escape dependency on free water, by way of internal fertilization, efficient kidneys, amniote eggs, and other adaptations. Examples are dinosaurs (extinct), crocodilians, snakes, lizards, and tuataras.

resource partitioning Of two or more species that compete for the same resource, a sharing of the resource in different ways or at different times, which allows them to coexist.

respiration [L. *respirare*, to breathe] Of all animals, exchange of environmental oxygen with carbon dioxide from cells (e.g., through integumentary exchange or a respiratory system).

respiratory surface Thin, moist epithelium that functions in gas exchange in animals.

resting membrane potential Of a neuron and other excitable cells, a steady voltage difference across the plasma membrane in the absence of outside stimulation.

restoration ecology Attempt to reestablish biodiversity in ecosystems that have become altered as a result of agriculture, mining, and other severe disturbances.

restriction enzyme One of a class of bacterial enzymes that cut apart foreign DNA injected into the cell body, as by viruses. Important tool of recombinant DNA technology.

reticular formation Low-level pathway of information flow through vertebrate nervous system. Mesh of interneurons that extends from the upper spinal cord, through the brain stem, and into the cerebral cortex.

reverse transcription Synthesis of DNA on an RNA template by using reverse transcriptase, a viral enzyme. Basis of RNA virus replication cycle and of cDNA synthesis in laboratory.

RFLP Short for restriction fragment length polymorphism. DNA fragments of different sizes, cleaved by restriction enzymes, that reveal genetic differences among individuals.

Rh blood typing Method of characterizing red blood cells according to a self-marker protein at their surface. Rh^+ cells have it; Rh^- cells do not.

rhizoid Simple rootlike absorptive structure of some fungi and nonvascular plants.

ribosomal RNA (rRNA) Type of RNA that combines with proteins to form ribosomes, on which polypeptide chains are assembled.

ribosome Structure composed of two subunits of rRNA and proteins. Has binding sites for mRNA and tRNAs, which interact to produce a polypeptide chain in translation stage of protein synthesis.

RNA Ribonucleic acid. Any of a class of single-stranded nucleic acids that function in transcribing and translating the genetic instructions encoded in DNA into proteins. A molecule of mRNA, rRNA, or tRNA.

rod cell Vertebrate photoreceptor sensitive to very dim light. Contributes to the coarse perception of movement across visual field.

root Plant part, typically belowground, that absorbs water and dissolved minerals, anchors aboveground parts, and often stores food.

root hair Threadlike extension of a specialized epidermal cell of a young root. Increases root surface area for absorbing water and minerals.

root nodule Localized swelling on a root of certain legumes and other plants. Develops when nitrogen-fixing bacteria infect the plant, multiply, and interact symbiotically with it.

rubisco RuBP carboxylase; an enzyme that catalyzes attachment of the carbon atom from CO_2 to RuBP and so starts the Calvin–Benson cycle of the light-independent reactions.

RuBP Short for ribulose bisphosphate. Organic compound that has a backbone of five carbon atoms that serves in carbon fixation and that is regenerated in the Calvin–Benson cycle in C3 plants.

salinization Salt buildup in soil by evaporation, poor drainage, and heavy irrigation.

saliva Glandular secretion that is mixed with food and starts starch breakdown in the mouth.

salt Compound that releases ions other than H^+ and OH^- in solution.

saltatory conduction Of myelinated neurons, a rapid form of action potential propagation. Excitation hops to nodes between jellyrolled membranes of cells making up myelin sheath.

sampling error Use of a sample (or subset) of a population, an event, or some other aspect of nature for an experimental group that is not large enough to be representative of the whole.

saprobe Heterotroph that obtains energy and carbon from nonliving organic matter and so causes its decay (e.g., many fungal species).

sarcomere (SAR-koe-meer) One of many basic units of contraction, defined by Z lines, that subdivide a myofibril (muscle cell). Every sarcomere shortens in response to ATP-driven interactions between its parallel arrays of actin and myosin components.

sarcoplasmic reticulum (sar-koe-PLAZ-mik reh-TIK-you-lum) System of membranous chambers threading around myofibrils that take up, store, and release the calcium ions required for cross-bridge formation.

Schwann cell Type of neuroglial cell that wraps around certain axons like a jellyroll. A series of these cells, separated only by small nodes, form a myelin sheath.

sclerenchyma (skler-ENG-kih-mah) Simple plant tissue that supports mature plant parts and commonly protects seeds. Most of its cells have thick, lignin-impregnated walls.

sea-floor spreading Ongoing event in which molten rock erupts from immense, continuous ridges on the ocean floor, flows laterally in both directions, and hardens to form new crust. Elsewhere, it forces older crust down into vast trenches in the seafloor.

second law of thermodynamics A law of nature stating that the spontaneous direction of energy flow is from forms organized to less organized forms. The total amount of energy in the universe is spontaneously flowing from forms of higher to lower quality; with each conversion, some energy becomes randomly dispersed in a form (heat, most often) not as readily available to do work.

second messenger Molecule within a cell that mediates a hormonal signal by initiating the cellular response to it.

secondary immune response Immune action against previously encountered antigen, more rapid and prolonged than a primary response owing to swift participation of memory cells.

secondary sexual trait A trait associated with maleness or femaleness but with no direct role in reproduction (e.g., distribution of body hair and body fat). The primary sexual trait is the presence of male or female gonads.

secondary wall A wall on the inner surface of the primary wall of an older plant cell that stopped growing but needs structural support. Contains lignin in older cells of woody plants.

secretion A cell acting on its own or as part of glandular tissue releases a substance across its plasma membrane, to the surroundings.

sedimentary cycle Biogeochemical cycle. An element having no gaseous phase moves from land, through food webs, to the seafloor, then returns to land through long-term uplifting.

seed Mature ovule with an embryo sporophyte inside and integuments that form a seed coat.

segmentation Of animal body plans, a series of units that may or may not be similar to one another in appearance. Of tubular organs, an oscillating movement produced by rings of circular muscle in the tube wall.

segregation, theory of [L. *se-*, apart, + *grex*, herd] Mendelian theory. Sexually reproducing organisms inherit pairs of genes (on pairs of homologous chromosomes), the two genes of each pair are separated from each other at meiosis, and they end up in separate gametes.

selective gene expression Control of which gene products a cell makes or activates during a specified interval. Depends on the type of cell, its adjustments to changing chemical conditions, which external signals it is receiving, and its built-in control systems.

selective permeability Of a cell membrane, a capacity to let some substances but not others cross it at certain sites, at certain times. The capacity arises as an outcome of its lipid bilayer structure and its transport proteins.

selfish behavior An individual protects or increases its own chance to produce offspring regardless of consequences to its social group.

selfish herd Social group held together simply by reproductive self-interest.

semen (SEE-mun) Sperm-bearing fluid expelled from a penis during male orgasm.

semiconservative replication [Gk. *hēmi*, half, + L. *conservare*, to keep] Mechanism of DNA duplication. The DNA double helix unzips, and a complementary strand is assembled on exposed bases of each strand. Each conserved strand and its new partner wind up together to form a double helix, thus being a half-old, half-new molecule.

senescence (sen-ESS-cents) [L. *senescere*, to grow old] Processes leading to the natural death of an organism or to parts of it (e.g., leaves).

sensation Conscious awareness of a stimulus.

sensory neuron Type of neuron that detects a stimulus and relays information about it toward an integrating center (e.g., a brain).

sensory system The "front door" of a nervous system; it detects external and internal stimuli and relays information to integrating centers that issue commands for responses.

sessile animal (SESS-ihl) Animal that remains attached to a substrate during some stage (often the adult stage) of its life cycle.

sex chromosome A chromosome with genes that influence primary sex determination (whether male or female gonads will develop in the new individual). Depending on the species, somatic cells have one or two sex chromosomes, of the same or different type. In mammals, females are XX and males XY.

sexual dimorphism Occurrence of female and male phenotypes among the individuals of a sexually reproducing species.

sexual reproduction Production of offspring by meiosis, gamete formation, and fertilization.

sexual selection A microevolutionary process. Natural selection favors a trait that gives the individual a competitive edge in attracting or keeping a mate, hence in reproductive success.

shell model Model of electron distribution in which all orbitals available to electrons of atoms occupy a nested series of shells.

shifting cultivation Cutting and burning trees in a plot of land, followed by tilling ashes into soil. Once called slash-and-burn agriculture.

shoot system Aboveground plant parts (e.g., stems, leaves, and flowers).

sieve-tube member One of the cells that join together as phloem's sugar-conducting tubes.

sign stimulus Simple environmental cue that triggers a response to a stimulus that the nervous system is prewired to recognize.

sink Any region of a plant where cells are storing or using food (e.g., roots).

sister chromatid Of a duplicated chromosome, one of two DNA molecules (and associated proteins) attached at the centromere until they are separated from each other during mitosis or meiosis. After separation, each is then called a chromosome in its own right.

six-kingdom classification scheme A recent phylogenetic scheme that groups all organisms into the kingdoms Eubacteria, Archaebacteria, Protista, Fungi, Plantae, and Animalia.

skeletal muscle An organ with hundreds to many thousands of muscle cells bundled inside a sheath of connective tissue, which extends past the muscle as tendons.

sliding filament model Explanation of how muscles contract. Myosin heads projecting from microtubules at the center of a sarcomere repeatedly bind to neighboring actin filaments (which project inward from the sarcomere's sides), and with short, ATP-driven power strokes make them ratchet toward the center of the sarcomere. The sarcomere shortens as the actin filaments slide toward its center.

small intestine Part of the vertebrate digestive system in which digestion is completed and most dietary nutrients are absorbed.

smog, industrial Polluted, gray-colored air that forms above industrialized cities during cold, wet winters.

smog, photochemical Polluted, brown, smelly air that forms above large cities with many gas-burning vehicles during warm weather.

social behavior Diverse interactions among individuals of a species, which display, send, and respond to shared forms of communication that have genetic and learned components.

social parasite Animal that exploits the social behavior of another species to assure its own survival and reproduction.

sodium–potassium pump Type of membrane transport protein that, when activated by ATP, selectively transports potassium ions across a membrane, against its concentration gradient, and passively allows sodium ions to cross it in the opposite direction.

soil Mixture of mineral particles of variable sizes and decomposing organic material; air and water occupy spaces between particles.

solute (SOL-yoot) [L. *solvere*, to loosen] Any substance dissolved in a solution. Spheres of hydration around charged parts of its ions and molecules keep them dispersed.

solvent Any fluid (e.g., water) in which one or more substances are dissolved.

somatic cell (SO-MAT-ik) [Gk. *sōmā*, body] Any body cell that is not a germ cell. (Germ cells are the forerunners of gametes.)

somatic nervous system Nerves leading from a central nervous system to skeletal muscles.

somite One of many paired segments in a vertebrate embryo that give rise to most bones, skeletal muscles of head and trunk, and dermis.

source Any plant part where photosynthetic cells are making organic compounds.

speciation (spee-see-AY-shun) The formation of a daughter species from a population or subpopulation of a parent species by way of microevolutionary processes. Routes vary in their details and in length of time before the required reproductive isolation is completed.

species (SPEE-sheez) [L. *species*, a kind] One kind of organism. Of sexually reproducing organisms, one or more natural populations in which individuals are interbreeding and are reproductively isolated from other such groups.

sperm [Gk. *sperma*, seed] Mature male gamete.

spermatogenesis (sper-MAT-oh-JEN-ih-sis) Formation of mature sperm from a germ cell.

sphere of hydration A clustering of water molecules around individual molecules or ions of a substance placed in water owing to positive and negative interactions among them.

spinal cord The part of the central nervous system in a canal inside the vertebral column; site of direct reflex connections between sensory and motor neurons; also has tracts to and from the brain.

spleen A lymphoid organ that is a filtering station for blood, a reservoir of red blood cells, and a reservoir of macrophages.

spore Reproductive or resting structure of one or a few cells, often walled or coated and adapted for resisting adverse conditions, for dispersal, or both. May be nonsexual or sexual (formed by way of meiosis). Sporozoans, fungi, plants, and some bacteria form spores.

sporophyte [Gk. *phyton*, plant] A vegetative body that grows, by mitotic cell divisions, from a plant zygote and that produces spore-bearing structures.

spring overturn Of many bodies of water (e.g., large lakes in temperate zones), a springtime downward movement of dissolved oxygen near the surface, and an upward movement of nutrients from bottom sediments to the surface that fans primary productivity.

S-shaped curve Type of diagrammatic curve that emerges when logistic population growth is plotted against time.

stabilizing selection Mode of natural selection by which intermediate phenotypes in the range of variation are favored and extremes at both ends are eliminated.

stamen (STAY-mun) A male reproductive part of a flower; usually a pollen-bearing structure (anther) on a single stalk (filament).

start codon Base triplet in mRNA that serves as the START signal for translation.

stem cell Self-perpetuating animal cell that stays unspecialized. Some of its daughter cells also are self-perpetuating; others differentiate into specialized cells (e.g., red blood cells that arise from stem cells in bone marrow).

steroid hormone Lipid-soluble hormone made from cholesterol that acts on a target cell's DNA by entering the nucleus alone or bound to intracellular receptor. Some act by binding to a receptor on a target's plasma membrane.

sterol (STAIR-all) Lipid with a rigid backbone of four fused carbon rings. Sterols differ in the number, position, and type of their functional groups. Cholesterol is one; it is a precursor of steroid hormones and occurs in animal cell membranes.

stigma Sticky or hairy surface tissue on upper part of a carpel (or fused carpels) that captures pollen grains and favors their germination.

stimulus [L. *stimulus*, goad] A specific form of energy (e.g., pressure, light, and heat) that activates a sensory receptor able to detect it.

stoma (STOW-muh), plural **stomata** [Gk. *stoma*, mouth] One of many gaps between two guard cells in leaf and stem epidermis. Opens and closes to control inward movement of carbon dioxide and outward movement of water vapor and oxygen, depending on whether conditions in the environment call for water conservation.

stomach A muscular, stretchable sac that mixes and stores ingested food, helps break it apart mechanically and chemically, and controls its expulsion (e.g., into the small intestine).

stop codon Base triplet in a strand of mRNA that serves as a STOP signal during translation; it blocks further additions of amino acids to a newly forming polypeptide chain.

strain One of two organisms with differences that are too minor to classify it as a separate species (e.g., *Escherichia coli* strain 018:K1:H).

stratification Stacked layers of sedimentary rock that resulted from a gradual deposition of volcanic ash, silt, and other materials over time.

stream A flowing-water ecosystem that starts out as a freshwater spring or seep.

stroma [Gk. *strōma*, bed] A semifluid matrix between the thylakoid membrane system and the two outer membranes of a chloroplast; a zone where sucrose, starch, cellulose, and other end products of photosynthesis are assembled.

stromatolite Fossilized mats of shallow-water, microbial communities (mainly cyanobacteria) of Archean to Precambrian times. Their tacky gel secretions blocked out ultraviolet radiation but trapped sediments; so new mats had to grow over older ones, like cake layers. Stromatolites are found on all continents; some are over a half mile thick and hundreds of miles across.

substrate Reactant or precursor for a specific enzyme-mediated metabolic reaction.

substrate-level phosphorylation The direct, enzyme-mediated transfer of a phosphate group from a substrate to a molecule, as when an intermediate of glycolysis gives up a phosphate group to ADP to form ATP.

succession, primary (suk-SESH-un) [L. *succedere*, to follow after] Ecological pattern by which a community develops in orderly progression, from the time pioneer species colonize a barren habitat to the climax community.

succession, secondary Ecological pattern by which a disturbed area of a community recovers and moves back toward the climax state. It is typical of abandoned croplands, forest burns, and storm-battered intertidal zones.

surface-to-volume ratio Mathematical relation in which volume increases with the cube of the diameter, but surface area increases only with the square. If a growing cell were simply to expand in diameter, its volume of cytoplasm would increase faster than the surface area of the plasma membrane required to service it. In general, this constraint keeps cells small, elongated, or with infoldings or outfoldings of its plasma membrane.

survivorship curve Plot of age-specific survival of a group of individuals in the environment, from the time of birth until the last one dies.

swim bladder Adjustable flotation device that changes in volume as it exchanges gases with blood; it helps many fishes maintain neutral buoyancy in water.

symbiosis (sim-by-OH-sis) [Gk. *sym*, together, + *bios*, life, mode of life] Individuals of one species live near, in, or on those of another species for at least part of life cycle (e.g., in commensalism, mutualism, and parasitism).

sympathetic nerve An autonomic nerve that deals mainly with increasing overall body activities at times of heightened awareness, excitement, or danger; also works continually in opposition with parasympathetic nerves to make minor adjustments in internal organ activities.

sympatric speciation [Gk. *sym*, together, + *patria*, native land] A speciation event within the home range of an existing species, in the absence of a physical barrier. Such species may form instantaneously, as by polyploidy.

synaptic integration (sin-AP-tik) Moment-by-moment combining of all excitatory and inhibitory signals arriving at the trigger zone of a neuron or some other excitable cell.

syndrome A set of symptoms that may not individually be a telling clue but collectively characterize a genetic disorder or disease.

systemic circuit (sis-TEM-ik) Of vertebrates, a cardiovascular route in which oxygen-enriched blood flows from the heart through the rest of the body (where it gives up oxygen and takes up carbon dioxide), then back to the heart.

T cell T lymphocyte; a type of white blood cell vital to immune responses (e.g., helper T cells and cytotoxic T cells).

taproot system A primary root together with all of its lateral branchings.

target cell Any cell with molecular receptors that can bind with a particular hormone or some other signaling molecule.

taxonomy Field of biology that deals with identifying, naming, and classifying species.

tectum Midbrain's roof. Coordinating center for most sensory inputs and initiating motor responses in fishes and amphibians. In most vertebrates (not mammals), a reflex center that relays sensory input to forebrain.

telophase (TEE-low-faze) Of meiosis I, the stage when one member of each pair of homologous chromosomes has arrived at a spindle pole. Of mitosis and of meiosis II, the stage when chromosomes decondense into threadlike structures and two daughter nuclei form.

temperature A measure of the kinetic energy of ions or molecules in a specified region.

tendon A cord or strap of dense connective tissue that attaches a muscle to bone.

territory An area that an animal is defending against competitors for mates, food, water, living space, other resources.

test A means to determine the accuracy of a prediction, as by conducting experimental or observational tests and by developing models. Scientific tests are made under controlled conditions in nature or the laboratory.

testcross Experimental cross to determine whether an individual of unknown genotype that shows dominance for a trait is either homozygous dominant or heterozygous.

testis, plural **testes** A primary reproductive organ (gonad) of some male animals; it produces male gametes and sex hormones.

testosterone (tess-TOSS-tuh-rown) A type of sex hormone necessary for the development and functioning of the male reproductive system of vertebrates.

tetanus (TET-uh-nuss) Of a muscle, a large contraction in which repeated stimulation of a motor unit causes muscle twitches to mechanically run together. In a disease by the same name, a toxin prevents muscles from being released from contraction.

thalamus (THAL-uh-muss) A forebrain region that is a coordinating center for sensory input and a relay station for signals to the cerebrum.

theory, scientific A testable explanation of a broad range of related phenomena, one that has been subjected to extensive experimental testing and can be used with a high degree of confidence. A scientific theory remains open to tests, revision, and tentative acceptance or rejection.

thermal inversion The trapping of a layer of dense, cool air beneath a layer of warm air. The inversion can result in an accumulation of air pollutants close to the ground.

thermophile A type of archaebacterium that is adapted to unusually hot aquatic habitats, such as hot springs and hydrothermal vents.

thermoreceptor Sensory cell or specialized cell next to it that detects radiant energy (heat).

thigmotropism (thig-MOE-truh-pizm) [Gk. *thigm*, touch] An orientation of the direction of growth in response to physical contact with a solid object, as when a vine curls around a fencepost.

threshold Of an excitable cell (e.g., a neuron or muscle cell), the minimum amount of change in the resting membrane potential that will trigger an action potential.

thylakoid Of chloroplasts, part of an internal membrane system folded repeatedly into a stack of disks. Such stacks (grana) have light-absorbing pigments and enzymes required to form ATP, NADPH, or both in photosynthesis. The stacks connect (by membranous channels) as a single functional compartment.

thymine Nitrogen-containing base; one of the nucleotides in DNA.

thymus gland A lymphoid organ that has endocrine functions. Lymphocytes of the immune system multiply, differentiate, and mature in its tissues; its hormone secretions affect their functioning.

thyroid gland Endocrine gland located in front of the trachea; its hormones have widespread effects on growth and development, and on the overall metabolic rates of warm-blooded animals.

tissue Of multicelled organisms, a group of cells and intercellular substances that function together in one or more specialized tasks.

tonicity (TOE-niss-ih-TEE) Relative solute concentrations of two fluids (e.g., cytoplasmic fluid relative to extracellular fluid).

touch-killing Mechanism by which cytotoxic T cells directly release perforins and toxins onto a target cell and cause its destruction.

toxin A normal metabolic product of one species with chemical effects that can hurt or kill individuals of a different species.

trace element Any element that represents less than 0.01 percent of body weight.

tracer Substance with a radioisotope attached, like a shipping label, that researchers can track after delivering it into a cell, body, ecosystem, or some other system. Laboratory devices detect emissions from the tracer as it moves through a pathway or reaches a destination.

trachea (TRAY-kee-uh), plural **tracheae** An air-conducting tube of respiratory systems. Of land vertebrates, the windpipe through which air passes between the larynx and bronchi.

tracheal respiration Of certain invertebrates (e.g., insects), respiration by way of finely branching tracheae that start at openings in the integument and dead-end in body tissues.

tracheid (TRAY-kid) One of two types of cells in xylem that conduct water and minerals.

tract A cordlike bundle of axons of sensory neurons, motor neurons, or both inside the brain or spinal cord. Comparable to a nerve.

transcription [L. *trans*, across, + *scribere*, to write] First stage of protein synthesis. An RNA strand is assembled on exposed bases of one unwound strand of a DNA double helix. The transcript's base sequence is complementary to that of the DNA template.

transfer RNA (tRNA) An RNA that binds with and delivers amino acids to a ribosome and that pairs with an mRNA codon during the translation stage of protein synthesis.

translation Stage of protein synthesis when an mRNA's base sequence becomes converted to a sequence of particular amino acids in a new polypeptide chain. rRNA, tRNA, and mRNA interact to bring this about.

translocation Of cells, a stretch of DNA that moved to a new location in a chromosome or in a different chromosome, with no molecular loss. Of vascular plants, a process by which organic compounds are distributed through the phloem.

transpiration Evaporative water loss from a plant's aboveground parts, leaves especially.

transport protein One of many kinds of membrane proteins involved in active or passive transport of water-soluble substances across the lipid bilayer of a cell membrane. Solutes on one side of the membrane pass through the protein's interior to the other side.

transposable element A stretch of DNA that can move at random from one location to another in the individual's genome. Often it inactivates the genes into which it becomes inserted and causes changes in phenotype.

triglyceride (neutral fat) A type of lipid that has three fatty acid tails attached to a glycerol backbone. Triglycerides are the body's most abundant lipids and its richest energy source.

trisomy (TRY-so-mee) The presence of three chromosomes of a given type in a cell rather than the two characteristic of a parental diploid chromosome number.

trophic level (TROE-fik) [Gk. *trophos*, feeder] Of an ecosystem, all organisms that are the same number of transfer steps away from the energy input into the system.

tropical rain forest A biome characterized by regular, heavy rainfall, an annual mean temperature of 25°C, humidity of 80 percent or more, and stunning biodiversity. Presently being obliterated in regions with fast-growing human populations but limited food, fuel, and lumber; projections are that most will disappear by 2035.

tropism (TROE-pizm) Of plants, a directional growth response to an environmental factor (e.g., growth toward light).

true breeding Of a sexually reproducing species, a lineage in which one version only of a trait shows up through the generations in all parents and their offspring.

tumor Tissue mass composed of cells that are dividing at an abnormally high rate. If benign, its cells remain in their home tissue; if malignant, they have metastasized.

turgor pressure (TUR-gore) [L. *turgere*, to swell] Internal fluid pressure on a cell wall when water moves into the cell by osmosis.

ultrafiltration Bulk flow of a small amount of protein-free plasma from a blood capillary when the outward-directed force of blood pressure is greater than the inward-directed osmotic force of interstitial fluid.

uniformity theory Early theory that the Earth's surface changes in gradual, uniformly repetitive ways (major floods, earthquakes, and other infrequent annual catastrophes were not considered unusual). Helped change Darwin's view of evolution. Has since been replaced by plate tectonics theory.

upwelling An upward movement of deep, nutrient-rich water along coasts; it replaces surface waters that move away from shore when the prevailing wind direction shifts.

uracil (YUR-uh-sill) Nitrogen-containing base of a nucleotide in RNA but not DNA. Like thymine, uracil can base-pair with adenine.

ureter One of a pair of tubes that conduct urine from the kidneys to the urinary bladder.

urethra Tube that conducts urine from the urinary bladder to an opening at the body's surface.

urinary bladder The distensible sac in which urine is stored before being excreted.

urinary excretion Mechanism by which excess water and solutes are removed from the body through the urinary system.

urinary system Organ system that adjusts the volume and composition of blood, and thereby helps maintain extracellular fluid.

urine Fluid consisting of any excess water, wastes, and solutes; it forms in kidneys by filtration, reabsorption, and secretion.

uterus (YOU-tur-us) [L. *uterus*, womb] Of a female placental mammal, a muscular, pear-shaped organ in which embryos are contained and nurtured during pregnancy.

vaccination Immunization procedure against a specific pathogen.

vaccine An antigen-containing preparation, swallowed or injected, designed to increase immunity to certain diseases by inducing formation of armies of effector and memory B and T cells.

vagina The part of the reproductive system of mammalian females that receives sperm, forms part of the birth canal, and channels menstrual flow to the exterior.

variable Of an experimental test, a specific aspect of an object or event that may differ over time and among individuals. A single variable is directly manipulated in an attempt to support or disprove a prediction; any other variables that might influence the results are identical (ideally) in both the experimental group and one or more control groups.

vascular bundle An array of primary xylem and phloem in multistranded, sheathed cords that thread lengthwise throughout the ground tissue system.

vascular cambium A lateral meristem that increases stem or root diameter.

vascular cylinder Arrangement of vascular tissues as a central cylinder in roots.

vascular plant Plant with xylem and phloem, and usually with well-developed roots, stems, and leaves.

vascular tissue system Xylem and phloem, the conducting tissues that distribute water and solutes through a vascular plant.

vein Of a cardiovascular system, any of the large-diameter vessels that lead back to the heart. Of leaves, one of the vascular bundles that thread through photosynthetic tissues.

venule A small blood vessel that serves as a transitional conducting tube between a small-diameter capillary and a larger diameter vein.

vernalization Environmental stimulation of flowering, by exposure to low temperatures.

vertebra, plural **vertebrae** One of a series of hard bones, with cushioning intervertebral disks stacked between them, that serve as a backbone and that protect the spinal cord.

vertebrate Animal with a backbone.

vesicle (VESS-ih-kul) [L. *vesicula*, little bladder] One of a variety of small, membrane-bound sacs in the cytoplasm that function in the transport, storage, or digestion of substances.

vessel member Type of cell in xylem; dead at maturity, but its wall becomes part of a water-conducting pipeline (a vessel).

vestigial (ves-TIDJ-ul) Applies to a small body part, tissue, or organ abnormally developed or degenerated and unable to function like its normal counterpart (e.g., vestigial wings of mutant fruit flies; "tail bones" of humans).

villus (VIL-us), plural **villi** Any of several fingerlike absorptive structures projecting from the free surface of an epithelium.

viroid An infectious particle consisting only of very short, tightly folded strands or circles of RNA. Viroids might have evolved from introns, which they resemble.

virus A noncellular infectious agent that is composed of DNA or RNA and a protein coat; it can become replicated only after its genetic material enters a host cell and subverts the host's metabolic machinery.

viscera All soft organs inside an animal body (e.g., heart, lungs, and stomach).

vision Perception of visual stimuli. Requires a focusing of light precisely onto a layer of photoreceptive cells that is dense enough to sample details of a light stimulus, followed by image formation in a brain.

visual signal An observable action or cue that functions as a communication signal.

vitamin Any of more than a dozen organic substances that an organism requires in small amounts for metabolism but that it generally cannot synthesize for itself.

vocal cord Of certain animals, one of the thickened, muscular folds of the larynx that help produce sound waves for vocalization.

water potential Sum of two opposing forces (osmosis and turgor pressure) that can cause a directional movement of water into or out of an enclosed volume (e.g., a walled cell).

water table Upper limit at which the ground in a specified land region is fully saturated with water.

watershed Any specified region in which all precipitation drains into one stream or river.

wavelength A wavelike form of energy in motion. The horizontal distance between the crests of every two successive waves.

wax Molecule having long-chain fatty acids packed together and linked to long-chain alcohols or to carbon rings. Waxes have a firm consistency and repel water.

white blood cell An eosinophil, neutrophil, macrophage, T or B cell, or other leukocyte which, with chemical mediators, counters tissue invasion and tissue damage. Different kinds take part in nonspecific and specific defense responses. Some (e.g., eosinophils) are fast-acting. Some (e.g., macrophages) take part in sustained immune responses.

white matter The portion of the brain and spinal cord with axons that have glistening white sheaths and that specialize in rapid signal transmission.

wild-type allele Allele that occurs normally or with the greatest frequency at a given gene locus among individuals of a population.

wing A body part that serves in flight, as among birds, bats, and many insects. A bird wing is a forelimb having feathers, strong muscles, and extremely lightweight bones. A bat wing is a modified forelimb; four thin, elongated digits are the framework for a thin integumentary membrane. An insect wing develops as a lateral fold of the exoskeleton.

X chromosome A type of sex chromosome. In mammals, an XX pairing causes an embryo to develop into a female; an XY pairing causes it to develop into a male.

X-linked gene Any gene that is located on an X chromosome.

X-linked recessive inheritance Recessive condition in which the responsible, mutated gene is located on the X chromosome.

xylem (ZYE-lum) [Gk. *xylon*, wood] A tissue having pipelines that conduct water and solutes through vascular plants. The pipelines are the interconnecting walls of cells that are dead at maturity.

Y chromosome Distinctive chromosome in males or females of many species, but not both (e.g., human males are XY and human females, XX).

yellow marrow A fatty tissue in the cavities of most mature bones that produces red blood cells when blood loss from the body is severe.

Y-linked gene Gene on a Y chromosome.

yolk sac An extraembryonic membrane. In most shelled eggs, it holds nutritive yolk; in humans, part of the yolk sac becomes a site of blood cell formation and some cells give rise to forerunners of gametes.

zero population growth A state in which the number of births in a population is balanced by the number of deaths over a specified period, assuming that immigration and emigration are balanced, also.

zooplankton A community of suspended or weak-swimming heterotrophs of freshwater or marine habitats. Most of its species are microscopic; commonly, rotifers and copepods are among the most abundant.

zygote (ZYE-goat) The first cell of a new individual, formed by fusion of a sperm nucleus with egg nucleus at fertilization; a fertilized egg.

CREDITS AND ACKNOWLEDGMENTS

This page is an extension of the copyright page. We thank the following authors, agents, and publishers for permission to use the material listed. We also thank graphics artists for the final rendering of original illustrations developed by Cecie Starr and, more recently, Lisa Starr, in collaboration with general advisers, researchers, and content reviewers.

TABLE OF CONTENTS *vi* (left) Gary Head, (right) model by Dr. David B. Goodin, The Scripps Research Institute / *vii* Larry West/FPG / *viii* Dr. Pascale Madaule, France / *ix* (left) From "Multicolor Spectral Karyotyping of Human Chromosomes," by E. Schrock, T. Ried, et al, *Science*, 26 July 1966, 273:495, (right) William E. Ferguson / *x* (left) Dr. Douglas Coleman, The Jackson Library, (right) Christopher Ralling / *xi* Jeff Hester and Paul Scowen, Arizona State University, and NASA, (right) Painting, Charles Knight (No. 2425), Courtesy Dept. of Library Services/American Museum of Natural History / *xii* (left to right) K. G. Murty/Visuals Unlimited; David M. Phillips/Visuals Unlimited; © Sherry K. Pittam; Edward S. Ross / *xiii* (left) Courtesy of Department of Library Services/American Museum of Natural History (Neg. K10273), (right) Gregory Dimijian/Photo Researchers / *xiv* (left to right) Kenneth Garrett/National Geographic Image Collection; Tom Till/Tony Stone Images, Inc.; Gary Head; David Lavagnaero /Peter Arnold, Inc.; David M. Phillips/Visuals Unlimited; John Lotter Gurling/Tom Stack and Associates / *xv* David Macdonald / *xvi* (left to right) Eric A. Newman; Lennart Nilsson © Boehringer Ingelheim International GmBH; Kenneth Garrett/National Geographic Image Collection; National Osteoporosis Foundation / *xvii* (left) From D. Waller *Physiology: The Servant of Medicine*, Hitchcock Lectures, University of London Press, 1910; (right) NSIBC/SPL/Photo Researchers / *xviii* (left to right) Stanley Flegler/ Visuals Unlimited; Lisa Starr; Ed Reschke; Gunter Ziesler/Bruce Coleman, Ltd. / *xix* (left) John H. Gerard; From Lennart Nilsson, *A Child Is Born*, © 1966, 1977 Dell Publishing Company, Inc. / *xx* (left) Stanley Sessions/Hartwick College; (right) Pieter Johnson / *xxi* Alan and Sandy Carey

CHAPTER 1 **1.1** Frank Kaczmarek / **1.2** Art, Lisa Starr / **1.3** Jack de Coningh / **1.4** © Y. Arthrus-Bertrand/ Peter Arnold, Inc. / **1.5** Art, Precision Graphics; photographs (above) Paul DeGreve/ FPG; (below) Norman Meyers/Bruce Coleman / **1.6** Art , Preface, Inc.; photographs (far left) Walt Anderson/Visuals Unlimited, (b,c) Gregory Dimijian/Photo Researchers / **1.7** (a) R. Robinson /Visuals Unlimited, (b) Tony Brain/SPL/Photo Researchers, (c) M. Abbey/Visuals Unlimited, (d), Edward S. Ross, (e) Gary Head, (f) Edward S. Ross, (g,h) Pat and Tom Leeson/Photo Researchers / **1.8** J. A. Bishop, L. M. Cook / **1.9** Courtesy Derrell Fowler, Tecumseh, OK / **1.10** Art, Preface, Inc. / Gary Head / **1.12** Art, Preface, Inc.; micrograph, CNRI/SPL/Photo Researchers, photographs Dr. Douglas Coleman, The Jackson Laboratory / **1.13** James Carmichael, Jr./ NHPA

PAGE 19 James M. Bell/Photo Researchers

CHAPTER 2 **2.1** Gary Head for Norman Terry, University of California, Berkeley / **2.2** Jack Carey / **2.3** Preface, Inc. / **2.4** Thyroid scans, Hank Morgan, Rainbow, computer-enhanced by Lisa Starr; photograph, Gary Head / **2.5** (a) Art, Raychel Ciemma; (b) Dr. Harry T. Chgani, M.D., UCLA School of Medicine / **2.6** Micrograph, Maris and Cramer / **2.7** Art, Preface, Inc., (b) Lisa Starr / **2.8** Art, Preface, Inc. / **2.9** Lisa Starr / **2.10** (a) Art, Preface, Inc., (b) Art, Lisa Starr; micrograph © Bruce Iverson / **2.11** Vandystadt/ Photo Researchers / **2.13** Art, Palay/ Beaubois / **2.13** Art, Raychel Ciemma; photograph © Kennan Ward/The Stock Market / **2.14** Art, Raychel Ciemma, photograph © Kennan Ward/The Stock Market / **2.15** (a) H. Eisenbeiss/Frank Lane Picture Agency; (b) Lisa Starr / **2.16** Lisa Starr / **2.17** Lisa Starr, using photographs © 2000 Photo Disc, Inc. / **2.18** Michael Grecco/Picture Group

CHAPTER 3 **3.1** Dave Schiefelbein / **p. 35** NASA / **page 36** Art , Precision Graphics and Lisa Starr / **3.3** Art, Lisa Starr; photograph, Tim Davis/Photo Researchers / **3.4 Lisa Starr / 3.5** Lisa Starr / **3.6** Art and photograph, Lisa Starr /**3.7** Lisa Starr /**3.8** David Scharf/Peter Arnold, Inc. / **3.9** Art, Precision Graphics / **3.10** Clem Haagner/Ardea, London / 3.11 (a,b) Art, Precision Graphics; (c) micrograph, Lewis L. Lainey; (d) Larry Lefever/ Grant Heilman; (e) Kenneth Lorenzen /**page 41** Lisa Starr / **3.12** (a) Ron Davis/Shooting Star; (b) Frank Trapper/Sygma / **3.13–3.16** Art, Precision Graphics / **3.17** Lisa Starr / **3.18** Art, Palay/ Beaubois / **3.19** (a) CNRI/SPL/Photo Researchers; (b) R. Demarest; (c) Lisa Starr / **3.20** (a) © Inga Spence/Tom Stack & Associates; inset photograph, Gary Head; (b–d) Art, Gary Head and Lisa Starr / (e) Sue Hartzell / **3.21** Art, Gary Head **/** Art, Precision Graphics

CHAPTER 4 **4.1** (a) Corbis/Bettmann; (b) Armed Forces Institute of Pathology; (c) National Library of Medicine; (e) Francis A. Countway Library of Medicine /**4.2** (a) Manfred Kage/Bruce Coleman Ltd., (b) George J. Wilder/Visuals Unlimited / **4.3** (a) Art, Raychel Ciemma, (b) George S. Ellmore, (c) R. Demarest / **4.4** (a,c) Art, Raychel Ciemma; (b) George S. Ellmore / **4.5** (a) Leica Microsystems, Inc., Deerfield, IL.; (b, d) Art, Gary Head; (c) © George Musil/Visuals Unlimited / **4.6** Art, Raychel Ciemma; micrograph, Driscoll, Youngquist, and Baldeschwieler/Cal Tech/ Science Source/Photo Researchers / **4.7** (a–d) Jeremy Pickett–Heaps, School of Botany, University of Melbourne / **4.8** Art, Raychel Ciemma and Precision Graphics / **4.9** Art, Raychel Ciemma and Precision Graphics; migrograph, M. C. Ledbetter, Brookhaven National Laboratory / **4.10** Art, Raychel Ciemma and Precision Graphics; micrograph G.L. Decker / **4.11** Stephen Wolfe / **4.12** (a, left) Don W. Fawcett/Visuals Unlimited, (a, right) A. C. Faberge, *Cell and Tissue Research*, 151: 403–415, 1974; (b) Lisa Starr / **4.13** Art, Raychel Ciemma and Precision Graphics / **4.14** Art, Raychel Ciemma; micrographs (a,b) Don W. Fawcett/Visuals Unlimited / **4.15** R. Demarest after a model by J. Kephart; micrograph, Gary Grimes / **4.16** Art, R. Demarest; micrograph, Keith R. Porter / **4.17** Art, Lisa Starr; micrograph, L. K. Shumway / **4.18** J. Victor Small, Gottfried Rinnerthaler / **4.19** Lisa Starr / **4.20** (a) © Lennart Nilsson; (b) CNRI/SPL /Photo Researchers/ **4.21** Art, Precision Graphics after Stephen Wolfe, *Molecular and Cellular Biology*, Wadsworth, 1993 / **4.22** Art, Precision Graphics / **4.23** Ronald Hoham, Dept. of Biology, Colgate University / **4.24** (a) Walter Hodge/ Peter Arnold; (c) Biophoto Associates/ Photo Researchers; art, Raychel Ciemma / **4.25** (a) George S. Ellmore; art, Raychel Ciemma; (b) Ed Reschke; art, Joel Ito / **4.26** Art, Raychel Ciemma / **4.27** (a) Lisa Starr; (b) Courtesy Dr. G. Cohen- Bazire; (c) K. G. Murti/Visuals Unlimited; (d) R. Calentine/ Visuals Unlimited; (e) Gary Gaard, Arthur Kelman / **4.28** Abraham Menashe

CHAPTER 5 **5.1** (a) Runk/Schoenberger/Grant Heilman; (b) Ingio Everson/Bruce Coleman Ltd. / **5.2** (a, b) Art, Precision Graphics; (c) art, Raychel Ciemma / **5.3** Art, Raychel Ciemma / **5.4** Art, Precision Graphics after Stephen Wolfe, *Molecular and Cellular Biology*, Wadsworth; photograph Charles D. Winters/Photo Researchers / **5.5** Art, Palay/Beubois; micrograph, P. Pinto da Silva and D. Branton, *Journal of Cell Biology*, 45:98, by permission of The Rockefeller University Press / **5.6** Art, Preface, Inc.; **5.7** Art, Precision Graphics / **5.8–5.10** Art, Raychel Ciemma / **5.11**, **5.12** Lisa Starr / **5.13** Art, Palay/Beaubois / **5.14** Art, Raychel Ciemma; micrographs, M. Sheetz, R. Painter, and S. Singer, *Journal of Cell Biology*, 70:193 (1976) by permission of The Rockefeller University Press / **5.15** Lisa Starr / **5.16–5.17** Art, Raychel Ciemma / **5.18** Art, Gary Head, micrographs, M. M. Perry and A. M. Gilbert / **5.19** M. Abbey/Visuals Unlimited / **5.20** Art, Raychel Ciemma /**5.22** Frieder Sauer/Bruce Coleman Ltd.

CHAPTER 6 **6.1** (a, above) © Oxford Scientific Films/Animals Animals, (a, below) © Raymond Mendez/Animals Animals; (c) Keith V. Wood / **6.2** C. Contag, *Molecular Microbiology*, November 1985, 18(4): 593–603. "Photonic Detection of Bacterial Pathogens in Living Hosts." Reprinted by permission of Blackwell Science / **6.3** Evan Cerasoli / **6.4** (above) NASA 97; (below) Manfred Kage/Peter Arnold, Inc. / **6.5** (a–d) Lisa Starr, using photographs © 1997, 1998 Jet Propulsion Laboratory, California Institute of Technology and NASA, 1972 / **6.6** Art, Gary Head / **6.7** Art, Nadine Sokol; photographs, NHPA/A.N.T. Photo Library / **6.8** Art, Raychel Ciemma and Gary Head / **6.9** Art, Preface Inc. / **6.10** Lisa Starr / **6.11** Lisa Starr / **6.12** Art, Precision Graphics / **6.13** Art, Gary Head / 6.14 Art, Precision Graphics / **6.15** Lisa Starr, using photographs © 1997, 1998 Jet Propulsion Laboratory, California Institute of Technology and NASA, 1972 / **6.16** (a,b) Thomas A. Steitz; (c) Lisa Starr / **6.17** (a) Art, Precision Graphics, (b) Douglas Faulkner/Sally Faulkner Collection / 6.18 Art , Precision Graphics / **6.19**, **6.20** Lisa Starr / **6.21** (a, b) Models by Dr. David B. Goodin, The Scripps Research Institute (c, d) © Gary Head / **page 111** Art, Gary Head

CHAPTER 7 **7.1** Art, Raychel Ciemma; algal micrograph, Carolina Biological Supply Company / **7.2** Art, Lisa Starr; photograph PhotoVault / **7.3** Lisa Starr / **pages 114, 115,** Lisa Starr / **7.4** Art, Gary Head and Lisa Starr with photograph from David Neal Parks; (b,c) Art, Precision Graphics; photograph Barker-Blakenship/FPG / **7.5** Lisa Starr after Stephen L. Wolfe, *Molecular and Cellular Biology*, Wadsworth / **7.6** Art, Precision Graphics / **7.7** Art, Precision Graphics / **7.8** Larry West/FPG / **7.9** Lisa Starr **/** 7.10, Lisa Starr / 7.11 Lisa Starr / 7.12 Lisa Starr / **7.13** E. R. Degginger / **7.14** Lisa Starr / **7.15** Lisa Starr / **7.16** Art, Raychel Ciemma and Precision Graphics; photographs (a) Grant Heilman, Inc, (b) Dick Davis/Photo Researchers; (c) (left) B. J. Miller, Fairfax, VA/BPS; (right) Martin Grosnick/Ardea London / **page 125** Steve Chamberlain, Syracuse University / **7.17** NASA / **7.18–7.20** Lisa Starr

CHAPTER 8 **8.1** Stephen Dalton/Photo Researchers / **8.2** Art, Preface, Inc. / **8.3** (a) Gary Head, (b) Paolo Fioratti, (c) Art, Preface, Inc. / **page 134** Lisa Starr / **8.4** (left) Art, Raychel Ciemma; (right) Lisa Starr after Ralph Taggart / **8.5** Art, Lisa Starr; micrograph, Keith R. Porter / **8.6** Lisa Starr / **8.7** Lisa Starr **/ 8.8,** 8.9, Lisa Starr / **8.10** (a) Lisa Starr; (b) Adrian Warren/Ardea London; (c) David M. Phillips/Visuals Unlimited / **8.11** William Grenfell/Visuals Unlimited / **8.12** Art, Precision Graphics; photograph, Gary Head / **page 144** R. Llewellyn/Superstock, Inc.

PAGE 147 © Lennart Nilsson

CHAPTER 9 **9.1** (left) Chris Huss, (right, above) Tony Dawson, (right, center) Chris Huss, (right, below) Tony Dawson / **9.2** Art, Raychel Ciemma; micrograph, C. J. Harrison, *Cytogenetics and Cell Genetics*, 35:21–27, © 1983, S. Karger, and A. G. Basel / **9.4** Micrographs, Andew S. Bajer, University of Oregon / **9.5** Art, Raychel Ciemma; photographs © Ed Reschke / **9.6** Art, Lisa Starr; micrographs (left) © R. Calentine/ Visuals Unlimited, (right) B. A. Palevita and E. H. Newcomb, University of Wisconsin/BPS/ Tom Stack & Associates / **9.7** Art, Raychel Ciemma, micrograph © D. M. Phillips/Visuals Unlimited / **9.8** (a–c, e) Lennart Nilsson from *A Child Is Born* © 1966, 1967 Dell Publishing Company, Inc.; (d) Lennart Nilsson from *Behold Man*, © 1974 by Albert Bonniers Forlag and Little, Brown & Company, Boston / **9.9** Art, Lisa Starr and Raychel Ciemma; micrographs (a) C. J. Harrison et al *Cytogenetics and Cell Genetics*, 35: 21–27 © 1983 S. Karger, A. G. Basel; (b) B. Hamkalo; (c) O. L. Miller, Jr., Steve L. McKnight / **9.9** Art, Preface, Inc. / **9.11** Photograph, Courtesy of The Family of Henrietta Lacks; micrograph, Dr. Pascal Madaule, France

CHAPTER 10 **10.1** (a) Jane Burton/Bruce Coleman Ltd.; (b) Dan Kline/Visuals Unlimited / **10.2** Art, Raychel Ciemma / **page 163** Art, Raychel Ciemma / **10.3** CNRI/SPL/Photo Researchers / **10.4** Art, Raychel Ciemma / **10.5** (c–f) Lisa Starr / **10.6** Art, Raychel Ciemma / 10.7 Art, Preface, Inc.; photograph David Maitland/Seaphot Limited / **10.8** Lisa Starr / **10.9** Art, Lisa Starr; micrograph, David M. Phillips/Visuals Unlimited / **10.10** Art, Raychel Ciemma / **10.11** Art, Precision Graphics / **10.12** © Richard Corman/Outline

CHAPTER 11 **11.1** (left) Frank Trapper/Sygma; (center) Focus on Sports; (right, above) The Moravian Museum, Brno; (right, below) Fabian/Sygma / **11.2** Art, Jennifer Wardrip; photograph Jean M. Labat/Ardea, London / **11.3** Lisa Starr / **11.5–11.7** Art, Hans & Cassady, Inc. / **11.8, 11.9** Art, Raychel Ciemma and Precision Graphic / **11.10** William E. Ferguson / **11.11, 11.12** Art, Preface, Inc.; micrographs, Stanley Flegler/Visuals Unlimited / **11.13** Art, Preface, Inc.; photographs (a, b) Michael Stuckey / Comstock, Inc., (c) Bosco Broyer, by Gary Head / **11.14** David Hosking / **11.15** Tedd Somes / **11.16** (above to below) Frank Cezus, Frank Cezus, Michael Keller, Ted Beaudin, Stan Sholik; all FPG / **11.17** (a) Art, Precision Graphics; photograph Daniel Fairbanks, Brigham Young University / **11.18** Art, D & V. Hennings; photograph Jane Burton/Bruce Coleman Ltd / **11.19** (left) William E. Ferguson; (right) Eric Crichton/Bruce Coleman Ltd. / **11.20** Evan Cerasoli / **11.21** Leslie Faltheisek/ Clacritter Manx / **11.22** Joe McDonald/Visuals Unlimited

CHAPTER 12 **12.1** From "Multicolor Spectral Karyotyping of Human Chromosomes," by E. Schrock, T. Ried, et al, *Science*, 26, July 1966, 273:495. Used by permission of E. Schrock and T. Reid and the American Association for the Advancement of Science; computer-enhanced by Lisa Starr / **12.2** Paul A. Thiessen / **12.3** Art, Raychel Ciemma; photograph, Charles D. Winters/Photo Researchers; micrograph, Omikron/Photo Researchers / **12.4** Art, Precision Graphics; photographs (left) Gary Head; (right) Lisa Starr / **12.5** (a) From Lennart Nilsson, *A Child Is Born*, © 1966, 1977 Dell Publishing Company, Inc.; (b) R. Demarest after Patten, Carlson & others; (c) R. Demarest with permission from M. Cummings, *Human Heredity: Principles and Issues*, p. 126 of third ed. © 1994 by Brooks/Cole. All rights reserved. / **12.6** Art, Raychel Ciemma and Preface Inc.; photograph, Carolina Biological Supply Company / **12.7-12.9** Art, Raychel Ciemma and Preface, Inc. / **12.10** Art, Precision Graphics / **12.11** (a,b) Art, Precision Graphics; photograph Dr. Victor A. McKusick; (c) Steve Uzzell / **12.12** Lisa Starr / **12.13** (a) Lisa Starr; (b) Giraudon/Art Resource, New York / **12.14** (a) Lisa Starr, (b) after V. A. McKusick, *Human Genetics*, 2nd Ed., © 1969. Reprinted by permission of Prentice-Hall, Inc., Englewood Cliffs, NJ; (b) Corbis-Bettmann / **12.15** C. J. Harrison / **12.16** Eddie Adams/AP Photo / **12.17** (a, b) Courtesy G. H. Valentine / **12.18** From "Multicolor Spectral Karyotyping of Human Chromosomes," by E. Schrock, T. Ried, et al, *Science*, 26, July 1966, 273:496. Used by permission of E. Schrock and T. Reid and the American Association for the Advancement of Science / **pages 206, 207** Art, Precision Graphics / **2.19** Art, Raychel Ciemma / **12.20** Micrograph, Cytogenetics Laboratory, The University of California, San Francisco; and photograph courtesy of Peninsula Association for Retarded Children and Adults, San Mateo Special Olympics, Burlingame, CA; (b) After Collman and Stoller, *American Journal of Public Health*, 52, 1962, photograph, by permission of Carole Lafrate / **12.21** Art, Raychel Ciemma / **12.22** Art, Lisa Starr; photograph courtesy of Lennart Nilsson from *A Child Is Born*, © 1966, 1977 Dell Publishing Company, Inc. / **12.23** (a) Fran Heyl Associates © Jacques Cohen, computer enhanced by © Pix Elation; (b) Art, Raychel Ciemma / **12.24** Art, Raychel Ciemma / **12.25** Carolina Biological Supply Company / **12.26** Bonnie Kamin/Stuart Kenter Associates / **12.27, 12.28** Art, Precision Graphics

CHAPTER 13 **13.1** A C. Barrington Brown © 1968 J. D. Watson / **13.2** Model by Dr. David B. Goodin, The Scripps Research Institute / **13.3**–13.5 Art, Raychel Ciemma / **13.4** (c) Lee D. Simon/Science Source/Photo Researchers / **13.6** Art rendered by Preface, Inc.; micrograph Biophoto Associates/SPL/Photo Researchers / **13.7** Lisa Starr, incorporating a model by Dr. David B. Goodin, The Scripps Research Institute / **13.8** Courtesy of the Cold Springs Harbor Laboratory Archives / **13.9, 13.10** Art, Precision Graphics / **13.11** (a) From Daniel Fairbanks; (b) PA News Photo Library

CHAPTER 14 **14.1** (above) Dennis Hallinan/ FPG; (below) © Bob Evan/Peter Arnold, Inc. / **14.3** Gary Head / **14.4** From Stephen Wolfe, *Molecular and Cellular Biology*, Wadsworth / **14.5** Art (a) Palay/Beaubois and (b,c) Precision Graphics / **14.8** (a) Art, Palay/Beaubois, (b–e) Lisa Starr / **14.9** Art, Preface, Inc. / **14.10, 14.11**, Art, Precision Graphics / **14.12** (a) tRNA model by Dr. David B. Goodin, The Scripps Research Institute; (b,c) Lisa Starr / **14.13** Lisa Starr / **13.14** Lisa Starr / **14.15, 14.16** Lisa Starr / **14.17** (left) Nik Kleinberg; (right) Peter Starlinger / **14.18** and **p. 238** Art, Raychel Ciemma and Lisa Starr / **14.19** Courtesy of the National Neurofibromatosis Foundation

CHAPTER 15 **15.1** (a) Ken Greer/Visuals Unlimited; (b) Biophoto Associates/Science Source/Photo Researchers; (c) James Stevenson /SPL/Photo Researchers; (d) Gary Head / **page 240** Lisa Starr / **15.2** Lennart Nilsson © Boehringer Ingelheim International GmbH / **15.3** Brian Matthews, University of Oregon / **15.4** Art, Palay/Beaubois and Hans & Cassady / **15.5** Art, Raychel Ciemma and Lisa Starr / **15.6** Art, Gary Head; micrograph, M. Roth and J. Gall / **15.7** Art, Palay/Beaubois / **15.8** (a) Art, Raychel Ciemma; micrograph, Dr. Karen Dyer Montgomery / **15.9** Jack Carey / **15.10** Art, Raychel Ciemma; micrograph, W. Beerman / **15.11** Art, Raychel Ciemma; photograph Frank B. Salisbury / **15.12** Art, Betsy Palay/Atemis / **15.13** Slim Films

CHAPTER 16 **16.1** Lewis L. Lainey / **16.2** Science VU/Visuals Unlimited / **16.3** (a) Dr. Huntington Potter and Dr. David Dressler; (b) Stanley N. Cohen/Science Source/Photo Researchers / **page 256**, art, Preface, Inc. / **16.4** Lisa Starr, photograph © 2000 PhotoDisc, Inc. / **16.5, 16.6** Art, Preface, Inc. / **16.7** (above) Damon Biotech, Inc.; (below) Cellmark Diagnostics, Abingdon, UK. / **16.8** Lisa Starr / **16.9** Lisa Starr / **16.10** Stephen Wolfe, *Molecular and Cellular Biology* / **16.11** Art, Lisa Starr; photograph, Herve Chaumeton, Agence Nature / **16.12** Keith V. Wood / **16.13** (a) Dr. Vincent Chiang, School of Forestry and Wood Products, Michigan Technology University; (b) Courtesy Calgene LLC / **16.14** R. Brinster, R. E. Hammer, School of Veterinary Medicine, University of Pennsylvania / **16.15** © Adrian Arbib/Still Pictures

PAGE 269 S. Stammers/SPL/Photo Researchers

CHAPTER 17 **17.1** Lisa Starr, with use of NASA satellite photographs / **17.2** Lisa Starr, with use of NASA satellite photographs / **17.3** (a) Jen & Des Bartlett/Bruce Coleman Ltd.; (b) Kenneth W. Fink/Photo Researchers; (c) Dave Watts/A.N.T. Photo Library / **17.4** Art, Raychel Ciemma / **17.5** (a) Courtesy George P. Darwin, Darwin Museum, Down House; (b) Heather Angel, computer-enhanced by Lisa Starr; (c) Christopher Ralling; (d) Art, L Morgan; (e) Dieter & Mary Plage/ Survival Anglia / **17.6** (a) © Philip Boyer/Photo Researchers; (b) Field Museum of Natural History, Chicago, and the artist, Charles R. Knight (Neg. CK21T) / **17.7** (a) Heather Angel; (b) David Cavagnaro; (c) Dr. P. Evans/Bruce Coleman, Ltd.; (d) Alan Root/Bruce Coleman Ltd. / **17.8** Courtesy of Down House and The Royal College of Surgeons of England / **17.9** Art, Raychel Ciemma; photograph, P. Morris/Ardea London

CHAPTER 18 **18.1** Elliot Erwitt/Magnum Photos, Inc. / **page 282** Photographs courtesy Derrell Fowler, Tecumseh, OK / **18.2** Alan Solem / **18.3** Art, Raychel Ciemma / **18.4** Art, Precision Graphics / **18.5** J. A. Bishop and L. M. Cook / **18.6** Art, Precision Graphics / **18.7** (a, b) Warren Abrahamson; (c) Forest W. Buchanan/Visuals Unlimited; (d) Kenneth McCrea and Warren Abrahamson / **18.8** Art , Precision Graphics / **18.9** (a, b) Thomas Bates Smith; (c) Art, Preface, Inc. / **18.10** Bruce Beehler / **18.11** Art, Precision Graphics after Ayala and others / **18.12** (above) David Neal Parks; (below) W. Carter Johnson / **18.13** Art, Precision Graphics after computer models developed by Jerry Coyne / **18.14** Art, Raychel Ciemma; photograph, David Cavagnaro / **18.15** Kjell Sandved/Visuals Unlimited / **18.16** Art, Precision Graphics using NIH data / **18.17** D. Avon/Ardea London

CHAPTER 19 **19.1** (a) © Photograph Gary Head; Snail courtesy of Larry Reed; (b) Adapted from R. K. Selander and D. W. Kaufman, *Evolution*, 29(3), December 31, 1975 / **19.2** Art, Jennifer Wardrip / **19.3** Art, Raychel Ciemma after C. T. Regan and E. Trewavas, 1932 / **19.4** Sound spectrograms, Aaron Bauer; photograph S. L. Collins and J. T. Collins; (a,b) *The Amphibians and Reptiles of Missouri*, by T. R. Johnson © 1987 by Conservation Commission, State of Missouri. Reprinted by permission / **19.5** After F. Ayala and J. Valentine, *Evolving*, Benjamin-Cummings, 1979 / **19.6** (a) Dr. Eugene Kozloff, (b) Mike Vasey, San Francisco State University / **19.7** Alvin E. Staffan/Photo Researchers / **19.8** G. Ziesler/ ZEFA / **19.9** John Alcock, Arizona State

University / **19.10** (a) Fred McConnaughey/ Photo Researchers; (b) Patrice Geisel/Visuals Unlimited; (right) Tom Van Sant/The Geosphere Project, Santa Monica, CA / **19.11** Art, Preface, Inc. / **19.12** Art, Raychel Ciemma / **19.12** Art, Precision Graphics based on U. Schliewen, et al., *Nature*, 368:629–632, 14 April 1994; (right) Tom Van Sant/The Geosphere Project, Santa Monica, CACA**19.14** Lisa Starr, after W. Jensen and F. B. Salisbury, *Botany: An Ecological Approach*, Wadsworth, 1972, and others / **19.15** Art, Preface, Inc.; photographs (left) © H. Clarke, VIREO/Academy of Natural Sciences; (right) Robert C. Simpson/Nature Stock / **19.16** Art, Precision Graphics / **19.17** After P. Dodson, *Evolution: Process and Products*, 3rd Ed., PWS; photographs (above) Jack Dermid, (below) Leonard Lee Rue III/FPG / **19.18** Jen & Des Bartlett/Bruce Coleman Ltd.

CHAPTER 20 **20.1** (left) Vatican Museums; (right) Martin Dohrn/SPL/Photo Researchers / **20.2** (a) Donald Baird, Princeton Museum of Natural History; (b) A. Feduccia, *The Age of Birds*, Harvard University Press, 1980; (c) H. P. Banks; (d) Jonathan Blair / **20.2** Art, Preface Inc. **20.4** / Art, Raychel Ciemma / **20.5** (above) Douglas P. Wilson/Eric and David Hosking; (center) Superstock, Inc.; (below) E. R. Degginger / **20.6** Art after E. Guerrant, *Evolution*, 36:699–712; photographs, Gary Head / **20.7** (b) From T. Storer et al., *General Zoology*, sixth ed., McGraw-Hill, 1979, reproduced by permission of McGraw -Hill, Inc. / **20.8** Art, Raychel Ciemma / **20.9** Art, Precision Graphics / **20.10** (left) Kjell B. Sandved/Visuals Unlimited; (center) Thomas D. Mangelsen/Images of Nature; (right) Jeffrey Sylvester/FPG / **20.11** (left to right) Larry Lefever/Grant Heilman, Inc.; R.I.M. Campbell/ Bruce Coleman, Ltd.; Runk and Schoenberger/ Grant Heilman, Inc.; Bruce Coleman, Ltd. / 20.12 Art, ACG / **20.13** Photographs, page 322 (left) C. B. & D. W. Frith/Bruce Coleman, Ltd., (right) W. A. Banaszewski/ Visuals Unlimited; page 323 (upper left © John Gurche 1989, (upper right) Hans Reinhard/ ZEFA; (center) Peter Scoones/ Seaphot Limited: Planet Earth Pictures; (below) Jane Burton/Bruce Coleman, Ltd. / **20.14** D. & V. Hennings / **20.15, 20.16** Art, ACG / **20.17** Art , Preface, Inc. / **20.19** Sandra Floyd, University of Colorado / **20.20** David Noble/FPG / **20.22** L. Morgan / **20.23** Lisa Starr after A. M. Ziegler, C. R. Scotese, and S. F. Barrett, "Mesozoic and Cenozoic Paleogeographic Maps," and J. Krohn and J. Sundermann (Eds.), *Tidal Friction and the Earth's Rotation II*, Springer–Verlag, 1983; photograph, Martin Land/Photo Researchers/ **20.24** (a) Kingsley R. Stern, (b) Chip Clark / **20.25** Art, Preface, Inc.

PAGE 333 © 1990 Arthur M. Greene

CHAPTER 21 **21.1** Jeff Hester and Paul Scowen, Arizona State University, and NASA / **21.2** William K. Hartmann / **21.3** (a) Chesley Bonestell, (b) Art, Raychel Ciemma / **21.4** Art, ACG / **21.5** (a) Sidney W. Fox; (b) W. Hargreaves and D. Deamer / **21.6** Art, ACG / **21.7** Art,y ACG and Raychel Ciemma / **21.8** Bill Bachman/Photo Researchers / **21.9** (a) Stanley M. Awramik; (b–f) Andrew H. Knoll, Harvard University / **21.10** P. L. Walne and J. H. Arnott, *Planta*, 77: 325–354, 1967 / **21.11** Art, Raychel Ciemma / **21.12** Robert K. Trench / **pages 344, 345** Maps, Lisa Starr, after Ziegler et al., *Tidal Friction and the Earth's Rotation II*, Springer–Verlag, 1983 / **21.13** (a, b) Neville Pledge/South Australian Museum; (c, d) Chip Clark; / **21.14** (a, b, d) Art, Raychel Ciemma; (c) Patricia G. Gensel / **page 346** Map, Lisa Starr after Ziegler et al., *Tidal Friction and the Earth's Rotation II*, Springer–Verlag, 1983 / **21.15** Painting by Megan Rohn courtesy of David Dilcher; art, Precision Graphics / **21.16** © John Gurche 1989 / **21.17** NASA Galileo Imaging Team / **21.18** Art, Raychel Ciemma / **page 349** Map, Lisa Starr after Ziegler et al., *Tidal Friction and the Earth's Rotation II*, Springer–Verlag, 1983 / **21.19** (a, b) Paintings © Ely Kish; (c) Field Museum of Natural History, Chicago, and artist Charles R. Knight (Neg. CK8T) / **21.20** (left) Maps, Lisa Starr after Ziegler et al., *Tidal Friction and the Earth's Rotation II*, Springer–Verlag, 1983; graph, Precision Graphics

CHAPTER 22 **22.1** (a–c) Tony Brain and David Parker/SPL/Photo Researchers / **22.2** Lee D. Simon/Photo Researchers / **22.3** Lisa Starr / **22.4** (a) Stanley Flegler/Visuals Unlimited; (b) P. Hawtin, University of Southhampton/SPL/ Photo Researchers; (c) CNRI/SPL/Photo Researchers / **22.5** Courtesy of K. Amako and A. Umeda, Kyshu University, Japan / **22.6** L. J. LeBeau, University of Illinois Hospital/BPS / **22.7** Art, Raychel Ciemma; micrograph L. Santo. / **22.8** Art, Raychel Ciemma / **22.10** (a) R. Robinson/Visuals Unlimited; (b) Jack Jones, *Archives of Microbiology*, 136: 254–261, 1983. Reprinted by permission of Springer–Verlag / **22.11** (a) © 2000 PhotoDisc, Inc., computer-enhanced by Lisa Starr; (b) Barry Rokeach; (c) NASA; (d) © Alan L. Detrick, Science Source/ Photo Researchers, Inc. / **22.12** (a) John D. Cunningham/Visuals Unlimited; (b) Tony Brain/SPL/Photo Researchers; (c) P. W. Johnson and J. MeN. Sieburth, Univ. Rhode Island/BPS / **22.13** T. J. Beveridge, University of Guelph/BPS / **22.14** © Ken Greer/Visuals Unlimited / **22.15** (a) Richard Blakemore; (b) Hans Reichenbach, Gesellschaft fur Biotechnologische Forschung, Braunschweig, Germany / **22.16** Art,, Raychel Ciemma after Stephen L. Wolfe and others / **22.17** (a) K. G. Murti/Visuals Unlimited; (b) George Musil/ Visuals Unlimited; (c, d) Kenneth M. Corbett / **22.18** Art, Palay/ Beaubois / **22.19** Art, Raychel Ciemma / **22.20** R. Regnery, Centers for Disease Control and Prevention, Atlanta / **22.21** E. A. Zottola, University of Minnesota

CHAPTER 23 **23.1** (a) Steven C. Wilson/Entheos; (b) Ronald W. Hoham, Dept. of Biology, Colgate University; (c) Edward S. Ross; (d) Gary W. Grimes and Steven L'Hernault / **23.2** (a) M. S. Fuller, *Zoosporic Fungi in Teaching and Research*, M. S. Fuller and A. Jaworski (eds.), 1987, Southeastern Publishing Company, Athens, Georgia; (b) Claude Taylor and University of Wisconsin Dept. of Botany; (c) Heather Angel; (d) W. Merrill / **23.3** (a) Art, Leonard Morgan; (b) M. Claviez, G. Gerish, and R. Guggenheim; (c) London Scientific Films; (d–f) Carolina Biological Supply Company; (g) courtesy Robert R. Kay from R. R. Kay, et al., *Development*, 1989 Supplement, pp. 81–90, © The Company of Biologists Ltd., 1989. / **23.4** M. Abbey/Visuals Unlimited / **23.5** (a) Neal Ericsson; (b) John Clegg/Ardea, London; (c) Courtesy Allan W. H. Bé and David A. Caron; (d) Dr. Howard J. Spero; (e, g) redrawn from V. & J. Pearse and M. & R. Buchsbaum, *Living Invertebrates*, The Boxwood Press, 1987. Used by permission; (f) G. Shih and R. Kessel/Visuals Unlimited; (h) T. E. Adams/ Visuals Unlimited / **23.6** (a) Micrograph Frieder Sauer/Bruce Coleman Ltd.; (b,c,e) Art redrawn from V. & J. Pearse and M. & R. Buchsbaum, *Living Invertebrates*, The Boxwood Press, 1987. Used by permission; (d) micrograph Sidney L. Tamm; (e) micrograph Ralph Buchsbaum / **23.7** Art, Raychel Ciemma after Eugene Kozloff / **23.8** Dr. Stan Erlandsen, University of Minnesota / **23.9** (a) © Oliver Meckes/Photo Researchers, Inc.; (b) Lisa Starr, after Prescott, *Microbiology*, third edition / **23.10** David M. Phillips/Visuals Unlimited / **23.11** Courtesy Saul Tzipori, Div. of Infectious Diseases, Tufts University School of Veterinary Medicine / **23.12** © A. M. Siegelman/ Visuals Unlimited / **23.13** Lauren and Homer, © Gary Head / **23.14** Art, Leonard Morgan; micrograph, Steven L'Hernault / **23.15** Art, Palay/Beaubois / **23.16** (a, b) Ronald W. Hoham, Dept. of Biology, Colgate University; (c) Greta Fryxell, University of Texas, Austin; (d) *Emiliania huxleyi* photograph by Vita Pariente; scanning electron micrograph taken on a Jeol T330A instrument at the Texas A&M University Electron Microscopy Center / **23.17** (a) S. Carty, Heidelberg College; (b, left) © S. Berry/Visuals Unlimited; (b right, and c, d) North Carolina State University, Aquatic Botany Lab / **23.18** (a) D. P. Wilson/Eric & David Hosking; (b) Douglas Faulkner/Sally Faulkner Collection / **23.19** Art, Raychel Ciemma / **23.20** (a) J. R. Waaland, University of Washington/ BPS; (b) Lewis Trusty/Animals Animals; art from T. Garrison, *Oceanography: An Invitation to Marine Science*, Brooks/Cole, 1993 / **23.21** (a) Herve Chaumeton /Agence Nature; (b) © Linda Sims/Visuals Unlimited; (c) Brian Parker/Tom Stack and Associates; (d, e) D. S. Littler; (f) Manfred Kage/ Peter Arnold, Inc.; (g, h) Ronald W. Hoham, Dept. Biology, Colgate University / **23.22** Art, Raychel Ciemma; photograph D. J. Patterson /Seaphot Limited: Planet Earth Pictures / **23.23** Carolina Biological Supply Company

CHAPTER 24 **24.1** © Sherry K. Pittam / **24.2** Robert C. Simpson/Nature Stock / **24.3** Robert C. Simpson/Nature Stock / **24.4** Art, Raychel Ciemma; micrograph Garry T. Cole, University of Texas, Austin/BPS / **24.5** (a) Jane Burton/ Bruce Coleman, Ltd.; (b) Thomas J. Duffy / **page 396** Micrograph J. D. Cunningham/ Visuals Unlimited / **24.6** Art, Raychel Ciemma, micrographs Ed Reschke / **24.7** (a) After T. Rost, et al., *Botany*, Wiley, 1979; (b) M. Eichelberger/ Visuals Unlimited; (c) Robert C. Simpson/ Nature Stock; (d) G. L. Barron, University of Guelph; (e) Garry T. Cole, University of Texas, Austin/BPS / **24.8** N. Allin and G. L. Barron / **24.9** (a) After Raven, Evert, and Eichhorn, *Biology of Plants*, 4th Ed., Worth Publishers, New York, 1986; (b) © Mark E. Gibson/Visuals Unlimited; (c) Gary Head, (d) Mark Mattock/ Planet Earth Pictures; (e) Edward S. Ross / **24.10** (a) Prof. D. J. Read, University of Sheffield; (b) © 1990 Gary Braasch; (c) F. B. Reeves / **24.11** (a) Dr. P. Marazzi/SPL/Photo Researchers; (b) Eric Crichton/Bruce Coleman, Ltd. / **24.12** John Hodgin

CHAPTER 25 **25.1** (a) Pat & Tom Leeson/Photo Researchers; (b, c) Edward S. Ross / **25.2** Art, Raychel Ciemma after E. O. Dodson and P. Dodson, *Evolution Process and Product*, third edition., p. 401, PWS / **25.3** Art, Precision Graphics / **25.4** (a) Craig Wood/ Visuals Unlimited; (b) Jane Burton/Bruce Coleman Ltd.; (c) art, Raychel Ciemma / **25.5** (a) Fred Bavendam/Peter Arnold; (b) John D. Cunningham/Visuals Unlimited / **25.6** (a, b) Kingsley R. Stern; (c) John D. Cunningham/ Visuals Unlimited / **25.7** (b) Kingsley R. Stern; (c) Ed Reschke/Peter Arnold; (d) William Ferguson; (e) W. H. Hodge; (f) Kratz/ZEFA / **25.8** Art, Raychel Ciemma; photographs (inset) A. & E. Bomford, Ardea, London, (below) Lee Casebere / **25.9** Art, Raychel Ciemma; (above) Brian Parker/Tom Stack & Associates; (below) Field Museum of Natural History, Chicago (Neg. 7500C) / **25.10** (a) Kathleen B. Pigg, Arizona State University; (b) George J. Wilder/Visuals Unlimited / **25.11** (a) Stan Elams/Visuals Unlimited; (b) Gary Head / **25.12** (a) Jeff Gnass Photography; (b) R. J. Erwin/Photo Researchers;

(c) Robert & Linda Mitchell; (d) Doug Sokell/ Visuals Unlimited / **25.13** (a) Robert & Linda Mitchell; (b) Ed Reschke / **25.14** (a) Joyce Photographics/Photo Researchers; (b) Runk/ Schoenberger/Grant Heilman, Inc.; (c) William Ferguson; (d) Kingsley R. Stern / **25.15** (a) Edward S. Ross; (b) William Ferguson; (c) Robert and Linda Mitchell Photography; (d) William Ferguson; (e) Biological Photo Service / **25.16** Art, Raychel Ciemma; photographs (left) Edward S. Ross; (above) Robert & Linda Mitchell Photography; (below) R. J. Erwin/ Photo Researchers / **25.17** (a) David Hiser, Photographers/Aspen, Inc.; (b) Terry Livingstone; (c) © 1993 Trygve Steen; (d) © 1994 Robert Glenn Ketchum / **25.18** (a) Art, Jennifer Wardrip; (b) M. P. L. Fogden/Bruce Coleman Ltd.; (c) Heather Angel; (d) L. Mellichamp/Visuals Unlimited; (e) Peter F. Zika/Visuals Unlimited / **25.19** Art, Raychel Ciemma / **25.20** (a) John Mason/Ardea, London; (b) Earl Roberge/Photo Researchers; (c) ZEFA-Rein; (d) George Loun/Visuals Unlimited; (e) Dick Davis/Photo Researchers; engravings after B. Simpson and M. Conner-Ogorzaly, *Economic Botany*, second edition, McGraw-Hill / **25.21** Bob Cerasoli / **25.22** (left, center) © 1989, 1991 Clinton Webb; (right) Gary Head

CHAPTER 26 **26.1** Courtesy of Department of Library Services, American Museum of Natural History (Neg. K10273) / **26.2** Art, Gary Head; photograph, Lisa Starr / **26.3** Art, L. Morgan / **26.4** Art, Raychel Ciemma / **26.5** (left) After Laszlo Meszoly in L. Margulis, *Early Life*, Jones and Bartlett, 1982 / **26.6** Bruce Hall / **26.7** (a–c) Art by Raychel Ciemma; (d, left) Art, Raychel Ciemma after Bayer and Owre, *The Free-Living Lower Invertebrates*, © 1968 Macmillan; (d, left) Preface, Inc., (d, right) Don W. Fawcett/ Visuals Unlimited; (e) Marty Snyderman/Planet Earth Pictures / **26.8** (a, b) Art, Raychel Ciemma; (c) F. Schensky; (d) Courtesy of Dr. William H. Hammer; (e) Kim Taylor/Bruce Coleman, Ltd. / **26.9** Art, Raychel Ciemma after Eugene Kozloff / **26.10** (a) Art, Raychel Ciemma, (b) Douglas Faulkner/ Sally Faulkner Collection; (c) F. S. Westmorland/Tom Stack & Associates / **26.11** Art, Precision Graphics after T. Storer et al., *General Zoology*, sixth Ed., © 1979 McGraw-Hill / **26.12** (a) Christian DellaCorte; (b) Douglas Faulkner/Photo Researchers; (c) Bill Wood/ Seaphot Limited: Planet Earth Pictures / **26.13** Andrew Mounter/Seaphot Limited: Planet Earth Pictures / **26.14** (a) Art, Preface, Inc.; (b) Kathie Atkinson/Oxford Scientific Films / **26.15** Art, Raychel Ciemma; photograph Kim Taylor/Bruce Coleman, Ltd. / **26.16** (a) Robert and Linda Mitchell Photography; (b) Cath Ellis, University of Hull/ SPL/Photo Researchers / **26.17** Kjell B. Sandved / **26.18, 26.19** Art, Raychel Ciemma / **26.20** Art, Raychel Ciemma; photograph Carolina Biological Supply Company / **26.21** (a) Lorus J. and Margery Milne; (b) Dianora Niccolini / **26.22** Art, Raychel Ciemma / **page 436** Art, Palay/Beaubois / **26.23** (a) Gary Head; (b) Alex Kerstitch; (c) Jeff Foott/Tom Stack & Associates; (d) Frank Park/A.N.T. Photo Library; (e) J. Grossauer/ ZEFA / **26.24** (a) Art, Palay/Beaubois, (b) Kjell B. Sandved, (c) Herve Chaumeton/ Agence Nature / **26.25** (a) Art, Raychel Ciemma, (b) Herve Chaumeton/Agence Nature / **26.26** (a) Art, Raychel Ciemma; (b) Bob Cranston; (c) A. Kerstitch, © 1992 Sea World of California. All rights reserved / **6.27** J. A. L. Cooke/Oxford Scientific Films / **26.28** (a) © Cabisco/Visuals Unlimited; (b) Jon Kenfield/ Bruce Coleman Ltd.; (c) after Eugene Kozloff, *Invertebrates* © 1990 by Saunders College Publishing. Reproduced by permission of the publisher; (d) adapted from Rasmussen, "Ophelia," Vol. 11 in Eugene Kozloff, *Invertebrates*, 1990 / **26.29** Art, Raychel Ciemma / **26.30** Jane Burton/Bruce Coleman, Ltd. / **26.31** (a) Jane Burton/Bruce Coleman, Ltd.; (b) Angelo Giampiccolo/FPG / **26.32** (a) P. J. Bryant, University of California, Irvine/BPS; (b) Redrawn from *Living Invertebrates*, V. & J. Pearse/M. &. R. Buchsbaum, The Boxwood Press, 1987. Used by permission / **26.33** (a) Herve Chaumeton; (b) Franz Lanting/Bruce Coleman, Ltd.; (c) Fred Bavendam/Peter Arnold, Inc.; (d) Agence Nature; Art, Raychel Ciemma / **26.34** After D. H. Milne, *Marine Life and the Sea*, Wadsworth, 1995 / **26.35** (a) Steve Martin/Tom Stack & Associates; (b) Z. Leszczynski/AnimalsAnimals / **26.36** Art, Raychel Ciemma / **26.37** From G. Pasteur, "Jean Henri Fabre," *Scientific American*, July 1994. © 1994 by *Scientific American*. All rights reserved / **26.38** (a,b,c,j) Edward S. Ross; (d) David Maitland/ Seaphot Limited: Planet Earth Pictures; (e,f,g,h) Edward S. Ross; (i) C. P. Hickman, Jr. / **26.39** (a) John H. Gerard; (b) Ken Lucas/Seaphot Limited: Planet Earth Pictures / **26.40** (a–d) Courtesy Pfizer Central Research; (e) Edward S. Ross, (f) Art, Gary Head / **26.41** B. Backes, Arizona State University / **26.42** (a) Marlin E. Rice, Iowa State University; (b,c) John Obermeyer, Purdue University, Dept. of Entomology / **26.43** (a) Chris Huss/The Wildlife Collection; (b) John Mason/Ardea, London; (c) Kjell B. Sandved; (d) Ian Took/Biofotos / **26.44** (a,b) Art, L. Calver; (c,d) Herve Chaumeton/ Agence Nature / **page 451** Inset, Jane Burton/ Bruce Coleman, Ltd. / **26.45** Art, Raychel Ciemma / **26.46** Walter Deas/Seaphot Limited: Planet Earth Pictures / **26.47** (a) J. Solliday/BPS; (b) Herve Chaumeton/Agence Nature

CHAPTER 27 **27.1** (a) Jean Phillipe Varin/Jacana/ Photo Researchers; (b) Tom McHugh/Photo Researchers / **27.2** Art, Gary Head / **27.3** (a) Rick M. Harbo; (b,c) Redrawn from *Living Invertebrates*, V. & J. Pearse and M. & R. Buchsbaum, The Boxwood Press, 1987. Used by permission / **27.4** (a) Art, Raychel Ciemma; (b) Runk & Schoenberger/Grant Heilman, Inc. / **27.7** Art. Preface, Inc. / **27.6** (a–c) Adapted from A. S. Romer and T. S. Parsons, *The Vertebrate Body*, sixth ed., Saunders College Publishing, 1986; (d) Al Giddings/Images Unlimited / **27.7** (a, b) After D. H. Milne, *Marine Life and the Sea*, Wadsworth, 1995; (c) Heather Angel /**27.8** (a) Erwin Christian /ZEFA; (b) Allan Power/Bruce Coleman, Ltd.; (c) Tom McHugh/Photo Researchers / **27.9** (a) Bill Wood/Bruce Coleman, Ltd.; (b) Art, Raychel Ciemma; (c) Robert and Linda Mitchell Photography ; (d) Patrice Ceisel/© 1986 John G. Shedd Aquarium; (e) © Norbert Wu/Peter Arnold, Inc.; (f) From T. Garrison, *Oceanography: An Invitation to Marine Science*, Wadsworth, 1993 / **27.10** After C. P. Hickman, et al., *Integrated Principles of Zoology*, 6th Ed., St. Louis: C. V. Mosby Co., 1979 / **27.11** (a) © Marianne Collins; (b,c) Art, Laszlo Meszoly and D. & V. Hennings / **27.12** (a) Jerry W. Nagel; (b) Stephen Dalton/ Photo Researchers; (c) John Serraro/Visuals Unlimited; (d) Juan M. Renjifo/Animals Animals / **page 463** Adapted from A. S. Romer and T. S. Parsons, *The Vertebrate Body*, sixth ed., Saunders College Publishing, 1986 / **27.13** (a) Art, Raychel Ciemma; (b) © 1989 D. Braginetz; (c) Z. Leszczynski/Animals Animals / **27.14** Art, D. & V. Hennings / **27.15** (a) Kevin Schafer/Tom Stack & Associates; (b) D. Kaleth/Image Bank; (c) art, Raychel Ciemma; (d) Andrew Dennis/ A.N.T. Photo Library; (e) art, Raychel Ciemma, photograph, Stephen Dalton/Photo Researchers; (f) Heather Angel / **27.16** (a) Lisa Starr; (b) Gerard Lacz/A.N.T. Photo Library; (c) Rajesh Bedi; (d) Thomas D. Mangelsen/Images of Nature; (e) J. L. G. Grande/Bruce Coleman, Ltd. / **27.17** (b) Art, Raychel Ciemma / **27.18** (a) Sandy Roessler/FPG; (b) Leonard Lee Rue III/FPG; (c) Art, Raychel Ciemma after M. Weiss and A. Mann, *Human Biology and Behavior*, 5th Ed., Harper Collins, 1990 / **27.19** Lisa Starr / **27.20** (a) D. & V. Blagden/ A.N.T. Photo Library; (b) courtesy The Zoological Society of San Diego, The San Diego Zoo; (c) S. Johnson / **27.21** Art, Raychel Ciemma / **27.22** (a) Clem Haagner/Ardea, London; (b) J. Scott Altenbach, University of New Mexico; (c) Douglas Faulkner/Photo Researchers; (d) Christopher Crowley; (e) Leonard Lee Rue III/FPG / **27.23** (a) Courtesy of Dr. Takeshi Furuichi, Biology, Meiji–Gakuin University-Yokohama; (b) Bruce Coleman, Ltd; (c) Tom McHugh/Photo Researchers; (d) Larry Burrows/Aspect Picture Library / **27.24** and page 475 Art, D. & V. Hennings / **27.25** Art, Precision Graphics / **27.26** © Time, Inc. 1965/ Larry Burrows Collection / **27.28** Art, Precision Graphics after *National Geographic*, February 1997, page 82 / **27.29** (a) Dr. Donald Johanson, Institute of Human Origins; (b) Louise M. Robbins; (c,d) Kenneth Garrett/National Geographic Image Collection / **27.30** Lisa Starr / **27.31** (a, b) Kenneth Garrett/ National Geographic Image Collection; (c) Jean Paul Tibbles / **27.32** (a–d) John Reader © 1981 / **27.33** Lisa Starr / **27.34** Art, Raychel Ciemma / **27.35** Art, Preface, Inc. / **27.36** Lisa Starr, with photograph from NASA / **27.37** Lisa Starr / **27.38** W. J. Weber/Visuals Unlimited

CHAPTER 28 **28.1** Tom Till/Tony Stone Images, Inc. / **28.2** Lisa Starr / **28.3** (a) Painting by Charles Knight, Department of Libary Services/ American Museum of Natural History (2425); (b) Mansell Collection/Time, Inc. / **28.4** Lisa Starr / **28.5** (a) Courtesy of Dan Stanford; (b) Lisa Starr, based on 1940–1984 reports from International Whaling Commission, (c) Steve Hillebrand, U.S. Fish & Wildlife Service / **page 490**, Douglas Faulkner/ Sally Faulkner Collection / **28.6** (left inset) Sea Studios/Peter Arnold, Inc.; (above) Jim Doran; (center, left to right) Jeff Rotman, Alex Kerstitch, Douglas Faulkner/Sally Faulkner Collection (below, and right insets) Douglas Faulkner/Sally Faulkner Collection / **28.7** Eric Hartmann, Magnum Photos / **28.8** Lisa Starr / **28.9** Lisa Starr, with photographs © 2000 PhotoDisc, Inc. / **28.10** Bureau of Land Management / **28.11** Lisa Starr

PAGE 497 © Gary Head

CHAPTER 29 **29.1** (a) R. Barrick/USGS; (b,c) © 1980 Gary Braasch; (d) Don Johnson/Photo Nats, Inc. / **29.2–29.4** Art, Raychel Ciemma / **29.5** James D. Mauseth, *Plant Anatomy*, Benjamin-Cummings, 1988 / **29.6** Biophoto Associates / **29.7** (a) Jan Robert Factor/Photo Researchers; (b) D. E. Akin and I. L. Risgby, Richard B. Russel Agricultural Research Center, Agricultural Research Service, U.S. Department of Agriculture, Athens, Georgia; (c) Kingsley R. Stern / **29.8** Art, Raychel Ciemma / **29.9** George S. Ellmore / **29.10** Art, D. & V. Hennings / **29.11** Art, Raychel Ciemma / **29.12** (a) Robert & Linda Mitchell Photography; (b) Roland R. Dute; (c) Gary Head / **29.13** Art, D. & V. Hennings; (a, left) Ray F. Evert; (a, right) James W. Perry; (b, left) Carolina Biological Supply Company; (b, right) James W. Perry / **29.14** Art, D. & V. Hennings / **29.15** (a) Heather Angel; (b) Gary Head / **29.16** (a) Art, Raychel Ciemma; (b) C. E. Jeffree, et al., *Planta*, 172(1):20–37, 1987. Reprinted by permission of C. E. Jeffree and Springer–Verlag; (c) Jeremy Burgess/SPL/Photo Researchers / **29.17** Art, Raychel Ciemma / **29.18** (a) Art after Salisbury and Ross, *Plant Physiology*, fourth edition, Wadsworth; (b) John Limbaugh/Ripon Microslides, Inc. / **29.19** (a) Micrographs Chuck Brown; (b) Carolina Biological Supply Company / **29.20** John Limbaugh/Ripon Microslides; sketch after T. Rost et al., *Botany: A Brief Introduction to Plant Biology*, second edition., © 1984, John Wiley & Sons / **29.21** (a) John Lotter Gurling/Tom Stack and Associates; (b) Art (left), Rost et al, *Plant Biology*, Wadsworth, (right) Raychel Ciemma; inset photograph, Biophot / **29.22** Art, Raychel

Ciemma / **29. 23** Art, ACG and Gary Head / **29.24** Lisa Starr / **29.25** H. A. Core, W. A. Cote, and A. C. Day, *Wood Structure and Identification,* second ed., Syracuse University Press, 1979 / **29.26** Art (a) Lisa Starr; (b,c) Raychel Ciemma / **29.27** Edward S. Ross / **page 515** Art, Lisa Starr, photograph © Stephen J. Krasemann/National Audubon Society Collection/Photo Researchers

CHAPTER 30 **30.1** (a,b) Robert & Linda Mitchell Photography; (c) © John N. A. Lott, *Scanning Electron Microscope Study of Green Plants,* St. Louis: C. V. Mosby Company, 1976; (d) Robert C. Simpson/Nature Stock / **30.2** David Lavagnaero /Peter Arnold, Inc. / **30.3** William Ferguson / **30.4** U.S. Dept. of Agriculture / **30.5** (a,c,d) Art, Leonard Morgan, micrograph Chuck Brown / **30.6** Micrograph Jean Paul Reve / **30.7** (a) Art, Jennifer Wardrip, (b) Mark E. Dudley and Sharon R. Long; (c) Adrian P. Davies/Bruce Coleman, Ltd.; (d) NifTAL Project, Univ. of Hawaii, Maui; Art by Jennifer Wardrip / **30.8** (a, left) Micrograph W. A. Cote of N. C. Brown Center for Ultra-Structure Studies, computer enhanced by Lisa Starr ; all other micrographs H. A. Core, W. A. Cote and A. C. Day, *Wood Structure and Identification,* 2nd Ed., Syracuse University Press, 1979 / **30.9** Art, Raychel Ciemma / **30.10** Frank B. Salisbury / **30.11** George S. Ellmore / **30.12** W. Thomson, *American Journal of Botany,* 57(3):316, 1970 / **30.13** Art, Palay/Beaubois / **30.14** T. A. Mansfield / **30.15** (a, b) Scanning electron micrographs, Jeremy Burgess/SPL/Photo Researchers; photograph, John Lawlor/FPG / **30.16** Art, Hans & Cassady, Inc. / **30.17** Martin Zimmerman, *Science,* 1961, 133:73–79, © AAAS / **30.18** Art, Palay/Beaubois / **30.20** James T. Brock

CHAPTER 31 **31.1** Courtesy Merlin D. Tuttle, Bat Conservation International / **31.2** (a) Robert A. Tyrrell; (b,c) Thomas Eisner, Cornell University / **31.3** Art, Raychel Ciemma and Precision Graphics; photographs © Gary Head / **31.4** John Shaw/Bruce Coleman, Ltd. / **31.5** (a,b) David M. Phillips/Visuals Unlimited; (c) David Scharf/ Peter Arnold, Inc. / **31.6** Art, Raychel Ciemma / **31.7** (a,b) Patricia Schulz; (c,d) Ray F. Evert; (e,f) John Limbaugh/Ripon Microslides, Inc., (g) Art, Raychel Ciemma / **31.8** (a) Gary Head, (b) B. J. Miller, Fairfax, VA/BPS; (c) Janet Jones; (e) Richard H. Gross / **31.9** (a) R. Carr; (b) Gary Head; (c) ZEFA-Rein / **31.10** John Alcock / **31.11** Russell Kaye, © 1993 The Walt Disney Co. Reprinted with permission of *Discover Magazine.* / **31.12** (a) Runk/Schoenberger/Grant Heilman, Inc.; (b) Kingsley R. Stern / **31.13** Gary Head

CHAPTER 32 **32.1** (a) Michael A. Keller/FPG; (b,c) R. Lyons/Visuals Unlimited / **32.2** Sylvan H. Wittwer/Visuals Unlimited / **32.3** From B. Bracegirdle and P. Miles, *An Atlas of Plant Structure,* Heinemann Educational Books, 1977 / **32.4** (a, b) Art, Raychel Ciemma (c) Herve Chaumeton/Agence Nature / **32.5** Art, Raychel Ciemma; (c) Barry L. Runk/Grant Heilman, Inc.; (d) from Mauseth / **32.6** Art, Raychel Ciemma / **32.7** Art, Raychel Ciemma; micrograph, Biophot / **32.8** (a) Kingsley R. Stern, (b) D & V Hennings / **32.9** (a,b) Micrographs courtesy Randy Moore from "How Roots Respond to Gravity," M. L. Evans, R. Moore, and K. Hasenstein, *Scientific American,* December 1986; (c) John Digby and Richard Firn / **32.10** Frank B. Salisbury / **32.11** Gary Head / **32.12** Cary Mitchell / **32.13** Art, Precision Graphics / **32.14** Frank B. Salisbury / **32.15** Art, Precision Graphics / **32.16** From Rost et al, Plant Biology, Wadsworth / **32.17** (a) Jan Zeevart; (b) Plant Industry Station, The Corps Research Division, Agricultural Research Service, U.S. Department of Agriculture / **32.18** N. R. Lersten / **32.19** Larry D. Nooden / **32.20** R. J. Down / **32.21** Lisa Starr; photographs from Eric Welzel, Fox Hill Nursery, 347 Lunt Road, Freeport, Maine 04032 / **32.22** Diana Starr / **32.23** Gary Head / **32.24** Grant Heilman, Inc.

PAGE 559 **559** © Kevin Schafer

CHAPTER 33 **33.1** David Macdonald / **33.2** (a) photograph, Focus on Sports, micrograph Manfred Kage/Bruce Coleman, Ltd.; art, Palay/Beaubois; (b, left to right) Lennart Nilsson from *Behold Man,* © 1974 by Albert Bonniers Forlag and Little, Brown and Company, Boston; Manfred Kage/Bruce Coleman, Ltd.; Ed Reschke /Peter Arnold, Inc. / **33.3** Art, Raychel Ciemma / **33.4** Art adapted from C. P. Hickman, Jr., L. S. Roberts, and A. Larson, *Integrated Principles of Zoology,* ninth ed., Wm. C. Brown, 1995; photo Gregory Dimijian/Photo Researchers / **33.5** Art, Lisa Starr; photographs (a–c,e) Ed Reschke, (d) Fred Hossler/Visuals Unlimited, (f, above) © P. Motta, Dept. of Anatomy, University "La Sapienza," Rome/SPL/Photoresearchers; (f, below) Ed Reschke / **33.6** Roger K. Burnard; (left) Art, Joel Ito, L. Calver / **33.7, 33.8** Ed Reschke / **33.9** R. Demarest / **33.10** (a) Lennart Nilsson from *Behold Man,* © 1974 Albert Bonniers Forlag and Little, Brown and Company, Boston; (b) Kim Taylor/Bruce Coleman, Ltd. / **33.11** Art, L. Calver / **33.12** Art, Palay/Beaubois / **33.14** Art, Precision Graphics; photograph Fred Bruemmer

CHAPTER 34 **34.1** © Warren Faidley/ Weather-stock, Inc. / **34.2** Art, Gary Head and Kevin Summerville / **34.3** Art, R. Demarest with micrograph, Manfred Kage/Peter Arnold, Inc. / **34.4–34.6** Art, Raychel Ciemma / **34.7** Art, Precision Graphics / **34.8** Art, K. Somerville; micrograph Ed Reschke / **34.9** Art, Lisa Starr; micrograph, Dr. Constantino Sotelo from *International Cell Biology,* p. 83, 1977. Used by copyright permission of Rockefeller University Press / **34.10** Art, Precision Graphics / **34.11** Art, R. Demarest; micrograph, R. G. Kessel and R. H. Kardon from *Tissues and Organs: A Text-Atlas of Scanning Electron Microscopy,* W. H. Freeman, 1979. Used by permission of R. G. Kessel / **34.13** (b) Art, R. Demarest / **34.14** Painting, Sir Charles Bell,1809, courtesy of Royal College of Surgeons, Edinburgh

CHAPTER 35 **35.1** Comstock, Inc. / **page 587** Art, Kevin Summerville / **35.2** Art, Raychel Ciemma / **35.3** Lisa Starr after Eugene Kozloff / **35.4** Art, Raychel Ciemma / **35.5** Art, Kevin Somerville / **35.6** Art, Precision Graphics / **35.7** Art. R. Demarest and Precision Graphics/ **35.8** Art, R. Demarest; micrograph Manfred Kage/ Peter Arnold, Inc. / **35.9–35.11** Art, Kevin Somerville / **35.10** (a) Kenneth Garrett/National Geographic Image Collection / **35.12** C. Yokochi and J. Rohen, *Photographic Anatomy of the Human Body,* 2nd Ed., Igaku-Shoin, Ltd., 1979 / **35.13** (b) Marcus Raichle, Washington University School of Medicine / **35.14** Art, Palay/Beaubois after Penfield and Rasmussen, *The Cerebral Cortex of Man,* © 1950 Macmillan Publishing Company, Inc. Renewed 1978 by Theodore Rasmussen; photograph Colin Chumbley/Science Source/ Photo Researchers / **35.15** Art, Lisa Starr and Kevin Summerville / **35.16, 35.16, 35.18** Art, Precision Graphics / **35.19** Art, Kevin Summerville / **35.20** Art, Precision Graphics / **35.21** PET scans from E. D. London et al., *Archives of General Psychiatry,* 47:567–574, 1990; photograph, Ogden Gigli/Photo Researchers / **35.22** (a) © David Stoecklein/The Stock Market; (b) © Lauren Greenfield/Corbis Sygma; (c) AP/ Wide World Photos / **page 604** C. Yokochi and J. Rohen, *Photographic Anatomy of the Human Body,* 2nd Ed., Igaku-Shoin, Ltd., 1979

CHAPTER 36 **36.1** (a) Eric A. Newman; (b) Merlin D. Tuttle, Bat Conservation International / **36.2** Lisa Starr, using photograph © 1997, 1998 Jet Propulsion Laboratory, California Institute of Technology and NASA / **36.3** From Hensel and Bowman *Journal of Physiology,* 23:564–568, 1960 / **36.4** Art, Palay/Beaubois after Penfield and Rasmussen, *The Cerebral Cortex of Man,* © 1950 Macmillan Publishing Company, Inc. Renewed 1978 by Theodore Rasmussen; photograph, Colin Chumbley/ Science Source/Photo Researchers / **36.5** Art, Raychel Ciemma / **36.6** Art, Gary Head / **36.7** Art, R. Demarest; micrograph Omikron/SPL /Photo Researchers / **36.9** Edward W. Bower/© 1991 TIB/West / **36.10** Art, Kevin Somerville / **36.11** Art, Precision Graphics / **36.12** Art, R. Demarest / **36.13** Robert E. Preston, courtesy Joseph E. Hawkins, Kresge Hearing Research Institute, University of Michigan Medical School / **36.14** Art, Raychel Ciemma; photographs Keith Gillett/Tom Stack & Associates / **36.15** Art after M. Gardiner, *The Biology of Vertebrates,* McGraw-Hill, 1972; photograph E. R. Degginger; / **36.16** G. A. Mazohkin-Porshnykov (1958). Reprinted with permission from *Insect Vision* © 1969 Plenum Press / **36.17** Art, Raychel Ciemma, photograph Chris Newbert / **36.18** Art, R. Demarest / **36.19** Chase Swift / **36.20, 36.21** Art, K. Somerville / **36.22** Art, Palay/Beaubois; micrograph Lennart Nilsson © Boehringer Ingelheim International GmbH / **36.23** Art, Precision Graphics / **36.24** Art, Palay/Beaubois after S. Kuffler and J. Nicholls, from *Neuron to Brain,* Sinauer, 1977 / **36.25, 36.26** Art, Kevin Summerville; photographs Gary Ellis/The Wildlife Collection / **36.27** Photograph Douglas Faulkner/Sally Faulkner Collection

CHAPTER 37 **37.1** Hugo van Lawick / **37.2** Art, Kevin Somerville / **37.3, 37.4** Lisa Starr / **37.5, 37.6** Art, R. Demarest / **37.7** (a) Mitchell Layton; (b) Syndication International (l986) Ltd.; (c) Courtesy of Dr. William H. Daughaday, Washington University School of Medicine, from A. I. Mendelhoff and D. E. Smith, eds., *American Journal of Medicine,* 1956, 20:133 / **37.8** Art, L. Morgan / **37.9** (a, b) Art, Raychel Ciemma; photograph, Corbis-Bettmann / **37.10** Art, Precision Graphics / **37.11** Biophoto Associates / SPL/Photo Researchers / **37.12** Art, L. Morgan / **37.13** (a) John S. Dunning/Ardea, London; (b) Evan Cerasoli / **37.14** The Stover Group/D. J. Fort / **37.15** (a,b) From R. C. Brusca and G. J. Brusca, *Invertebrates,* © 1990 Sinauer Associates. Used by permission; (c) Frans Lanting/Bruce Coleman, Ltd.; (d) Robert & Linda Mitchell Photography / **37.16** Roger K. Burnard

CHAPTER 38 **38.1** John Brandenberg/Minden Pictures / **38.2** Art, L. Calver / **38.3** (a) Art, R. Demarest; (b) CNRI/SPL/Photo Researchers; (c) Gregory Dimijian/Photo Researchers / **38.4** (a) M.P. L. Fogden/Bruce Coleman Ltd.; (b) Tom McHugh; (c) Chaumeton-Lanceau/Agence Nature; art after *The Life of Birds,* fourth ed., by L. Baptista and J. C. Welty, © 1988 Saunders College Publishing. Reproduced by permission of the publisher / **38.5** Ed Reschke / **38.6** Michael Keller/FPG / **38.7** Linda Pitkin/Planet Earth Pictures / **38.8** D. A. Parry, *Journal of Experimental Biology,* 1959, 36:654 / **38.9** (a) Stephen Dalton/Photo Researchers; (b) art, Raychel Ciemma / **38.10** Art, D. & V. Hennings / **38.11** Art, Joel Ito and Raychel Ciemma; micrograph Ed Reschke / **38.12** Art, K. Kasnot / **38.13** National Osteoporosis Foundation / **38.14** C. Yokochi and J. Rohen, *Photographic Anatomy of the Human Body,* second ed., Igaku-Shoin, Ltd., 1979 / **38.15, 38.16** Art, Raychel Ciemma / **38.17** N.H.P.A./A.N.T. Photo Library / **38.18** Art, R. Demarest / **38.19** Art, Raychel

Ciemma / **38.20** Art, R. Demarest; photographs (a) Dance Theatre of Harlem, by Frank Capri; (c,d) © Don Fawcett/Visuals Unlimited, from D. W. Fawcett, *The Cell*, Philadelphia; W. B. Saunders Co., 1966 / **38.21** Art, R. Demarest / **38.22** Art, Preface, Inc. / **38.23** Art, R. Demarest / **38.24** Art, Lisa Starr / **38.25** Art, Precision Graphics / **38.26** Michael Neveux / **38.27** © Sean Sprague/Stock Boston

CHAPTER 39 **39.1** (a) From A. D. Waller *Physiology: The Servant of Medicine*, Hitchcock Lectures, University of London Press, 1910; (b) Courtesy The New York Academy of Medicine Library / **39.2** Art, Precision Graphics / **39.3** Art (a) Precision Graphics, (b,d) Raychel Ciemma, (c) After M. Labarbera and S. Vogel, *American Scientist*, 1982, 70:54–60 / **39.4** Art, Precision Graphics / **39.5** Art, Palay/Beaubois / **39.6** CNRI/SPL/Photo Researchers / **39.7** Art, Raychel Ciemma / **39.8** (a,b) Lester V. Bergman & Associates, Inc.; (c) After F. Ayala and J. Kiger, *Modern Genetics*, © 1980 Benjamin-Cummings / **39.9** Art, Nadine Sokol after G. J. Tortora and N. P. Anagnostakos, *Principles of Anatomy and Physiology*, 6th Ed. © 1990 by Biological Sciences Textbooks, Inc., A & P Textbooks, Inc. and Elia-Sparta, Inc. Reprinted by permission of Harper Collins / **39.10** Art, Precision Graphics / **39.11** Art, Kevin Somerville / **39.12** Art, Raychel Ciemma; photograph, C. Yokochi and J. Rohen, *Photographic Anatomy of the Human Body*, second ed., Igaku-Shoin, Ltd., 1979 / **39.13** Art, Precision Graphics / **39.14** Art, Raychel Ciemma; micrograph Michael Abbey/Photo Researchers / **39.15** Art, Raychel Ciemma / **39.16** Art, R. Demarest based on A. Spence, *Basic Human Anatomy*, Benjamin-Cummings, 1982 / **39.17** Art, Precision Graphics / **39.18** Sheila Terry/SPL/Photo Researchers / **39.19** Lisa Starr / **39.20** Art, Kevin Somerville / **39.21** (a) Ed Reschke; (b) F. Sloop and W. Ober/Visuals Unlimited, (c) Art, L. Calver / **39.22** Art, L. Calver / **39.23** Art, Precision Graphics / **39.24** Lennart Nilsson © Boehringer Ingelheim International GmbH / **39.25, 39.26** Art, Raychel Ciemma / **39.27** Lennart Nilsson from *Behold Man*, © 1974 by Albert Bonniers Forlag and Little, Brown and Company, Boston

CHAPTER 40 **40.1** (a) The Granger Collection, New York; (b) Lennart Nilsson © Boehringer Ingelheim International GmbH / **40.2** Robert R. Dourmashkin, courtesy of Clinical Research Centre, Harrow, England / **40.3** Lisa Starr / **40.4** (a) NSIBC/SPL/Photo Researchers; (b) Biology Media/Photo Researchers / **40.5** Art, Raychel Ciemma / **40.6** Lisa Starr / 40.7 Art, Precision Graphics / **40.8** Art, Raychel Ciemma / **40.9** Lisa Starr; micrograph Dr. A. Liepins/SPL/Photo Researchers / **40.10** Lisa Starr; photograph courtesy Don C. Wiley, Harvard University / **40.11, 40.12** Art, Precision Graphics / **40.13, 40.14** Art, Palay/Beaubois after B. Alberts, et al., *Molecular Biology of the Cell*, Garland Publishing Company, 1983 / **40.15** Art, Precision Graphics / **40.16** (above) Lowell Georgia/Science Source/Photo Researchers; (below) Matt Meadows/Peter Arnold, Inc. / **40.17** (left) David Scharf/Peter Arnold, Inc.; (right) Kent Wood/Photo Researchers / **40.18** Ted Thai/*Time* magazine / **40.19** © Zeva Olbaum/Peter Arnold, Inc. / **40.20** (a) Art, after Stephen Wolfe, *Molecular Biology of the Cell*, Wadsworth, 1993; Micrographs Z. Salahuddin, National Institutes of Health / **40.21** Courtesy Dr. Wayne A. Hendrickson and Dr. Peter Kwong

CHAPTER 41 **41.1** Galen Rowell/Peter Arnold, Inc. / **41.2, 41.3** Art, Precision Graphics / **41.4** Art, Precision Graphics; photographs (a) Peter Parks/Oxford Scientific Films; (b) Herve Chaumeton/Agence Nature / **41.5** Art, Precision Graphics; micrograph Ed Reschke / **41.6, 41.7** Art, Raychel Ciemma and Preface, Inc. / **41.8** Lisa Starr / **41.9** Art, Raychel Ciemma, micrograph H. R. Duncker, Justus-Liebig University, Giessen, Germany / **41.10** Art, Kevin Somerville / **41.11** Modified from A. Spence and E. Mason, *Human Anatomy and Physiology*, fourth Ed., 1992, West Publishing Company / **41.12** (a) Lisa Starr; photograph © 2000 PhotoDisc, Inc.; (b–f) Lisa Starr; x-rays SIU/Visuals Unlimited / **41.13** From L. G. Mitchell, J. A. Mutchmor, and W. D. Dolphin, *Zoology*, © 1988 by The Benjamin-Cummings Publishing Company. Reprinted by permission / **41.14** (a) © R. Kessel/ Visuals Unlimited; (b,c) Lisa Starr / **41.15** Art, L. Morgan / **41.16** CNRI/SPL/Photo Researchers / **41.17** O. Auerbach/Visuals Unlimited / **41.18** Photograph Lennart Nilsson from *Behold Man*, © 1974 by Albert Bonniers Forlag and Little, Brown and Company, Boston / **41.19** Art, Precision Graphics; photograph Giorgio Gualco/Bruce Coleman, Ltd. / **41.20** Image Bank / **41.21** Christian Zuber/Bruce Coleman, Ltd. / **41.22** Courtesy of Ron Romanosky / **41.23** Art, Precision Graphics

CHAPTER 42 **42.1** Hulton Getty Collection/Tony Stone Images / **42.2** Art, Precision Graphics / **42.3** Art, Raychel Ciemma / **42.4** Adapted from A. Romer and T. Parsons, *The Vertebrate Body*, sixth ed., Saunders Publishing Company, 1986; photograph D. Robert Franz/Plant Earth Pictures / **42.5** Gunter Ziesler/Bruce Coleman, Inc. / **42.6** Art, Kevin Somerville / **42.7** Art, Nadine Sokol / **42.8** Art, R. Demarest; photograph Omikron/SPL/Photo Researchers / **42.9** After A. Vander, et al., *Human Physiology: Mechanisms of Body Function*, fifth ed., McGraw-Hill, 1990, used by permission; (c) redrawn from *Human Anatomy and Physiology*, fourth ed., by A. Spence and E. Mason, 1992, Brooks/Cole. All rights reserved / **42.10** Art, Lisa Starr after Sherwood and others.; microslide (b) courtesy Mark Nielsen, University of Utah; micrograph (e) © D. W. Fawcett/Photo Researchers / **42.11** Art, Raychel Ciemma / **page 737** After A. Vander, et al., *Human Physiology: Mechanisms of Body Function*, fifth ed., McGraw-Hill, 1990. Used by permission / **42.12** Art, Precision Graphics; photograph Ralph Pleasant/FPG / **42.13** Art, Precision Graphics / **42.14** Art, Precision Graphics; photograph Elizabeth Hathon/The Stock Market / **42.16** Gary Head / **42.17** Dr. Douglas Coleman, The Jackson Laboratory

CHAPTER 43 **43.1** (above left) David Noble/FPG; (below left) Claude Steelman/Tom Stack & Associates; (above right) Gary Head / **43.2** Art, Precision Graphics / **43.3, 43.4** R. Demarest / **43.5** (above) R. Demarest; (below) Art, Precision Graphics / **43.6** © Gary Head / **43.7** (a, b) From T. Garrison, *Oceanography: An Invitation to Marine Science*, Brooks/Cole, 1993. All rights reserved; (c) © Thomas D. Mangelsen/Images of Nature / **43.8** (a) Colin Monteath, Hedgehog House, New Zealand; (b) Bob McKeever/Tom Stack & Associates / **43.9** Everett C. Johnson / **43.10** Art, Precision Graphics / **43.11** David Jennings / Image Works / **43.12** Corbis-Bettmann

CHAPTER 44 **44.1** Art, Raychel Ciemma; photographs (a) Hans Pfletschinger, (c,f) John H. Gerard, (d,e) © David M. Dennis/Tom Stack & Associates, Inc. / **44.2** (a) Frieder Sauer/Bruce Coleman, Ltd; (b) Wisniewski/ZEFA; (c) L. L. Rue III; (d) Carolina Biological Supply Company; (e) Fred McKinney/FPG; (f) Gary Head / **44.3** Art, Palay/Beaubois and Precision Graphics / **44.4** Art, L. Calver; micrographs Carolina Biological Supply Company / **44.5** Art, R. Demarest / **44.6** Art, Precision Graphics / **44.7** Gary Head / **44.8** Art after V. E. Foe and B. M. Alberts, *Journal of Cell Science*, 1983, 61:32, © The Company of Biologists; micrograph J. B. Morrill / **44.9** (Photographic series) Carolina Biological Supply Company; (right photograph) Peter Parks/Oxford Scientific Films/Animals Animals / **44.10** (a) Art after B. Burnside, *Developmental Biology*, 1971, 26:416–441. Used by permission of Academic Press; (b) Lisa Starr / **44.11** After S. Gilbert, *Developmental Biology*, fourth ed., Sinauer / **44.12** Art, Lisa Starr; (c) Prof. Jonathon Slack / **44.13** Art, Preface, Inc.; micrographs from F. R. Turner / **44.14** Carolina Biological Supply Company / **44.15** Art, Palay/Beaubois and Preface, Inc. / **44.16** C. B. and D. W. Frith/Bruce Coleman, Ltd. / **44.17** Art, Lisa Starr; micrograph J. Sulston, MRC Laboratory of Molecular Biology / **page 778** Carolina Biological Supply Company / **page 779** Dennis Green/Bruce Coleman, Ltd. / **44.19** Art, Palay/Beaubois and Preface, Inc.

CHAPTER 44 **45.1** Lennart Nilsson from *A Child Is Born*, © 1966, 1977 Dell Publishing Company, Inc. / **45.2** Art, Raychel Ciemma, modified by Lisa Starr / **45.3** (a) Art, Raychel Ciemma; micrograph Ed Reschke / **45.4** Art, Raychel Ciemma / **45.5** Art, Precision Graphics / **45.6** Art, Raychel Ciemma / **45.7** Art, R. Demarest and Raychel Ciemma / **45.8** Art, Precision Graphics; photograph Lennart Nilsson from *A Child Is Born*, © 1966, 1977 Dell Publishing Company, Inc. / **45.9** R. Demarest and Preface Inc. / **45.10–45.13** Art, Raychel Ciemma / **45.14** Art, Raychel Ciemma; photographs Lennart Nilsson, *A Child Is Born*, © 1966, 1977 Dell Publishing Company, Inc. / **45.15** Modified from K. L. Moore, *The Developing Human: Clinically Oriented Embryology*, fourth ed., Philadelphia: W. B. Saunders Co., 1988 / **45.16** (a) From Lennart Nilsson, *A Child Is Born*, © 1966, 1977 Dell Publishing Company, Inc.; (b) James W. Hanson, M.D. / **45.17** Art, R. Demarest; photograph © Gary Head / **45.18** Art, Raychel Ciemma / **45.19** Adapted from L. B. Arey, *Developmental Anatomy*, Philadelphia, W. B. Saunders Co., 1965 / **45.20** Art, Preface Inc. / **45.21** CNRI/SPL/Photo Researchers / **45.22** (a) John D. Cunningham/Visuals Unlimited; (b) David M. Phillips/Visuals Unlimited; (c) © Kenneth Greer/Visuals Unlimited

PAGE 811 Alan and Sandy Carey

CHAPTER 46 **46.1** (a) Gary Head; (b) Antoinette Jongen/FPG / **46.2** Art, Precision Graphics / **46.3** (a) E. R. Degginger; (inset) Jeff Fott Productions/Bruce Coleman, Ltd.; (b) © Paul Lally/Stock, Boston / **46.4** Art, Precision Graphics; micrograph Stanley Flegler/Visuals Unlimited / **46.5, 46.6** Art, Precision Graphics / **46.7** Art, Precision Graphics; photograph E. Vetter/ZEFA / **page 820** Photograph Eric Crichton/Bruce Coleman, Ltd. / **46.8** Art, Precision Graphics; Photographs (a) Jonathan Scott/Planet Earth Pictures; (b) Wisniewski/ZEFA; (c) Fred Bavendam/Peter Arnold, Inc. / **46.9** Helen Rodd / **46.10** John A. Endler / **46.11** John A. Endler / **46.12** Art, Preface Inc. / **46.13** Art, Precision Graphics; photograph NASA / **46.14** Art after G. T. Miller, Jr., *Environmental Science*, 6th ed., Brooks/Cole, 1997. All rights reserved. / **46.15** Art, Gary Head; photograph United Nations / **46.16** Data from Population Reference Bureau after G. T. Miller, Jr., *Living in the Environment*, 8th ed., Brooks/Cole, 1993. All rights reserved / **46.17** Art, Precision Graphics; photograph, United Nations / **46.18, 46.19** After G. T. Miller, Jr., *Environmental Science*, sixth ed., Brooks/Cole, 1997. All rights reserved

CHAPTER 47 **47.1** (left) John Bova/Photo Researchers; (right) Robert Maier/Animals Animals / **47.2** (a) From L. Clark, *Parasitology Today*, 6 (11), Elsevier Trends Journals, 1990, Cambridge, U.K.; (b) Jack Clark/Comstock, Inc. / **47.3** (a) Eugene Kozloff; (b,c) Stevan Arnold / **47.4** Spectrograms, John Alcock; photograph Hans Reinhard/Bruce Coleman, Ltd.; (a,b) From G. Pohl-Apel and R. Sussinka, *Journal for Ornithologie*, 123:211–214 / **47.5** (a) Evan Cerasoli; (b) From A. N. Meltzoff and M. K. Moore, "Imitation of Facial and Manual Gestures by Human Neonate," *Science*, 1977, 198:75–78. ©1977 by the AAAS / **47.6** Eric Hosking / **47.7** Nina Leen in *Animal Behavior*, Life Nature Library / **47.8** Mark Bekoff / **47.9** (a) Edward S. Ross; (b) E. Mickleburgh/Ardea, London / **47.10** Art, D. &. V. Hennings / **47.11** John Alcock / **47.12** (left) David Fritts/Animals Animals; (right) Ray Richardson/Animals Animals / **47.13** Michael Francis/The Wildlife Collection / **47.14** Frank Lane Agency /Bruce Coleman, Inc. / **47.15** Fred Bruemmer / **47.16** John Alcock / **47.17** Timothy Ransom / **47.18** A. E. Zuckerman/Tom Stack & Associates / **47.19** Patricia Caulfield / **47.20** John Alcock / **47.21** Kenneth Lorenzen / **47.22** Gregory D. Dimijian/Photo Researchers / **47.23** Yvon Le Maho / **47.24** Lincoln P. Brower / **47.25** Eric Hosking / **47.26** F. Schutz

CHAPTER 48 **48.1** (left) Donna Hutchins; (right) Edward S. Ross / **48.2** (a,c) Harlo H. Hadow; (b) Bob and Miriam Francis/Tom Stack & Associates / **48.3** Art, Precision Graphics; photograph Clara Calhoun/Bruce Coleman, Ltd. / **48.4** Art after G. Gause, 1934 / **48.5** Stephen G. Tilley / **48.6** Art after N. Weland and F. Bazazz, *Ecology*, 56: 681–688, © 1975 Ecological Society of America / **48.7** Art, Precision Graphics; photograph John Dominis, *Life Magazine*, © Time, Inc. / **48.8** Art, Precision Graphics; photograph Ed Cesar/Photo Researchers / **48.9** (a) James H. Carmichael; (b) Edward S. Ross; (c) W. M. Laetsch / **48.10** Edward S. Ross / **48.11** (a,b) Thomas Eisner, Cornell University; (c) Douglas Faulkner/Sally Faulkner Collection; (d) Edward S. Ross / **48.12** (a) Pieter Johnson; (b) Stanley Sessions/Hartwick College; (c) © C. James Webb, Phototake, NY / **48.13** Stephen Dalton/Photo Researchers / **48.14** Data from P. Price and H. Tripp, *Canadian Entymology*, 1972, 104:1003–1016 / **48.15** (a–e, g) Roger K. Burnard; (f) E. R. Degginger / **48.16** Art, Precision Graphics; photographs (a,c) Jane Burton/Bruce Coleman, Ltd.; (b) Heather Angel; (d,e) Jane Lubchenco, *American Naturalist*, 112:23–29, © 1978 by The University of Chicago Press / **48.17** R. Slavin/FPG / **48.18** Peter Bird/Australian Picture Library/Westlight; (inset) John Carnemolla/Australian Picture Library/Westlight / **48.19** Angelina Lax/Photo Researchers / **48.20** After W. Dansgaard, et al, *Nature*, 364:218–220, 15 July 1993; D. Raymond et al., *Science*, 259:926–933, February 1993; W. Post, *American Scientist*, 78:310–326, July–August 1990 / **48.21** Art, after S. Fridriksson, *Evolution of Life on a Volcanic Island*, Butterworth: London, 1975; photograph Dr. Harold Simon/Tom Stack & Associates / **48.22** (a,b) After J. M. Diamond, *Proceedings of the National Academy of Sciences*, 69:3199–3201, 1972; (c) David Cavagnaro / **48.23** (a,b) Edward S. Ross / **48.24** Heather Angel/Biofotos

CHAPTER 49 **49.1** (a,b) Wolfgang Kaehler / **49.2** Art, Gary Head and Lisa Starr / **49.3** Art, Gary Head after R. L. Smith, *Ecology and Field Biology*, fifth edition; photograph Alan and Sandy Carey / **49.4, 49.5, 49.6** Lisa Starr, with photographs of tallgrass prairie, © Ann B. Swengel/Visuals Unlimited; bacterium, Stanley W. Watson, *International Journal of Systematic Bacteriology*, 21:254–270, 1971; cutworm, © Nigel Cattlin/Holt Studios International/Photo Researchers, Inc.; gopher and prairie vole, © Tom McHugh/Photo Researchers, Inc.; sparrow, © Rod Planck/Photo Researchers, Inc.; yellow spider, Michael Jeffords; frog, John H. Gerard; garter snake, Michael Jeffords; weasel, courtesy Biology Department, Loyola Marymount University; plover, © O. S. Pettingill Jr./Photo Researchers, Inc.; crow, © Ed Reschke; hawk, © J. Lichter/Photo Researchers, Inc.; all others © 2000 PhotoDisc, Inc. / **49.7** Art, Precision Graphics **/ 49.8** Hans Reinhard/Bruce Coleman, Ltd. / **49.9** Photograph © Gary Head / **49.10** Gene C. Feldman and Compton J. Tucker/NASA, Goddard Space Flight Center / **49.10** and **49.11** Art, Precision Graphics / **49.13** Art, Precision Graphics; photograph Gerry Ellis/The Wildlife Collection / **49.14** Courtesy United States Department of Agriculture / **49.15** Lisa Starr / **49.16** Art, Gary Head after G. E. Likens and F. H. Bormann, "An Experimental Approach to New England Landscapes," in A. D. Hasler (editor), *Coupling of Land and Water Systems*, Chapman & Hall, 1975; photographs (left) G. E. Likens from G. E. Likens and F. H. Bormann, *Proceedings First International Congress of Ecology*, pp. 330–335, September 1974, Centre Agric. Publ. Doc. Wagenigen, The Hague, The Netherlands; (right) G. E. Likens from G. E. Likens et al., *Ecology Monograph*, 40(1)23–47, 1970 / **49.17** (a) Lisa Starr after Paul Hertz; fox photographs © 2000 PhotoDisc, Inc.; (b) John Lawlor/FPG / **49.18** Lisa Starr, NASA photograph / **49.19** Lisa Starr; photographs NASA JSC Digital Image Collection / **49.20** Lisa Starr, compilation of historical records from Vostok and Law Dome ice core samples, Barnola, Raynaud, Lorius, Laboratoire do Glaciologie et de Geophysique de l'Environnement; N.I. Barkov, Arctic and Antarctic Research Institute, and Atmospheric CO_2 concentrations from the South Pole; Keeling andWhorf, Scripps Institute of Oceanography; Jones, Osborn, and Briffa, Climatic Research Unit, University of East Anglia; Parker, Hadley Centre for Climate Prediction and Research; Vostok ice core data set, Chapellaz and Jouzel, NOAA/NGDC Paleoclimatology Program / **49.21** (a) Lisa Starr, compilation of data from Mauna Loa Observatory, Keeling and Whorf, Scripps Institute of Oceanography; and photographs © 2000 PhotoDisc, Inc. / **49.21** (b) Lisa Starr, compilation of data from Prinn et al, CDIAC, Oak Ridge National Laboratory; World Resources Institute; Khalil and Rasmussen, Oregon Graduate Institute of Science and Technology, CDIAC DB-1010; Leifer and Chan, CDIAC DB-1019; photographs © 2000 PhotoDisc, Inc. and Eyewire, Inc.; (c, d) Lisa Starr, compilation of data from World Resources Institute; Law Dome ice core samples, Etheridge, Pearman, and Fraser, Commonwealth Scientific and Industrial Research Organisation; Prinn et al, CDIAC, Oak Ridge National Laboratory; Leifer and Chan, CDIAC DB-1019; photograph © 2000 PhotoDisc, Inc. / **49.22** Lisa Starr after P. Hertz; photograph © 2000 PhotoDisc, Inc. / **49.23** Heather Angel / **49.24** Lisa Starr; photograph © 2000 PhotoDisc, Inc. / **49.25** D. W. Schindler, *Science*, 184:897–899 / **49.27** Art, Gary Head; photograph Bruce Coleman, Ltd. / **49.28** Courtesy of Matthew Lazzara, Antarctic Meteorological Research Center, University of Wisconsin–Madison

CHAPTER 50 **50.1** (a, above) Edward S. Ross; (a, below) David Noble/FPG; (b, above) Edward S. Ross; (b, below) Richard Coomber/Planet Earth Pictures / **50.2** Edward S. Ross / **50.3** Art, Preface Inc.; photograph NASA / **50.4** (b) Art, L. Calver / **50.5** Art, Precision Graphics **/ 50.6** Lisa Starr, using NASA photograph / **50.7, 50.8** Lisa Starr / **50.10** Lisa Starr, after The World Wildlife Fund and other sources, and using NASA photograph / **50.11** Lisa Starr, with photographs © 2000 PhotoDisc, Inc. / **50.12** Art, D. & V. Hennings after Whittaker, Bland and Tilman / **50.13** Harlo H. Hadow / **50.14** (above) John C. Cunningham /Visuals Unlimited; (inset) AP/Wide World Photos; (below) Jack Wilburn/AnimalsAnimals / **50.15** (a) Kenneth W. Fink/Ardea, London; (b) Ray Wagner/Save the Tall Grass Prairie, Inc.; (c) Jonathan Scott/Planet Earth Pictures / **50.16** (left) © 1991 Gary Braasch; (right) Thase Daniel; (below left) Adolf Schmidecker/FPG; (below right) Edward S. Ross / **50.17** Thomas E. Hemmerly / **50.18** (a) Dennis Brokaw; (b) Jack Carey; (c) Nigel Cook/Dayton Beach News Journal/Sygma / **50.19** (a) Fred Bruemmer; (b) Doug Sokell/Visuals Unlimited / **50.20** D. W. MacManiman / **50.21** Art, Precision Graphics / **50.22** Modified after E. S. Deevy, Jr., *Scientific American*, October 1951 / **50.23** E. F. Benfield, Virginia Tech / **50.24** Art, L. K. Townsend / **50.25** (a) Ulli Seer/Tony Stone Images; (b) McCutcheon/ZEFA; (c) After T. Garrison; (d) J. Frederick Grassle, Woods Hole Institution of Oceanography (e) Robert Hessler, Woods Hole Institution of Oceanography / **50.26** (a) C. B. & D. W. Frith/Bruce Coleman, Ltd.; (b) Douglas Faulkner/Photo Researchers; (c) Douglas Faulkner/Sally Faulkner Collection; (d) From T. Garrison, *Oceanography*, third ed., Brooks/Cole, 2000. All rights reserved / **50.27** (above) Peter Scoones/Planet Earth Pictures; (below) from T. Garrison, *Oceanography*, third ed., Brooks/Cole, 2000. All rights reserved / **50.28** (a) Courtesy of Everglades National Park; (b) © Beth Davidow/Visuals Unlimited / **50.29** Art, D. & V. Hennings; photographs (left) E. R. Degginger; (right) © Gary Head / **50.30** (a) Courtesy J. L. Sumich, *Biology of Marine Life*, fifth ed., W. C. Brown, 1992; (b) © 1991 Gary Braasch / **50.31** Adapted from T. Garrison, *Essentials of Oceanography*, Brooks/Cole, 1995. All rights reserved / **50.32, 50.33** Lisa Starr / **50.34** NOAA/Laboratory for Satellite Altimerty / **50.35** (a) © Dennis Kunkel/Phototake NYC; (b) © Roland Birke/OKAPIA/Photo Researchers; (c,d) Courtesy Rita Colwell and A. Huq/UMBI and Byron Wood, Brace Lotitz/NASA; (e) Raghu Rai/Magnum Photos / **50.36** (a) Jack Carey

CHAPTER 51 **51.1** Maps, U. S. Geological Survey; photograph Gary Head / **51.2** (a–b) Art, Lisa Starr; (c) photograph © United Nations / **51.3** Art after G. T. Miller, Jr., *Environmental Science: An Introduction*, Brooks/Cole, 1986. All rights reserved; and the Environmental Protection Agency; photograph, USDA Forest Service / **51.4** Center for Air Pollution Impact and Trend Analysis (CAPITA), Washington University, St. Louis / **51.5** (a–c) NASA; (d) National Science Foundation / **51.6** Data from G. T. Miller, Jr. / **51.7** Art after G. T. Miller, Jr. *Living in the Environment*, eighth ed., Brooks/Cole, 1993. All rights reserved; photograph R. Bieregaard /Photo Researchers / **51.8** NASA / **51.9** William Campbell/*Time* magazine / **51.10** Gerry Ellis/The Wildlife Collection / **51.12** Agency for International Development / **51.13** Dr. Charles Henneghien/ Bruce Coleman, Ltd. / **51.14** Water Resources Council / **51.15** Ocean Arks International / **51.16** After G. T. Miller, Jr. *Living in the Environment*, tenth ed., Brooks/Cole. All rights reserved / **51.17** After M. H. Dickerson, "ARAC: Modeling an Ill Wind," in *Energy and Technology Review*, August 1987. Used by permission of University of California, Lawrence Livermore National Laboratory and U.S. Dept. of Energy; photograph AP/Wide World / **51.18** Alex MacLean/ Landslides / **page 939** J. McLoughlin/FPG / **51.19** © 1983 Billy Grimes / **51.20** After R. Noss, J. M. Scott, and E. LaRoe

INDEX *The letter* i *designates illustration; the letter* t *designates table.*

B

C

D

E

F

G

J

K

L

M

N

O

P

S

T

U

V

W

X

Y

Z

APPLICATIONS INDEX